ENCYCLOPEDIA *of* WORLD CLIMATOLOGY

Encyclopedia of Earth Sciences Series

ENCYCLOPEDIA OF WORLD CLIMATOLOGY

Volume Editor
John E. Oliver is Professor Emeritus at Indiana State University. He holds a BSc from London University, and an MA and PhD from Columbia University. He taught at Columbia University and then at Indiana State where he was formerly Chair of the Geography-Geology Department, and Associate Dean, College of Arts and Sciences. He has written many books and journal articles in Climatology, Applied Climatology and Physical Geography.

Aim of the Series
The *Encyclopedia of Earth Sciences Series* provides comprehensive and authoritative coverage of all the main areas in the Earth Sciences. Each volume comprises a focused and carefully chosen collection of contributions from leading names in the subject, with copious illustrations and reference lists.

These books represent one of the world's leading resources for the Earth Sciences community. Previous volumes are being updated and new works published so that the volumes will continue to be essential reading for all professional earth scientists, geologists, geophysicists, climatologists, and oceanographers as well as for teachers and students.

See the back of this volume for a current list of titles in the *Encyclopedia of Earth Sciences Series*. Go to www.eseo.com to visit the "Earth Sciences Encyclopedia Online" – the online version of this Encyclopedia Series.

About the Editors
Professor Rhodes W. Fairbridge has edited more than 30 Encyclopedias in the Earth Sciences Series. During his career he has worked as a petroleum geologist in the Middle East, been a WW II intelligence officer in the SW Pacific and led expeditions to the Sahara, Arctic Canada, Arctic Scandinavia, Brazil and New Guinea. He is currently Emeritus Professor of Geology at Columbia University and is affiliated with the Goddard Institute for Space Studies.

Professor Michael Rampino has published more than 100 papers in professional journals including *Science*, *Nature*, and *Scientific American*. He has worked in such diverse fields as volcanology, planetary science, sedimentology, and climate studies, and has done field work on six continents. He is currently Associate Professor of Earth and Environmental Sciences at New York University and a consultant at NASA's Goddard Institute for Space Studies.

ENCYCLOPEDIA OF EARTH SCIENCES SERIES

ENCYCLOPEDIA *of* WORLD CLIMATOLOGY

edited by

JOHN E. OLIVER
Indiana State University

A C.I.P. Catalogue record for this book is available from the Library of Congress.

ISBN-10 1-4020-3264-1 (HB) Springer Dordrecht, Berlin, Heidelberg, New York
ISBN-10 1-4020-3266-8 (e-book) Springer Dordrecht, Berlin, Heidelberg, New York
ISBN-13 978-1-4020-3264-6 (HB) Springer Dordrecht, Berlin, Heidelberg, New York
ISBN-13 978-1-4020-3266-0 (e-book) Springer Dordrecht, Berlin, Heidelberg, New York

---

Published by Springer
PO Box 17, 3300 AA Dordrecht, The Netherlands

Printed on acid-free paper

Cover photo: Camex 4 satellite image of Hurricane Fran in the Atlantic, 27 August 2004.
Photo credit: NASA.

Every effort has been made to contact the copyright holders of the figures and tables which have been reproduced from other sources. Anyone who has not been properly credited is requested to contact the publishers, so that due acknowledgement may be made in subsequent editions.

Printed and bound in Great Britain by MPG Books, Bodmin, Cornwall

Contents

Contributors

Edward Aguado
Department of Geography
San Diego State University
San Diego, CA 92182-4493, USA

Enric Aguilar
Climate Change Research Group
Geography Unit
Universitat Rovira I Virgili, Tarragona
Spain
e-mail: eaa@fll.urv.es

Eric J. Alfaro
Department of Physics
University of the West Indies
Mona, Kingston 7, Jamaica

Ted Alsop
Department of History and Geography
Utah State University
Logan, UT 84321, USA
e-mail: tjalsop@cc.usu.edu

A. John Arnfield
Department of Geography
Ohio State University
Columbus, OH 43210-1361, USA
e-mail: John.Arnfield@osu.edu

David L. Arnold
Department of Geography
Ball State University
Cooper Science Complex CL425
Muncie, IN 47306, USA
e-mail: darnold@bsu.edu

Ferdinand Baer
Department of Meteorology
University of Maryland
College Park, MD 20742, USA
e-mail: Baer@atmos.usmd.edu

G.W. Bailey
Department of Geography
Simon Fraser University
8888 University Drive
Burnaby, British Columbia V5A 156
Canada
e-mail: bailey@sfu.ca

Robert C. Balling
Department of Geography
Arizona State University
Tempe, AZ 85281, USA
e-mail: balling@imap1.asu.edu

Roger Barry
National Snow and Ice Data Center
CIRES
Campus Box 449
University of Colorado
Boulder, CO 80309, USA
e-mail: rbarry@Kryos.Colorada.edu

Susan Berta
Department of Geography, Geology and Anthropology
Indiana State University
Terre Haute, IN 47809, USA
e-mail: geberta@isugw.indstate.edu

Greg Bierly
Indiana State University
Department of Geography Geology and Anthropology
Terre Haute, IN 47809, USA
e-mail: gebierly@isugw.indstate.edu

G.R. Bigg
Department of Geography
University of Sheffield
Winter Street
Sheffield S10 2TN, UK
e-mail: Grant.Bigg@sheffield.ac.uk

Mark Binkley
Department of Geography
Rutgers University
New Brunswick, NJ 08903, USA

Keith Robert Boucher
Department of Geography
Loughborough University
Loughborough, Leicestershire LE11 3TU, UK
e-mail: k.r.boucher@lboro.ac.uk

Anthony J. Brazel
Arizona State University
Main Campus
College of Liberal Arts and Sciences-Geography
P.O. Box 870104
Tempe, AZ 85287-0104, USA
e-mail: abrazel@asu.edu

Michael J. Brewer
National Weather Service W/OS4
11325 East West Highway
Silver Spring, MD 20910, USA
e-mail: Michael.J.Brewer@NOAA.gov

Howard A. Bridgman
School of Environmental and Life Sciences
University of Newcastle
Newcastle, Australia 2308
e-mail: Howard.Bridgman@newcastle.edu.au

Ross D. Brown
Meteorological Service of Canada
2121 Trans Canada Highway
Dorval, QC, H9P 1J3, Canada
e-mail: Ross.Brown@ec.gc.ca

Rodger A. Brown
National Severe Storms Laboratory
1313 Halley Circle
Norman, OK 73069, USA
e-mail: Rodger.Brown@noaa.gov

L.J. Bruce McArthur
Meteorological Service of Canada
4905 Dufferin Street
Downsview, Ontario M3H 5T4, Canada
e-mail: bruce.mcarthur@ec.gc.ca

Reid Bryson
University of Wisconsin
Center for Climate Research
Madison, WI 53706-1491, USA
e-mail: rabryson@facstaff.wisc.edu

Adam Burnett
Department of Geography
Colgate University
Hamilton, NY 13346, USA
e-mail: Aburnett@mail.colgate.edu

W.J. Burroughs
Squirrels Oak
Clandon Road
West Clandon, Surrey GU4 7UW, UK
e-mail: billsqoak@aol.com

James E. Burt
Department of Geography
University of Wisconsin
Madison, Wisconsin 53706, USA
e-mail: jburt@geography.wisc.edu

Lisa M. Butler Harrington
Department of Geography
Dickens Hall
Kansas State University
Manhattan, KS 66506, USA
e-mail: lharrin@ksu.edu

Andrew M. Carleton
Department of Geography
The Pennsylvania University
University Park, PA 16802, USA
e-mail: AndrewMCarleton@aol.com

Randall S. Cerveny
Arizona State University
Department of Geography
Box 870104
Tempe, AZ 85287, USA
e-mail: cerveny@asu.edu

Stanley Changnon
801 Buckthorn Circle
Mahomet, IL 61853, USA
e-mail: schangno@uiuc.edu

David Changnon
Meteorology Program
Department of Geography
Northern Illinois University
DeKalb, IL 60115, USA
e-mail: dchangnon@niu.edu

Richard K. Cook
8517 Milford Avenue
Silver Spring, Maryland 20910, USA

William T. Corcoran
Southwest Missouri State University
Department of Geography, Geology and Planning
Springfield, MO 65804-0089, USA
e-mail: wtc928f@smsu.edu

Cameron Douglas Craig
Indiana State University
Department of Geography, Geology and Anthropology
Terre Haute, IN 47809, USA
e-mail: ccraig@mymail.indstate.edu

Heidi M. Cullen
The Weather Channel
300 Interstate North Parkway
Atlanta, GA 30339, USA

Christina Dando
Department of Geography-Geology
University of Nebraska at Omaha
Omaha, NE 68182-0119, USA
e-mail: cdando@mail.unomaha.edu

William A. Dando
Indiana State University
Senior Scholars Academy
Terre Haute, IN 47809, USA
e-mail: gedando@isugw.indstate.edu

S.K. Dash
Centre for Atmospheric Sciences
Indian Institute of Technology Delhi
Hauz Khas, New Delhi 110016
India
e-mail: skdash@cas.iitdernet.in

John A. Day
609 North Cowls
McMinnville, Oregon 97128, USA
e-mail: Cloudman@cloudman.com

K. Dewey
Department of Geography
University of Nebraska
Lincoln, Nebraska 98588, USA

Richard W. Dixon
Department of Geography
Southwest Texas State University
San Marcos, TX 78666, USA
e-mail: rd11@swt.edu

John Dodson
Faculty of Natural and Agricultural Sciences
University of Western Australia
35 Sterling Highway
Nedlands, Western Australia 6009
Australia
e-mail: johnd@geog.uwa.edu.au

Robert A. Duce
Department of Oceanography and Atmospheric Sciences
Texas A & M, 3146 TAMU
College Station, TX 77843-3146, USA
e-mail: rduce@ocean.tamu.edu

Rhodes W. Fairbridge
Box 801
Amagansett, NY 11930, USA

Robert W. Fenn
Air Force Cambridge Research Laboratories
Atmospheric Optic Branch
Optical Physics Laboratory
Bedford, Massachusetts 01730, USA

Harold C. Fritts
Indiana State University
Department of Geography, Geology and Anthropology
Terre Haute, IN 47809, USA
e-mail: gaspeer@isugw.indstate.edu

Joseph Gentilli (deceased)

T.W. Giambelluca
Department of Geography
University of Hawaii at Manoa
Honolulu, HI 96822, USA
e-mail: Thomas@Hawaii.edu

Brian Giles
3/50 Onepoto Road
Takapuna
Auckland, 1309
New Zealand
e-mail: gilesnz@ihug.co.nz

Michael H. Glantz
Environmental and Societal Impacts Group
National Center for Atmospheric Research
P.O. Box 3000
Boulder, CO 80306, USA
e-mail: glantz@ucar.edu

Vivien Gornitz
Center for Climate System Research
2880 Broadway
Columbia University
New York, NY 10025, USA
e-mail: vgornitz@giss.nasa.gov

A.S. Goudie
School of Geography and the Environment
Oxford University
Mansfield Road
Oxford 0X1 3TB, UK
e-mail: andrew.goudie@st-cross.oxford.ac.uk

Samuel N. Goward
Department of Geography
2181 LeFrak Hall
University of Maryland
College Park, MD 20742, USA
e-mail: sgoward@umd.edu

Orman E. Granger
Department of Geography
University of California-Berkeley
Berkeley, California 94720, USA
e-mail: Ogranger@socrates.berkeley.edu

David Greenland
Department of Biological and Agricultural Engineering
Louisiana State University Agricultural Center
Baton Rouge, Louisiana, USA
e-mail: greenlan@lsu.edu

John Griffiths (deceased)

John M. Grymes III
Department of Geography and Anthropology
Louisiana State University
Baton Rouge, LA 70803, USA

Brian Hanson
Department of Geography
University of Delaware
Newark, DE 19711, USA
e-mail: hanson@udel.edu

Jay R. Harman
Department of Geography
Michigan State University
Lansing, Michigan 48824, USA
e-mail: Harman@pilot.msu.edu

J.G. Harvey
17 Old Hall Gardens
Brooke, Norwich NR15 1JZ, UK
e-mail: jharvey@nr-15.fsnet.co.uk

John E. Hay
P.O. Box 102069
NSMC
North Shore City, New Zealand
e-mail: johnhay@ihug.co.nz

James Henry
Geosciences
Middle Tennessee State University
Murfreesboro, TN 37132, USA
e-mail: 07jhenry@bellsouth.net

John J. Hidore
5577 Lake Juno Road
Liberty, NC 27298, USA
e-mail: hidore@att.net

Jack Hobbs
Honorary Fellow, School of Human and Environmental Studies
University of New England
P.O. Box 343
Armidale, New South Wales 2350
Australia
e-mail: jhobbs@metz.une.edu.au

Jay Hobgood
Department of Geography
The Ohio State University
Derby Hall, 154 N. Oval Mall
Columbus, OH 43210, USA
e-mail: hobgood.1@osu.edu

Robert M. Hordon
Department of Geography
Rutgers University
54 Joyce Kilmer Avenue
Piscataway, NJ 08854, USA
e-mail: hordon@rci.rutgers.edu

S.A. Hsu
Department of Oceanography and Coastal Sciences
Louisiana State University
308 Howe-Russell Building
Baton Rouge, LA 70803, USA

Tim Hughes
PPM Energy
1125 NW Couch St
Portland, Oregon 97209, USA
e-mail: tim.hughes@ppmenergy.com

James W. Hurrell
NCAR
Climate Analysis
P.O. Box 3000
Boulder, CO 80307-3000, USA
e-mail: jhurrell@ucar.edu

Norys Jiminez
Urbani-Zacion
Los Bloques,
Maturui-Edo
Monagas, Venezuela

Elias Johnson
Department of Geography, Geology and Planning
Southwest Missouri State University
Springfield, Missouri 65804, USA
e-mail: Elj175F2smsu.edu

Laurence S. Kalkstein
Center for Climatic Research
University of Delaware
896 Banyan Court
Marco Island, FL 34145, USA
e-mail: larryk@udel.edu

Paul Kay
Department of Environmental and Resource Studies
University of Waterloo
Waterloo, Ontario N2L3G1
Canada
e-mail: pkay@fes.uwaterloo.ca

Mick Kelly
Climatic Research Unit
School of Environmental Sciences
University of East Anglia
Norwich NR4 7TJ, UK
e-mail: m.kelly@uea.ac.uk

Melissa L. Kemling
Nebraska Department of Environmental Quality
Remediation Section, 1200 N St.
Lincoln, NE 68509-8922, USA
e-mail: Melissa.Kemling@ndeq.state.ne.us

J. Kleypas
Environmental and Societal Impacts Group
National Center for Atmospheric Research
P.O. Box 3000
Boulder, CO 80307-3000, USA
e-mail: kleypas@atd.ucar.edu

George Kukla
Lamont-Doherty Earth Observatory
Columbia University
New York, USA

S. Ladochy
Department of Geography and Urban Analysis
California State University, LA
5151 State University Drive
Los Angeles, CA 90032-8222, USA
e-mail: sladoch@exchange.calstatela.edu

H.E. Landsberg (deceased)

James Lebeau
Center for the Study of Crime
Delinquency and Corrections
Southern Illinois University
Carbondale, IL 62901, USA
e-mail: lebeau@siu.edu

David Legates
Centre for Climatic Research
Department of Geography
University of Delaware
Newark, DE 19716, USA
e-mail: legates@udel.edu

G. Jay Lennartson
University of North Carolina-Greensboro
Geography Department
Greensboro, NC 27402-617, USA

J. Ben Liley
NIWA, Lauder
Private Bag 50061, Omakau
Central Otago
New Zealand
e-mail: b.liley@niwa.co.nz

John G. Lockwood
4 Woodthorne Croft
Leeds LS17 8XQ, UK
e-mail: jglockwood@clara.co.uk

Vivian Loftness
Department of Architecture
DH 1325
Carnegie Mellon University
Pittsburgh, PA 15213, USA
e-mail: loftness@andrew.cmu.edu

Abdel Maarouf
Institute for Environmental Change
University of Toronto
Canada
Meteorolgical Service of Canada
Downsview, Ontario Canada
e-mail: abdel.maarouf@ec.gc.ca

Michael C. MacCracken
6308 Berkshire Drive
Bethesda, MD 20814, USA
e-mail: mmaccrac@comcast.net

Glen A. Marotz
Associate Dean for Research
and Graduate Programs
University of Kansas
Lawrence, Kansas 66045, USA
e-mail: gama@ku.edu

John R. Mather (deceased)

Sandra Mather
13 Roosevelt Way
Avondale, PA 19311, USA
e-mail: mather@verizon.net

Michael McCracken
6308 Berkshire Drive
Bethesda, MD 20814, USA
e-mail: mmaccrac@comcast.net

Anthony J. McMichael
National Centre for Epidemiology and Population Health
M Block, Mills Road
Australian National University
Canberra ACT 0200, Australia
e-mail: tony.mcmichael@anu.edu.au

David H. Miller
Emeritus Professor
Geosciences
University of Wisconsin
Milwaukee, Wisconsin 53201, USA

Gerald Mills
Department of Geography
Arts Building
University College Dublin
Belfield, Dublin 4, Ireland
e-mail: gerald.mills@ucd.ie

J. Murray Mitchell (deceased)

Sir Patrick Moore
Farthings
West Street
Selsey, Sussex
UK

Michael D. Morgan
University of Wisconsin
2420 Nicolet Drive
Green Bay, WI 54311-7001, USA
e-mail: morganm@uwgb.edu

Benjamin Moulton
300 W. York Dr.,
Terre Haute, IN 47802
USA

Robert A. Muller
Department of Geography and Anthropology
Louisiana State University
Baton Rouge, LA 70803, USA
e-mail: wolfiandsonni@cox.net

R.E. Munn
Adaptation and Impacts Research Group
Meteorological Service of Canada
4905 Dufferin Street
Downsview, ON
M3H 5T4 Canada
e-mail: ted.munn@utoronto.ca

D.S. Munro
Department of Geography and Planning
University of Toronto
Toronto, Canada
e-mail: smunro@credit.erin.utoronto.ca

H. Neuberger (deceased)

James E. Newman
Department of Agronomy
Purdue University
West Lafayette, Indiana 47907, USA

Sharon E. Nicholson
Meteorology Department
Florida State University
404 Love Building
Tallahassee, FL 32302, USA
e-mail: sen@met.fsu.edu

Joel Norris
Scripps Institution of Oceanography
8810 Shellback Way
Room 440, Nierenberg Hall
La Jolla, CA 92037, USA
e-mail: jnorris@ucsd.edu

John E. Oliver
Geography, Geology, Anthropology
Indiana state University
Terre Haute
IN 47809, USA
e-mail: climatology1@yahoo.com

Hans Panofski (deceased)

Peitao Peng
CPC/NCEP/NOAA
5200 Auth Rd. Rm. 806
Camp Springs, MD 20746, USA
e-mail: peitao.peng@noaa.gov

Allen Perry
University of Wales Swansea
Department of Geography
Singleton Park
Swansea, SA2 8PP, UK
e-mail: A.H. Perry@Swansea.ac.uk

A.J. Pitman
Head of Department of Physical Geography
Macquarie University
North Ryde, NSW 2109
Australia
e-mail: pitman@penman.es.mq.edu.au

Dale Quatiocchi
Arizona State University
Main Campus
College of Liberal Arts and Sciences-Geography
P.O. Box 870104
Tempe, AZ 85287-0104, USA

Robert G. Quayle
1 Botany View Court
Asheville, NC 28805, USA
e-mail: rquayle@ncdc.noaa.gov

Elmar Reiter (deceased)

Neil Roberts
School of Geography
University of Plymouth
Plymouth, UK
e-mail: cnroberts@Plymouth.ac.uk

Scott M. Robeson
Department of Geography
120 Student Building
Indiana University
Bloomington, IN 47405, USA
e-mail: srobeson@indiana.edu

David A. Robinson
Department of Geography
Rutgers University
Piscataway, NJ 08854, USA
e-mail: drobins@rci.rutgers.edu

A.A.L.N. Sarma
Department of Meteorology and Oceanography
Andhra University
Waltair, Visaknopatnam, South India

Thomas W. Schlatter
Chief Scientist
NOAA Forecast Systems Laboratory
David Skaggs Research Center, Rm. 3B128
325 Broadway
Boulder, CO 80305-3328, USA
e-mail: tom.schlatter@noaa.gov

Thomas W. Schmidlin
Kent State University
413 McGilvrey Hall
P.O. Box 5190
Kent, OH 44242-0001, USA
e-mail: tschmidl@kent.edu

Robert M. Schwartz
Department of Geography
Ball State University
Cooper Science Building CL425
Muncie, IN 47306-1776, USA
e-mail: rmschwartz@bsu.edu

Mark D. Schwartz
Department of Geography, Bolton 410
University of Wisconsin-Milwaukee
P.O. Box 413
Milwaukee, WI 53201-0413, USA
e-mail: mds@uwm.edu

Mark C. Serreze
Coop Inst. Res/Envrm. Sci. Dir.
University of Colorado at Boulder
216 UCB
Boulder, CO 80309-0216, USA
e-mail: serreze@kyros.colorado.edu

James H. Shirley
Jet Propulsion Laboratory MS 183-601
Pasadena, CA 91109
e-mail: James.H.Shirley@jpl.nasa.gov

Akhtar H. Siddiqi (deceased)

Shanaka L. de Silva
Department of Space Studies
University of North Dakota
526 Clifford Hall
Grand Forks, MD 58202-9008, USA
e-mail: desilva@space.edu

Robert Mark Simpson
Department of Geology, Geography and Physics
215 Johnson EPS Building
University of Tennessee at Martin
Martin, TN 38238, USA
e-mail: msimpson@utm.edu

Richard Snow
Applied Aviation Sciences
Embry-Riddle Aeronautical University
600 S. Clyde Morris Boulevard
Daytona Beach, FL 32114-3900, USA
e-mail: snow4fc@erau.edu

Mary Snow
Embry-Riddle Aeronautical University
Aeronautical Science Department
600 S. Clyde Morris Boulevard
Daytona Beach, FL 32114-3900, USA
e-mail: snowm@erau.edu

James H. Speer
Indiana State University
Department of Geography, Geology and Anthropology
Terre Haute, IN 47809, USA

Rachel Spronken-Smith
Department of Geography
University of Canterbury
Private Bag 4800
Christchurch
New Zealand

Steve Stadler
207 Scott Hall
Geography Department
Oklahoma State University
Stillwater, OK 74078-4073, USA
e-mail: rroh69@okstate.edu

Stephen J. Stadler
Geography Department
Oklahoma State University
Stillwater, OK 74078-4073, USA
e-mail: rroh69@okstate.edu

Graeme L. Stephens
Department of Atmospheric Science
Colorado State University
Ft. Collins, CO 80523
e-mail: stephens@atmos.colostate.edu

W.C. Swinbank
Commonwealth Scientific and Industrial Research Organization
Division of Meteorological Physics
Aspendale, Victoria, Australia

Michael A. Taylor
Department of Physics
University of the West Indies
Mona, Kingston 7, Jamaica
e-mail: mataylor@uwimona.edu.jm

John E. Thornes
School of Geography and Environmental Science
University of Birmingham
Edgbaston
Birmingham B15 2TT, UK
e-mail: j.e.thornes@bham.ac.uk

L.M. Trapasso
Department of Geography and Geology
Western Kentucky University
Bowling Green, KY 42101, USA
e-mail: michael.trapasso@wku.edu

Donna Tucker
Department of Physics
Malott Hall
1251 Wescoe Hall Drive, Room 1082
University of Kansas
Lawrence, KS 66045-7582, USA
e-mail: dftwx@yahoo.com

Danny M. Vaughn
832E 3200N
North Ogden, UT 84414-1745, USA
e-mail: docdmv@quik.net

Claudio Vita-Finzi
Geological Sciences
University College
London, UK
e-mail: ucfbcvf@ucl.ac.uk

Claudia Wagner-Riddle
Department Land Resource Science
University of Guelph
Guelph, Ontario
Canada N1G 2W1
e-mail: cwagnerr@uoguelph.ca

H.J. Walker
Department of Oceanography and Coastal Sciences
Louisiana State University
308 Howe-Russell Building
Baton Rouge, LA 70803, USA

John E. Walsh
Atmospheric Sciences
University of Illinois
Urbana, IL 61801, USA
e-mail: walsh@atmos.uiuc.edu

Wayne Wendland
Climatology Section
Illinois State Water Survey
Champaign, IL 61820, USA
e-mail: wayne@uxl.cso.uiuc.edu

Donald A. Wilhite
National Drought Mitigation Center
International Drought Information Center
School of Natural Resources Sciences
239 L.W. Chase Hall
University of Nebraska
Lincoln, NE 68583-0749, USA
e-mail: dwilhite@uninotes.uni.edu

Cort J. Willmott
Department of Geography
University of Delaware
Newark, DE 19711, USA
e-mail: wilmott@udel.edu

Lewis Wixon
Department of Geography
St. Cloud State University
St. Cloud, MN 56301, USA
e-mail: Wixon@stcloudstate.edu

Rosalie E. Woodruff
National Centre for Epidemiology and Population Health
M Block, Mills Road
Australian National University
Canberra ACT 0200, Australia

Y. Xue
Departments of Geography and Atmospheric Sciences
1255 Bunche Hall, UCLA
Los Angeles, CA 90095-1524, USA
e-mail: yxue@geog.ucla.edu

M. Yanai
Departments of Geography and Atmospheric Sciences
1255 Bunche Hall, UCLA
Los Angeles, CA 90095-1524, USA

Yuk Yee Yan
Department of Geography
Hong Kong Baptist University
Kowloon Tong
Kowloon, Hong Kong
e-mail: yyan@hkbu.edu.hk

Masatoshi Yoshino
5-1-8-202 Komazawa
Setagayaku, Tokyo, 154-0012, Japan
e-mail: mtoshiyo@poplar.ocn.ne.jp

Peyman Zawar-Reza
Department of Geography
University of Canterbury
Private Bag 4800
Christchurch
New Zealand
e-mail: peyman.zawar-reza@canterbury.ac.nz

Introduction

World-wide events are creating an unprecedented growth of interest in regional and global climates. Extensive media coverage of droughts, floods, and very cold or exceptionally warm winters is exposing people from all segments of society to the significance of climate. The modern age of communications has permitted many formerly esoteric climatological terms, with El Niño providing an excellent example, to be widely known and frequently cited. The realization that climate, and its potential for change, can have profound influences on both the quality of life and the nature of the Earth's environment provides the basis for global actions. International conferences, publications, and agreements, many sponsored by the United Nations, are based upon the necessity of establishing international climate programs that assist nations in both understanding and responding to natural and human-induced climate processes and their implications. Results are widely reported and have become the basis for some controversies, especially regarding future climate scenarios.

Given that the public is becoming more aware of the importance of climate, it is necessary for the professional climatologist to respond to the public's needs. Research in both theoretical and applied climatology is beginning to provide a greater understanding of the actual and potential impacts of past, present and future climates, and it remains for the professional to provide the educated layperson with climatic information that is accurate, meaningful and readily comprehended. It is hoped that *The Encyclopedia of World Climatology* will be of value in this respect, for its content provides the basis for the understanding of many aspects of climate while, at the same time, providing information on recent advances in the field.

The content of this work reflects, in part, that of the *Encyclopedia of Climatology* that was published in 1986. However, there are many essential differences between the current volume and the earlier work, with the most obvious being the inclusion of modern developments in the field. Of importance, too, is the way in which past climates are treated. In this volume, apart from an overview of changes over geologic time, past climate articles deal with changes that have occurred in historic times and their impacts. This time restriction results from the explosive development of paleoclimatology in recent decades. So significant is this growth, that a companion volume, *Encyclopedia of Paleoclimatology and Ancient Environments*, is necessary. Such a volume is currently being composed and edited for the same *Encyclopedia of Earth Sciences Series*.

The transfer of paleoclimate articles to another volume permits the *Encyclopedia of World Climatology* to include many more articles dealing with impacts of climate, examined from both environmental and social points of view. For example, changes in atmospheric quality are discussed not only as an atmospheric process but also as a human hazard. Additionally, a number of biographies of climatologists who have played a significant role in the development of climatology are provided. Those included should be considered representative of the many scholars who contributed to the discipline at various stages of its development. It is, of course, impossible to include all those individuals whose research has added to the intellectual content of climatology.

The 1986 volume contained contributions by outstanding scholars who are now deceased. The skill with which such authors as Helmut Landsberg, Elmer Reiter, J. Murray Mitchell and Derek Schove presented their ideas and concepts was not something that I wished to lose. As a result, their works have been suitably and carefully revised and edited while retaining enough of the original to reflect the writings of the original contribution. Completing such a task is no easy undertaking, and the writers who completed those revisions are certainly to be commended.

Organization

Entries are in alphabetical order, with their length being related to the relative importance of the topic. In some cases the same topic is considered under separate headings, and there are a number of ways to find relevant information. Initially, to find a particular topic it is best to look for that subject in the alphabetical listing of entries. Beyond that, the index and article cross-references locate the required subject matter. The comprehensive index at the back of the volume will list, for a given name or term, every page where that item appears in the volume.

Each article is followed by a list of cross-references that locate related entries to that article. It should be noted that these cross-references are not intended to be exhaustive, for this would lead, in many instances, to a very long listing. Instead, the reader is guided to other topics which themselves are cross-referenced. In this way, the many aspects of a given topic may be meaningfully selected.

To further assist the reader to complete research, references are given at the end of each entry. The number of references usually varies in direct proportion to the length of the entry. It will be noted that some articles provide older citations. These are intended to allow the reader to examine the historical development of a topic if desired. For the most part, however, the references provide recent significant information.

Acknowledgements

Editing a multiauthored volume such as this is not without its problems. Ranging from selection of topics to be included to determining an appropriate length, the decisions I made were greatly helped by many others. I gratefully acknowledge the ideas and suggestions of my Advisory Board members who provided an international perspective upon selection of topics and authors:

Howard Bridgman, Newcastle University, Australia;
Henry Diaz, NOAA/CDC/OAR,USA;
Michael Glantz, ESIG, NOAA, USA;
John Hay, IGCI, University of Waikato, New Zealand;
Tim Oke, University of British Columbia, Canada;
John Thornes, University of Birmingham, United Kingdom.

At all stages, the sage advice of Rhodes Fairbridge was invaluable.

All of the major articles in this volume were peer-reviewed. In addition to the acknowledgements given following selected articles, I gratefully acknowledge the comments of each of the following:

Richard Allen, Eric Alfaro, Huug van den Dool, John Feddema, James Hansen, John Horel, John Hidore, Arun Kumar, Tim Osborn, David Robinson, Thomas Schmidlin, Andrew Sturman, Tsegaye Tadesse, Martin Wadley.

Additionally, the support of Dr. Susan Berta, Chair, and colleagues in the Department of Geography, Geology and Anthropology at Indiana State University is appreciated.

Given the magnitude of the task, compiling a work of this nature requires someone of infinite patience, organizational ability, and all office management skills. I am fortunate that my wife Loretta filled this role. Without her considerable effort and continuing support, this volume might never have been completed.

John E. Oliver
Terre Haute, Indiana, 2005

A

ACID RAIN

Acid rain is one of the major environmental issues confronting industrialized countries, and is a wet form of acid deposition. Acid deposition is composed of sulfuric acid, nitric acid and ammonium, and occurs as wet deposition (rain, snow, sleet, hail), dry deposition (particles, gases and vapor), and cloud or fog deposition. Understanding the nature of acid rain and its impact requires a working knowledge of the concept of acidity and means by which atmospheric processes affect the potential for acid deposition.

Acidity

The acidity of a material is associated with the relative abundance of free hydrogen ions (H^+) when that substance is in a water solution. The pH scale is a logarithmic scale, where a value of 7 indicates neutrality; decreasing values on the scale indicate an increase in acidity and increasing values represent alkalinity. A pH scale with representative examples is shown in Figure A1.

Absolutely pure water (distilled) has a pH of 7, but if it is allowed to react with clean air its pH will decrease to near 5.6. This is a result of the absorption of carbon dioxide in the atmosphere to form weak carbonic acid. Most rainwater will have this pH; acid rain occurs when the pH is less than 5.6. For this reaction to occur, the atmosphere must be composed of chemicals that provide an acidic source. The atmosphere must also be able to deposit the material at the surface. This deposition is accomplished through the process of atmospheric cleansing.

Atmospheric cleansing

Acid deposition is a result of the "wet deposition" process, which involves the removal of chemicals through precipitation. Impurities in the atmosphere are incorporated in the entire precipitation process, beginning with cloud droplet formation, and are deposited as part of the resulting precipitation. Cleansing can also happen via "dry deposition", which implies that substances in the atmosphere are deposited through gravitational settling. Large particles settle under their own weight; very small particles may be transported far from their source to eventually polymerize until they are large enough to fall under their own weight. These particles generally are returned to earth through the wet deposition process.

Wet and dry depositional processes do not include all aspects of deposition. Other forms include fog droplet interception (especially prevalent in high-altitude forests), dew, frost, and rime icing.

The current distribution of wet-deposition acid in the United States is shown in Figure A2. The eastern part of the country, centered on the northeastern states, clearly has higher acidity than the remainder of the area. In part, this distribution reflects the conditions in the drier west where the chemistry of precipitation is altered by alkaline substances contributed to the atmosphere.

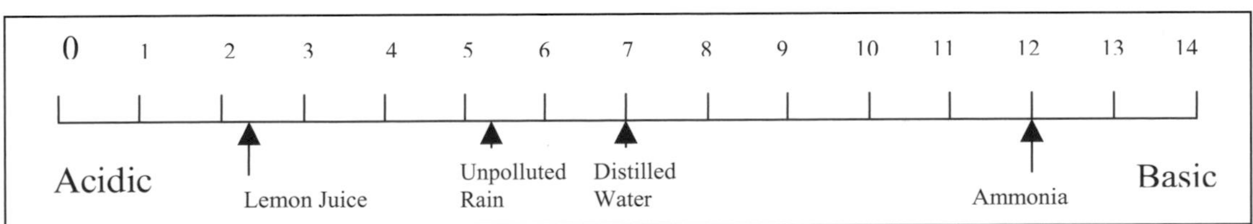

Figure A1 The pH scale denoting examples of some common substances.

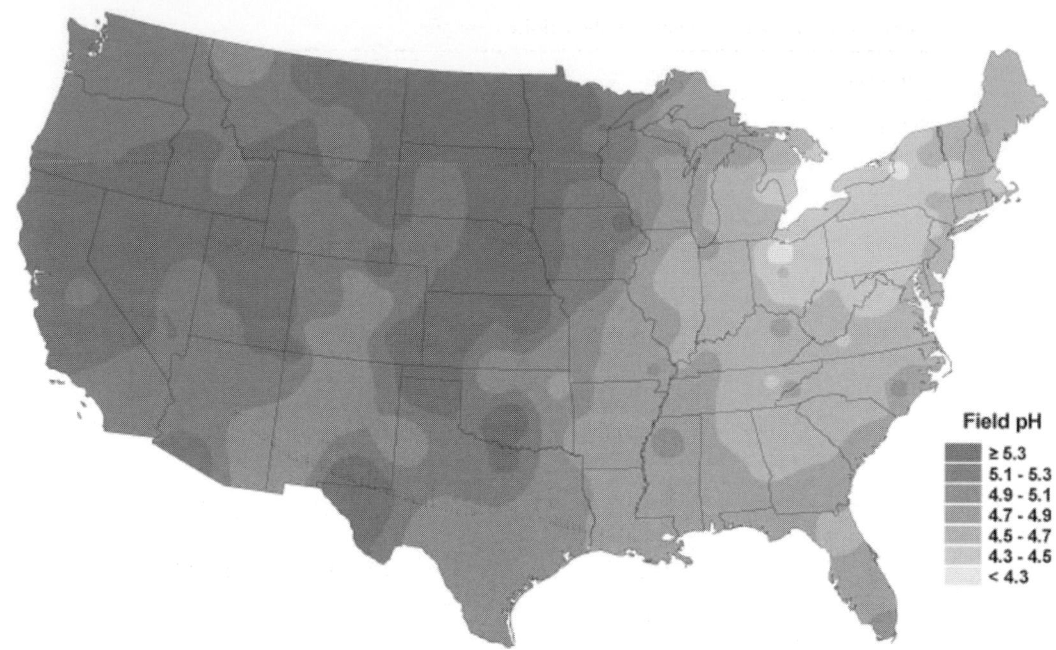

Figure A2 Distribution of acidity in the United States (National Atmospheric Deposition Program/NTN). Hydrogen ion concentration as pH from measurements made at the field laboratories, 2002.

Acid rain sources

Acid rain is primarily a result of the release of sulfur oxides (SO_2) and nitrogen oxides (NO_x) into the air by industrial and transportation sources. These oxides are transformed into sulfuric acid and nitric acid through oxidation and hydrolysis. Oxidants play a major role in several of these acid-forming processes. Rates of transformation are controlled by environmental conditions such as temperature, humidity, clouds, sunlight, and the presence of other select chemicals.

Table A1 summarizes all the major chemicals that influence the acidity of rain, with both natural and man-made sources identified. As noted above, man-made sources of SO_2 and NO_x contribute significantly to the development of acid rain. It is important to note that 95% of the elevated levels of NO_x are the result of human activities, including transportation sources (50%), electric utilities (26%), and industrial combustion (14%). The major sources of sulfur dioxide emissions were electric utilities (60%), industrial combustion (17%), and industrial processes (8%), measured in 1997.

Central to this conclusion is the question of a transport mechanism for air pollutants responsible for acid rain and the location of the source to areas that are currently experiencing acid deposition. Given that pollutants can return to earth instantaneously or remain airborne for a week or longer, then atmospheric circulation patterns and prevailing winds are significant. Examination of continental air currents demonstrates that seven states in the Midwest comprise the dominant source area for sulfur dioxide emissions that travel downwind to the Northeast (Driscoll et al., 2001). These states contributed 41% of the total national emissions of SO_2 in 1997. Furthermore, five of these states accounted for 20% of the total national emissions of NO_x.

Defining the location of actual sources contributing chemicals to acid rain is compounded by both political and economic circumstances. As mentioned above, there has been much controversy over the role of Midwestern industries and power plants using high-sulfur coal, thereby advancing acid rain in the northeastern United States. However, Phase I of the Clean Air Acts Amendment, Title IV (1995) has resulted in lower sulfate concentrations in precipitation in both the Eastern US and along the Ohio River Valley.

In 1980 the US Congress established the National Acid Precipitation Assessment Program to research the effects of acid deposition. Studies identified electric power generation as responsible for two-thirds of SO_2 emissions and one-third of NO_x emissions. Utilizing findings from these studies, Congress created the Acid Rain Program under Title IV (Acid Deposition Control) of the Clean Air Act Amendments (CAAA) of 1995. The Acid Rain Program set a goal of reducing NO_x by two million tons from 1980 levels, and requires a 50% decrease in SO_2 emissions from electric-generating facilities by 2010. By 2001, SO_2 levels have been reduced by 30% in the United States. However, nitrogen oxide emissions were not significantly affected. Research in the northeast United States has shown that acid rain is still a significant issue.

Effects of acid rain

In the 1960s and 1970s researchers in Scandinavia and the United States found that there was an increase in acidity and decrease in fish population in certain lakes and streams. Although acid rain can degrade water quality by lowering pH levels, the relationship is much more complex. Surrounding soils and vegetation contribute chemical ions that also modify acidity. Acid rain can also decrease the acid-neutralizing capacity and increase the aluminum concentrations of surface water.

Table A1 Chemicals that influence the acidity of rainfall

Ions in deposition	Source	Type	Significance
Sulfate (SO_4)	Swamps	Natural	
	Volcanoes	Natural	Minor
	Oceans	Natural	
	Power plants	Human activities	
	Industrial processes	Human activities	Major
	Smelters	Human activities	
Nitrate (NO_3)	Lightning	Natural	
	Soil biologic activity	Natural	Minor
	Industrial processes	Human activities	
	Transportation	Human activities	Major
Chloride (Cl)	Oceans	Natural	Minor
	Land surface	Natural	
	Industrial processes	Human activities	Minor
	Road salt	Human activities	
Ammonium (NH_4)	Biologic processes	Natural	
	Animals	Natural	Moderate
	Industrial processes	Human activities	
	Agriculture	Human activities	Minor
Calcium (Ca)	Land surface	Natural	Moderate
	Industrial processes	Human activities	Minor
	Agriculture	Human activities	
Sodium (Na)	Oceans	Natural	
	Land surface	Natural	Minor
	Road salt	Human activities	Minor
Magnesium (Mg)	Industrial processes	Human activities	
Potassium (K)	Land surface	Natural	Minor
	Agriculture	Human activities	

Acid rain is also thought to have a detrimental effect on forests. Trees may be injured by the direct action of acid on leaves or needles, or by changes in soil chemistry caused by the acid solution. By the early 1980s 20–25% of European forests were classified as moderately or severely damaged by acid rain. The decline of red spruce and sugar maple trees in the northeast United States has been occurring for the past four decades and has been linked to acid rain. Agricultural productivity may also be adversely impacted by acid rain.

Scientists have determined that, in soils impacted by acid rain, nutrients such as potassium, magnesium, and calcium are displaced by excess hydrogen ions, and that plant growth is retarded. Aluminum may also be released by excess hydrogen. Aluminum can be toxic to plants and, when released, can compromise the plant's ability to absorb water and nutrients. Acid rain also increases the accumulation of sulfur and nitrogen in the soil.

Buildings and monuments are also impacted by acid rain. Chemical reactions resulting from both wet and dry deposition have been recognized as a cause of extensive damage to historical buildings. Monuments such as the Acropolis in Athens and Jefferson Memorial in Washington DC show signs of damage from acid rain. Marble and limestone are susceptible due to the acids attacking the calcium carbonate in these materials. In addition to stone structures, acid rain also damages other building materials, particularly iron, steel, zinc, paint, and wood, and can cause the lifetimes of these structures to be shortened noticeably.

The chemicals comprising acid rain can also have an impact on human health, among other things. EPA has associated NO_x with photochemical smog, respiratory illness, ecosystem changes, visibility impairment, acidification of fresh-water bodies, eutrophication of estuarine and coastal waters, and increasing levels of toxins harmful to fish and other aquatic life. Likewise, high levels of SO_2 concentrations have been associated with increased respiratory difficulties.

The future

There are still many unanswered questions concerning acid rain. If acid rain proves to be as destructive as suggested by many writers, then it is certainly one of the major environmental concerns that must be dealt with in the near future.

Climatology will play an integral role in deriving many of the answers. Of importance is the transport of pollutants in the atmosphere, a problem that is associated with both meso- and macro-circulation patterns; a complete understanding of the chemistry involved in the wet deposition process; and comprehension of dry deposition processes. Additionally, politics, both national and international, will also play a role.

Melissa L. Kemling

Bibliography

Driscoll, C.T., Lawrence, G.B., and Bulger, A.J., et al., 2001. *Acid Rain Revisited: advances in scientific understanding since the passage of the 1970 and 1990 Clean Air Act Amendments*. Hubbard Brook Research Foundation. Science Links Publication, Vol. 1, no. 1.

Lynch, J.A., Bowersox, V.C., and Grimm, J.W., 1996. *Trends in Precipitation Chemistry in the United States, 1983–94: an analysis of the effects in 1995 of phase I of the Clean Air Act Amendments of 1990*, Title IV, US Geological Survey, Open File Report 96-0346.

National Atmospheric Deposition Program (NRSP-3)/National Trends Network, 2002. NADP Program Office, Illinois State Water Survey, 2204 Griffith Dr., Champaign, IL 61820.

Silver, C.S., and DeFries, R.S., 1990. *One Earth, One Future: our changing global environment*. Washington, DC: National Academy Press, pp. 131–144.

US EPA, 1999. *Progress Report on the EPA Acid Rain Program*. EPA 430-R-99-011.

US EPA, 2002. *Clearing the Air: the facts about capping and trading emissions*. EPA 430-F-02-009. Office of Air and Radiation. Clean Air Markets Division (6204N).

Cross-references

Aerosols
Air Pollution Climatology
Atmospheric Nuclei and Dust
Climatic Hazards
Precipitation

ADIABATIC PHENOMENA

An adiabatic process is one in which the system being considered does not exchange heat with its environment. The most common atmospheric adiabatic phenomena are those involving the change of air temperature due to change of pressure. If an air mass has its pressure decreased, it will expand and do mechanical work on the surrounding air. If no heat is taken from the surroundings, the energy required to do work is taken from the heat energy of the air mass, resulting in a temperature decrease. When pressure is increased, the work done on the air mass appears as heat, causing its temperature to rise.

The rates of adiabatic heating and cooling in the atmosphere are described as lapse rates and are expressed as the change of temperature with height. The adiabatic lapse rate for dry air is very nearly 1°C per 100 m. If condensation occurs in the air parcel, latent heat is released, thereby modifying the rate of temperature change. This retarded rate is called the pseudo-adiabatic lapse rate; it is not a constant for its value depends on the temperature at which the process takes place and the amount of water vapor in the air mass. However, for general descriptive purposes it is assumed as 0.5°C per 100 m.

The usefulness of the adiabatic approximation has led to the development of special thermodynamic diagrams that enable the results of adiabatic processes to be determined graphically. These charts are characterized by having two thermodynamic variables as coordinates and the values of various others represented by isopleths. Well-known examples of these "adiabatic diagrams" are the Tephigram, Stuve diagram, and Skew-T diagram.

Large-scale atmospheric motions are approximately adiabatic, and clouds and snow or rain associated with them are primarily adiabatic phenomena in that they result from cooling air associated with decreasing pressure of upward air motion. Simpler adiabatic phenomena occur on a smaller scale. A common example is that of rising "bubbles" of air on a warm day, leading to cumulus cloud forms. The growth of such cumulus clouds into thunderclouds is more complex but still a largely adiabatic phenomenon.

An example of the reverse effect, heating due to an increase in pressure, is the warm Foehn or Chinook wind that results from the descent of air from higher levels.

Rhodes W. Fairbridge

Bibliography

Cole, F.W., 1980. *Introduction to Meteorology*, 3rd edn. New York: Wiley.

McIlveen, R., 1992. *Fundamentals of Weather and Climate.* London: Chapman & Hall.

Cross-references

Boundary Layer Climatology
Cloud Climatology
Lapse Rate
Local Winds

AEROSOLS

Aerosols are technically defined as a relatively stable dispersion of solid or liquid particles in a gas. However, in common usage in the atmospheric sciences the term refers to the particles themselves. Aerosols are involved in a number of important atmospheric processes. These include atmospheric electricity; visibility; cloud and precipitation formation, including acidity of the precipitation; atmospheric biogeochemical cycles; and the radiative properties of the atmosphere. In all of these processes the size distribution of the aerosols is particularly important, as is their chemical composition.

Atmospheric aerosols are generated by a number of natural processes and as a result of human activities. Aerosols may be classified as "primary" or "secondary", depending on their production process. Primary aerosols are particles that are injected into the air directly, often by some mechanical process such as wind erosion of land surfaces (mineral aerosol), or wave breaking (sea salt aerosol). Secondary aerosols result from gas-to-particle conversion processes within the atmosphere that generate new particles. Table A2 presents the estimated source strengths for different types of primary and secondary aerosols. In terms of mass, sea salt and mineral aerosols (or dust) dominate the primary aerosols and also total global aerosol production. Industrial dust, various combustion processes and direct injection of particles from vegetation are also significant primary aerosol sources. The most significant secondary aerosols include sulfate, nitrate and organic particles, all resulting from the reaction of gaseous precursors in the atmosphere.

Table A2 Sources and radiative forcing of aerosols[a]

Aerosol type	Source strength (Tg/year)	Anthropogenic fraction (%)	Fine fraction (%)	Global mean direct radiative forcing[b] ($W\,m^{-2}$)
Primary				
Sea salt	3300	0	<1	—
Mineral dust	2200	~50	20	−0.60 to +0.40
Biomass burning	60	>90	~100	−0.20
Fossil fuel burning	35	100	~100	−0.10 or +0.20[c]
Biogenic	55	0	small	—
Industrial dust	100	100	~100	—
Total Primary	*~5800*	*~20*	*~10*	
Secondary				
Sulfate	200	~60	~100	−0.40
Nitrate	18	~80	~100	—
Organic compounds	17	~4	~100	—
Total Secondary	*~235*	*~60*	*~100*	
Global total	*~6000*	*~25*	*~15*	

[a] Numbers are derived from Houghton et al. (2001); 1 Tg = 1×10^{12} g.
[b] Values have uncertainties of a factor of 2 to 3. The global mean total *indirect* aerosol radiative forcing is estimated at 0 to $-2.0\,W\,m^{-2}$.
[c] Organic carbon is −0.10 and black carbon is +0.20.

Table A2 also provides information on the size of the particles from each of these sources. Typically aerosols are categorized according to size as "coarse (generally greater than ~1 μm radius) and "fine" (generally less than ~1 μm radius). Essentially all of the aerosol mass for secondary aerosols is in the fine category, whereas most of the mass of the primary aerosols is in the "coarse" category. The "fine" particles in turn have two components: the nuclei mode, from about 0.005 to 0.1 μm radius; and the accumulation mode, from about 0.1 to 1.0 μm radius. Particles in the nuclei mode are formed from the nucleation of atmospheric gases. The accumulation mode particles result from the coagulation of nuclei mode particles as well as from some condensation of gases onto existing particle surfaces.

The "fine" and "coarse" aerosols not only have very different sources, they are transformed differently, they have different chemical composition and optical properties, and they are removed from the atmosphere by different processes. For example, size has a significant impact on the atmospheric lifetimes of "coarse" and "fine" aerosols. The nuclei mode fraction of the "fine" aerosols has relatively short lifetimes (roughly hours or less) within that size range due to their rapid growth via coagulation with themselves or other particles into the larger accumulation mode aerosols. Accumulation mode aerosols can have atmospheric lifetimes of days to a week or more and can thus be transported great distances in the atmosphere. The primary removal mechanism for the accumulation mode aerosols is by precipitation. The "coarse" aerosols generally have atmospheric lifetimes of from hours to a day or two, with the lifetime decreasing as the particle size increases. Direct dry deposition is important for the coarse aerosol, and this becomes increasingly important as the particle size increases. Even "coarse" aerosols can be transported great distances, however. There have been numerous occasions when mineral dust from the Chinese deserts has been transported in the atmosphere to the east across the Pacific Ocean and North America to the Atlantic Ocean. Similarly, mineral dust from the Sahara Desert in Africa is often carried west across the tropical North Atlantic Ocean and Caribbean Sea to the eastern Pacific Ocean. Similar transoceanic transport of pollution-derived aerosols has also been observed on numerous occasions. Atmospheric sea salt aerosols from oceanic regions are also often observed in mid-continent locations.

The data in Table A2 indicate that about 20% of the primary aerosols and 60% of the secondary aerosols are derived from human-related, or anthropogenic, sources. This is particularly important for sulfate aerosols, which are often primarily responsible for visibility degradation and precipitation acidity in urban and near-urban regions. The predominant sources for the gaseous precursors for the secondary sulfate aerosols are industrial emissions and the combustion of fossil fuels. Similarly, fossil fuel combustion is the major source for the gaseous precursors of nitrate aerosols. Contrary to earlier estimates, anthropogenic sources do not appear to be a significant source of organic aerosols.

Table A3 provides information on typical number and mass concentrations of aerosols in different regions. In urban areas particle number concentrations of millions per cubic centimeter are common, whereas in the remote marine atmosphere concentrations can be less than 100 particles per cubic centimeter. In urban areas the mass concentrations can range up to several hundred micrograms per cubic meter, whereas under some

Table A3 Typical properties of aerosols in different regions*

Region	Total number (cm^{-3})	Mass <1 μm ($\mu g\,m^{-3}$)	Mass 1–10 μm ($\mu g\,m^{-3}$)
Urban (pollution)	1×10^5–4×10^6	30–150	70–150
Rural	2×10^3–1×10^4	3–8	5–30
Remote continental	1×10^2–1×10^4	0.5–3	1.5–7
Remote marine	1×10^2–4×10^2	1–4	10–35

* Derived from Seinfeld and Pandis (1998).

conditions in remote continental regions the mass concentrations can be as low as one microgram per cubic meter or less. In areas near major dust storms concentrations of 1 gram per cubic meter or more have been observed.

Aerosols have both a direct and an indirect role in the radiative forcing of climate. The direct forcing results because they can absorb and scatter both infrared and solar radiation in the atmosphere. Indirect forcing occurs because aerosols can affect and change the processes that control cloud and precipitation formation, which in turn can affect the radiative properties of the atmosphere. There has been strong interest in aerosol forcing. This is because the forcing is negative for most aerosol types, rather than the positive forcing associated with the major radiatively active trace gases, such as carbon dioxide, methane, nitrous oxide, halocarbons and tropospheric ozone. Once again, size distribution and chemical composition are critical factors in the efficiency of aerosols in affecting these forcings. While significant progress has been made in recent years in determining the importance of aerosols of various types in radiative forcing, the uncertainties are still quite large. The most recent Intergovernmental Panel on Climate Change (IPCC, Houghton et al., 2001) report has developed new estimates of aerosol forcing, and some of their results are shown in Table A2. Note that the numbers given represent the mean direct radiative forcing for the aerosols described (except for mineral dust, where only a range was given), but the uncertainties in these estimates range from a factor of 2 to 3. Aerosol types where the evidence shows negative forcing include sulfate aerosols, organic carbon aerosols, and aerosols from biomass burning. The sign of the net forcing for mineral aerosols is still uncertain. Black carbon aerosols appear to cause a positive forcing. Note also that the indirect forcing is estimated to be from 0 to $-2.0\,W\,m^{-2}$. The values for aerosol forcing in Table A2 can be compared with the best estimate for carbon dioxide forcing of $+1.46\,W\,m^{-2}$.

Robert A. Duce

Bibliography

Arimoto, R., 2000. Eolian dust and climate: relationships to sources, tropospheric chemistry, transport and deposition. *Earth Science Review*, **54**: 29–42.

Charlson, R. and Heintzenberg, J., eds., 1995. *Aerosol Forcing of Climate*. London: John Wiley & Sons.

Houghton, J.T., Ding, Y., Griggs, D.J., et al., 2001. *Climate Change 2001: the scientific basis* (The IPCC Third Assessment Report). Cambridge: Cambridge University Press.

Murphy, D.M., Anderson, J.R. et al., 1998. Influence of sea-salt on aerosol radiative properties in the Southern Ocean marine boundary layer, *Nature*, **392**: 62–65.

National Research Council, 1993. *Protecting Visibility in National Parks and Wilderness Areas*. Washington: NRC Press.

Seinfeld, J.H., and Pandis, S.N. 1998. *Atmospheric Chemistry and Physics*. New York: John Wiley & Sons.

Cross-references

Air Pollution Climatology
Albedo and Reflectivity
Cloud Climatology

AFRICA: CLIMATE OF

Africa covers an area of more than 30 million km^2 and is second in size only to Asia. Of all continents it is the most symmetrically located with regard to the equator, and this is reflected in its climatic zonation. The coastline is remarkably smooth and the continent has been called a giant plateau, since there is a relative absence of very pronounced topography, although some high mountains exist, especially in the East African region (Kilimanjaro, 5894 m; Mount Kenya, 5199 m; and the Ruwenzoris, 5120 m). Lake Victoria, astride the equator, covers an area of 70 000 km^2 and is exceeded in size only by Lake Superior among the world's fresh water lakes.

We will note below how certain evidence can be used to reconstruct the early climate of Africa, but climatic observations really only began with the European explorers of the late eighteenth and nineteenth centuries. Then, in the last few decades of the nineteenth century, meteorological services were formed that began systematic observations at a network of stations. In most cases the meteorological service followed the meteorological practices of the colonial or governing power and this characteristic has tended to persist, even after independence was obtained. A number of the countries have suffered from internal disturbance since independence, a fact often leading to a hiatus in the records. Nevertheless, the standard of observation generally has remained high at the first-order or synoptic stations, but care must be taken when using data from many of the cooperative or second-order stations.

Weather controls

As in other regions of the globe, the pattern of solar radiation is the major control of climate. However, the nature of the air–surface interface and topography also play important roles. Although ocean currents help to determine the climate of some narrow bands of land, it is by appreciating the nature of the air masses reaching a region that one can begin to understand the observed climatic pattern.

Most of the time the continent is affected by tropical air masses, often maritime (moist) in nature, but in certain areas and during certain months they can be of continental (dry) origin. At the extreme latitudinal boundaries of the continent, in the littoral region of the Mediterranean Sea, and in the area around and east of Cape Town, the effects of polar air masses cannot be ignored at the time of low sun. The terms high sun (summer) and low sun (winter) are often used when discussing reasonal variation in the topics.

Temperature, at a particular station, is a rather conservative element with a relatively small annual range, and wind speeds are normally low compared to areas in higher latitudes. Precipitation, mostly rainfall, is the significant feature of the African climate. For rainfall to occur two criteria must be met – an adequate amount of water vapor within the atmosphere and the initiation of a cooling mechanism.

The cooling mechanism is usually obtained through the ascent of a large parcel of air. Such uplifting is generally due either to topography or to horizontal convergence of the air, which is simply the coming together of air parcels or masses. The ways in which such horizontal convergence can occur over Africa are detailed in Johnson and Morth (1960).

There are four important phenomena that determine rainfall amounts and patterns over the tropical continents: (1) the intertropical convergence zone (ITCZ), (2) the equatorial trough (ET), (3) easterly waves, and (4) tropical cyclones. The latter two play relatively minor roles over Africa.

The ITCZ, defined as a surface discontinuity separating the trade winds of the two hemispheres, can be identified readily on climatic charts, but it is not easily found on daily weather maps. The confluence of convergence of the usually relatively moist air masses leads to rainfall patterns that reflect the seasonal migration of the sun, with a time lag.

The equatorial trough (ET), the zonal pressure minimum, is detected up to an average height of 500 mbar (5500 m) with a mean position near the equator at that height. At lower levels there is evidence of a pronounced shift in location with season.

Weather situations

To set the stage for an appreciation of the various climatic patterns experienced, it is helpful to understand the weather situations dominant during certain months.

January

A broad low-pressure region is noted north of the equator with only light winds in evidence. In the upper air the divergent northeasterlies act to suppress rainfall. The surface position of the ET is north of the rain belt in Central Africa whereas, in the southern sector, upper-level troughs cause heavy rains over the Angolan plateau. Frontal activity brings rain to the North African coast.

April

There has been a movement of the ET northward from the January situation. In West Africa the ET becomes identified more easily on the daily surface charts. Thompson (1965) considers the rainfall now to be the result of many complex interplays among synoptic processes and dismisses the concept of a continuous zonal belt of rainfall moving northward.

July

The position of the surface ET is now at about its furthest north, near 20°N. There is a meridional (longitudinal) pressure gradient extending from the high-pressure belt of the southern hemisphere to the intense heat lows of Arabia and North Africa. In East Africa the topography leads to periods of convergence above the 700 mbar (3000 m) level, giving rise to the wettest month, while there is subsidence at 850 mbar (1500 m), where little rainfall is reported.

October

The rain belt is now moving southward, while a new trough begins to develop over Somalia and the Arabian Sea. Like April, this is a transitional month between the extremes of January and July.

Continental patterns of important climatic elements

The best method of identifying analogous climatic zones is to consider aspects of each of the important elements separately and then to combine them to obtain the overall picture. The elements selected here are temperature, precipitation, humidity and radiation.

Temperature

The mean annual temperature range (MATR), the difference between the mean temperatures of the hottest and coldest months, is of small magnitude over most of the continent, being less than 6°C over about half the continent. Its minimum value is 1.4°C at Barumbu in northernmost Zaire whereas the greatest is 23–24°C in parts of the Algerian Sahara. The dependence of the MATR on the continentality of the station, as well as its latitude, is shown in Table A4.

The mean annual diurnal temperature range is extremely dependent on continentality, as shown in Figure A3. Nearly all the coastal regions exhibit values of below 10°C, whereas in the central Sahara the range reaches 20°C, one of the highest values for any region of the world.

Actually, the best measure of the temperature variation is the highest mean monthly maximum temperature (H) and the lowest mean monthly minimum temperature (L). The patterns of these two variables are given in Figures A4 and A5. Figure A4 shows that it is only north of the equator where values exceeded 35°C and only in parts of the foggy coastal strip of southwestern Africa where values less than 20°C were reported. In Figure A5 the effects of elevation and latitude are more evident than those of continentality. Values below 5°C are unusual and only at high altitudes (over 1000–1500 m) in Algeria, Morocco

Table A4 Mean annual temperature range (MATR)

	Latitude E	Longitude N	MATR (°C)
Port Harcourt	7°01′	4°46′	2.5
Lokoja	6°44′	7°48′	3.6
Kano	8°32′	12°32′	5.8
Zinder	9°00′	13°48′	9.4
Agadez	7°59′	16°59′	13.3
Tamanrasset	5°31′	22°42′	16.7
Ourgla	5°20′	31°54′	23.3
Biskra	5°44′	34°51′	22.7
Constantine	6°37′	36°22′	18.3
Philippeville	6°54′	36°52′	14.2

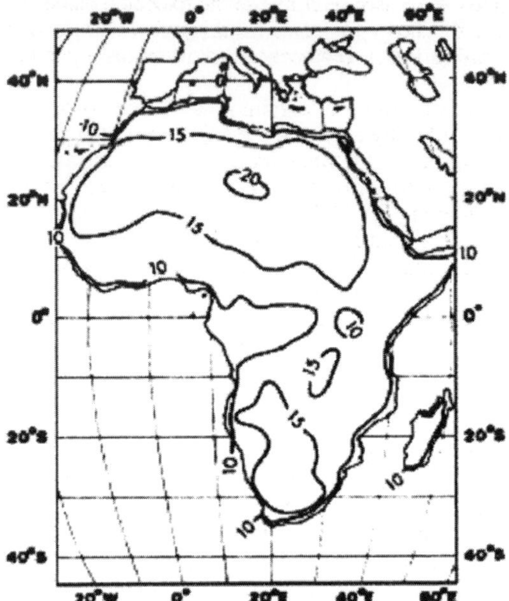

Figure A3 The mean diurnal temperature range (°C).

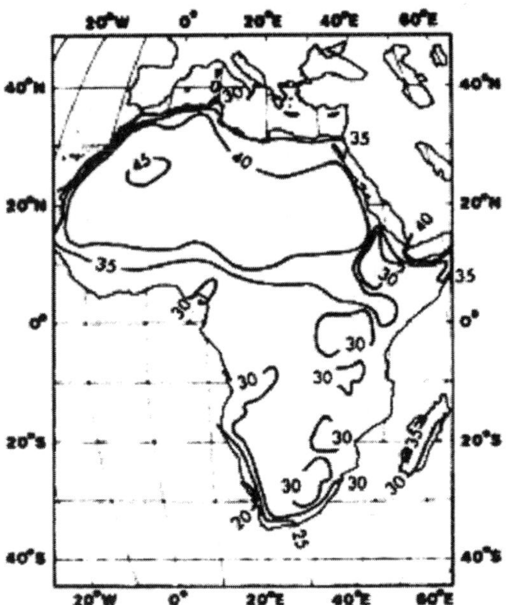

Figure A4 The highest mean monthly temperature (°C).

and South Africa does L go below 0°C. The largest value is 26°C noted at Dallol, Ethiopia. The mean annual temperature variation (MATV), defined as H − L, which is depicted in Figure A6, is a combined measure of both annual and diurnal range and shows a relationship with both latitude and continentality. Again, the maximum values (above 30°C) occur in drier areas, the greatest being at Adrar in western Algeria, with 42°C. As would be expected, the equatorial littoral yields the lowest values, reaching only 8°C in Liberia and Sierra Leone.

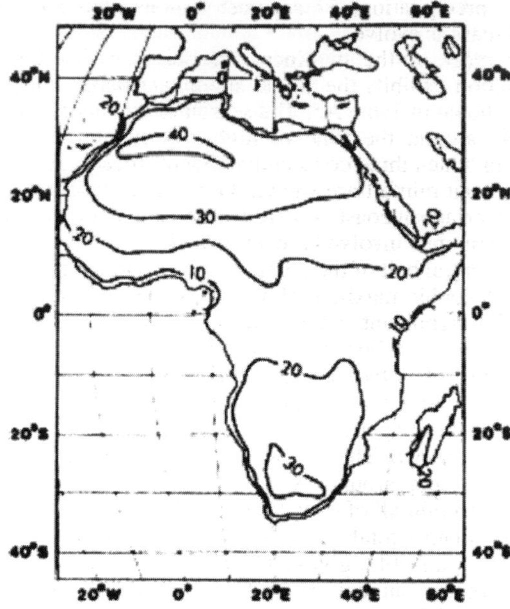

Figure A5 The lowest mean monthly temperature (°C).

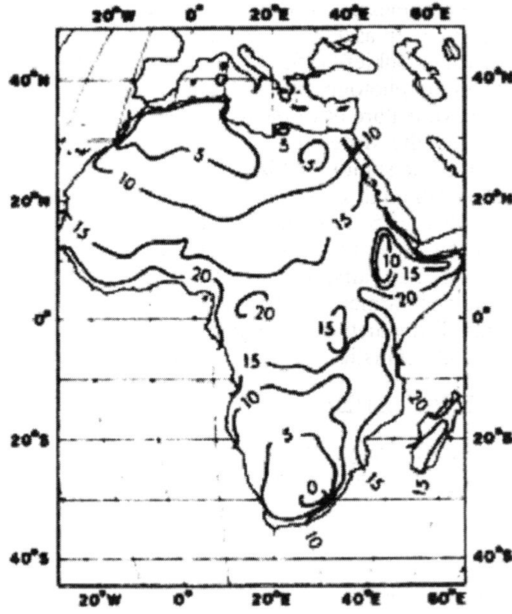

Figure A6 The difference between the highest mean monthly maximum temperature and the lowest mean monthly temperature.

Precipitation

Because both the ITCZ and the polar front lows exhibit large spatial movements, a basic seasonal pattern of precipitation can be identified on the continent. However, the complexities introduced by topography, upper-air conditions, ocean currents and inland lakes, among others, make the detailed pattern extremely complicated. An example of this is given by Griffiths (1972) for East Africa, whereby overlaying the spatial patterns of mean monthly precipitation, 52 separate regions with 30 different rainfall seasons evolve.

Over most of the continent the seasonal distribution of precipitation exhibits the single significant maximum pattern, such as shown in Table A5. The season of maximum amount is generally around the time of high sun. In Figure A6 those regions in which three consecutive months receive at least 50% of the annual rainfall are shown. Only in the eastern sector of the Mediterranean coast is there a maximum at the time of low sun, but amounts involved are very small.

In the central belt of the continent most stations exhibit some degree of double maxima. However, the areas in which there is a really significant double swing during the year are quite small (Figure A7). For this illustration, significance is defined as occurring when the difference between the secondary maximum and secondary minimum exceeds 5% of the annual mean (see example of Lagos in Table A5), and this criterion limits the regions to just two. The sector in the Horn of Africa is mostly semiarid, except around Nairobi, Kenya. For stations with annual mean rainfall of over 1000 mm, Kitui, Kenya, is unique; its mean monthly totals (mm) being 41, 24, 118, 244, 56, 5, 3, 5, 0, 82, 304 and 143, giving a 22% swing.

The average annual rainfall totals show a wide range (roughly 0–10 000 mm), exhibiting a decrease away from the equatorial regions to reach a minimum around 20–30° latitude, then showing a slight increase (Figure A8). Since snow and hail amounts are generally small, all precipitation can be considered as rain. Nevertheless, snowfalls have been recorded in the Sahara, as far south as 15°N. Some falls have been quite heavy and reference to Dubief (1959, 1963) will give fuller details and some interesting photographs.

A distinctive feature of tropical rainfall is its large variability, interpreted as the difference within monthly, seasonal and annual totals. The station of Makindu, Kenya, is outstanding in this respect. Although its mean annual value is 610 mm, it has recorded as low as 67 mm and as high as 1964 mm of precipitation. On the other hand, April, its wettest month (111 mm average), has had amounts ranging from 822 mm, which exceeds the annual mean, to 0 mm. In Figure A9 a measure of annual rainfall fluctuations is depicted. Use is made of the relative variability statistic, V_r, defined as mean deviation/mean:

$$V_r = \Sigma(X_i - \overline{X})/\Sigma X_i$$

where X_i is individual yearly amounts and $\overline{X}$ is the yearly mean. Values of V_r show dependence on the mean, $\overline{X}$, so data for 500 stations were used to compute the expected value of V_r as a function of $\overline{X}$, called $V_r(\overline{X})$, and comparing V_r for the station with its corresponding $V_r(\overline{X})$. Differences are given as a percentage of $V_r(\overline{X})$.

The great variability in certain areas of the continent can be illustrated further by two examples. Quseir, Egypt, has received 33 mm in a day – 11 times its average annual total; Lobito, Angola, had 536 mm of rain in one day – over 1.5 times its annual average fall of 330 mm.

Hail is not a common phenomenon on the continent, especially on the coast of the tropical regions. However, Maputo, Mozambique, experienced a very heavy fall in October 1977 that did considerable damage. Few places have more than five incidences annually, but a region around Kericho, Kenya, reports as many as 80 hailstorms per year. (Fresby and Sansom, 1967)

Thunderstorm days are frequent with over 20% of the continent reporting in excess of 100 annually. This band of 100 occurrences stretches from about Sierra Leone across the central area as far as Lakes Victoria and Malawi. There are a few locations where convective instability leads to annual values of more than 200, with Kampala, Uganda, 242; Bukavu, Zaire Republic, 221; and Calabar, Nigeria, 216 holding the top places.

Humidity

Relative humidity, as an expression of the atmospheric moisture condition, can be rather misleading because its impact on human comfort is dependent on the air temperature occurring at the same time. For this reason it is preferable to use the dew point temperature as an indicator since this shows little diurnal variation and can be related more readily to human comfort. Values in excess of around 21°C can be considered very sultry, and this isopleth is indicated by a thick line in Figures A10 and A11. Some scientists consider dew points above 18°C uncomfortable. With this threshold about 30% of the continent falls into this category in January and 25% in July.

Along the humid and hot coastal regions of Africa the trade winds and/or sea breezes provide reasonably comfortable conditions, contrary to most people's concepts of the humid tropics. When there is little wind, as is often the case inland, in cities or in wooded areas, the situation is quite enervating. For a good discussion of and information on human comfort conditions consult Terjung (1967).

Table A5 Seasonal distribution of precipitation

	Jan.	Feb.	Mar.	Apr.	May	June	July	Aug.	Sept.	Oct.	Nov.	Dec.	Annual
Single maximum													
Algiers, Algeria	116	76	57	65	36	14	2	4	27	84	93	117	641
Kano, Nigeria	0	1	2	8	71	119	209	311	137	14	1	0	873
Mbeya, Tanzania	199	165	161	116	17	1	1	1	3	15	52	152	883
Pretoria, South Africa	117	101	78	46	25	9	8	6	25	63	110	120	708
Wau, Sudan	0	4	20	69	132	170	199	234	179	130	8	0	1145
Double maxima													
Lagos, Nigeria	40	57	100	115	215	336	150	59	214	222	77	41	1625

Note: Double maxima significance (222 − 59)/1625 = 10%. See text.

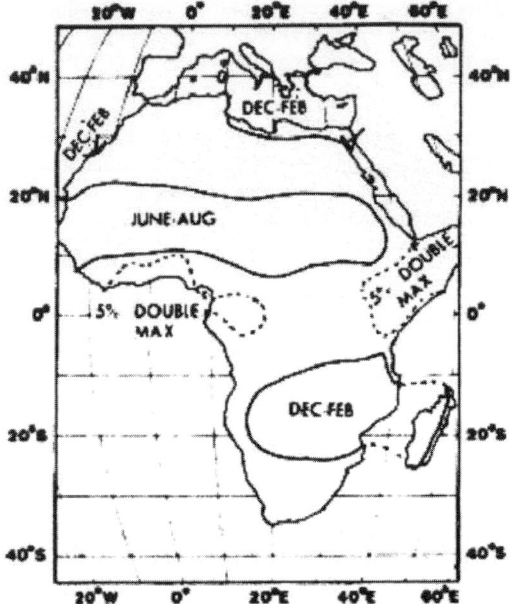

Figure A7 The 3-month period of maximum precipitation and those areas with a significant double maximum distribution.

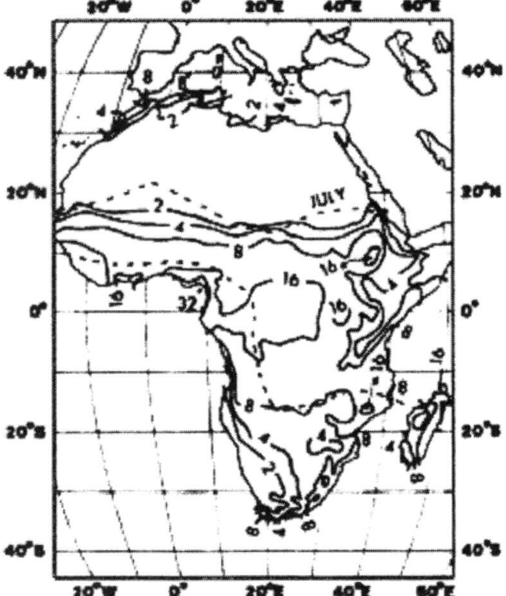

Figure A8 Mean annual precipitation (mm).

Radiation

Africa extends from 38°N to 35°S, so that the annual fluctuation of solar radiation at the top of the atmosphere is small compared with that in higher latitudes. Mean annual global radiation (solar radiation measured on a horizontal plane at the surface) varies from nearly 600 ly day^{-1} in the Sahara–Nubia area to something less than 400 ly day^{-1} around Gabon, the Algerian coast and East London, South Africa.

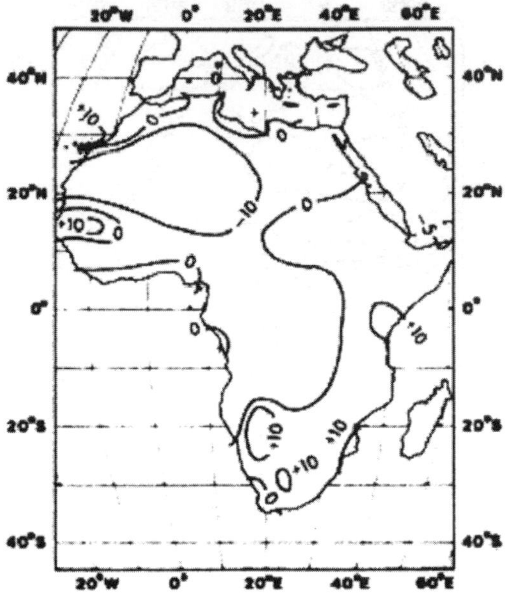

Figure A9 Variation of annual precipitation values in per cent (see text for details).

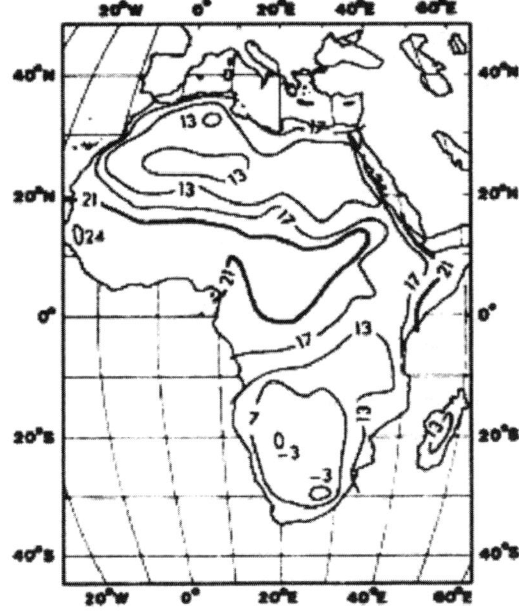

Figure A10 Mean January dewpoint temperature (°C).

Climatic zones

Using the findings of the earlier sections, it is possible to identify eight important climatic zones in Africa: (1) tropical wet; (2) tropical, short dry spell; (3) tropical, long dry spell; (4) tropical desert; (5) tropical highland; (6) subtropical desert; (7) subtropical, summer rain; and (8) subtropical, winter rain.

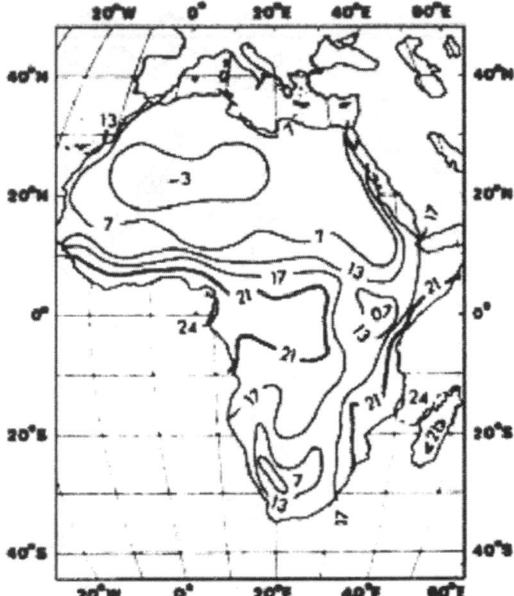

Figure A11 Mean July dewpoint temperature (°C).

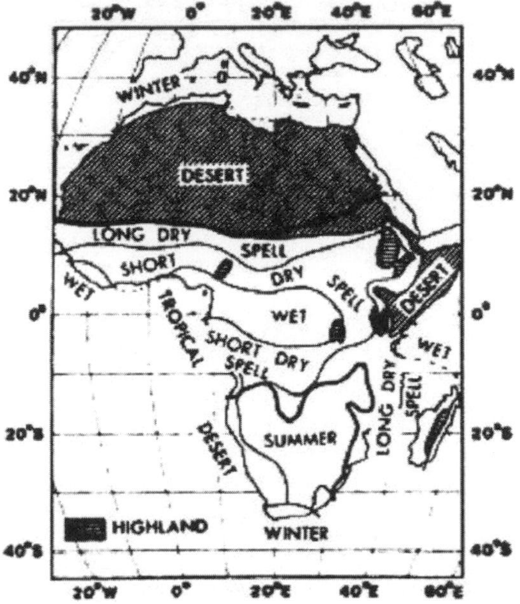

Figure A12 A simple climatic classification. Thick lines (N and S) indicate tropical boundaries (see text for details).

In addition, smaller zones of subtropical, uniform rain and subtropical highland can be found. The eight major zones are shown in Figure A12. For these purposes "tropical" designates that the mean temperature of each month is 18°C or greater; "desert" occurs when the mean annual rainfall (cm) is less than $16 + 0.9\overline{T}$, where $\overline{T}$ is the mean annual temperature (°C), and "highland" is where the altitude causes the region to be classified in a different thermal zone from what it would be if at sea level.

The tropical wet climate (Kisangani, Table A6) exhibits some rain in all months. Temperatures are uniformly high all year round and the conditions are very enervating, although sea breezes and/or trade winds can reduce the stress along the coast. The tropical rainforest is found in abundance in this zone.

Surrounding the first zone is the tropical, short dry spell climate (Kinshasa, Table A6). Here a period of 3–5 dry months is experienced. Precipitation and temperature are still high, but the annual temperature range tends to be larger than in the wet climate. Vegetation changes from forest near the boundary with the previous zone to deciduous woodland on the drier side although, because this is an important climate for agriculture, much clearing has taken place.

The tropical, long dry spell climate (Niamey, Table A6) is on the equatorial side of the desert regions and has low rainfall for at least 6 months. Rainfall amounts are less than in the two zones discussed above and temperatures show a much larger seasonal swing. The area is susceptible to drought and at such times the often marginal agriculture suffers tremendously. Vegetation is normally savanna and scrubland. The northern belt is referred to as the Sahel, a region in which famine has afflicted millions of inhabitants. The extreme variability of rainfall amount and frequency is a characteristic of the zone (Todorov, 1984).

Tropical desert climates (Obbia, Table A6) are not common, the biggest region being in the Horn of Africa where the prevailing winds, NE at low sun and SW at high sun, ensure that very few moist air masses reach the area.

The tropical highlands climate (Nairobi, Table A6) offers relief from the tropical heat, as well as a decrease in absolute humidity. Precipitation amounts can change quite rapidly in short distances as exposure to prevailing winds plays a dominant role. Generally, there is an increase in annual amount with height up to a belt of maximum rainfall, often around 2000 m or more, but changing according to the direction of slope. If the elevation is high enough (over about 5500 m) the region is permanently snowcapped. This transition from sea level to snowfield means that many vegetation belts are identifiable on the slopes.

The subtropical desert climate (Wadi Halfa, Table A6) is the most extensive of all zones on the continent. Summer temperatures in the Sahara are among the highest in the world, although they are not quite as great as in Namibia. Due to the low relative humidity and clear skies, diurnal temperature ranges can be extreme, with values in excess of 20°C often being reported. As may be expected, radiation and sunshine amounts are extremely large. Vegetation, while sparse, springs to life after any brief shower.

The subtropical, summer rain climate (Harare, Table A6) is found mainly in the southern plateau. Precipitation usually is so concentrated that about half the annual total falls in 3 months. Winters are generally very pleasant and comfortable.

The subtropical winter rain (or Mediterranean) climate (Cape Town, Table A6) is found at the extremities of the continent. These areas can experience extremely hot and dusty winds in summer from their adjacent deserts, but from fall to spring conditions are ideal. Vegetation is xerophytic, able to withstand the long dry spell.

Table A7 lists some climatic extremes for the continent. It is interesting to note how many of these are also world record extremes. Even in the precipitation class only two stations, Waialeale, Hawaii, with 1455 mm and Cherrapunji, India, with 10 820 mm, exceed Ureka's total. At another site near Cherrapunji a 5-year mean of 12 650 mm has been reported.

Table A6 Monthly temperature and precipitation data for representative stations

	Jan.	Feb.	Mar.	Apr.	May	June	July	Aug.	Sept.	Oct.	Nov.	Dec.	Average temperature or total precipitation for year
Kisangani, Congo, D.R.: 0°26′N, 25°14′E, 410m													
Mean maximum temperature (°C)	31.1	31.1	31.1	31.1	30.6	30.0	28.9	28.3	29.4	30.0	29.4	30.0	30.0
Mean minimum temperature (°C)	20.6	20.6	20.6	21.1	20.6	20.6	19.4	20.0	20.0	20.0	20.0	20.0	20.6
Precipitation (mm)	53	84	178	157	137	114	132	165	183	218	198	84	1703
Kinshasa, Congo, D.R.: 4°20′S, 15°18′E, 324m													
Mean maximum temperature (°C)	30.6	31.1	31.7	31.7	31.1	28.9	27.2	28.9	30.6	31.1	30.6	30.0	30.0
Mean minimum temperature (°C)	21.1	21.7	21.7	21.7	21.7	19.4	17.8	18.4	20.0	21.1	21.7	21.1	20.6
Precipitation (mm)	135	145	196	196	157	8	3	3	30	119	221	142	1355
Niamey, Niger: 13°31′N, 2°06′E, 215 m													
Mean maximum temperature (°C)	33.9	36.7	40.6	42.2	41.1	38.3	34.4	31.7	33.9	38.3	38.3	34.4	36.7
Mean minimum temperature (°C)	14.4	17.2	21.7	25.0	26.7	25.0	23.3	22.8	22.8	23.3	18.3	15.0	21.1
Precipitation (mm)	0	2	5	8	33	81	132	183	91	13	1	0	549
Obbia, Somalia: 5°20′N, 48°31′E, 15 m													
Mean maximum temperature (°C)	29.4	30.6	32.2	33.9	31.7	29.4	28.3	28.9	39.4	30.0	31.7	30.6	30.6
Mean minimum temperature (°C)	22.2	23.3	24.4	25.5	25.0	23.9	22.2	22.2	22.8	23.3	23.3	22.8	23.3
Precipitation (mm)	12	0	8	21	33	0	1	1	2	38	25	25	166
Nairobi, Kenya: 1°16′S, 36°48′E, 1820m													
Mean maximum temperature (°C)	25.0	26.1	25.0	23.9	22.2	21.1	20.6	21.1	23.9	24.4	23.3	23.3	23.3
Mean minimum temperature (°C)	12.2	12.8	13.9	14.4	13.3	11.7	10.6	11.1	11.1	12.8	13.3	12.8	12.8
Precipitation (mm)	38	64	124	410	157	46	15	23	30	53	109	86	1155
Wadi Halfa, Sudan: 21°55′N, 31°20′E, 125m													
Mean maximum temperature (°C)	23.9	26.1	31.1	36.7	40.0	41.1	41.1	40.6	38.3	36.7	30.6	25.6	34.4
Mean minimum temperature (°C)	7.8	8.9	12.2	16.7	21.1	23.3	23.3	23.9	22.2	19.4	14.4	9.4	16.7
Precipitation (mm)	T[a]	T	T	T	T	0	T	T	T	T	T	0	1
Harare, Zimbabwe: 17°50′S, 31°08′E, 1403m													
Mean maximum temperature (°C)	25.6	25.6	25.6	25.6	23.3	21.1	21.1	23.3	26.1	28.3	27.2	26.1	25.0
Mean minimum temperature (°C)	15.6	15.6	14.4	12.8	9.4	6.7	6.7	8.3	11.7	14.4	15.6	15.6	12.2
Precipitation (mm)	196	178	117	28	13	2	1	2	5	28	97	163	828
Cape Town, South Africa: 33°54′S, 18°32′E, 17m													
Mean maximum temperature (°C)	25.6	26.1	25.0	22.2	19.4	18.3	17.2	17.8	18.3	21.1	22.8	24.4	21.7
Mean minimum temperature (°C)	15.6	15.6	14.4	11.7	9.4	7.8	7.2	7.8	9.4	11.1	12.8	14.4	11.7
Precipitation (mm)	15	8	18	48	79	84	89	66	43	30	18	10	508

[a] Trace.

Table A7 Some climatic extremes for Africa

Temperature		
Absolute maximum	58°C[a]	Azizia, Libya (13 Sept. 1922)
Highest mean monthly maximum	47°C[a]	Bou-Bernous, Algeria (July)
Highest mean monthly	39°C[a]	Bou-Bernous, Algeria (July)
Highest mean annual	35°C[a]	Dallol, Ethiopia
Highest mean monthly minimum	32°C[a]	Dallol, Ethiopia
Highest mean of coldest month	31°C[a]	Dallol, Ethiopia
Highest absolute minimum	21°C[a]	Dallol, Ethiopia
Absolute minimum	−24°C (11°F)	Ifrane, Morocco (11 February 1935)
Precipitation		
Highest mean annual	10 450 mm	Ureka, Equat, Guinea
	10 300 mm	Debundscha, Cameroons
Lowest mean annual	0.5 mm	Wadi Halfa, Sudan
Miscellaneous		
Highest average dewpoint	29°C	Assab, Ethiopia (June afternoons)
Highest mean annual sunshine	4300 + h	Wadi Halfa, Sudan
Highest hourly radiation	113 langleys[a]	Malange, Angola
	112 langleys[a]	Windhoek, Namibia

[a] World record.

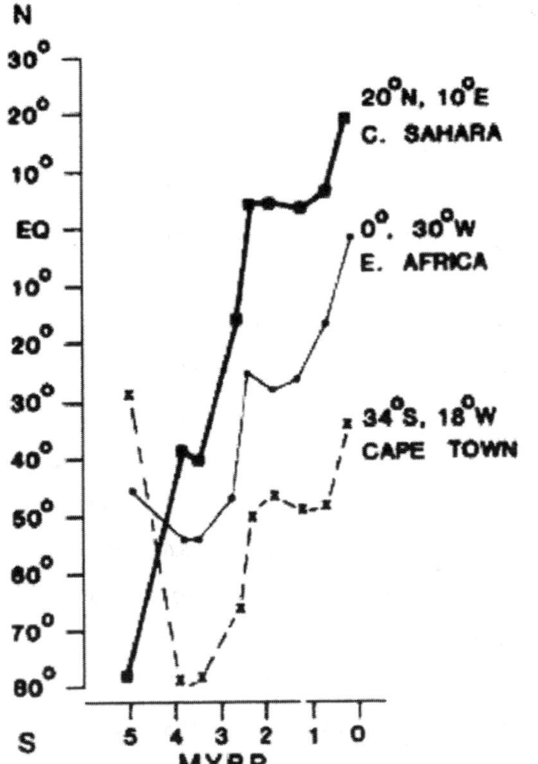

Figure A13 Latitudinal changes at three locations during the past 500 million years. (After Newell, 1974.)

Past climates

There has been relatively little study of past climates in Africa compared with studies of Europe and North America. However, it is known that the continent has occupied very different latitudes from that in which it is presently situated due to tectonic plate movements. In Figure A13 the latitudinal changes in the positions of three points on the continent are shown. From this alone it can be appreciated that in the period 450–200 million years before present (Ma BP) the Cape Town site, occupying a position within the Antarctic Circle, must have had a very cold climate, whereas now it is in relatively the same latitude as it was 500 Ma BP. The Central Saharan site has shown an almost steady progression from 80°S and it is likely that around 300 Ma BP it was also semiarid to arid, but from 200 to 50 Ma BP it was quite wet and humid. The East African location has not shown such extreme latitudinal variation but, nevertheless, must have experienced midlatitude and subtropical climates before reaching its present equatorial situation.

Some remark must be made concerning climate changes in the Saharan area. There have been many papers on this subject and the consensus of opinion is that from 20 000 to 12 000 years ago there was great aridity and the desert advanced southward. Following this period there were some very moist periods, while over the past 2000 years the rainfall has declined sufficiently to make the agriculture practiced in the time of Roman occupation no longer feasible (Carpenter, 1969). Murphey (1951) claimed that the cause is basically artificial and, even today, the growth of the desert must be attributed in great measure to anthropogenic influences. It is interesting to note that there are reports of ice on the Nile in the ninth and eleventh centuries (Oliver and Fairbridge, 1987).

For eastern Africa a more recent study (Hastenrath, 1984) suggests that there was a distinct retreat or disappearance of glaciers around 11 000–15 000 years ago, the deglaciation beginning at lower altitudes (*ca.* 3000 m). The most detailed studies of the historical climatology of Africa have been published by Nicholson (1976, 1978) and Nicholson and Flohn (1980). Nicholson finds, in the times of anomalous climate and climatic discontinuities, reasonable correlation between the sub-Saharan area and that of southern Africa. She identifies anomalous weather patterns in the 1680s and 1830s and a major rainfall change around 1800. Apparently the nineteenth century had greater snowfall than the twentieth century.

In the period 1870–1895, both the Sahara and eastern Africa had above-average rainfall, after which drier conditions set in and by the mid-1910s severe droughts were common in much of the tropics and subtropics. In the 1920s and 1930s there were indications of wetter conditions – Nile discharge up 35%, Lake Chad depth up 50%, and Sierra Leone reporting a third more rainfall than in the late nineteenth century.

Studies of African climate during the last 100 years or so are made problematic because of the vast areas for which the periods of record are very short. It is true that some temperature and precipitation measurements exist from the first half of the nineteenth century, such as in Tripoli and western Africa, but in general few reliable records exist before about 1890. Exceptions to this would include the island of Mauritius which has an almost unbroken record since around 1851.

A special project of the Global Climate Laboratory, part of the National Climatic Data Center, located in Asheville, North Carolina, is concerned with locating, extracting and digitizing data of monthly mean maximum and minimum temperatures and precipitation amount. In a few countries, including Egypt,

Nigeria and South Africa, there are enough stations to allow a regional investigation.

John F. Griffiths

Bibliography

Carpenter, R., 1969. Climate and history, *Horizon*, **11**(2): 48.

Desanker, P.V., and Justice, C.O., 2001. Africa and global climate change: critical issues and suggestions for further research and integrated assessment modeling. *Climate Research*, **17** (this issue is devoted to this topic).

Dubief, J., 1959. *Le Climat du Sahara*, Vol. 1. Algiers: University of Algeria.

Dubief, J., 1963. *Le Climat du Sahara*, Vol. 2. Algiers: University of Algeria.

Griffiths, J.F. (ed.), 1972. Climates of Africa, in *World Survey of Climatology*, Vol. 10. Amsterdam: Elsevier.

Hastenrath, S., 1984. *The Glaciers of Equatorial East Africa.* Dordrecht: Reidel.

Johnson, D.H., and Morth, H.T., 1960. Forecasting research in East Africa. In Bargman, D.J., ed., *Tropical Meteorology in Africa.* Nairobi: Munitalp Foundation, pp. 56–137.

Leroux, M., 2002. *The Meteorology and Climate of Tropical Africa.* Berlin: Springer Praxis Books.

Murphey, R., 1951. The decline of North Africa since the Roman occupation, *Association of American Geographers Annals*, **41**(2): 116–132.

Newell, R.E., 1974. The Earth's climate history, *Technol. Rev.*, pp. 30–45.

Nicholson, S.E., 1976. A climatic chronology for Africa: synthesis of geological, historical, and meteorological information and data. Unpublished PhD dissertation. Madison: University of Wisconsin.

Nicholson, S.E., 1978. Climatic variations in the Sahel and other African regions during the past five centuries, Journal of Arid Environments, **1**: 3–24.

Nicholson, S.E., and Flohn, H., 1980. African environmental and climatic changes and the general atmospheric circulation in Late Pleistocene and Holocene, *Climate Change*, **2**: 313–348.

Oliver, J.E., and Fairbridge, R.W., 1987. *The Encyclopedia of Climatology.* New York: Van Nostrand Reinhold, pp. 305–323.

Preston-Whyte, R.A., and Daughtrey Tyson, P., 2003. *The Weather and Climate of South Africa.* Oxford: Oxford University Press.

Terjung, W.H., 1967. The geographical application of some physioclimatic indices to Africa, *International Journal of Biometeorology*, **11**(1): 5–19.

Thompson, B.W., 1965. *The Climate of Africa.* London: Oxford University Press.

Todorov, A.V., 1984. The changing rainfall regions and the "normals" used for its assessment, *Journal of Climate and Applied Meteorology*, **24**: 97–107.

Cross-references

Airmass Climatology
Atmospheric Circulation, Global
Intertropical Convergence Zone
Mediterranean Climates
Rainforest Climates
Savanna Climate

AGROCLIMATOLOGY

Agroclimatology, often also referred to as agricultural climatology, is a field in the interdisciplinary science of agrometeorology, in which principles of climatology are applied to agricultural systems. Its origins relate to the foremost role that climate plays in plant and animal production. Formal references to the terms "agrometeorology" and "agroclimatology" date to the beginning of the twentieth century, but use of empirical knowledge can be traced back at least 2000 years (Monteith, 2000). Agroclimatology is sometimes used interchangeably with agrometeorology, but the former refers specifically to the interaction between long-term meteorological variables (i.e. climate) and agriculture. As such, they share common fundamental principles, methods and tools, but specific concepts are applied as described here.

Fundamental principles

Understanding the interactions between atmospheric variables and biological systems in agriculture, and applying this knowledge to increase food production and improve food quality, are the main goals of agrometeorology. Biological systems in agriculture are comprised of crops and forests, including the soil in which these grow; animals; and associated weeds, pests and diseases. Atmospheric variables that may affect these systems range from physical variables, such as solar radiation, precipitation, wind speed and direction, temperature, and humidity, to chemical variables, such as trace gas concentrations (e.g. CO_2, O_3). Agrometeorology is concerned with the characterization of these variables not only in the natural environment, but also in modified environments (e.g. irrigated areas, greenhouses, and animal shelters).

The fundamental principles used in the study of interaction between the atmosphere and agricultural systems are: (1) conservation of mass and energy, (2) radiation exchange, and (3) molecular and turbulent diffusion. The response of biological systems to these interactions draws on principles of soil physics, hydrology, plant and animal physiology, plant and animal pathology, entomology and ecology. Topics of research include water and radiation use efficiency by crops, animal comfort levels as affected by the physical environment, air pollution damage to crops, disease and pest development as a function of environmental conditions, and greenhouse gas emission by agricultural activities.

Methods and tools

Spatial scales in agroclimatology cover a wide range, from <0.1 m (e.g. response of fungi to leaf wetness) to regional and global scales (e.g. drought monitoring). Temporal scales may span past, present or future climate. Choice of instrumentation and measurement methods for weather and biological variables occurs according to the spatial and temporal scales of interest. Most often, agroclimatologists rely on long-term climate data provided by national meteorological services. In some countries weather stations originally were established in association with agricultural research institutes, attesting to the importance of weather and climate to agricultural production. Expertise in instrumentation, typically sensors for air temperature and humidity, solar radiation, precipitation, and wind measurement, is required from agroclimatologists in some applications, particularly those involving smaller spatial scales than provided by weather stations (i.e. microclimatological scales).

In all cases data describing the condition of the biological system are also needed. These include observations of developmental stages in crops, weeds, or insects; crop, milk or meat yield; grain or forage quality; and other physical and physiological measurements. Increasingly, remote sensing measurement techniques are being used to obtain regional and global scale estimates of variables, such as vegetation indices, or surface temperature.

Data handling, management, and processing procedures, spatial and temporal interpolation, statistical analyses, mathematical models that simulate the response of biological systems, and geographical information systems (GIS), decision support systems (DSS) are examples of tools used by agroclimatologists.

Concepts

A few concepts widely used in agroclimatology are presented here. The reader is referred to Griffith (1994) for additional background and examples of applications.

Degree-days and length of growing season

The concept of thermal time, or degree-days, was one of the earliest concepts developed, with wide application in agroclimatology. Climate records are matched to cardinal temperatures of crops, and daily mean temperatures above a minimum threshold, but below a maximum threshold, are accumulated over phenological phases. Stages in development for specific species or varieties occur once a certain degree-day value has been reached. Degree-day requirements for crops to reach maturity are then used to select suitable climatic regions for each crop or variety. Conversely, climatic records are used to calculate degree-days during the growing season. Provided water is available, length of growing season has its starting date defined by last mean spring frost date and ending date defined by first mean fall frost date. Probability distributions of historical degree-day records allow for selection of adequate crops and cultivars, those that will reach maturity for an accepted risk level.

The same concept can be applied to predict probability of occurrence of damaging development phases of pests in crops during certain times of the year, guiding the application of control measures.

Evapotranspiration and water balance

Water loss from the soil–plant system, through soil evaporation and plant transpiration, is termed evapotranspiration (ET). This is an energy-consuming process fueled mostly by net radiation absorption by the soil–plant surface. Soil water availability also determines the ET rate, with additional control provided by plant stomata. Under optimum water conditions plants do not need to control water loss, and carbon dioxide uptake also proceeds without restriction. Hence, crop yields are optimum under non-limiting water conditions. Numerous models have been developed for prediction of ET from crops given optimal and limiting soil water conditions. The Penman–Monteith equation, and the use of crop coefficients are commonly used (Allen et al., 1998) in agroclimatology to predict seasonal crop water requirements, and to determine irrigation needs of crops. This is accomplished through a soil water budget, where losses due to ET, runoff and deep percolation are compared to inputs such as precipitation and irrigation. Water stored, as determined by soil characteristics and rooting depth, decreases if outputs exceed inputs, and vice-versa. Limiting growth conditions, and reductions in yields, occur if soil water available to crops falls below a critical level, signaling the need for irrigation. A water budget applied using climatic data determines crop water requirements for optimal yields, and irrigation viability, contributing to long-range planning of land and water resource use.

Agroclimatic zoning

In contrast to climatic classifications, which are usually based on temperature and precipitation, agroclimatic classifications also take soil type, crop potential productivity and moisture deficiency, among other variables, into consideration (Bishnoi, 1989). Agroclimatic zoning allows for assessment of resources for agriculture, crop planning, and improvement in crop productivity. This concept was expanded to include environmental impact, resulting in the concept of agroecological zoning (AEZ) by the United Nations Food and Agriculture Organization (FAO). Recently, AEZ has been used to establish a global environmental resource database, including climatic, soil, terrain, and land cover, assessing the agricultural potential of 28 crops at three levels of farming technology (Fischer et al., 2001).

Animal comfort

The energy budget of homeothermic animals determines the level of comfort experienced, and if additional energy needs to be expended in keeping the body in a thermally comfortable zone. Excessive energy spent in thermal regulation results in reduced productivity, and may affect survival in extreme cases. Different livestock species and breeds have contrasting climatic requirements, and resistance to diseases typical to each environment. Consequently, climatic factors play a role in determining the level of success of the livestock enterprise. For example, a temperature–humidity index (THI), calculated from climate normals, has been used to indicate when temperate-evolved Holstein cows become less productive in various climate zones (Johnson, 1994).

Importance

Understanding of interactions between weather and agriculture leads to opportunities to use this knowledge for increased agricultural production, and also for minimizing agriculture's risks and impact on the environment. Thus, agrometeorology can play an important role in promoting sustainable development, through protection of the atmosphere and fresh water resources, desertification and drought control, sustainable agriculture, and education and training (Sivakumar et al., 2000). Knowledge obtained through research in agrometeorology is applied to guide: (1) strategic decisions in long-range planning; (2) tactical decisions in short-term planning, and (3) agrometeorological forecasts (WMO, 1981). Agroclimatology addresses needs in long-range planning, with typical examples of strategic decisions being selection of crops, livestock or forest species that match the existing climatic environment (agroclimatic characterization); and scheduling of agricultural operations (planting, pest control, etc.) that take into account long-term records and minimize risks. Hence, agroclimatology also provides tools needed in assessing the impact of climate change on agriculture, and strategies for farmers to adapt to a changing climate.

Claudia Wagner-Riddle

Bibliography

Allen, R.G., Pereira, L.S., Raes, D., and Smith, M., 1998. *Crop Evapotranspiration*. Food and Agriculture Organization, Irrigation and Drainage Paper No. 56.

Bishnoi, O.P., 1989. *Agroclimatic Zoning*. Commission of Agricultural Meteorology Report No. 30.

Fischer, G., Shah, M., van Velthuizen, H., and Nachtergaele, F.O., 2001. *Global Agro-ecological Assessment for Agriculture in the 21st Century*. Laxenburg, Austria: International Institute for Applied Systems Analysis.

Food and Agriculture Organization, 1993. *Agro-ecological Zoning Guidelines*. FAO Soil Bulletin 73.

Griffiths, J.F., 1994. *Handbook of Agricultural Meteorology*. Oxford: Oxford University Press.

Johnson, H.D., 1994. Animal physiology. In Griffiths, J.F., ed., *Handbook of Agricultural Meteorology*. Oxford: Oxford University Press, pp. 44–58.

Monteith, J.L., 2000. Agricultural meteorology: evolution and application. *Agricultural and Forest Meteorology*, **103**: 5–9.

Sivakumar, M.V.K., Gommes, R., and Baier, W., 2000. Agrometeorology and sustainable agriculture. *Agricultural and Forest Meteorology*, **103**: 11–26.

World Meteorological Organization, 1981. *Guide to Agricultural Meteorological Practices*. Geneva: WMO No. 143.

Cross-references

Applied Climatology
Climate Comfort Indices
Degree Days
Drought
Evapotranspiration
Spring Green Wave
Water Budget Analysis

AIRMASS CLIMATOLOGY

An airmass is a large body of air with relatively uniform temperature, humidity, and lapse rate characteristics across its horizontal extent. Especially in the midlatitudes, large seasonal and daily changes of weather commonly result from alternating dominance by airmasses derived from a variety of source regions. Specific types of these airmasses, typically covering hundreds of kilometers, are also responsible for stretches of days with similar weather, such as a cold polar outbreak in winter, or a hot–humid string of days in summer. Airmass climatology generally will describe the types of airmasses found in an area, will classify them according to their temperature and moisture characteristics, and will determine their seasonal or annual frequency of occurrence. This focus on temperature and moisture characteristics contrasts with weather-typing schemes that typically concentrate on pressure patterns and circulation features.

Airmass climatology grew out of the so-called "Norwegian" school of synoptic weather analysis. Bergeron (1928) formulated the concept of the airmass in the 1920s and proposed that a study be made of airmass characteristics and frequencies as a means of explaining recurring weather patterns (see Willet, 1931, for an English-language discussion). Petterssen (1940) first put together a northern hemisphere summary of airmasses. Strahler (1951) presented a worldwide synthesis showing general source regions and areas of dominance. Although too general for application to specific locations or during specific seasons, Strahler's model served well as a framework for other more detailed climatologies. Growing as it did out of midlatitude weather analysis, airmass climatology is most applicable where weather changes are frequent and airmasses numerous. The weather and climate of the tropics, poles, or other areas having a relative constancy of a single airmass type are generally less amenable to analysis by airmasses, since the significant weather patterns are caused by other mechanisms.

Classical airmass analysis is a manual exercise that requires subjective interpretations by the researcher that can be difficult to replicate, and often lack the precision necessary for successful applications. The availability of upper-air data and rise of numerical techniques, overshadowed the utility of the airmass as a weather forecasting tool by the late 1950s. Starting in the 1960s, dissatisfaction with the limitations of all manual classifications, coupled with increasing computer power, led to a dramatic reduction in their use. They were replaced by experimentation with many automated weather-typing techniques. None was widely adopted, nor did these schemes match the elegant simplicity and utility of the airmass concept for explaining midlatitude weather phenomena. However, neither did it appear possible to make classical-style airmass climatologies sufficiently rigorous for modern use.

Starting in the early 1990s, a new body of airmass work has grown from a commitment to a hybrid approach, combining manual and automated classification techniques to rehabilitate airmass climatology as a modern research tool (Schwartz, 1991; Kalkstein et al., 1996; Sheridan, 2002). Applications of the new approach have been successfully demonstrated in many areas, including describing the seasonal climate of an area, showing coherent upper-level circulation relationships to surface airmass distributions, interpreting precipitation intensity, assessing climate change, characterizing urban heat islands, identifying excessive heat-stress conditions, and even evaluating GCM output. Airmass climatology is again a fertile ground for discussion and research.

Airmasses

Formation

The critical properties of airmasses are temperature, humidity, and lapse rate. Of these, temperature and humidity are largely determined by the surface over which an airmass originates, whereas lapse rate is controlled by a number of factors. Usually, the geographic origin of an airmass within features of the general circulation (such as subtropical high pressure system or intertropical convergence zone) determines the lapse rate of the airmass.

In order for an extensive portion of the atmosphere to become recognizable as an airmass, it must reside for a time over a surface that is rather homogeneous across large expanses. The most effective way for this to occur is for a large anticyclone (high-pressure system) to remain stationary over an area for a number of days or weeks. Within the anticyclone, pressure gradients are weak, winds are light and slow subsidence induces radial drainage of air away from the center, which assures little mixing of foreign airstreams into the anticyclone. As the system stagnates, the atmosphere is slowly modified through radiative and convective processes to take on the temperature and humidity characteristics of the surface. For example, within a wintertime polar anticyclone the resident air will lose heat through radiative transfer to the surface, and the airmass will become cold and dry as water-holding capacity decreases with temperature. This airmass will have a relatively stable lapse rate because of cooling in the lower layers and adiabatic warming of subsiding air aloft.

Oceans, deserts, and extensive continental plains at high and low latitudes are the usual birthing places of the classical airmasses, as it is in these locations that large pieces of the atmosphere can reside for a long time over a relatively homogeneous surface. Mountainous areas and continental plains at midlatitudes are usually not considered source regions of airmasses, since the surface conditions of mountains are too heterogeneous, and midlatitudes continents do not often experience the stable weather conditions necessary for the formation of airmasses. On the other hand, midlatitude oceans typically are source regions for airmasses, since horizontal temperature variation in the oceans is usually rather small. The damp foggy conditions of midlatitude, west coast locations (marine type climate) are attributed to continual dominance by marine airmasses.

The subtropical high pressure systems (STH) near 30° latitude illustrate the complex interaction between general circulation features and surface conditions in influencing lapse rates of airmasses. These semipermanent high-pressure systems are strongest over subtropical oceans, and they are stronger to the east and weaker to the west. In addition, subtropical oceans usually have a relatively cold ocean current along their eastern portion and a warm ocean current in their western part. Airmasses formed within the strongly subsident eastern portion of a STH therefore have a very stable lapse rate aloft resulting from subsidence and adiabatic warming, and they also experience a slight cooling in the lower layers from contact with the cold ocean water. These airmasses are extremely stable; they typically have a surface temperature inversion and may also have a middle-level temperature inversion. With such a stable lapse rate it is almost impossible for these airmasses to produce rain. The aridity of coastal areas in eastern Australia, western North America, the Atacama Desert (South America) and the Kalahari Desert (Africa) is the result of the continual dominance by these stable airmasses that originate over oceans. Airmasses formed in the less strongly subsident western portion of a subtropical high originate over a warm water surface and are usually rather unstable. They often contain thunderstorms and are responsible for much of the precipitation received by the east coast of continents at subtropical latitudes (for example, southeastern United States or Japan).

Airmass types

Airmasses are traditionally classified based on temperature and humidity, although the location or latitude of the source region and lapse rate or stability are also important variables in classification. A typical qualitative air mass classification is shown in Table A8 (a good treatment of traditional manual airmass classification and weather patterns is provided in Trewartha and Horn, 1980; see also Pettersen, 1956). The temperature symbols in Table A8 (A, P, T, E) reflect both the midlatitude origin and evolution of airmass analysis since the original airmass meteorological work in Scandinavia, Canada, and the United States emphasized the role of cA, mP, cP, and mT airmasses in midlatitude weather systems. The gradual evolution of this terminology has caused some confusion, since polar airmasses, in reality, originate at latitude 45–66° and are more properly termed subpolar airmasses (James, 1970). The extremely frigid air originating in polar latitudes has come be called Arctic (or Antarctic) air. A symbol for equatorial airmasses (E) has also joined the list, primarily because airmasses originating from the unstable western limb of subtropical anticyclones near 30° latitude are usually classified as maritime tropical (mT) air, although their origin may not have been strictly in tropical latitudes. The mE classification, then, refers to actual tropical airmasses and mT air refers to the hot, moist, subtropical air experienced in summer by midlatitude locations.

In addition, other letters can be added to manual classifications to provide information on the lapse rate or the temperature of the airmass relative to the surface across which it moves. For

Table A8 Classical airmass types

Variable	Abbreviation	Source	Airmass characteristics
Temperature	A (Arctic)	Arctic	Frigid
	P (polar)	50°–70° Latitude	Cold
	T (tropical)	Subtropical	Hot
	E (equatorial)	Equatorial	Hot
Humidity	m (maritime)	Over water	Moist
	c (continental)	Over land	Dry
Symbol	**Airmass name**	**Airmass characteristics**	
mE	Maritime equatorial	Hot, moist	
mT	Maritime tropical	Hot, moist	
mP	Maritime polar	Cool, moist	
cT	Continental tropical	Hot, dry	
cP	Continental polar	Cold, dry	
cA	Continental Arctic	Frigid, dry	

The abbreviations in the upper part of the table are usually combined to give a specific airmass a two-letter identification symbol as shown in the lower part of the table.

Table A9 Integrated method airmass types (Schwartz, 1991)

Symbol	Airmass name	Airmass source
C	Continental	Central Canada
Pa	Pacific	Pacific, via Rocky Mountains
Po	Polar	C + Pa (summer only)
D	Dry tropical	SW USA (mostly summer)
dT	Dilute tropical	Modified T (non-summer)
T	Tropical	Gulf of Mexico
U	Unclassed	Transition and mixing

Table A10 Spatial synoptic classification airmass/weather types (Kalkstein et al., 1996)

Symbol	Airmass name	Airmass characteristics
DP	Dry polar	Similar to cP
DM	Dry moderate	No traditional source region
DT	Dry tropical	Analogous to cT
MP	Moist polar	Large subset of mP
MM	Moist moderate	Warmer, more humid than MP
MT	Moist tropical	Analogous to mT
TR	Transitional	Days one type yields to another

example, a cP airmass moving over a warm surface is colder than the surface, receives fluxes of energy and moisture, and is therefore unstable. This airmass would be given the classification cPku, where the k stands for cold (relative to the surface), and u stands for unstable. This type of airmass would be typical of conditions along the east coast of a midlatitude continent in winter, with cold, polar airmasses being advected across warmer ocean water, resulting in fog and frequent cyclogenesis.

Recent airmass climatologies using hybrid manual-automated techniques have started to move away from the classical definitions. Schwartz's (1991) methodology was not directly comparable to previous work, but tried to use generic names similar to the classical ones to recognize major airmass types in the north central United States (Table A9). Kalkstein et al. (1996) decided that the classical definitions were too limited for application to environmental problems, and developed a revised set of airmass/weather types for the spatial synoptic classification (SSC), for use in the conterminous USA (Table A10).

Airmass modification

One of the major difficulties in identifying and classifying particular airmasses is that their properties change over space and time. Schwartz (1991) and Kalkstein et al. (1996) employed differential station airmass criteria and new types to account for these effects. Central North America provides two excellent examples of these problems of identification, modification, and classification of airmasses. In the summertime, mT air from the Gulf of Mexico frequently streams northward across central North America, but convection and mixing with other non-mT air modifies the properties of these airmasses. Air that had a temperature of 32°C and a dewpoint temperature of 21°C over the Gulf may have a temperature of 28°C and a dewpoint temperature of 17°C by the time it reaches the Great Lakes. In the interim, midtropospheric dry air from the Rocky Mountains, subsiding beneath a ridge (or high-pressure system) in the upper air westerlies, has significantly diluted the mT air flowing northward at the surface.

Central North America also provides another example of a different style of airmass modification. Borchert (1950) found that, for over 6 months of the year, the central Great Plains are occupied by airmasses that originate over the Pacific Ocean and that are then advected over the western North American cordillera. In the process these Pacific airmasses lose most of their moisture over the mountains and gain latent heat. As these airmasses subside over the Great Plains they are much warmer and drier than when they originated. Borchert classified these airmasses as mP air, but he referred to them as modified mP air, recognizing that they are relatively warm and dry; in fact, mP would better refer to "mild Pacific" rather than "maritime Polar" air. Schwartz (1991) recognized this mP air as "Pacific" and Kalkstein et al. (1996) as "moist polar". These examples illustrate several of the problems associated with attempting a rigorous manual classification based solely on the origin of airmasses, and demonstrates some reasons why the new hybrid classification techniques are desirable.

Methods of airmass analysis

One of the central problems in airmass climatology is identification of particular airmasses. Classical airmass identification techniques may be grouped into two broad categories: those aimed at determining some characteristic of an airmass, and those that trace the path of movement of an airmass. Airmasses originate in core areas, and a single airmass ought to resemble closely the characteristics of the core area. Airmass identification for a given station usually begins, therefore, with an identification of the major airmass core areas and airmass types that might be expected to influence the study area. Further identification and classification of a single airmass occurring away from its core area is usually accomplished by comparing airmass characteristics such as temperature, relative humidity, dewpoint temperature, or lapse rate with characteristics of source regions. The weak link in this type of analysis is identification of source region characteristics; although these core areas are usually located based on widely agreed upon features of the general circulation, it is still a subjective judgement by the analyst who determines their temperature and moisture values. In North America, Brunnschweiler (1952) first attempted to classify climate based on airmass patterns, but his technique was not considered rigorous and remains unused. Other regional climatologies of note include Borisov (1965) for the USSR, Arakawa (1937) for Japan, Belasco (1952) for the British Isles, and Tu (1939) for China.

For rigorous scientific analysis some automated methods of airmass identification are desirable to reduce the possible bias introduced by analyst's assumptions. The partial collective method has been used to identify airmass dominance based on some property such as temperature. In the partial collective method, airmasses from a single source region are theorized to have temperatures that tend toward a modal value, and the deviation of the temperature of a specific airmass away from the mode is apparently caused by longer or shorter residence time in the source area. Theoretically, the daily temperature distribution at any weather station can be decomposed into a group of normal distributions (partial collectives of the whole distribution), each of which has a separate modal temperature representing multiple occurrences of a certain airmass. Bryson

(1966) used this method to determine North American airmass regions, but some of his airmass boundaries did not coincide with boundaries determined by Barry (1967), using frontal analysis methods. Ozorai (1963) did not find this method particularly useful for Hungary, since the temperature histogram of Budapest displayed many more partial collectives than existing airmass categories.

Taljaard (1969) attempted a southern hemisphere synthesis of airmass dominance, while Oliver (1970) concentrated on classifying Australia's climate based on airmass dominance. These two studies are notable since airmass literature concentrates strongly on northern hemisphere topics, and southern hemisphere climatologies are few in number. Oliver (1970) offered an alternative to the partial collective method. He assigned characteristics to airmasses in Australia based on thermohygric diagrams showing temperature, specific humidity, and relative humidity, as well as through theoretical identification of the presumed source areas. Monthly data were plotted and clusters of points were identified as airmass types. Oliver was able to generalize the analysis so that temperature and precipitation data could be used to identify regions of airmass dominance in Australia. His scheme for identifying the airmass dominance at a station (called monthly airmass identification, MAMI) remains largely untested for the rest of the world.

Trajectory analyses trace the path of an airmass as it moves and is modified. Again, the use of a trajectory method usually presupposes that source regions for airmasses are already known. An alternative type of this analysis attempts to trace airmass trajectories back to the source region. Wendland and Bryson (1981) determined such "airstream" regions for the northern hemisphere by tracing mean resultant near-surface wind flow, under the assumption that airmass trajectories follow wind flow. Source regions were identified as areas where mean wind flow diverged in many directions from a single point. Several areas were shown to be dominated by airstreams of relatively consistent and conservative properties, and these regions are analogous to airmass regions identified in other ways.

Schwartz (1991) explored the airmass classification problem with a hybrid manual – automated methodology, in order to benefit from the inherent advantages of each approach. The new scheme addressed the nonreplicability problem of manual classifications, the boundary identification problem of automated methods, and the difficulty in quantifying misclassification error common to all techniques. The manual portion of Schwartz's "integrated method" analysis identified maxima and minima of 850-hPa temperature and dewpoint values for each airmass type, at every station in the north central USA (NCUS) study area, during the four seasons (represented by the months of January, April, July, and October). The 850-hPa level was chosen because it is close enough to the ground to reflect surface conditions, but enough removed to be free of diurnal boundary layer processes, and fairly conservative of temperature and moisture characteristics. Trajectory analysis was used on a number of airmasses of known origin, until the accuracy of the specific numerical limits defining each airmass was ensured.

The automated portion of the integrated method used the partial collective technique to identify normal curve components in the temperature and dewpoint frequency distributions for each station-month, starting from initial estimates of each components' mean, standard deviation, and weight (percent of the total distribution), and the number of different air mass types present, based on the results of the manual analysis. The manual and automated analyses were combined to produce the final numerical limits and transition zones for each airmass type in the four seasons (Figure A14). Ideally, the transition zones were centered on the values where the component normal distributions crossed (had the same Z-score probability). In order to assess classification error, each airmass limit was assigned a probability value based on its relationship to the component normal distributions.

The results supported the validity of the normal component theory of airmasses, as average Z-score probabilities of the numerical limits were largest in the west-central portion of the NCUS. They then decreased consistently toward the outer edges of the study area, indicating stronger single airmass dominance and more distinct separation in airmass characteristics when moving closer to the respective Continental, Pacific, Dry

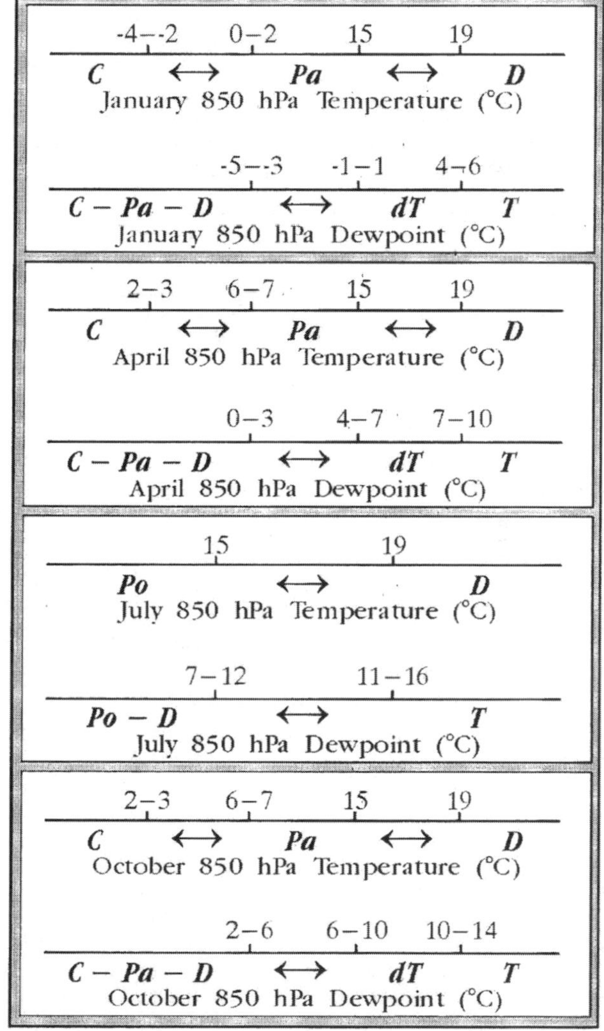

Figure A14 Integrated-method airmass criteria in January, April, July, and October for the north central United States (NCUS). Each airmass is defined by a range of temperature and dewpoint values except for dT and T airmasses which are distinguished by dewpoint alone. Double-ended arrow symbols show the U areas. A range of values indicates that limits vary depending on location within the NCUS. See Table A9 for names associated with airmass symbols (source: Schwartz, 1995).

Tropical, and Tropical air source regions (Table A9). Thus, the integrated method scheme is an easily interpreted approach to major airmass type classification that is numerically precise and allows assessment of misclassification error.

Kalkstein et al. (1996) continued development of hybrid techniques, by introducing the spatial synoptic classification (SSC) system, for the conterminous USA. Surface air temperature and dewpoint criteria were still central to the approach, but integration of multiple measurements each day, and the additions of other variables such as cloud cover, sea level pressure, and diurnal ranges of temperature and dewpoint, allowed for more detailed airmass definitions (Table A10). SSC's core was the manual selection of seed days for each airmass at every station location. A seed day is an actual day that displays the typical conditions for the given airmass at a selected station. Care was taken to ensure that adjacent stations have similar criteria for the same airmasses in order to create spatial continuity.

SSC used simultaneous discriminant analysis of all input variables to automate determination of the appropriate airmass type. Additionally, a parallel procedure identified and reassigned selected days as transitional situations (when one airmass yields to another). The result was a generally robust and replicable airmass classification scheme that has proven useful for many applications. However, an important limitation of SSC was its availability for only the winter and summer seasons.

Sheridan (2002) introduced enhancements to the SSC technique, with a revised approach labeled SSC2. Most notable were the creation of a new "sliding seed day" method that allowed the classification to be applied throughout the entire year, improvement in the spatial cohesiveness among stations, and expansion of the areas covered to include Canada, Alaska, and Hawaii, in addition to the conterminous USA.

Applications

One of the interesting applications of airmass climatology is ecology and paleoecology. Bryson (1966) showed that major vegetation boundaries in North America correlated very well with transition zones between major airmass types. He concluded that the location of each major vegetation regime was controlled by the dominance of a single airmass for more than 50% of a mean year. Barry (1967) questioned some of Bryson's methods and data, but Krebs and Barry (1970) found significant correspondence between vegetation transitions and mean frontal positions between airmass types in Eurasia. The implication that certain airmass types, through their influence on weather patterns, influence plant ecology, provides an exciting means of inquiry into the dynamics of plant distributions and the reasons behind the location of major natural regions.

Other researchers have used the correspondence between mean airmass dominance and vegetation regions to construct paleoairmass climatologies that describe the airmass climate of an area thousands of years ago. In this extremely complex analysis, researchers begin by making assumptions that climate (airmass dominance) largely determines the type of vegetation in a region, and the "pollen rain" from plants onto the surface of the earth is a reflection of the type of vegetation in an area. Although relatively few geologic deposits contain identifiable plants, certain ideal locations (peat bogs, lakes, ponds) have preserved a sediment accumulation for thousands of years, and from these sediments pollen found in ancient sediments presumably mirrors the types of vegetation grown in the area some time in the past and, from the pollen and radiocarbon dating of organic remains in the sediments, it is possible to reconstruct the type of vegetation that grew in an area thousands of years ago. It is assumed that vegetation regions are controlled by airmass dominance (for example, tundra vegetation exists in areas experiencing dominance by Arctic air for over 6 months of the year), then, from the records of many peat bogs or ponds across the continent, it is possible to piece together airmass climatologies for periods in Earth history from which we have no climatic records. For example, a profusion of grassland types of pollen found in 5000–7000-year-old sediments of peat bogs in currently forested areas of the eastern United States apparently records an expansion of the prairies of North America eastward as a result of expanded dominance by dry "mP" airmasses from the Pacific Ocean and Rocky Mountains (Wright, 1968). From today's correspondence between vegetation, pollen, and airmasses, along with the paleopollen record, it is possible to reconstruct a paleoclimatology of airmasses. These airmass paleoclimatologies provide an efficient means of understanding the character of past climates. (For example, see Bryson and Wendland, 1967; Webb and Bryson, 1972; Moran, 1973; and Sorenson and Knox, 1973.) However, these questions still need to be explored in greater detail.

A large number of applications have been undertaken in the broader field of synoptic climatology using automated "synoptic indices" that are similar to airmass climatologies. For example, the temporal synoptic index (TSI) developed by Kalkstein et al. (1987) has been applied to many problems, such as air pollution monitoring, glacier energy balance, and climate change detection (summarized in Kalkstein et al., 1990). Yarnal et al. (2001) provide an overview of developments in this and other areas of synoptic climatological research during the 1990s.

Applications have also been explored with modern hybrid airmass classification schemes since the mid-1990s. Schwartz and Skeeter (1994) showed that specific NCUS airmass distributions can be related to a small number of meaningful North American 500-hPa height and surface pressure patterns in all seasons, and also that key transects within these patterns can be used to describe changes in monthly airmass dominance from one year to another. Schwartz (1995) explored the use of airmass analysis as a tool for detecting structural climate change in the NCUS. Examinations of airmass temperature, dewpoint, and frequency changes over time, are superior to simple analyses of changes in monthly mean temperatures, as they may point to potential causes of change. Schwartz found increasing amounts of tropical air, coupled with decreasing polar air in summer over the 1958–1992 period, which suggested increases in the number of 500-hPa troughs over the western USA.

Schwartz (1996) used airmass analysis to assess the accuracy of a GCM control run. The results suggested that seasonal GCM performance was promising, but that substantial differences may exist between the control runs and observed climate data at the daily time scale. Greene (1996) used SSC to assess summertime precipitation intensity, and Sheridan et al. (2000) examined the urban heat island effect. Lastly, another important application of modern hybrid airmass analysis has been to identify excessive heat-stress conditions, and the development of health watch/warning systems (Sheridan and Kalkstein, 1998; Kalkstein, 1999). Sheridan (2002) summarized earlier work in this area, and recent studies incorporating SSC2 into systems designed for a number of cities in North America, Europe, and China.

Mark D. Schwartz and William T. Corcoran

Bibliography

Arakawa, H., 1937. Die Luftmassen in den Japanischen, *Am. Meteorol. Soc. Bull.*, **18**: 407–410.
Barry, R.G., 1967. Seasonal location of the Arctic front over North America. *Geog. Bull.*, **9**: 79–95.
Belasco, J.E., 1952. Characteristics of air masses over the British Isles. *Geophys. Mem.*, **11**.
Bergeron, T., 1928. Uber die dreidimensional veknupfende Wetteranalyse, Part 1: Prinzipelle Einfuhurung in das Problem der Luftmassen-und Frontenbildung, *Geofys. Publ.* (Oslo), **5**: 1–111.
Borchert, J.R., 1950. The climate of the central North American grasslands. *Assoc. Am. Geog. Ann.*, **40**: 1–39.
Borisov, A.A., 1965. *Climates of the U.S.S.R.*, 2nd edn. A. Ledwood (trans.), C.A. Halstead, (ed.). Edinburgh: Oliver & Boyd.
Brunnschweiler, D.H., 1952. The geographic distribution of airmasses of North America, *Naturf. Gesell. Zurich Viertel.*, **97**: 42–48.
Bryson, R.A., 1966. Airmasses, streamlines, and the boreal forest. *Geog. Bull.*, **8**: 228–269.
Bryson, R.A., and Wendland, W.M., 1967. Tentative climatic patterns for some late glacial and post glacial episodes in central North America. In Mayer-Oakes, W.J., ed., *Life, Land, and Water*. Winnipeg: University of Manitoba Press, pp. 271–299.
Greene, J.S., 1996. A synoptic climatological analysis of the summertime precipitation intensity in the eastern United States. *Phys. Geog.*, **17**: 401–418.
James, R.W., 1970. Air mass climatology. *Met. Rund.*, **23**: 65–70.
Kalkstein, L.S., 1999. Heat-health watch systems for cities. *World Meteor. Org. Bull.*, **48**: 69.
Kalkstein, L.S., Tan, G., and Skindlov, J., 1987. An evaluation of three clustering procedures for use in synoptic climatological classificiation. *J. Climate Appl. Meteorol.*, **26**: 717–730.
Kalkstein, L.S., Dunne, P.C., and Vose, R.S., 1990. Detection of climatic change in the western North American Arctic using a synoptic climatological approach. *J. Climate*, **3**: 1153–1167.
Kalkstein, L.S., Nichols, M.C., Barthel, C.D., and Greene, J.S., 1996. A new spatial synoptic classification: application to air-mass analysis, *Int. J. Climatol.*, **16**: 983–1004.
Krebs, J.S., and Barry, R.G., 1970. The Arctic front and the tundra–taiga boundary in Eurasia, *Geog. Rev.*, **60**: 548–554.
Moran, J.M., 1973. The late-glacial retreat of "Arctic" air as suggested by onset of *Picea* decline. *Prof. Geogr.*, **25**: 373–376.
Oliver, J.E., 1970. A genetic approach to climate classification, *Assoc. Am. Geogr. Ann.*, **60**: 615–637.
Ozorai, Z., 1963. An assessment of ideas in relation to airmasses. *Idojara*, **67**: 193–203.
Petterssen, S., 1940. *Weather Analysis and Forecasting*. New York: McGraw-Hill.
Petterssen, S., 1956. *Weather Analysis and Forecasting*, 2nd edn. New York: McGraw-Hill.
Schwartz, M.D., 1991. An integrated approach to air mass classification in the North Central United States. *Prof. Geogr.*, **43**: 77–91.
Schwartz, M.D., 1995. Detecting structural climate change: an air mass-based approach in the North Central United States, 1958–1992. *Assoc. Am. Geogr. Ann.*, **85**: 553–568.
Schwartz, M.D., 1996. An air mass-based approach to GCM validation. *Climate Res.*, **6**: 227–235.
Schwartz, M.D., and Skeeter, B.R., 1994. Linking air mass analysis to daily and monthly mid-tropospheric flow patterns. *Int. J. Climatol.*, **14**: 439–464.
Sheridan, S.C., 2002. The redevelopment of a weather-type classification scheme for North America. *Int. J. Climatol.*, **22**: 51–68.
Sheridan, S.C., and Kalkstein, L.S., 1998. Heat watch-warning systems in urban areas, *World Resource Rev.*, **10**: 375–383.
Sheridan, S.C., Kalkstein, L.S., and Scott, J.M., 2000. An evaluation of the variability of air mass character between urban and rural areas. In *Biometeorology and Urban Meteorology at the Turn of the Millennium*. World Meteorological Organization, pp. 487–490.
Sorenson, C.J., and Knox, J.C., 1973. Paleosols and paleoclimates related to Late Holocene forest/tundra border migrations: MacKenzie and Keewatin, N.W.T., Canada. In Raymond, S., and Schlederman, P., eds., *International Conference on the Prehistory and Paleoecology of Western North Arctic and Subarctic*. Calgary: University of Calgary Archeology Association, pp. 187–204.
Strahler, A.N., 1951. *Physical Geography*. New York: Wiley.
Taljaard, J.J., 1969. Airmasses of the southern hemisphere, *Notos*, pp. 79–104.
Trewartha, G.T., and Horn, L.H., 1980. *An Introduction to Climate*. New York: McGraw-Hill.
Tu, C.W., 1939. Chinese air mass properties, *Q. J. Roy. Meteorol. Soc.*, **65**: 33–51.
Webb, T., III, and Bryson, R.A., 1972. Late and post-glacial climatic change in the Northern Midwest, U.S.A. *Quatern. Res.*, **2**: 70–115.
Wendland, W.M., and Bryson, R.A., 1981. Northern hemisphere airstream regions. *Monthly Weather Rev.*, **109**: 255–270.
Willett, H.C., 1931. Ground plan for a dynamic climatology. *Monthly Weather Rev.*, **59**: 219–223.
Wright, H.E., 1968. History of the prairie peninsula. In Bergstrom, R.E., ed., *The Quaternary of Illinois*. Special Publication No. 14, University of Illinois College of Agriculture.
Yarnal, B., Comrie, A.C., Frakes, B., and Brown, D.P., 2001. Developments and prospects in synoptic climatology. *Int. J. Climatol.*, **21**: 1923–1950.

Cross-references

Atmospheric Circulation Global
Climatic Classification
Lakes, Effects on Climate
Local Winds

AIR POLLUTION CLIMATOLOGY

Air pollution is defined as an atmospheric condition in which substances (air pollutants) are present at concentrations higher than their normal ambient (clean atmosphere) levels to produce measurable adverse effects on humans, animals, vegetation, or materials (Seinfeld, 1986). Polluting substances can be noxious or benign, and can be released by natural and anthropogenic (human-made) sources. According to the World Health Organization an estimated 3 million people die each year because of exposure to air pollution (WHO, 2000). Air pollution climatology is concerned with the study of atmospheric phenomena and conditions that lead to occurrence of large concentrations of air pollutants and with their effects on the environment.

Air pollutants are typically classified into three categories: suspended particulate matter (SPM), gaseous pollutants (gases and vapors) and odors. SPM in the air includes PM_{10} (particulate matter with median diameter less than 10 μm), $PM_{2.5}$ (particulate matter with median diameter less than 2.5 μm), diesel exhaust, coal fly-ash, mineral dusts (e.g. asbestos, limestone, cement), paint pigments, carbon black and many others. Gaseous pollutants include sulfur compounds (e.g. sulfur dioxide (SO_2)), nitrogen compounds (e.g. nitric oxide (NO), ammonia (NH_3)), organic compounds (e.g. hydrocarbons (HC), volatile organic compounds (VOC) and polycyclic aromatic compounds (PAH), etc.). Known odorous agents are generally sulfur compounds such as hydrogen sulfide (H_2S), carbon disulfide (CS_2) and mercaptans.

The past few decades have seen much progress in the understanding of air pollution climatology and meteorology, due to rapid progress in pollutant measurement techniques, increased understanding of atmospheric dynamics, and computational capabilities, particularly for areas with complex terrain. At the time of the last edition of this *Encyclopaedia* (1986), the processes of air pollution dispersion were well understood for

Table A11 Typical spatial and temporal scales of air pollution problems

Air pollution type	Horizontal scale	Vertical scale	Temporal scale
Indoor	10–100 m	Up to 100 m	Minutes–hour
Local	100 m–10 km	Up to 3 km	Minutes–hours
Urban	10–100 km	Up to 3 km	Hours–days
Regional	100–1000 km	Up to 15 km	Hours–months
Global	40 000 km	Up to 50 km	Months–decades

Source: Modified after Stern et al. (1984).

urban areas over relatively simple terrain. However, knowledge gaps existed regarding the dispersion of pollutants in cities near coastlines or lake shores, or in areas of complex topography such as mountainous areas – all of which may be considered as "complex terrain". In addition, air pollution climatology now considers not only localized issues, but also global problems such as stratospheric ozone reduction. The role of long-range transport of pollutants in degradation of air quality in remote/distant regions has also been recognized. Over the past few decades a combination of field observational studies – such as the European Tracer Experiment (ETEX; Girardi et al., 1998) and the Across North America Tracer Experiment (ANATEX; Draxler et al., 1991) – in conjunction with different modeling approaches – has added to a growing knowledge base concerning the complexities of air pollution dispersion.

Natural air pollution has always occurred throughout Earth's history. Volcanic eruptions, natural fires and wind-blown dust are examples of natural phenomena that introduce pollutants into the atmosphere. Anthropogenic sources have only become a serious problem during the past 200 years with the growing population, increased urbanization and accelerating industrialization. It is estimated that 2 billion metric tons of air pollutants are released into the atmosphere world-wide (Arya, 1999). Air quality is affected not only by emissions into the atmosphere, but also by the pollution potential of air. Stern et al. (1984) provide a useful classification for the types and scales of air pollution problems (Table A11).

Earlier examples of important literature on air pollution climatology include Slade (1968), Williamson (1972), Stern (1976), Holzworth (1974) and Pasquill (1974). However, several introductory books that contain in-depth information on recent developments in this discipline are now available. Arya (1999) is an invaluable book that covers the physics of meteorology and dispersion, while Jacobson (2002) includes perspectives on history and science. The book by Seinfeld and Pandis (1998) also contains a wide range of topics on the physics and chemistry of air pollution. Those who require an extensive treatment of atmospheric chemistry (ambient and polluted) are encouraged to consult Finlayson-Pitts and Pitts (1986). A concise literature review is provided by Sturman (2000).

This account will focus on air pollution climatology by first introducing contemporary air pollution problems and issues; then by considering the components of air pollution, including emissions, atmospheric conditions and receptor response. The item continues by discussing air pollution dispersion models and air quality guidelines and standards.

Contemporary air pollution problems and issues

Air quality issues came to the forefront of scientific and societal attention during the 1950s. In 1952 London's "killer fog" caused a great increase in human fatality (4000 excess deaths), especially in people with a history of cardiopulmonary problems. The episode resulted in the implementation of air pollution mitigation measures by the authorities (Brimblecombe, 1987). The air pollution in London was due to the emission of smoke from the burning of coal and other raw materials into the foggy atmosphere; this type of pollution is known as the London-type smog (smoke plus fog). In the same decade another type of smog, which is formed by chemical reactions in the atmosphere – the Los Angeles photochemical smog – also gained notoriety. During the 1970s acid rain emerged as the top environmental concern. A decade later the reduction of stratospheric ozone due to emission of anthropogenic compounds became the top international concern.

Major air pollution problems facing humanity in the 21st century include urban smog, acid deposition, indoor air pollution, Antarctic stratospheric ozone depletion (global stratospheric ozone reduction), and the highly contentious issue of global warming due to alteration of the global longwave (infrared) radiation balance caused by emission of greenhouse gases (GHG) such as carbon dioxide, methane, and nitrous oxide. Table A12 lists some of the polluting substances present in the atmosphere.

Urban smog

This is characterized by build-up of harmful gases and particulates emitted from vehicles, industry, or other sources (primary pollutants) – that cause the London-type smog, or is formed chemically in the air from emitted precursors (producing *secondary* pollutants) – that cause the Los Angeles-type smog. London-type smog events have been observed in various places around the world, such as the Meuse Valley in Belgium, Pittsburgh in the United States, and Christchurch in New Zealand, to name just a few. Photochemical smog has been observed in most cities around the world; the problem is particularly severe in mega-cities such as Mexico City, Tokyo, Beijing and Tehran.

Transboundary pollution

This is characterized by transport of pollutants across political boundaries and/or vast distances. Examples of such pollution include acid deposition (also known as acid rain – a term introduced by Robert Angus Smith in the nineteenth century) and regional haze caused by forest fires. Acid deposition happens when sulfuric acid (H_2SO_4), hydrochloric acid (HCl), or nitric acid (HNO_3) is deposited on the ground in vapor form or dissolved in rainwater, fogwater, or particulates. This increased acidity in turn harms soils, lakes, vegetation and materials. Cowling (1982) provides an historical perspective on the discovery of and mitigation measures used to deal with acid rain. Natural examples of transboundary pollution include the 1998 dust clouds caused by intense storm events in western China, which carried vast quantities of particulate matter across the Pacific Ocean, contributing to increased concentrations in Vancouver, Canada (McKendry et al., 2001).

Indoor air pollution

Most people spend a large amount of time indoors where they are constantly exposed to indoor air pollution. This is caused from either the emission of gases and particulates in enclosed

Table A12 Gases and particulate matter components in air pollution

Outdoor urban air pollution	Indoor air pollution	Acid deposition	Stratospheric ozone depletion	Global climate change
Gases				
Ozone (O_3)	Nitrogen dioxide (NO_2)	Sulfur dioxide (SO_2)	Ozone (O_3)	Water vapor
Nitric oxide (NO)	Carbon monoxide (CO)	Sulfuric acid (H_2SO_4)	Nitric oxide (NO)	Carbon dioxide (CO_2)
Nitrogen dioxide (NO_2)	Formaldehyde (HCHO)	Nitrogen dioxide (NO_2)	Nitric acid (HNO_3)	Methane (CH_4)
Carbon monoxide (CO)	Sulfur dioxide (SO_2)	Nitric acid (HNO_3)	Hydrochloric acid (HCl)	Nitrous oxide (N_2O)
Ethene (C_2H_4)	Organic gases	Hydrochloric acid (HCl)	Chlorine nitrate ($ClONO_2$)	Ozone (O_3)
Toluene ($C_6H_5CH_3$)	Radon (Rn)	Carbon dioxide (CO_2)	CFC-11 ($CFCl_3$)	CFC-11 ($CFCl_3$)
Xylene ($C_6H_5CH_3$)			CFC-12 (CF_2Cl_2)	CFC-12 (CF_2Cl_2)
PAN ($CH_3C(=O)O_2NO_2$)				
Aerosol particle components				
Black carbon	Black carbon	Sulfate (SO_4)	Sulfate (SO_4)	Black carbon
Organic matter	Organic matter	Nitrate (NO_3)	Nitrate (NO_3)	Organic matter
Sulfate (SO_4)	Sulfate (SO_4)	Chloride (Cl)	Chloride (Cl)	Sulfate (SO_4)
Nitrate (NO_3)	Nitrate (NO_3)			Nitrate (NO_3)
Ammonium (NH_4)	Ammonium (NH_4)			Ammonium (NH_4)
Soil dust	Allergens			Soil dust
Sea spray	Asbestos			Sea spray
Tyre particles	Fungal spores			
Lead (Pb)	Pollens			
	Tobacco smoke			

Source: Modified after Jacobson (2002).

buildings or transport of pollutants from outdoors; but it is not as extensively researched as outdoor air pollution (Wanner, 1993). Health effects of indoor air pollutants can be severe and directly affect the respiratory and cardiovascular systems. Indoor air pollutants include radon gas, ammonia, volatile organic compounds (VOC), polycyclic aromatic hydrocarbons (PAH) and second-hand cigarette smoke. Radon gas has a long-term health effect and is naturally released by soil due to radioactive decay of radium (Bridgman et al., 2000). Indoor exposure to particulate matter increases the risk of acute respiratory infections, leading to an increase in infant and child mortality rates in developing countries. WHO reports that such exposure is responsible for between half and one million excess deaths in Asia, and 300 000–500 000 excess deaths in sub-Saharan Africa (WHO, 2000).

Antarctic ozone depletion

Antarctic ozone depletion and the reduction of global stratospheric ozone are due to emission of chlorine and bromine compounds such as chlorofluorocarbons (CFC). CFC break down by photolysis reaction (breakdown of molecules by solar radiation) after they have traveled to the upper atmosphere. The reduction of ozone increases the amount of ultraviolet (UV) radiation that reaches Earth's surface. UV radiation can damage genetic material of organisms and in humans it is known to cause skin cancer.

Scientific consensus supports the theory that human emission of carbon dioxide, methane, nitrous oxide and other gases may play a significant role in the cause of global warming (IPCC, 2001). Therefore global warming is also an air pollution issue.

Components of an air pollution problem

Air pollution problems usually have three components: (1) emission of polluting substances into the air; (2) the pollution potential of the atmosphere characterized by its ability to transport, diffuse, chemically transform and remove the pollutants; and (3) response of the receptors (e.g. people, animals, vegetation) to the exposed concentration. Each of these components is considered in this section.

Pollution emissions

Emission sources can be categorized as follows:

1. Urban and industrial. Industrial sources include power generation plants that use fossil fuel, mining, manufacturing, smelting, pulp and paper plants, and chemical industries. Major gaseous pollutants emitted are carbon monoxide (CO), carbon dioxide (CO_2), sulfur dioxide (SO_2) and volatile organic compounds.
2. Agricultural and rural. Agricultural areas can be sources of air contaminants due to decaying waste from animals and plants which can release ammonia, methane (CH_4, a powerful greenhouse gas) and other noxious gases. In addition, windblown dust due to plowing, tilting, and harvesting, and smoke and haze due to slash burning can lead to severe degradation in air quality. An interesting example of air pollution caused by agricultural practices is the regional haze and smog events that plagued the Southeast Asian countries in 1997 due to forest fires in Indonesia (Davies and Unam, 1999; Khandekar et al., 2000). Burning forests is a traditional way of clearing land for agricultural purposes in Indonesia (and indeed in many other parts of the world), but the fires became unmanageable due to prolonged drought and the delayed onset of the monsoon rains caused by the 1997 El Niño episode. A strong increase in tropospheric ozone concentrations occurred over tropical Southeast Asia, reaching as far as Hong Kong (Chan et al., 2001).
3. Natural. In addition to anthropogenic sources, many natural sources can contribute to air pollution. These include: volcanic

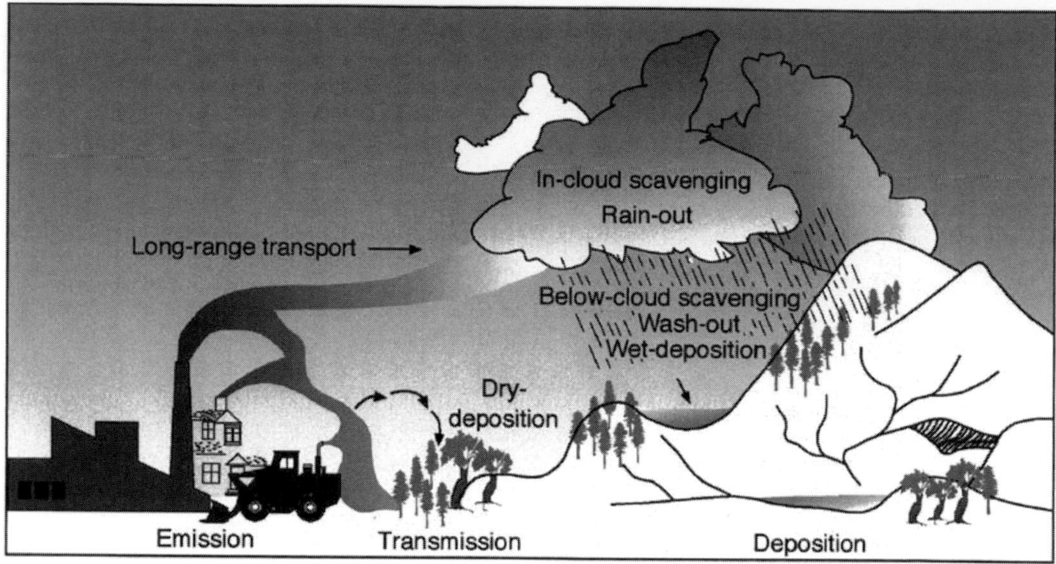

Figure A15 Schematic illustration of emission, transmission and deposition of air pollution.

eruptions that release vast quantities of particulate matter, CO_2, SO_2, and other gases, into the atmosphere; *biogenic* (from biological sources) emissions from forests and marshlands of compounds such as hydrocarbons (terpenes and isoprenes; Guenther et al., 1995), methane and ammonia; and soil microbial processes which contribute NO, CH_4 and H_2S.

Atmospheric conditions

The state of the atmosphere determines its pollution potential, which is affected by three important processes: *dispersion* which is the horizontal and vertical spread and movement of pollutants; *transformation* which involves chemical reactions that occur between pollutants or to pollutants under certain temperature and sunlight conditions; and *removal* of pollutants through such mechanisms as dry and wet deposition (Figure A15). Each of these processes will be considered in turn, but first the overriding control of synoptic conditions will be considered.

Synoptic controls

"Synoptic conditions" refers to an intermediate scale of atmospheric activity between global and regional scales and includes circulation systems such as anticyclones and cyclones. Synoptic circulation features have a spatial dimension of the order of thousands of kilometers and may last about a week on average. Atmospheric circulation at this scale has an important control on air pollution since it influences cloud (amount, thickness, height and type), temperature and relative humidity of the airmass, the type and amount of precipitation, and the wind speed and direction. These parameters in turn influence the vertical temperature structure of the atmosphere which determines the atmospheric stability. If the air temperature decreases rapidly with height (i.e. at a rate greater than the dry adiabatic lapse rate of 9.8°C km^{-1}) then the atmosphere is unstable. This means that vertical motion is promoted and any emissions will mix well into the boundary layer (the lowest layer of the atmosphere that is affected by the diurnal cycle of temperature, humidity, wind, and pollution). However, in stable conditions, which occur when temperature either decreases slowly with height (at a rate less than the dry adiabatic lapse rate) or increases with height (which is known as a temperature inversion), pollution may become trapped close to the emission source, resulting in increased concentrations. The layer through which the mixing of pollutants occurs is known as the mixed layer, and the height to which pollutants mix is called the mixing depth or mixing height.

Anticyclonic circulation is more conducive to pollution episodes since it usually involves weaker synoptic winds and clear skies. In summer, stronger surface heating under anticyclonic flow results in warmer temperatures in the boundary layer, and this warmth, together with high solar radiation levels and generally calm or light airflow, can lead to the formation of photochemical smog (discussed in more detail later). Since anticyclonic circulation involves subsidence of air towards the surface (and hence warming), this results in elevated inversions which can trap pollutants in a shallow mixed layer (Figure A16a).

In winter, clear skies and light winds during anticyclonic conditions lead to strong nocturnal radiative cooling of the surface which results in surface or radiative inversions (Figure A16b). During such conditions any pollutants emitted at the surface (e.g. through industrial emissions, domestic heating or traffic) become trapped very close to the surface and may result in high pollutant concentrations. This is the typical scenario that produces classic smog as a combination of smoke and fog. Figure A17 shows the severe degradation in air quality in Christchurch, New Zealand, as a result of emission of smoke from solid-fuel burning (wood) into a shallow surface inversion layer.

Dispersion

The dispersal of air pollutants is governed mainly by the mean wind speed and the characteristics of atmospheric turbulence. Together these factors determine the stability characteristics of the atmosphere. Turbulence consists of horizontal and vertical eddies that mix pollution with the surrounding air. As turbulence increases, so too does mixing, which results in good dispersion

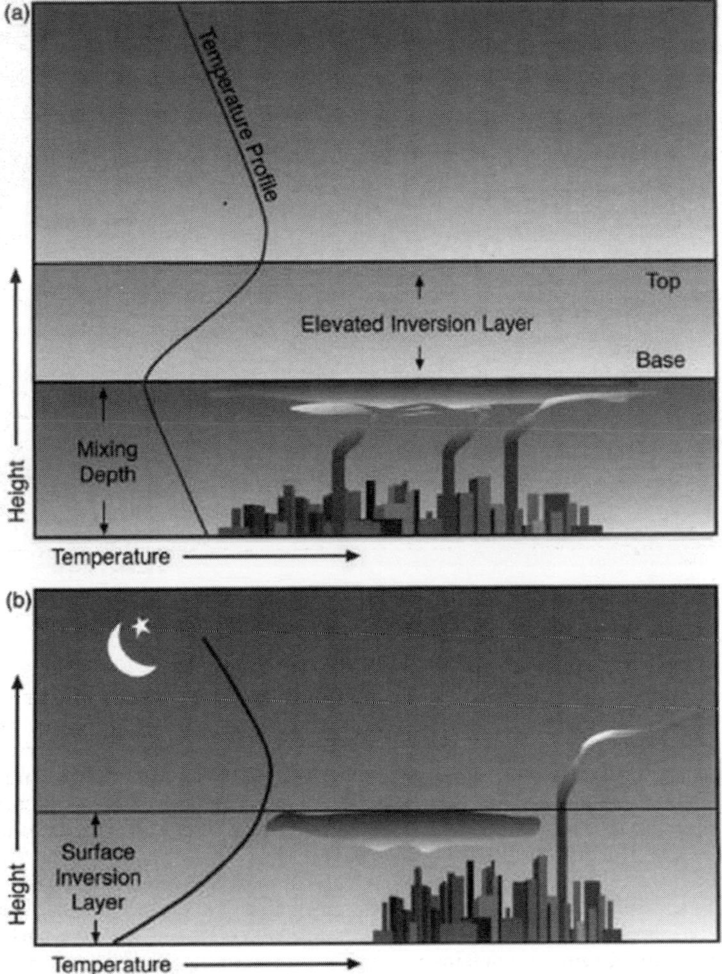

Figure A16 Air pollutants emitted at the surface can become trapped by (**a**) an elevated inversion layer and (**b**) a surface inversion layer (modified after Whiteman, 2000).

and lowered pollutant concentrations. Strong turbulence occurs in an unstable atmosphere in which vertical motion is enhanced. Conversely, when turbulence is weak, there is little propensity for mixing to occur and pollutant levels may increase. Conditions of weak turbulence are often typical of clear mornings in winter.

Turbulence can be generated through both thermal and mechanical means. Thermally generated turbulence results from heating of the surface, which leads to buoyancy-induced convective motion (i.e. vertical transport of heat away from the surface). With stronger heating of the surface, thermal turbulence will increase. Mechanically generated turbulence can result from wind shear as the wind blows across the Earth's surface. Mechanical turbulence increases with increasing wind speed and is greater over rougher surfaces (e.g. cities and forests). For a more in-depth explanation of the theory and mathematical representation of turbulence in the boundary layer refer to Stull (1998) or Garratt (1992).

A typical diurnal cycle of radiative heating and cooling in cloud-free conditions results in varying stability conditions throughout the day (Figure A18). In the early morning, before sunrise, the nocturnal temperature inversion is still present. Any pollutants emitted close to the ground may be trapped in this inversion, if the chimney stacks are lower than the inversion top, smoke plumes will take on a fanning shape (Figure A18a). After sunrise, surface heating results in the generation of convection which mixes the air close to the surface within the layer beneath the remains of the nocturnal inversion. This situation is one of the worst for pollutant concentrations at ground level as *fumigation* occurs (downward movement of pollutants from aloft because of a growing mixed layer), resulting in pollution originally emitted and trapped aloft being mixed to ground level, often in quite high concentrations (Figure A18b). As daytime heating continues the nocturnal inversion is completely eroded and replaced by a *lapse* profile (decrease of temperature with height; Figure A18c). This is typical of an unstable atmosphere which promotes vertical mixing. This mixing is evident in the behavior of smoke plumes which may take on looping behavior (Figure A18c). In late afternoon the atmosphere may become neutral, in which the environmental lapse rate equals the dry adiabatic lapse rate (Figure A18d). This results in coning of smoke plumes (Figure A18d). As the sun sets the surface begins

Figure A17 Christchurch, New Zealand, on a winter morning. The thick smoke at the surface is the result of poor ventilation due to a surface inversion layer that formed overnight. Smoke and particulate matter is emitted from home heating with solid-fuel burners (courtesy of The Christchurch Press).

to cool and the nocturnal inversion is again established (Figure A18e). If the smoke stack is above the inversion height, then lofting of the plume is observed (Figure A18e).

A common and relatively simple estimate of stability is the Pasquil–Gifford stability classification (Table A13). This scheme assumes that stability in the layers near the ground depends on net radiation as an indication of convective eddies, and on wind speed as an indication of mechanical eddies. At low wind speeds and with strong daytime insolation (i.e. radiative heating from the sun), the atmosphere will be extremely unstable (class A). As both cloud and wind speed increase, the atmosphere becomes less unstable (B, C) and may become neutral (D). Under light winds and clear skies at night the atmosphere may become stable (E and F).

Air pollution episodes are typically worse during stable atmospheric conditions since these restrict or limit the amount of vertical mixing. Temperature inversions, which are characteristic of a moderate to strongly stable atmosphere, can occur through several different mechanisms including surface cooling, warming of air aloft and *advection* (mean horizontal movement of air). Surface or radiation inversions occur during clear, calm nights as the ground cools rapidly due to the loss of longwave radiation to the outer atmosphere. Their strength and depth may be locally increased by cold air drainage. Surface cooling can also occur through evaporative cooling as water evaporates from a moist surface during the day in fine weather.

Inversions due to warming can occur through subsidence during anticyclones (as described earlier) or in the lee of mountains. In both cases the descending air warms adiabatically and creates elevated inversions which can effectively cap any mixing of pollution from below. In the lee of the Rocky Mountains in winter, radiative cooling over the interior of North America creates a very cold air mass which ponds against the mountains. Descending warm air spreads over the top of this stagnant pool and severely inhibits any upward dispersion (Oke, 1992).

Advection inversions can occur when warm air blows across a cooler surface such as a cold land surface, water body or snow cover. In this situation the cooling of the underside of the air mass creates an inversion with its base at the surface (Figure A19a). Existing inversion structures can also be modified during advection. For example, when a cool sea breeze (or rural air) flows over warmer land (or urban) areas, the inversion becomes elevated, and this can lead to serious fumigation problems if chimney stacks are located in the coastal zone (Figure A19b, c).

Complex terrain such as in coastal or alpine regions, or indeed anywhere that has significant variation in topography, can lead to complex airflows. In such areas under clear skies and light synoptic winds, local circulation systems often develop, such as land–sea breezes or mountain–valley winds. These circulation systems are not good pollution ventilators since they tend to have low wind speeds, are often closed systems and have a diurnal reversal of flow. Thus the pollution can be recirculated over the source area and may build up to high levels.

Pollutant transformation

Air pollutants may often undergo chemical transformations in the air to produce what are termed secondary pollutants. The damaging London smogs of 1952, which killed thousands of people, involved sulfuric acid fogs which remained stagnant for 4–5 days. These fogs occur when SO_2 (which is generated by the combustion of fuels – particularly from coal with high sulfur content) is oxidized in the air to sulfur trioxide (SO_3) which then reacts with water vapor (H_2O) in the presence of catalysts to form sulfuric acid mist. If the meteorological conditions are such that dispersion is poor, these toxic mists can have deadly consequences.

More common in recent decades is the formation of photochemical smog. There is a naturally occurring NO_2 photolytic cycle in which solar radiation results in the photo-dissociation

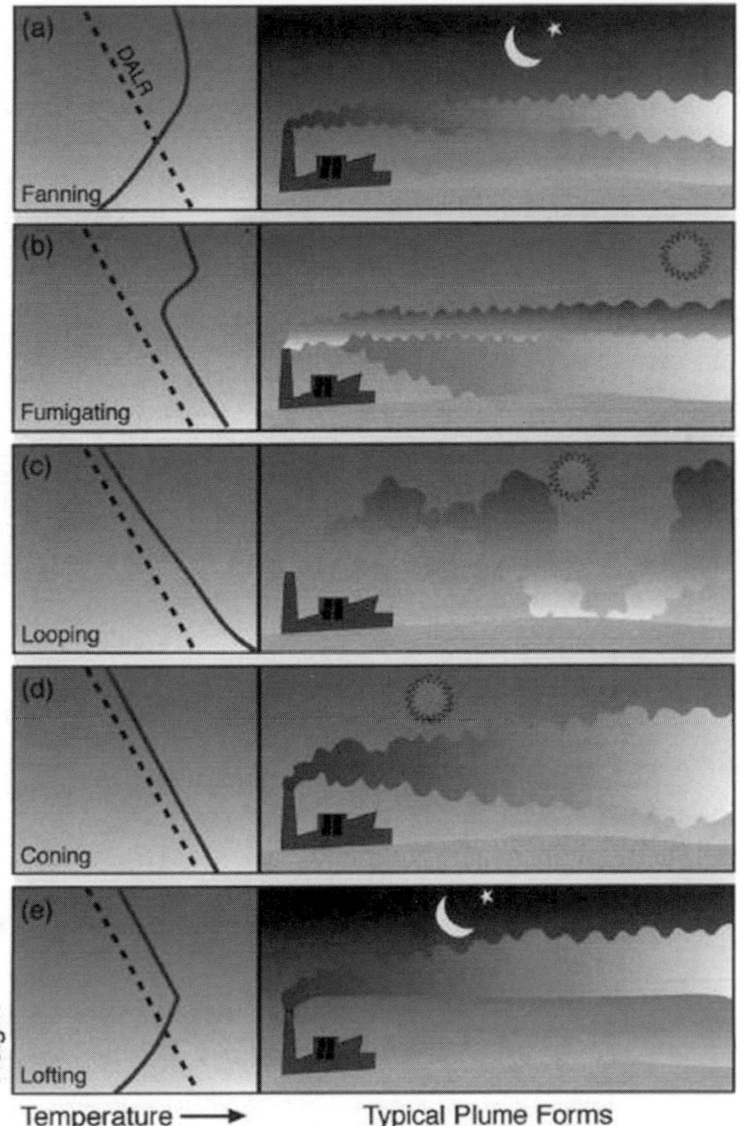

Figure A18 Stability changes throughout the day and effects on smoke plumes (modified after Whiteman, 2000). The dashed line (DALR) is the dry adiabatic lapse rate and the solid line is the environmental lapse rate.

Table A13 Pasquil–Gifford stability class

	Day-time insolation			Night-time cloudiness	
Wind speed at 10 m (ms^{-1})	Strong	Moderate	Slight	$\geq 4/8$ Cloudiness	$\leq 3/8$ Cloudiness
<2	A	A–B	B	—	—
2–3	A–B	B	C	E	F
3–5	B	B–C	C	D	E
5–6	C	C–D	D	D	D
>6	C	D	D	D	D

A, extremely unstable; B, moderately stable; C, slightly unstable; D, neutral, E, slightly stable; F, moderately stable. Insolation is the rate of solar radiation received per unit of Earth's surface and cloudiness is defined as the fraction (in octas) of sky above the local apparent horizon that is covered by clouds.

of NO_2 into NO and O (Figure A20). The oxygen atom (O) is highly reactive and combines with an oxygen molecule (O_2) to produce ozone (O_3). The ozone then reacts with NO to give NO_2 and O_2. However, the presence of reactive organic gases (ROG – which may be sourced from vehicles, but also naturally from vegetation) disrupts the cycle and can accelerate the production of ozone and other chemicals, such as aldehydes and ketones and peroxyacetyl nitrates (PAN) and even some particulates. Photochemical smog has a characteristic odor (partly due to the aldehydes), a brownish haze (due to NO_2 and light scattering by particulates) and may cause eye and throat irritation and plant damage. Figure A20 illustrates the sequence of chemical reactions that lead to formation of ozone as a by-product of photochemical smog following the emission of NO and ROG (the primary pollutants) into the atmosphere.

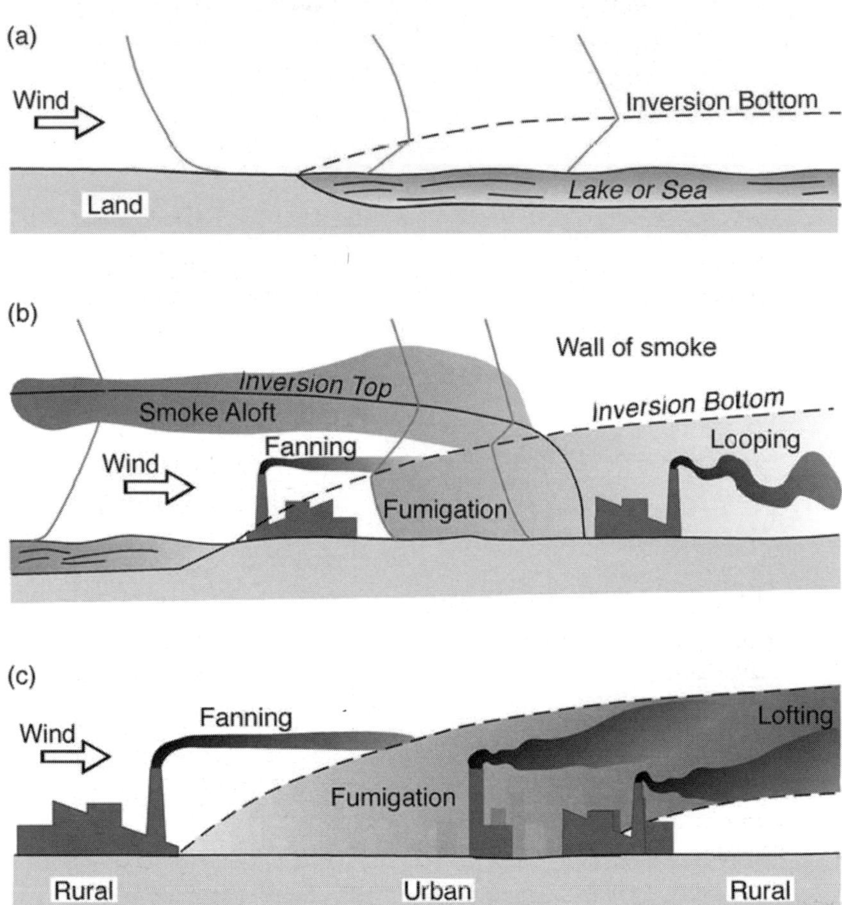

Figure A19 Examples of advection inversions. (**a**) Warm air flowing over cooler water, (**b**) sea breeze fumigation on a clear day and (**c**) fumigation over an urban area (modified after Oke, 1992 and Karl et al., 2000).

Organic peroxy radicals (RO_2) are formed by the break-up of ROG, which by reaction with NO form ozone. This type of smog has been reported from many cities around the world, including cities in Australia, Brazil, Britain, Canada, Greece, Netherlands, Iran, Israel, Japan and the United States.

Pollutant removal
As well as through chemical reaction, pollutants can be removed from the atmosphere by other processes, such as gravitational settling, and dry and wet deposition (Figure A15). Gravitational settling removes larger particulates, with the rate of removal related to the size and density of the particles and the strength of the wind. This process removes most particulates larger than 1 micrometer (μm) in diameter, but smaller particles become influenced by turbulence and may remain aloft for long periods. Dry deposition is a turbulent process in which there is a downward flux of pollutants to the underlying surface. The amount of deposition depends on the characteristics of turbulence, with increased rates under stronger turbulence.

Wet deposition involves the removal of air pollutants through absorption by precipitation elements (water droplets, ice particles and snowflakes) and consequent deposition to the Earth's surface during precipitation. This removal process includes the attachment of pollutants to cloud droplets during cloud formation, incloud scavenging (also known as *washout*), and through coalescence of rain drops with material below the cloud (*rainout*; Arya, 1999). Slin (1984) provides a comprehensive review of precipitation scavenging processes.

Receptor response

Air pollution is a concern due to its adverse effects on organisms and damage to materials. Research has demonstrated that air pollution can affect the health of humans and animals, damage plants, reduce visibility and solar radiation, and affect weather and global climate.

Effects on humans, animals and vegetation
During extreme air pollution episodes in urban areas the concentration of air pollutants can reach high levels for several hours or days. This can cause extreme discomfort, exacerbate illnesses and increase mortality rates among the most vulnerable part of the population (children, the elderly and the sick). Health-effect studies have revealed a number of adverse effects associated with common air pollutants. Peroxyacetyl nitrates (PAN), which are a

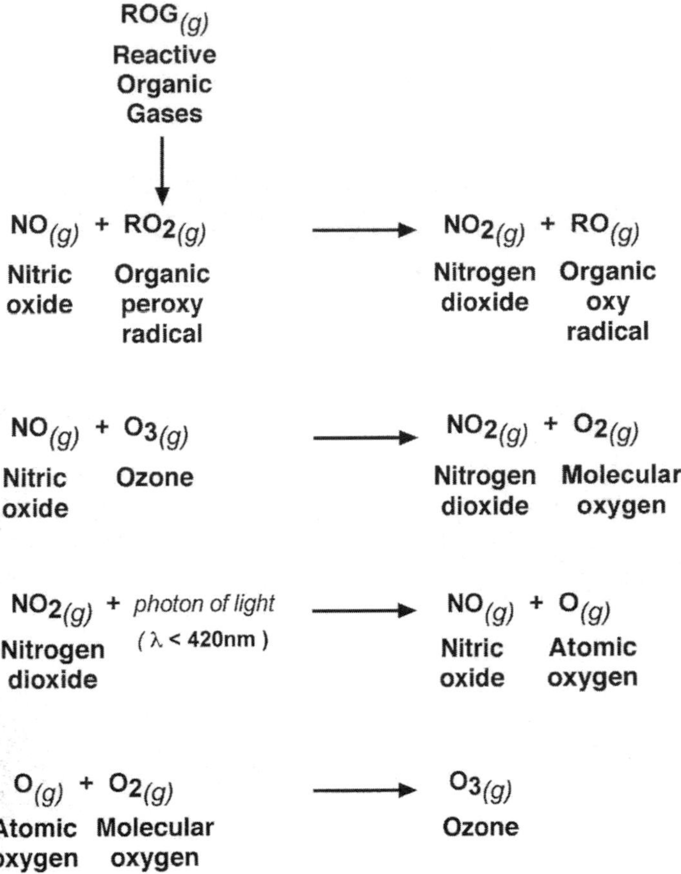

Figure A20 Chemical reactions in a polluted troposphere leading to photochemical ozone formation.

component of photochemical smog, can cause eye and throat irritation. Other health hazards of air pollutants include chronic bronchitis, pulmonary emphysema, lung cancer and respiratory infections. Even short-term exposure to particulate matter can cause an increase in the rate of asthma attacks.

Adverse effects of air pollution on vegetation include leaf damage, stunting of growth, decrease in size and yield of fruit, and severe damage to flowers. Domesticated animals such as cattle and dogs cannot only breathe in toxic air, but may also ingest pollution-contaminated feeds. The air pollutants that are of major concern are fluoride, lead, and other heavy metals and particulate matter.

Effects on materials
Air pollutants affect materials by chemical reactions and soiling. Extensive damage can occur to structural metals, building stones, fabrics, rubber, leather, paper, and other materials. For example, a well-known effect of photochemically produced ozone in the troposphere is cracking of rubber products. The number of cracks in automobile tyres, as well as their depth, has been related to ambient ozone concentrations.

Air pollution dispersion models

During the past two decades advances in computer technology have allowed the use of increasingly sophisticated air pollution dispersion models (also known as air quality models). Numerous books are now available that give an in-depth explanation of the theory and derivation behind the mathematical formulations used in air pollution models. For example, Zannetti (1990) provides information on a wide spectrum of dispersion models, while Jacobson (2000) offers an excellent and exhaustive treatise on the theory, the numerical techniques and chemistry of air pollution models. Arya (1999) also describes the diffusion/dispersion theories behind the modeling systems including detailed mathematical formulation. These models are used for both regulatory and research purposes.

Box models

The simplest air pollution models are single- and multi-box models (Lettau, 1970). Box models predict the concentration of primary pollutants inside a box using mass conservation principles, where the box represents a large volume of air over a city. A box model considers the emission source at the surface (lower boundary), advective inflow and outflow at the sides (lateral boundaries), and the evolution of the mixing depth and subsequent entrainment of pollutants from aloft. More complicated box models include chemical and photochemical transformations to predict concentrations of secondary pollutants like ozone (Demerjian and Schere, 1979; Dodge, 1977).

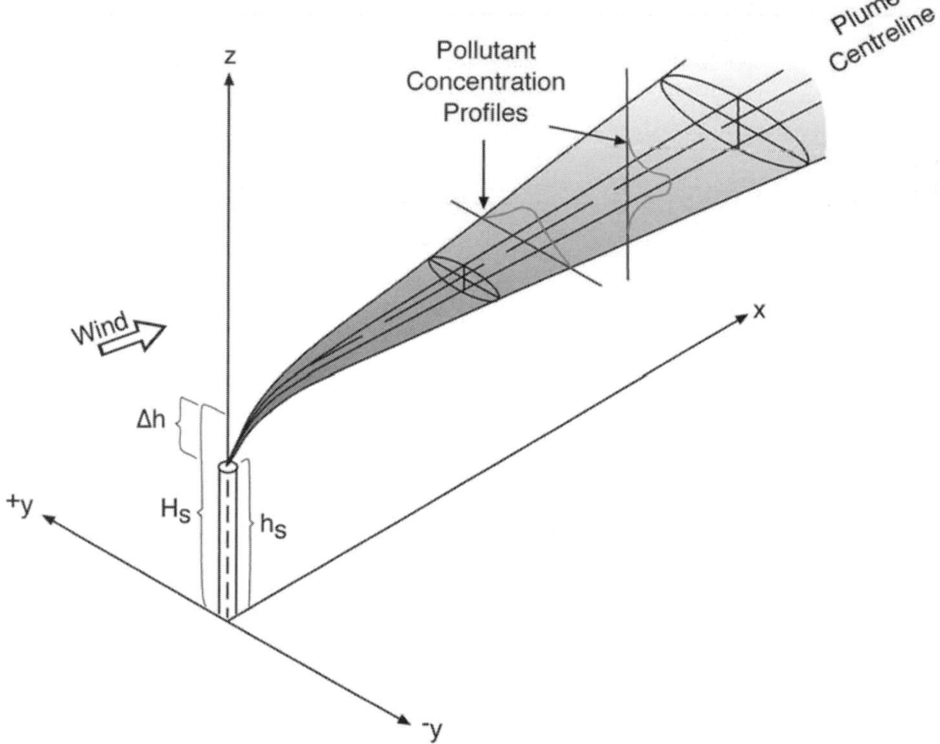

Figure A21 Schematic of plume dispersion from a stack with Gaussian distribution in horizontal and vertical (H_s is the height of the plume centerline and h_s is the stack height) (modified from Arya, 1999).

Gaussian plume models

Gaussian plume and puff models are used to estimate three-dimensional pollutant concentrations resulting from emissions generated at a point source, under stationary (unchanging) meteorological situations (Gifford and Hanna, 1973; Turner, 1964), although some models also incorporate non-stationary conditions. The underlying assumption in the Gaussian model is that with sufficiently long averaging periods (e.g. 1 hour) a bell-shaped (Gaussian) concentration distribution is produced in both horizontal and vertical directions (Figure A20). A plume model assumes a continuous point source with a uniform strong background flow, where release time and sampling time are longer than travel time. Puff models are used in cases where emission from a source is almost instantaneous and much shorter than travel time of the plume, such as accidental releases of hazardous material into the atmosphere. Gaussian models are commonly used in a regulatory framework because of their simplicity and cheap computational demands (relative to three-dimensional numerical models, which are described below). The severe limitation of the Gaussian models is that they generally do not work in calm conditions; in addition, the complexity of local terrain and variation of atmospheric conditions with time is ignored.

Numerical meteorological models

The most sophisticated and hence computationally demanding way of calculating pollutant concentrations is by simulating detailed meteorology (three-dimensional fields of wind, turbulence, temperature and water content) and coupling these fields to emission scenarios. The role of the meteorological component of the model is to provide input for the pollutant diffusion component. A numerical meteorological model uses numerical techniques to approximate meteorological equations that describe fluid flow – collectively known as the Navier-Stokes equations. These are highly complex and non-linear, and at present, no analytic solution is known for them. Stull (1998) and Pielke (2002) describe the numerical techniques and turbulence parameterizations behind these types of models. The Navier–Stokes equations take into account conservation of mass, heat, momentum and water. Air pollution scientists use many names for numerical models, but they are frequently referred to as grid-based and mesoscale.

Diagnostic models

Diagnostic models by definition do not provide a forecast for the state of the atmosphere. Diagnostic outputs are three-dimensional fields of meteorological variables obtained by interpolation and extrapolation of available meteorological measurements. They are computationally economical and appear to be effective analysis tools when the dominant factor driving the boundary layer airflow is the terrain. A major drawback is the requirement for sufficient observational data as input for the analysis. A three-dimensional diagnostic model such as CALMET – which is the meteorological component of the CALPUFF modeling system (Scire et al., 1990) – is widely used by regulatory/environmental organizations such as the US Environmental Protection Agency. The Earth Tech Corporation

Table A14 Summary of air quality standards from Canada, United States, New Zealand, and United Kingdom

Pollutant	Canada Air Quality Objectives and Standards[a]	United States Federal Primary Standards (NAAQS)	New Zealand Air Quality Guidelines[c]	United Kingdom National Standards
Carbon monoxide (CO)				
8-hour average	15 mg/m^3	10.5 mg/m^3	10 mg/m^3	11.6 mg/m^3 [d]
1-hour average	35 mg/m^3	40 mg/m^3	30 mg/m^3	—
Nitrogen dioxide (NO_2)				
Annual	100 μg/m^3	100 μg/m^3	—	—
24-hour average	200 μg/m^3	—	100 μg/m^3	—
1-hour average	400 μg/m^3	470 μg/m^3 [b]	200 μg/m^3	287 μg/m^3
Ozone (O_3)				
Annual average	30 μg/m^3	—	—	—
24-hour average	50 μg/m^3	—	—	—
8-hour average	—	160 μg/m^3	100 μg/m^3	100 μg/m^3 [e]
1-hour average	160 μg/m^3	235 μg/m^3	150 μg/m^3	—
Sulfur dioxide (SO_2)				
Annual average	60 μg/m^3	80 μg/m^3	—	—
24-hour average	300 μg/m^3	365 μg/m^3	120 μg/m^3	—
15 minute average	—	—	—	266 μg/m^3
PM_{10}				
Annual average		50 μg/m^3	20 μg/m^3	—
24-hour average		150 μg/m^3	50 μg/m^3	50 μg/m^3

[a] Values are maximum acceptable.
[b] California Standard.
[c] New Zealand has not legislated air quality standards yet.
[d] Running average.

freely distributes the code for the CALPUFF modeling system at http://www.calgrid.net/calpuff/calpuff1.htm.

Prognostic models

Prognostic models are able to forecast the evolution of the state of the atmosphere by numerically solving (both in time and space) the detailed set of approximations to the Navier–Stokes equations. Prognostic models take into account the evolution of synoptic scale winds and allow the formation of diurnally reversing local scale circulations such as sea/land breezes and valley/mountain flows (Whiteman, 2000). A great appeal of prognostic models, aside from their research applications, is that they can be the core of real-time air quality forecasting systems in case of accidental or intentional release of pollutants into the atmosphere. The forecast for the meteorological variables, in conjunction with appropriate turbulence statistics, is used to determine the probable spread and diffusion of the plume. Scientific centers that generate such forecasts and issue warnings use supercomputers, since expediency is important and prognostic models are computationally demanding. Examples of such organizations include the Atmospheric Research Division at the Commonwealth Scientific and Industrial Research Organization (CSIRO), which has implemented the Australian Air Quality Forecasting System (AAQFS), and the Lawrence Livermore National Laboratory in the United States which has the National Atmospheric Release Advisory Centre (NARAC). AAQFS routinely predicts daily levels of photochemical smog, atmospheric particles and haze, whereas NARAC is generally designed for emergency response purposes in case of accidental emissions of hazardous pollutants (e.g. radiological, chemical and biological substances).

Air quality guidelines and standards

Good health is (should be) a fundamental human right. The primary aim of air quality guidelines is to protect public health from the harmful effects of air pollution, and to eliminate or substantially reduce exposure to hazardous air pollutants. To achieve these aims, many countries use regulatory control through legislation to set legally enforceable air quality standards (AQS). These represent values that are the maximum average ground-level concentration allowed by law. AQS values are derived from air quality guidelines (AQG) determined purely from epidemiological/toxicological analysis. AQS have to consider technological feasibility; cost of compliance; and social, economic and cultural conditions. The maximum concentrations allowed are categorized by averaging periods, the most commonly used being hourly, daily and yearly. The air quality standards are different in each country. For example, in the United States they are referred to as National Ambient Air Quality Standards (NAAQS), while in Canada they are known as National Air Quality Objectives. Currently, the maximum allowable concentration for each pollutant (and for each averaging period) is prescribed differently in each standard. Table A14 shows maximum allowable concentrations for some common pollutants.

Peyman Zawar-Reza and Rachel Spronken-Smith

Bibliography

Arya, S.P., 1999. *Air Pollution Meteorology and Dispersion*. New York: Oxford University Press.

Bridgman, H., Warner, R., and Dodson, J., 2000. *Urban Biophysical Environment*. Melbourne: Oxford University Press.

Brimblecombe, P., 1987. *The Big Smoke*. London: Methuen.
Chan, C.Y., Chan, L.Y., Zheng, Y.G., Harris, J.M., Oltmans, S.J., and Christopher, S., 2001. Effects of 1997 Indonesian forest fires on tropospheric ozone enhancement, radiative forcing, and temperature change over Hong Kong Region. *Journal of Geophysical Research (D) Atmospheres*, **106**(14): 14875–14885.
Cowling, E.B., 1982. Acid precipitation in historical perspectives, *Environmental Science and Technology*, **16**: 110A–123A.
Davies, S.J. and Unam, L., 1999. Dec06 smoke-haze from the 1997 Indonesian forest fires: effects on pollution levels, local climate, atmospheric CO concentrations, and tree photosynthesis. *Forest Ecology and Management*, **124**(2–3): 137–144.
Demerjian, K.L. and Schere, K.L., 1979. Applications of a photochemical box model for O_3 air quality in Houston, Texas. *Proceedings of specialty conference on ozone/oxidants: interactions with the total environment*. Research Triangle Park, NC: US Environmental Protection Agency.
Dodge, M.C., 1977. Combined use of modelling techniques and smog chamber data to derive ozone precursor relationships. *Proceedings of the international conference on photochemical oxidant pollution and its control, Vol. II*. Research Triangle Park, NC: US Environmental Protection Agency, pp. 881–889.
Draxler, R.R., Dietz, R., Lagomarsino, R.J., and Start, G., 1991. Across North America Tracer Experiment (ANATEX): sampling and analysis. *Atmospheric Environment*, **25A**: 2815–2836.
Finlayson-Pitts B.J. and Pitts, J.N., 1986. *Atmospheric Chemistry: Fundamentals and Experimental Techniques*. New York: Wiley.
Garratt J.R., 1992. *The Atmospheric Boundary Layer*. Cambridge: Cambridge University Press.
Gifford, F.A. and Hanna, S.R. 1973. Modelling urban air pollution. *Atmospheric Environment*, **7**: 131–136.
Girardi, F., Graziani, G., van Velzen, D., et al., 1998. *ETEX – The European Tracer Experiment*. Luxembourg: EUR 18143 EN, Office for the official publications of the European communities.
Guenther, A., et al., 1995. A global model of natural volatile organic compound emissions. *Journal of Geophysical Research*, **100**: 8873–8892.
Holzworth, G.C., 1974. *Climatological Aspects of the Composition and Pollution of the Atmosphere*. Technical Note No.139. Geneva: World Meteorological Organization.
Intergovernmental Panel on Climate Change (IPCC), 2001. *Third Assessment Report. Climate Change 2001: The Scientific Basis*. (Houghton, J. et al., eds.) New York: Cambridge University Press.
Khandekar, M.L., Murty, T.S., Scott, D., and Baird, W., 2000. The 1997 El Niño, Indonesian forest fires and the Malaysian smoke problem: a deadly combination of natural and man-made hazards. *Natural Hazards*, **21**(2–3): 131–144.
Jacobson, M.Z., 2000. *Fundamentals of Atmospheric Modelling*. Cambridge: Cambridge University Press.
Jacobson, M.Z., 2002. *Atmospheric Pollution: History, Science and Regulation*. Cambridge: Cambridge University Press.
Lettau, H., 1970. Physical and meteorological basis for mathematical models of urban diffusion. *Proceedings of symposium on multiple source urban diffusion models*. Air pollution control official publication No. AP86, US Environmental Protection Agency.
McKendry, I.G., Hacker, J.P., Stull, R., Sakiyama, S., Mignacca, D., and Reid, K., 2001. Long-range transport of Asian dust to the lower Fraser Valley, British Columbia, Canada. *Journal of Geophysical Research – Atmosphere*, **106**(D16): 18361–18370.
Oke, T.R., 1992. *Boundary Layer Climates*. Canada: Routledge.
Pasquill, F., 1974. *Atmospheric Diffusion*, 2nd ed., Ellis Horwood Ltd., England: Chichester.
Pielke, R.A., 2002. *Mesoscale Meteorological Modelling*. San Diego: Academic Press.
Schnelle, K.B. and Dey, P.R., 2000. *Atmospheric Dispersion Modeling Compliance Guide*. New York: McGraw-Hill.
Scire, J.S., Insley, E.M., and Yamartino, R.J., 1990. *Model Formulations and User's Guide for the CALMET Meteorological Model*. Concord, MA: Sigma Research Corp.
Seinfeld, J.H., 1986. *Atmospheric Chemistry and Physics of Air Pollution*. New York: Wiley-Interscience.
Seinfeld, J.H. and Pandis, S.N., 1998. *Atmospheric Chemistry and Physics: From Air Pollution to Climate Change*. New York: Wiley-Interscience.
Slade, D.H., 1968. *Meteorolgy and Atomic Energy 1968*. U.S. Department of Energy, Technical Information Centre, Oak Ridge, TN.
Stern, A.C., 1976. *Fundamentals of Air Pollution*. New York: Academic Press.
Stern, A.C., Boubel, R.W., Turner, D.B., and Fox, D.L., 1984. *Fundamentals of Air Pollution*. San Diego: Academic Press.
Stull, R.B., 1998. *An Introduction to Boundary Layer Meteorology*. Boston: Kluwer.
Sturman, A.P., 2000. Applied Climatology. *Progress in Physical Geography*, **24**: 129–139.
Turner, D.B., 1964. A diffusion model for an urban area. *Journal of Applied Meteorology*, **3**: 83–91.
Wanner, H.U., 1993. Sources of pollutants in indoor air. *IARC Science Publications*, **109**: 19–30.
Whiteman, C.D., 2000. *Mountain Meteorology*. New York: Oxford University Press.
Williamson, S.J., 1972. *Fundamentals of Air Pollution*. MA: Addison-Wesley, Reading.
World Health Organization (WHO) (2000) http://www.who.int.
Zannetti, P., 1990. *Air Pollution Modelling: Theories, Computational Methods and Available Software*. New York: Van Nostrand Reinhold.

Cross-references

Acid Rain
Aerosols
Inversion
Lapse Rate
Turbulence and Diffusion
Winds and Wind System

ALBEDO AND REFLECTIVITY

Albedo

Albedo is the percentage of solar radiation reflected by an object. The term is derived from the Latin *albus*, white. A pure white object would reflect all radiation that impinges on it and have an albedo of 100%. A pure black object would absorb all radiation and have an albedo of 0%. Bright Earth features such as clouds, fresh snow, and ice have albedos that range from 50% to 95%. Forests, fresh asphalt, and dark soils have albedos between 5% and 20%. Table A15 presents representative albedos for a variety of objects. Knowledge of albedo is important because absorbed solar radiation increases the amount of energy available to the Earth's surface and atmosphere, whereas reflected radiation returns to space.

Appreciation of the relation between albedo and climate extends historically to at least classical Greek times. P. Bouguer and J. Lambert first formulated the principles and theories by which albedo and reflectivity may be explained and measured in the eighteenth century, but accurate measurements did not begin until the early twentieth century (Fritz and Rigby, 1957). The work of early investigators, including A. Ångström, C. Dorno, N. Katlin, F. Götz, H. Kimball, and others, rapidly developed an extensive body of knowledge concerning albedos that is still drawn on today (see annotated bibliography by Fritz and Rigby, 1957). One of the more interesting approaches to early observations of the Earth's planetary albedo employed measurements of earthshine and sunshine on the moon (Danjon, 1936, cited in Fritz and Rigby, 1957). Similar efforts continue today (Goode et al., 2001). Use of aircraft and spacecraft as observing platforms has significantly expanded albedo studies in recent decades (Barrett, 1974; Brest and Goward, 1987; Schaaf et al., 2002).

Table A15 Albedos for selected objects

Water surfaces	
Winter:	
0° latitude	6
30° latitude	9
60° latitude	21
Summer:	
0° latitude	6
30° latitude	6
60° latitude	7
Bare areas and soils	
Snow, fresh-fallen	75–95
Snow, several days old	40–70
Ice, sea	30–40
Sand dune, dry	35–45
Sand dune, wet	20–30
Soil, dark	5–15
Soil, moist gray	10–20
Soil, dry clay or gray	20–35
Soil, dry light sand	25–45
Concrete, dry	17–27
Road, black top	5–10
Natural surfaces	
Desert	25–30
Savanna, dry season	25–30
Savanna, wet season	15–20
Chaparral	15–20
Meadows, green	10–20
Forest, deciduous	10–20
Forest, coniferous	5–15
Tundra	15–20
Crops	15–25
Cloud overcast	
Cumuliform	70–90
Stratus (500–1000 ft thick)	59–84
Altrostratus	39–59
Cirrostratus	44–50
Planets	
Earth	34–42
Jupiter	73
Mars	16
Mercury	5.6
Moon	6.7
Neptune	84
Pluto	14
Saturn	76
Uranus	93
Venus	76
Human skin	
Blond	43–45
Brunette	35
Dark	16–22

Source: After Sellers (1965).

Reflectivity

Reflectivity is the capacity of an object to reflect solar radiation. It is described as a function of radiation wavelength and is determined by the physical composition of the object. The adjective "spectral" is frequently used in conjunction with reflectivity to indicate that reflectivity varies as a function of solar wavelength.

Representative spectral reflectivity measurements for common Earth surface features are given in Figure A22. Soil reflectivity in general increases monotonically with increasing wavelength to about 1.3 μm and then decreases with sharp dips at 1.4 μm and 1.9 μm because of absorption by soil water. Living green vegetation reflectivity is low in the visible portion of the spectrum (0.4–0.7 μm) as a result of absorption by chlorophyll and related pigments, high in the near infrared (0.7–1.3 μm) because of light scattering by internal leaf cellular structures, and decreases past 1.3 μm in a manner similar to soils due to absorption by water within leaves. Snow is highly reflective in the visible and decreases to low values in the infrared, again as a result of water absorption. Water reflects little radiation in any portion of the spectrum when solar elevation is high.

Spectral reflectivity varies significantly for each surface type as a function of physical condition and composition. Soil reflectivity varies because of variations in moisture content, particle size, organic matter content, surface roughness and mineral composition. Figure A23 presents the variation of a silty loam soil reflectivity due to changes in moisture content. Vegetation reflectivity varies with percentage ground cover, canopy geometry, leaf size, and area and plant growth stage. Snow reflectivity varies with crystal size, compaction, age, and liquid water content. Water reflectivity is affected by turbidity, depth, and phytoplankton concentrations. Also, because water in its pure form is a dielectric, its albedo increases as the angle of incidence of radiation decreases. Water albedo is lowest when the sun is near zenith and increases to near 100% when the sun is near the horizon. (For further discussion see chapter 4, Kondrotyev, 1973; chapter 8, Miller, 1981). Other factors may affect the reflectivity of these surfaces, such as lichen crusts, and other materials such as rocks, and man-made materials (e.g. asphalt and concrete) also display unique spectral reflectivity patterns.

Relation of reflectivity to albedo

Albedo is the integrated product of incident solar radiation spectral composition and the spectral reflectivity of the object. Outside the atmosphere, solar radiation spectral composition is relatively constant, peaking at about 0.5 μm, decreasing rapidly at shorter wavelengths to small amounts at 0.2 μm, and decreasing less rapidly at longer wavelengths to small amounts at about 4.0 μm.

The atmosphere selectively absorbs and scatters solar radiation. As a result, at the Earth's surface the spectral composition of solar radiation varies significantly as a function of atmospheric conditions (e.g. clouds, water vapor, and dust) and solar elevation (Robinson, 1966; Dickinson, 1983). The majority of albedo measurements have been carried out under clear-sky, high-sun elevation conditions (Table A15). Under cloudy conditions radiation is predominantly in the visible spectrum. This decreases the albedos of soils and vegetation but increases snow albedo (Miller, 1981). When atmospheric turbidity is high, or the sun is low in the sky, the spectral distribution of solar radiation shifts to the red and infrared portion of the spectrum. Soil and vegetation albedos increase and snow albedo decreases (Kondrotyev, 1973). This variability points out the need to know both the spectral reflectivity of objects and the spectral composition of incident radiation in order to evaluate earth albedos.

Surface and planetary albedos

Two global albedo measurements are of general interest to climatologists: surface and planetary albedos. Surface albedo is

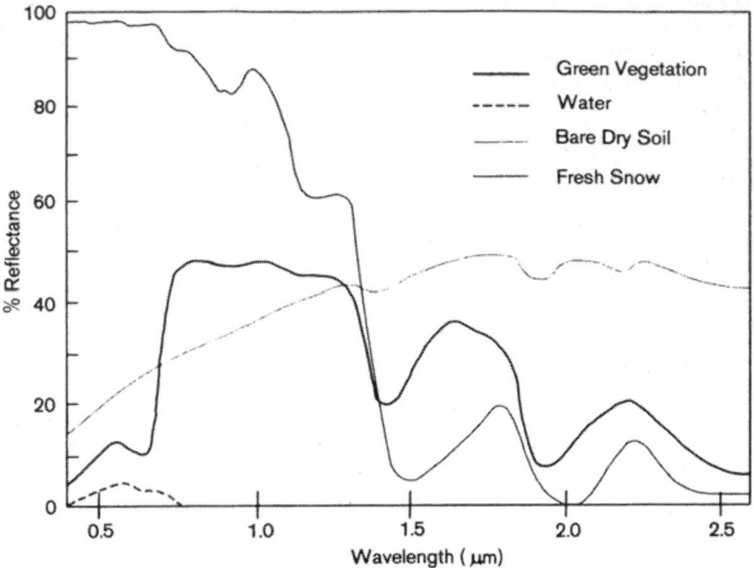

Figure A22 Spectral reflectance of selected Earth surface features.

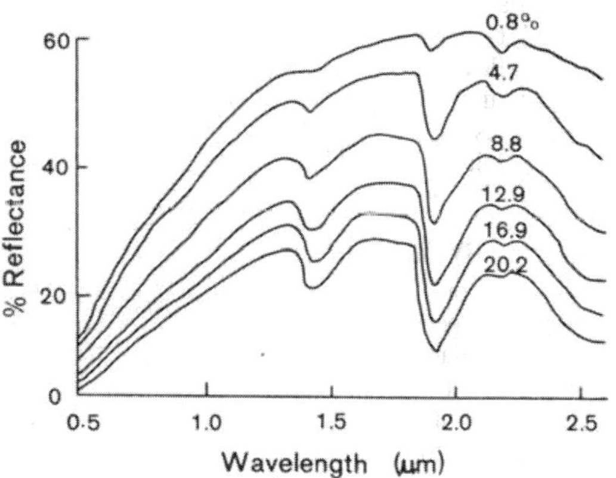

Figure A23 Variations of silty loam soil spectral reflectance as a function of water content (percentage water content shown by each plot) (after Kondrotyev, 1973).

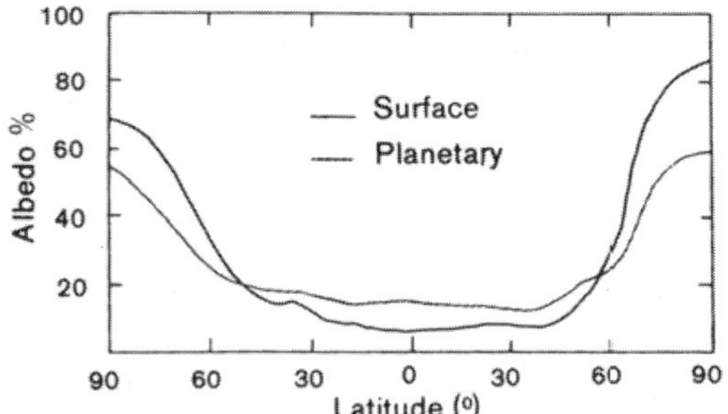

Figure A24 Latitudinal variations in surface and planetary albedo (after Hummel and Reck, 1979).

the ratio of incident to reflected radiation at the interface between the atmosphere and the Earth's land and water areas. Almost 75% of all solar energy absorbed by the Earth is absorbed at this interface; the remainder is absorbed in the atmosphere (Sellers, 1965). Any change in surface albedo will alter climate by significantly changing the amount of solar energy absorbed by the planet. Several studies have evaluated the Earth's surface albedo with resultant estimates ranging between 13% and 17% (Sellers, 1965; Kondrotyev, 1973; Hummel and Reck, 1979; Briegleb and Ramanathan, 1982). This range of results is suggestive of current limitations in knowledge of the distribution and reflectivity of Earth surface features.

Planetary albedo is the ratio between incident and reflected radiation at the top of the atmosphere, also referred to as "Bond" albedo, named after the astronomer who first described this metric. It includes the effects of reflection from the atmosphere, particularly clouds, and surface albedo. Only about 6% of the incident radiation is reflected by scattering in the atmosphere but, on average, 24% of incident radiation is reflected by clouds (Sellers, 1965). Changes in atmospheric turbidity or cloud cover can alter climate by changing the amount of solar radiation that reaches the Earth's surface. Studies of the Earth's planetary albedo have been carried out over the past 60 years (Barrett, 1974). Barrett notes that estimates have progressively decreased from 50% in early studies to current estimates between 30% and 35%, based on satellite estimates. He suggests that this trend is due to improved knowledge of global cloud cover. However, the possibility of inter-annual and longer-term variations in planetary albedo should not be overlooked (Rossow and Zhang, 1995).

Geographic patterns

Both surface and planetary albedos increase with distance from the equator (Figure A24). Surface albedo shows a slight minimum at the equator because of dense evergreen forest and a secondary minimum at 40°S latitude because of the large extent of open ocean at this latitude. Increases at 25° to 30° north and south result from the presence of subtropical deserts with relatively high albedos (>40%). At higher latitudes the seasonal or permanent occurrence of snow cover and sea ice raise average albedos, which are in excess of 60% in the polar regions. The latitudinal patterns of planetary albedo are less extreme than the surface trends. Two factors affect this difference. Tropical and midlatitude land areas that are vegetated are in latitudes of frequent cloud occurrence. Contrasts between vegetated and desert latitudes are thus less apparent in the planetary figures. In addition, interactions between surface albedo and atmospheric scattering and absorption tend to reduce albedo differences between the tropics and the poles. Where surface albedo is high, the reflected radiation passes back and forth through the atmosphere many times, increasing absorption both at the surface and in the atmosphere. Over regions of low albedo, scattering in the atmosphere increases planetary reflectance when compared to surface albedos (Hummel and Reck, 1979).

Measurements

Traditionally, two pyranometers, one pointed toward the sky, the other toward the surface, have been used to measure albedos. Pyranometers are solar radiation measurement devices that respond thermally to record the amount of radiation incident on the device. They are designed to absorb all wavelengths of solar radiation equally (Sellers, 1965). In the late 1960s, photoelectric detectors, such as silicon cells, became widely used for albedo and spectral reflectance measurements (Dirnhirm, 1968). The advantage of photon detectors is that they respond quickly to changes in incident radiation and thus permit high-resolution measurements, particularly from aircraft and spacecraft (Barrett, 1974; Justice and Townshend, 2002). One limitation of photon detectors is that they are sensitive to restricted spectral ranges. Silicon, for example, senses only wavelengths between 0.5 μm and 1.0 μm. Measurements must be either compensated for in those portions of the solar spectrum not observed, or two or more different detectors must be used. However, ease of use in the field and in aircraft and spacecraft has significantly increased their use for albedo measurements.

Human effects on Earth's albedo

Recently investigators have suggested that human modifications of the Earth's surface, accompanying continued expansion of urbanization, agriculture and forestry, may be altering the planet's albedo. For example, Otterman (1977) showed that overgrazing in desert regions can increase surface albedo by as much as 20%. Charney (1975) estimated that such changes may suppress rainfall, which would enhance the process of "desertification" that is occurring in sub-Saharan Africa. Sagan et al. (1979) proposed that extensive deforestation in tropical rainforests may significantly increase surface albedo and result in major climate changes. Such change may influence local climatic conditions, but global assessment suggests that such human changes to the Earth's land area contribute only slightly to possible global albedo changes (Henderson-Sellers and Gornitz, 1984).

Samuel N. Goward

Bibliography

Barrrett, E.C., 1974. *Climatology from Satellites*. London: Methuen.

Brest, C.L., and Goward, S.N. 1987. Deriving surface albedo measurements from narrow band satellite data. *International Journal of Remote Sensing*, **8**(3): 351–367.

Briegleb, B., and Ramanathan, V., 1982. Spectral and diurnal variations in clear sky planetary albedo. *Journal of Applied Meteorology*, **21**: 1160–1171.

Charney, J., 1975. Dynamics of deserts and drought in the Sahel, *Quarterly. Journal of the Royal Meteorological Society*, **101**: 193–202.

Dickinson, R.E., 1983. Land surface processes and climate-surface albedos and energy balance. *Advances in Geophysics*, **25**: 305–353

Dirnhirm, I., 1968. On the use of silicon cells in meteorological radiation studies. *Journal of Applied Meteorology*, **7**: 702–707.

Fritz, S., and Rigby, M., 1957. Selective annotated bibliography on albedo. *Meteorological. Abstracts and Bibliographies*, **8**: 952–998.

Goode, P.R., Qui, J., Yurchyshyn, V., et al., 2001. Earthshine observations of the earth's reflectance. *Geophysical Research Letters*, **28**(9): 1671–1674.

Henderson–Sellers, A., and Gornitz, V., 1984. Possible climatic impacts of land cover transformations, with particular emphasis on tropical deforestation. *Climatic Change*, **6**: 231–257.

Hummel, J.R., and Reck, R.A., 1979. A global surface albedo model. *Journal of Applied Meteorology*, **18**: 239–253.

Justice, C.O., and Townshend, J.R.G., 2002. Special Issue: The moderate resolution imaging spectrometer (MODIS): a new generation of land surface monitoring. *Remote Sensing of Environment*, **83**(1-2): 1–359.

Kondrotyev, K. Ya., ed., 1973. *Radiation Characteristics of the Atmosphere and the Earth's Surface*, V. Pondit (trans.). New Delhi: Amerind.

Miller, D.H., 1981. *Energy at the Surface of the Earth*. International Geophysics Series, vol. 27. New York: Academic Press.

Otterman, J., 1977. Anthropogenic impact on albedo of the Earth. *Climatic Change*, **1**: 137–155.

Petzold, D.E., and Goward, S.N., (1988) Reflectance spectra of Subarctic lichens. Remote Sensing of Environment, **24**: 481–492.

Robinson, N., 1966. *Solar Radiation*, New York: Elsevier.

Rossow, W.B., and Zhang, Y.-C., 1995. Calculation of surface and top of atmosphere radiative fluxes from physical quantities based on ISCCP data sets: 2. Validation and first results. *Geophysical Research*, **100**: 1167–1197.

Sagan, C., Toon, O.B., and Pollock, J.B., 1979. Anthropogenic albedo changes and the Earth's climate. *Science*, **206**(4425): 1363–1368.

Schaaf, C.B., Gao, F., Strahler, A.H., et al., 2002. First operational BRDF, albedo and nadir reflectance products from MODIS. *Remote Sensing of Environment*, **83**: 135–138.

Sellers, W.D., 1965. *Physical Climatology*. Chicago: University of Chicago Press.

Cross-references

Cloud Climatology
Energy Budget Climatology
Snow and Snow Cover

ALEUTIAN LOW

This is a low-pressure area over the Pacific Ocean that reflects the average position of a depression that formed on the western side of the North Pacific, developed into maturity over the Aleutian Islands, and then decayed over North America. It is located in the central part of the North Pacific and is associated with the subpolar low-pressure belt.

The Aleutian Low, in conjunction with its changing relationship with the North Pacific (Hawaiian) High in the eastern part of the Pacific Ocean, is the major feature of the general circulation controlling the weather over parts of North America. It is a semipermanent "center of action" of persistently lower pressure where barometric pressure of less than 1013 mb (29.92 in) prevails and where higher humidity and relatively higher temperatures for that latitude are common. In an analogous fashion to the Icelandic Low in the North Atlantic, the Aleutian Low is an area of average low pressure that reflects the frequent passage of cyclonic storms that migrate along the polar front from the continents that lie to the west (Akin, 1991). Consequently, the area is very stormy, with frequent strong winds, cloudy conditions, damp weather, and abundant precipitation.

The position of the Aleutian Low does change during the year. Beginning in November and continuing through February, the expanded low averages about 1002 mb (29.59 in). This causes migrating storms to move well into Mediterranean regions to the south more often as winter continues. In March the low diminishes slightly in intensity and migrating storms are less frequent by April. By July the low has virtually disappeared, showing up only as a trough extending from Kamchatka across central Alaska. By September the cycle starts again, as the low begins to form over the Bering Sea, leading to increasing disturbances.

Christoforou and Hameed (1997) studied the location of the Aleutian Low and found its location was highly correlated with mean annual sunspot numbers. When solar activity is at a minimum during its 11-year cycle the Aleutian Low migrates eastward.

Robert M. Hordon and Mark Binkley

Bibliography

Ahrens, C.D., 2003. *Meteorology Today: An Introduction to Weather, Climate, and the Environment*, 7th edn. Brooks/Cole (Thomson Learning).

Akin, W.E., 1991. *Global Patterns: Climate, Vegetation, and Soils*. Norman, Oklahoma: University of Oklahoma Press.

Barry, R.G., and Chorley, R.J., 2003. *Atmosphere, Weather, and Climate*, 8th edn. London: Routledge.

Christoforou, P., and Hameed, S., 1997. Solar cycle and the Pacific "centers of action". *Geophysical Research Letters*, Washington, DC: American Geophysical Union, **24**(3): 293–296.

Curry, J.A., and Webster, P.J., 1999. *Thermodynamics of Atmospheres and Oceans*. San Diego, California: Academic Press.

Graedel, T.E., and Crutzen, P.J., 1993. *Atmospheric Change: An Earth System Perspective*. New York: Freeman.

Hartmann, D.L., 1994. *Global Physical Climatology*. San Diego, California: Academic Press.

Lutgens, F.K., and Tarbuck, E.J., 2004. *The Atmosphere: An Introduction to Meteorology*, 9th edn. Upper Saddle River, New Jersey: Pearson Prentice Hall.

Nese, J.M., and Grenci, L.M., 1998. *A World of Weather: Fundamentals of Meteorology*, 2nd edn. Dubuque, Iowa: Kendall/Hunt.

Oliver, J.E., and Hidore, J.J., 2002. *Climatology: An Atmospheric Science*, 2nd edn. Upper Saddle River, New Jersey: Pearson Prentice Hall.

Robinson, P.J., and Henderson-Sellers, A., 1999. *Contemporary Climatology*, 2nd edn. Harlow, Longman.

Wallace, J.M., and Hobbs, P.V., 1997. *Atmospheric Science: An Introductory Survey*. San Diego, California: Academic Press.

Cross-references

Airmass Climatology
Atmospheric Circulation, Global
Centers of Action
Icelandic Low
Zonal Index

ANGULAR MOMENTUM, ANGULAR VELOCITY

An important physical concept appropriate to a rotating globe such as a planet, angular momentum is defined as the moment of the linear momentum of a particle about a point, thus:

$$M = \vec{r} \times m\vec{V}$$

where M is the angular momentum about a point 0, $\vec{r}$ the position vector from 0 to the particle, m the mass of the particle and $\vec{V}$ the velocity.

Angular momentum plays an important role in both the Earth's rotation and the coupling of the solid earth to the atmosphere. The speed of rotation (angular velocity) of the Earth varies only by small amounts, and these may be related in part to angular momentum lost to or received from the atmosphere. The atmospheric circulation patterns are modified by these energy transfers. The Earth's angular velocity is the speed of its rotation about the geographic pole, thus its pole of instantaneous rotation, and not with reference to a point in space. The angular velocity is approximately 2π/S or sidereal day of 86 164.09 seconds, roughly 24 hours, or 15.04106863 seconds of arc per second of time.

As Rossby (1940) explained it, the Earth's angular momentum can easily be visualized by taking a marble on the end of a string and rotating it on a smooth table. If the string is shortened the rate increases; if lengthened it slows. The speed is inversely proportional to radius, the product of the two remaining constant. This product, the angular momentum per unit mass, is a constant unless energy is added or subtracted (Newton's Law of Conservation of Momentum, his "first law of motion").

If part of the atmospheric envelope, rotating with the Earth at the equator at a speed of 465 m/s, moves poleward it will tend to speed up. At latitude 60° the radius (distance from axis of rotation) being reduced to half that at the equator, would indicate that the linear velocity of the Earth's surface would be reduced to 232 m/s. However, a frictionless atmosphere here (of the same mass, but halved radius) should be accelerated to double its original velocity, thus 930 m/s, but the relative speed would be higher, the Earth's surface linear velocity here being reduced by 232 m/s, the resultant relative ground speed of the atmosphere being 698 m/s (1560 mph); the direction would be easterly, since the Earth's sense of rotation is from west to east. The velocity would also be reduced owing to friction, etc. In the upper atmosphere high-velocity easterlies are often observed, but they are far from universal, being replaced by westerlies under the influence of frictional and differential heating, thermodynamic phenomena.

The easterlies have the effect of removing angular momentum from the Earth. If they were the only planetary winds they would cause a gradual deceleration of the Earth. The westerlies, however, compensate and, traveling faster than the Earth, impart angular momentum to it (through friction). The westerlies thus lose angular momentum to the Earth, and they would die out if they had no powerful source of energy. From one or possibly both sides of the westerly belts, large eddies such as cyclones and anticyclones set up by the shears of the westerlies, and in lesser degree by thermodynamic processes, bring in the necessary momentum to keep the westerlies in motion.

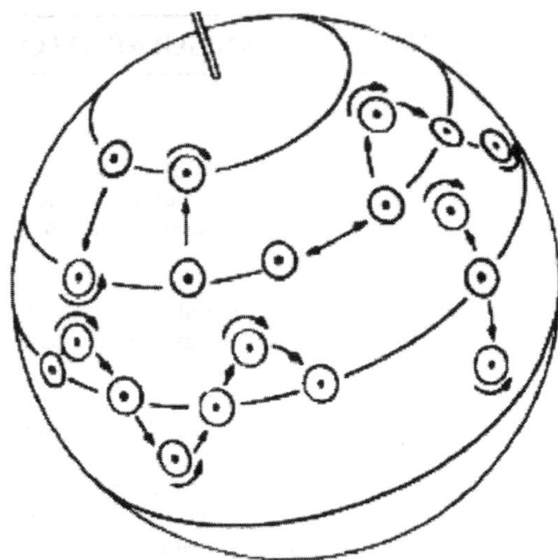

Figure A25 Planetary vorticity: angular momentum tends to be conserved in columns of constant height as they change latitude along different trajectories (from Von Arx, 1962; by permission of Addison-Wesley, Reading, MA).

An additional important principle, illustrated in Figure A25, is the conservation of angular momentum in any parcel of air (or water) as it changes latitude.

Rhodes W. Fairbridge

Bibliography

Barry, R.G. and Chorley, R.J., 1998. *Atmosphere, Weather, and Climate*, 7th edn. London: Routledge.

Clough, H.W., 1920. The principle of angular momentum as applied to atmospheric motion, *Monthly Weather Review*, **45**(8): 463.

Goody, R., 1995. *Principles of Atmospheric Physics and Chemistry.* New York: Oxford University Press.

McIlveen, R., 1992. *Fundamentals of Weather and Climate.* London: Chapman & Hall.

Nese, J.M., and Grenchi, L.M., 1998. *A World of Weather: Fundamentals of Meteorology*, 2nd edn. Dubuque, Iowa: Kendall Hunt.

Robinson, P.J., and Henderson-Sellers, A., 1999. *Contemporary Climatology*, 2nd edn. Harlow: Longman.

Rossby, C.G., 1940. Planetary flow patterns in the atmosphere. *Quarterly Journal Royal Meteorological Society*, **66**: 68–87.

Thompson, R.D., 1998. *Atmospheric Processes and Systems.* London: Routledge.

Von Arx, W.S., 1962. *An Introduction to Physical Oceanography.* Reading, MA: Addison-Wesley.

Wallace, J.M., and Hobbs, P.V., 1997. *Atmospheric Science: An Introductory Survey.* San Diego, CA: Academic Press.

Widger, W.K., 1949. A study of the flow of angular momentum in the atmosphere. *Journal of Meteorology*, **6**: 291.

Cross-references

Atmospheric Circulation, Global
Jet Streams
Vorticity
Zonal Index

ANTARCTIC CLIMATES

Introduction

Climate and the polar regions

Ice plays a critical role in polar climate (Bintanja, 2000). In contrast with lower latitudes, the high albedo and low thermal conductivity of snow cover and ice (sea ice, floating ice shelves, continental ice sheets) have associated negative values of the surface radiation budget (e.g. Nakamura and Oort, 1988). Accordingly, low temperatures occur at the surface and in the free atmosphere in polar regions, and the troposphere has low thickness and tends to be barotropic. The low average temperatures mean a low water-holding capacity of the air, realized as low precipitation amounts annually (i.e. "polar desert"). Because of the atmosphere's extreme static stability over ice-covered surfaces, clouds and precipitation tend not to be generated *in situ*; rather, they are advected in from lower latitudes by frontal cyclones. This poleward movement of moisture is a consequence of the inflow of energy into polar regions necessitated by the net-radiation deficit. Although these climatic characteristics are common to both polar regions on Earth, there are physical geographic major differences between the Arctic and Antarctic, manifest in their respective climates. Most notably, surface and free-air temperatures, along with cloud amounts, average lower in the Antarctic than the Arctic (Weller, 1982).

The Antarctic versus the Arctic

Whereas the Arctic is a landlocked ocean covered most of the year by sea ice, Antarctica is a polar continent upon which rests a thick ice sheet. The ice sheet rises to over 3000 m in East Antarctica (Figure A26), with a small area ("pole of inaccessibility") exceeding 4000 m. Moreover, Antarctica is surrounded by the Southern Ocean: the only interruption to the zonal configuration of the continent occurs in the Antarctic Peninsula (AP). The extreme continentality results from the high elevation of the ice sheet and the continent's isolation from other land masses (Carleton, 1992). In contrast, the Arctic is linked directly with middle latitudes via North America and Eurasia. The greater longitudinal variation of land and sea in boreal regions is expressed as a more meridional average pattern of the tropospheric waves and, accordingly, with enhanced transport of heat and momentum by the waves occurring between middle and high latitudes (van Loon, 1979). Over Antarctica the air is often poorly mixed, especially in austral winter and spring. This promotes a colder circumpolar vortex compared with that in the Arctic, and largely is responsible for the more extensive and intense stratospheric "ozone hole" of southern high latitudes.

Antarctica virtually doubles its own area between the early fall and spring because of the growth of sea ice (Figure A27). This large annual increase in sea ice occurs because, unlike the Arctic Ocean, the equatorward margin of Antarctica is unconstrained by land, and because atmospheric and oceanic mean circulations are relatively zonal (Carleton, 2003a). Interannual variations of Antarctic sea-ice conditions (its latitude extent, ice-water concentration) are most pronounced regionally (Parkinson, 1992), impacting the surface temperatures and precipitation. In addition to its negative feedback with surface

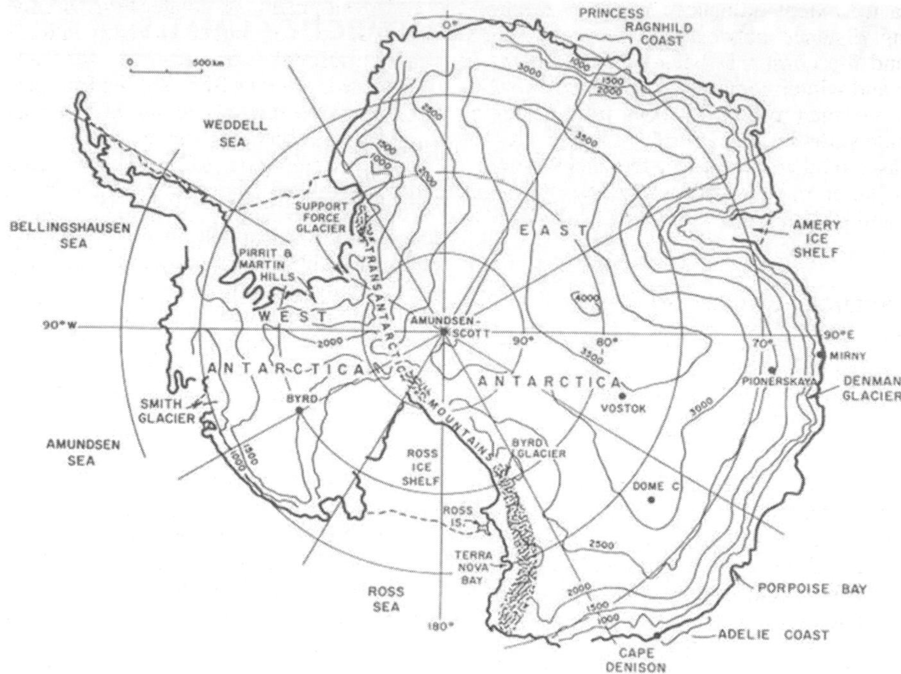

Figure A26 Map of Antarctica showing major physical geographic features (coastline, Antarctic seas, mountain ranges, ice-shelf margins, ice-sheet contours) and locations of selected permanent manned bases (named). (From Parish and Bromwich, 1991.)

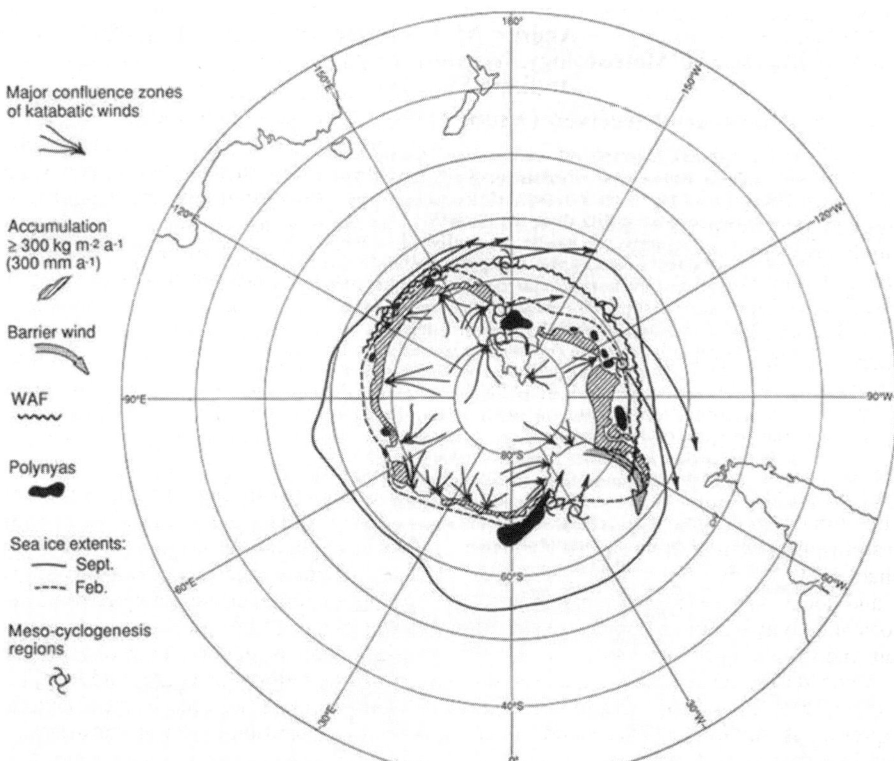

Figure A27 Schematic summarizing important features of Antarctic-region climate (side legend), as derived from multiple sources (map reoriented with respect to that appearing in Figure A26). (From Carleton, 1992.)

temperature, the sea ice extent influences snowfall in coastal areas: the increasing distance between the moisture source (Southern Ocean) and the coast that occurs as sea-ice extent increases during fall and winter, reduces snowfall amounts. Sea ice–climate interactions manifest fluctuations in atmospheric circulation that include shifts in the preferred "storm tracks" of synoptic cyclones, as modulated by lower-frequency climatic teleconnections. The latter involve particularly the "Antarctic Oscillation" (AAO) and El Niño Southern Oscillation (ENSO).

Developments in Antarctic climatology

Major advances have occurred in our understanding of Antarctic meteorology and climatology since the first *Encyclopedia of Climatology*, "Antarctic Climates" item was written 20 years ago. In that time the major developments have been as follows:

1. Improvements in the routine observational network. For the surface, automatic weather stations (AWS) provide data on pressure, temperature and winds, as frequently as every 15 minutes, and their spatial density has increased since 1980, especially in the Ross Sea area. AWS data have been used to improve the representation of 500 hPa height fields over the continent via the atmospheric thickness relationship (Phillpot, 1991); to characterize temperature–sea-ice extent associations locally (Carleton et al., 1998); and to better determine the surface energy balance (Reijmer and Oerlemans, 2002). Satellite retrievals of climate variables additional to those acquired by visible and infrared sensors, include scatterometer-derived near-surface winds over ice-free ocean, ice-sheet altitude and topography from radar altimetry, surface "skin" temperatures and sea ice extent and concentration, including open-water leads and polynyas, from microwave radiometers; stratospheric ozone from atmospheric sounders; and upper-ocean biological activity from shortwave narrow-band sensors. Crucial in these developments has been the establishment of satellite direct-readout facilities on the continent, and the ability to include Antarctic data in global real-time synoptic analyses. Thus, our understanding of Antarctic contemporary climates comes from an array of data sources; some conventional, some remotely sensed, yet with different lengths of record.
2. Advances in numerical modeling of Antarctic climates; not just using atmospheric General Circulation Models (GCM) but also coupled ocean-ice–atmosphere models (e.g. Hall and Visbeck, 2002). In particular, mesoscale meteorological models developed for middle-latitudes are now applied to polar environments (e.g. Polar MM5: Guo et al., 2003). More reliability can be placed in modeling the following features of Antarctic climate: the impacts of changes in sea-ice conditions on atmospheric circulations in the southern extratropics (Raphael, 2003a); the important role of ice-sheet topography and local sources of heat and moisture associated with coastal polynyas, for the surface katabatic winds and cold-air mesoscale cyclone systems (*mesocyclones*) generated in coastal regions of Antarctica (van den Broeke and van Lipzig, 2003); and understanding how the ENSO signal is expressed in Antarctica. These studies confirm that Antarctica interacts actively with lower latitudes, and even may influence climate in parts of the northern hemisphere at certain times of year (Hines and Bromwich, 2002).
3. Development of longer-term archives of daily synoptic analyses for higher southern latitudes, as compiled from operational analyses of the Australian Bureau of Meteorology (ABM) and the European Centre for Medium-Range Weather Forecasts (ECMWF), for studying the circulation climates of higher southern latitudes. More recently, the application of automated techniques for tracking cyclones and anticyclones in these data has greatly extended our understanding of Antarctic-region synoptic climatology (e.g. Simmonds et al., 2003). Reanalysis data sets – meteorological fields generated by an analysis/forecast model that is fixed rather than changing through time, and in which all available data are included – particularly are illuminating Antarctic-region circulation climate (Pezza and Ambrizzi, 2003). Composite studies of synoptic phenomena using reanalysis data have identified the typically occurring circulation environments of mesocyclones, explosively deepening cyclones, and extremely strong wind events (Carleton and Song, 1997; Murphy, 2003). However, reanalysis datasets are not equally reliable for climate change studies: the NCEP-NCAR reanalyses of SLP generally are considered inferior to those of the ERA (ECMWF reanalyses) over southern high latitudes.
4. Dedicated intensive observing periods and special monitoring programs to study Antarctic-region weather and climate. In the mid-1990s a concerted effort to better depict and understand Antarctic synoptic meteorology and climatology was realized with the FROST (First Regional Observing Study of the Troposphere) project (Turner et al., 1999). An outgrowth of FROST was the drive to develop an updated long-term climatology of surface and free-atmosphere conditions for the Antarctic manned stations as part of the SCAR (Scientific Committee for Antarctic Research) READER (Reference Antarctic Data for Environmental Research) project. Stringent quality control of the observation data acquired by the nations which are signatories to the Antarctic Treaty, is permitting the reliable determination of annual and seasonal trends and changes in climate. For longer time scales, recent scientific traverses, particularly the shallow ice cores drilled in West Antarctica as part of ITASE (International Trans-Antarctic Scientific Expedition), are permitting an assessment of climatic variability and changes over about the last two centuries at mesoscale spatial resolutions.
5. Improved determination of the variability (spatial, temporal) of Antarctic surface energy budgets, over both sea ice and continental ice. These have involved shipboard measurements in the sea ice zone, as well as satellite remote sensing (Wendler and Worby, 2001).
6. Refinements in depicting the spatial patterns and temporal variability of variables significant to the ice sheet mass balance (Vaughan et al., 1999); notably, precipitation and the near-surface katabatic winds (Guo et al., 2004).
7. Application of newer analytical and statistical techniques to determining the links between coarse-resolution climate data, such as mapped fields of geopotential height generated from reanalyses or GCM, and local climate conditions (i.e. "downscaling") (Reusch and Alley, 2002). These methods include artificial neural networks and "self-organizing maps".
8. Development of satellite-image based "climatologies" of mesocyclones for several genesis key regions around Antarctica. Moreover, there is a realization that these storms

are important for the snowfall and wind climatologies of coastal areas and embayments. Accordingly, attempts are being made to more accurately forecast these developments.

9. The addition of another 20 years' data has permitted identification of temporal trends and longer-term changes in Antarctic climate (atmospheric variables, sea ice conditions, ocean temperatures)). In particular, opposing trends of surface temperature between the AP and the rest of Antarctica have become evident (see "Antarctica and global change", below).
10. A realization that Antarctica is not isolated from the weather and climate processes of extratropical latitudes, or even the northern hemisphere. Rather, Antarctica is intimately connected with other places via atmospheric teleconnections, primarily the ENSO.
11. Identification and characterization of the dominant modes of variability (subseasonal, interannual, subdecadal) in the atmospheric circulation of higher southern latitudes, and links with the upper ocean circulation. These modes include particularly the ENSO, but also the AAO, the "Pacific–South America" (PSA) pattern, and an "Antarctic Circumpolar Wave" (ACW) that was discovered in the mid-1990s.

Accordingly, this revised item emphasizes particularly the above-noted developments in Antarctic climatology.

Antarctica and global climate change

Documenting Antarctic-region climate and its temporal variations has taken on increased urgency in recent decades, in the context of "global change". The following climate trends and changes, and their possible global associations, have spurred the increased monitoring of Antarctic weather, climate, biology, glaciology, and oceanography, particularly using satellites (e.g. Schneider and Steig, 2002).

1. "Global warming". This is likely to be evident earliest in polar regions, owing to "ice–albedo feedback". The latter involves a positive link between reduced snow/ice extent associated with warming, greater absorption of solar radiation by the surface, and continued warming. With respect to Antarctica, the following have been observed: a strong warming trend in the AP over the past several decades (Harangozo et al., 1997), especially in winter (van den Broeke, 2000a), that has been accompanied by reduced sea-ice extent west of the AP; the recent retreat and collapse of ice shelves such as Larsen-B on the eastern AP (Vaughan and Doake, 1996); an observed freshening of the upper ocean in the Ross Sea that suggests increased precipitation and/or melting from the West Antarctic ice sheet (Jacobs et al., 2002); and a subsurface warming concentrated within the Antarctic Circumpolar Current (ACC) (Gille, 2002). GCM studies (e.g. Wu et al., 1999) support a close link between recent climate warming and observed sea ice changes in Antarctica that mostly involve reduced ice–water concentration.
2. Stratospheric ozone depletion over the Antarctic in austral spring is a major contributor to global-scale reductions of ozone. Moreover, interannual variations in high-latitude circulation affect the configuration, spatial extent, intensity, and seasonal persistence of the ozone hole (e.g. Thompson and Solomon, 2002). The increased receipt of ultraviolet radiation in higher southern latitudes during spring and summer has deleterious consequences for organisms, including humans (e.g. Jones et al., 1998). In regions of Antarctica where cooling has occurred, a physical link with ozone depletion via the AAO has been proposed (e.g. Sexton, 2001).
3. The observed role of larger-scale atmospheric circulation patterns and their teleconnections for interannual variations and recent changes in climate over extrapolar as well as polar latitudes. These include surface temperature trends and summer melt periods linked to the AAO (Torinesi et al., 2003), and the sensitivity of the Antarctic ice sheet's mass balance to fluctuations in moisture transport, especially in the Pacific sector, where there is a strong ENSO signal (Bromwich et al., 2000).
4. Since the mid- to late-1970s the change in the dominant Semi-Annual Oscillation (SAO) of climate variables, especially SLP, temperature and precipitation in sub-Antarctic latitudes and coastal Antarctica, has been accompanied by warming in the tropical Pacific and cooling over large areas of the ice sheet (Hurrell and van Loon, 1994). For the same period the ozone hole worsened.
5. The strong interactions glimpsed between biological productivity in the Southern Ocean, and circulation–climate variations expressed as changes in sea surface temperature (SST), sea ice extent, and zonal wind speed.
6. The role of Antarctica in long-term global-scale climate changes (glaciations, deglaciations), via its influence on the deep-water ocean circulation. Moreover, contemporary climate processes in the Antarctic appear crucial influences on deep water formation; notably via the polynyas and leads of coastal and sea ice areas generated by strong offshore (katabatic) winds, and by the spin up of polar mesocyclones (see Figure A27). Mesocyclones extract large amounts of energy from the upper ocean, increasing the salinity and density of the water, which then sinks to great depths (Carleton, 1996).

Antarctic climate(s)

No single "Antarctic climate" can be identified (e.g. Bintanja, 2000). Regional ice–atmosphere interactions change markedly with latitude, elevation, season, and exposure to heat and moisture sources (either from open ocean or within the sea ice zone), and with the passage of synoptic systems (Schwerdtfeger, 1984). The seasonal sea ice zone (SSIZ) near maximum extent is considerably further equatorward than the Antarctic Circle, but it remains relatively close to the continent in longitudes south of Australia (Figure A27). The AP frequently exhibits a pattern of climate variability different to that of the rest of Antarctica (Rogers, 1983). The influence of the southern polar region in the global oceanic and climatic environment demands a broad definition, so here "Antarctica" is taken to include the continent and its ice shelves, the SSIZ, and the adjacent ocean area including the ACC (see Figure A26). A discussion of the radiation and energy climates of these higher southern latitudes, and their influence on the atmospheric circulation, is given below in the context of the general circulation, as well as synoptically and subsynoptically. Accordingly, regional variations in these climatic factors are also discussed.

Radiation climates and the energy balance

The radiation balance is the principal climate-causing factor (World Meteorological Organization, 1967) in the Antarctic, where there are strong seasonal variations in solar radiation as well as variations in elevation and surface characteristics. The high reflectivity of the surface (i.e. surface albedo) means that very little insolation is absorbed; thus keeping surface temperatures low all year. Satellite monitoring (visible, infrared, microwave) provides synoptic information on variables crucial to an accurate determination of the radiation balance, such as the extent, thickness and concentration of sea ice; accurate albedo measurements, and cloudiness. For modeling the surface energy balance, these satellite data augment ground-level observations.

Surface energy balance equation

The energy balance equation for unit horizontal area at Earth's surface is:

$$Q^* = K\downarrow (1-\alpha) + L^* = Q_H + Q_E + Q_G$$

Where $K\downarrow$ = global (direct plus diffuse) solar radiation; α = albedo, or fraction of shortwave reflected; L^* = net longwave radiation ($L\downarrow - L\uparrow$) (where $L\downarrow$ = atmospheric downward and $L\uparrow$ the terrestrial upward); Q_H = turbulent transfer of sensible heat; Q_E = turbulent transfer of latent heat (E = evaporation); Q_G = subsurface heat flux through ice, soil, or water. The conductive flux (Q_G) approximates to zero on an annual basis. Over snow- or ice-covered surfaces the radiative fluxes dominate the turbulent convective terms of Q_H and Q_E. However, over ice-free ocean or areas of low sea-ice concentration or small ice thickness within the pack, the convective terms become important, at least locally.

Solar radiation ($K\downarrow$)

The global radiation ($K\downarrow$) varies from virtually zero in midwinter south of the Antarctic Circle to values comparable to the highest on Earth over the ice sheet in December and January. The high summer insolation is due to lack of clouds, elevation of the Antarctic plateau, a highly transmissive (low aerosol and water vapor content) atmosphere, 24 hours of daylight, and the occurrence of perihelion. These high values are not reduced much even by periods of cloud cover; a decrease in the direct (clear-sky) component is more or less balanced by an increase in the diffuse radiation.

Annual total received solar radiation is relatively high over the continent, particularly for East Antarctica, but decreases over the pack ice zone to a minimum over the sub-Antarctic oceans. This gradient is due to the increasing cloudiness associated with traveling cyclonic systems and proximity to the Antarctic Circumpolar Trough, ACT (see "Climatic variables; Cloudiness", below).

Albedo

The Antarctic experiences an annual negative net radiation balance largely due to high surface albedo. Albedo values are highest over the permanent snow cover of the continental interior, where around 83% of the incident solar radiation is reflected, and lowest over the ocean (around 11%). Over the coastal sea ice and pack ice zones the albedo is strongly dependent on the presence of meltwater (low α), the presence of snowcover (increases α), and the amount of open water that includes substantial polynyas (Wendler and Worby, 2001). Wendler et al. (1997a) found that midsummer values of albedo for 10/10 ice cover averaged 59%, but with hourly values increasing to 76% where snow covered. Average values of albedo for 5/10 ice concentration were around 30%.

Although the surface albedo increases generally with latitude, the greater cloudiness over sub-Antarctic oceans modulates the effect of short-term variations in sea–ice concentrations on the planetary albedo. Over the continent, determination of cloud amount from space is difficult and leads to large discrepancies in computed solar radiation budget if not constrained using available surface observations (Hatzianastassiou and Vardavas, 2001). In summer the maximum variation in surface albedo occurs at about latitudes 60–70°S (78%), where it is dominated by the change from open water to ice cover. Seasonally, this constitutes about a 60% change in albedo. Between latitudes 80°S and 90°S the annual range does not exceed 10%.

Net longwave radiation (L^*) and the Antarctic inversion

The net longwave radiation (L^*) at the surface is a function of surface and atmospheric temperatures, and of the downward radiation from clouds, water vapor and aerosols. L^* is uniformly negative at the surface over southern high latitudes in winter. In summer there are large differences between the continent and peripheral oceans (50–60°S), due mainly to cloud cover effects. Increasing cloud cover tends to increase L^* over the ice sheet because $L\downarrow$ becomes more important over high albedo surfaces. Over sea ice, L^* varies due both to cloud amount and the ice–water concentration ($L\uparrow$ is reduced with increasing ice cover, Table A16). Satellite observations of the outgoing longwave radiation at the top of the Antarctic atmosphere show strong negative values in both summer and winter.

The intense radiational cooling and highly transmissive atmosphere produce a persistent low-level temperature inversion that reaches its greatest depth over the higher elevations of the ice sheet, where it is present almost the entire year (Schwerdtfeger, 1984). The inversion is no stronger at the end than at the beginning of the polar night, because an equilibrium is reached between $L\uparrow$ (decreases as surface temperature falls), and $L\downarrow$ from the atmosphere above the inversion layer, which changes relatively little with time. Combined with the slope of the ice surface, the semipermanent inversion is the source of the persistent katabatic winds (Connolley, 1996). A clearly defined time for the occurrence of the inversion maximum temperature is absent in the Antarctic, comprising the "coreless winter". This feature is evident for surface temperatures (T_s) at the Antarctic stations, including the AWS. This is because the rapid radiational cooling begun in the fall is arrested and, in many years, reversed in winter. Changes in the tropospheric longwave

Table A16 Mean outgoing longwave radiation ($L\uparrow$) and surface temperature (T_s) for different Antarctic sea-ice concentrations, from shipboard observations in the period 24 December, 1994 to 6 January, 1995

	Ice concentration in tenths		
	0–4	4–7	7–10
$L\uparrow$ (W m^{-2})	292	287	276
T_s (°C), for ϵ = 0.97[a]	−3.2	−4.4	−7.0

[a] ϵ = emissivity.
Source: After Wendler et al. (1997a).

Table A17 Mean fluxes of energy budget (EB) components for interior Adélie Land, Antarctica, in the period 20 November–22 December 1985. Values are expressed in $W\,m^{-2}$ and percent, and are positive (negative) toward (away from) the surface. Q_B replaces Q_G and is the snow heat flux. I = imbalance (3.4%), resulting from measurement errors

	EB components					
	K^*	L^*	Q_H	Q_E	Q_B	I
Heat balance: ($W\,m^{-2}$)	+72	−70	+9	−9	−5	+3
Percent	86	84	10	10	6	3

Source: After Wendler et al. (1988).

pattern between early winter and midwinter, and the consequent poleward movement of warmer air, help explain this temperature feature, which comprises part of the Semi-Annual Oscillation (SAO) of SLP and tropospheric height (see "Climatic variables; Temperature").

Net (all-wave) radiation (Q^*) and energy balance

For the South Pole, Carroll (1982) estimated that 85–90% of the winter deficits of Q^* (i.e. the net "all-wave" radiation, or net shortwave minus net longwave) are made up of sensible heat (Q_H) losses. Table A17 summarizes the energy budget over Adelie Land in East Antarctica, for the period 20 November–22 December 1985, emphasizing the importance of the radiation terms. In the coastal ablation zone, Q^* is positive from late September through late February. Q^* becomes slightly positive over much of the Antarctic interior around the time of maximum solar elevation, due to increased cloud cover and a downward flux of Q_H.

The annual deficit of Q^* necessitates a net influx of energy to Antarctica. This occurs primarily as eddy sensible heat advected by frontal cyclone systems migrating into high latitudes. Because the amplitudes of southern hemisphere planetary waves are reduced, on average, compared with their northern hemisphere counterparts, the waves themselves transport relatively little sensible heat into the Antarctic. Instead, the remainder of the heat transport is carried out by the ocean, whereby divergence (convergence) of air in the near-surface wind field occurs on the western (eastern) side of a standing tropospheric trough, promoting upwelling of cold (downwelling of warm) water.

Turbulent exchange and the role of sea ice in the energy balance

Seasonal and interannual variations in sea ice conditions (areal extent, ice–water concentration, presence/absence of snow cover, ice-edge latitude) profoundly influence the surface energy budget and climate. The sea ice cover modulates the oceanic fluxes of heat to the atmosphere, and thereby its own thickness (Wu et al., 2001). The average thickness of Antarctic sea ice is around 1 m or so in late winter, except where deformation occurs and it reaches thicknesses of up to 4 m. During the 1990s, which was broadly representative of the longer period 1978–2000, a trend to increased Antarctic sea–ice extent accompanied reduced ice–water concentration.

Close to the continent the sea ice tends to have higher concentration and a longer seasonal duration than further out. Moreover, when sea ice is advected equatoward (poleward) by winds and the ocean circulation, it undergoes divergence (convergence), which leads to increasing ice extent and decreasing ice concentration (decreasing ice extent and increasing ice concentration). These associations are evident particularly in the major embayments, notably the Weddell Sea, and on interannual time scales (Carleton, 1988; Turner et al., 2002b). Anomalies of this type appear related to the AAO and the ENSO). Additionally, the presence of low ice-concentration areas and open-water leads and polynyas can significantly influence the local to regional-scale climate (e.g. as increased cloud cover and precipitation). This occurs owing to the enhanced oceanic heat losses to the atmosphere as Q_H and Q_E, especially in winter. Then these fluxes can be at least one order of magnitude greater than those of adjacent areas of higher sea ice concentration.

Table A18 summarizes the radiation and energy fluxes for a ship transect through variable sea ice conditions in Antarctic longitudes of the Tasman Sea during summer (Wendler et al., 1997a). In addition to showing the very strong contribution of the radiative fluxes to the surface energy balance, and of the ocean as a heat sink, these results reveal that the fluxes of Q_H and Q_E are of approximately similar magnitude but opposite sign: Q_H is a source of energy to the surface, Q_E is a sink. In late winter, as sea ice concentration decreases below about 50%, the convective fluxes change relatively little compared with those from open ocean (Worby and Allison, 1991).

Climatic variables

Temperature

Setting the surface energy balance to zero determines the equilibrium T_s. Mean winter T_s of −60°C to −70°C occurs on the highest areas of the ice sheet. Maps of mean T_s (Figure A28) indicate that the cold pole is displaced into East Antarctica, both in January and July (Borisenkov and Dolganov, 1982). There is a general decrease of temperature with latitude, especially evident in the AP. Considerable interannual variability of

Table A18 Summary of radiation and energy fluxes ($W\,m^{-2}$) through the Antarctic sea ice for the period 24 December, 1994 to 6 January, 1995 (variable ice concentrations). Note the dominance of the radiative fluxes. As in Table A17, the fluxes are positive (negative) toward (away from) the surface. Q_B = ocean or ice heat flux or phase changes

	EB components						
	$K\downarrow$	$K\uparrow$	$L\downarrow$	$L\uparrow$	Q_H	Q_E	Q_B
Average fluxes	241	−93	252	−285	11	−17	−109
	$Q^* = 115$						
Percent of sources (+) or sinks (−)	+91.3%				+8.7%	−13.5%	−86.5%

Source: After Wendler et al. (1997a).

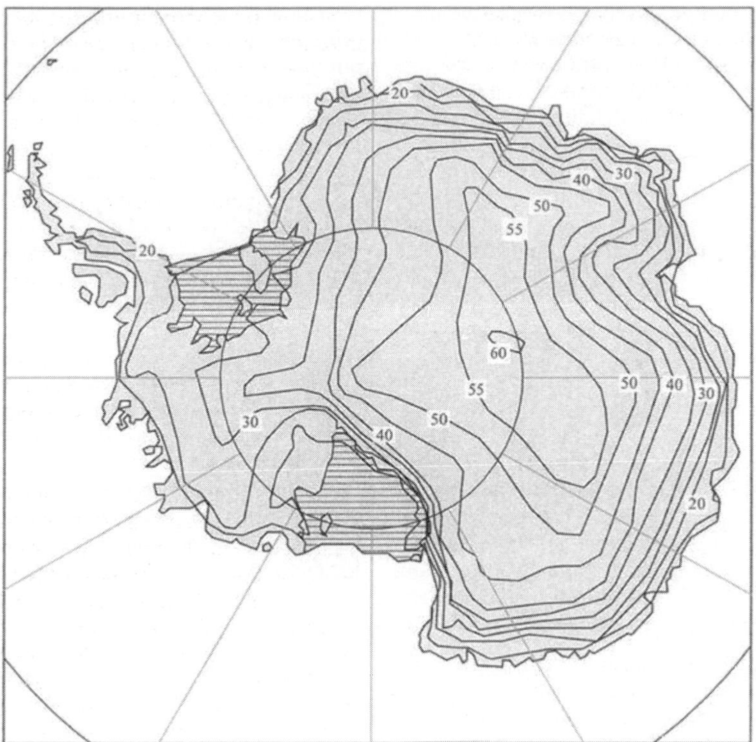

Figure A28 Map showing mean annual near-surface temperature field for Antarctica, synthesized from surface observations; isotherms in degrees below 0°C. (From Guo et al., 2004.)

T_s occurs, notably along the western AP and for stations south of the ACC (60–70°S) because of their position within the SSIZ (the temperature gradient in the latter zone increases to above 1°C/1° latitude in winter, compared with about 0.6°C/1° latitude between 30°S and 60°S). Some of the interannual variability of temperature is related to ENSO (Schneider and Steig, 2002). SST gradients are steepest in the vicinity of the ACC, comprising an Ocean Polar Front (OPF), but strong thermal gradients also border the continent, particularly in winter. On average the meridional strong gradients of temperature near the surface disappear in the mid-troposphere (van Loon et al., 1972).

Trends in Antarctic temperatures are evident over the past several decades. Although the sign and magnitude of recent temperature trends differ somewhat by study and the method of spatially extrapolating the sparse data (cf. Doran et al., 2002; Turner et al., 2002b), a strong warming of the AP region (winter on western side, summer on eastern side) is evident in many studies. The winter warming has been accompanied by decreases in sea ice extent and reduced concentration to the westward, suggesting the importance of ice–ocean–air positive feedbacks. Synoptically, warm (cold) winters in the AP are associated with a greater (reduced) frequency of northerly wind components as the Amundsen–Bellingshausen Sea mean low pressure intensifies (weakens). These SLP patterns also resemble those connected with the tropical ENSO and its teleconnection to higher southern latitudes occurring via the Pacific–South America (PSA) pattern (see "Circulation variations and climatic teleconnections; The El Niño Southern Oscillation (ENSO) in Antarctica").

In contrast to the recent warming of the AP, a cooling over much of the continent has coincided with lowering pressures in the ACT, strengthening westerlies over sub-Antarctic latitudes, and greater sea ice extent on average. These are associated with a trend toward more positive values of the AAO.

The dominant SAO in the annual march of temperature – and related changes in wind, pressure, geopotential height and precipitation – over sub-Antarctic latitudes and coastal Antarctica, produces gradient maxima in the equinoctial (March, September) months. The SAO results primarily from the tropospheric radiative imbalances between middle and high latitudes, which are at a maximum around the equinoxes. Also, a SAO has been detected in the latitude of the speed maximum of the ACC (Large and van Loon, 1989).

The atmospheric SAO is involved in the annual patterns of sea ice freeze-up and melt (Enomoto and Ohmura, 1990), as well as ice–water concentration (Watkins and Simmonds, 1999), via its interactions with the latitude location of the ACT and associated patterns of easterly (westerly) low-level winds occurring to the south (north) of this feature. In summer and early fall, when the ACT lies equatorward of the ice edge, easterly winds produce convergence of the pack, resulting in little increase in ice area despite the lowering surface temperatures. However, as the ACT shifts poleward of the ice edge in March, westerly winds to the north encourage ice divergence and a relatively rapid expansion of the pack through the winter and early spring. In December and January the ACT again lies north of the ice edge, which has melted back as temperatures have risen. The associated easterly winds encourage compaction of the ice pack and its retreat to higher latitudes.

There has been a climate change in the SAO since the mid-1970s to late 1970s (Hurrell and van Loon, 1994; Simmonds and Jones, 1998), involving primarily a shift in the springtime phase from September and early October into November. This change has coincided with strengthened latitude gradients of temperature originating mostly from a warming in the tropical Pacific, possibly associated with more frequent El Niño events (Mo, 2000); reduced sea-ice extent in the Amundsen–Bellingshausen seas (van den Broeke, 2000b), but significant cooling in coastal East Antarctica in early winter (van den Broeke, 2000a). At least part of the increased severity of the ozone hole over the past two decades likely is related to the longer isolation of the southern circumpolar vortex in austral spring, accompanying this change in the SAO.

Cloudiness

Satellite-observed cloud cover is at a maximum (80–100%) over ocean higher latitudes adjacent to the pack ice zone, varies little with the season, and is due to traveling cyclonic storms (see "Synoptic processes; Synoptic cyclone activity", below). Manned observing station data indicate a cloudiness minimum over the interior plateau (Figure A29 Borisenkov and Dolganov, 1982), especially in midwinter (less than about 25%), and generally low seasonal variability. Maximum cloud cover variability occurs at Antarctic coastal stations located between these two zones (van Loon et al., 1972), arising from higher rates of alternation of high- and low-pressure systems. The contribution of lower cloud to the total cloud amount reaches a maximum over coastal regions and the pack ice zone (cyclonic), and a minimum over interior East Antarctica where subsidence of air dominates. In recent decades, cloudiness in the spring and summer seasons has increased by about 15–20% at the South Pole. These increases appear to be associated with changes in the high-latitude circulation.

Precipitation

Snowfall dominates precipitation in the Antarctic. However, the contribution of clear-sky precipitation (e.g. "diamond dust") and blowing snow to the ice sheet mass balance, increases in importance over the higher, colder and dryer regions of the interior (Giovinetto et al., 1990). The mass balance is the result of precipitation, evaporation, direct deposition (rime, frost), drifting, and runoff from melting at low elevations (Figure A30). Snow drift represents a net loss from the continent to the coastal sea ice zone. Although our confidence in measurements of Antarctic continental T_s and pressure is reasonably high, it remains low for precipitation. This results from a lack of data and the difficulty of separating real precipitation from drifting snow, especially in katabatic wind-prone areas. Accordingly, indirect estimates of Antarctic precipitation have been more forthcoming. These compute the net surface mass balance as $P - E$, where P = precipitation and E = sublimation, from moisture fluxes directed toward the continent. Recently, model calculations of precipitation have been undertaken. The decrease of precipitable water, relative humidity, and precipitation with increasing latitude, together with increasing distance from the Southern Ocean moisture

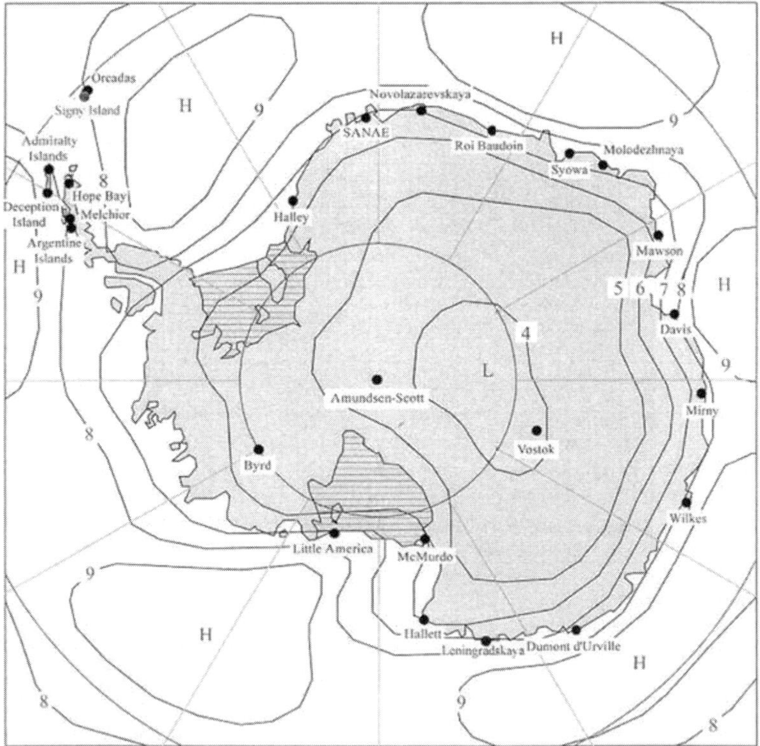

Figure A29 Map showing mean annual total cloud cover from station observations (dots show observation sites used for the compilation); cloud cover isopleths in tenths. (From Guo et al., 2004.)

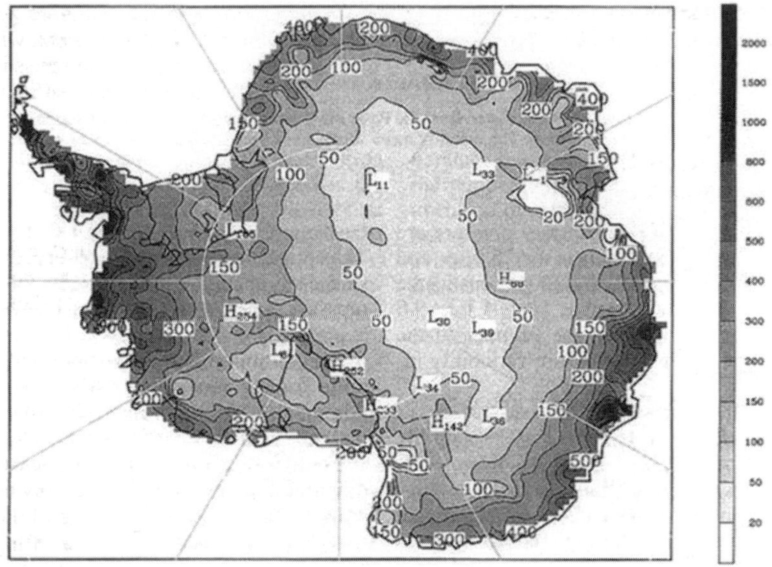

Figure A30 Map showing long-term accumulation distribution; units of $mm\,yr^{-1}$. (From Guo et al., 2004.)

source, means that annual precipitation on the high plateau is less than about $2\,cm\,yr^{-1}$ water equivalent. This increases by up to at least 2 orders of magnitude at the coast (Figure A30). The lower elevations of West Antarctica facilitate moisture transport into the interior, mostly by synoptic cyclones.

Interannual variations of coastal precipitation are high, in response to high variability of cyclonic activity, especially along the western AP where they manifest fluctuations in the intensity of the Amundsen–Bellingshausen mean low pressure. The latter has an ENSO component (Guo et al., 2004). There has been a statistically significant upward trend in precipitation averaged over the continent between 1979 and 1999, although weakly downward trends are evident regionally and in the interior.

Surface winds

The surface wind is a persistent katabatic (downslope) flow that is gravitational and thermal in origin, relatively shallow (100–200 m deep), and essentially decoupled from the synoptic gradient flow above. It dominates the annual wind speed and direction at many stations in West Antarctica and coastal East Antarctica (Borisenkov and Dolganov, 1982; Figure A31). At Cape Denison the mean annual windspeed exceeds $20\,m\,s^{-1}$, which is the strongest on Earth close to sea level (Wendler et al., 1997b). Extremely high wind speeds can persist for weeks or even a month or two, especially in winter. In summer, solar radiation absorption reduces the katabatic effect (van den Broeke and van Lipzig, 2003). Katabatic winds are important because they produce snowdrift and enhance coastal polynyas and sea ice production (Adolphs and Wendler, 1995), thereby affecting the surface mass and energy budgets. Also, katabatic winds promote mesoscale cyclogenesis in confluence areas near the coast (e.g. Heinemann, 1990; Bromwich, 1991; see Figure A27). Streten (1968) found that katabatic winds are best developed in coastal regions with little synoptic cyclone activity, but with generally lower pressure located to the eastward. However, individual synoptic events may enhance the katabatic winds at certain coastal locations. Modeling suggests that synoptic forcing may not be important in determining these winds over interior East Antarctica. There, they can be explained adequately from consideration of the topography (i.e. channeling effect) and the intensity of the wintertime inversion (Connolley, 1996). The katabatic winds are a critical component of the atmosphere's meridional circulation for southern higher latitudes and are linked to the intensity of the circumpolar vortex in the mid- to upper-troposphere. Katabatic winds may even participate in high latitude–tropical interactions on subseasonal time scales (Yasunari and Kodama, 1993).

Winds on the eastern side of the AP and the western Weddell Sea comprise a strong and persistent southwest to southerly flow, or *barrier wind* (see Figure A27). This wind is important for advecting ice equatorward into the westerly wind belt, thereby maintaining lower air temperatures in the southern South Atlantic than at comparable latitudes in the southeast Pacific (Schwerdtfeger, 1975, 1979). The barrier wind results from very cold and stable air that moves westward over the Weddell Sea, is dammed up against the eastern side of the AP, and sets up a thermal gradient similar to that forcing the katabatic wind. This explanation does not require the presence of a semipermanent low pressure in the Weddell, although such a feature often exists and may enhance the barrier wind (Figure A32). Interannual fluctuations in the intensity of the Weddell Sea low, some of which are ENSO-related (Yuan and Martinson, 2001), are strongly evident in the sea ice conditions of the embayment (Turner et al., 2002a).

Large-scale atmospheric circulation

Mean surface pressure and 500 hPa height fields

The essential features of the atmospheric circulation around the Antarctic have been known since at least the International Geophysical Year (IGY) of 1957–1958 (e.g. Lamb, 1959; van Loon et al., 1972). Mean monthly height, temperature, and wind

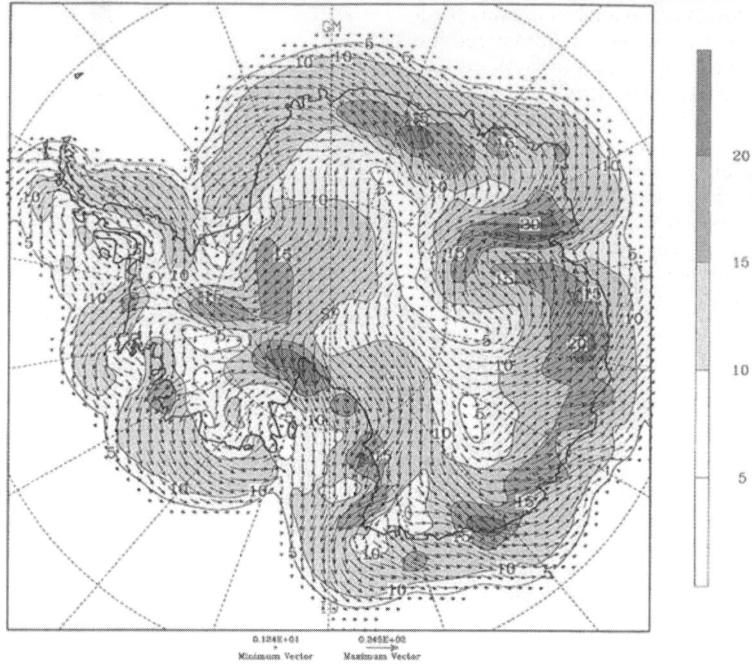

Figure A31 Antarctic near-surface winds for winter; units of m s^{-1}. (From Guo et al., 2004.)

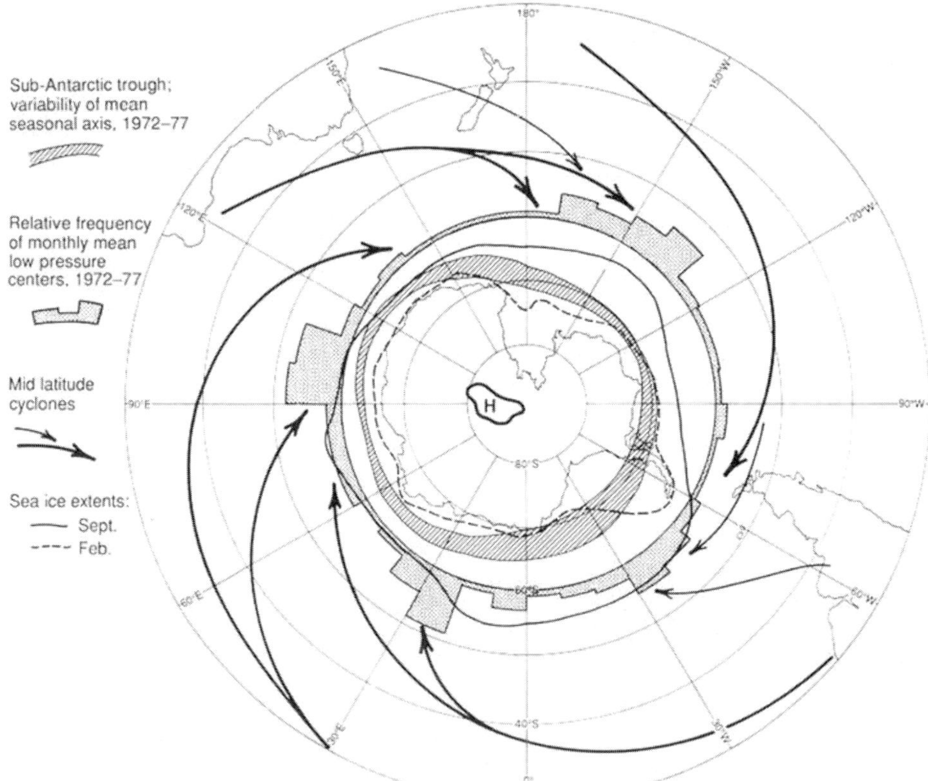

Figure A32 Schematic summarizing dominant features of the synoptic climatology of southern higher latitudes (side legend), compiled from multiple sources (map has similar orientation to that in Figure A27). (From Carleton, 1992.)

fields at different tropospheric levels are available in atlas form (see van Loon et al., 1972), and have been summarized by Schwerdtfeger (1984). Synoptic and dynamic climatologies of the higher latitude circulation derived from the ABM, ECMWF, and other datasets, largely confirm the IGY studies, although greater detail is now known for previously data-void ocean areas (the South Pacific). In addition, much more is known of the circulation variability on interannual and decadal time scales.

The dominant features of the SLP field of southern high latitudes are a thermal anticyclone over the continent (Figures A32 and A33), although this feature is somewhat artificial owing to the method of reducing pressures to sea level. An almost continuous belt of low pressure (the ACT) around the continent and pack ice zone has well-defined centers located near the major Antarctic embayments, and also off Wilkes Land in East Antarctica (Figure A33a,b). The Amundsen– Bellingshausen mean low (ASL) comprises a "pole of maximum variability" in the SLP field (Connolley, 1997). Modeling experiments suggest that the strong interannual variability of the ASL is explained by the displacement of Antarctica's highest elevation areas toward the Indian Ocean (Lachlan-Cope et al., 2001).

The mean lows of the ACT are the summation of individual synoptic cyclones moving in from middle latitudes (Figure A32), and represent areas of stagnation and cyclolysis (Carleton, 1979; Simmonds et al., 2003). However, recent satellite-based studies show that these "cyclonc graveyards" are often active sites for mesoscale cyclones forming in cold-air outbreaks just to the westward (e.g. Carleton and Song, 1997). Thc SLP has lowered over Antarctica since the mid-1970s to late 1970s as the surface and troposphere has cooled. This trend comprises part of the AAO teleconnection pattern.

Although the ACT undergoes little variation in intensity between summer and winter (Figure A33a,b), its latitudinal position changes markedly in connection with the SAO: more equatorward in the solsticial (January, July) months, and closer to Antarctica in the equinoctial (March, September) months. Also, the broad zonality of the Antarctic mean circulation undergoes quite marked meridionality on individual (daily) analyses, especially in the South Pacific. High-pressure ridges frequently interrupt the ACT (see Taljaard, in van Loon et al., 1972). Favored longitudes for ridges are in East Antarctica between 0–15°E, 50–60°E, and 140–150°E. The last location particularly is important in blocking events, which typically are more prevalent during ENSO warm phases (i.e. El Niño) (Marques and Rao, 2000).

In the free atmosphere the Antarctic is dominated by a circumpolar westerly vortex that expands equatorward in winter. Lowest heights (and layer thicknesses, or mean temperatures) occur over the ice sheet, and there is evidence that the intensity of this feature is coupled to the katabatic winds. On average the whole vortex is displaced slightly toward the Indian Ocean sector, or wave number 1 pattern. However, a three-wave pattern of tropospheric troughs and ridges also often occurs (Figure A32), especially during blocking events, with a trough over each of the three major ocean basins (e.g. Kiladis and Mo, 1998). In contrast to their counterparts in the northern hemisphere, the tropospheric waves have low amplitude and their positions do not change much from season to season. Interannual variations of wave number 1 occur preferentially toward either the Australian region or the Falkland Islands, with accompanying large variations in sea-ice conditions in the Scotia Sea (Carleton, 1989). This mode of variability is described by a Trans-Polar Index (TPI), which is based on the SLP anomalies at Hobart and Stanley (Pittock, 1984). However, the negative relationship of SLP between these two places is not stable in the long term (Carleton, 1989). Temporal changes in the tropospheric waves are confirmed from NCEP-NCAR reanalyses (Raphael, 2003b), showing that the largest changes have occurred since about 1975 in the southern late fall through early spring, and from the Indian Ocean eastward to the Weddell Sea.

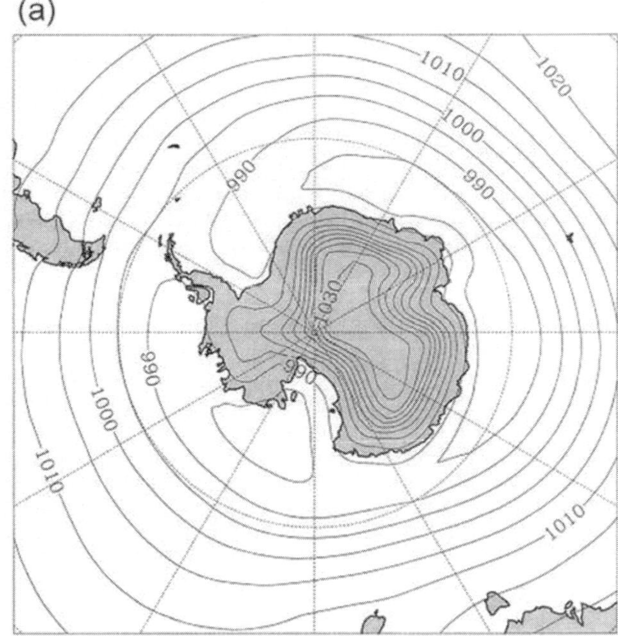

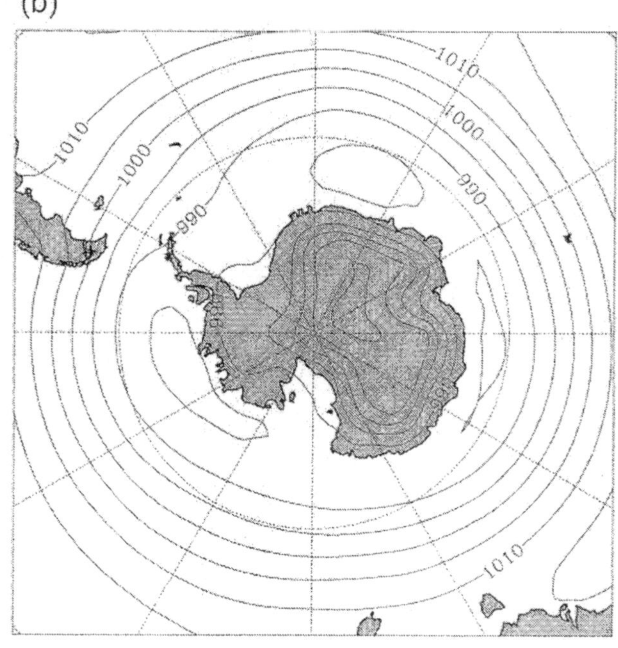

Figure A33 Mean sea level pressure (SLP, contour interval of 5 hPa) derived from NCEP-NCAR reanalyses for **(a)** winter, June through August, of 1979–1999; and **(b)** summer, December through February, of 1980–2000. (From Simmonds et al., 2003.)

Zonal circulation

In accordance with the limited summer-winter variation in the wave pattern (see "Mean surface pressure and 500 hPa height fields", above), there is little seasonal change in the intensity of the zonal westerlies at 500 hPa in middle and high latitudes (Trenberth, 1979). However, there is marked interannual variability, especially in the Pacific sector, where it is dominated by the ENSO – westerly zonal winds in adjacent broad latitude zones vary out-of-phase, or so-called "split jet" flow (Bals-Elsholz et al., 2001). During ENSO warm events the STJ (PFJ) tends to be stronger (weaker) (Figure A34); but in cold, or La Niña events, the STJ (PFJ) is weaker (stronger). These variations modify the momentum and heat transports, as well as storm tracks, over southern middle and higher latitudes. However, it is important to note that there is considerable variability of the above-noted patterns within a given phase of ENSO, especially the El Niño events.

Synoptic indexes long have been used to characterize the zonal atmospheric circulation in the southern extratropics (e.g. the TPI). These are constructed using SLP observations for station pairs across widely separated latitudes (zonal index) or longitudes (meridional index). Indexes of this type capture the SAO in near-surface SLP, and confirm the recent change in the AAO to a more positive (i.e. increased zonality) mode. The increased strength of the southern westerlies that occurs between fall and spring occurs concurrently with the expansion of the sea ice (Ackley, 1981); also, greater zonal wind variability tends to accompany increased variability of the ice. Similarly, between-year regional variations of the ice near maximum extent accompany changes in the zonal index which, in turn, are related to patterns of higher latitude cyclonic activity and the teleconnection patterns of AAO and ENSO (Carleton, 1989; Renwick, 2002).

Synoptic processes

Air masses and air streams

The Antarctic ice sheet is the source region for extremely cold, dry and stable Antarctic continental (cA) air year round (Wendland and McDonald, 1986). It is recognized as inversion air only over the continent and ice shelves because it transforms

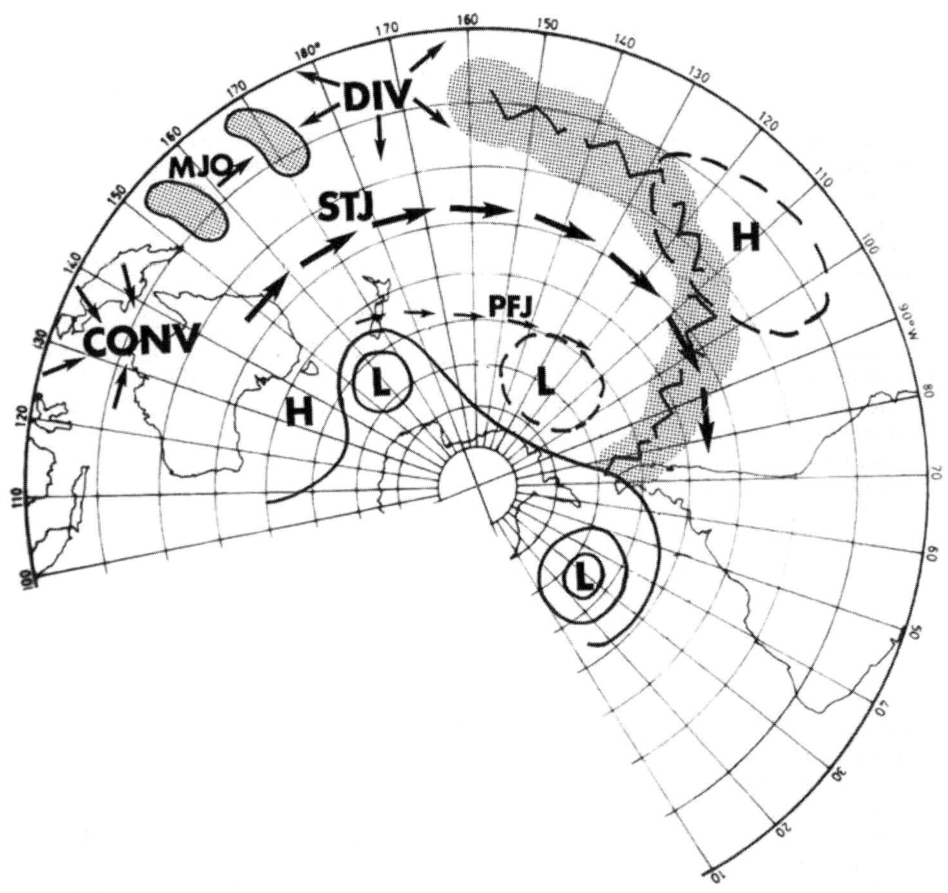

Figure A34 Schematic showing the dominant changes in mid- to upper-tropospheric circulation over the Pacific and southwest Atlantic sectors that accompany a warm (El Niño) event. Features over lower latitudes (MJO: Madden–Julian Oscillation, SPCB: South Pacific Cloud Band) mostly are representative of the summer season. Those over middle and higher latitudes (STJ, PFJ, pressure/height anomalies) apply to the winter season 4–8 months earlier. The solid (dashed) lines indicate that the represented feature is stronger (weaker) during El Niño than in La Niña, for the respective season. (From Carleton, 2003a.)

Table A19 Mean temperature T (°C), mixing ratio q (gm kg^{-1}), and pseudo wet-bulb potential temperature θ_{sw} (°C) of higher latitude air masses in summer (*s*, December-March) and winter (*w*, June-September) at selected stations

			Pressure Level (mb)											
			1000			850			700			500		
Station	Location	Season	T	q	θsw	T	q	θsw	T	q	θsw	T	q	θsw
Maritime Polar Air														
Macquarie I.	55°S, 159°E	*s*	5.7	4.5	4.1	−4.3	2.7	2.8	−12.5	1.3	4.8	−28.4	0.3	8.2
		w	3.7	4.1	1.8	−5.5	2.3	2.0	14.4	1.0	3.4	−32.7	0.3	5.5
Little America	78°S, 161°W	*s*	—	—	—	−8.3	2.0	−0.1	−15.2	1.5	3.6	−29.9	0.5	7.7
		w	—	—	—	−15.7	1.2	−6.4	−21.1	0.9	0.9	−37.2	0.3	2.9
Maritime Antarctic Air														
Marion I.	47°S, 37°E	*w*	0.3	2.9	−1.2	−9.6	1.6	−1.5	−19.4	0.8	−0.1	−34.3	0.2	4.3
Mirnyi	67°S, 93°E	*w*	—	—	—	−20.3	0.8	−10.6	−22.7	0.7	−2.2	−36.4	0.3	3.5
Little America	78°S, 161°W	*w*	—	—	—	−21.4	0.6	−11.7	−26.8	0.5	−5.5	−41.3	0.2	0.0
Continental Antarctic Air														
Mirnyi	67°S, 93°E	*s*	—	—	—	−12.3	1.2	−4.0	−19.8	0.8	−0.1	−32.1	0.3	6.0
		w	—	—	—	−26.8	0.4	−16.7	−28.2	0.3	−7.0	−40.3	0.1	0.7
Little America	78°S, 161°W	*s*	—	—	—	−12.7	1.1	−4.3	−21.0	0.6	−1.6	−35.1	0.2	4.2
		w	—	—	—	−28.9	0.3	−19.0	−31.9	0.2	−10.0	−44.0	0.1	−2.0

Source: Atter Taljaard, in van Loon et al. (1972).

strongly during equatorward excursions (Andreas and Makshtas, 1985). Taljaard (in van Loon et al., 1972) identified a transitional airmass (Antarctic maritime, mA) over the pack ice which is present only during winter. It is slightly warmer and less stable than cA (Table A19), but at times may be indistinguishable from that airmass.

Polar maritime (mP) air – sometimes also known as Southern maritime (mS) – dominates the middle and higher-latitude oceans that are ice free. Developed partly over the marked SST gradients of the OPF, mP air exhibits south–north variations in temperature and moisture (Table A19), being strongly destabilized on moving equatorward. Thus, the OPF is the site for many higher-latitude polar mesocyclone ("polar low") systems observed in satellite imagery and a high frequency of lower-level ("boundary-layer") fronts (Carleton, 1995). Modification of mP air also occurs when it returns poleward, typically in advance of frontal cyclones. In this case the air stabilizes and may penetrate far into the interior of Antarctica via the Pacific sector of West Antarctica.

Climatic fronts

The climatic frontal zones (and, hence, major cyclone tracks) of the southern hemisphere (see Figure A32) show a close association with the OPF at middle and high latitudes (Baker, 1979). These polar fronts are manifest as tropospheric thickness gradient maxima (TGM) in association with middle- and high-latitude jet streams (Taljaard, in van Loon et al., 1972). In summer there is some correspondence between the 1000–500 hPa TGM and the summer pack ice margin, implying frontal activity. In winter the higher-latitude frontal frequency maxima occur equatorward of the sea ice margin in the South Pacific, but there are no corresponding bands of TGM at this latitude owing to higher rates of alternation of high- and low-pressure systems (Taljaard, in van Loon et al., 1972). Instead, a well-defined TGM in the 1000–500 hPa layer lies poleward of the ice border along 65–70°S. This has been considered evidence of a secondary Antarctic (as opposed to Polar) Front (see Figure A27). In East Antarctica, TGM in the 500–300 hPa layer may correspond to an Antarctic Jet related to the topography of the ice sheet, as identified during the IGY and subsequently modeled (Mechoso, 1980). Interestingly, there has been little research on the climatology of atmospheric fronts over southern high latitudes in recent years. However, useful insights can be gleaned from satellite passive microwave retrievals of cloud water, column-integrated water vapor, and precipitation over ice-free ocean, as well as the summary maps of extratropical cyclone tracks derived using automatic tracking of systems in digital analyses.

Synoptically, fronts penetrating Antarctica often lose their identity on ascending the ice sheet, although they may temporarily disrupt the strong surface inversion. Warm fronts advancing southward ahead of mP or mA air provide overrunning precipitation, and frequently are involved in warming episodes associated with the coreless winter.

Synoptic cyclone activity

Automated cyclone-tracking routines applied to digital SLP data (Simmonds and Keay, 2000a,b) largely confirm the mean patterns of extratropical cyclones determined in the earlier satellite-based (cloud vortex) and IGY climatologies for the Antarctic and sub-Antarctic (Carleton, 1979; Figure A32). However, they add information on the variability (frequency, spatial locations) of these systems and also their characteristic intensities. Over the Southern Ocean, including the SSIZ, the circulation is dominantly cyclonic all year (see "Large-scale atmospheric circulation; Mean surface pressure and 500 hPa height fields", above), the mean pattern involving cyclogenesis over middle latitudes of the western South Atlantic, Indian Ocean, and mid-Pacific (Carleton, 1979). Synoptic cyclones bring clouds, precipitation and strong winds to coastal Antarctica. The maturity and dissipation (cyclolysis) of these systems occur at successively higher latitudes along well-defined climatic tracks that merge in the ACT. These "storm tracks" are evident as cloud bands in satellite VIS/IR imagery. They are also retrievable from digital daily analyses as zones of high variance of the v-wind (south–north) component and geopotential height, but low variance in the u-wind (west–east) component (Trenberth, 1987). However, there are some differences in the locations of storm tracks derived in this manner and those obtained from tracking cyclone systems (Jones and Simmonds, 1993). Relatively few lows invade interior East Antarctica, whereas many enter West Antarctica from the South Pacific. The latter region particularly shows an interannual variation in cyclone frequencies related to ENSO (Carleton and Song, 2000). Moreover, a trend to decreased (increased) cyclone frequencies over the sub-Antarctic (Antarctic) seas for the period 1960–1999 is consistent with numerical modeling studies employing anthropogenic forcings (Fyfe, 2003). These imply a poleward migration of the zone of maximum baroclinity under global warming. However, this change may depend, at least partly, on the intensity class of cyclone studied (Pezza and Ambrizzi, 2003).

Similar to earlier studies for the northern hemisphere, explosively deepening cyclones ("bombs") have been identified and studied for the southern extratropics. Those occurring closer to Antarctica have much in common with some "polar lows", in terms of their large-scale synoptic environments (cf. Carleton and Song, 1997, 2000; Simmonds et al., 2003).

The time-averaged centers of cyclolysis within the ACT are located just eastward of the longitudes of maximum variation of sea-ice extent; Carleton and Fitch, 1993; Yuan et al., 1999; cf. Figures A27 and A34). Therefore, the longitudinal frequency of depression centers influences ice distribution and advance/retreat patterns via temperature and dynamical (wind-induced) advection (Cavalieri and Parkinson, 1981). On average there is no clear connection between Antarctic sea ice extent and cyclone tracks, although closer associations can be found for certain time periods and regions (Yuan et al., 1999).

Mesoscale cyclone and "polar low" activity

High-resolution meteorological satellite imagery has shown that mesoscale cyclones, or polar-air mesocyclones (approximate length scales 100–800 km), occur frequently in Antarctic and sub-Antarctic latitudes (Carleton, 1992, 2003b). The improved detection of these systems likely is involved in a statistically significant recent increase in extratropical cyclone frequencies over the southern hemisphere that is not matched by significant changes in SLP (Sinclair et al., 1997). Mesocyclones comprise two main cloud-vortex types: inverted comma-shaped "polar lows", dominated by positive vorticity advection associated with short waves and jet stream maxima; and spiral-shaped (multi-banded) systems generated mostly by strong sea–air interactions of heat and moisture (Carleton, 1995, 1996). One zone of mesocyclone high incidence is located near the winter/spring sea-ice margin, especially in the South Pacific sector (see Figure A27). There, streams of cA or mA air are strongly destabilized in the vicinity of the ice edge and OPF. These storms often track east or northeast through Drake Passage to decay in the Weddell Sea area (Carleton and Song, 1997). The other major zone of meso-cyclogenesis is close to the Antarctic coastline and over the ice shelves; near the confluence areas for the katabatic winds (see Figure A27). There, "vortex stretching" as air descends the steep slope of the ice sheet, enhances cyclonic vorticity. Also, some mesocyclones develop in association with coastal polynyas because of the oceanic large heat losses to the atmosphere. These near-coastal mesocyclones may lack a cloud signature, being detectable only in the AWS data, or be composed mostly of low-level clouds.

The zonally averaged meso-cyclogenesis seems to be greatest in March and September, associated with the SAO (Carleton and Song, 1997). A number of regional-scale "synoptic climatologies" of mesocyclones in the Antarctic have been developed for time periods of varying length; for the Ross Sea, Marie Byrd Land and the Siple Coast, the Bellingshausen and Amundsen seas, and the Weddell Sea (refer to section 2.2.3 in Carleton, 2003b; also Carleton and Carpenter, 1990; Carrasco et al., 2003). In the Bellingshausen–Amundsen seas, strong differences in the ENSO-related SLP and SST anomalies (SSTA) for the southern winters 1988 and 1989 were reflected in major changes in mesocyclone activity (Carleton and Song, 2000). Mesocyclones were more (less) frequent in this sector in 1989 (1988), associated with decreasing (static) SSTA during the respective May to September periods.

Circulation variations and climatic teleconnections

Climate fluctuations in the Antarctic and sub-Antarctic occur interannually (i.e. climate variations), decadally (climate trends), and multidecadally (climate changes). Interannual variations primarily involve teleconnections to the tropical Pacific ENSO, which is, globally, the dominant coupled ocean–atmosphere interaction mode. On decadal time scales, Antarctic-region circulation–climate variations are most closely connected with an Antarctic Circumpolar Wave (ACW) of extratropical coupled anomalies of SST, upper-ocean salinity, SLP, meridional winds, and sea ice. At times the ACW may be linked to, and modulated by, the ENSO. Recent climate changes in the Antarctic, particularly as revealed in the retreat and collapse of ice shelves, and the ongoing warming of the western AP, may be evidence of "global warming". However, a more immediate explanation for these climate changes derives from long-term shifts in the atmospheric circulation; primarily, the zonally varying mode or Antarctic Oscillation (AAO).

The El Niño Southern Oscillation (ENSO) in Antarctica

The climate teleconnections to the ENSO phenomenon also are evident in Antarctica, especially in the Pacific sector (Kwok and Comiso, 2002). Interestingly, the pressure changes that occur over southern higher latitudes in the lead-up to an El Niño may be at least as large as those in the tropics, and often precede the changes there. (In an El Niño, tropical convection shifts eastward from the "maritime continent" to the central tropical Pacific, in association with the area of highest SST.) The associated weakening of the South Pacific subtropical high, a key "center of action" in ENSO, permits the semi-permanent South Pacific Convergence Zone (SPCZ) to move eastward (Figure A34). The SPCZ is a major trough ahead of which energy and moisture are advected rapidly southeastwards into extratropical latitudes. Consequently, its forward side is evident in satellite imagery as a quasi-meridional band of deep clouds that often energizes cyclonic circulations transiting the Pacific sector of the Southern Ocean. During El Niño the strengthened latitudinal pressure and temperature gradient resulting from the higher SST in the tropical central Pacific, helps intensify the STJ in that sector. At the same time the PFJ tends to weaken. These synoptic interactions are evident on monthly time scales as a standing wave train pattern of alternating anomalies linking the South Pacific and South Atlantic sectors, or so-called Pacific–South America (PSA) pattern (e.g. Mo, 2000). The PSA is reminiscent of the North Pacific PNA pattern, although the pressure/height anomalies generally are of lower magnitude. The PSA pattern during composite El Niño events is shown in Figure A34. Note particularly the out-of-phase relationship of the pressure anomalies between the Bellingshausen–Amundsen sector (Weddell Sea), where SLP anomalies are positive (negative). Similarly, this relationship is evident as out-of-phase anomalies of temperature and sea-ice extent. Thus, El Niño events tend to produce colder conditions, more frequent southerly winds and greater sea-ice extent east of the AP.

The out-of-phase temperature and related pressure anomaly pattern between the seas west and east of the AP comprise an Antarctic Dipole Index (ADP), which correlates well with tropical Pacific SST and surface air temperature, as well as the Antarctic sea ice edge (Yuan and Martinson, 2001). Thus, the ADP appears to link interannual climate variability in the Antarctic seas with the ENSO pattern. The signal is particularly strong in La Niña events, for which it may have some predictive value.

Over most of Antarctica, El Niño events tend to be accompanied by colder conditions than normal and negative anomalies of surface pressure (Smith and Stearns, 1993). The cooling of the troposphere helps intensify the katabatic winds, leading to stronger wind speeds near the coast; at least, as occurred during the major El Niño of 1982–1983. Accordingly, sea ice may be more extensive in those regions while it is reduced in others, such as the Ross Sea.

During La Niña, or ENSO "cold events", the regional pattern of coupled anomalies in pressure, meridional winds, temperature and sea ice over the Southern Ocean are more or less opposite those for El Niño (Sinclair et al., 1997). In the Pacific sector, La Niñas typically are expressed as a strengthened Amundsen Sea low and associated stronger PFJ (a weaker STJ), and a weaker Weddell Sea low (Yuan et al., 2001). Accordingly, the temperature anomalies along the western AP are mostly negative in La Niña, with greater sea-ice extent, in contrast to the eastern side of the peninsula.

As noted for climate variations in other parts of the world, the ENSO teleconnection is not stable through time. In the Pacific sector of West Antarctica a change in the relationship between ENSO and the poleward flux of moisture evidently occurred around 1991; from strongly positive before that time to significantly negative afterward. This change impacted regional precipitation amounts.

The Antarctic Circumpolar Wave (ACW)

Unlike the ENSO climate signal in Antarctica, the subdecadal ACW appears primarily to be of southern extratropical origin, although it likely interacts with the ENSO as a "slow teleconnection" (Peterson and White, 1998). In the atmosphere the ACW is a propagating wave number 2 pattern that makes a complete circuit about Antarctica in about 8 years (4 years per wave). It differs from the ADP, which is a quasistationary wave, although the two may interact in the Amundesen–Weddell Sea sectors (Yuan and Martinson, 2001). Because the western (eastern) sides of a trough are characterized by opposite anomalies of meridional wind and temperature, the patterns of upwelling and downwelling in the ocean – important for sea temperature and salinity differences – and of sea-ice extent, also are opposite across the wave (Motoi et al., 1998). East of a trough and west of a surface high pressure, the northerly winds advect mild air southward, leading to downwelling and decreased near-surface salinity, and a general retreat of the sea-ice extent. Conversely, west of a trough and east of a surface high pressure, the southerly winds advect cold air equatorward, producing upwelling and increased salinity, and advancing sea ice. Thus, the ACW in the upper ocean is evident as a slowly migrating (to the eastward) suite of anomalies that maintain their identity across seasons and through the different regions of the Southern Ocean. Interestingly, a number of modeling studies have successfully simulated an ACW-like phenomenon, yet its period and even wave number can be different from those evident in the data (e.g. Motoi et al., 1998). Similarly, the analysis of datasets for the periods before and after that analysed by Peterson and White (1998) suggests that the ACW is not always a prominent teleconnection (Kidson, 1999). This lack of temporal stability is reminiscent of the Antarctic teleconnection to ENSO.

The Antarctic Oscillation (AAO) and recent climate changes

The AAO, also known as the zonally varying mode, southern annular mode, and high-latitude mode, comprises the first EOF (empirical orthogonal function in a principal components analysis) of SLP and tropospheric height over southern middle and higher latitudes (Rogers and van Loon, 1982; Kidson, 1999). An intensification of the mean pressure centers respectively over middle latitudes and Antarctica results in stronger geostrophic westerly winds in the sub-Antarctic and PFJ, or the positive phase of AAO. Conversely, a weakening of the pressure meridional difference reduces the westerlies and PFJ, or negative phase of AAO. After the 1970s, the trend of the AAO has been one of increasing "positive polarity", partly linked to rising SST in the tropical Pacific and more frequent El Niño events. The stronger westerlies bring in relatively mild and moist air to the western side of the AP, providing at least a partial explanation for the observed winter warming and reduced sea ice there. During the same period the intensification of the polar vortex over the continent accompanied downward trends of temperature there, and reduced surface melt in the summer (Torinesi et al., 2003). Stronger westerlies in the AAO positive mode tend to be associated with a zonally averaged greater extent of the sea ice (Hall and Visbeck, 2002).

Although it is most marked in the troposphere, the AAO has links with the stratosphere, and may modulate the shape and intensity of the annually occurring Antarctic ozone hole (Thompson et al., 2000). In general, greater cooling over Antarctica accompanying positive AAO, better isolates stratospheric air from that in middle latitudes, permitting the formation of more polar stratospheric clouds upon which chlorine compounds accumulate, thereby accelerating the breakdown of ozone during austral springtime (Sexton, 2001). Thus, recent Antarctic-region climate changes involving human activities may be linked to changes in frequency of "natural" teleconnection patterns.

Andrew M. Carleton

Bibliography

Ackley, S.F., 1981. A review of sea-ice weather relationships in the southern hemisphere. In Allison, I., ed., *Sea Level, Ice, and Climatic Change*. International Association for Hydrological Sciences Publication 131. Canberra, Australia, pp. 127–159.

Adolphs, U., and Wendler, G., 1995. A pilot study on the interactions between katabatic winds and polynyas at the Adelie Coast, eastern Antarctica. *Antarctic Science*, **7**(3): 307–314.

Anderssen, E.C., 1965. A study of atmospheric long waves in the Southern Hemisphere, *NOTOS*, **14**: 57–65.

Andreas, E.L., and Makshtas, A.P., 1985. Energy exchange over Antarctic sea ice in the spring. *Journal of Geophysical Research*, **90**: 7199–7212.

Baker, Jr., D.J., 1979. Ocean-atmosphere interactions in high southern latitudes. *Dynamics of Atmospheres and Oceans*, **3**: 213–229.

Bals-Elsholz, T.M., Atallah, E.H., Bosart, L.F., Wasula, T.A., Cempa, M.J., and Lupo, A.R., 2001. The wintertime Southern Hemisphere split jet: structure, variability, and evolution. *Journal of Climate*, **14**(21): 4191–4215.

Bintanja, R., 2000. Surface heat budget of Antarctic snow and blue ice: interpretation of spatial and temporal variability. *Journal of Geophysical Research*, **105**(D19): 24387–24407.

Borisenkov, E.P., and Dolganov, L.V., 1982. Some results of climatic generalization of meteorological observations in the Antarctic. *Journal of Geophysical Research*, **87**(C12): 9653–9666.

Bromwich, D.H., 1991. Mesoscale cyclogenesis over the south-western Ross Sea linked to strong katabatic winds. *Monthly Weather Review*, **119**: 1736–1752.

Bromwich, D.H., Carrasco, J.F., Liu, Z., and Tzeng, R-Y., 1993. Hemispheric atmospheric variations and oceanographic impacts associated with katabatic surges across the Ross Ice Shelf, Antarctica. *Journal of Geophysical Research*, **98**: 13045–13062.

Bromwich, D.H., Monaghan, A.J., and Guo, Z., 2004a. Modeling the ENSO modulation of Antarctic climate in the late 1990s with the Polar MM5. *Journal of Climate*, **17**(1): 109–132.

Bromwich, D.H., Guo, Z., Bai, L., and Chen, Q-S., 2004b. Modeled Antarctic precipitation. Part I: Spatial and temporal variability. *Journal of Climate*, **17**(3): 427–447.

Bromwich, D.H., Rogers, A.N., Kållberg, P., Cullather, R.I., White, J.W.C., and Kreutz, K.J., 2000. ECMWF analyses and reanalyses depiction of ENSO signal in Antarctic precipitation. *Journal of Climate*, **13**: 1406–1420.

Carleton, A.M., 1979. A synoptic climatology of satellite-observed extratropical cyclone activity for the Southern Hemisphere winter. *Archiv für Meteorologie, Geophysik, und Bioklimatologie*, **B27**: 265–279.

Carleton, A.M., 1988: Sea-ice atmosphere signal of the Southern Oscillation in the Weddell Sea, Antarctica. *Journal of Climate*, **1**: 379–388.

Carleton, A.M., 1989. Antarctic sea-ice relationships with indices of the atmospheric circulation of the Southern Hemisphere. *Climate Dynamics*, **3**: 207–220.

Carleton, A.M., 1992. Synoptic interactions between Antarctica and lower latitudes. *Australian Meteorology Magazine*, **40**: 129–147.

Carleton, A.M., 1995. On the interpretation and classification of mesoscale cyclones from satellite infrared imagery. *International Journal of Remote Sensing*, **16**: 2457–2485.

Carleton, A.M., 1996. Satellite climatological aspects of cold air mesocyclones in the Arctic and Antarctic. *Global Atmospheric and Ocean System*, **5**: 1–42.

Carleton, A.M., 2003a. Atmospheric teleconnections involving the Southern Ocean. *Journal of Geophysical Research*, **107**, doi: 10.1029/2000JC000329.

Carleton, A.M., 2003b. The Antarctic, In: Rasmussen, E.A., and Turner, J., eds., *Polar Lows, Mesoscale Weather Systems in the Polar Regions*. Cambridge: Cambridge University Press, pp. 108–149.

Carleton, A.M., and Carpenter, D.A., 1990. Satellite climatology of "polar lows" and broadscale climatic associations for the Southern Hemisphere. *International Journal of Climatology*, **10**: 219–246.

Carleton, A.M., and Fitch, M., 1993. Synoptic aspects of Antarctic mesocyclones. *Journal of Geophysical Research*, **98**: 12997–13018.

Carleton, A.M., and Song, Y., 1997. Synoptic climatology and intrahemispheric associations of cold air mesocyclones in the Australasian sector. *Journal of Geophysical Research*, **102**(D12): 13873–13887.

Carleton, A.M., and Song, Y., 2000. Satellite passive sensing of the marine atmosphere associated with cold-air mesoscale cyclones. *Professional Geographer*, **52**: 289–306.

Carleton, A.M., John, G., and Welsch, R., 1998. Interannual variations and regionality of Antarctic sea-ice–temperature associations. *Annals of Glaciology*, **27**: 403–408.

Carrasco, J.F., Bromwich, D.H., and Monaghan, A.J., 2003. Distribution and characteristics of mesoscale cyclones in the Antarctic: Ross Sea eastward to the Weddell Sea. *Monthly Weather Reviews*, **131**(2): 289–301.

Carroll, J.J., 1982. Long-term means and short-term variability of the surface energy balance components at the South Pole. *Journal of Geophysical Research*, **87**(C6): 4277–4286.

Cavalieri, D.J., and Parkinson, C.L., 1981. Large-scale variations in observed Antarctic sea ice extent and associated atmospheric circulation. *Monthly Weather Review*, **109**: 2323–2336.

Connolley, W.M., 1996. The Antarctic temperature inversion. *International Journal of Climatology*, **16**: 1333–1342.

Connolley, W.M., 1997. Variability in annual mean circulation in southern high latitudes. *Climate Dynamics*, **13**: 745–756.

Doran, P.T., Priscu, J.C., and Lyons, W.B. et al., 2002. Antarctic climate cooling and terrestrial ecosystem response. *Nature*, **415**: 517–520.

Enomoto, H., 1991. Fluctuations of snow accumulation in the Antarctic and sea level pressure in the Southern Hemisphere in the last 100 years. *Climatic Change*, **18**: 67–87.

Enomoto, H., and Ohmura, A. 1990. The influences of atmospheric half-yearly cycle on the sea ice extent in the Antarctic. *Journal of Geophysical Research*, **95**: 9497–9511.

Fyfe, J.C., 2003. Extratropical Southern Hemisphere cyclones: harbingers of climate change. *Journal of Climate*, **16**(17): 2802–2805.

Gille, S.T., 2002. Warming of the Southern Ocean since the 1950s. *Science*, **295**(5558): 1275–1277.

Giovinetto, M.B., Waters, N.M., and Bentley, C.R., 1990. Dependence of Antarctic surface mass balance on temperature, elevation, and distance to open ocean. *Journal of Geophysical Research*, **95**: 3517–3531.

Guo, Z., Bromwich, D.H., and Comiso, J.J., 2003. Evaluation of Polar MM5 simulations of Antarctic atmospheric circulation. *Monthly Weather Review*, **131**(2): 384–411.

Guo, Z., Bromwich, D.H., and Hines, K.M., 2004. Modeled Antarctic precipitation. Part II: ENSO modulation over West Antarctica. *Journal of Climate*, **17**(3): 448–465.

Hall, A., and Visbeck, M., 2002. Synchronous variability in the Southern Hemisphere atmosphere, sea ice, and ocean resulting from the annular mode. *Journal of Climate*, **15**(21): 3043–3057.

Harangozo, S.A., Colwell, S.R., and King, J.C., 1997. An analysis of a 34-year air temperature record from Fossil Bluff (71°S, 68°W), Antarctica. *Antarctic Science*, **9**(3): 355–363.

Hatzianastassiou, N., and Vardavas, I., 2001. Shortwave radiation budget of the Southern Hemisphere using ISCCP-C2 and NCEP-NCAR climatological data. *Journal of Climate*, **14**(22): 4319–4329.

Heinemann, G., 1990. Mesoscale vortices in the Weddell Sea region (Antarctica). *Monthly Weather Review*, **118**: 779–793.

Hines, K.M., and Bromwich, D.H., 2002. A pole to pole west Pacific atmospheric teleconnection during August. *Journal of Geophysical Research*, **107**(D18): 4359, doi: 10.1029/2001JD001335, 2002.

Hurrell, J.W., and van Loon, H., 1994. A modulation of the atmospheric annual cycle in the Southern Hemisphere. *Tellus Series A*, **46**: 325–338.

Jacka, T.H., and Budd, W.F., 1998. Detection of temperature and sea-ice extent changes in the Antarctic and Southern Ocean, 1949–96. *Annals of Glaciology*, **27**: 553–559.

Jacobs, G.A., and Mitchell, J.L., 1996. Ocean circulation variations associated with the Antarctic Circumpolar Wave. *Geophysics Research Letters*, **23**: 2947–2950.

Jacobs, S.S., Giulivi, C.F., and Mele, P.A., 2002. Freshening of the Ross Sea during the late 20th century. *Science*, **297**(5580): 386–389.

Jones, A.E., Bowden, T., and Turner, J., 1998. Predicting total ozone based on GTS data: applications for South American high latitude populations. *Journal of Applied Meteorology*, **37**: 477–485.

Jones, D.A., and Simmonds, I., 1993. Time and space spectral analyses of Southern Hemisphere sea level pressure variability. *Monthly Weather Reviews* **121**: 661–672.

Kidson, J.W., 1999. Principal modes of Southern Hemisphere low-frequency variability obtained from NCEP-NCAR reanalyses. *Journal of Climate*, **12**: 2808–2830.

Kiladis, G.N., and Mo, K.C., 1998. Interannual and intraseasonal variability in the Southern Hemisphere. In Karoly, D.J. and Vincent, D.G., eds., *Meteorology of the Southern Hemisphere*. MA: American Meteorlogical Society, pp. 307–336.

Kwok, R., and Comiso, J.C., 2002. Southern Ocean climate and sea ice anomalies associated with the Southern Oscillation. *Journal of Climate*, **15**: 487–501.

Lachlan-Cope, T.A., Connolley, W.M., and Turner, J., 2001. The role of non-axisymmetric Antarctic orography in forcing the observed pattern of variability of the Antarctic climate. *Geophysics Research Letters*, **28**(21): 4111–4114.

Lamb, H.H., 1959. The southern westerlies: a preliminary survey; main characteristics and apparent associations. *Quarterly Journal of the Royal Meteorological Society*, **85**(363): 1–23.

Large, W.G., and van Loon, H., 1989. Large-scale low frequency variability of the 1979 FGGE surface buoy drifts and winds over the Southern Hemisphere. *Journal of Physical Oceanography*, **19**: 216–232.

Marques, R.F.C., and Rao, V.B., 2000. Interannual variations of blocking in the Southern Hemisphere and their energetics. *Journal of Geophysical Research*, **105**: 4625–4636.

Mechoso, C., 1980. The atmospheric circulation around Antarctica: linear stability and finite amplitude interactions with migrating cyclones. *Journal of Atmospheric Science*, **37**: 2209–2233.

Mo, K.C., 2000. Relationships between low-frequency variability in the Southern Hemisphere and sea surface temperature anomalies. *Journal of Climate*, **13**: 3599–3610.

Motoi, T., Kitoh, A., and Koide, H., 1998. Antarctic Circumpolar Wave in a coupled ocean-atmosphere model. *Annals of Glaciology*, **27**: 483–487.

Murphy, B.F., 2003. Prediction of severe synoptic events in coastal East Antarctica. *Monthly Weather Review*, **131**(2): 354–370.

Nakamura, N., and Oort, A.H., 1988. Atmospheric heat budgets of the polar regions. *Journal of Geophysical Research*, **93**: 9510–9524.

Parish, T.R., and Bromwich, D.H., 1991. Continental-scale simulation of the Antarctic katabatic wind regime. *Journal of Climate*, **4**(2): 135–146.

Parkinson, C.L., 1992. Interannual variability of monthly southern ocean sea ice distributions. *Journal of Geophysical Research*, **97**(C4): 5349–5363.

Peterson, R.G., and White, W.B., 1998. Slow oceanic teleconnections linking the Antarctic Circumpolar Wave with the tropical El Niño Southern Oscillation. *Journal of Geophysical Research* **103**: 24573–24583.

Pezza, A.B., and Ambrizzi, T., 2003. Variability of Southern Hemisphere cyclone and anticyclone behavior: further analysis. *Journal of Climate*, **16**(7): 1075–1083.

Phillpot, H.R., 1991. The derivation of 500 hPa height from automatic weather station surface observations in the Antarctic continental interior. *Australian Meteorology Magazine*, **39**: 79–86.

Pittock, A.B., 1984. On the reality, stability, and usefulness of southern hemisphere teleconnections. *Australian Meteorology Magazine*, **32**: 75–82.

Raphael, M., 2003a. Impact of observed sea-ice concentration on the Southern Hemisphere extratropical atmospheric circulation in summer. *Journal of Geophysical Research* 108(D22), 4687. doi: 10.1029/2002JD003308, 2003.

Raphael, M., 2003b. Recent, large-scale changes in the extratropical Southern Hemisphere atmospheric circulation. *Journal of Climate*, **16**(17): 2915–2924.

Reijmer, C.H., and Oerlemans, J., 2002. Temporal and spatial variability of the surface energy balance in Dronning Maud Land, East Antarctica. *Journal of Geophysical Research* 107(D24), 4759, doi: 10.1029/2000JD000110, 2002.

Renwick, J.A., 2002. Southern Hemisphere circulation and relations with sea ice and sea surface temperature. *Journal of Climate*, **15**(21): 3058–3068.

Reusch, D.B., and Alley, R.B., 2002. Automatic weather stations and artificial neural networks: improving the instrumental record in West Antarctica. *Monthly Weather Review*, **130**(12): 3037–3053.

Rogers, J.C., 1983. Spatial variability of Antarctic temperature anomalies and their association with the Southern Hemisphere atmospheric circulation. *Association of American Geographers Annals*, **73**(4): 502–518.

Rogers, J.C., and van Loon, H., 1982. Spatial variability of sea level pressure and 500 mb height anomalies over the Southern Hemisphere. *Monthly Weather Review*, **110**(10): 1375–1392.

Schneider, D.P., and Steig, E.J., 2002. Spatial and temporal variability of Antarctic ice sheet microwave brightness temperatures. *Geophysical Research Letters*, **29**(20): 1964, doi: 10.1029/2002GL015490, 2002.

Schwerdtfeger, W., 1975. The effect of the Antarctic Peninsula on the temperature regime of the Weddell Sea. *Monthly Weather Review*, **103**(1): 45–51.

Schwerdtfeger, W., 1979. Meteorological aspects of the drift of ice from the Weddell Sea toward the mid-latitude westerlies. *Journal of Geophysical Research*, **84**: 6321–6328.

Schwerdtfeger, W., 1984. *Weather and Climate of the Antarctic. Developments in Atmospheric Science*, vol. 15. Amsterdam: Elsevier.

Sexton, D.M.H., 2001. The effect of stratospheric ozone depletion on the phase of the Antarctic Oscillation. *Geophysical Research Letters*, **28**: 3697–3700.

Simmonds, I., and Jones, D.A., 1998. The mean structure and temporal variability of the semiannual oscillation in the southern extratropics. *International Journal of Climatology*, **18**: 473–504.

Simmonds, I., and Keay, K. 2000a. Mean Southern Hemisphere extratropical cyclone behavior in the 40-year NCEP-NCAR Reanalysis. *Journal of Climate*, **13**: 873–885.

Simmonds, I., and Keay, K., 2000b. Variability of Southern Hemisphere extratropical cyclone behavior, 1958–97. *Journal of Climate*, **13**: 550–561.

Simmonds, I., Keay, K., and Lim, E-P., 2003. Synoptic activity in the seas around Antarctica. *Monthly Weather Review*, **131**(2): 272–288.

Sinclair, M.R., Renwick, J.A., and Kidson, J.W., 1997. Low-frequency variability of Southern Hemisphere sea level pressure and weather system activity. *Monthly Weather Review*, **125**: 2531–2543.

Smith, S.R., and Stearns, C.R., 1993. Antarctic pressure and temperature anomalies surrounding the minimum in the Southern Oscillation Index. *Journal of Geophysical Research*, **98**: 13071–13083.

Streten, N.A., 1968. Some features of mean annual windspeed data for coastal East Antarctica. *Polar Record*, **14**(90): 315–322.

Streten, N.A., 1977. Seasonal climatic variability over the southern oceans. *Archiv für Meteorologie, Geophysik und Bioklimatologie*, **B25**: 1–19.

Streten, N.A., 1980a. Antarctic meteorology: the Australian contribution past, present and future. *Australian Meteorology Magazine*, **28**: 105–140.

Streten, N.A., 1980b. Some synoptic indices of the Southern Hemisphere mean sea level circulation 1972–77. *Monthly Weather Reviews* **108**(1): 18–36.

Streten, N.A., and Pike, D.J., 1980. Characteristics of the broadscale Antarctic sea ice extent and the associated atmospheric circulation 1972–1977. *Archiv für Meteorologie, Geophysik und Bioklimatologie*, **A20**: 279–299.

Streten, N.A., and Zillman, J.W., 1984. Climates of the South Pacific. In van Loon, H., ed., *Climates of the Oceans. World Survey of Climatology*, vol. 15. Amsterdam: Elsevier.

Thompson, D.W.J., and Solomon, S., 2002. Interpretation of recent Southern Hemisphere climate change. *Science*, **296**(5569): 895–899.

Thompson, D.W.J., Wallace, J.M., and Hegerl, G.C., 2000. Annular modes in the extratropical circulation, Part 2: Trends. *Journal of Climate*, **13**: 1018–1036.

Torinesi, O., Fily, M., and Genthon, C., 2003. Variability and trends of the summer melt period of Antarctic ice margines since 1980 from microwave sensors. *Journal of Climate*, **16**(7): 1047–1060.

Trenberth, K.E., 1979. Interannual variability of the 500 mb zonal mean flow in the Southern Hemisphere. *Monthly Weather Review*, **107**(11): 1515–1524.

Trenberth, K.E., 1987. The role of eddies in maintaining the westerlies in the Southern Hemisphere winter. *Journal of Atmospheric Science*, **44**: 1498–1508.

Turner, J., Leonard, S., Marshall, G.J., et al., 1999. An assessment of operational Antarctic analyses based on data from the FROST project. *Weather Forecast*, **14**: 817–834.

Turner, J., Harangozo, S.A., Marshall, G.J., King, J.C., and Colwell, S.R., 2002a. Anomalous atmospheric circulation over the Weddell Sea, Antarctica during the austral summer of 2001/02 resulting in extreme sea ice conditions. *Geophysical Research Letters*, **29**(24): 2160, doi: 10.1029/2002GL015565, 2002.

Turner, J., King, J.C., Lachlan-Cope, T.A., and Jones, P.D., 2002b. Communications arising, Recent temperature trends in the Antarctic. *Nature*, **418**: 291–292.

Turner, J., Lachlan-Cope, T.A., Marshall, G.J., Morris, E.M., Mulvaney, R., and Winter, B., 2002c. Spatial variability of Antarctic Peninsula net surface mass balance. *Journal of Geophysical Research*, **107**(D13): doi: 10.1029/2001JD000755, 2002.

Van Loon, H., 1979. The association between latitudinal temperature gradient and eddy transport. Part 1: Transport of sensible heat in winter. *Monthly Weather Review*, **107**: 525–534.

Van Loon, H., and Rogers, J.C., 1984. Interannual variations in the half-yearly cycle of pressure gradients and zonal wind at sea level on the Southern Hemisphere. *Tellus*, **36A**: 76–86.

Van Loon, H., Taljaard, J.J., Sasamori, T., et al., 1972. Meteorology of the Southern Hemisphere. In Newton, C.W., ed., *Meteorological Monographs*, **13**(35). Boston, MA: American Meteorological Society.

Van den Broeke, M., 2000a. On the interpretation of Antarctic temperature trends. *Journal of Climate*, **13**: 3885–3889.

Van den Broeke, M., 2000b. The semi-annual oscillation and Antarctic climate. Part 4: A note on sea ice cover in the Amundsen and Bellingshausen seas. *International Journal of Climatology*, **20**: 455–462.

Van den Broeke, M.R., and van Lipzig, N.P.M., 2003. Factors controlling the near-surface wind field in Antarctica. *Monthly Weather Review*, **131**(4): 733–743.

Vaughan, D.G., and Doake, C.S., 1996. Recent atmospheric warming and retreat of ice shelves on the Antarctic Peninsula. *Nature*, **379**: 328–331.

Vaughan, D.G., Bamber, J.L., Giovinetto, M., Russell, J., and Cooper, A.P.R., 1999. Reassessment of net surface mass balance in Antarctica. *Journal of Climate*, **12**(4): 933–946.

Vaughan, D.G., Marshall, G.J., Connolley, W.M., et al., 2003. Recent rapid regional climate warming on the Antarctic Peninsula. *Climatic Change*, **60**: 243–274.

Watkins, A.B., and Simmonds, I., 1999. A late spring surge in the open water of the Antarctic sea ice pack. *Geophysical Research Letters*, **26**: 1481–1484.

Weller, G., 1982. Polar problems in climate research: some comparisons between the Arctic and Antarctic. *Australian Meteorology Magazine*, **30**: 163–168.

Wendland, W.M., and McDonald, N.S., 1986. Southern Hemisphere airstream climatology. *Monthly Weather Review*, **114**(1): 88–94.

Wendler, G., and Worby, A., 2001. The surface energy budget in the Antarctic summer sea-ice pack. *Annals of Glaciology*, **33**: 275–279.

Wendler, G., Adolphs, U., Hauser, A., and Moore, B., 1997a. On the surface energy budget of sea ice. *Journal of Glaciology*, **43**(143): 122–130.

Wendler, G., Stearns, C., Weidner, G., Dargaud, G., and Parish, T., 1997b. On the extraordinary katabatic winds of Adélie Land. *Journal of Geophysical Research*, **102**(D4): 4463–4474.

Worby, A.P., and Allison, I., 1991. Ocean–atmosphere energy exchange over thin variable concentration Antarctic pack ice. *Annals of Glaciology*, **15**: 184–190.

World Meteorological Organization, 1967. *Polar Meteorology*, Technical Note No. 87. Geneva: World Meteorological Organization.

Wu, X., Budd, W.F., and Jacka, T.H., 1999. Simulations of Southern Hemisphere warming and Antarctic sea-ice changes using global climate models. *Annals of Glaciology*, **29**: 61–65.

Wu, X., Budd, W.F., Worby, A.P., and Allison, I., 2001. Sensitivity of the Antarctic sea-ice distribution to oceanic heat flux in a coupled atmosphere–sea ice model. *Annals of Glaciology*, **33**: 577–584.

Yasunari, T., and Kodama, S., 1993. Intraseasonal variability of katabatic wind over East Antarctica and planetary flow regime in the Southern Hemisphere. *Journal of Geophysical Research*, **98**: 13063–13070.

Yuan, X., and Martinson, D.G., 2001. The Antarctic Dipole and its predictability. *Geophysical Research Letters*, **28**(18): 3609–3612.

Yuan, X., Martinson, D.G., and Liu, W.T., 1999. Effect of air-sea-ice interaction on winter 1996 Southern Ocean subpolar storm distribution. *Journal of Geophysical Research*, **104**(D2): 1991–2007.

Cross-references

Albedo and Reflectivity
Arctic Climates
Atmosphere Circulation, Global
Energy Budget Climatology
Ocean Circulation
Oscillations
Sea Ice and Climate

APPLIED CLIMATOLOGY

Applied climatology has been the foundation upon which the world's weather-sensitive activities and infrastructure have been developed. Applications of climate data and information have likely contributed more to the development of most nations than any other function of the atmospheric sciences. Today weather forecasts are very useful and important but these only became available in the twentieth century, more than 150 years after applied climatology had been in service to the nation.

What is applied climatology? The answer may seem obvious, but the definition of applied climatology is elusive for several reasons. Activities falling under the umbrella of applied climatology are spread among many disciplines. The activities also have interfaces with many parts of the atmospheric sciences, forming a myriad number of interactions. The field is truly interdisciplinary, embracing climatologists and those of other disciplines including hydrology, agriculture, engineering, and business. This breadth of the field and its evolution over time make the history of applied climatology interesting.

Over 50 years ago, leaders in the field defined applied climatology as the scientific analysis of climate data in light of a useful application for an operational purpose (Landsberg and Jacobs, 1951). A recently published glossary of meteorological terms indicates that applied climatology includes agricultural climatology, aviation climatology, bioclimatology, industrial climatology, and others (AMS, 2000).

My interpretation is that applied climatology describes, defines, interprets, and explains the relationships between climate conditions and countless weather-sensitive activities. For example, if viewed in a business sense, applied climatology would embrace interactions with marketing, sales, customer services, research support, and the delivery arm for climatological data and information. The diversity of the field is one of its hallmarks.

Applied climatology does not include fundamental, basic studies of the climate system, but it does embrace causation as a functional part of explaining climatic relationships to other phenomena. For example, a climatological study of rainfall effects on streamflow may include an explanation of the atmospheric factors influencing the rainfall variability between days, months, or years.

The field of applied climatology has evolved over the past 60 years, into interactive groups embracing three functional areas. First is the inner core of applied climatology – focusing on instruments and data. Functions include data collection, transmission, quality assessment, archival, its representativeness in space and time, and access to it.

Functions found in the second group relate to the interpretation and generation of climate information generally based on interactions with users. Activities include statistical and physical analyses, performance of special interpretive studies, and generation of information (published or computer-based). Within the atmospheric sciences, applied climatology extends well into the areas of dynamic climatology, instrumentation, and climate data. Applied climatological research develops the informational products for the users, studies the climate–sector relationships, develops statistical techniques to express climate information effectively, contributes to weather data collection efforts, and develops databases to fulfill its mission. Today these functions are handled by state climatologists, university-based scientists, and staffs at regional climate centers, the National Climatic Data Center, the Climate Prediction Center, and the private sector that addresses climate services.

The third group consists of users of applied climatology products, and is easily the largest group. Users include scientists in many disciplines and decision makers in business and government. Effective relationships between applied climatologists and users require a two-way interaction to share needs and information. Applications of climate information fall into four classes: (1) design of structures and planning activities, (2) assessments of current and past conditions including evaluation of extreme events, (3) study of the relationships between weather–climate conditions and those in other parts of the physical and socioeconomic worlds, and (4) operation of weather-sensitive systems that employ climatic information in making decisions.

History

Applied climatology, as one sector of today's world of atmospheric sciences, was the first atmospheric sciences activity to serve humankind, and applied climatology remains a most successful contributor to the nation's well-being.

Persons such as Thomas Jefferson were among the nation's early users of climate information. Jefferson planned farming operations around local climate conditions and applied climate information in the equipment he designed and used and in building construction (Martin, 1952). Crop selection, construction styles, and building placement in the eighteenth century reflected awareness of climate conditions and their effects.

The beginning of weather observations, both privately and at military forts, helped further applications of climate data. The expansion and organization of the nation's weather data collection system, and science in general, occurred during the nineteenth century.

Early leaders in the emerging field of meteorology, such as James Espy and Elias Loomis, recognized regional differences in the nation's climate and identified how these differences created profoundly different effects on human endeavors (Fleming, 1990). A key scientist of the century, Joseph Henry, made invaluable studies in applied climatology, and in the 1850s analyzed the impacts of climate on US agriculture of the time. Koeppen (1885) and other scientists began associating climate conditions with various land uses and plant environments as a means to define climate regions in each continent.

Further growth in the understanding of climate interactions with various physical systems, such as water resources, had to await the systematic collection of data on climate as well as weather-sensitive conditions. As the nation entered the twentieth century data collection had been in progress sufficiently long to allow new, more definitive studies of how climate impacted activities. For example, Thiessen was a civil engineer who in 1912 defined how climate conditions were to be used in engineering designs. Famed hydrologist Robert Horton made basic discoveries regarding the components of the hydrologic cycle, and unraveled the complexities of climate–hydrology conditions of the Great Lakes in the 1920s (Horton, 1927). Early key studies of effects of climate on human health and behavior included those of C.F. Brooks (1925).

A national priority of the nineteenth century was to enhance agricultural production, and this led to the establishment of the US Weather Bureau in 1888, as part of the Department of Agriculture (Whitnah, 1961). Establishment of experimental farms around the nation ultimately produced the data needed to make definitive studies of how various components of the climate affected each crop, as well as livestock. Wallace (1920) made strides by developing the first climate–crop yield models. Thornthwaite (1937) pioneered modeling of the hydrologic system so as to derive measures of evapotranspiration and soil moisture and their effects on both plants and crops.

By 1940 many of the fundamental relationships between climate conditions and other physical systems had been sufficiently delineated to allow highly effective designs of structures, selections of regionally appropriate crop varieties, and wise management of water resource systems in the varying climatic zones of the nation (Landsberg, 1946). The pressures of World War II for climate data and expertise brought forth a new dimension to applied climatology (Jacobs, 1947). For example, studies focused on severe weather conditions such as tornadoes and hail, and their effects on crops, property, and human life (Flora, 1953). A flood of publications within the realm of applied climatology appeared during the 1940s and 1950s (Landsberg and Jacobs, 1951). This explosion of research and information generation was also tied to the creation of computers during World War II and use of punch cards as a means of digitizing historical weather data. By 1948 a national center housing all historical climate data had been established. The National Climatic Data Center (NCDC) staff began generating a myriad of publications giving potential users access to information never before widely available (Changnon, 1995).

Ever-improving computers and digitized data allowed major achievements in modeling of climate effects such as hydrologic models that related various climatic variables to streamflow behavior (Linsley et al., 1958). Heavy rainfall design information, critical to planning and designs to manage flooding, was generated (Hershfield, 1961; Huff, 1986). A highly useful national drought index was devised (Palmer, 1965), and sophisticated climate–crop yield models appeared, allowing accurate predictions of yield outcomes well before harvest (Thompson, 1964). Other studies addressed another important sector of applied climatology – the economic and environmental impacts of all forms of climate (Maunder, 1970; McQuigg, 1974). Major advances in statistical techniques for applied climate analysis occurred (Conrad and Pollack, 1944; Court, 1949). An area of applied climatology that has evolved over the past 100 years concerns urban climates and other human-induced changes to the landscape leading to altered local and regional climates (Landsberg, 1956; Changnon, 1973; Oke, 1973). The impacts of these changes on the physical world and socioeconomic sectors have also been assessed (Changnon, 1984; Hare, 1985) with climate guidance to urban planners (Changnon, 1979). Finally, textbooks that addressed applied climatology appeared (Oliver, 1973, 1981; Thompson and Perry, 1997), serving to enhance training and attention to the field.

By 1970 applied climatology had moved to a new level of recognition and ever-higher value to the user community. Atmospheric scientists within the field turned their attention to improvements to: weather-sensing instruments, data quality and its archival, access of data and climate information, and generation of user-friendly climate products.

The golden age

The era since about 1970 has seen a series of scientific and technological changes that have vastly enhanced the field of applied climatology. Coupled with these advances have been national and global economic conditions that acted to increase the demand for climate products. The golden age of applied climatology had begun.

The agricultural economy became global, and with this expansion came huge economic pressures. American firms searched for every activity that would give them an advantage. One of these was use of climate predictions. Firms that had previously ignored use of uncertain climate outlooks now shifted and became users. Other business sectors also became global, and the net effect was more use of climate data and information.

Another factor enhancing wide interest and use of climate information comprised major global climate anomalies of the 1970s and early 1980s and their severe impacts (Kates, 1980; Panel, 1981). This included the devastating Sahel drought, the record cold winters of 1976–1980 in the US, and the droughts of 1980, 1983, and 1988. Climate, and the problems it created, including escalation of federal relief payment for weather–climate disasters, attracted the attention of the federal government, and Congress passed the National Climate Program Act in 1978. This program fostered new climate institutions, enhanced applied research, and funded new data collection–transmission systems. However, at the federal level the program became overshadowed in the late 1980s by the rapidly expanding national climate change endeavors. Concerns over a climate change related to global warming became the new thrust enveloping most of the atmospheric sciences. An era of numerous weather extremes and large global losses during the 1990s has led applied climatologists to pursue climatological assessments (Changnon, 1999a,b). Some have assessed whether these increasing losses were due to the start of a climate change due to global warming, to increasing societal vulnerability to climate (Changnon et al., 2000), and/or to inadequate government policies (Changnon and Easterling, 2000). Recent major urban droughts in the US initiated new societal problems, indicating that the future of water supplies for major urban areas looks questionable (Changnon, 2000).

Development during the 1960s and 1970s of reasonably inexpensive computer systems capable of handling large volumes of climate data was another key factor in the recent growth of applied climatology. The systems allowed continual updates of information, and the development and delivery of near real-time climate information, coupled with wide use of PCs. Everyone could access a wealth of climate information quickly and at low cost (Kunkel et al., 1990). This enhanced use of climate information has helped create greater awareness of the value of applied climatology.

The above-mentioned fast access was facilitated by another critical step forward – the development of inexpensive means to quickly collect data, and to transmit climate data and information. This included satellites and the Internet. These allowed real-time transmittal of data and quick access to it, a huge step forward (Changnon and Kunkel, 1999). Closely coupled with this advance were the establishment of new climate service/research centers, which had been fostered by the National Climate Program (Changnon et al., 1990), as well as the growth of private sector providers of climate information (Vogelstein, 1998). The development of these new institutions with expertise and systems to serve the needs of users of climate data and information led to other advances in applied climatology.

Interdisciplinary research of climatologists and other physical and social scientists increased, and this led to a new level of sophisticated climate-effect models (crops, water, transportation, etc.). Further, these models, when fed with real-time data, allowed their use in operational settings. Thus, near real-time estimates of current and projected climate effects were generated for decision makers. These activities moved forward in the right arenas based on assessments of user needs that began during the 1970s, a form of market analysis for climate products (Changnon et al., 1988; Changnon, 1992).

Since the 1970s the nation and world have seen an increase in society's sensitivity to climate conditions and especially extremes. Population growth, coupled with demographic changes and wealth, have created greater vulnerability and hence higher costs from climate anomalies. These impacts in the US have further promoted the growth in the use of climate data and information to more effectively react, manage, and compete (Stern and Easterling, 1999). Climate information has taken on greater value. One reflection of this in the business world has been the development and use of "weather derivatives" during the 1990s, a means of insuring against climatological risk (Zeng, 2000). One company offers coverage for a fixed price against a climate outcome, say a cold winter, that a utility fears economically. If the cold winter occurs, the coverage firm pays the other for its losses, but if the cold winter does not occur, the coverage firm retains the original payment. Such risks are assessed using climate data.

One of the applied climate products long sought by weather-sensitive entities has been accurate long-term climate predictions. The past 15 years have seen major advances in climate prediction quality, related to a greater understanding of the climate system such as the effects of El Niño on the nation's climate (Kousky and Bell, 2000). Government predictions have improved, both in accuracy and formats needed by users (Changnon et al., 1995). Private firms now work more closely with firms to interpret predictions to meet specific corporate needs.

As stated, applied climatology has moved into its golden age in service to society. In a recent book Thompson and Perry (1997) provide a broad, all-encompassing view of the world of applied climatology. They and 27 other applied climatologists prepared chapters on wide-ranging topics such as climate effects on tourism, glaciers, fisheries, and air pollution. Hobbs (1997), in a sweeping assessment of applied climatology, points to the growing awareness of applied climatology and increasing use of climate information.

However, not all things are occurring at an optimum level. Some problems still face the field. Ironically, teaching of applied climatology is too limited and often not done at many colleges and universities. To be effective, quality instruction in applied climatology requires interdisciplinary training and experience (Changnon, 1995).

A second concern relates to the adequacy of weather instrumentation and data collection. Since the structure of weather data collection endeavors in the US is still dominated by the needs of forecasters, there are continuing problems with sustaining adequate spatial sampling of climate conditions and with the use of instruments that allow continuity with historical data (NRC, 1998). For example, the automated surface observation system installed during the 1990s for measuring many weather conditions at nation's first-order weather stations has led to alterations in the quality of certain data.

As noted above, the use of climate data and information has grown rapidly in the past 30 years, but sampling reveals many potential users are still not served and often unaware of applications (Changnon and Changnon, 2003). An outreach effort by government agencies involved in climate services and by private sector partners is needed to educate and demonstrate how to use climate information and the potential values apt to be realized from usage to manage climate risks.

There is no systematic collection of data on the impacts of climate extremes, and a national effort to begin such data collection is needed (NRC, 1999). Uncertainty over impacts under a changing future climate due to global warming is a current dilemma and one that will continue. Hare (1985) first noted the need to consider impacts of future climate change. Some have predicted future climate conditions exceed extremes sampled in the past 100 years, making use of existing climate-impact regression models as predictors of future impacts invalid. The implications of future climate changes remain a major challenge for applied climatologists. This is reflected in a recent book about applied climatology which includes four chapters addressing potential impacts of future climate change (Thompson and Perry, 1997).

Resolution of these four issues – better training, stabilization of weather measurements, better information on climate impacts, and effects of global warming induced climate change – needs to be accomplished to realize the full potential of applied climatology. Regardless, applied climatology is the oldest atmospheric sciences activity in service to society, and its most successful.

Stanley A. Changnon

Bibliography

American Meteorological Society, 2000. *Glossary of Meteorology*. Boston, MA.

Brooks, C.F., 1925. The cooling of man under various weather conditions. *Monthly Weather Review*, **53**: 28–33.

Changnon, D., 1998. Design and test of a "hands-on" applied climate course in an undergraduate meteorology program. *Bulletin of the American Meteorological Society*, **79**: 79–84.

Changnon, S.A., 1973. Atmospheric alterations from man-made biospheric changes. In *Weather Modification: social concerns and public policies*. Western Geographical Series, pp. 134–184.

Changnon, S.A., 1979. What to do about urban-generated weather and climate changes. *Journal of the American Planning Association*, **45**: 36–48.

Changnon, S.A., 1984. Purposeful and accidental weather modification: our current understanding. *Physical Geography*, **4**: 126–139.

Changnon, S.A., 1992. Contents of climate predictions desired by agricultural decision makers. *Journal of Applied Meteorology*, **31**: 1488–1491.

Changnon, S.A., 1995. Applied climatology: a glorious past, and uncertain future. *Historical Essays in Meteorology, American Meteorological Society*, **11**: 379–393.

Changnon, S.A., 1999a. Factors affecting temporal fluctuations in damaging storm activity in the U.S. based on insurance loss data. *Meteorological Applications*, **6**: 1–11.

Changnon, S.A., 1999b. Record high losses for weather hazards during the 1990s: how excessive and why? *Natural Hazards*, **18**: 287–300.

Changnon, S.A., 2000. Reactions and responses to recent urban droughts. *Physical Geography*, **21**: 1–20.

Changnon, D., and Changnon, S., 2003. Assessment of issues related to usage of climate predictions in the U.S. agribusiness sector and utility industry. *Proceedings, Workshop on Climate Predictions. American Meteorological Society*, Washington, DC.

Changnon, S.A., and Easterling, D.R., 2000. U.S. policies pertaining to weather and climate extremes. *Science*, **289**: 2053–2055.

Changnon, S.A., and Kunkel, K.E., 1999. Rapidly expanding uses of climate data and information in agriculture and water resources. *Bulletin of the American Meteorological Society*, **80**: 821–830.

Changnon, S.A., Sonka, S., and Hofing, S., 1988. Assessing climate information use in agribusiness. Part 1: Actual and potential use and impediments to use. *Journal of Climate*, **1**: 757–765.

Changnon, S.A., Lamb, P., and Hubbard, K., 1990. Regional climate centers: new institutions for climate services and climate impact records. *Bulletin of the American Meteorological Society*, **71**: 527–537.

Changnon, S.A., Changnon, J., and Changnon, D., 1995. Assessment of uses of climate forecasts in the utility industry in the central U.S. *Bulletin of the American Meteorological Society*, **76**: 711–720.

Changnon, S.A., Pielke, R.A., Changnon, D., Sylves, R., and Pulwarty, R., 2000. Human factors explain the increased losses from weather and climate extremes. *Bulletin of the American Meteorological Society*, **81**: 437–442.

Conrad, V., and Pollack, L., 1944. *Methods in Climatology*. Boston, MA: Harvard University Press.

Court, A., 1949. Separating frequency distributions into normal components. *Science*, **110**: 500–510.

Fleming, J.R., 1990. *Meteorology in America, 1800–1870*. Baltimore, MD: Johns Hopkins University Press.

Flora, S.D., 1953. *Tornadoes of the U.S.* Norman, OK: University of Oklahoma Press.

Hare, K.F., 1985. Future environments – can they be predicted? *Transactions, Institute of British Geographers*, **10**: 131–137.

Hershfield, D.M., 1961. *Rainfall Frequency Atlas of the U.S. Weather Bureau*. Washington, DC.

Hobbs, J.E., 1997. Introduction: the emergence of applied climatology and climate impact assessment. In *Applied Climatology Principles and Practice*. New York: Routledge, pp. 1–12.

Horton, R.E., 1927. *On the Hydrology of the Great Lakes*. Sanitary District of Chicago, Chicago, IL: Engineering Board of Review.

Huff, F.A., 1986. Urban hydrometeorology review. *Bulletin of the American Meteorological Society*, **67**: 703–711.

Jacobs, W.C., 1947. Wartime development in applied climatology. *Meteorology Monographs*, **1**(1): 1–52.

Kates, R.W., 1980. Climate and society: lessons from recent events. *Weather*, **35**: 17–25.

Koeppen, W., 1885. Zur Charaskteristik der regen in NW Europa and Nordaamerika. *Meteorologie Zhurnal*, **10**: 24.

Kousky, V.E., and Bell, G.D., 2000. Causes, predictions, and outcomes of El Nino 1997–1998. In *El Nino 1997–1998, the Climate Event of the Century*. New York: Oxford University Press, pp. 28–48.

Kunkel, K.E., Changnon, S.A., Lonnquist, C., and Angel, J.R., 1990. A real-time climate information system for the Midwestern U.S. *Bulletin of the American Meteorological Society*, **71**: 601–609.

Landsberg, H., 1946. Climate as a natural resource. *Scientific Monthly*, **63**: 293–298.

Landsberg, H., 1956. The climate of towns. In ed., *Man's Role in Changing the Face of the Earth*, Chicago, IL: University of Chicago Press.

Landsberg, H., and Jacobs, W.C., 1951. Applied climatology. In Malone, T.F., ed., *Compendium of Meteorology*. Boston, MA: American Meteorological Society, pp. 976–992.

Linsley, R.K., Kohler, M., and Paulhus, J., 1958. *Hydrology for Engineers*. New York: McGraw-Hill.

Martin, E.T., 1952. *Thomas Jefferson: Scientist*. New York: H. Schuman.

Maunder, W.J., 1970. *The Value of the Weather*. London: Methuen.

McQuigg, J.D., 1974. The use of meteorological information in economic development. Applications of meteorology to economic and social development. *World Meteorological Organization Technical Note*, **132**: 7–59.

National Research Council, 1998. *Future of the National Weather Service Cooperative Network*. Washington, DC: National Academy Press.

National Research Council, 1999. *The Costs of Natural Disasters: A Framework for Assessment*. Washington, DC: National Academy Press.

Oke, T.R., 1973. City size and the urban heat island. *Atmospheric Environment*, **7**: 769–779.

Oliver, J.E., 1973. *Climate and Man's Environment: An Introduction to Applied Climatology*. New York: Wiley.

Oliver, J. E., 1981. *Climatology: Selected Applications*. New York: Wiley.

Palmer, W.C., 1965. Meteorological drought. *U.S. Weather Bureau Research Paper* 45. Washington, DC.

Panel on the Effective Use of Climate Information in Decision Making, 1981. *Managing Climatic Resources and Risks*. Washington, DC: National Academy Press.

Stern, P., and Easterling, W., 1999. *Making Climate Forecasts Matter*. Washington, DC: National Academy Press.

Thompson, L.M., 1964. Our recent high yields – how much is due to weather? "Research in Water". *Soil Society of America*, **6**: 74–84.
Thompson, R.D., and Perry, A., 1997. *Applied Climatology Principles and Practice*. New York: Routledge.
Thornthwaite, C.W., 1937. The hydrologic cycle re-examined. *Soil Conservation*, **3**: 85–91.
Vogelstein, F., 1998. Corporate America loves the weather. *U.S. News & World Report*, 13 April, pp. 48–50.
Wallace, H.A., 1920. Mathematical inquiry into the effect of weather on corn yields in the eight corn belt states. *Monthly Weather Review*, **48**: 439–446.
Whitnah, D., 1961. *A History of the U.S. Weather Bureau*. Urbana, IL: University of Illinois Press.
Zeng, L., 2000. Weather derivatives and weather insurance: concept, application, and analysis. *Bulletin of the American Meteorological Society*, **81**: 2075–2082.

Cross-references

Art and Climate
Bioclimatology
Climate Hazards
Commerce and Climate
Crime and Climate
Hydroclimatology
Tourism and Climate

ARCHEOCLIMATOLOGY

The term archeoclimatology was coined by Reid A. and Robert U. Bryson, about 1990, to designate a particular approach to the estimation of past climates on the time and spatial scales appropriate for the use of archeologists (Bryson and Bryson, 1994). Because cultures change on less than the millennial scale, and because people live in relatively restricted areas, whatever data or model is used must be nearly site-specific and of high time-resolution. Ideally the method should be rather economical, since archaeology tends to be funded at a low level.

This subspecialty of paleoclimatology is largely concerned with bringing together various sources of estimation of past climates, models and proxy estimates from field studies, to provide the most consistent estimates of the past climatic environment at particular places and times. The study of proxy records has been well advanced by various other Earth and biological sciences, so archeoclimatology thus far has been concerned with climatic models designed with this particular end in sight rather than on studies of the atmosphere itself.

The definition of climate

The way a segment of the world is studied depends on how the topic is defined relative to the expertise of the scientist and the historical precedents of the scientist's discipline as much as on rational analysis. This may lead to definitions and biases that affect the study of the topic.

The usual definition of climate according to meteorologists is approximately that given in the *Glossary of Meteorology* (Huschke, 1959). Here climate is defined by reference to C.S. Durst, a meteorologist, as "The synthesis of the weather". Huschke then goes on: "More rigorously, the climate of a specified area is represented by the statistical collective of its weather conditions during a specified interval of time (usually several decades)."

Geographers tend to use a similar definition, witness Trewartha's "Climate . . . refers to a more enduring regime of the atmosphere; it is an abstract concept. It represents a composite of the day-to-day weather conditions, and of the atmospheric elements, within a specified area over a long period of time" (Trewartha and Horn, 1980). (How one can use facts, figures and equations to study an "abstract concept" is more than a little puzzling.) Lamb uses a similar definition of climate: "Climate is the sum total of the weather experienced at a place in the course of the year and over the years" (Lamb, 1972).

Viewing climate as essentially a summation of the weather has profound implications for the development of a body of climatological theory. Indeed, the preceding sentence might evoke disagreement from many atmospheric scientists for the common definition of climate indicates that there is no distinct body of theory separate from that of the weather, except perhaps statistical theory. With this viewpoint one might disagree, based on a quite different view of the relation of climate and weather.

The relation of climate and weather

An experienced meteorologist can identify the atmospheric circulation pattern on a weather map immediately as to whether it represents a summer pattern or a winter pattern. Usually the identification of the season can be even closer. The reason the array of weather patterns characteristic of a season differs from the array of another season is that the climate differs from season to season. This statement, of course, does not make sense if the climate is the summation of the weather. It does make sense if the thermodynamic status of the Earth–atmosphere–hydrosphere–cryosphere system *determines* the array of possible (and necessary) weather patterns. This status, which changes with time and season, *along with* the associated weather patterns, constitutes the climate.

The basis of the present discussion, and the research of which it is an elaboration, is that the climate is a consequence of external controls on the atmosphere–hydrosphere–cryosphere system. These "boundary conditions" force the state of the climate system, which in turn produces and requires sets of weather complexes, which differ as the climate differs from one time to another. For example, the climatic state thought of as summer has a different array of associated weather patterns than does the climatic state called winter (Bryson, 1997).

Modeling the climate on this basis utilizes what is here called a macrophysical approach. Relationships of large-scale nature are used, such as the Rossby long-wave equations, the thermal wind relationship, or the Z-criterion derived from the work of Smagorinsky (1963). The model on which archeoclimatology is based required first a model of global glacial volume, from which the glaciated area could be calculated and the ice albedo effect estimated (Bryson and Goodman, 1986). A hemispheric temperature model was then possible using the so-called Milankovitch-type variations in solar irradiance as modulated by volcanic aerosols (Bryson, 1988). This section of the overall model is essentially a "heat budget model" but with improved treatment of the volcanic modulation.

Going from hemispheric or global models to regional models requires that some broad-scale relationships be used. For example, one knows that the westerlies must become stronger with height up to about 10 km because the Earth is colder at the poles than at the equator, and the rate of increase is proportional to the magnitude of the south–north temperature gradient. Since equatorial temperatures change relatively little even from glacial to non-glacial times, the average hemispheric temperature and the temperature gradient must both depend primarily on high-latitude temperatures.

Assuming that the zonal component of the wind at 500 mb is proportional to the meridional temperature gradient, this concept was applied to North Africa, using the "Z-criterion" to calculate the latitude of the subtropical anticyclone and the latitude of the jet axis at 500 mb (Bryson, 1992). Empirical values of the appropriate gradients were derived from the inter-monthly variation, including the relation between the latitude of the anticyclone and the latitude of the intertropical convergence (ITC). Ilesanmi's model (1971) relating position of the ITC to rainfall south of it could then be used to estimate the rainfall in the Sahel at two- or five-century intervals for the late Pleistocene and Holocene. The match with palaeolimnological and archaeological field data was excellent. Ilesanmi's model is an example of the kind of synoptic climatology that constitutes the final calculation of the local climatic effect from the large-scale features of the atmospheric circulation.

The hierarchical steps and the fundamental assumptions

Step 1: Model glacier volume and area

Examination of the geological evidence tells us that the volume of glacier changes very slightly in the southern hemisphere from glacial to interglacial, and that most of the Pleistocene northern hemisphere glacier was in North America, with a large part of the rest being in northwestern Europe.

Synoptic climatology tells us that the snowiest winters in the major North American glacier accumulation areas are mild winters. Logic and experience tell us that snow melts slower in a cool, cloudy wastage season.

Assumption 1

Assume that the seasonal temperature of the hemisphere is related to the amount of radiation received and absorbed near the surface by the hemisphere in that season. (And secondarily, that the winter temperatures in the ice accumulation area of North America are essentially always well below freezing, even as now.)

It follows that accumulation rate minus wastage rate will be inversely related to the radiation seasonality of the hemisphere.

It also follows that the difference of these two rates is also a rate, so that this net accumulation rate must be integrated to simulate the total glacier volume at any particular time (arbitrary units).

The northern hemisphere irradiance seasonality may be computed using the "Milankovitch" periodicities in the usual Fourier series form, which may be readily integrated. This series contains no high-frequency terms (Bryson and Goodman, 1986).

Assumption 2

Assume that volcanic aerosol is the major source of modulation of the incoming radiation in ancient times.

Volcanic eruption chronologies may be extended back to about 30 000 years, however, using radiocarbon-dated eruptions. Using historically observed eruptions, Goodman was able to show that most of the variance of the observed aerosol loading can be explained by the volcanic record (Goodman, 1984).

Using the radiocarbon-dated volcanic eruption chronology it is possible to construct a proxy record of volcanic aerosol that extends back at least 30 000 years. This may be used to modulate the radiation for the integrated Milankovitch-based glacier volume modeling, and thus introduce higher-frequency terms into the glacier model.

The result of this modulation and the introduction of higher-frequency terms is a simulation in arbitrary units of glacial volume that follows the known geological record quite well, including such short-term features as the "Younger Dryas", and Neo-Glaciation. The arbitrary units may be readily calibrated into standard volume units by scaling (Bryson, 1988).

Step 2: Model hemispheric seasonal temperatures

Assumption 1

Assume that the largest source of variation in hemispheric albedo is due to variation in ice and snow cover.

Estimates of glacial volume from step 1 may be converted to glaciated area using the relationship developed by Moran and Bryson for typical glacial profiles (Moran and Bryson, 1969).

Measurements of typical albedos clearly indicate the dominance of ice as the source of variation in time for the total range of albedo for the other surfaces is less than even the range within various forms of ice cover (Kung et al., 1964).

Assumption 2

Assume equivalent opaque cloud cover. With the definition of equivalent opaque cloud cover developed by Dittberner and Bryson, the cloud amount drops out of the heat budget equations *but only for climatic scales* (Bryson and Dittberner, 1976).

Assumption 3

Assume that, throughout most of geological history, carbon dioxide and water vapor content of the atmosphere have been dependent variables, dependent on the temperature, and as a first approximation linearly so (Bryson, 1988).

Step 3: Model monthly hemispheric temperature, evaporation, and net radiation

Using the Lettau climatonomic method of obtaining a mutually consistent set of monthly surface heat budget components monthly temperatures may be obtained (Lettau and Lettau, 1975; Lettau, 1977, 1984).

Step 4: Calculate the position of the "centers of action" (large-scale features such as the subtropical anticyclones and the jet stream)

Assumption 1

Assume that the latitude of dynamic instability of the westerlies as calculated by Smagorinsky is a valid estimate of the latitude of the subtropical anticyclones. That this is valid was first suggested by Flohn, who used the term "Z-criterion". Tests with modern data give abundant evidence that the assumption is valid even regionally, and gives very close estimates (Flohn, 1965; Smagorinsky, 1963).

The Z-criterion calculation requires only the large-scale meridional and vertical temperature gradients.

Assumption 2
Assume that, since equatorial temperatures hardly vary during the course of the year and in time, that the meridional temperature gradient is a function of the mean hemispheric temperature and that the vertical temperature gradient is a function of season. These relationships may be determined empirically using modern data.

Assumption 3
Assume that the latitude of the jet stream is closely related to the outer edge of the westerlies as calculated with the Z-criterion (Bryson, 1992).

It has been shown that the latitude of the intertropical convergence is closely related to the latitude of the related subtropical high(s) (Bryson, 1973).

At this point one may have calculated the monthly latitude of the subtropical anticyclones, latitude of the jet stream and position of the intertropical convergence for each sector of the hemisphere for as far into the past as the volcanic aerosol record and Milankovitch calculations allow, and at intervals of centuries dictated by the quality of the volcanic record.

Step 5: Calculate local climate history from output of the above steps

At this stage, knowing the mean locations of the major anticyclones, mean latitude of the jet stream, and mean latitude of the intertropical convergence, sector by sector and month by month, one may use the well-established techniques of synoptic climatology to calculate the mean monthly local values of temperature, precipitation, rainfall frequency, potential evapotranspiration, etc. at one's station of interest.

For example, one might want to calculate the past precipitation amount for a location in the midwestern United States, such as Madison, Wisconsin. It is well known that tropical air from the North Atlantic cyclone brings warm-season rain when it is far north, but this is not a linear relationship. It is also known that the cyclonic storms associated with the storm track represented by the jet stream position bring the maximum rains of spring and early fall, and lesser amounts in winter. Other times, especially in winter, air from the Pacific anticyclone streams across the country, usually suppressing the rains in Madison. The problem then is to say how the mean monthly precipitation at Madison responds to the position of these circulation features. The synoptic climatology approach to this problem is empirical. By examination of the relationship of the precipitation at Madison to these circulation elements within the instrumental period, one can establish a quantitative relationship, which is

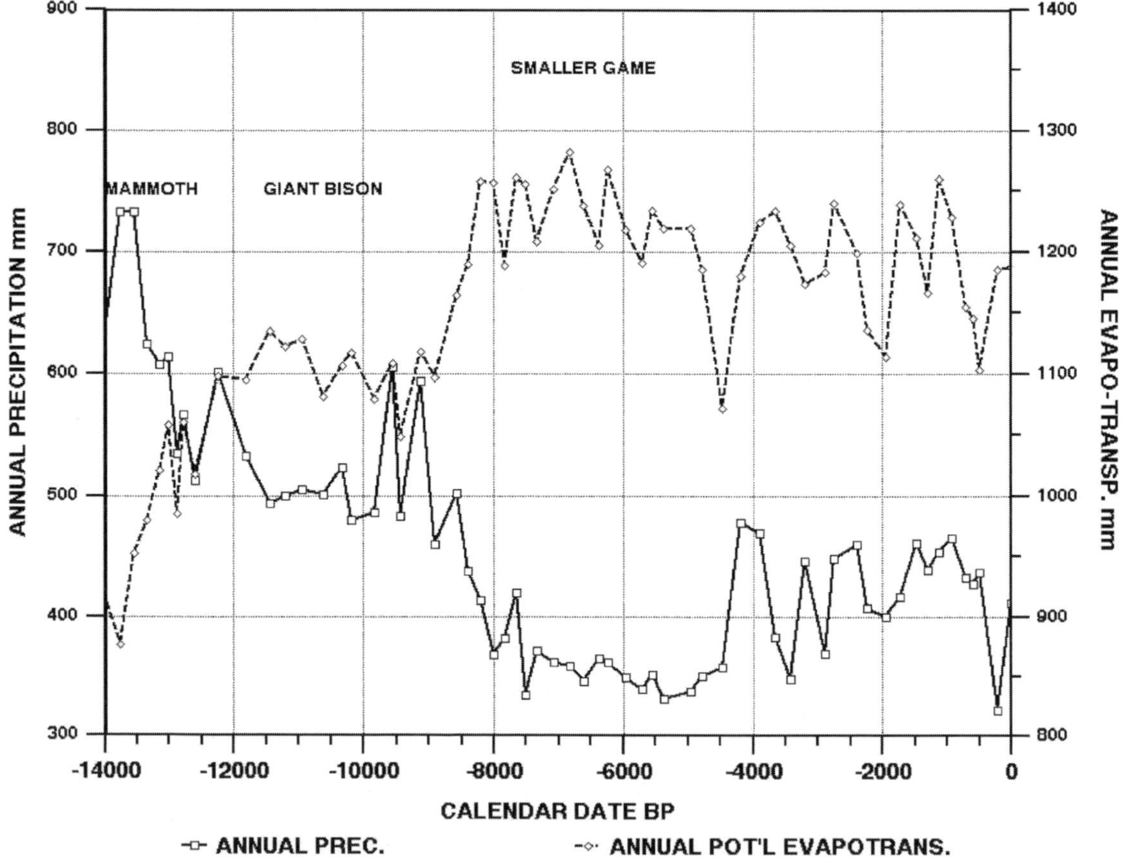

Figure A35 Modeled water supply history for Portales, New Mexico.

assumed to apply as far back as the general geographic pattern of waters and topography is relatively constant.

One virtue of the synoptic climatology approach is that the local and regional topography, etc. is implicitly included in the precipitation, temperature, etc. values that have been observed. For example, the data from Milwaukee, 75 miles east of Madison, differ from the data for Madison, in part because Miewaukee is on the shore of a great lake, and thus the relationship of the precipitation there to the large-scale circulation features would contain, implicitly, the effect of its location.

We can then say that the archeoclimatic model for North American precipitation at a particular place consists of the latitudes of the oceanic anticyclones, the jet stream and the intertropical convergence as inputs, and the calibrated response of the local mean values of the precipitation at that place as the output. Since the circulation features have been calculated far into the past, we can assume that the calibration was the same then, and calculate the local past climate. As a first approximation non-linear regression is used for calibration. Lacking a long series of observed values of all the needed data, the annual march of monthly positions of the circulation elements and of the precipitation are used, and in over 2000 trial cases seems to be an adequate first approximation.

The position of the intertropical convergence is used, even for northern locations, because it is related to the insertion of very moist air into the westerlies, sometimes in tropical cyclones, and hence is related to the precipitation even at higher latitudes.

On the other hand, the positions of the polar anticyclones are omitted because they are more dependent on topography and errors of pressure reduction to sea level than they are to dynamics of the atmosphere, and the latitude of the jet stream gives the outer limit of polar air.

Example 1: North America

Eastern New Mexico has been of particular interest to archeologists because it is a well-known locale for early big-game hunters in North America. These included hunters of mammoth and giant bison, the Clovis and Folsom cultures. Clovis, New Mexico, is the type locality for the distinctive projectile point of the Clovis people, a form found over a vast area of North America. The archeoclimatic model of the past climate of nearby Portales, NM yielded the next two graphs (Figures A35 and A36) of the precipitation, potential evapotranspiration and temperature history of the locale.

The graph of precipitation shows the high precipitation and lower evapotranspiration of latest Pleistocene time necessary to produce the rather lush vegetation required to support mammoths. This situation changes rapidly after the "Younger Dryas" peak precipitation to a quite different situation by 10 500 years before the present, and the typical big game changed to bison. With the advent of the Middle Holocene, after 8000 radiocarbon years ago, the environment only supports smaller game, and indigenous cultures no longer hunt big game. This pattern seems to agree with the field evidence.

Example 2: The Mediterranean

In 1965 Rhys Carpenter, a distinguished scholar of ancient Greece, proposed that the famous and very important Kingdom based at Mycenae, in the Peloponnesus, had

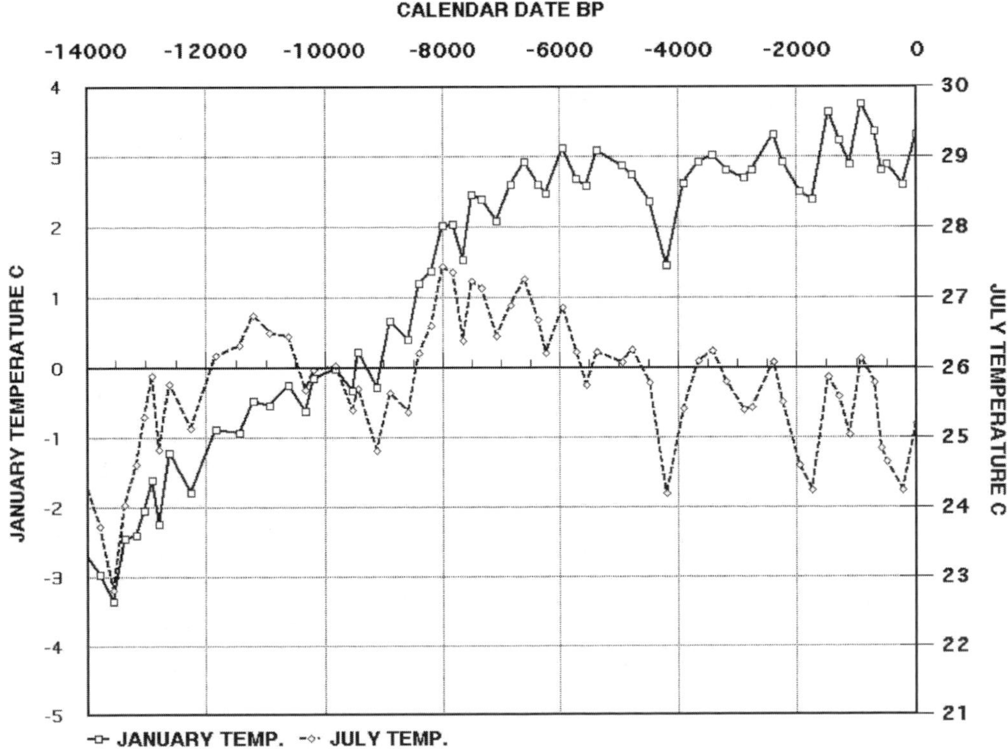

Figure A36 Modeled temperature history for Portales, New Mexico.

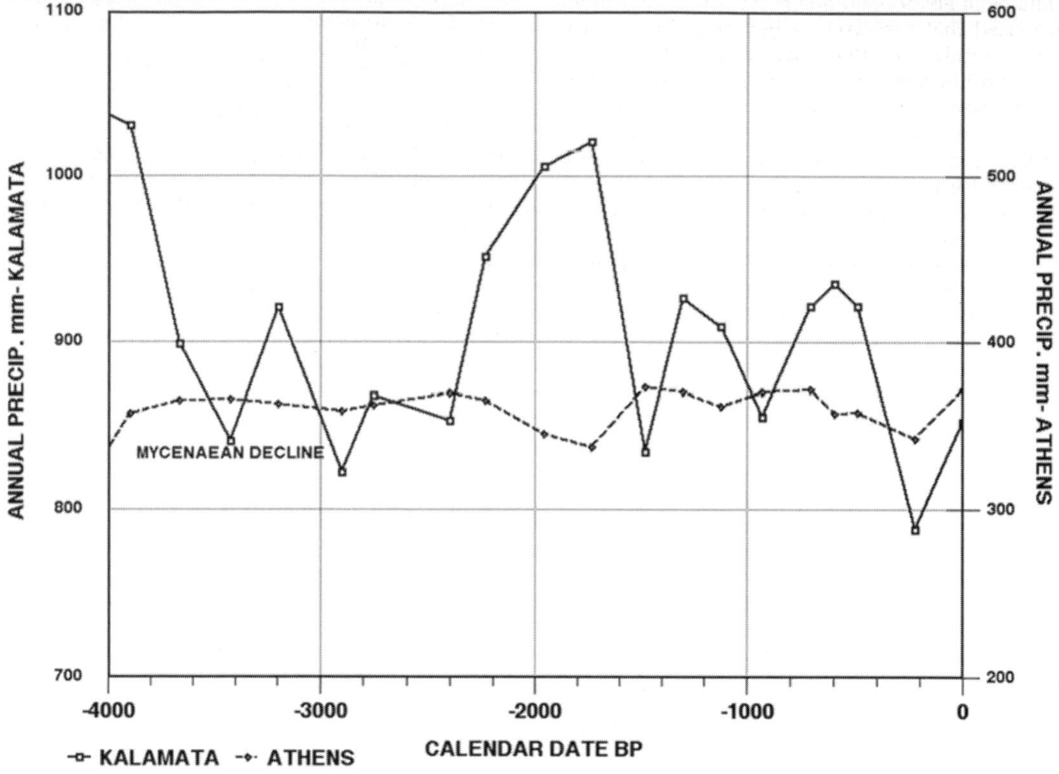

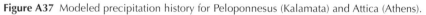

Figure A37 Modeled precipitation history for Peloponnesus (Kalamata) and Attica (Athens).

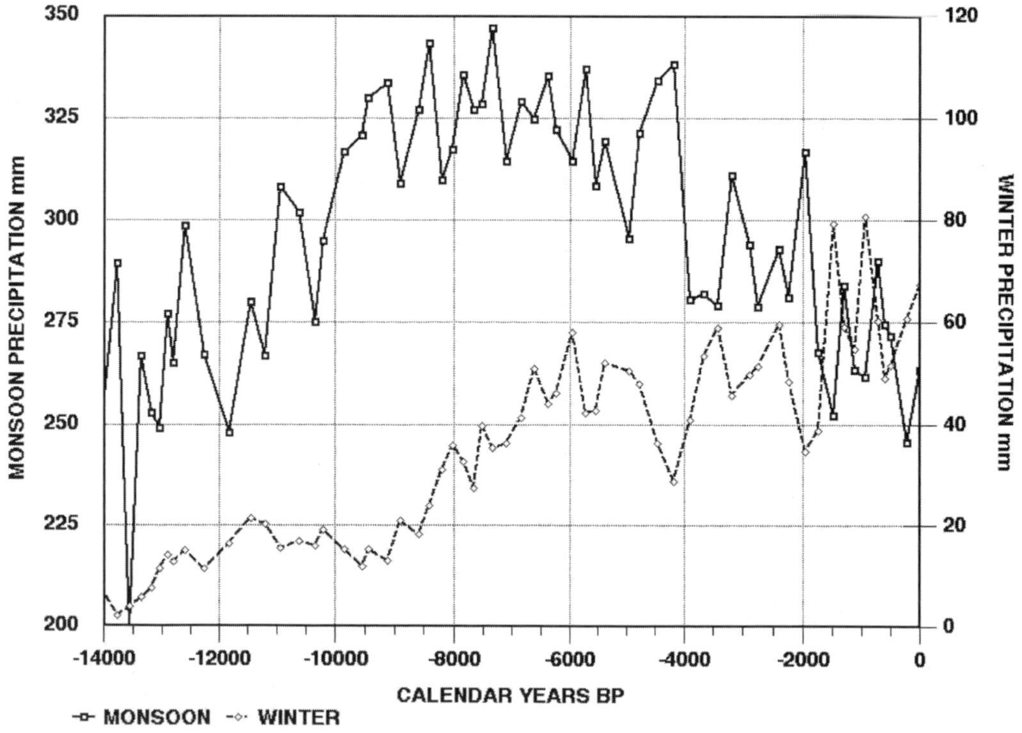

Figure A38 Modeled precipitation history for Harappa.

declined and had been abandoned due to a prolonged drought (1966). He proposed that the drought had been localized to regions by the terrain so that, while it affected the Peloponnesus, it did not affect the Attic Plain (Athens area). Using the archeoclimatic model we may compare the modeled history of these two neighboring areas to see whether such a pattern can be detected. Figure A37 shows the modeled history of precipitation for these two places.

This figure also shows the modeled precipitation for Kalamata in the southern Peloponnesus and for Athens on the Attic plain. Mycenae itself was nearer to Athens than Kalamata.

Some studies showed that Carpenter's drought hypothesis represented a possible, and even probable, pattern (Bryson et al., 1974). Archeoclimatic modeling also supports the hypothesis, and suggests that when the Peloponnesus became dry *ca* 1200 BCE, the Attic Plain did indeed not have a corresponding drought, even though the actual precipitation was usually less.

Example 3: Asia

In northwestern India and Pakistan, in an area now desert or semiarid, the extensive agriculture-based Indus culture existed between about 5500 and 3500 radiocarbon years ago. Studies of pollen in the northwestern Indian desert suggested that the termination of this culture was associated with a decrease in rainfall (Singh et al., 1974; Bryson and Swain, 1981). Figure A38, which shows the rainfall near Harappa, a major city of the Indus culture in the Indus valley, also suggests that there was a rapid change toward lower rainfall about 3500 years ago.

Reid Bryson

Bibliography

Bryson, R.A., 1973. Drought in Sahelia, who or what is to blame? *Ecologist*, **3**(10): 366–371.

Bryson, R.A., and Dittberner, G.J., 1976. A non-equilibrium model of hemispheric mean surface temperatures. *Journal of Atmospheric Sciences*, **33**: 2094–2106.

Bryson, R.A., 1988. Late Quaternary volcanic modulation of Milankovitch climate forcing. *Theoretical and Applied Climatology*, **39**: 115–125.

Bryson, R.A., 1992. A macrophysical model of the Holocene intertropical convergence and jetstream positions and rainfall for the Saharan region. *Meteorology and Atmospheric Physics*, **47**: 247–258.

Bryson, R.A., 1997. The paradigm of climatology: an essay. *Bulletin of the American Meteorological Society*, **78**(3): 449–455.

Bryson, R.U., and Bryson, R.A., 1994. A comparison of cultural evidence and simulated Holocene climates of the Pacific Northwest. Abstracts of the 59th Annual Meeting, Society for American Archaeology, Anaheim, CA, 20–24 April, p. 29.

Bryson, R.A., and Goodman, B.M., 1986. Milankovitch and global ice volume simulation. *Theoretical and Applied Climatology*, **37**: 22–28.

Bryson, R.A., and Swain, A.M., 1981. Holocene variations of monsoon rainfall in Rajasthan. *Quaternary Research*, **16**(2): 135–145.

Bryson, R.A., Lamb, H.H., and Donley, D.L., 1974. Drought and the decline of Mycenae. *Antiquity*, **XLVIII**, 46-50; also IES Report no. 20.

Carpenter, R., 1966. *Discontinuity in Greek Civilization*. Cambridge: Cambridge University Press.

Flohn, H., 1965. Probleme der theoretischen Klimatologie. *Naturwissenschaftliche Rundschau*, **10**: 385–392.

Goodman, B.M., 1984. The climatic impact of volcanic activity. Ph.D. thesis, Department of Meteorology, University of Wisconsin–Madison.

Huschke, R.E., ed., 1959. *Glossary of Meteorology*. Boston: American Meteorological Society.

Ilesanmi, Oluwafemi O., 1971. An empirical formulation of an ITD rainfall model for the tropics – a case study of Nigeria. *Journal of Applied Meteorology*, **10**: 882–891.

Kung, E.C., Bryson, R.A., and Lenschow, D.H., 1964. Study of a continental surface albedo on the basis of flight measurements and structure of the Earth's surface cover over North America. *Monthly Weather Review*, **92**(12): 543–564.

Lamb, H.H., 1972. *Climate: Past, Present, and Future*, Vol. 1: *Fundamentals and Climate Now*. London: Methuen.

Lettau, H.H., and Lettau, K., 1975. Regional climatonomy of tundra and boreal forests in Canada. In Wellers, G., and Bowling, S., eds., *Climate of the Arctic*. Fairbanks: University of Alaska, pp. 209–221.

Lettau, H.H., 1977. Climatonomic modeling of temperature response to dust contamination of Antarctic snow surfaces. *Boundary-Layer Meteorology*, **12**: 213–229.

Lettau, H.H., 1984. The Sauberer–Mahringer soilheat diffusion experiment: a re-analysis using force-response modeling equations. International Symposium in Memory of Dr Franz Sauberer, Universitat für Bodenkultur, Wien, pp. 21–28.

Moran, J.M., and Bryson, R.A., 1969. The contribution of Laurentide ice wastage to the eustatic rise of sea-level: 10 000 to 6000 Years BP. *Arctic and Alpine Research*, **1**(2): 97–104.

Singh, G., Joshi, R.A., Chopra, S.K., and Singh, A.B., 1974. Late Quaternary history of vegetation and climate of the Rajasthan Desert, India. *Philosophical Transactions of the Royal Society of London B*, **267**: 467–501.

Smagorinsky, J., 1963. General circulation experiments with the primitive equations, I: The basic experiment. *Monthly Weather Review*, **91**: 99–164.

Trewartha, G.T., and Horn, L.H., 1980. *An Introduction to Climate*, 5th edn. New York: McGraw-Hill.

Cross-references

Climate Change and Ancient Civilizations
Cultural Climatology

ARCHITECTURE AND CLIMATE

There are dramatic new building forms emerging around the world that have been generated by a growing understanding of climate and its importance to sustainability. From England to the US, from Malaysia to Malta, emerging architecture demonstrates the power of sun, wind, light, moisture and diurnal swing to generate innovative building forms, material detailing and systems integration for sustainability.

In an era of rediscovery, a growing number of architects and engineers begin their design efforts with regional climate graphics, study indigenous responses to climate, use climate-driven computer simulation tools for design development and specifications, and celebrate climatic variability through unique architectural results. These professionals will increasingly need accessible and consistent climate data worldwide, field testing of regional performance of climate-responsive buildings, regional design guidelines, and design simulation and modeling tools. The opportunities for improving environmental quality and reducing environmental costs could transform land-use and community development, building design and renovation, as well as speed the introduction of new materials and technologies – celebrating regional climates and sustainability in the pursuit of environmental security.

Climate change and architecture

In the face of climate change, surprisingly few global leaders have identified buildings and land-use as critical factors. Yet increasing levels of floods, hurricanes, quakes, and natural disasters are most clearly manifested in loss of property. Similarly, pollution, inversions, and warming are driven by combined forces of climate and building decision making. Even growing concerns about chemical and biological attack, and strategies for energy security, will only be addressed by a better understanding of climate and the built environment. International emphasis on avoiding environmental failure and improving environmental security must be fully tied to a rich understanding of regional climates, to ensure individual comfort and health, organizational effectiveness, and environmental sustainability.

Interface between climate and building performance

Total building performance

Today's design challenge is to "balance the high tech and the natural tech" to simultaneously ensure all building performance mandates – thermal comfort, acoustic comfort, visual comfort, air quality, spatial quality, and building integrity – for individual, organizational and environmental effectiveness.

To comprehend the breadth of this challenge it is important to begin with a complete definition of the building performance to be met assiduously by building policy makers, programmers, architects, engineers, contractors, owners, and managers.

First, there has been a fundamental mandate over the centuries for building-enclosure integrity, that is, protection from environmental degradation through moisture, temperature, air movement, radiation, chemical and biological attack, and environmental disasters (such as fires, floods, and earthquakes). Established by concerns for health, safety, welfare, resource management (energy, money), and image, the requirements for building integrity can be stated as limits of acceptable degradation and debilitation.

Second, there are a series of mandates relating to interior occupancy requirements (human, animal, plant, artifact, machine) and the elemental parameters of comfort, that is, thermal comfort, acoustic comfort, visual comfort, air quality, and functional comfort. Table A20 lists performance mandates of primary concern to the building industry, with admittedly lopsided emphasis given to building integrity (protection from degradation, debilitation, and devastation) reflecting the industry's focus on building failures in this area.

Table A20 Building performance mandates

- Building integrity
  - Moisture (rain, snow, ice, vapor)
    - penetration
    - migration
    - condensation
  - Temperature
    - insulation effectiveness
    - thermal bridging
    - freeze–thaw cycle
    - differential thermal expansion and contraction
  - Air movement
    - air exfiltrtion
    - air infiltration
  - Radiation and light
    - environmental radiation
    - solar radiation (e.g., ultraviolet)
    - visible light spectrum
  - Chemical attack
  - Biological attack
  - Fire safety
  - Disaster (earthquake, flood, hurricane, etc.)
- Thermal comfort
  - Air temperature
  - Radiant temperature
  - Humidity
  - Air speed
  - Occupancy factors and controls
- Acoustic comfort
  - Noise level and frequency
  - Reverberation
  - Speech privacy, articulation index
  - Vibration
  - Occupancy factors and controls
- Visual comfort
  - Ambient and task levels (artificial and daylight)
  - Contrast, brightness ratios (flare)
  - Color rendition
  - Occupany factor and controls
- Air quality
  - Ventilation rate: fresh air supply, circulation
  - Mass pollution (gases, vapors, microorganisms, fumes, smokes, dust)
  - Energy pollution (ionizing radiation, microwaves, radio waves, light waves, infrared)
  - Occupancy factors and controls
- Spatial comfort
  - Workstation layout (space, furniture surface, storage, seating, ergonomics)
  - Workgroup layout: adjacencies, compartmentalization, usable space; circulation/accessibility/way-finding/signage, indoor–outdoor relationships)
  - Conveniences, functional servicing (sanitary, electrical, security, telecommunication, circulation/transportation services)
  - Amenities
  - Occupancy factors and controls

Limits of acceptability

Each building performance mandate has a "comfort zone" that establishes the limits of acceptability for the type of occupancy concerned. These limits, often translated into standards and codes, budgets and guidelines, are established by the physiological, psychological, sociological, and economic requirements of a nation and its citizenry. The limits must be established in each country for the range of building functions and the range of occupancy types (human, artifact, plant, animal).

In regard to human occupancy, physiological requirements aim to ensure the health and safety of the building occupants, sheltering basic bodily functions from wear or destruction over time, against such conditions as fire, building collapse, poisonous fumes, high and low temperatures, and poor light. Psychological requirements aim to support individual mental health through appropriate provisions for privacy, interaction, access to the natural environment, clarity, status, change, and so forth. Sociological requirements (also referred to as sociocultural requirements) aim to support the effectiveness of the organization and the well-being of the community, relating the needs of the individual to those of the collective, including rising expectations for environmental quality. Finally, economic requirements aim to allocate resources – energy, water, land, air, materials – in the most efficient manner in the overall goal to serve user needs.

Although the distinction between climate and weather data is of major significance in the study of meteorology, the needs of the building industry span the range of *descriptive* climate data and *predictive* forecasted data without lucid differentiation. Consequently, both climate and weather data will be discussed under the broader term *climate*, with descriptive data referring to the long-term climate conditions and predictive or forecasted data referring to weather conditions.

Climate and thermal comfort

The climate database requirements for ensuring thermal comfort in buildings continues to be of great importance. Descriptive summaries of average temperatures are needed to estimate annual heat loss and heat gain, in order to establish building energy budgets. Record and extreme temperature conditions must be known to predict peak heat losses and gains, and to determine the size of mechanical conditioning equipment required in a given building. Forecasts of daily temperature and temperature swings must be given for setting mechanical system operating schedules and relieving day–night and seasonal conditioning imbalances. In addition, records of coincident wet-bulb temperatures or humidity conditions must be incorporated into air-conditioning system design, while predictive data will determine the level of humidification or desiccation needed. Annual, monthly, and typical-day solar data aids in estimating both solar heating contribution in winter and solar overheating potential in summer, with design responses to lower annual operating budgets and to allow the use of smaller mechanical conditioning equipment. Predictive solar data allow building operators to anticipate building thermal imbalances and to maximize the use of free heat. Finally, descriptive summaries of wind direction and speed, particularly coincident with temperature and humidity conditions, enable designers to prevent serious winter infiltration through vulnerable building components and to maximize the potential for natural ventilation.

The widespread shift to deeper section buildings that began in the mid-1970s, with less perimeter and more core, a move intended to reduce "climate-driven" conditioning loads, has instead dramatically compromised natural conditioning of buildings. Today, many commercial buildings have become so massive in plan and/or maintain such significant internal loads (from lights, equipment, and people) that energy costs include both high cooling loads in the core, as well as climate-related cooling and heating loads at the perimeter. The EU-funded PROBE studies identified 40–60% higher conditioning loads in newer commercial buildings than traditional, thinner buildings with identical functions (Bordass and Leaman, 2001).

Even in buildings that cannot take advantage of passive conditioning (daylighting, natural ventilation, solar heating, time lag cooling, etc.) attention to climate conditions is still critical for building energy conservation (Figure A39). Heating and cooling equipment size and consequent capital costs are set by peak heat loss and gain conditions, both of which are established at the perimeter. Annual heating and cooling costs (and associated personnel costs) are often significantly escalated by heating and cooling system "battles," resulting from the attempt to respond to daily shifts in temperature, sunshine, and wind infiltration in perimeter spaces. Energy reliability and utility rates are often set by peak demand, a condition that results from core internal loads being aggravated by poor climatic design at the perimeter (solar overheated roofs and facades in summer, cold air infiltration in winter). Finally, significant energy savings can be generated by conditioning buildings with fresh outside air through the use of the economizer cycle (passive conditioning with some energy cost). In the future we will need to shift away from massive, core-dominated buildings in order to provide thermal comfort through strategic combinations of passive conditioning and energy conservation, set climate by climate, as well as to reinstate the physiological and psychological "healthiness" of access to the natural environment.

Climate and lighting comfort

A significant descriptive and predictive database on sky cover conditions, daylight hours, and the quantity of direct, diffuse, and reflected radiation is needed for effective lighting and daylighting design. Artificial lighting effectiveness depends on absence of glare (caused most frequently by direct and diffuse sunshine), presence of adequate contrast to prevent eyestrain, and a controlled mix of wavelengths to provide effective color rendition. On the other hand, individual productivity, energy conservation, and a measurable amount of visual delight is achieved with effective daylighting. In the US more than 10% of all national energy use is for electric lighting in buildings, and most of that during the daytime (DOE/EIA 2002). A predominant number of new sustainable buildings, such as those achieving LEEDJ certification (US Green Building Council), clearly demonstrate that daylight can cost-effectively provide both ambient lighting for general way-finding and task lighting for the range of visual tasks in buildings. The use of daylight can offer major energy savings as well as improved health, productivity, and morale. Daylight, and all the climatic variables associated with daylight, also have the unique ability to provide "kinetic feedback" that gives the user a critical sense of time, place, and dynamic excitement (contributing to the emerging field of biophilia).

Climate and air quality

Wind speed and direction, relative humidity, and atmospheric composition are instrumental in maintaining or improving air quality within and around buildings. Interior pollution migration in large buildings is determined by variation in building pressurization and temperatures, which, in turn, depend on wind direction, wind speed, temperature differentials and possible solar-induced air flows. Controlling indoor pollution migration will also be critical to reducing chemical and biological risk. Anticipating pollution migration from neighboring sites also depends on the climate database, while avoiding pollution buildup in cities, commonly caused by temperature inversions, depends on access to a combination of wind, temperature, and atmospheric composition data sets. Moisture conditions also play a major role in building air quality. Requirements for humidification and dehumidification, linked to absolute and relative humidity conditions, remain one of the major conditioning challenges in building construction and operation. Water management, from driving rain and melting snow, and moisture management from infiltration and condensation on cold surfaces, is critical to avoiding long-term air quality risks from mold and mildew – demanding knowledge of climate data, weather events, and the interpretive power of building physics.

Figure A39 Atop a 100-year-old building, the dynamic enclosure of the Intelligent Workplace ™ at Carnegie Mellon University uses sunlight and natural ventilation to achieve an 80% reduction in heating, cooling and lighting loads of typical office buildings.

Figure A40 The glazed superstructure over the office space of Moncenis in Essen moderates the northern German climate to be more like Tuscany.

Climate and spatial comfort

One of the central activities of urban planners and building designers – the layout of spaces and functions for accessibility, way-finding, and effective use – also depends on climate data if the best use of the building site is to be ensured. Entering and exiting a building comfortably and safely depends on the designer's and operator's comprehension of local wind conditions, combined with occurrences of snow, rain, and ice. Servicing the building with water, electricity, and other transported goods also depends on reliable data of the averages and extremes of temperature, snow, ice, rain, and wind.

Moreover, planning for effective use of outdoor spaces and indoor–outdoor movement during comfortable periods, requires a full understanding of coincident climate conditions. In the design of outdoor spaces at least three microclimates can be created by the building itself: outdoor comfort without sun or wind; outdoor comfort with sun but without wind (extending the comfortable period into early spring and late fall); and outdoor comfort with wind but without sun (extending the comfortable period into the heat of the summer). In the past 10 years, architects and engineers have pursued a new generation of atrium spaces with passively moderated climates to support significant programmatic needs. Double envelopes have been developed for a number of landmark buildings in Europe, with modulated climate zones in the 1–5 meters between the inside and outside glazing. In one of the most dramatic projects the building's programmed spaces are laid out in a village configuration within a highly dynamic, climate-responsive glazed superstructure that moderates the northern German climate to be more like Tuscany (Figure A40). With knowledge of regional and local climate data, building designers can extend the use of outdoor spaces, enrich indoor–outdoor movement, create modified climates in buffer and atria spaces, and provide building occupants with the delight of daily, seasonal, and annual climate variability and manageability.

Climate and building integrity

The final and perhaps most touted area of overlap between climate and architecture is the contribution of climatic forces to building degradation, debilitation, and destruction. Humidity and precipitation contribute most heavily to these three failures in building integrity. Streaking, corrosion, efflorescence, and decay start as unsightly signs of age. With time this degradation leads to more serious debilitation, the inability to keep out the rain or to provide thermal comfort. However, at the point when the humidity, rain, snow, or ice begins to endanger structural integrity (through excessive loading, for example), the partial or complete reconstruction of the building may be inevitable. Wind and sun conditions also contribute to the loss of building integrity. Information about wind speed and direction and commensurate atmospheric conditions is needed to prevent corrosion (from sand and wind, for example), lifting, shear failure, cracking, and the popping in or out of claddings due to pressure differentials. Data on seasonal and daily solar intensity and position warn against potential fading, cracking, breaking due to expansion, or corroding due to dryness.

Each of these climatic stresses – precipitation, wind and sun – is aggravated by extreme temperatures or major temperature swings in a short period of time. The movement of the sun on extremely cold days and the action of changing solar gain values cause the greatest stress in materials and component assemblies. The combination of snow and freezing rain can cause the extreme structural loading that leads to failure. Consequently, descriptive and predictive database requirements for the prevention of building degradation, debilitation, and destruction are unique. In addition to predictive data as to climatic disasters (hurricanes, floods, tornadoes, and earthquakes), descriptive data on recent and repeated coincident climate conditions (average and extreme) are also needed. Finally, builders must track daily forecasts to ensure that potential material and component failures, leading to degradation, debilitation, or destruction, are not sealed into the building during the construction phase.

Rationale for action

These six performance mandates for buildings – thermal, acoustic, visual, air quality, spatial and building integrity – outline the breadth of the interchange needed between builders and climatologists. However, the rationale for action depends on the physiological, psychological, sociological, and economic limits society is willing to establish for *each* of the performance mandates. For a majority of developed nations these limits translate into five reasons for action.

Increasing environmental contact

Increasing environmental contact for passive conditioning was traditionally a prominent design criterion, but this climate-responsive design expertise was lost with the advent of air conditioning and the ability to mechanically condition year-round. With year-round conditioning, previously uninhabitable regions could be rapidly developed – especially desert and tropical regions. Designers could ensure comfort in any climate, with any building materials, packaged into any building image. However, the psychological unacceptability of this homogeneous and sealed-in approach to architecture has grown, despite its inherent development benefits. Indeed, pedestrianization (making outdoor spaces and facades more suitable for pedestrians), daylighting, natural ventilation, and outdoor space utilization have become novel foci for emerging design theory. It is on this basis that the newest rationale for climate data in building design is introduced – the desire to increase environmental contact.

Increasing environmental comfort

Even though the architectural and engineering community tout human comfort, the use of regional and local climate data to design for increasing environmental comfort within and around buildings has not been a primary step. The majority of urban planners, building designers, contractors, and managers rely heavily on deep, sealed buildings with mechanical conditioning to artificially provide thermal comfort, lighting comfort, and air quality. Indeed, little has been done to mitigate the liabilities of climate (cold wind infiltration and cold mean radiant temperatures in winter, glare and solar overheating in summer) or to use the assets of climate (solar gain and daylighting in winter, natural ventilation in summer) to provide a balanced environment *as a basis* for mechanical system design and operation. Yet thermal, visual and indoor air quality concerns have actually grown during this time due to an over-reliance on advanced technologies in sealed buildings that are oblivious to climate and energy consumption. Moreover, building energy consumption has also grown, for simultaneous heating and cooling, increased fan and pump energy, electric lighting during the daytime, and for ensuring adequate ventilation to purge indoor pollutants. The economic motivation for *energy conservation* is what puts emphasis on this second rationale for introducing climate data – to naturally increase environmental comfort within and around buildings.

Reducing environmental failure

The third rationale for action, reducing environmental failure, is well established in the building industry. Whether triggered by the desire to keep up appearances (psychological limits of acceptability) or by the need to prevent failure and consequent risk to life (physiological limits of acceptability), data on extreme climate conditions have become a part of most building design codes. Today, design for the safety and security of building occupants, in the face of earthquakes, hurricanes, tornadoes, and heavy snow loads, ensures that extreme climate conditions are well documented in the codes. It is the joint responsibility of the architectural, engineering and construction community and the climatologist to establish whether these data are of the appropriate amount and in the appropriate form to adequately ensure both the psychological and physiological limits of acceptable environmental stress on building integrity.

Reducing environmental costs

Reducing environmental costs is a fourth argument for including climate data in the building decision-making process. This rationale for action promotes the value of all resources: money, labor, materials, air, water, and energy in the face of carelessness and abuse. The introduction of climatic facts and thresholds into analyses of the renewability of air, water, soil, energy, and capital-intensive building materials has the potential to significantly reduce the environmental costs (from leaky roofs to pollution build-up) imposed by building development (see Table A21). Growing concerns about climate change, pollution build-up, water shortages and depleting resources should significantly

Table A21 Environmental costs of poor planning and design decision making

Construction category	Annual volume[a]		Potentially weather sensitive: Perishable material[a]		Potentially weather sensitive: On-site wages[a]		Potentially weather sensitive: Equipment[a]		Overhead and profit[a]		Total sensitive percent of annual volume	
	1964	1979	1964	1979	1964	1979	1964	1979	1964	1979	1964	1979
Residential	17.2	87.6	0.960	5.76	1.624	7.88	0.073	0.37	2.141	10.51	27.9	27.4
General building	29.7	86.9	1.928	5.21	4.029	12.17	0.222	0.87	2.670	7.82	30.0	30.0
Highways	6.6	9.5	1.666	2.38	1.633	2.38	0.773	1.14	0.727	1.05	72.7	73.0
Heavy and special	12.5	17.6	1.875	2.64	3.125	4.40	2.500	3.52	2.500	3.52	80.0	80.0
Repair and maintain	22.0	50.4	2.674	6.05	3.996	9.07	1.386	3.02	3.143	7.06	50.9	50.0
Totals	88.0	252.0	9.1	21.54	14.4	35.90	5.0	8.92	11.2	29.96	45.1	38.2

Sources: Original 1964 data and calculations after Russo (1966); 1979 construction volume from Bureau of Census and Bureau Business Development, US Department of Commerce; otherwise 1979 data calculated on same ratio as 1964 data.

[a] Billions of dollars.

boost the demand by the design community for climate data. It is important for the meteorological community to anticipate these demands and create interpretive databases on the vulnerability of regions and local ecosystems in relation to air, water, soil, natural energies, and material resources.

Increasing environmental security

In a parallel effort, it is important for the meteorological community to create interpretive databases on the opportunities of regions and local ecosystems to replenish air, water, soil, natural energies, and material resources. With growing international attention on energy and water security, the meteorological community has a major opportunity to emphasize the importance of sun, wind, hygrothermal and other data in the design and retrofit of buildings and communities. The best protection from chemical and biological attack within a building's mechanical system may be replacing central systems with natural conditioning. The best protection from brownouts and blackouts may be the effective use of solar electric, thermal and daylighting energies. The importance of climate data to environmental security – both reducing vulnerability and providing sustainable resources – cannot be overstated.

Climatic data and building decision making

Given these five rationales for action – to increase environmental contact, comfort, and security while decreasing environmental failure and costs – climate data form an essential resource for the building industry. By focusing on the steps in the building decision-making process, these data can be introduced in a form conducive to action. Table A22 outlines the decisions made in the early stages of planning and design and the eventual operational decisions that must be made in building construction and management.

Table A22 Building decision making

Land-use/town planning
Geological change (soils, vegetation, topography)
Density
Spacing
Collective massing
Height
Functional mix
Building design
Massing and orientation
Organization of spaces
Enclosure design
Opening design
Systems integration
Building construction
Worker comfort and stress
Worker safety and health
Material stress
Operational planning
Building management, maintenance, operation, and use
Occupant comfort and stress
Occupant safety and health
Building image over time
Facility management costs over time
Energy and environmental costs over time

Climate and land-use/town planning

Land-use development involves decisions concerning building density, spacing, height, collective massing, and functional mix. Land-use decision-making has a direct impact on air quality, pollution, thermal build-up, runoff and flooding potential, and water quality and availability, as well as changes in wind and sunlight conditions. Consequently, an understanding of a region's climatic profile and thresholds for concern are critical to the effective planning and design of building communities.

If regional and local wind conditions are to provide natural ventilation for thermal comfort, reduce pollution build-up, and improve use of outdoor spaces, then prevailing wind speeds and directions, as well as microclimate variations, must be known. Average and extreme wind conditions must also be addressed if liabilities are to be mitigated, including thermal discomfort through infiltration, acoustic discomfort through noise transportation, poor air quality through pollution transportation, and building degradation and destructive failure. Early design decisions regarding building density, massing, and functional mix dictate the availability or mitigation or regional wind conditions at the building microclimate level. According to Landsberg (1976):

> The major effect of urban areas on the wind circulation is a result of the increased roughness introduced by the building complexes. These induce many smaller and larger eddies which swirl around in the city. The arrangement of streets and avenues will cause channeling of the wind. The results may then be that strong currents flow through one set of streets, while relative calm prevails in the streets at right angles. The eddy currents that are caused by the buildings, especially the tall ones, can lead to very high wind gusts at street level. They can be so strong that it becomes difficult to walk on pavements and to open doors unless special protective overhangs are included in the design.

Sun conditions are equally affected by land-use decisions. If the benefits of solar heating, daylighting, and outdoor space utilization are to be realized, regional sunlight conditions and microclimate modifications must be optimized through careful control of density, height, and collective massing (Knowles, 1994). If the liabilities of solar overheating, thermal inversions causing poor air quality, and ultraviolet building degradation are to be reduced, regional sun position, intensity, variability, and potential microclimate shifts must be fully incorporated into site planning decisions.

Without a regional comprehension of overheated, underheated, and comfortable climate conditions, land-use decisions as to density, massing, and functional mix can aggravate a region's disadvantages, while losing the opportunity to make use of its climatic advantages. Still today, urban planning takes little advantage of the comfortable season for natural thermal conditioning or outdoor space utilization. In fact, town planning often shortens the natural season by poor manipulation of natural wind and sun conditions. Landsberg (1976) states that "on a local scale, the formation of an area of higher temperatures compared with rural environs is generally the first obvious manifestation of settlements, and can be directly related to the building density." Heat islands, which result from the complex actions of urban pollution, solar and heat build-up, and poor air flow, can be anticipated with climate information. In addition to increased thermal discomfort and reduced access to outdoor spaces, our poor understanding of regional temperature conditions and corresponding wind, sun, and humidity conditions leads to unsightly building degradation, unacceptable debilitation (inability to provide shelter), and, eventually, unsafe destruction.

Finally, regional humidity and precipitation data are needed in the effective planning of urban density, spacing, building height, and collective massing. To the designer's advantage, low humidity allows for effective evaporative cooling to provide thermal comfort and reduced conditioning costs. Design for the utilization of precipitation can provide for our water needs and improve air quality as well as reduce degradation of buildings due to pollution. However, comprehension of the serious liability presented by regional humidity and precipitation is also critical. Extreme humidity can be the major cost in conditioning for thermal comfort and air quality, as well as growing concerns about the health consequences of mold and mildew. While average precipitation contributes to building degradation and debilitation over time, extremes in precipitation, aggravated by extreme wind conditions, constitute the greatest cause for building and infrastructure debilitation and destructive failure.

Climate and building design

Although many design decisions regarding massing and orientation, organization of spaces, enclosure design, opening design, and systems integration are made without the benefit of regional and microclimate data, the importance of climatic conditions to thermal and lighting comfort, air quality, and building integrity (versus degradation and destruction) cannot be understated. Table A23 clarifies the advantages and disadvantages of various climatic forces on building performance as well as the building design decisions that are consequently affected.

Records of average wind speeds and directions are necessary if natural ventilation is to be provided, dictating appropriate building orientation, spatial organization, opening design, and systems integration. The liability of wind to thermal comfort, air quality, and building integrity must also be considered, demanding that building massing, organization, enclosure detailing, and opening design minimize the unacceptable wind forces around and through the building. Today, the most advanced architectural/engineering teams use computational fluid dynamic tools (CFD) to create innovative naturally ventilated buildings and even power-generating wind turbine designs integral with the building design (Hamzah et al., 2001).

The natural conditioning benefits of sunshine for solar heating, daylighting, and outdoor space utilization require a working knowledge of seasonal and daily sun position and intensity, along with their implications for building massing and orientation,

Table A23 Climate and building design decision making

	Massing and orientation	Spatial organization	Enclosure design	Opening design	Systems integration
Wind asset					
Thermal: natural ventilation	X	X		X	X
Air quality: pollution mitigation	X	X		X	X
Wind liability					
Thermal: infiltration	X	X	X	X	
Air quality: pollution migrations	X	X		X	
Building integrity: degradation/destruction	X	X	X	X	
Solar asset					
Thermal: solar heating	X	X		X	X
Lighting: daylighting	X	X		X	X
Spatial: outdoor utilization	X	X		X	
Solar liability					
Thermal: solar overheating	X	X	X	X	X
Lighting: glare	X	X		X	X
Building integrity: degradation	X		X	X	
Temperature asset					
Thermal: net comfort, time-lag cooling	X	X		X	X
Spatial: indoor–outdoor utilization	X	X		X	X
Temperature liability					
Thermal: heat loss/heat gain	X	X	X	X	
Building integrity: degradation/ debilitation/destruction	X		X	X	
Humidity/precipitation asset					
Thermal: evaporative cooling		X		X	X
Air quality: humidification, cleansing				X	X
Humidity/precipitation liability					
Thermal: dehydration, excess relative humidity				X	X
Spatial: building access, outdoot				X	X
utilization	X	X		X	
Building integrity: degradation, debilitation, destruction	X		X	X	

organization of spaces, opening design, and system integration. Average and extreme solar and sky conditions must also be considered by designers if solar overheating in summer, glare causing lighting discomfort, and building degradation are to be reduced. Today, the most advanced architectural/engineering teams use daylighting and solar thermal simulation tools to maximize passive solar heating, daylighting and power generation utilizing buildings rooftops and facades.

When temperature conditions offer significant periods of natural comfort, with or without the assistance of sunshine for warmth or wind for cooling, the massing of a building and its potential for indoor-to-outdoor movement through the appropriate organization of spaces and displacement of openings should be significantly affected. Often, however, the design of the building's mechanical system is not integrated effectively with climatic data, thereby eliminating the advantage of this natural conditioning potential. Today, the most advanced architectural/engineering teams use hour-by-hour simulation tools to design thermal mass and thermal distribution strategies – such as core-to-perimeter, E-S-W-N, and day to night load balancing systems to take advantage of diurnal energies in specific climates.

Knowledge of regional humidity conditions allows the designer to take advantage of evaporative cooling for thermal comfort and air quality, reducing the conditioning liability of excessively dry climates through careful opening design and systems integration. In humid climates the minimization of moisture build-up and the effective use of ventilation and solar "drying" or desiccant systems places equal demands on enclosure design and systems integration. Spatial comfort, including the ability to enter or exit a building and the usableness of outdoor spaces, is assured through an understanding of average and extreme rain, snow, and ice conditions, along with coincident wind and temperature data. Full knowledge of these coincident climatic liabilities is also necessary to prevent all three threats to building integrity: degradation; debilitation to such a degree that the building no longer provides comfort or shelter; and destructive failure. Tveit (1970) presents an illustrative discussion:

> Water may be brought into porous materials as liquid or as vapour. The driving force may originate from differences in hydrostatic pressure, osmotic pressure, vapour pressure, as well as from free surface energy. Considering rain penetration into building structures, this would be caused by (1) capillary action due to free surface energy; (2) kinetic energy of rain drops or water streams; (3) hydrostatic pressure due to gravity or to wind pressure on water films or menisci; or (4) air flow through the structures. The ability of a building structure to dry out and keep dry generally depends on the successful design in every detail of the building and its various structural components and draining systems above or under the ground as well as on certain physical properties of the applied materials and workmanship, all in relation to the ambient climate. Some of these principles are related to the design and construction of certain building components and are in exposed areas frequently reflected in a characteristic of local design and building practice.

Consequently, increased understanding of regional and local climatic assets, their combinations, and manageability could change significantly the present-day isolationist or "sealed box" tendencies of building designers. Setting regional priorities among these climatic assets and liabilities to ensure long-term thermal comfort, lighting comfort, air quality, spatial comfort, and building integrity, is the genius of climate-responsive building design. The regionalization of architecture is not a return to the past but a reconfirmation of the natural forces and resulting material availabilities that make survival and comfort in each location unique. The all-glass building in hot and humid Houston, the flat roof in rainy England, the pedestal building in windy Chicago all attempt to deny the very climate in which they sit. Today, armed with a full understanding of a building's climate and consequent microclimate, design priorities can be set that maximize the benefits of the various climate assets for natural conditioning while minimizing the liabilities – a delicate balance that will ensure building quality over time.

Climate and building construction

The building construction industry has several responsibilities that depend greatly on climate data. Worker comfort in relation to precipitation, hot or cold temperatures, biting or dehydrating winds, excessive solar gain, serious humidity and pollution conditions as well as protection against the stresses of heat stroke and frost bite, require that project managers keep a close watch on daily and hourly weather forecasts. The safety of the worker requires knowledge of climate conditions – wind speeds in high-rise construction, snow and ice conditions on various ground surfaces, excessive overheating or freezing of building materials as well as components that must be handled or walked on. The seriousness of these responsibilities, and the corresponding need for predictive weather information, is already an innate part of the standards and liability laws of the construction industry. However, material stress resulting from adverse climatic conditions (average, coincident, and extreme) is less enshrined in the working standards of the construction industry. Not only do extreme temperature, humidity, precipitation, wind, and sun conditions greatly affect the integrity of building materials, components, and assemblies on the job site, but historic average conditions also indicate the feasibility of using particular materials and assemblies in various climates. Without a working knowledge of the impact of average and extreme climate conditions on building integrity, the unsightly degradation of buildings, the debilitation of buildings to an extent where they no longer provide comfort or shelter, and the destructive failure of buildings will be initiated from the laying of the first foundation.

Balancing these climatic variables that dictate worker comfort and safety as well as material stress over time, against other construction variables such as worker availability, material delivery, financing, and occupancy schedules form the basis of operational planning on the construction site. However, because of the lack of descriptive and predictive climate data in a form readily usable by the construction industry, only sketchy climate information is used in construction planning, despite long-term implications of precise climate data for the performance of components and systems.

Climate and building management: maintenance, operation and use

The daily operation and use of a building for occupant comfort require data on present and near-term anticipated climate conditions. To provide thermal and visual comfort for the occupants, as well as the air quality they need, building operators (or automated operating systems) require informaton on outdoor temperature, humidity, and sun conditions. Energy load balancing from north to south, from day to night, and from season to season to reduce peak demands and annual energy costs requires descriptive average and extreme climate data, in addition to 24-hour weather forecasts. Table A24 shows that over 50% of

Table A24 Winnipeg office building: annual load breakdown and climate dependencies

	kWh/sq ft/ year	Operating dollars 1981	Climate impact annuals	Climate negotiable capital costs[a]
Process electricty (kitchen, power, computers)	4.0	22000	—	No
Electric conditioning				
Lights	5.0	28000	10000[b]	Yes
Fans for air quality[c]	2.0	11000	—	No
Fans for "core"/cooling				
Lights[d]	1.2	6600	2000	Yes
Equipment[e]	1.8	10000	—	No
Fans for perimeter cooling	1.0			
Air conditioning for perfimeter[f]	2.0	16000	16000	Yes
Nature gas conditioning				
Heating	13.0	72000	72000	Yes
How water	2.5	14000	—	No
Total	32.5	180200	100600[g]	

[a] Initial capital costs of equipment could be reduced if climate-responsive design.
[b] Only perimeter lighting can be reduced through daylight management.
[c] Majority of ventilation for air quality is provided through cooling and heating fanpower.
[d] 60% of total wattage is translated into cooling demand.
[e] 100% of total wattage is translated into cooling demand.
[f] Due to cold climate, the majority of air conditioning is provided through fan economizer (free cooling) cycles.
[g] Consequently 55% of total energy demand in this building is dependent on appropriate climate-responsive building design.
Source: Statistics from Public Works of Canada.

the annual operating costs in a 300000 square-foot office building in Winnipeg, Canada, depend on outdoor climate. These operating costs, and the occupant comfort that is associated with them, generate the demand for on-site weather monitoring equipment today. The innovative use of such equipment can yield tremendous improvement in the understanding of occupancy comfort and stress that is directly affected by climate and related building decision making.

Occupant safety, given adverse weather conditions and long-term building degradation due to these conditions, also constitutes a major demand for climate information. Wind, rain, snow, and ice can make certain parking areas, pedestrain areas, and entries unsafe. Material expansion and contraction, caused by swinging temperatures, fluctuating humidities, and sunshine, can make various building components unsafe to walk on or under. Material corrosion and degradation due to moisture migration and rain penetration can lead to building failures that are dangerous to occupants and passers-by. Disasters such as hurricanes, floods, tornadoes, or snowfalls followed by freezing rain can cause total building collapse, a phenomenon that should be anticipated in building design and upheld through building maintenance. Consequently, 24-hour and weekly forecasts are needed by building managers to anticipate building stress and occupant safety risks. Long-term historic data are needed to establish annual maintenance budgets for each building, while short-term historic data will compel building managers to undertake preventive maintenance programs for building materials and assemblies that have been under unexpected, but recorded, stress.

The image of the building – of interest to the occupants, the owner, and the community at large – is also a major factor in building maintenance programs. Before the building is occupied, the climatic liabilities of the region and site can be mitigated with careful planning, design, detailing, and construction. However, after the building is occupied, the stresses placed on the building fabric and site by climate will require continuous cycles of maintenance. Preventive maintenance, like preventive medicine, can slow the degradation of buildings and potentially prevent debilitation to the point where the building no longer provides comfort or shelter. For example, anticipating excessive rainfall, gutters can be cleared out and drainage channels opened, preventing the major roof leak that eventually demands millions of dollars of repair. In many countries, records of major snowfalls become mandates for maintenance crews to clear roofs, preventing excessive structural loading and eventual failure. Material and product manufacturers often establish preventive maintenance programs, given a range of average climate conditions. Table A25 lists the weather mechanisms that contribute to stone decay; factors that can be translated directly into building maintenance programs (Tombach, 1982). Eventually, these maintenance schedules, as well as the reactive maintenance required in the aftermath of major climatic events, will be computerized along with the descriptive, predictive, and historic climate data that necessitate the activities, providing each building owner with a specific climate-responsive maintenance program.

Methods of intervention

There is little question that the availability of climatic data is far greater than their use. Indeed, the willingness of climatologists to be of service to the building community is far greater than the demand for their services. Several theories have been advanced as to why this is so. John Page (1970) writes:

> "When an applied science is not applied in practice, it is always valuable to inquire why this situation should persist in spite of a host of apparently good scientific reasons why the situation should be otherwise. The aviation services use the

Table A25 Classification of mechanisms contributing to stone decay

Mechanism	Rainfall	Fog	Humidity	Temperature	Solar insolation	Wind	Gaseous pollutants	Aerosol
External abrasion								
Erosion by wind-borne particles						•		•
Erosion by rainfall	•							
Erosion by surface ice	•	•		•				
Volume change of stone								
Differential expansion of mineral grains				•			○	
Differential bulk expansion due to uneven heating				•	•			
Differential bulk expansion due to uneven moisture content	•	•	•	•	•	○	○	○
Differential expansion of differing material at joints				•				
Volume change of material in capillaries and interstices								
Freezing of water	•	•		•				
Expansion of water when heated by sun	•	•		•	•			
Trapping of water under pressure when surface freezes	•	•		•				
Swelling of water-imbibing minerals by osmotic pressure	•	•	•				○	○
Hydration of efflorescences, internal impurities, and stone constituents	•		•				○	○
Crystallization of salts			•	•	•	•	○	○
Oxidation of materials into more voluminous forms	•	•					○	
Dissolution of stone or change of chemical form								
Dissolution in rainwater	•			•		•	•	
Dissolution by acids formed on stone by atmospheric gases or particles and water	•	•	•	•			•	•
Reaction of stone with SO_2 to form water-soluble material	•	•		•			•	
Reaction of stone with acidic clay aerosol particles	•	•		•				•
Biological activity								
Chemical attack by chelating, nitrifying, sulfur-reducing, or sulfur-oxidizing bacteria			•	•			•	
Erosion by symbiotic assemblages and higher plants that penetrate stone or produce damaging excretions	○	○	•	•				

Source: From Tombach (1982).
Note: Solid circles denote principal atmospheric factors; open circles denotes secondary factors.

meteorological networks without persuasion, why not the architects, town planners, and civil engineers? There would seem to be three fundamental reasons for rejecting scientific information in any practical field of human activity: (1) the information provided is considered irrelevant by the potential user; (2) the information provided in the form presented is considered inapplicable by the potential user; and (3) the information provided is considered incomprehensible by the potential user. The first question implies a lack of understanding of the problem by the scientific worker, the third question a lack of understanding by the designer of the problem, and the second question shows a lack of communication between the two.

John Eberhard of the National Academy of Sciences (1986) adds:

> The resistance to change is a natural, and probably necessary, instinct in all people and consequently in human institutions. The organizations and institutions associated with the building industries of the United States are no exception – in fact they appear more resistant to change than many other sectors of the economy. One major indicator, and also contributor, to this lack of innovation is the relatively low volume of research and development. While industries such as electronics and aircraft are spending more than 20 percent of their revenue on R&D, all indications are that the industries of building spend less than one percent on Research and Development – in fact if quality control and product testing to meet government regulations is subtracted from the so-called R&D budgets of building industry organizations, the budget left to devote to new ideas and new knowledge is probably very small indeed.

These weaknesses in information transfer and in applied research are compounded by the unfamiliarity of the climatologist with the exact nature of individual building decisions and the uncertainty of the planner, designer, engineer, contractor, and building manager as to what climatology has to offer. Despite this overall lack of familiarity there are several concrete steps that can be taken to improve the contribution of climate data in the building decision-making process: accessible climate databases; climate data information packaging; algorithm development and information processing; model testing; field case studies and documentation; and advanced computer simulation tools.

Important climatic variables

For effective decision making in the early stages of land-use planning and building design and in the later stages of construction and building management, the building community needs a considerable quantity of average and extreme temperature, humidity, atmospheric content, wind, sun, and precipitation data. In the majority of industrialized nations this database exists to such an extent that data collection needs are insignificant in comparison to the needs for applied data packaging and processing in relation to the building sciences. The efforts of the European Community to consolidate solar and other climate databases for EC members in the publicly accessible web site – SoDa (2002) – will greatly enhance decision making in those climates.

However, there is a major gap in two areas of data collection and presentation: the availability of information on complex, coincident climate conditions often critical to building performance; and the development of micrometeorology. Lacy (1972) writes:

> Although in many multi-variate problems, an analysis of the simultaneous occurrence of various elements is usually required, there are others in which the sequence in which events occur is important. Ultra-violet radiation will only react with a material if the material has already been wetted by dew or rain. Frost will only damage porous materials if they have been wetted by rain a sufficiently short time before the frost occurs, so that there has been insufficient time for water loss by evaporation. Condensation of water on structures can be caused if a prolonged cold spell is followed rapidly by a mild one, so that the atmospheric dew-point rises more rapidly than the temperature of surfaces which have been chilled during the cold spell. In such cases the conventional statistical analyses of weather elements may be useless, and it is necessary to study the actual weather situations which cause such problems, and estimate their frequency of recurrence.

Only coincident climate data can ascertain the availability of wind when the temperatures are warm enough to merit natural ventilation, or the direction and speed of wind when it is raining to ensure appropriate enclosure detailing and opening design.

Equally important is the development of microclimate databases or scientific methods for the prediction of microclimates on specific building sites. Building design decision making will rely on whatever sketchy microclimatic databases exist, until the predictive science of micrometeorology is developed. Finally, the industrialized nations should be committed to establishing climate databases for the developing world as well, if climate responsive community and building design, construction and operation is to be achieved world-wide for environmental safety, comfort and security.

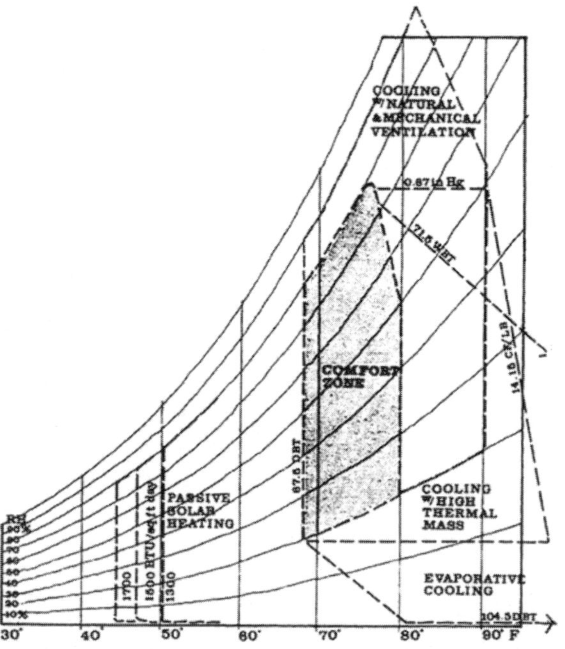

Figure A41 Evaluating passive conditioning potential (from Loftness, 1982).

Information packaging and climate graphics

Graphic interpretations of climate data are critical for professionals to understand the variations and coincident conditions that should significantly impact building and community design. From the early work of the H.E. Landsberg for the AIA (1950), the Olgyay brothers (1963), Loftness (1982), Brown and Dekary (2001), and a host of recent leading international designers (e.g. Short, Foster, Yeang) a commitment to graphically communicating climatic forces is key to climate-responsive design.

Based on the assumption that the greatest amount of information is learned in the doing, a how-to publication, *Climate/Energy Graphics*, was written by Loftness (1982) for the World Meteorological Organization. Loftness outlines 15 graphic exercises that can be completed by meteorologists, technicians, engineers, or architects using data from local weather stations. As shown in Figures A41 and A42, available climate data should be used to generate climate graphics and result in a summary translation with written design guidelines to ensure the climate-responsive design of building massing and orientation, spatial organization, enclosure and opening design, and systems integration (Figure A43).

Algorithms and climate information processing

To get beyond the elementary knowledge concerning regional and local climates that can be achieved through climate graphics, it is necessary to develop an understanding of the fundamental algorithms that relate climate to building success and failure. Boer (1970), in an article outlining a training and education syllabus, stipulates:

> Knowledge of the interaction between meteorological parameters and/or processes and building activities, of single structures and built-up areas, is an essential prerequisite for a functioning information system of building meteorology.

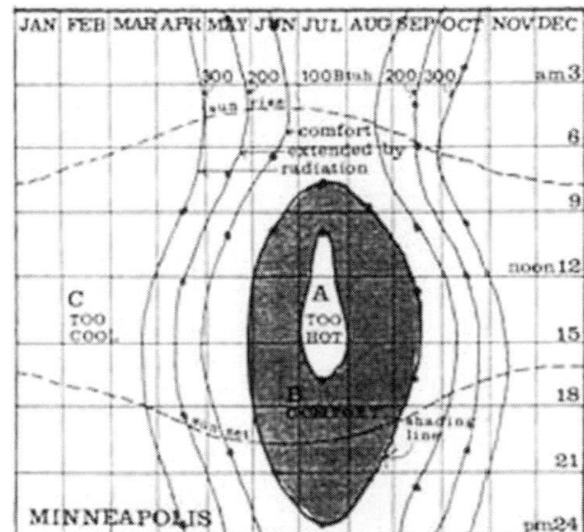

Figure A42 Climate design priorities for Minneapolis, Minnesota (from Loftness, 1982, based on Olgyay, 1963).

		RANK	A Bldg. & Landcaping	B Building Grouping	C Bldg. Shape & Configuration	D Bldg. Orientation	E Building Organization	F Envelope/ Enclosure Design	G Envelope/ Opening Design	H Systems Integration
HEATING										
Limit LIABILITY	Isolation from cold /hot temps	1	Group buildings, berm or site underground		Compact building form		Organize bldg. Into zones	High r-Value materials	Size & Type Windows Movable ins.	
	Isolation from cold winds	5	Protect bldg. By vegetation, grouping…		Form & orient bldgs to provide winter wind protection		Protect entries	Tight const. to limit infiltration	Place & detail windows vs. infiltration	
Use ASSET	Opening to solar heating	3	Site, landscape & group bldgs. For winter sun access			Orient bldg. To winter sun	Provide adequate collection & storage areas		Size opening & type glass	Radiant & convective DistRibution systems
COOLING										
Limit LIABILITY	Isolation from hot sun	3	Protect bldg. By vegetation Or berming	Self-shading forms and min. east & West exposures				Effective shading devices For walls & Windows exposed to summer sun		
	Isolation from high humidity	6								Prevent excess humidity creation
Use ASSET	Responding to diurnal temps.									
	Opening to cooling winds	2	Site & place bldgs. For natural ventilation			Orient bldg. For natural ventilation	Open spaces to each other		Size & orient opening for nat, ventil.	Induce & assist ventil.
	Opening to evap-cooling									
	Opening to radiant cooling									
COMFORTABLE										
Use ASSET	Opening to Comfort	4	Provide open spaces for winter, spring & Autumn uses					Provide connections w/ outside for winter, spring & autumn uses		

Figure A43 Developing design strategies from climate priorities (Charleston, from Loftness 1982).

The algorithms that explain the interaction between meteorological parameters and design decision making can be as simple as the snow loading that leads to failure of various structural systems and as complex as the relationship of town planning decisions to the alteration of regional wind patterns. While computational tools such as DOE 2.2 support the integration of on-line climate databases into energy-effective design decision making, there are numerous performance simulation challenges that remain, including the potential impact of time-lag, of radiant cooling, and the hygrothermal performance of materials and assemblies. In his book *Climate Considerations in Building and Urban Design* (1998), Baruch Givoni documents the importance of algorithms to describe the effect of mass in buildings with continuous ventilation, nocturnal ventilation, and night-time radiant exchange (Figure A44). In practice, Dr. Givoni continues to explore innovative cooling strategies using combined forces of wind, thermal stratification, and evaporative cooling, to create, for example, the dramatic Ashower cooling towers that captured the imagination of all who attended the 1992 World Expo in Seville, Spain (see Figure A45).

Laboratory and field case studies and documentation

Field studies of innovative indigenous and advanced buildings in a range of climates could aid in the development of performance algorithms linking climate and architecture, to enable designers to understand the regional appropriateness of climate-responsive design innovations. A range of diagnostic tools, equipment, and procedures should be further developed to determine the overall performance of building materials, assemblies and integrated systems in the full range of climates. According to Markus and Morris (1980):

> Historical studies of vernacular buildings are helping in the development of a new kind of planning and architectural theory – concerned not with monumental planning and design, but with the pattern of cities, settlements, and buildings as expressive of the structural relationship between technological, social, symbolic and natural forces – that is, a cultural theory of form. This theory attempts to unravel the meaning of patterns found in primitive and vernacular creations of the past and, by studying the processes and forms of still active authentic societies, to draw conclusions for design today.

Until now, field instrumentation has generally focused on the evaluation of building degradation, debilitation, and destruction. Indeed, there are no comprehensive approaches to evaluating the performance of materials, buildings or land-use alternatives in a range of climates for universities or professionals to decisively study indigenous and innovative climate-responsive architecture.

On the other hand, climate testing of scale models set in average and extreme wind or sun conditions has been pursued by a small number of designers for major projects or climatically challenging projects. Model wind testing is used on some large projects to visualize pedestrian safety and comfort, the integrity of cladding and structure against degradation or destruction, the viability of natural ventilation, and the predictability of pollution migration. Model sun/daylight testing is used to illuminate the solar availability in public spaces, the effectiveness of shading devices, the potential of solar heating, and daylight penetration. However, scale model testing could go much further. Model wind/snow testing could clarify the build-up of snow on roads, airports, building entries, roofs, and in recreational areas such as ski resorts. Model rain testing could be undertaken to demonstrate watersheds, storm runoff and site drainage effectiveness and the consequent impact on communities and regions.

Both laboratory and field studies of climate-responsive building materials, assemblies and integrated systems are critically needed for designers to regionally increase environmental contact, comfort, and security while decreasing environmental failure and costs. There should be a concerted effort to completely document field studies of both indigenous climate-responsive buildings and communities and recent innovations in a range of climates to inform the entire building community, from building owners to material suppliers, of the significance of climate in building decision making.

Advanced computer simulation tools

The computer plays a more active role in land-use planning and building design, construction, and management today, with climate data integral to performance simulation. However, more laboratory and field case studies linking the performance of materials, assemblies and integrated systems to regional climate conditions will help to mainstream the use of simulation tools in design and construction. When an adequate number of building climatological algorithms and descriptive case studies can be simplified into computer-simulation packages, the demand for climate data in building decision making will be clearly established. The environmental costs of inappropriate land-use and building and their corresponding change in soil, vegetation, and topography should be too graphic to ignore. The loss of environmental comfort through poor building orientation, organization, and opening design should be quickly simulated and alternatives chosen. The environmental stress placed on building materials in the face of anticipated and recorded climate conditions should be avoided through computerized construction management programs. Material manufacturers should be able to establish regional performance priorities and preventive maintenance schedules suited to the unique history of climate conditions surrounding the project. The acceptable averages and threshold of coincident climate conditions should be automatically processed through the algorithms of physiological, psychological, and economic comfort to continuously guide decision making in building design, construction, and use.

Although no attempt is made to establish which of these methods of intervention will generate the greatest impact – data packaging, data processing through algorithms, field studies, model testing, or computer simulation – each of them will reap results. Each method of intervention "describes historical and more recent notions of buildings as shelters; i.e., structures which intervene by acting as barriers and as responsive filters between the natural and urban environment, and the range of environments required for human activity" (Markus and Morris, 1980).

Vivian Loftness

Bibliography

AIA/House Beautiful, 1949–1952. *Regional Climate Analysis and Design Data: The House Beautiful Climate Control Project*. Ann Arbor, Mich.: Xerox University Microfilms.

Boer, W., 1970. Problems of distribution and effective exploitation of meteorological information for architecture and building Industry. In *Building Climatology, Technical Note 109*. Geneva: World Meteorological Organization, pp. 191–197.

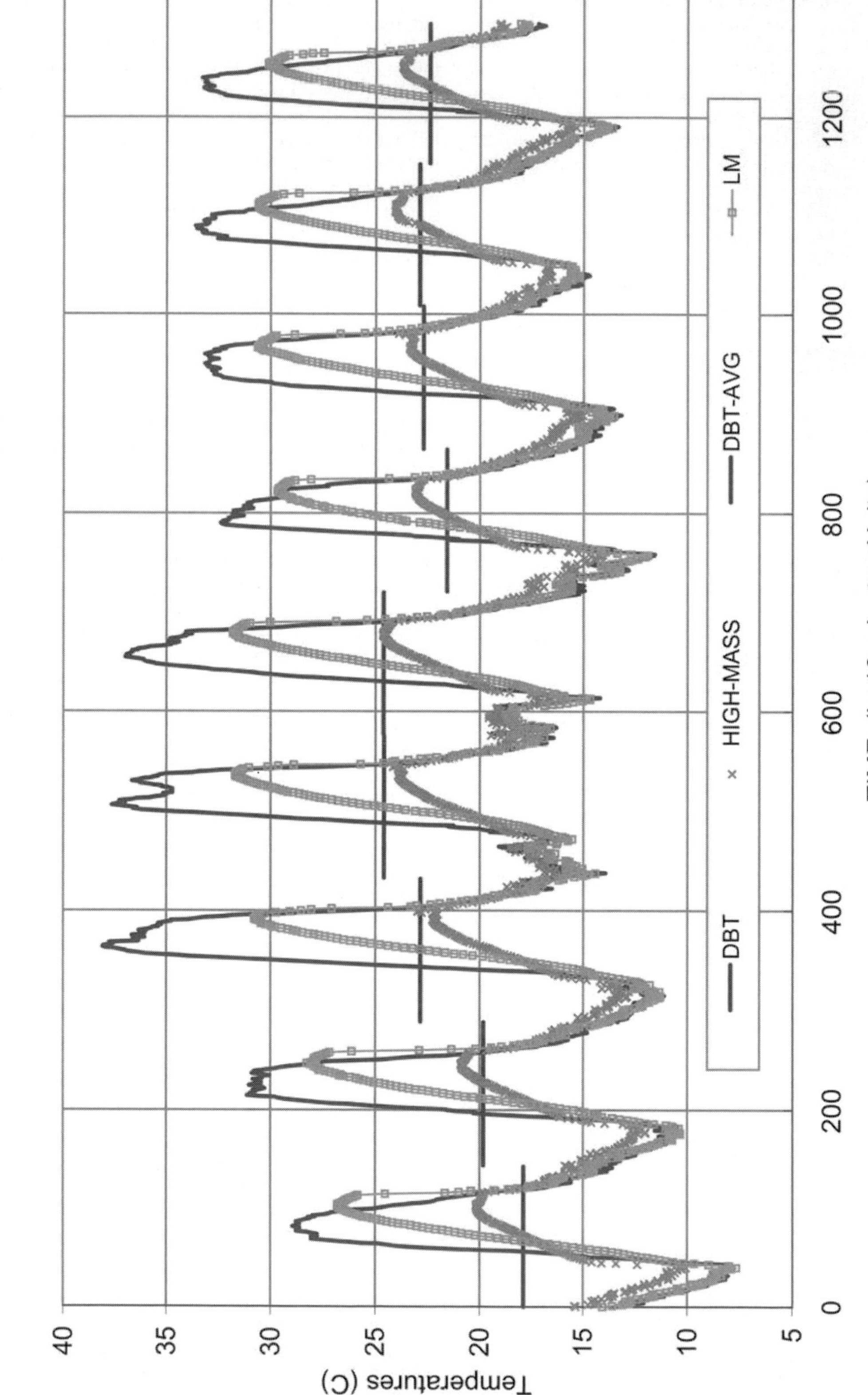

Figure A44 Dr. Baruch Givoni documents the importance of developing algorithms from field measurements to describe the effects of mass in buildings, in this case with night ventilation.

Figure A45 An evaporative cooling tower passively cooled an outdoor rest area at the 1992 EXPO in Seville, Spain with dramatic form.

Bordass, W., and Leaman, A. PROBE – Post occupancy Review of Buildings and their Engineering <http://www.usablebuildings.co.uk/probe/ProbeIndex.html> ESD Limited, Overmoor, Neston, Corsham, Wiltshire, SN13 9TZ, United Kingdom, 2004.

Briggs, R., Lucas, R., and Todd Taylor, Z., 2002. *Climate Classification for Building Energy Codes and Standards*. Technical paper Pacific Northwest National Laboratory.

Brown, G.Z. and Dekay, M., 2001. *Sun, Wind and Light: Architectural Design Strategies*. NewYork: John Wiley and Sons.

DOE2.2 (2002). PowerDOE, eQuest, www.DOE.com

Energy Information Agency, Building Energy Use Data, http://www.eia.doe.gov/

Givoni, B., 1998. *Climate Considerations in Building and Urban Design*. New York: Van Nostrand Reinhold.

Hamzah, T.R., Yeang, K., and Richards, I., 2001. *Ecology of the Sky*. Australia: Images Publishing.

Knowles, R., 1994. *Energy and Form: An Ecological Approach to Urban Growth*. Boston, MA: MIT Press.

Lacy, R.E., 1972. *Survey of Meteorological Information for Architecture and Building*. CIB Working Commission W4A.

Landsberg, H.E., 1976. Weather, climate and human settlements. Special Environmental Report 7. Geneva: World Meteorological Organization.

Leadership in Energy and Environmental Design of the US Green Building Council, http://www.usgbc.org/

Loftness, V., 1982. *Climate/Energy Graphics: Climate Data Applications in Architecture*. World Climate Program Publication WCP-30. Geneva: World Meteorological Organization.

Lloyd Jones, D. 1998, *Architecture and the Environment: Bioclimatic Building Design*. London: Laurence King.

Markus, T.A., and Morris, E.N., 1980. *Buildings, Climate and Energy*. London: Pitman.

National Academy of Sciences, Board on Applied Climatology, 1986. *Climate Data Management*. Final Report. Washington, DC: US National Research Council.

National Academy of Sciences, Building Research Advisory Board, 1950. *Weather and the Building Industry*. Conference Report 1. Washington, DC: US National Research Council.

Olgyay, V. 1963. *Design With Climate*. Princeton: Princeton University Press.

Page, J.K., 1970. The fundamental problems of building climatology considered from the point of view of decision making by the architect and urban designer. In *Building Climatology*. Technical Note 109. Geneva: World Meteorological Organization, pp. 9–21.

Page, J.K., 1976. *Application of Building Climatology to the Problems of Housing and Building for Human Settlements*. Technical Note 150. Geneva: World Meteorological Organization.

Page, J.K., 1992. Basic Climate Data in Energy in Architecture. In Goulding, J.R., Owen Lewis, J., and Steemers, T.C., eds., *The European Passive Solar Handbook*. EUR 13446. Brussels: Batsford, for the Commission of the European Communities, Chapter 2 and appendices.

Prior, M.J., and King, E.G., 1981. Weather forecasting for construction sites, *Meteorological Magazine*, **110**: 260–266.

SoDa, 2002. Solar Radiation Databases, http://soda.jrc.it for the Commission of the European Communities, Brussels.

Tombach, I., 1982. Measurement of local climatological and air pollution factors affecting stone decay. In *Conservation of Historic Stone Buildings and Monuments*. Washington, DC: National Academy of Sciences, pp. 197–210.

Tveit, A., 1970. Moisture absorption, penetration and transfer in building structures. In *Building Climatology*, Technical Note 109. Geneva: World Meteorological Organization, pp. 151–158.

Cross-references

Human Health and Climate
Urban Climatology

ARCTIC CLIMATES

The Arctic is the northern hemisphere heat sink that establishes latitudinal pressure gradients which drive the general circulation of the atmosphere. The Arctic typically conjures up mental images of a region dominated by extreme cold, snow cover and floating sea ice; but Arctic climates are quite diverse, both by season and region. This diversity reflects the pronounced seasonal cycle in solar radiation receipts, regional aspects of the atmospheric circulation and the contrasting thermal properties of different surface types. The most formal definition of the Arctic is the region north of the Arctic circle, approximately 66.5°N. At this latitude the sun does not rise above the horizon at the winter solstice and does not fall below the horizon at the summer solstice. However, climate conditions of Arctic "flavor" can be found south of the Arctic Circle, while surprisingly mild winter temperatures extend well north in the Atlantic sector. Other definitions of the Arctic include the region north of the 0°C mean annual isotherm, and the region north of the tree line.

Key physical features

The geography of the Arctic is in striking contrast to its southern counterpart. The Antarctic is characterized by a continental ice sheet surrounded by ocean. By comparison, most of the area north of latitude 70°N is occupied by the Arctic Ocean, which apart from the sector between about 20°E and 20°W longitude is almost entirely surrounded by land. The most striking feature of the Arctic Ocean is its floating sea ice cover. Total northern hemisphere sea ice extent ranges from about 14.8 million square kilometers in March to roughly half that value in September (Figure A46). Typical ice thicknesses in the Arctic Ocean are 1–5 m. The sea ice cover is not a solid slab, but contains roughly linear openings, known as leads, and more irregular areas of open water, termed polynyas. The ice cover is in near-constant motion due to winds and ocean currents. Snow cover overlies the sea ice and surrounding land for most of the year. Maximum snow depths range widely, especially on local scales, but values of 30–80 cm are typical.

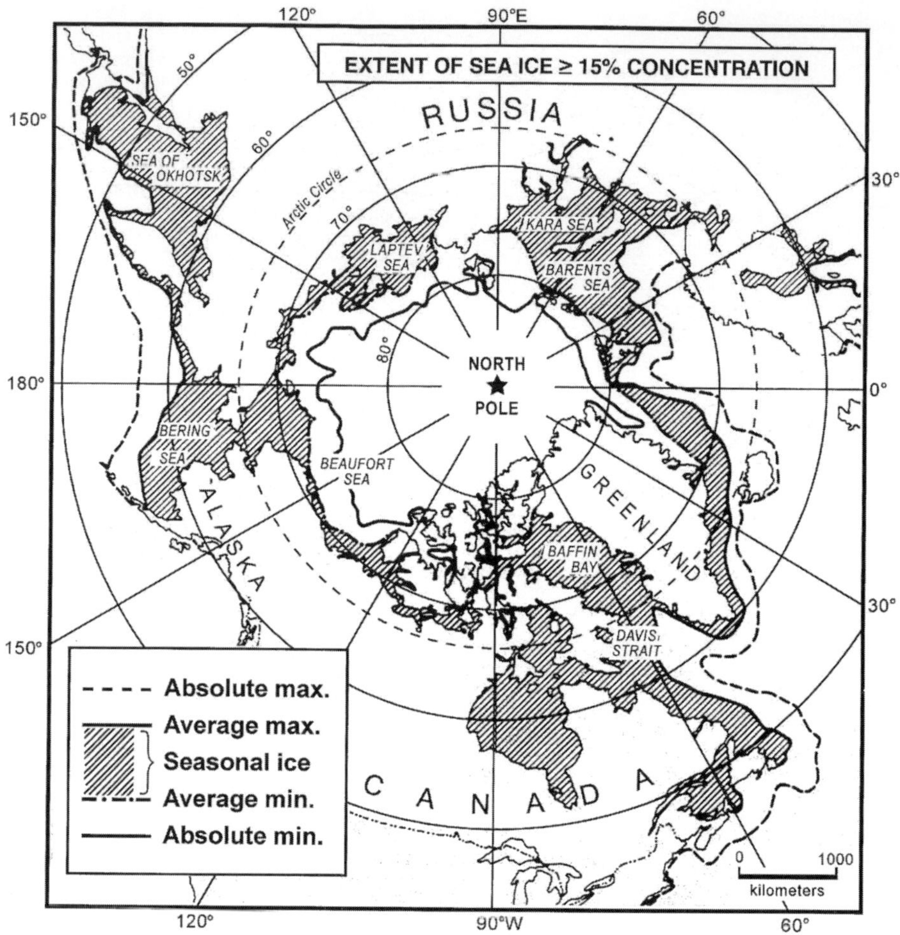

Figure A46 Geography of the Arctic region and limits of sea ice (adapted from Barry, 1983).

Apart from the Greenland ice sheet, which contains about 7.5 m of global sea level equivalent, ice caps and glaciers are primarily limited to the mountainous parts of the Siberian and Canadian Arctic Archipelagos, Svalbard and Iceland. However, most of the Arctic lands are underlain by perennially frozen ground, known as permafrost. The upper part of the ground in permafrost regions, termed the active layer, freezes and thaws seasonally, generally to depths of 100–200 cm. Permafrost acts as an impermeable barrier, which keeps moisture near the surface. Hence, many areas with poor drainage are covered by shallow thaw lakes in summer. In areas with more pronounced drainage, permafrost fosters rapid channeling of snow melt and precipitation into streams and river channels.

Much of the land surface is characterized as tundra. The most extreme and generally northernmost tundra landscape, polar desert, often has less than 5% vegetation cover. In lower latitudes, tundra commonly includes shrub vegetation of birch and willow, together with sedges and grasses. Vegetation covers 80–100% of the surface. At lower latitudes there is a forest–tundra transition, or ecotone, south of which is found the boreal forest.

Atmospheric circulation

The primary feature of the northern high-latitude mid-tropospheric circulation is the polar vortex, which in winter (Figure A47a) has major troughs over eastern North America and eastern Asia, a weaker trough over western Eurasia, and a strong ridge over western North America. The lowest winter pressure heights are located over northern Canada. These features are related to orography, land–ocean distribution and radiative forcing. The mean winter circulation at sea level (Figure A47b) is dominated by three "centers of action": (1) the Icelandic Low off the southeast coast of Greenland; (2) the Aleutian Low in the North Pacific basin; (3) the Siberian High over east-central Eurasia. The Icelandic and Aleutian lows are complex features. In the most fundamental sense they reflect position downstream of the major mid-tropospheric stationary troughs where cyclone activity is favored (they are hence part of the primary North Atlantic and East Asian cyclone tracks, respectively) and thermal effects of the relatively warm underlying ocean. The Icelandic Low is part of a broad area of low pressure extending into the Barents and Kara seas. This manifests open ocean waters and the penetration of cyclones (with associated strong horizontal heat and moisture transports) well into the Arctic Ocean. The Siberian High is a cold, shallow feature, driven largely by radiative cooling to space.

The polar vortex during summer is weaker and more symmetric (Figure A48a). The sea level Icelandic and Aleutian Lows essentially disappear and the Siberian High is replaced by a broad region of low pressure (Figure A48b). Mean low pressure is also found near the pole. Cyclone activity increases over land. Reflecting the mean low pressure, a summer cyclone

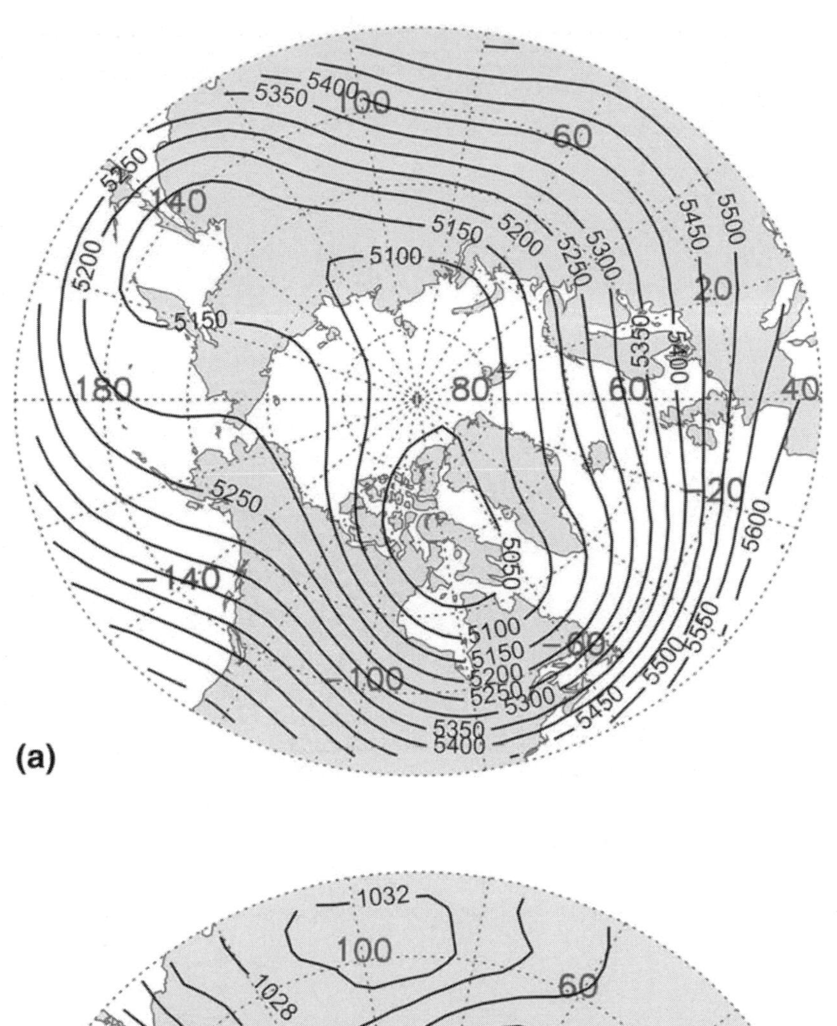

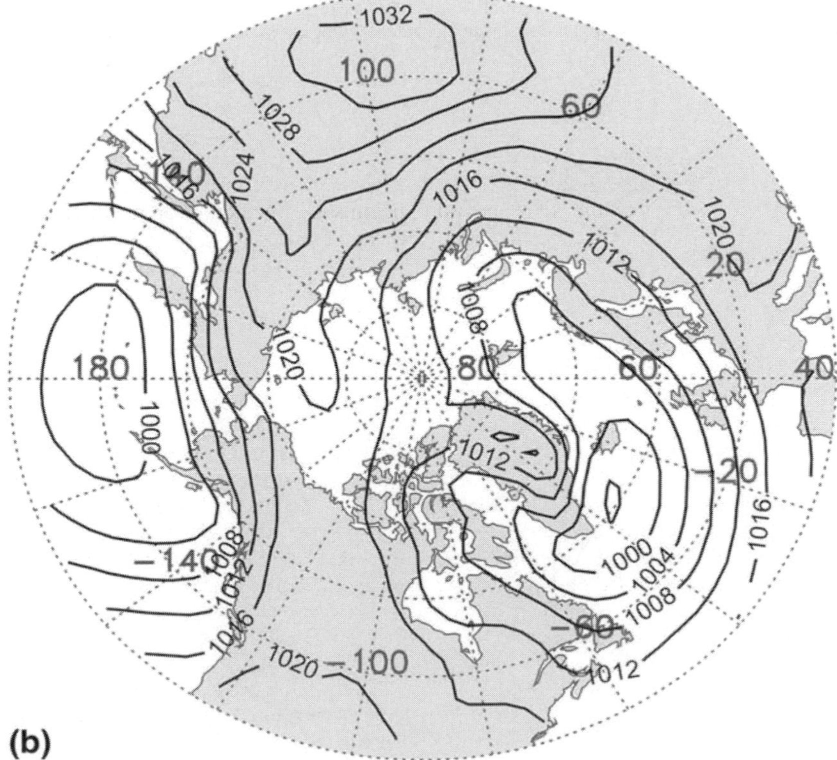

Figure A47 (**a**) Mean 500 hPa height (meters) and (**b**) sea level pressure (hPa) for January, based on National Center for Environmental Prediction/National Center for Atmospheric Research (NCEP/NCAR) reanalysis data for the period 1970–1999.

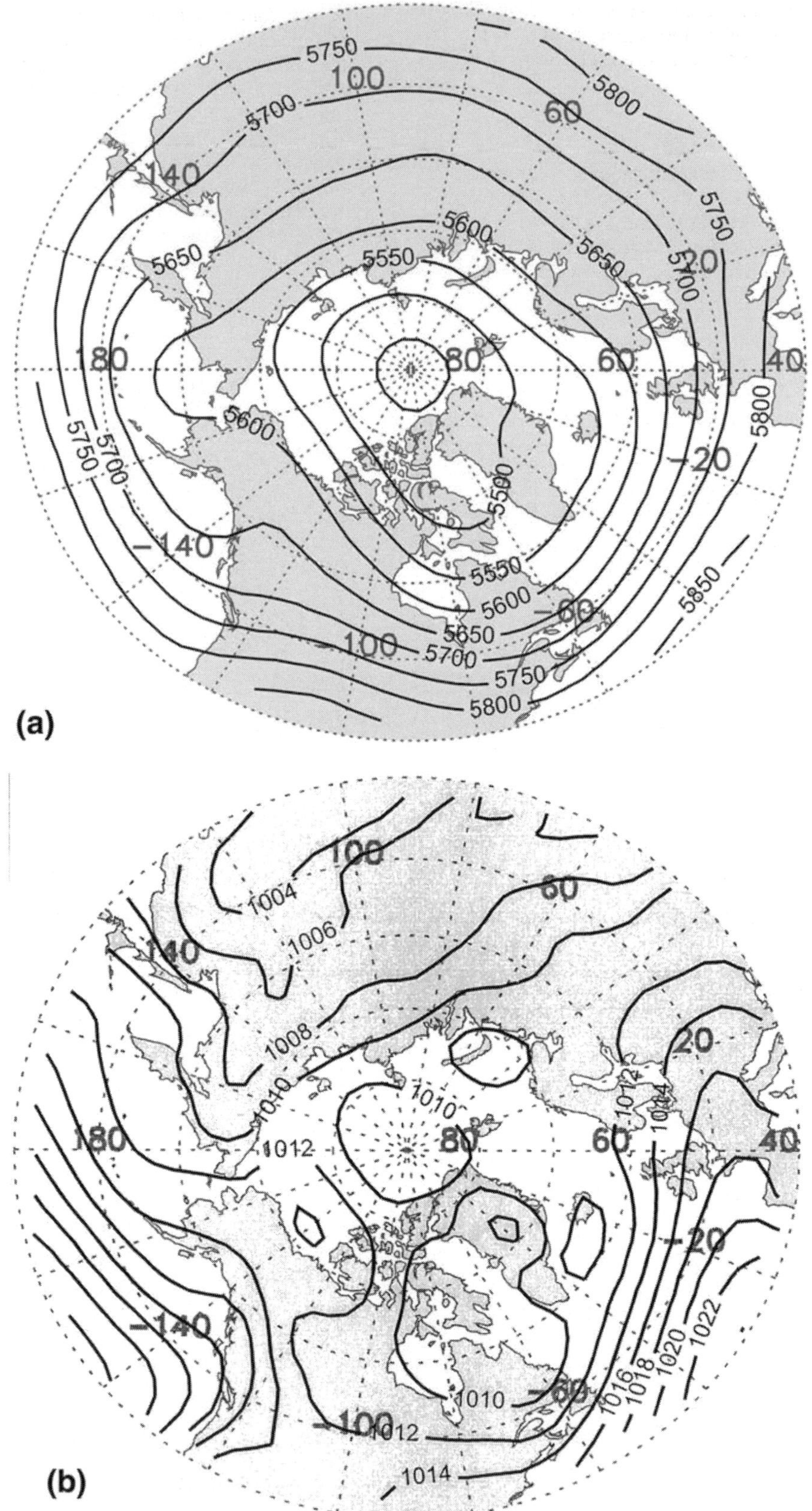

Figure A48 (**a**) Mean 500 hPa height (meters) and (**b**) sea level pressure (hPa) for July, based on NCEP/NCAR reanalysis data for the period 1970–1999. Note the different contour interval for sea level pressure (2 hPa) compared to the corresponding January plot (4 hPa) in Figure A47b.

maximum is also found over the central Arctic Ocean. This is largely due to the migration of cyclones generated over Eurasia and along the weakened North Atlantic track. Serreze (1995) provides further reading.

Surface air temperature

Figure A49 shows patterns of mean surface air temperature (at approximately the 2-m level) for the four mid-season months. January means near the North Pole are −32 to −33°C. The coldest conditions are found in northeastern Siberia in association with the winter Siberian High. Mean values of below −40°C are found in January. Locally, in valleys, temperatures are lower. Verkhoyansk, located along the Yana River valley at 67°N, is notorious for its extreme winter cold. The highest winter temperatures are found in the Atlantic sector. Building from previous discussion, this manifests warm, open water and horizontal transports of sensible and latent heat associated with the North Atlantic cyclone track. Over the ice-covered Arctic Ocean, leads and polynyas permit locally strong vertical heat fluxes from the ocean to the atmosphere. This prevents winter temperatures over the central Arctic Ocean from reaching the extremes observed in Siberia.

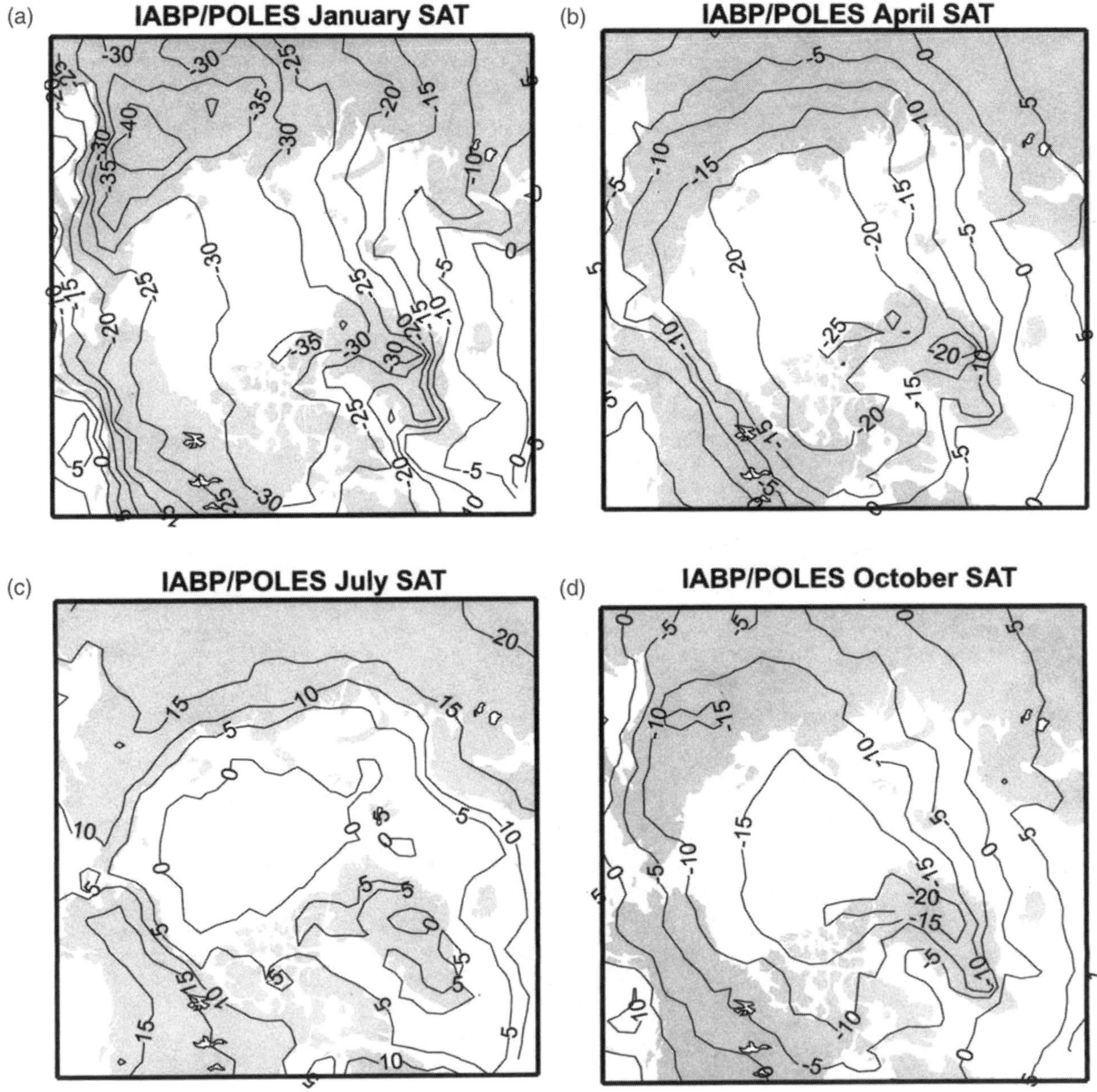

Figure A49 Mean surface air temperature (°C) for: (**a**) January; (**b**) April; (**c**) July; (**d**) October, based on the University of Washington International Arctic Buoy Programme/Polar Exchange at the Sea Surface (IABP/POLES) data set (updated and adapted from Rigor et al., 2000).

By April the Siberian temperature minimum has disappeared. The lowest temperatures are found over the Arctic Ocean and the Canadian Arctic Archipelago. Temperatures rise to the melting point over the coasts by the end of May and over most of the central Arctic Ocean by mid-June. The melt season over the central pack ice is about 60 days long. Summer temperatures over the central Arctic Ocean are constrained to hover near freezing point due to the presence of the melting sea ice cover. Because of strong differential heating between the Arctic Ocean and snow-free land, summer sees a sharp temperature gradient along the Eurasian and Alaskan coasts. Inland Arctic temperatures for July range from 10°C to 15°C. The sharp coastal gradient is not observed along the Canadian Arctic Archipelago due to the many ice-covered channels separating the islands.

A notable aspect of the Arctic environment is the presence of strong low-level temperature inversions (increasing temperature with height) during winter. These surface or near-surface-based features generally extend to about 1200 m during January–March, with a temperature difference from the inversion base to top of typically 11–12°C (Serreze et al., 1992). The inversion layer tends to be maintained by a radiative equilibrium associated with the different longwave emissivities of the surface (nearly a blackbody) and of the temperature maximum layer aloft, with northward heat advection balancing the outward radiation loss to space (Overland and Guest, 1991). Especially strong winter inversions are found inland in northwest Canada and eastern Siberia in inter-montane basins and valleys (e.g. at Verkhoyansk). In summer the Arctic is characterized by weaker, elevated inversions.

Precipitation

Mean annual precipitation totals across the Arctic (Figure A50a) vary by a factor of 20. The lowest annual totals of less than 100 mm are found over the northern Canadian Arctic Archipelago. Over the central Arctic Ocean the mean totals are around 250 mm. The largest annual amounts, locally exceeding 2000 mm, are found over the Atlantic sector. There is a strong seasonality in precipitation, illustrated by the January (Figure A50b) and July (Figure A50c) fields. The high annual totals in the Atlantic sector are largely driven by winter precipitation. This is a response to the strong Icelandic Low and North Atlantic cyclone track in this season. By contrast, most land areas and the central Arctic Ocean exhibit a summer maximum and winter minimum in precipitation. The winter minimum manifests the low moisture-holding capacity of the cold air in these regions and the relative infrequency of cyclone activity. The summer maximum over land areas is strongly associated with surface evaporation and more frequent cyclone activity. Convective precipitation is common over Alaska and Eurasia. The summer maximum for the central Arctic Ocean also manifests the summer cyclone maximum for this region.

Cloud cover

The Arctic is a cloudy place. Winter cloud fractions range from 40% to 70%, greatest over the Atlantic side where cyclone activity is frequent. Total cloud fractions rise to 70–90% in summer. Over the central Arctic Ocean there is a rapid increase between April and May, driven primarily by an increase in

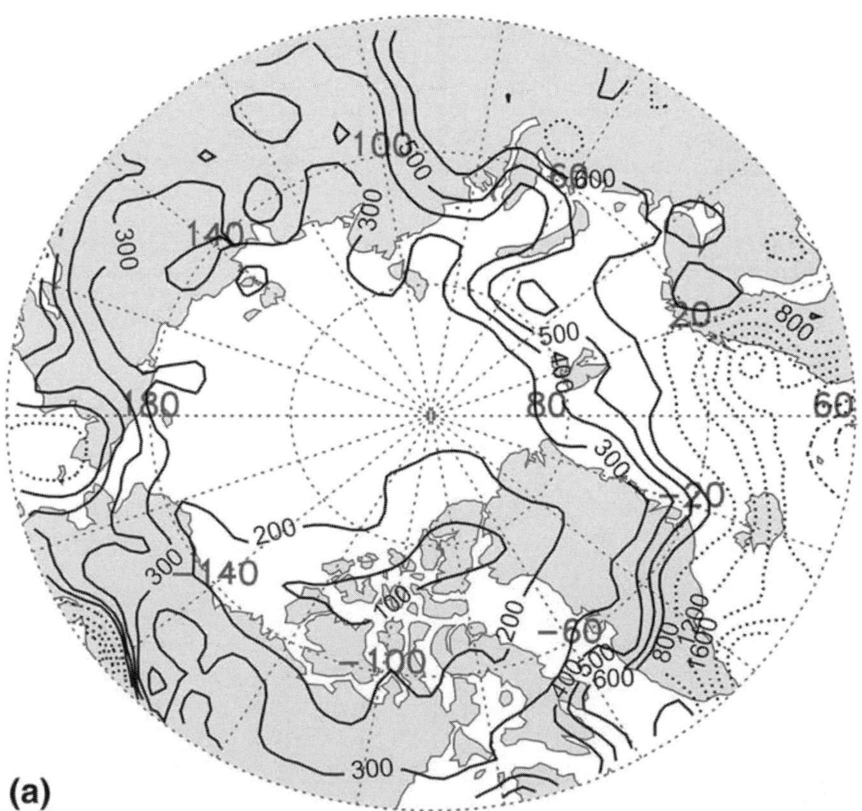

Figure A50 *(Continued)*

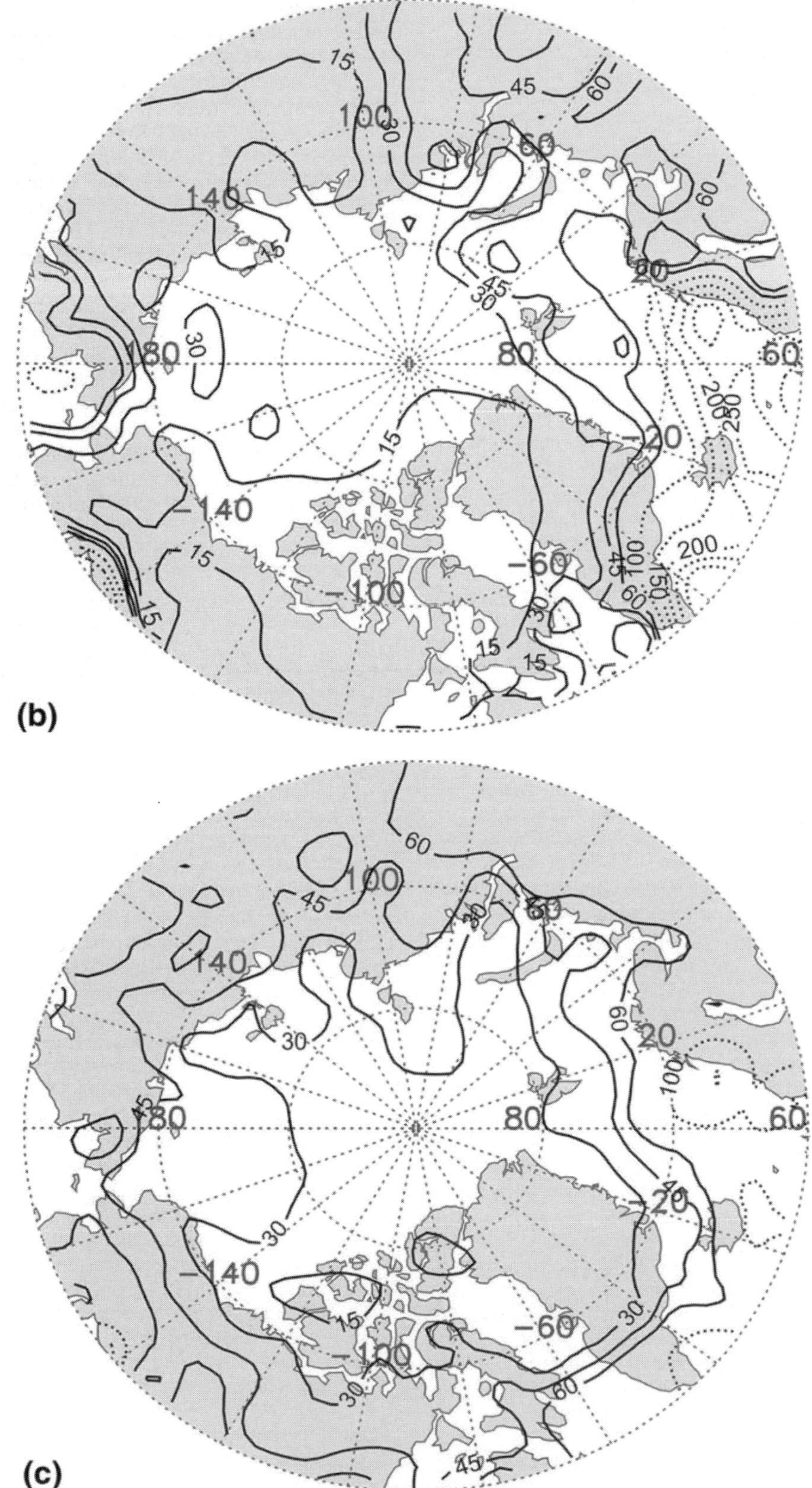

Figure A50 (**a**) Mean annual, (**b**) January and (**c**) July precipitation (millimeters) with estimated adjustments for wind-induced gauge undercatch, changes in instrument types and differences in observing methods. The annual map has isolines at every 100 mm up to 600 mm and at every 400 mm for higher amounts. The maps for January and July have isolines at every 15 mm up to 60 mm and at every 50 mm for higher amounts.

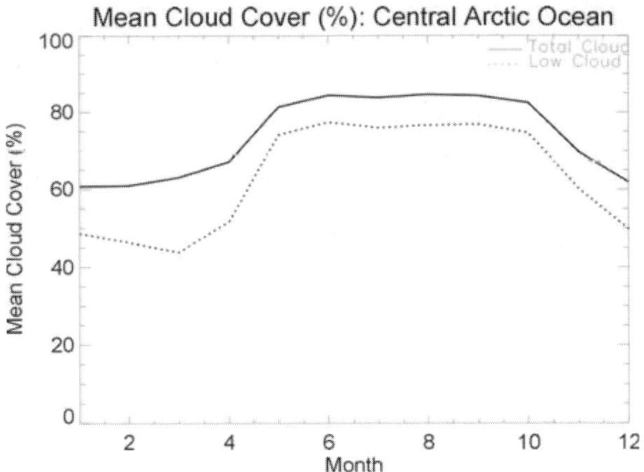

Figure A51 Mean monthly cloud cover (%) for the central Arctic Ocean based on available surface-based observations. Total (low-level) cloud cover is indicated by solid (dotted) line.

low-level stratus. While, traditionally, this summer increase is viewed as an airmass modification process whereby relatively moist air is chilled as it passes over the cold sea ice cover, the formative processes are not completely understood (Beesley and Moritz, 1999). The seasonal cycle of total and low-level cloud cover for the central Arctic Ocean is illustrated in Figure A51. While the dominance of low-level clouds is obvious, one can nevertheless observe many of the basic cloud types observed in middle latitudes. Perhaps surprisingly, convective cloud cover is frequent during winter over the Norwegian Sea – cold outbreaks from the north, when reaching the fairly warm, ice-free waters, result in destabilization of the air. Winter also finds low-level ice crystal clouds and "clear-sky" ice crystal precipitation (often termed "diamond dust") in stable boundary layers, along with ice crystal "plumes" emanating from open leads.

Surface energy budget

The surface net allwave radiation flux (the sum of shortwave and longwave exchanges) in the Arctic is negative for most of the year. Surface-based and satellite-derived sources indicate that, for the central Arctic Ocean, a net radiation deficit prevails from October through March, with winter monthly means of −20 to −35 Watts per square meter. There is a surplus from April through September, largest in July with typical means of +80 to +95 Watts per square meter. Winter deficits over land areas are broadly similar, but summer surpluses are higher, in the range of +100 to +130 Watts per square meter in July. That net allwave fluxes are small should come as no surprise. From October through March, solar radiation is either small or absent (6 months of polar darkness at the pole), such that the radiation budget is dominated by the longwave budget, which is almost always negative. Except in summer over land and ice-free waters, the high surface albedo (reflectivity in solar wavelengths) of snow and ice results in most of the incident solar flux being lost to space. The albedo of fresh snow exceeds 0.80 and is above 0.50 even for bare sea ice. Compared to other regions the Greenland ice sheet sees stronger net radiation losses in winter and smaller gains in summer. This is due to the high elevation, promoting stronger longwave losses, and the persistently high surface albedo.

With respect to the non-radiative terms, the fundamental difference between the surface energy budgets of the Arctic Ocean and tundra is the portion of net radiation used to melt snow and ice. Once the snow is melted from the tundra in summer, energy can be used in sensible heating and to evaporate water. By comparison, during summer, most of the available net radiation over sea ice is used for melt. This is consistent with Figure A49, which shows that July surface air temperatures over the central Arctic remain at about the freezing point. A major exception to the generally small turbulent heat exchanges over sea ice can be found in winter over leads and polynyas. Here, the exposure of warm, open water can yield sensible heat fluxes reaching 600 Watts per square meter.

A final point to be made is the strong role on the surface radiation budget played by cloud cover. For most of the year the cloud radiative forcing is positive in the Arctic, meaning that clouds have a warming effect at the surface. During polar darkness this can be fairly simply understood in that cloud cover represents an effective "blanket" that reduces longwave losses to space. When solar radiation is present the problem is more complex. One most consider the reduction in longwave losses in comparison to the reduction in the shortwave gains due to the cloud albedo, along with other factors such as shortwave absorption by clouds, the surface albedo, and multiple reflections between the surface and cloud base. Curry et al. (1996) review the problem.

Climate variability and change

The Arctic exhibits pronounced climate variability on interannual to decadal scales. A major source is the North Atlantic Oscillation (NAO). The NAO describes co-variability in the strengths of the Icelandic Low and Azores High (a semipermanent area of high pressure in the middle latitudes of the North Atlantic). When the Icelandic Low and Azores High are both strong, the NAO is in a positive state. Surface air temperatures are above normal over Eurasian Arctic lands and below normal over northeastern Canada. The North Atlantic cyclone track extends deeper into the Arctic Ocean and precipitation increases in the Atlantic sector of the Arctic. When the Icelandic Low and Azores High are both weak, the NAO is in a negative state, with roughly opposing climate signals. Climate signals associated with the NAO are best expressed in winter and have been recognized for centuries. In recent years many investigators have preferred to view the NAO as a component of a more fundamental northern hemisphere annular mode (NAM), popularly known as the Arctic oscillation, or AO (Thompson and Wallace, 1998).

From about 1970 through the late 1990s, the NAO and AO changed from primarily negative to positive states. Among other changes this has been attended by surface air temperature increases over most of the northern continental landmasses and the central Arctic Ocean (strongest in winter and spring), a downward tendency in summer sea ice extent, and changes in the circulation of the sea ice and ocean.

Global climate models predict that the effects of anthropogenic greenhouse warming will be amplified in the Arctic due to feedbacks in which variations in snow and sea ice extent, the stability of the lower troposphere and thawing of permafrost play key roles. There is some evidence that positive phases of the NAO and AO may be favored by increased greenhouse gas concentrations or stratospheric ozone loss. Interestingly, paleoclimate evidence suggests that Arctic temperatures of the late twentieth century were the highest of the past 400 years (Overpeck et al., 1997).

Mark C. Serreze

Bibliography

Barry, R.G., 1983. Arctic Ocean ice and climate: perspectives on a century of polar research. *Annals, Association of American Geographers*, **73**: 485–501.

Barry, R.G., Serreze, M.C., Maslanik, J.A., and Preller, R.H., 1993. The Arctic sea-ice climate system: observations and modeling. *Reviews of Geophysics*, **31**: 397–422.

Beesley, J.A., and Moritz, R.E., 1999. Toward an explanation of the annual cycle of cloudiness over the Arctic Ocean. *Journal of Climate*, **12**: 395–415.

Curry, J.A., Rossow, W.B., Randall, D., and Schramm, J.L., 1996. Overview of Arctic cloud and radiation characteristics. *Journal of Climate*, **9**: 1731–1764.

Overland, J.E., and Guest, P.S., 1991. The Arctic snow and air temperature budget over sea ice during winter. *Journal of Geophysical Research*, **96**(C3): 4651–4662.

Overpeck, J., Hughen, K., Hardy, D., et al., 1997. Arctic environmental change of the last four centuries. *Science*, **278**: 1251–1256.

Rigor, I.G., Colony, R.L., and Martin, S., 2000. Variations in surface air temperature observations in the Arctic, 1979–1997. *Journal of Climate*, **13**: 896–895.

Serreze, M.C., 1995. Climatological aspects of cyclone development and decay in the Arctic. *Atmosphere-Ocean*, **33**: 1–23.

Serreze, M.C., Kahl, J.D., and Schnell, R.C., 1992. Low-level temperature inversions of the Eurasian Arctic and comparisons with Soviet drifting station data. *Journal of Climate*, **5**: 615–629.

Serreze, M.C., Walsh, J.E., Chapin III, F.S. et al., 2000 Observational evidence of recent change in the northern high latitude environment. *Climatic Change*, **46**: 159–207.

Thompson, D.W.J., and Wallace, J.M., 1998. The Arctic Oscillation signature in the wintertime geopotential height and temperature fields. *Geophysical Research Letters*, **25**: 1297–1300.

Cross-references

Aleutian Low
Antarctic Climates
Asia, Climates of Siberia, Central and East Asia
Cloud Climatology
Energy Budget Climatology
North American High
Ocean Circulation
Siberian High
Snow and Snow Cover

ARID CLIMATES

The Earth's variable climate is a function of an unequal distribution of solar radiation. Average yearly net gains in insulation occur equatorward of 36° latitude, and net losses poleward this latitude. It follows that variations in air temperature and atmospheric pressure occur about the planet, which drive atmospheric circulation and ocean currents. Moisture content in the atmosphere is a function of air temperature and proximity to water sources. Warm tropical air rises, expands, cools adiabatically, condenses, forms clouds, and yields precipitation throughout much of the tropical zone. A rotating Earth causes air masses to migrate poleward where at about 20–40° north and south latitudes the air subsides, compresses, warms adiabatically, increases its moisture-holding tendency, and ultimately results in very warm, dry air at the surface. This middle tropospheric warming inhibits deep convection necessary for heavy rain to form. Arid climatic conditions ensure many of the world's deserts reside within these belts of high pressure that ring the planet. The Arctic and Antarctic contain polar deserts which will not be addressed here.

Aridity

Aridity refers to the dryness of the atmosphere. Aridity is a function of a continuum of environmental factors including temperature, precipitation, evaporation, and low vegetative cover. Aridity indexes are quantitative indicators of the degree of water deficiency at a given location (Stadler, 1987). Perhaps the best-known methods at identifying the arid zones of the world are classified by Köppen (1931), Thornthwaite (1931, 1948), and Meigs (1953), although the term aridity index specifically applies to Thornthwaite's work. A brief summary of these three works follows, but a more comprehensive treatment of aridity indexes may be found under Aridity Indexes (this volume).

Köppen devised a method for the identification of aridity that took into account the season of occurrence of precipitation. Rainfall occurring during a hot summer is less effective than the same amount of precipitation falling in winter. He classified arid environments as BW for arid deserts and BS for semiarid steppe regions. He also created a subcategory to distinguish between hot (h) and cold (k) dry climates. The B classification (dry climates in which potential evapotranspiration exceeds precipitation) is assigned to a region meeting one of the following criteria relating to mean annual precipitation (R in centimeters), and mean annual temperature in °C.

1. At least 70% of R occurs during the warmer 6 months, and $R < 2T + 28$.
2. If at least 70% of R occurs during the cooler 6 months, and $R < 2T$.
3. If neither half of the year includes more than 70% of R and $R < 2T + 14$.

Thornthwaite (1931, 1948) created indices based upon a percentage of water deficiency and water need during a growing season. The *Aridity Index* $AI = 100 \times d/n$, where d is water deficiency or the sum of the absolute values of monthly differences between rainfall and potential evapotranspiration when rainfall is less than potential evapotranspiration, and n is the water needed at a particular site calculated as the sum of the monthly values of potential evapotranspiration for months having a water deficiency. He also created a *Humidity Index* $HI = 100 \times s/m$, where s is excess water representing the sum of the monthly differences between rainfall and potential evapotranspiration for the months when precipitation exceeds evapotranspiration; and m is water needed at a site calculated as the sum of the monthly values for evapotranspiration during months having excess rainfall. The higher the Aridity or Humidity Index, the more arid or wet the region respectively.

Thornthwaite (1948) combined the Aridity Index with a Humidity Index to form a *Moisture Index* ($MI = HI - 0.60 \times AI$). This is a measure of precipitation effectiveness for plant growth that takes into account deficiency and excess water at a particular site. A climate classification was developed from this index where a MI less than 0 is assigned semiarid, and a MI less than -40 is assigned arid. Thornthwaite further considered regions arid when annual precipitation is less than 33% of potential evapotranspiration, and semiarid when annual precipitation is between 33% and 67% of potential evapotranspiration. Despite these indices there is little agreement on the complex number of factors leading to aridity, or how aridity can be compared using these indices.

Meigs (1953, 1960), under the auspices of UNESCO's Advisory Committee on Arid Zone Research, devised a system that delimited the arid zone and identified climatic differences within the zone. He identified arid climates (A) as those in which there is not enough precipitation for crop production. In addition, he identified an extremely arid zone (E) on the basis of a dry period of 12 months or more in which no precipitation fell (Thornthwaite's moisture index values of less than −40 can be used to make specific determinations of the boundaries of these zones).

Meigs followed Köppen in recognizing the importance of season of precipitation, and Thornthwaite in stressing heat as a factor essential for the growth and ripening of plants. Where water is available for irrigation in arid lands, temperature becomes a more important factor than moisture in the regional climate. Meig's maps of arid climate types include a number of subdivisions where the first digit represents the coldest month and the second digit the warmest month. A value of 03 indicates a winter month with temperatures below 0°C and 3 indicates a summer month with average temperatures of 20–30°C. Meigs distribution of arid climates is shown in the article on Deserts in this volume.

The arid environment

Arid environments occur at the poles, along the equator, on mountains, plateaus, below sea level, along coasts, and within mid-continents. They may be hot or cold climate types. Deserts grade imperceptibly into semiarid deserts forming expansive dry lands about the planet. Deserts are both arid and semiarid lands, and account for over 35% (>61 million km^2) of the Earth's land area (Mares, 1999). They are not capable of supporting a continuous cover of vegetation, and in the dryer portions of the arid zone, vast areas are without vegetation. Shortages of water characterize the arid climate. Water becomes a critical resource for the survival of plants and animals struggling to survive in this harsh environment. Arid zones are broad expanses (fuzzy boundaries) in which xerophytic vegetation composed of shrubs and succulents is widely scattered. It is along the desert margins that humans have tested nature by encroaching on areas that are often considered best left in their natural state.

The effect of the clear skies on temperature is readily apparent in deserts. Temperature ranges of 15°C to 22°C are common and objects in the shade are appreciably cooler than those in the sun. Some of the highest temperatures ever recorded have been measured in the arid lands. The highest temperature (58°C) was recorded at El Azizia about 40 km south of Tripoli (Kendrew, 1961). Temperatures in the tropical deserts are generally lower than this, but are usually above 40°C each day during the summer months. Nighttime temperatures are normally 16°C to 18°C lower. In winter, freezing temperatures at or near the surface are not unusual at night. Arid lands in higher latitudes experience temperature conditions comparable to those of adjacent humid environments. Advection of warm and cold air into these areas is more frequent and radiational cooling and heating of less significance.

Arid environments suffer a negative water balance in that the potential evaporating moisture exceeds the moisture supply provided by rainfall; therefore, potential evapotranspiration (the amount of moisture that would evaporate from soil and vegetation if a continuous supply of water were available) is directly proportional to temperature, wind speed, and relative humidity. Since transpiration is the most profuse of the two, if precipitation evaporates rapidly after falling, a much greater amount of moisture is required in the soil for the growth of plants than if precipitation evaporates slowly.

The results of climatic and human-induced arid conditions may initiate a threshold that spawns a physical transformation called desertification. The United Nations (1980) defines desertification as "a diminution or destruction of the biological potential of the land [which] can ultimately lead to desert-like conditions". In 1977 the United Nations announced that 35 million square kilometers were affected by desertification, and 35% of the Earth's surface at risk of similar changes. Disagreement exists on causes, processes, extent, and which changes are human-induced and what are natural. The question that begs to be answered is will the land that is desertified be reversible, and how much land is clearly at risk of change to deserts?

Identification and location of arid regions

In general terms arid regions (true deserts) typically receive less than 200–250 mm of irregular precipitation annually, with high evaporation rates. Hyperarid deserts receive less than 25 mm of precipitation annually, and have no rainy season and semiarid (steppe) regions receive greater than 250 mm and less than 600 mm of precipitation annually.

Deserts often form due to their proximity to cool, offshore ocean currents. Upwelling of colder water (Humboldt current off eastern South America, Benguela current off the southeast Africa coast) promotes cooler air at the lower contact with the ocean. Subtropical highs centered on the eastern half of oceans create equatorial movement of air along the western continental margins. This brings cooler, denser air to the warmer, subtropics enhancing the sinking motion associated with anticyclones. The relative humidity decreases, sharp temperature inversions develop that inhibit vertical convective activity, thus condensation and precipitation fail do develop. Little moisture is transferred from the colder waters flowing along the continental margins to air reaching land. The already dry, descending air mass becomes even dryer, resulting in some of the most arid deserts in the world (Sechura/Atacama Desert system of Chile and Peru 20–28° south latitude, western Sahara Desert of west Africa 18–30° north latitude, Namib Desert of southwest Africa 18–30° south latitude, and the Sonoran Desert of southern California and northern Mexico 30–33° north latitude (Lydolph, 1973).

Continental deserts form as moist air traveling from the oceans, inland, loses its moisture (Gobi Desert of central Asia and central Australian deserts 40–50° north latitude). Rainshadow deserts form when moisture is removed as air rises, cools, and condenses on the windward side of mountain barriers, descends the leeward side, is compressed, and warmed (Great Basin in the USA). In those areas subject to surges of tropical air where thunderstorms occur, intensities near the centers of the storms are high, but duration is generally short. In those arid lands reached by tropical cyclones, rainfall intensities are great, and the storms can persist for several days, resulting in areas of heavy precipitation. Such storms are of particular significance in North America and Australia. In other arid lands, unusually heavy precipitation is generally associated with isolated thunderstorms that produce flash floods in desert washes and wadis.

In desert areas where virtually no rain occurs, dew and fog drip represent the only forms of atmospheric moisture available. In the coastal deserts of Peru and Chile, moisture obtained from the fog that moves inland across the Coast Ranges sustains a moderately dense shrub forest. Because of the great amount of radiational cooling in the deserts, the formation of dew is a frequent occurrence. In the days of Roman occupation of North Africa, large rock piles above a cistern were used to collect dew for domestic water supply. Dew is an important contributor to the water budgets of many desert regions.

Perhaps the greatest amount of research into moisture sources in the arid regions has been directed at problems relating to reliability, frequency, duration, and intensity of precipitation. Generalizations about these matters are difficult because of the different kinds of atmospheric conditions that prevail in different parts of the world. For all of the arid lands it is safe to say that precipitation occurs infrequently.

Select references focusing on research organizations (Hunchinson and Varady, 1988; Hopkins and Jones, 1983), and discussions on world regional deserts (Mares, 1999; Ferrari, 1996; Bender, 1982; Petrov, 1976) are noteworthy.

World arid regions

One of the problems in delineating absolute boundaries to arid lands is that even the slightest climatic changes in these fragile environments may result in widespread changes in the natural and in human–environment relationships. Conversely, changes that humans make in the environment also result in changes in climate. General meteorological conditions are discussed for arid and hyperarid desert regions throughout the world. Table A26 lists many of the world's true deserts and hyperarid regions as classified by precipitation. Semiarid (steppe) regions are not included in this list.

African arid regions

The Sahara is a vast desert region that lies within the subtropical high-pressure belt. Across its northern margins, cyclonic storms that penetrate and cross the Mediterranean Sea from west to east drop precipitation, occasionally in the form of snow in the winter season. On the southern margins, precipitation is associated with the northward migration of the intertropical convergence zone. In the desert, dust storms associated with strong winds present problems for both sedentary agriculture and for migratory nomadic groups.

In southwest Africa, the Namib Desert along the coast and the Kalahari Desert inland are associated with the descending air masses and stable atmospheric conditions found at the eastern end of the South Atlantic subtropical high-pressure area. Portions of the Namib are extremely arid and devoid of vegetation. Stability of the air mass in the region is enhanced by cold upwelling water found on the coastal side of the Benguela Current. The outer desert of the Kalahari receives less than 250 mm of precipitation, while further into the interior precipitation amounts greater than 250 mm provide sufficient water to sustain a scrub forest.

Australian arid regions

Over half the continent is a desert with subtropical and tropical grasslands occupying much of the remainder. In the southern hemisphere summer the intertropical front dips southward into the continent, occasionally penetrating far into the desert interior. Runoff from heavy rains floods desert washes and creates vast playa lakes. Tropical cyclones affect the Queensland coast, and in the northwest the dreaded Willy Willies cause extensive destruction to coastal settlements. Winter precipitation is associated with the troughs of low pressure that lie between the migratory anticyclones that encircle the globe north of Antarctica. Cold fronts force moist marine air to rise over the low hills and mountains of the coastal zones. The easterly flow of air on the backsides of the migratory anticyclones is forced to rise along the slopes of the Great Dividing Range to produce precipitation on the eastern coast of New South Wales and Queensland.

Within the Great Australian Desert differences in the appearance of the landscape resulting from variation in environmental factors are reflected in local names for that portion of the desert. The Great Sand, Simpson, Gibson, Great Victoria, and Sturt deserts reflect the variety of landscapes to be observed within the continent.

North American arid regions

The arid zone of North America comes under the influence of the belt of subtropical high pressure. Portions lie behind mountain barriers, and much of it is remote from sources of moisture. The driest portion, the Sonoran Desert, is dominated by the Pacific High for most of the year. In winter, occasional cyclonic storms penetrate the region, bringing small amounts of precipitation, occasionally in the form of snow. However, summer is the season of maximum precipitation. Afternoon thunderstorms associated with surges of tropical air may drop copious amounts of moisture on limited areas, causing temporary flooding. By far the heaviest precipitation and the most serious flooding comes with tropical storms that migrate into the area from the southeast Pacific Ocean in late summer and early fall.

In the Great Basin, maximum amounts of precipitation are associated with the cyclonic storms of winter that cross the area. However, these storms have lost most of their moisture in crossing the Sierra–Cascade barrier, and rainfall is generally light. Snow falling in the surrounding mountains represents a major source of water used in the region. In the summer season, thunderstorms forming in tongues of moist tropical air occasionally produce heavy rain and runoff.

The Chihuahua Desert of the Rio Grande Valley and north central Mexico lie in an area that is protected from air masses from the Pacific and Atlantic oceans by mountain ranges. Aloft, stable descending air limits the formation of convective storms. Heavy rains occur only in summer, when tropical hurricanes and easterly waves from the Caribbean and Gulf of Mexico penetrate the area.

South American arid regions

The Peruvian and Atacama Deserts of the west coast of South America are among the driest areas on earth. On the north, the intertropical convergence zone only occasionally penetrates more than several degrees south of the equator. When it does, a warm ocean current – El Niño – appears offshore and heavy rains cause innumerable problems for the irrigated oases of northwest Peru. On the south, the cyclonic storms of winter rarely bring precipitation much farther north than 32°S. Offshore, the cold Humboldt Current enhances the stability of the air in the South Pacific High, and fog frequently blankets the hills of the coastal zone.

On the eastern flanks of the Andes lie the deserts and grasslands of Argentina, Bolivia, and Paraguay. Storms sweeping out of the Pacific Ocean drop their moisture on the western slopes

Table A26 Desert regions about the world. Some deserts appear in both arid and hyperarid columns because of varied climate

Continent	Arid region	Hyperarid region
Africa	Namib Desert (southwest coast, Namibia/Angola). Libya Desert (Sahara). Sahara Desert (Nubian Desert in Sudan and Ethiopia). Sinai Desert (Egypt).	Libya Desert (Sahara). Namib Desert (southwest coast, Namibia/Angola).
Australia	Great Sandy Desert. Tanami Desert. Gibson Desert. Simpson Desert. Great Victoria Desert. Nullarbor Plain.	
North America (Mexico and USA)	Chihuahuan Desert (USA/Mexico). Great Basin Desert (USA). Sonoran Desert (USA/Mexico). Mojave Desert (USA).	Mojave Desert (USA).
South America	Peruvian/Chilean Desert. Monte Desert (Argentina). Patagonia.	Sechura Desert (Peru). Atacama Desert (Chili). Peruvian/Chilean Desert.
Asia	Sistan depression (Sistan Desert, Afghanistan). Ust-Urt (midwest Asia). Kyzyl-Kum (Uzbekistan). Kara-Kau (Turkmenistan). Bet-Pak-Dala (Kazakhistan). Kara Kum (Irano-Turanian Region). Turanian Plain (Irano-Turanian Region). Dusht-e-Kavir Desert (Iran). Dasht-e-Naumid Desert (Iran). Ala-Shan Desert (China). Gobi Desert (Mongolia/China). Takla Makan Desert (China). Thar (India/Pakistan). Syrian Desert (Syria/Iraq/Jordan). Negev Desert (Israel, western Arabian Desert, Sinai Peninsula). Ordos (China).	Registan (Afghanistan). Dusht-e-Margo (Afghanistan). Dusht-e-Lut Desert (Iran). Tsaidam Desert (China).

of the Andes Mountains. The desiccated air flowing downslope on the eastern side is warmed by compression, and contributes to the aridity of the region by absorbing the available moisture as it crosses the plains.

Two unusual zones of aridity are located along the northern coast of Venezuela and in northeast Brazil. Both of these areas have been the subject of intensive investigations and are considered to be somewhat anomalous. The Brazilian arid zone, in particular, has been of great concern to that nation because of recurrent drought and forced migrations from the region because of the lack of food and water.

Asian arid regions

The Arabian, Iranian, and Thar Deserts fall under the influence of the subtropical high-pressure area of the northern hemisphere, but are also located far from the principal source of moisture in the storms that cross the area. In the winter season, cyclonic storms that originate over the Atlantic Ocean or Mediterranean Sea pass through the area. However, only small amounts of moisture are left by the time the storms reach these interior locations. Additionally, mountains and high pressure tend to block their movements and divert the paths of the storms to the north.

The Turkestan, Takla-Makan, and Gobi Desert regions lie at the interior of the Eurasian continent, remote from sources of oceanic moisture and shielded from tropical air masses by gigantic mountain systems. In winter the Asiatic High blocks the movement of cyclonic storms across the region. In the summer the moist stream of air associated with the summer monsoon is diverted around the southeastern corner of Asia by the mountain masses of Pakistan, India, China, and Malaysia.

Danny M. Vaughn

Bibliography

Bender, G.L., 1982. *Reference Handbook on the Deserts of North America.* Westport, CT: Greenwood Press. Westport, Conn.

Ferrari, M., 1996. *Deserts.* New York: Smithmark.

Hopkins, S.T., and Jones, D.E., 1983. *Research Guide to the Arid Lands of the World.* Phoenix, AZ: Oryx Press.

Hutchinson, B.S., and Varady, R.G., 1988. *Arid lands Research Institutions.* New York: Allerton Press.

Kendrew, W.G., 1961. *Climates of the Continents.* Oxford: Clarendon Press.

Köppen, W., 1931. *Grundriss der Klimakunde*, 2nd edn. Berlin: Walter de Gruyter.

Lydolph, P., 1973. *On the Causes of Aridity Along a Selected Group of Coasts*. In Amiran, D., and Wilson, A., eds., *Coastal Deserts: Their Natural and Human Environments*. Tucon, AZ: 1973.
Mares, M.A. (ed.), 1999. *Encyclopedia of Deserts*. Norman, OK: University of Oklahoma Press.
Meigs, P., 1953. *World Distribution of Arid and Semi-arid Homoclimates*. In *Reviews of Research on Arid Zone Hydrology*. Paris: UNESCO, Arid Zone Research, pp. 203–210.
Meigs, P., 1960. *Distribution of Arid and Semi-arid Homoclimates: Eastern Hemisphere; Western Hemisphere*. United Nations Maps No. 392 and No. 393, Revision 1. Paris: UNESCO.
National Science Board, 1972. *Drought: The Causes and Nature of Draught and Its Prediction*. In *Environmental Science*, Washington, DC.
Petrov, M.P., 1976. *Deserts of the World*. New York: Wiley.
Stadler, S.J., 1987. Aridity Indexes. In Oliver, J.E., and Fairbridge, R.W., eds., *The Encyclopedia of Climatology*. New York: Van Nostrand Reinhold, pp. 102–107.
Schneider, S.H. (ed.), 1996. *Encyclopedia of Climate and Weather*, Vol. 1. New York: Oxford University Press.
Thornthwaite, C.W., 1931. The climates of North America according to a new classification. *Geography Review*, **21**: 633–655.
Thornthwaite, C.W., 1948. An approach toward a rational classification of climate. *Geography Review*, **38**: 55–94.
United Nations, 1980. *Desertification*. Oxford: UN.

Cross-references

Aridity Indexes
Climate Classification
Desertification
Deserts
Rainshadow

ARIDITY INDEXES

Aridity indexes are quantitative indicators of the degree of water deficiency present at a given location. A variety of aridity indexes have been formulated, although the term *Aridity Index* specifically refers to the 1948 work of Thornthwaite. Aridity indexes have been applied at continental and subcontinental levels and are most commonly related to distributions of natural vegetation and crops. Critical values of the indexes have been derived from observed vegetation boundaries. For instance, Köppen's 1918 classification defines the desert/steppe boundary as the 200-mm annual isohyet in regions where there is no seasonality of rainfall and the mean annual temperature is 5–10°C.

Formulation of aridity indexes is not straightforward due to the nature of aridity. First, aridity is a function of the interplay between rainfall, temperature, and evaporation. Use of mean annual rainfall as an index of aridity ignores the importance of temperature and evaporation. Aridity indexes that have gained widespread acceptance directly or indirectly take into account all three factors. Second, the arid regions generally have been recognized as having a paucity of climatological data. Given the temporal variability of precipitation inherent in arid regions, the lack of climatological data has been detrimental in attempts to quantitatively define the boundaries of aridity. Third, aridity indexes must be considered from the standpoint of their eventual use. For example, the 1968 US Army World Desert Classification defines aridity with respect to military operations; application to world vegetation patterns would be inappropriate. A particular aridity index may serve several purposes, but no one index is appropriate for all uses. However, aridity indexes are often mathematically related and to some extent have been used interchangeably on a global scale.

Identification of the arid zones of the Earth has roots that can be traced two millennia. Classical Greek thought identified the latitudinally controlled torrid, temperate, and frigid zones of the world. Implicit in their thought was the concept that the torrid, low-latitude climates were arid. Not until long-term instrumental records and reliable world vegetation maps became available could true aridity indexes be developed. Thus, aridity indexes are a product of the twentieth century. Table A27 outlines the major developments regarding aridity indexes. For additional information, see Dzerdzeevskii (1958), Hare (1977), and International Crops Research Institute for the Semi-Arid Tropics (1980).

In 1900 Köppen originally qualitatively classified as arid those places that had desert vegetation. V.V. Dokutchaev in 1900 and A. Penck in 1910 qualitatively defined arid regions as places where annual evaporation exceeds precipitation. In 1905 both E.N. Transeau and G.N. Vyssotsky quantified this relationship. Yet this approach was not totally satisfactory because of the lack of reliable, worldwide evaporation measurements. Köppen's influential series of climatic classifications used mean annual temperature and precipitation combinations to define arid climates (1918, 1936). In a similar vein W. Lang's 1920 Rain Factor Index was a ratio between mean annual precipitation and mean annual temperature. Lang's index, and a modified version done by E. de Martonne in 1925, were widely used because their data requirements were minimal. However, their approach was limited in that the seasonality of temperature and precipitation were not addressed.

A. Meyer's 1926 Precipitation–Saturation Deficit Ratio was an attempt to obviate the need for dependable evaporation data. Meyer assumed the evaporation rate to be a function of the saturation deficit (saturation vapor pressure minus actual vapor pressure at a particular temperature). The Precipitation–Saturation Deficit Ratio was calculated from long-term temperature, precipitation, and relative humidity data and was found to be more reliable than temperature/precipitation-based indexes. Data availability limited the application of Meyer's ratio in that relative humidity data generally were not as available as were temperature and precipitation records.

Thornthwaite's work had an immense influence on the quantitative calculation of aridity. His Precipitation Effectiveness Index of 1931 is computed as ten times the sum of the monthly precipitation to evaporation ratio at a given location. Of practical importance was his accompanying empirical formula for deriving the Precipitation Effectiveness Index for stations recording only mean monthly temperature and precipitation. In 1948, and in subsequent revisions of his climatic classification, Thornthwaite employed the Aridity Index, which relates annual moisture deficit to annual potential evapotranspiration (see Water Budget Analysis). Weighted by 0.6 and subtracted from Thornthwaite's Humidity Index, the Aridity Index is a component of Thornthwaite's Index of Moisture. On the basis of the Index of Moisture, Thornthwaite categorized the world into nine moisture zones ranging from arid to perhumid. Evapotranspiration prominently figured in Thornthwaite's indexes, yet it was measured at only a handful of sites worldwide. So Thornthwaite devised a formula to estimate evapotranspiration through the use of a station's latitude and temperature.

Table A27 Selected summary of aridity indexes

Year	Author	Remarks	Formula
1900	W. Köppen	*Xerophytic* (arid and seimarid) climates qualitatively defined through presence of vegetative types. No formula used.	
1900	V.V. Dokutchaev	Defined aridity through comparison of annual precipitation with annual evaporation from a water surface. No formula used.	
1905	E.N. Transeau	Used ratio of annual precipitation to evaporation to describe aridity. Along with Vyssotsky, the first quantitative aridity index	$\frac{P}{E}$
1905	G.N. Vyssotsky	Used ratio of annual precipitation to evaporation to describe aridity. Along with Transeau, the first quantitative aridity index.	$\frac{P}{E}$
1910	A. Penck	Defined aridity through comparison of annual precipitation with annual evaporation from a water surface. An attempt to relate climate to landforms. No formula used.	
1911	E.M. Oldekop	Precipitation compared with potential evaporation. *E* computed by multiplying the saturation deficit of the air by a *coefficient of proportionality*.	$\frac{P}{E}$
1918	W. Köppen	Arbitrary climatic boundaries based on presumed vegetation boundaries. For example, desert and steppe were partitioned by 200 mm annual isohyet in areas where the mean annual temperature was 5–10°C; they were separated by the 320-mm isohyet where the mean annual temperature was 25°C. No formula used.	
1920	W. Lang	Rain factor. Mean annual precipitation (mm) and mean annual temperature (C) compared.	$\frac{P}{T}$
1922	W. Köppen	Precipitation compared formula at right. Several revisions of Köppen's scheme were formulated by the author himself and by others.	$2(T + 7)$
1926	E. de Martonne	Index of Aridity. A modification of Lang's Rain Factor Index.	$\frac{P}{T + 10}$
1926	A. Meyer	Absolute saturation deficit (mm of mercury) replaces evaporation.	$\frac{P}{D}$
1928	E. Reichel	Inserts the number of days with precipitation (*N*) in the formula of deMartonne.	$\frac{NP}{T + 10}$
1931	C.W. Thornthwaite	Precipitation effectiveness. Monthly precipitation to evaporation ratios determined, summed, and multiplied by 10 to eliminate fraction, (where *n* is an individual month, and *T* is the mean monthly temperature). For stations where evaporation data were not available, a formula using only precipitation and temperature data was provided. *(Note:* Formulae use English units.)	$\left\{\sum_{n=1}^{12} \frac{P_n}{E_n}\right\}10$
1932	V.B. Shostakovitch	*t* is the mean temperature during the growing period.	$\frac{P}{t10}$
1933	L. Emberger	An attempt to incorporate the effect of the seasonality of temperature on aridity. *M* is the mean maximum temperature of the warmest month and *m* is the minimum temperature of the coldest month.	$\frac{100P}{(M + m)(M - m)}$

Table A27 (Continued)

Year	Author	Remarks	Formula
1934	W. Gorozynski	Aridity coefficient. C is the cosecant of latitude, T_r is the difference between the means of the hottest and coldest months, and P_r is the difference between the greatest and least annual precipitation totals over 50 years. The coefficient increases with increasing aridity with its maximum value near 100. (*Note:* formula uses English units.)	CT_rP_r
1937	G.T. Selianinov	Effectiveness of precipitation in the growing season. Only mean monthly temperatures above 10°C are summed.	$\frac{P10}{\sum_{n=1}^{12} T_n}$
1941	N.N. Ivanova	Calculation of a precipitation "potential" evaporation ratio using the formula at right where t is the mean monthly temperature and a is the mean monthly relative humidity.	$\frac{P}{E}$ $E = 0.0018(25 + t)^2(100 - a)$
1942	E. de Martonne	Modification of earlier work incorporating a representation of the temperature (T_d) and precipitation of the driest month (P_d)'	$\frac{P}{T+10} + \frac{(12P_d/(T_d + 10))}{2}$
1947	N.V. Bova	Inclusion of soil moisture conditions in a precipitation/temperature ratio. H is the initial moisture content of the soil.	$\frac{H+P}{\sum_{n=1}^{12} T_n}$
1948	C.W. Thornthwaite	Represents a water balance approach to aridity where I_h is the Humidity Index, s is the surplus moisture in the humid season, n is the water deficiency in the dry season, I_a is the Aridity Index, and I_m is the Moisture Index. Later modifications were made to this work.	$I_h = 100\,s/n$ $I_a = 100\,d/n$ $I_m = I_h - 0.6I_a$ $I_m = \frac{100s - 60d}{n}$
1948	V.P. Popov	Index of Aridity. Σ_g is the annual effective precipitation, $t - t'$ is the mean annual wet bulb depression, and r is a factor based on daylength.	$\frac{\Sigma_g}{2.4(t - t')r}$
1949	J.A. Prescott	Refinement of earlier formulae using precipitation and saturation deficit. *Note:* This method uses English units.	$\frac{P}{0.7D}$
1950	A.A. Skvortsov	E_a, actual evaporation, compared to E_{st}, "standard" evaporation measured from a water surface.	$\frac{E_a}{E_{st}}$
1951	R. Capot-Rey	P and T refer to the mean annual precipitation and evaporation while p and t refer to the precipitation and evaporation of the wettest month.	$\frac{100\frac{P}{e} + 12\frac{P}{e}}{2}$
1951	M.I. Budyko	Radiational Index of Dryness. R is the mean annual net radiation and L is the latent heat of vaporization for water. This is the first index using a radiation balance approach.	$\frac{R}{LP}$
1952	S.J. Kostin	Precipitation versus potential evapotranspiration for the same period.	$\frac{P}{PE}$
1953	P. Meigs	Use of Thornthwaite's Moisture Index to classify and map the dry lands of the Earth. No formula used.	

Table A27 *(Continued)*

Year	Author	Remarks	Formula
1955	H. Gaussen	Classification based on the duration and severity of dry months. A dry month is defined by the conditions at right. Other factors considered by Gaussen's definition of aridity include number of rainy days, humidity, mist, and dew.	$P \leq 2T$
1957	F.R. Bharucha and G.Y Shanbhag	A reuse of the *P/E* index using the formula at right. *E* is the mean 24-hour evaporation in inches, *B* is the mean wind velocity in miles per hour, *h* is the mean relative humidity in percent, and *e* is the mean vapor pressure in inches of mercury.	$E = (1.465 - 0.0186B)$ $(0.44 + 0.11BW)\frac{100}{h} - 1e$
1960	P. Meigs	Revision of 1953 maps. This work has become the most widely used identification of the world's arid regions.	
1962	V.M. MeherHomji	Index of Aridity–Humidity. *S* is the "precipitation quantity factor" and *X* is the length of the day period.	$S + X$
1965	C. Troll	Defined arid climates on the basis of number of months that the expression at right holds true.	$P > PE$
1967	C.C. Wallen	Interannual variability (V_I), where *n* is a particular year in a series of *N* years. The second equation is an empirical one that describes the arid margin of dryland farming.	$V_I = \frac{100\sum(P_n - 1 - P_n)}{\overline{P}(N - 1)}$
1967	J. Cocheme and P. Franquin	Matches water availability to a crop's growth cycle through a comparison of the values of precipitation and evapotranspiration *(ET)*. Can be used for any growing period.	P vs. $\overline{ET}$ $V_I = 0.07\overline{P} + 22$
1968	US Army	World Desert Classification. Based on rainy days per month with a rainy day defined as any day with greater than 0.1 inch of precipitation. Months are categorized in four categories by the categorization of the cumulative number of wet months in a year. Used with respect to men and military equipment.	
1969	H. Lettau	Approaches aridity from the standpoint of surface energy and moisture fluxes. *B* is the Bowen Ratio (cf. the ratio of sensible to latent heat fluxes) and *C* is the annual water surplus divided by the precipitation. The formula is equivalent to Budyko's Radiational Index of Dryness.	$(1 + B)(1 - C)$
1970	W.K. Sly	The ratio of growing-season precipitation to total amount of water required by the crop if lack of water is not to limit production. *P* is the growing season precipitation, *SM* is the start of the growing season, and *IR* is the calculated irrigation requirement during the growing season.	$\frac{P}{P + SM + IR}$

Table A27 *(Continued)*

Year	Author	Remarks	Formula
1971	G.H. Hargreaves	Moisture Available Index (MAI). For a specified period the ratio of the monthly rainfall total expected with a 75% probability to the estimated potential evapotranspiration. Values of 1.00 to 0.00 were considered increasingly moisture-deficient.	$\frac{P_p}{PE}$
1979	UNESCO	Map of the World Distribution of Arid Regions. Based on the ratio of precipitation to evapotranspiration with evapotranspiration being determined by Penman's method. This work was intended to replace Meigs's 1960 work.	$\frac{P}{ET}$
1980	R.P. Sarker and B.C. Biwas	Modification of MAI to consider weekly periods, various levels of rainfall total probabilities so that P_A is the assured rainfall of a period and PE is the potential evapotranspiration for the same period.	$\frac{P_A}{PE}$

Notes: Dates given are first appearance in the literature. All formulae are in metric units unless otherwise noted. Symbols have been modified from original sources for purposes of comparison.

His concepts have gained wide use because of the simplicity of their data requirements and their general agreement with world vegetation patterns. However, some engineers and agriculturalists have criticized his methods as too general for use in specific applications. The formulae have been found to produce unreliable results in certain tropical locales.

Budyko (1951) offered a new approach by considering the heat and water balance equations of the Earth's surface. His Radiational Index of Dryness was the ratio of the mean annual net radiation (i.e. the radiation balance) to the product of the mean annual precipitation times the latent heat of vaporization for water. The warm dry conditions synonymous with arid regions are well characterized by Budyko's index. In practical terms the Radiational Index of Dryness is the number of times the net radiative energy income at the surface can evaporate the mean annual precipitation. Although a number of writers have preferred Budyko's method of calculating aridity, a major limitation in application is the lack of long-term radiation records at many observation stations.

Other indexes have tended to be refinements and hybridizations of the above notions. Of recent interest has been the use of aridity indexes to define the agricultural boundary between arid and semiarid climates. UNESCO, FAO, and WMO are in accord that the boundary should be drawn where lack of water makes dryland farming impossible. Thus, aridity indexes are gaining increased importance in the planning of water supplies for crops.

Meigs's maps (1953, 1960) have been the most widely cited classification of aridity. Meigs used Thornthwaite's Moisture Index to define aridity. The 1 : 2 500 000 Map of the World Distribution of Arid Regions (UNESCO, 1979) has been produced to refine Meigs's maps. In this latter work approximately one-third of the world's continental surface is classified as having some degree of aridity.

Although UNESCO's 1979 map continued use of the ratio of precipitation to evapotranspiration, evapotranspiration was calculated by the more-favored Penman method. Calculation of aridity indices usually includes an input of evapotranspiration. The present international consensus is to estimate reference evapotranspiration using the Penman–Monteith formula (Allen et al., 1998). This formula is state-of-the-art but needs relatively esoteric atmospheric and soil input such as solar radiation and ground heat flux; if such data are not available locally, they can be estimated. An excellent explanation and commentary of operational drought-related aridity indices was compiled by Heim Jr. (2002).

Traditionally, climatologists and agriculturalists have used long-term weather data from standard weather shelters to create maps of aridity at Earth's surface. The difficulties of interpolating between observation points have created profound uncertainties in areas where data are sparse. For instance, there has never been a worldwide map of Thornthwaite's Moisture Index. The advent of satellites with continuous coverage of immense areas over a course of years has had a profound impact in the use of aridity indices. Recent advances in the archiving and collation of data have allowed worldwide *monitoring* of aridity as opposed to hindsight assessments.

The most commonly used satellite measure is the Normalized Difference Vegetation Index (NDVI). Derived from the Advanced Very High Resolution Radiometer (AVHRR) on the NOAA polar orbiter series, the NDVI at a location is closely related to the proportion of photosynthetically absorbed radiation that can be calculated from a ratio of reflectances in visible (0.58–0.68 μm) and near-infrared (0.725–1.1 μm). AVHRR channels:

$$(\mathrm{CH2} - \mathrm{CH1}) / (\mathrm{CH2} + \mathrm{CH1})$$

Channel 1 (CH1) is the reflectance in the visible and Channel 2 (CH2) is the reflectance in the reflective infrared. CH1 is sensitive to chlorophyll's absorption of incoming radiation, and CH2 is in a portion of the spectrum in which the mesophyll structure in leaves causes great reflectance (Tucker et al., 1991). As the NDVI values increase, this infers increasing amounts of biomass. Active biomass is largely controlled by climate, so NDVI is essentially an aridity index when time series of values are calculated over area. Multiyear data sets are needed so as to sort the phonological effects of green-up and senescence from atmospheric variability.

Near-real-time monitoring of aridity to assess drought is now possible. Surface and satellite measures of short-term drought/aridity can be combined to provide operational assessments. An example of this sort of work is given by Svoboda et al. (2002). By using a blend of the Palmer Drought Severity Index, CPC Soil Moisture Model Percentiles, USGS Daily Streamflow Percentiles, Percent of Normal Precipitation, the Standardized Precipitation Index, and the Satellite Vegetation Health Index, a weekly Drought Monitor index value is calculated for each US climate division. The results are mapped and widely distributed.

As latter-day aridity monitoring and research continues, the climatological prospects are exciting – "intellectual descendants" of the aridity indices of the early 1900s are providing useful decision support to policymakers. As time progresses and time-series of these products achieve climatological proportions, academic analyses should offer unprecedented dynamism to our concepts of aridity.

Stephen J. Stadler

Bibliography

Allen, R.G., Pereira, L.S., Raes, D., and Smith, D., 1998. Crop evapotranspiration – guidelines for computing crop water requirements. FAO irrigation and drainage paper no. 56. Rome: FAO.

Budyko, M.I., 1951. O. Klimaticheskikh Factorakh Stoka (On climatic factors and runoff). *Problemyfiz. Geog.*, **16**: 41–48.

Dzerdeevskii, B.L., 1958. On some climatological problems and microclimatological studies of arid and semiarid regions in the U.S.S.R. In *Climatology and Microclimatology: Proceedings of the Canberra Symposium*. Paris: UNESCO, Arid Zone Research, pp. 315–323.

Hare, F.K., 1977. Climate and desertification. In Secretariat of the United Nations Conference on Desertification, ed., *Desertification: Its Causes and Consequences*. Oxford: Pergamon Press, pp. 63–168.

Heim, R., Jr, 2002. A review of twentieth century drought indices used in the United States. *Bull. Amer. Met. Soc.*, **83**(8): 1149–1165.

International Crops Research Institute for the Semi-Arid Tropics, 1980. *Climatic Classification: A Consultants Meeting*. Patancheru, India: IRCISAT.

Köppen, W., 1918. Klassification der Klimate nach Tempertur, Niederschlag and Jahreslauf. *Petermanns Geog. Mitt.*, **64**: 193–203, 243–248.

Köppen, W., 1936. Das Geographische System der Klimate. In Köppen, C.W., and Geiger, R., eds., *Handbuch der Klimatologie*, vol. 3. Berlin: Gebrüder Bornträger.

Meigs, P., 1953. *World Distribution of Arid and Semiarid Homoclimates*. Arid Zone Programme, vol. 1. Paris: UNESCO, pp. 203–210.

Meigs, P., 1960. *Distribution of Arid Homoclimates: Eastern Hemisphere: Western Hemisphere*. United Nations Maps No. 392 and No. 393, Revision 1. Paris: UNESCO.

Svoboda, M., LeComte, D., Hayes, M. et al., 2002. The drought monitor. *Bull. Am. Met. Soc.*, **83**(8): 1181–1189.

Thornthwaite, C.W., 1931. The climates of North America. *Geog. Rev.*, **21**(3): 633–655.

Thornthwaite, C.W., 1948. An approach toward a rational classification of climate. *Geog. Rev.*, **38**(1): 55–94.

Tucker, C.J., Newcomb, W.W., Los, S.O., and Prince, S.D., 1991. Mean and inter-year variation of growing-season normalized difference vegetation index for the Sahel 1981–1989. *Int. J. Remote Sens.*, **12**: 1113–1115.

UNESCO, 1979. *Map of the World Distribution of Arid Regions*, MAB Technical Note 7. Paris: UNESCO.

Cross-references

Arid Climates
Desertification
Deserts
Drought
Water Budget Analysis

ART AND CLIMATE

"Art and Climate" is the title of an article published by Richard Wagner, the famous German musician, in 1841. Wagner explains in the introduction that his opinions on the future of Art had been criticized for failing to take into account the *influence of Climate upon man's capacity for Art.* Broadly this criticism suggested that northern Europeans had a poorer capacity for art than those who were blessed with the Ionic skies of the warmer climates of the Mediterranean. Wagner was not impressed with this criticism and set out to prove in the article that "Everywhere, in every climate, will these works of Art be inspired by native skies: they will be beautiful alike and perfect". However, many of Wagner's ideas have been controversial and considered almost racial and linked to what Livingstone (2002) has called moral climatology. Hence we have to be careful and state that the quality of art in a region is not in any way determined by climate. Climatic determinism will undoubtedly influence the content but not quality of art. Climate will therefore influence art but can art influence climate? Oscar Wilde, writing in 1889, certainly thought so:

> At present, people see fogs, not because there are fogs, but because poets and painters have taught them the mysterious loveliness of such effects. There may have been fogs for centuries in London. But... They did not exist till Art had invented them (Wilde 1889: 925)

The climate of a region can be considered to be a restraining influence on the weather – keeping it within a season's allowable array. No such simple definition can be offered for "art", a term derived from the Latin *ars* that, like the equivalent Greek word, also means science, skill, craft, ruse, etc. Here, art is defined as the manifest expression of the human experience in response to stimuli from the outer and inner world.

These expressions assume a multitude of forms, but only those that have significant relationships to climatic features as stimuli are considered here. Nevertheless, it should be mentioned that in architecture, e.g., the thickness of the walls of old houses in Europe is a function of the continentality of the region (Landsberg, 1958); or a bridge over the Oreto river at Palermo, Sicily, built in AD 1113, indicates that it originally spanned a much larger river, i.e. during a moister period in the Mediterranean area than prevails in our time (Lamb, 1968). Outdoor sculpture, an art related to architecture, can suffer from the ravages of wind, hydrometeors, and air pollution.

Another powerful medium for expressing human experience is the "international language" of music, vocal and instrumental, which is not limited to titles or texts that refer to climatic features, such as "Seasons" (Vivaldi, Haydn, Tchaikowsky, Glazunov), "Nuages" (clouds) by Debussy, or Brahms' "Regenlied" (rain song). There is also musical imitation of weather sounds such as wind, thunder, even rain and lightning as, for example, in the fourth movement of Beethoven's sixth symphony (Neuberger, 1961; Burhop, 1994).

A much wider range of atmospheric imagery is used in language, i.e. written or verbal prose or poetry, including drama and comedy (Aristophanes, 423 BC) and pictorial art using various media, but excluding sculpture.

Language

The spoken word

When communicating with one another, people express not only their opinion about the topic of conversation, but in their language also express other experiences in symbolic form for the purpose of emphasis or as circumscription of situations, happenings, or emotions. Considering our continuing exposure to atmospheric properties and phenomena, it is not surprising that much of our idiomatic speech involves meteorological imagery. Phrases such as thunderous applause, lightning speed, whirlwind courtship, know which way the wind blows, to be under a cloud, hazy recollection, foggy notion, snow-white, thunderstruck, etc., involve the appropriate properties of the pertinent meteorological phenomena. Similar idiomatic expressions are also used in French, German, Italian, Turkish, and many other languages (Neuberger, 1961).

The written word

As in spoken language, meteorological imagery abounds in the literature of Europe, the Middle East and the Far East. Starting with the Bible, through classical Greek and Roman literature, and from the Renaissance (Heninger, 1968; Janković, 2000) to the present time, with parallels in the Asian literature, we find weather imagery and weather and climatic descriptions. They are used in prose and poetry, in drama and comedy, thereby providing, literally and figuratively, an "atmosphere" for the actions and events described or as substitute expressions of human emotions. Bone (1976) examines clouds in the poetry of Wordsworth, Byron, Shelley and Keats. Three anthologies of poetry (one in German) revealed that 1152 out of 2563 poems (i.e. 45%) contained references to meteorological elements.

It appears that the extent to which weather images or descriptions are used in writing, is linked to the severity of weather and climate; thus, one finds many more references to weather in the British, Scandinavian, German, and especially Russian literature than in French and that of the Mediterranean countries in which milder climates prevail. Heninger (1968), in his extensive analysis of a large number of meteorological references in the literature of England, also cites descriptions of, and literary reactions to, air pollution in London and its environs, resulting from the domestic and industrial uses of coal. The employment of coal as fuel started in the thirteenth century, and due to the resultant smoke has led to a higher frequency, greater density, and longer duration of fogs. Early writers have also remarked on the deleterious effects of smoke on people, plants, and buildings (Brimblecombe and Ogden, 1977; Brimblecombe, 1988, 2000). Monet painted 95 images of the London smog (Thornes and Metherell, 2003).

Literature written before the establishment of observational networks in the nineteenth century can also give valuable clues to the climate and its variations over time in different regions. Chronicles written by the ancient Egyptians and Babylonians always included special meteorological and climatological events such as advances or retreats of glaciers, lake or river freezes, floods, storms, hail, heavy rains, etc. (Lamb, 1967, 1968, 1995; Watson, 1984; Durschmied, 2000). In the literature of ancient Greece, the epics of Homer, the writings of Herodotus, of Hippocrates (460–377 BC), who could be called the first bioclimatologist, and of others indicate that colder and wetter periods than now occurred over most of Europe throughout the first millennium BC (Frisinger, 1977).

Hundreds of references to climatological and meteorological phenomena can be found in the works of Shakespeare. For example, in Hamlet, Act I, Scene I, line 63 reads: "He smote the sledded Polacks on the ice"; this undoubtedly refers to the deep freezes of the waters east and south of Denmark, as well as large portions of the Baltic Sea in 1589–1590. This was part of a longer cold epoch that also featured significantly during the Thirty Years' War (1618–1648).

An interesting point was raised by Bonacina (1939) when he compared written descriptions with paintings. Words can depict as many details of an atmospheric event as can paintings, but a description can also report changes or sequences of such events. Pictorial representations reproduce only a moment in an event; however, some Medieval and Renaissance artists tried to overcome part of this limitation by putting non-simultaneous aspects of a story on the same canvas.

Pictorial arts

General remarks

Although this category of the arts includes etchings, drawings, cave paintings, and pictorial decorations of pottery (especially those of classical Greece), only murals and paintings in the most usual sense of the term will be considered here. This is not to say that the other media cannot, or do not, represent, or at least give clues to, climatic features; but such is relatively rare. In drawings there are no colors by which to judge the condition of the sky or, in prints and pottery, the color range is very limited. Nevertheless, there is among the Tassili Frescoes in the caves of southeastern Algeria the depiction of pelican heads (Lhote, 1961) and of a canoe, probably used for hippopotamus hunting, that indicates the existence of lakes in about 3500 BC on what is now part of the African Sahara (Lamb, 1977). Also some decorated ancient Greek pottery shows ships with billowing sails, implying wind.

Excluded from consideration here are the murals of the ancient Greeks and Romans; there are too few of them from which to draw significant conclusions with respect to climate. Yet, according to the late art historian, Dr H.W. Janson, (personal communication, 1968), there is an

> almost complete lack of any clouds in ancient Roman painting. There are plenty of landscape backgrounds, but the skies are almost invariably cloudless. Suddenly, from the 4th c. AD on, clouds begin to appear in large quantity . . . the change from no-clouds to clouds occurs in a few decades.

This art–historic fact seems to correlate with the increase of nocturnal cloudiness in the zone between 30°N and 50°N latitudes from AD 300 to 500 as calculated from the frequency of discovery of comets in Europe and elsewhere. As in Roman paintings, the classical Greek and Etruscan frescoes and later the Byzantine school of painting contain only an occasional sky background. In general, Western art during the Middle Ages showed a preference for gold backgrounds in lieu of skies. It is only in the transition period to the Renaissance, starting perhaps with the turn of the fourteenth century, that artists seemed to become more aware of the pictorial opportunities of their natural environment; the awareness of nature being the hallmark of the Renaissance spirit.

Paintings

A visit to any art gallery, excepting those that exhibit abstract art exclusively, clearly shows that in every century many artists have chronicled almost everything in their own environment: fauna and flora, architecture, furniture, foods, fashions, weapons, utensils, tools, toys and games, sports equipment, and musical instruments.

Considering the powerful influence of the atmospheric environment on all life forms, we can expect that artists also became chroniclers of the climates they experienced. The expression of climatic features in art can be achieved in two mutually nonexclusive ways: (1) directly in pictorial representation of clouds, frozen ponds, snowy landscapes, etc.; or (2) indirectly by images of people wearing heavy clothes, indicating low temperatures, or of a flowing scarf, billowing sails, waves on bodies of water, and the drift of smoke, all indicating wind. An example of the first category is "Storm over Taos" by John Marin (1872–1953). This painting shows a forcefully executed sky with a towering thundercloud on the left, a descending shower in the middle, and zig-zag lightning on the right (Figure A52). This watercolor also shows that modern painters have not abandoned the representation of nature. An example of the second category is "Lady at the Harpsichord" by Jan Miense Molenaer (1610–1668), painted between 1635 and 1640. The lady's attire is made of very heavy materials; it must have been quite cold, indeed. In order to avoid contact with the cold tile floor her feet are propped up on a footstool with a ceramic brazier, presumably containing hot coals (Figure A53). Another interesting feature is the landscape with clouds painted on the inside of the spinet cover by which the artist brings the outdoors into the room. Many other artists similarly have painted indoor scenes with windows or open doors through which a view of sky and clouds was admitted.

It appears that artists have used atmospheric imagery in several ways: as a decorative or incidental feature, as a primary subject, and/or as a symbolic support of or contrast to the emotional content of the human scene.

Figure A52 "Storm over Taos", by John Marin (courtesy of the National Gallery of Art).

Figure A53 "Lady at the Harpsichord", by Jan Miense Molenaer (courtesy of Fotocommissie Rijkmuseum Amsterdam).

An example of the first category is seen in Molenaer's painting mentioned above; the landscape inside the harpsichord cover is merely a decorative element. The painting by John Marin is representative of the second category, as the title indicates. The title of a painting, however, cannot be considered the main criterion for the subject category. For instance, "Breezing Up", by Winslow Homer (1836–1910), while showing a large amount of the canvas covered with an almost overcast sky and an agitated ocean, is basically a scene of human activity, namely, three boys and a man in a sailboat (Figure A54). For the purpose of this article, any painting in which the sky and/or other atmospheric phenomenon is the dominant feature can be considered a meteorological or climatological painting. The "Skating Scene", by the American painter John Tooe (1815–1860), belongs clearly to the second category, as the overcast sky, the snow-covered landscape, and the frozen creek dwarf the skating scene proper (Figure A55). "Madonna and Child", by Giovanni Bellini (*ca.* 1430–1516), shows a sunlit strip of landscape in the middle and a few bright cumuliform clouds on the upper left; but then above the Madonna's head spreads an ominous dark thundercloud, a symbolic reminder of the future tragedy (Figure A56). This type of example can be found in innumerable religious paintings (Neuberger, 1961).

Many authors have written about the artist's perception of his or her climatic environment (Lamb, 1967; Botley, 1970; Gedzelman, 1989; Neuberger, 1970; Spink, 1970; Thornes, 1979, 1999; Burroughs, 1981; Walsh, 1991). A great deal of thought was given, not only to the esthetic aspects of clouds in landscapes, but also to the sudden and substantial increase of snowy landscapes in the sixteenth century, which coincided with the emergence of the "Little Ice Age". The first attempt at

Figure A54 "Breezing Up", by Winslow Homer (courtesy of the National Gallery of Art).

Figure A55 "Skating Scene", by John Toole (courtesy of the National Gallery of Art).

extracting quantitative information of climatic changes from paintings with respect to cloudiness was made by Lamb (1967); he demonstrated such changes by analyzing the amount of sky cover in 200 paintings of Dutch and British artists. He used only summer scenes painted between 1550–1568, 1590–1700, 1730–1788, 1790–1840, and 1930–1939; with such a small and biased sample the statistical results become uncertain, although they are supported by other evidence. In order to test the hypotheses that the average climatic features derived from the paintings of many artists living in the same climatic epoch should be different for different climatic regions, and that such differences should exist for different climatic epochs, Neuberger, in 1967, evaluated 12 284 paintings in 41 art museums in 17 cities of nine countries.

Figure A56 "Madonna and Child", by Giovanni Bellini (courtesy of the National Gallery of Art).

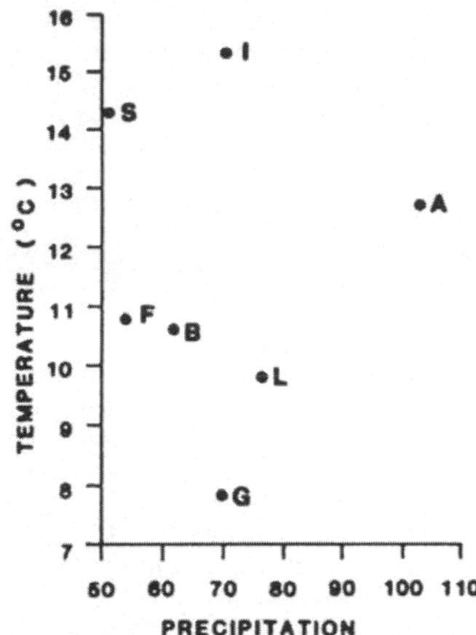

Figure A57 Mean regional annual temperatures and precipitation (in cm) for the period 1951–1960.

The regional climatic differences

For determining the location of the artists, the art–historic classification according to countries and schools was the only method available. This does not account for the travels of many artists or their move to different countries and climates. Vincent Van Gogh (1853–1890), though considered a Dutch painter, moved to England, Belgium, northern France, and finally southern France, where he produced a great many paintings during the last three years of his life. This fact, together with the differences in style and idiosyncrasies of individual artists living in the same region at the same time and in the same climate, will tend to diminish the average differences. Another "handicap" is the fact that the indoor climates of different art galleries, as well as the age of the paintings, may alter, to a lesser or greater degree, the colors of the skies represented. These and other factors that operate against the verification of the above hypothesis will result in minimizing the statistical differences (Neuberger, 1970).

The regions

Only the following regions (art schools) were involved in Neuberger's study. *American,* which is essentially the eastern seaboard from Richmond, Virginia, to Boston, Massachusetts, where 90% of the artists surveyed lived and worked. For this region the climate of New York City is considered representative, although the actual climate is different at both ends of the region. Finer detail of the climate did not seem to be warranted in view of the other factors mentioned above. *British,* with London representing the climate. *French,* with Paris representing the regional climate, although the same objection may be raised as in the case of New York City. *Low Countries,* combining Holland, represented by DeBilt, and Belgium, represented by Uccle: this combination is permitted in view of the great similarity of the climates as well as of the cultural aspects. *German,* including Austria, Switzerland and relatively few paintings from what is now the Czech Republic and Slovakia and adjacent countries – in other words, the central European region with Görlitz representing the regional climatic center. *Italian,* with Rome representing the climate. *Spanish,* represented by Madrid.

Figure A57 shows the regional differences of mean annual temperatures and precipitation for the seven regions, abbreviated with their initials, for the period 1951–1960. As will become evident in the evaluation of the paintings, the apparent similarity of the British and French climates is not always found in the paintings, largely because of the influence of the paintings by artists in southern France.

Climatic features in paintings

All paintings exhibited in each of the art galleries visited were evaluated with respect to the above-mentioned seven regions, the century and decade of the painting, the artist's name, the percent of canvas area covered by sky, the prevalent type and amount of clouds, the sky color (pale blue = 1; medium blue = 2; deep blue = 3), the visibility, i.e. the apparent transparency of the air painted in the picture (poor = 1; medium = 2; good = 3), and several other items. The qualitative values of sky color and visibility were subsequently converted into percentages by statistical means, so that the range from 0% to 100% would represent a very pale-blue to a very deep-blue sky and a dense fog to a very clear atmosphere, respectively.

With respect to sky color and visibility, the possible effects of aging of the painting, its illumination in the museum, and of the technique used by the painter were anticipated at the start. Pointillists, like the French artists Seurat (1859–1891) and Signac (1863–1935), would produce more or less hazy-looking atmospheres in their paintings. However, since these and other influences would tend to negate the hypothesis, they were considered tolerable. It should be noted that, for the statistical evaluations, 325 works painted before the year 1400 were eliminated, chiefly because they had gold backgrounds or showed no sky or other climatic features; thus, the total number of paintings was reduced to 11 959. Although this seems to be a substantial statistical population, in reality it is only a minuscule sample of the total output by all the artists of Western culture, which must amount to hundreds of thousands of paintings. A hundred collections in the United States alone hold 43 143 works of American painters.

Regional differences in blueness of sky

In Table A28 the average blueness in percentages and the ratios of frequencies of deep-blue (*d*) to pale-blue skies (*p*) are presented for a total of 4411 paintings; the ratio *d/p* amplifies the regional differences by disregarding the medium-blue category. There is no significant difference between regions A, L, and F, but region B has by far the fewest deep-blue skies in its paintings. Central European paintings (region G) have 30 more deep-blue than pale-blue skies, and the Mediterranean regions S and I show a large predominance of deep-blue skies.

Regional differences in visibility

The average visibilities in percentages are given by regions for a total of 5601 paintings in Table A29. The results are similar to those in Table A28 in that there is little difference between the regions A, L, and F. The British region has the greatest opacity of the painted atmosphere, and central and southern European paintings show the highest transparency of the air. The statistical significance levels are 0.1% for all pairs of regions, except for A-L and F-S, and no significance for the difference between G and S.

Regional differences in painted clouds

Of each of the 5805 paintings with clouds, only the dominant cloud type was recorded according to the four commonly used categories: high, middle, low, and convective clouds. In Table A30 the relative frequencies of each of the four cloud families are shown. The highest frequency for each region is in boldface type. The average frequencies for all regions combined are 8% for high, 35% for middle, 25% for low, and 32% for convective clouds, showing the artists' preference for middle and convective clouds. A notable exception is seen in the artists of the British region who show a significant preference for low clouds.

The cloud amounts were estimated in the four categories: (1) clear (no clouds); (2) scattered (one-half or less of the sky area covered by clouds); (3) broken (more than one-half of the sky area covered by clouds); and (4) overcast (no blue sky visible). Table A31 gives the relative frequencies (percentages) of 6483 paintings in the four categories by region.

Table A28 Average blueness of painted skies by region

	Region						
	A	B	L	F	G	S	I
Blueness (%)	40	35	44	45	54	79	70
Ratio *d/p*	0.5	0.3	0.6	0.7	1.3	8.1	3.9

Table A29 Average visibility by region

	Region						
	A	B	L	F	G	S	I
Visibility (%)	37	33	38	36	40	42	54

Table A30 Average frequencies of cloud families by region

	Region						
	A	B	L	F	G	S	I
High clouds	5	5	5	9	11	6	9
Middle clouds	**43**	23	**32**	**36**	**33**	**38**	**43**
Low clouds	30	**40**	27	27	27	28	14
Convective clouds	22	32	36	28	29	28	34

Table A31 Relative frequencies of cloudiness categories by region

	Region						
	A	B	L	F	G	S	I
Clear	5	0	7	9	16	13	12
Scattered	16	4	11	10	16	13	24
Broken	**47**	**48**	**44**	**49**	**42**	**38**	**42**
Overcast	32	**48**	38	32	26	36	22

All regions show a maximum in the "broken" category, except that British painters avoided clear skies and showed as many overcast as broken skies. All regional differences are significant at the 1% or better level.

The average cloudiness for each region was statistically converted into percent cloudiness from the percent frequencies in Table A31. The results again show that the British painters have the greatest preference for very cloudy skies, whereas the Mediterranean painters prefer skies with considerably less cloudiness, as is evident from Table A32. The differences among the regions are significant at better than the 1% level.

Painted versus observed visibilities by region

Although Table A29 established regional differences in visibility painted, the question arises concerning the extent to which these visibilities correspond to the climatically prevailing visibilities. Actual observations at the stations shown in Figure A57 are available only since 1850, so that, for comparison, only the painted data for the period 1850–1967 are usable. Because the painted visibilities were estimated in only three categories, the observed visibilities, which were recorded in different codes at the various stations, were also converted into three classes: 1 = visibilities less than 2.5 miles; 2 = those between 2.5 miles and 11 miles; 3 = those larger than 11 miles. Both sets of data were then expressed in percentages. Only 1677 paintings were available for the seven regions in the period 1850–1967. Table A33 shows that the observed data have a larger range and greater differences between regions. Both sets show minimum visibility for the Low Countries and maximum visibility for the Mediterranean regions. Nevertheless, considering the relatively small statistical sample and the coarseness of the estimates of painted transparency of the air, the fact that British painters paint a more transparent air than do the American and Low Countries artists seems to be supported by actual observations.

Painted versus observed cloudiness by region

For the comparison of painted with actually observed cloudiness, 1848 paintings were available for the period from 1850 to 1967. Figure A58 shows the observed versus the painted average cloudiness for each region. Although the artists exaggerate cloudiness, with the exception of the Italian painters, there is an excellent linear relationship between painted (P) and observed (O) averages with a correlation coefficient of 0.95 ± 0.03 and a regression equation $O_c = 0.5P + 26$ (O_c and P in percents). When we use the equation to compute O_c, we obtain the values in Table A34 in which the actually observed cloudiness averages (O) and the differences $O_c - O$ are given. These differences are well within the accuracy of the observations.

In summary, the quality of the relationships between the actually observed and the painted regional averages of visibility and cloudiness in the period from 1850 to 1967 permits the conclusion that the collective experience of climatic features by artists in a given region finds its expression in paintings that correspond well to the actually observed average values.

Table A32 Average percent cloudiness by region

	Region						
	A	B	L	F	G	S	I
Cloudiness	72	85	75	72	63	69	62

Table A33 Observed and painted visibilities in percent by region for the period 1850–1967

	Region						
	A	B	L	F	G	S	I
Observed	47	53	42	58	54	93	90
Painted	38	42	36	38	38	63	78

Table A34 Cloudiness (O_c) computed from the linear relationship between painted and actually observed (O) values

	Region						
	A	B	L	F	G	S	I
O_c (%)	60	68	65	62	61	51	44
O (%)	56	68	66	62	62	48	47
$O_c - O$	4	0	−1	0	−1	3	−3

Climatic changes in paintings

The last major ice age ended about 12 000 years ago. However, the warming trend during the postglacial period was not a steady, continuous process but had intermittant relapses into colder periods. The so-called Little Ice Age was the latest and probably the best-documented excursion of climate that affected the entire northern hemisphere (Lamb, 1967). There is disagreement as to when it started. It seems to have emerged slowly with the beginning of the sixteenth century, although there were some precursors of it in the preceding century. The end can be dated more easily, as a warming trend set in by 1850. In general, the Little Ice Age was an epoch of frequent wet and cold years. Southward displacement of storm tracks brought snow and ice into the Mediterranean regions and rapid advances of Alpine glaciers and other climatic disasters occurred all over Europe (Neuberger, 1970).

Lamb (1967) has found from his brief survey of Dutch and British paintings that cloudiness increased from around the middle of the sixteenth century. With the large volume of data used here in Tables A28, A29, A30, and A32, averages combining the data of all regions were calculated for three epochs: the preculmination period of the Little Ice Age from 1400 to 1549; the culmination period of the Little Ice Age from 1550 to 1849; and the postculmination period from 1850 to 1967. In Figure A59 these averages show that there was a substantial paling of the blue color of painted skies toward the culmination period, but no recovery afterwards. The same trend is true for visibility. The failure of sky blueness and visibility to improve in the last epoch is in part due to the effects of the techniques of impressionism, which created a hazy atmosphere by the uncertain contours resulting from short brush strokes. This method reached its extreme in pointillism. On the other hand, this trend should have been counteracted by expressionism with its bold colors and sharp contours. It seems reasonable to assume that in the postculmination period the progressive industrialization, with its attendant air pollution, diminished both the blueness of the sky and the transparency of the air. The combined frequencies of low and convective clouds, both of which may lead to

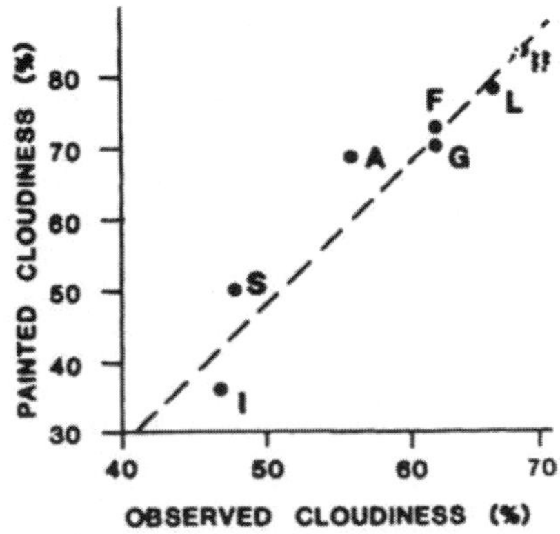

Figure A58 Relationship between the average regional cloudiness as observed and painted for the period 1850–1968.

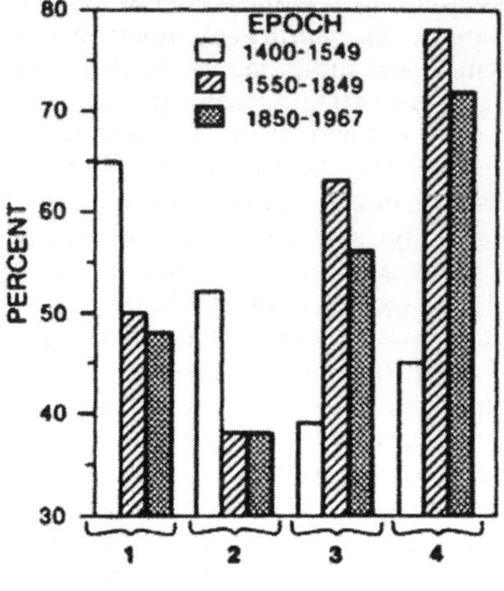

Figure A59 Averages for all regions of climatic features painted in the preculmination period, the culmination period, and the postculmination period of the Little Ice Age.

precipitation, show a strong increase from the first to the second epoch and a reduction in the third epoch. Similarly, the cloudiness has a large rise and a subsequent small decline, the latter probably being caused, at least to some extent, by human activities.

In conclusion, the collective climatic experience of artists in a given region at a given time is reflected in their paintings, notwithstanding the individual artist's techniques and intentions, or awareness of his or her surrounding climate. Subsequent physical and chemical effects of the climate on paintings are also reflected by the intensity and hue of the paint on the canvas. The importance of climate in the appreciation of landscape painting is now established and features in several recent art exihibitions such as "Impressionists in Winter" (Moffett, 1998); "Constable's Clouds" (Morris, 2000); "Holland Frozen in Time" (van Suchtelen, 2001).

Hans Neuberger and John E. Thornes

Bibliography

Aristophanes, 423 BC. *The Clouds*, ed. K. J. Dover. (Oxford: Clarendon Press, 1968).
Bonacina, L.C.W., 1939. Landscape meteorology and its reflection in art and literature, *Quarterly Journal of the Royal Meteorological Society*, **65**: 485–497.
Bone, J.D., 1976. Clouds in the poetry of four romantics. *Catalogue of the Cloud Watchers Exhibition*, Coventry City Art Gallery, Coventry, UK.
Botley, C.M., 1970. Climate in art. *Weather*, **25**: 289.
Brimblecombe, P., 1988. *The Big Smoke, A History of Air Pollution in London since Medieval Times*. London and New York: Methuen.
Brimblecombe, P., 2000. Aerosols and air pollution in art. *Proceedings of the Symposium on the History of Aerosol Science*, Vienna.
Brimblecombe, P., and Ogden, C., 1977. Air pollution in art and literature. *Weather*, **32**: 285–291.
Burhop, N.G., 1994. The Representation of Weather in Music. Unpublished MSc thesis, School of Geography, Earth and Environmental Sciences, University of Birmingham, UK.
Burroughs, W.J., 1981. Winter landscapes and climatic change. *Weather*, **36**: 352–357.
Durschmied, E., 2000. *The Weather Factor: How Nature has Changed History*. London: Coronet Books.
Frisinger, H.H., 1977. *The History of Meteorology to 1800*. History Monograph Series. New York: American Meteorological Society, Science History Publications.
Gedzelman, S.D., 1989. Cloud classification before Luke Howard. *Bulletin of the American Meteorological Society*, **70**: 381–395.
Heninger, S.K., 1968. *A Handbook of Renaissance Meteorology*. New York: Greenwood Press.
Janković, V., 2000. *Reading the Skies: a cultural history of English weather 1650-1820*. Manchester: Manchester University Press.
Lamb, H.H., 1967. Britain's changing climate. *Geography Journal*, **133**: 445–466.
Lamb, H.H., 1968. The climatic background to the birth of civilization. *Advances in Science*, **25**: 103–120.
Lamb, H.H., 1977. *Climate: Present, Past and Future*, vol. 2. New York: Barnes & Noble.
Lamb, H.H., 1995, *Climate, History and the Modern World*. London: Methuen.
Landsberg, H., 1958. *Physical Climatology*. DuBois, PA: Gray Printing.
Lhote, H., 1961. Rock art of the Maghreb and Sahara. In Bandi, H.G., et al. eds., *The Art of the Stone Age*. New York: Crown, pp. 99–152.
Livingstone, D.N., 2002. Race, space and moral climatology: notes towards a genealogy. *Journal of Historical Geography*, **28**: 159–180.
Moffet, C.S., 1998. *Impressionists in Winter*. Washington, DC: Philip Wilson.
Morris, E., 2000. *Constable's Clouds*. Edinburgh: National Gallery of Scotland.
Neuberger, H., 1961. Meteorological imagery in language–music–and art. *Mineral Ind.* **29**: 1–8.
Neuberger, H., 1970. Climate in art. *Weather*, **25**: 46–56, 61.
Spink, P.C., 1970. Climate in art. *Weather*, **25**: 289.
Suchtelen van, A., 2001. *Holland Frozen in Time*. Zwolle, Holland: Waanders.
Thornes, J.E., 1979. Landscape and clouds. *Geography Magazine*, **51**: 492–499.
Thornes, J.E., 1999. *John Constable's Skies*. Birmingham: University of Birmingham Press.
Thornes, J.E., and Metherell, G., 2003. Monet's "London Series" and the cultural climate of London at the turn of the twentieth century. In Strauss, S., and Orlove, B.S., eds., *Climate, Weather and Culture*. Oxford: Berg.
Wagner, R., 1841, *Kunst und Klima* (Art and Climate). *Samtliche Schriften und Dichtungen*, **3**: 207–221.
Walsh, J., 1991. Skies and reality in Dutch Landscape. In Freeberg, D., and de Fries, J., eds., *Art in History, History in Art*, ed by Chicago: University of Chicago Press.
Watson, L., 1984. *Heaven's Breath: a natural history of the wind*. London: Coronet Books.
Wilde, O., 1889. *The Decay of Lying*. In *The works of Oscar Wilde*, 1987 edition, Leicester: Galley Press.

Cross-references

Cultural Climatology
Literature and Climate

ASIA, CLIMATES OF SIBERIA, CENTRAL AND EAST ASIA

Central Asian, Siberian, and East Asian mega climatic regions, located on the world's largest and most physically diverse continent, are experiencing a period of marked climatic change. The impact of climatic change is affecting the lives of nearly two billion people. With more climatic types than any other continent, and with one-third of the world's land mass, any perturbation of Asian weather and climate is biophysically and socioeconomically important. Containing the highest point above sea level and the coldest inhabited place on earth, Asia's location and geomorphology contribute much to climatic diversity. Major variations of climatic types are defined by receipt of solar energy, moisture from the Atlantic Ocean in the west and the Pacific Ocean in the east, and the powerful Siberian Winter High Pressure Cell. A decrease in temperature occurs from south to north, with an increase in continentality from west to east. Latitudinal variations of climate are significant.

Asian climates

Using the Köppen climatic classification, north and east Asia may be divided into three major climatic realms: Siberia or Northeast Asia in the north (Dfb, Dfc, Dwa, Dwb, Dwc, Dwd, and ET); Central Asia or Desert Asia in the west and center (BS and BW); and East Asia or Monsoon Impacted Southeast Asia (Cfa and Cw). Siberia or Northeast Asia is the largest. A complex set of interactive climatic controls and many multifaceted physiographic features influence local climates and give distinctive regional climatic character to places (Figure A60). One of the most impacting and heat-determining factors is the length of day and night at different latitudes, particularly in Siberia and Central Asia (Table A35).

Siberia extends from the Ural Mountains in the west to the Pacific Ocean in the east. It is bounded on the north by the cold Arctic Ocean and on the south by Central Asia and South Asia. The climate of Siberia is one of the most continental on Earth, with great seasonal changes. The average frost-free period is only 75–82 days. Average annual precipitation ranges from 250 mm to over 2000 mm. During the winter Siberia is divided into two climatic parts: the northwest, dominated by frequent storms, high winds, and snow; and the southeast, dominated by clear skies and calm weather. A major part of Siberia is underlain by continuous permafrost, up to 1600 meters deep and extremely cold (−12°C). These factors impact temperature efficiency and moisture effectiveness in this region. The seasonal snow-melt layer is dynamical in space and time, impacts heat energetic conditions, and serves as diagnostic criteria for estimating global change or climate. Since the late 1960s a decrease in snow cover, a reduction in annual duration of lake and river ice cover, a temperature decrease, and significant changes in precipitation, cloud cover, temperature ranges, and drought frequency have occurred (Dando et al., 2003).

Table A35 Maximum lengths of day and night at different latitudes (excluding twilight)

North latitude	Daylight (in days)	Darkness (in days)
90°	189	176
85°	163	150
80°	137	123
75°	107	93
70°	70	55
67°38'	54	under 24 hours
60°	18 hours 27 minutes	18 hours 27 minutes
50°	16 hours 18 minutes	16 hours 18 minutes

Source: Kriuchkov, 1973 and Lounsbury, 1973.

Table A36 Number of days with urban fog in selected Siberian cities

1.	Anadyr	33	Evenly distributed annually
2.	Bratsk	97	Fall and winter primarily
3.	Irkutsk	102	Primarily in winter
4.	Novosibirsk	31	Primarily in late fall
5.	Petropavlovsk-Kamchatskii	48	Primarily in summer
6.	Salekhard	48	Primarily in winter
7.	Surgut	47	Primarily in winter
8.	Verkhoiansk	56	Primarily in winter
9.	Yakutsk	62	Primarily in winter

Source: Field work, interviews, archival research, and Kriuchkov, 1973.

Along with a change in Siberia's climate, and an increase in urban conglomeration, urban climates have impacted all biophysical processes. Increased atmospheric humidity, increased absolute humidity, and extensive use of firewood and low-quality coal for heating and for power have induced a phenomenon that is detrimental to all cultural and physical activities – urban fog (Table A36).

East Asia, recording atypical weather perturbations, including persistent drought in normally moist areas and devastating floods in normally dry regions, is experiencing an unusually strong manifestation of global warming. Changes in high-pressure cell positions and characteristics, shifts in low-pressure cell locations and intensities, and a variation of jet stream seasonal paths have modified weather patterns. Warming of the atmosphere has increased evaporation, increased specific humidity, intensified drought, increased the amount of moisture in air masses, and increased the amount of moisture and energy in the air for rainstorms, snowstorms, typhoons, and tornadoes. Concomitantly, extreme cold temperatures in early winter, followed by anonymously warm temperatures in late winter, have disrupted natural processes in eastern Mongolia and northeastern China. Changes in precipitation regimes in spring, summer, and autumn in China and Japan have enhanced the potential for flooding. A combination of human activities that modify climatic elements and controls and natural causes of climatic change could lead to large-scale weather-related disasters and to a redefinition of the traditional boundaries of East Asia's climatic regions (Figure A61).

Climatic changes

In the 1999–2002 period a severe drought persisted across a large part of Central Asia, portions of south and southwest Asia, including Afghanistan, Iraq, Iran, Saudi Arabia, Syria, Israel, Turkmenistan, Uzbekistan, Kazakhstan, Tajikistan, Pakistan, and parts of India (Lawton, 2002). At the time it was the largest contiguous region of severe drought in the world. The International Research Institute for Climatic Prediction considered this widespread drought a "grave humanitarian crisis". Scientists from the Ohio State University analyzing ice cores from glaciers in Central Asia noted a warming in this region over the past 50 years. At one site the recent warming and drying trend exceeded anything observed in the past 12 000 years. Current global warming has exceeded the normal range of climatic variations during the last 5000 years (Jones, 2001). Warming trends in Central and Desert Asia are not just natural fluctuations that will reverse quickly. Concomitant with the Pacific (SO) and North Atlantic Oscillations (NAS), humans have contributed to climatic changes in Central Asia. Drying of the Aral Sea is one of the best-known environmental disasters

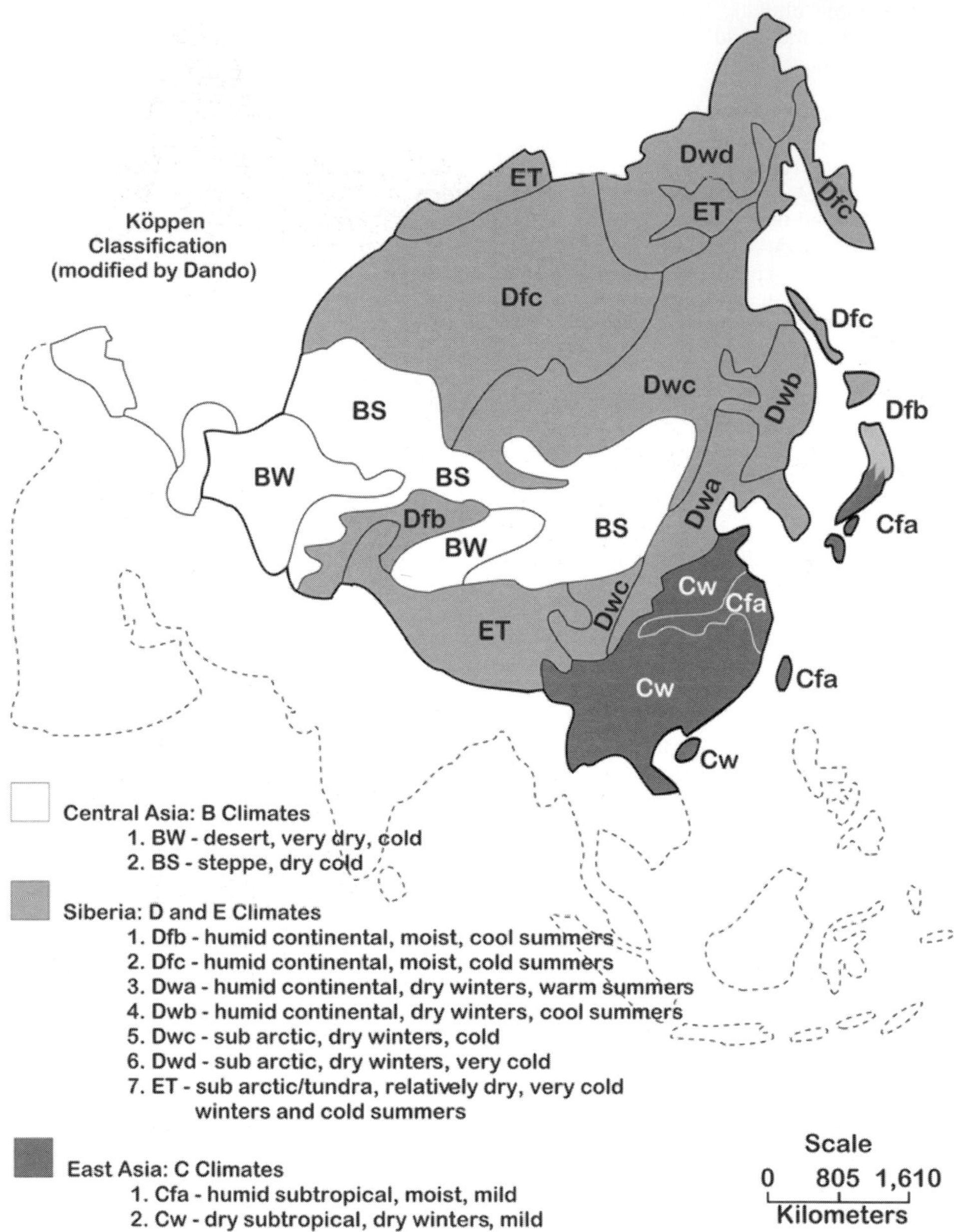

Figure A60 Asian climates (2003). Anna R. Carson 2003.

caused by humans in the past three or four decades. There is widespread agreement in the scientific community that, with the reduction in Aral Sea level, the regional climate has changed for the worst and become more extreme.

Climatic elements

The unequal distribution of solar radiation over Asia is the primary factor in Asia's multifaceted weather and climate. In Asia's tropical belt the sun remains high with little seasonal variation, and this accounts for continuous warm to hot year-round temperatures. In Asia's mid-latitudes, solar radiation receipts exhibit a strong seasonal maximum and a strong seasonal minimum which are reflected in greater seasonal variations in temperature than the tropical belt. And in Asia's high latitudes there is a period of limited to no solar radiation received at the surface of the Earth, resulting in a season with extremely low temperatures in the winter or low sun period (Figures A62 and A63). Total solar radiation received at the Earth's land – sea surface during the entire year in kcal/cm^2 per year ranges from 140 to 180 in Asia's tropical belt, 160 to 120 in Asia's mid-latitudes, and 100 to 60 in Asia's high latitudes. Maximum solar radiation received at the Earth's surface occurs in the steppe and desert regions of west-central Asia. Transformation of available solar

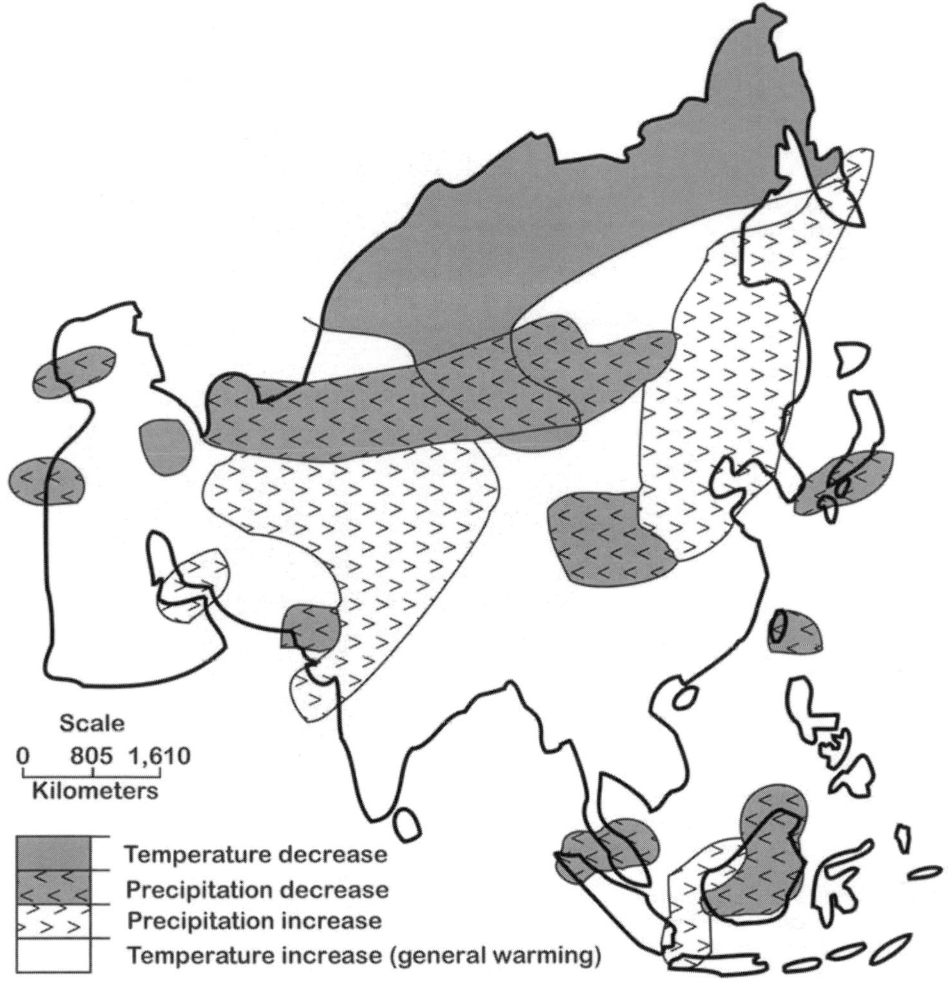

Figure A61 Trends in temperature and precipitation change: 1901–1998. Anna R. Carson 2003.

radiation is an essential ingredient of the process that produces Asia's climate – particularly temperature ranges (Borisov, 1965).

Temperature

Northern Asia, specifically Siberia, is climatically isolated from moist tropical air masses and records the greatest mean annual temperature range on earth (Table A37). At Verkhoyansk, in the valley of the Yana River, January temperatures average −49°C, July temperatures average +15°C, and the absolute temperature range is 103°C. Oymyakon, located in the same physiogeographic region, records an absolute temperature range of 104°C. Siberian winter low temperatures are proverbial, and for most of Siberia the January mean annual temperature is less than −25°C. In contrast, much of southwestern Asia is subject to moderate heat (Arakawa, 1969).

The temperatures of eastern China resemble those of the eastern United States. Winter cP and cA air masses from Siberia are more powerful and colder than those that move south from Canada. Summer mT air masses from the South

Table A37 Average annual temperatures at selected Siberian stations (in °C)

Station	Latitude (N)	Longitude (E)	Average annual Temperature
Salekhard	66°31′	66°35′	−6.6
Surgut	61°17′	72°30′	−4.0
Tiumen'	57°10′	65°32′	+1.0
Novosibirsk	54°58′	82°56′	−0.4
Igarka	67°27′	86°35′	−8.5
Eniseisk	58°27′	92°10′	−1.9
Bratsk	56°04′	101°50′	−2.7
Irkutsk	52°16′	104°19′	−1.1
Verkhoiansk	67°33′	133°25′	−16.0
Takutsk	62°01′	129°43′	−10.3
Baunak	54°43′	128°52′	−5.0
Nerchinsk	52°02′	116°31′	−3.8
Anadyr Petropavlovsk	64°27′	177°34′	−8.1
Kamchatskii	52°33′	158°43′	+0.6
Sovetskaia Gavan'	48°58′	140°17′	−0.5

Source: Nuttonson, 1950, appendices.

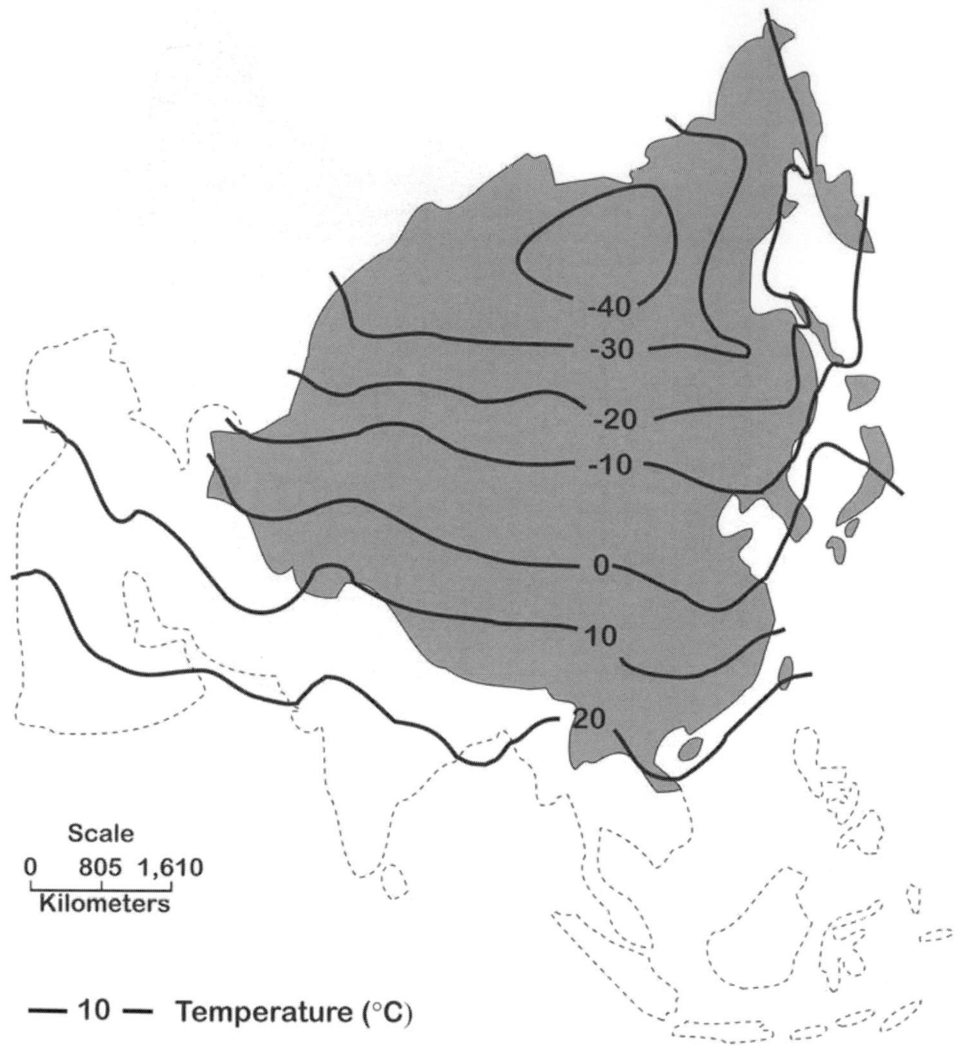

Figure A62 Average temperature: January. *Source: Geograficheski Atlas,* 1959, Moskva, p. 14; *Preroda i Resoursi Zemli,* 1999, Moskva and Vienna, p. 52. Anna R. Carson 2003.

China Sea are stronger than mT air masses from the Gulf of Mexico. China extends farther north to south than the United States and has a wider range of climates and temperature regimes. China extends from 18° to 53° north latitude. This is the equivalent distance from Puerto Rico to Labrador. Concomitantly, no simple summary can give adequate insights into Japan's temperature regimes. The Japanese islands extend for 1400 kilometers or more and are surrounded by temperature-modifying ocean and seas. Irregularities in island topography induce sharp vertical temperature contrasts. The winter monsoon typically brings cold air masses and low temperatures from Siberia, while the summer monsoon brings warm, moist maritime air masses and warm, mild temperatures to Japan. The normal January temperature gradient is slightly more than 2°C per degree of latitude; the normal July temperature gradient is less than 1/2°C per degree of latitude.

Asia's spatial temperature differences and ranges are greatest in winter, least in summer, and more extreme in all seasons within the land-locked core, rather than the south and eastern maritime periphery (Figures A62 and A63). Winter isotherms reflect the influence of ameliorating ocean currents, mountain barriers, altitude above sea level, high-pressure cells, and solar radiation. Summer isotherms reflect solar radiation and latitude, along with altitude (Global Change, A and B, 1999).

Precipitation

Precipitation receipts, in general, increase from north to south and from the southwest to the southeast. Throughout most of Siberia the annual average precipitation is scarcely 250 mm and in parts of Central Asia, the Tarim Basin, and the Gobi Desert, less than 100 mm. In east China there are regions that receive

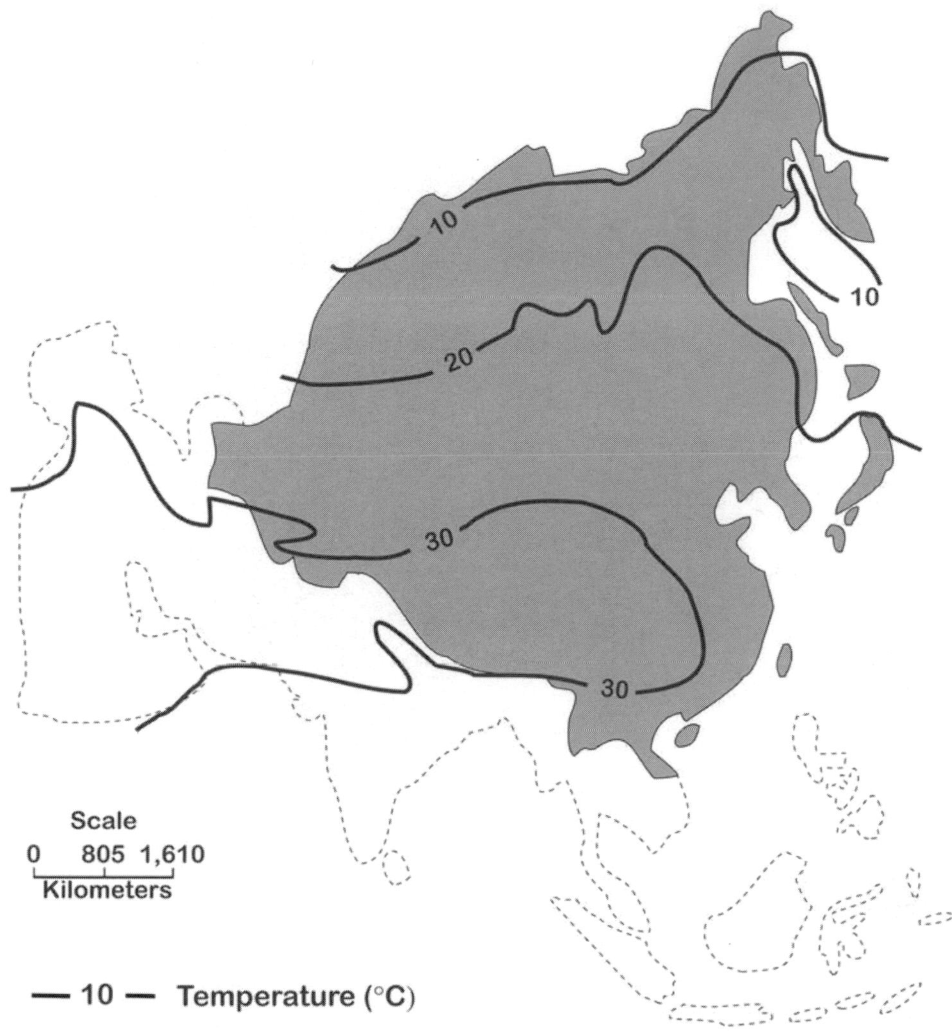

Figure A63 Average temperature: July. *Source: Geograficheski Atlas,* 1959, Moskva, p. 14; *Preroda i Resoursi Zemli,* 1999, Moskva and Vienna, p. 52. Anna R. Carson 2003.

more than 2000 mm per year (Figure A64). A small part of China, north of Nepal, receives over 3000 mm annually. Seasonal distribution of precipitation becomes of greatest importance, for the summer heat and continental character of Asia affect precipitation effectiveness and thermal efficiency. In insular and island-studded southeast and east Asia, where there is limited or no frost, rainfall is relatively evenly distributed throughout the year (Cfa and Dfb). Interior and rainshadow locations in south and east Asia experience a distinct dry season in the low-sun period or winter (CW, Dwa, and Dwb). Mediterranean Sea-influenced west Asia and the dry eastern coastal region of the Caspian Sea receive most of their precipitation in winter, at least three times as much precipitation in the wettest winter month as the driest month of summer (Cs). Precipitation in the dry realms of southwest Asia, Central Asia, Tarim Basin, and the Gobi Desert is minimal and erratic, but most of the precipitation is secured in summer from violent convective showers (BW) or a combination of violent convective showers and frontal activity (BS). And in the subarctic and arctic areas of Siberia (Dwc, Dwd, and ET), summer is the season of maximum temperatures, highest specific humidity, deepest penetration of maritime air masses under the influence of the summer monsoon, and is the period of maximum precipitation (Figures A62 and A63). Asia's central dry realm separates Asia's eastern warm-to-hot wet belt from Asia's north and northeastern cool-to-cold limited precipitation belt (Griffiths and Driscoll, 1982).

Barnaul – an example of a precipitation regime impacted by continentality

Barnaul is located in Siberia, west of the Ural Mountains, in the extreme eastern portion of the zone of the Russian forest steppe. Barnaul's maximum monthly precipitation falls in July that is also the warmest month of the year. Precipitation is

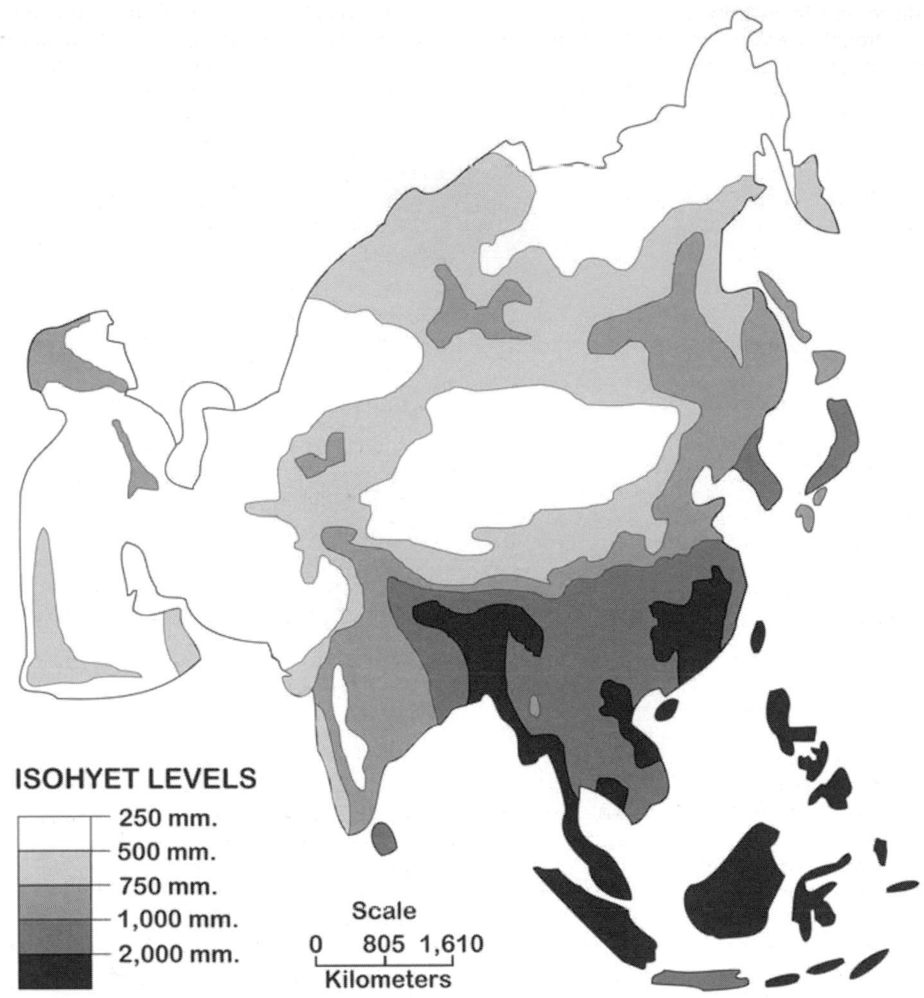

Figure A64 Asian precipitation. *Source: Geograficheski Atlas*, 1959, Moskva, p. 16; *Preroda i Resoursi Zemli*, 1999, Moskva and Vienna, p. 56; *Oxford Economic Atlas of the World*, 4th edn, 1972, p. 2. Anna R. Carson 2003.

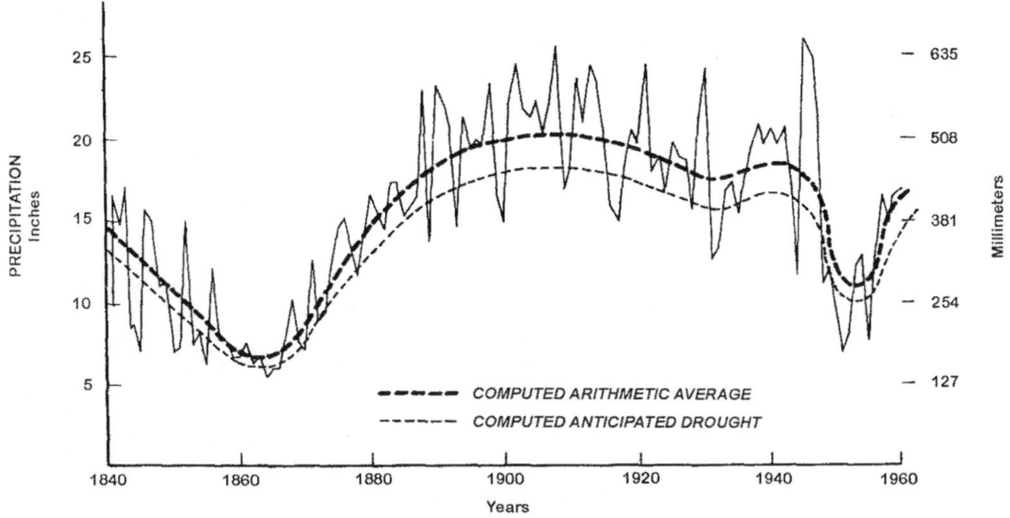

Figure A65 Barnaul – variation in precipitation, 1840–1960. *Source:* A.M. Shulyin, *Izv. AN USSR, serila geogr.*, 1963, no. 2, p. 168. Anna R. Carson 2003.

marginal for agriculture and for supplying moisture for economic development. Droughts are common, occurring once every 4 years on the average (Figure A65). A review of the climatic records for Barnaul from 1890 to 1966 shows that droughts were recorded in 1900–1902, 1904–1905, 1909–1910, 1917, 1920, 1923, 1927, 1929, 1931–1932, 1934, 1939, 1945, 1951–1952, 1955, and 1962–1963 (Figure A65). Analysis of precipitation data for Barnaul reveals a tendency toward the repetition of very moist years when all four seasons are wet or three seasons very wet and then a dry or very dry year or two followed by an average year or two.

Sukhovei – an Asian atmospheric enigma

Hot desiccating winds at times become associated with droughts, but these hot, dry winds or *sukhovei* can occur when there is no drought and can cause death to plants even when there are adequate reserves of moisture within the soil.

Sukhovei are hot, dry masses of air that envelop plants, increase evapotranspiration rates by factors of 2–3 to 20 times, and injure or destroy crops. Air in a *sukhovey* is so devoid of moisture that, in general, the relative humidity is less than 30% and, in many cases, less than 20% for an extended period of time. In a drought there are day and night variations in relative humidity, and at night dew may even form. In a *sukhovey* the relative humidity varies very little during the entire day/night period, and evaporation takes place continuously. The majority of cultivated plants cannot endure desiccation of their tissues; therefore the moisture lost to the atmosphere must be replaced by an inflow of moisture from the roots. Plants, unable to compensate for losses as a result of increased evaporation, lose turgescence, wilt, and die.

Sukhovei winds may reach speeds in excess of 70 miles per hour, but the usual velocity is from 7 to 11 miles per hour. Plant transpiration is increased by the most moderate wind speeds via the removal of water vapor from the apertures of the stomatal apparatus. Even wind with a velocity of 4–5 miles per hour increases plant transpiration threefold. When *sukhovei* winds achieve great speed they often turn into dust storms or so-called "black storms", which remove vast amounts of unconsolidated material and even small plants over great distances. As much as 5 inches of soil have been removed from various steppe regions in one year by dust storms. *Sukhovei* winds may last for an hour or two, or they may last for several days, but the end result is the same: dry winds increase evaporation and disrupt the water balance (the ratio of the water inflow from roots to that transpired to the atmosphere) of the plant tissue.

The origin of *sukhovei* remains an enigma, but there is a general agreement that both droughts and *sukhovei* develop under the south-central edge of displaced anticyclones, anticyclones 10° to 20° north from their usual summer position. Displaced anticyclones with diverging, clockwise air masses are characterized by little cloud cover, low relative humidity (dropping, in some instances, to 10–17%), and the absence of rainfall. Intense solar radiation and very high temperatures create moisture deficits and desiccation of the soil.

Basic atmospheric circulation

Air masses

The basic atmospheric circulation features involved in giving regional character to Asia's climates are the movement of air masses, transformation of air mass properties, and interactions between them along fronts. Asian air masses show great contrasts in temperature, moisture, and density. At any particular site the properties of air masses depend not only upon the nature of the source region but also the modification the air mass experienced en route from the source region. En route air mass modifications are of great importance in determining the nature of weather associated with an air mass (Lydolph, 1977; Takahashi and Arakawa, 1981). Some typical characteristics of air masses and their source regions are as follows:

1. Continental Arctic (winter) air masses are stable, intensely cold (−55° to −35°C), extremely dry, with inversions; cA air masses enter Siberia from the Arctic seas and at times penetrate to the Pacific Ocean; they are associated with clear skies, low temperatures, and dense air.
2. Continental polar (winter) air masses are very stable with strong temperature inversions and are cold (−35° to −20°C) and dry; they spread to the north and to the south of the "great ridge" of high pressure in Siberia and Outer Mongolia at approximately the 50th parallel and dominate the weather of the entire continent; generally cP air masses produce clear, cold, and cloudless weather.
3. Maritime tropical (winter) air masses are, for the most part, unable to penetrate eastern Asia in winter due to the intensity of the great Siberian High Pressure Cell; mT air masses that influence the weather in central Asia in winter circulate around the eastern edge of the Azores – Bermuda High Pressure Cell where upper level subsistence is strong; mT air masses are cool, dry, and stable in winter.
4. Maritime polar (winter) air masses have great difficulty penetrating deep into the Asian continent in winter, for the Siberian High Pressure Cell and strong outflowing winds in eastern Asia inhibit mP incursions. The mP air masses basically are conditionally stable, cool (0–10°C) and moist; they have significant influence upon the weather of maritime eastern Siberia, Manchuria, and Korea.
5. Continental polar (summer) air masses are, in general, stable or conditionally stable, cool (5–15°C) and somewhat moist; east and south Asia are so dominated by mT air in summer that cP air contributes little to the weather of that region.
6. Maritime polar (summer) air masses are a contributing factor to the summer weather of Manchuria, eastern Siberia, and Japan, and are, in most cases, stable to conditionally stable, mild (2–14°C), and humid; conflicts between mP and mT air masses in summer lead to the formation of a semi-stationary front in northeast Asia with associated overcast and drizzly weather.
7. Maritime tropical (summer) air masses initially are conditionally stable and very moist; mT air masses provide moisture for the summer monsoon and in spring invading mT air meets cP air in central and south China producing very active cyclogenesis; mT air masses dominate the weather of eastern Asia during the high-sun period.
8. Continental tropical (summer) air masses develop over southwest Asia in summer and are conditionally stable, dusty, very hot (30–42°C), and hold moisture; cT air flows north and northwest in summer into western Asia, advecting heat and absorbing moisture en route, plus contributing a characteristic opalescent haze to the local climates.

Asian air masses show great contrasts in temperature, moisture, and density. At any particular site, properties of air masses depend not only upon the nature of the source region but also the modification the air mass experienced en route from the source region. En route air mass modifications are of great importance in determining the nature of weather associated with an air mass (Lydolph, 1977; Takahashi and Arakawa, 1981).

Siberian high-pressure cell

General atmospheric circulation over Asia is controlled by centers or cells of high or low pressure whose axes are, in general, east–west and whose pressure centers vary drastically from winter to summer. In winter an extensive and well-developed high-pressure cell, centered over Mongolia, dominates the weather over most of east Asia (Figure A66). Triangular in shape, with the apex extending in the west to the Caspian Sea and the base anchored in the northeast near the Verkhoyansk Mountains of Siberia and in the southeast near the Chin Ling Mountains of east-central China, the Siberian High effectively blocks penetration of moisture-bearing, moderating maritime air masses in winter. Acting as a wedge forcing air masses to skirt northeasterly from the Black Sea across much of northern Siberia and to flow southerly along the Kamchatka Peninsula toward northern Japan, this intense high-pressure cell generates continental, land-trajectory, dry, cool, low-level air masses which surge from the north and northeast to the southwest across all of south and east Asia. The Siberian High, whose core and area of highest pressure is focused upon Lake Baikal, pulsates in intensity and breaks at times into smaller high-pressure cells of less intensity. February marks the height of the Siberian High's dominance of the winter circulation over Asia. The Siberian High weakens and shifts its center westward into a position over northeastern Central Asia in April, then dissipates in May. The weather map of Asia begins to be dominated by an intensive, thermally induced low-pressure cell over southwestern Asia and focused upon the tip of the Arabian Peninsula, the Iranian Plateau, and the Thar Desert of Pakistan (Davydova et al., 1966).

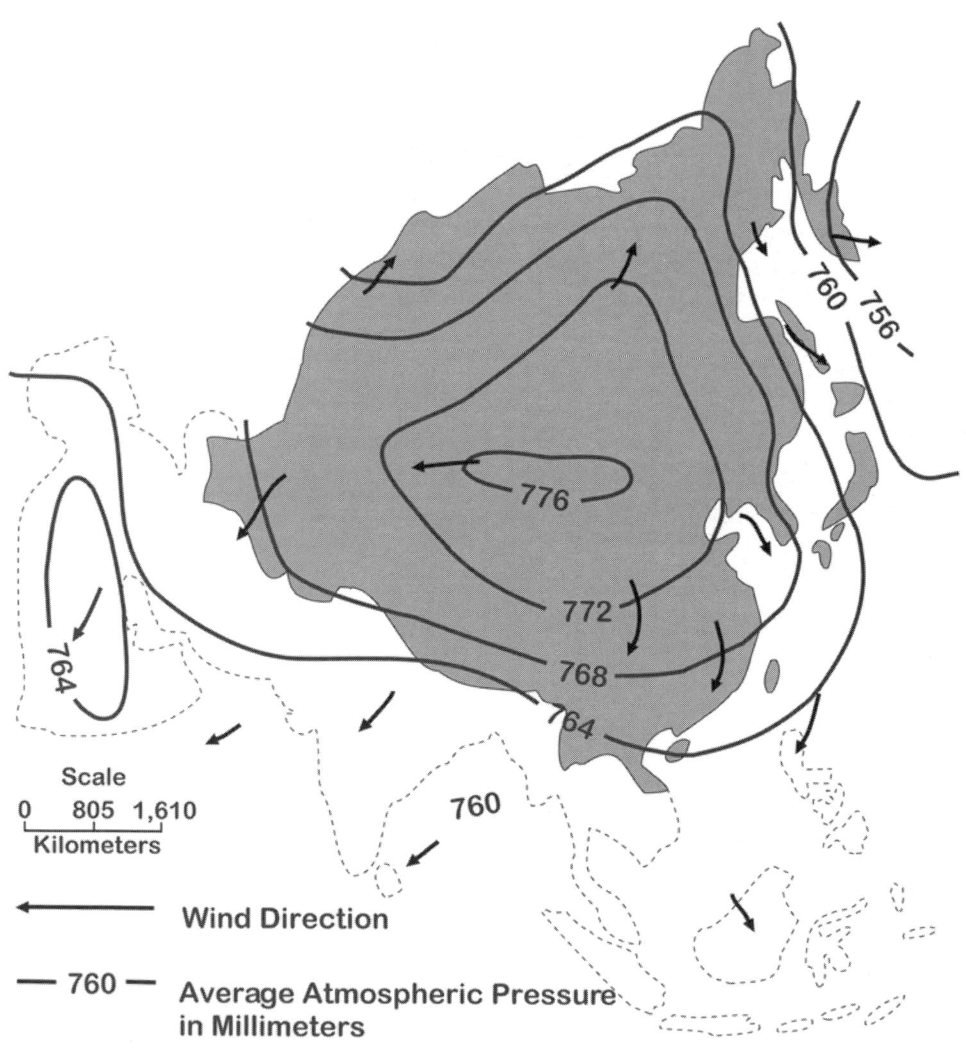

Figure A66 Atmospheric pressure and winds: January. *Source: Geograficheski Atlas*, 1959, Moskva, p. 15; *Preroda i Resoursi Zemli*, 1999, Moskva and Vienna, p. 51. Anna R. Carson 2003.

Southwest Asian low

In summer, east Asia's weather is dominated by a large, deep, thermally induced low-pressure cell that extends from the Arabian Peninsula to central China and from central India to Central Asia – centered approximately 30°N latitude and 70°E longitude (Figure A67). A complex cell, the Southwest Asian Low experiences east–west locational oscillations and occasional intense pressure deepening. This low in summer has the effect of interrupting the subtropical high-pressure system in the northern hemisphere by dividing the globe-girdling zonal band into two distinct large oceanic cells. One intense depression, the Southwest Asian Low, induces a radical change in prevailing winds and storm tracts during the high-sun period. Air masses from the stable eastern end of the Azores–Bermuda High Pressure Cell skirt this low from a north to northwesterly direction across the eastern rim of the Arabian Peninsula; less intense air masses from the northern quadrant of the Azores–Bermuda High sweep eastward across Turkey and into the northern extremities of Central Asia. Air masses and storms spawned under the western unstable quadrant of the Hawaiian High sweep from the south and southeast in a northerly trajectory across Japan, extreme eastern China, and the Russian Maritime Provinces. But the most constant and climatically significant air masses and storms are advected to this intense low-pressure cell as southwesterly trade winds spawned from a semipermanent high-pressure cell in the southern hemisphere over the Indian Ocean. In India and east Asia this modification of the general planetary wind system in summer constitutes the Asian monsoon (Tsuchija, 1964; Chang, 1967).

Frontal dynamics

Frontal zones and wind systems conform to the location and circulation of major air masses. In winter three major zones of cyclonic activity are distinguished over Asia. One zone is located

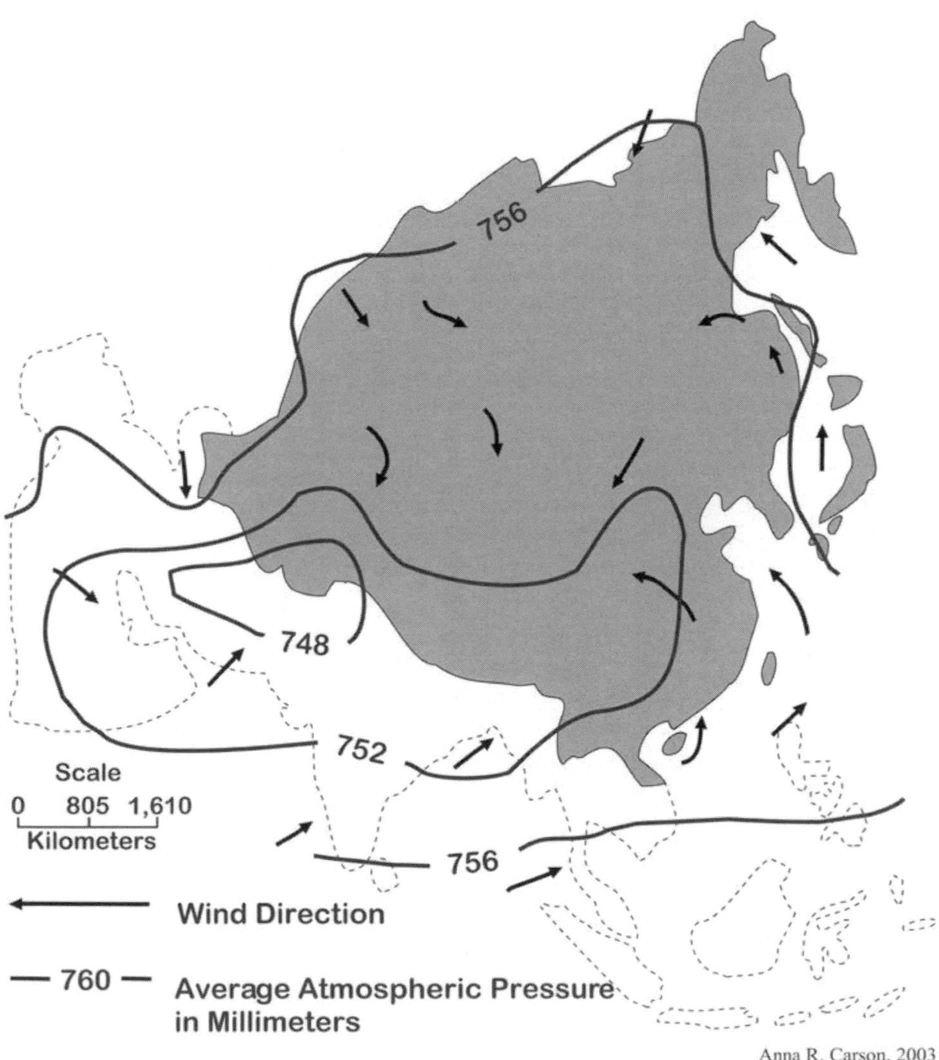

Figure A67 Atmospheric pressure and winds: July. *Source: Geograficheski Atlas,* 1959, Moskva, p. 15; *Preroda i Resoursi Zemli,* 1999, Moskva and Vienna, p. 51. Anna R. Carson 2003.

along the Asiatic Arctic Front, well above the Arctic Circle in northern Siberia, extending along the shores of the continent. This front fluctuates greatly and Arctic air masses, at times, penetrate east-central Asia. The second zone is along the Southwest Asian Polar Front that develops in winter over the Mediterranean Sea and extends to the Caspian Sea. The third zone is along the East Asian Polar Front that aligns itself in a northeasterly path from extreme south Asia toward Japan. Along these winter frontal zones, and moving at various directions and speeds, depressions and anticyclones impart to the climates of Asia that special character by which one area differs from another.

During summer, three major zones of pronounced cyclonic activity can be identified. The first is the southward-displaced Asiatic Arctic Front that at times extends east–west across northwestern Siberia along the 70°N parallel. A second zone of pronounced cyclonic activity is the Asian Polar Front that normally extends from the eastern tip of Lake Balkash in Central Asia, over Mongolia, to the northernmost bend of the Amur River along the 50°N latitude. This oscillating and at times southward-dipping Asiatic Polar Front zone has been called the "barometric backbone of Asia" in summer or the "great ridge", for a large number of anticyclones is observed each year between 50° and 55°N latitude. The third major zone is along the South Asian Intertropical Front. Extending over China at approximately 25°N latitude, the South Asian Intertropical Front is well defined in some locations, but in others it is weak or absent. In this belt, convergence of surface winds result in large-scale lifting of warm, humid, relatively unstable air, producing numerous weak, rain-generating disturbances (Trewartha, 1981). Along major frontal zones are found pronounced horizontal variations in temperature, humidity, and stability – usually jet streams. Strong, narrow jet stream currents, thousands of kilometers long, hundreds of kilometers wide, and several kilometers deep, are concentrated along a nearly horizontal axis in the upper troposphere or stratosphere, producing strong vertical and lateral shearing action.

Jet streams

Three distinct jet systems, the Polar Front Jet, the Subtropical Jet, and the Tropical Easterly Jet, have a major impact upon weather and climate in Asia. Latitudinal location of these jet systems, especially the Polar Jet, shifts considerably from day to day and from season to season, often following a meandering course. But in general, the Polar Front Jet gives rise to storms and cyclones in the middle latitudes of Asia, the Subtropical Jet noted for a predominant subsidence motion gives rise to fair weather, and the Tropical Easterly Jet is closely associated with the Indian monsoon. Over most of Asia the Polar Front Jet and the Subtropical Jet are most intense on the eastern margin of the continent and are best developed during winter and early spring. Jet stream wind speeds up to 500 km/h have been encountered over Japan. The Subtropical Jet, at times in winter, forms three planetary waves with ridges over eastern China and Japan. It frequently merges with the Polar Front Jet producing excessively strong jet wind speeds. During summer the Subtropical Jet loses its intensity and is mapped only occasionally.

East Asian monsoon

The East Asian monsoon is a gigantic multifactored, multifaceted, complex weather system, composed of diverse heat and moisture cycles intimately related to local topography, modified air masses, quasistationary troughs and ridges, and jet streams. Weather over most of eastern Asia during winter is dominated by the Siberian High-Pressure Cell and outflowing continental air masses. Low-level air flow is mainly from the north and is cold, dry, and stable. Successive waves of cold northeasterlies commence in late September and early October, progressing farther and farther southward, reaching the south China coast by late November or early December. In the higher mid-latitudes of Asia during the low-sun period, a steep north–south pressure gradient persists. During March and April the Siberian High gradually weakens, and incursions of moist maritime tropical air from the south and east replace the cold-to-cool northeasterlies, producing widespread stratus clouds, fog, and drizzle that may persist for days (Das, 1968).

In summer a combination of the deep and elongated Southwest Asian Low-Pressure Cell extending from the Arabian Peninsula to China, and the South Asian Intertropical Front, reaching its maximum poleward displacement, sets the stage for a marked seasonal reversal of air flow. Warm, moist, and conditionally unstable southwesterly maritime air masses from relatively cooler oceanic source regions eventually overcome blocking atmospheric conditions, and flow northwestward over China and Japan. Considerable convective activity develops over land, and heavy showers and thunderstorms contribute largely to the summer rainfall maximum of the region. Characteristics and attributes and onset and duration of the monsoon are site-specific. Duration of the summer monsoon in China varies between north and south. Dates of monsoon onset and percent of annual average rainfall received have been a focus of numerous studies in the past century. A synthesis of these studies is given in Table A38.

In all cases low-level wind patterns and resultant weather in summer are complicated by topography. Distance from air mass source regions, moisture content of air masses, and orographic barriers and atmospheric disturbances associated with cyclonic or inter-air mass convective activity determine distribution and quantity of precipitation. There is a pronounced difference between the East Asian and South Asian monsoonal weather. The East Asian winter monsoon is much stronger than the South Asian winter monsoon, and the East Asian summer

Table A38 Dates, duration and percent of rainfall by region

Location	Dates	Duration	Rainfall
1. South China	Mid-April; end October	6½ months	75–90%
2. South Central China	Mid-May; mid October	5½ months	60–80%
3. North Central China	June; end August	3 months	60–75%
4. North China/Mongolia	June; end August	3 months	Less than 60%

monsoon is much weaker than the South Asian summer monsoon (Trewartha and Horn, 1980).

Storms

Local winds

Owing to the great variety in relief and exposure, Asia is subject to numerous local winds and their associated weather. Ubiquitous in Asia are the warm, dry, gusty, downslope *foehn* winds generated, in most cases, by passing atmospheric disturbances in highland or mountainous areas. For example, in Central Asia and Siberia, *foehn*-like winds experienced along the Caspian Sea are locally called *germich*; in Uzbekistan, *Afghanets*; in Tadzhikistan, *harmsil*; in and near the great Fergana Valley, *ursatevskiy* and *kastek*; and along the Kazakh–Chinese border at the Dzhungarian Gate east of Lake Balkhash, *evgey*. Winds are given a medley of names, and the largest and most pronounced are a major contributing factor responsible for the broad aspects of Asia's climate (Critchfield, 1983).

For each season of the weather year, representative winds of local significance have been identified and named:

1. Summer – A hot, strong, and constant northerly summer wind carrying considerable dust and obscuring the atmosphere is called *karaburan* in the Tarim Basin of Sinkiang and northwestern China and *chang* in Turkmenia. At times, on the southern edge of a modified Arctic air mass that advects into southern Siberia and Central Asia, hot, dry moisture-absorbing *sukhovei* winds eliminate local cloud cover, permit intense solar radiation to strike the Earth and reduce relative humidity at the surface to a very low value both day and night.
2. Fall – Cold and often very dry northerly or northeasterly winds, preceded by a cold front, often blow with great strength and violent gusts down from the mountains and high plateaus of northern Siberia and the Caucasus. *Bora* is the term identifying this type of wind in the region between the Black and Caspian seas, and *sarma* and *kharanka* are bora-type winds in the Lake Baikal region of southern Siberia.
3. Winter – Very cold northerly or northeasterly gale-force winds, often blowing at temperatures below −20°C and accompanied by falling or drifting snow, are generated from the back side of winter depressions in Siberia and Central Asia. The sensation of cold temperatures is increased by the low wind-chill factor associated with a *buran* or a *purga*, and the break in the comparative calm associated with the Siberian High-Pressure Cell.
4. Spring – Continental depressions passing eastward over central China and the Yellow Sea in spring and early summer bring heavy overcasts, high humidity, and rain to China and Japan. *Bai-u*, *mai-yu*, or *plum rains*, as they are called, are an extended period of unstable weather caused by stagnation of the polar front.

Thunderstorms and tornadoes

Thunderstorms reach their maximum development in Asia over lowlands in summer. Most thunderstorms are ordinary convective cumulonimbus clouds within maritime tropical air masses that produce localized precipitation. Air mass thunderstorms, randomly scattered, are initiated primarily by daytime solar heating of land surfaces. Frontal and orographic thunderstorms have distinct patterns and movements, for they are triggered in a place or zone where unstable air is forced upward. Severe thunderstorms may produce hail, strong surface winds, and tornadoes. Convective thunderstorms develop more frequently here because of strong insolation and low wind velocities. Clouds of convective thunderstorms reach elevations of 12 000 meters or more, and rainfall associated with these cells is short in duration, intense, and localized. Very few thunderstorms are observed over the tundra regions of Siberia – less than five per year. Thunderstorm activity is a summer phenomenon in China, much of Central Asia, Singkiang, Mongolia, the Maritime Provinces of the Far East, and Siberia. They commence during the fall transitional period, are more prevalent in winter, and decline in number during the spring transitional period.

The most destructive spin-off of a thunderstorm is the tornado, an extremely violent rotating column of air that descends from a thunderstorm's cloud base and can cause great destruction along a narrow track. An Asian tornado travels in a 150–500 meter wide, approximately several kilometers long, straight track, at speeds ranging between 50 and 100 kph. Although one of the least extensive, it is the most violent of all Asian storms. As the whirling mass of unstable air gains force, a rotating column of white condensation is formed at the base of the cloud. Dirt and debris sucked into this whirlwind darken the column as the column of air reaches the ground. A notable feature of the climate of Asia is the relative infrequency of tornadoes, particularly as compared to central and eastern North America at similar latitudes. Although records are inadequate, there are sufficient data to conclude that tornadoes occur on the order of once every 3–5 years in the northern Caspian Sea area and Central Asia during May, June, and July and two to five or more annually in China and Japan during August and September.

Tropical cyclones

One of the most powerful and destructive types of cyclonic storms is the tropical cyclone (Hsu, 1982). Referred loosely to any pressure depression (near 1000 mb) originating above warm oceans in tropical regions, tropical cyclones form an important feature of the weather and climate of south and east Asia – particularly from July to October. Different terms are used worldwide to describe this tropical storm: *typhoon* in the western Pacific; *baquios* or *baruio* in the Philippines; *tropical cyclone* in the Indian Ocean; *willey-willeys* in Australia; *hurricane* in the eastern Pacific and Atlantic; *cordanazo* in Mexico; and *taino* in Haiti.

Organization and development of tropical cyclones are not fully understood, and they are under intensive study. Formation of this type of storm is associated with warm ocean surfaces not less than 27°C, located between 5° and 10°N latitude, light to calm initial winds, and waves or troughs of low pressure deeply embedded in easterly wind streams converging into an unstable atmospheric zone. Large quantities of latent heat released through condensation are converged and transferred to higher levels, deepening the pressure center and intensifying the storm. An almost circular storm of extremely low pressure, into which winds spiral with great speed, is formed. Asian tropical cyclones travel slowly at speeds of 16–48 kph, cut a destructive storm path 80–160 km wide, and winds in the wall cloud area achieve speeds in excess of 200 kph. Passage of a tropical cyclone over water and land is associated with strong winds and heavy rainfall. Storm tracks vary annually and no two recorded tracks have been exactly the same. Despite all irregularities, most tropical cyclones have a tendency to move westward, then poleward, finally turning eastward, toward higher latitudes under the influence of both internal circulation and external steering currents, penetrating into the belt of westerly winds. This awesome tropical storm contributes between 25% and 50% of the annual precipitation received in many tropical weather stations. Flooding, destructive wind force, and storm surge are responsible for much property damage and for human casualties.

Occurrence of tropical cyclones is restricted to specific seasons depending upon the geographical location of the storm-affected region. The Observatory of Hong Kong reports that 83% of the annual total recorded in Hong Kong occur between June and November. Approximately 50% of these severe tropical storms attain typhoon intensity. From mid-November to April, very few tropical storms pass over the coasts of China and Korea, but in the warm July through October period numerous tropical depressions, tropical storms, severe tropical storms, and tropical cyclones (typhoons) are experienced. In the 1884–1955 period at least 438 tropical cyclones crossed various sections of the Chinese and Korean coasts. Japan's tropical cyclone season begins in June and ends in November, reaching its peak occurrence in September. In the 1918–1947 period 85 typhoons were reported in Japan. The tropical cyclone season is slightly longer in the Philippines, extending from June to December. Almost 90% of all tropical cyclones in the 1948–1962 period were noted in the summer and fall seasons. Tropical cyclones usually weaken over land, and few penetrate and persist more than 500 km inland. The rise in sea level, when combined with high tides, accounts for more damage and loss of life in Asia than violent wind (Riehl, 1979).

Prospectus

Continentality, as a climatic control or factor, is best manifested in the extremes found in the climates of Central Asia, Siberia, and East Asia. The land-dominated interior climates present striking contrasts to the maritime-impacted coastal and island areas. Winters are much colder than corresponding latitudes in western Europe and North America; also, those regions where continentality is a climatic factor are the regions where climatic change is most noticeable. Vegetal, agricultural, and biotic zones have advanced 10 km northward in some climatic regions. This is best revealed in satellite remote sensing images of Central Asia and Siberia. In the past 20 years Asian temperatures have remained above the long-term averages (Figures 3 and 4). An annual change in temperature of only a few degrees centigrade affects the climates of Asia sufficiently to render marginal agricultural regions unacceptable for food production and to wreak havoc on local food supplies (Bryson and Murry, 1977). Moreover, a combination of human activities that modify climatic elements with natural causes of climatic change could lead to more frequent or more severe changes in Asia's climatic regions (Christianson, 1999). For those struggling to secure a meager existence from a hectare or two of land in Asia, and for the urban dweller seeking water to maintain bodily functions, any climatic change disrupts lifestyles – because humans and human institutions are adjusted to precisely the climate and weather that prevail (Budyko, 1977; Dando, 1980; Martens, 1999).

William A. Dando

Bibliography

Arakawa, H. (ed.), 1969. *Climates of Northern and Eastern Asia*. World Survey of Climatology, vol. 8. Amsterdam: Elsevier.
Borisov, A.A., 1965. *Climates of the U.S.S.R.* R. A. Ledward (trans.). Chicago: Aldine, pp. 1–28.
Bryson, R., and Murry, T., 1977. *Climates of Hunger*. Madison: University of Wisconsin Press, pp. 123–156.
Budyko, M.I., 1977. *Climatic Changes* (American Geophysical Union trans.). Washington, DC: American Geophysical Union, pp. 197–245.
Chang, J., 1967. The Indian Summer Monsoon. *Geographical Review*, **57**: 372–396.
Christianson, G., 1999. *Greenhouse: the 200-year story of global warming*. New York: Walker & Company.
Critchfield, H., 1983. *General Climatology*. Englewood Cliffs, NJ: Prentice-Hall, pp. 88–101.
Dando, W., 1980. *The Geography of Famine*. London: Edward Arnold.
Dando, W.A., Carson, A.R., and Dando, C.Z., 2003. *Russia*. Philadelphia: Chelsea House, pp. 30–35.
Das, P., 1968. *The Monsoons*. New York: St Martin's Press.
Davydova, M., Kamenskii, A., Nekiukova, N., and Tushinskii, G., 1966. *Fizicheskaia Geografiia SSSR*. Moskva: Prosveshchenie, pp. 57–82, 439–704.
Davydova, M., Kamenskii, A., Nekiukova, N., and Tushinskii, G., 1999A. Global change: a review of recent anomalies. *Global Change*, **5**(1): 6–7.
Davydova, M., Kamenskii, A., Nekiukova, N., and Tushinskii, G., 1999B. Russia plagued by heat and drought. *Global Change*, **5**(3): 7.
Griffiths, J., and Driscoll, D., 1982. *Survey of Climatology*. Columbus, Ohio: Charles E. Merrill, pp. 162–172.
Hsu, S.I., 1982. *Tropical Cyclone in the Western North Pacific*. Occasional Paper No. 24. Hong Kong: Chinese University of Hong Kong, Department of Geography, vol. 2.
Jones, P., 2001. "The Travails of a Warmer World." Presentation by Dr. Jones, University of East Anglia, Intergovernmental Panel on Climatic Change, Washington, DC.
Kriuchkov, V., 1973. *Krainii Sever: Problemy Ratsional'nogo Ispol'zovaniia Pirodnykh Resursov*. Moskva, Mysl, p. 15.
Lawton, J., 2002. Rebuilding in Afghanistan. *Saudi Aramco World*, **53**(6): 16.
Lounsbury, J.F., 1973. *A Workshop for Weather and Climate*. Dubuque: William C. Brown, p. 124.
Lydolph, P.E., 1977. *Climates of the Soviet Union*. World Survey of Climatology, vol. 7. Amsterdam: Elsevier.
Martens, P., 1999. How will climatic change affect human health, *American Scientist*, **87**(6): 534–541.
Nuttonson, M.Y., 1950. *Agricultural Climatology of Siberia*. Washington, DC: American Institute of Crop Ecology, appendices.
Nuttonson, M.Y., 1999. *Preroda i Resoursi Zemli*. Moskva and Vienna: Geographical Institute Ed. Holzel, pp. 51–56.
Riehl, H., 1979. *Climate and Weather in the Tropics*. London: Academic Press, pp. 459–496.
Takahashi, K., and Arakawa, H., (eds.), 1981. *Climates of Southern and Western Asia*. World Survey of Climatology, vol. 9. Amsterdam: Elsevier.
Trewartha, G.T., 1981. *The Earth's Problem Climates*, 2nd edn. Madison: University of Wisconsin Press, pp. 173–254.
Trewartha, G., and Horn, L., 1980. *An Introduction to Climate*. New York: McGraw-Hill, pp. 123–128.
Tsuchija, I., 1964. *Climate of Asia*. World Climatology, vol. 1. Tokyo: Kokon Shoin, p. 577.

Cross-references

Asia, Climate of South
Asia, Climate of Southwest
Climate Classification
Continental Climate and Continentality
Siberian (Asiatic) High
Taiga Climate
Tundra Climate

ASIA, CLIMATE OF SOUTH

South Asia, which mainly includes India, Pakistan, Bangladesh, Sri Lanka, Nepal, and Bhutan in this article, is located between about 5° and 37°N, in tropical and equatorial latitudes, with the Indian Ocean to the south, the Great Himalayas and Karakoram to the north, the Baluchistan highlands to the west, and the meridional chains of mountains in Indochina to the east. The distinct geographical features in South Asia restrict the horizontal exchange of oceanic and continental airmasses and modify the climate to a characteristic pattern. Besides great mountains and upland systems there are also lowland depressions. Numerous bays and gulfs enhance the influence of the ocean. In winter a warm, westward-flowing current reaches eastern Sri Lanka, from where it diverges northward, affecting the southeast coast of India. During the summer a warm current in the Indian Ocean reaches the west coast of India and flows southward; from south of Sri Lanka the current moves east (Martyn, 1992).

South Asia mainly consists of four climate zones. The northern Indian edge, as well as the upland and mountainous part of Pakistan, has a dry continental, subtropical climate. The far south of India and southwest Sri Lanka are located in the zone of equatorial climates. The rest of the areas lie in the tropical zone, with hot semi-tropical climate in northwest India, cool-winter hot tropical climate in Bangladesh, and semiarid tropical climate in the center of the peninsula. The last two tropical climate zones are greatly affected by monsoon circulations (Martyn, 1992; Rao, 1981; Papadakis, 1966). In addition, mountain climate occurs in the Himalayas. Climates there vary with elevation, from permanent ice caps at high elevation to subtropical climate at lower lands.

The most important climate feature in the region is the southwest monsoon, which lasts for 4 months, from June to September. In this southwest monsoon period the precipitation is one order of magnitude more than during the rest of the year. Because temperatures in the South Asian monsoon region drop dramatically in June and July, due to the monsoon onset, March to May becomes the warmest period.

In this article the observational data for precipitation are from the rain gauges station data over land and the data are interpolated to 1° resolution (Xie and Arkin, 1997), covering 1948–2003. The surface temperature data are from the National Center for Environmental Prediction (NCEP), the Climate Analysis Center, and the Climate Anomaly Monitoring System station data achieve (CAMS, Ropelewski et al., 1985); they include data from 1948 through June 2003. The NCEP/National Center for Atmospheric Research (NCAR) Global Reanalysis data from 1948 to 2003 (Kistler et al., 1999) are used for circulation and radiation figures.

Radiation

Although the sun's altitude is high during the entire year, and the amount of daylight varies little during the year for the area south of 30°N (Martyn, 1992), the solar radiation still exhibits high seasonality and great spatial variability (Figure A68). Annual mean solar radiation amounts decrease eastward, from 290 W m^{-2} in Baluchistan to less than 240 W m^{-2} in Bangladesh. Annual mean downward long-wave radiation at the surface decreases northward, which is consistent with the temperature gradient and water vapor quantity, with the two coasts having the highest long-wave radiation (about 390–400 W m^{-2}) due to cloudiness.

January solar radiation is about 240–250 W m^{-2} in the southern peninsula and western Sri Lanka, and drops to 220 W m^{-2} in the northern peninsula and eastern Sri Lanka. Much lower solar radiation is received in the northwest subcontinent and northeast Bangladesh, with north Pakistan having the lowest (Figure A68a). The long-wave radiation shows a clear zonal pattern, with high gradients in the Himalayas associated with topography (Figure A68c). In July the solar radiation decreases southeastward from northwest South Asia. Pakistan and northwest India receive the most solar radiation, between 300 and 360 W m^{-2}. In northeast India and Bangladesh the solar radiation decreases southward from 300 to 200 W m^{-2}. The southern continent has less solar radiation, especially along the west coast of India, due to the cloudiness in these regions (Figure A68b). There are two very high downward long-wave radiation centers over the Indian Ocean, located at both sides of the Indian peninsula and associated with heavy rain in these two areas. Apart from the areas influenced by these two centers, in other parts of the peninsula and Sri Lanka it is quite uniform, between 410 and 420 W m^{-2} (Figure A68d).

Circulation

A characteristic feature of the South Asian climate is the atmospheric circulation associated with the monsoon evolution. The extensive Eurasian continent and the Indian and Pacific Oceans play an important role in monsoon formation. In the winter the northeast monsoon prevails and sea-level pressure decreases to the south. The Intertropical Convergence Zone (ITCZ), associated with the equatorial low-pressure belt, runs across the entire Indian Ocean along 10°S latitude. The cold air from the Siberian High, which is centered at 45°N and 105°E, approaches eastern India and the Bay of Bengal starting in October. Since the barrier of Tibet and the Himalayas blocks the advance of cold polar air, low-level northeast winds only run across the Bay of Bengal and reach northeast India. The pressure gradient is strong to the north of the Himalayas but weak over the Indian Peninsula. From October to May the southern branch of the westerly jet stream, usually flowing between 28° and 30°N, crosses the Himalayas. This westerly wind in the upper troposphere is strongest in January and gradually slackens.

Meanwhile, in western South Asia, cold air from Iran and Afghanistan reaches Pakistan and northern India. Cyclonic disturbances are more frequent over the western Himalayas and Karakoram from early November to the end of May because of strong vertical wind shear in the basic current (Hamilton, 1979). They reach the Ounjab, Rajasthan, and even Patna on the Ganges Plain. They bring clouds and precipitation in November and December, and cold weather, thunderstorms, and precipitation in the normally hot, dry period from March to May (Lockwood, 1974; Martyn, 1992).

Development of low pressure due to increased heating over land starts for South Asia in March. Heat lows are usually found in an area extending from Somalia across southern Arabia to Pakistan and northwest India between May and September, and are most intense in July. Middle tropospheric subsidence is considered to affect the heat low (Hamilton, 1979). The distribution of sea-level pressure, as well as

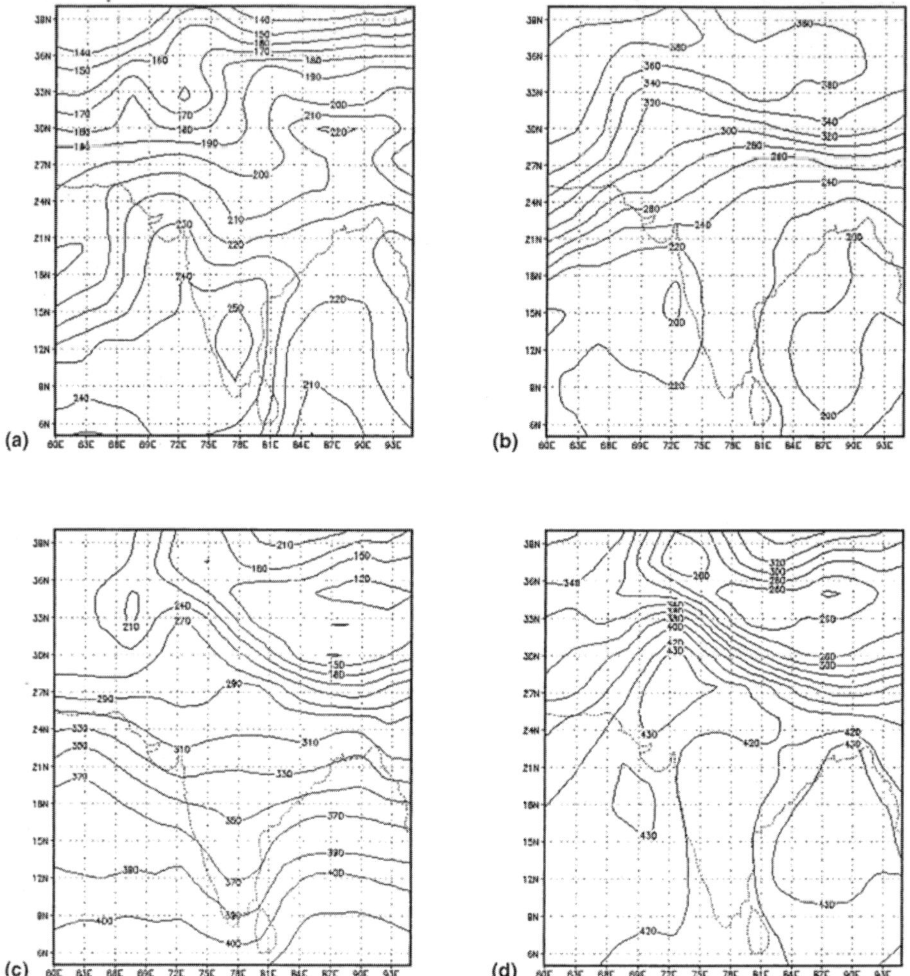

Figure A68 Monthly mean radiation flux at the surface (W m^{-2}): (**a**) January mean downward short-wave radiation; (**b**) July mean downward short-wave radiation; (**c**) January mean downward long-wave radiation; (**d**) July mean downward long-wave radiation.

low-level air flow over the Indian subcontinent, experiences a complete reversal from January to July (Figures A69a and A69b). Earlier northeasterlies are replaced by westerlies in April with the core of a subtropical jet stream at 200 mb near 27°N. This jet gradually weakens from an average speed of about 40 m s^{-1} in March to about 25 m s^{-1} in May (Hamilton, 1979; Keshavamurty and Rao, 1992).

In the summer the low pressures are centered over Pakistan, northwest India, and the Ganges Plains. The Indian Ocean high is strong and centered at 30°S and 60°E. At the beginning of June, as the jet stream disappears above the southern border of the Tibetan Plateau, the low over Pakistan deepens and the ITCZ moves in. The change in the direction of circulation to southwest occurs suddenly (called the burst of the monsoon) because of the very steep horizontal northward pressure gradient. The southwest monsoons, carrying enormous quantities of moisture from the Indian Ocean, reach the coast of India (Figure A69b). The monsoon circulation dominates South Asia. Because wind now blows off the sea, the summer monsoon contains far more water vapor than the winter monsoon. The depth of the summer monsoon over the Indian Peninsula decreases northwards (Martyn, 1992). Humidity, cloudiness, and precipitation all increase in the summer, but the increase of temperature is checked. When the monsoons change direction during the summer, cyclonic activity increases rapidly over very warm seas and oceans and is extremely violent over the Bay of Bengal.

Monsoon depressions form in the Bay of Bengal from June to September, move west/northwest, and reach at least the central parts of the subcontinent before weakening or disappearing. They are usually moderately vigorous, with surface winds below 20 m s^{-1} and a central pressure range from 2 to -10 mb below normal. Many depressions are probably of a baroclinic type (Koteswaram and George, 1958) and usually exist in the presence of a large zonal wind shear between the upper and lower troposphere. Monsoon depressions are an important summertime precipitation agency between 18°N and 25°N. They bring widespread rains in the southwest quadrant, with many heavy rainfall events. As a depression moves over land it usually weakens due to frictional dissipation of energy and a reduction of

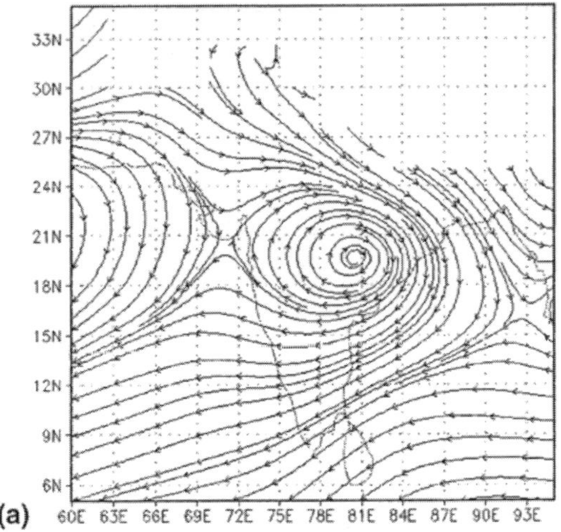

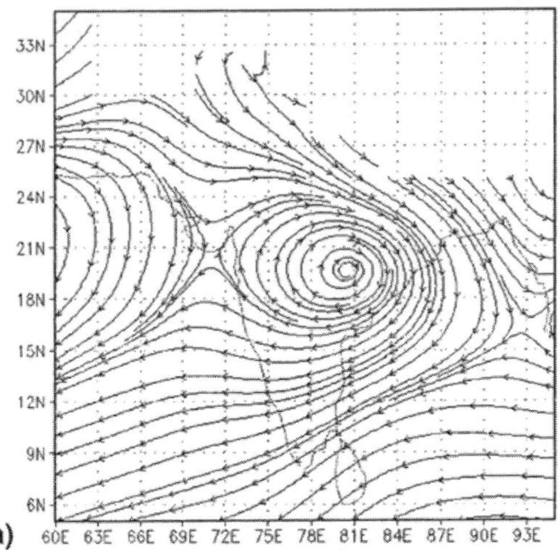

Figure A69 Monthly mean streamline at 850 mb: (**a**) January; (**b**) July.

moisture intake. Eventually, most depressions merge into the monsoon trough over northwestern India (Hamilton, 1979).

The transition of a summer circulation to a winter circulation is of shorter duration than the reversion from winter circulation to summer circulation. The transition comes suddenly over the Himalayas as the circumpolar westerlies begin to intensify and move south. A subtropical westerly jet stream develops at the northwestern end of the Himalayas, thereafter extending southeastward. By mid-October the jet is fully established south of the Himalayas. A winter circulation develops north of 25°N by early November (Yeh et al., 1959; Hamilton, 1979).

Surface temperature

The isotherms in January generally are parallel to the latitudes except near the coasts. January mean temperatures in northern South Asia are about 10°C, which is not as cold as most parts of Asia because the Himalayas and Sulaiman block the inflow of masses of cold air from the north. On the Indian Peninsula, south of 18–20°N in the east, the temperatures exceed 20°C and can reach 24–27°C on the equator (Figure A70a). In Kashmir the mean temperature can be 1°C at Srinnagar (1666 m) and −15.7°C at Dras (Martyn, 1992), where below-zero temperatures are usual from November to April.

Land becomes progressively heated after January. The warmest period of the year precedes the monsoon from March to mid-June. In the far north the warmest month is July Figure A70b), and the warm season begins earlier southward; for instance, in southern India it begins in April. During the warm period the mean temperatures in the northern Deccan exceed 35°C (and even reach 38–42°C), and in the interior of the peninsula they reach 33°C. Because of the greater moisture brought in with the sea breezes, coasts are not hotter in the afternoon hours than the peninsular interior. The west coast of India is about 5°C cooler than the east coast.

In July the southwest monsoon is established over India and the cloudiness is heavy. Monthly mean temperatures over much of Indian Peninsula are quite uniform, around 27°C (Figure A70b). However, temperatures of over 30°C are recorded in the Great Indian Desert, and may reach 32–36°C. Temperatures in the Himalayan mountain valleys are from 24°C at elevations of 1700 m to 17°C at 3500 m (Martyn, 1992). Absolute maximums in some interior regions even exceed 40°C. The summer monsoon season is the second cooler period in the year because clouds and rain prevent the temperature from rising too much during the day. Once the monsoon ends the temperatures rise and a second annual maximum is attained. This is not as high as the pre-monsoon maximum because of the greater quantity of moisture in the air and the sun's lower altitude. Monthly mean temperatures over the entire subcontinent in October are generally within the range of 27–29°C. This is the month of the most equable distribution of temperature on the Indian Peninsula. The range of annual temperature variation is about 1–5°C for most parts of the region.

Near-surface humidity and cloudiness

In January the relative humidity on the southern peninsula is higher than 50–60%, and even 70% on the coasts. A maximum relative humidity of over 80% in Sri Lanka and Hhasi-Janitia Hills in northeast India has been recorded. The driest areas are the Indus Plain and the adjacent areas of Pakistan and western India. In the western Himalayas the humidity is high when temperatures are very low, but becomes lowest in May–July. From March to mid-June (the warm months) the relative humidity is under 30–40% on the Indus Plain, in the Punjab, Rajasthan, and at the western end of the Ganges Plain, and under 50% on the Deccan Plateau and in central Burma (Martyn, 1992). In July the areas most affected by the monsoon have relative humidity of over 80%, while the southwest coasts of India, the Khasi-Janta Hills, and some sections of the Himalayas even reach above 90%. However, there are areas in Pakistan still with relatively low humidity. On the Coromandel Coast and in eastern Sri Lanka the relative humidity reaches the lowest value in a year (Martyn, 1992). In the driest regions of Rajasthan the relative humidity does not exceed 60%, and in central India it does not reach 80%. In the eastern Himalayas, where the effects of the monsoon are strong, there are two wet

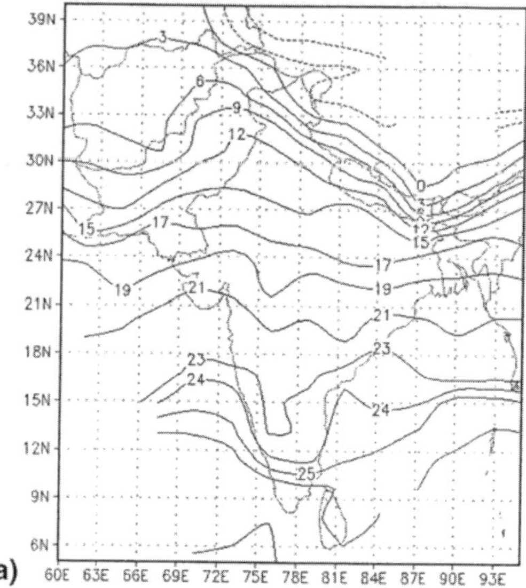

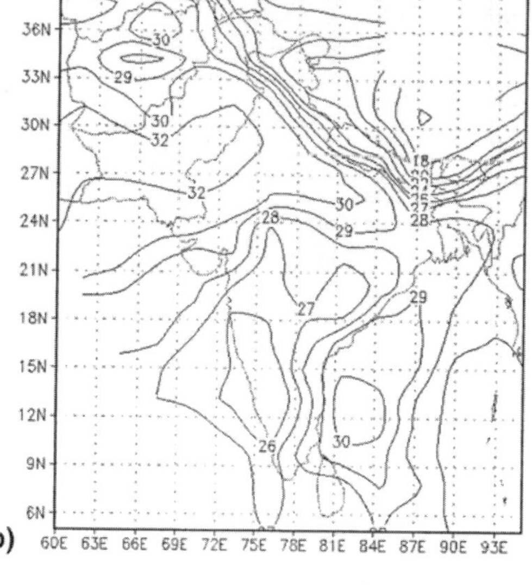

Figure A70 Monthly mean surface temperature: (**a**) January; (**b**) July.

periods, one in January and February and one from May through October, while two dry periods remain between these two wet periods.

The cloudiness and its annual variation are varied; they depend on atmospheric circulation and topography. With the arrival of the summer monsoon, which carries a huge amount of moisture from the Indian Ocean, South Asia becomes one of the most overcast parts of Asia, normally over 80%. Cloudiness even reaches 90% in western India in the upper reaches of the Godavari and the windward slopes of the Khasi-Jaintia Hills. The southern and northern parts of the region have greatest cloud cover from June to August, and from July to August, respectively. During the rest of the year the cloudiness is much less. In the west the cloudiness is 10–22% during the November–April period. In the east on the Bay of Bengal the lower cloudiness is about 22–45% during January–March. In January it is scanty (under 30% or even lower) in most parts of South Asia; but in the far north of India, where a westerly circulation is frequent, cloudiness is greatest (Martyn, 1992).

Precipitation

Precipitation attracts much more attention than any other aspect of Indian subcontinent meteorology. Precipitation in South Asia exhibits strong spatial variability. Pakistan and northwest India, where there is very little water vapor in the air and little precipitation due to limited evaporation and to dry air currents, have the most arid areas in South Asia. Annual precipitation is less than 250 mm in the driest parts of these areas (Figure A71a). Despite considerable low-level convergence into the low-pressure areas, there are very little clouds and precipitation because of extremely low humidity and/or because of stable inversion (Hamilton, 1979). From the east coast of India the precipitation decreases inland south of 17°N. The interior of the peninsulas, especially the lee side of the Western Ghats, has low precipitation, less than 1000 mm per year. Low precipitation is also found in the upper basin of the Bhima River (around 18°N) and its tributaries. In northern India there is a low precipitation strip from about 28°N and 75°E to 25°N and 87°E; but topographic effects and monsoons produce significant precipitation along the Himalayas.

In monsoon areas the precipitation exceeds 1000 mm per year. Maximums of precipitation coincide with higher elevations in the Ghats and the eastern Himalayas. In the Khasi-Jaintia Hills the precipitation can reach as much as 11 000 mm per year. On the Malabar Coast and at the southern end of the Western Ghats in India the value reaches more than 3000 mm per year. In Sri Lanka the annual precipitation is also affected by monsoons and is higher than 1300 mm per year; in the southwest it exceeds 2200 mm per year. The pattern of rainy days is similar to that of annual precipitation. The highest number of rainy days (200 per year) is recorded in southwest Sri Lanka. The rainy days number about 170 per year in the eastern Himalayas and about 150 on the Malabar Coast. The interior regions of India have rain on 5–80 days per year, while in the Great India Desert it falls on less than 5 days (Martyn, 1992; Rao, 1981).

The precipitation in South Asia is linked to seasonal changes in the atmospheric circulation. Rains develop over southern India and the southern Bay of Bengal in May–June (Webster et al., 1998). The abrupt northward jump appears with the onset of the South Asian summer monsoon. Based on analysis of various data sets it was found that the South Asian monsoon onset was associated with the reversal of the meridional gradient of upper tropospheric temperature south of the Tibetan Plateau (Flohn, 1957; Li and Yanai, 1996).

The monsoons bring most of the region's precipitation (Figure A71c). On the Indian Peninsula north of 13°N the rainy season begins in mid-June and lasts until September; in Bangladesh it lasts until October. The monsoon usually advances from two locations – one from the Bay of Bengal to central India and southern India in June and reaching northwest India by early July, and the other one over the Arabian Sea 1 month later. This progression of monsoons is associated with large-scale ascent through the troposphere with strong lower tropospheric convergence and upper tropospheric divergence. The periods of maximum thunderstorm frequency precede and

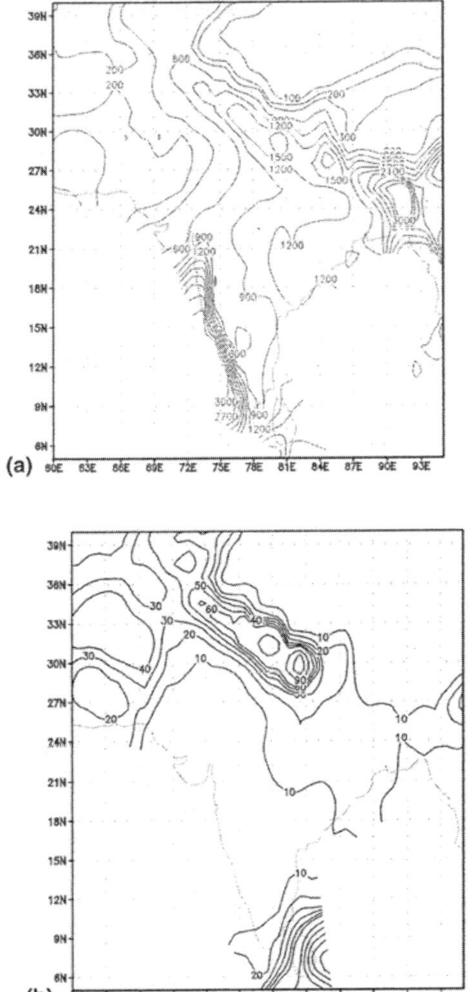

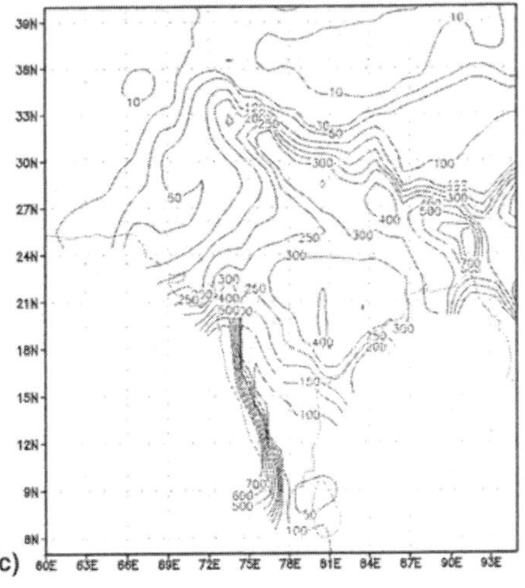

Figure A71 Precipitation: (**a**) Annual mean (mm year^{-1}); (**b**) January mean (mm month^{-1}); (**c**) July mean (mm month^{-1}).

follow the summer rains (Hamilton, 1979). In most parts of South Asia the spatial distribution of annual precipitation is similar to that in this period. The southwest monsoon rain provides water for agriculture in the world's very populous regions.

Orographic influence is dominant as onshore winds blow to the Western Ghats and the Khasi-Jaintia Hills. The main impact of the southwest monsoon is in a 250-km long stretch of the Himalayas between 87 and 90°E. Not only do the Himalayas receive heavy rains, but also the plains adjoining the foothills receive more rain than the plains further south. Even in north Pakistan the precipitation also ensues from monsoons from June to September. A trough (the monsoon trough) extends east-southeastward from Indo-Pakistan heat low to the northern part of the Bay of Bengal near 25°N in June and July and shifts a little further north in July–August, by which time it dominates weather over the Indian subcontinent. It retreats southward and disappears in September as a near-equatorial trough near 10°N. The position of this trough is an index of the activity of the monsoon over the subcontinent. If the trough is in a normal position it will be a normal monsoon year. When the trough shifts to the foothills of the Himalayas this indicates a break situation (Rao, 1981; Hamilton, 1979).

Temporal variations in precipitation are associated with shifts of position in large-scale upper tropospheric circulation, the low-level monsoon trough, and the development of synoptic disturbances. The precipitation maximum occurs in May–August (some areas extend to September). Rain is very heavy, especially at the start of the monsoon season and during tropical cyclones. Along the Coromandel Coast the rains begin in July, sometimes even in August. There is rain on more than 20 days in a month. In some areas, rainy days reach 25–28. However, it is very dry on the southeast tip of the peninsula. In the northwestern parts of the subcontinent the precipitation also decreases progressively westward.

A seasonal reversal of winds at lower levels, the displacement of the monsoon airmass by the continental airmass, and development of anticyclonic flow indicate withdrawal of the monsoon. From mid-September, as the surface temperature falls, the atmospheric pressure over the Indus and Ganges Plain rises rapidly and becomes higher than that over the Bay of Bengal. In October the ITCZ lies at 10°N and only affects southern India. Southern India and southwest Sri Lanka have two rainy seasons due to the movement of the ITCZ. The first one is from May to July, and the second one is from October to November. In addition, tropical cyclones have prolonged heavy rainfall on the Coromandel Coast in fall. The turn of the monsoons is the time when tropical cyclones develop over the Bay of Bengal. In the Bay of Bengal, based on 70-year records, 15% of cyclones arise in April and May, and 38% arise in October and November (Rao, 1981). Fall cyclones are much stronger than the pre-monsoon ones. Some are listed as severe cyclonic storms with winds of 48 knots or more. The most violent cyclones over the Bay of Bengal cross the central Deccan along the Krishnai river valley. Others turn to blow up the eastern coast toward Bangladesh. Considerable flood damage has occurred in the Ganges delta. Apart from this, the cyclones are accompanied by high winds, tidal waves, and heavy rain. There can be disastrous flooding.

The winter precipitation over South Asia is only a small fraction of the annual amount, except in Kashmir and its surroundings (Figure 71b). During winter in Kashmir, Karakoram, cyclonic precipitation is dominant. On the east coast of Sri Lanka the rainfall is still more than 30 mm per month, decreasing to 15 mm per month along the west coast.

Impact of the Himalayas/Tibetan Plateau on the South Asian climate

The northern edge of South Asia is bounded by the Himalayas, which run along the southern periphery of the Tibetan Plateau; it extends over the latitude–longitude domain of 70–105°E, 25–45°N, with a mean elevation of more than 4000 m above sea level. Strong contrast exists between the eastern and western parts of the Plateau in both land surface features and meteorological characteristics.

The sharp edge of the Himalayas and the vast and high Plateau force the large-scale airflow to go around rather than over it. This effect is clearly seen in the flow features well above the Plateau. The mid-tropospheric westerly jet stream is situated to the south of the Plateau before the onset of the summer monsoon, but it suddenly jumps to the north side with the monsoon onset. Earlier, the thermal influences of the Tibetan Plateau were inferred from the presence of a huge upper tropospheric anticyclone above the Plateau in summer (Flohn, 1957) and from the pronounced diurnal variations in the surface meteorological elements along the periphery of the Plateau (Yeh and Gao, 1979).

During winter the Tibetan Plateau is a cold source; it acts as a heat source from March until October. The elevated heat source above the Plateau acts to warm the mid-upper tropospheric air, and the temperature over the Plateau in summer exceeds that over the Indian Ocean (Yanai et al., 1992). This meridional gradient of temperature, appearing in May–June, is crucial for the onset of the South Asian monsoon. Concurrent with the reversal of the temperature gradient, the rain belt associated with the ITCZ over the Indian Ocean suddenly moves onto the Indian subcontinent and reaches the foot of the Himalayas

With the cooling of the Plateau surface the westerly jet stream retreats to the south side of the Plateau in late October; this is just after the end of the southwest monsoon in India and almost concurrent with the commencement of the northeast winter monsoon over southeast India (Matsumoto, 1992).

Y. Xue and M. Yanai

Bibliography

Flohn, H., 1957. Large-scale aspects of the "summer monsoon" in South and East Asia. *Journal of the Meteorological Society of Japan*, **75**: 180–186.

Hamilton, M.G., 1979. *The South Asian Summer Monsoon*. London: Edward Arnold.

Keshavamurty, R.N., and Rao, M.S., 1992. *The Physics of Monsoons*. New Delhi: Allied Publishers.

Kistler, E., Kalnay, E., Collins, W., Saha, S., et al. 1999. The NCEP/NCAR 50-year Reanalysis. *Bulletin of the American Meteorological Society*, **82**: 247–268.

Koteswaram, P., and George, C.A., 1958. On the formation of monsoon depression in the Bay of Bengal. *Indian Journal of Meteorology and Geophysics*, **9**: 9–22.

Li, C., and Yanai, M., 1996. The onset and interannual variability of the Asian summer monsoon in relation to land-sea thermal contrast. *Journal of Climate*, **9**: 358–375.

Lockwood, J.G., 1974. *World Climatology: an Environmental Approach*. London: Edward Arnold.

Matsumoto, J., 1992. The seasonal changes in Asian and Australian regions. *Journal of the Meteorological Society of Japan*, **70**: 257–273.

Martyn, D., 1992. *Climates of the World. Developments in Atmospheric Science*, vol. 18. Amsterdam: Elsevier.

Papadakis, J., 1966. *Climates of the World and their Agricultural Potentialities*. Buenos Aires: Cordoba.

Rao, Y.P., 1981. The climate of the Indian subcontinent. In Takahashi, K., and Arakawa, H., eds., *Climates of Southern and Western Asia*. Amsterdam: Elsevier, pp. 67–117.

Ropelewski, C.F., Janowiak, J.E., and Halpert, M.F., 1985. The analysis and display of real time surface climate data. *Monthly Weather Review*, **113**: 1101–1107.

Webster, P.J., Magana, V., Palmer, T.N., et al. 1998. Monsoons: processes, predictability, and the prospects for prediction. *Journal of Geophysical Research*, **103**(C7): 14,451–14,510.

Xie, P., and Arkin, P.A., 1997. Global precipitation: a 17-year monthly analysis based on gauge observations, satellite estimates and numerical model outputs. *Bulletin of the American Meteorological Society*, **78**: 2539–2558.

Yanai, M., Li, C., and Song, Z., 1992. Seasonal heating of the Tibetan Plateau and its effects on the evolution of the Asian summer monsoon. *Journal of the Meteorological Society of Japan*, **70**: 319–351.

Ye, D. (Yeh, T.C.), and Gao, Y.-X., 1979. *The Meteorology of the Qinghai-Xizang (Tibet) Plateau*. Science Press: Beijing (in Chinese).

Yeh, T.C., Dao, S.Y., and Li, M.T., 1959. The abrupt change of circulation over the northern hemisphere during June and October. *The Atmosphere and the Sea in Motion*. New York: Rockefeller Institute, pp. 249–267.

Cross-references

Asia, Climates of Siberia, Central and East Asia
Asia, Climate of Southwest
Climate Classification
Intertropical Convergence Zone
Monsoons and Monsoon Climate
Tropical and Equatorial Climates

ASIA, CLIMATE OF SOUTHWEST

The region commonly referred to as southwest Asia reaches from the mid-latitudes to the tropics and encompasses a swath of land stretching zonally from the eastern shores of the Mediterranean Sea to the Himalayan Mountains and meridionally from the Caspian Sea to the Arabian Sea (roughly 26°E to 70°E and 12°N to 42°N). In addition, this area includes the Gulf of Aden, the Arabian (Persian) Gulf, and the Red Sea. The

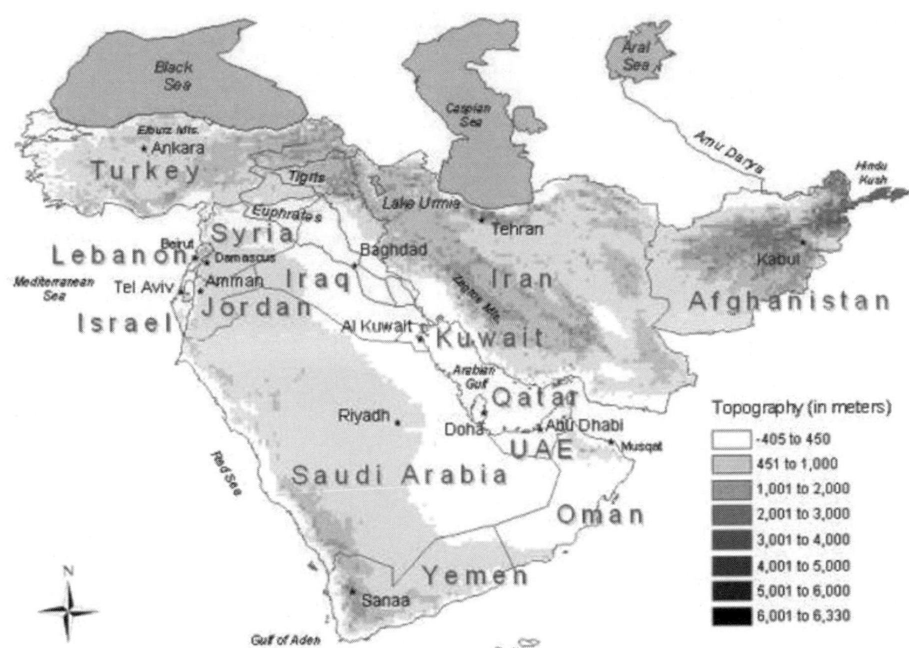

Figure A72 Topographic map of southwest Asia in meters above sea level. Topographic data were provided by the United States Geological Survey (USGS) Land Processes Distributed Active Archive Center (DAAC) established as part of NASA's Earth Observing System (EOS) Data and Information System (EOSDIS) and are available at http://edcdaac.usgs.gov/main.html. GTOPO30 is a global digital elevation model (DEM) with a horizontal grid spacing of 30 arc seconds (approximately 1 kilometer).

countries within this region include: Afghanistan, Bahrain, Iran, Iraq, Israel, Jordan, Kuwait, Lebanon, Oman, Qatar, Saudi Arabia, Syria, Turkey, the United Arab Emirates (UAE) and Yemen (Figure A72).

Location and topography

Southwest Asia is bordered by seven bodies of water: the Mediterranean, the Red, the Arabian, the Caspian and the Black Seas, the Gulf of Aden, and the Arabian (Persian) Gulf. There are three major plateaus:

1. The Iranian Plateau: with the Elburz mountain range bordering it to the north and the Zagros mountain range bordering it to the west.
2. The Anatolian Plateau: with the Septus and Pentus mountain ranges bordering in the north and the Toros mountain range bordering in the south.
3. The plateau of the Arabian Peninsula, which is characterized by a vast desert area including the Arabian, Syrian, and Iranian deserts.

Climatic classification

Southwest Asia is a region of diverse climates and is generally divided into three main climate types: arid, semiarid, and temperate. According to the Köppen classification system, a system of climate classification using latitude band and degree of continentality as its primary forcing factors, Central Asia is a predominantly B-type climate regime. This implies that 70% or more of the annual precipitation falls in the cooler half of the year (October–March in the northern hemisphere). Based on the Köppen–Geiger classification, which factors in seasonal distribution of rainfall and the degree of dryness/coldness of the season, southwest Asia can further be divided into five climatic types (Figure A73). These climatic types are are listed in Table A39.

General circulation

Southwest Asia lies at the boundary of three competing climate regimes. These three systems are: (1) the cold Siberian High in winter over Central Asia, (2) the monsoon Asian Low in summer over India , and (3) eastward propagating secondary low-pressure systems traveling through the Mediterranean and adjacent areas during non-summer seasons. Overall, the climate of northern southwest Asia is governed by extra-tropical synoptic scale disturbances while the climate of southern southwest Asia is governed by smaller-scale seasonal thermodynamics.

Airmasses

Polar continental air (cP)

Polar continental air (cP) is formed over Central Asia in winter and is characterized by extremely low temperatures in a shallow surface layer with a significant inversion reaching to 1500 m. The cP airmasses are stable in their source regions – the central parts of Asia and North Africa – but are modified upon entering southwest Asia. As an example, when cP air enters southwest Asia via the Caspian Sea the surface inversion is destroyed by heating from below, resulting in increased

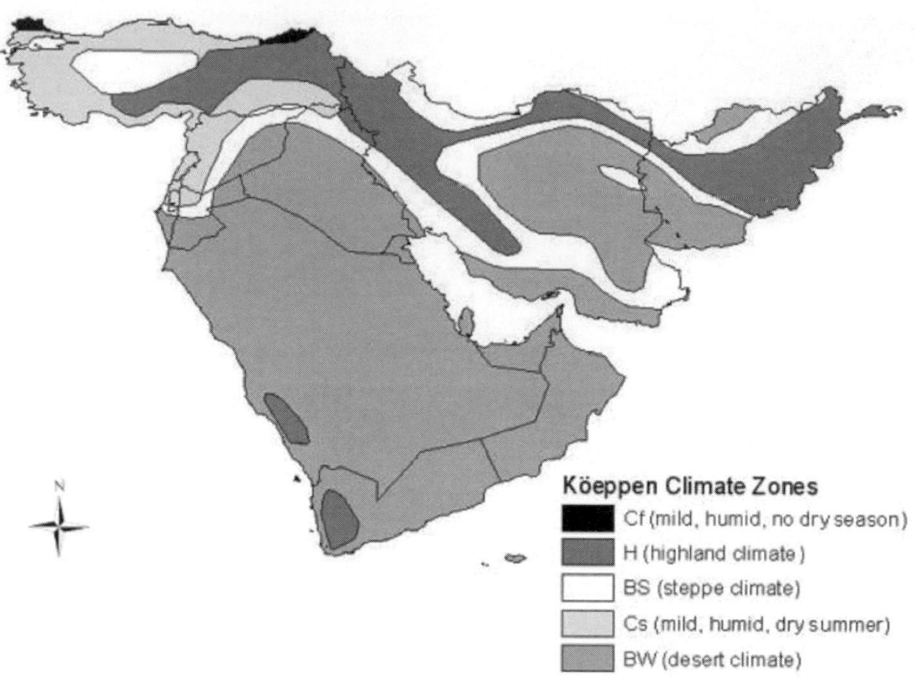

Figure A73 Köppen classification zones for southwest Asia. Köppen data were obtained through the Center for International Development at Harvard University and are available at http://www2.cid.harvard.edu/ciddata/geographydata.htm

Table A39 Köppen–Geiger climatic types for southwest Asia

Type	Winter	Summer	Location
Cs	Temperate, rainy	Hot, dry	Coast of Turkey, Syria, Lebanon and Israel
Cf	Temperate, rainy	Warm, rainy	Interior parts of the Anatolian Plateau
H	Temperate, rainy	Dry	Mountain ranges in southwest Arabian Peninsula
BW	Arid	Arid	Desert areas of Arabian Peninsula, Iran, Syria and Iraq
BS	Semi-arid	0.1/2.9	Arabian Peninsula, Oman, Jordan, Syria, Israel, Iraq and Iran

cloudiness due to the absorption of moisture. The cP airmasses are shifted north of 50°N in summer and hence do not influence the climate of the region.

Tropical maritime (mT)

The source region for tropical maritime (mT) airmasses is the southern portions of the oceans.

Polar maritime air (mP)

In winter polar maritime (mP) airmasses invade the Anatolian Plateau and the Black Sea and tend to be modified because of the long stretches across land. These airmasses tend to be more humid than cP airmasses and they follow the migrating low-pressure systems across Europe – losing most of their moisture upon orographic uplift. The mP airmasses are not present in the summer.

Tropical continental air (cT)

Formed over the dry land areas of central Asia and the African Sahara in summer, late spring, and early fall when surface heating is pronounced, tropical continental (cT) airmasses precede the "khamsin" dust storms occurring in late spring/early fall. These dust storms are very hot and dry and frequently proceed eastward through the Sinai or the northern Red Sea.

The characteristic airmasses of southwest Asia are cP, cT and mP in winter and cT and mT in summer – with each of these airmasses having a different source region.

Winter

The following section describes general climate conditions during the winter season in terms of sea-level pressure, temperature and precipitation. Average winter (December through February; DJF) total precipitation and average winter temperature are presented in Figure A74 and Table A40.

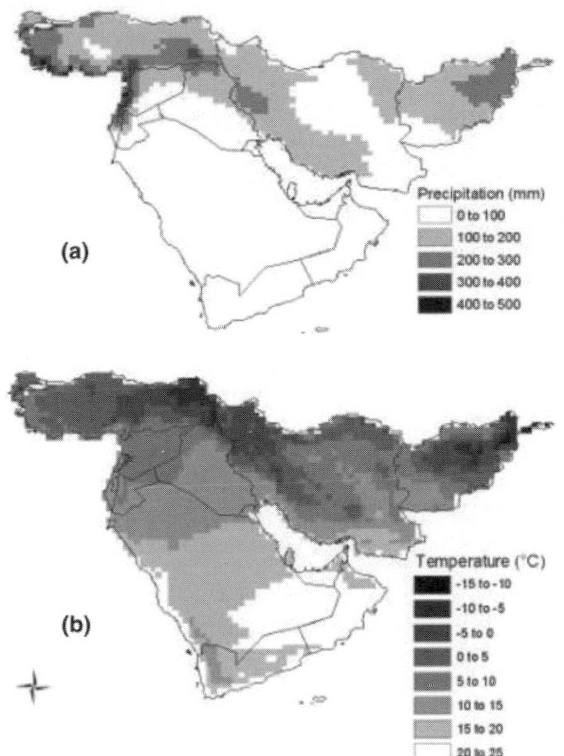

Figure A74 Average total winter precipitation is shown in (**a**) and average winter temperature is shown in (**b**). Monthly temperature and precipitation data (New et al., 2000) have a resolution of 0.5° × 0.5° and are averaged over the period 1901–1998.

Table A40 Mean average temperature (T) in °C and total precipitation (P) in mm/day for summer and winter (S/W)

Country	T (S/W °C)	P (S/W mm/day)
Afghanistan	24.1/3.3	0.8/5.9
Bahrain	NA/NA	NA/NA
Iran	27.4/7.4	0.3/3.6
Iraq	31.7/10.8	0.0/3.3
Israel	26.0/13.1	0.1/2.9
Jordan	26.0/10.4	0.1/2.8
Kuwait	35.7/14.7	0.0/1.4
Lebanon	23.9/9.8	0.0/10.1
Oman	29.0/21.4	1.2/1.5
Qatar	35.0/19.0	0.0/1.5
Saudi Arabia	31.7/17.5	0.40/1.0
Syria	27.6/8.0	0.02/4.3
Turkey	20.6/1.6	1.4/6.1
Yemen	26.7/19.3	2.6/1.3

Pressure

The general features of mean sea-level pressure in winter can be summarized as follows:

1. *Siberian and Azores High*: the northern portion of an anticyclonic ridge over the region is an extension of the Siberian High. This ridge results in northeast surface winds over the Black and Caspian Seas. The southern portion is related to the Azores High which extends over the southern half of the Arabian Peninsula and is associated with an easterly current thought to be related to the easterly trade winds or to the northeast monsoons which prevail over India in winter. The intensity of this ridge is directly related to the passage of mid-latitude winter storms entering into the Mediterranean – these will be discussed next.
2. *Mid-latitude storms*: these depressions enter into the Mediterranean from the Atlantic via different tracks. These tracks include (a) northern Spain, (b) southern France, and (c) straits of Gibraltar. Some are secondary low-pressure systems, meaning they further develop upon entering into the Mediterranean – mainly the Gulf of Genoa, over Cyprus and in the Adriatic Sea. Orographic depressions are created on the lee side of the Alps in Switzerland and Atlas Mountains over North Africa. In general the Mediterranean low-pressure systems travel only 20–30 miles inland; however, when these low-pressure systems come in contact with the Caspian Sea and Arabian Gulf they are strengthened and have enough energy to penetrate into Iraq, Iran, and Afghanistan – providing the major source of rain to this region. Overall, it is the Mediterranean depressions which carry most of the precipitation to the northern half of southwest Asia – the area of greatest precipitation being the coastal regions of the Black, Caspian, and Mediterranean Seas.
3. *Subtropical jet*: the 200 mn upper-level winds show an axis of strong westerly winds centered at approximately 23–27°N – this is the subtropical jet. Upper tropospheric divergence around the jet – together with low-level convergence – is considered a major factor in the development of the convective activity in non-summer seasons. The high cloud base and dryness over Saudi Arabia does not allow precipitation to fall to the surface; strong gusts, sandstorms and thunderstorms are observed. Upper tropospheric divergence, coupled with lower-level convergence, is also considered an important precursor for the development and maintenance of small-scale systems such as desert depressions.
4. *Polar jet:* along with the subtropical jet stream, a permanent feature of the upper tropospheric flow pattern in fall, winter, and spring, southwest Asia is also invaded by active branches of the polar jet which is associated with the passage of low-pressure systems generated in the Mediterranean.

Temperature and precipitation

In southwest Asia the precipitation primarily falls from winter storms moving eastward from the Mediterranean, with the high mountains of the region intercepting most of the water and the interior high plains left with large stretches of barren desert. This wintertime precipitation generally occurs between the months of November and April, with the peak between January and March. Much of the precipitation falls as snow in the higher elevations and the timing and amount of snowmelt is an important factor in the irrigated agriculture prevalent in the region.

Summer

The following section describes general climate conditions during the summer season in terms of sea-level pressure, temperature and precipitation. Average summer (June through August; JJA) total precipitation and average summer temperature are presented in Figure A75 and Table A40.

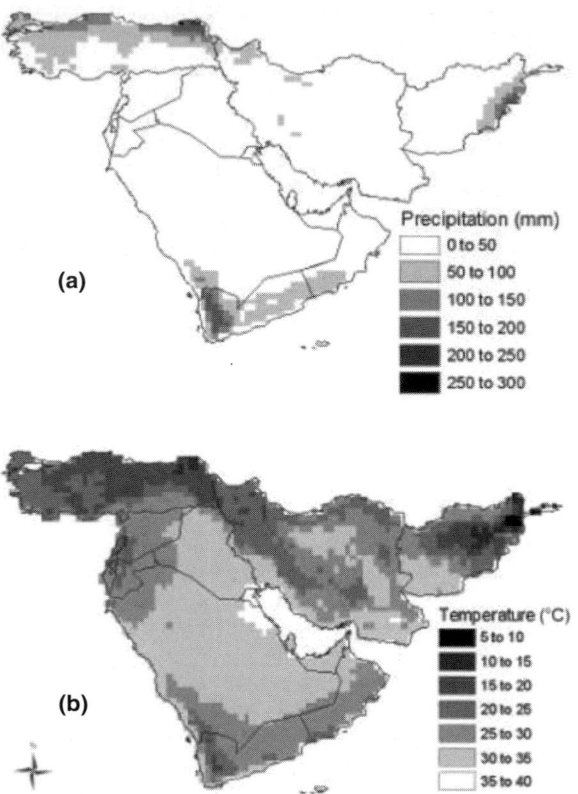

Figure A75 Average total summer (JJA) precipitation is shown in (**a**) and average summer temperature is shown in (**b**). Data used are same as in Figure A24.

Pressure

1. *Indian monsoon*: in summer the western branch of the Asian Low dominates southwest Asia which is approximately centered over 30°N. While appearing essentially fixed in time and space over the Arabian/Persian Gulf in the daily series of SLP, synoptic charts reveal zonal variability in the western branch – sometimes extending all the way into the eastern Mediterranean. During such an event, warm subtropical air will advect over the eastern Mediterranean region resulting in intense heat waves. The basic theory of these summer heat waves relies on the latitudinal elongation of the monsoon low over the Arabian/Persian Gulf or to an overall strengthening of the low.
2. *Quasi-stationary surfaces*: the summer season marks a time of generally stable conditions lacking significant daily variability. This allows for the emergence of quasi-stationary surfaces of separation between two unique air streams; namely the subtropical front and the intertropical front.

Temperature and precipitation

Very little rain falls in most of southwest Asia during the summer season. However, in eastern Pakistan the primary rainfall season is summer, associated with the northernmost advance of the Asian monsoon, which results in a summertime maximum in precipitation in the northern mountain regions of Pakistan but generally suppresses rainfall over Iran and Afghanistan. Dust storms occur throughout the year in the desert high plains. Such storms are prevalent through much of the region in summer, often associated with the "wind of 120 days", the highly persistent winds of the warm season which blow from north to south.

Observational records

Lack of meteorological observations continues to pose a problem for long-term climate monitoring aimed at "clarifying the climate picture", making it difficult to cleanly differentiate between different climatic regimes.

Global warming

According to the Third Assessment Report (Working Group II) of the Intergovernmental Panel on Climate Change continuing emissions of greenhouse gases are likely to result in significant changes in mean climate and its intraseasonal variability in Asia (IPCC, 2001). General circulation models (GCM) suggest that the area-averaged annual mean warming over Asia would reach about 3°C by 2050 and 5°C by 2080. When the cooling effects of sulfate aerosols are added to the models, surface warming is reduced slightly, reaching 2.5°C by 2050 and 4°C by 2080. Seasonally speaking, warming over Asia is expected to be higher during northern hemisphere (NH) winter. Both the annual mean and winter season minimum temperature is expected to rise more than the maximum temperature in Asia, leading to a decrease in the diurnal temperature range (DTR) over these time periods. During summer, however, DTR is expected to increase, suggesting a rise in average summertime maximum temperature. When compared with other regions, the summertime DTR increase is expected to be significantly higher.

All GCM project an enhanced hydrological cycle resulting in rainfall increases over Asia due to continued growth in the emissions of greenhouse gases. A 7% increase is projected by 2050 and an 11% increase is projected by 2080. When the cooling effects of sulfate aerosols are added to the models the projected increase in precipitation is reduced to 3% in 2050 and 7% in 2080. As with temperature, precipitation increases are expected to be greatest during the NH winter. A decline in summer precipitation is expected in southwest Asia. Because this area already receives so little precipitation, severe water-stress conditions resulting in expansion of the deserts are a possibility – when factoring in both the rise in temperature and loss of soil moisture. Increased temperatures can also affect the amount and timing of snowmelt and river flow in southwest Asia. In addition, global warming could affect the role the tropical oceans play in the climate of the region as well as the character of winter storms that currently supply the majority of cold season precipitation.

Overall, southwest Asia appears to be highly vulnerable to anticipated climate change projections. Water stress is expected to be high with significant negative impacts on the agricultural sector. Agricultural productivity in southwest Asia is likely to suffer severe losses due to increases in temperature, soil moisture depletion, and the increased potential for severe drought – ultimately increasing the potential for food insecurity and famine. Adaptation strategies identified for southwest Asia include: (1) transition from conventional crops to intensive greenhouse agriculture, (2) improve conservation of freshwater supply for times of enhanced water stress, and (3) protect lakes and reservoirs.

Bibliography

CID, 1999. CID Geography Datasets: Köppen–Geiger Climate zones, Center for International Development at Harvard University, Cambridge, MA, USA. Available online at http://www.cid.harvard.edu/cidglobal/economic.htm.

IPCC, 2001. Asia. In McCarthy, J.J., Canziani, O.F., Leary, N.A., Dokken, D.J., and White, K.S., eds., *Climate Change 2001: Impacts, Adaptation, and Vulnerability. Contribution of Working Group II to the Third Assessment Report of the Intergovernmental Panel on Climate Change.* Cambridge, New York: Cambridge University Press.

New, M.G., Hulme, M., and Jones, P.D., 2000. Representing twentieth-century space–time climate variability. Part II: Development of 1901–1996 monthly grids of terrestrial surface climate. *Journal of Climate*, **13**: 2217–2238.

Takahashi, K., and Arakawa, H., 1981. Climate of southern and western Asia. In Landsberg, H.E., ed., *World Survey of Climatology*, vol. 9. Amsterdam: Elsevier, pp. 183–256.

USGS, Global 30 Arc-second Elevation Dataset, Land Processes Distributed Active Archive Center (LPDAAC), EROS Data Center, Sioux Falls, SD, USA. Available online at http://edcdaac.usgs.gov.

Heidi M. Cullen

Cross-references

Airmass Climatology
Asia, Climate of South
Climate Classification
Deserts
Extratropical Cyclones
Mediterranean Climates
Monsoons and Monsoon Climate

ATMOSPHERE

The atmosphere is the envelope of air surrounding the Earth. It consists of a physical mixture of gases and particle matter. Based on its temperature structure, the atmosphere can be divided into several sections. Below about 12 km is the *troposphere*, where the majority of the world's weather occurs and the temperature broadly decreases from about 15°C at the Earth's surface to −54°C at the top. Almost all of the processes of vertical transfer of atmospheric properties through turbulence and mixing occur in the troposphere.

Above the troposphere the temperatures increase to a level of about 50 km, in the region called the *stratosphere*. Here the atmosphere is very stable and contains layers of gaseous and particle matter, mainly of volcanic origin. The troposphere and the stratosphere are separated by the *tropopause*, which is located where temperatures suddenly begin to increase with altitude. Above the stratosphere is the *stratopause*, which separates the stratosphere from the *mesosphere* (48–78 km), where temperatures decrease with altitude again. Above the mesosphere is the *thermosphere* (above 80 km), another stratum in which temperature increases with altitude, where the thin outer layers of the atmosphere are directly interacting with emissions from the sun. Both the mesosphere and thermosphere contain gaseous atoms and ions in very rarefied concentrations.

Table A41 Physical characteristics of the atmosphere at sea level

Descriptor	Value	Units
Pressure	1013.25	hPa
Temperature	283	K
Density	1.29×10^{-3}	g/cm
Molecular weight (dry air)	28.966	—

Table A42 Average dry air composition below 25 km (after Barry and Chorley, 1987)

Gaseous component	Chemical symbol	Percentage volume
Nitrogen	N_2	78.08
Oxygen	O_2	20.94
Argon	A	0.93
Carbon dioxide	CO_2	0.034*
Neon	Ne	0.0018
Helium	He	0.0005
Ozone	O_3	0.00006

* Variable and presently increasing in concentration on a global scale.

Table A41 presents the average physical characteristics of the atmosphere at sea level. This part of the atmosphere is where the greatest interactions with the Earth's surface occur, and is defined as the *boundary* (or *mixing*) *layer*. Above the surface both atmospheric density and pressure decrease logarithmically. Thus, approximately half the mass of the atmosphere is situated below an altitude of 5.6 km, and the pressure at this altitude is about 500 hPa.

The atmosphere itself is not a chemical compound, but contains a wide variety of chemical compounds and individual gas molecules as part of its composition. The composition is not constant, but is highly variable between locations and over time. In terms of content it is dominated by four gases: nitrogen, oxygen, argon, and carbon dioxide, as shown in Table A42. All of these gases are stable in concentration except carbon dioxide and ozone. Carbon dioxide is a major greenhouse gas, and is controversially linked to the rise in global air temperatures over the past century. Ozone occurs naturally as a layer in the stratosphere, between 10 and 50 km and peaking in concentration around 25 km. Its major role is to protect the Earth's surface from harmful ultraviolet radiation from the sun. The ozone layer is threatened by partial destruction as a result of chemical reactions with chlorofluorocarbon emissions from human activities.

The most important gas not included in Table A42 is water vapor. Concentrations range from almost 0% by volume over the driest regions of the Earth (deserts and polar ice caps) to about 4% in the hot tropical regions. A greenhouse gas, water vapor, is an essential part of the hydrologic cycle and is crucial to the maintenance of life on Earth. It is the only compound that can exist naturally in all three states at one location.

The troposphere and lower stratosphere also contain spatially inhomogeneous quantities of particles, dust, and aerosols. Particle matter in the stratosphere originates mainly from volcanic explosions, and through scattering of shortwave radiation may create spectacular sunsets and affect global temperatures for periods of up to 2 years. In the troposphere, particle matter originates from sea spray, windblown surface material, air pollution emissions, and chemical reactions. Particles form condensation nuclei for clouds, affect visibility through short wave radiation scattering, and are often responsible for pollution episodes in major cities around the globe.

Howard A. Bridgman

Bibliography

Barry, R.G., and Chorley, R.J., 1987. *Atmosphere, Weather and Climate*, 5th edn. London: Methuen.

Goody, R., 1995. *Principles of Atmospheric Physics and Chemistry*. New York: Oxford University Press.

Lutgens, F.K., and Tarbuck, E.J., 2001. *The Atmosphere*, 8th edn. Upper Saddle River, NJ: Prentice-Hall.

Wayne, R.P., 1999. *Chemistry of Atmospheres*, 3rd edn. New York: Van Nostrand.

Cross-references

Aerosols
Air Pollution Climatology
Atmospheric Nomenclature
Greenhouse Effect and Greenhouse Gases
Precipitation

ATMOSPHERIC CIRCULATION, GLOBAL

Global atmospheric circulation consists of the observed wind systems with their annual and seasonal variations, and is the principal factor in determining the distribution of climatic zones, while variations in the atmospheric and oceanic circulations are responsible for many of the observed longer-term fluctuations in climate. The two major controls on the global wind circulation are inequalities in radiation distribution over the Earth's surface and the Earth's rotation. Global radiation distribution, together with gravity, drives the global atmospheric circulation, whereas the Earth's rotation determines its shape. Basically, the mean surface circulation consists of easterly winds with equatorial components in the tropics and westerly winds with poleward components in middle latitudes, the corresponding meridional flows aloft being reversed. Weak surface easterlies are observed in polar regions, and extensive areas of calms in the equatorial and also the subtropical regions. Strong, westerly winds are found in the upper troposphere poleward of about 25°N and 25°S.

Causes of the global atmospheric circulation

Radiation distribution

The planet Earth receives energy from the sun in the form of short-wave radiation, while radiating an equal amount of heat to space in the form of longwave radiation. This balance of heat gain equaling heat loss applies only to the planet as a whole over several annual periods; it does not apply to any specific area for short periods of time. The equatorial region absorbs more radiation than it loses, while the polar regions radiate more heat than they receive. (The distribution of radiation over the Earth's surface is reviewed by Lockwood, 1979, 1985.) However, the equatorial belt does not become warmer during the year, nor do the poles become colder. This is because heat flows from the warm regions to the cold regions, maintaining the observed temperatures. An exchange of heat is brought about by the fluid motions in the atmosphere and oceans, thus forming the observed circulation of the atmosphere and oceans.

If the Earth's surface were homogeneous, and the planet was not rotating, the imposition of a theoretical latitudinal (north-south) heating gradient, with cold poles and a warm equator, would probably result in a single circulation cell in each hemisphere with an upward limb at the equator and downward limbs at the poles. Such cells would be energetically direct cells because they result from a direct transformation of potential to kinetic energy. They are often referred to as Hadley cells in tribute to the eighteenth-century scientist who first suggested their possible existence. In reality, on an Earth with one face always toward the sun, the equatorial segment of the sunward face would be extremely hot and the dark face extremely cold. The temperature gradient would be from the daylight to the dark face, not from equator to pole, with a corresponding circulation cell between the two faces. The Earth's rotation smears the region of strong solar tropical heating zonally around the whole globe and generates the observed equator – polar temperature gradients.

The Earth's rotation

The real Earth rotates, and this makes the existence of simple Hadley cells extending from equator to pole impossible, since the rotation generates east–west motions in the atmosphere. To an observer on the rotating Earth it appears as if a force is acting on moving air particles that causes them to be deflected from their original path, and this apparent force-per-unit mass is termed the Coriolis force. The Coriolis force turns the wind to the right in the northern hemisphere and to the left in the southern hemisphere. Therefore, in the northern hemisphere, the upper poleward current of the Hadley cell assumes a strong eastward component (westerly wind) and the lower equatorial current assumes a westward component (easterly wind). For a given heating gradient the turning of the wind tends to reduce the efficiency with which a single cell can transport heat poleward. For a sufficiently large rotation rate a balance cannot be maintained, and the imposed latitudinally dependent heating would cause the meridional temperature gradient to increase with time, while remaining circularly symmetric about the pole. However, there comes a time when small meridional displacements of a suitable east–west size can become dynamically unstable and grow, the process being an example of baroclinic instability.

Baroclinic instability

The growth of these wave disturbances under baroclinic instability is characterized by the ascent of warmer airmasses and descent of colder airmasses, causing a decrease of potential energy and an associated release of kinetic energy. Potential temperature increases with height; thus this ascent and descent of airmasses takes place in a manner described by the term slantwise convection. In the normal atmosphere equal potential temperature (isentropic) surfaces slope upward from lower to higher latitudes, and in slantwise convection the trajectories of individual air particles are tilted at an angle to the horizontal that is comparable with, but less than, the slope of the isentropic

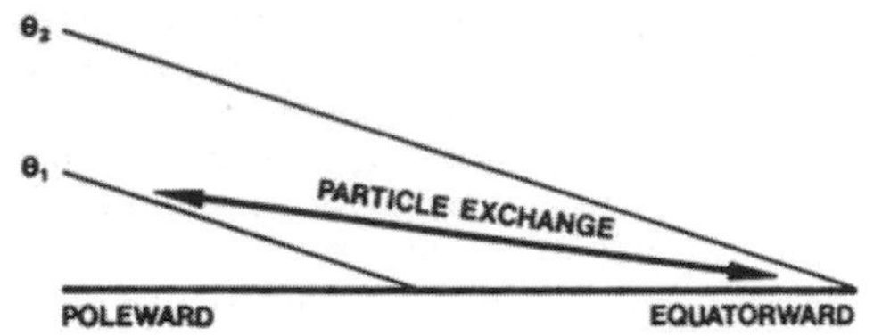

Figure A76 Height–latitude cross-section of the troposphere showing potential temperature θ increasing equatorward and upward. The system is stable for vertical convection, but energy can be released if particles are exchanged by slantwise convection along the thick solid arrow.

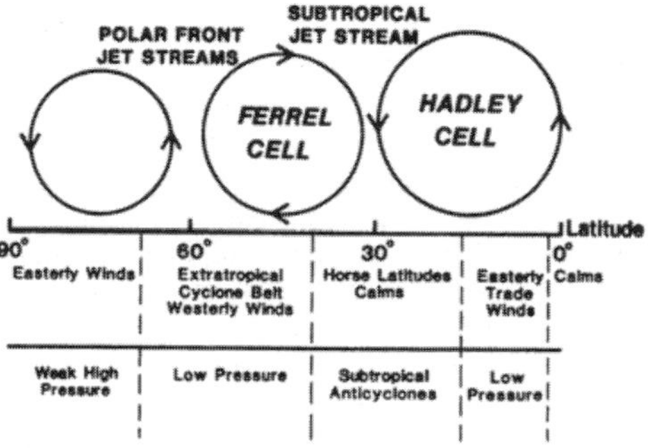

Figure A77 A schematic representation of the mean meridional circulation of the northern hemisphere.

surfaces (Figure A76). Therefore, while an air parcel may be prevented from rising vertically because the prevailing lapse rate is less than the appropriate adiabatic lapse rate, poleward travel brings it to an environment more dense than itself, enabling it to rise. Air at higher levels and higher latitudes may similarly descend by equatorward movement. Further details may be found in Lorenz (1967), Barry and Carleton (2001), and Houghton (2002).

Mean global atmospheric circulation

Mean meridional circulation

Disturbances with the dimensions of about five to eight waves about the pole will grow fastest under baroclinic instability. At the surface, anticyclones in this size range that develop will grow and ultimately move equatorward, finding final residence in the subtropics. Similarly, surface low-pressure systems develop and move poleward. The resultant mean meridional circulation is no longer simple. The single equator-to-pole direct cell postulated by Hadley to exist if the Earth did not rotate is contracted equatorward, so that the poleward descending limb coincides with the axis of the surface subtropical anticyclones (Figure A77). In the tropics baroclinic instability is virtually non-existent because of the weak Coriolis force, so that direct mean meridional circulation (the Hadley cell) is still the most efficient means for effecting a poleward heat transfer. In mid-latitudes unstable wave motion is the most efficient means for poleward heat transfer. An energetically weak indirect mean meridional circulation, known as the Ferrel cell, exists in middle latitudes.

Subtropical jet streams

A schematic representation of the mean meridional circulation in the northern hemisphere during winter is shown in Figure A77. The simple Hadley cell circulation is clearly seen south of 30°N. Eastward angular momentum is transported from the equatorial latitudes to the middle latitudes by nearly horizontal eddies, 1000 km or more across, moving in the upper troposphere and lower stratosphere. Such a transport, together with the dynamics of the middle latitude atmosphere, leads to an accumulation of eastward momentum between 30° and 40° latitude, where a strong meandering current of air, generally known as the subtropical westerly jet stream, develops. The cores of the subtropical westerly jet streams in both hemispheres and throughout the year occur at an altitude of about 12 km. The air subsiding from the jet streams forms subtropical anticyclones. More momentum than is necessary to maintain subtropical jet streams against dissipation through internal friction is transported to these zones of strong winds. The excess is transported downward to maintain the eastward-flowing surface winds of the middle latitudes against ground friction. The supply of eastward momentum to the Earth's surface in middle latitudes tends to speed up the Earth's rotation. Counteracting such potential speeding-up of the Earth's rotation, air flows from the subtropical anticyclones toward the equatorial regions, forming the so-called trade winds. The trade winds, with a strong component directed toward the west, tend to retard the Earth's rotation, and in turn gain eastward momentum.

Mean surface winds

An idealized mean surface wind circulation with associated pressure distribution, as shown in Figure A78, may be described as follows:

1. The equatorial trough. A shallow belt of low pressure on or near the equator with light or variable winds.
2. The trade winds. Between the equatorial trough and latitudes 30°N and 30°S are northeast winds in the northern hemisphere and southeast winds in the southern hemisphere.
3. The subtropical anticyclones. Ridges of high pressure between about latitudes 30–40°N and S, associated with light, variable winds.
4. The westerlies. Belts of generally westerly winds, southwest in the northern hemisphere and northwest in the southern hemisphere, between about latitudes 40–60°N and S.
5. The temperate latitude low-pressure systems. Variable winds converging into low-pressure belts at about 60°N and 60°S, the subpolar low-pressure area.
6. The polar anticyclones. Regions of outflowing winds with an easterly component, diverging from weak high-pressure systems near the poles.

Further detailed descriptions of mean atmospheric circulation patterns may be found in the textbooks listed in the Bibliography.

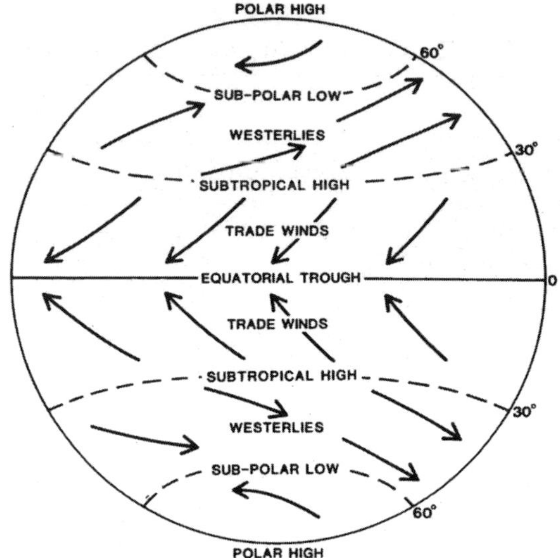

Figure A78 Idealized mean surface wind circulation.

Tropical circulation patterns

The mean surface winds in the tropics are directed predominantly toward the equator within latitudes 30°N and 30°S. Such motions can persist only if some force accelerates the air in that direction, and the only force in the atmosphere known to produce accelerations of large-scale windfields is the pressure force. Therefore, high pressure must exist in the subtropics and low pressure in the equatorial zone, with the resulting pressure gradients driving the mean winds. The average latitude of the equatorial low-pressure trough is about 5°S in the northern winter and 15°N in the southern winter. The annual mean latitude is about 5°N, so the meteorological southern hemisphere is larger than the northern hemisphere. This is largely a result of the differing ocean/continent configurations of the hemispheres (McGregor and Nieuwolt, 1998).

The trade winds

Between the equatorial low-pressure trough and the subtropical high-pressure belt lie two belts of tropical easterlies, each with an equatorial meridional flow. These wind systems are known as the northeast trade winds in the northern hemisphere and the southeast trade winds in the southern hemisphere. There tends to be a certain monotony about the weather of the trade winds since their extreme steadiness reflects the permanence of the subtropical anticyclones, which are inclined to be the most intense in winter, making the trade winds strongest in winter and weakest in summer. Figure A79 shows that all tropical oceans except the northern Indian Ocean have extensive areas of trade winds. Each trade wind region contains definite centers of resultant high wind speed that reach 6–8 m/s. In contrast, wind speeds are low in the subtropics and in the equatorial trough zone. According to Riehl (1979), when wind speed is averaged without regard to direction, an almost uniform value of 7 m/s appears everywhere except within the equatorial trough.

During the summer of 1856 an expedition under the direction of C. Piazzi-Smyth visited the island of Tenerife in the Canary Islands to make astronomical observations from the top of the Peak of Tenerife. On two of the journeys up and down the 3000 m mountain, Piazzi-Smyth carefully measured the temperature, moisture content, direction, and speed of the local trade winds. He found that an inversion was often present, and that it was not located at the top of the northeast trade regime, but was situated in the middle of the current; thus, it could not be explained as a boundary between two airstreams from different directions. He also noticed that the top of the cloud layer corresponded to the base of the inversion. These observations by Piazzi-Smyth have been confirmed many times and the trade-wind inversion is now known to be of great importance in the meteorology of the tropics. Broad-scale subsidence in the subtropical anticyclones is the main cause of the very dry air above the trade-wind inversion. The subsiding air normally meets a surface stream of relatively cool maritime air flowing toward the equator. The inversion forms at the meeting point of these two airstreams, both of which flow in the same direction, and the height of the inversion base is a measure of the depth to which the upper current has been able to penetrate downward.

Subsidence is most marked at the eastern ends of the subtropical anticyclonic cells; that is to say, along the desert cold-water coasts of the western edges of North and South America and Africa, and it is here that the trade-wind inversion is at its lowest. Normally, as the trade winds approach the equator, the trade inversion increases in altitude and conditions become less arid. Over the oceans the intense tropical radiation evaporates water that is carried aloft by thermals and eventually distributed throughout the layer below the trade inversion. The result is that the layer below the inversion becomes more moist as the trade wind nears the equator and the continual convection in the cool layer forces the trade inversion to rise in height.

The equatorial trough

Trade winds from the northern and southern hemispheres meet in the equatorial trough, a shallow trough of low pressure generally situated near the equator. Its position is clear-cut in January (except in the central South Pacific), and in July the trough can be located with ease over Africa, the Atlantic, and the Pacific to about 150°E. The equatorial trough shows a marked tendency to meander with longitude; in January it ranges from 17°S to 8°N, in July from 2°N to 27°N. This is largely the result of wide oscillations in the monsoon regions that cover the whole of southern Asia and North Africa in the northern summer, and southern Africa and Australia in the southern summer. The excursions into the southern hemisphere are smaller than those into the northern hemisphere, because the more constant westerly circulation of the southern hemisphere's middle latitudes constrains the equatorial trough to near the equator. Over southern Asia and the Indian Ocean the mean seasonal positions are around 15°N and 5°S, a difference of 20° latitude. In contrast, the trough is quasistationary over the oceanic half of the Pacific and Atlantic, where seasonal displacement is restricted to 5° latitude or less. Winds in the equatorial trough generally are calm or light easterly. Over some areas, such as the Indian

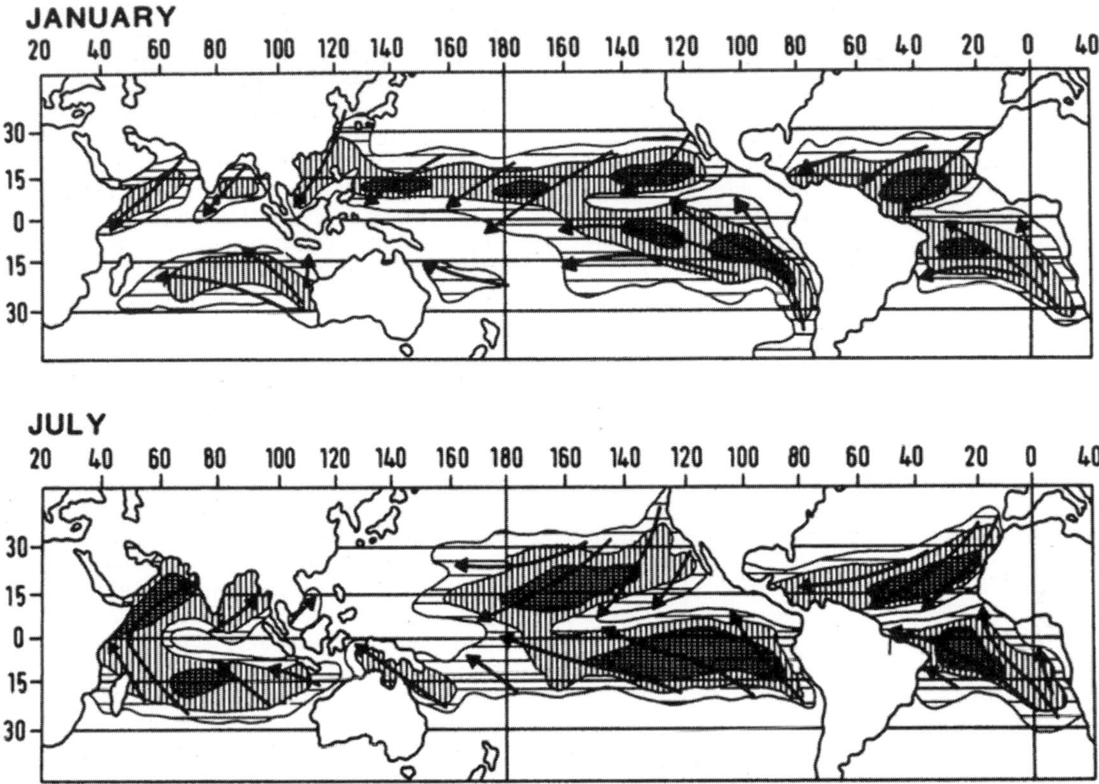

Figure A79 The trade-wind systems of the world in January and July. The isopleths are in terms of relative constancy of wind direction and enclose shaded areas where 50%, 70%, and 90% of all winds blow from the predominant quadrant with Beaufort force 3 or more (over 3.4 m/s) (after Crowe, 1971).

Ocean and Southeast Asia, the structure of the trough is complex and westerlies may be found.

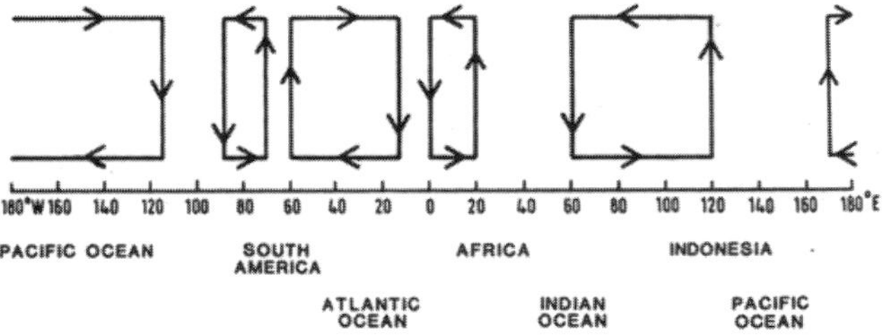

Figure A80 Schematic representation of the Walker circulation near the equator during the northern winter.

The Walker circulation and the Southern Oscillation

Dietrich and Kalle (1957) have mapped the difference between the sea-surface temperature and its average value along the part of each latitude circle situated over the oceans. If cold-water regions are defined as ocean areas with sea surface temperatures below the latitudinal average, by far the most extensive cold-water area is the South Pacific cold water stretching about 85° westward from the coast of South America. In contrast the South Atlantic cold water continues westward from the coast of Africa by only about 40° longitude.

Bjerknes (1969) believes that when the Pacific cold ocean water along the equator is well developed, the air above will be too cold to take part in the ascending motion of the Hadley cell circulation. Instead, the equatorial air flows westward between the Hadley cell circulations of the two hemispheres to the warm western Pacific, where, having been heated and supplied with moisture from the warmer waters, the equatorial air can take part in large-scale, moist adiabatic ascent. In Indonesia huge cloud clusters with a diameter of more than 600 m develop each day, giving an area-averaged rainfall of around 2200 mm per year. This is equivalent to a release of latent heat of about 170 W m^{-2}, much more than the net radiation near the surface. The resulting thermally driven circulation (Figure A80) between an equatorial heat center – the maritime continent of Indonesia – and a cooling area in the eastern Pacific is often known as a Walker circulation. As shown in Figure A80, the equatorial Atlantic is analogous to the equatorial Pacific in that the warmest part is in the west, at the coast of Brazil. West–east contrasts of water temperature in the Atlantic are much smaller than in the Pacific, so the potential for a Walker-type circulation is less. Nevertheless, in January a thermally driven Walker circulation may operate from the Gulf of Guinea to the Andes, with the axes of the circulation near the mouth of the Amazon.

Sir Gilbert Walker described in his papers in the 1920s and 1930s what he named the Southern Oscillation. The Southern Oscillation is dominated by an exchange of air between the Southeast Pacific subtropical high and the Indonesian equatorial low, with a period that varies between roughly 1 and 5 years. The Southern Oscillation may be defined in terms of the difference in sea level pressure between Darwin, Australia and Tahiti. Records are available, with the exception of a few years and occasional months, from the late 1800s. During one phase of this oscillation the trade winds are intense and converge into the warm western tropical Pacific, where rainfall is plentiful and sea-level pressures low. At such times the atmosphere over the eastern tropical Pacific is cold and dry. During the complementary phase the trade winds relax, the zone of warm surface waters and heavy precipitation shifts eastward, and sea-level pressure rises in the west while it falls in the east. The latter phase is the more unusual and in the eastern Pacific has become known as El Niño, while vigorous episodes of the former are often termed La Niña. The combined atmospheric/oceanic conditions that give rise to these changes in rainfall across the Pacific and neighbouring areas are referred to as El Niño Southern Oscillation (ENSO) events. The El Niño phenomena are associated with extreme negative Southern Oscillation values, but for much of the time the series exhibits continuous transitions from high to low values with most values being positive. ENSO is important climatologically (Glantz, 1996; Diaz et al., 2001) for two main reasons. First, it is one of the most striking examples of interannual climatic variability on an almost global scale. Secondly, in the Pacific it is associated with considerable fluctuations in rainfall and sea surface temperature, and also with extreme weather events around the world.

Monsoon climates of southern Asia

Originally the term monsoon was applied to the surface winds of southern Asia that reverse between winter and summer, but the word is now used for many different types of phenomena, such as the stratospheric monsoon, the European monsoon, etc. The "classic" characteristics of a monsoon climate are mainly to be found in the Indian subcontinent, where over much of the region the annual changes may conveniently be divided as follows:

1. The season of the northeast monsoon:
   (a) January and February, winter season
   (b) March to May, hot weather season.
2. The season of the southwest monsoon:
   (a) June to September, season of general rains
   (b) October to December, postmonsoon season.

Halley proposed the first explanation of the Asiatic monsoon in a memoir presented to the Royal Society in 1686. His theory is general and can be applied to all continental regions, not only Asia. During winter intense cooling over the cold continents leads to the establishment of thermal high-pressure systems, while pressure remains relatively low in the warmer, lighter air over the oceans. The surface air flow therefore has a component from the highs over the land toward the lows over the ocean. During summer the land is warmer than the sea and a reverse current circulates from the relatively cool sea toward the heated land. The average sea-level pressure charts for the two extreme seasons show a reversal of pressure over Asia and its maritime surroundings. In winter the Siberian anticyclone accumulates subzero masses of air (−40°C to −60°C) at an equivalent mean sea-level pressure of between 1040 and 1060 mb, while in summer the torrid heat (about 40°C) reduces the pressure to 950 mb over northwest India. Modern explanations of the Indian monsoon are much more complex (see Lockwood, 1974; Lighthill and Pearce, 1981; Pant and Rupa Kumar, 1997), but follow the basic idea of Halley.

According to Miller and Keshavamurthy (1968), the southwest monsoon current in the lower 5 km near India consists of two main branches; the Bay of Bengal branch, influencing the weather over the northeastern part of India and Burma, and the Arabian Sea branch, dominating the weather over the western, central, and northwestern parts of India. The low-level flow across the equator during the southwest monsoon is not evenly distributed between latitudes 40°E and 80°E as was previously thought, but has been found by Findlater (1969, 1972) to take the form of low-level, high-speed southerly currents that are concentrated between about 38°E and 55°E. A particularly important feature of this flow is the strong southerly current with a mean wind speed of about 14 m/s, observed at the equator over eastern Africa from April to October. The strongest flow occurs near the 1.5 km level, but it often increases to more than 25 m/s and occasionally to more than 45 m/s at heights between 1.2 km and 2.4 km. According to Findlater this high-speed current flows intermittently during the southwest monsoon from the vicinity of Mauritius through Madagascar, Kenya, eastern Ethiopia, Somalia, and then across the Indian Ocean toward India.

When upper winds are taken into account, the Asian monsoon is a fairly complex system (Pant and Rupa Kumar, 1997). During the northern winter season (Figure A81), the subtropical

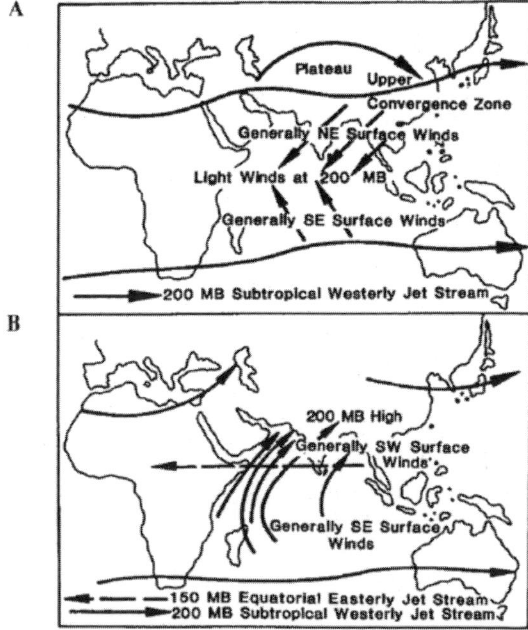

Figure A81 Schematic illustration of the major features of the Asian monsoon. **A:** northern winter; **B:** northern summer (after Lockwood, 1985).

westerly jet stream lies over southern Asia, with its core located at about 12 km altitude. It divides in the region of the Tibetan Plateau, with one branch flowing to the north of the plateau, and the other to the south. The two branches merge to the east of the plateau and form an immense upper convergence zone over China. In May and June the subtropical jet stream over northern India slowly weakens and disintegrates, causing the main westerly flow to move north into central Asia. While this is occurring, an easterly jet stream, mainly at about 14 km, builds up over the equatorial Indian Ocean and expands westward into Africa. The formation of the equatorial easterly jet stream is connected with the formation of an upper-level, high-pressure system over Tibet. In October the reverse process occurs; the equatorial easterly jet stream and the Tibetan high disintegrate, while the subtropical westerly jet stream reforms over northern India.

Middle and high latitude circulation patterns

The greatest atmospheric variability occurs in middle latitudes, from approximately 40–70°N and S, where large areas of the Earth's surface are affected by a succession of cyclones (depressions) and anticyclones or ridges. This is a region of strong thermal gradients with vigorous westerlies in the upper air, culminating in the polar-front jet stream near the base of the stratosphere. The zone of westerlies is permanently unstable and gives rise to a continuous stream of large-scale eddies near the surface, the cyclonic eddies moving poleward and the anticyclonic ones equatorward.

Compared to the Hadley cells, the middle-latitude atmosphere is highly disturbed and the suggested meridional circulation in Figure A77 is largely schematic. At the surface the predominant features are closed cyclonic and anticyclonic systems of irregular shape, while higher up, smooth wave-shaped patterns are the general rule. The dimensions of these upper waves are much larger than those of the surface cyclones and anticyclones which have dimensions of 1000–3000 km, and only rarely is there a one-to-one correspondence. In typical cases there are four or five major waves around the hemisphere, and superimposed on these are smaller waves that travel through the slowly moving train of larger waves. The major waves are called long waves or Rossby waves, after Rossby (1939, 1940, 1945), who first investigated their principal properties. These upper Rossby waves are important because they strongly influence the formation and subsequent evolution of the closed surface synoptic features.

Analyses of long time-series of climatogical data reveal large-scale correlations between fluctuations at remote locations. These fluctuations occur at the low frequency range of timescales, and they are called "teleconnections" to stress the correlation-at-a-distance aspect of their nature. The term "teleconnection" was introduced by Ångstrom (1935) in the context of patterns of climatic fluctuations; Bjerknes (1969) and Namias (1963, 1969) later used it to describe patterns of atmospheric responses to a remote surface forcing. Some teleconnections arise simply from natural preferred modes of the atmosphere associated with the mean climate state and the land–sea distribution, while several such as ENSO and the North Atlantic Oscillation are directly linked to sea surface changes.

In a simple atmosphere Rossby waves could arise anywhere in the middle latitude atmosphere, as is observed in the predominantly ocean-covered southern temperate latitudes. In contrast the northern temperate latitude Rossby waves tend to be locked in certain preferred locations. These preferred locations arise because the atmospheric circulation is influenced not only by the differing thermal properties of land and sea, but also by high mountain ranges and highlands in general. Influences on the propagation of Rossby waves through the atmosphere include zonal temperature asymmetries, transient synoptic weather systems and baroclinic effects.

Atmospheric thermal patterns

The basic thermal pattern of the lower half of the atmosphere in the temperate latitudes is partly controlled by prevailing mean surface temperatures. The mean thermal pattern may be usefully investigated by using the concept of thickness lines. The thickness of an atmospheric layer bounded by two fixed-pressure surfaces, 1000 mb and 500 mb, for example, is directly proportional to the mean temperature of the layer. Thus, low thickness values correspond to cold air and high thickness values to warm air. The geostrophic wind velocity at 500 mb is the vector sum of the 1000 mb geostrophic wind and the theoretical wind vector (the thermal wind) that blows parallel to the 1000–500-mb thickness lines with a velocity proportional to their gradient. This theoretical wind is known as the thermal wind and is shown in Figure A82.

The thermal wind is directed along the mean isotherms with lower temperature to the left in the northern hemisphere. Thus, in temperate latitudes the thermal wind will be westerly, and, since north–south horizontal temperature gradients are relatively steep, it will be strong. The upper winds are the vector sum of a rather weak surface wind field and a vigorous thermal wind field. This implies that, in the lower middle-latitude troposphere, winds will become increasingly westerly with altitude and that the upper wind field is strongly controlled by the thermal wind. Therefore, a parallel exists between the mean topography of the 500-mb level and the mean 1000–500-mb thickness patterns.

The mean January thickness patterns for the northern hemisphere show two dominant troughs near the eastern extremities of the two continental landmasses, while ridges lie over the eastern parts of the oceans. A third weak trough extends from northern Siberia to the eastern Mediterranean. Climatologically, the positions of the main troughs may be associated with cold air over the winter landmasses, and the positions of the ridges with relatively warm sea surfaces. As seen from Figure A83, the mean January 500-mb wind field, and therefore the mean Rossby wave locations, is similar to the thickness field.

In July the mean ridge in the thickness pattern found in January over the Pacific has moved about 25°W, and now lies over the warm North American continent, while there is a definite trough over the eastern Pacific. Patterns elsewhere are less marked, but a weak trough does appear over Europe, and may

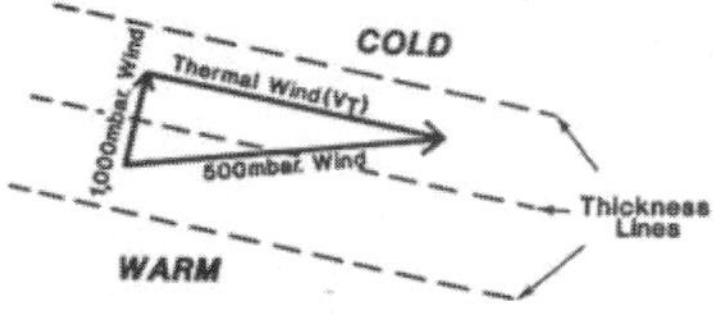

Figure A82 Definition of thermal wind. The thermal wind (VT) is the vector difference between the 1000-mb wind and the 500-mb wind.

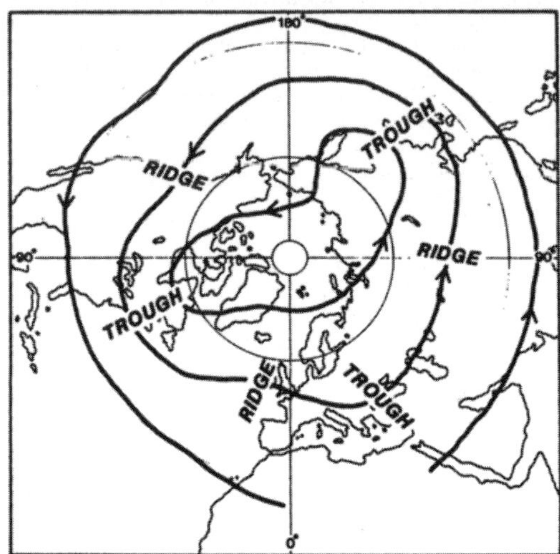

Figure A83 Selected mean contours of the 500-mb surface (about 5.5 km) in January. The upper winds blow along the contour lines.

perhaps be connected with the coolness of the North, Baltic, Mediterranean, and Black seas.

North Atlantic Oscillation

A major source of interannual variability in the atmospheric circulation over the North Atlantic and western Europe is the so-called North Atlantic Oscillation (NOA), which is associated with changes in the strength of the oceanic surface westerlies (Marshall et al., 2001). Its influence extends across much of the North Atlantic and well into Europe, and it is usually defined through the regional sea-level pressure field, although it is readily apparent in mid-troposphere height fields. The NAO's amplitude and phase vary over a range of time scales from intraseasonal to inter-decadal. The NAO is profoundly linked to the leading mode of variability of the whole northern hemisphere circulation, the annular mode or Arctic Oscillation. This suggests that Atlantic effects are more far-reaching and significant than previously thought (Marshall et al., 2001).

The NAO is often indexed by the standardized difference of December to February sea-level atmospheric pressure between Ponta Delgado, Azores (37.8°N, 25.5°W) or Lisbon, Portugal (38.8°N, 9.1°W) and Stkkisholmur, Iceland (65.18°N, 22.7°W) (Hurrell, 1995). Statistical analysis reveals that the NAO is the dominant mode of variability of the surface atmospheric circulation in the Atlantic and accounts for more than 36% of the variance of the mean December to March sea-level pressure field over the region from 20° to 80°N and 90°W to 40°E, during 1899 through to 1994. Marked differences are observed between winters with high and low values of the NAO. Typically, when the index is high the Icelandic low is strong, which increases the influence of cold Arctic airmasses on the northeastern seaboard of North America and enhances westerlies carrying warmer, moister airmasses into western Europe (Hurrell, 1995). During high NAO winters the westerlies directed onto northern Britain and southern Scandinavia are over 8 m/s stronger than during low NAO winters, with higher than normal pressures south of 55°N and a broad region of anomalously low pressure across the Arctic. In winter, western Europe has a negative radiation balance and mild temperature levels are maintained by the advection of warm air from the Atlantic. Thus strong westerlies are associated with anomalously warm winters, weak westerlies with anomalously cold winters, and NAO anomalies are related to downstream wintertime temperature and precipitation anomalies across Europe, Russia and Siberia (Hurrell, 1995; Hurrell and van Loon, 1996).

Overlying the interannual variability there have been four main phases of the NAO index during the historical record: prior to the 1900s the index was close to zero; between 1900 and 1930 strong positive anomalies were evident; between the 1930s and 1960s the index was low; and since the 1980s the index has been strongly positive. The recent persistent high positive phase of the NAO index, extending from about 1973 to 1995, is the most persistent and highest of the historical record. During each positive phase higher than normal winter temperatures prevail over much of Europe, culminating in the unprecedented strongly positive NAO index values and mild winters of 1989 and 1990. During high NAO index winters drier conditions also prevail over much of central and southern Europe and the Mediterranean, whilst enhanced rainfall occurs over the northwestern European seaboard (Hurrell, 1995).

Pacific atmospheric circulation oscillations

Interannual climate variability over the Pacific Ocean is dominated by ENSO (Salinger et al., 2001). This has the strongest SST signals of one sign along the equator over the central and eastern Pacific and a boomerang-shaped pattern of weaker SST signals of opposite sign extending over the middle latitudes of both hemispheres in the north and south Pacific. Salinger et al., (2001) comment that recently "ENSO-like" features in the climate system that operate on decadal to multidecadal time scales have been identified.

A recently identified pattern of longer-term variability is the Pacific Decadal oscillation (PDO) described by Mantua et al. (1997), which may be defined in terms of changes in Pacific sea-surface temperature north of 20°N latitude. Like the warm El Niño phase of ENSO, the warm or positive phase of PDO warms the Pacific near the equator and cools it at northern mid-latitudes; but unlike ENSO, PDO effects are stronger in the central and northern Pacific than near the equator, and it has an irregular period of several decades. It has been found that, during epochs in which the PDO is in its positive polarity, coastal central Alaska tends to experience an enhanced cyclonic flow of warm moist air which is consistent with heavier than normal precipitation. In these winters the mid-latitude depression track tends to split, with one branch carrying storms south to California and the other north to Alaska. The PDO was in its cool or negative phase from the first sea-surface temperature records in 1900 until 1925, then in warm or positive phase until 1945, cool again until 1977, and in warm phase until the 1990s.

An Inter-decadal Pacific Oscillation (IPO) has also been described by Power et al. (1998, 1999) and Folland et al. (1999), the time-series of which is broadly similar to the inter-decadal part of the North Pacific PDO index of Mantua et al. (1997). When the IPO is in a positive phase, SST over a large area of the southwest Pacific is cold, as is SST over the extra-tropical northwest Pacific. SST over the central tropical Pacific is warm, but unlike the ENSO situation it is less obviously warm over the equatorial far eastern Pacific. Also in contrast to

ENSO, warmth extends into the tropical west Pacific. Like the PDO the IPO shows three major phases this century: positive from 1922 to 1946 and from 1978 to at least 1998, with a negative phase between 1947 and 1976. A general reduction in sea-level pressure over the extra-tropical North Pacific Ocean during the winter (November to March) after about 1976 has been particularly evident. This appears as a deeper-than-normal Aleutian low-pressure system, accompanied by stronger-than-normal westerly winds across the central North Pacific and enhanced southerly to southwesterly flow along the west coast of North America.

Polar regions

Both polar regions are located in areas of general atmospheric subsidence, though the climate is not particularly anticyclonic; nor are the winds necessarily easterly. The moisture content of the air is low because of the intense cold, and horizontal thermal gradients are normally weak, with the result that energy sources for major atmospheric disturbances do not exist and are rarely observed. Vowinckel and Orvig (1970) suggest that the Arctic atmosphere can be defined as the hemispheric cap of fairly low kinetic energy circulation lying north of the main course of the planetary westerlies, which places it roughly north of 70°N. The situation over the south polar regions is more complex and the boundary of the Antarctic is not so clear.

John G. Lockwood

Bibliography

Ångstrom, A., 1935. Teleconnections of climate changes in present time. *Geography Annals*, **17**: 242–258.

Barry, R.G., and Carleton, A.M., 2001. *Synoptic and Dynamic Climatology*. London: Routledge.

Barry, R.G., and Chorley, R.J., 1976. *Atmosphere, Weather and Climate*. London: Methuen.

Berlage, H.P., 1966. The southern oscillation and world weather. *K. Meteorol. Inst.* (The Hague) Med. Verhand., **88**: 134.

Bjerknes, J., 1969. Atmospheric teleconnections from the equatorial Pacific. *Monthly Weather Review*, **97**: 163–72.

Corby, G.A., 1970. *The Global Circulation of the Atmosphere*. London: Royal Meteorological Society.

Crowe, P.R., 1971. *Concepts in Climatology*. London: Longman.

Diaz, H.F., Hoerling, M.P., and Eischeid, J.K., 2001. ENSO variability, teleconnections and climate change. *International Journal of Climatology*, **21**: 1845–1862.

Dietrich, G., and Kalle, K., 1957. *Allgemeine Meereskunde*. Berlin: Gebruder Borntraeger.

Findlater, J., 1969. A major low-level air current near the Indian Ocean during the northern summer. *Quarterly Journal of the Royal Meteorological Society*, **95**: 362–380.

Findlater, J., 1972. Aerial explorations of the low-level cross-equatorial current over eastern Africa. *Quarterly Journal of the Royal Meteorological Society*, **98**: 274–289.

Folland, C.K., Parker, D.E., Colman, A.W., and Washington, R., 1999. Large scale modes of ocean surface temperature since the late nineteenth century. In Navarra, A., ed., *Beyond El Niño: Decadal and Interdecadal Climate Variability*. Berlin: Springer, pp. 73–102.

Glantz, M.H., 1996. *Currents of Change: El Niño's Impact on Climate and society*. Cambridge: Cambridge University Press.

Hadley, G., 1735. Concerning the cause of the general trade-winds. *Royal Society of London Philosophical Transactions*, **29**: 58–62.

Halley, E., 1686. An historical account of trade-winds and monsoons observable in the seas between and near the tropics with an attempt to assign the physical causes of the said winds. *Royal Society of London Philosophical Transactions*, **26**: 153–168.

Houghton, J., 2002. *The Physics of Atmospheres*, 3rd edn. Cambridge: Cambridge University Press.

Hurrell, J.W., 1995. Decadal trends in the North Atlantic Oscillation: regional temperature and precipitation. *Science*, **269**: 676–679.

Hurrell, J.W., and van Loon, H., 1996. Decadal variations in climate associated with the North Atlantic Oscillation. *Climate Change*, **36**: 301–326.

Lamb, H.H., 1959. The southern westerlies. *Quarterly Journal of the Royal Meteorological Society*, **85**: 1–23.

Lamb, H.H., 1972. *Climate: Present, Past and Future*, vol. 1. London: Methuen.

Lighthill, J., and Pearce, R.P., 1981. *Monsoon Dynamics*. Cambridge: Cambridge University Press.

Lockwood, J.G., 1974. *World Climatology*. London: Edward Arnold.

Lockwood, J.G., 1979. *Causes of Climate*. London: Edward Arnold.

Lockwood, J.G., 1985. *World Climatic Systems*. London: Edward Arnold.

Lorenz, E.N., 1967. *The Nature and Theory of the General Circulation of the Atmosphere*. Geneva: World Meteorological Organization.

Lutgens, F.K., and Tarbuck, E.J., 1979. *The Atmosphere*. Englewood Cliffs, NJ: Prentice-Hall.

Mantua, N.J., Hare, S.R., Zhang, Y., Wallace, J.M., and Francis, R.C., 1997. A Pacific interdecadel climate oscillation with impacts on salmon production. *Bulletin of the American Meteorological Society*, **78**: 1069–1079.

Marshall, J., Kushnir, Y., Battisti, D., Chang, P., et al. 2001. North Atlantic climate variability: phenomena, impacts and mechanisms. *International Journal of Climatology*, **21**: 1863–1898.

McGregor, G.R., and Nieuwolt, S., 1998. *Tropical Climatology: An Introduction to the Climates of the Low Latitudes*. Chichester: Wiley.

Miller, F.R., and Keshavamurthy, R.N., 1968. *Structure of an Arabian Sea Summer Monsoon System*. Honolulu: East-West Center Press.

Namias, J., 1963. Interactions of circulation and weather between hemispheres. *Monthly Weather Review*, **91**: 482–486.

Namias, J., 1969. Seasonal interactions between the North Pacific and the atmosphere during the 1960s. *Monthly Weather Review*, **97**: 173–192.

Neiburger, M., Edinger, J.G., and Bonner, W.D., 1982. *Understanding Our Atmospheric Environment*. San Francisco: Freeman.

Nieuwolt, S., 1977. *Tropical Climatology*. New York: Wiley.

Pant, G.B., and Rupa Kumar, K., 1997. *Climates of South Asia*. Chichester: Wiley.

Piazzi-Smyth, C., 1858. An astronomical experiment on the Peak of Teneriffe. *Royal Society of London Philosophical Transactions*, **148**: 465–534.

Power, S., Tseitkin, F., Torok, S., Lavery, B., Dahni, R., and McAvaney, B., 1998. Australian temperature, Australian rainfall and the Southern Oscillation. 1910–1992: coherent variability and recent changes. *Australian Meteorological Magazine*, **47**: 85–101.

Power, S., Casey, T., Folland, C., Colman, A., and Mehta, V., 1999. Interdecadal modulation of the impact of ENSO on Australia. *Climate Dynamics*, **15**: 319–324.

Riehl, H., 1962. The tropical circulation. *Science*, **135**: 13–22.

Riehl, H., 1969. On the role of the tropics in the general circulation of the atmosphere. *Weather*, **24**: 288–308.

Riehl, H., 1978. *Introduction to the Atmosphere*. New York: McGraw-Hill.

Riehl, H., 1979. *Climate and Weather in the Tropics*. London: Academic Press.

Rossby, C.G., 1939. Relation between variations in the intensity of the zonal circulation of the atmosphere and the displacements of the semipermanent centres of action. *Journal of Marine Research,* **2**: 38–55.

Rossby, C.G., 1940. Planetary flow patterns in the atmosphere. *Quarterly Journal of the Royal Meteorological Society*, **66**: 68–87.

Rossby, C.G., 1945. On the propagation and energy in certain types of oceanic and atmospheric waves. *Journal of Meteorology*, **2**: 187–204.

Salinger, M.J., Renwick, J.A., and Mullan, A.B., 2001. Interdecadal Pacific Oscillation and South Pacific Climate. *International Journal of Climatology*, **21**: 1705–1721.

Troup, A.J., 1965. The southern oscillation. *Quarterly Journal of the Royal Meteorological Society*, **91**: 490–506.

Vowinckel, E., and Orvig, S., 1970. The climate of the north polar basin, In Orvig, S., ed., *Climates of the Polar Regions*. Amsterdam: Elsevier, pp. 129–252.

Walker, G.T., and Bliss, E.W., 1932. World weather V. *Royal Meteorological Society Memoirs*, **4**(36): 53–84.

Cross-references

Blocking
Centers of Action
Coriolis Effect
Hadley Cell
Jet Streams
Rossby Wave/Rossby Number
Trade Winds and the Trade Wind Inversion

ATMOSPHERIC NOMENCLATURE

Vertical temperature distribution is the most common criterion used in defining atmospheric regions. Three popular systems of nomenclature are based on the thermal stratification of the atmosphere. Figure A84 displays these three classification schemes and the vertical temperature structure of the atmosphere up to 110 km.

Chapman (1950) was the first to suggest the common four-layer division of the atmosphere. He classified the atmosphere into troposphere, stratosphere, mesosphere, and thermosphere. The upper boundary of each layer is given the suffix "pause"; for example, the upper boundary of the troposphere is referred to as the tropopause. Chapman's subdivisions of the mesosphere into the mesoincline, mesodecline, and mesopeak were based on the temperature maximum occurring near 50 km.

In 1954 Goody devised a different classification of the upper atmosphere, replacing Chapman's mesosphere. Goody suggested that Chapman's mesosphere should be called upper stratosphere, and the region above 80 km should be called ionosphere rather than thermosphere.

In 1960 the International Union of Geodesy and Geophysics recommended a nomenclature in an attempt to standardize terminology. In this classification the troposphere is the lowest 10–20 km of the atmosphere where temperature generally declines with height. The lower stratosphere is the relatively isothermal region immediately above the tropopause, and the upper stratosphere extends from the lower stratosphere to the maximum temperature level at 40–50 km (stratopause). The mesosphere is the region bounded by the stratopause and the temperature minimum occurring at 70–80 km (mesopause). The thermosphere is the region above the mesopause where temperature increases with height. Additionally, atmospheric physicists commonly often refer to the region from 15 km to 95 km as the middle atmosphere while the region above that is referred to as the upper atmosphere.

As opposed to thermal stratification, physical and/or chemical processes are also used to delineate discrete layers of the atmosphere. Today the term ionosphere is often restricted to the region around 70–80 km, where the existence of free electrons becomes significant. This region has importance in the transmission of radio signals over long distances. Conversely, the neutrosphere refers to the un-ionized region between the surface and the ionosphere. The ozonosphere, a subdivision of the neutrosphere based on appreciable ozone concentration, is located from 10 km to 50 km.

Other criteria have been suggested for classification of atmospheric layers. For example, the exosphere has been defined as the region near the top of the atmosphere where particles can move in free orbits, subject only to the Earth's gravitation. Another criterion, based on composition, defines the homosphere as the region from the surface to 80–90 km, where

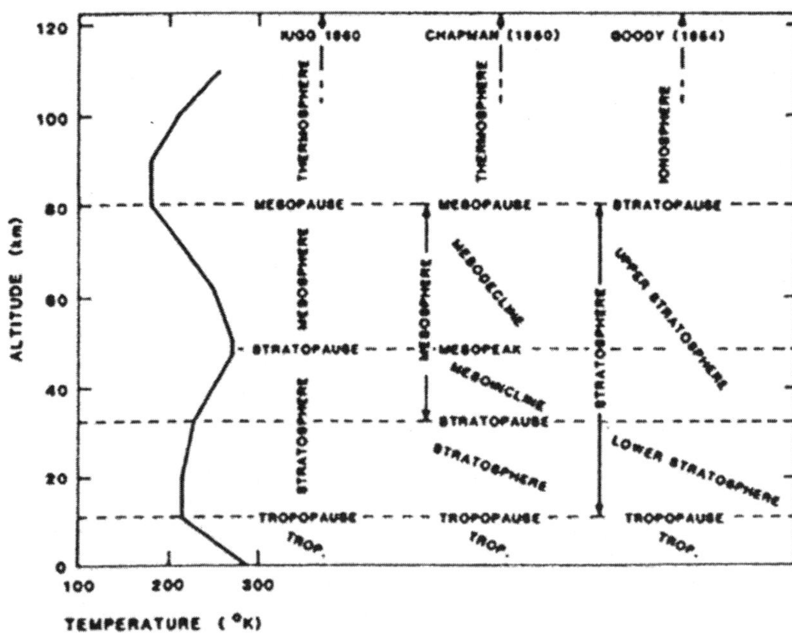

Figure A84 Vertical distribution of temperature up to 100 km, according to the US standard atmosphere, 1962, and three common systems of nomenclature (after Craig, 1965).

little photodissociation or gravitation separation occurs, and the heterosphere is the region above this, where atmospheric composition is not constant. The magnetosphere is defined as the region of the Earth's magnetic field which extends from the top of the ionosphere and is shaped by the solar wind. These and other variations in nomenclature have arisen in response to the needs of the various specialties in the atmospheric sciences.

Randall S. Cerveny

Bibliography

Brasseur, G., and Solomon, S., 1984. *Aeronomy of the Middle Atmosphere: chemistry and physics in the stratosphere and mesosphere*. Dordrecht, Holland: D. Reidel.

Chapman, S., 1950. Upper atmospheric nomenclature, *American Meteorological Society Bulletin*, **31**: 288–290.

Craig, R.A., 1965. *The Upper Atmosphere*. New York: Academic Press.

Goody, R.M., 1954. *The Physics of the Stratosphere*. London: Cambridge University Press.

Gordon, C.W., and Canuto, V., (eds), 1978: *The Earth, 1: The upper atmosphere, ionosphere and magnetosphere*. New York: Gordon & Breach.

Cross-references

Atmosphere
Inversion
Lapse Rate
Troposphere

ATMOSPHERIC NUCLEI AND DUST

Almost a century ago it was found that cloud and fog droplets form by condensation of water vapor from the atmosphere on small particles, so-called condensation nuclei. They were more systematically studied by Aitken toward the end of the nineteenth century. Condensation nuclei form part of the atmospheric aerosol content. These particles, aside from acting as condensation nuclei, also have an effect on the optical properties of the atmosphere and play an important role in atmospheric electricity. The light-scattering properties of these particles provided much information about their concentration and size distribution, especially in the early periods of investigation.

Nucleation process

In 1897 it was found that when a reasonably dust-free air sample, which was saturated with water vapor, was rapidly expanded so that adiabatic cooling occurred, droplet cloud formation would occur only if the expansion ratio was high enough to produce several hundred percent supersaturation. In the absence of nuclei a transition from the water-vapor phase to the liquid phase can occur only if the probability for aggregation of molecule clusters is high enough and supersaturation and temperature are adequate for such embryos to remain in equilibrium with their environment. However, if foreign particles are present in the air, condensation can occur at much lower supersaturations. The water-vapor molecules will attach themselves much more easily to such nuclei, thereby initiating the droplet-forming process. The effectiveness of a particle as a condensation nucleus largely depends on whether it is hygroscopic or hydrophobic (water attractive or repellent, respectively). Therefore, in the presence of condensation nuclei, condensation can occur at supersaturations on the order of a few tenths of a percent. This is in agreement with atmospheric observations. Similar observations apply to the formation of ice crystals. Small water droplets can be supercooled to temperatures as low as 40°C without freezing. At still lower temperatures, sudden formation of ice crystals occurs, i.e. spontaneous nucleation takes place. In the presence of freezing and sublimation nuclei, formation of ice crystals occurs at much warmer temperatures (−10°C or −15°C). The relative importance of freezing and sublimation nuclei remains to be determined. The term sublimation nucleus assumes that the transition from the vapor to the solid phase occurs directly by sublimation of molecules on the nucleus; in the case of freezing nuclei it is thought that the transition goes through the liquid water phase.

Properties and composition of atmospheric nuclei

Not all airborne particles (aerosol particles) act equally well either as condensation nuclei or as ice-forming nuclei. In the case of condensation nuclei, water affinity is an important factor in their efficiency. Supersaturation required to sustain a growing droplet is a function of the particle size; the smaller the particle (i.e. the greater the curvature of its surface), the higher the required saturation. From this consideration it follows that larger particles are more apt to act as condensation nuclei than small ones. In the case of freezing nuclei, particles whose crystal structure is similar to that of ice (pseudomorphic particles) are most likely to act as a base for building on the ice crystal lattice, and therefore form good freezing or sublimation nuclei. This concept does not seem to be in complete agreement with experimental evidence, however. For instance, quartz dust is not a good freezing nucleus; silver iodide is a good artificial freezing nucleus (cloud seeding).

Several conclusions may be drawn from this diagram: (a) In an isolated population of droplets containing equal masses of solute, the very small droplets must grow at the expense of the larger droplets. (b) Beyond a critical size the larger droplets may grow at the expense of the smaller. (c) If supersaturation of the vapor with respect to a flat surface is limited, say to about 0.1%, droplets must contain more than 10–15 gram of NaCl (or other similar solute) in order to reach critical size. (d) Small droplets may exist in equilibrium with moist air of relative humidity far below 100%.

According to Junge (1963), the size distribution of natural aerosols can be described by an inverse power law, according to which the number of particles is inversely proportional to the third or fourth power of their radius, although considerable deviations may occur in individual cases. Junge also defines three size groups, the Aitken particles (10^{-7} to 10^{-5} cm or 10^{-3} to 10^{-1} m radius); large particles (10^{-4} to 10^{-3} cm; see Table A43). Particles smaller than a few hundredths of a micron become unstable because their tendency to coagulate increases. Particles much larger than a few microns become more and more affected by gravitational forces and fall out more rapidly than smaller ones.

It can be concluded that approximately 60% in mass of the natural aerosol consists of inorganic, water-soluble material;

Table A43 Ranges of sizes of condensation nuclei and kerns occurring in the atmosphere[a]

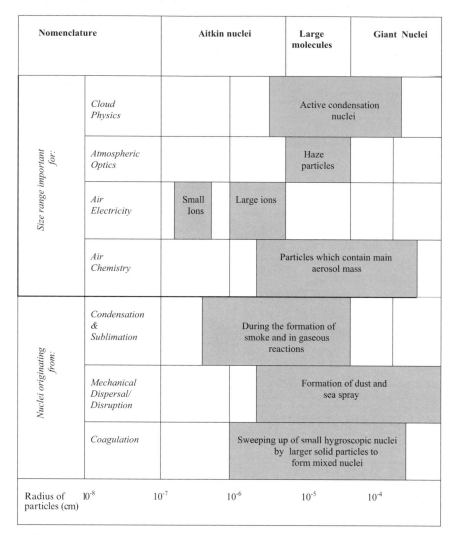

Nomenclature		Aitkin nuclei	Large molecules	Giant Nuclei
Size range important for:	*Cloud Physics*		Active condensation nuclei	
	Atmospheric Optics		Haze particles	
	Air Electricity	Small Ions; Large ions		
	Air Chemistry		Particles which contain main aerosol mass	
Nuclei originating from:	*Condensation & Sublimation*	During the formation of smoke and in gaseous reactions		
	Mechanical Dispersal/ Disruption		Formation of dust and sea spray	
	Coagulation		Sweeping up of small hygroscopic nuclei by larger solid particles to form mixed nuclei	
Radius of particles (cm)	10^{-8}	10^{-7}	10^{-6}	10^{-5} 10^{-4}

[a] Shading indicates size range at which labeled characteristics occur.

15% of organic material; and 25% of inorganic water-soluble material. The great majority of particles are mixed. In maritime airmasses, sea-salt particles are a major component in the natural aerosols. Such hygroscopic salt particles can act as excellent condensation nuclei. The continental aerosol originates partially from the ground itself (dust), but also from gaseous components. The inorganic nuclei seem to be formed to a large extent by the oxidation of sulfur gases (e.g. H_2S, SO_2) to sulfate compounds such as ammonium sulfate, etc. There are strong indications that a considerable portion of the nuclei in the size range below about 0.5 μm consists of organic materials and that these particles are extremely unstable. It is assumed that these are mostly particles formed from organic compounds with low vapor pressure. The vapors originate from certain types of vegetation and are changed to liquid or solid particles by photo-oxidation.

A group of cloud physicists believe that a correlation exists between the occurrence of meteor showers and precipitation. They assume the extraterrestrial meteors, by entering the atmosphere, disintegrate into fine dust particles that act as ice nuclei in the formation of cloud and precipitation particles.

Since 1947, when Vonnegut discovered that silver iodide crystals act as freezing nuclei at temperatures below about −4°C, much research has been done on artificial nuclei. It appears that nature, with few exceptions, provides enough nuclei for condensation and/or freezing. Adding artificial nuclei may change the size spectrum of cloud particles, leading to an increase in precipitation, but it may also result in a decrease (overseeding).

Robert W. Fenn

Bibliography

Aitken, L., 1888–1892. Related papers on the development and use of the dust counter. *Collected Scientific Papers (1923)*, pp. 187, 207, 236, 284.

Barry, R.G., and Chorley, R.J., 1992. *Atmosphere, Weather, and Climate*. New York: Methuen.
Eagleman, J.R., 1983. *Severe and Unusual Weather*. New York: Van Nostrand Reinhold.
Fleagle, R.G., and Businger, J.A., 1963. *An Introduction to Atmospheric Physics*. New York: Academic Press.
Goody, R., 1995. *Principles of Atmospheric Physics and Chemistry*. New York: Oxford University Press.
Junge, C.E., 1958. *Atmospheric Chemistry*. In *Advances in Geophysics*, **4**: 1–108.
Junge, C.E., 1963. *Air Chemistry and Radioactivity*. New York: Academic Press.
Ludlam, F.H., 1980. *Clouds and Storms*. State College, PA: Pennsylvania State University Press.
Meszaros, E., 1981. *Atmospheric Chemistry: Fundamental Aspects*. New York: Elsevier.
Peixoto, J.P., and Oort, A.H., 1992. *Physics of Climate*. New York: American Institute of Physics.
Roll, H.U., 1965. *Physics of the Marine Atmosphere*. New York: Academic Press.
Vonnegut, B., 1947. The nucleation of ice formation by silver iodide. *Journal of Applied Physics*, **18**: 593.
Woodcock, A.H., 1953. Salt nuclei of marine air as a function of altitude and wind force. *Journal of Meteorology*, **10**: 362–371.

Cross-references

Aerosols
Albedo and Reflectivity
Cloud Climatology
Fog and Mist
Precipitation
Radiation Laws

AUSTRALIA AND NEW ZEALAND, CLIMATE OF

The climates of Australia and New Zealand are strongly influenced by general circulation patterns prevalent in the SW Pacific. Critical to the strength of this circulation is the temperature gradient between Antarctica and the SE Asian tropics, and between Antarctica and the Australian continent. The ocean plays a key role, since there are no major land masses between the South island of New Zealand and Antarctica. More regional circulations, such as El Niño Southern Oscillation, affect the climate patterns in this region. The classic reference, which provides the first detailed summary of the climates of Australia and New Zealand, is Gentilli (1971).

Climates of Australia

Australia's climate depends on four major factors. First, the size and shape of Australia determine how the general circulation affects the continent. Australia covers 7 682 300 km^2, stretching from 10°41′S to 43°39′S and 113°09′E to 153°39′E at the extremes. Its greater longitudinal distance (4000 km) allows considerable airmass modification and a strong degree of continentality. Its limited latitudinal span creates moderate changes on a seasonal basis compared to larger continents such as North America.

Second, Australia's topography does little to obstruct air moving over the continent. The average altitude of the land mass is only 300 m, with 87% of the continent less than 500 m and 99.5% less than 1000 m. Only the eastern mountains, which extend north to south, have any major influence on synoptic flow.

Third, Australia is completely surrounded by water. Polar air originates from the Antarctic Ocean, and is basically cool and moist. The major influence of the ocean is coastal, and there is no significant upwelling of cold water. Australia's climate can be generally described as moderately continental with important ocean influences.

Fourth, Australia's geographical location places it under the global subtropical high-pressure zone. Much of Australia is under the influence of dry subsiding air and can be considered arid or semiarid, although not as dry as the west coasts of the other southern hemisphere continents. Particularly in summer, Australia experiences high sun angles and intense solar radiation under mostly clear skies. With the exception of Tasmania and much of the east coast, most of the continent receives over 3000 hours of sunshine per year, almost 70% of that possible (Bureau of Meteorology, 1988a).

Synoptic circulation patterns

In general, Australia is affected by midlatitude westerlies on the southern fringe, tropical convergence on the northern fringe in summer, and stable subsiding air under the subtropical anticyclone belt in between. The synoptic circulation patterns can best be described by comparing summer and winter seasons.

(a) Winter (May–October) (Figure A85)
The subtropical anticyclone belt covers the northern two-thirds of the continent. Dry stable SE trades dominate Australia north of 20°S, with greater than 50% frequency. This air originates over the Pacific, often loses some of its moisture due to orographic precipitation on the eastern highlands, and cools and stabilizes as it passes over the continental interior.

Frontal activity. South of 20°S the continent is under the influence of alternating anticyclones and cold fronts, associated with the northern edge of the midlatitude westerlies. Troughs and fronts from these systems extend into southern Australia (Figure A86), separating migratory anticyclones in 4–8-day intervals. These fronts move at an average speed of 35 km h^{-1} and can be divided into two types (Sturman and Tapper, 1996). Ana-fronts are more active and are associated with convection in the frontal zone. Kata-fronts are less active and are linked to subsiding air from anticyclones.

The impact of these fronts on the main continent is not usually strong, and only in spring extends into the northern half of the country (Smith et al., 1995). Tasmania is most severely affected, being closest in location to the midlatitude depressions. There is a striking absence of warm fronts because the frontal systems form well to the west of Australia and the warm front has swung south before the system reaches the continent.

Precipitation associated with winter frontal systems often occurs from the Northwest Australian cloud band (Sturman and Tapper, 1996), a source of moisture and deep convection which can stretch for distances of up to 5000 km between the Indian Ocean and S to SE Australia. Eighty percent of these cloud bands occur between April and September, with an average of about two per month.

Occasionally, secondary depressions or cut-off lows moving toward low latitudes can initiate severe storm situations.

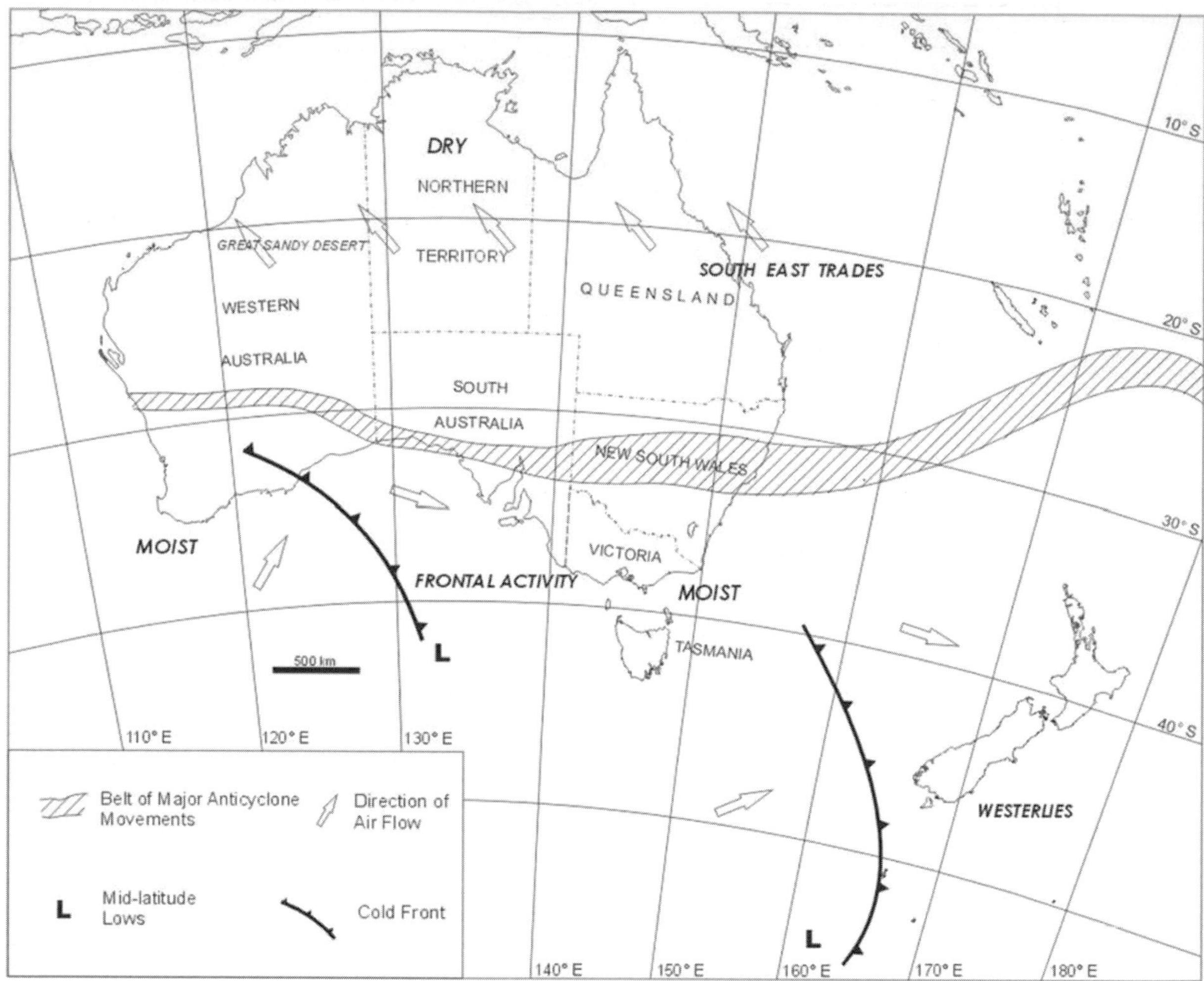

Figure A85 Simplified synoptic climatology of Australia, July surface circulation (Bridgman, 1987).

Development is better on the SE coast of Australia rather than on the SW. A slow-moving deep low-pressure system results, bringing strong winds and heavy precipitation for periods of 0.5–10 days (Tapper and Hurry, 1993). Such systems are more likely to develop in spring or fall (October–November or April–May) because the thermal gradient between airmasses is strongest. The cause is either the remnants of a tropical storm, or intense cyclonic development in an easterly dip in the isobaric pattern.

The interaction of the Indian Ocean air and the cooler maritime air south of the continent brings winter rains to S and SW Australia. In early winter these rains can be extensive and are related to mid-tropical convection and the strength of the westerlies. In late winter more showery precipitation occurs, associated with topographic and coastal convergence influences, and not with the general circulation (Allen and Haylock, 1993).

Winter anticyclones. Anticyclones dominate much of the continent in winter. On average, 40 anticyclones move over Australia in a year, with slightly less in fall–winter than spring–summer. Anticyclones enter the continent at about 115°E and 27°S on the west coast, dip to 35°S in S Australia, and exit about 30°S off the east coast (Figure A85). The mean anticyclonic track is furthest north in June–July except on the east coast of the continent, where the lag in ocean heating defers the maximum northward extent to August–September (Gentilli, 1971; Tapper and Hurry, 1993).

Over the continent the strength and persistence of the anticyclones can be intensified by wintertime cooling of the land. Anticyclones may remain stationary in the interior for several days, particularly if the high develops a meridional (longitudinal) orientation. This creates blocking, and is much more common east of 150°E, in the Tasman Sea. Eastward movement of weather systems is retarded and the convective motion associated with depressions is countered (Baines, 1990; Sturman and Tapper, 1996). Duration is up to 12 days, bringing dry, fine weather.

The presence of anticyclones in the winter enhances the effects of local topography on the continent, encouraging mesoscale inversions and cold air drainage to occur. It is not unusual to see nighttime floodplain temperatures 4–5°C cooler than those on the adjoining hilltop, caused by the downslope movement of cold air, and widespread frost and/or fog.

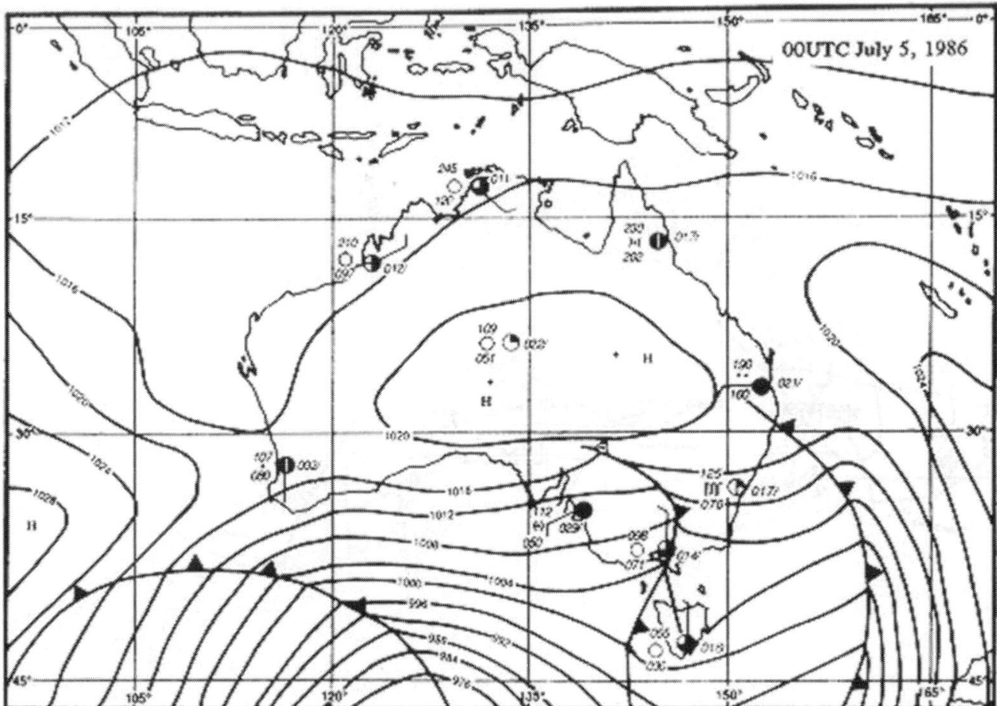

Figure A86 Representative synoptic meteorological situation in winter. Cold fronts parade across the southern third of the continent bringing cool cloudy weather and a light to moderate rainfall. Dry, cool conditions dominate the interior, with overnight frosts likely in many locations (Tapper and Hurry, 1983).

The subtropical jet stream. The trajectory and speed of movement of both anticyclones and fronts are strongly influenced by the subtropical jet stream, which is strongest over the southern part of the continent near 200–300 mb (10 000 m altitude) in the winter (Bridgman, 1998; Sturman and Tapper, 1996). The mean latitude of the jet is 26–32°S, but varies considerably from season to season and year to year. In winter its mean core speed is $70\,m\,s^{-1}$ and northerly meandering from about 25°S is most apparent. In summer the jet is much weaker ($30\,m\,s^{-1}$), with its mean axis position around 31°S. The jet is associated with the formation of upper air troughs and depressions which, when linked with a lower-level unstable airflow, can bring considerable rain to large areas of the southern half of the continent. Such troughs also provide the major medium for tropical moisture to reach the southern parts of the continent through the northwest cloud band. The strongest jets are also associated with strong cyclogenesis in the west Tasman Sea. It is not unusual in winter on the SE coast for a shallow surface anticyclone with surface pressures near 1020 mb to exist, with steady rain falling through the system from a mid-tropospheric trough.

When the jet is located between 20°S and 25°S it has a significant influence on rainfall systems occurring in late fall. Storms can bring strong rain and flooding to parts of the continental interior. They appear erratically in the general area from Geraldton–Port Hedland in Western Australia to Carpenteria–NE New South Wales in eastern Australia, providing secondary and sometimes primary rainfall maxima for some individual stations (Bridgman, 1998).

Anticyclonic shape and movements relate strongly to the area of upper tropospheric convergence on the equatorial side of the jet. If the jet meanders considerably across the continent, the anticyclones will be elongated longitudinally and will be slow-moving. If the jet has a strong zonal (east–west) component, the anticyclones will be zonally elongated and may move across the continent with considerable speed.

(b) Summer (November–April) (Figure A87)

Summer anticyclones. In summer the circulation systems are 5–8° further south, changing the climatic regimes affecting the northern and southern edge of the continent. Subsiding air from the subtropical high-pressure belt, now located poleward of Australia between 35°S and 45°S, covers most of the continent except for the northern tropical fringe and southern Tasmania. The descending air, originating at 10 000 m, warms adiabatically and subdivides into a series of traveling anticyclones (5–6 days periodicity) separated by troughs (Gentilli, 1971). The longitudinal extent of these anticyclones (Figure A88) is of the order of 2000–3000 km and the latitudinal extent 1000–2000 km. The counterclockwise circulation around the high during its passage ensures dry air from the hot interior moves coastward, limiting rainfall periods to troughs between anticyclonic systems or to onshore easterlies along the east coast of the continent. The pressure in the centre of the anticyclonic cells usually reaches 1030 mb, and falls to about 1005 mb in the trough between.

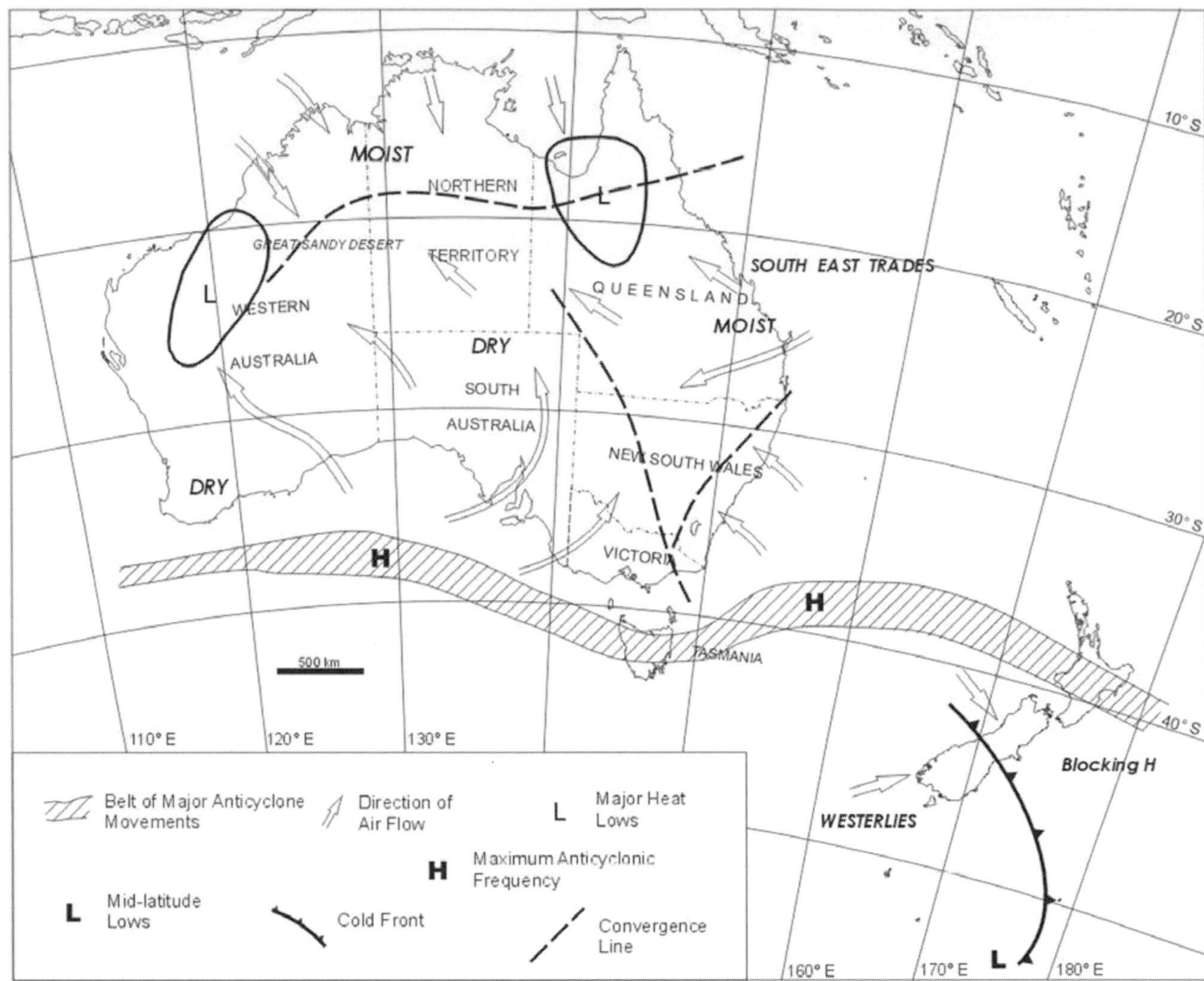

Figure A87 Simplified synoptic climatology of Australia, January surface circulation (Bridgman, 1987).

A typical weekly synoptic pattern for the southern half of Australia might be cooler S winds in the east side of a high, changing to E, NE and then NW winds as the high passes. Along the coasts, temperatures might warm from 22–24°C in the cooler Pacific maritime air to 36–38°C or higher in the continental air from the interior. The situation remains extremely hot and dry until the trough, or cool change between anticyclones, brings relief.

The summer cool change. In Western Australia the cool change is weak, often dry, and is mainly a change in temperature and wind direction (Reeder and Smith, 1992). It forms in the Cape Leeuwin–Kalgoorlie area as an incursion of Indian Ocean air, and is mainly controlled by the warm to hot air ahead, rather than the weak, cooler air behind. The change will often be sudden, bringing relief from exceedingly hot conditions, dropping temperatures by 10–15°C in the space of less than half an hour.

Cool changes, or Southerly Busters, reaching SE Australia, are often much stronger, with gusty winds and occasional rain, as they move east along the Victorian coast and then north along the NSW coast. They occur between November and March, and are channeled by the topography of the eastern highlands (Reeder and Smith, 1992; McInnis, 1993). Since the depth of the change is only about 3–5 km, this change in orientation is not reflected at 500 mb, but the buster may be associated with weak troughs at this level. The majority of the busters occur between 1300 and 2100, since their speed of movement strength is enhanced by a sea breeze. About half the cool changes are dry, but the change may be heralded by a roll cloud, gusty, squally winds (gusts up to 30 m s^{-1}) and thunderstorms. At passage a temperature drop of 10°C in a few minutes and up to 22°C in 24 hours may occur.

On occasion cool changes are associated with prefrontal troughs, which can bring severe weather, heavy precipitation, and flash floods. From 1957 to 1990, 69 meso-scale severe storms occurred in the Sydney region between November and

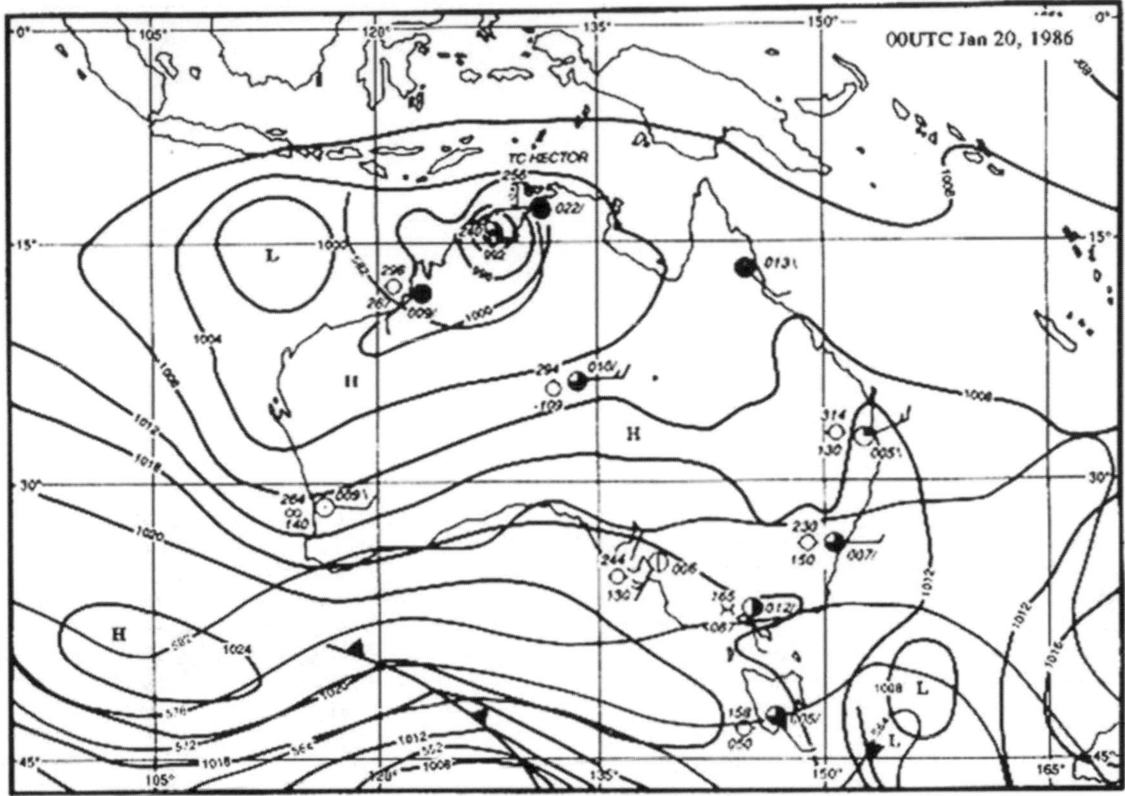

Figure A88 Representative synoptic meteorological situation in summer. A cool change separates an elongated high-pressure ridge. Hot NW winds from the continental interior invade SW Australia. Onshore SE winds affect the E coast, bringing cool, cloudy, and showery conditions. A tropical cycle (Hector) enters the continent near Darwin (Tapper and Hurry, 1993).

March (Speer and Geerts, 1994). The majority occurred between noon and 6 pm.

Regional and local circulations. The west coast trough develops between two strong highs centered south of the continent, as a dip in the isobaric pattern. The trough encourages hot strong NE winds from the interior to the coast, which continue until the trough moves inland and weakens (Tapper and Hurry, 1993).

The sea breeze brings cooling to the coast on summer days, as well as on occasional winter afternoons (Abbs and Physick, 1992). Established under clear skies and weak synoptic airflow, through land–water temperature contrasts of more than 7–8°C, the sea breeze can penetrate 80–90 km inland on average, depending on topography. A well-developed sea breeze may blow for 2–7 hours, depending on distance from the shore. Sea breezes are more strongly developed on the W and SW coasts than on the E coast. Bringing temperature relief of up to 15°C on a hot day, these breezes occur with daily frequencies of about 60% on the coast in summer. They often carry appropriate local names such as the Fremantle Doctor (Western Australia).

The morning glory is a spectacular wave structure that appears over the Gulf of Carpenteria just ahead of the wet season in Australia's north, created by a clash between two opposing sea breezes. It heralds strong wind squalls, long narrow roll clouds, and a pressure jump, but precipitation is rare (Collis and Whitaker, 1996).

The northern Australia rainy season, or monsoon. In northern Australia the shift of the subtropical high-pressure belts and subtropical jet allows a dynamically active low to form over the northern part of the continent (Figure A87). These encourage the entry of the intertropical convergence zone (ITCZ). The result is a moist NW air flow which meets the SE trades along a line approximately from Port Hedland (Western Australia) to Cairns (Queensland), which is commonly described as a monsoon (Joseph et al., 1991; Suppiah, 1992; Cook and Heedegen, 2001). Table A44 presents the range of dates of onset of the monsoon, illustrating the high variability in monsoon dynamics. There is a strong correlation with the previous season's monsoon rain in India, with

Table A44 Range of dates of onset and length of the North Australian summer monsoon (Suppiah, 1992)

	Minimum	Mean	Maximum	St Dev (days)
Onset	23 Nov.	24 Dec.	27 Jan.	15
Finish	1 Jan.	7 Mar.	6 Apr.	18
Length	11	74	119	25

below-normal rain creating a delay in monsoon onset in Australia.

The monsoonal flow generally extends to about 15°S, and is better developed east of 100°E. It brings the rainy season to northern Australia. Its onset can be abrupt, producing violent thunderstorms and tropical depressions during active phases. Rainfall is heaviest along the coast, decreasing quite rapidly inland. About 20% of the monsoon period consists of breaks, where moist humid air in the NW airstream is not triggered into precipitation (Sturman and Tapper, 1996).

To the south and east of the monsoonal trough SE winds dominate the eastern third of the continent. Low-level moisture from the Pacific Ocean is, to a large part, expended on the eastern highlands, but often reaches the interior. Convective and orographic activity results in rain through instability and thunderstorms, occasionally enhanced by an upper-level trough. The best rains occur when a mid-tropospheric trough extends southward from central Queensland, interacting with the unstable easterly flow (Tapper and Hurry, 1993).

Tropical cyclones. Australia is the only continent with the same incidence of tropical cyclones reaching either the W or E coast, due to the availability of warm seas to enhance formation and to the shape of the coastline. Ninety-five percent of tropical cyclones form in the latitudinal zone 9–19°S, in the shear line between the monsoon westerlies and the trade winds (Sturman and Tapper, 1996). The tropical cyclones season occurs between late November and early May. On average, eight to 10 cyclones form off the Australian coast every year, but there is considerable variation in frequency. About three on average reach NE Queensland each year, two or three reach Western Australia, and one reaches the Northern Territory (Figure A89). Landfall most often occurs around Port Hedland on the west coast, near Townsville on the east coast, and Wyndham, on the northeast coast of Western Australia. February and March are the months of highest frequency, particularly between 15°S and 20°S. The timing of the strongest tropical cyclones is linked to the active phases of the monsoon (Suppiah, 1992).

Tropical cyclones bring high winds and heavy rain to both coasts with considerable destruction potential. Flooding in much of Central Queensland and in Western Australia is inevitable if the cyclone moves inland. Much of the area is sparsely inhabited, however, and rains from tropical cyclones are often the main source of moisture for crops and grassland. Only on rare occasions does a tropical cyclone strike a heavily inhabited area. The most famous Australian cyclone is Tracy, which passed through Darwin on Christmas evening, 1974 and destroyed more than half the city.

Links to El Niño/Southern Oscillation. Tropical cyclones, the monsoonal flow, and precipitation in eastern Australia in general is very strongly influenced by the El Niño/Southern Oscillation (ENSO) circulation variations (Allen, 1988; Drosdowsky and Williams, 1991; Joseph et al., 1991), and the associated changes in sea-surface temperature, especially in spring and early summer. There is a significant correlation between the Southern Oscillation Index (SOI) and variations in precipitation, cloud cover, and the diurnal range of temperature. Lower SOI brings less rain (and often drought), lower cloud amounts, and greater diurnal temperature ranges. A positive SOI brings higher than average rainfall, deeper incursions of the monsoon into central Australia, and potential floods. Correlations (r) between precipitation and SOI reach 0.6 in western NSW, and are generally 0.4 over much of the eastern half of Australia (Sturman and Tapper, 1996). Correlations diminish greatly toward the SW of the continent.

The climatic elements

Tables A45 and A46 present means of various climatic elements for January and July and some climatic extremes. Two elements are of major concern to Australia: temperature and moisture.

Temperature

The range of annual average air temperature is from 28°C on the Kimberly coast (NW Western Australia) to about 4°C in the alpine mountain areas (>1500 m) in SE Australia. In January (summer) the average daily minimum is virtually equivalent to the annual average air temperature. In July (winter) minimums decrease from 21°C in the Darwin area to −20°C in the SE Australian mountains. Particularly in winter, the minimums are tempered on the coast by the ocean. Local cold air drainage and topography eliminates any regularity of pattern in the eastern highlands.

Maximum temperatures are much more important. Australia is known for its heat extremes rather than its cold (Bureau of Meteorology, 1988a, 1989). In January, 35°C is exceeded over most of the interior and 40°C regularly in NW Western Australia. Towns in this area, such as Marble Bar, exceed maxima of 40°C for several weeks at a time. Maxima drop to less than 20°C in Tasmania and in the higher altitudes of the SE Australian mountains. The coastal sea breezes create maximum temperature gradients in all directions from the interior, with the south and east coasts having maxima in the mid and upper 20s and the north and west coasts in the lower 30s. The maximum for single days under one synoptic system has reached 47.6°C in Adelaide, 46°C Melbourne, and 45.3°C in Sydney (11–14 January 1939).

In July, except for the eastern highlands, maximum temperatures follow a latitudinal pattern, decreasing from 30°C near Darwin and on the Cape York peninsula to 12°C on the south and Tasmanian coasts. July maxima in the mountains do not usually reach 5°C.

The sun and seasonal extent of cloud cover determines the month of highest temperature. Most of northern Australia, equatorward of 20°S, has highest maximum temperatures in November just before the onset of the monsoon, with some areas near Darwin and in the NW having the highest maximum in October. Just south of this area, to a line from Geraldton (W.A.) to Tennant Creek (N.T.) to Cooktown (Qld.), the highest maxima occur in December. Most of the interior and the NE coast has highest maxima in January, and the west and east coasts in February, due to the lag in ocean temperatures.

A combination of hot dry climatic conditions and strong winds from the center to the coast can create serious bushfire problems, particularly in late spring and early summer. Bushfires are enhanced when a blocking summer anticyclone occurs in the Tasman Sea and there is slow approach of a trough or depression from the southwest. Bushfire behavior is controlled by fuel availability, topography, pressure tendency and wind direction. Cumulative antecedent rainfall, or the total rainfall over several previous months or seasons, correlates strongly with bushfire frequency (Love and Downey, 1986). The region of highest fire hazard is the coastal zone of E Victoria and S

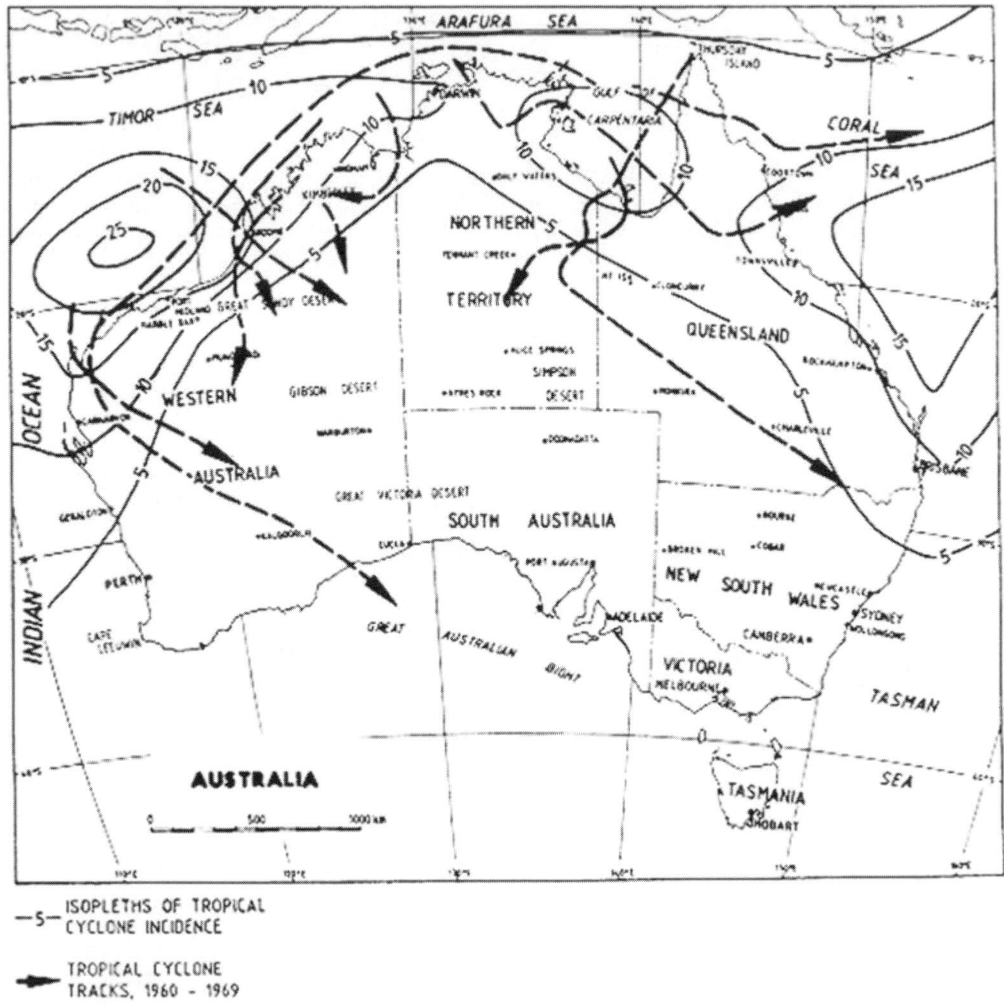

Figure A89 Average decadal incidence of tropical cyclone formation. Tropical cyclone tracks crossing the coast are shown to indicate representative storm paths (after Bridgman, 1998; Collis and Whitaker, 1996).

Table A45 Extreme weather events in Australia

Weather event	Location	Period	Value
Rainfall			
Hourly total	Deeral, Qld	13 Mar. 1936	330 mm
Daily total	Beerwah, Qld	3 Feb. 1893	907 mm
Monthly total	Bellanden Kerr, Qld Bellanden	Jan. 1979	5387 mm
Annual total	Kerr, Qld	1979	11251 mm
Highest annual mean	Babinda, Qld	32 years	4537 mm
Lowest annual mean	Troudaninna, SA	42 years	105 mm
Temperature			
Highest maximum	Cloncurry, Qld	16 Jan. 1889	53.1°C
Lowest minimum	Charlotte's Pass, NSW	29 Jun. 1994	−23°C
Longest heat wave[a]	Marble Bar, WA	30 Oct.−7 Apr. 1924	161 Days
Maximum annual average	Wyndham, WA	–	35.5°C
Wind			
Maximum gust	Mardie, WA	19 Feb. 1975	259 km/hr

[a] Defined as the number of days in a row the maximum temperature exceeded 37.8°C.
Data courtesy of the National Climate Centre, Australian Bureau of Meteorology.

Table A46 Mean climatic data for selected Australian capital cities, January and July

City (Lat./long.)	MSL alt (m)	Years of data	Pressure (hpa)	Wind speed (m/s)	Prevailing wind direction 9 a.m.	Prevailing wind direction 3 p.m.	Evap (mm)	Daily cloud (8ths)	Max. temp. (°C)	Min. temp. (°C)	Mean temp. (°C)	Hours of sunshine	9 a.m. relative humidity	Precipitation (mm) (days)	Fog days	Solar energy (MJ/m^2 per day)
Perth, WA (31°57′S/115°51′E)	19.5	42														
January			1012.6	4.9	E	SSW	280	2.3	31.5	16.7	23.5	10.5	51	7(3)	0.2	27.3
July			1018.8	3.9	NNE	W	58	4.5	17.7	9.0	13.2	5.3	76	164(18)	1.6	9.4
Darwin, NT (12°25′S/130°52′E)	31.0	45														
January			1006.2	2.6	W	NW	225	5.9	31.7	24.7	28.6	5.9	81	409(21)	0.0	18.4
July			1012.8	3.4	SE	E	229	1.3	30.3	19.2	25.1	9.8	62	1(1)	1.1	19.3
Sydney, NSW (33°56′S/151°10′E)	6.0	47														
January			1012.7	3.4	NE	NE	217	4.7	25.7	18.5	22.0	7.2	68	102(12)	0.3	22.5
July			1018.5	3.2	W	WSW	95	3.5	16.0	7.9	11.8	6.2	74	100(10)	2.1	10.4
Canberra, ACT (35°19′S/149°12′E)	571.0	47														
January			1012.0	1.8	NW	NW	251	4.1	27.7	12.9	20.3	8.9	60	60(8)	1.0	25.9
July			1020.2	1.2	NW	NW	54	4.4	11.0	-0.3	5.4	5.2	84	39(10)	7.9	9.5
Melbourne, Victoria (37°41′S/144°51′E)	132	16														
January			1012.8	3.6	S	S	228	4.1	26.0	13.4	19.9	8.1	68	42(8)	0.1	24.9
July			1202.2	3.6	N	N	47	5.2	12.8	4.9	9.5	3.7	81	33(14)	4.3	6.3
Hobart, Tasmonia (42°55′S/147°20′E)	4.0	28														
January			1010.6	3.5	NNW	SW	167	5.0	18.6	11.5	16.5	7.9	58	69(14)	0.3	23.2
July			1014.0	3.0	NNW	NNW	26	4.8	11.1	6.1	7.9	4.3	78	95(21)	1.4	5.5
Alice Spring, NT (23°49′S/133°53′E)	545	38														
January			1007.0	ND	ESE	SE	397	2.3	36.6	22.2	28.0	11.0	36	43(ND)	ND	26.7
July			1018.0	ND	ESE	SE	121	0.7	19.3	4.5	12.0	9.3	61	11(ND)	ND	16.0

After Bridgman (1987); Bureau of Meteorology (1988b).

New South Wales, where, due to the regular availability of good fuel grown during wet periods, large bushfires during dry periods occur every third year. In an area roughly from SE South Australia through Victoria and W New South Wales to the S Queensland coast, including S Tasmania near Hobart, one big fire occurs approximately every 10 years. Recent severe fires include the Ash Wednesday fires in Victoria and Southern Australia in February 1983, where 75 people were killed, more than 2000 houses destroyed, and property damage was $430 million (Tapper and Hurry, 1993); the east coast fires in the Sydney/Newcastle New South Wales area in January 1994; and the 2002 bushfires which created major destruction of houses in E New South Wales, and the Mt Stromolo astronomical observatory in Canberra.

Moisture (Figure A90). Australia is a dry continent with certain localized exceptions, and has periods of extreme rainfall variability (Haylock and Nicholls, 2000). Vapor pressure and atmospheric water vapor do not change much diurnally (or in the interior seasonally) except along the coast or where there is a significant rainy season. Potential evaporation regularly overshadows precipitation. Fifty percent of the continent has a median rainfall less than 300 mm y^{-1}, and 80% less than 600 mm y^{-1}. Highest rainfall occurs on the NE and N coasts, reaching 3800 mm y^{-1} in the eastern highlands, and lowest in the Simpson Desert (<150 mm y^{-1}) (Bureau of Meteorology, 1988a). Seventy-Five percent of the continent has an annual evaporation (class A pans) of greater than 2500 mm yr^{-1}. In the central and northwest sections of the continent, potential evaporation reaches 4500 mm yr^{-1}, more than 20 times the annual rainfall.

Rain days per year decrease from over 100 around Darwin and 120 or more on the NE Queensland coast, to less than 20 in the interior. In the area affected by the midlatitude westerlies, for example SW Tasmania, more than 150 rain days per year can occur.

Precipitation can be strongly seasonal, depending on location. Five main precipitation regimes dominate the continent (Bureau of Meteorology, 1988a, 1989; Stern et al., 2000) allowing an analysis by climatic zones (Figure A90):

1. In Northern Australia the monsoonal season (October to April) is markedly wet and the winter very dry. Most stations have more than 20 times the rainfall in summer compared to winter. The NE Queensland Coast has the highest annual rainfall in the country.

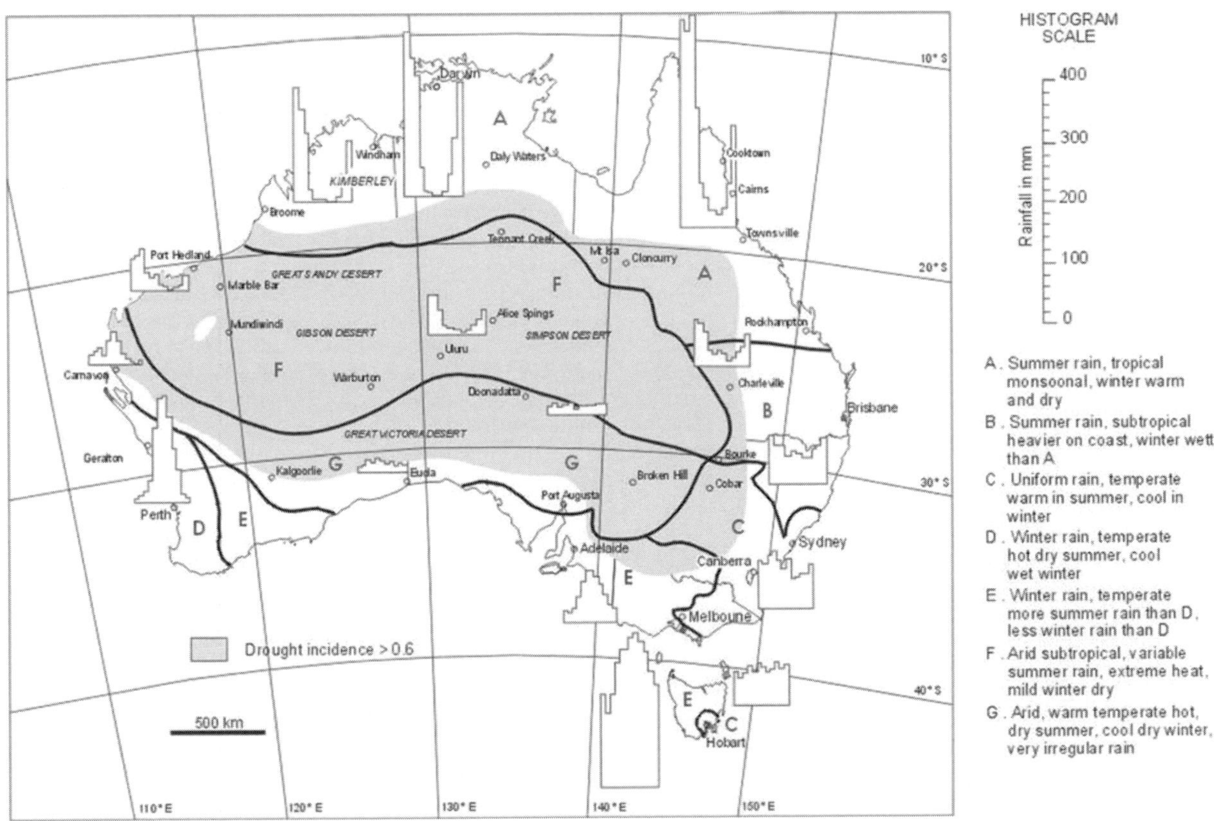

Figure A90 Representative monthly precipitation histograms (January begins at the left) for various parts of Australia. The data are averaged for meteorological districts and not representative of single stations. Included are climatic districts based on temperature and precipitation. The region of high to excessive drought incidence is shaded (Bridgman, 1998). A more detailed classification can be found in Stern et al. (2000).

2. In SE Queensland and NE New South Wales, summer is wet and winter is dry, but the difference between seasons is much less marked than under monsoon conditions. There is a tendency for a secondary peak in midwinter.
3. In SE Australia, including most of Victoria, E New South Wales and Tasmania, a relatively uniform precipitation regime exists. Victoria and Tasmania receive slightly more rain in winter than summer, and New South Wales the reverse, with the difference becoming greater closer to the equator.
4. SW Western Australia is winter wet (May to October), summer dry, the Mediterranean climate caused by the seasonal shifting of the subtropical high-pressure belt and the midlatitude westerlies. Climate here is strongly affected by changing conditions in the Indian Ocean (Smith et al., 2000).
5. More than half the continent, from NW Western Australia to the Great Australian Bight, is semiarid to arid, with a weak summer seasonal rainfall distribution in the west and north. This is the area of greatest rainfall variability, with heaviest rains associated with occasional tropical cyclones and severe local thunderstorms.

On a more local scale, severe thunderstorms associated with fronts and other convective triggers create flooding and damage (Kuleshov et al., 2002). Some of the strongest thunderstorms spawn tornadoes, particularly under extremely unstable conditions in the mid and upper troposphere. Tornadoes occur especially in the uplands of eastern Queensland and in the SW of Western Australia (Sturman and Tapper, 1996). The frequency of tornadoes may be 100–200 per year over the continent, occurring mainly outside the dry interior. In the northern half of the country, tornadoes are relatively rare and only develop in summer, as a result of strong local convection related to the wet season or sometimes associated with tropical cyclones. In the southern half, tornadoes in summer occur under convective activity associated with weak lows or troughs. In winter a secondary maximum occurs associated with cold fronts emanating from deep depressions in the midlatitude westerlies.

Tornado formation is strongly influenced by upper atmospheric airflow, the subtropical jet, and upper air trough location. Most occur within 6° latitude of the jet stream or within 14° longitude of an upper air trough. Diurnally tornadoes occur most often between 1530 and 1830. Winds occasionally exceed $50\,m\,s^{-1}$.

Snow, associated with winter cold fronts passing over elevated areas, is mainly confined to the mountains in SE Australia, particularly just south of Canberra, and on Tasmania's Central Plateau and uplands.

The lack of consistent rainfall over most of the continent creates one of Australia's most serious climatic problems, drought. Drought can be defined as a severe water shortage depending on the amount of water needed, or by the failure of rains at specific places. Where seasonal rainfall is critical, drought is defined as the failure of the wet season. The Australian Bureau of Meteorology (www.bom.gov.au) defines an index of drought to establish the concept of drought potential: 50th percentile of rainfall minus 10th percentile divided by the 30th percentile. High indexes (greater than 0.6) mean a strong potential for drought. More than two-thirds of Australia has high to extreme drought indexes (see Figure A90). The only areas escaping are the N, S and E coast and Tasmania.

Table A47 Annual average cost of weather-related disasters in Australia (Sturman and Tapper, 1996)

Type of disaster	Annual cost ($ million, 1989)	Percent of cost
Drought	303	24
Bushfire	68	5
Storm	202	16
Flood	386	31
Tropical cyclone	258	21
Other	33	3
Total	1250	100

Table A47 shows that drought, along with flood and tropical cyclones, create the highest cost from weather disasters in the country. The table emphasizes the importance of extreme weather to Australia.

New Zealand climates

New Zealand consists of two main islands and a number of smaller islands, located approximately 2500 km east of Australia. The country covers an area of $267\,000\,km^2$, and is aligned north to south between latitudes 34°S and 47°S (Figure A91a). The distance north to south is approximately 1930 km, but the width of New Zealand reaches a maximum of only 400 km. New Zealand's size, location, orientation, and dominance by mountains (especially on the South Island), create a very different climatic regime than Australia.

Four major physical factors dominate New Zealand's climate. First, the ocean surrounding on all sides strongly defines the type of air that reaches the country. Important airmasses include tropical maritime air from the Pacific, with warm temperatures, 18–21°C, and considerable moisture; and polar maritime air from the southern ocean, with temperatures approximately 7–13°C. Thus New Zealand is consistently under the influence of moist air masses, allowing considerable rainfall. The ocean influences bring cool summers and mild winters, although cold air advection from the Antarctic Ocean can bring snow and frosts.

Second, the mountain ranges on the western sides of the two main islands create a considerable influence on atmospheric circulation (see Figure A91a). Dynamically, the mountains block upstream airflow, creating orographic uplift and vertical mixing, which leads to lee troughs on the downwind side. Very large spatial differences in precipitation can occur over the short distance between the windward mountain slopes and the rainshadows a few kilometers to the east. The mountains also act as channels for winds, and have significant influences on local airflows.

Third, New Zealand's geographic position places most of the country in the westerly circumpolar wind belt. Subtropical influences only occur irregularly, in the northern part of the North Island in summer.

Fourth, the much larger Australian continent exerts a seasonal influence. In winter considerable surface cooling under clear skies creates stabilization in the lower atmosphere. This effect extends eastward in the circulation pattern, helping weaken the westerly airflows over New Zealand. In summer the

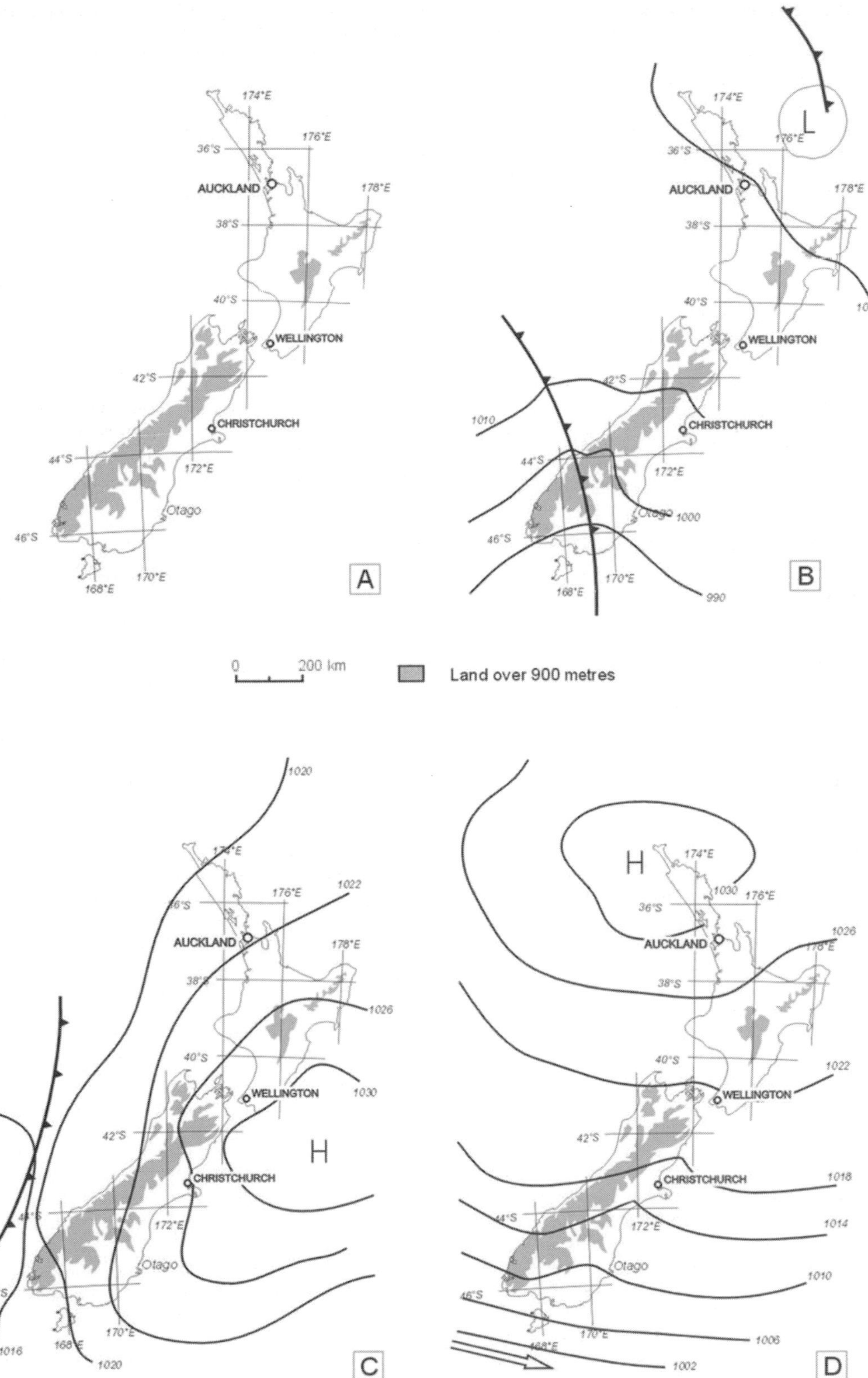

Figure A91 (**a**) New Zealand's physical features; (**b**) typical cold frontal passage over New Zealand; (**c**) typical blocking high situation to the east of New Zealand; (**d**) strong zonal circulation with high pressure to the north.

intense heating under clear skies over Australia creates a more convective atmosphere, which upon reaching New Zealand assists in creating stronger and more persistent westerly airflows.

The combination of these four factors has major influences on surface energy and water budgets. Unlike Australia, which has a water deficit over most of the continent and during most of the year, New Zealand's water budgets are generally positive. Latent heat has a greater influence than sensible heat, especially during daylight hours. The highly irregular terrain creates major spatial and temporal changes in the energy budget structure. For example, in the mountains, albedos can range from around 15% for alpine vegetation, to more than 60% for snow cover, over very small distances.

Synoptic circulation patterns

Figure A85 shows New Zealand's climatic circulation pattern in winter. The major band of anticyclonic movements extends to the north of the country. The westerlies, transporting fronts and depressions alternating with high-pressure systems, dominate. During this season the temperature gradient between Antarctica and locations further north is weakest, and therefore the strength of westerly circulation remains relatively weak.

In summer, as shown in Figure A87, the major belt of anticyclones now extends over most of the North Island. Here subsidence brings clear skies or extensive stratocumulus cloud sheets, over several days. The South Island remains under the influence of the westerlies. In spring and early summer the temperature gradient between Antarctica and the tropics, as well as the Australian continent, is strong. This is the season of strong winds and gales in the westerlies over southern New Zealand.

A more detailed evaluation of circulation in the New Zealand area reveals that three broad climate features provide the greatest influences:

1. *Troughs and depressions.* These are linked to cold frontal passages, occur regularly (approximately 40% of the time) during the year, although less frequently in fall. Depressions come from two main areas; those from the Southern Ocean are generally strong and move in a well-defined manner, guided by the westerly circulation. More infrequently, depressions form in the Tasman Sea. These can move slowly and erratically. Figure A91b shows a typical cold frontal passage. The isobaric pattern can be strongly deformed by the mountains. Frontal structure can be retarded or lost, and then re-formed once the system passes through the mountain area. Approximately half the fronts dissipate before reaching Cook Straight, the body of water between the two islands. Winds preceding the trough are generally northwesterly, and after frontal passage change to southwesterly. Higher elevations have a significant influence, causing wind direction deviations. Winds coming from a direction west of 220° can be prevented from reaching the eastern half of the country. The result is often heavy cloud and rain on the western sides of both islands, but partially cloudy to clear conditions to the east. Overall, precipitation is above average, but there is considerable spatial variation depending on geographical location.

   Troughs, depressions and fronts are usually steered by jet streams. In winter the jet stream is located between 25°S and 35°S, and can reach speeds of over 40 m s^{-1}. In summer the jet shifts to 45–50°S and is considerably weaker. There is considerable day-to-day variation.
2. *Anticyclonic or Blocking High.* Such conditions occur most frequently in summer and fall. Figure A91c presents an example. The frequency of blocking highs in this location is the highest in the southern hemisphere, and is linked to a split in the jet stream. Most frequently, blocking highs occur just to the east of New Zealand, and extend their influence over most of the country. They bring warmer temperatures than average over most of the country, higher precipitation to the northeast and lower precipitation to the southwest.

   Blocking highs bring settled weather, with light winds and mainly clear or partially cloudy skies. This encourages more local airflows to develop, such as sea breezes and katabatic winds (see discussion below). Fog can be prevalent inland.
3. Strong *zonal circulation* to the south with an anticyclone to the north is the third circulation regime. Figure A91d provides an example. This pattern creates drier and warmer than average conditions to the north, but higher rainfall to the

Table A48 Circulation periods affecting New Zealand climate during the twentieth century and impacts (after Salinger and Mullan, 1999; Salinger and Griffiths, 2001)

Period	Circulation	Temperature	Rainfall
1930–1950	More S to SW airflows	Lower over most of country; fewer days >25°C except NE North Island	Wetter to NE of South Island; drier in N and W of South Island
1951–1975	More E and NE airflow (higher frequency of blocking highs)	Higher over most of country; more days >25°C	Wetter in north of North Island; drier to SE of South Island
1976–1994	More W to SW airflows (more ENSO events)	Little change on average; fewer days >25°C except NE North Island	Wetter over South Island; drier in N of North Island

southwest. The enhanced westerlies to the south may result in somewhat milder conditions, especially in winter. This type of circulation regime is less frequent in summer than in other seasons of the year.

During the twentieth century there were changes in the circulation pattern which can be roughly classified into two-decade periods (Salinger and Mullan, 1999). Between 1930 and 1950 anomalously high frequencies of S and SW airflow occurred compared to the overall average. From 1951 to 1976 E and NE airflows were more frequent. From 1976 to 1998 the circulation reverted to more W and SW airflows. Table A48 summarizes the impacts of these three periods on New Zealand's climate.

Western Pacific regional influences on New Zealand circulation

There are three major interacting influences in the western Pacific that affect New Zealand's circulation pattern, and therefore climate variability (Kidson and Renwick, 2002). The enhancement of convection in the tropics in locations around the dateline decreases frequency of blocking and enhances zonal circulation. If convection is enhanced over the Indonesian area, blocking frequency near New Zealand is also enhanced, and zonal circulation reduced. The frequency and movements of troughs and depressions are not significantly affected.

Shifts in the location of the South Pacific Convergence Zone (SPCZ) affect the precipitation regime over New Zealand (Sturman and Tapper, 1996). When the SPCZ shifts eastward, circulation leading to drier conditions occurs over most of New Zealand except for the W and SW, where enhanced westerlies can bring more precipitation. A westward shift in the SPCZ creates the opposite effect, enhancing precipitation over most of the country except the W and SW coasts. These variations can be subtle and difficult to detect.

Both the strength and location of convection, and the location of the SPCZ, depend strongly on the Southern Oscillation, and whether an El Niño (ENSO) or La Niña (LNSO) event is occurring (Griffiths et al., 2003). During ENSO the SPCZ is shifted toward the NE, and convection is enhanced near the dateline. ENSO encourages high pressure with drought conditions in the north, with a stronger winter anticyclone. Enhanced meridional circulation leads to anomalous strength and frequency of southerly changes. During LNSO the SPCZ is shifted toward the SW, and convection is enhanced over the Indonesia area. Zonal circulation is enhanced, creating more rapidly moving troughs and depressions. NE airflow occurs at the expense of westerlies.

Local circulations

The interaction of the mountain terrain with synoptic circulation can create significant local airflows that affect New Zealand climate. The mountains both block airflow from more westerly sources and act to channel the airflow. The result is often a Foehn wind, which blows over the eastern parts of the country, creating warm dry conditions (Sturman and Tapper, 1996). On South Island a local name for this wind is the "Canterbury Northwester". Foehn winds are strong and gusty downslope winds from the lee of the mountain ranges, which warm through adiabatic compression. They often occur just before a cold frontal passage. Their strength and location depend on the atmospheric conditions to the windward of the mountain ranges, especially the vertical structure of the atmosphere. Foehn winds can create desiccation, very high temperatures, and damage to buildings, livestock and crops.

The Cook Straight acts a wind tunnel to encourage strong and gusty local wind passage from west to east of the country. These winds can recurve quite sharply, creating NE airflow on the E side of South Island.

Higher terrain can also create katabatic winds overnight over much of the eastern parts of New Zealand, when the synoptic situation is dominated by high pressure. The irregularity of the terrain creates complex downslope airflows, which eventually spill out over the glaciers, river valleys and plains. Low-level temperature inversions are common, especially in the winter, trapping air pollutants. A good example is wood smoke from home heating in the winter atmosphere in Christchurch, which creates highly elevated concentrations of fine particulate matter overnight.

Despite the narrowness of the islands, sea breezes occur on all coasts, especially in the summer. While these create some welcome relief from warm temperatures, their penetration inland is very much dependent on the terrain. Over very narrow areas, such as the very north of North Island, and in Auckland, sea breezes from both the Pacific Ocean and Tasman Sea can meet and interact together. Sea breeze depth ranges from 200 to 2000 m, and wind speeds can reach $15\,m\,s^{-1}$. They normally start in late morning and can extend into late evening (Sturman and Tapper, 1996).

Climatic elements (see Table A49)

The circulation variations over New Zealand, associated with changes affecting the circulation of the southwest Pacific region, significantly influence the temperature and precipitation distributions from year to year.

Temperature

Figure A92a shows the annual average temperature distribution. In summer the average temperatures range from about 18°C along the north coast of North Island to 14°C along the south coast of South Island. Particularly in South Island, the isotherm distribution is very strongly influenced by the terrain. Winter average temperatures range from 11°C in the very north of North Island to 4°C in the higher elevations below 45° latitude. Table A48 lists general changes from this average pattern as circulation periods change.

During ENSO periods the temperatures tend to be cooler over most of the country, affected by the increases in SW airflow. North Island is affected more strongly. Seasonally, this influence is weakest in summer (Kidson and Renwick, 2002). During LNSO periods the temperatures are generally milder in all areas, although the impact is less consistent on South Island. The S and W coasts are often cooler, associated with the stronger SW winds. Correlations between the South Oscillation and temperatures over New Zealand can reach 0.6 in the northern part of North Island, falling to below 0.4 along the south coast of the country and are significant.

Since 1950 the average temperatures over New Zealand have increased by about 0.5°C, in line with global increases

Table A49 Mean climatic data for selected New Zealand cities, January and July

City (Lat./long.)	Alt. (m)	Years of data	Max. temp. (°C)	Min. temp. (°C)	Mean temp. (°C)	Relative humidity 0900 (%)	Precipitation (mm)	Frost days	Daily sunshine (h)	Solar energy (MJ/m^2 per day)
Auckland (36.9°S/174.7°E)	41	25								
January			23.3	15.3	9.3	77.1	75	0	229	23.1
July			14.5	7.1	10.8	88.5	146	4	140	7.9
Wellington (41.3°S/174.8°E)	125	30								
January			20.3	13.8	16.9	80.3	72	0	246	23.6
July			11.4	6.3	7.5	86.3	136	3	117	5.7
Christchurch (43.5°S/172.6°E)	7	26								
January			22.5	12.2	17.4	72.9	42	0	230	21.9
July			11.3	1.9	6.6	87.3	79	16	124	5.1
Dunedin (45.9°S/170.5°E)	2	26								
January			18.9	11.5	15.2	73.1	72	0	178	18.5
July			9.8	3.2	6.5	80.2	69	16	101	4.5
Invercargill (46.5°S/168.3°E)	44	23								
January			18.6	9.4	14.0	80.8	114	1	180	20.4
July			9.5	0.9	5.2	88.9	88	18	93	4.3

Data from www.niwa.co.nz.

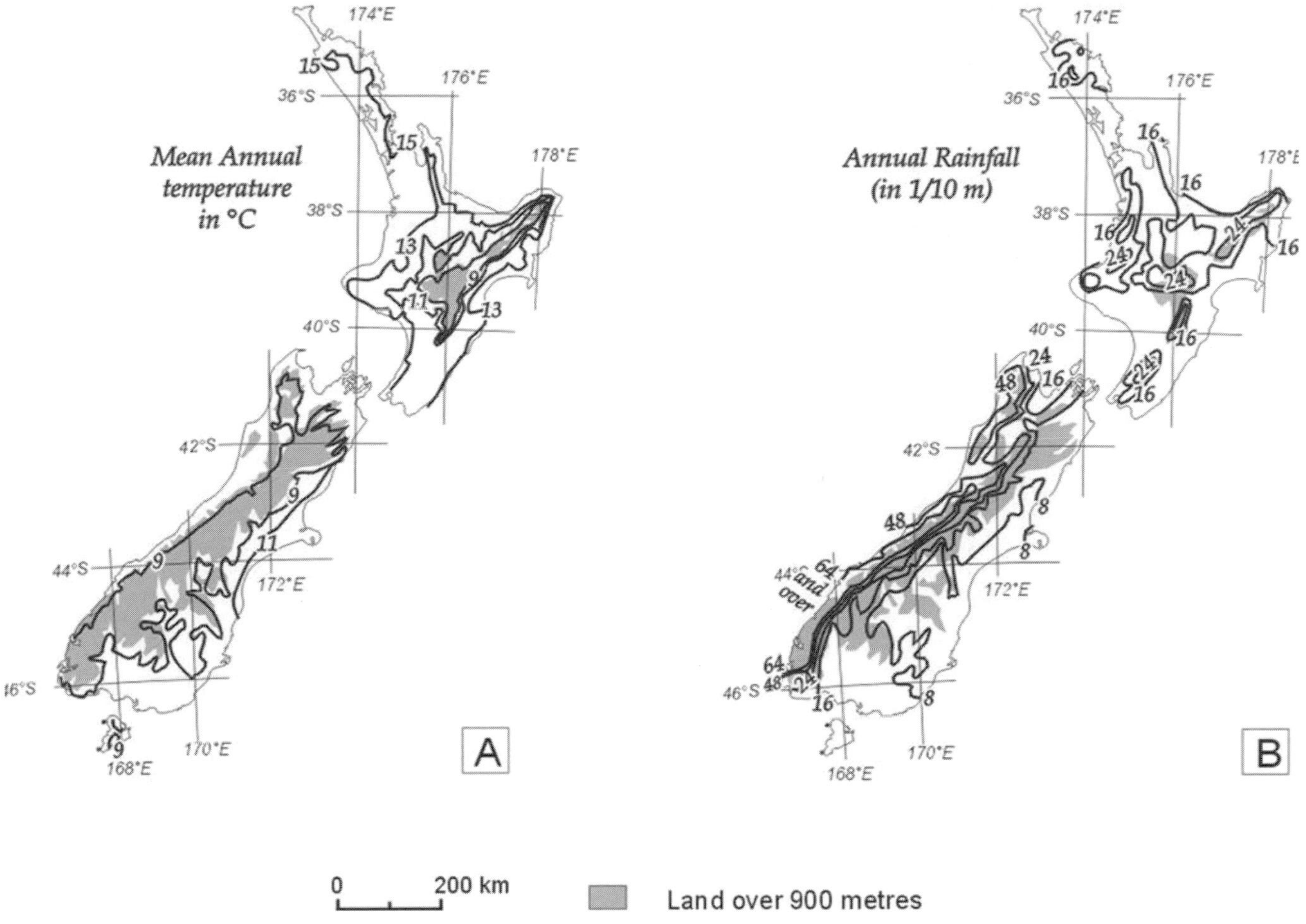

Figure A92 (**a**) New Zealand average yearly temperature distribution (°C); (**b**) New Zealand average yearly rainfall (cm divided by 10) (after Kirkpatrick, 1999).

associated with **greenhouse warming** (Salinger and Griffiths, 2001). There is little evidence of change in maximum temperatures or extremes, but minimum temperatures have increased significantly. There has been a decrease in the frequency of extremely cold nights. Between 1976 and 1998 the number of days with frost was lower by an average five to 15 in each year, compared to earlier periods.

Precipitation

The precipitation distribution pattern over New Zealand is highly irregular (Figure A92b), mainly due to terrain influences. Precipitation can range from more that 12 000 mm y^{-1} on the mountain tops of South Island to less than 600 mm y^{-1} on the eastern plains, for example the Canterbury area. Particularly on South Island, snow is an important component, covering 35% of the area on average but with high variability from year to year.

The greatest source of rainfall is fronts and troughs, particularly if they are associated with cold pools of air in the mid-to-upper troposphere. These can bring severe storms and create very heavy rain. Such storms are more likely to occur in winter and spring than in summer.

Overall, increased rainfall on a seasonal or yearly basis is associated with a higher frequency of N airflows, and thus linked to LNSO. Under this pattern rainfall to the S and W is often reduced, associated with weaker westerly circulation. This varies, depending on the time of year and the strength of the LNSO event.

ENSO periods bring drier weather to the N and E, and, compared to Australia, occasional much shorter periods of drought. Droughts rarely last more than 3 weeks. Increased westerlies during this period bring higher rainfall than average to the W and S coasts of especially South Island. The SOI does not correlate significantly with rainfall over New Zealand (TPE, 2001).

Changes in rainfall associated with the major changes in circulation in 1950 and 1976 are less well defined than for temperature (Salinger and Griffiths, 2001). The eastern part of South Island and the central area in North Island show significant decreases, by 15–25%. Increases occur on the south and west sides of both islands. Otherwise, there are no significant changes in rainfall distributions. The length of dry spells over New Zealand has been reduced by 1.5 to 4.5 days since 1950.

Table A50 Extreme weather events in New Zealand

Weather event	Location	Period	Value
Rainfall (mm)			
Hourly Total	Whenuapai	16 Feb 1966	107
Daily Total	Colliers Creek (S.I.)	21–22 Jan 1994	682
Monthly Total	Cropp River (S.I.)	Dec 1995	2927
Annual Total	Cropp River (S.I.)	1998	16617
Lowest Annual Total	Alexandra (S.I.)	Nov 1963–Oct 1964	167
Temperature (°C)			
Highest Maximum (N.I.)	Ruatoria	7 Feb 1973	39.2
Highest Maximum (S.I)	Rangiora & Jordan	7 Feb 1973	42.4
Lowest Minimum (N.I.)	Chateau Tongariro	7 Jul 1937	−13.6
Lowest Minimum (S.I.)	Ophir	3 Jul 1995	−21.6
Wind Gusts (km/h)			
Highest, North Island	Hawkins Hill	6 Nov 1959	248
Lowest, South Island	Mt John	18 Apr 1970	250

Data from http://www.metservice.co.nz
N.I. – North Island; S.I. – South Island.

Other climate elements
Thunderstorms are most often linked with fronts and troughs, and are higher in frequency on the west side of the islands, associated with orographic forcing. On the east coast of the islands, diurnal variations are typically a maximum in the afternoon and early evening, associated with a summertime peak in frequency and a thermal (convective) origin. On the west coast a higher frequency occurs at night, with an erratic seasonality pattern. Here wintertime thunderstorms occur more regularly, associated with friction and orographic effects. Along a north–south transect through the center of both islands, thunderstorm frequencies peak in the afternoon and early evening, but a secondary peak overnight is evident.

Hail frequencies increase from north to south, associated with the deep convection in cold fronts and troughs. New Zealand has approximately 25 tornadoes per year. North Island has an approximately three times higher frequency than South Island. Most tornadoes are relatively small by world standards, and are more frequent in the afternoon. Wind speeds can reach 190 km/h (Sturman and Tapper, 1996).

Tropical cyclones affect New Zealand rarely, creating strong winds, and heavy rain can occur on the northern and northeastern parts of North Island.

Table A50 lists extreme weather events for New Zealand. The maximum temperatures for both North and South Island are associated with a strong Foehn event, which occurred just before a cold frontal passage. Subsequently, the temperature then rapidly dropped to around 22°C. The lowest minimum temperatures occurred in snowfield stations in the winter. Short-term, high-rainfall events are associated with severe storms, and longer-term records with the higher altitudes on the western slopes of South Island mountain ranges. The lowest annual rainfall occurs at Alexandra on South Island, in the rainshadow east of the mountain ranges. In comparison with Australia (Table A45), New Zealand extremes show higher rainfall amounts but lower maximum and higher minimum temperatures.

By global standards New Zealand is basically one Köppen climate type, Cf (warm temperate climate with sufficient precipitation during all months). On a more detailed regional scale, however, a simplified climate classification of New Zealand produces eight categories, as shown in Figure A93.

Near-future climate change in Australia and New Zealand

The climate is varying continuously, and there are concerns about near-future changes associated with greenhouse warming in Australia and New Zealand. A series of reports focusing on the results of meteorological network measurements in both countries, and global and regional climate modeling, suggest the following scenario is likely for Australia and New Zealand (www.dar.csiro.com; www.niwa.co.nz).

Higher temperatures in general across Australia are probable, 0.4–2.0°C by 2030 and 1–6°C by 2070. The extra heat capacity of the oceans will delay and minimize warming along the coast. Most of the change will most likely occur in the minimum temperature, decreasing the overall daily temperature range. Evidence that this has already been occurring comes from Queensland (Lough, 1995) and other parts of the country (Bureau of Meteorology 2003). Warmer minimum temperatures would decrease the number of days below 0°C (by up to six per year) and less snow cover would exist. However, an increase in the number of extremely hot days in summer is expected, particularly in the desert areas of Western Australia.

Rainfall zones are likely to shift southwards, leading to lower rainfall in the south and east, especially in winter and spring, but higher rainfall in summer. Winter rainfall may increase slightly in central Australia, but decrease or remain the same in most other parts of the country. There has been a downward trend in rainfall in the SW since the mid-1960s (Allen and Haylock, 1993), which may continue. The climate record already shows an increase in Australian cloud cover during this century (Jones, 1991). There may be greater frequencies of high-rainfall events, and decreased numbers of low-rainfall events, as well as increased frequencies of drought, bushfires, and fine weather (Sturman and Tapper, 1996). Potential

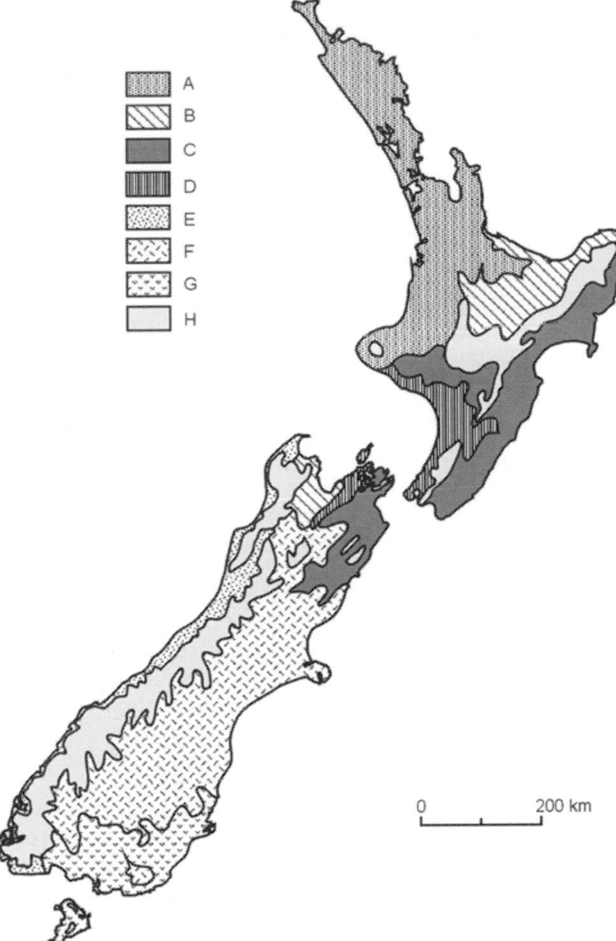

Figure A93 Simplified climate classification for New Zealand (modified from Sturman and Tapper, 1996). **A**, mild winter, warm moist summers, annual rainfall 1000–2500 mm, winter rainfall maximum, prevailing SW winds, occasional storms from E or NE. **B**, annual rainfall 1000–2000 mm, winter maximum, mild winters (but frequent frost), very warm summers, heavy rain from N or NE. **C**, very warm summers (strong Foehn effect), cool to moderate winters, annual rainfall 600–1500 mm (heavier in isolated places), winter maximum, lower and more unreliable summer rain. **D**, winds dominant from W and NW, frequent gales throughout year, annual rainfall 900–2000 mm, cooler temperatures. **E**, cool annual temperatures, range is small, prevailing SW winds, strong mountain influence on rainfall distribution (increase with altitude). **F**, cool winters, warm summers, frequent frosts, occasional snow (especially inland), annual rainfall 500–1500 mm, NE winds prevail along coast, NW winds inland. **G**, cool winters, warm summers, annual rainfall 500–1300 mm, slight winter maximum. **H**, extreme rainfall variations in mountains, cool to cold temperatures.

evaporation will increase. A greater intensity in extreme rainfall events and tropical cyclones is expected.

For New Zealand, the summer increase in temperature by the latter two decades of the 21st century may range from 1°C in the S to 1.8°C in the N. Winter increases are slightly higher, ranging from 1.6°C in the S to 2.2°C in the N. Model results suggest decreases in precipitation of around 5% on the east coast of both islands, and an increase of up to 15% on the west coasts. Winter precipitation may change quite dramatically, with reductions of 10–15% on the east coasts and increases of more than 30% in the mountains of the South Island. These changes reflect the increasing frequency and strength of SW and W airflows.

Readers are referred to the following Bibliography for further information on the climate of Australia and New Zealand.

Howard A. Bridgman

Bibliography

Abbs, D.J., and Physick, W.L., 1992. Sea-breeze observations and modelling, *Australian Meteorological Magazine*, **41**: 7–20.

Allen, R.J., 1988. El Niño Southern Oscillation influences in the Australasian region. *Progress in Physical Geography*, **12**: 313–348.

Allen, R.J., and Haylock, M.R., 1993. Circulation features associated with winter rainfall decrease in southwestern Australia. *Journal of Climate*, **6**: 1356–1367.

Baines, P.G., 1990. What's interesting and different about Australian meteorology?, *Australian Meteorological Magazine*, **38**: 123–146.

Bridgman, H.A., 1998. Australia's Climate and Water Resources. In Herschy, R., ed., *Encyclopaedia of Hydrology and Water Resources*. London: Chapman & Hall, pp. 98–109.

Bureau of Meteorology, 1988a. *Climatic Atlas of Australia*. Canberra: Department of Administrative Services, 67 pp.

Bureau of Meteorology, 1988b. *Climatic Averages, Australia*. Brisbane: Watson Ferguson.

Bureau of Meteorology, 1989. *Climate of Australia*. Canberra, Australian Government Publishing Service.

Bureau of Meteorology, 2003. *The Greenhouse Effect and Climate Change*. Melbourne: Commonwealth Bureau of Meteorology, Melbourne.

Castles, I., 1995. *Year Book Australia 1995*. Canberra: Australian Bureau of Statistics.

Collis, K., and Whitaker, R., 1996. *The Australian Weather Book*. Sydney: National Book Distributors.

Cook, G., and Heedegen, R., 2001, Spatial variation in the duration of the rainy season in monsoonal Australia, *International Journal of Climatology*, **21**: 1723–1732.

Drosdowsky, W., and Williams, M., 1991. The Southern Oscillation in the Australian region. Part I: anomalies at the extremes of the oscillation. *Journal of Climate*, **4**: 619–638.

Gentilli, J., 1971. Climates of Australia and New Zealand. In Landsberg, H., ed., *World Survey of Climatology*. Amsterdam: Elsevier.

Gentilli, J., 1972. *Australian Climatic Patterns*, Adelaide: Thomas Nelson.

Griffiths, G., Salinger, M., and Leleu, I., 2003. Trends in extreme daily rainfall across the South Pacific and relationship to the South Pacific Convergence Zone. *International Journal of Climatology* **23**: 847–869.

Haylock, M., and Nicholls, N., 2000. Trends in extreme rainfall indices for an updated high quality data set for Australia, 1910–1998. *International Journal of Climatology*, **20**: 1533–1542.

Hess, G.D., and Spillane, K.T., 1990. Characteristics of dust devils in Australia. *Journal of Applied Meteorology*, **29**: 498–507.

Hobbs, J., 1998. Present climates of Australia and New Zealand. In Hobbs, J., Lindesay, J., and Bridgman, H., eds., *Climates of the Southern Hemisphere Continents Past Present and Future*. Chichester: Wiley, pp. 63–106.

Holland, G.J., Lynch, A.H., and Leske, L.M., 1987. Australian east-coast cyclones. Part I. Synoptic overview and case study. *Monthly Weather Review*, **115**: 3024–3036.

Jones, P.A., 1991. Historical records of cloud cover and climate for Australia. *Australian Meteorological Magazine*, **39**: 181–190.

Joseph, P.V., Liebmann, B., and Hindon, H.H., 1991. Interannual variability of the Australian summer monsoon onset: possible influences of the Indian summer monsoon and El Niño, *Journal of Climate*, **4**: 529–538.

Kidson, J., 2000. An analysis of New Zealnad synoptic types and their use in defining weather regimes. *International Journal of Climatology*, **20**: 299–316.

Kidson, J. and Renwick, J., 2002. Patterns of convection in the tropical Pacific and their influence on New Zealand weather. *International Journal of Climatology*, **22**: 151–174.

Kirkpatrick, R., 1999. *Contemporary Atlas of New Zealand*. Auckland: Bateman.

Kuleshov, Y., deHoedt, G., Wright, W., and Brewster, A., 2002. Thunderstorm distribution and frequency in Australia. *Australian Meteorological Magazine*, **51**: 145–154.

Lough, J.M., 1995. Temperature variations in a tropical–subtropical environment: Queensland, Australia, 1910–1987. *International Journal of Climatology*, **15**: 77–96.

Love, G., and Downey, A., 1986. A prediction of bushfires in central Australia. *Australian Meteorological Magazine*, **34**: 93–102.

McInnis, K.L., 1993. Australian southerly busters. Part III: the physical mechanism and synoptic conditions contributing to development, *Monthly Weather Review*, **121**: 3261–3281.

Reeder, M.J., and Smith, R.K., 1992. Australian spring and summer cold fronts. *Australian Meteorological Magazine*, **41**: 101–123.

Salinger, M. and Griffiths, G., 2001. Trends in New Zealand daily temperature and rainfall extremes. *International Journal of Climatology*, **21**: 1437–1452.

Salinger, M., and Mullan, A., 1999. New Zealand climate: temperature and precipitation variations and their links with atmospheric circulation, 1930–1984. *International Journal of Climatology*, **19**: 1049–1072.

Salinger, M., Renwick, J., and Mullan, A., 2001, Interdecadal Pacific Oscillation and South Pacific climate. *International Journal of Climatology*, **21**: 1705–1723.

Smith, I., McIntosh, P., Ansell, T., Reason, C., and McInnes, K., 2000. Southwest Western Australian winter rainfall and its association with Indian Ocean climate variability. *International Journal of Climatology*, **20**: 1913–1930.

Smith, R.K., Reeder, M.J., Tapper, N.J., and Christie, D.R., 1995. Central Australian cold fronts, *Monthly Weather Review*, **123**: 16–28.

Speer, M., and Geerts, B., 1994. A synoptic-mesoalpha-scale climatology of flash-floods in the Sydney metropolitan area. *Australian Meteorological Magazine*, **43**: 87–103.

Stern, H., deHoedt, G., and Ernst, J., 2000. Objective classification of Australian climates. *Australian Meteorological Magazine*, **49**: 87–98.

Sturman, A.P., and Tapper, N.J., 1996. *The Weather and Climate of Australia and New Zealand*. Melbourne: Oxford University Press.

Suppiah, R., 1992. The Australian summer monsoon: a review. *Progress in Physical Geography*, **16**: 283–318.

Tapper, N., and Hurry, L., 1993. *Australia's Weather Patterns: An Introductory Guide*. Mt Waverly (Victoria): Dellastra.

TPE, 2001. (Sturman, A., and Spronken-Smith, R., eds.) *The Physical Environment A New Zealand Perspective*. Melbourne: Oxford University Press.

Cross-references

AirMass Climatology
Climate Classification
Drought
El Niño
Southern Oscillation
Synoptic Climatology
Vegetation and Climate

Web sites of interest

Australian Bureau of Meteorology, www.bom.gov.au
New Zealand Meteorological Bureau, www.metservice.co.nz
Division of Atmospheric Research, Commonwealth Scientific and Industrial Research Organisation, Australia, www.dar.csiro.au
National Institute of Water and Atmospheric Research (NIWA), www.niwa.co.nz/atmos/

AZORES (BERMUDA) HIGH

This is one of seven regions of year-round relatively high pressure where barometric values are typically higher than adjacent areas and where the air is drier, skies clearer, and surface wind speeds slower. The other six regions are the Siberian (Asiatic), Pacific (Hawaiian), and North American (Canadian) Highs in the northern hemisphere, and the Indian Ocean, South Atlantic (St Helena), and South Pacific Highs in the southern hemisphere.

The Azores (or Bermuda) High is a large subtropical anticyclone centered between 25°N and 35°N in the Atlantic Ocean. The high often extends westward as far as Bermuda; when this occurs it is known in North America as the Bermuda High. It extends to great elevations, with deep tropical easterlies on its equatorward side and midlatitude westerlies on its poleward side. This very persistent and quasistationary feature of the atmosphere's circulation is elongated in an east–west direction, with its center of highest pressure located over the eastern portion of the subtropical Atlantic in the general vicinity of the Azores. It is most pronounced in the summer, when land–water temperature contrasts are substantial, and has an extreme summertime position of about 34–38°N. The expanding high often gets large enough to affect hurricane trajectories, moving the storms westward toward the West Indies and the southeastern United States. In winter it is less pronounced and is located at somewhat lower latitudes.

Within the heart of the Azores High, winds are light and variable, and the high is characterized by divergence, subsidence, and fair weather. However, there is a marked contrast between the eastern and western portions of the high. The eastern side is characterized by strong subsidence and a dry climate, whereas the western side has little or no subsidence and a wetter climate. This contrast corresponds to the moisture and stability differences between the east and west sides of the oceanic trade winds.

The cause of this contrast is not clear. The longer trajectory of air over the warm waters of the subtropical Atlantic contributes to a more humid west-side climate. This circulation pattern results in hot and humid weather for the central and eastern United States in the summer. The ultimate cause is probably related to the inherent distribution of continents and oceans. Large east–west circulations exist in heating between subtropical continents and oceans. The strongly subsiding eastern portions of the Azores High coincide with the sinking branches of these east–west cells, whereas the western portion of the high is near the rising branch of the circulation cell.

The relative strength of the high is used as one variable to determine the North Atlantic Oscillation and its derived index.

Robert M. Hordon and Mark Binkley

Bibliography

Ahrens, C.D., 2003. *Meteorology Today: An Introduction to Weather, Climate, and the Environment*, 7th edn. Brooks/Cole (Thomson Learning).

Curry, J.A., and Webster, P.J., 1999. *Thermodynamics of Atmospheres and Oceans*. San Diego, CA: Academic Press.

Hartmann, D.L., 1994. *Global Physical Climatology.* San Diego, CA: Academic Press.

Lutgens, F.K., and Tarbuck, E.J., 2004. *The Atmosphere: An Introduction to Meteorology*, 9th edn. Upper Saddle River, New Jersey: Pearson Prentice Hall.

Nese, J.M., and Grenci, L.M., 1998. *A World of Weather: fundamentals of meteorology*, 2nd edn. Dubuque, IO: Kendall/Hunt.

Oliver, J.E., and Hidore, J.J., 2002. *Climatology: An Atmospheric Science*, 2nd edn. Upper Saddle River, New Jersey: Pearson Prentice Hall.

Robinson, P.J., and Henderson-Sellers, A., 1999. *Contemporary Climatology*, 2nd edn. Harlow: Longman.

Wallace, J.M., and Hobbs, P.V., 1997. *Atmospheric Science: an introductory survey*. San Diego, CA: Academic Press.

Cross-references

Airmass Climatology
Atmospheric Circulation, Global
Centers of Action
North Atlantic Oscillation
Oscillations
Pacific (Hawaiian) High
Zonal Index

B

BEAUFORT WIND SCALE

British Admiral (then Commander) Sir Francis Beaufort, KCB developed the Beaufort Wind Scale in 1805. The scale provided a practical method for estimating the force of the wind on a ship at sea by means of an arbitrary scale running from 0 (calm) to 12 (hurricanes) as shown in Table B1. In 1838 use of the Beaufort Wind Force Scale was made mandatory for log entries on all Royal Navy ships. In the scale's original form (which has been greatly altered in most current versions), there is no mention of wind velocity, it specifically being a force scale.

The scale is made up of three associated lists – a list of 13 integers (Beaufort force numbers) from 0 through 12; a list of common words for the strength of the wind; and a list that describes the state and behavior of a "well conditioned man-of-war" (Kinsman, 1969). Beaufort, and all seafaring individuals, recognized distinct levels in the behavior of a man-of-war of the early 1800s operating under conditions of "clean full" (probably close-hauled but held just off the wind to avoid any loss of speed due to luffing or shaking of the sails) and "in chase, full and by" (carrying all possible sails, all sails full, and heading as near to the wind as possible).

The Beaufort Wind Force Scale was developed at a time when instruments that measure wind speed were unavailable; but point wind speed information would not have been helpful on a sailing ship. Wind speed would have varied considerably from the height of the quarter deck (12 feet) to the crosstrees on the main mast (120 feet). What was really needed was information on the overall effect of the wind force at all

Table B1 Beaufort's Wind Force Scale, 1831 version (probably close to the original formulation of the scale)

0	Calm		
1	Light air	Or just sufficient to give steerage way	
2	Light breeze	Or that in which a man-of-war, with all sail set, and clean full, would go in smooth water from	1 to 2 knots
3	Gentle breeze		3 to 4 knots
4	Moderate breeze		5 to 6 knots
5	Fresh breeze		Royals and courses
6	Strong breeze	Or that to which a well-conditioned man-of-war could just carry in chase, full and by	Single-reefed top sails and top-gallant sails
7	Moderate gale		Double-reefed top sails, jib and courses
8	Fresh gale		Treble-reefed top sails and courses
9	Strong gale		Close-reefed top sails and courses
10	Whole gale	Or that with which she could scarcely bear close-reefed main top sail and reefed fore-sail	
11	Storm	Or that which would reduce her to storm stay-sails	
12	Hurricane	Or that which no canvas could withstand	

Source: From Fitzroy (1839), p. 40.

Table B2 Reasonable current specifications for the Beaufort Wind Force Scale

Beaufort	Descriptive term	Wind speed (knots) Mean	Limits	Effect on sea surface
0	Calm	0	<1	Sea like a mirror
1	Light air	3	1–4	Ripples with the appearance of scales, no foam crests
2	Light breeze	7	5–8	Small wavelets, crests have glassy appearance and do not break
3	Gentle breeze	11	9–12	Large wavelets, crests begin to break, perhaps scattered white horses
4	Moderate breeze	15	13–16	Small waves becoming longer, fairly frequent white horses
5	Fresh breeze	19	17–21	Moderate waves, many white horses, chance of some spray
6	Strong breeze	24	22–26	Large waves form, white foam crests extensive, probably some spray
7	Near gale	29	27–31	Sea heaps up and white foam from breaking waves blown in streaks
8	Gale	34	32–36	Moderately high waves of great length, edges of crests begin to break into the spin-drift, foam blown in streaks
9	Strong gale	39	37–42	High waves, dense streaks of foam, crests of waves begin to topple, tumble and roll over
10	Storm	45	43–48	Very high waves with long overhanging crests, sea surface takes white appearance, visibility affected by spray
11	Violent storm	52	49–55	Exceptionally high waves, sea completely covered by long white patches of foam, everywhere edges of wave crests blown into froth
12	Hurricane		>55	Air filled with foam and spray, sea completely white with driving spray, visibility seriously affected

Source: Wind speeds from Verploegh 1956; descriptive terms from Roll, 1965, pp. 24–25.

levels on the performance of the man-of-war operating at peak performance.

By the time anemometers were used to give wind speed information, men-of-war of the type utilized by Beaufort in the formulation of his scale were no longer available, so that there was no way to assign wind speed values to the Beaufort numbers. However, because of the widespread acceptance of the Beaufort Scale, later versions have included wind speeds and descriptive terms that refer to the effect of the wind on fishing smacks, on trees, and on the state of the sea (Table B2).

Wind speed values associated with specific Beaufort numbers vary considerably from scale to scale (Simpson, 1926; Verploegh, 1956). Beaufort force numbers were used as a telegraphic wind code until 1947 when the Directors of the International Meteorological Committee agreed to abandon the use of Beaufort numbers and to replace them with wind velocities in knots in meteorologic reports.

John R. Mather

Bibliography

Fitzroy, R., 1839. *Narrative of the Surveying Voyages of His Majesty's Ships, Adventure and Beagle*, vol. 2. London: Henry Colburn.

Kinsman, B., 1969. Who put the wind speeds in Admiral Beaufort's force scale, *Oceans* **2**(2): 18–25.

Roll, H.U., 1965. *Physics of the Marine Atmosphere*. New York: Academic Press.

Simpson, G.C., 1926. The velocity equivalents of the Beaufort Scale. *Air Ministry Professional Notes*, No. 44. London: Meteorological Office.

Verploegh, G., 1956. The equivalent velocities of the Beaufort estimates of the wind force at sea. *Koninkl. Nederlands Acad. Meteorol. Inst. Mededel. Verh.* Monograph No. 66, pp. 1–38.

Cross-references

Local Winds
Wind Power Climatology
Winds and Wind Systems

BIOCLIMATOLOGY

Bioclimatology (biometeorology) is the study of the relationships between climate (weather) and living organisms. The field is vast and brings together scientists from many disciplines. Bioclimatology is frequently divided into human, plant (agricultural and forest), and animal bioclimatology. Other subdivisions include aerobiology (the behavior of airborne living material), phenology, urban bioclimatology, air pollution bioclimatology, tourism and recreation bioclimatology, mountain bioclimatology, electromagnetic and ionization bioclimatology, and bioclimatological rhythms. However, no single classification system has been adopted universally. The American Meteorological Society, for example, has several committees with bioclimatological involvements: Agricultural and Forest Meteorology, Applied Climatology, Biometeorology and Aerobiology, and Meteorological Aspects of Air Pollution. The time intervals studied range from the daily cycle to geological times.

Bioclimatology is an interesting research field that has many important practical applications for human comfort, agricultural yields, regional land-use planning, forest management, building research, and so forth. A focus for bioclimatological studies is provided by the International Society of Biometeorology, which sponsors the *International Journal of Biometeorology* (Springer-Verlag Publishers, Heidelberg, Germany). However, research results appear in a wide range of journals.

Bioclimatologists customarily refer to an environmental stress that causes a biological response. The word stress does not necessarily imply a harmful effect; a brisk walk on a cold day may be considered invigorating by many people, for example. Over the centuries, humans have been increasing their ability to reduce unwanted environmental stresses through use of clothing, heated and air-conditioned buildings, irrigation, flood-control systems, fertilizers, and pesticides. By the year 1960 expectation was widespread that humans soon would have total control of their environment. However, beginning in the 1960s this utopian view began to be challenged, and since the 1980s it is generally believed that environmental stresses are necessary to retain the resilience of humans and supporting ecosystems. By using technology to shield living things from the rigors of the elements, the chances increase that the capacity to withstand extremes may become seriously impaired. For example, a severe ice storm in January 1998 in eastern Canada and northeastern United States caused serious damage to the power supply infrastructure, which led to electricity blackouts for several days (Higuchi et al., 2000). Communities in those regions that depended heavily on modern technology suffered more than those that were accustomed to power outages and had their own back-up alternatives. Bioclimatologists have a full agenda for the next few decades elaborating these issues, particularly in the face of global environmental change (see article in this Encyclopedia on Global Environmental Change: Impacts, by the same authors).

The field of bioclimatology is too broad to cover here; therefore, the approach taken is to discuss a few topics in sufficient detail to give the reader an idea of the flavor of bioclimatology. A list of further reading is added with respect to subjects not mentioned in the text.

Bioclimatological methods

Methods used in bioclimatological research are of three main types:

1. Statistical

Multivariate regressions or other statistical methods, in which a response characteristic is correlated with an array of possible indicators of stress. Although a significance test may indicate an association, it does not imply the existence of a cause– effect relationship.

2. Experimental

Controlled *laboratory studies* using human volunteers or greenhouses/phytotrons/biotrons in the case of plants, animals and whole ecosystems. An ethical problem with studies involving human volunteers is that the stress produced might be harmful (e.g. causing an asthma attack) for a few of the more sensitive people participating. Yet these individuals provide data for that part of the stress–response curve of most interest.

Field studies of stress–response relations, preferably under given sets of conditions, e.g. using only data when skies are clear, or when gradient winds are light.

3. Models

Mathematical statements synthesizing current understanding of stress–response relationships. For large systems, e.g. those used in integrated pest management, models can interconnect various kinds of information. Models can also be used to test the utility of various proposed management strategies.

Data sets selected for study

Biological data sets used by bioclimatologists are of four main types: (1) general biological data sets that happen to be available, e.g. census information, crop yields, insect populations; (2) data sets relating to biological indicators that are specific to conditions of extreme environmental stresses, e.g. studies of organisms existing at high elevations, studies of health effects during heat waves, cold waves, or air pollution episodes; (3) data sets relating to cycles, e.g. diurnal, annual, or life cycles; and (4) data sets that are model-dependent. In the initial stages of an investigation, e.g. with respect to electromagnetic effects, the stress–response mechanism may be obscure and only a time-series analysis will be possible.

Epidemiological (large-population) statistical regressions may suggest laboratory or field studies that should be undertaken, and models that should be tested. In the latter case the data sets must be selected with care, or new data must be collected, the specific goal being model-performance testing.

Meteorological data sets include hourly, daily and monthly records of variables such as temperature, humidity, precipitation, etc.

Some problems with bioclimatological studies

Bioclimatological studies are difficult to design for several reasons:

Lack of reproducibility

Even under controlled laboratory conditions the response of a living organism to environmental stress may vary greatly from time to time. The response depends on the time of day (most organisms display a circadian rhythm of about 24 hours) and the frequency with which the response is imposed. Sometimes *acclimatization* (short-term physiological adjustment) may take place and may lead to long-term *adaptation.*

Variation in population responses

The response of populations to an environmental stress can vary greatly. For example, not everyone suffers from hay fever, no matter how high the concentration of pollen. The World Health Organization makes a useful distinction between an *effect* (a biological reaction) and a *response* (the percentage of exposed organisms/people who react with a specific effect.)

Existence of time lags

An environmental stress may produce an almost immediate response (e.g. exposure to H_2S causes an instantaneously detected bad smell), or the response may be lagged by days, weeks, years, or decades (e.g. up to a 20-year latency period may be required before asbestos exposure causes lung cancer; or for UV-B radiation to cause skin cancer). In the meantime the population being studied may have changed its lifestyle, making it impossible to disentangle stress–response relationships.

Multiple stresses

Living organisms frequently are subjected to several stresses at the same time. For example, during heat waves, air quality becomes poor, but it is not possible to determine whether the resulting effects on health are due to high temperatures, poor air quality or both.

Estimation of exposures

Most living things move about, and it is not easy to characterize their exposure. Many people spend most of their time indoors where the environment may be quite different from that measured outdoors at weather-observing and air-quality stations. For this reason, studies of human exposure to air pollutants often use human volunteers who wear portable pollution samplers to obtain their daily exposures (see, e.g., Silverman et al., 1992).

Failure of standard statistical methods

Many responses are curvilinear, making standard statistical methods inappropriate. Furthermore, most data sets contain time correlations rather than being random samples, causing correlation coefficients between two variables to be inflated. For more information on biometeorological methods, see Munn (1970), Houghton (1985), and Kates et al. (1985).

Human bioclimatology

Atmospheric variables that may affect humans include heat, cold, wind, humidity, solar radiation (especially UV-B radiation), air pollution, pressure, negative ions, electromagnetism, and biorhythms. In the case of the first six factors the existence of stress–response relationships has been clearly demonstrated. However, with respect to the last four factors the results obtained are still controversial, even though studies began more than a hundred years ago, particularly at the great European health spas, where people went to seek relief from arthritis, respiratory ailments, and allergies.

Heat stress and cold stress continue to demonstrate the most obvious effects of weather and climate on people, and will be discussed in the following paragraphs as good examples of the state of the art in bioclimatology.

Warm-blooded mammals must keep their inner body temperature within a very narrow range (around 37°C) or irreparable harm ensues. The body gains heat by its own metabolism; gains (or loses) heat from (or to) its surroundings by radiation, conduction, and convection; and loses heat by evaporation (breathing and sweating). There are two types of metabolic heat stress: *basal metabolic heat* (released when the body is at rest) and *muscular metabolic heat* (released, in addition to basal metabolic heat, during periods of work or body exercise).

Basic heat transfer models can predict the heat balances of simple volumes such as spheres or cylinders. However, the human body has a complex shape and some parts of the body – such as nose, cheeks and earlobes – are fully exposed whereas other parts are covered with clothing. Therefore, the study of heat and cold stress is not a straightforward problem in thermodynamics. Recently, more advanced heat transfer models have been developed and are being tested for use in a wide range of environmental conditions (see, for example, Fiala et al., 2001; Huizenga et al., 2001; Tanabe et al., 2002; and the International Society of Biometeorology: http://www.biometeorology.org/).

Heat stress is associated with various combinations of the following conditions: (1) high muscular activity; (2) high solar radiation, particularly in the tropics at high elevations; (3) high infrared radiation, e.g. near a blast furnace in a steel mill; (4) air temperatures greater than body temperature (37°C), producing a net gain in body heat from convection; (5) high humidity, reducing the rate of evaporational cooling from the body; and (6) strong winds in combination with (4) and (5), increasing the convective heat gains of the body.

The *physiological* (involuntary) mechanisms used to cope with heat stress are (1) dilating of blood vessels near the surface of the skin, increasing the flow of blood near the skin and increasing the heat exchange from the body to its surroundings; (2) increased sweating, resulting in evaporational cooling; and (3) increased respiration (equivalent to panting of dogs). The *voluntary* mechanisms are (1) avoiding strenuous activity; (2) avoiding direct sunlight and strong infrared sources; (3) switching to lightweight clothing; (4) changing diet to reduce basal metabolic heat production; and (5) remaining in air-conditioned buildings.

An indicator of heat stress is *effective temperature*, the temperature at which motionless saturated air would induce the same degree of comfort/discomfort as that associated with ambient conditions of temperature, humidity, and wind. Although the words "comfort" and "discomfort" are subjective terms, some consensus on their meaning has been achieved through physiological studies of volunteers walking on treadmills in controlled conditions. During each experiment, measurements are made of skin temperature, sweat rate, and body weight, and the volunteers are asked about their degree of discomfort. In other kinds of studies the learning rates of students or work outputs of office employees are examined in relation to room temperature and humidity. The results of these types of studies have been summarized in tables of discomfort.

Table B3 Humidex from dry-bulb temperature and relative humidity readings

	RH (%)																
Temp (°C)	100	95	90	85	80	75	70	65	60	55	50	45	40	35	30	25	20
43													56	54	51	49	47
42												56	54	52	50	48	46
41											56	54	52	50	48	46	44
40										57	54	52	51	49	47	44	43
39									56	54	53	51	49	47	45	43	41
38							57	56	54	52	51	49	47	46	43	42	40
37					58	57	55	53	51	50	49	47	45	43	42	40	
36			58	57	56	54	53	51	50	48	47	45	43	42	40	38	
35		58	57	56	54	52	51	49	48	47	45	43	42	41	38	37	
34	58	57	55	53	52	51	49	48	47	45	43	42	41	39	37	36	
33	55	54	52	51	50	48	47	46	44	43	42	40	38	37	36	34	
32	52	51	50	49	47	46	45	43	42	41	39	38	37	36	34	33	
31	50	49	48	46	45	44	43	41	40	39	38	36	35	34	33	31	
30	48	47	46	44	43	42	41	40	38	37	36	35	34	33	31	31	
29	46	45	44	43	42	41	39	38	37	36	34	33	32	31	30		
28	43	42	41	41	39	38	37	36	35	34	33	32	31	29	28		
27	41	40	39	38	37	36	35	34	33	32	31	30	29	28	28		
26	39	38	37	36	35	34	33	32	31	31	29	28	28	27			
25	37	36	35	34	33	33	32	31	30	29	28	27	27	26			
24	35	34	33	33	32	31	30	29	28	28	27	26	26	25			
23	33	32	32	31	30	29	28	27	27	26	25	24	23				
22	31	29	29	28	28	27	26	26	24	24	23	23					
21	29	29	28	27	27	26	26	24	24	23	23	22					

Source: Data obtained from the Meteorological Service of Canada, Downsview, Ontario.

Table B4 Relation of Humidex with comfort

Humidex	Degree of discomfort
20–29	Comfortable
30–39	Varying degrees of discomfort
40–45	Almost everyone uncomfortable
46 and over	Many types of labor must be restricted

Source: Data obtained from the Meteorological Service of Canada, Downsview, Ontario.

An example of a simple empirical discomfort index (*Humidex*), used by the Meteorological Service of Canada, is shown in Table B3, and the associated discomfort ranges are given in Table B4. Humidex is derived from air temperature and humidity (Masterton and Richardson, 1979). For example, air temperature of 30°C and relative humidity of 90%, giving a Humidex value of 46, are an extremely uncomfortable combination.

Cold stress occurs during exposure to low temperature, strong winds, and thin or wet clothing. It is not always realized that severe cold stress can occur at temperatures well above freezing if clothing becomes saturated with moisture (see, for example, Pugh, 1966).

The *physiological* (involuntary) mechanisms activated to cope with cold are (1) contracting of blood vessels near the surface of the skin; and (2) shivering. The *voluntary* mechanisms are (1) switching to warmer and drier clothing; (2) moving indoors or to locations sheltered from wind and rain; (3) exercising; and (4) changing diet to increase basal metabolic heat production.

A widely used method of expressing cold stress is by calculating *wind chill*. One of the early empirical methods was derived from 89 sets of measurements made in Antarctica by Siple and Passel (1945), who recorded temperature, wind speed, and the time required to freeze a plastic cylinder of water. However, this method was found to exaggerate the effect of cold due to certain assumptions that do not apply for the human face (Bluestein and Zecher, 1999; Osczevski, 2000; Tikuisis and Osczevski, 2002). More complex but also more realistic approaches for calculating wind chill, as well as the full range of thermal stress, are based on heat budget models of the whole body (see, for example, Steadman, 1984; Hoeppe, 1999; Fiala et al., 2001; Huizenga et al., 2001; Laschewski and Jendritzky, 2002; Tanabe et al., 2002). However, these models have not yet been widely tested or used, and a coordinated international effort is currently under way to develop a universal thermal climate index (see International Society of Biometeorology: http://www.biometeorology.org/). A simpler approach has recently been developed and implemented in Canada and the United States, based on the cooling effect of wind on the human face, and was validated on a group of healthy volunteers (see Meteorological Service of Canada: http://www.msc-smc.ec.gc.ca/education/windchill/index.cfm). The data are shown in Table B5, together with explanatory notes concerning the risk of frostbite. The table is developed for the dry human face in the shade and may not represent the discomfort of a warmly dressed person, a person exposed to sunshine or a person with wet skin.

The thermal insulation of clothing is expressed in *clo* units, where 1 clo is the insulation that maintains comfort in a resting person indoors at an air temperature of 21°C with relative humidity of less than 50%. Experimentally, 1 clo has been estimated to be equal to 0.18°C m^2 h/kcal. The insulating efficiency of clothing can be measured, and the results can be used

Table B5 Wind chill, where T_{air} = Air temperature in °C, and V_{10} = Observed wind speed at 10-m elevation, in km/h

	T_{air}											
V_{10}	5	0	−5	−10	−15	−20	−25	−30	−35	−40	−45	−50
5	4	−2	−7	−13	−19	−24	−30	−36	−41	−47	−53	−58
10	3	−3	−9	−15	−21	−27	−33	−39	−45	−51	−57	−63
15	2	−4	−11	−17	−23	−29	−35	−41	−48	−54	−60	−66
20	1	−5	−12	−18	−24	−31	−37	−43	−49	−56	−62	−68
25	1	−6	−12	−19	−25	−32	−38	−45	−51	−57	−64	−70
30	0	−7	−13	−20	−26	−33	−39	−46	−52	−59	−65	−72
35	0	−7	−14	−20	−27	−33	−40	−47	−53	−60	−66	−73
40	−1	−7	−14	−21	−27	−34	−41	−48	−54	−61	−68	−74
45	1	−8	−15	−21	−28	−35	−42	−48	−55	−62	−69	−75
50	−1	−8	−15	−22	−29	−35	−42	−49	−56	−63	−70	−76
55	−2	−9	−15	−22	−29	−36	−43	−50	−57	−63	−70	−77
60	−2	−9	−16	−23	−30	−37	−43	−50	−57	−64	−71	−78
65	−2	−9	−16	−23	−30	−37	−44	−51	−58	−65	−72	−79
70	−2	−9	−16	−23	−30	−37	−44	−51	−59	−66	−73	−80
75	−3	−10	−17	−24	−31	−38	−45	−52	−59	−66	−73	−80
80	−3	−10	−17	−24	−31	−38	−45	−52	−60	−67	−74	−81

***Approximate thresholds*:**

Risk of frostbite in prolonged exposure: wind chill below	−25	
Frostbite possible in 10 minutes at	−35	Warm skin, suddenly exposed. Shorter time if skin is cool at the start.
Frostbite possible in less than 2 minutes at	−60	Warm skin, suddenly exposed. Shorter time if skin is cool at the start.

Note: Other factors such as wet skin (causing increased cold stress) or direct sunshine (causing warming) become important and are not accounted for in the table. (*Source:* Meteorological Service of Canada, Downsview, Ontario.)

in conjunction with joint climatological frequency distributions of temperature and wind speed to produce clo climatologies. For example, at Yellowknife in the Canadian Northwest Territories in January, the numbers of clos required to maintain comfort outdoors are 5.2, 4.1, 2.7, and 1.2 for metabolic rates of 80, 100, 150 and 300 $\mathrm{kcal\,m^{-2}h^{-1}}$, respectively (Auliciems et al., 1973).

That heat stress and cold stress indeed cause increases in morbidity and mortality has been amply demonstrated. See Macfarlane (1977, 1978) for reviews of relevant epidemiological studies on mortality. In one particularly interesting investigation, Clarke and Bach (1971) found that the number of deaths caused by heat in St Louis, Missouri, in July of 1966 could be fitted by a straight line when plotted against average temperature of the previous day; no deaths at 32°C (assumed to be the critical level) rising to 73 cases at 35°C. A more recent heat wave during 12–15 July of 1995 caused over 500 excess deaths in Chicago, Illinois (WMO, 1999; Klinenberg, 2002). The average daily maximum and minimum temperatures during the 4-day heat wave exceeded 38°C and 26°C, respectively, coupled with very high dewpoint temperatures during day and night (Kunkel et al., 1996). The high number of deaths was blamed on several factors, including an inadequate heat-wave warning system, power failure, inadequate health-care and air-conditioned facilities, an aging population, and poor home ventilation due to poverty or fear of crime (Changnon et al., 1996). A much more dramatic heat wave in Western Europe in August 2003 caused between 22,000 and 35,000 excess deaths (Schär and Jendritzky, 2004), over 14,000 of them in France alone (Allen and Lord, 2004).

This review of human bioclimatology admittedly is incomplete. However, the following list will lead the reader into the literature on specific topics:

Further reading

For a general overview, see Landsberg (1969), Jendritzky (1991), multi-author reviews published in *Experientia* (Birkhauser Verlag, Basel) in 1993, Hoeppe (1997), and WMO (1999).

For infectious diseases, see "Under the Weather: Climate, Ecosystems, and Infectious Disease," by the Committee on Climate, Ecosystems, Infectious Disease, and Human Health (CEIDH). Washington DC: National Academy Press, 2001.

For thermal insulating properties of clothing, see Cena and Clark (1978), Kaufman et al. (1982), Havenith (1999), and Chen et al. (2003).

For negative ions, see First (1980), Bissell et al. (1981), Yost and Moore (1981), Watanabe et al. (1997), and Nakane et al. (2002).

For electromagnetic effects, see Malin and Srivastava (1979), and NRC (1997).

For quantifying population exposures to air pollution, see WHO (1982, 1999).

For biorythms, see Wever (1979, 1986, 1989), and Min (2003).

For UV-B impacts, see UNEP (1998).

Plant bioclimatology (agricultural and forest)

Humans have been managing (or in some cases mismanaging) renewable resources for many centuries. The goal is to increase crop and timber yields without causing long-term degradation of soils and forests. Climate has a direct effect on plants; it provides the primary source of energy; it governs the temperature of plant organs; and it controls water loss by transpiration. The main climatological stresses that adversely affect vegetation are drought, flood, hail, strong winds, and frost. However, atmospheric processes are indirectly important in seed dispersal, pest outbreaks, air pollution concentration (including acidic rains), soil erosion, and soil moisture. Because many cash crops are grown outside their normal climatological ranges, they must be irrigated and protected from frost and strong winds. The bioclimatologist assists in this latter regard, either by modifying the landscape (e.g. with shelterbelts) or by providing advice on how to optimize the use of terrain irregularities (*topoclimatology*). For example, a forest clearing can create a frost pocket, making it difficult for young seedlings to survive. The forest industry therefore has a need to evaluate topoclimatological features in order to assure regeneration of clear-cut areas.

A single example, apple-growing in Britain, will illustrate some of the factors considered in a bioclimatological study (Landsberg, 1980). First, the life cycle of apple trees must be described: from seedlings to mature orchards and from bud production (late summer and early fall) through blossoming (spring) to fruiting (late summer). Next, the sensitivity of an orchard to environmental stresses must be evaluated with respect to each phenological stage of development of the tree and of the fruit. The main stresses are drought, unseasonably mild winters, spring frost, wet and cold summers, and weather-sensitive pest outbreaks. These several stresses have an effect on both quantity and quality of the harvest. Landsberg reviews the work that has been done to understand the effects of weather on the apple orchards of Britain. He asserts that part of the problem has been solved: the annual total dry matter produced by apple trees can be estimated satisfactorily, given the daily weather throughout the year being studied. But orchards are grown for their fruit, not for their biomass. An empirical relationship has therefore been developed that relates dry matter production to quantity and quality of apples.

Agricultural and forest bioclimatology has become increasingly significant in the recent climate change debate. Agricultural lands and forests sequester large amounts of carbon; therefore improved management practices are needed to partially offset the rising levels of atmospheric CO_2. Climate change may also alter the pattern of food and timber production, either for the better in some regions, or for the worse in others. Plant bioclimatology will therefore continue to play a major role in understanding the impacts of climate change, and in formulating mitigation and adaptation measures (see Global Environmental Change, in this volume).

For more information on agricultural and forest bioclimatology, consult the following references. For crop-weather relations, see Baier et al. (1976), Skjelvag (1980), Burt et al. (1981), Parry (1985), and Parry et al. (1986); for topoclimatology, see Skaar (1980), and Utaaker (1980); for shelterbelts, see Rosenberg (1975); for UV radiation, see Grant (1997); for ozone effects, see Krupa et al. (2001); for effects of nitrogen compounds, see a series of articles in *Environmental Pollution* (2002), vol. 118, issue no. 2, pp. 165–283. In general, see the journal *Agricultural and Forest Meteorology*, published by Elsevier Publishing Company, Amsterdam, and the book *Forest Microclimatology* (Lee, 1978).

Animal bioclimatology

Bioclimatologists seek to quantify the direct and indirect impacts of climate on animals, particularly domestic ones. In the case of farm animals the objective is to improve the quality and quantity of meats and dairy products, as well as to increase the work output of "beasts of burden". Direct effects involve heat exchanges between the animal and its environment, which are linked to air temperature, humidity, wind speed, and thermal radiation. These linkages affect animal health, growth, milk and wool production, reproduction, and performance in general. For example, heat and cold stress can have marked effects on milk yield of lactating cows, depending on breed, feed intake and degree of acclimatization. Conception rates of dairy cows are also sensitive to seasonal fluctuations. Climatic extremes such as droughts, floods, violent winds, heat waves, and severe winter storms can result in injury and death of vulnerable animals. Indirect effects include climatic influences on quantity and quality of feedstuffs such as pastures, forages, and grains, as well as the severity and distribution of livestock diseases and parasites.

Bioclimatologists also undertake studies on wild animals to determine the effects of atmospheric stresses such as acidic deposition, long-range transport of toxic chemicals, weather disasters, and more recently UV-B radiation and climate change. Ecologists need this information to help understand population changes, species diversity and ecosystem health. Biodiversity has become a fast-growing field of study linked in many ways to bioclimatology.

Animal bioclimatology is a vast field and only a few references are listed here. For general information, see Tromp (1980), and Johnson (1997). For heat transfer from animals, see Cena and Monteith (1975), and Tracy (1977); for bioclimatology of domestic animals, see Bianca (1976), Dragovich (1981), Christianson et al. (1982), Igono and Aliu (1982), and Starr (1983); for bioclimatology of wild animals, see Shkolnik (1971), and Picton (1979).

Phenology

Blooming wildflowers, falling leaves, migrating birds and insects, spawning fish, hibernating animals, freezing ponds and rivers, and the like, are all influenced primarily by climatic conditions. Bioclimatologists undertake phenological studies to understand the role of climate variables in the dynamics of plant and animal natural cycles. The relationships between climate variables and the timing of these phenophases provide useful information to farmers, gardeners, horticulturists, and beekeepers, e.g. first growth and flower dates, last frost of spring or first frost of fall, planting and harvesting dates, and appearance of insect or weed pest species. Wildlife managers also need information on bird and animal migration dates, growth stage dates of various plant and animal species, dates of critical lake and soil temperatures, and breeding activities and nesting/denning dates (see, e.g., Lieth and Schwartz, 1997).

Phenological studies have become increasingly valuable in recent years, because trends in phenology may serve as natural indicators of global climate change (IPCC, 2001; Lechowicz, 2001; Van Noordwijk, 2003).

Aerobiology

Airborne living material is transported by the wind, sometimes for thousands of kilometers. Field naturalists are interested in bird and insect migrations, and in large-scale movements associated with the life cycles of pollen, rusts, and spores, which are of economic and health significance. For example, pollen and other allergy-causing materials may be transported from countryside to city, or from state to state, causing health problems at considerable distances from the source regions. (A question often asked is how large an area surrounding a city should be cleared of ragweed in order to provide effective relief to hay fever sufferers?) Other examples include the potato blight, wheat rust, locusts, the gypsy moth, and spruce budworms. A final example is that many viruses and other substances that harm animals and/or people can also be transmitted by the wind, e.g. foot-and-mouth disease, rinderpest, influenza viruses, etc. In 2001 a severe outbreak of foot-and-mouth disease spread in the UK, leading authorities to destroy several million livestock. Computerized models are often used to predict the dispersion, movement and deposition of viruses and other biological agents.

Aerobiology contributes to understanding the atmospheric part of the life cycles of the pests mentioned above, and to an activity called *integrated pest management*. A good example is the study by Fleming et al. (1982), who examined the effect of field geometry on the spread of crop disease. Using reaction–diffusion models, rectangles of various dimensions were studied. The models suggest that the greater the perimeter-to-area ratio, the slower the increase of disease within a field. Assuming that a constant proportion of a region is allotted to a particular crop, the models also indicate that decreasing field size and elongating fields in the cross-wind direction will reduce disease losses.

There are, of course, many other types of aerobiological studies (see, e.g., Edmonds, 1979; Pedgley, 1982; and Isard and Gage, 2001).

Climatic adaptation

Finally, mention should be made of climatic adaptation. Because the time scale involved is of the order of centuries or longer, it is difficult to demonstrate stress–response relationships. Yet the presumption of climatic adaptation seems to be justified in many cases. For example, the shape of the Arctic igloo is optimal for minimizing heat losses. Similarly, small insects often have spherical or cylindrical bodies to reduce heat loss on cold nights. As another example, vegetation has "learned" to survive along the coastline of the Arctic Ocean at a site (the Smoking Hills) where SO_2 concentrations and soil pH values should be lethal; in this case the cliff has been burning naturally for at least several centuries (Havas and Hutchinson, 1983). Some of the species have been successfully transplanted on mine tailings near the Sudbury copper smelter.

Other examples of adaptation include desert plants and animals, which have minimized their water consumption and water losses through physiological mechanisms (Smith, 1978), and bird, insect, and animal migrations, which depend on the regularity of annual climatic cycles. Black people in the tropics have skin pigments that protect them from sunburn and skin cancer, while native people in the Arctic have adjusted diet and body fat to cope successfully with the long northern winters.

On the global scale, the *Gaia* theory of Lovelock (1979) asserts that the biosphere played an essential role in the evolution of the Earth over a geological time-scale. Once established, the primitive vegetation cover was able to control the oxygen and CO_2 concentrations of the atmosphere, the temperature of atmosphere and oceans, and the salinity and pH of oceans.

For a recent discussion of the *Gaia* theory, see a collection of papers in *Climatic Change* (2002), vol. 52, issue no. 4, pp. 383–509. See also the exchange of views between J. Lovelock and T. Volk in *Climatic Change* (2003), vol. 57, pp. 1–7.

Conclusion

Although the bioclimatologist has been studying stress–response relations for more than a century, nature unlocks its secrets slowly. Nevertheless, future prospects for meaningful research are good. The main problem to be overcome is the difficulty that specialists encounter when working with specialists in other fields. For example, medical doctors and climatologists have traditionally not communicated with one another except at the non-specialist level. It is hoped that interdisciplinary educational and research programs will be encouraged in all parts of the world.

A.R. Maarouf and R.E. Munn

Bibliography

Allen, M.R., and Lord R., 2004. The blame game. Who will pay for the damaging consequences of climate change? *Nature*, **432**: 551–552.

Auliciems, A., de Freitas, C.R., and Hare, F.K., 1973. *Winter Clothing Requirements for Canada*, Climatology Study No. 22. Downsview, Ontario: Atmospheric Environment Service.

Baier, W., Davidson, H., Desjardins, R.L., Ouellet, C.E., and Williams, G.D.V., 1976. Recent biometeorological application to crops. *International Journal of Biometeorology*, **20**: 108–127.

Bianca, W., 1976. The significance of meteorology in animal production. *International Journal of Biometeorology*, **20**: 139–156.

Bissell, M., Diamond, M.C., Ellman, G.L., Krueger, A.P., Orenberg, E.K., and Sigel, S.S.R.G., 1981. Air ion research. *Science*, **211**: 1114.

Bluestein, M., and Zecher, J., 1999. A new approach to an accurate wind chill factor. *Bulletin of the American Meteorological Society*, **80**(9): 1893–1899.

Burt, J.E., Hayes, J.T., O'Rourke, P.A., Terjung, W.H., and Todhunter, P.E., 1981. A parametric crop water use model. *Water Resources Research*, **17**: 1095–1108.

Cena, K., and Clark, J.A., 1978. Thermal insulation of animal coats and human clothing. *Physical Medicine and Biology*, **23**: 565–591.

Cena, K., and Monteith, J.L., 1975. Transfer processes in animal coats. *Royal Society of London Proceedings, ser. B*, **188**: 377–423.

Changnon, S.A., Kunkel, K.E., and Reinke, B.C., 1996. Impacts and responses to the 1995 heat wave: A call to action. *Bulletin of the American Meteorological Society*, **77**(7): 1497–1506.

Chen, Y.S., Fan, J., and Zhang, W., 2003. Clothing thermal insulation during sweating. *Textile Research Journal*, **73**(2): 152–157.

Christianson, L., Hahn, G.L., and Meador, N., 1982. Swine performance model for summer conditions. *International Journal of Biometeorology*, **26**: 137–145.

Clarke, J.F., and Bach, W., 1971. Comparison of the comfort conditions in different urban and suburban microenvironments. *International Journal of Biometeorology*, **15**: 41–54.

Dragovich, D., 1981. Thermal comfort and lactation yields of dairy cows grazed on farms in a pasture-based feed system in Eastern New South Wales, Australia. *International Journal of Biometeorology*, **25**: 167–174.

Edmonds, R.L. (ed.), 1979. *Aerobiology: the ecological systems approach*. Stroudsburg, PA: Dowden, Hutchinson & Ross.

Fiala, D., Lomas, K.J., and Stohrer, M., 2001. Computer prediction of human thermoregulatory and temperature responses to a wide range of environmental conditions, *International Journal of Biometeorology*, **45**: 143–159.

First, M.W., 1980. Effects of air ions. *Science*, **210**: 715–716.

Fleming, R.A., Marsh, L.M., and Tuckwell, H.C., 1982. Effect of field geometry on the spread of crop disease. *Prot Ecol*, **4**: 81–108.

Grant, R.H., 1997. Partitioning of biologically active radiation in plant canopies. *International Journal of Biometeorology*, **40**: 26–40.

Havas, M., and Hutchinson, T.C., 1983. The Smoking Hills: natural acidification of an aquatic ecosystem. *Nature*, **301**: 23–28.

Havenith, G., 1999. Heat balance when wearing protective clothing. *Annals of Occupational Hygiene*, **43**(5): 289–296.

Higuchi, K., Yuen, C.W., and Shabbar, A., 2000. Ice Storm '98 in south-central Canada and northeastern United States: A climatological perspective. *Theoretical and Applied Climatology*, **66**: 61–79.

Hoeppe, P., 1997. Aspects of human biometeorology in past, present and future. *International Journal of Biometeorology*, **40**(1): 19–23.

Hoeppe, P., 1999. The physiological equivalent temperature – a universal index for the biometeorological assessment of the thermal environment. *International Journal of Biometeorology*, **43**(2): 71–75.

Houghton, D.D. (ed.), 1985. *Handbook of Applied Meteorology*. Chichester: John Wiley.

Huizenga C., Hui, Z., and Arens, E., 2001. A model of human physiology and comfort for assessing complex thermal environments. *Building and Environment*, **36**(6): 691–699.

Hutchinson, T.C., Gizyn, W., Havas, M., and Zobens, V., 1978. Effect of long-term lignite burns on arctic ecosystems at the Smoking Hills, N.W.T. In Hemphill, D.D., ed., *Proceedings of Trace Substances in Environmental Health*. Columbia: University of Missouri, pp. 317–332.

Igono, M.O., and Aliu, Y.O., 1982. Environmental profile and milk production of Friesian-Zebu crosses in Nigerian Guinea savanna. *International Journal of Biometeorology*, **26**: 115–120.

IPCC (Intergovernmental Panel on Climate Change), 2001. Third Assessment Report – *Climate Change 2001*. Cambridge: Cambridge University Press.

Isard, S.A., and Gage, S.H., 2001. *Flow of Life in the Atmosphere: an airscape approach to understanding invasive organisms*. East Lansing, MI: Michigan State University Press.

Jendritzky, G., 1991. Selected questions of topical interest in human bioclimatology. *International Journal of Biometeorology*, **35**(3): 139–150.

Johnson, H.D., 1997. Aspects of animal biometeorology in the past and future. *International Journal of Biometeorology*, **40**(1): 16–18.

Kates, R.W., Ausubel, J.H., and Berberian, M., 1985. *Climate Impact Assessment SCOPE* 27. Chichester: John Wiley.

Kaufman, W.C., Bothe, D., and Meyer, S.D., 1982. Thermal insulating capabilities of outdoor clothing materials. *Science*, **215**: 690–691.

Klinenberg, E., 2002. *Heat Wave: a social autopsy of disaster in Chicago*. Chicago: University of Chicago Press.

Krupa, S., McGrath, M.T., Andersen, C.P., et al., 2001. Ambient ozone and plant health. *Plant Diseases*, **85**(1): 4–12.

Kunkel, K.E., Changnon, S.A., Reinke, B.C., and Arritt, R.W., 1996. The July 1995 heat wave in the midwest: a climatic perspective and critical weather factors. *Bulletin of the American Meteorological Society*, **77**(7): 1507–1518.

Landsberg, H.E., 1969. *Weather and Health*. Garden City, NY: Doubleday Anchor.

Landsberg, J.J., 1980. From bud to bursting blossom: weather and the apple crop. *Weather*, **34**: 394–407.

Laschewski, G., and Jendritzky, G., 2002. Effects of the thermal environment on human health: An investigation of 30 years of daily mortality data from SW Germany. *Climate Research*, **21**(1): 91–103.

Lechowicz, M.J., 2001. Phenology. *Encyclopaedia of Global Environmental Change*, vol. 2, Chichester: John Wiley, pp. 461–465.

Lee, R., 1978. *Forest Microclimatology*. New York: Columbia University Press.

Lieth, H., and Schwartz, M.D. (eds.), 1997. *Phenology in Seasonal Climates*. Leiden: Backhuys.

Lovelock, J., 1979. *GAIA: a new look at life on earth*. Oxford: Oxford University Press.

Macfarlane, A., 1977. Daily mortality and environment in English conurbations. I. Air pollution, low temperature and influenza in Greater London. *British Journal of Preventive Social Medicine*, **31**: 54–61.

Macfarlane, A., 1978. Daily mortality and environment in English conurbations. II. Deaths during summer hot spells in Greater London. *Environmental Research*, **15**: 332–341.

Malin, S.R.C., and Srivastava, B.J., 1979. Correlation between heart attacks and magnetic activity. *Nature*, **277**: 646–648.

Masterton, J.M., and Richardson, F.A., 1979. Humidex: A method of quantifying human discomfort due to excessive heat and humidity, Report No. CLI 1-79. Downsview, Ontario: Atmospheric Environment Service.

Min, H.T., 2003. The essence of circadian rhythms. *Journal of Biological Systems*, **11**(1): 85–100.

Munn, R.E., 1970. *Biometeorological Methods*. New York: Academic Press.

Nakane H., Asami, O., Yamada, Y., and Ohira, H., 2002. Effect of negative air ions on computer operation, anxiety and salivary chromogranin A-like immunoreactivity. *International Journal of Psychophysiology*, **46**(1): 85–89.

NRC (National Research Council), 1997. *Possible Health Effects of Exposure to Residential Electric and Magnetic Fields*. Washington, DC: National Academy Press.

Osczevski, R.J., 2000. Windward cooling: an overlooked factor in the calculation of wind chill. *Bulletin of the American Meteorological Society*, **81**(12): 2975–2978.

Parry, M., 1985. *The Sensitivity of Natural Ecosystems and Agriculture to Climatic Change*. In Parry, M., ed., *Climatic Change*, vol. 7, no. 1. Dordrecht: Reidel.

Parry, M., Carter, T., and Konijn, N., 1986. *The Assessment of Climate Impacts on Agriculture*, 2 vols. Dordrecht: Reidel.

Pedgley, D.E., 1982. *Windborne Pests and Diseases: meteorology of airborne organisms*. Chichester: Halsted Press.

Picton, H.D., 1979. A climate index and mule deer farm survival in Montana. *International Journal of Biometeorology*, **23**: 115–122.

Pugh, L.G.C., 1966. Clothing insulation and accidental hypothermia in youth. *Nature*, **209**: 1281–1286.

Rosenberg, N.J., 1975. Windbreak and shelter effects. In Tromp, S.W., ed., *Progress in Plant Biometeorology*, vol. 1. Amsterdam: Swets & Zeitlinger, pp. 108–153.

Schär, C., and Jendritzky, G., 2004. Hot news from summer 2003. *Nature*, **432**: 559–560.

Shkolnik, A., 1971. Diurnal activity in a small desert rodent. *International Journal of Biometeorology*, **15**: 115–120.

Silverman, F., Hosein, H.R., Corey, P., Holton, S., and Tarlo, S.M., 1992. Effects of particulate matter exposure and medication use on asthmatics. *Archives of Environmental Health*, **47**(1): 51–56.

Siple, P.A., and Passel, C.F., 1945. Measurements of dry atmospheric cooling in subfreezing temperatures. *American Philosophical Society Proceedings*, **89**: 177–199.

Skaar, E., 1980. Application of meteorological data to agroclimatological mapping. *International Journal of Biometeorology*, **24**: 3–12.

Skjelvag, A.O., 1980. Crop-weather analysis model applied to field bean. *International Journal of Biometeorology*, **24**: 301–313.

Smith, W.K., 1978. Temperatures of desert plants: another perspective on the adaptability of leaf size. *Science*, **201**: 614–616.

Starr, J.R., 1983. Animals and weather-pestilence, plague and productivity. *Weather*, **38**: 2–17.

Steadman, R.G., 1984. A universal scale of apparent temperature, *Journal of Climatology and Applied Meteorology*, **23**(12): 1674–1687.

Tanabe, S., Kobayashi, K., Nakano, J., Ozeki, Y., and Konishi, M., 2002. Evaluation of thermal comfort using combined multi-node thermoregulation (65 MN) and radiation models and computational fluid dynamics (CFD). *Energy and Buildings*, **34**(6): 637–646.

Tikuisis, P., and Osczevski, R.J., 2002. Dynamic model of facial cooling. *Journal of Applied Meteorology*, **41**(12): 1241–1246.

Tracy, C.R., 1977. Minimum size of mammalian homeotherms: role of thermal environment. *Science*, **198**: 1034–1035.

Tromp, S.W., 1980. *Biometeorology: the impact of the weather and climate on humans and their environment (animals and plants)*. London: Heyden.

UNEP (United Nations Environment Program), 1998. Environmental effects of ozone depletion: 1998 assessment. *Journal of Photochemistry and Photobiology*, **46**: 1–108.

Utaaker, K., 1980. Local climates and growth climates of the Sognefjord region. *International Journal of Biometeorology*, **24**: 13–22.

Van Noordwijk, A.J., 2003. Climate change – The earlier bird. *Nature*, **422**: 29.

Watanabe, I., Noro, H., Ohtsuka, Y., Mano, Y., and Agishi, Y., 1997. Physical effects of negative air ions in a wet sauna. *International Journal of Biometeorology*, **40**(2): 107–112.

Wever, R.A., 1979. *The Circadian System of Man*. Berlin: Springer-Verlag.

Wever, R.A., 1986. Characteristics of circadian-rhythms in human functions. *Journal of Neural Transmission*, 323–373 (Suppl. 21).

Wever, R.A., 1989. Light effects on human circadian-rhythms – a review of recent Andechs experiments. *Journal of Biological Rhythms*, **4**(2): 161–185 (Summary).

WHO (World Health Organization), 1982. *Estimating Human Exposure to Air Pollutants*, EFP/82-31. Geneva: World Health Organization.

WHO (World Health Organization), 1999. *Air Quality Guidelines*. Geneva: World Health Organization.

WMO (World Meteorological Organization), 1999. *Weather, Climate and Health*. Geneva: WMO – No. 892.

Yost, M., and Moore, A.D., 1981. Air ion research. *Science*, **211**: 1114–1115.

Cross-references

Agroclimatology
Applied Climatology
Climate Change and Human Health
Global Environmental Change: Impacts
Heat Index
Human Health and Climate
Spring Green Wave
Statistical Climatology
Wind Chill

BLIZZARD CLIMATOLOGY

The most extreme form of a winter storm is the blizzard, which combines strong winds with falling or blowing snow to cause low visibility, deep snowdrifts, and intense wind chill. This combination of strong winds and blowing snow causes a general cessation of routine societal activities, making the blizzard the most dangerous of winter storms (Schwartz and Schmidlin, 2002).

The National Weather Service (NWS) defines a blizzard as winds over $16\,m\,s^{-1}$ [35 miles per hour (mph), (30 knots)] and falling or blowing snow reducing visibility to less than 400 m (0.25 miles) lasting for at least 3 hours (Branick, 1997; NSIDC, 1999; NWS, 1999). An additional criterion for temperatures below 20°F (−7°C) was used for many years but was abandoned in the 1970s. Blizzards in Canada have regional-dependent definitions with one definition for the Northwest Territories and Nunavut and another for Alberta, Saskatchewan, and Manitoba (Lawson, 2002). Also, the Meteorological Office in Great Britain has another definition (Wild, 1997).

The term "blizzard" was first used by Henry Ellis along Hudson Bay in 1746 (Black, 1971) although Ellis's meaning at the time was ambiguous according to Wild (1997). Ludlum (1968) reported that the term was first used in the United States to describe the winter storm of March 1870 in Iowa; however, Wild (1997) reported use in South Dakota newspapers in 1867. By the 1880s the term was in general use according to Wild (1997).

Blizzards require interactions of meteorological processes that are vital for storm development. First, a cyclonic storm or low-pressure system that forms along frontal boundaries is usually the focal point for storm development and cyclogenesis. Important components include the position of the jet stream as the trough aloft often follows the 500 mb level flow. Cold polar air and a source of moisture are also necessary for the snow (Kocin and Uccellini, 1990). Deepening (strengthening) cyclones are another ingredient of a blizzard. The intensity is dependent on strength and positioning of the jet stream relative to the other upper-air disturbances along with the availability of moisture and strength of horizontal temperature gradients. Additionally, the upper-level trough/ridge pattern is important as this gives vorticity, upper-level divergence, and diffluence with a negatively tilted trough. The upper-level jet streaks or short waves increase divergence and vorticity, which gives the necessary uplift for heavy snow as described by Black, 1971; Dupigny Giroux, 1999; Ellis and Leathers, 1996; Kocin and Uccillini,1990; Leathers and Harrington, 1999; and Mote et al.,1997 (see references in Schwartz, 2001).

The physical geography of North America allows a favorable environment for blizzards as the Rocky Mountains run north–south, effectively isolating the Great Plains from comparatively warmer Pacific air (Rauber et al., 2002). Flat terrain in central North America provides an "open door" for cold polar and Arctic airmasses from the north, and moisture from the Gulf of Mexico is transported from the south. Some of the source regions for blizzards in North America originate in the Pacific Ocean (Westerly Lows), Alberta (Alberta Clippers), the Texas–Colorado Panhandle (Panhandle Hooks), the Louisiana Delta (Gulf Coast Low), and the Atlantic Ocean off the North Carolina coast (East Coast Low or Nor'easter). These were described by Doeskin and Judson, 1996; Goddard, 1998; Kocin and Uccellini, 1990; and Morgan and Moran, 1997.

There were 438 blizzards identified for the 41 winters from 1959/1960 to 1999/2000 in the conterminous United States. The annual number of blizzards per winter averaged 10.7 and ranged from 1 blizzard in 1980/1981 to 27 blizzards in 1996/1997 (Figure B1) (Schwartz, 2001; Schwartz and Schmidlin, 2002).

The frequency of blizzards by county over the period 1959–2000 is shown in Figure B2. Forty of the 48 states reported a blizzard during the study period. The number of states with blizzards was 26 in the 1960s, 37 in the 1970s, 31 in the 1980s, and 39 in the 1990s. Blizzards were most common in the northern Great Plains where 17 counties in North Dakota and eight counties in South Dakota had more than 60 blizzards during 1959–2000. Traill County, North Dakota, had 74 blizzards in the 41-year period, the most of any county in the conterminous United States. An apparent region described as the United States "blizzard zone" occupied all of North and South Dakota and 34 counties in western Minnesota. Each county in the blizzard zone had 41 or more blizzards in the 41 winters, for an average of one or more blizzards per winter (Schwartz, 2001; Schwartz and Schmidlin, 2002).

Outside of the blizzard zone there were 21–40 blizzards per county during 1959–2000 in the rest of Minnesota, northern and central Iowa, most of Nebraska, northwest Kansas, eastern Colorado, southeast Wyoming, and eastern Idaho. The blizzard frequency decreased to 11–20 per county in the northern and central Rockies, central Great Plains, and the western Midwest. Other areas with 11–20 blizzards during the period were in Erie County (Buffalo), New York; Garrett County, Maryland; all of Maine; and portions of southeastern New England. Hancock and Washington counties in Maine had 18 blizzards, the largest number in the eastern United States. There were 4–10 blizzards per county during 1959–2000 across the remainder of the northeastern United States, the Great Lakes region, the High Plains of Texas and New Mexico, and the Sierra Nevada Mountains (Schwartz, 2001; Schwartz and Schmidlin, 2002).

Monthly frequency of blizzards peaked during January in the blizzard zone and most other parts of the United States. However, blizzards were most frequent during December in the Sierra Nevada; during March in the central High Plains of western Nebraska, eastern Colorado, and Kansas; and during April in Montana. The blizzard frequency in western South Dakota showed a second peak during April, matching the January peak frequency. Early and late season blizzards generally occurred in the upper Plains and mountainous areas (Schwartz, 2001; Schwartz and Schmidlin, 2002).

The mean population was 2 462 949 people affected per blizzard. Additionally, the average total population affected

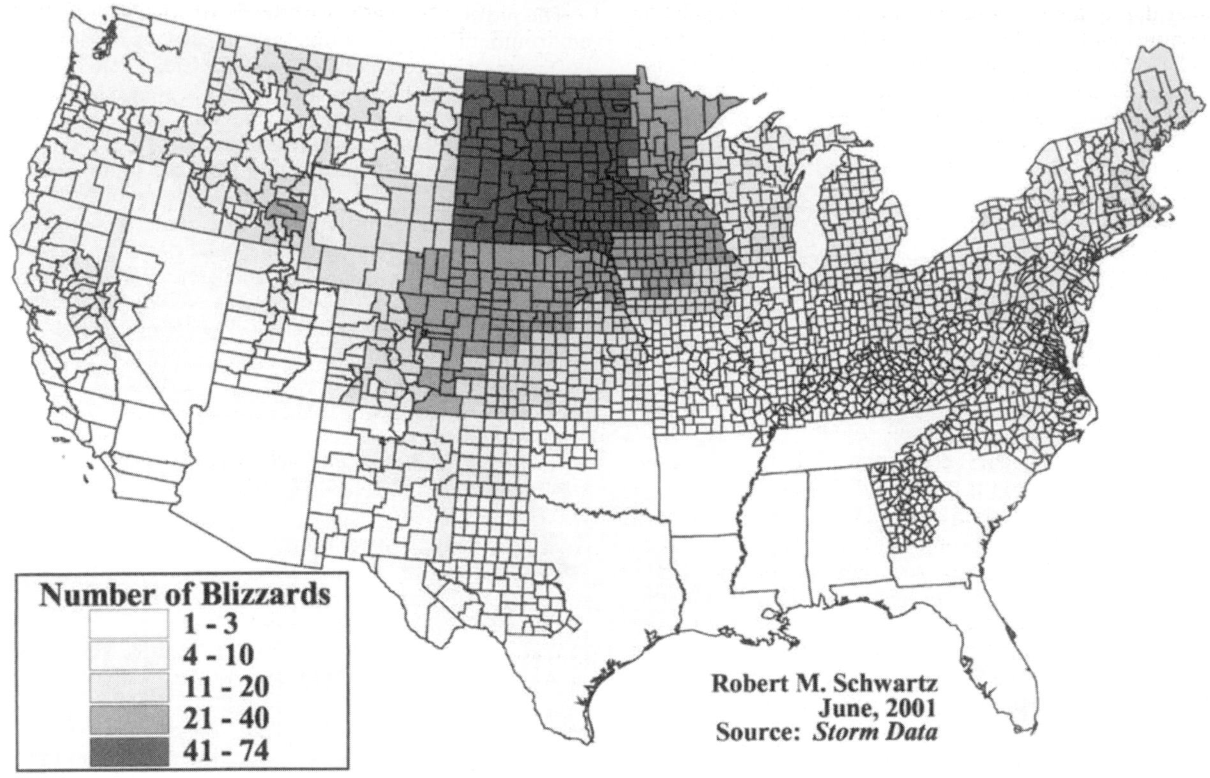

Figure B1 Number of blizzards in the United States from 1959/60 to 1999/2000 (courtesy of the American Meteorological Society).

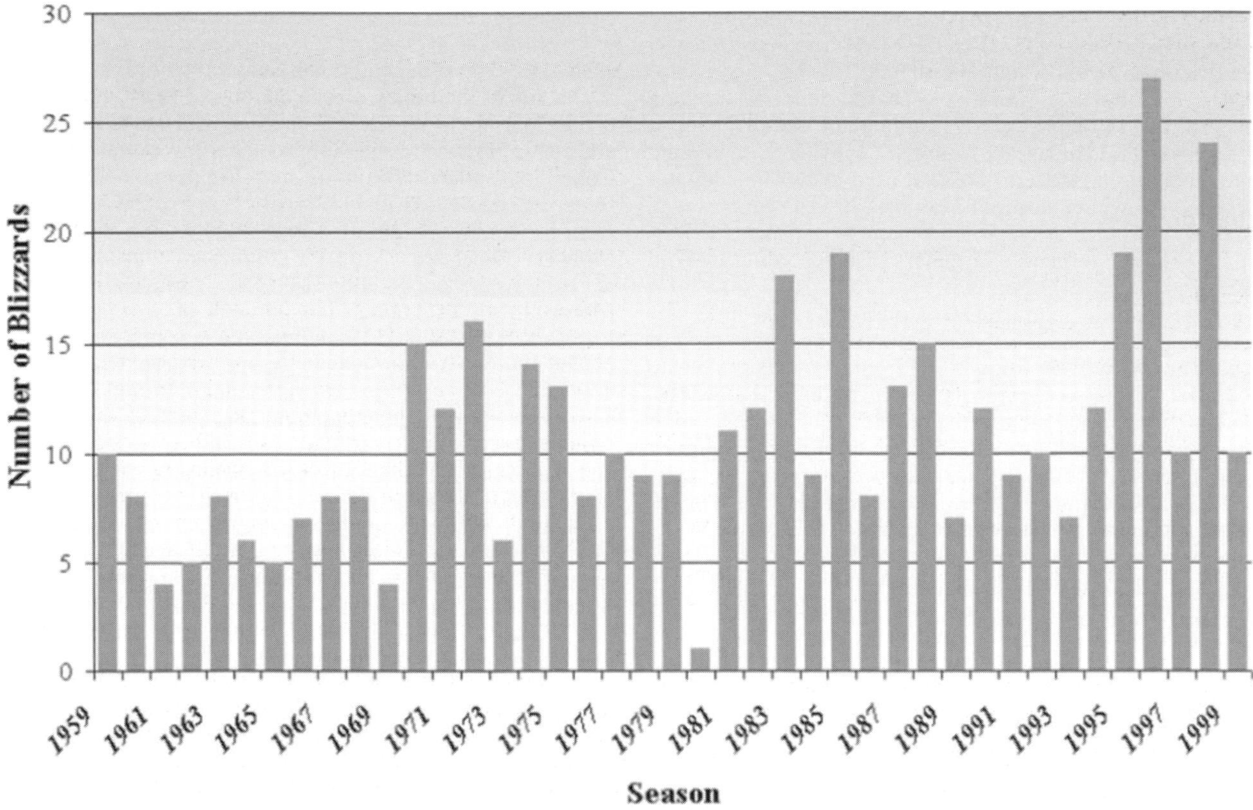

Figure B2 Blizzard frequency in the United States, 1959–2000 (courtesy of the American Meteorological Society).

per winter during the study period was 26 311 505. Population in the continental United States for 1960 through 1990 rose by 68 728 426 persons. In spite of the increased population, the population affected by blizzards did not increase. Although the number of blizzards increased, the affected population did not increase. The likely reason for no linear time trend is that the population decreased by 30 737 in the core blizzard region (Schwartz, 2001).

There were 679 reported fatalities directly associated with blizzards during the 41-year period, which is a mean of 1.55 fatalities per blizzard. Fatalities ranged from zero in several cases to 73 during the Midwest Blizzard of January 1978 (Schwartz, 2001).

During the study 2011 injuries were reported as direct results of blizzards. The mean was 4.59 per event and the minimum number was zero while the maximum was 426, with the Superstorm of March 1993 (Schwartz, 2001).

Total property damage adjusted to 2001 dollars was reported at $22 600 000 000 or a mean of $51 593 998 per blizzard. No damage was reported for several blizzards. The greatest property damage was $4 550 000 000 (unadjusted) or $12 300 000 000 (adjusted) from the Midwest Blizzard of January 1978 (Schwartz, 2001).

Even though many agricultural areas are affected by blizzards, crop damage was often not reported by *Storm Data*. In unadjusted dollars, $439 542 000 in crop damage was reported ($1 080 000 000 adjusted). The mean crop damage per blizzard was $1 003 520 unadjusted and $2 466 324 in adjusted dollars. Maximum reported crop damage was $227 250 000 (unadjusted) and $613 575 000 (adjusted) from the January 1978 Midwest Blizzard (Schwartz, 2001).

There were 25 disaster declarations associated with blizzards in the study period. The 1960s had no declared blizzard disasters while the 1970s had eight declarations, the 1980s had three, and the 1990s had 14 (Schwartz, 2001).

Robert M. Schwartz

Bibliography

Black, R.E., 1971. A synoptic climatology of blizzards on the North-Central Plains of the United States. *NOAA Technical Memorandum* NWS CR-39. Kansas City, MO.

Branick, M.L., 1997. A climatology of significant winter-type weather events in the contiguous United States, 1982–94. *Weather and Forecasting*, **12**: 193–199.

Doeskin, N.J., and Judson, A., 1996. *The Snow Booklet: a guide to the science, climatology, and measurement of snow in the United States.* Fort Collins, CO: Colorado State University Department of Atmospheric Science.

Dupigny-Giroux, L.-A., 1999. The influence of the El Niño-Southern Oscillation and other teleconnections on winter storms in the North American Northeast – a case study. *Physical Geography*, **20**(5): 394–412.

Ellis, A.W., and Leathers, D.J., 1996. A synoptic climatological approach to the analysis of lake-effect snowfall: potential forecasting applications. *Weather and Forecasting*, **11**(2): 216–229.

Goddard, D., 1998. *Dick Goddard's Weather Guide and Almanac for Northeast Ohio.* Cleveland, OH: Gray & Company.

Kocin, P.J., and Uccellini, L.W., 1990. *Snowstorms Along the Northeastern Coast of the United States: 1955 to 1985.* Boston: American Meteorological Society.

Lawson, B.D., 2002. Personal communication 6 March. Meteorologist, Environment Canada.

Leathers, N., and Harrington, J., Jr, 1999. Major snowfall events in the Great Plains: temporal aspects and relative importance. *Physical Geography*, **20**(2): 134–151.

Ludlum, D.M., 1968. *Early American Winters, II: 1821–1870.* Boston: American Meteorological Society.

Morgan, M.D., and Moran, J.M., 1997. *Weather and People.* Upper Saddle River, NJ: Prentice-Hall.

Mote, T.L., Gamble, D.W., Underwood, S.J., and Bentley, M.L., 1997. Synoptic-scale features common to heavy snowstorms in the Southeast United States. *Weather and Forecasting*, **12**(1): 5–23.

National Weather Service (NWS), 1999. *National Weather Service Says: know your winter weather terms.* Washington, DC: National Weather Service. http://www.nws.noaa.gov/om/winter/wntrtrms.htm (last accessed 13 September 2002).

NSIDC, 1999. The Blizzards of 1996. Online. http://nsidc.org/snow/blizzard/questions.html. (Last accessed 16 December 2002).

Rauber, R.M., Walsh, J.E., and Charlevoix, D.J., 2002. *Severe and Hazardous Weather.* Dubuque, IA: Kendall/Hunt.

Schwartz, R.M., 2001. Geography of blizzards in the conterminous United States, 1959–2000. Ph.D. dissertation, Kent State University.

Schwartz, R.M., and Schmidlin, T.W., 2002. Climatology of blizzards in the conterminous United States, 1959–2000. *Journal of Climate*, **15**: 1765–1772.

Wild, R., 1997. Historical review on the origin and definition of the word blizzard. *Journal of Meteorology*, **22**: 331–340.

Cross-references

Airmass Climatology
Climate Hazards
Precipitation
Snow and Snow Cover

BLOCKING

Middle-latitude circulation is classically represented by belts of high- and low-pressure systems that are approximately parallel to the equator. Winds associated with these systems also have their major components parallel to the equator and are said to be zonal. Periodically, this zonal flow breaks down and wind components perpendicular to the equator become more important. Such flow is said to be meridional. Meridional flow often results from the blocking action of a persistent anticyclone.

Blocking action is a diminution of the zonal circulation at all levels of the troposphere, with retardation of zonal flow tending to spread westward. The phenomenon may last from 3 days to a month and is often associated with a warm anticyclone in high latitudes and cold cyclonic circulation in lower latitudes. It may result in significant temperature and precipitation anomalies.

Various types of blocking patterns have been recognized. The most common is the omega block, resembling the Greek letter Ω. An example is shown in Figure B1, which provides a sequence of the 1000–500-mb thickness charts that illustrates the change from high-index zonal circulation to meridional circulation to the low-index pattern in which cut-off lows are separated by a deep circulation high – a blocking high. Note that, although pressure cells or ridges are usually cited as the blocking phenomena, cut-off lows can also act as a block to zonal circulations.

The Climate Prediction Center (CPC) utilize a blocking index (Tibaldi and Molteni, 1990) to consider the frequency of "blocked days". This index, shown as Table B6, has been used to calculate seasonal frequency and location of blocking highs for the northern hemisphere through reanalysis.

John E. Oliver

Table B6 Blocking index used by Climate Prediction Center (after CPC), northern hemisphere

For each longitude the southern 500 hPa geopotential height gradient (GHGS) and the northern 500 hPa geopotential height gradient (GHGN) are computed as follows:

$$\mathrm{GHGS} = \left[\frac{Z(\phi_o) - Z(\phi_s)}{\phi_o - \phi_s}\right]$$

$$\mathrm{GHGN} = \left[\frac{Z(\phi_n) - Z(\phi_o)}{\phi_n - \phi_s}\right]$$

where

$$\phi_n = 80°\mathrm{N} + \delta,\ \phi_o = 60°\mathrm{N} + \delta,\ \phi_s = 40°\mathrm{N} + \delta,\ \delta = -5°, 0°, 5°.$$

A given longitude is said to be blocked at a given time if the following conditions are satisfied for at least one value of δ:

(1) $\mathrm{GHGS} > 0$

(2) $\mathrm{GHGN} < -10$ m/deg latitude.

A 5-day running mean can be applied to the 500 hPa height field prior to calculating both GHGS and GHGN to isolate potential blocking episodes of sufficient duration.

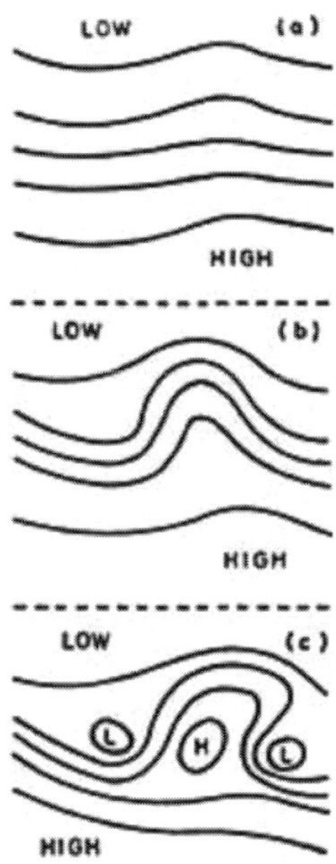

Figure B3 Typical successive 1000–500-mb thickness patterns during the formation of cut-off lows and blocking high. (**a**) High-index zonal circulation; (**b**) meridional circulation; (**c**) blocking high (after McIntosh and Thom, 1973).

Bibliography

Djuric, D., 1994. *Weather Analysis.* Englewood Cliffs, NJ: Prentice Hall. http://www.cpc.ncep.noaa.gov/products

McIntosh, D.H., and Thom, A.S., 1973. *Essentials of Meteorology.* London: Wykham.

Mullen, S.L., 1983. Computer simulation of atmospheric "blocking". *Weatherwise*, **36**(5): 232–233.

Stringer, E.T., 1972. *Foundations of Climatology.* San Francisco, CA: W.H. Freeman.

Tibalti, S., and Molteni, F., 1990. On the operational predictability of blocking. *Tellus*, **42A**: 343–365.

Cross-references

Centers of Action
Jet Streams
Reanalysis Projects
Rossby Wave/Rossby Number
Zonal Index

BOUNDARY LAYER CLIMATOLOGY

Near the Earth's surface the wind interacts with an endless variety of ground covers, each of which has distinctive surface properties that condition the air above, imparting to it distinctive characteristics. As the wind moves the air across the edge of a new area of ground, characteristics that were acquired upwind of the edge begin to change in response to new surface properties. The modified air grows in thickness as the distance downwind of the edge, or fetch, increases. The process continues until a boundary layer a few meters thick is established, its climate clearly linked to the nature of the surface boundary (Figure B4).

Boundary layer climatology is the study of the processes that link the surface of the Earth to the boundary layer, as well as the boundary layer response. The study of process is rooted in the field of micrometeorology, from which the theoretical basis is established. The more descriptive aspects of the response, such as wind speed, temperature and humidity profiles, are derived from the traditions of early investigators who studied microclimatology from the viewpoint of the case study (Geiger, 1966). This perspective still exists, but now the emphasis is directed more toward energy exchanges and their attendant processes (Bailey et al., 1997).

The term "boundary layer" was borrowed from the field of fluid mechanics by micrometeorologists who used it in their

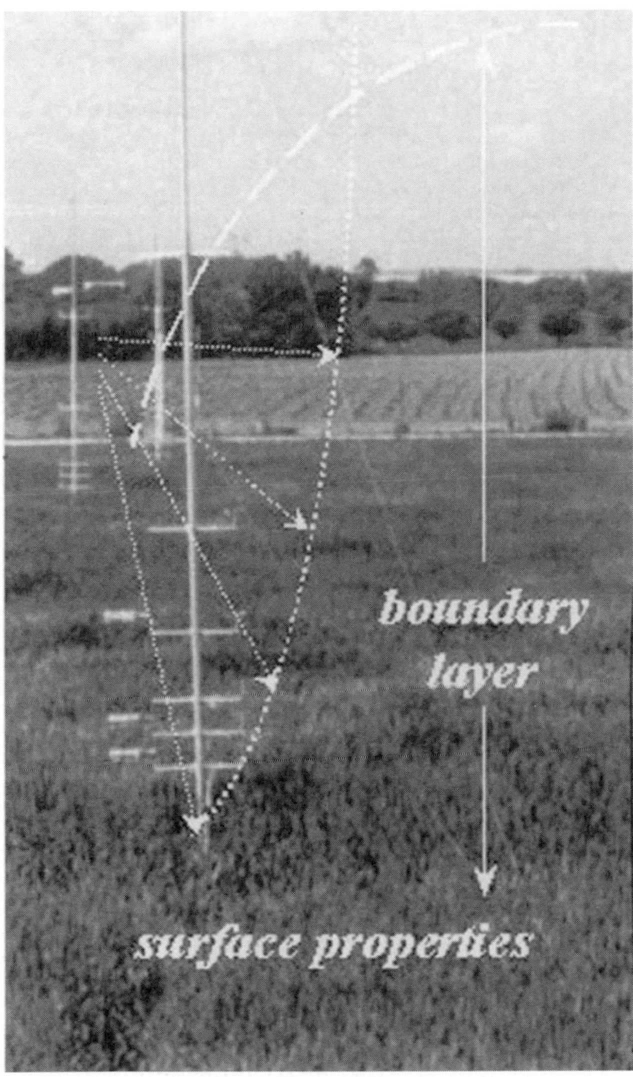

Figure B4 Boundary layer development over an instrumented wheat field, Simcoe, Southern Ontario, Canada. A new boundary layer begins where the wind (dotted arrows) encounters the wheat, after passing over upwind surfaces. The top of the boundary layer (broken line) marks the transition between upwind air and boundary layer air, where the degree of modification increases toward the base of the wind profile (dotted line).

investigations of the lower atmosphere. Some investigators focused attention on experimental studies of boundary layer growth downwind of a change in surface type (Elliott, 1958; Rider et al., 1963; Bradley, 1968). Such studies provided evidence that atmospheric data must be obtained at a considerable distance from the border if the data are to be linked accurately to energy exchanges at the surface. It has since become good practice to collect data at fetches that are at least 100 times the height of measurement above the surface.

Subsequent adoption of the term "boundary layer" in small-scale climatological work occurred gradually, as microclimatologists developed an increasing interest in processes that shaped the climate near the ground. The emphasis on process has become such that the casual observer may see little that differs from micrometeorology. However, the meteorological focus is clearly on improving the physical theory behind the processes at work. The climatological focus rests on the application of theory to gain a better understanding of how surface properties interact with the boundary layer to create distinctive surface *boundary layer climates* (Oke, 1987).

Energy balance at the boundary

The influence of the surface boundary is established by the way in which the surface responds to the transfer rate per unit area, or flux density, of global radiant energy supply, $K_{\downarrow}$ (Figure B5). Surface reflectivity initially reduces this to K_*, the net shortwave radiation supply, which in turn is reduced by L_*, the net longwave radiation loss, thus leaving the net radiative flux density, Q_*, to force a surface response (Figure B5a). The response to Q_* occurs in the form of other energy exchanges that are balanced against Q_* (Figure B5b):

$$Q_* = Q_H + Q_E + Q_G, \tag{1}$$

where Q_H refers to sensible heat transfer between air and ground, Q_E to latent heat transfer due to surface evaporation or condensation, and Q_G is heat transfer between the surface and the underlying medium, which changes the surface heat storage. The transfers are the result of radiative heating or cooling, which causes the boundary temperature to differ from that of the air above and the ground below, thus invoking Q_H and Q_G. It also establishes humidity differences across which the moisture transfer associated with Q_E takes place. The wind and the roughness property of the surface play an active role in the process, mixing the air as momentum is transferred to the surface.

The example in Figure B5 provides a representative picture of the major energy exchange components for a moist, vegetated boundary, where Q_H, Q_E and Q_G use the net radiation supplied during the day, when Q_* is positive, then compensate for net radiative loss at night. The energy flow induces atmospheric responses (Figure B5c), the most prominent features of which are the diurnal cycles of temperature, T, wind speed, u, and the persistent rise in atmospheric vapor pressure, e, until such time as the weather changes. Examples from other types of surfaces (Bailey et al., 1997; Geiger, 1966; Sellers, 1965) show considerable variation in the magnitudes and relative contributions of the energy balance components, as well as the need to modify some of the components or to add new terms to the equation. Thus, one finds that the calculation of Q_G in water may require the inclusion of horizontal heat transfer due to currents, that the energy used in melting may replace Q_G for melting ice and snow (Munro, 1990), or that anthropogenic heating may be important in cities (Oke, 1987). The negligibly small amounts of energy involved in photosynthesis are investigated as part of the increasingly important theme of carbon cycling, wherever the requirement of measuring atmospheric carbon dioxide concentrations can be met (Monteith and Unsworth, 1990; Soegaard, 1999).

Although geographic location and ambient weather conditions are recognized as major determinants of the overall magnitude of energy exchange, it appears that characteristics of the surface medium control the distribution of energy among the various components in equation (1). Water is well known for its relatively high amounts of heat storage and small diurnal temperature

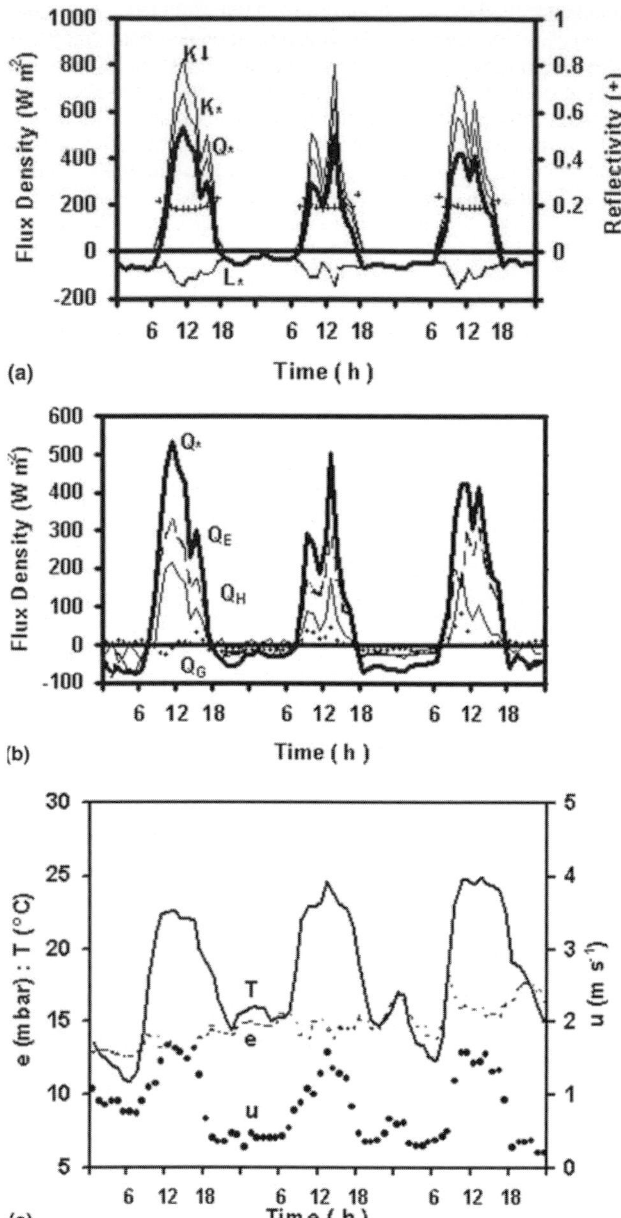

Figure B5 Hourly variations of (**a**) radiation terms, (**b**) energy balance components and (**c**) atmospheric properties for sawgrass marsh during the dry season, Everglades National Park, Florida (16–18 February 1999). Data source: author. (See Appendix for explanation of symbols and units.)

ranges in comparison with land. The proportion of net radiation used in evaporation over land tends to decrease as the amount of moisture in the soil decreases. Net radiation itself tends to be small for snow in comparison with its value for vegetation-covered or bare ground in similar locations. Such knowledge has been obtained by adapting micrometeorological theory to flux estimation, an exercise that often provides insights into the explicit role that surface properties play in energy transfer between the surface and the boundary layer.

Flow of energy at the boundary

Radiative exchange

The role of surface properties in determining net radiation at the boundary may be appreciated by separating Q_* into its shortwave and longwave radiation components in such a manner as to include the shortwave reflection coefficient, α, and the surface emissivity, ε_s:

$$Q_* = K_\downarrow(1 - \alpha) + \varepsilon_s(L_\downarrow - \sigma T_s^4), \quad (2)$$

in which $K_\downarrow$ is incoming shortwave (global) radiation from the sun, $L_\downarrow$ is incoming longwave radiation from the atmosphere, and T_s is the absolute temperature of the boundary. Part of the expression, σT_s^4, is the longwave radiation leaving the surface, where σ is the Stefan–Boltzmann constant. Because σT_s^4 is almost always greater than $L_\downarrow$, net longwave radiation exchange amounts to a perpetual energy loss, which ensures that Q_* is generally less than net shortwave radiation, $K_* = K_\downarrow (1 - \alpha)$, and is therefore negative during the night (Figure B5a).

The respective values of α and ε_s control the shortwave and longwave radiation exchanges. Their values must be derived experimentally (Monteith and Szeicz, 1961; Buettner and Kern, 1965), as there is no theoretical basis on which to make estimates. Tabulations of experimental results from various sources (Monteith and Unsworth, 1990; Oke, 1987) show little variation in ε_s, its value being close to 0.98 for most surfaces. On the other hand, α varies over a wide range, from less than 0.1 for water to nearly 0.9 in the case of freshly fallen snow. The value for most vegetated surfaces is close to 0.2, slightly less for forests, somewhat greater for bare soil and ice. Furthermore, α is known to increase rapidly with decreasing solar elevation over water (Nunez et al., 1972), a feature that is repeated to a lesser degree over land (Arnfield, 1975). Seasonal variation may also be important: the reflectivity of sawgrass during the dry season (Figure B5a), typical as it is of vegetated surfaces, is likely to be more characteristic of a water surface during the wet season.

Given reliable values of $K_\downarrow$, $L_\downarrow$, α, and ε_s, equation (2) might be used to calculate Q_* if it were possible to measure T_s. However, the measurement of surface temperature is difficult, except in the case of water, where thermometers submerged slightly below the surface have been shown to closely estimate T_s (Davies et al., 1971), or melting snow and ice, where T_s is the melting point temperature (Munro, 1990; Oerlemans, 2000). Since the invention of a reliable net radiometer, however, it has been possible to measure Q_* directly (Funk, 1959).

Heat storage in the boundary

Part of the difficulty in obtaining surface temperature arises from the fact that the surface boundary, though active as the principal site of energy exchange (Oke, 1987), is not a physical entity in itself. It is simply the point where the ground and the atmosphere meet. Moreover, the position of such a point is difficult to define, particularly over tall vegetation where individual leaves become the main sites of energy exchange activity, such that the soil surface no longer forms an appropriate zero reference level. In such cases the zero plane, d, is effectively displaced upward, so that Q_G now applies to the plant mass and intervening air space, as well as the soil. Following a customary rule of thumb (Oke, 1987), one may estimate $d \approx 2/3h$, where h is the vegetation height. This does not amount to much plant mass and air space heat storage for short vegetation covers, such as wheat and sawgrass. For tall, dense vegetation, such as is found in forested wetland, plant and air space heat storage may constitute 1/3 Q_G (Munro et al., 2000).

Although surface heating effects quickly diminish with increasing soil depth (Figure B6), they form an interesting topic for study (Monteith and Unsworth, 1990). Also, the diffusion of heat through soil and other components of the surface may be viewed as one of the necessary steps in the evaluation of $Q_* - Q_G$, the amount of energy available to balance Q_H and Q_E.

Heat transfer in soil is controlled by the thermal conductivity, k_s, the thermal diffusivity, κ_s, and the volumetric heat capacity, C_s, which are surface properties that apply to layers beneath the surface:

$$Q_G = -k_s \frac{\partial T'}{\partial z'}, \tag{3a}$$

$$\frac{\partial Q_G}{\partial z'} = -C_s \kappa_s \frac{\partial^2 T'}{\partial z'^2}, \tag{3b}$$

$$\Delta Q_G \approx -C_s \frac{\Delta T'}{\Delta t} \Delta z', \tag{3c}$$

where T' refers to soil temperature, $\Delta z'$ is the thickness of a particular layer at some depth, z', and t refers to time. It is clear from equation (3b) that Q_G undergoes substantial variation with depth; in other words a flux divergence, which accounts for the rapid reduction in temperature change with time as depth increases (Figure B6). The latter is incorporated into equation (3c), an approximation which constitutes a practical procedure for estimating heat transfer if a sufficient number of $\Delta z'$ layers is summed to calculate Q_G.

Although k_s, κ_s, and C_s are properties of the medium, soil is a mixture of solid, air, and water in which properties vary with depth and soil moisture conditions. Therefore, it has become standard procedure to follow the practice of Fuchs and Tanner (1968) in which equation (3a) is replaced by soil heat flux plates buried just below the surface, usually at a depth of 50 mm, and the remaining portion is obtained by employing equation (3c), inserting temperature measurements taken near the surface. Such an approach does not work in semitransparent media such as ice, snow, and water where shortwave radiative transfer penetrates the surface to act in conjunction with the conductive process (Schwerdtfeger and Weller, 1967). However, C_s is easily specified in such media, so equation (3c) can be applied to temperature profile data to obtain reasonably good results, particularly for water (Dutton and Bryson, 1962). Heat storage in vegetation can be estimated by means of suitable adaptations of equation (3c) (Monteith and Unsworth, 1990).

It is rare that Q_G amounts to more than 10% of net radiation for most surfaces, thus minimizing the significance of estimation errors. The notable exception is water, where the proportion may be substantially greater, depending on the season. This is due to a number of factors, notably its transparency to solar radiation and the large values that can be assigned to C_s and κ_s. The latter may be combined with density, ρ_s, to define the conductive capacity, or thermal admittance (Oke, 1987; Priestley, 1959): $\rho_s C_s \kappa_s^{1/2}$. Thermal admittance is generally greater for water than for land surfaces due to large C_s and mixing by wave action. Therefore, water boundaries are generally more effective than land boundaries in moving heat to and from storage.

Flow of energy through the boundary layer

Energy exchange and turbulence

Although boundary layer temperatures vary considerably over time (Figure B6), flux divergence due to heat storage is negligible

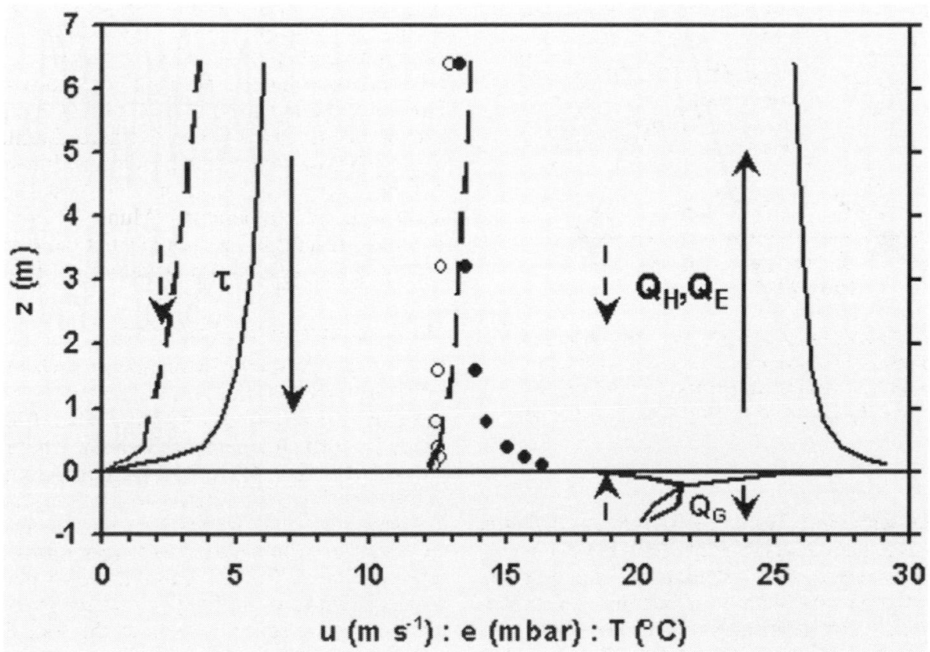

Figure B6 Profiles of wind speed (lines), vapor pressure (circles), air and soil temperature (lines) for late night (0235) and afternoon (1435), O'Neil, Nebraska (19 August 1953), where solid lines, circles and arrows denote afternoon profiles and flux directions. Data source: Great Plains Turbulence Field Program (Lettau and Davidson, 1957).

because the heat capacity of air is extremely small. Divergence due to horizontal gradients in u, T, and vapor pressure, e, poses a potentially greater problem, hence the interest in keeping measurement height, z, small in relation to fetch. The assumption of negligible flux divergence with height is crucial in developing procedures for estimating transfers from atmospheric data.

The development of transfer procedures lies at the heart of the rapidly evolving field of micrometeorology (Monteith and Unsworth, 1990; Priestley, 1959; Thom, 1975). A convenient starting point is to outline the convective transfer equations for Q_H, Q_E, and momentum, τ:

$$\tau = \rho K_M \frac{\partial u}{\partial z}, \quad (4a)$$

$$Q_H = -\rho c_p K_H \frac{\partial T}{\partial z}, \quad (4b)$$

$$Q_E = -\frac{\rho L_v \varepsilon}{p} K_V \frac{\partial e}{\partial z}, \quad (4c)$$

in which one finds ρ, the air density, c_p, the specific heat of air at constant pressure, L_v, the latent heat of vaporization, ε, the ratio of the weight of water vapor to dry air, and p, the air pressure. The quantities, K_M, K_H, and K_V are referred to as eddy diffusivities for momentum, heat, and water vapor, respectively, by analogy to the molecular diffusivities for conduction in solids. Although the conduction process is at work in air, its contribution to the task of transporting heat within the boundary layer is negligible in comparison with that of convection.

The use of the term "eddy" recognizes that convective transfer is a turbulent process. The nature of the turbulence may be appreciated by noting that air, being a fluid, is subject to mixing, and that the increase of wind speed with height imparts a circular motion to the flow (Figure B7a). The motion is sensed as an almost random series of upward and downward velocities, w', or eddies, which decrease in size but increase in frequency, as momentum passes down the wind profile toward the ground. The effect of the eddies is to cause small fluctuations, u', T', and e', from mean values of u, T, and e at any given height above the ground whenever the properties of air at adjacent levels are different, as is normally the case (Figure B6).

Schematically (Figure B7b), the pattern of u' or w' is similar in shape and sign to that of T' and e' at night. During the day the similarity in shape for T' and e' remains, but the sign is reversed, because temperature and humidity now decrease with height (Figure B6). Given sufficiently precise measurements it is thus possible to define an eddy correlation procedure for flux estimation:

$$\tau = -\rho \overline{w'u'}, \quad (5a)$$

$$Q_H = \rho c_p \overline{w'T'}, \quad (5b)$$

$$Q_E = \frac{\rho L_v \varepsilon}{p} \overline{w'e'}, \quad (5c)$$

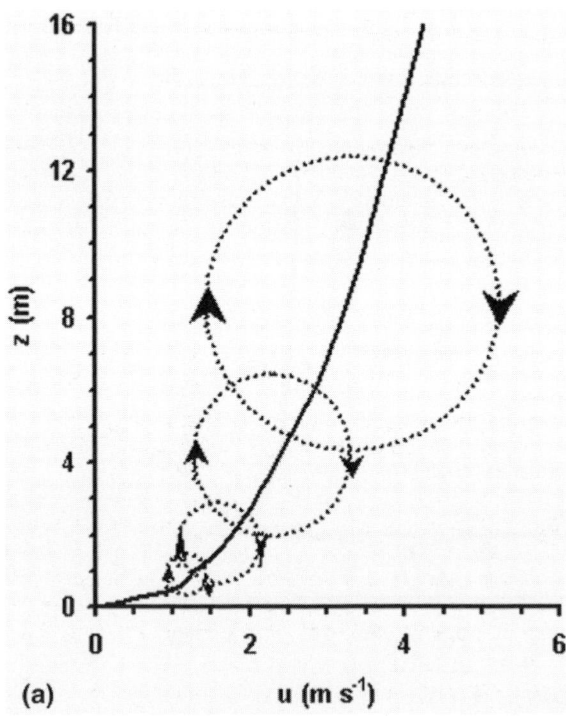

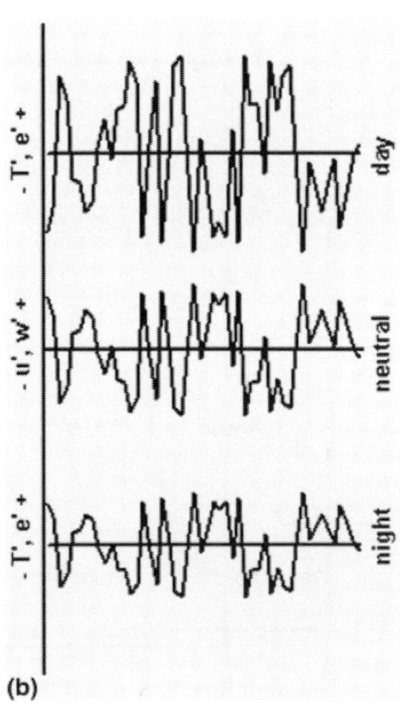

Figure B7 (**a**) Wind profile at Kerang, Victoria, Australia (17 February 1962), with representation of circular nature of eddy motion and scale in relation to height; (**b**) schematic depiction of variations from mean values (horizontal lines) due to turbulence, where velocity fluctuations expand beyond the neutral range during the day but contract at night, in accordance with T' and e'. Data source: Webb (1970).

where the overbars denote an average taken over a period of time. Since the time of successful field trials in Australia (Dyer, 1961), the method has been used in a variety of situations (Wesley et al., 1970; Yap and Oke, 1974; McKay and Thurtell, 1978). The great care and expense involved in the methodology once meant that the procedure was beyond the reach of most investigators. Recent developments in instrumentation have, however, given the procedure a much wider range of application (Campbell and Unsworth, 1979; Munro, 1989).

Eddy fluctuations are far more intensive during the day than they are at night (Figure B7b). The more intensive activity during the day is due to the increased amount of energy available at the surface, but only a part derives from the increase in wind speed. The remainder is obtained from the temperature gradient and, to a much lesser extent, the humidity gradient. Warm temperatures at the ground during the day lead to a strong lapse of temperature with height, causing the air to be relatively buoyant and therefore unstable. At night, cold temperatures near the ground invert the profile, resulting in relatively heavy, stable air. Consequently, instability enhances turbulence during the day but suppresses it at night, situations that can be visualized by rendering the circles in Figure B7a as a set of ellipses that have their long axes oriented vertically during unstable conditions; horizontally when the atmosphere is stable (Figure B8). The circular portrayal of eddy motion is appropriate for neutral conditions, when the temperature gradient is equivalent to that which would be expected from adiabatic cooling, a gradient that is virtually zero over small height intervals.

Stability effects are of minor concern in eddy correlation work, but they become important in understanding how fluxes are related to profiles. Therefore it is useful to obtain a numerical measure of stability such as the Richardson number, R_i, or the z/L stability parameter (Thom, 1975):

$$R_i = \frac{g}{T_a}\frac{\Delta T/\Delta z}{(\Delta u/\Delta z)^2}, \tag{6a}$$

$$\frac{z}{L} = \frac{g}{T_a}\frac{kzQ_H}{\rho c_p u_*^3}, \tag{6b}$$

where g denotes the gravitational acceleration, k is von Kármán's constant, and T_a is the absolute temperature of the air. The friction velocity, u_*, is defined by setting $u_* = (\tau/\rho)^{1/2}$. L, the Obukhov length scale, defines the height of the sub-layer of dynamic turbulence (Obukhov, 1971).

Equation (6) is negative in unstable conditions but positive in stable conditions, with zero signifying neutrality. Although equation (6a) is preferred for its simplicity, in that the gradients can be obtained from measurements at two elevations above the ground, it assumes that eddy diffusivities for heat and momentum are similar. The assumption is not required in equation (6b), which has the advantage of explicitly showing that stability effects increase with height above the boundary.

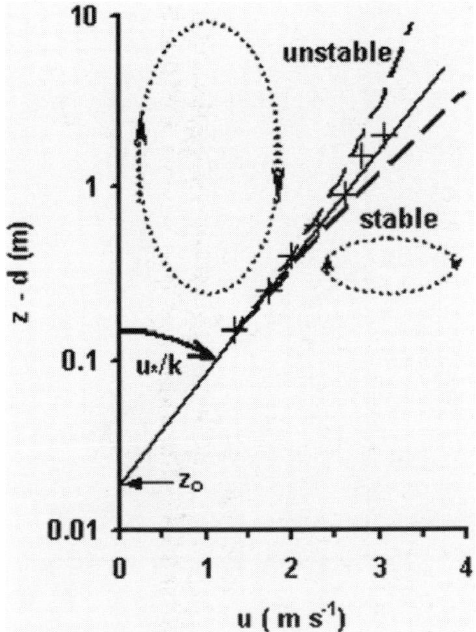

Figure B8 Logarithmic wind profile plot for Davis, California (9 May 1967), showing possible curvatures due to stability effects. Note the roughness length, z_0, and the height correction for zero-plane displacement, d. Data source: Morgan et al. (1971).

Calculation of fluxes from profiles

The calculation of fluxes from profiles requires the integration of equation (4) between any two measurement levels near the surface, in order that differences may be substituted for the gradients. The requirement is not simple because, unlike molecular diffusivities, the eddy diffusivities are properties of the turbulence. Therefore, they change according to height, wind speed, and stability. Taking the example of K_M, micrometeorological theory suggests a model for eddy diffusivity:

$$K_M = \frac{ku_* z}{\phi_M}, \tag{7}$$

where the stability correction factor, $\phi_M = f(z/L)$, is one in neutral conditions.

Taking $\tau = \rho u_*^2$ and substituting equation (7) into equation (4a) leads to a relationship that may be integrated across any height interval, $\Delta z = z_2 - z_1$, to give

$$\Delta u = \frac{u_*}{k}\left[\ln\left(\frac{z_2}{z_1}\right) + \Delta\psi_M\right], \tag{8}$$

in which z_1 and z_2 are the respective lower and upper limits of the height interval in question, and ψ_M is the integrated form of ϕ_M. For neutral conditions, $\Delta\psi_M = 0$ and equation (8) becomes the well-known logarithmic wind profile, where the slope of the profile is equal to u_*/k (Figure B8). The purpose of introducing $\Delta\psi_M$ into the equation is to correct for curvature due to stability effects (Figure B8).

The profiles may also include a correction for the zero-plane displacement, d. Failure to include the correction over any vegetated surface other than short grass introduces curvature to the logarithmic wind profile. Because height is plotted logarithmically, zero wind speed is found at a height, z_0, the roughness length. It tends to be an order of magnitude smaller than the vegetation height, and has often been regarded as an artifact of the analysis, but there is good reason to treat it as a physical property of the surface. Lettau (1969) showed how an independent z_0 calculation can be made from careful measurements of surface dimensions, a technique which works especially well over non-vegetated surfaces (Munro, 1989; Raupach, 1992; Smeets et al., 1999).

It is possible to rearrange equation (8) to estimate u_* and, therefore, τ by means of an aerodynamic approach that may be extended to Q_H and Q_E:

$$\tau = \frac{\rho k^2(\Delta u)^2}{[\ln(z_2/z_1) + \Delta\psi_M]^2}, \quad (9a)$$

$$Q_H = -\frac{\rho c_p \Delta u \Delta T}{[\ln(z_2/z_1) + \Delta\psi_M]^2}, \quad (9b)$$

$$Q_E = -\frac{\rho L_v \varepsilon k^2 \Delta u \Delta e}{p[\ln(z_2/z_1) + \Delta\psi_M]^2}, \quad (9c)$$

where all differences correspond to Δz. The expression for Q_E was first proposed by Thornthwaite and Holzman (1939), without the inclusion of ψ_M, and subsequently received much attention in evaporation studies. However, the inclusion of ψ_M in the denominator of each expression requires the assumption of similar eddy diffusivities for momentum, heat, and water vapor.

The similarity assumption was treated with suspicion by micrometeorologists, but experimental evidence remained vague until the publication of a stability analysis undertaken by Swinbank and Dyer (1967). They demonstrated that the functional form for K_M in unstable conditions is different from that for K_H and K_V, which were found to be equivalent quantities. Subsequent work (Dyer and Hicks, 1970; Webb, 1970) addressed the problem of integrating the stability corrections with height and demonstrated the numerical equivalence of R_i and z/L when the atmosphere is unstable.

In principle, the structure of equation (9) can be improved by means of suitable modifications to the denominators of equations (9b) and (9c). In practice this leads to a cumbersome structure that places great demands on measurement accuracy. Furthermore, investigators have produced a variety of different stability correction functions (Andreas, 2002; Dyer, 1974).

The finding that $K_H = K_V$ means that, regardless of the form that stability modifications might take, the denominators of equations (9b) and (9c) are equal. Thus, it becomes a simple matter to extract the Bowen ratio, $\beta = Q_H/Q_E$ from temperature and humidity data:

$$\beta = \gamma\frac{\Delta T}{\Delta e}, \quad (10a)$$

$$Q_E = \frac{Q_* - Q_G}{1 + \beta}, \quad (10b)$$

where equation (10b) is obtained by rearrangement of equation (1) and $\gamma = c_p p/L_V\varepsilon$ defines the psychrometric constant. This is the Bowen ratio, or energy balance approach, in which Q_H can be obtained as a residual in the surface energy balance, or as a different rearrangement of equation (1).

The energy balance approach has become very popular in boundary layer work because it reduces the problem of obtaining fluxes from profiles to the task of obtaining good measurements of ΔT and Δe. The task can be difficult where temperature and humidity differences are small, as for forest boundaries. Thus, effort has been directed toward the construction of reliable measurement systems for ΔT and Δe (Black and McNaughton, 1971; McNeil and Shuttleworth, 1975).

The Bowen ratio itself is a quantity of considerable climatological significance because it characterizes the surface partitioning of energy exchange with the boundary layer (Oke, 1987). As surface conditions become colder, β increases to values approaching 10, a value which is characteristic of dry desert surfaces. Also, β sensitivity to surface moisture content is well known, the value changing from a low of approximately 0.1 over water to a range of 0.5 to 4.0 over vegetated surfaces, where the higher values apply to drier boundary conditions.

Incorporating the boundary into boundary layer work

The incorporation of the boundary in flux calculation procedures is essentially a matter of substituting the surface for the lower measurement level in equation (9). The substitution is desirable on both operational and conceptual grounds. Operationally, it relaxes the stringent measurement requirements for profile measurements, thus allowing the use of standard measurements of wind speed, temperature, and humidity, such as might be obtained from a climatological station. Conceptually, it poses the stimulating problem of characterizing the boundary in such a way that it plays a useful role in the flux estimation scheme.

In attempting to characterize the boundary, one may consider the problem of defining surface temperature, T_s, and humidity, e_s, over a vegetative surface. One solution, proposed by Monteith (1963), is to plot temperature and humidity against windspeed, defining T_s and e_s at $u = 0$ (Figure B9). The approach was criticized by Tanner and Fuchs (1968) on the grounds of uncertainty regarding source levels for heat and

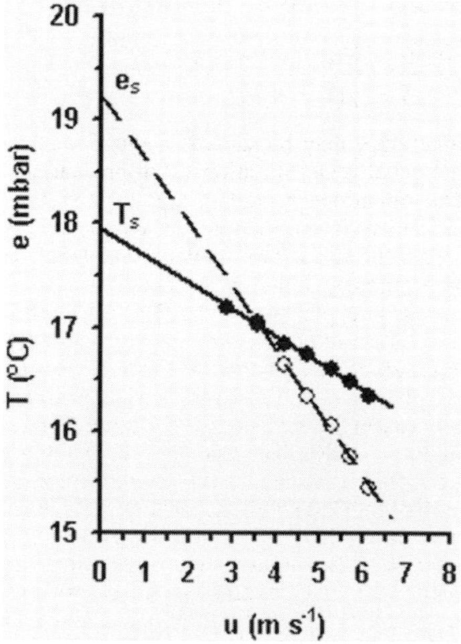

Figure B9 Plots of temperature and vapor pressure against wind speed to estimate surface values, Davis, California (9 May 1967). Data source: Morgan et al. (1971).

water vapor. Thom (1975) has since shown that reference levels for heat and water vapor differ from that for momentum, so z_0 is not the appropriate roughness length for temperature and humidity profiles. In fact, temperature and humidity roughness lengths may exceed z_0 by a factor of two for forest (Mölder et al., 1999), but are likely to be two to three orders of magnitude smaller than z_0 for ice and snow (Andreas, 2002).

Modifying the aerodynamic procedure to include the boundary in calculations of Q_H and Q_E causes equation (9) to take the form of a bulk transfer equation:

$$Q_H = -\frac{\rho c_p k^2 u(T - T_s)}{[\ln(z/z_0) + \psi_M][\ln(z/z_H) + \psi_H]}, \tag{11a}$$

$$Q_E = -\frac{\rho c_p k^2 u(e - e_s)}{[\ln(z/z_0) + \psi_M][\ln(z/z_V) + \psi_V]}, \tag{11b}$$

where the subscripts, H and V, apply to the appropriate roughness lengths and stability corrections for sensible heat and water vapor. Nevertheless, it is assumed that $\psi_H = \psi_V$ and $z_H = z_V$, pending strong experimental evidence to the contrary.

The difficulty in defining T_s and e_s severely restricts the use of a bulk transfer procedure over land, though there is a long history of its use over water (Deacon and Webb, 1962). The method is especially attractive for the special case of melting ice and snow on glaciers, where one may assume $T_s = 0°C$ and $e_s = 6.11$ mb (Munro, 1990; Oerlemans, 2000). In such cases, however, turbulence is generated by katabatic winds, such that alternative flux estimation approaches are required (Oerlemans and Grisogono, 2002). Also, regardless of surface cover type, $z_{H,V}$ is proportional to z_0, the proportion changing with the degree of turbulence (Andreas, 2002; Garratt and Hicks, 1973).

A more generalized approach to the problem of incorporating the boundary into flux computations arose from the initial work of Penman (1948) which, following subsequent work (Monteith, 1981), led to the development of the Penman–Monteith (PM) model for surface evaporation:

$$Q_E = \frac{S(Q_* - Q_G) + \rho c_p D_a/r_a}{S + \gamma(1 + r_c/r_a)}. \tag{12}$$

The air temperature may be used to determine S, the slope of the relationship between saturation vapor pressure, e', and temperature. The quantities r_a and r_c are respectively known as aerodynamic and surface (or canopy) resistances to water vapor movement from the surface to the boundary layer, while $D_a = e' - e$ is the saturation deficit of the air.

Monteith (1981) explains how the PM model combines diabatic, $S(Q_* - Q_G)$, and adiabatic, $\rho c_p D_a/r_a$, energy contributions to evaporation. The relative importance of each contributor depends upon the extent to which surface roughness results in *coupling* between the surface and the boundary layer (Jarvis and McNaughton, 1986). The adiabatic contribution is dominant where coupling is strong, as it is for tall, rough forest, such that Q_E may be modeled solely according to D_a and the resistances (Infante et al., 1997). The diabatic contribution prevails over smoother, weakly coupled short vegetation, such that D_a and the resistances may be expressed as a coefficient (Priestley and Taylor, 1972), or ignored altogether, leaving the equilibrium form of equation (12) which may be useful for cropped surfaces (Davies and Allen, 1973). The full PM model applies to most surfaces, however, because the degree of coupling is typically between these extremes.

The PM aerodynamic resistance is a function of turbulence, and may be obtained from the logarithmic wind profile, incorporating the correction for d:

$$r_a = \frac{1}{k^2 u}\left[\ln\left(\frac{z - d}{z_0}\right)\right]^2. \tag{13}$$

Some error may be introduced in ignoring a stability correction, and in using z_0 rather than z_H, but the error is likely to be small because r_a appears in both the numerator and the denominator of equation (12). Thus, if Q_* and Q_G are known, estimates of Q_E follow from measurements of T, D_a, and u above the ground provided the value of r_c can be defined.

Over water, r_c can be taken as zero, an assumption that has also been shown to work well for wet vegetation canopies (Stewart, 1977). Otherwise, it has been demonstrated that r_c is a physiologically sensitive parameter because r_c values obtained by rearranging equation (12), knowing the value of Q_E, closely follow patterns of stomatal resistance, r_s, obtained from direct measurements on the leaves of plants (Figure B10). The r_s to r_c ratio in Figure B10 reflects the leaf area index of the plant cover (Tan and Black, 1976), which suggests that r_c may be obtained from the rule of parallel resistances:

$$\frac{1}{r_c} = \sum_{i=1}^{n} \frac{\mathrm{LAI}_i}{r_{si}}, \tag{14}$$

in which i is any of n leaves or leaf layers. Conceptually, this compresses a multi-level plant cover into one *big* leaf, treating r_c to represent a full canopy which covers the whole surface.

Depending upon the plant root structure, r_c is sensitive to the availability of soil moisture (Figure B11), the sensitivity being strong for field crops, with their temporary root systems, but much weaker for the well-established root system of a forest (Szeicz and Long, 1969). Given the extensive work done with the PM model, it is now possible to generalize about r_c where soil water supply is not severely restrictive: ranges of 5 to 50 s m^{-1} for crops, 5 to 125 s m^{-1} for forests, where the low end of each

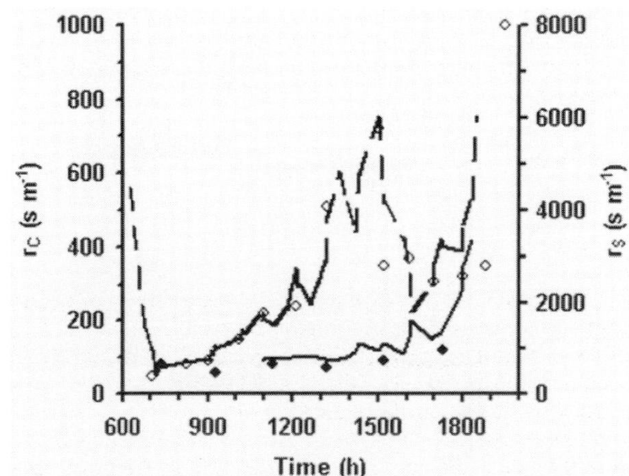

Figure B10 Hourly variations of surface (lines) and stomatal (circles) resistances for a Douglas fir forest, Courtenay, British Columbia, Canada, where solid lines and circles refer to relatively moist soil, broken line and open circles to drier soil (after Tan and Black, 1976).

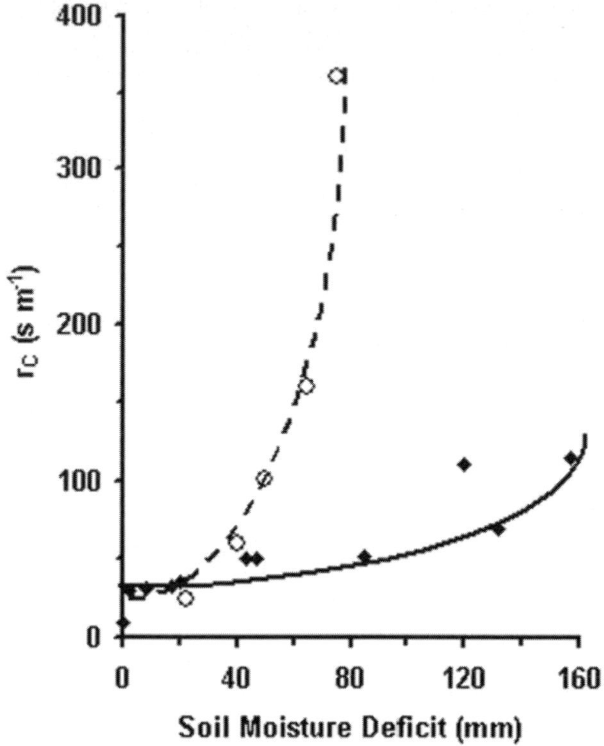

Figure B11 Surface resistance responses of grass–clover vegetation and forest vegetation (solid line and circles) to soil moisture deficits, where the deficit refers to the depth of water required to raise soil moisture to field capacity (from Szeicz and Long, 1969).

range applies to wet canopies (Oke, 1987). Thus, there is now sufficient experimental evidence to develop a climatology of r_c (Thom and Oliver, 1977), and to incorporate vegetation layers into simulation models of basin hydrology (Sellers and Lockwood, 1981).

Once the PM model had been well established for the full big leaf canopy, attention turned to evaporation from sparse canopies (Figure B12), where the ground appears between rows or patches of ground cover (Shuttleworth and Wallace, 1985). The approach taken here is to apply the PM model separately to each surface category, such that the ground itself has a surface resistance, r_g (Figure B12b). Thus Q_E is composed of canopy evaporation, Q_{Ec}, and ground evaporation, Q_{Eg}, each of which is obtained from the PM model (Wallace, 1995):

$$Q_E = f_c Q_{Ec} + f_g Q_{Eg}, \tag{15}$$

where the proportional contributions of each evaporation source, f_c and f_g, are functions of the aerodynamic and bulk resistances of canopy and ground.

Hence, the PM model will continue to be a powerful, adaptable operational tool and a conceptual stimulus in a number of environmental fields which involve boundary layer climatology. Operationally, it minimizes the need for sophisticated measurement procedures, such that basic weather station data are useful. Conceptually, it stimulates interest in the structure of the surface boundary, comprised as it may be of full or sparse vegetation cover, thus allowing scholars to consider greater variability in the surface properties which shape the many climates of the boundary layer.

D.S. Munro

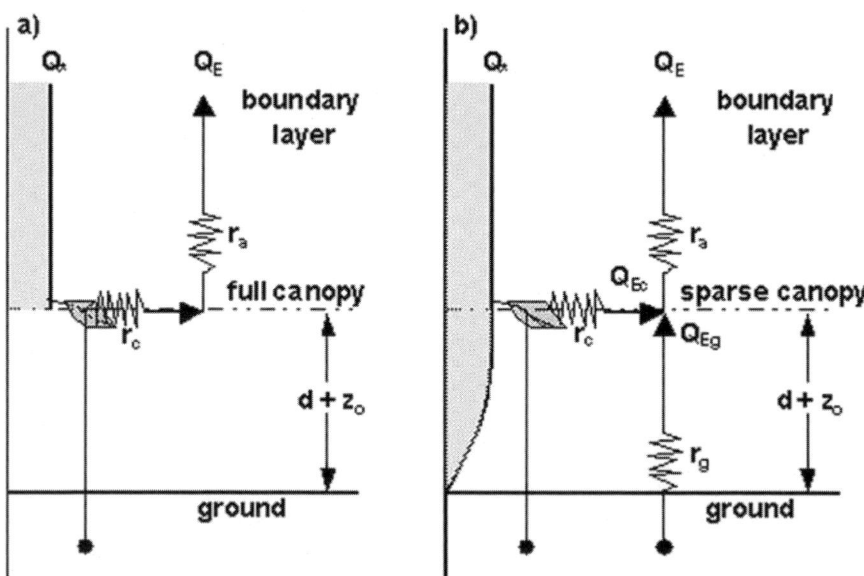

Figure B12 Schematic relationship of resistance pathways for evaporation from (**a**) full and (**b**) sparse canopies, where $Q_* = 0$ at the left margin of each diagram (adapted from Lafleur and Rouse, 1990; Wallace, 1995).

Appendix: Symbols and units

C_s = surface volumetric heat capacity ($J\,m^{-3}\,C^{-1}$)
D_a = vapor pressure deficit of air (mb, Pa)
$K_{M,H,V}$ = eddy diffusivities for momentum, heat, water vapor ($m^2\,s^{-1}$)
$K_\downarrow$ = incoming shortwave, or global, radiation ($W\,m^{-2}$)
L = Obukhov length (m)
L_v = latent heat of vaporization ($J\,kg^{-1}$)
$L_\downarrow$ = incoming longwave radiation ($W\,m^{-2}$)
LAI = leaf area index
Q_E = latent heat transfer ($W\,m^{-2}$)
Q_{Ec} = leaf canopy latent heat transfer ($W\,m^{-2}$)
Q_{Eg} = bare ground latent heat transfer ($W\,m^{-2}$)
Q_G = heat transfer to the surface medium ($W\,m^{-2}$)
Q_H = sensible heat transfer ($W\,m^{-2}$)
Q_* = net radiative transfer ($W\,m^{-2}$)
Ri = Richardson number
S = slope of the saturation vapor pressure curve ($Pa\,C^{-1}$)
T = air temperature (C, K)
T_a = absolute temperature of the air (K)
T' = soil or water temperature
T_s = surface temperature (C, K)
c_p = specific heat of air at constant pressure ($J\,kg^{-1}\,C^{-1}$)
d = zero-plane displacement (m)
e = vapor pressure of air (mb, Pa)
e' = vapor pressure of air (mb, Pa)
e_s = surface vapor pressure (mb, Pa)
$f_{c,g}$ = proportion of evaporation due to canopy, ground
g = acceleration due to gravity ($m\,s^{-2}$)
k = von Kármán's constant (≈ 0.4)
k_s = surface thermal conductivity ($W\,m^{-1}\,C^{-1}$)
p = air pressure (mb, Pa)
r_a = aerodynamic resistance ($s\,m^{-1}$)
r_c = surface, or canopy resistance ($s\,m^{-1}$)
r_g = ground resistance ($s\,m^{-1}$)
r_s = stomatal resistance ($s\,m^{-1}$)
t = time (s)
u = wind speed ($m\,s^{-1}$)
u_* = friction velocity ($m\,s^{-1}$)
z = height above ground (m)
z' = depth in soil (m)
$z_{0,H,V}$ = roughness length for wind speed, temperature, humidity (m)
α = reflectivity for shortwave radiation
β = Bowen ratio
γ = psychometric constant ($0.66\,mb\,C^{-1}$, $66\,Pa\,C^{-1}$)
ε = ratio of molecular weight of water vapor to dry air (0.622)
ε_s = surface emissivity
κ_s = thermal diffusivity of soil or water ($m^2\,s^{-1}$)
ϕ_M = gradient stability correction for momentum
σ = Stefan-Boltzmann constant ($5.67 \times 10^{-8}\,W\,m^{-2}\,K^{-4}$)
τ = momentum transfer ($N\,m^{-2}$)
ρ = air density ($kg\,m^{-3}$)
ρ' = ice, soil or water density ($kg\,m^{-3}$)
$\psi_{M,H,V}$ = integrated stability correction for momentum, heat, water vapor

Bibliography

Andreas, E.L., 2002. Parameterizing scaler transfer over snow and ice: a review. *Journal of Hydrometeorology*, **3**: 417–432.

Arnfield, A.J., 1975. A note on the diurnal, latitudinal and seasonal variation of the surface reflection coefficient. *Journal of Applied Meteorology*, **14**: 1603–1608.

Bailey, W.G., Oke, T.R., and Rouse, W.R., eds. 1997. *The Surface Climates of Canada*. Montreal and Kingston: McGill-Queen's.

Black, T.A., and McNaughton, K.G., 1971. Psychrometric apparatus for Bowen ratio determination over forests, *Boundary Layer Meteorology*, **1**: 246–254.

Bradley, E.F., 1968. A micrometeorological study of velocity profiles and surface drag in the region modified by a change in surface roughness, *Quarterly Journal of the Royal Meteorological Society*, **94**: 361–379.

Buettner, K.J.K., and Kern, C.D., 1965. The determination of infrared emissivities of terrestrial surfaces. *Journal of Geophysical Research*, **70**: 1329–1337.

Campbell, G.S., and Unsworth, M.H., 1979. An inexpensive sonic anemometer for eddy correlation. *Journal of Applied Meteorology*, **18**: 1072–1077.

Davies, J.A., and Allen, C.D., 1973. Equilibrium, potential and actual evapotranspiration from cropped surfaces in Southern Ontario. *Journal of Applied Meteorology*, **12**: 649–657.

Davies, J.A., Robinson, P.J., and Nunez, M., 1971. Field determinations of surface emissivity and temperature for Lake Ontario. *Journal of Applied Meteorology*, **10**: 811–819.

Deacon, E.L., and Webb, E.K., 1962. Small-scale interactions. In Hill, M., ed., *The Sea*, vol. 1. New York: Wiley, pp. 43–87.

Dutton, J.A., and Bryson, R.A., 1962. Heat flux in Lake Mendota. *Limnology and Oceanography*, **7**: 80–97.

Dyer, A.J., 1961. Measurement of evaporation and heat transfer in the lower atmosphere by an automatic eddy correlation technique. *Quarterly Journal of the Royal Meteorological Society*, **87**: 401–412.

Dyer, A.J., 1974. A review of flux–profile relationships. *Boundary Layer Meteorology*, **7**: 363–372.

Dyer, A.J., and Hicks, B.B., 1970. Flux–gradient relationships in the constant flux layer, *Quarterly Journal of the Royal Meteorological Society*, **96**: 715–721.

Elliott, W.P., 1958. The growth of the atmospheric internal boundary layer. *American Geophysics Union Transactions*, **39**: 1048–1054.

Fuchs, M., and Tanner, C.B., 1968. Calibration and field test of soil heat flux plates. *Soil Science Society of America Proceedings*, **32**: 326–328.

Funk, J.P., 1959. Improved polyetheylene-shielded net radiometer. *Journal of Scientific Instruments*, **36**: 267–270.

Garratt, J.R., and Hicks, B.B., 1973. Momentum, heat and water vapour transfer to and from natural and artificial surfaces, *Quarterly Journal of the Royal Meteorological Society*, **99**: 680–687.

Geiger, R., 1966. *The Climate Near the Ground*. Cambridge, MA: Harvard University Press.

Infante, J.M., Rambal, S., and Joffre, R., 1997. Modelling transpiration in holm-oak savannah: scaling up from leaf to the tree scale. *Agricultural and Forestry Meteorology*, **187**: 273–289.

Jarvis, P.G., and McNaughton, K.G., 1986. Stomatal control of transpiration: scaling up from leaf to region. In Ford, E.D., and Macfadyen, A., eds., *Advances in Ecological Research*, vol. 15, pp. 1–45.

Lafleur, P.M., and Rouse, W.R., 1990. Application of an energy combination model for evaporation from sparse canopies. *Agricultural and Forestry Meteorology*, **49**: 135–153.

Lettau, H., 1969. Note on aerodynamic roughness-parameter estimation on the basis of roughness element description. *Journal of Applied Meteorology*, **8**: 828–832.

Lettau, H.H., and Davidson, B., 1957. *Exploring the Atmosphere's First Mile*. London: Pergamon.

McKay, D.C., and Thurtell, G.W., 1978. Measurements of the energy fluxes involved in the energy budget of a snow cover. *Journal of Applied Meteorology*, **17**: 339–349.

McNeil, D.D., and Shuttleworth, W.J., 1975. Comparative measurements of the energy fluxes over a pine forest. *Boundary Layer Meteorology*, **9**: 297–313.

Mölder, M., Grelle, A., Lindroth, A., and Halldin, S., 1999. Flux-profile relationships over a boreal forest – roughness sublayer corrections. *Agricultural and Forestry Meteorology*, **98–99**: 645–658.

Monteith, J.L., 1963. Gas Exchange in Plant Communities. In Evans, L. T., ed., *Environmental Control of Plant Growth*. New York: Academic Press, pp. 95–110.

Monteith, J.L., 1981. Evaporation and surface temperature, *Quarterly Journal of the Royal Meteorological Society*, **107**: 1–27.

Monteith, J.L., and Szeicz, G., 1961. The radiation balance of bare soil and vegetation. *Quarterly Journal of the Royal Meteorological Society*, **87**: 159–170.

Monteith, J.L., and Unsworth, M.H., 1990. *Principles of Environmental Physics*. London: Edward Arnold.
Morgan, D.L., Pruitt, W.O., and Lourence, F.J., 1971. *Analyses of Energy, Momentum and Mass Transfers Above Vegetative Surfaces*, Research and Development Technical Report ECOM 68-G10-F. Fort Huachuca, AZ: US Army Electronics Command Atmospheric Sciences Laboratory.
Munro, D.S., 1990. Comparison of melt energy computations and ablatometer measurements on melting ice and snow. *Arctic and Alpine Resarch*, **22**: 153–162.
Munro, D.S., 1989. Surface roughness and bulk heat transfer on a glacier: comparison with eddy correlation. *Journal of Glaciology*, **35**: 343–348.
Munro, D.S., Bellisario, L.M., and Verseghy, D.L., 2000. Measuring and modeling the seasonal climatic regime of a temperate wooded wetland. *Atmosphere–Ocean*, **38**: 227–249.
Nunez, M., Davies, J.A., and Robinson, P.J., 1972. Surface albedo at a tower site in Lake Ontario. *Boundary Layer Meteorology*, **3**: 77–86.
Obukhov, A.M., 1971. Turbulence in an atmosphere with a non-uniform temperature. *Boundary Layer Meteorology*, **2**: 7–29.
Oerlemans, J., 2000. Analysis of a 3 year record from the ablation zone of Morteratschgletscher, Switzerland: energy and mass balance. *Journal of Glaciology*, **46**: 571–579.
Oerlemans, J., and Grisogono, B., 2002. Glacier winds and parameterization of the related surface heat fluxes. *Tellus*, **54A**: 440–452.
Oke, T.R., 1987. *Boundary Layer Climates*. London: Methuen.
Penman, H.L., 1948. Natural evaporation from open water, bare soil and grass, *Royal Society of London Proceedings, Ser. A*, **193**: 120–145.
Priestley, C.H.B., 1959. *Turbulent Transfer in the Lower Atmosphere*. Chicago: University of Chicago Press.
Priestley, C.H.B., and Taylor, R.J., 1972. On the assessment of surface heat flux and evaporation using large scale parameters, *Monthly Weather Review*, **100**: 81–92.
Raupach, M.R., 1992. Drag and drag partition on rough surfaces. *Boundary Layer Meteorology*, **60**: 375–395.
Rider, N.E., Philip, J.R., and Bradley, E.F., 1963. The horizontal transport of heat and moisture—a micro-meteorological study. *Quarterly Journal of the Royal Meteorological Society*, **89**: 507–531.
Schwerdtfeger, P., and Weller, G., 1967. The measurement of radiative and conductive heat transfer in ice and snow, *Archiv für Meteorologie, Geophysik, und Bioklimatologie*, **B15**: 24–38.
Sellers, P.J., and Lockwood, J.G., 1981. A computer simulation of the effects of differing crop types on the water balance of small catchments over long time periods. *Quarterly Journal of the Royal Meteorological Society*, **107**: 395–414.
Sellers, W.D., 1965. *Physical Climatology*. Chicago: University of Chicago Press.
Shuttleworth, W.J., and Wallace, J.S., 1985. Evaporation from sparse crops-an energy combination theory. *Quarterly Journal of the Royal Meteorological Society*, **111**: 839–855.
Smeets, C.J.P.P., Duynkerke, P.G., and Vugts, H.F., 1999. Observed wind profiles and turbulence fluxes over an ice surface with changing surface roughness. *Boundary Layer Meteorology*, **92**: 101–123.
Soegaard, H., 1999. Fluxes of carbon dioxide, water vapor and sensible heat in a boreal agricultural area of Sweden – scaled from canopy to landscape level. *Agricultural and Forestry Meteorology*, **98–99**: 463–478.
Stewart, J.B., 1977. Evaporation from the wet canopy of a pine forest. *Water Resources Research*, **13**: 915–921.
Swinbank, W.C., and Dyer, A.J. 1967. An experimental study in micrometeorology. *Quarterly Journal of the Royal Meteorological Society*, **93**: 494–500.
Szeicz, G., and Long, I.F., 1969. Surface resistance of crop canopies. *Water Resources Research*, **5**: 622–633.
Tan, A.S., and Black, T.A., 1976. Factors affecting the canopy resistance of a Douglas-fir forest. *Boundary Layer Meteorology*, **10**: 475–488.
Tanner, C.B., and Fuchs, M., 1968. Evaporation from unsaturated surfaces: a generalized combination method. *Journal of Geophysical Research*, **73**: 1299–1303.
Thom, A.S., 1975. Momentum, mass and heat exchange of plant communities. In Montieth, J.L. ed., *Principles*, vol. 1: *Vegetation and the Atmosphere*. London: Academic Press, pp. 57–109.
Thom, A.S., and Oliver, H.R., 1977. On Penman's equation for estimating regional evaporation, *Quarterly Journal of the Royal Meteorological Society*, **103**: 345–358.
Thornthwaite, C.W., and Holzman, B., 1939. The determination of evaporation from land and water surfaces, *Monthly Weather Review*, **67**: 4–11.
Wallace, J.S., 1995. Calculating evaporation: resistance to factors. *Agricultural and Forestry Meteorology*, **73**: 353–366.
Webb, E.K., 1970. Profile relationships: the log-linear range, and extension to strong stability, *Quarterly Journal of the Royal Meteorological Society*, **96**: 67–90.
Wesley, M.L., Thurtell, G.W., and Tanner, C.B., 1970. Eddy correlation measurements of sensible heat flux near the earth's surface. *Journal of Applied Meteorology*, **9**: 45–50.
Yap, D., and Oke, T.R., 1974. Sensible heat fluxes over an urban area – Vancouver, B.C., *Journal of Applied Meteorology*, **13**: 880–890.

Cross-references

Albedo and Reflectivity
Energy Budget Climatology
Microclimatology
Radiation Climatology
Urban Climatology

BOWEN RATIO

Net radiation at a surface – an expression of incoming less outgoing radiation – must be divided between energy that goes to sensible heat flow and to latent heat flow. The Bowen ratio (β) is used to express the partitioning of net radiation at a surface. Thus:

$$\beta = \frac{\text{Sensible heat loss to atmosphere } (C)}{\text{Latent heat loss to atmosphere } (LE)}.$$

Theoretically, in the absence of an atmosphere, β can vary from infinity (for a dry surface with no evaporation) to zero (for a wet surface with no sensible heat loss). Practically, measurements of surfaces with an atmosphere do not experience such extremes. Typical values are $\beta = 0.1$ for world oceans and $\beta = 5.8$ for desert–semidesert areas.

The derivation of values needed to determine the Bowen ratio has been outlined by Lockwood (1979), who notes that both the sensible and latent heat fluxes (C and LE, respectively) can be expressed in almost symmetrical form:

$$C = -\rho\, C_p K_H \frac{\partial T}{\partial z}.$$

$$LE = -\rho\, \frac{C_p}{\gamma} K_V \frac{\partial e}{\partial z}.$$

where ρ is the density of moist air; C_p is the specific heat of air at constant pressure; K_V and K_H are the eddy diffusivities for water vapor and heat, respectively, e/z and T/z are the vertical gradients of vapor pressure and temperature, respectively; and L is the latent heat of vaporization of liquid water.

The thermodynamic value of the psychometric constant, γ, is given by:

$$\gamma = \frac{C_p\, \rho}{0.621 L}.$$

If the following typical values are introduced into the above equation, γ becomes 0.66: $C_p = 0.240$ cal °C^{-1} g^{-1}; $L = 585$ cal g^{-1}; and $\rho = 1000$ mb (ρ = pressure).

Therefore, from the above equations it is possible to write for the Bowen ratio:

$$\beta = \gamma \frac{\partial T K_H}{\partial e K_V}.$$

Often the convenient simplification is made that $K_H = K_V$, giving:

$$\beta = \gamma \frac{\partial T}{\partial e}$$

The dimensionless Bowen ratio is a useful value in climatic analysis of energy balance studies. For example, it is interesting to compare β values of areas that have been modified through human activities. The change from a natural to an urban environment results in an increase in β, with the value for an urban area in humid climates approaching similar values to those of deserts. A decreasing ratio occurs in irrigated areas in dry regions where the β value is similar to that found in oases.

John E. Oliver

Bibliography

Lockwood, J.G., 1979. *Causes of Climate.* New York: Halsted Press.
Oke, T.R., 1987. *Boundary Layer Climates.* New York: Wiley.
Robinson, P.J., and Henderson–Sellers, A., 1999. *Contemporary Climatology.* Harlow: Longman.

Cross-references

Energy Budget Climatology
Evapotranspiration
Latent Heat
Sensible Heat
Water Budget Analysis

BUDYKO, MIKHAIL IVANOVICH (b. 1920)

Unquestionably, Mikhail Ivanovich Budyko was one of the greatest climatologists in the Soviet Union and is the most respected in Russia today. He has received worldwide recognition for his work. Budyko was born in 1920, at a time of civil war and famine, in Gomel, Belarussia (now Belarus). Educated in the Soviet system, he received his Master of Science degree from the Leningrad Polytechnic Institute in 1942. Budyko was assigned immediately to Leningrad Main Geophysical Observatory as a researcher in 1942. During the German Army siege of the city he experienced the travails of the famine of 1941–43. Then in 1951, Budyko received his PhD from the Observatory, became Deputy Director in 1954, and advanced to Director in 1972.

A respected and highly qualified scientist, Budyko was dismissed from his director's position in 1954 for not cooperating with the Communist Party's Leningrad Committee. Eventually he was reinstated. His brilliant research and major scientific contributions enabled him to be appointed Head of the Division of Physical Climatology at the Observatory in 1972, Head of the Division of Climatic Change Research at the State Hydrological Institute at Leningrad in 1975, and later, Academician of the Russian Academy of Science in 1992 (Pacchioli, 2001). To many scientists Budyko is the father of physical climatology, and he is noted for raising the level of precision of global climatic research.

Budyko became deeply involved in researching the Earth's heat budget and regional heat balances in the 1950s. The first systematic measurements of solar radiation had begun in the nineteenth century. Improved instrumentation, better communication between scientists, and exchange of data led to new insights into heat balance Arctic climatology during the Second International Polar Year in the 1930s. After World War II, investigation of the radiation balance of the Earth was stimulated by an expansion of actinometric observation and advanced computational methods. Dr Budyko postulated that the sun's energy was the most important factor in climatology. He calculated the heat balance of selected climatic regions, verified his numerical models with observational data, confirmed that his heat balance calculations were accurate, and published his conclusions in his seminal book, *The Heat Balance of the Earth's Surface*, in 1956. This book revolutionized the way climatology was taught, provided a base for research not considered before, and remolded climatology into a more quantitative and physical science. A few years later, in 1963, Dr Budyko's team of researchers produced a very sophisticated atlas, depicting the energy balance of the Earth as viewed from space (Budyko, 1963).

In 1970, supported by an excellent team of researchers and concerned about the role of biological organisms and human activities on climate, Dr Budyko began research on climatic change (Budyko, 1971). He believed that the Earth would enter a period of warming that would increase the global average temperature a few degrees Celsius. His research into the composition of the atmosphere in the geologic past indicated that a significant factor in global temperature changes in the past was the amount of carbon dioxide in the atmosphere. In 1973 Dr Budyko published a remarkable article entitled, "Atmospheric carbon dioxide and climate". He noted that a relatively small increase or decrease in the amount of carbon dioxide could change climatic conditions on Earth. Also, he observed that the atmosphere has retained approximately one-half of the carbon dioxide gas formed as a result of human activities (Budyko, 1973). Humans, through their actions, and in particular the burning of fossil fuels, were changing the climates of the Earth. This article stimulated much research throughout the world, and Dr Budyko continued his studies on the relationships between climate and human activity. He published a book entitled *Climate and Life*, in 1974. In this book he correlated climatic changes and the extinction of animal species, plus the effect of climate on all forms of life on earth (Budyko, 1974).

Realizing that any changes, whether large or small, in the stable and unstable components of the Earth's atmosphere impact world climate, Dr Budyko began to focus his attention on the atmospheric ramifications of a nuclear war. He postulated that if a nuclear exchange between the United States and the Soviet Union were to occur, large amounts of aerosol particles would be released into the atmosphere. Coordinating what was known about past climate and weather changes resulting from volcanic eruptions and collisions with meteorites, he concluded that a nuclear weapon exchange would threaten life in all forms on our Earth. His publications and presentations on a potential "nuclear winter" impacted Soviet military planners and decision-makers in the 1980s and are believed to have contributed to the decision of the Soviet Union and the United States to sign a treaty to reduce the number of nuclear missiles in their arsenals.

Recognized by the United Nations and the recipient of the Blue Planet Prize in 1998, Dr Budyko outlined in his Commemorative Lecture his charge for climatologists in the world. He firmly believes that there should be more research into climatic change, climate and global food problems, climatic change's effect upon natural zonality, global warming and permafrost, temperature increases and reduction of energy consumption in winter, hydrological response to climatic change, and a rise in the sea level. He has been the recipient of the Lenin National Prize in 1958, the Gold Medal of the World Meteorological Organization in 1987, the Professor R. Horton Medal of the American Geographical Union in 1994, the A. A. Grigoryev Prize of the Russian Academy of Science in 1995, and others. Mikhail Ivanovich Budyko personifies the best in atmospheric research, climatic change identification, and concerned scientist involvement in critical global issues. His research and publications represent major contributions to scientific work, focused upon seeking solutions to global environmental problems (AGF, 1999).

William A. Dando

Bibliography

AGF, 1999. *Profiles of the 1998 Blue Planet Prize Recipients*. Tokyo: Asaki Glass Foundation, pp. 1–2.

Budyko, M.I., 1956. *The Heat Balance of the Earth's Surface*. Leningrad: Gidrometeoizdat.

______, ed., 1963. *Atlas of the Heat Balance of the Globe*. Moscow.

______, 1971. "The energetics of the biosphere and its transformation under the influence of man", *Izvestia Akademii Nauk SSSR*, Geographic Series, No. 1.

______, 1973. "Atmospheric carbon dioxide and climate", *Gidrometeoizdat*, Leningrad, pp. 1–32.

______, 1974. *Climate and Life* (edited by D.H. Miller). New York: Academic Press.

Pacchioli, D., 2001. "Life in the extreme", *Astrobiology: The Search for Life in the Universe*, Penn State Online, vol. 22, Issue 1, January 2001, pp. 1–3; http://www.rps.psu.edu/0101/extreme.html

Cross-references

Bioclimatology
Energy Budget Climatology
Greenhouse Effect and Greenhouse Gases

C

CARBON-14 DATING

The introduction of the carbon-14 (^{14}C, radiocarbon) dating method in 1947 (for which Willard F. Libby received the Nobel Prize for Chemistry in 1960) transformed many aspects of environmental science by permitting numerical dating of fossils, artifacts and deposits whose age previously had to be estimated. Organisms and events could now be put into chronological order and correlated objectively, and the search for mechanisms of change placed on a sounder footing, leading to a better understanding of such matters as the viscosity of the Earth's mantle, the mechanisms of climatic change, processes of organic evolution and extinction, and climatic history, within the 70 000 or so years spanned by the method.

Three isotopes of carbon are present in the atmosphere in the ratio 100 : 1 : 0.01, of which two, ^{12}C and ^{13}C, are stable. The third, ^{14}C, is radioactive and thus subject to decay, but it is continually replenished by the action of cosmic rays, which interact with ^{14}N atoms in the upper atmosphere to form ^{14}C. The radiocarbon is oxidized to form CO_2, which is then incorporated into plants by photosynthesis or dissolved in the ocean and used to build carbonate structures by mollusks and corals. The current estimate of the half life ($t_{1/2}$) of ^{14}C is 5730 $\pm$30 years, but Libby's original value of 5568$\pm$30 is used in many date lists for consistency. The "Libby" age can be adjusted to the new value by multiplying by 1.03. His use of BP for Before Present (= 1950) also persists.

Originally Libby analyzed his samples as solid carbon. Nowadays in most radiocarbon laboratories ^{14}C content is measured by converting the sample into CO_2, whose radioactivity relative to a modern standard is counted in a gas-proportional counter or by synthesizing benzene (C_6H_6) from the gas and using a liquid scintillation counter. A third method, accelerator mass spectrometry (AMS), allows ^{14}C atoms to be counted directly and thus requires much smaller samples, typically 15 mg as opposed to 1 g of carbon. The $\pm$ value that follows the age is generally a statement of the counting error at 1 s.d., but some laboratories include analytical and other error estimates.

The method can be used for any organic material but some substances have proved less troublesome than others. Wood, charcoal and peat, suitably pretreated, are often favored, but bone collagen and unrecrystallized shell and coral yield reliable ages. Besides exercising great care in field attribution and handling, the collector can check the sample for contamination by old or young carbon by means of microscopy (both optical and SEM), X-ray diffraction and the $^{13}C/^{12}C$ (stable isotope) ratios measured by mass spectrometry. Corrections need to be made for ^{14}C contributed to the atmosphere by thermonuclear weapons testing after 1952 and by dead CO_2 produced from the burning of coal and oil fuels (the Suess effect).

Variations in the ^{14}C reservoir (the de Vries effect) can be allowed for by reference to calibration curves based on tree ring ages for sequoia (*Sequoia gigantea*) and bristlecone pine (*Pinus aristata*), which are available for the last 8000 years. The main sources of variation are the intensity of the cosmic ray flux, the strength of the Earth's magnetic field and changes in the Earth's carbon reservoir stemming from climatic changes. The solution of parochial dating errors is thus proving a source of information on environmental change both on Earth and on the sun.

Further information on carbon-14 dating can be gained from Bradley (1985, Ch. 3), Worsley (1981), and Raaen et al. (1968). Practical applications can be found in Ozer and Vita-Finzi (1986).

Claudio Vita-Finzi

Bibliography

Bradley, R.S., 1985. *Quaternary Paleoclimatology: Methods of Paleoclimatic Reconstruction*. Boston: Allen & Unwin.

Ozer, A., and Vita-Finzi, C. (eds), 1986. *Dating Mediterranean Shorelines*. Berlin: Gebruder Borntraeger.

Raaen, V.F., Ropp, G.A., and Raaen, H.P., 1968. *Carbon-14*. New York: McGraw-Hill.

Worsley, P., 1981. Radiocarbon dating: principles, applications and sample collection. In Goudie, A.S., ed., *Geomorphological Techniques*. London: Allen & Unwin, pp. 277–83.

Cross-references

Archeoclimatology
Climatic Variation: Historical
Tree-Ring Analysis

CENTERS OF ACTION

If there were no continents and seasonal variation on the Earth, an ideal atmospheric circulation system would be denoted by: (1) an equatorial belt of low pressure; (2) polar centers of high pressure; and (3) two intermediate belts, one of high pressure (the Horse Latitudes) at about 30 degrees and the other of low pressure in the vicinity of 60 degrees. The low-pressure zones would be centers of convergence and the high-pressure zones centers of divergence. Since the Earth is rotating, wave patterns would develop along the convergences and these would become the great storm-generating belts. However, the critical modification in this ideal pattern is the asymmetric presence of continents; meridional land masses and mountain belts block the 60 degree convergence zone in the northern hemisphere, but in the southern hemisphere this is a clear waterway – the Southern Ocean – all around the Earth.

Due to the different thermal characteristics of continents and oceans, semipermanent centers of high and low pressure tend to build up and remain along the zonal (east–west) pressure belts, with greater contrasts in the northern hemisphere as it contains most of the world's land mass. Over the oceans the sign of the pressure extremes remains constant in each pressure mass, but over continents it tends to alternate in summer (low) and winter (high), the so-called monsoon effect. Since the surface winds diverge from high-pressure areas, the semipermanent highs are known as source regions for airmasses. These semipermanent or regularly recurring pressure centers are known, as Rossby (1945) says, "by the somewhat misleading name" of centers of action. It is important to recognize that these centers are the result and not the cause of global circulation.

The semipermanent centers of action in the northern hemisphere are: (1) low pressure: Aleutian Low, Icelandic Low; and (2) high pressure: Azores or Bermuda High, Hawaiian or North Pacific High (Figure C1).

The continental highs in the northern hemisphere winter are the North American or Canadian High and the Siberian or Asiatic High; in summer these tend to be replaced by lows, but are more complicated by the zonal circulation, especially in the case of the smaller land mass of North America.

With the exception of the South Polar High over Antarctica, comparable centers of action are not as well developed in the southern hemisphere, though well-defined oceanic high-pressure centers occur: the South Atlantic or St Helena High, the Indian Ocean High, and the South Pacific High. In summer, lows develop over the continents and are replaced by highs in the winter; this is similar to, but not so marked as, the Siberian Low and High. These differences occur due to the continuous waterbody of the Southern Ocean symmetrically surrounding Antarctica, which is itself symmetrical about the South Pole, and because the other three southern continents occupy only minor fractions of the total area, the steady circulation is striking. Indeed, Gentilli (1949) found that the maximum zonal index was 6 (i.e. well-marked westerlies) and there was never a negative (easterly) reading between 35°S and 55°S; in contrast, the index varied from +15 to −5 over a 12-month period in North America.

It is known that large-scale phenomena such as the El Niño/Southern Oscillation (ENSO) and the North Atlantic Oscillation (NAO) affect the climate over a large part of the Earth on time scales that range from years to centuries. Kirov and Georgieva (2002) compared the century-long variability of ENSO and NAO with variations of solar activity, and suggest that the influence of solar activity on these phenomena is mediated by the atmospheric centers of action which react by changes in intensity and location.

Robert M. Hordon

Bibliography

Curry, J.A., 1999. *Thermodynamics of Atmospheres and Oceans*. San Diego, CA: Academic Press.

Gentilli, J., 1949. Air masses of the southern hemisphere. *Weather*, **4**: 258–261, 292–297.

Hartmann, D.L., 1994. *Global Physical Climatology*. San Diego, CA: Academic Press.

Kirov, B., and Georgieva, K., 2002. Long-term variations and interrelations of ENSO, NAO and solar activity. *Physics and Chemistry of the Earth*. Oxford: Pergamon-Elsevier Science, pp. 441–448.

Lamb, H.H., 1972. *Climate: Past, Present and Future*, vol. 1. London: Methuen.

Robinson, P.J., 1999. *Contemporary Climatology*. Harlow: Longman.

Rossby, G.C., 1945. The scientific basis of modern meteorology. In Berry, F.A., Bollay, E., an *Handbook of Meteorology*, New York: McGraw-Hill, pp. 502–529.

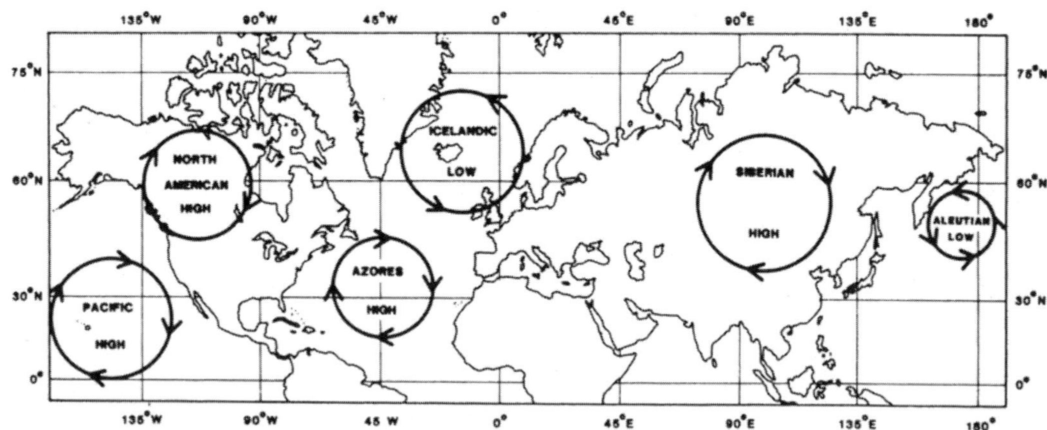

Figure C1 Idealized map showing locations of centers of action in the northern hemisphere.

Cross-references

Airmass Climatology
Aleutian Low
Atmospheric Circulation, Global
Azores (Bermuda) High
Icelandic Low
North American (Canadian) High
North Pacific (Hawaiian) High
Oscillations
Siberian (Asiatic) High
Teleconnections
Zonal Index

CENTRAL AMERICA AND THE CARIBBEAN, CLIMATE OF

Central America and the Caribbean span the deep tropics and subtropics. Because of the tropical maritime location temperature changes throughout the region are generally small, and rainfall is by far the most important meteorological element. In general the climate of the region is controlled by the migration of synoptic features, and the mean climate strongly reflects the annual cycle of these features.

The most dominant synoptic influence is the subtropical high of the north Atlantic. Subsidence associated with the spreading of the subtropical high from the north Atlantic to the north American landmass dominates during boreal winter, as do the strong easterly trades found on its equatorward flank. Coupled with a strong trade inversion, a cold ocean and reduced atmospheric humidity, the region is generally at its driest during the winter. With the onset of boreal spring, however, the subtropical high moves offshore and trade wind intensity decreases, with convergence characterizing their downstream. Especially for Central America, the variation in the strength of the trades is an important determinant of climate throughout the year. There is also a high and weak trade inversion, the ocean warms and atmospheric moisture is abundant. The region is consequently at its wettest in the northern summer half-year.

The contrast in summer and winter rainfall defines most climate classifications of the region. For example, Rudloff (1981) suggests that the climate of Central America and the Caribbean can be classified as dry-winter tropical. The contrast proves an important control for agricultural and tourism activity, water resource allocation, hydrological considerations and fishing – activities which are of utmost importance to the region.

The dry winter/wet summer regime, however, only *broadly* defines the climate of the region as orography and elevation are significant modifiers on the subregional scale. The region is one of complex and diverse topography including continental territories, island chains, and mountain ranges of varying orientations and elevations. The topography interacts with the large-scale circulation to produce local variations in the climate, including significant variations in annual rainfall totals, length of the rainy season, and the timing of maxima and minima. As examples, the windward slopes of the larger mountainous islands of the Greater Antilles have significantly higher rainfall totals than the smaller flatter islands in the eastern Caribbean Sea. Similarly, the Caribbean coastal stations of Honduras, Costa Rica and Panama possess a strikingly homogeneous rainfall regime (i.e. in stark contrast to the dominant dry winter/summer regime) due to strong interactions of low-level winds with the topography. The subregional variations make generalizations about the climate of the region difficult – a fact which must be borne in mind when considering the general overviews presented in the following sections.

Besides the subtropical high, other significant synoptic influences include: (a) the seasonal migration of the Intertropical Convergence Zone (ITCZ) – mainly affecting the Pacific side of southern Central America; (b) the intrusions of polar fronts of midlatitude origin (called "Nortes" in Spanish) which modify the dry winter and early summer climates of the northern Caribbean and north Central American; and (c) westward propagating tropical disturbances – a summer season feature associated with much rainfall especially over the Caribbean region.

Central America

Temperature

Temperature ranges during the course of the year in the Central America region are generally small. They may exceed 4°C to the north, from the Yucatan to Honduras, but drop to less than 2°C in the south and along the coastal zones. Maximum temperatures are dependent on altitude given the mountainous interior of the region. The common usage of the terms *tierra caliente* ("hot land"), *tierra templada* ("temperate land"), *tierra fria* ("cold land"), and *tierra helada* ("frozen land") suggests a longstanding recognition of the effect of altitude on temperatures. The terms loosely denote shifts in the mean temperature regime with increasing altitude, but their usage is relative (varies with location), with no consistent assignment of a temperature to elevation. Minimum temperatures similarly exhibit dependence on altitude, but also show the effect of the winter intrusions of cold air from North America, and can dip as low as 7°C.

Portig (see Schwerdtfeger, 1976) classifies the annual cycle of air surface temperature in Central America as tropical, predominantly maritime, with small annual changes and dependent on cloud cover and altitude. The dominant annual cycle of the Central American region (excepting the Atlantic coasts of Honduras and northern Nicaragua) is monsoonal, with highest temperatures occurring just before the summer rains (Figure C2). Temperatures are at their lowest during January, largely due to the cooling effect of the strong trade winds (Figure C3). Maximum temperatures occur during April and are associated with a decrease in trade wind strength, lower cloud coverture and higher values of solar radiation (Alfaro, 2000a). There is a temperature minimum in July which coincides with the onset of the midsummer drought (or in Spanish the *veranillos* or *caniculas* – see below). During this period the trade winds briefly increase in intensity (see again Figure C3), the subtropical ridge over the Caribbean intensifies, and there is a second minimum and maximum in cloud coverture and radiation respectively.

Precipitation

Mean annual rainfall totals vary over a wide range in Central America in keeping with the diversity in topographic conditions. Annual totals of less than 100 cm are typical of the plains of Guatemala, Honduras and northwestern Nicaragua, and portions of the Pacific coast of El Salvador, Honduras and

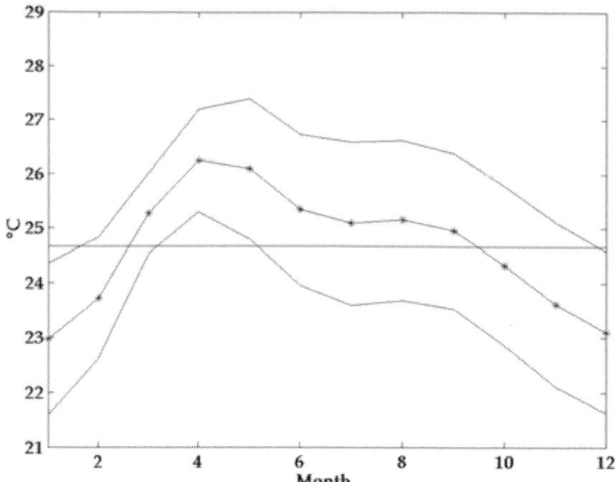

Figure C2 The central line (with asterisks) depicts the mean annual cycle of temperature over Central America as determined from principal component analysis. The upper and lower solid lines represent one standard deviation. The horizontal line is the annual mean of 24.7°C.

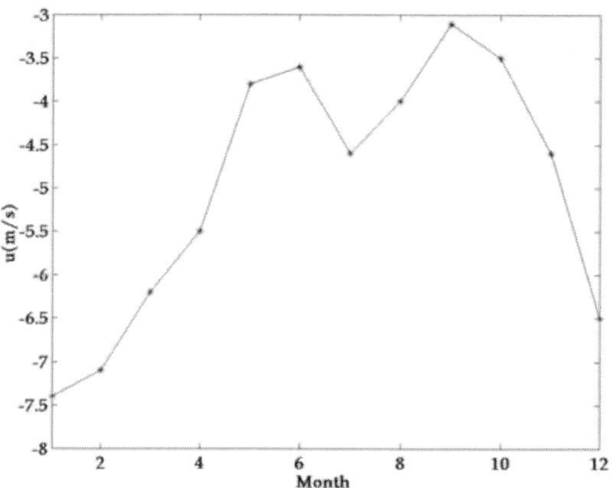

Figure C3 Annual cycle for zonal wind component (u) at the 900 hPa level. Monthly values are from Juan Santamaría synoptic station (10°00′N, 84°12′W, from Alvarado, 1999), period used: 1972–1989). The annual average value is $-5\,\mathrm{m\,s^{-1}}$. Notice that negative values are westward winds.

Panama. In contrast the northern and southern mountain ranges of Guatemala and the low mountains between Costa Rica and Panama receive large amounts of rainfall (in excess of 250 cm), as do the Atlantic coastal regions from Belize through Guatemala, the south coast of Costa Rica and a section of Panama's Atlantic coast. In general the Caribbean seaboard receives more rainfall than the Pacific side, reflecting the influence of tropical disturbances from the Caribbean Sea and the windward interaction of the trade winds with the mountains chains.

Fernandez et al. (1996) explored the variability of rainfall with altitude in Central America by examining the vertical rainfall distribution in Costa Rica along a topographic profile which crossed the country from the Pacific to the Caribbean coasts. The mountain profile, with highest peak of approximately 3000 m, is

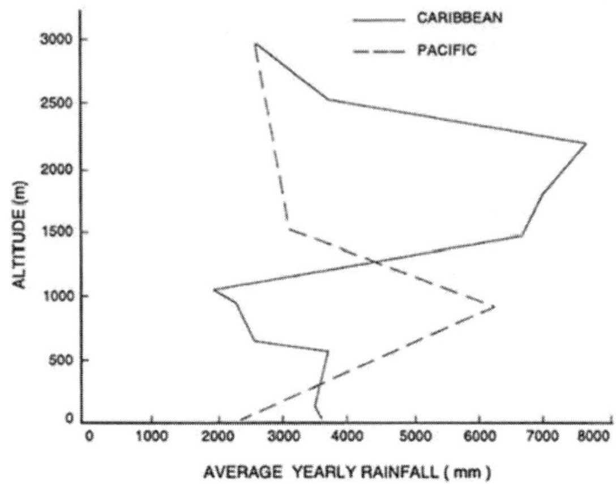

Figure C4 Distribution of average yearly rainfall with altitude on the Caribbean and Pacific sides of a topographic profile in Central America. The periods of analysis are 1966–1984 and 1981–1984 for the Caribbean and the Pacific sides respectively. (After Fernandez et al., 1996.)

oriented parallel to the prevailing large-scale northeasterly trade winds. Their analysis of rainfall amounts and the seasonal and diurnal variations at 14 rain-gauge stations located on or close to the topographic profile reveals considerable variation with altitude. Maximum rainfall on both the windward (Caribbean) and leeward (Pacific) sides of the main mountain range occur at intermediate altitudes rather than on the mountain tops. (See also the work of Chacon and Fernandez, 1985.) Average yearly maxima of 7735 mm on the windward side and 6692 mm on the leeward side were observed at about 2000 m and 800 m respectively (see Figure C4).

Analysis of the mean annual cycle of precipitation reveals two dominant modes (Figure C5). The first and more representative mode is characterized by two rainfall maxima in June and September, an extended dry season from November to May, and a shorter dry season in July–August. Figure C5a suggests this regime as characteristic (with a few exceptions) of the entire Central America. This regime is largely explained by the seasonal migration of the subtropical north Atlantic high and the ITCZ.

The dry season of winter and early spring accounts for less than 20% of the annual precipitation total of Central America. It is more intense on the Pacific slopes of the isthmus, possibly due to an additional drying effect caused by the seasonal reversal of the winds on the Pacific side which blow offshore during winter. The ITCZ is also at its maximum southeast position during February and March. The season is further characterized by strong Atlantic trades (Figure C3), high values of total radiation (direct plus diffuse radiation) and sunshine hours in the low troposphere levels, despite the fact that radiation at the top of the atmosphere is at minimum. Cortez (2000) shows that there is no evidence of deep convection during the period as mean OLR values are greater than $240\,\mathrm{W\,m^{-2}}$, and that there is poor humidity convergence over almost all the isthmus excepting the southeast.

The onset of the rainy season is in May. There is a latitudinal variation of onset dates with earlier onset (early May) in the south and late onset (late May) in the north. The latitudinal variation may be partially explained by the northward migration

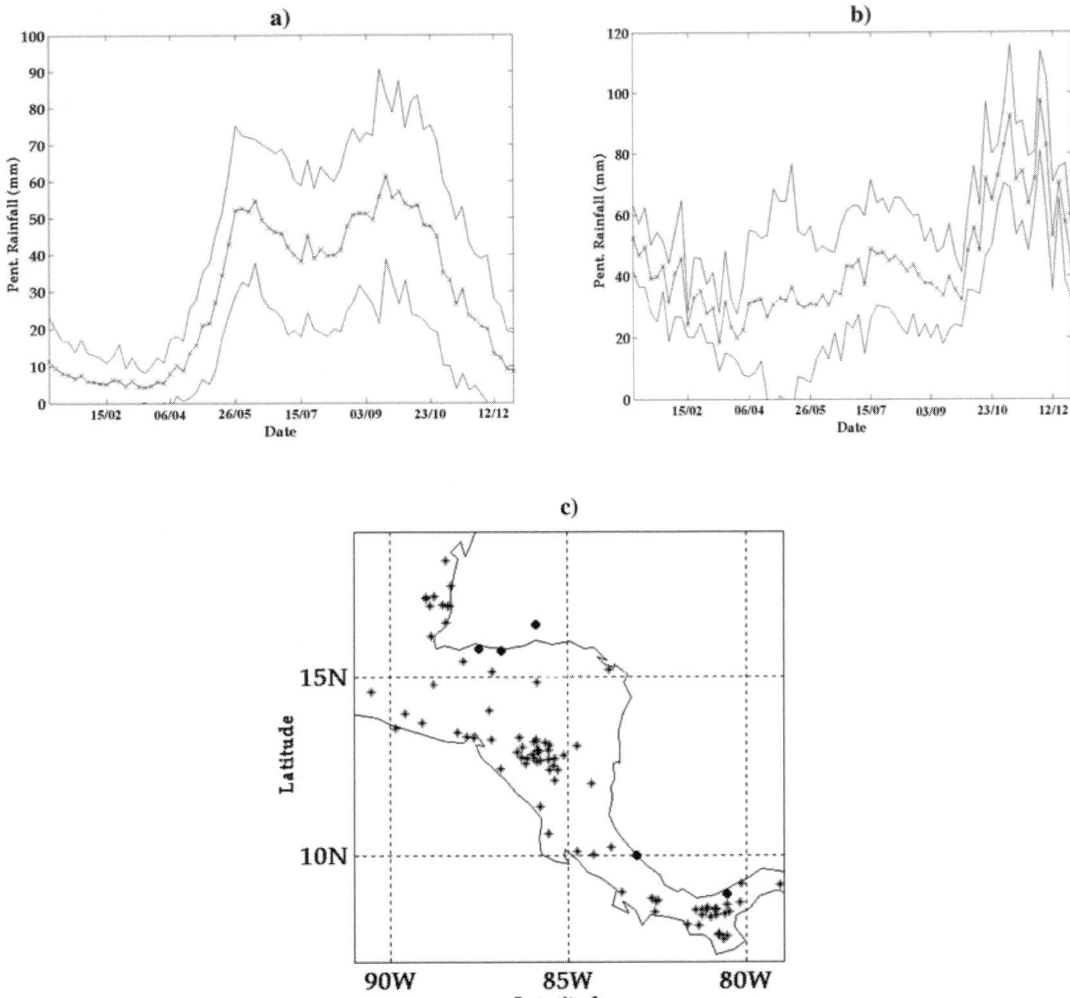

Figure C5 Lines with asterisks are the (**a**) first and (**b**) second pentad rainfall dominant annual cycle patterns, using empirical orthogonal function analysis of 94 gauge stations. Bands are given for one standard deviation (solid lines). (**c**) Asterisks (solid circles) are locations at which the first (second) mode dominates.

of the ITCZ that causes instability and humidity convergence in southern Central America from boreal late spring to early fall. As noted by Alfaro (2000b), however, the migration of the ITCZ cannot explain the generalized deep convection over the entire region during the rainy season as the ITCZ is not normally found at latitudes higher than 10–12°N. The first rainfall peak occurs in mid-June, and during this period (and for the entire rainy season), the trade winds are weak (Figure C3), allowing for the formation of mesoscale systems such as sea breezes, though mostly on the Pacific side.

By the middle of July there is a distinct decrease in rainfall over much of Central America. The brief dry period is known as the *veranillo, canícula,* or the midsummer drought (MSD, see Magaña et al., 1999). Again there is some latitudinal variation in the onset dates of the MSD, with earlier onset in the Southern-Pacific slope and later onset in the Northern-Caribbean slope. During the MSD there is an increase in the trade wind intensity associated with a brief retrogression of the north Atlantic subtropical high.

The second rainfall maximum occurs in mid-September/October, and is wetter than the first. The subsequent demise of the rainfall season also shows latitudinal variation with the tendency for earlier end dates in the north Pacific sector and later end dates in the south. This latitudinal behavior is only partially explained by the southward migration of the ITCZ, since latest end dates are found on the north Caribbean coast. The reason for such late end dates in the north is unclear. It could reflect the influence of cold fronts over the North-Caribbean Central American coast during boreal winter and/or the intensification of the winter trade winds which transport humidity and produce precipitation on the windward side.

The Caribbean coasts of Honduras, Costa Rica and Panama (Figure C5c) exhibit a second rainfall mode, different from the dominant dry-winter/wet-summer regime. It is difficult to define a dry season for this region as rainfall is nearly homogeneous between January and the middle of October, with typical accumulations of between 180 and 300 mm per month. The period accounts for approximately 60% of the annual total in this region. From mid-October onwards there is a marked increase in precipitation accumulation until the end of the year, with a maximum in the beginning of December. This is related to a maximum in humidity convergence at the end of the year in the southeastern region. Significantly, there is no rainfall minimum in July, with instead a relative maximum characterizing the mode 2 regions. This might result from the low-level jet influence and the interaction between topography and the low-level wind (Amador, 1998). Interestingly it is the same mechanism (i.e. the intensification of the trades) which both yields the MSD over most of Central America and makes the Caribbean slope wetter than the Pacific slope during July.

The Caribbean

Temperature

Temperatures on the islands of the Caribbean remain fairly constant throughout the year, with a small annual range of only 2–7°C (see Figure C6). As opposed to the dominant monsoonal type variation of Central America, the annual cycle of temperature for the Caribbean exhibits a summer–winter variation similar to that of more northern latitudes, but with high tropical temperatures and in the context of the small range. In the mean, temperatures steadily increase from May and peak in August–September. The summer months are the warmest with mean temperatures peaking in the upper 20 s Celsius, but rarely exceeding 37°C for any station. For the smaller islands of the Lesser Antilles, the summer heat is tempered by the easterly trades.

Coolest temperatures occur during boreal winter and early spring (December to April) with values generally in the low 20 s for sea level and coastal stations. Temperatures, however, never generally fall below 18°C. The summer–winter variation becomes more pronounced for territories at increasingly higher latitudes, which also come under the occasional influence of intruding polar fronts in winter. Inland, on the larger islands of the Greater Antilles, the cooling influence of the sea breeze is lost, though in some cases this is compensated for by increasing altitude.

Precipitation

The differences in size, shape, topography, and orientation with respect to the trade winds greatly influence the amount of rainfall received by the island territories of the Caribbean. The larger and more mountainous islands of the Greater Antilles (Cuba, Jamaica, Hispaniola, Puerto Rico) receive rainfall amounts of up to 160 cm per year, with in excess of 500 cm on the highest peaks. There is, however, a distinct rainshadow effect on their southern coasts, which are distinctly arid. To their east the northern Windward Islands (as far as Antigua) receive lower rainfall amounts, tending to be slightly drier. Precipitation increases with southerly latitude and the Windward Islands, Trinidad and Barbados tend to be well watered. Normal yearly rainfall is between 180 and 230 cm for the Windward Islands (from Montserrat through Grenada), 150 cm for Trinidad, and 125 cm for Barbados. The dry belt of the Caribbean is found over the southwestern islands of the Netherland Antilles (Aruba, Bonaire, Curacao), which possess a semiarid climate and receive less than 60 cm per year on average. In general the eastern sides and windward slopes of the Caribbean islands receive more rainfall because of the prevailing northeasterly trade winds, while the lee slopes and interior portions of the mountainous islands are drier.

Except for the Netherland Antilles, the Caribbean islands generally receive most of their yearly rainfall in the summer months (Figure C6). This reflects the influence of the migrating north Atlantic subtropical high, the warming of the Caribbean Sea, a reduction in trade wind strength, and the appearance in mid-June of easterly waves/tropical disturbances which leave the west coast of Africa and traverse through the Caribbean region. These easterly waves are convection centers carried by the trades that frequently mature into tropical storms and hurricanes under conducive atmospheric and oceanic conditions such as warm sea surface temperatures and low vertical shear. They represent the primary source of Caribbean rainfall, and their onset in June and demise in early November roughly coincide with the mean Caribbean rainy season. Unlike Central America, the southernmost Caribbean territories lie too far north (~11°N) to come under the direct influence of the ITCZ.

The northwestern Caribbean territories (Jamaica, eastern Cuba, and Hispaniola) possess a distinct midsummer break in rainfall similar to the MSD of Central America (see Figure C6). Onset, however, tends to be later than in Central America, with the MSD for the Caribbean territories occurring in mid to late July/early August. Consequently the annual cycle of rainfall for the northwest Caribbean is bimodal with an initial peak in May/June and a second greater maximum in October. The islands of the Lesser Antilles generally lack the earlier maximum, with rainfall totals increasing from the onset of the rainy season in May to a late fall rainfall peak. The eastern side of Dominica and Martinique has one long rainfall season with a peak in November, while Barbados has a broad minimum in February and March, and a broad maximum between September and November. This lack of an early rainfall season peak in the eastern Caribbean might be accounted for by the fact that only the western Caribbean Sea is warm enough to support convection (i.e. exceeds the 27°C convection threshold) at the onset of the rainy season (Wang and Enfield, 2001, 2003), and it is only the southwestern Caribbean which is conducive to tropical cyclogenesis at the start of the rainy season.

A slightly different rainfall regime characterizes the flat islands of the Netherland Antilles, which have their main rainy season from October through January, with the "small rains" from June to September. The islands are at their driest between February and May, with an absolute minimum in April.

Outside of the rainy season, rainfall in the Caribbean dry season can occur from the southern tip of intruding cold fronts form North America. The influence of the fronts is particularly noticeable in northern Caribbean territories (especially the Bahamas) but rarely extends below Jamaica or east of Puerto Rico.

Hurricanes

Tropical storms and hurricanes are seasonally common in the northern Caribbean and Gulf of Mexico. They merit brief discussion due to the significant loss of life, the extensive infrastructural damage, and the disruption to the Central American and Caribbean economies and way of life that occur with hurricane landfall or even passage nearby. In the mean eight hurricanes will pass near or through the Caribbean region in a year, but this number can vary significantly from year to year. Global climatic phenomena such as the El Niño Southern

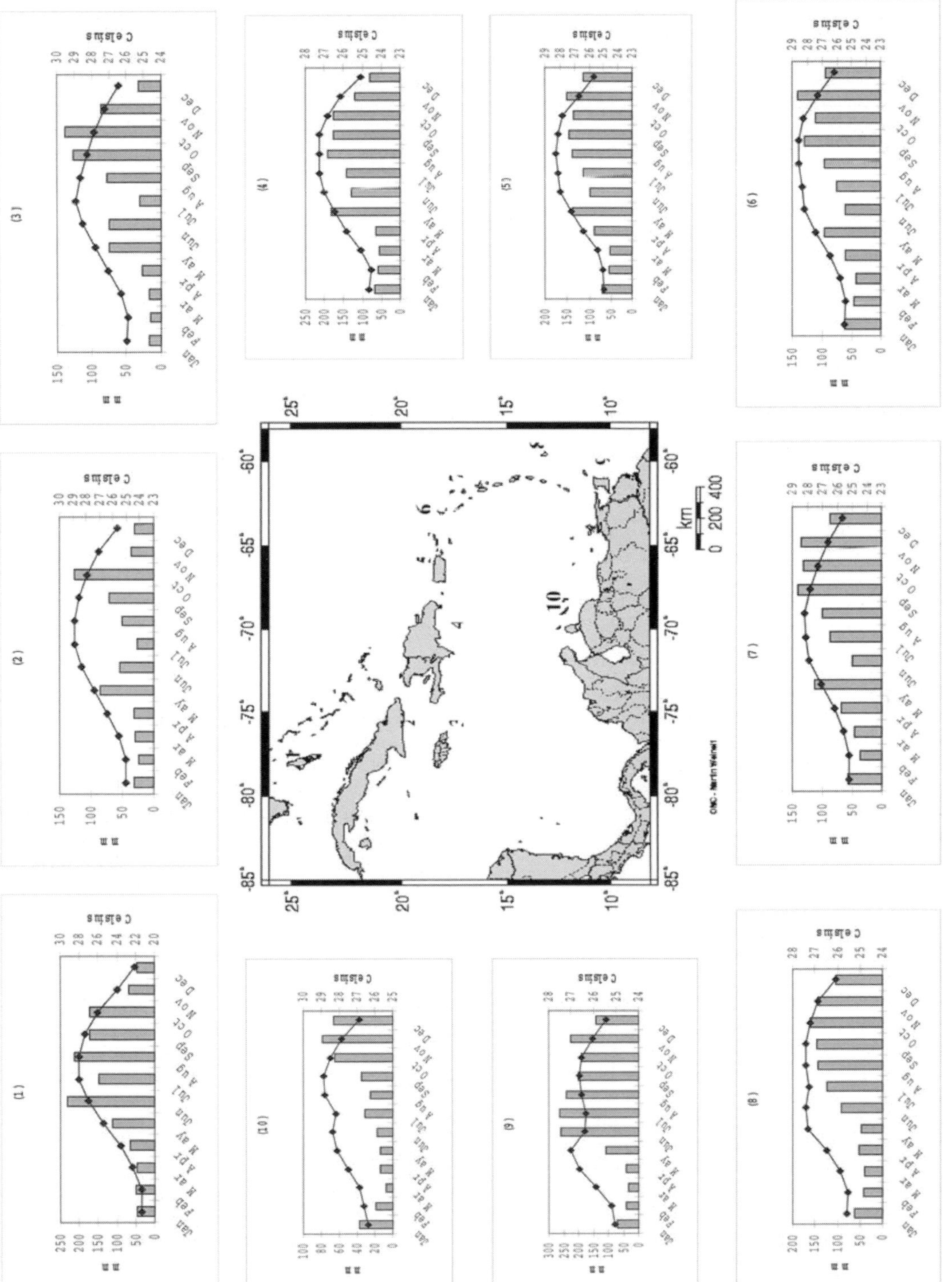

Figure C6 Precipitation and temperature climatology for 10 selected stations in the Caribbean. See Table C1 for station description. Bars denote precipitation in mm. The line graph denotes variation in temperature in °C.

Table C1 Information for the 10 Caribbean stations used to generate Figure C6. Values in column 1 correspond with a specific subplot in Figure C6. Values in column 6 indicate available monthly data used to calculate the climatology. Asterisked years indicate years with 6 months or more of missing data. Data source for stations 1, 2, 3, 4, 5, 6, 8 and 9 is the Global Historical Climate Network (GHCN). Data source for stations 7 and 10 is the respective meteorological services

Figure No.	Station name	WMO identifier	Coordinates	Elevation	Data range
1	Nassau International Airport (Bahamas)	78073	25.05N, 77.47E	7 m	1960–2000
2	Guantanamo Bay (Cuba)	78367	19.90N, 75.13E	23 m	1960–2000 *1992–1996
3	Kingston, Norman Manley International Airport (Jamaica)	78397	7.90N, 76.80E	14 m	1960–2000
4	Santo Domingo (Dominican Republic)	78486	18.43N, 69.88E	14 m	1960–2000 *91–95
5	San Juan International Airport (Puerto Rico)	78526	18.43N, 66.00E	19 m	1960–2000
6	Juliana Airport (St. Maarten)	78866	18.05N, 63.12E	9 m	1960–2000
7	V.C. Bird International Airport (Antigua)	78862	17.12N, 61.78E	8 m	1960–1995
8	Grantley Adams International Airport (Barbados)	78954	13.07N, 59.48E	56 m	1960–2000 *89–91, 93
9	Piarco International Airport (Trinidad and Tobago)	78970	10.62N, 61.35E	15 m	1960–2000 *81–83, 85–86, 95
10	Queen Beatrix International Airport (Aruba)	78982	12.50N, 70.02E	18 m	1971–2000

Oscillation (ENSO) seem to play a role in determining the number of storms which will develop and pass through the Caribbean. During the warm phase there is an apparent decrease in the frequency of tropical storms due to an increase in wind shear over the Caribbean during the hurricane season (Banichevich and Lizano, 1998). There is also evidence of decadal variation in storm activity with some decades on average being less active (1970s to 1990s) than others (1920s to 1960s) (Goldenberg et al., 2001).

The hurricane season runs from June to November with a peak in activity in September. The peak occurs prior to the rainfall maximum of the Caribbean, suggesting that although hurricanes are significant they are not the primary rainfall mechanism for the Caribbean territories, as is often thought, mainly because their effects over specific locations depend on the cyclone's relative position and their velocity over the Caribbean Sea. The coincidence of hurricane peak activity and maximum tropical wave activity is, however, not surprising given the importance of the waves to cyclogenesis. Most (though not all) of the tropical cyclones that ply the Caribbean Sea originate from tropical waves in the easterlies. A warm ocean and low vertical shear fuel their development. Development during the peak period generally occurs in the 10–20°N latitudinal band (termed the main development region or MDR) just east of the Lesser Antilles in the eastern north Atlantic. Early in the hurricane season, however (June), the development region resides in the Caribbean and Gulf of Mexico.

Hurricane development is suppressed in July leading to a bimodal distribution of the annual cycle of north Atlantic hurricanes. This suppression coincides with the previously noted intensification of the trades in July, which yields upwelling in the southwestern Caribbean and lower sea surface temperatures, high vertical shear and reduced convection.

Sometimes (as with Joan in 1988), hurricanes pass over Central America and continue their activity in the eastern tropical Pacific. The eastern tropical Pacific is itself an important region of hurricane development but these affect mainly the Pacific coast of Mexico, and occasionally the northern part of Central America (Guatemala and El Salvador).

Michael A. Taylor and Eric J. Alfaro

Bibliography

Alfaro, E.J., 2000a. Response of air surface temperatures over Central America to oceanic climate variability indices. *Topicos Meteorologicos y Oceanograficos*, **7**(2): 63–72.

Alfaro, E.J., 2000b. Some characteristics of the annual precipitation cycle in Central America and their relationships with its surrounding tropical oceans. *Topicos Meteorologicos y Oceanograficos*, **7**(2): 99–115.

Alvarado, L., 1999. Alteración de la atmósfera libre sobre Costa Rica durante eventos de El Niño. Degree thesis in Meteorology, School of Physics, University of Costa Rica, San José, Costa Rica.

Amador, J., 1998. A climate feature of the tropical Americas: the trade wind easterly jet. *Topicos Meteorologicos y Oceanograficos*, **5**(2): 91–102.

Banichevich, A., and Lizano, O., 1998. Interconexión a nivel ciclónico-atmosférico entre el Caribe y el Pacífico centroamericanos. *Revista de Biologia Tropical*, **46**(Suppl. 5): 9–22.

Chacon, R., and Fernandez, W., 1985. Temporal and spatial rainfall variability in the mountain region of the Reventazon river basin, Costa Rica. *J. Climate*, **5**: 175–188.

Cortez, M., 2000. Variaciones intraestacionales de la actividad convectiva en México y América Central. *Atmósfera*, **13**(2): 95–108.
Fernandez, W., Chacon, R., and Melgarejo, J., 1996. On the rainfall distribution with altitude over Costa Rica. *Revista Geofisica*, **44**: 57–72.
Giannini, A., Kushnir, Y., and Cane, M.A., 2000. Interannual variability of Caribbean rainfall, ENSO and the Atlantic Ocean. *Journal of Climate*, **13**: 297–311.
Gray, W.M., 1984. Atlantic seasonal hurricane frequency: Part I: El Niño and 30-mb quasi-bienniel oscillation influences. *Monthly Weather Review*, **112**: 1649–1668.
Goldenberg, S.B., Landsea, C.W., Mestas-Nuñez, A.M., and Gray, W.M., 2001. The recent increase in Atlantic hurricane activity: causes and implications. *Science*, **293**: 474–479.
Inoue, M., Handoh, I.C., and Bigg, G.R., 2002. Bimodal distribution of tropical cyclogenesis in the Caribbean: characteristics and environmental factors. *Journal of Climate*, **15**: 2897–2905.
Magaña, V., Amador, J.A., and Medna, S., 1999. The midsummer drought over Mexico and Central America. *Journal of Climate*, **12**: 1577–1588.
Martis, A., Van Oldenborgh, G.J., and Burgers, G., 2002. Predictign rainfall in the Dutch Caribbean – more than El Niño? *International Journal of Climatology*, **22**: 1219–1234.
Rudloff, W., 1981. *World-Climates, with Tables of Climatic Data and Practical Suggestions.* Wissenschaftliche Verlagsgesellschaft mbH Stuttgart.
Schwerdtfeger, W. (ed.), 1976. *Climates of Central and South America*, vol. 12: *World Survey of Climatology.* New York: Elsevier.
Taylor, M.A., Enfield D.B., and Anthony Chen, A., 2002. The influence of the tropical Atlantic vs. the tropical Pacific on Caribbean rainfall. *Journal of Geophysical Research*, **107**(C9) 3127, doi:10.1029/2001JC001097.
Wang, C., and Enfield, D., 2001. The tropical Western Hemisphere Warm Pool. *Geophysical Research Letters*, **28**(8): 1635–1638.
Wang, C., and Enfield, D., 2003. A further study of the tropical Western Hemisphere Warm Pool. *Journal of Climate*, **16**: 1476–1493.

Cross-references

Azores (Bermuda) High
Humid Climates
Intertropical Convergence Zone
Orographic Precipitation
Trade Winds and the Trade Wind Inversion
Tropical and Equatorial Climates
Tropical Cyclones

CLIMATE AFFAIRS

Climate affairs is a multidisciplinary concept that encompasses climate science, climate impacts on ecosystems and on societies, climate politics, policy, law and politics, climate economics, and climate ethics. While individual and societal concerns about climate have existed for millennia, the notion of climate affairs is new and is designed to meet the growing educational needs related to climate-related issues.

Climate in its various forms – variability, fluctuations, change, extremes – is appearing at the top of governmental lists of concern about global environmental and technological change. Aside from specific weather and climate phenomena that are of concern in their own right, there is a deep concern about what the adverse effects might be of those phenomena on human activities or on the resources on which societies depend: food production and security, water resources, energy, public health, public safety, economy and environment. Running through each of these categories are ethical and equity issues. One could argue that global change represents a version of a new Cold War; only this time the conflict is between societal activities and natural physical and biological processes, instead of between two nuclear superpowers.

Rich and poor societies alike have increasingly come to realize the extent to which their activities (e.g. industrialization processes and land-use practices) can affect the local and global atmosphere as well as be affected by it. In addition, an increasing number of government, individual, and corporate decisions are being made for which a knowledge of climate affairs is required. As a result, there is a growing awareness among educators in many disciplines of the need for a better understanding of just how climate variability, change and extremes affect the environment and the socioeconomic affairs of people, cultures and nations.

Climate issues have become increasingly important to governments, corporations, and the public. One could surmise that this sharp increase in interest in climate issues and weather has been a result of the end of the Cold War in the early 1990s. At the same time various storms and anomalies have been labeled as the "storm of the century", the worst hurricane, the most costly drought, the longest El Niño, and so forth, in history. Some observers have linked "blockbuster" weather and climate episodes to human-induced global warming of the atmosphere, while others have argued that these are random extreme events under normal global climate conditions.

Since the late 1980s, scientific and government interest in human-induced influences on the naturally occurring greenhouse effect has increased steadily and to new heights, leading up to serious international negotiations to control the emissions of greenhouse gases (carbon dioxide, nitrous oxides, CFCs, methane). The perception about the global climate regime is that climate anomalies have become more frequent, more costly and more deadly. The negative impacts of a changing climate on food, water, energy, health and public safety of a human-induced global warming loom large in the 21st century.

For our purposes, climate encompasses *variability* from season to season and from year to year, *fluctuations* on the order of decades, *change* on the order of centuries and beyond, and *extreme meteorological events*.

Climate variability

Although climate varies on several time scales from seasons to millennia, most people are most concerned about climate variability within a season, from one season to the next, and from one year to the next. These are the time scales that are perceived by people and societies to directly, as well as indirectly, affect them and resources such as food, water, energy, ecological and human health. People everywhere are aware of growing seasons, hunger seasons, rainy seasons, fire seasons, hurricane seasons, planting seasons, harvest time, and so on. Forecasters are now focused on producing reliable seasonal forecasts well in advance of the season, as these forecasts are most useful to societies and individuals, affording them time to take strategic evasive action. When the natural flow of the seasons is disrupted, social and economic problems arise for individuals as well as governments. Year-to-year climate variability is also of concern to society. The El Niño–Southern Oscillation (ENSO) cycle of warm–neutral–cold episodes operates on an annual basis with El Niño events recurring on the order of 2–10 years, often interspersed with the cold episodes.

There is a major difference in societal and environmental impacts in the way annual climate is sequenced. For example, if a dry year is followed by a wet year, societal (and individual) responses will differ from a dry year followed by a second and a third dry year. The same applies to wet years. As an example, in an arid area, societal coping strategies are designed to cope with one or two dry years but not with a third.

Climate fluctuations

Fluctuations refers to variability on the order of decades. For example, the frequency of hurricanes in the tropical Atlantic has changed over periods of decades: 1930–1960 was a very active period, whereas 1960–1995 was relatively inactive. Some researchers suggest that we are about to enter a multi-decadal period with an increase in the frequency of tropical storms on the order of the earlier (1930–1960) active hurricane period.

As another example, El Niño events were less frequent than La Niña between 1950 and 1975, but have become more frequent in the last quarter of the twentieth century. While this time scale is of interest to certain specialists dealing with long-range issues, such as engineers who deal with the planning of water resources and structures that have to last for many decades, they are apparently of much less concern to individuals and policy makers. The decadal scale is one about which societies should be more concerned and knowledgeable than they are at present.

Climate change (new global climate state)

Climate change is a euphemism for global warming. Scientific reports strongly suggest that human activities such as the burning of fossil fuels, tropical deforestation and the emission of other greenhouse gases are responsible for enhancing the naturally occurring greenhouse effect.

Political leaders in many countries have become directly involved in and concerned about the global warming issue. It relates to international energy politics as well as domestic energy debates. Governments equate the use of energy with economic progress. The popular source of energy derives from fossil fuel use (coal, oil and natural gas). These gases when emitted alter the radiative balance of the atmosphere, trapping outgoing longwave radiation in the Earth's atmosphere, which leads to global warming.

Furthermore, in order to prepare society to cope with the projected global and regional changes in climate, societies must evaluate how well they cope with variability and extremes today.

Extreme meteorological events (EMEs)

Societies everywhere worry about particular extreme weather, climate and climate-related events. Although each society faces its own subset of such extremes – droughts, floods, cyclones, heat waves, vector-borne infectious disease outbreaks, etc. – it does not have to face the same set of extremes as other societies with different climate regimes. While Floridians are concerned about hurricanes and fires, Californians are concerned about El Niño-related coastal storms and torrential rains in the southern part of the state. States in the Great Plains and the Gulf states are concerned about tornadoes and severe weather to a much greater extent than states in other parts of the US. There are, however, many similarities in the way that societies prepare for or cope with extreme events.

Researchers concerned about climate variability and those concerned about climate change are interested in EMEs, but for different reasons: the former community wants to forecast their onset and prepare society for their impacts, whereas the latter community of researchers wants to identify changes in frequency, intensity, duration or location of such extremes.

In order for a climate affairs activity (education, training, research) to be complete it must incorporate information about climate science, climate impacts, climate politics, policy and law, climate economics and climate ethics and equity.

Climate science

The objective of a climate science section of climate affairs is threefold:

1. To understand the climate system.
2. To understand its components.
3. To recognize society as a component.

It is important that climate affairs students and professionals understand the workings of the climate system. That system has numerous components. The shells of mollusks, for example, retain carbon for millennia. Termites produce methane, as do animals in feedlots, and so forth. The system's major components are sea ice, forests, oceans, deserts, clouds, the sun, topography, rangelands, and wetlands. Human activities have become a notable forcing factor as far as atmospheric processes are concerned and are now an integral part of the climate system along with its physical components. Social scientists are increasingly recognizing the importance of understanding the physical conditions under which societies they study must operate.

Climate impacts

Climate variability, change and extremes have positive and negative impacts on environment and society. Those impacts can vary over time even in the same location, depending on changes in the vulnerability of ecosystems and societies. The level of vulnerability or resilience of a society or an ecosystem is not just dependent on the intensity of a weather or climate anomaly.

Attributing a specific impact to a climate anomaly is not easy, as there are often several factors that have to be taken into consideration in doling out "blame". Often, the way attribution is done by the public is as follows: a major anomaly – drought, flood, frost, fire, severe storm, cyclone – occurs and most adverse impacts that happen to take place are identified; almost immediately they are attributed to it. However, a more considered assessment is necessary, so that appropriate safeguards can be developed to enable a society to cope with similar events in the future. Wrongly attributing climate-related impacts to a specific cause can prove to be wasteful of scarce resources. It can also lead to false expectations about the level of protection that a society might have from climate-related anomalies.

Climate impacts on terrestrial and marine ecosystems

For the most part the media concentrate their attention and coverage on visible climate and climate-related impacts on land-based ecosystems, both managed (irrigated, dryland, rangeland) and unmanaged (forested and other wilderness areas). Droughts and floods capture the most attention. More correctly, premature concern during dry spells about drought often fill the airwaves and the printed media, only to have that concern evaporate as a timely rain saves a crop. For example, in the midst of a devastating El Niño-related drought in 1997 in Australia, timely rains occurred over a few weeks and prevented a devastating agricultural production year.

Climate impacts on marine ecosystems receive much less attention for a variety of reasons: fishing activities take place where most people do not live; the fishing sectors in most countries involve a relatively small number of workers; it is an economic problem for the most part, rather than one of life and death; the relationship between the viability of living marine resources and variability in air–sea interactions in different parts of the world's oceans are much less studied and not well understood. There are notable examples, however, of climate–marine life interactions: the Peruvian anchoveta, Ecuadorian shrimp, Pacific Northwest Salmon, Pacific Sardine, among others.

The impacts of climate anomalies in industrialized countries are economic issues, for the most part, whereas in developing countries the impacts of anomalies on societies tend to create life–death situations for relatively large segments of their populations. In other words, rich countries have the economic (and therefore technological) wherewithal to minimize or mitigate climate impacts on their populations.

The public now knows that human activities can affect atmospheric chemistry and atmospheric processes at all levels from local (i.e. urban heat island effect) to regional (i.e. tropical deforestation, desertification, and SO_2 emissions), to global (i.e. ozone depletion as a result of CFC emissions and global warming as a result of greenhouse gas emissions). Research on human activities that can affect the atmosphere continues, as researchers seek to reduce scientific uncertainties surrounding their impacts on the atmosphere. Improved understanding of this interaction could provide a sound, non-controversial basis for societal actions to minimize human influences on the atmosphere at various spatial levels.

Attempts to identify human impacts on the atmosphere and the impact of atmospheric processes on society depend on both qualitative and quantitative methods of assessment. It is likely that there will always be some level of scientific uncertainty that surrounds society–atmosphere interactions and, as new methods are developed, a refining of our understanding of those interactions will take place. However, there may be useful qualitative measures to identify how societies might be affected by climate variability, change and extremes. Even the use of anecdotal information can be instructive about climate impacts in certain societies. Historical accounts of impacts and responses to climate-related problems can be very instructive.

One qualitative approach has been referred to as "forecasting by analogy". While the future impacts of a climate anomaly will not likely be a mirror image of a recent past anomaly, this approach can help decision makers identify the strengths and weaknesses in societal and especially institutional responses. The strengths can be maintained or enhanced, and the weaknesses can be addressed.

Climate politics, policy and law

Climate politics refers to the process to achieve (or not achieve) certain objectives by way of climate and climate-related policies and laws. The outcome of politics is climate policies and laws. There are many climate-related political issues in need of discussion and resolution. Politics related to climate issues occurs at local, national, regional and global levels.

Climate has become a highly charged (politically and economically) aspect of environmental change at local to global levels. Climate as a scientific concern is not only important to comprehend in its own right, but understanding it and its various aspects helps us to cope more effectively with other issues considered to be climate-sensitive (fisheries, health, history, war, economy, policy, disasters, institutional change, etc.).

Today, the international community is negotiating the development of what could be called a "Law of the Atmosphere", through the IPCC (Intergovernmental Panel on Climate Change) process. Coupling together international agreements related to the atmosphere that have been developed over time, a body of laws related to the chemistry of the atmosphere and atmospheric processes is being accumulated: the Vienna Convention, the Montreal Protocol, the Convention on Long-Range Transboundary Air Pollution, and the UN Framework Convention on Climate Change (UNFCCC) and subsequent Kyoto Protocol and the entire set of Conference of Parties (COP) conferences. In the process of developing what will essentially become a Law of the Atmosphere, negotiators continue to seek assistance from experts in the physical, biological, and social sciences. Thus, expertise on climate-related issues has been emerging within educational, governmental and non-governmental institutions around the world.

Actions have been taken since the mid-1970s to reduce and, later, to eliminate ozone-depleting chemicals from industrial use in both the industrialized and developing countries. The Montreal Protocol in 1987 (and later amended several times) has sharply reduced CFC emissions. Since the late 1980s to the present, there have been several international conferences on global warming. The most noteworthy is the UNFCCC and the various Conferences of Parties to develop a protocol for the effective control of greenhouse gas emissions.

Acid rain has also generated international discussions, conferences and agreements on the science and impacts. Transboundary acid rain concerns have led to a dilemma. The SO_2 output of industrialization processes has been found to counteract the warming of the atmosphere regionally caused by the burning of the suite of fossil fuels. Recently, considerable renewed concern has focused on water issues and how those issues are or can be affected by climate variability, fluctuations, change and extremes (e.g. the Water Forum in 2001 in Japan).

Climate economics

What was the cost of that storm, drought, flood, frost or El Niño event? When people think of climate economics, this is most likely what they will think about – cost or benefit of a climate-related impact. To do this properly, however, researchers have to get a better handle (e.g. understanding) on "attribution" or what impacts can appropriately be blamed on or associated with a weather or climate event. A climate anomaly affects a society that is also affected at the same time by other factors, such as poverty, poor land management, inappropriate land use, conflict, trans-boundary air or water disputes, bureaucratic rivalries, domestic political rivalries, and so forth. A researcher has to take great care in teasing out of this situation, what was the contribution of the climate anomaly. Sometimes the anomaly may not have been the most important factor by itself but became important when it combined with some of these other factors.

Financial loss or gain as the result of climate variability, change or extremes, in fact, is only one aspect of climate economics. Economic researchers are also concerned about the following, when it comes to assessing climate–society–environment interactions: risk analysis, discount rates, externalities – present vs. future generation welfare, forecast value, free market vs. government intervention, prevent, mitigate or adapt, climate variability and economic development.

Climate ethics

Of all the aspects of a climate affairs activity, the most neglected has been that of climate ethics and equity. This aspect encompasses a wide range of ethical and equity issues, such as the following, that are directly or indirectly affected by climate variability, change or extremes.

Inter- vs. intra-generational equity

Governments everywhere face the same question in the face of a climate issue: Who to help in the event of climate-related problems, present generations or future ones? For example, should the Brazilian government preserve Brazil's tropical rainforests, for example, for potential discovery of pharmaceuticals or cut them down to clear the land for cultivation and livestock and sell the timber? There is a desire also on the part of many to pass something of value on to future generations. As a result, they want to manage the Earth's resources (soil, vegetation, water, air) with the needs and welfare of future generations in mind.

Environmental justice

It is usually the poor and elderly in society who are most adversely affected by a drought, a flood or a heat wave. They are the ones whose living space has been relegated to precarious regions or areas or conditions: flood plains, steep slopes, arid lands, swampland, short growing seasons, and so forth. A forecast issued on the Internet gives those who have computers an advantage to receive that timely information before those who do have that access to the information highway. The rich in society have other options that the poor do not have available to avoid climate-related harm. Seldom are the upper classes in a society in harm's way to the same extent as the poor.

North–south cleavages and climate change

Today, there is considerable debate over responsibility for the global warming of the Earth's atmosphere. The developing countries argue forcefully and effectively that the industrialized countries were able to develop rapidly as a result of their burning of fossil fuels during *their* industrial revolution. As a result, the rich countries have saturated the atmosphere with radiatively active greenhouse gases. They argue that it is the responsibility of the industrialized countries to take the first steps – unconditionally – to reduce their greenhouse gas emissions. To counter, the industrialized and the coal- and oil-producing countries argue that the developing countries will be supplying the lion's share of greenhouse gases emissions in future decades.

There are several sayings that relate to the ethical aspects of climate, broadly defined: the "polluter pays principle", the "precautionary principle", and "common but differentiated responsibilities". Each of these notions generates concerns about equity among competing countries, companies and ideologies. Thus, climate ethics and equity is an integral part of climate affairs.

In sum, climate affairs is an integrator of disciplinary interests in issues that are directly or indirectly affected by the behavior (as well as perceptions of the behavior) and impacts of climate on environment and on society. It can also serve to stimulate new ways of thinking about traditional issues of disciplinary scientific research; conflicts over climate-related resources; political, economic, and technological development; famine; and political instability.

Michael H. Glantz

Bibliography

Burke, D. et al., 2001. *Under the Weather: Climate, Ecosystems, Infectious Diseases, and Human Health.* Washington, DC: Board on Atmospheric Sciences and Climate. National Research Council.

Burroughs, W. (ed.), 2003. *Climate: Into the 21st Century.* Rome: World Meteorological Organization.

Glantz, M.H., 2003. *Climate Affairs: A Primer.* Washington, DC: Island Press.

IPCC (Intergovernmental Panel on Climate Change), 2002. *Climate Change 2001: Impacts, Adaptation, and Vulnerability.* Contribution of Working Group II to the Third Assessment Report of the IPCC. Cambridge, UK: Cambridge University Press.

Kolstad, C.D., and Toman, M. 2001. *The Economics of Climate Policy.* Washington, DC: Resources for the Future, Inc.

Lamb, H.H., 1982. *Climate, History and the Modern World.* London: Methuen & Co. (2nd edn, 1995).

Larson, E., 1999. *Isaac's Storm: A Man, A Time, and the Deadliest Storm in History.* New York: Crown Publishers.

Spash, C.L., 2002. *Greenhouse Economics: Value and Ethics.* London: Routledge.

McMichael, T. (ed.), 1996. *Climate Change and Human Health.* Geneva: World Health Organization.

Cross-references

Applied Climatology
Climate Change Impacts: Potential Environmental and Societal Consequences
Climate Hazards
Climate Data Centers
Commerce and Climate

CLIMATIC CHANGE AND ANCIENT CIVILIZATIONS

Climatic change during the postglacial period

The Quaternary period is characterized as a glacial age. The coldest period of the last glaciation was about 18 000 years BP (based on uncalibrate ^{14}C estimate), when margins of ice sheets and glaciers began to retreat in northern Europe, North America, and regions of the Eurasian continent. The end of that glacial age brought great changes in the landscape, due not only to the retreat of ice sheets and glaciers but also to the rise in sea level caused by melting ice. The rise in sea level occurred not only in the postglacial period but also between 17 000 and 14 000 years BP and between 10 000 and 7000 years BP. The latter phase was drastic in northern Europe particularly rapid, and north America because of glacio-isostatic rebound. Worldwide rise in sea level from the last glaciation to the present is estimated to be 140 m in total.

The hypsithermal

The last glacial stage ended about 10 000 years BP, after which the climate became warmer and reached a peak period called the

Climatic Optimum or the Hypsithermal, about that lasted about 6000 to 4500 years BP. Lamb (1982a) reconstructed average latitudes of the lowest and highest air-pressure axes at sea level in the European sector of the northern hemisphere. In winter, during the Hypsithermal, the dry belt under the subtropical high was located at around 40–45°N (around 30°N today) and the wet belt under the subpolar low at around 57–58°N (about 50°N today). On the other hand, in summer the dry belt under the subtropical high was located at around 50°N during the Hypsithermal (around 33–35°N today); the wet belt under the subpolar low has not fluctuated greatly since 8000 years BP. Suzuki (2000) compiled the climatic change curves during the last 10 000 years all over the world, which were based on the continuous data analyzed, as shown in Figure C7. Even though the ranges of fluctuations and the local differences are different from place to place, and for the climatic elements, the general tendency mentioned above can be observed. Circulation patterns over Africa of 8000 years BP were reconstructed by Messerli (1980), as shown in Figure C8. The difference between Hypsithermal and present-day circulation patterns is strikingly clear in this illustration. Climatic fluctuations in the arid belt of the "Old World" since the last glacial maximum, taking examples in Africa, were discussed by Flohn and Nicholson (1980), Nicholson and Flohn (1980), Tyson (1986), and Tyson and Lindberg (1992).

In most parts of the world the climate between 7000 years BP or earlier and by 5000 years BP was warmer by 1–3°C than it is today (Harding, 1982). The middle latitudes enjoyed a warm climate and the Intertropical Convergence Zone (ITCZ) shifted northward in the northern hemisphere (Suzuki, 1975, 2000). One result of this shift was that the arid regions from the Sahel to northwestern India, which today lie in the Subtropical High-Pressure Zone (STHP) area, during the Hypsithermal experienced a moister regime south of the STHP. Europe experienced a warm, moist climate, and extensive beech and oak forests covered England and the Scandinavian peninsula. Overall mean temperatures in Europe and North America seem to have been up to 2°C higher than at present (Lamb et al., 1966; Lamb, 1977). In East Asia it was 5–8°C colder 10 000 years BP but 2–3°C warmer 5000–6000 years BP than today (Yoshino and Urushibara-Yoshino, 1978).

In the middle latitudes, glaciers on the mountains almost disappeared during the Hypsithermal. Ice in the Arctic Sea melted away and the ice sheets and glaciers in Greenland and Antarctica shrank in this period. The minimum reconstructed 1000–500-hPa-layer thickness of 5150–5200 gpm was found over Arctic North America (Lamb, 1974). By 6000 years BP thermal gradients had weakened and the area of steep gradient shifted north of its current position. The great cold trough had broadened, weakened, and shifted east, out over the Atlantic in summer. It appears that the trough was located farther east than where it had been over North America in winter at 8500 years BP (Lamb et al., 1966). In short, the circulation became weaker on the whole, more zonally oriented, with its action centers located farther north. The North American cold trough was displaced eastward and the wavelength was increased, associated with a spread of westerlies at surface level.

Post-Hypsithermal

After the Hypsithermal the climate deteriorated once more. Large forests disappeared from England and Scandinavia during 3500–3000 years BP. (These dates "BP" are all uncalibrated.) This tendency was observed not only in Europe but in all parts of the world. Suzuki (2000) concluded that there is evidence of a sharp decrease in air temperature around 3500 years BP in the Mediterranean region, East Asia, Australia, North America and South America. Of course, the decrease in temperature did not occur at the same time all over the world but it can be summarized that in 3500–3000 years BP temperature dropped by 2–3°C from the maximum of the Hypsithermal period. In the Near East this period was characterized by a mild dryness and in northwestern India a sharp dryness. Such severe desertification at the zone around 25–35°N was caused by the northward shift of the subtropical high-pressure zone in this period. The polar frontal zone also shifted poleward and wet conditions prevailed along both zones. There are other indications that the climate became slightly wetter in the tropical zone during this period, which may have been caused by the intensification of tropical westerlies. In short, in the period 3500–3000 yr BP the temperature decrease and the change in wet and dry conditions caused by the shifting of frontal zones and the subtropical high-pressure zone were striking features.

By 2500 years BP the renewed cooling trend of climates in the north had steepened the thermal gradient again, resulting in stronger circulation, longer wavelength, and more penetration of the westerlies over Europe (Lamb et al., 1966). Around 2850 years BP, glaciers in northern Europe and the Alps advanced suddenly and rivers were frozen in many parts of Europe due to the cold climate. East Asia was also cold during this period.

Suzuki (2000) compiled also the climatic change curves during the last 1100 years all over the world, as shown in Figure C9. The fluctuation ranges and phases are rather different region to region. Roughly speaking, however, the fourth to fifth centuries AD were colder and wetter in most regions. The evidence of climatic changes has been clarified recently (Bradley and Jones, 1992).

Ancient civilizations

After the climax stage of the last glaciations, the climate became warmer and reached its peak of the Hypsithermal as mentioned above. Until 12 000 years BP, human inhabitants in the northern part of the Near East had lived in caves and hunted wild game in mountain areas. With the climatic change to warmer conditions they came down to open living sites in the foothills where the ground was more favorable for cultivation (Wright, 1968). A northward shift in the tree line (tundra boundary) was seen in Europe. In accordance with the invasion of woodland-type vegetation, insects, birds, and fish extended their ranges.

During the warmest postglacial time, between 6500 and 5000 years BP, the Sahara was much more humid than today. Animals and humans could roam about and cross what is now the world's greatest desert. The moist condition of the Sahara is supported by the study of the change in water levels of Lake Chad and other lakes (Messerli, 1980). Although the levels fluctuated severely in the beginning of the postglacial time, they reached their highest levels about 8000 years BP, with a second peak about 5000 years BP.

Cooling first occurred in higher latitudes after about 5500 years BP. Especially after 4800 years BP the moist regime began to decline. The drier conditions confined human settlement and animals to oases and river valleys such as the Nile, Mesopotamia, Indus, and Hwang-ho (Yellow River). In this period, summer temperatures were 1–3°C higher than today's but winter temperatures were variable. The drying tendencies from about 5500 to 4800

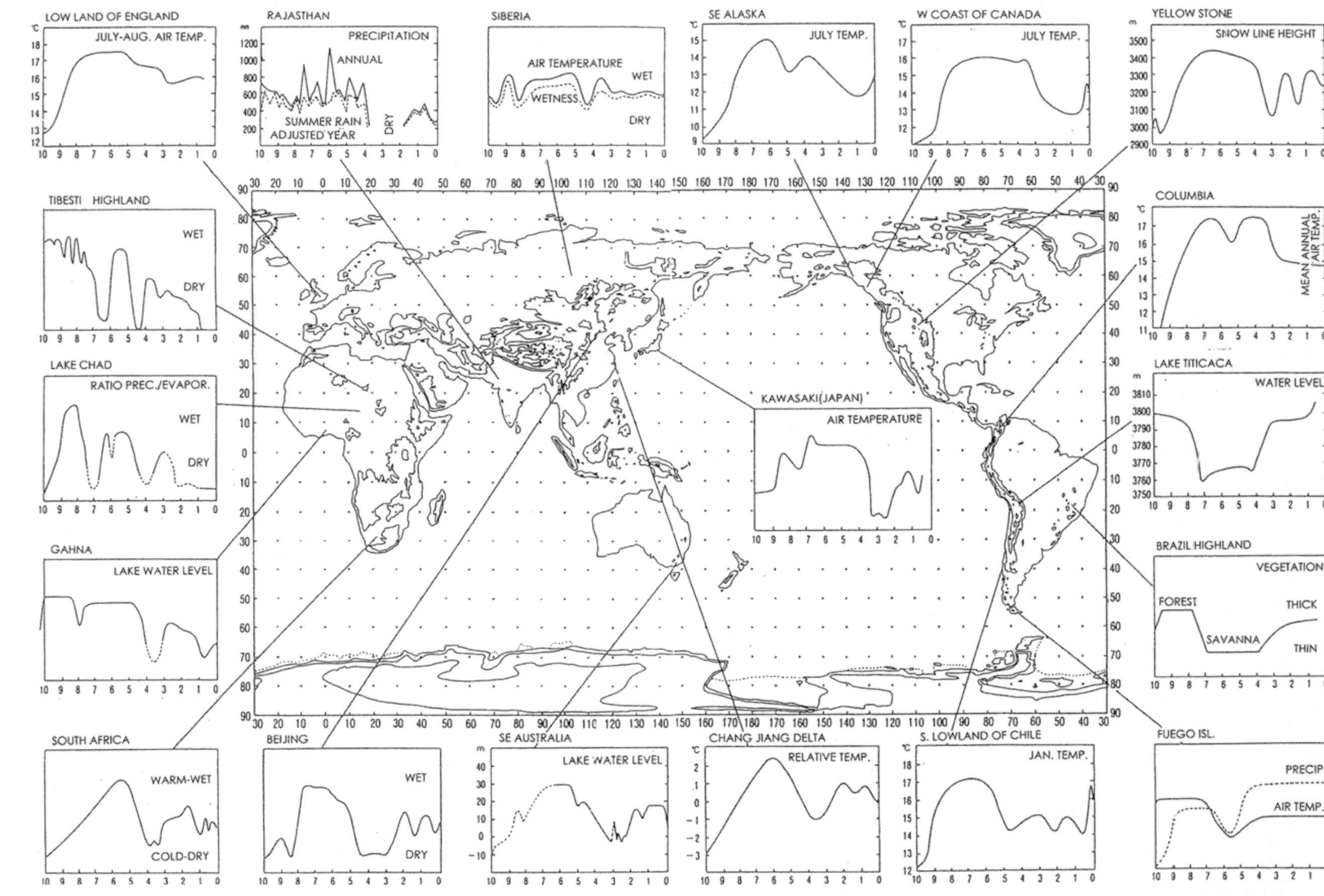

Figure C7 Climatic change curves during the last 10 000 years (after Suzuki, 2000, pp. 8–9).

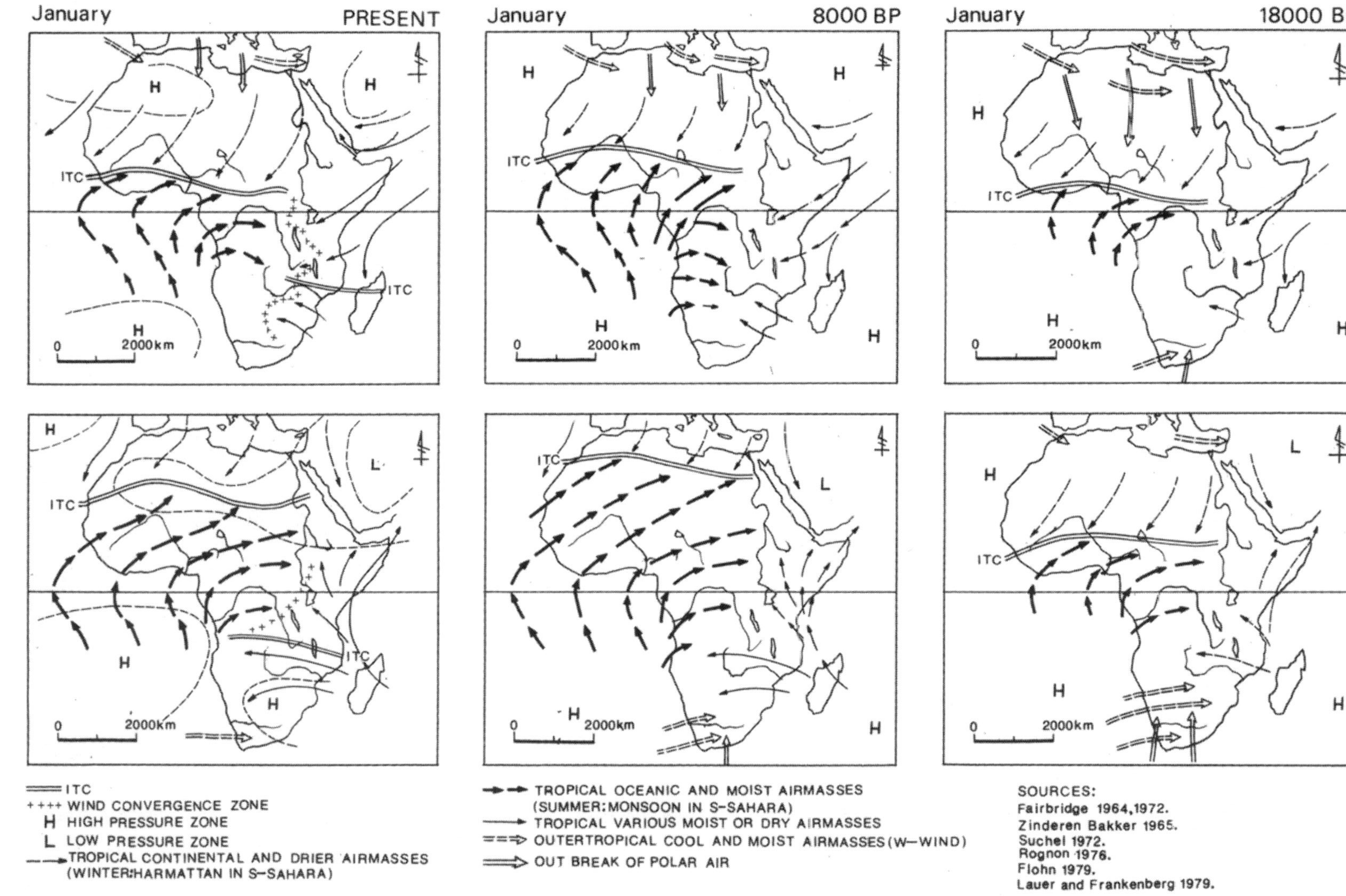

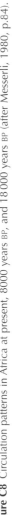

Figure C8 Circulation patterns in Africa at present, 8000 years BP, and 18 000 years BP (after Messerli, 1980, p.84).

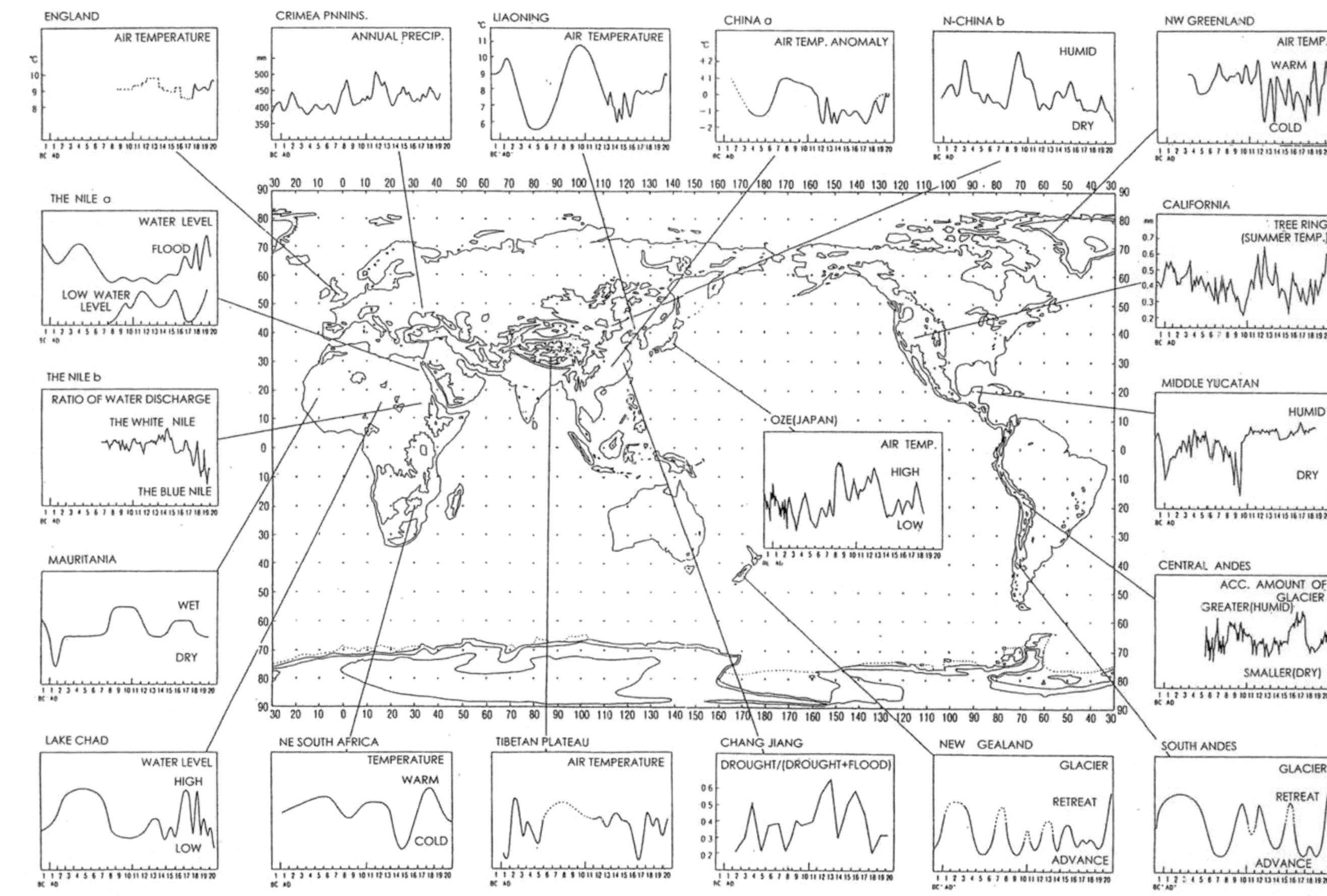

Figure C9 Climatic change curves during the last 2100 years (after Suzuki, 2000, pp. 126–127).

years BP onward seem to be have been related to a climatic development of hemispheric, and probably global, extent (Lamb, 1982a). Butzer (1966) also suggested that the Sahara recorded a moister subpluvial period between about 5500 and 2350 BC (with interruption by one or more dry spells) and a hyperarid oscillation between 2350 and 870 BC. This can also be observed on the curve of ratio for precipitation/evaporation in Lake Chad in Figure C7.

The dry and wet conditions also differed, from region to region during the Hypsithermal, as also shown in Figure C7. The regional change in cultural practices during the postglacial period is closely related to the climatic change, mainly due to the change of hunted animals. Irwin-William and Haynes (1970) reported changes in Paleo-Indian cultural areas in the southwestern United States since 11 500 years BP. After the last glacial period the dryness increased gradually and the Hypsithermal period was the driest in the Paleo-Indian region. Accordingly, their cultural area retreated owing to the decrease in the number of bison. Then, in the period between 5000 and 4500 years BP, a wet condition pervaded this region and a dramatic increase in population took place.

In Egypt, between the first and the second Dynasty (2152–2040 BC), and between the second and the third Dynasty (1552–1069 BC), periods of chaos prevailed. As one of the causes for confusion, Bell (1975) pointed out the anomalous changes of flood levels of the River Nile; which have a close relation to Egypt culture. These periods coincide roughly with the water-level changes of Lakes Faiyum and Moeris (Hassan, 1986), and the sea-level change of the Mediterranean (Fairbridge, 1965) as well as water-level changes of lakes in East Africa (Butzer, 1976) which were of course related to the monsoon rainfall and evaporation changes of the region.

In China, during the Hypsithermal period, the overall average temperature may have been about 2°C warmer and the midwinter temperature about 5°C higher than today. This enabled the cultures of the Yanshao and Yin-Hsu periods to flourish, because there were rich subtropical fauna and flora concentrated around old cities such as Sian (Xi'an) and Anyang (Chu, 1973). It is assumed that, at the beginning of the Chou Dynasty (1066–256 BC), the climate was warm enough to grow bamboo extensively in the Hwang-Ho valley. The cultivation was destroyed, however, and when protracted droughts followed, as well as freezing up of the Yangtze River region.

The first Chinese Neolithic agricultural civilization developed in the northern China plain. The sudden advance to a Bronze-Age culture and the cultivation with irrigation of wheat and millet seem to suggest some contact with the European culture across Central Asia, possibly during a moist period.

In Japan the Hypsithermal corresponded with the early and middle Johmon (Jomon) period (Fukui, 1977). The Johmon culture seems to have been established by the people who came from the west during the warmer Johmon period. After about 4000 years BP the climate tended to deteriorate; it was warm but unstable. In the second half of the latter stage of the Johmon period a cooler climate affected northern Japan: the wetter region retreated to southwestern Japan due to the southward shift of the polar frontal zone during the warm season (Yasuda, 1995). There is evidence that the Johmon culture developed first in the east, then migrated to the southwest. This may be attributed to the shifting of wetter regions in Japan (Yasuda, 1997).

The Yayoi period in Japan began about 2500 years bp, when the climate became cooler. This cooler period coincided with the above-mentioned cool period of the Chou Dynasty in China. There was a warm stage between 800 and 400 bc. Air temperature in the Yayoi period seems to have been 1.0–1.58C lower than today. In northern Japan it was cooler and wetter than today, with its peak in about 2500 years bp (Yasuda, 1978, 1995). Paddy rice cultivation spread from southwestern to northeastern and northern Japan and established the basis of Japanese culture today, even though the climatic conditions were not friendly in this period. It is interesting that even among the different cultures in different periods there are similar phenomena; namely, the coincidence in timing of the rise of a new culture with the decline of the prevailing favorable climate, as in the timing of the rise of the stone age culture in the Old World and its decline during the Hypsithermal.

A question is, why were flourishing civilizations not in the valleys of other great rivers such as the Ganges, the Yangtze, and the Mississippi, as they were in the Nile, Hwang Ho, etc.? At least one reason for this absence can be attributed to the climate. It seems that the drier conditions in the less inhabited valleys during the Hypsithermal did not experience the passage of the ITCZ or polar frontal zones in this period.

Summarizing the relationship between the beginning of ancient civilizations and their climates, we may conclude that the ancient major civilizations began around 5000 years BP – this was the time after the peak of the Hypsithermal. Strictly speaking, civilization began in the valleys along great rivers located in the marginal regions that started to experience cool and dry conditions. The rise of Egypt and the organized cultivation of the Nile valley by the use of early floods for irrigation may have been the necessary response to the increased food demand in a newly habitable region. In other words, civilization came about through the need to organize irrigation systems to produce food for the increased population, while the refugees presumably provided slave labor (Suzuki, 1979; Lamb, 1982a). The disruption of established ways, which the climatic events caused, provided the challenge and stimulus for undertaking deliberate cultivation and invention of new tools.

The decline of ancient civilization took place around 3500 years BP, which was a turning point in postglacial climatic history. It is obvious that the ancient civilizations were not able to continue in the valleys along great rivers because increasingly drier climates caused limited crop production.

The agricultural areas had been greatly reduced by 3500 years BP, and agricultural production was influenced by the decreased rainfall, e.g. several hundred millimeter decreases in rainfall in the Indus valley. Outside the agricultural areas, people had to shift to regions to live in accordance with the retreat and advance of grassland. The climate change was the cause of decline in the ancient civilizations and the migration of peoples.

In Roman times (Rome was founded in 753 BC) the Mediterranean world had a cooler climate with more winter rains (primarily from 600 to 200 BC). This period was one of great fertility for Greece, northern Africa, the Carthaginian, and later Roman croplands. For a few hundred years there was a warming tendency and increasing dry weather until about AD 400.

During the days of early Babylonia and Egypt there were two great inventions. One was the art of fashioning iron into tools with a cutting edge, and the other was the building of seagoing ships. The first invention meant that humans could now live quite comfortably in colder or more humid climates than those of Babylonia or Egypt. The second, shipbuilding, can be considered in relation to forest management. During the greatest days of Greece, the contrast between wet winters and dry summers was less marked than today. Such conditions were favorable for the growth of forests.

Viewpoints

There are several viewpoints on the relationship between climate and civilization (Pittock et al., 1978). In particular, the rise and fall of ancient civilizations seems to have been closely related to the change in climate. The ancient civilizations were built up on the basis of hunting and nomadism, which depended on the distribution of vegetation and animals. The flourishing of the ancient civilizations was supported by agricultural production and a definite form of organized village life.

Huntington (1945) wrote on the role of climate in developing civilizations in the Babylonian and Mesopotamian regions. He pointed out rightly that "the real problem [in developing civilizations] is to determine the exact nature of the [climatic] influence, its magnitude, and the extent to which its favorable and unfavorable aspects have counteracted one another". Because of his general deterministic treatment of the relationship between climate and human activities, including civilization, his descriptions are not accepted by many people today. However, at least as far as his account of the ancient, civilizations is concerned, his description is a sophisticated one. As pointed out by Spate (1952) and Oliver (1973), the historian Toynbee presented many similar ideas, although in a much more literary fashion.

It seems that the second half of the twentieth century was a time to discuss how climatic change may have influenced history (Manley, 1958; Carpenter, 1966; Lamb, 1968; Claiborn, 1970; Le Roy Ladurie, 1971; Singh, 1971; Chu, 1973; Dansgaard et al., 1975; Bryson, 1978; Suzuki, 1979, 2000; Wigley et al., 1981; Lamb, 1982a; Issar, 1998). Ranging from determinism to probablism, and possiblism to voluntarism, there can be various viewpoints on climate–civilization relationships (Oliver, 1973). Civilization is effected by economic determinism, political instability, racial invasion, decreasing activity, disaster by disease, population decreases etc., as well as climatic change (Yoshino et al., 1993). However, at the present stage, it can be concluded that even though the Holocene fluctuations are of relatively small magnitude, they have been sufficient to trigger cultural change in marginal situations. As Bryson (1978) wrote, cultural change includes changes in the economic base, such as agriculture, hunter–gathering, and herding. How humans responded to cultural change, i.e. in situational modification of lifestyle, migration, or literal disappearance of the people, must be determined locally.

Masatoshi Yoshino

Bibliography

Bell, B., 1975. Climate and the history of Egypt. *American Journal of Archaeology*, **79**(3): 224–269.

Bradley, R.S., and Jones, P.D., 1992. *Climate since A.D.1500*. London: Routledge.

Bryson, R.A., 1978. Cultural economic and climatic records. In Pittock, A.B., et al., eds., *Climatic Change and Variability*. Cambridge: Cambridge University Press, pp. 316–327.

Butzer, K.W., 1966. Climatic changes in the arid zones of Africa during early to mid-Holocene times. In *World Climate from 8000 to 0 B.C.* London: Royal Meteorological Society, pp. 72–83.

Butzer, K.W., 1976. *Early Hydraulic Civilization in Egypt*. Chicago: University of Chicago Press.

Carpenter, R., 1966. *Discontinuity in Greek Civilization*. Cambridge: Cambridge University Press.

Chu, Ko-chen, 1973. A preliminary study of the climatic fluctuations during the last 5000 years in China. *Scientifica Sinica*, **16**(2): 226–256.

Claiborn, R., 1970. *Climate, Man, and History*. New York: Norton.

Dansgaard, W., Johnson, S.J., Reeh, N., Gundestrup, N., Clausen, H.B., and Hammer, C., 1975. Climatic changes, Norseman and modern man. *Nature*, **255**: 24–28.

Fairbridge, R.W., 1965. Elszeitklima in Nordafrika. *Geologische Rundschau*, **54**: 385–398.

Flohn, H., and Fantechi, R., 1982. *The Climate of Europe: Past, Present and Future*, Dordrecht: Reidel.

Flohn, H., and Nicholson, S.E., 1980. Climatic fluctuations in the arid belt of the "Old World" since the last glacial maximum: possible causes and future implications. In *Palaeoecology of Africa*. Cape Town: Balkema, vol. 12, pp. 3–21.

Fukui, E., 1977. Climatic fluctuations, past and present. In Fukui, E., ed., *Climate of Japan*. Amsterdam: Elsevier, pp. 261–305.

Harding, A., 1982. *Climatic Change in Later Prehistory*. Edinburgh: Edinburgh University Press.

Hassan, F.A., 1986. Holocene lakes and prehistory settlements of the western Faiyun, Egypt. *Journal of Archaeological Sciences*, **13**: 483–501.

Huntington, E., 1945. *Mainsprings of Civilization*. New York: Wiley.

Irwin-William, C., and Haynes, C.V., 1970. Climatic change and early population dynamics in the southwestern United States. *Quaternary Research*, **1**: 59–71.

Issar, A.S., 1998. Climatic change and history during the Holocene in the eastern Mediterranean region. In Issar, A.S., and Brown, N., eds., *Water, Environment and Society in Times of Climatic Change*. Dordrecht: Reidel, pp. 113–128.

Lamb, H.H., 1968. Climatic changes during the course of early Greek history. *Antiquity*, **42**: 231–233.

Lamb, H.H., 1974. Climates and circulation regimes developed over the Northern Hemisphere during and since the last ice age. In *Physical and Dynamic Climatology*. Geneva: World Meteorological Organization, pp. 233–261.

Lamb, H.H., 1977. *Climate: Present, Past and Future*, vol. 2. London: Methuen.

Lamb, H.H., 1982a. *Climate, History and the Modern World*. London: Methuen.

Lamb, H.H., 1982b. Reconstruction of the course of postglacial climate over the world. In Harding, A., ed., *Climatic Change in Later Prehistory*. Edinburgh: Edinburgh University Press, pp. 11–32.

Lamb, H.H., Leweis, R.P.W., and Woodforrfe, A., 1966. Atmospheric circulation and the main climatic variables between 8000 and 0 B.C.: meteorological evidence. In *World Climate from 8000 to 0 B.C.* London: Royal Meteorological Society, pp. 174–217.

Le Roy Ladurie, E.L., 1971. History and climate. In Burke, P., ed., *Economy and Society in Early Modern Europe: Essay from Annales*. London: Routledge & Kegan Paul, pp. 134–169.

Manley, G., 1958. The revival of climate determinism. *Geographical Review*, **48**: 98–105.

Messerli, B., 1980. Die afridanischen Hochgebridge und die Klimageschichte Afrikas in den letzten 20 000 Jahren. In Oeschger, H., et al., eds., *Das Klima, Analysen und Modelle, Geschichte und Zukunft*. Berlin: Springer-Verlag, pp. 64–90.

Nicholson, S.E., and Flohn, H., 1980. African environmental and climatic changes and the general atmospheric circulation in Late Pleistocene and Holocene. *Climatic Change*, **2**: 313–348.

Oliver, J.E., 1973. *Climate and Man's Environment. An introduction to applied climatology*. New York: Wiley.

Pittock, A.B., Frakes, L.A., Jenssen, D., Peterson, J.A., and Zillman, J.W., 1978. The effect of climatic change and variability on mankind. In Pittock, A.B., et al., eds., *Climatic Change Variability*. Cambridge: Cambridge University Press, pp. 294–297.

Singh, G., 1971. The Indus Valley culture seen in the context postglacial climatic and ecological studies in Northwest India, *Archaeology and Physical. Anthropology of Oceana*, **6**: 177–189.

Spate, O.H.K., 1952. Toynbee and Huntington: a study in determinism. *Geographical Journal*, **118**: 406–428.

Suzuki, H., 1975. World precipitation, present and Hypsithermal. *University of Tokyo Department of Geography Bulletin*, **4**: 1–69 (in Japanese).

Suzuki, H., 1979. 3,500 years ago. *University of Tokyo Department of Geography Bulletin*, **10**: 43–58 (in Japanese).

Suzuki, H., 2000. *Kiko-henka to ningen (Climatic change and man)*. Tokyo: Taimeido (in Japanese).

Tyson, P.D., 1986. *Climatic Change and Variability in Southern Africa*. London: Oxford University Press.

Tyson, P.D., and Lindberg, J.A., 1992. The climate of the last 2000 years in southern Africa. *The Holocene*, **2**(3): 271–278.
Wigley, T.M.L., Ingram, M.J., and Farmer, G., 1981. *Climate and History: studies in past climates and their impact on man*. Cambridge: Cambridge University Press.
Wright, H.E., 1968. Natural Environment of early food production north of Mesopotamia. *Science*, **161**: 334–339.
Yasuda, Y., 1978. Prehistoric environment in Japan. *Tohoku University (Sendai) Science*, 7th Ser., **28**(2): 117–281.
Yasuda, Y., 1995. Climatic changes and development of Jomon culture in Japan. In Ito, S., and Yasuda, Y., eds., *Nature and Humankind in the Age of Environmental Crisis*. Kyoto: Kokusai Nihon Bunka Kenkyu Center.
Yasuda, Y., 1997. Environment of Jomon Civilization (jomon bunmei no kankyo). Tokyo, Kobunkan (in Japanese).
Yoshino, M.M., and Urushibara, K., 1978. Climatic change and fluctuation in South and Southeast Asia. *Climatological Notes*, **21**: 1–48.
Yoshino, M.M., and Urushibara-Yoshino, K., 1993. Monsoon changes and paleoenvironment in Southeast Asia. *Proceedings of the International Symposium on Global Environment (IGBP)*. Waseda University, Tokyo, 25–29 March, 1992, pp. 700–705.

Cross-references

Climatology, History of
Determinism, Climatic
Climatic Variation: Historical
Climatic Variation: Instrumental Record
Desertification
Tree-Ring Analysis

CLIMATE CHANGE AND GLOBAL WARMING[1]

Coal, oil, and natural gas that were formed from the fossilized remains of ancient plants and animals (leading to the term *fossil fuels*) provide 80–85% of the world's energy. This energy contributes to sustaining the world's standard of living and provides much of the power for transportation, generation of electricity, home heating, and food production. Compared to other sources of energy, fossil fuels are relatively inexpensive, transportable, safe, and abundant. At the same time their use contributes to environmental problems such as urban air pollution, acid rain, and elevated levels of mercury (particularly in the Arctic) that arise from trace elements in the fuels, and to long-term climate change that arises because combustion of these fuels releases carbon to the atmosphere that has been geologically sequestered for hundreds of millions of years. The health and environmental problems from the trace element emissions can be reduced by conventional emission control strategies. However, while some carbon can likely be sequestered in the deep ocean or in soils, sharply cutting overall carbon emissions will require substantial reductions in the combustion of fossil fuels for generating energy. The governments of the world began this process by negotiating the UN Framework Convention on Climate Change in 1992 and many of the developed nations are taking the first implementation step by adopting the Kyoto Protocol that was negotiated in 1997.

Concerns about the potential climatic influences of the burning of fossil fuels are not new. During the nineteenth century, initial interest focused on determining where the carbon went when coal was burned; analyses sought to determine if the atmospheric concentration of carbon dioxide (CO_2) would increase or if new carbonaceous fuels were being formed as rapidly as they were being combusted. In 1897 Swedish scientist Svante Arrhenius suggested that combustion of fossil fuels would cause the atmospheric CO_2 concentration to build up, and he calculated that this would lead to an increase in global temperatures because this gas absorbs and re-radiates infrared radiation. Just before World War II, British scientist G. S. Callendar reported observations indicating both a rising atmospheric CO_2 concentration and rising temperatures over much of the northern hemisphere. However, because analyses of ocean chemistry suggested that the oceans could take up the excess CO_2, and because northern hemisphere temperatures seemed to be cooling during the late 1940s and early 1950s, Callendar's results were only slowly accepted. In 1957 American scientists Roger Revelle and Hans Suess recognized that, because of ocean chemistry and the slow mixing of surface and deep ocean waters, the oceans would only be able to slowly moderate the excess CO_2 concentration that was being created by the combustion of fossil fuels. They coined the phrase "great geophysical experiment" to describe what was occurring. To provide the observations needed to confirm their findings, C. David Keeling established a monitoring station atop Mauna Loa in Hawaii to sample the CO_2 concentration in relatively clean and well-mixed air that was expected to represent the average conditions for the northern hemisphere. Over the next several years his observations of the rising CO_2 concentration, along with new calculations with atmospheric radiation models carried out by Syukuro Manabe and his colleagues at the Geophysical Fluid Dynamics Laboratory in Princeton, NJ, led to increasing acceptance of the suggestions of Arrhenius and Callendar. In a 1965 report of the US Government's Council on Environmental Quality, a panel of prominent scientists, led by Roger Revelle, reported to President Lyndon Johnson on the potential for global climate change and consequent impacts.

In the nearly four decades since this first report to government policy makers was submitted, much more information has become available on all aspects of this issue, which has come to be known as human-induced "global warming", even though this term greatly oversimplifies the complexity and diversity of the response of the climate system. Of particular importance have been the international scientific assessments prepared by the Intergovernmental Panel on Climate Change (IPCC), which was established in 1990 under the auspices of the United Nations (see listings of IPCC assessments in the Bibliography; references to the full scientific literature are provided in the technical chapters of these assessments). These assessment reports periodically summarize and evaluate the state of scientific understanding, and the IPCC findings have been unanimously accepted by representatives of the roughly 150 member nations, and endorsed by the leading national academies of science in the world.

Although uncertainties remain about many aspects of the chain of events from emissions to changes in concentration to climatic effects to societal and environmental impacts and options for addressing and responding to the issue, the broad outlines of the global warming situation are quite well established. This article summarizes the state of current understanding about changes in atmospheric composition and climate that are likely to occur as a result of past and future combustion of fossil fuels. An accompanying article provides an overview of the potential impacts of climate change and the challenge of adapting to the

[1] This paper is updated from MacCracken, M. C., 2001, Global warming: a science overview, in *Global Warming and Energy Policy*. New York: Kluwer/Plenum, Kluwer, pp. 151–159.

changes, and indicates the level of effort required and underway to limit long-term climate change.

Changes in atmospheric composition

That the composition of the atmosphere is changing is evident from many types of observations. For example, over the past several decades, dozens of observing stations have been established in pristine areas from the South Pole to Point Barrow, Alaska. Observations at all of these stations indicate that the global background concentrations of a number of gases present in the atmosphere at trace levels are increasing. In particular, the atmospheric concentrations of carbon dioxide (CO_2), methane (CH_4), nitrous oxide (N_2O), and of various halocarbons – including hydrofluorocarbons (HCFC) and, until very recently, chlorofluorocarbons (CFC) – are all increasing. Attention is focused on these gases, which are collectively referred to as "greenhouse" gases, because these gases can exert a warming influence that enhances the Earth's natural greenhouse effect through their absorption and downward re-emission of the infrared (heat) radiation emitted from the surface and lower atmosphere (see further explanation in next section).

Changes in greenhouse gas concentration

Atmospheric concentrations of greenhouse gases have been measured since 1957 and reconstructed for periods prior to when the instrumental record began. For example, by measuring the composition of air trapped in the bubbles in ice cores drilled through the Antarctic and Greenland ice sheets, the history of the CO_2 concentration now reaches back over 400 000 years. These records indicate that, except for a number of the halocarbons that were created in the early twentieth century for use as refrigerants, these gases were present in the natural (or preindustrial) atmosphere, but at much lower concentrations than at present. By looking at such factors as the history of emissions versus concentrations, analyses of carbon isotopes, tree rings, and other scientific measures, it is clear that human activities have caused the increase from the preindustrial level that has persisted since the end of the last glacial.

Taken together, these results indicate that, as of late 2004, the CO_2 concentration of about 380 parts per million by volume (ppmv) is about 35% above its preindustrial level of about 280 ppmv. In addition, the long record indicates that the present value has not been exceeded back through at least the past few glacial cycles (and likely as far back as several million years or more). The analyses also indicate that the increase in the CO_2 concentration since the mid-nineteenth century has been due primarily to the combustion of fossil fuels and secondarily to the release of carbon occurring in the clearing of forested land and the plowing of soils for agriculture (see Figure C10).

Similarly, the CH_4 concentration is up over 150% compared to its preindustrial value. The analyses indicate that its increase is due primarily to human-induced emissions, including particularly from rice agriculture, ruminant livestock, biomass burning, landfills, and fossil fuel development, transmission, and combustion. Analyses of past concentrations and emissions of halocarbons, however, indicate that many of these compounds were not present in the preindustrial atmosphere and are solely a result of human activities.

By comparing emissions and concentrations it is evident that the persistence (or lifetimes) of the excess contributions of these gases in the atmosphere range from decades (for CH_4) to centuries (for CO_2 and some halocarbons) to thousands of years (for some perfluorocarbons). For this reason, even in the absence of future emissions, the excesses of their concentrations above natural levels are expected to persist for many centuries. Because of their long lifetimes, the concentrations of these gases tend to be quite uniform around the world, with local variations occurring indicating strong source regions (and for CO_2, possibly strong seasonal uptake or release by vegetation).

Aerosols

Observations of atmospheric composition also indicate that human activities are contributing to an increase in the atmospheric concentrations of small particles (called aerosols), primarily as a result of emission of sulfur dioxide (SO_2), soot, and various organic compounds. A large fraction of the emissions of these human-induced aerosols results from the combustion of fossil fuels (primarily from coal combustion and diesel and two-stroke engines), with a somewhat smaller fraction resulting from biomass burning. Once in the atmosphere these compounds are transformed and combined in various ways. For example, SO_2 emissions are transformed into sulfate aerosols that create the whitish haze common over and downwind of many industrialized areas. Changes in land cover, especially where this leads to desertification, can also lead to increased lofting of dust particles into the atmosphere.

Of critical importance is that the typical lifetime of aerosols in the atmosphere is less than 10 days. Sulfate and nitrate compounds, for example, are often removed when they first encounter rain systems, causing the acidification of precipitation known popularly as acid rain. Because of their relatively short lifetime in the atmosphere compared to greenhouse gases, the radiative influence of aerosols is most often regional. As a result, large and sustained emissions must occur for aerosols to have a climatic influence that is comparable to the warming influences of the long-lived greenhouse gases.

Because aerosols have direct effects on human health and on visibility, control measures have tended to limit aerosol buildup in most developed nations; however, aerosol concentrations have become quite high in particular regions in many developing nations and in regions where biomass burning is extensive. Recent research is indicating that, in addition to the pollutant effects, the aerosols are causing regional disturbances of the climate. For example, aerosols lofted in southern Asia, primarily from two-stroke engines and inefficient combustion in homes and factories, may be contributing to the diminishment of the monsoon (Lelieveld et al., 2001). In addition, aerosols can have hemispheric effects when they happen to be carried across oceans by the large-scale atmospheric circulation and do not encounter precipitation systems.

Summary

Although natural processes can also affect the atmospheric concentrations of gases and aerosols, observations indicate that natural factors have not been an important cause of significant changes in atmospheric composition over the past 10 000 years. Combined with the evidence of changes in source emissions, this preindustrial stability in atmospheric composition makes it clear that human activities have been the major cause of the dramatic increases in greenhouse gas concentrations since the start of the Industrial Revolution.

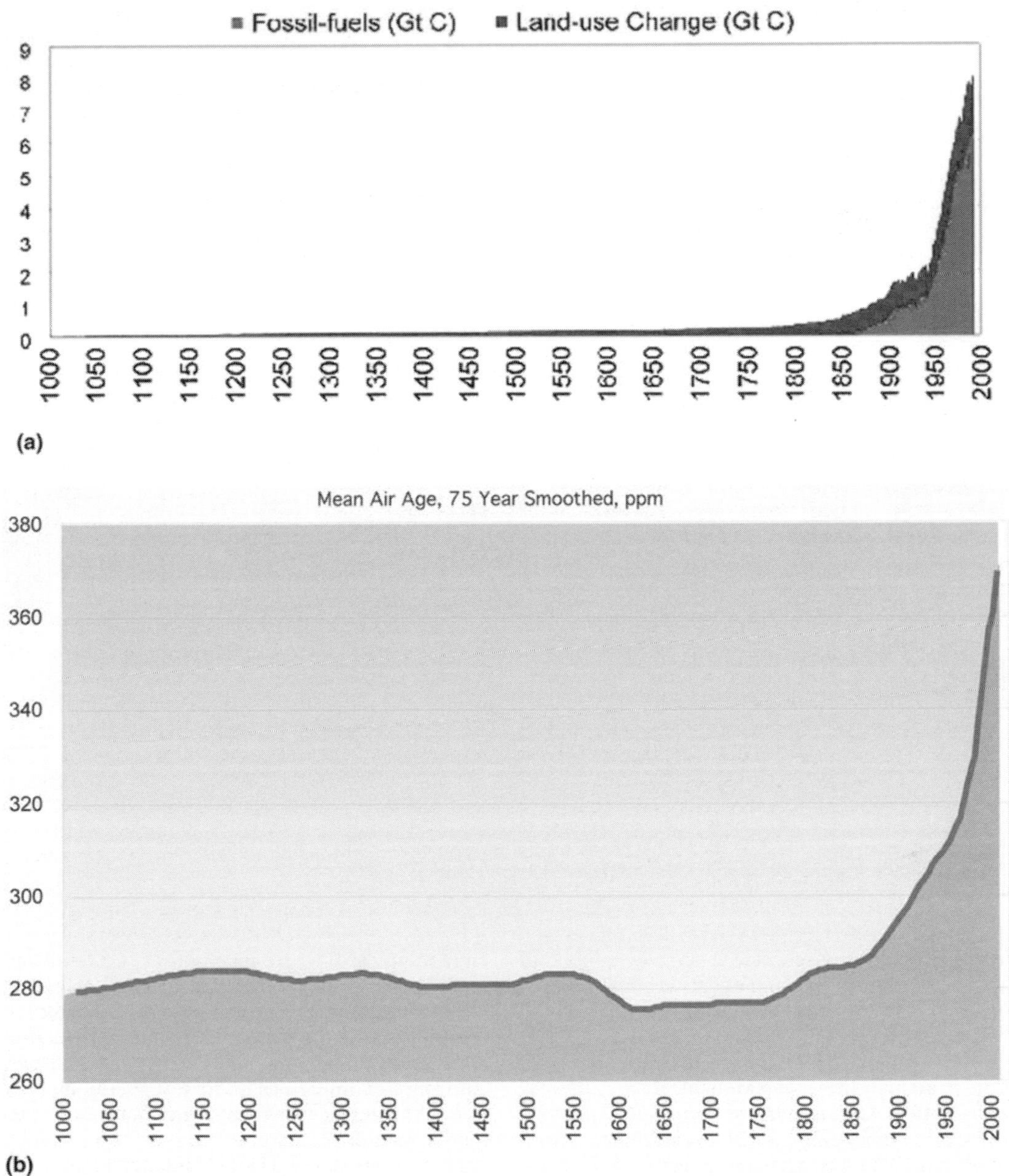

Figure C10 (**a**) Reconstructions of CO_2 emissions for the last 1000 years. Emissions of CO_2 since 1750 are based on compliations for land clearing and for combustion available from the Carbon Dioxide Information and Analysis Center (CDIAC, http//cdiac.esd.ornl.gov).
(**b**) Reconstruction of the CO_2 concentration for the last 1000 years. This record is based on the concentration of CO_2 in ice core bubbles up to the early twentieth century (Etheridge et al., 1998) and instrumental records from Mauna Loa since 1957 (Keeling, 1999).

Intensification of the greenhouse effect

From laboratory experiments, from study of the atmospheres of Mars and Venus, from observations and study of energy fluxes in the atmosphere and from space, and from reconstructions of past climatic changes and their likely causes, it is very clear that the atmospheric concentrations and distributions of radiatively active gases play a very important role in determining the surface temperature of the Earth and other planets. Figure C11 provides a schematic diagram of the energy processes and fluxes that determine the Earth's temperature.

Of the solar radiation reaching the top of the atmosphere, about 30% is reflected back to space by the atmosphere (primarily by clouds) and the surface; about 20% is absorbed in the atmosphere (primarily by water vapor, clouds, and aerosols); and about 50% is absorbed at the surface. For each part of a system to come to a steady-state temperature the energy absorbed in each component of the system must be balanced by the energy lost. For the planet as a whole, the radiation that is emitted away as infrared (or heat) radiation must balance the amount of solar energy absorbed. Were the Earth's

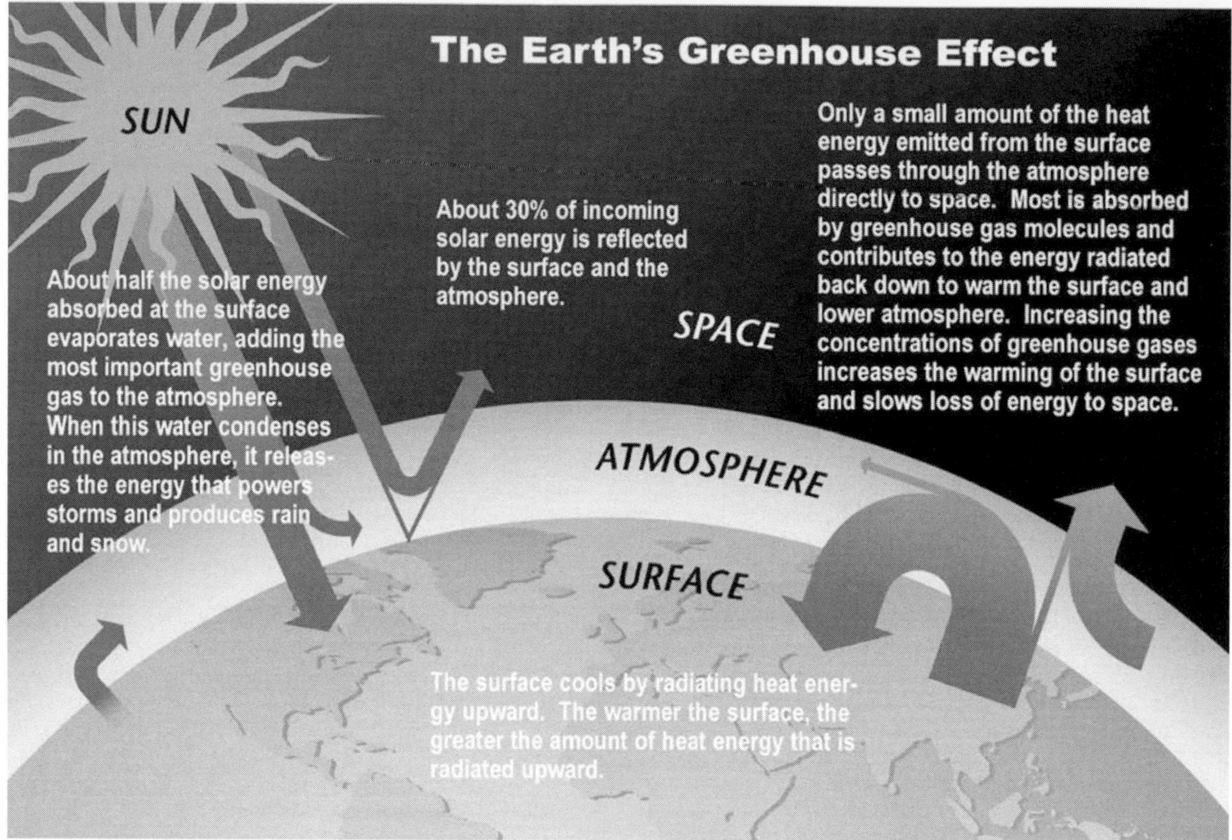

Figure C11 Schematic diagram of the energy fluxes that create the Earth's greenhouse effect. The width of the arrows is proportional to the amount of energy. Total incoming energy from the Sun averages 342 Wm^{-2} (NAST, 2000).

atmosphere transparent, and its surface a simple radiator of energy to space, the Stefan–Boltzmann equation could be used to calculate that the Earth's average surface temperature would equilibrate at close to 0°F (−18°C), given the current reflectivity of the Earth–atmosphere system. Such a temperature would be much too cold to sustain life as we know it.

However, the Earth's atmosphere is not transparent to infrared radiation; instead, a number of gases are able to absorb and emit infrared energy (to be radiatively active, molecules need to have at least three atoms so that various rotational and vibrational bands can be excited by the radiation). As a result of this property of particular atmospheric gases, a warming influence on the climate is created. This warming effect occurs because a large fraction of the infrared radiation emitted by the surface and by the greenhouse gases and low clouds in the atmosphere is absorbed by these radiatively active atmospheric gases before the energy is lost to space. For example, less than 10% of the infrared radiation emitted by the surface is at wavelengths that pass through the atmosphere directly to space without being absorbed. The energy that is absorbed in the atmosphere tends to warm it, in turn causing the radiatively active gases in the atmosphere and the clouds to emit infrared radiation. A significant fraction of this energy is radiated back toward the surface, providing additional energy to warm the surface. This radiation in turn causes the surface to warm, which raises its temperature and causes more radiation to be emitted upward, where much of it is again absorbed, providing more energy to be radiated back to the surface.

This emission–absorption–re-emission process that recycles energy back to the surface is popularly called the greenhouse effect, even though the processes are radiative and so are different from the limitations on loss of energy by convection and to freely flowing air that keep a greenhouse warm and humid. The Earth's natural greenhouse effect raises the average surface temperature from about −18°C (the temperature at which the Earth would radiate away as much energy as it absorbs from the Sun given its current albedo) to the observed value of about 15°C.

Radiative effects of greenhouse gases

The amount of this greenhouse warming is determined by the concentrations of greenhouse gases in the atmosphere. If the concentrations of greenhouse gases are increased, this causes an intensification of the natural greenhouse effect, inducing some warming. An additional warming influence results because the atmospheric temperature decreases with altitude up to the tropopause (about 10–12 km above the surface) before temperatures start to rise again in the stratosphere, which is warmed primarily as a result of solar absorption by ozone (O_3) molecules. When the concentrations of greenhouse gases are increased and the atmosphere becomes more opaque to infrared radiation, the atmosphere's vertical temperature structure (i.e.

its lapse rate) causes the absorption and reemission of infrared radiation to the surface to occur from lower and warmer layers in the atmosphere. Because the emission of infrared energy is proportional to the fourth power of temperature, this has the effect of increasing the downward-emitted radiation, tending to further enhance the natural greenhouse effect. Similarly, because of the resulting changes in the atmosphere's infrared opacity, increasing the concentrations of greenhouse gases tends to cause the emission of infrared radiation outward to space to occur from higher and colder layers; this leads to less radiation being emitted to space for a given average temperature of the lower atmosphere. With less infrared energy being emitted from the surface–atmosphere system to space, the excess of absorbed energy will cause the system to warm in order to re-establish the planetary energy balance.

The most important radiatively active (or greenhouse) gas is water vapor. Not only does water vapor absorb infrared radiation emitted from the Earth's surface and lower atmosphere, but it also absorbs infrared radiation from the Sun. In addition, under appropriate conditions, water vapor can condense and form clouds that absorb and emit infrared radiation as well as absorbing and scattering solar radiation. The amount of water vapor in the atmosphere is largely determined by the large-scale surface temperature of the Earth, with atmospheric circulation and precipitation processes causing its concentration to be highly variable throughout the atmosphere. Although most of the water vapor is found within about 2 km of the surface, the small amounts at higher levels in the troposphere have very strong influences on the fluxes of infrared radiation. Because of this, changes in atmospheric convection that lead to small changes in the amount of upper tropospheric water vapor could have important climatic effects.

Although human activities loft significant amounts of water vapor into the atmosphere, these additions tend to reduce the surface–atmosphere gradient of water vapor pressure. This tends to reduce the natural rate of evaporation of water vapor, thereby moderating the radiative effects of the injected water vapor. In addition, although there is a lot of water vapor in the atmosphere, the average lifetime of a water vapor molecule in the atmosphere is only about 8 days. This short lifetime makes it very difficult for the direct effects of human activities (e.g. by water vapor injection from cooling towers) to sustain a significant increase in water vapor's natural concentration. The only exception is for the case of water vapor injected by airplane exhausts at altitudes where water vapor molecule lifetimes are relatively long. Present levels of injection in this manner exert a relatively small warming influence that is balanced somewhat by the additional reflection of solar radiation from contrails.

Other than water vapor, the primary radiatively active gases are CO_2, CH_4, N_2O, and a number of HCFCs and CFCs. Because the concentrations of these gases are being directly affected by human activities, these greenhouse gases are usually referred to as the anthropogenic greenhouse gases (strictly speaking, their concentrations are being anthropogenically modified). In addition, the tropospheric and stratospheric concentrations of O_3, which is also a greenhouse gas, are being indirectly affected through chemical reactions caused by the emissions of gases such as non-methane hydrocarbons and nitrogen oxides, many of which are also tied to use of fossil fuels for energy.

Changes in the radiation balance

Changes in the concentrations of aerosols can also affect the radiation balance. Sulfate aerosols tend to be quite bright, and can exert a cooling influence on the climate by reflecting away a higher fraction of solar radiation than darker ground and ocean surfaces that typically lie below. Dust lofted by the winds also generally has a cooling influence on the climate. Because they are so dark and absorbing, soot aerosols by themselves tend to enhance absorption of solar radiation, thereby creating a warming influence, especially over very light surfaces. The various types of aerosols can also combine to form mixed-composition aerosols that can exert either warming or cooling influences, creating significant uncertainty in the overall effect of changes in aerosol loading.

Changes in land cover can also alter the energy balance of the Earth system. Although changes in vegetation from one year to the next can seem quite small, over time they can become quite extensive. For example, most of the northeastern US was forested in pre-settlement times, then mostly cleared by the early twentieth century due to the demands for farmland, wood for ships and buildings, and charcoal to make steel. This change tended to brighten the surface, exerting a cooling influence. Over the last 100 years, however, as farming shifted to the Great Plains and the Southwest, the forests have regrown and now cover the region, tending to darken it and exert a warming influence. Some studies indicate that such changes can have not only regional effects on the climate but, by altering the atmosphere's heat balance, these changes can also cause shifts in the atmospheric circulation that result in climatic changes elsewhere. With changes of land cover occurring in many regions around the globe (e.g. deforestation in Amazonia and in southern and western Asia), significant regional perturbations in climate are likely to be already occurring along with the growing increment in the global average temperature due to the rising concentration of greenhouse gases.

In addition to changes in the radiation balance occurring due to human activities, changes can result from natural processes. Most noticeable are the influences of volcanic eruptions, which can loft millions of tonnes of SO_2 and dust into the stratosphere, where it can spread out over much of the globe and persist for up to a few years. Major eruptions, such as Mount Pinatubo in the Philippines in 1991, can reduce incoming solar radiation by of order of 1% and cause cooling of order of 0.5°C for a year or more. Major volcanic activity has tended to be infrequent, and no long-term trend in the effects of volcanoes on the radiation budget has been found, although there have been some relatively active and inactive decades.

Variations in the Sun's output are a second natural source of changes in the Earth's radiation balance. Satellite observations have confirmed indications from the surface that the Sun's output varies with the cycle of sunspots, and other indications suggest that there may be multidecadal to multicentury variations. These variations are estimated to amount to perhaps a few tenths of 1% of the solar flux, and may have caused variations in global average temperature of roughly 0.5°C over the last few thousand years.

Over periods of many thousands of years, other natural forcings can also come into play. For example, the Yugoslavian scientist Milutin Milankovitch was the first to calculate the effect on the Earth's radiation balance resulting from its orbit about the Sun being elliptical rather than circular. Over the course of tens of thousands of years, changes in eccentricity (i.e. departure from circularity), tilt of the Earth's axis, and precession of the seasons (time of year of nearest approach to the Sun) cause a redistribution of the incoming solar radiation in latitude and season. That the timing of these orbital changes and the cycling of the glaciers coincide suggests that these changes have been the drivers of glacial cycling, especially over the last million years. The ice-core records of changes in atmospheric composition suggest that variations in greenhouse gas concentrations are likely one of several additional processes that have contributed to amplifying the climatic response to the orbital changes identified by Milankovitch.

The relative climatic influences of natural forcings and of changes in the atmospheric amounts of gases and aerosols are typically compared by using atmospheric radiation models to calculate their ability to change the fluxes of solar and terrestrial radiation at the tropopause; that is, to alter the amount of energy entering or leaving the surface–troposphere system. The change in these fluxes as a result of changes in the amounts of gases and aerosols is referred to as the "radiative forcing" that they exert. Based on the latest international assessment report of the IPCC (2001), Table C2 summarizes the best estimates of the changes in abundances of various gases and aerosols and the radiative forcings that are estimated to have resulted since

Table C2 Greenhouse gases and aerosols contributing to changes in climate (based on information in IPCC, 2001)

Greenhouse gas or aerosol or other forcing mechanism	Approximate abundance in 1750	Approximate abundance in 2000	Estimated radiative forcing from 1750–2000 ($W m^{-2}$)	Projected abundance in 2100 based on SRES emissions scenarios (IPCC, 2000)	Projected radiative forcing from 1750–2100 ($W m^{-2}$)
Carbon dioxide (CO_2)	278 ppmv	369 ppmv	1.46 (plus or minus 10 per cent for all well-mixed greenhouse gases)	549–970 ppmv (reference model)	3.64–6.69
Methane (CH_4)	700 ppbv	1745 ppbv	0.48	1574–3413 ppbv	0.42–1.09
Nitrous oxide (N_2O)	270 ppbv	314 ppbv	0.15	354–460 ppbv	0.27–0.57
Perfluoromethane (CF_4)	40 pptv	80 pptv	0.003	208–397 pptv	0.013–0.029
CFC-11 ($CFCl_3$)	0	268 pptv	0.07	45 pptv per Montreal Protocol	0.01
CFC-12 (CF_2Cl_2)	0	533 pptv	0.17	222 pptv per Montreal Protocol	0.07
Other anthropogenic halocarbons (generally contain C, H, and Cl, Br, or F)	0	Various	~0.1	Various	~0.06–0.33
Tropospheric ozone (O_3)	Increase of ~8 Dobson units		0.35 (0.2–0.5)	−18% to +5%	0.21–1.27
Stratospheric ozone (O_3)	Depletion by several percent since about 1970		−0.15 (−0.05 to −0.25) (1979–1997)	Recovery from halocarbon injections, but some depletion due to stratospheric cooling	Uncertain, but possibly less negative
Sulfate aerosol (SO_4) – direct effect		0.52 TgS	−0.4 (−0.2 to 0.8)	0.15–0.45 TgS	−0.12 to −0.35
Black carbon (BC)		0.26 Tg	0.2 (0.1–0.4)	0.13–0.68 Tg	0.2–1.05
Organic carbon (OC)		1.52 Tg	−0.1 (−0.03 to 0.3)	0.77–4.00 Tg	−0.88 to −1.32
Biomass burning aerosols			−0.2 (−0.07 to 0.6)		
Aerosol – indirect effects due to change in cloud reflectivity			0 to −2.0		Uncertain
Mineral dust			−0.6 to 0.4		
Contrails due to jet aircraft			0.02 (0.00 to 0.07)		
Aviation-induced cirrus			0 to 0.04		
Change in surface vegetation			−0.20 (0 to −0.4		
Solar irradiance			0.30 (0.1–0.5)		

[a] From 1993 to 2002, on assignment from the Lawrence Livermore National Laboratory to the Office of the US Global Change Research Program, serving as Executive Director of the Office from 1993 to 1997 and as Executive Director of the National Assessment Coordination Office from 1997 to 2001. Presently serving as Chief Scientist for Climate Change Programs at the Climate Institute, Washington DC, and president (2003–07) of the International Association of Meteorology and Atmospheric Sciences.

the beginning of the Industrial Revolution in about 1750. Although there is considerable uncertainty in the estimates of the radiative forcing of the various types of aerosols on the radiation balance, the net effect of all of the changes that have occurred is most likely to have been significantly positive, thereby exerting a warming influence on the global climate. Indeed, observations from space-based instruments since the 1970s clearly indicate that the rising concentrations of the anthropogenic greenhouse gases are indeed tending to enhance the natural greenhouse effect.

In addition to the direct effects of changes in greenhouse gas concentrations and aerosol loading on the radiation balance, there are several types of indirect influences on the radiative forcing of a gas or aerosol. For example, changes in the amounts and types of aerosols can affect the amounts and reflectivities of clouds that in turn alter the fluxes of solar and terrestrial radiation. There can also be indirect influences through various effects on atmospheric chemistry. For example, the chemical decomposition of methane in the stratosphere creates water vapor that has a powerful greenhouse effect, and the release of various halocarbons has depleted the stratospheric ozone layer, affecting its role in the infrared radiation balance. Estimates of these indirect influences are also included in Table C2.

Feedback mechanisms

The intensity of the Earth's natural greenhouse effect is also altered through various feedback mechanisms that can amplify or moderate the system's response to the changes in radiative forcing. For example, the warming influence caused by increases in the concentrations of CO_2, CH_4, and other anthropogenic greenhouse gases can be significantly amplified by a positive water-vapor feedback mechanism. This positive feedback occurs because more water vapor can be present in a warmer atmosphere, and an increase in atmospheric water vapor enhances the greenhouse effect and causes a further warming that in turn allows more water vapor to be present in the atmosphere. At the same time, however, changes in the amount of atmospheric water vapor (more specifically, in conditions affecting water vapor removal when the relative humidity exceeds saturation) and in atmospheric circulation can alter the extent and distribution of clouds. These changes in cloud cover and distribution can in turn affect the extent of the absorption and scattering of solar radiation and the absorption and re-emission of infrared radiation, thereby creating both amplifying and moderating feedbacks.

There are many other feedbacks and interactions. For example, warming tends to reduce the amount of snow and sea ice, which in turn reduces the reflectivity of the surface (i.e. the albedo), allowing increased absorption of solar radiation, and reinforces the original warming. Over the longer term, melting of glaciers and ice sheets can similarly enhance the warming influence. Changes in the climate can also affect the type of vegetation and its extent, affecting not only the surface albedo, but also its roughness, the rate of evapotranspiration from the surface, and the lofting of dust. The ocean circulation can also be affected by the climate, with wind changes affecting surface current and changes in ocean temperature and salinity causing global-scale changes in the thermohaline circulation that ventilates the deep ocean. Over thousands of years and longer, glaciological and isostatic processes can affect the height of the land and even the shapes of coastlines, which in turn can affect atmospheric and oceanic conditions. There are thus a great many processes and feedbacks that can potentially come into play once changes in the climate are initiated.

Climate sensitivity

Estimates of the climate's sensitivity to changes in radiative forcing have been derived from theoretical analysis, from reconstructions of climate changes in the geological past, and from simulations using global climate models. These approaches indicate that, at steady state, each increase in the radiative forcing by $1\,W\,m^{-2}$ seems to be associated with an increase in the global average temperature of about 0.4–1.2°C. This degree of sensitivity is also often expressed in terms of the warming that would result from the change in radiative forcing associated with a doubling of the atmospheric CO_2 concentration. When expressed in this manner, the radiative forcing from the doubling the CO_2 concentration (i.e. almost $4\ Wm^{-2}$) would lead to an increase in the global average temperature of 1.5–4.5°C. It is important to emphasize that these estimates are after a new steady-state climate is established, and that, while much of the readjustment would occur over a few decades, the full response would take many centuries because of the time it takes for the deep ocean temperature and feedback mechanisms involving glacial ice and ecological systems to come to a new equilibrium.

Overall, there is broad scientific agreement that increases in the atmospheric concentrations of the anthropogenic greenhouse gases will enhance the Earth's natural greenhouse effect and tend to raise the global average surface temperature by of order of a few degrees Celsius. The remaining uncertainties relate to pinning down by how much and how rapidly warming would occur.

Detecting and attributing past changes in the climate

With the evidence being clear-cut that the concentrations of greenhouse gases have risen significantly since the start of the Industrial Revolution, and that increasing the concentrations of greenhouse gases will induce a warming influence on the Earth's climate, then a key test of scientific understanding of the climatic influence of these changes is to determine if the time history and magnitude of historic changes in the climate match those expected to be occurring as a result of past changes in radiative forcing. Complications in this analysis arise for several reasons. First, there are shortcomings in the climate record, including a near-absence of observations at times before human-induced forcing began to have an influence on the global climate. As a result, *detecting* whether changes have actually occurred requires care to distinguish actual change from changes arising due to limitations in the data (e.g. due to the changing spatial coverage over time) and from the natural variability of the climate caused mainly by varying atmosphere–ocean interactions. Second, there are multiple influences contributing to changes in the radiative forcing; most are not definitively quantified and it is not even clear if all are yet recognized. As a result, *attributing* particular changes in the climate to particular factors requires very careful forensic analysis.

To have the best chance of both detecting and attributing the human influence, it is most useful to look at the longest records of the changing climatic state. Instrumental records of temperature go back only a few hundred years, and a reasonable network of observation stations was only established around the middle of the nineteenth century. Accounting carefully for changes in measurement technique and the local environment, records of how the monthly and annual average temperatures at each of thousands of locations have changed from their local baseline value can be used to construct a space and time history of such changes. These results can then be used to calculate an area-weighted global average of the change in local near-surface air temperature. This quantity is often referred to simply as the change in the global average temperature, even though this terminology really obscures how its value is determined and unfortunately implies that the global average temperature is actually measured in some direct way by some single instrument.

Detecting changes in climate

Spatially representative estimates of the change in global average temperature go back to about 1860. These records indicate a warming of over 0.6°C, mostly during the early and late twentieth century. Extensive proxy records (e.g. records derived from tree rings, ice cores, coral growth, etc.) have been used to attempt to reconstruct the amount and pattern of changes in surface temperature further back into the past. For the northern hemisphere, proxy records going back about 1000 years indicate that a slow cooling was under way for most of this period. These records also clearly indicate that the twentieth century was much warmer than earlier centuries and that the decade of the 1990s was almost certainly the warmest decade of the last 1000 years. This temporal history is clearly evident in Figure C12, which indicates that a sharp rise in the average temperature began during the late nineteenth century and has continued throughout the twentieth century. It is also clear that the twentieth-century warming has been much more persistent than earlier fluctuations, which were likely caused by the inherent natural variability of the ocean–atmosphere system (i.e. internally created variability) and natural variations in solar radiation and the occasional eruption of volcanoes (i.e. externally caused variability).

The global warming evident at the surface is also consistent with observations of increasing temperatures measured in boreholes (i.e. dry wells), retreating mountain glaciers, diminishing sea ice in the Arctic, increasing concentrations of atmospheric water vapor, rising sea level due to melting of mountain glaciers and thermal expansion in response to recent warming (augmenting the natural rise due to the long-term melting of parts of Antarctica), and related changes in other variables. As a whole, most records indicate that long-term changes in the climate have been occurring, even though some analyses of some shorter records indicate very little change (e.g. one analysis of changes in tropospheric radiance measured using microwave instruments on a succession of satellites indicates virtually no increase in the implied tropospheric average temperature since 1979, although another analysis and balloon observations of changes in tropopause height do indicate that tropospheric warming has been occurring).

Attributing climate change to human activities

A key question is whether these changes are mainly a natural fluctuation or whether human activities are playing a significant role. The twentieth-century warming is largely being attributed to

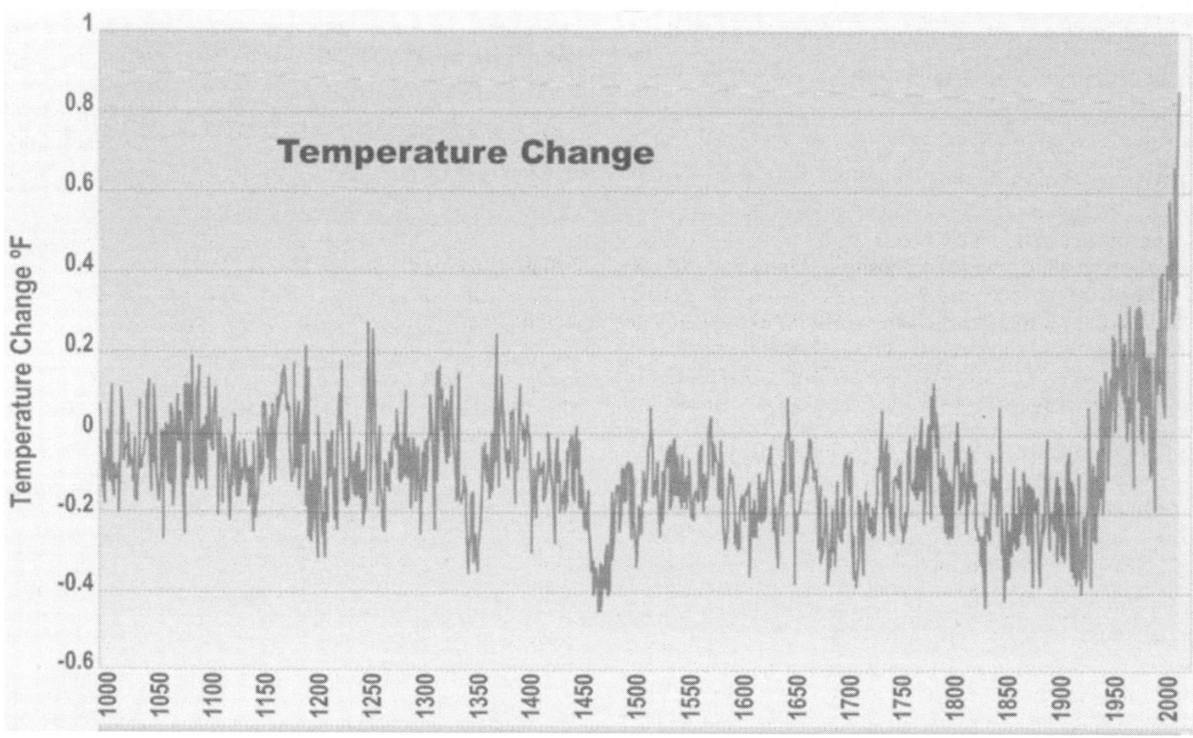

Figure C12 Reconstruction of the northern hemisphere average surface temperature for the last 1000 years. Up to the mid-nineteenth century the record is based on proxy records (e.g. tree rings analyzed by Mann et al., 1999). The estimated uncertainty for an individual year is several tenths of a degree. Since the mid-nineteenth century the record is based on instrumental data, for which the comparable uncertainty is about 0.2°F.

human activities because of the coincidence in timing with the warming influences exerted by the changes in greenhouse gas concentrations, the very large and unusual magnitude of the changes compared to past natural fluctuations, the warming of the lower atmosphere and cooling of the upper atmosphere (a sign of a change due to changes in greenhouse gas concentrations rather than changes in solar radiation), and the global pattern of warming (with larger changes in high than low latitudes). Some uncertainty is introduced because some of the warming occurred before the sharpest rise in greenhouse gas concentrations in the latter half of the twentieth century. For example, some analyses indicate that as much as 20–40% of the warming may be due to a coincidental increase in solar radiation, although other factors, such as changes in land cover or in soot emissions, may also have had an influence. In addition, some uncertainty has been introduced because the rise in tropospheric temperatures over the past two decades may have been a bit slower than the rise in surface temperature. Whether this difference is real or arises from, for example, calibration issues with the satellite instrumentation, natural variations in Earth-surface temperatures, or the confounding influences of ozone depletion, volcanic eruptions, and atmosphere–ocean interactions, is not yet clear.

Taking all of the scientific results into consideration, the IPCC concluded in its Second Assessment Report (IPCC, 1996) that "The balance of evidence suggests a discernible human influence on the global climate". This conclusion, in essence, is equivalent to the criterion for a civil rather than a criminal conviction in a court of law. The IPCC's Third Assessment Report (IPCC, 2001) documented even more clearly that the magnitude and timing of the warming during the twentieth century quite closely matches the changes that would be expected from the combined influences of human and known natural influences, concluding that "there is new and stronger evidence that most of the warming over the past 50 years is attributable to human activities".

Projections of future changes in emissions and atmospheric composition

The continued reliance on fossil fuels for generation of energy and the release of other greenhouse gases are sure to cause further changes in atmospheric composition. Along with the continuing adjustment of the climate system to past changes in atmospheric composition, these changes will far exceed changes in other contributions to radiative forcing, making them the driving influences for future climate change. While it is not possible to provide definitive predictions of how the world will evolve out for decades to centuries into the future, the IPCC (2000) reviewed projections of various trends for the future and, using a number of storylines about how population, energy use and energy technologies, economic development, and international cooperation might evolve, developed a set of plausible scenarios of future emissions of greenhouse gases and some types of aerosols and aerosol precursors. These scenarios indicate, for example, that annual emissions of CO_2 in the year 2100 could range from roughly 5 to over 30 GtC/yr (billions of tonnes of carbon per year) compared to present emissions of about 6 GtC/yr.

A rough indication of what the emissions might be can be derived by simply extending trends from current values. With the global population of 6 billion people and current use of fossil fuels, each person on Earth is presently responsible, on average, for emission of about 1 tC/yr (tonne of carbon per year). Per-capita use, however, varies widely across the world, reaching nearly 6 tC/yr in North America and about 3 tC/yr in Western Europe, but amounting to only about 0.5–1.0 tC/yr in developing countries such as China and India.

Central projections for the year 2100 foresee a global population of 8–10 billion. With the rising standard of living and the energy required to sustain it, and with continued reliance of fossil fuels for most of the world's energy, global average per-capita emissions of CO_2 seem likely to at least double, with most growth resulting from increased emissions in the highly populated, but currently underdeveloped, emerging economies. If this occurs, total annual emissions could more than triple, reaching roughly 20 GtC/yr or more. IPCC (2000) also considered the potential for variations of various kinds around an estimate such as this, with the possibility that intensified use of coal, which is among the least expensive and most available sources of energy in the developing world, could cause emissions to be even greater, whereas a drop in the rate of population growth and a greatly accelerated switch to non-carbon-based fuels could return emissions in 2100 to about current levels after higher levels through most of the twenty-first century.

If global emissions of CO_2 do gradually increase to about 20 GtC/yr by 2100, models of the Earth's carbon cycle that are verified against observations of past increases in concentration, project that the atmospheric CO_2 concentration would rise to just over 700 ppmv. This would be almost double its present concentration, and over 250% above its preindustrial value.

For the broader set of emission scenarios that are based on consideration of ranges in global population, energy technologies, economic development, and other factors, projections indicate that the atmospheric CO_2 concentration is likely to reach at least 500 ppmv and may even exceed over 900 ppmv in 2100. The projections also indicate that the CO_2 concentration would continue to increase during the twenty-second century unless global emissions have been reduced to 1–2 GtC/yr by 2100.

Depending on various control measures, the rise in the CO_2 concentration would be expected to be accompanied by increases in the concentrations of CH_4 and N_2O, as well as of sulfate and soot aerosols, although it would take an unrealistically large (and unhealthy) amount of aerosols to affect radiative forcing to an extent that would counterbalance the radiative forcing created by the increases in the concentrations of the anthropogenic greenhouse gases that are projected. Based on IPCC scenarios and analyses (IPCC, 2000, 2001), Table C2 provides estimates of the future abundances of the various gases and aerosols as well as an estimate of the total anthropogenic contribution to radiative forcing expected by the year 2100.

Because the historical temperature record includes natural fluctuations, and because the relative contributions to radiative forcing of the various greenhouse gases and aerosols are changing, extrapolation of historic trends in temperature into the future is not a particularly reliable means for projecting future trends. Similarly, although the geological record provides evidence that both the atmospheric composition and climate have changed in the past, only a rough indication can be provided of how the climate will change over the twenty-first century because the change in atmospheric composition is occurring so rapidly. Because of these limitations, and in order to get a more complete sense of how the many climatic variables would be expected to change, projections of the climatic effects of the various scenarios of changes in greenhouse gas and aerosol emissions must necessarily be based on the use of computer-based climate models.

Global climate models

Global climate models (also known as Earth system models) are assembled by coupling general circulation models of the atmosphere and oceans to land surface and cryospheric models. Each of these component models is constructed by applying the fundamental conservation laws of physics (i.e. for mass, momentum, and energy) and chemistry (i.e. for water and various other species, including chemical transformations) to many thousands of grid cells that are used to cover (or tile) the Earth. Although atmospheric and oceanic models with resolutions of order 10 km in horizontal resolution are starting to become available, developing reliable projections of climatic change requires running multiple simulations for hundreds to thousands of years. As a result of limits imposed by present computer speed and resources, a significantly coarser resolution (e.g. 100–200 km) is being used for most model simulations of climate change. Because this resolution is not adequate to represent the details of, for example, convective cloud development and wind drag caused by variable terrain, the effects of these processes are represented in the models in terms of larger-scale variables using parameterizations that are typically derived from the results of large-scale observations or extensive field investigations of how atmospheric or oceanic processes work. In that these parameterizations are only approximate, and because it is not possible to ensure that all processes have been represented, it is essential to test the models and to carefully evaluate their results.

To develop a sense of confidence in the models a wide range of tests is often performed. The most intensive efforts have been conducted under the auspices of the Program for Climate Model Diagnosis and Intercomparison, which has programs to compare the results of the various model components with observations and with the performance of other models (e.g. Gates et al., 1999). Comparisons of model simulations to the observed behavior of the climate typically cover recent decades, recent centuries, and even geological periods in the past. In general, the models do reasonably well at representing the latitudinal and seasonal distributions of the climate, but are still somewhat limited in their representations of climatic variability on scales from the El Niño/Southern Oscillation to the North Atlantic Oscillation. As indicated above, when driven by both natural and human-induced forcings, the models reasonably represent the changes in global average temperature observed over the twentieth century (see IPCC, 2001, chapter 8).

Projections of changes in climate and sea level

To generate projections of changes in the future climate, the models are typically initiated at conditions that have been reconstructed for the mid-nineteenth century before human activities were contributing substantially to radiative forcing. A control simulation with no change in forcing is then run to establish an estimate of the baseline climate in the absence of human activities. These results are then compared to the results of perturbation simulations that are driven by the scenarios of future emissions and/or changes in atmospheric composition, depending on the model and the type of forcing.

Based on the model simulations, and supported by theoretical analyses and extrapolations of recent trends, the set of IPCC emissions scenarios (IPCC, 2000) for the twenty-first century is projected to cause the global average temperature to increase by about 1.4–5.8°C over the period 1990–2100 (IPCC, 2001). A warming of this amount would be roughly 2–9 times as large as the warming during the twentieth century. Such a warming would mean that human activities would have increased global average temperature over its preindustrial average by from about 2°C to over 6°C. Model simulations are increasingly providing consistent indications of the warming to be expected in various regions of the Earth, indicating larger changes occurring in higher rather than lower latitudes and over land rather than the ocean (IPCC, 2001, chapter 10). Increases in the average temperatures projected for the end of the twenty-first century, even were they to be at the lower bound, would significantly exceed the temperatures that occurred over the last 1000 years, and most likely since the emergence from the last glacial maximum.

Many types of changes in the weather and climate would result. For example, shifts in storm tracks and an intensification of evaporation and precipitation cycles would be expected to alter the frequency and intensity of floods and droughts. IPCC (2001) also projects that the intensity of various other extreme weather and climate events would increase.

In addition, the projections indicate that the melting of mountain glaciers and thermal expansion of warming seawater would cause the rate of sea level rise to increase from the observed rate of 0.1 to 0.2 m/century during the twentieth century. Because of uncertainties relating to the fates of the Greenland and West Antarctic ice sheets, the estimated rate of rise is quite broad, ranging from about 0.1 to 0.9 m/century during the twenty-first century. The lower estimate of projected sea level rise reflects the possibility that, for perhaps the next century, increased snowfall could cause ice to accumulate over major areas of Greenland and Antarctica more rapidly than it would be lost at the edges. Higher rates of rise could occur during the latter part of the twenty-first century and are projected for the following centuries as the warming over Greenland (and perhaps West Antarctica) initiates a millennium-long melting that could add 0.5–1 m/century to the rate of future sea level rise.

As was the case for the sudden appearance of the Antarctic ozone hole, there is also a real possibility for surprising changes to occur, especially given the potential for thresholds and nonlinearities. One of the possibilities is the disruption of the Gulf Stream and the larger-scale deep ocean circulation of which it is a part. The weakening of the world-girdling thermohaline circulation apparently occurred during the Younger Dryas period about 11 000 years ago, which was an interruption in the recovery from the glacial conditions that had prevailed for the preceding 100 000 years. The weakening of the global ocean circulation, in that case likely caused by a sudden release into the ocean of glacial meltwaters, led to a strong cooling centered over Europe. Were a similar weakening of the thermohaline circulation to occur during the twenty-first century, the cooling temperatures would likely only be regionally significant, even though it might only moderate the influence of global warming. However, because such a change would reduce the rate of cold-water transport into the deep ocean, the rate of sea level rise would tend to increase, creating problems for coastal regions around the world.

An associated item in this volume (Climate Change Impacts) summarizes the expected types of impacts resulting from changes in climate of the magnitude discussed here. That item also indicates the magnitude of the emissions reduction that would be required to stabilize the atmospheric concentrations of greenhouse gases and eventually stabilize the climate, as has been set as an international objective in the Framework Convention on Climate Change that was negotiated in 1992, and has since been agreed to by most of the nations of the world.

Acknowledgments

The views expressed are those of the author and do not necessarily represent those of his former employer (the University of California) or those agencies supporting the Office of the US Global Change Research Program (now the Climate Change Science Program Office), where he was on assignment.

Michael C. MacCracken

Bibliography

Etheridge, M., Steele, L.P., Langenfelds, R.L., Francey, R.J., Barnola, J.M., and Morgan, V.I., Historical CO_2 records from the Law Dome DE08, DE08-2, and DSS ice cores. In *Trends: A Compendium of Data on Global Change*. Oak Ridge, TN: Carbon Dioxide Information Analysis Center, Oak Ridge National Laboratory, US Department of Energy.

Gates, W.L., et al., 1999. An overview of the results of the Atmospheric Model Intercomparison Project (AMIP I). *Bulletin of the American Meteorological Society*, **80**: 29–55.

IPCC (Intergovernmental Panel on Climate Change), 1996. *Climate Change 1995: The Science of Climate Change*, edited by Houghton, J.T., Meira Filho, L.G., Callander, B.A., Harris, N., Kattenberg, A., and Maskell, K., Cambridge: Cambridge University Press.

IPCC (Intergovernmental Panel on Climate Change), 2000. *Special Report on Emissions Scenarios*, N. Nakicenovic (lead author). Cambridge: Cambridge University Press.

IPCC (Intergovernmental Panel on Climate Change), 2001. *Climate Change 2001: The Scientific Basis*, edited by Houghton, J.T., Ding, Y., Griggs, D.J., et al. Cambridge: Cambridge University Press.

Keeling, C.D. and Whorf, T.P., 1999. Atmospheric CO_2 records from sites in the SIO air sampling network. In *Trends: A Compendium of Data on Global Change*. Oak Ridge TN: Carbon Dioxide Information Analysis Center, Oak Ridge National Laboratory, U.S. Department of Energy.

Lelieveld, J., et al., 2001. The Indian Ocean experiment: Widespread air pollution from south and southeast Asia. *Science*, **291**: 1031–1036.

Mann, M.E., Bradley, R.S., and Hughes, M.K., 1999. Northern Hemisphere temperatures during the past millennium: Inferences, uncertainties, and limitations. *Geophysical Research Letters*, **26**: 759–762.

NAST (National Assessment Synthesis Team), 2000. *Climate Change Impacts on the United States: The Potential Consequences of Climate Variability and Change. Overview Report*, US Global Change Research Program. Cambridge: Cambridge University Press (also see http://www.nacc.usgcrp.gov).

Cross-references

Aerosols
Climate Change Impacts: Potential Environmental and Societal Consequences
Climate Variations
Energy Budget Climatology
Greenhouse Effect and Greenhouse Gases
Models, Climatic

CLIMATE CHANGE AND HUMAN HEALTH

Global climate change is one of a larger set of large-scale environmental changes occurring in today's world. These changes reflect, in various ways, the increasing human domination of the ecosphere, as human numbers continue their unprecedented expansion and as human economic activities intensify (Vitousek et al., 1997). All of these changes – climate change, stratospheric ozone depletion, biodiversity loss, worldwide land degradation, freshwater depletion, disruption of the elemental cycles of nitrogen and sulfur, and the global dissemination of persistent organic pollutants – have great consequences for the sustainability of ecological systems, for food production, for human economic activities and, via those and other pathways, for human population health.

There is a growing realization that the sustainability of human health should be a central consideration in the public debate on sustainable development (McMichael et al., 2000). After all, society's ultimate, if unstated, objective is to maximize the wellbeing, health and survival of its population. The advent of climate change and other global environmental changes, as an indicator that our current developmental trajectory is unsustainable, is obliging us to take a more ecological view of population health. That is, population health is an index of the extent of our success in the longer-term management of social and natural environments.

A change in world climate will influence the structure and functioning of many ecosystems and the biological health of many plants and creatures. Indeed, it has emerged, in recent decades, that many non-human physical and biological systems have undergone alterations that are reasonably attributable to climate change. This includes the retreat of glaciers and sea ice, and the earlier occurrence of bird-nesting and insect migrations (IPCC, 2001; Root et al., 2003). Likewise, there would be various health impacts in human populations. These would differ by location since climate change will induce a range of environmental changes specific to the particular geographical setting. Besides, the vulnerability of each human population will vary as a function of locality, level of material resources, technological assets and type of governance.

Some of the anticipated health impacts would be beneficial. For example, milder winters would reduce the seasonal wintertime peak in deaths that currently occurs in temperate countries, whereas in currently hot regions a further increase in temperatures might reduce the viability of disease-transmitting mosquito populations. Overall, however, scientists assess that most of the health impacts of climate change would be adverse. The health impacts of climate change do not entail novel processes and unfamiliar disease outcomes (in contrast to the recent surprise appearances of HIV/AIDS and human "mad cow disease"). Rather, they entail climate-induced changes in the frequency or severity of familiar health risks – such as floods, storms and fires; the mortality toll of heatwaves; the range and seasonality of infectious diseases; the productivity of local agro-ecosystems; the health consequences of altered freshwater supplies; and the many repercussions of economic dislocation and population displacement.

The human species, via its social organization and cultural practices, is much better – and often intentionally – buffered against environmental stressors than are all other plant and animal species. Hence, *Homo sapiens* is likely to be affected less soon and less sensitively than most other species. Not surprisingly, therefore, at this early stage in the process of global climate change there is little empirical evidence that climate change has already begun to affect human health.

Climate and human health: an ancient struggle

> Whoever wishes to investigate medicine properly, should proceed thus: in the first place to consider the seasons of the year, and what effects each of them produces, for they are not all alike, but differ much from themselves in regard to their changes. [Hippocrates, in *Airs, Waters, and Places* (Hippocrates, 1978)]

The idea that states of human health and disease are linked with climatic conditions almost certainly predates written history. The Greek physician Hippocrates, around 400 BC, related epidemics to seasonal weather change. He wrote that physicians should have "due regard to the seasons of the year, and the diseases which they produce, and to the states of the wind peculiar to each country and the qualities of its waters". He exhorted them to observe "the waters which people use, whether they be marshy and soft, or hard and running from elevated and rocky situations, and then if saltish and unfit for cooking" and to note also "the localities of towns, and of the surrounding country, whether they are low or high, hot or cold, wet or dry... and of the diet and regimen of the inhabitants".

Two thousand years later, Robert Plot, Secretary to the newly-founded Royal Society in England, collated weather observations in 1683–84 and noted that if the same observations were made "in many foreign and remote parts at the same time" we would "probably in time thereby learn to be forewarned certainly of divers emergencies (such as heats, colds, dearths, plagues, and other epidemical distempers)".

In those intervening 2000 years many climatic disasters have affected communities and populations around the world, causing starvation, infectious disease, social collapse and, sometimes, the disappearance of whole populations. One such example is the mysterious demise of the two Viking settlements in Greenland in the fourteenth and fifteenth centuries, as temperatures in and around Europe began to fall (Pringle, 1997). These culturally conservative livestock-dependent settlements were established during the tenth century ad early in the Medieval Warm Period. They could not cope, however, with the progressive deterioration in climate that ensued from the late Middle Ages. Food production declined and food importation became more difficult as sea ice persisted. The Viking settlements eventually died out or were abandoned.

There are numerous historical accounts of acute famines occurring in response to climatic fluctuations (Bryson and Murray, 1977). Throughout pre-industrial Europe, food supplies were marginal, and the mass of people survived on monotonous diets of vegetables, grain gruel and bread. A particularly dramatic example in Europe was the great medieval famine of 1315–17. Climatic conditions were deteriorating, and the cold and soggy conditions led to widespread crop failures, food price rises, hunger and death. Social unrest increased, robberies multiplied, and bands of desperate peasants swarmed over the countryside. Animal diseases proliferated, contributing to the die-off of over half the sheep and oxen in Europe. This tumultuous event, and the Black Death that followed 30 years later, may have contributed to the weakening and dissolution of feudalism in Europe.

Over these and the ensuing centuries, average daily food intakes were less than 2000 calories, falling to around 1800 calories in the poorer regions of Europe. This caused widespread malnutrition, susceptibility to infectious disease, and low life expectancy. Recurrent famines culled the populations, often drastically. In Tuscany, for example, there were over 100 years of recorded famine between the fourteenth and eighteenth centuries. Meanwhile in China, where the rural diet of vegetables and rice accounted for an estimated 98% of caloric intake, in the 2000 years between around 100 BC and AD 1900 there were famines that involved at least one province in over 90% of years (Bryson and Murray, 1977).

The health impacts of climate change

It is important to stress, again, that there is yet little direct empirical evidence of climate change affecting human health over recent decades. In part this is because epidemiologists have had little interest in studying the relationship of climatic variations to human health. The low premium accorded to this topic has presumably reflected a general assumption that there are few opportunities for direct intervention to reduce adverse health impacts from climatic influences. Nevertheless, the position has changed markedly in recent years in the wake of newly recognized "anthropogenic" global climate change.

This upsurge in research activity falls into three categories: (a) studies seeking to broaden the empirical information base by examining how recent variations or trends in climatic conditions have related to changes in health outcomes, (b) studies seeking evidence of early impacts of global climate change on human health, and (c) studies that use existing empirical and theoretical knowledge to model the future impacts of given scenarios of climate change.

It is usual to distinguish several generic categories of health impacts due to climate change – first, those that arise from relatively direct impacts of alterations in temperature, precipitation and extreme weather events; second, those that occur in response to climate-induced changes in ecological processes and systems; and, third, those that result from the economic and demographic dislocation of human communities.

Direct effects

The more direct and immediate impacts include those due to changes in exposure to very cold and very hot weather extremes and those due to increases in extreme weather events (floods, tropical cyclones, storm-surges and droughts). Climate change would also directly increase the production of certain air pollutants, such as tropospheric ozone (the formation of which is affected by both temperature and level of sunlight), and various aeroallergens (spores and molds) involved in the causation of asthma, hay fever and other allergic disorders.

Populations display a characteristic pattern of daily deaths (and hospitalizations) in relation to daily temperature (Wilkinson et al., 2002). Typically, death rates increase both with greater heat and greater cold. In cool temperate countries, climate change would result in a decrease in winter mortality because of less severe winters, and this may offset the increases in summer heat-related mortality (Langford and Bentham, 1995; Rooney et al., 1998), at least initially. In warm temperate and tropical countries the overall number of temperature-related deaths is likely to increase. The net balance of future changes in hot and cold effects remains contentious and, anyway, varies substantially between different geographic regions, states of economic development, and between urban and rural populations.

The extent to which the regional frequency of extreme weather events will alter because of climate change remains somewhat uncertain. Climatologists, however, are becoming increasingly confident of their modeled regional predictions, as the downscaling of their climate models improves. Further, extreme weather events appear to have increased in tempo during the 1990s (e.g. McMichael et al., 2003; Milly et al., 2002), with commensurate risks to human life and limb. Nevertheless, despite the potentially great impact of such events on deaths, injuries and consequent diseases (infections, malnutrition and mental health disorders), estimating the future profile of health impacts is at best an indicative exercise. Hurricane Mitch, centered in Honduras in 1998, had a death toll of over 11 000 people, and was followed by tens of thousands of new cases of malaria, cholera and dengue fever.

Indirect effects (especially infectious diseases and food yields)

The indirect, second, category of health impact refers particularly to changes in the transmission patterns of infectious diseases and changes in regional food-producing systems. These impacts are, in essence, a result of ecologically mediated mechanisms. In the longer term it is probable that these indirect impacts on human health would have greater magnitude than the more direct impacts (McMichael et al., 1996, 2003).

For vector-borne infectious diseases the geographic range and abundance of vector organisms (and, in some cases, their intermediate hosts) are affected by various meteorological factors (temperature, precipitation, humidity, surface water and wind), biotic factors (vegetation, host species, predators, competitors, and parasites) and human interventions. The rate of maturation of the pathogen within the vector is typically sensitive to temperature, in a curvilinear fashion. This means, for example, that an increase of 1°C across different parts of the temperature range would yield very different increases in transmissibility.

There is some evidence that tick-borne encephalitis, a viral disease of humans, has increased its geographic range within Sweden over the past two decades, as winter temperatures have increased (Lindgren and Gustafson, 2001). Other studies have found some evidence that malaria has extended its range in parts of highland Eastern Africa in association with local warming (Patz et al., 2002) and that, in northeastern Australia, Ross River virus disease has increased its geographic distribution since 1990 (Tong et al., 2001).

Considerable recent effort has sought to develop better mathematical models for making scenario-based health impact projections (McMichael et al., 2003; Epstein, 1999; Martens, 1998). The models in current use have well-recognized limitations. For example, the transmission cycles of some vector-borne diseases are highly complex, involving multiple vector and host species. This can have the effect of limiting the geographical scale of predictive modeling (from the global to, say, the subnational level), as well as increasing the range of uncertainty around estimates of climate change impact on disease transmission (Woodruff et al., 2002). Nonetheless, mathematical modeling studies have provided useful indicative forecasts of the likely direction and magnitude of impacts, and are an important stimulus to further research.

Mathematical models have forecast that an increase in worldwide average temperature and associated changes in precipitation patterns would cause a net increase in the geographic range of malaria-transmitting mosquito species (Martens et al., 1999) – although some localized decreases may also occur in regions that become too hot or dry, such as in the Sahel region of Sub-Saharan Africa (McMichael et al., 1996; Martens, 1998). Further, temperature-related changes in the life-cycle dynamics of both the vector species and the pathogens (protozoa, bacteria and viruses) would increase the potential transmission of many vector-borne infectious diseases such as malaria (mosquito), dengue fever (another of the world's great vector-borne viral infectious diseases, transmitted by mosquito) and leishmaniasis (sand-fly) (Patz et al., 1996; Martens, 1998). Models indicate that dengue fever would extend its range and seasonality (Hales et al., 2002). In Australia, a combination of heavy rainfall and high temperatures have been used to predict epidemics of Ross River virus disease (transmitted by mosquitoes) with a likely increase in future transmission expected in some regions (Woodruff et al., 2002).

Climate change would also affect the transmission of water-borne infectious diseases via several mechanisms (McMichael et al., 2003), including intensification of rainfall episodes, leading to flooding that causes contamination of drinking water supplies. Bacterial water-borne illnesses such as gastroenteritis due to coliform bacteria, giardiasis, and cholera may thus be affected; so too would various protozoal water-borne infectious diseases such as cryptosporidiosis. Similarly, warmer temperatures would tend to increase the summer seasonal peaks of food-borne bacterial enteric infections, such as those due to *Salmonella* and *Campylobacter*. In low-income countries, with poor hygiene and resources, rates of serious child diarrheal disease would rise. The sensitivity of child diarrhea to variations in climatic conditions has been well demonstrated in studies in Lima, Peru and in Fiji (Singh et al., 2001; Checkley et al., 2000).

Another fundamental potential impact on human health would arise from any downturn in food availability. Climate change is likely to affect crop yields, livestock health and resultant food products, and fisheries. However, the processes are generally complex – and often entail seemingly stochastic events (e.g. the occurrence of new infectious diseases in plants and animals). Cereal grains account for around 70% of world food energy – both via direct consumption and via the feeding of grains to livestock for meat production. Various modeling studies have made estimates of the impact of climate change upon cereal grain yields – allowing for the expected "fertilization" effect of increasing concentration of atmospheric carbon dioxide. Globally, a slight downturn in grain production appears likely, especially in the latter half of the twenty-first century (Parry and Carter, 1998; McMichael et al., 2003). This downturn would be greater in the already food-insecure regions in South Asia, parts of Africa and Central America. Such downturns would increase the number of malnourished people in the world (currently an estimated 830 million) by at least several percent overall – with higher percentage decreases in Sub-Saharan Africa and South Asia. World cereal grain yields have become a little more unstable during the 1990s, displaying increased inter-annual variability: could this be (partly) due to changing climatic conditions?

Effects of economic, social and demographic disruption

The third category of health impact is somewhat speculative and difficult to model quantitatively. The situation of several small island states in the Pacific region typifies issues that are likely to be experienced, at some level, in other parts of the world. These islands are registering growing concern about sea-level rise and disturbance of natural and managed food systems, due to the depletion of freshwater stocks, arable land, and damage to coastal industries. The economic downturns, unemployment, and civil strife typically associated with such large-scale changes, and the resulting population displacement, are also conducive of a range of risks to health. Refugee populations typically experience mental health problems, malnutrition, infectious diseases, and the physical hazards of new and improvised living environments.

The anticipated health consequences of economic and social disruptions due to climate change may not become evident for several decades. A usual difficulty for researchers is in deciding whether an event, such as climate change, has caused such outcomes, given the many, often interrelated, causal influences involved. It will therefore be difficult to detect early climate impact "signals" against the considerable background "noise".

Population vulnerability, and adaptive responses

The magnitude of the impact of climate change upon health depends both on the extent of that change and on characteristics of the target population. Although changing climate conditions may make the environment more favorable for disease transmission and other negative health impacts, this does not necessarily mean these will occur, as long as there is the capacity to adapt to the changed conditions. For example, warmer temperatures are predicted to extend the distribution of many disease-carrying vectors into new areas. This increased risk does not have to mean a large increase in cases of disease, provided there is a continuing expansion of the public health response to the disease, additional quarantine efforts, etc.

A population's "vulnerability" is a function both of the sensitivity of the exposure–response function in that population and of the population's adaptive (i.e. impact-lessening) capacity (Parry and Carter, 1998). That adaptive capacity depends on factors such as population density, level of economic development, food availability, local environmental conditions, pre-existing health status, and the quality and availability of public health care. It also depends on various structural and politically determined characteristics, including social-cultural rigidity, international connectedness and political flexibility (Woodward et al., 2000). The level of risk to the population's health is therefore a function of that vulnerability and the amount of exposure to climatic or associated environmental factors.

The reduction of socioeconomic vulnerability remains a high priority. Poor populations will be at greatest health risk because of their lack of access to material and information resources and because of their typically lower average levels of health and resilience (nutritional and otherwise). The long-term improvement in the health of impoverished populations will require income redistribution, improved employment opportunities, better housing and stronger public health infrastructure – including primary health care, disease control, sanitation and disaster preparedness and relief.

Adaptation will be either reactive, in response to climate impacts, or anticipatory, in order to reduce vulnerability. These adaptive actions can be categorized as: (a) administrative or legislative, (b) engineering, or (c) personal (behavioral) (Patz, 1996). Legislative or regulatory action can be taken by government, requiring compliance by all (or by designated classes of) people. Alternatively, voluntary adaptive action may be encouraged via advocacy, education or economic incentives. The former type of action would normally be taken at a supranational, national or community level; the latter would range from supranational to individual levels. Adaptation can be undertaken at the international/national level, the community level and the individual level.

It would be shortsighted to expect that adaptation will provide a complete antidote to the problems of climate change. Reduction of greenhouse gas emissions remains a primary preventive health strategy, with appreciable 'collateral' health benefits in the short term. Transport policies to reduce dependence on the motorcar could lead, for example, to improved air quality and increased physical activity (linked to a reduction in obesity, diabetes and heart disease at the population level). Even so, we are already committed to some degree of global climate change over coming decades despite any mitigation actions we might take now. It is important to begin assessing, for any particular population, both its vulnerability and its adaptation options.

Conclusion

Although it is now generally agreed that human-induced climate change is under way, and will continue to occur, the magnitude and rate of change remain uncertain. Future levels of greenhouse gas emissions will be affected by population and economic growth, as well as by technological change, energy preferences, and social behavior. Our ability to explain the chaotic variations within the climate system is improving, but there will always be inaccuracies involved in modeling natural systems.

In addition to climatic change and increased variability, numerous other factors (i.e. social, behavioral, environmental) influence the risk of the diseases discussed in this item. Teasing out the climatic contribution is a major challenge to health researchers. Given the unavoidable uncertainties pertaining to the projected health impacts of climate change, social policy should be developed within the framework of the "precautionary principle" – that is, the principle of proceeding on the basis of the incomplete evidence that is available in relation to future processes that are never fully knowable in advance (McMichael et al., 2003).

Anthony J. McMichael and Rosalie E. Woodruff

Bibliography

Bryson, R.E., and Murray, T.J., 1977. *Climates of Hunger: Mankind and the World's Changing Weather.* Madison, WI: University of Wisconsin Press.

Checkley, W., Epstein, L.D., Gilman, R.H. et al., 2000. Effects of El Nino and ambient temperature on hospital admissions for diarrhoeal diseases in Peruvian children. *Lancet*, **355**(9202): 442–450.

Epstein, P.R., 1999. Climate and health. *Science*, **285**: 347–348.

Hales, S., de Wet, N., Maindonald, J., and Woodward, A., 2002. Potential effect of population and climate changes on global distribution of dengue fever: an empirical model. *Lancet*, **360**(9336): 830–834.

Hippocrates, 1978. Airs, waters and places. An essay on the influence of climate, water supply and situation on health. In Lloyd, G.E.R., ed., *Hippocratic Writings*. London, Penguin, p. 148.

IPCC, 2001. *Climate Change 2001*. Third Assessment Report. Cambridge: Cambridge University Press.

Langford, I.H., and Bentham, G., 1995. The potential effects of climate change on winter mortality in England and Wales. *International Journal of Biometeorology*, **38**: 141–147.

Lindgren, E., and Gustafson, R., 2001. Tick-borne encephalitis in Sweden and climate change. *Lancet*, **7**;358(9275): 16–18.

Martens, P., Kovats, R.S., Nijhof, S., et al., 1999. Climate change and future populations at risk of malaria. *Global Environmental Change*, **9**(99): S89–S107.

Martens, W.J.M., 1998. *Health and Climate Change: Modelling the Impacts of Global Warming and Ozone Depletion.* London: Earthscan.

McMichael, A.J., 2001. *Human Frontiers, Environments and Disease. Past Patterns, Uncertain Futures.* Cambridge: Cambridge University Press.

McMichael, A.J., Ando, M., Carcavallo, R., et al. 1996. Human population health. In Watson, R.T., Zinyowera, M.C. and Moss, R.H., eds., *Climate Change 1995: Scientific–technical analyses of impacts, adaptations, and mitigation of climate change.* New York: Cambridge University Press, pp. 561–584.

McMichael, A.J., Campbell-Lendrum, D., Ebi, K., Githeko, A., Scheraga, J., and Woodward, A. (eds.), 2003. *Climate Change and Human Health: Risks and Responses.* Geneva: World Health Organization.

McMichael, A.J., Smith, K.R., and Corvalan, C.F., 2000. The sustainability transition: a new challenge. *Bulletin of the World Health Organization*, **78**(9): 1067.

Milly, P.C.D, Wetherald, R.T., Dunne, K.A., and Delworth, T.L. 2002. Increasing risk of great floods in a changing climate. *Nature*, **415**: 514–517.

Parry, M.L., and Carter, T., 1998. *Climate Impact and Adaptation Assessment*. London: Earthscan.

Patz, J.A., 1996. *Adapting to Climate Change: An International Perspective*. Edited by Smith, J.B., et al. New York: Springer, pp. 440–464.

Patz, J.A., Epstein, P.R., Burke, T.A., and Balbus, J.M., 1996. Global climate change and emerging infectious diseases. *Journal of the American Medical Association*, **275**: 217–223.

Patz, J.A., Hulme, M., Rosenzweig, C., et al., 2002. Increasing incidence of malaria since 1970 parallels regional warming in East Africa. *Nature*, (In press).

Pringle, H., 1997. Death in Norse Greenland. *Science*, **275**: 924–926.

Rooney, C., McMichael, A.J., Kovats, R.S., and Coleman, M., 1998. Excess mortality in England and Wales during the 1995 heatwave. *Journal of Epidemiology and Community Health*, **52**: 482–486.

Root, T.L., Price, J.T., Hall, K.R., Schneider, S.H., Rosenzweig, C., and Pounds, J.A., 2003. Fingerprints of global warming on wild animals and plants. *Nature*, **421**: 57–60.

Singh, R.B., Hales, S., de Wet, N., Raj, R., Hearnden, M., and Weinstein, P., 2001. The influence of climate variation and change on diarrheal disease in the Pacific Islands. *Environmental Health Perspectives*, **109**(2): 155–159.

Tong, S., Bi, P., Hayes, J., Donald, K., and Mackenzie, J.S., 2001. Geographic variation of notified Ross River virus infections in Queensland, Australia, 1985–1996. *American Journal of Tropical Medicine and Hygiene*, **65**(3): 171–176.

Vitousek, P.M., Mooney, H.A., Lubchenco, J., and Melillo, J.M. 1997. Human domination of Earth's ecosystems. *Science*, **277**: 494–499.

Wilkinson, P., McMichael, A., Kovats, S., et al., 2002. International study of temperature and heatwaves on urban mortality in low-and-middle-income countries (ISOTHURM). Presented at the International Society of Environmental Epidemiology, Vancouver, August 2002. *Epidemiology*, **13**(4): S81.

Woodruff, R.E., Guest, C.S., Garner, M.G., et al., 2002. Predicting Ross River virus epidemics from regional weather data. *Epidemiology*, **13**(4): 384–393.

Woodward, A., Hales, S., Litidamu, N., Phillips, D., and Martin, J., 2000. Protecting human health in a changing world: the role of social and economic development. *Bulletin of the World Health Organization*, **78**: 1148–55.

Cross-references

Air Pollution Climatology
Applied Climatology
Climate Change Impacts: Potential Environmental and Societal Consequences
Climate Comfort Indices
Climatology, History of
Determinism, Climatic
Human Health and Climate
Seasonal Affective Disorder

CLIMATE CHANGE IMPACTS: POTENTIAL ENVIRONMENTAL AND SOCIETAL CONSEQUENCES[1]

With some changes in climate having already occurred, with the climate continuing to adjust to past changes in atmospheric composition, and with continuing greenhouse gas emissions certain to cause further climate change, there is no doubt that the world will be undergoing relatively large changes in climate that will be ongoing for centuries. Such changes will inevitably cause consequences for society and the environment.

[1] This item is updated from MacCracken, M.C., 2001, Global warming: a science overview, pp. 151–159 in *Global Warming and Energy Policy*. New York: Kluwer Plenum.

In developing estimates for these consequences, it is important to consider not only the impacts that would be expected were the projected changes in climate to affect systems as they exist at present, but also to carry out the analyses in the context of how these systems are changing (naturally and due to other human-induced stresses) and in recognition of the possible adaptation of strategies for responding and adapting to the projected combination of stresses. Because vulnerability is really the difference between the sensitivity to the impact and the ability to adapt, all considered in the context of other stresses on society and the environment and resources available to help respond, the vulnerability of each nation will depend particularly on its own situation and the local ability and commitment to respond.

The Second and Third Assessment Reports of the IPCC summarize present understanding of the types of consequences for continental-scale regions around the world (IPCC, 1996a,b, 2001b). Figure C13, taken from IPCC (2001b), provides a summary diagram of how the relative importance of the potential impacts is likely to increase as the amount of warming increases. This figure also makes clear that, because climate change will occur with differing intensities at different latitudes, and because different emissions scenarios will lead to different overall rates of change, different countries are likely to be in very different situations.

Recognizing this, many nations have begun to analyze the potential consequences that they anticipate having to deal with. Information for many nations on potential impacts and adaptive measures can be found in their national assessment reports (e.g. for the US, see NAST, 2000 and 2001; for Canada, see Environment Canada, 1998). Many of the nations of the world summarize their findings in reports called for by the United Nations Framework Convention on Climate Change (UNFCCC); for example, for the US, see Department of State (2002) and for other nations see "National Communications" section at http://unfccc.int/). Generally, it has been found that the impacts are likely to be relatively more important for developing than developed nations because developing nations must devote their more limited resources to more immediate challenges and so do not have the flexibility and resources to adapt that developed nation economies may have available. Because it is impossible to be comprehensive in a brief review, both the IPCC and the national assessment reports should be consulted to gain a fuller understanding.

The regional and temporal variations in projected consequences create significant difficulties for negotiators in determining criteria for meeting the central objective of the UNFCCC that was negotiated and internationally accepted in 1992. This objective is to achieve "stabilization of the greenhouse gas concentrations in the atmosphere at a level that would prevent dangerous anthropogenic interference with the climate system" while at the same time doing this in a way that would "allow ecosystems to adapt naturally to climate change,... ensure that food production is not threatened, and... enable economic development to proceed in a sustainable manner".

In thinking about the importance of potential impacts it is also important to remember that fossil fuels provide wide-ranging benefits to society, providing roughly 80% of the world's energy and so sustaining the world's economy and enabling the present standard-of-living of peoples around the world. Recognizing this, some argue that, if the use of the energy source that supports the global economy is to be sharply reduced, such impacts will have to be relatively certain and substantial. In contrast, others point out that we have only one "spaceship Earth", so

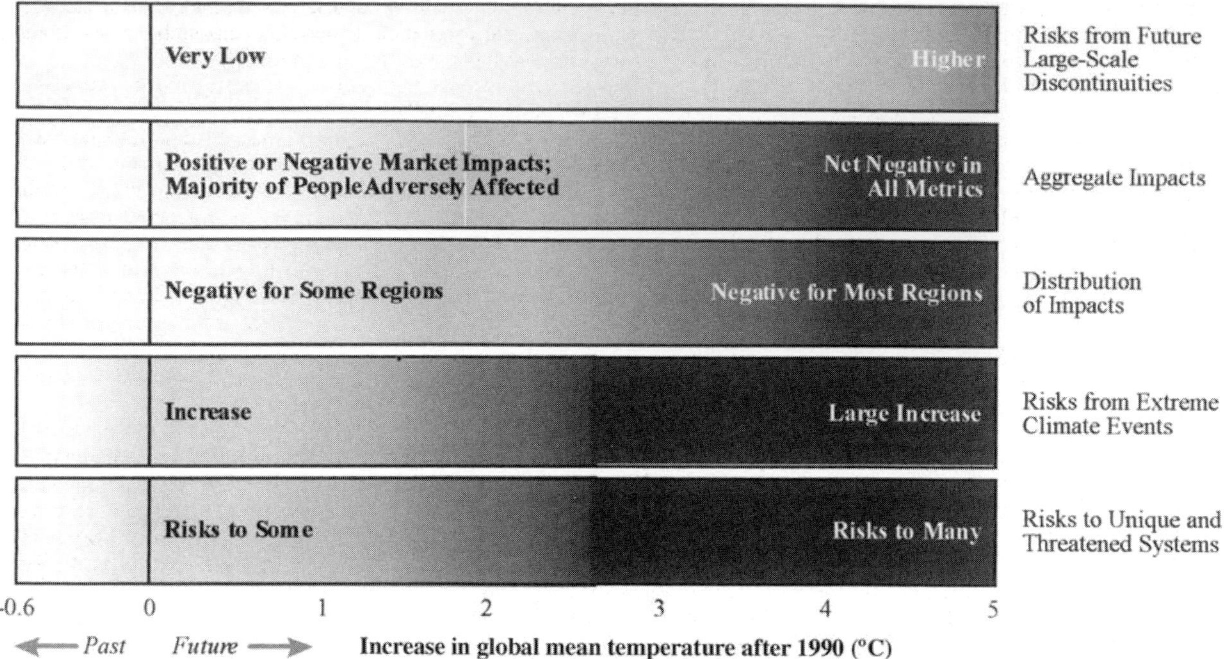

Figure C13 Estimated level of risk as a function of the change in global mean temperature from its value 1990 (in °C), with the areas to the left of the vertical line referring to the risk created by warming prior to 1990. Each horizontal bar refers to a different reason for being concerned. The more intense the shading, the greater the risk. Figure developed as part of IPCC's Third Assessment (IPCC, 2001b).

precaution suggests that the chance of significant or irreversible changes should be avoided. This article provides an overview of some of the important categories of potential impacts that have been identified, providing a sense of the types of impacts (for further details, see the various IPCC assessments).

Human health

Given the large number of people that could be affected, even a modest health impact could create very significant societal and economic consequences. For many regions, significant increases in summertime temperature (and in the absolute humidity that is often associated with such high temperatures) could lead to much more frequent and intense occurrence of dangerously high heat index conditions. While more extensive availability of air-conditioning and warning of dangerous events could limit an increase in heat-related mortality in developed nations, the threat for developing nations where such remedies might not be available could be quite significant.

With the poleward movement of wintertime frosts, mosquitoes and other disease vectors are likely to spread poleward; this has the potential to increase the incidence of infectious diseases if these influences cannot be offset by more attention to appropriate public health measures, enhanced building and community design, and new medicines. Again, those in developed nations are likely to be much better off than those in developing nations; for example, the 17-year total of cases of dengue fever, which is mosquito-borne, is about a thousand times as high in northeastern Mexico as in Texas, mainly due to the relative level of development. The increased intensity of extreme events such as hurricanes and typhoons is also likely to injure or kill more people (and disrupt communities) unless steps are taken to enhance risk-averse planning and construction and to provide better warning and recovery capabilities, and again, developing countries are much more vulnerable.

Food production

The UNFCCC objective specifically mentions the need to ensure that food production is not threatened, for if there is not enough food to survive into the future, the amount of climate change will not really matter. Fortunately, laboratory and field studies indicate that CO_2 concentrations are very likely to enhance the growth of many types of crops and to improve their water use efficiency. If this happens widely (i.e. if other constraints on agriculture do not arise), changes in climate would be expected to contribute to an increase in agricultural productivity (i.e. yield per hectare), thus working in the same positive direction as improvements in agricultural practices, seed strains, and technology. If these positive influences exceed the negative influences caused by deteriorating soil fertility and altered climatic conditions, the potential exists for increasing overall agricultural production while using less land for agriculture. Such an outcome would have the potential for reducing food costs for the public, and there are some studies indicating that the past increase in the CO_2 concentration has contributed to overall food production.

Given the varying soil, climatic, and agricultural capabilities around the world, however, the benefits and impacts are unlikely to be evenly spread. For example, the altered climatic conditions facing farmers in some low-latitude developing nations might well exceed the tolerances for present crops, or, for farmers used to growing only a few types of crops, could at least require a disruptive change in what is grown. For farmers in developed nations, many are facing economic hardships now because of the low prices that result, at least in part, from overproduction. This overproduction presently causes many governments to have to provide subsidies or to impose trade barriers of various kinds. Thus, if changes in climate lead to increases in productivity that cause additional downward pressure on commodity prices or to an increase in costs of production (e.g. to pay for additional irrigation), income for farmers in marginal growing areas is likely to decrease. Even though these farmers might benefit from some gain in overall crop productivity, it is likely to be less than the increase for farmers in the most productive areas, meaning that those in marginal areas are unlikely to remain competitive. This could cause economic and environmental problems in nearby rural communities unless other profitable crops are identified and farmers have the incentive and resources to tend and conserve the land resources.

Ecosystems and ecosystem services

As is clearly evident from a hike up a mountain or a vacation excursion across a continent, the prevailing vegetation is closely tied to the climate that a location experiences. Although the world's vegetation shifted dramatically as the climate warmed at the end of the last glacial maximum, the shifts took place over many millennia and were possible in part because the world's wildlife and plant species had been squeezed equatorward and so had a base from which to spread. For the future, the amount and rate of climatic change are projected to be much, much greater than has been experienced over the past 20 000 years. In addition, the ultimate climatic conditions are projected to reach levels that are well beyond existing conditions, thereby rapidly recreating climatic conditions that have not been experienced in the world for many millions, even tens of millions, of years. In that the ecosystems that make up our landscape provide such a range of services, including water and air purification, supply of food and fiber products, and recreational and esthetic enjoyment, the likelihood of significant changes in ecosystems prompted the provision in the UNFCCC objective that necessitates understanding how ecosystems are likely to change and adapt.

Ecosystem simulation models, which are calibrated based on past and present climate–vegetation relationships, project that changes are likely to be quite extensive. In general, temperature changes are likely to increase with latitude and to be larger over land areas than over ocean areas. Precipitation in low and high latitudes is expected to increase, but with an expanded subtropics where increased evaporation likely further reduces soil moisture. Wintertime precipitation in middle- and high latitudes is projected to increase (proportionately) more than summertime precipitation. Evaporation rates are likely to increase, especially in summer, causing reductions in soil moisture that could affect pest and fire disturbance frequencies for forests and grasslands. As the seasonal patterns of temperature and soil moisture change, the competition among various species will be changed, leading to changes in prevailing tree and grass types and then in the types, abundance, and migration of birds and other wildlife. Some, but not all, of these effects may be offset by the increased CO_2 concentration, which is expected to help many types of plants grow better if other factors, such as nutrients, are not limiting. To the extent that regions accumulate carbon in vegetation and dry out during persistent warm episodes, fire risk is likely to increase even though plant growth may overall be more vigorous. At the same time, increased precipitation over deserts caused by shifting atmospheric circulation patterns, even though perhaps seasonally concentrated, could lead to more vegetation in some regions.

Climatic changes will also lead to significant impacts on freshwater and marine ecosystems. Already, bleaching of coral reef ecosystems is being attributed, at least in part, to the warming oceanic conditions. The increasing CO_2 concentration also affects ocean chemistry, and for coral reefs the impact is to dramatically shrink the regions of the world where temperatures and ocean chemistry are conducive to generation of strong reef structures. Fish populations are also very sensitive to the temperatures of streams and ocean waters, and significant shifts are projected as the climate changes. Changes in ocean productivity also cause effects all the way up the food chain, ultimately affecting the resources, and so the populations, of species as varied as whales, seals, and penguins.

What is most important to understand is that the notion of ecosystem migration is a misconception – as climate changes, the dominant species in particular locations will change, and so the local ecosystem will change as species now dominant become, if they can, dominant elsewhere. However, because each species is distinct, this process is likely to lead to the deterioration of existing ecosystems and the creation of new ones. During this century the changes are likely to be relatively rapid, and the complexity and resilience of current systems is likely to be disrupted because there will not be sufficient time for adjustment and evolution to take place. As a result, the ecological services that society depends on seem likely to be diminished.

Water supplies

Water is absolutely vital to society, and water resource limitations are currently affecting many nations. By affecting the hydrological cycle in a variety of ways, climate change is likely to exacerbate the availability of water for peoples around the world. Changes in the location and timing of storms will alter the timing and amount of precipitation, soil moisture, and runoff. For example, the intensity of convective rainfall is projected to increase, creating the potential for enhanced flooding in watersheds that experience frequent rainfall. Conversely, increased summertime evaporation will reduce soil moisture and may also reduce groundwater recharge in areas such as the Great Plains, and cause lower levels in the lake and river systems, reducing water supplies for communities and agriculture as well as opportunities for shipping and recreation.

Warmer conditions are also likely to increase the rate of evaporation and to change precipitation in cold regions from snow to rain. These and other types of changes will thus affect the overall amounts and availability of water resources, as well as potentially intensifying both floods and droughts. Adapting to such changes will require changes in how river and reservoir systems are managed in order to satisfy the demands of water for communities, industry, hydroelectric power, irrigation, water levels, groundwater recharge, and preservation of fish migrations and aquatic ecosystems. The challenges will be especially severe in mountainous regions where the snowpack (and even mountain glaciers) presently serve as virtual water reservoirs, typically providing water during the spring and summer. Because wintertime precipitation is projected, on a relative basis, to increase the most, avoiding floods is likely to require an even greater lowering of reservoir levels in winter, even though this would reduce water availability in summer when demand is likely to be highest. In addition, because water resources are under increasing stress due to many non-climatic factors, the impacts of climate change are likely to be a significant exacerbation of these other influences.

Coastal endangerment

After the rapid rise of sea level at the end of the last interglacial, reconstructions of beach height, taking into account isostatic adjustments of the deformed land surface, suggest that sea level was relatively constant through most of the Holocene. Indications are that global sea level started rising during the nineteenth century, rose by about 0.1–0.2 m during the twentieth century, and will rise more rapidly in the future. If the rate of sea level rise triples (using mid-range estimates from IPCC, 2001a), this would put many coastal regions at significant risk of inundation. For regions currently experiencing subsidence of their coastlines (e.g. river deltas as sediments compact and are not renewed, regions just equatorward of ice sheet edges), there could be an even more significant acceleration in the rate of rise. Because there are few resources to protect sensitive natural areas such as wetlands and fish- and bird-breeding grounds, accelerating loss is very likely. The rate of loss will be greatest during coastal storms when storm surges (and therefore wind-whipped waves) will reach further inland and further up rivers and estuaries.

There are also many cities and significant infrastructure along coastlines. For highly developed areas, retreat would be very costly, and so strengthening of coastal protection is essential, not just to protect against sea-level rise, but also to reduce current vulnerability to the storm surge from coastal storms and hurricanes. For smaller communities and residential structures, protective structures such as levees are likely to be too expensive and so retreat will be required, or forced by events. Already, some coastal villages in Arctic regions are at risk because the wintertime retreat of sea ice is increasingly exposing these communities to high waves whipped up by winter storms.

With the likelihood that the rate of sea-level rise will continue at significant levels for many centuries, determining when near-term protective measures are best replaced by long-term relocation will be very difficult for many nations, especially with the increasing desire of affluent populations to live near the coastline and enjoy its scenic and recreational opportunities. However, it is essential that such analyses be done as the choice is important in determining how best to deal with rising sea level and its impacts over time (e.g. if the coastline is to be preserved, roads need to be parallel to the coast; if retreat is to occur, roads need to be perpendicular to the coast to allow phased movement).

Societal infrastructure

Although societal infrastructure is generally designed to withstand a wide variety of weather extremes, the changing climate is projected to create changes in the intensity and frequency of extreme events. For example, the climate model of the Canadian Climate Centre projects that the return period of 100-year storms could decrease by half or more over the next 100 years. Given the economic costs and psychological impacts of recovering from such intense and damaging storms and conditions, a phased response that leads to infrastructure designs that provide more resilience has the potential of limiting potential consequences. Although the incremental cost of providing more resilience is not likely to be high for any single project in developed nations, there is an extensive amount of infrastructure in these nations and so total costs could be quite significant, although mostly hidden within large budgets. For developing countries the cost of designing resilient infrastructure is likely to be a greater increment, but, in that old infrastructure is not having to be replaced, may be feasible. What is most important is that such precautionary investments (e.g. elevating new sewer plants to withstand projected changes in sea level) be made, starting as early as possible, rather than further investing in facilities that will soon have to be reconstructed.

Transportation systems represent a very large societal investment. Present intensities of severe weather and floods typically cause disruptive economic impacts and inconvenience that can become quite important for particular regions during particular extreme events. While information is only starting to emerge about how climate change might lead to changes in weather extremes, a range of possible types of impacts seem possible, including some that are location-dependent and some that are event-specific. Location-dependent consequences could include: lower levels in rivers and lakes used for transportation and as sources of water for communities and industry, relocation of coastal channels as a result of shifting sediments and barrier islands, and opening of Arctic shipping routes as sea ice melts back. Event-specific consequences could include: more frequent occurrence of heavy and extreme rains (a trend already evident during the twentieth century); reduced or shifted occurrence of winter snow cover that might reduce winter trucking and air traffic delays; altered frequency, location, or intensity of hurricanes and typhoons accompanied by an increase in flooding rains; and warmer summertime temperatures that raise the heat index and may increase the need for air pollution controls. By reducing the density of the air, warmer temperatures will also cause reductions in combustion efficiency, which would both increase costs and require longer runways or a lower load for aircraft, just as now occurs for airports in mountainous regions. Starting to consider climate variability and change now in the design of various types of infrastructure could well be a very cost-effective means of enhancing the short- and long-term resilience of society to weather extremes and climate change.

Air quality

Warmer temperatures generally tend to accelerate the formation of photochemical smog. The rising temperature and rising absolute (although perhaps not relative) humidity will raise the urban heat index significantly, potentially contributing to greater air pollution. Meeting air quality standards in the future is therefore likely to require reductions in pollutant emissions beyond those currently in force (although, of course, a move away from the combustion engine might make this change much easier). Increasing amounts of photochemical pollution could also increase impacts on stressed ecosystems, although the increasing concentration of CO_2 may help to alleviate some types of impacts by causing the leaf's stomata to close somewhat. In some regions, summertime dryness would be expected to exacerbate the potential for fire, creating the potential for increased amounts of smoke, whereas in other regions the climatic changes may make dust more of a problem.

International interactions

While it is natural to look most intently at consequences within a particular nation, each nation is intimately coupled to the world in many ways. For example, economic markets and investments are now coupled internationally and a nation's foods come from both within and outside its borders. Because of international travel for business and pleasure, health-related impacts are also international. Many resources, from water- and hydropower-derived electricity to fisheries and migrating species, are shared across borders, or move and are transferred internationally. Finally, people continue to relocate from one nation to another, becoming citizens of one area while tied by family and history to other areas, especially when disasters strike. Increasingly, all countries are connected to what happens outside their borders, and those in one country will be affected by the repercussions of the societal and environmental consequences experienced by others.

Summary of impacts

Given the variety of potential impacts and the varying capabilities and resources for responding to them, international assessments (e.g., IPCC, 1996b, 2001c) provide primarily a qualitative indication of the potential consequences. In addition to the scientific challenges of developing the estimates, issues of equity and cultural values introduce many more complications. For example, IPCC (1996b) struggled with the issue of whether the imputed value of a human life should be considered the same in a developed and a developing nation even though the lifetime earnings and economic evaluations of a life lost are quite different across the world.

Present assessments indicate that there are very likely to be important consequences, some negative and some positive, and that impacts are likely to be greater for developing than for developed nations. However, most analyses currently treat rather large regions and so average out the most costly and most beneficial outcomes, thereby perhaps obscuring what could be rather large implications for localized groups.

The mitigating effects of emissions reductions

In recognition of the potential for significant climatic and environmental change and consequent impacts, the nations of the world in 1992 agreed to the United Nations Framework Convention on Climate Change (UNFCCC). As indicated in the preceding section, this agreement set as its objective the "stabilization of the greenhouse gas concentrations in the atmosphere at a level that would prevent dangerous anthropogenic interference with the climate system". Because of difficulties in deciding upon what might be considered dangerous, and because uncertainties in understanding of the climate system create only a loose coupling to the changes in atmospheric emissions and composition and consequences that might occur, discussions about potential stabilization levels have so far been based mainly on a sense of the level of change in atmospheric composition that seems likely to have significant consequences, considered generally in the context of what might be technologically and economically possible. For example, considerable attention has been given to attempting to stabilize the atmospheric CO_2 concentration at double its preindustrial level (i.e. at about 550 ppmv). Although not discussed below, similar limitations would need to be imposed on the other greenhouse gases and aerosols if atmospheric composition is to be stabilized.

Based on current scientific understanding of the carbon cycle, limiting the atmospheric CO_2 concentration to 550 ppmv would require *the world as a whole* to remain, on average, at the present per-capita CO_2 emission level of about 1 ton of carbon (tC) per year throughout the twenty-first century. This contrasts with the expectation that per-capita emissions are likely to double over this period in the absence of agreements to control emissions. As context for considering the challenge that this would impose, the typical American is now responsible for emission of 5–6 tC/year and the typical European is responsible for emission of about 3 tC/year, ignoring any imputed contribution that might arise from considering the energy used to create imported products on which they depend. Maintaining the CO_2 concentration at 550 ppmv beyond the twenty-first century would require, for example, that global emissions for the twenty-second century drop to more than a factor of 2 below current global emissions levels (i.e. to less than half of the 6 billion tonnes of carbon now emitted each year). Such an emissions level would require that global per-capita emissions of carbon be about one-third of the emission level of developing countries today, or one-twentieth of the current per-capita level in North America. Such low levels would not mean that per-capita *energy* use would need to be reduced by this amount, only that net per-capita use of fossil fuels (so emissions minus sequestration) would need to be this low. The IPCC (2001c) suggests that the necessary transition in sources of energy could be accomplished by meeting most of the world's energy needs using renewables and nuclear, if there is significant effort to improve the efficiency of energy end uses, although this is controversial (Hoffert et al., 2002).

As perhaps a start toward this objective, the Kyoto Protocol has been negotiated. Even though its goal is relatively modest in a scientific sense (i.e. by 2008–2012, developed nation emissions of a set of greenhouse gases would be reduced on average to several percent below their 1990 emissions level), it is proving to be a political challenge to get approved and implemented. Even if the US were to participate in its implementation, the emissions limitation would only begin to limit the rate of increase in the global CO_2 concentration because emissions from developing nations would not be similarly controlled and are, not surprisingly, projected to increase significantly along with their economic growth, even though many of these countries are making voluntary efforts to limit growth in emissions. With full implementation of the Kyoto Protocol through the twenty-first century, the CO_2 concentration would be expected to rise to about 660 ppmv as compared to an expectation of a rise to 710 ppmv in the absence of emissions controls. Such analyses make it clear that reducing CO_2 emissions to achieve stabilization at 550 ppmv would thus require significant further steps, and these would need to involve all nations. Accomplishing such an emissions reduction would require much more extensive introduction of non-fossil energy technologies, improvement in energy generation and end-use efficiencies, and switching away from coal to natural gas (unless extensive carbon sequestration of carbon in forests, oceans, or underground can be accomplished).

Absent such efforts on a global basis, the CO_2 concentration is projected to rise to about 800–1100 ppmv (about 3–4 times the preindustrial level) during the twenty-second century. Were this to occur, climate simulations indicate that the resulting warming would be likely to induce such potentially dangerous long-term, global-scale impacts as the initiation of the eventual melting of the Greenland and the West Antarctic ice sheets (each capable of inducing a sea-level rise of up to about 5–7 m over the next several centuries), the loss of coral reef ecosystems due to warming and rapid sea-level rise, the disruption of the global oceanic circulation (which would disrupt the nutrient cycle sustaining ocean ecosystems), and extensive loss or displacement of critical terrestrial ecosystems on which societies depend for many ecological services (e.g. see O'Neill and Oppenheimer, 2002).

What seems most clear from present-day energy analyses is that there is no "silver bullet" that could easily accomplish the major emissions reduction that is needed to achieve the UNFCCC objective of stabilization of atmospheric composition, even in the face of significant potential impacts. Rather, moving toward the objective will require a multifaceted international approach involving a much more aggressive (although not unprecedented) rate of improvement in energy efficiency, broad-based use of non-fossil technologies (often selecting energy sources based on local resources and climatic conditions), and an accelerated rate of technology development and implementation.

Conclusion

A major reason for controversy about dealing with the climate change issue results from differing perspectives among and within nations about how to weight a wide range of influential factors. These factors include, among other aspects:

1. the need for scientific certainty versus making decisions in the face of uncertainty;
2. the flexibility and adaptability of the energy system that provides for the national and global standard of living;
3. the potential for improving efficiency and developing new technologies;
4. the potential risk to "Spaceship Earth" that is being imposed by this inadvertent and virtually irreversible geophysical experiment;
5. the economic and environmental costs and benefits of taking early actions to reduce emissions (including what factors to consider in the analysis and how to weight the relative importance of long-term potential impacts versus better defined near-term costs); and
6. the weight to give matters of equity involving relative impacts for rich versus poor within a nation, for developed versus developing nations, and for current generations versus future generations.

Moving toward an international consensus on these issues sufficient to negotiate international agreements will require that the publics and governmental officials of all nations become better informed about the science of climate change, about potential impacts and their implications, and about potential options for and costs of reducing emissions. Moving toward collective action will require finding approaches that tend to balance and reconcile these (and additional) diverse, yet simultaneously legitimate, concerns about how best to proceed.

Michael C. MacCracken

Acknowledgments

The views expressed are those of the author and do not necessarily represent those of his former employer (the University of California) or those agencies supporting the Office of the US Global Change Research Program (now, the Climate Change Science Program Office) where he was on assignment. Presently serving as Chief Scientist for Climate Change Programs at the Climate Institute, Washington DC, and president (2003–07) of the International Association of Meterology and Atmospheric Sciences.

Bibliography

Environment Canada, 1998. *Canada Country Study: climate change impacts and adaptation*. Ottawa, Canada.

Hoffert, M., et al., 2002. Advanced technology paths to global climate stability: energy for a greenhouse planet. *Science*, **298**: 981–987.

IPCC (Intergovernmental Panel on Climate Change), 1996a. *Climate Change 1995: the science of climate change*, edited by J.T. Houghton, L.G. Meira Filho, B.A. Callander, N. Harris, A. Kattenberg, and K. Maskell. Cambridge: Cambridge University Press.

IPCC (Intergovernmental Panel on Climate Change), 1996b. *Climate Change 1995: economic and social dimensions of climate change*, edited by E.J. Bruce, Hoesung Lee, and E. Haites. Cambridge: Cambridge University Press.

IPCC (Intergovernmental Panel on Climate Change), 2001a. *Climate Change 2001: the scientific basis*, edited by J.T. Houghton, Y. Ding, D.J. Griggs, et al. Cambridge: Cambridge University Press.

IPCC (Intergovernmental Panel on Climate Change), 2001b. *Climate Change 2001: impacts, adaptation, and vulnerability*, edited by J.J. McCarthy, O.F. Canziani, N.A. Leary, D.J. Dokken, and K.S. White. Cambridge: Cambridge University Press.

IPCC (Intergovernmental Panel on Climate Change), 2001c. *Climate Change 2001: mitigation*, edited by B. Metz, O. Davidson, R. Swart, and J. Pan. Cambridge: Cambridge University Press.

NAST (National Assessment Synthesis Team), 2000. *Climate Change Impacts on the United States: the potential consequences of climate variability and change, overview report*. U.S. Global Change Research Program. Cambridge: Cambridge University Press. (Also see http://www.nacc.usgcrp.gov).

NAST (National Assessment Synthesis Team), 2001. *Climate Change Impacts on the United States: the potential consequences of climate variability and change, foundation report*. U.S. Global Change Research Program. Cambridge: Cambridge University Press. (Also see http://www.nacc.usgcrp.gov).

O'Neill, B.C., and Oppenheimer, M., 2002, Climate change: dangerous climate impacts and the Kyoto Protocol. *Science*, **296**: 1971–1972.

Cross-references

Agroclimatology
Air Pollution Climatology
Bioclimatology
Coastal Climate
Cultural Climatology
Climate Change and Global Warming
Human Health and Climate
Vegetation and Climate

CLIMATE CLASSIFICATION

The philosophical base of climate classification follows the same logic as all classifications and can be dealt with accordingly. Classification is variously defined but is well described as the systematic grouping of objects or events into classes on the basis of properties or relationships they have in common. The function of classification is also given different interpretations but essentially concerns the needs:

(a) to bring structure, order and simplicity to a complex system,
(b) to provide an intellectual shorthand,
(c) to identify spatial limits and boundaries, and
(d) to provide practical as well as theoretical uses.

Each of these cited functional goals must be considered in the formulation of a useful climate classification scheme. However, the identification of spatial limits and boundaries has proved of special interest to climatologists. Much of the literature dealing with climatic classification deals with limits of a climate group rather than the climate itself.

Of the many ways in which to classify components of the world's climate, the simplest is through classification using a single variable. Each distributional map of any climatic variable is, by definition, a classification scheme. The long history of climatic classification has its roots in the orderly presentation of the distribution of a single variable.

Single-variable classification

The earliest writings providing a consideration of the different climatic environments over the Earth can be traced to the contributions of thinkers in the sixth century BC and to such scholars as Aristotle and Plutarch. The ancient Greeks had postulated a spherical Earth and, through knowledge of the travel of the sun, identified the five-zone system – a torrid zone, two temperate zones and two frigid zones – still identified today. Aristotle provided the first quantitative boundaries of the system by identifying the Tropics, a division based upon astronomy and geography. Following the earlier Greeks, Ptolemy (AD 90–168) used day length as the classification variable and identified seven climates based upon duration of the longest day.

The ideas of Greek and Egyptian scholars passed to the Arab schools and such writers as Ibn Hauqual (ninth century), al Biruni (tenth century) and Idrisi (twelfth century) reproduced and expanded upon the work of the ancients. This work carried over to the geographical writings of the Renaissance and, in 1650,Varenius presented a table, the climata, giving the day lengths at the solstices for zones of the Earth. From the findings of these early theorists it is apparent that day length zones rather than temperature zones provided the quantitative expression of climate.

Distributions other than those based upon day-length appeared soon after thermometric data were assembled in the early nineteenth century. In 1817 von Humboldt produced the first isothermal map; while in 1846 Dove prepared monthly temperature distributions and went on to compute mean temperature of latitudes and hence the concept of temperature anomalies. Thereafter contributions became more numerous. Supan (1879) developed the first map of climatic belts based upon mean annual temperature and temperature of the warmest month. Later, in 1884, he produced a system in which zones were identified by regional names.

Single-variable classifications also used precipitation data and it was here that Koppen, the best known of climatic classifiers, made his first major climatological contribution. Both Schimper (1898) and Herbertson (1905) drew upon Koppen's work to produce regionalized world precipitation maps. As more data became available, world maps and regions were constructed for most climatic variables, and classification passed to the use of multiple variables

Vegetation distribution and climatic classification

The early development of climatic classification is related closely to vegetation studies. Of significance in these early studies of vegetation was the approach used. Systematic botanists arranged plants of the world into classes, genera, etc. using what might be termed a floristic approach; plant geographers used a physiognomy of vegetation in which plant communities expressed in such terms as forests, meadows, moors, etc., were identified. Table C3 provides an example showing the relationship between climate and vegetation.

In 1874, De Candolle proposed a classification that was designed for tracing the development of plants through geologic time. Six subdivisions were initially used but one, the megistotherms (mean annual temperature above 30°C), while prevalent in earlier geologic times, are found today only in the vicinity of hot springs. The remaining five, megatherms, mesotherms, microtherms, hekistotherms and xerophiles, were designated with letters A through E, with B referring to the xerophiles. Given that De Candolle was identifying groups as parallel zones, the insertion of moisture-based unit (B) in a temperature-based grouping is not really illogical. It has led, however, to many misunderstandings in its transfer to the Köppen system.

The Köppen system

From the pioneer work on vegetation classification came climate organizational schemes based upon vegetation distribution. This realm of classification is dominated by the work of Wladimir Köppen, who published his first paper in 1868 and continued to be a productive scholar up to his death

Table C3 Correspondence between climatic and vegetation types

Climate	Vegetation type name	Vegetation
Rainy tropical	Malayan	Evergreen rain forest
Subhumid tropical	Nicaraguan	Deciduous or monsoon forest
	Timoran	Savanna forest or woodland
	Visayan	Tropic grassland
Warm semiarid	Tampicoan Tamaulipan	Thorn forest, thorn scrub Desert savanna, wetter parts
Warm arid	Tamaulipan Sonoran Tripolian	Desert savanna, drier parts Subtropic desert Short grass; desert grass
Hyperarid	Atacaman	"Barren" desert
Rainy subtropical	Kyushun	Warm temperate rain forest
	Argentinean	Prairie
Summer-dry subtropical	Mediterranean	Sclerophyll woodland and scrub
Rainy marine	Tasman	Subantarctic forest
Wet-winter temperate	Oregonian	Conifer forest
Rainy temperate	Virginian	Mixed deciduous and conifer forest
Cool semiarid	Patagonian	Cold desert, wetter parts
Cool arid	Patagonian	Cold desert, drier parts
Subpolar	Alaskan	Taiga forest
Polar	Aleutian	Tundra and polar barrens

After Putnam et al. (1960).

in 1940. Köppen's work in classification may be viewed as a climatic determinism of vegetation types, while his greatest contribution was to stress the fundamental unity of pattern in the location of climatic regions throughout the world.

Table C4 provides the outline of Köppen's 1900 classification. It is seen that the idea of providing letters (A – E) is introduced but not emphasized. Instead, Köppen retained the nineteenth-century approach of using plant and animal names for identified regions. Some of these, for example the Penguin Climates, are certainly interestingly informative. The classification also contained some quantitative boundaries, but the well-known characteristics of the classification known to us did not appear until 1918.

The 1918 Köppen classification provides the basis for that used today (Table C5, Figure C14). Each climatic type is given a distinctive upper- or lower-case letter, each of which has a specific meaning. Local regional and biological names are dropped. Over time, changes to the initial system have been both suggested and implemented. The main changes introduced

Table C4 Climatic regions identified in Köppen's (1900) classification

A. Megathermal or tropical Lowland climates; coldest months >18°C

1. Liana
2. Baobab

B. Xerophytic climates – arid and semiarid. Continuous scarcity of precipitation

I. Coastal deserts of low latitudes

3. Garua (Fog)

II. Lowland desert and steppe experiencing great summer heat

4. Date palm
5. Mesquite
6. Tragacanth
7. East Patagonian

III. Lowland desert and steppe with cold winters and short hot summers

8. Buran
9. Prairie

C. Mesothermal or temperate climates. Coldest month below 18°C, warmest month over 22°C

I. Subtropical climates with moist, hot summers

10. Camellia
11. Hickory
12. Maize

II. Subtropical climates with mild, wet winters and dry summers

13. Olive
14. Heather

III. Tropical mountain climates and maritime climates of middle latitudes

15. Fuchsia
16. Upland savanna

D. Microthermal or cool climates. Warmest month between 10°C and 22°C

17. Oak
18. Spruce
19. Southern beech

E. Hekistotherm or cold climates. Warmest month between 0°C and 10°C

20. Tundra
21. Penguin or Antarctic
22. Yak (Pamir)
23. Rhododendron
24. Ice cap

Quantitative values were given for each identified type.

Table C5 The Köppen classification of climate

A. *Temperature of the coolest month above 18°C (64.4°F)*

- *f:* rainfall of driest month at least 6 cm
- *m:* rainfall of driest month > (10 – annual rainfall/25) but < 6 cm
- *w:* rainfall of driest month, 6 cm and dry in low-sun season (but does not meet *m* criteria)
- *s:* rainfall in driest month, 6 cm and dry in high-sun season (but does not meet *m* criteria)

Also recognized:

- *w′:* maximum rainfall in summer
- *w″:* two rainfall maxima
- *i:* annual temperature range < 5°C
- *g:* warmest month precedes summer solstice

B. *Evaporation exceeds precipitation for the year*

BW (desert) and *BS* (steppe) determined as follows:
Substitute *r* (annual precipitation in cm) and *t* (annual average tempature °C) in
$r = 2(t + 14)$ when 70% rain is in summer 6 months
$r = 2t$ when 70% rain is in winter 6 months
$r = 2(t + 7)$ when evenly distributed or neither of above
If $r >$ right-hand side of equation then climate is not a desert
If $r <$ right-hand side of equation then climate is a desert (*BS* or *BW*)
If right-hand side of equation is $< r/2$ then *BW*, else it is *BS*

- *h:* average annual temperature > 18°C
- *k:* average annual temperature < 18°C

Also recognized:

- *k′:* average temperature of warmest month < 18°C
- *n:* high frequency of fog
- *s:* 70% of rainfall in winter 6 months (summer dry)
- *w:* 70% of rainfall in summer 6 months (winter dry)

C. *Coolest month average temperature < 18°C and above −3°C; warmest month average temperature >10°C*

- *f:* at least 3 cm precipitation each month
- *w:* at least 3 times as much precipitation in a summer month as in driest winter month
- *s:* at least 3 times as much precipitation in a winter month as in the driest summer month, with 1 month < 3 cm

- *a:* warmest month > 22°C
- *b:* warmest month < 22°C but at least 4 months > 10°C
- *c:* only 1–3 months > 10°C

Also recognized:

- *x:* rainfall maximum in late spring or early summer: dry late summer
- *n:* high frequency of fog
- *i:* annual temperature range < 5°C
- *g:* warmest month occurs before summer solstice
- *t′:* warmest month occurs in fall
- *s′:* maximum rainfall in fall

D. *Coolest month average temperature < −3°C; Warmest month >10°C*

d: coldest month < −38°C
Other categories same as *C* climates

E. *Warmest month average temperature <10°C*

- *ET*: average temperature of warmest month between 0°C and 10°C
- *EF*: average temperature of warmest month < 0°C

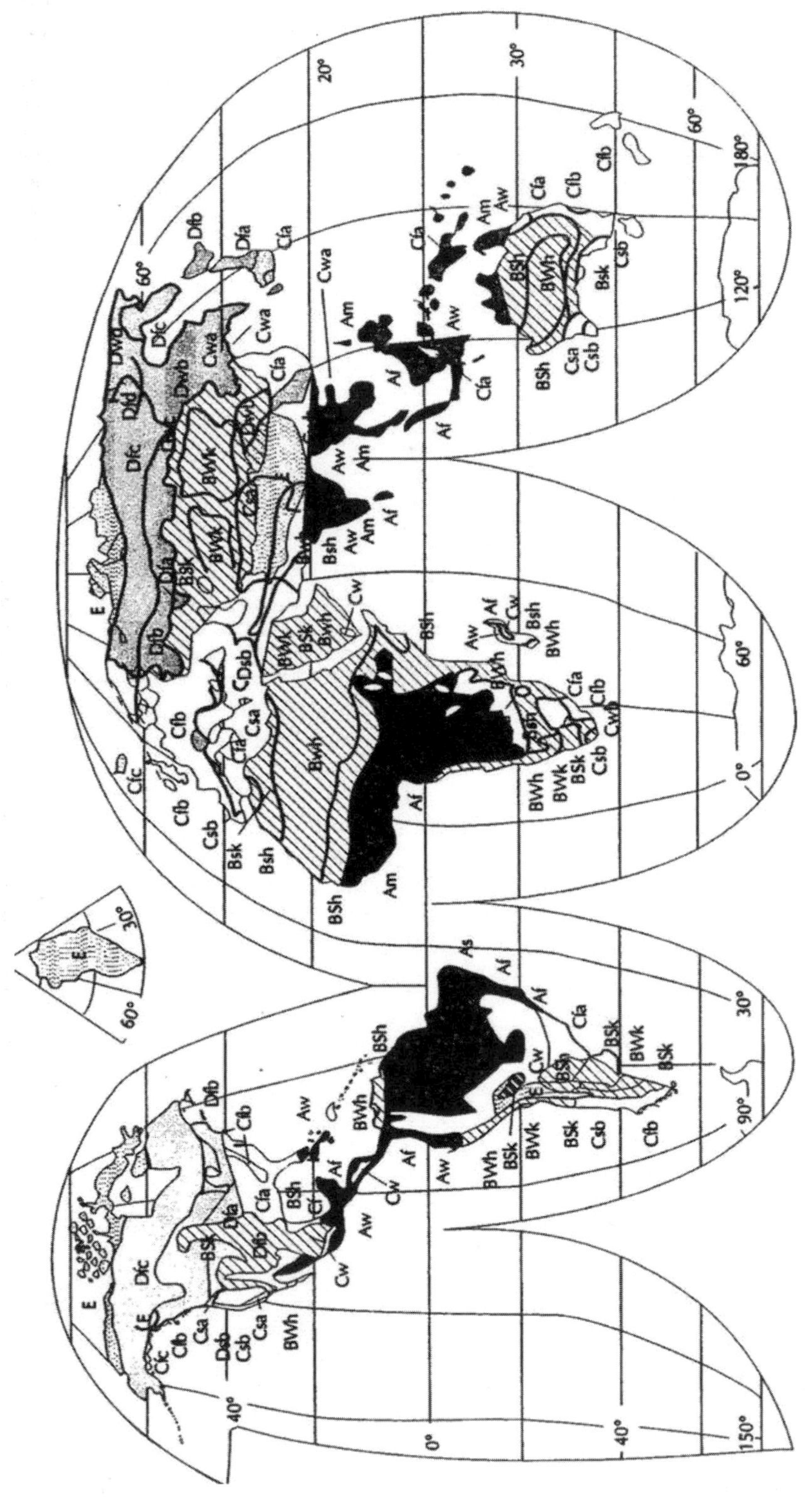

Figure C14 The Köppen classification of world climates.

by Köppen himself largely concerned the definition of the B climates. It remained for others to expound upon the work, and of these there were many. In a survey completed in 1952, Knoch and Schultz identify 70 works concerning the use of the Köppen system. Excellent reviews of the Köppen systems have been given by Thornthwaite (1943) and Wilcock (1968).

So dominant was the Köppen system that many fine alternate vegetation-based systems did not gain the widespread recognition. A system proposed by Vahl (1919), for example, overcomes the reliance upon average annual temperatures of the Köppen approach. Vahl based his classification upon four zonal climates that are subdivided by temperature limits that use both the warmest and coldest months. Another interesting approach was given by Troll (1958) who, by noting that "The life of plants, animals and humans . . . is subject to rhythm of climatic phenomena" produced a classification of the seasonal climates of the Earth. Graphical representation also plays an important role in the classification introduced by Peguy (1961).

A classification of considerable interest was proposed by Holdridge (1947), who explained its formulation. "While attempting to understand relationships between the mountain vegetation of an area of Haiti and other vegetation units of the island and surrounding regions, the literature was searched unsuccessfully for a comprehensive system which presented formation or vegetation units on a relatively equal or comparable basis." To meet this end he produced what is now called the Holdridge Model, which related temperature and precipitation to produce a series of nested hexagons. The ecological value of the system was ably demonstrated by Tosi (1964).

Table C6 Some empirical approaches to climate classification

Author(s)	Purpose	Base of System
Bagnouls and Gaussen (1957)	Biological climates	Duration of the dry season based upon the derived xerothermal index (*X*). Twelve major regions identified according to *X* values, temperature of coldest month, and frost/snow data.
Blair (1942)	An orderly description of world climates	Five main zonal climates distinguished: tropical (*T*), subtropical (*ST*), intermediate (*I*), subpolar (*SP*), and polar (*P*) Fourteen types and six subtypes distinguished using letter notation. Based upon precipitation and temperature data and related to vegetation types.
Brazol (1954)	Human comfort zones	Use of wet and dry bulb temperatures to establish comfort months. Twelve ranked classes ranked from no. 12 – lethal heat – to No. 1 – glacial cold.
Budyko (1958)	Distribution of energy in relation to water budget	Use of rational index of dryness to relate ratio of net radiation to energy required for vaporizing moisture. Moisture regions form basic unit.
Carter and Mather (1966)	Environmental biology	A modification of the 1948 Thornthwaite system.
Creutzberg (1950)	Climate–vegetation relationships	The annual rhythm of climate based on identification of *Isohygromen* (lines of equal duration of humid months) and *Tag-Isochione* (lines of equal daily snow cover duration). Four major zones differentiated, subdivided by monthly moisture values.
de Martonne (1909 and following years)	World regional (land-form) studies	Nine first-order divisions based upon temperature and precipitation criteria. Numerous subdivisions named for local areas in Europe. Considerable attention given to desert limits, but most boundaries derived nonquantitatively.
Emberger (1955)	Biologic (ecologic)–climate relationships	Two main climates differentiated: deserts and non-deserts. Differentiated and subdivided in terms of annual range of temperature and duration of light periods.
Federov	The "complex" method, utilization of day-to-day weather observations	An incomplete system that relies upon codification of daily weather events. For example, the first letter indicates character of prevailing wind; second letter character of temperature; third letter character of precipitation, cloudiness, humidity; fourth letter character various phenomena of atmosphere and state-of-ground surface.
Gorczynski (1945)	The decimal system	Ten "decimal" types associated with five main zonal climates. Emphasis on continental versus marine climates and definition of aridity.
Geiger-Pohl (1953)	World map of climate types	Modification of the Köppen system.
Köppen (1918 and following years)	See description in text	
Malmstrom (1969)	Precipitation effectiveness as a teaching scheme	Retention of the basic concepts of the Thornthwaite (1948) system but with arbitrary threshold values to express water needs of plants. Warmth index ($N-38\,m/100$) also used.
Miller (1931 and following years)	Temperature and precipitation used s main variables	A basic system for location of general world patterns.

Table C6 *(Continued)*

Author(s)	Purpose	Base of System
Papadakis (1996)	Agricultural potential of climatic regions	Use of "crop-ecologic" characteristics of a climate based upon empirically derived threshold values. Ten main climate groups recognized, each divided into subgroups which are themselves divided.
Passarge (1924)	Climate–vegetation relationships	Recognition of five climatic zones and their subdivision into 10 regions emphasizing vegetation distribution.
Penck (1910)	World climates in relation to studies in physical geography and physiography	Recognition of three main types of climates significant in determining weathering and erosion. Humid, Arid, and Nival. Each subdivided into two.
Peguy (1961)	See description in text	
Philippson (1933)	Climatic regionalization on the world, continental and regional levels	Based on temperature of warmest and coldest months and upon precipitation characteristics. Five climatic zones with 21 climatic types and 63 climatic provinces.
Putnam et al. (1960)	Coastal environments of the world, climate–vegetation characteristics	Fourteen types recognized. Climatic characteristics expressed as those occurring between the 25th and 75th percentile of the frequency distribution appropriate to each climatic type. Variables include mean maximum and minimum temperatures, mean annual precipitation and monthly precipitation frequency.
Terjung (1966)	Bioclimatic, based on humans	Use of comfort index and wind effect to identify physiological climates.
Thornthwaite (1948)	See description in text	
Trewartha (1954)	An orderly description of world climates	Modification of the Köppen system.
Vahl (1919)	World climates related to vegetation	Five zones appraised by temperature limits which are a function of the data for warmest and coldest months. Subdivision by precipitation expressed as a percentage of number of wet days in a given humid month.
von Wissman (1948)	World distribution of climate related to vegetation	Related to the Köppen approach. Five temperature zones subdivided by precipitation distribution and temperature regimes.

Table C6 provide a listing of some of empiric approaches to the classification of climate.

Applied empiric classifications

While the vegetation-based classifications, typified by the Köppen system, received most attention, many other authors proposed empiric schemes that might be classed as special-purpose. These applied empiric classifications were as diverse as the things upon which climate has an impact, but their nature can be demonstrated using selected examples.

In early published works using the applied empiric approach, climate was considered a component of the proposed classification rather than the base of the system itself. In the De Martonne (1909) classification the principal goal was to provide a basis for physiographic studies. This was accomplished by identifying nine main regions grouped upon a variety of parameters ranging from temperature thresholds, annual temperature range and annual average precipitation. De Martonne refined his approach in successive published books, many of which were translated into English (De Martonne, 1927). Another physiographer, A. Penck (1910), used an already-existing formula (rainfall/evaporation ratio) to suggest that humid, dry and transitional climates could form the bias of a world classification of utility in physical geographic studies.

The more recent applied empiric systems relate climate to a wide range of applied areas. Terjung (1968) formulated a bioclimatic classification, while in his text *World Climates,* Rudloff (1981) applies a bioclimatic aspect to the Köppen system. Of systems relating climate to crops, the classification outlined by Papadakis (1966) is of interest. In some ways such an approach is an extension of the crop analog studies completed by Nuttenson (1947).

The moisture balance

In a paper published in 1931, C. W. Thornthwaite proposed a climatic classification that appeared as a marked departure from preexisting systems. Unlike most classifications available at the time Thornthwaite based his system on the concept of precipitation and thermal efficiency. Of more significance at present, having superseded the earlier system, is Thornthwaite's 1948 classification (Figure C15). This shows a radical departure from the 1931 system because it makes use of the important concept of evapotranspiration. While the earlier system had been concerned with the loss of moisture through evaporation, the new approach considers loss through the combined process of evaporation and transpiration. Plants are considered as physical mechanisms by which moisture is returned to the air. The combined loss is termed evapotranspiration, and when the amount of moisture available is non-limiting, the term potential evapotranspiration is used. As with any widely used system, the 1948 classification has been subject to criticism. Many of the criticisms relate to the empiric formula used to express evapotranspiration, and to the

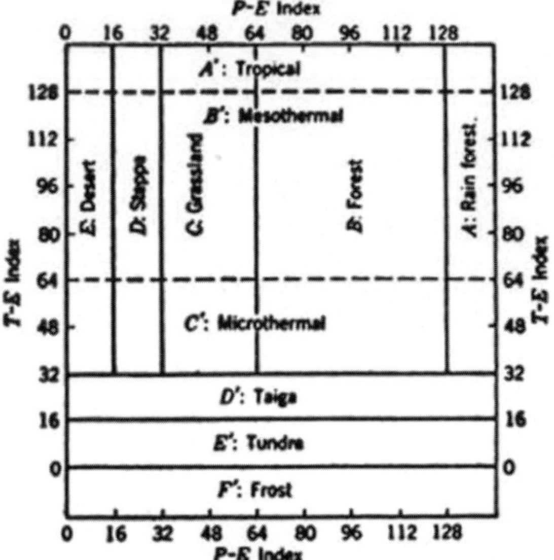

Figure C15 Temperature and humidity provinces of the Thornthwaite classification.

way in which the water budget of a station is manipulated. Its present status is well described by Muller and Oberlander (1984), who note that because its computations are time-consuming, the Thornthwaite system is not often used to define climatic regions on a global continental scale. At the same time it is an invaluable tool in water budget analyses.

Airmass and synoptic climatology

Just as climates may be classified according to the effect of climate upon environmental systems, so may they be grouped according to factors that contribute toward their cause. These causative effects relate to the origin (or genesis) of the climates concerned and resulting systems may be termed the genetic classifications. The development of synoptic meteorology and the concept of airmasses led to enormous improvements in comprehending weather. It seemed that a similar methodology could provide a suitable base for the study of genetic climatology for, as Thornthwaite (1943) suggested "Climatic types are to climate what airmasses are to weather". Subsequent to the basic delineation of airmass types, a number of significant climatological studies based upon airmasses were published. Willett (1933) described the properties of airmasses of the United States, while Showalter (1939) produced airmass identification tables. Despite these advances no comprehensive climatic classification based upon airmasses was formulated and, even at present, such systems are still only first approximations of the climatic complex of the Earth. Valuable maps of seasonal airmasses were presented by Berry and Bollay (1945), while Oliver (1970) used an airmass identification chart to classify world climates using set theory. But of the airmass groupings available perhaps the most utilitarian is that given by Strahler (1951 and following years) in his popular text *Physical Geography* (Figures C16 and C17). Essentially, three main groups of climates are differentiated:

1. Group I. Climates dominated by equatorial and tropical airmasses all the year.
2. Group II. Climates that occur between groups I and III and that are influenced by the interaction between tropical airmasses (group I) and polar airmasses (group III).
3. Group III. Climates controlled by polar airmasses.

Dynamic/synoptic classification

Today, some the most significant classification schemes are developed as part of dynamic/synoptic climatology. The forerunners to the most recent systems are those that were based upon identified physical determinants. One of the earliest workable systems of classifying climate by cause was introduced by Hettner (1930), with that by Flohn (1950) exemplifying a later approach. The nature of more recent work in synoptic classification, and the concepts of grosswetterlagen (large-scale weather patterns) is well described by Barry and Perry (1973). These authors note the various approaches that can be used in synoptic classification, and differentiate the static pattern (the features of circulation including pressure patterns) and the kinematic, in which streamline analysis is of importance. Of note, however, is that many of the studies using such approaches are purpose-oriented classifications and differ in objective and character from the classic world schemes. Barry (in Oliver et al., 1989) has described the significant advances in classification procedures in synoptic climatology made in recent years. He cites the progress in evaluating existing objective procedures (e.g. Yarnal and White, 1984), and the value of clustering techniques. Many of the techniques used are an integral part of the methods used in numerical classification.

Numerical classification

Of singular importance in the developmental history of climatic classification is the introduction of numerical classification or numerical taxonomy. Classifications may be derived either by the logical subdivision of a population or by agglomeration of similar individuals within that population. Most of the classification systems described thus far use logical subdivision as the methodological choice. By contrast, the use of any one of the discriminatory techniques now available, permit classification by agglomeration. The advances in electronic computing permit univariate and multivariate systems of climate to be generated from enormous databases and sophisticated techniques have been incorporated in existing classification methodologies. In reality the numerical taxonomic method supplies a rational mathematical approach to the basic needs of a classification. Willmott (1977) points out that decision making in deriving any classification must concern (a) the number of regions to be identified, (b) the identification of boundary criteria, (c) the variables selected to represent the climate, and (d) the methods in which the selected variables may be summarized. Prior to the utilization of taxonomic theory, the availability of extensive databases and electronic computing devices, decisions concerning these were highly arbitrary. Using the newer methodology the decisions obviate arbitrary choices. McBoyle (1971), following Steiner (1965), produced one of the earliest numerical classifications of climate, and his methodology illustrates the nature of the approach. Using 20 variables for 26 stations in Australia, McBoyle completed a factor analysis to produce three factors that together accounted

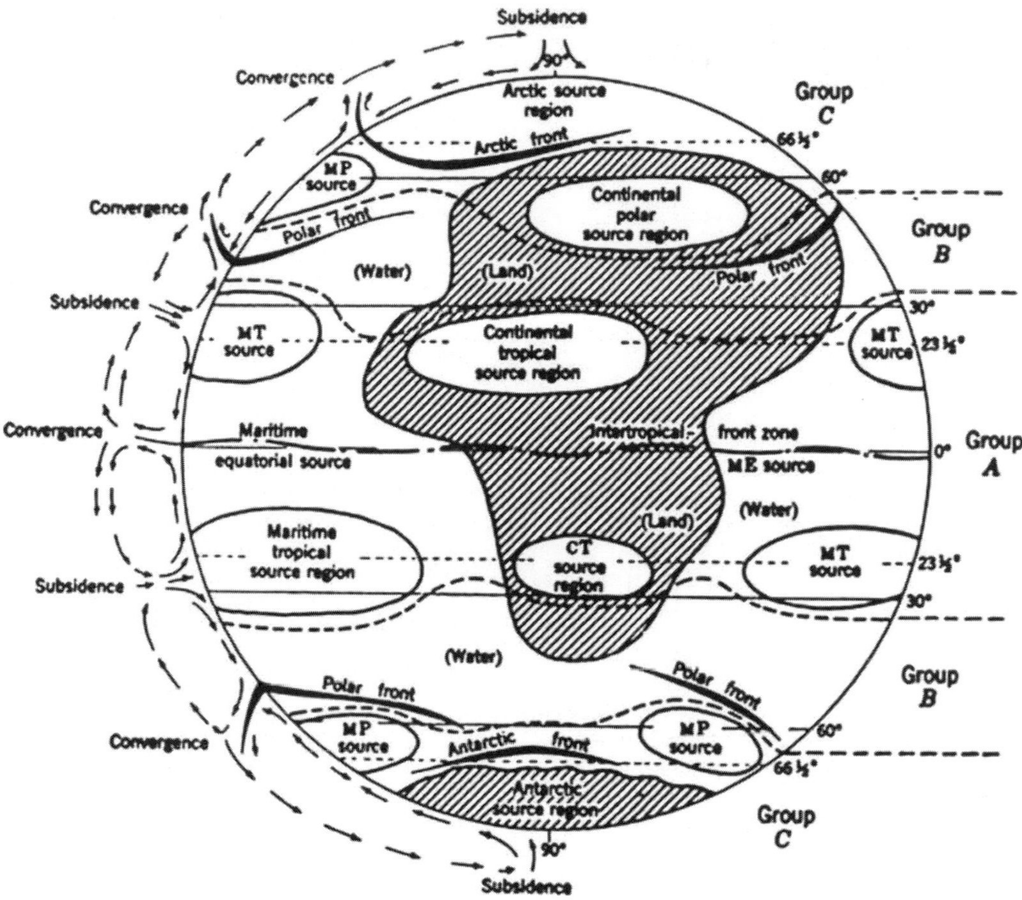

Figure C16 Global diagram showing the basis for three major climate divisions in the Strahler classification (from Strahler, 1951, by permission of John Wiley & Sons).

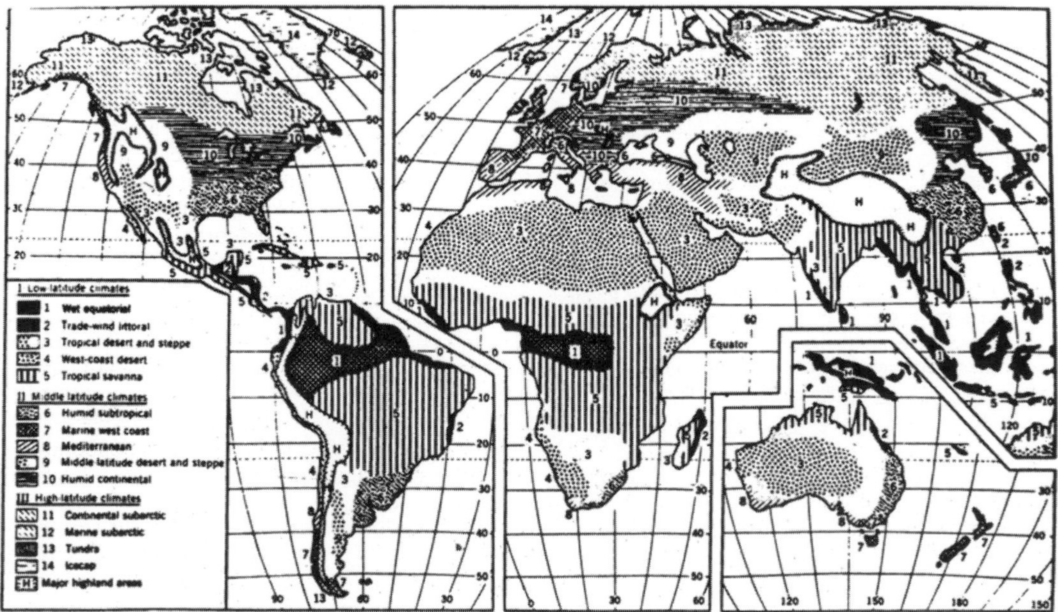

Figure C17 Generalized and simplified map showing distribution of climates according to the Strahler system (from Strahler, 1951, by permission of John Wiley & Sons).

Table C7 Numerical procedures applicable to climatic classification[a]

Similarity matrices
Correlation coefficients
Distance measures
Principal component transformation
Grouping
Taxonomic structure evaluation cophenetic correlation
Cluster selection and testing canonical correlation

[a] Derived from Balling (1984), who provides details of procedures.

for almost 90% of the total variance. Then, through linkage analysis, he showed how a grouping of Australian climates could produce a number of regions most applicable to the needs for which the classification was completed. This latter point is of considerable importance for, ultimately, classification rests with the goals of the user. This example is but one of many numerical classifications now available. In reviewing available grouping methods, Balling (1984) provides a most useful inventory of both the techniques and appropriate examples of their use. Table C7 outlines the approaches he identifies

The introduction of numerical taxonomy has opened new vistas for the regionalization of climate, and perhaps it is this realm that represents the future of climatic classification. In effect, it seems that, for general pedagogic purposes, users appear satisfied with world climatic systems that are available. It is possible that other general-purpose world classification systems may be proposed, but given the history of non-use of some excellent classifications, the likelihood of widespread use and acceptance is minimal. In contrast, the increasing significance of climate impact studies, applied climatic research and scenario analysis suggests that special-purpose classifications will be used, and that these will largely be based upon numerical classification.

John E. Oliver

Bibliography

Abbe, C., 1905. A first report on the relations between climates and crops. *U.S. Weather Bureau Bulletin*, vol. 36.

Alissow, B.P., 1954. *Die Klimate der Erde.* Berlin: Deutch. Ubers.

Ayoade, J.O., 1977. On the use of multivariate–climatic classification and regionalization. *Geophys. Biokl.* Series B, **24**: 257–267.

Bagnouls, F., and Gaussen H., 1957. Les climates ecologiques et leur classification. *Annals de Geographie*, **66**: 193–220.

Bailey, H.P., 1960. Toward a unified concept of the temperate climate. *Geographical Review*, **54**(4): 516–545.

Balling, R.C., 1984. Classification in climatology. In Gaile, G.L. and Willmott, C.J., eds., *Spatial Statistics and models.* Dordrecht: Reidel.

Barry R.G., and Perry, A.H., 1973. *Synoptic Climatology: methods and applications.* London: Methuen.

Berry, B.L., 1958. A note concerning methods of classification. *Annals of the Association of American Geographers*, **48**: 300–303.

Berry, F.A., and Bollay, E., 1945. Airmasses. In Berry, F.A., Bollay, E., and Beers, N.R., eds., *Handbook of Meteorology.* New York: McGraw-Hill, pp. 604–637.

Bunge, W., 1962. *Theoretical Geography.* Lund: Lund Series Geography, series C: General and Mathematical Geography.

De Candolle, A., 1874. *Geographie Botanique Raisonnee.* Geneva.

De Candolle, A., 1855. *Geographie botanique raisonee, ou exposition des faits principaux et des lois concernant la distribution geographigue des plantes de Ilepogue actuelle.* 2 volumes. Paris and Geneva.

Carter, D.B., and Mather, J.R., 1966. *Climatic Classification for Environmental Biology.* Elmer, NJ: C.W. Thornthwaite Associates, Laboratory of Climatology Publications, 19(4).

Carter, D.B., 1967. Farewell to the Köppen climate. *Proceedings of the Association of American Geographers*, 1967.

Chang, J-Hu, 1959. An evaluation of the 1948 Thornthwaite classification. *Annals of the Association of American Geographers*, **49**: 24–30.

Christenson W.J., and Bryson, R.A., 1966. An investigation of the potential of computer analysis for weather classification. *Monthly Weather Review*, **94**(12): 697–709.

Epperson, D.L., Davis, J.M., Watson, G.F., and Monahan, J.F., 1989. Bivariate normal classification of North American air masses using the Normix Program. *Journal of Climate*, **9**: 527–543.

Gentilli, J., 1958. *A Geography of Climate.* Perth: University of Western Australia Press.

Gilmour, J. S. L., 1952. Taxonomy and philosophy. In Huxley, J., ed., *The New Systematics.* London: Oxford University Press.

Grigg, D., 1965. The logic of regional systems. *Annals of the Association of American Georgraphers*, **55**(3): 264–278.

Gorczynski, W., 1942. Climatic types of California according to the decimal scheme of world climates. *Bulletin of the American Meteorological Society*, **28**: 161–165.

Gregory, S., 1954. Climatic classification and climatic change. *Erdkunde*, 246–252.

Hare, F.K., 1951. Climatic classification. In Stamp, L.D., and Wooldridge, S.W., eds., *Essays in Geography.* London: Longman, pp. 111–134.

Herbertson, A.J., 1905. The major natural regions: an essay in systematic geography. *Geographical Journal*, **1**: 300–312.

Hettner, A., 1930. *Die Klimate der Erde.* Leipzig: B.G. Teubner.

Holdridge, L.R., 1947. Determination of world plant formations from simple climatic data. *Science*, **105**: 367–368.

Johnston, R.J., 1968. Choice in classification: the subjectivity of objective methods. *Annals of the Association of American Geographers*, **58**: 575–539.

Kalkstein, L.S., Tan, G., and Skindlov, J.A., 1987. An evaluation of three clustering procedures for use in synoptic climatological classification. *Journal of Climate and Applied Meteorlogy*, **26**: 717–730.

Kendrew, W.G., 1936. *The Climates of the Continents.* New York: Oxford University Press.

Kesseli, J.E., 1942. The climates of California according to the Köppen classification. *Geography Review*, **32**(3): 476–480.

Köppen, W., 1918. Klassifikation der Klimate nach Temperatur, Niederschlag und Jahreslauf. *Petermanns Mitteilungen*, **64**: 193–203.

Köppen, W., 1923. *Die Klimate der Erde: Klimakunde.* Berlin and Leipzig: Grundriss.

McBoyle, G.R., 1971. Climatic classification by computer. *Australian Geographical Studies*, vol. **9**: 1–14.

Micklin, P.P., and Dickason, D.G., 1981. The climatic structure of the Soviet Union: a factor analytic approach. *Soviet Geographer*, **22**(4): 226–239.

Miller, A.A., 1931. *Climatology.* London: Methuen.

Newnham, R.M., 1968. A classification of climate by principal component analysis and its relationship to tree species distribution. *Forest Science*, **14**: 226–264.

Oliver, J.E., 1970. A genetic approach to Climatic Classification, 637. *Annals of the Association of American Geographers*, **60**(4): 615–637.

Oliver, J.E., 1991. The history, status and future of climatic classification. *Physical Geography*, **12**: 235–251.

Oliver, J.E., Barry, R.G., Brinkmann, W.A.R., and Rayner, J.N., 1989. Climatology. In Gaile, G.L., and Willmott, C.J., eds., *Geography in America.* Columbus: Merrill.

Oliver, J.E., Siddiqi, A.H., and Goward, S., 1978. Spatial patterns of climate and irrigation in Pakistan: A multivariate statistical approach. *Arch. Met. Geophys. Biokl.*, Ser B, **25**: 345–357.

Preston-Whyte, R.A., 1974. Climatic classification of South Africa: A multivariate approach. *South African Geological Journal*, **65**: 79–86.

Schimper, F.W., 1903. *Plant Geography Based upon a Physiological Basis* (trans. W. R. Fisher). Oxford: Clarendon Press.

Shear, J.A., 1964. The polar marine climate. *Annals of the Association of American Geographers*, **54**: 310–317.

Steiner, D., 1965. A multivariate statistical approach to climatic regionalization and classification. *Tiidshrift van het Koninklilk Nederlandsch Aadrilkskundig Genootschap*, **82**: 329–347.

Strahler, A.N., 1951. *Physical Geography.* New York: Wiley.

Terjung, W.H., 1970. Toward a climatic classification based upon net radiation. *Proceedings of the Association of American Geographers*, **2**: 140–142.

Thornthwaite, C.W., 1931. The climates of North America according to a new classification. *Geography Review*, **38**: 655.

Thornthwaite, C.W., 1948. An approach toward a rational classification of climate. *Geography Review*, **38**: 55–94.
Wilcock, A.A., 1968. Köppen after fifty years. *Annals of the Association of American Geographers*, **58**: 12–28.
Willmott, C.J., 1976. A component analytic approach to precipitation regions in California. *Arch. Met. Geophys. Biokl.*, Ser. B, **24**: 269–281.
Willmott, C.J., 1977. Toward a new regionalization of the contiguous United States. *Proceedings, Second National Solar Radiation Workshop*, Johnson Environmental and Energy Center, Huntsville, DI-DII.
Yarnal, B., and White, D.A., 1987. Subjectivity in a computer assisted synoptic climatology. *Journal of Climate*, **7**: 119–128.

Cross-references

Airmass Climatology
Antarctic Climates
Arctic Climates
Arid Climates
Atmospheric Circulation, Global
Climatology
Evaporation
Evapotranspiration
Humid Climates
Maritime Climate
Middle Latitude Climates

CLIMATE COMFORT INDICES

Climate has profound influence on human comfort and health. Weather elements can act singly or in combination to bring about effects on human bodies. Assessment of the influence of atmospheric environment on human comfort becomes an area of great concern in the context of human health and diseases, recreation, migration, tourism, heating and ventilating industry and architecture.

To obtain relative comfort the human body has to achieve a thermal equilibrium with the surrounding environment. The basis of human thermal response is the energy balance equation, which is expressed by

$$M \pm R \pm C - E = \pm S \tag{1}$$

where M is metabolic rate, R and C are heat exchange through radiation and convection, E is heat loss through evaporation, and S is heat storage in the body. Positive S indicates an energy gain for the body whereas negative S shows an energy loss. Physiologically, least thermal stress is experienced when S equals zero. To maintain a constant core body temperature near 37°C, the human body acts as an energy exchange surface and a balance is conserved through heat exchange with the environment and heat gain from metabolism.

Human thermal comfort is the result of the combined effect of several atmospheric variables. To quantify comfort or discomfort it is necessary to devise biometeorological indices to predict the human responses to weather stress and to assess the physiological strain. These indices incorporate a combination of atmospheric parameters depending upon their applications. Also, the terms in the energy balance equation given earlier may have to be simplified or detailed. In cold environments, evaporative–expired air energy loss is assumed to be constant, and it is presumed that metabolic rates need to increase to make up the losses in the energy balance. On the contrary, in warm environments, sweating, that leads to elevated metabolic rate and inefficiency of cooling due to sweat covering the body surface, makes calculation extremely complicated.

Empirical indices

Table C8 presents representative comfort indices, largely from Driscoll (1985), Beshir and Ramsey (1988), Auliciems (1997) and Pepi (1999). The listing is not inclusive.

The Effective Temperature (ET) combines temperature and humidity into a single index; and is defined as the temperature of still-saturated air that produced the same thermal sensation as the actual atmosphere. ET is criticized for overestimation of the effect of humidity at low air temperature and underestimation at higher air temperature. ET can be modified to include air movement. Its derivative corrected effective temperature (CET) is simple and easy to use, and is considered as a useful index for engineers. Gagge et al. (1971) introduced the new effective temperature (ET*). This revised ET* employs reference conditions of 50% relative humidity and 0.6 clo of clothing insulation instead of 100% humidity and 1.0 clo. ET* can be applied in environments in high altitudes and underground mines.

The Discomfort Index (DI) was developed by the US Weather Bureau (now the National Weather Service). It gives the equivalent temperature at 100% humidity. It was renamed the Temperature–Humidity Index (THI). The index does not include radiation and wind speed.

The major criticism of the THI is the use of 100% saturation in the computation. A new index called the Summer Simmer Index (SSI), which is calculated as the THI in the usual manner, but 10% humidity is employed, was developed. Ten percent humidity is selected because it is a value experienced at typical temperature extremes in dry climates of the United States. So the SSI is an index expressing how hot one feels relative to a dry climate. A new SSI was proposed in 1999 using dewpoint of 2°C (35°F) as the dry equivalent base. This new index is confirmed by subjects testing at Kansas State University.

Humiture (H) incorporates temperature and vapor pressure to characterize warm and hot environment. The Atmospheric Environment Service of Canada modified and changed the units in the index from degree Fahrenheit to degree Celsius and named the new index Humidex (Masterton and Richardson, 1979), that is used to inform the public as to the heat stress.

The Wet-bulb Globe Temperature Index (WBGT) is an index of heat stress and incorporates air temperature, wet bulb temperature and black globe temperature. The WBGT is a better index than the old ET as it shows a good correlation with sweat rate. The index is commonly used by the US Marine Corps to control drill activity outdoors, and by industries to estimate heat stress potential in industrial environments.

The Relative Strain Index (RSI) includes temperature, humidity, air movement, the insulating effect of clothing and net radiation of heat of the body. It assumes that a person, dressed in a light business suit, walking at a moderate pace in a very light air motion, has a metabolic rate of 3.2 km/h.

The Predicted 4-hourly Sweat Rate Index (P_4SR) adds the rate of heat produced from physical work to the WBGT. The index assumes to assess the amount of sweat perspired by a physically fit young man in the conditions under review for a period of 4 hours. It is commonly utilized in hot climates to estimate sweat loss and the required water intake.

The Belding–Hatch Heat Index also involves the heat rate produced by physical work. It compares the amount of sweat

Table C8 Representative comfort indices

Name of index	Description or comment
Effective Temperature (ET)	Physiological principles are not considered. Overestimates the effect of humidity at low temperature.
Corrected Effective Temperature (CET)	A modification of ET. Combines temperature, humidity, air velocity and radiation. Underestimates the effect of humidity and low air movement. Used in the British armed forces.
New Effective Temperature (ET*)	Used by ASHRAE for indoor comfort. Good indicator of physiological strain and warmth discomfort. Difficult to apply and complicated instruments are needed.
Temperature–Humidity Index (THI); formerly called the Discomfort Index (DI)	Gives the equivalent temperature at 100% humidity.
Summer Simmer Index (SSI)	Same as THI but 10% humidity is used. To indicate how hot one feels to a dry climate. New SSI proposed in 1999, also related to a dry environment and a dewpoint base of 2°C (35°F) is used.
Humiture (H)	Incorporates air temperature and vapor pressure. The Atmospheric Environment Service of Canada modified the index and named it the Humidex, that is more sensitive to humidity change.
Wet-Bulb Globe Temperature Index (WBGT)	Good correlation with sweat rate. Practical for industrial purposes. Estimation gets poorer under low humidity conditions. Metabolic workload is not considered.
Relative Strain Index (RSI)	Includes temperature, humidity, air movement, clothing and radiation of heat of the body. Applicable to assess heat stress of manual workers under shelter at various metabolic rates.
Predicted 4-hourly Sweat Rate Index (P_4SR)	Same as WBGT with addition of heat rate produced by physical work. Estimates sweat loss and required water intake.
Belding–Hatch Heat Index	Same as WBGT with addition of heat rate produced by physical work. Evaluates hot working environment. Underestimates the adverse effect of low wind speed and hot humid environment.
Apparent Temperature (AT)	To quantify sultriness.
Heat Index (HI)	A multiple regression model converted from AT. Used by the US NWS to alert the public of heat stress.
Wind Chill	Most widely used cold stress index. A new formula was implemented by the US NWS on 1 November 2001.

needed to maintain thermal equilibrium to the maximum amount of sweat that can be evaporated. It incorporates air temperature, globe temperature and the wet bulb temperature.

Steadman (1979a,b) studied sultriness that led to the development of the Apparent Temperature (AT). AT was an attempt to quantify sultriness. The index incorporated temperature, humidity, clothing and human physiology. The algorithms were simplified and regression equations were provided for indoor, shaded and sunny conditions (Steadman, 1984). The US National Wealth Service (NWS) converted AT into a multiple regression model called Heat Index (HI), which is used to alert the public to heat stress. Table C9 presents the apparent temperatures of HI, and the heat stress categories and their associated symptoms based on the severity of impact on humans.

Wind Chill (WCT), a measure of the combined effects of low temperature and wind, was devised by Siple and Passel (1945) on the basis of experiments examining the freezing rate of water in a cylinder at various temperatures and wind speeds in Antarctica. The index and derived wind chill equivalent temperature are widely used, especially among the media; meanwhile they have also received criticisms over the past five decades. The recent reviews of wind chill are given by Dixon and Prior (1987) and Brauner and Shacham (1995). Steadman (1971) introduced a new wind chill formula that included the effect of clothing and all forms of heat loss. The model was justified in human conditions. However, it is less popular due to its complex equation. Because of the dissatisfaction of the WCT, the Office of the Federal Coordinator for Meteorological Services and Supporting Research (OFCM) formed a special group called the Joint Action Group for Temperature Indices (JAG/TI), that included the Meteorological Services of Canada (MSC), the US NWS and several members of the academic community, to develop a new WCT formula. Besides mathematical modeling, human experiments were also involved. Volunteers, with temperature sensors on their faces, were exposed to various thermal conditions in a wind tunnel. The

Table C9 The Heat Index in terms of air temperature and relative humidity

Air temperature (°C)	Relative humidity (%) 0	10	20	30	40	50	60	70	80	90	100
20	16	17	17	18	19	19	20	20	21	21	21
21	18	18	19	19	20	20	21	21	22	22	23
22	19	19	20	20	21	21	22	22	23	23	24
23	20	20	21	22	22	23	23	24	24	24	25
24	21	22	22	23	23	24	24	25	25	26	26
25	22	23	24	24	24	25	25	26	27	27	28
26	24	24	25	25	26	26	27	27	28	29	30
27	25	25	26	26	27	27	28	29	30	31	33
28	26	26	27	27	28	29	29	31	32	34	36
29	26	27	27	28	29	30	31	33	35	37	40
30	27	28	28	29	30	31	33	35	37	40	45
31	28	29	29	30	31	33	35	37	40	45	
32	29	29	30	31	33	35	37	40	44	51	
33	29	30	31	33	34	36	39	43	49		
34	30	31	32	34	36	38	42	47			
35	31	32	33	35	37	40	45	51			
36	32	33	35	37	39	43	49				
37	32	34	36	38	41	46					
38	33	35	37	40	44	49					
39	34	36	38	41	46						
40	35	37	40	43	49						
41	35	38	41	45							
42	36	39	42	47							
43	37	40	44	49							
44	38	41	45	52							
45	38	42	47								
46	39	43	49								
47	40	44	51								
48	41	45	53								
49	42	47									
50	42	48									

The US National Weather Service has guidelines for the following ranges of heat index:

32–40°C Heatstroke, heat cramps, or heat exhaustion are possible with prolonged exposure and/or physical activity.
41–53°C Heat cramps or heat exhaustion are likely, and heatstroke possible with continued exposure.
≥54°C Heatstroke is highly likely with continued exposure.

NWS implemented the new WCT on 1 November 2001. The new formula is shown in equation (2).

$$\mathrm{WCT} = 13.13 + 0.62T - 13.95V^{0.16} + 0.486TV^{0.16} \quad (2)$$

where T is the air temperature (°C) and V is wind speed (m/s).Using the new WCT formula, frostbite may occur in 30 min or less according to Table C10.

Energy balance models

A large variety of models have been developed to include all energy exchange processes and outdoor environments. These models are extremely complex. The major models are presented in Table C11.

Fanger (1970) devised a model that comprises all forms of heat exchange. The model was based on the results of laboratory experiments involving nearly 1400 subjects wearing light clothing, and the relationship between the computed energy exchange and the predicted mean vote (PMV) of comfort sensation was derived. The PMV of comfort sensation is rated on a seven-point scale ranging from +3 (hot) to −3 (cold). The PMV equation represents the human physiological responses to particular environments. The model was designed for indoor conditions and has been modified to incorporate the complex outdoor radiation conditions, being known as the Klima Michel model (Jendritzky et al., 1979).

Höppe (1984) proposed the MEMI model (Munich Energy-balance Model for Individuals) that took into account the basic thermoregulatory processes, such as constriction or dilation of peripheral blood vessels and sweat rate. The MEMI consists of three equations. They are the energy balance equation of the total body, the equation of heat flux from the body core to the skin, and the equation of heat flux from the skin through clothing layers to the surface of the clothing. The model is the basis for the calculation of the physiological equivalent temperature (PET), that is equivalent to the air temperature at which, in a typical room indoors, the heat budget of the human body is balanced with the same core and skin temperatures as under the outdoor conditions being assessed.

Blazejczyk (1994) proposed the MENEX model (Man–environment Heat Exchange Model), that is the evolution of the model proposed by Budyko (1974). The model also takes into account human thermo-physiological considerations. Its applications include forecasting human thermal conditions outdoors, and evaluations of bioclimates and heat load at work.

De Freitas (1985) developed the STEBIDEX (Skin Temperature Energy Balance Index) and the HEBIDEX (Heat

Table C10 The new Wind Chill Temperature (WCT) chart

	T_{air}											
V_{10}	5	0	−5	−10	−15	−20	−25	−30	−35	−40	−45	−50
5	4	−2	−7	−13	−19	−24	−30	−36	−41	−47	−53	−58
10	3	−3	−9	−15	−21	−27	33	−39	−45	−51	−57	−63
15	2	−4	−11	−17	−23	−29	−35	−41	−48	−54	−60	−66
20	1	−5	−12	−18	−24	−31	−37	−43	−49	−56	−62	−68
25	1	−6	−12	−19	−25	−32	−38	−45	−51	−57	−64	−70
30	0	−7	−13	−20	−26	−33	−39	−46	−52	−59	−65	−72
35	0	−7	−14	−20	−27	−33	−40	−47	−53	−60	−66	−73
40	−1	−7	−14	−21	−27	−34	−41	−48	−54	−61	−68	−74
45	−1	−8	−15	−21	−28	−35	−42	−48	−55	−62	−69	−75
50	−1	−8	−15	−22	−29	−35	−42	−49	−56	−63	−70	−76
55	−2	−9	−15	−22	−29	−36	−43	−50	−57	−63	−70	−77
60	−2	−9	−16	−23	−30	−37	−43	−50	−57	−64	−71	−78
65	−2	−9	−16	−23	−30	−37	−44	−51	−58	−65	−72	−79
70	−2	−9	−16	−23	−30	−37	−44	−51	−59	−66	−73	−80
75	−3	−10	−17	−24	−31	−38	−45	−52	−59	−66	−73	−80
80	−3	−10	−17	−24	−31	−38	−45	−52	−60	−67	−74	−81

Approximate thresholds:

Risk of frostbite in prolonged exposure: wind chill below	−25	
Frostbite possible in 10 minutes at	−35	Warm skin, suddenly exposed. Shorter time if skin is cool at the start.
Frostbite possible in less than 2 minutes at	−60	Warm skin, suddenly exposed. Shorter time if skin is cool at the start.

T_{air} = air temperature in °C and V_{10} = observed wind speed at 10 m elevation, in km/h; courtesy of the Meterological Service of Canada.

Table C11 Major energy balance models

Name of model	Description or comment
Predicted Mean Vote (PMV)	First model including all forms of heat exchange. Mean skin temperature and sweat rate are quantified as comfort values, being only dependent on activities.
MEMI	Thermo-physiological heat balance model. Basis for the calculation of PET.
MENEX	Thermo-physiological heat balance model.
Klima Michel Model	Modification of Fanger's PMV to incorporate complex outdoor radiation.
HEBIDEX and STEBIDEX	To identify the relationship between the environmental stress being experienced and the condition of mind, being expressed in thermal sensation.
STOEC	Based on the fall of body core temperature. Applied in cold environment.

Budget Index). The adequacy of both indices was tested by empirical study of human thermal response outdoors on the beach. The results indicated that the STEBIDEX model could provide a more reliable estimate of thermal sensation.

De Freitas and Symon (1987) developed the STOEC (Survival Time Outdoors in Extreme Cold). The STOEC incorporates temperature, wind, solar radiation, clothing and the human energy budget, which is based on the rate of fall of core temperature form 37°C to 27°C. It provides the calculations for the shortest, longest, mean and indefinite survival times. It is useful in estimating the duration of research and rescue operations.

Conclusion

The trends in the development of comfort indices reflect the move from the conventional static approach to environmental assessment. This change has been toward the use of dynamic energy exchange models, that synthesize the interaction between the human thermo-physiological processes and the atmospheric environments. However, it is important for an index to provide realistic meaning and simple interpretation of abstract index value to the lay public.

Yuk Yee Yan

Bibliography

Auliciems, A., 1997. Comfort, clothing and health. In Thompson, R.D., and Perry, A., eds., *Applied Climatology: Principles and Practice*. New York: Routledge, pp. 155–174.

Beshir, M.Y., and Ramsey, J.D., 1988. Heat stress indices: a review paper. *International Journal of Industrial Ergonomics*, **3**: 89–102.

Blazejczyk, K., 1994. New climatological-and-physiological model of human heat balance outdoor (MENEX) and its applications in bioclimatological studies in different scales. In Blazejczyk, K., and Krawczyk, B., eds., *Bioclimatic Research of the Human Heat Balance*. Warsaw: Polish Academy of Sciences, Institute of Geography and Spatial Organization, pp. 27–58.
Brauner, N., and Shacham, M., 1995. Meaningful wind chill indicators derived from heat transfer principles. *International Journal of Biometeorology*, **39**: 46–52.
Budyko, M.I., 1974. *Climate and Life*. New York: Academic Press.
De Freitas, C.R. 1985. Assessment of human bioclimate based on thermal response. *International Journal of Biometeorology*, **29**: 97–119.
De Freitas, C.R., and Symon, L.V., 1987. A bioclimatic index of human survival times in the Antarctic. *Polar Record*, **23**: 651–659.
Dixon, J.C., and Prior, M.J., 1987. Wind-chill indices – a review. *Meteorological Magazine*, **116**: 1–15.
Driscoll, D.M., 1985. Human health. In Houghton, D.H., ed., *Handbook of Applied Meteorology*. New York: John Wiley, pp. 778–814.
Fanger, P.O., 1970. *Thermal Comfort*. New York: McGraw-Hill.
Gagge, A.P., Stolwijk, J.A.J., and Nishi, Y., 1971. An effective temperature scale based on a simple model of human physiological regulatory response. *ASHRAE Transactions*, **77**: 247–272.
Höppe, P., 1984. *Die Energiebilanz des Menschen* (dissertation). Wiss Mitt Meteorological Institute, University of München, p. 49.
Jendritzky, G., Sönning, W., and Swantes, H.J., 1979. *Ein objektives Be-wertungsverfahren zur Beschreibung des thermischen Milieus in der Stadt- und Landschaftsplanung (Klima-Michel-Modell)*. Beiträge der Akademic für Raumforschung and Landesplanung. Hanover: Hermann Schroedel, p. 28.
Masterton, J., and Richardson, F., 1979. *Humidex: A Method of Quantifying Human Discomfort due to Excessive Heat and Humidity*. Downsview, Ontario: Atmospheric Environment Service, CLI 1-79.
Pepi, J.W., 1999. The new Summer Simmer Index: a comfort index for the new millennium, from http://www.idad.com/iss/
Siple, P.A., and Passel, C.F., 1945. Measurements of dry atmospheric cooling in sub-freezing temperatures. *Proceedings of the American Philosophical Society*, **89**: 177–199.
Steadman, R.G., 1971. Indices of windchill of clothed persons. *Journal of Applied Meteorology*, **10**: 674–683.
Steadman, R.G., 1979a. The assessment of sultriness, Part I: A temperature–humidity index based on human physiology and clothing science. *Journal of Applied Meteorology*, **18**: 861–873.
Steadman, R.G., 1979b. The assessment of sultriness, Part II: Effects of wind, extra radiation and barometric pressure on apparent temperature. *Journal of Applied Meteorology*, **18**: 874–885.
Steadman, R.G., 1984. A universal scale of apparent temperature. *Journal of Climate and Applied Meteorology*, **23**: 1674–1687.

Cross-references

Architecture and Climate
Heat Index
Seasonal Affective Disorder
Human Health and Climate
Wind Chill

CLIMATE DATA CENTERS

With recent advances in computer storage capability, and the proliferation of the Internet, climate data for virtually every country in the world are becoming easily obtainable for researchers and the general public alike. Many countries have set up, within the organizational structure of the various departments of meteorology and climatology, systems for gathering, editing and archiving the volumes of climate information that are available. While the following is not a comprehensive list of the climate centers around the world, it represents places of international renown that service the research and public domain communities by providing raw, edited, or aggregated datasets and have important links to modeling and research centers.

Japan

Tokyo Climate Center

The Tokyo Climate Center (TCC) was established in April 2002 to provide climate data and services to the Asian-Pacific environment and to assist in the mission of the Japan Meteorological Agency and the National Meteorological and Hydrological Service of Japan. The activities of the center include monthly reporting of climate events of a global nature and investigations in the climate systems that may affect the weather of the Asian-Pacific realm, including, and especially reporting on, the El Niño phenomenon and to produce ensemble monthly forecast models and verification of them.

The TCC is involved in assisting with the technical expertise needed to design the delivery of climate services throughout the Asia-Pacific region and to provide climatology data to the region. It also works in concert with other organizations in nations within the region to help facilitate the delivery of climate data, services and forecasts by the meteorological and climatological agencies in other Asia-Pacific nations.

United Kingdom

British Atmospheric Data Center (BADC)

The BADC is the main archive for atmospheric data in the UK. It is one of seven data centers designated by the National Environmental Research Center and is housed within the Space Science and Technology Department of the Rutherford Appleton Laboratory in Oxfordshire. The BADC produces datasets derived from NERC-funded projects, and serves as an efficient link to data provided by such services as the European Center for Medium-Range Weather Forecasting Office (the ECMWF) and the UK Meteorology Office.

The BADC was established in 1994 building upon the work of the former Geophysical Data Facility (GDF). The BADC presently holds up to 70 datasets of a variety of atmospheric variables and is the archive for the ECMWF reanalysis dataset. In addition, research work by BADC staff includes projects delegated to three main groups: atmospheric dynamics, cloud physics and data assimilation. BADC staff are also engaged in research efforts with the RAL SSTD Atmospheric Modeling and Data Interpretation Group (AMDI).

United States

Carbon Dioxide Information Analysis Center, CDIAC, Oak Ridge, Tennessee

The CDIAC, operating out of the Oak Ridge Laboratory, compile data and information related to greenhouse gases, their rates of emission, and the cycle of these gases between the oceans, the biosphere and the atmosphere. They also have surface climate data, edited for selected stations that are part of the United States Historical Climate Network and historic mid-tropospheric height data. The CDIAC also houses surface data for the People's Republic of China. The World Data Center for Atmospheric Trace Gases is housed in the CDIAC.

The Climate Diagnostics Center (NOAA – CIRES, CDC), Boulder, Colorado

The CDC is a research and archival unit designed to analyze climate events. It archives the National Center for Environmental Protections Re-Analysis Dataset, as well as time series of sea-surface temperatures, various El Niño and Southern Oscillation indices, as well as the northern hemisphere teleconnection indices.

US National Snow and Ice Data Center (NSIDC), Boulder, Colorado

The NSIDC, under a joint agreement with the University of Colorado, operates houses the World Data Center for Glaciology. The center compiles, archives and maintains data related to snow cover, snow depth, snow pack, sea-ice extent and depth, freshwater ice and glacier extent.

US National Climate Data Center (NCDC), Asheville, North Carolina

The NCDC has the largest climate and meteorology database in the world. It also houses the World Data Center for Meteorology. Designed originally to house weather records obtained by National Weather Service offices and cooperative stations, it now archives 99% of the data obtained by departments of the National Oceanic and Atmospheric Administration (NOAA). The center also maintains and archives radar data, satellite imagery from 1960, radiosonde,

Table C12 Selected members of the World Data Center network directly related to research in climatology

World Data Centers	Affiliation and location	Data emphasis
Atmospheric Trace Gases	Carbon Dioxide Information Center, Oak Ridge National Laboratory, Oak Ridge, Tennessee	Data related to atmospheric trace gases that affect and contribute to the Earth's energy budget.
Glaciology – USA	National Snow and Ice Data Center, Boulder, Colorado	Snow cover, snow pack, sea-ice extent, sea-ice thickness, images and pictures of historic glacial extent.
Glaciology – Geocryology, China	Lanzhou Institute of Glaciology and Geocryology, Chinese Academey of Sciences. Lanzhou, China	Glacial atlas of China, snow cover, glacial extent and variation, periglacial data and hydrologic data.
Glaciology, UK	The Royal Society and the Scott Polar Research Institute, University of Cambridge	Data related to glaciers, periglacial processes, satellite imagery, snow and ice chemistry.
Airglow – Japan	National Astronomical Observatory, Tokyo, Japan	Airglow data, solar radiation data.
Meteorology – China	Climate Data and Applications Office, Beijing, China	Real-time synoptic data, historical surface climate data, dendrochronology data, glacial data and atmospheric chemistry data.
Meteorology – Russia	Federal Service of Russia for Hydrometeorology and Monitoring of the Environment, Obninsk, Russia	Both surface observations and gridded surface and upper air meteorology data, marine ship observation data, and aerology data.
Meteorology – USA	National Climate Data Center, Ashville, North Carolina	Archives of data from many national and international research projects and experiments including data from the IGY 1957–1958 and International Quiet Sun Year, 1964–1965, among many others.
Paleo-climatology – USA	National Geophysical Data Center (NGDC), Boulder, Colorado	Dendrochronology data, ice-core data, sea-floor sediment cores, Coral data, proxy data on climatic forcing, including volcanic aerosol data, ice volume, atmospheric composition, etc. Numerical model simulation experiments data, climate reconstructions and maps.

rawinsonde, rocketsonde and other advanced technology sources. The center also houses weather data from diaries, notes and other sources. The center also services a network of regional and state climatology data centers designed to deliver climate data of regional and local interest.

International Organizations

The Intergovernmental Panel on Climate Change Data Distribution Center (DCC)

The DCC acts as a gateway to many datasets used to assess changes in the global climate. The DDC acts in concert with the Deutsches Klimareschenzentrum (DKRZ) in Germany and the Climate Research Unit (CRU) in the United Kingdom to archive and distribute much of the data that the IPCC provides. The DDC maintains links to locations that have data in the public domain. Output from Global Circulation Model (GCM) runs are available at the DKRZ distribution site. The DDC also archives the CRU Global Climate Dataset that has monthly surface variables from 1905 to 1995 and is arranged on a 0.5° by 0.5° latitude–longitude grid. This includes all land grid cells except for Antarctica. The DDC also maintains another dataset arranged by country and containing both averages (1961–1990) and month-by-month variations (1901–1995).

World Climate Data Centers

The World Climate Data Center system is a network of data centers that was originally designed to archive the data derived during the International Geophysical Year (IGY) but has subsequently expanded to include climatological, meteorological, astronomical, oceanograhical as well as geophysical datasets. The centers are usually housed within other organizations. Each of the following regions, United States, Europe, Australia, China, Russia, Japan and India, have several centers specializing in one of the areas of geophysical data and research. Table C12 lists those centers that have data directly related to research in climatology in particular.

Robert Mark Simpson

Sources

Carbon Dioxide Information Analysis Center (CDIAC), Oak Ridge National Laboratory, 2003. http://cdiac.ornl.gov/home.html

Intergovernmental Panel on Climate Change (IPCC) Data Distribution Center (DDC), 2003. http://ipcc-ddc.cru.uea.ac.uk/

Japan Meteorological Agency, 2002. http://www.jma.go.jp/JMA_HP/jma/jma-eng/contents/meet-cli/meeting_e.html

Kaiser, D. et al., 2003. Climate Data Bases of the People's Republic of China 1841–1988. http://cdiac.ornl.gov/ndps/tr055.html

US Department of Commerce, NOAA, Climate Diagnostics Center, 2003. http://www/cdc.noaa.gov

US Department of Commerce, NOAA, National Climate Data Center, 2003. http://www.ncdc.noaa.gov/whatisncdc.html

Tokyo Climate Center, 2003. http://okdk.kishou.go.jp/about.html

World Data Center – Airglow, Tokyo, Japan. http://www.ngdc.noaa.gov/wdc/wdcc2/wdcc2_airglow.html

World Data Center – Atmospheric Trace Gases. http://www.ngdc.noaa.gov/wdc/wdca/wdca_atmosgas.html

World Data Center – Glaciology, Langshou, China. http://www.ngdc.noaa.gov/wdc/wdcd/wdcd_glaciology.html

World Data Center – Glaciology, Cambridge, United Kingdom. http://www.ngdc.noaa.gov/wdc/wdcc1/wdcc1_glaciology.html

World Data Center – Meteorology, Beijing, China. http://www.ngdc.noaa.gov/wdc/wdcd/wdcd_meteorology.html

World Data Center – Meteorology, Obninsk, Russia. http://www.ngdc.noaa.gov/wdc/wdcb/wdcb_meteor.html

World Data Center A – Meteorology, United States. http://www.ngdc.noaa.gov/wdc/wdca/wdca_meteor.html

World Data Center – Paleoclimatology, Boulder, Colorado. http://www.ngdc.noaa.gov/wdc/wdca/wdca_paleo.html

World Data Center System. http://www.ngdc.noaa.gov/wdc/wdcmain.html

Cross-references

Climate Modeling and Research Centers
Models, Climatic
Oscillations
Teleconnections

CLIMATE HAZARDS

This review of climate hazards outlines the great variety of such events and their impacts. It illustrates some of the present concerns of and future challenges for the climatologist.

Nature and variety of the threats

Potentially hazardous atmospheric phenomena include tropical cyclones, thunderstorms, tornadoes, drought, rain, hail, snow, lightning, fog, wind, temperature extremes, air pollution, and climatic change. Hazards may arise from single-element extremes, such as excessively high temperatures causing physiological heat stress; or from various combinations of elements, such as tropical cyclones with high wind, torrential rain and storm surge, all posing threats to people and their property. Estimates of annual global economic losses due to meteorological disasters showed a fourfold increase from the 1960s to the early 1990s to nearly $90 billion, and insured losses increased about tenfold to over $50 billion (Bruce, 1994).

Hazards producing disasters with economic, political, and social repercussions have always been newsworthy. The public view of climatic hazards has been nurtured by frequent media reports of events such as the tropical cyclone that killed about 300 000 people in the former East Pakistan in 1970; the severe frost damage to coffee crops in Brazil in 1975, followed by huge increases in world coffee prices; the storm that devastated the Fastnet yacht race in 1980; the exceptional storm in 1988 which left the most severe damage for many generations in southern England; another tropical cyclone in Bangladesh in 1991 which killed about 139 000 people; Yangtze floods which destroyed over four million dwellings in 1991; Hurricane Andrew which produced damage estimated at over $25 billion in Florida and Louisiana in 1992; ice and snow storms in eastern North America in 1993, 1994 and 1998, with total damages of several billions of dollars; the heatwave in the Midwest and eastern USA in 1995; El Niño-related weather events in 1997–1998 around the world; Hurricane Mitch in Central America in 1998; and in 2002 Typhoon Rusa, the deadliest storm in South Korea in over 40 years, as well as widespread flooding in Europe.

Perhaps Leonardo da Vinci was the first to recognize the downburst associated with severe thunderstorms (Gedzelman, 1990), but there is little evidence of research on climatic hazards before the nineteenth century, with early emphasis mainly on the cataloguing of events, such as tropical cyclones in a particular area (e.g. Knipping, 1893). The work of Barrows (1923) on human structuring of, and adjustment to, environment provided an identifiable start for a more balanced research tradition, the development of which may be traced through studies such as those by Visher (1925) on tropical cyclones in the Pacific Ocean and by Foley (1957) on drought in Australia. Interest and emphasis have shifted to a much greater concern with the nature and alleviation of impacts, as reflected in the wide range of natural, physical, and behavioral scientists, and economists working on frequencies, magnitudes, causes, behavior, and impacts of climatic hazards, as well as with studies of perception, preparedness, planning, mitigation and control. This diversity of interests, and the directions in which research has been heading, is exemplified in works such as those by White (1974), Heathcote and Thom (1979), Hewitt (1983, 1997), Burton et al. (1993) and Smith (2001).

Identification of hazardous events is not always easy, although certain criteria are usually present. They include property damage; economic loss, such as loss of income or a halt in production; major disruption of social services, communications and transportation; excessive strain on essential services such as police, fire, hospitals, and public utilities; and psychological stress, injuries, and fatalities (e.g. Changnon, 1989, 1999; Barker and Miller, 1990; Morison and Butterfield, 1990; Hoque et al., 1993; Brugge, 1994a, b; Kalkstein, 1995; Fink et al., 1996; Curran et al., 2000; Dupigny-Giroux, 2000; Palecki et al., 2001; Ulbrich et al., 2001; Pielke and Carbone, 2002).

It can be difficult to distinguish between the atmospheric and nonatmospheric factors producing climatic hazards. Hot, dry winds may promote fire disasters but do not cause them. An avalanche depends on the quality and quantity of snow and on the timing of a thaw, but it is unlikely to happen without certain slope characteristics. Increasing losses over time point to the significance of socioeconomic factors in exacerbating the vulnerability of communities to hazard events. More than half of the world's population lives within 60 km of the ocean. Known flood plains, hurricane- and drought-prone regions have experienced development pressures and increased volumes and values of property at risk (e.g. Riebsame et al., 1986; Pielke and Pielke, 1997; Kunkel et al., 1999; Changnon et al., 2000; Easterling et al., 2000). Climate change is also seen by some as creating potential new threats (e.g. Obasi, 1994; Woodhouse and Overpeck, 1998; Yarnal et al., 1999; Parry et al., 2001). There are already very large disparities in the nature and magnitude of hazard losses in different parts of the world; differences mainly attributable to socioeconomic factors (Degg, 1992).

Atmospheric, and socioeconomic, factors fulfill a variety of roles in the development of a hazardous situation. A broad distinction can be made between phenomena such as tropical cyclones or severe local storms, and their associated weather extremes, which involve the sudden impact of very large amounts of energy discharged over relatively short periods; and those features that become hazards only if they exceed tolerable magnitudes within or beyond certain limits (Gentilli, 1979). In the latter category can be included heat waves; cold spells; flood-producing rains; frosts; fogs; droughts; high winds, snow and ice associated with extratropical low-pressure systems; and the effects of climatic change. Some climatic hazards result from human activity. Under the broad umbrella of air pollution, these include hazards to human health, the possible dangers of inadvertent modification of climatic patterns, and the effects of acid rain on natural ecosystems.

The treatment of climatic hazards that follows is not exhaustive, nor is it the only approach possible, but it illustrates various types of threats and their impacts. Examples of alternative approaches include Changnon and Changnon (1992), who group 11 causes of storms into four classes; and Smith (1997, 2001), who classifies climatic hazards into four groups: single-element extremes from common and less common hazards, and compound-element events from primary or secondary hazards.

Sudden-impact hazards

Tropical cyclones

Tropical cyclones can be the most dangerous and deadly storms on Earth. They are usually very mobile and relatively unpredictable. The main dangers to people and property arise from three distinct hazards: violent winds, storm waves and surges, and torrential rain. Winds in Hurricane Gilbert in the Caribbean in 1988 gusted to over 320 km/h (Eden, 1988), and in Hurricane Linda in the East Pacific in 1997 reached about 350 km/h (Brugge, 1998). Sustained winds approaching 280 km/h were reported for Hurricane Gilbert, but most tropical cyclones do not reach such intensity. Death and injury may result from structural collapse or from the impact of flying objects. Rainfall associated with tropical cyclones may total over 1000 mm in 24 h. Cyclone Hyacinthe dumped 6433 mm of rain on Reunion in 14 days in January 1980, with more than 1000 mm falling on each of two successive days (Smithson, 1993). During Hurricane Mitch in October 1998 parts of Honduras received 698 mm in 41 h, leading to widespread flooding and landslides (Hellin and Haigh, 1999). Flooding can be severe when such heavy rain falls in restricted catchments, or when run-off combines with storm wave and storm surge effects. The northern Bay of Bengal suffers a particularly serious storm surge problem because of a combination of large astronomical tides, a funneling coastal configuration, low and flat terrain, and frequent severe tropical storms. The November 1970 storm in this region may have been the deadliest ever to devastate a coastal area in recent times, with a storm surge of over 6 m and about 300 000 people killed (Frank and Husain, 1971), although tropical cyclones in the same region in 1737 and 1876 may have rivaled this (Sensarma, 1994).

Most of the 80–100 tropical cyclones each year form between latitudes 5° and 25°, 60–70% in the northern hemisphere. Tropical cyclone tracks sometimes reach beyond 401 latitude. The area immediately affected by the full force of a storm is typically about 1° by 1° latitude, corresponding to approximately 100 km of coastline. On this basis there are about 800 separate prime target areas around the globe, each with peculiar local conditions, especially with respect to storm surges.

Tropical cyclones are responsible for an annual average of about 20 000 deaths and over $6 billion in damage globally (Obasi, 1994). In some cases, like the 1991 Bangladesh storm which destroyed over 500 000 dwellings, millions of residents can be directly affected (Haque and Blair, 1993). Three types of most vulnerable areas can be identified: densely populated, fertile coastal plains and deltas (e.g. the Ganges); island groups dependent on agricultural economies (e.g. Oceania, the Caribbean, the Philippines); and highly populated coastal regions developed as residential resorts (e.g. Florida, the

Queensland Gold Coast) or for industry (e.g. the Texas Gulf Coast, Japan) (Stevens, 1991).

Regional variation in the socioeconomic impact of tropical cyclones is related to factors such as the geographic vulnerability of communities, their experience with severe storms, population density, coastal and inland topography, land use, social organization, property and infrastructure at risk, and warning and response capabilities. Extensive delta regions in developing countries present the greatest potential for loss of life.

Evacuation is one form of emergency response in developed countries. However, as population densities increase, evacuation plans may prove inadequate, so that evacuation times may exceed what is feasible given the number of people to be moved and likely lead times from existing forecast capabilities (American Meteorological Society, 1993, 2000b). Costs of evacuation, estimated at about $0.6 million per kilometer of coastline (Pielke and Carbone, 2002), could in some cases at least match those resulting from storm damage. Research after cyclone Tracy hit Darwin has suggested that mass evacuation may pose more problems than it solves, and that greater use of the victims' resources might speed up reconstruction and reduce psychological stress (Western and Milne, 1979). Research on the impact of tropical cyclones in such areas, however, cannot adequately represent impacts in countries with labor-intensive rural economies. Hurricane Gilbert produced damage in Jamaica estimated at $800–1000 million, and estimates of reinsurance inflows were $650 million; even the lower value exceeded the country's annual foreign exchange earnings from exports (Barker and Miller, 1990). Shortfalls in domestic food production had to be met by imports, virtually the whole banana crop and most of the coffee crop were wiped out. In Bangladesh in 1991 an estimated 51 000 ha of crops were completely destroyed and over 150 000 ha were partially damaged (Haque and Blair, 1993).

An industrialized society may plan for a low rate of building failures. This requires a high degree of engineering attention to housing design and construction, together with stringent building regulations. However, a developing country may elect to accept a higher level of building failures, adopting a strategy of temporary housing during and after a tropical cyclone. A variety of public buildings would then be strengthened and designed as refugee or reception centers. These contrasting approaches are exemplified by the decisions made by authorities in Darwin after cyclone Tracy in 1974 and by those in Sri Lanka after a rare tropical cyclone in 1978.

Most building damage is caused by the effects of wind on buildings that are not properly engineered, such as domestic houses and small, low-rise industrial and commercial structures. The effects of extreme winds on buildings are well understood, and wind engineering technology in theory can be applied anywhere. The impact of tropical cyclones is greatest when the population is rendered homeless, so housing design in vulnerable areas should include consideration of the effects of wind and water. Areas of infrequent tropical cyclone occurrence (e.g. the Queensland Gold Coast, the Atlantic Coast of the United States north of Cape Hatteras, parts of Mexico, India, and Japan) tend to have greater building vulnerability; whereas communities in areas battered relatively frequently by tropical cyclones have learned to cope with the hazard and tend to be less visible as disaster areas (e.g. Mauritius, Reunion Island, Guam, Northwest Australia, some South Pacific islands).

Improvements in warnings and in community preparedness have decreased the death toll from tropical cyclones, but the average annual inflation-adjusted hurricane losses in the United States have grown from about $5 billion in the 1940s to more than $40 billion in the 1990s (Easterling et al., 2000). In developing countries, where some of the expected increase in world population will be in regions subject to tropical cyclone impact, a growing death toll and damage may be unavoidable, partly due to limitations on national resources to apply effective mitigation measures and partly due to a lack of understanding of the vulnerability of certain areas to the effects of tropical cyclones. In all parts of the world there is a continuing need for better community education about the potentially damaging hazards and about long-term mitigation planning (American Meteorological Society, 2000b).

The warning system is a primary feature of organization for disaster preparedness. Key elements in a warning system, apart from early detection and accurate forecasting, are the efficiency of the dissemination process and the reaction of the community. Education is vital to ensure that warnings are understood and that people are aware of the necessity to heed such warnings. Hazard warning involves a sensitive balance between overwarning, resulting in unnecessary and sometimes expensive preparations and leading ultimately to complacency and apathy, and underwarning, which gives insufficient time for adequate protective measures to be taken. The benefits of improvements in tropical cyclone detection technology and in behavior prediction are lost without corresponding progress in ability to utilize the information in planning, organizing, and acting for protection and convenience (Jagger et al., 2002). McAdie and Lawrence (2000) note that tropical cyclone track forecasts for the Atlantic basin improved from 1970 to 1998, but Powell and Aberson (2001) suggest that no statistically significant change is apparent for landfall position forecasts during the last 25 years. In most other parts of the world the situation is no better. In Taiwan, for example, hit by an average of nearly four typhoons per year, forecasting, and hence warning, is difficult because of lack of data over the North Pacific (Wu and Kuo, 1999). Despite some recent possible reductions in track forecast errors there has been little improvement in forecasts of storm intensity or structure, including overall size; and, while there have been advances in basic understanding, accurate predictions of tropical cyclone genesis are still some way off (American Meteorological Society, 2000b).

There were many false warnings in East Pakistan prior to the storm of November 1970, mainly because of lack of facilities to distinguish between killer and nonkiller cyclones. An estimated 90+% of the people in the disaster area knew about the storm, yet less than 1% sought refuge in substantial buildings. Most residents had never experienced a storm surge like that predicted, and thus felt no urgency to leave their homes. In addition, few had the means to move, or anywhere to go in the time available. A comparable situation occurred in Darwin with cyclone Tracy when the potential benefits of technically good predictions were not realized. Records show several near-misses for Darwin, including cyclone Selma only 3 weeks before Tracy, and at least six tropical cyclones which had seriously affected the community. However, the last severe event had been in 1937 and was remembered by few residents in 1974; Christmas Eve as Tracy approached, and Christmas Day when it struck, were not ideal times to muster enthusiasm for effective action.

Severe local storms

Severe local storm hazards are widespread, relatively unpredictable, seemingly impossible to prevent, and often costly in lives and property damage (Atlas, 1976). At any instant there may be about 2000 active thunderstorms around the world (Dudhia, 1996). As separate cells, or as organized line squalls, thunderstorms develop cold downdraughts with high velocities that are capable of causing severe localized damage. The squall of a thunderstorm can gust to 185 km/h, and its effects are often compounded by intense rainfall, large hail, or lightning. Individual storms usually affect only small areas, but there may be many such storms at any one time in a particular region. Their association with flash floods, downbursts, strong winds, tornadoes and lightning makes accurate forecasting vital.

High-intensity, localized thunderstorms may produce flash floods. A storm in August 1988 at Khartoum, with daily rainfall of about 200 mm being more than twice the previous fall on record, left tens of thousands of homes destroyed and vast areas around the city inundated (Hulme and Trilsbach, 1989). An estimated 1.5 million were made homeless, and there were over 100 deaths and hundreds of injuries. Most of the homeless had nowhere to go; even areas which had not suffered damage were under water and there were few areas dry enough for temporary camps to be erected. A flash flood, produced by rainfall approaching a 500-year return period event, at Fort Collins in Colorado in July 1997, caused five deaths, 62 injuries and more than $250 million in property damage (Weaver et al., 2000). Lessons from that event, with a systematic effort to improve awareness, communication and warning, meant that another flash flood in April 1999 caused significantly less damage.

Hailstones are usually less than 10 mm in diameter and cause little damage, but they are occasionally 100 mm or more in diameter and can cause serious damage to crops, buildings, and motor vehicles. About 20 000 severe thunderstorms occur annually in the USA, with damage to property and agriculture of up to $3 billion (Bentley et al., 2002). Hailstones up to softball size were reported in a storm causing about $350 million damage in Denver in June 1984 (Blanchard and Howard, 1986), and there have been at least two reports from South Africa of coconut-sized stones (Perry, 1995). In the period 1982–1989, 250 people were reported killed by hail in India (Nizamuddin, 1993). Giant hailstones weighing 2–3 kg were reported in February 1988 in parts of Orissa state.

Thunderstorms are the most common cause of air traffic delays and play a major role in weather-related aircraft accidents (Bromley, 1977). Turbulence, hail, and wind shear within storm clouds have damaged many aircraft, with wind shear recognized as a major aviation hazard, particularly in the airport environment. The gust front ahead of the cold outflow from thunderstorm downdraughts is particularly hazardous because of associated large surface wind shears and because of its very localized nature. The strongest downdraughts, those most likely to be hazardous to aircraft during takeoff and landing, are called downbursts or microbursts (Fujita and Caracena, 1977; McCarthy and Serafin, 1984; Dudhia, 1997). Very high-resolution modeling of a developing thunderstorm has confirmed reports from aircraft of relatively narrow regions of turbulence in layers more than 1 km deep above the cloud (Lane et al., 2002). Thunderstorm downdraughts may also be a critical factor in driving some bushfire fronts; sometimes the fires themselves having been started by lightning strikes.

Lightning is one of the major causes of fatalities in the USA and is a significant hazard to outdoor activity during the summer months (Watson and Holle, 1996; Holle et al., 1999; Curran et al., 2000). It is probable that lightning deaths and injuries are considerably underestimated (Lopez et al., 1993), with most people displaying inappropriate behavior during thunderstorms and not realizing the range of possible medical implications from a lightning strike, including paralysis, external burns, severe headaches, hearing and memory loss, and many others (Shearman and Ojala, 1999).

Some thunderstorms and the peripheral circulation of some tropical cyclones are accompanied by tornadoes, which are among the smallest but most destructive features of atmospheric circulation. Tornado wind speeds can exceed 350 km/h. The Tri-State tornado in March 1925 traveled about 350 km from Missouri to Indiana at speeds of 91 km/h to 109 km/h. About 80 km^2 of land were totally devastated, 689 people were killed, nearly 2000 people were injured, and over 11 000 people were left homeless (Flora, 1953). Over 2 days in April 1974 an outbreak of 148 tornadoes in the Midwest and southeast USA left 350 dead (Brugge, 1994c). On a single day in November 1981 an outbreak of 102 tornadoes struck parts of England and Wales (Rowe and Meaden, 1985). Just one tornado in May 1996 is reported to have killed at least 400 people in Bangladesh (Snow and Wyatt, 1997).

The global distribution of tornadoes is difficult to determine accurately (Perry and Reynolds, 1993). Most reports of tornadoes come from the United States, but they occur in many parts of the world including most of Europe, northern India and Bangladesh, Australia and New Zealand, Japan, Uruguay, and southern Africa (Fujita, 1973). Frequency data for tornadoes are unreliable for most parts of the world. During the 1950s about 200 tornadoes were reported annually in the United States, but by the late 1990s around 1200 were noted each year (American Meteorological Society, 2000a). Most of the increase may be accounted for by more frequent reports of weak tornadoes, probably because of growing public awareness, rather than any meteorological factors. Numbers of tornadoes reported from other parts of the world are substantially less, perhaps related to lack of awareness and sparse weather-observing networks. A study of tornadoes on the Indian subcontinent identified only 51 events between 1835 and 1977, but the path lengths and widths were larger than those characteristic of the United States, so many smaller tornadoes may have passed unreported (Peterson and Mehta, 1981). Elsewhere, for example, 191 tornadoes were reported in Argentina from 1930 to 1979; 42 in Taiwan from 1951 to 1978; 87 in Japan in 5 years from 1968; and 273 in France from 1680 to 1998 (Snow and Wyatt, 1997; Paul, 1999).

Tornadoes exhibit a considerable range of intensity, size, and duration. Typical New Zealand tornadoes have been reported to have damage paths only 10–15 m wide. The typical or median path of tornadoes in the United States has been given as 3.2 km long by just under 50 m wide (National Severe Storm Forecast Center Staff, 1980), but there have also been reports of tornadoes with paths ranging from less than 50 m to over 400 km. Damage associated with tornadoes is very localized, including severe structural damage to buildings (roofs lifted, walls and windows collapsed) and mobile homes, and tops screwed or snapped off trees, which may also be uprooted.

Tornado-generated missiles, ranging from gravel to semi-trailer trucks, present engineers with major design problems. In tornado-prone areas of the United States, where protection of people in buildings is important, the tornado missile can be the controlling design factor. Buildings such as hospitals, fire stations, and emergency operating centers, where critical functions must be maintained, are particularly susceptible to missile damage (McDonald, 1976). Small outdoor equipment and larger objects such as utility poles can become tornado missiles. A tornado at Lubbock, Texas, in 1970 was responsible for moving a cylindrical tank (3.35 × 12.5 m) weighing over 11 tonnes about 1.21 km; and three 40-passenger school buses apparently became airborne in a tornado at McComb, Mississippi, in 1974.

Increasing public awareness, emphasis on tornado preparedness, improved warning systems, better understanding of tornado formation, the use of tornado drills, spotter groups and other measures at local and state levels have contributed to the decreasing death toll from tornadoes in the United States. However, a single tornado in Bangladesh may kill several hundred people because none of these conditions or measures exists, and because of the very dense population.

The broad environmental conditions leading to the development of severe thunderstorms and tornadoes are reasonably well recognized and can be forecast with some skill, but there is still no reliable method for predicting the development of a specific severe storm (Hoium et al., 1997; Snow and Wyatt, 1998). Technological advances such as Doppler radar, high-resolution satellite imagery, and acoustic sounders have improved prospects for forecasting severe local weather. In the USA there have been marked improvements in the remote detection of severe local storms and flash floods, accompanied by better warnings of such events (Polger et al., 1994; Vasiloff, 2001); but in most regions of the world the observing network and the available technology are inadequate for detection and measurement of most small-scale events.

Some local storms occur in areas where warnings, if they are given, will not help very much. A severe dust storm followed by heavy rain, for example, hit Karachi in May 1986 causing extensive property damage and dislocation of services (Middleton and Chaudhary, 1988). Thousands of bamboo huts and improvised houses with tin or asbestos roofs were blown away; telephone and electricity wires were snapped, giant trees uprooted and vehicles overturned; and more than half the city was plunged into darkness. There is no doubt that improved short-term forecasts can greatly diminish the impact of some local storms, in areas where such forecasts are available and can be readily disseminated. Warnings during the May 1999 tornado outbreak in Oklahoma and Kansas are credited with saving hundreds of lives, and Doppler radar provided significant warning lead time during the tornado outbreak in Georgia in February 2000 (Vasiloff, 2001).

Cumulative hazards

Many atmospheric disasters result from an accumulation of events, that singly would not be hazardous. One dry day or even one dry year does not necessarily constitute a drought, but a succession of abnormally dry years can have disastrous effects on the environment and its inhabitants. Similarly, hot days are common in many parts of the world, but a succession of many very hot days can prove lethal, especially in areas not normally accustomed to heat waves combined with high humidities.

Drought

In the period from 1967 to 1991 droughts were estimated to have affected 50% of the 2.8 billion people who suffered from weather-related disasters (Kogan, 1997). It is difficult to find a generally accepted definition of drought. Drought clearly involves a shortage of water, but realistically can be defined only in terms of a particular need (Linsley, 1982). Drought is not just a physical phenomenon; it results from interplay between a natural event and demands placed on water supply by human use systems (American Meteorological Society, 1997). The absence of a precise definition adds to confusion about whether or not a drought exists and, if it does, its severity. The effects of a drought accumulate slowly over long periods and may linger for years after termination of the event. The reporting of drought occurrence may be overestimated or underestimated, because the material well-being of the reporter may be affected (Heathcote, 1979).

Four types of drought are usually recognized: meteorological or climatological, agricultural, hydrological and socioeconomic. Droughts impact both surface and groundwater resources. They can lead to reductions in water supply, diminished water quality, crop failure, reduced power generation, disturbed riparian habitats, suspended or curtailed recreation activities and a variety of other associated economic and social activities (Riebsame et al., 1991; Woodhouse and Overpeck, 1998). Long-term impacts on plant and animal life have received relatively little attention, and many questions are unanswered concerning the role of drought in ecosystems.

The impact of drought on human activities is usually described in terms of reduced water supplies and economic losses throughout the community. The evaluation of such impacts is complicated, with many factors needing to be taken into consideration. Several uncertainties can be involved, such as shortfalls in expected yields tending to inflate the value of actual yields on which the value of lost production is calculated, yet there being no guarantee that lost production could have been sold anyway.

Frequent droughts around the world, and interest in their possible links with phenomena such as El Niño, keep the hazard in evidence even for the casual observer. The 1975–1976 drought in Western Europe had widespread effects on agriculture, domestic and industrial water supplies, and on river and canal traffic. The drought in Britain, particularly southern England, was the worst for about 250 years. In some parts of the country, water supplies to domestic consumers were cut for up to 17 hours per day, and production of root and vegetable crops was down by as much as 40%. Droughts of the early to mid-1980s in Africa affected more than 40 million people. The Canadian Prairie Provinces are particularly sensitive to rainfall shortages and associated soil losses due to wind erosion (Wheaton and Chakravarti, 1990). In 1984 conditions rivaled those of the "Dust Bowl" years of 1936–1937, with farmers losing up to half of their grain crop to the value of about $2.5 billion (Sweeney, 1985). The 1988 drought in the USA was rated as one of the worst in 100 years, with an estimated impact on the economy of $40 billion (Kogan, 1997). An unexpected impact in this instance was on barge traffic on the lower Mississippi river, the industry suffering an income loss of about 20% (Changnon, 1989). The El Niño event of 1997–1998 was linked to drought in Central America and southeast Asia, with major impacts on vegetable quantity and quality in the former and on coffee and palm oil in the latter, both leading to reduced exports and increased prices on global markets (Changnon, 1999). It has been suggested that worldwide disasters triggered by droughts are twice as frequent during year two of ENSO warm events as during other years, particularly in southern Africa and southeast Asia (Dilley and Heyman, 1995).

Several factors may be implicated as potential causes of drought: ENSO, abnormal sea surface temperature patterns in areas other than the equatorial eastern Pacific, soil moisture desiccation, and nonlinear behavior of the climate system (Orville, 1990). It is tempting also to suggest that climate is changing and that droughts are becoming more frequent and/or more severe. However, there have always been droughts, and records show that events such as those mentioned are within the realm of statistical expectations (Landsberg, 1982). Examination of the paleoclimatic record for the Great Plains suggests that the droughts of the 1930s, 1950s and 1980s were eclipsed several times by droughts within the last 2000 years, and that more severe droughts could occur in the future (Woodhouse and Overpeck, 1998).

The massive Australian drought of 1895–1903, which followed rapid growth of rural enterprises in the 1870s and 1880s, and the great American drought of the 1930s, are good examples of lack of understanding of the environment leading to unwise land use. Farmers in southeastern Australia before 1893 believed that the climate was on their side, and in South Australia, the leading wheat producer, there was a belief that rain followed the plow. Notions that plowing and tree planting could bring rain were widespread, enticing farmers into marginal areas. The number of sheep in Australia fell from an estimated 106 million in 1891 to 54 million in 1902, cattle numbers were almost halved to about 7 million, and dust storms were common as a result of the vast expansion of land plowed for wheat. In areas such as the Sahel, where nomadism and intermittent grazing have been prevalent and more or less in balance with environmental conditions, more intensive exploitation has had disastrous results for social systems and ecosystems when drought has struck.

Drought is a common feature in many countries but is often regarded as an unfortunate and irregular abnormality of the environment. It would be more appropriate to consider drought as part of the normal sequence of events. Society must be prepared to cope with the effects of drought at any time. Impacts in the past have been exacerbated by absence of coping mechanisms, with too little preparation during non-drought periods.

Heat and cold

Excessive heat possibly contributes to more illness and mortality than any other direct, weather-related cause, certainly in regions better equipped to cope with the more violent hazards (Kalkstein, 1995). Most heat-related deaths occur in midlatitude cities, in northern India and China, eastern and Midwestern USA and Western Europe, with infrequent but extreme heat waves. Subtropical and tropical cities with higher mean summer temperatures seem less vulnerable, partly due to acclimatization of residents and perhaps due to more efficient cooling of houses. Heat-related deaths, averaging about 1000 per year, appear to be on the increase in the USA, exceeding those caused by other hazardous weather conditions (Changnon et al., 1996). Heat waves may also be associated with increases in the incidence of rioting, violence and homicide.

Heat-related death rates are usually higher in urban areas than in rural areas. This is almost certainly the result of climatic modification and heat retention due to urbanization, the heat island effect, plus the pollution trapping and concentrating effects of stagnant atmospheric conditions of heat waves, adding to those of heat stress. Death rates are also higher among the aged, especially as a result of aggravating effects on pre-existing conditions such as heart disease or cancer.

An exceptionally severe heat wave in parts of the southwestern USA in June 1994 produced the highest temperatures ever recorded in four states (Brugge, 1995). At the same time authorities in southern Ontario advised residents with heart or respiratory problems to rest, and also requested people to avoid driving, and avoid the use of aerosols that could worsen ground-level ozone. A short but intense heat wave in mid-July 1995 caused over 800 deaths in the USA, over 500 in Chicago alone (Kunkel et al., 1996; Karl and Knight, 1997). About 70 daily maximum temperature records were set during this heat wave at locations from the central and northern Great Plains to the Atlantic coast (Livezey and Tinker, 1996).

Many places in England and Wales experienced record high temperatures during a heat wave in August 1990. Apart from the usual impacts upon those unaccustomed to temperatures over 37°C, tar on roads melted, leading to closures, one runway at Heathrow airport was closed when newly laid tar failed to set, the entire stock of a Liverpool chocolate factory melted, a life-sized waxwork knight at a castle in Essex melted into a puddle, and there was a spate of drownings as people tried to keep cool by swimming (Brugge, 1991).

Increased demands for air-conditioning and refrigeration can produce overloading of power supply systems during heat waves, leading to power restrictions and breakdowns, tending to aggravate the heat-stress situation. Assessment of the causes of death in the 1995 Chicago heat wave included factors such as inadequate warning systems, insufficient time to acclimatize, the heat island effect, an aging population, an inadequate ambulance service, and the inability of many residents to properly ventilate homes due to fear of crime or lack of resources for fans or air-conditioning (Changnon et al., 1996). During a subsequent heat wave in Chicago in July 1999 the death toll was much lower, because lessons had been learned from the previous event. Heat wave plans included timely warnings, activation of cooling shelters, frequent broadcasting of information and the ready availability of helplines (Palecki et al., 2001). In any such event, indeed in almost any hazard event, there are winners. In this case air-conditioner sales increased, sales of ice creams and ice set records, private ambulance operators were busier than normal, utilities not experiencing major equipment failures made record profits, people went to shopping malls and movie theaters to escape the heat in air-conditioned facilities, and merchants at lakeside outlets benefited from record attendances at the height of the heat wave. 'Properly spaced green areas are the most effective and aesthetically pleasing means of controlling urban temperature excess by improving ventilation and circulation and reducing heat storage capacity. Building materials with lower heat conductivity and storage properties, and water bodies within or close to cities help to keep maximum temperatures down and encourage mixing and ventilation resulting from enhanced temperature differentials. Adequate surveillance systems are needed, to alert the public and authorities that potentially dangerous weather is imminent. A promising approach is based on the identification of high risk air masses historically associated with increased mortality (Kalkstein, 1995; Kalkstein et al., 1996). A similar approach, using a weather type classification scheme developed for North America, has been used in several heat stress warning systems worldwide. It has been incorporated in systems developed for Rome, Shanghai, Toronto, Phoenix, New Orleans and Cincinnati (Sheridan, 2002).'

Excessive cold directly causes death through the effects of exposure and is indirectly responsible for deaths from causes such as fatal heart attacks while clearing snow and asphyxiation in stranded vehicles. The annual average mortality rate attributable to winter storms and cold in the USA up to the mid-1990s was between 130 and 200 (Changnon et al., 1996). The number of cold-related fatalities may be increasing, perhaps because an aging population becomes more sensitive to the effects of cold. Over 600 people died from cold-related accidents in each of the North American winters of 1977–1978 and 1978–1979, with temperatures averaging 6°C below normal from December 1978 to February 1979.

The impact of a severe winter extends beyond increased fatalities. Production losses in industry, crop losses, transportation losses in revenue and damage to roads and bridges, losses in retail sales, and losses resulting from increased energy consumption in the winter of 1976–1977 in the United States were at least $40 billion. Unemployment rose by 2 million, the consumer price index had its largest 2-month jump in 3 years, the inflation rate rose, the number of business failures increased, and the balance of payments dropped by $1.4 billion (Hughes, 1982). Record low temperatures were registered in January 1994 across central and eastern USA and eastern Canada, down to −31°C at Akron, Ohio, −30°C at Pittsburgh, −36°C in parts of Minnesota, −41°C in New York State and below −45°C in parts of Ontario (Brugge, 1994b). A wind-chill figure of −70°C was reported from the Midwest. Impacts of the severe cold included school closures; interrupted train services; closed airports; closed public services; record power consumption and electricity blackouts; water mains frozen for up to a week; ice-making machines frozen; canceled garbage collections because of frozen equipment; many weather-related injuries, including a man who set himself on fire trying to light a fire; a halt to brewery production in Milwaukee; and a crack in a gasoline pipeline leaking over 450 000 L, much into the Mississippi. To impacts like these should be added elements such as personal inconvenience, worry and stress, extra work, injuries, higher taxes to cover costs of repairs, and decreased tax income through lost work-days.

Frost presents a hazard to crops around the world. Freezing temperatures in southern Brazil, in 1979 and 1985 at around 20°S, damaged coffee trees and severely reduced coffee production. Severe frost, although infrequent, is also a major climatic hazard in the near equatorial latitudes of the highlands of Papua New Guinea (Brown and Powell, 1974). Risk of ground frost in Papua New Guinea starts at about 1 500 m, increasing with elevation and in valleys and basins. The worst frost in memory occurred in the early 1940s, accompanied by a prolonged drought. Gardens as low as 2000 m were wiped out by weeks of frosts, with sleet and snow at higher levels destroying gardens, food-bearing trees, domestic pigs, and wild animals. Food shortages forced extensive migrations to lower areas, leading to severe social conflict and occasional fighting. In March–April 1990 much of the UK was affected by unusually severe ground frosts and prolonged air frosts, leading to widespread damage to tree fruit crops in flower at the time and to young nursery stock (Morison and Butterfield, 1990). There were also reports of substantial damage to winter-sown cereal crops, with from 10% to 90% of ears in barley and wheat crops being killed. The damage was exacerbated because of previously mild conditions throughout the preceding winter and the corresponding advanced development of many of the plants.

There is growing recognition of the potential value of meteorological information in decision making. Deciduous fruit trees are very susceptible to frost damage in spring, with loss of fruit yield and even permanent tree damage. Several protective devices are available, such as wind machines, sprinklers, and heaters, but these are expensive. Accurate minimum temperature forecasts can help the grower in making decisions whether or not to protect on a particular occasion, and so may help to save money (Katz et al., 1982).

Snow and ice

Snow hardly worries an alpine village; indeed it may be welcomed as improving prospects for the tourism industry, but the same depth of snow may bring panic and total disruption to a lowland city unaccustomed to it. The nature of the snow hazard is influenced by other weather conditions, particularly wind, by topography and by road surface materials. On crowded roads even a shallow snow cover of 20 mm may be sufficient to halt or hinder traffic flow, leading to increased costs in work losses, late delivery of goods, losses of consumer sales and receipt losses by public transportation, apart from increasing accident and casualty rates (Perry, 1981). Snow has widespread impacts on construction, merchandising, manufacturing, agriculture, power supply, communications, recreation, and public health and safety services and can induce heavy financial losses either in terms of direct damage and disruption or in costs of mitigation (Rooney, 1967). However, snow can be beneficial if it covers crops or vegetation and protects them from air frost.

There are numerous examples of the impacts of major snow storms on large urban areas (e.g. Speakman, 1994; Brugge, 1944a; Wild et al., 1996). The severe winter storm of March 1993 along the east coast of North America illustrates the possible impacts of snow on communities. Property damage was estimated at $1 billion; snow clearance costs were estimated at $100 million; and total damage was put at about $4 billion. Structural damage was widespread from Florida to Canada; hundreds of roofs collapsed under snow loads; and over 3 million customers were without electrical power due to fallen trees and high winds, which reached 210 km/h on Cape Breton Island. Over 200 people died from weather-related causes and 48 were lost at sea. Many deaths were attributed to heart attacks clearing snow, others died in fires, drowned or suffered carbon monoxide poisoning when trapped in cars or buildings (Brugge, 1994a).

A record early snow storm which hit parts of the High Plains and Midwest of the USA in October 1997 and an extremely severe ice storm in the northeast and in eastern Canada in January 1998 were both attributed to the El Niño event of 1997–1998, as were many other climatic hazards at that time (Changnon, 1999).

Ice storms can cause major economic and social disruption (Assel et al., 1996). Aircraft icing, through the accretion of supercooled water, has been identified as a cause of many aircraft accidents during winter storms (Rasmussen et al., 1992). The 1998 ice storm was remarkable for its spatial extent and duration, as well as in terms of the severity of its impacts on the region (DeGaetano, 2000; Dupigny-Giroux, 2000). Total storm damages in the USA and Canada exceeded $2 billion, over $800 million to the hydroelectric installations in Quebec alone, where well over 1 million people were at times without power. In Montreal the entire commercial sector was closed down for a week. Deaths occurred from falling ice, house fires, hypothermia and carbon monoxide poisoning. Thousands of farmers in the dairy industry were particularly hard hit, with losses of cattle and milk production; and the long-term consequences of milk shortages affected ice-cream makers, cheese suppliers and all relying on dairy products. The livelihoods of maple syrup producers were placed in jeopardy because of extensive ice damage to trees.

Telecommunications and electrical services were tested beyond design limits by ice accumulations, radio and television stations were put off the air, airports were closed and rail services suspended. Damage to hiking and snowmobile trails had impacts on the leisure industry, and it was estimated that most Christmas trees intended for the 1998 season were damaged beyond salvage.

Fog

Fog is a hazard to land, sea, and air transportation. Many multiple-vehicle accidents occur when visibility is severely restricted on highways. In 1974 fog was estimated to have cost over £12 million on roads in the United Kingdom (Perry, 1981). The worst aviation accident on record occurred in March of 1977 in the Canary Islands when two wide-bodied jets collided on a runway in heavy fog, killing 583 people. Possibly the worst fog disaster occurred in May of 1914 when the *Empress of Ireland* sank after a collision in the St Lawrence estuary with the loss of 1078 lives (Whittow, 1980). The collision between the *Stockholm* and *Andrea Doria*, and subsequent sinking of the latter, in sea-fog off Massachusetts in July 1956, is another example of a dramatic impact of what can appear as an innocuous phenomenon.

The aviation industry loses millions of dollars each year when fog causes aircraft diversions and delays, inconveniencing thousands of passengers and incurring increased operating costs. Costly adjustments, such as automatic aids for aircraft, are possible and many fog dispersal techniques have been tried. Supercooled fog can be modified but accounts for only a small proportion of fogs. Warm fog can also be modified using thermal energy, but high costs of installation and operation of dispersal systems and the problem of pollution have discouraged their use (Kunkel, 1980).

Visibility is one of the most difficult of meteorological phenomena to forecast. Several objective fog forecasting techniques have been tried, but there is still no entirely reliable method. Publicity of the hazards and driver education remain the surest ways of reducing the dangers of fog on highways.

Air pollution

Many studies have pointed to an association between air pollution and ill-health, particularly for diseases such as bronchitis and lung cancer, and less so for cardiovascular ailments and nonrespiratory tract cancers (e.g. Lave and Seskin, 1970).

The first noticeable effect of photochemical pollution is a stinging of the eyes due to peroxyacetylnitrate gas. Ozone, also formed in photochemical reactions, damages lung tissue, increases death rates as a result of swellings in lung passages, and reduces athletic performance. Sulfur dioxide can lead to infections of the lower respiratory tract, especially among the elderly, the very young, and those already weakened by illness. Carbon monoxide may cause drowsiness in low concentrations, extending to severe headaches, nausea, and collapse in high concentrations. Large cities such as Beijing, Bombay, Cairo, Jakarta and Mexico City are now the most susceptible to hazards arising from air pollution.

Air pollution impact can be reduced, chiefly by not locating pollution sources in badly ventilated areas known to experience frequent low-level temperature inversions. A common approach to air pollution from motor vehicles is emission control, either through the fuel or through exhaust abstraction systems, although the efficiency of these is debatable and their public acceptance often poor. Design of buildings and layout of city streets can help promote greater turbulence and better diffusion of pollutants. Good traffic flow promotes better urban air quality, but traffic engineering and air pollution control aims are frequently at variance. Implementation of fuel conservation policies, use of alternative fuels, and pollutant-emission regulations are steps toward minimization of total pollution emissions and thus of the pollution hazard. On the other hand, it may be better and cheaper to achieve clean air from the outset through sensible planning rather than through remedial steps later. Transport of air pollution by atmospheric circulation means that the problem is one for legislators at the national or continental level.

Acid deposition from the atmosphere is a major worldwide environmental concern, particularly in parts of northern and western Europe, North America and China (Mason, 1990). It originates from the release of sulfur and nitrogen oxides by industry and transport. Oxidation and hydrolysis of the oxides produce sulfuric and nitric acids or related sulfates and nitrates, which are transported in and eventually removed from the atmosphere in rain, snow, dew, frost, fog, gases, or particles. There is evidence that acid rain may be responsible for long-term adverse effects on the environment. These effects may include acidification of rivers, lakes, and groundwater, with damage to all components of the aquatic ecosystems; acidification and demineralization of soils; changes in agricultural and forest productivity; corrosion damage to buildings, monuments and water supply systems; degradation of water supplies; and reductions in biodiversity, with acid-tolerant species flourishing in severely stressed ecosystems (Bridgman, 1997).

Biomass burning for forest clearing, fossil fuel burning and "slash-and-burn" agricultural practices in the tropics of southeast Asia, the Amazon Basin and central Africa create a wide range of pollutants, including smoke components and dust. Such emissions can be transported long distances downwind. The Asian brown haze, largely composed of particulates and sulfate aerosols from south, southeast and east Asia, significantly reduces solar radiation and has possible impacts including a reduction in rainfall, reduction of light available for photosynthesis and adverse health effects (Anon, 2002). Cooling of the land surface; a stronger and more frequent thermal inversion trapping more pollutants; and an overall reduction in evaporation and precipitation, with implications for water quality and availability, are possible indirect effects of the brown haze.

Climatic changes

The importance of climatic changes to the world's future and of human dependence on a stable climate to maintain present patterns of agriculture has been highlighted by many authors (e.g. Schneider, 1976; Bryson and Murray, 1977; Roberts and Lansford, 1979; Lamb, 1982).

Climatic inputs provide the essential resource base for agriculture; even with present technology the world is not immune to the effects of climatic variation. In 1972, when the global climate was particularly unfavorable for food production, millions starved. The increasing demands of a growing population mean that there may be more frequent imbalances between food supply and need, especially in regions of increasing pressure for cultivation of climatically marginal land. The results could be depletion of grain reserves, malnutrition, starvation, and political unrest. Parry et al. (2001) have examined the possible impacts of climate change in some key areas of risk: hunger, water shortage, exposure to malaria transmission, and coastal flooding. They concluded that millions more people will become at risk and that it will be necessary to find a blend of mitigation, to buy time, and adaptation, which in turn can raise thresholds of tolerance.

There has been much speculation about what might happen to frequencies, intensities and distribution of climatic hazards in the future, although not all changes in risk would necessarily be attributable to climate change; societal shifts such as population growth, demographic moves to more at-risk locations, and the growth of wealth have all made for greater vulnerability (Changnon et al., 2000). Some projections suggest an increase in the risk from droughts, floods, heat waves, tropical cyclones and storms (e.g. Changnon and Changnon, 1992; Bruce, 1994; Yarnal et al., 1999). There is evidence that diseases such as malaria and dengue fever, carried by mosquitoes, are undergoing resurgence and redistribution, possibly in response to global warming (Epstein et al., 1998). Mosquito-borne diseases are being reported at higher elevations and higher latitudes, consistent with a spread of warmer, wetter conditions. The distribution of agricultural pests could also shift, with implications for food security and pointing to the need for the public and policy makers to be aware of the possible biological consequences of climate change (Harrington and Woiwood, 1995).

There is still considerable uncertainty about the nature of climatic trends, much depending upon the rate at which greenhouse gas concentrations increase and on the particular models and scenarios used to predict future climate. It has been suggested, for example, that there is no clear evidence of long-term trends in global tropical cyclone activity, with indications that the broad geographic regions of cyclogenesis and of regions affected by tropical cyclones will not change significantly (Henderson-Sellers et al., 1998). Increasing trends in weather-related losses in recent decades have led to a popular view, nurtured by greater media coverage, that hazard frequency and intensity are increasing, but societal changes may be the primary cause (Kunkel et al., 1999). The lack of high-quality long-term data makes it difficult to determine changes in extremes, and observations of the impact of global warming are based on very short records for analysis (Easterling et al., 2000).

Climatologists can assist in many ways to plan for severe weather, now and in the future (Hunt, 1990): for example, with advice at design and planning stages of projects to reduce possibilities of weather-related disasters; with advice on land-use planning and design of structures to minimize risks from climatic hazards; and with improving forecast and warning capabilities and awareness and coping strategies. Whatever changes may occur in the future, and whatever adjustments they may necessitate, there is a need for greater recognition by the public, policy makers and planners that change is possible and preparation is necessary (e.g. Glantz, 1979; Wilson, 1981; Robinson and Hill, 1987; Changnon et al., 1995).

Jack Hobbs

Bibliography

American Meteorological Society, 1993. Policy statement: hurricane detection, tracking and forecasting. *American Meteorological Society Bulletin*, **74**(7): 1377–1380.

American Meteorological Society, 1997. Policy statement: meteorological drought. *American Meteorological Society Bulletin*, **78**(5): 847–849.

American Meteorological Society, 2000a. Policy statement: tornado preparedness and safety. *American Meteorological Society Bulletin*, **81**(5): 1061–1065.

American Meteorological Society, 2000b. Policy statement: hurricane research and forecasting. *American Meteorological Society Bulletin*, **81**(6): 1341–1346.

Anon, 2002. Asian brown cloud impacts regional climate. *Australian Meteorol. and Oceanographic Society Bulletin*, **15**(4): 73.

Assel, R. A., Janowiak, J.E., Young, S., and Boyce, D., 1996. Winter 1994 weather and ice conditions for the Laurentian Great lakes. *American Meteorological Society Bulletin*, **77**(1): 71–88.

Atlas, D., 1976. Overview: the prediction, detection and warning of severe storms. *American Meteorological Society Bulletin*, **57**(4): 398–401.

Barker, D., and Miller, D., 1990. Hurricane Gilbert: anthropomorphising a natural disaster. *Area*, **22**(2): 107–116.

Barrows, H.H., 1923. Geography as human ecology. *Association of American Geographers Annals*, **13**: 1–14.

Bendel, W.B., and Paton, D., 1981. A review of the effect of ice storms on the power industry. *Journal of Applied Meteorology*, **20**(12): 1445–1449.

Bentley, M.L., Mote, T.L., and Thebpanya, P., 2002. Using Landsat to identify thunderstorm damage in agricultural regions. *American Meteorological Society Bulletin*, **83**(3): 363–376.

Bergen, W.R., and Murphy, A.H., 1978. Potential economic and social value of short-range forecasts of Boulder windstorms. *American Meteorological Society Bulletin*, **59**(1): 29–44.

Blanchard, D.O., and Howard, K.W., 1986. The Denver hailstorm of 13 June 1984. *American Meteorological Society Bulletin*, **67**(9): 1123–1131.

Bridgman, H.A., 1997. Air pollution. In Thompson, R.D., and Perry, A., eds., *Applied Climatology: Principles and Practice*. London and New York: Routledge, pp. 288–303.

Bromley, E., 1977. Aeronautical meteorology: progress and challenges – today and tomorrow. *American Meteorological Society Bulletin*, **58**(11): 1156–1160.

Brown, M., and Powell, J., 1974. Frost and drought in the highlands of Papua New Guinea. *Journal of Tropical Geography*, **38**: 1–6.

Bruce, J.P., 1994. Natural disaster reduction and global change. *American Meteorological Society Bulletin*, **75**(10): 1831–1835.

Brugge, R., 1991. The record-breaking heatwave of 1–4 August 1990 over England and Wales. *Weather*, **46**(1): 2–10.

Brugge, R., 1994a. The blizzard of 12–15 March 1993 in the USA and Canada. *Weather*, **49**(3): 82–89.

Brugge, R., 1994b. The record-breaking low temperatures of January 1994 in the USA and Canada. *Weather*, **49**(10): 337–346.

Brugge, R., 1994c. The Alabama tornado outbreak of 27 March 1994 – an example of tornado formation. *Weather*, **49**(12): 407–411.

Brugge, R., 1995. Heatwaves and record temperatures in North America, June 1994. *Weather*, **50**(1): 20–23.

Brugge, R., 1998. East Pacific Hurricane Linda. *Weather*, **53**(1): 19–20.

Burton, I., Kates, R.W., and White, G.F., 1993. *The Environment as Hazard*, 2nd edn. London and New York: Guildford Press.

Bryson, R.A., and Murray, T.J., 1977. *Climates of Hunger*. Canberra: Australian National University Press.

Changnon, S.A., 1979. How a severe winter impacts on individuals. *American Meteorological Society Bulletin*, **60**(2): 110–114.

Changnon, S.A., 1989. The 1988 drought, barges and diversion. *American Meteorological Society Bulletin*, **70**(9): 1092–1104.

Changnon, S.A., 1999. Impacts of 1997-98 El Niño-generated weather in the United States. *American Meteorological Society Bulletin*, **80**(9): 1819–1827.

Changnon, S.A., and Changnon, J.M., 1992. Temporal fluctuations in weather disasters: 1950–1989. *Climatic Change*, **22**: 191–208.

Changnon, S.A., Changnon, J.M., and Changnon, D., 1995. Uses and applications of climate forecasts for power utilities. *American Meteorological Society Bulletin*, **76**(5): 711–720.

Changnon, S.A., Kunkel, K.E., and Reinke, B.C., 1996. Impacts and responses to the 1995 heat wave: a call to action. *American Meteorological Society Bulletin*, **77**(7): 1497–1506.

Changnon, S.A., Pielke, R.A., Jr, Changnon, D., Sylves, R.T., and Pulwarty, R., 2000. Human factors explain the increased losses from weather and climate extremes. *American Meteorological Society Bulletin*, **81**(3): 437–441.

Curran, E.B., Holle, R.L., and Lopez, R.E., 2000. Lightning casualties and damages in the United States from 1959 to 1994. *Journal of Climate*, **13**: 3448–3464.

DeGaetano, A.T., 2000. Climatic perspective and impacts of the 1998 northern New York and New England ice storm. *American Meteorological Society Bulletin*, **81**(2): 237–254.

Degg, M., 1992. Natural disasters: recent trends and future prospects. *Geography*, **77**: 198–209.

Dilley, M., and Heyman, B.N., 1995. ENSO and disaster: droughts, floods and El Niño-Southern Oscillation warm events. *Disasters*, **19**(2): 181–193.

Dudhia, J., 1996. Back to basics: thunderstorms: Part 1. *Weather*, **51**(11): 371–376.

Dudhia, J., 1997. Back to basics: thunderstorms: Part 2 – Storm types and associated weather. *Weather*, **52**(1): 2–7.

Dupigny-Giroux, L.-A., 2000. Impacts and consequences of the ice storm of 1998 for the North American north-east. *Weather*, **55**(1): 7–15.

Easterling, D.R., Evans, J.L., Groisman, P.Ya, Karl, T.R., Kunkel, K.E., and Ambenje, P., 2000. Observed variability and trends in extreme climate events: a brief review. *American Meteorological Society Bulletin*, **81**(3): 417–425.

Economic and Social Commission for Asia and the Pacific, League of Red Cross Societies, and World Meteorological Organization, 1977. *Guidelines for Disaster Prevention and Preparedness in Tropical Cyclone Areas*. Geneva/Bangkok: published by the authors.

Eden, P., 1988. Hurricane Gilbert. *Weather*, **43**(12): 446–448.

Epstein, P.R., Diaz, H.F., Elias, S., et al., 1998. Biological and physical signs of climate change: focus on mosquito-borne diseases. *American Meteorological Society Bulletin*, **79**(3): 409–417.

Fink, A., Ulbrich, U., and Engel, H., 1996. Aspects of the January 1995 flood in Germany. *Weather*, **51**(2): 34–39.

Flora, S.D., 1953. *Tornadoes of the United States*. Norman: University of Oklahoma Press.

Foley, J.C., 1957. *Droughts in Australia: Review of Records from Earliest Years of Settlement to 1955*. Melbourne: Commonwealth of Australia, Bureau of Meteorology.

Frank, N.L., and Husain, S.A., 1971. The deadliest tropical cyclone in history? *American Meteorological Society Bulletin*, **52**(4): 438–444.

Fujita, T.T., 1973. Tornadoes around the world. *Weatherwise*, **26**: 56–62, 78–83.

Fujita, T.T., and Caracena, F., 1977. An analysis of three weather-related aircraft accidents. *American Meteorological Society Bulletin*, **58**(11): 1164–1181.

Gedzelman, S.D., 1990. Leonardo da Vinci and the downburst. *American Meteorological Society Bulletin*, **71**(5): 649–655.

Gentilli, J., 1979. Atmospheric factors in disaster: an appraisal of their role. In Heathcote, R.L., and Thom, B.G., eds., *Natural Hazards in Australia*. Canberra: Australian Academy of Science, pp. 34–50.

Glantz, M., 1979. A political view of CO_2. *Nature*, **280**: 189–190.

Gribbin, J., 1976. *Forecasts, Famines and Freezes*. London: Wildwood House.

Haque, C.E., and Blair, D., 1993. Vulnerability to tropical cyclones: evidence from the April 1991 cyclone in coastal Bangladesh. *Disasters*, **16**(3): 217–229.

Harrington, R., and Woiwood, I.P., 1995. Insect crop pests and the changing climate. *Weather*, **50**(6): 200–208.

Heathcote, R.L., 1979. Drought in Australia: Some problems for future research. In Heathcote, R.L., and Thom, B.G., eds., *Natural Hazards in Australia*. Canberra: Australian Academy of Science, pp. 290–296.

Heathcote, R.L., and Thom, B.G., (eds.), 1979. *Natural Hazards in Australia*. Canberra: Australian Academy of Science.

Hellin, J., and Haigh, M.J., 1999. Rainfall in Honduras during Hurricane Mitch. *Weather*, **54**(11): 350–359.

Henderson-Sellers, A., Zhang, H., Berz, G., et al., 1998. Tropical cyclones and global climate change: a post-IPCC assessment. *American Meteorological Society Bulletin*, **79**(1): 19–38.

Hewitt, K. (ed.), 1983. *Interpretations of Calamity, from the Viewpoint of Human Ecology*. Boston: Allen & Unwin.

Hewitt, K., 1997. *Regions of Risk: A Geographical Introduction to Disasters*. London: Longman.

Hoium, D.K., Riordan, A.J., Monahan, J., and Keeter, K.K., 1997. Severe thunderstorms and tornado warnings at Raleigh, North Carolina. *American Meteorological Society Bulletin*, **78**(11): 2559–2575.

Holle, R.L., Lopez, R.E., and Zimmerman, C., 1999. Updated recommendations for lightning safety – 1998. *American Meteorological Society Bulletin*, **80**(10): 2035–2041.

Hoque, B.A., Sack, R.B., Siddiqui, M., Jahangir, A.M., Hazera, N., and Nahid, A., 1993. Environmental health and the 1991 Bangladesh cyclone. *Disasters*, **17**(2): 143–152.

Hughes, P., 1982. Weather, climate and the economy. *Weatherwise*, **35**: 60–63.

Hulme, M., and Trilsbach, A., 1989. The August 1988 storm over Khartoum: its climatology and impact. *Weather*, **44**(2): 82–90.

Hunt, R.D., 1990. Disaster alleviation in the United Kingdom and overseas. *Weather*, **45**(4): 133–138.

Jagger, T.H., Niu, X., and Elsner, J.B., 2002. A space–time model for seasonal hurricane prediction. *International Journal of Climatology*, **22**(4): 451–465.

Kalkstein, L.S., 1995. Lessons from a very hot summer. *Lancet*, **346**: 857–859.

Kalkstein, L.S., Jamason, P.F., Greene, J.S., Libby, J., and Robinson, L., 1996. The Philadelphia Hot Weather–Health Watch/Warning System: development and application, summer 1995. *American Meteorological Society Bulletin*, **77**(7): 1519–1528.

Karl, T.R., and Knight, R.W., 1997. The 1995 Chicago heat wave: how likely is a recurrence? *American Meteorological Society Bulletin*, **78**(6): 1107–1119.

Katz, R.W., Murphy, A.H., and Winkler, R.L., 1982. Assessing the value of frost forecasts to orchardists: a dynamic decision-making approach. *Journal of Applied Meteorology*, **21**(4): 518–531.

Knipping, E., 1893, *Die Tropischen Orkane der Sudsee*. Hamburg: Archives, Deutschen Seewarte.

Kogan, F.N., 1997. Global drought watch from space. *American Meteorological Society Bulletin*, **78**(4): 621–636.

Kunkel, B.A., 1980. Controlling fog. *Weatherwise*, **33**(3): 117–123.

Kunkel, K.E., Changnon, S.A., Reinke, B.C., and Arritt, R.W., 1996. The July 1995 heat wave in the Midwest: a climatic perspective and critical weather factors, *American Meteorological Society Bulletin*, **77**(7): 1507–1518.

Kunkel, K.E., Pielke, R.A., Jr, and Changnon, S.A., 1999. Temporal fluctuations in weather and climate extremes that cause economic and human health impacts. *American Meteorological Society Bulletin*, **80**(6): 1077–1098.

Lamb, H.H., 1982. *Climate, History and the Modern World*. London: Methuen.

Landsberg, H.E., 1982. Climatic aspects of droughts. *American Meteorological Society Bulletin*, **63**(6): 593–596.

Lane, T.P., Sharman, R.D., and Clark, T.L., 2002. A modeling investigation of "near-cloud" turbulence. Paper presented to 10th Conference on Aviation, Range and Aerospace Meteorology, Portland, Oregon.

Lave, L.B., and Seskin, E.P., 1970. Air pollution and human health. *Science*, **169**: 723–733.

Linsley, R.K., 1982. Social and political aspects of drought. *American Meteorological Society Bulletin*, **63**(6): 586–592.

Livezey, R.E., and Tinker, R., 1996. Some meteorological, climatological, and microclimatological considerations of the severe U. S. heat wave of mid-July 1995. *American Meteorological Society Bulletin*, **77**(9): 2043–2054.

Lopez, R.E., Holle, R.L., Heitkamp, T.A., Boyson, M., Cherington, M., and Langford, K., 1993. The underreporting of lightning injuries and deaths in Colorado. *American Meteorological Society Bulletin*, **74**(11): 2171–2178.

Mason, B.J., 1990. Acid rain – causes and consequences. *Weather*, **45**(3): 70–79.

McAdie, C.J., and Lawrence, M.B., 2000. Improvements in tropical cyclone track forecasting in the Atlantic Basin, 1970–98. *American Meteorological Society Bulletin*, **81**(5): 989–997.

McCarthy, J., and Serafin, R., 1984. The microburst: hazard to aircraft. *Weatherwise*, **37**: 120–127.

McDonald, J.R., 1976. Tornado-generated missiles and their effects. In Peterson, R.E., ed., *Proceedings of the Symposium on Tornadoes: assessment of knowledge and implications for man. 22–24 June 1976*, Lubbock, Texas. Lubbock: Texas Technical University, pp. 331–348.

Macfarlane, A., and Waller, R.E., 1976. Short term increases in mortality during heatwaves. *Nature*, **264**: 434–436.

Middleton, N.J., and Chaudhary, Q.Z., 1988. Severe dust storm at Karachi, 31 May 1986. *Weather*, **43**(8): 298–301.

Morison, J.I.L., and Butterfield, R.E., 1990. Cereal crop damage by frosts, spring 1990. *Weather* **45**(8): 308–313.

National Severe Storm Forecast Center Staff, 1980. Tornadoes, when, where, how often? *Weatherwise*, **33**(2): 52–59.

Nizamuddin, S., 1993. Hail occurrences in India. *Weather*, **48**(3): 90–92.

Obasi, G.O.P., 1994. WMO's role in the International Decade for Natural Disaster Reduction. *American Meteorological Society Bulletin*, **75**(9): 1655–1661.

Orville, H.D., 1990. AMS statement on meteorological drought. *American Meteorological Society Bulletin*, **71**(7): 1021–1023.

Palecki, M.A., Changnon S.A., and Kunkel, K.E., 2001. The nature and impacts of the July 1999 heat wave in the Midwestern United States: learning from the lessons of 1995. *American Meteorological Society Bulletin*, **82**(7): 1353–1367.

Parry, M., Arnell, N., McMichael, T., et al., 2001. Millions at risk: defining critical climate change threats and targets. *Global Environmental Change*, **11**(3): 1–3.

Paul, A.H., 1980. Hailstorms in southern Saskatchewan. *Journal of Applied Meteorology*, **19**(3): 305–314.

Paul, F., 1999. An inventory of tornadoes in France. *Weather*, **54**(7): 217–219.

Perry, A.H., 1981. *Environmental Hazards in the British Isles*. London: Allen & Unwin.

Perry, A.H., 1995. Severe hailstorm at Grahamstown in relation to convective weather hazards in South Africa. *Weather*, **50**(6): 211–214.

Perry, A.H., and Reynolds, D., 1993. Tornadoes: the most violent of all atmospheric phenomena. *Geography*, **79**: 174–178.

Peterson, R.W., and Mehta, K.C., 1981. Climatology of tornadoes in India and Bangladesh, *Archiv für Meteorologie, Geophysik, una Bioklimatologie*, ser. B, **29**: 345–356.

Pielke, R. Jr, and Carbone, R.E., 2002. Weather impacts, forecasts, and policy. *American Meteorological Society Bulletin*, **83**(3): 393–403.

Pielke, R.A., Jr, and Pielke, R.A., Sr, 1997. *Hurricanes: their nature and impacts on society*. Chichester and London: John Wiley.

Polger, P.D., Goldsmith, B.S., Przywarty, R.C., and Bocchieri, J.R., 1994. National Weather Service warning performance based on the WSR-88D. *American Meteorological Society Bulletin*, **75**(2): 203–214.

Powell, M.D., and Aberson, S.D., 2001. Accuracy of United States tropical cyclone landfall forecasts in the Atlantic Basin (1976-2000). *American Meteorological Society Bullelin*, **82**(12): 2749–2768.

Rasmussen, R., Politovich, M., Marwitz, J., et al., 1992. Winter icing and storms project (WISP). *American Meteorological Society Bullelin*, **73**(7): 951–974.

Riebsame, W.E., Changnon, S.A., and Karl, T.R., 1991. *Drought and Natural Resources Management in the United States: impacts and implications of the 1987-89 Drought*. Boulder: Westview.

Riebsame, W.E., Diaz, H.F., Moses, T., and Price, M., 1986. The social burden of weather and climate hazards. *American Meteorological Society Bulletin*, **67**(11): 1378–1388.

Roberts, W.O., and Lansford, H., 1979. *The Climate Mandate*. San Francisco: Freeman.

Robinson, P.J., and Hill, H.L., 1987. Toward a policy for climate impacts. *American Meteorological Society Bulletin*, **68**(7): 769–772.

Rooney, J.F., 1967. The urban snow hazard in the United States: an appraisal of disruption. *Geog. Review*, **57**: 538–559.

Rowe, M.W., and Meaden, G.T., 1985. Britain's greatest tornado outbreak. *Weather*, **40**(8): 230–235.

Scheider, S.H., 1976. *The Genesis Strategy: Climate and Global Survival*. New York: Plenum.

Sensarma, A.K., 1994. The great Bengal cyclone of 1737 – an enquiry into the legend, *Weather*, **49**(3): 90–96.

Shearman, K.M., and Ojala, C.F., 1999. Some causes for lightning data inaccuracies: the case of Michigan. *American Meteorological Society Bulletin*, **80**(9): 1883–1891.

Sheridan, S.C., 2002. The redevelopment of a weather-type classification scheme for North America. *International Journal of Climatology*, **22**(1): 51–68.

Smith, K., 1982. How seasonal and weather conditions influence road accidents in Glasgow. *Scottish Geography Magazine*, **98**: 103–114.

Smith, K., 1997. Climatic extremes as hazards to humans. In Thompson, R.D., and Perry, A., eds., *Applied Climatology: Principles and Practice*. London and New York: Routledge.

Smith, K., 2001. *Environmental Hazards: Assessing Risk and Reducing Disaster*, 3rd edn. London and New York: Routledge.

Smithson, P., 1993. Tropical cyclones and their changing impact. *Geography*, **78**: 170–174.

Snow, J.T., and Wyatt, A.L., 1997. Back to basics: the tornado, Nature's most violent wind: Part 1 – World-wide occurrence and categorization. *Weather*, **52**(10): 298–304.

Snow, J.T., and Wyatt, A.L., 1998. Back to basics: the tornado, Nature's most violent wind: Part 2 – Formation and current research. *Weather*, **53**(3): 66–72.

Speakman, D., 1994. Ambulances adrift: the impact of the snowstorms of the winter of 1990/91 upon the services of the City of Birmingham. *Weather*, **49**(3): 96–101.

Stevens, L.P., 1991. Tropical storm Marco flirts with Florida's flourishing west coast. *Weather*, **46**(8): 239–244.

Sweeney, J., 1985. The 1984 drought on the Canadian Prairies. *Weather*, **40**(10): 302–309.

Ulbrich, U., Fink, A.H., Klawa, M., and Pinto, J.G., 2001. Three extreme storms over Europe in December 1999. *Weather*, **56**(3): 70–80.

Vasiloff, S.V., 2001. Improving tornado warnings with the Federal Aviation Administration's Terminal Doppler Weather Radar. *American Meteorological Society Bulletin*, **82**(5): 861–874.

Visher, S.S., 1925. *Tropical Cyclones of the Pacific*. Honolulu: Bernice P. Bishop Museum.

Watson, A.I., and Holle, R.L., 1996. An eight-year lightning climatology of the southeast United States prepared for the 1996 summer Olympics. *American Meteorological Society Bulletin*, **77**(5): 883–890.

Weaver, J.F., Gruntfest, E., and Levy, G.M., 2000. Two floods in Fort Collins, Colorado: learning from a natural disaster. *American Meteorological Society Bulletin*, **81**(10): 2359–2366.

Western, J.S., and Milne, G., 1979. Some social effects of a natural hazard: Darwin residents and cyclone Tracy. In Heathcote, R., and Thom, B.G., eds., *Natural Hazards in Australia*. Canberra: Australian Academy of Science, pp. 488–502.

Wheaton, E.E., and Chakravarti, A.K., 1990. Duststorms in the Canadian prairies. *International Journal of Climatology*, **10**(8): 829–837.

White, G.F., 1974. *Natural Hazards: Local, National, Global*. New York: Oxford University Press.

Whittow, J., 1980. *Disasters: the anatomy of environmental hazards*. Harmondsworth: Penguin.

Wild, R., O'Hare, G., and Wilby, R., 1996. A historical record of blizzards/major snow events in the British Isles, 1880–1989. *Weather*, **51**(3): 82–91.

Wilson, L.S., 1981. The world's changing climate – some issues for planners. *Long Range Planning*, **14**(5): 83–89.

Woodhouse, C.A., and Overpeck, J.T., as before 1998. 2000 years of drought variability in the central United States. *American Meteorological Society Bulletin*, **79**(12): 2693–2714.

Wu, C.-C., and Kuo, Y.-H., 1999. Typhoons affecting Taiwan: current understanding and future challenges. *American Meteorological Society Bulletin*, **80**(1): 67–80.

Yarnal, B., Frakes, B., Bowles, I., Johnson, D., and Pascale, P., 1999. Severe convective storms, flash floods and global warming in Pennsylvania. *Weather*, **54**(1): 19–25.

Cross-references

Acid Rain
Air Pollution Climatology
Applied Climatology
Bioclimatology
Desertification
Drought
Lightning
Thunderstorms
Tornadoes
Tropical Cyclones

CLIMATE MODELING AND RESEARCH CENTERS

The following provides an overview of the main world centers where climate research and climate modeling are carried out.

Australia

Bureau of Meteorology Research Centre, Melbourne (BMRC) – Department of the Environment and Heritage, Commonwealth of Australia

The BMRC has six research groups focusing on different aspects of climatic and meteorological modeling, including model development, model evaluation, data assimilation, weather and climate forecasting and marine and ocean forecasting. Attention is given to the Australasian region in the development of numerical models, long-range forecasting, tropical meteorology and meteorological observation systems.

The research objectives of the center are generally designed to support the operations of the Bureau of Meteorology as well as to produce research that would be of interest to the scientific community. The center maintains close ties to the World Meteorological Organization (WMO), the Commonwealth Scientific and Industrial Research Organization (CSIRO) and is involved with the Cooperative Research Centre program of the Commonwealth Government.

Commonwealth Scientific and Industrial Research Organization (CSIRO), Melbourne – Atmospheric Research

The CSIRO is the major science research institution in Australia. It has several divisions that produce research in all aspects of the environment, of which the Atmospheric Research Division (ARD) is but one. The ARD operates three laboratories, located in Canberra, Victoria and Aspendale, and is grouped into three major programs: Earth systems modeling program; pollution program; and measurements, processes and remote sensing programs. Most of the research conducted within this organization is applied and covers topics that are specific to Australia's interests. The CSIRO collaborates with the BMRC, as well as participating in such programs as the International Geosphere–Biosphere program and the World Climate Research Program.

Canada

Canadian Centre for Climate Modelling and Analysis (CCCma), Victoria

The CCCma is a division of the Climate Research Branch of the Meteorological Services of Canada and is located on the campus of the University of Victoria. This center specializes in modeling coupled sea-ice and atmospheric modeling, climate variability, climate predictability and the carbon cycle, among other areas. They developed the Atmospheric General Circulation Model (AGCM2), which models equilibrium climate change simulations using variations of sulfur aerosols. This model has been used in paleoclimate studies, passer tracer studies, and predictability studies. The model has also been used to produce operational seasonal forecasts for the Canadian Meteorological Centre and by other collaborators. A newer model (AGCM3) has been developed and became operational as of January 2003.

The CCCma is involved with many international projects such as the Program for Climate Model Diagnosis and Intercomparison Project (PCMDI), hosted by the Lawrence Livermore Laboratory in the United States. CCCma has contributed to the Intergovernmental Panel on Climate Change (IPCC) reports of 1990, 1995 and 2001 and presently participates in the Climate Variability Predictability Program (CLIVAR) and the Scholarly Publishing and Academic Resources Coalition (SPARC).

France

Institut Pierre-Simon Laplace des sciences de l'environment (IPSL) (Campus du Jussieu)

The IPSL is composed of several laboratories that model different aspects of the environment and have working relationships with other institutions within France, throughout Europe and the rest of the world. Under the auspices and collaborative efforts of all of the laboratories in their institute, they have developed their own ocean-climate models (e.g. the IPSL model, the LMD and LMDZ, OPA-ICE). All of the IPSL laboratories are associated with the National Center for Research in the Sciences (CRNS) of France. Below is a survey of three laboratories related directly to climate modeling. All of them are affiliated with the IPSL.

1. Laboratoire de Meteorologie Dynamicque du CNRS, Paris. The Laboratoire de Meteorologie Dynamicque du CNRS (LMD) is part of the IPSL. It encompasses three sites: l'École Polytechnique in Palaiseau, l'École Normale Supérieure, in Paris and l'Université Pierre et Marie Curie, also in Paris. The LMD has four areas of research: climate modeling, remote sensing, experimental data measurements, and theoretical activities. The research is divided into several different teams located at the three university sites. They have developed the LMD and the LMDZ models of the general circulation of the atmosphere and presently collaborate with numerous laboratories, research centers and universities to develop linkages to study a wide variety of climate issues.
2. Laboratoire d'Optique Atmospherique (LOA) Lille. This laboratory, housed at the University of Lille, is involved with modeling the radiative properties of the atmosphere, especially as relating to its optical properties and its role in the global energy budget. The LOA focuses on the interaction of solar and terrestrial radiation with the atmosphere in the context of the global climate and climate change. It initially began as a research project on solar ultraviolet irradiance and radiative transfer theory, then shortly thereafter concentrated on air–sea interactions and on the remote sensing of the oceans. In the 1970s the laboratory focused its research on the radiation budgets of Venus and Mars. In the 1980s it applied its modeling efforts to the radiative forcing of the Earth–climate system, particularly to the development of radiative models involving the oceans, cloud cover and other aerosols, and the greenhouse gases.
3. Laboratoire d' Océanographie Dynamique et de Climatologie (LODYC). Housed in the University of Pierre et Marie Curie, the LODYC works to foster understanding of the Earth–climate system and the oceans' contribution to it. They have developed the OPA-ICE model that simulates oceanic circulation that can be coupled with other models (like the LMDZ) through the research at IPSL. They are also interested in modeling marine ecosystems in order to understand the role of the oceans in the exchange of carbon between the ocean and the atmosphere.

European Center for Research and Advanced Training in Scientific Computation (CERFACS), Toulouse, France

CERFACS has a set of research teams devoted to various aspects of scientific modeling, including climate modeling and global change, and modeling in fluid dynamics. They developed the OASIS model that is used as the coupler between general circulation models of the atmosphere and ocean circulation models in the research done by the laboratories of IPSL.

Germany

Alfred Wegener Institute for Polar and Marine Research (AWI – Bremerhaven, Germany)

The AWI is an institute devoted to modeling the Arctic and Antarctic environments and for using and developing general circulation models for polar and marine research. They also devote some attention to modeling for research in middle-latitude regions. The AWI has four main units, one of which is devoted to the climate system. In the climate unit, they have sections focusing on research in large scale and regional circulations, and on the physical and chemical processes of the atmosphere.

Deutsches Klimareschenzentrum (DKRZ) – German Climate Research Center, Hamburg

The DKRZ is a limited corporation funded in part by the Federal Ministry for Research and Technology, and is a service center for those participating in climate research in Germany. It was founded in 1987 at the University of Hamburg and is responsible for maintaining the infrastructure necessary for basic and applied research in climatology. It also cooperates with the European Climate Computer Network to allow for climate research to be pursued in both the larger and smaller nations of Europe.

Max Planck Institute for Meteorology (MPI-Met), Hamburg

Located in the university district of Hamburg, the MPI-Met contributes research into climate model development, investigates new methodologies of observation and measurement of the Earth–climate system and examines human–climate interactions that affect policy-making decisions. The institute is composed of three scientific divisions: biogeochemical systems, climate processes and physical systems, and one research group which itself has two scientific divisions: a regional climate group and a socioeconomic modeling group. The scientific divisions' activities focus on model formulation and application and are at present developing procedures to couple the models produced by the individual scientific departments and research groups with the goal of establishing a comprehensive model of the Earth–climate system.

United Kingdom

UK Universities Global Atmospheric Modelling Programme (UGAMP)

UGAMP is a network of university research centers that are primarily focusing on the development of large-scale numerical global climate models. The Natural Environmental Research Council (NERC), who fund the program, initiated UGAMP in 1987. There are nine affiliated university sites including London, Cambridge, Reading, Southampton, Rutherford (Appleton Lab), Oxford, East Anglia, Leicester and Edinburgh. Two universities in particular house the major climate and modeling centers while the other universities contribute to their work. These two centers are: the Center for Global and Atmospheric Modelling (CGAM) located at the University of Reading and the Atmospheric Chemistry Modelling Support Unit (ACMSU) located at the University of Cambridge. The CGAM is the main group that supports the UGAMP network of research sites. It emphasizes model development for large-scale atmospheric circulations and has partnerships with modeling centers and activities in the rest of Europe, the United States, and Japan, as well as with their colleagues in the United Kingdom. They are responsible for maintenance and diagnostics of the General Circulation Models formulated by the UGAMP community. ACMSU essentially works to provide the chemical data and codes that the rest of the UGAMP community will need in model development.

Hadley Center for Climate Prediction and Research (HC) – London

The Hadley Center for Climate Prediction and Research is the climate research arm of the UK Meteorology Office. It uses a variety of models including GCM, coupled ocean–atmosphere models, regional climate models and atmospheric chemistry models. The research is divided into teams focusing on the following themes: model development, model paramaterization, atmospheric chemistry and ecosystems, global and regional climate prediction, quantifying model uncertainties, rapid climate change, environmental stress, extreme events, climate monitoring, database development and interannual and interdecadal forecasting.

United States

Geophysical Fluid Dynamics Laboratory (GFDL), Princeton, New Jersey

The GFDL is a modeling center that is funded through the US Department of Commerce and the National Oceanic and Atmospheric Administration (NOAA). It has six research groups in it: Climate Dynamics, Hurricane Dynamics, Atmospheric Processes, Mesoscale Dynamics, Climate Diagnostics and Oceanic Circulation. There is also a technical support division that maintains the computing environment for the laboratory. They also maintain the Atmospheric and Oceanic Sciences Program (AOSP), a collaborative effort with Princeton University and funded by NOAA. This center is interested in developing atmospheric models at small and regional scales, as well as large-scale global circulation models. They have produced several models, notably the GFDL R15 and GFDL R30, which are coupled general circulation models of different spatial resolutions and the Modular Ocean Model, the most widely used global oceanic circulation model.

Lawrence Livermore Laboratory (LLL), Livermore, California

The LLL, operated under the US Department of Energy, includes the Program for Climate Model Diagnostic and Intercomparison Project (PCMDI) in which various atmospheric (Atmospheric Model Intercomparison Projects, AMIP), coupled (Coupled Model Intercomparison Projects (CMIP) and Paleoclimate models (PMIP)) are analyzed and compared in order to improve model development and reliability. These projects facilitate a community-based protocol that allows comparison and diagnostics of these models to be made. Models formulated by researchers in research and modeling centers throughout the world participate.

MIT Center for Global Change Science Climate Modeling Initiative (CMI), Cambridge

The CMI, developed as an outgrowth of the existing MIT Center for Global Change Science to concentrate efforts in climate modeling in order to contribute to understanding the larger context of global change science. The main attention initially was to improve climate prediction by ascertaining the limits of predictability. Funding has come from both private and public sources.

One of the objectives of the CMI is to use a hierarchy of models to simulate different portions of the Earth–climate system so that certain specific research goals can be met. They are also interested in addressing the chaotic behavior of the climate system in order to reduce the errors of prediction and to minimize errors in setting initial conditions for the models. Currently the CMI is developing a new coupled-atmosphere–ocean general circulation model that will incorporate and enhance an existing model developed by the Max Planck Institute (MPI-Met) with their collaboration.

NASA Goddard Institute for Space Studies (GISS), New York

GISS is a research institute designed to research a number of disciplines in Earth and space sciences. The institute works closely with Columbia University's Earth Institute and Lamont–Doherty Earth Observatory. Many GISS scientists are also closely involved with Columbia's Center for Climate Systems Research. The primary objective of GISS in the atmospheric sciences is to produce research leading to the understanding of global climate change. The research tends to revolve around the following themes: climate modeling development, climate forcings and impacts, Earth observations, paleoclimatology, radiation studies, atmospheric chemistry and planetary atmospheres.

Most of Goddard's research in climate modeling involves the development and use of Global Circulation Models and coupled GCM and ocean models. Some research is also conducted using energy balance and radiative transfer models. The research primarily focuses on using models geared toward understanding climate sensitivities to a variety of potential perturbations, such as greenhouse gas emissions, aerosols and solar variability. The research is also applied to understanding the possible impacts to society that climate changes associated with these possibilities may produce.

National Center for Atmospheric Research (NCAR): Climate and Global Dynamics Division (CGD), Boulder, Colorado

NCAR is a center devoted specifically to all areas of the atmospheric sciences. It is directed under the auspices of the University Corporation for Atmospheric Research (UCAR) and funded through the National Science Foundation. The major climate modeling division of NCAR is the Climate and Global Dynamics Division (CGD). This division is responsible for developing global circulation models of varying resolutions and scope. The Climate System Model (CSM), which consists of coupling models of the atmosphere, oceans, the land surface and sea-ice without the benefit of artificially providing flux adjustments, is the primary model that is being currently used. The CGD division of NCAR works closely with other divisions in NCAR, with the university research community and with other national and international laboratories to develop simulations using the CSM for use in other projects.

Other modeling and research institutions

There are many modeling centers and research institutions that are smaller in size, narrower in scope, or are part of larger organizations who do not have, as their primary function, climate or atmospheric research. The following is a list of laboratories and institutes and their locations that have climate modeling and research teams or divisions. Their contributions to model development and diagnosis cannot be underestimated.

Ames Research Center (NASA), Moffett Field, California
Argonne National Laboratory (Energy), Chicago, Illinois
Brookhaven National Laboratory (Energy), Upton, New York
The Jet Propulsion Laboratory (NASA), Pasadena, California
The Pacific Northwest National Laboratory, Richland, Washington
The Scripps Institute for Oceanography, San Diego, California
Woods Hole Oceanographic Institute, Woods Hole, Massachusetts

Robert Mark Simpson

Bibliography

AMIP Project Office, Lawrence Livermore Laboratory: 2003. *What is AMIP*. http://www-pcmdi.llnl.gov/

Atmospheric Chemistry Modeling Support Unit, National Environmental Research Center, 2003. http://www.acmsu.nerc.ac.uk/acmsu.html

Brix, H., ed. Alfred Wegener Institute for Polar and Marine Research 2003. http://www.awi.bremerhaven.de/AWI/index.html and http://www.awi-bremerhaven.de/Modelling/GLOBAL/

Bureau of Meteorology Research Center, M. Manton, Director. 2003. http://www.bom.gov.au/bmrc/basic/rol_hp.htm

Centre for Global Atmospheric Modelling, National Environmental Research Center, 2003. http://cgam.nerc.ac.uk/

Commonwealth Scientific and Industrial Research Organization, Geoff Garrett, Chief Executive Officer, 2003. http://www.csiro.au/

Delworth, T.L., et al., 1999. *Coupled Climate Modeling at GFDL*. Newsletter of the Climate Variability and Predictability Programme, vol. 4, no. 4; http://www.clivar.ucar.edu/publications/exchanges/ex14/exchv4n4p1.htm#DELW

European Center for Research and Advanced Training in Scientific Computation, (CERFACS) Paris, France. 2003. http://www.cerfacs.fr/globc/

GFDL, A. Leetma, Director, 2003. *Geophysical Fluid Dynamics Laboratory Brochure*. http://www.gfdl.noaa.gov/brochure/Cover.doc. html

Hadley Center for Climate Prediction and Research, UK Meteorology Office, 2003. http://www.met-office.gov.uk/research/hadleycentre/

Helhummer, H., Director. Alfred Wegener Institute for Polar and Marine Research: Climate System Section, 2003. http://www.awi-bremerhaven.de/Climate/index.html

Institut Pierre-Simon Laplace des Sciences de l'Environment. Publication Directors. Jouzel, J., Mégie, G., Senior, C., and Wilmart, E., 2003. http://www.ipsl.jussieu.fr/

Laboratoire d'Optique Atmospherique, Lille, France, 2003. http://loasys.univ-lille1.fr/activite_generale_gb.html, and http://loasys.univ-lille1.fr/divers/histoire_gb.html

Laboratoire Meteorologie Dynamique, H. Le Treut, Director. http://www.lmd.jussieu.fr/en/LMD_General.html, and http://www.lmd .jussieu.fr/~lmdz/PUBLIS/publis.html)
Luthardt, H. Deutsches Klimareschenzentrum, 1995. http://www.dkrz.de/dkrz/institution-eng.html
Massachussetts Institute of Technology, Center for Global Change Science Climate Modeling Inititiative, Peter Stone, Director, 2003. http://web.mit.edu/cgcs/www/cmi.html
Max Planck Institute for Meteorology, Hamburg. Hartmut Graßl, Director of the Climate Processes Division, 2003. http://www.mpimet.mpg.de/en/web/
National Center for Atmospheric Research, Tim Kileen, Director, 2003. http://www.ncar.ucar.edu/ncar/about.html
National Aeronautics and Space Administration, Goddard Institute for Space Studies, James Hansen, Director, 2003. http://www.giss.nasa.gov/about/ and http://www.giss.nasa.gov/research/modeling/
Program for Climate Model Diagnostics and Intercomparison, Lawrence Livermore Laboratory, 2003. http://www-pcmdi.llnl.gov/
Universities Global Atmospheric Modelling Programme, National Environmental Resarch Center, 2003. http://ugamp.nerc.ac.uk/
White, R. 1995. *Geophysical Fluid Dynamics Laboratory.* http://www.gfdl.noaa.gov/overview.html

Cross-reference

Climate Data Centers

CLIMATE VARIATION: HISTORICAL

The basis of history is documentation, and given that database we can attempt to draw some conclusions; one might even try to use this historical record, in combination with the instrumental material of the last century or two, to anticipate the future. Experimental science should, ideally, provide the basis for prediction. In closed systems, like the chemists' test tube, prediction can be made with some degree of certainty. But in the open systems of the planet Earth–its atmosphere, ocean, magnetic fields, moon, sun and other planets–all these bodies and systems are interacting.

The trouble with the meteorological records is that they are too short to capture the "big picture." Instrumental data provide a numerical, quantitative basis for working out the dynamic equations of atmospheric motion. They also offer a quantitative way of comparing day-to-day variables, the contrasts between winter and summer: a definite improvement on our intuitive, sensory, and memory-based impressions of hot and cold, moist and dry. In combination with computer-based storage facilities and sophisticated methods of statistical analysis, some hitherto unknown, or unproven, features are now beginning to emerge from the meteorological materials, such as the 14-month Chandler wobble pulse and the 18.6-year lunar nodal cycle. But what of the longer periodicities and peculiarities of the climatic machine? The agreed base for a standardized climatic average for any observing station, by convention of the World Meteorological Organization, is only 30 years. The climatic variation of the geologic record should indicate clearly that the day–month–year basis of instrumentally based climatology only scratches the surface. But the geological picture has a tendency to drift (sublimely) away from the human scale of things. This is where the historical record helps to fill in an essential gap. It takes us from the instrumental field of "hard numbers" to ancient diaries, tax records, narratives, and eventually to the geological data sources, where the quantitative values must be calculated, albeit by highly sophisticated analytic techniques, not least the measurement of the time factors.

Definition

Historical climatology is seen in a variable context by different workers. (A sampling may be seen in several collections: Rotberg and Rabb, 1981; Wigley et al., 1981; Mörner and Karlén, 1984; Roberts, 1989, 1998; Burroughs, 1992; Shindell et al., 2001). Let us define historical climatology as the study of climate through the time-range of civilized *Homo sapiens*, during the period in which humans have developed the arts of writing and the construction of permanent dwellings and other structures relating to their maintenance and culture. Harding (1982) has provided a useful volume on climate in a context of Holocene archeology.

That time-range varies from region to region. In Britain it is about 3500 years (Stonehenge), but in Egypt it is more like 6000 years. The world's oldest continually inhabited town is said to be Jericho, in the Jordan Valley, which was established 10 000 years ago, or 8000 BC. This date, roughly is also the conventionally adopted date for the Pleistocene–Holocene boundary, agreed by a commission of INQUA–the International Union for Quaternary Research (Olausson, 1982; Fairbridge, 1983). Thus, the 10 000 years of the Holocene Epoch become the logical and ultimate time frame of historical climatology.

Pfister (1980) recognized two categories of data that are appropriately integrated to construct a historical record of climate for any region and interval of time within the designated range: (a) *documentary proxy data* and (b) *field data*.

(a) The use of documentary proxies has been made a special, life-long study by Hubert Lamb (1972, 1977), who integrated them closely with the instrumentally based meteorological standards. The documents provide varied types of information and are of varied quality. They range from diaries, which usually have the virtue of homogeneity (a single observer, living in a given place, and dated with reasonable accuracy), to records of particular events (often found in monastic documents, such things as eclipses, comets, auroras, floods, droughts, killing frosts), and eventually to folklore or sagas (weak in chronology but often helpful if used in conjunction with field data). The diaries may contain actual weather observations (temperatures, wind, precipitation), or they may generate proxy indicators (such as crop planting, harvest times, with yields and costs). Tax records are very illuminating, when interpreted by a skilled specialist (e.g. Pettersson, 1912, 1914, 1930).

(b) Field data are mostly geological, botanical, or glaciological, where the chronology is established on an incremental basis, usually year by year or perhaps century by century. The time-series may go backward from the instrumental era or it may be "hanging" chronology, intended to be integrated "some day." The source material is generally stratigraphically layered (following the *Law of Superposition*), as in the seasonally varved bedding of a lake, or ice layers in a glacier, or the speleological accumulations of a limestone cave. It may be concentrically banded as in the inorganic growth rings of stalactites and stalagmites or in tree rings (*dendrochronology*). Event indicators include such things as volcanic ash layers (tephra, hence *tephrochronology*, frost or burn rings in dendrochronology, and coarse sediments in thick varves, indicating major flood years) or thickness increase in the dark layers of varves (thus higher organic productivity). Seismic events (earthquakes) may be recorded by rock falls in cave deposits (archeological data often record such rude interruptions to the human occupancy), or by "neotectonic" faults or slumps in stratigraphic sequences.

World-class standard time-series

In this entry are listed selected examples of world-class historical standards within the Holocene Epoch that deserve maximum attention from climatologists. If twentieth-century climatologists ignore these mutually interlocking records, they do so at their own risk. History does not repeat itself, and *time's arrow* flies only in one way, but those who ignore history are condemned to repeat its errors. The mutual interlocking of chronologies has, up till recently, been a matter of hypothesis. But now the sheer volume of data is enough to prove that certain events and trends are of global significance, and cannot be dismissed with a shrug or sneering aspersions as to accuracy of dating or whatever. The error bars are often large, but the ever-narrowing ranges of those error bars now permit conclusions based on the convergence-of-evidence line of reasoning (Windelius and Carlborg, 1995).

As to accuracy of chronology, one may point to the vast accumulation of tree-ring data that is now being extended to the whole world (Stockton et al., 1985). Rings are counted year by year, and some sequences reach back to 13 000 BP and more. Errors can be made but, with multiple samples and multiple workers, the margin of error is gradually being refined to the ±1-year level or better. As will be noted below, de Geer's counting of annual varves, which started a century ago, was strongly attacked during his lifetime, but now a Swedish team of devoted workers has repeated his more than 10 000-year record. Small refinements and corrections have been made, to be sure, but the basic chronology is correct (Cato, 1985). More recently, ice cores have furnished a third system of year-by-year counting (Dansgaard et al., 1971). Again errors can occur, ablation can skim off a layer or two, but there are over 100 ice cores over 100 m deep from both hemispheres, which cannot all be wrong.

All three proxy calendars are recorded in *sidereal years*, documenting the seasonal regularity of the Earth's revolution about the sun. The invaluable radiocarbon-dating method, for all its built-in potential for errors, is now calibrated to the sidereal record, and so can be used for bringing "floating" chronologies into the firm records (Klein et al., 1982; Stuiver, 1982). With time measured in sidereal years we can now use astronomical parameters to retrodict planetary–solar–lunar events (Fairbridge, 1984a).

Ten world-class categories of examples of standard time-series are now discussed: ice cores, dendrochronology, lake varves, palynologic sequences, fluvial sediments, beach ridges, marine varves, the volcanic signal, the paleomagnetic signal, and the documentary archives.

Standard time-series

1. Arctic and Antarctic ice cores

One of the most promising fields of research in the scope of historical proxies is the drilling and study of glacial ice cores. The drilling reached 2164 m at Byrd Station (Antarctica) and 2035 m at Dye 3 (Greenland) and dates events back to more than 100 000 (Barry, 1985). The first analyses generated an invaluable ^{18}O–^{16}O isotope curve that is believed to reflect the temperature (and salinity) of the evaporated sea-surface waters (Dansgaard et al., 1971; Lorius et al., 1979). As remarked by Hecht (1985, p. 20), the reproduction of the "Dansgaard profile" in Flint's textbook (1971), "became a standard climate series against which other climatic records could be compared."

An examination of cores both in the Arctic and Antarctic showed that particulate contamination was very much higher during the glacial phases than during interglacials. At first it led to the idea of volcanic forcing of the major glaciations (Bray, 1976). A closer look at the particles, however, showed that they were dominated by desert dust, reflecting the worldwide aridity and greater meridional upper air flow of the glacial phases.

Volcanic activity, however, is recorded in the ice cores (Hammer et al., 1980). The invention of a rapid pH-measuring device opened the way to identifying specific annual layers of high acidity, reflecting volcanogenic SO_2 and CO_2 injections into the upper atmosphere. Precision counting of the ice layers, year by year for the last few millennia, has been rewarded by the discovery that within 12 months of almost all the major historic eruptions there is an acidic *spike* in the cores. Its intensity is presumed to be more or less proportional to the gas production in the eruption. Some eruptions generate only a little gas. Numbers of such spikes already have been found that lack known volcanic events, which only goes to show how little is really known of the Holocene sequences in many parts of the globe (See the AD 535 eruption in the Sunda Strait: Keys, 1999).

Acidity and particulate pollution are not the only variables that can be measured in the ice cores, in addition to the basic ^{18}O signal. Chloride, nitrate, and sulfate have been analyzed by Herron and Herron (1983) from Dye 3. The isotope beryllium-10 (^{10}Be), which is comparable to radiocarbon insofar as it is generated in the upper atmosphere by cosmic ray collisions, it may well serve to measure, inversely, the solar-activity modulation of the geomagnetic field.

Within the ^{18}O ice-core records, there is a prominent cyclicity, on the order of 90, 180, and 360 years. Unfortunately the sampling accuracy is not yet sufficiently refined to give a precise year-by-year variance. Nevertheless, these periodicities are dramatically close to those recorded in the beachridge studies (see subsection 6 below), as well as in sunspots and in known planetary cycles (Fairbridge, 1984a).

Precise attention to the microparticle deposition over the last millennium at the South Pole has shown that accumulation was maximal from AD 1450 to 1860 during the Little Ice Age (Mosley-Thompson and Thompson, 1982). It also provided a basis for calculating annual increments of snow. The Little Ice Age also was marked by the minimal rate of snow accumulation (1657–1686), whereas the warm phases of 1057–1086 (Little Climatic Optimum) and 1867–1896 were marked by maximal snow accumulation. On Greenland the pronounced warming of the Little Climatic Optimum (or Medieval Warm Period) made colonization feasible for the Viking voyagers, and the demonstrated cooling of the fifteenth century brought those colonies to their tragic end (Dansgaard et al., 1975). Even their Icelandic settlements nearly collapsed (see Ogilvie in Wigley et al., 1981).

A newly developed infrared laser spectroscopic technique has made it possible to analyze tiny gas bubbles in the ice and their CO_2 values (Neftel et al., 1982). This has led to the very important discovery (in the light of the present fossil-fuels controversy), that during the last glacial the CO_2 level in the atmosphere was 210 ppm, as compared to the modern 340 ppm. Several specialists regard CO_2 fluctuation as a biological consequence of climate change, not a cause (e.g. Newell and Hsiung, 1984); because of increased upwelling and phytoplanktonic productivity, larger quantities of CO_2 are withdrawn from the atmosphere (and hydrosphere) during cold epochs.

2. The California dendrochronology

Counting tree rings has been carried from a fine art to a sophisticated science in the US Southwest, and is now being extended to the world (Stockton et al., 1985). The crowning achievement has been the radiocarbon analyses of the dendrochronologically dated Bristlecone Pine series from the Sierra Nevada by Hans Suess and a large team of associates. Over eight millennia of records now document the fluctuations of the ^{14}C flux rate through the entire period. This provides an inverse quantification of solar emissions (see Sunspots), as discussed by Damon et al. (1978) and Eddy (1977). The actual ring width (and density), as an indicator of favorable growing climate, correlates precisely with the radiocarbon value in selected areas (Sonett and Suess, 1984). Refined studies have been accorded the rings of the last 2000 years (Stuiver, 1982), and later about 10 000 years (Stuiver and Braziunas, 1993) which permits correlation with the long-term climate proxies of northwest Europe. For a long time tree-ring counts in northwest Europe were in hanging chronologies, that is to say, not interconnected, but the whole sequence has now been unified (Hughes et al., 1982).

The main feature about tree-ring analyses for climatic interpretation (*dendroclimatology*) is that although one particular tree provides only a record of its own locality a regional picture can be constructed when a considerable number of samples are analyzed and integrated. Excessively large sample areas, however, are likely to smooth out the interesting signals, so that large numbers of individual tree-growth areas should be treated separately (Currie, 1984).

Tree-ring analysts are now extending their work to all continents. All latitudes, however, cannot be treated, because certain climatic belts do not furnish distinctive annual rings, as in the year-round wet regimes of the equatorial regions. In some settings the tree grows its thickest rings in the wettest years, but in others it grows them in the warmest years. In some low-latitude mountain areas the vertical zonation (and thus the meaning of the signal) will change dramatically several times in the course of the Holocene Epoch. Nevertheless, the dendroclimatology of the Holocene offers the most universally useful and potentially productive source of long-term climatic data and deserves the most energetic investigations.

3. The Swedish varve chronology (and other varves)

Nearly a century ago the Baron Gerard de Geer (1858–1943) began measuring the thickness of the annual sediment layers in Swedish clay pits (used for the brick industry). The clays were laid down in late glacial and postglacial lakes in distinctive couplet layers (varves, from the Swedish name), a dark band representing winter freezeover and a light band the summer meltwater input. The thickness of the light material is a measure of the melt season in the watershed area, and of the dark layer the organic productivity, and thus a good climatic proxy. Some layers contain volcanic ash impurities from eruptions in Iceland or elsewhere and can thus be used for teleconnections.

De Geer completed an integrated count of thousands of lake deposits, and thus provided a chronology for the entire Holocene, a time scale that for a long time represented the world's only absolute (year-by-year) chronology. In northwestern Europe all other stratigraphic systems, from pollen analysis to archeology, to the stratigraphy of the Baltic, were dated by de Geer's method. His methodology was strongly criticized during his lifetime, but subsequently the Swedish Geological Survey has resurveyed the entire sequence, essentially confirming the old standard, but introducing small corrections here and there (Cato, 1985). It was de Geer's chronology that made it possible to date the sea-level fluctuations represented by the raised beach (strandline) deposits of southern Sweden. Those same sea-level changes were then identified in the South Pacific and dated by de Geer's chronology; subsequently, shells from those beaches were radiocarbon-dated and (after isotopic correction) found to be identical with Swedish standard (Fairbridge, 1961, 1981).

Varved lake clays also are found in Finland, Russia (Lake Saki), Iran (Lake Van), Patagonia, Canada, and the United States. Teleconnections between such far-removed areas are fraught with difficulties, but provided that an approximate correlation can be demonstrated, perhaps within a 100-year error range, then a "vernier tuning" can be applied, as demonstrated by Schove (1978, 1983). The method is based on the twentieth-century observation that the well-known 26-month QBO or quasi-biennial climate cycle sometimes breaks down, especially during sequences of weak solar cycles, to become a quasi-triennial cycle. The length and timing of such breakdowns are globally synchronous so that when they are isolated in hanging chronologies (tree rings, varves, ice cores, or whatever), the century error can be reduced to 1 or 2 years. With the contemporary interest shown by meteorologists and oceanographers in the QBO and ENSO (El Niño–Southern Oscillation), this is an attractive way of extending their statistical data base to 10 000 or more years by using the historical proxies.

Varved marine layers (see subsection 7 below) are much less well known than the fresh-water accumulations, because they are mostly in deep, isolated basins, requiring very costly core-sampling expeditions.

4. Palynology of peat and lake sequences

The science of palynology, usually treated as a branch of botany, treats with the collection, sampling, identification, correlation, and chronology of pollen and spores. Stratigraphic treatment of cored sections provides a continuous paleoecologic record for a specific region that can be compared with actual data (natural contemporary floras and their growth patterns). This comparison permits standardization of results in terms of climatic parameters (Birks and Birks, 1980).

Pioneer studies of this sort were carried out in the nineteenth-century by two Scandinavian botanists, A. Blytt (a Norwegian) and R. Sernander (a Swede), who subdivided the 10 000 years of Holocene time into a series of biozones: *Preboreal, Boreal, Atlantic, Subboreal,* and *Subatlantic*. For convenience they are often identified by the Danish system using Roman numerals: respectively, IV through IX (zones I–III are late glacial Pleistocene).

Mangerud et al. (1974) proposed that, because the local Scandinavian terms of Blytt and Sernander were so useful in designating paleoclimatically determined divisions of Holocene time elsewhere, they should be employed as global stratigraphic time labels, as *chronozones*, and they have been formally defined in terms of radiocarbon years.

Probably the most significant climatic event in the Holocene (Nilsson, 1983), terminating the warmer *Hypsithermal* or

Climatic Optimum phase (Atlantic chronozone, with its mixed-climax forests), was first identified from the palynological data as the Ulmus (Elm) Decline about 5000 ^{14}C-years BP (5800 sidereal years BP. All across northwest Europe this dramatic decrease in the elm forests has been attributed by some botanists either to Neolithic farming or the disease (like the one affecting elms in the twentieth-century). However, the same decline has now been traced by Soviet palynologists all across Asia as well (in Velichko, 1984) and evidence of a simultaneous climatic deterioration and neoglacial advances (Denton and Karlén, 1973) in Arctic Canada, Greenland, and southern hemisphere lands combine to show that it was a global phenomenon. In the temperate belt of eastern North America, according to Davis (1983), the time boundary 5000 ^{14}C-years BP. (5730 Cal. Yr) is marked by a widespread hemlock *(Tsuga)* decline, with a concurrent rise of hickory *(Carya)*. In the western United States, in the central Rockies, from 5000 ^{14}C-years BP on, the tree line became distinctly lower, and is further evidence of the cooling climate. These examples are simply citations documenting a global event of climatic nature. A vast literature has already built up, documenting this one and many others during the Holocene.

A remarkable site is a peat swamp at Ozegahara moor, Japan, where a 4–5 m core has been obtained covering a continuous time-series for 7600 sidereal years (Sakaguchi, 1982). The closely sampled and sharp fluctuations offer a unique record that correlates well with the better-known history of Europe (Roberts, 1998).

Of importance for climatology has been the rising level of confidence in recent decades, that shows an appreciation of the reality of these fluctuations documented by the palynological stratigraphy. Instead of regarding them as local events, experimental error, or whatever, the chronological precision is becoming steadily more refined, and from the repeatability of the pollen diagrams it is becoming quite clear that we are looking at finely tuned paleoclimate records. Care must be taken, especially in the more densely populated areas, to avoid anthropogenic influences, but such problems hardly arise in such remote and sparsely inhabited areas as Finnish Lapland, eastern Siberia, northern Canada, and Patagonia. Yet in all those regions the strongly fluctuating palynologic records proclaim the varied climatic history of the last 10 000 or so years.

5. Nile floods

A unique data source is provided by the documentary record of the Nile floods, which have two peaks each year: the high summer peak reflects the northern hemisphere summer monsoonal rains in Ethiopia; the low, winter peak reflects the equatorial rains in East Africa and the overflow of Lake Victoria. Inasmuch as Egypt's agricultural wealth has always been gauged by the height of the annual Nile floods, and thus for tax reasons, the records have been carefully kept since the foundation of Islam (AD 622), with some gaps caused by social upheavals and invasion. Nevertheless, the record is enormously helpful, providing not only a tropical climatic standard, but also some global signals (Fairbridge, 1984b; Currie, 1995). Historical events such as persistent droughts have been correlated with it all across North Africa from Senegal to Ethiopia. One remarkable example was experienced during a succession of very low Nile floods in Egypt from 827 to 848, but in 828–832 the Nile is reported to have frozen over (presumably in the delta); this correlation of dry cycles with low temperatures is frequently noted in the subtropics. In some epochs the dry cycles match sunspot minimums, but in others there is a phase shift to match the sunspot maximums.

The lowest Nile flood of the twentieth century (the year 1913) corresponded to the highest lunar declination in a millennium; similar low extremes were reported all across Africa. Particularly interesting is the fact that in some centuries there is a strong correlation with the lunar nodal cycle (18.6 years), but in others the solar (11 years) signals appear. This question has been discussed by Currie (1984, 1995). In the Senegal River flood–drought sequence Faure and Gac (1981) found a 30-year periodicity that could represent a beat frequency of these two, or perhaps the 31-year perigee–syzygy cycle of the moon.

Prior to the Islamic era the Nile records are more scattered, but pharaonic data are quite informative (Bell, 1975), especially when taken in conjunction with stratigraphic material. The latter can be dated by an integration of several methods: radiocarbondating of freshwater shells (mollusca) and charcoal (from fireplaces) and by association with dated archeological cultures or with established structures such as temples.

Evidently the pharaohs, for all their interest in both taxes and immortality, had little idea of how much variance the annual flood could offer. Some of the temple walls show little white bands (of calcium carbonate) marking flood levels up to 3 m *above* the flood plain on which they were built. At Semna (above the Second Cataract) the writer measured the present maximum flood height at 8 m *below* a high watermark inscribed about 1800 BC during the reign of Amenemhat III. Such fluctuations correlate with the ups and downs of world sea level (Fairbridge, 1962), itself a proxy for sea-surface temperature (SST) and for glacier advance and retreat (glacial eustasy).

The Nile deposits (with their radiocarbon and historical data series) provided the first quantitative demonstration of the falsity of a long-accepted climatic premise: that chronologically the glacial epochs equaled pluvials (wet or snowy cycles), and interglacials equaled interpluvials (i.e. dry periods). The delusion goes back in the past century to biblical scholars who observed the wadi gravels of the Dead Sea region and elsewhere, associating them with the great rains of the Noachian flood. High lake levels in East Africa and high terraces on the Nile were easily drawn into the same net. On logical, climatological grounds, objections were raised in 1913 and 1928 by Albrecht Penck (citations and discussion in Fairbridge, 1962). A cold glacial-phase ocean did not lead to more evaporation, but less. Besides, it was eustatically smaller, and large areas were covered by sea ice, a 13% reduction in effective sea surface. Net evaporation may have dropped 40%. East African lakes were in fact low during glacial phases. The middle Nile valley became a desiccated inland delta because of the reduced discharge, and the high terraces seen today are relics of hyperarid phases when discharge failed to reach the Mediterranean; indeed, the middle valley filled up to 40 m with silt. Glacials were *not* pluvials; there was ice-age aridity in Africa (Fairbridge, 1962, 1976b).

A similar inland delta has been discovered on the Okawango of northern Botswana and another on the Niger, in Mali, upstream from Timbuktou. In the Sudan and Egypt the aridity ended with the overflow of Lake Victoria into the Nile around 13 000 BP, and this can be matched by similar pluvial activity all across North Africa to India and northern Australia. The monsoon, which had been brought to an end during the glacial phase by the year-round high-albedo and high-pressure system established over Central Asia by the growth of ice caps, began to reassert itself around 13 000 BP. Heavy seasonal rains, a real pluvial, set in as part of the usual monsoonal pattern. Oxygen-18 isotopic studies and planktonic indicators in the surroundings of the Arabian sea confirm the story from the oceanographic side (Prell, 1984).

During the Holocene, from time to time, throughout the entire tropical region, there also were strong fluctuations of the climatic regime (Fairbridge, 1976b), just as is indicated by the Nile record. They are particularly interesting from the human point of view, because this same belt also happens to include the Cradle of Civilization. Climatic forcing of human migrations seems as real here as in central Asia (Huntington and Visher, 1922), although sociologically oriented historians seem to have a marked distaste for climatic determinism. There is no doubt that the nomadic Neolithic people were forced out of the north African savannas by the advance of the climatic (*not* human-induced) desertification of the Sahara. It was surely not a question of overpopulation or politics around 6000 BP! At this same time began the intensive settlements along the Nile that were associated with a meteoric rise in building, engineering, and irrigation skills. Equally well, during certain catastrophic climatic deteriorations, such as the desiccation of the Indus Valley around 3000 BP, one sees the total obliteration of a major culture, in this case Mohenjodaro (Bryson and Murray, 1977). Care needs to be taken, of course, to avoid psychological and viewpoint errors (Bryson and Padoch, 1980), but the large number of investigators is a fairly healthy insurance against jumping to conclusions.

6. Coastal beach ridges of Hudson Bay and elsewhere

Around the coastlines of the world certain areas are favored by gentle offshore gradients and large supplies of gravel, sand, or silt so that a prograding shore tends to develop. In regions of crustal subsidence such as the Mississippi or Rhine deltas, those shoreline deposits are progressively drowned and buried. In rather stable regions such as the shores of Western Australia, West Africa, and Brazil, the prograding systems develop sheaves of beach ridges across flat coastal plains. In uplift regions such as the glacioisostatic rebound coasts of Hudson Bay, the Canadian Arctic, Spitsbergen, Scotland, and Scandinavia those beach ridges are progressively uplifted, so that in protected places they form dramatic staircases of shore terraces. This third category is the best for the study of historical time-series, because the crustal uplift is a slow, steady, non-fluctuating motion and relatively easy to measure. The residual value in the elevation, which with the width ratio of the terraces, is a measure of past sea levels and degree of storminess.

The staircase beach ridges of Richmond Gulf on eastern Hudson Bay (Quebec) have been measured in the field in numerous traverses (thesis work of Hillaire-Marcel), and associated formations have been dated by radiocarbon with many cross-checks, so that chronology is considered well established (Fairbridge and Hillaire-Marcel, 1977; also discussion in Fairbridge, 1983). The area is near the former center of the maximum ice loading of the Laurentide (New Quebec) ice sheet and today is the site of maximum deglacial isostatic uplift. The oldest Marine Limit of 8300 BP is uplifted to around 315 m elevation. The Earth's crust here is still rising at about 8 mm per year. The rising eustatic sea level of the last post-glacial (Flandrian) phase caused sea water to intrude the Hudson Bay beneath the continental ice, leading to buoyancy and a very rapid calving, so that within a few hundred years of 8300 ^{14}C years BP the entire bay was emptied of icebergs and normal marine deposits began to accumulate. Concordant dates for this event are obtained all around the bay. A shallow-water marine inland sea of very large size (520 000 km^2) was created, at first several hundred meters deep but today with an average depth of only about 100 m. It is thus hydrographically rather comparable to the Baltic, and subject to strong fluctuations of water level due to atmospheric pressure, wind, and seiche effects; with the addition of tides (of the order of 1 m) the annual water level fluctuation may thus be as much as 2.5 m.

The staircase of raised beach ridges is for the most part uniform, with 185 steps distributed from the present beach up to 315 m. There is a gradually increasing individual height toward the top, a phenomenon due to the initially rapid isostatic rebound, now gradually decreasing. The mean age difference between beaches is 45 ± 1–5 years. On average, they are 1 m high and 5 m wide, but exceptional ones are 3–4 times higher. The individual ridges represent relatively short periods (3–10 years) when the degree of storminess was high, those episodes being separated by longer intervals when only fine sands were moved by the waves and relative sea level was rather stable. The Hudson Bay, at the present time, freezes over for 7–9 months of the year, so it is only summer storminess that is involved. Meteorological and tide-gauge records provide a guide to this alternation between stormy and calm periods during the present century, but the duration of the ice-free to frozen-in states is documented rather well, back to the early 1700s, thanks to content analyses of the old fur-trading log books of the Hudson Bay Company (Catchpole and Ball, 1981).

On the south western Hudson Bay, a more gentle slope permitted the preservation of intermediate beach ridges at 22 and 11-yr intervals (Grant, 1993). Prior to about 3000 BP the higher isostatic emergence rate amplified the contrasts (Figure C18).

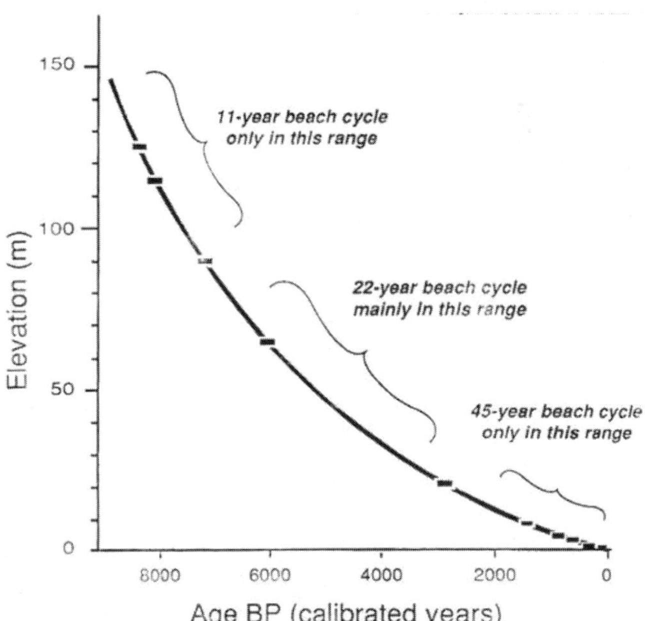

Figure C18 Holocene beach-emergence cycles in southwestern Hudson Bay, showing gradual decrease of glacio-isostatic recovery over the last 8000 years (courtesy: Douglas R. Grant, Geological Survey of Canada). Note: Progressive increase of storminess cycles from 11-year prominence, in the early stage, before 7000 yr BP (calibrated), through 22-yr before 3000 BP, down to 45 yr in last 3000 years. In the eastern coast, the 45 yr and longer cycles occur throughout the sequence (Fairbridge & Hillaire-Marcel, 1977).

Maximum beach building corresponds to longest periods of summer ice-free conditions and to the incidence of frontal storm systems across the bay. When the jet stream lies well to the south, the Arctic high-pressure cells move sluggishly across northern Canada from northwest to southeast, often remaining stationary over northern Quebec, causing cold easterly to northerly winds on the bay. In contrast, when the jet trajectory runs farther north, the frontal systems cross the bay, bringing in warm, wet, southerly and southwesterly air; a considerable storm fetch and wave set-up is then generated along the eastern margin of the bay.

From studies of pollen indicators and dune sand outbreaks in the same region, Filion (1983, 1984) discovered a correlation between the cold-dry climatic cycles, forest fires, dune building, and anomalous negative sea-level fluctuations that have a cyclicity of about 360 years, as noted earlier in the beach ridges by Fairbridge and Hillaire-Marcel (1977). Both amplitude and period are very similar to the Rocky Mountain tree-ring record of LaMarche (1974).

For convenient identification the ridges are numbered from no. 1 to no. 185, with 1 tentatively dated from tide-gauge and meteorological data as AD 1962. Beach 2 would be 1917, and so on. To calculate sidereal BP dates we create an imaginary beach 0 at 2007 which is 57 years AP (after present, the arbitrary astronomic and geochemical datum being 1950). Then for any beach number (n) we calculate: $(n \times 45) - 57 =$ date (in sidereal years BP)

The most astonishing feature of these beach ridges is their regularity. No stochastic variance would seem possible for such a uniform cycle of 45 years. When plotted as a simple graph, with the isostatic component removed, it is evident that there is also a further modulation into sets of somewhat higher and somewhat lower ridges. The length of this secondary long-term oscillation seems to have at least two components, of the order of a half and a third of a millennium. What manner of cycles are these?

In view of the dramatic demonstrations by a number of workers (Currie, 1984 and others) that both solar and lunar cycles are clearly recorded in long-term historical series, we need to compare them with both the standard oscillations (i.e. the lunar nodal precession at 18.6 years and the solar activity cycle of 11–22 years, together with their harmonics) and various higher-order astronomic alignment periodicities. In the first paper (Fairbridge and Hillaire-Marcel, 1977), a double-Hale solar cycle of 45 years was suggested, and it is indeed observed that 11-year solar cycles are commonly arranged in groups of four (with increasing sunspot numbers). A 45-year periodicity is seen in eastern North American summer temperatures, in sea levels, and in the secular drift of the geomagnetic field. A careful analysis of Canadian Arctic winter temperature and pressure data (meteorology plus proxies) by Guiot (1985) show a phase coherence with Currie's solar and lunar results, as well as a volcanic dust signal. As mentioned earlier, the 18.6-year lunar tidal effect not only appears in sea-surface temperatures and sea tides but also triggers some earthquakes and volcanicity. Currie's work shows that its phase fluctuates by 180° regionally and over long time-series, and this was confirmed by Guiot for the northern Arctic region (including the area of the north magnetic pole). When the solar signal is strong the lunar one is often weak, and vice-versa. Maximum winter temperatures often correspond to the even-numbered (south positive) solar cycles that are most often weaker than the next cycle. Over the last centuries the solar (22-year) signal has been strongest over the Hudson Bay. However, after having established the most probable twentieth-century dates (tide- and weather-based evidence) for the Richmond Gulf beaches 1 (1962) and 2 (1917), it is noticed that 1962 matches the lunar perigee–syzygy cycle (31-year) peak and 1917 matches the 18.6-year lunar–solar tidal maximum. The relationships of the lunar cycles to sea level and storm-flooding events are described by Wood (1978, 2001).

Figure C19 Emerged beach ridges facing the Barents Sea in northern Russia (Møller et al., 2002), the dates ranging from late Pleistocene to early Holocene, and the storminess frequency is less than on Hudson Bay. The "Marine Limit" (ML) here is probably Alleröd, and the Younger Dryes (YD) is dated 9190 BC (Cal. 11,140 BP).

A systematic exploration of Arctic coasts was undertaken by the writer and associates (Fletcher et al., 1993), with comparable results except that in northern Norway and Russia (Møller et al., 2002) ice-borne boulders marked cold cycles with advance of Arctic sea ice (Figure C19).

Referring now to the long-term climatic oscillations on the Hudson Bay shown by Fairbridge and Hillaire-Marcel (1977), it is curious that the solar–planetary influences (178-year and 356-year cycles) seem to be most prominent since about 3000 BP, but prior to that the 556-year (50 $\times$ sunspot cycle), and 558-year lunar cycle (28 $\times$ nodel cycle and precession of lunar perigee) seem to be associated with larger fluctuations. From evidence in the Baltic history, Pettersson (1912, 1914, 1930) reached the conclusion that the long-term lunar cycles played a leading role in the regional oceanographically forced climatic behavior but recognized that they also were gravitationally tied-in to the sun. A long-term correlation with the Baltic herring fishery introduced an interesting economic overtone.

In nonglacial areas, dated series of prograding beach ridges have been described in Alaska, Mexico, and South Australia. Several hundred-odd beach ridge series north and south of Perth, Western Australia, again disclosed a 45-year periodicity with a larger fluctuation of *ca.* 360 years (Woods and Searle, 1983). This is remarkably interesting from the climatological viewpoint, because they are evidently responding to southern hemisphere weather systems as well as northern triggers.

7. Marine varves

Normally all deep-sea deposits are homogenized to about a 3000-year smoothing by bioturbation (burrowing worms and so on), but in a few isolated basins anoxic conditions isolate the bottom waters and annual layers are found, e.g. the Adriatic Sea, Red Sea, Cariaco Trench, and Gulf of California. Well-studied records were those from the Santa Barbara Basin (Pisias, 1978). By analyzing the climatogenic significance of plankton carried by the California Current, he established sharp fluctuations in sea-surface temperatures over the last 8000 years. The major cycles are of the order of 500–1000 years, and disclose, for example, a variance of the mean February (midwinter) temperature of as much as 15°C. This astonishing range is apparently due to sort of northern hemisphere El Niño effect, whereby the normally cold south-setting California Current is replaced in some periods by warm tropical water. Pollen analyses disclose that the southern California landscape has been correspondingly vegetated at cold times by a dry, chaparral cactus and coastal sage scrub flora–the Sonora Desert type, which today exists 500 km farther south in Baja California. At warm times this flora was replaced by a mixed deciduous and pine forest, reflecting warm–wet conditions with southerly winds and summer rains.

Most significantly, the weakening of the California Current appears to match northward shifts in the Kuro Shio in the western Pacific, indicated in Japan by northerly appearance of reef-building corals in the littoral terraces, assumed to be partly eustatic (Taira, 1980). Chronologically, they seem to approximate those dated by Fairbridge (1961) and others from Australia and elsewhere in tropical seas. Climatologically, it appears that during the major warm cycles, the subtropical highs expand and migrate somewhat poleward, accompanied by a reduction in the equator–pole thermal gradient. Trade winds weaken and the equatorial Pacific westerly air flow increases. Sea-surface temperature rises and we have a super-El Niño effect. The major warm cycles were most prominent during the mid-Holocene (Fairbridge, 1993). In the coastal shell middens of Peru a remarkable shift from warm to cold-water mollusca is observed in the late Holocene.

8. Volcanic signal

As developed by Lamb (1972, 1977) and much favored by meteorologists and journalists in the present century, the concept of violent, dust- and gas-producing volcanic eruptions is important climatologically because they often distribute a dust veil into the stratosphere that effectively screens incoming solar radiation, thus cooling the Earth's surface. In the four largest of such eruptions of the recent historical era, Asama (Japan) in 1783, Tambora in 1815, Krakatau in 1883, and Agung in 1963, the effect may have been a planetary cooling that lowered the mean global temperature by as much as 0.5°C for up 2 years (Rampino and Self, 1982). Tambora dramatically made 1816 the ill-famed "Year without a Summer" (Stommel and Stommel, 1983). No longterm climatic effects have been demonstrated, but sociological effects were, in places, profound, ranging from forced migrations from New England, to famine in central Europe and Bengal. Even more dramatic results followed a Sunda Strait eruption in AD 535 (Keys, 1999), although its effects were mainly felt in AD 536.

Volcanic eruptions have been traced back and catalogued as far as possible for the whole 10 000-year interval of the Holocene Epoch (Simkin et al., 1981), but their climatic potential has been variously appraised. They have even been elevated to the role of controlling the primary cooling in the 100 000-year glacial–interglacial cycles (but see criticism by Huntington and Visher, 1922; Schwarzbach, 1950, 1974). Evidence for that hypothesis seems to be entirely in the wishful-thinking category. Volcanism, as a source of CO_2, might even lead to warm cycles, as suggested long ago by Fritz Frech (see discussion in Schwarzbach, 1974, p. 291). Seen in a global context, volcanic eruptions are too brief and discontinuous to play a major role in long-term climatic cycles–either heating or cooling, although their effect on regional populations can be apocalyptic (Winchestor, 2003).

Although they clearly modulate climatic fluctuations on a less-than-decadal basis, the major volcanic eruptions over the last few centuries, for which we possess good chronological data, all seem to have taken place during but not at the beginning of cooling cycles. Lag times are of the order of 10–20 years. This anomalous reversal of possible cause and effect was commented on long ago by Huntington and Visher (1922), who pointed out that a month-by-month global study by Arctowski showed that the great eruption of Katmai, in June 1912, coincided with a general cooling cycle. The volcanic school submit that only equatorial volcanoes should count because of the lower tropopause in low latitudes, but neither Agung, nor Krakatau, nor Tambora–all near the equator–had lasting effects.

Tambora, in 1815, was one of the largest eruptions and it was certainly associated with cooling, but the whole of the preceding decade was one of global cooling. Pettersson (1914) noted that the three coldest winters ever recorded in Stockholm were in 1805, 1809, and 1814. In 1809 a Russian army attacked Sweden by crossing over the Baltic on the ice from Finland, the general reporting to the czar an air temperature measured at −36°C. Napoleon's retreat from Moscow was in December of 1810. For several winters even the southern Baltic was frozen over as far as the Kattegat. The East River (a salt-water channel) and lower harbor at New York City also froze over. The years leading up to 1815 were marked by one of the worst droughts in East Africa, subtropical droughts generally being associated with worldwide cooling phases. The year 1810 also marks the lowest point in mean sunspot numbers in the last two centuries.

An alternative hypothesis has therefore been offered: "Can rapid climatic change cause volcanic eruptions?" (Rampino et al., 1979). Climatic stress, notably atmospheric angular momentum for short pulses, and ice build-up and decay for longer cycles, is proposed as a trigger for crustal stress. Sudden accelerations in the atmosphere's angular momentum (Rosen et al., 1984) are accompanied by changes in spin rate as well as in the rate of secular drift and intensity of the geomagnetic field. It seems hardly a random coincidence that the two largest changes in spin rate and geomagnetic field intensity in the twentieth century match the declination maximums of the lunar nodal cycle (1913, 1969) when that cycle correlates precisely with either sunspot maximums or minimums (Courtillot et al., 1982).

Long ago, Köppen 1914 proposed a correlation between air temperatures, sunspots, and volcanicity. Köppen had noticed as early as 1873 that a temperature–sunspot correlation could be seen in low-latitude records but not elsewhere. The cumulative factors of lunar periodicities combine to give 1913 the maximum crustal stress build-up in more than one millennium. The cool cycle at this time and the volcanic eruptions (of which Katmai in June of 1912 was only one) do not seem to be random occurrences. A statistical survey of large earthquakes in California (Kilston and Knopoff, 1983) shows a clear link to lunar periodicities, including the 18.6-year cycle, probably through tidal loading on the continental shelf. A similar periodicity is observed in oceanic tidal height, and in temperature (SST), reflecting enhanced water exchange. Russian geologists have also observed the same 18.6-year period in the volcanic eruptions on Kamchatka (Shirokov, 1983).

During the Maunder Minimum (the sunspot dearth of 1645–1715) the incidence of global volcanic activity was twice as high as during the interval 1715–1800, and in the two decades 1680–1700 it was six times higher than for 1740–1760 (Fairbridge, 1980, p. 390). The coldest decades indicated by the Greenland ice-core oxygen-18 proxies were 1660–1680, preceding the volcanicity peak (see also Little Ice Age).

None of these complex relationships demonstrates conclusively a cause-and-effect situation. If the lunar declination cycle is locked into the sun's barycentric orbit and to solar radiation, as urged by Pettersson (1914), the 18.6-year period in historical climate proxies may also involve a solar pulse. The volcanic signal in long-term historical proxies thus deserves more far-reaching investigations, not so much as the cause of important climatic fluctuations but as a consequence, and also for their taphrochronological value, where they have a potential of providing a "golden spike" or a "silver thread" that runs through multiple proxy records, not least being the human observations of the actual eruption. Archeological documentations of the burial of villages or towns should be integrated into the record.

9. Paleomagnetic signal

The Earth's magnetic field is just beginning to be recognized as a major link between solar particulate radiations and their role in modulating terrestrial climates. This important geophysical parameter has been carefully studied and documented over a longer period than any other, initially because of its use in navigation. In 1692 Edmund Halley recognized that there was a secular westward drift of the magnetic field of some 0.2° per year, involving not only declination but also changes in inclination and total intensity. Because of the eccentric arrangement of the dipole and quadripolar fields, the actual readings of these variables at the Earth's surface are quite complicated and constantly changing; for navigational use they needed to be continuously monitored and mapped. Most significantly, the north and south magnetic poles (nonantipodal) are found to migrate some 5–10 km per year.

Prior to the documented historical period, i.e. the last five centuries or so, the record has been extended back by a technique known as *archeomagnetism*, whereby the former field characteristics may be measured in bricks or fireplace clays of archeologically established ages that have been heated and had cooled through their Curie point, thus preserving a fossil relic of the magnetic field of that time. The key observation was made in 1906 by a French geophysicist, Bernard Brunhes, studying the cooling of modern bricks (see Imbrie and Imbrie, 1979). Pioneer work on historical bricks was done much later by the Czech Geophysical Institute in Prague (Bucha, 1970). Recent studies with refined techniques confirm the general pattern, making the values stratigraphically useful.

The most striking result of this work was the discovery of a slow secular rise and fall of the total intensity field during the Holocene. At about 6000 BP (the Climatic Optimum of the Holocene Epoch), the dipole field moment strength was about 10% weaker than its present value. This variable is particularly interesting in connection with the flux rate of radiocarbon as determined from dendrochronological series (Suess, 1981), magnetic intensity being inversely proportional to radiocarbon value. That is to say, when the magnetic field intensity is weak, the influx of the weakly radioactive ^{14}C isotope is high, and vice-versa.

The Earth's spin rate is influenced by both exogenetic and endogenetic variables. Most emphatic of these variables is the mass transfer from glacial to interglacial mode and vice-versa; ice build-up near the poles shifts the moment of inertia and increases the spin rate, whereas deglaciation slows it down. Angular momentum is conserved by differential motions of the Earth spheres and by changes in the lunar orbit, thus affecting ocean tides. Changes in spin rate affect the angular momentum of all fluid systems, from the outer core, and asthenosphere, to the hydrosphere and atmosphere. The last is the most fluid and reacts most rapidly. The atmosphere is also most subject to extraterrestrial energy input and thus is most likely to disclose evidence of the two disturbing factors, external and internal (see also subsection 8 above).

The Earth's principal magnetic field, the dipole field, appears to be generated largely by the dynamo-like turbulence in a 100-km layer of low strength in the semiliquid outer core. Inasmuch as the spin rate of the mantle and crust are repeatedly subject to sudden changes, a differential torque is generated. The nearest familiar analog in human experience is the automatic transmission and gearshift of the family car.

The geomagnetic field appears, in fact, to be modulated by three principal phenomena, generating cycles or fluctuations of variable length and intensity. These are still subject to much research but appear to be as follows: orbital variations, magnetic pole migration and westerly drift, and solar wind modulation of the Earth's magnetic field. All three appear to play distinctive roles in the Earth's climate either by affecting the geomagnetic field intensity or by the geographic location of the magnetic pole. The relevant climatic link is through their control over the amplitude and location of incoming solar particulate radiation, which modifies the atmospheric chemistry and dynamics. The three control mechanisms are briefly summarized here.

Orbital variations.

Most important in this context is the variation of the obliquity of the ecliptic (*ca.* 41 000 years; Williams, 1993), together with some input from the eccentricity cycle (*ca.* 96 000 years) and the precession cycle (*ca.* 23 000 years and 19 000 years). These orbital variations, as already mentioned, correlate very precisely with geological proxies (Imbrie and Imbrie, 1979), but it has been questioned whether insolation alone has the ability to produce the large glacial–interglacial fluctuations. The question is usually skirted by pointing to the powerful feedback mechanisms that come into play, but the geomagnetic factor should also be included in the model.

Studies of the paleomagnetic intensity of Pleistocene deep-sea cores suggest higher fields in the cold, high spin-rate intervals, but the glacial-cycle sediments themselves carry more terrigenous (and magnetic) particles, which weakens that argument. Nevertheless, cold episodes during the Holocene also correlate to higher intensity signals, without contamination, so that the interrelationship is probably nonetheless true. What triggers the reversals in the geological past is still a controversial problem, but it seems that rapid changes in the spin rate and tilt must accompany the intervals of rapid crustal break-up and redistribution. Long periods of no magnetic reversal seem to coincide with times of mild climates (e.g. the late Cretaceous). Conversely, the times of disturbed crustal distribution and cycles of ice-age glacier loading must affect both tilt and spin.

During the course of the Holocene Epoch there have been appreciable variations in the secular orbital parameters. In particular, the tilt angle is now nearly 2° less than at the beginning of the Holocene (Berger, 1978). This secular change may be the principal component in Bucha's archeomagnetic variation. Mörner (Mörner and Karlén, 1984, p. 489) has presented the concept of a paleogeoid, linked to a phase transition at the core-mantle boundary, that is paralleled during the last 10 000 years by the curves of archeomagnetism and ^{14}C production.

Magnetic pole migration and westerly drift.
Within a period of about 1800 years the eccentric dipole field migrates westward, as does the weaker quadripole field, but more slowly, giving it a relatively eastward component, having a period of about 2800 years. The magnetic pole migrates at about 5–10 km per year. Bucha (1984) illustrates the position of the geomagnetic north pole over the last 18 000 years based on values obtained in Czechoslovakia, Japan, and the United States, its trajectory appearing to follow a quasi-ovoid form over the period of historical measurements, with a four-leafed clover form in its longer records.

To obtain more detailed records of total intensity, declination, and inclination, a vigorous campaign of lake coring has been undertaken in Europe, North America, and Japan (e.g. Creer, 1981; King et al., 1983). Records are pushed back to beyond 15 000 BP. They show regular fluctuations of all three variables but, because of the field asymmetry, the pattern varies slightly from one region to the next and with a general lag corresponding to west drift. Eight inclination cycles occurred during the last 10 000 years.

It was proposed by Fairbridge and Hillaire-Marcel (1977) from evidence of alternating calm and storminess cycles in the Hudson Bay that the more westerly sites of the magnetic pole generated a higher continentality situation (cold and calm) in the Canadian Arctic, in contrast to the more easterly pole situations (more oceanic sites around Spitsbergen), which coincided with warmer temperatures, less summer ice, and increased stormwave action (Fletcher et al., 1993).

Bucha (1984) believes that during brief but highly energetic solar flares the site of the magnetic pole and its associated auroral oval favors the development of low-pressure systems and storminess, which in turn helps to maintain milder temperatures, and that this condition applied to the North Pacific–Aleutian area during the late Pleistocene (18 000–12 000 BP) This would explain the lack of ice sheets in eastern Siberia and Alaska at the time. In contrast, according to Bucha, the remoteness of the North Atlantic region from the north magnetic polar area during the last glaciation maximum permitted a stable high-pressure situation favoring meridional air flow and the build-up of major continental ice sheets in eastern North America and northwestern Europe. If these hypotheses are correct, then not only would historical Holocene climatic fluctuations but also the glacial–interglacial relationships have a geomagnetic component in the 1800-year range and its subharmonics.

When the geomagnetic field is in an extended weak cycle, as during the mid-Holocene, and is modulated by unusually weak, short-period solar cycles (Damon et al., 1978), there will be an enhanced effect of galactic cosmic radiation, marked by prominent spikes in the ^{14}C flux record (Suess, 1981; Stuiver and Braziunas, 1993). The cosmic rays lead to the nucleation of high cirrus clouds (Roberts and Olson, 1973). This stratospheric cloud screen would furnish a high-albedo, cooling screen for the midlatitudes and Arctic.

During magnetic excursions, field intensity drops away dramatically and the result can be catastrophic cooling events as shown by evidence from the Pleistocene Lake Biwa, Japan (Kawai et al., 1975). In contrast, when the long-term magnetic field is strong, as seems to be the case during glacial cycles when the spin rate is high, the solar signal will be largely suppressed.

Solar wind modulation of the Earth's magnetic field.
This exogenetic factor provides a variable modulation of the internal field that can be identified in two independent ways. On a day-to-day basis, eruptions of giant solar flares (see Sunspots), observed astronomically, can be traced through a series of magnetosphere and ionosphere interactions, ultimately to their role in atmospheric chemistry.

The major flare events trigger upper atmosphere chemical reactions, particularly in the magnetic pole area, that in the troposphere at the 500-mb level lead to a dramatic warming (20–30°C) within 24 hours, accompanied by a drop in air pressure and the initiation of giant frontal weather systems that bring in a flow of warm humid air from as far away as the North Pacific. Such events trigger displacements of the jet stream and related atmospheric systems throughout the high- and mid-latitude belt of the northern hemisphere (Bucha, 1984).

One such solar–terrestrial event will not, of course, generate a climatic change, but the rate of incidence and magnitude of major solar flares is a variable that correlates roughly with the 11-year solar cycle. Landscheidt (1984) relates this variable to the crescendos of the torque stress developed by the spin and orbital positions of the planets. It is, therefore, an astronomically predictable quantity.

The most complete long-term record reflecting geomagnetic–climatic relations, now extending over almost the whole 10 750-year length of the Holocene, is provided by the radiocarbon flux-rate measurements made from dendrochronological series (Suess, 1981; Damon et al., 1978; Stuiver and Braziunas, 1993). These represent the ^{14}C isotope product of cosmic particle impacts on nitrogen in the stratosphere. The actual flux rate depends on two principal factors: (1) variation of the geomagnetic field strength in an inverse relationship–that is, when the field is weak, the flux rate is high; and (2) variation in the strength of the solar wind, which is modulated by planetary motion dynamics on the sun's photosphere (see Sunspots). A third but less frequent phenomenon is the modulation of galactic cosmic radiation by supernova events. Damon et al. (1978) commented on a very important aspect of these interrelationships already noted above in the subsection "Magnetic pole migration and westerly drift." When the internal geomagnetic field is weak, as in the mid-Holocene about 6000 BP, the amplitude of ^{14}C flux-rate departures is up to five times higher than their level during the high-field intensity intervals. These large fluctuations are noted at about 500–1000-year intervals (Figure C20). A similar oscillation may be seen in many of the standard climatic time-series noted in this entry: sea-level fluctuations, palynological diagrams, ice-core data, hydrological indicators, and so on.

From all these indications it is evident that the solar (i.e. exogenetic) component is most important in terrestrial climate series in the 500–1000-year time frame. However, on the 10 000–100 000-year scale the endogenetic components (controlled by spin-rate and tilt factors) are paramount.

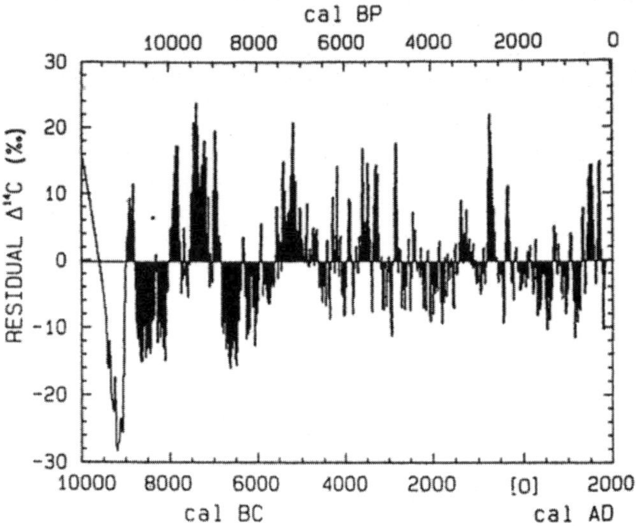

Figure C20 Solar variability, indicated by C-14 flux (inverse to solar activity) based on tree rings, and partly on corals (Stuiver and Braziunas, 1993). Maxima correspond to temperature minima in the Greenland ice cores. The post-Younger Dryes (cold spell, not shown) was followed by the dramatic warming of the Pretoreal and its several oscillations. Glaciers tend to respond rather sluggishly, so that eustatic sea level shows a retardation (Fairbridge, 1961).

10. Climate–aurora–sunspots documentation of ancient China

A gratifying feature of the new cultural orientation in modern China is the liberal approach to science and an appreciation that the written archives of that country contain the largest storehouse of proxy climatic data outside of Europe (Chu, 1973; Wang, 1979). Studies of the aurora and sunspot records were undertaken by Schove (1955), assisted by expatriate Chinese scholars, and in this way the first long-term (2600-year) reconstruction of solar history became available. Subsequent improvements and reprints of the key literature have been published in Schove (1983). The aurora have been separately treated by Siscoe (1980). Spectrum analyses by Jelbring (1995) broadly confirm the Schove model.

A preliminary study of the long-term variables in China's climate has been offered by Ren and Li (1980), who came to the conclusion that its cyclicity was in phase with the All-Planet Synod, an approximately 178-year alignment of planets in quadrature. At such times, as demonstrated by Newton, the sun is displaced as much as 1.5×10^6 km from the barycenter (center of mass) of the solar system, its spin rate accelerates (up to 5% from the usual 27 days) and its photospheric convection is retarded, resulting in a decrease in solar emissions (see Sunspots). On Earth these sunspot-dearth periods are associated with cooling spells (see Little Ice Age).

A study of diary and yearbook records for the period AD 1470–1979 by Wang and Zhao (1981), using content analysis, has disclosed six spatial patterns for droughts and floods, using 25 stations in eastern China, arranged in east–west belts in more or less meridional distribution. The first eigenvector relates to long-term variations of the Southern Oscillation Index, which is linked to the intensity of the western Pacific subtropical high. The second is linked to the same high, specifying whether its influence extends north or south of the Yangtze River. The third relates to the high-frequency variations of the East Asian Circulation Index.

The linkage to the Southern Oscillation makes these figures of global significance, especially for the Indo-Pacific subtropics and regions affected by the El Niñ Southern Oscillation, which in the recent example (1982–1983) means practically the whole world. When the Southern Oscillation and Walker Circulation indices are both high, in the mid-Pacific there is low sea-surface temperature (SST), the Pacific Hadley cell weakens and the summer monsoon is strong in southeast Asia (Yoshino and Xie, 1983), and the Yangtze discharge level is high.

In an investigation of winter thunder from 250 BC to AD 1900 by Wang (1980), it was possible to show a correlation with cold airmasses associated with unusual winter front development. A close match was found with temperature records collected by Chu (1973). It is significant that the patterns also correlate with similar historical records from Japan based on the dates of the blossoming of cherry trees (an important ceremonial event) and the freezings of Lake Suwa (Arakawa, 1957).

Application of modern computer technology and statistical theory to the Chinese data for the Beijing (Peking) are was begun by Hameed et al. (1983), who found in the proxy drought–flood records since AD 1470 signals of both the 18.6-year lunar nodal periods, probably a tidally induced modulation, and the 11-year solar emission cycle. These conclusions were dismissed by Clegg and Wigley (1984), but in a study by Currie and Fairbridge (1986) the results of Hameed et al. were supported, and, in addition, the curious bistable phase reversals found elsewhere in the world (Currie, 1984) appeared once more. Currie employs the maximum entropy method (MEM) to develop power spectra, and then constructs waveforms for the principal cycles as indicated. This permits a visual comparison of phase and amplitude, which can then be conveniently compared with historical events.

Testing and modeling

In order to derive a useful scientific product from the vast store of historical data, documentary and field, evidence that are either available at the present time or merely potential and unexploited, it is desirable to develop programs of testing and modeling. This rich area of research endeavor has hardly been explored up to the present moment, except in the most sporadic or generalized way. Intuitive matching has been the usual, nonrigorous methodology, and the lack of adequate statistical procedures has been brought out by Pittock (1978, 1979, 1983), Currie (1984), and others. It is petulant and childish however to make fun of the pioneer investigations, many of which rested on the most limited or crudest of databases. Yet, almost miraculously, many of their preliminary conclusions have turned out to have more than a grain of truth, as may be seen from a careful reading of Lamb (1972, 1977), for example.

In the preceding sections dealing with standard time-series, 10 examples of historical data types covering the last 10 millennia are presented. They have been derived conceptually from a short contemporary base line of instrumentally observed processes, furnished by meteorology, oceanography, geology, botany, astronomy, and astrophysics. Building up on this minuscule chronologic base line we are able to construct an array of time-series of varying quality but with sufficient common threads that the researcher gains confidence.

The last 10 millennia are today not a sea of unknowns. During the approximately 100 years that have elapsed since Blytt and Sernander constructed the first climatologically oriented chronology (covering the last 15 000 years), based on Scandinavian pollen analysis, we have now come to possess three independent time scales that have the potential of providing year-by-year chronologies and that are mutually verifiable. Probably their error bars are still in the range of ±10 years, but the three methodologies have the potential of refinement and correction; they also can be extended to other (favorable) regions around the globe. An immense field of potential research is thus open and still largely unexploited.

The first and most ambitious attempt to develop a global model of the glacial–interglacial climatic fluctuation generated the CLIMAP (1976), which, for all its faults, was a remarkable first attempt. It exploited a largely oceanographic–paleontologic database to derive a broad picture of an instant in time during a glacial-phase maximum. Further endeavors have been made for modeling distinctive intermediate (Holocene) stages. Using the three year-by-year chronologies, supplemented by input from the other seven signals, an exciting vista is now open for establishing what *did* happen. The geologists have a pet phrase: What did happen, can happen! From that base comes the potential of explanation and prediction.

Summary of historical climatic epochs

In the space of one encyclopedia entry it is impossible to review the entire 10 000 years of climate history on a world basis. Here we concentrate on the best-known Scandinavian and northwest European data series with only brief references to events elsewhere. No book exists that summarizes the record, but several useful references deal with specific fields of research. Three excellent multiauthored volumes cover the Holocene history of coastal changes and deposits around the Baltic (Gudelis and Königsson, 1979), the North Sea (Oele et al., 1979) and the Irish Sea (Kidson and Tooley, 1977). The general vegetational and climatic history of Europe has been summarized by Frenzel (1973). A single-volume, multiauthored work has been dedicated to the Holocene of the United States (Wright, 1983) and the Holocene is included in a companion volume on the Quaternary of the USSR (Velichko, 1984). Detailed investigations are covered in the journal *The Holocene* (vol. 1, no. 1, 1991).

There is now an abundance of professional papers but, alas, the much-needed syntheses are for the most part lacking. The explanation for this lack of what might appear to be a key area of scientific and humanistic literature is associated in part with the inherent costs and problems of dating, and in part with the multidisciplinary training that would be needed to study the whole field. The specialists are botanists, or malacologists, or archeologists or whatever; unfortunately very few scientists are trained today outside of their specialized fields. In as much as the Holocene Epoch is a segment of geologic time, there is primarily a problem in geological chronology. Its climate history is a derivative product that must be extracted from a proxy record that basically is stratigraphical and preserved in geologic strata, or ice, or tree rings.

Reasons of priority dictate that, all things being equal, the first classification worked out by any team of scientists for an interval of geologic time is the one used as a basis for future workers to build on. In the case of the Holocene it may appear at first sight to be anomalous that the Blytt–Sernander botanical classification is commonly adopted (see earlier discussion under subsection 4: Palynology). This unusual action becomes understandable when it is remembered that the geologic basis for chronologic ordering is, first, the superposition rule, second, the fossil content of strata, and third, any absolute (geochemical) method of dating that may be appropriately applied to the formations in question. In this case it was the integration of pollen studies by Blytt and Sernander, in stratigraphic sequences of lake or swamp deposits, and provided with the varve counting by de Geer (see subsection 3), that furnished the first standard geologically based record of the last 10 000 years of Earth history.

This serendipitous combination of distinctive botanical fossils, sequential strata, and a precise year-by-year chronology placed Scandinavia in an extraordinarily advantageous position for presenting a world-standard historical record (Mangerud et al., 1974). Thanks to written documents and early scientific instrumental records, it would be additionally very helpful if this region could be selected for various stratotypes, i.e. chosen areas that contain well-preserved and well-analyzed records. It is unquestionably true that several other parts of the world *could* provide detailed, continuous stratigraphic records of the last 10 000 years of the Earth's history, but up till now no area has been proposed that has had the same rich, multidisciplinary background of study. Continuously cored lake or swamp deposits are now available from Britain, France, Nova Scotia, Minnesota, Japan, India, Colombia, and elsewhere. In time it is hoped much of the world will be brought into this chronological framework.

The smallest units of geological time, derived in the case of the Holocene from the original Scandinavian pollen-based stratigraphy, are categorized as *chronozones*. Five of these units, unequal in duration, are used to classify the epoch. In different parts of the world these boundaries are not always conveniently applicable to local records, and local terminologies need to be established. The purpose of a set of chronologic labels is simply that it is helpful to have a reference standard that is universally known to scientists. The local records should never be forced into a straitjacket of any assumed universality. Many natural boundaries are diachronous; this is particularly true when we are dealing with a eustatically rising sea level, where the beginning of a transgression may be hundreds of years earlier than its culmination. Ocean water warms or cools slowly; glacier melting involves latent heat and other delaying factors; again there will be lags or retardation. The purpose of agreed time planes is simply to coordinate our data within an internationally understood framework of time intervals, often called *epochs* or *ages* in an informal sense. (It may be noted, according to the internationally agreed stratigraphic code, that each term has a specified meaning in the chronostratigraphical heirarchy; and *epoch* is thus longer than an *age*, and shorter than a *period*, e.g. Holocene Epoch, Quaternary Period.)

The Blytt–Sernander scheme of pollen stratigraphy was applied to every chronozone back to the first formations to emerge from the melting ice fronts of northwest Europe; that is, from about 14 000 BP, ranging up to the present. The earlier zones (14 000–10 000 BP) are classified as late Pleistocene (Weichselian or Wisconsinan glaciation). In this entry we only consider the Holocene, which includes five principal units. Several are now subdivided, giving 10 units in all.

The latest Pleistocene corresponds to the first part of the very rapid deglacial process, when the midlatitude glaciers were melting at an astonishing rate, peaking in two warm phases (Bölling: 12 500–12 000 BP; and Alleröd, 11 800–11 000 BP) that were separated by three cooler fluctuations, known as the *Tundra* ages (Oldest, Older, and Younger Dryas). Heavy snowfalls consequent upon the oceanic warming (the Simpson principle), were probably responsible in part for glacial readvances in the mountains during those colder intervals. In the last one, the Loch Lomond Readvance in Scotland, glaciers actually reached down to sea level. Even in the tropical oceans signs of these cool fluctuations are seen in the oxygen-18 records.

Thus the stage was set for the beginning of the Holocene, the boundary being established, by international agreement, as the time when warm marine fossils replaced cold ones in the uplifted deposits of southwest Sweden (Olausson 1982; Fairbridge, 1983) the time was roughly 10 000 BP (by radiocarbon) ±300 years. (This time equivalent in sidereal years is believed to be approximately 8000 BC, but precise calibration is still controversial; Cato, 1985.) In other parts of the world another boundary might have seemed better. In tropical regions the dramatic postglacial warming was generally about 13 500 BP. In northern Canada or Alaska about 8000 BP or even 6000 BP might seem more appropriate. The local boundaries are always diachronous, so that only the somewhat arbitrary chronostratigraphic boundary should be shown as a horizontal plane on our diagrams, a "time line" that has been agreed to by a commission of the International Union for Quaternary Research (Olausson, 1982).

Holocene chronozones

We will now briefly review the 10 chronozones of the Holocene. The ages will be given first in sidereal (calendar) years, expressed as BC or AD, then in BP equivalent (before 1950), and finally in radiocarbon years, indicated as b.p. (lower case). Calibration tables for conversion of one to the other were provided by Klein et al. (1982), and subsequent authors, simplified in Roberts (1998). When precise astronomic equivalents are required, one should remember there is no year zero in the AD–BC system, so that the astronomical year − 100 is 101 BC and conversion to BP made by adding 1949 to the BC date, thus 2050 BP. Inasmuch as radiocarbon dates normally carry at least a ±100-year range of error, it is usual to round off calendar equivalents of such dates to the nearest 10 or 20 years. The chronozones are listed with abbreviated initials and with the Danish pollen zones (Roman numerals). Botanical aspects are treated in Birks and Birks (1980), for the North American setting by Davis (1983), for the Scandinerian setting by Berglund (1986).

Preboreal (PB, IV) 9350–7500 BC (11 300–8900 BP, 10 400–9650 b.p., calibration doubtful)

Glaciers still covered most of North America north of the St Lawrence and the Great Lakes, most of Scandinavia north of Stockholm and all of the central Alps. Sea level, at minus 40–45 m, had reached about halfway across the continental shelves in mid- to low-latitudes, introducing maritime air and leading to a rapidly rising humidity. The Baltic area was invaded by the Yoldia Sea. Marine formations of this age are isostatically uplifted today in the Oslo Fjord to 221 m and to lower elevations in southwest Sweden. In Finland the Yoldia strandline reaches 220 m. Tundra vegetation was quickly replaced by birch and pine forests across western Europe. In North America, boreal forests (mainly spruce) followed closely on the retreating ice but in the northern plains was replaced by deciduous forest by 9500 b.p. The Champlain Sea, which had invaded the St Lawrence lowlands somewhat earlier, brought maritime influences to the lower Great Lakes; whale bones and other marine fossils are found from Quebec to Michigan. (Note: radiocarbon dates given for the early Holocene are not precise.)

Early Boreal (BO-1, V) 7500–6900 BC (9450–8750 BP, 9650–8550 b.p.)

Glaciers were in universal retreat. The *climatic optimum* (Europe) or *hypsithermal* (North America), a diachronous rise of air temperatures to higher-than-present levels, reached southern Sweden, Britain, and much of North America except for the still-ice-covered areas of Keewatin and Quebec. Vegetation in northern Europe now became dominated by pine forests, but in favored areas in Sweden the hazel was flourishing, a thermal indicator requiring 2°C to 2.5°C warmer summers than today. Isostatic uplifts isolated the Baltic (the Ancylus Lake) and emptied the St Lawrence Lowland largely of the Champlain Sea, leaving the Goldthwait Sea in its lower valley. In the plains of the United States, prairie grasslands began to replace deciduous forests.

Late Boreal (BO-2, VI) 6800–6200 BC (8750–8150 BP, 8550–7700 b.p.)

Glaciers in Scandinavia now shrank to the principal mountain axis but the Ancylus Lake persisted until 8300 BP. In North America the Keewatin Ice Sheet and Laurantide (New Quebec) Ice Sheet became separate, with Hudson Bay open to the sea. Ice remnants still remained on the high ground of the Maritime Provinces and Maine. A strong zonal (westerly) air flow began to affect most of Europe and North America and northern Mexico, with Gulf air limited to the southeast United States. Alaska and most of Canada were influenced by a strong Arctic airmass. In northern Europe, pines and spruce were progressively replaced by mixed oak forests.

Early Atlantic (AT-1, VII-a), 6200–5500 BC (8150–7450 BP, 7700–6600 b.p.)

Glaciers became limited to high mountains in Scandinavia and to residual patches in northern Quebec and Keewatin. Sea level rose to within 20 m of its present level, and a warm-water marine transgression invaded the Baltic (the Littorina Sea). Progressive isostatic uplift has left former Littorina strandlines (high eustatic oscillations) in parallel belts around the coasts. Mixed oak forests spread across most of Europe and North America. Strong westerly air flows maintained highly maritime climates. Even in northern India winter rains were almost as important as the summer monsoon. The AT-1–AT-2 boundary is marked in Greenland ^{18}O ice cores by a distinctive cooling indication, and likewise in the California tree-ring ^{14}C series.

Late or Main Atlantic (AT-2, VII-b), 5500–4200 BC (7450–6150 BP, 6500–5300 b.p.)

The last ice patch disappeared from northern Quebec. Only mountain glaciers and polar latitude glaciers persisted for the rest of the Holocene. Around 6500–6000 BP eustatic sea level reached its present height and has only fluctuated somewhat

(+1 m to +3 m) since then. The late Atlantic is marked by the 3- m *Early Peron* shore terrace in the tropic and the *Nizza* terrace in the Mediterranean. Successive uplifted Littorina strandlines rim the Baltic. Climatic optimum warmth now reached its farthest northerly influence. In northern Canada the boreal (spruce) forest–tundra boundary almost reached the Arctic shore. In eastern North America and Europe mixed oak–elm forests reached climax states. From the Swedish flora, in 1910, Andersson determined that the mean temperature was 2.5°C above present. This value is about 20% of the total glacial–interglacial range for this latitude. In the tropics the East African rains were much heavier than today and the Nile floods were on average three times above today's. The Early Kingdom pharaonic culture was established in Egypt and the Mohenjodaro civilization developed along the Indus valley.

Early Subboreal (SB-1, VIII-a), 4200–2300 BC (6150–4250 BP, 5300–4000 b.p.)

In most latitudes this chronozone marks the beginning of the fall in global temperatures from the peak of the climatic optimum or hypsithermal. Its beginning is marked all across Eurasia, from Britain to Siberia by the *Ulmus* (elm) decline, with a dramatic change in the climax mixed forests. The progressive cooling is well shown by the gradual reduction in tree-ring widths; in the Bristlecone Pine series (White Mountains, California), they reached an initial low point about 2800 BC (matched by a high level of ^{14}C flux, evidence of low solar activity). This corresponds to the *Bahama Emergence* (−3 m) on tropical coasts. In the upper Midwest (Elk-Lake core, Minnesota), the same interval is marked by a 300-year period of reduced varve thickness and drop in diatom (*Melosira*) frequency; it was followed by an abrupt rise in varve thickness to a peak around 2000 BC (late Subboreal). Mountain glaciers showed repeated advances, corresponding to sherp ^{14}C flux peaks.

Late Subboreal (SB-2, VIII-b), 2300–900 BC (4250–2850 BP, 3800–2900 b.p.)

What is sometimes called a *mid-Subboreal* mild phase generated warmer climates for about five centuries in Scandinavia and elsewhere. Then from 1400 to 1000 BC the Bristlecone Pine and most pollen series show a major climatic deterioration, marked by a high ^{14}C flux; it was paralleled by the *Crane Key Emergence* in the tropical seas, followed shortly after by the *Pelham Bay Emergence* about 800 BC, both being certainly glacioeustatic in view of important neoglacial readvances in the interval 1350–450 BC (Denton and Karlén, 1973). In northern Canada the forest–tundra border retreated. Culturally this epoch was marked in Europe by the Bronze Age.

Early Subatlantic (SA-1, IX-a), 900 BC–AD 175 (2850–1775 BP, 2775–1700 b.p.)

In northern Europe this chronozone was characterized by cool to mild and moist conditions, accompanied by the general spread of beech forests. The Mediterranean saw the rise of the Etruscans, and of the Roman Empire. Sea level dropped from a high around +2 m in the classical Greek period, to a low around −1 m around the first century AD, followed by a progressive rise (late and post-Roman transgressions). Culturally and economically this sequence had a significant impact: Roman port facilities all around the Mediterranean are today at least partly submerged and commonly built over by Medieval structures. The coastal salt-pan economy was ruined by the rising sea level and salt mines were developed inland (Spain, Austria, Germany, Dead Sea), coastal farmers, particularly in France and the North Sea area, were forced inland or started to build dikes.

Mid-Subatlantic (SA-2, IX-aa), AD 175–750 (1775–1200 BP, 1700–1150 b.p.)

Corresponding to the late Roman/Byzantian cultural phase, the coasts of northern Europe were extensively inundated of the Mediterranean region was plagued by droughts and famines, which in the Middle East favored the rise of Islam.

Late Subatlantic (SA-3, IX-b), AD 750–present (1300–1 BP, 1300–1 b.p.)

In broad terms the youngest chronozone is divided into three: the *Medieval Warming* or *Little Climatic Optimum* (q.r.) from AD 950 until about 1250, the *Little Ice Age* (with two maximums, divided by a slight fifteenth-century warming (about 1390–1540), and finally the late nineteenth–twentieth-century warming that brings us up till today. The ^{14}C flux according to the California dendrochronological series remained fairly stable until about AD 1250, after which a precipitous decline is registered with minimums about AD 1350 and 1700, after which, with a minor low around 1810–1820, it rose to modern levels; after 1950 it becomes grossly contaminated by nuclear bomb pollution, followed by anthropogenic burning of fossil fuels and deforestation.

Sea level rose during the Medieval Warming to about 0.5 m above present, as may be seen from raised coral platforms and shell banks in the tropics (Rottnest Submergence in Australia), and from anomalously high Viking beaches in Sweden. It fell to at least −0.5 m during the Little Ice Age, after which its fluctuating upward trace is recorded by the Amsterdam tide gauge installed in 1682. With decadal fluctuations of around 5 cm, it rose to a peak around 1770 and then sank again to a low about 1820, since when it has been mostly rising (after correction for subsidence; Fairbridge, 1961, p. 104). The Little Ice Age regression has been called the *Paria Emergence*, from South American data.

As mentioned in the entry on the Little Ice Age, the most striking development of that interval was the enormous expansion of sea ice in the Greenland Sea, which around 1600 reached a peak of 26 weeks of ice cover on the north coast of Iceland each winter. The effect on the albedo was appreciable and long-term average temperatures in western Europe fell by 1–2°C. Around 1600 the north magnetic pole came to lie northeast of Iceland, and the low solar activity phases like the Wolf, Spörer and Maunder minimums led to excessive fluxes of cosmic radiation (shown by high ^{14}C and ^{10}Be values). According to the Roberts and Olson (1973) model, the geomagnetic screen is weakened during those solar activity minimums and the amplified cosmic ray flux created frequent high-altitude cirrus cloud cover, thus generating a primary stratospheric albedo.

Summary

Why study the historical geological record of the last 10 000 years? To the climatologist, trained on the straightforward hard data of meteorological instrument observations, who would perhaps ask that question, it might be useful to summarize some key points:

1. The current Quaternary Ice Age (of the last 2 million years or so) alternates rather abruptly between two modes: glacial and interglacial. We are living in one of the latter, the Holocene Epoch of geology, its lower boundary being 10 750 ± 300 BP.
2. During these 10 millennia, sea level has achieved its present level glacioeustatically, that is, mainly by the melting of midlatitude continental ice sheets. The last traces of the latter were gone by 6000 BP.
3. Following the Milankovitch theory of orbital control of terrestrial insolation, the last peak of effective radiation was about 9000 BP, around which time the sea-surface and low-latitude land temperatures (in areas not dominated by the residual ice sheets) a *climatic optimum or hypsithermal* climatic regime developed. Gradual deterioration (global Cooling and desiccation) began after about 5000 BP.
4. The last 5000 years have been marked by remarkably non-linear climate, glacial, and sea-level record. Fluctuations are mostly in saw-toothed patterns of various scales, progressive cooling with abrupt deviations to a generally warm mode; the upper thermal bound is episodically interrupted by cold spikes which vary in length from 1 year to a century or more. The amplitude of the fluctuations has oscillated or even increased, not decreased toward the present (Figure C21).

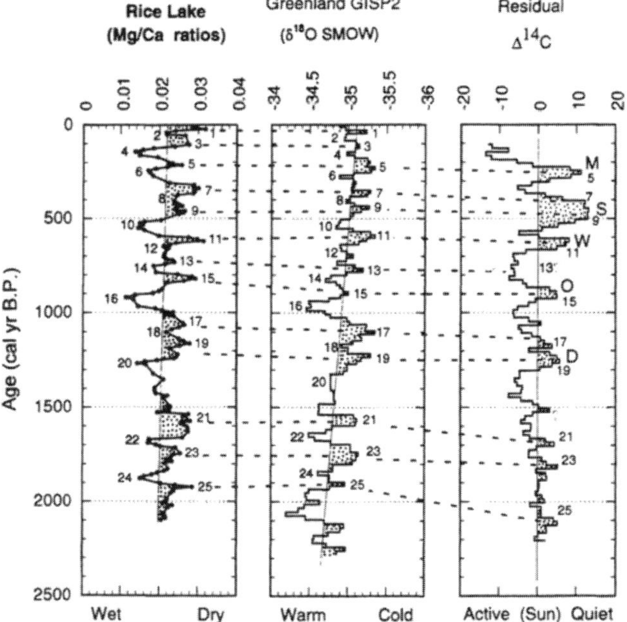

Figure C21 Three examples of cyclic climatic signals spanning the last 2000 years. Left: a lake core in western Canada (Rice Lake). Middle: an ice from Greenland (GISP-2). Right: bidecadal residual C-14 from tree rings (Stuiver & Reimer, 1993). With slight modifications for simplicity (R.W.F.). Abbreviations: **M** (Maunder Minimum: AD 1645–1715); **S** (Spoerer minimum: AD 1420–1530); **W** (Wolf minimum: AD 1280–1340); **O** (Oort minimum: AD 1010–1050); **D** (Dark Age minimum: AD 660–740). Each represents low levels of solar activity (sunspots), corresponding high O–18 levels in ice cores, i.e. cold episodes, and cool-dry climates in western Canada.

5. The nature of the cooling episodes is highly varied and subject to considerable research effort. The longer ones clearly relate to solar radiation (cf. ^{14}C data). More or less random spikes of 1–2 years correlate with specific incidents of volcanism, particular explosions leading to a gross loading of the *dust veil index* (DVI).
6. More or less regular quasi-biennial or quasi-triennial climate cycles relate to equatorial statospheric winds and the Walker circulation; the El Niño and Southern Oscillation are well-recognized expressions of these systems, which are controversial in origin, variously explained as stochastic (endogenetic thermodynamics), or as produced by solar–lunar forcing (exogenetic: radiational and tidal).
7. Time-series of millennial and multimillennial length furnished by ice cores, varves, and tree rings provide a sound statistical basis for exploring the nature of longer climate cycles. The solar (11- and 22-year) and lunar nodal (18.6-year) periods are now well established, but are associated with *noise* generated by complex feedback processes, ranging from magnetic and electrical fields to oceanic circulation and lagging sea-surface temperatures.
8. Long solar–lunar periodicities of the order of 45, 79, 93, 180, 360, 558, 1112, 1850 and more years have been studied only in a very preliminary way. Clearly, there is a strong astronomical pulse recognizable in all the long time-series but far more investigation is still needed.
9. The role of CO_2 in climate change, both long-term and authropogenic, is complex and remains to be resolved.
10. Finally, a word about the future. We cannot yet confidently make predictions, even for a few months. For the curious and the dedicated, however, the historical record offers a vast reservoir of information that is simply waiting to be tapped. The stratigraphic and other natural repositories have only just begun to be explored.

Rhodes W. Fairbridge

Bibliography

Arakawa, H., 1957. Climate change as revealed by the data from the Far East. *Weather*, **12**: 46–51.

Barry, R.G., 1985. Snow and ice data. In Hecht, A., ed., *Paleoclimate Analysis and Modeling*. New York: Wiley-Interscience.

Bell, B., 1975. Climate and history of Egypt. *American Journal of Archeology*, **79**: 223–269.

Berger, A.L., 1978. Long term variations of caloric insolation resulting from the Earth's orbital elements. *Quaternary Research*, **9**: 139–167.

Berglund, B.E., ed., 1986. *Handbook of Holocene Palaeoecology and Palaeohydrology*. Chichester: J. Wiley, 869p.

Birks, H.J.B., and Birks, H.H., 1980. *Quaternary Palaeoecology*. London: E. Arnold.

Bray, J.R., 1976. Volcanic triggering of glaciation. Nature, **260**: 414–415.

Bryson, R.A., and Murray, T.J., 1977. *Climates of Hunger*. Madison, WI: University of Wisconsin Press.

Bryson, R.A., and Padoch, C., 1980. On the climate of history. *Journal of Interdisciplinary History*, **10**: 583–597.

Bryson, R.A., and Swain, A.M., 1981. Holocene variations of monsoonal rainfall in Rajasthan. *Quaternary Research*, **16**: 135–145.

Bucha, V., 1970. Influence of the Earth's magnetic field on radiocarbon dating (with discussion), In *Radiocarbon Variations and Absolute Chronology*. Nobel Symposium, 12th proceedings, pp. 501–511.

Bucha, V., 1984. Mechanism for linking solar activity to weather-scale effects, climatic changes and glaciations in the northern hemisphere, In Mörner, N.A., and Karlén, W., eds., *Climatic Changes on a Yearly to Millennial Basis*. Dordrecht: Reidel, pp. 415–448.

Burroughs, W.J., 1992. *Weather Cycles: Real or Imaginary?* Cambridge, England: Cambridge University Press, 201p.

Catchpole, A.J.W., and Ball, T.F., 1981. Analysis of historical evidence of climatic change in western and northern Canada. In *Syllogeous 33: Climatic Change in Canada*. Ottawa: National Museum of Natural Science, pp. 96–98.

Cato, I., 1985. The definitive connection of the Swedish geochronological time scale with the present, and the new date of the zero year in Doviken, northern Sweden. *Boreas*, **14**: 117–122.

Chu, K.C., 1973. A preliminary study on the climatic fluctuations during the last 5,000 years in China. *Scientia Sinica*, **16**: 226–256.

Clegg, S.L., and Wigley, T.M.L., 1984. Periodicities in precipitation in north-east China, 1470–1979. *Geophysical Research Letters*, **11**: 1219–1222.

CLIMAP, 1976, The surface of the ice-age Earth. *Science*, **191**: 1131–1137.

Courtillot, V., Le Mouel, J., Ducruix, J., and Cazenave, A., 1982. Magnetic secular variation as a precursor of climatic change. *Nature*, **297**: 386–387.

Creer, K.M., 1981. Long-period geomagnetic secular variations since 12,000 yr. BP. *Nature*, **292**: 208–212.

Currie, R.G., 1984. On bistable phasing of 18.6 year nodal induced flood in India. *Geophysical Research Letters*, **11**: 50–53.

Currie, R.G., 1984. Periodic (18.6 year) and cyclic (11 year) induced drought in western North America. *Journal Geophysical Research*, **89**: 7215–7230.

Currie, R.G., 1995. Bariance contribution of M_n and S_c signals to Nile River data over a 30-8 year bandwidth. *Journal Coastal Research*, Special Issue **17**: 29–38.

Currie, R.G., and Fairbridge, R.W., 1986. Periodic 18.6-year and cyclic 11-year induced drought and flood in northeastern China. *Quaternary Science Review*, **4**(2): 109–134.

Damon, P.E., J.C. Lerman, and Long, A. 1978. Temporal fluctuations of atmospheric ^{14}C: causal factors and implications. *Annual Review of Earth and Planetary Sciences* **6**: 457–494.

Dansgaard, W., Johnson, S.J., Clausen, H.B., and Langway, C. C., Jr, 1971. Climatic record revealed by the Camp Century ice core. In Turekian, K.K., ed., *The Late Cenozoic Glacial Ages*. New Haven, CT: Yale University Press, pp. 37–56.

Dansgaard, W., Johnsen, S., Reeh, N., Cundestrup, N., Clausen H.B., and Hammer, C.U., 1975. Climatic changes, Norsemen and modern man. *Nature*, **255**: 24–28.

Davis, M.B., 1983. Holocene vegetational history of the eastern United States. In Wright, H.E., Jr, ed., *Late-Quaternary Environments of the United States*, vol. 2: *The Holocene*. Minneapolis; MN: University of Minnesota Press, pp. 166–181.

Denton, G.H., and Karlén, W., 1973. Holocene climatic variations–their pattern and possible cause. *Quaternary Research*, **3**: 155–205.

Eddy, J.A., 1977. Climate and the changing Sun, *Climate Change*, **1**: 173–190.

Fairbridge, R.W., 1961. Eustatic changes in sea level. In *Physics and Chemistry of the Earth*, vol. 4. London: Pergamon Press, pp. 99–185.

Fairbridge, R.W., 1962. New radiocarbon dates of Nile sediments. *Nature*, **196**: 108–110.

Fairbridge, R.W., 1976a Effects of Holocene climatic change on some tropical geomorphic processes. *Quaternary Research*, **6**: 529–556.

Fairbridge, R.W., 1976b. Shellfish-eating preceramic Indians in coastal Brazil. *Science*, **191**: 353–359.

Fairbridge, R.W., 1980. Prediction of long-term geologic and climatic changes that might affect the isolation of radioactive waste. In *Underground Disposal of Radioactive Wastes*, vol. 2. Vienna: International Atomic Energy Agency, pp. 385–405.

Fairbridge, R.W., 1981. Holocene sea-level oscillations. *Striae*, **14**: 131–141.

Fairbridge, R.W., 1983. The Pleistocene–Holocene boundary. *Quaternary Science Review*, **1**: 215–244.

Fairbridge, R.W., 1984a. Planetary periodicities and terrestrial climate stress. In Mörner, N.A., and Karlén, W., eds., *Climatic Changes on a Yearly to Millennial Basis*. Dordrecht: Reidel, pp. 509–520.

Fairbridge, R.W., 1984b. The Nile floods as a glacial climatic/solar proxy. In Mörner, N.A., and Karlén, W., eds., *Climatic Changes on a Yearly to Millennial Basis*. Dordrecht: Reidel, pp. 181–190.

Fairbridge, R.W., 1993. El Niño is not unpredictable as coastal hazard: luni-solar synergy. *Journal Coastal Research*, Special Issue **26**: 52–63.

Fairbridge, R.W., and Hillaire-Marcel, C., 1977. An 8,000-yr palaeoclimatic record of the "Double-Hale" 45-yr solar cycle. *Nature*, **268**: 413–416.

Faure, H., and Gac, J.Y., 1981. Sahelian drought to end in 1985? *Nature*, **291**: 475–478.

Filion, L., 1983. Dynamique holocène des système éoliens et signification paléodynamique (Québec nordique), PhD, dissertations Québec: Université Laval; summary in *Nature*, **309**: 543–546 (1984).

Filion, L., 1984. A relationship between dunes, fire, and climate recorded in the Holocene deposits of Quebec. *Nature*, **309**: 543–546.

Fletcher, C.H. III, Fairbridge, R.W., Møller, J.J. and Long, A.J., 1993. Emergence of the Varanger Peninsula, Arctic Norway, and climate changes since deglaciation. *The Holocene*, **3**: 116–127.

Flint, R.F., 1971. *Glacial and Quaternary Geology*. New York: Wiley.

Frenzel, B., 1973. *Climatic Fluctuations of the Ice Age*. Cleveland, OH: Case Western Reserve University Press.

Gudelis, V., and Königssson L.K., eds., 1979. *The Quaternary History of the Baltic*. Uppsala: Acta Universite Uppsala.

Guiot, J., 1985. *Réconstructions des champs thermiques et barométriques de la région de la Baie d'Hudson depuis 1700*. Downsview: Environment Canada SEA.

Hameed, S., Yeh, W.M., Li, M.T., Cess, R.D., and Wang, W.C., 1983. An analysis of periodicities in the 1470 to 1974 Beijing precipitation record. *Geophysical Research Letters*, **10**(6): 436–439.

Hammer, C.U., Clausen, H.B., and Dansgaard, W., 1980. Greenland ice sheet evidence of postglacial volcanism and its climatic impact. *Nature*, **288**: 230–235.

Harding, A.F., ed., 1982. *Climatic Change in Later Prehistory*. Edinburgh: Edinburgh University Press.

Hecht, A.D., 1985. *Paleoclimate Analysis and Modeling*. New York: Wiley-Interscience.

Herron, M.M., and Herron, S.L., 1983. Past atmospheric environments revealed by polar ice core studies. *Hydrological Sciences Journal*, **28**: 139–153.

Hughes, M.K., Kelly, P.M., Pilcher, J.R., and LaMarche V.C., eds., 1982. *Climate from Tree-Rings*. London: Cambridge University Press.

Huntington, E., and Visher, S.S., 1922. *Climatic Changes: Their Nature and Causes*. New Haven; CT: Yale University Press.

Imbrie, J., and Imbrie, K.P., 1979. *Ice Ages*. New York: Macmillan.

Jelbring, H., 1995. Analysis of sunspot cycles phase variations: based on D. Justin Schove's proxy data. *Journal Coastal Research*, Special Issue **17**: 363–369.

Kawai, N., et al., 1975. Paleomagnetism of Lake Biwa sediments. *Rock Magnetics and Paleogeophysics*, **3**: 24–37.

Keys, D., 1999. *Catastrophe*. New York: Ballantine Books, 343p.

Kidson, C., and Tooley M.J., eds., 1977. The Quaternary history of the Irish Sea, *Geological Journal*, (*Liverpool*), special issue **7**.

Kilston, S., and Knopoff, L., 1983; Lunar–solar periodicities of large earthquakes in Southern California. *Nature*, **304**: 21–25.

King, J.W., Bannerjee, S.K., Marvin, J., and Lund, S., 1983. Use of small amplitude paleomagentic fluctuations for correlation and dating of continental climatic changes. *Palaeogeography, Palaeoclimatology, Palaeoecology*, **42**: 167–183.

Köppen, W., 1914. Luftltemperatur, Sonnenflecken und Vulkanausbrüche, *Meterologische Zeitschrift*, **35**: 305–328.

Klein, J., Lerman, J.C., Damon, P.E., and Ralph, E.K., 1982. Calibration of radiocarbon dates: Tables based on the consensus data of the workshop on calibrating the radiocarbon time scale. *Radiocarbon*, **24**: 103–150.

LaMarche, V.C., Jr, 1974. Paleoclimate inferences from long tree-ring records. *Science*, **183**: 1043–1048.

Lamb, H.H., 1972. *Climate: Present, Past and Future*, vol. 1. London: Methuen.

Lamb, H.H., 1977. *Climate: Present, Past and Future*, vol. 2. London: Methuen.

Landscheidt, T., 1984. Cycles of solar flares and weather. In Mörner, N.A., and Karlén, W., eds., *Climatic Changes on a Yearly to Millennial Basis*. Dordrecht: Reidel, pp. 473–481.

Lorius, C., Merlivat, L., Jouzel, J., and Pourchet, M., 1979. A 30,000 yr isotope climatic record from Antarctic ice. *Nature*, **280**: 644–648.

Mangerud, J., et al., 1974. Quaternary stratigraphy of Norden, a proposal for terminology and classification. *Boreas*, **3**: 109–127.

Mörner, N.A., and Karlén, W., eds., 1984. *Climatic Changes on a Yearly to Millennial Basis*. Dordrecht: Reidel.

Møller, J.J. et al., 2002. Holocene raised beachridges and sea-ice-pushed boulders on Kola Peninsula, northwest Russia: indicators of climate change. *The Holocene*, **12**(2): 169–176.

Mosley-Thompson, E., and Thompson, L.G., 1982. Nine centuries of microparticle deposition at the South Pole. *Quaternary Research*, **17**: 1–13.

Neftel, A., Oeschger, H., Schwander, J., Stauffer, B., and Zumbrunn, R., 1982. Ice core sample measurements give atmospheric CO_2 content during the past 40,000 years. *Nature*, **295**: 220–223.

Newell, R.E., and Hsiung, J., 1984. Sea surface temperature, atmospheric CO_2 and the global energy budget: some comparisons between the past and present. In Mörner, N.A., and Karlén, W., eds., *Climatic Changes on a Yearly to Millennial Basis*. Dordrecht: Reidel, pp. 533–561.

Nilsson, T., 1983. *The Pleistocene. Geology and Life in the Quaternary Ice Age*. Stuttgart: Enke Verlag.

Oele, E., et al. (eds.), 1979. *The Quaternary History of the North Sea*. Uppsala: Acta Universite Uppsala.

Olausson, E., ed., 1982. *The Pleistocene/Holocene Boundary in South-western Sweden*. Stockholm: Sveriges Geologiska Undersokning Arsbok.

Pettersson, O., 1912. The connection between hydrographical and meteorological phenomena, *Royal Meteorological Society Quartely , Journal*, **38**: 173–191.

Pettersson, O., 1914. Climatic variations in historic and prehistoric time. *Svenska Hydrogr.-Biol. Komm Skriften*, **5**.

Pettersson, O., 1930. The tidal force. *Geographische Annaler*, **12**: 261–322.

Pfister, C., 1980. The Little Ice Age: thermal and wetness indices for Central Europe, Journal of *Interdisciplinary History*, **10**: 665–696.

Pisias, N.G., 1978. Palaeo-cean-nography of the Santa Barbara Basin during the last 8000 years, *Quaternary Research*, **103**: 366–384.

Pittock, A.B., 1978. A critical look at long-term sunweather relationships, *Review of Geophysics and Space Physics*, **16**: 400–420.

Pittock, A.B., 1979. Solar cycles and the weather: successful experiments in autosuggestion? In Mc Cormac, B.M., and Seliga, T.A., eds., *Solar–Terrestrial Influences on Weather and Climate*. Dordrecht: Reidel, pp. 181–191.

Pittock, A.B., 1983. Solar variability, weather and climate: an update, Royal Meteorological *Society Quarterly Journal*, **109**: 23–55.

Prell, W.L., 1984. Monsoonal climate of the Arabian Sea during the late Quaternary: a response to changing solar radiation. In Berger, A. et al., eds., *Milankovitch and Climate*. Dordrecht: Reidel.

Raisbeck, G.M., et al., 1982. Cosmogenic B-10 concentrations in Antarctic ice during the past 30,000 years. *Nature*, **293**: 825–826.

Rampino, M.R., and Self, S., 1982. Historic eruptions of Tambora (1815), Krakatau (1883), and Agung (1963), their stratospheric aerosols, and climatic impact. *Quaternary Research*, **18**: 127–143.

Rampino, M.R., Self, S., and Fairbridge, R.W., 1979. Can rapid climatic change cause volcanic eruptions? *Science*, **206**: 826–829.

Ren, Z., and Li, Z., 1980. Effect of motions of planets on climate changes in china. *Kexue Tongbao*, **25**: 417–422.

Roberts, W.O., and Olson, R.H., 1973. New evidence for effects of variable solar corpuscular emission on the weather, *Reviews of Geophysics and Space Physics*, **11**: 731–740.

Roberts, N., 1989/98. *The Holocene: An Environmental History*. Oxford: Blackwell Ltd., 2nd ed., 316p.

Rosen, R.D., Salstein, D.A., Eubanks, T.M., Dickey, J.O., and Steppe, J.A., 1984. An El Niño signal in atmospheric angular momentum and earth rotation, *Science*, **225**: 411–414.

Rotberg, R.I., and Rabb, T.K., 1981. *Climate and History*. Princeton, NJ: Princeton University Press.

Sakaguchi, Y., 1982. Climatic variability during the Holocene Epoch in Japan and its causes, *Tokyo University Department of geography Bulletin*, **14**: 1–27.

Schove, D.J., 1955. The sunspot cycle, 649 BC to AD 2000. *Journal of Geophysical Research*, **60**: 127–146.

Schove, D.J., 1978. Tree ring and varve scales combined, *c.* 13,500 B.C. to A.D. 1977. *Palaeogeography, Palaeoclimatology, Palaeoecology*, **25**: 209–233.

Schove, D.J., ed., 1983. *Sunspot Cycles*. Benchmark Papers in Geology, vol. 68. Stroudsburg, PA: Hutchinson & Ross.

Schwarzbach, M., 1950. *Das Klimate der Vorzeit*. Stuttgart: Enke Verlag.

Schwarzbach, M., 1974. *Das Klimate der Vorzeit*, 3rd en. Stuttgart: Enke Verlag.

Shindell, D.T., Schmidt, G.A., et al., 2001. Northern hemisphere winter climate response to greenhouse gas, ozone, solar, and volcanic forcing. *Journal Geophysical Research*, **106**(D-7): 7193–7210.

Shirokov, V.A., 1983. The influence of the 19-year tidal cycle on large-scale eruptions and earthquakes in Kamchatka, and their long-term prediction. In Fedotor, S.A., and Markhinin, Y.K., eds., *The Great Tolbachik Fissure Eruption*. Cambridge: Cambridge University Press.

Simkin, T., Siebert, L., McClelland, L., Bridge, D., Newhall, C., and Latter, J.H., 1981. *Volcanoes of the World*. Stroudsburg, PA: Hutchinson & Ross.

Siscoe, G.L., 1980. Evidence in the auroral record for secular solar variability. *Review of Geophysical and Space Physics*, **18**: 647–658.

Sonett, C.P., and Suess, H.E., 1984. Correlation of bristlecone pine ring widths with atmospheric C-14 variations: a climate–Sun relation. *Nature*, **307**: 141–143.

Stockton, C.W., Boggess, W.R., and Meko, D.M., 1985. Climate and tree rings, In Hecht, A., ed., *Paleoclimate Analysis and Modeling*. New York: Wiley-Interscience, pp. 71–140.

Stommel, H., and Stommel, E., 1983. *Volcano Weather: The Story of 1816, the Year Without a Summer*. Newport, RI: Seven Seas Press.

Stuiver, M., 1982. A high-precision calibration of the A.D. radiocarbon time scale. *Radiocarbon*, **24**: 1–26.

Stuiver, M., and Braziunas, T.F., 1993. Modelling atmospheric 14C influences and 14C ages of marine samples back to 10,000 BC. *Radiocarbon*, **35**: 137–189.

Stuiver, M., and Reimer, P.J., 1993. Extended 14C database and revised CALIB radiocarbon calibration program. *Radiocarbon*, **35**: 215–230.

Suess, H.E., 1981. Solar activity, cosmic-ray produced carbon-14, and the terrestrial climate. In *Proceedings of the Conference on Sun and Climate, 1980*. Toulouse: Center Nat. D'Et. Spatiale, pp. 307–310.

Taira, K., 1980. Holocene events in Japan: Palaeo-oceanography, volcanism and relative sea-level oscillation, *Palaeogeography, Palaeoclimatology, Palaeoecology*, **32**: 69–77.

Velichko, A.A., ed., 1984. *Late-Quaternary Environments in the Soviet Union*. Minneapolis; MN: University of Minnesota Press.

Wang, P.K., 1979. Meteorological records from ancient chronicles of China, *American Meteorological Society Bulleting* **60**: 313–318.

Wang, P.K., 1980. On the relation between winter thunder and the climatic change in China in the past 220 years, *Climate. Change*, **31**: 37–44.

Wang, S.W., and Zhao, Z.C., 1981. Droughts and floods in China, 1470–1979. In Wigley, T.M.L., Ingram, M.J., and Farmer, G., eds., *Climate and History*. Cambridge: Cambridge University Press, pp. 271–288.

Wigley, T.M.L., Ingram, M.J., and Farmer G., eds., 1981. *Climate and History: Studies of Past Climates and Their Impact on Man*. Cambridge: Cambridge University Press.

Williams, G.E., 1993. History of the Earth's obliquity. *Earth Science Reviews*, **34**, 1-45.

Winchester, S., 2003. *Krakatos: The Day The World Exploded*. New York: Harper Collins, 432p.

Windelius, G., and Carlborg, N., 1995. Solar orbital angular momentum and some cyclic effects on Earth systems. *Journal Coastal Research*, Special issue **17**: 383–395.

Wood, F.J., 1978. The strategic role of Perigean spring tides. In *Nautical History and North American Coastal Flooding, 1635–1976*. Washington, DC: National Oceanic and Atmospheric Administration, US Department of Commerce.

Wood, F.J., 1985. *Tidal Dynamics: Coastal Flooding and Cycles of Gravitational Force*. Dordrecht: Reidel.

Wood, F.J., 2001. *Tidal Dynamics* (third edition). West Palm Beach: Coastal Education and Research Foundation, 2 vols.

Woods, P.J., 1983. Evolution of, and soil development on, Holocene ridge sequences, West Coast, Western Australia. PhD, dissertation, University of Western Australia.

Wright, H.E., ed., 1983. *Late-Quaternary Environments of the United States*, vol. 2: *The Holocene*. Minneapolis MN: University of Minnesota Press.

Yoshino, M.M., and Xie, S., 1983. A preliminary study on climatic anomalies in East Asia and sea surface temperatures in the North Pacific. *Tsukuba University Ins. Geoscience and Science, Reports*, **A4**: 1–23.

Cross-references

Climate Change and Ancient Civilizations
Cycles and Periodicities
El Niño
Climate Change and Global Warming
Ice Ages
Little Ice Age
Southern Oscillation
Sunspots
Tree-Ring Analysis

CLIMATE VARIATION: INSTRUMENTAL DATA

Climate is often regarded as the average state of weather. Although this definition is not incorrect, it can lead (incorrectly) to an interpretation of climate as a constant, unchanging phenomenon. On the contrary, the climate system is characterized by constant change. Climatic variations result in the modification of the average, variability, and frequency of extreme values of temperature, precipitation, pressure and other meteorological variables. Such changes, which may happen in all temporal and spatial scales beyond those of individual weather events, are defined by the Intergovernmental Panel on Climate Change (IPCC) as *climatic variability*. Statistically significant changes in the mean state of climate or its expected variability, which persist for an extended period (decades or longer), are referred to as *climatic change* (IPCC, 2001).

Causes of climatic variability and change include: natural internal processes such as dynamic and thermodynamic interactions between the atmosphere and oceans; volcanic eruptions; external forcing, like e.g. changes of intrinsic solar radiation, or persistent anthropogenic changes in the composition of the atmosphere, or changes in land use. Such factors may affect climate directly by altering the local atmospheric heat budget and/or indirectly by modifying the planetary wind and ocean current circulation by which local climate to a large extent is regulated. The effects of climatic variation are numerous and often affect socioeconomic and natural systems.

Geographically, climatic changes are so extensive that they deserve to be treated as an integrated planetary phenomenon. Indeed, a hemispheric or global viewpoint is required in order to evaluate cause-and-effect relationships. On a worldwide basis, variations of the mean state of surface climate can be described with good confidence from instrumental observations since the mid-nineteenth century, although analyses of surface variability and extremes, and the climate of the upper atmosphere, are more limited.

Climatic analyses require homogeneous data, and this is particularly important for analyses of climate change and variability. A homogeneous climatic time series is defined as one where variations are caused only by variations in climate. Unfortunately, most long-term climatological time series have been affected by a number of non-climatic factors that make these data unrepresentative of the actual climate variation occurring over time. These factors include changes in: instruments, observing practices, station locations, formulae used to calculate derived quantities, and station environment, among others. Some changes cause sharp discontinuities while other changes, particularly a change in the environment around the station, can cause gradual biases in the data. All of these inhomogeneities can bias a time series and lead to misinterpretations. Climate researchers must carefully remove all these non-climatic biases or inhomogeneities or at least determine the possible error they may cause before drawing conclusions from data (Aguilar et al., 2004).

The available instrumental record broadly covers the period of anthropogenically induced increases in CO_2 and other greenhouse gas concentrations and the contemporary global warming. Global warming has led to a more extreme climate, an increase in global precipitation and atmospheric water vapor content, an increase in ocean heat content and a sea level rise, a reduction in mountain glacier extent and snow coverage, a shortening of the seasons of lake and river ice and a systematic decrease in spring and summer ice in the Arctic regions, among other effects.

Observational data contribute to the knowledge of paleoclimatic variations, helping to calibrate the relationship between proxy data (tree rings, ice cores, boreholes, sediments, corals, historical documents) and meteorological variables. Finally, instrumental data are needed too for the climate models which can predict the future evolution of climate and assess the goodness-of-fit of their results. The next section will focus on the analysis of temperature and precipitation evolution.

Climatic variations in the instrumental record

Global temperatures can be reconstructed through instrumental data back to the mid-nineteenth, century, as shown in Figure C22 (IPCC, 2001). The average global land surface temperature has increased by 0.6°C since 1860 (Folland et al., 2001). Warming is found to be significant for the aforementioned period annually and in all seasons. Independently compiled, homogenized and analyzed datasets (Peterson and Vose, 1997; Hansen et al., 1999; Jones et al., 2001) agree and show two remarkable warming phases, 1910–1945 and 1976 to present, as the most distinctive features of the observational record. Sea surface temperatures have experienced similar pulses, although variations and trends have been slightly less pronounced. As for land data, different compilations of oceanic data agree on similar results (Quayle et al., 1999; Jones et al., 2001; Reynolds et al., 2002).

The late nineteenth century was the coldest period in the instrumental record. This is true not only for the global estimates extending back to 1860, but also when some longer temperature records for central England, Fennoscandia and central Europe, going back to the eighteenth century, are considered (Jones, 2001). The early twentieth century showed positive anomalies, mainly concentrated in the North Atlantic and nearby regions, with a contrary cooling tendency prevailing in parts of North America, southern Eurasia, and the southern hemisphere.

The ongoing warming is qualitatively consistent with some climate model projections driven by anthropogenic forcing (Mitchell et al., 2001), and has a nearly global extent, although some oceanic areas and Antarctica (excluding the Antarctic Peninsula) are still cooling. Global temperatures – including land and ocean – have been rising at about 0.165°C per decade between 1977 and 2001. Trends are larger over land than over oceans and, as a consequence, the northern hemisphere warming (0.223°C per decade) is much larger than that observed for the southern hemisphere (0.106°C per decade). The highest regional year-round warming rates are found over Europe and the Arctic, and – on a seasonal basis – winter and spring over North America and Eurasia show the largest increase (Jones and Moberg, 2003), probably in relation to the enhanced westerly flow caused by the maintained positive phase of the North Atlantic/Arctic Oscillation.

The described warming trend places the 1990s decade as the warmest in the observational record, and probably the warmest (at least in the northern hemisphere) of the last 1800 years as derived from multi-proxy data reconstructions (Mann and Jones, 2003). This happened even with the short-lived cooling influence of the 1991 eruption of the Pinatubo volcano (Parker et al., 1996). The year 1998, affected by the strong 1997/1998 El Niño, is, up to this writing (late 2003),

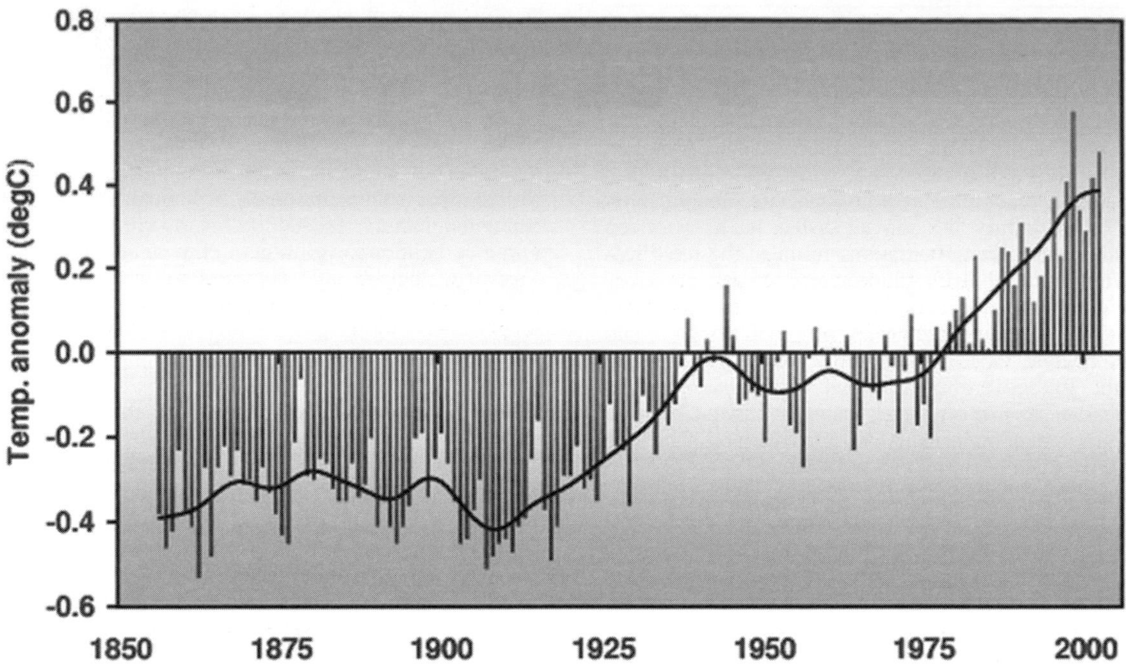

Figure C22 Global air temperature (HadCRUT2v), 2002 anomaly +0.48°C (second-warmest on record). Reproduced by kind permission of the Climatic Research Unit, University of East Anglia, Norwich, U.K., to whom any queries should be directed (cru@uea.ac.uk). Copyright: Climatic Research Unit.

the warmest year on record, with an anomaly of roughly + 0.5°C with respect to the 1961–1990 normal period. The El Niño – Southern Oscillation (ENSO) phenomenon has shown an increased occurrence, persistence, and intensity of its warm phase (El Niño) in recent decades, which enhances the global positive anomalies of the climate. The intensities of the 1982/1983 and the 1997/1998 El Niños were unprecedented in the observational record. The most recent event led to much warmer global temperatures, in spite of the offsetting effect of the 1982 El Chichon volcanic eruption, in part because of a warmer starting point in 1998. The initial decade of the twenty first century, continues with the warming trend started in the mid-1970s. The year 2002 was the second warmest in the instrumental record and the warmest for latitudes poleward of 30°N (Waple and Lawrimore 2003). If the current trend is maintained a future El Niño event will easily lead to a new record high global temperature.

During the second part of the twentieth century the DTR (diurnal temperature range or difference between daytime and nighttime temperature) has been reduced on a global basis by 0.1°C per decade. This happened, in agreement with an observed increase in global cloudiness, because three-quarters of the observed warming over land areas has been attributed to nighttime (usually daily minimum) temperatures and the remaining to daily maxima (Easterling et al., 1997). Nevertheless, several regions do not follow the planetary pattern – among them, eastern North America, middle Canada, portions of Europe, parts of Southern Africa, India, Nepal, Japan, New Zealand, the western tropical Pacific Islands, Australia, Antarctica and Spain display no change or slight increase of the day–night temperature differences (Karl et al., 1994; Horton, 1995; Brunet et al., 2001).

In the 50 years of available radiosonde data, upper air tropospheric temperatures (roughly the lowest 10–15 km of the atmosphere) have experienced warming, in agreement with the surface observations, and also in agreement (for thermodynamic reasons) with the observed stratospheric cooling. The extra-tropical northern hemispheric troposphere concentrated its warming since the mid-late 1970s, while the austral mid to high-latitude troposphere started warming earlier in the record (Lanzante et al., 2003).

Increases in global temperatures are very likely to be reflected in changes in precipitation and atmospheric moisture via induced changes in the atmospheric circulation; a more active hydrological cycle; increases in the water vapor holding capacity throughout the atmosphere (Folland et al., 2001); and changes in the leading modes of interannual variability, e.g. ENSO, the Arctic Oscillation, and the North Atlantic Oscillation. Different rain-gauge datasets (Peterson and Vose, 1997; Rudolf et al., 1999; New et al., 2001) and satellite measurements, available since the late 1970s (Xie and Arkin, 1996; Huffman et al., 1997; Doherty et al., 1999; Huffman et al., 2000) can be combined to produce global and regional estimates of changes in precipitation (New et al., 2001). The later authors reported a global precipitation increase of 0.89 mm per decade, mainly the result of an increase of 40 mm from 1901 to the mid-1950s, when global precipitation peaked. This trend was not globally uniform. Between the late nineteenth century and the mid-twentieth, precipitation in the tropics and along the east coasts of the continents decreased by roughly 10%. Some investigators believe this decrease occurred quite abruptly around the turn of the century. Pressure data point to a slight, gradual weakening of the general circulation during the same period, in the northern hemisphere at least, and to a contraction of the zonal westerly wind belt in temperate latitudes toward higher latitudes (analogous to its annual contraction from winter to summer).

The magnitude of the century-long trend (9 mm) appears to be modest when compared to the interdecadal and interannual variability of precipitation. Spatially, larger increases came from the mid and high northern hemisphere latitudes (40–80°N). The year 1998 was the wettest year on record for latitudes exceeding 55°N (Folland et al., 2001). Precipitation in the Northern Subtropics did not show significant trends and was rather characterized by strong subdecadal variability. Finally, areas between 20°N and 40°N experienced a decrease of 6.3 mm per decade, most of it after the mid-1950s. The southern hemisphere tropics and midlatitude precipitation remained steady in the last 100 years, when the austral subtropics showed an increase of 3.6 mm per decade (New et al., 2001).

Besides the analysis of the mean state of climate and its variability, it is crucial to understand the evolution of the extreme values. Extremes tend to have a notorious impact on natural and socioeconomic systems. Relatively modest changes in the mean state of climate can lead to large changes in the frequency of extremes (Katz, 1999). Despite a good degree of spatial differences, and the lack of data for significant parts of the southern hemisphere, it is safe to say that a significant proportion of the global land area was increasingly affected by significant changes in climatic extremes during the second half of the twentieth century. During this period, days with cold temperatures have diminished and, consequently, the number of frost days has been reduced; the frequency of extremely cold days has been reduced, and the number of days with extremely warm temperatures seems to be increasing (a result of increased variances), as well as the number of days with heavy rainfall and the total amount of precipitation coming from wet spells (Frich et al., 2002).

Causes of climatic variation

The immediate meteorological causes of observed local climatic variation involve changes of the general atmospheric circulation pattern. In this respect many regional anomalies in the world-average variation can be understood. However, it is unlikely that the net world-average variations of temperature and precipitation are traceable to circulation changes alone; rather, they are caused by more fundamental environmental changes that can affect both climate and circulation. Ultimate causes of climatic variation have not yet been identified with certainty, but they are thought to be several. The most likely among these fall into the first four of the following five main categories.

1. *Air–sea interaction.* Since the atmosphere is closely coupled with the oceans both dynamically and thermally, the relatively much slower rates of mixing and circulation of the oceans, together with their enormous heat-storage capacity, open up many possibilities for the oceans to stimulate long-term variations of climate. Air–sea interactions are, in fact, a probable source of much of the observed variability of climate on all scales of time from years to centuries to millennia. In general, the longer the time scale of variation involved, the deeper are the oceanic depths involved in the interaction. On interannual time scales a principal mode of air–sea interaction is the ENSO phenomenon, centered in the tropical Pacific Ocean, but whose effects on climate are of worldwide extent (Rasmusson and Wallace, 1983). The strong El Niño event of 1982–1983 was accompanied by extremely wet weather in parts of the Americas and, concurrently, by severe drought in parts of Australasia.
2. *Explosive volcanic activity.* In modern times, explosive volcanic eruptions like that of Krakatoa in 1883, violent enough to inject large amounts of fine ash and sulfurous gases into the midstratosphere (altitudes of 20–30 km), have occurred with an average frequency of several per century. Some such eruptions have produced veils of ash and sulfuric acid aerosols that have spread worldwide in the stratosphere with lifetimes of 2 or more years. Because such volcanic veils scatter and absorb significant amounts of incoming solar radiation while having little effect on outgoing terrestrial radiation, they are capable of cooling worldwide climate by a fraction of a degree Celsius over a period of years (Newell and Deepak, 1982). The warming of world climate during the 1920s and 1930s can be attributed in part to the absence of such volcanic eruptions at that time, and the cooling that culminated in the 1960s can be partly attributed to a renewal of eruptive activity that included the major eruption of Agung in 1963. The more recent major eruption of El Chichon in Mexico in the spring of 1982 was expected to result in a climatic cooling in the years following. This cooling, however, is not readily apparent in the temperature record (see Figure C22), a fact that may be explained by the contrary warming effect of the great El Niño event of 1982–1983 in which enormous quantities of heat were transferred to the atmosphere from the warmer-than-normal tropical Pacific Ocean.
3. *Changes of atmospheric composition.* Changes of the gaseous composition of the atmosphere may alter the terrestrial heat balance if the gases are selective absorbers of radiation such as water vapor, carbon dioxide (CO_2), and ozone (O_3). Carbon dioxide has accumulated in the atmosphere in the past two centuries, partly owing to vast deforestation for agriculture and fuel wood and sharply increasing use of fossil fuels. The net accumulation of CO_2 in the atmosphere has by now increased from preindustrial levels of about 280 parts per million by volume to levels of about 360 ppmv (as of 2003). Other trace gases of industrial origin, such as nitrous oxide and chlorofluorocarbons as well as methane, are also increasing. These other gases behave in the manner of CO_2, to cause a "greenhouse" warming of climate. Their combined effect on climate is thought to have been a warming since 1880 of about 0.6°C (0.9°F); further warming at an accelerating pace will likely continue into the future as long as these gases continue to accumulate as expected (IPCC, 2001). Changes of "greenhouse" gas concentrations to date are a likely cause of much of the net warming of climate observed in the past century (see Figure C22).
4. *Solar radiation changes.* Solar radiation may vary in several respects, including the total solar constant, its ultraviolet component, and solar wind intensity. Long-term changes of total solar constant, if real, would clearly affect climate but evidence of such changes other than those of a fraction of 1% in the past few years of direct satellite measurement is as yet unverified (Foukal, 2002). Possibly larger changes may have occurred over the 80–90-year Gleissberg cycle of solar activity or in the course of longer historical solar disturbances such as the Maunder Minimum, a time of unusual solar inactivity between about 1650 and 1710. Changes of ultraviolet emissions are known to parallel the 11-year solar cycle, and may influence surface climate indirectly via their effect on upper atmospheric conditions such as ozone amount. Changes of solar wind intensity may likewise be capable of altering the behavior of atmospheric circulation through subtle

effects on upper-atmospheric temperatures and winds and the "ducting" of Rossby-wave energy in the troposphere (Geophysics Study Committee, 1982). To date, however, such presumed solar–climate relationships lack adequate observational and theoretical confirmation.

5. *Impacts by large asteroids*. Though extremely rare (with time scales of hundreds of thousands to hundreds of millions of years), impacts by comets or asteroids have the potential to cause massive climatic changes via mechanisms similar to volcanic eruptions. Such impacts are thought to have resulted in a few mass extinctions in the geological record. No such event has ever been recorded in the instrumental record.

Enric Aguilar, Robert G. Quayle and J. Murray Mitchell, Junior

Bibliography

Aguilar, E., Auer, I., Brunet, M., Peterson, T.C. and Wieringa, J., 2004. *Guidance on Metadata and Homogenization*. Geneva: Commission on Climatology, World Meteorological Organization.

Brunet, M., Aguilar, E., Saladié, O., SigróJ., and López, D., 2001. A differential response of northeastern Spain to asymmetric trends in diurnal warming detected on a global scale. In Brunet, M., and López, D., eds., *Detecting and Modeling Regional Climate Change*. Berlin: Springer, pp. 95–107.

Doherty, R.M., Hulme, M., and Jones, C.G., 1999. A gridded reconstruction of land and ocean precipitation for the extend tropics from 1974 to 1994. *International Journal of Climatalogy*, **19**: 119–142.

Easterling, D.R., Horton, B., Jones, P.D. et al., 1997. Maximum and minimum temperature trends for the globe. *Science*, **277**: 364–367.

Folland, C.K., Karl, T.R., Christy, J.R. et al., 2001. Observed climate variability and change. In Houghton, J.T., et al., eds., *Climate Change 2001: the scientific basis*. Contribution Working Group I to the Third Assessment Report of the Intergovernmental Panel on Climate Change. Cambridge: Cambridge Univerity Press.

Foukal, P., 2002. A comparison of variable solar total and ultraviolet irradiance outputs in the 20th century. *Geophysical Research Letters*, **29**: 10.1029/2002GL015474.

Frich, P., Alexander, L.V., Della-Marta, P. et al., 2002. Observed coherent changes in climatic extremes during the second half of the twentieth century. *Climate Research*, **19**: 193–212.

Geophysics Study Committee, Geophysics Research Board, 1982. *Solar Variability, Weather, and Climate*. Washington, DC: National Academy Press.

Hansen, J., Ruedy, R., Glascoe, J., and Sato, M., 1999. GISS analysis of surface temperature change. *Journal of Geophysical Research*, **104**(D24): 30997–31022.

Horton, B., 1995. Geographical distribution of changes in maximum and minimum temperatures. *Atmospheric Research*, **37**: 101–117.

Huffmann, G.J., Adler, R.F., Morrisey, M.M., et al., 2000. Global precipitation at one degree daily resolution from multi-satellite observations. *Journal of Hydrometeorology*, **2**(1): 36–50.

Huffmann, G.J., Adler, R.F., Arkin, P., et al., 1997. The Global Precipitation Climatology Project (GPCP) Combined Precipitation Dataset. *Bulletin of the American Meteorological Society*, **78**: 5–20.

IPCC, 2001. *Climate Change 2001: the scientific basis*. Contribution Working Group I to the Third Assessment Report of the Intergovernmental Panel on Climate Change, Houghton, J.T., et al., (eds.) Cambridge: Cambridge University Press, pp. 99–181.

Jones, P.D., 2001. Instrumental temperature change in the context of the last 1000 years. In Brunet, M., and López, D., eds., *Detecting and Modeling Regional Climate Change*. Berlin: Springer, pp. 55–68.

Jones, P.D., Osborn, T.J., Briffa, K.R., et al., 2001. Adjusting for sampling density in grid box land and ocean surface temperature time series. *Journal of Geophysical Research*, **106**: 3371–3380.

Jones, P.D., and Moberg, A., 2003. Hemispheric and large-scale surface air temperature variations: an extensive revision and an update to 2001. *Journal of Climate*, **16**: 206–223.

Karl, T.R., Easterling, D., Peterson, T.C., et al., 1994. An update of the asymmetric day/night land surface warming. *Proceedings of the Sixth Conference on Climate Variations*, American Meteorological Society, pp. 170–173.

Katz, R.W., 1999. Extreme value theory for precipitation: sensitivity analysis for climate change. *Advances in Water Resources*, **23**: 133–139.

Lanzante, J.R., Klein, S.A., and Seidel, D.J., 2003. Temporal homogenization of monthly radiosonde temperature data. Part II: Trends, sensitivities, and MSU comparison. *Journal of Climate*, **16**(2): 241–262.

Mann, M.E., and Jones, P.D., 2003. Global surface temperatures over the past two millennia. *Geophysics Research Letter*, **15**, 1820.

Mitchell, J.F.B., Karoly, D.J., Hegerl, G.C., Zwierse, F.W., Allen, M.R., and Marengo, J., 2001. Detection of climate change and attribution of causes. In Houghton, J.T., et al., eds., *Climate Change 2001: the scientific basis*. Contribution Working Group I to the Third Assessment Report of the Intergovernmental Panel on Climate Change. Cambridge: Cambridge University Press, pp. 000–000.

New, M., Todd, M., Hulme M., and Jones, P.D., 2001. Precipitation measurements and trends in the twentieth century. *International Journal of Climatology*, **21**: 1899–1922.

Newell, R.E., and Deepak A., (eds.), 1982. *Mount St. Helens Eruptions of 1980, Atmospheric Effects and Potential Climate Impact*. Washington, DC: NASA (see chapter 6).

Jones, P.D., and Moberg, A., 2003. Hemispheric and large-scale surface air temperature variations: an extensive revision and an update to 2001. *Journal of Climate*, **16**(2): 206–223.

Parker, D.E., Wilson, H., Jones, P.D., Christy, J., and Folland, C.K., 1996. The impact of Mount Pinatubo on climate. *International Journal of Climatology*, **16**: 487–497.

Peterson T.C., and Vose, R.S., 1997. An overview on of the global historical climatology network temperature database. *Bulletin of the American Meteorological Society*, **78**: 2837–2849.

Quayle, R.G., Peterson, T.C., Basist, A.N., and Godfrey, C.S., 1999: An operational near real-time global temperature index. *Geophysical Research Letters*, **26**: 333–225.

Rasmusson, E.M., and Wallace, J.M., 1983. Meteorological aspects of the El Niño/Southern Oscillation. *Science*, **222**: 1195–1202.

Reynolds, R.W., Rayner, N.A., Smith, T.M., Stokes, D.C., and Wang, W., 2002. An improved in situ and satellite SST analysis for climate. *Journal of Climate*, **15**(13): 1609–1625.

Rudolf, B., Gruber, A., Adler, P., Huffman, G., Janowiak, J., and Xie, P.P., 1999. GPCP precipitation analyses based on observations as a basis for NWP and climate model verification. *Proceedings of WRCP Second International Reanalysis Conference*. Reading, UK, 23–27 August 1999. WCRP report.

Waple, A.M., and Lawrimore, J.H., 2003. State of the climate in 2002. *Bulletin of the American Meteorological Society*, **84**:(6) S1–S68.

Xie, P.P., and Arkin, P.A., 1996. Analyses of global monthly precipitation using gauge observations, and satellite estimates and numerical model prediction. *International Journal of Climatology*, **34**: 1143–1160.

CLIMATE VULNERABILITY

To be vulnerable is to be "susceptible of receiving wounds or physical injury" (*Oxford English Dictionary*). The phrase "climate vulnerability" raises an immediate question: what is it that might be injured? Climate might be affected by the actions of or changes in natural and social systems. Or natural and social systems experience impacts from changes in climate. Or the phrase may mean both of these interpretations.

Vulnerability of climate

Climate – the assemblage of atmospheric properties and behaviors, in the mean and with their inherent and characteristic variances – responds to the state, and changes of state, of the hydrosphere, cryosphere, and biosphere. Conversely, and reflexively, those natural systems are conditioned by the state of global climate. As a simple example, the poles are frozen due to the fundamentals of global climate; the extent of Arctic pack ice responds to episodes of hemispheric warming or cooling; hemispheric climate responds to the state of the surface of the Arctic basin. Most models of climate consider coupled natural systems, describing sensitivities and feedbacks, but not "vulnerabilities" in the colloquial sense of the term. In the case of social systems, where there are direct and indirect effects from alterations in atmospheric composition and of biosphere, it does make sense to think of climate being vulnerable. Much research in global climate change at the turn of the millennium focuses on these processes, patterns, and implications (see, for example, the IPCC Working Group I assessment; Houghton et al., 2001).

Strongest among the evidence that human activity can effect change in the climate system is the relationship of global temperature and "greenhouse gases". Concentrations of carbon dioxide, methane, nitrous oxides, and other gases have increased exponentially during the industrial era, i.e. since the beginning of the eighteenth century. Theory about atmospheric chemistry and radiative forcing of climate suggests that those changes should produce atmospheric warming, and models suggest a good fit of observed global surface temperatures with the records of gas concentrations. Models of likely socio-economic futures, greenhouse gas emissions and resultant concentrations in the atmosphere, suggest that, even with marked reductions in greenhouse gas production, the atmosphere and thus the climate system will remain perturbed for a century or more (cf. the summary by Albritton and Meira Filho, 2001, and the more detailed reviews within Houghton et al., 2001).

The notion that human alteration of surface features at a regional scale might change the climate is not new (Brooks, 1970, originally published 1926). In late nineteenth-century America, for example, a popular idea was that "rain follows the plow"; conversion of the vast central grasslands to agriculture would increase precipitation of the area, to the benefit of agriculture (Lawson, 1974). In the latter half of the twentieth century, speculations about the effects of other major landscape alterations were made, often without aid of global circulation models: redirection of northward-flowing rivers in Asia; Amazon deforestation; expansion of desertification in Africa. Models are less successful at – or perhaps have paid less attention to – reproducing the effects of land surface changes on climate, perhaps because of the complexity of the climate system and because most attention has focused on atmospheric chemistry and the impacts of climate change on natural and human systems.

Vulnerability to climate

The dictionary definition cited at the head of this article implies a straightforward model of impacts: the characteristics of an entity (or system) are changed in a deleterious sense due to the action of outside forces. Application of the concept in natural sciences such as ecology led to notions of degree of vulnerability. Entities might, for example, not exhibit any evidence of impact until external actions (uniquely or cumulatively) surpassed some threshold beyond which the entity could not continue as it had been. If the entity returned to its original condition after an impact had occurred, it was said to be resilient. In some usage, resilience is the ability to withstand change, to persist. Thus, vulnerability and resilience are related concepts, although they are not – as might be the usage in colloquial speech – antonyms. More on this point follows in the paragraphs below.

The concept of vulnerability has broadened considerably in the work on climate change and adaptations that emerged toward the end of the twentieth century. In the Third Assessment Report of the IPCC, Working Group II specifically addressed climate impacts, adaptations and vulnerability (McCarthy et al., 2001). The 19 sectoral and regional chapters of this volume provide the most comprehensive review of the topic. The overview chapter (Schneider and Sarukhan, 2001), the final chapter that synthesizes vulnerability to climate change and reasons for concern (Smith et al., 2001), and the extensive bibliographies therein, are of particular relevance to this item. The other chapters and their bibliographies should not be ignored. Much of what follows in this article is drawn from this volume.

Definitions

The IPCC's working definition of climate vulnerability is "the degree to which a system is susceptible to, or unable to cope with, adverse effects of climate change, including climate variability and extremes" (McCarthy et al., 2001, p. 995). That is, vulnerability is "the extent to which a natural or social system is susceptible to sustaining damage from climate change" (Schneider and Sarukhan, 2001, p. 89). These definitions are notably more epansive than the classical concept stated at the outset of this article, incorporating fundamental ideas from earlier definitions in the hazards and development (poverty) literatures. Whereas the simple definition contained the idea of exposure to a hazard and resultant impacts, the IPCC's definition integrates hazard, exposure, consequences, and adaptive capacity. In this sense it is akin to the concept of "risk" in the risk-assessment literature.

What makes an entity or system vulnerable to climatic impact? Vulnerability is a function of both the climate change or variability to which the system is exposed, and the nature of the system in terms of its sensitivity to climatic control and its adaptive capacity. Exposure is a function of the climatic event or change – its character or direction, its magnitude, its rate of onset, and its duration.

Sensitivity is "the degree to which a system will respond to a given change in climate, including beneficial and harmful effects" (Schneider and Sarukhan, 2001, p. 89). Effects may be direct, such as crop yield responses to changes in temperature or precipitation; or they may be indirect, via intermediary actions, such as coastal flooding by sea-level rise due to climate change. Thus impacts might be sequential, nested, or hierarchical, and vulnerability may be immediate or delayed, and muted to lesser or greater degree (cf. Kates, 1986).

Adaptive capacity is "the ability of a system to adjust to climate change (including climate variability and extremes) to moderate potential damages, to take advantage of opportunities, or to cope with the consequences" (McCarthy et al., 2001, p. 982). Adaptations in social systems may be made by adjustments in practices (institutions), structures (infrastructure), or processes (mitigation). Many natural systems, or elements thereof, may be limited in adaptive capacity because of human alteration of the globe. For example, ecosystems will not adjust en masse, but as the suite of species responses; some species may be unable to adjust fast enough, or to find suitable habitat because of fragmentation and destruction of ecosystems that has already occurred.

"[A] highly vulnerable system would be a system that is very sensitive to modest changes in climate, where the sensitivity includes the potential for substantial harmful effects, and for which the ability to adapt is severely constrained" (Schneider and Sarukhan, 2001, p. 89). Many researchers believe "[s]ocial systems generally are more resilient than natural systems because of the potential for deliberate adaptation" (Schneider and Sarukhan, 2001, p. 91). Natural systems cannot access coping mechanisms or strategies beyond the autochthonous, whereas social systems can draw upon or invent purposeful adaptations. Synergies among purposeful adaptations may also yield additional coping capacity.

A system with low vulnerability is often termed resilient. Resilience results from adaptive measures, whether in reaction to actual impact or in anticipation of potential impact. The Resilience Alliance (2002) goes so far as to state that "[t]he antonym of resilience is often denoted vulnerability". Adger (2003) calls resilience "the ability to persist and the ability to adapt". A system's resilience is determined by: the magnitude of perturbations it can absorb without changing its overall function; the system's capacity for self-organization; and the system's capacity for learning and adapting (Adger, 2003, p. 2). It is not correct, however, to think of persistence merely as stasis, such that the system returns to the pre-existing state that prevailed prior to impact. Such was the classic systems view of the steady state. Rather, the current concept of resilience recognizes change as an essential characteristic, as the system adapts to new conditions, incorporating learning and adjustment.

Implications for assessment

The first generation of climate change assessments looked like an impacts model with (simple) feedback (see Kates, 1986, for impact and interaction models). Adaptation comes at the end of a causal chain between climate and impact, and is not an integral part of the system from the start. Such assessments, as in the Second Assessment Report of the IPCC (Watson et al., 1995), addressed the following questions: How will climate change? What first-order impacts will there be on natural systems? What will be the effects of those impacts on humans? How might human society adapt? How might the adaptations alter the impacts? (Leary, 2003).

Climate change assessments of the second generation, as the Third Assessment Report typifies, resemble adaptation models with synergies between natural and social systems (cf. Kates, 1986). Vulnerability and adaptation occupy a central place in these assessments. The latest generation of assessments builds upon the earlier in four major ways. First, vulnerability and adaptation to climate change are assessed: who is vulnerable, the source and nature of that vulnerability, and capacities for coping with current variability and extremes (and implications for the future). Second, stakeholders are integrated with scientists in the process: who needs to know what, and how can they be involved in production of credible and relevant information? Third, risk-assessment approaches are utilized. Fourth, scenarios of future environmental and socioeconomic conditions are integrated with climate change scenarios (Leary, 2003).

The emphasis on vulnerability permeates the Third Assessment Report to a much greater degree than previous reports. This focus can be seen by the attention paid to: adaptation to climate change; links between climate change, sustainable development, and equity; vulnerability to changes in climate variability; discontinuous responses to climate change; emphasis on ranges of outcomes using subjective probabilities; and other matters.

Vulnerability of agriculture

Agriculture is arguably the social system in which vulnerability might be easiest to define and earliest to be noted. Food production is of course important to existence and well-being, as well as livelihood and social structure. Agriculture exists at the interface of environmental and social systems, and forms a nexus with water resources not easily subdivided into components.

Agricultural production is, at base, affected by water and nutrient availability, and temperature. Human factors – genetic manipulation whether by traditional breeding or biotechnological engineering; artificial inputs; and so on – modify but do not replace these natural controls. Global temperature increases may allow cultivation in subpolar areas now inhospitable because of short growing seasons, if soil conditions in those areas are amenable. In other areas, however – tropics, drylands – heat or drought stress might increase, limiting productivity. Many of the least-developed countries are in the tropics, and have limited adaptive capacities for socioeconomic and political reasons. At moderate global warming (*c.* 1°C), subpolar and midlatitudes may benefit; above *c.* 2.5°C that productivity is modeled to diminish, and tropical decreases are more severe. Thus, it is possible to imagine an increased disparity in food production between developed and developing nations under scenarios of climate change for the twenty-first century (Schneider and Sarukhan, 2003; Smith et al., 2003).

Water resources are controlled to a great extent by precipitation and (temperature-related) evaporation. Climate scenarios suggest precipitation increases in high-latitude and some equatorial regions, which would decrease water stress, but decreases in many midlatitude, subtropical, and semiarid areas, which would increase water stress. These changes imply respective positive and negative impacts for rain-fed and irrigation agricultural productivity in these regions. With growing population, urbanization, and improving living standards, there will be increased demand for urban and municipal uses of water, thus increasing competition for scarce water at the same time that agricultural demand increases. There is marked disagreement in the literature about prospects for meeting food demands. Some see growing populations as necessitating great expansion of agricultural lands into marginal areas (and climate change may also require shifts in agricultural areas or make expansion infeasible). Others say intensification of activity, with technological improvements, will meet demand, satisfy growth in nutritional standards, and save land for ecosystem functions (but socioeconomic factors mitigating against agricultural lifestyles, growing urbanization, and competition for water may confound these prospects) (Smith et al., 2003).

Differential vulnerability

Some physical systems, by virtue of their size, location, or sensitivity to climate factors, are highly vulnerable to anticipated direction and magnitude of climate change. Among those so identified are: tropical glaciers, such as in the Andes Mountains of South America; and small lakes of interior drainage, such as in central Asia. Similarly, unique and threatened biological systems include: montane systems, where high-elevation species have nowhere higher to go; prairie wetlands and remnant native grasslands of North America, because of extensive alteration and fragmentation; coldwater fish habitat of northern latitudes; ecosystems overlying permafrost in circum-Arctic locations; ice-edge ecosystems of polar latitudes; neotropical cloud forests; coral reefs and mangroves ecosystems of tropical latitudes; and ecotones between existing ecosystems (Schneider and Sarukhan, 2001).

On a global scale some human systems are thought to be particularly vulnerable by virtue of poverty, isolation, size, or dependence on ecosystems of limited extent or diversity. Among those so identified are: small island states, especially atoll nations in the Pacific and Indian Oceans; and indigenous communities with low levels of technological development. Within developed nations the hazards literature indicates differential vulnerability between and within populations. Socially or economically disadvantaged groups – the young, the elderly, females, ethnic minorities, the poor, among other social strata – are found to be particularly at risk; that is, vulnerable. Further, vulnerability may be situational; that is, it may vary depending on the kind, character, or combination of hazards (Bohle et al., 1994; Buckle et al., 2000).

In general, "[s]ystems that are exposed to multiple pressures (synergistic effects) usually are more vulnerable to climate changes than systems that are not" (Schneider and Sarukhan, 2001, p. 91). Such pressures include current and projected demands on resources, unsustainable management, pollution, and fragmentation of natural areas.

Has climate made a difference?

There is a large literature of long standing about impacts of and adaptations to extreme weather events. Extreme weather, however, does not constitute climate, although scenarios of future climate suggest that present-day extremes may be more frequent (Albritton and Meira Filho, 2001). Assessments of vulnerability to recent climate variability are hampered by the limited range of climatic experience in the recent record. While responses to events such as those anticipated in the future – drought, say – may provide some indication of impacts and adaptations, we just have not had the magnitude and duration of changes that scenarios suggest might be in our future. However, the IPCC reviews suggested that many ecosystem changes that had been reported during the twentieth century were either associated with observed climatic changes, or in the directions expected as responses to directions of climate change (Schneider and Sarukhan, 2001; Smith et al., 2001).

Because climate change, system vulnerability, and adaptation involve changes in natural and social systems over time, as well as the synergistic and cumulative effects of multiple events, there is a need to aggregate over time and to accumulate over occurrences. There is a smaller body of literature that looks to the historical record for evidence of impacts of climate on society (Wigley et al., 1981; Rotberg and Rabb, 1981; Jones et al., 2001). Some papers in these sources concern themselves with the search for appropriate methodologies. How are we to know that social changes over historical time are the results of impacts by and adjustments to climate, apart from social, political, and economic developments? One methodological example is Parry's (1978) multi-stage "postdictive" approach: climate sensitivity of agricultural production was modeled by relating crop yields to various climatic factors; a history of climate was produced from the paleobotanical and other records, independent of the record of human activity; impacts on agriculture were postdicted by applying the sensitivity analysis to the climate record; the reconstructed record of impacts was interpreted in terms of implications for human settlement at the margins of sustainable agriculture; and the historical record of settlement was then compared to the postdicted impacts for verification.

The historical and archeological evidence suggests that societies, especially those at the margins of sustainability, have exhibited vulnerabilities to climate variability, yet have also learned to adapt in some ways (among the adaptations selected by pre-modern societies has been abandonment of inhospitable areas; cf. Burton et al., 1983, for the categorization of choices available to society as adaptations to hazards). More research using historical materials, with refined methods based on the above postdiction model, should combine with the more quantitatively detailed modeling of interactions from recent records, to produce a fuller understanding of vulnerabilities.

Paul Kay

Bibliography

Adger, W.N., 2003. Building resilience to promote sustainability. *IHDP Newsletter*, 02/2003, 1.

Albritton, D.L., and Meira Filho, L.G., 2001. Technical summary. In Houghton, J.T., et al., eds., *Climate Change 2001: The Scientific Basis*. Cambridge: Cambridge University Press, pp. 21–83.

Bohle, H., Downing, T., and Watt, M., 1994. Climate change and social vulnerability. *Global Environmental Change*, **4**: 37–48.

Brooks, C.E.P., 1970 (1926). *Climate Through the Ages: A Study of the Climatic Factors and Their Variations*. New York: Dover.

Buckle, P., Mars, G., and Smale, S., 2000. New approaches to assessing vulnerability and resilience. *Australian Journal of Emergency Management*, **15**(2): 8–14.

Burton, I., Kates, R.W., and White, G.F., 1993. *The Environment as Hazard*, 2nd edn. New York: Guilford Press.

Houghton, J.T., Ding, Y., Griggs, D.J., et al., eds., 2001. *Climate Change 2001: The Scientific Basis*. Contribution of Working Group I to the Third Assessment Report of the Intergovernmental Panel on Climate Change. Cambridge, UK: Cambridge University Press.

Jones, P.D., Ogilvie, A.E.J., Davies, T.D., and Briffa, K.R., eds., 2001. *History and Climate: Memories of the Future?* New York: Kluwer/Plenum.

Kates, R.W. 1986. The interaction of climate and society. In Kates, R.W., Ausubel, J.H., and Berberian, M., eds., *Climate Impact Assessment*. Chichester: John Wiley, pp. 3–36.

Lawson, M.P., 1974. *The Climate of the Great American Desert: Reconstruction of the Climate of Western Interior United States*. Lincoln, NE: University of Nebraska Press.

Leary, N., 2003. AIACC: Contributing to a Second Generation of Climate Change Assessments. AIACC Reports and Publications. http://www.aiaccproject.org/publications_reports/Pub_Reports.html

McCarthy, J.J., Canzlani, O.F., Leary, N.A., Dokken, D.J., and White, K.S., eds., 2001. *Climate Change 2001: Impacts, Adaptation, and Vulnerability*. Contribution of Working Group II to the Third Assessment Report of the Intergovernmental Panel on Climate Change. Cambridge: Cambridge University Press.

Parry, M.L., 1978. *Climatic Change, Agriculture and Settlement*. Hamden, CT: Archon Books.

Resilience Alliance. 2002. What is Resilience? http://www.resalliance.org/ev.php

Rotberg, R.I., and Rabb, T.K., eds., 1981. *Climate and History: Studies in Interdisciplinary History*. Princeton, NJ: Princeton University Press.

Schneider, S., and Sarukhan, J., 2001. Overview of impacts, adaptation, and vulnerability to climate change. In McCarthy, J.J., Canzlani, O.F., Leary, N.A., Dokken, D.J., and White, K.S. eds., *Climate Change 2001: Impacts, Adaptation, and Vulnerability*. Cambridge: Cambridge University Press, pp. 913–967.

Smith, J.B., Schellnhuber, H.-J., and Mirza, M.M.Q., 2001. Vulnerability to climate change and reasons for concern: a synthesis. In McCarthy, J.J., Canzlani, O.F., Leary, N.A., Dokken, D.J., and White, K.S., eds., *Climate Change 2001: Impacts, Adaptation, and Vulnerability*. Cambridge: Cambridge University Press, pp. 75–103.

Watson, R.T., Zinyowera, M.C., and Moss, R.H., eds., 1995. *Climate Change 1995: Impacts, Adaptations and Mitigation of Climate Change: Scientific–Technical Analyses*. Cambridge: Cambridge University Press.
Wigley, T.M.L., Ingram, M.J., and Farmer, G., eds., 1981. *Climate and History: Studies in Past Climates and their Impact on Man*. Cambridge: Cambridge University Press.

Cross-references

Agroclimatology
Archeoclimatology
Climate Change Impacts: Potential Environmental and Societal Consequences
Desertification
Greenhouse Effect and Greenhouse Gases
Models, Climatic

CLIMATE ZONES

A climate zone is a world area or region distinguished from a neighbor by a major physical climatic characteristic that is of global scale. Climate zones are bounded by limits that parallel lines of latitude to form "belts" that mostly extend around the globe; the word "zone" is actually derived from the Greek word meaning belt, and it is from classical Greek times that the concept of zones is derived.

History of the zonal concept

Beginning in the sixth century BC, Greek scholars identified zones of the Earth based upon astronomical knowledge and position of the overhead sun as related to changes in length of daylight. These illumination zones were called *klimata* (a word that is the origin of our word climate) and differentiated (1) a torrid zone (where the noonday sun was never far from overhead), (2) a temperate zone, and (3) a frigid polar zone. It was in the fourth century BC that Aristotle identified the parallels that gave the identified zones actual boundaries. He named the Tropics of Cancer (23½°N), and Capricorn (23½°S), and the Arctic (66½°N) and Antarctic (66½°S) Circles as lines separating the zones. Aristotle, however, also adhered to the viewpoint that the Torrid Zone was too hot and the Polar Zone too frigid for human habitation.

Claudius Ptolemy (AD 100–170) made use of earlier Greek ideas and, as part of his extensive geographic writings, divided the Earth into seven zones based upon length of daylight.

Ptolemy's zones of the Earth remained in effect for many centuries. Although their utility was questioned by such Arabic scholars as Ibn Hauqal in the ninth century, and modified by Idrisi in the twelfth century, they still formed the basis of the highly popular work of the Dutch Renaissance geographer Bernhardus Varenius (1622–1650). He presented a system of geography in which a table of zones relating day length at the summer solstice to latitude was titled *climata,* climates to the English-speaking world.

Temperature zones

Edmond Halley was perhaps the first to suggest that heat rather than day-length hours provided a more meaningful way to identify climatic zones; in his 1686 work he came to the conclusion that winds were caused by solar heating. Halley's suggestion to identify zones by temperature could not be undertaken until actual temperature data for world areas became available. This did not occur until 1817 when Alexander von Humboldt used observed data to draw the first isothermal map. By 1848 H.W. Dove was able to prepare maps of average temperatures of individual months. A new zonality was born; that based upon heat rather than day length.

The first modern map of climatic zones was devised by A. Supan in 1879. The division was based upon mean annual temperature and the temperature of the warmest month (Table C13a). Shortly after, in 1884, Wladimir Köppen (whose climatic classification remains widely used today) gave a detailed listing of temperature zones based upon the number of months above or below selected thresholds. As Table C13b shows, this introduces the concept of duration of a given climatic factor, quite different from the use of mean annual values.

Numerous other authors suggested alternate temperature zones of the Earth, but all show a similarity in that hot,

Table C13 Temperature zones: selected examples

(a) Supan's 1879 thermal zones

	Annual mean temperature		Mean temperature of warmest month	
	°C	°F	°C	°F
Hot	Over 20	Over 68	Over 10	Over 50
Temperate	Under 20	Under 68	Over 10	Over 50
Cold	Under 20	Under 68	Under 10	Under 50

(b) Köppen's 1884 thermal zones

	Duration in months of critical temperatures		
	Above 20°C (68°F)	10–20°C (50–68°F)	Below 10°C (below 50°F)
Tropical	12	—	—
Subtropical	4–11	1–8	—
Temperate	Less than 4	4–12	Less than 4
Cold	—	1–4	8–11
Polar	—	—	12

(c) Herbertson's 1905 thermal zones and economic belts

Economic belt	Temperature limits (°C)
No Crops	<10
Very few crops	0–10
Temperate Climate Crops	10–20
Tropical Crops	<20

(d) Miller's 1951 thermal zones

	Number of months 43°F or warmer
Warm temperate	12
Cool temperate	6–12
Cold	3–5
Arctic	1–2
Polar	0

temperate and cold zones are identified. In some instances, as in the case of Herbertson's 1905 thermal regions, the identified zones were used to delineate other variables, including economic zones as shown in Table C13c. As part of the development of climatic zones, many new concepts were introduced. For example, in 1951 A.A. Miller used the idea of month–degree and accumulated temperature. This identifies the length of the growing season for broadleaf trees and extends the relationships between zonal climates and vegetation growth. The idea of relating climatic zones to vegetation zones was not new; Köppen, for example, had relied heavily upon the work of plant geographers in devising his climatic zones and classification.

While the identification of thermal zonation initially relied upon the new availability of data, it soon became apparent that the zone designated "temperate" contained some of the most extreme conditions on Earth, and was in fact highly "intemperate" in terms of temperatures. This led to a number of variations in the way in which zonal maps were prepared. For example, Figure C23, taken from a 1901 publication by Dryer, is based upon the following rationale: "The tropics and polar circles do not divide the face of the earth into zones of temperature but of insolation. The true temperature zones are bounded by isotherms. By drawing isotherms of 70° and 30° for January and July upon one map, we obtain a set of zones which are not shifting but fixed, and reveal in striking manner the temperature conditions of the globe." Despite attempts to provide a more meaningful delineation of the temperate zone, it is now more appropriate to call the area between the tropics and the Arctic and Antarctic circles the mid- or middle-latitude climatic zone when dealing with anything other than the Greek concept of temperate.

Circulation zones

The availability of actual climatic data, that permitted the construction of thermometric zones in the late nineteenth century, also permitted identification of other zonal concepts, especially those based upon atmospheric circulation. In 1735 Hadley had postulated a simple atmospheric circulation and later the general motion he identified became known as a Hadley Cell. The Hadley Cell was incorporated into models of global circulation by, for example, Maury (1855), Ferrell (1856 and 1889), and Thompson (1857). While in such representations the use of the term "zone" is limited, the term "belt" is used and, essentially, each of the other circulation characteristics is a zone; thus, the Horse Latitude calms, the Trade Wind zone, and the Westerly Wind zone are zonal concepts based upon wind patterns.

The simplicity of the zonal concept of circulation, while attractive, had great limitations. During the twentieth century, meteorologists and climatologists have striven for more realism in model construction and the broad zones identified by the three-cell zonal concept proved unacceptable. New ideas, in-depth studies and, in some cases, actual hostility to the zonal concept demoted its representation to highly general introductory books about the atmosphere. In meteorology today, the term "zonal" invariably implies a zonal circulation type in

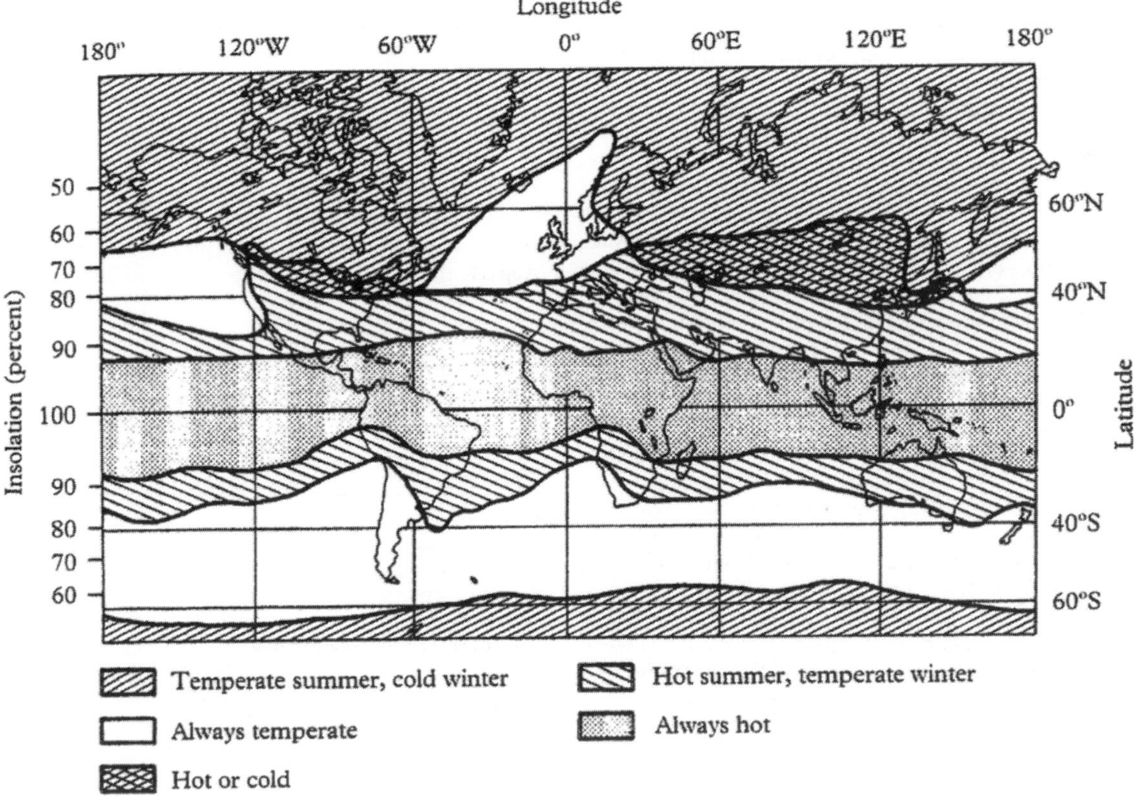

Figure C23 Redrawn from a 1901 map by Dryer, the map represents an earlier attempt to provide meaningful limits to temperature zones using winter and summer isotherms.

Table C14 de Candolle's (1874) plant zones

Symbol	Physiologic plant group	Temperature limit (°F)
*	Megistotherms	Over 86
A	Megatherms	68–86
B[a]	Xerophiles	
C	Mesotherms	59–68
D	Microtherms	32–59
E	Hekistotherms	Below 32

* Found only in high temperatures of geologic past or in hot springs.
[a] Plants adapted to dry conditions.

which flows, especially those in the upper air westerlies, were essentially in an east–west pattern. Nonzonal flow, or meridional, indicates a more significant north–south component.

Related zonal systems

The nineteenth-century development of climatology was strongly influenced by botanists. Köppen's boundaries for climatic zones were initially based upon vegetation zones identified by de Candolle, who had published his work in 1874. As Table C14 indicates, his physiologic plant groups are based upon the same temperature limits as those used by Köppen in his classification of climates.

The concept of vegetation zonation was, at least in part, emphasized in the monumental work of the German botanist Schimper, published in 1898. Schimper noted that certain plant orders and families are limited by high or low temperature limits, as found in the tropics and polar realms. As a result, the plant cover of the Earth is arranged in more or less parallel zones of different systematic characteristics. He further noted, however, that temperature zones cannot be equated to floristic and ecological characteristics, and these are identified regionally rather than zonally.

A similar zonal relationship exists in relation to the identification of soils. The Russian soil scientist Dokuchaev is credited, in a 1898 publication, with identifying soil zones of the northern hemisphere. These are (a) the Boreal or Tundra zone, (b) the Taiga or Forest Zone, (c) the Black-Earth Zone, (d) the Dry Subtropical Zone and (e) the Red-earth or Laterite Zone. This organization of hemispheric soil types resulted in a much-used soil classification based upon climatic processes. It is only in recent years, with the production of the soil taxonomy system, that it has been replaced.

John E. Oliver

Bibliography

Bailey, H.P., 1964. Toward a unified concept of the temperate zone. *Geographical Review*, **54**: 516–545.
Crowe, P.R., 1971. *Concepts in Climatology*. London: Longman Group
Dryer, C.R., 1901. *Lessons in Physical Geography*. New York.
Gentilli, J., 1958. *A Geography of Climate*. Perth.
Oliver, J.E., 1991. History, status and future of climatic classification. *Physical Geography*, **12**: 231–251.
Peattie, R., 1932. *New College Geography*. Boston, MA: Ginn and Co.

Cross-references

Atmospheric Circulation, Global
Climatic Classification
Climatology, History of
Doldrums
Horse Latitudes
Temperature Distribution
Zonal Index

CLIMATOLOGY

Climatology is the science of climate, which in turn is the composite of all weather events. It should be noted at the outset that climate fluctuates on all time scales: monthly, yearly, decadally, centennially, and millennially. Thus, climate is a statistical collective. It has often been described in terms of mean values of particular climatic elements, but it encompasses a wide range of values, including occasional extremes. Although atmospheric conditions and variations follow well-known physical laws, the day-to-day variations and those for longer intervals can, for many purposes, be treated like quasirandom variables. They can be represented by statistical distributions applicable to stochastic universes.

Climatology has multiple aims. In its historical development first efforts were directed toward a geographical inventory of climates, which led to attempts at climate classification. As a physical science, developments led to theories of the causation of global as well as regional and local climates. Because of the influence of climate on human activities, much effort has been devoted to applying the knowledge gained about climate for human welfare, including attempts to develop schemes for prediction of future climatic conditions. A major motivation for this has been to assess the potential for anthropogenic alterations of climate.

Areas of study

The wide range of the physical controls and the impacts of climate have led to the development of specialized realms of study in climatology. The major realms are considered here.

Physical climatology

The incorporation of physical principles in explaining climate is examined under the title of physical climatology, a subject area that deals with mass and energy exchanges at or close to the surface of the Earth. The focus is on energy and water balances and budgets and upon boundary layer studies.

Energy budget studies concern the availability of solar energy at a location, the transformations of that energy and its transfer through the atmosphere. Because the amount of solar radiation striking the Earth varies over both time and space, and because the Earth's surface is made up of materials of varying properties, then the physical climatology of any particular surface is unique. The patterns and exchanges are complicated by the special physical attributes of water and the exchanges of energy when water changes from one form to another.

These energy and water exchanges are dealt with in terms of boundary layer studies. Differing surfaces of the Earth result in highly variable boundary characteristics and hence different types of energy/mass exchanges and eventually different climates. Studies concern natural surfaces as well as those modified through human activities. The most modified type of surface occurs in cities; as a result, urban climatology has become an important component of the study of physical climatology.

Dynamic and synoptic climatology

The term dynamic climatology was first used by Tor Bergeron in 1929 when he outlined how the concept of airmasses and fronts might be adapted to develop a comprehensive dynamic climatology. Since then, the approach used by dynamic climatologists has evolved into studies based upon atmospheric motion characteristics and the thermodynamic processes that produce them. Of major importance today is modeling the dynamics of selected systems.

Most modeling is used to study the reaction of climate to forced changes in various physical parameters; these are the sensitivity studies. The most popular of these at the present time are models related to the increase of greenhouse gases and the attendant change of global temperatures. At the same time important studies in dynamic climatology concern such topics as sea surface temperatures and the role of mountains in determining dynamics of the atmosphere.

Synoptic climatology may be defined as the study of climate from the viewpoint of atmospheric circulation with emphasis upon the connections between circulation patterns and climatic differences. The term was first used in the 1940s when military services became concerned with the climatology of weather types and surface/air transport. While dynamic climatology is often global in scope, synoptic largely concerns hemispheric and local climatologies. In relation to the former, hemispheric teleconnections, such as ENSO events, have been investigated; at the local scale many types of studies, such as the relationships between circulation features and severe weather, are of interest.

Regional climatology

As its name implies, this concerns the climate of regions. The size of the region can vary but is often synonymous with identified geographic regions; the Great Plains, the Paris Basin, Southern China, providing apt examples. The manner in which the climatology is presented can range from highly descriptive, such as in the classic climatography of the early twentieth century, or can be based purely upon derived climatic indices, such as in component analysis.

To compare the climate of regions and to derive analog climates, regional climatologies often use a climatic classification. Earliest classifications were often based upon a single variable, such as temperature or daylight hours. By the nineteenth century increased data availability permitted a more quantitative approach and the use of more than one variable. Vegetation distribution strongly influenced the development of climatic regions in classification systems, with that devised by Wladimir Köppen proving the most durable. The Köppen approach provides a unique shorthand system based upon numeric rules for classifying climates, and the system remains in wide use today. Many other climatic classification systems have been devised, most for special-purpose uses such as human comfort or crop distribution. Such systems, based upon observed data, are termed empiric classifications. Regionalization by cause is dealt with in genetic classification systems, often based upon airmass climatology. To demonstrate the nature of classification, Table C15 shows a 13-category scheme with some prototypes shown as examples.

Applied climatology

Applications of climatology have risen rapidly in recent years. Aside from agriculture, the construction industry makes major use of climatic data. Temperature, wind, and humidity statistics permit rational design of buildings and dwellings, as well as their heating and cooling plants. Siting of factories and power plants requires analysis of wind and vertical atmospheric stability parameters to judge the pollution potential of effluents. In the case of nuclear power plants, climatological analysis is mandated. For major structures such as bridges, broadcast towers, and electric transmission towers, data on extreme wind speeds, low-level icing, and exposure to lightning are needed, although they must often be inferred from locations other than the building site. The transference of climatic information from an observing site to a "silent" area has become an important part of applied climatology. A combination of the use of interpolated data from a climatological observing network and site-specific interpretations from a series of synoptic weather maps is an applicable technique.

Climatological information also enters into the housing problems of farm animals. Climate affects the design of clothing and its seasonal distribution and marketing. The effect of climatic conditions has had a major impact on transportation and packaging and storage of goods, especially foodstuffs.

A major role for combined climatic–synoptic analysis is in the planning of dam construction, both for water-resource conservation and for flood control. Floods still remain a major hazard and their probability of occurrence has entered into calculations for setting insurance premium rates. Estimates of extreme values of precipitation leading to floods, as well as for extreme values of other damaging weather events, has resulted in the use of a variety of statistical techniques. These are attempts to simulate the frequency distribution of past rare events and high values. There are a number of such extreme value distributions. Such extreme value probability analyses can be used for predictions without date. They can inform a planner or construction engineer as to what extreme value of a particular climatic element can be expected at least once in 50 or 100 years. Such information, while not indicating when this value will occur, will at least permit the necessary safety factors to be included in the design. Aside from wind speeds, maximum precipitation amounts, and extreme temperatures, this type of analysis is also particularly useful for estimates of high snow loads, especially on places of public assembly.

The causation of climate

In regard to the causation of climate, the principal factor is the astronomical position of the Earth with respect to the sun, which is the source of all energy. The distance between the Earth and the sun governs the total energy received from the sun. There are seasonal differences in that distance and

Table C15 Basic climatic elements at selected stations

Climatic category	Location	Temperature (°C) W	C	Y	Mean precipitation (cm) J	F	M	A	M	J	J	A	S	O	N	D	Y
Glaciated	Ice cap 77°56′W Greenland	−7	−35	−23													<10
	South Pole, Antarctica	−28	−59	−49													<15
	Barrow, Alaska	4.3	−27.7	−12.2	0.41	0.38	0.31	0.25	0.33	0.71	2.11	2.03	1.40	1.32	0.69	0.51	10.4
Tundra	Ostrov Dikson, USSR	4.8	−27.5	−12.3	2.0	1.3	1.7	0.9	1.1	2.3	3.2	4.6	4.2	2.1	1.4	1.8	26.6
	Nome, Alaska	9.7	−15.3	−3.3	2.6	2.4	2.2	2.0	1.8	2.4	5.8	9.7	6.8	4.3	2.9	2.5	45.4
Short summer with frost	Salekhard, USSR	13.8	−24.4	−6.7	2.4	2.0	2.4	3.2	3.9	5.1	5.7	5.7	5.4	4.6	3.1	2.9	46.4
	Bismarck, North Dakota	22.3	−12.8	5.4	1.1	1.1	2.0	3.1	5.0	8.6	5.6	4.4	3.0	2.2	1.5	0.9	38.5
Cold winter, adequate precipitation	Moscow, USSR	17.8	−10.3	3.6	3.1	2.8	3.3	3.5	5.2	6.7	7.4	7.4	5.8	5.1	3.6	3.6	57.5
	Cincinnati, Ohio	24.9	0.9	12.9	9.3	7.1	9.9	9.2	9.7	10.6	9.1	8.3	6.9	5.7	7.5	7.0	100.3
Distinct winter, moist	Lyon, France	20.7	2.1	11.4	5.2	4.6	5.3	5.6	6.9	8.5	5.6	8.9	9.3	7.7	8.0	5.7	81.3
	San Francisco, California	16.7	10.4	13.8	11.6	9.3	7.4	3.7	1.6	0.4	0.1	0.1	0.6	2.3	5.1	10.8	52.9
Mild, winter rain	Rome, Italy	24.7	6.9	15.6	7.6	8.8	7.7	7.2	6.3	4.8	1.4	2.1	7.0	12.8	11.6	10.6	88.1
	Elko, Nevada	20.9	−5.2	7.4	2.9	2.3	2.1	2.1	2.4	1.8	1.0	0.8	0.9	1.9	2.3	2.6	23.1
Cool, arid	Semipalatinsk, USSR	22.1	−16.2	3.1	1.4	1.5	1.7	1.9	2.2	3.0	3.2	2.3	2.1	2.2	2.7	2.2	26.4
	Singapore, Malaya	24.7	22.9	23.9	25.2	17.5	20.0	19.6	17.4	17.1	16.7	19.1	17.9	20.8	25.1	26.6	243.0
Hot, rainy	Mlänje, Zambia	24.5	17.5	22.0	31.3	31.9	30.8	12.5	5.9	6.4	3.8	3.3	3.0	5.8	18.3	26.8	179.6
	Barra do Corda, Brazil	27.4	24.2	25.8	19.0	20.8	21.4	14.4	6.0	1.6	0.7	0.7	2.3	4.1	7.0	11.7	109.7
Hot, seasonal rain	Kinshasa, Zaire	27.0	17.0	26.5	32.1	13.9	18.1	20.9	13.4	0.5	0.0	0.4	3.3	13.7	23.6	17.1	137.8
	Salta, Argentina	21.5	10.6	16.8	17.6	14.9	9.4	2.5	0.6	0.3	0.2	0.4	0.5	2.5	6.1	12.1	67.1
Warm, seasonal rain	Windhoek, Namibia	23.5	13.0	19.0	7.7	7.3	8.1	3.8	0.6	0.1	0.1	0.0	0.1	1.2	3.3	4.7	37.0
	Guaymas, Mexico	31.2	17.9	25.0	1.2	0.7	0.3	0.1	0.0	0.3	3.9	6.0	5.1	2.2	0.7	1.7	22.1
Warm, semiarid	Karachi, Pakistan	31.4	17.7	25.9	0.7	1.3	0.5	0.2	0.1	0.9	10.1	4.7	2.3	0.8	0.3	0.6	22.1
	Yuma, Arizona	35.3	13.0	23.2	1.0	0.7	0.5	0.3	0.1	0.0	0.5	1.1	0.6	0.7	0.6	0.9	6.8
Warm desert	Cairo, Egypt	29.0	14.0	22.0	0.4	0.3	0.5	0.2	0.2	0.1	0.0	0.1	0.1	0.2	0.5	0.1	2.7
	Mount Washington, New Hampshire	9.3	−15.0	4.6	13.0	16.6	14.2	13.9	14.8	16.5	17.2	18.3	16.2	15.5	19.5	17.9	193.5
Mountain type	Sonnblick, Austria	1.6	−13.2	−6.0	11.5	10.8	11.2	15.3	13.6	14.2	15.4	13.4	10.4	11.8	10.8	11.1	149.5

Notes: W = mean temperature of warmest month; C = mean temperature of coldest month; Y = mean temperature of the year.
* Precipitation not measurable because of drifting snow.

long-term changes in the eccentricity of the Earth's orbit. Of equal importance is the inclination of the Earth's axis with respect to the ecliptic. This too is subject to cyclical changes; it is presently 23.5°. This inclination causes the seasons, with the poles during the respective summer seasons of the hemispheres inclined toward the sun and away from the sun in winter. Thus the polar regions change from complete daylight to complete darkness during the course of the yearly revolution of the Earth around the sun. The axis tilt therefore governs the amount of solar radiation as the seasons change. This fact was well known in ancient times and is reflected in the Greek word *klima*, which means slope.

The cyclical elements in the astronomical position of the Earth – namely, the *precession* of the spring point (~21 000 years), the tilt of the Earth's axis (~41 000 years), and the change of the *eccentricity* of the Earth's path (~97 000 years) – lead to different radiation conditions at various latitudes. Determination of these elements has led to formulation of a model of climatic changes by Milutin Milankovitch. The model permits calculation of radiation energy received as a time series prior to the present. For the past half million years, during which continents and oceans have been in their present position, the radiation curve coincides with the glaciations and interglacials derived from geologic and isotopic evidence. Whether or not the model is adequate to explain the initiation of glaciation remains controversial. Evidence for changes in solar energy output exists, but quantitative measurements have been available only in recent years from satellites; prior to that, sunspot observations were used as corollaries to climatic fluctuations since the eighteenth century, but statistical tests indicate that less than 10% of the variance can be accounted for, both for temperature and precipitation.

If the sun ceased radiating energy to the Earth, it has been estimated that atmospheric motions would cease in about 2 weeks. Satellite determinations of the energy received from the sun at the boundary of the atmosphere gives values between ~1360 and ~1378 $W m^{-2}$. Partitioned over a spherical surface this fixes the solar energy amount entering the atmosphere at ~340–344 $W m^{-2}$.

As solar radiation enters the atmosphere and penetrates to the surface, a complex system of interaction comes into play. Atmospheric gases, clouds, and conditions of the surface partition the solar radiative flux. At the surface reflectivity plays a key role; it varies from the high reflectivity of areas covered with snow and ice to intermediate reflectivity (albedo) of desert and vegetated land areas, and low albedos of water surfaces. The high albedo of the frozen surfaces contributes to their maintenance. The low albedo of water leads to large heat absorption in the oceans, which become the principal secondary energy source of the atmosphere. These effects have led to the designation of climate as an atmospheric–oceanic–cryospheric system. In the energy exchanges the interception of outgoing long-wave radiation from the Earth by atmospheric gases, principally water vapor and carbon dioxide, keeps the Earth at temperatures favorable for biota. The depletion of the solar radiation in various spectral regions is shown in Figure C24, which also shows the wavelengths in which radiation escapes from the Earth to space. The atmospheric gases are nearly opaque to the longwave infrared radiation from the Earth, except for a few "windows" in the spectral absorption of water vapor and carbon dioxide. It is these few escaping infrared fluxes that make satellite scanning of atmospheric and surface temperatures possible.

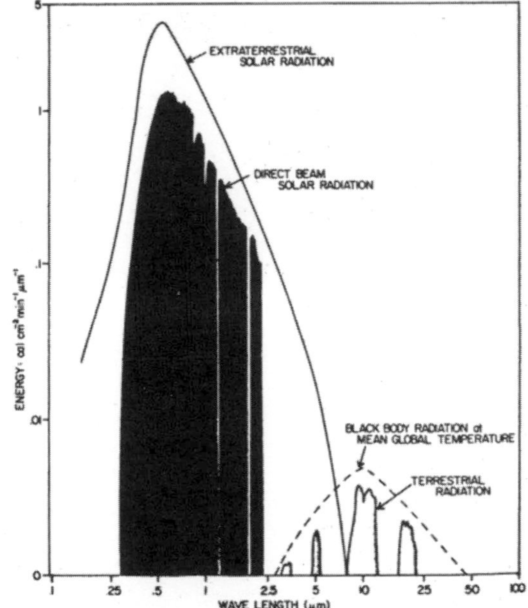

Figure C24 Spectral distribution of solar beam (left); shaded area showing flux reaching the Earth's surface, depletion caused by absorbing gases; terrestrial radiation to space (right); shaded areas showing the only fluxes escaping through the so-called atmospheric "windows" in the water vapor and carbon dioxide absorption bands.

A relative partitioning of the various radiative fluxes of Earth and atmosphere (in percent of the incoming solar radiation) are shown in Figure C25. This diagram also shows clearly that more radiative energy transactions take place in the atmosphere than at the surface. Radiation is the primary element in the climate-forming processes. For this reason energy balance models have gained a dominant place in modern climatology. They sum up all energy fluxes of an area or locality and incorporate incoming solar radiation, albedo, longwave outgoing radiation, as well as the terrestrial processes of evaporation, which adds latent heat to the atmosphere, and sensible heat transfers. On a hemispheric scale, some of these components are shown in Figure C26 for hemispheric summer and winter as zonal averages. High-energy incomes are indicated for summer, but no solar energy is received in polar regions during winter. The net energy income is positive for the whole summer hemisphere but negative for most extratropical regions in winter. The latent heat values stay at the high level throughout, and hence play a role as a major heat source in the atmosphere.

Global circulation

Satellite observations have given better insight into the net radiation balance of various parts of the globe. Not only the areas of high latitudes show net heat loss, but large areas at high altitude in Asia as well. Some of the North African and Arabian desert regions show radiation balance deficits. The tropics and major ocean areas are the areas of positive

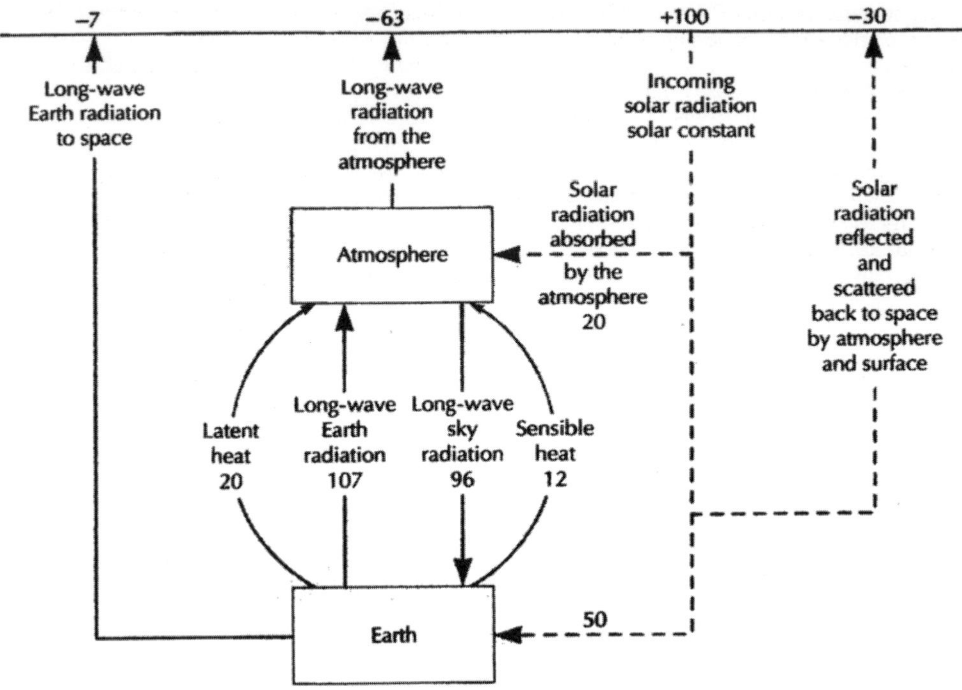

Figure C25 Radiative fluxes in and through the Earth's atmosphere, in percent of the solar radiation received at the boundary of the atmosphere.

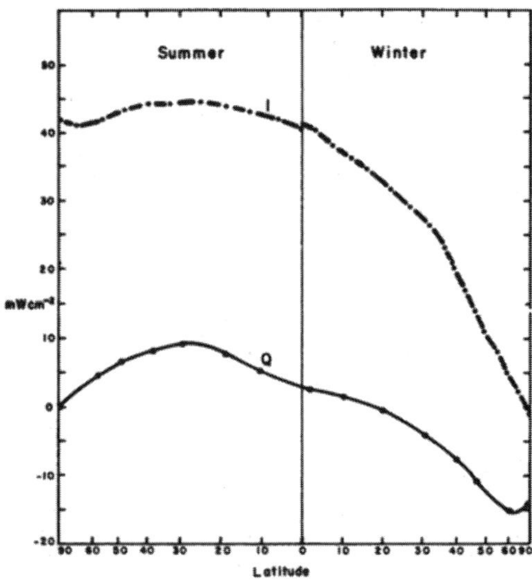

Figure C26 Zonal distribution of incoming solar radiation flux I and net radiation (gain or loss) Q, for hemispheric summer and winter.

radiation balance. These differences are the driving force of the global circulation in atmosphere and oceans. Changing air pressures transport heat from the areas of surplus to those of deficit and try to smooth out the differences. Radiation imbalance is the principal ingredient of climate. About half of the heat transfer from low to high latitudes takes place in the ocean. Oceanic currents are very important in climatogenesis of many areas of the globe. In the atmosphere the heat contrasts lead to a multicellular circulation, which follows the principles of fluid dynamics as modified by the rotation of the Earth. This leads to the existence of three meridional circulation cells in each hemisphere, causing different dynamic processes. Between the equator and the tropics, or subtropics, is a direct circulation of air from warmer to cooler areas, the so-called Hadley cell, with low pressure near the equator and high pressure in the subtropics. The resulting wind system produces the trade winds over the oceans, northeasterly in the northern hemisphere, southeasterly in the southern hemisphere. Near the equator in an area of weak winds – the doldrums – the trade winds create a convergence zone with cloud bands and convective precipitation. This intertropical convergence zone migrates somewhat, following the seasonal changes.

In the polar regions another direct circulation cell exists. The cold air forming there breaks out in irregular bursts toward the warmer areas. This is delineated by a notable discontinuity, the polar front, toward the lower latitudes. Its thermodynamic interaction with the warmer air forming in the lower latitudes, usually generously endowed with latent heat from evaporated water vapor, causes migrating pressure systems in an indirect cell, the Ferrel cell. These bring various midlatitude areas under rapidly changing weather regimes with alternations of warm and cold airmasses and wet and dry spells. A schematic representation of the atmospheric circulation is shown in Figure C27, which shows another globally important circulation that is zonal in character. This is imposed over the other circulations in the tropical area and is rather weak and slow, but it has important temporal effects on both atmospheric and oceanic currents. This so-called Walker circulation affects the intertropical convergence zone and the monsoonal circulations.

Although the general atmospheric and oceanic circulations affect the broad-scale distributions of climate over the globe, the fine fabric is created by the differentiation of the surface. Mountains and their orientation with respect to the general circulation are extremely influential. Windward and leeward sides have radically different climates. Proximity to water surfaces, both oceanic and inland, are other factors causing climatic differences. The presence or absence of vegetation and the land use by humans also cause notable climatic differences. Anthropogenic effluents are the cause of changes in atmospheric composition. Their local influences have been explored for decades, but there are potential global repercussions if the concentration of heat-absorbing gases becomes sufficiently large to cause measurable alterations of climate. An uncertain, sporadic factor in the climatic conditions of individual years, and perhaps longer intervals, are volcanic eruptions. These can cause changes in the radiation balance of the globe. Even a limited analysis of climatogenic factors shows so many feedback mechanisms that one can begin to visualize the diversified mosaic of climate in space and time. In Figure C28 some of the various elements of climatogenesis and some of the mutual interactions are reflected. The many agents having effects on climate make their exact representation in climatic models difficult, and hence these remain currently inadequate to represent the rich fabric of terrestrial climates.

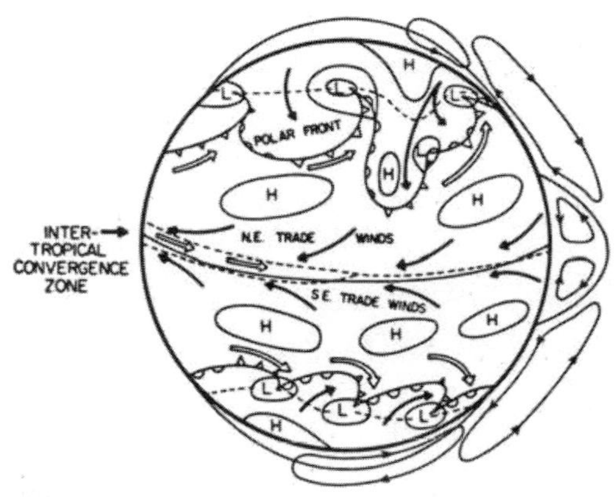

Figure C27 Schematic representation of the global atmospheric circulation showing principal frontal systems, high (H) and low (L) pressure areas and wind systems.

Global patterns

Systematic instrumental observations of the surface climate of the Earth have been carried on in most land areas for over a century, and in some localities for two centuries. In central England over 300 years of temperature observations are available. Time series of the various climatic elements such as pressure, temperature, precipitation, sunshine, cloudiness and wind show notable fluctuations from year to year. Even longer intervals can show marked differences, yet the mean value of

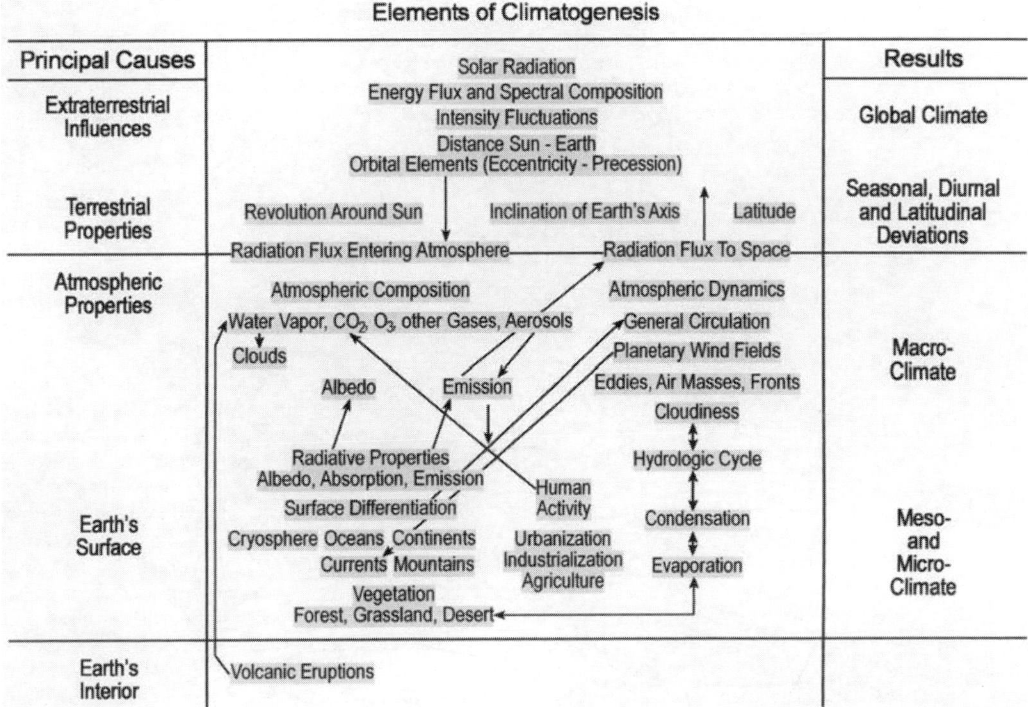

Figure C28 Principal climatogenetic factors on Earth with limited indications of feedback mechanisms.

the various elements has traditionally been used to represent climate. Although this has shortcomings, it permits broad-scale comparisons of geographical areas. The mean pressure distribution, as shown for January and July in Figures C29 and C30, shows the great differences in flow patterns between the two hemispheres caused by the greatly different distribution of land and sea. In January the two great centers of action are the Aleutian and the Icelandic low-pressure systems. These reflect the steady stream of cyclonic disturbances along the polar front bringing alternating outbreaks of cold air and intrusions of warmer subtropical air, generally accompanied by precipitation, to the middle northern latitudes. The sunny subtropical high-pressure cells have migrated southward, as has the intertropical convergence zone, which is now south of the equator. The strong Asiatic anticyclone not only coincides with intense Siberian cold, but its outflow controls the dry winter monsoon in south and east Asia. In the southern hemisphere there are some vestiges of heat lows over the continents and some inflow of moist air from the oceans, but in the great southern west wind belt the winds keep roaring at the 40th parallel.

In July the southern hemisphere has not changed much. The subtropical high-pressure cells have strengthened. Sunshine, and often drought, prevails over the continents. The Intertropical Convergence Zone has migrated north of the equator. The oceans show higher pressure than the continents and oceanic air influx to the land brings monsoonal rains. The classical monsoonal region is east Asia, especially the Indian subcontinent. Even though there are large year-to-year fluctuations, the broad general patterns have persisted.

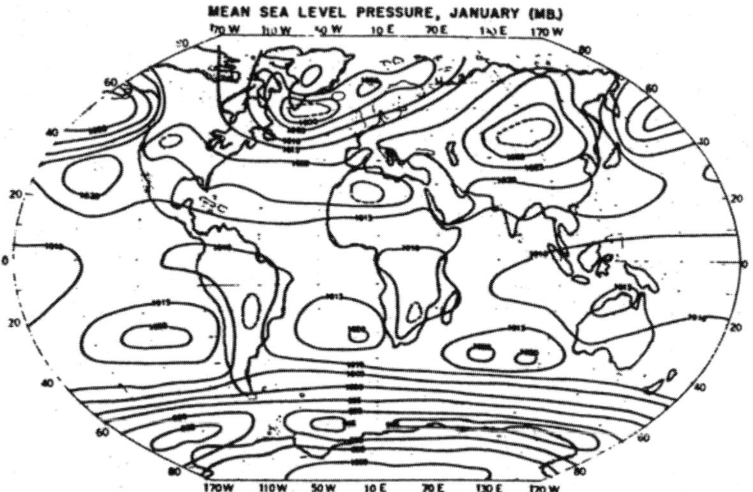

Figure C29 Mean sea-level pressure in January, in millibars (hPa).

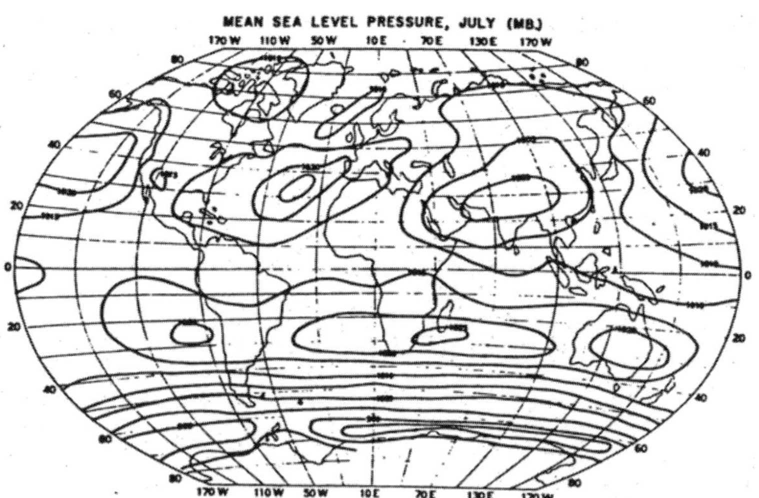

Figure C30 Mean sea-level pressure in July, in millibars (hPa).

These circulation patterns are reflected in the meridional distribution of cloudiness and precipitation (Figure C31). The cloudiness shows three major belts: the subpolar latitudes in both hemispheres and the equatorial zone. The precipitation amounts are somewhat less strikingly distributed latitudinally, but the zone just north of the equator shows a notable maximum. Equally pronounced are the subtropical dry zones where the prevalent high-pressure belts produced by the subsiding branches of the Hadley circulation create vast desert areas, especially in the northern hemisphere. But within the prevailing structure of global circulation, numerous other controls shape and modify the climate. Among these are proximity to the ocean or other large water bodies, and position with respect to mountain ranges. Oceanic currents exercise profound influences. Thus the North Atlantic Gulf stream and its continuation, the North Atlantic Drift, create a notable asymmetry of latitudinal temperature distribution between the east coast of North America and west coast of Europe. Whereas along the American Atlantic shore there is a wide annual range of temperature and cold winters prevail, the west coast of Europe far north into Scandinavia has a much smaller range and mild winters.

Other coastal climates are equally influenced by oceanic currents. Some of them are tied to the broad flow patterns of the general circulation, others are produced by more local air flow. The interaction of wind and currents can produce coastal upwelling of cold water from greater depths at sea. Such coastal cold waters not only have ecological benefits for fisheries, but also modify the regional climate. Cold currents cause coastal dry zones or even deserts and often cause high fog frequency. In perspective, the arid and semiarid areas of the globe cover 48×10^6 km^2 or 31% of the land areas, excluding the Antarctic continent. If one adds the 12% glaciated areas, which usually have very little precipitation, to the 31% of arid areas, the great contrasts of climate on a planet covered 70% by ocean becomes clear.

Precipitation is the climatic element that shows the greatest variability in time and space. Some extremes, together with those of other elements, are shown in Table C16. As the time

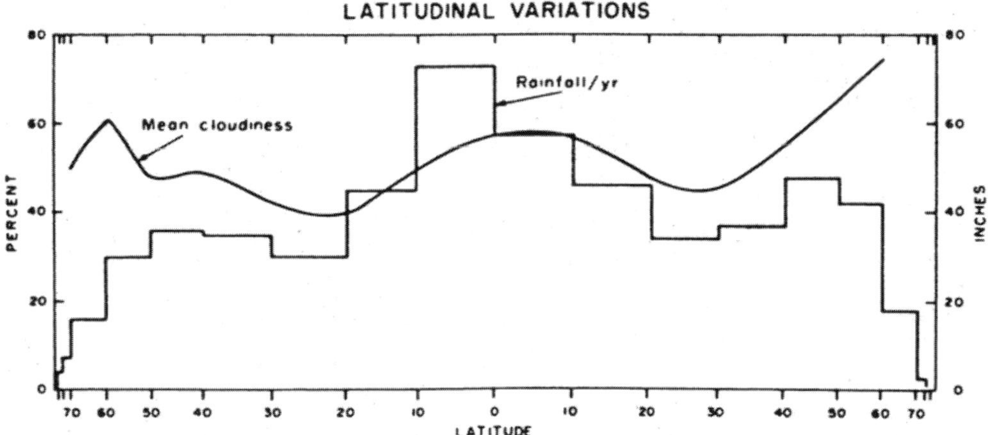

Figure C31 Zonal distribution of mean rainfall (in right scale) and mean cloudiness (percent, left scale) on Earth.

Table C16 Some extremes of climatic elements at the surface of the Earth (observed through 2000)

Element	Measurement	Globe location	Measurement	US location
Temperature (°C)				
Highest	58	Azizia, Libya	57	Death Valley, California
Lowest	−89	Vostok, Antarctica	−63	Snag, Yukon, Canada
Pressure (mb)				
Highest	1084	Agata, USSR	1068	Barrow, Alaska
Lowest	870	Typhoon Tip, Ocean	892	Matecumbe Key, Florida
Precipitation (cm)				
1 minute	3.12	Unionville, Maryland	3.12	Unionville, Maryland
24 hours	187.0	Cilaos, La Reunion, Indian Ocean	109.2	Alvin, Texas
1 month	923.0	Cherrapunji, Assam, India	271.8	Kukui, Maui, Hawaii
1 year	2647	Cherrapunji Assam, India	271.8	Kukui, Maui, Hawaii
Greatest annual mean	1168.4	Mt. Waialeale, Kauai, Hawaii	1168.4	Mt. Waialeale, Kauai, Hawaii
Windspeed (km/h)	372	Mt Washington, New Hampshire	372	Mt. Washington, New Hampshire
Longest rainless period	14 years	Arica, Chile	993 days	Bagdad, California

interval of observations and their spatial coverage increase, new extremes can be expected. It might be noted here that areas subjected to tropical cyclones are particularly apt to experience rainfall and wind extremes. Most of these storms occur in the northern hemisphere, with a peak frequency in fall. Much of their energy is derived from the latent heat of water vapor condensation, which derives from evaporation at the ocean surface, using the large energy income from the warm season. The tropical storms balance the water vapor excess in the northern hemisphere and adjust the vapor pressure to the seasonal temperatures.

Of all the features of global circulation, the seasonal winds called "monsoons" have the greatest climatic significance. About half the population of the globe is affected by monsoonal currents. These are most marked in the northern hemisphere and are most conspicuous in India and Southeast Asia. There are, however, monsoonal currents in West Africa, Australia, Western Europe and southern North America. The basic cause is the seasonal contrast of temperatures between land and adjacent oceans due to changes in radiation balance between the different types of surfaces. This creates high pressure over land and lower pressure over the sea in winter. The reverse is true in summer. Thus the winter monsoon creates outflow of air from the continent and the summer monsoon, inflow into the continent. This flow pattern results in dry winters and rainy summers. In India and Southeast Asia the rain tendency in summer is reinforced by the migration of the Intertropical Convergence Zone to almost 20°N latitude. The early spring heating of the Tibetan Plateau also plays a major role in initiating the summer monsoon. The seasonal shift in wind directions is another principal character of the monsoons, additional to the precipitation patterns (Figure C32).

Airmass frequencies are useful adjuncts to describe climatic conditions other than by numerical values of climatic elements. Prevalence of a given airmass indicates a stable, uniform climate whereas frequent airmass changes indicate a turbulent climate. Higher frequency of maritime airmasses over continental airmasses indicates a marine influence and probability of a rainy climate. If the continental types dominate, a dry climate is likely to prevail. Similarly, if the polar airmasses dominate, a cool or cold climate is indicated. High frequency of tropical airmasses will be associated with a warm climate. Seasonal changes in the frequency of various airmasses readily indicate the annual course of climate.

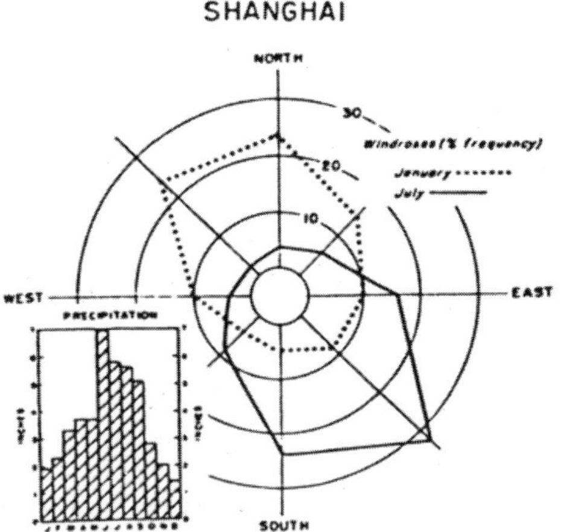

Figure C32 Wind roses for midwinter and midsummer showing the monsoonal wind shift in east Asia at Shanghai and the associated rainfall distribution (lower left).

Scale and time in climatology

It need be appreciated that climatology can be considered at almost any spatial scale and over any time period that is long enough to establish a climatic record.

Scale

The smallest area scale is considered as microclimatology, which is characterized, for example, by the climate that might occur in an individual field or around a single building. A microclimate may extend horizontally from less than 1 m to 100 m, vertically from the surface to 100 m.

A local climate comprises a number of microclimates that make up a distinctive surface such as a forest or a city. The size of a local climate may extend horizontally from 100 m to 10 000 m and vertically to 100 m. A mesoclimate may contain a number of individual landscape types but often having a similar physiographic component. Thus the neighboring states within the Great Plains may be part of the same mesoclimate, with a vertical extent ranging to 6000 m. Macroclimatology concerns the climates of continents and, eventually, global climates.

Time

Climatology considers many time scales. In terms of the reconstruction of past climates, studies vary depending upon the type of information available. Given that reliable and extensive climatic data have been available for little more than 100 years, climatologists rely upon proxy data for interpreting past climates. Table C17 provides a listing of the common types of proxy data used. The level of sophistication of interpretation of proxy data is a function of time. The more recent the event, the more information that is available. Thus, the representation of climates of, for example, the Paleozoic is much more general than that of say, the Pleistocene. When people began to observe the conditions around them, and eventually describe them in writing, proxy data had a new dimension added. The type of information used to reconstruct conditions of, for example, the Little Ice Age, illustrates how such data are used. Contributing to reconstruction of such climates are descriptions of unusual events, at least in terms of today's climate; the cited frequency of river freezings, reported crop failures, writings describing "years without summers", contemporary paintings and etchings showing wintry conditions, are some examples.

Timed predictions of climatic developments have been attempted for decades, so far with only limited success. Nearly all of the schemes are based on persistence, lag correlations, and teleconnections of climatic elements. The emphasis has been on temperature and precipitation projections for a month or season. Some skill is shown for temperatures but very little for precipitation. Relations to solar activity have proved elusive and no reliable prediction models for energy emission from the sun have so far been found. Periodicities have also been ardently searched for. There are some indications that a small amount of the variance in both temperature and precipitation that might be explained by solar fluctuations in long climatic

Table C17 Palcoclimatic data sources

Data Source	Variable measured	Potential geographical coverage	Period open to study (years BP)	Climate Inference
Ocean sediments (cores, accumulation rate of < 2 cm/ 1000 years)	Isotopic composition of planktonic fossils; benthic fossils; mineralogic composition			Sea-surface temperature, global ice volume; bottom temperature and bottom-water flux; bottom-water chemistry
Ancient soils	Soil type	Lower and midlatitudes	1 000 000	Temperature, precipitation, drainage
Marine shorelines	Coastal features, reef growth	Stable coasts, oceanic islands	400 000	Sea level, ice volume
Ocean sediments (common deep-sea cores, 2–5 cm/1000 years)	Ash and sand accumulation	Global ocean (outside red clay areas)	200 000	Wind direction
Ocean sediments (common deep-sea cores, 2–5 cm/1000 years)	Fossil plankton composition	Global ocean (outside red clay areas)	200 000	Sea-surface temperature, surface salinity, sea-ice extent
Ocean sediments (common deep-sea cores, 2–5 cm/1000 years)	Isotopic composition of planktonic fossils; benthic fossils; mineralogic composition	Global ocean (above $CaCO_3$ compensation level)	200 000	Surface temperature, global ice volume; bottom temperature and bottom-water flux; bottom-water chemistry
Layered ice cores	Oxygen-isotope concentration (long cores)	Antarctica; Greenland	100 000+	Temperature
Closed-basin lakes	Lake level	Lower and midlatitudes	50 000	Evaporation, runoff, precipitation, temperature
Mountain glaciers	Terminal positions	45°S to 70°N	50 000	Extent of mountain glaciers
Ice sheets	Terminal positions	Midlatitudes, high latitudes	25 000 to 1 000 000	Area of ice sheets
Bog or lake sediments	Pollen-type and concentration; mineralogic composition	50°S to 70°N	10 000+ to 200 000	Temperature, precipitation, soil moisture
Ocean sediments (rare cores, > 10 cm/ 1000 years)	Isotopic composition of planktonic fossils; benthic fossils; mineralogic composition	Along continental margins	10 000+	Surface temperature, global ice volume; bottom temperature and bottom-water flux; bottom-water chemistry
Layered ice cores	Oxygen-isotope concentration, thickness (short cores)	Antarctica; Greenland	10 000+	Temperature, accumulation
Layered lake sediments	Pollen type and concentration (annually layered core)	Midlatitude continents	10 000+	Temperature, precipitation, soil moisture
Tree rings	Ring width anomaly, density, isotopic composition	Midlatitude and high-latitude continents	1000 to 8000	Temperature, runoff, precipitation, soil moisture
Written records	Phenology, weather logs, sailing logs, etc.	Global	1000+	Varied
Archeological records	Varied	Global	10 000+	Varied

After various sources.

time series, is generally less than 10% of the total. In the Western United States a coherence of the extent of drought areas, as found by tree ring analysis with the double sunspot rhythm (Hale cycle), has been found; it has only limited prediction value. Similarly, the persistent quasibiennial oscillation found in many time series of climatic elements seems to wax and wane in time and space to prevent practical predictive use. The Southern Oscillation, a large-scale pressure see-saw in the South Pacific Ocean, is repetitious on a 5–7-year time scale. It influences the intensity of the Indian summer monsoon and the occasional warm water current El Niño off the Peruvian coast.

Of considerable interest to climate researchers are the identification of climatic analogs, periods of climate in the past that can be considered analogous to those of the present or of the possible future. Thus warm climates of the past, such as the Eemian interglacial (125 000–130 000 years BP) or the Holocene "climatic optimum" (5000–6000 years BP) have been used in attempts to assess future greenhouse climates. For the limited period for which observed climatic data are available, climatologists use numerous statistical methods to search for trends, periodicities or iterations. The specter of global warming, as promulgated in the 1980s, has resulted in extensive examination of observed data with attempts to standardize and validate them.

The forecasting of climate is a difficult task. At present there are two main methods of attempting to determine what lies ahead. The first uses climatic analogs which, as earlier noted, are time periods in the past that resemble conditions as they may soon be if the Earth's climate were to warm. Second, climate models are generated. Climatologists use various types of models but of particular importance at present are general circulation models (GCM). Modern GCM are a product of computers' calculating capability for GCM, attempt to solve equations of motion, thermodynamics and conservation for moisture and mass in a defined global area through various atmospheric levels. A number of models have been run, and while problems still exist, the output of such models is becoming more acceptable.

Climatic data

All of the understanding within the science of climatology must rest upon the availability of data. For climates preceding the instrumental period actual measurements are not available, and proxy data are used. Since about 1850, especially for land surfaces in the northern hemisphere, monitoring and data collection have gradually improved. Today, routine measurements of the atmosphere are accumulated for many world locations and are stored in archives. For most climatic analysis the statistics derived for a 30-year period are adequate. Unless otherwise noted, the cited average data of most locations represents the most recent 30-year block, ending in a decennial year. Thus 1951–1980 observations are replaced by those of 1961–1990 and so on.

For examination of climatic trends and periodicities, periods longer than 30 years are needed. Often, errors enter into older data from such things as change of observing location or modified instruments. Such is true in many urban locations where the development of an urban heat island causes the observed data to differ from surrounding areas. Climatologists check historic data carefully before assuming it is error-free. Despite the gains made in collection, quality control and archiving of climatic data, there should be improvements especially over the oceans and in underrepresented global areas. Climate is a basic global resource, and for managing and comprehending this resource, climatologists must have good data. Certainly, without such a base, the important task of predicting future climates becomes even more difficult.

H.E. Landsberg and J.E. Oliver

Bibliography

(The following list deals with broad segments of climatology and serve as guides to the immense periodical literature).

Aguado, E., and Burt, J.E., 2001. *Understanding Weather and Climate*. Upper Saddle River, NJ: Prentice-Hall.

Ahrens, D.L., 2000. *Meteorology Today: An Introduction to Weather, Climate, and the Environment,* 6th edn. Pacific Grove, CA: Brooks & Cole.

Akin, W.E., 1991. *Global Patterns: Climate, Vegetation, and Soils.* Norman, OK: University of Oklahoma Press.

Arya, S. Pal, 1998. *Air Pollution Meteorology and Dispersion.* New York: Oxford University Press.

Barry, R.G., and Chorley, R.J., 1998. *Atmosphere, Weather, and Climate*, 7th edn. London: Routledge.

Barry, R.G., and Carleton, A.M., 2001. *Synoptic and Dynamic Climatology.* London: Routledge.

Berger, A. (ed.), 1981. *Climatic Variations and Variability: facts and theories.* Dordrecht: Reidel.

Bluthgen, J., 1966. *Allgemeine Klimageographie,* 2nd edn. Berlin: Walter de Gruyter.

Boucher, K., 1975. *Global Climates*. New York: Halstead Press.

Brasseur, G.P., Orlando, J.J., and Tyndall, G.S. 1999. *Atmospheric Chemistry and Global Change.* New York: Oxford University Press,.

Bryant, E., 1997. *Climate Process and Change*. Cambridge: Cambridge University Press.

Budyko, M.I., 1982. *The Earth's Climate: Past and Future.* New York: Academic Press.

Burroughs, W.J., 2003, *Weather Cycles: Real or Imaginary*, 2nd edn. Cambridge: Cambridge University Press.

Conrad, V., and Pollak, L.W., 1950. *Methods in Climatology,* 2nd edn. Cambridge, MA: Harvard University Press.

Curry, J.A., and Webster, P.J., 1999. *Thermodynamics of Atmospheres and Oceans.* San Diego, CA: Academic Press.

Flohn, H., 1969. *Climate and Weather.* New York: World University Library/McGraw-Hill.

Gedzelman, S.D., 1980. *The Science and Wonders of the Atmosphere*. New York: John Wiley.

Geer, I.W., 1996, *Glossary of Weather and Climate*. Boston, MA: American Meteorological Society.

Glickman, T., 2000. *Glossary of Meteorology,* 2nd edn. Boston, MA: American Meteorological Society.

Graedel, T.E., and Crutzen, P.J., 1993. *Atmospheric Change: an earth system perspective*. New York: Freeman.

Griffiths, J.F., and Driscoll, D.M., *Survey of Climatology*. Columbus, OH: Merrill.

Hansen, J.E., and Takuhashi, T., (eds.), 1984. *Climate Processes and Climate Sensitivity*. Geophysical Monograph 29. Washington, DC: American Geophysical Union.

Hartmann, D.L., 1994. *Global Physical Climatology.* San Diego, CA: Academic Press.

Hidore, J.J., 1996. *Global Environmental Change.* Upper Saddle River, NJ: Prentice-Hall.

Holton, J., 1992. *An Introduction to Dynamic Meteorology,* 3rd edn. New York: Academic Press.

Houghton, J.T. (ed.), 1984. *The Global Climate.* Cambridge: Cambridge University Press.

Intergovernamental Panel on Climatic Change (IPCC), 2001. *Summary for Policy Makers.* Online. www.ipcc.ch/

Kopec, R.J. (ed.), 1976. *Atmospheric Quality and Climatic Change.* Studies in Geography No. 9. Chapel Hill, NC: University of North Carolina.

Lamb, H.H., 1967. *Climate: Present, Past and Future*, vol. 1: *Fundamentals and Climate Now*. London: Methuen.

Lamb, H.H., 1977. *Climate: Present, Past and Future*, vol. 2: *Climatic History and the Future.* London: Methuen.

Landsberg, H.E. (ed.), 1969–1984. *World Survey of Climatology*. Amsterdam: Elsevier.

Landsberg, H.E., 1981. *The Urban Climate*. New York: Academic Press.

Licht, S., 1964. *Medical Climatology.* New Haven, CT: Elizabeth Licht.

Lutgens, F.K., and Tarbuck, E.J., 2001. *The Atmosphere*, 8th edn. Upper Saddle River, NJ: Prentice-Hall.

Mather, J.R., 1975. *Climatology: fundamentals and applications.* New York: McGraw-Hill.

McIlveen, R., 1992. *Fundamentals of Weather and Climate*. London: Chapman & Hall.

Miller, D.H., 1981. *Energy at the Surface of the Earth.* New York: Academic Press.

Moran, J.M., and Morgan, M.D., 1997. *Meteorology: the atmosphere and the science of weather,* 5th edn. Upper Saddle River, NJ: Prentice-Hall.

Morgan M.D., and Moran, J.M., 1997. *Weather and People*. Upper Saddle River, NJ: Prentice Hall Kendall/Hunt.
National Aeronautics and Space Administration, NASA, www.nasa.gov/
National Climate Data Center, www.noaa.gov/ncdc.html
National Oceanographic and Atmospheric Administration, www.noaa.gov/
Oliver, J.E., 1973. *Climate and Man's Environment: an introduction to applied climatalogy.* New York: Wiley.
Oliver, J.E., and Fairbridge, R.W., 1987. *Encyclopedia of Climatology*. New York: Van Nostrand Reinhold.
Oliver, J.E., and Hidore, J.J., 2002. *Climatology: an atmospheric science.* Upper Saddle River, NJ: Prentice Hall.
Orme, A. (ed.), 2000. *The Physical Geography of North America.* New York: Oxford University Press.
Pittock, A.B., Frakes, L.A., Jennssen, D., Peterson, J.A., and Zillman, J.W., (eds.), 1978. *Climatic Change and Variability – a southern perspective.* Cambridge: Cambridge University Press.
Robinson, P.J., and Henderson-Sellers, A., 1999. *Contemporary Climatology,* 2nd edn. Harlow: Longman.
Sellers, W.D., 1965. *Physical Climatology.* Chicago, IL: University of Chicago Press.
Thompson, R.D., 1998. *Atmospheric Processes and Systems*. London: Routledge.
Tooley, M.J., and Sheail, G.M., (eds.), 1985. *The Climatic Scene*. London: George Allen & Unwin.
Trewartha, G.T., 1954. *An Introduction to Climate*, 2nd edn. New York: McGraw-Hill.
Turco, R.P., 1996. *Earth Under Siege: from air pollution to global change.* New York: Oxford University Press.
Wallace, J.M., and Hobbs, P.V., 1997. *Atmospheric Science: an introductory survey*. San Diego, CA: Academic Press.
Warrick, R.A., Barrow, E.M., and Wigley, T.M., 1993. *Climate and Sea Level Change.* New York: Cambridge University Press.
Wayne, R.P., 1999. *Chemistry of Atmospheres*, 3rd edn. New York: van Nostrand Reinhold.
Yoshino, M.M., 1975. *Climate in a Small Area*. Tokyo: University of Tokyo Press.

Cross-references

Applied Climatology
Bioclimatology
Cycles and Periodicities
Dynamic Climatology
Climatology, History of
Hydroclimatology
Microclimatology
Scales of Climate
Synoptic Climatology
Water Budget Analysis

CLIMATOLOGY, HISTORY OF

Since the mid-1980s the content and research topics of climatology have broadened considerably. Less-developed topics, such as teleconnections and periodicities, have attained new significance. Additionally, during this period the ready availability of data and information via the worldwide web and the enhancement of data processing and modeling have also produced new research directions. This more recent history is dealt with in many pages of this volume. As a result the historical perspective provided here provides information up to the time prior to the explosive development of paleoclimatology and the paradoxes of global warming. It deals with the history of climatology to the mid-1980s.

Introduction

In one sense climatology has no special history. Outdoor people – farmers, foresters, sailors – have always been aware of their environment, its physical and chemical dimensions, and its biological effects. As Brooks (1948, p. 155) said: "Anyone who has lived here [Charlottesville, Virginia] a decade or two can give you a pretty clear picture of the climate", but records and instrumental readings are needed when comparisons or explanations are wanted. Measurements of state, e.g. temperature of air or soil, and of fluxes, such as solar radiation, are required for these purposes: (1) to picture the variability of a climate and to assess risks of extreme conditions for personal, agricultural, or economic planning; (2) to compare one place with another or to describe a remote, perhaps never-visited place, and answer the perennial question, what is it like there? (3) to understand how climate is formed, how it relates to plants, soil, landforms, and water bodies and the rest of the environment; and (4) to define the roles that solar energy or marine influences play, and what is the outcome of the eternal conflict between local, indigenous processes, and inflows of energy or water from outside. These kinds of understanding make it possible to apply science to models of crop growth, moisture stress, and stream-flow, and for climatology to contribute to its sister sciences of hydrology and meteorology, ecology, forestry and agronomy, the geophysical and geological sciences, as well as to the management of regional resources and national economic health.

Early observations

The Greeks, who lived under cloudless summer skies, were aware of how the height of the daytime sun differed in the desert to the south of them and in the cold lands to the north. Their zoning of climate (= angle of inclination of the sun to the surface of the Earth) is reasonable, emphasizing the role of the Earth–air interface (Tweedie, 1967). Later expressed as day length, this zonation of the globe persisted into the period of Arabian science (Blüthgen and Weischet, 1980). Meanwhile, the Chinese were living where a cloudy, wet summer is the productive season of the year, and they thought of their environment in terms of abundance or failure of the rains. In about the same period as the Greeks, they apparently were recording variations in their environment in terms of the rainfall that determined harvests and tax revenues. They recorded floods from the second century, dry periods from the seventh century, and severe winters from the eleventh century (Needham and Ling, 1959).

The Mediterranean center emphasized the sun and latitude, which seemed to be associated with the range from the warmth of the deserts of northern Africa to the cold of Scythia and to the miserable winters beyond the Alps where the legionnaires huddled around the warm springs of Baden. The effect of the sea was recognized, especially at places exposed to the steady winds used in navigation. During the same time period that the Greeks built the Tower of the Winds, the Chinese began to record wind direction, which was associated with air streams delivering water; they measured rainfall even earlier than the thirteenth-century records in Korea (Needham and Ling, 1959), measured snowfall, and apparently organized networks that reported these readings to the central government. The Greeks gave the winds names and qualities (Wright, 1925), Cicero spoke of the marine climate of England (Khrgian, 1959), and the general effects of the sea and of mountain barriers against polar winds were well known (Wright, 1925). During the late

Middle Ages weather diaries and phenological observations provided climatological information, much of it as yet unused.

When fifteenth-century navigators moved out of the Mediterranean into the lower latitudes of the eastern Atlantic, they encountered the steady trade winds, which differed from the seasonal winds of the Mediterranean or the monsoons of the South Asian seas (Landsberg, 1985). When continuing southward they encountered another belt of trade winds, the low-latitude elements of the global circulation. Halley, who had been studying the relation between land elevation and barometric pressure, turned in 1686 to the winds associated with pressure readings and recognized a thermal driving force for the trades (Frisinger, 1977), and in 1735 G. Hadley developed the first general circulation model (Lorenz, 1983).

As settlers moved westward from Europe and eastward from Russia into new lands, they encountered more severe climates. The cold winters and overheated summers of North America forced changes in the pattern of farming and subsistence (US Department of Agriculture, 1941), and seventeenth-century reports on Siberia discussed whether or not grain could be grown (Khrgian, 1959). Concern about the effects of the climatic environment on plant and human health promoted the use of instruments, especially the barometer and thermometer, which were just being developed in Western Europe. These were first employed by scientists in different fields who shared an interest in their environment; some of the earliest observations were made at astronomical observatories (Blüthgen and Weischet, 1980), and it may be recalled that many eighteenth-century discoveries in chemistry arose out of interest in the atmosphere. These early measurements, made carefully and systematically, gave climatology numerical data to enjoy centuries before geology, geomorphology, ecology, or other field sciences had such materials. In fact, the word *climatology* came to be used for the ordering of any variable displaying an annual regime. We have a climatology of Atlantic shelf water, of wind stress on the oceans, of clouds, and of upper-atmosphere conditions. Time series of early measurements have been recently constructed by Landsberg (1985).

The nineteenth century

At the beginning of the nineteenth century A. von Humboldt, during his expeditions in the New World, noted differences in climates, mapped them by isotherms, and recognized the effect of altitude on air temperature in the Andes (Leighly, 1949a; Khrgian, 1959). This work brought the end of the mathematical formalisms that had so oversimplified climatic distributions, although Fourier series continued to be published as summaries of diurnal and annual regimes at individual places (Leighly, 1949a).

As records at particular places accumulated, it became possible to describe the climate of a city, as L. Howard did for London in 1818 (Landsberg, 1981), E. Renou for Paris in 1855, and M.F. Spasskii for Moscow in 1847; some pointed out urban–rural differences, others the variability from year to year. Single-station observations of the seventeenth and eighteenth centuries expanded into networks (Rigby, 1965), the system in the Rhineland beginning in 1780 was "decisive for the history of climatology" (Landsberg, 1985, p. 58). Not much later (1818), the Surgeon General of the United States began receiving and publishing regular climatic reports from military posts (Landsberg, 1964). This network and those in several states, which later merged into the Smithsonian network, served as models for a Russian system that started in 1839 (Landsberg, 1964). The value of these networks can be seen in early handbooks on climate (Hellman, 1922), which dealt with plants (1822, 1827), forestry and agriculture (1840, 1853), that were generally associated with air temperature, as by H.W. Dove (Khrgian, 1959). The climate of mountains, e.g., Humboldt in 1831 and 1843, introduced the question of sea-level reduction of shelter temperature readings, which continued, parallel to the question of the decrease in temperature of the free air with height, through much of the century. The measurement networks provided data for national surveys of climate like those of L. Blodget for the United States (Leighly, 1949a) and K.S. Veselovskii for Russia (Khrgian, 1959), which represented a high degree of literary skill as well as concern for practical applications. Both were published in 1857.

Extension to the worldwide scale had already taken place for temperature with Humboldt's isotherm maps of 1817, L.F. Kämtz's (1831–1836) discussions of temperature and wind patterns and Dove's polar and tropical air streams (Khrgian, 1959). Humboldt's research in the 1840s on the mountains and climates of central Asia led him to a concept of a solar climate (Flohn, 1971) that was a comprehensive expression of physical climatology. World maps of the winds, pressure, and rainfall followed (Leighly, 1949a); temperature maps became less schematic. Many world maps were products of the Smithsonian Institution in the years before 1870, when it still operated both climatological and synoptic services (Leighly, 1954).

Observations from all over the world, on sea as on land, flowed into the scientific community. J.F. Maury had enlisted the maritime nations in standardized observations at sea in the 1850s and sketched a general circulation model. Later, W. Köppen, at the Deutsche Seewarte in Hamburg, had at hand worldwide flows of data of interest to seamen and merchants. J. von Hann, an editor in Vienna for decades, developed a worldwide network of correspondents whose reports appeared in his journals. A.I. Voeikov, who had traveled widely and worked in many scientific institutions such as the Smithsonian, had similar global contacts and experiences, especially of the influences of climate on agriculture and settlement in the Americas and east Asia. Voeikov, Köppen, and Hann studied the general circulation and, with the scientists of the Smithsonian and other organizations, created the classic period of climatology that occurred during the middle third of the century from about 1855 to 1885 (Leighly, 1949a).

While Maury (1806–1873) preceded the others, it is notable that three of these founders of climatology were born within a few years of one another. Hann lived from 1839 to 1921, Voeikov from 1842 to 1916, and Köppen from 1846 to 1940 (within our own times). Similarly, V.V. Dokuchaev lived from 1846 to 1903 and studied global soils in their climatic context.

By midcentury the most easily measured climatic elements on land, shelter temperature, and rainfall were observed widely; cloudiness, wind, and humidity were observed at fewer places; radiation, notwithstanding its recognized role in the energetics of climate, still remained beyond reach. As observations accumulated, the global circulation of the atmosphere expanded and deepened into a global view of climate. Cartographic analyses, one of the major strands of climatology in the nineteenth century (Leighly, 1949a), drew attention to problems of causation, but much effort still went into futile attempts to reproduce the zonation of shelter temperature and to cast global energy budgets. These were an improvement over the earlier trigonometrical formulas for global temperature but failed because of lack of data on the thermodynamic properties of the underlying surface, which either went unrecognized or were incorrectly assumed, as was albedo for many years (Miller, 1969). Physical laws of climate stated by Spasskii included absorption of radiation, heat conduction, evaporation, and so on (Khrgian, 1959) but could not be investigated because of the lack of accurate measurements of solar radiation and the late discovery of longwave radiation.

Local climates, though not explained, could be described from phenological observations in both their year-to-year variability and their spatial patterns, as well as by tables, maps, and verbal discussion of such variables of state as shelter temperature. These descriptions make up a large part of the classic books that essentially defined the field of climatology (Hann, 1883; Voeikov, 1884). These two gave rise to books in English and Italian (Hellmann, 1922); Voeikov's papers from his travels in Japan influenced the development of climatology in that country (Fukui, 1977). Explanatory power increased as time went on; for example, Voeikov's ideas on the effect of snow cover on climate were introduced in the second edition of Hann's handbook (1897), and near the end of the century a few rough data on radiation began to appear.

New directions in energetics

Maury (1861) outlined the components of ocean surface energetics, and Spasskii and Voeikov those of the land interface. The gaps mentioned in radiation data remained and augmented the lack of knowledge of properties of the interface, which influences not only the atmosphere as a whole, but particularly its lowest meters in which biological activity and physical interactions are concentrated. As a result, the effects of climate on agriculture, for instance, in any but gross features remained inadequate, and basic understanding of climatic genesis was absent. Voeikov's book and papers pointed out the importance of this interface for climatic energetics, the bookkeeping of solar energy receipts, and included such indicators of a regional climate as the temperatures of lakes and rivers, presence and effects of a snow cover on air and soil climate, and mutual interactions between vegetation cover and the lower air. Following Russian observations of soil freezing and temperature, Homén's heat-budget measurements (1897) in three sites in Finland centered on heat fluxes in the substrates. A number of investigators subsequently concerned themselves with heat storage in soil and water bodies. Radiation observations at the surface of the Earth – although considered as early as L.W. Meech's mathematical work in 1857 (Miller, 1969) – were just appearing. Nineteenth-century radiation work had been chiefly a matter of instrument development (Khrgian, 1959) on both sides of the Atlantic, much of it near the 60th parallel north (St Petersburg–Leningrad, Helsinki, Stockholm, and later Tartu). Radiation studies in Vienna entered climatology after the work of J. Stefan and L. Boltzmann, and a long period of productive work in that city and the Alps began. By the 1920s the capability to measure radiation fluxes converged with studies on the survival and growth of plants, especially in central Europe (Munich, the Alps) because plant growth could not be explained by shelter temperature. Field studies in dissected terrain by foresters, ecologists and meteorologists resulted in data that, in other hands, could sometimes be given a physical interpretation. In doing this R. Geiger (1927) invented the discipline of microclimatology, which in subsequent decades proved invaluable to climatology at larger spatial scales (Miller, 1969). Another biological connection was made by A.E. Douglass's work on tree rings (Fritts, 1976), in collaboration with archeologists trying to extend the climatic record back in time. Brooks (1926) associated climate changes with drops in energy gained from solar radiation resulting from volcanism or high polar albedo.

Also at this time, the feeling that recent changes in climate might be measured in terms of changes in glaciers led Scandinavian scientists to examine mass balances of snowfields, where summer melting rates express both solar and advected energy. Energy budgets were developed by A. Ångström in the 1920s and subsequently by H.U. Sverdrup (1942), H.W. Ahlmann, and C. C. Wallén (1948–1949).

Sverdrup also did notable work on turbulent processes at the interface, which had been studied by Schmidt (1925) and others, completing the roster of energy fluxes in climate. Leighly (1938, 1942) applied energetics reasoning to assess the influence of advection on the regional climates of California and the Great Lakes area. By 1940 Albrecht had calculated budgets for several parts of the Earth, including one from detailed observations brought back from an expedition into Central Asia. Unfortunately, cities remained refractory subjects for energetic analysis, but their mosaics of climates were described in detail in terms of state variables – principally air temperature (Kratzer, 1937), radiation and microscale interface temperature (courtyards, streets, walls) by Viennese climatologists.

Energetics, which is central in the calculations at each grid point in large-scale climate models and provides a tool to explain the genesis of climate, forms in addition a link between the climate as environment and the environed organisms. Energy budgets were cast for plant leaves in the 1950s and for animals in the 1960s. For the first time such physically significant characteristics of the Earth's vegetation cover as the geometry, mass, radiative properties, and ventilation of crop and forest canopies were measured and introduced into energy budgets (Gates, 1962; Rauner, 1972; Ross, 1975). Such budgets were able to assess the effect of forest on the atmosphere, and also to explain and predict the way living organisms utilize time and space. Embodying the same radiative and turbulent fluxes that figure in the interface budget, they defined the true impact of environment on an organism's thermoregulation, productivity, water and nutrient needs, behavior during the diurnal and annual cycles, and even its life strategy and evolution. Climatology took on a new immediacy and precision.

The water flux

The fluxes of water in climate, difficult to approach on a global basis because of lack of data over the oceans, never gave rise to the studies of merely state variables that had slowed progress in climatic energetics. Perhaps European scientists, like Dove, considered water a secondary factor, and this attitude persisted through much of the nineteenth century. It had been known since the late seventeenth century that rainfall supplied enough water to support rivers, not only, as E. Halley thought (Middleton, 1965), to mountains in the center of land masses, but even in lowlands like the basin of the Seine (Biswas, 1970). However, without knowledge of the partitioning and conversions of water at the interface, late nineteenth-century efforts to evaluate the true role of moisture in climate and to bring it into climatic typing and classification were in vain, even after it had become clear by 1900 that rainfall by itself was an inadequate variable. Trials of many combinations of rainfall and shelter temperature were made to make various schemes of classifying climates fit vegetation better (Leighly, 1949a). The most notable of these were the successive stages Köppen went through from 1874 up until the 1930s (Leighly, 1949a; Khrgian, 1959). None of these classifications, however, turned out to have lasting physical relevance; partly, at least, because of the error in the assumption that the kind of observations established in the first half of the nineteenth century provided sufficient data for solving the problems of climate (Leighly, 1954).

Agricultural climatology was, from the beginning, an important part of the subject and was well summarized in 1941 by *Climate and Man*, a yearbook of the US Department of Agriculture. This compendium discussed agricultural settlement of major climatic regions along with chapters on corn, cotton and other crops affected by the conventional elements of shelter temperature and rainfall, especially as measured by the network of cooperative stations, then at its peak. Because administrative shifts later separated this work from interface and synoptic climatology, a hiatus followed. *Climate and Man* marked, along with the Köppen-Geiger *Handbuch der Klimatologie* of the 1930s and especially Conrad's volume (1936) on the dependence of climatic elements on terrestrial influences, a summing-up of the early twentieth century.

The lag in the hydrologic dimension of climate behind energetics and microclimatology ended in the 1930s when flood-mitigation programs and droughts in the United States revealed the deficiencies of direct rainfall-to-runoff transfers. R.E. Horton and others emphasized soil moisture as a dimension of drought and a condition of infiltration during rainstorms. Evaluating soil moisture, in turn, called attention to evapotranspiration. This long-sought convergence of water and energy burst into prominence in 1948, when important papers were published by C.W. Thornthwaite, H.L. Penman, and M.I. Budyko (see references in Miller, 1977). These three papers, which recognized the role of energetics, made it possible for the first time to provide an adequate representation of moisture in climate (Tweedie, 1967). Their energetics aspect stimulated research on all the exchanges between the interface and the atmosphere (Thornthwaite, 1961), especially radiation. Lettau's partitioning (1952) of sensible heat between substrate and atmosphere supplemented this work on latent heat and rounded out the theory for interface effects on climate, despite the fact that observational data remained scanty in most kinds of terrain. These methods have been adapted to faster computation methods in such applications as drought indices, probabilities of soil-moisture stress, streamflow modeling, flood frequencies, irrigation scheduling, and crop-growth models. Water-budget methods developed in Thornthwaite's Laboratory of Climatology were applied to reforested uplands, coastal marshes and estuaries, stream flow, and biological phenomena such as decomposition of litter (Muller, 1972; Mather, 1978). Ocean climatology used the methods of interface energetics to outline large-scale patterns of evaporation and hence salinity as a factor in ocean dynamics. Sverdrup (1942) discussed radiation, energy budgets, and sensible-heat exchanges with the air, all related to evaporation.

Circulation

Large-scale climatology benefited from advances in meteorology following 1920, when frontal and air-stream analyses (Namias, 1983), combined with upper-air observations, sparked research in a science stagnant for half a century (Bergeron, 1959) that had been a drag on climatology whenever the two were associated in governmental weather-forecasting services. Some studies of airmass formation, e.g. the classic by Wexler (1936), added to knowledge of climate genesis by virtue of developing interface energy budgets; Flohn (1971) applied energy budgets of the Arctic vs. Antarctic to explain the strength and steadiness of the circulation of the southern hemisphere and budgets of Tibet to evaluate the effects of the injection of heat into the middle troposphere. Many investigators, especially in England and Germany, found air streams useful entities for typing climates; calendars of synoptic air flow pursued the beginnings that Köppen in 1874 (Leighly, 1949a) and others had made toward stratifying climatic events by direction and vorticity of air flow. These became a comprehensive synoptic climatology in World War II (Jacobs, 1947; Hare, 1955; Court, 1957) and served as frameworks in regional climatologies like that of the Paris Basin (Pédelaborde, 1957), or baseline climatic data for New Orleans (Muller, 1977). On a larger scale, circulation analyses provided indexes related to meltwater generation in anticyclonic conditions and to glacial accumulation (Leighly, 1949b). Circulation types over the northern hemisphere (Dzerdzeevskii, 1968) and vapor transport in monsoon climates (Yoshino, 1971) were evaluated. They provided a basis for climatic classifications (Flohn, 1971), revisions in general circulation models (Lorenz, 1983), world maps (Hendl, 1963), and study of climatic genesis and fluctuations (Flohn, 1971; Lamb, 1972, 1977). Fluctuations of large magnitude presumably reflected circulation changes, which could be associated with profiles of pollen and other sediments and statistical properties of tree-ring measurements (Fritts, 1976).

Evaluation of atmospheric vapor transport (Benton and Estoque, 1954) provided an independent check on assessments of evapotranspiration at the underlying interface and not only had hydrologic value but also showed the ties of a regional climate to other parts of the world. Such studies assessed the strength of advection from other regions *vis-á-vis* local factors, as for example, the Andean antiplano, a nearly autochthonous climatic region (Hendl, 1963) that receives little advection (Gutman and Schwerdtfeger, 1965). Postcirculations are sketched on the basis of plate tectonic data on locations of the continents as far back as the Cambrian (National Research Council, Geophysics Study Committee, 1982; Budyko, 1982).

Expanding applications of climatology

Applications of climatology increased greatly during World War II, particularly in the US Air Force. The worldwide distribution and diverse nature of these applications, including the design of equipment and clothing to meet climatic conditions anywhere in the world, had important effects on the whole field. Among other things, they showed the inadequacy of the old hope that a single classification could meet all needs of science (Landsberg, 1958). Networks expanded, archives grew and were put into computable form and machine methods for sorting and summarizing data became standard. This effort was aided by a series of initiatives by the World Meteorological Organization's Committee on Climatology, first chaired by C.W. Thornthwaite. Large-scale statistical programs that had started in the late 1930s with marine data and research on drought, developed machine methods that made possible frequency analyses on an unheard-of scale, as well as correlations, principal-component analysis and other ways to identify major features of a regional climate. Statistical techniques were transferred from agricultural and biological research and developed within climatology (Court, 1951). As storage and computing techniques grew more powerful, however, the gritty problem of assuring the quality of field measurements, especially those not immediately and directly utilized in forecasting the weather, persisted, and in spite of efforts of the World Meteorological Organization, records have lost homogeneity.

Concern in the 1970s about variations in climate that threatened world food production revealed the fact that serious questions of uniformity of series and representativeness of stations had grown up, questions that had been thoroughly examined a century before but forgotten in many countries. The struggles to maintain even a skeleton network of benchmark stations to detect changes in temperature of the kind Mitchell (1963) investigated revealed the difficulty of these historic problems of reliability. The question of change was also examined from the 56-year record of circulation types (Dzerdzeevskii, 1962) and upper-air records. The study of climatic changes at longer time scales, i.e. in the preinstrumental periods, which had been stimulated by nineteenth-century research on glaciation, was again aided by geological research beginning in the 1960s on oceanic sediments (CLIMAP Project Members, 1976) and plate tectonics (Frakes, 1979; National Research Council, Geophysics Study Committee, 1982; Budyko, 1982). For example, the movement of the Australian plate into the latitude of the subtropical anticyclones caused a decrease in rainfall, changes in the structure and fire regime of the vegetation cover and its microclimatic habitats. These geological and geochemical studies indicate that carbon dioxide and water vapor have been components of the atmosphere since the earliest times and stabilize the planet's energetics against "greenhouse" overheating or a cooling to a condition of complete glaciation or a "white earth" (Budyko, 1982, p. 131). The Earth has a long climatic history, only glimpsed in part at the present time.

A large program in energetics at the Voeikov Main Geophysical Observatory in Leningrad (Budyko, 1956, 1971) began with physical research on energy and water fluxes at the interface, tested the results in expeditions to desert, forest and mountain locations, and applied them to calculate national and global patterns of distribution (Budyko, 1963) and to analyze changes (Budyko, 1982). This expansion to global scale paralleled the climatologically based soil science of V. V. Dokuchaev (Khrgian, 1959). The results of the energetics program provided better data for research because of the foundation in adequate measurements of physical exchanges at the interface, a consideration lacked by some earlier research on global climatology (Miller, 1969). This program was supplemented by energetics research in several regional hydrometeorological institutes, e.g. that at Tashkent (Aizenshtat, 1960), where soil climate had been measured for half a century and radiation for several decades; at Tbilisi, another old observatory, at Kiev, and at Tartu, for radiation (Ross, 1975); and in a program developed by Dzerdzeevskii (1962) and Rauner (1972) to examine the influence of forested regions on the atmosphere. Many of the global patterns became inputs to large-scale computer models of the circulation that examined possible changes caused by alterations in atmospheric composition, interface characteristics such as sea ice and continental snow cover or albedo, remotely sensed over extensive areas or in the intake of solar energy. Examining the annual march of climate affords a useful approach to its physical causes, a method used earlier (Leighly, 1938) and well adapted to quantitative computation techniques exploring solar forcing of climatic fluctuations.

Meteorological observations that expanded during and after World War II into the low latitudes and in the 1950s into the polar regions, filled in gaps in the patterns of world climates and added variables desired since Hann and Voeikov, but not earlier measured in routine programs. The results are summed up in the many volumes of Landsberg's *World Survey of Climatology* (1969–1984) and made possible worldwide water budgeting in several countries. Intensified observations of radiation and other heat fluxes in the USSR and Western Europe served for both applied purposes and definition of large-scale patterns, which were also shown by improved upper-air observations, the measurement of sea ice, snow cover, and interface temperature at sea by remote sensing, with its potential for synchronous and macroscale surveillance. Short field programs over the low-latitude oceans in the 1970s might have climatological significance for the light they cast on the phenomenon that is the major exception to the astronomical regimes that otherwise dominate climatology: the Southern Oscillation and El Niño, with their wide teleconnections.

The climatology of the energy budget influenced ecology and agriculture; one-third of Chang's text (1968) is devoted to radiation and photosynthesis and more than a third to evapotranspiration; shelter temperature receives only part of a chapter. Agricultural climatology advanced in practicality in most countries, including China, where much use of radiation data is made (Huang, 1981), and in many states of the United States, which developed their own observational networks to support these methods. A phase of this work was the growth of "near real-time" climatology, which called on the relevant observations of antecedent atmospheric and soil conditions to make direct recommendations on the management of rainfed cropland, fire hazards in wildlands, irrigation of crops, resources of polar lands of Canada, and to predict crop harvests. This time dimension of application was expressed most clearly in mesoscale and regional climatology, though the basic physical and biological relations were generally microscale.

Because most applications of climatology, as distinguished from attempts to improve weather forecasting, dealt with management of water resources or crop production, which operate at regional scales, information at this scale was most useful. For example, new data on mountain climates in practical energy and water terms came from programs in snow hydrology in three mountain ranges of the western United States, to provide data for reservoir design and scheduling, and from investigations at the timberline in Austria and Switzerland to determine how to regenerate forests in avalanche source areas. These programs included multidisciplinary work on the soil, topographic, and biological conditions associated with the energetics of the water and biomass yields of mountain climates.

Lettau's climatonomy (1969) centered on the energy budget, using as forcing functions rainfall and solar radiation, which are independent at the regional scale, to develop the regimes of the state variables of soil moisture and interface temperature and of the flux of latent heat. These quantities are basic in agricultural climatology as it developed in several midwestern states of the United States, making use of observational programs in radiation, soil temperature, and soil moisture, as well as a dense network of recording rain gauges. Concern in several states about energy supplies in the 1970s encouraged research in wind and solar energy climatology at regional and mesoscales. Basic factors in small-scale climatology were classified by Yoshino (1975) as topography, local air streams and atmospheric disturbances, and interface characteristics of open land, forests, and cities. Urban regions received attention beyond the old heat- island idea as their pollution thickened, but while descriptions of urban climates included more physical considerations (Landsberg, 1981) it was generally in spatially averaged terms, although patterns of rain and pollutants were examined in the Metromex program in 1971–1976 (Changnon, 1981).

Addition of chemistry to the usual energetics and hydrology of climatological analysis caught observation programs and climatologists unaware, and much research on diffusion of pollutants neglected important climatological considerations. Some pollutants, such as ozone, were found to be transported on a regional scale, and some, like acid rain, following C.G. Rossby's revival of interest in precipitation chemistry on even larger scales; modeling these transports and studying their effects entered the area of global climatology. Concern about possible effects on climate of increased atmospheric concentrations of carbon dioxide and other radiatively active substances was expressed in the 1950s, but served less to bring about better observations of, say longwave radiation, than to justify computer-assisted models of global circulation. While these models portrayed zonal averages better than regional climates, where water resources and food production are most likely to be affected by fluctuations, they are expected to help study the climates of the past and foreshadow those of the future.

Summary

Seven strands were identified by Leighly (1949a) throughout the development of climatology since 1800: empirical formulation, classification, physical explanation, synoptics, cartographic analysis, description, and variability. Empirical formulations have nearly disappeared as an important activity of climatologists, as has classification of climates. They were compensated for by a strengthening of physical science, especially energetics and hydrology, which were compatible with the newer concerns of chemical and ecological climatology and with more sophisticated applications. Interface exchanges and tropospheric synoptics shape climate, and both increased with advances in meteorology after about 1920. Climatic, water, and agricultural atlases shifted to portray the elements specific to the needs of particular applications of climatology, and as new physical relations became known the relevant observations (e.g. solar energy) were made, duly mapped, and given verbal description. Many descriptions were written to explain applications in water management, food production, and construction, both in design and planning, and in operation. Climatic fluctuations became better defined with the lengthening of the instrumental record, more accurate deductions from information in the preinstrumental period, and from geology; this knowledge has gained importance with concern over the frequencies of severe winters and multiyear droughts, as well as the possibility of permanent shifts resulting from human activity.

David H. Miller

Bibliography

Aizenshtat, B.A., 1960. *The Heat Balance and Microclimates of Certain Landscapes in a Sandy Desert.* G.S. Mitchell (Trans.), Washington DC., U.S. Weather Bureau.

Albrecht, F., 1940. Untersuchungen über den Wärmehaushalt der Erdoberfläche in verschiedenen Klimagebieten. Reichsamt Wetterdienöt. *Wissenschaffliche Abhandlungen*, **8**(2).

Benton, G.S., and Estoque, M.A., 1954. Water-vapor transfer over the North American continent. *Journal of Meteorology*, **11**: 462–477.

Bergeron, T., 1959. Methods in scientific weather analysis and forecasting. An outline in the history of ideas and hints at a program. In Bolin, B., ed., *The Atmosphere and the Sea in Motion.* New York: Rockefeller Institute Press, pp. 440–474.

Biswas, A.K., 1970. *History of Hydrology.* Amsterdam: North-Holland.

Blüthgen, J., and Weischet, W., 1980. *Allgemeine Klimageographie*, 3rd edn. Berlin: Walter de Gruyter.

Brooks, C.E.P., 1926. *Climate Through the Ages.* London: Benn.

Brooks, C.F., 1948. The climatic record: Its content, limitations, and geographic value. *Association of Geographers Annals*, **38**: 153–168.

Budyko, M.I., 1956. *Teplovoi Balans Zemnoi Poverkhnosti.* Leningrad: Gidrometeoizdat. (Translated by N. A. Stepanova, 1958 as *The Heat Balance of the Earth's Surface*. Washington, DC: US Weather Bureau.)

Budyko, M.I. (ed.), 1963. *Atlas Teplovogo Balansa Zemnogo Shara.* Moscow: Mezhduvedomstrenny Geofizicheskii Komitet pri Prezidiume Akademii Nauk SSSR i Glavnaia Geofizicheskaia Observatoriia im A. I. Voeikova.

Budyko, M.I., 1971. Klimat i Zhizn. Leningrad: Gidrometeoizdat. (Translated by D. H. Miller, 1974 as *Climate and Life.* New York: Academic Press.)

Budyko, M.I., 1982. *The Earth's Climate: past and future.* International Geophysics Series 29. New York: Academic Press.

Chang, J.H., 1968. *Climate and Agriculture: An Ecological Survey.* Chicago, IL: Aldine.

Changnon, S.A., Jr (ed.), 1981. Metromex: a review and summary. *Meteorological Monographs,* vol. 18, no. 40. Boston, MA: American Meteorological Society.

CLIMAP Project Members, 1976. The surface of the ice-age Earth. *Science*, **191**: 1131–1137.

Conrad, V., 1936. Die klimatologischen Elemente und ihre Abhängigkeit von terrestrischen Einflüssen. In Köppen, W., and Geiger, R., eds., *Handbuch der Klimatologie.* Berlin: Gebruder Borntraeger.

Court, A., 1951. Temperature frequencies in the United States. *Journal of Meteorology,* **8**: 367–380.

Court, A., 1957. Climatology: complex, dynamic, and synoptic. *Association of American Geographers Annals*, **47**: 125–136.

Dzerdzeevskii, B.L., 1962. Fluctuations of climate and of general circulation of the atmosphere in extratropical latitudes of the Northern Hemisphere and some problems of dynamic climatology. *Tellus*, **14**: 328–336.

Dzerdzeevskii, B.L., 1968. *Tsirkuliatsionnye Mekhanizmy v Atmosfere Severno Polushariia v XX Stoletii.* Moscow: Mezhduvedomstrennyi Geofizicheskii Komitet pri Prezidiume Akademii Nauk SSSR i Institut Geografii.

Flohn, H., 1971. *Arbeiten zur Allgemeinen Klimatologie.* Darmstadt: Wissenschaftliche Buchgesellschaft.

Frakes, L.A., 1979. *Climates Throughout Geologic Time.* Amsterdam: Elsevier.

Frisinger, H.H., 1977. *The History of Meteorology: to 1800.* American Meteorological Society Historical Monographs Series. New York: Science History Publications.

Fritts, H.C., 1976. *Tree Rings and Climate.* London: Academic Press.

Fukui, E. (ed.), 1977. *The Climate of Japan.* Tokyo: Kodansha.

Gates, D.M., 1962. *Energy Exchange in the Biosphere.* New York: Harper.

Geiger, R., 1927. Das Klima der bodennahen Luftschicht. Ein Lehrbuch der Mikroklimatologie. Braunschweig, Germany: Vieweg (2nd edn published in 1942, 3rd edn in 1950, 4th edn in 1961. (Translated in 1957 by M. N. Stewart as *The Climate Near the Ground.* Cambridge, MA: Harvard University Press.)

Gutman, G.J., and Schwerdtfeger, W., 1965. The role of latent and sensible heat for the development of a high pressure system over the subtropical Andes, in the summer, *Meteorologische Rundschau,* **18**: 69–75.

Hann, J. von, 1883. *Handbuch der Klimatologie*. Stuttgart: Engelhorn (2nd edn published in 1897, 3 vols; 3rd edn in 1908–1911, 3 vols).

Hare, F.K., 1955. Dynamic and synoptic climatology. *Association of American Geographers Annals*, **45**: 152–162.

Hellmann, G., 1922. Entwicklungsgeschichte des klimatologischen Lehrbuches. *Beitrage Geschichte Meteorologie*, **11**: 1–14.

Hendl, M., 1963. *Einfuhrung in die physikalische Klimatologie*, Bank II: Systematische Klimatologie. Berlin: Deutscher Verlag der Wissenschoff.

Homén, T., 1897. Der tägliche Wärmeumsatz im Boden und die Warmestrahlung zwischen Himmel und Erde, *Acta Societtis Scientiarum Fennicae*, **23**(3).

Huang, P.W., 1981. Environmental factors and the potential agricultural productivity of China: an analysis of sunlight, temperature, and soil moisture. In Ma, L.J.C., and Nobel, A.G., eds., *The Environment: Chinese and American views*. London: Methuen, pp. 45–71.

Jacobs, W.C., 1947. *Wartime Developments in Applied Climatology*. Boston, MA: American Meteorological Society.

Kamtz, L.F., 1831–1832. *Lehrbuch der Meteorologie,* Halle, 2 vols. (Translated by C. V. Walker, 1845 as *A Complete Course of Meteorology*. London: Bailliere.)

Khrgian, A. Kh., 1959. *Ocherki Razvitiia Meteorologii,* 2nd edn. Leningrad: Gidrometeoizdat. (Translated in 1970 as *Meteorology: A Historical Survey*. Jerusalem: Israeli Program for Scientific Translation.)

Köppen, W., and Geiger, R., (eds.), 1930–1939. *Handbuch der Klimatologie*. 5 vols. Berlin: Gebruder Borntraeger.

Kratzer, A., 1937. *Das Stadtklima*. Braunschweig, Germany: Vieweg (2nd edn published in 1956).

Lamb, H.H., 1972. *Climate: Present, Past and Future*, vol. 1: *Fundamentals and Climate Now*. London: Methuen.

Lamb, H.H., 1977. *Climate: Present, Past and Future*, vol. 2: *Climatic History and the Future*. London: Methuen.

Landsberg, H.E., 1958. Trends in climatology. *Science*, **128**: 749–758.

Landsberg, H.E., 1964. Early stages of climatology in the United States. *American Meteorological Society Bulletin*, **45**: 268–275.

Landsberg, H.E. (ed.), 1969–1984. *World Survey of Climatology*, 15 vols. Amsterdam: Elsevier.

Landsberg, H.E., 1981. *The Urban Climate*. New York: Academic Press.

Landsberg, H.E., 1985. Historic weather data and early meteorological observations. In Hecht, A.D., ed., *Paleoclimate Analysis and Modeling*. New York: Wiley, pp. 27–70.

Leighly, J., 1938. The extremes of the annual temperature march with particular reference to California. *California University Publications in Geography*, **6**(6): 191–234.

Leighly, J., 1942. Effects of the Great Lakes on the annual march of air temperature in their vicinity. *Michigan Academy of Science, Arts and Letters Papers*, **27**: 377–414.

Leighly, J., 1949a. Climatology since the year 1800. *American Geophysical Union Transactions*, **30**: 658–672.

Leighly, J., 1949b. On continentality and glaciation. *Geographische Annaler*, **31**: 133–145.

Leighly, J., 1954. Climatology. In James, P.E., amd Jones, C.F., eds., *American Geography Inventory and Prospect*. Syracuse: Syracuse University Press, pp. 334–361.

Lettau, H., 1952. *Synthetische Klimatologie. Berichte des Deutsches Wetterdienstes in der US Zone*, **38**: 127–136.

Lettau, H., 1969. Evapotranspiration climatonomy. I. A new approach to numerical prediction of monthly evapotranspiration runoff, and soil moisture storage, *Monthly Weather Review*, **97**: 691–699.

Lorenz, E.N., 1983. A history of prevailing ideas about the general circulation of the atmosphere, *American Meteorological Society Bulletin*, **64**: 730–734.

Mather, J.R., 1978. *The Climatic Water Budget in Environmental Analysis*. Lexington, MA: Lexington Books.

Maury, J.F., 1861. *The Physical Geography of the Sea, and its Meteorology*, 8th edn. Edited by J. Leighly, 1963. Cambridge, MA: Belknap Press.

Middleton, W.E.K., 1965. *A History of the Theories of Rain and Other Forms of Precipitation*. New York: Watts.

Miller, D.H., 1969. Development of the heat budget concept. In Court, A., ed., *Eclectic Climatology*. Corvallis, OR: Oregon State University Press, pp. 123–144. Also as Yearbook Pacific Coast Geographical Association, **30** (1968): 123–144.

Miller, D.H., 1977. *Water at the Surface of the Earth: an introduction to ecosystem hydrodynamics*. New York: Academic Press.

Mitchell, J.M., Jr, 1963. On the world-wide pattern of secular temperature change. In *Changes of Climate. Arid Zone Series* 20. Paris: UNESCO, pp. 161–181.

Muller, R.A., 1972. Application of Thornthwaite water balance components for regional environmental inventory, *Publications in Climatology*, **25**(2): 28–33.

Muller, R.A., 1977. A synoptic climatology for environmental baseline analysis: New Orleans, *Journal of Applied Meteorology*, **16**: 20–33.

Namias, J., 1983. The history of polar front and air mass concepts in the United States—an eyewitness account. *American Meteorological Society Bulletin*, **64**: 734–755.

National Research Council, Geophysics Study Committee, 1982. *Climate in Earth History*. Washington, DC: National Academy Press.

Needham, J., and Ling, W., 1959. *Science and Civilization in China*, vol. 3: *Mathematics and the Sciences of the Heavens and the Earth*. Cambridge: Cambridge University Press.

Pédelaborde, P., 1957. *Le Climat du Bassin Parisien. Eassai dùne methode rationnelle de climatologie physique,* 2 vols. Paris: Editions Genin, Librairie de Médicis.

Rauner, Iu. L., 1972. *Teplovoi Balans Restitelnugo Pokrova*. Leningrad: Gidrometeoizdat.

Rigby, M., 1965. Evolution of international cooperation in meteorology 1654–1965. *American Meteorological Society Bulletin*, **46**: 630–633.

Ross, Iu. K., 1975. *Radiatsionnyi Rezhim i Arkhitektonika Rastitel'nogo Pokrova*. Leningrad: Gidrometeoizdat.

Schmidt, W., 1915. Strahlung und Verdunstung an freien Wasserflächen; ein Beitrag zum Wärmehaushalt des Weltmeers und zum Wasserhaushalt der Erde, Ann. *Calender Hydrographie und Maritimen Meteorologie*, **43**: 111–124, 169–178.

Schmidt, W., 1925. *Der Massenaustausch in freier Luft und verwandte Erscheinungen*. Hamburg: Grand.

Sverdrup, H.U., 1942. *Oceanography for Meteorologists*. New York: Prentice-Hall.

Thornthwaite, C.W., 1961. The task ahead. *Association of American Geographers Annals*, **51**: 345–356.

Tweedie, A.D., 1967. Challenges in climatology. *Australian Journal of Science*, **29**: 273–278.

U.S. Department of Agriculture, 1941. Climate and Man. *Yearbook of Agriculture 1941*. Washington, DC.

Voeikov, A.I., 1884. *Klimaty Zemnogo Shara, v Osobennosti Rossii*. St. Petersburg. (Reprinted in 1948 by Izdatel'stvo Akademii Nauk SSSR.) (Translated in 1887 as *Die Klimate der Erde,* 2 vols. Jena: Costenoble.)

Wallén, C.C., 1948–1949. Glacial–meteorological investigations on the Kårsa Glacier in Swedish Lappland 1942–1948. *Geographische Annaler*, **30**: 451–472.

Wexler, H., 1936. Cooling in the lower atmosphere and the structure of polar continental air. *Monthly Weather Review*, **64**: 122–136.

Wright, J.K., 1925. *The Geographical Lore of the Time of the Crusades. A Study in the History of Medieval Science and Tradition in Western Europe*. New York: American Geographical Society.

Yoshino, M.M. (ed.), 1971. *Water Balance of Monsoon Asia: a climatological approach*. Honolulu: University of Hawaii Press.

Yoshino, M.M., 1975. *Climate in a Small Area: an introduction to local meteorology*. Tokyo: University of Tokyo Press.

Cross-references

Applied Climatology
Archeoclimatology
Bioclimatology
Climate Classification
Climatology
Microclimatology

CLOUD CLIMATOLOGY

A cloud is an assemblage of liquid water particles and/or ice particles in the atmosphere. Almost all clouds form when moist air is cooled until it becomes supersaturated, i.e. the relative humidity is above 100%. If the temperature is above freezing, supersaturated water vapor then condenses onto cloud condensation nuclei to produce liquid particles. If the temperature is below freezing, the supersaturated water vapor deposits on ice nuclei to form ice particles. If there are not enough ice nuclei present, supercooled liquid particles will form even though the temperature is below 0°C. When the temperature is below −40°C, ice particles will form without ice nuclei and supercooled liquid particles will spontaneously freeze. Under certain conditions cloud particles will combine and/or grow large enough to fall as precipitation.

Clouds play a key role in the climate system. They are intimately connected to precipitation and have a large influence on the transfer of energy within the atmosphere. Latent heat is released when water vapor is converted to liquid water or ice and when liquid water is converted to ice. This added heat increases the temperature, and hence buoyancy, of the air in which a cloud forms. Correspondingly, the temperature of the air decreases due to conversion into latent heat when cloud or precipitation particles melt or evaporate. Latent heating and cooling associated with clouds and precipitation drive atmospheric motions, especially in the tropics. Clouds also have a strong influence on the transfer of solar and terrestrial radiation within the atmosphere. Because they are generally brighter than the surface of the Earth, clouds reflect more solar radiation back to space. This decreases the amount of energy absorbed by the Earth, thus causing it to be cooler than would be the case were it cloudless. On the other hand, clouds are generally colder than the surface and consequently emit less thermal infrared radiation than does the surface. This decreases the amount of energy emitted to space, thus causing the Earth to be warmer than would be the case were it cloudless. The greatest reduction in emitted thermal radiation occurs for clouds high in the atmosphere, because they are coldest. Globally, the reduction in absorbed solar radiation by the Earth caused by clouds outweighs the reduction in emitted thermal radiation by the Earth caused by clouds. This result is a net cloud cooling effect.

Cloud types

The most basic categorization of clouds is by the altitude range in which they occur. Low-level clouds have bases less than 2 km above the surface. Mid-level clouds occur between approximately 2 km and 6 km above the surface. High-level clouds typically occur more than 6 km above the surface. Each altitude range is further divided into genera, or types of clouds, according to their morphology. Low-level genera are stratus, cumulus, and stratocumulus. A stratus cloud exists as a uniform grayish layer with little apparent structure. Cumulus clouds develop vertically as mounds or towers separated from one another. Stratocumulus cloudiness occurs as a layer with distinct clumps. Mid-level genera are altostratus and altocumulus. Both altostratus and altocumulus clouds occur in layers, but differ in that altocumulus has distinct cloud elements and altostratus has a more uniform appearance. High-level genera are cirrus, cirrostratus, and cirrocumulus. Cirrus clouds are detached from one another and appear white and often feathery. A cirrostratus cloud occurs in a whitish and transparent layer. Cirrocumulus cloudiness exists as a whitish layer composed of very small but distinct cloud elements. Two cloud types that produce rain or snow are cumulonimbus and nimbostratus. A cumulonimbus cloud, typical of thunderstorms, resembles a cumulus cloud and may grow very high, but is nevertheless classified as a low-level cloud type. Cumulonimbus clouds sometimes have tops shaped like anvils and occur with thunder and lightning. Nimbostratus cloudiness exists as a gray and frequently dark layer and is distinguished from stratus cloudiness by the occurrence of rain or snow. Although nimbostratus clouds are thick, and can occupy a large range of altitude, they are often classified as a midlevel cloud. Table C18 lists these 10 basic cloud types. Fog, essentially a cloud that touches the ground, can be identified as an additional cloud type. Because specific meteorological processes produce different types of clouds, cloud morphology qualitatively describes the local atmospheric environment. The global distributions of different types of clouds are closely related to the locations of various climate zones of the Earth.

Table C18 Cloud types

Altitude range	Genera (cloud type)	Abbreviation
Low	Stratus	St
	Cumulus	Cu
	Stratocumulus	Sc
	Cumulonimbus	Cb
Middle	Altostratus	As
	Altocumulus	Ac
	Nimbostratus	Ns
High	Cirrus	Ci
	Cirrostratus	Cs
	Cirrocumulus	Cc

Cloud observations

The primary sources of cloud data are visual observations made by people at the surface of the Earth and satellite observations made from orbit. Human observers report cloud fraction, the fraction of sky-dome covered by clouds, and what cloud types are present. They have a clear view of low-level clouds but often cannot see higher clouds because lower clouds obscure them. Satellites have an unobscured view of high clouds but not lower clouds. Since human observers view the hemispheric sky-dome and include cloud sides as part of cloud cover, the cloud fraction they report is generally not identical to the cloud fraction reported by satellite. The difference is usually greatest for cumuliform clouds (cumulus and cumulonimbus) and least for stratiform clouds (stratus, stratocumulus, and fog). People have made regular visual observations of clouds for many decades in connection with weather forecasting, and it is possible to use these data to study past changes in cloud fraction and cloud type. Unfortunately human observers are not everywhere at all times on the Earth, and there are large regions of the globe, especially over the ocean, where no surface observations of clouds are available. However, a small number of satellites can observe clouds over every part of the globe several times a day. The International Satellite Cloud Climatology Project (ISCCP) (Rossow and Schiffer, 1999) provides cloud fraction and other information from July 1983 onwards. Unlike human visual observations, satellite observations cannot unambiguously identify morphological cloud type. Satellites instead report the reflectivity of clouds and the amount of emitted thermal radiation. This information is useful for determining how clouds affect the transfer of visible and infrared radiation within the atmosphere.

Satellites directly measure radiances, the amounts of visible, infrared, or microwave radiation upwelling in the direction of the satellite from various locations on the Earth. Meteorological satellites typically cannot resolve, i.e. distinguish, features less than several kilometers in size. The minimum area resolved by a satellite is called a pixel. Clouds smaller than the size of a pixel cannot be distinguished; the pixel is assumed either to be completely clear or completely cloud-filled. A common method for identifying clouds by satellite makes use of the fact that clouds are generally brighter and colder than the underlying surface. Consequently more visible radiation will be reflected and less thermal infrared radiation will be emitted by a cloudy pixel than by a cloud-free pixel. The threshold method used by ISCCP identifies pixels as cloudy if their visible radiance is greater by some threshold, and/or if their thermal infrared radiance is less by some threshold, than the expected visible and thermal infrared radiances of a cloud-free pixel at that location. Increasing the threshold will cause some true cloudy pixels to be identified as cloud-free, and decreasing the threshold will cause some true cloud-free pixels to be identified as cloudy. Clouds are also difficult to identify over snow and ice using the threshold method since the surface can be brighter and colder than the clouds. Satellite cloud fraction within a particular area is defined as the number of cloudy pixels divided by the number of all pixels. Cloud reflectivity is obtained from visible radiances using additional information about surface and other atmospheric properties at the location of the cloudy pixel. Because the amount of thermal infrared radiation emitted by a cloud is quantitatively related to its temperature, cloud top temperature is obtained from thermal infrared radiances, also using additional information about surface and atmospheric properties. Matching cloud top temperature to a profile of atmospheric temperature as a function of altitude provides cloud top height. Unlike the case for clouds observed from the surface, clouds detected by satellite are classified into altitude ranges according to the height of cloud top. ISCCP arbitrarily defines middle-level clouds as those with tops between approximately 3.2 km and 6.5 km above sea level and high-level clouds as those with tops more than 6.5 km above sea level.

Geographical distributions of cloud types

Figure C33 shows annual mean cloud fraction of all clouds over the entire globe, as reported by ISCCP. There is more cloudiness at middle latitudes than at low latitudes, except near the equator. The geographical variations in cloud fraction are largely related to the large-scale circulation of the atmosphere. Oceanic low-level trade winds flowing from the northern and southern hemispheres come together near the equator in the Intertropical Convergence Zone (ITCZ). Conservation of mass requires upward motion to balance the converging trade winds, and the upward motion causes air to cool and produce substantial cloudiness and precipitation. Latent heat release provides buoyancy that strengthens the upward motion. The climatological ITCZ can be identified by the relatively large fraction of middle- and high-level cloud near the equator, as seen in Figure C34. The ITCZ is broad over the Indian Ocean and western Pacific Ocean and is narrow and slightly north of the equator over the Atlantic Ocean and eastern Pacific Ocean. Over continents, the area of equatorial precipitation and middle- and high-level cloudiness is very widespread. Enhanced cloudiness marks tropical rainforests in South America and Africa. Air moving upwards in the ITCZ diverges and flows out toward the subtropical latitudes of the northern and southern hemispheres and then downward to complete the circuit of the trade winds, also called the Hadley circulation. The downward-moving air warms, thus decreasing its relative humidity. Figure C34 shows that very little middle- and high-level cloudiness exists over most regions between 10°

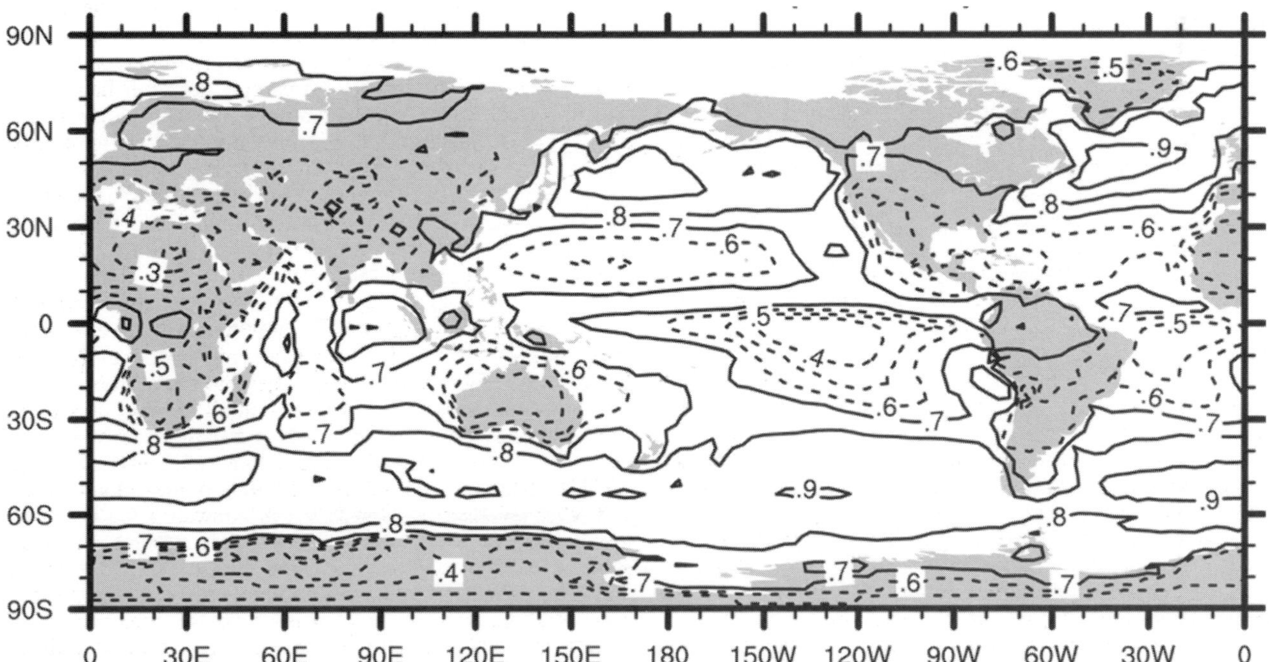

Figure C33 Annual mean cloud fraction of all (total) cloudiness obtained from the International Satellite Cloud Climatology Project and averaged from the years 1984–1999. Contour interval is 0.1, and values equal to 0.6 and lower are dashed.

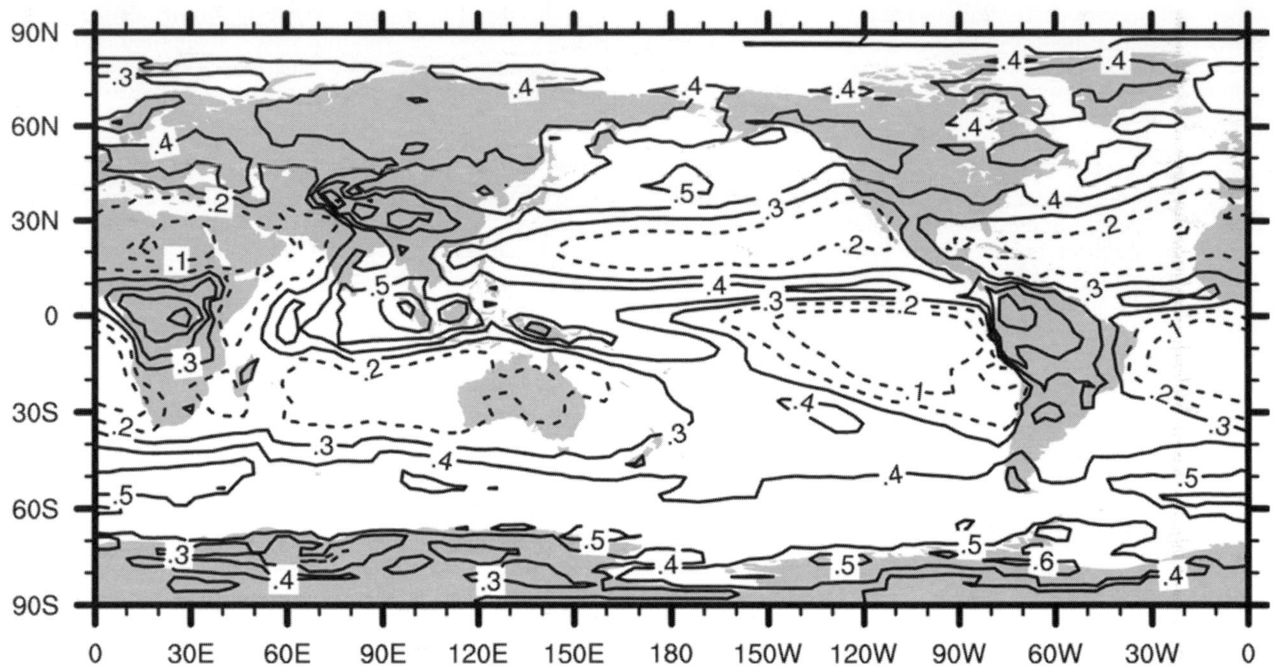

Figure C34 Annual mean cloud fraction of middle- and high-level cloudiness obtained from the International Satellite Cloud Climatology Project and averaged from the years 1984–1999. Contour interval is 0.1, and values equal to 0.2 and lower are dashed.

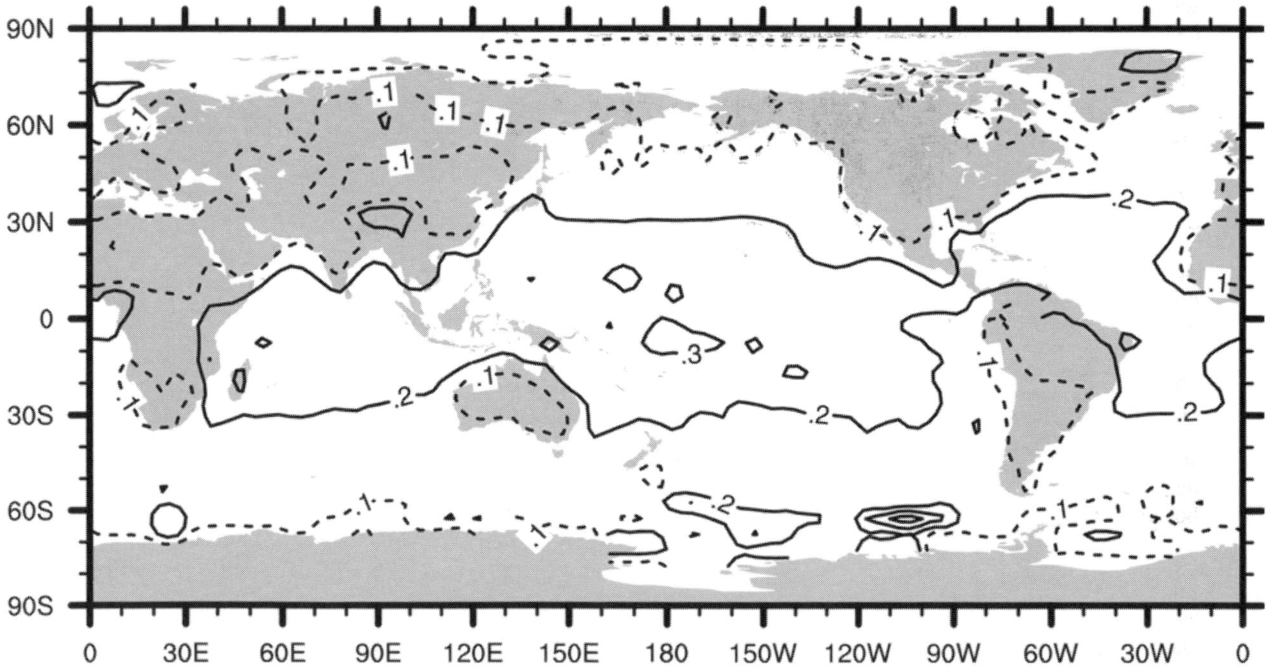

Figure C35 Annual mean cloud fraction of low-level cumulus and cumulonimbus cloudiness obtained from surface observations in the Extended Edited Cloud Report Archive and averaged from the years 1971–1996. Contour interval is 0.1, and the value equal to 0.1 is dashed.

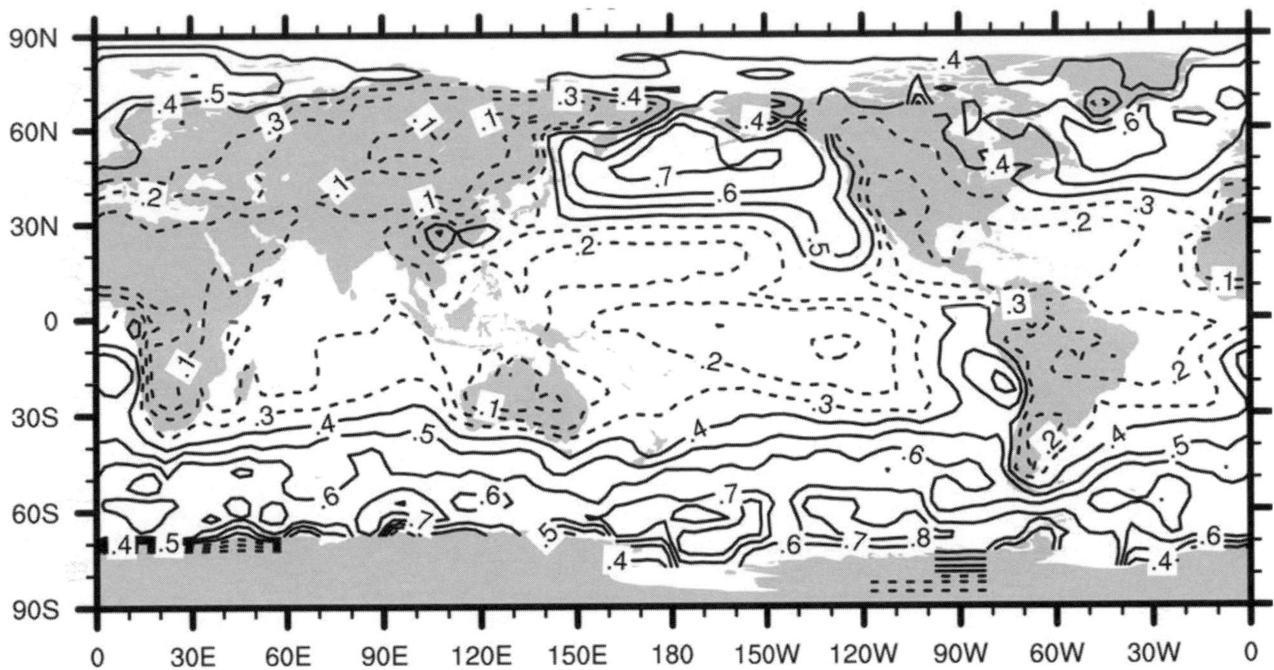

Figure C36 Annual mean cloud fraction of low-level stratus and stratocumulus cloudiness and fog obtained from surface observations in the Extended Edited Cloud Report Archive and averaged from the years 1971–1996. Contour interval is 0.1, and values equal to 0.3 and lower are dashed.

and 35° latitude. Middle- and high-level cloud fraction is greater poleward of 35°, especially over oceans. This is the latitude zone of the storm tracks, which are stronger over ocean than over land. Upward motion associated with extratropical cyclones passing through the storm track produces large amounts of cloudiness at middle and high levels.

Figures C35 and C36 show global distributions of low-level cumulus-type cloud fraction and low-level stratus-type cloud fraction, respectively. The source of cloud data for these plots was the Extended Edited Cloud Report Archive (Hahn and Warren, 1999), a collection of visual observations made by humans on ships or at land meteorological stations. Due to greater moisture availability, more low-level clouds occur over oceans than over continents. Cumulus-type clouds primarily occur equatorward and stratiform clouds primarily occur poleward of 30° latitude. A notable exception to this rule is the presence of large amounts of persistent stratocumulus clouds over eastern subtropical oceans, particularly off the coasts of California, Namibia, and Peru. In these regions the downward motion of the Hadley circulation is stronger, and the sea surface temperature is colder, than elsewhere in the subtropics. The marine boundary layer, the layer of the atmosphere directly affected by the ocean, is kept shallow by the strength of the descending branch of the Hadley circulation and kept cool and moist by the relatively cold underlying water. These conditions are conducive for the production of horizontally extensive layer clouds. Outside of stratocumulus regions, the tropical and subtropical ocean is warm and generates plumes of buoyant moist air that intermittently rise up from the surface. The plumes cool as they rise and form cumulus or cumulonimbus clouds when saturation is reached. Stratiform clouds occasionally occur over the tropical ocean in association with large cumulonimbus cloud systems. The cold oceans at middle and high latitudes are favorable for the production of stratiform clouds. Widespread uplift associated with storms also produces many stratiform clouds. Cumulus clouds sometimes occur during winter over middle-latitude oceans when cold continental air flows offshore over relatively warmer water and generates many buoyant plumes.

Joel Norris

Bibliography

Hahn, C.J., and Warren, S.G., 1999. *Extended Edited Synoptic Cloud Reports from Ships and Land Stations over the Globe, 1952–1996*. Oak Ridge, TN: Carbon Dioxide Information Analysis Center, Oak Ridge National Laboratory.

Norris, J.R., 1998. Low cloud type over the ocean from surface observations. Part II: geographical and seasonal variations. *Journal of Climate*, **11**: 383–403.

Rossow, W.B., and Schiffer, R.A., 1999. Advances in understanding clouds from ISCCP. *Bulletin of the American Meteorological Society*, **80**: 2261–2287.

Warren, S.G., Hahn, C.J., London, J., Chervin, R.M., and Jenne, R.L., 1986. *Global Distribution of Total Cloud Cover and Cloud Type Amounts over Land*. Boulder, CO: National Center for Atmospheric Research.

Warren, S.G., Hahn, C.J., London, J., Chervin, R.M., and Jenne, R.L., 1988. *Global Distribution of Total Cloud Cover and Cloud Type Amounts over the Ocean*. Boulder, CO: National Center for Atmospheric Research.

World Meteorological Organization, 1987. *International Cloud Atlas*, vols I and II. Geneva: World Meteorological Organization.

Cross-references

Adiabatic Phenomena
Coastal Climate
Condensation
Lapse Rate
Latent Heat
Thunderstorms

COASTAL CLIMATE

For some distance on both sides of the 450 000-km-long world shoreline, meteorological processes and resulting climates differ from those farther seaward and landward. The coast is in essence a transitional zone between marine on one side and terrestrial on the other, a transitional zone or band that varies in total width as well as in the relative width of its two parts, land and water. The distinctiveness of this band varies with the amount and rate of meteorologic and climatic change across it. Although most of its characteristics are a blend of marine and terrestrial phenomena, there are a few characteristics, such as land and sea breezes and coastal fogs, that can be truly labeled coastal.

The climatic characteristics of the coastal zone most often have been considered a landward invasion of adjacent sea air and have been referred to as oceanic, marine, maritime, littoral, and coastal. The first three designations have been loosely used, and the last two appear only rarely in the literature. Maritime is the term that comes closest to our usage here of coastal (see Fairbridge, 1967). With the rapid evolution of the concept of the coastal zone, and especially because of the recent tendency to include the seaward segment of the coastal zone as one of its integral parts, the terms coastal meteorology and coastal climatology are now more appropriate than any of the others.

Climatic controls

Three of the most important climatic controls on Earth are variation of insolation with latitude, distribution of land and water, and surface configuration. Coastlines cross virtually every parallel on Earth except those immediately surrounding the two poles. Although coasts do not occur south of 78°S latitude because of the presence of Antarctica and its ice shelves, nor are they present north of 82°N latitude because of the presence of the Arctic Ocean, they nonetheless have a wide range of latitudinally influenced characteristics.

Whereas the juxtaposition of any two environments (e.g. forest/grassland, mountain/plain, and city/country) give rise to climatic modification, the land/water combination is the most distinctive on a worldwide basis. There are marked variations in the heat and moisture budgets of these two surfaces because of contrasts in albedo, evaporation rate, transparency, surface mobility, heat capacity, and sensible heat flux. Moreover, there are important modifications in circulation patterns because of variations in surface irregularities across the interface, including variations in the actual relief as well as in the mobility and spatial characteristics of roughness forms. Possibly the two most important differences are in wetness and roughness – the sea is almost always wetter and smoother than the land.

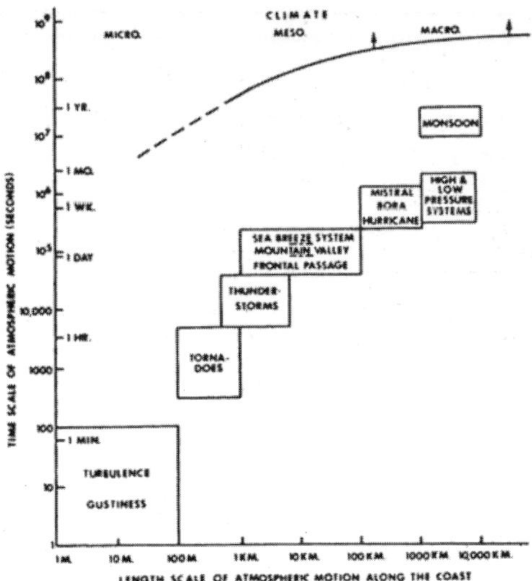

Figure C37 Time-length scales of atmospheric motions.

These controls are especially important in affecting atmospheric motion in the coastal zone. Pertinent motions have a great range in scale both spatially and temporally (Figure C37), and although all of these phenomena have some coastal expression, only a few have actually been considered from that standpoint. Studies to date most often have dealt with the impact of these levels of atmospheric motion (especially turbulence, sea breezes, and hurricanes) on the coastal environment rather than with the motion itself as a coastal phenomenon. Such an emphasis reflects the fact that coastal meteorology and climatology are integral parts of the total systems approach to the study of coastal environments.

Boundary-layer meteorology (microscale and mesoscale meteorology)

The boundary layer is the region of the atmosphere that is directly affected by friction caused by interaction with the Earth's surface. Transport in this layer in nearshore and estuarine environments is important from several points of view. For example, the wind stress or momentum flux is one of the most essential driving forces in shallow-water circulation. Heat and convection are the origins of some localized coastal weather systems. Sensible heat and water vapor fluxes are necessary elements in radiation and heat budget considerations, including the computation of salt flux for a given estuarine system. Experiments in these environments ranging from the tropics to the Arctic have produced a large number of conflicting drag and bulk transfer coefficients for both deep and shallow waters. There are several reasons for the discrepancies, but the most important is that early methods did not take into account the simultaneous contribution of wind, dominant waves, and atmospheric stability. This difficulty was removed by the development of a wind/wave interaction method of determining wind stress from commonly available wind and wave parameters (Hsu, 1976).

Aerosol transport

Atmospheric particles, particularly sea salts, have become an increasingly important subject for investigation in recent years.

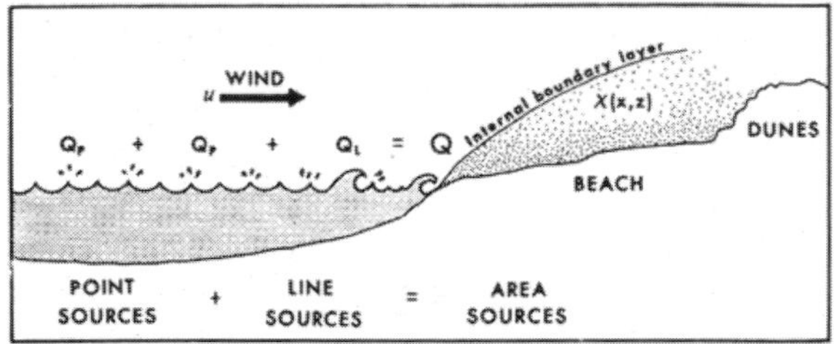

Figure C38 Schematic representation of point and line sources for sea salt and the effect of the internal boundary layer (from Hsu, 1977).

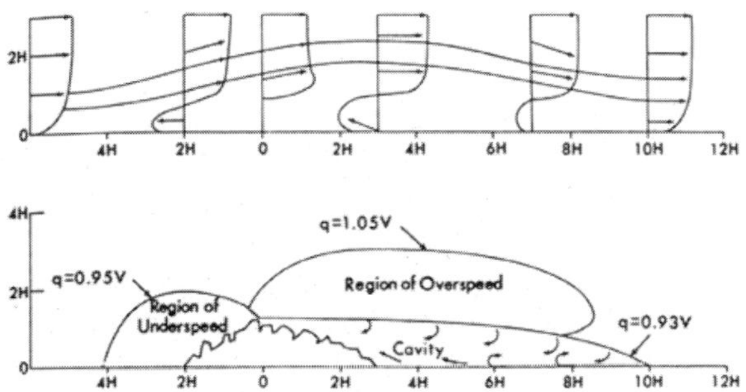

Figure C39 A basic flow model and a similarly shaped ice ridge. q is the local resultant mean velocity and V is the reference velocity in the uniform stream (from Hsu, 1977).

Sea-salt aerosols are a significant source of condensation nuclei. The quantity and quality of sea-salt particles deposited on land may be important in determining the physical and chemical characteristics of coastal soils and plants. The generation of aerosols depends on many meteorological and oceanographic factors. Among those in the coastal region are wind speed, direction, duration, fetch (which govern sea state and whitecap distribution), and subaqueous bathymetry (which controls the breaking wave condition in the surf zone). Aerosol contribution from these point sources (whitecaps) and line sources (surf zone) can be considered as originating from an area source (Figure C38). Experiments measuring the vertical distribution of sea salt in the atmospheric boundary layer over beaches have shown that on a coast influenced by synoptic onshore winds, such as a beach–dune complex, mixing depths are less thick than they are under sea breeze conditions (Hsu, 1977).

Sand dune and ice ridge air flow

When air passes over an obstacle it must adjust to a new set of boundary conditions. Major characteristics associated with this adjustment are the displacement zone, the wake, and the cavity. Measurements of the wind field in the region of coastal sand dunes in the tropics and ice ridges in the Arctic were combined with data from laboratory experiments in the preparation of a basic flow model. This model (Figure C39) is aiding the study of the modification of the structure of airflow over selected coastal topographies and is furthering the understanding of low-altitude atmospheric dispersion characteristics in the coastal zone.

Sea breeze

The sea breeze, a coastal local wind that blows from sea to land, occurs when the temperature of the sea surface is lower than that of the adjacent land (Defant, 1951; Hsu, 1970). It usually blows on relatively calm, sunny, summer days, and alternates with an oppositely directed, usually weaker, nighttime land breeze. As a sea breeze regime progresses, coriolis deflection causes the development of a wind component parallel to the coast. Coastal air circulations of this kind have been known and recorded since the time of Aristotle, about 350 BC, and have been studied more or less scientifically since the seventeenth century. Of all coastal meteorological phenomena, these have received by far the greatest amount of attention and are continuing to do so. For example, during the First Conference on Coastal Meteorology in 1976 about 40% of the papers dealt in some way with the sea breeze.

Synoptic meteorology and its coastal impact (macrometeorology)

Associated with the readjustment of airflow at the sea/land interface are other parameters such as temperature and humidity. As considered above at the microscale and mesoscale, these are quite distinctive. Although their effect is less obvious at the larger scale, they nonetheless have a significant impact on synoptic and planetary circulation systems.

Data support for the conduct of studies at this scale is scarce. Few meteorological data are routinely obtained at coastline stations; the major exception is at the all too few US Coast Guard stations where observations of temperature, precipitation, and wind are made. The most useful data, however, come from coastal airports, even though these are usually located several kilometers inland.

Such data have been used in the analysis of the weather input and environmental response in the coastal zone in Louisiana (Muller, 1977). This approach classified Louisiana coastal weather into eight inclusive synoptic types: Pacific High (PH), Continental High (CH), Frontal Overrunning (FOR), Coastal Return (CR), Gulf Return (GR), Frontal Gulf Return (FGR), Gulf Tropical Disturbance (GTD), and Gulf High (GH). Because each type can be identified on weather maps, climatic calendars were devised that are useful in the analysis of mean properties by months and seasons, climatic variation from year to year, and geographical comparisons of one coastal region with another (Muller and Wax, 1977).

Synoptic weather type changes have been found to be related to water level and salinity changes in the coastal marshes and estuaries of Louisiana (Borengasser et al., 1978). When the effects of astronomical tides were "filtered" from water-level data, it was found that water levels tended to drop during the 24-hour periods associated with a progression from the Frontal Gulf Return type to the Pacific High, Continental High, or Frontal Overrunning types. In contrast, water levels were shown to rise during other types of progression. Such analyses are only beginning but appear to hold promise for other non-tropical coastal areas as well.

In recent years there has been increasing focus on coastal storms, classification (Dolan and Davis, 1992), synoptic patterns (Davis et al., 1993), and some of the environmental and economic impacts (Davis and Dolan, 1993). Very important environmental issues still not well resolved focus on questions about increasing or decreasing frequencies and magnitudes of

Table C19 Climatic characteristics related to coastal landscapes

Climatic type and percent of world coastline	Typical locality	Climatic conditions[a]				
		Temperature (°C)		Precipitation		
		Mean maximum warmest month (tm)	Mean minimum coldest month (tm)	Mean annual depth, cm (P)	Mean annual no. days ⩾0.004 cm (FP)	Winder concentration of precipitation[b] (R)
Rainy tropical (20%)	Inner tropics	30–32	19–23	198–310	134–185	14–49
Subhumid tropical (10%)	Border tropics	30–33	15–21	104–140	61–114	17–38
Warm semiarid (2%)	Tamaulipas, Venezuela	32–34	13–19	53–71	42–60	9–40
Warm arid (5%)	Horn of Africa, Sonora	33–37	11–20	13–25	10–32	36–94
Hyperarid (4%)	Cool-water coasts of subtropics	24–34	9–14	<5	1–4	42–100
Rainy subtropical (6%)	East coasts, lat. 20–35°	29–32	6–9	114–147	93–142	29–49
Summer-dry subtropical (7%)	Mediterranean	27–31	6–9	43–69	54–103	74–87
Rainy marine (1%)	W. coasts, Tasmania, New Zealand	17–20	4–7	109–206	166–187	51–69
Wet-winter temperate (2%)	Oregon, Washington	17–22	0–6	99–170	120–198	67–78
Rainy temperate (9%)	NE United States, W. Europe	20–27	−7–2	66–112	127–188	41–54
Cool semiarid (1%)	Bahia Blanca	19–31	−3–9	30–53	45–87	37–52
Cool arid (2%)	Patagonia	21–26	1–7	10–15	24–41	54–88
Subpolar (6%)	Gulfs of Alaska, Bothnia	15–23	−23–−13	46–104	106–184	32–50
Polar (25%)	Arctic Sea border	9–13	−34–−13	18–66	91–131	30–49

Source: From Bailey (1976).

[a] Each pair of numerals in the body of the table refers to data from climatic stations at the 25th and 75th percentiles of the frequency distribution appropriate to the climatic type and element. As only long-period means have been entered into the frequency distributions, the data in the table above show the spread in average conditions of climate in the most representative parts of the several climatic regions. Because approximately equal spacing was employed in the station network, it is also true that the data illustrate, for a given climatic type, conditions in about 50% of the aggregate length of coastlline affected by that climatic type.

[b] The winter concentration of precipitation is defined as the percentage of the mean annual total that falls in the winter half-year. October through March in the northern hemisphere, April through September in the southern hemisphere. The computation was not carried out for those places where the difference between the mean monthly temperatures of the warmest and coldest month was less than 3°C.

severe and extreme events in terms of global change. For tropical storms and hurricanes there is clear evidence that regional frequencies of events vary significantly along the Atlantic coast of the United States and appear to be associated with ENSO and North Atlantic Oscillation indices (Muller and Stone, 2001).

Coastal climates

Whereas the atmospheric processes thus far considered are measured in intervals of seconds to weeks, there remain longer time periods to be considered – periods representative of climate (Figure C39). Climate, as a synthesis of atmospheric events, is unique for each location on Earth. Yet vast areas have climates that are sufficiently similar to justify consideration together. In most classifications continental types are extended directly to the shore, so that their coastal ramifications are indistinguishable (Wilson, 1967).

The major exception to date is the work of Bailey (1960, 1976), who developed a classification of coastal regions through the utilization of data from coastal stations. Defining his climatic types so that they closely approximated the distribution of coastal vegetation, Bailey (1976) arrived at 14 types (Table C19). The relative length of coastline represented by each type varies from 1% and 2% for five types to 20% and 25% for rainy tropical and polar climates, respectively. Only about one-fourth of the coastline of the world has a temperate climate.

The longitudinal gradients of the 14 coastal climates in Bailey's classification are generally weak. In contrast, the climatic gradients across the coastal zone are frequently steep. That there is a distinction between oceanic, coastal, and continental conditions has been demonstrated by Bailey (1976) through comparison of the data in Table C19 with that from non-coastal areas. When the mean annual range of temperature is related to mean annual temperature, the intermediate nature of the coastal zone is evident (Figure C40). There are decided interhemispheric contrasts in the curves, contrasts that are especially evident in the seasonal swings (A of Figure C40) of temperature.

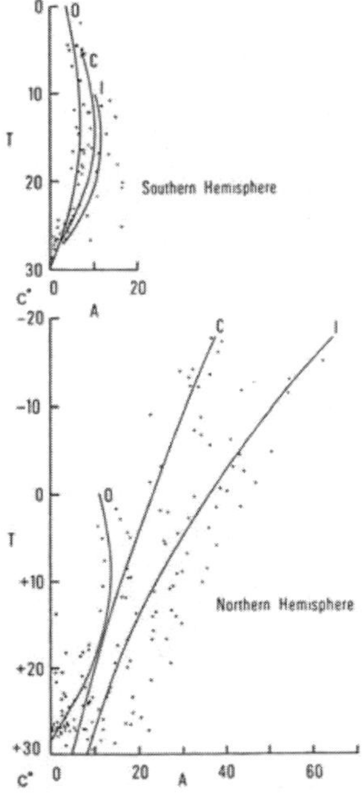

Figure C40 The relation between mean annual range of temperature (A) and mean annual temperature (T) in each hemisphere for oceanic (O), coastal (C), and interior (I) places (from Bailey, 1976).

Bailey's classification of coastal climates, like most climatic classifications, is based on temperature and precipitation. Dolan et al. (1972), however, used airmass climatology as their basis for separating the coastal climates of the Americas. The airmass characteristics they considered significant are: airmass seasonality, airmass source region, nature of the surface over which the airmass moves, and the confluence of airstreams at the coast. Further conditions strengthening this approach are: each airmass possesses a set of secondary characteristics, including the traditionally utilized meteorological elements; the airmass is modified upon crossing the coast; and the boundary between airmasses (i.e. a front) is distinct and separates natural climatic complexes. Although the two approaches are different, the mapped distributions of coastal climates are nonetheless similar in both.

The future

Coastal meteorology and coastal climatology are in their infancy as research disciplines, but the future for both looks bright. Many research organizations are beginning to place greater emphasis on the meteorology and climatology of coasts, albeit an emphasis that is still related mostly to the demand for data by other coastal sciences. In addition to the topics briefly discussed here, research is under way on storm surges, upwelling, tidal stirring, coastal fog, coastal dunes, and coastal frontogenesis, among others.

R.A. Muller, H.J. Walker and S.A. Hsu

Bibliography

Bailey, H.P., 1960. Climates of coastal regions. In Putnam, W.C., Axelrod, D.I., Bailey, H.P., and McGill, J.T., eds., *Natural Coastal Environments of the World*. Berkeley and Los Angeles, CA: University of California Press, pp. 59–77.

Bailey, H.P., 1976. Coastal climates in global perspective. In Walker, H.J., ed., *Geoscience and Man*, vol. 15: *Coastal Research*. Baton Rouge, LA: Louisiana State University School of Geoscience, pp. 65–72.

Borengasser, M., Muller, R.A., and Wax, C.L., 1978. *Barataria Basin: Synoptic Weather Types and Environmental Responses*, Sea Grant Publication No. LSU-7-78-001. Baton Rouge, LA: Louisiana State University Center for Wetland Resources.

Davis, R.E., and Dolan, R., 1993. Northeasters. *American Scientist*, **81**: 428–439.

Davis, R.E., Dolan, R., and Demme, G., 1993. Synoptic climatology of Atlantic Coast north-easters. *International Journal of Climatology*, **13**: 171–189.

Defant, F., 1951. Local winds. In Malone, T.F., ed., *Compendium of Meteorology*. Boston, MA: American Meteorological Society, pp. 655–672.

Dolan, R., and Davis, R.E., 1992. Rating northeasters. *Mariners Weather Log*, Winter, pp. 4–11.

Dolan, R., Hayden, B.P., Hornberger, G., Zieman, J., and Vincent, M.K., 1972. *Classification of the Coastal Environments of the World. I: The Americas*. Technical Report 1. Charlottesville, VA: Department of Environmental Sciences, University of Virginia.

Fairbridge, R.W., 1967. Maritime climate: oceanicity. In Fairbridge, R.W., ed., *The Encyclopedia of Atmospheric Sciences and Astrogeology*. New York: Reinhold, pp. 546–549.

Hsu, S.A., 1970. Coastal air circulation system: observations and empirical model, *Monthly Weather Review*, **10**: 187–194.
Hsu, S.A., 1976. Determination of the momentum flux at the air–sea interface under variable meteorological and oceanographic conditions: further applications of the wind-wave interaction method. *Boundary-Layer Meteorology*, **10**: 221–226.
Hsu, S.A., 1977. Boundary-layer meteorological research in the coastal zone. In Walker, H.J., ed., *Geoscience and Man*, vol. 18: *Research Techniques in the Coastal Environment*. Baton Rouge, LA: Louisiana State University School of Geoscience, pp. 99–111.
Muller, R.A., 1977. A synoptic climatology for environmental baseline analysis, New Orleans. *Journal of Applied Meteorology*, **16**: 20–33.
Muller, R.A., and Stone, G.W., 2001. A climatology of tropical storm and hurricane strikes to enhance vulnerability prediction for the southeast US coast. *Journal of Coastal Research*, **17**: 949–956.
Muller, R.A., and Wax, C.L., 1977. A comparative synoptic climatic baseline for coastal Louisiana. In Walker, H.J., ed., *Geoscience and Man*, vol. 18: *Research Techniques in the Coastal Environment*. Baton Rouge, LA: Louisiana State University School of Geoscience, pp. 121–129.
Wilson, L., 1967. Climatic classification. In Fairbridge, R.W., ed., *The Encyclopedia of Atmospheric Sciences and Astrogeology*. New York: Reinhold, pp. 171–193.

Cross-references

Coral Reefs and Climate
Lakes, Effects on Climate
Land and Sea Breezes
Local Climatology
Maritime Climate
Ocean Circulation
Sea Level Rise

COMMERCE AND CLIMATE

The saying that "everyone talks about the weather but no one can seem to do anything about it" may seem like an appropriate statement for the vast majority of people even today. In many sectors of the nation's economy the effect of weather and climate variations are well known yet the "weather" variable is the least understood in terms of all factors involved in weather-sensitive decisions (Maunder, 1970). Enhanced use of climate information has come about in recent years through improved understanding of the weather and climate, enhanced technology, and an interdependent global community that is ever more reliant on decisions impacted by weather and climate issues.

The "value" associated with the use of climate information in aspects of "commerce" (herein related to retail sales, the commodities market, and insurance sector) has been difficult to determine over time due to the lack of available data and investigations by either climatologists or businesses. Since the 1970s atmospheric scientists in three sectors – government agencies, universities, and commercial meteorology firms – have become increasingly engaged in activities with business decision makers (Smith, 2002). What triggered this increased interest in knowing more about the benefits and costs associated with weather and climate? First, as climate data and information became more timely and accessible, users could easily incorporate them into timely weather-related decisions. Second, since the 1950s weather forecasts at all time scales, hours to seasons, have improved and become more accurate. Third, weather-related losses in the United States and around the world have increased dramatically, forcing government officials and insurance risk managers to re-examine climate risks and adjust accordingly. Recent studies (Dutton, 2002; Murnane et al., 2002) have indicated that the value of weather and climate information has increased and will continue to do so in this highly urbanized, industrial, and interdependent world. These new challenges have created new areas of growth for atmospheric scientists.

For the United States, Dutton (2002) estimated that nearly one-third (approximately $3.0 trillion in 2000) of all private industry activities were weather-sensitive. Commercial activities (retail trade) accounted for $894 billion and he estimated that nearly 100% of those activities were weather-sensitive. Changnon (1999) noted that United States retail sales increased nearly $5.6 billion as a result of warmer than average conditions across the northern tier of states during the El Niño winter of 1997–1998. There is uncertainty in the economic value of weather in commercial activities but it is clearly significant.

Increased understanding of oceanic/atmospheric interactions and their relationship to regional climate variability around the world is considered one of the major climate advances since the 1970s. As a result, monthly and seasonal climate outlooks, which provide users with probabilistic forecasts of temperature and precipitation anomalies, are now being integrated into financial strategies used to hedge weather risks in energy, retail, and other industries (Dutton, 2002).

Retail sector

When examining the influence of weather/climate on the retail sector two time scales have been considered. Many studies have examined whether short-term day-to-day sales of a specific item are influenced (and to what degree) by weather conditions (whether it rains or not). For example, Zeisel (1950) examined the relationship of summer hot periods to beer sales, while Linden (1959) analyzed the influence of precipitation events on umbrella sales. Longer-term seasonal sales have generally been related to seasonal climate (either temperature or precipitation) anomalies such as those associated with El Niño (Changnon, 1999).

In any retail assessment it is difficult to document non-weather factors such as whether a shopper has the resources (money) to make purchases. To minimize this uncertainty in order to draw a clearer picture of how retail sales associated with one item are related to specific weather events or climate anomalies, studies have examined the retail–weather relationship over a number of years. This approach provides a range of outcomes and allows one to examine sales during warm versus cold periods and wet versus dry periods (Linden, 1962).

When examining the weather influences on all retail sales it is also important to consider that regional differences in weather across a country as large as the United States may in fact balance out regional profits in one area and losses in another. For example, during the 1997–1998 El Niño winter retail sales for the United States were $4 billion above average; however, the gains were not experienced nationwide. Warmer and drier winter conditions along the northern tier of the United States helped to enhance retail sales as people were more likely to venture out of their homes to shop, whereas retail sales from southern California eastward along the Gulf Coast and into Florida sagged due to cooler, cloudier, and wetter winter conditions (Changnon, 1999). Similar regional increases in retail sales were experienced in the eastern two-thirds of the United States during the record warm winter of 2001–2002 (Changnon and Changnon, 2002). On the other end of the scale, the record cold and snowy winters of the late 1970s in the Midwest and Northeast United States had a largely negative impact on retail sales (exceptions being winter clothing) in these regions (Changnon and Changnon, 1978). In summary, when examining national retail sales, one must keep in mind that there will often be regional "winners" and "losers" as some aspects of the retail sector are enhanced by pleasant weather while are others hurt by a specific weather event or climate anomaly.

Commodities sector

This sector has witnessed some of the most dramatic changes in how it values weather and climate information. The commodities market involves trading of futures commodities such as oil, grains, meat, etc. (purchase now for something expected at a later date). Agricultural commodities have always been directly impacted by two conditions: present and forecasted weather events (Dutton, 2002). Mitigative strategies, such as contracts between different parts of the decision chain (resource, producer, processor, distributor, consumer) and a third party, were developed as a type of insurance plan for traders (Figure C41). Any commodities trader knows that the "weather factor" is ever changing, and that because commodities are grown and traded worldwide, one must examine weather issues and impacts around the globe (Davis, 2002). In the United States the weather-sensitive sector dominating commodities trading in the mid-1980s was agriculture. However, due to the deregulation of oil, natural gas, and electricity industries in the early 1990s, the need for weather and climate information in the United States energy sector has since increased and is now equal to agricultural needs.

Many decision makers in weather-sensitive industries realize that in this interdependent world they must better manage the weather and climate "risk" to reduce profit volatility and deal with increased competition. These efforts are becoming more comprehensive and require quantitative answers (Dutton, 2002). As part of the commodities sector, the weather risk market has generated considerable interest in new approaches to management of the financial aspects of weather and climate risk. The recent development of weather "derivatives", as a financial instrument to deal with seasonal climate anomalies, requires the use of very accurate climate information (Dischel, 1999, 2002; Dutton, 2002).

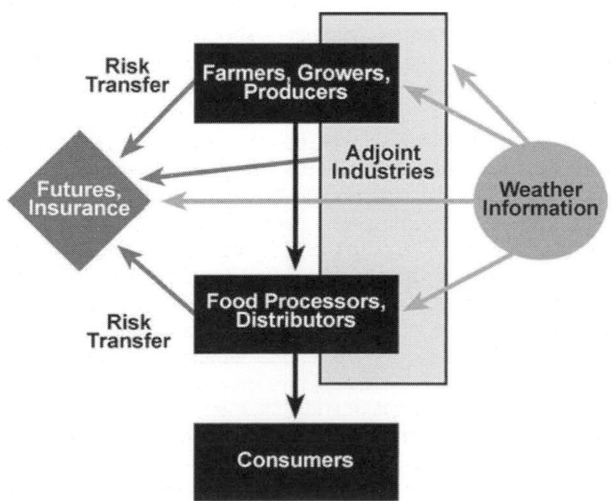

Figure C41 Flows of agricultural products, weather information, and transfers of financial risk (Dutton, 2002).

How does a weather derivative (contract) work? For instance, you own a Chicago company that is negatively impacted by very cold winter temperatures and desire protection. Once the risk associated with a particular seasonal climate anomaly (e.g. average winter temperatures in Chicago at 6°F below average) is determined, a contract can be established between you and another party. That is, one party concerned about this weather risk purchases an insurance-type policy that will pay (cover losses) if the average winter temperature is ≥6°F colder than average. According to the Weather Risk Management Association (WRMA, 2002) the total national value of seasonal weather derivatives executed between parties over-the-counter was about \$2 billion per year in 1998–2000 and \$4 billion in 2001. This climate-based tool has dramatically altered how those involved with the commodities markets manage weather and climate risks and, as a result, it has opened new and extremely important opportunities for atmospheric scientists (Dutton, 2002).

Insurance sector

Since the late 1950s insurance companies have asked atmospheric scientists to assist them with the difficult task of determining weather-related risks to both crops and property. Initially, crop-hail insurance companies, which had experienced severe and extremely variable losses during the 1950s, wanted to gain a greater knowledge of the hail risk they faced at the local and regional scale. Using a statewide index of annual hail frequency and the annual "loss cost" value calculated for each state by the insurance sector, the hail risk calculations were completed (Changnon and Stout, 1967). These initial analyses, which were updated in the late 1990s (Changnon and Changnon, 2000), showed that, except in the western Great Plains and northern Texas, the hail risk had generally decreased or not changed in most agricultural regions of the United States (Figure C42). The research conducted by these meteorologists has assisted insurance decision makers in establishing the degree of risk and hence the premiums to be levied. Importantly, atmospheric scientists can help explain the causes of weather-related losses.

When examining property and casualty-insured losses related to weather catastrophes since the 1950s, one is impressed with the steady increase in total losses (Figure C43) and may have initially associated this increase with global warming. However, when losses for each catastrophe were adjusted for changing conditions (changes in property values, cost of repairs, growth in the size of the fixed property market, and relative change over time in the share of the total property market that was insured against weather perils), the trend in losses looked very different (Figure C44). Years with high values occurred when a major tropical storm/hurricane hit the United States (e.g. estimated losses for Hurricane Andrew in 1992 adjusted to 1996 dollars were \$30.5 billion).

Assessment of climate extreme data (hail, tornadoes, thunderstorms, floods, hurricanes, etc.) and insurance data led to important findings. Although annual insurance losses have

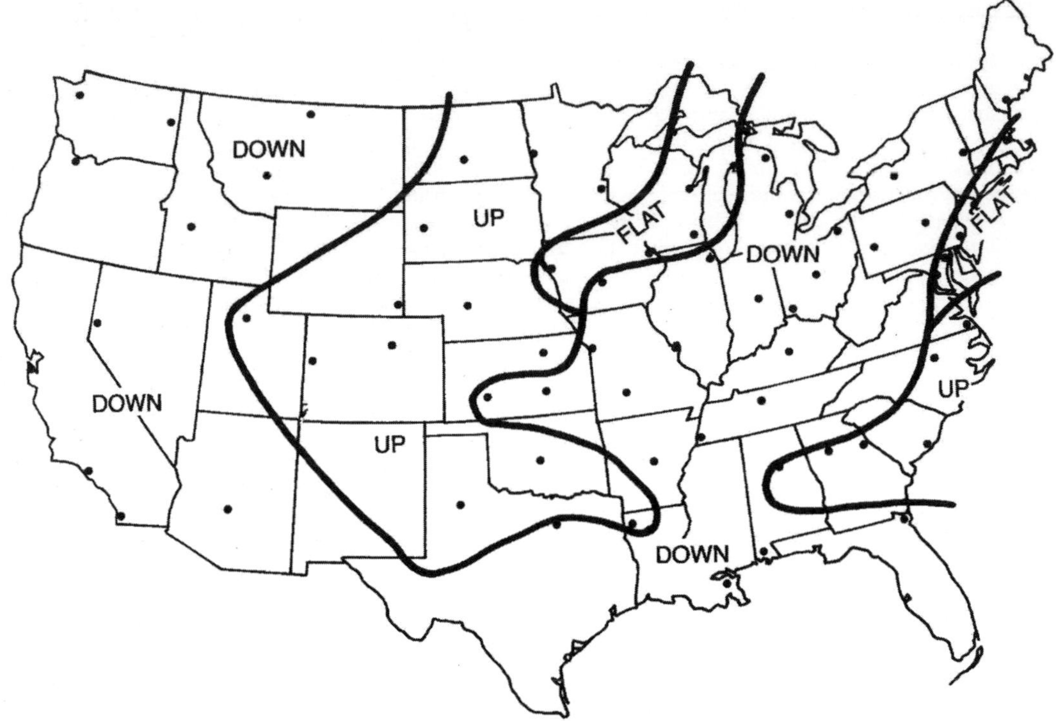

Figure C42 Regions defined based on linear trends of the 100-year hail-day values. Up is for increasing trends significant at the 5% level, flat is essentially no trend up or down, and down is for decreasing trends significant at the 5% level (Changnon and Changnon, 2000).

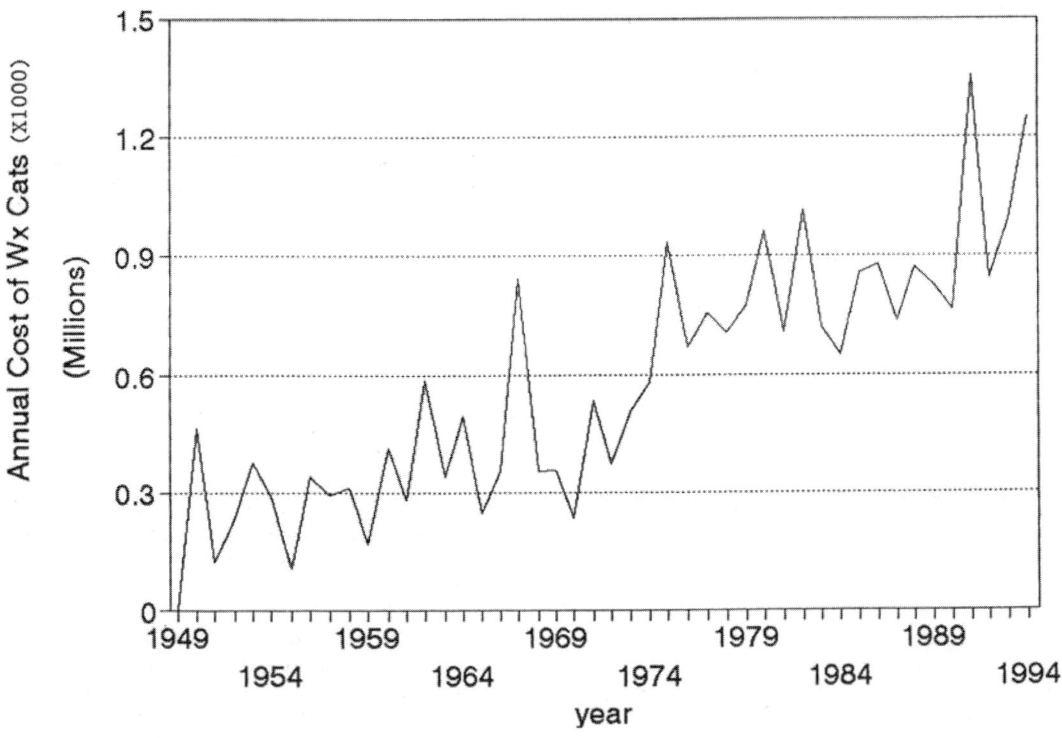

Figure C43 Annual losses produced by $10 million to $100 million catastrophes during 1949–1994 (Changnon and Changnon, 1998).

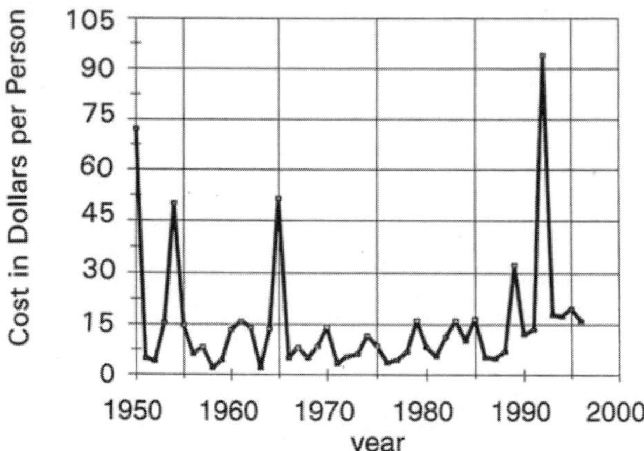

Figure C44 Annual losses caused by catastrophes causing <$10 million in insured losses normalized by dividing annual losses by the annual US population during the 1950–1996 period. Values are dollars per person (Changnon and Changnon, 1999).

increased, and will continue to do so into the future, the changes during the past 50 years appear to be more closely related to societal changes than to changes in weather extremes (Changnon et al., 1999). Most climate extremes in the United States do not exhibit steady, multidecadal increases over time, as found in the insurance losses. Societal (demographic) changes – moving to coastlines, cities, and mountains) – explain some of the loss, along with growth of population and wealth. These societal shifts suggest that the "targets" for damaging weather have changed and have become more vulnerable to huge financial losses (Changnon and Changnon, 1999). Furthermore, more climate "impact" data are needed if climatologists working with insurers are to continue to understand and forecast insured losses.

Summary

There are a number of messages that need to be learned by climatologists if they are to become more effective in relating weather and climate information to commercial activities described herein. Many of the fundamental points listed below have been identified earlier by other atmospheric scientists (Maunder, 1970, 1973). However, given increased knowledge of the atmospheric sciences, enhanced technology, and increased loss amounts related to weather catastrophes, a number of these fundamental points have been modified to reflect current issues. The overall theme of this list is that interactions and research conducted with and for commercial business decision makers (Figure C45) should be "user-centered" (Changnon, 1998). The concept that atmospheric scientists totally understand (without user interaction) and solve the weather-related issues facing the commercial sector must be tossed aside. It is time climatologists and business decision makers got together. Following are five lessons to enhance commerce in the future.

1. Climatologists must work closely with the user in this changing world.
2. Climatologists should understand the value of climate information and develop relevant tools for commercial decision makers.

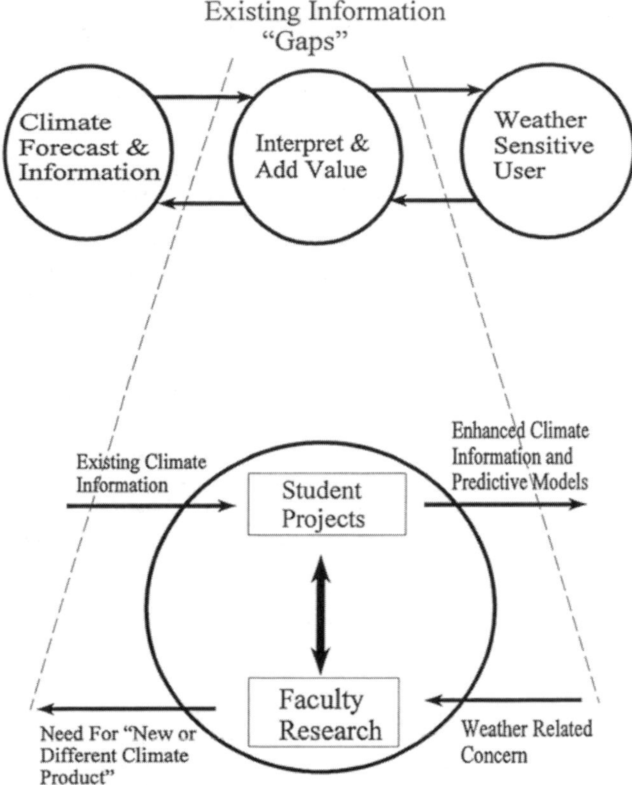

Figure C45 Bridging the information "gap" between climatologists and users (Changnon, 1998).

3. Climatologists need to separate climate variations from those of nonclimate factors involved in decision models.
4. Climatologists need to understand that uncertainty exists in any climate-related decision and this must be explained to users.
5. Climatological products need to change as the commercial world evolves.

Increased resources (public and private) for maintaining important climate databases and information are necessary if long-term benefits are to occur. Due to the dynamic nature of both the atmosphere and commerce, true benefits will be found only through long-term relationships. Basic reference unbiased climate datasets need to be maintained by the government to help users determine and minimize weather risks (Murnane et al., 2002).

David Changnon

Bibliography

Changnon, D., 1998. Design and test of a "hands-on" applied climate course in an undergraduate meteorology program. *Bulletin of the American Meteorological Society*, **79**: 79–84.

Changnon, D., and Changnon, S.A., 1998. Climatological relevance of major USA weather losses during 1991–1994. *International Journal of Climatology*, **18**: 37–48.

Changnon, S.A., 1999. Impacts of 1997–98 El Niño-generated weather in the United States. *Bulletin of the American Meteorological Society*, **80**: 1819–1827.
Changnon, S.A., and Changnon, D., 1978. Winter storms and the record-breaking winter in Illinois. *Weatherwise*, **31**: 218–225.
Changnon, S.A., and Changnon, D., 1999. Record-high losses for weather disasters in the United States during the 1990s: how excessive and why? *Natural Hazards*, **18**: 287–300.
Changnon, S.A., and Changnon, D., 2000. Long-term fluctuations in hail incidences in the United States. *Journal of Climate*, **13**: 658–664.
Changnon, S.A., and Changnon, D., 2002. Unusual winter weather of 2001–2002 caused major impacts on the nation's economy. *Changnon Climatologist, Mahomet, IL.*
Changnon, S.A., and Stout, G.E., 1967. Crop-hail intensities in central and northwest United States. *Journal of Applied Meteorology*, **6**: 542–548.
Changnon, S.A., Pielke, R.A., Changnon, D., Sylves, R., and Pulwarty, R., 2000. Human factors explain the increased losses from weather and climate extremes. *Bulletin of the American Meteorological Society*, **81**: 437–442.
Davis, J.B., 2002. Personal communication; Vice-President in charge of Meteorology, Salomon Smith Barney, Inc., Chicago, IL.
Dischel, R., 1999. Weather risk at the Frozen Falls Fuel Company. *Chicago Mercantile Exchange*.
Dischel, R., 2002. *Climate Risk and the Weather Market: Financial Risk Management with Weather Hedges.* Risk Books.
Dutton, J.A., 2002. Opportunities and priorities in a new era for weather and climate services. *Bulletin of the American Meteorological Society*, **83**: 1303–1311.
Linden, F., 1959. Weather in business. *Conference Board Business Record*, **16**: 90–94.
Linden, F., 1962. Merchandising weather. *Conference Board Business Record*, **19**: 15–16.
Maunder, W.J., 1970: *The Value of the Weather.* London: Methuen.
Maunder, W.J. 1973. Weakly weather and economic activities on a national scale: an example using United States retail trade data. *Weather*, **28**: 2–12.
Murnane, R.J., Crowe, M., Eustis, A., et al., 2002. The weather risk management industry's climate forecast and data needs: A workshop report. *Bulletin of the American Meteorological Society*, **83**: 1193–1198.
Smith, M.R., 2002. Five myths of commercial meteorology. *Bulletin of the American Meteorological Society*, **83**: 993–996.
WRMA, 2002. Second annual industry survey. www.wrma.org.
Zeisel, H., 1950. How temperature affects sales. *Printers Ink*, **233**: 40–42.

Cross-references

Applied Climatology
Agroclimatology
Climate Affairs
Climate Hazards
Tourism and Climate

CONDENSATION

Condensation is the phase change of water from its vapor to liquid. When moist air is cooled sufficiently, it attains a temperature at which saturation occurs – the dewpoint. Further cooling produces supersaturation, a condition in which relative humidity exceeds 100%. Supersaturation seldom occurs in the atmosphere, therefore condensation is the result of cooling. The condensed water may appear as dew or frost on the ground or as a suspension of water droplets in the air, fog or cloud.

Table C20 Heat transfers associated with phase changes of water

Phase change	Heat transfer	Type of heat
Liquid water to water vapor	540–590 cal absorbed	Latent heat of vaporization
Ice to liquid water	80 cal absorbed	Latent heat of fusion
Ice to water vapor	680 cal absorbed	Latent heat of sublimation
Water vapor to liquid water	540–590 cal released	Latent heat of condensation
Liquid water to ice	80 cal released	Latent heat of fusion
Water vapor to ice	680 cal released	Latent heat of sublimation

Condensation is brought about in three ways: (1) adiabatic expansion, which occurs when air rises to progressively lower pressure levels; (2) contact with a cold surface, for example, if the temperature of the surface is below the dew point, condensation may occur on that surface; and (3) mixing, for example, when two moist airmasses of very different temperatures are mixed, condensation may result.

If the air were perfectly clean, supersaturation of several hundred percent could occur. Air, however, contains impurities that act as nuclei on which condensation can occur. These condensation nuclei can be conveniently considered in three categories:

1. Aitkin nuclei with radii between 5×10^{-3} and 2×10^{-1} μm.
2. Large nuclei with radii between 0.2 and 1 μm.
3. Giant nuclei with radii greater than 1 μm.

The sizes and concentration of the condensation nuclei influence such factors as visibility and the precipitation process.

The condensation process, as a phase change, is responsible for the transfer of heat energy in the atmosphere. When water melts or in vaporized heat is absorbed. When the reverse process – condensation – occurs heat is released. Table C20 provides a summary of the energy exchanges that occur.

John E. Oliver

Bibliography

Meteorological Office, 1981. *Elementary Meteorology.* London: Her Majesty's Stationery Office.
Neiberger, M., Edinger, J.G., and Bonner, W.D., 1982. *Understanding Our Atmospheric Environment.* San Francisco, CA: W.H. Freeman.
Oliver, J.E., and Hidore, J.J., 2002. *Climatology: an atmospheric science.* Upper Saddle River, NJ: Prentice Hall.
Thompson, R.D., 1998. *Atmospheric Processes and Systems.* London: Routledge.

Cross-references

Adiabatic Phenomena
Atmospheric Nuclei and Dust
Evaporation
Evapotranspiration
Latent Heat
Phase Changes

CONTINENTAL CLIMATE AND CONTINENTALITY

A continental climate has characteristics associated with areas within a continental interior as opposed to coastal locations under the moderating influence of water. Among the most significant of these climatic properties are large annual temperature ranges, which can be attributed to the distinctive thermal differences between land and water. Heat capacity is the ratio of the amount of heat energy absorbed by a substance to its corresponding temperature increase. Specific heat refers to the amount of energy necessary to raise the temperature of an exact mass of a substance by a certain quantity. It takes the energy of 4190 joules to increase the temperature of pure water by one degree Celsius. Thus, water has a high specific heat causing it to cool and heat slowly. Conversely, land has a low specific heat (five to six times less than that of water) causing it to cool and heat faster than water. Additionally, land surfaces lack the evaporative cooling effect of water and, unlike water bodies that distribute heat energy both horizontally and vertically, the solar radiation received by land heats only a shallow surface layer. As a result of these characteristics, continental locations tend to have cold winters and warm to hot summers.

Climatic characteristics

Continental climates also have large diurnal temperature ranges. Because these regions tend to be located in the interior of a continent, they are far removed from sources of moisture, reducing the availability of water vapor and subsequent cloud development. Clear skies allow insolation to quickly heat the land surface by day, causing temperatures to rise, while at night the absence of clouds enhances radiational cooling as temperatures rapidly decrease. Reno, Nevada, is an example of a continental location with a large diurnal temperature range. The mean maximum temperature in July is 92°F; however, the average minimum temperature during the same month is 47°F, yielding a daily range of 45°F.

A lack of significant moisture sources also causes continental climates to have relatively low annual precipitation amounts. Maritime airmasses that originate over distant water bodies typically lose much of their moisture before arriving at a continental interior. However, higher temperatures during summer increase potential evapotranspiration, thus contributing water vapor to the regime and improving the possibility of precipitation. Table C21 lists the mean annual temperature, annual temperature range, and annual precipitation of selected continental stations.

Continentality

Continentality refers to the degree to which a place on the planet is affected by the temperature and moisture characteristics of a large, interior landmass. With the exception of latitude and its effect on solar radiation receipt, continentality is the most meaningful climatic control, and it is practical, for certain purposes, to separate every climate into a maritime or continental category (Driscoll and Fong, 1992). Further, Kopec (1965) asserts that it is possible to characterize a location based on its proportion of continentality, that is, its rank between perfect continentality and the total lack of continentality. D'Ooge (1955) notes that continentality quantifies the degree to which a region is influenced by continental characteristics. In the classical sense of the term, Stamp defines continentality simply as a climate with a wide range of diurnal and annual temperatures, as in the geographical center of a continent (Oliver, 1970).

The degree of continentality at a particular location is typically measured through the use of a derived climatic index. For example, the index of continentality at Prairie du Chien, Wisconsin, is 50%, which implies that the effect of continentality on the local climate is 50% greater at Prairie du Chien that on the island statio of Pago Pago (Samoa) where the index of continentality is 0%. The converse of continentality is "oceanicity." A value of 0 represents. the climate on an Earth completely covered with water (Kopec, 1965).

Continentality indices

Several climatic indices that account for the daily or annual range of temperature and which include some allowance for latitude have been derived. One of the earliest attempts to assign a numerical value based on geographic location and temperature came from Forbes (1859), who correlated the dependent variable, temperature, with the independent variable, latitude, in a formula expressed as:

$$T = -17.8^\circ + 44.9^\circ \cos^2(\phi - 6^\circ 30')^\circ\mathrm{C}$$

where T represents temperature and ϕ represents latitude.

Table C21 Temperature, precipitation, and continentality of selected stations

Station	January mean temperature (°C)	July mean temperature (°C)	Annual range (°C)	Continentality value (%)	Annual precipitation (cm)
Beijing, China	−04.4	26.1	30.5	54	62.23
Calgary, Canada	−10.6	16.8	27.4	39	42.42
Fargo, North Dakota	−13.9	21.7	35.6	58	47.45
Fort Simpson, Canada	−27.2	17.2	44.4	65	33.27
Irkutsk, Russia	−21.1	15.6	36.7	57	37.59
Moscow, Russia	−10.0	18.9	28.9	40	62.99
Sverdlovsk, Ukraine	−17.8	16.8	34.6	50	42.41
Verkhoyansk, Siberia	−50.0	13.3	63.3	96	13.46

In 1905 Kerner developed a measure of the maritime influence on a location, or oceanicity, which he referred to as the thermoisodromic ratio. Kerner's formula is expressed as:

$$O = 100[(T_o - T_a)/A]$$

where O represents oceanicity, T_o represents the monthly mean temperature for October, T_a represents the monthly mean temperature for April, and A is equal to the average annual temperature range (Oliver and Fairbridge, 1987).

Among the earliest indices specific to continentality is that of Gorczynski (1920), whose formula is expressed as:

$$K = 1.7(A/\sin\phi) - 14$$

where K represents continentality, A is the average annual temperature range, and ϕ is latitude.

Conrad (1946) later amended Gorczynski's equation, suggesting that the sine of the latitude plus a constant of 10 would allow for use of the index at low latitudes. Conrad's index represents a reliable formula, which is widely accepted and expressed as:

$$K = 1.7[A/\sin(\phi + 10)] - 14$$

where K represents continentality, A is the difference between the mean temperature of the warmest and coldest month in degrees Celsius, and ϕ is station latitude in degrees.

Conrad's formula is perhaps the most frequently used of all continentality indices. Based on temperature records published in *Climatological Data* for 1952, Fobes (1954) applied Conrad's index in an assessment of the continentality of New England. Conrad personally assisted in the study, and the researchers concluded that northern stations in Maine possess the highest continental index values in the region with the effect of continentality diminishing as one approaches the coasts. Index values for New England range from 30% along the coasts to 50% in the north.

D'Ooge (1955) examined the western United States using Conrad's index of continentality and temperature data from 1952. He found the maritime influence to be much greater than in New England with the 40% continentality values extending approximately 500 miles inland. The highest values are in eastern Montana and North Dakota (60%). However, the values are still less than the 100% calculated by Conrad (1946) for Verkhoyansk, Siberia, which represents the maximum continentality value on the planet.

Trewartha (1961) applied Conrad's index to an analysis of North America and found continentality to be much weaker on the West Coast than on the East Coast, with continentality values of 5–10% for the Pacific Coast, 25–30% along the Gulf Coast, and 30–40% along the Atlantic Seaboard. The core area of maximum continentality (60%) is located west of Hudson Bay in central Canada. Figure C46 depicts continentality values of North America as derived by Conrad's index.

Mapping continentality

Using data from the United States Weather Bureau for the period from 1931 through 1952, and Canadian data for the period from 1921 through 1950, Kopec (1965) analyzed continentality around the Great Lakes. Conrad's index reveals that the highest values for the region (60%) can be found northeast of the Great Lakes. As expected, the Great Lakes act as a moderating influence on the surrounding region with the lowest continentality values located along the shores. Kopec (1965) explains that the Great Lakes delineate a zone of reduced

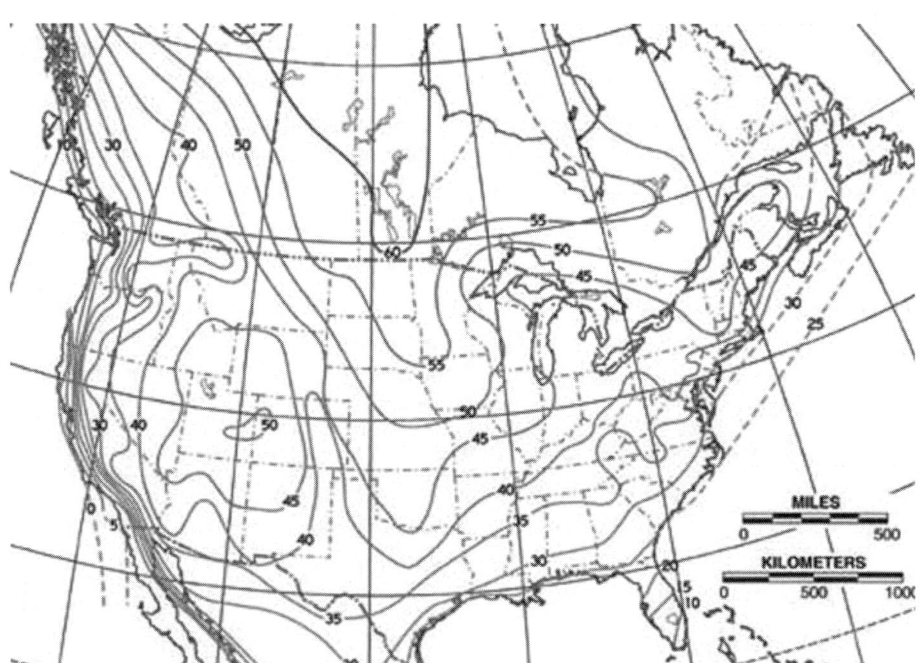

Figure C46 Continentality values of North America (after Trewartha, 1961).

continentality values situated within a region that would otherwise represent the core of continentality for North America. In Siberia a comparable role is played by Lake Baikal. Even small lakes, as in Switzerland, a mild effect is noticeable.

More recently, Conrad's continentality index is among the formulas selected for a major study conducted by the Climate Criteria Examination Team (CET). The CET acts under the auspices of the Atmospheric Radiation Measurement Program (ARM), a multi-laboratory, interagency program created in 1989 with funding from the US Department of Energy (DOE). The CET is tasked with identifying and characterizing the geographic regions of the world based on the climatic attributes of latitude, seasonality, and continentality.

Despite its widespread use, the inclusion of the sine of latitude in continentality indices has been criticized because it assumes that contrasts in seasonal radiation receipts are approximately distributed such that dividing by the sine of the latitude negates the influence. This leaves aridity and continentality as the only remaining controls over annual temperature range. However, Driscoll and Fong (1992) insist the technique is flawed since seasonal radiation does not increase monotonically and have derived their own formula based on residuals from the regression line of annual temperature range on latitude. Their index yields a scale for North America that ranges from -10 along the coasts to $+10$ in the heart of the continent. However, the researchers acknowledge that, for regional studies in which the latitudinal area is limited, the use of sine ϕ is essentially correct, and the results are not necessarily invalid (Driscoll and Fong, 1992).

Others have attempted to quantify continentality without including the sine of the latitude in the formula. Oliver (1970) applied an airmass evaluation to the concept of continentality using the formula:

$$K = L \cos A$$

where K is continentality, L is the length of the long axis (in millimeters), and A is the angular deviation away from the vertical on a climograph. Oliver concluded that a continental climate is one that experiences a large daily and annual range of temperature due to the proportionate preeminence of continental airmasses. Despite the ambiguity often associated with the classification of climatic regimes, continentality remains a viable concept that has inspired a wealth of research.

Richard Snow

Bibliography

Conrad, V., 1946. Usual formulas for continentality and their limits of validity. *American Geophysical Union Transactions*, **27**: 663–664.

D'Ooge, C., 1955. Continentality in the Western United States. *Bulletin of the American Meteorological Society*, **36**: 175–177.

Driscoll, D.M., and Fong, J.M. Yee, 1992. Continentality: a basic climatic parameter re-examined. *International Journal of Climatology*, **12**: 185–192.

Fobes, C., 1954. Continentality in New England. *Bulletin of the American Meteorological Society*, **35**: 197.

Forbes, J.D. 1859. Inquiries about terrestrial temperature. *Transactions of the Royal Society Edinburgh*, **22**: part 1, 75.

Gorczynski, W., 1920. Sur le calcul du degrée de continentalisme et son application dans la climatologie. *Geografiska Annaler*, **2**: 324–331.

Kopec, R., 1965. Continentality around the Great Lakes. *Bulletin of the American Meteorological Society*, **46**: 54–57.

Oliver, J.E., 1970. An air mass evaluation of the concept of continentality. *Professional Geographer*, **22**: 83–87.

Oliver, J.E., and Fairbridge, R.W., eds. 1987. *The Encyclopedia of Climatology*. New York: Van Nostrand Reinhold.

Trewartha, G., 1961. *The Earth's Problem Climates*. Madison: University of Wisconsin Press.

Cross-references

Airmass Climatology
Asia, Climates of Siberia, Central and East Asia
Coastal Climate
Maritime Climate
Temperature Distribution

CORAL REEFS AND CLIMATE

Coral reef ecosystems are found in tropical marine environments, and extend between about 30° north and south of the equator. Optimal reef development occurs in clear, nutrient-poor waters where winter temperatures remain above 18°C. Coral reefs are unique in having a strong geological component (calcium carbonate, or limestone) as well as an ecological one (the coral reef ecosystem). Coral reef structures are built by the coral reef ecosystems themselves. Corals, calcareous algae, and many other organisms secrete limestone skeletons that progressively accumulate into complex three-dimensional structures that can attain tens of meters in thickness over a period of several thousand years. The coral reef ecosystem exists as a living veneer over these structures.

Although corals are animals, their success in reef-building is largely founded on their strong symbiotic relationship with microscopic algae, called zooxanthellae. Almost all reef-building corals require this symbiosis to survive. The zooxanthellae grow directly within the otherwise transparent coral tissue, fueled by photosynthesis and benefiting from nutrients provided from the coral animal. The coral animal in turn benefits from the carbohydrates produced by the zooxanthellae. This internal recycling allows corals to thrive in an otherwise nutrient- and food-poor environment. Corals which have these symbiotic algae grow and secrete calcium carbonate much faster than those that do not.

Early predictions of how climate change would affect coral reefs were optimistic. Because coral reefs thrive in warm waters, an expansion of the tropics during global warming was considered a probable boon for coral reefs, allowing them to migrate poleward with the 18°C isotherm. Over the last few decades the negative effects of climate change on coral reefs have dwarfed this optimism.

The most immediate problem facing coral reefs in a high-CO_2 world is global warming, which has so far proved to be damaging to corals and many other reef dwellers. The phenomenon of temperature-induced coral bleaching is the most serious climatic threat to coral reefs. Bleaching is a natural stress response of corals and other organisms that host zooxanthellae, whereby the host organism expels the algae and becomes colorless. A rule of thumb is that corals and many other organisms bleach once temperatures exceed the normal maximum temperature by 1–2°C. Many corals can and do recover from bleaching if the offending conditions are short-lived and of moderate intensity. Until the 1980s, bleaching events were rare and isolated events. In

1982–1983, mass coral bleaching was observed in the eastern Pacific, which resulted in nearly total coral mortality on Galápagos reefs. Mass bleaching events have become increasingly common and widespread during the past two decades, and nearly all have been associated with elevated temperatures. Secondary climate factors in coral bleaching and mortality include water circulation and light. Increased light penetration, particularly of ultra-violet radiation, intensifies bleaching. Vigorous mixing by winds, tides, and currents tends to keep sea surface temperatures in check, and to reduce light penetration. Bleaching in many areas has often been associated with several climatic patterns that occur during El Niño events: decreased cloud cover and increased insolation; increased sea surface warming; reduced winds; and reduced upwelling. In 1997–1998 an estimated 16% of the world's coral reefs were destroyed by bleaching.

Global warming also leads to other changes in the climate system that can affect reefs. Cyclones (hurricanes, typhoons) can destroy coral reefs, and although there is little evidence of an increase in storm frequency, there is some evidence that maximum winds will intensify. Increased precipitation on land can lead to greater sedimentation on many reefs, particularly those near deforested areas. Sea level rise is a consequence of climate change that can benefit some reefs and harm others. Because of the prolific production of calcium carbonate by coral reef organisms, many of the coral reefs that we see today were able to keep pace with the sea level rise of 100 m or more that occurred between about 18 000 and 3000 years ago. Sea level rise is predicted to be less than 1 m over the course of the twenty-first century. This relatively small rise could affect reefs near land if flooding of the coastal zone releases nutrients and sediments that degrade water quality. However, coral reefs that occur away from major landmasses are not considered threatened by a 1–2 m sea level rise.

An additional consequence of increasing atmospheric CO_2 concentration is ocean acidification. Ocean acidification in the tropics is a predictable process, driven directly by rising atmospheric CO_2 concentration rather than by climate change *per se*. The increase in partial pressure of atmospheric CO_2 drives more CO_2 into the ocean, some of which combines with water to form carbonic acid (H_2CO_3), which lowers the pH. Over this century, predicted decreases in surface ocean pH (about 0.25–0.33 pH units) are not likely to cause direct coral mortality, but constitute a creeping environmental problem that affects skeletal formation in corals and reef-building algae, and the overall calcium carbonate accumulation on a reef. Under increased acidity, corals secrete less calcium carbonate, and reef limestone dissolves more readily. It is likely that these effects will cause corals to secrete their skeletons more slowly, which would leave them less competitive for space; or less densely, which would leave them more fragile.

Predicting the effects of future climate on coral reefs depends on how well we can predict climate change, and on how well we can predict coral reef ecosystem response to this change. Corals may adapt to increases in temperature through a variety of mechanisms, including natural selection, and an interesting phenomenon whereby a bleached coral becomes repopulated with a more heat-tolerant variety of zooxanthellae. The recent increase in coral bleaching indicates that these mechanisms are not yet keeping pace with temperature rise. There is no evidence that coral calcification can adapt to changes in seawater chemistry. However, some coral species calcify more rapidly with temperature increase, and for those that survive the negative effects of future warming, a decrease in calcification due to ocean acidification may be offset by increases in calcification due to warmer temperature. Projections of the state of coral reef ecosystems have ranged from moderate to total degradation; all predict significant disruption of ecosystem functioning. Current conservation efforts aim toward removing the controllable stresses on coral (e.g. overfishing, sedimentation, pollution), while encouraging a reduction in fossil fuel emissions.

J. Kleypas

Bibliography

Birkeland, C. (ed.) 1997. *Life and Death of Coral Reefs*. New York: Chapman & Hall.

Buddemeier, R.W., Kleypas, J.A., and Aronson, R.B., 2004. *Coral Reefs & Global Climate Change: potential contributions of climate change to stresses on coral reef ecosystems*. Prepared for the Pew Center on Global Climate Change.

Hughes, T.P., Baird, A.H., Bellwood, D.R., et al., 2003. Climate change, human impacts, and the Resilience of coral reefs. *Science*, **301**: 929–933.

Spalding, M.D., Ravilious, C., and Green, E.P., 2001. *World Atlas of Coral Reefs*. UNEP-WCMC, University of California Press, Berkeley.

Cross-references

Coastal Climate
Climate Change and Global Warming

CORIOLIS EFFECT

Any moving object above the Earth's surface tends to deflect from its course because of Earth's rotation. This deflection is known as the Coriolis force, named after a French engineer and mathematician Gaspard Gustave de Coriolis (1792–1843) in 1835. Given that this deflection is an apparent force, it is perhaps more correctly known as the Coriolis effect.

The Earth rotates eastward and has the same rotational velocity. However, places at different latitude have various linear velocities. A point near the equator goes around at 1000 miles an hour, while one near the pole moves only a few miles an hour. An object launched from the equator to the north will maintain the eastward component of velocity of other objects at the equator. When this object travels away from the equator it will be heading east faster than the ground beneath it. Similarly, an object moving to the equator from the north will be moving more slowly than the ground beneath it, and deflects to the right of its true path (apparently deflects to the west).

In fact there is no actual force involved in these displacements, and the ground is simply moving at a different speed than its original ground speed. Therefore this deflection (the Coriolis force) will only affect the direction and not the speed of the moving body. An object moving a north–south path on Earth will deflect to the right in the northern hemisphere and left in the southern hemisphere.

The magnitude of Coriolis deflection is directly related to the velocity of the object and its latitude, and can be expressed as:

$$F_c = 2(\Omega \sin \phi)v,$$

where: F_c is magnitude of Coriolis force per unit volume,
Ω is Earth's angular velocity,
ϕ is latitude and
v is velocity of the object.

The Coriolis effect is zero at the Equator and maximum at the poles.

The Coriolis effect is significant in the dynamics of atmosphere, in which it affects large-scale atmospheric circulation and development of storms. Global scale winds blow in the paths commanded by the Coriolis effect. Wind deflects to the right in the northern hemisphere and left in the southern hemisphere. Stronger winds will have a greater amount of deflection than weaker winds. Wind blowing near the poles will deflect more than that with the same speed near the equator.

Yuk Yee Yan

Bibliography

Neiburger, M., Edinger, J.G., and Bonner, W.D., 1982. *Understanding Our Atmospheric Environment*, San Francisco, CA: W.H. Freeman.

Persson, A., 1998. How do we understand the Coriolis Force? *Bulletin of the American Meteorological Society*, **79**: 1373–1385.

Van Domelen, D.J., 2000. Getting around the Coriolis force. In http://www.physics.ohio-state.edu/~dvandom/Edu/newcor.html

Cross-references

Atmospheric Circulation, Global
Wind, Principles

CRIME AND CLIMATE

Scientific inquiries into the relationship between climate and crime began during the early nineteenth century. Usually climate has referred to the monthly or seasonal mean temperatures and crime has been classified into two types: crimes against the person (or violent crimes such as homicide, rape, assault, and robbery); and crimes against property (such as burglary, larceny–theft, and motor vehicle theft).

During the nineteenth century A.M. Guerry in France and A. Quetelet in Belgium, in two independent studies, observed the coincidence of higher frequencies of crimes against persons in warmer areas and during warmer months. Moreover, crimes against property increased in winter months and in cooler areas. Quetelet conceptualized these observations into the *Thermic Law of Delinquency* (Cohen, 1941; Harries, 1980). Neither Guerry nor Quetelet felt that climatic conditions were the principal causes of delinquency. Both criminologists are credited with starting the geographic school of criminology where the ecological facts of crime are pursued – namely, the relationship between crime and social conditions (Gibbons, 1978).

Other European scholars, during the late nineteenth and early twentieth centuries, corroborated Guerry's and Quetelet's findings and agreed that climate was not the principal cause of crime (Cohen, 1941). Yet studies asserting a direct causal linkage emerged and persisted.

E.G. Dexter, studying 40 000 assault cases in New York City between 1891 and 1897, suggested that temperature was the most significant condition aggravating the emotions and erupting into fighting (Cohen, 1941; Harries, 1980). A stronger and more enduring endorsement of the causal relationship came from Ellsworth Huntington (1945) who claimed physiological and psychological conditions vary with climate and weather, hence the propensity for riots and assaults to increase with temperature. Huntington's vehement advocacy of climatic determinism as the cause of crime overshadowed the works of others including Cohen, whose work in 1941 contradicted climatic determinism.

During 1941, using the Federal Bureau of Investigation's Uniform Crime Reports, Cohen observed the expected highs for violence during the midsummer and property crimes during the midwinter. However, the most important finding was that regional variations of frequencies and types of crimes were more pronounced and significant than the seasonal variations. Cohen believed that the regional variations were the surrogates for significant social forces (1941). Despite findings like these the academic community's reaction to Huntington's environmental determinism was so strong and negative that inquiries into the relationship between climate and social phenomena in general, and climate and crime in particular, became almost taboo (Harries and Stadler, 1983).

Gradually, studies began emerging disproving the direct climate–crime linkage or including climatic variables with a host of other independent variables in causal analyses. An example of the former is a study by Lewis and Alford (1975) testing the *Thermic Law of Delinquency* by examining 3 years of monthly assault data across 56 US cities. Beginning with the winter months, the expectation was that the number of assaults would oscillate seasonally with the march of the sun on a south to north vector. Such an orderly progression was not found and, regardless of region, the transition from lower to higher levels of assaults was rather abrupt (Lewis and Alford, 1975). An example of the latter is a study by Van de Vliert et al. (1999) associating riots and armed attacks in 136 countries between 1948 and 1977 with temperature, population size and density, degree of democratic government, socioeconomic development, and a cultural masculinity dimension. They conclude that the culture mediates the effect between temperature and collective violence.

The fast pace of changes in computing and information systems technology since the mid-1980s has decreased the number of inquiries into the climate–crime relationship. Easy access to weather and crime information of varying temporal and spatial scales has spawned numerous studies on the relationship between crime or police activity with weather elements or events (see LeBeau and Corcoran, 1990; Rotton and Cohn, 2000). Despite the changes in focus the *Thermic Law of Delinquency* remains an integral guide for these and other studies.

The relationship between climate and crime is more suggestive than definitive, but climate is viewed as indirectly providing enhanced opportunities for the commission of different crimes. Finally, regardless of the empirical evidence, the public may feel there is a definitive relationship between the two; especially since many residential real-estate companies on their Internet websites list a range of factors describing the quality of life in a place and on the list climate is followed by crime.

James Lebeau

Bibliography

Cohen, J., 1941. The geography of crime. *Academy of Political Science and Social Science Annals*, **217**: 29–37.
Gibbons, D.C., 1978. *Society, Crime, and Criminal Behavior*, 4th edn. Englewood Cliffs, NJ: Prentice-Hall.
Harries, K.D., 1980. *Crime and Environment*. Springfield, IL: Charles C. Thomas.
Harries, K.D., and Stadler, S.J., 1983. Climatic determinism revisited: assault and heat stress in Dallas, 1980. *Environment and Behavior*, **15**(2): 235–256.
Huntington, E., 1945. *Mainsprings of Civilization*. New York: Wiley.
LeBeau, J.L., and Corcoran, W.T., 1990. Changes in calls for police service with changes in routine activities and the arrival and passage of weather fronts. *Journal of Quantitative Criminology*, **6**(3): 269–291.
Lewis, L.T., and Alford, J.J., 1975. The influence of season on assault. *Professional Geographer*, **27**: 214–217.
Rotton, J., and Cohn, E., 2000. Violence as a curvilinear function of temperature in Dallas: a replication. *Journal of Personality and Social Psychology*, **78**(6): 1074–1081.
Van de Vliert, E., Schwartz, S., Huismans, S., Hofstede, G., and Daan, S., 1999. Temperature, cultural masculinity, and domestic political violence, *Journal of Cross Cultural Psychology*, **30**(3): 291–314.

Cross-references

Applied Climatology
Determinism, Climatic

CULTURAL CLIMATOLOGY

Climate and life are intimately linked. Without life the composition of the Earth's atmosphere today would be totally different and the climate of the Earth would be more like that of Venus – the Earth's atmosphere would be 98% CO_2 (Morrison and Owen, 1996). This intimate link between life and the atmosphere on Earth has led to claims that the atmosphere is effectively managed by the biosphere, leading to the "Gaia hypothesis" (Lovelock, 1979). No-one is sure how life developed on Earth about 3850 million years ago, but after continuous evolution humans emerged about 5 million years ago, and civilization and society as we know it began about 5000 years ago. Hence it is impossible to study climate in isolation from the biosphere and society, and any changes to the Earth's atmosphere caused directly or indirectly by society should be firmly within the remit of climatology. Since the industrial revolution of the nineteenth century society has threatened to change our climate, ironically to increase CO_2 concentrations and other so-called greenhouse gases, the need for effective climate management has never been greater.

Today the links between climate and society are explored under the headings of applied climatology where the climate is seen as both a resource and a hazard that needs to be effectively managed by society. Climate change cannot be managed under the heading of applied climatology as the issue has global consequences and climatologists do not have the social science skills to effectively, manage climate change and social scientists do not understand the vagaries of climate sufficiently. Hence it is time that climatologists were trained to understand how to manage the climate effectively, otherwise it will be left to politicians and economists to decide the future of our climate. This has never been a more important issue than in the assessment and management of climate change. Hence the need for cultural climatology to emphasize this requirement for climatologists who understand how society works, as well as the climate.

Cultural climatology is a new branch of climatology concerned with the study of physical and dynamical processes in the atmosphere at various space and time scales in conjunction with evaluating and understanding climate–society interactions and feedbacks. Climate is considered to be an integral part of culture and conversely culture is an integral part of climate. This climate–culture dialectic gives meaning and value to the study of climatology that has been missing in the past. Given that the ultimate goal of climatology is to apply the field's knowledge to the solution of both environmental and socioeconomic problems, a cultural turn is taking place as climatologists enter the post-normal phase of their science. The nature of this cultural turn, and what constitutes cultural climatology, is discussed in detail elsewhere, together with a consideration of climate and society relationships as these are at the heart of cultural climatology (Thornes and McGregor, 2003).

Culture is a complex and overused word, and is used to encapsulate everything from the "total way of life" of a people (language, dress, religion, music, values, etc.) through to works of art and advertising. Mitchell (2000, p. 14) gives six possible definitions of culture, but the simplest definition of all is: "Culture is all that is not nature". Although the definition of nature is not as simple as it might seem (Macnaghten and Ury, 1998) we can fruitfully examine the relationship between climate, as part of nature, and culture.

The summons to cultural climatology should not be misconstrued as a call for climatologists to abandon the research mainstays of synoptic, dynamic and physical climatology. On the contrary we expect cultural climatologists to be concerned with issues such as global climate system change, establishing what the principal drivers of the climate system are, assessing how the climate system will respond to natural and human-induced changes, answering how society might respond to the opportunities and threats posed by climate change, and evaluating to what extent the changes expected in the climate system can be predicted. This call to cultural climatology should be seen more as a signal to climatologists that opportunities await us at the interface between science and society, an area which climatologists on the whole have felt great apprehension with in the past.

Cultural climatology is just the ticket for catching the climate and society boat of opportunity. A passage on this boat will bring climatologists closer to understanding the physical and societal mechanisms underlying the complex interactions between components of the climate–human system. All students and purveyors of climatology courses need to look beyond the learning and teaching of straight climate processes by considering the multitude of ways in which climate and society may interact. Such a broadening of horizons into the realms of cultural climatology will not only will provide us with learning and research opportunities, but, will provide society with a better understanding of the meaning of climate.

John Thornes

Bibliography

Lovelock, J.E., 1979. *Gaia: A New Look at Life on Earth*. Oxford: Oxford University Press.
Macnaghten, P., and Urry, J., 1998. *Contested Natures*. London: Sage.
Mitchell, D., 2000. *Cultural Geography*. Oxford: Blackwell.

Morrison, D., and Owen, T., 1996. *The Planetary System.* Reading, MA: Addison–Wesley.
Thornes, J.E., and McGregor, G.R., 2003. Cultural Climatology. In Trudgill, S., and Roy, A., eds., *Contemporary Meanings in Physical Geography.* London: Arnold, pp. 173–197.

Cross-references

Art and Climate
Climatology, History of
Folklore, Myths and Climate
Literature and Climate

CYCLES AND PERIODICITIES

Throughout recorded history people have been aware of periodic meteorological events that have influenced their well-being. Because so much of our lives is governed by seasonal rhythms, it is natural that there has been a search for long-term order in the physical world. Today the study of cycles, which may be defined as a series of recurring events that repeat with some sense of periodicity, is a well-defined area of research.

The search for cycles

The search for cycles in weather records has been a constant endeavor of a lot of meteorologists for many years. Whether the product of natural variability of the climate system or the result of interactions between the Earth's atmosphere and oceans and external agencies, including periodic variations of solar activity and the astronomical motions of the Moon and the planets, there has been a long and largely fruitless search for clear evidence of predictable cycles.

Interest in weather cycles has fluctuated over time, but has recently experienced an upsurge in activity for two principal reasons. First, the record-breaking El Niño of 1982/1983 stimulated huge interest in the possibility of events in the tropical Pacific Ocean having an impact on the weather around the globe (Diaz and Markgraf, 2000). The realization that the El Niño was part of atmosphere–ocean interactions that can set up quasi-cyclic behavior in the climate has led to a burgeoning study of such "oscillations". Second, satellite measurements since 1980 have shown that solar irradiance varies in phase with the 11-year cycle in the number of sunspots on the sun (Figure C47). This has led to renewed interest as to how changes in solar activity might affect the climate (Haigh, 2000).

The quasi-biennial oscillation (QBO)

Prior to the 1980s the one oscillation that had established as being a real feature of the climate was the periodic reversal of the equatorial winds in the stratosphere (Baldwin et al., 2001). On average the cycles in these winds have a period of around 27 months, hence the name quasi-biennial oscillation (QBO). First observed in the 1950s, their period has varied from over 3 years to well under 2 years (Figure C48). The QBO shows well-defined characteristics. The wind regime propagates downward as time progresses. The amplitude is greatest at an

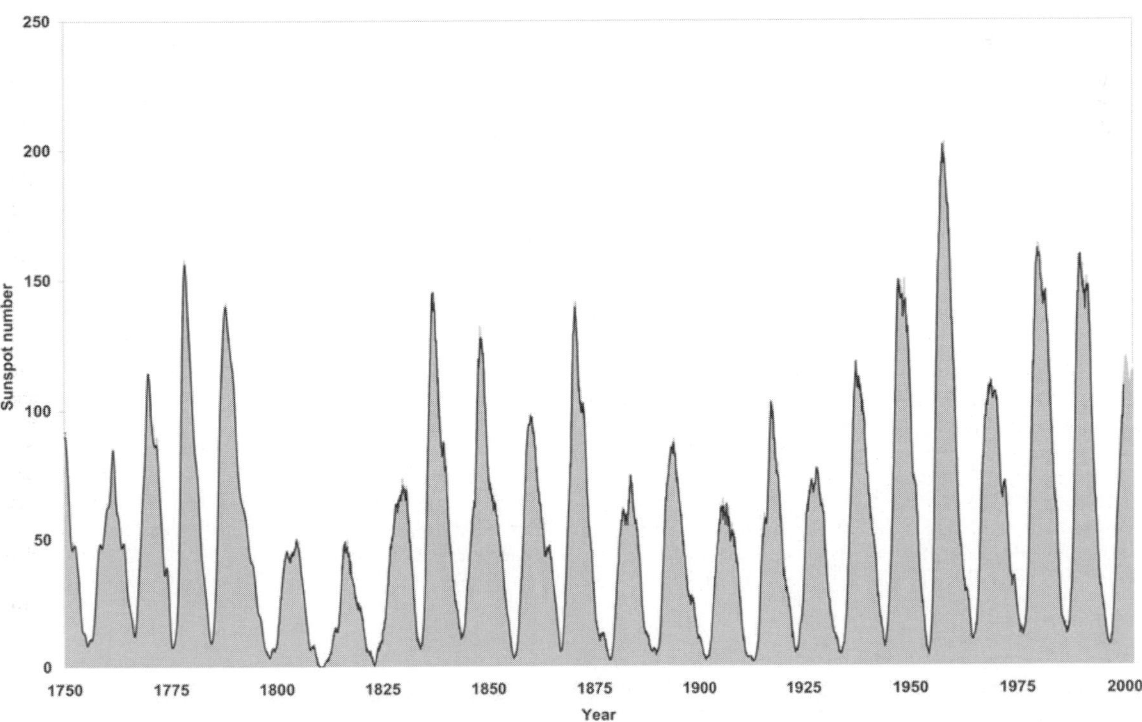

Figure C47 The observed variations in the number of sunspots since 1750, showing the 13-month running mean of monthly numbers. (Data from National Geophysical Data Center, NOAA.)

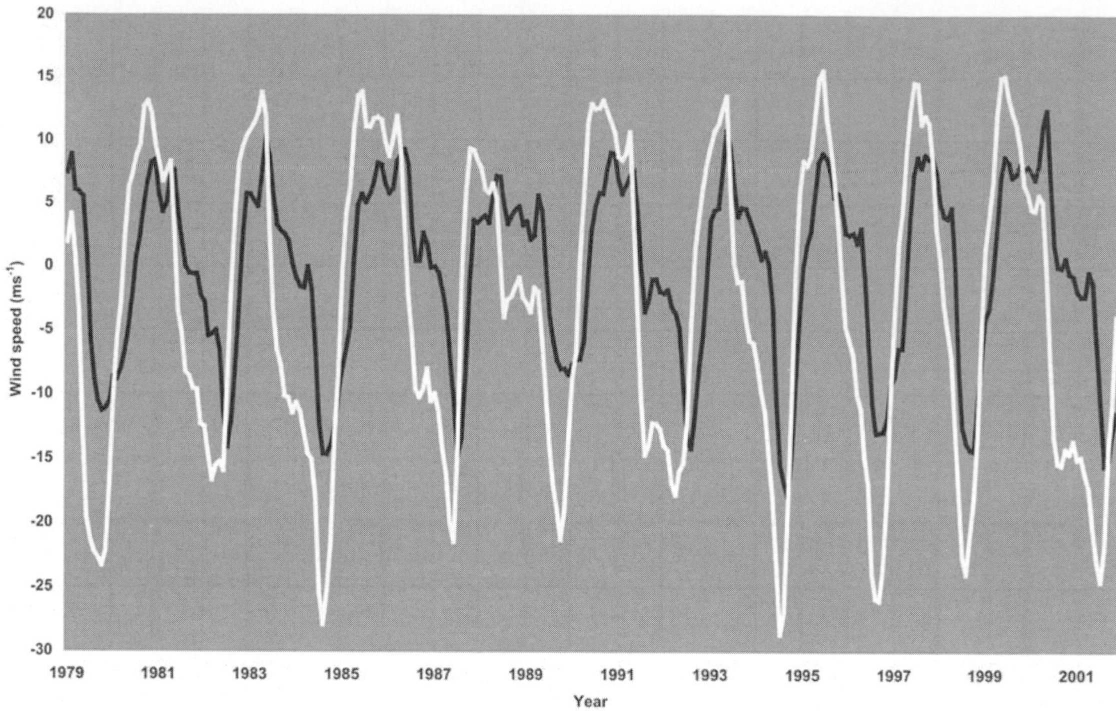

Figure C48 The quasi-biennial oscillation of the winds in the stratosphere over the equator shows a pronounced periodic reversal. The scale of this oscillation is greatest at around 30 hPa (white line) and reduces at lower levels, as shown by the values at 50 hPa (black line) as the periodic feature migrates downward over time, so that the peaks occur first at high levels. (Data from NOAA, available on websites: ftp://ftp.ncep.noaa.gov/pub/cpc/wd52dg/data/indices/qbo.u30.index and ftp://ftp.ncep.noaa.gov/pub/cpc/wd52dg/data/indices/qbo.u50.index.)

altitude of 30 km (a pressure of 30 hPa). The easterly phase of the QBO is stronger than the westerly phase.

The accepted explanation for reversal of the winds in the equatorial stratosphere involves the dissipation of upward-propagating Kelvin and Rossby waves in the stratosphere. These waves originate in the troposphere and lose momentum in the stratosphere by the process of radiative damping. This involves rising air cooling and radiating less than the air that is warmed in the descending part of the wave pattern. Westerly momentum is imparted by decaying Kelvin waves. Rossby waves perform a similar function in respect of the easterly phase. These two sets of waves combine to produce a regular reversal of the upper atmosphere winds.

The tropospheric biennial oscillation (QBO)

The tropospheric biennial oscillation (TBO) is the less well-defined cousin of the QBO. The TBO appears in many meteorological series, but it is nowhere near as distinct as the QBO. Moreover, both within individual series and between different series, its period varies from around 2.1 to 2.6 years in a manner that cannot be directly linked to the QBO. In part this disconnection may be related to the dominant role of the annual cycle in many meteorological series, and especially to the extremes of winter and summer. Because the QBO and TBO move in and out of phase with the annual cycle there is possibility that the weaker oscillation will tend to "phase lock" with the annual cycle and then skip a beat every third year or so. This potentially chaotic behavior has compounded the difficulty of establishing whether the QBO and TBO are manifestations of a common meteorological phenomenon or reflect separate physical processes.

El Niño Southern Oscillation (ENSO)

The El Niño Southern Oscillation (ENSO) is a major contributor to interannual fluctuations in the tropical climate. It is best known for its warm (El Niño) episodes, when much of the eastern tropical Pacific has well above normal sea surface temperatures (SST). These are interspersed with either more normal conditions or cool (La Niña) episodes. This coupling between the atmosphere and the ocean across the equatorial Pacific shows a quasi-cyclic periodicity centered around 3–5 years with the strength of both warm and cool events varying appreciably over time. Detailed analysis of SST for the central equatorial Pacific since 1950 shows that much of the variance is made up of two periodicities: the first being the TBO (period 26 months) and a quasi-quadrennial oscillation (QQO) with a period of 53 months (Figure C49). Here again there is evidence of "phase locking" between ENSO and the annual cycle, as warm (El Niño) episodes tend to peak around the end of the calendar year, thereby making their spacing erratic and difficult to predict.

Other interannual oscillations

The insights provided by modeling ENSO have led to renewed interest in other forms of atmosphere–ocean oscillations around the world. In particular the North Atlantic Oscillation (NAO),

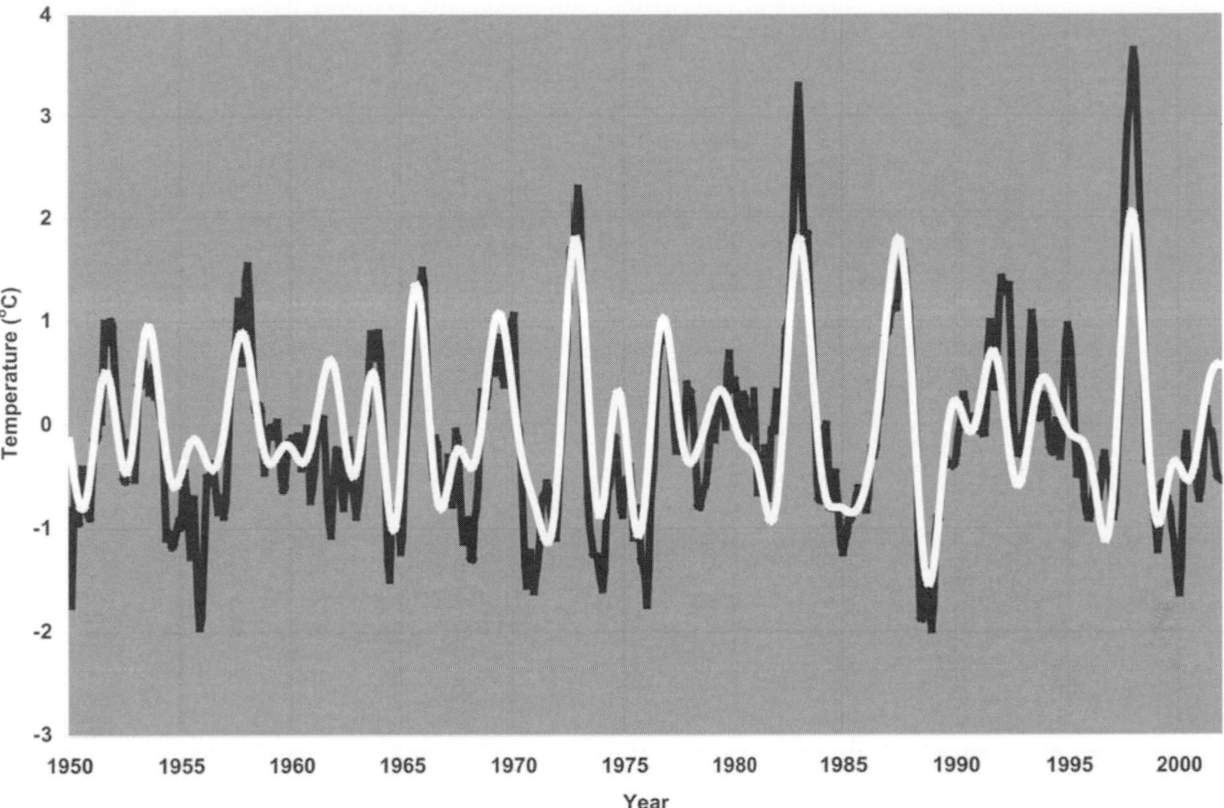

Figure C49 A comparison of the observed bimonthly temperature anomalies in the central equatorial Pacific Ocean (black line) and a reconstruction of this series using a combination of two filtered versions of the data (white line): one centered on 24 months and the other centered on 48 months. This result shows that the data can be largely represented by the combination of a tropical–biennial oscillation (TBO) and a quasi-quadriquennial oscillation (QQO) that is effectively "phase locked" to the warm events peaking at around the end of the calendar year.

the Pacific Decadal Oscillation (PDO) and Antarctic Oscillation have attracted most interest. In certain cases these oscillations were first identified in interannual fluctuations in the pressure difference patterns between mid- and high latitudes. So the NAO is usually measured in terms of the pressure difference between Iceland and the Azores during the months of December to March and is a measure of the strength of the winter circulation of the midlatitudes of the northern hemisphere. Although the NAO shows some evidence of periodicities in the region of 7–9 years and around 20 years, these do not show real statistical significance. The PDO tends to occur with a periodicity of around 20 years.

Decadal oscillation

Many meteorological records show evidence of periodicities around 10–12 years. Often these have been linked with solar activity, (see section on Solar activity), but the case for the link is not always convincing as they are inclined to shift in phase with respect to the 11-year sunspot cycle. So, although seen in both large-scale temperature patterns across the northern hemisphere, notable over North America, and in the oscillations in the Pacific Ocean, including ENSO, these various manifestations do not show any particular coherence.

Bidecadal oscillation

A cycle with around 20 years frequently appears in meteorological records, and in some instances is closely linked with the double-sunspot cycle (known as the Hale cycle). One of the most convincing associations is with the incidence of drought in the western United States. An alternative hypothesis however, is that the 18.6-year lunar tidal cycle may be the cause of this variation. Other impressive examples of bidecadal oscillations are found in Greenland ice-core data, central England temperature records and extratropical Pacific SST, where it is part of the case for the PDO.

Multidecadal oscillations

There is considerable evidence of 65–90-year oscillation in both temperature records and tree-ring data. In particular, for the North Atlantic and its bounding continents, there is a basin-wide pattern of sea surface temperature anomalies. This is probably related to the NAO, but it has now become known as the Atlantic Multidecadal Oscillation (AMO). This periodicity has also shown up in tree-ring data in northern Eurasia, sea level measurements in northern Europe and rainfall observations in the United States. It appears also to be part of a wider phenomenon, as there is evidence of an oscillation in the global climate system of period 65–70 years, although in a record of only 135 years this is the subject of some doubt.

Longer-term cycles

The best-established cycles in the climate are associated with the waxing and waning of the ice ages over the last million years or so. These have periodicities centered on 100 000, 41 000 and 23 000 years (100, 41 and 23 kyr). These cycles appear to be the result of variations in the Earth's orbital parameters, which lead to the amount of solar radiation falling at different latitudes at different times of the year varying in a cyclic manner. These variations result in the amount of snow and ice in polar regions rising and falling with a characteristic 100-kyr sawtooth pattern, which is modulated by the 41- and 23-kyr cycles (Figure C50).

In addition, within the last ice-age cycle there is evidence of a roughly 1500-year cycle that has been linked with variations of the ocean circulation in the North Atlantic. While the most striking evidence of this cycle is during the last ice age, it appears to have played a part in less dramatic climatic changes during the last 10 kyr.

Ocean–atmosphere coupling

The most obvious explanation for the many oscillations observed in the global climate is that they are the product of feedback processes between the atmosphere and the oceans. Because the atmosphere and the oceans respond to various physical stimuli on different timescales, it is possible for them to interact nonlinearly to produce quasiperiodic behaviour. Indeed, it is possible to produce nonlinear models of the atmosphere alone that are capable of producing substantial interannual variability. When combined with the fact that the oceans are more likely to exhibit longer-term fluctuations, it is possible for coupled atmosphere–ocean computer models to produce periodic fluctuations on every timescale from the interannual to the millennial.

Physical explanations of ENSO involve large-scale atmosphere–ocean interactions. During warm (El Niño) episodes the combination of the warming of the ocean off the coast of South America, the associated strong convection and the reduced trade winds all contrive to maintain the abnormal conditions. Similarly, during cool (La Niña) episodes the strong convection in the western Pacific and the strong trade winds should maintain the below-normal SST in the eastern equatorial Pacific. So, in principle, the system should stay in one or other state indefinitely. The reason this does not happen is because of changes in the depth of the thermocline in the ocean due to Kelvin and Rossby waves in the ocean.

Close to the equator downwelling and upwelling Kelvin waves in the depth of the thermocline move rapidly eastwards across the Pacific basin every 2 months or so. Farther from the equator wind fields generate Rossby waves moving westward whose speed depends on the latitude. These speeds are much slower than those of the equatorial Kelvin waves. Near the equator they cross the Pacific basin in about 9 months. Toward the poles the time increases rapidly, to be about 4 years at 12°N

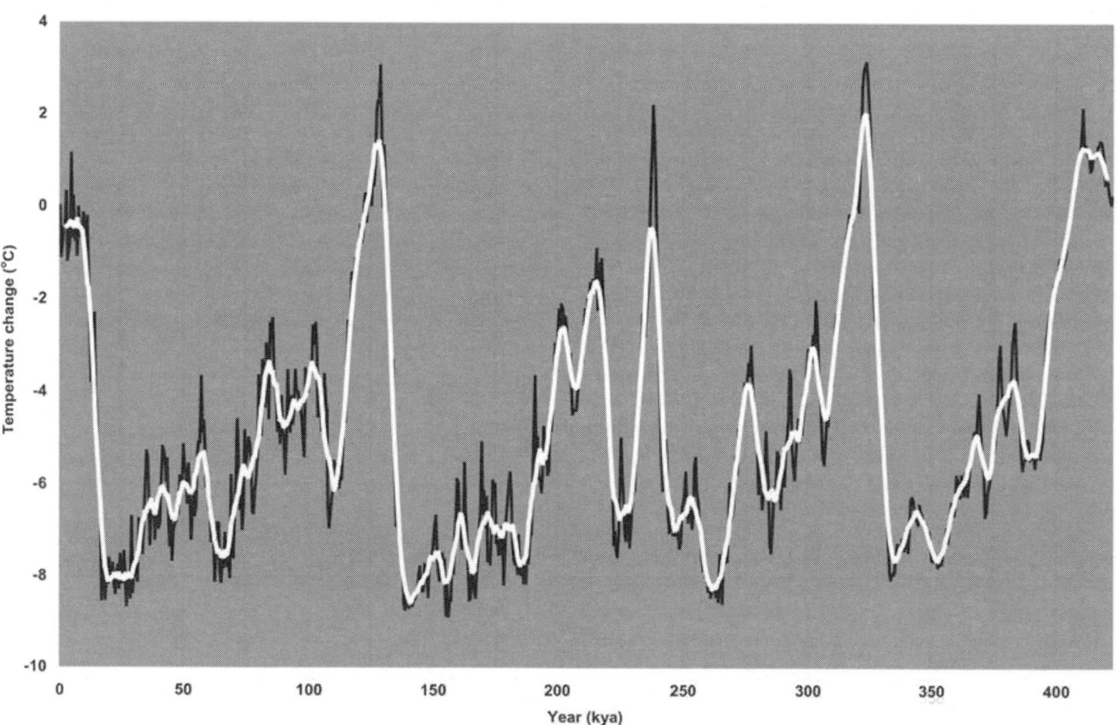

Figure C50 The estimate of the temperature change over Antarctica using measurements of the hydrogen isotope ratio (H/D) in an ice core drilled at the Vostok research station, going back to over 400 000 years ago (400 kya). This shows how the record is dominated by a 100-kyr periodicity over the last four ice ages, while during the glacial periods there is evidence of shorter periodicities of around 20 and 40 kyr. (Data archived at the World Data Center for Paleoclimatology, Boulder, Colorado, USA.)

and S. The combination of these effects leads to the quasiperiodic appearance of warm water in the eastern Pacific (El Niño) every 3–5 years, interspersed with cool (La Niña) episodes.

The unresolved question about atmosphere–ocean coupling is whether the atmosphere or the ocean is the driving force behind interannual periodicities. In principle, the greater thermal inertia of the oceans should exert a dominant influence, and hence make the prediction of behavior months and years ahead a practical proposition. In practice, both the models and experimental forecasting suggest that, with both ENSO and the NAO, the chaotic influence of the atmosphere exerts an overriding influence.

In the longer term changes in the thermohaline circulation of the world's ocean currents may play a part in the 1500-year periodicity seen within the ice age cycle. In particular the pattern of currents in the North Atlantic appears to be especially sensitive to shifts in temperature and salinity of the surface water up around Iceland. Computer models of the oceans show that changes in evaporation, precipitation and freshwater runoff from the surrounding continents can produce sudden switches in circulation patterns. It is proposed that, during the last ice age, such changes, combined with surges of icebergs from the northern ice sheet, led to dramatic swings in temperatures around the northern hemisphere.

Solar activity

Since the seventeenth century, observations on the sun's photosphere have revealed sunspots whose number, size and duration vary over time in a more or less regular manner with a period of about 11 years (Figure C49). The variation in number during each cycle is more than two orders of magnitude greater than for any shorter period. It ranges from virtually no spots during the minimum of solar activity to just over 200 in the most active cycle that peaked in 1957. The instrumental record has now accumulated reliable data since around 1750, and is now in its twenty-third cycle of activity, which peaked at around 120 spots in 2000.

Another measure of solar activity is the number of faculae, which is closely associated with sunspots: they brighten as the number of dark sunspots increases. Their increased brightness is the dominant factor in changing solar output rather than sunspot darkening. A series of satellites have made measurements that provide unequivocal observations of how the sun's output varies with the 11-year cycle in solar activity (Figure C47). These results show that the level of total solar irradiance (TSI) is greatest around the solar maxima and the amplitude of the solar cycle variation is slightly less than 0.1%.

Models of the TSI that combine changes in sunspot number and faculae brightness produce a good fit with satellite observations of the TSI. These models explain not only the observed changes in solar output since 1980, but also the longstanding hypothesis that the cold period known as the Little Ice Age may be related to changes in solar activity. In particular, the colder weather of the seventeenth century appears to have been the result of an almost complete absence of sunspots during this period, known as the "Maunder Minimum".

The small change in the TSI during the last two solar cycles is, however, on its own, an order of magnitude too small to explain observed correlation between solar activity and global temperature trends since the late nineteenth century. Predicted changes in TSI since the seventeenth century, which include the sun's magnetic, solar cycle length, solar cycle decay rate, solar rotation rate and various empirical combinations of all of these, suggest greater changes in the TSI (Haigh, 2000). These models produce widely different estimates of changes since the Maunder Minimum, ranging between 1 and 15 $W m^{-2}$ although most estimates lie in the range 2.5 to 5 $W m^{-2}$. Even so, this may be sufficient to explain much of the observed correlation between global climate change and solar variability during the last three to four centuries.

The TSI also varies with wavelength. The greatest changes are in the ultraviolet (UV) region. These wavelengths are largely absorbed in the atmosphere and this could enhance the impact on the weather. In particular, the amount of UV affects the amount of ozone formed in the stratosphere. This, in turn, alters the radiative balance of both the stratosphere and lower levels of the atmosphere. Computer models, which include realistic changes in UV and ozone, show a significant effect on simulated climate. Solar cycle variability may therefore play a significant role in regional surface temperatures, even though its influence on the global mean surface temperature is small. So it is possible, by both radiative and dynamical means, that variations in solar UV output can produce disproportionate changes in the atmosphere.

Possible links between the sun's magnetic field, solar activity and the Earth's weather involve a complex set of mechanisms. The magnetic polarity of sunspots alternates between successive cycles. Known as the Hale magnetic cycle, this 22-year cycle could be the key to the amplification process. Furthermore, the sun's magnetic flux at the Earth rose by 130% between 1901 and around 1990. While it exhibits periodic fluctuations apparently linked to 11-year sunspot cycle, these are not precisely in phase with it and are less regular.

The direct effect of fluctuations of the sun's magnetic field on the weather may relate to the quantity of energetic particles emitted from the sun. Second, the overall strength of the sun's magnetic field alters the Earth's magnetic field, and with it the amount of cosmic rays (energetic particles from both the sun and from elsewhere in the universe) that are funneled down into the atmosphere.

The impact of variations in the solar magnetic field depends on the energy of the cosmic rays as they enter the Earth's atmosphere. Cosmic rays of low energy have their origin in the sun and are absorbed high in the atmosphere. Galactic cosmic rays (GCR) are of much higher energy and have an appreciable impact on the troposphere. When the sun is more active GCR are less able to reach the Earth and so their impact on the lower atmosphere is inversely related to solar activity.

Cosmic rays produce various chemical species, which alter the concentrations of radiatively active molecules such as ozone (O_3), nitrogen dioxide (NO_2), nitrous oxide (N_2O) and methane (CH_4). These species are most likely to be seen in the stratosphere where their impact will be similar, but less significant to the changes caused by solar UV variations. In addition, the formation of ions will affect the behavior of aerosols and cirrus clouds that have a direct radiative impact and also alter the amount of water vapor throughout the atmosphere. These changes could lead to shifts in the radiative balance of both the stratosphere and troposphere and so produce long-term fluctuations in the temperature.

There is also a question of whether GCR alter cloud cover at lower levels. An analysis of satellite observations of global total cloud cover, between 1984 and 1991, suggested a correlation with cosmic ray flux. Increased GCR flux appears to cause total cloud amounts to rise, and this cools the climate. The interannual variations in cloudiness are, however, difficult to distinguish from parallel changes caused by warm and cold ENSO events.

There is, however, a possibility that charged particles are more efficient than uncharged ones in acting as cloud condensation nuclei (CCN), even though only a small proportion of aerosol particles are capable of acting as ice nuclei, depending on chemical composition or shape. Recent work has provided the first observational evidence of cosmic ray-induced aerosol formation in the upper troposphere. In theory, the higher level of charge carried by such aerosols can enhance the formation of ice nuclei. By changing precipitation rates or radiative balance, the changes in the clouds then affect atmospheric dynamics and temperature.

Tidal forces

The other obvious periodic influences on the weather tidal forces, which can have a direct effect on the movement of the atmosphere, the oceans and even the Earth's crust, are the Earth's tides. The nature of these links varies in complexity. Tidal effects in the atmosphere are relatively predictable and measurable, but tiny compared with normal atmospheric fluctuations. In the oceans the broad effects can be calculated, but estimating changes in the major currents is much more difficult. In particular, recent satellite altimetry studies have come up with some interesting results that may shed new light on the climatic implications of the dissipation of tidal energy. Prior to this work it was widely assumed that this energy, representing the effect the moon receding from the Earth at a rate of some four centimeters a year, was dissipated in the shallow waters of the continental shelves around the world. It is now reckoned that about half this energy is fed into deeper water where it exerts a significant effect on the strength of the major ocean currents and hence the transport of energy from the tropics to polar regions. So the 18.61-year lunar cycle affects the strength of the principal ocean currents.

As for movements of the Earth's crust, the problems are compounded by possible links with solar activity. The direct influence of the tides could influence the release of tectonic energy in the form of volcanism. Since there is considerable evidence that major volcanic eruptions have triggered periods of climatic cooling, this would enable small extraterrestrial effects to be amplified to produce more significant climatic fluctuations. In addition, there is evidence that intense bursts of solar activity interact with the Earth's magnetic field to produce measurable changes in the length of the day. Such sudden tiny changes in the rate of rotation of the Earth could also trigger volcanic activity. It should be noted that, while there is no evidence that their occurrence was in any way related to this effect, the three climatically important volcanoes in the second half of the twentieth century (Agung in 1963, El Chichon in 1982 and Pinatubo in 1991) were spaced in such a way as to have a confusing impact on the interpretation of any solar or tidal effects.

The Earth's orbital parameters

The explanation of the periodicities in the ice ages relies on the Earth's orbital parameters producing latitudinal and seasonal variations in incident solar radiation. In particular, the variations in the amount of sunlight falling in summer at high latitudes in the northern hemisphere appear to affect the amount of build-up of ice sheets over the continental landmasses. These changes are controlled by the precession of the equinoxes (the 19- and 23-kyr periodicities) and the variations of the tilt of the Earth's axis (the 41-kyr periodicity) and the 100-kyr periodicity due to the eccentricity of the Earth's orbit. There is, however, a problem with this proposal. While the first two periodicities are probably sufficient to trigger significant climatic changes, the 100-kyr eccentricity periodicity is the weakest of the orbital effects. Nevertheless, the 100-kyr ice age cycle is the strongest feature of the climatic record in the last 800 000 years.

A solution to this problem is to use a differential model in which the rate of change of the climate is a function of both the orbital forcing and the current state of the climate (Imbrie et al., 1992, 1993). Not only is this a more realistic representation of climatic behavior, but it also contains a nonlinear relationship between the input and the output that has important physical consequences. For instance, if the model is sensitive to changes in ice volume that are reflected in the time constants for the growth and decay of the ice sheets and the lag between the changes in solar radiation falling in summer at high latitudes of the northern hemisphere, it can be shown to achieve a reasonable representation of the observed long-term behavior. Various models have tuned the parameters to achieve the best fit between the calculated ice-volume changes and the oxygen-isotope record to give a reasonable match with past changes. The dependence on empirically derived time constants, which are broadly linked to the physical behaviour of the ice sheets, is a measure of the nonlinear response of the global climate to changes in the orbital parameters.

A more sophisticated approach is to draw on the fact that there is evidence that the North Atlantic appears to have exhibited three different modes of circulation during the last ice age (Paillard, 1998). As a result of this behavior it appears that the global climate is capable of switching between three distinct regimes (e.g. interglacial, mild glacial and full glacial). Switches between these regimes could well be controlled by a combination of changes in insolation, resulting from the Earth's orbital parameters, combined with changes in ice sheet volume. By defining the conditions for the transition between the three regimes a model has been produced that is capable of reproducing with remarkable accuracy the changes in ice sheet volume over the last million years.

W.J. Burroughs

Bibliography

Baldwin, M.P., Gray, L.J., and Dunkerton, T.J., 2001. The quasi-biennial oscillation. *Reviews of Geophysics*, **32**: 179–229.

Burroughs, W.J., 2003. *Weather Cycles: real or imaginary"*, 2nd edn. Cambridge: Cambridge University Press.

Diaz, H.F., and Markgraf, V. (eds), 2000. *El Niño and the Southern Oscillation*. Cambridge: Cambridge University Press.

Haigh, J.D., 2000. Solar variability and climate. *Weather*, **55**: 399–405.

Imbrie, J., Boyle, E.A., and Clemens, S.C., 1992. On the structure and origin of major glaciation cycles. 1. Linear responses to Milankovitch forcing. *Paleoceanography*, **7**: 701–738.

Imbrie, J. Berger, A., and Boyle, E.A., 1993. On the structure and origin of major glaciation cycles. 2. The 100 000 year cycle. *Paleoceanography*, **8**: 699–735.

Paillard, D., 1998. The timing of Pleistocene glaciations from a simple multiple-state model. *Nature*, **391**: 378–381.

Cross-references

El Niño
Maunder, Edward Walter and Maunder Minimum
North Atlantic Oscillation
Ocean–Atmosphere Interaction
Oscillations
Sunspots

D

DEGREE DAYS

The degree-day is an index number originally developed by heating, ventilation, and air conditioning engineers (Houghton, 1985). The heating degree-day was developed to assess energy needs necessary to heat business buildings and private homes. The heating degree-day index assumes that people will begin to use their furnaces when the mean daily temperature drops below 65°F. It is computed by subtracting the mean temperature for a day from 65°F to determine the total number of heating degree-days for a given day. When the mean daily temperature is above 65°F there are no heating degree-days. The greater the range in a mean daily temperature from 65, the greater the number of heating degree-days; hence, more fuel must be expended (Ahrens, 2001).

The cooling degree-day is used to assess energy needs required to cool buildings and homes. When the mean daily temperature exceeds 65°F people will begin to employ measures to cool their environment by switching on their air-conditioners and cooling pumps. The cooling degree-day index is computed by subtracting 65°F from the mean daily temperature. If the two temperatures are the same, there are no cooling degree-days. The greater the range in a mean daily temperature from 65, the greater the number of cooling degree-days; hence, more power must be expended to maintain a comfortable environment. Cooling degree-days provide a guide for builders when planning the size and type of cooling units to install in structures. Power companies may use these data to predict energy demands throughout a summer season. When combined, the heating and cooling degree-day indices provide a practical way of assessing energy requirements for a region throughout the year.

The freezing or thawing degree-day is a measure of the combined duration and magnitude of below or above freezing temperatures during a freezing season. It is designated as an air freezing/thawing index when taken for air temperatures 4.5 ft above the ground, and surface freezing/thawing index for temperatures immediately below the surface. The freezing/thawing degree-day (FDD) is used to assess conditions related to problems with highway construction, soil disturbance involving mass movements, and foundation assessment due to groundwater through flow and sapping processes resulting from freeze/thaw conditions.

Table D1 Base temperatures for selected crops and insects. By reviewing the climate data for a region and comparing mean daily temperatures, growing degree-days can be computed for a crop

Crops and insects	Base temperatures
Wheat, barley, rye, oats, flaxseed, lettuce, asparagus	40° F
Sunflower, potato	45° F
Sweet corn, corn, sorghum, rice, soybeans, tomato	50° F
Corn rootworm	44° F
Alfalfa weevil	48° F
Black cutworm, European corn borer	50° F
Green cloverworm	52° F

The growing degree-day is used by farmers as a guide for planting and estimating when a crop will be ready for harvest; and closely associated with crop production, the peak time of a season in which crop-destroying insects occur. The growing rate or biological development in plants and insects is based in part upon the amount of heat absorbed during a growing season of a plant or the life cycle of an insect. Each species, whether crop, weed, insect, or disease organism, is adapted to grow over a certain minimum (base) temperature, and decline in growth at a maximum temperature. The quantity of heat absorbed as a requirement of sustained growth and development is often referred to as a heat unit or degree-day. By subtracting the base temperature (Table D1) for a given crop from the daily mean temperature, the number of growing degree-days may be computed. When growing degree-days are totaled, farmers can establish an estimate of the minimum number of days of growth required for a crop until it is ready for harvest. While this index provides a quick guide for crop development, soil moisture properties, precipitation amounts, and other climatic variables are not considered in this general format.

Discussion

The *Handbook of Applied Meteorology* (Houghton, 1985) discusses the degree-day on the basis of heating or cooling requirements, growing degree-days, or freezing/thawing degree-days. Degree-days are generally quantified from a threshold or base temperature determined for each condition, i.e. cooling, heating, growing, or freezing/thawing degree-days. Threshold temperatures include:

1. 65°F (18°C) for cooling energy requirements generating cooling degree-days (CDD), and expressed as, CDD = $\Sigma(\check{T}_i - A)$ where, $\check{T}_i$ is the mean daily temperature for a day i, A is a threshold temperature, and Σ is the summation over a period of days, such as a week, month, or season. If $\check{T}_i$ is < 65°F, CDD is 0.
2. 65°F (18°C) for heating energy requirements generating heating degree-days (HDD) with the equation written as, HDD = $\Sigma(65 - \check{T}_i)$. If $\check{T}_i$ is > 65°F, HDD is 0.
3. Variable base temperatures (B) for growing degree-days (GDD) dependent upon crop type where, GDD = $\Sigma(\check{T}_i - B)$, and B is a base temperature for a given crop type. If $\check{T}_i < B$, GDD is 0. Growing degree-days are similar to cooling degree-days, only the base temperature can vary with different plant types in computing growing degree-days, while the threshold temperature for cooling degree-days generally remains constant at 65°F.
4. 32°F (0°C) for freezing degree-days (FDD) where, FDD = $\Sigma(32 - \check{T}_i)$. If $\check{T}_i$ is > 32°F, HDD is 0. By extension, thawing degree-days (TDD) are computed as, TDD = $\Sigma(\check{T}_i - 32)$. If $\check{T}_i$ is < 32°F, TDD is 0.

Degree-day values have been used to show that climate change models may not adequately explain the energy expenditures that would be expected with reported increases in mean temperatures, increases in extreme high temperatures, and increases in the frequency, duration, and magnitude of summer heat waves. Hansen et al. (1998) developed a common-sense climate index using degree-days and reported that parts of North America are showing changes consistent with climate model predictions given increasing levels of greenhouse gases. A graph for New York City's LaGuardia airport shows a decrease in heating and cooling degree-days for the past few decades. Balling (1999) analyzed temperature data derived from NOAA's Daily Historical Climatology Network for the conterminous United States from 1950 to 1995. He computed and plotted the heating, cooling, and total degree-days for this time period. His graph paradoxically shows no statistically significant trends throughout this time. He reports a 0.2% decrease in heating degree-days, 5.7% decrease in cooling degree-days, and a 1.0% decrease in total degree-days. When considering climate change and energy needs, this slight decrease in cooling degree-days would indicate a reduction in the energy needs to cool buildings, and is contrary to what would be expected with warming trends. The period of time reviewed in this study was one of substantial buildup of greenhouse gases throughout the Earth's atmosphere. Numerous climate models for global warming suggest that a decrease in heating degree-days and increase in cooling degree-days would follow rises in these greenhouse gases, yet the analysis of degree-day patterns from the historical climate record for this 35-year period does not provide empirical evidence to support these climate model simulations.

A climatic classification using degree-days has also been reported. The somewhat generalized nature of this classification does not appear to provide a significant contribution toward the enhancement of the more widely accepted climate classification systems since it only addresses degree-days with respect to ordinal-scaled (e.g. very severe, severe, moderate, mild, warm, or hot) seasonal characteristics. Figure D1 (Houghton, 1985) illustrates this classification for the eastern United States.

Phenology models

The degree-day concept is fundamentally simple in design, but it has limits when used to determine growing degree-days for agricultural use. Phenology models predict time of growth events for organisms (insects and plants). Plants and insects require a minimal amount of heat to develop throughout their life cycles. An accumulation of heat units is called physiological time, and it is measured in degree-days (°D). Degree-day methods provide scientists with quantitative methods used to assess the physiological time of organisms. Estimating heat units or accumulated degree-days provides a more biologically accurate snapshot of an organism's stages of development than using calendar days based upon variations in yearly climate records.

Organisms (plant and insect) develop faster when temperatures are higher, although this does not imply there is a yield or quality benefit in warm seasons. While organisms will grow faster in warmer temperatures, and they are exposed to greater heat for fewer days, the net accumulation of heat units (degree-days) required for development is about the same as for organisms developing under cooler conditions for more days. Since temperatures fluctuate from cool to hot during a growing season, it is the total heat accumulation derived between the lower and upper threshold temperatures for each plant species that determines the time to complete development. Plant development ceases when a temperature falls below a lower threshold, and also begins to decline and ultimately stop development when the temperature exceeds an upper threshold. In some cases the maximum temperature can exceed the base temperature for a plant, thus 0 growing degree-days are accumulated, e.g. if a maximum temperature is 60°F, a minimum temperature is 35°F, then the average daily temperature is $(60 - 35)/2 = 47.5$°F. If the base temperature for this plant = 50, then $47.5 - 50 = -2.5 = 0$ degree-days. This example results in 0 growing degree-days (an underestimation), although the average temperature used in the calculations is above the lower threshold for the plant.

When temperatures become excessively hot, the degree-day system will give heavy weighting to these temperature extremes that can actually be detrimental to the plant. A modified growing degree-day provides an adjustment for high temperature extremes. If a temperature exceeds a threshold high (Figure D2) for a given plant, it is adjusted back to the threshold. Alternatively, if the lowest temperature is below the threshold low (Figure D2) for a plant, it is adjusted up to the threshold. Once the temperatures have been adjusted, the average daily temperature is computed and used to compute growing degree-days. Threshold temperatures are used for certain crops (e.g. corn) and based on the assumption that development is limited once a threshold low or high is exceeded.

While degree-days are generally based upon a mean daily temperature (maximum daily temperature − minimum daily temperature/2), the influence of diurnal patterns is lost when only considering these temperature extremes. Computing growing degree-hours (GDH) may provide a more representative way of assessing variability in the daily temperatures. Mimoun and DeJong (1999) have documented that accumulated

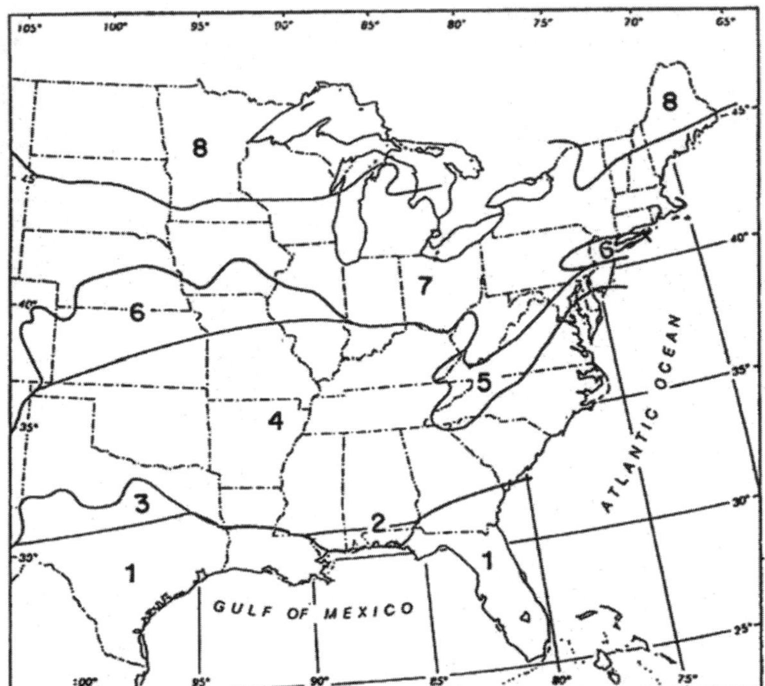

Figure D1 A climatic classification of the United States using heating (HDD) and cooling (CDD) degree days. Key:

	Winter	HDD	Summer	CDD
1	Mild	<2000	Hot	>2000
2	Mild		Warm	1000–2500
3	Moderate	2000–4000	Hot	
4	Moderate		Warm	
5	Moderate		Mild	<1000
6	Severe	4000–8000	Warm	
7	Severe		Mild	
8	Very severe	>8000	Mild	

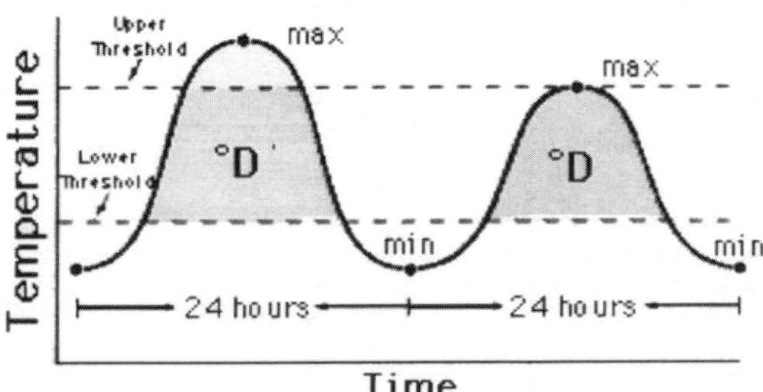

Figure D2 Thresholds (dashed lines) and area representing degree-days (shaded) for 24-hour time periods (from the University of California, Pet Management Program, 1990).

temperatures (in growing degree-hours) for 30 days after bloom are highly correlated with yearly differences in harvest date for peaches, plums, and nectarine cultivar. A general use would be to compare records of bloom dates (when 50% of the flowers on a tree or orchard are fully open) and harvest dates for previous years, look for a year with a comparable GDH accumulation (at 30 days after bloom), and expect the number of days from bloom to harvest will be similar to that year.

Seasonal variances and climatic region can also impact how accurately a degree-day method represents actual degree-days. Several refinements in computing degree-days have been reported (Allen, 1976; Wilson and Barnett, 1983; Zalom et al., 1983). These methods include (from the simplest to the most mathematically complex): single triangle, double triangle, single sine, double sine, and Huber's method (a modification of the single sine method). Each has been applied to determine heat units required for insect development, and by extension to agricultural development since insects can have a pronounced and detrimental affect on annual crop yield. These methods are based on the area under a diurnal temperature curve and between critical threshold temperatures.

The degree-day computational techniques including the cut-off methods that establish parameters for the upper and lower threshold temperatures are discussed in University of California, 1990, Zalom et al. (1983), and at the web site for University of California, Davis, (http://www.ipm.ucdavis.edu/WEATHER/ddconcepts.html).

Summary

The degree-day was initially designed by heating and cooling engineers to assess the cooling and heating needs of business buildings and homes. A number of procedures for computing degree-days have been presented in this discussion. The more robust mathematical models (triangle and sine methods) tend to provide more credible results than those models that only use the mean daily temperature. Growing degree-days have been used to study pest management and further applied in agricultural studies. An ordinal climatic classification has also been proposed, although it does not provide any significant contribution to the more widely accepted climate classification systems.

Danny M. Vaughn

Bibliography

Ahrens, C.D., 2001. *Essentials of Meteorology: An Invitation to the Atmosphere*. Pacific Grove, CA: Brooks/Cole.

Allen, J.C., 1976. A modified sine wave method for calculating degree days. *Environmental Entomology*, **5**(3): 388–396.

Balling, R.C., 1999. Analysis of Trends in United States Degree Days: 1950-1995. Report from the Office of Climatology, Arizona State University, Tempe, Arizona (URL: http://www.greeningearthsociety.org/Articles/1999/trends.htm).

Hansen, J.M., Sato, J.G., and Ruedy, R., 1998. A common-sense climate index: is climate changing noticeably? *Proceedings of the National Academy of Sciences*, **95**: 4118–4120.

Houghton, D.D. (ed.), 1985. *Handbook of Applied Meteorology*. New York: John Wiley, pp. 98–100.

Mimoun, M.B., and DeJong, T.M., 1999. Using the Relation Between Growing Degree Hours and Harvest Time to Estimate Run-Times for PEACH: a tree growth and yield simulation model. Department of Pomology, University of California, Davis (URL: http://fruitsandnuts.ucdavis.edu/weather/GDHpaper.pdf).

University of California, 1990. DDU Degree-Day Utility User's Guide and Software, Statewide Integrated Pest Management Program. University of California, Davis. URL: http://www.ipm.ucdavis.edu/WEATHER/ddconcepts.html. *Degree Day Concepts*, University of California, Davis.

Wilson, L.T., and Barnett, W.W., 1983. Degree-days: an aid in crop and pest management. *California Agriculture*, **37**: 4–7.

Zalom, F.G., Goodell, P.B., Wilson, L.T., Barnett, W.W., and Bentley, W.J., 1983. Degree-days: the calculation and use of heat units in pest management. University of California, Division of Agriculture and Natural Resources Leaflet, 21373.

Cross-references

Agroclimatology
Climate Change and Human Health
Spring Green Wave
Vegetation and Climate

DESERTIFICATION

The prolonged multiyear drought in the early 1970s in the Sahelian zone, stretching from West Africa to the Horn of Africa, precipitated a sharp increase in death and morbidity of humans and livestock, as well as widespread environmental deterioration. Stark satellite images capturing the degradation of the land surface throughout the region were backed up by ground-truth photographs of the landscape and human suffering.

Although many articles, papers, and reports from various countries begin with comments on the role of the Sahelian drought in the growing interest in desertification (e.g. Glantz, 1977; UN Secretariat of the Conference on Desertification, 1977; Quintanilla, 1981; Zonn, 1981), that drought was neither the first manifestation of the desertification phenomenon nor the only reason for scientific interest in it. In fact, Aubreville (1949), a French scientist, was the first to use the term "desertification" in his report, and others (e.g. Le Houérou, 1962) have discussed the phenomenon since the late 1950s. Desertification received immediate and widespread public attention ever since the early 1970s, as witnessed by the creation of the United Nations Conference on Desertification (UNCOD) in Nairobi in 1977. It was clear that the conference was convened mainly as a result of the devastating impacts of drought in the West African Sahel earlier in the decade.

This conference brought the concept of desertification and its demographic aspects into the spotlight (UN Secretariat of the Conference on Desertification, 1977). The conference report described desertification in the following way:

> The diminution or destruction of the biological potential of the land ... can lead ultimately to desert-like conditions. It is an aspect of the widespread deterioration of ecosystems, and has diminished or destroyed the biological potential, i.e., plant and animal production, for multiple use purposes at a time when increased productivity is needed to support growing populations in quest of development. Important factors in contemporary society – the struggle for development and the effort to increase food production, and to adapt and apply modern technologies, set against a background of population growth and demographic change – interlock in a network of cause and effect. Progress in development, planned population growth and improvements in all types of biological production and relevant technologies must therefore be integrated. The deterioration of productive ecosystems is an obvious and serious threat to human progress. In general, the quest for ever greater productivity has intensified exploitation and has carried disturbance by man into less productive and more fragile lands.
>
> Overexploitation gives rise to degradation of vegetation, soil and water, the three elements which serve as the natural foundation for human existence. In exceptionally fragile ecosystems, such as those on the desert margins, the loss of biological productivity through the degradation of plant, animal, soil and water resources can easily become irreversible, and permanently reduce their capacity to support human life. Desertification is a self-accelerating process, feeding on itself, and as it advances, rehabilitation costs rise exponentially. Action to combat desertification is required

urgently before the costs of rehabilitation rise beyond practical possibility or before the opportunity to act is lost forever (Reining, 1978, p. 3).

The 1977 Nairobi conference also served to bring together representatives of countries whose arid and semiarid landscapes had been directly or indirectly affected by desertification processes. More than a score of governments produced and shared their national case studies about desertification processes within their borders. The actual usefulness of this conference varied from country to country. It became quite clear that desertification was a local to national-level problem with transboundary impacts. It is listed among major global problems because of its occurrence and interest worldwide.

For some countries, such as the People's Republic of China, the conference drew the attention of national policy makers toward arid lands research and helped to elevate such research to a sustained national priority. It remains a high priority more than 25 years after UNCOD. The former Soviet Union's State Committee for Science and Technology (GKNT) established, with the United Nations Environment Program (UNEP) support, international training courses on various aspects of desertification (e.g. overgrazing, salinization, woodcutting) and on ways to identify and combat them (Plit et al., 1995; Zonn, 1981). Today, almost all of the arid parts of the former Soviet Union are within independent countries of Central Asia. Nevertheless, the Russian Federation contains in its Kalmykia Republic the only desert in Europe (Saiko and Zonn, 1997).

In the United States the attention of national policy makers was directed at the time toward environmental degradation in arid areas, and a plan of action was drawn up to assess desertification in a national context (Sabadell et al., 1982). However, the US government refused to admit that desertification processes were taking place on its soil. It believed that desertification was a Third World problem, even though the US rangelands were being overgrazed, dust storms were occurring, and salinization and waterlogging of the soils could be witnessed.

Desertification is a complex multifaceted phenomenon, requiring the expertise of researchers in disciplines such as climatology, soil science, meteorology, hydrology, range science, agronomy, ecology, veterinary medicine, as well as geography, political science, economics, and anthropology. It has been defined in different ways by researchers in these and other disciplines, as well as from national and bureaucratic (institutional) perspectives, with each perspective emphasizing different aspects of desertification of concern to it. Scores of definitions exist for the concept, so it is important to know what one is referring to when discussing desertification processes.

At one time desertification as a process was of particular interest to climatologists in their attempts to understand climate variation and change on both short and long time scales (e.g. Charney, 1975; Hare, 1976). With increasing pressure on government decision makers to allow populations to move into climatically marginal areas, the implications of natural variations in climate have become even more important in decisions relating to the use by society of its land in desertification-prone regions. There will always be climatic deserts on the Earth, although their locations can shift. Human-induced expansion of these desertlike conditions can take place in areas where they had not previously existed.

Desertification literature shows a great diversity (and confusion) among definitions (e.g. Carder, 1981). The wide range of meanings attributed to the concept leads to misunderstandings *among* physical and social science researchers and policy makers, as well as *between* researchers and policy makers (see International Geophysical Union Working Group on Desertification, 1975, *passim*). An analysis of the definitions of desertification helps to improve understanding of the phenomenon, of how it is viewed from different disciplines, countries, and bureaucratic units, and whether progress in combating it has in fact been as slow as many observers suggest (e.g. Glantz and Orlovsky, 1986; UN General Assembly, 1981). Scientific interest in desertification remains high, because, as Reynolds and Smith (2002) noted,

> There is disagreement concerning the causes and processes of land degradation and its importance; the extent to which land changes are natural (climate-driven) versus anthropogenic; the role of "grass roots" abatement efforts versus scientific and technological ones; how to determine the amount of land affected or at risk; and whether or not desertification is reversible.

Various reviews of the progress to combat desertification processes worldwide exposed limited successes in isolated situations. Some of the scientists who in the 1970s and early 1980s had written about those adverse conditions were, by the end of the 1980s, suggesting that natural causes of desertification in the West African Sahel were dominant. Thus, scientific papers were beginning to challenge the hypotheses about the irreversibility of desertification and about the dominance of influence on the fragile environment of human activities in the region. In other words, desertification is, under certain conditions, reversible in spite of human pressures on fragile arid, semiarid, and dry subhumid ecosystems.

According to the UNCCD (UN Convention to Combat Desertification) website, at the Earth Summit (the UN Conference of Environment and Development, or UNCED) in Rio de Janeiro in 1992, delegates supported a

> new, integrated approach to the problem, emphasizing action to promote sustainable development at the community level. It also called on the UN General Assembly to establish an Intergovernmental Negotiating Committee to prepare, by June 1994, a convention to combat desertification, particularly in Africa. In December 1992, the General Assembly agreed and adopted Resolution 47/188 (www.unccd.int).

The UNCCD entered into force in December 1996, after the fiftieth ratification by a government was received. The Conference of Parties, the Convention's supreme governing body, held its first session in October 1997 in Rome, Italy. The UNCCD also noted that

> The Convention aims to promote effective action through innovative local programs and supportive international partnerships. The treaty acknowledges that the struggle to protect drylands will be a long one – there will be no quick fix. This is because the causes of desertification are many and complex, ranging from international trade patterns to unsustainable land management practices. Difficult changes will have to be made, both at the international and the local levels (www.unccd.int).

In the following sections these definitions are used as the basis for discussion, as they are often what is seen and used by decision makers. In the last section it is argued that the concept of desertification also applies to higher rainfall regions than those cited by contemporary arid lands researchers.

Desertification: a process or an event?

Some people consider desertification to be a *process* of change (degradation), whereas others view it as the *end result* of a process of change (a desertlike landscape). This distinction underlies one of the main disagreements on what constitutes desertification. Desertification-as-process has generally been viewed as a series of incremental (sometimes stepwise) cumulative adverse changes in biological productivity in arid, semiarid, and dry subhumid ecosystems. It can encompass decline in yield of the same crop or, more drastically, the replacement of one (maybe equally productive or equally useful) vegetative species by another, or as a decrease in the density of the existing vegetative cover. Desertification-as-event is the creation of desertlike conditions, where perhaps none had existed in the recent past, as the end result of a long-term process of cumulative adverse changes. Many find it difficult to accept incremental changes, whether natural or human-induced, as a manifestation of desertification processes.

In fact, both perspectives represent different aspects of a broader overarching concept of desertification. Thus, seemingly different statements such as "the creation of desertlike conditions in areas once green", "the encroachment of desertlike conditions", or "the intensification of a desertlike landscape", as well as less drastic projections such as "changes in soil structure or in climate" or "the land is becoming less fit for range and crops", can be encompassed by the concept of desertification.

Form of change

Within the dozens of existing definitions of desertification, many words are used to describe the phenomenon. Some of them complement each other, whereas others appear to be contradictory. A point on which they all agree, however, is that desertification is an adverse environmental process. The negative descriptors used in these definitions of desertification include: deterioration of ecosystems (e.g. Reining, 1978), degradation of various forms of vegetation (e.g. Le Houérou, 1975), destruction of biological potential (e.g. UN Conference on Desertification, 1978), decay of a productive ecosystem (e.g. Hare, 1977), reduction of productivity (e.g. Kassas, 1977), decrease of biological productivity (e.g. Kovda, 1980), alteration in the biomass (e.g. UN Secretariat of the Conference on Desertification, 1977), intensification of desert conditions (e.g. Meckelein, 1980; World Meteorological Organization, 1980), and impoverishment of ecosystems (e.g. Dregne, 1976). More recent definitions continue to fall into one or another of the above categories.

Each of these terms suggests a change in landscape from a favored or preferred state (with respect to quality, societal value, or ecological stability) to a less preferred one. Depending on the particular definition, each has been used to describe the condition of vegetation, soils, moisture availability, or atmospheric phenomena. Other descriptors used in these definitions connote an expansive movement or a transfer of characteristics of a desertlike landscape into an area where such characteristics had not previously existed: extension, encroachment, acceleration, spread, and transformation. If one combined each of these negative and transfer descriptors with all the other factors cited in the existing definitions, desertification would encompass most kinds of environmental changes related to biological productivity (see Rozanov, 1981). UNEP (1992) produced the *World Atlas of Desertification*, which identified the location as well as threats of desertification processes around the globe (Figure D3).

What is being changed?

Different definitions have focused on changes either in soil (e.g. salinization), vegetation (e.g. reduced density of biomass), water (e.g. waterlogging), air (e.g. increased aridity), or land surface (e.g. increased albedo). Most of them, regardless of primary emphasis, also describe changes in biological productivity, with comments related to the type, density, and even socially determined value of vegetation.

Type-of-vegetation comments center on changes from desired (or accepted) species to less desired (or less accepted) ones. Such comments include a reduction in the proportion of preferred species having an economic or societal value, the lowering of yields of an existing preferred species, or a major ecological change such as species replacement.

Change in the density of the vegetative cover is an important factor acknowledged in many definitions of desertification. As density decreases, for example, the risks of wind erosion, water erosion, and the adverse effect of increased solar radiation on bare soils are increased dramatically. Surface albedo (reflectivity), also enhanced by a reduction in the vegetative cover, is a major contributor to desertification processes.

With respect to the value of vegetation, researchers have also referred to "lower useful productivity" (Johnson, 1977, p. 317), "reduced productivity of desirable plants" (Dregne, 1976, p. 12), "sustained decline in the yield of useful crops" (UN Secretariat of the Conference on Desertification, 1977, p. 17), and "loss of primary species" (Rapp et al., 1976, p. 8). However, the concern with the value of vegetation in desertification processes is not shared by all researchers. Some researchers have dismissed the value concern, suggesting that any type of vegetation that holds the soil in place is of value in the combat against desertification, whether or not it has an economic benefit for a given society.

As a final comment on what desertification is, it is important to note that disciplinary and institutional biases might appear in any given definition of desertification. For example, a meteorological bias might require that a change take place in the meteorological parameters of a given region, so that they become similar to those for a desert region (e.g. high evaporation rates, aridity, increased rainfall intensity, etc.). As another example, Meckelein (1976), cited in Kharin and Petrov (1977), alluded to the disciplinary bases for desertification when he wrote that desertification could be characterized by the following components: *climate:* increasing aridity (diminishing water supply); *hydrological processes:* runoff becoming more irregular; *morphodynamic processes:* intensification of distinct geomorphological processes (accelerated soil erosion by wind and water); *soil dynamics:* desiccation of soils and accumulation of salt; *vegetation dynamics:* decline of vegetation. It is important to note that, while one aspect of desertification processes could be arrested, others may appear (Prince, 2002).

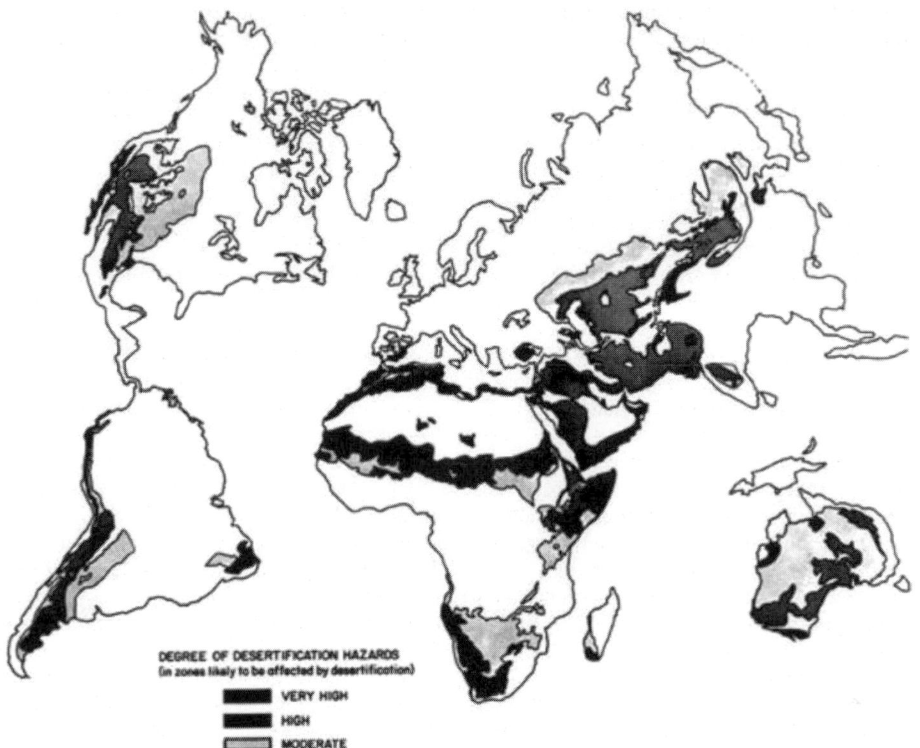

Figure D3 Degree of desertification hazards (in zones likely to be affected by desertification) (after *United Nations Map of World Desertification,* 1977; see also for comparison UNEP's *World Atlas of Desertification,* 1997).

Location of change

There is no agreement on where desertification can take place. Many researchers identify arid, semiarid, and sometimes dry subhumid regions as the areas in which desertification can occur or where the risks of desertification are highest. Others imply that the areas prone to desertification might not be restricted to arid, semiarid, or subhumid regions, by using such descriptive words as extension, encroachment, and spread of desert characteristics into nondesert regions. Still others (e.g. Mabbutt and Wilson, 1980) refer to the intensification of desertlike conditions, suggesting that desertification can occur in desertlike areas. Many oppose this view, however, contending that desertlike conditions cannot be created in a desert. They assert that desertification can only occur along the desert fringes. According to Le Houérou (1975), for example, desertification can occur only in the 50–300 mm isohyet zone.

Reversibility

Few definitions explicitly refer to whether or not desertification is permanent. Le Houérou (1975) briefly explained the conditions under which desertification (he called it "desertization") might be reversible. Others have implied reversibility with reference to the higher costs of rehabilitation of desertified areas (as opposed to prevention). For example, Adams suggested that the "reversibility of desertification was a function of technology and the cost of rejuvenating an area.... Irreversibility should refer to a situation in which the costs of reclamation were greater than the return from a known form of land use" (International Geophysical Union Working Group on Desertification, 1975, p. 138). Still others implied irreversibility by referring to the *end result* of desertification as the creation of desertlike conditions.

Two additional important considerations relating to the *in-situ* permanence of desertification are: (1) when desertification (as a process or event) might be reversed (i.e. the "time" factor); and (2) under what conditions (i.e. the "how" factor).

With respect to the time factor, some observers consider desertification to be irreversible during periods of up to several seasons but reversible on the order of decades and, if not decades, perhaps centuries. Peel "saw great danger in the concept of irreversibility because it has no time limitations whatever" (International Geophysical Union Working Group on Desertification, 1975, p. 138). One author has drawn a distinction between temporary and permanent desertification (Kove, 1982). Is it possible to distinguish between temporary desertification and, for example, seasonal environmental changes? Some have addressed this question by defining desertification as a sustained (as opposed to temporary) decline in biological productivity (e.g. Sabadell et al., 1982; UN Secretariat of the Conference on Desertification, 1977). Le Houérou, commenting on "what is temporary?", noted that, while temporary fluctuations may be interspersed with more favorable conditions, such a condition of successive crises does involve a progressively deteriorating situation, possibly past a threshold of irreversibility (International Geophysical Union Working Group on Desertification, 1975, p. 27).

How can desertification be reversed? Reversal might occur naturally, once the natural contributing causes have been removed. Otherwise, human intervention would be required (e.g. Kassas, 1977) if there is a desire on the part of decision makers to reverse it in less time than might be required to do so naturally. Chinese scientists have actively pursued programs for the past few decades designed to reverse desertification in China. In a 1990 speech the Executive Secretary of the UN Convention to Combat Desertification used China as an example of the threats imposed by desertification. His comments echoed those made during the early 1970s and the prolonged drought in the West African Sahel, when he suggested that "the desert has moved to within 100 miles of Beijing, and in the long term this desertification will have serious effects on food scarcity and people's health and will force people to migrate" (D'Aleo, 2000). He also suggested that the desertlike condition was approaching Beijing at a rate of 1.2 miles per year.

The Chinese government has become increasingly concerned about another sign of desertification – the enhanced frequency and intensity of dust storms originating in its western provinces and blanketing the nation's capital, Beijing (Royston, 2001), as well as dust storms originating in northern China. The government is especially intent on arresting such dust storms in advance of the 2008 Olympics to which they are hosts.

Under specific circumstances, desertification is totally irreversable. For example, in mountainous areas of West Africa, once the higher elevations are denuded of soil (wind and water erosion), it will require millions of years to reform. Former sand dunes, now vegetated, if deprived of that protection, will be remobilized, and desert dune encroachment will result. Those desert sands cannot be easily stabilized, nor made fertile (Fairbridge, 1968, p. 1134 et seq.). Examples can be seen in Senegal, Maci, Niger, Sudan, as well as in India and Australia.

Desertification: why does it occur?

Ever since the mid-1970s, researchers have been divided over whether to blame the climate system or human activities for desertification. Some researchers consider climate to be the major contributor to desertification processes, with human factors playing a relatively minor supporting role. Other researchers reverse the significance of these two factors. For example, Le Houérou (1959) concluded that "on its edges the Sahara is mainly made by man; climate being only a supporting factor" (quoted in Rapp, 1974, p. 32). A third group blames climate and humans more or less equally. For example, Grove (1974, p. 137) has noted that "desertification or desert encroachment can result from a change in climate or from human action and it is often difficult to distinguish between the two". Each of these views can be shown to be valid, at least at the local level and on a case-by-case basis. Thus, there is a region-specific bias to perceptions about desertification, one that spills over to the definitions.

Climate

References to climate in these definitions relate either to interannual climate variability, climate fluctuations, climate change, or drought. *Climate variability* (a term that is usually overlooked in these definitions) refers to the natural variations that appear in the atmospheric statistics for a designated period of time, usually on the order of months to years. Variations can occur in any or all of the atmospheric variables (such as precipitation, temperature, wind speed and direction, relative humidity, evaporation, etc.). Those variations could alter an ecosystem, which would eventually affect human activities in adverse ways, activities that had been designed to exploit the productivity of that ecosystem.

It is important to note that during the annual dry season the characteristics of the atmosphere in an arid or semiarid area are like that of a desertlike region (low precipitation, high evaporation, high solar radiation, etc.), and if the land is improperly used during this period, degradation results (Aubreville, 1949). Thus, short-term variations in climatic factors as well as seasonal dry periods, when combined with improper land-use practices, can give the appearance of the impact of a regional climate change (e.g. global warming) when none may have occurred at all. *Climate fluctuation* refers to variations in climate conditions that occur on the order of decades. Climate and hydrologic regimes are known to fluctuate between extended wet and dry periods that can last up to several decades. *Climate change* refers to the view that the statistics that represent the average state of the weather for a relatively longer period of time are changing, and that desertification is primarily the result of such natural climate change. For example, there has been an obvious multidecadal trend beginning in the late 1960s toward increased aridity in the West African Sahel; a natural desiccation of the region that humans can do nothing to stop. Usually cited as evidence for long-term climate changes in that area in the past are the fossil dune fields near the West African coast far from the active dunes close to the desert. The debate over the possible impacts of long-term global climate change on the climate conditions in the West African Sahel (and on other arid lands around the globe) continues.

Nevertheless, paleoclimatologists, point out that since the "Little Ice age", which reached its nadir in the 17th century, there has been a systematic global warming (of natural causes), associated with aridification of the Sahara and other high pressure belts.

Drought episodes are also considered to be a major cause of high-pressure desertification. Especially during multiyear meteorological droughts, desertification becomes relatively severe, widespread, as well as more visible, and its rate of development increases sharply (e.g. Grainger, 1984). As the probability of drought increases as one moves from the humid to the more arid regions under present-day global climate conditions, so too, does the likelihood of severe desertification. Land forms, soils, and vegetation are often transformed, sometimes irreversibly, during such extended drought periods.

Tree-ring analysis, lake sedimentation, Nile floods and other proxies prove conclusively that drought cycles have been recurrent for at least the last few thousand years. Man-made pollution (CO_2), however, has amplified the late 20th century and early 21st centuary intervals.

The view held by the general public of what constitutes climate change exposes a misunderstanding. It is not just a matter of a summer warmer than last year, or less snow in winter than in the preceding decade. Scientists are referring to a profound climate change (a global warming of a few degrees Celsius), the magnitude of which societies have not witnessed for millennia. For reasons of clarity for the public, this profound type of change might be called "deep climate change". Many countries are now concerned about the possible impacts of global warming of the atmosphere on the rate and irreversibility, as well as the location, of desertification processes in future decades (Ci et al., 2002).

Human activities

Cultivation, herding, and wood-gathering practices, as well as the use of technology and even the occurrence of conflict (Timberlake and Tinker, 1984), have been cited in the definitions as major causes of, or contributors to, the desertification process in arid, semiarid, and, subhumid areas (e.g. Swedish Red Cross, 1984). *Cultivation practices* that can lead to desertification include irrigation, land clearing, deforestation, cultivation of marginal climatic regions, cultivation of poor soils, woodcutting for firewood and construction, and inappropriate cultivation tactics such as reduced fallow time, improper tillage, drainage, and water use. Areas that might support agriculture on a short-term basis may be unable to do so on a long-term sustained basis. Even land surfaces that are considered suitable for cultivation of some sort may become degraded, if they are managed in a way that is inappropriate to their ecological and climatic setting.

Rangeland abuse leading to desertification includes excessively large herds in relation to existing range conditions (e.g. overgrazing and trampling) and herd concentration around human settlements and watering points. Government policies toward their pastoral populations can also indirectly lead to desertification by, for example, not pursuing payment policies that encourage herders to cull their herds, by arbitrarily putting a minimum sale price on grain and a ceiling on prices that pastoralists might receive for their livestock, and so forth.

Gathering firewood by itself or in combination with overgrazing or inappropriate cultivation practices can create conditions that expose the land to existing "otherwise benign" meteorological factors (such as wind, evaporation, precipitation runoff, solar radiation on bare soil, etc.), thereby contributing to desertification.

The *use of technology* in arid, semiarid, and subhumid environments is often a result of the policy-makers' desire for economic growth and development. Thus, deep wells, irrigation and cash crop schemes, even the reduction of livestock diseases, each in its own way, increases the risks of desertification processes, if the technology is not properly applied. Desertification can result from road building, industrial construction, geological surveys, ore mining, settlement construction, irrigation facilities, and motor transport (Rozanov, 1977).

In sum, most researchers accept that both human intervention and climate are involved in the desertification process, with a few observers noting that the two factors are so entwined that to separate them as to primary and secondary contributors would be a fruitless endeavor.

Return to Aubreville

Aubreville discussed desertification at great length in his 1949 report entitled *Climats, Forets et Desertification de l'Afrique Tropicale*. His work, when compared to the scores of contemporary definitions, raises the issue about *where* desertification can take place. Most of the contemporary definitions relate desertification to what broadly speaking might be viewed as the desert fringes: the arid, semiarid, and dry subhumid areas. Aubreville explicitly referred to the dry tropical forest of Africa, noting that "these are real deserts that are being born today, under our eyes, in the regions where the annual rainfall is from 700 mm to 1500 mm" (1949, p. 332). Interestingly, Aubreville's original view of desertification would likely have no place in desertification studies today because it fails to meet the criteria identified in most contemporary definitions.

Aubreville viewed desertification primarily as a process but also referred to it as an event. He described how forested regions were transformed into savanna and savanna into desertlike regions. One of Aubreville's central concerns was the rate of destruction, resulting from human activities, of Africa's dry tropical forests. He noted that cultivation, deforestation, and erosion were so entwined as to lead to the destruction of the vegetative cover and soils in the forested regions of tropical Africa where "the desert always menaces, more or less evident, but it is always present in the embryonic state, during the dry and hot season" (Aubreville, 1949, p. 331). Savanna would result. Continued disregard for the fragility of the savanna would result in the creation of desert-like conditions.

Aubreville's original research findings and observations are still relevant to contemporary efforts to identify, understand, and combat desertification processes. The resurrection of his research on desertification is not a call to discard other definitions, but a call to broaden thinking about what constitutes desertification as a process and where on Earth that process might occur. If desertification can be identified by some of its component subprocesses, such as soil erosion, deforestation, overgrazing, or cultivation in marginal areas as defined, for example, by soil characteristics or by the amount of rainfall, then there is a great deal of research activity under way that relates directly as well as indirectly to desertification, even though the word desertification does not appear in its title (e.g. Riquer, 1982).

A broad view of desertification would shed a different light on progress in the understanding and combating of desertification. There is much research under way on soil erosion, range management, deforestation, increasing biological productivity in arid and semiarid lands. Only in this broader conceptualization of desertification can we develop a more accurate assessment of how nations are really doing in their national "war" against desertification.

Michael H. Glantz

Bibliography

Adams, R., Adams, M., Willens, A., and Willens, A., 1979. *Dry Lands: Man and Plants*. New York: St Martin's Press.

Aubreville, A., 1949. *Climats, Forets et Desertification de l'Afrique Tropicale*. Paris: Societe d'Editions Geographiques, Maritimes et Coloniales.

Carder, D.J., 1981. Desertification in Australia – a muddled concept. *Search*, **12**(7): 218–221.

Charney, J.G., 1975. Dynamics of deserts and drought in the Sahel. *Quarterly Journal of the Royal Meteorological Society*, **101**: 193–202.

Ci, L.J., Yang, X.H., and Chen, Z.X., 2002. The potential impacts of climate change scenarios on desertification in China. *Earth Science Frontiers*, **9**: 287–294.

D'Aleo, J., 2000. Expanding deserts are the latest signs of a changing climate. On line 15 November at www.intellicast.com/DrDewpoint/Library/1174/

Dregne, H.E., 1976. Desertification: symptoms of a crisis. In Paylore, P., and Haney, R., eds., *Desertification: process, problems, perspectives.* Tucson, AZ: University of Arizona, Office of Arid Lands Studies, pp. 11–24.

Fairbridge, R.W. (ed.), 19?? *The Encyclopedia of Geomorphology*. New York: Van Nostrand-Reinhold.

Glantz, M.H. (ed.), 1977. *Desertification: environmental degradation in and around arid lands*. Boulder, CO: Westview Press.

Glantz, M.H., and Orlovsky, N., 1986. Desertification: anatomy of a complex environmental process. In Dahlberg, K.A., and Bennett, J.W., eds., *Natural Resources and People: conceptual issues in interdisciplinary research*. Boulder, CO: Westview Press, pp. 213–229.

Grainger, A., 1984. *Desertification.* London: International Institute for Environment and Development.

Grove, A.T., 1974. Desertification in the African environment. *African Affairs*, **73**: 137–151.

Hare, K., 1976. *Climate and Desertification.* Toronto: Working Group on Climate and Human Response, Institute for Environmental Studies, University of Toronto.

Hare, K., 1977. Connections between climate and desertification. *Environmental Conservation*, **4**(2): 82.

International Geographical Union Working Group on Desertification, 1975. *Proceedings of the IGU Meeting on Desertification.* Cambridge: University of Cambridge, 22–26 September.

Johnson, D.L., 1977. The human dimensions of desertification. *Economic Geography*, **53**(4): 317–318.

Kassas, M., 1977. Arid and semi-arid lands: problems and prospects. *Agro-Ecosystems*, **3**: 186.

Kharin, N.G., and Petrov, M.P., 1977. *Glossary of Terms on Natural Conditions and Desert Development.* Materials for the UN Conference on Desertification. Ashkhabad, USSR: Desert Research Institute.

Kovda, V., 1980. *Land Aridization and Drought Control.* Boulder, Co: Westview Press.

Kove, N.G., 1982. WMO Draft. Geneva: World Meteorological Organization.

Le Houérou, H.N., 1962. *Les paturages naturels de la Tunisie aride et desertique*. Paris: Inst. Sces. Econ. Appl. Tunis.

Le Houérou, H.N., 1975. The nature and causes of desertization. In *Proceedings of the IGU Meeting on Desertification*. Cambridge: University of Cambridge, 22–26 September 1975. (Reprinted in Glantz, 1977.)

Mabbutt, J.A., and Wilson, A.W. (eds.), 1980. *Social and Environmental Aspects of Desertification.* Tokyo: UN University.

Meckelein, W., 1980. The problem of desertification within deserts. In Meckelein W., ed., *Desertification in Extremely Arid Environments.* Stuttgart: Stuttgarter Geograpisches Studien.

Plit, F., Plit, J., and Zakowski, W., 1995. *Drylands Development and Combating Desertification: bibliographic study of experiences in countries of CIS*. Paper no. 14, Environment and Energy. Rome: UN Food and Agriculture Organization.

Prince, S.D., 2002. Spatial and temporal scales for detection of desertification. In Reynolds, J.F., and Stafford-Smith, D.M., eds., *Global Desertification: do humans cause deserts?* Berlin: Dahlem University Press, pp. 23–40.

Quintanilla, E.G., 1981. Regional aspects of desertification in Peru. In *Combating Desertification through Integrated Development.* United Nations Environment Program/USSR Commission for UNEP International Scientific Symposium, Abstract of Papers, Tashkent, USSR, pp. 114–115.

Rapp, A., 1974. *A Review of Desertization in Africa: water, vegetation and man.* Stockholm: Secretariat for International Ecology.

Rapp, A., Le Houérou, H.N., and Lundholm, B., (eds.), 1976. *Can Desert Encroachment Be Stopped? A study with emphasis on Africa.* Stockholm: Swedish Natural Science Research Council.

Reining, P. (comp.), 1978. *Handbook on Desertification Indicators.* Washington, DC: American Association for the Advancement of Science.

Reynolds, J.F., and Smith, M.S., 2002. An integrated assessment of the ecological, meteorological and human dimensions of global desertification. Dahlem Konferenzen, DWR 88. In Reynolds, J.F., and Smith, M.S., eds., *Global Desertification: do humans cause deserts*? Berlin: Dahlem University Press.

Riquer, J., 1982. A world assessment of soil degradation. *Nature and Resources*, **18**(2): 18–21.

Royston, R., 2001. China's dust storms raise fears of impending catastrophe. National Geographic News, 1 June. On line at news.nationalgeographic.com/news/2001/06/0601_chinadust.html

Rozanov, B.G., 1977. Degradation of arid lands in the world and international cooperation in desertification control. *Pchvovedenie*, **8**: 5–11.

Rozanov, B.G., 1981. Principles of Desertification diagnostics and assessment. In *Combating Desertification through Integrated Development.* United Nations Environment Program/USSR Commission for UNEP, International Scientific Symposium, Abstract of Papers, Tashkent, USSR, pp. 24–26.

Sabadell, J.E., Risley, E.M., Jorgenson, H.T., and Thornton, B.S., 1982. *Desertification in the United States: status and issues*. Washington, DC: Bureau of Land Management, Department of the Interior.

Saiko, T.A., and Zonn, I.S., 1997. Europe's first desert. In Glantz, M.H., and Zonn, I.S., eds., *Scientific, Environmental, and Political Issues in the Circum-Caspian Region.* Dordrecht: Kluwer, pp. 141–144.

Swedish Red Cross, 1984. *Prevention Better than Cure*. Stockholm and Geneva: Swedish Red Cross.

Timberlake, L., and Tinker, J., 1984. *Environment and Conflict*. Earthscan Briefing Document 40. London: International Institute for Environment and Development.

UN Conference on Desertification, 1978. *Round-up, Plan of Action and Resolutions*. New York: United Nations.

UN Environment Program, 1992, 1997. *World Atlas of Desertification*, 1st and 2nd edn. London: Edward Arnold.

UN Environment Program, 1975. *Overviews in the Priority Subject Area Land, Water and Desertification.* Nairobi: UNEP.

UN General Assembly, 1981. Development and International Economic Co-operation: Environment. *Study on Financing the Plan of Action to Combat Desertification.* Report A/36/141. New York: Report of the Secretary General, United Nations.

UN Secretariat of the Conference on Desertification, 1977. *Desertification: an overview.* In *Desertification: its causes and consequences*. New York: Pergamon Press.

World Meteorological Organization, 1980. *Expert Meeting on Meteorological Aspects of Desertification.* Geneva: World Meteorological Organization.

Zhenda, Z., and Shu, L., 1981. Desertification and desertification control in northern China. *Desertification Control*, **5**: 13–19.

Zonn, I.S. (ed.), 1981. *USSR/UNEP Projects to Combat Desertification.* Moscow: Centre of International Projects GKNT.

Cross-references

Arid Climates
Aridity Indexes
Deserts
Drought

DESERTS

The world's deserts are some of its most extreme and spectacular environments. The incredible dune systems of the Sahara or the Namib are but two examples; the bountiful succulents fields of the American southwest are another. Desert environments globally are highly diverse, a result of the complexity of the climatic conditions that can produce aridity and the underlying surface materials. However, all deserts share one essential commonality: a relatively scarcity of precipitation.

The definition of a desert depends on the aspect of environment of principal concern, such as vegetation, terrain, culture, climate or resources; hence there is no simple way to define or delineate desert environments. Even climatic boundaries of deserts, and therefore geographic boundaries, are hard to establish because there is a gradual and continual transition between arid and semiarid environments.

In basic climatic terms a desert can be defined as an area which receives little or no rainfall and experiences no season of the year in which rain regularly occurs. However, rainfall alone insufficiently defines climatic boundaries of deserts because it varies dramatically from year to year and because aridity is not determined solely by rainfall. What distinguishes deserts from non-deserts is the amount of moisture available to the ecosystem, or the difference between the rainfall a region receives and that lost through evaporation. According to estimates based on this concept, true deserts make up roughly 20% of the Earth's land surface and another 15% is semiarid.

Deserts are concentrated in Asia, Africa, and Australia. The largest expanse of arid land lies in an almost continuous area stretching nearly half a hemisphere across northern Africa and Asia, an inhospitable barrier separating most of Europe from southeastern Asia. Africa and Asia respectively contain 37% and 34% of the world's arid lands. The principal deserts of the world include (see Figure D4) the Kalahari–Namib, Somali–Chalbi and Sahara on the African continent; the Arabian, Iranian and Turkestan Deserts of the Middle East; the Thar, Takla-Makan and Gobi Deserts of Asia; the Monte–Patagonian and Atacama–Peruvian Deserts of South America; the Australian Desert; and the North American Desert, including the Great Basin and the Sonoran, Mojave and Chihuahuan Deserts. Vast tracts of semiarid land, such as the American Great Plains and the African Sahel, border each of these.

The deserts shown in Figure D4 tend to be situated in subtropical latitudes or in the lee of major mountain ranges, as the factors which promote aridity are common to these locations. Deserts are also found at high altitudes and high latitudes, but these differ in two basic respects: the low rainfall is primarily a result of cold conditions and evapotranspiration is inherently low and of minor importance in determining water availability.

Classification of deserts

Although sparse or negligible rainfall is a common characteristic of all deserts, aridity is better defined by moisture availability. This a residual in the balance between water availability through precipitation and water loss via evapotranspiration. The latter quantity represents the maximum amount of water that could be evaporated by solar energy or transpired by plants under conditions of constant moisture supply. Potential evapotranspiration can be as high as 3000 mm in some regions. In deserts, however, water is limited and hence this maximum is never achieved. Nevertheless, this parameter can be considered as "water demand" and climatic characteristics such as temperature or incoming radiation are used to approximate this demand. Rainfall, in turn, represents water availability.

Water demand, whether assessed by net radiation, temperature or potential evapotranspiration, varies greatly from one region to another. It tends to decrease with increasing latitude and it is higher in summer than winter. Therefore the amount of rainfall at the desert margin tends to decrease with increasing latitude, and it is higher for deserts with summer rainfall. Within a geographical region, water demand is more constant and desert boundaries are often delineated regionally on the basis of mean annual rainfall. In Australia, for example, the generally accepted limit for the desert is 400–500 mm of rainfall in the north and 250 mm in the more temperate climate of the southern border.

A comparison between supply and demand forms the basis of most definitions of deserts. All systems of assessments are in reasonable agreement on global space scales, but show significant differences along the margins at regional scale. The best-known

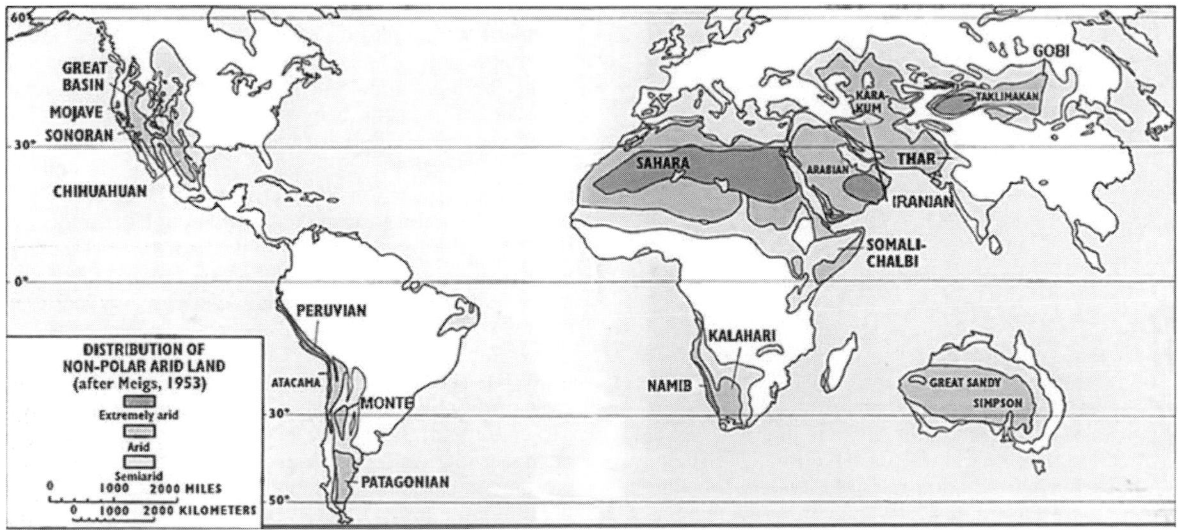

Figure D4 Location of the world's major deserts, according to Meig's classification.

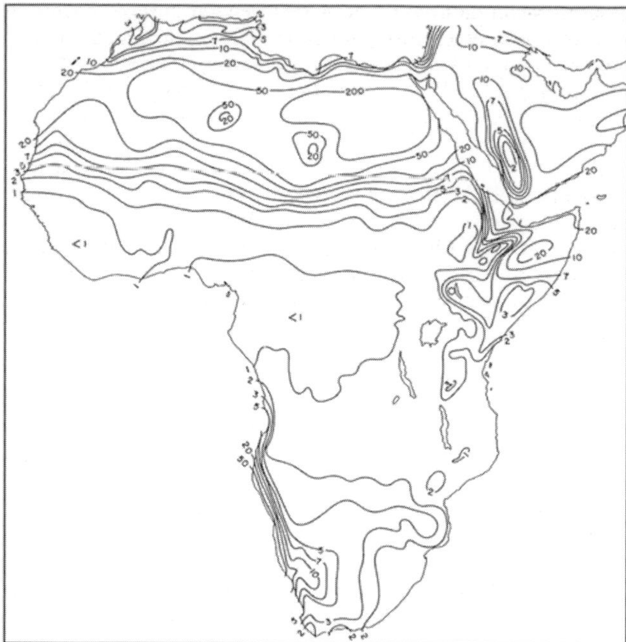

Figure D5 Budyko's radiational index of dryness over Africa.

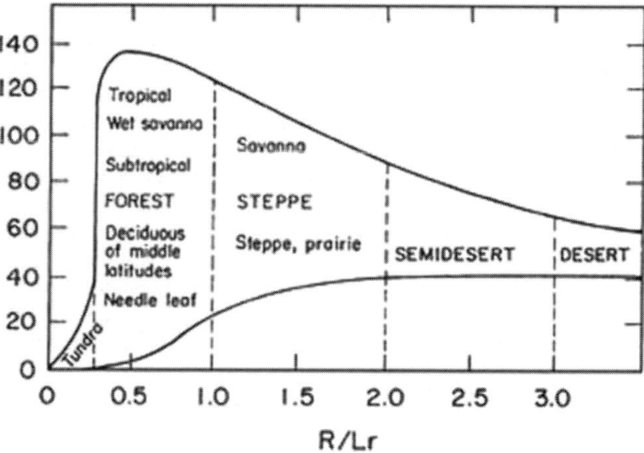

Figure D6 Budyko's concept of geobotanical zonality: the principal ecosystems as a function of net radiation and the dryness ratio (ratio between annual net radiation and annual average precipitation multiplied by the latent heat of condensation).

classification scheme for arid lands, that of Peveril Meigs, assesses "demand" by way of potential evapotranspiration. An example of Meigs' classification for Africa is shown in Figure D4. The classification scheme of the Russian climatologist Budyko represents it by the net radiation received at some location, i.e. the difference between the net heat gain from the sun and the net heat loss by radiation from the ground (Figure D5).

The Budyko classification has some advantages, in that it allows for a simple quantitative comparison of the degree of aridity of various deserts and its aridity index is readily interpreted physically. The index is a parameter termed the "dryness ratio". To produce it, net radiation is converted to energy units and the index is defined as the ratio of mean annual rainfall to the amount of rainfall which this energy suffices to evaporate. In Budyko's classification the boundary between deserts and semiarid regions is where the ratio is 3 (Figure D6). This ratio is as high as 200 in the central Sahara, but is generally below 10 in the North American deserts.

Causes of aridity

The formation of rainfall requires moisture supply, unstable atmospheric conditions and ascending air. Ascent can be produced by extreme heating of the surface, by a convergent pattern of air flow, or orographically forced by mountain barriers. Aridity is promoted by the opposite conditions: a lack of atmospheric moisture, stable air, subsidence (i.e. descending air motion), and a divergent pattern of air flow (i.e. air streams spreading apart from each other). Another cause is distance from the main tracks of major weather systems.

Except in polar and high-altitude deserts, a lack of sufficient moisture is generally not the overriding cause of the arid climate. The case of the Sahara Desert clearly demonstrates this. In summer the atmosphere above the Sahara is enriched with about as much moisture as that over the wetter regions of the southeastern United States. Likewise, the Namib Desert of southwestern Africa is a very humid environment, although it is one of the driest (i.e. most rainless) deserts on Earth.

The other factors, stable and subsiding air and divergent flow, are common to two meteorological situations linked to the occurrence of dry climates: the semipermanent high-pressure systems which prevail in the subtropical latitudes and the rainshadows on the leeward sides of mountain barriers. These are the most common locations of deserts.

Descending currents are particularly effective in promoting aridity. They originate at high altitudes, where the air is dry and, as they sink, they undergo intense compressional heating. Consequently, the air is very hot and dry when it reaches the surface. The warming also produces a temperature inversion (temperature increasing with elevation), which stabilizes the lower atmosphere and suppresses the uplift required to produce clouds and precipitation. Air descends in the core of high-pressure cells because of the pattern of surface air flow; it descends also in the lee of mountains.

Near the high-pressure cells, aridity is further enhanced by a temperature inversion produced by the descending air and by cold water on the eastern flanks of the highs. The occurrence of this unusually cold water is linked to the surface wind flow around the high, which produces the upwelling of cold water and advects cold polar water into relatively low latitudes. Along the coast of South America the upwelled water is as cold as 12°C, compared to over 20°C in the mid-ocean.

Both the subtropical highs and mountain barriers further promote aridity by blocking the passage of major weather systems. Mountain barriers also act as a barrier to the inland penetration of moist, maritime airmasses. The transformation of airmasses crossing the mountains produces a "rainshadow". Air approaching a mountain range is forced to rise, promoting cloud formation and rainfall on the windward side and the peaks. The air which reaches the leeward side has dropped its moisture at higher elevations and is relatively dry. As it sinks in the lee, compressional heating accentuates its dryness.

Most of the world's largest deserts are under the influence of the subtropical highs: the Australian Desert, the Peruvian–Atacama Desert of South America, the southwestern United States, the Namib Desert and Kalahari of southern Africa, and the Sahara. Aridity is strongest on the western sides of continents in these latitudes, where the influence of the highs is greatest and where the coastal deserts lie.

The deserts of the midlatitudes are generally located in the lee of major mountain ranges: the Patagonian Desert of South America, many of the deserts of the Middle East and Asia, and the intermontane region of the western United States. Topography also plays a role in the Somali–Chalbi Desert of Ethiopia and the Horn of Africa and, to some extent, the Thar desert of India.

In most cases, however, complementary factors also play a role in producing aridity. For example, the influence of the subtropical highs cannot suffice to explain the full extent of the Sahara and its continuation into Arabia and the Middle East. There, neither the cold-season temperate latitude weather systems to the north nor the warm-season tropical disturbances on its equatorward side penetrate to its interior. The deserts of Asia and the Middle East are far from prevailing storm tracks. The latter two regions are also quite distant from any moisture sources. In the Somali-Cholbi Desert, one factor is local patterns of wind flow, jet streams lying near the surface. A coastal, low-level jet stream enhances the aridity of the Peruvian Desert, the most latitudinally extensive desert in the world.

Thus, the causes of desert climates are quite complex and often regional influences must be taken into account. The complexity and the regionalization of these factors gives a nearly unique climatic identity to each of the Earth's major dryland regions, despite their many common climatic characteristics.

Surface environments of the desert

The vision of the sandy desert is applicable to only a portion of the world's arid lands. Other surface types include bedrock (often termed hammada), stone pavements (termed reg and serir), depositional flats, and desert crusts. Sand fields and dunes are common in the Sahara and in parts of the Australian Desert, but the Syrian and Gobi deserts are largely hammada and reg types. As the desert gives way to semiarid landscape, soils become more common and widespread.

Overall sand dunes occupy less than 20% of the surface area of the world's deserts. They cover about 28% of the Sahara but only 1% of the desert surface in North America. Classically, two types of dunes are distinguished: longitudinal (or linear) dunes, which are roughly parallel to the prevailing winds, and transverse dunes, which are aligned roughly normal to the wind. In reality, a great variety and complexity of dune forms exist. The simpler dune forms result from simple wind regimes, i.e. one or two prevailing directions. As the complexity of the prevailing winds increases, so does the complexity of dunes.

Many deserts sustain a rich vegetation cover, which can include grasses, trees and shrubs (although fewer in number than in the wetter environments). In the true deserts arboreal species (trees and tall shrubs) are rare, as are perennial grasses. The dominant vegetation in most deserts includes dwarf shrubs and low shrubs (such as *Artemisia* or sagebrush) and succulents (cacti and euphorbia). Annual grasses may be present in the years with abnormally high rainfall.

Desert vegetation consists of two general classes, based on the primary way to resist moisture stress: perennials which *avoid* drought, and annuals or ephemerals, which *evade* it. The latter grow profusely during periods of favorable precipitation and produce seeds which lie dormant during drier episodes, until wetter conditions return. The perennials are phreatophytes, which develop long taproots penetrating downward to the water table, or xerophytes, which adapt to low water supply and high salinity. The xerophytes include many dwarfed, woody species, with reduced water demand, and succulents, that resist drought by storing water in their leaves, roots and stems. The succulents include the cacti of the American deserts and the Old World euphorbia.

Diversity and extremes of desert climates

Most desert environments share several common climatic characteristics. These include meager rainfall which is highly variable in time and space, and often concentrated in very small, localized areas; low atmospheric relative humidity; an extreme thermal environment with large fluctuations of temperature between day and night and during the year; and generally low cloudiness and high insolation. Individual deserts differ markedly with respect to the amount of rainfall and the degree of aridity. The thermal characteristics, especially mean annual temperature and its seasonal variations, are determined to a large extent by latitude.

Consequently, several types of deserts are distinguished on the basis of climate: warm, cold-winter and foggy or coastal deserts (Figure D7). In the warm deserts of the lower latitudes, temperatures are high year-round and freezing is rare or absent. These may occasionally receive incursions of frigid air from higher latitudes, producing frost or even snow. The midlatitude deserts generally experience cold winters, often with seasonally occurring freezing temperatures. Both are characterized by low relative humidity year-round.

In contrast, coastal deserts are relatively humid and the temperatures relatively mild in both winter and summer, a consequence of the influence of the coastal waters. Compared to inland deserts, both the diurnal and annual temperature ranges are reduced. Also, they tend to be frequently cloud-covered and fog is a common occurrence.

The rainfall conditions in deserts are quite diverse. Some, such as the Sahara, have virtually rainless sectors with less than 10 mm of rainfall annually. Rainless stretches of 10 years have been recorded in in parts of the Sahara and at locations such as Swakopmund (Namibia) and Lima (Peru) in the coastal deserts of Southern Africa and South America. Few deserts are this dry. In the Sonoran, Mojave, Iran and Arabian Deserts and in the Takla-Makan, mean annual rainfall is on the order of tens of millimeters in their driest cores. It exceeds 80–100 mm everywhere in the Thar, Gobi, Patagonian and Australian Deserts and in the Kalahari. In contrast, it scarcely falls below 200 mm in the Chihuahuan Desert and Great Basin of North America.

The temperature conditions are also diverse because they depend on several factors, including latitude, elevation, distance from coast and degree of aridity (Figure D8). Except in coastal deserts mean daily maximum temperatures during the summer months are generally on the order of 30–45°C. The mean daily winter minima are usually above −10°C, except in the interior of Asia, where they may be as low as −20 or −30°C. There is a general tendency for annual temperature

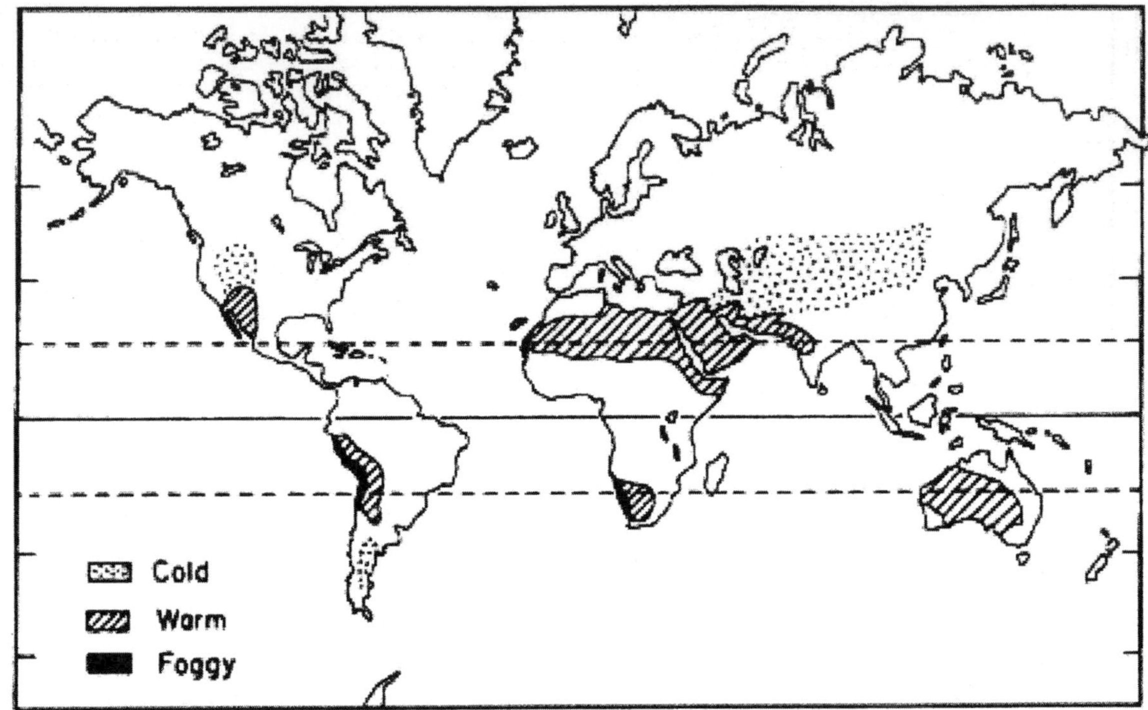

Figure D7 Classification of desert regions into cold, warm and foggy deserts (after Schmida).

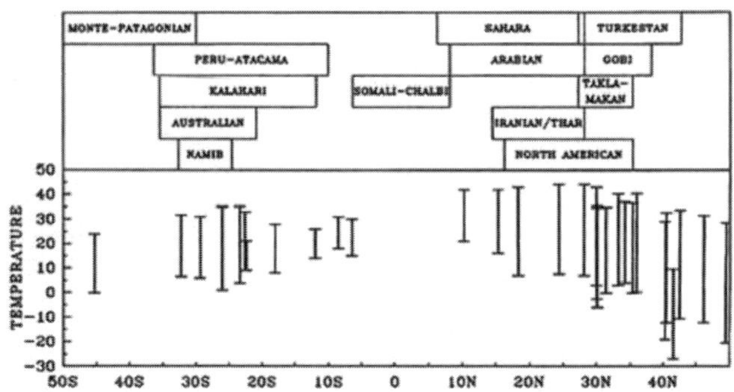

Figure D8 Annual temperature range vs. latitude for arid and semiarid regions. Bars extend from mean temperature of the warmest month to mean temperature of the coldest month. Note the smaller temperature range in the southern hemisphere, a consequence of the large ratio of water to land.

range to increase with latitude, varying from about 15°C in low-latitude deserts to 20–35°C in the cold-winter deserts. The mean diurnal range is likewise diverse, varying from about 4°C in coastal deserts to 22°C in high latitudes.

These mean conditions of rainfall and temperature do not fully illustrate the extreme range of thermal conditions experienced in deserts. In Australia and Asia the absolute maximum air temperatures are on the order of 48–50°C, but temperatures of 57°C and 58°C have been recorded in the Sahara, 57°C at Death Valley, California. Surface temperature commonly fluctuates by over 30°C between day and night in the Kara Kum Desert of Asia and by over 40°C in the parts of the North American Deserts. Some deserts of the Soviet Union experience absolute temperature ranges in excess of over 100°C, with air temperatures as low as −58°C having been recorded.

Ground temperatures are even more extreme, with temperatures in excess of 70°C having been recorded at several locations. At Port Sudan on the Red Sea, a sand temperature of 83.5°C was recorded. The deserts also experience extreme cold: at Repetek, in the Kara Kum desert of Asia, where a sand temperature of 79.4°C was once recorded, the ground temperature can drop to about −40°C.

The precipitation regime is similarly extreme. In some areas many years pass without a drop of rain, but when rain does fall it is often torrential. Many times the mean annual rainfall can fall within hours or days. At Lima, where the mean annual rainfall is 46 mm, 1524 mm fell during one storm in 1925. In the Sahara at Nouakchott (Mauritania), over twice the annual mean, 249 mm, fell in 1 day; at Tamanrasset (Algeria), the mean annual rainfall of 48 mm once fell in hour.

Thermal regime

The subtropical location of many deserts and the scant cloud cover of most produce a regime of strong solar insolation and high surface temperatures. Temperature is enhanced by the dryness of the soil and the lack of a dense vegetation cover to absorb and redistribute the solar radiation near the ground. The heat is concentrated at the surface because the dry ground transports little heat to deeper layers of soils; temperature decreases rapidly with depth into the ground and with height above the

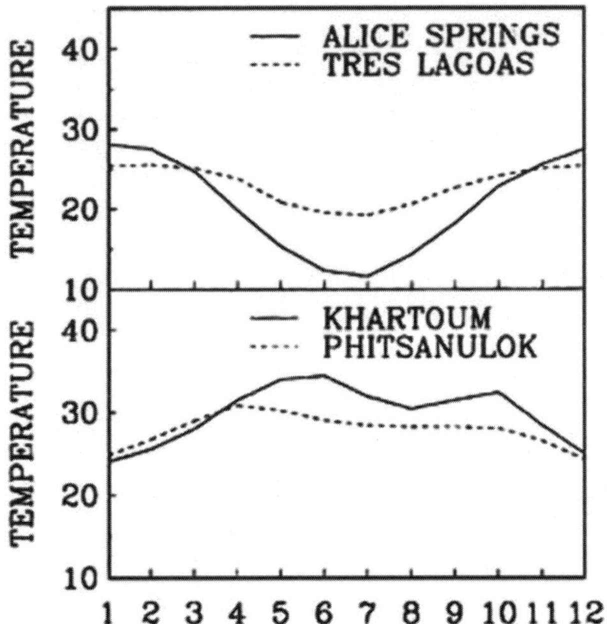

Figure D9 Mean monthly temperature at pairs of stations at comparable latitudes but differing precipitation regimes. Alice Springs and Khartoum are desert locations; Tres Lagoas and Phitsanulok are regions of tropical forests.

surface. In the Kara Kum Desert of Asia daytime temperature typically drops over 15°C within the first few centimeters and 20–30°C within the first 10 cm below the surface. Air temperature can drop by more than 10°C within the first 10 cm above the surface.

The concentration of heat at the surface means that there is no subsurface thermal reservoir. As a consequence of this, and the sparse vegetation cover, the surface cools extremely rapidly and efficiently at night. The generally clear skies and the dryness of the air near the ground allow most of the heat accumulated by day to escape to the upper atmosphere by night (90%, compared to 50% in humid regions). This accentuates the nighttime cooling.

The result is a large daily range of both ground and air temperature. At Khartoum, Sudan, at 15°N latitude in the eastern Sahara, the daily air temperature range is on the order of 18°C in the dry season and 11°C in the wet season. At Alice Springs, at 23°S in central Australia, the diurnal range is about 15°C. The daily fluctuations at humid stations of comparable latitude would be about half as great during the wet season. At a station in the Sahara, during the course of a day the air temperature fell from a daytime maximum exceeding 37°C to −1°C at night. At Death Valley, California, a daily range of 41°C was observed in August 1891; at Tucson, Arizona, the record is 56°C.

The lack of a surface heat reservoir in deserts tends to lead also to a high annual range of temperature. To a large extent the annual range is dependent on latitude, increasing with latitude as seasonal contrasts become important. Nevertheless, the annual temperature range at a desert location will tend to be greater than that at a humid location at a comparable latitude. The annual ranges for Alice Springs (24°S) and Khartoum (16°N) are 16°C and 10°C respectively, compared with approximately 6°C at Tres Lagoas (Brazil) and Phitsanulok (Thailand), at similar latitudes but with humid climates (Figure D9). In deserts near the equator the annual range is often much less, so that the daily temperature range is several times larger than the annual range.

Hydrologic regime

Much of the hydrologic character of a desert is dependent on its latitudinal location. Tropical rainfall prevails at low latitudes, such as in the southern Sahara or the northern sector of the Australian Desert. A midlatitude rainfall regime prevails at higher latitudes, such as the poleward sectors of these deserts or in the deserts of North America and much of Asia. The tropical and midlatitude regimes differ with respect to both the nature and seasonality of the precipitation and, consequentially, with respect to the loss of moisture through runoff and evapotranspiration.

Tropical rainfall is produced by convective processes (i.e. cloud formation related to localized surface heating and dynamic processes). In the midlatitudes most rainfall (especially that which occurs in winter) is linked to large-scale warm or cold fronts. Convective rainfall tends to be much more localized and intense than frontal rainfall. As a result, in areas where convective rainfall prevails, whether deserts or humid regions, rainfall is highly variable in space, confined to raincells on the order of 10–50 km, or less. The monthly rainfall at two locations a few kilometers apart might be quite different. For this reason rainfall in deserts was often thought to be a completely random and localized occurrence. Satellites have shown, however, that although this is true of individual raincells, these cells tend to occur in organized, large-scale, meteorological disturbances.

Convective rainfall tends to be of both short duration and high intensity. A typical rain might last for a few minutes or hours, compared to days for frontal rainfall. The high intensity of the convective rainfall also affects its effectiveness, since the ground quickly becomes saturated and much of the rainfall is lost to runoff. Because there are few drainage channels in deserts, the runoff may form a thin sheet of water, transforming the barren ground to an enormous lake within minutes. This runoff produces flashflooding in many deserts, especially when rain falls over barren rock at higher elevations. As little as a few millimeters of rain can produce flow in dry desert wadis; the flow may emerge suddenly, peak rapidly, but last only a few hours. Less than 100 mm can produce catastrophic floods. In the Tadmait Plateau of the Sahara a rain of approximately 16 mm generated a flood with peak discharge of about 1600 m^3/s.

Not only the rainfall in individual storms, but also the mean distribution in a desert, can be a mosaic, particularly in the rainshadow deserts. In the state of Washington, in the western United States, mean annual rainfall varies from 3000 mm on the peaks of the Cascades to less than 200 mm in the leeside valleys only 40 km away. Isolated mountains, such as Tibesti or the Hoggar in the Sahara, dramatically enhance rainfall. Peaks may routinely receive 100 mm or more per year, compared to a few millimeters in the surrounding regions.

Although by some definitions a true desert is a location without a regular rainy season, in most desert regions there is a preference for concentration in either the warm or cool season (Figure D10). The subtropical deserts represent a transition

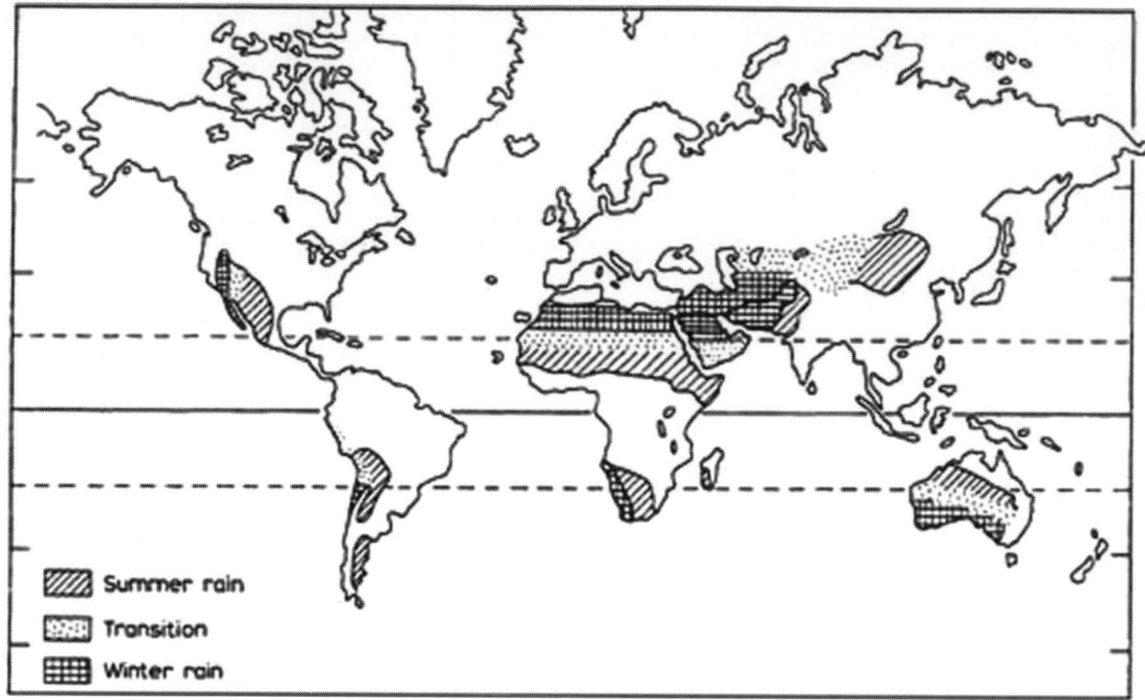

Figure D10 Areas of hot deserts with summer rain, winter rain and rainfall during the transition seasons (from Schmida).

Figure D11 Saguarro cactus in the Sonoran Desert of Arizona.

from tropical, summer rainfall, which prevails along their equatorward margins, to midlatitude, winter rainfall on their poleward borders. Both the seasonality and amount of rainfall decrease toward their center.

In the cold-winter deserts of the midlatitudes, fewer generalizations can be made about the seasonal occurrence of precipitation. For example, in the western Great Plains of the United States, precipitation is concentrated in summer, but the desert Southwest tends to receive both summer and winter rainfall, with greater aridity in the transition seasons. In some deserts of the Middle East the maximum is in spring. In most midlatitude deserts summer rainfall has the characteristics of tropical rain, falling in intense but brief bursts which promote high runoff. Winter rainfall is usually linked to large frontal systems, of low to moderate intensity but persisting for long periods.

For these reasons, and because potential evapotranspiration is higher in summer than winter, summer precipitation is less effective for vegetation growth. The seasonality of the meager precipitation can produce vastly different surface conditions. The Mojave Desert of California is distinguished from the adjacent Sonoran Desert of Arizona and northern Mexico on the basis of the percent of precipitation fall during the winter season. The latter is characterized by succulent vegetation, such as the giant saguarro cactus (Figure D11); while the former features dwarf shrubs and the uniquely-formed Joshua tree.

Snow occasionally falls in the poleward margins of the subtropical deserts. At oases such as Ouarghla and Ghardaia in the northern Sahara, snow falls as often as 1 in 10 years. At Laghouat snow fell nearly every year toward the end of the nineteenth century and now occurs every 2–5 years. The traditional housing is not meant to withstand such occurrences, which may lead to the collapse of roofs. In the higher latitudes snow is much more common, especially in deserts where precipitation is concentrated in the winter months. These include the Takla-Makan and Iranian Deserts.

The distribution of rainfall in a desert is erratic in both space and time, especially in regions of tropical rainfall. The amount of rainfall varies tremendously from year to year; many years may pass without a drop. Usually rain occurs only a few days within the year and most of the rain which falls occurs within short periods, sometimes as briefly as a few hours. Several times the mean annual rainfall may occur within one day. In Helwan (Egypt), where the mean annual rainfall is about 20 mm, seven storms produced a quarter of the rain that fell during an entire 20-year period. At Nouadibou (Mauritania), with mean annual rainfall of 32 mm, annual totals have ranged from 0 to 301 mm; on one day in 1909, 140 mm fell. At Biskra (Algeria) (mean annual rainfall of 140 mm), annual totals range from 32 to 638 mm, but 299 mm fell in September of 1969 (210 of it in 2 days). The same storm system brought nearly 800 mm to Sidi bou Zid (Tunisia) during September and October, months in which the mean rainfall is on the order of 10–20 mm.

Coastal deserts

Deserts are common along the western coasts of continents in the subtropical latitudes, where high pressure prevails. These deserts have certain climatic characteristics, such as a high frequency of fog, which distinguish them from other desert regions. Their origin lies in three main factors: the aridifying influence of the subtropical high-pressure cells over the oceans and adjacent coasts; the cold water which exists along these coasts; and a frictional effect of the shoreline on the coastal winds. Aridity is greatest at the coast, with rainfall increasing inland. Many local factors serve to accentuate the coastal aridity, such as near-surface jet streams constrained by coastal mountain chains.

The extent of the coastal deserts ranges from 25° of latitude for the Peruvian–Atacama Desert of western South America to less than 5° of latitude for western Australia. The latitude of their equatorward margin, with summer rainfall, is highly variant, but the poleward border with winter rainfall is in all cases at about 25°N or S of the equator. In those of Africa and South America, rainfall approaches zero in the arid core.

The climatic conditions of coastal deserts are less extreme than those of inland deserts. Many of the coastal deserts are relatively moist environments, with a high frequency of fog and relatively high atmospheric humidity. In some, more precipitation is received as fog-water than as rainfall. The temperature regime is generally quite moderate, as the water, with its high thermal capacity, dampens both diurnal and annual fluctuations. In the Namib, for example, the daily and annual range are about 6°C at the coast, compared to 15–20°C just 100 km inland.

These more moderate temperatures, together with the moisture provided by the fogs, create a more favorable environment for plants and animals than interior deserts. In the Peru–Atacama and Namib Deserts some plant and animal species have special adaptations which allow them to utilize fog-water. Certain plants can absorb water directly through the leaf surface. A species of beetle in the Namib builds trenches on the dunes to trap water as it condenses on the sand. Another "basks" on the dune with its head pointed downward and body to the wind, causing fog droplets to condense on its back and roll downward into its mouth.

Many coastal deserts are affected by a phenomenon called El a major change of temperature and wind patterns in the Pacific that has global climatic consequences. During El Niño years the cold water along the South America Desert coast disappears, establishing conditions which promote intense rainfall. In March and April of 1965 an El Niño brought 600 mm of rainfall to areas of coastal Peru, that receive on average about 80 mm during those months. The El Niño often renders similar changes along the coasts of Southern California, southwestern Africa and in other coastal deserts.

Winds, sand and dust

The desert surface conditions, barren ground and extreme heat, interact with the prevailing regional scale winds to produce a local wind regime which is strong and turbulent. The sparse vegetation cover means little surface friction to dissipate wind near the ground, so mean wind speeds are quite high in deserts. In the afternoon steady gale-force winds may prevail for hours, but at night, when the surface cools and the air becomes stable, winds are often calm. The desert winds are hot and dry and often blow from the desert to surrounding regions of more humid climate. Some, like the *Harmattan* of the Sahara, may provide welcome relief to stifling humidity.

The hot desert ground and rapid temperature drop above the ground produce unstable conditions which create gusty, turbulent winds. These are very effective in lifting particles from the surface. These winds are reflected in the patterns of sand dunes and surface erosion. They also produce sand and dust storms and smaller and shorter-lived dust devils. The heavy sand does not stay aloft very long, but the smaller dust particles do. The dust layer over the Sahara, for example, extends to over 5 km in altitude, producing vivid red colors in the clouds at this height.

When a dust storm occurs in a desert its effects can be instant and dramatic. Within minutes bright sunshine changes to an eerie, dusk-like ambience of red–brown haze and the temperature can drop more than 15°C. One particular type of dust storm, called a *haboob* in North Africa and the Southwestern United States, originates as a strong, turbulent downdraft in a thunderstorm. The dust is kicked up by what is called a density current, cold air which sinks to the ground from high altitude. As it hits the surface it spreads laterally, churning up dust in violent gusts that may exceed 60 mph. The visibility can rapidly approach zero, producing near darkness in mid-afternoon.

Microclimates in the desert

There are a number of microenvironments, or habitats, in deserts which offer shelter from the harsh conditions. It is here where life thrives. Some, such as parts of sand dunes and desert plants, modulate the thermal extremes. Others, such as oases and riverine environments, offer more favorable moisture conditions. Even dry riverbeds can support a virtual forest. Favorable habitats can also be dictated by topography or soils; the arid surface conditions are accentuated by stone pavements but moderated in lowlands near high relief. On the other hand, desert depressions, such as the Qattara depressions of Egypt or the Chott el Jerid of Tunisia, are the hottest places on Earth. The heat of the walls of the depression (Figure D12) creates a circulation cell of hot air that continues to heat up via conduction from the hot surfaces. At Tozeur, in the Chott el Jerid, the mean daily maximum temperature exceeds 42°C in June and July. In Death Valley, in the United States, the mean daily maximum exceeds 46°C in July.

Even an environment as small as a dune contains a number of microhabitats, as each part of the dune is affected differently by sun and wind. The side facing the morning sun will heat up first and by 8 a.m. may be 10°C warmer than elsewhere. In the afternoon the windy dune crest may be the

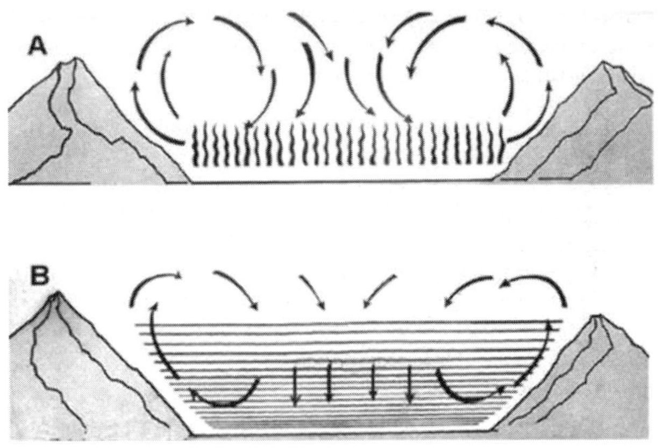

Figure D12 Heating of a desert depression.

Figure D13 The *Welwitschia,* an unusual plant of the Namib Desert, absorbs the fog water through its leaves. The long, flat, low surfaces of its leaves promote condensation.

coolest location by more than 10°C. Characteristic plant and animal communities reside at different parts of the dunes and in the interdune areas.

Water accumulates deep inside sand dunes, because the surface dries out first. This deep water reservoir helps to support plant life. Moisture also varies among the dune habitats. In coastal deserts, for example, where sea breezes bring fogs inland, moisture is most efficiently captured where wind is strongest.

Temperature and surface moisture are also moderated by plant cover. The temperature in litter underneath and within the leaves of a *Welwitschia* plant (Figures D13 and D14) may be as much as 20–30°C cooler than on the surrounding, exposed ground surface. The soil moisture content can be twice as high as in the exposed ground. Insects and small animals take refuge within the plant cover.

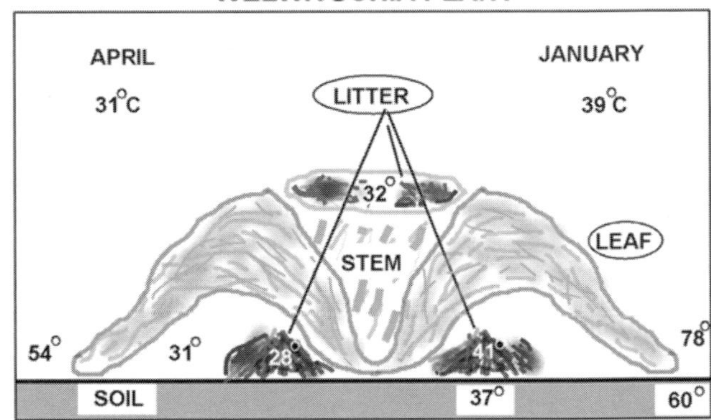

Figure D14 The relatively moderate microclimate of the *Welwitschia* plant provides a refuge for small animals and insects.

Climatic change

The Earth's deserts have undergone numerous climatic fluctuations on time scales of tens to tens of thousands of years. The fluctuations in these regions are roughly synchronous with major changes over the Earth as a whole. Those of higher latitudes have experienced fluctuations of both temperature and rainfall, but the major changes in the low-latitude deserts involve mainly rainfall. In recent times generally only the semiarid margins have been affected, but over the last 30 000 years the global extent of deserts has changed markedly.

During the last glacial maximum, about 18 000 years ago, the low-latitude deserts generally expanded, with the Sahara advancing nearly 10° toward the equator and its dunes overriding the lakes and rivers south of its previous border. The expansion of the arid zone into the tropics was so complete that the rainforests nearly vanished, their species taking refuge in a few highland habitats. At the same time "pluvial", or humid, conditions prevailed in many of the midlatitude deserts. Great lakes covered much of the current states of Utah, Nevada and California. Many dried up at the end of the Ice Age, about 10 000 years ago, and the left behind huge beds of fine sediments, like the Bonneville Salt Flats of Utah, where the ultra-smooth surface has permitted a vehicle to set a land-speed record of 622 mph.

The conditions which commenced about 10 000 years ago are in stark contrast, with a low latitude "pluvial" period reaching a maximum some 5000 years ago. Then, the low-latitude deserts were generally reduced to their hyperarid cores. There is little evidence anywhere of active sand dunes at that time. In parts of the Sahara, Neolithic man herded cattle and animals grazed on savanna vegetation; fish hooks uncovered in archeological sites attest to the presence of lakes in, and human occupation of, now-hyperarid regions.

A more recent period in which aridity has waxed and waned was around the Middle Ages, some 600–1200 years ago. Considered to be a global warm epoch, the core of this period was one of wetter conditions along the margins of many deserts. This was probably the case in the Peruvian Desert, the southwestern United States, Australia, the Mediterranean, the Middle East and the southern Sahara. Major civilizations, such as the Mali empire, thrived in presently semiarid regions of Africa. Caravan routes traversed now-waterless plains of the

Sahara and several towns flourished along these routes. On the other hand, drier conditions prevailed in some deserts and semi-desert regions of higher latitudes, such as the Great Plains of the central United States.

In recent centuries similar fluctuations of climate have affected desert regions, although less extreme. These are an inherent characteristic of the desert environment, and one to which life in the deserts has adapted. A quick look at the major deserts suggests no systematic change in recent decades, with one exception: throughout most of Africa there is a trend toward more arid conditions and expansion of the total area occupied by deserts. This is a significant percentage of the global arid environments and is cause for concern.

Sharon E. Nicholson

Bibliography

Cooke, R.U., and Warren, A., 1975. *Geomorphology in Deserts*. London: Batsford.

Evenari, M., Noy-Meir, I., and Goodall, D.W., (eds.), 1985. *Hot Deserts and Arid Shrublands. Ecosystems of the World*, **12**. Amsterdam: Elsevier.

Goudie, A. and Wilkinson, J., 1977. *The Warm Desert Environment*. Cambridge: Cambridge University Press.

Seely, M.K., (ed.), 1994. *Deserts: the illustrated library of the Earth*. Sydney Weldon-Owen.

West, N.E., (ed.), 1985. *Temperate Deserts and Semi-Deserts. Ecosystems of the World*, **5**. Amsterdam: Elsevier.

Cross-references

Asia, Climate of Southwest
Desertification
Drought
Vegetation and Climate

DETERMINISM, CLIMATIC

Determinism is the doctrine that all happenings and existences are the inevitable outcome of a set of preexisting conditions, and specifically that human actions are not based on free will, but are determined by environmental influences. In its widest context such a philosophy is encompassed within the framework of environmental determinism. Climatic determinism is a subset of environmental determinism, with climate considered the major influence or control.

Environmental determinism has its roots in classical antiquity but in its modern terms can be traced to the influence of Charles Darwin, whose ideas made it inevitable that social scientists should see natural laws operating in the different cultures and activities of people. The formal statements of such a relationship are often attributed to Friedrich Ratzel (1844–1904), although many of the ideas presented in his work *Anthropogeographie* are not as naive as sometimes suggested by those who have referred to it. In the United States, Ellen C. Semple, a student of Ratzel's, became the foremost proponent of environmental determinism. Despite the avoidance of the word "determinism" in her work, she expressed her basic thesis in the declaration "man is a product of the earth's surface" (Semple, 1935).

Reaction to many of the simplifications of environmental determinism led other writers to proposed modifications of the role of environment in human endeavors. Thus "stop-and-go" determinism was proposed by Griffiths Taylor, who wrote:

> Man is able to accelerate, slow or stop the progress of a country's development. But he should not, if he is wise, depart from the direction as indicated by the natural environment. He is like the traffic controller in a large city who alters the rate and not the direction of progress (Taylor, 1955).

Similarly, the ideas of environmental possibilism and probabilism reflect a spectrum of the role of environment in determining the possible and probable outcome of human endeavors.

Within this flow of ideas it was to be expected that the role of climate as a significant part of the human environment should be given special emphasis. Climatic determinism has a developmental history of its own.

The early concepts

Variations from climatic conditions that existed in the heart of their world caused classical Greek philosophers and writers to postulate many ideas concerning the role of climate in the nature of people. In his discussion *On Airs, Waters and Places*, Hippocrates (*ca.* 420 BC) contrasted easy-going Asiatics with penurious Europeans by suggesting that the latter had to be much more active to ameliorate their environment. A similar theme occurs in the work of Aristotle (*ca.* 350 BC), who commented "The inhabitants of colder countries in Europe are brave but deficient in thoughts and technical skills" (see Tatham, 1957).

Such concepts were also used by Roman writers. Strabo (*ca.* 64 BC–AD 20), for example, suggested that the rise and strength of Rome was due in part to the climate, physical makeup, and position of Italy. With the decline of Rome, however, explanations of the role of humans in their environment diminished, and in medieval times writers were not encouraged to deal with items that did not completely conform to biblical teachings.

While this was true in the Christian world, Arabic writers did contribute to the climate–human relationship. The great historian–geographer Ibn Khaldun (1332–1406) divided the hemisphere into seven climatic zones. The extremes were uninhabitable and only the middle zone provided the climate in which people could excel in wisdom and were neither too stolid nor excessively passionate. He said that

> the human inhabitants of these zones are well proportioned in their bodies, color, character qualities and general conditions. The inhabitants of the zones that are far from temperate are also farther removed from being temperate in all their conditions. Thus the inhabitants of the middle zone have all the natural conditions necessary for a civilized life (Khaldun, 1958, vol. 1).

The sixteenth and seventeenth centuries produced writers who once again resurrected the classical ideas. Of particular note was the work of the French social writer Montesquieu (1689–1755). In his *The Spirit of Laws* he sought to determine the effects of climate and soils on the character of people and, in his opinion, climate was of particular importance.

Table D2 Selected concepts on climatic determinism

Author	Concepts
Hippocrates (*ca.* 420 BC)	Developed ideas on environmental influence on human behavior.
Aristotle (384–322 BC)	The varying habitability of the earth with differences of latitude: the inhabitants of extreme climate conditions are deficient in thought and technical skill, whereas living in the intermediate region combined the best quality of living.
Strabo (*ca.* 64 BC–AD 20)	Attempted to explain how climate affected the rise and strength of Rome.
Ibn Khaldun, A.R. (1332–1406)	Argued that environmental factors, such as climate, provide their due influence on such social factors as cohesion, occupation, and wealth.
Bodin, Jean (1530–1596)	Presented ideas of environmental influence on human life: the habitability of a place was a function of its latitude.
Carpenter, Nathanael (1589–1628)	Human character is determined by climate.
Montesquieu, Charles Louis de Secondat (1689–1755)	Presented ideas about the influence of climate on politics.
Kant, Immanuel (1724–1804)	Animals and men gradually change by their environment: effect of temperatures on nerves and veins, making the movements stiff and unsupple.
Von Humbolt, Alexander (1769–1859)	Developed the relations of latitude air, temperature, vegetation and agriculture in tropical mountains: made investigations with a view to seeking answers to questions about interconnections among the phenomena grouped together in rich diversity on the face of the Earth. He cautiously pointed out that atmospheric conditions merely favored that which had been called for by mental qualifications, and by the contact of highly gifted races with more civilized neighbors.
Ritter, Carl (1779–1859)	Some of his basic ideas concerning the influence of the Earth's major features on the course of history were developed. He sought to understand the interconnections, the causal interrelations, that make the areal associations cohesive. He was interested in the effect of the Earth on humans, the reciprocal action of humans on the Earth was to him equally significant.
Darwin, Charles (1809–1882)	Mechanism of natural selection of the cause of evolution, rather than need or use.
Spencer, Herbert (1820–1903)	Pointed out close resemblance of human societies to animal organisms.
Buckle, H.T. (1821–1862)	Physical agents have powerfully influenced the human race. These are climate, food, soil, and general aspects of nature. Contended that climate, generally working indirectly, is a major environmental influence.
Haeckel, E.H. (1834–1919)	Humans are only one of the organisms to be studied, and was equally in the grip of the surrounding forces; a proponent of Darwin's theory.
Ratzel, Friedrich (1844–1904)	The application of Darwin's biological concepts of human societies: human beings must struggle to survive in particular environments much as plant and animal organisms must do, known as Social Darwinism. He recognized climate as a chief environmental influence but pointed out that its effects are often indirect, affecting man by way of its influence on plants, animals, and soil.
Davis, William Morris (1850–1934)	Set forth his concept of geography as an explanatory description of the physical character of the Earth's surface, and of the human response to the physical surroundings. Began to seek cause and effect generalizations, usually between some elements of inorganic control and some elements of organic response. Attempted to bring cohesion to geography by introducing Spencer's ideas of Social Darwinism.
Demolins, Edmond (1852–1907)	Argued the role of climate produces grass vegetation; and the presence of grass determines a uniform mode of work – the art of pastrolism. Sought the "laws" which govern human actions.
Semple, Ellen Churchill (1863–1932)	Very careful to make the point that the environment does not control human activity: only that under certain circumstances there is a tendency for people to behave in predictable ways.
Dexter, E.G. (1868–1938)	Compared the daily temperature, atmospheric pressure, humidity, wind, sunshine, and precipitation in respect to the behavior condition investigated, and found statistically significant correlations between weather and conduct, suicide, and other crimes.
Patterson, S.O. (1868–1941)	Noted the human effects of the period of climatic instability that culminated the fourteenth century.

Table D2 *(Continued)*

Author	Concepts
Moore, H.L. (1869–1958)	Presented the idea of economic cycles as influenced by the climatic conditions.
Brunhes, Jean (1869–1930)	It is not influences that are sought but geographical relations between physical facts and human destinies; humans are creations of habit and habit once established becomes a part of their environment and exerts considerable influence on their later development.
Huntington, Ellsworth (1876–1947)	Refined the climatic hypothesis as related to climatic causation.
Bowman, Isaiah (1878–1950)	There is widespread evidence that there are climatic limits to profitable agricultural settlement in spite of the production of quicker maturing crops for hardier cattle.
Taylor, Griffith (1880–1963)	Humans are not free agents. If humans are wise they should not depart from the directions as indicated by the natural environment.
Visher, Stephen Sargent (1888–1967)	Of all the geographical influences to which humans are subjected. climate seems to be the most potent. Despite the fact that most climatic influences are "permissive", "not deterministic". various aspects of climate are so highly restrictive as to be in effect controls.
Toynbee, Arnold J. (1889–1975)	Was influenced by Huntington's ideas about the relation between human beings and their physical environment; is not a determinist.
Mills, Clarence A. (1891–1974)	A professor of experimental medicine; he put forward many ideas in his book *Climate Makes the Man*, a title that expresses clearly his philosophy.
Markham, S.F. (1897–1975)	Related the relative

Accordingly, he noted that people who live in cold climates are stronger, more courageous, less suspicious and less cunning than those in hot climates, who tend to be like old men and are timorous, weak in body, indolent and passive. A similar theme was presented by the philosopher Immanual Kant (1724–1804), who also added that outside of temperate climes the inhabitants of both hot and cold lands were stiff and unsupple because too much or too little perspiration makes blood thick and viscous while too much heat and cold dry out nerves and veins.

Many other thinkers and writers, such as Ritter (1799–1859) and Humboldt (1767–1835) expressed opinions that, although more cautious, exhibited a similar outlook. Table D2, while far from complete, provides a list of writers who contributed toward the ideas of climatic determinism.

The twentieth century

The two names most closely associated with determinism in the twentieth century are Ellen Churchill Semple (1863–1932) and Ellsworth Huntington (1876–1947), with Semple being closely associated with environmental determinism in all its aspects and Huntington being best known as the archetypal climatic determinist. The work of Huntington is dealt with here.

Huntington's early work was influenced by his travels in Asia and by James Geikie's book *The Great Ice Age and its Relation to the Antiquity of Man*. Arising from these early influences was his book *The Pulse of Asia* (1907), in which he developed the idea that climate in postglacial time was becoming drier and that in Central Asia not only the habits but also the character of the people were molded by the environments resulting from climatic change. In effect, climatic change is one of the greatest factors in determining the course of human progress.

While continuing to work on climatic change, Huntington also developed the idea that there is a climatic optimum for the human being regardless of race or background. Studies on human productivity in New England led to the conclusion that temperature is the most significant variable and that the people in Bridgeport and New Haven are physically most active when the average temperature is between 60°F and 65°F, and mentally most active when outdoor temperature is about 38°F. Further, people do not work well when the temperature is fairly constant or extremely changeable. The ideal conditions are moderate changes, especially a cooling of the air at frequent intervals. These ideas were further developed in articles such as "The handicap of the tropics" (1913) and "The adaptability of the white man in tropical America" (1914).

These two main themes – climatic change and history and climate and the human response – are represented in much of Huntington's further work. About the former, Martin (1973) has written:

> His bold scheme, imaginatively wrought, involving tree rings, lake levels, valley terraces, dunes, disappearing rivers, wilderness ruins and the whole apparatus for climatic change, was applied to history . . . it revealed an ever-changing climatic mileau explaining nomadic migrations, the fortunes of Kings, the downfalls of empires, outbursts of intellectual advance and lesser matters concerning rhythms.

His work on climate and human activities was marked by less scientific evidence and oversimplification of complex variables. Nonetheless, Huntington did recognize, long before it was popular, many aspects of applied climatology (see Fonaroff, 1965).

Current thought

There are few climatologists today who would admit to being climatic determinists. The sweeping generalizations, backed by minimal evidence, of Huntington and other writers cannot be accepted at face value. However, the extreme stance of determinists of the midtwentieth century led to a backlash effect in which climate and environment, when considered in any aspect of human activities, became an area of nonresearch. Voluntarism, the other end of the spectrum of thought, became viable.

Fortunately, other schools of thought, taking a midpath between the two extremes, were introduced and today a more rigorous approach to human–climatic relationships and the rise and fall of civilizations has again made it a respectable area of study.

Akhtar H. Siddiqi and John E. Oliver

Bibliography

Buckle, H.T., 1887. *History of Civilization in England*. New York: D. Appleton.

Brunhes, J., 1904. *La Geographic*, **10**: 104.

Dexter, E.G., 1904. *Weather Influences: an empirical study of mental and physiological effects of definite meteorological conditions*. New York: Macmillan.

Dickinson, R.E., 1969. *The Makers of Modern Geography*. London: Routledge & Kegan Paul.

Fonaroff, L.S., 1965. Was Huntington right about human nutrition? *Association of American Geographers Annals*, **55**(3): 365–376.

Hartshorne, R., 1966. *Perspective on the Nature of Geography*. Chicago, IL: Rand McNally.

Huntington, E., 1915. *Civilization and Climate*. New Haven, CT: Yale University Press.

Huntington, E., 1945. *Mainsprings of Civilization*. New York: Wiley.

James, P.E., and Martin, G.J., 1981. *All Possible Worlds*. New York: Wiley.

Jones, E., 1956. Causes and effect in human geography, *Association of American Geographers Annals*, **46**: 369–377.

Khaldun, I., 1958. *The Muqaddimah: An Introduction to History*, 3 vols., (F. Rosenthal, trans.). Bollinger Series XLIII. New York: Pantheon Books.

Lewthwaite, C.R., 1966. Environmentalism and determinism: a search for clarification. *Association of American Geographers Annals*, **56**: 1–23.

Markham, S.F., 1944. *Climate and the Energy of Nations*. New York: Oxford University Press.

Martin, G.J., 1973. *Ellsworth Huntington: his life and thought*. New York: Archon Books.

Mills, C.A., 1942. *Climate Makes the Man*. New York: Harper.

Platt, R.S., 1948. Environmentalism versus geography. *American Journal of Sociology*, **53**: 351–355.

Ratzel, F., 1882–1891. *Anthropogeographie*, 2 vols. Stuttgart: Englehorn.

Ritter, C., 1822–1859. *Die Erdkunde*, 19 vols. Berlin: Reimer.

Semple, E.C., 1935. *Influences of the Geographical Environment*. London: Constable.

Tatham, G., 1957. Environmentalism and possibilism. In Taylor, G., ed., *Geography in the 20th Century*. London: Methuen, pp. 128–162.

Taylor, G., 1955. *Australia*. London: Methuen.

Taylor, G. (ed.), 1957. *Geography in the 20th Century*. London: Methuen.

Toynbee, A.J., 1955. *A Study of History*, 10 vols. London: Oxford University Press.

Visher, S.S., 1957. Climatic influences. In Taylor, G., ed., *Geography in the 20th Century*. London: Methuen, pp. 196–220.

Wanklyn, H., 1961. *Friedrich Ratzel: a bibliographical memoir and bibliography*. Cambridge: Cambridge University Press.

Cross-references

Climatic Change and Ancient Civilization
Crime and Climate
Cultural Climatology
Human Health and Climate

DEW/DEWPOINT

Dew is any water that is condensed onto the ground, rocks, grass, and so on. It generally forms at night due to radiational cooling (most effective on clear, still, cool nights), which causes the temperature of the air to fall below its dewpoint.

Dewpoint is the temperature at which saturation of the air occurs, i.e. the temperature at which the observed vapor pressure is equal to the saturation vapor pressure. If a parcel of air is hypothetically held at constant pressure and vapor content, the temperature at which it must be cooled to reach saturation is called its dewpoint or, if below 0°C, its frost point. Provided that any change in temperature of a parcel of air occurs without change in pressure or vapor content, the dewpoint will always be constant, i.e. conservative. On the other hand, if the parcel rises adiabatically, the dewpoint is quasi-conservative, because the dewpoint of moist air drops only at about one-fifth the rate of the dry adiabatic lapse rate.

A dewpoint hygrometer may be used to determine surface dewpoint. A refrigerated metal surface is brought in contact with air, causing condensation when it is slightly below the temperature of the thermodynamic dewpoint. More usually dewpoint is simply determined with a psychrometer (wet and dry bulb thermometers), used with appropriate tables (Table D3).

If the temperature is below freezing, hoar frost will form, but if the temperature drops below freezing *after* the dew has formed the result is white dew. Dew is especially effective when the ground layer of air has a high relative humidity, as along river valleys, near swamps, etc. Dew is often the only form of moisture available to plants and animals in extreme deserts.

An optical effect of dew is known as heiligenschein or Cellini's halo, brought about by an early-morning sun producing a shadow over the observer's head. Reflection from the dew-covered surface produces a "saintly" halo, said to have been described first by Cellini, who found it a divine sign.

John E. Oliver

Bibliography

Moran, M.J., and Morgan, M.D., 1997. *Meteorology: the atmosphere and the science of weather*, 5th edn. Upper Saddle River, NJ: Prentice-Hall.

Wallace, J.M., and Hobbs, P.V., 1997. *Atmospheric Science: an introductory survey*. San Diego, CA: Academic Press.

Cross-references

Humidity
Phase Changes

Table D3 Dewpoint temperature (1000 mb)

Dry-bulb temperature (°C)	Saturation vapor pressure (mb)	Wet-bulb depression ($T_d - T_w$)																					
		1	2	3	4	5	6	7	8	9	10	11	12	13	14	15	16	17	18	19	20	21	22
−20	1.2540	−33																					
−18	1.4877	−28																					
−16	1.7597	−24																					
−14	2.0755	−21	−36																				
−12	2.4409	−18	−28																				
−10	2.8627	−14	−22																				
−8	3.3484	−12	−18	−29																			
−6	3.9061	−10	−14	−22																			
−4	4.5451	−7	−11	−17	−29																		
−2	5.2753	−5	−8	−13	−20																		
0	6.1078	−3	−6	−9	−15	−24																	
2	7.0547	−1	−3	−6	−11	−17																	
4	8.1294	1	−1	−4	−7	−11	−19																
6	9.3465	4	1	−1	−4	−7	−13	−21															
8	10.722	6	3	1	−2	−5	−9	14															
10	12.272	8	6	4	1	−2	−5	−9	−14	−28													
12	14.017	10	8	6	4	1	−2	−5	−9	−16													
14	15.977	12	11	9	6	4	1	−2	−5	−10	−17												
16	18.173	14	13	11	9	7	4	1	−1	−6	−10	17											
18	20.630	16	15	13	11	9	7	4	2	−2	−5	−10	−19										
20	23.373	19	17	15	14	12	10	7	4	2	−2	−5	−10	−19									
22	26.430	21	19	17	16	14	12	10	8	5	3	−1	−5	−10	−19								
24	29.831	23	21	20	18	16	14	12	10	8	6	2	−1	−5	−10	−18							
26	33.608	25	23	22	20	18	17	15	13	11	9	6	3	0	−4	−9	−18						
28	37.796	27	25	24	22	21	19	17	16	14	11	9	7	4	1	−3	−9	−16					
30	42.430	29	27	26	24	23	21	19	18	16	14	12	10	8	5	1	−2	−8	−15				
32	47.551	31	29	28	27	25	24	22	21	19	17	15	13	11	8	5	2	−2	−7	−14			
34	53.200	33	31	30	29	27	26	24	23	21	20	18	16	14	12	9	6	3	−1	−5	−12	−29	
36	59.422	35	33	32	31	29	28	27	25	24	22	20	19	17	15	13	10	7	4	0	−4	−10	
38	66.264	37	35	34	33	32	30	29	28	26	25	23	21	19	17	15	13	11	8	5	1	−3	−9
40	73.777	39	37	36	35	34	32	31	30	28	27	25	24	22	20	18	16	14	12	9	6	2	−2

DOLDRUMS

Doldrums are the calms of the Intertropical Convergence Zone. The regions are characterized by low atmospheric pressure, high humidity, and thunderstorms. They are often associated with the source of tropical hurricanes, the sites of water spouts, and windlessness, sometimes alternating with sharp squalls. They are not to be confused with the calms of the horse latitudes.

Doldrums is an old English word meaning an unpleasant, depressed feeling. According to the *Oxford English Dictionary* there apparently arose a misunderstanding of the phrase "in the doldrums" used by some sailors, the human physiological state or condition of being transferred to a geographic belt or locality. Maury in 1855 used the term in this sense, and the usage became widespread. The colloquial French for doldrums is *potau-noir*, literally the boot-polish jar. Figuratively, *être dans le pot-au-noir* means exactly "to be in the doldrums", hence the extension of the meaning to the equatorial calms later on. Most other European languages have no special name for these equatorial calms.

While the term doldrums is now seldom used in the literature, descriptions are common in older works. As pointed out by Brunt (1939), the doldrums are not by any means continuous throughout the year and do not form an uninterrupted equatorial belt. A sharp discontinuity occurs in the equatorial Pacific where the northeast and southeast trades meet; the latter rise over the former, accompanied by heavy rainfall. In the Pacific area, therefore, true doldrums only occur near the continental shores; while in the Atlantic they fluctuate both in width and in position (Durst, 1926).

Over the continents the equatorial belts are regions of calms and are subject to diurnal instability and thermal systems with heavier rainfall than over the oceans.

Joseph Gentilli

Bibliography

Brunt, D., 1939. *Physical and Dynamical Meteorology*. London: Cambridge University Press.

Durst, C.S., 1926. The doldrums of the Atlantic. *Geophysics Memoirs*, **28**(8): 228–238.

Lorenz, E.N., 1967. *The Nature and Theory of the General Circulation of the Atmosphere*. Geneva: World Meteorological Organization.

Lorenz, E.N., 1983. A history of prevailing ideas about the general circulation of the atmosphere. *America Meteorological Society Bulletin*, **7**: 730–734.

Reihl, H., 1978. *Introduction to the Atmosphere*, 3rd edn. New York: McGraw-Hill.

Cross-references

Atmospheric Circulation, Global
Intertropical Convergence Zone
Trade Winds and the Trade Wind Inversion
Tropical Cyclones

DROUGHT

Drought is an insidious natural hazard that results from a deficiency of precipitation from expected or "normal" that, when extended over a season or longer period of time, is insufficient to meet the demands of human activities. As the world's population increases, societies are placing greater and greater pressure on finite water supplies and other limited natural resources – thus potentially increasing our vulnerability to extended periods of drought. In addition, growing concern over the impact of human activities on climate has heightened awareness of our sensitivity to climatic extremes such as drought. The results from scientific investigations indicate that the frequency and severity of drought may increase for some regions in the future as a direct result of increasing concentrations of greenhouse gases in the atmosphere.

Scores of definitions of drought exist, reflecting different climatic characteristics from region to region and sector-specific impacts. Although droughts are usually classified as meteorological, agricultural, hydrological, or socioeconomic, all types of drought originate from a deficiency of precipitation that results in water shortage for some activity or some group. Drought must be considered a relative, rather than absolute, condition. The ultimate results of these precipitation deficiencies are, at times, enormous economic and environmental impacts as well as personal hardship. Impacts of drought appear to be increasing in both developing and developed countries, a clear indication of nonsustainable development in many cases. Lessening the impacts of future drought events will require nations to pursue development of drought policies that emphasize a wide range of risk-management techniques, including improved monitoring and early-warning systems, preparedness plans, and appropriate mitigation actions and programs.

Drought concepts and definitions

Drought differs from other natural hazards in several ways. First, drought is a slow-onset, creeping natural hazard. Its effects often accumulate slowly over a considerable period of time and may linger for years after the termination of the event. Therefore, the onset and end of drought is difficult to determine. Because of this slow-onset characteristic it is difficult to recognize the onset of drought, and scientists and policy makers continue to debate the basis (i.e. criteria) for declaring an end to a drought. Second, the absence of a precise and universally accepted definition of drought adds to the confusion about whether or not a drought exists and, if it does, its degree of severity. Realistically, definitions of drought must be region- and application- (or impact-) specific. This is one explanation for the scores of definitions that have been developed. Third, drought impacts are nonstructural and spread over a larger geographical area than are damages that result from other natural hazards. Quantifying the impacts and providing disaster relief are far more difficult tasks for drought than they are for other natural hazards. These characteristics of drought have hindered the development of accurate, reliable, and timely estimates of severity and impacts and, ultimately, the formulation of drought preparedness plans.

Many people consider drought to be largely a natural or physical event. Like other natural hazards, drought has both a natural and social component. The risk associated with drought for any region is a product of both the region's exposure to the event (i.e. probability of occurrence at various severity levels) and the vulnerability of society to the event. The natural event (i.e. meteorological drought) is a result of the occurrence of persistent large-scale disruptions in the global circulation pattern of the atmosphere. Exposure to drought varies spatially and there is little, if anything, that we can do to alter drought occurrence. Vulnerability, on the other hand, is determined by social factors such as population changes, population shifts (regional and rural to urban), demographic characteristics, technology, government policies, environmental awareness, water-use trends, and social behavior. These factors change over time and thus vulnerability is likely to increase or decrease in response to these changes. Subsequent droughts in the same region will have different effects, even if they are identical in intensity, duration, and spatial characteristics, because societal characteristics will have changed.

Defining drought

Drought is the consequence of a natural reduction in the amount of precipitation received over an extended period of time, usually a season or more in length, although other climatic factors (such as high temperatures, high winds, and low relative humidity) are often associated with it in many regions of the world and can significantly aggravate the severity of the event. Drought is also related to the timing (i.e. principal season of occurrence, delays in the start of the rainy season, occurrence of rains in relation to principal crop growth stages) and the effectiveness of the rains (i.e. rainfall intensity, number of rainfall events). Thus, each drought event is unique in its climatic characteristics, spatial extent, and impacts. The area affected by drought is rarely static during the course of the event. As drought emerges and intensifies, its core area or epicenter shifts and its spatial extent expands and contracts throughout the duration of the event.

Because drought affects so many economic and social sectors, scores of definitions have been developed by a variety of disciplines. In addition, because drought occurs with varying frequency in nearly all regions of the globe, in all types of economic systems, and in developed and developing countries alike, the approaches taken to define it also reflect regional differences. Impacts also differ spatially and temporally, depending on the societal context of drought. A universal definition of drought is an unrealistic expectation.

Many disciplinary perspectives of drought exist, often causing considerable confusion about what constitutes a drought. Research has shown that the lack of a precise and objective definition in specific situations has been an obstacle to understanding drought, which has led to indecision and/or inaction on the part of managers, policy makers, and others. It must be accepted that the importance of drought lies in its impacts. Thus definitions should be region- and impact- or application-specific in order to be used in an operational mode by decision makers.

Drought is normally grouped by type as follows: meteorological, hydrological, agricultural, and socioeconomic. Meteorological drought is expressed solely on the basis of the degree of dryness (often in comparison to some normal or average amount) and the duration of the dry period. Thus, intensity and duration are the key characteristics of these definitions. Meteorological drought definitions must be considered as region-specific since the atmospheric conditions that result in deficiencies of precipitation are climate regime-dependent.

Agriculture is usually the first economic sector to be affected by drought because soil moisture supplies are often quickly depleted. Agricultural drought links various characteristics of meteorological drought to agricultural impacts, focusing on precipitation shortages, differences between actual and potential evapotranspiration, and soil water deficits. A plant's demand for water is dependent on prevailing weather conditions, biological characteristics of the specific plant, its stage of growth, and the physical and biological properties of the soil. A definition of agricultural drought should account for the variable susceptibility of crops at different stages of crop development.

Hydrological droughts are associated with the effects of periods of precipitation shortfall on surface or subsurface water supply (e.g. streamflow, reservoir and lake levels, groundwater) rather than with precipitation shortfalls. Hydrological droughts are usually out of phase or lag the occurrence of meteorological and agricultural droughts. More time elapses before precipitation deficiencies are detected in other components of the hydrological system (e.g. reservoirs, groundwater). As a result, impacts are out of phase with those in other economic sectors. Also, water in hydrological storage systems (e.g. reservoirs, rivers) is often used for multiple and competing purposes (e.g. power generation, flood control, irrigation, recreation), further complicating the sequence and quantification of impacts. Competition for water in these storage systems escalates during drought, and conflicts between water users increase significantly.

Finally, socioeconomic drought associates the supply and demand of some economic good or service with elements of meteorological, hydrological, and agricultural drought. Socioeconomic drought is associated directly with the supply of some commodity or economic good (e.g. water, hay, hydroelectric power) that is the result of precipitation shortages. Increases in population can alter substantially the demand for these economic goods over time. This concept of drought supports the strong symbiosis that exists between drought and its impacts and human activities. Thus, the incidence of drought could increase because of a change in the frequency of meteorological drought, a change in societal vulnerability to water shortages, or both.

Drought characteristics and severity

Droughts differ from one another in three essential characteristics: intensity, duration, and spatial coverage. Intensity refers to the degree of the precipitation shortfall and/or the severity of impacts associated with the shortfall. It is generally measured by the departure of some climatic index from normal and is closely linked to duration in the determination of impact. Another distinguishing feature of drought is its duration. Droughts usually require a minimum of 2–3 months to become established, but then can continue for months or years. The magnitude of drought impacts is closely related to the timing of the onset of the precipitation shortage, its intensity, and the duration of the event.

Droughts also differ in terms of their spatial characteristics. The areas affected by severe drought evolve gradually, and regions of maximum intensity shift from season to season. In larger countries, such as Brazil, China, India, the United States, or Australia, drought would rarely, if ever, affect the entire country. During the severe drought of the 1930s in the United States, for example, the area affected by severe and extreme drought never exceeded 65% of the country. However, because of its size and diverse climatic regimes, drought occurs somewhere in the country each year. From a planning perspective the spatial characteristics of drought have serious implications. Nations should determine the probability that drought may simultaneously affect all or several major crop-producing regions within their borders, and develop contingencies if such an event were to occur. Likewise, it is important for governments to know the chances of a regional drought simultaneously affecting agricultural productivity in their country as well as adjacent or nearby nations on whom they are dependent for food supplies.

The impacts of drought

The impacts of drought are diverse and often ripple through the economy. Thus, impacts are often referred to as direct or indirect. Because of the number of affected groups and sectors associated with drought, its spatial extent, and the difficulties connected with quantifying environmental damages and personal hardships, the precise determination of the financial costs of drought is an arduous task. It has been estimated that the average annual impacts of drought in the United States are $6–8 billion, but drought years often occur in clusters.

The impacts of drought can be classified into three principal areas: economic, environmental, and social. Economic impacts range from direct losses in the broad agricultural and agriculturally related sectors, including forestry and fishing, to losses in recreation, transportation, banking, and energy sectors. Environmental losses are the result of damages to plant and animal species, wildlife habitat, and air and water quality; forest and range fires; degradation of landscape quality; and soil erosion. Although these losses are difficult to quantify, growing public awareness and concern for environmental quality has forced public officials to focus greater attention on these effects. Social impacts mainly involve public safety, health, conflicts between water users, and inequities in the distribution of impacts and disaster relief programs. As with all natural hazards, the economic impacts of drought are highly variable within and between economic sectors and geographic regions, producing a complex assortment of winners and losers with the occurrence of each disaster.

Drought preparedness

Drought planning is defined as actions taken by individual citizens, industry, government, and others in advance of drought for the purpose of mitigating some of the impacts and conflicts associated with its occurrence. Because drought is a normal part of climate variability for virtually all regions, it is important to develop plans to deal with these extended periods of water shortage in a timely, systematic manner as they evolve. This planning process needs to occur at various levels of government and be integrated between levels of government.

The purpose of a drought policy and plan is to reduce the impacts of drought by identifying the principal sectors, groups, or regions most at risk and developing mitigation actions and programs that can reduce these risks in advance of future drought events. Generally, drought plans have three basic components: monitoring, early warning, and prediction; risk and impact assessment; and mitigation and response. Plans will also improve coordination within agencies of government and between levels of government. Before developing a preparedness plan, government officials should first define, in consultation with principal stakeholder groups, the goals of the plan.

The awareness of the need for drought planning has increased dramatically in both developed and developing countries. For example, in the United States, the number of states with drought plans has increased from three in 1982 to 34 in 2002. This trend demonstrates an increased concern about the potential impacts of extended water shortages and the complexity of those impacts. Drought plans are at the foundation of improved drought management, but only if they emphasize risk assessment and mitigation programs and actions.

Summary

Drought is an insidious natural hazard that is a normal part of the climate for virtually all regions. It should not be viewed as merely a physical phenomenon. Rather, drought is the result of an interplay between a natural event and the demand placed on water supply by human-use systems. Drought should be considered relative to some long-term average condition of balance between precipitation and evapotranspiration.

Many definitions of drought exist; it is unrealistic to expect a universal definition to be derived. The three characteristics that differentiate one drought from another are intensity, duration, and spatial extent. The impacts of drought are diverse and generally classified as economic, social, and environmental. Impacts ripple through the economy and may linger for years after the termination of the drought episode.

It appears that societal vulnerability to drought is escalating in both developing and developing countries, and at a significant rate. It is imperative that increased emphasis be placed on mitigation, preparedness, and prediction and early warning if society is to reduce the economic and environmental damages associated with drought and its personal hardships. This will require improved coordination within and between levels of government and the active participation of stakeholders.

Donald A. Wilhite

Bibliography

Benson, C., and Clay, E., 1998. *The Impact of Drought on Sub-Saharan African Economies*. World Bank Technical Paper No. 401. Washington, DC: World Bank.

Bruins, H.J., and Lithwick, H., 1998. *The Arid Frontier: interactive management of environment and development*. Dordrecht: Kluwer.

Glantz, M.H., (ed.), 1994. *Drought Follows the Plow: cultivating marginal areas*. Cambridge: Cambridge University Press.

National Drought Policy Commission, 2000. *Preparing for Drought in the 21st Century*. Washington, DC: United States Department of Agriculture.

Vogt, J.V., and Somma, F., (eds.), 2000. *Drought and Drought Mitigation in Europe: advances in natural and technological hazards research*. Dordrecht: Kluwer.

White, D.H., (ed.), 1994. Climate and risk. *Agricultural Systems and Information Technology*, **6**(2), Barton, ACT, Australia: Bureau of Resource Sciences.

Wilhite, D.A., (ed.), 1993. *Drought Assessment, Management, and Planning: theory and case studies*. Natural Resources Management and Policy Series. (Dinar, A., and Zilberman, D., Series editors). Dordrecht: Kluwer.

Wilhite, D.A., (ed.), 2000. *Drought: a global assessment* (vols 1 and 2). *Hazards and Disasters: A Series of Definitive Major Works*. London: Routledge.

Wilhite, D.A., Hayes, M.J., Knutson, C., and Smith, K.H., 2000. Planning for drought: moving from crisis to risk management, *Journal of the American Water Resources Association*, **36**(4): 697–710.

Wilhite, D.A., Sivakumar, M.K.V., and Wood, D.A., (eds.), 2000. *Early Warning Systems for Drought Preparedness and Drought Management*. Proceedings of an Expert Group Meeting. Lisbon, Portugal, 5–7 September. Geneva: World Meteorological Organization.

Cross-references

Agroclimatology
Arid Climates
Deserts
Climate Hazards
Desertification
Palmer Index/Palmer Drought Severity Index

DYNAMIC CLIMATOLOGY

Dynamic climatology, now frequently termed "climate dynamics", is an attempt to study and explain atmospheric circulation over a large part of the Earth in terms of the available sources and transformations of energy (Court, 1957). Hare (1957) notes that dynamic climatology is the prime approach for explanation of world climates as integrations of atmospheric circulation and disturbances. It also attempts to derive circulation types at the regional scale.

Agreement on the scope of dynamic climatology is not universal, and differences among dynamic, complex, and synoptic climatology are not always well defined. For example, Bergeron (1930) and others considered what is now synoptic climatology to be dynamic climatology. Interest in weather patterns among synoptic climatologists has also obscured contrasts with complex climatology. A history of the origin and content of dynamic climatology has been given by Rayner et al. (1991) while a modern survey of the field is available in a text by Barry and Carlton (2001).

Principles

The dynamic climatological approach is based on motion characteristics and the thermodynamic processes that produce them. Dynamic meteorologists have as their principal interest the development and interpretation of relationships that describe these latter items; their usage separates the climatologist from the meteorologist, although the boundary between the disciplines is a tenuous one. An excellent introduction to dynamics and thermodynamics as they apply to the atmosphere is contained in Hess (1959) and Atkinson (1981).

Moving air is the result of forces brought about by thermodynamic processes that originate primarily at the Earth–air interface. The driving force for all processes is the radiant energy supplied to the Earth–atmosphere system by the sun.

A small fraction of available solar energy is absorbed directly by atmospheric constituents. About three times as much energy reaches the surface of the Earth and is absorbed in differing amounts, depending on the geometry and physical properties of the surface. Various unevenly distributed energy sources and sinks are thus created. Energy is defined as the capacity to do work, to move an object (mass) through a distance by application of a force. Disposition of thermal energy in this case alters temperature, density, and pressure values, thereby yielding unbalanced vertical and horizontal forces that cause air to move. Heat, moisture, and momentum transfers result from such motions at various scales. Ultimately, the Earth–atmosphere system returns thermal energy to space through radiation in an amount equivalent to the solar energy originally received, thus balancing long-term gains and maintaining the atmospheric heat engine in a steady state.

The fundamental equations (often called the Primitive Equations) that describe processes and motion in the atmosphere are the Ideal Gas Law, the First Law of Thermodynamics, the Equation of Motion, the Equation of Moisture Conservation, and the Equation of Continuity (Table D4). They are used to describe the interactions and characteristics of the basic building blocks comprising the Earth–atmosphere system. Manipulations of the basic equations allow us to decipher relationships among thickness between layers in the atmosphere, temperature, the wind field, and to develop stability–change equations, pressure–tendency equations, and the vorticity equation, among others. Some concepts that follow from these relationships include geopotential and geopotential height, isobaric and constant level representations, streamlines and trajectories, circulation, and divergence.

Imperfect characterization

If the entire atmosphere could be confined to a laboratory setting, application of the equations in Table D4 would be a straightforward problem. Application to the real, unconfined atmosphere presents several difficulties. For example, data on the state of the atmosphere are not available at every point at all possible moments. This means that climatologists must work with average, or smoothed, data among sampled points. Unfortunately, such smoothing often eliminates small-scale density or temperature patterns that may have a significant but unknown effect at some location and time. Ramage (1978) points out that it is small-scale factors embedded within the atmosphere that prevent numerical weather forecasting efforts from being more accurate.

Lack of continuous, pervasive sampling also requires that boundaries be fixed on the edges of an area of interest; the west coast of the United States serves as one such large boundary because data on conditions over the Pacific Ocean are not widely available. Changes at the boundary become difficult to determine and ultimately filter through all equations used to characterize the atmosphere.

Turbulent eddies (small-scale air currents produced by frictional interaction by juxtaposed, unlike airmasses) also contribute substantially to the status of the atmosphere at any time. Again, unfortunately, the ability to describe energy transfers by

Table D4 Fundamental variables and equations

Variable	Equation
Pressure	Ideal Gas Law
Temperature	First Law of Thermodynamics
Density	Equation of Continuity
Moisture	Equation of Conservation of Moisture
Velocity	Equation of Motion

Source: After Panofsky (1956).

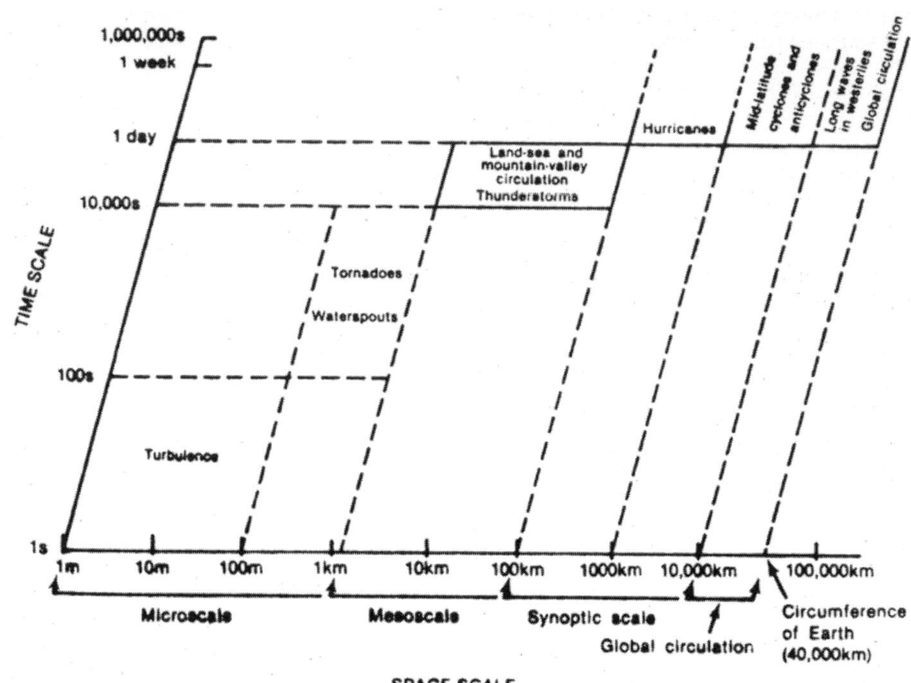

Figure D15 Time and space scales of atmospheric motion (after Anthes et al., 1981).

eddy conduction, diffusion and friction within the friction layer (the lowest 1500 m of the atmosphere) is limited; such transfers are neglected.

Finally, the basic equations used to describe the workings of the atmosphere are nonlinear. This means that, even at large scales, irregular fluctuations above and below mean values of variables do not cancel. Equations can be made to perform as though they were linear by using an artifice that considers all turbulent contributions as perturbations on the mean pattern. Given the contribution of turbulence to the mean state of the atmosphere, this introduces errors, is unsatisfactory, but is all that can be done at present without the use of numerical methods.

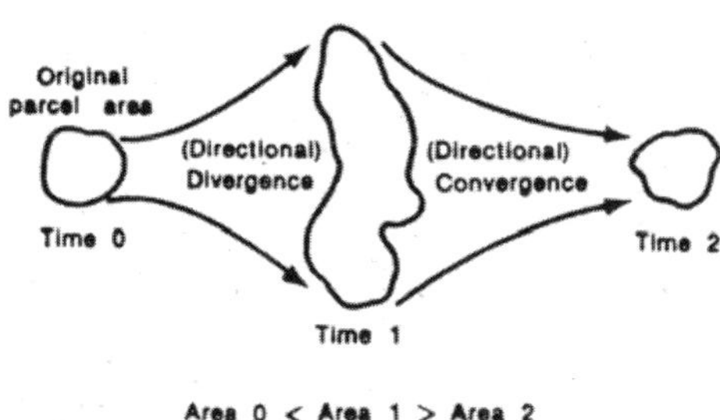

Figure D16 Horizontal divergence.

Description of motion

Although forces producing motion are interwoven in the atmospheric fabric, some equation terms and variables can be excluded at certain scales, because their contribution to an event or process is minimal (Figure D15). For example, at the level of the global circulation, vertical motion can be neglected, but not heat transfer, because at the global scale motion is overwhelmingly horizontal. On the other hand, certain microscale processes allow neglect of all thermodynamic equations and the variables temperature, moisture, and density if the atmosphere is considered incompressible. Thus, complexity is decreased without decreasing understanding of important processes and the results.

The Equation of Motion listed in Table D4 describes a vector and thus has three parts, each expressing Newton's Second Law of Motion (force = mass × acceleration) for three directions in a Cartesian system. Manipulations of this general equation yield the collective forces that must be specifically expressed; namely, gravity, pressure gradient, Coriolis force, and friction. Stated as a vector in words:

> Changes in velocity per unit time = Coriolis force + gravitational force − pressure gradient force + effect of friction.

Motion that takes place on a plane (two equations describe motion on a plane) above the friction layer between straight, parallel isobars (i.e. accelerations are not produced by the pressure field), is called the geostrophic wind, and is expressed as a balance between the pressure gradient and Coriolis forces. Consideration of curved isobars and the effect of friction yields other types of flow regimes, such as gradient (observed around low- and high-pressure systems), cyclostrophic (typical of tornadic circulation), and cross-isobaric motion due to the effect of friction on the velocity component of the geostrophic wind. Subtraction of geostrophic winds at two levels yields the thermal wind, which provides ties among thickness between layers, virtual temperature, and warm or cold air advection.

The third equation of motion applies to movement along a vertical axis. The Coriolis and frictional forces are neglected because their vertical components are small. Vertical motion then represents an interplay between an upward-directed pressure gradient force and gravity as a restoring force; this relationship is called the Hydrostatic Equation.

Manipulations of the three motion equations yield other concepts that are used in the dynamic climatological approach. For example, divergence and its opposite convergence occur when flow leads to increasing or decreasing area changes through time, as in Figure D16. Mass is neither created nor destroyed in the Earth–atmosphere system, and this principle (the First Law of Thermodynamics) stated in the form of the Equation of Continuity, ties changes in the horizontal with vertical motion (Figure D17). As the atmosphere contracts horizontally (converges), it expands vertically (stretches), and vice-versa. If we consider the troposphere as the region where these compensating mechanisms operate, then a general pattern like Figure D18 results. These patterns can also be related to vertical motion and typical convergence–divergence patterns in an eastward-moving upper-air wave system (Figure D19).

Finally, vorticity (which can be derived from the two horizontal equations of motion) area changes and vertical motion

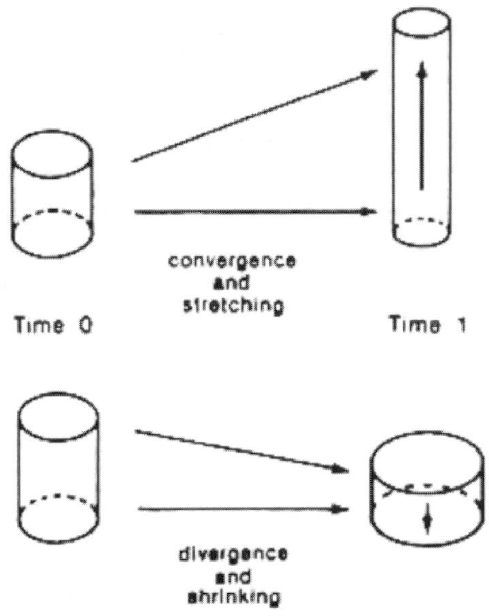

Figure D17 Convergence and stretching (upward vertical motion); divergence and shrinking (downward vertical motion).

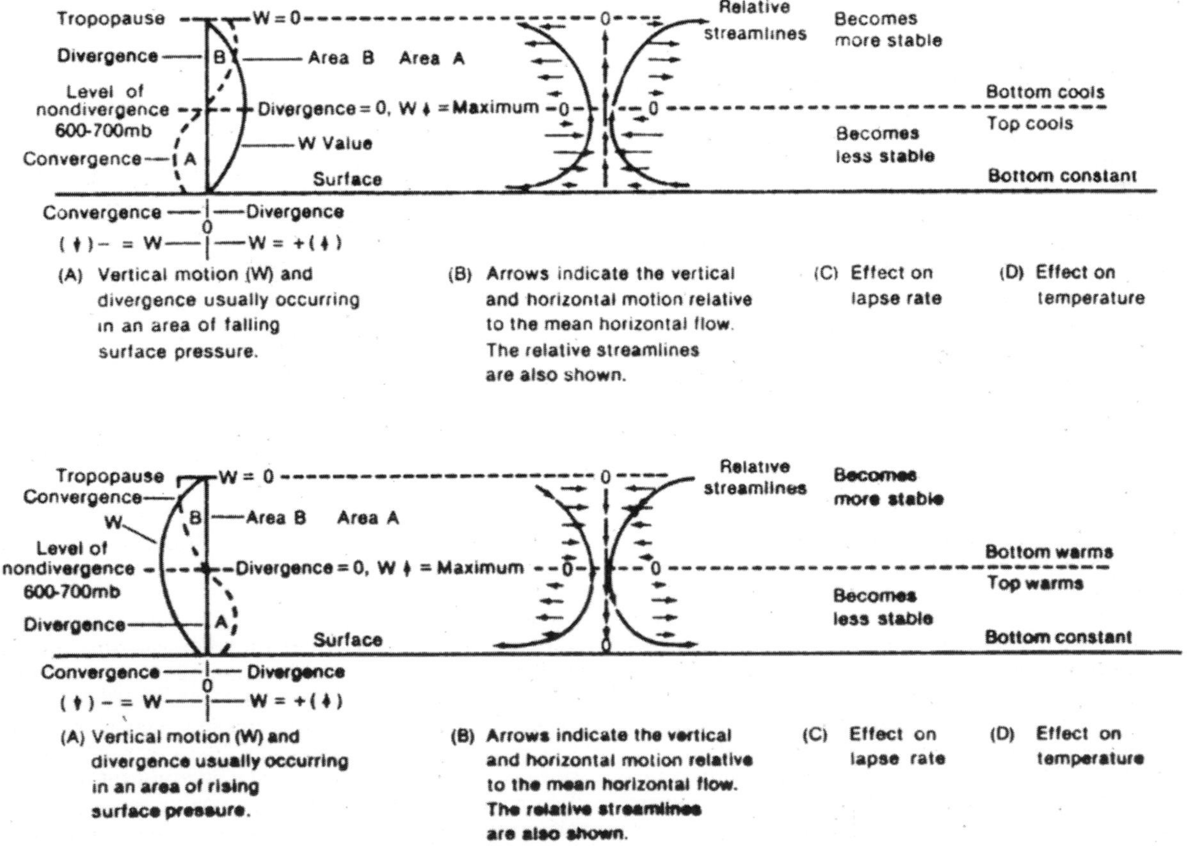

Figure D18 Convergence, divergence, and vertical motion in a typical atmospheric cross-section in the midlatitudes.

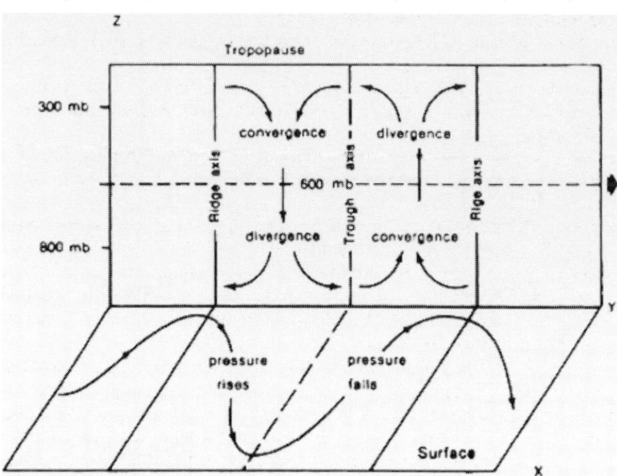

Figure D19 Organized patterns of convergence and divergence in an eastward-moving system in the midlatitudes.

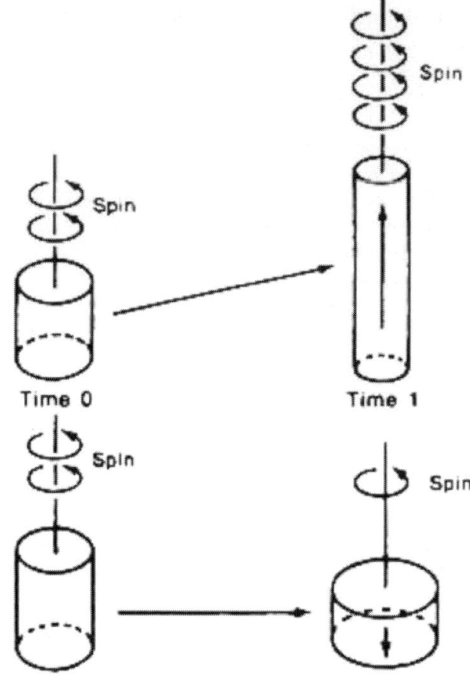

Figure D20 Relationship among vorticity, vertical motion, and area.

are related through the vorticity equation. Vorticity is defined as circulation (velocity along a closed path times the length of path) per unit area; since circulation for most processes is relatively constant through time, absolute vorticity (composed of relative plus Earth vorticity) times area is a constant. Again, equation of continuity is used to relate area changes to depth changes; i.e. decreases in area are related to convergence, upward motion, and positive vorticity (counterclockwise spin). Positive vorticity is normally associated with weather system development. Negative vorticity (clockwise spin) is associated with divergence, downward vertical motion, and lack of significant system development.

An example of the interplay among area, depth, and vorticity changes is shown on Figure D20. Figure D21 shows these latter items as they change when topographic changes are encountered. Lower-level air moving perpendicular to a topographic barrier is forced to ascend as it approaches. Decreases in absolute vorticity below the value that should exist at the given latitude occur as the cylinder reaches the mountain crest, causing an increase in divergence, subsidence, and ultimately anticyclonic deflection of the air parcel. As the column leaves the crest, the lower levels are free to expand, thereby increasing upward vertical motion and positive spin. The topographically forced condition depicted in the figure often is the situation prevailing to the lee of the Rocky Mountains, and may lead to development of large cyclonic storm systems.

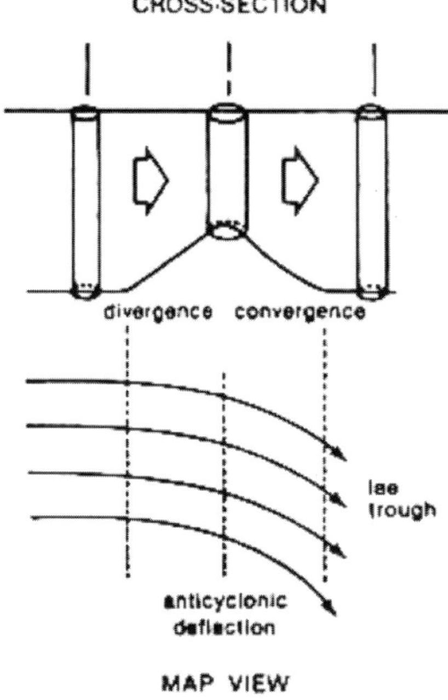

Figure D21 Patterns of convergence and divergence in an air current crossing a topographic barrier.

Thermodynamic processes

The use of thermodynamic principles shows how changes in heat content of the atmosphere, which behaves as a fluid under most circumstances, affect its dynamic characteristics. Volume and pressure changes within the atmosphere are produced by heat addition to, or subtraction from, an air parcel or by work done on or by the parcel within its surroundings. Thermodynamic laws describe what occurs among pressure, volume, and heat content as changes take place.

For example, the First Law of Thermodynamics states that heat added to a parcel can be used to increase its internal energy or to do work on its surrounding environment. From the First Law we can derive the adiabatic relationship, which states that all temperature changes within a parcel are due to expansion or compression if no external heat is added or taken away. External heat may come from radiation, eddy heat conduction, evaporation, or condensation. The adiabatic relationship allows stability to be determined and tells us what will happen to parcel air temperatures as parcels move about in the atmosphere.

The state of the atmosphere at any moment can be described by application of the Ideal Gas Law, which relates pressure, density, virtual temperature, and a gas constant for air. The Moisture Equation allows determination of changes in water vapor content of air parcels through time. The equation relates changes in specific humidity (grams of water vapor per unit mass of air including moisture) to occurrence of condensation, evaporation, molecular diffusion and eddy diffusion of vapor between unlike air masses.

Glen A. Marotz

Bibliography

Anthes, R., Cahir, J., and Panofsly, H., 1981. *The Atmosphere*, 3rd edn. Columbus, OH: Merrill.

Atkinson, B.W. (ed.), 1981. *Dynamical Meteorology: An Introductory Selection*. New York: Methuen.

Barry, R.G., and Carlton, A.M., 2001. *Synoptic and Dynamic Climatology*. London: Routledge.

Barry, R., and Perry, A., 1973. *Synoptic Climatology*. London: Methuen.

Bergeron, T., 1930. Richtlinien einer dynamischen Klimatologie. *Meteorologische Zeitung*, **47**: 246–262.

Charney, J., 1975. Dynamics of deserts and drought in the Sahel, *Royal Meteorological Society Quarterly Journal*, **101**: 193–202.

Court, A., 1957. Climatology: complex, dynamic and synoptic. *Association of American Geographers Annals*, **47**: 125–136.

Fu, C., and Fletcher, J.O., 1985. The relationship between Tibet-tropical ocean thermal contrast and interannual variability of Indian monsoon rainfall. *Journal of Climatology and Applied Meteorology*, **24**: 841–847.

Hare, K., 1957. The dynamic aspects of climatology. *Geographische Annaler*, **39**: 87–104.

Harnack, R., 1980. An appraisal of the circulation and temperature pattern for winter 1978–79 and a comparison with the previous two winters. *Monthly Weather Review*, **108**: 37–55.

Hess, S., 1959. *Introduction to Theoretical Meteorology*. New York: Holt, Rinehart & Winston.

Namias, J., 1952. The annual course of month-to-month persistence in climatic anomalies. *American Meteorological Society Bulletin*, **33**: 279–285.

Namias, J., 1954. Further aspects of month-to-month persistence in the mid-troposphere. *American Meteorological Society Bulletin*, **35**: 112–117.

Namias J., 1978. Multiple causes of the North American abnormal winter, 1976–77. *Monthly Weather Review*, **106**: 279–295.

Panofsky, H.A., 1956. *Introduction to Dynamic Meteorology*. University Park, PA: College of Mineral Industries, Pennsylvania State University.

Peixoto, J.P., and Oort, A.H., 1992. *Physics of Climate*. New York: American Institute of Physics.

Ramage, C.S., 1978. Future outlook – hazy. *Bulletin American Meterological Society*, **59:** 18–21.

Rayner, J.N., Hobgood, J.S., and Howarth, D.A., 1991. Dynamic climatology: its history and future. *Physical Geography*, **25**: 207–219.

Reiter, E., 1978. Long-term wind variability in the Tropical Pacific, its possible causes and effects. *Monthly Weather Review*, **106**: 324–330.

Smagorinsky, J., 1979. Global atmospheric modeling and the numerical simulation of climate. In Hess, E., ed., *Weather and Climate Modification*. New York: Wiley, pp. 633–686.

Cross-references

Atmospheric Circulation, Global
Climatology
Coriolis Effect
Jet Streams
Rossby Wave/Rossby Number
Synoptic Climatology
Vorticity
Winds and Wind Systems

E

EASTERLY WAVES

On the western sides of oceans in the trade wind belt, surface winds vary little and heavy rainfall occurs mostly as periodic rather than continuous events. This rainfall is most frequently associated with a wavelike disturbance of the normal isobaric pressure pattern, a disturbance called an easterly wave.

In the North Atlantic Ocean, between 5 and 15 degrees north, the waves are first seen in April or May and continue until October or November. With a wavelength of about 2000–2500 km, and a period of 3–4 days, they move at approximately 18–36 km/h. Passing from the Africa onto the cool Eastern Atlantic Ocean, the waves generally decay, but some survive to regenerate over the Western Atlantic and Caribbean. About 10% of all easterly waves survive to develop into gale-force tropical storms or hurricanes.

Ahead of the wave, the east or northeast trade wind backs to northeast or north, the pressure falls, and the trade wind inversion lowers, producing fine weather. Low-level divergence and high-level convergence are accompanied by subsidence and suppressed convection. About 320 km (200 miles) ahead of the wave trough, the moist layer reaches a minimum, often at 1500 m (5000 ft). As the wave trough approaches, the depth of the moist layer increases. Behind the wave, low-level convergence and upper divergence result in ascending motion. The pressure begins to rise, heavy cumulus and cumulonimbus develop as the moist layer destroys the inversion and reaches a maximum height of about 7500 m (25 000 ft). Shower and thunderstorm activity, if present, reaches a maximum intensity some 320 km (200 miles) behind the surface trough (Figures E1 and E2).

Easterly waves vary widely among themselves with respect to stability and the amount of rainfall produced. Weak waves produce no rain over the open sea but usually intensify as they pass over islands. The variation in precipitation amount depends on whether the divergence or convergence extends to 850 mb or to 700 mb. Even with small divergence in weak disturbance, the precipitation amount may vary greatly in different parts of the waves.

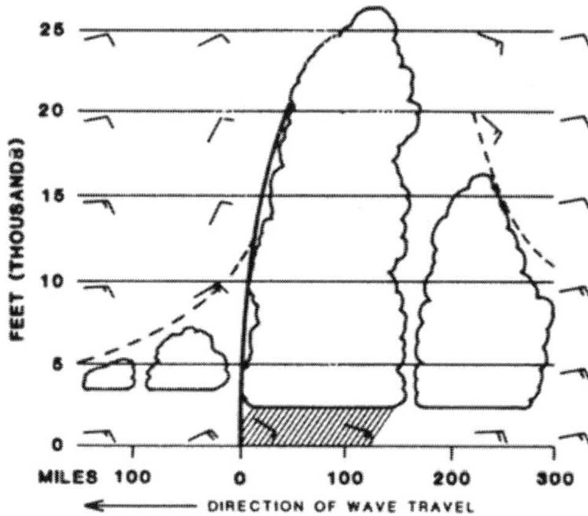

Figure E1 Cross-section of an easterly wave (after Malkus, 1957).

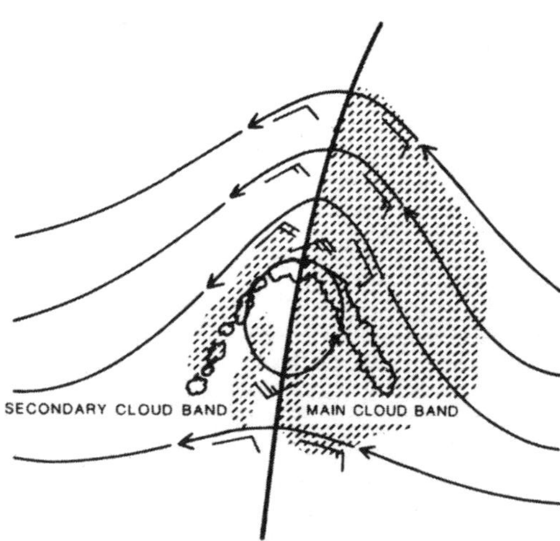

Figure E2 A deep easterly wave. The shaded area represents extensive rain and cloud amounts.

It has also been observed that when an upper-air polar trough penetrates deeply into low latitudes its tropical portion sometimes breaks off from the parent body in the westerlies. The fractured trough then may reverse its direction to move westward in the form of a wave in the easterlies.

Yanai (1961) suggested that easterly waves receive their energy from horizonal shear in the zonal flow and that the cold cores result from dynamically forced lifting associated with weak temperature gradients. The required horizontal shear may be provided by an Intertropical Convergence (ITC) that has become displaced some distance away from the equator. In general, the zonal shear is nearly three times as large as the meridional shear while the contribution of the curvature term to vorticity is negligible. Such unstable barotropic waves caused primarily by north–south shear are common in the eastern Pacific Ocean during the summer season.

On those occasions when the easterly wave develops rapidly, convection becomes organized into a definite pattern, pressure falls and a deep, intense tropical storm forms. These provide the potential genesis for hurricane formation.

John E. Oliver

Bibliography

Case, B., 1990. Hurricanes: strong storms out of Africa. *Weatherwise*, 23–29.

Chang. J.-H., 1972. *Atmospheric Circulation Systems and Climate*. Honolulu: Oriental Publishing Co.

Malkus, J., 1957. The origin of hurricanes. *Scientific American*, **197**: 3–9.

Yanai, M., 1961. Dynamic aspects of typhoon formation. *Journal of the Meteorological Society of Japan*, **39**: 282–309.

Cross-references

Atmospheric Circulation, Global
Intertropical Convergence Zone
Tropical Cyclones
Tropical and Equatorial Climates

EKMAN (SPIRAL) LAYER

The lower layers of the troposphere include a transition zone between the free atmosphere above and the surface boundary layer close to the Earth's surface. This zone, extending from perhaps 100 m above the surface to the geostrophic wind level, is sometimes considered as the spiral layer. The upper free atmosphere behaves as an ideal fluid with air motion responding to pressure and Coriolis forces; near the Earth's surface the shearing stress may be considered constant. Between these two the spiral layer needs be considered separately.

In this intermediate layer there is ideally an *Ekman spiral*, in which there is decreasing stress. Here the coefficient of eddy viscosity and density are assumed to be constant (for ease of calculation), but in nature considerable departures from the ideal are recorded. The spiral was originally worked out by W.F. Ekman in 1902 for an ocean current condition and was named for him (Figure E3). Shortly thereafter its applications to the atmosphere were identified.

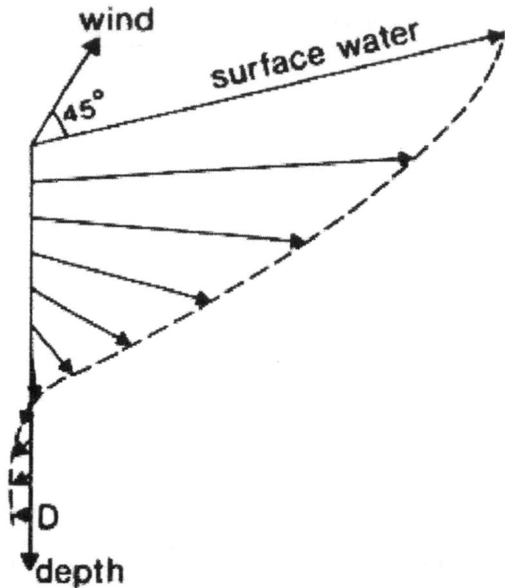

Figure E3 The Ekman spiral applied to increasing water depth.

In the spiral layer the wind blows across the isobars toward the lower pressure, as a so-called *secondary flow*, being at a maximum angle near the surface (less than 45°); with rising altitude, the wind bears to the right (veers) in the northern hemisphere, but bears left (backs) in the southern hemisphere. The Ekman formula provides the horizontal wind components u and v in the x and y directions, x being parallel to the pressure gradient and y normal to it for any altitude level z, thus in Cartesian coordinates, as follows:

$$u = v_g e^{-\beta} \sin \beta$$

$$v = v_g(1 - e^{-\beta} \cos \beta)$$

where v_g is the geostrophic wind speed, e the vapor pressure, and $\beta = z\sqrt{f/2K_M}$ with f as the Coriolis parameter and K_M the eddy viscosity. The wind vector in the spiral layer deviates from the geostrophic wind by an amount that diminishes upward at an exponential rate. An ideal graphical solution of this spiral can be seen on a polar-coordinate plot of wind direction tangentially against wind speed radially for successive altitudes.

Rhodes W. Fairbridge

Bibliography

Curry, J.A., and Webster, P.J., 1999. *Thermodynamics of Atmospheres and Oceans*. San Diego, CA: Academic Press.

Hess, S.L., 1959. *Introduction to Theoretical Meteorology*. New York: Holt.

Holton, J., 1992. *An Introduction to Dynamic Meteorology*, 3rd edn. New York: Academic Press.

Sutton, O.G., 1953. *Micrometeorology*. New York: McGraw-Hill.

Cross-references

Coriolis Effect
Dynamic Climatology
Ocean Circulation

ELECTROMAGNETIC RADIATION

Electromagnetic radiation is energy propagated through space (as from the sun or the cosmos) or through various solid media in the manner of a vibrating electric and magnetic field. According to James Clerk Maxwell, light was to be explained not by a transverse wave motion in an elastic medium, but by electric and magnetic charges, vibrating at right angles to each other and to the direction of propagation. The electric charges were supposed to transmit energy as an electromagnetic wave. According to the quantum theory these charges behave as particles, with mass and momentum.

Electromagnetic radiation (or simply radiation, in common practice) is characterized by wavelength (or frequency) and amplitude (or flux density). Wavelength (in one or two words) is generally given as λ, being related to frequency f and phase speed ν by $\lambda = \nu/f$. The reciprocal of wavelength is wave number, or the number of waves per unit distance in the direction of propagation. Wavelength is usually measured from midpoint (going from negative to positive) in one cycle to similar midpoint on the next; however, in oceanography, oscillating sea waves are measured from crest to crest. A phase is a hemicycle. In an oscillation of uniform amplitude, two cycles are said to be 100% out of phase if the positive part of the one cycle exactly matches the negative part of the other.

The propagation speed of an electromagnetic radiation is given as

$$\frac{c}{\sqrt{\mu K}}$$

where the constant c (the speed of light) $= 2\,997\,930 \times 10^{10}$ cm/s, the magnetic permeability is μ and the specific inductive capacity is K. The speed through vacuum, and roughly speaking through air, approximates c.

The electromagnetic spectrum covers a vast range, from about 20 km down to microns (10^{-6} meters), angstroms (10^{-10} meters), and less (see Figure E4). Of longer wavelength than 20 km is the audio-frequency range (sound waves, of about 20 kc to 20 cycles). One may note the relative narrowness of the visible light band. Solar radiation covers the spectrum from about 1500 Å (ultraviolet) to over 40 000 Å (in the infrared).

Frequencies are expressed in cycles per second, or in the radio wavelengths in kilocycles and megacycles. Since it was the German physicist Heinrich Rudolf Hertz (1857–1894) who studied Maxwell's theory, but actually first demonstrated the nature of electromagnetic wave propagation, measuring wavelength and frequency, the unit hertz, one complete cycle per second, is used. Radiowaves are often expressed in kilohertz and megahertz. The relationship of wavelength λ in meters to frequency (f) in kilocycles (or kHz) is: $\lambda \times 3 \times 10^5/f$, or 1 kc $=$ 30 000 m, e.g. a 6-Mc radio transmission uses the 50-m band.

Rhodes W. Fairbridge

Bibliography

Gates, D.M., 1966. Spectral distribution of solar radiation at the Earth's surface. *Science*, **151**: 523–529.

Goody, R., *Principles of Atmospheric Physics and Chemistry*, New York: Oxford University Press.

Hartman, D.L., 1994. *Global Physical Climatology*. San Diego, CA: Academic Press.

McIlveen, R., 1992. *Fundamentals of Weather and Climate*. London: Chapman & Hall.

Cross-references

Energy Budget Climatology
Terrestrial Radiation
Radiation Laws
Ultraviolet Radiation

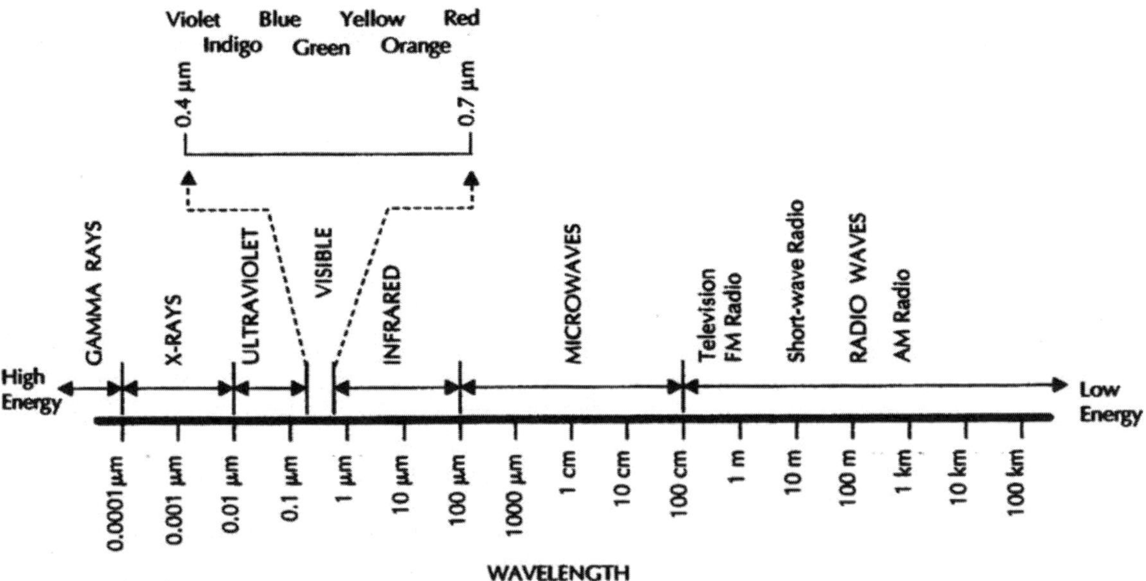

Figure E4 The electromagnetic spectrum.

EL NIÑO

Newspaper headlines are increasingly highlighting stories about "severe", "bizarre", "crazy", "extreme", and "anomalous" weather patterns that have disrupted human activities in many parts of the world and have caused devastation to life and property on a local, national, and regional scale. Much of this unusual weather – and even an infinitesimal slowing of the Earth's rotation – has been blamed on an oceanic–atmospheric interaction known as El Niño.

The eastern equatorial Pacific is usually dominated by coastal upwelling – the upsurge of deep, nutrient-rich cold water to the ocean's surface. Such waters provide excellent breeding grounds for various levels of the food chain from phytoplankton to anchovies. Exploitation of this fish population made Peru the number one fishing nation in the world by the early 1970s. However, every Christmas season the winds that blow along the coast from the southeast slacken, the strength of the upwelling weakens, and the surface waters begin to warm and the fish population is less available to the fishing boats. This warming trend usually lasts for a few months and by March or so the strong upwelling resumes again. Occasionally the warm water continues to accumulate in the eastern and central equatorial Pacific, resulting in a large area of warm surface water.

This appearance along the coast of a large mass of warm water is associated with sharply reduced catches off the coast of Peru and with destructive rains that occur in normally dry areas of Peru and Chile. It has also been associated with the development of drought in Bolivia and northeast Brazil, and other impacts such as outbreaks of malaria and cholera in various parts of South America.

El Niño can be defined as an invasion of warm surface water from the western part of the equatorial Pacific into the normally cooler waters of the central and eastern part of the Pacific Basin off the western coast of South America – mainly Peru and Ecuador.

El Niño is usually described as an anomaly, an unusual or abnormal interaction between air and sea in the equatorial Pacific that is not part of the normal regional or global climate system. While we can talk of sea surface temperatures in the eastern and central Pacific Ocean departing from a mathematically identified average condition, we should not view an El Niño-related departure from the average as being abnormal and unexpected. In fact, El Niño is a normal and to-be-expected part of it.

Under El Niño conditions the warm water heats the air above it, and if the right conditions exist aloft, the rising water vapor condenses and forms clouds. Heavy rains then fall over normally dry areas, and as this warm mass of air eventually moves out of the region, sinking in the western Pacific, it inhibits rain in this usually wet area. Many atmospheric scientists believe that the atmospheric processes set in motion during an El Niño event also influence climate in regions outside the tropics.

Although El Niño originally referred to local conditions off the coast of Peru and Ecuador, the use of the term has been broadened by many scientists to represent all surface temperature warming in the eastern and central Pacific. Because this warming process typically begins around Christmastime, Peruvian fishermen called it El Niño, which is Spanish for the Christ Child.

Recurrent phenomenon

El Niño is only the warm phase of a larger cycle called ENSO, or El Niño–Southern Oscillation. The Southern Oscillation refers to a seesaw-like shift in sea level pressure systems across the Pacific Basin. When sea level pressure is high at Darwin, Australia, it tends to be low at Tahiti, and vice versa. ENSO includes a cold phase – La Niña. Extreme weather events around the globe have also been associated with La Niña, though many policy makers are unaware of its existence or its potential impacts. Some scientists suggest that, in various regions, La Niña-related extreme events are the opposite of those related to El Niño. For example, in southern Africa drought accompanies El Niño, while rain is associated with La Niña. In the United States, some researchers claim that La Niña's impacts are more costly that those of El Niño. Researchers are just beginning to realize that they should focus more attention on the cold part of the cycle to enhance their overall forecast possibilities.

El Niño (a warm event), like its counterpart La Niña (a cold event), is an integral part of the global climate system. It is a recurrent phenomenon with an average return period of 4½ years, but can recur as little as 2 or as much as 10 years apart. Such events have occurred for millennia, and can be expected to continue to occur in the future. Indeed, to go through a decade or two without an El Niño would be truly unusual. Since El Niño does not represent unusual global climate behavior, we can attempt to prepare for it.

Although El Niño events have occurred throughout the twentieth century, the 1982–1983 event was the one that first sparked serious interest. Decision makers in water resources, agriculture, energy and health realized that they could benefit from advance warnings of weather conditions. That is why there is so much interest in El Niño.

Researchers and policy makers, in particular in Australia, Brazil, Chile, Ecuador, Peru, Southern Africa and the United States of America, became interested in developing an improved understanding of El Niño and its impact on weather systems.

The onset of a strong El Niño in early 1997, later called "the El Niño of the century" because of its costly impacts, intensified general interest in the phenomenon, especially in the potential value of planning to mitigate its direct effects. However, still needed is a better understanding of the ways El Niño events can develop and evolve. Also, societies have not yet seen all of the ways that they can affect society and ecosystems.

Policy makers, governments and scientists are increasingly focusing on El Niño as a tool for forecasting climatic conditions months in advance, particularly for identifying the potential for droughts or floods. Droughts can have devastating impacts on communities, affecting food and energy production and, in arid zones, the very survival of humans and animals. Floods can destroy property and kill people. In January 1998, in northern Peru and Kenya, for example, heavy rains and flooding had serious consequences for their built environments, including communications and transportation networks.

Timely warning of such extreme events can enable civil defense experts to plan for evacuation of at-risk populations. If reservoir managers are given a reliable forecast of seasonally above-average rainfall, they can lower reservoir water levels to make storage room for the expected excess water.

El Niño-associated flooding has many adverse effects: it can disrupt farming along river borders; drinking water can become contaminated during excessive rainfall and flooding; and local water delivery systems can be damaged or destroyed, with negative consequences for human health.

Effects on water resources

Knowing that an El Niño is due can enable managers of reservoirs supporting irrigation schemes in the affected region to alter their decisions about when and how much water to allocate and to whom, if the El Niño effect is likely, for example, to bring drought. It could perhaps have helped planners in southern Africa during the 1982–1983 El Niño-induced drought, when lack of water led to the closure of factories, with workers laid off for many months.

That same El Niño brought severe flooding to northern Peru and southern Ecuador. There had been no warning. In November 1997 the northern Peruvian city of Piura, which typically receives about 6 mm of rain in November, received almost 600 mm of rain. This time, however, the Government of Peru took the El Niño forecast seriously and was able to mobilize engineering and civil defense resources to mitigate some of the worst impacts of flooding in the north.

Not every weather anomaly throughout the world that occurs during an El Niño year is caused by the El Niño itself, though in 1997 there was an increasing tendency to blame it for everything. For example, floods in the Midwest United States of America during 1993 were blamed on that year's El Niño, even though scientific evidence for this conclusion was very weak. On the other hand, research did support the conclusion that the severe drought in the same region in the summer of 1988 was caused by La Niña and not necessarily global warming as some scientists had suggested at the time.

Extreme record-setting weather events are occurring around the globe, even in non-El Niño years. Only some parts of the globe are directly influenced by El Niño and even these areas are not influenced in the same way by each event.

Teleconnections, or long-distance effects between El Niño and regional climate anomalies around the globe, have been identified statistically and by observations of direct links between warm surface water in the central and eastern equatorial Pacific and distant anomalies such as drought in northeast Brazil or Ethiopia.

El Niño can have positive and negative aspects at the same time. For example, during an El Niño the number of hurricanes and tropical storms along the Atlantic and Gulf coasts of North America and in the Caribbean is noticeably reduced. During the 1997 El Niño, the hurricane season was almost non-existent and no devastating storm occurred. For many this was good news, but not for everyone. The absence of tropical storms also reduces the amount of freshwater that tropical islands normally receive from passing storms. The rain from these storms usually fulfills their water needs until the onset of the next year's hurricane season.

The impact of global warming on El Niño is not yet known, though there are hundreds of studies on the possible effects of global warming on the global hydrological cycle and what these effects might mean for regional and local hydrological cycles. Some researchers suggest that dry areas will become drier with global warming and wet areas wetter; others have suggested the opposite.

Forecasting

Knowing when an El Niño will strike, however, is different from forecasting the effect it will have in different areas. Although some kinds of impacts tend to recur in the same places, each El Niño can cause a different set of impacts (droughts, floods, fires). Predicting its arrival is likely to prove relatively easy compared with forecasting what will actually happen to disparate locations around the globe.

But forecasting can provide useful insights. Certain teleconnections do have a tendency to occur when an El Niño has been forecast. In areas where the link seems reliable, authorities can begin to prepare for seasonal changes in water resources.

El Niño as a spawner of natural hazards

El Niño's characteristics fit well within the analytical definition of a natural hazard. However, it seems that hazards researchers prefer to argue why El Niño should *not* be viewed as a natural hazard. Perhaps it is time for El Niño to be viewed as a legitimate hazard (in this broader sense).

El Niño is a natural process that has been associated with various kinds of hazards. It is only since 1969 that it was discovered to have such impacts around the globe. The hazards it spawns include, but are not limited to, droughts, floods, frosts, fires, and landslides. Perhaps if researchers can link what it is that spawns natural hazards (e.g. El Niño) more closely to the potentially spawned hazards, societies can shift their short-term responses toward proaction (i.e. prevention and mitigation) and away from reaction (i.e. adaptation and cleanup). Doing so would tend to push the early warning of specific El Niño-related hazards further "upstream", thereby providing more lead time for societal coping mechanisms to come into play. This would make El Niño forecasts the earliest possible warning that a government might get of climate-related impacts they wish to avoid.

During El Niño events, almost all the Pacific islands suffer from drought and, as a result, water shortages are highly likely to follow. Water is a scarce commodity in these island nations even under "normal" conditions (and it is growing scarcer). Expanding population, combined with agricultural and industrial growth, are increasing the pressure on limited freshwater supplies. So now El Niño has become synonymous with drought in the North and South Pacific island states. The only exception is Kiribati, which receives considerable rain during El Niño and suffers drought during La Niña events.

Disaster managers in this region clearly have their hands full. This has always been the case, even before they became aware of El Niño and La Niña. What is different now is that the onset of drought or flood can be linked to the ENSO cycle. Therefore, El Niño and La Niña forecasts can convert disaster responses from reaction to proaction. Based on the adage that "forewarned is forearmed", disaster managers in these island nations can prepare their countries to reduce the damage of El Niño-related adverse impacts.

The intensity of an El Niño event can vary from weak to moderate to strong to extraordinary. A strong event is more likely to influence the climate conditions far from the Pacific Basin, whereas weak events are likely to have their strongest impacts in Pacific Rim countries. El Niño hotspots are regions that are likely to be affected by an El Niño event. A list of likely El Niño hotspots might include, among others, the following:

- Drought in Zimbabwe
- Drought in Mozambique
- Drought in South Africa
- Drought-related food shortages in Ethiopia
- Warm winter in the northern half of the United States and southern Canada
- Heavy rains in southern Ecuador and northern Peru
- Drought in northeast Brazil (a region known as the Nordeste)

- Flooding in southern Brazil
- Drought and fires in Indonesia
- Drought in the Philippines in the tropical Atlantic
- Coral bleaching worldwide
- Droughts in various South Pacific island nations
- Drought in eastern Australia
- Heavy rains in southern California

Often the actions that a society needs to take to mitigate or avoid the worst impacts of an El Niño-related societal or environmental impact are those from which society would benefit anyway: cleaning up our dry river channels so torrential rainwater can pass to the sea; repairing leaky roofs; shoring up bridges, rail lines, and roads that are in poor condition.

Clearly, the science and art of forecasting El Niño are in their early stages. For some locations around the globe, and for some socioeconomic activities worldwide, El Niño information can provide a "heads up" to decision makers in climate–and weather sensitive regions and sectors. Such information provides power to those who choose to use it.

Michael H. Glantz

Bibliography

Carredes, C.N., 2001. *El Niño in History: storming through the ages.* Gainesville, FL: University Press of Florida.

Davis, M., 2001. *Late Victorian Holocausts: El Niño, famines, and the making of the Third World.* New York: Verso Books.

Diaz, H.F., and Markgraf, F., (eds.), 1992. *El Niño: historical and paleoclimatic aspects of the Southern Ocean.* Cambridge: Cambridge University Press.

Fagan, B.M., 2000. *Floods, Famines and Emperors: El Niño and the fate of civilizations.* New York: Harper Collins.

Glantz, M.H., 2001. *Currents of Change: El Niño and La Niña impacts on climate and society.* Cambridge: Cambridge University Press.

Suplee, C., 2000. *Milestones of Science.* Washington, DC: National Geographic Society.

Cross-references

Cycles and Periodicities
Pacific North American Oscillation
Ocean–Atmosphere Interaction
Oscillations
Southern Oscillation
Teleconnections

ENERGY BUDGET CLIMATOLOGY

The energy budget of a surface or an object is the amount of heat energy, in any form, that arrives at or departs from the surface of the object in a specified time period. Energy budget climatology most commonly is the climatology of the energy budgets of the Earth's surface that are studied at a variety of scales. The term may also be broadened to include the energy budget of the atmosphere as a whole or any part of it. We will concentrate here mainly on the energy budget climatology of the Earth's surface.

The energy budget of a surface is an application of the law of conservation of energy for the Earth's surface. The most simple form of the law may be expressed as:

$$Q^* = H + LE + G$$

where Q^* is net radiation, H is sensible heat (enthalpy), LE is latent heat (L being the latent heat of vaporization of water and E the amount of water evaporated), and G is the flow into or from the submedium. Each term can be positive or negative. The net radiation term may be further broken into the component terms of the radiation budget. The form of the energy budget described by this equation is one-dimensional and refers to flows of heat in the vertical direction to and from the surface of the Earth. In some applications of the concept, such as when advective influences are important, it is useful to take into account the two horizontal space dimensions as well.

According to Miller (1965), it was during the nineteenth century that George Perkins Marsh in the United States and Alexander Voeikov in Russia independently suggested that estimation of the energy budget of the Earth's surface and its parts would greatly enhance our understanding and explanation of climate. However, with a few exceptions, and owing to a lack of measurement technology, the idea was not really developed until the middle part of the twentieth century. At that time ongoing work in the western hemisphere was stimulated by the translation and publication of Mikhail Budyko's *The Heat Balance of the Earth's Surface* and its associated atlas. Since that time it has been realized that energy budget climatology not only provides insights into the causal explanations of the Earth's climates, but it has many other advantages as well. In general, these include its interdisciplinary function in linking applications in hydrology, glaciology, and bioclimatology, its link with the water budget through the evaporation term, its use for quantifying human influences that alter the surface of the Earth, and its application to the phenomenon of climatic change. Some of the most comprehensive sources detailing energy budget climatology, its uses and advantages include the works of Budyko (1958, 1974), Miller (1965), Sellers (1965), Oke (1987), Gates (1980), Rosenberg et al. (1983), Campbell and Norman (1998) and the continuing series of reports of the Intergovernmental Panel on Climate Change (e.g. Houghton et al., 2001).

A brief description of how surface energy budgets are measured and modeled is provided below. The energy budgets of different surfaces are then described for the micro- and mesoscale. Consideration is then given to the energy budget climatology of the regional (synoptic) and global scale. The application of energy budget climatology to climatic change is discussed. Finally, recent trends in energy budget climatology are noted.

Measurement of the energy budget at the surface

There have been many advances in the measurement of the energy budget at the surface since the mid-1950s. These advances range from relatively low-cost surface flux eddy correlation measurement systems to increasingly accurate satellite measurements. In a typical instrumented site at any point on land on the Earth's surface, net radiation and the ground heat flow would normally be measured directly by a net radiometer and one or more soil heat flux plates. The turbulent flows of latent and sensible heat are still sometimes estimated from measurements of temperature and water vapor pressure gradients in the vertical direction with adjustments made for the stability of the air, which may vary over time. The profile measurement of the turbulent fluxes assumes no major changes of the wind field or the radiation budget over a short measurement period, no vertical divergence of the fluxes, and the equivalence of the diffusion coefficients for momentum, sensible heat, and water vapor. Since these assumptions are not always fulfilled, the most common turbulent flux estimation methods based on profile measurements, usually known as the Aerodynamic method and the Energy Balance or Bowen ratio method, are not always the most accurate. Greater accuracy may be achieved at larger expense. Direct measurements of the turbulent fluxes may be made, for example, by the eddy correlation method, which directly measures the fluctuating part of the turbulent heat fluxes using highly sensitive, fast-response sensors and a real-time computer to separate the instantaneous deviations from the time-averaged mean. The eddy correlation technique has become more frequently used with the emergence of commercially available equipment. Networks of Flux Tower measurements are also being formed; the best known of these being organized by the Ameriflux and Euroflux programs as well as a worldwide program called Fluxnet. Sensitive, large weighing lysimeters that track the loss or gain of the weight of water from a soil monolith are an accurate though expensive way of directly measuring the latent heat flux. Oke (1987) presents a clear description of the detail of these common methods of energy budget instrumentation.

On a larger geographic scale, satellite data are increasingly being used to estimate surface energy budget values. A number of important multidisciplinary national and international programs have concentrated on research to find out how best satellite measurements may be used to estimate surface energy budget and other parameter values. Simultaneous measurements are made at the Earth surface and from space-borne and airborne sensor platforms. Some of the more important of these programs include the First ISLCP Field Experiment (FIFE) for grasslands in Kansas (Sellers and Hall, 1992), and a program for the Boreal forests of Canada (BOREAS) (Sellers et al., 1995). Remote sensing results have also played a role in the debate over desertification in West Africa and elsewhere. The bright star for remote sensing and the derivation of even better surface energy budget values is the Earth Observing System (EOS) that is beginning to give us huge amounts of highly relevant and fascinating data.

There are a number of difficulties in making actual observations of the surface energy budget. These include the inherent complexity of the energy flux processes, especially those dealing with the turbulent fluxes, and the need for continual attention to the instrumentation, particularly in severe environments. Gradually these difficulties are being overcome. A prime example is the flux tower at the Harvard Forest Site in central Massachusetts. At this site energy balance values and other fluxes have been measured continuously for about a decade (Waring et al., 1995). More and more actual accurate surface energy budget observations are being made. Eugster et al. (2000) report 38 different sets of measurements for just the Boreal forest and Arctic Tundra surfaces alone. The energy budgets of most major surface types of the Earth have now been investigated in detail, including that of the urban surface (Grimmond and Oke, 1995, 2002). Russian scientists used energy budget measurements in the middle of the last century to estimate the effects of human activities such as the flooding or draining of land areas. Similar uses of the energy budget approach are still being made today. In California, for example, investigators are using the technique to monitor the effects of restoring wetlands in the San Joaquin river delta area (Anderson et al., 2002).

Modeling energy budgets

Historically, several important works used standard meteorological data to estimate the surface heat budget. Probably the most significant is that of Budyko (1958) in his initial establishment of the heat budget of the Earth. Vowinckel and Orvig (1968) did pioneering work to estimate the heat budget of the Canadian Arctic. The *Heat Budget Atlas of the Tropical Atlantic and Eastern Pacific Ocean* by Hastenrath and Lamb (1978) and the comparable volume for the Indian Ocean (Hastenrath and Lamb, 1979) at one time represented the state of the art in these endeavors. Between about 1960 and 1990, when actual measurements of the surface energy balance were difficult to make, considerable attention was given to modeling such values. With greater access to surface and satellite measurements, modeling activities have become a little less pronounced. However, modeling still remains important, especially where large areas are concerned. Estimates of surface heat balance and other climate changes accompanying deforestation of the Amazon Basin are a prime example. Most of these estimates show potential increases in albedo value and a reduction in Q^* and LE, as well as a decrease in "recycled" precipitation (Hahmann and Dickinson, 1997). Another conceptual area where modeling remains important is in the field that has become known as Land Surface Parameterization (LSP). This field encompasses the estimation of the exchange of surface energy, momentum, water, heat, and carbon between the land surface and the atmosphere. The generation of LSP developed in the late 1990s combine energy, water and momentum balance values with the biophysics of photosynthesis and recognize vegetation as part of the climate system (Stocker, 2001, p. 442).

Further insights into energy budget climatology may be gained by examining the actual budgets of different surfaces.

The local scale

An examination of energy budgets of different surfaces demonstrates the manner in which the energy budget concept quantitatively explains the complex two-way interaction between the surface and the atmosphere. Oke (1987) has gathered together data on the energy budgets of many different surfaces and most of the following description is based on his efforts. One of the most simple surfaces in terms of energy budgets is bare ground. Table E1 lists the mean daily energy budget values for such a surface in the Turkestan mountains (41°N) for the month of September. Note that for the horizontal surface the largest proportion of the heat received as net radiation (Q^*) is expended into the atmosphere as sensible heat (H). Equally small amounts are lost to evapotranspiration (LE) and into the submedium (G). The data also show the effect of aspect in helping to determine the amount of net radiation received, and thus the amount of heat energy available for heating the air and the ground on slopes facing in different directions. Note, however, that the proportioning of use of the available net radiant energy is not too different on the two slopes from that on the horizontal surface. The bare-ground situation is in contrast to a vegetated surface. The energy budget for one day over a barley crop in southern England, for example, shows that almost all energy available in the form of net radiation is used for evapotranspiration

Table E1 Energy budgets for different surfaces

Surface	Q^*	H	LE	G	Data source
Bare soil; Turkestan Mountains					
Horizontal surface	14.4	9.4	2.1	2.1	Aisenshtat, 1966
North-facing slope	6.0	3.5	1.7	0.7	
South-facing slope	17.6	—	3.1	1.9	
Barley crop; southern England	15.7	−1.1	12.2	1.2	Long et al., 1964
Ocean surface; Tropical Atlantic	22.2	1.2	16.2	4.8	Holland, 1971
Melting glacial ice; Peyto Glacier, Alberta	11.4	−3.7	−0.9	16.0	Munro, 1975
Urban building air volume; Vancouver (10 h period only)	16.9	5.7	7.0	4.2	Yap and Oke, 1974

Note: Values are in MJ/m^2 per day.

(Table E1). In fact the sensible heat flow exhibits negative values, indicating heat was passed down from the atmosphere to the ground. The high amount of *LE* is not unusual for a vegetated surface, especially when the surface is irrigated. Under these conditions, and with the horizontal influx (advection) of warm air, *LE* often exceeds Q^*. The irrigated vegetated surface in many respects acts like an ocean surface. The ocean surface energy budget, as might be expected, shows that most of the heat income is used in evaporation. The energy budget for this surface does not balance in Table E1, however, because some heat is involved in a seasonal heat storage change in the water and in advective heat exchange by the water currents. Heat used at the surface is also important in the case of the energy budget of a melting ice surface of a glacier. Here, downward flows of *H* and *LE* (in the form of condensation on the ice surface) are used to supplement Q^* to sustain the amount of heat used in melting the ice. The presence or absence of snow on a surface makes a large difference to the net radiation received because of the higher reflectivity of the snow surface. Ohmura (1982) has documented this for the Arctic tundra surface. The energy budgets of urban surfaces show that some of these surfaces are not as arid as might be expected. The energy budget values for an urban building air volume in Vancouver (Table E1) actually show a larger amount of latent heat loss than sensible heat loss. This is due partially to the presence of urban vegetation and irrigated lawns (Suckling, 1980). The energy budget values of a suburban Vancouver surface have been found to be intermediate between those for truly urban and surrounding rural vegetated surfaces (Kalanda et al., 1980).

Finally, the effect of human manipulation of a surface may be discussed. One of the most common manipulations is the flooding or draining of a land surface. Either procedure greatly alters the ratio of surface heat loss between *H* and *LE*. Surface energy budgets related to these activities and to certain kinds of vegetation management have been discussed by Budyko (1958, 1974). The effect of another common kind of surface manipulation may be demonstrated by quoting the energy budgets of a bare soil plot and the same soil covered by mulches of different types (Table E2). Net radiation is strongly influenced by the reflectivity (albedo) of the surface. Lower albedos, usually associated with darker, less reflective surfaces, give rise to a higher absorption of solar radiation and thus higher values of net radiation. Consequently, compared to the bare soil, the black plastic mulch has higher Q^* values and the paper surface has lower Q^* values. The different surfaces have varying permeabilities to water vapor flow. This is reflected in the different amounts of latent heat loss with no loss for the black plastic surface and increasingly greater losses, respectively, with the paper, hay, and bare soil surfaces. The opposite effect is seen with the sensible heat loss. All of the mulches decrease the flow of heat into the ground, but the effect is least with the black plastic, partly because the surface receives more net radiation. Studies are now paying attention to the manipulation of surfaces at larger geographic scales. Bonan (1999) used simulation modeling to indicate a regional cooling of 1–2°C over the last century in the east and midwest of the United States due to enhanced *LE* values and increased winter albedo associated with the agricultural cultivation of large areas. This brings us to the regional and global scale.

Table E2 Energy budgets of bare soil and soil with various mulches, Hamden, Connecticut

Surface	Q^*	H	LE	G
Bare soil	642	363	195	84
Black plastic	711	635	0	42
Paper	432	349	42	35
Hay	607	488	84	35

Note: Units are W/sq m. Observations taken near noon in June.
Source: After Waggoner et al. (1960).

The regional and global-scale

The value of the energy budget approach in comparing the climates of different regions of the Earth is demonstrated by examining the variation of the values of the energy budget components over the annual period for stations representative of different climatic environments. Data for six stations whose annual energy budgets were calculated and first presented by Budyko (1958) are shown in Figure E5. On this figure the usually negative values of *LE* and *P* (*H*) are shown with a reversed sign (i.e. positive) in order to save space on the diagram. The normal convention is to regard heat traveling towards the Earth's surface (either from above or below) as positive and vice-versa.

The equatorial, continental climate of Manaos, Brazil, is notable for the high proportion of heat used in evapotranspiration. It is also significant that, owing to cloudiness, Q^* is relatively small, being surpassed in some seasons by that in the midlatitude climate of Paris, France, and equaled even, during the summer time, by the values in the subarctic climate of Turukhausk, Russia (also spelled Turukhansk). Energy budget values at Thanh Pho Ho Chi Minh (listed on Figure E5 as Saigon), Viet Nam, manifest the importance of a hot dry season in a monsoonal regime. During this season *LE* is reduced with a corresponding rise in *H* (for which the symbol *P* is used on

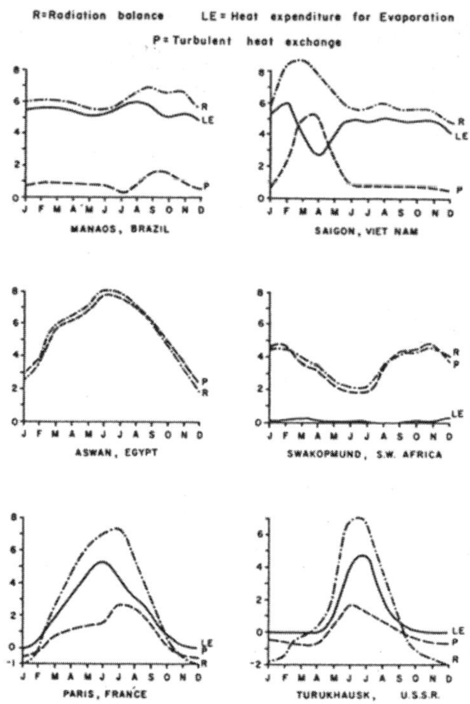

Figure E5 Heat balance regimes for different climates. Units are in kilocalories per square centimeter per month (after Budyko, 1958).

Figure E5), whereas in the wet season the energy budget is similar to that of Manaos. A distinct lack of surface moisture available for evapotranspiration is shown by the heat budget values for the tropical continental climate of Aswan, Egypt. It is significant here, however, that the highest value of net radiation is lower than that of Saigon owing to the greater albedo of the desert surface. Swakopmund (Namibia), in South West Africa, is the only southern hemisphere station presented here. Its energy budget is similar to that of Aswan, although the available heat (Q^*) is generally lower because of the higher latitude. A midlatitude maritime example is given by the energy budget of Paris, France, where Q^* and *LE* values follow their expected course. A noteworthy point arises here, in that, during the winter months, a net radiation loss is registered. This situation, if prolonged, would lead to very low air temperatures, but in this case it is ameliorated by the downward flow to the surface of sensible heat as shown by the negative values in the graph. This extra source of heat in the air is originally derived from the adjacent, warm ocean current. Thus the mild winters of the northwest European type of climate are guaranteed as long as the ocean remains warm. Such a state of affairs may be contrasted with that of the continental, subarctic climate exemplified by the energy budget of Turukhaust in Siberia. Here, although there is some downward transfer of sensible heat, the net radiation loss in the winter is even greater due to the high latitude of the location and the continual snow cover. The result is the severe winters of this continental climate characterized by very low air temperatures.

Although only a few features of the energy budgets of this selection of stations have been discussed, it is possible to see the value of the energy budget approach in comparing the climates of different areas. Possibly more important is the demonstration of many of the causal factors of these climates by placing emphasis on the physical processes at work in the fundamental link between the Earth's surface and the atmosphere. These links are of vital importance on a global basis as well. Budyko (1958, 1974) has published energy budget values for the whole globe on an annual basis and for selected months. The classic nature of this work is demonstrated by the fact that the pattern identified by Budyko is reconfirmed by later authors (e.g. Hartmann, 1994). The principal features of the global energy budget are now described briefly.

Global values of net radiation show a rather zonal (latitudinal) distribution as might be expected. Values over land are generally lower than those over water. The maximum difference is in the tropics, with the discrepancy becoming less as latitude increases. A distinction between land and sea surfaces is also exhibited in latent heat flux. Over land *LE* values are a maximum near the equator, decrease in the subtropics, increase in midlatitudes and decrease again at high latitudes. Over the oceans *LE* values are at a maximum in the area under the subtropical high-pressure zones, remain fairly high in midlatitudes, especially where there are warm ocean currents, and then decline in high latitudes. Maps of sensible heat flux show maximum values over land in the subtropics and decreasing values toward both higher and lower latitudes. Over the sea this flux increases steadily with latitude.

The global heat energy budget is the link between the surface of the Earth and the atmosphere. Together with an understanding of the general circulation of the air flow across the globe, the concept helps not only to explain present-day climates but also provides a framework for discussion of possible past and future changes of the climate of the Earth.

Energy budget climatology and climatic change

Studies of the planetary energy budget have played an important role in the development of climate models because exchanges of energy and matter at the Earth's surface are so important to explaining the climate of the planet. These models have, in turn, been used to further our understanding of the Earth–atmosphere climate system and its possible changes over time. Simple one- or two-dimensional energy balance models are still used for approximate estimations of possible climate change as well as for instructional purposes. However, most current estimation of possible climate change is made with General Circulation Models (GCM) that couple the ocean and atmosphere dynamics and are run in a transient mode (Houghton et al., 2001). Surface energy budget climatology could be said to be at the heart of these models because it is through surface processes that the coupling between land and water surfaces and the atmosphere is achieved. However, the surface energy budget climatology is not as "visible" in the GCM as it used to be in the simpler models.

Trends in energy budget climatology

The fundamental importance of energy and mass exchanges between the Earth's surface and the atmosphere is now so well acknowledged that no other type of climatology can ignore it. There are several channels for research endeavors in the field. Some of the most productive are: (1) further development of land-surface parameterization (LSP) techniques for complex surfaces, (2) the further refinement and application at the mesoscale of three-dimensional surface energy budget models, and (3) the incorporation of satellite data into energy budget modeling systems.

Current LSP efficiently combine biology and atmospheric physics. Stocker (2001) describes how the models and parameterizations provide more opportunities for parameter calibration, constraint, process submodel validation, and "upscaling" algorithms through comparison with observations. These models may be used with global datasets of vegetation type at monthly time steps and a spatial resolution of about one degree of latitude and longitude. The vital importance of the two-way interaction between vegetation and atmosphere has now been recognized at all spatial scales and it is likely that future work will provide further revelations related to this interaction. Three-dimensional mesoscale models have seen a rapid development and are now being used for climate as well as meteorological purposes (e.g. Liston and Pielke, 2001). These models have also shown the importance of the nature of the ground surface in affecting such things as the location of the development of convective storms. Finally, the rapid development of satellite technology has revolutionized energy balance climatology by providing formerly unavailable estimates of energy balance components over large areas. The technology has also, to a certain extent, caused greater attention to be given to certain energy budget factors, such as albedo and surface resistance, in ways that were not previously contemplated. Undoubtedly, further huge advances will occur in all of the above areas, and many more, as the twenty first century continues.

David Greenland

Bibliography

Anderson, F.E., Snyder, R.L., Kyaw Tha Paw, U., and Drexler, J., 2002. An energy budget evaluation of a restored California delta ecosystem using the eddy-covariance method in comparison with the surface renewal method. Preprints of the 15th American Meteorological Society Conference on Biometeorology and Aerobiology. Boston American Meteorological Society, pp. 401–402.

Aisenshat, B.A., 1966. Investigations on the heat budget of Central Asia, in *Sowremennye Problemy Klimatologii* (M.I. Budyko, ed.). Leningrad: Meorol. Gidrol. pp. 83–129.

Bonan, G.B., 1999. Frost followed the plow: impacts of deforestation in the climate of the United States. *Ecological Applications*, **9**: 1305–1315.

Budyko, M.I., 1958. *The Heat Balance of the Earth's Surface*, (N. Stepanova, trans.). Washington, DC: Office of Technical Services, US Department of Commerce.

Budyko, M.I., 1974. *Climate and Life*, (D. H. Miller, ed.). New York: Academic Press.

Campbell, G.S., and Norman, J.M., 1998. *An Introduction to Environmental Physics*. New York: Springer.

Eugster, W., Rouse, W.R., Pielke, R.A., Sr., et al., 2000. Land–atmosphere energy exchange in Arctic tundra and boreal forest: available data and feedbacks to climate. *Global Change Biology* **6**(Suppl. 1): 84–115.

Gates, D.M., 1980. *Biophysical Ecology*. New York: Springer-Verlag.

Grimmond, C.S.B., and Oke, T.R., 1995. Comparison of heat fluxes from summertime observations in the suburbs of 4 North-American cities. *Journal of Applied Meteorology*, **34**(4): 873–889.

Grimmond, C.S.B., and Oke, T.R., 2002. Turbulent heat fluxes in urban areas: observations and a local-scale urban meteorological parameterization scheme (LUMPS). *Journal of Applied Meteorology*, **41**(7): 792–810.

Hahmann, A.N., and Dickinson, R.E., 1997. RCCM2-BATS model over tropical South America: applications to tropical deforestation. *Journal of Climate*, **10**: 1944–1964.

Hartmann, D.L., 1994. *Global Physical Climatology.* San Diego, CA: Academic Press.

Hastenrath, S., and Lamb, P.J., 1978. *Heat Budget Atlas of the Tropical Atlantic and Eastern Pacific Oceans.* Madison, WI: University of Wisconsin Press.

Hastenrath, S., and Lamb, P.J., 1979. *Climatic Atlas of the Indian Ocean.* Part II: *The Oceanic Heat Budget.* Madison, WI: University of Wisconsin Press.

Holland, J.Z., 1971. Interim report on results from the BOMEX core experiment, *Bomex Bulletin*, **10**: 31–43

Houghton J.T. et al. (eds). 2001. *Climate Change 2001: the scientific basis.* Cambridge: Cambridge University Press.

Kalanda, B.D., Oke, T.R., and Spittlehouse, D.L., 1980. Suburban energy balance estimates for Vancouver, B.C., using the Bowen ratio-energy balance approach. *Journal of Applied Meteorology*, **19**(7): 791–802.

Liston, G.E., and Pielke, R.A., 2001. A climate version of the Regional Atmospheric Modeling System. *Theoretical and Applied Climatology*, **68**: 155–173.

Long, I.F., Montieth, J.L., Penman, H.L., and Sziecz, G., 1964. The plant and its environment. *Meteorol. Rundsch*, **17**: 97–101.

Miller, D.H., 1965. The heat and water budget of the Earth's surface, *Advances in Geophysics*, **11**: 176–302.

Munro, D.S., 1975. Energy exchange on a melting glacier, Ph.D. diss. Hamilton, Ontario: McMasters University.

Ohmura, A., 1982. Climate and energy balance on the alpine tundra. *Journal of Climatology*, **2**: 65–84.

Oke, T.R., 1987. *Boundary Layer Climates*, 2nd edn. London: Methuen.

Rosenberg, N.J., Blad, B.L., and Verma, S.B., 1983. *Microclimate: the biological environment*, 2nd edn. New York: Wiley.

Sellers, P.J., and Hall, F.G., 1992. FIFE in 1992: results, scientific gains, and future research directions. *Journal of Geophysical Research*, **97**(D17): 19,091–19,109.

Sellers, P.J., Hall, F.G., Margolis, H., et al., 1995. The boreal ecosystem–atmosphere study (BOREAS): an overview and early results from the 1994 field year. *Bulletin of the American Meteorological Society*, **76**(9): 1549–1577.

Sellers, W.D., 1965. *Physical Climatology.* Chicago, IL: University of Chicago Press.

Stocker, T.F., 2001. (Lead author). Physical climate processes and feedbacks. In Houghton, J. et al., eds., *Climate Change 2001: The Scientific Basis*. Cambridge: Cambridge University Press, pp. 417–470.

Suckling, P.W., 1980. The energy balance microclimate of a suburban lawn. *Journal of Applied Meteorology*, **19**(5): 606–608.

Vowinckel, E., and Orvig, S., 1968. *A Method for Calculating Synoptic Energy Budget*. Publications in Meteorology, No. 93. Montreal: Arctic Meteorology Research Group, Department of Meteorology, McGill University.

Waggoner, P.E., Miller, P.M., and De Roo, H.C., 1960. *Plastic Mulching – Principles and Benefits*. Bull. No. 634. New Haven, Conn.: Connecticut Agricultural Experimental Station.

Waring R.H., Law, B.E., Goulden, M.L. et al., 1995. Scaling gross ecosystem production at Harvard Forest with remote-sensing – a comparison of estimates from a constrained quantum-use efficiency model and eddy-correlation. *Plant Cell and Environment*, **18**(10): 1201–1213.

Yap, D., and Oke, T.R., 1974. Sensible heat fluxes over an urban area – Vancouver, B.C., *Journal of Applied Meteorology*, **13**: 880–890.

Cross-references

Albedo and Reflectivity
Atmospheric Circulation, Global
Bioclimatology
Boundary Layer Climatology
Climatology
Dynamic Climatology
Greenhouse Effect and Greenhouse Gases
Radiation Laws
Solar Radiation

EUROPE, CLIMATE OF

Analysis of climatic data

Most climate data are assembled and kept in databanks held at regional climate centers or in research centers. The data are then analyzed using powerful statistical tools that enable the researcher to search for patterns and correlations. An example of one such data set is NACD – the North Atlantic Climatological Dataset containing temperature and precipitation series. Regional or subregional datasets are then compiled and adjusted to build a homogeneous dataset (Tuomenvirta, 2001). Other regional datasets are compiled for use in climate analysis computational programs covering a wide range of aspects from global warming through to hydrological management. This item has been compiled from a combination of descriptive sources and of analytical studies. In places there is some difficulty in linking together the climatic ensemble and this must be borne in mind by the reader. Many datasets are now available on websites.

Factors affecting the climate of Europe

The continent of Europe extends from the mid-Atlantic islands of the Azores (29°W) across the open North European Plain to the Ural Mountains of the Soviet Union (60°E) and from the islands of the Mediterranean Sea in the south (35°N) to the islands of the Barents Sea, which lie well within the Arctic Circle. Europe, along with its islands and peninsulas, is the most maritime of the continents. Large areas to the west of the 20°E longitude experience a temperate oceanic climate where extremes of temperature and rainfall are rarely experienced. The chief factors promoting this comparatively unique assemblage of climates are:

1. The predominant west-to-east movement of weather systems.
2. The extensive area of abnormally warm surface waters of the North Atlantic.
3. The virtual absence of north–south-aligned mountain ranges between latitudes 45°N and 60°N that might otherwise alter the nature of the westerly flow.
4. The presence of large inland seas such as the warm Mediterranean, the cooler Baltic, and the smaller, less warm, Caspian and Black seas.

Elements of the climate of Europe

Energy

In common with other midlatitude regions, Europe experiences a net deficit in radiation except at its southern margins. Values derived from satellite information for the Earth–atmosphere system indicate that annual deficits increase from $-20\,\mathrm{W\,m^2}$ over central Italy to around $-80\,\mathrm{W\,m^2}$ over southern Scandinavia and northern Norway (Raschke et al., 1973). The deficit cannot be attributed principally to the excessive loss of heat from the land surface, though this loss may be significant in cold snowy winters over central and eastern Europe. Compensation for this deficit in radiant energy is effected by the transfer of sensible and latent heat from lower latitudes. Radiative cooling is at its maximum at mid-tropospheric levels around latitude 60°N, and it is in this region that maximum cooling takes place in winter (Figure E6). The effects of this cooling on European climates are profound. In the long term such cooling produces latitudinal zones of temperature gradient in the atmosphere. These zones are associated with high winds and disturbed weather conditions so frequently experienced over Europe.

Temperature

The influence of the warm waters of the Atlantic Ocean is most clearly seen in the winter pattern of surface temperatures (Junge and Stephenson, 2003). Average January values of 7°C over western Ireland decline steadily eastward to around minus 10°C in the vicinity of Moscow, producing a north–south thermal pattern between latitudes 45°N and 60°N. In summer the heating of the continental area leads to a more zonal thermal pattern with isotherms tilted WSE–ENE (see Figure E7). The 17°C isotherm for July (2005) now lies through southern England across to the southern Baltic Sea toward Leningrad. Summers in north European Russia, though warm, are usually short and are constantly being threatened by outbreaks of cold air of Arctic origin.

A measure of the increasing warmth from north to south through European Russia has been visualized using summer daily air temperatures that exceed 10°C. Accumulated summer temperature values of 750°C occur at around 70°N, increasing to 2000°C at 55°N and exceeding 3000°C around the Black Sea

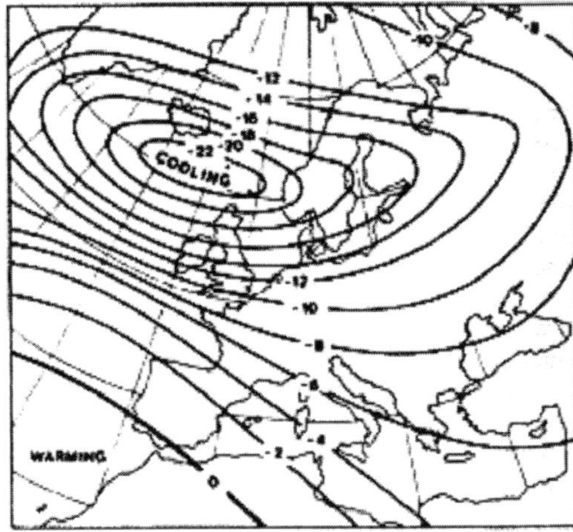

Figure E6 Change in the heat content of the troposphere from mid-November through mid-December in W/m². Average values for 1955–1959. Values are relatively small with maximum cooling representing 1.8% of the solar constant (after Crowe, 1971; Tucker, 1965).

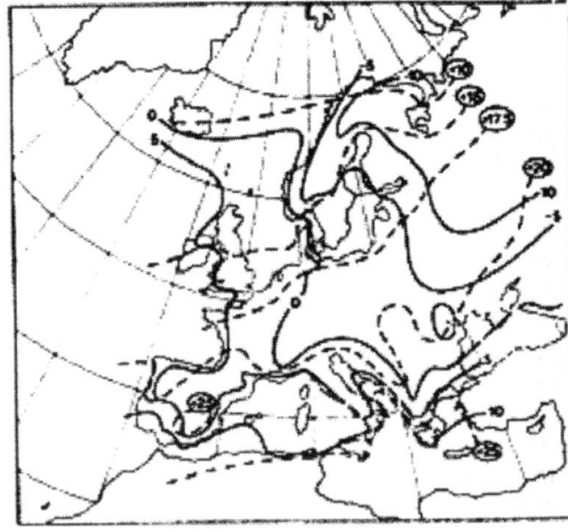

Figure E7 Mean temperature for January (solid lines) and July (dashed lines) in °C.

at 45°N (Budyko, 1974). In the cloudier climate of the British Isles these values are usually lower. They range from below 500 in northern England (55°N) and approach 1000 in the Channel Islands (49°N). Using this method the equivalent value for Moscow would be 722. Analysis of temperature data by Gerasimov (1964) showed that average daily temperatures rose above 10°C by 21 April in the Crimea, but not until July along the coastlands of the Barents Sea. The average date when temperatures have fallen below 10°C in the fall have been 15 of October for the Crimea and 11 August for North Russia. Using a slightly different definition of the growing season, Mitchell and Hulme (2002) show that for Central England there has been an increase in length of the growing season over the twentieth century of 28 days, much of this in the period 1980–2005. Recent research by Zhou et al. (2001) and Menzel and Fabian (1999) has confirmed that this is also true for most of Europe. A more detailed daily temperature dataset has been compiled for the ECA (European Climate Assessment) at De Bilt in The Netherlands by Klein Tank et al. (2002), and shows a good correlation with existing gridded data sets for Europe.

Frost-free period

Closely associated with the growing season is the average length of the frost-free period. Over the open plains of European Russia this increases from at least 75 days in the north to around 200 along the northern shore of the Black Sea. Further west, across the European Plain, the length of the frost-free period becomes less easy to determine because of the alternating airflows from cold and warm sources (Lednicky, 1985). Over lowland France mid-twentieth-century values have ranged from 160 days in Alsace in eastern France to 240 days along the coastlands of the Bay of Biscay (Garnier, 1954). In contrast, the determination of the frost-free period in mountainous countries such as Austria is largely a function of elevation. In the Austrian Alps the frost-free season at 800 m above mean sea level lasts for about 155 days per year. At a height of 1600 m this is reduced to 104 days, whereas above 2500 m spells of below freezing temperature may even be expected in cold summers. In the Iberian Peninsula the frost-free period is governed both by altitude and proximity to the Mediterranean Sea and Atlantic Ocean. Much of interior Spain may experience less than 250 days frost-free. Only the extreme southeast and southwest of the peninsula are likely to be entirely free of frost during most years.

Extreme values

Extreme values of temperature tend to reflect the local controls of topography as well as distance from the North Atlantic Ocean. Very low temperatures show a much greater variation in time and space than high temperatures. Very dry, cold air with little wind is required for temperatures to fall below −15°C. Such an occurrence is far more common in winter over countries such as Finland, where temperatures below −30°C are recorded nearly every winter. The lowest temperature indicated in the standard climatic tables is −51°C at Syktyvkar (61°40′N, 50°51′E) in northeast European Russia. Moscow, along with many stations in northern Russia and Finland, has experienced temperatures of −40°C at screen height. The penetration of cold air into parts of eastern Europe also means that very low temperatures have occurred at relatively low latitudes. This may be seen from cities such as Iasi (Rumania) at 47°N, recording −30°C and Sofia (Bulgaria) 42°30′N, −27.5°C. There were also reports of −26°C from several locations in Thrace (Greece) in December 1972. Only the southernmost part of Spain and a part of the Mediterranean island of Crete have absolute minimums at or above 0°C. Comparatively, maritime countries such as Spain experience frost in severe winters. Madrid recorded −10.1°C during the period 1901–1930, and Rome (Italy) reported −7.4°C. Over the British Isles, minimums of −10°C have been recorded along the western seaboard. Inland values of −25°C are very rare, but were registered in the Dee Valley of the Grampians in Scotland and in the West Midlands of England during the winter 1981–1982. An extensive climatology of extreme temperatures has been assembled by Yan et al. (2002).

In contrast, there is little variation in absolute maximums. Over much of northern Europe, temperatures have occasionally climbed into the low 30s (Jenkinson, 1985) whilst over the remainder of Europe temperatures between 35°C and 39°C have sometimes occurred, as in the summer of 2003 (Eden, 2003). Apart from isolated events, temperatures in excess of 40°C have only been recorded south of latitude 42°N. Absolute maximums of 45°C (Palermo, Sicily) and 47°C (Malaga, Spain) testify to occasional incursion of very warm desert air from the Sahara. Trends in minimum temperatures (Weber et al., 1997) over the period 1901–1990 for Central Europe, show significant increases across the plains, whereas little change was found across the mountains.

Continentality

Various indices have been formulated to express the degree of continentality over Europe. In a general sense the indices make use of the range of the mean temperature between the coldest and warmest months of the year, and this is recorded as the annual range of mean temperature. The range increases from the Atlantic coastlands eastward across Europe into Russia (Figure E8). A relatively simple index has been used by Tsenker (in Borisov, 1965), who proposed that $K = A/\phi$ where K is the index, A is the annual range and ϕ the latitude. Similar values are obtained from Conrad's index (Conrad, 1946). More complex formulae appear in Russian literature, an example being that of Ivanov (1959).

Figure E8 Annual range of mean monthly air temperature in °C.

Using Conrad's equation a value of 4 is obtained for western Ireland, 12.5 for London, 24 for Berlin, and 40 for Moscow. Indices of over 30 are also found in parts of the Mediterranean countries, especially where inversions of temperature occur in winter such as in North Italy (Milan: 32). A different index may be calculated for an individual year to describe the degree of Atlantic influence, such as by adding the number of days of westerly flow either using Lamb's weather types or the North Atlantic Oscillation Index (Jones and Hulme, 1997).

Precipitation

Annual precipitation

Values of annual lowland rainfall show little spatial variation in the zone stretching from the English Midlands through to the Urals of the USSR (Figure E9). The passage of fronts usually ensures moderate precipitation across the European Plain, excessive falls being rare and usually associated with slow-moving disturbances. A slow decline from 640 mm to 500 mm can be observed eastward, and is evidence of the ease with which moist Atlantic air can penetrate across Europe. A few areas such as Bydgoszcz (northwest of Warsaw) and coastal regions of Lithuania record annual values less than 500 mm. As a general rule, annual rainfall over lowlands declines both northward toward the Arctic and southward toward the Mediterranean. Everywhere the existence of hills, mountains, enclosed basins, and peninsulas complicates this simple pattern. Highest annual values are found in mountainous areas where valleys are open to the southwest or westerly airflow. A few localities have recorded in excess of 5000 mm mean annual precipitation for the period 1961–1990, notably south of Nordfjord (Norway) and Sprinkling Tarn (English Lake District) and possibly inland from Boka Kotorska near the Yugoslav–Albanian border. A value of 4000 mm may have occurred at Monchsgrat in the Bernese Alps, although heavy winter snow makes measurement difficult. It is probable that one or two localities in the Bavarian Alps (West Germany) and in the Sierra de Gredos (west of Madrid, Spain) receive 3000 mm, whereas 2000 mm would seem feasible for the highest parts of the Central Massif and Jura mountains in France.

Seasonal pattern

Invasions of moist air across Europe are most frequent when the circulation is vigorous and depressions are moving quickly east across Scotland and Scania into the Baltic region. Such a situation is most likely to occur in early winter and is least likely in spring or early fall when anticyclones are usually present over some parts of Europe. In fact, over much of Europe the percentage of annual precipitation falling in spring is remarkably constant, lying between 16% and 21% (Table E3). A winter maximum in rainfall might well be expected to occur fairly widely. That this is not so indicates the many other factors that influence the seasonal distribution. In particular, higher summer temperatures over much of central and eastern Europe can lead to enhanced convection. This shower activity may be random if pressure gradients are weak, or may be organized along cold fronts as incursions of cool maritime polar air from the Atlantic travel southeast over much of the region. As may be seen from five of the stations listed in Table E3, a summer maximum occurs over eastern England and central, eastern, and northern Europe. A zone of fall maximum occurs from Portugal through

Figure E9 Mean annual precipitation in cm.

Table E3 Seasonal precipitation as a percentage of the annual precipitation (1931–1960)

	Latitude	Winter (DJF) (%)	Spring (MAM) (%)	Summer (JJA) (%)	Fall (SON) (%)	Annual total (in mm)
Northern Europe						
Bergen (west coast)	60°23′	27	17	22	34	1958
Stockholm (central)	59°17′	22	16	33	29	555
Arkhangelsk (northeast)	64°40′	18	18	33	31	539
Central Europe						
Valentia	51°56′	31	19	20	30	1398
Brussels	51°51′	26	17	29	27	817
Warsaw	52°13′	19	19	39	22	471
Moscow	55°45′	17	21	37	25	575
Southern Europe						
Madrid	40°25′	28	30	12	30	436
Rome	41°54′	35	20	6	39	749
Athens	37°58′	42	21	7	30	402

western France, central Britain, the coastlands of the Low Countries and Norway (see Bergen, Table E3). In a climatic context this maximum is the result of relatively high sea surface temperatures leading to upward transfer of water vapor into the boundary layer. The moisture may then be ingested into the circulation of midlatitude depressions via a conveyor belt mechanism (Barry and Chorley, 2003) These depressions become more vigorous and active from September through November across central and northern Europe.

The Mediterranean region with its mountains, peninsulas, islands, and inland sea presents a more complex pattern of seasonal precipitation (Figure E10). It is more convenient to state the driest period of the year, which is summer, when extensions of the subtropical high pressure lie over the region. During the remainder of the year, although frontal depressions can sometimes be identified, many disturbances are ill-defined and rather transitory. Moreover, their activity in terms of producing rain varies on both a seasonal and annual basis. One or two very wet Octobers can lead to an imprint on the climatic record that can be misleading if arithmetic averages are employed. Bearing this in mind, it is generally true that a fall maximum is more likely in the western and northern Mediterranean and that a winter maximum is more evident in the southern and eastern Mediterranean (see Athens, Greece, Table E3). A bimodal maximum (spring, fall) is a feature of much of Spain, at least when the westerly flow is relatively strong. Comprehensive analysis of seasonal rainfall frequency in the Mediterranean was carried out by Reichel and Huttary (in Trewartha, 1981). More recent investigations have employed indices and statistical analyses in attempts to describe present rainfall patterns and to search for regular periodicities (see Fukui, 1966; Tabony, 1981). It should also be noted that there is a tendency for the general characteristics of a rainfall regime to change over periods of 30 years. Thus fall may be considered "wet" in one decade, spring in the next (Boucher, 1994). More detailed gridded analysis of precipitation has recently been carried out for a 25 km grid over the entire region of the European Alps and then extended over the time period 1901–1990. This reveals an increase in winter precipitation of 20–30% per 100 years over the western part of the Alps and a decrease of 20–40% fall precipitation in the southern part of the Alps (Schmidli, 2002).

Types of rainfall and duration of heavy falls

Attempts have been made to classify rainfall into three categories of cyclonic (predominantly frontal uplift), convective (thunderstorm type), and orographic (uplift over mountains). It is not easy to distinguish clearly between these categories, though convective thunderstorms contribute significant amounts to summer rainfall over much of Europe. Precipitation over mountainous areas is enhanced by uplift of airmasses on windward slopes. Extreme maximum precipitation over periods of less than 24 hours is highest at lowland stations where convection is dominant. Continuous heavy rain depends on steady inflow of moisture into a region usually associated with slow-moving low-pressure systems. Recent analysis of heavy rainfall events in Central Europe (Ulbrich et al., 2001, 2003a,b) has shown that extreme rainfall periods have return periods of between 300 and 1000 years with rainfall totals over 5–10 days exceeding the monthly rainfall by 300%. Data from Ulbrich and others are shown in Table E4. Such widespread rainfall usually combines three components: uplift of moist air over mountains, vigorous wave activity along a quasi-stationary frontal zone, and convective instability. Two examples are given in Table E5, in which over 600 million cubic meters of rain fell over a limited area.

Thunderstorms and related phenomena

Over much of continental Europe, thunderstorms are typically a summer phenomenon. Within European Russia about 50% of the thunderstorms are associated with cold fronts, 22% with warm fronts, and 28% with convergence within airmasses. Thunderstorm frequency increases with the rise in surface temperatures in early summer and reaches a maximum in June in parts of southern European Russia; in July over much of central and northern Europe, including northern Italy; in August in parts of Sweden; in September and October in the western

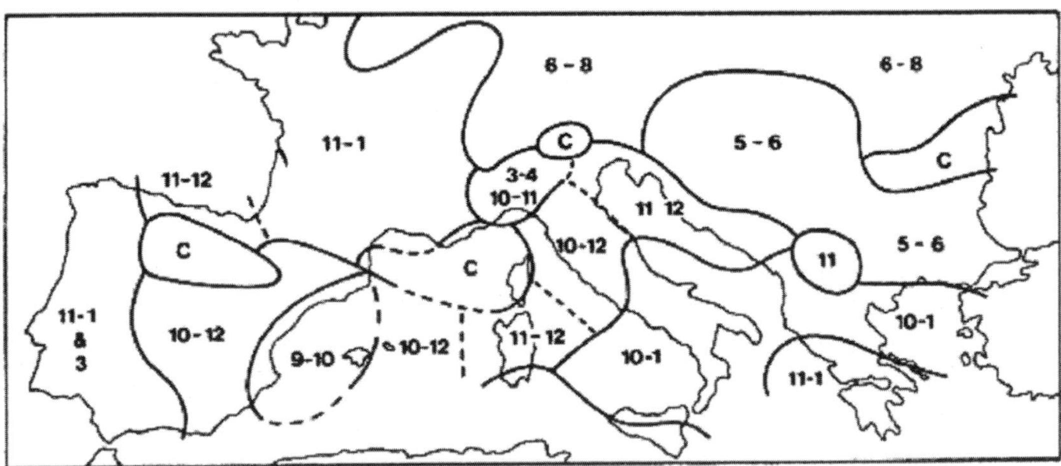

Figure E10 Seasonal distribution of maximum rainfall. The figure shows months of maximum rainfall over southern Europe based on World Meteorological Organization Data 1961–1990. Months are numbered from January (1) through December (12). Complex regimes with no distinct seasonal maximum are indicated by C.

Table E4 Maximum precipitation sample extreme values – Europe 1960–2005

Event	Location	Amount	Time period/date	Notes
1	Northampton, England	48 mm	15 hrs/9 April 2002	Frontal
2	Mid-Thames, England	80 mm	27 Dec–3 Jan. 2002–3	Frontal systems
3	Lower Elbe, Austria	610 mm	6–7 Aug. 2002	Low moving East
4	Devon, England	300–500 mm	15 Aug. 1952	Thunderstorm
5	Udine-Rivolto, Italy	44 mm	4 Aug. 2002	Small mesoscale convection
6	North Wales, UK	417 mm	30 Jan.–5 Feb. 2004	6 days of cyclonic rain
7	Majorca, Spain	200 mm	5 Sept. 1989	Cyclonic storm–Cold pool

Table E5 Data relating to two notable rainfall events with return periods in excess of 100 years

Central and northern Rumania	12–14 May 1970	Duration: 48 hours
Isohyet	Area enclosed	Rainfall in m^3 × 10^9
25 mm	50 000 km^2	2.5
100 mm	6000 km^2	0.66
Scottish Highlands	26–27 March 1968	Duration: 36 hours
Isohyet	Area enclosed	Rainfall in m^3 × 10^9
100 mm	12 500 km^2	1.96
200 mm	2700 km^2	0.64

Sources: After *Doneaud (1971)*, p. 30, and *Bleasdale (1975)*, p. 225.

Mediterranean and in winter over southern Italy as well as parts of northwestern Britain.

The occurrence of hail is most closely associated with air-mass thunderstorms, especially where these develop in hilly or mountainous regions. They may be accompanied by squally winds that are best attributed to storm downdraughts. Small-scale tornadoes occasionally accompany storms, but nowhere have they been reported to attain the magnitude and frequency of those in the United States. Over western Europe they may be closely associated with cold fronts at any time of the year. Their tracks rarely exceed 1 km. There have been fairly frequent sightings of water spouts over the western and central Mediterranean during thundery weather. Small-scale whirlwinds and tornadoes occur fairly regularly during summer months.

Snowfall

The relatively high sea surface temperatures in winter mean that snowfall in winter is variable over low ground in western Europe. Snow cover may persist for more than 7 days only in the coldest winters. As a general rule, monthly average temperatures must lie below 1°C for snow cover to persist. The number of individual days with complete snow cover increases northeast across the European Plain from 5 days in the Midlands of England to 40 over central Poland, 135 days around Moscow and up to 200 along the Arctic fringe. The mean maximum depth of snow cover increases from 20 cm in southwest Russia to about 75 cm on the lowlands to the west of the Urals. Most mountain areas are liable to receive heavy snowfall, though aspect and character of individual winters along with recent rise in temperatures have reduced snow accumulations. Prolonged snowfall exceeding 25 cm over parts of western Europe is only likely to the north of an active frontal zone. Such a situation leads to very cold continental air being drawn west over northern Germany toward central Britain, while humid air is forced to ascend over the cold easterlies. Snow enhancement occurs if small disturbances run east, typically along the English Channel into central Germany. Polar lows occasionally bring blizzards to Scotland, and eastern parts of the North Sea, but their activity depends on unstable cyclonic disturbances developing in cold northerly airflows crossing warm ocean waters.

Drought

Lack of precipitation is a normal part of the seasonal climatic cycle within the Mediterranean in summer. An appropriate index is used to describe the rainless period. Aridity indices, such as that of De Martonne, have traditionally been employed by Italian and French climatologists to describe summer drought (Pinna, 1957). The existence of a rainless period is closely associated with midtropospheric subsidence. Therefore the existence and maintenance of summer drought depend on the strength and position of the Azores anticyclone and its eastward extension. When this anticyclone is well developed, drought is complete and may last for up to 4 months in southern parts of Spain, Italy, and Greece. Occasionally during summers such as those of 1959, 1976, 1992 and 2003 the belt of high pressure may be located farther north over northern France and the North Sea, bringing drought to areas usually well supplied with precipitation and delivering summer rainfall to the Mediterranean (Perry, 1976). Such annual variations are of deep concern, not only to countries such as Spain and Portugal (Estrela et al., 2000), but also to the grain-growing areas of Poland and Russia where the failure of adequate snowmelt and spring–summer rain may seriously reduce yields, as in 1972 and 2003. Hydrological droughts (low streamflow) have continued to be cause for anxiety (EEA, 1998) but show both spatial and temporal variations (Lloyd-Hughes and Saunders, 2002). Increasing drought deficit volumes have occurred in the eastern sector of East Europe and large parts of the UK, but decreasing deficits have characterized many parts of central Europe (Hisdal et al., 2001).

Severe droughts are associated with rainfall deficiencies of 50% over periods of 6 months or more. The impact of such droughts is greatest if they occur during the soil moisture recharge period (Hounam et al., 1975). As elsewhere, the intensity of the drought is increased by high winds, low humidity, and high temperatures. Areas likely to suffer such conditions lie along the southern border of Europe and across the steppes of the Ukraine and Kazakhstan (north Caspian Sea). In the latter area the steppe climate is characterized by extreme annual range of temperature approaching 80°C, frequent droughts, and desiccating, dust-laden winds known as *sukhovei*. During the period 1880–1950, serious drought occurred on average every 3.5 years

(Poltarus, in Borisov, 1965). The effect of one year's drought is magnified by the tendency of one drought year to follow another. Such was the case over much of European Russia from 1889 through 1892, and in southern Spain from 1917 to 1919.

Within the midlatitude belt of Europe the occurrence of drought is closely associated with the position, height and persistence of ridges in the middle part of the troposphere. Persistence of high pressure tendency over a number of years may produce general deficits in precipitation. The droughts of the early 1970s in European Russia prompted a number of studies (Buchinskiy, 1975; Chistyakova, 1975). In an earlier investigation Davitaya (1958) found that droughts over southern Russia reflected shifts in the midtropospheric ridge–trough pattern. A thorough investigation by Rauner (1980) into the simultaneous recurrence of drought in different regions of Europe revealed that drought affected the whole of the grain zone of the USSR in 5 years during the period 1891–1975. On a regional basis droughts were found to have affected the Ukraine in 43 out of 106 years (1870–1976) and the Volga region, a 1000 km further east, in 41 out of 96 years (1880–1976) of which 19 years were common to both regions. The occurrence of drought simultaneously in European Russia and Western Europe evidently has taken place in 20 years during the period 1700–1976, three of which occurred consecutively from 1747 to 1749.

Climate change

Rising surface temperatures in the European sector have been shown by Parker and Alexander (2002) whilst their annual analysis of rainfall reveals high variability on time scales of 10 years with no easily discernible trends.

Synoptic climatology

Weather types

Regional assessment of climate has often been carried out on the basis of a classification of air movement and associated weather conditions. British research (Lamb, 1972) has favored a less rigorous system than a number of German and Russian climatologists (see Hess and Brezowsky, 1969; Chubukov, 1977). Despite its limitations, an example of such an annual regime is included. Other circulation indices include that of Namias (1950) and some variants used by Dole and Gordon (1983). Research has also focused on indices that affect part of, or all of, the northern hemisphere, such as the North Atlantic Oscillation (NAO). This fluctuation in pressure gradient between the Icelandic low (65°N) and the Azores high (40°N) is believed to influence weather patterns over much of Europe in winter (van Loon and Rogers, 1978). Since 1989 the NAO has been mostly positive, leading to warmer winters over many parts of Europe (Barry and Chorley, 2003).

Spells of weather

The term "singularities" has been applied to periods of distinctive weather lasting between 7 and 14 days that have a tendency to recur in most years on or about the same dates. Trewartha (1981) associates this calendar of weather episodes with the *Grosswetterlagen* of German climatologists. For a certain region of Europe a calendar sequence of probable types of weather may be assembled. In essence, singularities represent adjustments of the general circulation to various distributions of energy within the boundary layer and in the troposphere above. A singularity such as an anticyclonic spell may be regarded as a statistical probability of a circulation pattern occurring for a given period of the climatic record. An example of such a calendar is presented in Table E6, which refers chiefly to the British Isles. Most singularities shown also occur over Germany; however, the region of European Russia lies too far downstream in the westerlies for statistical correlations to be significant.

Due to the ridge–trough nature of the air flow at 5000 m periods of anticyclonic activity over western Europe are likely to be associated with cyclonic activity over eastern Europe. It might appear from Table E6 that periods of cyclonic activity alternate

Table E6 Abbreviated calendar of singularities

Dates	Singularity	Probability of occurrence over British Isles	Comments for Germany
11/22–12/10	Early winter storms	>50% (1)	zonal westerlies
12/17–12/24	Continental anticyclones	25%	>60%
12/26–1/12	Storms of midwinter	westerly 60%	westerlies 72%
1/19–1/25	Continental anticyclones	25% (5)	>75%
1/27–2/4	Renewed winter storms	variable (1)	precipitation peak 70%
2/7–2/22	Anticyclones	25%	>60%
2/26–3/9	Cold spell	50%	no regular features
3/12–3/22	Anticyclones	35% (1)	>60%
3/30–4/15	Atlantic depressions	westerly 40% (5)	rainfall peak
4/16–5/20	Spring northerlies	25%	3 days 75% (1)
6/10–6/30	Increase in Westerlies	westerly 52% by 20th	westerlies 80%
7/15–8/15	Unstable cyclonic	30–35%	westerlies >80%
8/17–9/2	End of summer	westerly 50%	rainfall peak
9/6–9/19	Anticyclones	30–40%	>70%
9/21–9/30	Atlantic storms	25%	anticyclones >70%
9/30–10/15	Anticyclones	peak 40% (5)	part > 70%
10/24–11/13	Late autumn rains	westerly 35%	anticyclones >60%
11/15–11/24	Anticyclones	30% (1)	anticyclones >70%

Notes: Statistically tested figures exceeding the level of 5% probability of occurrence are indicatd by (5) and 1% probability by (1) usually for part of the period indicated in column 1. Length of the record in total was from 1873 to 1961. Some figures refer to parts of this period.
Source: After *Lamb (1964)*, pp. 176–188.

with anticyclones, but such a deduction can be misleading. The singularity table is not a forecasting tool. For reasons not well understood, certain singularities become statistically significant for limited periods of the record before declining, even disappearing, for a number of years only to re-emerge at a later period. It is interesting to postulate that such changes form part of a climatic signal in the atmospheric system, indicating fundamental adjustments of the general circulation (see Lamb, 1982).

The general circulation

Pressure patterns at the surface

The mean centers of both anticyclonic and cyclonic activity lie over the Azores islands, west of Portugal, and southwest of Iceland respectively – outside continental Europe. In periods when the westerly flow is well established, ridges and troughs move eastward away from these centers across central Europe. These may be associated with closed isobaric highs and lows at the surface. On the average monthly mean pressure charts for 1951–1966 (Meteorological Office, 1975), the shape of the North Atlantic low changes, being at its most intense in December (997 mb); thereafter it declines until July when there is no easily identifiable center of low pressure close to Europe. The Azores anticyclone appears weakest in March (1020 mb) and lies equatorward of 30°N. It reasserts its dominance and is at its most extensive in July (1025 mb) with ridges extending east and northeast across Europe (Figure E11). It has been noted that July is also the wettest month over much of Europe, indicating that breakdowns in the high pressure are frequent in some summers. The greatest monthly variation in pressure during the period 1951–1966 occurred west of Ireland in February and November, with standard deviation values of 11 mb and 8 mb, respectively. This indicates marked changes in flow patterns in these months from one year to another, indicating the variable nature of West European climate (Tables E7 and E8).

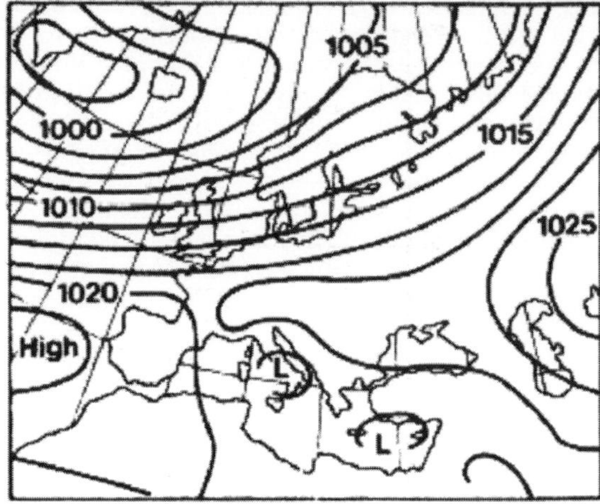

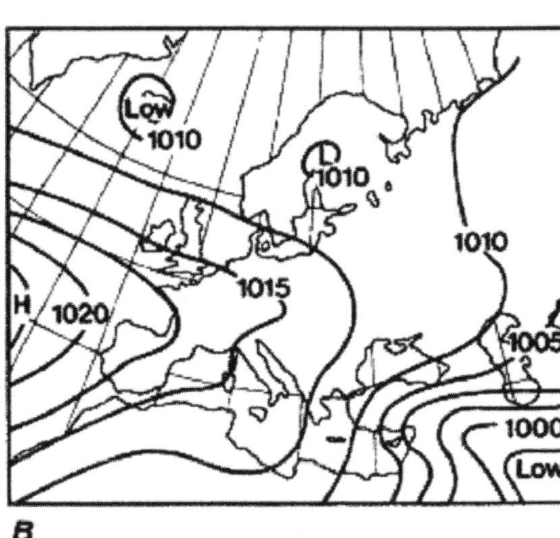

Figure E11 Normal sea level pressure in mb for (***A***) January and (***B***) July (after Meteorological Office, 1975, pp. 110–111).

Table E7 Meridional mean pressure difference 40°N to 60°N in mb at the surface

	20°W	0	20°E	40°E
January	15.2	9.0	3.7	3.0
April	11.2	2.5	−1.0	−4.0
July	11.5	6.2	2.1	−0.3
October	14.5	7.0	3.7	3.0

Note: Positive values represent high pressure at 40°N.

Table E8 Meridional mean difference in the height of the 500 mb level in geopotential meters between 40°N and 60°N

	20°W	0	20°E	40°E
January	140	82	68	76
July	92	86	66	60

Note: Positive values indicate westerly flow.

The jet flow across europe

Mean airflow over Europe in the troposphere is predominantly from the west between high pressure to the south and low pressure to the north. Imposed on this pattern are waves of large amplitude having a wavelength of several thousand kilometers. Their position, movement, and intensity produce an ever-changing pattern of airflow. Embedded in this flow lie narrow zones of fast-moving air known as jet streams that extend downstream for several thousand kilometers. They may be about 400 km wide, and are associated with airmass boundaries of the Polar Front. They appear and disappear with the changes in the circulation pattern. Within the core of the Polar Front jet, winds may exceed values of 60 m/s over Europe.

It is not possible to represent a mean position of the Polar Front jet, but some idea of its strength and location at 700 mb is given in Figure E12, which shows the axes of maximum wind at around 3000 m. The direction and movement of midlatitude depressions in any one month are closely associated with the mean position of the Polar Front jet during that month. The Subtropical jet is far more persistent. Its mean winter position lies over the central Sahara. Figure E13 shows cross-sections of the average zonal wind components in January and July: (1) 40°E; (2) 15°E; and (3) 10°W. The position of the Polar

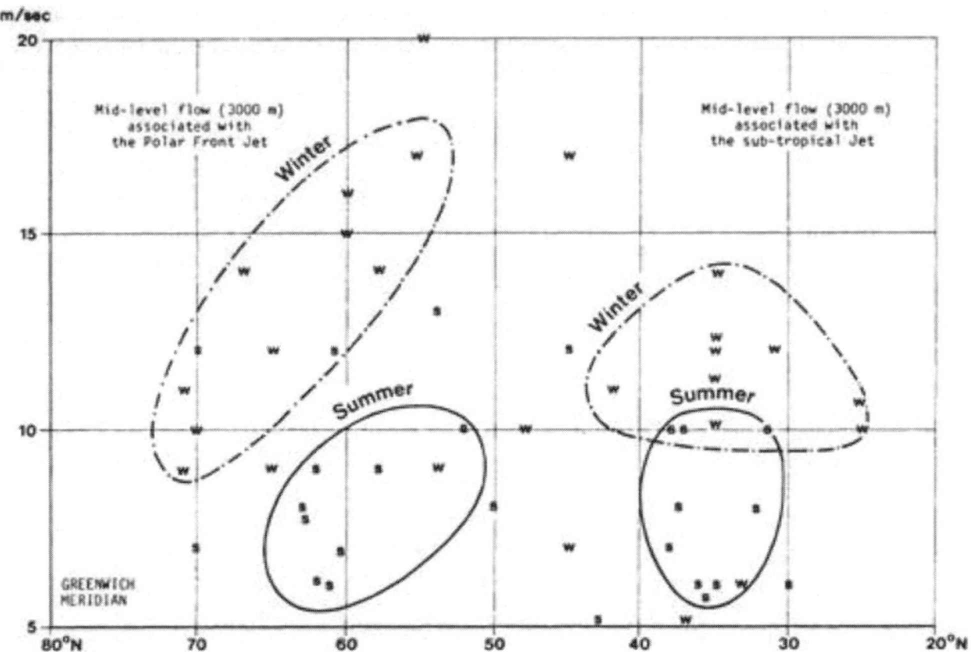

Figure E12 Envelopes showing the position and strength of the monthly mean maximum geostrophic wind at 700 mb for winter (December to February) and summer (June to August) along the Greenwich Meridian. Plots for individual months are also shown for 1969–1975 (partial record).

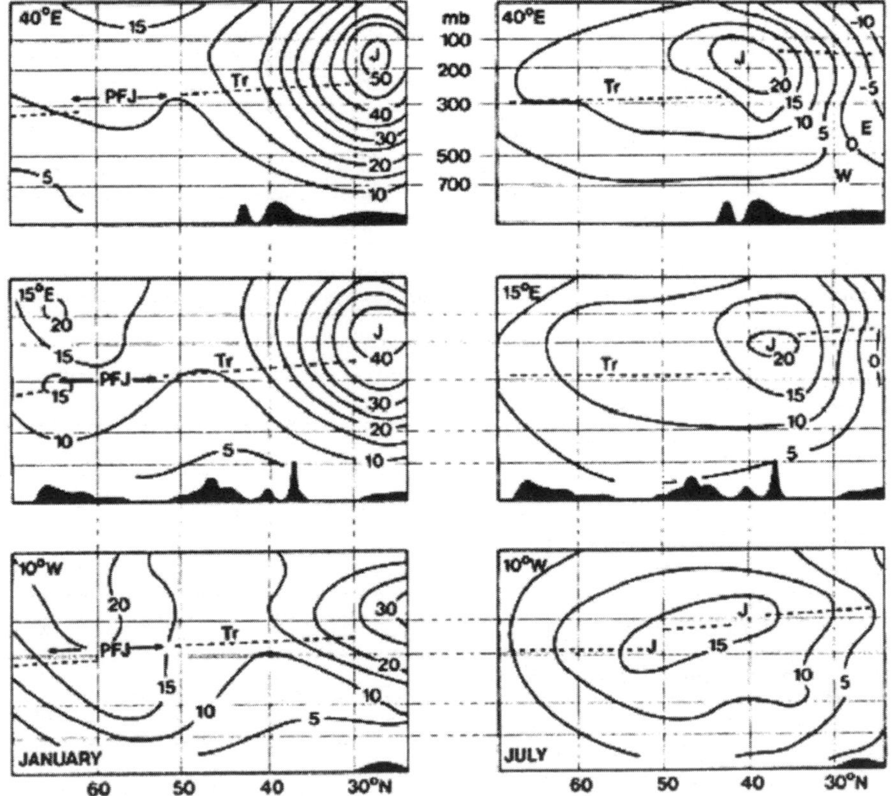

Figure E13 Meridional cross-sections at 10°W, 15°E, and 40°E showing the westerly (zonal) component of the wind in meters per second. Positive speeds represent westerly wind, negative values easterly wind. Schematic position of the tropopause (Tr) is also shown (after Meteorological Office, 1975, p. 121). PFJ indicates the likely position of the Polar Front Jet.

Front jet is purely schematic. In winter it may be found anywhere between 30°N and 70°N, but it is too ephemeral to imprint its presence on average maps of zonal flow. In summer only one jet flow is recognizable at 300 mb over Europe even if there is a suggestion of two axes at 700 mb.

Blocking patterns

One of the most frequent interruptions to the westerly flow over Europe results from "blocking" high-pressure patterns developing either upstream in the mid-North Atlantic or over Europe. The presence of a large anticyclone, persisting on average for 16 days, causes disruption of and breakdown in the mid-level flow. This blocking may be dominant in some years (such as 2003) or only weakly developed in most other years. The effect of blocking on the distribution of precipitation is potentially important. Abnormally low temperatures in winter may also result, as in February 1986. Much attention has been given to these features in the literature. Rex (1950) has provided a comprehensive review of the climatic data relating to blocking over Scandinavia.

Obstruction to westerly flow below 3000 m also results from the development and persistence of a cold anticyclone or thermal high over snow-covered Siberia. The westward extension of this high into European Russia causes continental temperate air to flow around the southern margin of the high in winter. On average, air associated with the Siberian high is found on 24 days in January over the lower Volga Plain, decreasing to 12 days in the vicinity of Leningrad (Lydolph, 1977).

Surface anticyclones and depressions

Anticyclones

Throughout the year, "cells" may break away from the semi-permanent Azores anticyclone over the east central North Atlantic and travel slowly northeast, close to the English Channel and across the European Plain or central Europe into southern European Russia. Two to four new anticyclones a year tend to form along this axis. Another favorite center for anticyclogenesis lies over Scandinavia, while high values over the Mediterranean represent the relocation and development of the Azores high. Annual mean speeds of eastward (progressive) moving highs are of the order of 6° of longitude per day at 55°N. Somewhat surprisingly, anticyclones moving westward (retrogressing) away from Europe traveled at 7° per day on average (Meteorological Office, 1975) and accounted for 40% of mobile anticyclones. There was little significant seasonal pattern evident. Figure E14 shows the total number of days in which the center of a high was located within a 5° × 5° grid. Counts were made daily. Climatological interest in this empirical aspect of climatology has since declined.

Midlatitude depressions

In European literature the term "cyclones" usually is applied to tropical storms. Midlatitude cyclonic disturbances are referred to as depressions or lows. Frontal depressions affect much of Europe north of 45°N throughout the year, and parts of the Mediterranean during the winter half. A wide variety of non-frontal depressions also occur and include heat lows (thunder lows), polar lows (Businger, 1985), and lee depressions. Most of these are associated with infrequent synoptic weather patterns. Major depressions that develop central pressures below 980 mb originate over the western North Atlantic and attain maximum intensity before reaching land. This is evident from Figure E15, which shows the diminishing influence of low pressure southeastward into central Europe. A feature of great importance over the sea areas bordering northwestern Europe is the ability of some lows to suddenly deepen, creating high winds and storm surges as in 1987.

Figure E16 shows the areas most likely to experience cyclogenesis. It should be noted that the map does not convey any dynamic information about the development, movement, and intensity of the lows (Campins et al., 2000). For example, the center over northern Italy reflects the formation of many shallow but active lee depressions (see Zenone and Lecce, in Wallén, 1977). Other "centers" are less geographically fixed and do not appear on all map interpretations of the data (see Trewartha, 1981; Borisov, 1970).

The movement of depressions over the northern hemisphere has been investigated by Klein (1957) and others, and is shown in Figure E17. The continuous lines denote main tracks and the

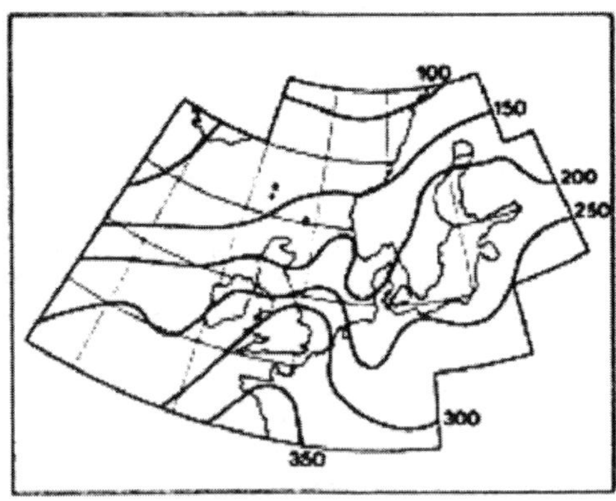

Figure E14 Total number of days with anticyclonic centers at 1230 GMT, 1899–1938 (after Meteorological Office, 1975, p. 63).

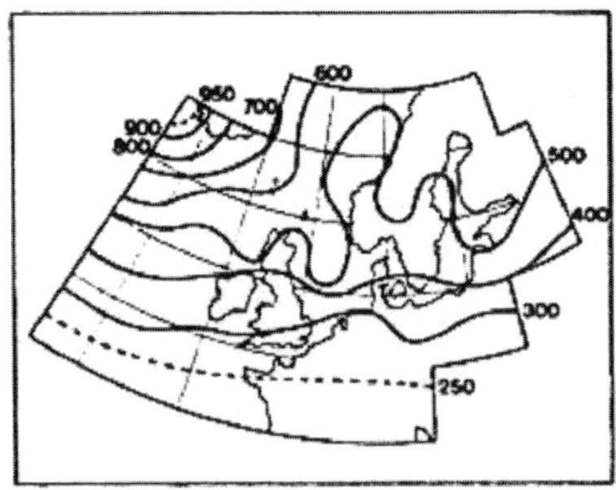

Figure E15 Total number of days with low pressure centers at 1230 GMT, 1899–1938, adjusted to unit area size (after Meteorological Office, 1975, p. 46).

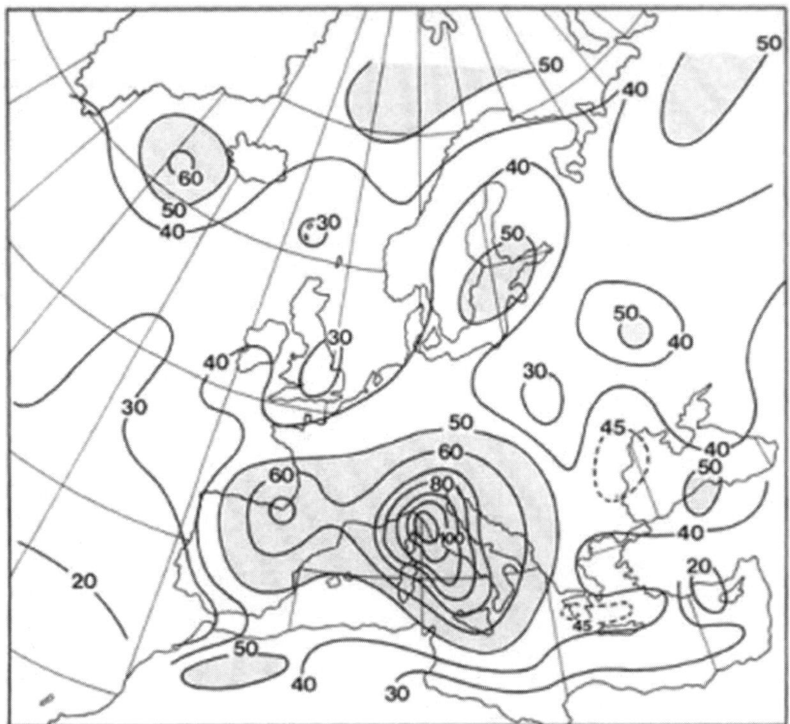

Figure E16 Twenty-year frequency distribution of cyclogenesis, 1909–1914 and 1924–1937 (after Meteorological Office, 1975, p. 57 and Trewartha, 1981).

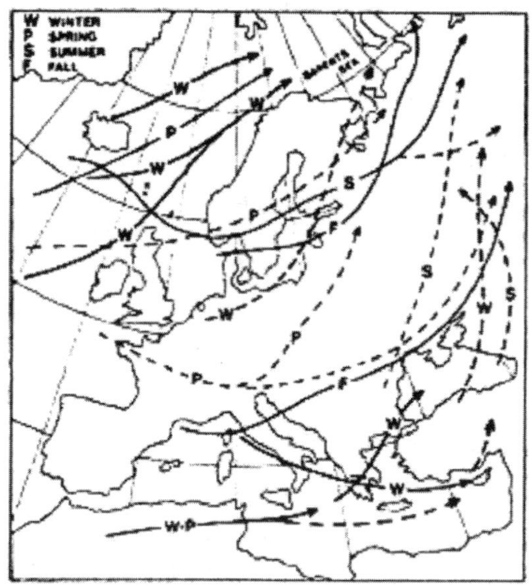

Figure E17 Main tracks of traveling depressions (after Klein, 1957, p. 15).

dashed lines indicate less frequent routes. The main Icelandic low is renewed by depressions originating further west. Tracks across northwestern Europe are usually followed by subsidiary lows circulating around the parent depression, although some do break away beneath the westerly jet and travel across the Barents Sea. Most European lows either follow the northern or southern boundary of the continent in winter, whereas in summer a central route across the North and Baltic seas has been established. Over European Russia, tracks are difficult to identify in winter when high pressure predominates, but low pressure is more in evidence in summer. This is particularly so in central European Russia – a region that stretches from the western borders with Poland eastward toward the southern Urals. In this region there are twice as many cyclonic centers in summer as in winter. Some of these lows will be associated with fronts and outbreaks of cool polar air. Others are more easily identified with zones of marked baroclinicity (Ulbrich et al., 2001) sometimes associated with unusual pressure features (Pearce et al., 2001; Barry and Chorley, 2003).

Depressions in the mediterranean

It is unusual for the whole of the Mediterranean, which extends 3700 km eastward from Gibraltar, to be free from depressions at any one time, even in summer. When depressions form within or travel through the Mediterranean, there is often a quasistationary anticyclone over central or western Europe. Other factors contributing to cyclogenesis in the Mediterranean are the presence of a baroclinic or frontal zone, lee effects in strong airflows, and instability in airmasses (Meteorological Office, 1975). Analysis of upper-air charts shows that depressions travel in the direction of the flow at 200 mb, whereas thunderstorms are more likely to follow the flow pattern at 700 mb. On the other hand, Perry (1981) has drawn attention to the role of sea surface temperature anomalies in promoting cyclonic activity within the Mediterranean.

Cold pools

During periods of low index and meridional flow masses of cold air may be transported southward over Europe and may become entirely surrounded within the troposphere by relatively warm air. The severing of the tip of this tongue of cold air produces a tropospheric cold pool. When this occurs over the Mediterranean it becomes the source of instability, thunderstorms, and abnormal precipitation (Boucher, 1982). One notable cold pool and accompanying storm developed in late January 1986 with central pressure falling below 980 mb.

Conclusion

The ever-changing pattern of the climate of western Europe provided the stimulus for the development of the science of meteorology in the early part of the twentieth century – notably the establishment of the Bergen School founded in 1918 by the Norwegian physicist Vilhelm Bjerknes (Jewell, 1981). This day-to-day variability has posed a challenge for climatologists who seek to establish order out of seeming chaos (see Volkert, 1985 and Lamb, 1985). Others have turned their attention to archived records across Europe in an attempt to place the climate records across Europe into a meaningful sequence (New et al., 2001). Elsewhere the emphasis on classification and indices has enabled greater understanding of climatic processes to be combined with issues such as that of global warming. Even so, the quantitative expression of the regional characteristics of climate remains a daunting task for climatologists and statisticians alike (Trewartha, 1981).

Keith Robert Boucher

Bibliography

Barry, R.G., and Perry, A.H., 1973. *Synoptic Climatology*. London: Methuen.

Barry, R.G., and Chorley, R.J., 2003. *Atmosphere, Weather and Climate*, 8th edn. London: Routledge.

Belasco, J.E., 1952. Characteristics of air masses over the British Isles. *Geophysics Memoirs*, **11**(87): 1–34.

Borisov, A.A., 1965. *Climates of the USSR*. London: Oliver & Boyd.

Borisov, A.A., 1970. *Klimatografiya Sovetskogo Soyruza*. Leningrad: Leningrad University.

Boucher, K.R., 1982. Tropospheric cold pool development over the Western Mediterranean. In *Proceedings of the First Hellenic–British Climatological Congress*, Athens, 1980, pp. 139–154.

Boucher, K., 1994. *Loughborough and its Region*. Climate 68–77, Loughborough University.

Boucher, K., 1999. Global warming. In Pacione, M., ed., *Applied Geography: principles and practice*. London: Routledge.

Buchinskiy, I.Ye., 1975. Droughts in the Ukraine (in Russian), Moscow Vses. *Geog. Obshch. Izv.* **107**: 207–213.

Budyko, M.I., 1974. *Climate and Life*. New York: Academic Press.

Businger, S., 1985. The synoptic climatology of polar low outbreaks, *Tellus*, **37A**: 419–432.

Campins, J. et al., 2000. A catalogue and a classification of surface cyclones for the Western Mediterranean. *International Journal of Climatology*, **20**: 969–984.

Chistyakova, Ye.A., 1975. April precipitation deficits and surpluses in the southern half of Soviet Europe (in Russian), Leningrad Gidromet, *Nauc. Issled. Tsent. SSSR T. Vyp.* **166**: 17–29.

Chubukov, L.A., 1977. Graphical methods of representing climatic weather types, Moscow. *Akad. Nauk Mezhd. Geof. Kom.* **2**: 81–88.

Conrad, V., 1946. Usual formulas of continentality and their limits of validity. *American Geophysical Union Transactions*, **27**: 663–664.

Crowe, P.R., 1971. *Concepts in Climatology*. London: Longman.

Davies, D.R., 1978. Blocking anticyclones. *Weather*, **33**(1): 30–32.

Davitaya, F.F., 1958. Studies of drought and sukhovei (in Russian). *Izv. Akad. Nauk SSSR Ser. Geog.* **5**: 131–136.

Dole, R.M., and Gordon, N.D., 1983. Persistent anomalies of the extratropical Northern Hemisphere wintertime circulation: geographical distribution and regional persistence characteristics. *Monthly Weather Review*, **111**: 1567–1586.

Eden, P., 2003. Weather log, August 2003. *Weather*, **58**: 390 (insert).

Estrela, M.J. et al., 2000. Multi-annual drought episodes in the Mediterranean (Valencia Region) from 1950–1996. A spatio-temporal analysis. *International Journal of Climatology*, **20**: 1599–1618.

European Environment Agency, 1998. *Europe's Enviroment: the second assessment*. European Communities, Elsevier.

Fukui, E., 1966. Numerical expression for the development of the Mediterranean climate with its regional varieties. *Tokyo Geographical Papers*, **10**: 149–173.

Garnier, M., 1954. Contribution à l'étude des gelées en France. *La Météorologie*, **35**: 369–378.

Gerasimov, I.P. (ed.), 1964. *Fiziko-Geograficheskiy Atlas Mira*. Moscow: Academia Naukita.

Hess, P., and Brezowsky, H., 1969. Katalog der Grosswetterlagen Europas. *Ber. Dtsch. Wetterd (Offenbach)* **15**(113).

Hisdal, H., et al., 2001. Have streamflow droughts in Europe become more severe or frequent? *International Journal of Climatology*, **21**: 317–333.

Hounam, C.E., et al., 1975. *Drought and Agriculture*. Technical Note No. 138. Geneva: World Meteorological Organization.

Ivanov, N.N., 1959. Belts of continentality on the globe. *Izv. Vses. Geogr. Obshch.* **91**: 419–423.

Jenkinson, A.F., 1985. Hot spells in central England. *Weather*, **40**: 127–128.

Jewell, R., 1981. *Tor Bergeron's First Year in the Bergen School*. Basel: Birkhäuser Verlag.

Jones P., and Hulme, M., 1997. The changing temperature of Central England. In Hulme, M., and Barrow, E., eds., *Climates of the British Isles*. London: Routledge.

Junge, M.M., and Stephenson, D.B., 2003. Mediated and direct effects of the North Atlantic Ocean on Winter Temperatures in North West Europe. *International Journal of Climatology*, **23**: 245–261

Klein, W.H., 1957. *Principal Tracks and Mean Frequencies of Cyclones and Anticyclones in the Northern Hemisphere*. Research Paper No. 4. Washington, DC: US Weather Bureau.

Klein Tank, A.M.G., et al., 2002. Daily dataset of 20th-century surface air temperature and precipitation series for the European Climate Assessment. *International Journal of Climatology*, **22**. 1441–1453.

Lamb, H.H., 1964. *The English Climate*. London: English Universities Press.

Lamb, H.H., 1972. British Isles weather types and a register of the daily sequence of circulation patterns 1861–1971. *Geophysics Memoirs*, **16**(116): 1–85.

Lamb, H.H., 1982. *Climate, History and the Modern World*. London: Methuen.

Lamb, H.H., 1985. *Climate and its Variability in the North Sea–Northeast Atlantic Region*. Stavanger: North Sea.

Lednicky, V., 1985. Climatological characteristics of a summer period index. *Hydromet. Ustav: Met. Zpravy*, **38**: 94–95.

Lloyd-Hughes, B., and Saunders, M.A., 2002. A drought climatology for Europe. *International Journal of Climatology*, **22**: 1571–1592.

Lydolph, P.E., 1977. *Climates of the Soviet Union*. London: Oliver & Boyd.

Menzel, A., and Fabian, P., 1999. Growing season extended in Europe. *Nature*, **397**: 659.

Menzel, A., et al., 2003. Variations of the climatological growing season (1951–2000) in Germany compared with other countries. *International Journal of Climatology*, **23**: 793–812.

Meteorological Office, Great Britain, 1975. *Weather in Home Waters*, vol. II, part I. London: Her Majesty's Stationery Office.

Mitchell, T.D., and Hulme, M., 2002. Length of the growing season. *Weather*, **57**: 196–198.

Namias, J., 1950. The index cycle and its role in the general circulation. *Journal of Meteorology*, **7**: 130–139

Namias, J., 1964. Seasonal persistence and recurrence of European blocking during 1958–1960. *Tellus*, **16**: 394–407.

New, M., et al., 2001. Precipitation measurements and trends in the twentieth century, *International Journal of Climatology*, **21**: 1889–1922.

Parker, D.E., and Alexander, L.V., 2002. Global and regional climate in 2001. *Weather*, **57**: 328–340.

Pearce, R. et al., 2001. The post-Christmas 'French' storms of 1999. *Weather*, **56**: 81–91.
Perry, A.H., 1976. Mediterranean downpours in 1976. *Journal of Meteorology*, **2**: 10–11.
Perry, A.H., 1981. Mediterranean climate – a synoptic reappraisal. *Progress in Physical Geogrphy*, **5**(1): 107–113.
Pinna, M., 1957. *La carta dell'indice di aridita par l'Italia*. Attidel XVII Congresso Geografico Italiano.
Raschke, E., Vonderhaar, T.H., Bandeen, W.R., and Pasternak, M., 1973. The radiation balance of the earth–atmosphere system during 1969–70 from Nimbus 3 measurements. *Journal of Atmosphere Science*, **30**: 341–364.
Rauner, Yu.L., 1980. The synchronous recurrence of droughts in the grain growing regions of the Northern Hemisphere. *Soviet Geography*, **21**(3): 159 179.
Rex, D.F., 1950. Blocking action in the middle troposphere and its effects upon regional climate. *Tellus*, **2**: 196–211, 275–301.
Schmidli, J., et al., 2002. Mesoscale precipitation variability in the region of the European Alps during the 20th century. *International Journal of Climatology*, **22**: 1049–1074.
Tabony, R.C., 1981. A principal component and spectral analysis of European rainfall. *Journal of Climatology*, **1**: 283–294.
Trewartha, G.T., 1981. *The Earth's Problem Climates*, 2nd edn. Madison, WI: University of Wisconsin Press.
Tucker, A., 1965. The distribution and annual cycle of local heating rate throughout the troposphere in the northern hemisphere, *Meteorology Magazine*, **94**: 205–214.
Tuomenvirta, H., 2001. Homogeneity adjustments of temperature and precipitation series – Finish and Nordic data. *International Journal of Climatology*, **21**: 495–506.
Ulbrich, U., et al., 2001. Three extreme storms over Europe in December 1999. *Weather*, **56**: 70–80.
Ulbrich, U., et al., 2003(a). The central European floods of August 2002: Part 1 – Rainfall periods and flood development. *Weather*, **58**: 371–377.
Ulbrich, U., et al., 2003(b). The central European floods of August 2002: Part 2 – Synoptic causes and considerations with respect to climate change. *Weather*, **58**: 434–442.
van Loon, H., and Rogers, J. C., 1978. The see-saw in winter temperatures between Greenland and Northern Europe, Part 1. General description. *Monthly Weather Review*, **106**: 296–310.
Volkert, H., 1985. On the mesoscale variability of meteorological fields – the example of Southern Bavaria. *Contributions in Atmospheric Physics*, **58**: 498–516.
Wallen, C.C., 1977. *Climates of Central and Southern Europe, vol. 6: World Survey of Climatology*. Amsterdam: Elsevier.
Weber, R.O. et al., 1997. 20th-century changes of temperature in the mountain regions of Central Europe. *Climate Change*, **36**: 327–344.
Yan, Z., et al., 2002. Trends of extreme temperatures in Europe and China based on daily observations. *Climatic Change*, **53**: 355–392.
Zhou, L.M., et al., 2001. Variations in northern vegetation activity inferred from satellite data of vegetation index during 1981–1999. *Journal of Geophysical Research, Atmospherics*, **106**: 20069–20083.

Cross-references

Airmass Climatology
Continental Climate and Continentality
Extratropical Cyclones
Maritime Climate
Winds and Wind Systems

EVAPORATION

Natural evaporation is the process whereby water at the Earth's surface, either in liquid or solid form, is converted to vapor and transferred into the atmosphere. It is therefore the reverse component to precipitation in the global water cycle and, in total over the whole surface of the Earth, the two must balance on the average. Furthermore, since the storage capacity of the atmosphere for water in all phases is equivalent to only a few centimeters depth of liquid water, whereas the average evaporation for the whole Earth approaches 100 cm year^{-1}, this balance must hold over comparatively short intervals of time.

It should, however, be stressed that the balance is achieved only on a global reckoning, the world patterns of evaporation and precipitation showing gross differences. The consequent transfer of water vapor from the source of evaporation to the area of precipitation, with the associated transfer of latent heat, provides the most important single factor in the overall redistribution of heat over the Earth's surface. It is for this reason that knowledge of the distribution of evaporation over the whole Earth is basic to an understanding of the general circulation of the atmosphere, while the economic importance of evaporation over the continents is being increasingly recognized as water usage grows rapidly. Over the oceans the latter aspect is, of course, of no consequence, but the combined study of evaporation and precipitation in these regions will lead to increased knowledge of drift currents in and between the oceans, which are already known to be considerable.

Measurement of evaporation

The measurement of evaporation presents difficult problems, the difficulty varying with the nature and extent of the evaporating surface. Here we shall discuss briefly the principal methods that have been developed to meet different requirements, indicating the applicability and limitations of each.

Hydrological balance

When all other factors contributing to the hydrological or water balance for a given area are known, evaporation may be estimated as a residual. The hydrological balance may be expressed as

$$P = R + S + E$$

P being precipitation, R the surface runoff of water, S the seepage of water into the ground, which may increase the local storage or percolate to deeper levels and be removed as underground currents, and E the evaporation. Adequate measurement of P and R presents problems in sampling over the area in question. S is obviously a difficult term to assess, although in certain circumstances (when it is known from the local geological structure that underground currents out of the area do not exist) it can, over a sufficient time interval, be assumed to be zero. This method of estimating evaporation must evidently be very approximate in general, although in certain favorable circumstances, for example catchment areas of easy topographical form, results of acceptable accuracy may be obtained.

In certain studies, e.g. in agricultural research, the gravimetric method of determining evaporation is of prime importance. This technique, which is a specific refinement of the hydrological balance method, consists of isolating in an impermeable container *in situ* a soil block of appropriate size, whose water balance is known in terms of water applied and water lost by successive weighings of the block. This method is capable of high accuracy and, depending on the precision of the weighing equipment, can be used for the measurement of evaporation over periods less than 1 h when the evaporation rate is high. There are, however, obvious limitations to the size of such installations and the method is unsuitable for the study of evaporation when the terrain carries irregular cover.

Soil moisture profile

Other methods of measuring the water loss from soil by evaporation may be mentioned briefly. They are all based on the determination of the vertical distribution of moisture in the soil, the "soil moisture profile", and its change with time, and include measurement of the moisture content by sampling and weighing; measurement of electrical conductivity of materials placed in the soil, whose conductivity is moisture-dependent (gypsum, fiberglass, nylon, etc.); the measurement of the thermal conductivity of soil, which is also moisture-dependent; and the determination of the scattering of neutrons by the hydrogen atoms in water. Each is subject to limitations of one sort or another (for details and reference to original work, see review by Deacon et al., 1958).

Pan evaporation

The measurement of water loss from pans and tanks has occupied a central position in experimental studies of evaporation. The results from such observations are usually presented by means of a formula due originally to Dalton (Shaw, 1993):

$$E = C(e_w - e_a)$$

where E is the rate of evaporation, e_w is the saturation water vapor pressure at the surface temperature of the water, e_a is the vapor pressure in the air, and C is a factor that incorporates the effects of wind speed, barometric pressure and other variables such as exposure. Much of this work was originally carried out in the United States, in particular by Rohwer (1931) at Fort Collins, Colorado. Rohwer summarized the results of his studies, made from a tank 3 ft (0.9 m) square, in the formula

$$E = 0.771(1.465 - 0.0186B)$$
$$(0.44 + 0.118w)(e_w - e_a)$$

where E is measured in inches per day; B (the barometric pressure), e_w and e_a, in inches of mercury; and w (the wind speed near the ground), in miles per hour.

Such measurements, if they can be related to water losses from larger water bodies such as dams and reservoirs and from freely transpiring vegetation, are of obvious economic importance. What is wanted, ideally, is a single conversion factor (pan coefficient) applicable at all times under any conditions. Consideration of the basic dependence of pan evaporation on the size, design, and exposure of the pan and its comparatively rapid response to changing atmospheric conditions (whereas the larger body of water will be more or less thermally inert depending on its size) emphasizes the difficulty of achieving this factor. This is clearly brought out in the results of evaporation studies over Lake Superior. Figure E18 shows the monthly evaporation rates from Lake Superior and a nearby small shallow lake, both calculated from Rohwer's formula. The physical reasons for the curves being 6 months out of phase are obviously connected with the relative thermal inertia of the two lakes. This case is admittedly extreme, but it does emphasize the fact that each water body will have its own pan coefficient, and that this is likely to vary with pan or tank type and season. Nevertheless, such instruments, particularly if they are standardized, have a useful role to play in assessing the comparative evaporative needs of different regions, even though their quantitative indications must be interpreted with caution.

Other methods

A direct method of measuring evaporation has been developed in recent years, which stems from the recognition of evaporation as a process of turbulent diffusion. It can be shown that the product, averaged over any time interval, of the instantaneous values of vertical air velocity and specific humidity near the evaporating surface represents the average evaporation for the period. Instrumentation has been designed to measure the fine structure of the air in respect of these properties (Deacon et al., 1958) and the technique has been verified in the field. This method has the advantage that it can be used for the measurement of evaporation from any type of uniform surface and is subject to none of the restrictions that limit many other techniques. Though up to the present time its use has been confined to research studies of evaporation, it is capable of development for routine use (Shaw, 1993).

Evaporation in the climatic context

Each of the above methods finds its particular use in specific studies of evaporation over relatively small areas, the interest being chiefly in the economy of water storages and in the water needs for agriculture. None is suitable in the larger context of the delineation of evaporation on the climatic scale, although

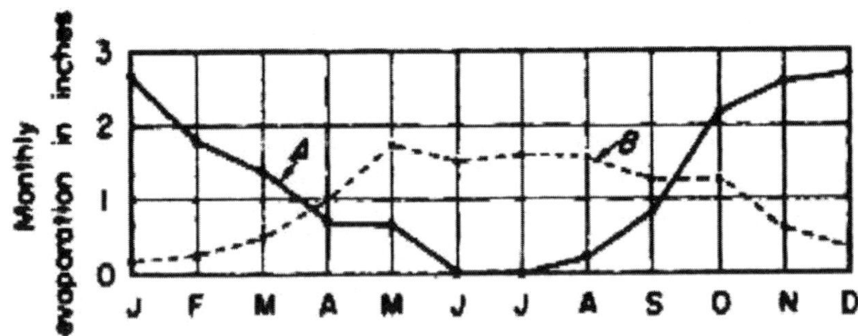

Figure E18 Estimated monthly evaporation rates for (**A**) Lake Superior and (**B**) nearby shallow lake (from Wisler and Brater, 1959).

the hydrological balance method in favorable circumstances provides an exception to both of these statements.

For such surveys, where it is sought to estimate evaporation in terms of climatological variables, there are two principal approaches. In the first, which is usually known as the aerodynamic method, the rate of evaporation E is related to the difference in water vapor content of the air at two levels above the evaporating surface and some function $f(u)$ of the wind speed in the diffusion formula

$$E = f(u)(q_1 - q_2)$$

where q_1 and q_2 represent the specific humidity of the air at the levels z_1 and z_2.

The quantity $f(u)$ is determined empirically, or may be derived from the theory of atmospheric turbulence as

$$f(u) = \frac{\rho k^2 u_2 (1 - u_1/u_2)}{(\ln z_2/z_1)}$$

where u_1 and u_2 are the wind speeds at the levels z_1 and z_2, ρ is the air density and k is von Karman's constant ($\sim$0.4). It must be stressed that this formulation for $f(u)$ is valid only when the air is in or near neutral equilibrium. However, over much of the Earth's oceanic surface the assumption of neutral stability is justified, and then u_1/u_2 has a constant value so $f(u)$ reduces to a constant factor times the wind speed, and

$$E = A(q_1 - q_2)u$$

For the empirical determination of $f(u)$ it is necessary for the evaporation to be known independently, either from consideration of the water balance or from energy conservation. In the latter method, which can be used in its own right for the determination of evaporation and is the second principal approach referred to above, the latent heat expended on evaporation can be estimated as a residual when all other components in the energy balance at the evaporating surface are known. This balance is represented as

$$R_N = H + LE + S$$

where R_N is the net radiant energy reaching the surface, which can be estimated from empirical formulas in terms of climatological variables or can be measured; H is the sensible heat conduction from the surface to the air; L is the latent heat of vaporization of water; and S is the rate of heat storage below the evaporating surface. Over land, S is a relatively small component and, over long periods, may be neglected. However, over the oceans this is not so, and for this reason among others, the aerodynamic method has been preferred over these regions, as we shall see below.

The mechanism of transfer of sensible heat into the air is similar to that of water vapor, and so the problem of determining evaporation from the energy balance resolves into that of correctly apportioning the available energy $R_N - S$ (or, nearly enough, R_N) between sensible and latent heat transfer. For this purpose the Bowen ratio technique is available. In this method, identity of the transfer mechanisms of heat and water vapor is assumed, and the ratio of the transfer of sensible heat H to latent heat LE (the Bowen ratio) may be written

$$\frac{H}{LE} = \frac{c_p\,(T_1 - T_2)}{L(q_1 - q_2)} = \beta$$

where T_1 and T_2 are the temperatures of the air at the levels z_1 and z_2, and c_p is the specific heat of air at constant pressure.

Hence, from the equation representing energy balance, E may be expressed as

$$E = \frac{R_N}{L(1 + \beta)}$$

Although each method may, in principle, be used for the measurement of evaporation, each requires great accuracy in the determination of the small differences in temperature and specific humidity that usually occur. The instrumentation for achieving this on a routine network basis is not yet available. Over the oceans, however, the difficulty is relieved by the fact that differences in temperature and vapor pressure between the surface and the air are relatively large, and a knowledge of the surface temperature automatically provides the value of the vapor pressure there. Over the land no such simplification is generally available. Furthermore, departures from neutrality that call into question the assumptions of equality of transfer coefficients, referred to above, are much more pronounced over the land than over sea. For reasons such as these, evaporation surveys over land areas have, up to the present, had to rely largely on estimates made from the hydrological balance and, being subject to the inherent errors of this method, can at best be only very approximately correct.

One of the most extensive surveys of evaporation yet published is that due to Budyko (1956). Evaporation is estimated by the aerodynamic method, the appropriate transfer coefficient being determined from a variety of evidence including energy balance, studies of lake evaporation, etc. The formula used by Budyko is

$$E = 2.4 \times 10^{-6} u(q_s - q_a)$$

where E is in $\mathrm{g\,cm^{-2}\,s^{-1}}$, u is the wind speed in $\mathrm{cm\,s^{-1}}$ at the usual height for shipboard observation, and q_s and q_a are the specific humidity of the air at the sea surface and at shipboard level, respectively: q_s is assumed to correspond to saturation at the sea surface temperature.

Since the world totals of evaporation and precipitation must balance, and since the estimates of either must be made on grossly inadequate observations, there is considerable latitude for interpreting and reconciling the observations of precipitation from coastal and island stations with estimates of evaporation from formulae whose coefficients are only approximately known and that necessarily use observational material whose quality may often be questionable. Observations of wind speed, temperature, and humidity can only properly be made on shipboard if special precautions are taken, and the measurement of sea surface temperature is another source of error.

Although estimates of evaporation show a variation in excess of 20%, and thus emphasize the approximate knowledge of the level of oceanic evaporation, there is nevertheless reasonably good agreement in the evaporation pattern on the broad scale. A notable feature is the distribution of evaporation rates in the subtropical high atmospheric pressure belts, with high rates over oceanic areas reaching a maximum in the central Indian Ocean interrupted by low values over land areas. The equatorial minimum reflects the comparatively high cloudiness of that region. The marked contrast in the precipitation–evaporation relationship between land and ocean surfaces is brought out in Figure E19. Over the land areas, where evaporation relies largely on local precipitation, the patterns of the two quantities are in good agreement, in strong distinction to the precipitation minima and evaporation maxima over the oceans in the trade wind belts. Table E9 presents final estimates of average annual

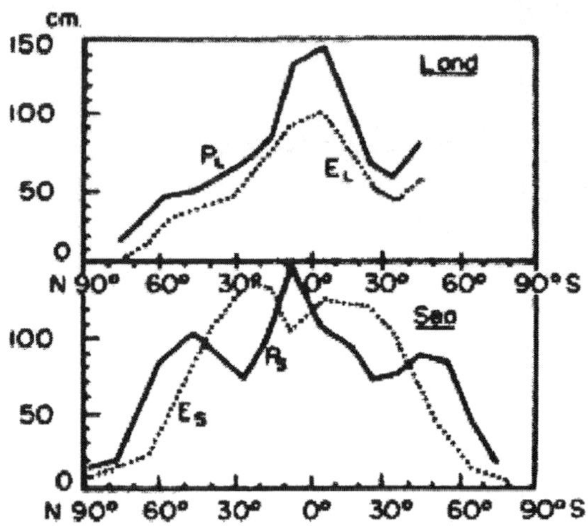

Figure E19 Mean annual precipitation (*P*) and evaporation (*E*) for land and ocean surfaces (cm).

Table E9 Average annual rates of evaporation for oceans and continents (cm $year^{-1}$) (after Budyko, 1956)

Oceans		Continents	
Atlantic,	104	Europe,	36
Indian,	138	Asia,	39
Pacific,	114	North America,	40
Arctic,	12	South America,	86
		Africa,	51
		Australia,	41

rates of evaporation from the oceans and the continents, based on the work of Budyko, 1956. The approximate nature of the figures shown in the must be borne in mind.

W.C. Swinbank

Bibliography

Budyko, M.I., 1956. *The Heat Balance of the Earth's Surface* (N.A. Stepanova, trans.). Washington, DC: Office of Technical Services; US Department of Commerce.

Deacon, E.L., Priestley, C.H.B., and Swinbank, W.C., 1958. Evaporation and the water balance. In *Climatology, Reviews of Research; Arid Zones Research*, No. 10. Paris: UNESCO.

Kessler, A., 1985. *Heat Balance Climatology*. New York: Wiley.

Rohwer, C., 1931. *Evaporation from Free Water Surfaces*. US Department of Agriculture Technical Bulletin 271. Washington, DC.

Shaw, E., 1993. *Hydrology in Practice*. London: Chapman & Hall.

Wisler, C.O., and Brater, E.F., 1959. *Hydrology*. NewYork: Wiley.

Cross-references

Bowen Ratio
Energy Budget Climatology
Evapotranspiration
Latent Heat
Phase Changes
Water Budget Analysis

EVAPOTRANSPIRATION

Moisture is returned directly to the atmosphere through a number of processes. The change in state from solid or liquid form to gaseous water vapor comprises the process of evaporation and sublimation. Evaporation occurs when input of energy onto an evaporating surface causes water molecules to pass from that surface to the atmosphere; this will occur when the vapor pressure of the air is below its saturation value. The rate of evaporation is governed by the state of a number of variables including water vapor, temperature and air motion, and several formulae are available to determine the rate at which it occurs. Perhaps the oldest of these is the Dalton equation, given as

$$E_0 = (e_s - e)\,f(u)$$

where e_s is the vapor pressure of the evaporating surface, e is the vapor pressure at some height above that surface, and $f(u)$ is a function of the horizontal wind speed. Since these parameters vary widely over the Earth's surface, the rates of evaporation vary enormously.

Transpiration

Just as moisture is returned to the atmosphere through evaporation, it is also returned by the process of transpiration. Transpiration is the term applied to the loss of moisture by plants, such losses occurring through stomata on the exposed leaf surface, with the rate of loss of water being controlled by guard cells on the leaf. During the day these are usually open to expose the moist leaf interior, with resulting high transpiration; at night they are generally closed. Some question exists concerning the role of transpiration in plant growth and development, but it is generally agreed that transpiration rates control the movement of moisture through plants, and this is obviously related to the transport of materials through the plant. The relative amounts of moisture lost through evaporation and transpiration obviously vary appreciably, depending on the nature of the ground surface. But, as shown in Table E10, transpiration rates are often significantly higher than evaporation rates over densely vegetated areas.

While the study of both evaporation and transpiration is significant in itself, it is convenient to treat them as a single process in applied climatic studies. The loss of water through the combined process is termed evapotranspiration.

Evapotranspiration measurement

The rate at which evapotranspiration occurs depends on both meteorological and botanical characteristics, but it can be

Table E10 Evaporation and transpiration losses for selected forest types

Forest	Forest age (years)	Evaporation (*e*) (mm/growing season)	Transpiration (*T*) + interception (*I*) (mm/growing season)	Ratio *E*/(*T*+*I*)
Aspen	20	78	314	1:4.0
Aspen	60	84	282	1:3.4
Scots pine	20	48	363	1:7.6
Scots pine	60	87	340	1:3.9

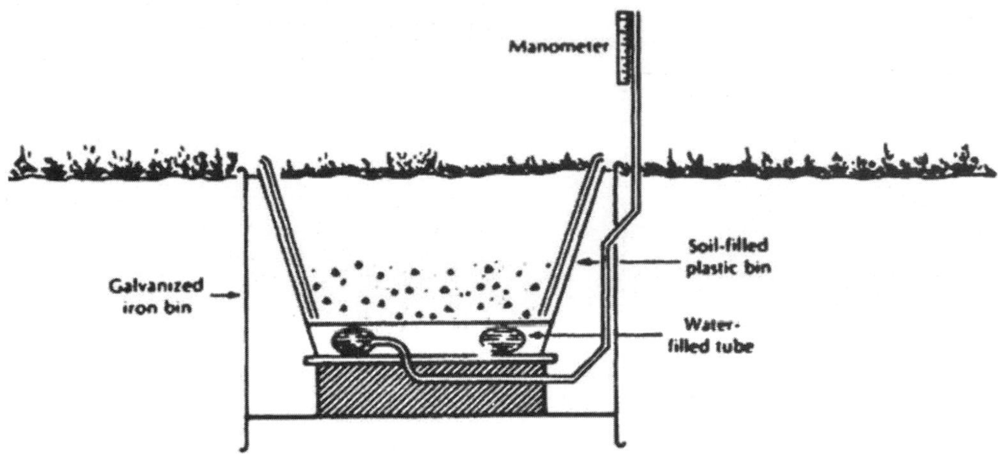

Figure E20 Section through a simple weighing lysimeter.

assumed that, under a given set of conditions, an upper limit will exist. Clearly, to satisfy this maximum rate there must be sufficient moisture available. If moisture is in limited supply the loss of water will be lower than the maximum rate. It is therefore necessary to recognize two evapotranspiration rates. Potential evapotranspiration is the maximum amount of water lost, assuming all moisture requirements can be met; actual evapotranspiration is the observed amount that, if moisture is limited, will be lower than the potential rate.

Problems exist in determining both the evaporation rate and the transpiration rate. It follows then that the estimation of evapotranspiration rates is equally problematic. In measuring the potential rate, water can be added at known quantities and its disposal can be determined more readily. A number of instruments have been designed to measure both rates. The evapotranspirometer, since it is measuring potential evapotranspiration, gives fairly reliable results. It consists of a number of tanks that are filled with a soil of similar constituency to that of the surrounding area and covered by a continuous vegetation cover. Since all of the moisture entering the system is accounted for by evapotranspiration or by percolation into the collecting jars, then the moisture consumed by evapotranspiration can be calculated. The measurement of the actual rate is open to larger experimental error. An example of the type of instrument used, in this case the floating lysimeter, is shown in Figure E20.

The emplacement, location and operation of lysimeters require great care and expenditure. Clearly, they must be large enough to provide a representative sample of a given area and deep enough so that they do not significantly alter the natural profile of soil moisture. Such instruments are rarely found outside of experimental stations and thus provide standards and supplements to evapotranspiration rates derived using other methods.

With the lack of observed data, investigators have turned to other methods of estimating evapotranspiration losses. Essentially, these concern analysis of the variables that influence moisture losses: derived formulae include such factors as solar radiation, temperature, plant cover, and wind speed. Three approaches are available: the empiric in which derived equations are based on observation, the aerodynamic where the physics of the atmospheric processes responsible for evapotranspiration are evaluated, and the energy budget approach, which estimates the amount of energy available to cause moisture transfer back to the air. Numerous formulas have been derived using each approach. Empirical methods are represented by the work of Thornthwaite (1948), Blaney and Criddle (1950) and Makkink (1957). The aerodynamic approach is illustrated by the method of Thornthwaite and Holzman (1942), whereas the energy budget approach has been investigated by Bowen (1926), Budyko (1956) and Penman (1963). In a call for standardization, the FAO suggest use of the Penman–Monteith method.

Estimating methods

Penman's method uses the combined influences of turbulent transfer and the energy budget. His derived equation for the determination of evapotranspiration uses vapor pressure, net radiation and the drying power of air at a given temperature. To derive this formula, Penman had first to obtain an expression that allowed determination of both net radiation and the drying power of air. Such a formula is necessarily quite complex, as is the expression for the evaluation of the drying power of air. The formula was derived through research in the UK, but results obtained from its use appear to hold true for many other parts of the world. As shown in Table E11, it appears to provide fairly reliable results for the Australian region in which it was tested. Investigators looking into the agricultural potential of most regions would do well to consider the Penman approach in estimating moisture requirements.

Thornthwaite's method is probably the best known and most widely used for estimating potential evapotranspiration in the United States. Working in the eastern part of the country, Thornthwaite devised a formula that essentially is based on the availability of temperature data. His method uses mean monthly temperature and an empiric heat index, which is itself an exponential function of temperature, as inputs. The derived unadjusted potential evapotranspiration is corrected by using actual daylight hours and number of days in the month in question.

While the method is widely used, it has been criticized. Perhaps the fault most often cited is the fact that temperature,

Table E11 Methods for estimating potential evapotranspiration from short grass at Aspendale. Australia (after Sellers, 1965)

		Estimated potential evapotranspiration (mm/day^{-1})					
Month	T(°C)	ET_g^a (mm day^{-1})	Budyko and Penman	McIlroy	Thornthwaite	Blaney and Criddle	$0.2T^b$
January	23.3	7.76	6.86	7.57	4.47	4.91	4.66
February	21.1	5.62	5.44	6.12	3.51	4.34	4.22
March	19.6	4.21	3.76	4.22	2.80	3.82	3.92
April	17.2	2.94	2.76	3.16	2.04	3.24	3.44
May	12.6	1.31	1.48	1.67	1.09	2.52	2.52
June	10.9	0.99	1.01	1.19	0.80	2.26	2.18
July	10.0	0.93	1.17	1.34	0.75	2.25	2.00
August	11.0	1.37	1.60	1.76	0.89	2.55	2.20
September	13.0	2.30	2.64	2.88	1.50	3.02	2.60
October	15.8	4.07	4.00	4.28	2.09	3.64	3.16
November	17.8	5.26	4.76	5.27	2.73	4.17	3.56
December	10.1	5.99	5.99	6.56	3.47	4.62	4.02
Annual total		1296	1260	1398	793	1257	1170

[a] ET_g based upon observed values of seven lysimeters.
[b] A simple empiric statement, where T is mean monthly temperature, which appears to provide a reasonably good estimate.

which is the major variable used in the system, is not the best indicator of evapotranspiration rates; radiation values probably provide a more precise guide. Chang (1959) gives a number of examples where, because of the time lag between incoming radiation and temperature maxima, the Thornthwaite method gives imprecise results. Furthermore, since the formula was based on lysimeter data observed in watersheds in the eastern United States, the method does not always give good results elsewhere in the world. Criticism is also made of the fact that Thornthwaite assumes that evapotranspiration ceases at temperatures below 0°C.

Despite such criticism, there is little doubt that the Thornthwaite method is a useful and valuable approach, particularly when monthly data are used (shorter-term results are more questionable). Its value is enhanced since the only data required to estimate evapotranspiration are temperatures, and these variables are readily available for many stations throughout the world. The numerous publications by the Thornthwaite Associates (in 1950 and following years) are also an asset to its application.

Other estimates, in which actual plant types are considered, are also available. The method proposed by Blaney and Criddle (1950) uses the equation

$$U = KF = kf$$

where U = the consumptive use (evapotranspiration) in inches, F = the sum of the monthly consumptive use factors (the sum of f, the product of mean monthly temperature and monthly percentage of daytime hours) and K = the empiric coefficient for the plant in question (k being the monthly coefficient). Estimates of the plant coefficient (K) are derived from observed data for crops in arid and semiarid regions. Typical values are alfalfa 0.85, corn 0.80, citrus trees 0.06 and rice 1.20. When used in humid regions the coefficient values need to be decreased by 10%.

Papadakis (1966) has used evapotranspiration under different climatic regimes to formulate a climate classification based on agricultural potential. Although the classification he proposes is somewhat complex, it contains much that is interesting regarding climate and water needs. For example, in evaluating evapotranspiration he uses saturation vapor pressure corresponding to monthly temperatures. This is given by

$$E = 0.5625(e_{ma} - e_d)$$

where E is the monthly potential evapotranspiration in centimeters, e_{ma} is the saturation vapor pressure corresponding to the average daily maximum temperature, and e_d is vapor pressure corresponding to the mean monthly temperature. In tests of various estimates of potential evapotranspiration the Papadakis formula appears to hold up well.

Calls for standardization

Given the high variability in results derived from the calculation methods, there have been calls for selection of a single approach. In particular the FAO suggested that the Penman–Monteith method be used as a standard for derivation of evapotranspiration from meteorological data.

The Penman–Monteith equation is:

$$\lambda ET = \frac{\Delta(R_n - G) + \rho_a c_p \dfrac{(e_s - e_a)}{r_a}}{\Delta + \gamma\left(1 + \dfrac{r_s}{r_a}\right)}$$

where:
R_n is the net radiation,
G is the soil heat flux,
$(e_s - e_a)$ represents the vapor pressure deficit of the air,
ρ_a is the mean air density at constant pressure,
c_p is the specific heat of the air,
Δ represents the slope of the saturation vapor pressure–temperature relationship,
γ is the psychrometric constant, and
r_s and r_a are the (bulk) surface and aerodynamic resistances.

A full explanation of the equation and methods of calculation are provided by Allen et al. (1998).

John E. Oliver

Bibliography

Allen, R.G., Pereira, L.S., Raes, D., and Smith, M., 1998. *Crop Evapotranspiration – Guidelines for Computing Crop Water Requirements*. Rome: FAO Irrigation and Drainage Paper 56.

Blaney, H.F., and Criddle, W.D., 1950. *Determining Water Requirements in Irrigated Areas from Climatological Data*. Soil Conservation Service Technical Publication No. 96. Washington, DC: US Department of Agriculture.

Bowen, I.S., 1926. The ratio of heat losses by conduction and by evaporation from any water surface. *Physics Review*, **27**: 779–787.

Budyko, M.I., 1956. *The Heat Balance of the Earth's Surface* (N.A. Stepanova, trans.). Washington, DC: Office of Technical Services, US Department of Commerce.

Chang, J.-Hu., 1959. An evaluation of the 1948 Thornthwaite classification. *Association of American Geographers Annals*, **49**: 24–30.

Itier, B. 1996. Measurement and estimation of evapotranspiration. In Pereira L.S., et al., eds., *Sustainability of Irrigated Agriculture*. Dordrecht: Kluwer.

Makkink, G.F., 1957. Testing the Penman formula by means of lysimeters. *Journal of the Institute of Water Engineers*, **11**: 277–288.

Mather, J.R., 1992. Evaporation and Evapotranspiration. In *Encyclopedia of Earth Sciences,* vol. 2. New York: pp. 187–196.

Monteith, J.L. 1981. Evaporation and surface temperature. *Quarterly Journal of the Royal Meteorological Society,* **107**: 1–27.

Papadakis, J., 1966. *Climates of the World and their Agricultural Potentialities*. Buenos Aires: published by author.

Penman, H.L., 1963. *Vegetation and Hydrology.* Commonwealth Bureau of Soils Tech. Communication No. 53. Farnham Royal: Commonwealth Agricultural Bureaux.

Sellers, W.D., 1965. *Physical Climatology*. Chicago, IL: University of Chicago Press.

Thornthwaite, C.W., 1948. An approach toward a rational classification of climate. *Geographical Review*, **38**: 55–94.

Thornthwaite, C.W., and Holzman, B., 1942. *Measurements of Evaporation from Land and Water Surfaces*. Technical Bulletin No. 817. Washington, DC: US Department of Agriculture.

Thornthwaite Associates, 1963–1964. *Average Climatic Water Balance Data of the Continents*, vols 16 and 17. *Publications in Climatology.* Centerton, NJ.

Ward, R.C., 1967. *Principles of Hydrology.* New York: McGraw-Hill.

Cross-references

Agroclimatology
Bowen Ratio
Evaporation
Hydroclimatology
Water Budget Analysis

EXTRATROPICAL CYCLONES

Extratropical cyclones are among the largest disturbances of atmospheric circulation found on Earth. These giant low-pressure systems of the middle latitudes typically span the entire depth of the troposphere vertically, frequently involve lower stratospheric air, and extend 1000–3000 km in horizontal diameter. Central pressure values within extratropical cyclones vary markedly and are typically less than those for smaller cyclonic systems, ranging between 990 and 1000 mb, achieving much lower values in intense systems. Like all organized low-pressure systems, extratropical cyclones rotate in sympathy with the Earth's spin in the hemisphere of occurrence. This cyclonic motion, forced by the interaction of pressure gradient force, centripetal acceleration and Coriolis effect, is disrupted near the Earth's surface by friction, inducing low-level convergence. Thus, midlatitude cyclonic storms are marked by large values of absolute and relative vorticity.

The circulation of air within midlatitude cyclones induces a complex pattern of rising and sinking motion and generates atmospheric discontinuities in temperature, moisture and wind shear, known as fronts. The geometry and evolution of circulation and fronts exert a powerful influence on the temperature, precipitation and severe weather climatology of a region. Cyclone behavior is therefore of keen interest to forecasters, as it forms a central component of weather and climate in the midlatitudes.

In addition to influencing the weather elements of the regions where they occur, extratropical cyclones play a key role in the global energy balance. Essentially eddies within the westerly circulation of the midlatitudes, cyclones serve to deplete energy imbalances that exist in the vicinity of the baroclinic zones of the so-called polar front, where regional temperature variations can be quite large, particularly in the cold season of a hemisphere (Palmén, 1949a, 1951a, b). Cyclones move heat and momentum along "storm tracks", modifying the large-scale flow in which they are imbedded (Holopainen, 1984).

Although models describing the dynamics of midlatitude cyclones have existed for well over a century, meteorologists' understanding of cyclone dynamics did not experience major growth until the second half of the twentieth century. Since the late 1940s the increased availability and sophistication of upper-air observations has revealed the three-dimensional structure of cyclones, particularly the apparent feedbacks that occur between upper and low tropospheric circulation during development. More recently, exponential growth in the ability of computers to process vast quantitites of data continues to improve the ability to simulate and investigate cyclone behavior.

Cyclogenesis

Midlatitude cyclones form initially where the westerlies are perturbed and a pre-existing upper tropopheric disturbance, a short wave or isotach maximum, is amplified, often by interaction with a low-level baroclinic zone. Consequently, the initial development of a surface low-pressure feature, termed cyclogenesis, tends to occur east (downwind) of major mountain ranges, over warm water and along semipermanent frontal zones. Cyclones have been classified according to their key mode of development, i.e. orographic, surface front or upper level disturbance; however, most cyclones couple two or more of these mechanisms during their development (Pettersen and Smeybe, 1971).

Orographic cyclogenesis

Orographic, or lee cyclogenesis, as discussed by Buzzi and Tibaldi (1978), Newton (1956) and others, is a mechanism whereby the existing potential and absolute vorticity of a westerly current are modified by passage over a topographic barrier. A zonal current that flows normal to a north–south-oriented mountain range undergoes diminished relative vorticity as air parcels diverge over the region of elevated topography. In the mountain lee, where tropospheric thickness increases, relative vorticity rises via convergence. In addition, because westerly flow is diverted equatorward by its initial vorticity decrease, it must gain additional relative vorticity in the mountain lee to conserve absolute vorticity (Rossby, 1940; Ertel, 1942). This requisite increase in relative vorticity amplifies the short wave in the westerlies, generating a likely parent trough for a surface cyclone.

Occasionally, an existing cyclonic circulation is masked by divergence as it crosses a mountain range, only to reappear in the lee equatorward of its initial latitude. The portion of the cyclonic circulation equatorward of the storm center descends the mountain range in the direction of flow, undergoing the vorticity adjustments described above and intensification. Easterly airstreams poleward of the storm center ascend the mountain range and become anticyclonic as their relative vorticity is reduced. The new position of the lee cyclone is no longer in synchrony with the old low center, but now coincides with that equatorward portion of the original circulation that was cyclonically enhanced by its passage over the mountain barrier. In North America, cyclonic parent systems enter the Rocky Mountain complex from the northwestern United States, only to form a lee cyclone in the southern Great Plains.

Frontal cyclones and upper level development

Cyclones also form and strengthen as a result of rising air and the destabilization of geopotential height surface inherent in baroclinic zones. Airflow across a strong surface front, such as that found in midlatitude east coast regions during winter, induces powerful geopotential height changes and vertical motions that produce or amplify a disturbance in the troposphere. Amplification of the wave disturbance increases kinetic energy and rotation in the upper atmosphere and activates an unstable cascade of positive physical feedbacks that enable a cyclone to grow.

The interaction between an upper tropospheric short wave and a surface cyclone can be described in terms of the advection of temperature and vorticity. A parent trough, by definition, is a local anomaly of reduced geopotential height and elevated relative vorticity. As air flows through a geopotential trough, it experiences a relative vorticity increase as it approaches the trough axis, followed by a relative vorticity decrease as it moves downstream. Thus, upwind of the trough axis, in a zone of negative vorticity advection, air parcels undergo convergence in the upper troposphere, while downstream, in the zone of positive vorticity advection, air parcels diverge as they exit the trough (Figure E21). According to the principle of continuity, divergent horizontal motion in the upper troposphere (above the level of non-divergence) is accompanied by ascent, which reduces pressure and forces convergence at the surface below – cyclogenesis (Palmén and Holopainen, 1962). Once a low-pressure feature has developed, the surface pressure gradient produces a low-level circulation that advects cold and warm air across the initially stationary temperature gradient.

Quasigeostrophic theory defines two key elements of cyclone intensification, geopotential height falls and upward vertical velocity, in terms of differential vorticity and temperature advection with height. Geopotential height falls, which define the amplification of upper tropospheric disturbances, occur in zones of decreasing cold air advection with height and positive vorticity advection. The primary location for upper-level development is downstream of the trough axis, in the regions overlying and behind the surface cold front. In cases of rapid, or explosive, cyclogenesis, upper-level intensification is enhanced by intrusions of stratospheric air and warm air advection aloft (Uccellini et al., 1985).

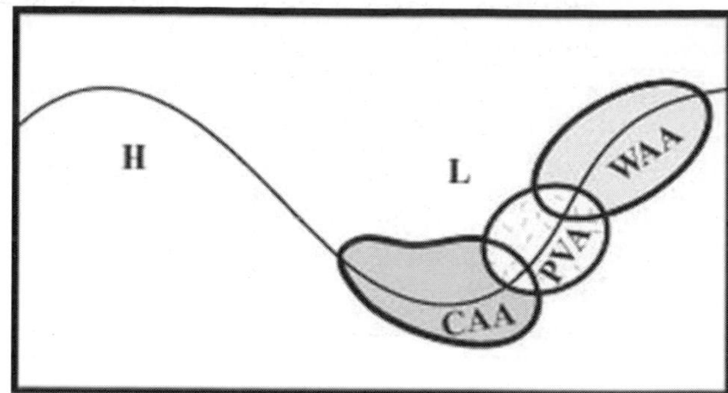

Figure E21 Idealized vorticity and temperature advection patterns in relation to an upper-level baroclinic wave.

Rising air defines the location and strength of the surface low by forcing low-level pressure falls, thereby enhancing the pressure gradient and wind speed. Such upward motion is preferred in locations where increasing positive vorticity advection aloft and warm air advection coincide (Bjerknes and Holmboe, 1944; Hoskins et al., 1978; Sutcliffe, 1947; Trenberth, 1978). Because the optimum dynamics for rising air occur downstream of the upper tropospheric geopotential height minimum, midlatitude systems tilt westward with height. This tilted configuration is distinct from other types of low-pressure system, including cold core lows that strengthen with height (e.g. polar lows) and warm core lows that weaken with height (e.g. tropical cyclones).

The mutual enhancement of upper and lower tropospheric pressure features through the advections described above is essential to the process known as baroclinic instability (Charney, 1947; Eady, 1949). Motions that increase surface pressure falls and upper tropospheric geopotential height falls also serve to strengthen existing horizontal temperature gradients, enhancing baroclinicity (Eliassen, 1966; Sawyer, 1949). A more baroclinic environment is conducive to strong upper-level winds, increased advection and more powerful vertical motion.

As a cyclone strengthens and moves, its low-level convergent circulation entrains moisture, which is then lifted across a broad expanse of the storm's interior. Massive condensation and latent heat release result, further destabilizing the atmosphere and contributing to ongoing vertical motion. Through vertical motion, which strengthens the surface pressure gradient and drives low-level advection, moisture entrainment becomes an additional component of the positive feedback system. The cyclone's intensification can theoretically proceed until its low-level temperature gradient is eliminated or no longer effectively oriented to the advecting flow, a phase known broadly as occlusion.

Cyclone structure, weather and climate

The classical cyclone model of the Norwegian school describes the evolution of surface pressure and fronts in a developing low,

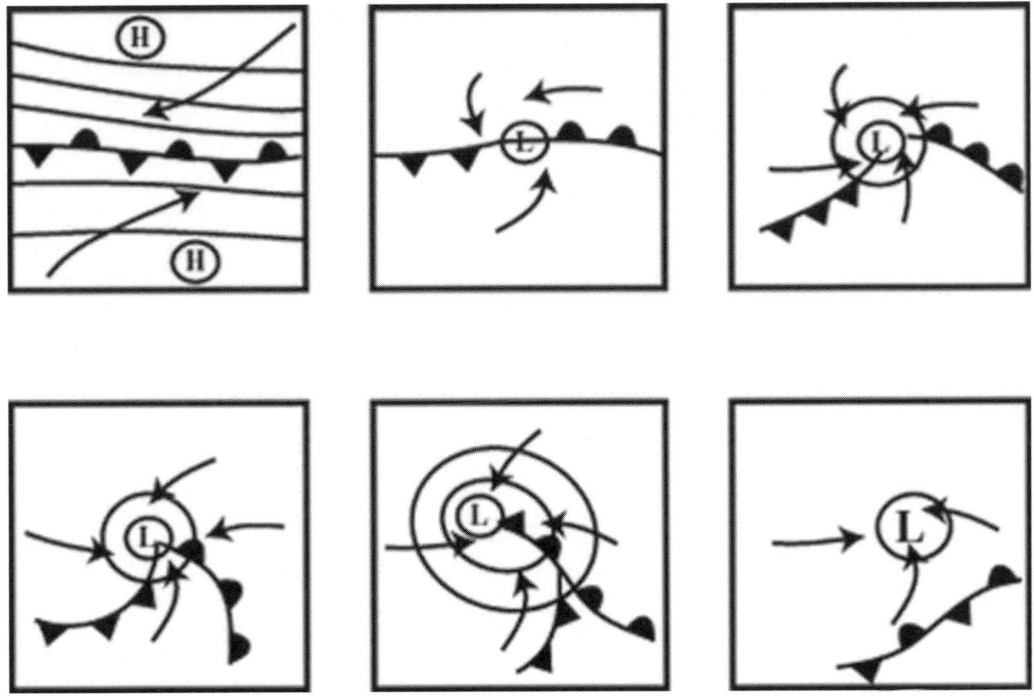

Figure E22 Classical Norwegian cyclone model (after Bjerknes and Solberg, 1922).

based on northern hemisphere observations (Bjerknes and Solberg, 1922). Despite an absence of upper air data during its formulation, this model still captures and generalizes the principal phases of the cyclone life, from cyclogenesis to occlusion.

In the classical model a cyclone originates as a low-pressure disturbance on an existing stationary front. The frontal boundary deforms as cyclonic circulation develops, with active fronts leading the cold air north and west of the low and the warm air equatorward of the low in a cyclonic migration about the center of the low. During the mature phase of the cyclone the surface storm is characterized by regions of markedly different temperature and humidity. Warm air is advected poleward east of the low from within the system's warm sector, where it ascends passively along slantwise trajectories. Cold air sweeps southeastward upstream of the low center, forcing aloft warm air in its path. The high density of the cold air north and west of the low center, coupled with the strong pressure gradient there, enable the cold front to propagate more rapidly than its warm counterpart, a trend that gradually erodes the region of warm air within the storm.

Occlusion, the source of much confusion in the analysis of cyclone evolution, is seen in surface analysis as the general separation of the cyclone center from its cold and warm fronts (Wallace and Hobbs, 1977). In an idealized model of occlusion, polar air behind the cold front appears to overtake the warm front, eliminating or reducing the horizontal temperature gradient that was initially present at the surface (Bergeron, 1937). An alternative form of occlusion, known as instant occlusion, arises when a comma-shaped cloud in cold air poleward and upstream of a polar front overtakes and merges with the front and its attendant cloud-field. The process of occlusion often marks a turning point in the cyclone's ability to generate the rising motion and geopotential height falls necessary for survival.

The cloud and precipitation fields within a cyclonic system are complex, but can be explained in terms of the vertical motion of airstreams and their frontal boundaries (Browning and Harrold, 1969; Harrold, 1973). The broad corridor of stratiform cloud cover and imbedded precipitation downstream of the low center may be attributed, in part, to slantwise ascent along and in the advance of the warm front. Similarly, the rapid, assertive motion of the cold front frequently corresponds to sharp linear bands of cumuliform cloud cover and shower-type precipitation. Between the fronts, warm, frequently moist air supports general convective cloud development. Fronts correspond spatially to cloud and precipitation edges and marked transitions in humidity and wind direction.

The basic features of the Norwegian model, the surface low, fronts and their evolution toward occlusion, are adequately described by quasigeostophic energetics. When the synergy between upper-level jet streaks and surface fronts is coupled to this model, much of the typical cyclone behavior is explained. However, precipitation and cloud features are often more complex than suggested by the dynamics of isobaric models and do not correspond directly with the traditional lifting mechanisms of the polar front model. To describe these characteristics, an alternative approach to cyclone description, known as conveyor belt theory, may be employed.

Because three-dimensional flow in the atmosphere does not follow isobaric surfaces, estimation of vorticity and thermal advection along surfaces of constant pressure can provide only a crude measure of storm evolution. Such estimation is adequate locally, for point determinations of geopotential height

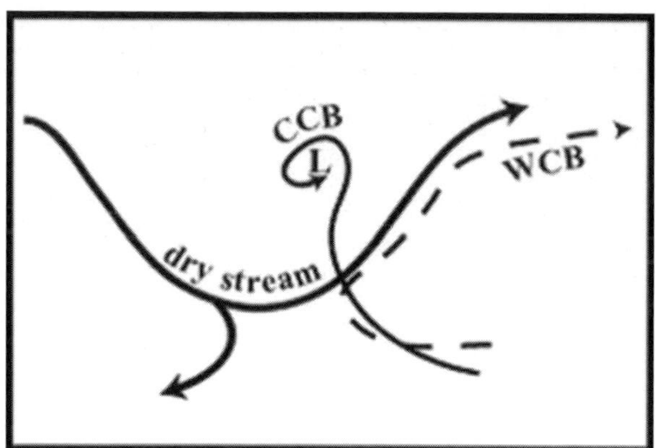

Figure E23 Idealized conveyor belt airflow through a midlatitude cyclone.

change and vertical motion, but is difficult to extrapolate over large areas. Consequently, the quasigeostrophic description of an extratropical cyclone is accurate at each point of assessment, but simplistic in overall scope.

Isentropic analysis provides a more accurate representation of actual airflow and a more intuitive one (Green et al., 1966). Unsaturated air flows along isentropic surfaces (surface of constant potential temperature), which ascend toward cold air and descend in warm regions. Alternatively, the motion of saturated air is well-approximated by the contours of equivalent potential temperature surfaces. When the circulation of a short wave is mapped on representative isentropic surfaces and placed in a relative-motion framework (the translation of the system is subtracted from the environmental winds), several common wind features emerge (Figure E23).

For a northern hemisphere system, two distinct corridors of air dominate cyclone airflow aloft. Arriving from the upper troposphere/lower stratosphere, a dry stream of air flows from north and west of the low, descending, curving cyclonically, and then flowing downstream while ascending. The dry stream closely resembles the position of the shortwave as seen on an isobaric map. However, the isentropic depiction reveals the descent and ascent of this airstream along its path, as well as a significant bifurcation that results south and west of the surface low.

The dry stream interacts with a second airstream, often termed the warm conveyor belt (WCB) or polar front conveyor belt. The WCB approaches the cyclone from the south and east, curves anticyclonically and ascends parallel to the dry stream, in stark contrast to its temperature and humidity properties. Because the dry stream originates in cold, dry environs, its confluence with the warm, moist WCB is a natural frontal boundary, induced by relative motion within the moving system. Examination of a vertical profile of the dry stream–WCB confluence reveals that this upper front tilts forward with height. Thus a front structure exists at many elevations downstream of the traditional model of the surface cold front and located above the warm sector of the cyclone. Advection of cold dry air above the low-level warm sector constitutes generation of static instability. The slantwise ascent of the moist WCB and its western boundary with the dry stream correspond well spatially with portions of the classic "comma cloud" pattern in cyclones, the comma tail and baroclinic leaf features.

Beneath the dry stream and WCB and their confluent boundary flows a third airstream, termed the cold conveyor belt (CCB) by Carlson (1980) and the cyclonic moist airstream by Bierly and Winkler (2001) (they found the airstream to originate in air as warm as, or warmer than, that of the WCB during early phases of the storm life) and polar trough conveyor belt. The CCB, like the WCB, originates in the low troposphere and ascends as it approaches the cold air north and west of the cyclone center. Although contradictory, most analyses of the CCB suggest that it wraps cyclonically around the low center, where it then descends to near the surface (Schultz, 2001). The CCB migrates from a source area near that of the WCB origin shortly after cyclogenesis to a cooler region of origin at high latitude later during the cyclone life span. The CCB delivers moist air beneath the dry airmass domain of the dry stream, elevating the potential instability of the atmosphere near the comma cloud head about the low center.

According to conveyor belt theory, the cloud regions of the familiar comma pattern are explained in terms of relative-wind airstream ascent along isentropic surfaces, confluence between airstreams of different origin and the static and convective instability created by superposition of airstreams of different origin. Temperature, humidity and wind direction aspects of fronts all derive from the airstream confluence model. Finally, upper and low-level fronts are naturally defined through examination of relative wind confluence patterns at multiple levels. Occlusion is seen as the cessation of temperature advection through the onset of airflow parallel to isohypses on surfaces of constant potential temperature.

Cyclone climatology

The geographical distribution of dynamic controls on cyclone development leads to spatial and seasonal patterns in their occurrence. In the northern hemisphere winter, extratropical cyclones develop most frequently over western midlatitude ocean basins and along east coast margins, particularly the eastern Asia–western Pacific Ocean region and the eastern North America–western Atlantic Ocean (Klein, 1957; Pettersen, 1956; Whittaker and Horn, 1982). In both cases, warm ocean currents, the Kuroshio current of the Pacific and the Gulf Stream–North Atlantic Drift of the Atlantic, flow poleward along the cold margin of a large continental landmass. Disturbances moving through these comparable areas amplify from the combined effects of strong baroclinicity and latent heat release over the warm water surface. As thermal contrasts between land and sea diminish toward the warm season, coastal cyclogenesis frequency wanes.

Cyclogenesis in the lee of high elevations in the Coastal Range of British Columbia (Alberta cyclones), southern US Rocky Mountains (Colorado cyclones) and the Alps (Genoa cyclones) is of moderate importance during winter; Rocky Mountains lee cyclones grow in significance during the spring transition of the westerlies. Lee cyclogenesis is less common during the warm season as baroclinicity in the westerlies weakens and contracts poleward, but resumes during fall, particularly Alps–Adriatic cyclogenesis, which peaks in November. The frequency of lee cyclogenesis reflects the passage of short wave disturbances during periods of hemispheric wave number change and proximity of high-elevation disturbance zones to the strongest baroclinicity within the circumpolar vortex of the westerlies.

During the northern hemisphere warm season the origin zones and paths of cyclones move northward with the position of the now-weakened polar front. Storm development during summer is focused in the Hudson–James Bay region and to a lesser extent in the northwest Atlantic Ocean.

Cyclone activity in the southern hemisphere has received considerably less attention than that of the northern hemisphere. Despite fewer and weaker continent–ocean temperature contrasts in the southern hemisphere, cold season cyclogenesis frequency is most pronounced along the Australian east coast and in the lee of the Andes of southern Argentina (Sinclair, 1995). Both locations are characterized by strong SST gradients. In summer, southern hemisphere cyclone activity decreases, with most storm initiation focused east of Argentina and in the Indian Ocean southeast of Madascar.

Greg Bierly

Bibliography

Bergeron, T., 1937. On the physics of fronts. *Bulletin of the American Meteorological Society*, **18**: 265–275.

Bierly, G.D., and Winkler, J.A., 2001. The structure and evolution of cold-season Colorado cyclones. *Weather and Forecasting*, **16**: 57–80.

Bjerknes, J., and Holmboe, J., 1944. On the theory of cyclones. *Journal of Meteorology*, **1**: 1–22.

Bjerknes, J., and Solberg, H., 1922. Life cycle of cyclones and the polar front theory of atmospheric circulation. *Geophys. Publik.*, **3**(1): 1–18.

Browning, K.A., and Harrold, T.W., 1969. Air motion and precipitation growth in a wave depression. *Quarterly Journal of the Royal Meteorological Society*, **95**: 288–309.

Buzzi, A., and Tibaldi, S., 1978. Cyclogenesis in the lee of the Alps: a case study. *Quarterly Journal of the Royal Meteorological Society*, **104**: 271–287.

Carlson, T.N., 1980. Airflow through mid-latitude cyclones and the comma cloud pattern. *Monthly Weather Review*, **108**: 1498–1509.

Charney, J.G., 1947. The dynamics of long waves in a baroclinic westerly current. *Tellus*, **42A**: 28–40.

Eady, E.T., 1949. Long waves and cyclone waves. *Tellus*, **1**(3): 33–52.

Eliassen, A., 1966. Motions of intermediate scale: fronts and cyclones. In Hurley, P.M., ed., *Advances in Earth Science*, Cambridge, MA: MIT Press, pp. 111–138.

Ertel, H., 1942. Ein Neuer hydrodynamischer Wirbelsatz. *Meteorologische Zeitschrift*, **59**: 271–281.

Green, J.S.A., Ludlam, F.H., and McIlveen, J.F.R., 1966. Isentropic relative-flow analysis and the parcel theory. *Quarterly Journal of the Royal Meteorological Society*, **92**: 210–219.

Harrold, T.W., 1973. Mechanisms influencing the distribution of precipitation within baroclinic disturbances. *Quarterly Journal of the Royal Meteorological Society*, **99**: 232–251.

Holopainen, E.O., 1984. Statistical local effect of synoptic-scale transient eddies on the time-mean flow in the Northern Hemisphere winter. *Journal of Atmospheric Science*, **41**: 2505–2515.

Hoskins, B.J., Draghici, I., and Davies, H.C., 1978. A new look at the ω-equation. *Quarterly Journal of the Royal Meteorological Society*, **104**: 31–38.

Klein, W.H., 1957. *Principal Tracks and Mean Frequencies of Cyclones and Anticyclones in the Northern Hemisphere*. Resource Paper No. 40, US Weather Bureau. Washington, DC: US Government Printing Office.

Newton, C., 1956. Mechanism of circulation change during lee cyclogenesis. *Journal of Meteorology*, **13**: 528–539.

Palmén, E., 1949b. Meridional circulations and the transfer of angular momentum in the atmosphere. *Journal of Meteorology*, **6**: 429–430.

Palmén, E., 1951a. The aerology of extratropical disturbances. In *Compendium of Meteorology*, T.F. Malone (ed.), American Meterological Society, 599–620.

Palmén, E., 1951b. The role of extratropical disturbances in the general circulation. *Quarterly Journal of the Royal Meteorological Society*, **77**: 337–354.

Palmén, E., and Holopainen, E.O., 1962. Divergence, vertical velocity and conversion between potential and kinetic energy in an extratropical disturbance. *Geophysica*, **8**: 89–113.

Pettersen, S., 1956. *Weather Analysis and Forecasting*, 2nd edn. vol. 1. New York: McGraw-Hill.

Petterson, S., and Smeybe, S.J., 1971. On the development of extratropical cyclones. *Quarterly Journal of the Royal Meteorological Society*, **97**: 457–482.

Rossby, C.-G., 1940. Planetary flow patterns in the atmosphere. *Quarterly Journal of the Royal Meteorological Society*, **66**: 68–87.

Sawyer, J.S., 1949. The significance of dynamic instability in atmospheric motions. *Quarterly Journal of the Royal Meteorological Society*, **75**: 364–374.

Schultz, D.M., 2001. Reexamining the cold conveyor belt. *Monthly Weather Review*, **129**: 2205–2225.

Sinclair, M.R., 1995. A climatology of cyclogenesis for the Southern Hemisphere. *Monthly Weather Review*, **123**: 1601–1619.

Sutcliffe, R.C., 1947. A contribution to the problem of development. *Quarterly Journal of the Royal Meteorological Society*, **73**: 370–383.

Trenberth, K.E., 1978. On the interpretation of the diagnostic quasi-geostrophic omega equation. *Monthly Weather Review*, **106**: 131–137.

Uccellini, L.W., Keyser, D., Brill, K.F., and Walsh, C.H., 1985. Presidents' Day cyclone of 18–19 February 1979: Influence of upstream trough amplification and associated tropopause folding on rapid cyclogenesis. *Monthly Weather Review*, **112**: 962–988.

Wallace, J.M., and Hobbs, P.V., 1977. *Atmospheric Science: An Introductory Survey*. New York: Academic Press.

Whittaker, L.M., and Horn, L.H., 1982. *Atlas of Northern Hemisphere Extratropical Cyclone Activity, 1958–1977*. Madison, WI: University of Wisconsin Press.

Cross-references

Airmass Climatology
Climatology
Europe, Climate of
Vorticity
Mediterranean Climates
North America, Climate of
Synoptic Climatology
Vorticity

F

FERREL CELL

A Ferrel Cell is a circulation pattern named for William Ferrel (1817–1891), an American meteorologist who discovered the effects of the Earth's motion in wind systems. Although George Hadley had recognized the convective nature of the air in 1735, and M. F. Maury, in 1855, had proposed a good model for the atmospheric circulation, both lacked an appreciation of the nature of the westerlies. Ferrel brought a fundamental knowledge of mechanics and mathematics to the subject (Clayton, 1923) and first sketched the diagram of the planetary surface winds that appears in almost every geography textbook (Figure F1). He pointed out that the total movement of air to the east must balance the total movement to the west or else the Earth's rotation would be accelerated or retarded. His model also included a profile of the upper air winds, which incorporated Hadley Cells (simple convection) north and south of the equator and at the poles; in the latter a polar anticyclone operated by centrifugal force was postulated. A reversed circulation was shown in the temperate belts (30–60 lat.), thereby recognizing the component of the westerlies.

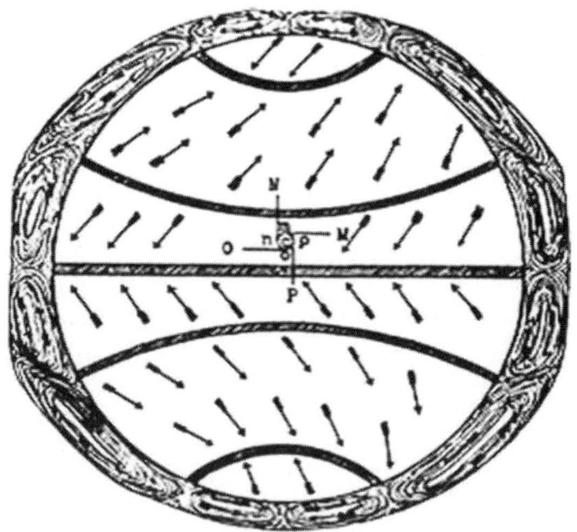

Figure F1 Atmospheric circulation (from Ferrel, 1856).

The warm air of the tropical cell sinks along the high-pressure belts and in parts moves into the northerly setting westerly circulation. This circulation component is now known as the Ferrel Cell. Even though modern models reject the polar anticyclonic cell, the Ferrel and Hadley cells are now fairly well confirmed by observations at all levels. The Ferrel Cell plays a major part in the poleward energy (mainly heat) transport. North of about latitude 38°N in the northern hemisphere the precipitation exceeds evaporation, so that water vapor (representing latent heat) must be transported northward. Beyond this contribution, in an 1856 paper Ferrel also helped expand the convection cell interpretation of vertical motion in the oceans, proposed by A. von Humboldt in 1814, and later developed by M. F. Maury.

Ferrel's Law concerns the principle that relates the deflection of winds by the Earth's rotation, now generally known as the Coriolis Effect. Ferrel stated the law "If a body moves in any direction upon the earth's surface, there is a deflecting force arising from the earth's rotation, which deflects it to the right in the northern hemisphere, but to the left in the southern hemisphere" (Ferrel, 1856).

Rhodes W. Fairbridge

Bibliography

Clayton, H.H., 1923. *World Weather*. New York: Macmillan.

Ferrel, W., 1856. The winds and currents of the oceans. *Nashville Journal of Medicine and Surgery*.

Ferrel, W., 1889. *A Popular Treatise on the Winds*. New York.

Cross-references

Atmospheric Circulation, Global
Coriolis Effect
Hadley Cell
Maury, Matthew Fontaine
von Humboldt (Baron), Friedrich Alexander
Winds and Wind Systems

FLOHN, HERMANN (1912–1997)

Hermann Flohn was born on 19 February 1912 in Frankfurt, Germany. Considered by many as one of the world's greatest climatologists, he began his lifelong learning in medical meteorology and bioclimatology when studying the weather-related spread of influenza in Germany. Pursuing an academic career, at the age of 28 he worked at the regional meteorological offices in Leipzig and Hamburg. Later he was the climatologist at Potsdam for the German air force. This career guided him into the understanding of the climate of the Earth and an interest in synoptic climatology.

Under the Nazi regime in 1941 he was charged to investigate the upper-air flow of three routes over Asia (Siberia, India, and Burma) to Japan, which led to several publications after the Second World War concerning the atmospheric character over Siberia ("Zum Klima der freien Atmosphäre über Sibirien. I. Temperatur und Luftdruck in der Troposphäre über Jakutsk", *Meteorologische Zeitschrift*, 1944), India's monsoons ("Witterungs-Singularitäten im Monsunklima Indiens", *Annals of Hydrography and Marine Meteorology*, 1943), and the equatorial westerlies over the Indian Ocean ("Studies on trade-wind circulation and equatorial westerlies", in *Procès-Verbaux des Séances de l'Association de Météorologie. Union Géodesique et Géophysique Internationale.* IV. Assemblée Générale, August 1951).

Following his release as a German prisoner in 1946, he became involved in establishing a new German weather service, of which he became Head of Research. In 1961 he became the Founding Director and then head of the department at the Institute of Meteorology of Bonn University. He retired as Professor Emeritus in 1977. During this period he was inspired by the birth of his first grandchild and began focusing his research on climate change. The result of his former research interest in medical meteorology and climate change culminated into: "How does human activity influence the climate?" Flohn's enthusiasm for his new avenue of research resulted in his participation in the first international meeting on climate modification in 1971 and contributed to several conferences on the subject: the First World Climate Conference (1979), the Second World Climate Conference (1990), and he held the position of advisor on the Enquete Commission of the German parliament, *Protecting the Earth's Atmosphere*. His concentration on climate modification produced about 360 publications and membership in numerous scientific societies such as the Bavarian Academy, the Leopoldina in Saxonia, and the Royal Belgium Academy.

Cameron Douglas Craig

Major contributions

Flohn, H., 1957. Large-scale aspects of the "summer monsoon" in South and East Asia. *Journal of the Meteorological Society of Japan*, 75th annual volume, **11**: 180–186.

Flohn, H., 1969. *Climate and Weather*. World University Library. New York: McGraw-Hill.

Flohn, H., Henning, D., and Korff, H.C., 1974. Possibilities and limitations of a large-scale water budget modification in the Sudan–Sahel belt of Africa. *Meteorologische Rundschau*, **27**: 100–109.

Flohn, H., Kapala, A., Knoche, H.R., and Michel, H., 1990. Recent changes of the tropical water and energy budget and of midlatitude circulations. *Climate Dynamics*, **4**: 237–252.

Flohn, H., Kapala, A., Knoche, H.R., and Michel, H., 1992. Water vapor as an amplifier of the greenhouse effect: new aspects. *Meteorologische Zeitschrift*, **122**: 122–138.

Reference

Obituary, January 1998. Klaus Fraedrich Hermann Flohn. *Quarterly Journal of the Royal Meteorological Society*, **124**(546), Part B, p. 653.

FOG AND MIST

Fog

Fog is a stratus cloud that lies on, or very close to, the surface of the Earth. The horizontal visibility in fog is reduced to less than one kilometer, according to international definition. Fog is an aggregate of very small water droplets in a size range of 10–50 microns, typically in concentrations of 10–100 per cc. The air in fog usually feels wet because the humidity is very high, often but not necessarily above 95%, and the observer is experiencing contact with many small droplets.

The atmosphere always contains an adequate number of condensation nuclei, though in certain circumstances there will be differing mixes of types of nuclei, e.g. along an ocean shoreline there will be a greater than normal number of salt nuclei, which are highly water-loving, or hygroscopic. The haziness of beaches with breaking surf is accounted for by the fact that salt particles start to take on water at relative humidities as low as 60–70%.

A critical relative humidity beyond which condensation will be initiated, and fog form, can be achieved in four ways: (1) addition of water vapor to the volume of space in question; (2) cooling of the volume by contact with a colder surface; (3) cooling by infrared radiation from the volume itself; (4) expansional cooling due to ascent of the airmass. Of these four ways, the first two are of major importance in the formation of fog.

Fogs may be broadly classified as those that form within airmasses and those that form at the boundaries between different airmasses, i.e., in conjunction with fronts. In outline form:

- I Airmass fogs
  - A Advection types
    - 1. Transport of warm air over a cold surface
      - (a) Land and sea breeze fog
      - (b) Sea fog
    - 2. Transport of cold air over a warmer, wet surface
      - (a) Steam fog
      - (b) Arctic "sea smoke"
  - B Radiation types
    - 1. Ground fog (ice fog if particles are ice crystals)
    - 2. High inversion fog
  - C Expansional cooling for (upslope fog)
  - D Combinations of A, B, and C
- II Frontal fogs
  - A Prefrontal warm front fog
  - B Frontal passage fog

Advection implies primarily horizontal transport, though vertical transport may be significantly present in certain circumstances. When relatively warm air is carried over a cooler surface its temperature is lowered by contact cooling (conduction) and the relative humidity rises. After the critical value is exceeded, condensation on the cloud condensation nuclei (CCN) begins and fog forms. The cooling of the surface layers is carried to higher layers by turbulent mixing, which results from wind drag over the surface. If the air becomes slightly unstable because of this mixing the fog may have a low ceiling, as is common with the California coastal stratus.

When cold, dry air moves over a warmer, wet surface, rapid evaporation takes place and saturates the colder air. The resultant condensation is the steam fog (steam smoke) seen in the Arctic Ocean areas.

Radiation fog results from a different set of conditions. If skies are clear and the air is relatively still, nocturnal cooling of the ground by escaping terrestrial radiation chills the surface layers of air. Fog forms at ground level when the air temperature is lowered to the dewpoint. Slight turbulent mixing increases the depth of cooling and the thickness of the fog. Ground fog tends to form in the late night or early morning hours; it then "burns off" in the later morning hours when solar shortwave radiation penetrates the shallow cloud, warms the ground, lowers the relative humidity, and causes the water droplets to evaporate.

Upslope fog forms when a stable airmass moves slowly up over higher terrain, cools by expansion, and finds its temperature lowered to the dewpoint. This fog is characteristic of the Western Plains states in the United States.

Frontal fogs form primarily when additional water vapor from evaporating precipitation elevates the dewpoint temperature, and evaporative cooling lowers the temperature of the air. Such fogs are most characteristic of warm fronts.

Mist

According to international definition, mist consists of an aggregate of microscopic sized droplets (~10 microns) producing a thin, grayish veil over the landscape, reducing visibility to a lesser extent than fog. Mist is intermediate in all respects between damp haze and fog. However, in the United States the term *mist* has come to have a popular usage of a hydrometer that is intermediate between fog and drizzle. *Oregon mist* or *Scotch mist* are terms used to describe the occurrences of very light, "misty" precipitation. Mist particles range in size from 50 to 500 microns, the latter large enough to fall from the cloud.

Trees and other objects such as grasses collect moisture from drifting heavy fog or mist, as sometimes occur in the Redwood Forests in Northern California. Fog-drip can collect as much as 0.05 inch of water in a single night – the equivalent of a light shower. This phenomenon prevents excessive aridity in the coastal forests during the rainless California summers.

John A. Day

Bibliography

Day, J.A., and Sternes, G., 1970. *Climate and Weather*. Reading, MA: Addison-Wesley.

Aguado, E., and Burt, J.E., 2001. *Understanding Weather and Climate*. Upper Saddle River, NJ: Prentice Hall.

Nese, J.M., and Grenci, L.M., 1988. *A World of Weather: Fundamentals of Meteorology*, Dubuque, IO: Kendall/Hunt.

Schaefer, V.J., and Day, J.A., 1981. *A Field Guide to the Atmosphere*. Boston, MA: Houghton Mifflin.

Cross-references

Aerosols
Air Pollution Climatology
Atmospheric Nuclei and Dust
Cloud Climatology
Dew/Dewpoint
Precipitation

FOLKLORE, MYTHS AND CLIMATE

Why do humans feel the need to personify or deify the weather? Many academics have attempted to address that need through analysis of the human psyche, ranging from the landmark psychological works of Sigmund Freud and Carl Jung to the anthropological analyses of J.G. Frazer and Joseph Campbell. Some works have linked these myths and legends to deep-seated sexual or cultural motivations in our societies. Others have attempted to associate the diverse mythologies to a basic human trait to create explanations for our surroundings. Yet, despite spirited debate, all can agree that weather is a prevalent aspect of all myths and folklore of the world's myriad societies.

In order to explain the various atmospheric phenomena that early societies and civilizations experienced and suffered, these cultures often depicted higher-order beings, such as gods and other supernatural beings, who were given the various attributes of weather. For the Caribbean islands the god Hurrican was the supernatural personification of the oft-deadly tropical cyclones that strike the area – and, indeed, his name has become the term used to refer to the deadliest of these tropical storms in the Atlantic Ocean, the hurricane. In ancient Europe, Jupiter (in Roman myth) or Zeus (in Greek myth) was a god who is often depicted as a god of weather whose weapon of choice was a lightning bolt. For example, in Homer's masterpieces, the *Iliad* and the *Odyssey*, lightning strikes ofhumans or their structures were considered as vehicles for Zeus's divine justice.

Because weather was the property of the gods, generally imitation of weather was the height of sacrilege. As Virgil recounted, Salmoneus, king of Elis, discovered the consequences of such "hubris" first hand. Salmoneus desired to be worshipped as a god by his people and to this end he rode his chariot over a bronze bridge to show how he could make thunder, while he hurdled torches hurled on either side of his chariot to demonstrate that he could produce lightning as well. The all-mighty Zeus was not amused and decided to throw the real thing down at Salmoneus. The pretender to divinity subsequently was burnt to a crisp and sentenced to Hades for his blasphemy.

Priests of many early cultures attempted to divine the future and the favor of the gods based on atmospheric signs and portents. In Rome, for example, divination was in the hands of the College of Augurs. Lightning was one of the greatest of the potents for which the augurs scanned the skies. Of great interest to the augurs was the specific direction that lightning took as it crossed the sky .If the flash progressed from left to right, the gods favored the current state of affairs in Rome. If, however, it passed from right to left, the augur was duty-bound to proclaim that the gods did not favor the current actions in the Roman government.

Yet, surprisingly, many of these myths and legends are rooted in accurate observations. Ancient civilizations consistently displayed a strong interest in observing the weather. For example, Homer's epic poem, the *Odyssey*, consistently shows a credible set of weather observations that indicates the tragedies experienced by the ancient Greeks may have been caused by a cyclonic storm crossing the area in the early summer. It is amazing that, given the narrative restrictions of the media imposed on Homer, the *Odyssey* contains vivid and apparently accurate descriptions of weather. Indeed, it may be possible to interpret an episode near the saga's end in the light of an only recently defined meteorological phenomenon, a microburst, a thunderstorm's small- scale blast of air (up to 100 miles per hour) that strikes the ground and spreads out:

> "That said, he [Poseidon] massed the clouds, and, as he gripped his trident, whipped the surge and urged all winds to blow at will . . . Eurus [the East Wind], Notus [the South Wind], and voracious Zephyr [the West Wind] and Boreas [the North Wind], who's born in the bright ether, attacked together" *(V, 293-5)*. . . . The force of all winds crushes me *(V, 304)*.

Similarly, research in the area of the Black Sea by Walter Pitman and William Ryan of Lamont-Doherty Laboratory indicates that the Great Flood chronicled in the Bible as "Noah's Flood", and which many people thought mythical, may actually have occurred as a result of climate change producing an inundation of the Black Sea from the Mediterranean.

As Joseph Campbell has noted, myths often change as civilizations change. With the eventual prominence of monotheism, medieval societies often depicted the elements of weather as aspects of fundamental evil or divine displeasure. For example, by the 1500s the practice of ringing church bells had become established as a means of dispersing lightning. As a result, many medieval bells still bear the inscription "Fulgura Frango" (I break up the lightning). Of course, ringing metal bells posed a certain danger to the bell-ringer. A medieval scholar wrote an interesting treatise on the subject entitled "A proof that the ringing of bells during thunderstorms may be more dangerous than useful", in which he stated that, in 33 years, lightning had struck 386 church towers and killed 103 bell-ringers. Conversely, in the year 1360, when his army was battered by an incredible hailstorm that killed 1000 soldiers and 6000 horses, English King Edward III, sensing apparent divine displeasure with his army, turned to the Cathedral of Chartres and "vowed to the Virgin that he would conclude a treaty".

Even after the industrial revolution, weather myths persisted. By the mid-nineteenth century the weather gods and goddesses of early cultures, for example, had morphed into the heroic superman myths of the American West – the Great Lakes' Paul Bunyan, Texas's Pecos Bill or the more obscure Febold Feboldson of Nebraska. These characters tangled with, and normally tamed, the violent weather of the Great Plains and Midwest in a manner reflective of the tough character of the people of that time and place. When Pecos Bill needed a stronger beast than a wild horse to break, he saddled up a tornado and rode it into submission. It was so cold in Paul Bunyan's logging camp one bitter winter that people's words froze and they had to be stored until the spring thaw to hear what had been said. Febold Feboldson once had to hire English fog-cutters to come and cut a Nebraska fog, which he then laid in strips along the dirt roads where it continues to drip out even today, especially in spring. These stories served to re-enforce and strength the "can-do" spirit of pioneers of the west.

Do weather-related myths still exist today? Many people would immediately answer no, citing that today's civilization is too scientific to still cling to old-fashioned weather myths and folklore. Yet many modern people still believe that washing their cars will somehow lead to the appearance of a thunderstorm or that a groundhog's shadow in Pennsylvania on February 2 in some way forecasts the coming of spring. And elements of weather, such as the still-elusive "ball lightning", are still linked in popular society to supernatural causes, such as extraterrestrial aliens and "UFOs".

As our human history has shown, because of their unique personal impact on every individual of the planet, weather and climate continue to hold a fascination. That enthrallment has led to their prominent positions in our myths, folklore and legends, and helped to form the basic foundations and cultures of our societies.

Randall S. Cerveny

Bibliography

Campbell, J., 1949. *The Hero with a Thousand Faces*. Princeton NJ: Princeton University Press.

Frazer, J.G., Sir, 1913. *The Golden Bough: a study in magic and religion*, London: Macmillan, 8 vols.

Ryan, W.B.F., and Pitman, W., 1998: *Noah's Flood: the new scientific discoveries about the event that changed history*. New York: Simon & Schuster.

Cross-references

Cultural Climatology
Literature and Climate

FROST

Frost may be considered as (1) a form of mineral; (2) a climatic condition; and (3) an economic situation. Each of these is dealt with sequentially.

Frost as a mineral

Frost is a solid phase of water. As a mineral it crystallizes according to the hexagonal system featuring six-sided plates, needles, clusters, and columns. The crystals are subject to innumerable variations of twinning and dendritic or arborescent growth. None of the variations is unusual in the mineral world, but the melting point of water limits observations, whereas similar crystallization can be found in commonly seen metallic and non-metallic minerals. The complex crystals are the object of a variety of platitudinal visual imitations in the forms of snow crystals drawn by artists and in reality as seen by frost on the windows.

Climatology of frost

From the climatologists' point of view, the frost of specific interest is *hoar frost* or white frost that accumulates on surfaces in places with appropriate temperatures. The white color is a product of small air bubbles in the ice, cutting down on the transparency of the ice crystal, the reaction of poor sky light and the lack of sunshine. At the critical temperature at which frost forms, sunshine would prove disastrous in a very short time. The "hoar" implies a gray or grayish tone often seen on objects covered with frost. Hoar frost or the veneer of ice crystals is indicative of three conditions:

1. Surfaces on which the frost forms must be 0°C or below.
2. The surrounding air is saturated at 0°C or slightly below.
3. Nuclei are present so that the process of sublimation can take place.

Each of these three conditions is part of the natural environment from time to time.

Surfaces suitable for the accumulation of frost are cooled by outward radiation and advection, so that the ambient temperature of 0°C or less is reached. The temperature of the atmosphere has already been reduced to the near-freezing condition by normal processes. The loss of heat to or below the freezing level by the atmosphere and surface objects in dry air and without reaching saturation creates a frost known as *black frost*. Vegetation, when exposed to freezing conditions, upon thawing will turn dark or black. The water that is part of the cell structure of the plant solidifies, cell walls burst, and the plant materials deteriorate.

Saturated air at the appropriate temperature begins to give up its moisture. If this is in the range above 0°C, the product is dew and, should the temperature drop, the dew particles solidify to coat the object with a veneer of ice, which is amorphous. Thus the saturation-level temperature is most critical in the formation of frost.

Nuclei are as essential for crystal formation as they are for dew, raindrop, or cloud formation. The surfaces on which frost accumulates depend on the presence of these nuclei. Nuclei can be dust particles, irregularities in the configuration of leaves, plant hairs, or a host of other features. It is difficult to conceive of a situation where an adequate number of nuclei would not be present.

Although the literature is sprinkled with terms related to frost such as frost-wedging, ice-wedging, frost-buckling, and others implying solid ice, under scrutiny many of these terms are not frost at all but gelification of water as a liquid. Innumerable citations imply that ice formed by sublimation in pore space and microcellular space in rocks, soils, and other material should be classified as frost. Again it is probably interstitial and interfacial water that has crystallized or solidified. There is no recognition in the literature that water vapor is found free in cellular or pore space of such consequence that sublimation could occur to form frost to the extent of creating expansion of material and the destruction of, for example, plant cells.

Frost as a hazard

Protoplasm in plants functions at varying temperatures for each plant within restricted ranges. Frost, as defined here, implies near 0°C temperatures, but some plants cease protoplasm functioning at temperatures higher or lower than 0°C. Thus a plant may suffer "chilling" injury but not as a result of frost or freezing.

It is conceivable in certain types of cavities and caverns that frost accumulates over a period of time and a repetition of the conditions may permit consolidation of this frost into ice masses. The condition is simulated in large cold-storage or deep-freeze rooms. Thus, large and small cavities opened to the atmosphere and near the surface experience frost activities. The results are such phenomena as ice-filled sink holes or ice caves.

Whatever the process of formation or product, frost is a symptom of a climatic situation in which temperatures have been reduced through radiation or advection to the freezing stage. Thus frost, particularly hoar frost, is a symptom of or forerunner to plant or crop loss.

Because frost can freeze plant tissue, which marks the end of growth for the plant, it is significant that this can occur at either end of the growing season. Since destruction is involved, it is evident that frost is a natural hazard along with other climatic hazards such as excessive precipitation, tornadoes, hail, etc. Most citations to frost overlook this hazard factor, perhaps because it is not as spectacular or forceful as high winds, high water, large hailstones, or heavy snowfalls. The agriculturist could well proclaim the facts by playing the game and trying to produce a crop that fits in between the last killing frost of the growing season in spring and the first killing frost at the end of the growing season in fall. In doing so, the farmer recognizes frost as a hazard.

The accumulated experiences of frost as a hazard leads to a considerable discussion in many climatology texts of techniques for frost or freeze prevention. These are generally beyond what the botanist and geneticist have already done to produce plants that mature in shorter periods of time, or by the climatologist who has defined the perimeters of frost incidence for any particular area. Indeed the greatest incidence of activity in frost prevention is in the marginal or fringe areas where, for reasons of possible economic gain, the farmer or orchardist is willing to take a chance on planting in the hope of being able to produce a crop.

Frost damage prevention techniques can be classified into several groups by procedures and practices:

1. Identification of the critical and perhaps typical physical site of low or high frost incidence in a general region. These can be defined as wind shelters, topographic favorable regions, or shore position, etc.
2. Development of physical equipment such as fans, heaters, brushes, sprinklers, and plant shields to modify temperatures or reduce radiation.

Whatever method or combination of methods is chosen to prevent destruction by frost, the choice is usually to modify temperatures a few degrees, usually not more than four or five. The techniques involve reducing radiation, improving wind circulation, discouraging sublimation, or creating a fog or smoke cover. The physical modification of the environment in the immediate situation of a frost or freeze hazard does not represent all that can be done.

Long before agriculture is considered for a particular area or crop, a climatic record or history would reveal the probabilities of frost or freeze incidence and the chance of success if the correct plans and methods are used. Additional terrain analysis helps to identify local microclimatic conditions in basins, alluvial fans, slopes, plains, or other topographic conditions. Peri-shore or valley sites could and should be evaluated.

In the competition for the early harvest–high price crop, farmers for many years have tried to protect young plants with shields of plastic or paper. Commercial outlets have provided the devices in large numbers. As is true of other procedures, the shield protects within limited temperature ranges. The shield in addition may speed growth and thus "harden" the plant against chill.

Another shield device is the drowning or flooding of crops, especially cranberries, to prevent low temperatures from damaging tender berries near the end of the growing season.

Benjamin Moulton and John E. Oliver

Bibliography

Critchfield, H.J., 1974. *General Climatology*, 3rd edn. Englewood Cliffs, NJ: Prentice-Hall.
Geiger, R., 1966. *The Climate Near the Ground*, rev. edn. Cambridge, MA: Harvard University Press.
Goody, R., 1995. *Principles of Atmospheric Physics and Chemistry.* New York: Oxford University Press.
Schaefer, V.J., 1964. Preparation of permanent replicas of snow, ice, and frost. *Weatherwise*, **17**: 278–287.
van Loon, G.W., and Duffy, S.J., 2000. *Environmental Chemistry.* New York: Oxford University Press.
Wallace, J.M., and Hobbs, P.V., 1997. *Atmospheric Science: An Introductory Survey.* San Diego, CA: Academic Press.

Cross-references

Agroclimatology
Climatic Hazards
Dew/Dewpoint
Phase Changes
Snow and Snow Cover

G

GENTILLI, JOSEPH

Dr Joseph Gentilli was born in the small Italian village of San Daniele de Friuli in 1913 and died in Australia in August 2000. Educated in Italy and the United States Dr Gentilli had a long and distinguished career of over 60 active years at the University of Western Australia. He was an outstanding scholar, teacher and researcher and he contributed over 150 works across a range of areas in geography including: migration studies, economic development in arid lands, regional planning, biogeography and refugees in Australia. His greatest contribution was in the field of climatology, and for this he was made a fellow of the Royal Meteorological Society and the Royal Geographical Society; and a member of the American Meteorological Society. Dr Gentilli's distinguished research career saw him prepare the first accurate climate and rainfall maps for Australia, Western Australia and Tasmania. He was also the first to recognize that a regional or "pseudo-monsoon" system operated over northwestern Australia and that there were summer heat lows with mean positions over the Pilbarra region of Western Australia and Cloncurry in Queensland. These features are now recognized as crucial for understanding climate patterns across northern and central Australia, but their impact extends further. Indeed, when troughs form along the west coast these bring moist humid air and high humidity into the otherwise dry hot conditions generally experienced in Perth, for example, over summer.

The prominent ocean current called the Leeuwin brings warm water in winter south along the west coast of Australia and into the Southern Ocean. Dr Gentilli was the first to describe this and comment on its significance for marine life and fisheries in Western Australia. He also pointed out that its width and distance offshore made it unlikely that it had major climatic significance. In 1972 he wrote about the importance of factors in the Indian Ocean leading to the onset of weather patterns in other parts of the world. Joe was one of the first to begin looking for evidence of the "enhanced greenhouse effect" in climate records. While he was unable to demonstrate any clear signal it is typical that he was a trendsetter in this most modern areas of climatic work.

His synthesis work on climate is particularly well known. His *Climates of Australia and New Zealand* was published in 1971, yet it remains a major work in the field. In preparing this he was led toward comparative work on climate systems of the southern continents, and from this he was able to show that the southern hemisphere was not a simple analog of the northern hemisphere. He took this further by asking questions on why even in the southern hemisphere there were interesting differences across the continents, and he began, and one might say pioneereed, in this region the search for teleconnections of climate patterns.

In looking back over Dr Gentilli's work it must be remembered that all this was done by carefully comparing published observations of rainfall and daily weather maps. It took a finely honed mind to search for patterns visually across seasons, years and geography.

John Dodson

Bibliography

Gentilli, J., 1941. Wheat yields and variability in West Australian districts, 1929–39. *Journal of the Department of Agriculture, Western Australia*, **18**: 62–65.

Gentilli, J., 1946. *Australian Climates and Resources*. Melbourne: Whitcombe & Tombs.

Gentilli, J., 1948. Present-day volcanicity and climate change. *Geological Magazine*, **85**: 172–175.

Gentilli, J., 1949. Air masses of the Southern Hemisphere. *Weather*, **4**: 258–261, 292–297.

Gentilli, J., 1961. Quaternary climates of the Australian region. *Annals of the New York Academy of Sciences*, **95**: 465–501.

Gentilli, J., (ed.), 1971. *Climates of Australia and New Zealand* (*World Survey of Climatology*, vol. 13). Amsterdam: Elsevier (Gentilli was author of chapters 1, 2, 4–7).

Gentilli, J., 1972. Thermal anomalies in the eastern Indian Ocean. *Nature, London*, **238**: 93–95.

Gentilli, J., 1979. Atmospheric factors in disasters: an appraisal of their role in Australia. In Symposium volume on *Natural Calamities*. Canberra: Australian Academy of Science.

Gentilli, J., 1991. Homologous peri-oceanic west coast climates in the southern hemisphere. *Journal of the Royal Society of Western Australia*, **74**: 15–33.

Gentilli, J., 1993. Floods in the desert. Heavy rains in the dry regions of Western Australia. *Western Australia Naturalist*, 19: 201–218.

Cross-reference

Australia and NewZealand, Climate of

GEOMORPHOLOGY AND CLIMATE

The driving forces that influence climate act as significant agents in the exogenic processes that operate at or near the Earth's surface. "The landforms of the earth are the result of the interplay between internal, or endogenetic processes, and surface, or exogenetic, processes" (Derbyshire, 1997, p. 89). The exogenic processes are driven by gravity and by solar radiation, therefore the climate of a region and the variability of climate in a region contribute to geomorphic variability.

Scale is a factor to be considered when examining climate and geomorphologic relationships. Surface variability interacts with atmospheric variables and results in energy movement and transfers. Five topoclimatic variables that affect incoming radiation are:

1. aspect – the direction an object faces;
2. slope – influences the temperature characteristics that will develop;
3. relief – direct radiation of the sun on a surface area;
4. color – impact on albedo; and
5. texture – smooth versus rough impacts heat transfer through the surface.

Each of these variables influences the geomorphic mechanics of a locale allowing for microclimates to be monitored and small anomalies to be accounted for within a given region.

Climatic geomorphology assumes that the geomorphic mechanics of a locale vary in type and rate according to the particular climatic zone in which they function. Classical climatic geomorphology is rooted in the investigation of landforms as products of particular climatic environments. Agassiz (1840) identified distinctive suites of landforms, such as moraines, as unique products of glaciers that existed in contemporary glacial environments and also beyond that climatic regime. Davis (1899, 1900, 1905) identified distinctive landforms characteristically associated with arid climates and additional features uniquely associated with humid, midlatitude regions. The landforms serve as indicators of dominant climatic influence in a region; in the past and/or present. In 1950 Peltier proposed his concept of "morphogenetic regions" (Figure G1). Climatic morphogenesis is the approach to geomorphology that emphasizes the shaping of landforms in response to various climatic conditions (Birot, 1960, 1968; Tricart and Cailleux, 1965, 1972; Andrews, 1975; Budel, 1982). The more acceptable terminology for areas characterized by landforms associated with a particular climatic environment are morphoclimatic zones (or regions). Tanner (1961) proposed a model that used potential evapotranspiration instead of temperature (Figure G2) to convey erosion rates as a function of effective precipitation. Tanner's model identified a total of four classes: arid, selva, moderate, and tundra-glacial. Wilson (1968) introduced a morphogenetic classification called the climate–process system (CPS). This system identifies sets of temperature–precipitation values that tend to drive (or dominate) geomorphic processes under those climatic conditions (Figure G3). The climate–process system considers climatic seasonality as well as airmass dominance as factors in defining the dominant geomorphic processes. Landscape characteristics (both erosional and depositional)

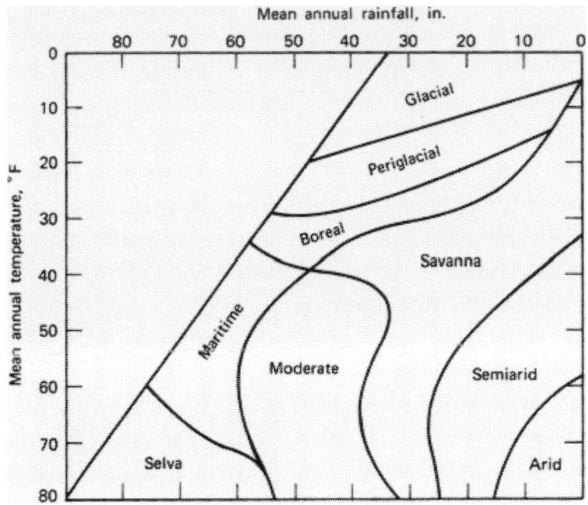

Figure G1 Morphogenetic regions distinguished by Peltier (after Peltier, 1950).

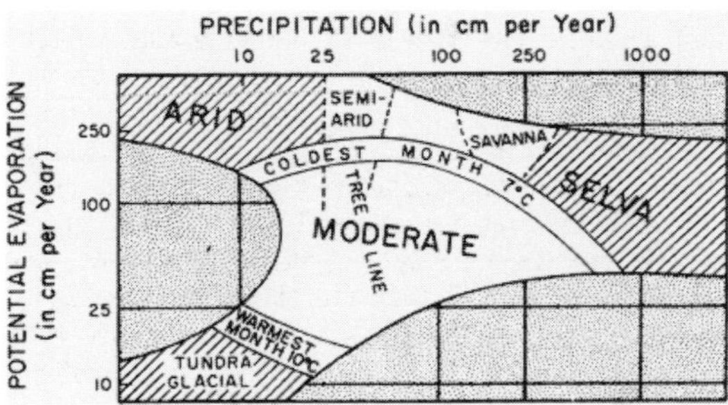

Figure G2 Tanner's morphogenetic regions (after Tanner, 1961).

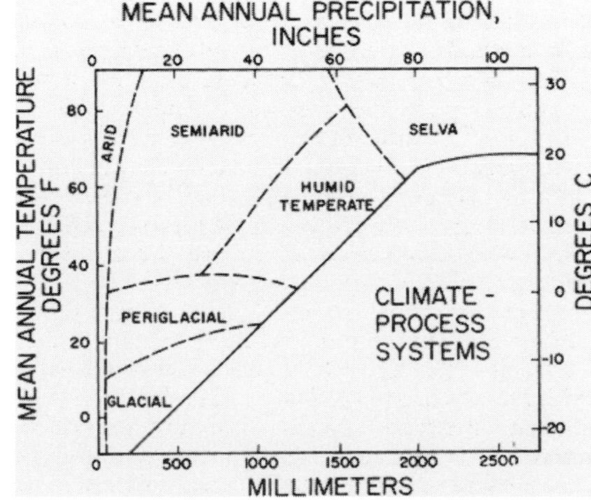

Figure G3 Wilson's climate-process systems (after Wilson, 1968).

Table G1 Simple morphogenetic systems and their landscapes (after Wilson, 1968)

System name	Dominant geomorphic processes[a]	Landscape characteristics[b]
Glacial	Glaciation Nivation Wind action (freeze-thaw)	Glacial scour Alpine topography Moraines, kames, eskers
Periglacial	Frost action Solifluction Running Water	Patterned ground Solifluction slopes, lobes, terraces Outwash plains
Arid	Desiccation Wind action Running Water	Dunes, salt pans (playas) Deflation basins Cavernous weathering Angular slopes, arroyos
Semiarid (subhumid)	Running water Weathering (especially mechanical) Rapid mass movements	Pediments, fans Angular slopes with coarse debris Badlands
Humid temperate	Running water Weathering (especially chemical) Creep (and other mass movements)	Smooth slopes, soil covered Ridges and valleys Stream deposits extensive
Selva	Chemical weathering Mass movements Running water	Steep slopes, knife-edge ridges Deep soils (laterites included) Reefs

[a] Processes are listed in order of relative importance to landscape (not in order of absolute magnitude). List is abbreviated.
[b] Both erosional and depositional forms are included. List is neither comprehensive nor definitive; merely suggestive.
Source: Wilson (1968).

Table G2 Earth's major morphoclimatic zones (after Summerfield, 1991, based on various sources)

Morphoclimatic zone	Mean annual temperature (°C)	Mean annual precipitation (mm)	Relative importance of geomorphic processess
Humid tropical	20–30	>1500	High potential rates of chemical weathering; mechanical weathering limited; active, highly episodic mass movement; moderate to low rates of stream corrosion but locally high rates of dissolved and suspended load transport.
Tropical wet-dry	20–30	600–1500	Chemical weathering active during wet season; rates of mechanical weathering low to moderate; mass movement fairly active; fluvial action high during wet season with overland and channel flow; wind action generally minimal but locally moderate in dry season.
Tropical semi-arid	10–30	300–600	Chemical weathering rates moderate to low; mechanical weathering locally active especially on drier and cooler margins; mass movement locally active but sporadic; fluvial action rates high but episodic; wind action moderate to high.
Tropical arid	10–30	0–300	Mechanical weathering rates high (especially salt weathering); chemical weathering minimal; mass movement minimal; rates of fluvial activity generally very low but sporadically high; wind action at a maximum.
Humid midlatitude	0–20	400–1800	Chemical weathering rates moderate, increasing to high at lower latitudes; mechanical weathering activity moderate with frost action important at higher latitudes; mass movement activity moderate to high; moderate rates of fluvial processes: wind action confined to coasts.
Dry continental	0–10	100–400	Chemical weathering rates low to moderate; mechanical weathering, especially frost action, seasonally active; mass movement moderate and episodic; fluvial processes active in wet season; wind action locally moderate.
Periglacial	<0	100–1000	Mechanical weathering very active with frost action at a maximum; chemical weathering rates low to moderate; mass movement very active; fluvial processes seasonally active; wind action rates locally high
Glacial	<0	0–1000	Mechanical weathering rates (especially frost action) high; chemical weathering rates low; mass movement rates low except locally; fluvial action confined to seasonal melt; glacial action at a maximum; wind action significant.
Azonal mountain zone	Highly variable	Highly variable	Rates of all processes vary significantly with altitude; mechanical and glacial action become significant at high elevations.

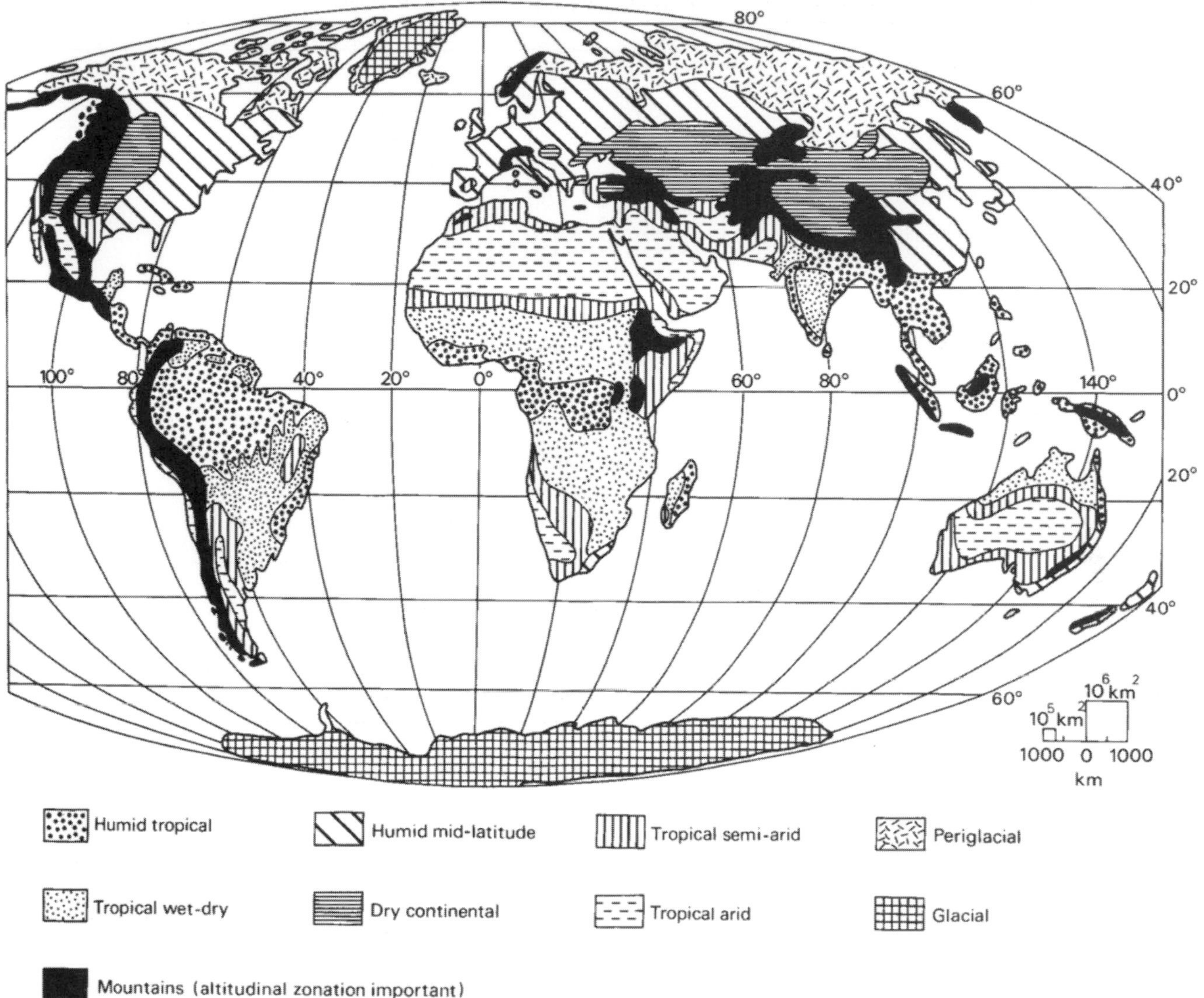

Figure G4 Global distribution of morphoclimatic zones (from Summerfield, 1991, after Tricart and Cailleux, 1972). See Table G2 for descriptions.

may be classified using this system (Table G1). Summerfield (1991) has revised the morphogenic classification to incorporate the Earth's major morphoclimatic zones as defined by mean annual temperature, mean annual precipitation and relative importance of geomorphic processes (Table G2).

In every proposed morphoclimatic analysis solar radiation is considered the most fundamental input to the Earth's climate system. A change in solar radiation input results in responses in the atmosphere, oceans, hydrologic cycle, vegetation cover, and glaciers and ice sheets (Benn and Evans, 1998). The rate of geomorphic activity, when taken into consideration with climate and landform morphology, seems to be a more dominating element in determining whether or not a landscape is formed by the prevailing climate or, in areas with a generally low rate of geomorphic activity, are dominated by relict landforms developed under quite different climates from the current regime (Summerfield, 1991).

Susan Berta

Bibliography

Agassiz, L., 1840. *Etudes sur les glaciers*. Neuchatel.

Andrews, J.T., 1975. *Glacial Systems: an approach to glaciers and their environments*. North Scituate, MA: Duxbury Press.

Benn, D.I., and Evans, D.J.A., 1998. *Glaciers and Glaciation*. London: Arnold.

Birot, P., 1960. *Le cycle d'erosion sous les differents climates*. Rio de Janeiro: Univ. do Brasil Centro de Pesquisas de Geografia do Brasil, Rio de Jaeiro.

Birot, P., 1968. *Cycle of Erosion in Different Climates* (trans. C.I. Jackson and K.M. Clayton). Berkeley, CA: University of California Press.
Budel, J., 1982. *Climatic Geomorphology* (trans. L. Fischer and D. Busche). Princeton, NJ: Princeton University Press.
Davis, W.M., 1899. The geographical cycle. *Geographical Journal*, **14**: 481–504.
Davis, W.M., 1900. Glacial erosion in France, Switzerland and Norway. *Proceedings of the Boston Society of Natural History*, **29**: 273–322.
Davis, W.M., 1905. The geographical cycle in an arid climate. *Journal of Geology*, **13**: 381–407.
Derbyshire, E., 1997. Geomorphic process and landforms. In Thompson, R.D., and Perry, A., eds., *Applied Climatology: principles and Practice*. London: Routledge, chap. 8.
Peltier, L.C., 1950. The geographical cycle in periglacial regions as it is related to climatic geomorphology. *Annals of the Association of American Geographers*, **40**: 214–236.
Summerfield, M.A., 1991. *Global Geomorphology: an introduction to the study of landforms*. Harlow: Longman Scientific and Technical.
Tanner, W.F., 1961. An alternate approach to morphogenetic climates. *Southeastern Geologist*, **2**: 251–257.
Tricart, J., and Cailleux, A., 1965. *Introduction a la geomorphologie climatique*. Paris: Soc. D'edition d'enseignement superieur.
Tricart, J., and Cailleux, A., 1972. *Introduction to Climatic Geomorphology* (trans. by C.J.K de Jonge). London: Longman.
Wilson, L., 1968. Morphogenetic classification. In Fairbridge, R.W., ed., *Encyclopedia of Geomorphology*. New York: Reinhold, pp. 717–729.

Cross-references

Arctic Climates
Arid Climates
Climate Classification
Desertification
Paleoclimatology
Tropical and Equatorial Climates

GLOBAL ENVIRONMENTAL CHANGE: IMPACTS

In preparing the entry "Bioclimatology" for this volume, the authors found that many of the articles they wished to cite appeared in literature pertaining to the impacts of global environmental change on humans and the biosphere. To provide a guide to that literature it was decided that a separate entry would prove of value. However, the literature relating to the impacts of global environmental change on humans and the biosphere is developing so rapidly (probably due to the beneficial influence of the IPCC on the world community of environmental scientists) that it would be quite impossible to survey the field in a few pages. Instead, an extensive Bibliography is given.

"Global environmental change" has become a widely discussed topic in recent years. Of particular concern is the issue of climate change (commonly referred to as global warming), which has become a priority policy question, both nationally and internationally. However, there are other kinds of global environmental change (e.g. land-use changes), some of which have feedbacks into the global climate system. Table G3 lists several important types of global environmental change.

The environment is constantly changing, of course, but the changes are now occurring at an accelerating pace. Examples include the continuing urbanization in many parts of the world, the increasing emissions of greenhouse gases and, since about 1970, growing evidence of human-induced greenhouse-gas warming (e.g. pronounced winter and spring warming over most of the northern high latitudes, a decrease in the extent of sea ice in the Arctic Ocean and a northward extension of the tree-line (Serreze et al., 2000)). By the end of the twenty-first century the world's environment is likely to be quite different from that of today.

For articles on global climate change, the main overview references are Volume 1 of the *Encyclopaedia of Global Environmental Change* (2001) and the *Third Assessment Report of the Intergovernmental Panel on Climate Change* (IPCC, 2002). Additionally, appropriate volumes in the *Encyclopedia of Earth Science Series* provide an important source.

Among the journals that deal particularly with global change issues are *Climatic Change, Global Change Biology, Global Change and Human Health, Global Environmental Change, Mitigation and Adaptation Strategies for Global Change, AMBIO*, and *Science of the Total Environment*. Nevertheless, it should be noted that research papers on global environmental change are dispersed across these and many other scientific journals. This makes it difficult for research bioclimatologists to keep abreast of the field.

Table G3 Types of global change that may affect climate, and thus, the health and welfare of humans, and the resilience of ecosystems

Increasing concentrations of greenhouse gases:
Warming, especially in polar regions
Melting of glaciers and of polar snow and ice cover
Increasing evaporation, leading to increasing water shortages and desertification
More frequent air pollution stagnation episodes
Sea level rise
Spread of tropical diseases (e.g. malaria) into the temperate zone
Destruction of the stratospheric ozone layer

Disruption of the global biogeochemical cycles:
Acid rain, leading to forest dieback and fish kills in sensitive lakes
Nitrogen overload, leading to increasing pollution loadings and changes in biodiversity

Land-use changes:
Urbanization and industrialization
Irrigation, diverting water from rivers and causing salinization and waterlogging
Soil erosion and desertification
Habitat loss and changes in biological diversity

Over-consumption of renewable and non-renewable resources:
Over-consumption of fresh water
Deforestation
Over-fishing
Increasing energy consumption, due to, e.g., increasing numbers of automobiles
Population growth in some developing countries accelerating the consumption of natural resources and desertification

New technology:
Historical examples include the invention of the automobile; and development of chlorofluorocarbons (causing stratospheric ozone depletion)

Environmental consequences of war:
Destruction of ecosystems
Nuclear winter

Impacts of global environmental change: general

The main impacts of global environmental change on humans and on the biosphere can be classified as physical/chemical, physical/biological, demographic, economic and sociological. In the physical/chemical and physical/biological domains, the environmental impacts are direct, e.g. increased frequency and intensity of heat waves or air pollution episodes lead to increased rates of morbidity and mortality. In other cases the environmental impacts may be indirect, e.g. war may lead to a decline in agricultural production or to a closure of trade routes, in both cases causing malnutrition and even migrations of affected peoples.

When environmental stress factors are considered one at a time, the effects on humans and on ecosystems are reasonably well predicted. But in the case of the boreal forest, for example, where many different factors are likely to change over the next 50 years (climate, frequencies of forest fires and pest outbreaks, increasing nitrogen overload downwind of cities, and changing forest management practices), predictions of the net effects on a forest contain considerable uncertainty. Furthermore, trends in the socioeconomic drivers of environmental change are uncertain (rates of change in emissions of greenhouse gases, rates of human consumption of water, etc.). In summary, single-species and/or single-stressor models are useful diagnostic tools but they are often poor predictors of the behavior of whole ecosystems. See, for example, Fleming (2000) and Fleming et al. (2002).

Thus scenarios must be constructed, usually for several possible futures, and policy-makers must adopt the *precautionary principle* whenever the net losses of not taking action are very great (O'Riordon, 2001). A problem with this approach is that estimates of the benefits of taking action and the disbenefits of not taking action are both subject to uncertainty.

Impacts of global environmental change on humans

Table G4 provides a list of the research fields in which the impacts of climate change/warming on human health and welfare are currently being studied. Some of the major references in these fields are:

Human health:
General: Hancock (2001), Hammad (2002), IPCC (2002), McMichael (2001).
Infectious diseases: Epstein (2001), Guest (2001), NRC (2001), Pimentel (2001).
UV-B: UNEP (1998).

Socioeconomic impacts:
Tourism: Burger (2001), Lise and Tol (2002), Stadelbauer (2001), Unis (2001), Wong (2001).
Economics: Goodstein (2001), Mendelsohn (2001), Norgaard (2001).
Global land use: Douglas (2001a).
Sea-level rise: IPCC (2002).

Table G4 Current research fields relating to the impacts of global change/climate warming on human health and welfare

Direct impacts on health of climate warming
Summer: increasing frequency and intensity of heatwaves
Winter: decreasing frequency of extreme cold and wind chill episodes
Increasing UV-B, causing more frequent cases of melanoma
More frequent weather disasters
Indirect impacts on health of climate warming
More frequent stagnation episodes generally in valleys, and in temperate zones and the Arctic, causing greater air pollution potential
Spread of tropical diseases into temperate zones
Fresh water shortages
Regional disparity in food yields and incidence of malnutrition
Increasing abundance of aeroallergens, particularly pollen
Socioeconomic impacts of climate warming
Impacts on fisheries
Impacts on forestry
Impacts on agriculture
Impacts on transport
Impacts on tourism
Impacts on the energy sector
Coastal zone damage from sea-level rise

Impacts of global environmental change on the biosphere

Table G5 lists the root causes of global change, and the resulting impacts on climate, and gives a summary of the types of biosphere responses to these impacts. Some of the major references on biosphere responses to global change are:

Table G5 Causes of global change, resulting impacts on climate, and types of biosphere responses

Increases in human consumption and in human population leading to:
Increasing concentrations of greenhouse gases, leading to climate warming
Land-use changes, e.g. deforestation and other losses of habitat
Increasing pollution (e.g. acidic deposition, nitrogen overload, persistent organic pesticides, stratospheric ozone destruction)
Types of climate change:
Warming at the Earth's surface, particularly in polar regions
Decay of glaciers and polar ice
Increasing evaporation
Increasing frequencies of weather-related extreme events, e.g. droughts and floods
Rising sea level
Changes in the atmospheric general circulation, producing changes in the behavior of the Asian monsoon, El Niño, the Gulf Stream and possible "rapid" large-scale climate flip-flops to new regimes
Types of biosphere responses:
Longer growing seasons in polar and temperate zones
Poleward shifts of biomes
Changes in biodiversity regimes, e.g. nitrogen overloading may cause a severe decline in the Boreal forest
Oceanic shifts in the migration patterns of fish
Biological invasions and increasing numbers of species extinctions

General:
IPCC (2002), McCarthy et al. (2001), Parmesan and Yohe (2003), Root et al. (2003), Walther et al. (2002).

Land organisms and ecosystems:
Plants: Bazzaz and Catofsky (2001), Catofsky and Bazzaz (2001), Gregg et al. (2003), Howden and Barnett (2001), Hungate and Marks (2001), Lechowicz (2001), Korner (2001), Paruelo (2001), Poorter and Perez-Soba (2001), Reich and Frelich (2001), Rustad and Norby (2001), Waring (2001), Wookey (2001).
Ecotones: Risser (2001).

Insects: Fleming (2000), Fleming et al. (2002), Harrington (2001), Parmesan et al. (1999), Percy et al. (2002)
Animals: Howden and Barnett (2001), Hungate and Marks (2001), Kingsolver (2001), Martin and Nagy (2001).

Marine ecosystems:
Coral reefs: Gattuso and Buddemeier (2001).
Fisheries: Caddy and Regier (2001), Douglas (2001b), Fleming and Jensen (2001), Regier (2001), Welsh (2001).

Arctic ecosystems:
ACIA (2004).

Biodiversity:
Adams and Wall (2001), Drake et al. (1989), Levêque (2001), Staley and Fuhrman (2001).

Land-use trends:
Douglas (2001a).

Sustainability:
Baskin (1997), Caddy and Regier (2001), Duinker (2001).

Biological invasions:
Clark et al. (2001), Drake et al. (1989), Lonsdale (2001), McNeely et al. (2002), Wittenberg and Cock (2002).

Extinctions:
Brooks (2001).

Ecological economics:
Daily and Ellison (2002), Norgaard (2001).

R.E. Munn and A.R. Maarouf

Bibliography

ACIA, 2004. *Impacts of Warming Arctic: Arctic Climate Impact Assessment*. Cambridge: Cambridge University Press.

Adams, G.A., and Wall, D.H., 2001. Biodiversity in soils and sediments: potential effects of global change. *Encyclopaedia of Global Environmental Change*, Vol. 2. Chichester: John Wiley, pp. 52–159.

Baskin, Y., 1997. *The Work of Nature: how the diversity of life sustains* us. Washington D.C.: Island Press, 263 pp.

Bazzaz, F.A., and Catofsky, S., 2001. Plants – from cells to ecosystems: impacts of global change. *Encyclopaedia of Global Environmental Change*, Vol. 2. Chichester: John Wiley, pp. 94–111.

Brooks, T.M., 2001. Extinctions (contemporary and future). *Encyclopaedia of Global Environmental Change*, Vol. 2. Chichester: John Wiley, pp. 301–307.

Burger, J., 2001. Tourism and ecosystems. *Encyclopaedia of Global Environmental Change*, Vol. 3. Chichester: John Wiley, pp. 597–609.

Caddy, J., and Regier, H., 2001. Policies for sustainable and responsible fisheries. *Encyclopaedia of Global Environmental Change*, Vol. 4. Chichester: John Wiley, pp. 343–351.

Catofsky, S., and Bazzaz, F.A., 2001. Plant competition in an elevated CO_2 world. *Encyclopaedia of Global Environmental Change*, Vol. 2. Chichester: John Wiley, pp. 471–481.

Clark, J.S., Beckage, B., Hillerislambers, J. et al., 2001. Plant dispersal and migration. *Encyclopaedia of Global Environmental Change*, Vol. 2. Chichester: John Wiley, pp. 81–93.

Daily, G.C., and Ellison, K., 2002. *The New Economy of Nature*. Washington D.C.: Island Press.

Douglas, I., 2001a. Global land cover and land use trends and changes. *Encyclopaedia of Global Environmental Change*, Vol. 3. Chichester: John Wiley, pp. 13–15.

Douglas, I., 2001b. Changes of world marine fish stocks. *Encyclopaedia of Global Environmental Change*, Vol. 3. Chichester: John Wiley, pp. 238–241.

Drake, J.A., Mooney, H.A., di Castri, F. et al., (eds), 1989. *Biological Invasions: a global perspective*. Chichester: SCOPE/John Wiley.

Duinker, P.N., 2001. Policies for sustainable forests: examples from Canada. *Encyclopaedia of Global Environmental Change*, Vol. 4. Chichester: John Wiley, pp. 351–356.

Epstein, P.R., 2001. Infectious diseases. *Encyclopaedia of Global Environmental Change*, Vol. 2. Chichester: John Wiley, pp. 357–363.

Fleming, I., and Jensen, A.J., 2001. Fisheries: effects of climate change on the life cycles of salmon. *Encyclopaedia of Global Environmental Change*, Vol. 3. Chichester: John Wiley, pp. 309–312.

Fleming, R.A., 2000. Climate change and insect disturbance regimes in Canada's boreal forests. *World Resources Review*, **12**: 521–555.

Fleming, R.A., Candau, J.-N., and McAlpine, R.S., 2002. Landscape-scale analysis of interactions between insect defoliation and forest fire in central Canada. *Climatic Change*, **55**: 251–272.

Gattuso, J.-P., and Buddemeier, R.W., 2001. Coral reefs: an ecosystem subject to multiple environmental threats. *Encyclopaedia of Global Environmental Change*, Vol. 2. Chichester: John Wiley, pp. 232–241.

Goodstein, E., 2001. Economics and global environmental change. *Encyclopaedia of Global Environmental Change*, Vol. 5. Chichester: John Wiley, pp. 25–36.

Gregg, J.W., Jones, C.G., and Dawson, T.E., 2003. Urbanization effects on tree growth in the vicinity of New York City. *Nature*, **424**: 183–187.

Guest, C., 2001. Viral diseases and the influence of climate change. *Encyclopaedia of Global Environmental Change*, Vol. 3. Chichester: John Wiley, pp. 690–694.

Hammad, A.El B., 2002. Policy responses to public health issues relating to global environmental change. *Encyclopaedia of Global Environmental Change*, Vol. 4. Chichester: John Wiley, pp. 47–64.

Hancock, T., 2001. Theories of health and the environment. *Encyclopaedia of Global Environmental Change*, Vol. 5. Chichester: John Wiley, pp. 492–502.

Harrington, R., 2001. Insect pests and global environmental change. *Encyclopaedia of Global Environmental Change*, Vol. 3. Chichester: John Wiley, pp. 381–386.

Howden, S.M., and Barnett, G.B., 2001. Natural systems: impacts of climate change, *Encyclopaedia of Global Environmental Change*, Vol. 2. Chichester: John Wiley, pp. 67–70.

Hungate, B.A., and Marks, J.C., 2001. Terrestrial and freshwater ecosystems: impacts of global change. *Encyclopaedia of Global Environmental Change*, Vol. 2. Chichester: John Wiley, pp. 122–134.

IPCC, 2002. *Climate Change 2001*. Synthesis Report (R.T. Watson, ed.), Cambridge: Cambridge University Press.

Kingsolver, J.G., 2001. Impacts of global environmental change on animals. *Encyclopaedia of Global Environmental Change*, Vol. 2. Chichester: John Wiley, pp. 56–66.

Körner, Ch., 2001. CO_2 enrichment: effects on ecosystems. *Encyclopaedia of Global Environmental Change*, Vol. 2. Chichester: John Wiley, pp. 215–224.

Lechowicz, M.J., 2001. Phenology. *Encyclopaedia of Global Environmental Change*, Vol. 2. Chichester: John Wiley, pp. 461–465.

Levêque, C., 2001. Biodiversity in freshwater. *Encyclopaedia of Global Environmental Change*, Vol. 2. Chichester: John Wiley, pp. 146–152.

Lise, W., and Tol, R.S.J., 2002. Impact of climate on tourist demand. *Climatic Change*, **55**: 429–449.

Lonsdale, W.M., 2001. Biological invasions. *Encyclopaedia of Global Environmental Change*, Vol. 2. Chichester: John Wiley, pp. 11–19.

Martens, P.J., and McMichael, A.J., 2002. *Environmental Change, Climate and Health*. Cambridge: Cambridge University Press.

Martin, K., and Nagy, K., 2001. Animal physiology and global environmental change, *Encyclopaedia of Global Environmental Change*, Vol. 2. Chichester: John Wiley, pp. 136–139.

McCarthy, Canziani, O.F., Leary, N.A., Dokken, D., and White, K.S. (eds), 2001. *Climate Change 2001: impacts, adaptation and vulnerability*. Cambridge: Cambridge University Press.

McMichael, A.J., 2001. Environmental change and human health: extending the sustainability agenda. *Encyclopaedia of Global Environmental Change*, Vol. 3. Chichester: John Wiley, pp. 130–145.

McNeely, J.A., Mooney, H.A., Neville, L.E., Schei, P.J., and Waage, J.K., 2002. *Global Strategy on Invasive Alien Species*. Wallingford: CABI.

Mendelsohn, R. (ed.), 2001. *Global Warming and the American Economy*. Cheltenham, Glos., UK: Edward Elgar.

Norgaard, R.B., 2001. Ecological economics. *Encyclopaedia of Global Environmental Change*, Vol. 5. Chichester: John Wiley, pp. 37–48.

NRC, 2001. *Under the Weather: climate, ecosystems and infectious disease*. National Research Council. Washington, DC: National Academy Press.

O'Riordan, T., 2001. Precautionary principle. *Encyclopaedia of Global Environmental Change*, Vol. 5. Chichester: John Wiley, pp. 455–457.
Parmesan, C., and Yohe, G., 2003. A globally coherent fingerprint of climate change impacts across natural systems. *Nature*, **421**: 37–42.
Parmesan, C., Ryrholm, N., Stefanescu, S. et al., 1999. Poleward shifts in geographical ranges of butterfly species associated with regional warming, *Nature*, **399**: 579–583.
Paruelo, J.M., 2001. Temperate grasslands. *Encyclopaedia of Global Environmental Change*, Vol. 2. Chichester: John Wiley, pp. 569–574.
Percy, K.E., Awmack, C.S., Lindroth, R.L., et al., 2002. Altered performance of forest pests under atmospheres enriched by CO_2 and O_3. *Nature*, **420**: 403–407.
Pimentel, D., 2001. Malnutrition, infectious diseases and global environmental change. *Encyclopaedia of Global Environmental Change*, Vol. 3. Chichester: John Wiley, pp. 440–446.
Poorter, H., and Perez-Soba, M., 2001. Plant growth at elevated CO_2. *Encyclopaedia of Global Environmental Change*, Vol. 2. Chichester: John Wiley, pp. 489–496.
Regier, H., 2001. Freshwater fisheries. *Encyclopaedia of Global Environmental Change*, Vol. 2. Chichester: John Wiley, pp. 327–331.
Reich, P.B., and Frelich, L., 2001. Temperate deciduous forests. *Encyclopaedia of Global Environmental Change*, Vol. 2. Chichester: John Wiley, pp. 565–569.
Risser, P., 2001. Ecotones. *Encyclopaedia of Global Environmental Change*, Vol. 2. Chichester: John Wiley, pp. 283–288.
Root, T.L., Price, J.T., Hall, K.R., Schneider, S.H., Rosenzweig, C. and Pounds, J.A., 2003. Fingerprints of global warming on wild animals and plants. *Nature*, **421**: 57–60.
Rustad, L.E., and Norby, R.J., 2001. Temperature increase: effects on terrestrial ecosystems. *Encyclopaedia of Global Environmental Change*, Vol. 2. Chichester: John Wiley, pp. 575–581.
Serreze, M.C., Walsh, J.E., Chapin III, F.S., et al., 2000. Observational evidence of recent change in the northern high-latitude environment. *Climatic Change*, **46**: 159–207.
Stadelbauer, J., 2001. Tourism: climate change and tourist resorts. *Encyclopaedia of Global Environmental Change*, Vol. 4. Chichester: John Wiley, pp. 623–627.
Staley, J.T., and Fuhrman, J., 2001. Microbial diversity. *Encyclopaedia of Global Environmental Change*, Vol. 2. Chichester: John Wiley, pp. 421–425.
UNEP, 1998. Environmental effects of ozone depletion: 1998 assessment. *Journal of Photochemistry and Photobiology*, **46**: 1–108.
Unis, E., 2001. Policies to achieve sustainable tourism. *Encyclopaedia of Global Environmental Change*, Vol. 4. Chichester: John Wiley, pp. 362–366.
Walther, G.-R., Post, E., Convey, P. et al., 2002. Ecological responses to recent climate change. *Nature*, **416**: 389–395.
Waring, R., 2001. Temperate coniferous forests. *Encyclopaedia of Global Environmental Change*, Vol. 2. Chichester: John Wiley, pp. 561–565.
Welsh, D., 2001. Fisheries: Pacific coast salmon. *Encyclopaedia of Global Environmental Change*, Vol. 3. Chichester: John Wiley, pp. 314–316.
Wittenberg, R., and Cock, M.J.W., 2002. *Invasive Alien Species: a toolkit of best prevention and management practices*. Wallingford: CABI.
Wong, P.P., 2001. Tourism as a global driving force for environmental change. *Encyclopaedia of Global Environmental Change*, Vol. 3. Chichester: John Wiley, pp. 609–623.
Wookey, P.A., 2001. Tundra. *Encyclopaedia of Global Environmental Change*, Vol. 2. Chichester: John Wiley, pp. 593–602.

Cross-references

Precipitation
Air Pollution Climatology
Bioclimatology
Climate Vulnerability
Greenhouse Effect and Greenhouse Gases
Human Health and Climate
Ozone
Reanalysis Projects
Sea Ice and Climate
Teleconnections
Urban Climatology

GREENHOUSE EFFECT AND GREENHOUSE GASES[1]

The greenhouse effect is a natural phenomenon that warms the Earth by about 33°C. It is caused by greenhouse gases in the atmosphere (Table G6). The natural greenhouse effect is currently being enhanced through human activity via the release of additional greenhouse gases into the atmosphere. The consensus of the scientific community, voiced through reports written by the Intergovernmental Panel on Climate Change (IPCC), is that this enhancement of the natural greenhouse effect has caused global warming over the last 150 years and that this warming is very likely to accelerate in the future.

A simple explanation of the greenhouse effect is straightforward, although the full mechanics, processes and feedbacks that exist within the climate system are extremely complex. All things with a temperature above −237.16°C (known as absolute zero) emit electromagnetic radiation. A key property of electromagnetic radiation is that the wavelength of this radiation is related to the temperature of the emitting object. This relationship, described by Planck's Law, means that relatively cool objects emit radiation at longer wavelengths than relatively hot objects.

The sun has a temperature of about 6000°C) and thus emits shortwave radiation (also known as solar radiation). Figure G5 shows a curve (known as a Planck or black-body curve) for the sun's emissions (dashed line, left-hand side). The actual wavelengths of radiation emitted by the sun vary across a wide range

Table G6 List of major greenhouse and non-greenhouse gases in the atmosphere

Gas	Concentration
Non-greenhouse gases	
Nitrogen	78.1% volume mixing ratio
Oxygen	20.9% volume mixing ratio
Argon	0.93% volume mixing ratio
Greenhouse gases	
Water vapor	Variable – of order 1% volume mixing ratio
Carbon dioxide	Trace gas
Methane	Trace gas
Nitrous oxide	Trace gas
Ozone	Trace gas
Hydrofluorocarbons	Trace gas
Perfluorocarbons and sulfur hexafluoride	Trace gas

[1] This entry is sourced primarily from Working Group 1 of the 2001 report (Houghton et al., 2001) of the Intergovernmental Panel on Climate Change (IPCC) which provides the scientific basis for greenhouse-related science. This entry therefore represents the consensus opinion of 122 lead authors and 515 contributing authors. The substantive information reported below has been reviewed by several hundred scientists from expert groups, governments and non-government organizations. In writing this entry I have had to simplify and condense the IPCC report and I would strongly recommend that interested readers refer to the Summary for Policy Makers and Technical Summary of Working Group 1 reports available at http://www. ipcc.ch/pub/spm22-01.pdf and http://www.ipcc.ch/pub/wg1TARtechsum.pdf.

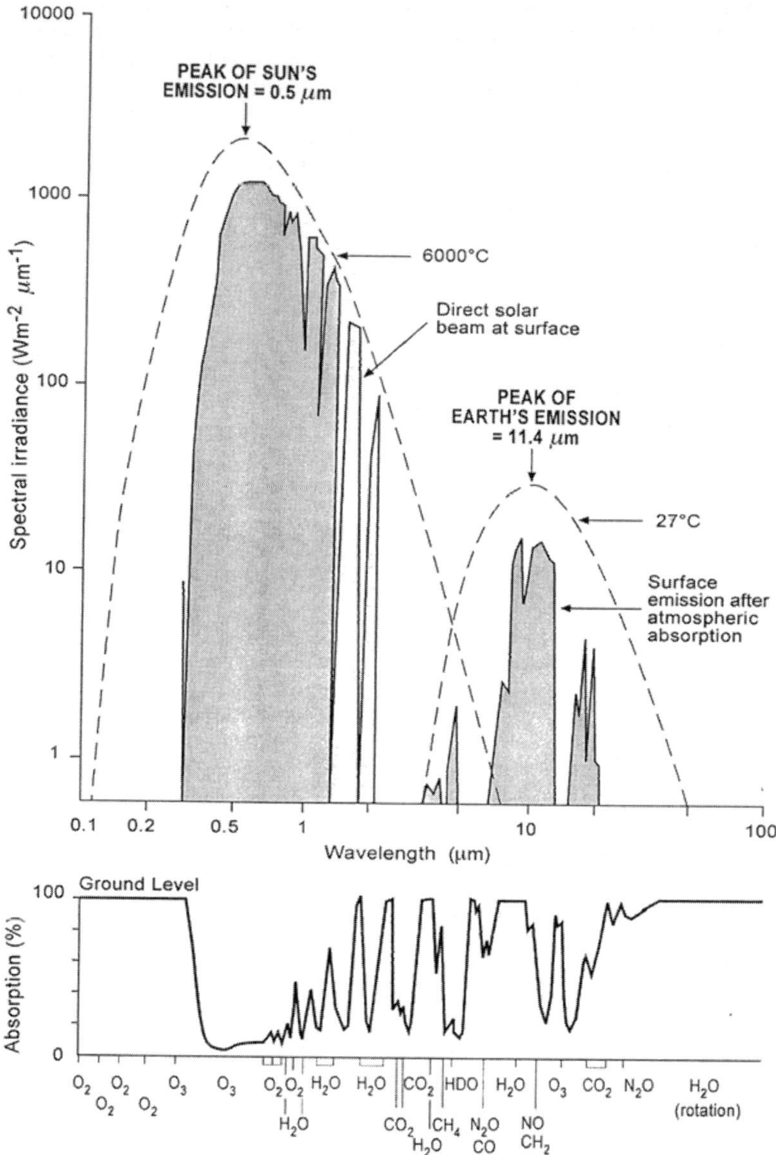

Figure G5 The top pane shows the Planck curves (the range and amounts of electromagnetic radiation) emitted by the sun and Earth's surface (dashed lines). The shaded areas show the amount of the sun's radiation the reaches the Earth's surface and the amount of the longwave (terrestrial) radition the escapes to space through atmospheric windows. the lower panel shows the percentage of electromagnetic radiation absorbed at these wavelengths. Note the horizontal scale on both panels is the same.

but most is between 0.15 and 5 micrometers. Radiation at these short wavelengths can pass through the atmosphere relatively unimpeded as shown by the lower panel in Figure G5 where the percent of absorption of wavelengths around the peak of the sun's emissions (0.5 micrometers) is close to zero. Figure G5 shows a second curve (solid line, left-hand side) which shows the radiation from the sun that actually reaches the Earth's surface. With the exception of wavelengths shorter than about 0.3 micrometers and wavelengths above about 2 micrometers, the amount of radiation emitted by the sun that reaches the top of the atmosphere is very similar to the amount that reaches the Earth's surface. The substantial difference in the sun's emissions of wavelengths of less than 0.3 micrometers and the receipt of this radiation at the Earth's surface is critical to life on Earth, since these wavelengths are known as ultraviolet radiation and are highly dangerous. The relative lack of this radiation reaching the Earth's surface is the result of absorption in the upper atmosphere principally by ozone (Figure G5, lower panel).

As the solar radiation passes through the atmosphere, some is reflected by clouds and atmospheric aerosols and some is absorbed by the atmosphere. Of the 342 W m^{-2} of shortwave solar radiation received at the top of the atmosphere, 198 W m^{-2}

reaches the Earth's surface. Since the overall Earth's surface is relatively non-reflective, only about 30 $W m^{-2}$ is reflected and thus 168 $W m^{-2}$ is absorbed at the Earth's surface.

To maintain a surface energy balance, the Earth's surface emits some of this absorbed energy as longwave or terrestrial radiation. Figure G5 shows the overall emissions by the Earth and includes emissions by the atmosphere. Since the Earth's surface is very much cooler than the sun, it emits radiation at longer wavelengths (mainly 3–50 micrometers, see Figure G5, right-hand side). The atmosphere is relatively more opaque (it has a higher absorptivity) to this terrestrial radiation because of greenhouse gases. Thus a significant fraction of the radiation emitted by the Earth's surface is absorbed in the atmosphere and re-emitted. This is particularly clear in Figure G5 (right-hand side) where the difference between what the Earth should emit to space (dashed line) is very different from what is actually emitted (shaded region). It is the presence of greenhouse gases in the atmosphere that causes this difference. As a result of this increased opacity (Figure G5, lower panel) more longwave radiation is absorbed and re-emitted, and this acts to warm both the atmosphere and the Earth's surface. The various greenhouse gases absorb longwave radiation at particular wavelengths (Figure G5, lower panel). If a greenhouse gas is released that happens to absorb radiation at a wavelength that is already strongly absorbed in the atmosphere then the impact of increasing this gas is likely to be relatively small (but not zero). In contrast, the release of a greenhouse gas that absorbs in a wavelength region poorly absorbed is likely to have a much larger effect. These wavelength regions, where the atmosphere is a poor absorber of longwave radiation, are known as atmospheric windows, and several of these windows are being partially "closed" through the emission of key greenhouse gases.

This role of greenhouse gases in the atmosphere is a very ancient natural process. The greenhouse effect is the main mechanism to explain how the Earth remained non-frozen during the Archean period (2.5–3.8 billion years ago) when the sun's luminosity is believed to have been 30% less than present (this is known as the faint-sun paradox). It is currently believed that methane generated a sufficiently strong greenhouse effect to counteract the faint sun. Through geological time the amounts of greenhouse gases in the atmosphere have varied, driven by long timescale changes in the sources and sinks of the natural greenhouse gases. For example, over the last 160 000 years carbon dioxide has varied naturally from below 180 parts per million by volume (ppmv) (some 40 000 years ago) to above 300 ppmv (some 140 000 years ago).

Human activity began to affect the Earth when humans began to use fire. However, the impact of humans began to accelerate when large-scale burning of fossil fuels began in Europe at the beginning of the Industrial Revolution (around 1750). Burning carbon-based fossil fuels in an oxygen-rich atmosphere releases carbon dioxide (CO_2), and since the 1750s the concentration of CO_2 has increased from about 280 ppmv to 375 ppmv (2003) at the Mauna Loa observing station which provides a very high quality and continuous data set (see Figure G7) (http://cdiac.esd.ornl.gov/trends/co2/sio-mlo.htm). Similarly, methane, nitrous oxide and a suite of human-made chemicals (the chlorofluorocarbons (CFC) and hydroflurocarbons (HFC)) have increased in the atmosphere. At the same time industrial activity releases sulfate aerosols into the atmosphere.

Before examining the details of the greenhouse effect, some details on these greenhouse gases will be provided, since these form one of the foundations for the concerns over the enhancement of the natural greenhouse effect by human activity. Basic data on major greenhouse gases are provided for particular greenhouse gases compared to an emission of a kilogram of carbon dioxide. The global warming potentials calculated for various time periods show the effects of the atmospheric lifetime of the different gases. Note that the global warming potentials are on the basis of kilograms of each greenhouse gas. While billions of tonnes of carbon dioxide are released annually, the amount of HFC and CFC released is far smaller: HFC-134a is of order $3.2 \times 10^{-5} Gt y^{-1}$ or 1/170 000 of the amount of carbon dioxide. Thus, even 500 years into the future, carbon dioxide will remain far more important that the HFC or CFC for net impact on radiative forcing.

Greenhouse gases and aerosols

Carbon dioxide has increased in the atmosphere by 32% between 1750 (280 ppmv) and 1998 (365 ppmv) (Figure G6a and Table G7); this increase is due to human activity. The current level of CO_2 has not been exceeded during the past 420 000 years and it is unlikely that it has been exceeded in the past 20 million years. The current rate of increase over the last century is unprecedented, at least over the last 20 000 years. Most of this increase in CO_2 (about 75%) comes from fossil fuel burning; the rest is largely the result of deforestation (Table G8). The sinks of CO_2 into the land and ocean currently take up about

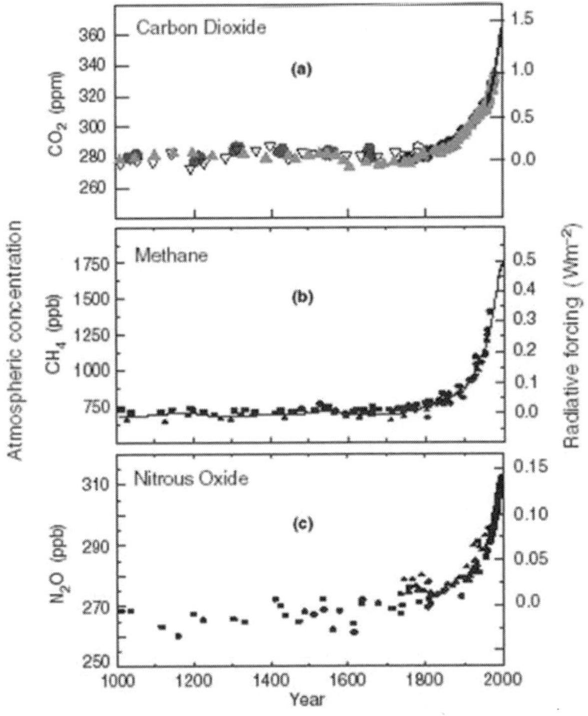

Figure G6 The variation in (**a**) carbon dioxide, (**b**) methane, (**c**) nitrous oxide since AD 1000. These data are derived from a combination of measurements including ice cores. In the last few decades observations have been taken directly form the atmosphere. The left-hand scale shows atmospheric concentration and the right-hand scale shows radiative forcing. This figure is taken from the Summary for Policy Makers Report of the Intergovernmental Panel on Climate Change, available at *http://www.ipce. ch/pub/spm22-01.pdf*

Table G7 Various data for key greenhouse gases

Greenhouse gases	Concentration (preindustrial)	Concentration (1998)	Approximate lifetime (years)	Radiative forcing ($W m^{-2}$)	Global warming potential (years into the future)		
					20	100	500
Carbon dioxide	270 ppmv	365 ppm	5–200	1.46	1	1	1
Methane	700 ppbv	1745 ppb	12	0.48	62	23	7
Nitrous oxide	270 ppbv	314 ppb	114	0.25	275	296	156
Hydrofluorocarbon-23 (HFC-23)	0 ppt	14 ppt	260		9400	12 000	10 000
Chlorofluorocarbon (CFC-11)	0 ppt	−1.4 ppt	45	0.34		4600	
Perfluoromethane (CF_4)	40 ppt	80 ppt	50 000		3900	5700	8900

The global warming potentials are relative to carbon dioxide and are an index for estimating relative warming contribution due to an emission of a kilogram of a particular greenhouse gas compared to an emission of a kilogram of carbon dioxide. The global warming potentials calculated for various time periods show the effects of the atmospheric lifetime of the different gases. Note that the global warming potentials are on the basis of kilograms of each greenhouse gas. While billions of tonnes of carbon dioxide are released annually, the amount of HFC and CFC released is far smaller: HFC-134a is of order 3.2×10^{-5} Gt yr^{-1} or 1/170 000 of the amount of carbon dioxide. Thus, even 500 years into the future, carbon dioxide will remain far more important that the HFC or CFC for net impact on radiative forcing.

Table G8 Global carbon dioxide budgets. Positive values are fluxes to the atmosphere, negative values represent uptake from the atmosphere (from table 3.1 and associated text in Prentice et al., 2001). See also Figure G7

Carbon dioxide	Size (1980s)
Sources	
Fossil fuels emissions	5.3 ± 0.3 GtC yr^{-1}
Cement production	0.1 GtC yr^{-1}
Land-use change	1.7 (0.6 to 2.5) GtC yr^{-1}
Sinks	
Oceans–atmosphere exchange	-1.9 ± 0.6 GtC yr^{-1}
Atmospheric increase	3.2 ± 0.1 GtC yr^{-1}
Terrestrial sink	−1.9 (−3.8 to 0.3) GtC yr^{-1}

Table G9 Major anthropogenic sources and major sinks for the global methane budget, derived from table 4.2 and associated text in Prather et al. (2001). See also Figure G7

Methane (CH_4)	Size (1980s)
Sources	
Energy production	75–110 Tg $CH_4 yr^{-1}$
Landfills	35–73 Tg $CH_4 yr^{-1}$
Ruminants	8–150 Tg $CH_4 yr^{-1}$
Rice production	25–100 Tg $CH_4 yr^{-1}$
Biomass burning	23–55 Tg $CH_4 yr^{-1}$
Total (anthropogenic)	300–500 Tg $CH_4 yr^{-1}$
Sinks	
Soils	10–44 Tg $CH_4 yr^{-1}$
Tropospheric OH	450–510 Tg $CH_4 yr^{-1}$
Stratospheric loss	40–46 Tg $CH_4 yr^{-1}$
Total	460–580 Tg $CH_4 yr^{-1}$

50% of the CO_2 released by human activity, the other 50% is accumulating in the atmosphere. CO_2 contributes about 60% of the enhancement to the natural greenhouse effect of globally well-mixed greenhouse gases. This enhancement is measured by the impact on radiative forcing (the amount of radiation available to drive the climate system). In the case of CO_2 the increase in radiative forcing is 1.46 $W m^{-2}$. A molecule of CO_2 released into the atmosphere remains for between 5 and 200 years (the lifetime or residence time), depending on the nature of the eventual sink.

Methane (CH_4) is also increasing in the atmosphere, by about 150% between 1750 (700 ppbv) and 1998 (1745 ppbv) (Table G7, Figure G6). CH_4 has an approximate lifetime in the atmosphere of 12 years. About half of the release of CH_4 is the result of human activity (mainly use of fossil fuels, agriculture, landfills etc) (Table G9). CH_4 contributes approximately 20% of the enhancement to the natural greenhouse effect of globally well-mixed greenhouse gases (an increase in radiative forcing of 0.48 $W m^{-2}$). Methane is removed from the atmosphere by chemical reactions. Table G7 shows the global warming potential of CH_4 relative to CO_2. A single kilogram of CH_4 emitted into the atmosphere will have a global warming potential of 62 times a kilogram of CO_2 20 years into the future (Table G7). This global warming potential is reduced to 23 times at 100 years into the future and 7 times at 500 years into the future. This change in global warming potential reflects the atmospheric lifetime of CH_4 being quite short, while the overall greater global warming potential compared to CO_2 reflects the role of CH_4 in trapping longwave radiation in atmospheric windows. Note that, since the total anthropogenic emission of CH_4 is approximately 5% of the emissions of CO_2 (Figure G7), the actual impact of CO_2 will remain much larger than CH_4.

Finally, the large family of halocarbons are a group of very long-lived greenhouse gases. As a consequence of the Montreal Protocol, the concentration of some of these gases is falling (e.g. CFC-11) or increasing more slowly (e.g. CFC-12) than in the 1990s. The family of halocarbons contributes approximately 14% of the enhancement to the natural greenhouse effect of globally well-mixed greenhouse gases and adds 0.34 $W m^{-2}$ to radiative forcing. These gases are removed from the atmosphere when the molecules are broken down in the stratosphere under the influence of ultraviolet radiation. Unfortunately, this leads to the release of bromine and chlorine, which contribute to the breakdown of ozone in the stratosphere. This breakdown of these molecules is typically slow; and hence many of these chemicals have long atmospheric lifetimes. Typically, the HFC range between a few months (HFC-161) to a few centuries (HFC-23). Some gases (known as fully fluorinated species) have atmospheric lifetimes of thousands of years and perfluoromethane exceeds 50 000 years. This leads to very large global warming potentials for these gases (Table G7), both because they will remain in the atmosphere for long periods and because they commonly absorb longwave radiation at wavelengths where the atmosphere is presently relatively transparent.

In terms of the enhanced greenhouse effect, the key measure of the importance of a greenhouse gas is the role it plays in the radiative forcing of climate (the energy available to drive the climate system). Very many factors have changed since 1750, and these all need to be considered as potential influences on the climate system. If radiative forcing increases, then warming is likely to occur, and conversely a reduction in radiative forcing is likely to cool the planet.

Figure G8 shows changes in radiative forcing between 1750 and the present. The main contribution to increases in radiative forcing is from the increases in globally well-mixed greenhouse gases (CO_2, CH_4, N_2O and the halocarbons) that combine to contribute about $2.5\,W\,m^{-2}$ of radiative forcing. Figure G8 shows many other aspects of change within the climate system between 1750 and the present that contribute to radiative forcing. Most of these are unlikely to be a significant factor in future climate in comparison to the greenhouse gases. However, there are a series of issues worthy of note. The changes in ozone

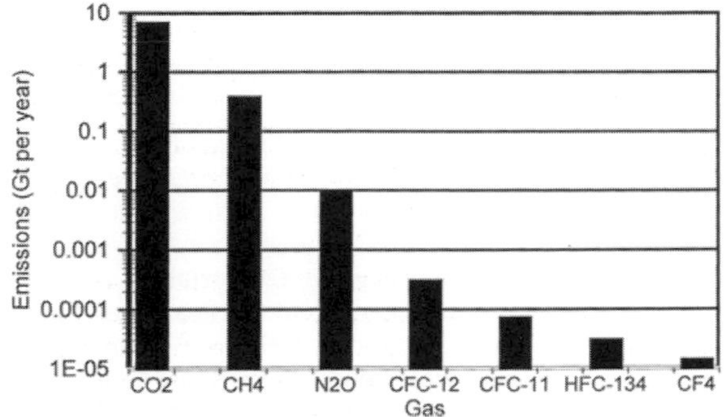

Figure G7 Total anthropogenic emission of various greenhous gases (converted into $Gt\,yr^{-1}$). Note the log scale on the vertical axis. HFC-134 is hydrofluorocarbon-134 and CF_4 is perfluromenthane. These data are from various year in the 1990s.

Table G10 Major anthropogenic sources and all major sinks for the global nitrous oxide budget, derived from table 4.4 and associated text in Prather et al. (2001). See also Figure G7

Nitrous oxide (N_2O)	Size (1980s)
Sources	
Agricultural soils	$0.6–14.8\,Tg\,N\,yr^{-1}$
Biomass burning	$0.2–1.0\,Tg\,N\,yr^{-1}$
Industrial sources	$0.7–1.8\,Tg\,N\,yr^{-1}$
Cattle and feedlots	$0.6–3.1\,Tg\,N\,yr^{-1}$
Total (anthropogenic)	$2.1–20.7\,Tg\,N\,yr^{-1}$
Sinks	
Stratospheric loss	$9–16\,Tg\,N\,yr^{-1}$

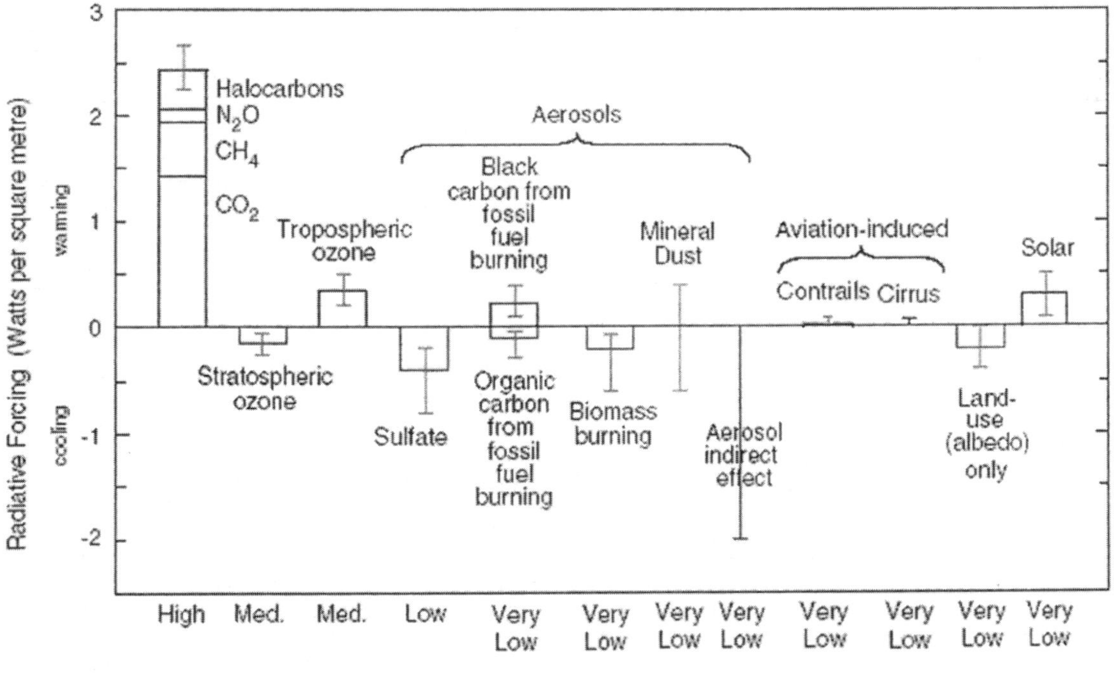

Figure G8 The global mean change in radiative forcing of the climate system between 1750 and 2000. Positive values imply warming and negative values imply cooling. This figure is taken from the Summary for Policy Markers Report of the Intergovernmental Panel on Climate Change available at *http://www.ipcc.ch/pup/spm22-01.pdf*

(O_3) have cooled the stratosphere (where the majority of ozone depletion has occurred), which may have lessened the observed tropospheric warming caused by greenhouse gases. The role of aerosols is also very important but quite poorly understood. There are two ways in which aerosols affect radiative forcing: (a) the direct effect, whereby aerosols scatter and absorb solar and thermal longwave radiation; and (b) the indirect effect, whereby aerosols modify the microphysical and hence the radiative properties and amounts of cloud. A major source of atmospheric aerosols are the emissions of sulfates by industry. These reflect incoming solar radiation and thus generate a negative radiative forcing that partially counteracts the effect of increasing greenhouse gases. It is believed that this may have led to a reduction in the amount of warming observed since 1860. This is not good news, however, since these sulfate aerosols affect air quality and are the cause of acid deposition. Overall, the best estimate of the role of the direct and indirect effect of aerosols is shown in Figure G8, but it should be emphasized that these estimates remain very uncertain.

In summary, the changes in the well-mixed greenhouse gases since 1750 are extremely well known. Their impact on radiative forcing is also very well known. The impact of the increases in CO_2, CH_4, N_2O and the halocarbons is to enhance the natural greenhouse effect. There is no doubt that this enhancement, as a direct result of human activity, will lead to ongoing climate change. There are questions, however, regarding how large these changes might be.

Impacts 1750 to present day

The Earth's surface is warming. The global average surface temperature (with measurements taken over land and ocean surfaces) has increased since 1861 (the earliest year where reliable measurements are available) (Figure G9). Over the twentieth century the global average surface temperature has increased by 0.6 ± 0.2°C. It is very likely that the 1990s was the warmest decade, and 1998 the warmest year, in the instrumental record since 1861. The global average surface temperature in 2002 was the second warmest since 1860, approximately 0.5°C above the 1961–1990 average. Remarkably, while the warmest year since 1860 remains 1998, 2002 was the second warmest, 2001 the third warmest and the 10 warmest years since 1860 have all occurred since 1987. However, it is not just the amount of warming that has concerned climate scientists. It is likely that the rate and duration of warming over the twentieth century is larger than any other time during the last 1000 years, even if uncertainties in data are taken into account.

It is also possible to explore the warming trend geographically decade by decade. An analysis of annual trends in the observed temperature record highlights an apparent acceleration in warming in recent decades. Between 1901 and 2000 most of the Earth shows warming of approximately 0.2°C per decade. Warming was more rapid between 1910 and 1945. A mixed pattern of warming and cooling was apparent between 1946 and 1975 followed by very rapid warming, particularly over the continental surfaces of the northern hemisphere, between 1976 and 2000) (see http://www.ipcc.ch/pub/wg1TARtechsum.pdf, Figure G7).

This pattern of regional change is one piece of a vast collection of evidence showing that the Earth is warming. Measurements of ocean heat content, measurements from weather balloons, observations of precipitation, atmospheric moisture, snow cover, land-ice, sea-ice, sea level, changes in terrestrial ecosystems and observed changes in some ocean circulation patterns and possibly climate variability and extreme weather and climate all point to a warming world. No single piece of evidence is safely taken in isolation; however, when taken collectively, looking at the spatial and temporal evolution of all these aspects of the climate, the evidence is genuinely convincing that the enhanced greenhouse effect is leading to significant climate change.

Thus, convincing evidence exists that the Earth is warming and theoretical evidence that the changes in greenhouse gases lead to an increase in radiative forcing also exists. Can we put these two pieces of evidence together and *attribute* the observed warming to human activity? An alternative question is, can we explain the observed changes in climate in other ways; ways that do not involve the observed increases in radiative forcing caused by greenhouse gases?

The process of demonstrating that an observed change in the climate system is statistically significantly different from

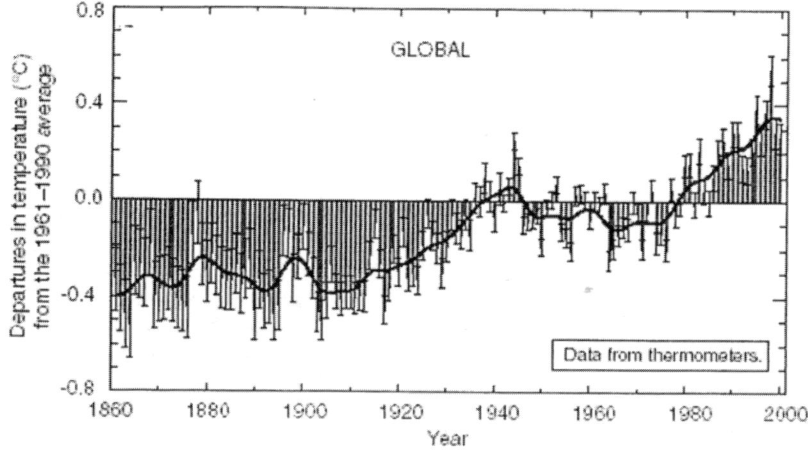

Figure G9 Combined annual land-surface air and sea surface temperature anomalies (°C) 1861 to 2000, relative to 1961 to 1990. Two standard error uncertainties are shown as bars on the annual number. This figure is taken from the Technical Summary of the Working Group 1 Report of the Intergovernmental Panel on Climate Change available at *http//www.ipcc.ch/pub/wg1TARtechsum.pdf*

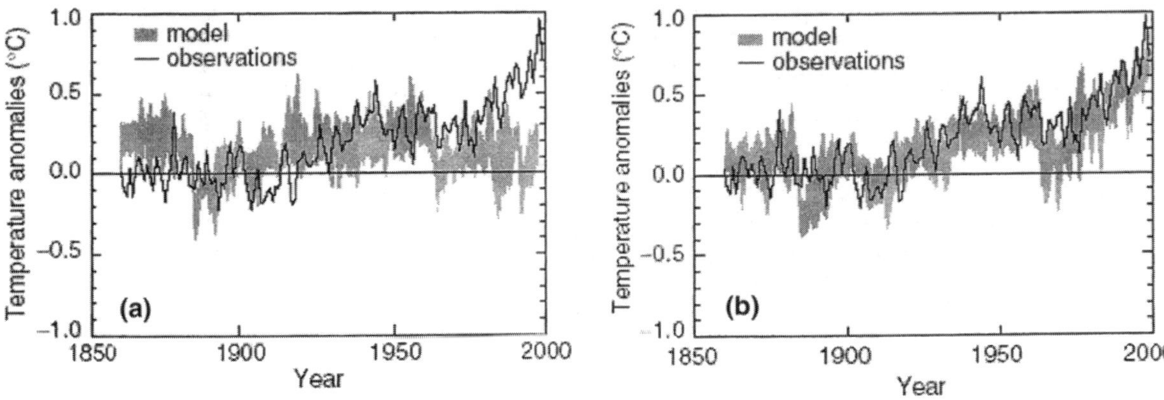

Figure G10 Simulating the Earth's temperature variations using a climate model without including the effects of greenhouse gas increases (**a**) and including these effects (**b**). Both simulations include the effects of solar variation and volcanic avtivity.

natural changes is called *detection*. The process of linking a cause (say the enhanced greenhouse effect) to an effect (say climate change) at a defined level of confidence is called *attribution*. A substantial effort has been invested in the past 5 years in attempts to detect the fingerprint of human activity on climate, and to then attribute these observed changes in climate to human activity.

The climate varies naturally, and has done so for billions of years. This natural climate variability is commonly called "noise", and the key question is whether the currently observed changes in climate are just "noise" or whether they are a human-induced "signal" that is statistically unusual. It is also important to show that any "signal" is caused by human-induced changes, and that the signal is the approximate size expected given the changes in greenhouse gases in the atmosphere.

At the present time (2003), it appears that the temperature changes observed over the last 100 years are unlikely to be entirely natural in origin – recent globally averaged temperatures exceed reconstructions of temperature over the last 1000 years, even taking into account the uncertainties in those reconstructions. Evidence from climate models also shows that the warming observed since 1850 is very unlikely to be due to natural variability – because climate models do not warm at the rate that has been observed since 1860 for the length of time the observations show the Earth has warmed. This is illustrated in Figure G10a, where a climate model is used with no changes in greenhouse gases or aerosols. Under these circumstances the climate model responds to natural variability (forced by solar and volcanic activity) and shows no trend in temperature, in contrast to the observed record (Figure G10a) which shows the same upward trend seen in Figure G9. If, however, these natural processes are combined with changes in greenhouse gases and aerosols, the climate model shows a remarkable ability to simulate the observed changes in temperature since 1850 (Figure G10b). Climate models never show a warming trend which matches the observed since 1850 *unless* greenhouse gases are increased within the model. This is not proof that observed warming is caused by the increase in greenhouse gases, but it is another piece of strong supporting evidence that it is human activity that has caused the increase in surface temperature.

In summary, no one piece of evidence is totally convincing that humans have caused the observed changes in climate. However, when taken together the case is becoming convincing that most of the observed warming over at least the last 50 years is likely to have been due to the increase in greenhouse gas concentrations caused by human activity. A further piece of evidence is worthy of note. It is very hard to dispute the reliability of the observed record, which is derived rigorously from many independent sources. Our understanding for the enhanced greenhouse effect can largely explain this record. We do not have an alternative explanation for the changes that have been observed. There is no known explanation for all the changes that are taking place in the terrestrial ecology – sea level, glaciers, snow extent, temperature, rainfall, etc. – *except* the enhanced greenhouse effect. It is valid scientifically to critique the impact of the enhanced greenhouse effect, but if we are to replace the enhanced greenhouse effect as the cause of the observed changes in climate, then an alternative theory that explains the observed phenomenon will have to be developed.

A.J. Pitman

Bibliography

Prather, M., Ehhalt, D., Dentener, F., et al., 2001. Atmospheric chemistry and greenhouse gases. In Houghton, J.T., Ding, Y., Griggs, D.J., et al., eds., *Climate Change 2001: the scientific basis*. Contribution of Working Group I to the Third Assessment Report of the Intergovernmental Panel on Climate Change. Cambridge: Cambridge University Press, pp. 239–287.

Houghton, J.T., Ding, Y., Griggs, D.J., et al., eds., 2001. *Climate Change, 2001: the scientific basis*. Contribution of Working Group I to the third assessment report of the Intergovernmental Panel on Climate Change. Cambridge: Cambridge University Press.

Prentice, I.C., Farquhar, G.D., Fasham, M.J.R., et al., 2001. The carbon cycle and atmospheric carbon dioxide. In Houghton, J.T., Ding, Y., Griggs, D.J., et al., eds., *Climate Change 2001: the scientific basis*. Contribution of Working Group I to the Third Assessment Report of the Intergovernmental Panel on Climate Change. Cambridge: Cambridge University Press, pp. 183–237.

Cross-references

Aerosols
Climate Change Impacts: Potential Environmental and Societal Consequences
Global Environmental Change: Impacts
Models, Climatic
Radiation Climatology

H

HADLEY CELL

In 1735 George Hadley proposed a cellular model to explain the primary circulation of the atmosphere. Based upon pressure differences produced by uneven heating of the earth, Hadley postulated that cold air sinking at the poles would flow equatorward to be replaced by warm air rising at the equator and flowing poleward aloft. This would result in two large Hadley circulation cells in each hemisphere (Figure H1a). This model was shown to be greatly oversimplified and to disregard the Earth's rotation, but it offered a very important first approximation and is reproducible in dishpan experiments. Hadley's model was refined over time and gradually altered to a three-cell model for each hemisphere as in Figure H1b.

Today, the low-latitude circulation between the equatorial low-pressure and the semipermanent high-pressure belts is referred to as the Hadley Cell, although now it is considered more complex than a simple thermal cell.

John E. Oliver

Bibliography

Gedzelman, S.D., 1980. *The Science and Wonders of the Atmosphere*. New York: John Wiley.

Geer, I.W., 1996. *Glossary of Weather and Climate.* Boston, MA: American Meteorological Society.

Lutgens, F.K., and Tarbuck, E.J., 2001. *The Atmosphere*, 8th edn. Upper Saddle River, NJ: Prentice-Hall.

McIlveen, R., 1992. *Fundamentals of Weather and Climate*. London: Chapman & Hall.

Oliver, J.E., and Hidore, J.J., 2002, *Climatology: an atmospheric science.* Upper Saddle River, NJ: Prentice-Hall.

Robinson, P.J., and Henderson-Sellers, A., 1999. *Contemporary Climatology*, 2nd edn. Harlow: Longman.

Thompson, R.D., 1998. *Atmospheric Processes and Systems*. London: Routledge.

Cross-references

Atmospheric Circulation, Global
Ferrel Cell

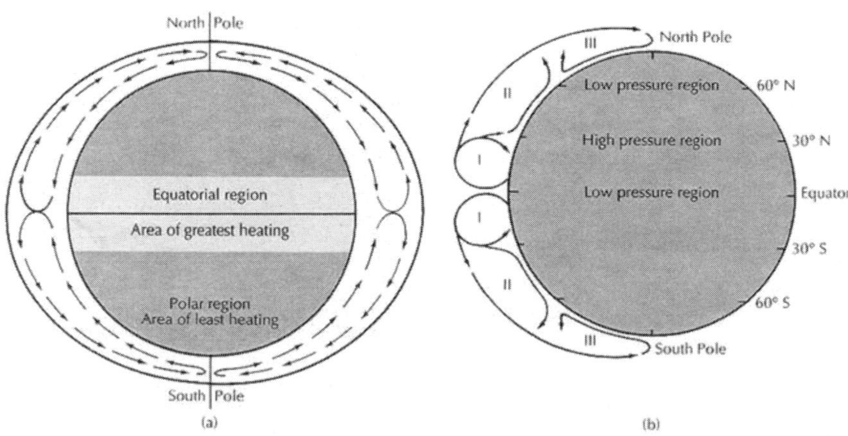

Figure H1 (**a**) A single cell in each hemisphere provides the fundamental air flow based upon temperature differences. (**b**) A three-cell model, for each hemisphere, that results from differential heating and Earth's rotation. The Hadley Cell is labeled 1.

HADLEY, GEORGE, 1685–1768

George Hadley was born in London, England, and educated at Oxford University where he trained in law. He also developed an interest in atmospheric physics and produced meteorological observations for the Royal Society of London. Hadley built on the 1686 work of Edmond Halley, who presented basic ideas about the rising of heated equatorial air and its replacement with cooler air flowing in from higher latitudes. Halley had not been able to describe a cause for the tropical easterlies, or tradewinds, however.

Hadley's key publication, "Concerning the cause of the general trade winds", appeared in 1735. In this paper he described variable speeds of rotation with latitude and deflection of airflow dependent on differences in Earth circumference between the equator and the tropics. Hadley combined this with an understanding of rising warm air in the tropics and sinking cold air in higher latitudes to explain the easterly direction of the tradewinds. He also described upper air motion in the opposite direction, from the equatorial region toward the higher latitudes, sinking and moving as westerlies beyond the tropics. Hadley described tropical circulation as two huge cells of tropical rising air caused by solar heating, poleward-moving air higher in the atmosphere, sinking cooled air, and return flow to the tropics at the surface. In doing so, he also attempted to explain the westerlies, as well as the northeast tradewinds in the northern hemisphere and southeast trades in the southern hemisphere. Although his understanding of circulation and deflection were incomplete, his view was generally accepted until better wind measurements in the mid-1800s led to modifications, with some authors showing two or three circulation cells in each hemisphere. Tropical circulation, however, follows the general pattern suggested by Hadley, and continues to bear the name "Hadley Cell".

Lisa M. Butler Harrington

Bibliography

Daintith, J., and Gjertsen, D., eds., 1999. *A Dictionary of Scientists*. Oxford: Oxford University Press.

Hadley, G., 1735. Concerning the cause of the general trade-winds. *Philosophical Transactions of the Royal Society of London*, **39**: 58–62.

Harrington, J. Jr, and Oliver, J.E., 2000. Understanding and portraying the global atmospheric circulation. *Journal of Geography*, **99**(1): 23–31.

Lau, W.K.M., 1996. General circulation. In Schneider, S.H., chief ed., *Encyclopedia of Climate and Weather*, Vol. 1. Oxford: Oxford University Press, pp. 330–334.

Monmonier, M., 1999. *Air Apparent: how meteorologists learned to map, predict, and dramatize weather*. Chicago, IL: University of Chicago Press.

HAIL

Hail is a type of precipitation often composed of concentric spheres of alternating clear and opaque ice, having a diameter of up to 50 mm or more. The largest single hailstone on record to fall in the United States struck Coffeyville, Kansas, on 3 September 1970; this hailstone weighed 1.7 lb (766 g), was 5.7 in (144 mm) in maximum diameter and fell with a calculated velocity of over 100 mph (see Figure H2).

Three forms of hail have been recognized: graupel, small hail, and true or severe hail. *Graupel*, commonly called "soft hail", is composed of loosely compacted ice crystals, and is roughly spherical with a tendency to fracture upon striking the ground. Having a diameter of 5 mm or less, it is common for graupel to melt before it reaches the Earth, especially if it falls during the summer. *Small hail* is about the size of graupel, but is semitransparent with a translucent or milky-white center. Frequently, graupel serves as a nucleus for small hail, allowing a thin layer of ice to form around it. This ice sphere gives small hail a glazed appearance and causes it to be less compressible than graupel. When small hail strikes hard surfaces, often in conjunction with rain, the hailstone remains intact. *True hail* or *severe hail* is composed of hailstones greater than 5 mm that often cause extensive damage.

It is not known for certain how hailstones form. Traditionally, multiple-incursion theories had wide acceptance. According to these theories, ice embryos form as liquid water droplets from the warmer, lower portion of a cloud by being carried aloft by strong vertical currents, and subsequently freezing upon encountering the subfreezing temperatures associated with the upper portion of the cloud prior to falling out of the freezing portion of the cloud. Where a raindrop may require a million cloud droplets to form, a hailstone the size of a golfball needs approximately 10 billion cloud droplets to form. Additional water is acquired from the warmer, lower portion and, with each repetition of the process, a sphere of transparent ice is added to the embryo. Repeated circulation from the nonfreezing to the freezing level of the cloud contributes to the onionlike structure of the concentric spheres of the hailstorms. As many as 25 concentric rings have been identified on some hailstones. Serious questions have been raised, however,

Figure H2 Photograph of the largest officially recognized hailstone to fall in the United States (© National Center for Atmospheric Research/University Corporation for Atmospheric Research/National Science Foundation; used with permission).

regarding the attempts of the multiple-incursion theories to account for the frequency and velocity of the essential updrafts required to create the concentric banding of the hailstone.

Gaining wider acceptance are theories that attribute growth of the hailstone to steady nonviolent updrafts within the cloud. According to these theories, the structure of the ice in the hailstone depends on the rate of deposition. If the hailstone acquires water too quickly to freeze completely, the resulting ice will be transparent, as dissolved air diffuses and escapes from the relatively slow-forming ice surface. If, however, multitudes of air bubbles become trapped as small droplets and freeze rapidly on impact, milky-white opaque ice will form. Hailstorms may, therefore, be a product of two different modes of growth, specifically "wet" growth resulting in a sphere of clear ice and "dry" growth depositing the opaque layer. The freezing level for hail in a cumulonimbus is often 3 miles (5 km) above the ground.

Large clouds, particularly the cumulonimbus, experience subsiding cold air from high levels that later create an intense, violent squall with rapid updrafts immediately ahead of the cold air. These updrafts pick up small protohailstones and carry them aloft. Most of these hailstones fall from the cloud and melt upon descent. Some of these hailstorms fall from the clouds, whereas others may fall back into the clouds. The size of those re-entering the cloud may be large enough to be slowly uplifted by the updraft and thus grow at a rate sufficient to almost match the speed of the updraft. Eventually these hailstones move forward near the cloud top and fall from the cloud. Occasionally, they may briefly pass through the updraft on their descent. Thus, in the same cloud, hailstones may be growing and rising while being very close to the condition in which hailstones are just becoming wet. Following these theories, slight variation in either the cloud-water concentration or in the velocity of the updraft may cause fluctuation back and forth between wet and dry growth stages, also resulting in opaque or clear ice being formed even in nearly-steady updrafts. Since these updrafts may be tilted in the cloud, the hailstone may traverse the cloud "sideways", experiencing wide variation in the irregular distribution of humidity with height.

Analysis of the structure of hailstones has provided insights concerning the environmental conditions under which they form. The interpretation of the process involved in hail formation is ambiguous and still a matter of controversy.

Distribution of hail

Hail occurs most often in deep continental interiors within the midlatitudes and occurs less frequently toward the poles and the equator. Hail is frequently associated with large, highly developed clouds, most notably the cumulonimbus. Clouds found in cold climates are seldom well developed enough to produce hail or are lacking water–ice concentrations necessary to facilitate the growth of large hailstones. Hailstones are rare in tropical regions because the steep horizontal temperature gradients and vertical wind shears conducive to hail formation fail to adequately develop. Likewise, hail seldom falls over temperate oceans, due in part to the absence of intense surface heating.

Maximum frequency of hail often occurs on the drier leeward side of high mountains. Representative of this distribution are the hailstorms in the pampas of western Argentina, the Po Valley of Italy, southern France, eastern New Zealand, South Africa, and the Caucasus regions of Russia. North America's principal hail area exists on the leeward side of the Rocky Mountains from New Mexico to Alberta. Probably the stronger vertical currents associated with the mountains enhance hail formation. No portion of the United States is immune from hail but frequently hail occurs only once per year along the Atlantic–Gulf Coastal Plains.

More is known about the climatology of hail in the United States than in any other country. Across the United States the average number of days experiencing hail ranges from over 8 per year to less than 1 per year. The highest frequency region, known as "Hail Alley", is located where the borders of Wyoming, Nebraska, and Colorado meet. Great variability exists in the number of hailstorms experienced annually and varies considerably by size of hailstone: small (0.25–0.5 inches or 6–12 mm); moderate (0.5–1 inch or 12.5–25 mm); large (greater than 1 inch or 25 mm). In some years localized regions in eastern Colorado may record 20–30 days with hail. In the United States hail is recognized as a small-scale phenomenon with relatively infrequent occurrence at any specific site. Small hail is the most common classification recorded throughout the United States and often is correlated with regions experiencing relatively infrequent hail. Hail of moderate size is concentrated along the Rocky Mountains wedging through the Great Plains toward the Midwest. Large hail mostly falls in Wyoming and Colorado, comprising 20% of the number of days experiencing hail in this region (see Figure H3).

It is hail intensity that produces damage, and intensity is determined by the size and number of hailstones experienced, the velocity of their fall, and the accompanying winds. Wide variations in intensity likewise occur across the United States. Hail in the western Great Plains may be up to 18 times more intense than the hail falling farther to the east in the Midwest. The largest hailstorms in the United States tend to fall on the leeward side of mountains, whereas small hailstones are more representative of New England, Indiana, Illinois and the Pacific Northwest.

Hailstorms occur in any season and at any time of day. East of the Great Plains, hail activity is greatest in spring with the southern states being most vulnerable in March. The northern states are most susceptible to hailstorms in May. States located on the leeward side of mountains on the Great Plains record hail mainly in the summer, averaging over 25 episodes in July, whereas the west coast experiences maximum hail activity in late winter or spring. The Great Lakes region is the only area that records peak hail activity in the fall. Most hail occurs in the afternoon hours when the Earth's surface has achieved maximum ground heating. Regions close to the Rocky Mountains tend to experience most hailstorms between 3 and 6 p.m., whereas in areas farther east the peak hours for hail are between 6 and 9 p.m., suggesting a west-to-east storm sequence. Hailstorms that occur near mountains typically last 10–15 minutes, while those in the Midwest last 3–6 minutes. The average size of a *hailstreak*, the path of the ground completely covered by hail, is 5 square miles, and there may be several hailstreaks in a single storm. In the Midwest, half of the thunderstorms associated with warm fronts and low pressure produce hail, while 75% of cold fronts or stationary fronts associated with thunder produce hail. In Iowa, Illinois, and Colorado, hail is produced on 70% of the days during which thunderstorms occur.

Fourteen hail regions, based on meteorological origin, frequency, peak season, and regional intensity, have been delineated in the United States (Figure H4). Note that hail intensity differentiates the Rocky Mountain and Great Plains region from the upper Midwest region.

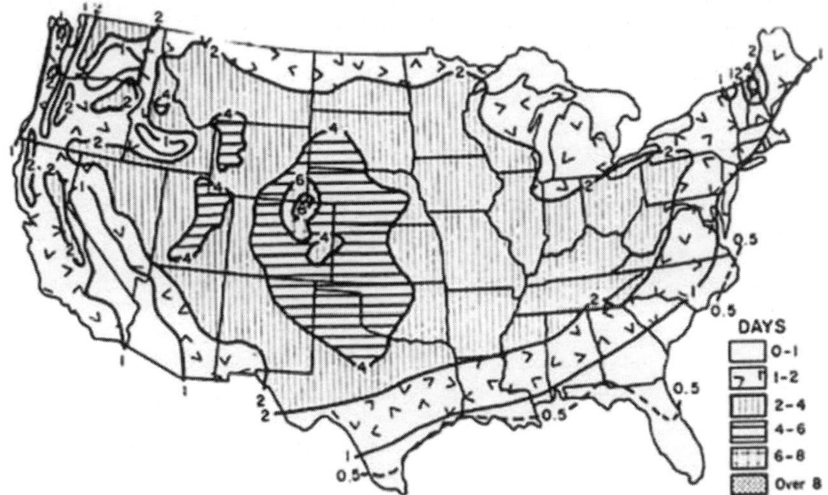

Figure H3 Average annual number of days with hail based on date for 1901–1994 (from *Thunderstorms Across the Nation: an atlas of storms, hail and their damages in the 20th century* (© 2001 Stanley A. Changnon; used with permission).

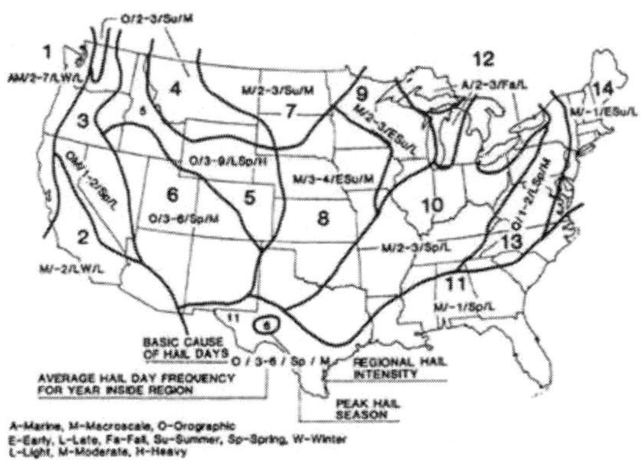

Figure H4 Major hail climate regions of the United States, based on hail frequency, cause of storms, hail intensity, and the peak hail season (from *Thunderstorms Across the Nation: an atlas of storms, hail and their damages in the 20th century* © 2001 Stanley A. Changnon; used with permission).

Hail damage

Accounts of devastation by hail have been recorded since ancient times. Hail has been known to kill cattle, horses, and even on rare occasion humans. One of the most deadly hailstorms to date occurred in India in 1888, killing 230 people at Moradabad and 16 others 50 miles away at Bareilly. The *South Carolina Gazette* reported, in what is likely the first recorded account of death by hail in the United States, that eight people died in Winnsborough, South Carolina on 8 May, 1784, from pieces of falling ice 9 inches in diameter. Even though hail has been cited as cause of death in Uvalde, Texas in 1909 and Windsor, North Carolina in 1931, the National Weather Service officially acknowledge only two deaths from hail during the twentieth Century. The first was on 13 May, 1939 in Lubbock, Texas, of a 39 year-old-farmer, and on 30 July, 1979, a 3-month-old boy (Jolene Kappelman) whose skull was crushed while in his mother's arm, resulting in his death from the injury 8 days later. Although fatalities from hail are rare in the United States, the United States leads the world in total hailstorm damage.

Crops are most vulnerable to hail damage. In terms of total dollar value of damage sustained, the most susceptible crops in the United States are wheat, corn, soybeans, and tobacco. In 1997 hail destruction to crops was estimated at $1.3 billion annually, representing up to 2% of the annual value of all crops in the United States. Property damage by hail in 1997 almost paralleled that of crop damage that approached $1 billion. To lessen the economic consequences of hail, one out of every six farmers in the United States takes out hail insurance. In high-hail regions, insurance rates may amount to 30% of the crop value, with up to 6% of crops in the Great Plains experiencing loss. Annually, 25–30% of hail loss to crops annually in the United States is covered by hail insurance, with wide regional and temporal variations.

Hail suppression

One potential means of reducing losses due to hail is to reduce the fall of damaging hail. With cloud seeding an attempt is made to increase ice nuclei as a means of increasing the number of hail embryos competing for the supercooled water. The object of such seeding is to reduce the accumulation of large amounts of supercooled water by supplying the cloud with numerous freezing nuclei capable of converting water to ice, thus preventing the growth of large hailstones. It is thought that, as a result of these efforts, the hail reaching the Earth will be smaller and hence less damaging or that, under favorable conditions, the hail will melt completely during descent.

Hail suppression programs have been undertaken by over 15 nations in the past 40 years. Of the early projects the most ambitious were those of Russia, focusing on regions near the Caucasus Mountains. Russian scientists claimed a 50–80% reduction in crop hail losses by using rockets and artillery shells to supply the clouds with silver iodide nuclei. Statistical evaluation of Russian projects has not been possible, leaving their claims open to interpretation.

Stimulated by the Russian claims, the United States initiated the National Hail Research Experiment (NHRE) in 1972, eventually monitoring about 30 hailstorms (15 used as control storms) near the Colorado–Nebraska border. In 1975 an analysis of the observations of the NHRE failed to support statistically significant increases in the number of hailstorms associated with small hail or an increase in rain. In the 18-year period from 1958 to 1975, 137 cloud-seeding projects were attempted in the United States, and of these approximately 17% (61) dealt with some aspect of hail suppression. Controversy continues in assessing the possibility of significantly reducing hail. Cloud-seeding may increase the number of small hailstones under certain atmospheric combinations of temperature, velocity of updraft, and amount of supercooled water available, whereas other combinations may yield an increase in the number of large hailstones as well. Although there has been much interest and activity in hail suppression, controversy continues, since no adequate standard exists to determine the degree of success. Research into hail suppression has resulted in much additional information on hail formation, but with suspension of most of the federal funding research for suppression in the 1980s, and with declining public interest as to its effectiveness, investigations focusing specifically on hail suppression have declined. The formation, distribution, and suppression of hail continues to be a major topic of interest to climatologists and meteorologists.

Lewis G. Wixon

Bibliography

Baldwin, J.L., 1973. *Climates of the United States.* Washington, DC: Environmental Data and Information Service.

Battan, L.J., 1965. *Recent Studies on Hail and Hail Modification in the Soviet Union.* Scientific Report No. 21. Tucson, Az: Atmospheric Physics.

Changnon, S.A., Jr, 1977. The scales of hail, *Journal of Applied Meteorology*, **16**: 626–648.

Changnon, S.A., Jr, Davis, R.J., and Farhar, B.C. et al., 1977. *Hail Suppression Impacts and Issues.* Urbana, IL: Illinois State Water Survey.

Changnon, S.A., 1997. Trends in hail in the United States. *Workshop on the Social and Economic Impacts of Weather.* Boulder, Colorado, 2–4 April.

Changnon, S.A., 2001. Thunderstorms across the nation: an atlas of storms, hail, and their damages in the 20th century. *Changnon Climatologist.*

Fassig, O.L. (ed.), 1889. *Bibliography of Meteorology, Part II: Moisture.* Washington, DC: United States Signal Office.

Flora, S.D., 1956. *Hailstorms of the United States.* Norman, OK: University of Oklahoma Press.

Frisby, E.M., and Sansom, H.W., 1967. Hail incidence in the tropics, *Journal of Applied Meteorology*, **6**: 339–354.

Hughes, P., and Wood, R., 1993. The white plague. *Weatherwise*, **46**: 16–21.

Lane, F.W., 1965. *The Elements Rage.* Philadelphia, PA: Chilton Books.

Ludlam, F.H., 1961. The hailstorm. *Weather*, **16**: 152–162.

Moran, J.M., Morgan, M.D., and Pauley, P.M., 1997. *Meteorology: the atmosphere and the science of weather.* New York: Prentice-Hall.

Oliver, J.E., 1981. *Climatology: selected applications.* London: F. H. Winston & Sons, Edward Arnold. Schleusener, R.A., 1967. The 1959 hail suppression effort in Colorado, and evidence of its effectiveness, *Nubila*, **5**: 31–59.

Skaggs, R., 1974. Severe hail in the United States. *Association of American Geographers Proceedings*, **6**: 43–46.

Stout, G.E., and Changnon, S.A., Jr, 1968. *Climatology of Hail in the Central United States.* Research Report No. 38. Chicago, IL: Crop Hail Insurance Actuarial Association.

Cross-references

Climatic Hazards
Cloud Climatology

HARE, F. KENNETH (1919–2002)

In his as-yet-unpublished autobiography, Ken Hare concludes with:

> two conflicting self assessments . . . my students, colleagues and collaborators have done most of the real work. . . . I should have concentrated on a narrower range of subjects. I diffused my efforts – partly because of the strange cocktail of opportunities that I had to capitalize, but mainly because I am like my own pet beagle: I can't resist following any interesting scent that crosses my path.

The first comment hints at two key attributes of an individual who made important contributions during key times for the evolution of climatology – his typical selfdeprecation, and his personally inspiring and enabling nature. The second self-assessment is poignant – it hints at a person who has left this world feeling his contributions could have been more substantive, and given greater recognition. But his contributions were not only substantial but also pioneering and perceptive. As a result of Hare's many and diverse contributions to climatology, and its cognate disciplines, the world is decidedly the better off for the somewhat eclectic (but far from esoteric) nature of his interests and contributions. The evidence is summarized in the paragraphs that follow. Ken Hare did deserve greater recognition, though in truth that was never his motivation. But the reality is that the world is more willing to recognize the innovative work of the reductionist scientist than the equally important contributions of the interdisciplinary researcher who explodes our understanding of policy-relevant science. Hare is a relatively unsung hero of climatology, but in 1989 he was awarded the International Meteorological Organization Prize from the World Meteorological Organization. He held 11 honorary degrees and was a Companion of the Order of Canada, recipient of the Order of Ontario, a Fellow of the Royal Society of Canada and President of the Royal Meteorological Society. Bailey, Oke and Rouse chose to dedicate their book *The Surface Climates of Canada* to Ken Hare, and accorded him the fitting honour of writing the opening chapter (see Bibliography, below). Ken Hare's important contributions to the University of Toronto were recognized by his prestigious appointment as University Professor. For several years Hare was Chairman of the Canadian Climate Planning Board for Environment Canada.

After graduating from the University of London (UK) with a first-class honours BSc, (Special) in geography, and with geology as a subsidiary subject, Hare's further studies and early career were disrupted by the Second World War. He was called up to join the British Meteorological Office, where he was trained and served as a weather forecaster. Thus began a distinguished career which linked many disciplines and continents. Hare's initial interests in landforms, landscapes and land-use were married with his expertise in dynamic meteorology, to forge an enduring interest in global climate and its stability. This is turn was often manifest by his studies into the exchanges and equilibria between the Earth's surface and its atmosphere, in order to understand the mutual interactions between soil and vegetation patterns and micro- and macro-climates. Hare's first postwar position was as an assistant professor of geography at McGill University, a formative start to an illustrious academic career that straddled the Atlantic, including

Dean of Arts and Science at McGill, Master of Birkbeck College, University of London, President of the University of British Columbia, Director of the Institute for Environmental Studies at the University of Toronto, Provost of that University's Trinity College, and Chancellor of Trent University.

Much of Hare's lifetime research focused on relating the characteristics of the northern boreal forest to the physical, dynamic and micro climates of the Canadian Arctic and sub-Arctic. Inspired by the work of such luminaries as Thornthwaite, Penman and Budyko, one of Ken Hare's major interests and research thrusts was heat and water balance climatology. Influenced also by the work of Lettau, he showed that all the familiar annual balance parameters of the hydrologists and climatologists – including the Bowen and run-off ratios, the Thornthwaite indices, Priestley–Taylor's alpha, and Budyko's dryness index – are simple linear or logarithmic transforms of one another, provided there is no change in annual water storage. Moreover, any three or four of these hydroclimatological measures form a conjugate set, definable for any locality, thereby characterizing uniquely the mean annual heat and water balance regime. Applications were many, including contributing to the debate on desiccation of the Sahel, and arid and semiarid climates more widely, by clarifying the parameterization of aridity. But perhaps the most seminal contribution was demonstrating that the depictions of mean annual precipitation and runoff for Canada were irreconcilable, to the extent that in some places mean annual runoff exceeded precipitation. This resulted in a comprehensive revision of the fields for precipitation, runoff and lake evaporation. Hare's specific and wider contributions on these topics are captured in such contributions as *Climate Canada* and *The Climate of Canada and Alaska* (see below).

In his later years, and again on both sides of the Atlantic, Hare was involved in numerous policy-related enquiries, reflecting not only his stature as an interdisciplinary scientist but also his integrity, objectivity, diplomacy and ability to think holistically, laterally and creatively. The enquiries covered such diverse themes and issues as nuclear waste disposal, transboundary air pollution and acid deposition, nuclear winter, global climate change, lead pollution, nuclear reactor safety and asbestos hazards.

But Ken Hare's most lasting legacy to climatology is the large number of his students who now lead and inspire others in their climatological and related investigations and other endeavors, throughout the world.

John Hay

Selected Bibliography

Hare, F.K., 1968. The Arctic. *Quarterly Journal of the Royal Meteorological Society*, **94**: 439–459.

Hare, F.K., 1997. Canada's climate: an overall perspective. In Bailey, W.G., Oke, T.R., and Rouse, W.R., eds., *The Surface Climates of Canada*. Montreal: McGill-Queen's University Press, pp. 3–20.

Hare, F.K., and Hay, J.E., 1971. Anomalies in the large-scale annual water balance over northern North America. *Canadian Geographer*, **15**: 79–94.

Hare, F.K. and Hay, J.E., 1974. The climate of Canada and Alaska. In Bryson, R.A., and Hare, F.K., eds., *The Climates of North America*. World Survey of Climatology, Vol. 11. Amsterdam: Elsevier, pp. 49–192.

Hare, F.K., and Thomas, M.K., 1979. *Climate Canada*, 2nd edn. Toronto: Wiley.

HEAT INDEX

High moisture content in the air adds to human discomfort when the temperature is high. Many indices have been developed over the years that are intended to approximate how hot it feels when both temperature and humidity are high (Quayle and Doehring, 1981). For example, when the temperature is 35°C and the dewpoint is 25°C, the air feels oppressively hot; whereas if the dewpoint is only 5°C, a temperature of 35° can seem almost tolerable.

The index that finds current favor in the U.S. is the Steadman heat index or simply heat index. It stems from the work of R.G. Steadman (1979a,b), who investigated the perception of heat under a variety of meteorological conditions using extensive data on human physiology. Steadman's tables give apparent temperature as a function of air temperature and relative humidity (Table H2) or, equivalently, air temperature and dewpoint (Table H1). The US National Weather Service quotes the heat index each summer.

Steadman's calculations assume standard sea-level barometric pressure (101.3 kPa, where kPa is kilopascal, given that 1 Pascal is a force of 1 Newton applied over an area of 1 square meter), a vapor pressure of 1.6 kPa (equivalent to a sea-level dewpoint of 14°C), a wind speed of 2.5 m s^{-1}, and no direct solar radiation. The person experiencing the heat is assumed to be walking.

In the table, note that if the dewpoint is less than 14°C, the apparent temperature is less than the actual air temperature. The reason is that evaporating sweat causes cooling on the surface of the skin and gives the sensation that the air is somewhat cooler than the actual temperature would suggest. If the dewpoint is greater than 14°C, sweat evaporates less readily, collects on the skin, and makes the subject more uncomfortable than the air temperature alone would suggest.

Steadman (1979b) notes that wind greater than 2.5 m s^{-1} generally lowers the apparent temperature because of the increased evaporation of sweat. An exception occurs, however, when the air temperature approaches or exceeds body temperature (37°C) and the humidity is low. A hot, dry wind of 15 m s^{-1} can raise the apparent temperature by 1–4°C. Finally, exposure to direct sunlight can raise the apparent temperature by as much as 7°C.

Thomas W. Schlatter

Bibliography

Quayle, R., and Doehring, F., 1981. Heat stress: a comparison of indices. *Weatherwise*, **34**(3): 120–124.

Steadman, R.G., 1979a. The assessment of sultriness. Part I: A temperature–humidity index based on human physiology and clothing. *Journal of Applied Meteorology*, **18**(7): 861–873.

Steadman, R.G., 1979b. The assessment of sultriness. Part II: Effects of wind, extra radiation and barometric pressure on apparent temperature. *Journal of Applied Meteorology*, **18**(7): 874–885.

http://www.zunis.org/index.html contains much information about heat-related stress on the human body, background information on two heat indices, including Steadman's, and guidelines for the prevention of heat-related injury during sporting events.

Cross-references

Climate Comfort Indices
Wind Chill

Table H1 Figures from Steadman (1979a), which give the heat index in the form of an apparent temperature as a function of the actual temperature and dewpoint

Air temperature (°C)	Dewpoint (°C)															
	0	2	4	6	8	10	12	14	16	18	20	22	24	26	28	30
20	18	18	18	19	19	19	20	20	21	21	21	–	–	–	–	–
22	20	20	21	21	21	22	22	22	22	23	23	24	–	–	–	–
24	22	23	23	23	23	24	24	24	24	25	25	26	26	–	–	–
26	24	25	25	25	25	25	26	26	26	27	27	28	29	30	–	–
28	26	27	27	27	27	27	28	28	29	29	30	31	32	33	36	–
30	28	28	28	29	29	29	30	30	31	31	32	33	35	36	38	45
32	30	30	30	31	31	31	31	32	33	33	34	36	37	39	41	50
34	32	32	32	33	33	33	33	34	35	36	37	38	40	42	45	–
36	33	33	34	34	34	35	35	36	37	38	39	41	43	45	48	–
38	35	35	35	36	36	37	37	38	39	41	42	44	46	49	52	–
40	36	37	37	38	38	39	39	40	41	43	45	47	49	52	–	–
42	38	38	39	39	40	40	41	42	43	45	47	49	52	–	–	–
44	40	40	41	41	42	42	43	44	45	47	49	52	–	–	–	–
46	42	42	42	43	44	44	45	46	47	49	51	–	–	–	–	–
48	43	44	44	45	45	45	46	48	49	51	–	–	–	–	–	–
50	45	45	46	46	47	47	48	50	52	–	–	–	–	–	–	–

Table H2 Figures also from Steadman (1979a), which give the heat index in terms of air temperature and relative humidity

Air temperature (°C)	Relative humidity (%)										
	0	10	20	30	40	50	60	70	80	90	100
20	16	17	17	18	19	19	20	20	21	21	21
21	18	18	19	19	20	20	21	21	22	22	23
22	19	19	20	20	21	21	22	22	23	23	24
23	20	20	21	22	22	23	23	24	24	24	25
24	21	22	22	23	23	24	24	25	25	26	26
25	22	23	24	24	24	25	25	26	27	27	28
26	24	24	25	25	26	26	27	27	28	29	30
27	25	25	26	26	27	27	28	29	30	31	33
28	26	26	27	27	28	29	29	31	32	34	36
29	26	27	27	28	29	30	31	33	35	37	40
30	27	28	28	29	30	31	33	35	37	40	45
31	28	29	29	30	31	33	35	37	40	45	
32	29	29	30	31	33	35	37	40	44	51	
33	29	30	31	33	34	36	39	43	49		
34	30	31	32	34	36	38	42	47			
35	31	32	33	35	37	40	45	51			
36	32	33	35	37	39	43	49				
37	32	34	36	38	41	46					
38	33	35	37	40	44	49					
39	34	36	38	41	46						
40	35	37	40	43	49						
41	35	38	41	45							
42	36	39	42	47							
43	37	40	44	49							
44	38	41	45	52							
45	38	42	47								
46	39	43	49								
47	40	44	51								
48	41	45	53								
49	42	47									
50	42	48									

The US National Weather Service has guidelines for the following ranges of heat index:

32–40°C Heatstroke, heat cramps, or heat exhaustion are possible with prolonged exposure and/or physical activity.
41–53°C Heat cramps or heat exhaustion are likely, and heatstroke possible with continued exposure.
≥ 54°C Heatstroke is highly likely with continued exposure.

HEAT LOW

A heat low is a region of low barometric (atmospheric) pressure caused by intense local heating of the earth's surface. These systems are not frontal in nature. The air in direct contact with the heated surface will warm, become more buoyant, and rise. Rising air causes atmospheric pressure at the surface to decrease, and a "low" is created. This upward movement will eventually create an upper-level outflow (a "high" aloft) that induces more air to flow into the surface low. Heat lows (also called thermal lows) are commonly found over tropical and subtropical continental regions during the summer months. If moisture is available the thermal low may enhance convective rainfall or thunderstorm activity. Thermal lows are, at least, partially responsible for the movement of certain monsoonal winds.

L.M. Trapasso

Bibliography

Glickman, T., 2000. *Glossary of Meteorology*, 2nd edn. Boston, MA: American Meteorological Society.

Holton, J., 1992. *An Introduction to Dynamic Meteorology*, 3rd edn. New York: Academic Press.

Cross-reference

Pressure, Surface

HOLOCENE EPOCH

The Holocene, or "wholly recent", Epoch is the youngest phase of Earth history. It began when the last glaciation ended, and for this reason is sometimes also known as the postglacial period. In reality, however, the Holocene is one of many interglacials which have punctuated the late Cainozoic Ice Age. The term was introduced by Gervais in 1869 and was accepted as part of valid geological nomenclature by the International Geological Congress in 1885. The International Union for Quaternary Research (INQUA) has a Commission devoted to the study of the Holocene, and several IGCP projects have been based around environmental changes during the Holocene. A technical guide produced by IGCP Subproject 158B (*Palaeohydrological Changes in the Temperate Zone*) represents one of the most comprehensive accounts so far of Holocene research methods (Berglund, 1986). Since 1991 there has also existed a journal dedicated exclusively to Holocene research (*The Holocene*, published by Edward Arnold).

During the Holocene, the Earth's climates and environments took on their modern, natural form. Change was especially rapid during the first few millennia, with forests returning from their glacial refugia, the remaining ice sheets over Scandinavia and Canada melting away, and sea levels rising to within a few meters of their modern elevations in most parts of the world. By contrast, during the second half of the Holocene, human impact has become an increasingly important agency in the modification of natural environments. A critical point in this endeavor was when *Homo sapiens* began the domestication of plants and animals, a process which began in regions such as the Near East and Mesoamerica very early in the Holocene, and which then spread progressively to almost all areas of the globe. For short histories of the Holocene, see Roberts (1989) and Bell and Walker (1992). Although there are different schools of thought about how the Holocene should be formally defined (see Watson and Wright, 1980), the most common view, and one which is supported by INQUA, is that the Holocene began 10 000 radiocarbon (^{14}C) years ago. But ^{14}C chronologies count AD 1950 as being the "present day" and also underestimate true, or calendar, ages by several centuries for most of the Holocene. None the less, there is evidence of a global climatic shift remarkably close to 10 000 ^{14}C yr BP (years before present), often involving a sharp rise in temperature (see Atkinson et al., 1987).

Various attempts have been made to subdivide the Holocene, usually on the basis of inferred climatic changes. Blytt and Semander, for instance, proposed a scheme of alternating cool–wet and warm–dry phases based on shifts in peat stratigraphy in northern Europe. Some researchers believe there is evidence of a "thermal optimum" during the early-to-mid part of the Holocene. During the 1980s the Cooperative Holocene Mapping Project (COHMAP) members established a comprehensive paleoclimatic database for the Holocene (Wright et al., 1993), and showed that variations in the Earth's orbit were the principal cause of differences in climate between the early Holocene and the present day. For this reason the early Holocene is unlikely to provide a good direct analog for a future climate subject to greenhouse-gas warming (Street-Perrott and Roberts, 1993).

Neil Roberts

Bibliography

Atkinson, T.C., Briffa, K.R., and Coope, G.R., 1987. Seasonal temperatures in Britain during the past 22,000 years reconstructed using beetle remains. *Nature*, **325**: 587–92.

Bell, M., and Walker, M.J.C., 1992. *Late Quaternary Environmental Change: physical and human perspectives*. London: Longman.

Berglund, B. (ed.), 1986. *Handbook of Holocene Palaeoecology and Palaeohydrology*. New York: Wiley.

Roberts, N., 1989. *The Holocene. An Environmental History*. Oxford: Blackwell.

Street-Perrott, F.A., and Roberts, N., 1993. Past climates and future greenhouse warming, In Roberts, N., ed., *The Changing Global Environment*. Oxford: Blackwell.

Watson, R.A., and Wright, H.E. Jr, 1980. The end of the Pleistocene: a general critique of chronostratigraphic classification. *Boreas*, **9**: 153–63.

Wright, H.E., Kutzbach, J.E., Webb, T., Ruddiman, W.F., Street-Perrott, F.A., and Bartlein, P.J., eds., 1993. *Global Climates for 9000 and 6000 Years Ago*. Minneapolis, MN: University of Minnesota Press.

Cross-references

Archeoclimatology
Climatic Variation: Historical
Ice Ages
Vegetation and Climate

HORSE LATITUDES

Horse latitudes is a term used to describe north and south Subtropical High atmospheric pressure belts. The term is now used only rarely in modern studies. Marked by dry sinking air, these Subtropical High atmospheric pressure belts are centered around 35°N and 30°S, extending approximately 25° to 40°, respectively, N and S, and shifting seasonally – to the north in the northern summer and to the south in the southern summer. Characterized by light winds and fine, clear weather, where not modified by continental factors, they are sometimes known as the *Calms of Cancer* and *Calms of Capricorn*. In the northern Atlantic the Calms of Cancer include the Sargasso Sea, the lack of wind combining with the gyratory current system to make this an exceptionally difficult region to cross in the days of the sailing ships. According to the *Oxford English Dictionary,* the origin of the name "horse latitudes" is uncertain. However, it has been suggested that sailing ships carrying horses to the West Indies, if becalmed unduly in the Sargasso Sea, occasionally had to jettison their live cargo as fodder ran out.

The two belts merge equatorially into the belts of the Northeast Trade Winds and the Southeast Trade Winds. Poleward they merge into the westerly, zonal wind belts. Over the oceans the horse latitudes correspond approximately to the areas of maximal evaporation. Over land the air is exceptionally dry.

Rhodes W. Fairbridge

Bibliography

Huschke, R.E., 1959. *Glossary of Meteorology.* Boston, MA: American Meteorological Society.

Riehl, H., 1954. *Tropical Meteorology.* New York: McGraw-Hill.

Cross-references

Atmospheric Circulation, Global
Azores (Bermuda) High
Centers of Action
Pacific (Hawaiian) High

HOWARD, LUKE (1772–1864)

Luke Howard was born in 1772 in London, England. Although by profession he was a manufacturing chemist and pharmacist, he was also an amateur meteorologist who is credited with the first successful and logical classification of clouds based on appearance. Howard was educated at a Quaker school in Burford, Oxfordshire and became a devout Quaker and family man – Mariabella (Eliot) Howard, his wife, and three children. Although his professional life was that of a chemist, he stated that "meteorology was my real penchant".

In his attempt to classify clouds, Howard believed that the terms should describe the appearance of the clouds and use a common language that Europeans and non-Europeans could understand – Latin – rather than the vernacular which was attempted by Jean-Baptiste Lamarck of France. Lamarck's classification of clouds was not unlike Howard's basic idea that clouds have distinct appearances. However, there were three reasons that Lamarck's scheme failed – first, the use of the scheme which did not use the Linnaean style of nomenclature; second, the language was French not Latin, which would only be useful to French-speaking people; and third, it was too difficult to understand. Howard's success of classifying clouds came about when he used a similar method of classifying plants and animals by the Swedish taxonomist Carl von Linné (Linnaeus). The style of nomenclature used identified the genus and further specified the species. Howard's scheme was widely accepted for two reasons: similar clouds occur globally, which allowed the use of an internationally accepted language, and the proper use of nomenclature.

In his book, *On the Modification* (classification) *of Clouds* (1804), Howard presented three distinct basic forms of clouds: *stratus* meaning layered, *cumulus* having a stacked appearance, and *cirrus* regarding the appearance of an angel's hair. In addition, four cloud forms were introduced: *cirro-cumulus*, *cirro-stratus*, *cumulo-stratus*, and *cumulo-cirro-stratus*. Howard believed, from his study of cloud formations, that one could predict the weather through observing clouds.

Howard's classification scheme and detailed descriptions from republished essays inspired a few Romantic painters such as John Constable and Joseph M. W. Turner in the accuracy of illustrating clouds in their works of art (Thornes, 1984; Heidorn, 1999).

In addition to Howard's book, *On the Modification of Clouds*, he further presented a range of papers and lectures on meteorological topics, including *Average Barometer* (1800), and *Theories of Rain* (1802). In 1806 Howard began a register of meteorological observations, which were regularly published in the *Athenaeum Magazine* in 1807. A series of *Seven Lectures in Meteorology* (presented in 1817) that Howard gave were later published as a textbook on meteorology in 1837. Howard began a study of London's urban climate that culminated into two volumes, *The Climate of London* (1818–1819). In 1821 Howard became a member of the Royal Society, an organization of scientists and inventors, and later published his final work, *Barometrographia* (1847).

Although not much has been written about Howard's life and his classification of the clouds, Richard Hamblyn's book, *The Invention of the Clouds: how an amateur meteorologist forged the language of the skies*, provides a history into the life of Luke Howard and his creation of the classification of clouds. On 17 April 2002 English Heritage of England awarded a plaque to be placed on the home of Howard to honor his achievements to the study of meteorology and climatology. Although Howard's original classification has been modified and further cloud types added, such as *cumulus congestus*, *cumulus fractus*, *cumulonimbus calvus*, and *stratocumulus undulates*, his concept is still used around the world in meteorological and climatological studies.

Cameron Douglas Craig

Bibliography

Hamblyn, R., 2002. *The Invention of the Clouds: how an amateur meteorologist forged the language of the skies.* New York: Picador.

Heidorn, K.C., 1999. Luke Howard: the man who named the clouds. *Weather People and History*. www.islandnet.com/~see/weather/history/howard.htm.
"Luke Howard Met Office Press Release", www.meto,govt.uk
Pedgley, D.E., 2003. Luke Howard and his clouds. *Weather*, **58**: 51–55.
Thornes, J.E., 1984. Luke Howard's influence on art and literature in the early nineteenth century. *Weather*, **39**: 252–255.

Cross-reference

Cloud Climatology

HUMAN HEALTH AND CLIMATE

The impact of climate on human health has commanded increasing attention in light of growing research indicating a strong relationship and because of the very real possibility of a global-scale climate change. Human physiological adaptive responses, such as sweating, are largely determined by climatic factors. Even though modifications of the indoor environment, such as heating and air-conditioning, may influence considerably the atmospheric conditions to which humans are exposed, it is still clear that the human population is significantly impacted by the vagaries of weather. In addition, individual lifestyles, such as behavioral modifications and clothing habits, also influence the exposure levels of people, even if they live in the same habitat.

Climate also acts on human health indirectly through its effects on ecosystems, on the hydrologic cycle, on food species, and on disease agents and vectors. Studies on climate's indirect effects upon human health are less intensive than on direct effects, but it is clear that many causes of illness and death occur through climate's impact on ecosystem modification.

Climatic thresholds

Within certain ranges of tolerance, human biology can handle most variations in climate, whether these relate to rate of change or degree of change. But large short-term fluctuations in weather can cause acute adverse effects, often indicated by increased death rates, upswings in hospital admissions, and increases in the number of individuals complaining of mental stress, such as depression (WHO/WMO/UNEP, 1996). In fact, the notion of a "temperature threshold", which represents the temperature beyond which human health declines precipitously, is common in numerous studies, and in many midlatitude cities in summer it is apparent that elevated mortality occurs only during the hottest 10–15% of days with temperatures above the threshold (Auliciems et al., 1998). Even more interesting is the notion that these threshold temperatures are *relative* rather than *absolute*; they actually vary from locale to locale, depending upon their frequency. For example, the temperature exceeds 36°C in St Louis with approximately the same frequency that it exceeds 32°C in Detroit, values which represent threshold temperatures for these cities (Mills, 2004). This strongly suggests that the notion of a "heat wave" is relative on an interregional scale, and is dependent upon the frequency of times that the local threshold is exceeded.

A most interesting finding related to heat- and cold-induced mortality which potentially could have a profound influence on such deaths in a warmer world is the large degree of variation in response on an interregional level. Recent studies on summer mortality demonstrate that many cities in temperate regions, where hot weather is severe but irregular in occurrence, show a sharp rise in total mortality during unusually hot weather conditions (IPCC, 1996; Kalkstein and Greene, 1997). In some cases daily mortality can be more than double baseline levels when weather is oppressive (Figure H5). Conversely, cities in more tropical locales seem less affected by hot weather. One possible explanation for this interregional disparity may involve the variance in summer temperatures across regions. In the cities of the northern and midwestern US, for example, the very hot days or episodes are imbedded within periods of cooler weather. Thus, the physiological and behavioral "shock value" of very high temperature is considerable. In tropical cities the hottest periods are less unique, as they do not vary much from the mean. Low variability in the fluctuation of thermal meteorological variables seems to lead to a diminishing impact of a very hot episode on human health. Thus, less people die in Phoenix when the temperature is 45°C than in New York City when the temperature is 35°C. Although the concept is not intuitive, weather variability thus plays as important a role in affecting human health as does the intensity of individual weather events.

Similar findings occur with cold-related mortality and morbidity, but the prevalence of infectious agents in the winter, and their rapid spread because of indoor confinement, renders winter health research a more daunting task. However, there is evidence that the impact of weather on human health in winter is also relative in nature, and weather variability also plays a large role in creating winter weather health problems (Keatinge et al., 1997).

Direct weather-related health problems

Clearly, direct exposure to hostile weather elements, such as lightning strikes and hurricanes, can kill large numbers of people very rapidly. However, the largest direct weather-related killer is extreme heat and cold. The combination of temperature, wind, and humidity produce an "apparent temperature", which is the perceived temperature to the human body (Steadman, 1984). Healthy persons have efficient heat-regulatory mechanisms which cope with increases in apparent temperature up to the threshold condition. When exposed to heat the body can increase radiant, convective, and evaporative heat loss by methods such as vasodilation (enlargement of blood vessels) and perspiration (Diamond, 1991). In addition, acclimatization to oppressive conditions can occur during several days with continuing exposure (Kilbourne, 1992). However, there are obvious indications that the body cannot cope with oppressive conditions indefinitely, as indicated in Figure H5 (Kalkstein and Smoyer, 1993). The rapidity of deteriorating conditions beyond the threshold is startling (WMO, 2003).

Some recent studies have utilized a "synoptic climatological approach" to identify meteorological conditions which exceed the threshold. The synoptic approach identifies "airmasses" which comprise the entire umbrella of air affecting individuals. Thus, the interaction of numerous meteorological factors is evaluated simultaneously, and their impact on human health can be assessed (Kalkstein, 1999). Each day is assigned to a particular airmass, allowing for the identification of "offensive airmasses", which contain days with mortality that exceeds the

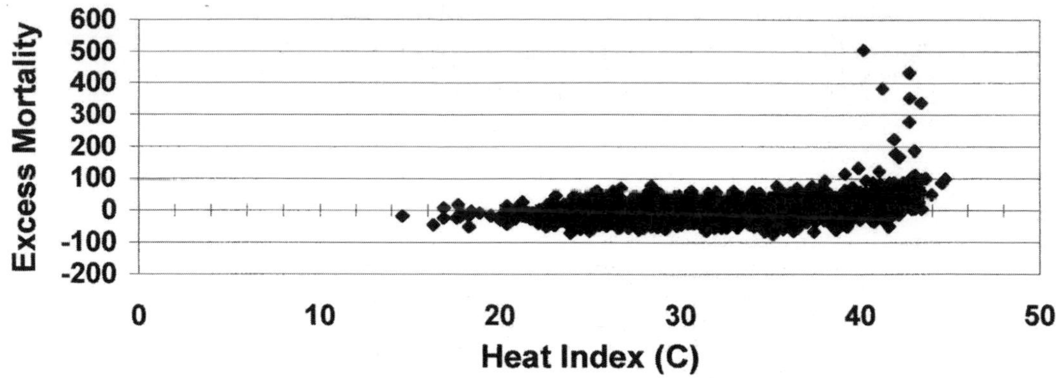

Figure H5 Shanghai daily mortality.

baseline for a locale. For example, in Shanghai, an airmass which included days with high afternoon temperatures, low wind speeds, and low humidity was deemed offensive (Tan et al., 2004).

The elderly and very young appear to be disproportionately affected by offensive airmasses, as they possess a limited ability to adapt to stressful conditions. There is evidence that individuals living in poverty, as well as urban populations in developing countries such as China and northern India, are particularly vulnerable to heat stress. Poor housing conditions, the exacerbation of stress because of the "urban heat island", and the lack of access to air-conditioning are frequently cited as primary causes (Kilbourne, 1989; Mestel, 1995).

Although the most direct impact of heat stress on the human body is the onset of heat exhaustion or heat stroke, it appears that morbidity and mortality increases associated with hot weather are related to a variety of causes. For example, deaths from cardiovascular, respiratory, and immune system disorders, as well as accidents, appear to increase during stressful weather conditions and; surprisingly, heat stroke and heat exhaustion represent only a small proportion of the mortality increase (IPCC, 1996).

The relationship between weather and cold-related mortality is less clear, and disagreements appear in the literature. In many countries the death rate in winter is 10–25% higher than during the rest of the year. However, the existence of a threshold temperature is less evident in winter than in summer (Jendritzky, 1995). Although deaths from cardiovascular disease are important in both seasons, influenza, pneumonia, and increased mortality from a variety of accidents appear to be of greater significance in winter than in summer (Langford and Bentham, 1995). In some cold-weather locales, such as Montreal and Minneapolis, the increase in mortality with decreasing winter temperature is slight (Frost and Auliciems, 1993). However, in less extreme climates, such as those in Europe, mortality rates seem to rise linearly with decreasing wintertime temperatures (Keatinge et al., 1997). People in temperate environments may be less accustomed to extreme cold, while in colder climates, behavioral responses such as cold avoidance may be a dominant thermoregulatory process when very cold temperatures occur.

There is a major discussion among biometeorologists who are exploring the possibility that a global climate change might exacerbate health problems from direct causes. The IPCC clearly expects that a climate change including increased warming will produce increases in direct weather-related summer mortality, primarily because of the existence of greater numbers of days above threshold conditions and the inability of the population to adapt to the increased heat (IPCC, 1996, 2001). Although most research points to worsening conditions for direct weather-related human health problems if the globe warms, there are large question marks. First, will possible decreases in cold-related mortality compensate for increases in hot-weather mortality in a warmer world? Some authors believe that this will be the case, especially in moderate European climates (Langford and Bentham, 1995). Others believe that the increased respiratory infections in winter will not be decreased significantly in a warmer world, since such infections partially depend on aerosol transmission in poorly ventilated places. The relatively small rise in winter temperatures will probably not encourage outdoor activities to a level where risk is reduced (Mills, 2004), and to support this stand, there appears to be a poor correlation between winter temperature and annual influenza outbreaks (CDC, 1994).

However, it is not certain how much summer mortality will increase, even if the world warms (Davis et al., 2004). For example, if climate change is accompanied by a decrease in summer temperature variability, it is possible that heat-related mortality will not increase substantially, even in a warmer world. However, if variability is similar to the present, or if it increases, the likelihood of increased summer mortality is much more likely (WHO/WMO/UNEP, 1996; Mills, 2004).

Mitigating measures to lessen direct-weather-related health problems

A number of strategies are possible to reduce the impact of weather on present and future weather-related mortality. For example, the introduction of protective technologies (e.g. improving insulation, extension of air-conditioning use) would be helpful (WHO/WMO/UNEP, 1996). In addition, there are a number of cost-effective options to lower summer temperatures in dwellings while saving billions of dollars in energy costs. Widespread use of reflective paint and innovative roofing materials might decrease in-home temperatures sufficiently to lessen heat-related mortality (WMO, 2003).

A pressing requirement is for an adequate surveillance system to alert the public and city agencies responsible for public health that dangerous weather is predicted. Such a system is in place at a number of municipalities around the world, and recent forecast improvements have probably saved many lives (Kalkstein et al., 1996; Tiesberg et al., 2004). The present

TORONTO HEAT HEALTH ALERT SYSTEM
Afternoon Forecast
Issued 8/7/2001 15:13:49
Forecast for 8/8 – 8/9/2001

8/8: HEAT EMERGENCY
Conditions oppressive – with a 97% chance of excess mortality

8/9: HEAT EMERGENCY
Conditions oppressive – with a 92% chance of excess mortality

DAY				08/08				08/09
HOUR	05	11	17	23	05	11	17	23
TEMPERATURE	23	31	35	29	25	29	31	25
DEW POINT	22	22	23	23	22	23	23	22
CLOUDINESS				4				5
AIR MASS				MT+				MT+
DAY IN ROW				3				4

Forecast data provided by Meteorological Service of Canada – Ontario Region
Click here for the latest 5-day Public Forecast and latest observation at Pearson Airport

SYSTEM LEVELS

HEAT EMERGENCY

The likelihood of weather-related excess mortality occurring exceeds 90 percent.

HEAT ALERT

The likelihood of weather-related excess mortality occurring exceeds 65 percent.

ROUTINE MONITORING

An oppressive air mass is forecast, although conditions do not suggest excess mortality is likely

Figure H6 The Toronto heat health alert system.

systems in Philadelphia and Toronto are considered by many as state-of-the-art (Figure H6). Besides possessing a high degree of accuracy, these cities has instituted a large number of intervention techniques that are novel, including the use of numerous city volunteers to deal with vulnerable citizens within their community. The number and sophistication of these systems are increasing rapidly, especially in the US, where a nationalization of standardized systems is presently under way.

Infectious disease and climate

Since the introduction of vaccines and improved nutrition and sanitation, such once-devastating infectious diseases as influenza are no longer among the leading causes of death in developed nations. However, other infectious diseases, once considered to be in broad retreat, are now resurgent in many parts of the world. This situation reflects a global convergence of a number of factors, including rapid population growth and density, greater human mobility, inappropriate use of pesticides and antibiotics, and regional climate disturbances (WHO/WMO/UNEP, 1996). Many of the biological organisms linked to the spread of infectious diseases are especially influenced by fluctuations in climate, notably thermal and moisture variables. A majority of these fluctuations are part of normal climatic variability, as indicated by the seasonality of the diseases. For example, onchocerciasis (river blindness), an important vector-borne disease in tropical countries which is transmitted by a small black fly, is largely tied into the monsoonal climates of these regions. Wet conditions during the high-sun season increase transmission dramatically, while dry conditions initiate great reductions (Mills, 1995).

Climate affects the vector biology in many ways. For example, warmer temperatures accelerate vectors' metabolic processes, consequently affecting their nutritional requirements. Blood-feeding vectors need to feed more frequently. Their biting rate therefore increases, which can lead to increases in egg production. Temperature changes can also affect the distribution of many arthropod vectors, since this is limited geographically by ranges of tolerance of the organisms (Dobson and Carper, 1993). High humidity prolongs the survival of most arthropods, while low humidity might cause some vectors to feed more frequently to compensate for dehydration. Precipitation is important, especially for insects that have aquatic larvae and pupae stages, as it is precipitation that determines the presence or absence of breeding sites. In short, climate is a major factor in vector development and subsequent disease transmission.

Of a number of infectious diseases which are closely tied to climate, there are at least three that may become particularly troublesome if the climate warms. One of them, malaria, presently kills millions annually, and is transmitted by a number of species of *Anopheles* mosquitoes. The infectious agent of malaria is any of four species of parasitic microorganisms of the genus *Plasmodium*. The malarial parasite requires a temperature of at least 15°C to complete its development within the mosquito, while the mosquito vector requires a temperature of at least 16°C and prefers a relative humidity above 60% (WHO/WMO/UNEP, 1996). The parasitic agent develops much more quickly with elevated temperatures; at 30°C its reproductive rate is more than twice that at 20°C. Thus, abundance of both the parasite and the vector could increase considerably under warmer conditions. Moreover, increased use of irrigation, which is possible if moisture stress increases under climate change, has the potential for creating numerous new breeding grounds for mosquitoes. Malaria is already showing a range increase; whether this is due to climate change is uncertain, but the potential is great for more widespread appearances of this dangerous disease (Carcavallo et al., 1996).

Tripanosomiasis (sleeping sickness) is a major disease of humans and their domestic animals in Africa. It is particularly important because its presence may preclude human habitation from areas where wild animals act as a reservoir for the disease. Its agents are various species of parasitic microorganisms of the genus *Trypanosoma*, which is transmitted by an insect vector, the tsetse fly. Research has indicated that mortality rates of the tsetse fly are closely related to humidity, and to a lesser extent to temperature. Information on the climatic requirements of the insect has been used to construct a possible distribution, given a 2°C increase in temperature in sub-Saharan Africa. The analysis suggests that the tsetse fly would become less frequent near the north edge of its range (because of drier conditions adjacent to desert landscapes) but would become more frequent toward the south (Rogers and Williams, 1993). Studies show that a very small change in temperatures can alter the range of the tsetse fly significantly.

Dengue fever is widespread in Asia, Oceania, Australia, tropical America, and the Caribbean. It is caused by one of four distinct viruses (known as "arboviruses", because they are transmitted by arthropods) which are transmitted by a mosquito, *Aedes aegypti* or *Aedes albopictus*. The current range of these insects is limited by cold weather, which kills both larvae and adults. Dengue is a particularly insidious disease that is characterized by the abrupt onset of fever, severe headache, muscle and bone pain (the disease is commonly called "breakbone fever"), and sometimes hemorrhaging of blood vessels. Some forms can be fatal if treatment is not available.

One of the unique challenges to climatologists involving dengue is that the insect vector breeds frequently in old tires and containers. Thus, many climatologists have taken a water-budget approach to evaluate the likelihood that the vector is prevalent (Cheng et al., 1998). At present a number of models are available to evaluate and estimate mosquito populations that transmit dengue, and are being used to control pesticide application. However, this disease is a major menace, and can continue to spread if the climate warms.

There are a number of vector-borne and water-borne diseases that are sensitive to climate. Until the recent past, research on these diseases was limited to the medical community, and finally, climatologists are becoming more involved in modeling the spread and transmission of these diseases.

Other weather-related/health issues

The relationship between climate and human health is multi-faceted and complex beyond any potential discussion here. However, some other key issues of interest include:

1. *The potential synergism between weather/pollution/human health*. It is clear that pollution has a sizable negative impact on human health, but the synergistic impacts of oppressive weather and unhealthy pollution episodes are still not fully understood. Some research indicates that weather is a confounding factor in strengthening the impact of pollution on human health (Samet et al., 1998). Other studies indicate that, when the weather is already oppressive (especially in summer), many extra deaths will occur no matter what the pollution concentration (Kalkstein, 1995). Studies of this type acknowledge that high pollution concentrations are dangerous to humans, but the real acute impact of these concentrations is generally noted when the weather is benign. However, there seems to be general agreement that weather and pollution can combine synergistically to create a negative environment for asthma sufferers. With this in mind, there are attempts to develop asthma warning systems, to advise sufferers that the combination of meteorology, pollution, and aeroallergens is such that there is a high likelihood of discomfort among asthma patients (Jamason et al., 1997).
2. *Stratospheric ozone depletion*. Although not exactly a weather/health issue, there has been considerable research describing the impacts of ultraviolet exposure on the human body. These analyses assume that the thinning ozone layer over parts of the Earth has created environments where humans are more vulnerable to direct sun exposure. Some of

the most common conditions associated with extended exposure to UV-B radiation include skin cancers, cataracts and eye damage, a reduction of immune system function, and large negative impacts to terrestrial and aquatic species of animals and plants (WHO/WMO/UNEP, 1996).

3. *Sea level rise.* Many climate change advocates have developed extensive research on how sea level rise can negatively impact human health. Some of these assume that a long-term climate warming can increase sea level by up to a meter over the next 100 years (IPCC, 2001). The potential impacts of sea level rise on human health are multifaceted, and include population displacement, increase in vector-borne diseases, lessening of food availability, and greater probabilities of coastal flooding (WHO/WMO/UNEP, 1996).

Considering the numerous links between meteorology and human health, the World Meteorological Organization, the World Health Organization, and the United Nations Environmental Programme have developed an initiative to encourage interdisciplinary collaboration to achieve greater integration between the human dimensions and biophysical dimensions of this issue. Some of the recommendations include: expansion of electronic data exchange to enhance climate–health researchers, development of integrated process-oriented models to understand the complex linkages, strengthening the role of international agencies (such as the WMO, WHO, and UNEP) to provide technical support and policy advice to various governments, and creation of multidisciplinary scientific advisory boards to address long-term environmental and health issues. The heightened awareness of climate-induced health problems should permit meteorological and health practitioners to move aggressively in meeting these ends.

Laurence S. Kalkstein

Bibliography

Auliciems, A., deDear, R., Fagence, M., et al. (eds.), 1998. *Advances in Bioclimatology, 5: Human Bioclimatology*. New York: Springer.

Carcavallo, R.U., and Curto de Casas, S.I., 1996. Some health impacts of global warming in South America: vector-borne diseases. *Journal of Epidemiology*, **6**(4): S153–S157.

CDC, 1994. Update influenza activity – United States and worldwide, 1993 season and composition of the 1994-95 influenza vaccine. *Morbidity and Mortality Weekly Reports*, **43**(10): 179–183.

Cheng, S., Kalkstein, L.S., Focks, D.A., and Nnaji, A., 1998. New procedures to estimate water temperatures and water depths for application in climate-dengue modeling. *Journal of Medical Entomology*, **35**(5): 646–652.

Davis, R.E., Knappenberger, P.C., Michaels, P.J., and Novicoff, W.M., 2004. Changing heat-related mortality in the United States. *Environmental Health Perspectives* (In press).

Diamond, J., 1991. *The Rise and Fall of the Third Chimpanzee*. London: Radius.

Dobson, A., and Carper, R., 1993. Health and climate change: biodiversity. *Lancet*, **342**: 1096–1099.

Frost, D.J., and Auliciems, A., 1993. Myocardial infarct death, the population at risk and temperature habituation. *International Journal of Biometerology*, **37**: 46–51.

IPCC (Intergovernmental Panel on Climate Change), 1996. *Climate Change 1995: impacts, adaptations and mitigation of climate change: scientific-technical analysis*. Watson, R.T., Zinyowera, M.C., and Moss, R.H., eds. New York: Cambridge University Press.

IPCC (Intergovernmental Panel on Climate Change), 2001. *Impacts, Adaptation and Vulnerability*. Contribution of Working Group II to the Third Assessment Report of the IPCC. McCarthy, J.J., Canziani, O.F., Leary, N.A., Dokken, D.J., and White, K.S., eds. Cambridge: Cambridge University Press.

Jamason, P.F., Kalkstein, L.S., and Gergen, P., 1997. A synoptic evaluation of asthma hospital admissions in New York City. *American Journal of Respiratory Diseases and Critical Care Medicine*, **156**: 1–8.

Jendritzky, G., 1995. Umweltfaktor Klima. In Beyer A., and Eis, D., eds., *Praktische Umweltmedizin*. Berlin: Springer.

Kalkstein, L.S., 1995. Effects of weather and climate on human mortality and their roles as confounding factors for air pollution. *Air Quality Criteria for Particulate Matter*, Vol. III. Research Triangle Park, NC: National Center for Environmental Assessment, Office of Research and development, US EPA. EPA/600/Ap-95/001cF, Washington, DC, Chapter 12.

Kalkstein, L.S., 1999. Predicting heat worldwide. *Environmental Health Perspectives*, **107**(5): A238–A244.

Kalkstein, L.S., and Greene, J.S., 1997. An evaluation of climate/mortality relationships in large U.S. cities and the possible impacts of a climate change. *Environmental Health Perspectives*, **105**(1): 84–93.

Kalkstein, L.S., and Smoyer K.E., 1993. The impact of climate change on human health: some international implications. *Experientia*, **49**: 469–479.

Kalkstein, L.S., Jamason, P.F., Greene, J.S., Libby, J., and Robinson, L., 1996. The Philadelphia hot weather-health watch/warning system: development and application, Summer 1995. *Bulletin of the American Meteorological Society*, **77**(7): 1519–1528.

Keatinge, W.R., Donaldson, G.C., Bucher, K., et al., 1997. Cold exposure and winter mortality from ischaemic heart disease, cerebrovascular disease, respiratory disease, and all causes in warm and cold regions of Europe. *Lancet*, **349**: 1341–1346.

Kilbourne, E.M., 1989. Heatwaves. In Gregg, M.B., ed. *The Public Health Consequences of Disasters 1989*. Atlanta, GA: US Department of Health and Human Services, pp. 51–61.

Kilbourne, E.M., 1992. Illness due to thermal extremes. In Last, J.M., and Wallace, R.B., eds. *Public Health and Preventive Medicine*, 13th edn. Appleton, Norwalk: Lange, pp. 491–501.

Langford, I.H., and Bentham, G., 1995. The potential effects of climate change on winter mortality in England and Wales. *International Journal of Biometeorology*, **38**: 141–147.

Mestel, R., 1995. White paint. *New Scientist*, 25 March, pp. 34–37.

Mills, D.M., 1995. A climatic water budget approach to blackfly population dynamics. *Publications in Climatology*, **XLVIII**(2).

Mills, D.M., 2004. *Final Report for US EPA on Impacts of Climate Change Upon Human Health.* Boulder, Co: Stratus Consulting (In press).

Rogers, D.J., and Williams, B.G., 1993. Monitoring trypanosomiasis in space and time. *Parasitology*, **106**: S77–S92.

Samet, J.M., Zeger, S.L., and Kalkstein, L.S., 1998. Does weather confound or modify the association of particulate air pollution with mortality? An analysis of the Philadelphia data 1973–1980. *Environmental Research, A*, **77**: 9–19.

Steadman, R.C., 1984. A universal scale of apparent temperature. *Journal of Climatology and Applied Meteorology*, **23**: 1674–1687.

Tan, J., Kalkstein, L.S., Huang, J., Lin, S., Yin, H., and Shao, D., 2004. An operational heat/health warning system in Shanghai. *International Journal of Biometeorology* (In press).

Teisberg, T.J., Ebi, K.L., Kalkstein, L.S., Robinson, L., and Weiher, F., 2004. Heat watch/warning systems save lives: estimated costs and benefits for Philadelphia 1995-1998. *Bulletin of the American Meteorological Society* (In press).

WHO/WMO/UNEP, 1996. *Climate change and human health*. McMichael, A.J., Haines, A., Slooff, R., and Kovats, S., eds. Geneva.

WMO, 2003. *Climate into the 21st Century*. Cambridge: Cambridge University Press.

Cross-references

Air Pollution Climatology
Climate Change and Human Health
Heat Index
Climate Comfort Indices
Urban Climatology
Wind Chill

VON HUMBOLDT (BARON), FRIEDRICH HEINRICH ALEXANDER (1769–1859)

Alexander von Humboldt was born in Berlin in 1769 and died there in 1859. He is widely regarded as a scientific "renaissance man" who was able to observe and integrate a variety of natural phenomena. He attended the universities of Göttingen, Frankfurt-on-Oder, and Freiburg, with fieldwork emphasizing botany and geology. His travels with botanist Aimé Bonpland to South America, Mexico, Cuba, and the United States (1799–1804) led to development of 30 volumes, *Voyage de Humboldt et Bonpland (Personal Narrative* in the English edition), published from 1807 to 1817. He sought to understand the interaction of natural forces, the influences of geography on living things, and "the unity of nature".

In 1817 Humboldt became the first to use temperature observations to construct an isothermal map, evidently based on earlier isogonic maps by Edmond Halley. During his travels he had identified the Peruvian current, which sometimes is called the Humboldt current. Humboldt's isotherm maps were adjusted to sea level, depicting the effects of latitude, but also showing ocean current effects. Although his isothermal map removed the effect of topography on temperature, his experiences in South America, particularly the Andes, led to recognition of the similar effects of altitude and latitude on vegetation.

In 1829 Humboldt spent time in Russia, with 6 months of exploration funded by Tsar Nicholas I. Humboldt was able to use extensively collected temperature data to develop the idea of continentality. He worked with mathematician Karl Gauss to establish a worldwide network of weather and magnetic observatories. A number of European countries were convinced to do so, including Russia; this created "the first large-scale international collaboration".

Humboldt's second great series of volumes, *Kosmos*, was published 1845–1862. In its five volumes Humboldt attempted to describe the universe as it was then known. He spent a quarter-century writing this series; the last volume was completed after his death, based on his notes. *Kosmos* was Humboldt's attempt to present "the unity of nature".

Lisa M. Butler Harrington

Bibliography

Agassiz, L., 1869. *Address Delivered on the Centennial Anniversary of the Birth of Alexander von Humboldt, under the Auspices of the Boston Society of Natural History*. Boston, MA: Boston Society of Natural History.

Axelrod, A., and Phillips, C., 1993. *The Environmentalists: a biographical dictionary from the 17th century to the present*. New York: Facts on File, (pp. 112–113).

Lee, J., 2001. Alexander von Humboldt. *Focus*, **46**(3): 29–30.

Monmonier, M., 1999. *Air Apparent: how meteorologists learned to map, predict, and dramatize weather.* Chicago, IL: University of Chicago Press.

Oliver, J., 1996. Climatic zones. In Schneider, S.H., chief editor. *Encyclopedia of Climate and Weather*, Vol. 1. New York: Oxford University Press, pp. 141–145.

Oliver, J.E., 1991. The history, status and future of climatic classification. *Physical Geography*, **12**(3): 231–251.

Stringer, E.T., 1972. *Foundations of Climatology: an introduction to physical, dynamic, synoptic, and geographical climatology*. San Francisco, CA: W.H. Freeman.

HUMID CLIMATES

Humid climates are commonly considered to be those that support forests as the typical vegetation cover (Huschke, 1970); thus the term forest climate often is applied. However, several alternate definitions, from broad generalizations to detailed specifications, also exist.

One alternate definition specifies that a humid climate is one in which annual precipitation exceeds evaporation. This definition encompasses all areas that are not desert or semiarid, although the icecap climates of Antarctica and Greenland, where annual precipitation is generally less than 130 mm, often are also excluded. According to this interpretation, humid climates cover approximately 74% of Earth's land surface (approximately 64% if the icecaps are excluded).

Other approaches to delineating humid regions were developed by Thornthwaite (Griffiths and Driscoll, 1982; Oliver and Hidore, 2002; Thornthwaite, 1931, 1948), as part of general climatic classifications. A central assumption of Thornthwaite's earlier (1931) classification is that to quantitatively determine whether a region is truly humid or not, the precipitation of the region must be compared with what has been termed the climatic demand for water. This concept was incorporated through the use of a moisture index. Humid climates have positive index values and dry climates have negative values. Table H3 shows the basic groups of the moisture index. Climate types A (perhumid) and B (humid) are generally referred to collectively as the humid climates. Together they comprise approximately 18% of Earth's land. The generalized global distribution of the perhumid and humid climates is shown in Figure H7. The perhumid climate (group A in Table H3) is concentrated in tropical regions, such as Malaysia, southern Philippines and the west coast of India. It is also found on the coast of west-central Africa, throughout much of Central America, especially the eastern portion, on parts of the northern coast of South America, and on west coasts in higher middle latitudes, especially northwestern United States and southwestern Chile.

Group B climate (humid group in Table H3), which is slightly less wet than the perhumid group, is extensive in tropical and middle latitudes, particularly the Amazon Basin, southeastern Brazil, equatorial Africa, eastern and northwestern United States, Japan, southeast Asia, eastern Australia, New Zealand, and central and western Europe.

Interestingly, the extensive taiga (boreal) forests centered at about 60°N latitude are not included in Thornthwaite's humid (groups A and B) climates. In this region, as well as in more poleward areas, temperature efficiency, rather than precipitation effectiveness, is the most critical influence on plant distribution; temperature provinces, instead of moisture groups, are the basis for Thornthwaite's climatic divisions in these areas. Soil moisture is relatively abundant in the taiga zone, although it is often frozen for a portion of the year.

Because humid climates occupy such diverse global positions, there are few universally applicable characteristics. However, in comparison with arid areas, humid regions, which are generally dominated by forests, utilize a much larger portion of solar radiation in the process of evapotranspiration. Thus there is less energy available to heat the atmosphere; consequently humid regions are less likely to have the

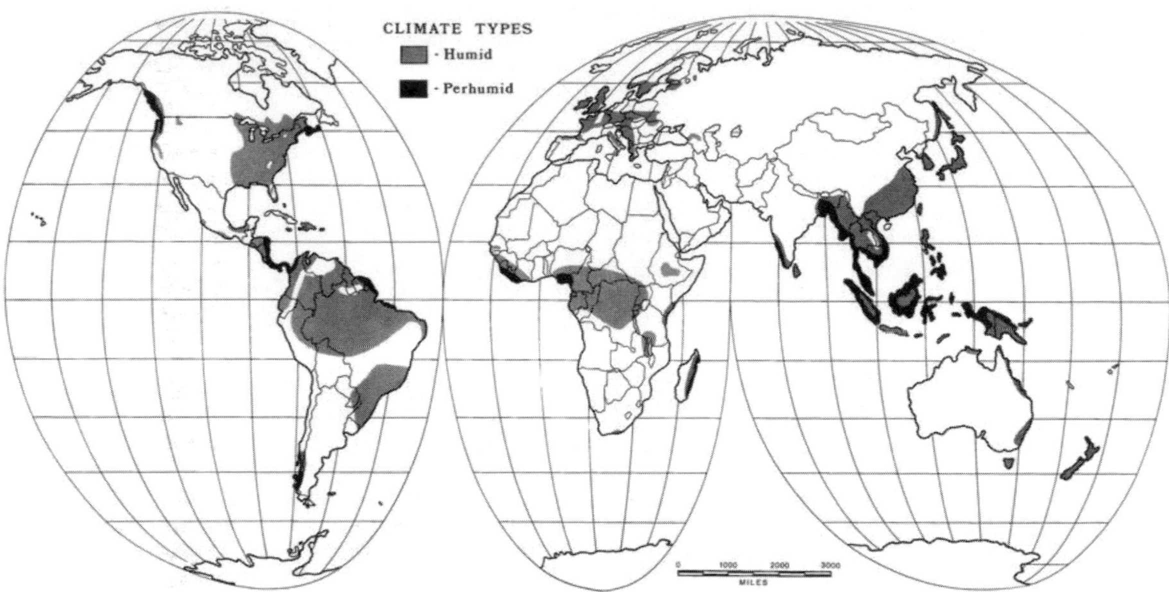

Figure H7 Thornthwaite's perhumid and humid moisture regimes (after US Department of Agriculture, 1941; Griffiths and Driscoll, 1982).

Table H3 Moisture index groups

A	Perhumid (Wet)
B	Humid
C	Subhumid
D	Semiarid
E	Arid

very high air and ground temperatures often experienced in arid areas.

James A. Henry

Bibliography

Griffiths, J.F., and Driscoll, D.M., 1982. *Survey of Climatology*. Columbus, OH: Charles E. Merrill.

Oliver, J.E., and Hidore, J.J., 2002. *Climatology: an atmospheric science*. New York: Prentice-Hall.

Huschke, R.E., 1970. *Glossary of Meteorology*. Boston, MA: American Meteorological Society.

Thornthwaite, C.W., 1931. The climates of North America according to a new classification. *Geographical Review*, **21**: 327–336.

Thornthwaite, C.W., 1948. An approach toward a rational classification of climate. *Geographical Review*, **38**: 55–94.

US Department of Agriculture, 1941. *Climate and Man: yearbook of agriculture*. Washington, DC: US Government Printing Office.

Cross-references

Arid Climates
Climate Classification
Maritime Climate
Middle Latitude Climates
Monsoons and Monsoon Climate
Tropical and Equatorial Climates

HUMIDITY

Humidity refers to the water vapor content of air. Several ways of expressing humidity are:

1. *Absolute humidity* (also known as vapor concentration or vapor density). For a moist parcel of air the ratio of the mass of water vapor to the volume of the parcel is called the absolute humidity, i.e. the density of the water vapor. It is expressed in grams of water vapor per cubic meter, or by engineers (as in air-conditioning) in grains per cubic foot (1 grain = 0.0648 g and 1 cubic foot = 28 317 cm^3). For meteorologists the absolute humidity is not very useful since it neglects questions of pressure.
2. *Specific humidity*. This is the ratio of the mass of water vapor to the mass of air, expressed as grams of vapor per kilogram of air or as grains per pound (in air-conditioning). This figure may be quoted regardless of temperature and volume (related to the atmospheric pressure), therefore for meteorology it is a much more useful expression than absolute humidity. For most purposes the specific humidity may be approximated by the mixing ratio (the ratio of mass of water vapor to mass of dry air) thus:

$$q = \frac{w}{(1 + w)}$$

where q is specific humidity and w the mixing ratio. The terms specific humidity and mixing ratio thus are used interchangeably because the amount that they differ by is less than the error involved in measuring either one of them.
3. *Relative humidity*. This is what is commonly meant by "humidity". It is the dimensionless ratio of vapor that a given quantity of air can contain for a given temperature,

expressed as a percentage. Perfectly dry air has a relative humidity of 0%; totally saturated air, 100%. Thus:

$$r = \frac{w}{w_s}$$

where w_s is the saturation mixing ratio.

Humidity measurement, hygrometry and hygrometers

The techniques of humidity determination are known as hygrometry and the class of instruments as hygrometers. The term hygroscopic pertains to a substance altered on contact with water vapor. In certain clouds, fog, and haze, condensation may be initiated on hygroscopic nuclei (NaCl) even when relative humidity may be only 75%. Of the hygrometric instruments there are several types, operating on varied principles:

1. *Hair hygrometer* and related devices depend on the absorption of moisture and change of physical dimensions of a particular substance. This substance may be a strand of human hair or other animal hair, which may be appropriately treated so that it varies in length (it increases approximately logarithmically) with a relative humidity increase of 20–100%. Its lag time at low temperatures is serious and essentially infinity at −40°C. The torsion hygrometer is simply a variety where the hair is twisted to cause a rotation (indicated on a dial) during expansion or contraction. Another is the goldbeater's-skin hygrometer, using the expansion characteristics of a length of skin from the large intestine of the ox, such as is used in goldbeating.
2. *Electrical absorption hygrometers* depend on change of electrical properties during absorption of water vapor. A simple form, electrically operated, uses a film of carbon black dispersed in a hygroscopic binder on a plastic strip. Another, called the humidity strip, often used in radiosondes, uses the electrolytic principle on a plastic strip coated with lithium chloride or other hygroscopic compound. Absorption of moisture changes the resistance of the circuit, thus indicating humidity.
3. *Diffusion hygrometer* measures the diffusion of water vapor through a porous membrane, to be absorbed in a container by a hygroscopic medium, the drop in pressure in the container being measured by a manometer.
4. *Spectral hygrometer*, an electronic–optical class of instruments that measure the absorption spectra of water vapor, i.e. the attenuation of radiant energy due to the absorption bands of water vapor.
5. *Psychrometer or wet- and dry-bulb hygrometer*, the most widespread and generally useful instrument, a thermodynamic procedure, utilizing two matched mercury thermometers, one of which has damped muslin wrapped around the bulb, the muslin being wetted in distilled water before making the reading. A psychrometric chart is a nomograph for graphical determination of relative humidity, absolute humidity, and dewpoint from wet- and dry-bulb thermometer readings. Alternatively, a circular slide rule may be used as a psychrometric calculator.

Humidity in climatology

Humidity is one of the basic concepts of climatology, essential for considering comfort, the habitability of a region, the agricultural potential, and the pedologic characteristics. It has to be considered alongside an aridity index.

Humidity coefficient, proposed by A. J. Angstrom to indicate the precipitation effectiveness of a region, recognizes the exponential relationship of temperature to plant growth, and is expressed as $P/(1.07)^t$, where P is precipitation in centimeters and t is mean temperature in degrees Celsius for the period.

Humidity index, devised by C. W. Thornthwaite (1948) as a measure of water surplus over water need at a given point, and given by $100s/n$, with s being water surplus (the sum of monthly differences between precipitation and potential evapotranspiration for those months when the former exceeds the latter, and n being water need, determined by the sum of monthly potential evaporation for those months of surplus. The humidity index is used as a component in Thornthwaite's moisture index and as a basis for classifying climates in general.

Humidity province, a regional unit in Thornthwaite's classification of 1931, characterized by climax vegetation (rainforest, tundra, etc.). Five classes of humidity province were proposed (bounded by values of precipitation effectiveness index): (A) wet or rain forest; (B) humid or forest; (C) subhumid or grassland (D) semiarid or steppe; (E) arid or desert. In his 1948 system Thornthwaite removed the biologic consideration (vegetation climax) and used values of the moisture index for identifying rather similar regions: (A) perhumid; (B) humid; (C) subhumid; (D) semiarid; (E) arid. Moisture index can be calculated by the formula:

$$I_m = \frac{(100 \times \text{water surplus}) - (60 \times \text{water deficit})}{\text{potential evapotranspiration}}$$

which becomes

$$I_m = \frac{100s - 60d}{n}.$$

where I_m is the moisture index, s the water surplus, d the water deficiency, and n the water need.

John E. Oliver and Rhodes W. Fairbridge

Bibliography

Bemis, A.C., 1951. Aircraft meteorological instruments. In Malone, T. (ed.) *Compendium of Meteorology*. Boston, MA: American Meteorological Society, pp. 1223–1231.

DeFelice, T.P., 1998. *Meteorological Instrumentation and Measurement*. Upper Saddle River, NJ: Prentice-Hall.

Ference, M., Jr, 1951. Instruments and techniques for meteorological measurements. In Malone, T. (ed.) *Compendium of Meteorology*. Boston, MA: American Meteorological Society, pp. 1207–1222.

Hess, S.L., 1959. *Introduction to Theoretical Meteorology*. New York: Holt.

Middleton, W.E.K., and Spilhaus, A.F., 1953. *Meteorological Instruments*, 3rd edn. Toronto: University of Toronto Press.

Thornthwaite, C.W., 1948. An approach toward a rational classification of climate, *Geographical Review*, **38**: 55–94.

Cross-references

Aridity Indexes
Climate Classification
Dew/Dewpoint
Evapotranspiration
Precipitation
Relative Humidity

HYDROCLIMATOLOGY

Hydroclimatology was defined by Langbein (1967) as the study of the influence of climate upon the waters of the land. It includes hydrometeorology as well as the surface and near-surface water processes of evaporation, runoff, groundwater recharge, and interception. The total hydrologic cycle, then, is the basis for a discussion of hydroclimatology.

The water budget equation based on conservation of mass is given as:

$$P = R + ET + U \pm \Delta S_s \pm \Delta S_g + I$$

where P is precipitation, R is streamflow, ET is evapotranspiration, U is subsurface underflow, ΔS_s is the change in soil moisture, ΔS_g the change in groundwater storage, and I is interception. When the equation is evaluated using annual mean data, the changes in groundwater storage, soil moisture and subsurface underflow are assumed to be zero; and if the interception is sufficiently small, all four terms may be deleted. Although the components of the total budget are easy to list, quantification of each of the components over various time scales is always difficult, and sometimes impossible to measure or to estimate correctly.

Observations of the standard meteorological parameters have been made continuously since the late 1880s at some 250 National Weather Service (NWS) First Order Stations (major airports in the US). Daily observations of temperature, precipitation and snowfall are routinely made at about 6000 NWS so-called Second Order sites. These provide the necessary means for the study of climate and climate change. During the last 20 years or so many states have inaugurated networks of automated observing sites where the standard parameters are continuously sensed and recorded. In addition, these data are generally on line and hence available on a near real-time basis.

Measurement of water budget components

Precipitation

Measurement of precipitation is typically accomplished by catching the rain or snow in a flat-bottomed, vertically sided canister, and simply measuring the depth, either after each precipitation event or continuously by an automatic sensor and recorder. Catchment of solid precipitation (particularly under windy conditions) often underestimates, in that snow is easily blown over the gauge opening rather than falling in. Liquid precipitation and the water equivalent of freezing precipitation, too, are subject to depletion by wind.

A different, but equally important problem, relates to the measurement of the spatial distribution of rainfall rates and amounts. Low-latitude and most warm-season precipitation in summer is convective in nature, i.e. precipitation emanating from cumuliform clouds. Individual clouds typically exhibit a horizontal extent of only a few kilometers. Therefore the spatial distribution of precipitation, even over flat terrain or the ocean, can be quite variable. In the extreme case, rainfall at the rate of a few tens of millimeters per hour can occur in one location with no precipitation occurring only a few kilometers distant. Rain gauge densities sufficient to measure small-scale anomalies of, say, 1-km scale have only been maintained for research, rather than for operational networks. Changnon (1979) describes 14 meso-networks or rain gauges in Illinois operated by the Illinois State Water Survey during the last three decades. The networks ranged from 0.03 km^2 (18 gauges) to 3800 km^2 (250 gauges). The former was operated during 20 warm seasons, the latter for four. Some of the precipitation anomaly patterns are discussed by Changnon and Vogel (1981).

The spatial distribution of precipitation is typically determined from measurements made at several rain gauges within a given area. On scales of a few tens of kilometers, spatial correlation of rainfall is directly related to latitude, and it is greater during the cold season. For example, summer rainfall for sites with 5 km separation typically correlates about +0.2 to +0.3 in Florida, and about +0.4 to +0.6 in Illinois (Jones and Wendland, 1984).

Although the spatial distribution of precipitation can be inferred from radar imagery, the calibration of the radar echo intensity to rainfall rate can only be fulfilled if rain guages are sufficient in number and appropriately located, a condition not always possible (Browning, 1979). However, Browning foresees precipitation data from gauges, overlapping radars, and geostationary satellites contributing to a meso-scale numerical model to produce areal precipitation forecasts up to *c.* 6 h. Currently, the decrease in precipitation data quality from gauges to radars to satellites will require much human interaction to integrate the complex data sets.

Surface runoff

Runoff is composed of several potential components: immediate runoff from rainfall and snow-melt, lagged runoff from snow- and ice-melt, and runoff contributed for interflow if reentering a stream. Each of the above lags the precipitation event by substantially different intervals, the lag being a function of the rate of liquid precipitation, depth of snowpack-ice layer, slope and aspect of the hydraulic conductivity and its vertical variation in the subsurface material, and the presence of ground frost. The length of time that runoff lags the precipitation event varies inversely with the rate of liquid precipitation, and may increase greatly if the precipitation is of a solid form; it is substantially diminished when the underlying soil is relatively porous. From monthly observations from several Illinois sites streamflow and shallow groundwater responded to anomalous precipitation within the month following the event (Wendland, 2001). The response of soil moisture, however, was not as clear, in that wet months were followed by either increases, decreases or no change depending upon the season (growing or fallow) and the magnitude of evaporation, whereas dry months were either followed by declining soil moisture or no change.

The severe drought in the Upper Midwest during the growing season of 1988 drew down all surface and groundwater resources, delaying any recovery until fall and early winter. The heavy precipitation of the growing season of 1993 was of sufficient magnitude so that none of the hydrologic responses showed any decline through the usually desiccating growing season.

Interestingly, infiltration of liquid precipitation or snow melt into essentially frozen soil under a snowcover can represent in excess of 95% of the precipitation (Price et al., 1978). Runoff from surface ice and snow is substantially reduced if ablation (i.e. snow or ice loss through either melting or evaporating) occurs via sublimation during clear sunny days with temperatures less than 0°C.

The rate of runoff largely determines the degree of erosion from a surface, although erosion also depends on surface cover, slope, aspect, etc. Sediment yield in a stream (including eroded material) has been found to be highly related to the coefficient of variation (standard deviation/mean) of precipitation (Harlin, 1978).

Evapotranspiration

The rate of evaporation from a surface into the air is a function of the availability of an energy source, the vapor pressure gradient between the soil or vegetation surfaces and that of the ambient air, and windspeed. Evapotranspiration is the total moisture loss through the soil–air interface, and that which transpires through the vegetation.

In the United States, evaporation is measured by the water loss from a Class A evaporation pan (25 cm deep, 1.2 m diameter) initially filled with water to 5 cm from the rim. Because of turbulence to air flow caused by the structure itself, and heat transfer through the sides of the container, pan evaporation tends to be greater than that over an open water surface (Budyko, 1974). This difference from actual evaporation can be reduced by sinking the pan so that the evaporating surface is near the level of the undisturbed ground surface (World Meteorological Organization, 1971). Reasonable estimates of evaporation from a cropped surface can be obtained from a lysimeter (a container several meters in diameter and 1 or 2 m in depth), filled with soil and growing vegetation. Water loss is usually determined by noting weight loss of the lysimeter. That water loss can vary greatly is seen from the observation that evapotranspiration over a pine forest was found to exceed that from a water surface by about 10% over six warm seasons (Holmes and Wronski, 1981).

Estimates of evaporation and evapotranspiration can be calculated. For example, Thornthwaite (1948) presented an equation to calculate potential evapotranspiration (evapotranspiration from an unlimited supply of water) as a function of daily mean air temperature and day length. Penman (1963) related evapotranspiration to vapor pressure, wind speed, surface temperature, and the shortwave radiation available at the surface. Estimates in Illinois using Penman's method are typically greater than those calculated by the Thornthwaite equation (Jones, 1966). Monteith (1965) suggested an improved evapotranspiration calculation, whereby evapotranspiration is determined as a function of energy (usually radiation), temperature, humidity, wind speed, leaf resistance, and the number and distribution of stomata. Bavel (1966) derived a relationship between potential evapotranspiration and net radiation, ambient air properties, and surface roughness. Priestley and Taylor (1972) demonstrated that latent and sensible heat fluxes could be closely estimated using parameterizations of energy over land and of bulk aerodynamic type over the sea. The latter three relationships, based on a greater number of physically related independent variables, yield good estimates of evapotranspiration under varying conditions. However, the greater number of independent variables limits their use to sites where sufficient observations are available, i.e. primarily micrometerological research sites.

On a larger scale, Rasmusson (1966) presented the spatial distribution of mean monthly water vapor transport over North America, determined from 2 years of twice-daily aerological data. These analyses exhibit a reasonable pattern of flux divergence and inferred evapotranspiration. On the whole, however, calculated evaporation rates without corroborating observations must be considered suspect.

Groundwater

Groundwater includes two components: subsurface underflow and groundwater storage. These two components are difficult to measure separately and are therefore often treated as one quantity and evaluated as a residual in the water budget equation, i.e. groundwater recharge, that quantity representing the change to subsurface underflow and groundwater storage. Rehm et al. (1982) review methods to estimate groundwater recharge: calculations based on hydrograph records, Darcy flux calculations (where vertical flow is a function of permeability, hydraulic gradient, and porosity), and flow partitioning from an analysis of groundwater flow. Two examples of the first method are found in Schicht and Walton (1961) and O'Hearn and Gibb (1980). Depth to shallow groundwater is relatively easy to measure, but is routinely done at only a few locations due to cost of infrastructure and continued measurements.

Soil moisture

Soil moisture is moisture in the unsaturated zone, originating from surface water that enters the soil through percolation. Soil moisture can be measured by physically removing a soil column. It is then weighed, heated, and weighed again. The weight difference is attributed to water from the soil mass. Obviously, disturbing the soil precludes further measurement at the same site.

In-situ measurements of soil moisture can be made with a calibrated neutron probe. A neutron source/sensor is lowered into a pipe extending downward from the soil surface, obtaining measurements at any level. These sites can be remeasured as necessary.

Other than "casual" observations of soil moisture in terms of "adequate", "short", etc., quantified observations were few and far between in the US until the late 1980s, when routine observations began at several locations. Even today, however, soil moisture is routinely observed at only a few score sites, e.g. on a bi-weekly basis during the growing season and monthly thereafter at some 18 sites in Illinois.

Models have been developed to estimate the change in soil moisture from an earlier time, that time often being saturated conditions in early spring. Soil moisture estimates are updated as a function of temperature and precipitation. Dale et al. (1982) describe a budget technique that estimates soil moisture under different field and drainage conditions.

Interception

Interception is that quantity of precipitation that impinges on vegetation and other structures above the soil surface, some of which is either assimilated into the intercepting medium or evaporated prior to reaching the surface. In long-term determinations interception is usually assumed to be minimal. However, interception can be a significant moisture source to vegetation. For example, in coastal Peru where precipitation is minimal, vegetation obtains moisture by interception from surface-based clouds that drift through the vegetation. Cloud droplets settle on the leaves, drip to the soil surface and are available to the tree roots. Interception in an Australian plantation forest with closed canopy was found to be 1.8 mm by Holmes and Wronski (1981).

Worldwide distribution of evaporation and runoff ratios

Maps of annual and seasonal measured or (as necessary) estimated precipitation and evaporation are presented in most climatology texts and are not reproduced here. The mean annual distribution of the above two variables and runoff are given in Sellers (1967), and Budyko (1974).

Two parameters helpful in delineating hydroclimatological regions are evaporation ratio and runoff ratio. Evaporation ratio is the heat consumed by evaporation (E) during a given time interval expressed as a percent of the total net radiation (R_n) received during the same interval. Runoff ratio is the total runoff (RO) during a given time interval expressed as a percent of total precipitation (P) accumulated during the same interval. The raw data required for these ratios (i.e. heat expended for mean annual evaporation, net radiation, and precipitation) were obtained from Barry (1969), Budyko (1956), and Geiger (n.d.), respectively. Additional heat budget detail from tropical oceans was determined from data presented by Hastenrath and Lamb (1978, 1979). Runoff was calculated from precipitation minus evaporation estimates, and the appropriate ratios constructed. Resolution of detail is limited to distances of about 1000 km.

Evaporation ratio

Budyko (1956) and Sellers (1967) have discussed the concept of evaporation ratio. The ratio (E/R_n 100) differentiates between those areas where evaporation can be totally supported by net radiation (indicated by evaporation ratios less than 100%) and those where the energy required for evaporation comes in part from sensible energy sources, i.e. atmospheric advection, Earth's surface, etc. (indicated by values greater than 100%). Figure H8 shows the large-scale annual mean patterns of evaporation ratio over oceans and over continents.

Ratios greater than 100% dominate the midlatitude oceans of both hemispheres. Note the relatively large ratios along the western oceanic margins, where the sea surface is relatively warm, and diminished ratios over eastern oceans. Maximum values in excess of 250% are found over the Gulf Stream. The tropical oceans and eastern margins of the Atlantic and Pacific oceans typically exhibit evaporation ratios of 50–60%.

Ratios less than 50% dominate the continents, including the southwestern United States, southern South America, Africa, southwestern and central Asia, and Australia. Sizeable areas with evaporation ratios less than 10% are found over northern Africa and interior Australia, where minimal precipitation limits evaporation. Maximum ratios are found over northern hemisphere continents poleward of about 60°N, exhibiting values in excess of 150%. Midlatitude continental ratios are between 50% and 100%. For the continents as a whole, net radiation is sufficient to support estimated evaporation.

Evaporation and net radiation estimates are undoubtedly less accurate over oceans and high-latitude continents, as are therefore the evaporation ratios. However, it is unlikely that the estimates are sufficiently in error to significantly change the patterns of the evaporation ratio.

Runoff ratio

Budyko (1956) and Sellers (1967) have also discussed another hydrologic parameter, i.e., runoff ratio ($RO/P \times 100$) where runoff is defined as mean annual precipitation minus evaporation. The pattern of runoff ratio is shown over the continents as broken lines in Figure H9. Substantial areas of the continents exhibit ratios near

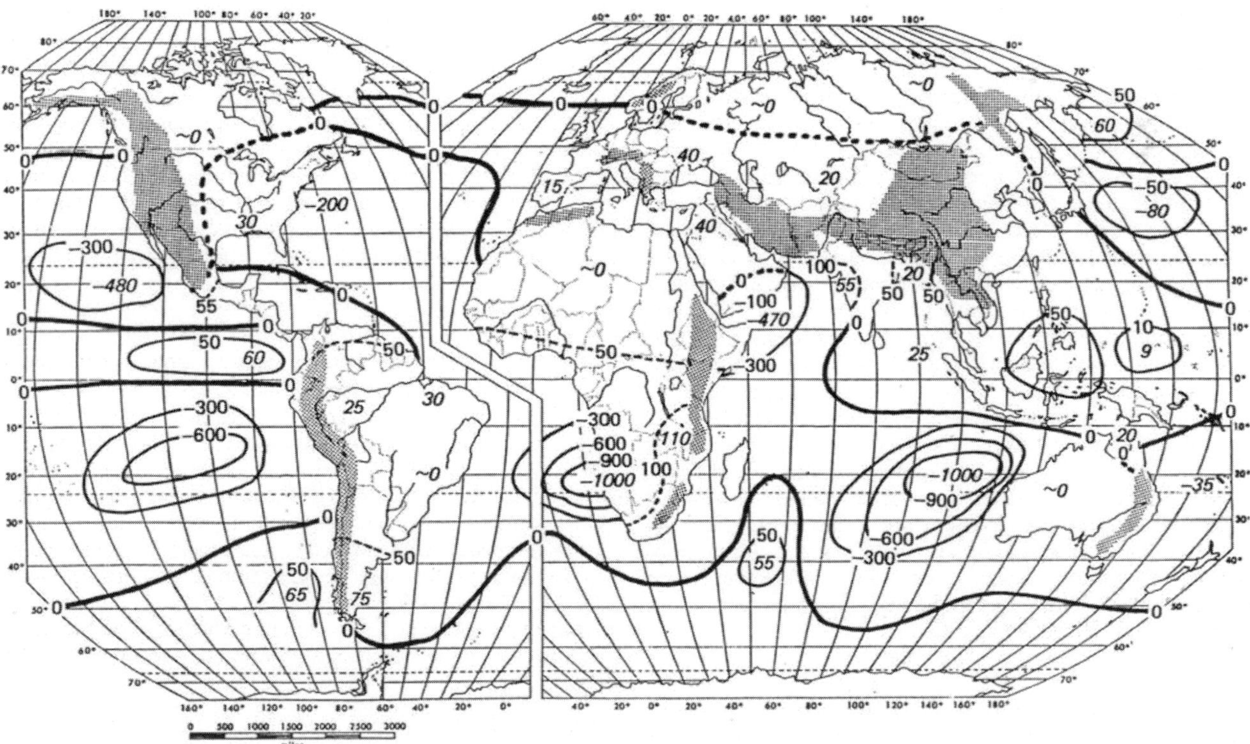

Figure H8 Mean annual evaporation ratios (evaporation/net radiation ×100). High-altitude areas shown by shading; continental ratios analyzed by dotted lines; those over oceans by continuous lines. Maximum and minimum values are italicized.

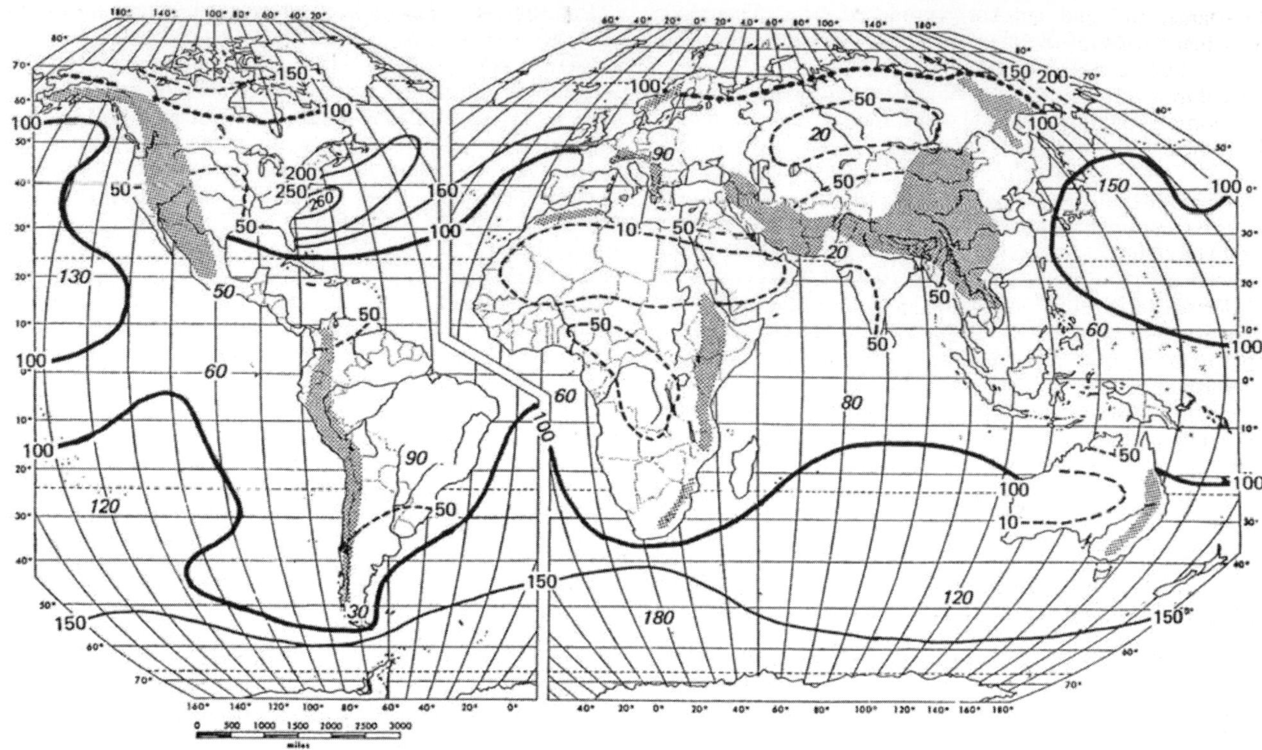

Figure H9 Mean annual runoff ratios (runoff/precipitation ×100) over the continents, indicated by dotted analysis. Mean annual effective precipitation ratios (precipitation−evaporation/precipitation ×100) are shown by continuous lines over oceans. High-altitude areas are shown by shading. Maximum and minimum values are italicized.

zero, including northern and western North America, east central South America, Australia, northern Africa, and northern Eurasia. Evaporation equals precipitation in these areas. The remainder of the continents typically exhibit ratios between 30% and 50% except southern South America and western India (>75%), and southeastern Africa (>100%). Only in these latter two areas does runoff exceed precipitation, at least at this scale of analysis.

The solid lines of Figure H9 (over oceans) show the distribution of effective precipitation ratio, i.e. $[(P - E)/P \times 100]$. Low-latitude oceans typically exhibit effective precipitation ratios less than zero, due to the considerable evaporation under subtropical anticyclones. The ratios tend to be slightly positive near the equator and over high-latitude oceans, where cloud cover tends to dominate due to the intertropical convergence and the subpolar cyclones, respectively. Lowest effective precipitation ratios are found over the eastern margins of southern hemisphere oceans.

Again, because evaporation and precipitation estimates are less likely to be accurate over oceans and high latitudes, runoff and effective precipitation ratios may be in error. However, it is unlikely that any errors are sufficiently large to change the patterns shown in Figure H9.

Human influences on hydroclimate

Purposeful and inadvertent impacts have been exerted on the hydroclimate in limited areas and times by human activity. The effect of large metropolitan areas on the urban and downstream precipitation distribution has been demonstrated by data from, and downstream of, St Louis, Missouri, by Changnon et al. (1981).

Reducing water loss by evaporation from lakes, and evapotranspiration from vegetation by the addition of monomolecular films to the surfaces, has been shown to be feasible (Roberts, 1961).

Atmospheric carbon dioxide concentrations have been increasing since observations began about 40 years ago. Though water vapor is a more effective greenhouse gas its atmospheric concentration is minimal and relatively stable, fluctuating only with changes in airmass and advection. The concentration of carbon dioxide, on the other hand, is increasing near-exponentially, generally attributed to the burning of fossil fuels and forest clearing, i.e. human activity. The presence of atmospheric carbon dioxide warms the lower atmosphere by reducing terrestrial longwave losses to space. It is anticipated that the present atmospheric carbon dioxide concentration will double by the mid-twenty-first century, resulting in mean annual warming of 1.5–3°C (Schneider, 1990; Houghton et al., 2001). This warming may decrease atmospheric stability affecting precipitation, and will further modify the hydrologic budget since relative humidities are expected to remain constant, with a concomitant increase in absolute humidity (Watts, 1980).

The damming and channeling of the Mississippi River to improve navigation and flood control has changed the river's flow rate and sediment-carrying capacity. The change imposed on discharge and residence time of water on its journey to the Gulf of Mexico has impacted the local hydrologic budget. Reversing the direction of flow of the Snowy River in southeastern Australia modified local water budgets by draining regions that formerly received runoff and delivering water to others that formerly exported runoff.

On a larger and speculative scale, Borisov (1973) discusses changes that would result from diverting warm Pacific water to the Arctic Ocean, thereby freeing Arctic Ocean coastal areas from ice during much of the year. In addition to improving shipping opportunities for currently ice-locked ports during much of the year, the open ocean may lead to increased precipitation over nearby continental regions. This would increase runoff of fresh water to the Arctic Ocean, modifying the distribution of density and thus having an impact on ocean circulation.

The hydrologic budget may also be modified by purposeful cloud seeding, in which a nucleating agent is injected into a cloud. These agents promote ice crystal formation either by cooling cloud temperatures or by providing a crystalline structure conducive to ice crystal formation (briefly reviewed by Kerr, 1982). Positive results have been observed in some locations and in some cases, but cloud seeding does not produce consistent results, suggesting that the process is more/less efficient at some locations and times because of topography, ground cover, stability etc., or because of differences in application technique. Clearly, the process is not yet well understood and should not be adopted *a priori* as a strategy for drought mitigation.

Conclusions

The components of the water budget exhibit spatial and temporal patterns over the surface of the Earth in response to various climatic forcing functions and surficial constraints, e.g. components of atmospheric general circulation, atmospheric stability and moisture, radiation, topography, surface albedo, and surface water. As any one component changes, so does one or more component of the hydrologic budget.

The occurrence and frequency of unusual hydrologic events prompts interest in the concept of climatic change and what the future may hold. Recent examples of such events include the intensity and frequency of droughts in northern Africa. Lamb (1985) has shown that the sub-Saharan drought has persisted from 1968 through 1985, a relatively long string of moisture-deficient years relative to the *ca.* 45-year record. Recently, the upper Midwest experienced two substantial precipitation anomalies, a monumental drought in 1988 and massive flooding in 1993, events which exhibited 50–100-year recurrence intervals. The occurrence of extreme-magnitude events, or several such events in a relatively short time, may merely reflect natural variance or may signal a change from one stable climatic regime to another.

That climate, and the hydrological responses, change with time is clearly shown from the paleoclimatic record. Continued monitoring and research of atmospheric and hydrologic parameters will provide a database from which better dynamic relationships can be developed.

Wayne M. Wendland

Bibliography

Barry, R.G., 1969. The world hydrologic cycle. In Chorley R.J., ed., *Water, Earth and Man*. London: Methuen, pp. 11–29.

Bavel, C.H.M., van, 1966. Potential evaporation: the combination concept and its experimental verification. *Water Resources Research*, **2**: 455–467.

Borisov, P., 1973. *Can Man Change the Climate?* Moscow: Progress Publishers.

Browning, K.A., 1979. The FRONTIERS plan: a strategy for using radar and satellite imagery for very short range precipitation forecasting. *Meteorological Magazine*, **108**: 161–184.

Budyko, M.I., 1956. *The Heat Balance of the Earth's Surface* (N.A. Stepanova, trans.). Washington, DC: Office of Technical Services US Department of Commerce.

Budyko, M.I., 1974. *Climate and Life* (D.H. Miller, trans.). New York: Academic Press.

Changnon, S.A., 1979. The Illinois climate center. *American Meteorological Society Bulletin*, **60**: 1157–1164.

Changnon, S.A., and Vogel, J.L., 1981. Hydroclimatological characteristics of isolated severe rainstorms. *Water Resources Research*, **17**: 1694–1700.

Changnon, S.A., Semonin, R.G., Auer, A.H., Braham, R.R., and Hales, J.M., 1981. METROMEX: a review and summary. *Meteorological Mons*. **18**: 1–181.

Dale, R.F., Nelson, W.L., Scheeringa, K.L., Stuff, R.G., and Reetz, H.F., 1982. Generalization and testing of a soil moisture budget for different drainage conditions. *Journal of Applied Meteorology*, **21**: 1417–1426.

Geiger, R., n.d., *Mean Annual Precipitation (Chart)*. Darmstadt: Justus Perthes Publications.

Harlin, J.M., 1978. Reservoir sedimentation as a function of precipitation variability. *Water Resources Bulletin*, **14**: 1457–1465.

Hastenrath, S., and Lamb, P.J., 1978. *Heat Budget Atlas of the Tropical Atlantic and Eastern Pacific Oceans* Madison, WI: University of Wisconsin Press.

Hastenrath, S., and Lamb, P.J., 1979. *Climatic Atlas of the Indian Ocean*, Pt II. Madison, WI: University of Wisconsin Press.

Holmes, J.W., and Wronski, E.B., 1981. The influence of plant communities upon the hydrology of catchments. *Agriculture and Water Management* **4**: 19–34.

Houghton, J.T., Ding, Y., Griggs, D.J., et al. (eds), 2001. *Climate Change 2001: the scientific basis* Cambridge: Cambridge University Press.

Jones, D.M.A., 1966. *Variability of Evapotranspiration in Illinois*. Circular 89. Champaign, IL: Illinois State Water Survey.

Jones, D.M.A., and Wendland, W.M., 1984. Some statistics of instantaneous precipitation. *Journal of Climatology and Applied Climatology*, **23**: 1273–1285.

Kerr, R.A., 1982. Test fails to confirm cloud seeding effect. *Science*, **217**: 234–236.

Lamb, P.J., 1985. Rainfall in Subsaharan West Africa during 1941–1983, *Zeitschrift für Gletscherkunde und Glazialgeologie*, **21**: 131–139.

Langbein, W.G., 1967. Hydroclimate. In Fairbridge, R.W., ed., *The Encyclopedia of Atmospheric Sciences and Astrogeology*, New York: Reinhold, pp. 447–451.

Monteith, J.L., 1965. Evaporation and environment. *Symposium of the Society for Experimental Biology*, **19**: 205–234.

O'Hearn, M., and Gibb, J.P., 1980. *Groundwater Discharge to Illinois Streams*. Contract Report 246. Champaign, IL: Illinois State Water Survey.

Penman, H.L., 1963. *Vegetation and Hydrology*, Technical Communication No. 53. Farnham: Commonwealth Bureau of Soils, Commonwealth Agricultural Bureau.

Price, A.G., Hendrie, L.K., and Dunne, T., 1978. Controls on the production of snowmelt runoff. In Colbeck, S.C., and Ray, M., eds., *Modeling of Snow Cover Runoff*. Hanover, NH: U.S. Army Cold Regions Research and Engineering Laboratory, pp. 257–268.

Priestley, C.H.B., and Taylor, R.J., 1972. On the assessment of surface heat flux and evaporation using largescale parameters. *Monthly Weather Review*, **100**: 81–92.

Rasmusson, E.M., 1966. *Atmospheric Water Vapor Transport and the Hydrology of North America*. Report A. Cambridge, MA: Massachusetts Institute of Technology, Department of Meteorology.

Rehm, B.W., Morgan, S.R., and Groenwold, G.H., 1982. National groundwater recharge in an upland area of central North Dakota, U.S.A. *Journal of Hydrology*, **59**: 293–314.

Roberts, W.J., 1961. Reduction of transpiration. *Journal of Geophysics Research*, **66**: 3309–3312.

Schicht, R.J., and Walton, W.C., 1961. *Hydrologic Budgets for Three Small Watersheds in* Illinois. Report of Investigation 40. Champaign, IL: Illinois State Water Survey.

Schneider, S., 1990. The global warming debate heats up: an analysis and perspective. *Bulletin of the American Meteorological Society*, **71**(9): 1291–1304.

Sellers, W.D., 1967. *Physical Climatology*. Chicago, IL: University of Chicago Press.

Thornthwaite, C.W., 1948. Approach toward a rational classification of climate, *Geographic Review* **38**: 55–94.

Watts, R.G., 1980. Climate models and CO_2-induced climatic changes. *Climate Change*, **2**: 387–408.

Wendland, W.M., 2001. Temporal responses of surface-water and groundwater to precipitation in Illinois. *Journal of the American Water Reserach Association*, **37**(3): 685–693.

World Meteorological Organization, 1971. *Guide to Meteorological Instruments and Observing Practices*, 4th edn. Geneva: World Meteorological Organization.

Cross-references

Acid Rain
Climate Hazards
Cloud Climatology
Evaporation
Evapotranspiration
Precipitation
Water Budget Analysis

HURRICANES – DATA AND DATA SOURCES

The word hurricane is a regionally specific name for a strong tropical cyclone in the North Atlantic Ocean, the Northeast Pacific Ocean east of the dateline, or the South Pacific Ocean east of 160°E. A tropical cyclone is the generic term for a non-frontal synoptic scale low-pressure system over tropical or subtropical waters with organized cyclonic surface wind circulation.

The details of hurricane formation are provided in the item on "Tropical cyclones". Given the enormous damage, human suffering and cost of the hurricanes season in Florida and the Caribbean, there is considerable interest in hurricane information. This entry provides some basic data and sources applicable to hurricanes.

Tables H4 and H5 list names selected for Atlantic and Eastern Pacific storms. A partial listing of the most intense and most deadly hurricanes influencing the US is provided in Tables H6 and H7.

John E. Oliver

Table H4 Names for Atlantic storms

2001	2002	2003	2004	2005	2006
Allison	Arthur	Ana	Alex	Arlene	Alberto
Barry	Bertha	Bill	Bonnie	Bret	Beryl
Chantal	Cristobal	Claudette	Charley	Cindy	Chris
Dean	Dolly	Danny	Danielle	Dennis	Debby
Erin	Edouard	Erika	Earl	Emily	Ernesto
Felix	Fay	Fabian	Frances	Franklin	Florence
Gabielle	Gustav	Grace	Gaston	Gert	Gordon
Humberto	Hanna	Henri	Hermine	Harvey	Helene
Iris	Isidore	Isabel	Ivan	Irene	Isaac
Jerry	Josephine	Juan	Jeanne	Jose	Joyce
Karen	Kyle	Kate	Karl	Katrina	Kirk
Lorenzo	Lili	Larry	Lisa	Lee	Leslie
Michelle	Marco	Mindy	Matthew	Maria	Michael
Noel	Nana	Nicholas	Nicole	Nate	Nadine
Olga	Omar	Odette	Otto	Ophelia	Oscar
Pablo	Paloma	Peter	Paula	Philippe	Patty
Rebekah	Rene	Rose	Richard	Rita	Rafael
Sebastien	Sally	Sam	Shary	Stan	Sandy
Tanya	Teddy	Teresa	Tomas	Tammy	Tony
Van	Vicky	Victor	Virginie	Vince	Valerie
Wendy	Wilfred	Wanda	Walter	Wilma	William

Names are recycled every 6 years. Influential hurricanes have their names retired.

Table H5 Names for Eastern Pacific Storms

2001	2002	2003	2004	2005	2006
Adolph	Alma	Andres	Agatha	Adrian	Aletta
Barbara	Boris	Blanca	Blas	Beatriz	Bud
Cosme	Cristina	Carlos	Celia	Calvin	Carlotta
Dalila	Douglas	Dolores	Darby	Dora	Daniel
Erick	Elida	Enrique	Estelle	Eugene	Emilia
Flossie	Fausto	Felicia	Frank	Fernanda	Fabio
Gil	Genevieve	Guillermo	Georgette	Greg	Gilma
Henriette	Hernan	Hilda	Howard	Hilary	Hector
Ivo	Iselle	Ignacio	Isis	Irwin	Ileana
Juliette	Julio	Jimena	Javier	Jova	John
Kiko	Kenna	Kevin	Kay	Kenneth	Kristy
Lorena	Lowell	Linda	Lester	Lidia	Lane
Manuel	Marie	Marty	Madeline	Max	Miriam
Narda	Norbert	Nora	Newton	Norma	Norman
Octave	Odile	Olaf	Orlene	Otis	Olivia
Priscilla	Polo	Patricia	Paine	Pilar	Paul
Raymond	Rachel	Rick	Roslyn	Ramon	Rosa
Sonia	Simon	Sandra	Seymour	Selma	Sergio
Tico	Trudy	Terry	Tina	Todd	Tara
Velma	Vance	Vivian	Virgil	Veronica	Vicente
Wallis	Winnie	Waldo	Winifred	Wiley	Willa
Xina	Xavier	Xina	Xavier	Xina	Xavier
York	Yolanda	York	Yolanda	York	Yolanda
Zelda	Zeke	Zelda	Zeke	Zelda	Zeke

These names are recycled every 6 years.

Table H6 Ten most intense USA (continental) hurricanes; 1900–2003 (at time of landfall with landfall area); updated from Hebert et al. (1997)

				Pressure	
Ranking	Hurricane (landfall)	Year	Category	Millibars	Inches of Hg
1	"Labor Day" (FL Keys)	1935	5	892	26.35
2	Camille (MS,SE LA,VA)	1969	5	909	26.84
3	Andrew (SE FL,SE LA)	1992	5	922	27.23
4	(FL Keys, South TX)	1919	4	927	27.37
5	"Lake Okeechobee" (S FL)	1928	4	929	27.43
6	Donna (FL,Eastern U.S.)	1960	4	930	27.46
7	(New Orleans, LA)	1915	4	931	27.49
8	Carla (North and Central TX)	1961	4	931	27.49
9	Hugo (SC)	1989	4	934	27.58
10	"Great Miami" (FL,MS,AL, NW FL)				

Table H7 Deadliest USA (continental) hurricanes, 1900–2003; updated from Hebert et al. (1997)

Ranking	Hurricane (location)	Year	Category	Deaths
1	"Galveston" (TX)	1900	4	8000+
2	"Lake Okeechobee" (S FL)	1928	4	2500+
3	Unnamed (Fl Keys,S TX)	1919	4	600+
4	"New England" (NY,RI)	1938	3	600
5	"Labor Day" (FL Keys)	1935	5	408
6	Audrey (SW LA,N TX)	1957	4	390
7	"Great Atlantic" (NE U.S.)	1944	3	390#
8	"Grand Isle" (LA)	1909	4	350
9	Unnamed (New Orleans, LA)	1915	4	275
10	Unnamed (Galveston, TX)	1915	4	275

Source for Tables: http://www.nhc.noaa.gov

Bibliography

Data of the type given here are available from web sites of NOAA including http://www.nhc.noaa.gov and http://www.aoml.noaa.gov and other sites including http://www.unisys,com/hurricane.

Dvorak, V.F., 1984. Tropical cyclone intensity analysis using satellite data. *NOAA Technical Report NESDIS 11*.

Gentry, R.C., 1966. Nature and scope of hurricane damage. American Society for Oceanography. *Hurricane Symposium, Publication Number One*.

Hebert, P.J., Taylor, J.G., and Case, R.A., 1984. Hurricane experience levels of coastal county populations – Texas to Maine. *NOAA, Technical Memorandum* NWS-NHC-24.

Hebert, P.J., Jarrell, J.D., and Mayfield, B.M., 1997. The deadliest, costliest and most intense United States hurricanes of this century (and other frequently requested hurricane facts). *NOAA, Technical Memorandum* NWS-TPC-1.

Jarrell, J.D., Hebert, P.J., and Mayfield, B.M., 1992. Hurricane experience levels of coastal county populations – Texas to Maine. *NOAA, Technical Memorandum* NWS-NHC-46.

Neumann, C.J., Jarvinen, B.R., McAdie, C.J., and Hammer, G.R., 1999. Tropical cyclones of the North Atlantic Ocean, 1871–1998. *NOAA, Historical Climatogy Series* 6-2.

Simpson, R.H., 1974. The hurricane disaster potential scale. *Weatherwise*, **27**: 169–186.

US Weather Bureau. *Climatological Data and Storm Data*, various volumes, various periods, National and State Summaries (National Weather Service 1971–1998).

US Weather Bureau. *Monthly Weather Review*, 1872–1970 (National Weather Service 1971–1973, and American Meteorological Society 1974–2001).

Cross-references

Tropical Cyclones

I

ICE AGES

An ice age, in its broad sense, is any cold period in geologic history when glaciers extended over mountainous areas and, in high latitudes, covered continent-sized regions. At their maxima the land area covered by ice would exceed 45 million km^2 and its volume 80 million/km^3. During the last 4 billion years of geologic time, ice ages have occupied less than 10%; at other times more equable climates prevailed and major glaciation was absent, except in high mountains (John, 1979). During that history the movements of continents due to plate tectonic processes brought those land masses into paleolatitudes which at certain times favored ice-age initiation (Fairbridge, 1973). At other times, equatorial distribution of land areas made it unlikely (Schwarzbach, 1963).

Ice ages of the past have generally lasted 10–20 million years, and have a recurrence time of about 150–250 million years (Fairbridge, 1987; Williams, 1980). This "cyclicity" may be related to the passage of the solar system through the two spiral arms of the Milky Way galaxy. Within any ice age there is a powerful cyclicity with terms around 20 000, 41 000, and 93 000 years, which is related to the variations of orbital motions (the so-called "Milankovitch periods", Berger et al., 1984; Milankovitch, 1941) of the moon around the Earth, and of the Earth around the sun, that change the eccentricity, the tilt angles and the precessional seasonality (Mörth and Schlamminger, 1979). These variables create an alternation of "glacial" and "interglacial" stages. Modern mankind is living in an interglacial (i.e. mild) stage, following a glacial interval that had its climax about 20 000 years ago. In popular use, it is the latter that is known as *the* Ice Age, a time characterized by Paleolithic ("Stone Age") Man, and various extinct creatures such as the woolly mammoth, mastodon, saber-tooth tiger and Irish elk. (Only in science fiction did early man coexist with the dinosaurs, which became extinct some 65 million years earlier.)

The present ice-age condition has lasted about 2 million years, and is identified as the "Quaternary Period". (The name is a survivor from an obsolete terminology, first introduced in 1750 with "Primary", "Secondary" and "Tertiary" for labeling divisions of geologic time.) All of the glacial and interglacial intervals have received local names in different regions, but there is little consensus on an acceptable international nomenclature. In deep-sea deposits, which contain the only continuously stratified record of the Quaternary Period, there is evidence of about 28 glacial-type climatic oscillations. In rare instances there is a land record, e.g. in 28 coastal dune ridges in South Australia; also, in the loess region of northwest China, there is a comparable record with desert-dust accumulation during the cold–arid phases, alternating with red soil layers marking the warm–humid interglacials.

Of the ice-covered areas of the present day, Antarctica is the largest (over 12 million km^2), and it has been glaciated throughout the Quaternary. It displays signs of only minor interglacial melting from time to time. In contrast, the northern hemisphere landmasses of North America and northwestern Europe were repeatedly buried by ice and then "deglaciated"; each melting phase was completed within about 10 000–15 000 years. The melting ice left behind swaths of rock debris and piled-up soil (moraines) that marked the outer limits of "ice-push" or standstill episodes during the ice retreat. Giant boulders (erratics), some bigger than a house, are included in that debris and were the source of much speculation in the nineteenth century, before the modern glacial theory was developed (in the 1840s). One early theory attributed erratics to sea-ice flotation during the biblical flood. Discoveries of pre-Quaternary ice age indicators in the second half of the nineteenth century were to have a profound effect on geological thinking and philosophy. Most striking were finds in India, Australia, South Africa and Brazil in rocks, shown by the fossils, to be of Permian age (about 250 million years old), at latitudes so close to the present equator that various "nonactualistic" theories were proposed to account for them. The paradox was partially solved by the development of plate tectonics (Fairbridge, 1973; Tarling, 1978), but the overall problem of cycles of global cooling remains a scientific challenge. A still-earlier ice age of 450 million years ago (the Ordovician Period) was actually discovered (in 1970) in the middle of the Sahara Desert and later found to extend from West Africa to Saudi Arabia. Although predicted on paleontological evidence by Grabau (1940) and paleomagnetic data (Fairbridge, 1970a), it was only later confirmed in the field (Beuf et al., 1971; Fairbridge, 1970b, 1973). After plate-tectonic closing of the Atlantic, the late Ordovician ice sheets may once have extended from Brazil to the Persian Gulf. In a review by Crowell (1999) it was largely ignored (see also Royer (2004).

Ice ages that are still older, around 600, 750 and 900 million years, or even earlier, have introduced further paradoxical problems that hint at changes in the Earth's spin axis (the "obliquity"). Such changes cannot be explained by ordinary astronomic variables, but might be caused during the early history of the Earth by catastrophic collisions with large asteroids which would be expected from the evidence of such impacts on the moon and Mars. Talk of a "snowball Earth" at one stage has not gained universal support.

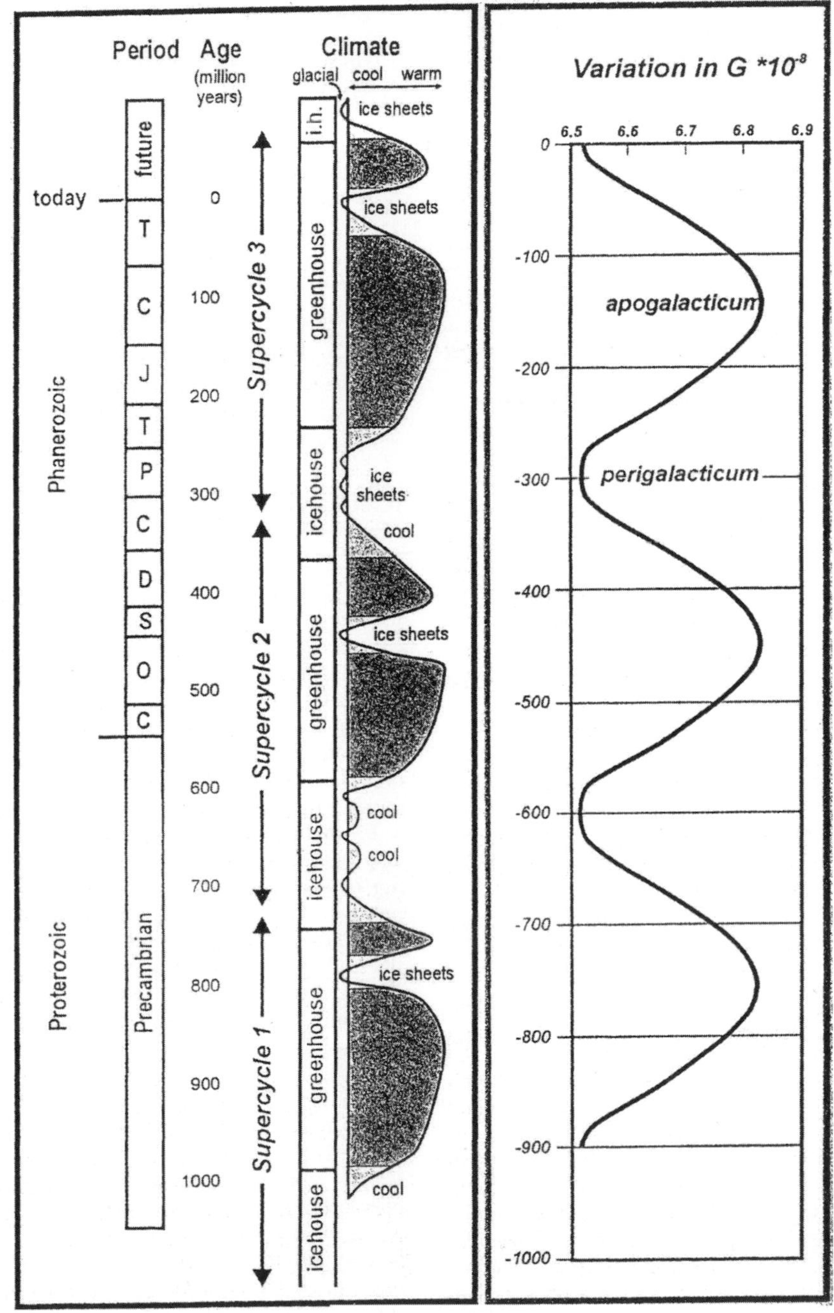

Figure I1 Steiner's concept of climate changes over the last billion years, as adapted by Veevers (1990) and others, postulating three supercycles, with "icehouse" to "greenhouse" alterations. These correspond approximately to a postulated variation of the gravitational constant **G** in the Milky Way galaxy, where the "apogalacticum" matches peaks of greenhouse conditions, and "perigalacticum" the icehouse maxima. Additional cooling stages are presumed to match the "Simpson Effect". Over the same billion-year period the cosmic-ray flux shows the same broad cycles, but in an inverse relationship to temperature (Shaviv and Veizer, 2003).

The primary "driver" of ice ages is variously attributed to CO_2, solar dynamics and the rotation of the galaxy (Figure I1). In general, temperature seems to be inversely related to cosmic ray flux (Shaviv and Veizer, 2003).

For the environmentalist, what important messages are contained in the history and dynamics of the Earth's ice ages? To begin with they show that, although the mean history of the Earth has been "uniformitarian" and there is other evidence to show that the mean temperature of the Earth over the last 4 billion years has been in the order of 20 ± 5°C, this equable-to-subtropical world was sometimes seriously disturbed by intervals of multiple cold cycles that, individually, during their extreme phases, may each have lasted 5000–10 000 years (Budyko et al., 1987). Such episodes of climatic extremes would have been sufficient to trigger numerous evolutionary extinctions, so-called "punctuated evolution" (Clube, 1990). These concurrent hydrologic exchanges, of water to ice and back, also led to eustatic changes of sea level, themselves introducing several stress conditions that would also be liable to cause extinctions. By the same token, Darwinian adaptations to those stresses would accelerate positive ("intelligent") changes; one such "nudge" may be modern human evolution to Homo sapiens sapiens from the Neanderthals, perhaps around 30 000–20 000 years ago.

Much speculation has appeared on the subject of "the next ice age", meaning the next glacial stage of the Quaternary Ice Age (Imbrie and Imbrie, 1979). Close studies of the lat interglacial (about 100 000 ± 20 000 years ago) furnish a basis for modeling. An astronomically forced reduction of solar radiation can lower mean global temperatures by 3–5°C within a few decades (Öpik, 1965). If events of this sort augment long-term trends (paleogeographic shifts, Milankovitch cycles), a glacial cycle is predictable. But when? Much more research is still needed.

Rhodes W. Fairbridge

Bibliography

Berger, A., Imbrie, J., Hays, G., Kukla, G., and Saltzman, B., 1984. *Milankovitch and Climate*. New York: Reidel.

Beuf, S., Biju-Duval, V., De Charpal, O., Rognon, P., Gariel, O., and Bennacef, A., 1971. *Les Grès du Paléozoique inférieur au Sahara*. Paris: Technip.

Budyko, M.I., and Ronov, A.B., 1987. *History of the Earth's Atmosphere*. Berlin: Springer-Verlag.

Clube, S.V.M., (ed.), 1990. *Catastrophes and Evolution: astronomical foundations*. Cambridge: Cambridge University Press.

Crowell, J.C., 1999. Pre-Mesozoic ice ages. *Geological Society of America Memoirs*, vol. 192.

Fairbridge, R.W., 1970a. An ice age in the Sahara. *Geotimes*, **15**(6): 18–20.

Fairbridge, R.W., 1970b. South Pole reaches the Sahara. *Science*, **168**: 878–881.

Fairbridge, R.W., 1973. Glaciation and plate migration. In Tarling, D.H., and Runcorn, K., eds., *Implications of Continental Drift to the Earth Sciences*, vol. 1. London: Academic Press, pp. 503–515.

Fairbridge, R.W., 1987. Ice Age theory. In Oliver, J.E., and Fairbridge, R.W., eds., *The Encyclopedia of Climatology*. New York: Van Nostrand Reinhold, pp. 503–514.

Grabau, A.W., 1940. *The Rhythm of the Ages*. Peking: Henri Vetch (reprinted New York: Krieger, 1978).

Imbrie, J., and Imbrie, K.P., 1979. *Ice Ages: solving the mystery*. Short Hills, NJ: Enslow.

John, B. (ed.), 1979. *The Winters of the World*. Newton Abbot: David & Charles.

Milankovitch, M., 1941. Kanon der Erdbestrahlung und seine Anwendung auf das Eiszeitenproblem. *Royal Serbian Academy*, Special Publication 133 (English translation 1969 as *Canon of Insolation and the Ice-Age Problem*, by *Israel Program, Scientific Translations*), Washington, DC: US Department of Commerce.

Mörth, H.T., and Schlamminger, L., 1979. Planetary motion, sunspots and climate. In McCormac, B.M., and Selliga, T.A., eds., *Solar-Terrestrial Influences on Weather and Climate*. Dordrecht: Reidel, pp. 193–207.

Öpik, E.J., 1965. Climatic change in cosmic perspective. *Icarus*, **4**: 289–307.

Royer, D.L., 2004. CO_2 as a primary driver of Phanerozoic climate. *GSA Today*, **14**(3): 4–10.

Schwarzbach, M., 1963. *Climates of the Past* (transl. Muir, R.O.). London; Van Nostrand.

Shaviv, N.J., and Veizer, J., 2003. Celestial driver of Phenerozoic climate? *GSA Today*, July, 4–10.

Tarling, D.H., 1978. The geological–geophysical framework of ice ages. In Gribbin, J., ed., *Climatic Change*. Cambridge: Cambridge University Press, pp. 3–24.

Veevers, J.J., 1990. Tectonic-climatic cycle in the billion year plate-tectonic eon. *Sedimentary Geology*, **68**, 1–16.

Williams, G.E., 1980. *Megacycles: long-term episodicity in Earth and planetary history*. Stroudsburg, PE: Dowden, Hutchinson & Ross (Benchmark, vol. 57).

Cross-references

Geomorphology and Climate
Little Ice Age
Milankovitch, Milutin
Paleoclimatology

ICEBERGS AND HEINRICH EVENTS

Icebergs are seasonal phenomena in modern oceans, but are commonly associated with cooling events during climatic history, as seen on 10^3 to 10^9-yr time scales (Hambrey and Glasser, 2003). However, when considered in century or decadal terms the facts are paradoxically opposite. Glacial erratics scattered about the landscape in northern Europe when first noticed were often attributed to iceberg transport during Noah's Flood – another misconception doomed to be discarded.

From the geological record it is observed that around 2.7 Myr ago there was a sudden increase in bipolar sea-ice and ice-rafting of sedimentary debris ("IRD", or ice-rafted debris). This pattern fluctuated from glacial to interglacial stages, but at glacial extremes ice-rafting took the IRD, even large erratics, as far south in the Atlantic as 30° or even 25°S. In the Southern Ocean, ice transport reached the coasts of South Africa and Australia. As a biological proxy, various species of penguins, riding as "passengers" on the ice, were able to colonize the low-latitude Galapagos Islands, and their bones are found in Australian paleosols as far north as 28°.

The nature of IRD tends to reflect the source areas. Thus, the shale blocks with Paleozoic fossils, dredged up off the Canary Islands on a nineteenth-century oceanographic expedition, could be traced immediately to the British Isles. Granitic boulders found by Ewing near the northern edge of the Bahama plateau were mistakenly attributed to tree-trunk transport from Brazil, but eventually were traced to iceberg carriers from the Laurentian ice-sheet. An extensive literature now treats with the provenance of the North Atlantic IRD deposits. Besides the lithology, i.e. composition, of the debris, in the case of pebbles and small boulders it is sometimes possible to use shape as a guide (Middleton, 2003, p. 580). Following the Cailleux rule, those of ovoid shape reflect intertidal, planar abrasion (such as found in ice-foot or "glaciel" coastal settings, while those of three-dimensional to subspherical form point to the tumbling dynamics of subglacial meltwater torrents (and therefore more likely to be carried by icebergs).

Actualistic, or "MAT" modern-analog, observations show that ice-shelf (tabular) bergs are typically developed in the Southern Ocean, off Antarctica, whereas the irregular blocky bergs are more likely to be generated at the valley-glacier snouts of Greenland, Spitzbergen, southern Alaska, New Zealand and southern Chile. The tabular bergs generated around 1985–2005 in the Weddell Sea or Ross Sea are often of giant size (as "big as the state of Connecticut", i.e. 5000 miles^2, or 13 000 km^2) before they disintegrate. The accelerating production of icebergs in recent decades is demonstrably related to two climatic variables, the rising MSL and rising SST, both aspects of twentieth-century "global warming". The sea-level rise increases the glacier-ice buoyancy, thus its distal areas float rather than rest on the shelf. Secondly, the warming water leads to melting from below, even when atmospheric temperature is well below zero. Surface melting of the glacier during warm summers feeds meltwater into crevasses, thence to subglacial rivers, which further increases its general mobility.

Historically, the longest record of North Atlantic sea ice can be derived from Icelandic evidence, which began with the first Norse settlers in AD 865. Schell (in Fairbridge, 1967, p. 858) provides a century-by-century synthesis, expressed as weeks of coastal ice severity:

865–1100: 0.1–0.7
1101–1200: zero
1201–1300: 7.5
1301–1600: 3.2–6.0
1601–1800: 22.6–25.3
1801–1900: 40.8
1901–1950: 8.6

From this evidence it seems that the Medieval Warming Period was indeed most favorable for the settlers, while the Little Ice Age is known to have brought great privations and the abandonment of many farms in Iceland and to the total loss of the Greenland colonies. However, it is curious to note that the maximum sea ice off Iceland was not reached until the nineteenth century (probably due to a shift in the Irminger Current). At the present time the ice cover of the Arctic Ocean may exceed 10 Mkm^2 in winter, calving and melting away to 800 000 km^2 or so in late summer. In the Pacific the winter ice cover of the Sea of Okhotsk today may even beset northern Japan, and the winter ice of the Bering Strait still links Siberia and Alaska, though all is gone by early summer. Closely associated with global climate, heavy ice years are often matched by a southerly shift of the jet stream over the North Atlantic (corresponding to the NAO). Even the monsoons are affected by these global cycles, as shown by the flood years of the Nile (Currie, 1995), as well as India's succession of drought/flood years.

Icebergs and tides

The relationship of iceberg production to oceanic tides was first demonstrated by the Swedish scientist, Otto Pettersson (1914a), with respect to SST in the Norwegian Sea. Pettersson also recognized that the periodicity of the temperature and tidal oscillations were linked primarily to the luni-solar forcings. Normal fortnightly tidal cycles were greatly amplified by the long-term effects of internal waves coming in across the Continental Shelf, at times near the shore achieving amplitudes in excess of 20 m. (A discussion of internal waves by Cucchione and Pratson, 2004, treats some of the oceanographic and geological aspects involved.)

The master study of tidal dynamics by Wood (2001) brings out both short- and long-term tidal cycles which tend to have additive effects. Thus the perigee's closest approaches to the Earth, enhancing the tides, he called the "proxigee". Over the long term the proxigee–perihelion–syzygy precession tides return at only some 558-yr intervals; their last epoch was 1912. Around southern Greenland the SST was then 2–3°C above normal. Small wonder that "armadas" of icebergs were reported drifting out into the northwest Atlantic. It was just an incredible misfortune that this event in the early summer of 1912 coincided with the maiden voyage of the RMS *Titanic* and its disastrous fate (Wood, 2001).

Precisely one declination cycle earlier, in 1893–1894, coinciding with the apsides–perihelion cycle, there was a gigantic outburst of icebergs in the Antarctic, even reaching 40°N (Pettersson, 1914a,b), and forcing the Cape-Town to Australia shipping to shift from their usual (great circle) course. Five years previously no icebergs at all were sighted along that route.

The luni-solar tidal acceleration (or "slow-down") of the 18.6-yr nodal and nutation (declination) cycles are a striking feature of all the geostrophic ocean currents. The two periodicities recur at 18.6 yr intervals but with a nodal lag of 4 yr, which is often disregarded (Currie, 1995). Thus not only the Gulf Stream and Kuro Shio are accelerated, but also the return flows of the Canaries and California currents, as well as those of the southern hemisphere. The 18.6-yr effect in the North Atlantic was discovered by Maksimov and Smirnov (1965) of the Russian Oceanology Institute in Saint Petersburg from the evidence of warming SST. This was confirmed by the Canadian workers in the Maritime provinces, and by MSL rise in New England (Kaye and Stuckey, 1973). On a global basis Currie (1987, 1995) recognized the 18.6-yr effect in MSL, SST, rainfall and other proxies in his lifelong researches.

Economic aspects were recognized by Pettersson (1914a, b) inasmuch as the incoming megatides brought tremendous catches of herring and cod to the Swedish west coast. In the Middle Ages these tides even reached into the southern Baltic, as shown by tax records in the Hanseatic League ports dating back to the fourteenth century. Fisheries are also affected in the western Atlantic, as reported by Currie et al. (1993).

Heinrich events

Evidence of extensive late Pleistocene sea-ice and iceberg transport of IRD was discovered during sediment coring operations in the northwest Atlantic, particularly around 45°N, but later to 60°N and beyond (Heinrich, 1988; Grousset et al. 1993), and notably in the Labrador Sea (Bond et al., 1994; Hillaire-Marcel, 1994). These crescendos here become known as "Heinrich Events" and are numbered from "H-0" back to "H-11". The principal IRD provenance was the North American Laurentide ice sheet (Alley and MacAyeal, 1994), but Greenland and the Canadian Arctic also contributed. The IRD composition was particularly distinctive for its carbonate enrichment when derived from Arctic Canada, but almost devoid of carbonates in the northeast Atlantic. Unlike the IRD of the northwest Atlantic with its carbonates traced to the Paleozoics of "Canada", cores in the northeast are non-calcareous, and the IRD can be associated with fluctuations of the Fennoscandian ice-sheet (Fronval et al., 1995).

From various temperature proxies covering the span of the late Pleistocene (oxygen isotope stages 2–4) a sawtooth pattern is evident. While not exactly cyclical they have a distinctive regularity (Bond et al., 1992, 1993, 1997, 1999) and are accordingly identified as "Bond Cycles", with Heinrich Events following each nadir. Their association with ice-melting was shown by decrease in salinity.

Measured over the last 100 000 yr or so, Bond et al. (1993) showed that the fluctuations in 0–18 in the Greenland ice cores were indicators of temperature-cycle proxies in the sea-floor sediments of the northwest and northeast Atlantic. In a comparison of North Atlantic cores in general, it can be seen that the largest Heinrich/Bond oscillations were in core K 708-1 located at 25°W, a site due west of Ireland.

A secondary feature of the Bond Cycles with their asymmetric sawtooth pattern are the Dansgaard–Oeschger events that were first recognized in 0–18 trace in the Greenland ice cores (Dansgaard et al., 1993). These are abrupt warming events of 5–8°C, not associated with a sine curve of temperature proxies, but with abrupt rises and falls (sometimes likened to the pattern of skyscrapers on a city skyline). They can be matched with the pattern of the ^{14}C flux record in Holocene tree rings (Stuiver and Braziunas, 1988), with only 10–40 yr or so difference between the start of the rise and the irregular "plateau" at the top. After a few centuries this is followed by an equally abrupt fall. They are much more prominent in Greenland than in Antarctica at Vostok.

Solar and lunar links

Global distribution of Heinrich Event teleconnections far away from the glacial regions has been suggested in several papers, but most clearly in a well-dated lake core from Florida (Grimm et al., 1993).

To calculate the ages of these events, as a working hypothesis, one can use the luni-solar tidal periodicities, established in the Santa Barbara Basin (Behl and Kennett, 1996; Berger et al., 2004), with AD 1943 as a convenient index year. (That year marks the continentality peak for Stockholm and summer temperature crescendo.) It is also a solar spin-rate minimum, which marks a weakening of the geomagnetic field, favoring a high influx of solar particles to the Earth. Behl and Kennett identified 17 anoxia horizons since 59.6 ka (see Table I1) which perfectly match the Greenland ice-core record of warming events of Bond et al. (1994). Resonance of two major planetary periodicities, 18.6134 yr (L_N) and 11.17 yr (SSC) occurs ideally at 156.392-yr intervals, respectively 15 × L_N and 14 × SSC, the so-called "Libby Cycle" (Libby, 1955), but due to varying amplitude and length of the solar cycle, however, it is not

Table I1 Tentative chronology of Heinrich Events and climatic proxies

No.	(a)	(b)	(c)	(d)
			7.6	Faiyum/Fromentine
H-0	(11.67 ka)	11.84 ka	1.1 ka	End Younger Dryas
H-1	16 ka (15.84 ka)	14.81 [14.0]	14.1	Allerød Interstadial
		[16.0]		Bølling Interstadial
Late Weichselian–Wisconsinan Glacial Maximum (21-17 ka) OIS-2				
H-2	23 ka (22.93)	21.0	20.6	
		[24.5]		OIS-3 (60-21 ka)
H-3	29 ka (29.19)	26.0 [28.3]	27.1	
		[30.7]		
		[31.4]	33.6	
		[33.6]		
		[38.0]		
	(39.2)	[40.3]	40.1	Denekamp Interstadial
		45.5 [45.5]	46.6	Hengelo Interstadial
		59.8		
H-6	52 ka (52.12)	[51.5]	53.1	Glinde Interstadial
		[56.4]	59.6	
		[60.9]		
H-6	65 (65.051)	64.5 [67.3]	66.1	OIS-4 (75-60 ka)
		[70.35]	72.6	
Early–middle Weichselian–Wisconsinan Boundary (74 ka)				
H-7	(83.82)	83.9 [79.5]	79.1	Odderade Interstadial
			85.6	MIS-5(a)
H-8	(102.16)	102.2 [88.7]	105.1	Brørup Interstadial
				MIS-5(c)

(a) following Mayewski et al. (1994), based on calcium and other ions in Greenland ice core "Summit", in kyr BP; in parenthesis is suggested the 208.5/417-yr solar cycle based on tree-ring ^{14}C flux with an index year of AD 1943. (b) Following Bond et al. (1997) are shown estimates of the start (warmest) phase of "Bond Cycles" with prominent Dansgaard–Oeschger events in brackets. (c) Cyclic returns of 6672 yr period (337 × 18.0334 yr lunar tide cycle; 600 × 11.12 yr solar cycle; 32 × 208.5 yr ^{14}C flux cycle). (d) Notes. "ka" indicates chronology in kiloyears (kyr) BP, AD 1950. Oeschger events [in brackets]: e.g. 70.35 − 24.5 ka = 45 850 yr (208.5227 × 220 = 45 875 yr). OIS = oxygen isotope stage (also given as "MIS").

always quite regular. Several other periodicities are found to cluster around AD 1900. The 31.01-yr lunar perigee syzygy extreme falls in 1900 with L_N in 1898. The SSC minimum came in 1901. Astronomically, it is interesting that 1900 marked the sun–barycentric radius maximum (greatest inertial swing of the sun from the system center of gravity) and minimum in its orbital velocity, a time characterized by fewer solar flares with their climatic impact. On Earth there was a spin-rate maximum, associated with global outbreaks of volcanicity.

The previous, closely calculated event of this nature was in 1340, some 558 yr before 1898, and also the largest tidal pulse during the past millennium (Wood, 2001). This value can be used for estimating the approximate ages of the various Heinrich events which at present are only stated in radiocarbon years that need to be calibrated to astronomic time in order to relate their iceberg outbreaks to tidal and climatic events. With the guidance of ice-core 0–18 values and the tree-ring ^{14}C values, both dated in sidereal time, it can be suggested that the Heinrich incidents can each spread over some 500 yr or somewhat more.

Analysis of calcium and other indicator ions in a Greenland ice core ("Summit": Mayewski et al., 1994) show that their amplitude corresponds very closely during the late Pleistocene glaciation to a GCI or "general circulation index" for the Polar Vortex system. The larger the vortex, the colder the high-latitude climates. At such times stronger (easterly) winds would transport more dust particles to the ice sheet. In China and Mongolia such winds would be from the northwest (the Gobi and Maklakan deserts), producing gigantic loess deposits.

Massive iceberg discharges during Pleistocene cycles are not confined in their climatic effects to a single region, the North Atlantic, but appear to have global repercussions. Broecker (1994) has suggested they act as triggers for worldwide climate fluctuations, mediated by the thermohaline circulation. Very abrupt transitions are apparent. It is clear that the abrupt termination of each Bond Cycle is marked by a dramatic reduction of salinity over a broad surface of the ocean, which in turn is accompanied by the warming atmospheric temperature, a rise that must result in extensive melting of glacier ice. However, there will also be an increase in evaporation and thus the snowfall over the ice sheets, which accounts for the brief duration of the Dansgaard–Oeschger events. Where precise dating is available it is also evident that each Heinrich Event and its high IRD supply continues until the climax of the steep warming phase that follows each Bond Cycle. Iceberg production then decreases during the extended cooling phase that follows.

With its abrupt change following the Younger Dryas and during the Holocene to globally warm climates the GCI is inversely related to ^{14}C flux (which is high during cold intervals, e.g. during the Younger Dryas). Progressively over the last 10 kyr there has been a decrease in the amplitude, both positive and negative, of the ^{14}C flux (Stuiver and Braziunas, 1988) that has probably been related to the decrease in the extremes of the Earth–moon orbital cycles. Since about 6000 BP there has been also a progressive fall in the cold-cycle temperatures, reaching their nadir in the last stage of the Little Ice Age.

To sum up: water-level glaciers and grounded sea ice are made buoyant by rise of sea level such as at peak tides in AD 1912, an 18.6-yr cycle, with crescendos at 558-yr intervals. These events are marked by "armadas" of icebergs in both hemispheres. Icebergs and sea ice increase generally during major cold cycles, but their production accelerates greatly during brief warming events marked by an abrupt rise in temperature and a rise in MSL. This rise is caused by outbursts in solar emissions, with the synergy provided by lunar cycles.

Rhodes W. Fairbridge

Bibliography

Alley, R.B., and MacAyeal, D.R., 1994. Ice-rafted debris associated with binge/purge oscillations of the Laurentide ice sheet. *Paleoceanography*, **9**: 503–511.

Behl, R.J., and Kennett, J.P., 1996. Brief interstadial events in the Santa Barbara Basin during the past 60 kyr. *Nature*, **379**: 243–245.

Berger, W.H., et al., 2004. Tidal cycles in the sediments of Santa Barbara Basin. *Geology*, **32**(4): 329–332.

Bond, G., et al., 1992. Evidence for massive discharges of icebergs into the North Atlantic Ocean during the last glacial period. *Nature*, **360**: 245–249.

Bond, G., et al., 1993. Correlations between climate records from North Atlantic sediments and Greenland ice. *Nature*, **365**: 143–147.

Bond, G., et al., 1997. A pervasive millennial-scale in North Atlantic Holocene and glacial climates. *Science*, **278**: 1257–1266.

Bond, G., et al., 1999. The North Atlantic's 1-2kyr climate rhythm: relation to Heinrich events, Dansgaard-Oeschger cycles and the Little Ice Age. In Clark, P.U., et al., eds., *Mechanisms of Global Climate Change at Millennial Time Scales*. Washington DC: AGU, Geophysics Monographs, **112**: 35–58.

Broecker, W.S., 1994. Massive iceberg discharges as triggers for global climatic change. *Nature*, **372**: 421–424.

Broecker, W.S., et al., 1992. Origin of the North Atlantic Heinrich events. *Climate Dynamics*, **6**: 265–273.

Cucchione, D.A., and Pratson, L.F., 2004. Internal tides and the continental slope. *American Scientist*, **92**(2): 130–137.

Currie, R.G., 1987. Examples and implications of 18.6- and 11-yr terms in world weather records. In Rampino, M.R., et al., eds., *Climate: history, periodicity, and predictability*. New York: Van Nostrand Reinhold, pp. 378–403.

Currie, R.G., 1995. Variance contribution of M_n of and S_c signals to Nile River data over a 30.8 year bandwidth. *Journal of Coastal Research*, special issue 17, 29–38.

Currie, R.G., Wyatt, T., and O'Brien, D.P., 1993. Deterministic signals in European fish catches, wine harvests, sea level, and further experiments. *International Journal of Climatology*, **13**: 655–687.

Dansgaard, W., and Oeschger, H., 1989. Past environmental long-term tercords from the Arctic. In Oeschger, H., and Landway Jr, C.C., eds., *The Environmental Record in Glaciers and Ice Sheets*. Chichester: John Wiley, pp. 287–317.

Dansgaard, W., Johnsen, S.J., Clausen, H.B. et al., 1993. Evidence for general instability of past climate from a 250-kyr ice-core record. *Nature*, **364**, 218–220.

Fairbridge, R.W., ed., 1967. *The Encyclopedia of Atmospheric Sciences and Astrogeology*. New York: Reinhold.

Fronval, T., et al., 1995. Oceanic evidence for coherent fluctuations in Fennoscandian and Laurentide ice sheets on millennium timescales. *Nature*, **374**: 443–446.

Grimm, E.C., et al., 1993. A 50 000-year record of climatic oscillations from Florida and its temporal correlation with the Heinrich Events. *Science*, **261**: 198 200.

Grousset, F.E., et al., 1993. Patterns of ice-rafted detritus in the glacial North Atlantic (40-55°N). *Paleoceanography*, **8**: 175–192.

Hambrey, M.J., and Glasser, N.F., (2003). Glacial sediments, processes, environments and facies. In Middleton, G.V., ed., *Encyclopedia of Sediments and Sedimentary Rocks*. Dordrecht: Kluwer, pp. 316–331.

Heinrich, R., 1988. Origin and consequences of cyclic ice rafting in the northeast Atlantic Ocean during the past 1 30 000 years. *Quaternary Research*, **29**: 142–152.

Hillaire-Marcel, C., (ed.), 1994. The Labrador Sea during the late Quaternary. *Canadian Journal of Earth Sciences*, **31**(1): 1–158.

Kaye, C.A., and Stuckey, G.W., 1973. Nodal tidal cycle of 18.6 yr: its importance in sea-level curves of the east coast of the United States and its value in explaining long-term sea-level changes. *Geology*, **1**: 141–144.

Maksimov, I.V., and Smirnov, N.P., 1965. A contribution to the study of the causes of long period variations in the activity of the Gulf Stream. *Oceanology, Moscow*, **5**(2) (translation by American Geophysics Union).

Mayewski, P.A., et al., 1994. Changes in atmospheric circulation and ocean ice cover over the North Atlantic during the last 41,000 years. *Science*, **263**, 1747–1751.

Middleton, G.V., (ed.), 2003. *Encyclopedia of Sediments and Sedimentary Rocks*. Dordrecht: Kluwer.

Pettersson, O., 1912. The connection between hydrographical and meteorological phenomena. *Royal Meteorological Society Quarterly Journal*, **38**: 173–191.

Pettersson, O., 1914a. Climatic variations in historic and prehistoric time. *Svenska Hydrogr. Biol. Komm., Skriften*.

Pettersson, O., 1914b. On the occurrence of lunar periods in solar activity and the climate of the Earth, a study in geophysics and cosmic physics. *Svenska Hydrogr. Biol. Komm., Skriften*.

Williams, M., et al., 1998. *Quaternary Environments*, 2nd edn. London: Arnold.

Wood, F.E., 2001. *Tidal Dynamics*, 2 vols. *Journal of Coastal Research*, special issue 331.

Cross-references

Climate Variation: Historical
Evaporation
Ice Ages
Sea Ice and Climate
Teleconnections

ICELANDIC LOW

The Icelandic Low is a semipermanent low-pressure cell or trough located in the North Atlantic between Iceland and southern Greenland at 60–65°N (Trewartha and Horne, 1980). This subpolar low is one of the major "centers of action" in the northern hemisphere.

Daily synoptic weather charts (i.e. weather maps drawn from observations taken simultaneously from a network of stations over a large area, thereby yielding a general view of weather conditions) for the northern hemisphere show many high-and low-pressure systems associated with the disturbances along the polar front between the cold and warm airmasses. These moving disturbances can be eliminated cartographically by constructing mean charts that delineate rather clearly their semipermanent character (Rossby, 1945). Thus, the mean Icelandic Low is a result of numerous traveling depressions alternating with shorter periods of anticyclonic weather.

The center of the Icelandic Low has a January mean sea-level pressure of 996 mb (29.41 in), 17 mb (0.51 in) below the normal sea-level pressure of 1013 mb (29.92 in). By comparison, the center of the Aleutian Low has a January mean sea-level pressure of 1002 mb (29.59 in), 6 mb (0.18 in) higher than the Icelandic Low.

Although both the Aleutian and Icelandic Lows are considered subpolar in latitude, the latter is centered further poleward, tending to stretch along the poleward edge of the Gulf Stream. In contrast, the Aleutian Low is displaced somewhat equatorward by the effect of the Aleutian Islands in restricting the movement of warm Pacific waters to the Bering Sea (Willett and Sanders, 1959).

The Icelandic Low expands during the low-sun period (winter in the northern hemisphere) when the ocean is relatively warmer than the cold continents. The steady pressure gradient associated with this zone of converging cyclonic activity is largely responsible for the severe gales and winds common to the North Atlantic in winter (Ahrens, 2003). Winds of 88–121 km/h (55–75 mph) are common in winter cyclones. In the summer the low greatly decreases in size as the land temperature becomes warmer and the water relatively (but not actually) cooler.

During periods of strong circulation, both the Icelandic and Aleutian Lows cover large areas and form single well-formed centers. Conversely, the centers may be split into two separate cells of smaller size during periods of weak circulation, often due to blocking. At this time the Icelandic Low is often displaced westward and southward of Iceland.

The Icelandic Low is a major component of the North Atlantic Oscillation (NAO), a periodic fluctuation in the distribution of pressure between the low and the Azores High that strongly affects weather in Europe. Visbeck et al. (2001) note that the phase of the NAO has been shifting from mostly negative values to mostly positive values during the past three decades, a change that may be attributed to anthropogenic activities. Recent changes in Arctic Ocean sea ice motion appear to be related to the intensity of the Icelandic Low, as this center of action influences the sea-level pressure distribution over a large part of the Arctic Ocean and contiguous waters (Kwok, 2000).

Robert M. Hordon

Bibliography

Ahrens, C.D., 2003. *Meteorology Today: An introduction to Weather, Climate, and the Environment*, 7th edn. Brooks/Cole (Thomson Learning).

Akin, W.E., 1991. *Global Patterns: climate, vegetation, and soils*. Norman, OK: University of Oklahoma Press.

Barry, R.G. and Chorley, R.J., 2003. *Atmosphere, Weather, and Climate*, 8th edn. London: Routledge.

Curry, J.A., and Webster, P.J., 1999. *Thermodynamics of Atmospheres and Oceans*. San Diego, CA: Academic Press.

Graedel, T.E., and Crutzen, P.J., 1993. *Atmospheric Change: an earth system perspective*. New York: Freeman.

Hartmann, D.L., 1994. *Global Physical Climatology*. San Diego, CA: Academic Press.

Oliver, J.E., and Hidore, J.J., 2002. *Climatology: An Atmospheric Science*, 2nd edn. Upper Saddle River, New Jersey: Pearson Prentice Hall.

Kwok, R., 2000. Recent changes in Arctic Ocean sea ice motion associated with the North Atlantic Oscillation. *Geophysical Research Letters*, **27**(6): 775–778.

Robinson, P.J., and Henderson-Sellers, A., 1999. *Contemporary Climatology*, 2nd edn. Harlow: Longman.

Rossby, C.G., 1945. The scientific basis of modern meteorology. In Berry, F.A., Bollay, E., and Beers, N.R., eds., *Handbook of Meteorology*, New York: McGraw-Hill, pp. 502–509.

Trewartha, G.T., and Horne, L.H., 1980. *An Introduction to Climate*, 5th edn. New York: McGraw-Hill.

Visbeck, M.H., Hurrell, J.W., Polvani, L., and Cullen, H.M., 2001. The North Atlantic Oscillation: past, present, and future. *Proceedings of the National Academy of Sciences*, **98**(23): 12876–12877.

Wallace, J.M., and Hobbs, P.V., 1997. *Atmospheric Science: an introductory survey*. San Diego, CA: Academic Press.

Cross-references

Airmass Climatology
Aleutian Low

Atmospheric Circulation, Global
Centers of Action
Oscillations
Zonal Index

INDIAN SUMMER

The term "Indian Summer" is applied in North America to an autumn period, usually in late October or November, when delightfully warm, but not really hot, weather dominates for a period of days. It occurs in the midst of cooler days after the first frost has occurred and the first signs of winter have made their appearance. While there is no formal meteorological definition of Indian Summer, it is usually assumed that, subsequent to a hard frost, warmer days, clear skies, and calm or light winds prevail for at least 3 days.

This late autumn warming period is also identified elsewhere. In some parts of Europe it sometimes may be referred to as "Saint Martin's Summer", a time of mild weather during late fall. In England it is also known as St Martin's Festival, a time that coincides with "All-hallows Summer".

A number of explanations have been given to account for the name Indian Summer. The most commonly accepted are those applying to American Indians. Early settlers in the United States noted that the Indians recognized this summerlike weather as a time for storing late crops and readying their lodges for the coming winter. Brooks (1924) suggests other versions including that "early settlers attributed the blue haze which is characteristic of the season to the smoke of fires set by Indians on the western prairies, and bestowed the name for that reason. Yet another suggested possibility is that the name was bestowed by some early traveler who, familiar with the dry, hazy weather of India, recognized a similar type of weather here." The versions could go on, but, irrespective of its origin, the name Indian Summer is a beautiful description of a delightful weather event.

Indian Summer is a "weather singularity", a weather event that recurs around a specific calendar date each year. The "January Thaw" is another well-established North American weather singularity.

The idea of Indian Summer dates back to the late eighteenth century and has been an oft-quoted literary subject. For example, in the 1855 poem *Hiawatha*, Longfellow wrote:

Brought the tender Indian Summer
To the melancholy north-land,
In the dreary Moon of Snow-shoes.

It is a most appropriate description.

John E. Oliver

Bibliography

Brooks, C.F., 1924. *Why the Weather*? New York: Harcourt Brace.

Cross-reference

Singularities

INTERGOVERNMENTAL PANEL IN CLIMATE CHANGE (IPCC)[1]

The work of the IPCC is guided by the mandate given to it by its parent organizations the World Meteorological Organization (WMO) and the United Nations Environment Programme (UNEP). IPCC Principles contain specific procedures for the preparation, review, acceptance, approval, adoption and publication of IPCC material. The three main classes of IPCC material are IPCC Reports, which include Assessment, Synthesis, Special and Methodology Reports; Technical Papers; and Supporting Material such as proceedings of IPCC workshops. The procedures provide also guidance on the tasks of authors and reviewers.

The IPCC has three working groups. Working Group I assesses scientific aspects of climate systems and climate change; Working Group II studies the vulnerability of human and natural systems to climate change and options for adapting to them; Working Group III assesses options for limiting greenhouse gas emissions and economic issues. Also, a task force studies national greenhouse gas inventories.

John E. Oliver

Cross-references

Climate Change and Global Warming
Climate Change and Human Health
Climate Change Impacts: Potential Environmental and Societal Consequences
Climate Variation: Historical
Climate Vulnerability
Global Environmental Change: Impacts
Greenhouse Effect and Greenhouse Gases

INTERTROPICAL CONVERGENCE ZONE (ITCZ)

The intertropical convergence zone (ITCZ) is an east–west-oriented low-pressure region near the equator where the surface northeast and southeast trade winds meet. As these winds converge, moist air is forced upward, producing cumulus clouds and heavy precipitation. These clouds occur in scattered masses about 100 km (60 miles) across. Among these randomly spaced clouds are synoptic-scale waves that move from east to west; they are known as the easterly waves. Within the ITZC, winds are often light and variable in both speed and direction. It was known as the doldrums by sailors because of the lack of horizontal air movement.

The ITCZ is most clearly developed in the tropical oceans (in the eastern Pacific and Atlantic). It is the most prominent climatic feature in the tropics and plays principal roles in tropical climate by interacting with the planetary-scale circulations of the atmosphere and ocean.

[1] This account is drawn from www.ipcc.ch/ which is the home page of the IPCC. The site provides a wealth of information about organization and publications.

The positions of the ITCZ

The ITCZ is usually found within 350 km of the equator; about 80–300 km wide and 5° north and south from the equator. The position of the ITCZ varies seasonally (Figure I2), being drawn toward the areas of the most intense solar heating. The largest seasonal migration occurs over land. In January the ITCZ lies about at 15°S over South America and Africa. In July it lies at about 25°N over Africa and about 30°N over Asia. The ITCZ is less mobile over the oceans, where it holds a stationary position just north of the equator. It was discovered that the equatorial asymmetric sea surface temperature (SST) distribution and the ocean–atmospheric feedback mechanisms, including the Bjerkness feedback and wind-evaporation-SST (WES) feedback, cause the oceanic ITCZ to stay to the north of the equator (Xie, 1998).

Double ITCZ

The double ITCZ refers to two ITCZ, one at each side of the equator. Double ITCZ experience spatial and temporal variations. The most noticeable double ITCZ are usually found over the eastern Pacific during boreal spring. Over the western and central Pacific, signals of double ITCZ can be seen during June to September. Infrequent weak signals of double ITCZ are also observed over the Indian Ocean during November (Zhang, 2001). However, the QuikScat images show that the double ITCZ exist in the Atlantic and eastern Pacific all year round. For most of the time the south ITCZ is weaker than the north ITCZ. The stronger north ITCZ is created when the northeast and southeast trade winds converge over the warm waters, resulting in intense convection and formation of vigorous cumulonimbus clouds that produce heavy precipitation. The south ITCZ is formed when the southern trade winds blow over the cold upwelling waters near the equator, that cause decrease in vertical mixing and deceleration of surface winds (Liu and Xie, 2002). The south ITCZ contains only the southern wind blowing to the equator, and does not have the cloud creation capability of the north ITCZ.

Models and theories for the explanations of the ITCZ

The ITCZ is commonly viewed as the ascending branch between the Hadley cells, that is a deep circulation pattern extending from the surface to the top of the troposphere (Figure I3a). Heating due to latent heat release in deep convection plays a crucial role in the tropical meridional circulation. Fletcher (1945) proposed a twin equatorial cell model (Figure I3b). Latent heat release first heats up the atmosphere. However, radiational cooling from the cloud top and evaporative cooling at the lowest layer of the cloud will create a directional reversal in the solenoidal field. Thus, the equator becomes a cold source from which air flows outward into the two hemispheres, resulting in two small new circulation cells and two convergence zones.

Asnani (1968) suggested another model with a single equatorial cell separating the two Hadley cells (Figures I3c and I3d). In the single cell the air flows toward the equator in the lower layers and flows away from the equator aloft, producing upward motion on one side and downward motion on the other. Near the equator, subsidence occurs. There are two regions of absolute vorticity maximum, one on each side of the equator; one coincides with the ITCZ that is an area of maximum absolute vorticity where a kinematic tough is formed.

Hastenrath (1968) stated that Asnani's three-cell model could be explained by varying the vertical distribution of absolute vorticity, frictional force and vertical zonal wind components in the vorticity equation. However, for Fletcher's twin equatorial cell model, the requirement of the eastward-directed frictional force in the lower layers to decrease from the equator poleward, within the subsiding portion of either of

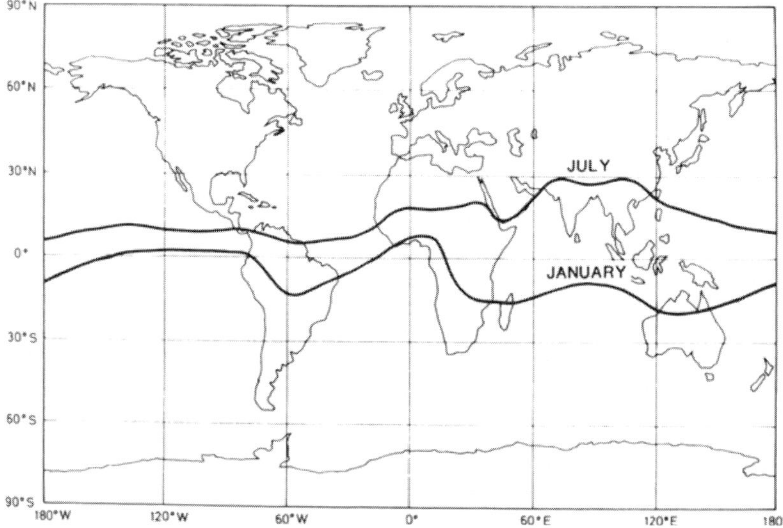

Figure I2 Mean positions of the Intertropical Convergence Zone (ITCZ) in January and July (after Henderson-Sellers and Robinson, 1991).

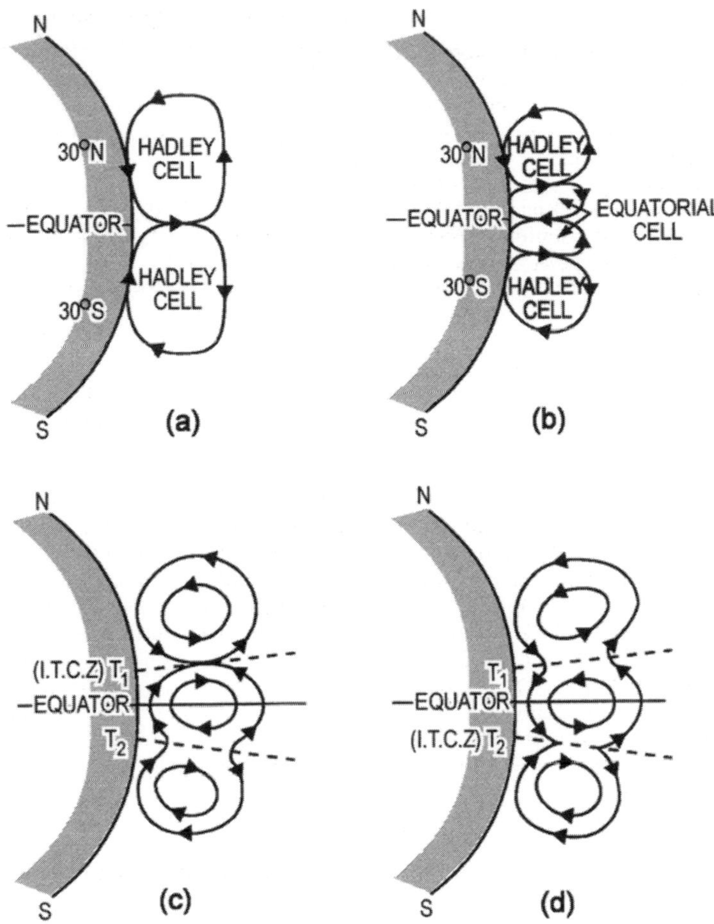

Figure I3 Models of tropical meridional circulation. (**a**) Hadley cells on two side of the equator; (**b**) two equatorial cells separating the two Hadley cells (adopted from Fletcher, 1945); (**c**) and (**d**) a single equatorial cell separating the Hadley cells (adopted from Asnani, 1968). (*Source:* http://orca.rsmas.miami.edu/~czhang/Research/ITCZ/research.itcz).

the two equatorial cells, can only be met over the oceans but not the continents.

There are numerous theories and numerical modeling studies for the explanation of the ITCZ. Charney (1971) considered that conditional instability of the second kind (CISK) and stated that the position of the ITCZ depends on Ekman pumping efficiency and moisture availability. Holton et al. (1971) stated that the easterly waves were associated with the convergence at off-equatorial location of the ITCZ. Lindzen (1974) employed a wave-CISK argument to explain a convective maximum off the equator.

Schneider and Lindzen (1977) discovered that zonally symmetric convective heating was crucial in determining the location of the ITCZ. Tomas et al. (1999) stated that a zonally symmetric cross-equatorial pressure gradient would result in a convergent flow at the ITCZ latitudes. Some modeling studies reveal that the ITCZ follows the SST maximum. Goswami et al. (1984) discovered a steady ITCZ occurred over the SST maximum. Hess et al. (1993) and Numaguti (1993) found that the ITCZ location following SST maximum depended heavily on the cumulus parameterization used.

Attempts to address the ITCZ position have been conducted using coupled ocean–atmosphere models. Pike (1971) found that the ITCZ was displaced because of equatorial oceanic upwelling. Manabe et al. (1974) explained that maximum precipitation tending to occur off the equator was a consequence of oceanic upwelling at the equator. Xie and Philander (1994) argued that the processes such as mixing and evaporation were as important as dynamic oceanic upwelling.

There are many numerical modeling studies examining the association of the ITCZ with atmospheric internal dynamics (Waliser and Somerville, 1994), energy transport by transient waves (Kirtman and Schneider, 2000), moist static energy budget (Neelin et al., 1987), and radiative convective instability (Raymond, 2000).

Despite the many ongoing numerical modeling studies, no single theory can explain the formation and variations of the ITCZ as various competing mechanisms are involved.

Yuk Yee Yan

Bibliography

Asnani, G.C., 1968. The equatorial cell in the general circulation. *Journal of Atmospheric Sciences*, **25**: 133–134.

Chang, J.H., 1972. *Atmospheric Circulation Systems and Climates*. Honolulu, Hawaii: Oriental Publishing Co.

Charney, J.G., 1971. Tropical cyclogenesis and the formation of the Intertropical Convergence Zone. In Reid, W.H., ed., *Mathematical Problems of Geophysical Fluid Dynamics. Lectures in Applied Mathematics*, Vol. 13. American Mathematics Society, pp. 355–368.

Fletcher, R.D., 1945. The general circulation of the tropical and equatorial atmosphere. *Journal of Meteorology*, **2**: 167–174.

Goswami, B.N., Shukla, J., Schneider, E.K., and Sud, Y.C., 1984. Study of the dynamics of the Intertropical Convergence Zone with a symmetric version of the GLAS climate model.

Hastenrath, S.L., 1968. On mean meridional circulations in the tropics. *Journal of Atmospheric Sciences*, **25**: 979–983.

Henderson-Sellers, A., and Robinson, P.J., 1991. *Contemporary Climatology*. NY: John Wiley.

Hess, P.G., Battisti, D.S., and Rasch, P.J., 1993. Maintenance of the Intertropical Convergence Zones and the large scale tropical circulation on a water covered earth. *Journal of Atmospheric Sciences*, **50**: 691–713.

Holton, J.R., Wallace, J.M., and Young, J.A., 1971. On the boundary layer dynamics and the ITCZ. *Journal of Atmospheric Sciences*, **28**: 275–280.

Kirtman, B.P., and Schneider, E.K., 2000. Spontaneously generated tropical atmospheric general circulation. *Journal of Atmospheric Sciences*, **57**: 2080–2093.

Lindzen, R.S., 1974. Wave-CISK in the tropics. *Journal of Atmospheric Sciences*, **31**: 156–179.

Liu, W.T., and Xie, X., 2002. Double intertropical convergence zones – a new look using scatterometer. *Geophysical Research Letters*, **29**(22): 291–294.

Manabe, S.D., Hahn, D.G., and Holloway, J.L., Jr, 1974. The seasonal variation of the tropical circulation as simulated by a global model of the atmosphere. *Journal of Atmospheric Sciences*, **31**: 43–48.

Neelin, J.D., Held, I.M., and Cook, K.H., 1987. Evaporation-wind feedback and low-frequency in the tropical atmosphere. *Journal of Atmospheric Sciences*, **44**: 2341–2348.

Numaguti, A., 1993. Dynamic and energy balance of the Hadley circulation and the tropical precipitation zones: significance of the distribution of evaporation. *Journal of Atmospheric Sciences*, **50**: 1874–1887.

Pike, A.C., 1971. Intertropical convergence zone studied with an interacting atmospheric and ocean model. *Monthly Weather Review*, **99**: 469–477.

Raymond, D.J., 2000. The Hadley circulation as a radiative–convective instability. *Journal of Atmospheric Sciences*, **57**: 1286–1297.

Schneider, E.K., and Lindzen, R.S., 1977. Axially symmetric steady-state models of the basic state for instability and climate studies. Part I: Linearized calculations. *Journal of Atmospheric Sciences*, **34**: 263–279.

Tomas, R.A., Holton, J.R., and Webster, P.J., 1999. The influence of cross-equatorial pressure gradients on the location of near-equatorial convection. *Quarterly Journal of the Royal Meteorological Society*, **125**. 1107–1127.

Waliser, D.E., and Somerville, R.C.J., 1994. Preferred latitudes of the intertropical convergence zone. *Journal of Atmospheric Sciences*, **51**: 1619–1639.

Xie, S.P., 1998. What keeps the ITCZ north of the equator? An interim review. *UCLA Tropical Meteorological Newsletter*, No. 22, 25 May. (http://iprc.soest.hawaii.edu/~xie/ITCZ.html)

Xie, S.P., and Philander, S.G.H., 1994. A coupled ocean-atmosphere model of relevance to the ITCZ in the eastern Pacific. *Tellus*, **46A**: 340–350.

Zhang, C., 2001. Double ITCZs. *Journal of Geophysical Research* **106**(D11): 11785–11792; http://orca.rsmas.miami.edu/~czhang/Research/ITCZ/research.itcz

Cross-references

Atmospheric Circulation, Global
Doldrums
Monsoons and Monsoon Climate
Trade Winds and the Trade Wind Inversion
Tropical and Equatorial Climates

INVERSION

An inversion occurs when there is an anomaly in the normal positive lapse rate. Such a condition may occur for the vertical

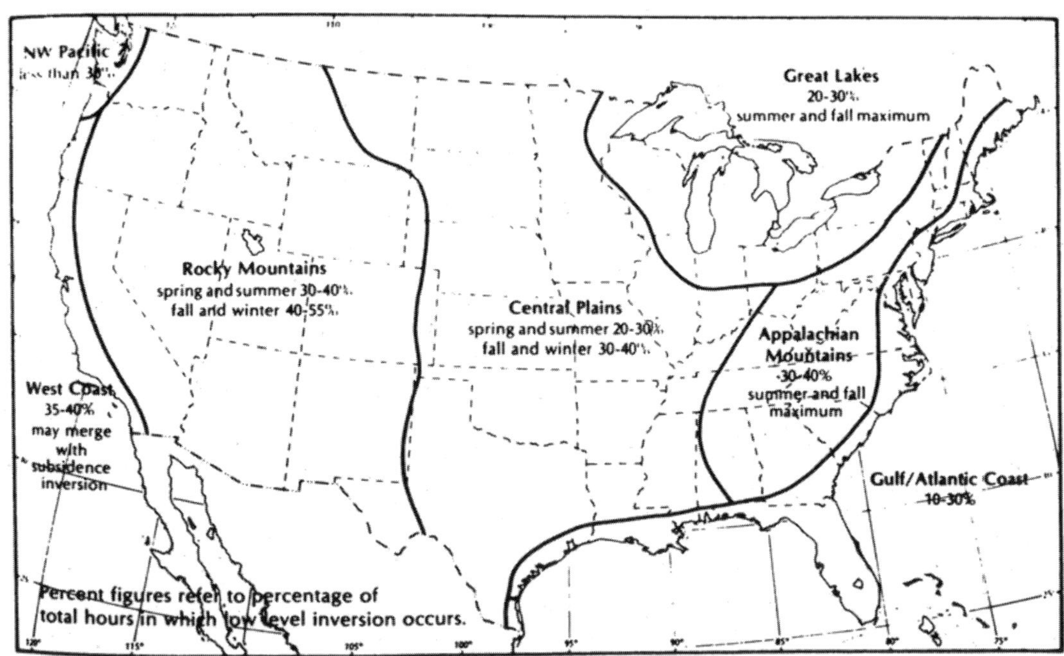

Figure 14 Frequency of low-level inversions over the United States.

variation of any atmospheric property, but the unqualified term usually implies a temperature inversion, i.e. a situation in which temperature increases rather than decreases with height. Temperature inversions can be found at or above the surface.

Surface inversions generally occur in highly stable air with minimum turbulence to allow the lower levels of air to be cooled. Such a situation often occurs during early-morning hours and gives rise mist and fog.

Inversions above the surface level occur as part of large-scale circulation systems in which subsiding air causes temperatures aloft to be higher than those closer to the surface. Such subsidence inversions are caused by the adiabatic heating of descending air that is often associated with anticyclonic circulation. Upper-level inversions can also result when cold anticyclones pass over warm surfaces. A winter flow of cP air over the Great Lakes is one case in point; if vertical mixing of air below the inversion level takes place, inversion stratus may form at the base of the inversion layer to give light snowfall.

Given the highly stable conditions under which inversions form, they are associated with poor ventilation conditions and a resulting propensity for adverse air pollution levels. Because of this, the spatial distribution of inversion frequency is of particular concern. Figure I4 provides a generalized map of the temporal and spatial distribution of low-level inversions in the United States. Note the high incidence of such inversions in the mountainous states.

Subsidence inversions are of particular importance in determining the climate of a number of locations. The aridity of coastal areas on the eastern side of the subtropical high-pressure cells (e.g. California in summer) is partly a response to stable maritime air, mTs. Similarly, the Trade Wind Inversion plays an important role in the climatology of locations in the trade wind belt.

John E. Oliver

Bibliography

Curry, J.A., and Webster, P.J., 1999. *Thermodynamics of Atmospheres and Oceans*. San Diego, CA: Academic Press.

Hartmann, D.L., 1994. *Global Physical Climatology*. San Diego, CA: Academic Press.

Thompson, R.D., 1998. *Atmospheric Processes and Systems*. London: Routledge.

Cross-references

Adiabatic Phenomena
Airmass Climatology
Air Pollution Climatology
Fog and Mist
Lapse Rate
Trade Winds and Trade Wind Inversion

"ISO" TERMS

"Iso" is the Greek prefix meaning equal; therefore, "iso" terms are usually applied to some sort of contoured map or diagram showing equal distributions. Stamp (1963) compiled a listing of the many "iso" terms in use, and the following lists those used in climatic studies:

Isallobar: lines of equal pressure tendency, marking similar changes within a given time.
Isanomaly: lines or contours of equal anomalies, or departures from normal.
Isarithm: lines of constant value (of any sort); similar to isopleth (*q.v.*).
Isamplitude: lines of equal amplitude of variation of any feature or phenomenon.
Isobar: lines of equal barometric pressure.
Isobase: lines of equal land uplift.
Isobath: lines of equal depth (generally below sea or lake level).
Isobathytherm: lines of equal temperature at given ocean depths.
Isocheim: lines on a map through places of equal mean winter temperature.
Isochrone: lines on a map joining points that can be reached in equal time from a given origin. Isochronous also refers to anything related by equal time.
Isocryme: lines on a map connecting places of equal lowest mean temperature for specified periods, e.g. for the coldest month.
Isodynamic line: indicating points of equal intensity of terrestrial magnetism.
Isogeotherm: a surface within the Earth connecting points of equal temperature.
Isogram: lines of equal values of any sort (similar to isarithm and isopleth; not recommended, as of mixed derivation).
Isohaline: lines of equal salinity, in the ocean or lakes.
Isohel: lines of equal sunshine.
Isohydrics: lines or surfaces (within a water body) of equal pH or hydrogen-ion concentration.
Isohyet: lines of equal rainfall.
Isyhyomene: lines of equal wet months.
Isohypse: lines of contour, generally of topography; also used for level of ground water.
Isoikete: lines of equal habitability.
Isokeraunic: lines of equal thunderstorm incidence.
Isokrymene: lines of equal minimum temperature in the ocean.
Isoline: lines of equal anything (not recommended, as a mixed derivation; better, see isopleth).
Isomer: lines of equal average monthly rainfall, expressed as percentage of annual average.
Isometric: lines of equal value, specifically applied to a method of projection, often used in block diagrams, the plane of the projection being equally inclined to the three principal axes of the object. Isometric is also a system in crystallography.
Isoneph: lines indicating equal degree of cloudiness.
Isonif: lines indicating equal snowfall.
Isophene/Isophane: lines of equal seasonal phenomena, e.g. in botany, flowering season.
Isophenomenal: lines on a map connecting places of equal phenomena of any sort.
Isophotic: lines or planes of equal light emission or penetration, e.g. in the ocean or lakes.
Isophyte: lines of equal height of vegetation.
Isophytochrone: lines of equally long growing season.
Isopleth: lines of equal values, a general descriptive term.
Isopone: lines of equal annual change in magnetic variation.
Isopotential: lines, levels, or planes to which artesian water may rise.
Isoryme: lines of equal frost incidence.

Isostasy: equilibrium within the Earth's crust and mantle obtained by vertical adjustment to mass and density.
Isotade: lines of equal significant dates.
Isotalantose: lines connecting places with equal range between the mean temperatures of the hottest and coldest months.
Isoterp: lines of equal comfort (bioclimatology).
Isothere: lines of equal mean summer temperature.
Isotherm: lines of equal temperature in air or sea.
Isothermombrose: lines of equal summer rainfall.
Isotope: term of chemistry, indicating elements of similar chemical character, but of differing electron numbers.
Isotype: a type of organism common to different regions.

John E. Oliver and Rhodes W. Fairbridge

Bibliography

Monkhouse, F.J., 1965. *A Dictionary of Geography*. Chicago, IL: Aldine.
Stamp, L.D., 1963. *A Glossary of Geographic Terms*. London: Longmans.

J

JET STREAMS

The term jet stream primarily refers to concentrated, narrow atmospheric currents in the upper troposphere or stratosphere, associated with large wind shears and single or multiple velocity maxima. According to the World Meteorological Organization, the typical jet stream is "thousands of kilometers in length, hundreds of kilometers in width and some kilometers in depth". The vertical shear of wind is of the order of 5–10 m/s per kilometer and the lateral shear is of the order of 5 m/s per 100 km. An arbitrary lower limit of 30 m/s is assigned to the speed of the wind along the axis of the jet stream. Ocean currents, such as the Gulf Stream and Kuroshio current, follow the same physical laws and could be considered jet streams, although their velocities are considerably less than atmospheric examples.

Formation of jet streams

The uneven distribution of available energy over the Earth's surface, marked by surplus solar radiation absorption at low latitudes and excessive radiative loss at high latitudes, necessitates a heat transport from equator to pole. If the Earth did not rotate, this transport would occur as a simple Hadley cell (Figure J1).

In a rotating system, in addition to heat, absolute angular momentum is transported poleward with the meridional circulation generated by differential heating. In the absence of external forces, absolute angular momentum is conserved along the path of displacement, resulting in an increase in westerly velocity.

According to Figure J1, air rising over the equator has been in frictional contact with the Earth's surface, acquiring angular momentum characteristic for these latitudes. As the air travels poleward beneath the tropopause it turns – under conservation of angular momentum, from a southerly into a westerly direction. At the same time its westerly velocity increases with the distance moved meridionally. Although relatively small displacements are sufficient to generate a strong westerly jets stream, wind velocities often fall short of those derived from angular momentum conservation because of the counteracting effect of friction and pressure forces.

In a slowly rotating system the meridional circulation extends close to the pole, accompanied by a high-latitude jet stream. This position is possible because only small amounts of absolute angular momentum need be transported at a slow rate of rotation. When rotation is more rapid, horizontal wind shear and turbulence near the pole (as jet velocities drop to zero directly over the pole) force the jet stream position to shift equatorward.

Beyond a critical rate of rotation the meridional circulation is no longer capable of transporting heat and angular momentum in required amounts. The jet stream breaks down into a series of symmetric waves or eddies. Figure J2 shows a simulated

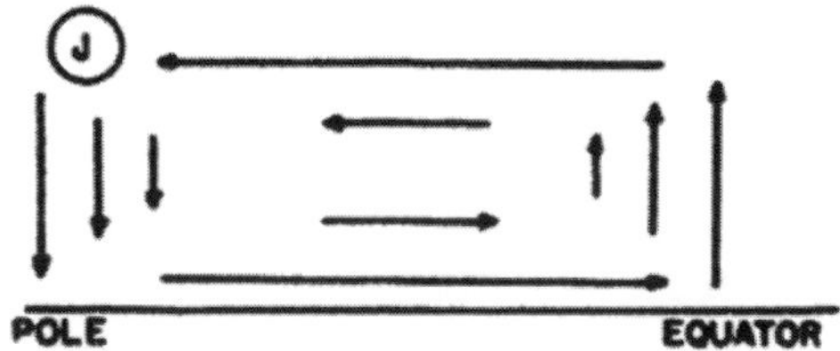

Figure J1 Hadley cell on a non-rotating Earth.

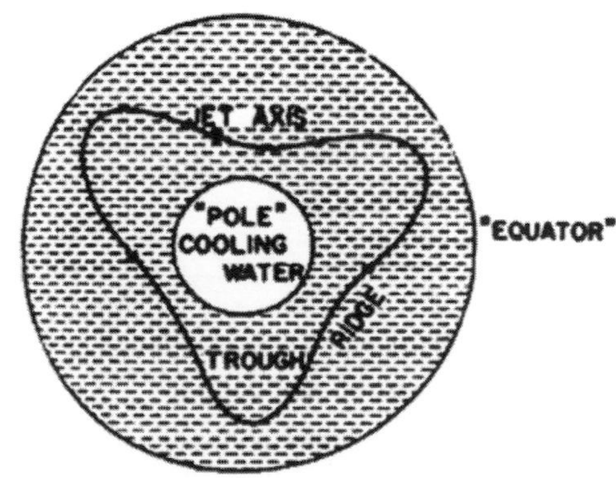

Figure J2 Three symmetric Rossby waves in a rotating dishpan.

three-wave arrangement in a rotating dishpan. Such waves were first theoretically described by Rossby. With still-increasing rotation rates the number of such waves increases. As the rotation rate continues relative to the magnitude of the surface heating imbalance within the system, irregular waves form, resembling those observed on daily hemispheric weather maps. These waves have a strong tilt of their axes (Figure J3) permitting more westerly momentum and heat to flow northward along the east side of troughs than southward on the west side of troughs.

Atmospheric conditions

Conditions in Earth's atmosphere are more complex than those considered in classical jet stream simulations (e.g. dishpan experiments). The Coriolis parameter (f) varies with latitude, rendering the deflection of airflow under conservation of absolute angular momentum greater in polar than in equatorial regions. As f increases with latitude, so does the maximum allowable anticyclonic wind shear, causing high-latitude jet streams to be more pronounced in terms of horizontal wind shear than their low-latitude counterparts.

In addition, the presence of large topographic disruptions to the depth of the troposphere, such as the Tibetan Plateau and the Rocky Mountains, cause deflections in the jet stream flow that form quasistationary planetary (Rossby) waves. The variable heating effects of land and ocean also exert a strong influence on planetary long-wave patterns. Superimposed on the long-wave patterns are migratory Rossby waves, observed in jet stream simulations under relatively high rotation rates.

Because of large momentum transports resulting from the relatively fast effective rotation rate of the Earth at high latitudes, the Hadley cell of Figure J1 is confined to latitudes equatorward of 30°. Impossibly high wind speeds would result from a Hadley cell more meridionally extensive. Rather, the circulation model shown in Figure J4 is observed, revealing multiple jet streams.

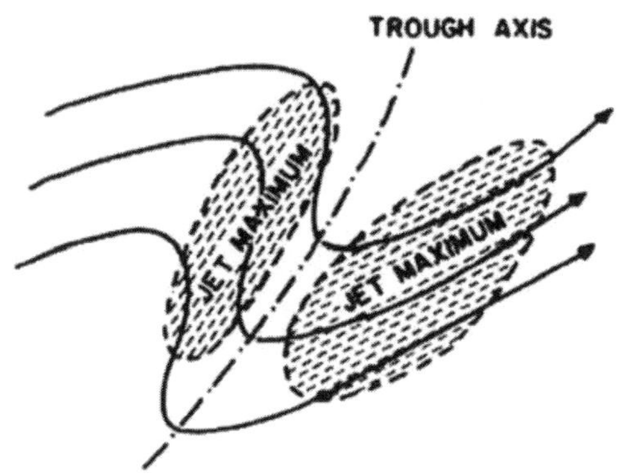

Figure J3 Streamlines and jet streams in a tilted low-pressure trough.

Jet stream systems

Although absolute angular momentum is not strictly conserved in airflow because of external pressure and frictional forces, a tendency toward such conservation leads to jet streams in regions where considerable meridional circulation occurs. Such meridional displacements typically reach extreme values at characteristic atmospheric interfaces, prescribed by the thermal structure of the atmosphere and designated "pauses" (Figure J5). Jet streams are observed at the tropopause, stratopause and mesopause.

Tropopause jet streams

The Subtropical Jet Stream (STJ) is found near 30°N during the northern hemisphere winter as a continuous corridor around the hemisphere with three waves superimposed and jet maxima center near 70°W, 40°E and 150°E (Krishnamurti, 1959; Chen et al., 1988). The maximum at 150°E shows the strongest mean winds (e.g. >150 knots during the 1955–1956 winter). The STJ draws its high speeds from the meridional circulation of the Hadley cell. Momentum transfer is accomplished via both the meridional component of the mean circulation and horizontal eddy transport associated with the three-wave pattern. During the summer the STJ is less pronounced, particularly over Asia, where it gives way to an eastlery jet stream during the monsoon (see Riehl, 1962).

The core of highest winds in the STJ is located near the 200 mb level (about 12 km); the tropopause is disrupted on its poleward side. A zone of enhanced baroclinicity (horizontal temperature gradient on a surface of constant pressure) occurs beneath the STJ but does not extend below the 400 mb level. Below this shallow baroclinic zone the atmosphere is barotropic; wind shears and horizontal temperature gradients are small and naturally dissipated by the subsidence and surface divergence associated with the subtropical high-pressure zones.

The Polar Front Jet Stream (PFJ) is located above the polar front of the middle latitudes near the 300 mb level (about 10 km). The PFJ interacts strongly with developing extratropical cyclones, commonly inducing frontogenesis and precipitation via low-level convergence and upslope flow along the frontal zone. In addition to their occurrence above the baroclinic environment of the polar front, jet streams are also associated with continental and maritime Arctic fronts.

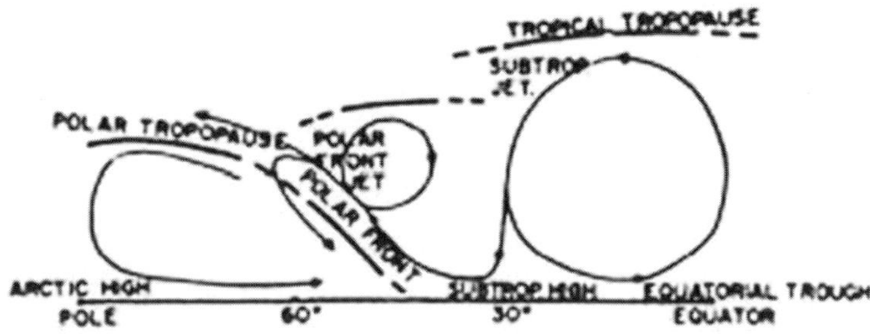

Figure J4 Schematic diagram of mean meridional circulation during winter, northern hemisphere (after Palmén, see Reiter, 1963).

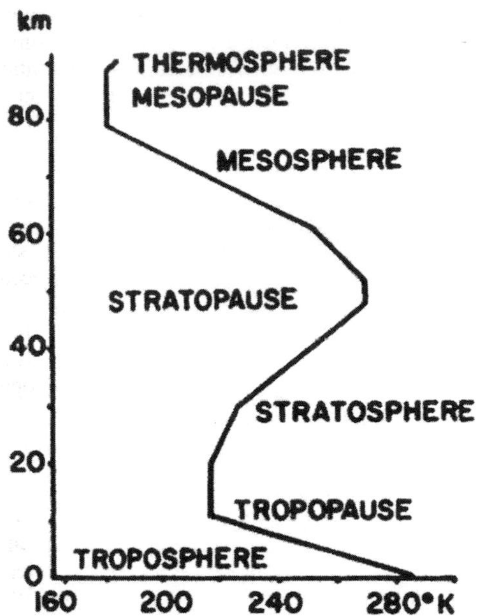

Figure J5 Thermal structure and nomenclature of the standard atmosphere.

Although diversion of air poleward along the polar front supports conservation of absolute angular momentum as a mechanism for jet stream formation, atmospheric conditions near the PFJ are more complex than those of the STJ. Individual jet maxima develop along the bases of deep low-pressure troughs, rather than at the crests of high-pressure ridges, as suggested by angular momentum theory and exhibited by the STJ. In the vicinity of jet maxima, pressure forces and potential energy released through sinking (rising) cold (warm) air exert considerable influence on the formation and maintenance of the PFJ.

Stratopause and mesopause jet streams

The stratopause located near the 50-km level harbors a strong westerly jet stream during winter, with winds in excess of 150 knots near 70° (Lee and Godson, 1957). A two-wave pattern dominates, with mean troughs centered over Siberia and Hudson Bay, suggesting the strong orographic influence of the Himalayas and Rocky Mountains, even at high latitudes. Large-scale horizontal and vertical transport processes induce anomalous seasonal ozone concentrations, which are not accounted for by the radiation budget of these latitudes. During the northern hemisphere warm season, continuous insolation over the pole reverses the hemispheric temperature and pressure gradients and the jet stream gives ways to moderate easterly winds. Stratospheric flow patterns in the southern hemisphere show similar seasonal trends, although the summer easterlies are weaker than in the northern hemisphere.

The breakdown of the winter circulation may be rather abrupt. Whereas in the southern hemisphere it falls close to the period of the equinoxes, in the northern hemisphere large variations occur from one year to another. As the orientation and intensity of the stratospheric jet changes, stratospheric temperatures below the jet may experience "explosive warming" of more than 20°C within a few days. The transition to the warm season solar radiation regime is marked by a shift from a prominent standing wave pattern in winter to summer easterly winds positioned symmetrically about the pole. According to Charney, planetary waves are trapped as they encounter an atmospheric layer in which the wind speed is equal to their phase velocity. In winter, with westerly winds (westerly velocity $u > 0$) in the troposphere and stratosphere, standing waves (phase velocity $= 0$) propagate freely into the stratospheric jet stream. During summer, standing waves will be trapped as they encounter easterly winds ($u < 0$) and are thus prevented from influencing the stratosphere.

Jet streams of the mesopause are still poorly understood. Because they occur in the ionosphere, and consist of drifts of charged-particle clouds, they are sometimes referred to as electrojets. Electromagnetic forces must be taken into account when considering air motions at these altitudes (>80 km).

Low-level jet streams

Low-level Jet Streams (LLJ) tend to form along *pauses* in the lower troposphere, i.e. along inversions on top of the planetary boundary layer. The precise depth of this layer is dependent upon convective mixing and nocturnal cooling and stabilization of air near the ground. Bonner and Paegle (1970) determined that the LLJ is influenced by differential heating along an inclined surface (e.g. the Great Plains between the Mississippi River and Rocky Mountains). The diurnal heating and cooling of the planetary boundary layer induce changes in the horizontal pressure field, to which the wind adjusts with a time lag dictated by the period of inertial oscillation. The inertial oscillation is contingent upon the Coriolis parameter, which varies by latitude. At 30° latitude the 24-h period of inertial oscillation matches the period of the diurnal temperature flux and it is here that LLJ develop most frequently.

LLJ phenomena are also associated with synoptic-scale forcing in extratropical low-pressure systems. In particular, LLJ have been observed imbedded within the southerly flow downstream from transient shortwave axes in northern hemisphere systems. Whether motivated by the diurnal heating and cooling cycle or the circulation with a midlatitude cyclone, LLJ can carry abundant moisture from adjacent seas (Gulf of Mexico, South China Sea) and thus contribute to severe weather.

Effects of jet streams

On a surface of constant pressure the vertical component of relative vorticity

$$\zeta = V/r - \delta V/\delta n$$

(where V is the velocity of the air current, r is the radius of curvature, n is the coordinate normal to the flow of direction) varies strongly to either side of the jet axis. These strong shears, coupled with the effects of curvature within the airflow, generate positive vorticity poleward and negative vorticity equatorward of a westerly jet maximum. Poleward of the jet axis, air flowing through this vorticity pattern will undergo convergence as it enters the jet maximum, then divergence as it passes downstream of the jet center. Equatorward of the jet axis, air diverges as it enters the wind maximum, then converges downstream of the center. Figure J6 shows schematically the divergence field around a jet maximum, superimposed on straight flow. Relative vorticity adjustments from curvature have been shown to modify the divergence field as a jet maximum migrates through a geopotential trough.

Since divergence at tropopause level will result in a mass deficit in the vertical air column, surface pressure falls should result beneath the regions of divergence. If a frontal zone is present at low levels in the zone of pressure falls, cyclogenesis may occur, as shown by Pettersen, 1956; Newton, 1954, and others). Thus, stormy weather patterns, particularly cloudiness, precipitation and cyclones, are intimately linked to jet streams at tropopause level. These weather patterns in turn modify the kinematics of the jet streak environment (see Wolf and Johnson, 1995).

The divergence distribution at jet-stream level also forces significant vertical mass displacements that induce mixing of tropospheric and stratospheric air. These processes are illustrated schematically in Figure J7 (Shapiro et al., 1980). Stratospheric air, characterized by high ozone concentrations and large potential vorticity values (the product of thermal stability and absolute vorticity) subsides within the stable layer beneath the jet maximum, the so-called jet-stream front. Deep and sustained intrusions of stratospheric air to the middle troposphere and below are associated with explosive cyclogenesis.

Conversely, tropospheric air intrudes into the stratosphere in a large wedge between the stability tropopause and the potential vorticity tropopause. In contrast to the stratospheric intrusion, the tropospheric air is recognizable from low ozone concentrations and smaller potential vorticity values. Low-pressure troughs are thus characterized by high total ozone content, due to the sinking motion in the upper troposphere and lower stratosphere. Low-latitude tropospheric air is transferred poleward in high-pressure ridges, reducing the ozone values there.

Jet stream winds also have an important effect on airline operations and the planning of their economy. Flight times may be improved in transcontinental and transoceanic flights by riding the jet stream, while headwind flights are diverted around the jet stream path when possible, as required by Air Traffic Control regulations. Strong shears in the jet stream region may cause clear air turbulence that may, under serious conditions, lead to material fatigue and even aircraft failure.

Elmar Reiter and Greg Bierly

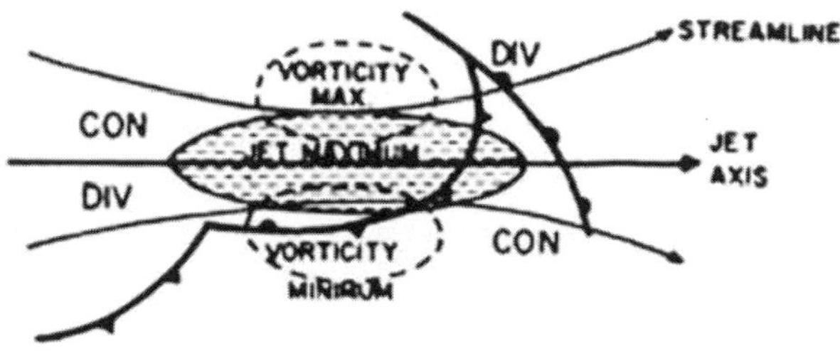

Figure J6 Schematic position of convergence and divergence at jet stream level and fronts at the Earth's surface.

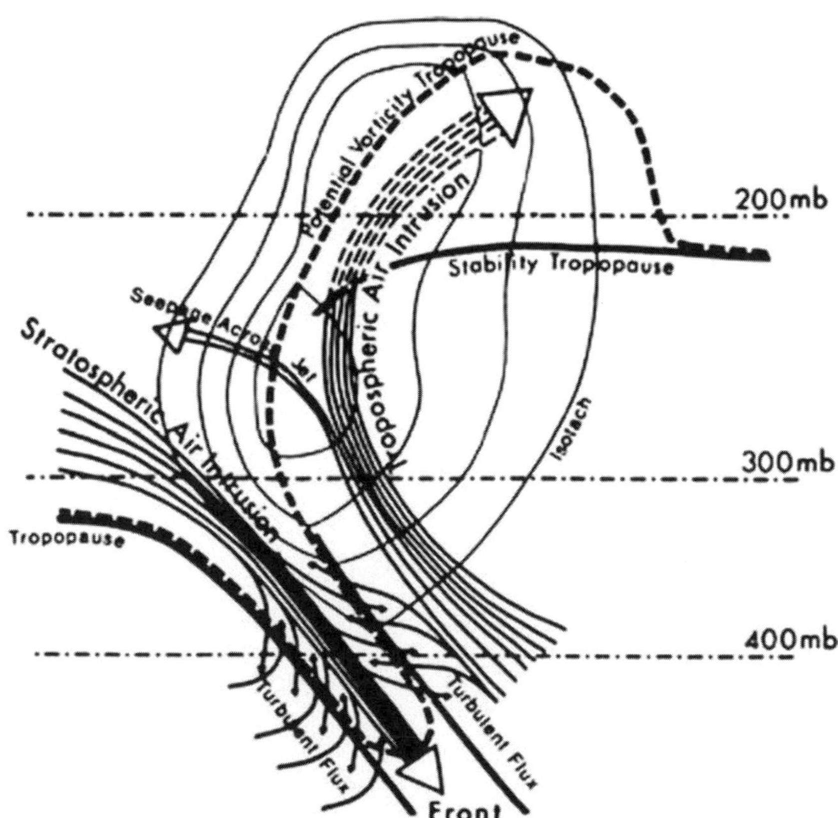

Figure J7 Schematic diagram of stratospheric–tropospheric exchange in the vicinity of a cyclogenetically active jet maximum (after Shapiro et al., 1980).

Bibliography

Bonner, W.D., 1968. Climatology of the low level jet. *Monthly Weather Review*, **96**: 833–850.

Bonner, W.D., and Paegle, J., 1970. Diurnal variations in boundary layer winds over the south-central United States in summer. *Monthly Weather Review*, **98**: 735–744.

Browning, K.A., and Pardoe, C.W., 1973. Structure of low-level jet streams ahead of mid-latitude fronts. *Quarterly Journal of the Royal Meteorological Society*, **99**: 619–638.

Charney, J.G., 1947. The dynamics of long waves in a baroclinic westerly current. *Journal of Meteorology*, **4**(5): 135–162.

Chen, T.-C., Teng, R.-Y., and van Loon, H., 1988. A study on the maintenance of the winter subtropical jet streams in the northern winter. *Tellus*, **40A**: 392–397.

Danielsen, E.F., 1968. Stratospheric-tropospheric exchange based on radioactivity, ozone and potential vorticity. *Journal of Atmospheric Science*, **25**: 502–518.

Keyser, D., and Shapiro, M.A., 1986. A review of the structure and dynamics of upper level frontal zones. *Monthly Weather Review*, **114**: 452–499.

Krishnamurti, T.N., 1959. *The Subtropical Jet Stream of Winter*. Research Report, Contract No. N6ori-02036, NR 082-120, University of Chicago.

Lee, R., and Godson, W.L., 1957. The arctic stratospheric jet stream during the winter of 1955–56. *Journal of Meteorology*, **14**(2): 126–135.

Murray, R., and Daniels, S.M., 1953. Transverse flow at entrance and exit to jet streams. *Quarterly Journal of the Royal Meteorological Society*, **79**: 236–241.

Newton, C.W., 1954. Frontogenesis and frontolysis as a three-dimensional process. *Journal of Meteorology*, **11**: 449–461.

Palmén, E., 1948. On the distribution of temperature and wind in the upper westerlies. *Journal of Meteorology*, **5**: 20–27.

Pettersen, S., 1956. *Weather Analysis and Forecasting*. New York: McGraw-Hill.

Reiter, E.R., 1963. *Jet Stream Meteorology*. Chicago, IL: University of Chicago Press.

Reiter, E.R., 1975. Stratospheric–tropospheric exchange processes. *Reviews of Geophysical and Space Physics*, **13**: 459–474.

Riehl, H., 1962. *Jet Streams of the Atmosphere*. Atmospheric Scientific and Technical Report No. 32. Colorado State University.

Rossby, C.G., 1940. Planetary flow patterns in the atmosphere. *Quarterly Journal of the Royal Meteorological Society*, **66**: 68–87.

Shapiro, M.A., Reiter, E.R., Cadle, R.D., and Sedlacek, W.A., 1980. Vertical mass and trace constituent transports in the vicinity of jet streams. *Archiv für Meteorologie, Geophysik und Bioklimatologie*, **B28**: 193–206.

Uccellini, L.W., 1980. On the role of upper tropospheric jet streaks and leeside cyclogenesis in the development of low-level jets in the Great Plains. *Monthly Weather Review*, **108**: 1689–1696.

Uccellini, L.W. and Johnson, D.R., 1979. The coupling of upper and lower tropospheric jet streaks and implications for the development of severe convective storms. *Monthly Weather Review*, **107**: 682–703.

Uccellini, L.W., Kocin, P.J., Petersen, R.A., Walsh, C.H., and Brill, K.F., 1984. The President's Day cyclone of 18–19 February 1979: synoptic overview and analysis of the subtropical jet streak influencing the precyclogenetic period. *Monthly Weather Review*, **112**: 31–55.

Uccellini, L.W., Keyser, D., Brill, K.F., and Walsh, C.H., 1985. The President's Day cyclone of 18–19 February 1979: influence of upstream trough amplification and associate tropopause folding on rapid cyclogenesis. *Monthly Weather Review*, **113**: 962–988.

Wolf, B.J., and Johnson, D.R., 1995. The mesoscale forcing of a midlatitude upper-tropospheric jet streak by a simulated convective system. Part I: Mass circulation and ageostrophic processes. *Monthly Weather Review*, **123**: 1059–1087.

Cross-references

Atmospheric Circulation, Global
Dynamic Climatology
Extratropical Cyclones
Hadley Cell
Rossby Wave/Rossby Number

K

KATABATIC (GRAVITY) WINDS

Katabatic winds occur when air is cooled from below over sloping terrain. Such cooling causes a shallow blanket of air adjacent to the surface to become colder and therefore heavier than the atmosphere above, thus forming a thermally distinct layer that exchanges little energy with the overlying air. The slope of the ground imparts a component of gravitational acceleration to the surface layer, thus allowing the air to drain downslope.

Radiative cooling of the air plays a dominant role in most katabatic wind systems, particularly those that occur in mountain valleys where katabatic flow forms one component of a local wind system referred to as the mountain and valley breeze (Defant, 1951). In such locations the katabatic wind, or mountain wind, is a nighttime event, spurred on by cooling in the higher elevations. The situation is reversed during the day when radiative heating at the same elevation causes air next to the ground to move upslope in a flow known as the anabatic wind. Either case is accompanied by a return flow in the air above the surface layer.

Katabatic winds are more persistent where the ground is covered by ice, provided the air remains warmer than the ground. Such conditions are found in warm air advected over melting valley glaciers in summer where a katabatic flow, sometimes referred to as the glacier wind, occurs over the ice throughout the day. However, far more favorable conditions are found over major ice caps, where the air may be warmed through large-scale subsidence. Radiative cooling in the lowest layers of such air gives rise to the major katabatic systems that dominate the wind regime of Antarctica (Ball, 1957).

Various forces govern the katabatic wind, the importance of each depending upon local conditions and the scale of the flow (Mahrt, 1982). The most important is the force generated by gravity acting on cold air at the ground, a force that drives the wind with increasing strength as the slope of the surface increases. It may be substantially modified by forces arising from other accelerations that occur over time and space (Kottmeier, 1986). Local pressure gradient forces arising from differential surface heating or cooling may also be at work, either reinforcing or weakening the katabatic wind. As time and space scales become large, the Coriolis force assumes growing importance, deflecting the wind so that it flows at an angle to the slope azimuth.

Scale determines the thickness of the katabatic wind layer, and the speed at which the air moves. The relatively small scale of valley wind systems, where cooling might occur over distances of 5 km or less, tends to restrict katabatic layers to thicknesses that rarely exceed 10 m, and wind speeds on the order of 1 to 5 $m\,s^{-1}$. Antarctic katabatic winds, which can be generated over distances of hundreds of kilometers, may extend through the first 0.5 km of the atmosphere and achieve speeds in excess of 25 $m\,s^{-1}$. They are especially well developed near coastal Antarctic locations such as Cape Dension where, at mean monthly speeds approaching 90 $km\,h^{-1}$, they rank among the world's strongest winds.

D.S. Munro

Bibliography

Ball, F.K., 1957. The katabatic winds of Adélie Land and King George V Land. *Tellus*, **9**: 201–228.

Defant, F., 1951. Local winds. In Malone, T.F., ed., *Compendium of Meteorology*. Boston, MA: American Meteorological Society, pp. 655–672.

Kottmeier, C., 1986. Shallow gravity flows over the Ekström ice shelf. *Boundary Layer Meteorology*, **35**: 1–20.

Mahrt, L., 1982. Momentum balance of gravity flows. *Journal of the Atmospheric Society*, **39**(12): 2701–2711.

Cross-references

Adiabatic Phenomena
Antarctic Climates
Local Winds
Orographic Precipitation
Rain Shadow

KEPLER'S LAWS

Johann Kepler (1571–1630) was one of the founders of "classical" astronomy. By painstaking calculations from the observations of Tycho Brahe he was able to obtain empirically the three fundamental laws of celestial mechanics, two in 1609 and a third in 1619. At first it was assumed with Copernicus that planets were in circular motion about the Sun, in short the Heliocentric Theory, which rejected the Ptolemaic geocentric concept of the solar system. Kepler tried to compute circular orbits from Tycho Brahe's data, but eventually recognized that planetary orbits were elliptical, and this opened the way to his famous laws.

The first law

The orbit of each planet is an ellipse with the Sun as one of its foci. The longest axis of an ellipse is called its major axis (AB) and half this length, defining the center point (C), is the semimajor axis. The ratio of the distance from C to the sun (S) over the distance SA is called the eccentricity. The ellipse may then be specified by these two distances. Geometrically the orbit may be plotted mechanically by taking a piece of string twice the length of SA and fixing the ends at point S and S' the second focus, an equal distance from B as S is from A; a pencil placed in the loop of string, maintaining it taut, will then move in the ellipse as defined. The distance from the planet to the sun, or radius vector, thus varies constantly. The complete planetary circuit in such an ellipse, or orbit, is its revolution, the time occupied being the revolutionary or sidereal period. The planet when nearest to the Sun is said to be in perihelion; when farthest away aphelion. If the eccentricity were zero we would have a circle, but according to probability reasoning the chance of evolving such an orbit is zero.

The second law – law of areas

For a given period of time as the planet moves in its orbit, the radius vector sweeps out an equal geometric area. Plots of equal area along such an ellipse evidently intersect the orbit in arcs of varying length, longest near perihelion and shortest near aphelion. It follows that the orbital (and angular) velocity of the planet is highest when closest to the sun and lowest when farthest away. The mean angular velocity or mean motion through 360° is then easily determined by dividing the revolution period by 360, and is usually expressed in degrees, minutes, and seconds of arc per day. Since the angular momentum of a planet (J) relative to the sun is given by its mass (m) times its velocity (v) times the distance (y) at any moment, i.e. $J = mvy$, it can be shown that the angular momentum remains constant or conserved.

The third law – the harmonic law

The cube of a planet's semimajor axis (its mean distance from the sun) divided by the square of the sidereal period is the same for every planet. The Earth's mean distance from the sun is known as the astronomical unit (AU). In Kepler's day the actual distance was not well established, so by taking the AU as unity and the unit of time as 1 year, the quotient of the third law formula for the Earth was likewise 1; thus for any other planet the square of its sidereal period (expressed in years) provides the cube of its mean distance from the sun. In this way the periods, easy to determine, gave the clue to the mean orbital characteristics of the planets even though the precise scale was not known.

Rhodes W. Fairbridge

Bibliography

Motz, L., and Duveen, A., 1966. *The Essentials of Astronomy*. Belmont, CA: Wadsworth.

Poincare, J.H., 1905–1910. *Lecons de mecanique celeste*. Paris.

Wintner, A., 1941. *Analytical Foundations of Celestial Mechanics*. Princeton, NJ: Princeton University Press.

KÖPPEN, WLADIMIR PETER

As noted by C. W. Thornthwaite, the most important name in the history of climatology, and to many the father of modern climatology, is Wladimir Peter Köppen (Thornthwaite, 1943). Köppen published his first significant paper in 1868 and was researching, writing and publishing at the time of his death. He was born on 25 September 1846, in St Petersburg, the capital of the Imperial Russian Empire, and died in Graz, Austria, on 22 June 1940. Raised in an intellectual academic environment of German, French, and Russian scholars, Köppen developed a deep sense of environmental perception, a wonder for the varied vegetal zones of Imperial Russia, and an appreciation of the effects that climate has upon life on Earth. His grandfather was a German medical doctor invited to Imperial Russia by Empress Catherine the Great. He became a personal physician to the royal family. His father, Peter von Köppen, was a geographer who served on the faculty at the Imperial Academy of Science. Appointed to the highest academic rank in Imperial Russia by Tsar Alexander II, Peter was granted a seaside estate on the southern coast of the Crimea. As a child, Köppen explored the vegetal zones on the low mountain ranges along the Black Sea coast. Variations in vegetation, soil, and land use stimulated his interest in the geography of plants and the relationship of plants to climate.

Köppen began his formal academic studies when he was accepted as a student at the University of St Petersburg in 1864. His undergraduate major was in botany. Between academic terms he traveled from St Petersburg to the family estate in the Crimea. The vegetal changes he noted on the trip south and then living in the beautiful subtropical littoral of the Crimea broadened his biogeographic perspectives. Köppen transferred to the University of Heidelberg in 1867. He received a PhD from Heidelberg in 1870. His doctoral dissertation was focused upon the relationship between plant growth and temperature. Köppen returned to St Petersburg in 1871 to a position at the Central Physical Laboratory. In 1874 he returned to Hamburg, Germany, as head of a newly established division of weather telegraphy, storm warning systems, and marine meteorology at the German Naval Observatory. Köppen became the meteorologist of the observatory in 1879, and in 1884 published a world map of temperature belts according to the number of months above or below a certain mean temperature. At the time this was considered a very significant publication. He delimited six principal temperature belts – tropical, subtropical, summer-hot-temperate, winter-cold-temperate, cold, and polar (*Briefnachlass*, 1932).

Köppen's map stimulated other physical science researchers. Oscar Drude adopted the six principal belts and created a world map of floral kingdoms. A.F.W. Schimper expanded Drude's seminal work on plant geography and published his work in 1898. Schimper's classification of vegetal zones was similar to Drude's in that the vegetal zones or regions graded from humid to arid. Two years after Schimper published his study, Köppen published his first of many studies on climatic classification; it was a mathematical system of climatic regional differentiation (Köppen, 1900). He contended that plants serve as natural meteorological instruments integrating various climatic elements. Köppen did not see Schimper's publication, and envisioned De Candolle's plant regions of the world to be climatic regions. He focused his research upon determining numerical values for the boundaries of plant regions while retaining De Candolle's use of the letters A, B, C, D, E to represent climatic regions. Realizing the problems associated with his 1900 map, Köppen published a revised climatic map of the world in 1918. He had difficulties in finding numerical means to denote effective precipitation, particularly in delimiting desert and steppe boundaries (Leighly, 1926). Recognizing the lack of worldwide temperature and precipitation data, and realizing the use of mean values restricted precision, Köppen presented at minimum four separate formulas or means to define climatic regions in 1918, 1923, and 1928. In a series of publications on climatic classification, he created the classification scheme shown in Table K1.

Köppen retired from his position at the German Naval Observatory in 1924. He moved to Graz, Austria, a beautiful university town 87 miles southwest of Vienna. Working with Rudolph Geiger he began research on a five-volume *Handbook of Climatology*. It was nearly completed when he died (vol. 1, Part C, was published in 1936). The basic Köppen classification system divided the world into five main zones. These zones were subdivided according to temperature and precipitation, seasonal variations in precipitation, and the effect of temperature and precipitation on vegetation. He ignored other climatic controls that also affect climate such as planetary winds and pressure cells, frontal systems, and ocean currents (Van Royen, 1927). Still Köppen's basic map of world climates and his later revisions have been recognized as a great scientific contribution by most physical scientists, and its simplicity has contributed to its acceptance and longevity. Many problems in defining the climatic regions, and in the formulas or indexes, existed in Köppen's works. Yet his system has been modified, its many problems rectified, and it does give a reasonably acceptable representation of the world's climates.

Wladimir Peter Köppen published more than 500 items between 1868 and 1939. Fluent in Russian and German, he read much, kept well informed on scientific topics, and was receptive to new ideas and new concepts. Reflecting the background from which he came in Imperial Russia, and from his family, Köppen devoted much time and energy to improving the nutrition of the underprivileged, chronic alcoholism, school education reform, and conservation. He was a founder of the Eimsbutteler Boys Home in Hamburg, worked to secure a peaceful Europe, and provided help to refugees. A modest, small, dignified scholar, he made a very significant contribution to climatology (AWI, 2002).

Table K1 Climatic regions classification

Full humid	Semihumid	Semiarid	Arid
Af Cf Df	As/Aw Cs/Cw Ds/Dw	BS	BW

William A. Dando

Bibliography

AWI, 2002. Wladimir Köppen. Stiftung Alfred-Wegener-Institute fur Polar- und Meeresforschung, http://www.awi-bremerhaven.de/AWI/wladimir-d.html, p. 1.

Briefnachlass, 1932. Wladimir Peter Köppen (1846–1940), meterologe, klimatologe. Universittsbibliothek Graz (all).

Köppen, V., 1900. Vetsch einer Klassifikation der Klimate. *Geographische Zeitschrift*, **6**: 593–611.

Leighly, J.B., 1926. Graphic studies in climatology: graphic representation of a classification of climates. *University of California Publications in Geography*, **2**(3).

Thornthwaite, C.W., 1943. Problems in the classification of climate. *Geographical Review*, **33**(2): 1.

Van Royen, W., 1927. The climatic regions of North America. *Monthly Weather Review*, **55**: 315–319.

KRAKATOA WINDS

When the giant volcanic eruption of Krakatoa (also Krakatau) occurred in the Sunda Strait, between Java and Sumatra, Indonesia, on 27 August 1883, its ash was carried up to the lower stratosphere (18–34 km). The volcano is located at 6°9′S, 105°22′E, and at the time of the eruption there was an easterly equatorial air flow, so that the ash initially at the 9 mb (32 km) level, after some weeks settled down at about 25 km and was carried around the globe. This particulate loading affected the cloudscapes in the troposphere, particularly at sunrise and sunset, at times creating dramatic coloration which persisted in some places, e.g. London and Paris, for up to 3 years.

Information about the eruption and its effects was gathered over several years by the Royal Society of London, and published. The local geology arid environmental effects were published in Dutch by Verbeek. In the century that followed, smaller eruptions have occurred from time to time, and a new island, Anak Krakatau ("Child of Krakatoa") has begun to form. A semi-popular book-length review of the entire history was published in the year 2003 by a geologically trained journalist and in 2004 it was presented on television by the BBC in its science series.

The name "Krakatoa Winds" was first used by Van Bemmelen (this and other references in Fairbridge, 1967). From the meteorological point of view the Krakatoa Winds are most interesting, because the year after the eruption, 1884, the ash clouds were seen to reverse, to create a westerly stratospheric air flow, sometimes called the "Von Berson Westerlies" (Palmer, 1958). These westerlies are actually a lower-level air flow, which rises and falls in a cyclical manner (Bjerknes, 1969; Reed et al., 1961; Wexler, 1951). Furthermore, this biennial reversal is not quite regular and has sometimes become triennial (Schove, 1983). It has subsequently been correlated with the QBO or Quasibiennial Oscillation. This QBO periodicity has been found to have a long-term average of 2.172 yr, which appears in the spectra of sunspots (Currie, 1973) as well as in many terrestrial time series (Schove, 1983). Evidently these stratospheric winds are externally forced, but their physics still leave much to be explained. Sir Gilbert Walker (1924) proposed they belonged to equatorial upper-level cells of what is now called the "Walker Circulation" (Piexoto and Oort, 1995, p. 419).

The velocity of these winds is up to 25–50 m/s (i.e. 50–100 knots). The mean velocity of the initial dust cloud was 60 knots (Wexler, 1951). Subsequent to the 1883 eruption, several equatorial or low-latitude volcanic eruptions, e.g. Agung and Pinatubo, together with the nuclear bomb tests in the Marshall Islands, have added information, developed further by instrumental testing and satellite studies. An analogous and even greater eruption was that of AD 535, which has been described in a book by Keys (2001). It is believed to have also been located in the Sunda Strait, but has not yet been thoroughly investigated.

Rhodes W. Fairbridge

Bibliography

Bjerknes, J., 1969. Atmospheric teleconnections from the equatorial Pacific. *Monthly Weather Review*, **97**: 163–172.

Currie, R.G., 1973. Fine structure in the sunspot spectrum −2 to 70 years. *Astrophysics and Space Science*, **20**: 509–518.

Fairbridge, R.W., 1967. Albedo and reflectivity. In Fairbridge, R.W., ed., *The Encyclopedia of Atmospheric Sciences and Astrogeology*. New York: Reinhold, pp. 12–13.

Keys, D., 2001. *Catastrophe*. New York: Ballantine Books.

Palmer, C.E., 1958. The general circulation between 200 mb and 10 mb over the equatorial Pacific. *Weather*, **9**: 341–349.

Piexoto, J.P., and Oort, A.H., 1992. *Physics of Climate*. New York: American Institute of Physics.

Reed, R.J., et al., 1961. Evidence of a downward propagating annual wind reversal in the equatorial stratosphere. *Journal of Geophysics Research*, **66**: 813–818.

Schove, D.J. (ed.), 1983. *Sunspot Cycles* (*Benchmark Papers in Geology*, vol. 68). Stroudsburg: Hutchinson Ross.

Walker, G.T., 1924. *World Weather, I and II*. Indian Meteorological Department, memoir 24(4).

Wexler, H., 1951. Spread of the Krakatoa volcanic dust cloud as related to the high-level circulation. *Bulletin of the American Meteorological Society*, **32**: 48–51.

Cross-references

Hadley cell
Quasi-Biennial Oscillation
Volcanic Eruptions and their Impact on the Earth's Climate
Walker Circulation

KYOTO PROTOCOL

In 1992, at a meeting in Rio de Janeiro, the United Nations Framework Convention on Climate Change (UNFCCC) was adopted. Its purpose was an attempt to combat global warming by stabilizing the emissions of greenhouse gases. The Kyoto Protocol or, more properly, the Kyoto Protocol to the United Nations Framework Convention on Climate Change, is a proposed amendment that was adopted in 1997 at a Conference of Parties (COP) in Kyoto, Japan.

The Kyoto Protocol commits developed countries to reduce emissions of carbon dioxide, methane and nitrous oxides and to phase out hydrofluorocarbons, sulfur hexafluoride, and perfluorocarbons over a 30-year period. It reaffirms the idea that developed countries must supply technology to other countries in climate-related studies and related projects.

The Protocol was endorsed by 160 countries and will become binding provided that 55 countries, including developed nations responsible for most global emissions, ratify the accord. By 2002 some 104 countries had signed the agreement. Signing the agreement is essentially symbolic, and for the Protocol to become fully effective the UNFCCC requires that parties complete their "instruments of ratification, acceptance, approval or accession".

Since the Kyoto meeting other COP have attempted to resolve problems and issues, not always with success. Over time, and with appropriate political activity, full ratification may occur.

John E. Oliver

Bibliography

Grubb, M., Vrelijk, C., and Brack, D., 2002. *The Kyoto Protocol: a guide and assessment*. London: Royal Institute of International Affairs. http://unfcc.int

Houghton, J.T., Meira Filho, L.G., Griggs, D.J., and Maskell, K. (eds), 1997. *Implications of Proposed CO_2 Emissions and Limitations*, IPCC Technical Paper IV. Geneva: IPCC.

Singer, S.F., 2000. *Climate Policy – from Rio to Kyoto*. Hoover Institute, Stanford University.

Cross-references

Climate Change and Global Warming
Climate Change Impacts: Potential Environmental and Societal Consequences
Global Environmental Change: Impacts
Greenhouse Effect and Greenhouse Gases

L

LAKES, EFFECTS ON CLIMATE

Bodies of water modify the atmospheric environment in their vicinity. Small bodies of water of a few hectares or less cause small, local modifications that are generally not significant. Larger bodies of water, such as lakes, cause more significant effects on climate at scales ranging from the microscale to the synoptic scale. The effects of lakes on the atmosphere vary with areal extent, depth, and configuration of the lake, the velocity and direction of the winds, whether winter ice forms on the lake, and the regional climate of the lake. Large, deep lakes cause more pronounced modifications than small, shallow lakes. Most of this discussion will focus on the effects of large lakes on climate.

The effects of the Great Lakes of North America on climate have been studied extensively, as have the climatic effects of other large lakes of the world. Studies have also been conducted on artificial lakes, such as reservoirs. Some studies of the effects of lakes on climate have been observational with collection of field data for a few months or several years. Other studies have been computer models of the expected effects.

Lakes affect the climate through alterations of the atmospheric boundary layer due to: (1) the thermal lag of lake surface temperatures compared to the adjacent land areas, (2) the availability of open water over lakes for evaporation, and (3) alterations of winds by lakes as a result of contrasts in surface roughness between the lake and the land surfaces.

Large lakes, such as the Great Lakes of North America, contribute large amounts of latent and sensible heat to winter storm systems. Large lakes cause passing synoptic scale cyclones to accelerate toward the lakes and rates of intensification increase over the lakes during the ice-free, unstable season of September–November. Passing cyclones during the ice-covered, unstable season of December–April and during the stable season of May–July accelerate less than during the ice-free September–November season and rates of intensification do not change over the lakes. Computer simulations in general circulation models show that the Great Lakes cause an intensified temperature gradient north of the lakes in fall and winter. This intensifies the mean jet stream core and displaces it northward. Such synoptic scale forcing is unlikely from lakes smaller than the Great Lakes.

Air temperature

Effects of lakes on the air temperature climate are due primarily to the thermal lag of lakes. This is most pronounced for lakes in climates with a large seasonality in temperature. Water has a heat capacity that is about three times greater than the heat capacity of soil and, with mixing in water, lakes warm up slowly in spring and summer and cool down slowly in fall compared to land surfaces. The effects on climate are greatest for the largest lakes, such as the Great Lakes ($>25\,000\,km^2$) but are significant also for lakes and artificial reservoirs of less than $200\,km^2$.

Lakes create a modified marine climate along their shores, especially on the prevailing downwind shore. Summer temperatures are cooler over and near large lakes and winter temperatures are warmer. Extremes of heat and cold temperatures are reduced by lakes. The moderating effects of lakes on winter air temperatures are reduced, but not eliminated, if the lake surface freezes.

The start of the growing season in spring is delayed by the chilling effects of large lakes, but the first freeze in fall is also delayed near lakes. The freeze-free season is generally lengthened by several weeks near large lakes, although growing season heat availability for plants is reduced near the lake due to cooler summer temperatures. This effect on climate extends 10–30 km inland for the largest lakes. The moderated air temperatures and longer freeze-free seasons near lakes have led to the cultivation of fruit crops along the shores in climates where they may not otherwise be suited.

Moisture and precipitation

Moisture fluxes over a lake are dependent upon the vapor pressure difference between the lake water surface and the overlying air. Evaporation is at a maximum in late fall and early winter when the vapor pressure difference is greatest. Evaporation is diminished when the vapor pressure difference reaches a minimum in spring and early summer. Water vapor will condense onto a lake, the reverse of evaporation, if the water temperature is cooler than the dewpoint of the air.

Lakes are commonly colder than prevailing air temperatures in spring and summer due to the thermal lag discussed previously. A stable marine layer develops in the lower atmosphere under these conditions. This causes a suppression of convective cloud development and suppression of afternoon rain shower and thunderstorm activity over the lake and adjacent land downwind. On the other hand, during fall and winter when the lake temperature is commonly warmer than the air temperature, lakes release large amounts of sensible heat and latent heat to the atmosphere. This causes an unstable lower atmosphere and stimulates convection, condensation, clouds, and precipitation over the lakes and downwind – a process called lake-effect clouds and precipitation. The number of cloudy days in winter is 36% greater downwind of Lake Michigan, USA, than upwind of the lake, while the number of cloudy days in summer is 15% less downwind. Fogs may occur over and adjacent to lakes in spring and summer when the lake is cold with respect to the air, and in winter when steam fogs form in cold air over warmer lake waters.

In regions where winter temperatures are below freezing, lake-effect snowfall is common over and downwind of lakes, causing two to three times as much snowfall as occurs upwind of the lake. Lake-effect snowfall is diminished if the lake surface freezes. Lake-effect rainfall occurs from early autumn until temperatures fall below freezing. The enhanced snowfall downwind of large lakes causes additional expenses for snow removal along transportation routes and requires additional strength in buildings to support the roof snow loads. The enhanced snowfall also provides opportunities for skiing and other winter recreation.

Wind patterns

Wind speeds over lakes are higher than over land due to less friction over water surfaces. Data from the Great Lakes show wind speeds over the lakes are 30% higher than inland stations in summer and 100% higher in winter. On land adjacent to lakes, wind speeds are greatest near the shore and diminish farther inland where friction slows the wind.

A lake breeze may develop on spring and summer days with weak synoptic scale flow and a lake that is much cooler than the land and prevailing air temperatures. A cool, moist onshore air flow from the lake across the shore toward land develops several hours after sunrise and persists through the afternoon. The lake breeze may extend several kilometers inland and cause air temperatures to be 5–10°C cooler along the shore than inland. The zone of convergence at the lake breeze front can initiate convection and clouds.

Thomas W. Schmidlin

Bibliography

Angel, J.R., and Isard, S.A., 1997. An observational study of the influence of the Great Lakes on the speed and intensity of passing cyclones. *Monthly Weather Review*, **125**: 2228–2237.

Bates, G.T., Giorgi, F., and Hostetler, S.W., 1993. Toward the simulation of the effects of the Great Lakes on regional climate. *Monthly Weather Review*, **121**: 1373–1387.

Changnon, Jr, S.A., and Jones, D.M.A., 1972. Review of the influences of the Great Lakes on weather. *Water Resources Research*, **8**: 360–371.

Eichenlaub, V.L., 1970. Lake effect snowfall to the lee of the Great Lakes: its role in Michigan. *Bulletin of the American Meteorological Society*, **51**: 403–412.

Eichenlaub, V.L., 1979. *Weather and Climate of the Great Lakes Region*. Notre Dame, IN: University of Notre Dame Press.

Enger, L., and Tjernstrom, M., 1991. Estimating the effects on the regional precipitation climate in a semiarid region caused by an artificial lake using a mesoscale model. *Journal of Applied Meteorology*, **30**: 227–250.

Kristovich, D.A.R., and Laird, N.F., 1998. Observations of widespread lake-effect cloudiness: Influences of lake surface temperature and upwind conditions. *Weather and Forecasting*, **13**: 811–821.

Lofgren, B.M., 1997. Simulated effects of idealized Laurentian Great Lakes on regional and large-scale climate. *Journal of Climate*, **10**: 2847–2858.

Miner, T.J., and Fritsch, J.M., 1997. Lake-effect rain events. *Monthly Weather Review*, **125**: 3231–3248.

Niziol, T.A., Snider, W.R., and Waldstreicher, J.S., 1995. Winter weather forecasting throughout the eastern United States. Part IV: Lake effect snow. *Weather and Forecasting*, **10**: 61–77.

Wilson, J.W., 1977. Effect of Lake Ontario on precipitation. *Monthly Weather Review*, **105**: 207–214.

Cross-references

Continental Climate and Continentality
Land and Sea Breezes
Maritime Climate

LAMB, HUBERT H. (1914–1997)

The British scientist Hubert Horace Lamb was one of the twentieth-century pioneers of modern climate research. With a handful of peers, scattered across the inhabited continents, he transformed the subject of climatology from the rather dry record-keeping exercise it had become during the first half of the twentieth century into the dynamic study of climate variability that it is today. Hubert Lamb's particular contribution stemmed from his broad interdisciplinary interests and expertise. He established an international reputation as a scientist able to cross subject boundaries freely, making use of often fragmentary data to weave a rich story of climate change and its impact on human affairs.

Lamb studied history and languages before entering Cambridge University, then transferred from the natural sciences to geography whilst an undergraduate. Thus, he was well prepared to handle the wide range of evidence of climate change in the recent and more remote past. His two-volume work, *Climate: Present, Past and Future*, published in the 1970s, provides a fitting testament to his ability to analyze, evaluate and synthesize data ranging from observations from the modern meteorological network, through anecdotal evidence from the most recent millennium, to biological and other proxy indicators over geological timescales. That he earned the respect of practitioners in a wide array of disciplines contributed in no small way to the widespread acceptance of climate change as a subject warranting serious academic study during the 1950s and 1960s.

Lamb's interdisciplinarity was coupled with a deep understanding of the workings of the climate system, stemming from his years as a bench forecaster. Joining the Meteorological Office in 1936, he was transferred to the Irish Meteorological Office during the 1940s, having refused to accept an instruction to work on the meteorology of gas spraying. In Ireland, his main responsibility was forecasting for the new trans-Atlantic passenger flights. In 1945, Lamb resigned from this post after disagreeing with the Director of the Irish Meteorological Service over the effect of increasing workloads on aircraft safety, and rejoined the British Meteorological Office the following year. Shortly afterwards he headed south to the Antarctic as expedition meteorologist on the whaler *Balaena*. It was during this time that Lamb, on the basis of his own observations in southern waters, began to doubt that climate was quite as constant as generally believed. An empirical scientist at heart, he pursued this realization by developing new indicators of the state of climate and the atmospheric circulation and extending these records back in time. It was a short step to studying the potential causes of climate change.

July 1970 saw the publication by the Royal Society of one of Hubert Lamb's most significant contributions to climate research, the monograph "Volcanic dust in the atmosphere; with a chronology and assessment of its meteorological significance". The landmark report contained not only the well-known chronology of volcanic eruptions and dust veil index but also an exhaustive review of the scientific basis of the volcanic forcing mechanism and a preliminary analysis of the observational evidence linking climate change and volcanic events. It had long been recognized that the pollution ejected by a major, explosive volcanic eruption could affect weather and climate. Lamb's achievement was to draw together existing material from a number of disparate fields, initiating currents in climatological research that remain evident today.

In 1971 he left the Meteorological Office and established the Climatic Research Unit at the University of East Anglia as a base for interdisciplinary work on climate change. Director of the Unit for 6 years before retirement, he again broke new ground, creating a research group that remains recognized as a world authority on natural climate variability and anthropogenic change. After retirement, Lamb published two editions of his highly regarded account of climate and human affairs, *Climate, History and the Modern World*, returning to his historical roots.

During his later years Lamb was sceptical of the more strident claims regarding the dangers posed by global warming. As he observed in 1994, "there has been too much theory and not enough fact in predicting the future". He died on 28 June 1997, aged 83, having lived to see the United Nations Framework Convention on Climate Change come into force. Climate change was now not only a subject of academic concern but was firmly on the political agenda, a development that Hubert Lamb played no small part in making possible.

Mick Kelly

Selected publications

Lamb, H.H., 1966. *The Changing Climate.* London: Methuen.

Lamb, H.H., 1970. Volcanic dust in the atmosphere, with a chronology and assessment of meteorological significance. *Philosophical Transactions of the Royal Society of London, A: Mathematical and Physical Sciences*, **266**(1178): 425–533.

Lamb, H.H., 1972. *Climate: present, past and future*, Vol. 1: *Fundamentals and Climate Now.* London: Methuen.

Lamb, H.H., 1977. *Climate: present, past and future*, Vol. 2: *Climatic History and the Future.* London: Methuen.

Lamb, H.H., 1995. *Climate, History and the Modern World*, 2nd edn. London: Routledge.

Lamb, H.H., and Johnson, A.I., 1966. Secular variations of the atmospheric circulation since 1750. *Geophysical Memoirs No. 110.* London.

LAND AND SEA BREEZES

Land and sea breezes develop because of differential heating and cooling of adjacent land and water surfaces. Water has a greater heat capacity than land, i.e. land absorbs and emits radiation more efficiently and faster. The diurnal temperature differentials between land and water create a distinctive wind pattern – an example of thermal circulation.

During the day the air over land is heated and expands more rapidly than that over water, resulting in pressure changes and a thermally driven circulation with rising air over land and sinking air over water (Figure L1a). At the shoreline the cool sea air blows inland and is known as a sea breeze.

A sea breeze usually begins several hours after dawn and intensifies in the afternoon. The inland-moving sea breeze causes a decrease in temperature and a rise in humidity. The cooling effect on land is crucial for human comfort, particularly in the tropics (the Europeans called the sea breeze "the doctor", as it helped overcome the tropical heat in Africa); its influence can penetrate to about 30–40 km inland. This air is quite shallow, and ranges to several kilometers deep at its maximum development. A sea breeze front may develop between the warm inland air and the cool sea air. This front is responsible for the abundant precipitation in Florida. Sea breezes here develop on both sides of the Florida peninsula and converge over central Florida, producing strong thunderstorms.

At night the land cools more rapidly than water, resulting in a higher pressure relative to that over water. The circulation reverses and wind blows offshore; this is known as a land breeze (Figure L1b). A land breeze has a smaller magnitude than a sea breeze because the water surface does not gain heat but retains the heat absorbed during daytime, thus producing a smaller temperature difference between land and water at night. A land breeze can penetrate seaward to about 10 km.

Similar thermal circulation may develop over lakes. The inland-moving wind is called a lake breeze. Lake breezes form in regions around the Great Lakes and are common in later spring and summer.

Yuk Yee Yan

Bibliography

Houghton, D.D. (ed.), 1985. *Handbook of Applied Meteorology*. New York: Wiley.

Lutgens, F.K., and Tarbuck, E.J., 1995. *The Atmosphere: an introduction to meteorology*. Upper Saddle River, NJ: Prentice-Hall.

http://www. islandnet.com/~see/weather/elements/seabrz.htm

Cross-references

Coastal Climate
Local Winds
Ocean Circulation
Wind and Wind Systems

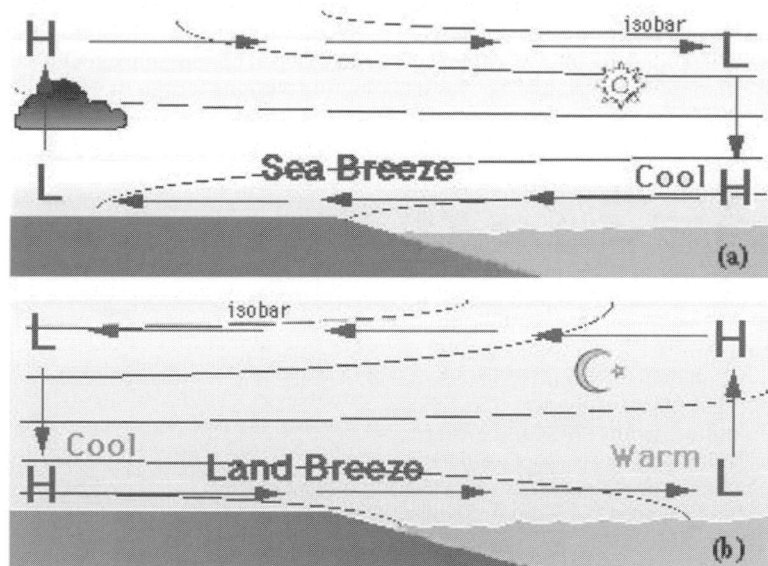

Figure L1 Land breeze and sea breeze.

LANDSBERG, HELMUT E. (1906–1985)

Helmut E. Landsberg died on 6 December 1985, in Geneva, Switzerland, while attending the ninth session of the World Meteorological Organization (WMO) Commission for Climatology. He was at that time a professor emeritus of meteorology at the University of Maryland, College Park.

As a renowned climatologist of the twentieth century, Landsberg advanced our knowledge of climatology through his own research, that of his students, and that of his many co-workers. He was in the forefront of teaching at universities, guiding international organizations, and directing national agencies. He developed regional and national applied climatological networks. He integrated atmospheric science into human affairs through political input and public documents. He stimulated professional organizations to interact and clarify scientific thought, and he interwove government, university, and private practitioners into a creative medium for scientific progress.

Landsberg was born in Frankfurt am Main, Germany, on 9 February 1906. He spent his student years at the University of Frankfurt, taking a basic curriculum of physics, mathematics, and geophysics, and completed his thesis dissertation in seismology in 1930 at the Geophysics Institute. Landsberg developed an intense interest in the atmosphere, principally from an observational viewpoint. He turned to practical forecasting by joining the Taunus Observatory, where he stayed until 1934, ultimately becoming its chief. The observatory provided extensive experience and broadened Landsberg's perspective. Indeed, it was at Taunus that his love for libraries blossomed.

Landsberg was recruited to teach geophysics at Penn State in 1934. As the first meteorologist at that institution he set up an observatory with the help of some students and began a teaching program that evolved into the present meteorology department. In 1941 he joined the faculty at the University of Chicago, where he developed a field course for cadets who were in attendance during that period.

Shortly before the end of the Second World War, Landsberg moved out of academia to the federal government and into the world of science administration. He first was a consultant to the US Air Force and then became executive director of the Commission on Geophysics and Geography of the Research and Development Board, a position he held from 1948 to 1951. From 1951 to 1954 he was director of the Geophysics Directorate of the Air Force Cambridge Research Center. There he had occasion not only to develop in-house research with young protégés but also to fund high-quality university research. He then returned to Washington to direct the Office of Climatology of the United States Weather Bureau (USWB), at which position he remained until reorganization of the Weather Bureau into the Environmental Science Services Administration (ESSA) in 1965. At that time he became head of the Environmental Data Services of ESSA.

Landsberg returned to the academic world in 1967 and spent his remaining years there. He joined the University of Maryland as a research professor and concurrently held several administrative posts. He was acting director of the Institute for Fluid Dynamics and Applied Mathematics and subsequently became director of the newly founded Meteorology Program. In 1976 he retired from the university as a professor emeritus but continued to work with no loss of intensity.

Landsberg was elected to membership in the National Academy of Engineering in 1966, and was an honorary life member of the New York Academy of Sciences. He was a fellow of the Royal Meteorological Society, the American Academy of Arts and Sciences, the American Association for the Advancement of Science, the American Geophysical Union (AGU), the Meteoritical Society, the American Meteorological Society (AMS), and the Washington Academy of Sciences. In addition, he was a member of the German Meteorological Society, the American Institute for Medical Climatology, the

International Society of Biometeorology, the Mount Washington Observatory, the Sneckenberg Society of Natural History, the Society of Sigma Xi, the Society of Sigma Pi Sigma, the Society of Sigma Gamma Epsilon, and he was an honorary member of Phi Beta Kappa.

He served as president of the WMO Commission for Special Applications of Meteorology and Climatology from 1969 to 1978; he was a member of a WMO Advisory Working Group from 1978 to 1981; and he was a member of its Commission for Climatology from 1981 until his death. He served the National Academy of Engineering as a member of its Awards Committee in 1974 and 1975. At the National Research Council (NRC) he served on the Geophysics Research Board as chairman of the Geophysical Predictions Panel in 1977 and 1978, and as a member of its Panel on Energy and Climate from 1975 to 1978. For the NRC Division of Physical Sciences, he was a member of its Climatic Impact Committee from 1972 to 1975. He served the government as a member of the National Advisory Committee on Oceans and Atmospheres from 1975 to 1977. He served the universities as trustee to University Corporation for Atmospheric Research (UCAR) from 1968 to 1972 and assisted the private sector as a certified consulting meteorologist of the AMS. He served the AMS as councilor from 1952 to 1960, as vice-president in 1963–1964, and as chairman of the Awards Committee in 1974–1975. In support of the American Association for the Advancement of Science (AAAS) he was vice-president of Section E in 1972. He served the AGU as vice-president, Section on Meteorology (1953–1956), as president of that section (1956–1959), as vice-president of the Union (1966-1968), and finally as its president (1968–1970).

Landsberg loved books and acquired a unique and unequalled collection of rare historical books on meteorology, which he subsequently donated to various libraries. He was an associate editor of the *Journal of Meteorology* (1950–1961), the editor of *Advances* in *Geophysics* (1952–1977), the editor-in-chief of the *World Survey of Climatology* (1964–1985), and the chairman of the Publications Committee of the International Society of Biometeorology (1960–1985).

For his extensive professional contributions he was rewarded by numerous acknowledgments. He received the Exceptional Meritorious Service Award from the Department of Commerce in 1960. The AMS bestowed on Landsberg their Award for Outstanding Achievement in Bioclimatology in 1964, the Charles Franklin Brooks Award for Outstanding Services to the Society in 1972, and the Cleveland Abbe Award for Distinguished Service to Atmospheric Sciences by an Individual in 1983. The Deutsche Meteorologische Gesellschaft EV, the German Meteorological Society, awarded him the Alfred Wegener Medaille in 1980. He was the recipient of the William Bowie Medal in 1978 from the AGU. In 1979 he received the International Meteorological Organization Prize from the WMO. The W.F. Peterson Foundation Gold Medal was awarded to him in 1983, and in 1985 he was honored with the Solco W. Tromp Memorial Award by the Enviroscience Foundation. Ultimately he was bestowed the National Medal of Science by President Reagan in 1985.

Landsberg's scientific productivity was astounding and his breadth of interest and involvement remarkable. He left almost four hundred written published documents, including several books, and his written contributions to collective documents from his many committee assignments would substantially augment that total. His research explorations commanded numerous topics, including seismology, geography, geology, climatology, weather forecasting, bioclimatology, urban climate, climate history, and climate services among many others. He had the perception and serenity that unfailingly led to meaningful solutions for the most complex of problems.

Ferdinand Baer

Selected publications

For a complete listing of Landsberg's published works, see Baer, F., Canfield, N.L., and Mitchell, J.M., 1991. *Climate in Human Perspective: a tribute to Helmut E. Landsberg*. Boston, MA: Kluwer.

Landsberg, H., 1958. *Physical Climatology*, 2nd edn. Dubois, PA: Gray Printing Co.

Landsberg, H., 1946. Climate as a natural resource. *Science Monthly*, **63**: 293–298.

Landsberg, H., 1962. Goals for climatology. *Journal of the Washington Academy of Scienees*, **52**(4): 85–91.

Landsberg, H., 1969. *Weather and Health*. Garden City, NY: Doubleday Anchor.

Landsberg, H., 1976. Weather, Climate and Human Settlements. WMO Special Environmental Report 7.

Landsberg, H., 1981. *The Urban Climate*. International Geophysics Series, **28**, New York: Academic Press.

Landsberg, H., 1982. Climatology: now and henceforth. *WMO Bulletin*, **31**(4): 361–368.

Landsberg, H., 1985. *The History of Geophysics and Meteorology: an annotated bibliography*. New York: Garland (with S. G. Brush).

Landsberg, H., 1985. The value and challenge of climate predictions. *Invited Scientific Lectures Presented at the Ninth World Meteorology Congress*, WMO No. 614, pp. 20–32.

LAPSE RATE

The lapse rate is generally understood to be the decrease of temperature with height (or depth in oceanography), defined symbolically as

$$\gamma = \frac{\partial T}{\partial z}$$

where γ is the lapse rate, T is temperature and z height. In meteorology and oceanography two distinct types of lapse rate are used.

1. *Environmental or lapse rate in situ* is the temperature decrease with ascent through the atmosphere; it reflects the actual temperature change with altitude. The temperature instrument samples successively different parts of the air to determine the temperature conditions in the environment, commonly by a radiosonde instrument sent up by balloon.
2. *Process lapse rate* is the temperature of the same sample of air recorded as it changes with vertical motion. In this case the rate is determined by the character of the sample. Two contrasting air types may be considered:

Unsaturated air, called dry air, cools adiabatically due to expansion as it is lifted. This is called the *dry adiabatic lapse rate*, which has a numerical value of nearly 1°C/100 m or 5.5°F/1000 ft, derived from the first law of thermodynamics and the equation of state for adiabatic conditions. Thus, one can determine the dry adiabatic lapse rate for any planetary atmosphere in general and substitute particular values of the constants. Following Byers (1959), to obtain the adiabatic rate of temperature change with height, one may substitute *dp* from the hydrostatic equation into the differential form of the adiabatic equation with the result

$$C_r \frac{dT}{T} + \frac{R\rho g dz}{mp} = 0.$$

and since, from the equation of state $p = \rho m/RT$,

$$C_r \frac{dT}{T} = -\frac{mp}{RT}\frac{Rg\,dz}{mp}$$

and

$$\frac{dT}{dz} = -\frac{g}{C_r}.$$

This gives an expression for the rate of change of temperature with height in an adiabatic process, the negative sign indicating that the temperature would decrease with altitude. The term g/C_p has the value of approximately 10^{-4}°C/cm, or 1°C/100 m, as mentioned above.

Moist air is cooled by lifting to the dewpoint, after which further cooling causes condensation of the water vapor to liquid or ice (sublimation), which releases the *latent heat*. Thus the *moist adiabatic lapse rate* (or *pseudo-adiabatic* since it is not truly adiabatic) is less than the dry rate due to the added heat. Following Byers (1959):

$$dT = \frac{-L}{C_r} dw.$$

Evaporating an amount of liquid water $-dw$ produces an amount of vapor $+dw$. Since this is a mixture, it is more convenient to use the specific humidity change dq in place of the change in mixing ratio dw. The new temperature, then, is

$$T'' = T_m + dT = T_m - \frac{L}{C_r} dq$$

$$= T_m - \frac{L}{C_r}(q'' - q_m).$$

where T'' and q'' are the actual temperature and specific humidity, respectively, resulting in the cloud. The numerical value varies from 0.3–1°C/100 m or 2–5°F/1000 ft, depending on other conditions (pressure, temperature), but it is usually taken as near 6°C/100 m or 3°F/1000 ft.

The *average lapse rate* is the mean *in situ* lapse rate for many observations, usually given as 0.5°C/100 m or 3°F/1000 ft.

The lapse rate normally increases slightly with altitude. Near the tropopause the observed lapse rate approaches the dry adiabatic lapse rate, which is 9.8°C/km. The stratosphere has a lapse rate nearly zero (i.e. it is *isothermal*) until *ozone* begins to affect it. Reversals or negative lapse rates result from the overlap of contrasting air masses or, near ground, from nighttime high net radiation, and are then called *inversions* (see Figure L2).

When an environmental lapse rate exceeds the dry adiabatic rate, and the *potential temperature* decreases with height, it is said to be a *superadiabatic lapse rate*. Such a situation would develop on a microclimatic scale over a warmed land surface such as a desert or a paved road as the airmass warmed. Lapse rates over 1°C/100 m are superadiabatic.

There usually is also a *diurnal variation in lapse rate*, owing to daytime warming and nighttime radiative cooling. Daytime heating may steepen the lapse rate, which may exceed the adiabatic rate, causing instability and convection. Cooling at night

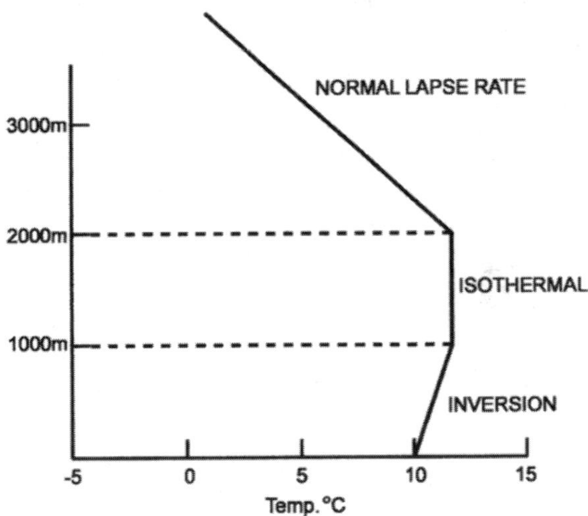

Figure L2 Normal lapse rate, isothermal condition, and inversion.

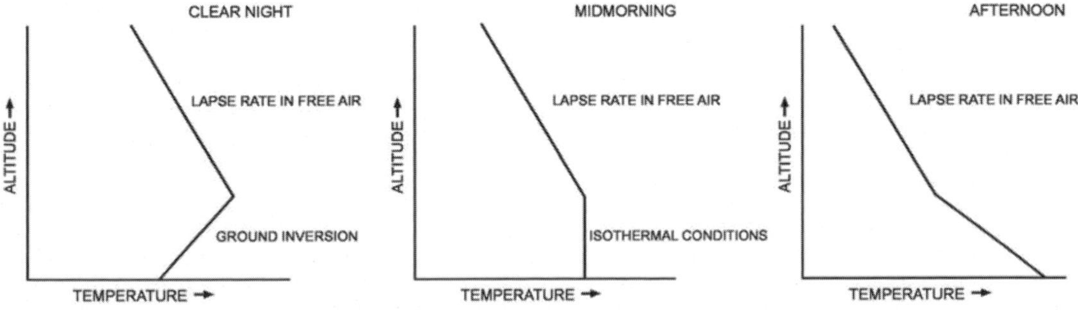

Figure L3 Diurnal variation of lapse rates over land in clear weather.

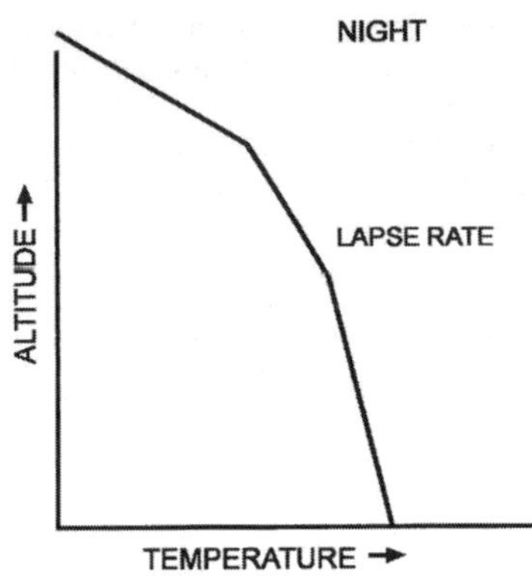

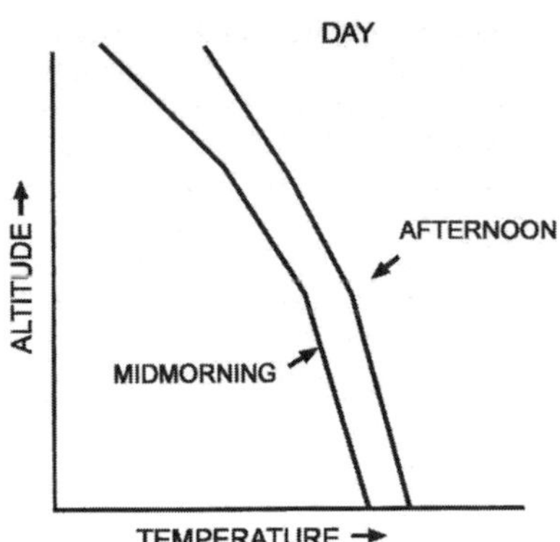

Figure L4 Lapse rates over the oceans. The temperature of the ocean remains nearly constant both day and night. Since there is little surface heating or cooling, air temperature in lower levels remains virtually constant.

reduces the lapse rate, increases *stability*, and in the early morning may cause inversion (see Figure L3, left-hand curve). Over a warm sea, which neither absorbs nor radiates heat very readily, there rarely is an inversion, but nocturnal instability may occur (see Figure L4).

Rhodes W. Fairbridge and John E. Oliver

Bibliography

Aguado, E., and Burt, J.E., 2001. *Understanding Weather and Climate.* Upper Saddle River, NJ: Prentice-Hall.

Ahrens, D.L., 2000. *Meteorology Today: an introduction to weather, climate, and the environment*, 6th edn. Pacific Grove, CA: Brooks & Cole.

Byers, H.R., 1959. *General Meteorology*, 3rd ed. New York: McGraw-Hill.

Her Majesty's Stationery Office, 1981. *A Course in Elementary Meteorology.* Meteorology 0.911. London: AMSO.

Lutgens, F.K., and Tarbuck, E.J., 2001. *The Atmosphere*, 8th edn. Upper Saddle River, NJ: Prentice-Hall.

Wallace, J.M., and Hobbs, P.V., 1997. *Atmospheric Science: an introductory survey*. San Diego, CA: Academic Press.

Cross-references

Cloud Climatology
Microclimatology
Standard Atmosphere

LATENT HEAT

Latent heat is a form of internal or potential energy stored by evaporated or melted water. As ice melts or liquid water evaporates, the molecules change state – from a solid to a liquid, from a liquid to a gas, or from a solid directly to a gas. The energy required to melt ice or to evaporate water does not result in a change in its temperature; rather, it is stored through the phase change of the water molecules and can be subsequently released through condensation of the water vapor (through either dew or frost formation or precipitation) or refreezing of liquid water. This energy is termed *latent heat* since it does not result in a change in the temperature of the water molecules – thermometers cannot measure this energy.

A total of 334 J of energy are required to melt 1 g of ice at 0°C, which is called the *latent heat of melting*. At 0°C, liquid water has 334 J g^{-1} more energy than ice at the same temperature. This energy is released when the liquid water subsequently freezes, and it is called the *latent heat of fusion*. Similarly, 2501 J of energy are required to evaporate 1 g of liquid water at 0°C, and this energy is called the *latent heat of vaporization*. This energy is released when the water vapor subsequently condenses, where it is called the *latent heat of condensation*. As before, water vapor at 0°C has 2501 J g^{-1} more energy than liquid water at 0°C. Thus, latent heat changes are symmetric with respect to the direction of a phase change, as is required through the Law of Conservation of Energy.

By contrast, to raise the temperature of 1 g of water by 1°C requires only 4.19 J of energy. Note that it takes almost six times as much energy to evaporate 1 g of water at 0°C than to raise its temperature to 100°C!

Water vapor can change directly to ice without ever becoming a liquid, as during the formation of frost, for example. Similarly, ice can change directly to water vapor without becoming a liquid, as occurs with dry ice (frozen carbon dioxide) or with ice cubes in a freezer. Since a fundamental law of physics is that energy must be conserved, this *latent heat of sublimation* must equal 2835 J g^{-1} at 0°C – the sum of the latent heats of melting/fusion and vaporization/sublimation at that temperature. Thus, water vapor at 0°C has 2835 J g^{-1} more energy than ice at 0°C.

Latent heat can be thought of as the energy required to break the bonds between the molecules at each change of phase. In ice, water molecules are strongly bound together in crystalline form. When 334 J of energy are added to 1 g of ice at 0°C, these bonds are loosened, producing liquid water at 0°C. If a further 2501 J of energy are added to this gram of liquid water at 0°C, the bonds are broken entirely, producing water vapor at 0°C. As a result, water vapor can be thought of as having "stored" energy as latent heat because condensing water vapor will release far more heat into sensible heat than can be stored by temperature changes in the water vapor alone.

In the above discussion the latent heats have been specified for a temperature of 0°C. At warmer temperatures the water molecules contain more kinetic energy (i.e. their *sensible heat* is greater) and thus less energy is required for evaporation to occur. Thus, the latent heat of vaporization decreases with increasing temperature. For example, the latent heat of vaporization equals 2442 J g^{-1} at 25°C and only 2264 J g^{-1} at 100°C. Similarly, molecules of water contain less internal energy at colder temperatures so that less energy is released when they freeze. For example, the latent heat of fusion equals only 289 J g^{-1} at −20°C. Latent heats, therefore, are inversely proportional to temperature. Bolton (1980) gives an equation to describe the relationship between the latent heat of vaporization/condensation, L_v, in Joules per gram of water, and temperature, T, in °C as

$$L_v = (2501 - 2.37T) \text{ for } -35°\text{C} \leq T \leq 35°\text{C}.$$

Storms in both the tropics and midlatitudes, including tropical storms (hurricanes) and thunderstorms, are really latent-heat engines, fueled by energy released by the condensing moisture. As it takes a comparatively large amount of energy to evaporate water, a considerable amount of energy is released when that moisture condenses. Consequently, the advection of a large quantity of water vapor enhances the severity of the storm and the loss of this water vapor advection often contributes to the storm's demise. This is one of the reasons why hurricanes often dissipate shortly after landfall.

David R. Legates

Bibliography

Bolton, D., 1980. The computation of equivalent potential temperature. *Monthly Weather Review*, **108**(7): 1046–1053.

Wallace, J.M., and Hobbs, P.V., 1977. *Atmospheric Science: an introductory survey*. Orlando, FL: Academic Press.

Cross-references

Bowen Ratio
Phase Changes
Sensible Heat

LIGHTNING

At any one time some 2000 thunderstorms are at work (Uman, 1987) illuminating the skies and striking the Earth with lightning between 25 and 300 times per second – more than 8.6 million times a day. This contributes significantly to the electrical energy balance of the Earth and its atmosphere.

Lightning is the discharge of static electricity within thunderstorm clouds, between clouds and the ground or into the surrounding air. Dust storms, fires and volcanic eruptions can also provide the necessary turbulent atmospheric environment for lightning to occur.

The necessary condition preceding any lightning flash is a separation of charge within the associated cloud or clouds. That is, electrons and positive ions must somehow be transported to different regions of the cloud to form an electrical dipole. The generally accepted model for a typical cumulonimbus lightning cloud has a region of negative charge just above its base, with the positive region just below its top. The voltage difference between charged centers may reach several million volts. Once the voltage difference exceeds the insulating capacity (dielectric strength) of the air, a lightning stroke occurs. While a satisfactory, quantitative description of charge separation in the turbulent environment of a thunderstorm is still lacking, most atmospheric physicists are in agreement that the process involves the interaction, by collision, between supercooled water droplets and ice particles at and above the freezing level in the cloud. This explanation unfortunately leaves out occurrences of lightning in "warm clouds" as in the tropics, although warm cloud (entirely above 0°C) lightning has never been completely substantiated.

The lightning flash

What we see as a single cloud-to-ground lightning flash is actually several separate discharges in rapid succession, moving both from cloud to ground and from ground to cloud, as many as 30–40 strokes, each lasting only 0.00005 s (50 microseconds) to 0.01 s. A moving camera, as early as 1903, was used to show those properties and later the speed of each lightning stroke.

The cloud-to-ground discharge begins with a virtually invisible stepped leader stroke that moves downward at about 1.5×10^5 m/s in regular distinct steps, typically of 50 m length, seeking the most conductive path to the ground. As the stepped leader approaches the ground, an upward discharge occurs from the ground to meet the leader. When a conductive channel is thus established, a highly luminous return stroke travels upward from the ground at about 3×10^7 m/s. This may be followed by other leaders, called dart strokes or dart leaders, using the same channel, each followed by return strokes, with speeds intermediate to the stepped leader and original return stroke.

Cloud-to-ground discharges typically lower negative charge to Earth (the negative CG or −CG). CG discharges are sometimes initiated by positively charged leaders, with the subsequent return strokes lowering positive charge to the ground (the +CG). Once thought to be quite rare, the +CG is now thought to make up about 10% of all CG (Lyons, 1999).

Only one-fifth of all lightning bolts are cloud-to-ground; intracloud (IC) lightning is the most common. Air discharges, where conducting channels from the cloud do not reach the ground, and cloud-to-cloud strokes are less frequent. Figures L5 and L6 illustrate the CG and IC forms.

Figure L5 Cloud-to-ground lightning flash under the base of a cumulonimbus cloud over Mesquite, Arizona (photo courtesy of William T. Reid).

Figure L6 Intracloud lightning, also known as crawler lightning, explodes across the Great Plains skies (photo courtesy of William T. Reid).

Types of lightning flashes

Lightning flashes are also typed by their appearance. Highly branched cloud-to-ground discharges are called fork lightning, whereas streak lightning has very few branches. Sheet lightning is an intracloud discharge where the channel and branches are not visible; the discharge appears as a white sheet against the cloud. If the lightning is so distant that only some light near the horizon is seen, it is called heat lightning.

Air discharges come out from the bottom of the cloud branching into the air but do not reach the ground. These discharges are more common in dry areas where the cloud base is normally much higher. Air discharges have been seen shooting up vertically from tops of thunderstorm clouds even into the stratosphere. The discharges may extend as far as 20 km from the thunderstorm before finally reaching the ground. What may appear to be coming from clear sky and not from a distant cloud is a "bolt from the blue".

On rare occasions, if a strong wind is blowing, the separate strokes of a nearby flash can be seen as ribbon lightning, the appearance produced as the channel is blown sideways. Bead, or chain, lightning appears when the return stroke breaks into luminous beads, as the light intensity of the channel decays, some portions cooling faster than others.

Ball lightning, or *kugelblitz*, is a rare and unusual form that is not well understood. It appears near a lightning discharge as a ball of luminosity with a diameter of about 20 cm, usually red, orange, or yellow in color, making a hissing sound and leaving a foul-smelling odor. As most frequently reported, it usually lasts only a few seconds, floating down to the ground, where it strikes something and either explodes or simply dissipates quietly. A related phenomenon, corona or "St Elmo's Fire", discharged at the ground under electrified clouds, can form a luminous glow of light from the mast of ships, airplane wings, or church steeples when they become extremely positively charged.

There are also forms of lightning moving skyward from tops of thunderstorms. Sprites, which are enormous red-colored aurora-like features, lasting 100 ms or less, can span the stratosphere and even parts of the mesosphere. They appear to be a response to large peaked positive CG (Boccippio et al., 1995). Another high-altitude response to energetic +CG are elves, which are brief (500 μs) disks of light found above thunderstorms at about 90 km. Blue jets are rare powerful beams of blue light traveling upward out of thunderstorm tops at speeds of 100 km/s up to 50 km elevations (Wescott et al., 1995). Upward-moving "superbolts", over 100 times brighter than typical lightning, have also been recorded traveling into the stratosphere from thunderstorm cloud tops (Thurman, 1977).

Lightning climatology

Lightning climatic data have been traditionally linked with thunderstorms, so that lightning days and thunderstorm days should correspond, although one relies on sight, the other on hearing. In the US, as well as in most other countries, thunder must be heard by a trained observer in order for an official thunderstorm to occur. As most airport weather stations are quite noisy, and many weather offices are located in closed buildings, comparisons between lightning networks and official thunder reports show thunderstorms may be underreported at many sites.

Most lightning networks only record CG lightning, only a fraction of the total flashes that may result in thunderstorm records. Various correlations between thunderstorm days and actual flash density, expressed as flashes/km^2 per year, have resulted in estimates of 1–2 flashes/km^2 per year per 10 thunderstorm days (Lyons, 1999). However, regional variability in this relationship shows that this estimate can be far off the mark. Two major differences are shown to exist between the average thunderstorm patterns and those of lightning in the US (Changnon, 1993). In the intermountain west the number of thunderstorm days and events is relatively high, while CG lightning flash frequency is low. Here the flash counts per thunderstorm event are also much lower than in regions east of the Rockies. The second disparity occurs along the East Coast from the Carolinas to New York, where high flash frequencies accompany not so high thunderstorms. Average CG flashes in the west run between 10 and 25 per thunderstorm event. Florida, the Midwest–High Plains and the northeast average more than 125 CG flashes per event, while the mid-Atlantic region is highest with averages of 200 CG flashes per thunderstorm event. Over North America, central Florida is known as the "Lightning Capital" of the United States. Recent data from the US National Lightning Detection Network (NLDN) indicate that central Florida most likely has the highest long-term CG flash density with over 11 flashes/km^2 per year (Orville and Huffines, 2001). However, year-to-year variations have placed the maxima over the Midwest or southern Louisiana in some years. Comparisons between NLDN and satellite data indicate IC to CG ratios over the Great Plains to be nearly 6:1, while coastal Gulf stations are nearly 2.5:1 (Christian et al., 2002).

Globally, about 78% of all lightning flashes (IC and CG) occur between 30°N and 30°S latitudes, with those over land or their immediate vicinity ten times more frequent than those over oceans, depending on season (Christian et al., 2002). Recent satellite measurements put the global lightning flash rate at 44 ($\pm$5) flashes each second, with nearly 1.4 billion flashes/year over the Earth. Concentrations of activity over interior South America and equatorial Africa are well documented, as well as the overall maximum activity around the Indonesian Islands. The Congo Basin is the planet's hot spot, with mean annual flash density of 80 flashes/km^2 per year, including an area over 3 million km^2 with flash densities above 30, that of central Florida. The annual cycle is dominated by the Northern Hemisphere's summer peak. Seasonal differences follow heating patterns and the shift in the Intertropical Convergence Zone. Over the oceans, higher rates are shown year-round in the western Atlantic and western Pacific basins where cold air passing over warm water provides instability. Lightning minima are found in polar regions, subtropical deserts and oceans, and along western, cold-water coasts of continents, in general.

Lightning measurements

The total energy produced by a lightning flash depends on its magnitude and duration. An average return stroke carries a current of 30 000 amperes, which at 125 million volts may develop a peak power in its brief lifetime to raise a large ship 6 feet. One survey of 60 million lightning flashes found 23% had peak currents >75 kA, with the largest positive cloud-to-ground (CG) reaching 580 kA and the largest negative 960 kA (Lyons et al., 1998). "Cold" lightning of very short duration but high current usually causes explosions when it strikes, while "hot" lightning of longer duration but lower current is known to burn things and cause forest fires. Most of the lightning flash's energy is expended as heat. Spectroscopic measurements have shown that the temperature of a lightning channel peaks near 30 000 K in a few microseconds. The air around the channel is heated so rapidly it expands at a supersonic rate, producing the shock waves heard as thunder. The global frequencies of lightning strikes were previously derived from thunderstorm or thunderstorm days information. More recently, local, regional and national lightning frequencies are estimated using lightning detectors, based on spherics (atmospheric electric or magnetic fields) research and satellite observations. Lightning detection networks are based on two distinct methods used to detect and locate CG lightning. One is based on magnetic direction finding (MDF), while the second uses time-of-arrival (TOA) techniques. The current US National Lightning Detection Network (NLDN) is made up of a hybrid system using the best of both methods (Cummins et al., 1995). The NLDN provides data on the time, location, polarity and peak current of the first stroke in a flash, also stroke multiplicity and estimate of location accuracy. Operating continuously since 1989, flash detection efficiency ranges between 80% and 90% (Cummins et al., 1995). Other national and regional networks have developed in many other areas, including Canada, Brazil, Japan, Europe, Asia, and Australia. Global measurements became possible with the use of optical sensors carried by satellites. An interesting dataset was acquired by a military satellite that observed local midnight lightning, on a global scale, for 1 year (Orville and Henderson, 1986). In April 1995 NASA's Optical Transient Detector (OTD) was launched. The polar-orbiting satellite passes overhead twice a day. Although not continuous, the observation intervals vary throughout the day for each location, making a more uniform sampling of lightning over much of the Earth. The LIS (Lightning Imaging Sensor), which has been updated on the Tropical Rainfall Measuring Mission (TRMM) spacecraft in 1997, has provided global data for both IC and CG lightning for several years. A need for a geostationary sensor for continuous coverage seems obvious for warnings of potentially dangerous lightning storms. Lightning deaths in the United States continue to fall; however, annual fluctuations seem to be linked to climatic features, such as thunderstorm days and average surface temperature values (Lopez and Holle, 1998).

S. Ladochy

Bibliography

Boccippio, D.J., Williams, E.R., Heckman, S.J., Lyons, W.A., Baker, I.T., and Boldi, R., 1995. Sprites, ELF transients, and positive ground strokes. *Science*, **269**: 1088–1091.

Changnon, S.A., 1993. Relationships between thunderstorms and cloud-to-ground lightning frequencies. *Journal of Climate*, **2**: 897–921.

Christian, H.J., Blakeslee, R.J., Boccippio, D.J., et al., 2003. Global frequency and distribution of lightning as observed from space by the Optical Transient Detector. *Journal of Geophysics Research* 108 (D1), 4005, doi: 10.1029/2002JD002347, 2003.

Cummins, K.L., Bardo, E.A., Hiscox, W.L., Pyle, R.B., and Pifer, A.E., 1995. NLDN'95: a combined TOA/MDF technology upgrade of the US National Lightning Detection Network. *Proceedings, International Aesospace and Ground Conference on Lightning and Static Electricity*, Williamsport, VA.

Golde, R.H., 1977. *Lightning*, 2 vols. London: Academic Press.

Lopez, R.E., and Holle, R.L., 1998. Changes in the number of lightning deaths in the United States during the twentieth century. *Journal of Climate*, **11**: 2070–2077.

Lyons, W.A., 1999. Lightning. In Pielke, R., Jr, and Pielke, R., Sr, eds., *Storms*, Vol. II. New York: Routledge, pp. 60–79.

Lyons, W.A., Uliasz, M., and Nelson, T.E., 1998. Large peak current cloud-to-ground lightning flashes during the summer months in the contiguous United States. *Monthly Weather Review*, **126**: 2217–2233.

Orville, R.E., and Henderson, R., 1986. Global distribution of midnight lightning: September 1977 to August 1978. *Monthly Weather Review*, **114**: 2640–2648.

Orville, R.E., and Huffines, G.R., 2001. Cloud-to-ground lightning in the United States: NLDN results in the first decade, 1989–1998. *Monthly Weather Review*, **129**: 1179–1193.

Thurman, B.N., 1977. Detection of lightning superbolts. *Journal of Geophysics Research*, **97**: 2566–2568.

Uman, M.A., 1987. *The Lightning Discharge*. International Geophysics Series, vol. 39. Orlando, FL: Academic Press.

Wescott, E.M., Sentman, D., Osborne, D., Hampton, D., and Heavner, M., 1995. Preliminary results from the Sprites94 aircraft campaign: 2. Blue jets. *Geophysics Research Letters*, **22**: 1209–12.

Cross-references

Cloud Climatology
Thunder
Thunderstorms

LITERATURE AND CLIMATE

Below, the cracked, brown earth,
  Like ancient earthen-ware,
Spreads out its dusty, worn
  Old surface, baked and bare.

Above, the polished blue
  Of a burnished August sky
Is an inverted bowl
  Of every drop drained dry (Converse, 1927).

In these few lines and sparse verse are contained a vivid description of the Great Plains landscape and the experience of drought. Literature has had a long relationship with climate as the human species attempts to capture its home environment and convey a sense of it to others, develop an understanding of why landscapes vary from one place to another, and communicate as well as preserve our environmental knowledge. Literature can be any writings of universal interest, carefully crafted in form and expression. Literature can be in the form of fiction, writing from the imagination, as in novels, short stories, drama, or poetry. Or it can be non-fiction, writing from "real-life" situations, such as essays, histories, or scholarly works. The term literature can refer also to writings dealing with a certain subject, period, language, or location. Climate is the prevailing atmospheric conditions of a location or region, its annual patterns of warmth and cold, moisture and dryness, cloudiness and clear, determined after an extended period of observation, usually over many decades. The term can also be used to describe a region characterized by a given climate. Today contemporary scholars associate climatology with the sciences and literature with the humanities, but these categorizations are fairly recent. Learned individuals historically practiced both arts and sciences, such as Leonardo da Vinci. Despite the seeming dichotomy, there remains today a strong reciprocal relationship between literature and climate: climatologists utilize literature and literary writers utilize climate. In exploring this relationship, examples from the American Great Plains will prove illuminating.

The Great Plains are a region, while readily brought to mind's eye with its flat landscape and enormous sky, most often defined by the climate. The Plains stretch from south of the Rio Grande to the edge of the Boreal Forest in the north, including vast stretches of the United States and Canada. The western boundary of the Plains generally is viewed as the foothills of the Rocky Mountains. The eastern boundary is more of a challenge: there is no sharp physical feature to delineate it. Climatic features are most often invoked as a result, such as the 20-inch isohyet or the tall-grass/short-grass prairie boundary, both pointing to the margin between humid and subhumid climatic regions.

On the Plains, climate not only serves to define the region, it is also central to life. As environmental historian Donald Worster points out:

> The Great Plains of North America offer a revealing place to study the challenges of rural communities trying to cope with climate and that chaotically changing envelope of gases we call the atmosphere. Despite the seeming monotony of flat, immutable land meeting big, unchanging sky, the plains are in fact the most volatile place on the North American continent. Their complexity lies not in land forms but in climate. Nowhere else do Americans confront such extremes of cold and hot or such rapid oscillations around the crucial point that divides wet from dry (Worster, 1999).

The distinctive environment of the Plains results in a distinctive sense of place communicated in its literature. Some of the earliest uses of Plains climate in literature can be found in the traditional stories of Plains Native American tribes, such as the drought that led to the creation of the Great Medicine Dance (called the Sun Dance by other Plains tribes), according to the Cheyenne:

> Long ago, when the earth and the people dwelling upon it were young, our tribe was starving. The earth itself was starving, for no rain was falling. Plants and trees wilted. Many rivers dried up. The animals were dying of hunger and thirst (Erdoes and Ortiz, 1984).

By accepting and performing the ritual, Maheo, the Creator, promised that "the rains will fall again". In this case, both the humans and the Earth are at the mercy of climate, foreshadowing the European–American experience on the Plains.

Climatologists and literature

Literature has served climatology for centuries. The roots of climatology in Ancient Greece, as well as Ancient China, took the form of written descriptions of climate based on observation but also at times imagination. While instrumentation and standardization of climate measures are the contemporary norm, scholars investigating the Earth's climatic past rely on "weather diaries" and other available weather descriptions (Lamb, 1995). By combining descriptive reports with known climatic events and proxy data it is possible to produce sound studies that advance our understanding of past and present climate. Chu K'o-Chen, considered the father of Chinese meteorology, utilized a wide range of writings to construct the climatic history of China. Carvings on oracle bones, descriptions in regional gazetteers, poetry and literary descriptions of birds, flowers, and trees were used to supplement the available records and assess climatic change (Chiao-Min Hsieh, 1976). Tyrell (1987) utilized historic travel accounts of Kenya to reconstruct nineteenth-century vegetation patterns, creating a snapshot of the Kenyan physical environment prior to colonization. Climatologists have also utilized literature to humanize climate, to communicate the human experience of climatic conditions. Roots, in introducing work on glacial mass balance measurements, reminds his readers that evocative language can be an effective communicator of science:

> On a more poetic and less scientific plane, the nineteenth-century naturalist-poet Clarence King, describing a valley glacier as a great white dragon, alluded to the concept of mass balance when he told how the dragon crawled down the valleys to flee the sun, but slept and grew fat during the cold dark days of winter. It would be hard to describe glacier behaviour much better today (Roots, 1984).

Besides employing literature as source material or to richly convey climatological information to readers, climatologists have also used climate data to support literary works as well as to speculate on the employment of climate data in constructing imaginary landscapes. In "Anne Frank's diary – a meteorologist's view" (1995), Persson draws on weather data to confirm Frank's text, correlating Frank's depiction of weather with the recorded weather. Pike's "An Appreciation of the Weather in *The Lord of the Rings*" (2002) examines the principal weather scenes of the novel and ties them to typical British climate patterns.

Correspondingly, in Great Plains research, historical writings have been employed in a variety of ways to advance our understanding of its climate. Rannie and Blair (1995) examined an "unusual" 1993 weather pattern on the upper Plains, involving a trough and strong negative pressure, producing record precipitation and severe flooding. Noting a similar pattern occurred in 1958, they wondered if a similar event had occurred prior to instrumental records. Using an account of a military expedition and settlement accounts, Rannie and Blair established an analogous pattern had likely occurred in 1849, suggesting that the pattern may be more common than originally thought.

"Desert myth and climatic reality" explores the concept of the Plains as "Great American Desert" from a climatological perspective (Lawson and Stockton, 1981). Authors Lawson and Stockton reconstructed Plains drought history with tree-ring indices and available records, mapped the reconstructed moisture conditions according to the early expedition routes, and then compared them to diaries, journals and excursion records.

Major Stephen Long's expedition of 1819–1820, responsible for the introduction of "Great American Desert" on American maps, was found to have passed through a region experiencing extreme drought, thought to rival that of the 1930s. Long and his expedition's perception of a desert may not have been an "illusion" but rather an experienced "reality".

Finally, *The Dreaming Is Finished: human perception of the Great Plains through poetry, 1890–1950*, examined poetry written about the Plains, by individuals on the Plains, to assess whether the extreme drought conditions of the 1930s impacted how the Plains were perceived and described (Dando, 1992). While elements of Plains climate including drought did appear in the poetry, overwhelmingly the imagery of the Plains was as a verdant garden with ample precipitation and favorable conditions for agriculture. The poets tended to focus their writings on the rich potential of the Plains yet did convey a sense of frustration in an environment they could neither control nor predict. Drought did not appear to impact human perception of the region: what was more significant was the passage of time. Early Plains settlers wrote of transforming the landscape as well as the climate of the Plains. By the end of the study there was recognition of human transformation of the landscape, positive as well as negative, as well as acknowledgment of humans adapting their methods to the Plains climate.

Whether we are considering the Great Plains or any other part of the world, climatologists have applied literature to their subject: to convey the "feeling" of climate, to capture data not otherwise documented, or even to express their own experiences. A document on the National Oceanic and Atmospheric Administration's (NOAA) website is entitled "Weather Man Poems" and comprises poems written by George Mindling, "Official in Charge" of the Atlanta Weather Bureau Office in 1939. The 12 poems are written from the "weather man's" perspective and can only be truly understood and appreciated by someone in similar situations. Climatologists are, after all, human and "the aim of artist and scientist alike is to communicate a new and valuable way of regarding the natural world around us" (Stephens, 2003).

Literature and climate

To write about places and environments is to call them into being, whether they are real or imaginary. Yi-fu Tuan writes: "Naming is power – the creative power to call something into being, to render the invisible visible, to impart a certain character to things" (Tuan, 1991). Writing about climate makes concrete human experiences and observations, rendering them in a form that can be compared and contrasted with other descriptions. It can be the beginning of climatology or an entryway to another world existing solely in our imaginations. While words are the basis of climatic descriptions, climate can shape literature. Mitchell contends "It is my conviction that literature, like all art, is ultimately a reflection and illustration of the landscape that produced it" (in Mallory and Simpson-Housley, 1987). Geography, being composed of the topography, the climate, and the human landscape, has an immense influence in shaping society and can be "read" in the literature of the region.

Literary scholars and human geographers have examined writers' use of climate and its impact on their literary constructions, such as Pérez and Aycock's *Climate and Literature: reflections of environment* (1995), Mallory and Simpson-Housley's *Geography and Literature* (1987), and Pocock's *Humanistic Geography and Literature* (1981). Great Plains literature, rich in description conveying a sense of the region's climate, has a well-established body of literary scholarship (Keahy, 1998; Quantic, 1995, 1997; Thacker, 1989). All writers utilize climate, together with place, in a variety of ways: as in establishing the setting, indicating the passage of time, serving as a symbol or metaphor, or character development (Lutwack, 1984; Pérez and Aycock, 1995).

The most straightforward use of climate in literature is as spatial location. If an actual place name is used, a realistic depiction of the place and climate is typically expected from its readers who may have first-hand experience of the location. In the case of the Plains, the landscape is known more for its absences: lack of water, physical features, and trees. With such absences, writers focus on capturing the sense of space of land and sky, the constant wind rippling the grasses, the sharp contrasts in weather (Quantic, 1995). Quantic (1997) points out "in a place with no familiar geographical features, men and women had to learn to *see* the Great Plains". Many enjoyed what they saw: "Even against our will, the bigness and peace of the open spaces were bound to soak in.... We could not but respond to air that was like old wine.... Never were moon and stars so bright" (Kohl, 1938). The "bigness" and "peace" of the Plains is often likened to an ocean: ocean imagery is found throughout Plains literature, capturing the expanse but also the "sameness", the "waves" and the silence.

Specific environments, at times, become stereotyped with certain types of writing: such as rainforests (jungles) or arid lands (deserts) with adventures or the arid American West with the Cowboy Western. The Great Plains environment is largely associated with pioneer dramas, pitting humans against their environment, such as Ole Rølvaag's *Giants in the Earth* (1927) or Willa Cather's *O Pioneers!* (1913). The environmental conflict is ultimately against the Plains climate, as the settlers struggle to survive in the face of erratic moisture patterns, drought, hail, blizzards, and tornadoes. Many of the classic Plains characters, such as Rølvaag's Per Hansa or Cather's Alexandra Bergson, must come to terms with the Plains landscape and conquer it or be defeated by it (Quantic, 1995). The environment at times is virtually a literary character, providing the conflict that drives the action. Louise Erdrich paints a vivid portrait of a tornado in her novel *Tracks* (1988):

> The odd cloud became a fat snout that nosed along the earth and sniffled, jabbed, picked at things, sucked them up, blew them apart, rooted around us as if it was following a certain scent, then stopped behind us at the butcher shop and bored down like a drill.

In Erdrich's description, the tornado becomes a malevolent beast, rooting and snorting around until it attacks. In Rolvaag's *Giants in the Earth* the Plains environment provides the tension but is personified:

> Monsterlike the Plain lay there – sucked in her breath one week, and the next week blew it out again. Man she scorned; his works she would not brook.... She would know, when the time came, how to guard herself and her own against him! (Rølvaag, 1927).

Climate, abstract, remote but with very real human consequences, is personified or bestialized, allowing climate to occupy space, perhaps even to have conscious thoughts. But the personification of Plains climate has not all been negative. The Plains landscape has often been symbolized as a woman, with the representation depending on the historical period. In writings from early European settlement (or representing this time), the Plains are a fertile young bride, waiting for her husband, the settler, to consummate the marriage and bring forth new life. Imagery from the settled Plains suggests a nurturing mother, providing home and sustenance. In Plains as woman imagery, the settlers must come to terms with the Plains, establish a "relationship" with it, if the "marriage" is to work.

Climate can be used to indicate the passage of time. Time passing can be expressed through the cycle of seasons:

> Gestating in the mid-year
> Seldom did it show.
> Youthful green in springtime.
> White, and gaunt with snow (McRae, 1986).

McRae uses Plains as woman imagery, likening the changing seasons to women's fertility over their lifespan. Evoking past landscapes can also suggest the passage of time:

> The place is as dry as a crater cup,
>   Yet you hear, as the stars shine free,
> From the barren gulches sounding up,
>   The lap of a spawning sea... (Clark, 1915).

The evocation of dinosaurs and oceans (referring to the Plains once being ocean floor) captures the timelessness of the Plains, while pointing to past climates that were markedly different.

Climatic shifts also can indicate spatial movement, as in Canadian poet Robert Kroetsch's *Advice to my Friends*,

> Early yesterday, driving west alone
> Across the prairies, I thought of swinging
> south off Number One to visit shortgrass
> country. Down there, everything is real,
> even the emptiness. The buttes are bare.
> Trees are only a memory... (R. Kroetsch, 1985).

Space and distance are common themes and imagery in Plains literature. Particularly in land not "domesticated", expanses of grasslands are depicted to stretch on forever, unless broken into management segments by roads, fields, and farmsteads, constrained by human actions. The literary journeys of settlers are depicted as crossing climatic regimes while also symbolizing spiritual journeys.

Climate may be symbolic, serving both geographical and literary ends. The repeated association of certain conditions with specific values has resulted in what Lutwack (1984) terms "archetypal place symbolism". Dark clouds on the horizon are seen as "ominous", foreshadowing dark events to come. The sun coming out is a hopeful sign. Lush vegetation is indicative of life, evoking the Garden of Eden. Deserts, lacking vegetation, are akin to death. Most relevant to the Plains is the desert/garden imagery that is pervasive in its literature (Quantic, 1995). Garden imagery presents the Plains as a cornucopia of richness, everything lush and productive, vibrant with life:

> Across bronze farmlands binders click their way,
> While lathery horses nibble at a spray
> Of tender grasses near the raveled edge
> Of broken loam. Elusive quail in sedge
> And covey call their wandering broods.... (Hoover, 1940).

With the Edenic imagery comes the implied position of humans in charge of the garden, of humans conquering the Plains. The climate is obviously ideal for human habitation, even conducive to great success. With the desert imagery, the Plains are stripped of vegetation, the sun appears overpowering, and there is no refuge for any species:

> Beneath me the fissured earth powders barren in my hands;
> my taut, dry body knows the earth's pain.
>
> There in the vertical glare of the sun
> the bald hill lifts a blanching alkali spot;
> my soul is like that in its desertness (Crafton, 1927).

The desert is both in climatic region and in spirit: everything is stripped bare, exposed. The desert/garden split is reflective of the wet/dry climatic pattern of the Plains. While a firm temporal pattern has not been established, there is evidence that the Plains alternate between wet and dry periods (Worster, 1999).

In the dry periods hope comes only in the form of rain, which at times supported the belief of positive climatic change. The notion of "rain follows the plow", can be found, not only in essays and historical accounts, but also in its imaginative literature, such as "Quivera – Kansas". The poem describes Coronado's search for Quivera, the departure of the Spanish, and the arrival of American settlers. While acknowledging the Spanish, it has the Plains unresponsive until the arrival of the "blue-eyed Saxon race", weaving an impression of European exploration but also of American Manifest Destiny:

> And it bade the climate vary;
>   And awaiting no reply
>   From the elements on high,
>   It with plows besieged the sky –
> Vexed the heavens with the prairie.
>
> Then the vitreous sky relented,
>   And the unacquainted rain
>   Fell upon the thirsty plain,
>   Whence had gone the knights of Spain,
> Disappointed, discontented (Ironquill, 1899).

"Quivera – Kansas" not only captures the notion of plowing transforming climate but also the environmental determinism that tied landscape conquest with race, here suggesting that the "Latin race" could not work the Plains-desert but the "blue-eyed Saxons" altered the climate and made it productive.

Today, the appreciation of the Plains' sparse landscape has been termed "Prairie Zen" and continues to inspire writers, 'such as Kathleen Norris' *Dakota: a spiritual geography* (1993). While some are repelled by the Plains' climatic extremes and sparse landscape, others are drawn to it, appreciating the subtle textural changes in its topography and vegetation, enjoying the ever-changing panorama offered by its expansive sky. The nature of the plains as garden and desert can be seen as embodying the cycle of life, death and rejuvenation. To appreciate the Plains is to accept them as they are; their landscape, their climate, their sense of place:

> The prairie is never really a thing or even a group of things. This absence leaves us with nothing to stand against, nothing to be a subject toward. We cannot play the role of detached evaluator. We can only accept the gentle onslaught of prairie, the sterilizing light and the desiccation of hubris.... The prairie is an experience, not an object – a sensation, not a view. The prairie is a way of being and not a thing at all (Evernden, 1983).

Reflecting on climate and literature

Plains literature has been shaped by its climate while also reflecting it. We could select almost any world region and illustrate the relationship between its climate and its literature. Climate figures prominently in the work of Colombian novelist and Nobel Prize laureate Gabriel García Márquez, whose work is set in his home region, the Caribbean coast of Colombia (Elbow, 1995). Russian author Ivan Turgenev's writings reflect his roots on the Russian Plain (Paul, 1987). All humans experience climate. It is an essential element of our life on this Earth. Humans also desire to share and connect with others, whether they are climatologists or writers. At the heart of climate and literature is the simple human desire to communicate.

Christina Dando

Bibliography

Cather, W., 1913. *O Pioneers!* Boston: Houghton Mifflin.
Chiao-Min Hsieh, 1976. Chu K'o-chen and China's Climatic Changes. *Geographical Journal*, **142**(2): 248–256.
Clark, C., 1915. The bad lands. In Coursey, O.W., ed., *The Literature of South Dakota*. Mitchell, SD: Educator Supply Company.
Converse, M., 1927. Drought. In Hoopes, H.R., ed., *Contemporary Kansas Poetry*. Lawrence, KS: Franklin Watts – The Book Nook.
Crafton, J., 1927. Alien. In Hoopes, H.R., ed., *Contemporary Kansas Poetry*. Lawrence, KS: Franklin Watts – The Book Nook.
Dando, C.E., 1992. *The Dreaming Is Finished: human perception of the Great Plains through poetry, 1890–1950*. Master's thesis: Department of Geography, University of Wisconsin–Madison.
Elbow, G., 1995. Creating an atmosphere: depiction of climate in the works of Gabriel García Márquez. In Pérez, J., and Aycock, W., eds., *Climate and Literature: reflections of environment*. Lubbock, TX: Texas Tech University Press.
Erdoes, R., and Ortiz, A., (eds), 1984. *American Indian Myths and Legends*. New York: Pantheon Books.
Erdrich, L., 1988. *Tracks*. New York: Henry Holt.
Evernden, N., 1983. Beauty and nothingness: prairie as failed resource. *Landscape*, **27**(3): 1–8.
Hoover, G., 1940. The land comes back. In Davis, A.S., ed., *Davis' Anthology of Newspaper Verse for 1940*. New York: Henry Harrison.
Ironquill [Eugene Ware]. 1899. Quivera – Kansas. In *Rhymes of Ironquill*. Topeka, KS: Crane.
Keahy, D., 1998. *Making It Home: place in Canadian prairie literature*. Winnipeg: University of Manitoba Press.
Kohl, E.E., 1986. *Land of the Burnt Thigh*. St Paul: Minnesota Historical Society Press.
Kroetsch, R., 1985. *Advice to My Friends: a continuing poem*. Don Mills: Stoddart.
Lamb, H.H., 1995. *Climate, History and the Modern World*, 2nd edn London: Routledge.
Lawson, M.P., and Stockton, C.W., 1981. Desert myth and climatic reality. *Annals of the Association of American Geographers*, **71**(4): 527–535.
Lutwack, L., 1984. *The Role of Place in Literature*. Syracuse, NY: Syracuse University Press.
McRae, W., 1986. The Land. In *It's Just Grass & Water*. Spokane, WA: Oxalis Group.
Mallory, W.E., and Simpson-Housley, P., (eds), 1987. *Geography and Literature: a meeting of the disciplines*. Syracuse, NY: Syracuse University Press.
Mindling, G.W., 1939. Weather Man Poems. Documents, National Oceanic and Atmospheric Administration. http://www.history.noaa.gov/poetrycorner.html
Norris, K., 1993. *Dakota: a spiritual geography*. New York: Ticknor & Fields.
Paul, A., 1987. Russian landscape in literature: Lermontov and Turgenev. In Mallory, W., and Simpson-Housley, P., eds., *Geography and Literature: A Meeting of the Disciplines*, Syracuse, NY: Syracuse University Press, pp. 131–151.
Pérez, J., and Aycock, W., (eds), 1995. *Climate and Literature: reflections of environment*. Lubbock, TX: Texas Tech University Press.
Persson, A., 1995. Anne Frank's Diary – a meteorologist's view. *Weather*, **50**(6): 218–220.
Pike, W.S., 2002. An appreciation of the weather in *The Lord of the Rings*. *Weather*, **57**(12): 439–446.
Pocock, C.D., (ed), 1981. *Humanistic Geography and Literature: essays on the experience of place*. London: Croom Helm.
Quantic, D.D., 1995. *The Nature of Place: a study of Great Plains fiction*. Lincoln, NE: University of Nebraska Press.
Quantic, D.D., 1997. The Midwest and the Great Plains. In *Updating the Literary West*. (Sponsored by The Western Literature Association) Fort Worth, TX: Texas Christian University Press, pp. 641–649.
Rannie, W.F., and Blair, D., 1995. Historic and recent analogues for the extreme 1993 summer precipitation in the North American Mid-Continent. *Weather*, **50**(6): 193–200.
Rølvaag, O., 1927. *Giants in the Earth: a saga of the prairie*. Translated from the Norwegian; English text by Lincoln Colcord and the author. New York: Harper.
Roots, E.F., 1984. Glacier mass balance measurements – an honourable past, an important future. *Geografiska Annaler*, **66A**(3): 165–167.
Stephens, G.L., 2003. The useful pursuit of shadows. *American Scientist*, **91**(5): 442–449.
Thacker, R., 1989. *The Great Prairie Fact and Literary Imagination*. Albuquerque, NM: University of New Mexico Press.
Tuan, Y., 1991. Language and the making of place: a narrative–descriptive approach. *Annals of the Association of American Geographers*, **81**(4): 684–696.
Tyrell, J.G., 1987. Reconstructing 19th-century vegetation patterns in Kenya. *Geographical Review*, **77**(3): 293–308.
Worster, D., 1999. Climate and history: lessons from the Great Plains. In Conway, J.K., et al., eds., *Earth, Air, Fire, Water: humanistic studies of the environment*. Amherst, MA: University of Massachusetts Press, pp. 51–77.

Cross-reference

Cultural Climatology

LITTLE ICE AGE

Subject to varying definitions, the "Little Ice Age" (LIA) is here taken to be the dramatic climatic deterioration that followed the "Medieval Warm Period" (of the high Middle Ages, *q.v.*). A recent book on the subject by Brian Fagan (2000) bears the subtitle *How Climate Made History* (AD) *1300–1850*. Lamb (1977) described the catastrophic cooling in Britain after about AD 1310, many villages being completely wiped out. On the European continent conditions were even worse and in Bohemia (modern Czech Republic) at one stage up to half the population died of starvation. Alexandre (1987) has cataloged a year-by-year documentation, and Barbara Tuchman (1978) has presented a socioeconomic picture of the "calamitous" fourteenth century. "The Great European Famine" lasted no less than three consecutive years (1315, 1316, 1317). The first overall treatments were by LeRoy Ladurie (1971) and by Jean Grove (1988), though now both are a little dated. The coldest episodes were around 1315, 1530, 1600–1616 (Fernau neoglacial) and 1800–1825 (Napoleonic era).

Glaciers are not always the best indicators of climate, because they tend to vary with elevation and show sluggish response behavior. The famines of the fourteenth century were mainly the result of heavy year-round precipitation, but as temperatures began to fall, increasing snows led to glacial advances. It was not until the seventeenth century that the maximum glacier advances in the Alps were documented. However, in Greenland the first advances began in the thirteenth century, and the Norse settlers began their progressive withdrawal; the last survivor was finally frozen to death in the mid-fifteenth century (McGovern, 1981).

In North America the best-known human reactions were among the Anasazi people of the southwest, where progressive desiccation led to failures of their staple crop, maize, and their retreat to upland canyons, where cliff-dwellers survived for a while, before their final extermination. In the Pacific, which had been extensively colonized by the Polynesians before the first millennium AD, the atoll settlements were partially threatened by droughts, resulting in migrations to the "high islands" that were mostly well watered. Isolated settlements such as Easter Island became overpopulated, and by about 1350 moai carving was abandoned as deforestation and droughts led to internecine warfare.

South America, with its topographic variety providing a constant buffer against climatic fluctuations, did not fare badly in the LIA, although the periodic incidence of "super-El Niños" led to catastrophic floods along the major river valleys of Peru, the coastal belt of which is normally semiarid. The glaciers in Patagonia, as also in New Zealand, appear to have fluctuated more or less in phase with those of the northern hemisphere, although some areas alternate (Carter and Gammon, 2004). Eustatically, the LIA was marked by the "Paria Emergence" (Fairbridge, 1961).

In Africa, apart from the Nile Valley and other major waterways, the subsistence farming of most of the people favored migratory solutions to periodic droughts. This was highly effective in the days before colonial boundary lines. In the well watered "inland delta" of the upper Niger a powerful empire developed.

In the Nile Valley and the northern rim of Africa, after AD 622, the Islamic hegemony introduced a new culture that was largely based on the use of sub-Saharan slaves who could be sold off during droughts. The price of slaves in the markets of Morocco and the other North African centers correlates quite closely with the LIA rainfall indicators (from tree rings). In the Nile Valley the general fertility (and economy) depended largely on the regularity of the summer flood that was fed by the monsoonal rains in Ethiopia. The flood records were generally maintained with great care, and closely monitored because they provided the taxation basis. Power spectrum analyses have provided evidence of both the 11/22-year solar and 18.6-year lunar cycles (Currie, 1987, 1995).

In Asia, particularly the west-central parts (now Kazakhstan, Uzbekistan, etc.), the first inkling of a former general desiccation came from Ellsworth Huntington (see Huntington and Visher, 1922), an American geographer and explorer, who discovered that there had been a general aridification during the LIA. The great river, the Oxus or Amu Darya, during classical times, had flowed from the northern borders of Afghanistan and Damir Mountains westward to the Caspian Sea (near Krasnovodsk), but decreasing volume and encroaching Kara Kum dunefields (in AD 1559) deflected its flow into the Uzboi Channel and the Aral Sea. Here a vast inland sea, the "Oxian Lake" existed, as noted by Alexander the Great, Strabo and Marco Polo.

Glaciers in the Tieni-Shan, Kuan Lun, Karakoram and other mountains in western China and northern Himalaya, fed oases that traditionally serviced the Old Silk Road and other trade routes in the time of Greece and Rome. General desiccation forced the abandonment of many sectors and rerouting to more favorable localities. People have pointed to the parallelism of the desiccations with the low eustatic levels in the sea-level curve of Fairbridge (1961).

The level of the Caspian Sea is largely dependent upon the latitude of the jet stream and rainfall in the Volga Basin. It is therefore out of phase with the level of the Black Sea, which is eustatically linked to that of the world ocean. During much of the twentieth century its level was falling, in contrast to the general rise of the world trend, but in the last decade or so, precipitation in the Volga basin has been rising, and likewise in the Caspian. Environmentalists had claimed erroneously that the fall in the earlier part of the century was due to dam construction. On the other hand, the Aral Sea fluctuations prior to the modern irrigation diversions were in phase with the world ocean.

In China there are two distinctive climatic indicators: the southern sector with the Asian monsoon and the northern one with the yellow dust (loess) transport from the region of the Gobi Desert, together with the yellow mud of the Hwang-Ho and Gulf of Pohai. As shown by an abundant literature touching on climate matters in China, the weaker monsoons led to summer rainfall deficiencies, a feature of the LIA, together with increasing northeasterly winds bringing dust storms in the winter season. Short-term modulations reflect both the sunspot cycle and the 18.6-year lunar period (Currie and Fairbridge, 1985).

In Southeast Asia the monsoon is the dominant factor, but flooding in the great river valleys during the LIA reflects the heavier snowfalls in the upper waters, all of which are located on the north face of the Himalayas or in eastern Tibet, but debouch over a wide span – from west to east – the Indus, Ganges, Brahmaputra, Irrawaddy, Salween, Mekong and Yangtzi. The LIA flooding left a particularly impressive result in the region of Angkor (Cambodia) where the annual flood causes the lake of Tonle Sap to back up. However, this LIA back-up was repeated on a vast scale and the entire temple complex was eventually buried under more than 10 m of mud.

A very generalized picture of the LIA is thus presented, but with enough contrast to make it clear than global averaging, almost universally espoused by modeling climatologists, is not a reasonable procedure. Regional compartmentalism, recognizing airmass systems and circulations, seems more actualistic and logical than the indiscriminate stirring of a giant stewpot with its inevitable inhomogeneous data sources. Historical studies (e.g. Lamb, 1977) show occasional overlaps of one compartment to another occur, creating for example the freezing of the Thames in England, snowfall in Ethiopia, heavy rains in central Australia, and so on. They call for explanation, but do not excuse homogenization.

Luni-solar forcing

In as much as most of the span of the LIA lies outside the range of modern instrumental records, a variety of proxy evidence is employed, based on:

1. Documentary material such as diaries, historical events, e.g. the El Niño cycles, the Scandinavian auroras, and the Schove sunspot cycle reconstruction.
2. Ice cores, Greenland, Canadian Arctic, central Andes, western China and Antarctica.
3. Tree rings, particularly where analyzed for ${}^{14}_{7}C$ flux and ${}^{18}_{7}O$ evidence.
4. Lake varves and fluviatile flood levels, the lake varves being treated primarily in Sweden, the marine varves in the Santa Barbara Basin (California), the flood levels (in annual records) on the Nile.

5. Beach-ridge series in sites where chronologic calibration is feasible, as in the Hudson Bay and northern Scandinavian and Russian littorals.

Power-spectrum analyses, covering worldwide sources and an extensive variety of geophysical indicators, disclose that both the 11–22-year sunspot cycle and the 18.6-year lunar cycle are universally present (Currie, 1987, 1995). From lacustrine proxies in western Canada, Campbell et al. (1998) conclude that "postglacial climatic anomalies such as the LIA and the Younger Dryas were at least in part periodic phenomena rather than the result of unique, aperiodic events".

It should not be forgotten that there are two distinct avenues that link solar activity with Earth systems, i.e.:

1. Electromagnetic radiation (optical, UV etc.), focused on the intertropical belt.
2. Particulate magnetic radiation (in the solar wind), focused on the magnetic poles.

The principal lunar cycles also have two critical areas, i.e.:

1. The Indonesian–New Guinea constriction, where ocean tides modulate heat transfer to the atmosphere and the orographic uplift carries it to the stratosphere.
2. The geostrophic currents, notably the Gulf Stream, are accelerated by lunar declination, bringing sensible heat to the North Atlantic and Arctic Ocean (Murmansk Current). Positive feedback is added by the melting of sea ice and reduced albedo.

Of overwhelming importance are the magnetic cycles that are related to the solar inertial orbits of the systemic barycenter (described in detail in Fairbridge and Sanders, 1987; and expanded by Charvátová, 1997). On Earth the incoming solar particles are funneled into the magnetic poles and displayed above the stratosphere in the auroras. The entire LIA interval was covered by Scandinavian diaries (mainly by the Lutheran pastors, who used the auroras as messages from the deity). A classical analysis of the 800-year record by Ekholm and Arrhenius (1898) discovered not only the sunspot cycle, but also the lunar cycle. The latter, primarily seen as an Earth-oriented tidal phenomenon, also involves the sun, and was recognized by Pettersson (1914), although it is often ignored (Sanders, 1995).

Four intervals of high geomagnetic activity are recognized, coinciding with sunspot minima, of varied length and importance. They are named after solar pioneers, in chronologic order:

1. Wolf Minimum.
2. Spoerer Minimum.
3. Maunder Minimum.
4. Sabine or Dalton Minimum.

Their importance and duration is indicated by the values of ^{14}C flux derived from tree-ring analyses (Stuiver and Braziunas, 1989), but subject to the sampling intervals and instrumental "errors" of the technology (about $\pm$ 5 years, on average). The dendrochronology in this interval has been so often repeated and checked that the dates obtained are not subject to any correction.

The pattern of $^{14}_{7}C$ flux suggests bimodal states, fluctuating from length intervals of an "adagio" mood (gentle oscillations: Windelius and Carlborg, 1995) to brief but violent phases ("staccato" mood). Comparable bimodality was independently observed by Charvátová (1997). The key parameter was the radius of the sun's orbital loops (Fairbridge and Sanders, 1987), and their symmetry, the simplest form being compared with a three-leafed clover (Charvátová's "trefoil"). The whole orbital symmetry cycle (OSC) has a length of 178.73 years, being controlled by the beat frequency of the two largest planets Jupiter and Saturn (19.8593 years $\times$ 9), each clover-leaf loop being one-third of this.

Superimposed on this is the sunspot cycle (SSC mean: 11.121 years) and its magnetic reversal (SMR: 22.24 years), but that is not quite in phase with the 178.73-year orbital symmetry cycle. A second cycle of 177.94 years is set up by alignments of Jupiter, the heliocenter and the barycenter (the "King–Hele Cycle"), which is 16 $\times$ SSC, and 8 $\times$ SMR.

During the late twentieth century worldwide concern regarding atmospheric pollution by CO_2, CH_4 and other "greenhouse gases" has led to an interest in "natural", i.e. preindustrial, climatic patterns. On several different time scales there is a systematic tendency for a sawtooth pattern to develop. The scale may be 100 000, 10 000, or 1000 years, or even shorter. Time marches on, unidirectionally. Graphs should be drawn with the x (ordinate) parameter, i.e. time, always advancing from left to right, so that the viewer perceives the future, or unknown area, to the right. Thus a sawtoothed curve of the last 10 000 years will show the Medieval Warm Period (*q.v.*)terminating at about AD 1300, and the LIA beginning at about that point. Temperatures will fluctuate, but the coldest extremes will appear gradually lower and lower, with the maximum cold indicators of the Maunder Minimum about AD 1680. A steep rise is then initiated to reach a preliminary peak around AD 1760. A ratio of 380:80 (or 4.75) defines the "cut" of the sawblade. Since 1760 the anthropogenic rise destroys the natural pattern.

All climatic factors and different regional aspects should not be expected to conform to such a simplistic model, due to multiple delaying or feedback mechanisms (latent heat in large water bodies, albedo, ice-melting rates, forest growth, etc.). Mann (2002) points out the value of multiple proxies. Other writers have suggested developing climatic patterns over the LIA and its aftermath (Free and Robock, 1999; Jones et al., 2001; Rind, 2002). A consensus, however, is beginning to emerge: the sun must play the essential role (Shindell et al., 2001).

Acknowledgment

This brief entry is based on a lifelong study, aided by many friends worldwide. Of outstanding material help has been Jim Hanson of the NASA Goddard Institute for Climate Studies.

Rhodes W. Fairbridge

Bibliography

Alexandre, P. 1987. Le climat en Europe au Moyen Age. Contribution á l'historie des variations climatiques de 1000 á 1425, d'aprés les sources narratives de l'Europe occidentale. Ecole des Hautes Etudes en Sciences Sociales, Paris, 825 pp.

Briffa, K.R., and Osborn, T.J., 2002. Blowing hot and cold. *Science*, **295**: 2227–2228.

Carter, R.M., and Gammon, P., 2004. New Zealand maritime glaciation: millennial-scale southern climate change since 3.9 Ma. *Science*, **304**: 1659–1662.

Campbell, I.D., Campbell, C., Apps, M.J., Rutter, N.W., and Bush, A.B.G., 1998. Late Holocene approximately 1500 year climatic periodicities and their implications. *Geology*, **6**(5): 471–473.

Charvátová, I., 1997. Solar–terrestrial and climatic phenomena in relation to solar inertial motion. *Surveys in Geophysics*, **18**: 131–146.

Currie, R.G., 1987. Examples and implications of 18.6- and 11-yr terms in world weather records. In Rampino, M.R., Sanders, J.E., Newman, W.S., and Konigsson, L.K. (eds.). *Climate: history, periodicity, and predictability*. New York: Van Nostrand Reinhold, pp. 378–403.

Currie, R.G., 1995. Variance contribution of M_n and S_c signals to Nile River data over a 30.8 year bandwidth. *Journal of Coastal Research*, Special issue **17**: 29–38.

Currie, R.B., and Fairbridge, R.W., 1985. Periodic 18.6-year and cyclic 11-year induced drought and flood in northeastern China, and some global implications. *Quaternary Science Reviews*, **4**(2): 109–134.

Currie, R.G., Wyatt, T., and O'Brien, D.P., 1993. Deterministic signals in European fish catches, wine harvests, sea level, and further experiments. *International Journal of Climatology*, **13**: 655–687.

Ekholm, N., and Arrhenius, S.A., 1898. Ueber den Einfluss des Mondes auf die Polarlichter und Gewitter. *Kongl. Svenska Vetenskaps Akademiens Handlingar* (Stockholm), **31**(2): 77–156.

Fagan, B., 2000. *The Little Ice Age*. New York: Basic Books.

Fairbridge, R.W., 1961. Eustatic changes in sea level. In Ahrens, L.H. et al., eds., *Physics and Chemistry of the Earth*. London: Pergamon Press, Vol. 4, pp. 99–185.

Fairbridge, R.W., and Sanders, J.E., 1987. The Sun's orbit, AD 750–2050: basis for new perspectives on planetary dynamics and Earth–Moon linkage. In Rampino, M.R. et al., eds., *Climate: history, periodicity, and predictability*. New York: Van Nostrand Reinhold, pp. 446–471 (bibliography, 475–541).

Free, M., and Robock, A., 1999. Global warming in the context of the Little Ice Age. *Journal of Geophysical Research*, **104**: 19 057–19 070.

Grove, J.M., 1988. *The Little Ice Age*. London: Methuen.

Haigh, J.D., 1996. The impact of solar variability on climate. *Science*, **272**: 981–984.

Huntington, E., and Visher, S.S., 1992. *Climatic Changes*. New Haven, Yale University Press.

Jones, P. et al., 1998. High-resolution palaeoclimatic records for the last millennium: interpretation, integration and comparison with General Circulation Model control-run temperatures. *Holocene*, **8**: 455–471.

Lamb, H.H., 1977. *Climate History and the Future*. London: Methuen.

Lean, J., Beer, J., and Bradley, R., 1995. Reconstruction of solar irradiance since 1610: implications for climatic change. *Geophysical Research Letters*, **22**: 3195–3198.

LeRoy Ladurie, E., 1971. *Times of Feast, Times of Famine: A History of Climate Since the Year 1000,* translated by B. Bray, Garden City, NY, Doubleday.

McGovern, T.H., 1981. The economics of extinction in Norse Greenland. In: Wigley, L.M.L., et al. (eds.). *Climate and History*. Cambridge University Press, 404–433.

Mann, M.E., 2002. The value of multiple proxies. *Science*, **297**: 1481–1482.

Pettersson, O., 1914. On the occurrence of lunar periods in solar activity and the climate of the Earth. *Svenska Hydrogr. Biol. Komm. Skriften*.

Rampino, M.R., Sanders, J.E., Newman, W.S., and Konigsson, L.K. (eds), *Climate: History, Periodicity, and Predictability*. Van Nostrand Reinhold.

Rind, D., 2002. The sun's role in climate variations. *Science*, **296**: 673–677.

Sanders, J.E., 1995. Astronomical forcing functions: from Hutton to Milankovitch and beyond. *Northeastern Geology and Environmental Sciences*, **17**(3): 306–345.

Schove, D.J. ed., 1983. *Sunspot Cycles* (Benchmark Papers in Geology, vol. 68). Stroudsburg, PA: Hutchinson Ross.

Shindell, D.T. et al., 2001. Solar forcing of regional climate change during the Maunder Minimum. *Science*, **294**: 2149–2152.

Stuiver, M., and Braziunas, T.F., 1989. Atmospheric 14-C and century-scale solar oscillations. *Nature*, **338**: 405–408.

Tuchman, B.W., 1978. *A Distant Mirror: The Calamitous 14th century*. New York: A. Knopf.

Windelius, G., and Carlborg, N., 1995. Solar orbital angular momentum and some cyclic effects on Earth systems. *Journal of Coastal Research*, Special issue **17**: 383–395.

LOCAL CLIMATOLOGY

The first use of the term "local climate" is unknown; however, German climatologists began to use it frequently as a synonym for *Kleinklima* (small climate) in the early 1930s. Since then local climate has appeared in many monographs and textbooks (Fukui, 1938; Landsberg, 1941; Sapozhnikova, 1950; Yoshino, 1961, 1975; Berényi, 1967; Barry, 1981). However, R. Geiger, a founder of microclimatology, did not use the term local climate; instead, he used *orographisches Mikroklima* (orographical microclimate) and *Kleinklima* (small climate) in his early works (Geiger, 1927, 1929). He continued to use these terms in the revised editions of his book (Geiger, 1961, 1965). These terms are now used as synonyms for microclimate, topoclimate and mesoclimate.

Although any region of the Earth's surface can be either small or large depending on the point of view for the classification, the expanse of a region can be determined based on the climatological phenomenon under consideration. Local climate is the climate in an area where local conditions of the Earth's surface are clearly different from those in the nearby surrounding areas; for instance, it is the climate in and around mountains, hills, rivers, lakes, coasts, forests, and cities. Different definitions of local climate and different sizes of an area of local climate have been used by different research workers; for example, Landsberg (1941) used local climate for a climate scale between the spot climate and the regional climate; Mörikofer (1947) first defined the horizontal scale of local climate as an order of 100–1000 m; Flohn (1959) and Barry (1970) defined it as an order of 100 m to 10 km and the vertical scale an order of 10 cm to 1 km (Yoshino, 1975). A schematic illustration of local climate is given in Figure L7.

Elements, factors, and conditions of local climate

Elements of local climate are similar to those of global climate, i.e. air temperature, precipitation, wind, humidity, sunshine, insolation, and so on; but air pressure, which is one of the most important elements for global climate, does not play an important role in local climate. Factors affecting local climate are topography, land and sea distribution, ground cover, and so on; but latitude, which is the most important factor affecting global climate, has no bearing on local climate.

To understand the physical processes that influence the formation of local climate, location, exposure, and different conditions of the ground surface – such as color, apparent density, heat capacity, moisture content, and permeability of the soil, characteristics of vegetation cover, albedo, and roughness of the ground surface – should be taken into consideration. These are all factors that influence the heat, moisture, and momentum exchange on the ground surface (Thornthwaite, 1953). Accordingly, the investigation of local differences in these factors is necessary on the open, flat ground. In a hilly area it is also necessary to investigate thoroughly the influence of terrains on the air current and to discuss the distribution of the climatic elements caused by this influence. Here, the word terrains means the inclination and the direction of main valleys and ridges, the extent of mountain masses, and so forth.

Most local climates are distinguishable primarily under the condition of weak winds and usually clear skies. Therefore, in Germany, a nickname *Schön-Wetter Klimatologie* (fair-weather climatology) was once given to local climatology, because local climatological observations were performed only under fair-weather conditions. This means that the wind, one of the elements of local climate, plays an important role in determining the local climate distribution. It also implies that a study in the synoptic–climatologic framework is desirable. For instance, in the investigation of local climatological features such as frost

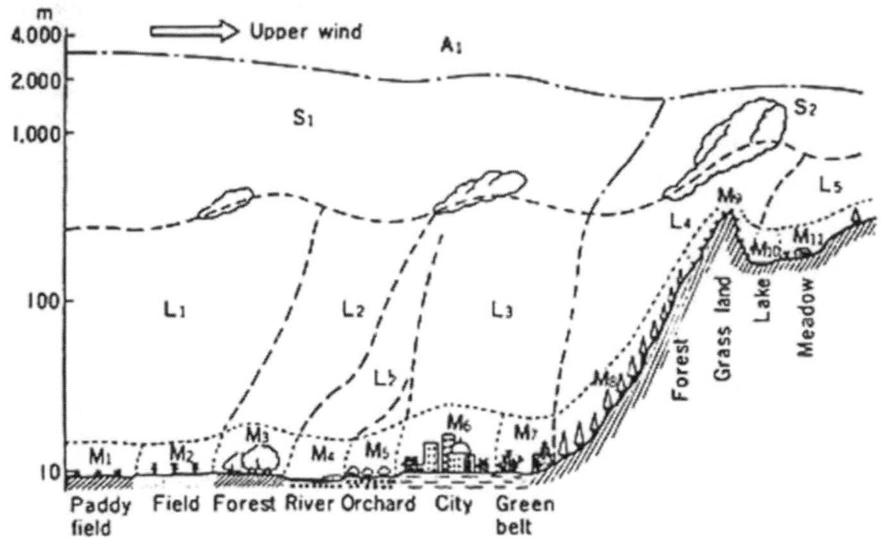

Figure L7 Schematic illustration of micro- (M1–M11), local (L1–L5), meso (S1–S2) and macro (A1) climatic phenomena (Yoshino, 1975).

distribution, occurrence of strong local winds, existence of heat islands in cities, and precipitation maximums on mountain slopes, the synoptic climatologic conditions are important for the consideration of local climatic phenomena.

A detailed classification of meteorological and geographical factors producing local climates was presented (1958), as shown in Table L1.

History of local climatic description

Many local climatic events have been recognized since ancient times, even though the term local climate was not used. Marcus Vitruvius (75–26 BC), a famous Roman engineer and architect, made remarks concerning local climatic conditions for planning cities. He dealt with the direction of slopes and stressed that the distance to swamps, lakes, and beaches must be considered from the standpoint of local climatic conditions. In ancient India it was pointed out in the Mānasāra Silpásāstra that the direction of major roads should be decided by referring to the prevailing wind directions. Also in India, the Arthastra, a book on politics written in the fourth century BC by Chanakya, has a chapter that deals with local differences of rainfall, probably measured by a rain gauge.

In Japan, *Harima-Fudoki* (Regional geography for Harima), written in the early eighth century, recorded the grades of soil that may represent the overall environment suitable for the agricultural production in an area 40 × 60 km. Although the judgement of productivity seems to have been based on subjective criteria, this record is noteworthy because it presents not only a map showing the conditions 1200 years ago, but also a map based on observed results on the local climatic conditions with grades in a small area.

Sir John Evelyn (1620–1706), a famous British diarist, wrote in his book, *Fumifugium, or the Inconvenience of the Aer and Smoke of London dissipated*, published in 1661, that "the weary Traveller at many Miles distance, sooner smells, than sees the City to which he repairs" (Landsberg, 1981). This is a keen observation on the dissipation of polluted air from London from the standpoint of local climatology. He also reported an observation on the result, due to local differences, of the severe frost damage in the century that occurred in the winter of 1683–1684.

Table L1 Geographical factors producing local climates

Factor	Items to Be Mapped
Type of surface	
Rock	Type, color, thermal conductivity.
Soil	Type, texture, color, air and moisture content, thermal conductivity.
Water	Surface area, depth, movements.
Vegetation	Type, height, density, color, seasonal change.
Agricultural	Fallow land; type, height, and color of crops; seasonal change.
Urban and industrial	Material (concrete, tarmac, wood, metal, etc.) color, thermal conductivity; sources of heat, moisture, pollution, etc.
Properties of surface	
Geometrical shape	Flat, convex, concave, etc.
Energy supply	Latitude and altitude, degree of screening of natural horizon, aspect, slope, exposure.
Exposure	Shelter provided by macro- and micro-orographic features; shelter provided by buildings, trees, etc.
Topographic roughness	Rural areas: extent of woodland, grassland, arable; location of windbreaks and hedges; degree of agglomeration or dispersal of individual buildings.
	Urban areas; distribution and average height of different types of built-up zones; orientation and exposure of streets, blocks and individual buildings; density of parks, gardens and other open spaces; vertical profiles across area.
Albedo	Type of surface.
Radiating capacity	Type and maximum temperature of surface; observed earth radiation.

Source: From Stringer (1958), p. 384.

Mountain climates were observed in connection with the development of modern alpinism. Horace Bénédicte de Saussure (1740–1799) was one of those who were dedicated to mountain climbing and the science of mountains. He contributed to the understanding of local differences of mountain climates.

The forest influences on local climate were also observed in many places of the world in the seventeenth and eighteenth centuries in connection with the construction of windbreaks and shelter belts.

People in the seventeenth century were already well aware that urban areas have unique climates as compared to rural districts. Howard first compiled in 1818 various records on the climatic elements for London and released them in 1833 after arranging the observed records for the years between 1797 and 1831 (Howard, 1833). Heat island, fog, wind, insolation, sunshine hours, and air pollution were the main subjects of local climatological recognition in the nineteenth century.

Agricultural microclimatology has been developing since the beginning of the last century. For example, in Russia it started at the Main Geophysical Observatory in Leningrad since 1932 (Gol'tsberg, 1970). Many researchers studied local climatic conditions for agriculture from subtropics, arid areas to boreal forest regions. A local climate subdivision of the climatology division was established at the Observatory in 1949–1950. A book entitled *Microclimate and Local Climate* was written by Sapozhhnikova (1950), a leader of the subdivision. An English translation of this book was widely read in the USA.

From the standpoint of air pollution, it has been pointed out that air as a storage medium, mainly in a local scale, could no longer sufficiently purify itself on a global scale since the 1950s (Wanner, 1997).

After modern climatology and meteorology became established disciplines, year-long records by instrumental observation, air temperature lapse rate on mountain slopes, change of precipitation amount with increasing altitude on slopes, air temperature at seashores and lakeshores, land and sea breezes, mountain and valley breezes, ground temperature distribution on small hills, and other such measurements became the main topics of local climatology.

Development of research methods, summarizing the previous studies (Geiger, 1961; Geiger et al., 1995; Oke, 1978; Yoshino, 1975) could be chronologically classified into the following three periods:

1. The period when analyses were made on the effect of local conditions at a certain climatological station. This was accomplished by comparing mean values of adjacent stations. The climax period of this research method was from the nineteenth century to the beginning of the twentieth century.
2. The period when analyses were made in as much detail as possible on the horizontal distribution of the climatic elements in small areas. In Vienna, air temperature distribution was observed by driving a car to various locations in the city on 12 May 1927. After this study, cars became indispensable for so-called mobile observation in small areas.
3. The present period when climatic phenomena in small areas are examined by experiments such as wind tunnels and mathematical models using supercomputers. Remote sensing techniques are introduced also in this period after the 1970s.

The above chronological classification is not necessarily included in individual research of local climatology. The research methods of the first and second stages are still significant today. All three methods should be used together in small areas. One example of research that used all three methods was a project on the local wind Bora along the Adriatic Coast of Kroatia, and in the Ajdovscina Basin of Slovenia (Yoshino, 1976).

Local climatic phenomena

Insolation and sunshine

Insolation is affected by the elevation and slope inclination of a mountain. The higher a mountain, the shorter is the distance for solar rays to travel through the atmosphere. They also become stronger at higher altitudes, because the amount of water vapor and dust in the atmosphere decreases. The sun rises earlier and sets later on high mountains, so the annual variation of insolation becomes smaller than on lowlands.

As a rule, it is expected that sunshine hours become greater with increasing elevation but they decrease distinctively where fog occurs frequently. Figure L8a shows the relationship between the annual total sunshine hours and the height above sea level and a maximum at 400–500 m. On the other hand, the relationship between the annual and August total number of fog days and the height above sea level is given in Figure L8b. These curves show marked maximums at 1500 m above sea level, which correspond to the minimum of sunshine hours at the same elevation.

These phenomena are related to other local climates in the mountains. For instance, the vertical change of the annual range of minimum sunshine hours and the maximum at 1100–1200 m on the mountains in central Japan. However, the heights of

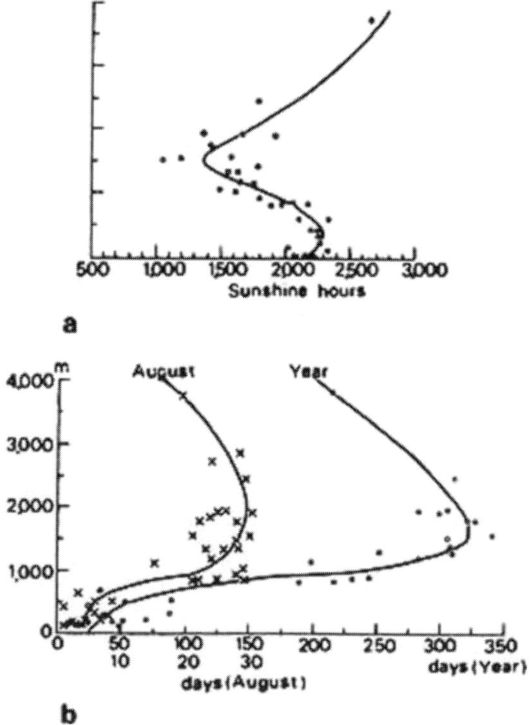

Figure L8 (**a**) Vertical distribution of annual sunshine hours in Japanese mountains; (**b**) vertical distribution of number of days with fog in the Japanese mountains (from Yoshino, 1975).

minimum sunshine hours and the maximum fog (cloud) occurrence change from region to region in accordance with macroclimatic conditions. In particular, the height of condensation level changes from lower latitude to higher latitude except for the subtropical high-pressure zone, which experiences a different height for maximum fog (cloud) occurrences on mountains.

On the valley slopes in the higher mountain areas, contrast of climates between sunny side and shadow side is quite obvious, in particular, in the middle and high-latitude regions. Not only air temperature, sunshine hours, and other climatic elements, but also landscape such as remaining snow in spring, vegetation, land use and human settlements show clear contrast as caused by these local climate differences. Since human life is affected by these conditions, place names are reflected sometimes by these conditions (Yoshino, 2001): Sonnenseite (Sunny-side) or Sonnenheim (Sunny-village) in Austria, Adrey (in the Roman language), Adritto (in the Italian language) or Poele chaud (in the French language) in Switzerland, Sol, Solana, and Soula (sun in the Catalonian language) in Spain, and Hinata, Himuki or Hyuuga (Facing sun) in the Japanese language are the examples of sunny side slopes. Contrarily, Schattenseite (Shadow-side) or Winterhadle (Winter-slope) in Austria, Ubae, Opaco, and Pierre gelee in Switzerland, Baga, Ubach, and Ubago in Catalonia, and Hikage, or Kage in Japan are the examples of place names of shadow side slopes.

Precipitation on mountains

The amount of precipitation increases with height and reaches a maximum at a certain height on the slope of a mountain because of the lifting of moist air caused by the orographic effect. The mechanism of this rainfall is air ascending along a slope, water vapor condensing into water droplets through the adiabatic change and falling as raindrops at a certain height. Precipitation decreases at higher slopes, above the height of the rainfall maximum. Therefore, precipitation on high mountains is scarce. The distribution of annual precipitation on both sides of Mount Everest (8848 m) shows a maximum at Zhangmu on the southern foot, caused by prevailing southerly, moist winds, and decreases toward the summit as shown in Figure L9. On the northern side, precipitation is only 200–300 m because of the so-called rainshadow effect and the high elevation of the Tibetan Plateau. For example, annual precipitation at Tingri (28°40′N, 87°00′E) is 236.2 mm.

Lauscher (1976) distinguished five types of vertical distribution of precipitation on mountains of the world: (1) in the tropics there is a marked maximum at 1.0–1.5 km; (2) in the equatorial climates there is a general decrease in precipitation with increasing height; (3) the subtropics show a transitional type with a slight increase at increased height; (4) in the middle latitudes there is a sharp increase with height; and (5) in the polar regions there is a very weak decrease with height. A more detailed examination, however, would show that there are many local and regional deviations. For instance, Lauer (1976) found the maximum belts of annual total precipitation on the Cameroon Mountains, Mexico, depending on height above sea level, as shown in Figure L10. In general, the drier a mountain base, the higher the maximum belt of annual precipitation totals, as shown in Figure L10. Lauer came to the conclusions that: (1) in the monsoon circulation regions within the tropical westerly winds, the maximum belt descends to sea level as shown for the examples in the Cameroon Mountains; (2) on the Caribbean slopes of the Mexican Meseta, the maximum belt appears between 600 and 1400 m with weak secondary maximums around 3000 m; (3) in the desert of marginal tropics, such as those in the Sahara, convection by intense heating is set off within disturbances in the tropical easterlies giving a maximum precipitation at 2500 m in the Hoggar Mountains.

Thermal belt on mountain slopes

It is said that Thomas Jefferson described the thermal belt, calling it a *verdant zone*, in 1781. The stratification of air temperature in the valleys and the enclosing slopes was brought to the attention of Silas McDowell of Franklin, North Carolina, a learned farmer, in connection with fruit growing. He mentioned the concept of thermal belts in 1858 (Dunbar, 1966).

Below the thermal belt, air temperature inversion is intense near the ground on clear calm nights; hence, the air temperature is lowest at the lowest level. On the contrary, above the thermal belt, air temperature decreases with height at a normal lapse rate.

A detailed study of the distribution of air temperature around Mount Tsukuba (876 m, 70 km NNE of Tokyo) showed the thermal belt at heights about 200–300 m above sea level, as shown in Figure L11.

According to the comparative study with previous results, the following can be summarized on the thermal belt (Yoshino, 1975):

1. The height of the center of the thermal belt appears at a height of 100–400 m above a valley floor situated below a mountain. In most cases it is 200–300 m.
2. The position of the thermal belt depends on the cross-section shape of the valley and mountain slopes.
3. The thermal belt is higher and more distinctive on a clear, calm night.
4. The thermal belt is higher in winter than in summer.
5. Climatologically, the appearance frequency of the thermal belt depends on the occurrence of synoptic conditions which bring clear, calm nights.

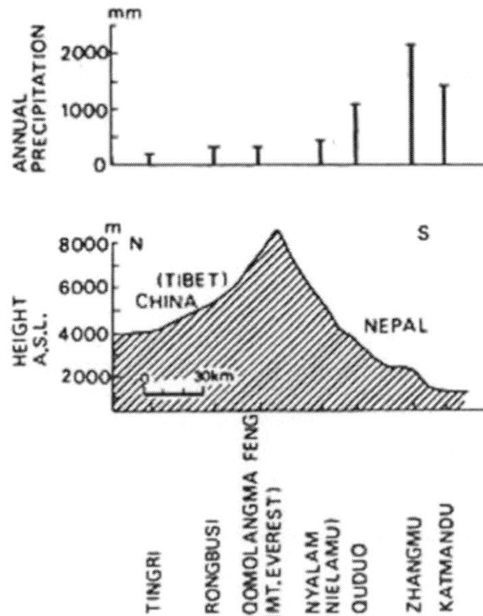

Figure L9 Annual precipitation along the north–south cross-section of Mount Everest, Himalaya.

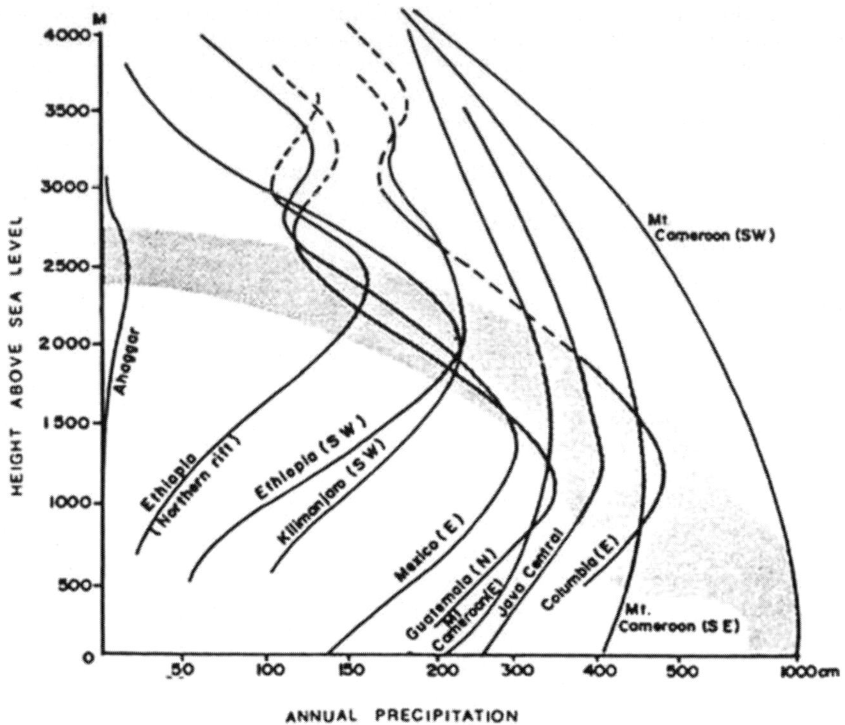

Figure L10 Vertical distribution of annual precipitation in the tropics, showing the maximum belt (shaded areas) dependent upon annual totals (from Lauer, 1976).

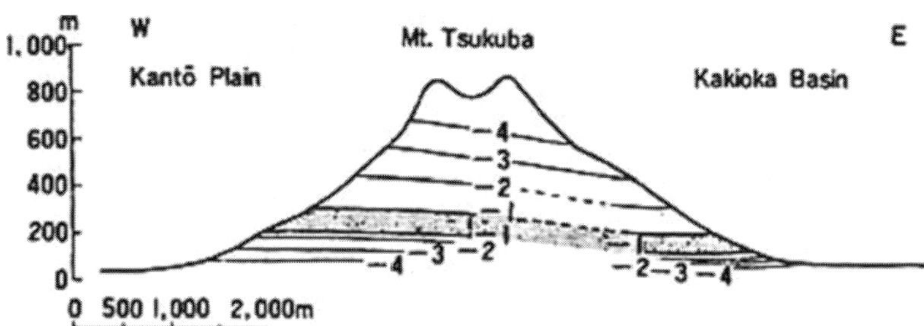

Figure L11 Vertical zonation of monthly mean minimum temperature along the west–east cross-section of Mount Tsukuba, Japan, January 1955 (from Yoshino, 1975).

There are many examples of fruit orchards, such as orange, peach and pear, and of human settlement and plants that originally were grown in lower-latitude regions and are found at the thermal belt of mountain slopes in the middle latitudes, because of the favorable thermal conditions. Thermal belts provide very good examples for the use of local climatic conditions to benefit human life.

Cold-air lake, valley inversion, and inversions on plains

The accumulation of cold air in a basin or in a valley on a clear, calm night is called a *cold-air lake*. The diameter of cold-air lakes varies from a few meters to several kilometers in most cases. Large cold-air lakes, several tens of kilometers in diameter, are sometimes found. Although not commonly used at present, a large cold-air lake has been called a *cold-air sea*, and a small cold-air lake a *cold-air pond*.

The layer of cold-air accumulation is an inversion stratification: that is, temperature is the lowest at the ground surface. This inversion layer is formed not only by strong cooling by the nocturnal outgoing long-wave radiation at the basin or the valley floor, but also is intensified by the inflow of cold air from the surrounding slopes.

Summarizing the results and schematic illustrations obtained from many parts of the world, the following conclusions can be drawn:

1. The thickness of the cold-air lake or valley inversion is 0.2–0.3% of the relative height between the bottom of the basin or valley and the surrounding mountain ridges.

2. Over the cold-air lake, a downslope wind or cold-air drainage flows down from surrounding slopes, whereas its compensating current, an upwind, occurs in the upper layers. These winds form a circulation system.
3. The level of compensating currents coincides with the thermal belt on the above-mentioned slopes. Therefore, the cold-air lake is generally deeper in the colder season than in the warmer season.
4. Some cold air from the slope flows into the cold-air lake as an important source of cold air to the lake.
5. Above this circulation system, general upper winds flow. A model of these wind systems is shown in Figure L12 (Yoshino, 1984).

At the open ground surface on a plain, the air layer near the ground begins to cool rapidly about 1 hour before sunset under a calm, clear sky condition. This is due to the greater radiation loss of heat at and near the ground surface. As a result of this process, an increase in temperature with height begins at ground level. This is called ground inversion, surface inversion, low-level inversion, or radiation inversion.

The inversion layer plays an important role in the local distribution of weather and climate. In particular the wind direction, velocity, and pollutant diffusion are much different under the influence of the nocturnal inversion layer than under the daytime, unstable conditions. In fall and winter a dense but local fog layer is formed near the ground, frequently due to the cooling of air near the ground.

In the cold-air lake, therefore, the fog is dense; it is called cold basin fog or valley fog. Fog disappears with the dissipation of the inversion layer in the morning hours. This means, in turn, that the appearance of this local fog is a sign of dominant anticyclonic conditions synoptically, and one can anticipate fair weather in the daytime.

Local climate in coastal regions

Climate in coastal regions is characterized by a smaller diurnal and annual air temperature range and the noticeable periodic change of wind direction, i.e. a land breeze at night and a sea breeze during the day.

There are both direct and indirect causes for the smaller diurnal and annual range in temperature. The direct cause is the difference between the physical characteristics of land and water. Since the warming caused by insolation in the daytime is smaller above the water surface than above the land surface, because of the higher specific heat of water, water turbulence, transparency, and increased evaporation from water, the maximum air temperature above water is lower and the cooling by long-wave radiation above the water at night is smaller. Therefore, the diurnal variation is smaller on the coast than inland.

The indirect effect of smaller temperature changes is caused by wind. Generally speaking, winds are stronger on the coast than inland because of the smaller roughness on the coast. Strong winds cause lowering of the temperature in the air layer nearest the ground during the daytime hours and an increase in temperature near the ground at night. Thus, the diurnal range of air temperature is small on the coast. The onshore sea breezes on the coast further counteract a rise in air temperature in the afternoon in the warm season. Conversely, synoptic scale winds on the coast lessen the extreme lowering of air temperature at night during the cold season, because they do not allow development of an intense inversion layer under the stronger wind condition. This may be one reason for the smaller annual range in air temperature in the coastal region.

An extreme example of marine climate is seen on small islands and capes; they have low frequency of frost occurrence but higher humidity and less air pollution. For instance, the monthly mean minimum air temperature in January at Cape Ashizuri (32°43′N, 133°01′E), Shikoku, Japan is 7.3°C, but 1.6°C at Tosashimizu (31°47′N, 132°58′E). The horizontal distance between these two stations is only about 10 km and the horizontal temperature gradient is 0.6°C/km. Although wind is considered an element of local climate, it also acts as a climatic factor for influencing the distribution pattern of other elements of local climates. This phenomenon is clearly seen in the coastal region.

Local front, local cyclone, and local anticyclone

In local climates, fronts, cyclones and anticyclones are important from the dynamic standpoint because they cause

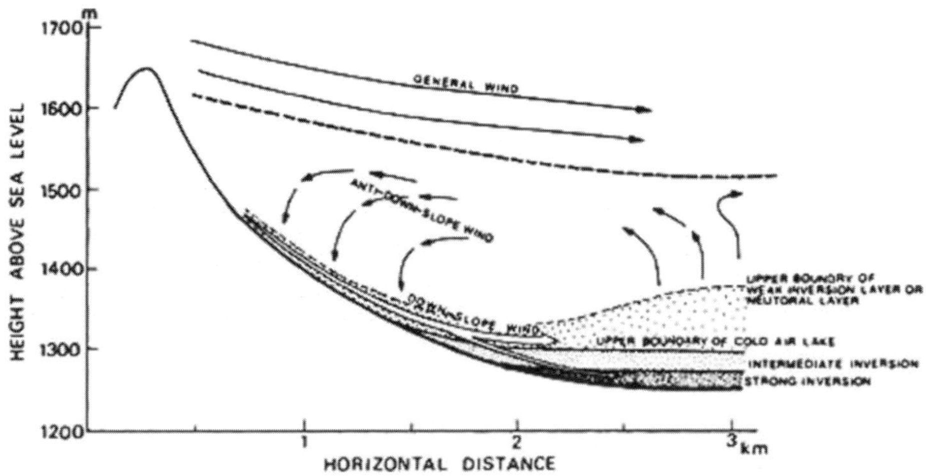

Figure L12 Schematic illustration of cold-air lake and cold-air drainage in a basin on a clear, calm night.

differences in local weather. In general meteorology, a front or frontal zone means the boundary between two airmasses of different densities. On the other hand, the local front can be distinguished not only by the difference in air temperature, but also by the discontinuities of wind direction, moisture, clouds, precipitation, visibility and pollutants. In most cases air pressure differences are small along the local front.

A sea breeze front sometimes becomes a smog front because pollutants stagnate on the ocean side of the sea breeze front. This situation occurs frequently in California (Kauper, 1960). Along the local front two converging air streams sometimes create a severe local storm, causing intense precipitation during a short period. On the local front between a cold mountain wind or land breeze and the warmer synoptic-scale wind such as a monsoon, precipitation occurs at night or in the early morning. Such precipitation is generally not heavy but occurs very frequently in some places. In this sense it is important climatically. A typical example may be the morning maximum precipitation found at the mouths of Himalayan valleys.

Local cyclones are formed thermally over the broad flat land areas under the influence of intense insolation during the warmer season and in the relatively warmer inland seas, lakes, and bays during the colder season. Mechanically, local cyclones are formed on the lee side of mountain ranges (Atkinson, 1981). A local cyclone itself seldom causes heavy precipitation, but it develops very quickly in some cases, bringing strong precipitation.

Local anticyclones often develop in winter over broad basins and land areas surrounded by relatively warmer water, whereas in summer they are formed over the broad water surfaces surrounded by relatively warmer land areas. The former is typically created over central Japan under the weak winter monsoon situation (Nomoto, 1975) and the latter over the Great Lakes in the summer (Strong, 1972).

Land-use/cover change, human activities and local climate

The IGBP (International Geosphere, Biosphere Program) contains many core projects, most of which are concerned with (a) conditions on Earth's ground surface, and (b) analyses of the impacts of land cover change on the planet's biogeochemistry, climate and ecology and their complex relationships. On the other hand, the IHDP (International Human Dimensions Program) deals with interaction among land-use change and global environmental change under the influence of human activities. Linkage between the human activities and physical dimension of global environmental change is shown in Figure L13. The special dimensions of global, regional, local and micro-scales are noted on the right-hand side of the figure. The driving forces (upper left-hand side) are population, level of affluence, technology, political economy, political structure, attitudes and values. Developments of local climatology related to agricultural land-use are: (a) division of area according to local climates, (b) establishment of practical application of local climatological knowledge, (c) statistical analysis of small local differences caused by land-use/cover and microtopographical conditions, (d) simulation by modeling at a local scale, (e) analyses by remote sensing methods, and (f) presentation of local-scale phenomena on maps or through other visual methods (Yoshino, 1997).

Masatoshi Yoshino

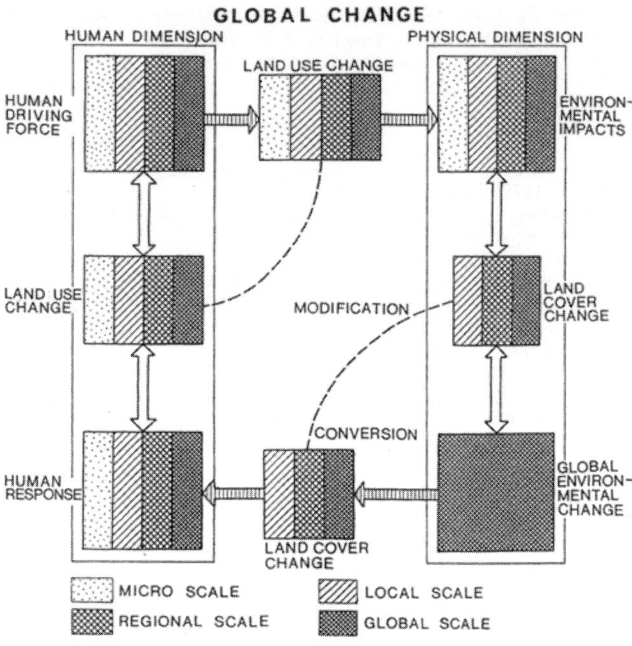

Figure L13 Linkage between the human activities and physical dimension of global environmental change.

Bibliography

Atkinson, B.W., 1981. *Meso-scale Atmospheric Circulation.* London: Academic Press.

Barry, R.G., 1970. A framework for climatological research with particular reference to scale concepts. *Institute of British Geographers Transactions*, **49**: 61–70.

Barry, R.G., 1981. *Mountain Weather and Climate.* London: Methuen.

Berényi, D., 1967. *Mikroklimatologie.* Stuttgast: Gustav Fischer Verlag.

Dunbar, G.S., 1966. Thermal belts in North Carolina. *Geography Review*, **56**(4): 516–526.

Flohn, H., 1959. Bemerkungen zum Problem der globalen Klimaschwankungen. *Archiv für Meteorologie, Geophysik, und Bioklimatologie*, **9**(3): 1–13.

Fukui, E., 1938. *Climatology.* Tokyo: Kokonshoin (in Japanese).

Geiger, R., 1927. *Das Klima der bodennahen Luftschicht.* Braunschweig: Friedrich Vieweg.

Geiger, R., 1929. Die vier stufen der Klimatologie, *Meteorologische Zeitschrift*, **46**: 7–10.

Geiger, R., 1961. *Das Klima der bodennahen Luftschicht*, 4th edn. Braunschweig: Friedrich Vieweg.

Geiger, R., 1965. *Climate near the Ground.* Cambridge, MA: Harvard University Press.

Geiger, R., Aron, R.H., and Todhuntes, P., 1995. *The Climate Near the Ground.* Braunschweig: Friedrich Vieweg.

Gol'tsberg, I.A., 1970. Research in agro- and micro-climatology. In Budyko, M.I., ed., *Voeikov Main Geophysical Observatory 1917–1967.* Jerusalem: Israel Program for Scientific Translation, pp. 84–97.

Howard, L., 1833. *Climate of London deduced from Meteorological Observations*, 3rd edn, 3 vols. London: Harvey & Darton.

Kauper, E.K., 1960. The zone of discontinuity between the land and sea breezes and its importance to southern California air pollution studies, *American Meteorological Society Bulletin*, **41**: 410–422.

Kratzer, A., 1956. *Das Stadtklima*, 2nd edn Buaunschweig: Friedrich Vieweg.

Landsberg, H.E., 1981. *The Urban Climate.* New York: Academic Press.

Landsberg, H.E., 1941. *Physical Climatology.* State College, PA: Pennsylvania State College.

Lauer, W., 1976. Zur hygrischen Hoehenstufung tropischer Gebirge. In Schmithuesen, J., ed., *Neotropische Oekosysteme, Biogeographica*, **7**: 169–182.

Lauscher, F., 1976. Weltweite Typen der Hoehenabhaengigkeit des Niederschlages. *Wetter und Leben*, **28**: 80–90.
Mörikofer, W., 1947. Die Bedeutung lokalklimatischer Einfluss fuer die Kurortplannung. *Annalen der Gesellschaft für Balneologie und Klimatologie*, **38**: 31–38.
Nomoto, S., 1975. A synoptic climatological analysis on local anticyclones, cyclones and fronts in Central Japan. *Geographical Reviews, Japan*, **48**(6): 424–437 (in Japanese with English abstract).
Oke, T.R., 1978. *Boundary Layer Climates*. London: Methuen.
Sapozhnikova, S.A., 1950. *Microclimate and Local Climate* Leningrad: Gidrometeorologicheske Izdatelistvo (in Russian).
Stringer, E.T., 1958. Geographical meteorology. *Weather*, **13**(11): 377–384.
Strong, A.E., 1972. The influence of a Great Lake anticyclone on the atmospheric circulation. *Journal of Applied Meteorology*, **11**(4): 598–612.
Thornthwaite, C.W., 1953. Topoclimatology In *Proceedings of the Toronto Meteorological Conference*, 15–17 September 1953. Boston: American Meteorological Society, pp. 227–232.
Wanner, H., 1997. Air pollution: a local problem becomes a global problem. In Yoshino, M., Domrös, M., Douguédroit, A., Paszynski J., and Nkemdirim, L.C., eds., *Climates and Societies – a climatological perspective*. Dordrecht: Kluwer, pp. 375–380.
Yoshino, M., 1961. *Local Climate*. Tokyo: Chijin-Shokan (in Japanese).
Yoshino, M., 1975. *Climate in a Small Area*. Tokyo: University of Tokyo Press.
Yoshino, M., 1976. *Local Wind Bora*. Tokyo: University of Tokyo Press.
Yoshino, M., 1984. Thermal belt and cold air drainage on the mountain slope and cold air lake in the basin at quiet, clear night. *GeoJournal*, **8**(3): 235–250.
Yoshino, M., 1997. Agricultural land use and local climate. In Yoshino, M., Domrös, M., Douguédroit, A., Paszynski, J., and Nkemdirim, L.C., eds., *Climates and Societies – a climatological perspective*. Dordrecht: Kluwer, pp. 381–400.
Yoshino, M., 2001. *Kiko-chimei shuusei* (climatic place names). Tokyo: Kokonshoin (in Japanese).

Cross-references

Applied Climatology
Coastal Climate
Lakes, Effects on Climate
Local Winds
Microclimatology

LOCAL WINDS

In many areas of the world, regional conditions give rise to winds that have been identified by the inhabitants as having a special effect or as occurring at a definite season. The recurrence of these winds caused them to be seen by the people living in an area as having a uniqueness particular to that region. Accordingly, these local winds were identified by names applicable to a given location.

The effects of these local winds depend on their thermal and humidity characteristics and the velocities they attain. Warm winds, such as the Santa Ana of California and the Sukhovey of the USSR, are extremely dry and cause widespread desiccation of plants. As indicated by the frequency of forest fires in California, the prevalence of such winds and resulting tinder-dry vegetation is a hazard. Such warm winds, however, can also be highly beneficial. The Chinook of Canada, for example, often hastens the melting of snow in spring to permit early planting activities.

In analyzing the causes of local winds, it is evident that most of them are either the Föhn-type wind, resulting from adiabatic heating of descending air, or of a wind that is associated with a regional pressure system that causes a channeling of air by topography. Figure L14 provides an example of the Föhn-type wind and Figure L15, the Mistral, a wind that results from the relative positions of pressure systems.

A comprehensive listing of local winds has been compiled by Forrester (1982) and the following descriptions are drawn largely from his account.

Abroholos

A squally wind on the southeast coast of Brazil, more frequent from May through August.

Austru

A west wind blowing over the lower Danube lands in the winter, usually bringing dry, clear, and cold weather.

Bad-i-sad-o-bistroz

A violent downslope wind affecting Afghanistan and adjacent areas, blowing from the northwest from May to September. Sometimes called *the wind of 120 days*.

Bali

A strong east wind that blows over the eastern end of Java across the Java Sea.

Barat

A squally, occasionally violent, and damaging wind blowing across the Celebes Sea to the northeast coast of the island of Celebes. The barat occurs most frequently from December to February.

Barber

A term used in some sections of the United States and Canada to describe a strong wind carrying precipitation that freezes on contact with objects, especially the beard and hair.

Barine

Unusual winds from the west that blow over eastern Venezuela.

Bayomo

A violent gusty wind usually associated with thunderstorms forming on the windward side of the Sierra Maestra mountain range in southern Cuba. The winds travel with the storms southward from the land over the south coast of Cuba.

Belot or Belat

A strong land wind from the north and northwest that blows along the southeast areas of Saudi Arabia during the period from December to March. Sand picked up from the interior by this wind often imparts a hazy appearance to the atmosphere.

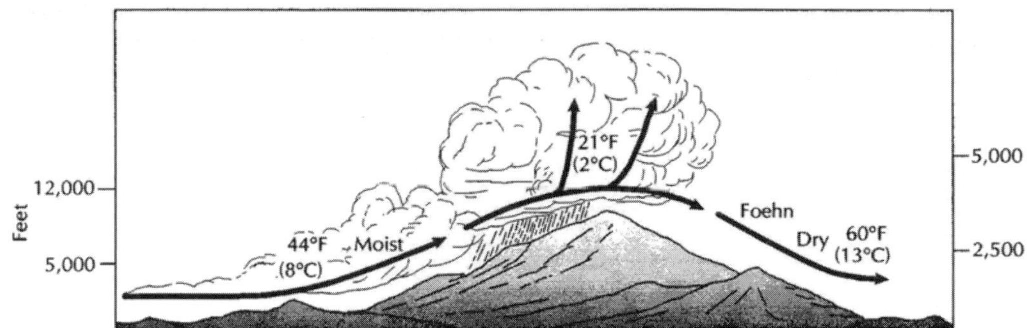

Figure L14 An example of the formation and characteristics of the Föhn wind. Elevations in feet (left axis) and meters (right axis).

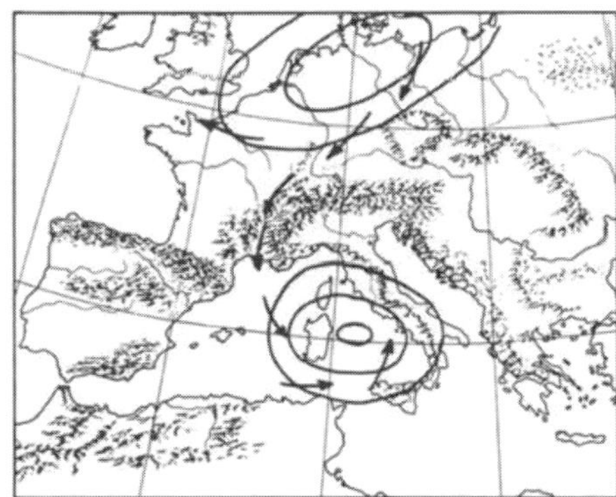

Figure L15 Synoptic conditions that give rise to the Mistral.

Bentu

An east wind blowing in the Mediterranean along the coast of Sardinia.

Berg

Hot, dry, and occasionally dust-laden winds blowing from interior portions of a large part of the southern African continent. Most frequent in the winter season, the berg generally flows in a southerly direction and can bring high temperatures for a period of several days all the way to the southern coast areas of Africa.

Bhoot

In India, a term describing a relatively small-scale counterclockwise whirling of air filled with loose dust, somewhat similar to a dust-devil.

Bise

A strong outbreak of cold dry air from the north that sweeps over the mountainous regions of southern France and Switzerland.

Blizzard

A term used in Canada and the northern United States to describe a howling, cold, piercing wind, usually of gale force, out of the north or northwest. These winds are intensely cold and punishing. They sweep southward in strong surges, moving from areas between Hudson Bay and Alaska. They are laden with blinding, powdery snow, mostly picked up from the ground.

Bohorok

A dry, warm, downslope type of wind that blows frequently from May to September from the leeward side of the Barison Mountains in the southwest to the northwest coast of Sumatra. The wind comes from the Indian Ocean, but loses much of its moisture on the windward side of the backbone mountains in Sumatra.

Bora

Excessively cold wind that moves rapidly southward from the valleys of the Karst and Dinaric Alps across the Adriatic Sea, affecting the Dalmatian Coast from Trieste to Albania. The bora's icy wind sometimes exceeds 100 mph. Also in the gulf of Finland (USSR and South Finland) for a cold northeast wind from Siberia.

Borasco

Winds associated with violent thunderstorms, especially in the Mediterranean.

Bornan

One of the many valley winds in the Swiss Alps, blowing over the central part of Lake Geneva from the valley of the Drance.

Brave West

The strong, often stormy winds from the west and northwest – the prevailing belt of westerly winds of the Temperate Zone of the Southern Hemisphere.

Breva

A valley wind blowing over Lake Como in northern Italy.

Brickfielder or Brickfelder

A hot, dry, and dust-laden wind blowing across the south of Australia from the deserts of the interior during the summer. The name originated in Sydney because of the dust the hot winds raised from the brickfields south of the city.

Brisa

Trade winds that blow from the northeast over the coast of South America or from the east over Puerto Rico.

Brisote

A northeast wind, part of the trades, blowing at a stronger than normal rate over Cuba.

Broboe

A dry wind prevailing from the east from June to October blowing over the southwest part of the island of Celebes.

Brubu

A squally wind in Indonesia.

Bruscha

A cold northwest wind in the Besgell Valley, Switzerland.

Bull's-eye Squall

A sudden squall forming in apparently fair weather. This type of squall is characteristic over the ocean off the Cape of Good Hope, South Africa. Its name is derived from the appearance of a small and isolated cloud seen at the beginning of the squall marking the top of an otherwise invisible vortex of the storm.

Buran

A dreaded, wildly violent, intensely cold wind that breaks out throughout Siberia and into south Russia from the northeast. Similar to the blizzard of the United States and Canada, the buran is dangerous because of its snow-filled character.

Burga

A strong windstorm in Alaska, usually attended by snow or sleet.

Cacimbo

A cooling sea breeze blowing from the southwest to the port of Lobito on the coast of Angola in western Africa. The refreshing breeze, occurring almost daily in July and August, starts about 10 a.m. and lasts through most of the day. It is cooled by its passage over the Buenguela water current lying to the west of Lobito in the South Atlantic.

Cat's paw

A light puff of wind in America just barely noticeable – enough to cause a patch of ripples on water. It is a very light breeze that affects a small area.

Challiho

A strong southerly wind experienced in parts of India during the spring months. It is the forerunner of the southwest monsoon wind that prevails during the summer months. The latter wind, moving from the Indian Ocean, is moisture-laden, and as it moves northward into the mountainous areas, yields enormous amounts of rain on an almost daily basis.

Chergui

An intrusion of hot air into Morocco in northwest Africa from the Sahara Desert areas to the east.

Chili

A hot wind from the deserts of North Africa and Saudi Arabia flowing into central and southern portions of the Mediterranean. Chili is the name applied when this wind passes off Tunisia. It is a dry wind, carrying considerable amounts of dust and sand, occurring more frequently in the spring months.

Chinook

A dry, warm wind blowing from a westerly direction down the east slopes of the Rockies in North America. This föhn-type wind loses its moisture on the windward sides of the Rockies and heats up as it slides down the lee sides. The chinook sometimes raises the air temperature as much as 25°C in less than a half-hour, causing a rapid melting of snow; thus, the chinook is sometimes called the *snow eater*.

Chubasco

A violent squall-type wind associated with severe thunderstorms, frequently occurring on the western coastal sections of Central America and Mexico between Costa Rica and Point Eugenio (Lower California), most frequent during the rainy season from May to November.

Churada

A fierce rain squall in the Mariana Islands, occurring mostly during January, February, or March.

Cockeyed bob

A squall wind associated with thunderstorms on the northwest coast of Australia, most frequent from December to March.

Collada

Strong northerly winds blowing over the Gulf of California.

Contrastes

Winds that blow from opposite directions even though a short distance apart. The western Mediterranean area frequently is subjected to these contrasting winds in the spring and fall seasons when airmasses may move toward each other from the European continent and the desert regions of Africa.

Coromell

A nighttime offshore breeze occurring with great regularity in the La Paz area at the southern part of Lower California from

November to May. At these times the land area cools rapidly at night and the overlying air flows toward the relatively much warmer water areas around La Paz.

Coronazo

Strong south winds blowing from the west coast of Mexico. These winds are usually the eastern peripheries of tropical storms located well offshore to the west.

Criador

A west wind in northern Spain, usually associated with traveling atmospheric disturbances that bring rain.

Crivetz

A cold wind blowing from the northeast over the lower Danube lands from the Russian interior.

Datoo

A westerly sea breeze that blows over Gibraltar from the adjacent water of the Atlantic Ocean.

Doctor

A term originating in England to describe the cooling sea breezes occurring in the tropics. The term *Cape doctor* refers to a strong southeast wind blowing on the South African coast.

Dust devil

A rapidly rotating column of air about 100 ft to 300 ft or higher that picks up dust, straw, leaves, or other light material. It usually develops on calm hot afternoons with clear skies, mostly in desert regions and has no relationship to a large-scale dust storm. In the southwestern part of the United States it is called a *dancing devil*; in India a *devil*; and in South Africa a *desert devil*. In Death Valley, California, it is sometimes called a *sand auger*. It also is called a *dust whirl*.

Elephanta

A strong wind blowing from the south or southwest along the Malabar coast at the extreme southwest end of India during September and October. It heralds the beginning of the dry season and marks the ending of the rainy southwest monsoon.

Etesian

Outbreaks of pleasantly cool air blowing over the eastern Mediterranean, particularly over the Aegean Sea. These outbreaks are most frequent in July and August, generally are moderate to strong, occasionally attaining gale force, although the stronger velocities are limited to the northerly portions.

Feh

A gentle breeze of the Shanghai region.

Föhn

A wind characteristic of many mountainous regions of the world and called by different names in different countries (chinook, Santa Ana, etc.). The term Föhn is more specifically applied in many Alpine valleys, notably in the upper Rhine, the Reuss and upper Aar in central Europe. It is a warm dry wind blowing down the lee side of a hill, mountain, or mountain range after it has risen over the windward side where it has lost its moisture. Such winds may have a marked effect on the climate of a region, sometimes creating an oasis of warm pleasant weather in what would normally be a cold region. In the Reuss Valley of Switzerland, such winds blow on an average of 48 days a year, mostly in March, April, and May.

Gallego

In Spain, a cold wind from the north.

Gharbi

Occasionally, winds from the Sahara Desert blow northward over the northern and eastern Mediterranean. As these hot winds move northward, they occasionally pick up moisture en route and arrive on the north coasts of the Mediterranean as strong winds, warm and damp. These winds are called *gharbi* in the Adriatic and Aegean sea regions. They bring heavy rains, especially on mountainous coasts. Some of the dust of the Sahara is often mixed with these rains, causing "red rain".

Ghibli

A dry hot wind in Tripoli, flowing from the hot deserts of North Africa.

Gregales

A strong polar outbreak of air from the northeast sweeping over the northeastern coasts of the Mediterranean, usually in spring and autumn, and associated with extremely variable weather conditions.

Haboob

Extremely severe dust storms that occur mostly in the summer in large regional areas of northwest Africa, including Egypt. When such squall winds blow in these areas, walls of loose dust advance with the wind, rising to heights of several thousand feet, about 15 miles along a front and moving at speeds of about 35 mph. It is mostly a dry squall caused by the interplay of southwest monsoon winds at the surface and dry hot winds from the northeast aloft.

Harmattan

This wind is a continental part of the globe-encircling trade winds. It dominates the Sahara Desert and impresses its extremely dry and warm characteristics on a huge area of north and northwest Africa. In the summer the parching winds blow moderate to strong from the Mediterranean Sea southward and eastward to about latitude 17°N. In the winter, the wind blows from latitude 30°N to coastal areas of West Africa, sometimes penetrating to the African equator.

Helm

A strong wind in the Pennine Chain in north-central England. One portion of the mountain chain, east of Westmorland County, runs northwest–southeast for about 10 miles, with an average steep-descending southwest slope of about 2000 ft. Extremely strong winds blow from the northeast over the Pennines to the Westmorland–Cumberland regions on the central west coast of England. These helm winds are sometimes associated with a roll or series of rolls of clouds that overhang the crest of a wind wave to leeward of the hills. The mountaintop clouds that form in the windstorm are called *helm clouds*.

Howling fifties

A term probably originating with whalers and others of the nineteenth century who sailed the oceans of the southern hemisphere. These sailors found the winds to be punishing over their routes southward, unhampered by land or mountains. The different latitudes of the general area over which these strong prevailing westerly winds blow (40–50°S of the equator) were given names that reflected their impression of these battering winds; thus, the howling fifties and the roaring forties.

Imbat

A refreshing sea breeze that tempers the heat of the North African coastal areas.

Kai

A balmy south wind in China.

Kapalilua

A prevailing type of sea breeze in Hawaii.

Karaburan

From early spring to late summer, these gale-force winds form daily in the Gobi Desert and surrounding regions in the heart of Asia. Blowing from the east-northeast, they carry clouds of dust from the desert. The blowing sand often darkens the air and is the reason for the karaburan sometimes being called the *black storm*. Lighter dust from the stirred sands carries far beyond the desert areas and provides a characteristic summer haze. These winds rage by day only; at night the desert air calms and skies clear rapidly. In this respect the karaburan of the Gobi and the harmattan of the Sahara are somewhat similar.

Kaus

A wind blowing from the southeast over the Persian Gulf during the winter. It is usually temporary in character and associated with low-pressure areas moving over the gulf region from across the Mediterranean. The wind usually brings cloudy skies, some rain, and above-normal temperatures.

Khamsin

In the winter season, Egypt is invaded alternately by outbreaks of cool and hot winds. Khamsin is the name given to a hot wind sometimes pulled into Egypt from Arabia, the Gulf of Aden and possibly the Arabian Sea far to the south and east. The khamsin, besides being quite hot, is extremely dry and so hazy with fine dust that lights are sometimes required at midday. It usually continues for a few days and then is swept away by an invasion of cold air moving from the northwest behind a cold front. As the cold air arrives, dust and sand is raised, the sky becomes cloudy and sometimes showers fall. Khamsin winds are usually moderate in force but may reach gale force. The mean frequency is about three a month in February, March, and April.

Kharif

A strong, often gale-force wind blowing from the southwest in the Gulf of Aden. It is called the kharif on the Somaliland coast on the south shore of the gulf where the wind descends sand-laden and uncomfortably hot from the African interior.

Knik

A strong southeast wind in the vicinity of Palmer, Alaska.

Kohilo

A light, pleasant breeze of Hawaii.

Kona

A somewhat oppressive and sultry wind of Hawaii, from the southwest.

Koshava

A stormy northeast wind in Yugoslavia, carrying snow from Russia.

Leste

A hot, dry wind blowing from the east and south sections of the north-central African desert regions to Madeira and the Canary islands.

Leung

A cold wind from the north blowing over the China coast.

Levanter

A strong east wind frequently blowing through the Strait of Gibraltar from the Mediterranean. When the wind is excessively strong and stormy, it is sometimes called *llevantades*.

Levanto

A hot wind from the southeast that blows over the Canary Islands; similar to the leste wind.

Leveche

One of the many hot and dry winds that originate in the hot deserts of North Africa and Saudi Arabia, affecting areas of the middle and south Mediterranean. All of them may be grouped under the general term *sirocco*, but they are so important in different areas that many of these sirocco-type winds have been given different names. Leveche is such a wind, blowing from the south and southeast over Spain.

Maloja

A warm, föhn-type wind flowing down from the high Maloja Pass in Switzerland.

Marin

A sirocco-type wind, blowing with strong intensity from the southeast in the Gulf of Lions and the neighboring shorelines of southeast France. It is quite warm and oppressive, usually attended by cloudy skies and heavy rains.

Matsukazee

A soft, pleasant breeze in some agricultural areas of Japan.

Mistral

This well-known wind, the "masterful" north wind of the Gulf of Lions, surges southward in outbreaks from polar regions over north and central Europe to affect a wide area over the northwest coast of the Mediterranean. It is a cold and dry wind that can cause damage, and is particularly noted in the winter in the lower Rhone Valley. Although it blows throughout the year, it is most prominent in the winter and spring. Marseilles is exposed to the mistral for about 100 days a year. Occasionally, the mistral is wide enough in extent to affect the coast between Barcelona and Genoa, the Gulf of Lions (with squalls reaching over hurricane force), and the Balearic Sea. It has, on some occasions, crossed the Mediterranean to the African coast.

Monsoon

In winter a consistent cool-to-cold dry wind flowing southward across India from the Tibetan plateau. In summer it is a warm, humid flow of air northward over India from the Indian Ocean, and precipitates torrential rainfalls in many mountainous areas of India.

Myatel

A stormy, northeast wind in northern Russia.

Naalehu

A wind that occasionally blows from the arid interior of Hawaii.

Narai

A cold wind in Japan, blowing from the northeast and polar regions of the Asiatic land mass.

El Norte

An outbreak of polar air, originating in north-central United States or the Canadian basin, which sweeps as far south as the Gulf of Mexico and parts of Central America. The term also refers to cold winds from the north that flow over eastern Spain in the winter.

Northeaster

A wind that blows from moderate to strong force from the northeast over the New England coastal regions. This wind, moving in from the North Atlantic, is generally moist and often chilly or cold. It is frequently accompanied by cloudiness and precipitation. The "nor'easter," usually a portion of a deep migrating low-pressure area, occasionally can have a severe impact on coastal communities.

Norther

A strong cold wind from the north, originating in the polar regions of Canada, that moves rapidly southward as far as the Gulf of Mexico and eastern coasts of Mexico. Similar to el norte.

Northwester

A moderate to strong wind from the northwest bringing cool to cold temperature over broad regions east of the Rockies. The name also is applied to frequent gale winds that batter the Cape region of South Africa from the northwest, attended by overcast skies and heavy rain in the winter.

Oe

A localized type of whirlwind that occurs off the coast of the Faeroe Islands in the northeastern Atlantic.

Orkan

A high gale of Norway.

Pampero

A violent squall that attends cold fronts as they move from the southwest to the northeast in the pampas of Argentina and Uruguay. It resembles the norther of the United States Plains section in the sense that it is an outbreak of air from polar latitudes. Behind the violent frontal squall the wind eases, blowing moderate and steady. The sky clears and the air is cool to cold. Buenos Aires has about 12 pamperos a year, mostly in spring and summer; Montevideo about 16; the River Plate area has about 20 a year.

Papagayo

A violent wind from the north that invades the Gulf of Papagayo on the northwestern coast of Costa Rica. It is akin to the norther of the United States and el norte of Mexico, resulting from large southward surges of polar air from the North American continent.

Ponente

A westerly wind over the Mediterranean, particularly as a refreshing sea breeze on the western Italian coastine.

Purga

Another name for the dreaded buran of the tundra regions in northern Siberia in the winter. The purga sweeps down from the north with extraordinary violence throughout Siberia and sometimes to south Russia, particularly violent over the open plains sections. The air is filled with snow picked up from the snow-covered tundra areas and this often cuts visibility to zero. The purga is quite similar to the North American blizzard.

Reshabar

A strong wind blowing from the northwest over the Caucasus Mountain range between the Black and Caspian seas.

Roaring forties

The winds prevailing in areas of the oceans between 40° and 50° south latitude where, day after day, the wind force may exceed 40–50 mph from the west. The winds are the prevailing westerly belt of winds that circle the Earth in the southern hemisphere as part of the Earth's primary atmospheric circulation.

Roc

A cold wind of Iceland.

Santa Ana

A föhn-type wind named after a community southeast of Los Angeles in the coastal area of southern California. During the winter, when north and east winds blow from the deserts and plateaus of lower eastern California, they cross the coast ranges and descend through such passes as the Cajon and Santa Ana to reach the coast as hot and dry winds, often laden with piercing particles of dust.

Seistan

Another name for the "wind of 120 days" or the bad-i-sad-o-bistroz that blows strongly from the north in summer over the Seistan basin in eastern Iran. Known for its consistency over a long period of time.

Shamal

A northwest wind blowing down from the Mesopotamian corridor over Iraq and the Persian Gulf in the summer. It often is strong during the day, carrying clouds of dust and sand, but decreases in intensity at night.

Sharki

A wind from the southeast occasionally blowing over the Persian Gulf.

Siffanto

A warm south wind blowing from the "heel" of Italy.

Simmoom or Simoon

A sirocco-type wind, hot, dry, and dust-laden, blowing over the middle and southern portions of the Mediterranean. It is called a simmoom when it is abnormally strong over the southeast part of the Mediterranean. The wind blows northward, originating in the hot desert sections of north-central Africa. The Turkish version of this wind is the samuel.

Sirocco

A warm wind of the Mediterranean area usually sweeping northward from the hot and dry Sahara or Arabian deserts. The sirocco usually invades the Mediterranean shores in the spring when many atmospheric low-pressure areas move over the Mediterranean. As the storms move, their forward sectors pull air northward that originated hundreds of miles to the south. Frequently, therefore, the sirocco is dusty. It often picks up moisture as it crosses the Mediterranean and may arrive on the north Mediterranean shores as a warm and damp wind.

Sno

Cold, swift-moving currents of air that fill the Scandinavian valleys in the winter from the highlands, attaining considerably strong velocities in the fiords.

Solano

An oppressively warm and dusty east wind that blows over Gibraltar and southeast Spain.

Sonora

A warm wind crossing Arizona from Mexico and California.

Southeaster

Strong winds from the southeast, sometimes of gale force, that blow near the extreme southwest end of the Cape of Good Hope, South Africa. They usually occur in the winter and often in advance of a cold spell. Clear skies attend the southeaster at the Cape, but the surface layer of air often carries a whitish haze of salt particles and sea spray.

Southerly burster

Cold winds that move in from polar zones northward over Australia. The air moves behind cold fronts that are attended by strong gale-force winds accelerating and intensifying over the highlands of New South Wales on the southeast coast. About 30 bursters a year occur over southern and southwest Australia, most of them in spring and summer.

Steppenwind

A cold northeast wind that occasionally blows over Germany from the steppe regions of Russia.

Stikine

A strong and gusty northerly wind of extreme southern coastal areas of Alaska near Wrangell, named for the Stikine River or Stikine Mountains to the northeast of Wrangell.

Sudestades or Suestado

Strong to gale-force winds from the southeast that affect coastal areas of Uruguay, Argentina, and Brazil. They are part of frequent migrating cyclonic (low-pressure) circulations in these regions and are accompanied by considerable cloudiness and rain.

Suhaili

A strong wind from the southwest blowing over the Persian Gulf, bringing thick clouds and rain.

Sumatra

Strong thunderstorm squalls blowing over the Malacca Straits from the southwest during the southwest monsoon season. These squally winds, attended by severe lightning and thunder, move along fronts that sometimes are about 100 miles long. They usually last for a few hours, and occur mostly at night. They appear to be surges in the southwest monsoon given extra push by the mountain ranges of Sumatra, which lie parallel to the Malacca Straits.

Surazos

Strong cold polar winds of the Andes Plateau in Peru, with intensifying velocities as they sweep through the mountain passes. The surazos usually bring temperatures below freezing with clear skies.

Taku

A strong wind from the east or northeast in the vicinity of Juneau, Alaska. The name is taken from the Taku River, the mouth of which is in Alaska but that flows over the Alaskan line from Canada. At the mouth of the Taku near Juneau, the wind sometimes reaches near hurricane force.

Tehuantepecer

A violent north wind in the region around the Gulf of Tehuantepec on the extreme south coast of Mexico. In the winter, polar winds sometimes pour southward in great surges from the North American continent (norther or el norte). The winds roar southward across the Gulf of Mexico and reach the Gulf of Campeche in the area just southeast of Vera Cruz. The wind then enters a low-level pass on the Isthmus of Tehuantepec in the mountain chain of the Central American Cordillera. It is intensified by this funnel effect and sweeps southward over the Gulf of Tehuantepec at gale strength. The wind is a scourge to boatmen in the gulf because there are few, if any, precursor signs.

Thalwind

A light pleasant valley breeze in Germany.

Tramontana

A pleasantly cool wind blowing from the north or northeast over the Mediterranean. The name is applied more specifically to this wind when it blows off the western coast of Italy.

Vardarac

Cold, dry polar winds blowing in the winter over the north Aegean Sea. They blow down from the Vardar River Valley in southeastern Yugoslavia.

Vento coado

An erratic, gusty wind blowing over hillside communities of Portugal.

Vento de baixo

A sea breeze along the coast of Portugal

Virazon

A consistent and prominent sea breeze blowing from the Pacific Ocean to the coast of Chile. The virazon is particularly strong during summer afternoons at Valparaiso, affecting harbor activities. The opposite-blowing land breeze in the region is called the *terral*.

Viuga

Stormy northeast winds of southern Siberia, associated with passing low-pressure areas.

Waff

In Scotland, a slight puff of air or gentle breeze, quite similar to the cat's paw of the United States.

Warm braw

A warm, dry föhn-type wind of New Guinea in the South Pacific. Crossing the Nassau and Orange Mountains of New Guinea, the air loses its moisture on the southern side of the ranges during the southwest monsoon season and sweeps northward as a warm and dry flow.

Whirly

A small but violent storm in the Antarctic. The whirling winds of the storm may cover an area of up to 100 yards or so in diameter. They occur most frequently near the time of the equinoxes.

Williwaw

(Not to be confused with Australia williwaw or willie-willie.) Strong, gusty wind in southwest Alaska coming down through the valleys and off the glaciers at speeds of 60–100 knots.

Wisper

A well-defined Rhine valley wind.

Yamo Oroshi

Another föhn-type warm, downslope wind blowing in the steep valleys of Japan.

Zephyr

A soft, gentle breeze, mostly from the west, of the Mediterranean regions (note the similarity of the name to zephyrus, the name given to the west wind in Greek mythology).

Zonda

A strong and dry west wind blowing over the western region of Argentina. The wind is hot, dry, and dusty and occurs most frequently in the spring. It acquires its dry character from the föhn action as it descends to Argentina from the Andes Mountains lying to the west.

Bibliography

Griffiths, J.F., and Driscoll, D.M., 1980. *Survey of Climatology*. Columbus, Ott: Charles E. Merrill.

Forrester, F., 1982. Winds of the world. *Weatherwise*, **35**(5): 204–210.

Oliver, J.E., 1981. *Climatology: selected applications*. London: Edward Arnold.

Cross-references

Adiabatic Phenomena
Katabatic (Gravity) Winds
Land and Sea Breezes
Mountain and Valley Winds

M

THE MADDEN–JULIAN OSCILLATION (MJO)

The MJO is a low-latitude intraseasonal oscillation, meaning that it passes through an identified cycle in a period of 60–90 days, quite unlike the annual, biennial or decadal cycles of other oscillations. It is so named for Roland Madden and Paul Julian of NCAR, who discovered the wave in the early 1970s. Its identification and possible forecasting is of considerable importance in long-range predictability of tropical and subtropical weather as well as short-term climate variability.

The MJO is a feature of the tropical atmosphere–ocean system that plays a significant role in precipitation variability. It is characterized by anomalous rainfall conditions that can be either enhanced or suppressed. The beginning of the cycle, the anomalous rainfall event, usually appears first over the Indian Ocean and Pacific Ocean. It remains identifiable as it moves over the very warm water of the western and central parts of the Pacific Ocean. On meeting the cooler waters of the eastern Pacific Ocean it usually becomes less defined, only to reappear over the tropical Atlantic Ocean and Indian Ocean. As noted this cycle can last 1–2 months.

The higher or lower than normal precipitation of the event are associated with both surface and upper-air conditions as they relate to ascending and descending air. A knowledge of the cycle characteristics would enable a clearer understanding of tropical rainfall variability, the role of the cycle in relation to Pacific Ocean and Atlantic Ocean tropical cyclones and the impact of the oscillation upon middle-latitude precipitation. In their study of the MJO's role in hurricanes in the Caribbean Sea and Gulf of Mexico, Maloney and Hartmann (2000) found a distinct relationship. They noted that hurricanes are four times more likely to occur when, in the ascending phase, the rising air and surface westerly winds are conducive to formation of the storm.

The Climate Prediction Center (CPC) of NOAA (http://www.cpc.ncep.noaa.gov/) shows how the tropical oscillation impacts rainfall in the Pacific northwest of North America. The evolution of this event is often referred to as the "pineapple express", so named because a significant amount of the deep tropical moisture traverses the Hawaiian Islands on its way toward western North America.

The MJO is thought to play a significant role in the formation and frequency of hurricanes. If an easterly wave, the initial formation feature of Atlantic hurricanes, meets the cloudy sky phase of the MJO the conditions for hurricane formation improve. The MJO is, of course, moving in the opposite direction to the easterly wave. Unfortunately, although hurricane researchers of the National Hurricane Center have been informally using the MJO, the linkage is currently insufficient for accurate modeling and forecasting.

John E. Oliver

Bibliography

Ångstrom, A., 1935. Teleconnections of climate changes in the present time. *Geografiska Analer*, **17**: 242–258.

Bjerknes, J., 1966. A possible response of the atmospheric Hadley circulation, to equatorial anomalies of ocean temperature. *Tellus*, **8**: 820–829.

Madden, R.A., and Julian, P.R., 1971. Detection of a 40–50 day oscillation in the zonal wind of the tropical Pacific. *Journal of Atmospheric Sciences*, **28**: 702–708.

Maloney, E.D., and Hartmann, D.L., 2000. Modulation of hurricane activity in the Gulf of Mexico by the Madden–Julian Oscillation. *Science*, **284**: 2002–2004.

Slingo, J.M., Rowell, D.P., Sperber, K.R., and Nortley, F., 1999. On the predictability of the interannual behavior of the Madden–Julian Oscillation and its relationship to El Niño. *Meteorology and Geophysics Abstracts*, **51**: 20264.

Cross-references

Cycles and Periodicities
El Niño
North Atlantic Oscillation
Oscillations
Pacific North American Oscillation
Quasi-Biennial Oscillation
Southern Oscillation
Teleconnections

MARITIME CLIMATE

A maritime climate occurs in regions whose climatic characteristics are conditioned by their position close to a sea or an ocean. Such regions, also known as oceanic climates or marine climates, are considered the converse of continental climates.

Land and sea differences

About 72% of the Earth is covered by water. The continents and oceans basins are unevenly distributed over the Earth's surface, with the northern hemisphere containing some 65% of the total continental surface. The vast area covered by water and its relative distribution play a very important role in the climates that occur not only at sea, but also, to varying degrees, on land.

Properties of water and land

The climate at any location can be determined by evaluating the amount of energy the surface receives and the way in which it is budgeted. This energy budget for the oceans differs from that of land and, as a result, produces a different climatic regime. The basic physical parameters that give rise to the differences are given in Table M1.

Water has a much greater capacity for absorbing and storing heat energy than does other Earth materials. The basis of this difference is the high specific heat of water. Specific heat is defined as the amount of heat required to heat a unit mass by a unit of temperature. In the example given in Table M1, the temperature of water would be raised only one-fifth that of the same amount of a given soil upon applying the same unit of temperature.

The reflectivity of a surface, its albedo, varies over both land and water. Generally, the lighter, smoother, less transparent the surface, the higher the albedo (a perfect reflector has an albedo of 100%). Land surface albedos vary because of the great range of surface types; a desert surface may have an albedo of 25% while an asphalt parking lot may be only 5%. The major difference in albedo over water is the angle of inclination of the sun. If the sun is high in the sky very little reflection occurs, but as the angle increases so the albedo increases. At a sun inclination angle of 90°, water albedo is less than 2%; at 10° it becomes 35%.

Thermal conductivity is the amount of heat that can pass through a given thickness of material in a given time (using standardized units). It is roughly proportional to the density of the material, causing the surfaces of dense material to heat slowly. If transparency is considered along with conductivity, then a marked difference is seen between land and water. Transparency is a measure of how much radiant energy (such as sunlight) can penetrate a surface. Since opaque a materials, granite rocks for example, have no transparency, radiant energy is transformed to heat energy at the surface. The transparency of water permits energy to penetrate to various depths with most infrared energy absorbed close to the surface; with increasing depth other portions of the electromagnetic radiation energy are absorbed until, at about 10 m, the unabsorbed blue portion of the spectrum is reflected back. This property gives the sea its blue color.

Table M1 Selected properties of air, soil and water

Material	Specific heat	Albedo (%)	Thermal transparency	Conductivity
Air	0.17	6	Very low	High
Soil	0.2	5–20	Variable	Low
Water	1.0	5–10	Medium	Medium

The heat budget

To these physical parameters must be added the dynamic elements that make up the heat budget of a surface. These include:

1. The amount of energy used to evaporate water is much greater over the ocean than the continent. This means that a large part of available energy goes to change water to water vapor. This latent heat is later released when water vapor changes back to water.
2. The large portion of available energy over the ocean that is used for evaporation, means less is available for sensible heat, the heat energy that is distributed through the atmosphere through turbulent transfer. This causes air over the land to be warmer than that over the ocean.
3. As noted, oceans absorb radiation to a substantial depth; over land the penetration of energy is limited to often less than a few feet. The oceans are mobile and the absorbed energy may be advected to other areas through currents and drifts.

The combined result of the physical and dynamic differences cause temperature variations in oceans to be much less than those of a land surface. The ocean is an equable environment that absorbs, transports and eventually releases stored heat to the atmosphere. Oceans thus warm up more slowly than landmasses but, because they conserve the heat, they cool down more slowly. Temperatures in the oceans are conservative, those over landmasses, which warm and cool quickly, are extreme. This maritime effect will be most clearly seen in extratropical regions where seasonal extremes are expected. In tropical latitudes the seasonality is more a function of precipitation than of temperature.

The influence of oceans

Surface ocean currents

Surface ocean currents are wind-driven and, like the wind, are subject to considerable variation. The actual pattern of world ocean currents can be related to the global circulation of winds and the shape of landmasses. Just as the subtropical high-pressure zones are a dominant feature of the atmospheric circulation so are the oceanic gyres (oceanic circulation cells) associated with them. On viewing a generalized map of ocean currents of the world the dominance of the subtropical gyres is seen.

The direction in which currents flow is important. Those flowing poleward, the warm currents, transport heat energy from low to high latitudes, while the cold currents, flowing from high to low latitudes, bring cool water to warmer areas. The relative impact of this movement upon temperature, particularly coastal temperatures, is illustrated by temperature distribution over the North Atlantic Ocean. The 10°F (−12°C) isotherm is located at about 50°N on the east coast of North America but extends to almost 70°N in western Europe. This is largely the result of the influence of the warm waters by the North Atlantic Drift.

Upwelling

The top waters of the ocean, perhaps down to 100 m, are mixed by motion of the wind and waves. Below this mixing layer there is a general decline in temperature until the coldness of the bottom waters is reached. The strong temperature contrast between top and bottom waters inhibits vertical mixing unless cold water is forced up to the surface. This upwelling occurs in a number of locations, often where surface water is swept away from the coast by divergence or offshore winds.

The upwelling process, while locally important in influencing ocean temperatures, is really part of a centuries-long circulation of the oceans that involves both surface and deep currents. Upwelling represents the return of cold waters to the surface circulation while downwelling is the return of surface water to the depths.

The influence of upwelling, together with cold ocean currents, will affect the rainfall regime of some coastal regions. By inhibiting convection, cold offshore waters result in desert climates along some coastlines. Although the dominant air is maritime, precipitation is minimal. The Atacama and the desert of Baja California provide examples.

Maritime airmasses

Air derives its thermal and moisture properties from the surface over which it originates. Air originating over the oceans, maritime tropical (mT) or maritime polar (mP) air, is generally characterized by (a) relatively small changes in the annual range of temperature in the source region and (b) a high moisture content derived from its origin over the oceans. These characteristics are, of course, reflective of the ocean properties already outlined. The relative frequency of maritime air received by a location will play a significant role in determining its climate.

Precipitation

Maritime precipitation in the westerly belts is marked by the cyclonic activity characteristic of the polar fronts. These are most intense in the winter months, but the fall season is likely to be wettest. The amount of evaporation into the maritime airmasses is related to both water and air temperature over the ocean. Accordingly, since the warmest conditions of the winter season occur in the fall, the latter is marked by the heaviest precipitation. In midwinter and spring a cold continental airmass may become more important.

There is marked diurnal variation in maritime precipitation, having its maximum at night or in the early morning, in contrast to the continental type, which tends to occur in the afternoon or early evening (Haurwitz and Austin, 1944), although admittedly many variants are known. There may, for example, be a maritime type in winter and a convectional, continental type in summer (especially in the lower-latitude stations). In the tropics there is greater radiation over the ocean at night, although the sea surface remains warm; thus the nighttime temperature gradient is steeper than that of the daytime, and convectional rain is common at night.

Humidity

Over the oceans humidity will reach a maximum in the afternoon and a minimum in the early morning, reflecting the diurnal air temperature cycle. The vapor pressure is controlled in part by the air temperature and in part by the turbulent transport that tends to disperse the humid layer upward, but the latter is less important over sea than over land, where a double cycle tends to develop over the 24-hour period. Relative humidity is lowest in the high-pressure belts of each hemisphere. In the completely maritime stations (all-year) relative humidity is likely to be over 80% at all seasons

Climates near the oceans

The discussion of the factors that influence climates over the oceans can now be related to the climates that occur in regions close to the oceans. It has been emphasized that the oceans are conservative in terms of temperature changes; such is also the characteristic of maritime climates.

The equable maritime climate is found in small, isolated oceanic islands that are most representative of the influence of maritime airmasses and oceanic climate controls. Table M2 shows examples.

There is an appreciable climatic difference between east and west coast climates in middle latitudes. In continental areas the influence of the oceans on climate is experienced in the west coast locations. This is a result of the general circulation of the atmosphere which, outside of the tropics, is dominated by a westerly air flow. Thus the most frequent airmasses that influence the west coast of middle latitudes are from maritime source regions. Those on the east coasts experience air that moves from a continental source.

Ocean currents, which themselves reflect atmospheric circulation, influence both temperature and rainfall in some west coast locations. The cool air of the offshore water lowers annual temperatures while the resulting stable air provides little rainfall.

Continentality and oceanicity

To measure the effect of a continental landmass on climate (i.e. a minimum impact of oceans), climatologists use the concept of continentality. As early as 1888 it was suggested that, by measuring the average annual range of climate and adjusting for latitude, continentality could be quantified. Later workers used this idea as illustrated by the formula used by Gorzynsky:

$$K = [1.7(A/\sin\varphi)] - 14$$

where K is continentality, A is annual thermal amplitude and φ is latitude.

Many modifications of this formula have been suggested. Using the formula where continentality is given by

$$[1.7A/\sin(\varphi + 10)] - 14$$

Conrad (1946), showed that Thorshaven in the Faeroe Islands would have a continentality index of 0.

Table M2 Climatic data for some maritime stations

Island	Warmest month	Coolest month	Annual range	Precipitation (mm)
Fiji	26.6	22.9	2.7	3026
Midway	25.9	18.4	7.5	1176
Waitangi	14.9	7.8	7.1	851
Dutch Harbor	11.7	−0.3	12.0	1443

While maritime climates may be expressed by low values of continental indices, oceanicity indices have also been proposed. The first, in 1905, is given by

$$O = 100[(T_o - T_a)/A]$$

where O is oceanicity, T_o and T_a are the mean monthly temperatures for October and April, and A is mean annual range of temperature.

Although the scientific validity of the empiric measures has been questioned (Driscoll and Fong, 1992), they do provide a general comparison of the oceanic and continental influences.

John F. Oliver

Bibliography

Conrad V., 1946. Usual formulas for Continentality and their limits of validity. *American Geophys. Union Transactions*, **27**: 663–664.

Currey, D.R., 1974. Continentality of extra-tropical climates. *Association of American Geographers Annals*, **64**: 268–280.

Driscill, D.M., and Fong, J., 1992. Continentality: A basic climatic parameter examined. *International Journal of Climatology*, **12**: 185–192.

Hartmann, D.L., 1995. *Global Physical Geography*, San Diego, CA: Academic Press.

Haurwitz, B., and Austin, J.M., 1944. *Climatology*. New York: McGraw-Hill.

Kerner, F., 1905. Thermisodromen, versucheiner Kartographischen Dartstellung des jährlichen Ganges der Lufttemperatur. K. K. *Geographische Gesellschaft, Wien*, **6**(3).

Köppen, W., 1936. Das Geographische System der Klimate. *Handbuch der Klimatologie*, **1**: 1–44.

Landsberg, H., 1958. *Physical Climatology*, 2nd edn, Dubois, PA: Gray Publishing Co.

Lockwood, J.G., 1974. *World Climatology*. New York: St Martin's Press.

Schroedor, M.J., 1967. Maritime air invasion of the Pacific Coast: A problem analysis. *American Meteorological Society Bulletin*, **48**: 802–808.

Thompson, R.D., 1998. *Atmospheric Processes and Systems*. London: Routledge.

Thornthwaite, C.W., 1948. An approach toward a rational classification of climate. *Geographical Review*, **38**: 55–94.

Trewartha, G.T., 1961. *The Earth's Problem Climates*. Madison, WI: University of Wisconsin Press.

Trewartha, G.T., and Horn, L.H., 1980. *An Introduction to Climate*. New York: McGraw-Hill.

Cross-references

Climatology
Continental Climate and Continentality
Evapotranspiration
Lakes, Effects on Climate
Land and Sea Breezes
Wind and Wind Systems

MATHER, JOHN R. (1923–2003)

John Russell Mather was born in Boston on 9 October 1923. After serving as a weather forecaster in World War II, he earned his BA (1945) at Williams College and a BS (1947) and MS (1948) in Meteorology from MIT.

The earliest days of his long distinguished career were spent at the C. Warren Thornthwaite Laboratory of Climatology as one of the first group of graduate students to become intimately involved in the research of the laboratory. The laboratory, which began operation in 1946, was originally located at Seabrook Farms in New Jersey and, in the spring of 1954, was relocated to Centerton, New Jersey. Mather reveled in the research being conducted there, writing that he enjoyed the job because he could go into the woods and not sit at a desk all day. His investigations while at the laboratory included topics such as micrometeorology, evapotranspiration, soil tractionability, soil moisture, climatic classification, and the movement of radioactive strontium in the soil.

Mather is probably best known for his work with the water budget, defined as the daily or monthly accounting of moisture inflows, outflows and storages at a particular place or over a geographical area for any period of time. He felt that the water budget was the basic tool of applied climatology, providing quantitative information on such things as moisture surplus, deficit, retention and runoff.

Mather earned his PhD from Johns Hopkins University in 1951 and remained at the laboratory for 16 years. When Thornthwaite died in 1963, Mather assumed the presidency of the laboratory, and continued his research there for 9 more years.

It has been said that Thornthwaite had the ideas but Mather put them into practice. In many instances this was true; but Mather took those ideas and refined them, making them his own. He expanded Thornthwaite's 1940 version of the water budget, and in 1955 he co-authored with Thornthwaite, *The Water Balance*, which made the ideas more useful for a wide range of soil and vegetative conditions. Thornthwaite had planned to write a book entitled *Physical and Applied Climatology*. The book was never written; it remained for Mather to do it in 1974. His book, *Climate Fundamentals and Applications*, was the fulfillment of Thornthwaite's plan. In it Mather hoped to show how humans could adjust their activities to the atmospheric environment. Chapters focused on hydrology, water resources, agriculture, clothing, human comfort, human health, air pollution, architecture, and industry.

Mather began teaching part-time at the University of Delaware in 1961. When Thornthwaite died that same year, Mather assumed the presidency of the laboratory and remained as president until 1972 when he left to devote his full efforts to the department at the University of Delaware. Thornthwaite's plans for his Laboratory of Climatology were outlined in a 1943 publication. They included, among other things, the training of climatologists on the graduate level. Students, because of the collaborative nature of the research, would need a broad background in all of the sciences. Thornthwaite's ideas became the foundation on which Mather, as chair, and the faculty built the Geography Department at the University of Delaware.

Mather felt that the best career decision he made was to move to the University of Delaware. He saw creating the new department and hiring new faculty as the most challenging job he faced in his life. As Mather wrote in the department history, three decades after Thornthwaite had written about an institution for climatological research, one had been created at the University of Delaware with the rigor in mathematics and the sciences he envisioned.

Permeating Mather's teaching and research was his desire to make climatology come alive, be dynamic and relevant. He hoped his students would be able to apply the principles of climatology to solving problems concerning humans and their environment. The descriptive, encyclopedic approach to climatology, he felt, was wrong; it was "dry as dust."

Mather's list of publications while at the University of Delaware is extensive, touching a wide variety of topics. He studied large- and small-scale applications of the water budget on climate, soils, vegetation, sea level, stream flow, irrigation, agricultural yield, forest products; the effect of land-use changes on crop yield; the effect of urbanization on the water table; the impact of society on the water budget of a particular area; and natural hazards such as coastal storms. His book, *The Genius of C. Warren Thornthwaite, Climatologist–Geographer*, co-authored with Marie Sanderson, is a major contribution to the history of climatology. In it he and Sanderson detail the life of Thornthwaite and his contributions to the discipline.

Mather's own service to the discipline is extensive: American Geographical Society Council member from 1981 to 2000 and secretary of the Council from 1982 to 2000; President of the Association of American Geographers in 1991–1992; Director of the Center for Climatic Research at the University of Delaware from 1972 to 1991; and State Climatologist for Delaware from 1978 to 1991, to name just a few examples.

Mather received many awards: the University of Delaware's Excellence in Teaching Award in 1989; the Commander's Award for Public Service, US Department of the Army, in 1990; the Francis Alison Award for distinguished scholarship at the University of Delaware in 1991; the Association of American Geographers Career Achievement Honors Award in 1998; the Association of American Geographers Climate Specialty Group's Lifetime Achievement Award in 1999; and the American Geographical Society's Charles P. Daly Award in 1999 for "valuable distinguished geographical labors."

Mather retired from the University of Delaware in the fall of 2000 as *professor emeritus*. He died on 3 January 2003, at the age of 79.

Mather was a "people" person who enjoyed interacting with colleagues and students. His legacy is huge; his ideas living on in his students and publications. He called Thornthwaite the "Father of Applied Climatology" and the Laboratory of Climatology the "mecca" for climatologists from around the world. Because of his position as a valued partner of Thornthwaite, Mather spread the gospel of the water budget to succeeding generations. He was eulogized as "an able leader, whose contributions to geography and especially physical geography have rarely been equaled."

Sandra F. Pritchard Mather

Selected publications

Mather, J.R., and Thornthwaite, C.W., 1955. The water budget and its use in Irrigation. *1955 Yearbook of Agriculture*, US Department of Agriculture.

Mather, J.R., Thornthwaite, C.W., 1955. *The Water Balance. Publications in Climatology*. Laboratory of Climatology, Vol. VIII, No. 1.

Mather, J.R., 1960. The laboratory of climatology – an institute for climatic research. *Weather*, **XV**(5).

Mather, J.R., 1972. Thornthwaite's legacy to climatology. *Publications in Climatology*, Laboratory of Climatology, Vol. XXV, No. 2.

Mather, J.R., 1974. *Climatology: fundamentals and applications*. New York: McGraw-Hill.

Mather, J.R., 1980. Hydrologic cycle, Evapotranspiration, Thornthwaite climatic classification. Three articles prepared for and published in *American Academic Encyclopedia*. Princeton, NJ: Arete.

Mather, J.R., 1985. The water budget and the distribution of climates, vegetation and soils. *Publications in Climatology*, Laboratory of Climatology, Vol. XXXVIII, No. 2.

Mather, J.R., and Sanderson, M., 1991. *The Genius of C. Warren Thornthwaite: climatologist–geographer*, Tucson, AZ: University of Arizona Press.

Mather, J.R., 1966. Water budget climatology. In Hanson, S., ed., *Geographic Ideas That Have Changed the World*. New Brunswick: Rutgers University Press.

MAUNDER, EDWARD WALTER (1851–1928), AND MAUNDER MINIMUM

Maunder was the son of a Wesleyan minister, best remembered for two things: his solar research, and as the founder of the British Astronomical Association. He attended King's College in London, and later secured a post as photographic and spectroscopic assistant at the Royal Greenwich Observatory; he initiated a long series of daily photographic sunspot records there. Eventually he was able to prepare a famous diagram, known as Maunder's "butterfly diagram", linking the latitudes of sunspot groups with the state of the 11-year solar cycle. He also carried out spectroscopic work. He publicized the apparent dearth of sunspots between the years 1645 and 1715, coinciding with a very cold spell in Europe; this is now formally known as the "Maunder minimum". His photographic records of sunspots were meticulous, and are still of great value today.

The sun was by no means his only interest. He observed the planets, and in a famous experiment demonstrated that the "canals" on Mars, claimed by Lowell and other observers, were optical illusions. He also made careful studies of the zodiacal light. He took part in various eclipse expeditions, notably the Royal Greenwich Observatory expedition to Mauritius in 1901, during which he obtained an excellent series of photographs. On other expeditions in 1896 and 1898 the weather was unkind. Between 1877 and 1887 Maunder was coeditor of the *Observatory Magazine*; for many years he served on the Council of the Royal Astronomical Society, and was joint Honorary Secretary from 1892 to 1897. In 1890 he founded the British Astronomical Association, and subsequently served as President. He officially retired from Greenwich in 1913, but was recalled during World War I to carry on the sunspot record.

His first wife died in 1888. He then married Miss Annie Russell, herself a brilliant mathematician and an astronomer on the staff of the Royal Greenwich Observatory, who collaborated with him in much of his later work in connection with the sun. He wrote several books, including a history of the Royal Observatory (1900), *Astronomy Without a Telescope* (1901) and *Astronomy of the Bible* (1908).

Maunder minimum

The Maunder minimum refers to a period from about 1645 to 1715 when sunspot activity virtually disappeared (Eddy, 1976; Schove, 1983).The Maunder minimum is one of a number of protracted periods of low sunspot activity; among others identified are the Oort minimum (1010–1050), the Wolf minimum (1280–1340) and the Spörer minimum (1420–1530). It was following the work of Gustav Spörer that Maunder first called attention to the interruption of the normal course of sunspot activity during the second half of the seventeenth and first part of the eighteenth centuries. Maunder published his findings in

1894 but the research appeared to have created little academic interest. In 1922 Maunder published another paper with the same title as that produced earlier (A prolonged sunspot minimum), but it remained for more recent astronomers and astrophysicists to verify and discuss it. Eddy (1976) noted that the sunspot dearth, 1645–1715, is supported by direct accounts in contemporary literature of the day. That the low sunspot activity was real, and not caused by a limitation in observing capability, is illustrated by drawings made of the sun at the time. Additionally, a tree-ring anomaly spanning the same period shows evidence of a concurrent drought in the US Southwest (Douglass, 1928) and when a ^{14}C lag time is assumed, there is a high degree of agreement in time between major excursions in world temperatures and excursions of solar behavior in the records of ^{14}C. The coincidence of Maunder's prolonged sunspot minimum with the coldest excursion of the Little Ice Age has been noted by many researchers examining possible relations between solar activity and terrestrial climate (e.g. Schneider and Mass, 1975; Fairbridge, 1987; Grove, 1988). Given the time relationship between the Little Ice Age and the Maunder minimum, the role of solar variability and sunspot activity has also been examined in terms of recent and future climatic change (e.g. Kelly and Wigley, 1990) and as it may be related to possible changes in the sun's brightness (Baliunas and Jastrow, 1990).

Patrick Moore and John E. Oliver

Bibliography

Anon, 1928. E.W. Maunder. *Monthly Notes of the Royal Astronomical Society*, **89**: 318.

Anon, 1928. E.W. Maunder. *Journal of the British Astronomical Association*, 38, 231.

Anon, 1926/7. A.S.D. Russell (Mrs Maunder). *Journal of the British Astronomical Association*, **57**: 238.

Baliunas, S., and Jastrow, R., 1990. Evidence for long-term brightness changes in solar-type stars. *Nature*, **348**: 520–523.

Douglass, A.E., 1928. *Climatic Cycles and Tree Growth*, vol. 2. Washington, DC: Carnegie Institute.

Eddy, J.A., 1976. The Maunder minimum. *Science*, **192**: 1189–1202.

Fairbridge, R.W., 1987. Little Ice Age. In Oliver, J.E., and Fairbridge, R.W., eds., *The Encyclopedia of Climatology*. New York: Van Nostrand Reinhold, pp. 547–551.

Grove, J.M., 1988. *The Little Ice Age*. London: Methuen.

Kelly, P.M., and Wigley, T.L.M., 1990 The influence of solar forcing trends on global mean temperature since 1861. *Nature*, **347**: 460–462.

Maunder, E.W., 1894. A prolonged sunspot minimum. *Knowledge*, **17**: 173.

Maunder, E.W., 1922. A prolonged sunspot minimum. *Journal of the British Astronomical Association*, **32**: 140.

Schneider, S.A., and Mass, C., 1975. Volcanic dust, sunspots and temperature trends. *Science*, **190**: 741–746.

Schove, D.J., 1983. *Sunspot Cycles* (*Benchmark Papers in Geology*, vol. 68). Stroudsburg: Hutchinson Ross.

Warner, D.J., 1974. Maunder, Edward Walter. *Dictionary of Scientific Biography*, vol. 9, pp. 183–185.

Cross-references

Little Ice Age
Sunspots

MAURY, MATTHEW FONTAINE (1806–1873)

Matthew Fontaine Maury was born in Spottsylvania County, Virginia in 1806 and died in Lexington, Virginia in 1873. Maury worked his way from midshipman to Lieutenant in the US Navy. From 1825 to 1834 his voyages took him to Europe, the Pacific coast of South America, and around the world. He showed special interest in navigation, publishing a *Treatise on Navigation* in 1836. Although his sea career was ended by a leg injury in 1839, Maury was appointed as superintendent of the Depot of Charts and Instruments in 1842. His job title changed to director of the US Naval Observatory and Hydrographical Office in 1844, when the Depot was moved and renamed. In this post Maury collected both ocean current and meteorological data, recorded in specially designed logbooks, from sea captains. He began publishing *Wind and Current* pilot charts of the North Atlantic in 1847, helping to reduce sailing times. He also proposed cooperative weather observation stations on land. Maury used two research vessels after 1849 to collect ocean temperature data and sea-floor samples. Data collected led to his publication of a bathymetrical map of the North Atlantic Basin in 1854, which depicted the sea floor between Yucatan and Cape Verde.

Maury wrote *The Physical Geography of the Sea* in 1855, recognized as one of the earliest oceanographic books, if not the earliest. Maury is widely considered to be "the father of oceanography". Several editions of *The Physical Geography of the Sea* were published, with later iterations taking on the title *The Physical Geography of the Sea, and its Meteorology*. With the exception of polar movement, Maury's 1855 depiction of global circulation appears to be remarkably close to current understandings. As important as his contributions were, however, many of his ideas rightly received criticisms on scientific grounds (e.g. tradewind control by magnetic forces).

Maury resigned and became commander of the Confederate Navy during the Civil War, then Imperial Commissioner for Immigration under Emperor Maximilian of Mexico. With the collapse of the Mexican Empire he moved to England, returning to the US in 1868 as professor of meteorology at Virginia Military Institute. His professorship there lasted until his death in 1873.

Lisa M. Butler Harrington

Bibliography

Daintith, J., and Gjertsen, D., eds., 1999. *A Dictionary of Scientists*. New York: Oxford University Press, pp. 365–366.

Maury, M.F. (edited by John Leighly), 1963. *The Physical Geography of the Sea and its Meteorology.* (Based on the 8th edn, 1861.) Cambridge, MA: Belknap Press.

Oliver, J., 1996. Climatic zones. In Schneider, S.H., chief editor. *Encyclopedia of Climate and Weather*, Vol. 1. New York: Oxford University Press, pp. 141–145.

Stringer, E.T., 1972. *Foundations of Climatology: an introduction to physical, dynamic, synoptic, and geographical climatology*. San Francisco, CA: W.H. Freeman, p. 5.

MEDIEVAL WARM PERIOD OR "LITTLE CLIMATIC OPTIMUM"

Histories in western Europe have long recognized a "post-Carolingian" climatic amelioration, approximately since the death of Charlemagne (AD 814), the first of the "Holy Roman Emperors" of the Middle Ages. Long-term indicators (sea level, glaciers, etc.) suggest it may have begun around AD 550, i.e. the "post-Roman Rise". The expression "Little Climatic Optimum" (LCO), a term believed to have been coined by Bryson, is essentially synonymous with "Early Medieval Warm Epoch" (Lamb, 1977), our "MWP", and also with the "Sub-Atlantic" (SA-2) Biozone of some palynologists.

Lamb (1977, p. 35) identifies a broad long-term "summer wetness index" (rising) and compares it with a "winter severity index" (falling), based on 50-year means. Both display a systematic lengthening of the upper westerlies' wavelength across northern Europe, reaching its peak around AD 1100. A comparable warming swing marked the twentieth century, but is now complicated by anthropogenic pollution. The overall cycle would appear to be about 900 years. The two were separated by the "Little Ice Age" (LIA) from 1300 to 1750. The warmest interval of the MWP in Europe was 1150 to 1300. A study of Canadian borehole temperatures (mentioned by Lamb, 1977, pp. 104–105) suggests that the mean MWP was up to 1.6°C warmer than the average for the last millennium. An interesting humanistic comparison of the MWP and the LIA is made by Fagan (2000).

Sea-surface temperatures in the high latitudes were higher, and sea-ice coverage was appreciably reduced, thus introducing a positive feedback with the decreasing albedo of open water. These factors played an important role in the occupation of Iceland by Viking (Norse) colonists in the ninth century (AD 860). In the North Atlantic there were frequent summers with calm seas and easterly winds that greatly favored their particular type of ships (shallow draft and square sails). Eventually there came the explorations by Eric the Red and Lief Ericsson, the colonization of southern Greenland by the Norsemen (Dansgaard et al., 1975), and their brief settlement (AD 986) at Anse aux Meadows in Newfoundland ("Vinland"), the first European foothold in North America.

In the same MWP Arctic warming came a widespread development of Eskimo cultures, reaching as far north as Ellesmere Land (AD 900). From the Bering Sea they spread out to northern Alaska and in Siberia, as far north as the New Siberian Islands. In the eleventh century a new wave of migrants brought the "Thule Culture" that spread from Alaska to northern Greenland.

In North America native peoples spread northward up the valleys of the Mississippi and Missouri, bringing an agricultural economy into Wisconsin and Minnesota, where pollen studies have shown there were reliable summer rains (Griffin, 1961). Eventually they reached the northern Rockies and even Utah. The Hohokam people were meanwhile developing settled agriculturally oriented communities in New Mexico and Arizona.

In Asia the behavior of the Caspian Sea provides hydrographic evidence of the MWP climate. From the ninth to fourteenth centuries the rise of the Caspian by 18 m flooded vast areas of the

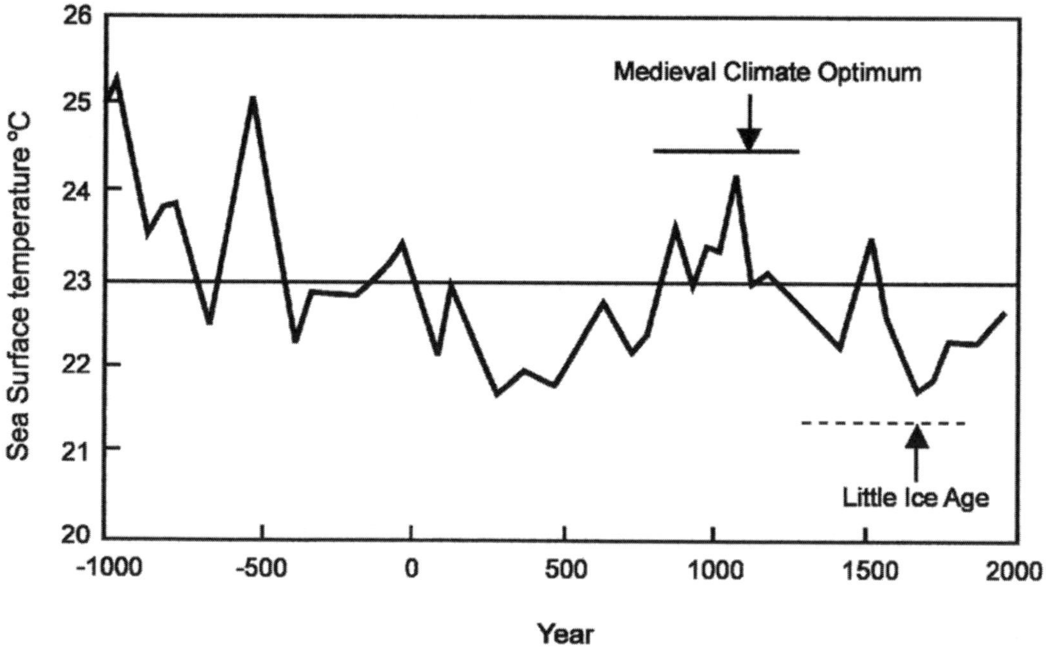

Figure M1 Trace of Sargasso Sea temperature, over 3000 years, based on benthic foraminifera oxygen-18 isotopes (adapted from Keigwin et al., 1994). The positions of the Medieval Warm Period and the Little ice Age have been expanded by the writer (R.W.F.) to conform to the other climatic proxies identified in the text.

Volga delta, reflecting a general increase in precipitation over northern Russia. Clearly this also reflects increased evaporation over the warming Gulf Stream and North Atlantic in general. The populations of central Asia showed a general expansion, but this reached a crescendo in the early thirteenth century, which launched the celebrated conquests of Genghis Khan and his "Mongol Hordes" (AD 1205–1225).

It was the exploration of central Asia in the first decades of the twentieth century that provided the database for Ellsworth Huntington's concept of climatic determinism (Huntington, 1924) that set up a great deal of controversy at a time when climate was assumed (by the "Establishment") to be a changeless aspect of the environment. A long sediment core obtained from Lake Saki (Crimea) illustrates the higher precipitation levels of the LCD also in southern Russia. The thickness of clay varves suggests a heavier rainfall at this time than for 2300 years before (Lamb, 1977, p. 408; Xanthakis et al., 1995).

The MWP warm cycle was registered also in Scandinavia and the Alps by glacier retreats (Röthlisberger, 1986; Karlén et al., 1995). Particularly in Sweden such glacier melts have been dated most precisely by sedimentation measurements in the corresponding glacial lakes. Worldwide response was marked by eustatic sea-level indicators including higher-than-normal beach ridges in many different regions, ranging from the Arctic coast of Alaska, to the coasts of the Gulf of California, and southwestern Florida (Fairbridge, 1992). The first radiocarbon dates of this high sea-level stage were obtained in Western Australia, on Rottnest Island, and became thus the "Rottnest Stage" (Fairbridge, 1961). In Scandinavia it is often referred to as the "Viking Stage" from the abundant relics displayed by the isostatic emergence around Stockholm and Uppsala. In northern Germany, the Netherlands and Belgium (Bennema, 1954), it is partly drowned by tectonic subsidence, but after correction its level was about 0.5 m above present MSL.

Inundation of low-lying coastal areas completely changed the geography in some areas. There was frantic dike building in northern Germany, the Netherlands, Belgium, France and England. The critical moment when man-made dikes were overtopped in particular storms is recorded by historic disasters when tens of thousands of people and livestock were drowned while farms and whole villages were destroyed (Bakker, 1957). In the lower Rhine valley river floods amplified the oceanic factors (Berendsen, 1995). Outstanding events included the flooding of the lower Ijssel valley in the Netherlands with the creation of the Zuider Zee (1250/1251, 1287), and in Britain with the creation of the Norfolk Broads (Lamb, 1977, p. 433). Other areas were affected including western France, the Rhone and Po deltas.

The MWP ("Rottnest") rise of sea level had an important demographic effect inasmuch as it drove coastal dwellers inland, creating endless strife with the people (mostly farmers and pastoralists) already occupying the areas. In the southern Baltic (modern Poland and northeast Germany), southern Denmark and the Frisian belt in northwest Germany, postglacial tectonic subsidence amplified the eustatic rise of the MWP. An economic side-effect of the MWP sea-level rise was in the salt industry. During periods of low sea level, as during much of the Roman era, there were coastal salt pans in which sea-water ponds were organized in series to permit progressive evaporation. This was a major industry because common salt (NaCl) is an essential dietary condiment, as well as being in widespread demand for food preservation, leather tanning, etc. With the MWP rise, the salt pans were repeatedly inundated and in many cases could not be expanded inland because of low bluffs, dunes or cliffs. Industrial disaster was alleviated by a wholesale shift inland to "fossil" salt deposits, mostly of Triassic or Permian age that had been long exploited since the Bronze Age or earlier. These deposits are known in Britain, Spain, France, Germany, Austria, Poland and Russia (Moores and Fairbridge, 1997). Many geographic names in central Europe incorporate the roots Salz, Sal, etc., e.g. Salzburg, Salzkammergut, Salzach, etc.

Metallic ores were extensively mined in mountainous areas of Europe when glaciers receded and exposed dry outcrop areas. With the ensuing Little Ice Age (after 1300), there were neoglacial advances and flooding of mines, leading to closures of the high passes in the Alps (Lamb, 1977, pp. 273–274).

Lamb (1967, 1977, pp. 276f.) was fascinated by the history of grape vines and viticulture in Medieval Britain, a country not particularly celebrated today for its home-grown wines. There were widespread vineyards south of latitude 53°N in Medieval times, some 500 km north of present-day limits in France and Germany. To quote from William of Malmesbury writing in 1150 (translated from Latin) of the vale of Gloucester "here you may behold highways and public roads full of fruit-trees, not planted, but growing naturally.... No county in England has so many or so good vineyards as this, either for fertility or for sweetness of the grape...." The export of British wines to France and Germany was a major factor in international trade, and even led to diplomatic friction. It should be recalled that in 1066 William of Normandy had led an invasion of Saxon Britain, at that time considered a remarkable prize. But then came the Little Ice Age, and soon after 1300, within 10 years the entire wine industry collapsed, the vines surviving only in rare cases under the protection of walled gardens.

The eastern parts of North America and Greenland seem to have shared equally with the European experience of the MWP, as did New Zealand and other parts of the southern hemisphere. However, much of the Pacific region, China, Japan and the western part of North America experienced the exact opposite. In Japan the mean blooming dates of the cherry trees of the imperial gardens in Kyoto have long been monitored as the harbingers of spring and the occasion of joyful festivals. For the MWP, however, the mean blossoming date was a full 10 days later than the millennial average, or 11 days comparing the mean for the twelfth century with that for the fifteenth century. To Lamb (1977, p. 400) "this suggests an eccentric position of the circumpolar vortex over the northern hemisphere in Europe's High Middle Ages, with the climatic zones persistently displaced north over the Atlantic sector and south over the whole Pacific sector and the Far East".

Archeomagnetic studies employing the paleomagnetic orientation of baked bricks or fireplace clays indicate that the North Magnetic Pole at the time lay in the eastern hemisphere (Bucha, 1984, 1988). Naked-eye sunspot observations and auroras (closely correlated to solar activity) were widely reported from China, Korea and Japan, which were then much closer to the North Magnetic Pole (Schove, 1983).

In the Mediterranean latitudes there were commonly heavy summer rains during the MWP. In central America these had an important role in bringing about the downfall of the Mayan civilization in Yucatan and elsewhere, where the maize (sweet corn) economy requires less humid seasons (Brooks, 1949), which needed to be nicely adjusted to their rainy seasons. Suffering repeated droughts, the population in the central valley of Mexico reached a minimum around AD 1000–1100.

The same delicate adjustment of seasons may also help to explain the wonderful surge in culture and religious buildings at Angkor in Cambodia, which reached a crescendo around AD 1000. During the Little Ice Age these structures were seasonally inundated by the rise of the Mekong and backfilling of Tonle-Sap that eventually buried the entire complex in 10 m or more of sediment.

Basing his estimates mainly on botanical proxies (palynology), Lamb (1977, p. 404) presented a comparison between the major climatic intervals for Britain of the last 10 000 years. The MWP discloses the warmest annual mean temperatures since the "Atlantic Biozone" 6000 years ago and 0.8°C above the warmest decades of the twentieth century. It was also wetter, except for the two summer months, July and August, which were characteristically dry, and thus good for grain harvesting, haymaking, thatching and so on (besides viticulture, noted earlier).

A series of northern hemisphere-oriented maps is presented by Lamb (1977, pp. 444/5) for selected centuries depicting troughs of the upper westerlies across Europe and the typical frontal depressions. For summer months during the MWP climax (AD 1000–1099), the mean polar jet swung from southern Greenland across northern Scandinavia to the Arctic shore of Siberia. Meanwhile the major frontal depressions radiated from northern Greenland to east-coast North America, from near Spitsbergen to Sicily, from near Lake Baikal to western China, and from Alaska to about Wake Island. For the same areas and season during the LIA (1550–1599) the polar jet ran across Scotland and southern Sweden, while the principal European frontal depressions ran from western Spain across the British Isles and the west of Norway.

For the winter seasons the MWP polar jet crossed Quebec and swung north of Iceland to the Barents Sea while its southern branch crossed Spain and the Mediterranean to Syria. The latter brought plentiful winter rains and prosperity to the kingdom of Palmyra and the peoples of Palestine and Mesopotamia. Five centuries later in the LIA the winter jet unified and crossed Iceland to head for southern Russia, bringing frequent droughts to the Middle East.

These mean jet trajectories help explain the antiphase relationships between climate proxies in western Europe and the Far East. The inferred position of the polar vortex caused swings to and fro, to eastern or western hemisphere; this appears to be analogous, on a greatly amplified (century) scale, with the "Atlantic Oscillation" (AO) which operates on a decadal basis. Biological proxies are very interesting. In central Europe history reports incidents of locust (*Locusta migratoria*) plagues in the drought summers as follows: ninth century, 8 seasons; tenth to twelfth centuries, none or next to none; fourteenth century, 15 seasons; fifteenth century, none; sixteenth century, 6; seventeenth century, one only; eighteenth century, 6; nineteenth century, 4 (Lamb, 1977). From recent data in North Africa Landscheidt (1987) believed there was a correlation between locust plagues and solar radiation.

Summary

1. The MWP, or LCD (Little Climatic Optimum), is a roughly 450-year climatic cycle, about AD 850–1300. Based on historical documentation this was a warming cycle in Europe.
2. Based on (a) CO_2 and temperature indicators in the Greenland ice cores, and (b) ^{14}C flux (inverse) levels in the dendrochronology, this interval is confirmed as a warming cycle, at least 1°C on average above millennial means.
3. Based on the astrochronology (using the 45.4 year resonance interval of the three major planets, Jupiter, Saturn and Uranus), there were three warm peaks: "Post-Carolingian" (AD 852), "Dunkerque-3" (944), "Viking/Rottnest" (1125). Two cool peaks (marked by extended interruptions caused by low sunspots and reduced solar emissions) occurred at the "Normanian" (peak: AD 1034) and the "Ottoman" (AD 1307).
4. In general, precipitation rose in step with the mean temperature oscillations, and glaciers tended to retreat. However, in exceptional areas glaciers showed temporary advances.
5. Mean sea level rose and fell eustatically and in more or less stable crustal areas it caused coastal retreat matching eustatic rise and vice-versa. However, in areas of crustal subsidence or rapid compaction, this relationship was reversed.
6. General atmospheric circulation (prevailing westerlies, trade winds, etc.) slackened, but in the sub-Arctic latitudes the 45-year summer storminess cycle persisted. With weakening trade winds, El Niño cycles became more important (but detailed monitoring is not available). However, the associated warm coastal currents reached farther south in Peru, and farther north in California.
7. The Gulf Stream and Kuro Shio were shifted northwards, with their characteristic ameliorating effects in maritime areas (increased oceanicity).
8. Sea ice coverage in the North Atlantic and Barents Sea decreased. Due to the longer ice-free seasons in areas of abundant sediment supply the beach ridge buildup was in places anomalously high.
9. In sub-Arctic lakes and swamp deposits the level of pine (or conifer) pollen rose in contrast to that of tundra species. Farther south the mixed forest boundaries moved northwards.
10. Equatorial lakes displayed little change in level, as did rainfall, but savanna (monsoon) boundaries tended to shift poleward by some hundreds of kilometers.

Acknowledgment

Appreciation is due to Dr James Hansen and colleagues at the Goddard Institute for Space Studies (NASA).

Rhodes W. Fairbridge

Bibliography

Bakker, J.P., 1957. Relative sea-level changes in northwest floods in the Netherlands in recorded time. *Deutsches Geographiches Verhandlung*, **31**: 233–237. (Translation in Leatherman, S.P., ed., *Overwash Processes: Benchmark*, vol. 58. Stroudsburg, PA: DHR, pp. 51-56).

Bennema, J., 1954. Holocene movements of land and sea-level in the coastal areas of the Netherlands. *Geologie en Mijnbouw*, **63**: 343–350.

Berendsen, H.J.A., 1995. Holocene fluvial cycles in the Rhine delta? *Journal of Coastal Research*, special issue 17, pp. 103–108.

Brooks, C.E.P., 1926/1949. *Climate Through the Ages*. London: Ernest Bonn; 2nd edn. (1949); New York: McGraw-Hill (Dover reprint, 1970).

Bucha, V., 1984. Mechanism for linking solar activity to weather-scale effects, climatic changes and glaciations in the northern hemisphere. In Moerner, N.-A., and Karlen, W., eds., *Climatic Changes on a Yearly to Millennial Basis*. Dordrecht: Reidel, pp. 415–448.

Bucha, V., 1988. Influence of solar activity on atmospheric circulation types. *Annals of Geophysics*, **6**(5): 513–524.

Dansgaard, W., et al., 1975. Climatic changes, Norsemen and modern man. *Nature*, **255**: 24–28.
Fagan, B., 2000. *The Little Ice Age*. New York: Basic Books.
Fairbridge, R.W., (ed.), 1961. Solar variations, climatic change, and related geophysical problems. *New York Acad Science, Annals*, **95**: 1–740.
Fairbridge, R.W., 1992. Holocene marine coastal evolution of the United States. *SEPM Special Publications*, **48**: 9–20.
Griffin, J.B., 1961. Some correlations of climate and cultural change in eastern American prehistory. *Annals of the New York Academy of Sciences*, **95**(1): 710–717.
Huntington, E., 1924. *Civilization and Climate*, 3rd edn. New Haven, CT: Yale University Press, (reprinted 1977; Hamden, CT: Archon Books).
Karlén, W., Bodin, A., Kuylenstierna. J., et al. 1995. Climate of northern Sweden during the Holocene. *Journal of Coastal Research*, special issue 17, pp. 49–54.
Keigwin, L.D., et al., 1994. The role of the deep ocean in North Atlantic climate change between 70 and 130 kyr ago. *Nature*, **371**: 323–326.
Lamb, H.H., 1967. Britain's changing climate. *Geographical Journal*, **133**(4): 445–468.
Lamb, H.H., 1977. *Climate History and the Future*. London: Methuen.
Landscheidt, T., 1987. Long-range forecasts of solar, cycles and climate change. In Rampino, M.R., et al., eds., *Climate: history, periodicity, and predictability*. New York: Van Nostrand, pp. 421–445.
Moores, E., and Fairbridge, R W., (eds.), 1997. *Encyclopedia of European and Asian Regional Geology*. London: Chapman & Hall.
Röthlisberger, F., 1986. *10 000 Jahre Gletchergeschichte der Erde*. Aarau: Verlag Sauerlaender.
Schove, D.J. (ed.), 1983. *Sunspot Cycles*. Stroudsberg: Hutchinson & Ross (Benchmark **68**).
Xanthakis, J., Liritzis, I., and Poulakos, C., 1995. Solar-climatic cycles in the 4190-year Lake Saki mud layer thickness record. *Journal of Coastal Research*, special issue 17, pp. 79–86.

MEDITERRANEAN CLIMATES

The Mediterranean climate is distinct for its sunny and dry, mild-to-hot summers, and mild, wetter winters. The archetypal example of this climate is found in the area adjacent to the Mediterranean Sea – hence the name Mediterranean climate. Because its summer season is markedly drier than the other seasons of the year, it is sometimes referred to as the summer-dry climate type. The Mediterranean climate is unique amongst Earth's climates for its pronounced winter maximum of precipitation.

Locations

Besides its occurrence in the Mediterranean Basin, it is found on the west coasts of the continental landmasses, between 25 and 40 degrees latitude. Its proximity to the ocean/sea has a significant influence on its temperature, precipitation, and humidity characteristics – particularly in near-coastal areas. The largest example – in geographic area – of the Mediterranean climate, occurs in the area adjacent to the Mediterranean Sea. Other major areas with this climate type include the west coast of North America, from the Mexico–United States border northward to the Pacific Northwest region of the United States, and southwestern and southern Australia. Smaller examples of Mediterranean climate occur in South Africa (Cape Province) and central Chile. In North America the north–south-trending Cascade and Sierra Nevada Mountains limit the eastern extent of Mediterranean climate. In total, the Mediterranean climate type covers only a very small portion – just 2% – of the continental landmass.

Climate characteristics

The semipermanent subtropical anticyclones in the Atlantic and Pacific oceans are largely responsible for the unique characteristics of the Mediterranean climate type. The strong subsidence and resultant adiabatic warming that occurs on the eastern limb of these anticyclones, produces the pronounced and distinct summer-dry period of the Mediterranean climate. The subtropical highs also act to keep summer cyclones poleward of the region, further reducing the opportunity for precipitation in the summer months.

As an example of just how dry the summer months of the Mediterranean climate can be, Santa Monica, California – located toward the equatorward margin of this climate type in North America – has never recorded more than a trace of precipitation during the months of June, July, and August (Oliver and Hidore, 2002). Winter is the distinctly wetter period for the Mediterrranean climate type because the subtropical anticyclones sag equatorward, allowing midlatitude cyclones embedded in the westerlies to bring rain to the area. In Perth, Australia, 85% of the annual precipitation comes during the winter half of the year. In Istanbul, Turkey, the 6 winter months of the year account for 70% of the precipitation that falls annually. San Diego, California, receives 90% of its total precipitation for the year during the 6 months from November through April (Oliver and Hidore, 2002).

Total annual precipitation amounts range between 40 and 80 cm, and characteristically increase poleward because of the diminished influence of the subtropical anticyclones, and thus greater frequency of penetration of midlatitude cyclones. Precipitation totals also increase in the interior highlands as orographic precipitation falls on the windward slopes of mountain ranges. In southern California, Los Angeles and San Diego average 37 and 26 cm of precipitation a year, respectively. However, in the adjacent mountains, annual precipitation totals are four to five times those amounts – with most of the precipitation falling as snow (Lydolph, 1985). Summers in the Mediterranean Basin are not as dry as they are along the west coast of North America because of the greater instability of the atmosphere in the area.

While the winter half of the year is the wetter and stormier half, the amount of precipitation that falls in any given year is quite variable. Moreover, precipitation events are generally shortlived and the period between events is characterized by fair weather. Thus, even during the winter months, pleasant conditions predominate. One caveat to this rule, however, is required along the California coast in El Niño years. During El Niño events the jet steam over the Pacific steers storms into California, bringing wetter than normal conditions to the region.

Along the west coast of North America, as winter progresses the polar front gradually moves southward, bringing precipitation to the more equatorward areas by late in the winter season. Maximum precipitation occurs during December in Seattle and Portland, in January in San Francisco, and in February in Los Angeles and San Diego (Lydolph, 1985). In the highland areas away from the coast, strong thunderstorms can occur and copious amounts of rain may fall in a short period of time – leading to flooding and hazardous mass wasting events.

Sea breezes and coastal upwelling moderate coastal temperatures in the summer. Daily maximum temperatures rarely get much higher than the low 20s (Celsius). In the winter, because of the moderating influence of the ocean, temperatures remain quite mild along coastal areas. Inland however, temperatures are warmer in the summer and colder in the winter. Sacramento, California, just 120 km to the northeast of San Francisco, has an average temperature for the month of July that is 9°C higher than that of San Francisco (Ahrens, 2002). Inland areas may even experience frost on occasion – which puts economically valuable citrus and other crops at risk.

Coastal areas adjacent to the Mediterranean Sea are warmer than their coastal counterpart in North America because upwelling does not occur in the Mediterranean Sea. Upwelling is quite strong, however, off the California coast during the summer months of the year. Air temperatures all along the length of the California coast, in July, are some 10–11°C below normal for their respective latitudes (Lydolph, 1985). Upwelling tends to be strongest just west of San Francisco Bay, making the area one of the coolest along the entire west coast of North America in the summer months. As moisture-laden air from the Pacific passes over the very cool water close to the coast in the San Francisco Bay area, it is cooled to its dewpoint and fog is produced. This fog is then advected inland. San Francisco, California, is renowned for its numerous days of fog each summer. Fog is common elsewhere along the California coast as well.

Inversions

The subsidence and adiabatic warming that occur on the southeastern edge of the Pacific subtropical high, in the summer, produce strong inversions in southern California's two largest cities, Los Angeles and San Diego. The base of the inversion typically lies only just a few hundred meters above sea level (Lydolph, 1985). Convection from surface heating in the moist marine layer below the inversion mixes emissions from stationary and mobile sources upward, and concentrates them just beneath the base of the inversion, where they collect and form a reddish-brown haze that is a signature of highly polluted urban atmospheres. Mountains to the north and east of the city do not allow for the horizontal dispersion of air, thus letting pollution build to very dangerous levels. Los Angeles, because of its geography and large number of automobiles, is infamous worldwide for its exceptionally high levels of air pollution.

Because inversions are such a common feature of Mediterranean climates, high levels of air pollution are characteristic of many large cities in Mediterranean climates. During the winter months the inversion is not as pronounced because of the equatorward movement of the Pacific High. Moreover, when migrating cyclones pass through the region they dramatically alter the thermal structure of the atmosphere and break down the inversions.

Summary

The abundant warmth, sunshine, high number of rainfree days, and cool sea breezes in coastal areas, make the Mediterranean climate type a favorite vacation destination. Its long, dry summers, however, do not allow for the growth of lush vegetation. Instead, scrubby, low-growing woody plants and trees predominate. In North America this type of vegetation is referred to as chaparral. In the Mediterranean basin such vegetation goes by the name *maquis*. Citrus, grapes (for wine-making), and other fruits and vegetables are grown in Mediterranean climates. Irrigation of these crops is required, however (Lydolph, 1985).

G. Jay Lennartson

Bibliography

Aguado, E., and Burt, J.E., 2001. *Understanding Weather and Climate*. Upper Saddle River, NJ: Prentice Hall.

Ahrens, C.D., 2003. *Meteorology Today: an introduction to weather, climate, and the environment*. Pacific Grove, CA: Brooks & Cole.

Lydolph, P.E., 1985. *Climates of the Earth*. Totowa, NJ: Rowan & Allanheld.

Moran, J.M., and Morgan, M.D., 1994. *Meteorology, the Atmosphere and the Science of Weather*. New York: Macmillian.

Oliver, J.E., and Hidore, J.J., 2002. *Climatology, an Atmospheric Science*. Upper Saddle River, NJ: Prentice Hall.

Cross-references

Art and Climate
Australia and New Zealand, Climate of
Europe, Climate of
Climate Classification
Precipitation Distribution

MICROCLIMATOLOGY

Microclimatology, the scientific study of microclimates, is concerned with the atmospheric layer that extends from the surface of the Earth to a height where the effects of the features of the underlying surface can no longer be distinguished from the local climate (American Meteorological Society, 2000). Microclimates vary in response to, and in turn are superimposed upon, larger-scale climates (the mesoclimate and macroclimate). Microclimates span a spatial scale (Figure M2) that includes the envelope of air that surrounds an insect or a spore on a plant leaf to the climatic conditions on a hillslope that permit or preclude the growth of specialty agricultural crops such as grapes for wine production. Hence, the horizontal extent of microclimates is often not rigidly defined, although several millimeters to 1 kilometer is often employed (Oke, 1987). Local climate or topoclimate terminology (Oliver and Hidore, 2002) is often employed between microclimates and mesoclimates to represent spatial scales between several hundred meters and 10 kilometers. Microscale atmospheric phenomena usually have temporal spans of less than a few minutes. Microclimate phenomena and their cumulative interactions, feedbacks and impacts may range from seconds to seasons. The thickness of the air layer of concern in microclimatology is also not rigidly defined (Figure M2). Sometimes it can extend from below the surface (the soil or substrate regime is often considered) to a height where recognizable microclimatic signatures in the atmosphere become essentially indistinguishable. During the night this might extend into the atmosphere for several hundred meters and, during the daytime, for over one kilometer. Usually, a much thinner air layer, in close proximity to the surface, receives particular attention. Another important aspect in the definition of microclimates and their relevance and importance is the perspective that is brought by those observing, studying or utilizing them. Human perspective and its acknowledged subjectivity often play a profound and fundamental role in delimiting the spatial and temporal scales of a microclimate as well as the scope of the controlling mechanisms.

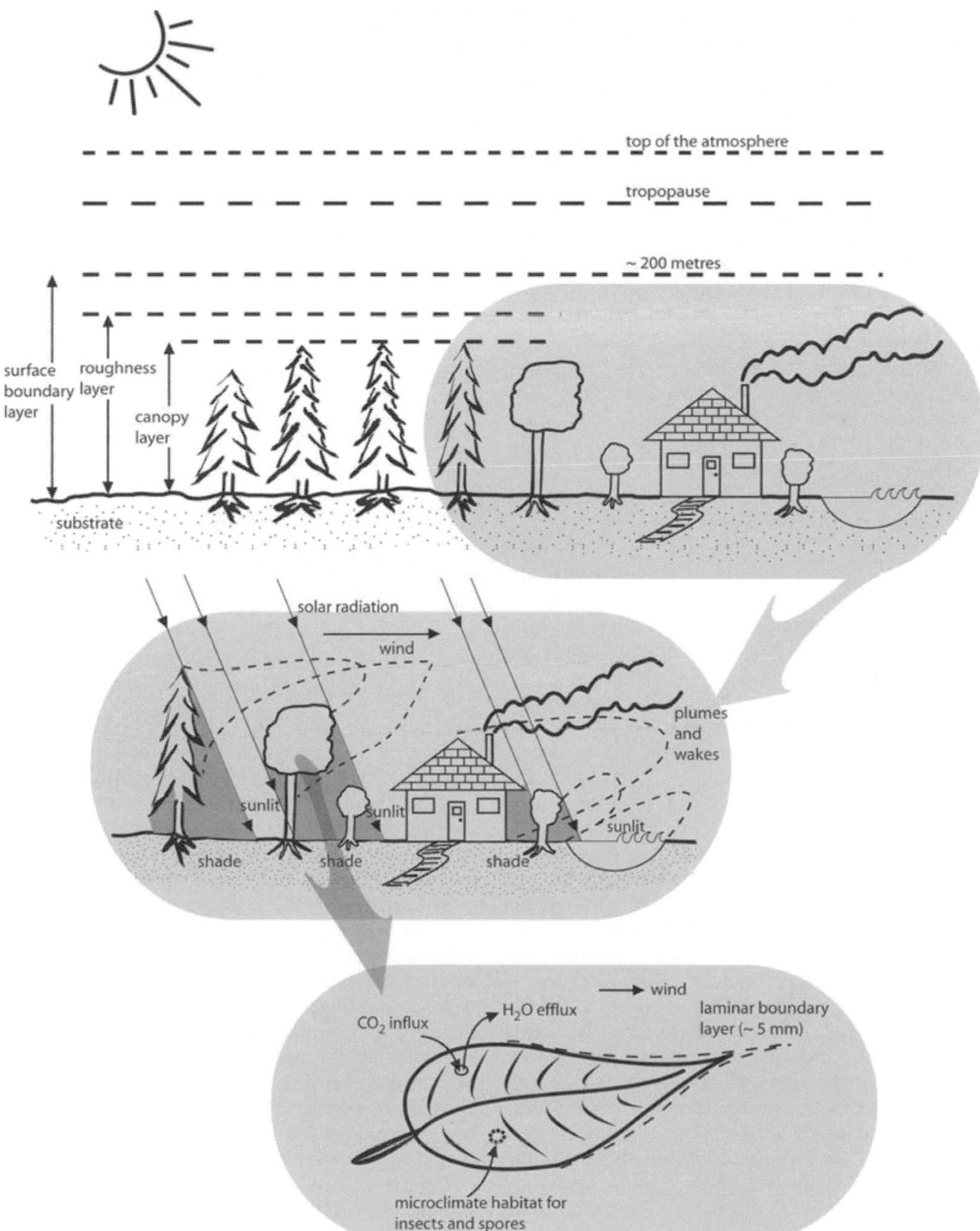

Figure M2 Schematic representation of a vertical cross-section of the Earth's atmosphere. Microclimates extend from the surface, or just below it, to a height where the effects of the surface features cannot be distinguished from the local climate. Numerous microclimates (forest, forest canopy, horticultural plants, building, lake edge, soil surface, insect habitat and others) can be readily identified.

Whereas microclimatology is a subdivision of climatology based primarily on spatial scale attributes, microclimates themselves are often subdivided to consider specific surfaces or habitats. Such subdividing, for example, gives rise to: the microclimatology of plant and animal environments in which the climatic resources for individuals and communities is examined (Bonan, 2002; Jones, 1992; Lowry and Lowry, 2001; Rosenberg et al., 1983; Stoutjesdijk and Barkman, 1992); agricultural microclimatology (Rose, 1966) and forest microclimatology (Lee, 1978) that focus more specialized treatments of particular surface types; forest canopy microclimatology that details biophysical and biochemical processes in foliage environments (Lowman and Nadkarni, 1995; Mynemi and Ross, 1991); urban microclimatology that focuses on human habitation and examines cities, building structures and materials (Brown and DeKay, 2001; Lowry, 1988), air quality (Bell and

Treshow, 2002) and landscape design (Brown and Gillespie, 1995). Additionally, microclimate investigations of many novel environments have been undertaken: cryptomicroclimatology which examines the microclimatology of confined spaces; the microclimatology of cultural heritage which examines the influence of atmospheric variables on cultural objects in both outdoor and museum environments (Camuffo, 1998; Thompson, 1990); animal microclimatology which examines the interrelationship between animals and microclimate variables in both natural and artificial environments (Albright, 1990; Johnson, 1987; Scientific Committee on Animal Health and Animal Welfare, 1999; Unwin and Corbet, 1991). Treatments on microclimatology can be found in general volumes on the atmosphere (for example Oliver and Hidore (2002) and many others) or for specific environments (i.e. mountains (Whiteman, 2000)), in specialized volumes that commence with first principles and cover both theory and application (Campbell and Norman, 1998; Geiger et al., 1995; Lowry and Lowry, 1989; Monteith and Unsworth, 1990; Oke, 1987; Rosenberg et al., 1983; Stoutjesdijk and Barkman, 1992) and in presentations for more focused readerships (i.e. microclimate and spray dispersion (Bache and Johnstone, 1992); extreme heat waves, societal response and human mortality (Klinenberg, 2002); and countless scientific journal publications). Additionally, other treatments have focused on the synthesis of microclimate information for various landscape types within political borders (Bailey et al., 1997).

Human recognition of, and influence by, climate phenomena predate recorded history (Calvin, 1991, 2002). Early human adaptation to the environment, migration and the utilization of resources for both survival and progress attest to an interrelationship with microclimates that parallels the long course of civilization and societal development (Brown, 2001; Butzer, 1971; Jones et al., 2001). Cumulative human impacts across a range of spatial scales have led to marked changes in the Earth's surface, atmosphere and in the concomitant atmosphere–surface interactions. Activities such as deforestation, agriculture, fossil fuel and biomass burning, industrialization and transportation, which occur at the microscale, have led to documented changes that are now being realized at local, regional and global scales (IPCC, 2001a,b,c). In turn, mesoscale and macroscale influences then impact the mosaic of microclimates that blanket the Earth's surface. Although human–climate interaction has been long recognized, the first major synthesis of microclimate research was not published until 1927. *Das Klima der bodennahen Luftschicht* by Rudolf Geiger (1894–1981) was a pioneering contribution, and four German editions appeared before *The Climate Near the Ground* was published in English (Geiger, 1966). It was the most exhaustive work on microclimatology until that time, and a modern edition (Geiger et al., 1995), to celebrate the 100th anniversary of Geiger's birth, now carries the volume's influence into the twenty-first century. Since the end of World War II, considerable advancements in the study of the Earth's environment, and associated with this microclimatology, have been achieved. A number of interrelated factors have played significant roles: the widespread quest for scientific understanding of the environment and the atmosphere has led to substantial increases in overall knowledge; the theoretical understanding of the air layer in contact with the surface, in terms of energy and mass exchange, has been greatly advanced; the development, refinement and extensive application of instrumentation, data recording and computational technology; the deployment of extensive spatial monitoring networks and the widespread availability of data (i.e. worldwide web and many other sources); the lure of and the opportunity to research unknown, challenging and remote locations on the Earth's surface; and the application of remote sensing technologies across a range of spatial and temporal scales for documenting patterns and trends (Kramer, 2002). Advancements have come from many persons in many countries, and contributions can be viewed as examples of interdisciplinary achievements with much microclimatic research being reported in agronomy, architecture, ecology, engineering, forestry, hydrology, physical geography, zoology and other communications.

The creation and control of microclimates

Microclimates arise in response to the external forcings of energy, precipitation and wind, and the magnitude of these forcings establishes the boundaries and character of the microclimates found. Transformations in microclimates arise from changes in the forcings and changes in the responses that occur at the surface to the forcings. Hence, a mosaic of microclimates is readily created over short distance scales.

The surface of the Earth is warmed by solar energy during daylight hours, swept by winds that are driven by interlocked large-scale circulation (macroclimates) and synoptic systems (mesoclimates), and is moistened by precipitation that arrives in either liquid (rain) or solid (snow) forms. Some of the inputs that create the climate of a surface are cyclical in character, for example the diurnal and seasonal variations in solar energy receipt and the associated thermal regimes. Others are quasi-periodic or perhaps close to random in character, for example the seasonal movement of storm systems, and the spatial and temporal occurrence of precipitation. For any geographical setting a general climatology will be created from the integrated influences of latitude, altitude, continentality and location in relation to mesoscale synoptic flows. Within this, an integrated blend of radiative, aerodynamic, thermal and moisture attributes establishes the microclimate for all surfaces (Table M3).

Table M3 Surface attributes important in the creation of microclimates. Adapted after Oke (1997)

Radiative	-Surface albedo, surface emissivity, surface temperature, geometric positioning of the surface and the surrounding environment that will influence radiant energy receipt and loss
Aerodynamic	-Surface roughness length, zero plane displacement, presence of elements upwind that obstruct or channel wind flow
Thermal	-Thermal conductivity, heat capacity, thermal diffusivity, thermal admittance
Moisture	-The surface character (vegetation, soil, etc.) that impacts plant transpiration and/or surface evaporation; the moisture status of the substrate and its availability for evaporation and/or transpiration

Solar radiation (in the wavelength range 0.15–4 micrometers) that reaches a surface has both direct beam and diffuse components. A portion reaching the surface is reflected, and the reflected solar radiation is dependent on the surface albedo. The result is net solar radiation (Figure M3), and this provides a direct contribution to surface heating. The albedo is an explicit surface control of the microclimate. Albedos vary from about 5–10% for water (midday), coniferous forest and dark soils, to

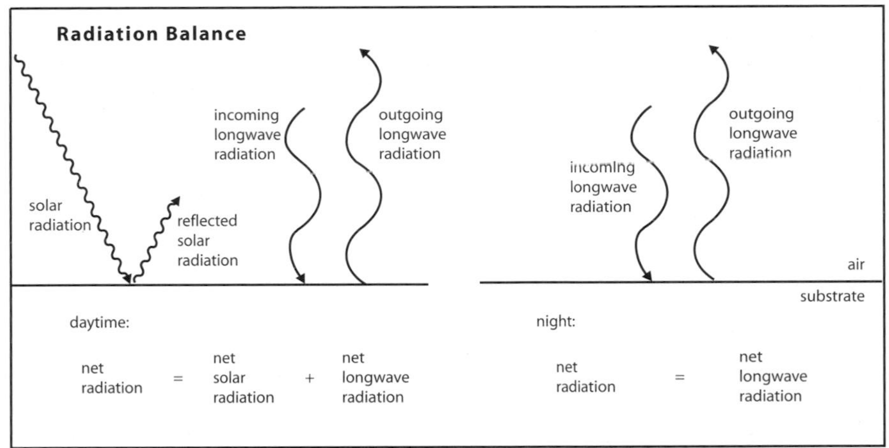

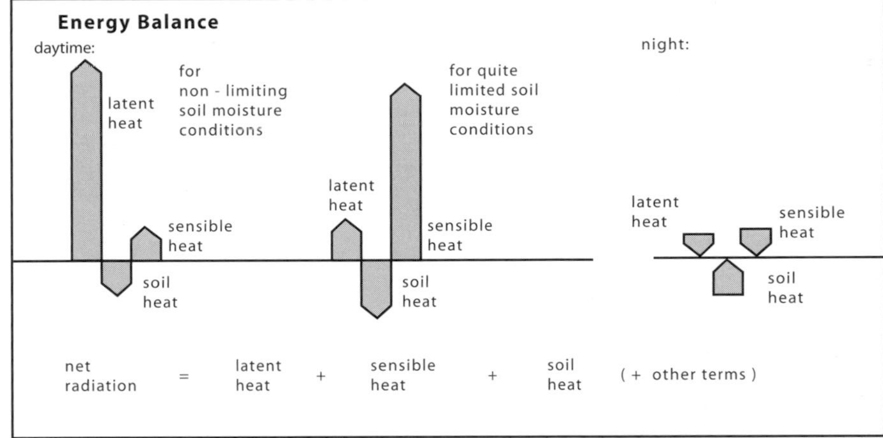

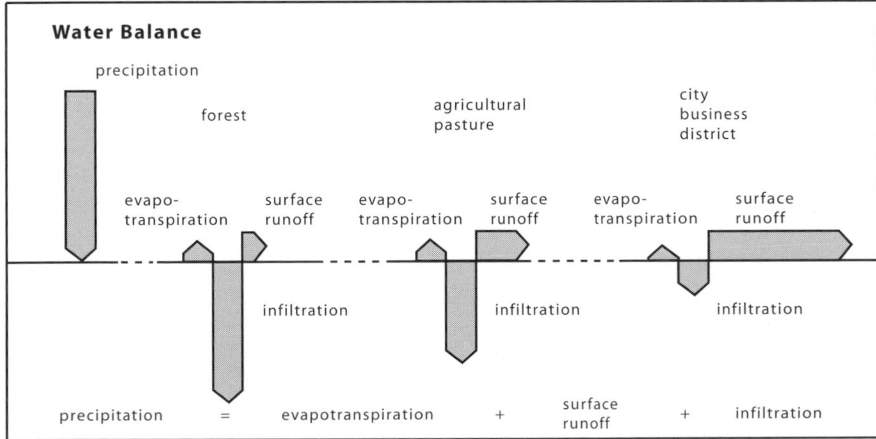

Figure M3 Schematic representation of the radiation, energy and water balances. The radiation balance is presented for both daytime and night periods. The energy balance is presented for both daytime and night periods, with the role of non-limiting and quite limiting surface moisture availability illustrated for daytime periods. For the water balance the response to precipitation input is illustrated for three distinct surfaces (forest, agricultural pasture and a city business district).

20–25% for agricultural crops and other vegetated surfaces, to 80–90% for fresh fallen snow. For three-dimensional surfaces, such as plant canopies, radiation is transmitted from the upper plant canopy toward the ground surface, with absorption and reflection of radiation by the foliage elements occurring. Solar radiation also penetrates into water bodies, snowpacks and glacier ice. Of the solar radiation reaching the Earth's surface, approximately 50% is in the visible spectral range. This also coincides with the approximate wavelength range that plants can biochemically harvest for photosynthesis. All elements in the Earth–atmosphere system also emit radiation, but at longer wavelengths (approximately 5–100 micrometers). The surface is both a recipient of the longwave radiation that is emitted by the atmosphere, as well as an emitter of longwave radiation to the atmosphere. Longwave radiation is governed by the Stefan–Boltzmann law, where the radiation is a function of the emitting surface's temperature raised to the fourth power and its emissivity. As Earth surface temperatures usually exceed the radiating temperature of the atmospheric column, the surface's net balance of longwave radiation is usually an energy loss. The net balance of solar and longwave radiation is known as net radiation (Figure M3). Daytime microclimates respond to the combined influences of solar and longwave radiation. At night the net radiation is solely a longwave radiation regime. The influence of diurnal and seasonal cycles in radiant energy input is pronounced in most microclimates (Figures M3 and M4).

The radiant energy surplus or deficit at the surface obeys the law of conservation of energy through the energy balance. In the most simple of treatments (Figure M3), net radiation is dissipated into latent heat, sensible heat and substrate heat. During the daytime, sensible and latent heat are energies employed in the convective transfer of heat and water into the air. Substrate heat is the energy lost by conduction into the underlying surface. The partitioning of available energy is primarily affected by surface moisture (Figure M3). At night, heating from the atmosphere and condensation often lead to surface-directed sensible and latent heat exchanges, although these are modest when compared to the magnitudes of the daytime regime. The substrate often serves as an energy source at night as conductive energy can flow toward the surface. Vertical profiles of temperature and humidity demonstrate the continuous coupling that exists between the surface and atmosphere in terms of energy and mass exchange (Figure M5). Simple treatments of the energy balance can be readily adapted to incorporate other factors. These could include the energy stored in surface vegetation, the energy used by plants in photosynthesis, the energy used in snow melt, the energy stored in the water bodies such as lakes, the energy contributed by anthropogenic sources, the horizontal transfer of sensible and latent energy from upwind environments, and the lateral transfer of energy to downwind environments. All of these can play significant roles in the creation and character of specific microclimates.

The input of precipitation to the surface and its routing is governed by the hydrological cycle. In the water balance the precipitation to the surface is lost to the atmosphere through evapotranspiration, to surface runoff by horizontal flow and to infiltration. The infiltrated water replenishes surface soil storage and thereafter the drainage flows horizontally or vertically into fluvial or groundwater systems. The character of the surface and substrate plays a pronounced role in the routing of precipitation (Figure M3). Precipitation input is not continuous, and the amount and timeliness is critical to the nature of the hydrological cycle across the range of spatial scales. Additionally, precipitation falling as snow may well introduce important temporal lags in the response of the hydrological system. The water balance is closely linked with the energy balance, as the transpiration from plant surfaces and evaporation from the soil are the water mass volumes represented in the energy balance by latent heat, the energy utilized in the change of phase of water from liquid to gaseous form. Analogous to the water balance, biogeochemical cycles can be defined and studied for various microclimates, with a prime example being carbon dioxide.

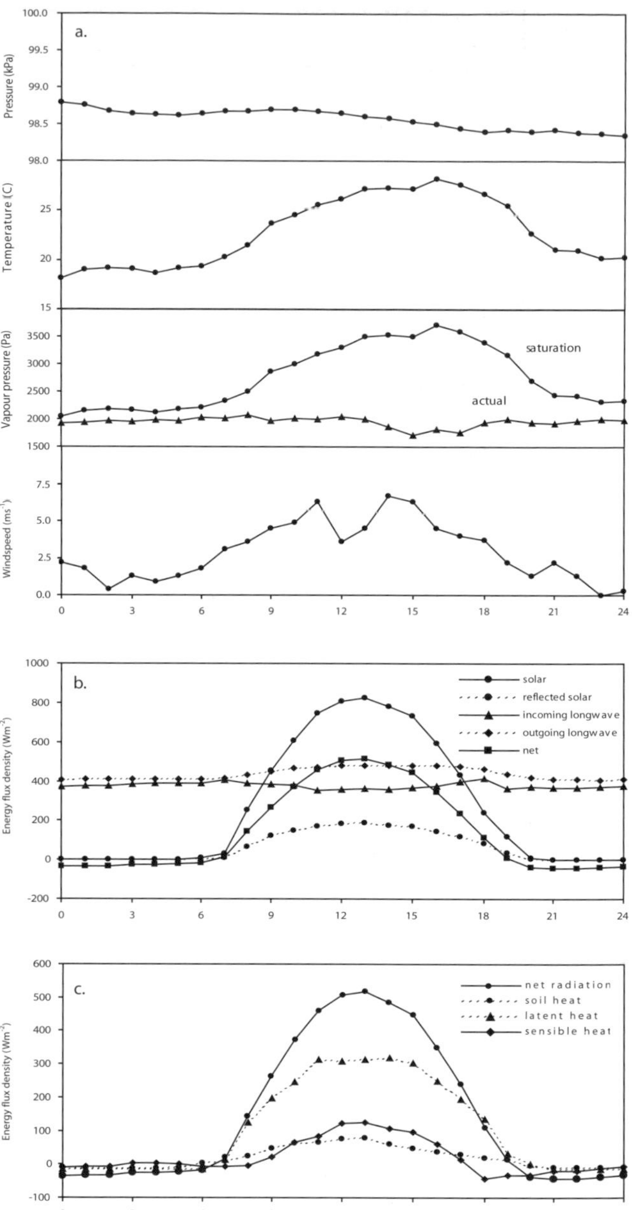

Figure M4 The diurnal trend of midsummer atmospheric variables monitored at a meteorological station and the radiation and energy balance components of a cropped surface located nearby. The day was almost cloudless and available moisture to the surface becomes slightly limiting during the afternoon period.

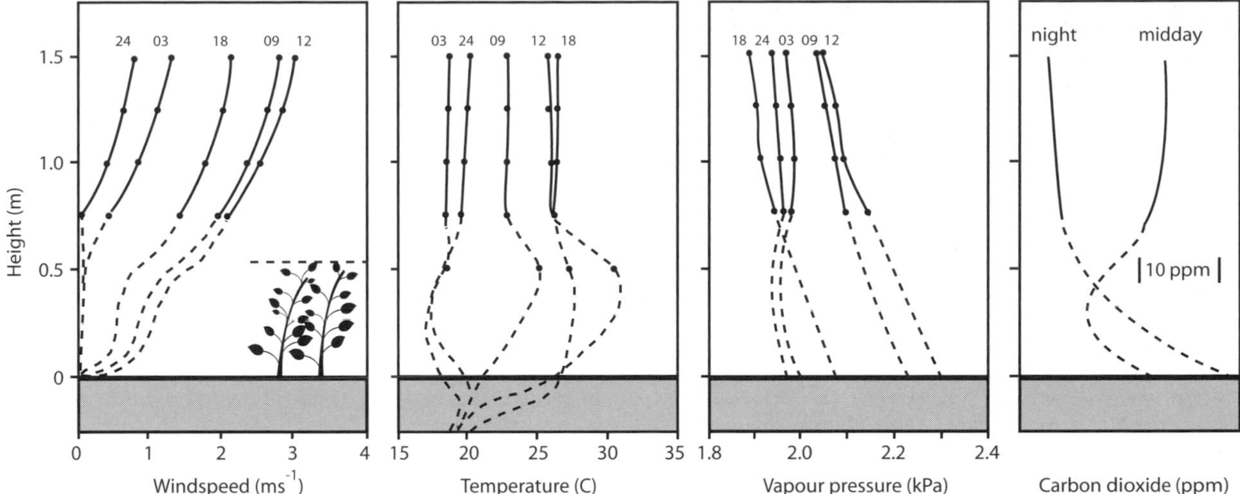

Figure M5 Hourly average vertical profiles of windspeed, temperature, vapor pressure and carbon dioxide for the near-surface layer for selected periods throughout the day (03, 09, 12, 18 and 24 hours). The profiles were measured on the day documented in Figure M5 (cropped surface in midsummer). Profiles within the plant canopy are denoted by dashed lines.

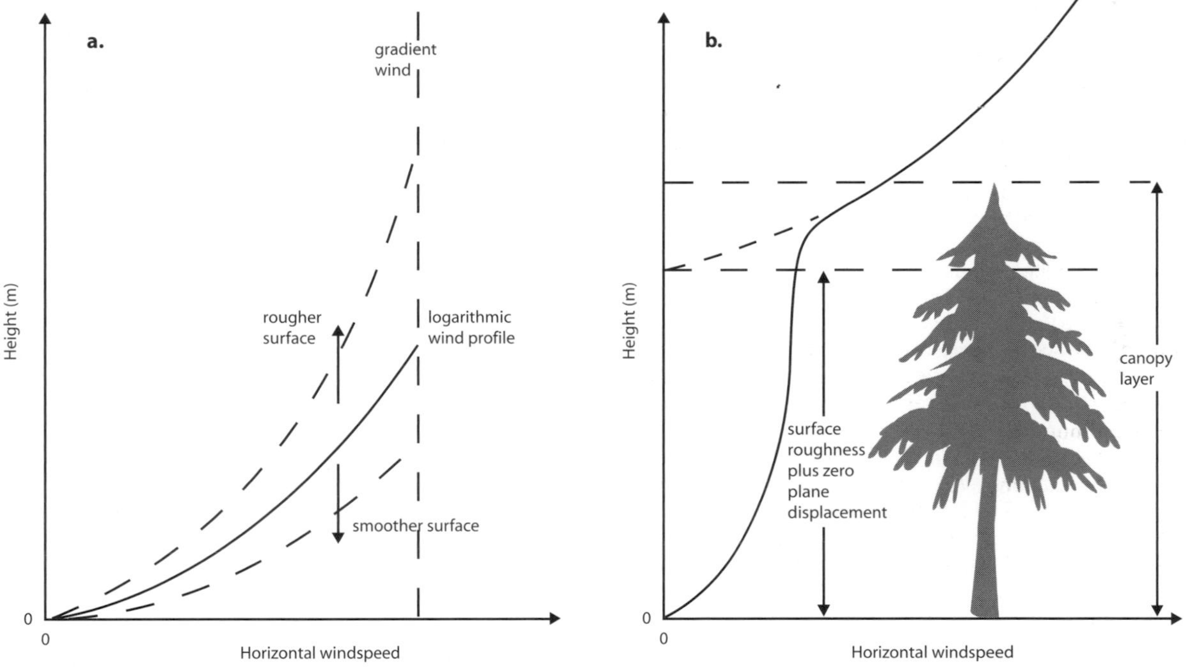

Figure M6 (**a**) The characteristic profile of windspeed near the Earth's surface when thermal enhancement or suppression is absent. Increasing surface roughness, from smooth (surfaces such as snow, calm water and desert sand) to rough (forests, cities), changes the profile of windspeed in the near surface layer. (**b**) The profile of windspeed through a plant canopy. For such three-dimensional surfaces a zero plane displacement must be subtracted from the height scale to preserve the characteristics of the logarithmic wind profile.

The flow of air above the surface, from regions of high pressure to low pressure, illustrates the combined influence of driving and steering forces. As the surface is approached, the wind velocity is reduced in response to friction effects. When no thermal gradients are present, the height variation of windspeed above the surface is logarithmic in character (Figure M6). Smooth surfaces such as water, snow and desert sand are relatively windy, but the effect of the surface extends only a short distance above the surface. Rough surfaces such as tall forests and cities generate more turbulence and affect a much deeper layer of air. The surface roughness parameter denotes the theoretical height where windspeed goes to zero, and it is typically 10% of the average surface element height (Figure M6). If a vegetation canopy covers the surface, the effective surface height is adjusted upward to account for this. This zero plane displacement is approximately 70% of vegetation height (Figure M6).

The fundamental property of flow for a surface is momentum. The momentum aloft is extracted from the flow by the friction effects of terrain roughness. Mixing of air by turbulence results in the creation of eddies. These parcels of air are of varying sizes and, at any instant, may exhibit almost random motion. Each eddy carries the characteristics of heat, mass substances and momentum. The vertical profiles of temperature, water vapor, other gases and windspeed exhibit fundamental linkages between the surface and the atmosphere (Figure M5). Turbulence in the lower atmosphere arises from buoyancy forces (free convection) and mechanical forces (forced convection). Departures from isothermal atmospheric conditions create changes in stability; these demonstrate profound roles in turbulent mixing. During daylight hours, unstable conditions predominate with strong lapse conditions existing (temperatures decreasing with height). Turbulence and, therefore, mixing is enhanced by buoyancy. At night, stable atmospheric conditions prevail with inversion conditions existing (temperatures increasing with height). Buoyancy is minimal and mixing is suppressed.

Few surfaces on the Earth are horizontal and smooth, and microclimate variation arising from topographic influence is commonplace. Topography transforms patterns of solar radiation, temperature, wind and precipitation. North, south, east and west slopes of hills or mountains, or the sides of valleys, can have markedly different microclimates (Figure M7), with solar energy receipt and wind patterns being major contributors to these differences. In northern midlatitudes, south-facing surfaces receive more solar radiation than do other orientations. North-facing slopes may receive no direct beam solar radiation for months, depending on latitude, declination of the sun, slope and orientation of the surface and slope position (valley bottom for example). In rugged terrain, adjacent topographic features may cast shade for portions of the day or year, and some places (valleys and steep poleward-facing slopes for example), may never receive direct beam solar radiation. As a result, each facet of the terrain has a unique radiation balance and a microclimate that differs from adjacent facets (Figure M7).

The contour of the landscape also creates winds that are superimposed on large-scale circulation patterns. Anabatic winds develop along slopes that have been heated by solar radiation. Associated cloud development throughout the daytime hours often provides a visible account of the intensity of the upslope flow. These winds are best developed during anticyclonic conditions, and when coupled with subsiding air in the valley bottom, a larger local circulation system is realized. Nighttime cooling results in an increase in the density of surface air on slopes and downslope katabatic (gravitational) flow results. Such topographically induced patterns of air flow can result in marked differences in near-surface temperature. In spring and fall, frost pockets can develop where cold air drainage is impeded by topographic features (valley bottoms, earthen berms, walls of buildings, rows of trees). In these pools of cold air, near-surface air temperatures at night may be many degrees less than temperatures a very short distance away. In valleys, nocturnal temperature regimes generally increase upslope as the impacts of the frost pockets diminish with elevation. Above the height of these zones of relative warmth, adiabatic cooling prevails.

Precipitation onto any surface is determined by regional synoptic conditions, the slope and aspect of the surface, and the nature of the raindrops themselves. Changes in the trajectory of raindrops from the vertical results from wind effects, and small raindrops with lower terminal velocities are most affected. Locations with higher wind velocities have more perturbed precipitation patterns than protected sites. The intensity of precipitation at the surface depends on the angle of the wind-induced trajectories relative to the aspect and slope of the surface. A surface that is parallel to the trajectory of the raindrops, for example a wall during calm conditions, may receive no precipitation. Maximum intensities will be received on a surface that is perpendicular to the raindrop trajectory. Hence, in complex terrain and around architectural structures, significantly different precipitation amounts can occur on various slopes, and the differences can be accentuated or moderated over long periods. In winter, snow is subject to resuspension, transport and redistribution by wind. Snow accumulation zones do not necessarily coincide with the original pattern of the snowfall. Skillful design and positioning of snow fencing is a prime example of microclimate modification. The reduction of wind and turbulence around the barrier can enhance snow deposition and minimize the drifting across nearby transportation corridors.

Examples of microclimates

Examples of microclimates are plentiful and contribute significantly to the makeup of the environments that surround us: the warming of one's face as it becomes aligned with the sun's beam on a cool morning; the seeking of shade by animals and people as a respite from the harsh afternoon sun; the distinct mosaic in surface temperatures experienced during summer outings to the beach or seaside; the creative use of sunlight and shade areas for ornamental plants; the cradling of snow by conifer branches in forests; and the fresh-fallen snow in open clearings that is swept into drift patterns that are repetitive from storm-to-storm and from year-to-year. Recent advancements in the study of microclimatology have sought not only to observe the features of microclimates, but also to investigate and explain the forcings and surface responses that give rise to the distinctiveness that is found over short distances.

With high solar energy, the microclimate energy forcing in tropical forest ecosystems (Sluiter and Smit, 2001) is considerable (Figure M8a). In such environments most of the solar radiation is absorbed by the upper levels of the forest canopy, and energy availability near the forest floor is very much reduced. In adjacent forest clearings (gaps), available energy is quite abundant in the near-surface layer. Diurnal temperature and windspeed observations document the distinct differences between these forest and gap environments (Figure M8b,c,d). Additionally, the size and nature of the gaps themselves play determinate roles in the microclimates created. An abundance of solar radiation and the availability of surface moisture provide microclimate resources for forest regrowth along the gap edges and within the gaps themselves. Ecological changes in response to microclimate resource availability will not only occur, but will in turn play an active role in modifying the surface microclimates.

The spatial pattern of surface thermal and moisture regimes is usually quite complex, and the transition across the boundary between different surfaces can display sharp discontinuities. Soil temperature and moisture transects across an open clear-cut in a midlatitude forest document both of these aspects (Figure M9). Variation within the forest and clear-cut surface regimes is considerable. Generally, soil temperatures are higher in the clear-cut than the forest, and differences are enhanced during cloudless conditions. The moisture of the uppermost surface materials is lower in the clear-cut due to the strong drying regimes present when compared to the shadier forest floor environment. Transitions across the borders of the clear-cut

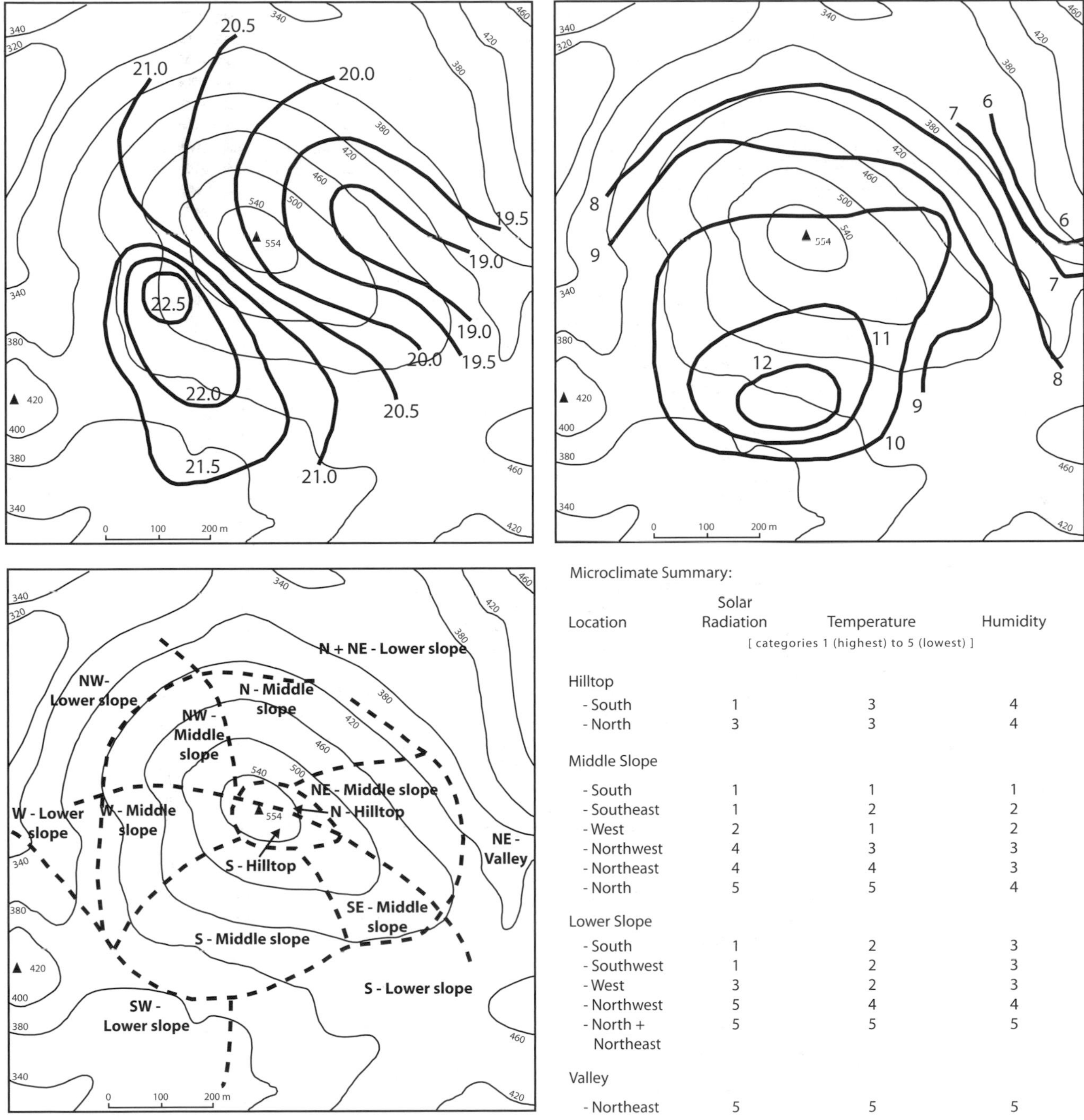

Microclimate Summary:

Location	Solar Radiation	Temperature	Humidity
	[categories 1 (highest) to 5 (lowest)]		
Hilltop			
- South	1	3	4
- North	3	3	4
Middle Slope			
- South	1	1	1
- Southeast	1	2	2
- West	2	1	2
- Northwest	4	3	3
- Northeast	4	4	3
- North	5	5	4
Lower Slope			
- South	1	2	3
- Southwest	1	2	3
- West	3	2	3
- Northwest	5	4	4
- North + Northeast	5	5	5
Valley			
- Northeast	5	5	5

Figure M7 Topographic control of maximum and minimum near-surface air temperature for forested terrain in Germany. Isotherms are based on observations made during a series of sunny days in the month of June. Solar radiation, temperature and humidity regimes for each topographic zone are categorized from 1 (highest values) to 5 (lowest values). Adapted after Geiger (1961).

illustrate the role of south-facing and north-facing orientations in radiant energy receipt, with the latter demonstrating a sharper gradient of change.

Caves provide a means for both the assessment and interpretation of paleoclimates. Cave sediments have yielded significant archeological finds, and artistic renderings on the walls of caves provide profound insights into early humans and their activities. In a cave with a single opening (Figure M10), the cave entrance area can be sunlit and the air relatively hot. Upon entry, passageway narrowing diminishes the effectiveness of turbulent mixing, and the temperature cools and the air becomes moister. Further from the entrance, the temperature

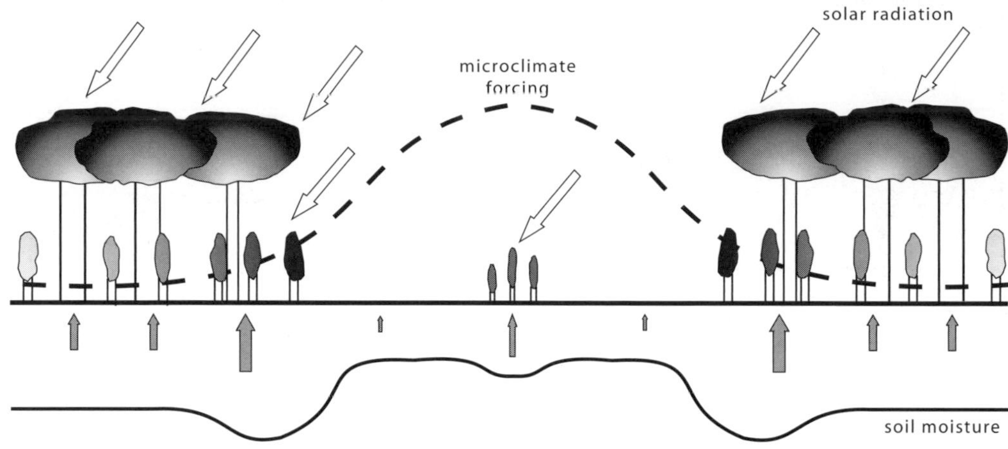

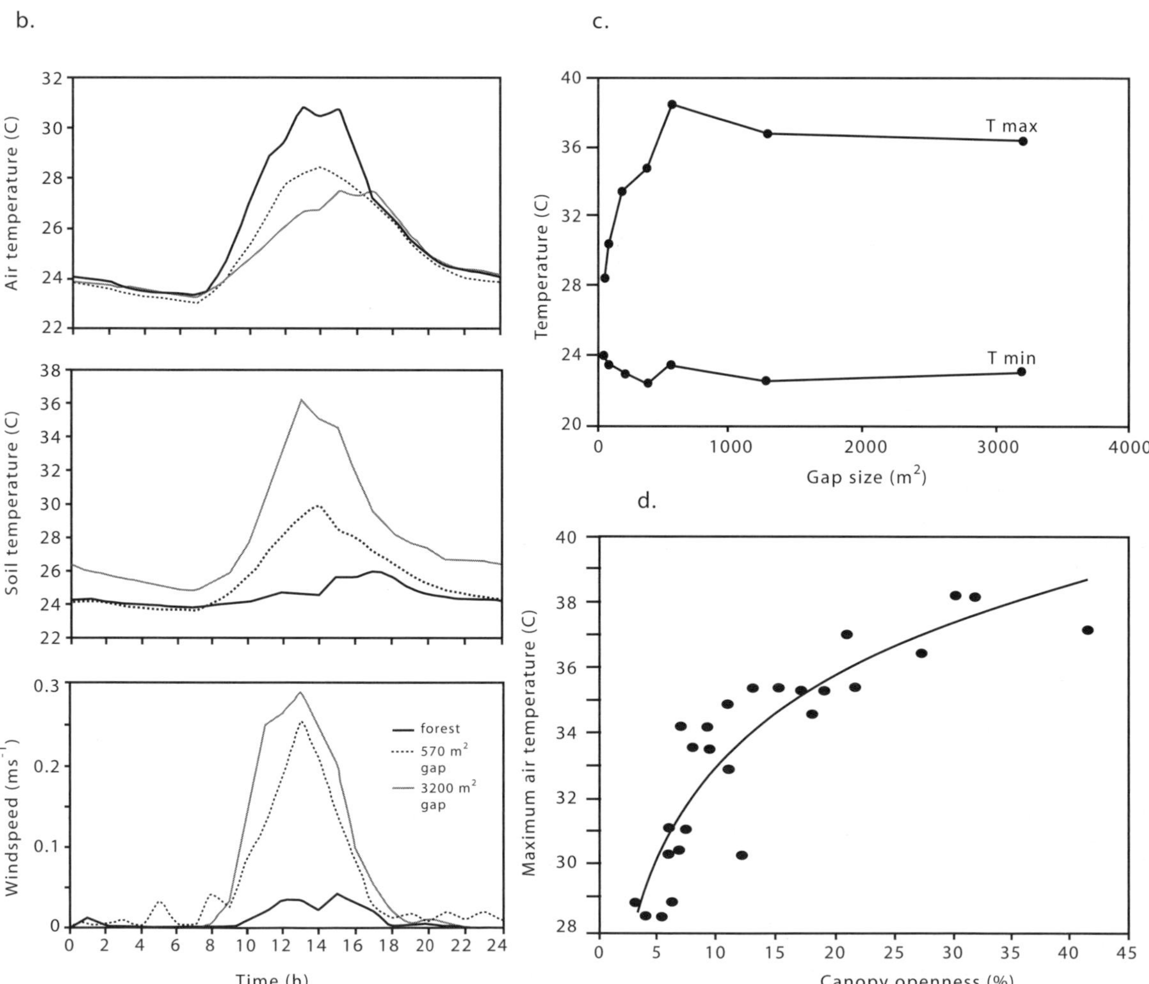

Figure M8 (**a**) A schematic representation of near-surface microclimate forcing in a tropical forest and clearing environment. The availability of soil moisture for the rooting zone is also indicated. (**b**) Diurnal air temperature, soil temperature and windspeed observations within a forest and in the center of two gaps with different areas in Guyana. (**c**) The influence of gap size on maximum and minimum air temperature. (**d**) The relationship between maximum air temperature and canopy openness for tropical forests. Adapted after Sluiter and Smit (2001).

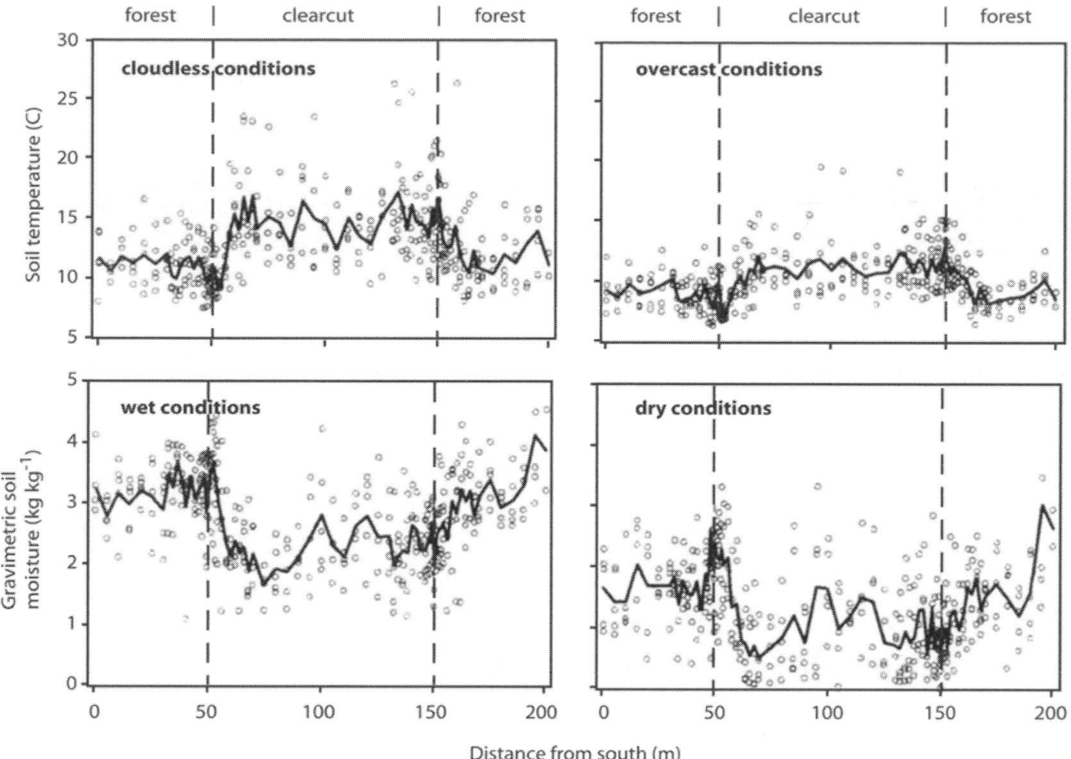

Figure M9 Transects of soil temperature and surface moisture across a 1 hectare clear-cut opening in a subalpine fir and spruce forest in British Columbia, Canada. Both average transect trends and individual site observations are presented. For soil temperature (20 mm below the surface), a comparison between cloudless and cloudy conditions is also displayed. For surface layer moisture a comparison between wet and dry conditions is exhibited. Adapted after Redding et al. (2003).

falls to near-constant values and the humidity approaches saturation. For caves with multiple openings, air circulation may develop on diurnal or seasonal timescales. Such unique cryptomicroclimates have long been used by humans for shelter and security. Knowledge of such subterranean environments also has importance in understanding animal habitats, the characteristics of past and present human subsurface or hillslope dwellings, and the nature of the thermal and humidity regimes in below-surface storage facilities.

The survival and functioning of living organisms is dependent on an interrelationship between an organism and its surrounding environment. Organisms whose temperature is almost totally dictated by the surrounding environment are known as poikilotherms (so called "cold-blooded" animals such as fish, reptiles and insects). The body temperatures of poikilotherms exhibit a strong tracking and correlation with environmental temperature. Organisms that maintain near-constant deep-body temperatures are known as homeotherms (so called "warm-blooded animals" such as birds and mammals, including humans). Homeotherms exhibit near-constant body core temperatures over wide environmental temperature ranges. All animals respond to their microclimate, and heat production by animals can be separated into two distinct components: minimal heat production and regulatory heat production (Figure M11). With favorable environmental temperatures an animal will be in a thermoneutral zone and only minimal heat production will be required. Conditions will be near optimal, but perhaps with slight coolness or warmth. As environmental temperatures deviate from this zone, heat or cold stress dictates increases in regulatory heat production. If environmental conditions further deteriorate, and regulatory heat production is unable to stabilize the thermal regime of the organism, threshold levels can be exceeded. Mortality can then result. For some organisms daily existence and survival is dependent on such microclimate circumstances (Blumberg, 2002). Aspects of animal care, husbandry and transport (Albright, 1990; Johnson, 1987; Scientific Committee on Animal Health and Animal Welfare, 1999) are but a few of the practical acknowledgements of these principles.

Human activities that reshape the landscape give rise to many microclimate regimes. The thermal and moisture conditions in a cultivated field vary markedly from furrow crest to bottom. In many ways these are analogous to features found in larger landscapes (i.e. the topographic control exhibited in Figure M7). The excavation and landscaping of slopes or embankments with orientations that favor higher solar radiation receipt usually realize greater warmth and dryness (Brown and Gillespie, 1995). Humans have created built environments for shelter, security and a host of other reasons. The walls and roofs of buildings are extreme examples of the effects of orientation on solar radiation receipt. In the northern hemisphere, east walls receive direct beam solar radiation in the morning and west walls in the afternoon. Radiant energy receipt on north walls is minimal at most times of the

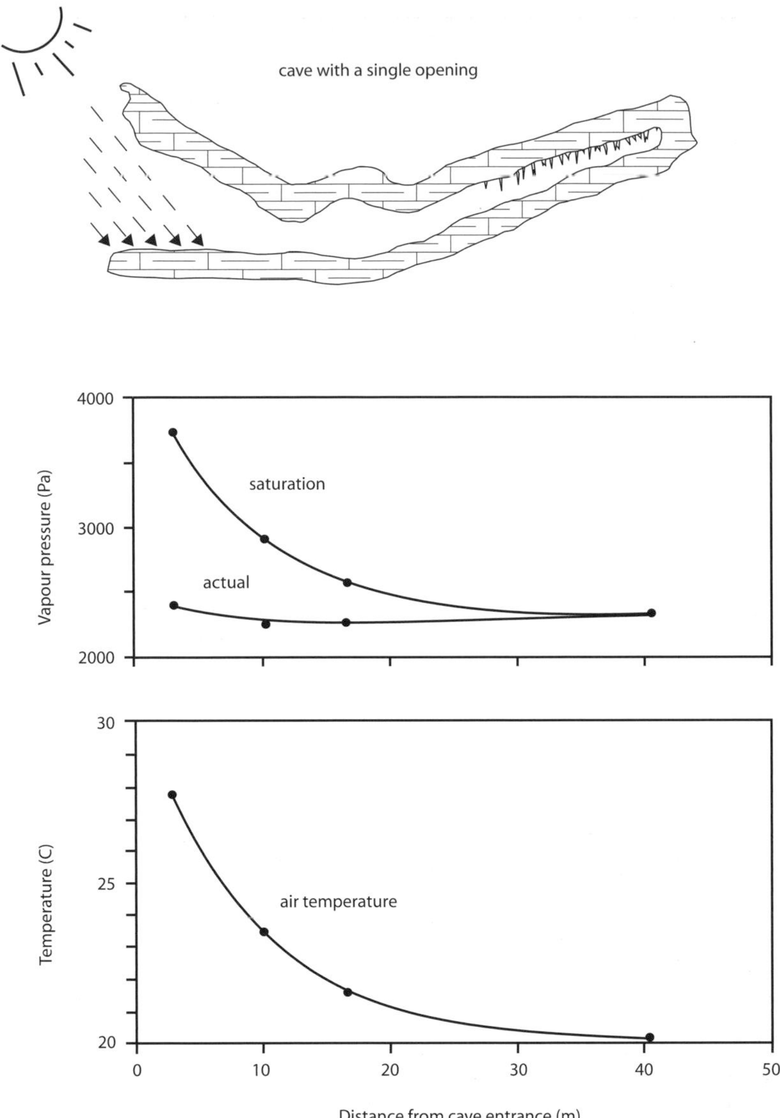

Figure M10 The temperature and humidity regime of a cave with a single opening in the eastern Mediterranean region. Adapted after Geiger (1961).

year and south walls may receive higher solar radiation during the late fall, winter and early spring. Microclimates within buildings may be linked to outside conditions (Brown and DeKay, 2001), or may have temperature, humidity and ventilation cycles that are artificially controlled. Concerns about indoor air quality and "sick buildings" are well documented, and provide challenges for architects and engineers. Humans themselves play distinct roles in the modification of the microclimates within buildings. The inhalation of oxygen and the exhalation of carbon dioxide and water vapor modify the atmospheric environment of rooms (Figure M12). The design of the best microclimate environment possible for buildings and cities, as well as a means for allowing people to modify the environment as needed, presents numerous and ongoing challenges.

The microclimatology of cultural heritage represents a unique blend of seemingly independent disciplines (Camuffo, 1998). However, knowledge of macroclimate, mesoclimate and microclimate regimes is an important factor in the location and architectural design of museums and other facilities that house cultural heritage. Within museums the monitoring and control of microclimate variables (radiation, temperature, humidity and air quality) is fundamental to the preservation of the collections as well as for the comfort of visitors (Thompson, 1990). Outdoor statues and monuments face additional complications as they are continually subject to microclimate influences, for example diurnal and seasonal heating and cooling, and dry and wet deposition arising from pollution. Pisa's leaning tower has been the focus of tourism, public awareness, and scientific and engineering study for many hundreds of years. In addition to

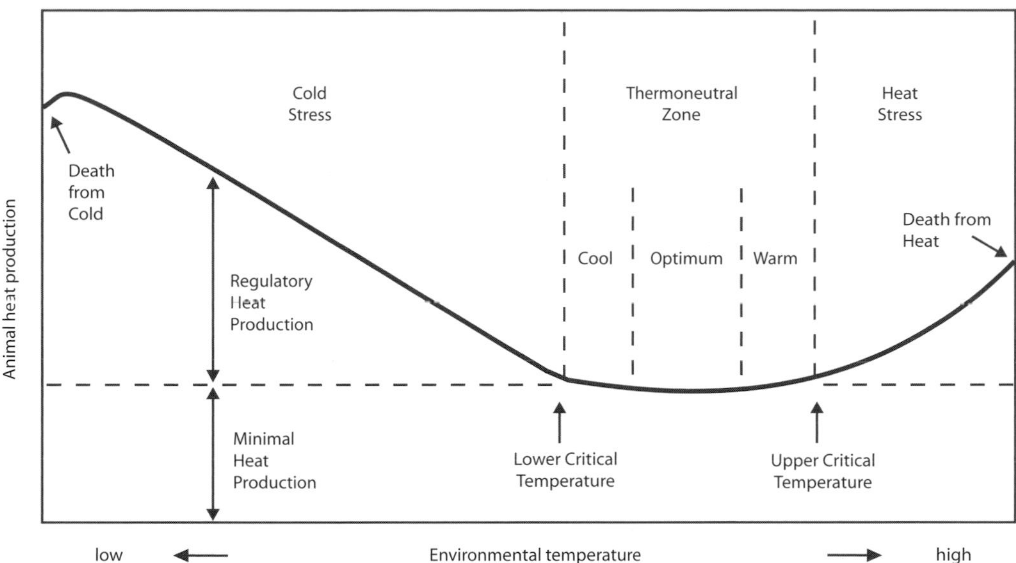

Figure M11 Schematic representation of heat production and the thermal zones of animals. Temperatures falling below the lower critical temperature results in cold stress (hypothermia), whereas temperatures rising above the upper critical temperature results in heat stress (hyperthermia). When critical thresholds are exceeded, mortality can result. Adapted after Scientific Committee on Animal Health and Animal Welfare (1999).

the tower's well-documented ballet with gravity, microclimate consequences are relentless (Camuffo et al., 1999) (Figure M13). The outer wall temperature regimes of La Tower Pendente di Pisa are closely linked to diurnal and seasonal trends in solar radiation. Throughout the day, the warmest wall temperatures move from the east-facing, to the south-facing, and finally to the west-facing walls. Expansion and contraction of the tower, and the associated diurnal twisting that is associated with this thermal warming of the stone, tracks the apparent motion of the sun in the sky. In summer, the warmed stone surfaces can exceed air temperatures by more than 10°C, and the north side of the tower is often colder than the surrounding air. The tower's splendor is continually disfigured by airborne particulate matter deposition, and a complex balance exists between deposition, the tilt and material of the tower, rainfall washout and surface runoff. Micropores on the stone surface play significant nanoclimate and picoclimate roles. The smallest micropores are always filled with water and constitute a reservoir for microbiological life, whereas condensation occurs only occasionally in the largest pores. Micropores between 4 and 5 nanometers undergo more than 250 condensation–evaporation cycles per year, and are influential in the dissolution of the material matrix of the stone and the migration of dissolved salts. Particles that are deposited on the tower can be washed away by rainfall. The washing is governed by the wind direction and the shape of the tower. Patterns in the surface washing demonstrate the direction of storms, the shielding provided by the tilt of the tower and the tower height. Rainfall washing is more effective at the top of the tower. Here airflow is less perturbed by the roughness of the surrounding urban landscape, and the trajectory of the raindrops has a greater horizontal component.

The Great Sphinx of Giza in Egypt is a globally recognized monument. Over past millennia it has been partially submerged in sand and subject to serious erosion about the neck and shoulder region. Excavation around the Sphinx has resulted in an advantageous microclimate modification as a result of altered airflow patterns (Figure M14). Wind that now sweeps across the Sphinx creates vortices on each flank of the monument, and the trajectory of abrasive granules has been altered with most falling into the excavation. As a consequence, one of the many threats to the preservation of the monument has been reduced (Camuffo, 1992, 1998).

New frontiers in the study of microclimatology are continuing to evolve. The study of the microclimatology of spacecraft, orbiting space stations and the surfaces of other planets has already commenced. In the latter, the integration of information gained from pioneer space probes, theoretical understanding and computer modeling holds the potential to yield remarkable advancements. The identification of severe microclimates on Earth, and their use as training areas for future planetary landings, is in its infancy. This attests to bold frontiers ahead for humans and their technology in the microclimates that will be visited far from Earth.

The challenges ahead

The impetus behind the advancements made in microclimatology knowledge during the twentieth century will continue into the twenty-first century. Knowledge about many environments exists, but only scant details are available for others. Demanding questions will continue to be posed, and the quest for insight and answers will persist. Global climate change is driven by processes that often commence with microscale

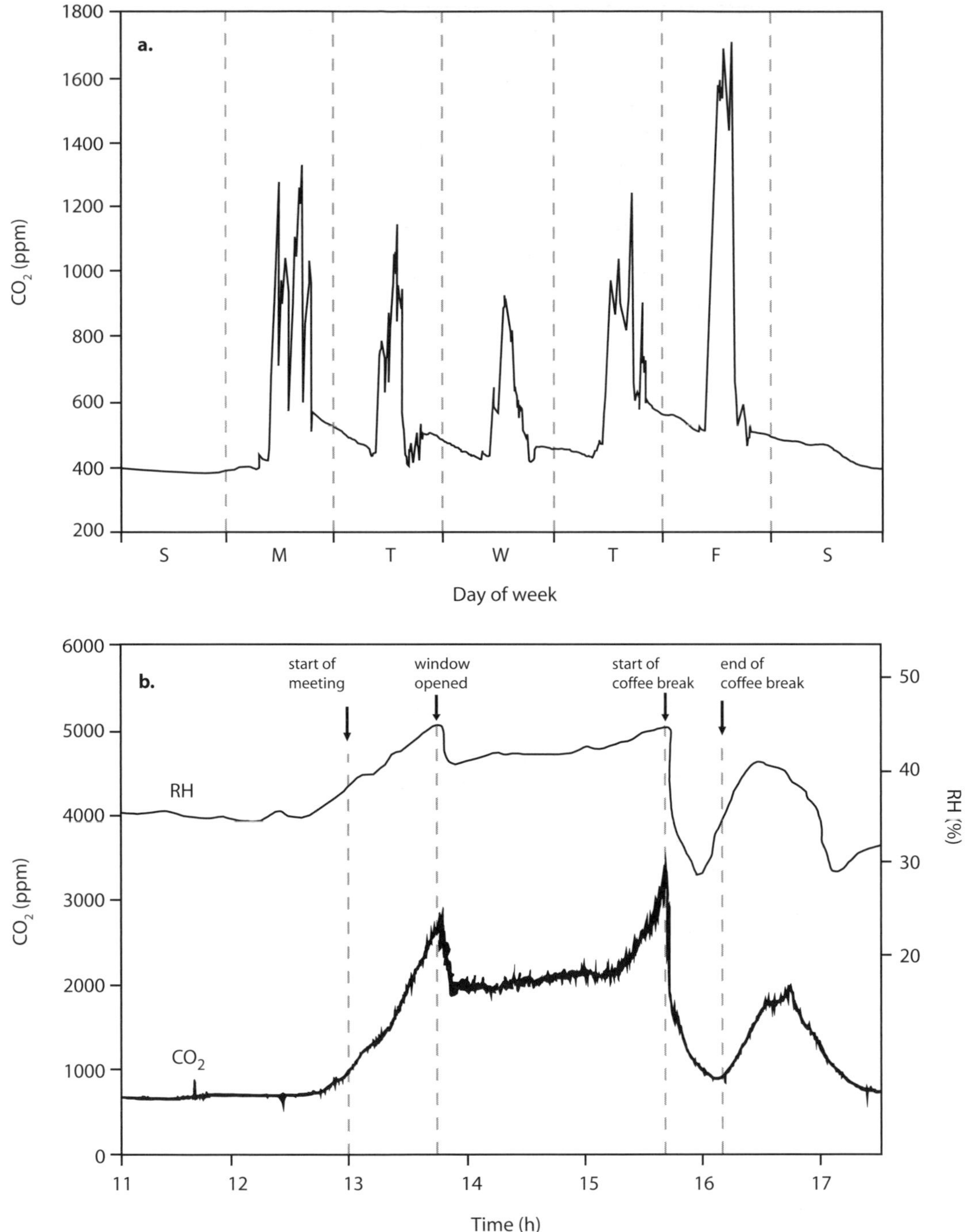

Figure M12 Variations in carbon dioxide in indoor environments. As CO_2 levels exceed 1000 ppm, tiredness and difficulties with concentration arise. Threshold values of CO_2 are approximately 5000 ppm. (**a**) The time-series of CO_2 in an office environment (maximum of three persons present) throughout a weekly period. Decreases in CO_2 occur during the daytime in response to windows being opened to allow the entry of fresh air. (**b**) The time-series of CO_2 and relative humidity during a meeting. Decreases in CO_2 and humidity occur due to window opening and the coffee break activities. Adapted after Endres (2001).

activities (IPCC, 2001c). Adequate representation and the scaling-up of important microclimate influences in numerical modeling approaches, for local and regional weather forecasting to climate change prediction and scenarios, will be an important contribution to the advancement of overall climate science (IPCC, 2001b). Forcings that are imbedded in natural and anthropogenic climate change will impact all microclimates. The response to these changes, in many environments,

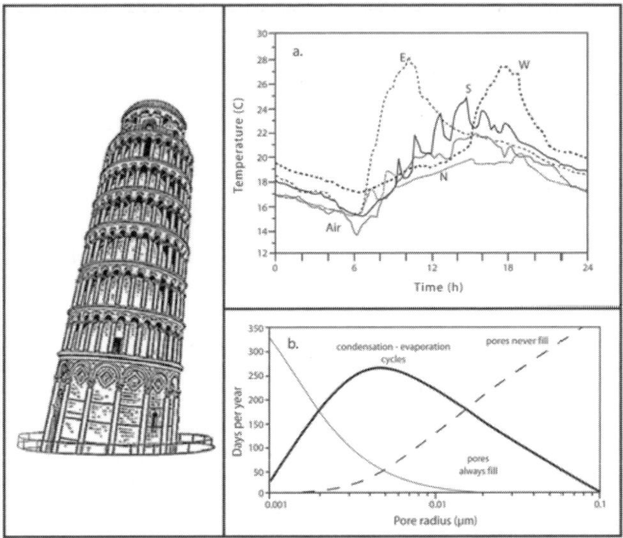

Figure M13 (**a**) The diurnal trend of surface temperatures on the outside wall surface of the central body of La Torre Pendente di Pisa (built from 1173 to 1370). Temperatures are given for positions facing the four cardinal directions. Air temperature was observed on the north side of the tower. (**b**) The influence of micropore size on the number of days each year in which condensation–evaporation cycles occur and the number of days that occur in which micropores are filled or never fill. Adapted after Camuffo et al. (1999).

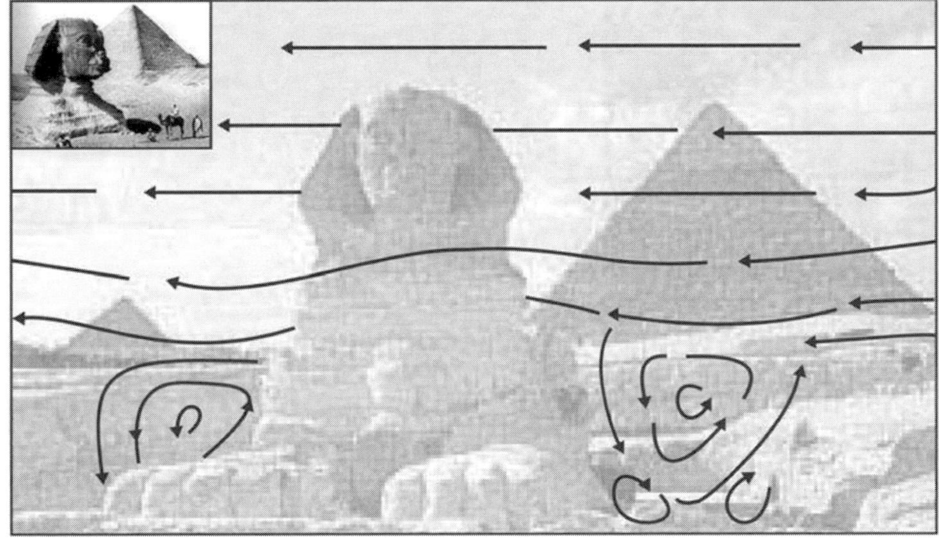

Figure M14 A schematic representation of airflow patterns across the Great Sphinx of Giza. Over the past millennia (insert), the Sphinx was often partially submerged in sand and subject to severe erosion in the neck and chest region. Excavation has resulted in the creation of a major vortex along each flank of the monument. The resultant trajectory of abrasive granules is such that most now fall into the excavation.

will have profound impacts on the organisms that inhabit them, perhaps their very survival (IPCC, 2001a).

Microclimatology is, and will remain, an inexact science. The profound challenges ahead will be the utilization of microclimate resources in the wisest and most prudent manner, and the minimization of deleterious modifications to microclimate regimes.

W.G. Bailey

Bibliography

Albright, L.D., 1990. *Environment Control for Animals and Plants*. St Joseph, MO: American Society of Agricultural Engineers.

American Meteorological Society, 2000. *Glossary of Meteorology*. Boston, MA: American Meteorological Society.

Bache, D.H., and Johnstone, D.R., 1992. *Microclimate and Spray Dispersion*. New York: Ellis Harwood.

Bailey, W.G., Oke, T.R., and Rouse, W.R., 1997. *The Surface Climates of Canada*. Montreal: McGill–Queen's University Press.

Bell, J.N.B., and Treshow, M., 2002. *Air Pollution and Plant Life*. New York: John Wiley.
Blumberg, M.S., 2002. Body Heat: temperature and life on Earth. Cambridge, MA: Harvard University Press.
Bonan, G.B., 2002. *Ecological Climatology: concepts and applications*. Cambridge: Cambridge University Press.
Brown, G.Z., and DeKay, M., 2001. *Sun, Wind and Light: architectural design strategies*. New York: Wiley.
Brown, N., 2001. *History and Climate Change*. London: Routledge.
Brown, R.D., and Gillespie, T.J., 1995. *Microclimate Landscape Design*. New York: John Wiley.
Butzer, K., 1971. *Environment and Archeology*. Chicago, IL: Aldine Atherton.
Calvin, W.H., 1991. *The Ascent of Mind: Ice Age climates and the evolution of intelligence*. New York: Bantam Books.
Calvin, W.H., 2002. *A Brain for All Seasons: human evolution and abrupt climate change*. Chicago, IL: University of Chicago Press.
Campbell, G.S., and Norman, J.M., 1998. *An Introduction to Environmental Biophysics*. Heidelberg: Springer-Verlag.
Camuffo, D., 1992. Controlling the aeolian erosion of the Great Sphinx. *Studies in Conservation*, **38**: 198–205.
Camuffo, D., 1998. *Microclimate for Cultural Heritage*. Amsterdam: Elsevier.
Camuffo, D., Sturaro, G., and Valentino, A., 1999. Urban climatology applied to the deterioration of the Pisa Leaning Tower, Italy. *Theoretical and Applied Climatology*, **63**: 223–231.
Endres, H.E., 2001. Air quality measurement and management. In Gassmaan, O., and Meixner, H., eds., *Sensors in Intelligent Buildings*. Weinheim: Wiley-VCH, pp. 85–102.
Geiger, R., 1961. *Das Klima der bodennahen Luftschicht*. Braunschweig: Vieweg.
Geiger, R., 1966. *The Climate Near the Ground*. Cambridge, MA: Harvard University Press.
Geiger, R., Aron, R.H., and Todhunter. P., 1995. *The Climate Near the Ground*. Braunschweig: Vieweg.
IPCC, 2001a. *Climate Change 2001: impacts, adaptation, and vulnerability*. Contribution of Working Group II to the Third Assessment Report of the Intergovernmental Panel on Climate Change. Cambridge: Cambridge University Press.
IPCC, 2001b. *Climate Change 2001: synthesis report*. A Contribution of Working Groups I, II, and III to the Third Assessment Report of the Intergovernmental Panel on Climate Change. Cambridge: Cambridge University Press.
IPCC, 2001c. *Climate Change 2001: the scientific basis*. Contribution of Working Group I to the Third Assessment Report of the Intergovernmental Panel on Climate Change. Cambridge: Cambridge University Press.
Johnson, H.D., 1987. *Bioclimatology and the Adaptation of Livestock*. Amsterdam: Elsevier.
Jones, H.G., 1992. *Plants and Microclimate*. Cambridge: Cambridge University Press.
Jones, P.D., Ogilvie, A.E.J., Davis, T.D., and Briffa, K.R., 2001. *History and Climate: memories of the future*. New York: Kluwer.
Klinenberg, E., 2002. *Heat Wave: a social autopsy of disaster in Chicago*. Chicago, IL: University of Chicago Press.
Kramer, H.J., 2002. *Observation of the Earth and its Environment. Survey of Missions and Sensors*. Berlin: Springer.
Lee, R., 1978. *Forest Microclimatology*. New York: Columbia University Press.
Lowman, M.D., and Nadkarni, N.M., 1995. *Forest Canopies*. San Diego, CA: Academic Press.
Lowry, W.P., 1988. *Atmospheric Ecology for Designers and Planners*. McMinnville, OR: Peavine.
Lowry, W.P., and Lowry, P.P., 1989. *Fundamentals of Biometeorology*, vol. I: *Interactions of Organisms and the Atmosphere*. McMinnville, OR: Peavine.
Lowry, W.P., and Lowry, P.P., 2001. *Fundamentals of Biometeorology*, vol. II: *The Biological Environment*. St. Louis, MO: Peavine.
Monteith, J.L., and Unsworth, M.H., 1990. *Principles of Environmental Physics*. London Edward Arnold.
Mynemi, R.B., and Ross, J., 1991. *Photon–Vegetation Interactions*. Berlin: Springer-Verlag.
Oke, T.R., 1987. *Boundary Layer Climates*. London: Routledge.
Oke, T.R., 1997. Surface climates processes. In Bailey, W.G., Oke, T.R., and Rouse, W.R., eds., *The Surface Climates of Canada*. Montreal: McGill–Queen's University Press, 21–43.
Oliver, J.E., and Hidore, J.J., 2002. *Climatology: an atmospheric science*. Upper Saddle River, NJ: Prentice-Hall.
Redding, T.E., Hope, G.D., Fortin, M.J., Schmidt, M.G., and Bailey, W.G., 2003. Spatial patterns of soil temperature and moisture across subalpine forest–clearcut edges in the southern interior of British Columbia. *Canadian Journal of Soil Science*, **83**: 121–130.
Rose, C.W., 1966. *Agricultural Physics*. New York: Pergamon Press.
Rosenberg, N.J., Blad, B.L., and Verma, S.B., 1983. *Microclimate: the biological environment*. New York: Wiley-Interscience.
Scientific Committee on Animal Health and Animal Welfare, 1999. *Standards for the Microclimate Inside Animal Transport Road Vehicles*. European Commission, Health and Consumer Protection Directorate-General, Directorate B, Scientific Health Opinions.
Sluiter, R., and Smit, N., 2001. Gap size effects on microclimate and soil moisture. In van Dam, O., ed., *Forest Filled with Gaps: effects of gap size on water and nutrient cycling in tropical rain forest*. Tropenbos-Guyana Series 10. Amsterdam: PrintPartners Ipskamp, pp. 49–66.
Stoutjesdijk, Ph., and Barkman, J.J., 1992. *Microclimate: vegetation and fauna*. Knivsta: Opulus.
Thompson, G., 1990. *The Museum Environment*. London: Butterworth–Heinemann.
Unwin, D.M., and Corbet, S.A., 1991. *Insects, Plants and Microclimate*. Slough: Richmond Publishers.
Whiteman, C.D., 2000. *Mountain Meteorology*. Oxford: Oxford University Press.

Cross-references

Albedo and Reflectivity
Bioclimatology
Climatology
Local Climatology
Scales of Climate
Vegetation and Climate

MIDDLE LATITUDE CLIMATES

The middle latitudes are regions of great atmospheric restlessness and variability, dominated at the surface and in the upper atmosphere by westerly winds (Hare, 1960). The climate is controlled at the surface by a succession of cyclones and anticyclones, normally moving from west to east, that are steered by the upper flow. The fall of temperature toward the poles occurs not in a uniform manner, but with the strong thermal gradients concentrated in one or more narrow latitudinal bands, or fronts. Baroclinic conditions lead to the development of jet streams just below the tropopause (Barry and Carleton, 2001).

The equatorward limit of middle latitude climates is often taken as the surface subtropical high-pressure belt. The poleward limit is more diffuse and variable, although it is often marked by the subpolar lows. The latitudinal extent of the climatic zone will vary from month to month and year to year, depending on changes in the position of the bordering centers of action, but most frequently it occupies the zone between 35 and 56 degrees N and S. As the overall equator-to-pole temperature contrast decreases from winter to summer, so does the strength of the westerlies. Day-to-day weather at a particular location is much affected by the latitude and strength of the westerly current. The upper westerlies vary between two extreme states. At times the upper flow is little deformed by waves, and the flow is parallel to the lines of latitude. At such times of high zonal index the westerlies blow strongly over large longitudinal zones and sweep a succession of depressions eastward at high speeds with frequent rain and gales in middle latitudes. At other times the Long or Rossby waves in the upper westerlies become greatly amplified, resulting in a meridional flow, a condition described as low zonal index. If the tips of the waves become cut off, leaving cold upper pools with associated low pressure at the surface in low latitudes and warm upper highs and associated surface anticyclones at high latitudes, blocking develops.

Blocks have preferred seasonal and geographical incidences, being most common in spring and in the eastern Atlantic and over eastern Asia. Having once formed, blocks frequently are very persistent and may dominate the circulation for several weeks and on occasion for a whole season, introducing large temperature and precipitation anomalies into middle latitudes. Evidence exists of annual variations in blocking frequency and persistent anomalous climatic conditions, e.g. the mid-1970s European drought could have been caused by persistent or repetitive blocking episodes.

The strength of the circulation in any season is often expressed in relation to oscillations in the strength and position of the main centers of action, e.g. the Icelandic Low, and the Azores High. Such pressure oscillations include the North Atlantic Oscillation (NAO) and North Pacific Oscillation (NPO). In recent years the NAO, in particular, has often been strongly positive, implying strong westerlies in the Atlantic-European sector and a succession of mild winters in much of Europe (Perry, 2000).

Great differences occur in the geographical extent of land and sea areas in the two hemispheres. In the northern hemisphere midlatitudes contain the large landmasses of North America and Eurasia that lead to prominent contrasts in regional climate with important longitudinal variations. West coast locations are exposed to the predominant westerly winds and have equable temperature regimes, often with abundant precipitation throughout the year. The interiors of the continents have a more continental climate with greater annual temperature ranges and lower average precipitation totals, most of which falls in the summer season as a result of convective processes. In the southern hemisphere the much greater extent of sea area ensures that the westerlies blow more consistently and strongly (being known by a variety of names, including the Roaring Forties), and the circulation is less disturbed by periods of blocking than in the northern hemisphere. The kinetic energy of the southern westerlies is about 60% larger than the northern hemisphere westerlies. Smith (1967) has suggested that continentality and oceanicity are the principal criteria in defining subdivisions of climate in middle latitudes with general temperature levels forming a secondary subdivision. The adjective "temperate" is often applied to midlatitudes or used as a proxy term (e.g. Trewartha, 1968) but is certainly not suited to all climates in these latitudes as Bailey (1964) has shown.

Marine midlatitude climates

Along the western margins of the continents in the middle latitudes of both hemispheres, the climate is under a considerable maritime influence. In Western Europe the topography allows penetration of maritime airmasses deep into the continent in contrast to the situation in both North America and South America. A distinction is normally made between warm temperate or Mediterranean and cool temperate climates.

Mediterranean type

The approximate latitudinal extent of this type is on the order of 5–10 degrees between latitudes 30/35–40. A large proportion of the total precipitation falls in the winter when day-to-day weather is controlled by the behavior of the westerlies (Perry, 1981). During the hot dry summers the subtropical high-pressure area and its attendant ridges take control. In the type area (the Mediterranean Basin), substantial winter precipitation totals occur during low zonal index phases, when a meandering jet stream with a major trough exists over the Mediterranean, favoring cyclogenesis. Lee depressions form in the Gulf of Genoa and south of the Atlas Mountains and move in a generally eastward direction through the basin, often becoming reinvigorated in the east near Cyprus. Whereas the sea acts as a heat sink in summer, in winter it represents a heat source and cold airmasses entering the basin quickly become unstable after passage over the relatively warm water. The high intensity of rainfall is reflected in the small number of rain-days, even in the wetter areas. Regional winds, e.g. the cold Mistral and Bora, are related to meteorological and topographic factors, whereas the persistent northerly winds of summer in the eastern Mediterranean known as the etesians give this area a distinct climatic subtype. High annual sunshine totals are a feature of Mediterranean climates, with totals exceeding 3000 hours being quite common. In the northern hemisphere Southern California has this type of climate, as does the southern hemisphere coast of Chile around Santiago, the West Australian coast around Perth, and the South African coast around Cape Town.

Cool temperate

Poleward of the Mediterranean climate, on the west side of the continents, the climate is changeable throughout the year with well-distributed precipitation, brought by a series of depressions and their associated fronts and a predominance of maritime air masses. In northwestern Europe and on the Oregon and Washington coasts of North America, large positive temperature anomalies occur in winter compared with the average for the latitude, as maritime airmasses cross the warm-water currents in the North Atlantic and North Pacific. Since summers are relatively cool for the latitude, annual temperature ranges of less than 20°C are common. Because depressions are deeper and more vigorous in winter near the coast, there is normally a fall or winter precipitation maximum, whereas spring and early summer have a minimum of precipitation, reflecting both the increased frequency of blocking anticyclones in these seasons and the lower moisture content of maritime airmasses at this period due to the lower sea temperatures. High precipitation totals are a feature of all mountain areas and large lapse rates give such areas a short growing season and a cloudy, damp and often raw climate. A high degree of changeability from day to day is a characteristic of these climates as rapid alternations of airmasses occur, although occasionally persistence of a particular synoptic situation leads to more settled conditions. The coldest winter weather and the warmest summer spells develop when continental airmasses replace, for a time, the more usual airflow from oceanic sources. In the southern hemisphere sizeable belts of this climatic type occur in Chile, Tasmania, and the South Island of New Zealand.

Continental midlatitude climates

In both North America and Scandinavia the transition from maritime to continental climates is rapid due to the barrier effect of the mountains imposed on the invasion of surface maritime airmasses, but across the European Plain the transition is much more gradual. Meridional airmass movement is a particular characteristic of the North American climate and rapid changes of temperature level can occur, especially in winter, as frontal depressions cross the continent. Very severe winters occur at times with disrupting snowstorms, and are the result of strong amplification of the long waves with ridging over western North America and a deep trough in the east. This allows deployment of Arctic airmasses into the eastern states, while often in the west weather is mild and dry. Blizzard conditions develop on the cold polar side of traveling lows. In summer severe heatwaves can develop when warm air from the Gulf of Mexico is advected northward into the central and eastern states.

Over the USSR the winter circulation is dominated by the intense Siberian anticyclone, although the position and the intensity of this cold anticyclone can cause considerable departures of temperature values from the normal in individual years. Mean January temperatures below −40°C occur in parts of Siberia, accompanied by dry sunny weather. During the short summers, temperatures can rise to 35°C on occasion, even as far north as the Arctic Circle. In these continental climates the transition seasons of spring and fall are very short. Although the seasonal variation of precipitation in the interior and eastern parts of both North America and Asia can be complex, there is normally a summer maximum brought on by instability showers and thunderstorms. Local variations in precipitation totals reflect such factors as the presence of large lakes, e.g. the Great Lakes of North America enhance snowfall totals on their eastern shores. In the United States the 51 cm annual isohyet follows approximately the 100° West Meridian, and in the dry area between the Rocky Mountains and this longitude occasional drought years occur, such as those resulting in the Dust Bowl conditions of the 1930s.

Allen Perry

Bibliography

Bailey, H.R., 1964. Towards a unified concept of the temperate climate. *Geographical Review*, **54**: 516–545.

Barry, R.G., and Chorley, R.J., 1998. *Atmosphere, Weather and Climate*, 7th edn. London: Methuen.

Barry, R.G., and Carleton, A.M., 2001. *Synoptic and Dynamic Climatology*. London: Routledge.

Hare, F.K., 1960. The westerlies. *Geographical Review*, **50**: 345–367.

Perry, A.H., 1981. Mediterranean climate: a synoptic reappraisal. *Progress in Physical Geography*, **5**: 107–113.

Perry, A.H., 2000. The North Atlantic Oscillation: an enigmatic see-saw. *Progress in Physical Geography*, **24**: 289–294.

Smith, D., 1967. Middle latitude climates. In Fairbridge, R.W., ed., *The Encyclopedia of Atmospheric Sciences and Astrogeology*. New York: Reinhold, pp. 604–605.

Trewartha, G.T., 1968. *An Introduction to Climate*, 4th edn., New York: McGraw-Hill, pp. 305–339.

Cross-references

Airmass Climatology
Climate Classification
Climatology
Continental Climate and Continentality
Europe, Climate of
Jet Streams
Mediterranean Climate
Maritime Climate
North America, Climate of

MILANKOVITCH, MILUTIN (1879–1958)

Milutin Milankovitch developed the theory of orbital control of terrestrial insolation and climate change. Milankovitch was of Serbian origin, born in Dalj (Slavonia) in what is now Croatia. He obtained his PhD in 1904 in Vienna and worked as a civil engineer for 5 years before going to the University of Belgrade as Professor of Applied Mathematics. Captured during World War I, but allowed to work at the Hungarian Academy of Sciences, Milankovitch completed the insolation theory for the Earth and had also worked out a climate history for Venus and Mars.

The "Milankovitch parameters" involved in his theory are: (1) the eccentricity (or ellipticity) of the orbit e, which measures the departure of the Earth's circumsolar orbit from a circle (with a period of about 90 000–100 000 years); (2) the obliquity of the ecliptic (or tilt), which is the angle between the equator and the plane of the orbit (principal period about 41 000 years); and (3) the precessional parameter, related to the longitude of the perihelion, which is conveniently expressed as the angular distance of the spring equinox point from the perihelion. It has two principal terms, about 19 000 and 23 000 years. This is related to the "general" *precession* that has been refined to 25 694 years (Berger, 1992).

After World War I, Wladimir Köppen, the famous climatologist, was working on a book with his son-in-law Alfred Wegener (of continental drift fame). This became a standard textbook on climate (1924) and carried the Milankovitch message. Unfortunately it only reached a German-speaking audience. However, a German geologist and archeologist from Breslau, Frederick Zeuner, applied the insolation theory to the central European geological record, and after moving to London in the 1930s Zeuner explained to a world audience the Milankovitch ideas (Zeuner, 1959). The "establishment" of the day responded to the Milankovitch theory with severe criticism, an outrage almost paralleled by the hostile reception of Wegener's continental drift theory. During World War II Milankovitch worked on a complete revision of his radiation theory, which was published (in German) by the Royal Serbian Academy of Sciences in 1941 as the *Kanon der Erdbestrahlung*.

In the post-World War II decades, however, dramatic discoveries were being made at sea thanks to sediment coring and deep-sea drilling. The sea-level record paralleled the last phase of the Milankovitch radiation curve for 65°N. During the next two decades, dating systems were expanded, deep-sea cores were obtained worldwide, and the results were conclusive: there was an exact match with the Milankovitch pattern (Fairbridge, 1967), and it was recognized as "the pacemaker of the ice ages" (Hays et al., 1976).

Milankovitch was destined to become one of those rare scientists who developed truly pivotal ideas. His inspirational conversion from engineering to orbital dynamics is elegantly told by Imbrie and Imbrie (1979), and may be followed with details and bibliography in an autobiographical work (1957). In 1979, to mark the 100th anniversary of the birth of Milankovitch, a symposium was organized in Belgrade by the Serbian Academy of Sciences, and a conference convened at the Lamont–Doherty Geological Observatory (Palisades, NY). The evidence was overwhelming (see symposium volume, edited by Berger et al., 1984).

The acceptance of the Milankovitch theory was far more than a quantification of dynamic change for the last ice age. Geologists are now applying it to the whole of Earth history. For climatologists, long-term modeling is now feasible, and for meteorologists it carries a crucial message: the terrestrial climate machine is neither chaotic nor unpredictable – it is forced by extraterrestrial agencies. The same message must apply also to other planets.

George Kukla and Rhodes W. Fairbridge

Bibliography

Berger, A.L. (ed.), 1981. *Climatic Variations and Variability: facts and theories*. Dordrecht: Reidel.
Berger, A.L., 1992. Astronomical theory of paleoclimates and the last glacial–interglacial cycle. *Quaternary Science Review*, **11**: 571–81.
Berger, A.L., and Loutre, M.F., 1991. Insolation values for the climate of the last 10 million years. *Quaternary Science Review*, **10**: 297–317.
Berger, A., Imbrie, J., Hays, J. et al. (eds), 1984. *Milankovitch and Climate*. Dordrecht: Reidel.
Berger, A., Loutre, M.F., and Laskar, J., 1992. Stability of the astronomical frequencies over the Earth's history for paleoclimate studies. *Science*, **255**: 560–566.
Broecker, W.S., 1968. Milankovitch hypothesis supported by precise dating of coral reefs and deep sea sediments. *Science*, **159**: 297–300.
Emiliani, C., 1966. Isotope paleotemperatures. *Science*, **154**: 851–857.
Fairbridge, R.W., 1960. The changing level of the sea. *Scientific American*, **292**(5): 70–79.
Fairbridge, R.W., 1961. Eustatic changes in sea level. In Ahrens, L.H., et al., eds., *Physics and Chemistry of the Earth*, vol. 4. London: Pergamon Press, pp. 99–185.
Fairbridge, R.W., 1967. Ice-age theory. In Fairbridge, R.W., ed., *The Encyclopedia of Atmospheric Sciences and Astrogeology*. New York: Reinhold, pp. 462–467.
Hays, J.D., Imbrie, J., and Shackleton, N.J., 1976. Variations in the Earth's orbit: pacemaker of the ice ages. *Science*, **194**: 1221–1332.
Imbrie, J., and Imbrie, K.P., 1979. *Ice Ages: solving the mystery*. Short Hills, NJ: Enslow.
Köppen, W., and Wegener, A., 1924. *Die Klimate der Geologischen Vorzeit*. Berlin: Gebr. Borntraeger.
Kukla, G., and Gavin, J., 1992. Insolation regime of the warm to cold transition. In Kukla, G., and We, E., eds., *Start of a Glacial*. NATO ASI Series I, 3, pp. 307–339.
Kutzbach, J.E., 1985. Modeling of paleoclimates. *Advances in Geophysics*, **28A**: 159–196.
Liu, H.-S., 1995. A new view on the driving mechanisms of Milankovitch glaciation cycles. *Earth and Planetary Science Letters*, **131**: 17–26.
Milankovitch, M., 1941. Kanon der Erdbestrahlung und seine Anwendung auf das Eiszeitenproblem. Royal Serbian Academy, Special Publication 133. (English translation 1969 as *Canon of Insolation and the Ice-Age Problem*, by Israel Progr. Sci. Transl. Washington, DC: US Department of Commerce.)
Milankovitch, M., 1957. Astronomische Theorie der Klimaschwan-kungen: ihr Werdegang und Widerhall. *Serbian Academy of Sciences Monography*, **280**: 1–58 (incl. comprehensive bibliography).
Sanders, J.E., 1995. Astronomical forcing functions: from Hutton to Milankovitch and beyond. *Northeastern Geology and Environmental Sciences*, **17**(3): 306–347.
Zeuner, F.E., 1959. *The Pleistocene Period*. London: Hutchinson.

Cross-references

Climatic Variation: Historical Record
Paleoclimatology

MILITARY AFFAIRS AND CLIMATE

Today, the military is called upon for a variety of operations ranging from war fighting to humanitarian or peacekeeping missions (military operations other than war, MOOTW) to peacetime operations (for example training and research) similar to any other industry. Post-September 11 the military has also been asked to increase its role in domestic security issues and anti-terrorism special operations. All of these complex missions require timely weather and climate information. For the military, climate and weather are not distinct areas of study; rather, they flow along the entire knowledge continuum embracing strategy and tactics. Knowledge of the climatic conditions of an area is necessary in developing the strategic plans for operating in that area. Training as well as development, test, and evaluation of equipment also requires detailed climatic knowledge. On timescales up to a week, knowledge of the weather is critical in developing the tactics used during a mission. As demonstrated in Winters (1998), failure to effectively exploit this climatic knowledge, or a chance encounter with weather conditions representative of extreme departures from anticipated conditions, have resulted in disastrous defeats throughout military history. This essay will highlight many of the climatic elements that impact on the military as they perform their various missions. It will conclude with a brief look at "climates" of space and the oceans which also impact military operations. Readers interested in other areas of military geography should consult the numerous references in Collins (1998) and Plaka and Galgano (2000).

Impacts of climatic elements

A climatic element is some component of the climate system such as temperature or precipitation. Information on these elements is measured using conventional weather equipment. The (near) instantaneous values of these components is termed weather. The synthesis of these values, including not only average values but also their variability, is termed climate. Each has critical military significance. This section will relate the major climatic elements to some of their military impacts.

Temperature is the most obvious climatic element. Humans are homeotherms; they attempt to maintain a core body temperature within narrow limits. Extreme hot or cold temperatures impact the body thermoregulation system and degrade performance. In hot environments, armies will consume more fluids, and troops must be monitored to prevent heat illness. Extreme cold increases the logistical need for clothing and food to ensure good health. Equipment as well is impacted by extreme temperatures. Hot ambient conditions reduce the efficiency of engine radiators and may lead to overheating. Lubricants become less efficient at extreme temperatures causing increased wear or damage. In very cold conditions machinery may freeze if not continuously tended. In the free atmosphere temperature inversions may cause areas of anomalous propagation of electromagnetic radiation. The resulting degraded radar performance may prevent the acquisition or tracking of targets or threats. Jet engine efficiency also is a function of air temperature. Low-flying cruise missiles may have their effective range shortened in hot environments.

Humidity also impacts human comfort and performance. High humidity combined with high temperatures may produce dangerous heat stress conditions. At low humidity, drying effects can influence not only human comfort but fluid evaporation rates. A particular concern is the build-up in static charge which can make handling of fuels and munitions dangerous. Temperature and humidity determine air density, which impacts projectile performance. In tropical environments, high humidity contributes to the biofouling and deterioration of equipment. It is often necessary to artificially heat confined spaces to discourage the growth of mold or provide rust-resistant coatings to metal surfaces.

Precipitation has inspired many a soldier's lament. Heavy rains often result in trafficability problems as roads become impassable due to mud or debris flows. Floods are also a problem. Flash floods may imperil troops or equipment while riverine floods may cause delays in bridging operations. Snow or heavy rain restricts visibility and limits aircraft operations. The in-flight build-up of ice can affect aircraft control surfaces. Sensors or communication systems operating at microwave frequencies are particularly susceptible to interference from heavy precipitation.

Wind impacts include lowering of visibility due to blowing dust or sand. High winds may preclude the use of smoke or other obscurants to hide troop movements or impact the dispersal of chemical or biological weapons. Radar-absorbing chaff clouds, used as an electronic warfare countermeasure, may be quickly dispersed in high winds, thus reducing their effectiveness. High winds may also impact aircraft operations. Helicopter maneuvering in gusty winds is often problematic and launch and recovery of unmanned aerial drones is complicated by unsteady winds. Cross-winds and turbulence can impact aircraft providing low-level combat air support.

Many impacts on visibility have been listed under other climatic elements. Fog, blowing dust, and the smoke created by battle significantly impact low level visibility. For aircraft, cloud layers may hinder target identification and weapons delivery. Many of these problems have technological solutions beyond the scope of this item. One historically important aspect of visibility has been all but eliminated by technology: night versus day. Night vision equipment and associated sensors which allow pilots and ground personnel to operate in near-total darkness have resulted in a radical change in battlefield tactics. War fighting is no longer a "sunrise to sunset" affair.

A final climatic element of military concern is atmospheric pressure. Aircraft must have oxygen systems for prolonged high-altitude flight. Troops operating in mountainous terrain need time to acclimatize to the lower pressures to avoid rapid fatigue, or in extreme cases altitude sickness. Even the often-vilified army cook must be aware of the impact of high elevation on cooking times and the boiling point of water!

Impacts of ocean and space "climates"

Most of this item has been directed toward the ground and lower troposphere. However, naval and marine units encounter another set of impacts peculiar to the ocean and the land/water interface. High winds and wave heights can make naval operations and shipkeeping extremely difficult to the point of structural damage. Personnel operating in and under the water must be protected from the effects of hypothermia. Amphibious landings must contend with tides, currents, and breaking waves. Submarine warfare is especially dependent on knowledge of sound-velocity profiles which are largely determined by seawater temperature. Fog and structural icing are also concerns.

With the development of space-based satellite observation and communication systems, the realm of "climate effects" has been extended to the exosphere. Periods of intense solar activity produce storms of charged particles which impact satellite operations. Multi and hyper-spectral sensors are used to minimize the effects of clouds and water vapor on the satellite geometry.

Military weather and climate monitoring

The military maintains weather-monitoring stations at many of its facilities around the world. Part of their mission is to provide forecasting support to operational units. The US Air Force and Navy maintain detachments at the National Climatic Data Center. Here climatic data collected from weather stations around the world are used to produce climatic summaries for areas of interest to the military. During times of crisis these detachments can produce specific mission-directed products. Operational units are typically deployed with personnel trained in analyzing these weather and climate products.

Lessons learned from historical engagements are combined with advances in technology at a series of national laboratories operated by the Army, Air Force, and Navy. Research results and products from these laboratories are tested and evaluated at a series of field laboratories before being approved for use by operational forces. The testing must be done under similar climatic conditions to those in which the products will used; therefore field laboratories are located in tropical, temperate, and Arctic locations. In addition, troops must be trained to operate in a variety of climatic conditions prior to their deployment. The goal is to exercise personnel and machinery prior to employment to maximize effectiveness and minimize loss. Climate is one consideration in this important process.

Conclusion

Weather and climate have played an important role in historical war fighting. The role continues today along with increased responsibilities for MOOTW and peacetime operations. Climate impacts all these functions. From training to development of new weapon systems to operational deployment climate considerations abound. Strategic planners and logisticians make use of climate data in deciding what mix of forces will be used in a particular area and how they will be equipped and supplied. Operational commanders can prepare tactics based on climatic knowledge and be ready to adapt those tactics to take advantage of changing weather conditions. Climatologists work with engineers and other technicians to provide new equipment and climatological summaries relevant to the evolving missions of today's military.

Richard W. Dixon

Bibliography

Collins, J., 1998. *Military Geography for Professionals and the Public*. Dulles, VA: Brassey's.

Plaka, E., and Galgano, F., 2000. *The Scope of Military Geography: across the spectrum from peacetime to war*. New York: McGraw-Hill.

Winters, H., 1998. *Battling the Elements: weather and terrain in the conduct of war*. Baltimore: Johns Hopkins University Press.

Cross-references

Climate Comfort Indices
Cultural Climatology
Human Health and Climate

MODELS, CLIMATIC

Climatic models are abstractions of climate that are used for one or more of three overlapping purposes: to explore and describe the interrelationships within and between the observed elements of climate, to predict important climatic or climate-forced events, or to explain the fundamental biophysical mechanisms that control climate. A widely accepted, comprehensive definition of climate and, subsequently, of climatic models has not been published although many climatologists would agree that climate is a synthesis of weather in which the time-period of integration often exceeds the length of a weather forecast. Climatic models, in fact, have been developed and used to investigate and describe climatic processes which are manifested over tens, hundreds and even thousands of years of Earth history as well as over but a few days. Equally variable is the geographic scale on which climate occurs and is modeled (Steyn et al., 1981).

Microclimatic models, for instance, may attempt to characterize climatic processes that function within only a few cubic meters, whereas macroclimatic models often strive to describe climate as it extends over hundreds or thousands of square kilometers and vertically from the surface to the bottom of the stratosphere and perhaps beyond (Schneider and Dickinson, 1974). From an anthropocentric vantage point, however – regardless of scale – all climatic processes and phenomena are manifested within the context of a four-dimensional (the three space dimensions plus time), circum-surface climatic environment (Figure M15). Models of climate, it follows, are chiefly formulated for and used within this context, even though application-specific requirements result in a wide variety of model forms.

A common purpose

A number of climatologists would qualify the above characterization by emphasizing that climatic models are distinct from meteorological models only when they attempt to couple terrestrial systems or oceanic processes to the lower atmosphere. Since the lower atmosphere and Earth are coupled across their contact surface, the energy and moisture crossing that interface control the equilibrium of atmosphere–Earth energy and mass exchange and, thus, have primary climatological importance. Climate at the surface also is primarily what humans experience, influencing a wide range of human activities. Climatic models, as a consequence, are mainly developed and used to explore, describe, predict or explain the integrated surface or primary interface (Figure M15) temperature and moisture states of the world's oceans, lakes, rivers, snowfields, soil, flora and fauna, as well as of people and their edifices.

As the integrated temperature and moisture states of any terrestrial or aquatic surface(s) of interest, that is climate, can only be evaluated by accounting for the fluxes of energy and mass (predominantly moisture) to and from the surface (i.e. by solving the surface "energy [and mass] budget"), the search for improved solutions to the budget – for the myriad of disparate space and time scales as well as environments – provides the *raison d'etre* for much of the development, refinement and use of climatic models. A primary reason for climatic modeling, in other words, is the establishment of good, quantitative solutions to all or parts of

$$Q(1-\alpha) + I\downarrow - I\uparrow = H + LE + G + F \tag{1}$$

where Q is solar irradiance ($\mathrm{W\,m^{-2}}$), α is the integrated reflectivity of the surface to solar (shortwave) irradiance, $I\downarrow$ is longwave radiation from the surrounding environment and the atmosphere which is absorbed by the surface ($\mathrm{W\,m^{-2}}$), $I\uparrow$ is longwave radiation emitted by the surface ($\mathrm{W\,m^{-2}}$), H is the rate of sensible heat exchange between the surface and lower atmosphere ($\mathrm{W\,m^{-2}}$), L is the latent heat of vaporization ($\mathrm{J\,kg^{-1}}$), E is the rate of water vapor exchange between the surface and lower atmosphere ($\mathrm{kg\,m^{-2}\,s^{-1}}$), G is the rate of heat exchange by conduction between the surface and underlying ground or water ($\mathrm{W\,m^{-2}}$) and F is the heat convected by the ocean, lake or river from its surface ($\mathrm{W\,m^{-2}}$). Consistent with convention [$Q(1-\alpha)$] and $I\downarrow$ are positive fluxes that contribute to surface warming whereas $I\uparrow$ is a positive, cooling term. On the other hand, H, LE, G and F are positive when the direction of transfer is away from the surface (cooling) and negative when they contribute to surface warming.

In order to solve the surface energy budget (equation 1), climate models must account for the ambient or representative energy and moisture states of the near-surface air, soil, snow, vegetation and so on, as well as for the states of the surface, because the magnitudes of most of the fluxes depend upon temperature, humidity and momentum gradients that exist between the surface, the atmosphere above and the terrestrial or aquatic environ below. That is, climatic models must provide supportive, quantitative characterizations of the near-surface state variables, such as air temperature, vapor pressure and soil moisture, before equation (1) can be adequately solved. Concern for the ambient states of the near-surface, in turn, leads to the related problem of evaluating those convective and radiative fluxes that take place across the climatic environment's secondary interfaces (Figure M15), for they – like the exchanges across the atmosphere–Earth interface – affect the states of the near-surface which then influence the surface energy budget. Thus, equation (1) is fundamentally nonlinear, in that the surface energy and moisture states are simultaneously required by the equation as input.

Evaluating the energy, mass and momentum fluxes into and out of the climatic environment entails modeling atmospheric dynamics and physics at the global scale. Over the last three decades, work on the development of global-scale climate models has increased dramatically, and become the main emphasis within the climate-modeling community. Our discussion, in turn, focuses on the history and status of global climate models. Although modern versions of these global climate models are both three-dimensional and time varying, important contributions to the history of global climate modeling and climatology have also been made with simpler models.

Global energy balance models (EBMs)

Global energy-balance models (EBMs) use the basic radiation laws and conservation of energy as their physical principles.

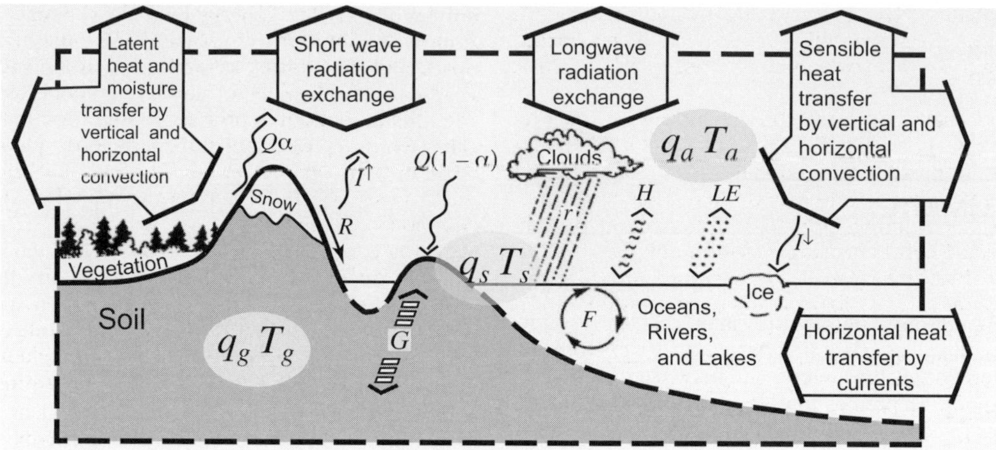

Figure M15 Schematic representation of the climatic environment represented in models. Primary fluxes occur to and from the surface (solid line), which also is referred to as the primary or atmosphere–Earth interface, and they are labeled with symbols for energy budget components (defined in text). Large symbols in oval represent temperature (*T*) and moisture (*q*) states of the atmosphere (q_a T_a), surface (q_s T_s), and near-surface ground (q_g T_g) respectively. Precipitation and total runoff are noted as *r* and *R*, respectively. Horizontal and vertical heat, moisture and momentum exchanges between the climatic environment and external environments are conceptually depicted by phrases within arrows crossing the secondary interfaces (dashed lines).

The balance implied is between income and outgo. On climatic timescales, whatever sunlight is intercepted by the Earth must be either reflected back to space or absorbed and reradiated as infrared radiation. A prototype for a zero-dimensional EBM is used in nearly every introductory weather and climate textbook in which the shortwave and longwave energy balances are presented for the Earth as a whole. In the textbook version, solar input is distributed between surface and atmospheric energy balances, with terms for latent heat, sensible heat, greenhouse gas warming, and albedo. All that is needed to make that into a climate model is to turn some of the ratios between energy flows into variables instead of constants, translate the percentage values into actual energy flux units, and constrain the model with the three energy balances it requires (surface, atmosphere, and space). Such a simple model can be used to provide heuristic calculations of the relative importance of the major climate forcing components, such as infrared radiation trapping (greenhouse effect) or changes in the vertical structure of the atmosphere.

Simple global EBMs reduce equation (1) to

$$Q[1 - \alpha(T,f)] = I(T,f) \qquad (2)$$

in which Q is the average solar flux at the top of the atmosphere, T is temperature, f is cloud-cover fraction, and α and I are functions representing the variation of planetary albedo and infrared output to space, respectively, with temperature and cloud-cover fraction. Simple linear relationships for α and I – derived from satellite data (Bintanja, 1996; Graves et al., 1993) – allow equation (2) to be inverted to solve for temperature.

In the most famous use of a zero-dimensional climate model, Arrhenius (1896) explored the role of carbon dioxide in the Earth's radiation balance. As early as the end of the nineteenth century, Arrhenius noted that carbon dioxide levels were rising from human use of fossil fuels, and he projected that a doubling of carbon dioxide levels in the atmosphere could produce a 1.5–3°C warming in the average surface temperature of the Earth. A century later, Arrhenius's calculation is still within the mainstream of climate model projections – neither alarmist nor contrarian.

Global energy-balance models have been usefully extended to higher dimensions, most notably to one-dimensional models. These zonal EBMs add a latitudinal component to the energy balance. In addition to the vertical energy balance implied by equation (2), a meridional transport term must be added, as in

$$QS(\varphi)[1 - \alpha(T(\varphi),f(\varphi),\varphi)] - I(T(\varphi),f(\varphi)) = F(T(\varphi),f(\varphi),\varphi) \qquad (3)$$

in which φ is latitude, S is a distribution function for solar energy, and F is a function representing the net meridional flux of energy, and all the other functions now vary with latitude. A direct dependence on latitude is included for the albedo and meridional flux functions, in addition to their indirect dependence on latitude via the latitude-dependence of temperature and cloud cover. The direct dependence of albedo on latitude allows for variation of vegetative cover and continental extent with latitude. The direct dependence of horizontal flux convergence on latitude reflects the fact that models of this simplicity do not adequately simulate the high meridional transport in the Hadley circulation without some numerical assistance.

Two particular versions of equation (3), published nearly simultaneously, hold a place of high importance in the history of climate models. Budyko (1969) and Sellers (1969) independently presented simple versions of (3) with slightly different formulations for F and a similar formulation for $\alpha(T)$ in which the albedo of the land surface jumped sharply if the zonal temperature went below −10°C, representing the onset of a snowline. Both found that a small decrease in solar input, amounting to just a few percent, would raise the planetary albedo sufficiently to cause runaway cooling. This runaway positive feedback raised considerable interest in this kind of modeling through the 1970s. Through that decade, considerable knowledge of the Earth's stability, sensitivity, and feedback systems was gained via exploration of zonal EBMs (North et al., 1981).

Schneider and Dickinson (1974) reviewed a wide range of modeling activities and concluded that a hierarchy of climate models ranging from zero-dimensional, steady-state models through fully three-dimensional general circulation models had a role to play in understanding climate. Within a short time after that, however, zonal EBMs had been found to be too sensitive to describe global sensitivity to solar radiation variations, too insensitive to produce a realistic ice-age cycle from orbital parameter variations, too parameterized to produce a useful greenhouse-gas-change simulation, and generally not competitive for cutting-edge research. Today, zonal EBMs remain an important heuristic tool: they are well worth using as exploratory tools in an introductory course, and building a simple model can illustrate a variety of climate principles and numerical modeling techniques for an advanced student. A few research problems make good use of simple models when coupled with ocean models (Egger, 1999) or glacier models (Sakai and Peltier, 1999) for long-term climate experiments. Beyond these specialized uses, EBMs have mostly faded from view in research.

General circulation models (GCMs)

General circulation models (GCMs) share with global EBMs the idea that physical principles drive the system, rather than atmospheric data. Beyond that commonality, GCMs are both genetically and morphologically different from EBMs. The core of the modern GCM is the atmospheric GCM (AGCM).

AGCMs grew out of weather-forecasting models. The early history of numerical weather prediction has been told elsewhere (Thompson, 1978). The first useful weather simulations were developed in the years immediately after World War II as a project led by John von Neumann at Princeton, the inventor of the stored-program computer. The size and complexity of weather and climate models has been closely tied to improvements in computing technology ever since (Hack, 1992, figure 9.1). The transition from weather model to climate model occurs when a model's statistical behavior approaches the statistical behavior of weather over a long enough period that we may consider climate statistics from the model output to be comparable to climate statistics from weather data. Phillips (1956) ran a baroclinic model long enough to produce a climate simulation in what is often called the first general circulation model.

Six partial differential equations define an atmospheric general circulation model: three for conservation of momentum (one for each dimension of physical space), two for conservation of mass (one each for air and for water), and one for conservation of energy. The six independent field variables in this system are the three components of atmospheric velocity, temperature, specific humidity, and pressure. Thus, nominally, we have a system that can be solved. In practice, the effort required to create such a model is huge.

Because of the complexity and scale of AGCM building, most state-of-the-art models bear institutional names, such as the NCAR CAM (the National Center for Atmospheric Research Community Atmospheric Model, Boulder, Colorado), the GFDL GCM (Geophysical Fluid Dynamics Laboratory, Princeton, New Jersey), or ECHAM model (European Center for Medium Range Weather Forecasting – Hamburg, Germany). Each of these has an evolving version history from over 30 years of continuous development involving hundreds of scientists and programmers. The most-developed models from large institutions coexist with a plethora of other models, but most share some common origins. What AGCMs have in common is the concept of simulating weather using a grid. The entire global atmosphere is broken up into boxes, both in the horizontal and vertical dimensions. Within each vertical column of boxes, all of the *subgrid-scale* processes must be evaluated. Between vertical columns of the model, most horizontal transfer processes proceed by advection, which requires a prediction of wind. For this prediction the equations of motion are solved approximately.

More difficult to understand, but currently more popular in use, are *spectral* methods (Bourke et al., 1977). In these methods, the fields that require horizontal derivatives are subjected to a generalized Fourier transform, so that they may be represented by a set of coefficients of the spherical harmonics. Some of the advantage of this method arises because the truncated Fourier series of spherical harmonics used to represent the horizontal fields is intrinsically smooth.

Grid scales have been decreased as computational power has increased. Up to a limiting point that we are far from approaching, decreasing the grid size so that fewer things need to be relegated to the subgrid-scale parameterizations will always be an improvement. However, halving the grid size requires an increase in computing power of more than a factor of 100, so grid sizes have not decreased at a rate that might be expected from a simple understanding of how fast computers have improved in recent decades. A common grid scale now would be roughly 3°latitude/longitude boxes, 20 layers in the vertical, and a 20-minute time step.

Climate system models (CSMs)

Climate system models (CSMs) are AGCMs within which boundary conditions at the land and ocean surfaces respond to the changing atmospheric conditions. In the earliest AGCM climate experiments a land model may have consisted of a simple parameterization of soil-moisture availability beneath each grid box and a specified vegetation albedo that could be raised if snowcover were predicted. A number of more realistic land-surface formulations have been developed since then, and they incorporate important vegetation and soil characteristics, as well as human-modified aspects of the land surface (Dickinson, 1984; Sellers et al., 1986). These land-surface models have been widely applied within CSMs, and they have improved significantly CSMs estimates of land-surface albedo, evapotranspiration and soil moisture. Dickinson (1984), in particular, raised the level of land modeling from a parameterization to a significant submodel with the Biosphere–Atmosphere Transfer Scheme (BATS). Continued development of BATS and similar models has led to land models which can be considered independent models in themselves, to be run either in conjunction with an AGCM or in standalone mode using atmospheric data (Bonan et al., 2002).

Just as land-surface boundary conditions have become more interactive and sophisticated, oceanic boundary conditions have matured. From "swamp" models in which the oceans were flat surfaces with unlimited thermodynamic capacity and moisture availability, we now recognize that for realistic long-term climate simulations, an oceanic general circulation model (OGCM) must be coupled to an AGCM. Such coupled experiments still have not been entirely successful in reconciling the energy and moisture fluxes between the atmospheric and oceanic components into a stable equilibrium. Usually, a flux correction field must be applied, a nonconservative difference between the energy leaving the top of the ocean and that entering the bottom of the atmosphere. Models have come close enough to equilibrium to avoid using the flux correction, but these still show long-term drift (Bryan, 1998).

Another boundary condition model required for coupling into a CSM is one for sea ice. Sea ice has an importance in climate systems well out of proportion to its mass, both because of its sharp albedo effect and because its role as an insulator with respect to both thermal and moisture fluxes between the oceans and the atmosphere. Although the thermal prediction of freezing or melting is fairly simple, and the simulation of the effects of sea ice on the atmosphere–ocean fluxes is also straightforward (Curry et al., 1995), sea ice moves in a complicated response to wind action, modulated by ocean currents. Errors in sea-ice prediction may be the largest identified, systematic problem with current CSM simulations.

Models of everything

Climate modelers have long dreamed of complete models of the climate system – models of "everything" which couple all of the related components of the climate system. The coupling of human activities to climate (Terjung, 1976), however – owing to its complexity, will continue to remain indirect for the foreseeable future. Nonetheless, all of the components of new CSM suites are available for simulating decadal to century-scale climatic variations. Specific improvements in nearly every aspect of atmospheric, oceanic, and land-surface models are being discussed widely and implemented. Even with dramatic improvements in these components, however, climate simulations will remain restricted to medium-term variations, perhaps up to the scale of a millennium, because the simulated climate system is still not complete.

The largest climatic changes of the last several million years have involved the advance and retreat of large ice sheets from North America and northern Eurasia, on time scales for which 1000 years is a high-speed fluctuation and 100 000 years appears to characterize a complete cycle. The pioneering ice-sheet model of Weertman (1976) demonstrated that such cycles could be controlled primarily by the lagged response within the ice sheet. The eventual model of everything will need to include an ice GCM for resolving problems on that scale. Ice-sheet models in current use are not quite ready for full coupling. The best continent-scale ice models are still map-plane models that have sub-grid scale weaknesses with important problems such as ice streams (Fastook and Prentice, 1994), whereas the fully three-dimensional ice sheet models that can simulate ice stream convergence cannot readily be scaled up to continental scales (Hanson, 1995; Blatter, 1995). Nevertheless, some interesting uncoupled simulations using ice-model-generated boundary conditions in an AGCM have been attempted (Bromwich et al., 2001). Full coupling of an ice-sheet model into a CSM is probably a few model generations away, but the path to that eventual coupling is at least being followed.

Another component that will eventually need to be dealt with is the vegetational response. When dealing with the last 20 000 years of climate, going from full glacial conditions to the present, we see significant vegetational responses. For example, the current Amazon rainforest was probably mostly savanna during the last ice age, with a few refugia of forests, from which the current rainforest was created by slow species migration as the climate changed. One of the first things specified into any paleoclimate simulation, immediately after the pattern of continents and oceans, is the vegetation types that will cover each grid box. No model currently presupposes the ability to modify those vegetation types in response to climate change, nor should models bother trying so long as a CSM is restricted to a relatively short simulation. We expect that such a vegetation model will be needed before the model of everything is truly complete, and we expect that building it will be a very difficult problem.

Outlook for the future

There is every reason to believe that the sophistication of our climate models will continue to increase over the next decades. Ongoing improvements in our computational resources will allow climatic models to incorporate increasingly realistic levels of detail at the same time that a burgeoning corps of climatologists is uncovering those essences of climate which will provide the bases for the models of the future. Vast and continuing increases in our stores of climatic and climate-related data – owing significantly to advances in remote sensing technology – will contribute many of the terrestrial and oceanic observations necessary to determine boundary and initial conditions for future models. With these advances in climatic theory and computational resources, as well as in the type and quantity of available data, climatologists will develop, refine and apply a hierarchy of CSMs. These CSMs will couple the oceans, vegetation, ice and sea ice to the atmosphere, and they will be used to describe, predict and explain climatic processes and phenomena over a wide range of geographic scales.

Brian Hanson and Cort J. Willmott

Bibliography

Arrhenius, S., 1896. On the influence of carbonic acid in the air upon the temperature of the ground. *Philosophical Magazine, Series 5*, **41**(251): 237–276.

Bintanja, R., 1996. The parameterization of shortwave and longwave radiative fluxes for use in zonally averaged climate models. *Journal of Climate*, **9**: 439–454.

Blatter, H., 1995. Velocity and stress fields in grounded glaciers: a simple algorithm for including deviatoric stress gradients. *Journal of Glaciology*, **41**(138): 333–344.

Bonan, G.B., Oleson, K.W., Vertenstein, M., et al., 2002. The land surface climatology of the community land model coupled to the NCAR Community Climate Model. *Journal of Climate*, **15**(22): 3123–3149.

Bourke, W., McAvaney, B., Puri, K., and Thurling, R., 1977. Global modeling of atmospheric flow by spectral methods. In Chang, J., ed., *Methods in Computational Physics 17: General circulation models of the atmosphere*. New York: Academic Press, pp. 268–324.

Bromwich, D.H., Cassano, J.J., Klein, T., et al., 2001. Mesoscale modeling of katabatic winds over Greenland with the Polar MM5. *Monthly Weather Review*, **129**: 2290–2309.

Bryan, F., 1998. Climate drift in a multicentury integration of the NCAR Climate System Model. *Journal of Climate*, **4**(6): 1455–1471.

Budyko, M.I., 1969. The effect of solar radiation variations on the climate of the earth. *Tellus*, **21**: 611–619.

Crowley, T.J., and Baum, S.K., 1997. Effect of vegetation on an ice-age climate model simulation. *Journal of Geophysical Research*, **102**(D14): 16463–16480.
Curry, J.A., Schramm, J.L., and Ebert, E.E., 1995. Sea ice-albedo climate feedback mechanism. *Journal of Climate*, **8**: 240–247.
Dickinson, R.E., 1984. Modeling evapotranspiration for three-dimensional global climate models. In Hansen, J.E., and Takahashi, T., eds., *Climate Processes and Climate Sensitivity*. Washington, DC, American Geophysical Union Geophysical Monograph 29, pp. 58–72.
Egger, J., 1999. Internal fluctuations in an ocean–atmosphere box model with sea-ice. *Climate Dynamics*, **15**: 595–604.
Fastook, J.L., and Prentice, M., 1994. A finite-element model of Antarctica: sensitivity test for meteorological mass balance relationship. *Journal of Glaciology*, **40**(134): 167–175.
Graves, C.E., Lee, W.-H., and North, G.R., 1993. New parameterizations and sensitivities for simple climate models. *Journal of Geophysical Research*, **93**(D3): 5025–5036.
Hack, J.J., 1992. Climate system simulation. basic numerical and computational concepts. In Trenberth, K.E., ed., *Climate System Modeling*. Cambridge: Cambridge University Press, pp. 283–318.
Hanson, B., 1995. A fully three-dimensional finite-element model applied to velocities on Storglaciären, Sweden. *Journal of Glaciology*, **41**(137): 91–102.
North, G.R., Cahalan, R.F., and Coakley, J.A., Jr, 1981. Energy balance climate models. *Reviews of Geophysics and Space Physics*, **19**(1): 91–121.
Phillips, N., 1956. The general circulation of the atmosphere: a numerical experiment. *Quarterly Journal of the Royal Meteorological Society*, **82**: 123–164.
Sakai, K., and Peltier, W.R., 1999. A dynamical systems model of the Dansgaard-Oeschger oscillation and the origin of the Bond cycle. *Journal of Climate*, **12**(8): 2238–2255.
Schiller, A., Mikolajewicz, U., and Voss, R., 1997. The stability of the North Atlantic thermohaline circulation in a coupled ocean-atmosphere general circulation model. *Climate Dynamics*, **13**: 324–347.
Schneider, S.H., and Dickinson, R.E., 1974. Climate modeling. *Reviews of Geophysics and Space Physics*, **12**: 447–493.
Sellers, P.J., Mintz, Y., Sud, Y.C., and Dalcher, A., 1986. A simple biosphere model (SIB) for use within general-circulation models. *Journal of the Atmospheric Sciences*, **43**(6): 505–531.
Sellers, W.D., 1969. A global climate model based on the energy balance of the earth-atmosphere system. *Journal of Applied Meteorology*, **8**: 392–400.
Shukla. J., and Mintz, Y., 1982. Influence of land-surface evapotranspiration on the Earth's climate, *Science*, **215**: 1498–1501.
Steyn, D.G., Oke, T.R., Hay, J.E., and Knox, J.L., 1981. On scales in meteorology and climatology. *Climatological Bulletin*, **30**: 1–8.
Terjung, W.H., 1976. Climatology for geographers. *Annals of the Association of American Geographers*, **66**: 199–222.
Thompson, P.D., 1978. The mathematics of meteorology. In Steen, L.A., ed., *Mathematics Today: twelve informal essays*. New York: Springer-Verlag, pp. 127–152.
Weertman, J., 1976. Milankovitch solar radiation variations and ice age ice sheet sizes. *Nature*, **261**: 17–20.

Cross-references

Boundary Layer Climatology
Energy Budget Climatology
Water Budget Analysis
Statistical Climatology

MONSOONS AND MONSOON CLIMATE

Monsoon regions

The monsoon broadly refers to an atmospheric phenomenon in which the mean surface wind reverses its direction from summer to winter. However, the monsoon is popularly used to denote the rains without reference to the winds. The term "monsoon" has its origin from an Arabic word meaning season. In English, the original Arabic word is spelled in several ways, such as, "mausam", "mausem", "mausim", "mawsim" and "mausin". The term was used by seamen, several centuries ago, to describe southwesterly wind during summer, and northeasterly wind during winter over the Arabian sea. Along with progress in meteorological sciences, more regional circulations have been categorized as "monsoonal"; they are based on both wind and rainfall characteristics. Monsoonal regions over the globe are generally identified by certain characteristics of surface circulations in January and July as laid down by C.S. Ramage. These include a shift in wind direction by at least 120°, average frequency of prevailing wind directions exceeding 40%, mean wind strength in at least one of the months greater than 3 m/s and fewer than one cyclone–anticyclone alternation in either month in every 2 years. Although these monsoon criteria do not include the rainfall explicitly, the seasonality in rainfall is the most important manifestation of the monsoon circulation. Most of the countries lying between 35°N and 25°S and between 30°W and 170°E (Figures M16a,b) satisfy the criteria of monsoons and, hence, are widely accepted as a part of the single most important monsoon domain on Earth. The monsoon is primarily an Asian phenomenon. However, numerous studies have established that monsoon circulation occurs in some other parts of the world, such as western Africa, northern Australia and North America at varying intensities. In the monsoon regions, wind blows inland from the cooler oceans toward warm continents in summer and from cold continents toward the warm oceans in winter (Figures M16a,b). Thus, broadly speaking the summer monsoons of both the hemispheres are very wet (Figures M17a,b) and winter monsoons are dry. Comparison of seasonal wind directions in Figures M16a and M17b demonstrates that the monsoon conditions are best developed in east and south Asia, with winds from southwest in summer and from the northeast in winter. Accordingly, those are known as the southwest and northeast monsoons respectively. The Asian summer monsoon consists of the Indian monsoon, and the east Asian monsoon, both of which are responsible for abundant summer rainfall in the region shown in Figure M17a. The Indian monsoon is effectively separated from the east Asian monsoon by the massive Himalayan mountain range. About 80% of the annual rainfall over India occurs during the southwest monsoon from June to September. Consequently, the summer monsoon is very important for the economy of India, which is predominantly an agricultural country. Usually, in eastern Asia, the winter monsoon wind is stronger than the summer monsoon wind while the opposite happens in south Asia.

West Africa also experiences a wind reversal to some extent, from southwesterly in summer (Figure M16a) to northeasterly in winter (Figure M16b). In the west coast of Africa there is heavy rainfall from June to August, although the rains actually commence in March–April. In some parts of eastern Africa near the equator there are two rainy seasons, one during March to May, popularly know as the "long rains", and the other from October to November, which is termed as "short rains". These rainy periods fall between the two African monsoon circulations.

Over north Australia there is a northwesterly flow of humid maritime air in the southern hemisphere summer (Figure M16b) from December to February, and the southeast trades in the southern hemisphere winter (Figure M16a). The east Asian winter monsoon and the north Australian summer monsoon are

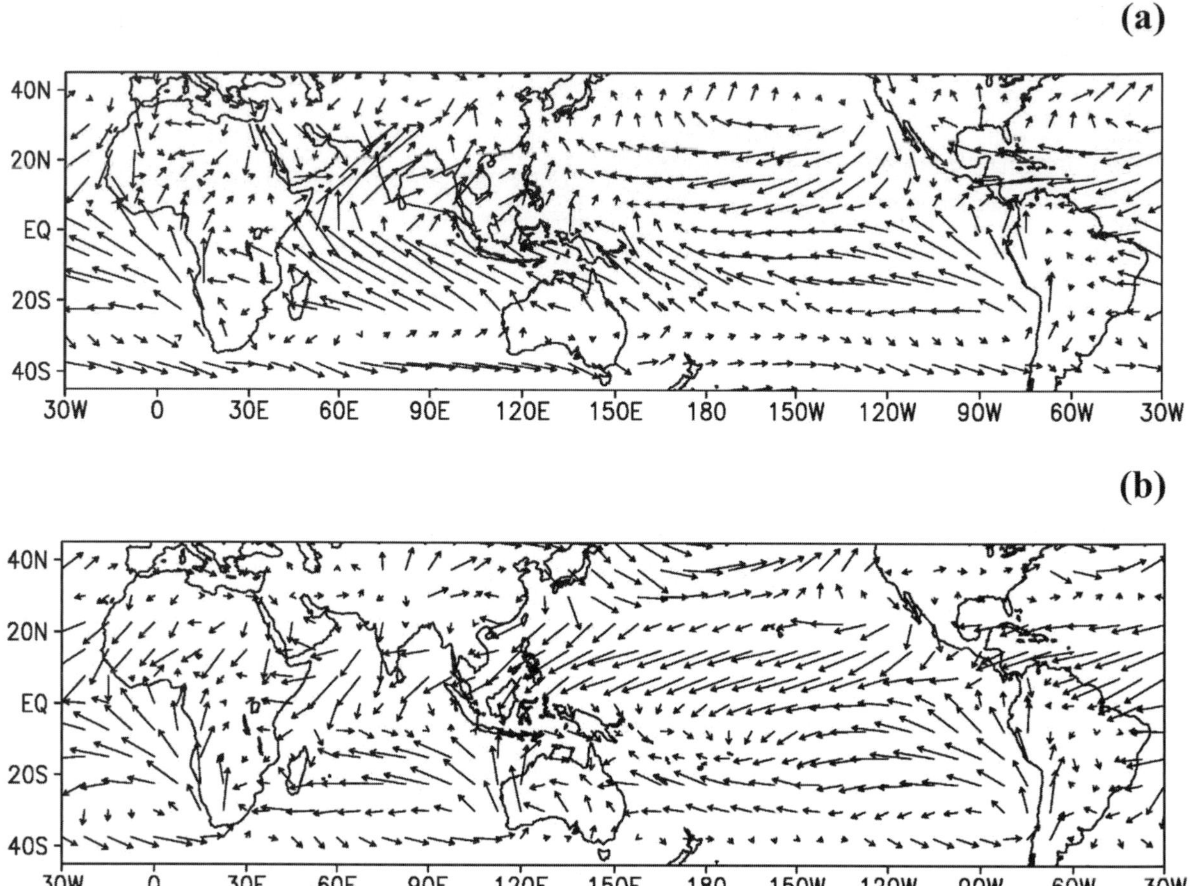

Figure M16 Surface wind (m/s) in the tropical belt (**a**) June, July and August average; (**b**) December, January and February average based on NCEP/NCAR reanalysis.

intermingled because the dry winter air of the northern hemisphere flows across the equator toward the southern hemisphere continents, picking up moisture from the warm tropical oceans to become the wet monsoon over north Australia.

The southwestern part of North America is also considered to be monsoonal. Here the surface zonal wind is observed to change from an easterly in January to westerly in July. There is a pronounced increase in rainfall over large areas of southwestern North America and southern Mexico from June to July, although the summer rains may last till September. This region also experiences significant winter rainfall.

Based on recent data, studies justify the existence of summer monsoon circulation over the subtropical south American highland. In southern hemisphere summer, distinct rainfall enhancement occurs over the central Andes and the southern parts and north coast of Brazil.

Important features of planetary scale monsoon

In 1686 Edmund Halley explained the Asiatic monsoon as resulting from thermal contrasts between the continent and oceans. In simple terms one may compare the monsoon circulation to a land–sea breeze occurring on larger spatial and temporal scales. In summer the continents surrounding the Arabian sea begin to receive large amount of heat, due both to solar insolation and the heat emitted from the Earth's surface; consequently, by the end of May, a trough of low pressure develops over a large area from Somalia to Pakistan and northwest India. This is usually known as the heat low. It is accompanied by subsidence and hence fine weather in general. The southerly wind, after crossing the equator, turns to the east under the influence of the Coriolis force and blows over the Indian subcontinent as a southwesterly monsoon wind (Figure M16a) in summer. This current appears to originate in the southeast trades. George Hadley in 1735 incorporated the concept of the Coriolis force, which arises due to the rotation of the Earth around its own axis from west to east.

In the northern hemisphere summer, prior to the arrival of monsoon, a low-pressure zone forms around 5°N and 5°S on either side of the equator. This is referred to as a near-equatorial double trough. During the Asian summer monsoon the trough is located north of 15°N and is associated with cyclonic vortices. Surface air converging into these troughs ascends, accompanied by moist convection, and gives rise to unsettled weather, mainly clouds and rain. Climatologically, the heat low and the near-equatorial trough in the northern hemisphere form a continuous low-pressure belt, although the two systems give rise to opposite weather. Hence, the location of the low-pressure

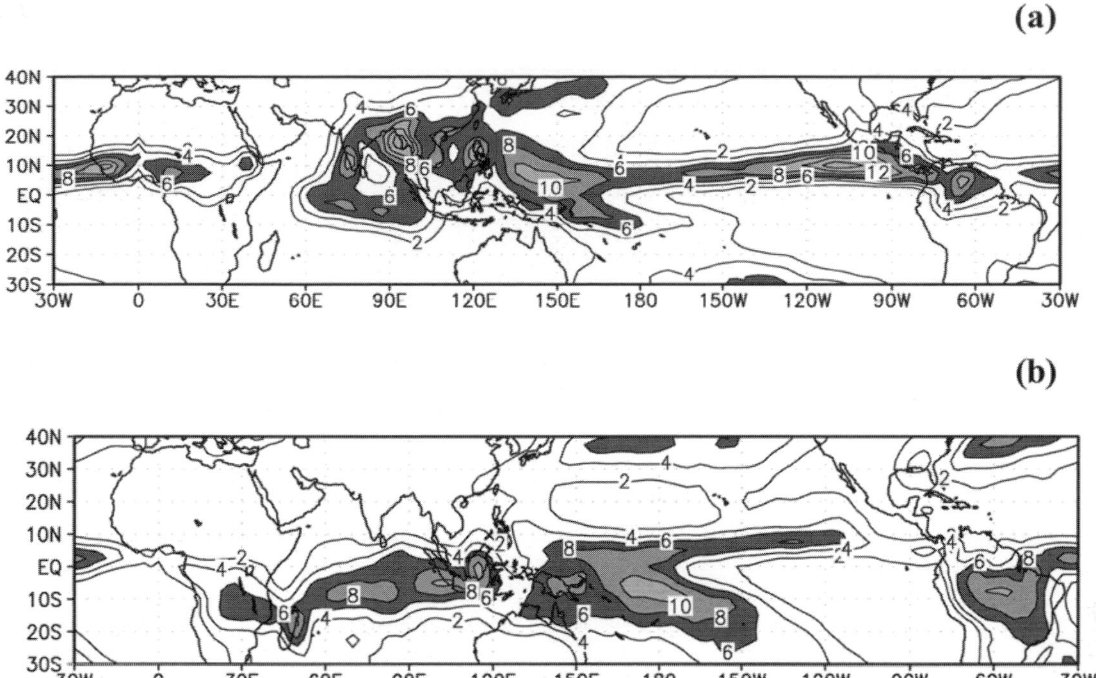

Figure M17 Accumulated rainfall (mm/day) in (**a**) northern (southern) hemisphere summer (winter) monsoon and (**b**) southern (northern) hemisphere summer (winter) monsoon based on Global Precipitation Climatology Project (GPCP) data updated by NCEP/NCAR.

belt over the Asian continent does not always coincide with maximum cloud and rains. Nevertheless, the huge land-mass of the Asian continent with wide east–west extension in the north and, primarily, an ocean to the south makes it geographically the most favorable for dominant summer monsoon flow compared to the other monsoons of the world.

The difference in the global distribution of heating gives rise to planetary-scale quasistationary waves in the tropics and subtropics, which form the most important component of both the summer and winter monsoons. The convective latent heat released forms the most important part of the differential heating, and globally the course of maximum rainfall follows that of convective heating. As shown in Figures M17a and M17b, the maximum precipitation bands occur at around 10°N and 10°S in the northern and southern hemisphere summers respectively. The monsoonal regions receive most of the rainfall due to the seasonal migration of the Intertropical Convergence Zone (ITCZ). This is a relatively narrow low-latitude zone in which the northeast and southeast trade winds originating in the northern and southern hemispheres respectively converge. The mean position of the ITCZ is somewhat north of the equator and it moves northward in the northern summer and southward in the southern summer. The ITCZ may not be the zone of organized moist convection always, especially over the Asian region. In this context it is convenient to use the term maximum cloud zone (MCZ) which is a bright cloud band of about 10° longitude width spreading along the latitude circles. There are two favorable locations of the MCZ; one within 15°N and 25°N latitudes over the land and another over the equatorial region (Figure M17a) at about 5°N have been identified. Successive generations of northward epochs of MCZ in the equatorial region pushes it to about 20°N in July–August from its mean position near 5°N. The northward progress of MCZ in surges is akin to that of monsoon rains over India.

Convection is best represented by low values of outgoing longwave radiation (OLR). A comparison of Figures M18a and M18b indicates the seasonal variation of low values of OLR, which is large over the monsoon regions. Specifically in the case of the Asian summer monsoon between the meridians 70°E and 110°E, the belt of low OLR extends over a large area from 5°S to 35°N, while for the rest of the monsoonal region it is confined to a narrow latitude belt (Figure M18a).

Each monsoon appears to have a dominant circulation either in a north–south or in the east–west direction, with a rising branch located near a heat source, and is associated with upper-level divergent circulation. Similarly, the descending branch lies over a heat sink. The north–south and east–west dominant circulations are popularly called the Hadley and Walker circulations respectively. Generally, the velocity potential field is used to depict divergence and convergence centers in the atmosphere. As shown in Figure M19a, in July the upper-level divergence circulation is prominent over the Asian monsoon region accompanied by convergence centers on both sides in the east–west direction as a part of the Walker circulation. This center of divergence is responsible for spillover of mass from the Asian summer monsoon region in all directions. Thus the Asian summer monsoon is supposed to have a profound impact on the global atmospheric circulation. In the northern hemisphere winter the center of divergence shifts to the south over the north Australian region, as shown in Figure M19b. In the northern hemisphere the Walker and Hadley type overturnings are observed to be prominent in summer (Figure M19a) and winter (Figure M19b) respectively.

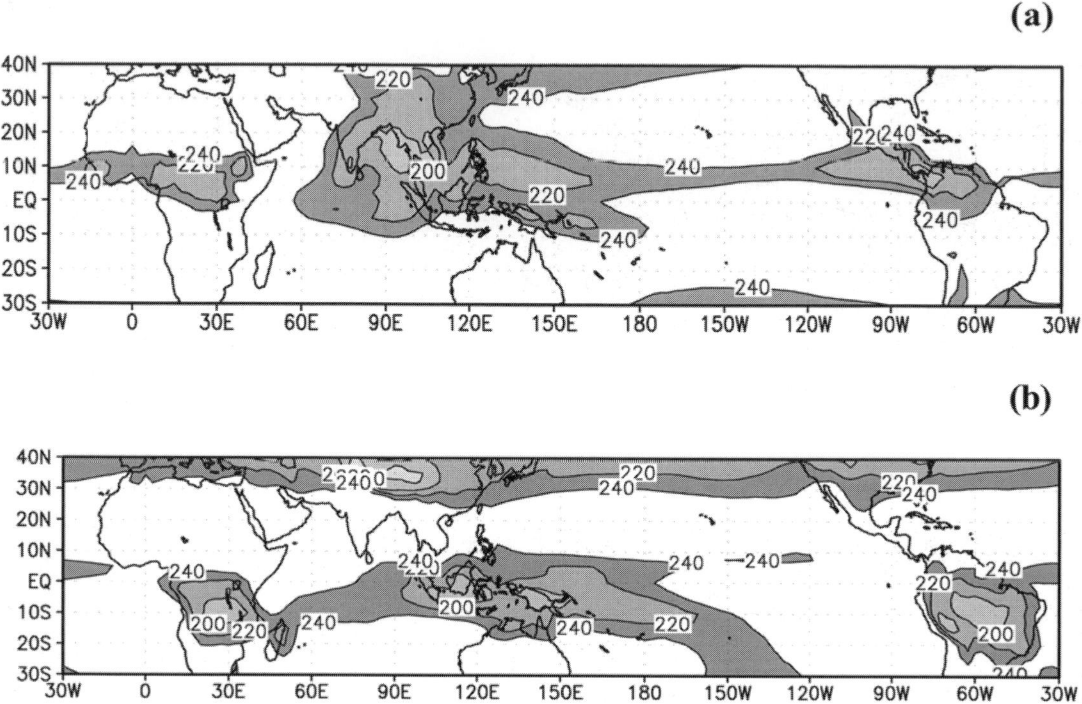

Figure M18 Same as in Figure M16 except for outgoing longwave radiation (W/m^2).

Important features of regional monsoons

The principal components of the Indian summer monsoon are the monsoon trough, the cross-equatorial low-level jet over coastal Africa, the Mascarene High, the Tibetan anticyclone and the tropical easterly jet stream at the upper level, monsoon disturbances and the cloud cover which gives rise to rainfall. The Western Ghats and the Himalayan mountains help in the ascent of moist air, and this gives rise to heavy orographic rains. Around July the monsoon activity is most pronounced, and the low-pressure area is intense, extending from north Africa to northeast Siberia. At this time a trough lies over north India with its axis extending from a place in northwest India to the northern tip of the Bay of Bengal running almost parallel to the Himalayas. This is referred to as the *monsoon trough*. The pressure gradient is strong south of this trough. Simultaneously, in the southern hemisphere, off the coast of Madagascar, a region of high pressure persists giving rise to anticyclonic circulation. This is known as the *Mascarene High*. The *cross-equatorial low-level jet* is most pronounced at a height of about 1.0–1.5 km and its major part penetrates into east Africa during May. Subsequently, it crosses the Arabian Sea and reaches the Indian west coast. Due to significant boundary layer exchange processes between the ocean and the atmosphere, there is an abundant supply of moisture to the Indian summer monsoon. The Tibetan plateau acts as a source of heat in summer and as a sink in winter. The heating of this elevated landmass in summer leads to the development of an intense anticyclone in the upper troposphere, which is popularly known as the *Tibetan Anticyclone* with strong east-northeasterly flow over north India. Here deep convection exists along with strong vertical wind shear. Monsoonal circulation is also accompanied by a variety of cyclonic disturbances; especially over the Indian and adjoining oceanic region one observes the *onset vortex, mid-tropospheric cyclones, offshore vortices, lows, monsoon depressions and other cyclonic disturbances of higher intensity*. Observations show that most of the remnants of tropical cyclones from the South China Sea under favorable atmospheric conditions intensify to monsoon depressions. Studies also show that the horizontal and vertical wind shears of the mean monsoon flow in the presence of deep convection supply energy for the growth of monsoon disturbances.

The east Asian monsoon consists of a monsoon trough in the South China Sea and western Pacific, a cross-equatorial flow around 100°E, the cold anticyclone in Australia, the subtropical high in the western Pacific, the upper-level northeasterly flow, the convection along the monsoon trough and the midlatitude disturbances. In the middle of June the ITCZ is normally prominent over India, corresponding to an active phase of the Indian monsoon, and at the same time the subtropical convergence, which is called Mei-yu in China or the Baiu frontal zone in Japan, is most active, especially around Japan.

For the west African monsoon, the major components are the low-level southwesterly wind over the coastal areas of the Gulf of Guinea flanked by northeasterly winds to its north. The African easterly jet and tropical easterly jet are the dominant middle- and upper-level atmospheric flows in this region. With the change of season from summer to winter, there is no complete reversal of wind direction over the west coast of Africa (Figures M17a and M17b) as in the case of the Indian monsoon. The boundary between the southwesterly and northeasterly winds is known as the intertropical discontinuity in west Africa. The major rainfall in the east African monsoon is due to the north–south movement of the ITCZ.

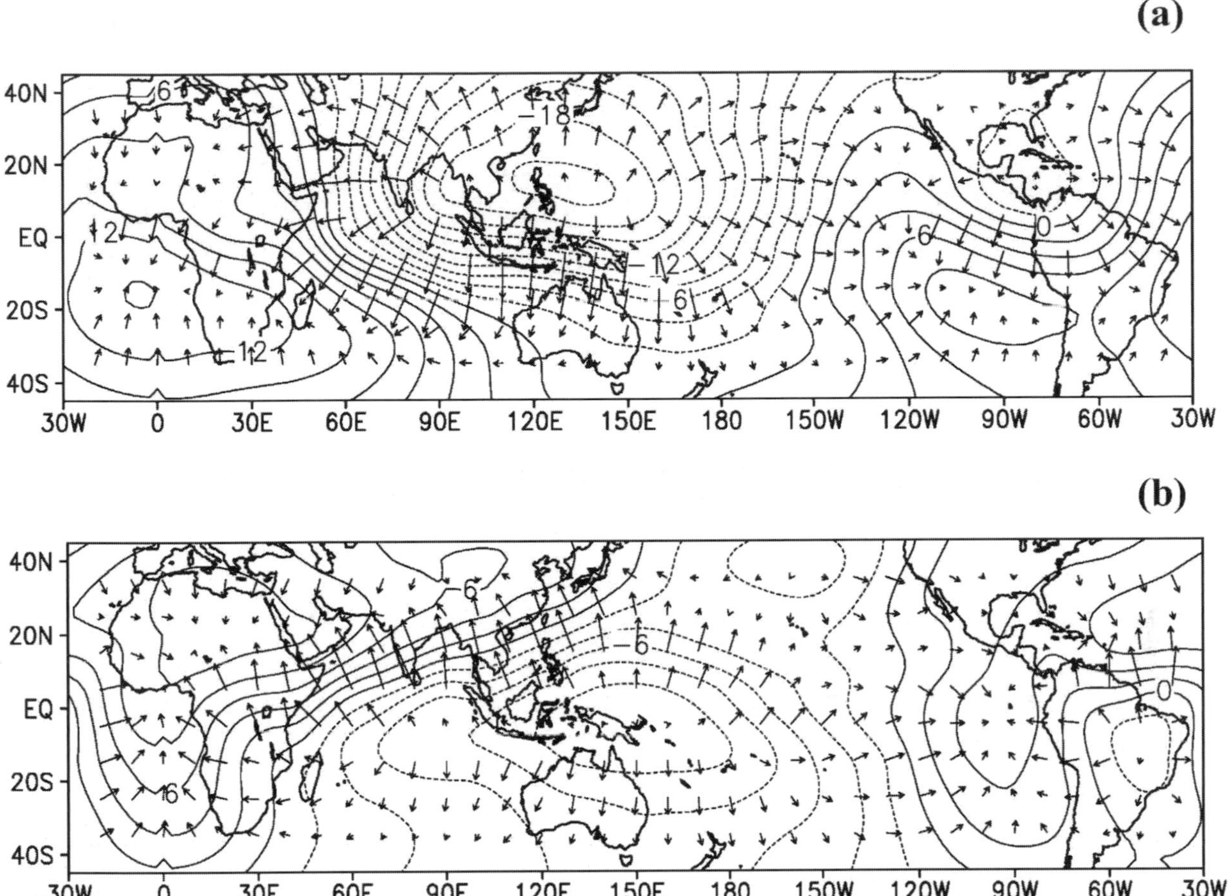

Figure M19 Divergent wind and contours of monthly mean velocity potential ($10^6\,m^2\,s^{-1}$) at 200 hPa (**a**) July and (**b**) January based be NCEP/NCAR reanalysis.

There are many similarities between the Asian summer monsoon and the northern Australian monsoon, although in general the latter is less intense. The ITCZ across northern Australia, near-equatorial upper-level easterlies and lower-level westerlies, cross-equatorial flow, tropical cyclones and monsoon depressions are the important components of the Australian monsoon. The southern hemisphere near-equatorial low lies to the south of the ITCZ over northern Australia. This low-pressure area is referred to as the monsoon trough and the corresponding westerlies are the north Australia monsoon winds.

The North American monsoon domain is considered to be large, as it spreads over much of the southwest United States, northwest Mexico and even Central America. The eastern north Pacific ITCZ between 120°W and 80°W migrates northwards (Figures M17a and M17b) in summer and the corresponding zonal wind becomes westerly at the lower level. At the upper level the zonal wind is easterly. The opposite happens in the winter. Over the North American Arctic region there is a seasonal reversal of wind direction and technically it may be considered to be monsoonal.

The occurrence of South American summer monsoon circulation can be shown by considering the difference between the prevailing surface wind and its annual mean. In the southern hemisphere summer this anomaly flow from the sub-Saharan region, after crossing the equator, becomes northwesterly along the eastern side of the Andes. In winter the anomaly flow reverses its direction. The seasonality of surface wind over this region is not normally noticed because of the fact that the easterly trade winds prevail over the tropical Atlantic throughout the year (Figures M16a and M16b).

Intraseasonal and interannual variabilities

The monsoon passes through different phases, such as its onset, active periods, breaks and withdrawal, which do not have any regularity. There are annual variations in the dates of onset and withdrawal of the monsoon at a place, and also in the duration and intensity of the active and break conditions of monsoon. The normal duration of the Indian summer monsoon is about 4 months starting from 1 June. It begins to withdraw from the northwest part of India by the middle of September. The extreme dates of onset of India summer monsoon is found to be as early as 11 May and as late as 15 June. Based on the temporal rainfall distribution at a particular place, there is a statistical method for delineating the normal dates for onset and withdrawal of monsoon at that place. The date of onset of monsoon over the southern tip of India every year is based on a

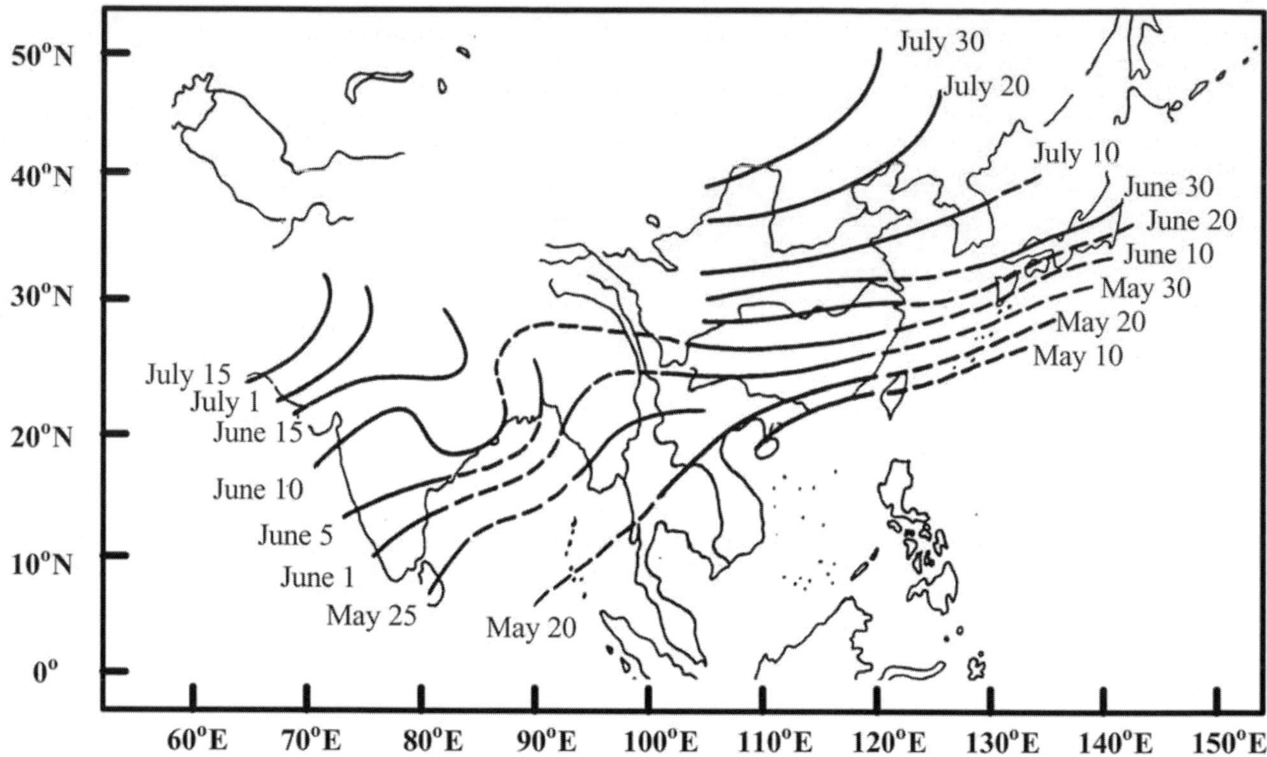

Figure M20 Mean dates of onset of Asian summer monsoon (based on Figures 3.1 and 3.9 in Rao (1976) and Chang and Krishnamurti (1987) respectively).

working rule by taking into account the continuously accumulated rainfall for certain days over certain meteorological stations in the west coast of southern India. Figure M20 shows the normal dates of onset of the summer monsoon at different places along with the northward progress of the rain belt over the Asian landmass. By mid-July the whole of India is observed to be covered by the summer monsoon. The withdrawal of the monsoon from India is slower than the onset; in fact it is difficult to know when southwest monsoon ends and the northeast monsoon begins over the extreme southern part of India.

The entire duration of the monsoon does not exhibit any uniformity or regularity in weather systems. During a season there are periods of active rainfall and also there are continuous days of no rainfall, called a "break". Normally, monsoon breaks are accompanied by cessation of rainfall in central and the adjoining parts of India, and an increase in rainfall over northeast India and foothills of the Himalayas. Observations show that, during breaks, the monsoon trough usually shifts northward to the foothills of the Himalayas and surface pressures are above normal over central parts of India. Prolonged monsoon breaks may lead to severe droughts and hence have disastrous effects. Thus, during a monsoon season, the characteristics and duration of weather systems and the amount and distribution of rainfall vary. Such intraseasonal variations in the Indian summer monsoon circulation and rainfall are the most important aspects of the monsoon.

The monsoon of a year is not identical with the monsoon of another year in all aspects. There are year-to-year variations in the strength of circulation and also in the associated rainfall at a place. It is observed that the mean rainfall over India during June to September in a year is about 88 cm with a coefficient of variation of 10%. Thus the amount of Indian summer monsoon rainfall (ISMR) equal to or more than 110% and less than 90% of the mean value are taken as excess and deficient rain respectively, and that equal to or more than 90% and less than 110% is normal. This annual variation in ISMR averaged over the whole of India may seem to be small, but its impact is high because of large spatial variations in the annual rainfall. The coefficient of variation of annual rainfall between different parts of India varies between 12% and 45%. There are occasions when some parts of the country experience deficient rain and drought, although the mean monsoon rainfall over the country as a whole is normal or in excess. The opposite also happens. Such temporal and spatial variations in the rainfall make the agriculture and the economy of India most vulnerable to the summer monsoon.

Variabilities of the Indian summer monsoon in the interannual and intraseasonal scales are scientifically intriguing. The exact reasons for such variations are not yet known. It is well known that changes in weather occur due to atmospheric instabilities. It is also known that the dynamic processes occurring in the atmosphere are nonlinear in nature, involving interactions between different spatial scales starting from few kilometers to hundreds of kilometers. Thus, "internal dynamics" play a very important role in the interannual variations of the seasonal mean circulation and rainfall. Secondly, the slowly varying "surface boundary conditions", such as the sea surface temperature (SST), soil moisture, sea ice, snow extent and depth, and land surface conditions play very important roles in the interannual variations of the ISMR. The exact physical processes responsible for the interannual variation of ISMR are not yet known. General circulation models (GCM) are increasingly used as convenient scientific tools to design and conduct sensitivity experiments on state-of-the-art computer systems for understanding the nonlinear processes leading to such variations. Large-scale field experiments using sophisticated sensors fitted to ships, aircraft, balloons and weather satellites yield good-quality meteorological data which help in understanding the evolution of weather systems. They are also needed for preparing reasonably accurate initial data for numerical weather prediction (NWP) models.

The Asian summer monsoon circulation is very different from other monsoons in its intensity, accompanying rainfall and its intraseasonal and interannual variabilities. A detailed analysis of satellite imagery indicates that the Asian summer monsoon is peculiar in having the double MCZ (one over the continents and the other over the seas) and also in the large meridional extent of low OLR spanning over 30° to 40° latitudes. It is believed that understanding the relationship between the continental MCZ and the oceanic MCZ may help a great deal in understanding the interannual and intraseasonal variations of ISMR.

Forecasting the Indian summer monsoon rainfall

Although the Indian summer monsoon is the most dominant feature of the atmospheric circulation, and although its annual appearance is certain, the most challenging task is forecasting its onset dates at different places, for active and break periods, withdrawal dates and above all the amount and spatial distribution of rainfall. Forecasts of rainfall at all time scales, such as short-range (up to 2 days), medium-range (between 3 and 10 days) and an extended range (a month to a season) are important. These involve the detailed understanding of the interaction between different physical processes, vast network for data collection, improved data assimilation procedures, sophisticated coupled ocean atmosphere models and the availability of immense computing power. The accuracy of the initial atmospheric variables is very important for the short-range forecasts, while both the initial atmospheric conditions and changes in the surface boundary conditions play very crucial roles in the forecasting of monsoons in the medium-range. In the long-range forecasts the variations in the slowly varying boundary conditions matter the most. Efforts are going on to forecast the Indian summer monsoon and its associated rainfall with reasonable accuracy by numerical weather prediction models at least a week in advance. As the mesoscale phenomena are very crucial, high-resolution regional models are nested to global models. They are best suited for forecasting monsoon rainfall up to few days with greater skill.

Forecasting the seasonal mean monsoon rainfall a couple of months ahead is very useful for farmers and planners as well. Based on the information of favorable and unfavorable predictors, a parametric model is generally used for a qualitative forecast of seasonal mean rainfall. Other statistical models, such as multiple-power regression models, dynamic stochastic transfer models, principal component regression models and models based on neural network technique are being tried for quantitative forecasting of ISMR for the whole of India, as well as for some of its homogeneous regions, such as northeast, northwest and peninsular India. These models depend on the long time relationship of some well-identified predictors such as the El Niño, Southern Oscillation Index, Eurasian and Himalayan snow covers, South Indian Ocean and Arabian Sea SST, Indian Ocean Equatorial Pressure, Central India temperature, northern hemisphere temperature and pressure, etc. with ISMR. The statistical models play a very important role in the long-range forecast of ISMR because the interannual variation of ISMR has not been successfully simulated by most of the NWP models. However, statistical models have their limitations, especially when the predictors are large in number as in case of ISMR. Changes in the global and regional circulation patterns may generate temporal variations of several predictors and hence affect their relationship with ISMR.

Because of the chaotic nature of the atmosphere there is a limitation of about 2 weeks in the deterministic prediction of weather. However, the seasonal mean monsoon circulation is potentially more predictable because of the slowly varying surface boundary conditions. The internal dynamics also affects the monsoon circulation. This makes the prediction of Indian summer monsoon very difficult. Intercomparison of the results of a number of GCM indicates that the simulation of ISMR is more difficult than the rainfall over the rest of the tropics. This is attributed to the occurrence of the double MCZ over the Indian summer monsoon region, one above the continent and the other over the Indian ocean. The successive fluctuations of the MCZ between these two favorable positions is found to be a major limitation for most of the GCM.

Oceans cover about three-fourths of the Earth's surface, and water has a much greater capacity for absorbing and storing heat energy than any other material on Earth. Thus the oceans have large memory. Hence, the interactions between the ocean and atmosphere have a crucial role in determining the state of the atmosphere. Results of sensitivity experiments show that the Pacific SST influences monsoon circulations more than other surface boundary conditions. Therefore, coupled ocean–atmosphere models are very essential for extended-range monsoon forecast by numerical methods. Such models also need proper parameterization of the land surface processes and cloud-radiation feedbacks. Ensemble simulations obtained from a combination of different GCM may help to yield reasonable forecasts of Indian summer monsoon.

The influence of global warming on the monsoons in general, and the Indian summer monsoon in particular, is not very clear, since the former occurs in the time-scale of a century. In general, global warming is expected to intensify the hydrological cycle and hence the monsoon rainfall. However, considering the limitations of GCM in simulating ISMR and the large time-scale of model integration, the different projections of Indian summer monsoon arising due to global climate change should be interpreted intelligently.

S.K. Dash

Bibliography

Asnani, G.C., 1993. *Tropical Meteorology*. G.C. Asnani.

Das, P.K., 2002. *The Monsoons*. New Delhi: National Book Trust of India.

Das, P.K., 1986. *Monsoons*. World Meteorological Organization, WMO-No.613.

Chang, C.-P., and Krishnamurti, T.N. (ed.), 1987. *Monsoon Meteorology*. New York: Oxford University Press.

Fein, J.S., and Stephens, P.L. (ed.), 1987. *Monsoons*. New York: John Wiley & Sons.

Gadgil, S., Sajani, S., and participating AMIP Modelling Groups, 1998. *Monsoon precipitation in Atmospheric Model Intercomparison Project runs*. WCRP-100, WMO/TD-No.837.
Keshavamurty, R.N., and Sankar Rao, M., 1992. *The Physics of Monsoons*. New Delhi: Allied Publishers.
Krishnamurti, T.N. (ed.), 1978. *Monsoon Dynamics*. Basel: Birkhauser Verlag.
Lighthill, J., and Pearce, R.P. (ed.), 1981. *Monsoon Dynamics*. Cambridge: Cambridge University Press.
Ramage, C.S., 1971. *Monsoon Meteorology*. New York: Academic Press.
Rao, Y.P., 1976. *Southwest Monsoon*. India Meteorological Department.

Cross-references

Asia, Climate of South
Ocean–Atmosphere Interaction
Trade Winds and the Trade Wind Inversion
Tropical and Equatorial Climates
Winds and Wind Systems

MONTREAL PROTOCOL

An international agreement for the protection of the ozone layer was initiated in 1985 in Vienna, Austria. Details were defined in the *Montreal Protocol on Substances that Deplete the Ozone Layer* that was signed in 1987 and ratified in 1989.

The substances defined are CFC, HCFC, HBFC, halons, carbon tetrachloride, methyl chloroform, and methyl bromide. The phase-out time for these was defined in the Montreal Protocol, with appropriate changes made at Vienna in 1995, Copenhagen in 1996, and Beijing in 1999. In all cases different schedules were applied to developed and developing countries.

The agreement to reduce production of ozone-depleting substances has been successful. Public concerns, political action, and industrial innovation, including the identification and production of appropriate alternative chemicals, has, in most cases, seen the phase-out meet or exceed the provisions of the Montreal Protocol.

The complete text of the Protocol including adjustments and amendments is available at the UN Environment Program site listed below.

John E. Oliver

Bibliography

Editor's note: The most recent data and status concerning ozone depletion are available from websites including:
http://www.unep.org/ozone/Montreal-Protocol/MontrealProtocol2000.shtml
http://www.afeas.org.montreal_protocol.html
http://www.epa.gov.ebtpages.intestratosphericozone.html

Cross-references

Climate Change and Human Health
Global Environmental Change: Impacts
Kyoto Protocol
Ozone

MOUNTAIN AND VALLEY WINDS

There is a special category of local or tertiary winds that are directly related to the topography. Sir Harold Jeffreys called them *antitriptic winds* because of the dominance of friction in their thermodynamics. Known as moutain and valley winds or breezes (Figure M21), they are diurnal in character and most obviously operative in otherwise calm, clear weather.

At night on the high mountain slopes, the air cools rapidly by radiation and thus becomes denser; it flows downhill into the valleys, gaining velocity and momentum, as a *katabatic wind*. A similar cold wind often flows down the surface of a glacier. This nightly drainage of cool air from mountain slopes may lead to important temperature inversion in the high plains, e.g. of Colorado or central Switzerland, leading to extremely cold nights. In high plains of the central Rockies the basins at 2500–2800 m elevation are treeless and known as *parks*, although the actual tree line is much higher (3000–3400 m), and the slopes carry forests of conifers and aspen: the cold layer below the inversion predicates what might be called an *inverted tree line*.

In the daytime the mountain slopes are warmed by insolation and the breeze begins to flow upslope, usually beginning about half an hour after sunrise and continuing until half an hour before sunset. It may reach a maximum of 6 m/s (12 knots) up sunny slopes but much less on the northern slopes. In depth the air flow may exceed 150 m increasing uphill. This type of flow is called an *anabatic wind*.

According to Defant (1951), an additional factor was suggested by Wagner. A pressure gradient from the plain to the valley must exist during the day, whereas a reverse gradient must develop at night. Equalization of the pressure differences must

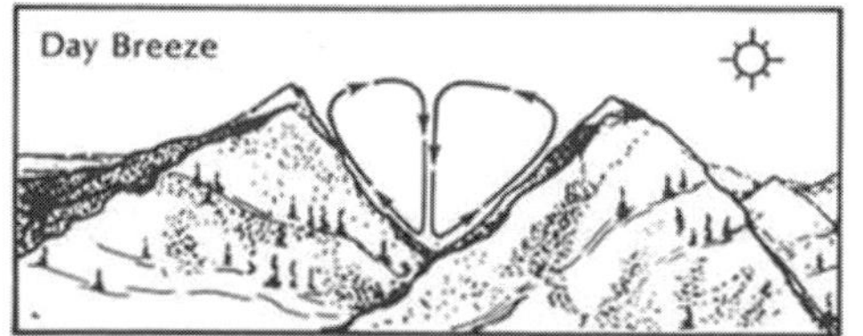

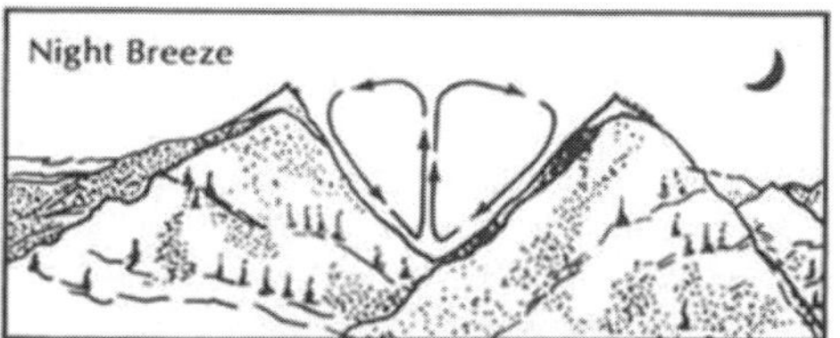

Figure M21 Schematic diagram illustrating moutain wind (breeze) and valley wind (breeze).

obtain at effective ridge altitudes, so that largest differences can be expected in valley floors. The pressure gradient is then due to the combined effects of the valley bottom slope and the effective ridge altitude; in practice one has two wind systems, a thermal slope circulation and a valley wind

The following sequence of mountain and valley winds may be identified:

1. Sunrise: onset of upslope winds; continuation of mountain wind. Valley cold, plains warm.
2. Forenoon (about 0900); strong slope winds, transition from mountain wind to valley wind. Valley temperature same as plains.
3. Noon and early afternoon; diminishing slope winds, fully developed valley wind. Valley warmer than plains.
4. Late afternoon; slope winds have ceased, valley wind continues. Valley continues warmer than plains.
5. Evening; onset of downslope winds, diminishing valley wind. Valley only slightly warmer than plains.
6. Early night; well-developed downslope winds, transition from valley wind to mountain wind. Valley and plains at same temperature.
7. Middle of night; downslope winds continue, mountain wind fully developed. Valley colder than plains.
8. Late night to morning; downslope winds have ceased, mountain wind fills valley. Valley colder than plains.

John E. Oliver and Rhodes W. Fairbridge

Bibliography

Barry, R.G., and Chorley, R.J., 1998. *Atmosphere, Weather and Climate*, 7th edn. London: Routledge.

Defant, F., 1951. Local winds. In Malone, T. (ed.) *Compendium of Meteorology*. Boston, MA: American Meteorological Society, pp. 655–672.

Jeffreys, H., 1922. On the dynamics of wind. *Royal Meteorological Society Quarterly Journal*, **48**: 29–46.

Lutgens, F.K., and Tarbuck, F.J., 2002. *The Atmosphere*, 8th edn. Upper Saddle River, NJ: Prentice-Hall.

Cross-references

Local Winds
Microclimatology
Mountain Climates
Winds and Wind Systems

MOUNTAIN CLIMATES

Mountain climates show great variation in the values of climatic elements over a short distance due to elevational differences, complex topography, and the physical presence of the mountains themselves. Whereas mountain climates are usually characterized by their spatial variety, the same causes often lead to large temporal variation as well.

It has been estimated that mountains or high plateaus occupy 20.2% of the Earth's land surface (Louis, quoted by Barry, 1992) and influence the life-support systems of 400 million people (Ives, 1981). Messerli (personal communication, 2001) claimed that 10–26% of the world's population lived on or within 50 km of mountains. The most prominent mountain areas include the Cascade–Sierra Nevada and Rockies of North America, the Andes of South America, the Alpine system of Europe, the Himalayas and Tibetan Plateau and Caucasus of Asia, the east African Highland, and the mountain backbones of Central America, Borneo, New Guinea, and New Zealand. There are many less prominent highland areas that should be recognized, such as the ice plateau of Greenland and Antarctica and the mountain ranges in both, together with elevated regions of older geologic materials, exemplified by the Appalachians of North America and the uplands of Great Britain and Scandinavia.

The variety of climates in mountain areas leads to confusion in climatic classification, therefore such areas are usually either excluded or grouped together under a broad category such as "highland climate". Modern versions of the Köppen classification, for example, use the letter H for mountain climates (Christopherson, 2002); the H standing for the German *hochgebirge* (high mountain). No quantitative definition of mountain climates has ever been given but, if one were needed, it would include the characteristic of rapid spatial change of climate classes within complex terrain. Such a characteristic would include not only areas currently classified as H but also areas such as the central highlands of Madagascar, which sometimes are attributed to other classes in the Köppen system.

Although humans have noted some of the effects of mountain climates whenever they have had contact with elevated areas, the scientific study of them dates only from the eighteenth century. The investigations of H.B. De Saussure, started in 1787, assured him of recognition as the first mountain meteorologist (Barry, 1978). Numerous mountain meteorological observatories were established in the nineteenth century, many of which were later closed due to lack of funds or for other reasons. Barry (1992) lists 30 of the most important observing stations, many of which have long records and continue to collect data through to the present time. The continuing importance of mountain climates is demonstrated by the appearance of excellent reviews of the subject such as those of Barry (1992) and Whiteman (2000). Additionally, there are more frequent conferences on the subject both in Europe and in North America (American Meteorological Society, 2000), and a number of national and international research programs into mountain meteorology are currently operating or planned.

Variety in mountain climates

Geographic variety in mountain climates is seen both on the global scale and on meso- and local scales. On the global scale the nature of mountain climate is determined by the latitude of the site as well as other factors that influence most climates, such as proximity to the ocean or features of the general circulation of the atmosphere. Table M4 gives some idea of the range of temperature and precipitation values found at high-altitude stations at different latitudes. La Paz at 16°S has a very small annual temperature range, whereas the range at Lhasa (29°N) is rather larger and similar to the midlatitude station at Longs Peak, Colorado. The Antarctic station of Vostok (which might also be regarded as a high-latitude climate) has an even larger annual temperature range but is more noteworthy for its extremely low temperatures. The precipitation regime at La Paz is influenced by the southward movement of the Intertropical Convergence Zone giving a maximum of rainfall between December and February. The same phenomenon in conjunction with the Asian monsoon system is

Table M4 Monthly mean temperature (°C) and total precipitation (mm) at selected high-altitude stations

Station	Altitude	Jan.	Feb.	Mar.	Apr.	May	June	July	Aug.	Sept.	Oct.	Nov.	Dec.	Year
La Paz, Bolivia 16°30′S 68°08′W	3658 m	11.7° 165.1 mm	11.7° 106.7 mm	11.7° 66.0 mm	11.4° 33.0 mm	10.3° 12.7 mm	8.9° 7.6 mm	8.6° 10.2 mm	9.5° 12.7 mm	10.6° 27.9 mm	11.7° 40.6 mm	12.5° 48.3 mm	12.0° 94.0 mm	10.8° 574.0 mm
Lhasa, Tibet 29°40′N 91°07′E	3685 m	−1.7° 2.5 mm	1.1° 12.7 mm	4.7° 7.6 mm	8.1° 5.1 mm	12.2° 25.4 mm	16.7° 63.5 mm	16.4° 121.9 mm	15.6° 88.9 mm	14.2° 66.0 mm	8.9° 12.7 mm	3.9° 2.5 mm	0.0° 0.0 mm	8.3° 406.4 mm
Longs Peak, Colorado 40°15′N 105°35′W	2729 m	−5.0° 17.8 mm	−5.0° 27.9 mm	−2.8° 50.8 mm	−1.1° 68.6 mm	5.0° 61.0 mm	10.0° 43.2 mm	12.8° 91.4 mm	12.8° 55.9 mm	8.9° 43.2 mm	3.9° 43.2 mm	−1.1° 22.9 mm	−5.0° 22.9 mm	2.8° 558.8 mm
Vostok, Antarctica 78°27′S 106°52′E	3420 m	−33.6°	−43.9°	−53.9°	−63.1°	−63.4°	−66.7°	−67.0°	−70.6°	−67.3°	−58.4°	−63.9°	−32 2°	−55.6°
						No precipitation data								

Source: After Liljequist (1970), pp. 483–515 and Critchfield (1983), pp. 421–429.

responsible for the maximum of precipitation at Lhasa between June and September. Longs Peak shows a summer maximum of precipitation owing to the frequency of thunderstorms at this season. There is very little measurable precipitation in Antarctica where the annual value is believed to be less than 6 in (150 mm; Trewartha and Horn, 1980).

On smaller geographic scales mountain climates are affected by the rapid change of temperature with height, aspect, particular wind systems, and the barrier effect of mountains as they act with respect to precipitation values. The complexities of mountain climates are represented by surface energy budget values. All of these factors will be discussed below.

Characteristic features of mountain climates

Radiation

The amount of global solar radiation received at the Earth's surface increases with altitude due to the decreased radiation through the atmosphere and atmospheric water vapor that is concentrated at the lower part of the atmosphere. Sauberer and Dirmhirn (quoted by Barry and Van Wie, 1974) note an increase in global solar radiation in the Austrian Alps from 650 ft (200 m) to 9840 ft (3000 m) of 21% in June and 33% in December. The increase is approximately exponential and occurs in both cloudy and clear skies. However, varying cloud amounts in mountain areas can sometimes disturb this pattern (Greenland, 1978). The increase in radiation is most marked in the shorter wavelengths. Reiter and Munzert (1982) observed that total ultraviolet radiation could be increased 1.4 times in January and 1.5 times in June, accompanying an altitude increase from 2296 ft (700 m) to 9840 ft (3000 m), using a 5-year observing period in the northern European Alps. The large variety of slope orientations and gradients give rise to a whole spectrum of different aspects in mountain areas, and it is the varying degree of radiation receipt on these slopes that essentially leads to the importance of aspect in mountain climates. At latitude 50°, for example, there is a fourfold difference in the radiation receipt on 45° slopes facing north and south (Barry and Van Wie, 1974). Variation of radiation receipt due to aspect can overshadow that due to any other cause, and leads to variation in temperatures, evaporation rates, and processes of the soil and vegetation growth. The specific location of long-lasting snow and ice patches is also associated with aspect.

Temperature

Since the atmosphere is heated principally by infrared radiation from the Earth's surface, air temperature usually decreases with altitude. The decrease varies in the free atmosphere from about −2.7°F/1000 ft (−5°C/km) for saturated air to −5.5°F/1000 ft (−10°C/km) for dry air and averages about −3.2°F/1000 ft (−6°C/km). The air above mountain slopes is also influenced by latent and sensible turbulent heat flows and so the temperature changes with altitudes, or lapse rates, on slopes are often different from those in the free air (Coulter, 1967). Average lapse rates in mountain climates have also been shown to vary with season, global climatic zone and with airmass frequency (Barry, 1992).

The rapid change of temperature with height leads to a climatic zonation that, in some parts of the world, has been given a special nomenclature. Trewartha and Horn (1980) point out that in tropical Latin America four zones are recognized; these are Tierra Caliente (hot lands), Tierra Templada (temperate lands), Tierra Fria (cool lands), and Tierra Helada (land of frost). Each of these zones is associated with different types of vegetation or agricultural crops. Figure M22 displays a similar kind of zonation in the New Zealand Southern Alps. From the perpetually snow-covered slopes of Mount Tasman one descends through rainforest vegetation before reaching sea level.

Figure M22 The west coast of the South Island of New Zealand from the top of Mount Tasman. The snow-covered mountain peaks give way to subtropical rainforest before sea level is reached.

Another important feature of temperature in mountain climates is related to the sinking downslope of radiationally cooled, relatively dense cold air that accumulates at the bottom of valleys and in mountain basins. The colder air at the lower elevations, and the relatively warmer air above, result in increasing temperatures with height, a situation known as a temperature inversion. Inversions are commonly formed at night but can last through the whole day, especially in winter, when radiative heating is not strong enough to disperse them. Stagnating cold air in valley bottoms can have the effect of trapping air pollutants.

The frequency of temperature inversions and the consequent higher temperatures at some distance up the slope can be so great as to cause generally warmer areas in midslope. These are called thermal belts and have been used to advantage by agriculturalists in many parts of the world (Dunbar, 1966). The center of the thermal belt in the European Alps is found to be between 330 and 1310 ft (100–400 m) above the valley in areas of relief under 1650 ft (500 m). In higher mountains the thermal belt is centered 1150 ft (350 m) above the valley floor in summer and 2300 ft (700 m) above it in winter (Barry, 1992).

Precipitation

A general global pattern of precipitation change with altitude has been noted by Lauscher (quoted by Barry, 1992). There is a general decrease of precipitation with elevation in equatorial latitudes, which is also seen, to a lesser degree, at polar stations open to a maritime influence. In the tropics, however, there is a maximum of precipitation at about 3000 ft (900 m). In midlatitudes there is a general increase of precipitation up to the highest altitudes at which observations are made at about 12 000 ft (3660 m).

This global pattern is often superceded by the effect of mountain barriers on the vertical flow of air and subsequent increase or suppression of precipitation formation processes. When air is forced to higher altitudes in its passage over a mountain, the air is cooled and condensation of water vapor often gives rise to precipitation formation on the windward side. This is sometimes called orographic precipitation. Descending air on the leeward side of the barrier is accompanied by warming of the air, an increase in potential for evaporation, a decrease in relative humidity, and a decrease in the likelihood for precipitation formation. Dry areas on the lee side of mountain barriers are called rainshadow areas. These phenomena are found in many of the mountain areas of the world. The South Island of New Zealand is a good example of such orographic effects. On the windward side of the island at Milford Sound, annual precipitation is 245 in (6233 mm), whereas on the lee side in the interior plateau of central Otago the value decreases to 13 in (330 mm) at Alexandra (Garnier, 1958).

A complicating but important feature of mountain precipitation is the fact that much of it falls in the form of snow. The difficulty of measuring snowfall makes values of precipitation from the higher elevations of mountain climates less accurate than their lowland flat-terrain counterparts. In both environments it is hard to design a shield for precipitation gauges that suppresses local wind eddy currents and permits representative assessment of snowfall but, in addition, mountain observational sites are seldom representative of larger geographic areas. As a result it is often expedient to sample the water content of the mountain snowpack over snow courses set up in a systematic manner. As much accuracy in snowpack measurements as possible is necessary because many lowland areas depend on this water source for agricultural, industrial, hydroelectric energy, and domestic purposes. The frozen form of precipitation in mountain areas also adds immensely to their esthetic qualities in terms of such features as long-lasting snow cover and glaciers.

Wind

The decrease of the effect of friction between the Earth's surface and the movement of air in the free atmosphere causes wind velocities to increase with altitude. This effect is also noted from wind observations at high mountain stations although Wahl (quoted by Barry, 1992) found that for the European Alps mountain summit wind velocities are about half those at corresponding altitudes in the free atmosphere. Nevertheless, mountain areas exhibit some of the highest wind velocities anywhere on the surface of the planet – a gust of over 231 mph (103 m/s) having been recorded on Mount Washington, New Hampshire, where the annual average wind speed exceeds 31 mph (15.7 m/s: Smith, 1982).

More important than high wind velocities is the topographic influence of mountain barriers on winds. These barriers impose wavelike motion on airflow both in the horizontal and in the vertical planes. On the global scale, downwind of major mountain barriers, horizontal waves with wavelengths on the order of 3000 miles (4800 km) are established with their troughs to the lee of the mountain range. This is most pronounced in the northern hemisphere, resulting from the presence of the Rockies in North America and the Himalayas in Asia. A similar smaller-scale effect is noticed, however, east of the Andes in South America and in the New Zealand Southern Alps. In all cases the presence of the mountains has a marked effect on the climates downwind of them, particularly in causing rainshadow areas and, further downwind, regions of cyclogenesis or formation of cyclones. Waves are also imposed on airflow over mountains in the vertical plane. These vertical undulations or waves created by gravity acting on local variations in air density are sometimes called gravity waves (Whiteman, 2000). The effects of these waves can sometimes be noticed when lee wave, or

Figure M23 Lee wave clouds near Aviemore, South Island, New Zealand.

lenticular, clouds are formed on their crests, as illustrated in Figure M23. Wind phenomena in the lee of mountains are not limited to lee waves, however. Another interesting feature is a rotor effect that can give air movement at the surface back toward the mountain. This effect sometimes carries snow that can feed alpine glaciers and maintain their existence at latitudes that would otherwise be impossible (Johnson, 1980).

High-velocity winds in the lee of mountain ranges are also common features of mountain climates. These downslope winds may be warm or cold. The warm variety are called chinook winds in North America and föhn winds in Europe, and their warmth is due to compressional heating as the air moves downward to more dense parts of the atmosphere. "Chinook" is a North American Indian term meaning "snow eater", and signifies the large amounts of snow that can be melted by the wind, especially when it occurs in the spring (Brinkman, 1970). The wind storms at Boulder, Colorado, originally studied by Brinkman, have been found to be difficult to predict because of lack of upstream observational data but also because more than one synoptic-scale pressure situation can give rise to the storms (Leptuch et al., 2000). As in the case of the chinook winds, particular synoptic-scale pressure patterns are required for the establishment of the cold winds. These winds that have several local names are exemplified in Europe by the Bora winds of Croatia and Slovenia (parts of the former Yugoslavia). Besides a strong pressure gradient, their formation occurs when there is a damming up of cold air east of the mountains in the type area (Petkovsæk and Paradiû, quoted by Barry, 1992). Locally, under certain conditions, downslope winds can give rise to intense windstorms, causing much damage in populated areas (Miller et al., 1974). A windstorm in 1972 caused $2.5 million damage in Boulder, Colorado (Whiteman, 2000).

The modification of wind flow direction is another important aspect of wind in mountain climates. Modifications include the lateral movement of air around obstacles, the channeling of air through topographically formed tunnels, and the altering of wind direction between that found in valleys and that above the ridge line. The fact that it is possible to have different wind regimes in different parts of mountainous topography is another noteworthy feature of mountain winds. When two or more regimes influence each other they are said to be coupled, but when they act as separate entities they are said to be uncoupled.

Some of the most frequent kinds of mountain winds are thermally induced. Downslope movement of cold air at night is called katabatic flow, whereas upslope movement of a warm air during the day is termed anabatic flow. In the absence of strong synoptic-scale winds, mountain-valley wind systems can develop. The classic description of these involves the combined results of katabatic nighttime winds, giving rise to a downvalley wind best marked in the early morning. This is gradually stilled when radiative heating of slopes gives rise to anabatic winds. Air to feed these is drawn up the valley, leading to an upvalley wind system that is most noticeable in the early afternoon. More recent observational studies show the phenomenon to be very complex and involving compensatory winds at different levels and characteristic fluctuations in wind velocity (Barry, 1992). The theory of these wind systems has yet to be completely worked out but Whiteman (in American Meteorological Society, 1981) has made an important contribution to it in his study of the breakup of valley temperature inversions. Whiteman (2000) distinguishes four components of the overall mountain wind system; namely slope winds, along-valley winds, cross-valley winds, and mountain-plain winds. He comments that it is difficult to study the pure form of any one of these components because they interact with one another in various combinations and also with the winds of the larger-scale synoptic situation.

Surface energy budgets

The energy budget of the surface is an estimation of heat flow to and from the surface by radiation and by sensible and latent heat flows to and from the atmosphere, and by conduction of heat into and out of the submedium. The estimation of surface energy budgets often helps our understanding of a climate under consideration. Reiter (1982) has drawn attention to the importance on global climate particularly of sensible heat fluxes from the Tibetan Plateau, where a resulting large-scale wind circulation may be established. Reiter has described how anomalies of heat flowing from the Tibetan Plateau could set off significant interannual variability of monsoon circulation systems. Complete energy budgets of snow- and ice-free mountain area surfaces are difficult to estimate for any length of time and hence very few exist. Barry (1992), for example, while giving information on several others, tabulates only six studies, and all of these are for short periods. One of the few long-term studies (daily values for 1 year) for the New Zealand Southern Alps shows a positive net radiation to be used more or less in equal parts by sensible and latent heat flows, whereas ground heat flow is rather small (Greenland, 1973). These results, which relate to a shrub and grass surface, have been paralleled by shorter-term studies in similar environments such as Alaska (Wendler, 1971). However, the effect of aspect plays a large role in determining the radiation receipt, and therefore the energy budgets, of mountain areas, as has been demonstrated by Brazel and Outcalt (1973) for another Alaskan site. Cline (1997a) performed pioneering work on measuring the energy balances of melting snow surfaces at Niwot Ridge, Colorado. He found that, besides net radiation, surface flowing turbulent fluxes were an important source of energy for snow melt and could provide 25–54% of the energy needed. He has also developed a point energy and mass balance model of a snow cover (called SNTHERM) for use in alpine conditions (Cline, 1997b).

The energy budgets of snow and glacial surfaces of mountain areas are better known partly because their details are important for practical hydrological purposes. A summer study of the Peyto glacier in the Canadian Rockies indicates that heat flows are generally toward the ice surface. These heat flows are used in melting the ice; 71% of the ice melt was from net radiation, 23% from the downward transfer of sensible heat, and 6% was from the downward transfer of latent heat in the form of condensation (Munro, quoted by Oke, 1987). These values appear to be generally representative as judged by the reviews presented by Patterson (1994). The difficult problems of aspect and surface heterogeneity still remain, however, and because of this several investigators have turned their attention to attempting to model the energy budgets of such surfaces. Dozier and Outcalt (1979) performed pioneering work in this area. With the advent of satellite data the heterogeneity of mountain areas becomes more approachable. Duguay (1994) has used remote sensing data to map the growing season radiation of the Niwot Ridge alpine site in Colorado. Researchers at the Institute for Computational Earth System Science at the University of California, Santa Barbara, are leaders in handling the large datasets of energy budget values and other data that result from the remotely sensed products. In another advance, Kumar et al. (1997) have extended the process of modeling topographic variation in solar radiation for use within a GIS environment. It is with respect to their surface energy budgets that mountain climates show one of their greatest degrees of variability.

Applied aspects of mountain climates

The increasing use of mountain areas by humans has stimulated the field of applied climatology in mountain regions. Studies have taken a wide variety of forms. These forms include the study of climate change and tourism, hydro-electric power potential, and marginal agriculture (Beniston, 1994), investigations of weather-related mountain hazards such as avalanches (Mock and Birkeland, 2000) and lightning (Peterson, quoted by Barry, 1992), examination of forest fire weather and smoke management, air pollution dispersion, and aerial spraying (Whiteman, 2000).

One topic of increasing importance is the development of systems to interpolate and extrapolate values of climatic variables in mountain environments. Such an endeavor is extremely useful in almost all applied aspects related to mountain climates. This development is particularly important because there will never be enough actual meteorological observations in mountain areas. Indeed there are fears that increasingly mountain observatories are being closed down for budgetary reasons (Beniston et al., 1997). Pioneering work has been performed on interpolation and extrapolation by researchers at the University of Montana who first developed a methodology called MT-CLIM and later refined it into a system called Daymet. These methodologies permit the interpolation and extrapolation of such variables as air temperature, solar radiation, and surface humidity from sparsely distributed observation sites to other parts of the terrain (Running et al., 1987; Glassy and Running, 1994; Kimball et al., 1997; and Thornton and Running, 1999). This work has been complemented by the development of a methodology, called PRISM (Parameter-elevation Regressions on Independent Slopes Model). PRISM is an analytical tool that uses point data, a digital elevation model, and other spatial datasets to generate gridded estimates of monthly, yearly, and event-based climatic parameters, such as precipitation, temperature, and dewpoint for interpolating precipitation values in mountain (and other) terrain (Daly et al., 1997). Ollinger et al. (1998) have also used regression techniques to establish the spatial variation of temperature and precipitation in the complex terrain of the northeastern United States.

Research programs in mountain climatology

Our knowledge of the characteristics of mountain climates has been achieved by the painstaking research of many individuals. However, in recent decades, as in many branches of the atmospheric sciences, it has been realized that significant further progress could be made by employing large integrated research programs in addition to the work of individuals and small groups.

Some noteworthy early programs of the 1980s included ALPEX that was focused on the European Alps and ASCOT that was developed in the United States in order to investigate the potential of mountain areas for air pollution. In more recent years the Special Observation Period of the Mesoscale Alpine Programme (MAP) in the European Alps was completed in the fall, 1999. MAP was possibly the most comprehensive field program to date documenting the influence of mountains on the atmosphere. MAP is likely to provide data and the development of new theory in years to come. The following are the goals of MAP as outlined on the website of the programme

> (1a) To improve the understanding of orographically influenced precipitation events and related flooding episodes involving deep convection, frontal precipitation and runoff. (1b) To improve the numerical prediction of moist processes over and in the vicinity of complex topography, including interactions with land-surface processes. (2a) To improve the understanding and forecasting of the life-cycle of Föhn-related phenomena, including their three-dimensional structure and associated boundary layer processes. (2b) To improve the understanding of three-dimensional gravity wave breaking and associated wave drag in order to improve the parameterization of gravity wave drag effects in numerical weather prediction and climate models. (3) To provide data sets for the validation and improvement of high-resolution numerical weather prediction, hydrological and coupled models in mountainous terrain.

Another recent field program was the Intermountain Precipitation Experiment (IPEX). This program, focused in the US Rocky Mountains, also had detailed and complex goals as described by the program leaders.

> IPEX is a research program designed to improve the understanding, analysis, and prediction of precipitation over the complex orography of the Intermountain West of the United States. The goals of IPEX are to advance knowledge of the structure and dynamics of Great Salt Lake-effect and orographic precipitation, especially in and adjacent to the Wasatch Mountains of northern Utah; to better understand the relationships between orographic circulations and cloud microphysics; to verify and improve data assimilation, numerical weather prediction, and radar-derived quantitative precipitation estimates over the Intermountain West; to explore the electrical structure of continental winter storms; and to raise awareness of mountain meteorology and the associated scientific and forecasting challenges at the public, K-12, undergraduate, and graduate levels (Schultz et al., 2002).

Programs such as MAP and IPEX clearly demonstrate the areas of concern and depth of investigation in modern-day studies of mountain climatology.

The future outlook

In mountain climatology, as with many other fields, the more that is discovered, the more new questions arise. In addition, recent discoveries in mountain climatology have highlighted the increasing importance of the topic. For example, evidence suggests that late twentieth-century global-warming signals may be amplified in mountain climates. Temperature increases in parts of the European Alps of up to 2.0°C have been recorded. It has also been shown for the European Alps that temperature extremes shift by a factor of 1.5 for a unit shift in the global means of the twentieth century (Beniston et al., 1997). Beniston and colleagues have outlined the most pressing needs for further activities in the field. They advocate the encouragement of field studies including monitoring programs, an increase in the depth and breadth of the paleo-database, and more attention to the downscaling of climate models. The future appears to continue to be closely tied to the increasing human populations in and near mountains and the particular kinds of uses to which mountain lands are being put. The main activities include: (1) recreation; (2) timber production and watershed management; and (3) mining for mineral and energy production. Each of these activities carries with it particular, although sometimes interrelated, problems of mountain climatology.

Recreationists demand clean air and this, together with clean air legislation in some countries, provides the incentive for air-quality studies that examine pollutant transport, chemical reaction, and visibility studies. Avalanche potential prediction is another item of interest to recreationists. Mountain forest land management requires knowledge of mountain microclimate and climatic factors affecting growth rates and forest fire. Overall, whereas projection of future climates is a difficult task in itself, the task becomes even more complicated in mountain areas because of the complex nature of mountain climates. Increasing population growth, however, continues to raise the importance of understanding mountain climates.

David Greenland

Bibliography

American Meteorological Society, 1981. *Second Conference on Mountain Meteorology*, 9–12 November. Steamboat Springs, Colorado. Preprints. Boston, MA: American Meteorological Society.

American Meteorological Society, 2000. *Ninth Conference on Mountain Meteorology*. Aspen, Colorado. Preprints. Boston, MA: American Meteorological Society.

Barry, R.G., 1978. DeSaussare H.B.: the first mountain meteorologist. *American Meteorological Society Bulletin*, **59**: 702–705.

Barry, R.G., 1992. *Mountain Weather and Climate*, 2dn edn. London: Methuen.

Barry, R.G., and Van Wie, C.C., 1974. Topo- and microclimatology. In Ives J.D., and Barry, R.G., eds., *Alpine Areas in Arctic and Alpine Environments*. London: Methuen, pp. 73–84.

Beniston, M. (ed). 1994. *Mountain Environments in Changing Climates*. London: Routledge.

Beniston, M., Diaz, H.F., and Bradley, R.S. (eds), 1997. Climatic changes at high elevation sites: an overview. In Diaz, H.F., Beniston, M., and Bradley, R.S., eds., *Climatic Changes at High Elevation Sites*. Dordrecht: Kluwer, pp. 1–19.

Brazel, A.J., and Outcalt, S.I., 1973. The observation and simulation of diurnal evaporation contrast in an Alaskan alpine pass. *Journal of Applied Meteorology*, **12**(7): 1134–1143.

Brinkmann, W.A., 1970. The chinook at Calgary (Canada). *Archiv für Meteorologie, Geophysik und Bioklimatologie*, **B18**: 269–278.

Cline, D.W., 1997a. Effect of seasonality of snow accumulation and melt on snow surface energy exchanges at a continental alpine site. *Journal of Applied Meteorology*, **36**(1): 22–41.

Cline, D.W., 1997b. Snow surface energy exchanges and snowmelt at a continental, midlatitude Alpine site. *Water Resources Research*, **33**(4): 689–701.

Christopherson, R.W., 2002. *Geosystems*. Upper Saddle River, NJ: Prentice-Hall.

Critchfield, H., 1983. *General Climatology*, 4th edn. Englewood Cliffs. NJ: Prentice-Hall.

Coulter, J.D., 1967. Mountain climate. *New Zealand Ecological Society Proceedings*, **14**: 40–57.

Daly, C., Taylor, G., and Gibson, W., 1997. The PRISM approach to mapping precipitation and temperature. *10th Conference on Applied Climatology*. Reno, NV: American Meteorological Society, pp. 10–12.

Dozier, J., and Outcalt, S.I., 1979. An approach toward energy balance simulation over rugged terrain. *Geographical Analysis*, **11**(1): 65–85.

Duguay, C.R., 1994. Remote sensing of the radiation balance during the growing season at the Niwot Ridge Long-Term Ecological Research site, Front range, Colorado, U.S.A. *Arctic and Alpine Research*, **26**(4): 393–402.

Dunbar, G.S., 1966. Thermal belts in North Carolina. *Geographical Review*, **56**(4): 516–526.

Garnier, B.J., 1958. *The Climate of New Zealand*. London: Edward Arnold.

Glassy, J.M., and Running, S.W., 1994. Validating diurnal climatology of the MT-CLIM model across a climatic gradient in Oregon. *Ecological Applications*, **4**(2): 248–257.

Greenland, D., 1973. An estimate of the heat balance of an alpine valley in the New Zealand Southern Alps. *Agricultural Meteorology*, **11**: 293–302.

Greenland, D., 1978. Spatial distribution of radiation in the Colorado Front Range. *Climatological Bulletin*, **14**: 1–14.

Ives, J.D., 1981. Editorial, *Mountain Research and Development*, **1**(1): 3–4.

Johnson, J.B., 1980. Mass balance studies on the Arikaree Glacier. In Ives, J.D., ed., Geoecology of the Colorado Front Range. *Boulder, Co: Westview Press,* pp. 209–213.

Kimball, J.S., Running, S.W., and Nemani, R., 1997. An improved method for estimating surface humidity from daily minimum temperature. *Agricultural and Forest Meteorology*, **85**: 87–98.

Kumar, L.A., Skidmore, K., and Knowles, E., 1997. Modelling topographic variation in solar radiation in a GIS environment. *International Journal for Geographical Information Science*. **11**(5): 475–497.

Leptuch, P., Brown, J.M., Bluestein, H.B., Thaler, E., and Richman, M.B., 2000. Forecasting downslope windstorms at Boulder, Colorado: the empirical–statistical approach revisited. *Preprints of Ninth Conference on Mountain Meteorology*. 7–11 August 2000, Aspen, Colorado. American Meteorological Society, pp. 105–108.

Liljequist, G.H., 1970. *Klimatologi*. Stockholm: Generalstabens Litografiska Anstalt.

Messerli, B., 2001. Opening remarks to International Year of Mountains 2002. http://www.iisd.ca/linkages/sd/mountains/

Miller, D.J., Brinkmann, W.A.R., and Barry, R.G., 1974. Windstorms: a case study of wind hazards for Boulder, Colorado. In White, G.F., ed., *Natural Hazards: local, global and national*. Oxford: Oxford University Press, pp. 80–85.

Mock, C., and Birkeland, K.W., 2000. Snow avalanche climatology of the western United States mountain ranges. *Bulletin of the American Meteorological Society*, **81**: 2367–2392.

Oke, T.R., 1987. *Boundary Layer Climates*, 2nd edn, London: Methuen.

Ollinger S.V., Aber, J.D., and Federer, C.A., 1998. Estimating regional forest productivity and water yield using an ecosystem model linked to a GIS. *Landscape Ecology*, **13**: 323–334.

Patterson, W.S.B., 1994. *The Physics of Glaciers*, 3rd edn, Oxford: Elsevier.

Reiter, E.R., 1982. Where we are and where we are going in mountain meteorology. *American Meteorological Society Bulletin*, **63**(10): 1114–1122.

Reiter, R., and Munzert, K., 1982. Values of UV and global radiation in the Northern Alps. *Archiv für Meteorologie, Geophysik und Bioklimatologie*, **B30**: 239–246.

Running, S.W., Nemani, R.R., and Hungerford, R.D., 1987. Extrapolation of synoptic meteorological data in mountainous terrain and its use for simulating forest evaporation and photosynthesis. *Canadian Journal of Forest Research*, **17**: 472–483.

Schultz, D.M., Steenburgh, W.J., Trapp, R.J., et al., 2002. Understanding Utah winter storms: The Intermountain Precipitation Experiment. *Bulletin of the American Meteorological Society*, **83**: 189–210.

Smith, A.A., 1982. The Mount Washington Observatory – 50 years old. *American Meteorological Society Bulletin*, **63**(9): 986–995.

Thornton, P.E., and Running, S.W., 1999. An improved algorithm for estimating incident daily solar radiation from measurements of temperature, humidity, and precipitation. *Agricultural and Forest Meteorology*, **93**: 211–228.

Trewartha, G.T., and Horn, L.H., 1980. *An Introduction to Climate*, 5th edn., New York: McGraw-Hill.

University Corporation for Atmospheric Research, 1980. ALPEX: upcoming GARP mountain subprogram. *UCAR Newsletter*, **4**(3): 1.

Wendler, G., 1971. An estimate of heat balance of a valley and hill station in central Alaska. *Journal of Applied Meteorology*, **10**: 684–693.

Whiteman, C.D., 2000. *Mountain Meteorology: fundamentals and applications*. New York: Oxford University Press.

Cross-references

Mountain and Valley Winds
Precipitation Distribution
Rainshadow

N

NATIONAL CENTER FOR ATMOSPHERIC RESEARCH*

The National Center for Atmospheric Research (NCAR) was established in 1960 to serve as a center for atmospheric and related science problems. It major sites are in Boulder, CO, at the I.M. Pei Mesa Laboratory and the Foothills Laboratory. It is operated by University Corporation for Atmospheric Research (UCAR) under a cooperative agreement with the National Science Foundation (NSF).

Anon

Cross-references

Climate Data and Research Centers
Climate Modeling and Research Centers

* The article is derived mainly from *www.ncar.ucar.edu*, where much additional information is given.

NATIONAL CLIMATE DATA CENTER/DATA SOURCES*

The National Climatic Data Center (NCDC) is part of the Department of Commerce, National Oceanic and Atmospheric Administration (NOAA), and the National Environmental Satellite, Data and Information Service (NESDIS).

The Weather Bureau, Air Force and Navy Tabulation Units in New Orleans, LA were combined and formed into the National Weather Records Center in Asheville, NC in November 1951. Authority to establish the joint Weather Records Center was granted under section 506(c) of the Federal Records Act of 1950 (Public Law 754, 81st Congress). The Center was eventually renamed the National Climatic Data Center. Data are received from a wide variety of sources, including satellites, radar, remote sensing systems, NWS cooperative observers, aircraft, ships, radiosonde, wind profiler, rocketsonde, solar radiation networks, and NWS Forecast/Warnings/Analyses Products.

The NWS's Cooperative Network, composed mainly of 8000 volunteer observers, has been recording daily records since the 1880s. Ships at sea have also been observing the weather essentially the same way for over 100 years. However, new observing systems have presented themselves as technology has advanced. As aircraft began to fill the skies, information on the upper atmosphere was needed. Balloon-borne instruments radioed data; radars began to probe clouds; rockets reached the fringes of the atmosphere; and weather satellites, both geostationary and polar orbiting, now continuously watch the weather. These data are all archived by the National Climatic Data Center.

Global weather and climate

The global weather and climate community has established a communications network to flash observations around the world in a matter of seconds. These data are supplied to NCDC by the National Weather Service's National Meteorological Center, the Air Force's Global Weather Center, and the Navy's Fleet Numerical Oceanography Center. NCDC also maintains World Data Center for Meteorology, Asheville. The four World Centers (US, USSR, Japan, China) have created a free and open atmosphere in which data and dialogue are exchanged.

The World Meteorological Organization, through its Resolution 35, has established the mechanisms whereby marine observations are exchanged among participating countries. This exchange is perhaps the prime example of country-to-country cooperation over several decades. However, to supplement all of the exchanges already listed, NCDC also maintains cooperative agreements with individual countries. In exchange for US

* This article is derived from information provided by the NCDC. The web source www.ncdc.noaa.gov provides much additional information.

data and information, countries are glad to exchange their own data and information.

Publications

Climatological publications have been produced and disseminated for over 100 years and include:

Local Climatological Data (LCD) is produced monthly and annually for some 270 cities. The *LCD* contains 3-hourly, daily, and monthly values. The annual issue contains the year in review plus normals, means and extremes.
Climatological Data (CD), also produced monthly and annually, contains daily temperature and precipitation data for over 8000 locations. The *CD* is published by state or region (New England), with a total of 45 issues produced each month.
Hourly Precipitation Data (HPD) is produced monthly. It contains data on nearly 3000 hourly precipitation stations, and is published by state or region.
Storm Data (SD) documents significant US storms and contains statistics on property damage and human injuries and deaths.
Monthly Climatic Data for the World (MCDW) provides monthly statistics for some 1500 surface stations and approximately 800 upper air stations.

In addition to routine publications, NCDC also generates many nonperiodicals including normals, probabilities, long-term station and state summaries, and several atlases covering the land areas, coastal zones, and oceans of the world.

John E. Oliver

Cross-reference

Climate Data Centers

NORTH AMERICA, CLIMATE OF

The climate of North America ranges from the frost-free tropical of southernmost Florida to the perennial ice and snow of the northernmost islands of the Canadian Archipelago, from the rain-drenched mountains of the northwest coast to the drought-ridden deserts of the southwestern United States. Within the limits set by these extremes is a variety of climates that results from the interplay of atmospheric systems with the complex geography of this large landmass. Its great size (about 25 million km^2), the span of latitudes from about 25° to 80°, "the configuration and alignment of its shorelines, plus the distribution of major landform regions have generated a diversity of climates that very closely resemble those of the much larger continent of Eurasia within the same latitudes.

The scattering of persistent ice fields in the Canadian Arctic from the northern Ellesmere southward to Baffin Island may be viewed as a fragmented extension of the north polar ice cap that dominates most of Greenland's elevated mass to the east and the surface of the Arctic Ocean to the north. These ice fields are only a negligible component of North America's climatic array, but it is a reminder that past climates have permitted continental glaciers to spread over much of North America during earlier geological epochs.

Organization rationale

Given the vast amount of climatic literature pertaining to the continent, it is possible to take a number of approaches to describe North American climates Certainly, the best description of climate would be a genetic one that relates each climate to a particular set of causes. One approach that assesses climates by cause draws upon airmass climatology (e.g. Oliver and Hidore, 2002). In this approach source regions are identified (Figure N1) and their characteristics are related to prevailing conditions (Table N1). Airmass climatology is examined in related articles in this volume. Similarly, reasons for the distribution of temperature (Figure N2) and precipitation (Figure N3) and other climatic elements are also dealt with in other entries.

It is also possible to delineate the climates according to a descriptive classification scheme, such as that of Kp̈pen, and use the identified regions as the basis for discussion. This approach has been used in a number of textbooks (e.g. Trewartha and Horn, 1980), and it is the approach adopted herein. Figure N4 identifies the location of the regions described on the following pages, along with a number of climographs showing mean monthly temperature and precipitation for a few selected stations. The stations in Figure N4 were selected to illustrate gradients in climatic conditions that are not always reflected by the categorization of the Kp̈pen system. Some stations may be mentioned in discussions of climates different from the climate class in which they reside simply because the climograph illustrates an important trend. For example, Amarillo is used to illustrate the westward decrease in precipitation in C climates, yet it is a BS climate. Similarly, generalization has eliminated some geographic variation and Kp̈pen subtypes.

Tundra climate (ET)

The northernmost of the continent's major climates is ice-free only during the summer, a region of rolling, virtually treeless plains stretching eastward from the shores of the Bering Sea to the coast of Labrador. Long, severely cold winters alternate with short, cool summers. Sunless darkness prevails for several weeks before and after the winter solstice around 22 December. Lowest temperatures are normally reached in February, except near the sea after oceanic ice has attained its maximum extent, where March is the coldest month. February averages between −15° and −30°C (Inuvik, Figure N4). From time to time, values drop to −50°C or lower. At this season precipitation diminishes to about 1 cm a month over the interior and falls mainly in the form of fine, light, dry snow, brought by the fairly rapid passage of small, intense frontal storms from the north Pacific. Occasionally, when clear, transparent air is calm and intensely cold, showers of minute ice crystals fall from the cloudless sky. Normal yearly precipitation throughout the tundra is usually less than 25 cm, although coastal stations may average up to 50 cm or more.

Summer is a season of prolonged daylight when temperatures average above freezing for 3–4 months at most stations. The warmest month is commonly July when readings are above 5°C but not above 10°C. Comparatively high temperatures are not unknown at numerous inland sites where values exceeding 30°C have been reached. However, it is the sporadic occurrence of subfreezing temperatures from June to September that characterizes the luminous summer of the tundra. This is the season of maximum precipitation for most of the region, mainly falling as light, steady drizzling rain of frontal origin. Monthly values are typically around 5 cm. Thunderstorms are rare. By

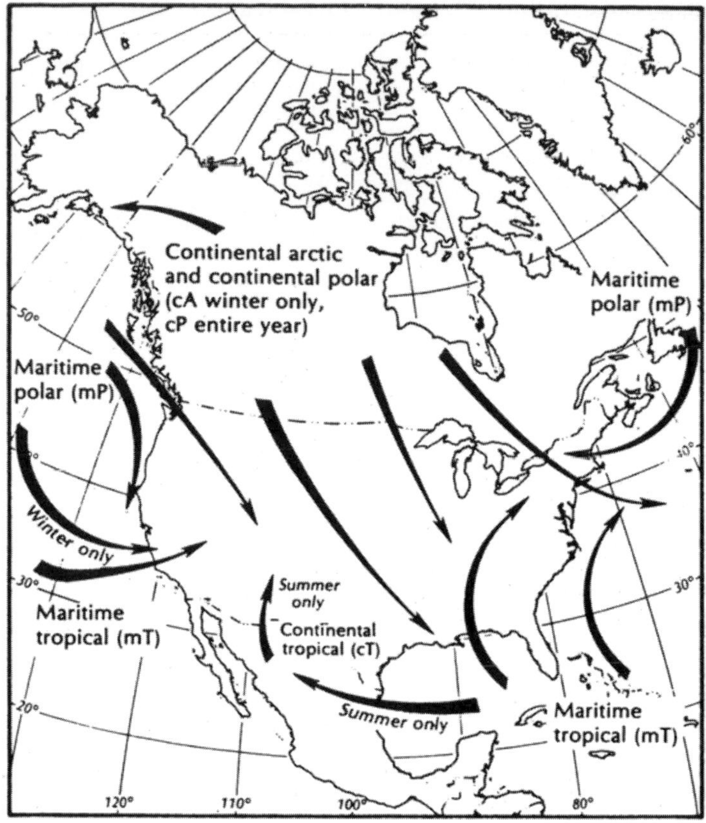

Figure N1 Source regions and paths of airmasses influencing the North American continent.

midsummer, landscapes are released from the grip of frost, streams flow, lakes and ponds are ice-free, and the low-lying plant life engages in the year's brief period of vegetative growth. Very few plant species tolerate the lengthy dormancy of winter and the short, cool, frost-threatened growing season. The assemblage of mosses, lichens, a few flowering herbs and ligneous shrubs that spread laterally close to the surface, plus isolated clumps of hardy spruces, a few meters high in sheltered depressions, are the commonest plant forms making up the tundra vegetation and giving this climate its unique name.

Snow climates (Dfa, Dfb, Dfc)

Along the southern margins of the tundra, treeless barrens give way gradually and unevenly to forest. Southward through the interior of the continent, these forests become more lush and the trees larger, changing in character from the coniferous boreal forest to mixed types, and eventually to subtropical species along the Gulf of Mexico. The largest portion of this forest contains the D or snow climates, those that have one or more months with average temperatures below −3°C. The Kp̈pen system's various modifiers succinctly indicate that all seasons of the year receive significant moisture (the "f" in Dfa), and the length and warmth of summer decrease from south to north (Dfa, Dfb, Dfc). It is this climate that offers the widest annual temperature range as well as the greatest geographic variation in temperature in North America.

Variability is the hallmark of the snow climates. The northern portion of the D climates is in the Arctic, and the southern portion verges on the subtropics. Besides the extreme variation in solar radiation caused by this latitudinal extent, the variability is also caused by the seasonal variation in the location of the polar front and its attendant airmasses and cyclones. During the summer the westerly jet stream retreats to the margin of the Arctic, while in the winter it may frequently reach to the Gulf of Mexico. The resulting storm tracks (Figure N5) are somewhat seasonal, with the Alberta track being occupied largely in summer and the Gulf Coast track in the winter. The Colorado storm track may see activity in all seasons, but cyclone activity as a whole decreases during the summer months.

The length and severity of winter are noteworthy characteristics of the Snow Climates. In the Dfc, (Anchorage) deep cold settles over the land from October through April. January is usually the coldest month when normal temperatures average near −10°C, and thermometer readings often plunge to less than −50°C. In northwest Canada, interior stations have recorded values below −60°C. The high frequency of anticyclonic weather is chiefly responsible for lengthy, intensely cold periods, when the calm, dry, crystal-clear atmosphere settles over the land permitting rapid radiative heat loss to proceed from the surface with exceptional effect.

Farther south (Dfb, Winnipeg), winter extends from November through March. But monthly and daily temperatures may be every bit as extreme as in Dfc, since the relatively flat,

Table N1 Characteristics of North American airmasses

Airmass	Source region	Temperature and moisture characteristics in source region	Stability in source region	Associated weather
cA	Arctic basin and Greenland ice cap	Bitterly cold and very dry in winter	Stable	Cold waves in winter.
cP	Interior Canada and Alaska	Very cold and dry in winter	Stable entire year	Cold waves in winter. Modified to cPk in winter over Great Lakes bringing "lake-effect" snow to leeward shores.
		Cool and dry in summer		
mP	North Pacific	Mild (cool) and humid entire year	Unstable in winter	Low clouds and showers in winter. Heavy orographic precipitation on windward side of western mountains in winter.
			Stable in summer	Low stratus and fog along coast in summer; modified to cP inland.
mP	Northwestern Atlantic	Cold and humid in winter	Unstable in winter	Occasional "northeaster" in winter.
		Cool and humid in summer	Stable in summer	Occasional periods of clear, cool weather in summer.
cT	Northern interior Mexico and southwestern US (summer only)	Hot and dry	Unstable	Hot, dry, and clear, rarely influencing areas outside source region. Occasional drought to southern Great Plains.
mT	Gulf of Mexico, Caribbean Sea, western Atlantic	Warm and humid entire year	Unstable entire year	In winter it usually becomes mTw moving northward and brings occasional widespread precipitation or advection fog. In summer, hot and humid conditions, frequent cumulus development and showers or thunderstorms.
mT	Subtropical Pacific	Warm and humid entire year	Stable entire year	In winter it brings fog, drizzle, and occasional moderate precipitation to NW Mexico and SW United States. In summer it occasionally reaches western US providing moisture for infrequent conventional thunderstorms.

Source: After Lutgens and Tarbuck (1982).

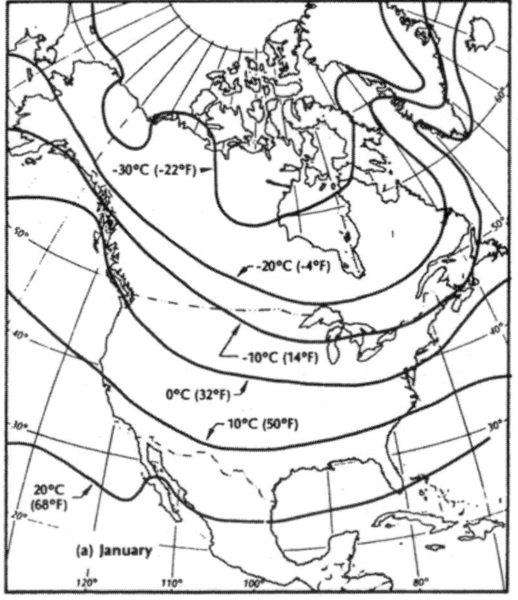

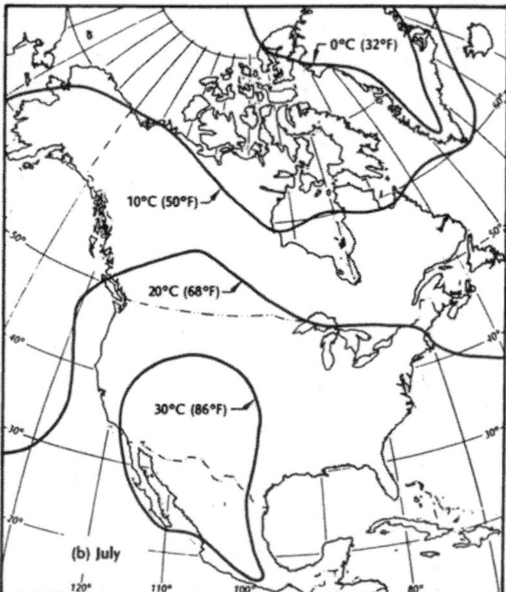

Figure N2 Mean January and July temperatures over North America.

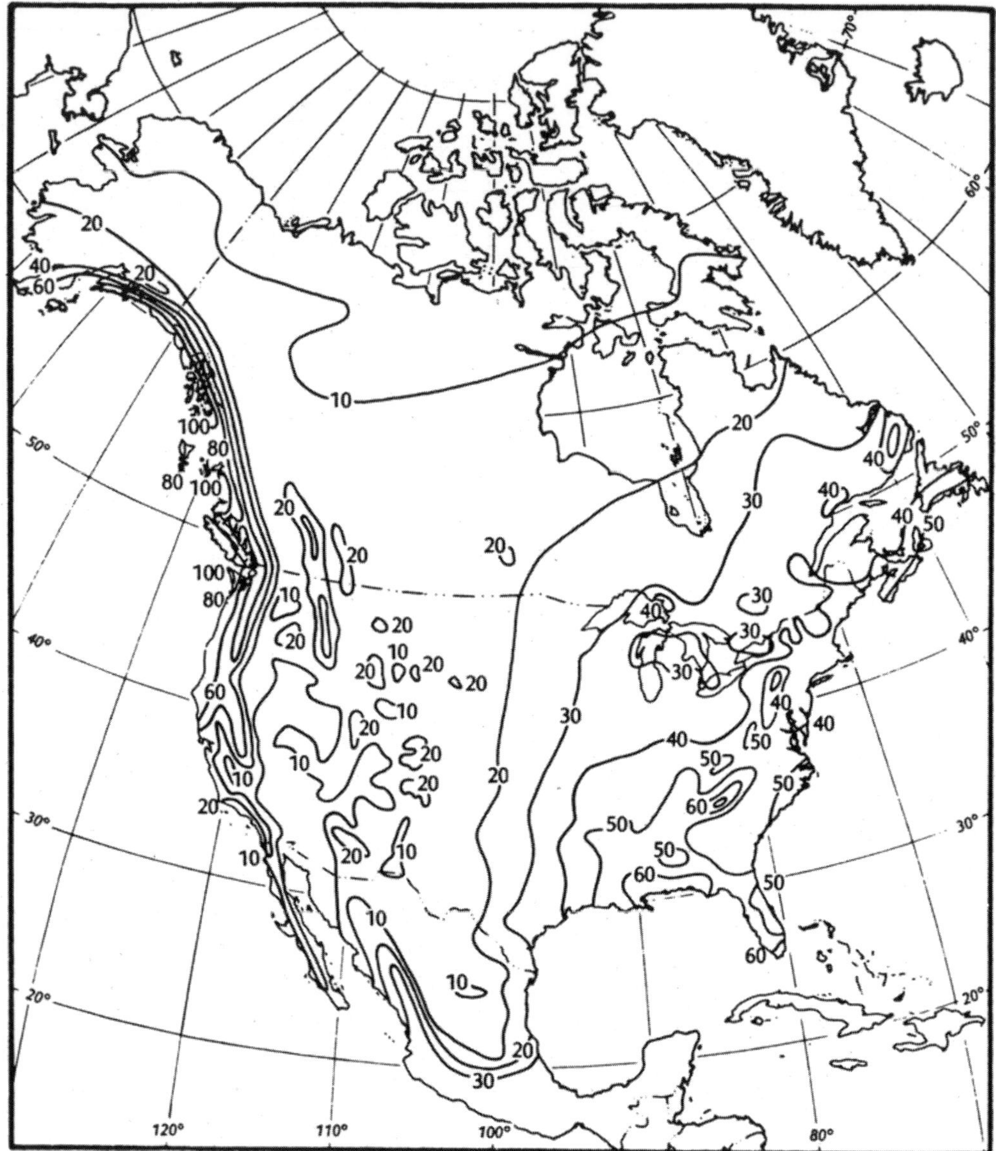

Figure N3 Distribution of precipitation in North America.

snow-covered continental interior offers a breeding ground and easy transit for cold Arctic and polar air masses. By the time these air masses reach the Dfa region, however, they warm considerably, as snow cover is usually less continuous and the sun is more intense. Average winter temperatures in the Dfa region are therefore nearer to 0°C (Springfield), and the cold season persists only from December through February.

Winter is the season of diminished precipitation when most frontal disturbances move rapidly over the surface at about twice the speed of summertime. The seasonal decrease is largely controlled by the expansion of drier air polar and Arctic masses in winter, and it becomes more pronounced as latitude increases. Longitudinal variation also occurs, as winter storms increase in frequency from west to east In the Dfc climate the storms yield a snow that is light, fine, and dry, while along the southern boundary of Dfa, wet, heavy snow combined with rain and sleet is common. Within the Dfc climate the incidence of snowfall rises from about 50 days per year in the Yukon Valley to over 100 days in northern Quebec and Labrador. Mean yearly snowfall increases from around 100 cm in central Alaska to over 500 cm in easternmost Labrador. Around the Great Lakes in the Dfb climate, snowfall values average between 200 and 300 cm, except for lake effect snow which may double or triple that amount for a region within a few miles of the lakes. Snowfall values are between 50 and 100 cm for most of the Dfa region but the geographic and annual variation can be extreme during any particular year.

The retreat of the polar front jet stream to the north in the summer allows maritime tropical airmasses from the Gulf of Mexico and continental tropical airmasses from the southwest

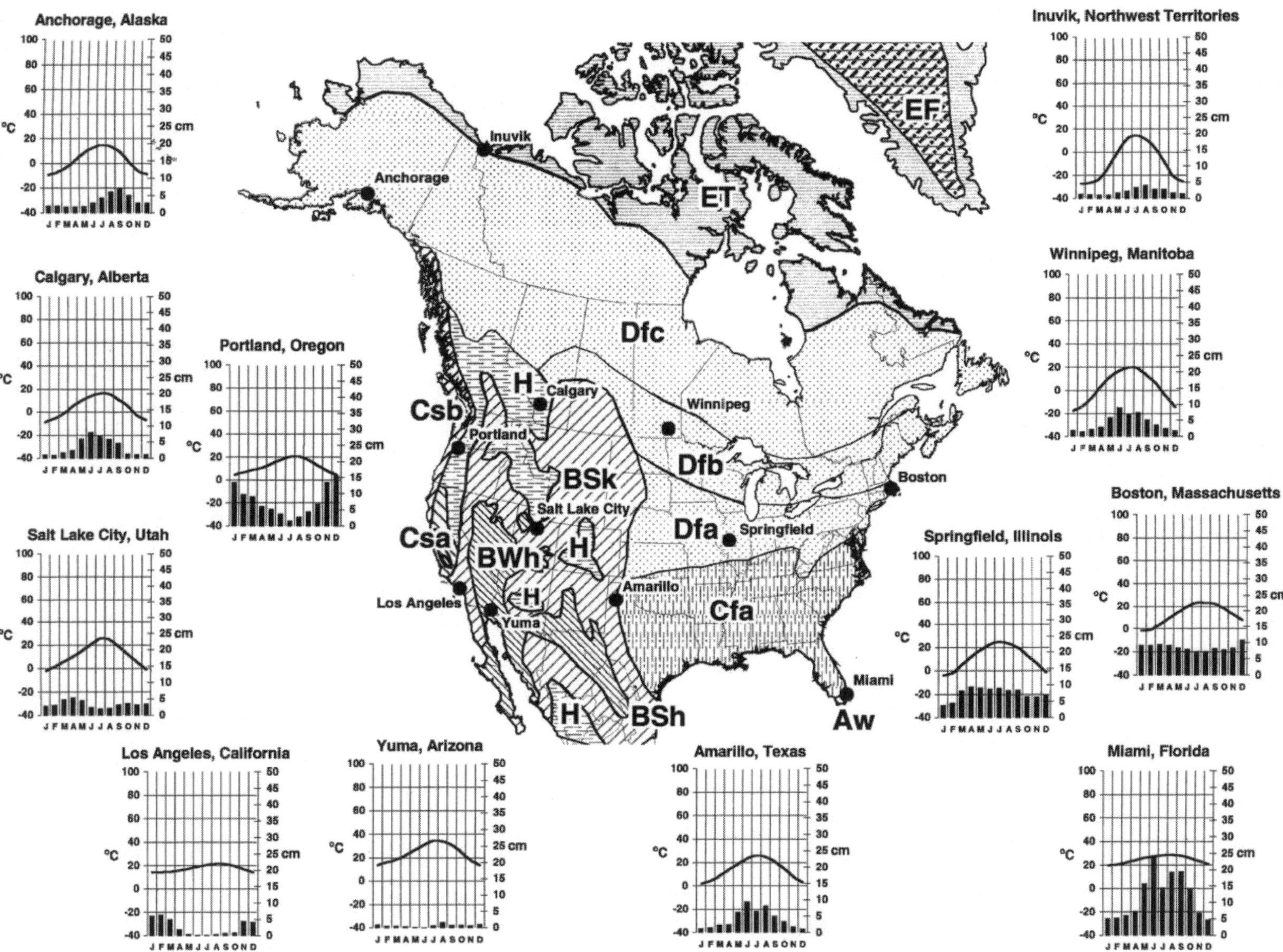

Figure N4 Köppen climate distribution.

Figure N5 Major storm tracks over North America.

to expand into the D climates bringing heat, humidity, and a decrease in extratropical cyclone activity. Summer in the D climates brings a growing season which varies from about 75 days in the northern portion of the Dfc to about 180 days on the southern margin of the Dfa. Although frost may occur sporadically in the north, mean monthly temperatures in the Dfc are above 10°C for 1–3 months. July is commonly the warmest month with average readings above 18°C but never above 22°C. Extreme temperatures in excess of 38°C have been recorded at some inland stations, although prolonged hot spells do not occur. Summers are longer and warmer in the Dfb climates: 4 months have average temperatures over 10°C, although none exceeds 22°C, and prolonged hot spells are common. Long, hot summers are the norm in the Dfa climate. July temperatures average over 22°C, and at least 4 months have average temperatures over 10°C. Daily maximum temperatures over 38°C, sometimes for days at a time, have been recorded in the Dfa climate.

Summer is the season of maximum precipitation when about two-thirds of the normal yearly amounts are received, except along the northeast coast which records a winter maximum (Boston). Most summer precipitation is produced by slow-moving extratropical cyclones and fronts. In the Dfc climate light, steady rains are the norm, but farther south the increased humidity allows fronts to trigger convective showers. Annual thunderstorm frequency varies from 5 to 10 days in the Dfc to as many as 50 days per year in the Dfa. The increase in convective activity also leads to tornadoes, and although the causes and distribution of tornadoes are quite complex, the southwest corner of the Dfa climate in North America occupies the northern portion of "Tornado Alley". Within this portion of the Dfa, annual tornado frequency may reach seven tornadoes per 27 000 km^2. May, June, and April, respectively, are the months of highest tornado frequency.

Significant regional variability in this picture is introduced by the Great Lakes, mountainous terrain and proximity to oceans. The Appalachian mountains in the east and the various ranges in the west all impart thermal and precipitation gradients dependent on altitude and location relative to prevailing winds on a rather fine scale. Similarly, within a few miles of the Great Lakes, thermal and precipitation gradients can be severe, most notably in the form of lake-effect snow in the winter. Location near the ocean generally causes warmer, wetter winters and cooler summers in the D climates, which can be seen by comparing, for example, Winnipeg and Boston, or Anchorage and Inuvik.

Humid subtropical climate (Cfa)

The polar front jet stream frequently inflicts winter storms on the Cfa climate, sometimes bringing rain, sometimes snow, and sometimes a mix of rain, snow, sleet, thunder, lightning, and tornadoes. Arctic airmasses following behind cyclones bring days of temperatures below 0°C, and then give way to warmer maritime tropical airmasses from the Gulf of Mexico ahead of the next winter storm. As the polar front retreats in the spring, the trade winds blowing out of the Bermuda extension of the Azores subtropical high-pressure system, and southerly winds from the Gulf of Mexico, blow warm, humid airmasses over the eastern portion of the region, providing the heat and humidity so typical of this climate. Peak summer temperatures are usually somewhat cooler in the east because the increased presence of water and lush vegetation forces the use of solar radiation for evaporation and transpiration rather than heating Along with the heat and humidity come convective storms, some triggered by extratropical cyclones and fronts, some accompanying tropical cyclones, and some just random, convective storms. Any convective storm may initiate tornadoes, and the Cfa climates bear the distinction of having both the highest frequency of tornadoes recorded in North America and the highest frequency of damage from hurricanes.

The four seasons become much less distinct farther south in all climates, a result mainly of increasing nearness to the equator. Annual possible sunshine increases from about 40% in southeastern Canada to over 65% in Florida: mean daily solar radiation rises from less than 300 Langleys in the north to more than 450 in the south, and the growing season lengthens from under 80 days in northern Canada to more than 350 days along the Gulf Coast. This change is particularly evident within the Cfa climate, as the growing season ranges from 180 days in the north to 350 days in the south. January temperatures during the coldest winter months remain substantially above freezing, and mean monthly temperatures in winter are all above −3°C (Amarillo). Summers are long and hot, with daily temperatures over 38°C common, and for months at a time the daily maximum temperature may reach 32°C. In the western portion of the region (Amarillo), drier continental tropical airmasses may persist for days, bringing temperatures over 45°C.

In the subtropical region snow is expected on fewer than 10 days and does not normally exceed 10 cm. But the gradient is a steep one, as snow is common in the northern portion of the Cfa but a rarity in the southern part. Winter precipitation may increase during El Niñ years as a result of increased high-level moisture advected into the area by the subtropical jet stream. Because of the tendency for El Niñ events to displace the polar front jet stream farther north over much of North America, this precipitation generally falls as rain rather than snow.

Mean annual precipitation increases southward and eastward from about 100 cm in the north to over 160 cm along the coast, largely because of greater atmospheric moisture content the closer proximity to the Gulf of Mexico or Atlantic Ocean. Because of these humid airmasses, summer experiences significantly more precipitation than winter even though winter has the higher frequency of synoptic-scale storms. Thunderstorms account for a much larger percentage of yearly precipitation in this climate, occurring on 50–100 days per year, and reaching a maximum between May and October. This convective precipitation is frequently intense and of a short duration; throughout the region, between 70% and 90% of the annual precipitation falls from storms dropping greater than 12.7 mm of water per day.

June through October is the peak season of tropical cyclone activity which accounts for 15% or more of annual precipitation along the Gulf Coast and the Atlantic seaboard. Many tropical cyclones develop into hurricanes, and the landfall of any one of these intense tropical storms is usually a disaster. In North America the greatest proportion of hurricane landfalls occurs in the Cfa climate.

Tropical climate (Aw)

The southern tip of the Florida peninsula contains the only tropical climate in North America (Miami). The average temperature of the coldest month (January) is over 18°C, and freezing temperatures do not occur, as the intensity of sunlight, and the proximity of the warm ocean waters, keep the area hot all year. Summer daily temperatures are usually in the range of 27–30°C. The summertime maximum in precipitation is the result of frequent and intense airmass thunderstorms. Approximately 90–100 thunderstorms per year bring monthly precipitation values over 15–20 cm, and an annual average precipitation of 190 cm. For much of the summer a daily thunderstorm is the rule. In the winter, though, drier continental polar airmasses spread south, and precipitation drops to 5 cm per month or less for 4 months.

Dry climates – steppe (BSk and BSh)

In the United States and southern Canada the westward decrease in precipitation eventually produces the dry climates, where potential evaporation exceeds precipitation. Potential evaporation is defined as the amount of water that could potentially evaporate, assuming it was available. On the average, then, the dry climates experience a water deficit. Within the dry climates, BS refers to steppe-type climates, where grasses were the originally the dominant vegetation, and BW is desert. The boundary between steppe and desert is where potential evaporation is double the annual precipitation. The modifiers "h" and "k" indicate whether the average annual temperature is above 18°C (h) or below 18°C (k), and their presence on the map generally reflects the north–south temperature gradient across the continent.

The greatest expanse of dry climates is largely the result of three factors: distance from oceans, prevailing winds, and rainshadow of mountains. The prevailing westerly winds across the United States, combined with the rainshadow of the large, western cordillera, creates a large expanse of dry land interfingered between mountain ranges in the west. Dry climates can be found even on the coast of the Gulf of Mexico because of the prevailing westerly winds. On the west coast (Los Angeles), however, the situation is somewhat different, as an area of steppe and desert climate reaches all the way to the Pacific ocean. In this region the proximity of the subtropical high-pressure system, the freshly subsided airmasses, and the cold California current combine to create very stable airmasses in the summer. This situation is similar to the west coast of Africa and South America at the same latitude.

The steppe climates are distinctly continental in character, featuring four well-defined seasons, a large annual temperature range, and a pronounced summer precipitation increase. From 65% to 80% of yearly precipitation occurs in the period April through September, maximum amounts commonly falling early in the growing season. Summer rains are mainly in the form of thermal convective showers of short duration and highly variable distribution. Summer convection is initiated by the intense solar heating, but most of the region is far from a water source, and so the precipitation coverage is spotty. Tornadoes can accompany these storms.

Skies during summer are predominantly cloudless or partly cloudy, and from April through September over 60% of possible sunshine is recorded, with a maximum of 70% in July and August. During July and August thermometer readings above 32°C occur on more than 15 days per month in the central portion of the region, more than 25 days per month in Texas, but fewer than 10 days in the Canadian provinces. Evaporative moisture loss to the atmosphere is notably high in the steppe, especially along the drier western margins, attaining values in excess of 250 cm per year in southern sections and over 140 cm per year in the north. The growing season varies in length from about 240 days in central Texas to fewer than 120 days in the prairie provinces of Canada. July is the warmest month as a rule, averaging above 27°C in the south, diminishing to about 22°C in North Dakota, and to less than 18°C in Alberta.

Winter is dominated by airmasses originating in the higher latitudes, most of which are continental in character and thus are often very cold and very dry. Moisture arrives mainly from the north Pacific, and frontal storms bring most of the season's precipitation. Snow, usually fine, light, and dry and easily drifted, is the chief form, occurring on over 60 days per year in the north and amounting to about 125 cm in depth. Snow is expected on fewer than 10 days per year in southern sections and rarely amounts to more than 25 cm. The cold wave, blizzard, and chinook are also characteristic of winter in the steppe. The cold wave arrives when a very cold, dry airmass moves rapidly southward, causing a temperature drop in 24 hours of more than 12°C to a value of lower than −18°C. When deep cold, dry air moves rapidly into a developing low-pressure system, a blizzard results. This is usually when wind speeds within the frontal system reach between 15 and 20 m s^{-1}. Accumulations of more than 60 cm from a single storm have been reported, with drifts to more than 300 cm, the open plains allowing winds to attain speeds of up to 130 km h^{-1}. Much of the surface is laid bare by the winters' winds and ground frost may reach depths of 250 cm in northern areas.

The chinook provides one of the more dramatic phenomena of winter's weather. It is essentially a downslope wind, variable in strength and persistence, but capable of attaining speeds exceeding 130 km h^{-1}. Moving rapidly down the east slopes of the Rockies, it may bring a temperature rise of more than 20°C in 3 hours and usher in a period of higher than normal temperatures lasting for several days. Through sublimation, snow on the ground may diminish at the rate of 2.5 cm h^{-1}. The chinook may develop at any hour of the day or night throughout the year, but it mainly appears in January and March. January is the coldest month of winter and the chinook offers an ameliorating effect, especially since January temperatures average −20°C or lower in northern sections and may drop to −50°C.

Dry climates – desert (BWh and BWk)

Desert in North America is geographically discontinuous, as mountain ranges disrupt the apparent continuous coverage on Figure N4. Deserts above 1500 m elevation such as in the region known as the Colorado Plateau may be cold desert (BWk). Adjoining this region to the south, and at a somewhat lower elevations, are fringes of the Sonoran Desert and the Chihuahua Desert, both hot deserts (BWh). The detailed distribution of actual desert climates is greatly complicated by the extremely variable topography of the intermountain west which causes a widespread scattering of steppe and the forested highland climates (H).

Clear, cloudless skies from dawn until dusk are the overriding atmospheric condition of the desert climate. In summer most of the western deserts receive more than 80% of all possible sunshine from day to day; in southern California more than 90% is the rule. During the winter months clear skies are less frequent, although even then southern deserts receive more than 70% of possible sunshine. Under clear skies the sun's rays at dawn rapidly heat the bare ground and the overlying air begins to warm at once. Air temperatures rise to a maximum in midafternoon, falling more slowly with the approach of evening. At Yuma, Arizona, on a typical August day, air temperature reached a minimum of 26°C during the calm interval at dawn, but by 1000 hours had risen to 32°C and by 1500 hours reached a maximum of 44°C. The dry soil at the same location registered 64°C. Relative humidity was 8% at noon and remained below 10% during the early afternoon. High temperatures and low humidities after midday are strongly characteristic of desert climates.

Extremes of summer heat are well known in desert situations of low altitude. At Greenland Ranch (el. −54 m) in Death Valley the mean July temperature is 38°C, at Yuma (el. 61 m) it is 35°C and at Brawley (el. −30 m) in the Imperial Valley it is 34°C. Extreme summer temperatures are frequently above 38°C. At Greenland Ranch the thermometer has reached 56.5°C, at Yuma 49°C and at Las Vegas, 47°C. Winter temperatures tend to remain above freezing except at higher elevations.

Mean yearly precipitation is generally less than 25 cm over most North American desert regions. Of far greater importance than the normals, however, are the degree of variability, the kinds of precipitation, and the seasons of their occurrence. In the northern portions, and in southern California, precipitation falls mainly in winter and is produced by cyclonic storms from the Pacific. In the southern deserts of Arizona and New Mexico, snow falls infrequently and in trifling amounts. Summer experiences higher precipitation values in a monsoon-type circulation: the intense desert heat causes widespread convection leading to the development of a thermal low pressure on the synoptic weather map. Humidity, perhaps in the form of clouds, is advected in from the Gulf of California, and the resulting convective showers are the chief rain-generating mechanisms for these deserts. In the areas of summer maximum, departures from normal are over 20%. In winter precipitation regions, variability averages between 25% and 35%. The random scattering of precipitating clouds is a common cause of variability. A notable phenomenon, known as *virga*, is the appearance of showers falling from a cloud base that fail to reach the surface because of the low relative humidity.

Mountain climates (H)

The western reaches of North America are dominated by high mountain ranges and intermontane plateaus and basins that create a climatic pattern of great complexity totally different from the rest of the continent. The Rocky Mountains extend from west Texas northwestward some 4500 km to Alaska. The Pacific mountain system is a series of narrower ranges reaching southward from the Alaska archipelago to include the Cascades of Washington and Oregon and the Sierra Nevada of eastern California. Less prominent coastal ranges overlook the Pacific from Mount Olympus in Washington to the peninsula of Baja California in Mexico. Between the Pacific system and the Rockies are broad tablelands such as the Colorado Plateau and extensive elevated depressions such as the Great Basin of Nevada and Utah.

Among the western mountains, maximum climatic diversity is encountered within short distances. Differences in altitude alone impose pronounced changes in atmospheric properties. With increasing altitude, moisture and dust content diminish, as well as atmospheric density and barometric pressure. The atmosphere's heat capacity also decreases, resulting in lower air temperatures. Along the Front Range in Colorado, for example, the mean annual temperature at Colorado Springs (el. 1860 m) is 8.9°C, whereas at Pike's Peak (el. 4300 m) about 15 km west, it is −7.2°C. As density decreases, there is an increase in the intensity of solar radiation, particularly in the ultraviolet range of the spectrum. Wind speed also increases with altitude as well as the proportion of precipitation that falls as snow.

Perhaps the chief effect on the atmosphere of mountain features is the barrier they interpose on the movement of airstreams over the Earth's surface. Airflow deflected upward is cooled adiabatically and encounters increasingly cooler atmospheric strata in the process. The visible result of these movements is the development of cloud formations along windward mountain slopes that offer one of the more striking features of the skies over mountain terrain. Precipitation is commonly produced by such deflection and is known as orographic rain or snow. Thus precipitation increases with altitude, reaching a maximum at heights that vary with latitude, but in western North America range from 1200 m to 2400 m. Convective motions that generate the cumulus clouds of mountain regions are often intensified and prolonged sufficiently to produce spectacular cumulonimbus forms that yield extremely heavy showers. Such cloudbursts, added to meltwater runoff from higher snow fields, often cause sudden, destructive freshets to rush swiftly down through the lower stream courses with disastrous consequences to life and property.

With increasing altitude, most of the year's precipitation falls as snow. As latitude increases, snow fields are seen at ever lower elevations, approaching sea level in the coastal ranges of southern Alaska. Snow is without doubt one of the more significant products of atmospheric activity in mountain country. With its esthetic value is coupled its usefulness in winter sports and in the annual release of meltwater to mountain streams in spring and summer. Mountain runoff is indeed the chief source of water supply for streams in arid regions of the west. On the other hand, mountain snows can bring sudden death and destruction through rapid accumulation during unusually severe winter storms or through the sudden, swift downward rush of an avalanche. On mountain tops above 3000 m snow has been observed during every month of the year.

The greatest amounts of snow are normally recorded in the coastal mountains of British Columbia, the Cascades, and the Sierra Nevada. In the state of Washington, at levels from 1200 m to 1700 m, normal wintertime snows amount to between 1000 cm and 1500 cm during the period from November to June. In Oregon, amounts average between 760 cm and 1400 cm at levels from 1400 m to 1800 m, and in California mean amounts are around 1000 cm in the central Sierra Nevada between 1800 m and 2400 m. Similarly high values are reported from British Columbia and are several times greater than the normal amounts reported outside the mountainous west. Some exceptional amounts have been measured during a single winter season. At Tamarack, California (el. 2438 m) 1200 cm (over 22 m) fell during the winter of 1906–1907. At the Paradise Ranger station (el. 1676 m) on Mount Rainier in Washington, 2540 cm were recorded in 1953–1956. At Silver Lake (el. 2600 m) about 35 miles northwest of Denver, 220 cm of fresh snow fell in 27½ hours in April of 1921.

Atmospheric turbulence is commonly intensified in mountain regions as a product of the barrier effect on normal air movement. Hazardous turbulence not infrequently occurs in clear, cloudless air over mountain terrain and is most noteworthy over middle and upper slopes. This is partly due to wind speeds increasing with altitude. The levels at which maximum wind speeds are attained vary widely, but one example suggests this common tendency. At the summit of Old Glory Mountain (el. 2347 m), British Columbia, average wind speeds during April have been 26 km h^{-1}, with a maximum for 1 h of 100 km. During the same period at Carmi (el. 1245 m) average wind speed was 8 km h^{-1}, with a maximum for 1 h of 71 km.

West coast climates (Csa, Csb)

From northwestern Mexico to southern Alaska the coastal climates of western North America are uniquely elongated and narrowly confined to within 80 km to 160 km of the sea. Within this narrow region exists a large north–south temperature gradients (although not as great as in the continental interior), and an extreme north–south precipitation gradient unlike any other on the continent. In summer the subtropical high-pressure system immediately offshore expands northward, bringing warm, dry, stable air on to the margins of the continent. As the polar front jet stream moves poleward, frontal systems develop less frequently and are less vigorous, and precipitation is substantially reduced. In winter the anticyclonic subtropical circulation contracts and moves southward, allowing more humid air from the Pacific to encroach. Cyclones and frontal systems develop much more frequently, and are usually larger and more intense. They pursue paths that lie much farther south than in summer and dispense the greater part of the year's precipitation. This is the only part of North America where winter precipitation is such a pronounced climatic characteristic.

The feature common to both of these climates is the precipitation regime of dry summers and wet winters. The dry summer becomes more pronounced farther south (Los Angeles), and less visible farther north (Portland), while the winter becomes wetter farther north. Csa is the classic Mediterranean climate with summers that are long and hot, whereas the companion climate, Csb, has long, cool summers.

In the Csb, cool, damp summers alternate with moderate, cloudy winters with much fog and frequent rain and snow. In summer, mean temperatures for the warmest month are 13°C in Eureka, California,. Extreme maximums have rarely exceeded 30°C. Inland, sheltered stations are somewhat warmer. Portland records a July mean of 19°C, Vancouver 18°C, and Seattle 17°C. In the Puget Sound–Willamette Valley most stations have recorded extreme maximums of more than 38°C. The growing season is about 240 days from Vancouver to Eureka.

Even though summer is the dry season, some rain falls in every month of summer. During the 3 months June through August, normal amounts increase northward, with less than 2.5 cm at Eureka, Portland 6.6 cm, Seattle 7 cm, and Vancouver 14 cm.

The increased frontal activity of winter brings a higher number of depressions to the Gulf of Alaska than anywhere else in the northern hemisphere at any time of the year. December is thus the wettest month at nearly all stations in the region. Although most of the year's precipitation occurs in winter, the actual percentage decreases northward. From October through March, Eureka receives 83%, Portland 77%, and Vancouver 73. Snowfall accounts for an increasing proportion of winter's precipitation. However, at sea-level stations it is usually less than 25 cm. Greater amounts of precipitation fall on higher elevations. In general, drizzling rain is the chief source of annual precipitation, which averages over 200 cm from northern California to southern Alaska. On seaward slopes just a short distance inland, amounts may range from 250 cm to 500 cm. The annual departure from normal averages about 15%.

Winter temperatures are unusually high for the latitudes: Eureka averages 8°C and Seattle about 7°C in January. Occasionally in winter a mass of cold, dry continental air from the deep-frozen interior spills out over the sea, producing very low thermometer readings at low-lying stations. Portland has reported −19°C and Vancouver −17°C.

The Mediterranean climate (Csa) is one of warm to hot, dry summers and mild, wet winters, with an abundance of cloud-free, sunny skies. Mean annual precipitation averages between 40 and 90 cm. An important climatic characteristic is that about 95% of the normal yearly amount falls during the 7-month period from October through April. Most stations record only negligible amounts the rest of the year. In southern areas the rainy season lasts for only 5 months.

Winter precipitation is produced almost entirely by frontal disturbances off the Pacific. Thunderstorms are rare, seldom occurring more than three or four times yearly, and are usually small, weak and of short duration. Winter rains ordinarily begin before the end of October and increase to a maximum for the season in December at most stations, although February is the wettest month around Los Angeles. During El Niñ years the subtropical jet stream may contribute increased moisture and energy to polar front storms and enhance precipitation greatly. Average amounts for December are largest toward the north. Redding, far up the Sacramento Valley, averages 20 cm; the Russian River Valley, north of San Francisco, averages between 20 cm and 23 cm; and Placerville (el. 575 m) in the Sierra Nevada foothills also records 20 cm. Elsewhere, normal December amounts are about 10 cm. From year to year the total amount of winter's precipitation may vary widely around the arithmetic mean, especially during El Niñ years. Red Bluff, in the northern interior, averages 64 cm, but during a recent year received 175 cm, followed a few years later by a total of 26 cm. San Francisco records a normal of 52 cm, but has received as little as 23 cm and as much as 90 cm. Variability rises to more than 40% in southern sections. During the 1997–1998 El Niñ, San Francisco and Los Angeles received over 200% of their normal annual rainfall.

Snow seldom falls in the Mediterranean region, except on the coastal mountains where it is a common occurrence at altitudes above 1200 m. At Sandberg (el. 1377 m) less than 80 km northwest of Los Angeles, winter snowfall averages about 70 cm. High in the Sierra Nevada, however, heavy snows accumulate each winter, creating an annual reserve of water that is released in spring and summer, adding meltwater to the rain-fed streamflow.

Despite the high frequency of precipitating disturbances during winter, the sun shines much of the time. About one-third of the days of each winter month are sunny, dry, and pleasant. Under clear skies, nocturnal temperatures sometimes drop well below freezing. Red Bluff in the north has recorded −8°C in both December and January, Sacramento −6°C in January, and San Diego −2°C. January is without exception the coldest month and mean values are usually well above 2°C in the north and above 4°C elsewhere in the region. In February winter begins to wane as frontal activity diminishes and precipitating storms retreat northward.

Summer weather is controlled for the most part by the expanded oceanic high-pressure fields. Except for the coast, most of southern California is under clear, dry, cloudless skies from May to October. More than 90% of possible sunshine prevails from June through August. Temperatures rise rapidly after sunrise each day, reaching close to 30°C by midafternoon. July is the warmest month, when mean temperatures are largely in the mid to upper 20s Celsius. Red Bluff averages 28°C, Sacramento 24°C, and Bakersfield 29°C. The thermometer rises above 32°C in July on 28 days at Red Bluff, 20 days at Sacramento, and 30 days at Bakersfield. Relative humidity by midafternoon during July drops to low values, usually less than 30%. The average at Red Bluff is 18% and at Sacramento 28%. In the drier southern sections of the central valley, under the desiccating influence of hot, dry winds, called Santa Ana, values often drop to between 5% and 10%. The period May through October accounts for over 75% of annual evaporation, which amounts to about 180 cm in the north and over 230 cm in the south. The growing season at interior locations averages between 240 and 270 days, whereas along the coast it increases to more than 330 days.

The summer weather of coastal California is cooler, more cloudy, and more humid than areas that are sheltered from chilly air off the Pacific, even though many are only a few miles inland. The south-flowing California Current, coupled with upwelling deep water, brings sea surface temperatures of about 15°C past San Francisco, and up to 20°C south of Los Angeles, from June through August. The warmest month is usually either August or September at most seaside stations. San Francisco averages 17°C in September, Santa Cruz 17°C from July through September, and Santa Barbara 19°C in July and August.

Warm air off the open Pacific, 130 km to 160 km offshore, passing over the cool coastal waters, produces frequent low stratus clouds and dense fog. Only about 60% of possible sunshine is recorded at shoreline stations, and most observe more than 60 days per year with dense fog. Cool, humid maritime air penetrates inland through the Golden Gate providing lower temperatures and higher humidities over much of the San Francisco Bay area. A similar effect is seen in the Los Angeles Basin, which is exposed to the sea. In both areas combustion effluents frequently combine with stable, humid air to develop into California's notorious *smog*.

William T. Corcoran and Elias Johnson

Bibliography

Bryson, R.A., and Hare, F.K., (eds.), 1974. *Climates of North America*. New York: Elsevier.

Hare, F.K., 1950. Climate and zonal divisions of the Boreal Forest in eastern Canada. *Geographical Review*, **40**: 615–635.

Harman, J.R., 1991. *Synoptic Climatology of the Westerlies: process and patterns*. Washington, DC: Association of American Geographers.

Houghton, J.G., 1969. *Characteristics of Rainfall in the Great Basin*. Reno, NA: University of Nevada Press.

Kendrew, W.G., 1961. *Climates of the Continents*, 6th edn. Oxford: Clarendon Press.

Kendrew, W.G., and Currie, B.W., 1955. *Climate of Central Canada*. Ottawa: Queen's Printer.

Lutgens, F.K., and Tarbuck, E.J., 1982. *The Atmosphere*, 2nd edn. Englewood Cliffs, N.J.: Prentice Hall.

Oliver, J.E., and Hidore, J., 2002. *Climatology: an atmospheric science*. Upper Saddle River, NJ: Prentice-Hall.

Pyke, C.B., 1971. Some meteorological aspects of the distribution of precipitation in the western United States and Baja California. *Water Resources Contribution*, No. 139. Davis, CA: University of California Desert Research Center.

Rumney, G.R., 1968. *Climatology and the World's Climates*. New York: Macmillan.

Schonher, T., and Nicholson, S.E., 1989. The relationship between California rainfall and ENSO events. *Journal of Climate*, **2**: 1258–1269.

Thomas, M.K., 1964. *Snowfall in Canada*. Toronto: Department of Transport, Meteorological Branch.

Trewartha, G.T., and Horn, L.H., 1980. *An Introduction to Climate*, 5th edn. New York: McGraw-Hill.

US Weather Bureau, 1968. *Climatic Atlas of the United States*. Washington, DC: Environmental Sciences Services Administration.

Visher, S.S., 1954. *Climatic Atlas of the United States*. Cambridge, MA: Harvard University Press.

Cross-references

Airmass Climatology
Atmospheric Circulation, Global
Centers of Action
Climate Zones
Continental Climate and Continentality
Dynamic Climatology

Local Climatology
Mediterranean Climates
Middle Latitude Climates
Taiga Climate
Tundra Climate

NORTH AMERICAN (CANADIAN) HIGH

The North American High consists of one or two semipermanent cells of high pressure that develop over the interior of the continent during winter. The Siberian (Asiatic) High represents an analogous situation in the northeastern interior of Asia.

The influence of the North American High is rarely felt above 2500 m (8200 ft), as it is made up of cold, dense air. At 3000 m (9800 ft) the high loses its identity (Petterssen, 1969).

The weaker thermal influence of the smaller North American landmass (24 063 000 km^2; 9 291 000 miles^2) as contrasted to the much larger landmass of Asia (44 134 000 km^2; 17 040 000 miles^2) results in a weaker and divided anticyclone. The southern center, the Plateau High of the western United States, represents a moderate poleward displacement of the subtropical high-pressure belt. The northern center, the Canadian or Alaskan High, is primarily caused by continental cooling.

As shown in Figure N6, the Canadian High has a January mean sea-level pressure of 1020 mb (30.12 in) in contrast to the mean pressure of 1035 mb (30.56 in) of the Siberian High. The Canadian High is located just to the east of the Rocky Mountains, where it is most protected from heat transfer from the Pacific. The highs generally move southeastward and then eastward toward the Atlantic coast. Coming over the warm Atlantic, the high loses its identity and becomes absorbed in the Azores High. The Canadian High also tends to build a bridge between the polar high and subtropical belt of high pressure. Continental cooling is so strong that the subpolar low-presure belt, which would occur on a uniform Earth, is almost entirely eliminated (Petterssen, 1969).

During periods of strong circulation the Plateau (or Great Basin) High that is located over the southern portion of the Rocky Mountain states increases in size. A rapid inflow of relatively mild Pacific airmasses occurs north of this high-pressure area, and rapid air movements eastward prevail over the northern United States. Thus, maritime Pacific air from the southwest characterizes weather conditions during the time of strong circulation. The Pacific coast can then expect to receive some precipitation.

The Canadian High develops and the Plateau High tends to disappear as the zonal circulation of middle latitudes weakens and reaches a minimum value. The strong continental anticyclone extends a wedge of high pressure southward into the United States. There is now little air motion in the surface layers from west to east across North America. Alaska and British Columbia may now experience an outflow of cold, dry continental air from the east.

Continental heating during the summer causes North America to become an area of relatively low pressure. Figure N7 clearly shows a small low-pressure center with a July mean of 1011 mb (29.86 in) in the hot desert regions of southern California, Arizona, and northern Mexico. Note how the low breaks the subtropical high-pressure belt existing over the oceans.

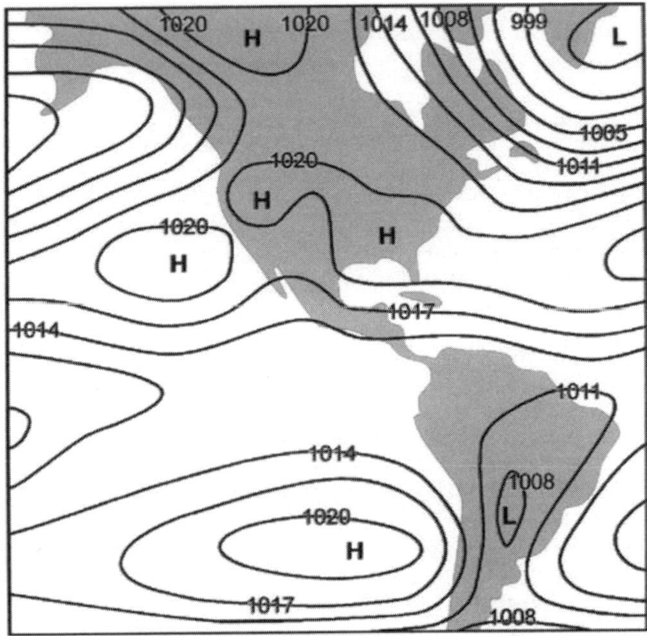

Figure N6 Mean January sea-level pressure (mb).

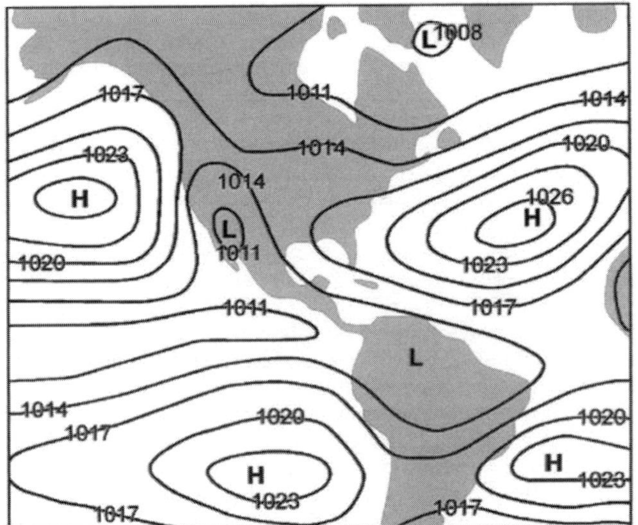

Figure N7 Mean July sea-level pressure (mb).

Robert M. Hordon

Bibliography

Ahrens, C.D., 2003. *Meteorology Today: An Introduction to Weather, Climate, and the Environment*, 7th edn. Brooks/Cole (Thomson Learning).

Akin, W.E., 1991. *Global Patterns: climate, vegetation, and soils*. Norman, OK: University of Oklahoma Press.
Barry, R.G., and Chorley, R.J., 2003. *Atmosphere, Weather, and Climate*, 8th edn. London: Routledge.
Curry, J.A., and Webster, P.J., 1999. *Thermodynamics of Atmospheres and Oceans*. San Diego, CA: Academic Press.
Graedel, T.E., and Crutzen, P.J., 1993. *Atmospheric Change: an earth system perspective*. New York: Freeman.
Hartmann, D.L., 1994. *Global Physical Climatology*. San Diego, CA: Academic Press.
Lutgens, F.K., and Tarbuck, E.J., 2004. *The Atmosphere: An Introduction to Meteorology*, 9th edn. Upper Saddle River, New Jersey: Pearson Prentice Hall.
Nese, J.M., and Grenci, L.M., 1998. *A World of Weather: fundamentals of meteorology*, 2nd edn. Dubuque, IO: Kendall/Hunt.
Oliver, J.E., and Hidore, J.J., 2002. *Climatology: An Atmospheric Science*, 2nd edn. Upper Saddle River, New Jersey: Pearson Prentice Hall.
Petterssen, S., 1969. *Introduction to Meteorology*. 3rd edn. New York: McGraw-Hill.
Robinson, P.J., and Henderson-Sellers, A., 1999. *Contemporary Climatology*, 2nd edn. Harlow: Longman.
Wallace, J.M., and Hobbs, P.V., 1997. *Atmospheric Science: an introductory survey*. San Diego, CA: Academic Press.

Cross-references

Airmass Climatology
Atmospheric Circulation, Global
Centers of Action
Continental Climate and Continentality
North America, Climate of
Siberian (Asiatic) High
Zonal Index

NORTH ATLANTIC OSCILLATION

Introduction

Simultaneous variations in weather and climate over widely separated points on Earth have long been noted in the meteorological literature. Such variations are commonly referred to as "teleconnections". In the extratropics, teleconnections link neighboring regions mainly through the transient behavior of atmospheric planetary-scale waves. Consequently, some regions may be cooler than average, while thousands of kilometers away warmer conditions prevail. Though the precise nature and shape of these structures vary to some extent according to the statistical methodology and the dataset employed in the analysis, consistent regional characteristics that identify the most conspicuous patterns emerge.

Over the middle and high latitudes of the Northern Hemisphere (NH), many teleconnection patterns have been identified, although there are only a few that are truly independent (Kerr, 2004). One of the most prominent is the North Atlantic Oscillation (NAO), which refers to changes in the atmospheric sea-level pressure difference between the Arctic and the subtropical Atlantic (Hurrell et al., 2003). Although it is the only teleconnection pattern evident throughout the year in the northern hemisphere, the climate anomalies associated with the NAO are largest during the boreal winter months (November–April) when the atmosphere is dynamically the most active.

A time-series (or index) of nearly 150 years of wintertime NAO variability and the spatial pattern of the oscillation are shown in Figures N8 and N9. In the positive index phase, higher than normal surface pressures south of 55°N combine with a broad region of anomalously low pressure throughout the Arctic. Because air flows counterclockwise around low pressure and clockwise around high pressure in the northern hemisphere, this phase of the oscillation is associated with stronger-than-average westerly winds across the middle latitudes of the Atlantic onto Europe, with anomalous southerly flow over the eastern United States and anomalous northerly flow across western Greenland, the Canadian Arctic, and the Mediterranean.

Impacts of the NAO

Swings in the NAO produce changes in wind speed and direction over the Atlantic that significantly alter the transport of heat and moisture. During positive NAO index winters, enhanced westerly flow across the North Atlantic moves relatively warm and moist maritime air over much of Europe and far downstream across Asia, while stronger northerlies carry cold air southward and decrease land and sea surface temperatures over the northwest Atlantic (Figure N10). Temperature variations over North Africa and the Middle East (cooling), as well as North America (warming), associated with the stronger clockwise flow around the subtropical Atlantic high-pressure center are also notable.

This pattern of temperature change is important. Because the heat storage capacity of the ocean is much greater than that of land, changes in continental surface temperatures are much larger than those over the oceans, so they tend to dominate average northern hemisphere (and global) temperature variability. Given especially the large and coherent NAO signal across the Eurasian continent from the Atlantic to the Pacific, it is not surprising that NAO variability accounts for about one-third of the year-to-year changes in average northern hemisphere winter surface temperature. Moreover, the long-term changes in atmospheric circulation associated with the upward trend in the NAO index have contributed substantially to the winter warming of the northern hemisphere in recent decades (Thompson et al., 2000). Changes in the mean circulation patterns over the North Atlantic associated with the NAO are also accompanied by changes in the intensity and paths of storms (atmospheric disturbances operating on time scales of about a week or less). During boreal winter a well-defined storm track connects the North Pacific and North Atlantic basins, with maximum storm activity over the oceans. Positive NAO index winters are associated with a northeastward shift in the Atlantic storm activity, with enhanced action from southern Greenland across Iceland into northern Europe and a modest decrease in activity from the Azores across the Iberian Peninsula and the Mediterranean. Positive NAO winters are also typified by more intense and frequent storms in the vicinity of Iceland and the Norwegian Sea (Hurrell et al., 2003).

The ocean realizes the effects of storms in the form of surface waves, so that it exhibits a marked response to long-lasting shifts in the storm climate. The recent upward trend toward more positive NAO index winters, for example, has been associated with increased wave heights over the northeast Atlantic and decreased wave heights south of 40°N (Kushnir et al., 1997). Such changes have consequences for the operation and safety of shipping, offshore industries, and coastal development.

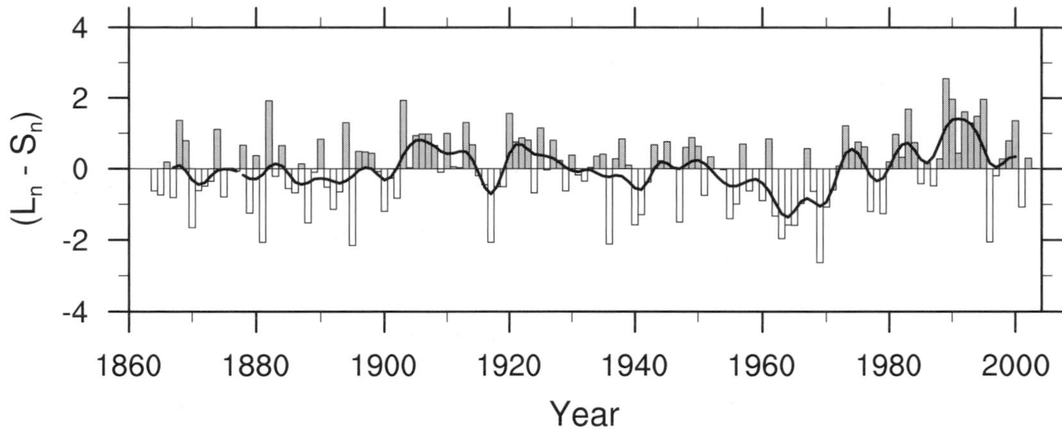

Figure N8 Index of the boreal winter (December–March) mean NAO constructed as the difference in sea-level pressure between Lisbon, Portugal and Stykkisholmur/Reykjavik, Iceland from 1864 through 2003. The mean winter sea-level pressure data at each station were normalized by division of each seasonal pressure by the long-term mean (1864–1983) standard deviation. The heavy solid line represents the index smoothed to remove fluctuations with periods less than 4 years. The indicated year corresponds to the January of the winter season (e.g. 1990 is the winter of 1989/1990). See http://www.cgd.ucar.edu/~jhurrell/nao.html for updated time-series.

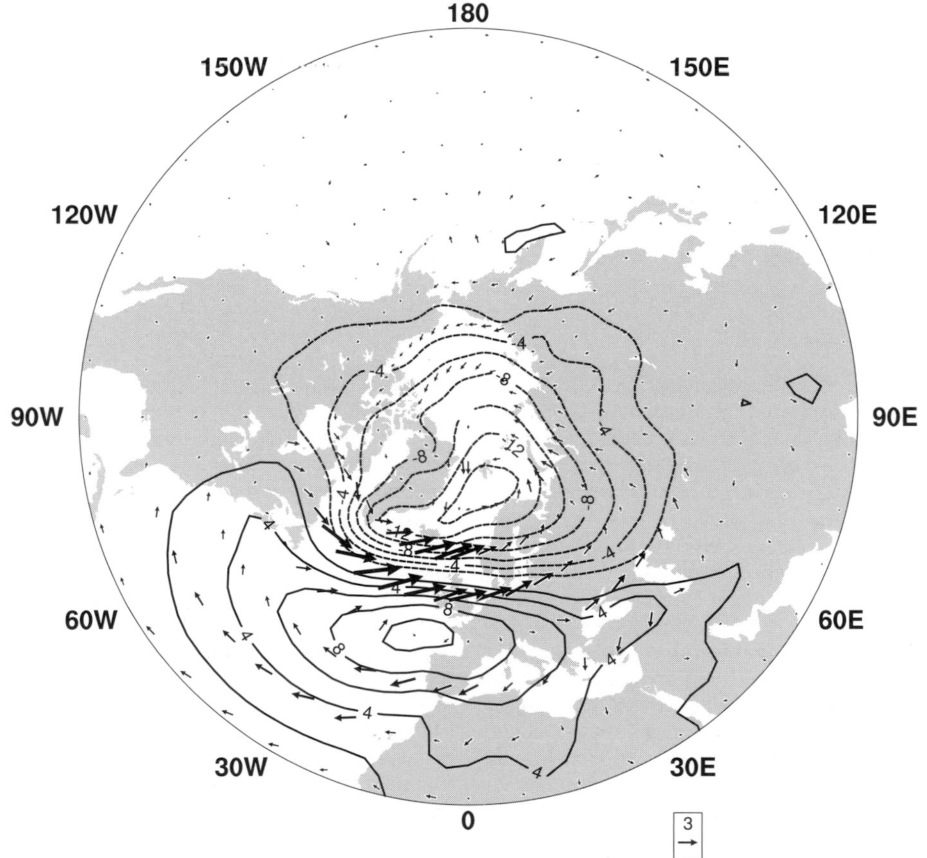

Figure N9 The difference in boreal winter (December–March) mean sea-level pressure and surface vector winds between positive (hi) and negative (lo) index phases of the NAO. The composites are constructed from data when the magnitude of the NAO index exceeds one standard deviation. Nine winters are included in each composite. The contour increment for sea-level pressure is 2 hPa; negative values are indicated by the dashed contours, and the zero contour has been excluded. The scaling vector is 3 m s^{-1}.

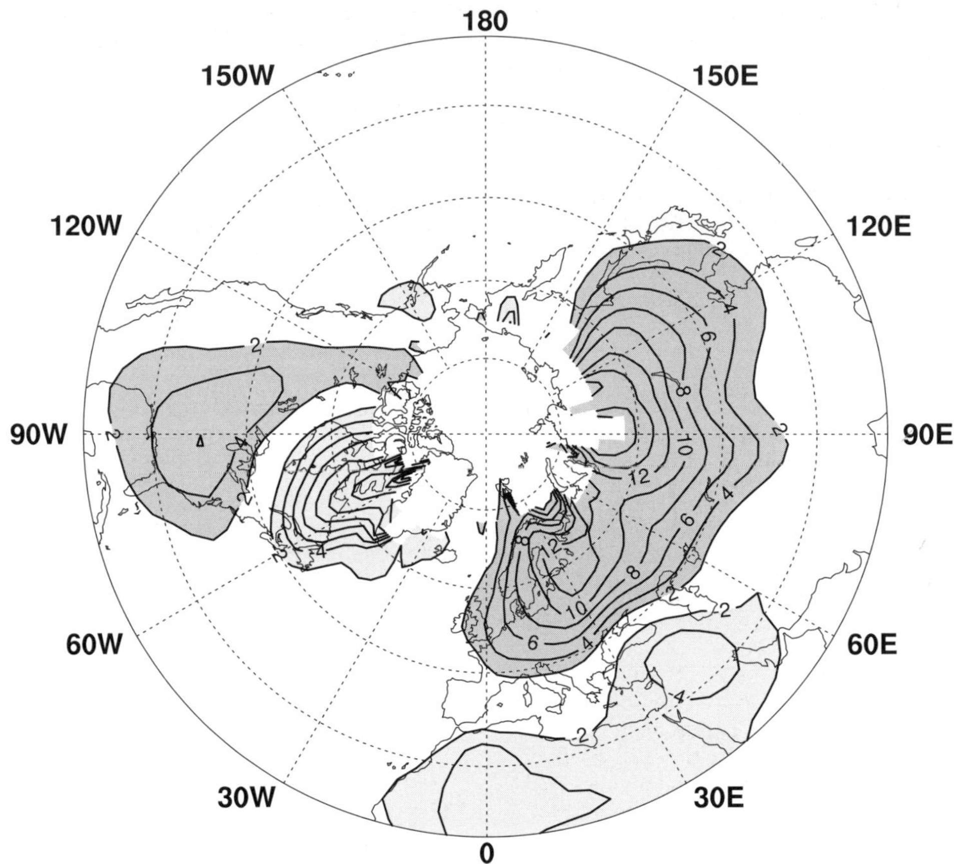

Figure N10 Changes in mean winter (December–March) land surface and sea surface temperatures ($\times 10^{-1}$°C) corresponding to a unit deviation of the NAO index over 1900–2003. The contour increment is 0.2°C. Temperature changes >0.2°C are indicated by dark shading, and those <−0.2°C are indicated by light shading. Regions of insufficient data (e.g. over much of the Arctic) are not contoured, and the zero contour has been excluded.

Changes in the mean flow and storminess associated with swings in the NAO are also reflected in pronounced changes in the transport and convergence of atmospheric moisture (Hurrell et al., 2003). Winters tend to be drier than average over much of Greenland, the Canadian Arctic, much of central and southern Europe, the Mediterranean and parts of the Middle East during positive NAO index winters, whereas more precipitation than normal falls from Iceland through Scandinavia. This pattern, together with upward trend in the NAO index since the late 1960s, is consistent with recent observed changes in precipitation over much of the Atlantic basin. One of the few regions of the world where glaciers have not exhibited a retreat over the past several decades is in Scandinavia, where more than average amounts of precipitation have been typical of many winters since the early 1980s. In contrast, over the Alps, snow depth and duration in recent winters have been among the lowest recorded, and the retreat of Alpine glaciers has been widespread.

In addition to these impacts, significant changes in ocean surface temperature and heat content, ocean currents and their related heat transport, and sea ice cover in the Arctic and sub-Arctic regions are also induced by changes in the NAO. For example, the oceanic response includes changes in the distribution and intensity of winter convective activity in the North Atlantic. The convective renewal of intermediate and deep waters in the Labrador Sea and the Greenland–Iceland–Norwegian Seas contribute significantly to the production and export of North Atlantic Deep Water and, thus, helps to drive the global thermohaline circulation. The intensity of winter convection at these sites is characterized not only by large interannual variability, but also interdecadal variations that are synchronized with variations in the NAO (Dickson et al., 1996). Likewise, the strongest interannual variability of Arctic sea ice occurs in the North Atlantic sector and is primarily driven by changes in the NAO. A seesaw in ice extent between the Labrador and Greenland Seas characterizes the ice variations (Deser et al., 2000).

Mechanisms

Considering the significant impact the NAO exerts on the climate (Hurrell et al., 2003), the economy (Visbeck et al., 2002) and ecosystems (Stenseth et al., 2002) of the northern hemisphere, understanding the mechanisms that determine its structure and variability in time is of central importance. It is well known that the basic structure of the NAO results from processes internal to the atmosphere (Hurrell et al., 2003). As such, the month-to-month and even year-to-year changes in the phase and amplitude of the NAO are largely unpredictable. But that forces external to the atmosphere might nudge the NAO over a particular month or season is important: even a small amount of predictability could be useful, and a better understanding of how the NAO responds to external forcing is crucial to the current debate on climate variability and change.

A number of different mechanisms that could influence the detailed state of the NAO have been proposed. Within the atmosphere itself, changes in the rate and location of tropical rainfall have been shown to influence the atmospheric circulation over the North Atlantic and, in particular, the NAO. Tropical convection is sensitive to the underlying distribution of ocean surface temperature, which exhibits much more persistence than sea surface temperature variability in middle latitudes. This might lead, therefore, to some predictability of the NAO phenomenon.

Interactions with the lower stratosphere are a second possibility. This mechanism is of interest because it might also explain how changes in atmospheric composition influence the NAO. For example, changes in ozone, greenhouse gas concentrations and/or levels of solar output affect the radiative balance of the stratosphere that, in turn, modulates the strength of the winter polar vortex in the lower stratosphere. Given the relatively long time-scales of stratospheric circulation variability (anomalies persist for weeks), dynamic coupling between the stratosphere and the troposphere could yield a useful level of predictive skill for the wintertime NAO.

A third possibility is that the state of the NAO is influenced by variations in heat exchange between the atmosphere and the ocean, sea-ice and/or land systems. A significant amount of numerical experimentation has been done to test the influence of tropical and extratropical sea surface temperature anomalies on the NAO, and these experiments are beginning to lead to more conclusive and coherent results.

James W. Hurrell

Acknowledgments

The National Center for Atmospheric Research is sponsored by the National Science Foundation. The author thanks Dr. Clara Deser for reviewing the item, and Adam Phillips, who produced the figures and assisted in preparing the copy.

Bibliography

Deser, C., Walsh, J.E., and Timlin, M.S., 2000. Arctic sea ice variability in the context of recent atmospheric circulation trends. *Journal of Climate*, **13**: 617–633.

Dickson, R.R., Lazier, J., Meincke, J., Rhines, P., and Swift, J., 1996. Long-term coordinated changes in the convective activity of the North Atlantic. *Progress in Oceanography*, **38**: 241–295.

Hurrell, J.W., Kushnir, Y., Ottersen, G., and Visbeck, M., 2003. *The North Atlantic Oscillation: climatic significance and environmental impact.* J. W. Hurrell, Y. Kushnir, G. Ottersen, and M. Visbeck, Eds. *Geophysical Monograph*, **134**. Washington, DC: American Geophysical Union.

Kerr, R.A., 2004. A few good climate shifters. *Science*, **306**: 599–601.

Kushnir, Y., Cardone, V.J., Greenwood, J.G., and Cane, M., 1997. On the recent increase in North Atlantic wave heights. *Journal of Climate*, **10**: 2107–2113.

Stenseth, N.C., Mysterud, A., Ottersen, G., Hurrell, J.W., Chan, K.-S., and Lima, M., 2002. Ecological effects of large-scale climate fluctuations. *Science*, **297**: 1292–1296.

Thompson, D.W.J., Wallace, J.M., and Hegerl, G.C., 2000. Annular modes in the extratropical circulation, Part II: Trends. *Journal of Climate*, **13**: 1018–1036.

Visbeck, M., Cherry, J.E., and Cullen, H.M., 2002. The North Atlantic Oscillation and energy markets in Scandinavia. *Climate Report*, **3(4)**: 2–8.

Cross-references

Arctic Climates
Global Environmental Change: Impacts
Atmospheric Circulation, Global
Ozone
Sea Ice and Climate
Teleconnections
Volcanic Eruptions and their Impact on the Earth's Climate

O

OCEAN–ATMOSPHERE INTERACTION

The oceans form 70.8% of the atmosphere's lower surface, and are therefore of major significance in structuring both weather and climate over the globe. The actual circulation of the ocean and the interactions between sea-ice and the atmosphere are largely dealt with elsewhere, although feedbacks between the ocean and atmospheric circulation will figure below. The emphasis here is on the ocean as a boundary, permeable to a number of climatically important gases but directly affecting the atmosphere thermodynamically through its sea surface temperature (SST).

Air–sea exchanges

The ocean exchanges heat, moisture, momentum and various gases and particles with the atmosphere. Many of these have an impact on weather and climate and we will return to consideration of the climatically important gas transfers later. However, the predominant exchange of climatic importance is of heat. The mean exchange of heat helps determine the global climate; anomalies in this exchange lead to changes in weather patterns through to longer-term climatic changes.

Heat is exchanged across the air–sea interface predominantly through four physical processes. Solar radiation enters the ocean and is rapidly absorbed below the surface. The ocean radiates back longer wavelength infrared energy; 60–80% of this is returned to the ocean via the greenhouse effect. Turbulent exchange of sensible heat occurs in the planetary boundary layer. This is proportional to the air–sea temperature gradient and the wind speed, with atmospheric instability amplifying the constant of proportionality (the Dalton number). Lastly, heat is (normally) lost to the atmosphere through evaporation and latent heat transfer. This is again proportional to the wind speed, though in a nonlinear way at higher wind speeds, and the specific humidity gradient in the boundary layer.

Typically, the latent heat flux is the largest and most variable heat transfer term, being particularly pronounced above western boundary currents in winter where, even in a monthly average, latent heat transfer to the atmosphere can be several hundred $\mathrm{W\,m^{-2}}$. As will be seen below, this has important implications for both mean and anomalous climates, through feedback between storm tracks and upper ocean circulation. A fuller discussion can be found in Bigg (2003).

The ocean and atmosphere also exchange water, with a global mean excess of evaporation over precipitation over the oceans of $36 \times 10^{15}\,\mathrm{kg\,yr^{-1}}$ ($\sim 100\,\mathrm{mm\,yr^{-1}}$). This provides approximately a third of the precipitation that falls over land. Regionally, however, depending on the balance between dry, windy conditions and moist, cool weather, there may be an oceanic excess of either evaporation or precipitation. In polar regions the oceanic freshwater budget is complicated by the impact of sea-ice and iceberg transport, and subsequent melting. While under normal circumstances these can be neglected for atmospheric studies, we will see later that exceptional, but real, anomalies in iceberg fluxes can lead to extreme climate change.

The ocean and atmosphere also exchange climatically important gases such as carbon dioxide (CO_2), dimethyl sulfide (DMS), hydrogen sulfide (H_2S), nitrous oxide (N_2O), methane (CH_4), carbon monoxide (CO) and methyl iodide (CH_3I). Many of these play a role in the greenhouse effect, and thus influence longer-term climate change, while some (DMS, H_2S and CH_3I) lead to the atmospheric formation of radiatively active particles and cloud condensation nuclei through oxidation. The sea is also an important source of sea salt particles, a major natural seeding source for the coalescence process within clouds, and smaller organic particles.

Monitoring

Sea surface data used in preparing weather forecasting analyses, climate forecasts or merely for climatological studies are obtained in a number of ways. For many years the primary source was merchant ships providing SST through bucket (largely pre-1940) or engine intake measurement, with a small number of fixed-location weather ships taking more extensive measurements. Since the mid-1970s these have been increasingly supplemented by expendable bathythermographs (XBT) for upper ocean temperature profiles, satellite advanced very high resolution radiometers (AVHRR) for SST, microwave scatterometers for surface winds and waves, and floating and pop-up buoys. Thus over the last 30 years data availability has gone from being heavily biased to shipping routes (and hence the mid-latitude North Atlantic and North Pacific) to global.

There are still problems with more recent datasets. For example cloud cover restricts AVHRR SST temporally and spatially, making polar latitudes very hard to observe. Buoy locations are determined by ocean currents. Scatterometer wind directions are not always uniquely determined. Nevertheless, we now have a much better idea of both mean surface ocean conditions and their temporal variability.

Mean climatic effects of the ocean

The ocean's supply of heat to the atmosphere, and its large heat capacity acting as a thermal moderator to the more variable atmosphere above it, has a significant impact on the geographical pattern of the climate. Figure O1, showing the mean SST for February and August, is a guide to these links.

The basic SST pattern in both seasons and hemispheres is predominantly zonal, as is the atmospheric temperature variation. This is as expected from the basic need for the planet to redistribute heat from the tropics to the poles to maintain thermal equilibrium. However, there are several important exceptions to this zonality. Firstly, in summer (northern hemisphere August and southern hemisphere February), the SST tends to be colder than the nearby land temperature, even where the SST pattern is largely zonal. For example the mean February air temperature in inland subtropical Australia is close to 30°C, yet the adjacent SST is in the low 20s. The opposite is true in winter (northern hemisphere February and southern hemisphere August), when SST tends to be warmer than the nearby land. Thus, Las Vegas in Nevada has a February air temperature just below 10°C while the SST in both the Atlantic and Pacific at a similar latitude is closer to 15°C. The high heat capacity of the ocean, and restriction of heat exchange to its surface only, means that the ocean has a strong moderating effect on climate. Climates of coastal locations are therefore significantly milder than in continental interiors. Note, however, that in regions adjacent to winter or permanent sea-ice the ameliorating effect of the ocean is lost.

Along the western boundaries of midlatitude and subpolar oceans winter temperatures are anomalously warm for their latitude, due to the presence of intense, but narrow, western boundary currents in the upper ocean. These very efficiently transport heat poleward. Indeed, toward the subtropics the ocean provides over half of the total meridional heat transport needed to balance the planetary heat budget (Bryden et al., 1991). These poleward oceanic heat transports are evident in each major ocean basin in Figure O1, but especially in the North Atlantic and North Pacific in winter. The Northeast Atlantic, stretching into the Norwegian Sea, shows a particularly striking signature of warm SST associated with the North Atlantic Drift.

The ocean can provide a net cooling of the atmosphere. Along eastern coasts of the oceans in the midlatitude and subtropics, and to some extent along the equator, cooler than expected SST can be seen in Figure O1. These are caused by upwelling, where equatorward alongshore winds (or easterly along-equator winds) cause water to diverge from the coast (or equator), allowing cooler, subsurface water to upwell. This is most noticeable along the eastern coasts of the Pacific and Atlantic, and in the equatorial Pacific, where the atmosphere above is cooled.

These land–sea thermal contrasts through the seasons have an impact on mean atmospheric pressure fields. Thus, the subtropical high-pressure zones of the North Pacific, South Pacific and South Atlantic are several hPa stronger during the summer than in the winter, due to the relatively cool water at these subtropical latitudes relative to the nearby land. The presence of the warm western boundary currents tends to displace all the subtropical highs to the eastern part of their respective basins. Another direct impact of the SST on atmospheric circulation is that tropical cyclones, or hurricanes, are only formed over off-equatorial waters warmer than about 27°C.

Anomalies in SST are directly linked to both seasonal and inter-annual climate variability so the next four sections will briefly discuss several such climate signals.

Long-range forecasting

For some time it has been known that SST anomalies are related to seasonal-scale climate variability, not just locally but possibly hundreds to thousands of kilometers downstream (Ratcliffe and Murray, 1970). The strongest links between SST and seasonal weather are found in the tropics, and there has been some success in forecasting the onset and strength of El Niñ events since the mid-1980s (see next section). However, there is evidence that similar, if weaker, links are present in other parts of the globe. Statistical techniques were first used to exploit these teleconnections and still remain a powerful tool, particularly in tropical regions. More recently, dynamical models of varying complexity up to full-scale coupled ocean–atmosphere–sea ice climate models have been used, increasingly on a routine basis, to forecast both tropical and midlatitude seasonal trends.

Seasonal forecasts have a rather different appearance compared with more familiar daily weather forecasts. As the link between weather and SST is manifested in long-term weather averages, then forecasts are given for conditions averaged over 3-month periods and are often stated in terms of geographical maps of probability. As initial conditions are uncertain, an ensemble of model runs is produced. This allows probabilities to be obtained. Thus, to use an example quoted on the Meteorological Office web-site (http://www.metoffice.com/research/seasonal/background.html) for seasonal forecasting over Europe, if 19 of 27 forecasts indicate above normal seasonal temperature over Europe, and eight of the forecasts indicate below-normal temperature, the chance of above-normal temperature is 70%, while the chance of below-normal temperature is 30%.

El Niño

The see-saw in pressure across the tropical Pacific between the Indo-Australian Convergence Zone and the southeastern Pacific, known as the Southern Oscillation (Figure O2), is the largest and most geographically widespread annual scale climatic interaction between the ocean and atmosphere. In the mean, surface pressure is low over the western Pacific and high in the southeast Pacific (the Walker circulation). When these "normal" conditions are accentuated then there is often a La Niñ climate anomaly, but when the pressure gradient across the Pacific is reduced there tends to be an El Niñ climate anomaly. Some relaxation of east–west pressure gradient occurs annually late in the calendar year, causing warmer than normal water to appear in the equatorial eastern Pacific, as

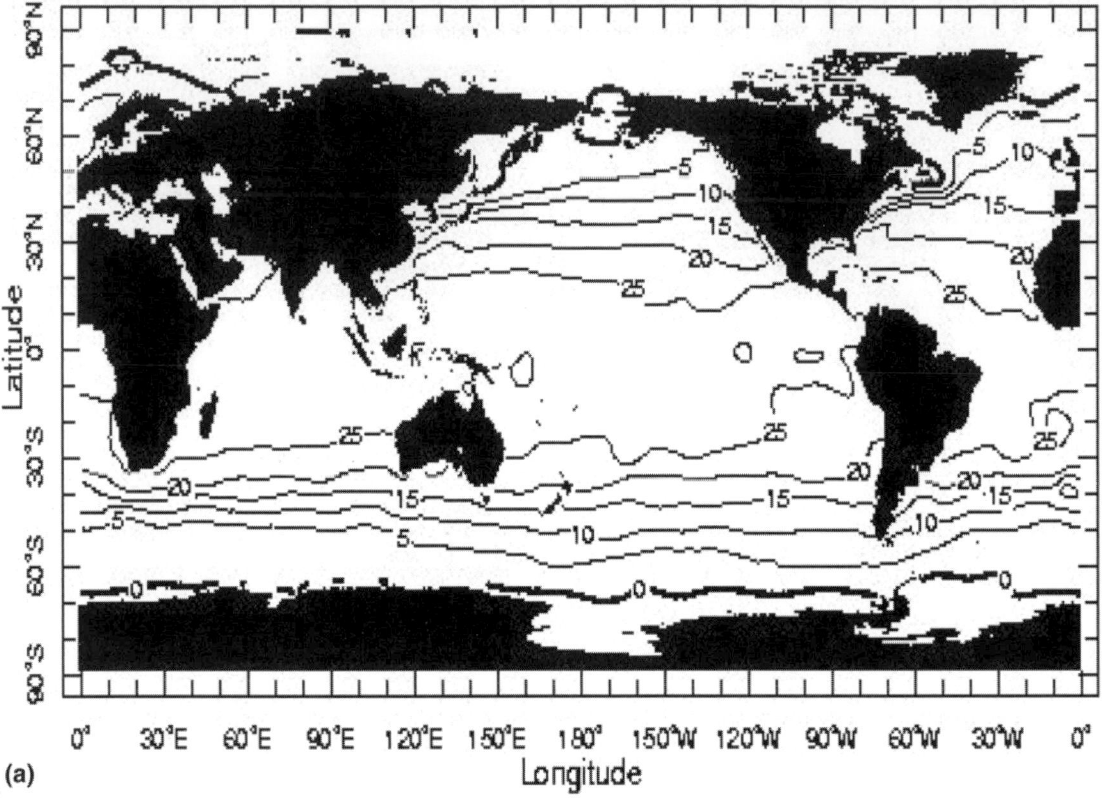

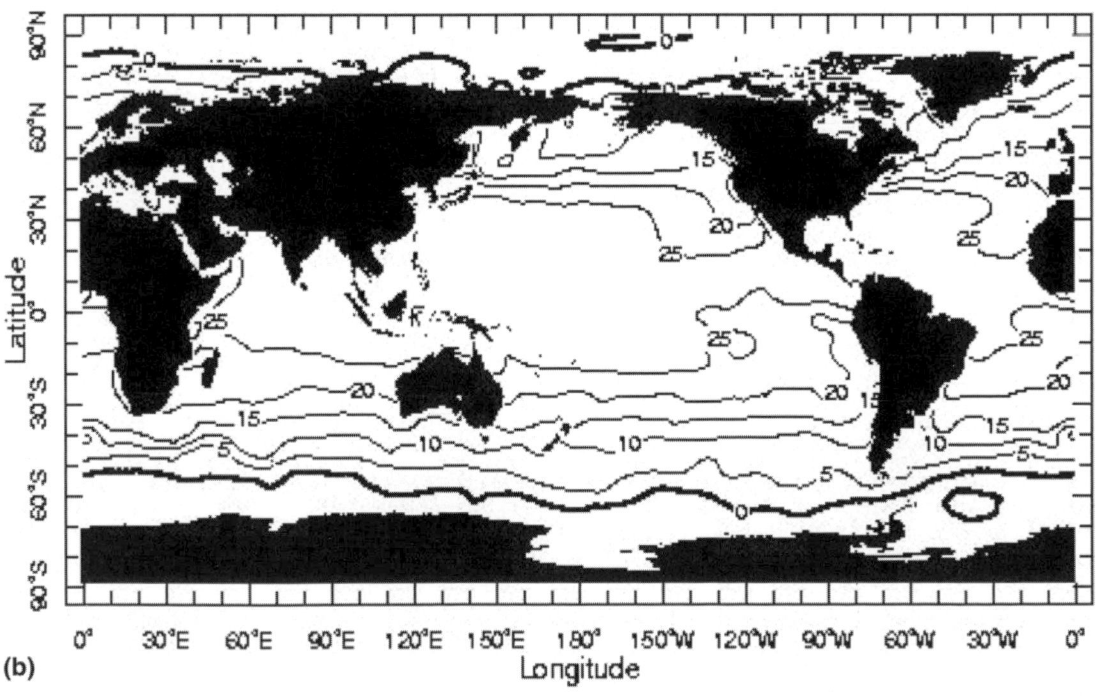

Figure O1 Mean SST for (**a**) February and (**b**) August; from the Lamont–Doherty Data Library.

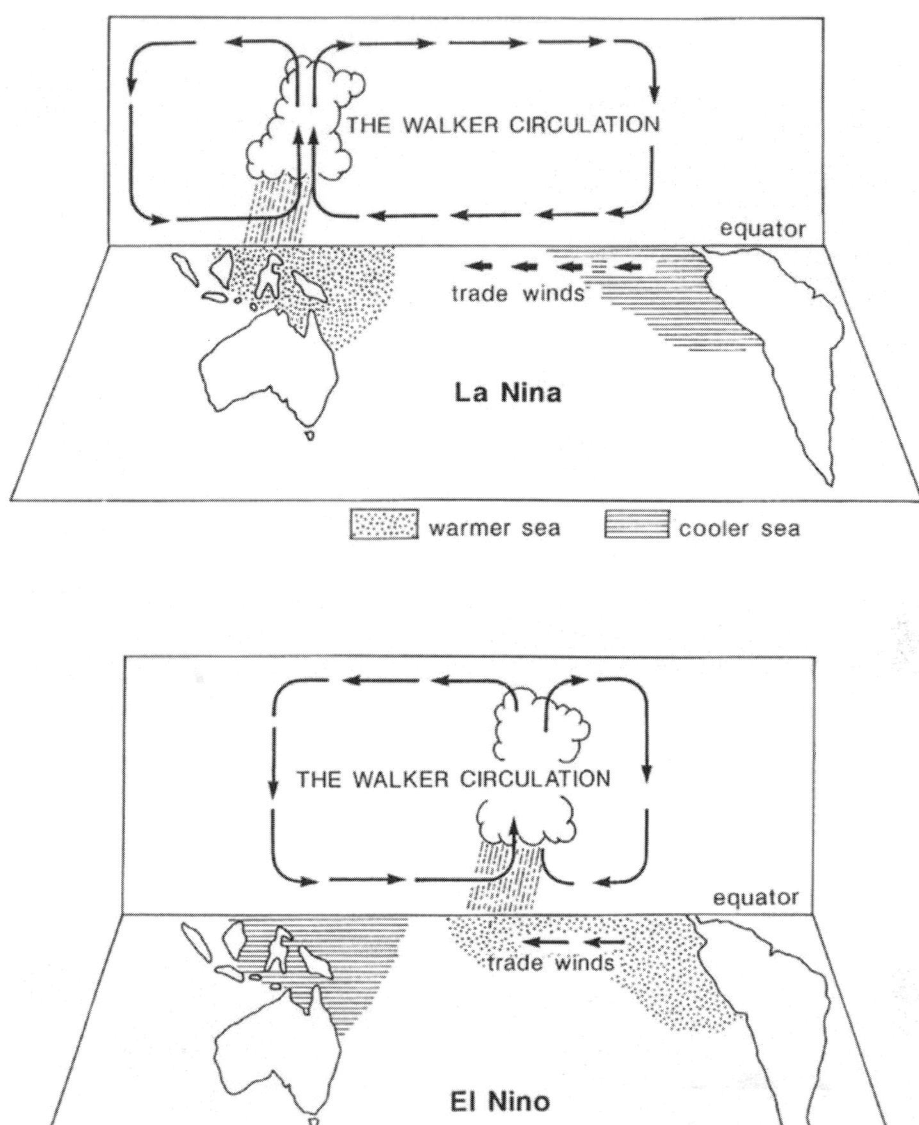

Figure O2 Schematic of the Walker circulation. Adapted from the Australian Bureau of Meteorology online tutorial.

upwelling decreases, and for a warm current to appear along the South American coast where there is usually a cold, northward flow. This current was given the name El Niõ (Christ Child) by local fisherman as it usually appeared around Christmas time. Every 3–4 years on average a much warmer and longer-lasting current forms, which is now known to be directly linked, through air–sea interaction, to the sudden relaxation of the east–west pressure gradient along the equator. Such events have therefore become known as El Niõ events, with the opposite, cold, events named for a girl child, La Niã (Philander, 1990).

While the existence of the Southern Oscillation has been known of since Sir Gilbert Walker's work in the early decades of the twentieth century, its connection to El Niõ events was not fully realized until Bjerknes' work in 1969. An explanation of the mechanisms underlying such an event had to wait until the mid-1980s (Graham and White, 1988). Even now, the mechanisms leading to variability in the frequency of events are not clear. At some times El Niõ events do not occur for a decade (1930s), while at others events can recur annually (early 1940s and early 1990s). Sometimes a La Niã immediately follows an El Niõ (1998 following 1997/8), but more often it does not. The character of individual events also varies significantly. Nevertheless, common features do exist and these form the basis of the following discussion.

The trigger for El Niõ is a prolonged period of westerly wind anomalies in the strong easterly trade wind belt in the western Pacific. This can come about in a number of ways: a random anomaly, part of the organized convection signal of the eastward passage of the intra-seasonal Madden–Julian Oscillation (Madden, 1986), a linking of the inflow from hurricanes on opposite sides of the equator. Whatever the cause, the westerly anomalies alter the upper ocean, sending a Kelvin wave along the equatorial waveguide. This pulse carries a signal of a relaxing sea level, and thermocline, gradient at speeds of several meters per second towards the eastern Pacific, suppressing upwelling and leading to the upper ocean warming experienced by fisherman off South America. The suppression of upwelling to the east of the original wind anomaly causes the SST to rise, encouraging convection to move eastwards toward it. The westerly wind anomaly propagates eastward with it, initially amplifying and then sustaining the signal in the ocean and atmosphere.

The movement of the Indo-Australian Convergence Zone into the central Pacific has ramifications for the atmospheric circulation well away from the tropical Pacific. Because the Hadley Cell allied with this convergence zone also moves eastward, and an associated Rossby wave is initiated in the atmosphere, a wave train is set off in the upper troposphere, arcing poleward and eastward. This leads to enhancement and southward movement of the Aleutian Low, bringing storms to California, colder and drier conditions over central Canada and wet conditions over the southeast United States. These anomalies often peak in the boreal winter. An extreme signal in the Pacific can lead to further, but weaker, effects downstream over Europe.

Associated with the change in the tropical Walker circulation is disruption to the tropical circulation over the Atlantic, leading to fewer Atlantic hurricanes and drought over west Africa. The Indian Monsoon circulation system can also be affected by a weakening of the summer convection over south Asia, although independent air–sea interaction in the Indian Ocean can limit or eliminate this signal, as was the case in the 1990s (Saji et al., 1999).

Typically, these widespread anomalies last for several months to a year before the oceanic anomaly in the tropical Pacific is eroded by the repeated impact of reflected equatorial long waves from the western Pacific. Once the oceanic heat source sustaining the convection is removed, the convection zone can return to its more normal anchoring in the western Pacific, and the wind field returns to normal. La Niã occurs if this return to normal convection overshoots and the equatorial easterlies become anomalously strong, leading to a strengthening of "normal" atmospheric and oceanic circulation – and air–sea feedback – for a similar time to the preceding El Niõ.

The very clear ocean–atmosphere feedback processes set off in El Niõ mean that there is the potential for its forecasting to achieve a useful measure of skill. Particularly once the initial signal has begun, climate prediction one or two seasons in advance is possible, and forecasters using a mix of statistical and dynamical methods achieved considerable success in the extreme 1997–1998 event – the so-called climate event of the century (Changnon, 2000) – and also correctly forecast the development of the weaker 2002 event.

The North Atlantic Oscillation

Another region where a pressure see-saw between two semi-permanent pressure centers has considerable climatic impact is in the North Atlantic, where oscillation in the gradient between the Icelandic Low and the Azores High is known as the North Atlantic Oscillation (NAO; Wallace and Gutzler, 1981). This oscillation operates on a number of different time-scales. It is most pronounced in boreal winter, when the average pressure gradient is at its greatest. Fluctuation in the amplitude of the NAO can also occur from synoptic through seasonal to decadal time-scales. It is sometimes thought to be part of a geographically wider atmospheric pattern called the Arctic Oscillation (Thompson and Wallace, 1998). However, the air–sea interaction involved is clearest in the North Atlantic, so here we will restrict our consideration to the NAO.

The most important effect of changing the pressure gradient between the Icelandic Low and Azores High is in the strength of the westerly winds across the North Atlantic and into western Europe. At times when the pressure difference is enhanced (positive NAO) the westerlies are stronger, leading to stormier but milder conditions in western Europe. In contrast, a reduced pressure gradient (negative NAO) corresponds to weaker atmospheric flow, and so drier, colder conditions over western Europe. Movement in the centers of action is not well captured by the basic pressure difference describing the NAO Index, and can result in significant changes in the regional impact of this climatic signal. There is some evidence that decadal variability in the NAO is related to movement of the key pressure centers.

Stronger winds in positive NAO periods cause more evaporation, and hence more latent heat transfer to the atmosphere from the ocean. This latent heat flux encourages storm formation, and leads to warmer conditions downstream. A negative NAO will reduce latent heat transfer from the ocean, although note that it is easier for continental weather systems to affect the eastern Atlantic in such conditions. There is therefore a strong interaction with the ocean, which contains the potential for a negative feedback: latent heat transfer cools the ocean reducing the heat flux to the atmosphere and discouraging cyclogenesis. The thermal inertia of the ocean means that any such feedback will only operate on seasonal or longer time-scales, and indeed persistent multi-year signals in the SST of the North Atlantic Drift have been observed to propagate eastwards (Sutton and Allen, 1997). The position of the Drift, and hence the source of the latent heat transfer into the atmosphere, is also coupled to the winds, allowing feedback between ocean and atmosphere.

Many environmental parameters have been found to correlate well with the NAO, because of the strong climatic impact in its variation. The change in western European climate between the cold 1960s and the warm 1990s can be seen to be strongly linked to a persistent long-term change from negative to positive NAO over this time. Whether this change was a consequence of global warming or a natural fluctuation coordinating with this signal is not yet known (Stocker et al., 2001).

Pacific Decadal Oscillation

The Pacific Decadal Oscillation (PDO) is a pattern of Pacific climate variability with an El Niõ-like spatial fingerprint, but a very different time-scale. Steven Hare described the climate signal in 1996 while researching connections between Alaskan salmon and Pacific climate (Mantua et al., 1997). The PDO differs from phases of the Southern Oscillation in two main ways. The time-scale is much longer: twentieth-century PDO "events" persisted for several decades, while typical El Niõ events last for around a year or less. Spatially, the PDO events are strongest in the North Pacific/North American sector, while secondary signatures exist in the tropics – the opposite is true for El Niñ/La Niñ events.

There appear to have been only two full PDO cycles in the past century: "cool" PDO regimes prevailed from 1890 to 1924 and again from 1947 to 1976, while "warm" PDO regimes dominated from 1925 to 1946 and from 1977 through (at least) the mid-1990s. The terms "cool" and "warm" refer to the tropical SST signature, while the North Pacific SST tends to be opposite to this. Major changes in northeast Pacific marine ecosystems have been correlated with phase changes in the PDO: warm eras have seen enhanced coastal ocean biological productivity in Alaska and inhibited productivity off the west coast of the contiguous United States, while cold PDO eras have seen the opposite north–south pattern of marine ecosystem productivity.

The causes of the PDO are not currently known, although the time-scale of the oscillation is suggestively similar to that of an oceanic midlatitude Rossby wave. Some coupled climate simulation models produce PDO-like oscillations. These can be similar to the NAO, involving interaction between advective SST anomalies in the western boundary current (the Kuroshio in this case) and variations in the westerlies and latent heat transfer. However, stochastic forcing cannot yet be discounted, because of the long time-scales of oceanic Rossby wave propagation.

Thermohaline circulation

The thermohaline circulation of the ocean is the oceanic equivalent of the meridional circulation in the atmosphere. Surface waters become sufficiently dense to sink, through cooling (North Atlantic) or sea-ice-induced salinification (Antarctic). These dense, deep waters then return to the surface through turbulent mixing near rough topography over the whole globe, and also through mixing in the strong zonal current, known as the Antarctic Circumpolar Current (ACC), underlying the westerlies in the Southern Ocean, at around 50°S. Surface flow closes the circuit, through entrainment of part of the ACC into the subtropical gyre of the South Atlantic, and through the generation of northwestward-moving eddies from the Agulhas Current of the Indian Ocean as it retroflects at the southern tip of Africa (Figure O3).

The overturning time-scale of the thermohaline circulation is of the order of 1000 years, so the ocean has memory of atmospheric conditions at the millennial time-scale. However, because the strong poleward movement of warm water in the North Atlantic forms part of the thermohaline circulation disruption of this flow on time-scales of only decades (e.g. through catastrophic freshwater inputs damping the ocean's ability to convect) gives the potential for abrupt climate change over years to decades. Reducing the supply of warm water to the northern Atlantic through suppressing convection reduces the ability of the ocean to provide large quantities of latent heat to the atmosphere. The downstream impact of this can be a reduction in western European temperature of several degrees. Particularly during glacial periods and deglaciation, such changes in northern Atlantic climate have left signs in the paleoclimate record, most dramatically associated with the release of either icebergs, or flood waters from ice dam breaks, from the North American Laurentide ice sheet. The iceberg releases are known as Heinrich events, after the German scientist who, in the 1980s, first identified the ice-rafted debris signal from such events (Heinrich, 1988). Heinrich events are believed to have occurred when the ice sheet over Hudson Bay and central Canada became sufficiently deep that the basal layer became fluid and much of the ice drained into the sea over a few centuries (MacAyeal, 1993). This freshened northern waters,

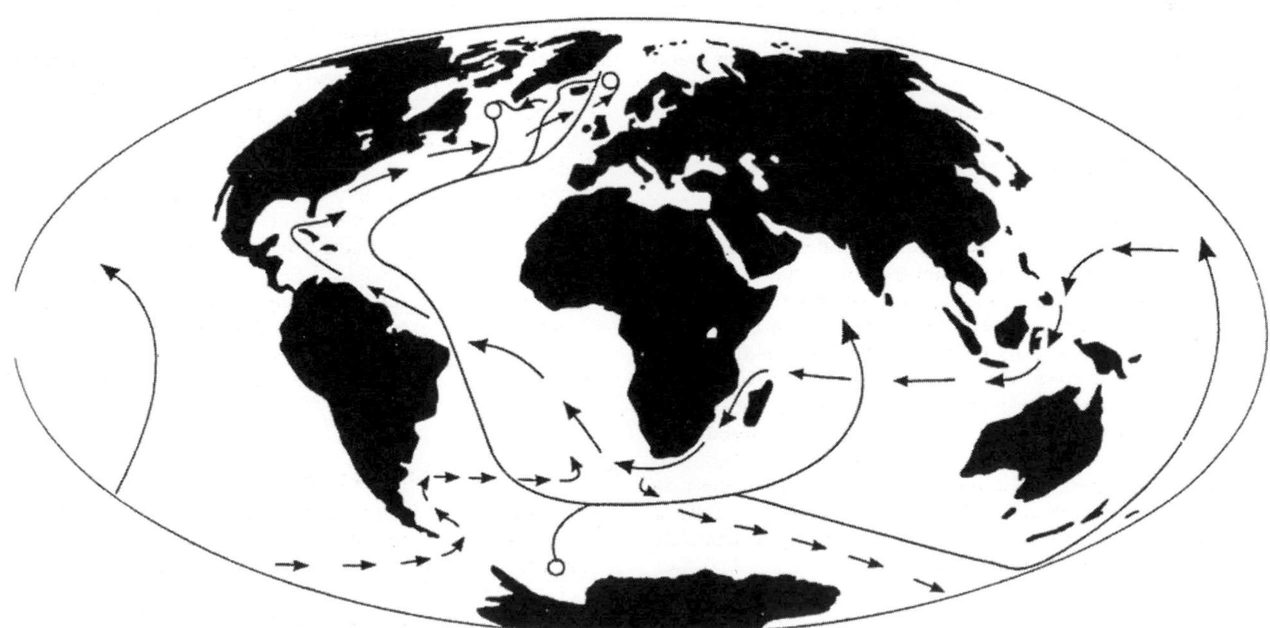

Figure O3 Schematic of the conveyor belt circulation. Circles indicate where deep water is formed; the solid lines show deep-water flow, and the dashed lines show near-surface flow (from Fig. 1.14 of Bigg (2003); reproduced with permission of Cambridge University Press).

stopping convection and plunging the North Atlantic and much of the northern hemisphere into a cold period. The rebounds to warm conditions some centuries later seem to have occurred over a few decades, as convection suddenly switched back on. The most recent such climate oscillation, around 12 000 years ago, is known as the Younger Dryas, although the cause of this is more likely to be related to ice-sheet meltwater release than icebergs. The Heinrich events proper recurred at varying intervals of the order of 10 000 years apart.

Between these extreme climate oscillations another cycle of about 1400 years exists – the Dansgaard–Oeschger Oscillations (Dansgaard et al., 1984) in the records of the Greenland ice cores, or the Bond cycle of ocean floor sediments (Bond et al., 1993). These appear to be less extreme convection transitions, but with similarly temporally abrupt endings and slightly less abrupt beginnings, a climatic signal largely confined to the North Atlantic, and probably involve air–sea interaction at the thermohaline time-scale (Alley, 2001).

Biological feedback

A major change to the way air–sea interaction has been perceived in the last two decades is via the rising profile of potential feedbacks between marine phytoplankton and atmospheric climate. It has been known for many years that the ocean was important in the carbon cycle, and hence in determining the level of atmospheric carbon dioxide, an important greenhouse gas. However, measurements of atmospheric carbon dioxide levels inside ice cores have now shown that as far back as half a million years ago there were tight couplings between global temperature and carbon dioxide concentration. The terrestrial biosphere no doubt contributes to this, but many scientists believe that much of the oscillation between the low levels of atmospheric carbon dioxide in glacial periods and the high levels in interglacials occurs due to feedbacks with the ocean carbon store, and most probably with phytoplankton stocks (Watson et al., 2000). Phytoplankton consume carbon dioxide during photosynthesis, and so atmospheric levels would be reduced if plankton were more abundant. Fertilizing currently iron-poor reaches of the surface ocean in the Southern Ocean and equatorial waters in particular through windblown dust is a possible mechanism forcing higher phytoplankton numbers. The ocean is also a source for other greenhouse gases, in particular methane and nitrous oxide, through reducing environments in sediments or low oxygen environments.

An important, but short-lived, gas produced by the ocean is dimethyl sulfide (DMS). This gas is produced by phytoplankton, probably as part of their regulation of cell ionic strength. Amounts vary widely according to species, however. The ocean is a net source of this gas, which has a lifetime in the atmosphere of only a day or so. Its importance lies in the rapid oxidation of DMS to sulfate particles in the atmosphere, a major source of cloud condensation nuclei. Changes in phytoplankton abundance can therefore directly influence cloud coverage and properties, and hence the Earth's radiation budget. The CLAW hypothesis (Charlson et al., 1987) proposes that increased phytoplankton, through elevated DMS levels, leads to increased concentrations of smaller cloud condensation nuclei, which tend to reflect solar radiation more efficiently and so cool the planet. Depending on the effect of cooling on the marine biosphere, which is complex and not fully resolved, this effect might be a positive or negative feedback on global temperature.

G.R. Bigg

Bibliography

Alley, R.B., 2001. *The Two-mile Time Machine*. Princeton, NJ: Princeton University Press.

Bigg, G.R., 2003. *The Ocean and Climate*, 2nd edn. Cambridge: Cambridge University Press.

Bjerknes, J., 1969. Atmospheric teleconnections from the equatorial Pacific. *Monthly Weather Review*, **97**: 163–172.

Bond, G., Broecker, W., Johnsen, S. et al., 1993. Correlations between climate records from North-Atlantic sediments and Greenland ice. *Nature*, **365**(6442): 143–147.

Bryden, H.L., Roemmich, D.H., and Church, J.A., 1991. Ocean heat-transport across 24-degrees-N in the Pacific. *Deep-Sea Research, A*, **38**(3): 297–324.

Changnon, S.E. (ed.), 2000. *El Niño 1997–1998*. Oxford: Oxford University Press.

Charlson, R.J., Lovelock, J.E., Andreae, M.O., and Warren, S.G., 1987. Oceanic phytoplankton, atmospheric sulphur, cloud albedo and climate. *Nature*, **326**: 655–661.

Dansgaard, W., Johnsen, S.J., Clausen, H.B. et al., 1984. North Atlantic climatic oscillations revealed by deep Greenland ice cores. In Harsen, J., and Takahashi, T. eds., *Climate Processes and Climate Sensitivity*. Washington, DC: American Geophysical Union, pp. 288–298.

Graham, N.E., and White, W.B., 1988. The El-Niñ cycle – a natural oscillator of the Pacific-ocean atmosphere system. *Science*, **240**(4857): 1293–1302.

Heinrich, H., 1988. Origin and consequences of cyclic ice rafting in the Northeast Atlantic-Ocean during the past 1 30 000 years. *Quaternary Research*, **29**(2): 142–152.

MacAyeal, D.R., 1993. Binge/purge oscillations of the Laurentide Ice-Sheet as a cause of the North-Atlantic's Heinrich events. *Paleoceanography*, **8**(6): 775–784.

Madden, R.A., 1986. Seasonal variations of the 40–50 day oscillation in the Tropics. *Journal of Atmospheric Science*, **43**(24): 3138–3158.

Mantua, N.J., Hare, S.R., Zhang, Y., Wallace, J.M., and Francis, R.C., 1997. A Pacific interdecadal climate oscillation with impacts on salmon production. *Bulletin of the American Meteorological Society*, **78**(6): 1069–1079.

Philander, S.G., 1990. *El Niño, La Niña, and the Southern Oscillation*. San Diego, CA: Academic Press.

Ratcliffe, R.A.S., and Murray, R., 1970. New lag associations between North Atlantic sea temperatures and European pressure, applied to long-range weather forecasting. *Quarterly Journal of the Royal Meteorological Society*, **96**: 226–246.

Saji, N.H., Goswami, B.N., Vinayachandran, P.N., and Yamagata, T., 1999. A dipole mode in the tropical Indian Ocean. *Nature*, **401**: 360–363.

Stocker, T.F. Clarke, G.K.C., and Le Treut, H., et al. Physical climate processes and feedbacks. In Houghton, J.T., Ding, Y., Griggs, M. et al., eds., *Climate Change 2001: the scientific basis*. Contribution of working group I to the third assessment report of the Intergovernmental Panel on Climate Change. Cambridge: Cambridge University Press, pp. 415–470.

Sutton, R.T., and Allen, M.R., 1997. Decadal predictability of North Atlantic sea surface temperature and climate. *Nature*, **388**(6642): 563–567.

Thompson, D.W.J., and Wallace, J.M., 1998. The Arctic Oscillation signature in the wintertime geopotential height and temperature fields. *Geophysics Research Letters*, **25**(9): 1297–1300.

Wallace, J.M., and Gutzler, D.S., 1981. Teleconnections in the geopotential height field during the Northern Hemisphere winter. *Monthly Weather Review*, **118**: 877–926.

Watson, A.J., Bakker, D.C.E., Ridgwell, A.J., Boyd, P.W., and Law, C.S., 2000. Effect of iron supply on Southern Ocean CO_2 uptake and implications for glacial atmospheric CO_2. *Nature*, **407**(6805): 730–733.

Cross-references

Cycles and Periodicities
El Niño
Maritime Climate
North Atlantic Oscillation
Quasi-Biennial Oscillation

OCEAN CIRCULATION[1]

Most ocean currents other than tidal currents derive from either the stress of the wind on the water surface or the uneven distribution of mass due to variations in temperature and salinity. In general, the currents in the upper layers of the ocean are attributable to wind stress, whereas those below are part of the thermohaline circulation resulting from the distribution of temperature and salinity, but this distinction is often unclear, and the two are connected by vertical water movements. Just as winds in the atmosphere are rarely steady, so ocean currents are subject to considerable variation. This variation, which is associated particularly with eddy-like features with length scales of the order 100 km, has been the subject of much research in recent years. Here we will be concerned essentially with the time-mean or net circulation on which this variability may be considered to be superimposed.

Wind-induced currents

The wind blowing across a water surface exerts a frictional stress in the direction in which it is blowing, as well as generating waves. There is some forward movement of water particles in wave motion, and the magnitude of the stress that the wind exerts depends on the roughness of the water surface, so the two processes are closely interrelated.

The simplest situation in which to consider the nature of the currents resulting from wind stress is that in which a wind blows at constant velocity over a deep, infinite, and homogeneous ocean, so that the water movement is not impeded by any continental barriers and the coefficient of (eddy) viscosity remains constant with depth. We could reasonably assume that in such a situation the speed of the resulting current would be greatest at the surface and would decrease with increasing depth, and that below some depth at which the current became negligibly small we could ignore friction. Considering the layer above this depth as a whole, and assuming that the wind remains steady for a period long enough to allow a steady current to become established with no acceleration, we can equate the magnitude of the wind stress (τ) with the Coriolis Effect and conclude that the mean current ($\bar{u}$), and therefore the net transport of water in this layer, is 90° *cum sole* (*cum sole* is sometimes used to mean to the right in the northern hemisphere and to the left in the southern hemisphere) from the wind direction. Thus

$$\tau = D\rho f\bar{u},$$

where D is the depth of the layer so that $D\rho$ is the mass of the layer per unit surface area. Hence

$$\bar{u} = \frac{\tau}{D\rho f}.$$

At any one depth, however, the water is subject to three forces – the stress of the overlying water (or, at the surface, the wind), the stress of the underlying water, and the Coriolis Effect. The effect of this is that the current deviates increasingly *cum sole* with increasing depth. Ekman, who examined this point theoretically, found that the surface current deviates 45° from the wind direction and that the speed of the current decreases exponentially with depth as well as turning further *cum sole* until, at the depth D, it has about 4% of the surface speed and is directed exactly opposite to the surface current. This current structure is known as the *Ekman Spiral* (Figure O4), and the depth D, which varies with the eddy viscosity and the latitude, is called the *depth of frictional influence* and is typically between 50 m and 200 m. The speed of the surface current depends on the same factors as D and on the wind stress, and is usually between 1% and 3% of the wind speed.

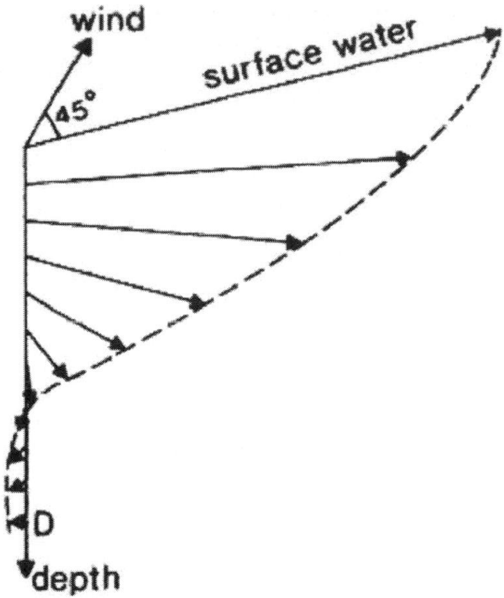

Figure O4 The Ekman spiral, showing variation of current velocity with depth in the Ekman layer.

Other atmospheric parameters in addition to wind can affect water movement. The sea surface appears to react as an inverted barometer to atmospheric pressure variations. An atmospheric low is thus accompanied by high sea level, and if the winds associated with the low cause water to move toward a coast against which it piles up, particularly high water levels can be experienced. Both midlatitude depressions and tropical cyclones are moving systems; if their speed is appropriate, such high water levels or *storm surges* can travel with them as long waves. This occurs fairly often in the North Sea where a storm surge, accompanying a depression moving rapidly east and with its center passing just to the north of Britain, travels around the coast of Scotland and then follows an anticlockwise path around the edge of the North Sea. Its effects are greatest where the North Sea narrows and shoals in the south, and by the time the storm surge reaches there the northerly winds in the cold sector of the depression may well be causing further water to flow southward in the North Sea.

Surface circulation of the oceans

The Atlantic and Pacific oceans are essentially similar in shape, extending northward from the Southern Ocean and narrowing

[1] This article is considerably revised from an original entry in the *Encyclopedia of Climatology* that was based upon Harvey, J.G., 1976. Oceanic circulation. In *Atmosphere and Ocean: our fluid environments*. Sussex: Artemis Press, pp. 119–126.

at their northern ends. The winds between about 10° and 50° latitude are essentially anticyclonic around the subtropical high. This leads to water in the Ekman layer above the depth of frictional influence being transported toward the center of the ocean. The convergence here depresses the main thermocline, but it also leads to the water level sloping downward from the center of the ocean outward, and this gives rise to a gradient current that is anticyclonic, i.e. running in the same direction as the wind (Figure O5).

This *subtropical gyre* is asymmetrical, particularly in the northern oceans, its center being displaced to the western side of the ocean, so that, for example, the Gulf Stream in the North Atlantic and the Kuroshio in the North Pacific are much stronger than any currents on the eastern side of these oceans. The reason for this involves vorticity. If we simplify the situation appreciably by ignoring vertical motion, we can consider the factors that bring about changes in vorticity as the water moves around the subtropical gyre. Throughout the gyre the water is acquiring negative vorticity in the northern hemisphere from the anticyclonic wind stress at the surface. The water on the eastern side is moving equatorward, and its wind-induced negative vorticity may be just sufficient to fit it to the lower positive planetary vorticity at lower latitudes, so that its absolute vorticity is conserved. On the western side, water is moving poleward so that its vorticity becomes negative relative to the Earth and its absolute vorticity becomes increasingly negative. Some braking action is needed to prevent vorticity increasing indefinitely, and this can only be supplied by friction, either at the lateral boundaries or at the sea floor, or by viscosity within the water. This requires much higher velocities on the western side, as friction is proportional to something like the square of the water speed. The result is that the speeds are typically some ten times greater in the warm western boundary current than in the cool eastern boundary current, and the western boundary current extends to greater depths; but it is less than one-twentieth the width of the ocean and water flows equatorward in most of the remainder, so that the same amounts of water are being transported in both directions. The asymmetry of the subtropical gyres south of the equator is less well marked, and in the South Pacific the cold Peru current on the eastern side is perhaps the dominant feature. The different pattern of distribution of land and sea in the southern hemisphere appears to be responsible for this contrast.

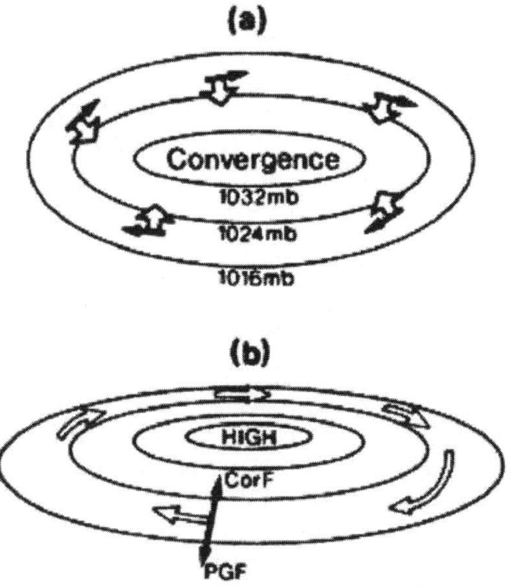

Figure O5 Water movements associated with anticyclonic winds in the northern hemisphere: (**a**) atmospheric pressure and winds, and associated Ekman transports to the right: (**b**) resultant topography of the sea surface and associated gradient currents below the depth of frictional influence.

The North Atlantic and North Pacific Ocean currents

The surface currents which most inhabitants of North America are familiar with are those off our East and West coasts. These currents are part of the larger circulation of the North Atlantic and North Pacific. A detailed examination of the currents in the North Atlantic Ocean serves as an illustration of where these currents flow, and their effect on weather and climate. North of the equator in the Atlantic Ocean is a current which moves from east to west. Here the trade winds drag the surface water westward in a slow-moving stream called the North Equatorial Current. The current splits upon reaching South America, with most of the westward-moving water flowing northwest along the coast into the Caribbean Sea and northward into the Atlantic off the North American coast.

Between the island of Cuba and Florida there is a strong current of very warm water flowing out of the Gulf of Mexico and into the Atlantic Ocean. This warm water from the Gulf of Mexico merges with the northward-flowing water from the Caribbean Sea near the West Indies and forms the Gulf Stream. The Gulf Stream, along with the other poleward-flowing currents, transport a tremendous amount of heat from tropical regions toward the poles. While some 60% of the poleward movement of heat is accomplished by the winds of the general circulation, about 40% of the heat is moved by ocean currents. The Gulf Stream is an extremely large current transporting warm water from the tropics, and hence heat, northward off the east coast of the United States. This current is a major factor in weather along the east coast. The current provides a lot of latent heat and water to the atmosphere through evaporation that plays a part in storm activity. Hurricanes often track northward along the Gulf Stream due to the heat that is available to maintain the storms. Because of the size of these storms, even if they remain offshore over the current they can have major impact on coastal areas of the Carolinas. In winter the warm water often provides enough heat for midlatitude lows to form, which then move north along the coast. Some of these lows become well developed nor'easters which inflict major damage to beaches and coastal property. How far northward the warm water flows, and how close to shore the current flows, varies with the season and from year to year.

The Gulf Stream is one of a number of currents that flow along the western side of ocean basins. In the North Pacific Ocean is the Kuroshio Current. It has its origin near the equator, then flows northward toward the islands that make up Japan, and then turns eastward into the Pacific Ocean. These warm currents are not limited to the northern hemisphere; they are found in the Indian Ocean and in the South Atlantic Ocean. These currents flow away from the equator, as do the Gulf Stream and Kuroshio Currents of the northern hemisphere. The Agulhas Current flows southward along the east coast of Africa. All of these currents flow fairly quickly, and move a very large volume of water. They all affect the regional climate.

In the general vicinity of New England the westerly winds direct the Gulf Stream away from North America. Much of the Gulf Stream water re-circulates to the west of the mid-Atlantic Ridge. That which crosses the Ridge spreads out as it goes eastward and is known as the North Atlantic Drift. The current remains warm as it flows toward northern Europe, which makes winter conditions there unusually mild for those latitudes. Some of the warm water flows north between Iceland and the British Isles and along the Scandinavian Peninsula (Figure O6).

Part of the North Atlantic Drift splits and flows southward along the west coast of Spain and Portugal as the Canary Current, named for the Canary Islands. The water in this current is relatively cool as it flows south after having crossed the North Atlantic. It is cool only relative to the water temperature in mid-ocean. Many of the characteristics of the Canary Current are seen in its equivalent current off North America, the California Current.

The California Current flows southward roughly parallel to the west coast of North America. Since the water making up this current comes from the North Pacific it is quite cool. All along the west coast of the United States the near-shore water is quite cool As the current flows south from Washington to California it is to be expected the water would warm as it receives more solar radiation. However, upwelling cold water keeps the surface water rather cold all the way south past the Mexican border. This upwelling provides a cold surface over which the onshore winds cross before reaching land. The cold surface tends to stabilize the onshore moving air, increasing the aridity along some coasts. This is particularly the case along southern California and Baja California. This cool current and upwelling keeps summer water temperatures off the California beaches fairly cool, and much cooler than the water at the same latitudes off the east coast. Temperatures in the California current are typically below 20°C as surfers and bathers are only too aware.

Another region which is exceptionally dry as a result of cool stabilizing water offshore is along the coast of Chile. Here precipitation seldom falls except during times of El Niñ. Details of this well-known ocean current, and the Southern Oscillation with which it is associated, are discussed in various items in this volume.

Currents of low and high latitudes

In the trade wind zone, water is transported across the oceans to the western side, and thus there is a slope of the sea surface from west to east. Along the intertropical convergence zone, where winds are light, water is able to flow back downhill to the east, and as it is so near to the equator Coriolis has little effect on it. This current is known in each ocean as the *Equatorial Counter Current*. At the equator itself, where the Coriolis Effect changes direction, there is divergence at the water surface. This brings the thermocline nearer to the surface and also promotes vertical mixing in it, which leads to the water above the thermocline being *denser* at the equator than on either side, but that below the thermocline being *less dense* at the equator than on either side. This causes the pressure gradients at the depth of the thermocline (about 100 m) to slope away from the equator to either side, which in turn causes geostrophic currents toward the east, both immediately north and south of the equator. Further, on either side of the equator, the directions of the wind-induced currents at a depth of 100 m will be nearly opposite those of the surface currents and, although very weak, they will lead to convergence at the equator and will contribute to the flow there with an easterly component.

At latitudes greater than 50° there is a marked contrast between the northern and southern hemispheres. In the North Atlantic and North Pacific, water movement is obstructed by continental barriers, but in the southern hemisphere it is able to travel right around the globe in the Antarctic Circumpolar

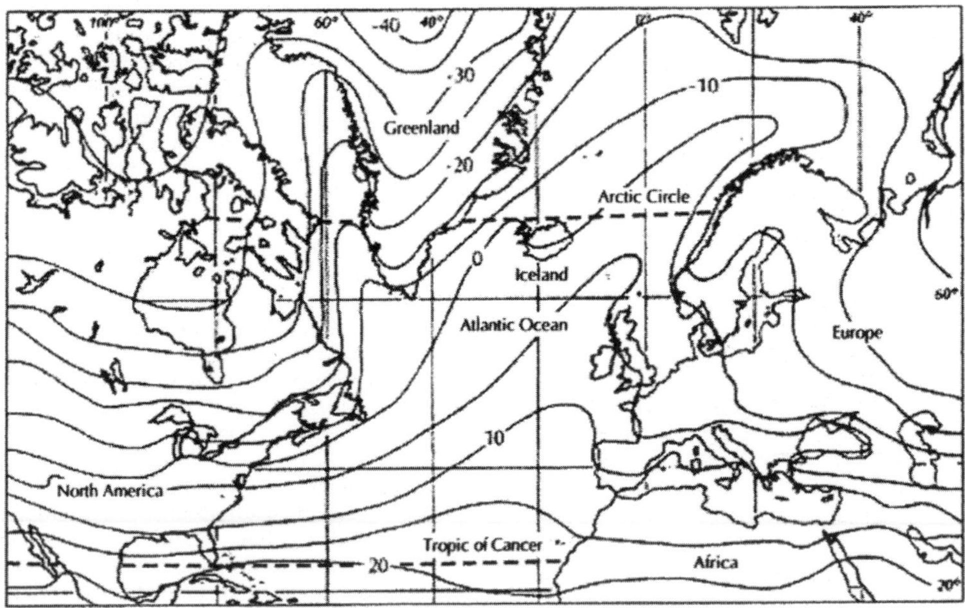

Figure O6 Average January air temperatures across the North Atlantic and neighboring lands. Note the comparative position of the freezing isotherm over North America and northern Europe.

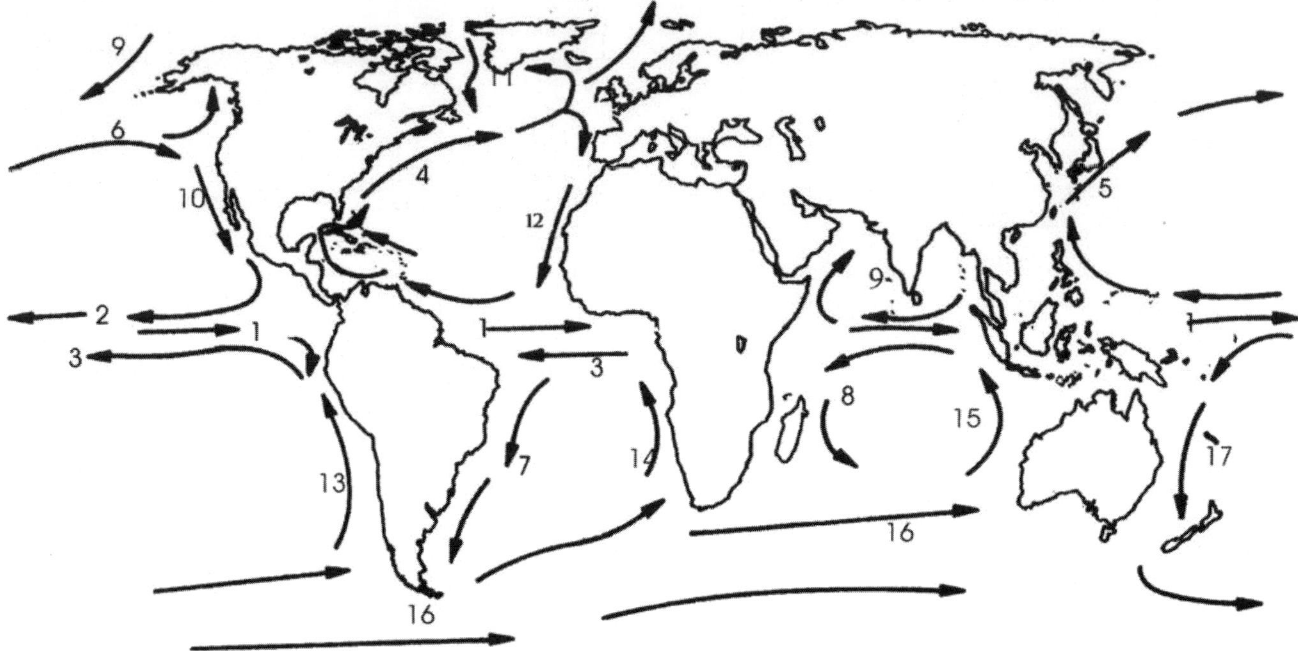

Figure O7 Generalized map of world surface ocean currents (see Table O1 for details).

Current. Over the Southern Ocean the winds are essentially westerly, giving net water transport away from the Antarctic continent. This causes the sea surface to slope upward toward the equator, and the associated gradient currents run toward the east, again in the same direction as the wind is blowing. There is, however, a very important convergence zone surrounding Antarctica, lying generally somewhere between 50°S and 60°S and known as the *Antarctic Convergence*. There is a marked increase in sea surface temperature as one proceeds northward across this convergence, and it is here that water sinks to form *Antarctic Intermediate Water*. Its location has been related to the zone of strongest winds over the Southern Ocean, but its position is too constant for it to be attributable entirely to the wind field, and there are various theories relating it to the circulation of subsurface water.

In the North Pacific there is a *subpolar gyre* comprising the Alaska and Oyashio currents that is comparable to the subtropical gyre to the south of it, but in the opposite sense associated with the cyclonic wind stress in latitudes 50–70°. In the North Atlantic the equivalent subpolar gyre is interrupted by Greenland and exists in two parts, one on either side of Greenland. The East Greenland Current carries pack ice and icebergs southward from the glaciers that reach the east coast of Greenland. Off Cape Farewell this current converges with the warm Irminger Current, and most of the ice soon melts, though some may persist in the cool inshore part of the northerly-flowing West Greenland Current. To this is added a considerable input of icebergs from glaciers that reach the West Greenland coast, particularly in North-East Bay and Disko Bay. These continue in the subpolar gyre of the Labrador Sea and Baffin Bay, and in due course travel southward in the Labrador Current to reach the Grand Banks off Newfoundland. The greatest numbers are encountered here in spring and early summer, having been released perhaps two years earlier from a fjord in the spring thaw and then having spent their last winter frozen in the ice between Baffin Land and Labrador. Off Newfoundland the larger icebergs will ground and gradually break up and melt in this region of convergence between the warm Gulf Stream and the cold Labrador Current. The Indian Ocean extends less far to the north than the Atlantic and Pacific oceans, and is subject to the major seasonal wind reversal of the monsoons.

Figure O7 provides a simplified view of world surface ocean currents.

Table O1 Ocean currents shown in Figure O7[a]

1. Equatorial Counter Current	10. California Current
2. North Equatorial Current	11. Labrador Current
3. South Equatorial Current	12. Canary Current
4. North Atlantic Drift	13. Peru Current
5. Kuroshio Current	14. Benguela Current
6. North Pacific Drift	15. West Australian Current
7. Brazil Current	16. West Wind Drift
8. Agulhas Current	17. East Australian Current

[a] Only major currents are named. Area marked 9 indicates seasonably variable conditions of Indian Ocean.

Upwelling

Zones of convergence and divergence in the flow of the upper layers of the ocean have already been noted. Divergence occurs particularly where the wind-induced transport is offshore, and it requires upwelling from below. This brings water rich in nutrients to the photic zone, and hence areas of upwelling are characterized by considerable biological productivity. As previously noted, the main regions to which this applies are the cool

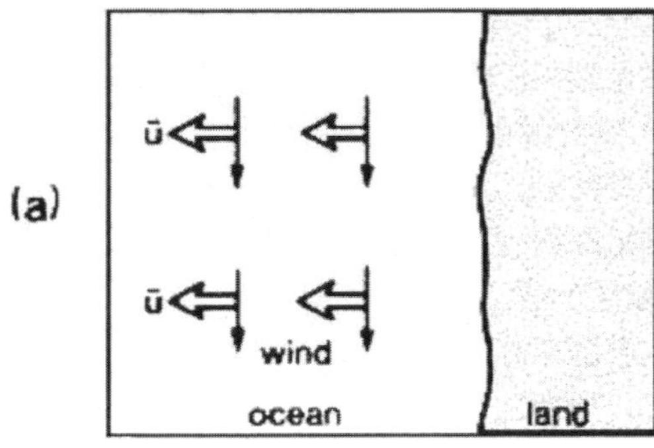

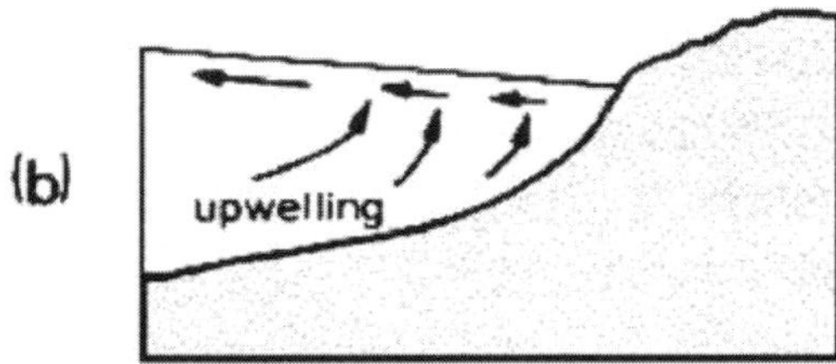

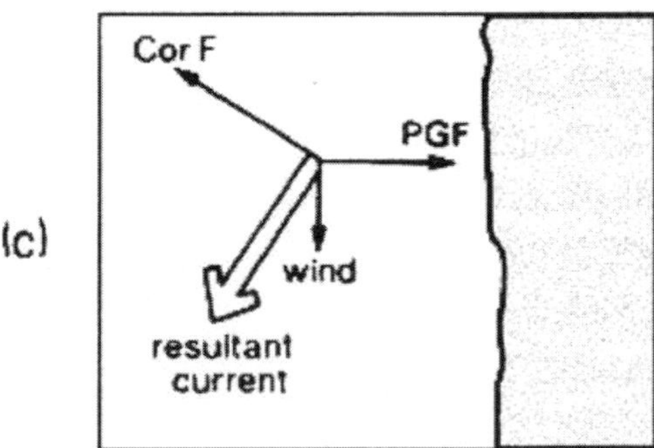

Figure O8 Upwelling in eastern boundary currents in the northern hemisphere: (**a**) plan view showing transport resulting from the wind stress; (**b**) profile showing associated vertical water movements and sloping sea surface; (**c**) possible balance of forces and resultant current.

eastern boundary currents of the subtropical gyre (Figure O8). The anomalously low temperatures of these currents are depressed further by the upwelling of cool water, thus advection fogs are often encountered in such regions. The vertical water movements associated with the convergences and divergences so far discussed are chiefly limited to the water above and within the main thermocline.

Thermohaline circulation

Figure O9 provides the elements of flow at various ocean levels. Figure O9(b) is a schematic model showing how surface flow from equatorial regions is countered by sinking in the polar realm. Between the spreading deep water and the surface flow a thermocline occurs. This is a layer in which a vertical temperature gradient, one that is appreciably greater than the gradients above or below, occurs. Figure O9(a) provides a diagrammatic interpretation of this as applied to the Atlantic Ocean from north to south.

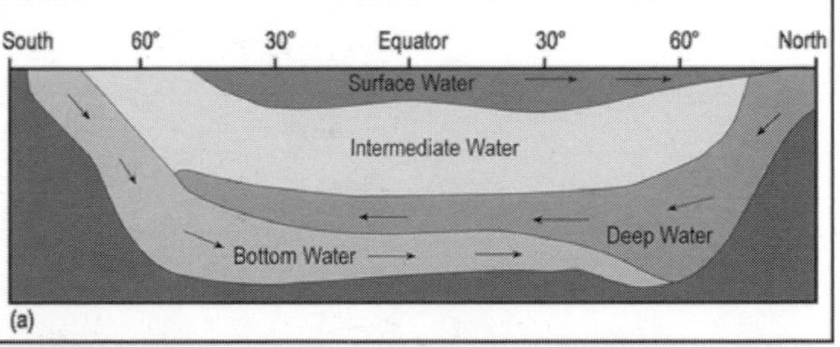

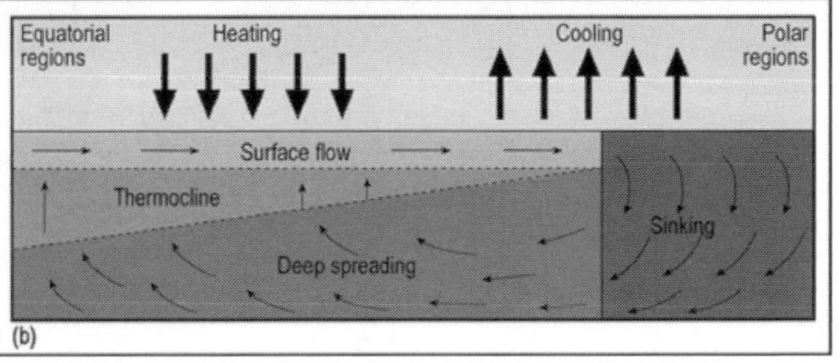

Figure O9 (**a**) A cross-section of the Atlantic Ocrean showing the variations in water with depth. (**b**) A schematic model showing the relationship between surface flow, sinking, and deep spreading of ocean waters between low and high latitudes.

The surface waters become sufficiently dense to sink, through cooling (North Atlantic) or sea-ice induced salinification (Antarctic). Bigg (2003) notes that these dense, deep waters then return to the surface through turbulent mixing near rough topography over the whole globe, and also through mixing in the strong zonal current, known as the Antarctic Circumpolar Current (ACC), underlying the westerlies in the Southern Ocean, at around 50°S. Surface flow closes the circuit, through entrainment of part of the ACC into the subtropical gyre of the South Atlantic, and through the generation of northwestward-moving eddies from the Agulhas Current of the Indian Ocean as it retroflects at the southern tip of Africa. As shown in Figure O10, this is often referred to as the great ocean conveyor belt.

Paleoclimatologists are particularly interested in the overturning time-scale of the thermohaline circulation. It is of the order of 1000 years, so the ocean has memory of atmospheric conditions at the millennial time-scale. However, because the strong poleward movement of warm water in the North Atlantic forms part of the thermohaline circulation disruption of this flow on time-scales of only decades gives the potential for abrupt climate change over years to decades. Reducing the supply of warm water to the northern Atlantic through suppressing convection reduces the ability of the ocean to provide large quantities of latent heat to the atmosphere. The downstream impact of this can be a reduction in western European temperature of several degrees. In his account of ocean–atmosphere interactions in this volume, Bigg provides further analysis of the effects of such an impact. Much additional information about deep-water circulation and climatic change is also to be found in the *Encyclopedia of Paleoclimatology and Ancient Environments*.

J.G. Harvey and J.E. Oliver

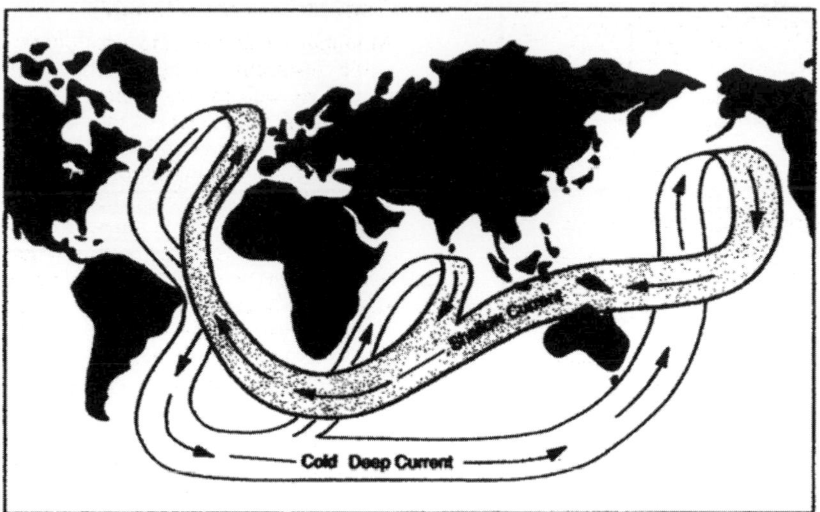

Figure O10 The great ocean conveyor belt (after various sources including Broeker, 1996).

Bibliography

Bigg, G.R., 2003. *The Ocean and Climate*, 2nd edn. Cambridge: Cambridge University Press.

Bond, G., Broecker, W., Johnsen, S. et al., 1993. Correlations between climate records from North-Atlantic sediments and Greenland ice. *Nature*, **365**(6442): 143–147.

Broeker, W., 1996. Impacts, adaptations and mitigation of climate change: scientific–technical analysis. Contribution of Working group II to the Second Assessment Report of the Intergovernmental Panel on Climate Change, UNEP and WMO. Cambridge: Cambridge University Press.

Bryden, H.L., Roemmich, D.H., and Church, J.A., 1991. Ocean heat-transport across 24-degrees-N in the Pacific. *Deep-Sea Research A*, **38**(3): 297–324.

Changnon, S.E. (ed.), 2000. *El Niño 1997–1998*. Oxford: Oxford University Press.

Dietrich, G.L., 1963. *General Oceanography: an introduction*. New York: Wiley-Interscience.

Hill, M.N. (ed.), 1963. *The Sea*, vol. 2. New York: Wiley-Interscience.

Knauss, J.A., 1978. *Introduction to Physical Oceanography*. Englewood Cliffs, NJ: Prentice-Hall.

Mantua, N.J., Hare, S.R., Zhang, Y., Wallace, J.M., and Francis, R.C., 1997. A Pacific interdecadal climate oscillation with impacts on salmon production. *Bulletin of the American Meteorological Society*, **78**(6): 1069–1079.

Philander, S.G., 1990. *El Niño, La Niña, and the Southern Oscillation*. San Diego, CA: Academic Press.

Pickard, G.L., and Emery, W.J., 1982. *Descriptive Physical Oceanography*, 4th edn. Oxford: Pergamon Press.

Pond, S., and Pickard, G.L., 1983. *Introductory Dynamical Oceanography*, 2nd edn. Oxford: Pergamon Press.

Ratcliffe, R.A.S., and Murray, R., 1970. New lag associations between North Atlantic sea temperatures and European pressure, applied to long-range weather forecasting. *Quarterly Journal of the Royal Meteorological Society*, **96**: 226–246.

Stommel, H., 1958. The circulation of the abyss. *Scientific American*, **199**: 85–90.

Stommel, H., 1965. *The Gulf Stream*, 2nd edn. Los Angeles; CA: University of California Press.

Sutton, R.T., and Allen, M.R., 1997. Decadal predictability of North Atlantic sea surface temperature and climate. *Nature*, **388**(6642): 563–567.

Wallace, J.M., and Gutzler, D.S., 1981. Teleconnections in the geopotential height field during the Northern Hemisphere winter. *Monthly Weather Review*, **118**: 877–926.

Warren, B.A., and Wunsch, C. (eds.), 1981. *Evolution of Physical Oceanography*. Cambridge, MA: MIT Press.

Worthington, L.V., 1976. *On the North Atlantic Circulation*. Baltimore, MD: Johns Hopkins University Press.

Cross-references

Coastal Climate
Continental Climate and Continentality
El Niõ
Icebergs and Heinrich Events
Land and Sea Breezes
Maritime Climate
Ocean–Atmosphere Interaction
Tropical Cyclones

OROGRAPHIC PRECIPITATION

Orographic precipitation is caused or enhanced by one or more of the effects of mountains on the atmosphere. These effects include the upward or lateral motions of air directly caused by mountains acting as a barrier, as well as the thermal effects of the mountains which cause elevated heat or cold sources. In addition, mountains influence the atmospheric circulation and precipitation processes indirectly on a range of scales from the production of lee waves downstream from isolated peaks to the development of cyclones downstream from major mountain ranges.

Stratiform precipitation

Mountains can generate both stratiform precipitation, which takes place in a statically stable atmosphere, and convective precipitation, which results from the release of static instability. The most obvious effect of mountains is that they can cause the

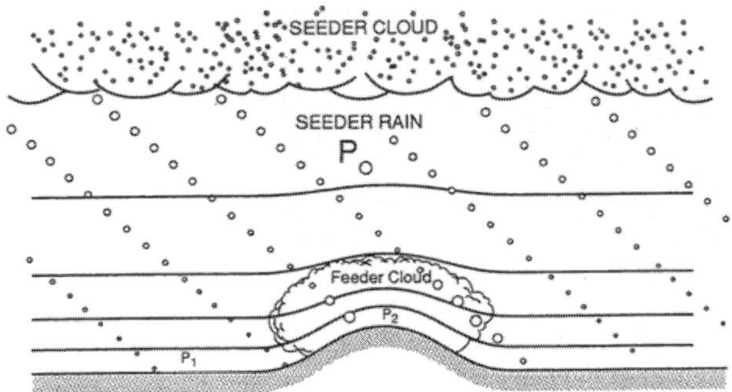

Figure O11 Conceptual model illustrating orographic enhancement of rain by the seeder–feeder mechanism. The size of a droplet falling from the cloud indicates the precipitation intensity at that point. Here P_0 indicates the intensity of the seeder precipitation entering the top of the feeder cloud, P_1 is the intensity of the rain at the surface upwind of the hill and P_2 is the intensity of the rain at the surface in the orographic maximum near the top of the hill (from Banta, 1990).

air encountering them to rise. Rising air cools adiabatically and if it is sufficiently humid condensation and perhaps precipitation can occur. It is widely recognized that the slope on the upwind side of the prevailing wind (the windward side), is more likely to have precipitation than the leeward side. Precipitation formed by this mechanism is widely referred to as upslope precipitation. In contrast, some of the world's deserts are on the leeward side of mountain ranges.

Upslope precipitation may be enhanced by the seeder–feeder mechanism. A low-level stratus cloud (feeder cloud) forms near the top of a mountain. Its temperature is below freezing but warm enough so that it lacks ice nuclei. Ice crystals fall from a higher (and colder) seeder cloud into the feeder cloud. These ice crystals grow at the expense of the water droplets in the feeder cloud by the Bergeron process. The large ice crystals then precipitate out of the feeder cloud to the Earth's surface (Figure O11). This process increases the precipitation efficiency of the feeder cloud because the moisture from small water droplets, which otherwise would not have reached the Earth's surface, evaporates and is deposited on the ice crystals. People have been inspired to encourage this process artificially by seeding the feeder cloud with excellent ice nuclei such as silver iodide. Under the appropriate conditions enhanced precipitation may result.

Upslope precipitation can be displaced upwind of the mountain by blocking. Air which is colder than its environment resists vertical displacement. As it is pushed against the mountain, a high-pressure area is created. Flow approaching the barrier encounters this high-pressure area and decelerates. Convergence and rising motion result, possibly producing precipitation. Convergence can also be generated by a mountain valley which grows narrower in the direction toward which the wind is blowing (channeling) or by air currents that are deflected around the mountain coming together on the mountain's lee side (lee side convergence) (Figure O12). Orographically generated waves can also modify upslope flow and produce areas of enhanced upward motion. Thus, there is no general rule about where upslope flow will produce the maximum precipitation amount along the slope. This place depends on the wind, stability, and local topography as well as the prevalence of blocking and mountain waves.

Convective precipitation

Mountains can generate precipitation through the release of static instability by providing a lifting mechanism for air parcels as well as by orographic processes that destabilize the air column. Precipitation can be generated by three categories of processes: orographic lifting, thermal effects and obstacle effects. Some examples of these processes are illustrated in Figure O12. Orographic lifting can bring air to the level of free convection (i.e. the level at which it becomes positively buoyant). If the atmosphere has a sufficiently thick layer above this level which is statically unstable, precipitation may result. Lifting of an air column will make it more statically unstable. An atmospheric layer which is potentially unstable is one which will become statically unstable if it is lifted. The upward motion generated by mountains may release this potential instability.

The thermal effects of mountains occur because orographic features can serve as elevated heat or cold sources. During the day the slope is heated by the sun. It warms the air next to it which then becomes warmer than the atmosphere at the same level away from the mountain over lower land. Thus, there is low pressure next to the mountain toward which the air moves – producing upslope flow. This heating can be modified by surface conditions. Bare rock will transfer more sensible heat to the atmosphere than snow cover. At night the mountain surface cools, which causes the air above it to cool and results in air flows down the slopes. Air moving across the mountain top during the day is heated by the mountain. This heating helps to destabilize the air column. In addition, convergence may be produced at the ridge crest as air flows up the slope from opposite directions. The combination of the heating and the heating-induced convergence can generate thunderstorms. At night, downslope winds from mountains on opposite sides of a valley can converge in the valley and encourage the generation of thunderstorms as well.

Mountain ranges and high-plateaus can have larger-scale thermal effects on precipitation. During the summer, especially during the day, large areas of high elevation act as an elevated heat source. The air over them is warmer than the air at the same height in the surrounding lowlands. Thus, low-pressure forms near the surface over high-elevation areas. Convergence into this low-pressure area can encourage the formation of thunderstorms. Since heating is stronger during the day, the location of the lowest pressure varies diurnally. The changes in this location can affect the time of day that is most favorable for precipitation to form for a particular place in or near a mountainous region.

Mountains also act as obstacles to air flow, and in so doing can encourage the development of convective precipitation. Channeling and lee side convergence, that favor stratiform precipitation, may also favor convective precipitation if the vertical profile of moisture and temperature are favorable. These processes can cause moisture convergence as well, and static stability is quite sensitive to atmospheric humidity.

Under stable static stability conditions, mountains may generate buoyancy waves – commonly known as gravity waves. These waves propagate downstream (and sometimes upstream as well) of the mountain. The parts of the wave which contain upward-moving air may aid in the development of convective storms, especially if they interact with other mountain-generated circulations such as the diurnal mountain-valley winds.

Orographic variations can contribute to the development of convective storms in their surrounding areas as well as in the

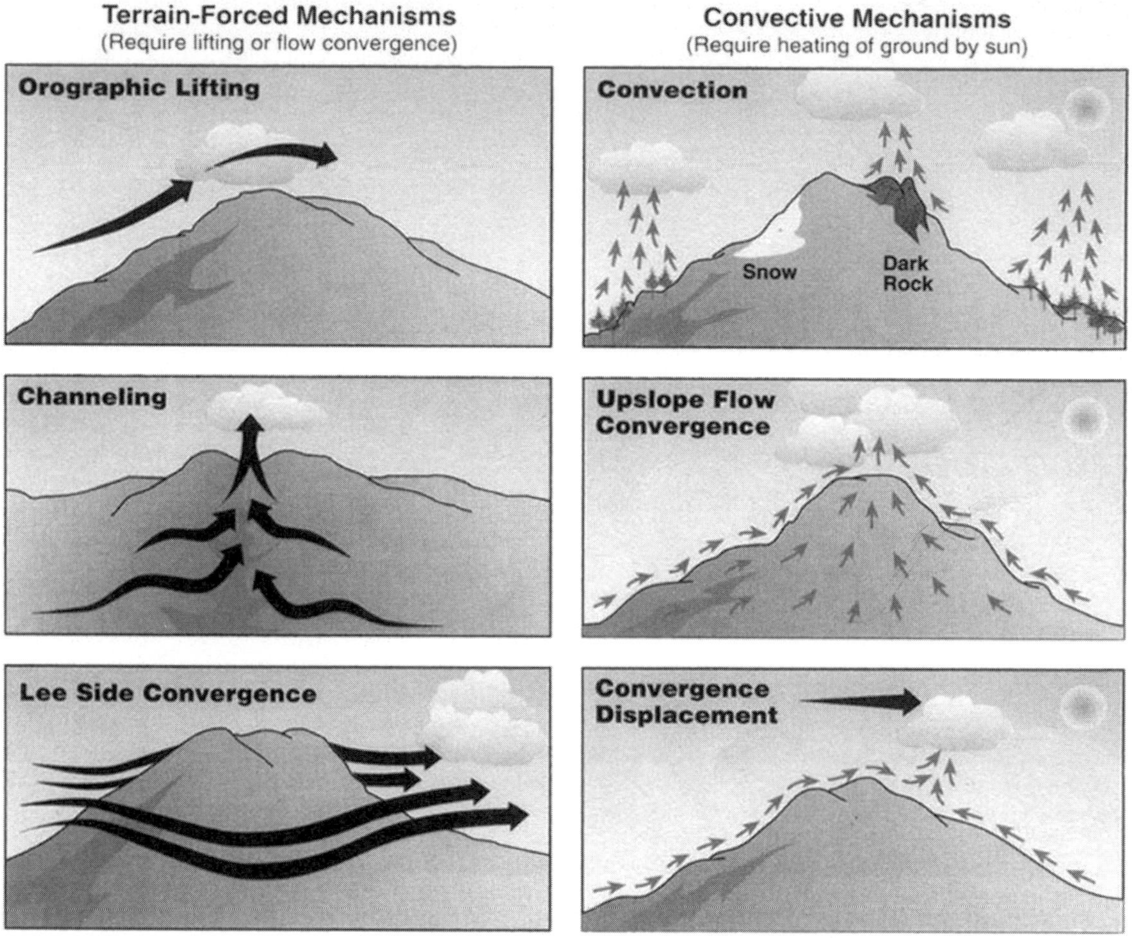

Figure O12 Terrain-related mechanisms that can produce lifting motions and that may trigger thunderstorms (from Whitman, 2000).

mountainous area itself. As they interact with the large-scale wind field, mountains may generate downstream eddies in the flow under some conditions. An example of such an eddy is the so-called "Denver cyclone". Such a cyclonic circulation can have areas of convergence and vorticity which encourage the development of convective storms. Outflow from thunderstorms generated within the mountains may initiate convective storms in the surrounding lowlands.

Another obstacle effect which encourages convective precipitation is mountain blocking of cold advection. In this situation cold advection is occurring to a height which reaches above the top of the mountain. The mountain blocks and retards the cold advection at its base. Above the peak, however, the cold advection proceeds unhindered. The increase in the rate of cold advection with height above the mountain's base destabilizes the layer – perhaps enough to make it suitable for the development of convective storms. This process does not occur often; most cold airmasses are shallow, not even as high as some mountain ranges. In addition, temperature advection in the atmosphere tends to be stronger near the Earth's surface because of stronger temperature gradients there. Thus, this process occurs only when the cold airmass is unusually deep and the winds above the surface are unusually strong.

The large number of processes that can contribute to the initiation of thunderstorms over mountainous areas would imply that in many cases two or more of these processes are involved in storm generation. Many of these processes depend on a particular topographic feature or a process that occurs during one part of the day. Thus, one would expect that some places in mountainous regions are more likely to generate thunderstorms than others. Indeed such places have been observed but they vary, even within one day, depending on the winds and atmospheric static stability. It has also been observed that most places in and around mountainous regions are more likely to receive convective precipitation at some times of day than at others. Furthermore, the time of day which is most favorable can change drastically over a horizontal distance of a few tens of kilometers.

Influence on extratropical cyclones

Extratropical cyclones can generate upslope flow which is enhanced by baroclinicity and may produce considerable precipitation amounts. Mountains also exert influence on the extratropical cyclones themselves. Since these weather systems often produce precipitation, this is an indirect orographic influence on precipitation. An eastward-moving extratropical cyclone approaching a north–south barrier will often not actually cross the barrier but move northward. A secondary cyclone then develops to its southeast on the lee side of the mountain range. This process is known as lee cyclogenesis and it is why the lee sides of mountain ranges are often known as places where cyclogenesis is common.

Donna Tucker

Bibliography

Banta, R.M., 1990. The role of mountain flows in making clouds. *Meteorology Monographs*, **23**: 229–283.

Barry, R.G., 1992. *Mountain Weather and Climate*, 2nd edn. London: Routledge.

Carlson, T.N., 1991. *Mid-Latitude Weather Systems*. London: Routledge.

Cotton, W.R., and Anthes, R.A., 1989. *Storm and Cloud Dynamics*. San Diego, CA: Academic Press.

Reiter, E.R., and Tang, M.C., 1984. Plateau effects on diurnal circulation patterns. *Monthly Weather Review*, **112**: 638–651.

Tucker, D.F., 1993. Diurnal precipitation variations in south-central New Mexico. *Monthly Weather Review*, **121**: 1979–1991.

Whitman, C.D., 2000. *Mountain Meteorology: fundamentals and applications*. New York: Oxford University Press.

OSCILLATIONS

Any phenomenon that tends to vary above or below a mean value in some sort of periodic way is properly designated as an *oscillation*. It may be recognized eventually as a predictable *cycle*, but this term should not be used unless the period has a recognizable regularity; thus the sunspot cycle has a variable amplitude and period (8–13 years from peak to peak) but after several centuries it is a predictable event, even though the parameters are not simple. An oscillation is sometimes used for the swing from one extreme to the other, that is a half-cycle.

Several types of oscillation are recognized: (a) *damped oscillation*, one with constantly decreasing amplitude; (b) *neutral, persistent* or *undamped oscillation*, maintaining constant amplitude; (c) *unstable oscillation*, growing in amplitude and then breaking down; (d) *stable oscillation*, generic class including (a) and (b); (e) *forced oscillation*, one set up periodically by an external force; (f) *free oscillation*, a motion established externally but which then receives no further external energy.

The major oscillations of the planetary atmospheric pressure fields were proposed by Sir Gilbert Walker, who first described and named the seesaw pressure readings of the southern oceans as the Southern Oscillation. Walker's work went relatively unnoticed until recent decades. In a 1966 paper Bjerknes recognized the contributions of Walker and identified the Walker Circulation.

Increased research on teleconnections has identified other oscillations. A brief description of the main oscillations is given here; more detailed information may be derived from cross-referenced items.

The *Southern Oscillation* (SO) is a pressure anomaly over the Indian and South Pacific Oceans. It has a slightly variable period averaging 2.33 years and is often analyzed as part of an ENSO (El Niñ-SO) event. A large amount of literature is available for this oscillation with a good overview, from both the scientific and social aspects, provided by Glantz (2001).

The *North Atlantic Oscillation* (NAO) concerns the stability of the Icelandic low-pressure cell and the Azores–Bermuda high-pressure cell. The NAO has a marked influence upon the climates of Western Europe.

The *North Pacific Oscillation* (NPO) is also known as the Pacific Decadal Oscillation (PDO). It is a long-lived phenomenon and is defined by surface ocean temperatures in the northeast and tropical Pacific Ocean. The NPO has a long-term influence on the climate of North America.

The *Madden–Julian Oscillation* (MJO) is characterized by an eastward progression of tropical rainfall. Anomalous rainfall patterns, which may be enhanced or suppressed, are first identified over the western Indian Ocean and remain viable across the warm waters of the western and central Pacific Ocean.

The *Pacific North American Oscillation* (PNA) is an alternating pattern between pressures in the central Pacific Ocean and centers of action over western Canada and the southeastern United States. It is expressed as an index that is both ocean- and land-based.

The *Arctic Oscillation* (AO) is an oscillation in which atmospheric pressure, at polar and midlatitude locations, fluctuates between defined positive and negative phases. It is computed as an index by comparing pressure in the polar region with pressure at 45°N.

The *Antarctic Oscillation* (AAO) is represented by an oscillation in values of mid- and high-latitude surface pressure systems in the southern hemisphere. It is quantified by the Antarctic Oscillation Index (AAOI). This is the monthly zonal mean sea level pressure at 45°S in relation to the same variable calculated for 65°S.

The *Quasibiennial Oscillation* (QBO) is a low-latitude oscillation that is longer than the dominant annual cycle and whose identification rests with sophisticated computer modeling. Associated with fluctuations of the ITCZ, it is a periodic reversal of winds in the lower stratosphere at elevations between 20 and 30 km.

Most oscillations are involved with the strength and location of the major planetary highs and lows (*centers of action*) and sea surface temperatures, and are derived statistically from long-range pressure observation series. For some, and to a lesser extent, rainfall and temperature series can be employed. Much information about the calculation and use of the oscillations may be derived from NOAA, Climate Prediction Center (http://www.ncep.gov/).

John E. Oliver

Bibliography

Bjerknes, J., 1966. A possible response of the atmospheric Hadley circulation, to equatorial anomalies of ocean temperature. *Tellus*, **8**: 820–829.

Bond, N.A., and Harrison, D.E., 2000. The Pacific Decadal Oscillation, air–sea interaction and central north Pacific winter atmospheric regimes. *Geophysics Research Letters*, **27**(5): 731–734.

Glantz, M.H., 2001. *Currents of Change: El Niño and La Niña impacts on climate and society*. Cambridge: Cambridge University Press.

Malony, E.D., and Hartmann, D.L., 2000. Modulation of hurricane activity in the Gulf of Mexico by the Madden–Julian Oscillation. *Science*, **284**: 2002–2004.

Mantua, N.I., Hare, S.R., Zhang, Y., Wallace, I.M., and Francis, R.C., 1997. A Pacific decadal climate oscillation with impacts on salmon. *Bulletin of the American Meteorological Society*, **78**: 1069–1079.

Maruyama, T., 1997. The Quasi-Biennial Oscillation (QBO) and equatorial waves – a historical review. *Meteorology and Geophysics*, **48**: 1–17.
Pekeris, C.L., 1937. Atmospheric oscillations, *Royal Society (London) Proceedings*, **A158**: 650–671.
Rogers, J.C., 1984. The association between the North Atlantic Oscillation and the Southern Oscillation in the northern hemisphere. *Monthly Weather Review*, **112**: 1999–2051.
Thompson, D.W.J., and Wallace, J.M., 1998. The Arctic Oscillation signature in the wintertime geopotential height and temperature fields. *Geophysics Research Letters*, **25**(9): 1297–1300.
Trenberth, K.E., 1990. Recent observed interdecadal climate changes in the northern hemisphere. *Bulletin of the American Meteorological Society*, **71**: 988–993.
Walker, G.T., 1923–1924. World weather, I and II. *Indian Meteorology Department Memoir*, **24**(4): 9.
Zhang, Y., Wallace, J.M., and Battisti, D.S., 1997. ENSO-like interdecadal variability; 1900–1903. *Journal of Climate*, **10**: 1004–1020.

Cross-references

Cycles and Periodicities
El Niõ
The Madden–Julian Oscillation
North Atlantic Oscillation
Pacific North American Oscillation
Quasi-Biennial Oscillation
Southern Oscillation
Teleconnections

OZONE

Electrical discharges in air produce a distinctive smell, once thought to be that of electricity itself. This was reported in 1785 by Van Marum, who observed that oxygen gas in which electricity had sparked would then tarnish mercury. In 1840 Schöbein noted the smell again in oxygen from electrolysis of acidulated water and concluded it was a gas, which he named from the Greek *ozon*, "smelling". Five years later, de la Rive formed ozone from pure oxygen gas (O_2), proving de Marignac's idea that ozone was an allotrope, or alternate form, of oxygen, but it was not until 1865 that Soret found, from volumetric analysis, the correct formula of O_3.

Ozone is present in natural air, as first shown by Houzeau in 1858. Schöbein's paper strips saturated with starch-iodide solution, which turn blue in the presence of ozone, provided an inexpensive test, and paper ozonometry became popular in nineteenth-century Europe. Concentrations were higher in coastal areas and bright sunshine, and depleted in smoky rooms, mines, and industrial areas, so ozone was associated with clean air and good health. As a powerful oxidant it deodorizes air, and it is also a potent disinfectant. Used instead of chlorine in many water-treatment plants, it leaves only oxygen as a residue, and from the early twentieth century ozone found use for some medical treatments. The idea that ozone is good for health is heard to this day, but proof was always elusive. Like many supposed tonics its effects are damaging in some circumstances or at higher concentration, and the truth is both more complicated and more interesting.

Ozone condenses below −112°C to a (highly explosive) deep ink-blue liquid, and a 10-cm cell of pure gas is also visibly blue due to absorption of red light. Ozone even more strongly absorbs ultraviolet (UV) light, especially at wavelengths between 230 and 290 nm for which even 2–3 mm of the gas is opaque. This discovery by Hartley in 1881 enabled him to account for a sharp cut-off below 300 nm in the spectrum of sunlight, as observed by Cornu. The natural ozone concentrations of 20–50 ppb (molecules per 10^9 air molecules) at ground level could not account for the amount of absorption, so Hartley concluded that much greater ozone concentrations must occur at higher altitude. Improved spectroscopy, first by photographic techniques and subsequently by photoelectric techniques, provided dramatic confirmation in the 1920s. The clearest evidence, and approximate profiles of ozone concentration, came with the 1929 discovery by Götz of the Umkehr effect; relative absorption at two ultraviolet wavelengths reverses as light from the rising or setting sun takes different paths through layers of the atmosphere. In 1934 Götz located the maximum ozone density at 22 km altitude, and in the same year the first balloon soundings gave similar results. Exact profiles of ozone and temperature change measurably by the hour, but they all look similar for given latitude and season.

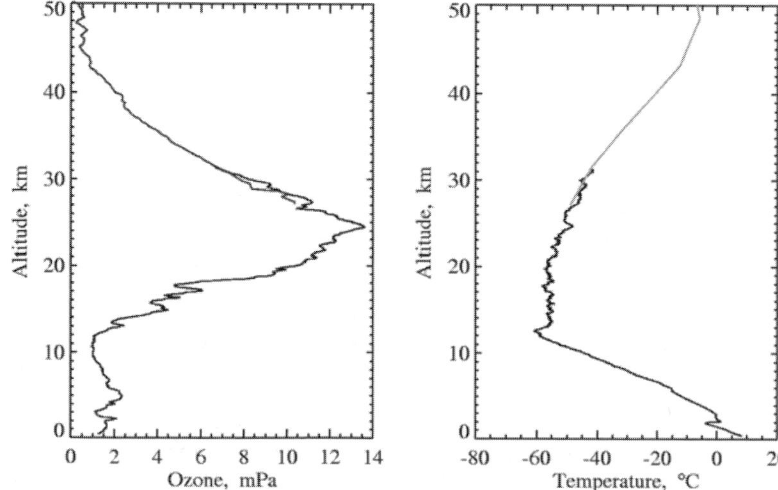

Figure O13 Profiles of ozone concentration (partial pressure) and temperature at 45°S, January 2004, measured by electrochemical sonde and lidar. The ozone layer from ~12 to 50 km altitude defines the stratosphere and maintains the temperature profile.

In 1926 Dobson built the first of the ozone spectrophotometers that bear his name. Individually numbered, and now regularly intercalibrated, over 90 of these instruments provide routine measurements of column ozone around the globe. Their stability and precision has played a pivotal role monitoring atmospheric change and in comparison with later instruments, including the satellite-borne systems which now give true global coverage. Dobson's name also persists in the unit of measurement for column density of ozone and other trace gases, the Dobson Unit (DU) or milli-atmosphere-centimeter.

A typical ozone column of 300 DU is equivalent to a layer of pure gas at standard temperature and pressure (STP) just 3 mm thick. As thin as this seems, it is sufficient to block over 95% of the solar UV that causes eye, skin, cell and DNA damage in living things, and photochemical breakdown in many materials. Life on Earth has evolved to depend on ozone for protection from damaging UV radiation, and it is threatened by reduction in the ozone layer.

Relative uniformity and stability of ozone column amounts was an early surprise from the network of ozone spectrometers. Photochemical production was expected to enhance ozone by day, and produce greater column amounts at the equator and in summer. Instead there is little diurnal change, there is more ozone over high latitudes than above the tropics, and the seasonal pattern is complex and differs between hemispheres. As found by Dobson, global ozone distribution depends mostly on atmospheric circulation, in which long-lived ozone produced 30 km above the tropics is carried toward the poles, and at all latitudes is mixed only slowly downward.

The Chapman equations

In 1930 Chapman provided a mechanism for the formation and stability of the ozone layer. At altitudes above about 20 km, UV radiation of wavelength $\lambda < 243$ nm causes photodissociation of oxygen molecules

$$O_2 + h\nu \rightarrow O + O \qquad (1)$$

Ozone is formed when the atomic oxygen, O, combines with O_2. The newly formed bond is unstable unless a third body carries away excess energy, so the process is written

$$O + O_2 + M \rightarrow O_3 + M \qquad (2)$$

The additional body, M, is a molecule of atmospheric nitrogen, N_2, or oxygen. In its turn, ozone absorbs UV at longer wavelengths to photodissociate

$$O_3 + h\nu \rightarrow O_2 + O \qquad (3)$$

Although (3) appears to destroy ozone, the atomic oxygen can again participate in (2), so together these reactions just partition "odd oxygen", the total of O and O_3. Destruction of ozone is by reaction between the odd oxygen species

$$O + O_3 \rightarrow O_2 + O_2 \qquad (4)$$

At its peak, ozone mixing ratio is about 8 ppm, or less than one ozone molecule to every 100 000 oxygen molecules, so (4) predicts only slow removal of ozone by chemistry.

The Chapman model explained the vertical profile of ozone in the middle atmosphere, and its persistence; it also explains much of the dynamics. In the lower atmosphere, temperature decreases with height, allowing warm or moist air to remain buoyant over large vertical motions and enabling the normal weather processes of cloud formation and precipitation. This region of convective motion is the "troposphere" (Greek *tropos* = turning), extending to the "tropopause" at about 17 km altitude above the tropics and about 10 km over the poles.

At higher altitude, absorption of solar UV by O_2 and O_3 in reactions (1) and (3), and some absorption by ozone of infrared radiation from below, heats the atmosphere so that temperature increases with altitude, suppressing convection." heats the atmosphere so that temperature increases with altitude, suppressing convection. Air movement is confined within layers, giving the name "stratosphere" to the region up to about 50 km altitude. Though it contains less than 10% of the total atmospheric mass, the stratosphere includes more than 90% of the ozone, whence its alternate name, the "ozone layer". There is very little mixing across the tropopause, or vertically within the stratosphere, so that air rising over the tropics carries ozone slowly outward to the poles. As a tracer of stratospheric motion, ozone measurements were used to improve weather prediction, and a global network of ozone stations was proposed for the International Geophysical Year in 1958. The World Meteorological Organization (WMO) assumed responsibility for standard procedures and the Global Ozone Observing System (G030S) was established.

The dispersal of radioactive debris from nuclear weapons tests in the 1960s gave a more detailed picture of stratospheric circulation, and with more accurate values for chemical reaction rates it became clear that the Chapman model underestimated ozone loss. Reaction (4) is supplemented with several catalytic cycles of the form

$$\begin{aligned} X + O_3 &\rightarrow XO + O_2 \\ XO + O &\rightarrow X + O_2 \end{aligned} \qquad (5)$$

The first-proposed catalyst X was the hydroxyl radical, OH, produced from water vapor. In 1970 Crutzen and Johnston discovered that oxides of nitrogen, with X = NO in (5), played a much greater role. At that time fleets of supersonic aircraft flying in the lower stratosphere were seen as the future of commercial aviation, and jet engines generate large amounts of "active nitrogen", or NO_x (NO and NO_2). Consequent damage of the ozone layer could increase the incidence of skin cancers and other health risks of UV light. In an era of increasing concern about anthropogenic pollution of terrestrial and aquatic environments, the stratosphere was now also seen to be at risk, and soon a much larger threat appeared.

Chlorofluorocarbons

Chlorofluorocarbons (CFCs) had been discovered in 1928 to solve the many problems in refrigeration with toxic and corrosive gases such as ammonia. With very good thermodynamic characteristics and nil toxicity, CFCs seemed the perfect choice, and their usage steadily increased. Through the 1950s they found use in air-conditioning, as propellants and foaming agents, and as solvents. Their chemical inertness made release to the environment seem harmless.

With the 1961 invention by Lovelock of the electron-capture detector, air samples from the remotest places on Earth were found to contain measurable, and steadily increasing, amounts of CFCs. Breakdown of CFCs in the troposphere is negligible, and they are well mixed through a hemisphere within months of release. Complete mixing between hemispheres takes 1–2 years, and transport into much of the stratosphere takes 2–5 years. It is only high in the stratosphere, exposed to solar radiation at shorter UV wavelengths, that CFCs breakdown occurs, releasing atomic chlorine. This life history of CFCs was determined in 1974 by Molina and Rowland, and they further showed that Cl atoms are very effective as the catalyst X in (5). Chlorine compounds are common in the troposphere, both from sea salt and volcanic HCl, but their high solubility in cloud water effectively traps them below the tropopause. Molina and Rowland established that stratospheric chlorine was almost entirely from manufactured CFCs. Over a century it could reduce the ozone layer by 5%.

The finding was widely publicized, and some consumer resistance to CFCs followed. A ban on CFCs in spray cans and other inessential uses was promoted, and it was enacted by the United States, Canada, Norway and Sweden in 1978. National and international agencies undertook reviews of the science and state of global ozone. In collaboration with WMO's newly constituted Global Ozone Research and Monitoring Project, the United Nations Environment Program (UNEP) established a Plan of Action on the Ozone Layer in 1977. Every 3–4 years from 1982 its reports would summarize the global consensus on ozone depletion science, environmental impacts, and technical advances. In March 1985, after 4 years of negotiation, the Vienna Convention for the Protection of the Ozone Layer was adopted by 21 nations. Though unspecific, and committing nations to do little more than encourage research, cooperate and exchange information, it was a milestone. Nations agreed in principle to tackle a global environmental problem before its effects were felt, or even scientifically proven. They did not have to wait long for that.

Antarctica

Unusually low (~200 DU) total ozone at Syowa, Antarctica, in October 1982, was reported at the Ozone Commission Symposium in 1984, but its significance was not recognized. In 1985 Farman, Gardiner and Shanklin discovered the "ozone hole" over Antarctica from Dobson spectrophotometer data. Reanalyzed satellite data from the late 1970s onward showed its extent; ozone across the continent was massively depleted every spring as sunlight returned to air trapped in the winter polar darkness.

Such a large effect had not been foreseen, but the previous concerns had primed the research community for a quick response with ground-based, *in-situ*, and satellite-borne measurements, and detailed models of stratospheric chemistry, radiation and dynamics. The mechanism of the Antarctic ozone hole was elucidated by Solomon et al. in 1986. At lower latitudes, catalytic destruction of ozone by chlorine is limited by NO, which reduces the chlorine monoxide (ClO) needed to complete the sequence. Much of the reactive chlorine is in the form of HCl or $ClONO_2$, reservoir species which do not themselves contribute to ozone destruction. In polar winter, temperatures below −78°C lead to the formation of clouds in the stratosphere, which is otherwise too dry to form clouds. Polar stratospheric cloud (PSC) particles occur as both solid and liquid, in various mixtures of nitric and sulfuric acids and water according to temperature and formation sequence. Surface reactions on PSCs catalyze the reaction

$$HCl(s) + ClONO_2 \rightarrow Cl_2 + HNO_3(s) \tag{6}$$

where the (s) denotes attachment to the surface. While temperatures remain cold enough, the PSC particles fall out of the stratosphere on a time-scale of weeks to months, removing water and active nitrogen but leaving chlorine gas. As sunlight returns in the spring it is rapidly photolysed, initiating the sequence:

$$\begin{aligned} &Cl + O_3 \rightarrow ClO + O_2 \\ &ClO + ClO \rightarrow (ClO)_2 \\ &(ClO)_2 + h\nu \rightarrow ClOO + Cl \\ &ClOO \rightarrow Cl + O_2 \end{aligned} \tag{7}$$

The first reaction is the same as in (5) for X = Cl, but it now works with the products of both the third and fourth reactions in (7). The net effect is very rapid destruction of ozone, and in the 13–20 km altitude range ozone is completely destroyed over the entire Antarctic continent. Stratospheric circulation again plays a very large part, as the extreme cold of the Antarctic stratosphere maintains a strong polar vortex isolated from mid latitude air. Bromine acts in a similar manner to, and also in concert with, the much more abundant chlorine, especially through the reactions:

$$\begin{aligned} &Br + O_3 \rightarrow BrO + O_2 \\ &BrO + ClO \rightarrow Br + Cl + O_2 \end{aligned} \tag{8}$$

As the Antarctic ozone hole steadily worsened through the 1980s and early 1990s, a massive international research effort sought to fully understand the phenomenon, monitor and model its effects, and determine its influence on global stratospheric ozone. Of particular concern was whether there would also be an ozone hole over the Arctic, which would affect a substantial proportion of the world's population, especially in the wealthy industrialized countries. With warmer temperatures and more variable winds, the Arctic vortex is much less stable than its Antarctic counterpart, and it is not confined to the winter polar night. Nonetheless ozone depletion over the Arctic was observed from about this time, and it continued to worsen even after the Antarctic ozone hole stabilized in its annual extent (Figure O14). There would indeed have been a substantial threat to northern hemisphere peoples but for the action

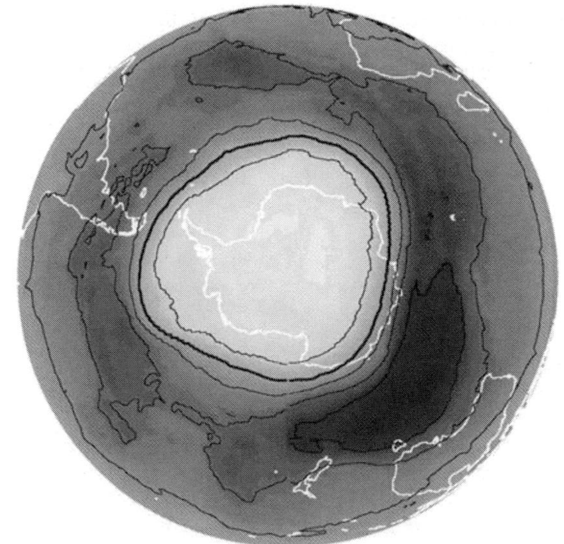

Figure O14 Southern hemisphere ozone from NASA satellite on 24 September 2003. Contours are at 60 DU intervals, with a thicker contour at 220 DU to mark the boundary of the Antarctic ozone hole. In surface area this was the second largest hole on record.

prompted by the earlier Antarctic phenomenon.

Protocols

The severity of Antarctic ozone depletion greatly hastened ratification of the Vienna Convention, and its implementation, so that in 1987 the landmark Montreal Protocol on Substances that Deplete the Ozone Layer was concluded by UNEP. It was much tougher than had been thought possible only a few months before.

The Protocol allows for special circumstances, especially developing countries with low consumption rates, but it sets the elimination of ozone-depleting substances as its final objective. When the Montreal Protocol came into force, on time, on 1 January 1989, 29 countries and the EEC, representing approximately 82% of world consumption, had ratified it. Since then several other countries have joined.

The Protocol was only a first step, but it was intentionally flexible, to be tightened or modified as the scientific evidence developed. Indeed it stipulates regular assessment of all relevant scientific, technological, environmental and economic research to guide its revision and implementation. Through international effort, understanding of stratospheric ozone chemistry and its consequences advanced rapidly, and it showed that very much tighter and greater controls would be needed. Governments and industry moved with surprising speed and determination to meet this challenge.

By 1988 a decrease of ozone concentrations by $\sim$10% per decade in the mid-latitude lower stratosphere was documented, and airborne Antarctic research proved conclusively that active chlorine and bromine byproducts of human activities are the cause of the springtime Antarctic ozone hole. The 1990 London Amendment strengthened the Montreal Protocol by phasing out all CFC production and consumption by 2000. The 1991 WMO/UNEP Ozone Assessment confirmed ozone decline not only in winter and spring but all year round, and everywhere except over the tropics; and very large concentrations of ClO measured in the Arctic increased concern for potentially stronger ozone decline. The 1992 Copenhagen Amendment sought to phase out CFCs by the end of 1995, and added controls on other compounds. Further amendments were agreed in Montreal (1997) and Beijing (1999).

In June 1991 Mount Pinatubo in the Philippines erupted massively, injecting around 20 million tonnes of sulfur dioxide directly into the stratosphere, where it oxidized to sulfate aerosol and spread over the globe. This aerosol cloud caused the greatest single disturbance to Earth's radiative balance in a century, and catalyzed ozone destruction by halogens in a manner similar to PSCs. Satellite measurements show that, in 1992–1993, global total ozone reached a minimum, not surpassed since, when it dipped to 5% below the 1964–1980 average.

Though it was the immediate cause of the ozone minimum, the volcanically derived aerosol had a large impact only because of human influence on the chemical state of the stratosphere. There is some similarity in the story of stratospheric bromine. The Montreal Protocol included, from its inception, brominated CFCs called halons that were used in fire extinguishers and suppressants. With atmospheric longevity of several decades, halons were known to be a major source of stratospheric bromine. By 1994 bromine was found to be 50 times as destructive of ozone as chlorine per atom, largely because its deactivation to reservoir species was very slow. At the same time methyl bromide, manufactured for use as a fumigant, was found to live long enough to reach the stratosphere and contribute to ozone destruction, so its production and use was restricted. Methyl bromide is produced naturally in the ocean, which is the largest source but also a sink for the gas. It was also found to be a product of biomass burning, a regular seasonal activity, especially in the tropics, over centuries. Once again, the rapid change in atmospheric chemistry in recent times had created such instability that even natural processes could trigger damage.

Between 1992 and 1994 the Antarctic ozone hole spread over more than 24 million square kilometers during spring, and column amounts less than 100 DU were recorded. Whether measured by total area, ozone minimum, or amount of "missing" ozone, the Antarctic ozone hole was at its worst from the 1990s into the early 2000s. Arctic ozone destruction became routine through this period, but with much greater variability as noted earlier.

Computational models of stratospheric chemistry and dynamics predict slow recovery of polar and global ozone through the first century of the new millennium, but at the time of writing it is too early to detect such recovery. Controls on manufacture and use of ozone-depleting substances seem to be working: total tropospheric concentrations of halogenated compounds, weighted by their ozone depletion potential (ODP), have fallen gradually since the peak in 1992–1994. The peak in stratospheric loading of halogens was delayed by mixing times to the end of the millennium, with reductions expected to be detectable by the end of the first decade.

The early beneficial image of ozone was well supported by understanding of its role in the stratosphere, which protects the surface environment from damaging, especially carcinogenic, UV radiation. At the same time, the image of tropospheric ozone took a different turn.

Urban Smog

From the late nineteenth into the mid twentieth century northern cities such as London suffered serious smog, as fog on damp winter nights became laden with smoke containing soot and sulfur dioxide from burning coal, oil, and wood. Such smog causes severe irritation of the respiratory system; it is very debilitating and can be fatal. As automobile usage increased through the mid-twentieth century, air pollution in Los Angeles, and later other cities, also caused respiratory problems and stinging eyes. It was again described as smog, but this is misleading. It is worst in sunlit conditions, with low humidity, and there was another important difference.

In traditional smog, ozone is depleted by reaction with airborne particles. By contrast Haagen-Smit showed in the early 1950s that high ozone concentrations are a feature of Los Angeles smog, which results from photochemistry in a mixture of NO_x and hydrocarbons. The basis of this process is:

$$\begin{aligned} &NO_2 + h\nu \rightarrow NO + O \\ &O + O_2 + M \rightarrow O_3 + M \\ &O_3 + NO \rightarrow O_2 + NO_2 \end{aligned} \tag{9}$$

This reaction series is rapid in sunny conditions, where the lifetime of the NO_2 molecule is just a few minutes, but alone it does not raise ozone concentration because of destruction in the third reaction, necessary to reform NO_2 and sustain the cycle. With hydrocarbons present, NO is converted to NO_2 without removal of ozone. Using R to represent a carbon chain (alkyl group), the main reactions can be summarized as:

$$RH + OH \rightarrow R + H_2O$$
$$R + O_2 + M \rightarrow RO_2 + M \quad (10)$$
$$RO_2 + NO \rightarrow RO + NO_2$$

and, rewriting R as $R'CH_2$:

$$R'CH_2O + O_2 \rightarrow HO_2 + R'CHO$$
$$HO_2 + NO \rightarrow NO_2 + OH \quad (11)$$

Carbon monoxide also contributes, through the reactions:

$$CO + OH \rightarrow CO_2 + H \quad (12)$$
$$H + O_2 + M \rightarrow HO_2 + M$$

With alkyl peroxy (RO_2) and hydroperoxy (HO_2) radicals converting NO to NO_2, as in (10) and (11), ozone produced in (9) can increase to high concentrations in photochemical smog, and it is the cause of eye and bronchial irritation as well as plant and material damage. Contrary to its past status as a sign of clean air, ozone is recognized now as an unwelcome product of industrialized society. Its concentration is an index of air pollution, and regulations in many countries seek to reduce it by controlling emissions of the NO_x and hydrocarbon precursors. The main sources of both in the developed world are combustion of fossil fuels – oil, coal, and gas – in a tradeoff with combustion temperature. Inefficient burning results in more hydrocarbon and CO emissions, while high temperatures and efficient burning produce more NO_x. There is some irony that the same jet aircraft which threaten life-supporting ozone by flying in the lower stratosphere can increase the production of ozone near the ground where it is a health hazard. Motor vehicles, however, are a much greater source of the precursors.

Understanding of these reactions in polluted environments prompted Crutzen in 1973 to another insight in global ozone chemistry. As carbon monoxide, methane, and other hydrocarbons are dispersed across the globe, reactions (9)–(12) produce ozone in the background atmosphere. Tropospheric ozone had been thought to come from exchange across the tropopause, but production *in situ* is of comparable magnitude. The industrial sources of NO_x and hydrocarbons are supplemented by rural burning of biomass, especially widespread in the tropical dry season, and so it also contributes substantially to tropospheric ozone production.

Future impacts

Again and again, the story of ozone has demonstrated the long reach of human influence on the chemistry of Earth's atmosphere. It has also shown that such influence can be controlled, through concerted international effort directed by detailed scientific study. For their part in understanding the formation and decomposition of atmospheric ozone, Paul Crutzen, Mario Molina and Sherwood Rowland were awarded the 1995 Nobel Prize in Chemistry.

Even with success in atmospheric ozone chemistry, major questions remain. Destruction of ozone has caused an increase in solar UV radiation reaching Earth's surface, where it damages living tissue. The incidence of ocular cataract and skin cancer has increased worldwide, though more as a consequence of migration and lifestyle changes than ozone depletion. Damaging as it is, UV has long been known to be essential for life through production of vitamin D, deficiency of which causes rickets and even, ironically, increased cancer risk. Human skin color has evolved over millennia from humanity's African origins so that natural pallor increases strongly with distance from the equator. Such evolution cannot keep pace with recent rapid migration, and concomitant changes to more indoor or outdoor lifestyles, so that many people are exposed to too much UV at certain times even as others may receive too little. Ozone change may compound this problem, but it has also fostered research to understand UV variation and especially its biochemical effects.

Climate change through enhanced greenhouse gas concentrations is now the main focus of worldwide concern about the effects of atmospheric chemistry on transmission of solar radiation. Here too the science of ozone depletion has made a large contribution. Many participants in ozone chemistry are greenhouse gases, including CFCs and HFCs, hydrocarbons, water vapor, and ozone itself. Accurate models of stratospheric chemistry, dynamics and radiation are a valuable adjunct to their much more complex counterparts in the troposphere. Very significantly, the success in addressing and solving human-induced ozone destruction serves as a model for the United Nations Framework Convention on Climate Change and its implementation through the Kyoto Protocol.

J. Ben Liley

Bibliography

Brasseur, G., and Solomon, S., 1986. *Aeronomy of the Middle Atmosphere: Chemistry and Physics of the Stratosphere and Mesosphere*. Dordrecht: Reidel.

Chapman, S., 1930. A theory of upper atmospheric ozone. *Memoirs of the Royal Meteorological Society*, **3**: 103–125.

Crutzen, P.J., 1970. The influence of nitrogen oxides on the atmospheric ozone content. *Quarterly Journal of the Royal Meteorological Society*, **96**: 320–325.

Farman, J.C., Gardiner, B.G., and Shanklin, J.D., 1985. Large losses of total ozone in Antarctica reveal seasonal ClO_x/NO_x interaction. *Nature*, **315**: 207–210.

Finlayson-Pitts, B.J., and Pitts, J.N., Jr., 2000. *Chemistry of the Upper and Lower Atmosphere: Theory, Experiments, and Applications*. San Diego: Academic Press.

IPCC, 2001. Climate Change 2001: *The Scientific Basis*. Contribution of Working Group I to the Third Assessment Report of the Intergovernmental Panel on Climate Change. Cambridge: Cambridge University Press.

Johnston, H., 1971. Reduction of stratospheric ozone by nitrogen oxide catalysts from supersonic transport exhausts. *Science*, **173**: 517–522.

Molina, M.J., Molina, L.T., and Kolb, C.E., 1996. Gas-phase and heterogeneous chemical kinetics of the troposphere and stratosphere. *Annual Review of Physics and Chemistry*, **47**: 327–367.

Rowland, F.S., and Molina, M.J., 1975. Chlorofluoromethanes in the environment. *Reviews of Geophysics and Space Science*, **13**: 1–35.

Schmidt, M., 1988. Pioneers of Ozone Research: A Historical Survey. Max-Planck-Institute for Aeronomy, Gttingen.

Solomon, S., Garcia, R.R., Rowland, F.S., and Wuebbles, D.J., 1986. On the depletion of Antarctic ozone. *Nature*, **321**: 755–758.

UNEP, 2002. *Environmental effects of ozone depletion and its interactions with climate change: 2002 assessment*. Nairobi: United Nations Environment Programme.

Warneck, P., 2000. *Chemistry of the Natural Atmosphere*. San Diego: Academic Press.

World Meteorological Organization (WMO), 1986. *Atmospheric Ozone 1985*. Global Ozone Research and Monitoring Project, Report no. 16. NASA/FAA/NOAA/UNEP/WMO/CEC/BMFT.

World Meteorological Organization (WMO), 1995. *Scientific Assessment of Ozone Depletion: 1991*. Global Ozone Research and Monitoring Project, Report no. 25. NASA/NOAA/UKDOE/UNEP/WMO.

World Meteorological Organization (WMO), 1995. *Scientific Assessment of Ozone Depletion: 1994*. Global Ozone Research and Monitoring Project, Report no. 37. NOAA/NASA/UNEP/WMO.

World Meteorological Organization (WMO), 1999. *Scientific Assessment of Ozone Depletion: 1998*. Global Ozone Research and Monitoring Project, Report no. 44. NOAA/NASA/UNEP/WMO/EC.

World Meteorological Organization (WMO), 2003. *Scientific Assessment of Ozone Depletion: 2002*. Global Ozone Research and Monitoring Project, Report no. 47. NOAA/NASA/UNEP/WMO/EC.

Cross-references

Air Pollution Climatology
Antarctic Climates
Energy Budget Climatology
Human Health and Climate
Montreal Protocol

P

PACIFIC (HAWAIIAN) HIGH

The (North) Pacific (Hawaiian) High is a semipermanent cell of high pressure centered in the eastern Pacific from 35 to 45°N. It is one of the principal "centers of action" in the northern hemisphere, expanding in summer and contracting in winter.

As shown in Figure P1, the Pacific High has a July mean sea-level pressure of 1026 mb (30.3 in), analogous in pressure and dimensions to the Azores High, its Atlantic counterpart. The subtropical highs are one of the key elements of the Earth's surface pressure. Their origin is not wholly understood, but it is believed that dynamic rather than direct thermal causes are foremost (Trewartha and Horne, 1980).

The Pacific waters are relatively cooler than the continents in the summer, thereby permitting the expansion of the Pacific High and saving the subtropical high-pressure belt from elimination by continental heating. It is extensive enough in the summer to spread its influence into the middle latitudes. Air tends to subside on the eastern sides of subtropical highs, heating adiabatically and leading to warm, dry weather conditions. As the Pacific High expands along the Pacific coast of North America it causes a marked reduction in rainfall. Thus, the lengthy summer drought of Mediterranean-type climates (southern California) is attributed to the effect of the expanding subtropical high.

The subtropical belt of high pressure is rather irregular at sea level, but becomes fairly uniform at higher elevations. The axis of the belt tilts southward as the warm air is to the south. At

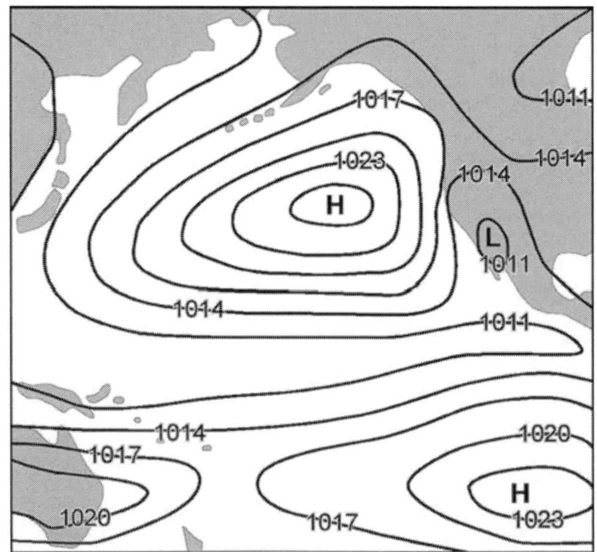

Figure P1 Average sea-level pressure in July.

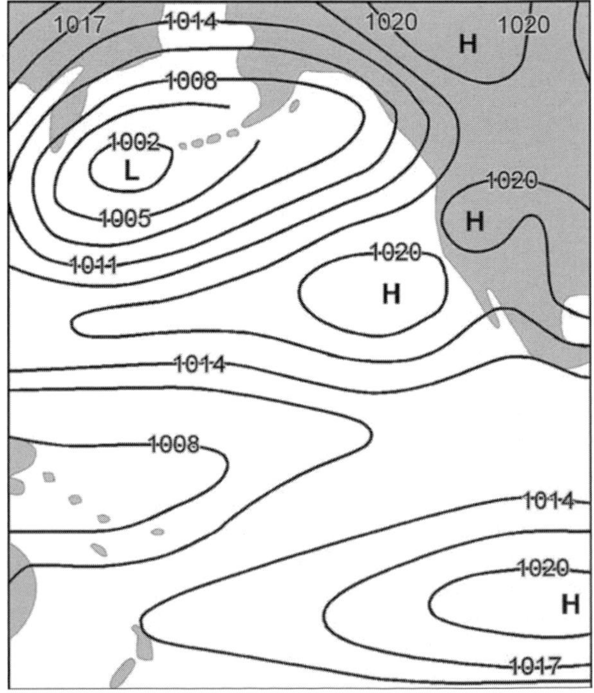

Figure P2 Average sea-level pressure in January.

3000 m (10 000 ft) the belt is located 10–15°S of its sea-level position (Petterssen, 1969).

During the winter the relative warmth of the Pacific Ocean in subtropical latitudes is insufficient to eliminate completely the dynamic subtropical belt of high pressure. Thus, the Pacific High contracts to become a weaker, smaller, but still noticeable anticyclonic cell (Willett and Sanders, 1959). As shown in Figure P2, the January mean sea-level pressure is 1020 mb (30.12 in), compared to the July mean of 1026 mb (30.3 in). Also, the Pacific High shifts equatorward during the winter, centering around 30°N.

In their study of the Pacific "centers of action", Christoforou and Hameed (1997) found that the location of the Pacific High was highly correlated with mean annual sunspot numbers. When solar activity is at a minimum during its 11-year cycle, the Pacific High migrates southward.

Robert M. Hordon

Bibliography

Ahrens, C.D., 2003. *Meteorology Today: An Introduction to Weather, Climate, and the Environment*, 7th edn. Brooks Cole. (Thomson Learning)

Akin, W.E., 1991. *Global Patterns: Climate, Vegetation, and Soils.* Norman, OK: University of Oklahoma Press.

Barry, R.G., and Chorley, R.J., 2003. *Atmosphere, Weather, and Climate*, 8th edn. London: Routledge.

Christoforou, P., and Hameed, S., 1997. *Solar cycle and the Pacific "centers of action". Geophysical Research Letters*, **24**(3): 293–296.

Curry, J.A., and Webster, P.J., 1999. *Thermodynamics of Atmospheres and Oceans.* San Diego, CA: Academic Press.

Graedel, T.E., and Crutzen, P.J., 1993. *Atmospheric Change: An Earth System Perspective.* New York: Freeman.

Hartmann, D.L., 1994. *Global Physical Climatology.* San Diego, CA: Academic Press.

Lutgens, F.K., and Tarbuck, E.J., 2004. *The Atmosphere: An Introduction to Meteorology*, 9th edn. Upper Saddle River, New Jersey: Pearson Prentice-Hall.

Pettersen, S., 1969. *Introduction to Meteorology*, 3rd edn. New York: McGraw-Hill.

Robinson, P.J., and Henderson-Sellers, A., 1999. *Contemporary Climatology*, 2nd edn. Harlow: Longman.

Trewartha, G.T., and Horne, L.H., 1980. *An Introduction to Climate*, 5th edn. New York: McGraw-Hill.

Wallace, J.M., and Hobbs, P.V., 1997. *Atmospheric Science: An Introductory Survey.* San Diego, CA: Academic Press.

Willett, H.C., and Sanders, F., 1959. *Descriptive Meteorology*, 2nd edn. New York: Academic Press.

Cross-references

Airmass Climatology
Atmospheric Circulation, Global
Centers of Action
Ocean–Atmosphere Interaction
Zonal Index

PACIFIC NORTH AMERICAN OSCILLATION (PNA)

The Pacific North American oscillation (PNA) is an alternating pattern between pressures in the central Pacific Ocean and centers of action over western Canada and the southeastern United States. The PNA is associated with a Rossby wave pattern and refers to the relative amplitudes of the ridge over western North America and the troughs over the central North Pacific and southeastern United States. Such a pattern tends to be most pronounced in the winter months.

There are two phases of this teleconnection. A positive phase occurs when deeper than normal troughs occur over the eastern United States and the region of the Aleutians. A negative phase (sometimes called the reverse Pacific North American pattern) is characterized when the troughs are filled and a ridge over the Rockies is lowered. The positive phase is associated with meridional upper air flow and the negative by zonal upper air flow.

According to Wallace and Gutzler (1981) the PNA is a "quadripolar" pattern of pressure height anomalies. Anomalies with a similar signs are located south of the Aleutian Islands and the southeastern United States and those with opposite signs (to the Aleutian center) are in the vicinity of Hawaii and the intermountain region of Canada.

To identify PNA patterns the Climate Prediction Center (CPC) of NOAA uses empirical orthogonal functions (EOF) applied to the 500-hPa height anomalies in the northern hemisphere. Daily and monthly PNA values are constructed and used in climate prediction. Figure P3 provides an example of the plotted data that the CPC provide.

In a two-part study of the role of the PNA in the climate of the United States Leathers et al. (1991) found a number of meaningful relationships. They demonstrated that regional temperatures and precipitation are highly correlated to a derived PNA index across the United States, especially in winter. Correlations for precipitation were less extensive than those of temperature, but clear relationships were obtained. Studies have also found that the PNA index is strongly linked to moisture variability in an area extending from southeast Missouri to Ohio, with a very strong correlation in southern Indiana. It has been observed that cyclones in the Great Lakes region of North America originate more from the northwest in those months when PNA-positive occurs, while cyclones from the west and southwest occur more often with a PNA-negative index. From these studies, and others like them, it is clearly evident that the PNA plays a significant role in the climates of North America.

John E. Oliver

Bibliography

Coleman, J.S.M., and Rogers, J.C., 1995. Ohio River Valley winter moisture condition associated with the Pacific-North American teleconnection pattern. *Journal of Climate*, **16**: 969–981.

Latif, M., and Harnett, T.P., 1996. Decadal climate variability over the North Pacific and North America: dynamics and predictability. *Journal of Climate*, **9**: 2407–2423.

Leathers, D.J., Yarnal, B., and Palecki, M.A., 1991. The Pacific/North American teleconnection pattern and United States Climate. Part I: Regional temperature and precipitation associations. *Journal of Climate*, **4**: 517–528.

Leathers, D.J., Yarnal, B., and Palecki, M.A., 1991. The Pacific/North American teleconnection pattern and United States Climate. Part II: Temporal characteristics and index specification. *Journal of Climate*, **4**: 707–716.

Minobe, S., 1997. A 50–70-year climatic oscillation over the North Pacific and North America. *Geophysical Research Letters*, **24**: 683–686.

Rogers, J.C., 1984. The association between the North Atlantic Oscillation and the Southern Oscillation in the northern hemisphere. *Monthly Weather Review*, **112**: 1999–2051.

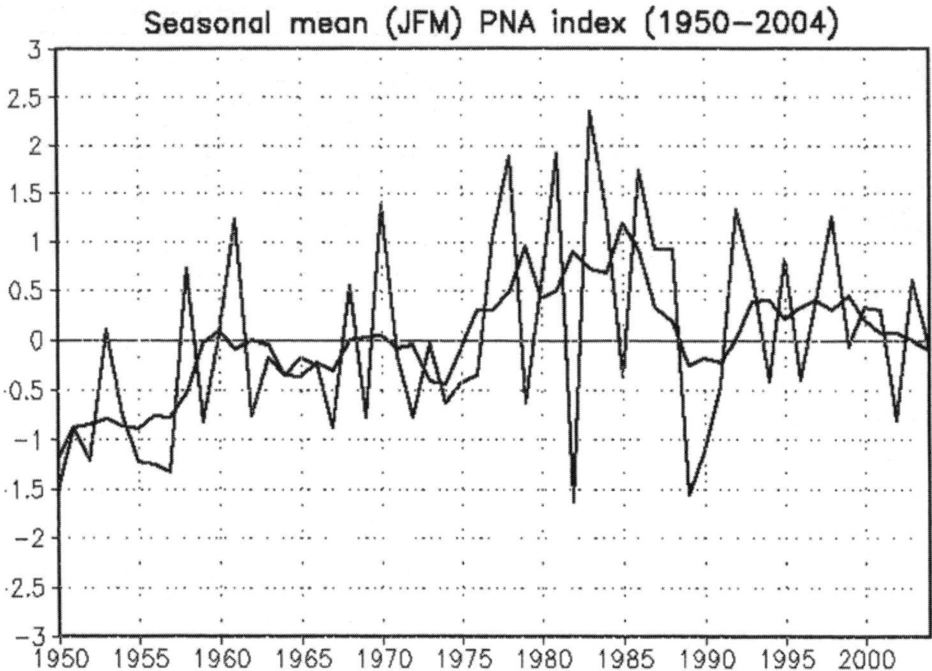

Figure P3 The Climate Prediction Center of NOAA produce a PNA index for various time periods. Here the seasonal winter mean for the period 1950–2000 shows the changing pattern of positive and negative phases. (*Source*: www.cps.noaa.gov.)

Wallace, J.M., and Gutzler, D.S., 1981. Teleconnections in the geopotential height field during the northern hemisphere winter. *Monthly Weather Review*, **109**: 784–812.

Cross-references

Centers of Action
North Atlantic Oscillation
Oscillations
Teleconnections

PALEOCLIMATOLOGY

Climate has changed through geological time, and paleoclimatology is the study of such changes (Frakes, 1978; Ruddiman, 2001). The changes have taken place at a variety of scales, and range from minor fluctuations within the instrumental record (with durations of the order of a decade or decades) to major geological periods (with durations of many millions of years). The shorter-term changes include events such as the period of warming that took place in the first decades of the twentieth century. The changes with durations of hundreds of years were characteristic of the Holocene and include various phases of glacial advance and retreat such as the Little Ice Age between 1500 and 1850. The fluctuations within the Pleistocene Ice Age (Bowen, 1978), consisting of major glacials and interglacials, lasted for 10 000–100 000 years, whereas the Pleistocene itself, termed *a minor geological variation*, consisted of a group of glacial and interglacial events that lasted in total approximately 2 million years. Such major phases of ice age activity appear to have been separated by about 250 million years. Figure P4 shows the nature of these different orders of change.

At the beginning of the nineteenth century it was still widely believed that the world was created in 4004 BC, and that its surface had been molded by catastrophic events such as the biblical Noah's flood. Given such a short time span for Earth history, and given the invocation of the flood to explain changes in the sedimentary record, it is not surprising that at that time there were few grounds for believing that the Earth's climate had undergone major changes through an extended time period (Goudie, 1992). However, as geologists examined the sedimentary and paleontological record, the evidence for the Earth having a more lengthy history became apparent, and the idea that climate and other aspects of the environment had fluctuated during this enlarged span of time resulted initially from the discovery that Norwegian and Alpine glaciers had formerly extended farther than the bounds of their current limits. After the 1820s the concept of the Ice Age became accepted by more and more geologists, notably by Louis Agassiz. By the 1860s it was evident that the Pleistocene ice age had consisted of multiple glaciations, and Penck and Brükner introduced their highly influential fourfold model of the Güz, Mindel, Riss, and Wüm glacials from work in Bavaria. Early studies of continental drift involved an appreciation of how climates in particular areas might have changed over longer time-scales, thereby explaining the presence of Dwyka glacial deposits in central Africa in the Permo-Carboniferous (see, for example, Wegener, 1967). Major developments in paleoclimatological studies occurred after World War II when isotopic methods for paleotemperature assessment became available, dating techniques such as radiocarbon and paleomagnetism enabled sequences to

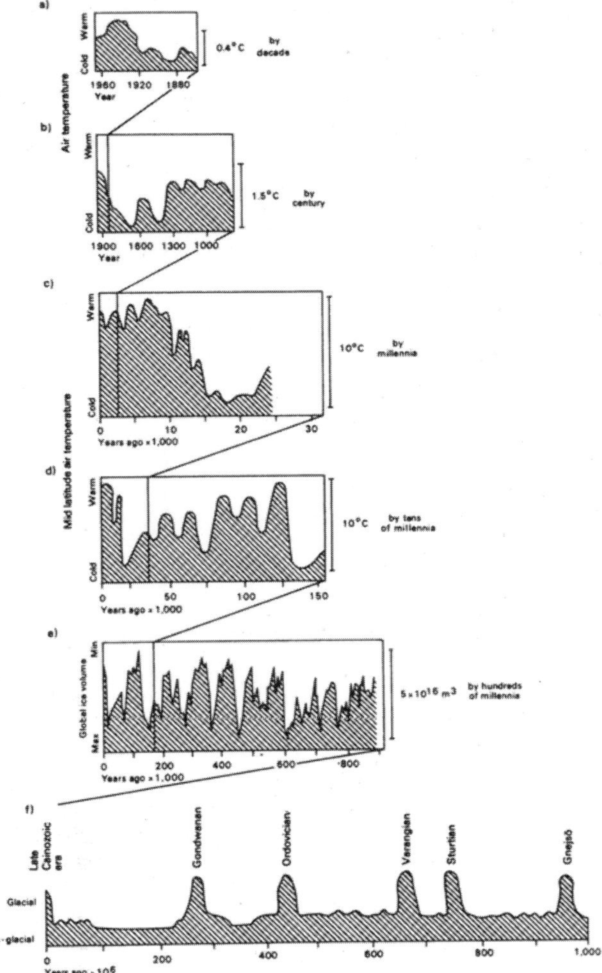

Figure P4 Different scales of climatic change. (**a**) Changes in the 5-year average surface temperature over the region 0–80°N; (**b**) winter severity index for eastern Europe; (**c**) generalized northern hemisphere air-temperature trends, based on fluctuations in alpine glaciers, changes in tree-lines, marginal fluctuations in continental glaciers, and shifts in vegetation patterns revealed in pollen spectra; (**d**) generalized northern hemisphere air/temperature trends, based on midlatitude sea-surface temperature, pollen records, and worldwide sea-level records; (**e**) fluctuations in global ice-volume recorded by changes in isotopic composition of fossil plankton in a deep-sea core; (**f**) spacing of ice ages in geological time.

be established more firmly, and improved techniques involving piston corers enabled lengthy, relatively undisturbed sedimentary sequences to be recovered from the ocean floors (Turekian, 1971; Bradley, 1999).

Techniques of paleoclimatological analysis

The various lines of evidence used to infer past climates are listed in Table P1. The most reliable information comes from instrumental observations made since the late seventeenth century. Climatic information from preinstrumental times is provided by various historical sources, for many types of material have survived: accounts of severe climatic events, information about recurring phenomena such as crop yields and harvest dates, and statements concerning weather-related events such as the dates of freezing of lakes and rivers. However, historical accounts, though useful (see, for example, Lamb, 1982), are subjective and generally qualitative.

Over longer time spans more indirect evidence for climatic change has to be employed (Lowe and Walker, 1997). Any variable that responds in an identifiable manner to climatic change and that can be dated can be used. Such data vary greatly in terms of the facet of climate that they indicate, their sensitivity, their speed of response and the reliability with which they can be dated. In Table P1 these methods are divided into certain main classes: the faunal and floral, the sedimentological, and the geomorphological.

Of all these different techniques, the ones that must be singled out for special attention because of their importance are those associated with the use of deep-sea cores (Kukla, 1977). Although the ocean floors are by no means completely stable (burrowing organisms, solution, and currents create problems), they offer a more continuous and lengthy stratigraphic record than most terrestrial sections, so that cores of sediment taken from them can be used and interpreted in a variety of ways.

Table P1 Sources of information on climatic change

Source
Instrumental records since the late seventeenth century
Historical sources, especially since AD 1000
Faunal and floral information
Terrestrial plants (e.g. palms, mangroves, etc.) in fossil form
Large reptiles, probably indicating warmth
Corals, indicating warmth
Giant shelled mollusca, indicating warmth
Coal, indicating humidity
Peat remains, indicating vegetation types
Pollen analysis of sediments, indicating vegetation types
Temperature-sensitive Foraminifera in deep-sea cores, etc.
Coleoptera
Tree rings, indicating warmth and moisture
Sedimentological and stratigraphic information
Ironstones and bauxites, indicating tropical moisture
Reed beds, indicating tropical moisture and aridity
Evaporites, indicating hot (?) and dry conditions
Eolian sandstones, indicating desert conditions
Limestones, thick, pure sequences indicating warm seas
Phosphates, e.g. collophane, indicating arid coast upwelling
Tillites, lithified glacial material, indicating cold conditions
Iceberg rafted debris in deep-sea cores, indicating cold conditions
Eolian dust on land (loess) and in ocean cores, indicating aridity
Oxygen isotope composition of fossils, indicating temperature
Chemical composition of lake cores, indicating precipitation levels
Oxygen isotopic and chemical composition of deep ice cores, indicating temperature
Fossil soils, e.g. calcrete crusts, indicating semiaridity
Geomorphological information
Pingos, palsen, ice wedge polygons, etc., indicating permafrost
Cirques, indicating snow line positions
Closed lake basin shorelines, indicating precipitation
Fossil dunes, indicating aridity and wind direction
Tufa mounds, indicating higher groundwater
Caves, alternations of solution, indicating moisture, and precipitation, indicating aridity
Angular screes, indicating frost action
Misfit valley meanders, indicating higher stream discharge levels
Eolian fluted bedrock and shaped deflation basins, indicating aridity, deflation, and wind direction

They can be dated by radiometric means, paleomagnetic epochs, whether normal or reversed, can be examined, and the lithological characteristics of the sediments within the cores can be determined with a view to finding out about changes in terrigenous sediment sources (Table P2). One of the most productive ways of examining the cores has been the study of changes in the frequency of particular "sensitive species" of Foraminifera that are thought to reflect changes in the temperature of ocean waters (e.g. *Globorotalia menardii*). Another way in which the Foraminifera can be used is by the measurement of the O^{18}/O^{16} ratios in their calcitic tests, for this ratio reflects the temperature of the water in which the beasts lived and also the volume of water stored up in the ice caps. During glacial episodes immense ice sheets of isotopically light ice (i.e. depleted in O^{18}) accumulated in North America and over Scandinavia. When this occurred, the oceans diminished in volume, became slightly more saline, and isotopically more positive (i.e. enriched in O^{18}). This enrichment is recorded in the isotopic composition of the Foraminifera. The changing composition thus provides a record of glacials and interglacials. In addition, deep-sea cores may contain varying proportions of aeolian debris, fluvial sediments, and iceberg rafted debris that indicate varying inputs of material from terrestrial sources.

Early climate history of the Earth

Little is known about the very early climatic history of the Earth, for rocks deposited before about 2800 million years ago are generally so metamorphosed that it is extremely difficult to extract climatic information from them. Indeed, the only way of deducing the climatic history from the formation of the Earth (4500 million years ago) until 2800 million years ago is by numerical modeling, with certain assumptions made about the gaseous content of the atmosphere. If the carbon dioxide and ammonia contents of the early atmosphere were considerably elevated over present levels, then it is likely that Earth surface temperatures would also have been elevated because of a marked greenhouse effect.

In the early Proterozoic (around 2700–1800 million years ago) evidence from Canada suggests that the first known glaciation occurred. It was probably a complex event with recurring glacial events rather than a continuous glaciation, and it was followed by a relatively warm phase that persisted to around 950 million years ago. During the late Precambrian at least three major glacial episodes occurred, each lasting about 100 million years and centering on about 615 million, 770 million, and 940 million years ago.

Toward the end of the Proterozoic, the record of climate improves because the sedimentary record is better preserved and animals with shells that could be preserved as fossils had spread through the oceans. In general climatic conditions between 570 million and 225 million years ago were relatively warm, though in the Sahara there is evidence of a brief glacial event in the late Ordovician around 430 million years ago. The Permocarboniferous was characterized by a long glaciation from 330 to 250 million years ago, and ice caps were extensive on the supercontinent of Pangaea.

During the succeeding Mesozoic era the evidence of deep-sea cores becomes of greater utility. Temperatures ranged from 10–20°C at the poles to 25–30°C at the equator. There were, however, fluctuations in temperatures, with slight cooling taking place at the start of the Jurassic and with marked high-latitude warming taking place during the first half of the Cretaceous. Global cooling took place toward the end of the Cretaceous, however, and a long-term cooling trend commenced at the start of the Eocene epoch some 55 million years ago. The Cenozoic suffered various marked short-term cooling episodes: at the mid-Palaeocene (*ca.* 60 million years ago), in the Middle Eocene (*ca.* 45 million years ago); at the Eocene–Oligocene boundary (*ca.* 38 million years ago); and in the Middle Oligocene (*ca.* 28 million years ago). There is some evidence that the Antarctic glaciers may have come into existence in the Eocene and that about 10–15 million years ago mountain glaciers came into existence in the northern hemisphere.

There is also evidence for climatic deterioration from floral remains, and the general decline in temperatures during the Tertiary is shown in Table P3. In the early Tertiary the North Atlantic region was characterized by a widespread, tropical rainforest vegetation. For example, in the London–Hampshire basin of southern England there was the Malaysian *Nypa* palm and *Avicennia* (mangrove) swamp. In the pre-Miocene Bovey Tracey beds in western England are the remains of many tropical plants including *Osmunda, Calamus, Ficus, Symplocos*, and *Lauras*. However, by Pliocene times the tropical vegetation of the North Atlantic Region was being replaced by a largely deciduous, warm temperate flora including *Sciadopitys, Tsuga, Sequoia, Taxodium*, and *Carya*. The last appearance of palms north of the European Alps is in the late Miocene flora at Lake Constance.

The Tertiary may also have been a time of increasing aridity. Studies of aeolian material in deep-sea cores from the Pacific Ocean (Leinen and Heath, 1981) have demonstrated that there were low rates of dust deposition 50–25 million years ago. This, Leinen and Heath believe, reflects both the temperate, humid environment that was seemingly characteristic of the early Tertiary and the lack of vigorous atmospheric circulation at that time. From 25 million to 7 million years ago the rate of

Table P2 Information to be gained from deep-sea cores

Indicator	Environmental information
O^{18}/O^{16} ratios	Temperature, ice volume
Coarse debris	Iceberg rafting
Eolian dust	Aridity
Sensitive species Foraminifera	Temperature
Total species Foraminifera	Temperature
Clay minerals	Weathering regimes on land
Fluvial sediments	River inputs
Coccolith carbonates	Temperature
Aeolian sand turbidites	Aridity

Source: From Goudie (1992), p. 20.

Table P3 Tertiary mean annual temperatures (°C)

Geological period	Northwest Europe	Western United States	Pacific Coast of North America
Recent	—	—	10
Pliocene	14–10	8–5	12
Miocene	19–16	14–9	18–11
Oligocene	20–18	18–14	20–18.5
Eocene	22–20	25–18	25–18.5

Source: From data in Butzer (1972), p. 19.

aeolian accumulation on the ocean floor increased, but it became greatly accelerated from 7 million to 3 million years ago. However, it was around 2.5 million years ago, at the end of the Tertiary, that the most dramatic increase in aeolian sedimentation occurred. This accompanied the onset of northern hemisphere glaciation.

The causes of the Cenozoic climate decline, illustrated in Figure P5, are still not fully understood, but the trend seems to be associated with the breakup of the ancient supercontinent *Pangaea* into the individual continents we know today. About 50 million years ago Antarctica separated from Australia and gradually shifted southward into its present position, centered over the South Pole. At the same time the continents of Eurasia and North America moved toward the North Pole. As more and more land became concentrated in high latitudes, ice caps could develop, surface reflectivity could increase, and, as a consequence, climate probably cooled all over the world. In addition, as the southern continents drifted away from Gondwanaland, India collided with Eurasia to produce the Tibetan Plateau, a feature so large and high that it changed the whole circulation of the northern hemisphere and drew cold Arctic air into Europe and Canada. Moreover, the closure of the straits of Panama in the Pliocene caused the ocean circulation to become less zonal and more meridional, thereby modifying the crucial thermohaline circulation.

Pleistocene and Holocene

The climatic deterioration of the later Tertiary eventually led to the glacial events of the Pleistocene. The classic model for the

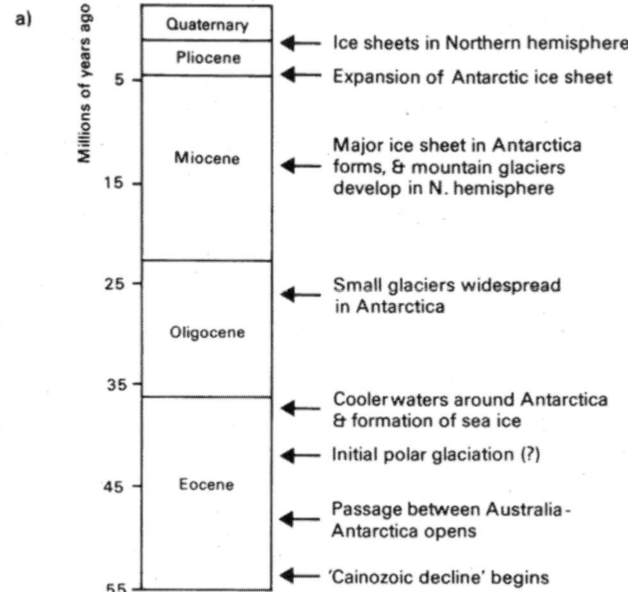

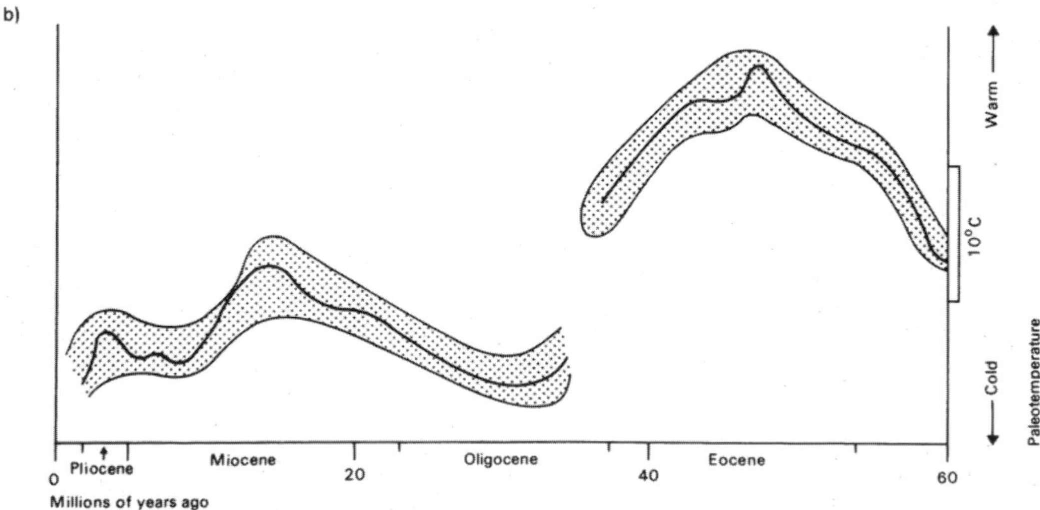

Figure P5 Cenozoic climatic decline. Generalized outline of significant events: temperature changes calculated from oxygen isotope values of shells in the North Sea (*source:* Goudie, 1983).

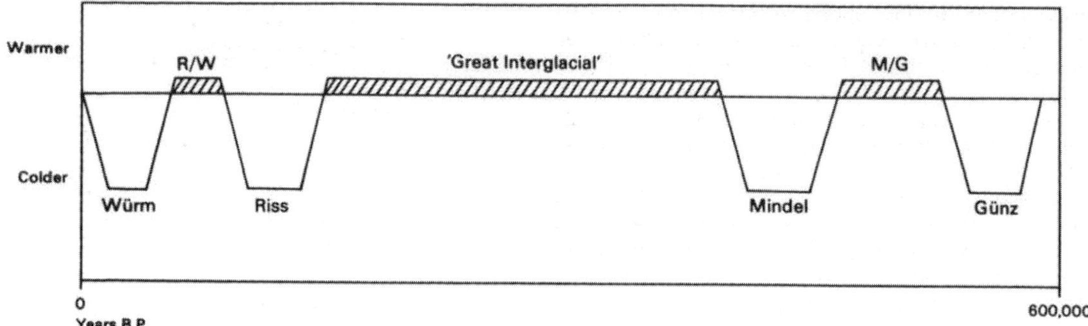

Figure P6 Simple Penck and Brückner model of Pleistocene glaciations showing the four main glacials. Note the extent of the "Great Interglacial".

Quaternary glaciations was proposed by Penck and Brükner between 1901 and 1909 in a major three-volume work *Die Alpen in Eiszeitalten*. By studying what was left of old moraines, they were convinced that there had been four great glaciations of different intensities in the alpine region of Europe. They also proved, from the preservation of plant remains at some sites such as Hötting, that some of the intervening periods were fairly mild. They named the four glacial periods after valleys in which evidence of their existence occurred: the Güz, Mindel, Riss, and Wüm. Of particular note was their belief that the interglacial between the Mindel and the Riss lasted a longer time than any other. In the classic model this event was therefore termed the *Great Interglacial*. Their model is summarized in Figure P6.

The various terrestrial sequences that have been laboriously compiled by classic stratigraphic techniques have demonstrated beyond doubt that the Quaternary was a period of remarkable geomorphological and ecological instability. However, because of the incomplete nature of the evidence created by hiatuses in the depositional record occasioned by erosion, it is not possible to use this information to construct a correct, long-term model of environmental changes for this period. For this to be achieved it is necessary to look at the more complete record of deposition preserved in the ocean core sediments.

A Pacific Ocean piston core, V-28–238, taken from near the equator at a depth of over 3000 m, has been used as the type locality of the O^{18}/O^{16} record. The upper 14 m of the core have been subdivided into 22 oxygen isotopic stages numbered in order of increasing age. Odd-numbered units are relatively deficient in O^{18} (and thus represent phases when temperatures were relatively high and the ice caps small), whereas the even-numbered units are relatively rich in O^{18} and represent the colder phases when the ice caps were more extensive. Boundaries separating especially pronounced isotopic maximums from exceptionally pronounced minimums have been called *terminations*. They are in effect rapid deglaciations and are conveniently numbered by Roman numerals in order of increasing age. The segments bounded by two terminations are called glacial cycles. The youngest glacial cycle, which starts with isotopic stage 1 and follows Termination I, is not yet completed.

The results of the analysis of this major type core, illustrated in Figure P7, can be summarized as follows. There have been eight completed glacial cycles (named *b* to *i*) and nine

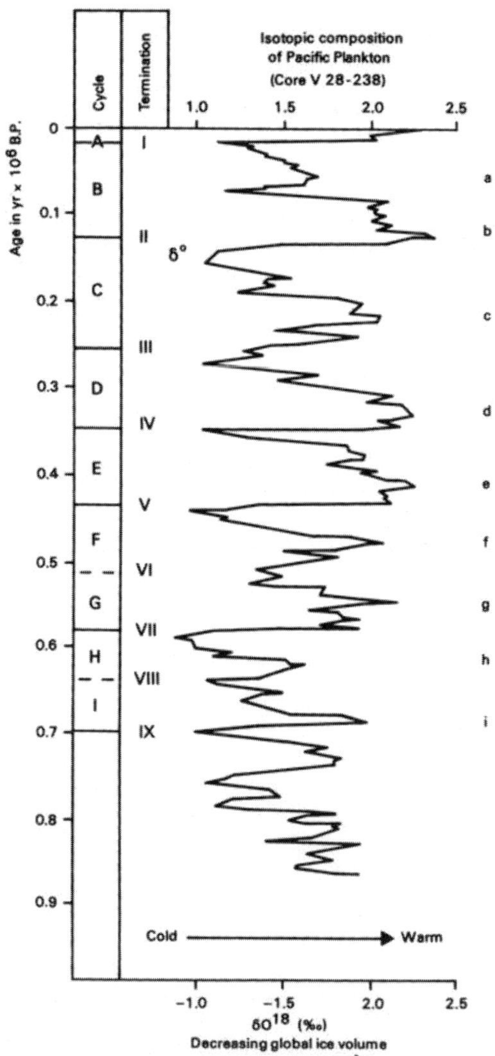

Figure P7 History of the last 900 000 years as revealed in isotopic composition of a Pacific core (V-28-238).

terminations (I-IX) in the last epoch of normal polarity (the Brunhes), which lasted just over 0.7 million years. There are 10 completed glacial cycles and 11 terminations in the 1 million years following the Jaramillo normal polarity event. Oscillations of the ice cap volume within each glacial cycle seem to follow a sawtooth pattern, progressing from an early minimum to a late maximum. The number of secondary fluctuations in isotopic composition within each glacial cycle varies, commonly between four and six. Over the last million years the general shape and amplitude of the last nine completed glacial cycles shows no great variation. However, cold stages 2, 6, 12, 16, and 22 are marked on the average by deeper and/or longer-lasting isotopic highs than the remaining cold stages, with cold stage 14 seeming to be one with an exceptionally low ice volume. The warm peaks (except stage 3) all seem to approach a similar level, indicating that global ice volume and sea-surface temperatures in peak interglacials were similar to those of the present day. During the last 0.9 million years there have been nine episodes with global climate comparable to today; in other words, there have been nine interglacials. The interglacials only constitute about 10% of the time, and seem to have had a duration of the order of 10^4 years, whereas a full glacial cycle seems to have lasted on the order of 10^5 years. Conditions such as those experienced in the twentieth century have thus been relatively short-lived and atypical of the Pleistocene as a whole. There appears to be scant evidence in the deep-sea cores for the Great Interglacial of Penck and Brückner.

The oxygen isotope record in the ocean sediments reveals a number of cycles or pacings related to the orbital variations of the Earth around the sun (the Milankovitch cycles) at 100, 41, 23, and 19 thousand years. There are also many shorter changes of an abrupt nature that have been identified both in the ocean cores and in ice cores (Pettit et al., 1999). The rapid temperature oscillations that hav been identified from ice cores are known as Dansgaard–Oeschger events. Dansgaard et al. (1993), for example, documented no less than 24 interstadials (warmer phases) from the GRIP ice core in the last glacial period. In ocean cores, high-frequency abrupt changes have been termed Bond cycles (Oppo et al., 1998). Also within the ocean core record are deposits interpreted as the result of deposition by massive armadas of icebergs (Bond et al., 1992). These are called Heinrich events and represent stadials (colder phases) of short duration (less than 1000 years) (Andrews, 1998). The causes of such abrupt short changes may include changes in the thermohaline circulation of the oceans and various feedback processes (Adams et al., 1999).

Going further back, the record seems less clear. Shackleton and Opdyke (1977) found from their study of Core V23–179 that glacial–interglacial cycles had been characteristic of the last 3.2 million years but that the scale of glaciations increased around 2.5 million years ago. In all there have been about 17 cycles in the Pleistocene *sensu stricto* (i.e. in the last 1.6 million years).

A fairly precise comprehension of the nature of one glacial cycle, the last, has been gained from other relatively continuous depositional sequences (Figure P8), including the polar ice caps and some large bogs such as Grand Pile in France (Woillard, 1979). From about 116 000 years ago until 60 000–70 000 years ago there was a relatively warm interval. The onset of the last glacial was about 60 000 BP, but the cold period following the initial fall in temperature lasted for only a few thousand years and was followed by a relatively clement period (an interstadial) that lasted until about 30 000 BP. A further cold period, the Glacial Maximum, followed. This lasted about 10 000 years and the lowest temperature and the greatest extent of ice sheets of this entire glacial episode were attained at about 20 000–18 000 BP. Widespread deglaciation began abruptly about 14 000 BP, so that by 10 000 BP the North American Cordilleran Ice Sheet had disappeared.

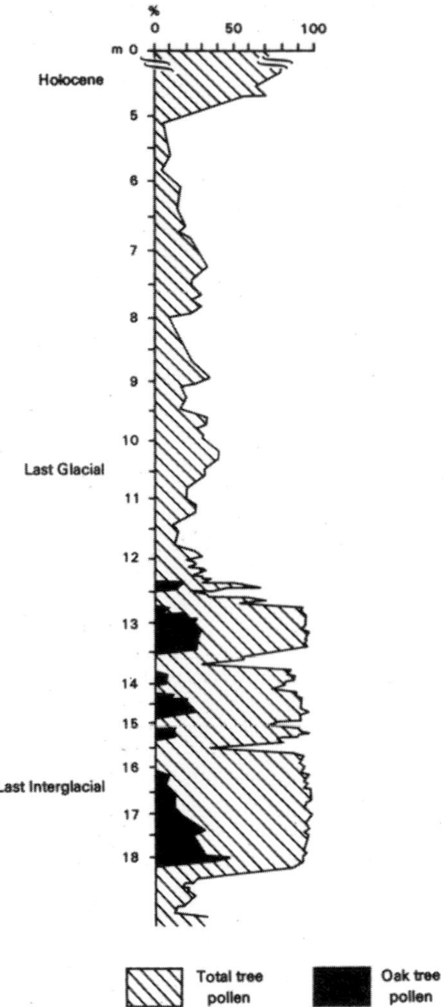

Figure P8 Pollen record from Grande Pile, France.

The Holocene has been a time of relative warmth – an interglacial – though it has also seen some substantial changes in climate. Glaciers have oscillated during a series of Neoglacial advances and retreats (Grove, 1988), and there is evidence that there have been some warmer phases than the present, such as the so-called *Climatic Optimum* of the mid-Holocene when temperatures may have been 1–3°C higher than now. There have also been a number of abrupt events at *ca.* 8200 BP and between 3500 and 3900 BP (Alley et al., 1997; Anderson et al., 1998).

Low-latitude conditions

The glacials and interglacials of the Pleistocene were complemented in lower latitudes by changes in precipitation

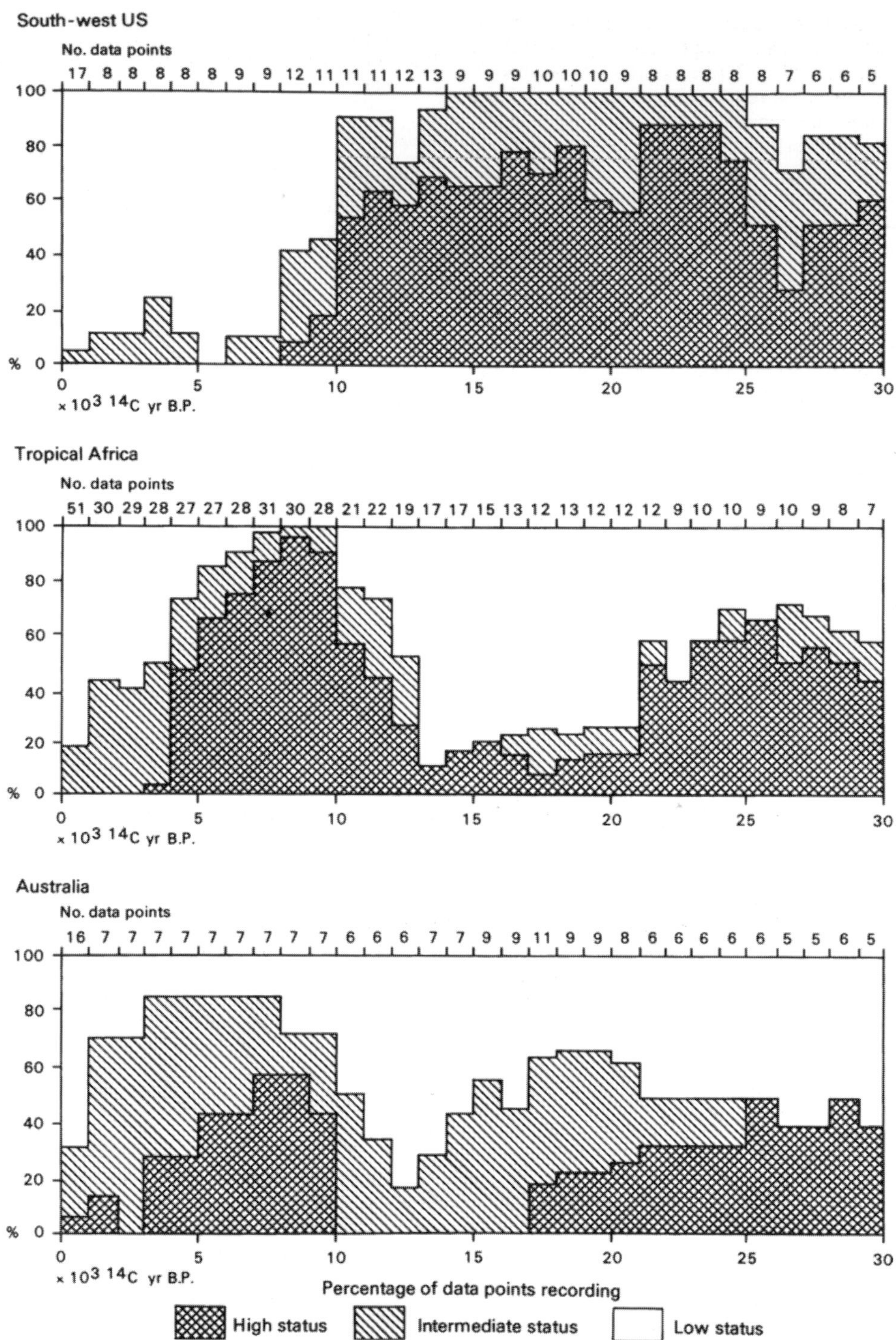

Figure P9 Histograms showing lake-level status for 1000-year time periods from 30 000 BP to the present day for three areas: southwestern United States, intertropical Africa, and Australia (*source:* Goudie, 1983).

(Street, 1981). At certain times there appears to have been greater humidity (pluvials or lacustrals), whereas at others aridity was increased so that the major deserts of the world expanded. In many parts of lower latitudes the last Glacial Maximum (*ca.* 18 000 years ago and slightly after) was a time of low rainfall, and it has been calculated that deserts then characterized 50% of the land area between 30°N and 30°S, forming two vast belts. In between, tropical rainforests and adjacent savannas were reduced to a narrow corridor. The prime evidence for this aridity is provided by fossil dune systems (ergs) (Goudie, 2002). There is also dramatic evidence for wetter conditions in the form of ancient lake shorelines. The greatest concentration of pluvial lakes in the western hemisphere, and possibly in the world, occurred in the Basin and Range Province of the United States. Lake Bonneville, for example, as its maximum extent was almost the size of Lake Michigan but now contains only 2600–6500 km^2 of water (the Great Salt Lake in Utah). The dates for such high lake levels vary in different parts of the world (Figure P9). In the southwest United States the Glacial Maximum appears to have been a time of high lake levels compared with the Holocene, whereas in tropical Africa and Australia, lake levels seem to have been low (especially about 14 000 years ago) and then became very high in the early Holocene (*ca.* 9000 years ago). At that time the Sahara became green, had through-flowing rivers, large populations of herbivores and pastoralists, and many freshwater lakes.

A.S. Goudie

Bibliography

Adams J., Maslin, M., and Thomas, E., 1999. Sudden climate transitions during the Quaternary. *Progress in Physical Geography*, **23**: 1–36.

Alley, R.B., Mayewski, P.A., Sowers, T., Struiver, M., Taylor, K.C., and Clark, P.V., 1997. Holocene climate instability: a prominent widespread event 8200 years ago. *Geology*, **25**: 483–486.

Anderson, D.E., Binney, H.A., and Smith, M.A., 1998. Evidence for abrupt climate change in northern Scotland between 3900 and 3500 calendar years BP. *The Holocene*, **7**: 97–103.

Andrews, J.T., 1998. Abrupt changes (Heinrich events) in late Quaternary North Atlantic marine environments. *Journal of Quaternary Science*, **13**: 3–16.

Bond G., Heinrich, H., and Broecker, W., et al., 1992. Evidence for massive discharges into the North Atlantic during the last glacial period *Nature*, **360**: 245–249.

Bowen, D.Q., 1978. *Quaternary Geology*. Oxford: Pergamon.

Bradley, R.S., 1999. *Paleoclimatology: reconstructing climates of the Quaternary.* London: Academic Press.

Buchard, B., 1978. Oxygen isotope palaeotemperatures from the Tertiary period in the North Sea area, *Nature*, **275**: 121–123.

Butzer, K.W., 1972. *Environment and Archeology – An Ecological Approach to Prehistory*. London: Methuen.

Dansgaard, W., Johnson, S.J., and Clausen, H.B., et al., 1993. Evidence for general instability of past climate from a 250 kyr Ice-Core record. *Nature*, **364**: 218–220.

Frakes, L.A., 1978. *Climates Throughout Geological Time*. Amsterdam: Elsevier.

Goudie, A.S., 1992. *Environmental Change*, 3rd edn. Oxford: Oxford University Press.

Goudie, A.S., 2002. *Great Warm Deserts of the World. Landscapes and Evolution.* Oxford: Oxford University Press.

Grove, J.M., 1988. *The Little Ice Age.* London: Methuen.

Kukla, G.J., 1977. Pleistocene land–sea correlations. I. Europe. *Earth-Sciences Review*, **13**: 307–374.

Lamb, H.H., 1982. *Climate, History and the Modern World.* Cambridge: Cambridge University Press.

Leinen, M., and Heath, G.R., 1981. Sedimentary indicators of atmospheric activity in the Northern Hemisphere during the Cenozoic. *Palaeogeography, Palaeoclimatology, Palaeoecology*, **36**: 1–21.

Lowe, J.J., and Walker, M.J.C., 1997. *Reconstructing Quaternary Environments*, 2nd edn. London: Addison-Wesley–Longman.

Oppo, D.W., McManus, J.F., and Cullen, J.L., 1998. Abrupt climate events 500 000 to 340 000 years ago: evidence from sub-polar North atlantic sediments. *Science*, **279**: 1335–1338.

Penck, A., and Brükner, E., 1909. *Die Alpen in Eiszeitalten*. Leipzig: Tauchnitz.

Pettit, J.R., Jouzel, J., and Raynaud, D., et al., 1999. Climate and atmospheric history of the past 420 000 years from the Vostok Ice Core, Antarctica. *Nature*, **399**: 429–436.

Ruddiman, W.F., 2001. *Earth's Climate: past and future*. San Francisco, CA: W.H. Freeman.

Shackleton, N.J., and Opdyke, N.D., 1973. Oxygen isotope and palaeomagnetic stratigraphy of Equatorial Pacific Core V28–238: oxygen isotope temperatures and ice volumes on a 10^5 year and 10^6 year scale. *Quaternary Research*, **3**: 39–55.

Shackleton, N.J., and Opdyke, N.D., 1977. Oxygen isotope and palaeomagnetic evidence for early Northern Hemisphere glaciation. *Nature*, **270**: 216–219.

Street, F.A., 1981. Tropical palaeoenvironments. *Progress in Physical Geography*, **5**: 157–185.

Turekian, K.K., 1971. *The Late Cenozoic Glacial Ages*. New Haven, CT: Yale University Press.

Wegener, A., 1967. *The Origins of Continents and Oceans*. London: Methuen.

Woillard, G., 1979. Abrupt end of the last interglacial S.S. in north-east France. *Nature*, **281**: 558–562.

Cross-references

Climatic Variation: Historical
Ice Ages
Milankovitch, Milutin
Volcanic Eruptions and their Impact on the Earth's Climate

PALMER INDEX/PALMER DROUGHT SEVERITY INDEX

In 1965 Wayne C. Palmer, a scientist in the Office of Climatology, Weather Bureau, US Department of Commerce, Washington, DC, published a research paper entitled "Meteorological drought". Applications of the analytic procedures set forth in this paper have become widely known as the Palmer Index or the Palmer Drought Severity Index (PDSI). It is a hydrologic or persistent climatological drought index, since it attempts to quantify the scope, severity, and frequency of prolonged periods of abnormally dry weather. It works reasonably well for this purpose.

The Palmer indexing procedure was developed to help fulfill the need to define and quantify droughts. This need is very real to agencies at all levels of government since drought means various things to people, depending on their specific problems related to it. To a farmer drought means a shortage of soil moisture in the rooting zone of crops. To the hydrologist or urban water system engineer it means low levels in groundwater, reservoirs, lakes, and stream flow. To the economist or businessperson it means a water shortage that has adversely affected the economy. The only common denominator among these many concerns is that drought is caused by a prolonged weather anomaly of less than expected precipitation.

Basically, the Palmer Index is an index of relative moisture deficiency under a wide range of climatic conditions. Its general concept is one of supply vs. demand. The supply is represented by precipitation and stored water in the form of soil moisture, groundwater, lakes, and reservoirs. The demand is the combination of potential evapo-transpiration and the amount of water needed to recharge soil moisture, plus the runoff needed to keep lakes, reservoirs, and stream flow at a normal level. It is a water balance accounting procedure that results in a positive or negative anomaly estimate weighted by the climatic conditions in time and space. The final product is an index that expresses the abnormality for a given place over a particular time period. These calculations produce an index value that can be compared across a wide variety of climates. In practice the time period increment is added to a portion of the previous time period value. In this manner the index value is weighted for the duration of the anomaly.

The generalized equation for the moisture-balance accounting assumes a given soil surface under natural climatic conditions. It is assumed that evapotranspiration takes place at the potential rate from the surface until all available moisture is removed. Likewise, it is assumed that there will be no recharge to the underlying soil body until the surface has been brought to normal holding capacity under the pull of gravity. It is further assumed that water loss from the underlying soil body depends on the initial soil moisture content as well as on computed potential evapotranspiration and the available capacity of the total soil system. Further, it is assumed that no surface runoff or runon occurs. Therefore:

$$L_{\mathrm{sm}} = (PE - P - L_{\mathrm{ss}}) \frac{S_{\mathrm{sm}}}{ASM} L_{\mathrm{sm}} \leq S_{\mathrm{sm}}$$

where L_{sm} is moisture loss from soil body, PE is potential evapotranspiration, P is precipitation, L_{ss} is surface soil moisture loss, S_{sm} is available surface moisture, and ASM is available soil moisture. These generalized accounting procedures produce quantitative estimates of the moisture balance from one time to another. In this manner estimates of positive or negative anomalous values can be produced. These anomalous moisture balance values are then assigned an index scale of severity with reference to that expected for the area and the season of the year. It is these scale values that have become known as the Palmer Index.

Palmer arbitrarily selected the classification scale of moisture conditions based on his original study areas in central Iowa and western Kansas (Table P4). Ideally, the Palmer Index is designed so that a −4.0 in South Carolina has the same meaning in terms of the moisture departure from a climatological normal as a −4.0 in Idaho. The Palmer Index has been calculated on a monthly basis, and a long-term archive of the monthly PDSI values for every Climate Division in the

Table P4 PDSI classifications for dry and wet periods

4.00 or more	Extremely wet
3.00 to 3.99	Very wet
2.00 to 2.99	Moderately wet
1.00 to 1.99	Slightly wet
0.50 to 0.99	Incipient wet spell
0.49 to −0.49	Near normal
−0.50 to −0.99	Incipient dry spell
−1.00 to −1.99	Mild drought
−2.00 to −2.99	Moderate drought
−3.00 to −3.99	Severe drought
−4.00 or less	Extreme drought

United States exists with the National Climatic Data Center from 1895 through the present. In addition, weekly Palmer Index values are calculated for the Climate Divisions during every growing season and are available in the *Weekly Weather and Crop Bulletin*. This is jointly prepared by the US Department of Commerce, National Oceanic and Atmospheric Administration, and the US Department of Agriculture. The bulletin is published weekly on Tuesdays. These are also available on the worldwide web from the Climate Prediction Center.

The Palmer Index is popular and has been widely used for a variety of applications across the United States. It is most effective in measuring impacts sensitive to the soil moisture conditions, such as agriculture. It has also been useful as a drought monitoring tool and has been used to start or end drought contingency plans. There are three positive characteristics of the Palmer Index that contribute to its popularity: (1) it provides decision makers with a measurement of the abnormality of recent weather for a region; (2) it provides an opportunity to place current conditions in an historical perspective; and (3) it provides spatial and temporal representations of historical droughts. It does, however, work best east of the Continental Divide.

Another index used to evaluate moisture is the Crop Moisture Index (CMI). Developed by Palmer subsequent to his development of the PDSI, the CMI is designed to indicate normal conditions at the beginning and end of the growing season. It uses the same levels as the PDSI but differs from it by placing less weight on the data from previous weeks and more weight on the recent week. As a result, the CMI responds more rapidly than the Palmer Index and can change considerably from week to week. It is thus more effective in calculating short-term abnormal dryness or wetness affecting agriculture.

Other indices include the Standardized Precipitation Index (SPI), a relatively new drought index, based only on precipitation. The SPI can be calculated for a variety of time-scales. This flexibility allows the SPI to be of value in both short-term agricultural and long-term hydrological applications. The index is based on data from the National Climatic Data Center, and on analysis by the Western Regional Climate Center and by the National Drought Mitigation Center. Also used as a drought indicator are values for Percent of Normal Rainfall for the past 30 and 90 days. These are updated daily and are derived from comparisons with data from earlier long-term data.

The PDSI has become a management tool for decision making in government, industry, and agriculture (Strommen, 1981). It is used as a standard of comparison by government for defining drought intensity throughout the United States. Industry and agriculture use the index in water resource planning as well as in estimation of current water supplies. Perhaps the most significant evidence of its place in meteorological practice is its adoption by foreign national meteorological services and agricultural ministries. This is particularly true among the nations in the third world within the semiarid tropics.

James E. Newman and John E. Oliver

Bibliography

National Oceanic and Atmospheric Administration, National Climatic Data Center, 1985. *Atlas of Monthly Palmer Moisture Anomaly*

Indices (1895–1930) for the Contiguous United States. Historical Climatic Series 3–8. Asheville, NC: NOAA, National Climatic Data Center.
National Oceanic and Atmospheric Administration, National Climatic Data Center, 1985. *Atlas of Monthly Palmer Moisture Anomaly Indices 1931–1984 for the Contiguous United States*. Historical Climatic Series 3–9. Asheville, NC: NOAA, National Climatic Data Center.
Palmer, W.C., 1965. *Meteorological Drought*. Weather Bureau Research Paper No. 45. Washington, DC: US Department of Commerce.
Palmer, W.C., 1968. Crop moisture index. *Weekly Weather and Crop Bulletin*, **55**(20): 8–9.
Strommen, N.S., 1981. The Palmer drought index – a management tool. *Weekly Weather and Crop Bulletin*, **68**(6): 10.
US Department of Agriculture, 1976. Droughts that made U.S. history. *Weekly Weather and Crop Bulletin*, **62**(27): 15–17.

Cross-references

Arid Climates
Aridity Indexes
Drought
Evaporation
Evapotranspiration
Hydroclimatology

PHASE CHANGES

Most substances can exist in three phases or states of matter – solid, liquid, and gaseous – and changes from one state to another can occur. In terms of atmospheric sciences, such phase changes are most significant in relation to water, whose phases include ice (solid phase), water (liquid phase), and vapor (gaseous phase).

Table P5 provides a listing of the reversible changes of the phases of water. Under normal conditions ice melts at a temperature of 0°C, its melting point. Below this temperature ice is the stable phase of water. Above 0°C the stable phase is water. Although ice cannot normally exist above 0°C, water can exist in the liquid phase at temperatures below 0°C and even to a temperature of −40°C. In such a state the water is said to be supercooled. This is a metastable phase, meaning that, although the supercooled water is stable, it is not the most stable phase and has a tendency to change to ice. The presence of supercooled water in the atmosphere is highly significant in the precipitation process and in potential rainmaking operations. The significance rests with the fact that when metastable supercooled water is in contact with ice it will freeze.

Table P5 Reversible changes of the phases of water

Initial phase	Final phase	Name of phase change
Solid	Liquid	Melting, or fusion
Liquid	Solid	Freezing, or crystallization
Liquid	Vapor	Evaporation
Vapor	Liquid	Condensation
Solid	Vapor	Sublimation
Vapor	Solid	Deposition

Table P6 Heat transfers associated with phase changes of water

Phase change	Heat transfer	Type of heat
Liquid water to water vapor	540–590 cal absorbed	Latent heat of vaporization
Ice to liquid water	80 cal absorbed	Latent heat of fusion
Ice to water vapor	680 cal absorbed	Latent heat of sublimation
Water vapor to liquid water	540–590 cal released	Latent heat of condensation
Liquid water to ice	80 cal released	Latent heat of fusion
Water vapor to ice	680 cal released	Latent heat of sublimation

Phase changes involving water vapor can occur up to a maximum of 100°C, the boiling point of water. Below this temperature vapor can exist in contact with both ice and water but above 100°C, vapor is the only stable phase. Under set conditions water can exist above 100°C, superheated water, but in the atmospheric sciences this condition can be ignored.

The values of both melting and boiling points of water depend on pressure exerted at their surfaces. Within the normal confines of surface atmospheric pressure, the 0°C and 100°C values can be considered operative. It is mostly in relation to increased elevation that the values, especially the boiling point, are significantly changed.

When phase changes occur there is an exchange of energy. Table P6 provides a summary of the amount of heat transferred. As noted in other entries (see cross-references) this exchange plays a very important part in the transfer of energy over the globe.

John E. Oliver

Bibliography

Aguado, E., and Burt, J.E., 2001. *Understanding Weather and Climate*. Upper Saddle River, NJ: Prentice-Hall.
Ahrens, D.L., 2000. *Meteorology Today: an introduction to weather, climate, and the environment*, 6th edn. Pacific Grove, CA: Brooks & Cole.
McIlveen, R., 1992. *Fundamentals of Weather and Climate*. London: Chapman & Hall.
Meteorological Office, 1981. *A Course in Elementary Meteorology*. London: Her Majesty's Stationery Office.
Oliver, J.E., and Hidore, J.J., 2002, *Climatology: an atmospheric science*. Upper Saddle River, NJ: Prentice-Hall.

Cross-references

Condensation
Evaporation
Evapotranspiration
Latent Heat
Precipitation

PRECIPITATION

Precipitation refers to solid or liquid phase water particles that are formed in the atmosphere and that fall and reach the Earth's surface. It also refers to the quantity of such particles that have reached the Earth's surface at a given location over a given period of time. This quantity is generally measured in linear units of liquid water (e.g. millimeters or inches) – with an assumption that this measurement is per unit area.

Formation of precipitation

Condensation of water vapor in the atmosphere may take place when the air is saturated. Air is typically brought to saturation by cooling and this cooling is generally accomplished by upward motion and the resulting adiabatic expansion of the air. The saturation vapor pressure over a curved surface is much higher than that over a flat surface. To form a stable water droplet would require much larger supersaturations than are observed in the atmosphere unless the water condensed on some other substance. Condensation on another substance reduces the radius of curvature of the water droplet. Thus, water vapor requires solid and liquid particles known as *Condensation nuclei* on which to condense. The concentrations of these nuclei vary greatly in the atmosphere from the order of 10 to 1000 cm^{-3}. Larger values are found over the continents and urban areas. Some condensation nuclei, known as *hygroscopic nuclei*, have a particular affinity for water and water may condense on them at relative humidities which are less than 100% (with respect to a flat surface of pure water).

Deposition of ice in the atmosphere requires that the atmosphere be saturated and also that the temperature be below the freezing point of water. At temperatures above −40°C deposition requires, in addition, that suitable particles known as *ice nuclei* are present. The number of ice nuclei in the atmosphere varies with temperature – more ice nuclei activate at lower temperatures – but averages about 1 per liter at −20°C. Thus, ice nuclei are considerably scarcer than condensation nuclei.

Water droplets and ice crystals fall under the influence of gravity but their fall is impeded by air resistance. Therefore, they do not continue to accelerate indefinitely but reach a terminal velocity. The terminal velocities for water droplets and ice crystals are given in Figures P10 and P11, respectively. Since cloud droplets are typically 2–50 μm in diameter, it can be seen that they must grow considerably larger to reach the Earth's surface in a reasonable period of time. Ice crystals must grow larger still.

Water droplets can grow larger by additional condensation. Condensation is favored on larger droplets as well as on droplets containing dissolved substances. Water droplets can also grow by encountering and merging with other cloud droplets in a process known as collision and coalescence which also favors the growth of larger droplets as opposed to smaller ones. Precipitation can and does develop in clouds whose temperatures are not low enough for ice crystals to form. Such conditions are more likely to occur in the tropics and subtropics than in middle or polar latitudes.

Ice crystals can also grow larger by additional deposition of water vapor. But since there are relatively few ice nuclei in the atmosphere, ice crystals and supercooled water droplets often coexist in clouds whose temperature is below freezing. The saturation vapor pressure over water is greater than that over ice. Thus, conditions can exist where the water droplets are subsaturated and the ice crystals are supersaturated. The ice crystals grow at the expense of the water droplets in what is known as the *Bergeron* process. The *Bergeron* process is responsible for the vast majority of precipitation (including rain) in middle and higher latitudes as well as in tropical thunderstorms. Ice crystals are always six-sided but may take a variety of forms: columns, needles, plates and the elaborate branching structures, known as dendrites, of the stereotypical snowflake. The structure of the ice crystal is dependent on the temperature and humidity of the air in which it forms. Ice crystals may collide and merge in a process called *aggregation*. In addition ice crystals may encounter water droplets that freeze on impact with the ice crystal. This process is known as *riming*.

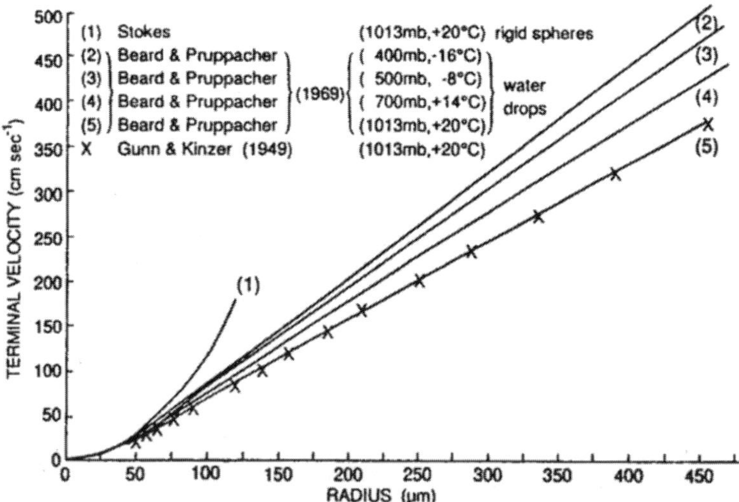

Figure P10 Variation with size of the terminal fall velocity of water drops smaller than 500 μm radius in air (from Pruppacher and Klett, 1997).

Types of precipitation

Precipitation in the liquid form when it reaches the Earth's surface may be divided into two categories: *rain* and *drizzle*. The distinction between them is entirely based on size – drizzle drops are less than 0.5 mm in diameter and rain drops are at least 0.5 mm in diameter. Drizzle typically falls from fog or low stratus clouds and generally produces small quantities of precipitation. Rain will fall from nimbostratus or cumulonimbus clouds.

Likewise, frozen precipitation that has experienced little or no riming and has not previously melted may come in the form of *snow* or *snow grains*. Snow is larger in size having often experienced aggregation, especially if the temperature is near freezing. Snow grains are the frozen equivalent of drizzle and typically fall from fog or low stratus clouds. They are small, white and opaque with diameters less than 1 mm. Snow crystals that are moderately rimed are called graupel, snow pellets or soft hail. They are white due to their core, which was formed by deposition. They are typically 2–5 mm in diameter and may bounce when they hit the ground.

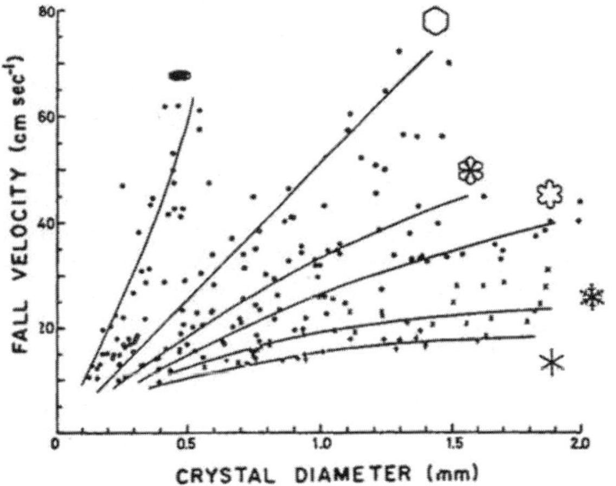

Figure P11 Variation with size of observed terminal fall velocities (symbols) and velocities computed from drag data (lines) for ice crystals of various shapes, −10°C, 1000 mb (from Pruppacher and Klett, 1997).

When ice crystals fall though a thick enough layer that is above freezing they will melt. If they then fall through another fairly thick layer that is below freezing before they reach the ground, they will refreeze. The resulting precipitation is called ice pellets or sleet (in the United States). (In British English, sleet refers to a mixture of rain and snow.) Ice pellets are transparent or translucent because they were formed by freezing liquid and not deposition of vapor. They are less than 5 mm in diameter. If the layer which is below freezing next to the Earth's surface is very thin, a falling water drop may not have time to freeze before it hits the ground. In this case the liquid water drop will freeze on contact with the ground, creating a thin sheet of ice. The precipitation is then known a freezing rain. Figure P12 illustrates the environment in which the various types of precipitation form. Freezing rain (and to a lesser extent, ice pellets) can be extremely hazardous to ground transportation because friction with a surface coated with freezing rain is even less than for a surface coated with snow. Freezing rain can also coat trees and power lines, unlike snow which will fall off both if accumulations are high enough. Heavy coatings can break tree branches and power lines, resulting in substantial damage.

Hail may be considered an extreme case of rimed ice crystals. Hail is greater than 5 mm in diameter. Hail typically has had several coatings of liquid water freeze on it. On larger hailstones these layers can be seen if the hailstone is cut open. To collect enough water to grow to even minimum size, the embryo hailstone must remain in the cloud longer than other types of precipitation. Although there is some disagreement as to all the processes involved in forming a hailstone, it is generally agreed

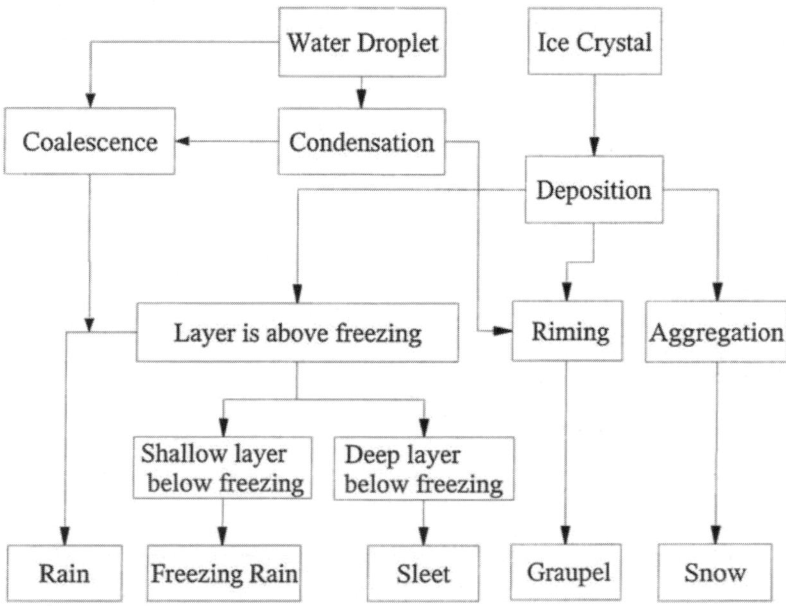

Figure P12 Processes forming different types of precipitation.

that the storm must have at least a moderately strong updraft to prevent the precipitation from falling out too soon. Thus, hail falls almost exclusively from cumulonimbus clouds. The shape and thickness of the thunderstorm updraft may be important factors as well as its orientation with respect to the downdraft.

Measurement of precipitation

Precipitation has traditionally been measured directly at the Earth's surface with a container into which the precipitation falls. Although the depth of frozen precipitation can be measured, its water equivalent varies greatly and the depth of snow changes depending on the length of time since the snow fell. Therefore, the liquid equivalent of the frozen precipitation has typically been a more useful measurement.

The standard rain gauge has a wide, usually circular, opening converging to a narrower inner tube. The cross-sectional area of the inner tube is generally one-tenth that of the outer opening. The depth of precipitation is measured manually with a specially calibrated ruler. This method can be economically advantageous because of the low cost and durability of the equipment. Its disadvantage lies in the difficulty of determining the rate at which the precipitation fell.

A weighing gauge can record the rate of precipitation accumulation. It measures the change in weight as liquid accumulates in a container. The liquid equivalent of frozen precipitation can also be measured if a known weight of antifreeze solution is in the container. The antifreeze can also reduce evaporative losses of accumulated precipitation. The minimum amount of precipitation recorded depends on the sensitivity of the gauge. This type of gauge is more costly and needs to be calibrated often.

Remote readings of precipitation accumulation are easier with the tipping bucket type of rain gauge. This instrument has a small container which is divided in two. The container pivots around its center in a seesaw manner such that only one side of the container accumulates precipitation at a time. When a known weight of precipitation accumulates in one side of the container, it will tip, empty its load and allow the other side to accumulate precipitation. The number of tips is recorded electronically. This type of gauge makes no record of less than some minimum precipitation accumulation – i.e. less than one tip's worth. It may have difficulty keeping up in very heavy rain and it tends to have more splash errors than other types of direct measurements.

Overall these direct measurements are subject to a number of errors. One is that precipitation in the gauge may evaporate before it is recorded. Another is that moisture may adhere to the collecting tube and not be measured. The most serious errors occur because catch decreases with increasing wind speed. Gauges are usually mounted as close to the Earth's surface as possible, to minimize this error. Some gauges may be fitted with a wind shield to help reduce this bias. It should be kept in mind that places with high average wind speeds may have precipitation measurements which are consistently lower than the actual amount which fell. Even outside of windy areas, errors in gauge measurements trend more towards underestimating rather than overestimating the amount of precipitation that actually fell.

Precipitation displays extreme areal variability and it is not usually feasible to have enough gauges to measure most of the variability. Generally the horizontal resolution of observations is very uneven. Thus, there is considerable incentive to estimate precipitation amounts with techniques yielding more spatial resolution.

Satellite measurements of infrared radiation (IR) in a narrow wavelength band have been related to precipitation. These estimates have been based on the assumption that higher clouds emitting less IR are more likely to be generating precipitation. This method works better in regions such as the tropics, where most precipitation is convective in nature. The total outgoing IR radiation (i.e. wide band of wavelengths) has also been related to precipitation as it is also sensitive to soil moisture. Microwave radiation is more sensitive to cloud water and ice and thus easier to relate to precipitation directly. But the best algorithm for doing so varies by location and season.

Radar reflectivity has also been used to estimate precipitation amounts. To use this technique a drop size distribution must be assumed. Estimates of precipitation may be poor if drop size distribution in a particular storm is unusual. This method cannot directly determine how much of the precipitation in the air will reach the ground without evaporating. This technique is more difficult to apply to frozen precipitation as radar is less sensitive to it. Obviously, it is only applicable in parts of the world with reliable radar measurements.

Perhaps the most promising way of determining how much precipitation has fallen is to combine remotely sensed and gauge data. These methods use the gauge data to reduce the impact of the assumptions which must be made in remote-sensing estimates. They are therefore able to take advantage of the high resolution of the remotely sensed data as well as the greater (but not perfect) accuracy of the gauge data.

Donna Tucker

Bibliography

Houze, R.A., 1993. *Cloud Dynamics*. San Diego, CA: Academic Press.

New, M.M., Todd, M., Hulme, M., and Jones, P., 2001. Precipitation measurements and trends in the twentieth century. *International Journal of Climatology*, **21**: 1899–1922.

Pruppacher, H.R., and Klett, J.D., 1997. *Microphysics of Clouds and Precipitation*, 2nd edn. Dordrecht: Kluwer.

Rogers, R.R., and Yao, M.K., 1989. *A Short Course in Cloud Physics*. New York: Pergamon Press.

Sumner, G., 1988. *Precipitation Process and Analysis*. New York: John Wiley & Sons.

Cross-references

Acid Rain
Cloud Climatology
Hail
Hydroclimatology
Precipitation Distribution
Water Budget Analysis

PRECIPITATION DISTRIBUTION

The amount of precipitation received at any place on the Earth's surface and over any given time period depends on a variety of factors and results from several complex causes and interacting processes that are themselves time- and space-dependent. Atmospheric water vapor, the basic element in precipitation, varies geographically and temporally. For example, the mean precipitable water content of the atmosphere at any given moment is 25 mm with a maximum near the equator of

44 mm and a minimum in the polar regions of 2–8 mm depending on the season. Between 40° and 50° latitude, precipitable water ranges above 20 mm in the summer and drops to 20 mm in the winter. However, the presence of atmospheric moisture is a necessary but not a sufficient condition for precipitation. Precipitation totals depend not only on amount of precipitable water in the atmosphere but also on lifting mechanisms that causes the moisture to precipitate. Except for irregularities induced by rugged topography, the greatest amount of precipitation occurs along portions of the Intertropical Convergence Zone (ITCZ). Any discussion of precipitation distribution, therefore, must of necessity address variations both in time and space. This can be accomplished by examining the major features of global patterns of precipitation and the factors that combine to influence those patterns, by looking at the nature and causes of seasonal, diurnal and other temporal variations in precipitation distribution to discern major temporal regimes and their forcing mechanisms, and finally, by examining distribution from a statistical perspective rather than from a purely climatological one. In this last case, distribution characteristics such as variability, persistence, frequency and intensity become important.

The distribution pattern of precipitation is considerably more complex than either temperature or global radiation because of the influences of several factors which can be classified into those influencing vertical motion in the atmosphere and those relating to the nature of the atmosphere itself. Of the latter group, the stability and instability of the atmosphere and its thermal and moisture characteristics, which in turn are determined by the nature of the source region of the air masses and their subsequent trajectory, are particularly noteworthy. Practically all precipitation results from the adiabatic cooling of ascending moist airmasses. The areas of most frequent and rapid airmass ascent are the zones of convergent horizontal air flow in the equatorial belt and areas affected by the midlatitude cyclonic disturbances, as well as along the windward side of high mountain ranges adjacent to extensive sources of moisture (Figure P13). Additionally, the distribution patterns of precipitation may be influenced by the distribution and density of the data collection network and the accuracy of the precipitation measuring instruments. Most measuring stations tend to be in highly populated areas with few stations over the oceans and in sparsely peopled regions. Moreover, gauge measurements tend to be underestimates of the true precipitation, largely because of wind-induced turbulence at the gauge orifice and wetting losses on the internal walls of the gauge. Monthly estimates of the bias often vary from 5% to 40%, are larger in winter than in summer, and over landmasses in the northern hemisphere, increases to the north due largely to the deleterious effects of wind on snowfall.

Geography of annual distributions

The mean annual precipitation for the Earth is about 1000 mm but its spatial distribution is variable (Figure P13). Several factors combine to influence the global distribution of precipitation. Mean annual totals are greater in the equatorial belt, where the general circulation of the atmosphere is characterized by the rapid ascent of warm moist air in the tropical zone of converging winds, than they are in the subtropical area of dominant subsidence and high surface pressure which inhibit precipitation and lead to much lower totals. The tropical convergence zone with its widespread uplift of warm, humid airmasses is the most effective precipitation-producing part of the general circulation of the atmosphere. From this zone of heavy precipitation, amounts decrease irregularly toward the poles, and at high latitudes the lower moisture-carrying capacity of the colder atmosphere, fewer incursions of moist warm air and less active thermal convection result in limited precipitation. The minimum in precipitation at latitudes between 20° and 30° is caused by subsidence in the high-pressure zones and the adjacent stable portions of the trade wind inversion. In midlatitudes, precipitation increases again because of synoptic storm systems there. The forced ascent of moist surface air in midlatitude cyclones and orographic uplift in the westerly flow give rise to heavy precipitation. Although the distribution of annual precipitation shows a strong latitudinal zonation, there are non-latitudinal differences, especially in the tropics. Mount Waialeale on the island of Kauai, the northwesternmost of the large islands of the Hawaiian chain, averages 11 675 mm of rain annually; Cherrapunji, at an elevation of 1313 m on the southern slope of the Khasi Hills in northeastern India, averages 11 419 mm; and Debundscha, near the base of the Cameroon Mountain just north of the equator in western Africa, averages 10 279 mm annually. Areas with the least rainfall are Arica in northern Chile, which over a 59-year period averaged 0.75 mm, and Wadi Haifa, Sudan, averaged 3.0 mm over 39 years. Most of the coast of Peru and northern Chile averages less than 25 mm, as do portions of the southwest coast of Africa. In some places years may pass with no measurable precipitation and then a rare shower will account for the long-term average.

In middle and high latitudes the interannual differences in precipitation distribution patterns that lead, for example, to extensive areas of drought existing concurrently with other and equally extensive areas of severe flooding, can be traced to interannual differences in the configuration and intensity of the mid-tropospheric jet stream that is responsible for steering storms and storm systems through a series of troughs and ridges, the wavelengths and amplitudes of which vary from year to year and place to place. This upper-air steering mechanism is directly related to the magnitude of the internal energy of the atmospheric system, which in turn is related to the amount of solar energy absorbed by radiatively active surfaces in the boundary layer and its apportionment into sensible heat energy, latent heat energy and energy stored in the submedia modulated by the heat capacity and the response time of each storage medium. Interannual variability of precipitation over a large part of the tropics and subtropics is related to the El Niõ/Southern Oscillation (ENSO) phenomenon.

The distribution of land and water greatly influences the precipitation patterns in middle and high latitudes. Regions in the interior of continents remote from oceanic sources of moisture receive less precipitation than coastal areas. Windward coasts tend to receive more precipitation than leeward coasts in regions where favorable prevailing winds transport moisture toward the land. Ocean currents or drifts may introduce further modifications since warm currents such as the North Atlantic Drift increase the moisture content of the air in contact with it, and in some cases induce atmospheric instability that may lead to precipitation. Cold currents, on the other hand, decrease precipitation over the adjacent land areas by cooling the lowest layers of the atmospheric column, increasing atmospheric stability, inhibiting vertical mixing and hence precipitation. The cooling process in the lower atmosphere may lead to fog or low stratus but no precipitation. Large lakes and lake systems in the middle and high latitudes produce extensive lake-effect

Figure P13 Mean precipitation (mm) over the Earth.

snowfall on their downwind edge in winter (Dewey, 1970; Brahan and Dungey, 1984). Cold moist air, in its trajectory over the lake, becomes unstable as a result of heating from below. This instability and uplift, accentuated by convergence in the lower layers as the airmass moves from an aerodynamically smooth lake surface to a comparatively rough land surface, result in comparatively higher snowfall totals from given airmasses on the downwind edge of the lake.

Forced ascent of moist air by topographic barriers concentrates precipitation on the windward slopes but creates a rainshadow to leeward, by causing descending motion and hence heating of the already wrung-out air. Apart from tropical regions, the wettest places are on the windward side of mountains exposed to winds with long fetches across warm ocean surfaces. Mountains therefore tend to contribute to the spatial inequality in precipitation distribution. Examples of topographically induced inequalities in precipitation distribution are mountainous islands in the trade wind belt, the southern Andes along the Chilean–Argentine border, the Southern Alps of New Zealand and the coastal ranges of western North America (Figure P14). A cogent example is afforded by the state of Washington in the western United States (Figure P15). Along the Pacific coast where topographic variations are small the rainfall averages about 1525 mm; with increasing topographic variations annual totals may increase by 500 mm to totals of 2050 mm. On the western slopes of the Olympic Mountains up to 3800 mm are recorded. However, just to the east in the Puget Sound–Williamette Valley region, totals average around 1020 mm. Farther east, the western slopes of the Cascades receive amounts of 1525–1780 mm. Across the Cascade Divide there is a sharp transition spanning the range from 180 mm in Yakima to about 510 mm around Spokane. Orographic barriers also affect the vertical distribution of precipitation such that totals tend to increase with elevation on the windward slopes up to a certain level. Very high totals due to orographic lifting also occur in monsoonal areas and where tropical cyclones occur frequently, such as the western coast of India, Burma, Sumatra and Borneo in the first case and in the Caribbean region in the second case.

The major features of global patterns of annual precipitation distribution ascertainable from the preceding discussion and Figure P13 are (1) large totals in the equatorial zone and moderate to heavy amounts in the middle latitudes; (2) comparative dryness in the subtropical belts and in the regions around the poles; (3) within the subtropics, the west coasts of continents tend to be dry whereas east coasts tend to be wet; (4) in higher latitudes, west coasts tend to be wetter than east coasts; (5) in mountainous regions, especially where the mountains are transverse to prevailing winds, precipitation tends to be abundant on windward slopes but sparser on leeward slopes; and (6) maritime locations receive more precipitation than interiors of continents that are remotely located with respect to sources of moisture in the oceans, except perhaps in innermost Amazonia where prevailing winds, extensive horizontal convergence, large evapotranspiration rates and orographic effects combine to produce large amounts in a continental interior (Lettau et al., 1979).

Seasonal distribution of precipitation

Variations in the spatial distribution of annual precipitation and the accumulated totals over a given year are quite important climatically, but total annual precipitation is an insufficient measure of moisture availability because it does not take into account how precipitation is distributed over time. Equally significant from the point of view of agriculture, hydroelectric power generation and water resources planning are the seasonality, the interannual dependability, month-to-month variations and the frequency and intensity. Precipitation-inducing mechanisms and processes do not operate in the same way throughout any given year. Belts of convergence shift poleward in summer and equatorward in winter, causing a shift in the associated precipitation belts. In these convergence belts, rainfall is comparatively well distributed throughout the year, but a few degrees of latitude poleward, where alternating convergence and subsidence predominate, marked dry and wet seasons exist. Near the equator two precipitation maximums are discernible, one occurring approximately 1 month after the equinoxes. With

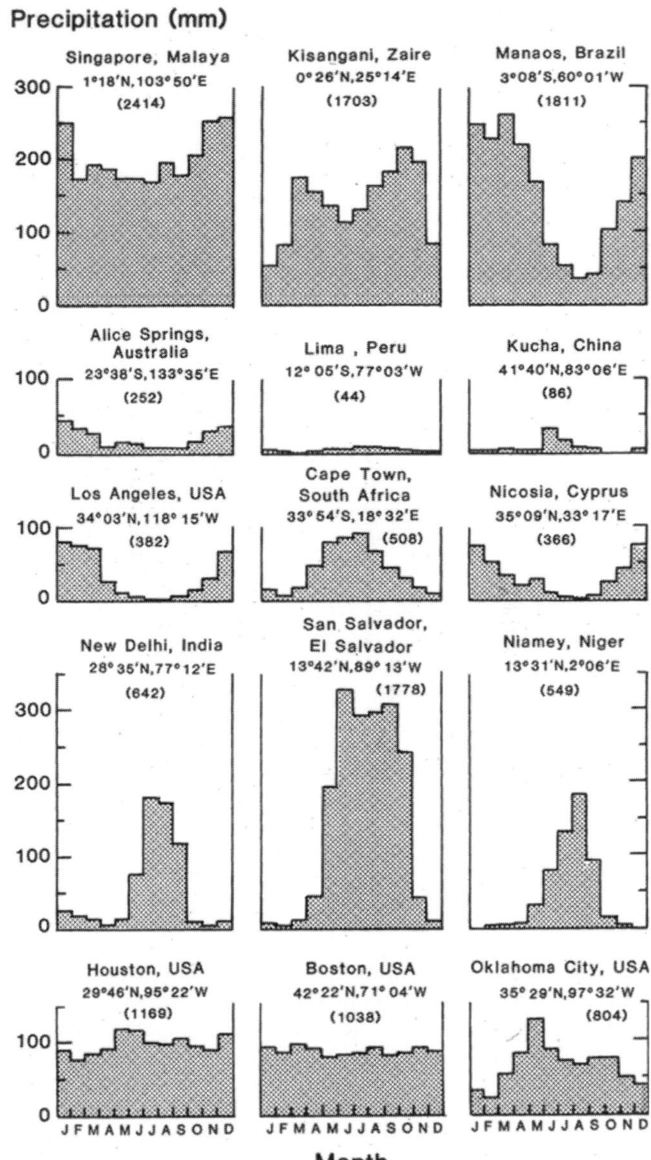

Figure P14 Seasonal precipitation distribution at selected groups of stations.

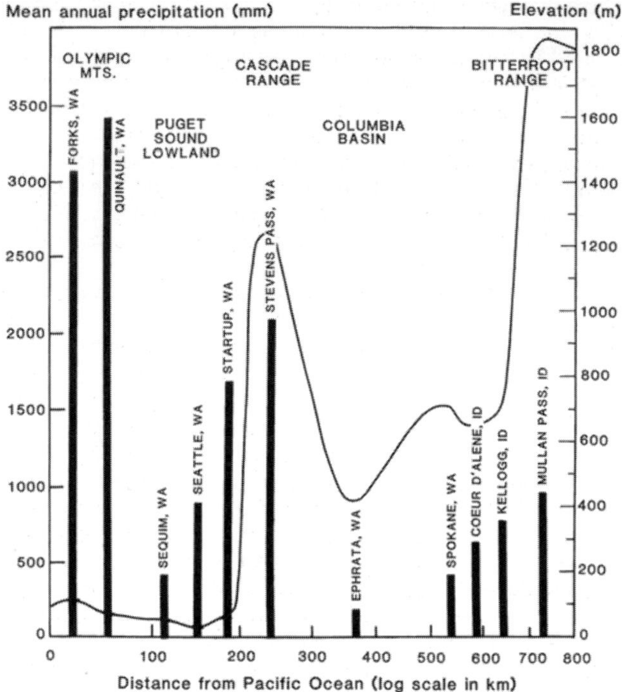

Figure P15 Orographic influences on precipitation exemplified by a transect of mean annual precipitation distribution along the 48°N parallel in western North America.

increasing distance from the equator the two maxima move closer together in time, reducing one minimum in length while the other dry season becomes longer and more intense. The total amount of annual rainfall is less here than closer to the equator. Between 15° and 20° latitude the short dry season disappears completely and the two maximums merge. North and south of this wet–dry precipitation regime, the influence of the subtropical quasistationary high leads to dry summers, whereas wet winters generally follow with the increasing influence of the westerlies and migrating cyclones and frontal disturbances. In India and Southeast Asia, monsoonal variations result in even more striking differences in the seasonal distribution of precipitation. Monsoons are seasonally prevailing winds that blow from one direction in summer and an approximately opposite direction in winter. These winds are best developed in India and Southeast Asia and are the result of differences in heat capacities of land and water, seasonally unequal heating and the pressure differentiation it produces, and the influence of mid-tropospheric circulation patterns, especially the subtropical jet and its interaction with the Himalayas. Northern Australia experiences a similar, though less well-developed, monsoon circulation, which is also related to seasonal migration of components of the general circulation. Along the east coast of Asia and across central Africa there are monsoon effects that merge into other shifting patterns of the general circulation. The southeast monsoon of India, which blows from the Arabian Sea across much of India from June to October, brings with it general rains that, along the west coast and the foothills of the Himalayas, are among the heaviest on Earth. In monsoon regions there is usually a marked summer maximum in precipitation (Figure P14), and delays in the onset of the west monsoon season have disastrous consequences for agriculture and water supplies. In East Africa, agricultural practices are predicated on two maximums, the larger of which is monsoonal whereas the smaller is due to the shifting tropical convergence zone. Over the tropical continents, seasonal rainfall distributions are further complicated by intense convectional processes that increase the amounts of rainfall locally.

The seasonal distribution of precipitation is as important as the total distribution over large portions of the Earth's surface. In many parts of the tropics precipitation is strongly seasonal and the times of initiation, duration and end of the rainy season control agricultural activities and water resources allocations. In the tropics, seasonal rainfall distribution is so important that it forms the basis of most classifications of tropical climates. This does not imply, however, that the seasonal distribution of precipitation is not important in middle latitudes; tropical rainfall is effective for plant growth no matter what time of year it falls, but in middle latitudes only that part of the annual precipitation that falls during the non-freezing season may be effective. In middle latitudes there is a fairly pronounced summer maximum of convective activity in continental interiors, where the summer maximum tends to lag later into summer the farther poleward one goes because of the lateness of snowmelt and ground thaw that delay convective activity.

In general terms, it is possible to discern the following seasonal regimes: (1) abundant rainfall, largely convectional, occurring year-round in equatorial regions; (2) largely convectional rainfall occurring mainly in summer months; (3) low rainfall totals in all seasons; (4) mainly cyclonic precipitation in midlatitudes occurring in winter whereas summers are dry; (5) cyclonically induced precipitation more abundant in winter than in summer; (6) mainly summer precipitation in continental regions; (7) abundant precipitation on east coasts due to onshore winds in low latitudes and inflowing warm moist airmasses in summer and traveling cyclones in winter in middle latitudes; and (8) low levels of precipitation in polar regions with a maximum in summer.

Diurnal distribution

Diurnal rainfall distribution is generally not as important as seasonal distribution except in low latitudes. The diurnal regime influences transportation systems, many outdoor activities and evapotranspiration losses. Daytime precipitation input is subject to very heavy evaporation losses. Diurnal precipitation distributions are more regular in low latitudes than in higher latitudes because of the differences in the dominant rain-producing mechanisms. In the tropics the convectional processes that predominate are forced by the diurnal march of global radiation input and are therefore considerably more regular than the midlatitude disturbances and fronts responsible for a large proportion of the diurnal variations in middle and high latitudes, especially in winter and spring. The nocturnal maximum in the midwestern United States as compared with the late afternoon maximum in the eastern part of the country indicate that, even within midlatitude regions, there can be distinguishable diurnal variations, particularly in summer. At almost all latitudes diurnal regimes are also influenced by the effects of local factors, such as topography, on small-scale atmospheric processes, such as thunderstorms, and by the form and physiographic configuration of the coastline. Two types of diurnal regimes are recognizable in low latitudes: (1) the inland or continental type characterized by a late morning or afternoon maximum attendant on convection initiated by surface heating of the land; and (2) the maritime or coastal type with a maximum during the night or early hours of the morning caused by severe nighttime instability and convection predicated on the steepened lapse rate attendant on radiative cooling of the upper troposphere while the lower layers remain warm due to its thermal coupling with the warm ocean surface (Kraus, 1963; Malkus, 1964). Ramage (1952) questioned the validity of this explanation but substituted no generally accepted theory. In the Caribbean region both types coexist, leading to the expectation of showers soon after solar noon, especially during the wet season. Equally expected, although not with the same regularity, are nocturnal severe thunderstorms occurring late at night or in the early hours of the morning. Some of the most intense precipitation, rivaled only by that accompanying hurricanes, occurs in these nighttime episodes.

The manner in which precipitation falls is also important. If it falls in infrequent cloudbursts it may cause floods and heavy erosion. On the other hand, prolonged drizzle effectively moistens the soil and provides high-humidity conditions. Europe receives less than 750 mm of precipitation per year but in many areas this is spread over more than 150 days. Under the cool cloudy, humid conditions that prevail there, the rainfall is quite adequate. In some parts of the world precipitation falls nearly every day of the year. Bahia Felix, Chile, averages 325 days per year with measurable precipitation, so that there is an 0.89 probability that it will rain or snow on any given day of the year. Buitenzorg, Java, averages 322 days per year with thunderstorms. Conversely, Arica, Chile, averages about one rain day annually; at nearby Iquique, Chile, there was no measurable rain between 1899 and 1913.

Statistical distribution

It is evident from the foregoing that the character of the spatial and temporal distribution of precipitation can be very important and enlightening but, in order to adequately analyze past sets of precipitation measurements so that conclusions about the future can be drawn, and in order to test the significance of physical experiments and check the validity of hypotheses concerning the distribution of precipitation, we need to know something about the statistical characteristics of the precipitation datasets. Statistical theory is based on the concept that a set of realizations is assumed to be representative of their population and therefore deductions can be made from that sample concerning the nature of the population. In order to draw inferences about a population, the values within the data sample must be random, independent and homogeneous. The precipitation datasets we work with, however, rarely meet all these criteria at once. In the first place there can be very little control over the selection of a sample, and whatever nature provides over a period of time must be used; second, the degree of independence varies with the precipitation-inducing mechanisms, season and location, and, in some cases and places, persistence is prevalent. In many regions, sample data lack homogeneity because station sites have shifted due to urbanization or development and because separate and distinct causal mechanisms contribute to the precipitation sample. Moreover, precipitation is not an infinite continuum in both directions; it cannot be less than zero. Therefore, variations on the dry side of the average cannot be as great in magnitude as variations on the wet side, and a few wet years can balance a great number of comparatively dry years. Despite all this, precipitation samples tend to follow one of several types of statistical distributions. The most important of these is the normal or Gaussian distribution, which is a bell-shaped curve that is symmetrical about the mean value for the sample. If the sample is not normal it can be normalized by various methods such as using the logarithms or cube roots of the variates in the sample. Other types of distributions include the gamma distribution, the Pearson types I, II and III distributions and, as already alluded to, logarithmically transformed lognormal, logextremal and truncated lognormal distributions. Each of these distributions has special characteristics that allow researchers to extract the maximum amount of information from the many and diverse precipitation records available throughout the world.

The characteristics of the statistical distribution of a precipitation sample are described by statistical parameters. The most important of these relate to measures of central tendency, variability and skewness. The parameters that generally represent measures of central tendency are the mean, the median and the mode. The mean can be arithmetic, geometric or harmonic. The arithmetic mean is the most familiar and generally used. The geometric mean is the nth root of the product of n terms, so that if one takes the logarithm of the individual values of a precipitation record and takes the mean, the logarithm of the geometric mean will result. The harmonic mean is the reciprocal of the mean value of the reciprocals of the individual values (Chow, 1964; Munn, 1970; Panofsky and Brier, 1968). The arithmetic mean is used more often than any other measure of central tendency because of its computational simplicity and, in general, its greater sampling stability. However, in extremely skewed precipitation distributions such as would be found in samples taken in arid and semiarid regions, the mean may be misleading because, although it is based on sound mathematics, weight is given to each occurrence according to its magnitude so that extreme values are excessively stressed in comparison with middling values. In a distribution that approximates the normal, this is of minor importance at most. The median is the middle value or the variate that divides the frequencies in a distribution into two equal parts so that all variates greater or less than the median always occur half the time. In the distribution of discrete variables the variate that occurs most frequently is the mode. In a distribution of continuous variables the mode is the variate with the maximum probability density. In many stations, particularly in the tropics, precipitation distributions are not unimodal but bimodal, and in that case it becomes difficult to decide exactly where the mode is, especially since the actual value arrived at may in part result from a subjective choice of groupings. Furthermore, the mode does not possess any true mathematical quality, having at best only a generalized relationship to the average.

The parameters representing variability or dispersion are the standard deviation, the variance, the range and the coefficient of variation. At many stations the frequency distribution of precipitation totals for individual years is positively skewed – the negative departures from the annual mean are more numerous than the positive ones. Such skewness is generally strongest where rainfall totals are low and where correct estimation of the rainfall that can be expected is most critical but where the annual mean is inflated by a few very high annual totals. For predictive purposes, medians and other probabilistic estimates such as quartiles, quintiles and deciles are more reliable despite the computational complexities and the requirement of long records. The standard deviation is the square root of the mean squared deviation of individual measurements from their mean, whereas the variance is the square of the standard deviation. The curve representing a particular normal or Gaussian distribution is completely determined by its standard deviation. Generally, for such a curve the standard deviation is the distance on either side of the center of the point where the slope is steepest. The coefficient of variation is the standard deviation divided by the mean, whereas the range is the difference between the largest and smallest values. Finally, skewness is a measure of the lack of symmetry of a distribution. The simplest measure of variability is the range but, because it is based on two observations only, there are large variations from sample to sample, making the range very unstable.

Precipitation intensity, which relates the total amount of precipitation to its duration over some specified finite time period, is another descriptor of precipitation. It may be taken as the amount of precipitation occurring on an average rainy day of a certain period. It is an important measure because it controls the probability and seriousness of local floods and is critical in planning dams, reservoirs and drainage canals. In addition, when intensity exceeds the maximum infiltration rate of the soil, surface runoff results, so that intensity can be related to erosion and sedimentation rates in lakes and reservoirs. In addition to intensity, the frequency distribution of different amounts of precipitation is also important. For example, in the tropical belt, in the wet season as well as in the dry season, the frequency distribution is represented surprisingly enough by a decay curve, demonstrating that days with no or very light rain are the most frequent and days with increasing amounts occur more rarely. Also, in the dry period nearly 85% of all days, and even in the wet season 35% of all days, have no or little rain.

The frequency distribution is used to organize precipitation data so their characteristics may be easily and quickly summarized. Also, an estimate of the frequency with which a given magnitude of precipitation may be exceeded in the future is based on the frequency with which it has been exceeded in the past. The parameters of the frequency distribution can be used in a probabilistic framework to calculate recurrence intervals of specified magnitudes of precipitation events with varying durations – hourly, daily or seasonal.

Precipitation distribution can be viewed from many different perspectives, each of which possesses both diagnostic and prognostic components. The time–space nature of that distribution, and the many and varied causal mechanisms and relationships, render the topic extremely complex, but at the same time immensely useful for an understanding of climate and its spatial and temporal variability and how these interact with the human-use system. At a time when the activities of human societies are modifying the climate system and reducing the margin between water supply and demand for agriculture, hydroelectric power generation and domestic use by an ever-expanding population in many parts of the world, it is imperative that we not only know where the areas of precipitation deficits and surpluses are, and the time periods in which they are most likely to exist, but also how the distribution patterns change over longer time periods.

Precipitation distribution in a warmer world

There is strong evidence in the instrumental records that global temperatures have been increasing over the last century and that the increase has been largest in the last two decades. The consensus, based on the results of general circulation models and higher-resolution regional models, is that most of that warming is due to increasing concentrations of trace gases in the atmosphere, and at the present rates of greenhouse warming, significant global environmental change will ensue. Among these changes are increased air and ocean surface temperatures, sea-level rise, an increase in the intensity of tropical storms, and a probable increase in the frequency of extreme climatic events. If temperatures change because of trace gas-induced warming, changes in other elements will occur also. Precipitation patterns are likely to be altered. Total precipitation over the globe will increase as the water-holding capacity of the air and evapotranspiration increase, but it is difficult to predict precisely how the additional moisture will be distributed either spatially or temporally. The IPCC estimates that summer precipitation will decline by 5–10% and soil moisture by 15–20%. The American Mid-West will become drier in summer. Confidence in these estimates is low, however (Mitchell et al., 1990). Precipitation would be less frequent over most of Europe in summer and fall and throughout the year in the south. According to some projections, more rainfall is possible in parts of Africa and Southeast Asia. In parts of Africa such as the Sahel, winter rainfall will decline by 5–10% and summer precipitation will increase by as much as 5% (Wigley et al., 1986; Kellogg, 1987). Hurricane intensity is expected to increase, as is the precipitation and the flooding associated with such storms. Whatever the magnitude and specific distribution of precipitation that a warmer atmosphere and ocean will foster, global warming is sure to change the distribution of precipitation in ways that may be beneficial to some regions of the world but detrimental to others. Finally, the knowledge about and understanding of precipitation and its distribution should culminate in an improvement in our predictive capability if it is to serve as a useful input into decision-making processes in agriculture, water resources and precipitation-related disaster mitigation, if only in a probabilistic form.

Orman E. Granger

Bibliography

Braham, R.R. Jr, and Dungey, J., 1984. Quantitative estimates of the effect of Lake Michigan on snowfall. *Journal of Climatology and Applied Meteorology*, **23**: 940–949.

Chow, V.T. (ed.), 1964. *Handbook of Applied Hydrology*. New York: McGraw-Hill.

Dewey, K.F., 1970. An analysis of lake-effect snowfall. *Illinois Geographical Society Bulletin*, **12**: 27–42.

Gilman, C.S., 1964. Rainfall. In Chow, V.T., ed., *Handbook of Applied Hydrology*. New York: McGraw-Hill, pp. 955–956.

Kellogg, W.W., 1987. Mankind's impact on climate: the evolution of an awareness. *Climate Change*, **10**: 113–136.

Kraus, E.B., 1963. The diurnal precipitation change over the sea. *Journal of Atmospheric Science*, **20**: 551–556.

Lettau, H., Lettau, K., and Molion, L.C., 1979. Amazonia hydrologic cycle and the role of atmospheric recycling in assessing deforestation effects. *Monthly Weather Review*, **107**(3): 227–238.

Malkus, J.S., 1964. Tropical convection: progress and outlook. In *Proceedings of the Symposium on Tropical Meteorology*. Boston; MA: American Meteorological Society, pp. 247–277.

Mitchell, J.F.B., Manabe, S., Tokioka, T., and Melishko, V., 1990. Equilibrium climate change. In Houghton, J.J., Jenkins, G.J., and Ephraums, J.J., eds., *Climate Change: the IPCC scientific assessment*. Cambridge: Cambridge University Press, pp. 131–172.

Munn, R.E., 1970. *Biometeorological Methods*. New York: Academic Press.

Panofsky, H.A., and Brier, G.W., 1968. *Some Applications of Statistics to Meteorology*. University Park, PA: Pennsylvania State University.

Ramage, C.S., 1952. Diurnal variation of summer rainfall over East China, Korea, and Japan. *Journal of Meteorology*, **9**: 83.

US Department of Commerce. National Oceanic and Atmospheric Administration, *Climatological Data: Annual Summary*. Asheville, NC: National Climatic Center.

Wigley, T.M.L., Jones, P.D., and Kelly, P.M., 1986. Empirical climate studies. In Bolin, B., Doos, B.R., Jager, J., and Warrick, R.A., eds., *The Greenhouse Effect, Climate Change and Ecosystems, SCOPE 29*. New York: John Wiley & Sons.

Cross-references

Arid Climates
Cloud Climatology
Hydroclimatology
Orographic Precipitation
Precipitation
Water Budget Analysis

PRESSURE, SURFACE

The atmosphere consists of a variety of gases, collectively called air, that form an envelope around the Earth and that are held there by gravity. The molecules of air are attracted toward the Earth and become more densely packed as sea level is approached because each layer of the atmosphere is being compressed by the mass of air above it. If this mass is multiplied by the force of gravity, the result is the weight of the air, which is described in terms of the area over which it is measured (i.e. pounds per square inch, grams per square centimeter). This weight of air is the atmospheric pressure, first mentioned by Robert Boyle in 1660, only 17 years after Torricelli invented the barometer. During the next 150 years barometric measurements gradually accumulated, mainly in Europe but also in scattered locations in North America and Siberia. Some of these were collected and published during the eighteenth century by the Royal Society of London (1732–1742) and by the Palatine Meteorological Society of Mannheim (1781–1792), the latter including data from Cambridge, Massachusetts.

Measurements and corrections

The need to correct barometer readings for variations in the temperature of the mercury was first noted by Halley in 1693 and recommended by Amontons in 1704. Similarly, Laplace, in 1805, pointed out that, since barometric pressure measured the weight of the atmosphere, it would vary with gravity at different latitudes; therefore, readings required a gravity correction before they could be used for comparative purposes.The variation of pressure with altitude was demonstrated by Perier in 1648, but it was not until the middle of the nineteenth century that a need to compensate for station altitude was generally accepted. Other corrections noted and allowed for in the eighteenth and early nineteenth centuries were due to the minor variations in the manufacture of barometers. These were overcome by comparing barometers with standard instruments held at main national laboratories. Various sets of tables giving these corrections were published in Europe during the eighteenth century, together with additional tables that allowed results to be reported in relation to standard atmospheric conditions. Unfortunately these standard conditions also varied from place to place. For example, the Palatine Meteorological Society followed De Luc (1755) who adopted 10°R (54.5°F) as a standard; the Royal Society of London adopted 50°F (1776), and Cotte (1788) decided on freezing point. By the middle of the nineteenth century the ideal of freezing point at sea level, at latitude 45°, was agreed on as the standard condition.

Units

These early measurements were reported in the form of the length of the mercury column in the barometer tube, which naturally balanced the weight of the atmosphere. Consequently, a variety of units of length were used. Conversion tables were produced during the nineteenth century that allowed comparisons of barometric readings originally published in English inches, European millimeters, or Paris lines. Foremost in this respect were the tables produced for the Smithsonian Institution by Guyot, beginning in 1852. During the nineteenth century attempts were made by the scientific community to define internationally acceptable units. The first system was the centimeter–gram–second (cgs), agreed on in 1870, in which the unit of force is the dyne and the derived unit of pressure is dynes per square centimeter. Since this unit is extremely small, for several years (until the beginning of the twentieth century) a much larger unit, the megadyne (= 1 million dynes), was used to report atmospheric pressures in addition to inches or millimeters of mercury.The more familiar millibar (103 dynes cm^{-2}) came into regular use early in the twentieth century following the introduction of the terms bar, decibar, and centibar by V. Bjerknes in 1906 and the proposal by W. Köppen at a conference in Monaco in 1909 that the term bar be used to indicate megadynes per square centimeter. The meter–kilogram–second (mks) system was initially suggested by Giorgi in 1906 but did not become widely used until the 1930s. The unit of force was the newton (N), but this was never popular with meteorologists, who preferred to retain the millibar (mb or mbar), which was defined as $102\,N\,m^{-2}$. The mks system was refined in 1954 and in its new form (Le Système International or SI) was internationally approved in 1960. Again the newton was the unit of force, but a derived unit of pressure – the pascal (pa = $1\,N\,m^{-2}$) – was introduced. The American Meteorological Society accepted the SI system in 1974 but agreed that the mb (= 1 hPa = 100 Pa) should be retained while it was used by the international meteorological community.

Periodic variation: diurnal

Once comparable measurements of pressure were collected together, it became possible to analyze variations of pressure at different time-scales. The first major periodic variation to be noted was the diurnal oscillation with maximum pressures occurring around 10 a.m. and 10 p.m. local time and minimum pressures at about 4 a.m. and 4 p.m. local time. The decrease in the amplitude of the diurnal pressure wave away from the equator, and its slight variation with the season, was shown by Bouvard in 1829 and by Forbes in 1832. As data from polar stations became available during the nineteenth century, it was found that in these high latitudes the diurnal oscillation did not occur at the same local time but varied longitudinally instead. From the evidence now available it is clear that this diurnal variation in pressure consists of three elements:

1. A progressive 12-hour zonal oscillation that travels from east to west, decreasing in intensity from the equator (3 hPa) to the poles (0 hPa).
2. A standing 12-hour sectorial oscillation that oscillates north and south between the equator and the poles (0.1 hPa).
3. A progressive 8-hour tesseral oscillation traveling from east to west with a maximum intensity of about 01.1 hPa at 30°N and S. Three highs and three lows appear in each hemisphere, a high in one balanced by a low in the other. The pattern in each hemisphere is valid only for a solstice and reverses at the equinoxes when these tesseral oscillations disappear.

The first and second oscillations are undoubtedly caused by the tidal influences of the sun and moon and are amplified by the natural 12-hour resonance of the atmosphere. The third oscillation can only be thermally induced since no 8-hour tidal influence is known. The result is that oscillations 1 and 3 produce the diurnal wave of maximum and minimum pressures that occurs at fixed local times between the equator and about 508 latitude. Oscillation 2 produces the pressure increases and decreases that occur at the same Greenwich meantime in polar areas. The composite result in the diurnal pressure varies from about 3.5 hPa at the equator to about 0.3 hPa at the poles. Buchan (1889) showed the geographical distribution of the composite oscillation in July. Maximums (3.4 hPa) appeared over the tropical continental areas and minimums (0.3 hPa) over the North Atlantic and the Arctic regions. His isopleths did not extend into the Pacific Ocean. He also provided tables of the diurnal variation for 146 stations throughout the world, of which 13 are shown to illustrate the diurnal curve and its geographical variation (Figure P16).

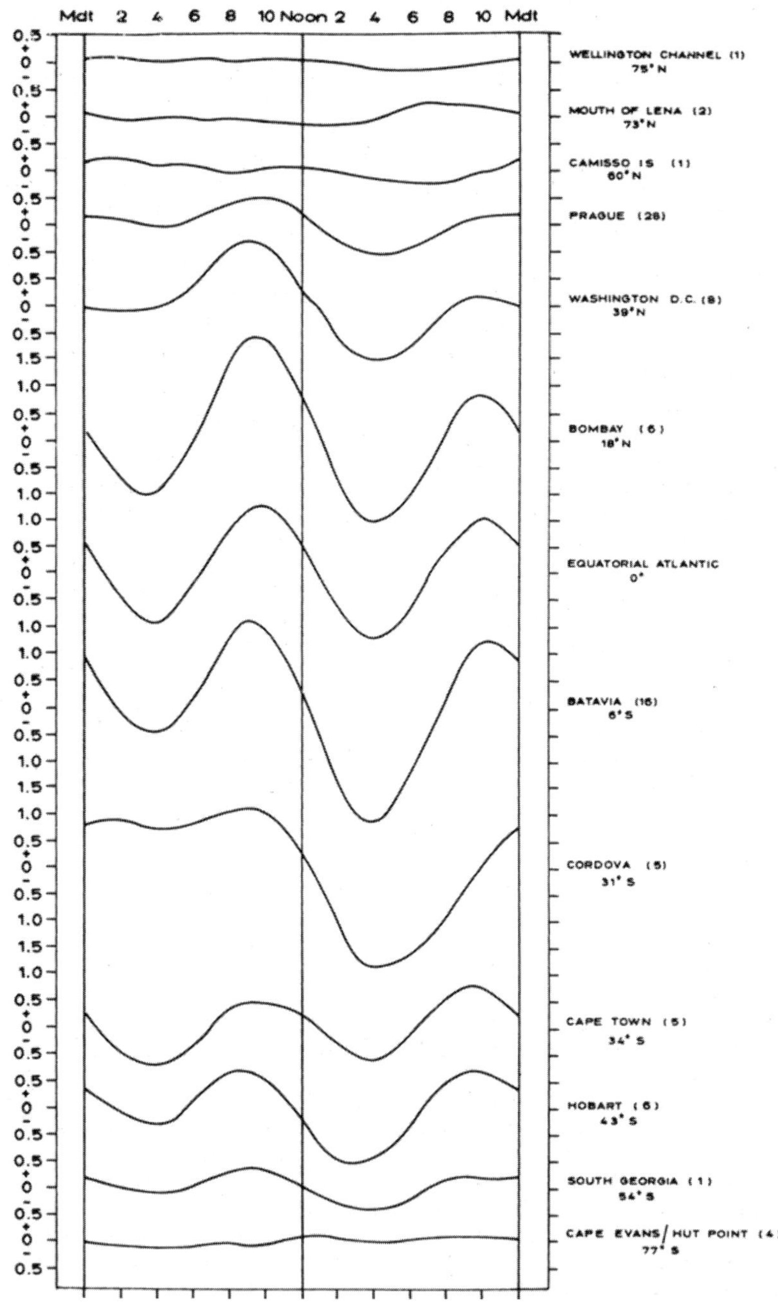

Figure P16 The diurnal variation of pressure of 13 stations situated at varying latitudes between 75°N and 77°S. The gradual disappearance of the normal double wave in high latitudes is clearly visible.

Periodic variation: monthly and seasonal

The second major periodic variation was the annual change considered on a world scale. The proliferation of hourly and daily observations allowed learned societies and various national organizations to publish monthly means of pressure. Many of these were collected together by Buchan and the results published in three papers. The first (1868) described the worldwide variation in sea-level pressure in January and July, and for the year. He described (but did not publish) the three charts "showing, by isobarometric lines, the mean pressure over the globe... drawn for every tenth of an English inch" with 30 in (762 mm or 1015.9 mb) considered to be the world average. These charts were based on the records of 358 stations: 167 in Europe, 63 in North America, 51 in Asia, 35 in Latin America, 22 in Africa, and 20 in Australia and the Southern Ocean. They enabled Buchan to define the major pressure belts of the world, which can be found in virtually the same form in many textbooks (Oliver, 1979, p. 153; Moran and Morgan, 1994, p. 219; Sturman and Tapper, 1996, p. 87) or atlases. He described what Humboldt had surmised 50 years earlier from much more limited data, that pressure was lower at the equator than in temperate latitudes and that it might also be low at the poles. Buchan clearly identified the low-pressure belt at the equator, the two subtropical high-pressure belts (of which the southern hemisphere one is more continuous), and the low pressures around the two poles. Buchan says "the region of low pressure round the south pole... is subject to little variation throughout the year. The depression in the neighborhood of the north pole is divided into two distinct centres... these are the north part of the Atlantic and the north part of the Pacific oceans." He also noted the major reversal of surface pressure that occurs in Siberia between January and July. Buchan followed this pioneering work by publishing in 1869 a much-extended paper containing mean pressure charts for each month of the year, as well as an annual chart based on 478 stations, most of the additional data coming from the southern hemisphere. This paper (1869) included an in-extenso table giving the mean monthly and annual pressure for each station in English inches (and thousandths) as well as the 105 sources he used. In 1882 Buchan was invited to write the meteorological report of the *Challenger* Expedition. The task of analyzing the *Challenger* data (439 pages) and incorporating it with information published during the 1870s took Buchan 7 years. The result was a monumental work (Buchan, 1889) of 342 pages and 52 plates of the world that included, from the point of view of surface pressure, the data from 1366 stations. For each month and for the year there were two mean pressure charts, one on a rectangular projection and the other on a polar projection for the northern hemisphere. This was the first time a polar projection had been used to illustrate the mean pressure distribution. (Equivalent southern hemisphere charts did not appear until 1915 when Meinardus and Mecking published the results of the German South Polar Expedition, 1903–1904.) The isobars of the *Challenger* charts were once again in English inches (and hundredths) with color being used to differentiate values above or below 29.95 in. The isobaric pattern for each month was described in detail with the main areas of high and low pressure noted. The January and July charts were updated by J. Hann in 1887 (isobars at 2 mm intervals and again color was used to differentiate areas above and below 760 mm) and again in 1901. For the next 20 or 30 years these major compilations were used in textbooks and atlases with either or both inch and millimeter as units. Although millibars had been in use since before World War I on daily charts (see below), their use on mean monthly charts in textbooks was very slow. Even in the late 1950s some US textbooks continued to give world mean pressure distribution charts in inch units. Generally, European textbooks adopted millibar on mean charts in the 1920s and 1930s and US textbooks followed suit, using Weather Bureau data, in the 1940s and 1950s. A much-simplified version of the surface pressure patterns in January and July is shown in Figure P17. These maps indicate the great extent of the subtropical high-pressure areas in the summer season when they almost encircle their respective hemispheres; in winter they withdraw to the oceanic areas. In contrast, the major low-pressure areas differ in each hemisphere. In the north they are clearly seen in the Aleutian and Icelandic areas in winter, but they virtually disappear in the summer when the main low-pressure area is found in southern Asia. In contrast, in the southern hemisphere there is a low that encircles the Antarctic continent in both summer and winter apart from a major break in the southeast Pacific in the vicinity of Ellsworth Land. In summary, the subtropical oceans have permanent high-pressure areas whereas the continents (apart from South America) generate high-pressure areas in winter. Low-pressure areas are found in temperate oceanic areas, particularly in winter, and in southwest Asia in summer. These semipermanent pressure areas were given the name of "des grandes centres d'action de l'atmosphere" or centers of action, by Teisserenc de Bort in 1881 (Mossman, 1913).

Aperiodic variations

There is some controversy concerning the first weather map or synoptic chart. Although not now in existence, it is most likely that Brandes drew, at some time in the first few years of the nineteenth century, the first weather chart for 6 March 1783, using data collected by the Palatine Meteorological Society. The first published weather map was presented by Elias Loomis in a paper to the American Philosophical Society in 1846. Both Brandes and Loomis depicted surface pressure in terms of deviations from a normal pressure. The pattern of using surface pressure charts to illustrate particular days or weather features continued until railroads and the telegraph enabled many reports to be collected and published quickly. The first daily weather report was published in the *British Daily News* on 31 August 1848 and was based on reports sent by train to London from 50 railway stations. In 1849 both Henry in the United States and Glaisher in England organized telegraphic reports and portrayed the results. The first printed daily weather maps were on sale at the Great Exhibition in London from 8 August to 11 October 1851 and included pressures in inches and hundredths from numerous English stations. Commercial daily weather maps appeared in England in 1861 and weather services began producing them in France (Paris Observatory, 5 September 1861), in the United States (Army Signal Service, 1 January 1871), and in England (Meteorological Office, 11 March 1872). The units varied with country of origin, but slowly mb became the standard in the early part of the twentieth century. Lines of equal pressure, or isobars, replaced lines of equal pressure variation, or deviation, in the middle of the nineteenth century, although isobars had been used by Berghaus in his *Physical Atlas* of 1845. The first patterns to be noted on the daily charts were the low- and high-pressure areas studied so assiduously in the United States by Elias Loomis. The low-pressure areas were initially considered as storms, but slowly the term cyclone became common, and by 1887 Abercrombie was able to define seven fundamental shapes of isobars: three low-pressure

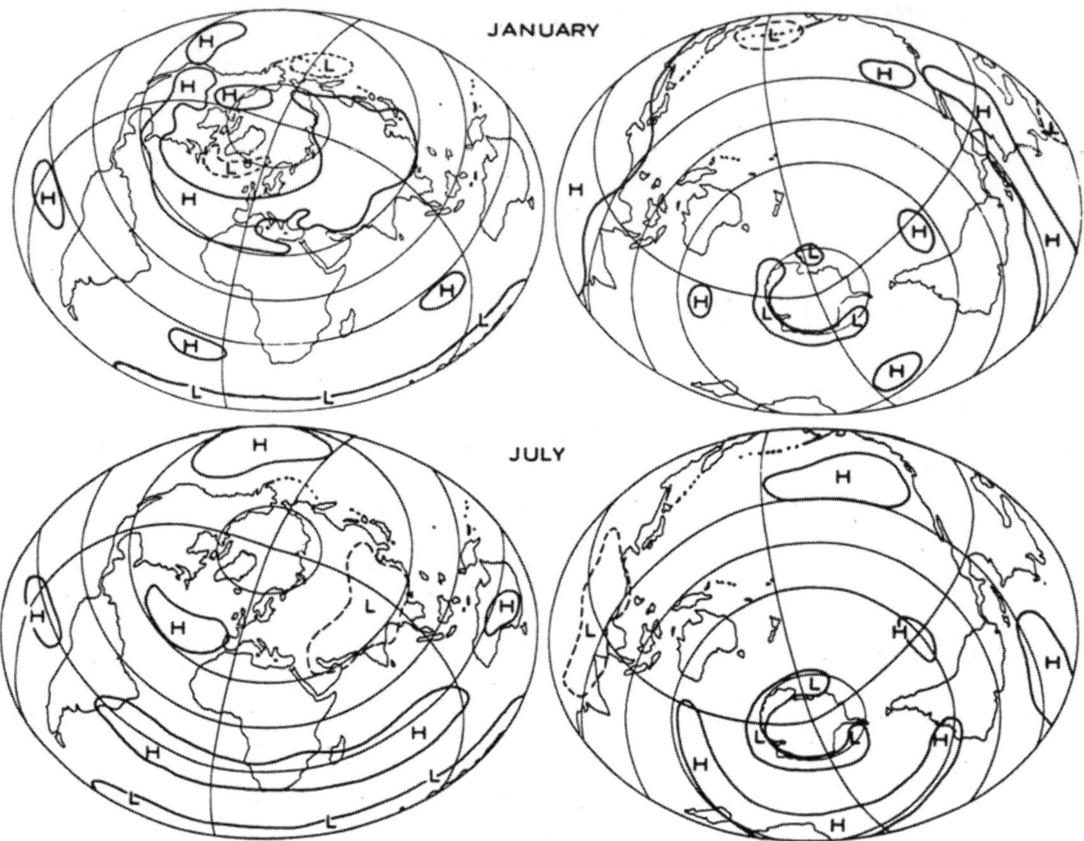

Figure P17 The main areas of high and low sea level pressure in January and July (after *Pergamon World Atlas*, 1968).

(cyclone, secondary cyclone, V-shaped depression); two high-pressure (anticyclone, wedge); the col and straight isobars. During the last quarter of the nineteenth century, English and French writers began to use depression synonymously for cyclone, and today depression is the normal expression (apart from the United States) for a large area of low pressure. The word is simply a contraction of "areas in which there is a depression of the barometer". In contrast, the word anticyclone was deliberately coined by Galton in 1863 and has remained in the literature since. These aperiodic isobar patterns that form and dissipate on the daily weather charts remain in vogue today. The major addition has been the frontal depression, which arrived with the model of fronts published by Bjerknes and Solberg in 1922.

Once long series of daily weather charts were available and collated, it was possible for climatologists to use them in a variety of ways – a development known as synoptic climatology (Barry and Perry, 1973). Thus W. Kṗpen (in 1880) and Van Bebber (in 1891) noted and analyzed the tracks of high- and low-pressure centers over the North Atlantic and Europe, while Elias Loomis, in a series of papers in the *American Journal of Science and Arts* between 1874 and 1888, did the same for tracks across the United States. Such studies on a hemispheric scale began in the 1950s in the northern hemisphere and in the 1960s in the southern hemisphere. Pressure pattern classification developed from this work, particularly in Europe, and resulted in the weather types of Lamb (1950) in the UK and Muller (1977) in the USA, and in Europe the *grosswetterlagen* of Baur (1936). With the advent of computers and advanced statistical methods a second generation of automated classification procedures appeared in the 1970s and 1980s. These are fully explained with worked examples by Yarnal (1993).

Extremes of surface pressure

The highest sea-level pressures have been reported at the centers of continental anticyclones, whereas the lowest pressures have been measured in the centers of hurricanes or typhoons. Table P7 gives the sea-level pressure extremes for the world, North America, and the British Isles.

World sources of surface pressure data

Historical and current-sea level or station-level data are available either as tabular compilations or mean pressure charts. The seven major tabulations

1. *Monthly Climatic Data for the World* (US Weather Bureau).
2. *Climatological Normals (CLINO) for Climat and Climat Ship Stations for the Period 1931–1960* (World Meteorological Organization, 1971).
3. *Climatological Normals (CLINO) for the period 1961–1990* (World Meteorological Organization, 1996).
4. *World Weather Records* (Smithsonian Miscellaneous Collections, Vols 79, 90, and 185; US Weather Bureau and

Table P7 Sea-level pressure extremes

	World	North America	British Isles
High	1083.8 mb 32.00 in Agata, Siberia 31, December 1968	1068 mb 31.43 in — 24, January 1897	1054.7 mb 31.15 in — 31, January 1902
Low	856 mb 25.55 in Off Okinawa 16, September 1945	892.3 mb 26.35 in Matecumber Key, Florida 2, September 1935	925.5 mb 27.33 in Tayside, Scotland 26, January 1884

US Department of Commerce and the US National Climate Data Center (NCDC).

5. The comprehensive ocean–atmosphere dataset (COADS) produced by the US National Oceanic and Atmospheric Administration (NOAA) which is available on the NOAA-CIRES Climatic Diagnostics Center website (www.cdc.noaa.gov/coads); it is described by Woodruffe et al., 1987, 1998.
6. The global mean sea-level pressure dataset which is a blend of existing gridded data sets and a variety of station level data banks (Allan et al., 1996).
7. There are also the two marine atlases produced by the US Navy (1955–59, 1995).

The International Geophysical Year (1957–1958) resulted in a special series of daily weather charts covering the northern and southern hemispheres and the tropics, published in monthly volumes by the United States, Republic of South Africa, and German Federal Republic, respectively. Mean charts of the northern hemisphere have been produced by both the US Weather Bureau (Technical Paper 21, 1952) and the Freie University of Berlin (Meteorologische Abhandlungen, 100, 1969) and for the southern hemisphere in a series of articles in the South African journal (*Notos*). Over the last 20 years there have been a number of gridded, rather than station-based, datasets produced by governmental and university research groups. These are easier to utilize in mainframe and laptop computers. Examples include the following datasets:

(a) United Kingdom Meteorological Office (UKMO) historical, 1873–1994.
(b) National Center for Atmospheric Research (NCAR), USA, 1899–1994.
(c) Climate Research Unit (CRU), University of East Anglia, UK, 1951–1985.
(d) Scripps Institute of Oceanography (SIO), USA, 1951–1993.
(e) CSIRO Division of Atmospheric Research (DAR), Australia, 1871–1989 (Allan et al., 1996).

Full details on climatic datasets including pressure for each member country are given in the World Meteorological Organization's world climate data and monitoring program, especially the INFOCLIMA project (www.wmo.ch). In the 1990s several reanalysis projects began which aimed at producing homogeneous global analyses of atmospheric fields.

Brian D. Giles

Bibliography

Allan, R., Lindesay, J., and Parker, D., 1996. *El Niño, Southern Oscillation and Climatic Variability*. CSIRO, Australia.

Barry, R.G., and Perry, A.H., 1973. *Synoptic Climatology: methods and applications*. New York: Harper & Row.

Baur, F., 1936. Wetter, Witterug, Grosswelter und Weltwetter. *Zeitschrift angewandt Meteorologie*, **53**: 377–381.

Bjerknes, J., and Solberg, H., 1922. Life cycles of cyclones and the polar front theory of atmospheric circulation. *Geophys. Publik.* (*Oslo*), **3**(1): 1–18.

Buchan, A., 1868. The mean pressure of the atmosphere over the globe for the months and the year. Part I: January, July, and the year. *Royal Society of Edinburgh Proceedings*, **6**: 303–307.

Buchan, A., 1869. The mean pressure of the atmosphere and the prevailing winds over the globe, for the months and for the year. Part II. *Royal Society of Edinburgh Transactions*, **25**(2): 575–637.

Buchan, A., 1889. Report on atmospheric circulation. In *Report on the Scientific Results of the Voyage of H.M.S. Challenger*, vol. 2, part 2. London: Eyre & Spottiswoode.

Godske, C.L., Bergeron, T., Bjerknes, J., and Bundgaard, R.C., 1957. *Dynamic Meteorology and Weather Forecasting*. Boston, MA: American Meteorological Society and Carnegie Institution.

Gordon, A.H., 1953. Seasonal changes in the mean pressure distribution over the world and some inferences about the general circulation. *American Meteorological Society Bulletin*, **34**(8): 357–367.

Hann, J., 1901. *Lehrbuch der Meteorologie*. Leipzig.

Lamb, H.H., 1950. Types and spells of weather around the year in the British Isles: Annual trends, seasonal structure of the year, singularities. *Royal Meteorological Society Quarterly Journal*, **76**: 393–429.

Lamb, H.H., 1972. British Isles weather types and a register of the daily sequence of circulation patterns, 1861–1971. *Geophysics Memoirs*, **116**.

Loomis, E., 1846. On two storms which were experienced throughout the United States in the month of February 1842. *American Philosophical Society Transactions*, **9**: 161–184.

Meinardus, W., and Mecking, L., 1915. Mittlere Isobarenkarten der Hheren sdlichen Breiten, von Oktober 1901 bis Mäz 1904. Deutsche Sudpolar-Exped., 1901–03. *Meteorological Atlas*. Berlin.

Moran, J.M., and Morgan, M.D., 1994. *Meteorology. The atmosphere and the science of weather*. New York: Macmillan.

Mossman, R.C., 1913. Southern hemisphere seasonal circulations. *Symon's Meteorological Magazine*, **48**: 2–7, 44–47, 82–85, 104–105, 119–124, 160–163, 200–207, 226–229.

Munn, T., 2001. *Encyclopedia of Global Environmental Change*, 5 vols. Chichester: Wiley.

Newton, C.W., 1972. Meteorology of the Southern Hemisphere. *Meteorological Monographs*, **13**(35):

Oliver, J.E., 1979. *Physical Geography: principles and applications*. North Scituate, MA: Duxbury Press.

Oort, A.H., 1983. *Global atmospheric circulation statistics, 1958–1973*. NOAA Professional Paper No. 14. Washington, DC: NOAA.

Shaw, Sir N., 1928. *Manual of Meteorology*, vol. 2: *Comparative Meteorology*. Cambridge: Cambridge University Press.

Sturman, A.P., and Tapper, N.J., 1996. *The Weather and Climate of Australia and New Zealand*. Melbourne: Oxford University Press.

Teisserenc de Bort, L., 1881. Etude sur l'hiver de 1879–80 et recherches sur la position des centres d'action de l'atmosphée dans les hivers anormaux. *Annals Bureau Central Météologie de France*, Part 4, 17–62.

US National Climate Data Center. 1986–1994. *World Weather Records, 1971–80*, 6 vols. Washington, DC (www.ncdc.noaa.gov).
US National Oceanic and Atmospheric Administration. *Monthly Climatic Data for the World*. Asheville, NC National Climatic Center (monthly) (www.noaa.gov/climate/climatedata.html).
US Navy, 1955–1959. *Marine Climatic Atlas of the World*, 5 vols. NAVAIR 50-1C-528/, . . . Washington, DC: Chief of Naval Operations (www.ncdc.noaa.gov/products/paper/navair.html).
US Navy, 1995. *Marine Climate Atlas of the World* (MCA). Fleet Numerical Meteorological and Oceanographic Detachment, Asheville, NC (www.navy.ncdc,noaa.gov/products/cdrom.html).
Woodruffe, S.D., Slutz, R.J., Jenne, R.I., and Steurer, P.M., 1987. A comprehensive ocean–atmosphere data set. *Bulletin of the American Meteorological Society*, **68**: 1239–1250.
Woodruffe, S.D., Diaz, H.F., Elms, J.D., and Wooley, S.J., 1998. COADS Release 2 data and metadata enhancements for improvements of marine surface flux fields. *Physics and Chemistry of the Earth*, **23**(5–6): 517–526.
World Meteorological Organization, 1971. *Climatological Normals (CLINO) for Climat and Climat Ship Stations for the Period 1931–1960*. Geneva: World Meteorological Organization (www.wmo.ch).
World Meteorological Organization, 1996. *Climatological Normals (CLINO) for the period 1961–1990*. Geneva: World Meteorological Organization (www.wmo.ch).
Yarnal, B., 1993. *Synoptic Climatology in Environmental Analysis. A Primer*. London: Belhaven Press.

Cross-references

Adiabatic Phenomena
Atmospheric Circulation, Global
Centers of Action
Climatology, History of
Land and Sea Breezes
Mountain and Valley Winds
Pressure, Upper Air
Reanalysis Projects
Synoptic Climatology
Vorticity

PRESSURE, UPPER AIR

The decrease of atmospheric pressure as one leaves the Earth's surface was first noted by Blaise Pascal on ascending the 52-m-high Tour St Jacques in Paris in 1643. It was verified by his brother-in-law Péier in 1648 in an experiment that demonstrated that the pressure at the top of the Puy de Dôme (elevation 1467 m) in Clermont, France, was 3 Parisian inches, 1½ lines (8.5 cm) lower than the base. Pascal therefore concluded that a barometer could be used to determine the difference in the altitude between places. The idea was taken up by Mariotte who used both his own experiments at the Paris Observatory and those of others to show in 1679 that: (1) pressure decreased by about 1 line in every 60 ft (1 mm in 8.6 m); and (2) pressure drop with height decreased in a geometric progression; therefore, the atmosphere was 8×10^6 times more tenuous at 30 leagues (133 km) than at the Earth's surface.

During the next 100 years, several formulas were derived that related barometric pressure and height. A complete solution was proposed by P.S. Laplace (1805, pp. 289–293), which superceded all others. Laplace's formula was used as the basis for the tables published by A. Guyot (1862) in the *Smithsonian Miscellaneous Collections*, No. 153 (1859) that were republished as part of vol. 1 of the *Smithsonian Miscellaneous Collections*. This collection of tables also included a set produced by the American, Elias Loomis, for heights in feet and temperatures in degrees Fahrenheit.

Relationship of pressure and height

The relationship between pressure and height can be expressed in the form of the hydrostatic equation or the barometric equation, which shows that the change of pressure (Δp) with the change of height (Δh) is equal to gravity (g) and air density (ρ):

$$\frac{\Delta p}{\Delta h} = -g\rho$$

or

$$-\Delta p = g\rho\Delta h$$

with the minus sign indicating the decrease of pressure upward. Density is difficult to measure directly, but it can be computed using the gas law relationship between pressure, density, temperature, and volume. In this way, by substitution and integration, the height interval ($h_2 - h_1$) between pressure levels (p_2 and p_1) can be calculated using:

$$h_2 - h_1 = 64.7T\,(\log p_1 - \log p_2)$$

where h is in meters, p is in millibars and T is in Kelvin.

The decrease of pressure with altitude was used in the nineteenth century to ascertain the heights both of mountains and the levels reached by balloons. The first unmanned balloon ascents were made in France in 1782 (hot air) and in Scotland in 1766 and 1783 (hydrogen). The first manned balloon ascent was by Jeffries in November 1784 from London, and this was followed by several from Hamburg, St Petersburg (now Leningrad), and Paris in 1803 and 1804. These ascents were mainly concerned with temperature and humidity changes and the barometers were instruments for indicating the elevations reached. There was a lull in such flights until the middle of the nineteenth century when a considerable number were made in England by James Glaisher (who made 28 flights between 1862 and 1866, reaching heights of up to 5 miles), and in France by Flammeron and Tissander in the 1870s. In the last quarter of the nineteenth century there was a move toward captive balloons, kites, and balloons-sondes (now known as radiosondes), which gave strong impetus to improvements in self-recording instruments. The first continuously recording instrument was flown with a kite at the Blue Hill Observatory in 1894. A full contemporary account can be found in Gold and Harwood (1910).

The main thrust of this research was toward temperature and wind variations with height, the latter calculated from barometers. By 1904 Teisserenc de Bort was able to draw a map of the northern hemisphere showing isobars at the level of 4000 m (Figure P18), although these were calculated using surface pressures and the temperatures between sea level and 4000 m. This set a precedent and until the end of World War II it was usual to draw charts at fixed heights (in kilometers or in thousands of feet) on which measured pressures and hence isobars were drawn. Thus linear height was a fixed unit and pressure was a variable unit. Shortly after World War II the format was reversed for mainly aeronautical reasons, and current upper air charts are constant-pressure charts (pressure is the fixed unit) on which variations in height (the variable unit) are plotted; the consequent isopleths were known as contours, i.e. lines of equal height (in decameters or hundreds of feet).

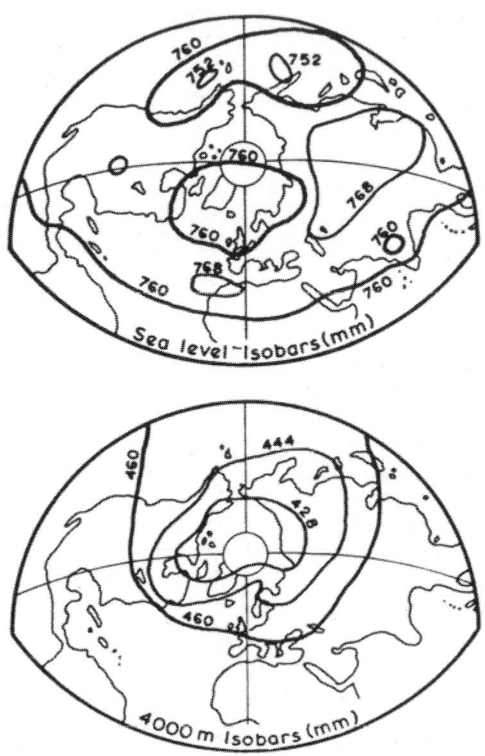

Figure P18 Isobars at 4000 m in the Northern Hemisphere (after de Bort, 1904).

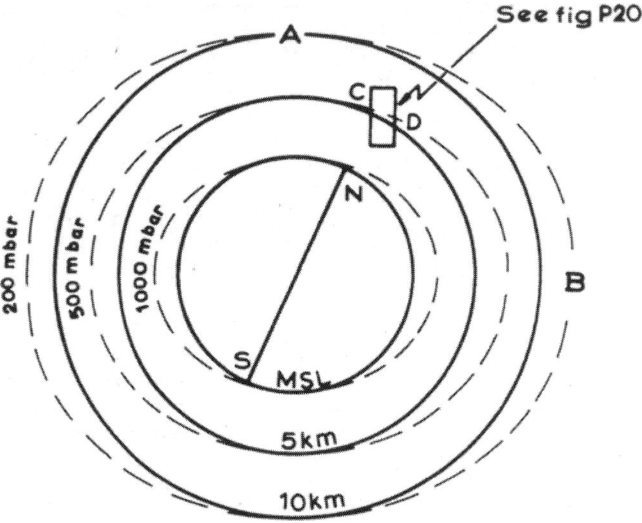

Figure P19 Relationship between surfaces of equal height or equal pressure that surround the Earth (see text).

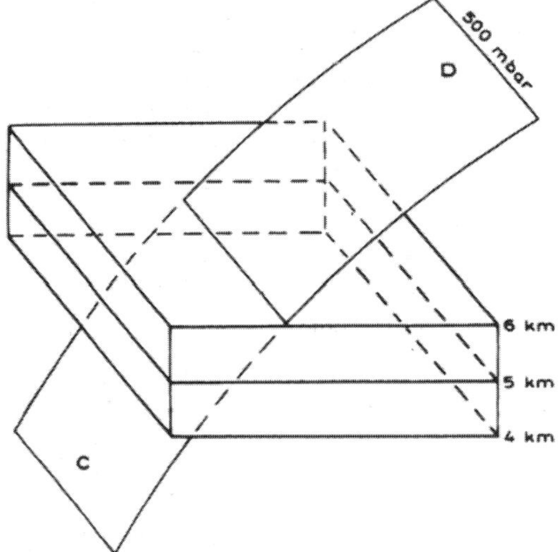

Figure P20 Intersection of a pressure surface and a height surface, enlarged from Figure P19.

Units

Units describing upper-air pressure have had a far less checkered career than their surface equivalents. Millibars (mb) were usual on fixed height charts (although the earliest ones were in millimeters of mercury) and meters or feet on fixed pressure charts. Millibars have now been replaced by hectopascals (hPa); 1 mb = 1 hPa. The need for fixed pressure charts grew out of the theoretical work of V. Bjerknes who, in 1898, formulated a circulation theorem using the distribution of pressure and specific volume. This and other circulation theorems are beyond the scope of this item, but the point must be made that the theory required that the elevations or heights on such charts took gravity into account.

The atmosphere can be considered in two ways:

1. As a series of spheres at fixed distances from the Earth's surface. On each sphere the pressure will vary because of the changes in temperature, humidity, density, and so on.
2. As another series of spheres that represent decreasing pressures. On each of these spheres the height above mean sea level varies. This is shown diagrammatically in Figure P19, where the encircling spheres (of either unit) can be considered as surfaces of height or pressure. In the figure, the points A and B are 10 km above the Earth's surface, but the pressure at A is less than, and the pressure at B is greater than, 250 hPa. Similarly, points C and D are both at a pressure of 500 hPa but C is less than 5 km above the Earth's surface whereas D is more than 5 km above. This can be easily visualized in Figure P20, an enlargement of part of Figure P19, in a three-dimensional form. A volume of atmosphere between 4 km and 6 km is shown through which the 500 hPa pressure surface passes. This clearly shows that D is much higher above the Earth's surface than C. An upper-air pressure surface should be thought of as a sheet of rubber that is distorted out of a flat (horizontal) plane, and the consequent measure of this distortion is the variable height of the sheet above sea level. This three-dimensional conceptualization of a particular pattern is shown in detail by Giles (1976) for a series of upper-air pressure surfaces.

In a very simplified form Figure P19 indicates that pressure surfaces tend to be further from the Earth's surface in equatorial and tropical regions and much closer in polar regions. This is taken into account when a standard atmosphere (q.v.) is defined. A skeleton version of the US Supplemental Atmosphere 1976

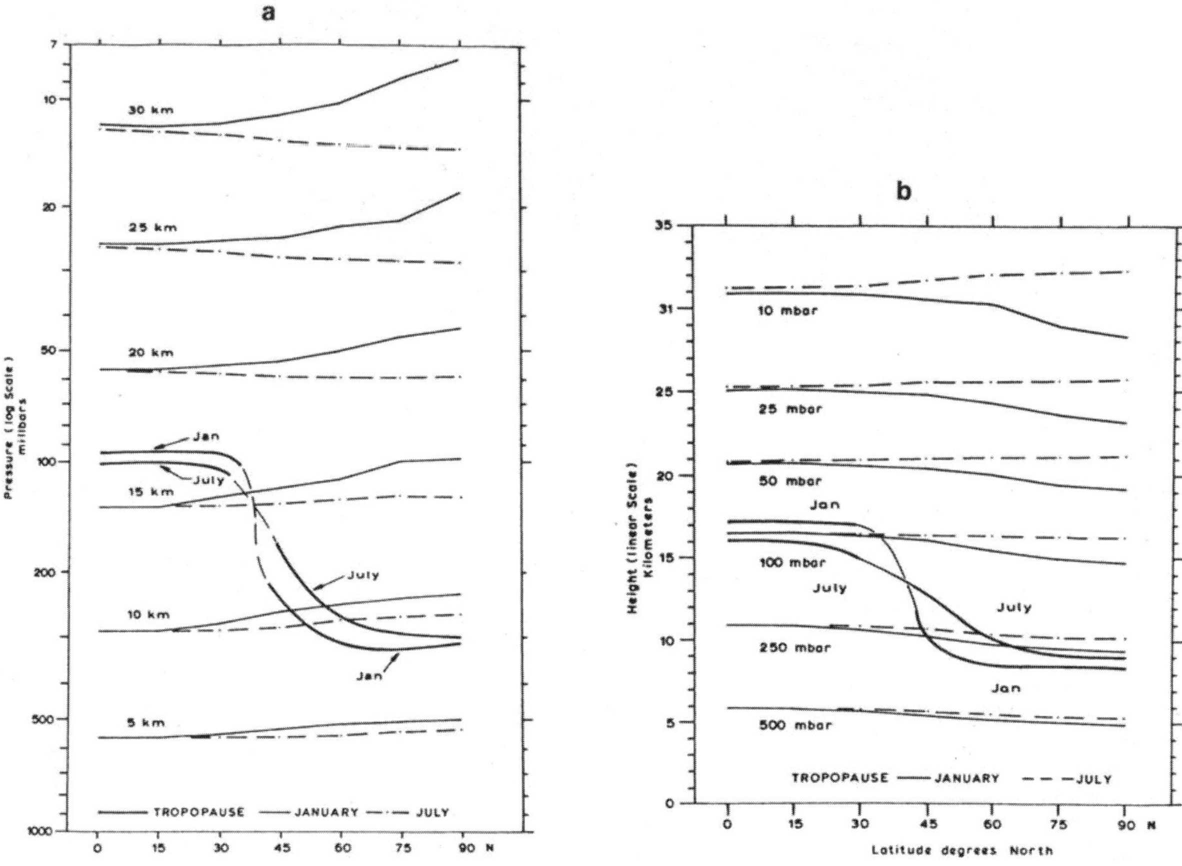

Figure P21 Latitudinal and seasonal variation of pressure with height (**a**) and height with pressure (**b**) of the U.S. Supplemental Atmosphere, 1976 (based on data in Cole and Kantor, 1978).

has been drawn in Figure P21, using data given by Cole and Kantor (1978), to illustrate the latitudinal and seasonal variation of pressure with height (Figure P21a) and of height with pressure (Figure P21b) in the northern hemisphere. In the troposphere (q.v.), pressure surfaces decrease in height from equator to pole but are generally higher in summer than in winter. In the stratosphere (q.v.) they increase in height in tropical areas and then decrease toward the poles in winter (January). However, in summer (July) there is an increase in the height of pressure surfaces toward the pole, which becomes greater the farther one goes into the atmosphere. The seasonal difference in pressure surface height increases both poleward and upward (see Figure P21a). Because pressure decreases logarithmically with height, the latitudinal and seasonal variation of the pressure found at fixed heights seems much greater (see Figure P21b). Whereas the pressures at 5 km vary from 560 hPa at the equator to 500 hPa at the pole in January, at 30 km the pressure increases from 12.3 hPa at the equator to 13.8 hPa at the North Pole in summer but decreases from a maximum of 12.5 hPa at 15°N to a minimum of 8.8 hPa at the pole (Figure P21b). Figure P21 also includes the envelope of the tropopause, which shows the marked charge in this tropospheric boundary between summer and winter, particularly in the subtropics between 30° and 45°N. The COSPAR International Reference Atmosphere (CIRA) provides data including pressure from 0 to 120 km and from 80°N to 80°S (see Fleming et al., 1988; Rycroft et al., 1990).

Charts

It was pointed out above that upper air charts generally were of the constant level type until the end of World War II when they were replaced by constant pressure charts. Upper air charts were drawn only for research purposes until 1933 when upper air constant level charts were produced on a daily basis in the United States and England. At the same time, daily constant pressure chart analysis began in Germany. The British Meteorological Office changed from constant level to constant pressure in 1941. US meteorologists continued to use constant level charts in North America, but conformed to European practice when they were working in Europe after 1942. The end of constant level charts occurred in 1945 when the Joint Meteorological Committee adopted constant pressure charts on an international basis. Today a series of standard pressure levels (for which charts are drawn) has been agreed on: 1000, 850, 700, 500, 400, 300, 250, 200, 150, 100, 70, 50, 30, 20, 10, 7, 5, 3, 2, and 1 hPa. However, not all these levels are analyzed and drawn routinely, and different countries have individual procedures. Generally speaking, the 500 hPa chart is found worldwide, but the 700 hPa chart is additionally found in American journals.

The main reasons for the comparatively recent arrival of daily upper air charts were:

1. The great expense involved in both the radiosondes and the necessary ground-receiving equipment.

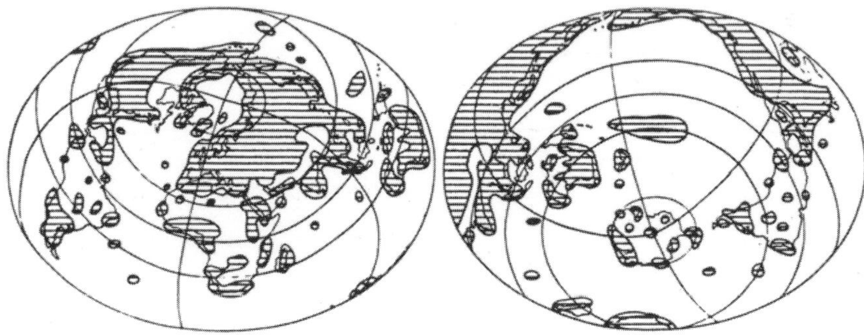

Figure P22 Worldwide upper air coverage during the 1970s.

2. Consequently, there was a much smaller number of upper air stations with wide gaps in areal coverage.
3. The large amount of data that had to be collected and transmitted from each station and analyzed at meteorological centers.

For the First Global Atmospheric Research Program (GARP) Global Experiment (FGGE) in 1978–1979 it was estimated that the exchange of information from 1300 upper air reports for the standard levels up to 30 hPa would constitute about 4.5 × 105 characters each day. Figure P22 shows the worldwide upper air coverage typical of the 1970s. The poor coverage in most of the southern hemisphere and in the North Pacific and North Atlantic oceans is obvious. However, the derived information can be enhanced by satellite-derived information.

Diurnal variations

Because upper air observations of pressure are usually only taken twice a day (at noon and midnight GMT), there has been little work on diurnal variations in the heights of pressure surfaces. From the small amount of data that has been published it is clear that there is a diurnal variation with maximum heights found shortly after local noon and minimum heights just after local midnight. No detailed analysis of the daily periodic variation is possible because of the fixed 12-hourly times of observation based on Greenwich rather than local time. In any case the variation is very small and heavily masked by aperiodic synoptic variations, which are the result of the pressure systems that are continuously changing and modifying the contour patterns found on the upper air charts. These patterns are generally in the form of smooth waves – the long waves that were an important item in the Rossby regime (q.v.) and that are described in more detail in Atmospheric Circulation, Global. In both hemispheres these waves are the result of the effect of topographic barriers such as the Rockies and Andes. The consequence is a series of troughs and ridges extending eastwards from the longitudinal mountain barriers around both hemispheres in extratropical latitudes. Usually there are between one and five such waves around a hemisphere with smaller waves embedded in them. Since the 1970s use has been made of constant level balloons which have been used to show the day-to-day variations of a particular pressure level. These balloons are usually flown at upper tropospheric and stratosperic levels.

Seasonal variations, vertical

The positions of the ridges and troughs vary on daily charts, but they have preferred longitudes that clearly appear if means over several years are calculated. Figures P23 and P24 are based on data given in Heastie and Stephenson (1960), and show the average heights of the 500 hPa and 200 hPa pressure levels at various latitude circles. On each diagram the mean positions of the ridges and troughs are marked on the height curves. At 500 hPa in January (Figure P23), three ridges and troughs can be seen at 50°N with the troughs found over central North America, European Russia, and the western North Pacific. At 80°N only two waves are apparent with troughs over North America and Central Asia. In the subtropics (30°N) the three troughs are found at longitudes somewhat to the west of their midlatitude positions. The intensity of the waves is indicated by the difference in height of the pressure surface at a given latitude. Thus in January at 500 hPa, the largest fluctuations in height are at 50°N. In July the intensity diminished and three waves are found only in midlatitudes – poleward and equatorward only two waves appear. In the southern hemisphere in both January and July, the height of the 500 hPa pressure surface varies to a smaller extent, and a three-wave pattern occurs only at 50°S in winter. In this hemisphere mean troughs occur over the oceans and ridges are loosely associated with areas to the south of the three major landmasses. Figure P24 is a similar diagram for the 200 hPa level. The height fluctuations are greater in the winter hemisphere than in the summer hemisphere and are largest in the northern hemisphere winter.

By comparing the same hemisphere and season at each level it is clear that the longitudinal position of the mean ridges and troughs does not vary much with height. This means that the two pressure surfaces (and all those between) undulate more or less in unison around the globe, and only over the poles and the equator do the undulations flatten out. In the former case this is simply because there is insufficient distance around the hemisphere to accommodate them. A measure of these undulations at different levels in the troposphere and lower stratosphere is given in Table P8. This shows the difference in decameters between the high equatorial and lower polar mean heights for each hemisphere in summer and winter. The variations in the heights of the pressure surfaces increase upward in summer and winter in both hemispheres to the region of the lower stratosphere, beginning to decline above 200 hPa in the northern hemisphere and above 150 hPa south of the equator. In summer at all levels up to and including 150 hPa in the northern hemisphere, mean height differences are about half their winter values. This is not true of the southern hemisphere, where summer values are only 25% lower than winter values. This indicates a smaller seasonal

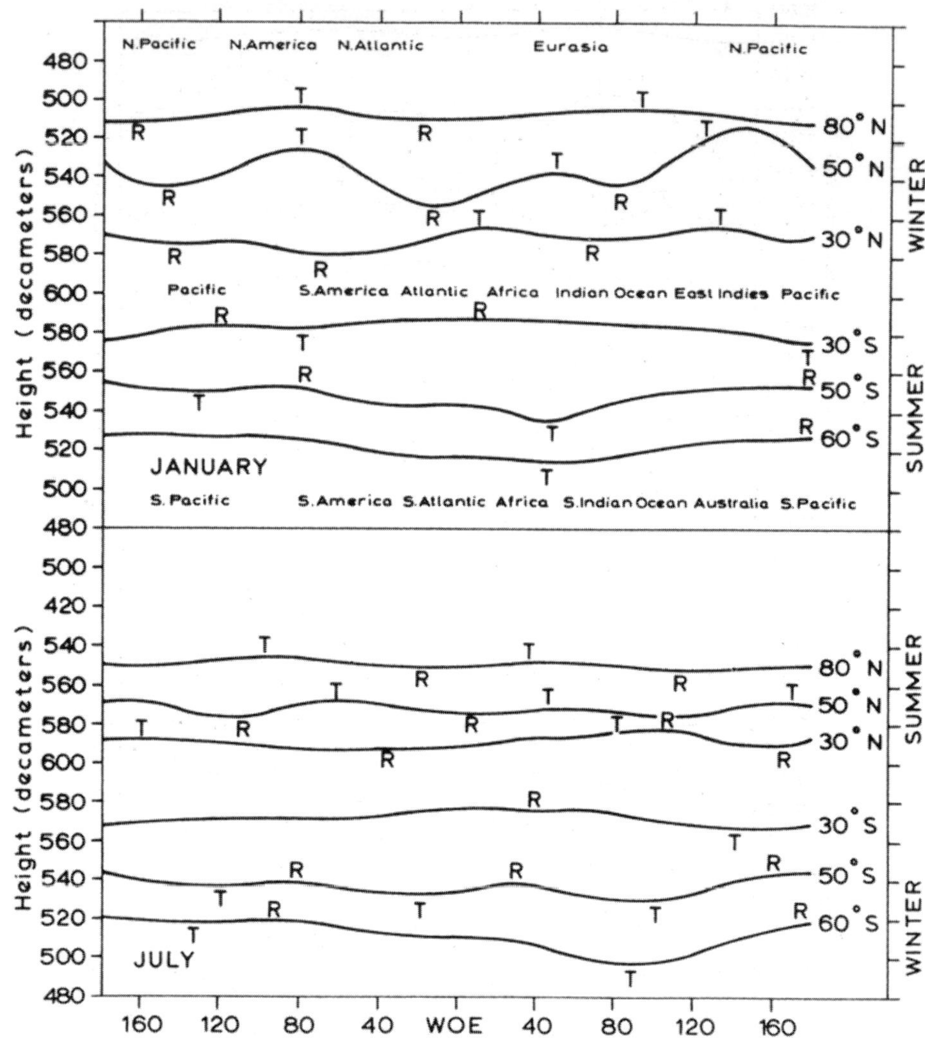

Figure P23 Average height of the 500 hPa pressure level at various latitudes in January and July.

variation in the southern hemisphere than in the northern hemisphere.

Seasonal variations, horizontal

The areal distribution of the pressure surfaces is shown in Figures P25 and P26 for 500 and 200 hPa, these being typical of mid- and upper-tropospheric patterns. Because of the projection used, the ridges and troughs do not stand out as clearly as in Figures P23 and P24. However, the upper air "centers of action" (see Pressure, Surface) clearly show areas of low heights in the regions of the geographic poles. The low in the northern hemisphere is eccentrically placed toward the Canadian Archipelago and Hudson Bay. In the southern hemisphere the low lies over the lower land of West Antarctica (i.e. the part of the continent adjacent to the Pacific Ocean). Thus the major low pressures of the surface charts disappear rapidly upward in the troposphere to be replaced by polar vortexes. In the tropics the pressure surfaces are generally at their highest. In January the greatest heights straddle the equator, but in July they move northward to the vicinity of the Tropic of Cancer. The exact pattern of the tropical highs varies from pressure surface to pressure surface, and indeed at 100 hPa the greatest heights in January are actually (for the first time) in the southern hemisphere. They go over the Tropic of Capricorn, extending from Madagascar across southern Africa and the South Atlantic to South America.

The marked difference in the mean height of the 500 hPa pressure surface in the two hemispheres and at the solstices (Figure P25) should be noted. In the northern hemisphere the winter polar low is over 40 dam nearer to the Earth's surface than the summer low, whereas the tropical high is some 200 dam higher. In the southern hemisphere there is less difference in height between the summer and winter polar lows.

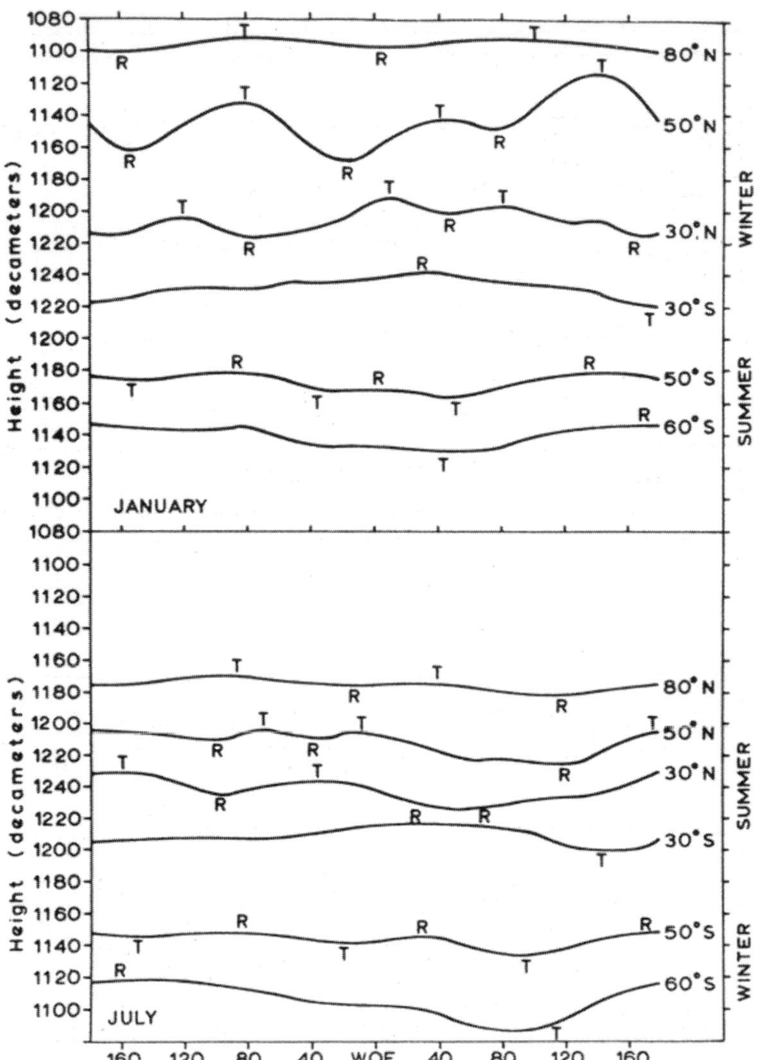

Figure P24 Average height of the 200 hPa pressure level at various latitudes in January and July.

Table P8 Summer and winter differences (decameters) between high equatorial and low polar mean heights for each hemisphere

Pressure level (mb)	Northern hemisphere		Southern hemisphere[a]	
	Winter	Summer	Winter	Summer
700	55	31	60	43
500	92	46	90	75
300	142	74	133	110
200	163	86	157	118
150	160	77	161	105
100	137	49	150	77

[a] Southern hemisphere analysis extends only to 60°S.

The pattern around the troposphere–stratosphere boundary is shown in Figure P26, which is for the 200 hPa surface. In the northern hemisphere the mean height difference of the polar low between winter and summer has increased to more than 1 km; in the southern hemisphere it is over 50 dam. In the tropics the major high lies athwart the equator, over the Indian Ocean, in January and moves north in July to cover an area of the Tropic of Cancer extending from North Africa across southern Asia to the China Sea. The North American equivalent appears over Arizona and New Mexico in July.

Data sources

Mean monthly values of upper air data, including heights of the main pressure surfaces, are published by the US National Oceanic and Atmospheric Administration (NOAA) as *Monthly Climatic Data of the World* under the sponsorship of the World Meteorological Organization. This is based on Climat Temp reports from over 400 upper air stations and comprises the raw

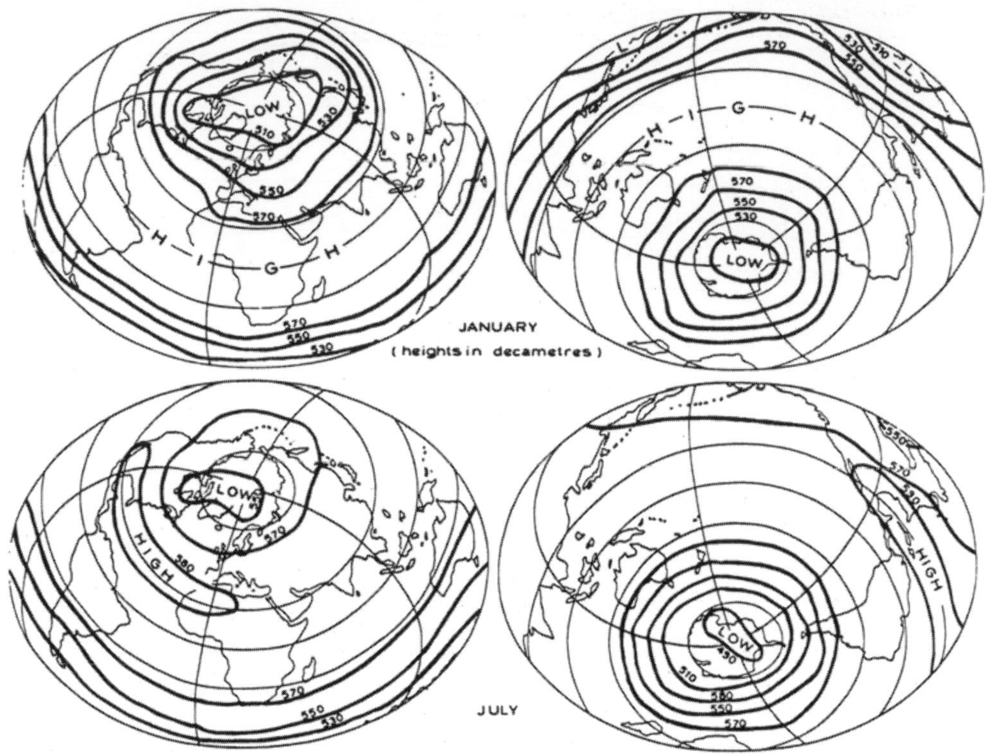

Figure 25 World pattern of the 500 hPa pressure surface in January and July.

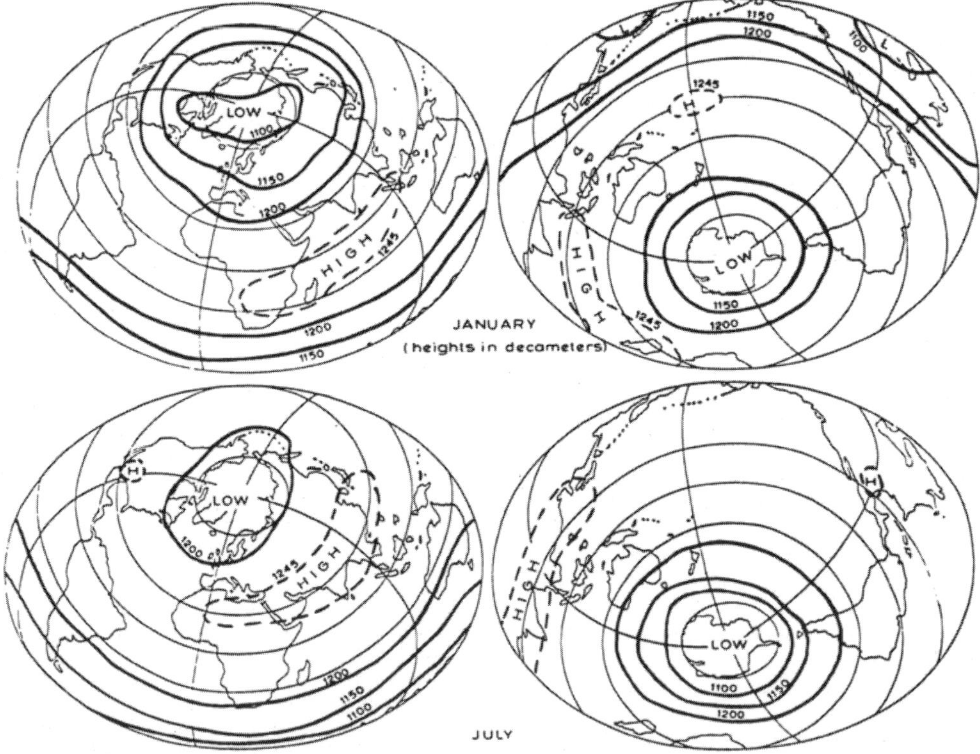

Figure P26 World pattern of the 200 hPa pressure surface in January and July.

Table P9 Data available for various pressure levels

Pressure level (hPa)	A	B	C	D	E	F	G	H
850	X			X	X	X	X	X
700	X	X	X	X	X	X	X	X
500	X	X	X	X	X	X	X	X
300	X	X		X	X	X	X	X
200	X	X		X	X	X	X	X
150		X						X
100	X	X			X	X	X	X
50	X							X
30	X							X

Notes: A: US National Oceanic and Atmospheric Administration, monthly
B: Heastie and Stephenson, 1960
C: US Weather Bureau, 1952
D: US Navy, 1955
E: Dubenov and Davidova, 1964
F: Taljaard et al., 1969
G: Crutcher and Meserve, 1970
H: Fleming et al., 1988

data from which a variety of climatological means at latitude and longitude grid points have been calculated. Data in the form of grid point values are more easily stored and useful than individual station values; they are derived from analyzed computer and statistical checks. Data in this form, including mean heights of pressure surfaces, are considered as the basic dataset for the upper air and are available in magnetic tape and/or CD-rom formats from the National Climatic Data Center. Similar datasets are also available from the National Space Science Data Center (NSSDC) as binary files (ftp://nssdcftp.gsfc.nasa.gov/models/atmospheric/cira/cira1986) or in ASCII format (ftp://nssdcftp.gsfc.nasa.gov/models/atmospheric/cira/cira86ascii). In addition Heastie and Stephenson (1960) published tables giving average heights of six pressure levels for January, April, July, and October at various latitude–longitude intersections from the North Pole to 60°S. The data used in this compilation were for the period 1949–1953. Such large amounts of data are better portrayed in chart form, and a number of publications give the mean contour patterns at various levels (Table P9). One of the earliest was by the US Weather Bureau (1952), which included mean contour charts for 700 and 500 hPa for the northern hemisphere. This was followed by the US Navy (1955–1959) whose *Marine Climatic Atlas* had contour charts for five pressure levels for the four seasons over the North Atlantic and North America (vol. 1) and the North Pacific (vol. 2). The publication drew, where possible, on 8 years' data (1946–1954) but in the southern hemisphere, in particular, it was usually for shorter periods and sometimes consisted of data derived solely from the International Geophysical Year (1957–1958). This shortage of data precluded the drawing of contours over the southern hemisphere oceans in volumes 3–5. This sparse data network did not prevent a Russian atlas (Dubencov and Davidova, 1964) from appearing with southern hemisphere contour charts for six pressure levels for the mid-season months of January, April, July, and October. Only 300 copies of the atlas were printed so it is not readily accessible. It used observations from a sparse network of about 70 stations reporting over the 5-year period 1956–1960 in addition to various other compilations published in the Soviet Union.

The next major step was made in the late 1960s when the results of two major research efforts were published. For the northern hemisphere, Crutcher and Meserve (1970) produced mean charts for six pressure levels using 1950–1964 data. Their analyzed charts were used as the basis for the production of grid point values held on magnetic tape in the National Weather Record Archives. A similar but more difficult exercise for the southern hemisphere culminated in the four-volume work by Taljaard et al. (1969). Only vol. 1 is relevant to this discussion since it contains the mean monthly contour charts for six levels (see Table P9) together with a full report of the data used – data that were widely scattered in both space and time. Finally, in the 1990s the US Navy (1995) updated the *Marine Climate Atlas of the World* (MCA); the latest version of CIRA86 has been made available on the internet (ftp://nssdcftp.gsfc.nasa. gov/models/atmospheric) and reanalysis projects (Kalnay et al., 1996) produce homogeneous global analyses of atmospheric fields including pressure.

Brian D. Giles

Bibliography

Cole, A.E., and Kantor, A.J., 1978. Air Force Reference Atmospheres. *Air Force Surveys in Geophysics*, No. 382, AFGL-TR-78-0051.

Crutcher, H.L., and Meserve, J.M., 1970. *Selected Level Heights, Temperatures and Dew Points for the Northern Hemisphere*, rev. NAVAIR-50-1C-52. Washington, DC: Chief of Naval Operations.

Dubencov, V.R., and Davidova, N.G., 1964. *Atlas of Aeroclimatic Charts and Charts of the Paths and Frequencies of Depressions and Anticyclones in the Southern Hemisphere*. Moscow: Chief Office of the Hydrometeorological Services, Central Forecasting Institute (in Russian).

Fleming, E.L., Chandra, S., Shoeberl, M.R., and Barnett, J.J., 1988. Monthly mean global climatology of temperature, wind, geopotential height and pressure for 0–120 km. *NASA Technical Memorandum* 100697, Washington, DC: NASA.

Giles, B.D., 1976. On isobars, isohypses and isopachs or pressure, contour and thickness charts. *Weather*, **31**(4): 113–121.

Gold, E., and Harwood, W.A., 1910. The present state of our knowledge of the upper atmosphere as obtained by the use of kites, balloons and pilot balloons. *British Association for the Advancement of Science*, **79**: Winnipeg, 1909. London: BASS, pp. 71–124.

Guyot, A., 1862. *Tables, Meteorological and Physical Prepared for the Smithsonian Institution*, 3rd edn. Smithsonian Miscellaneous Collections, vol. 1. Washington, DC: US Government Printing Office.

Heastie, H., and Stephenson, P.M., 1960. *Upper Winds over the World*, Parts I and II. Geophysical Memoirs, No. 103. London: Meteorological Office.

Kalnay, E., Kanamitsu, M., Kistler, R., et al. 1996. The NCEP/NCAR 40-year reanalysis project, *Bulletin of the American Meteorological Society*, **77**: 437–471.

Laplace, P.S., 1805. *Traite de Mecanique Céleste*, vol. 4.

Ryecroft, M.J., Keetmay, G.M., and Rees, D. (eds), 1990. Upper atmosphere models and research. *Advances in Space Research*, **10**(6).

Taljaard, J.J., van Loon, H., Crutcher, H.L., and Jenne, R.L., 1969. *Climate of the Upper Air: southern hemisphere, temperatures, dew points, and heights at selected pressure levels*, vol. 1. NAVAIR-50-1C-55. Washington, DC: Chief of Naval Operations.

Tiesserenc de Bort, L., 1904. Etudes sur les Dépressions Barometriques à Diverses Hauteurs. *British Association for the Advancement of Science*, **73**: Southport, 1903. London: BASS, pp. 549–555.

US National Oceanic and Atmospheric Administration. *Monthly Climatic Data of the World*. Asheville, NC: National Climatic Center (monthly) (www.noaa.gov/climate/climatedata.html).

US Navy, 1955–1959. *Marine Climatic Atlas of the World*, vol. 1, *North Atlantic Ocean*; vol. 2, *North Pacific Ocean*; vol. 3, *Indian Ocean*; vol. 4, *South Atlantic Ocean*; vol. 5, *South Pacific Ocean*. NAVAIR-50-1C-528. . . . Washington, DC: Chief of Naval Operations (www.ncdc.noaa. gov/products/paper/navair.html).

US Navy, 1995. *Marine Climatic Atlas of the World* (MCA). Fleet Numerical Meteorological and Oceanic Detachment, Asheville, NC (www.navy.ncdc.noaa.gov/products/cdrom.html).

US Weather Bureau, 1952. *Normal Weather Charts of the Northern Hemisphere*. Technical Paper No. 21. Washington, DC: US Government Printing Office.

Cross-references

Adiabatic Phenomena
Atmospheric Circulation, Global
Dynamic Climatology
Rossby Wave/Rossby Number
Reanalysis Projects
Vorticity
Zonal Index

Q

QUASIBIENNIAL OSCILLATION

Following the American Meteorological Society *Glossary*, Sir Gilbert Walker defined "oscillation" in a way acceptable to climatology as "a single number, empirically derived, which represents the distribution of pressure and temperature over a wide ocean area" as in Southern Oscillation, North Atlantic Oscillation, for example. In the German literature, von Rudloff (1967) used "Klimaschwankungen" for fluctuations longer than 30 yr, and "Klimapendelungen" for short-term oscillations (Schuurmans, p. 91, in Flohn and Fantechi, 1984). Von Rudloff's instrumentally based time-series go back to the seventeenth century and the Dutch records at De Bilt started in 1634. Three stations go back to before 1579 (Groveman and Landsberg, 1979). Earlier indications have to be based on proxies such as tree rings and harvests (Lamb, 1977, p. 534, etc.).

In American temperature data a 2-yr oscillation was first recognized by Clayton in 1884, but after several returns it jumps 1 year and hence the name "Quasibiennial Oscillation" (Schove, 1983). It was described in detail by Reed (1961) and even more thoroughly by Ebdon (1975) in its association with tropospheric patterns.

Most significant is the alternation of easterly and westerly equatorial upper-air winds, which have been systematically observed, with increasing precision, since the famous eruption of the Krakatoa (Krakatau) volcano in the East Indies in August of 1883. Its ash reached the lower stratosphere drifting in alternate years either easterly or westerly. The former were labeled the "Krakatoa Easterlies" and the latter the "Berson Westerlies" (see "Krakatoa Winds" in Fairbridge, 1967, p. 524). The alternation has a mean periodicity of about 27 months, corresponding to the QBO link.

Solar forcing mechanism

A key paper on the subject was by Berson and Kulkarni (1968), who related the QBO to the Sunspot Cycle (SSC). Labitzke and Van Loon (1990) found that the correlation was only positive during the west phase of the QBO in the north polar temperatures, but antiphase in the easterly condition. As remarked by Peixoto and Oort (1992), "if confirmed… this will be an important new… factor in climate". The trouble with this linkage is the variable length and amplitude of the SSC (11 ± 5 yr). Short cycles correspond to high-energy phases, and long ones to weak phases.

Furthermore, there are solar emissions (of several varieties, from UV to neutron flux) that disclose periodicities of the order of 2.172 yr (793.3 days). Spectral analyses of SSC by Currie (1973) show a prominent spike at 2.17 yr. The mean high-energy phase, about 8.6 yr ($4 \times$ QBO), contrasts with the low-energy mean of about 13 yr ($5 \times$ QBO). Schove (1983), picturesquely, likens the QBO to the period of a child on a swing, which every fourth or fifth return receives a nudge from the SSC. The analogy does not appear to have been rigorously explored, for the good reason that the long-term cycles that modify solar radiations grossly exceed the range of precise instrumental observations.

For a meeting convened by the Royal Society in London in 1990, a review of solar variability based on astronomic observations was prepared by Elizabeth Ribes (1990), the French specialist. In a series of astrolabe measurements, ranging 1977–1987, and disclosing solar radius (her fig. 3), when compared with Earth's stratospheric winds at the 10 mb level (Naujokat, 1986), a quasibiennial oscillation in the sun's diameter is clearly disclosed. Further observations emerged from an Arizona conference the next year (Ribes et al., 1991). The question of solar dimensions has long been controversial, but the Ribes evidence seems to be unarguable.

As noted above, the radiations of the sun are subject to long-term variations, based on the evidence of reliable proxies (ice cores, tree rings, geological history of glaciations and ice ages). Although a sun–planetary link is unavoidable in terms of astrophysics, it has been difficult to prove to universal satisfaction. However, the relationships presented by Fairbridge and Sanders (1987) have never been systematically refuted, and in fact are supported by Charvátová and Střeštík (1991), Windelius and Carlborg (1995), and others. It will therefore have been assumed below.

The planetary motions in question are based on the NASA-JPL ephemeris and calculations that span time-scales of the order of 10^5 yr. Individual planetary periods provide a basis for applying the Helmholtz formula for beat frequencies. The most important of these is that for the two largest planets, Jupiter and Saturn, a period of 19.8593 ± 0.8 yr, the so-called "heartbeat" of the solar system. All of the planets become geodetically arranged in approximately 90° quadrants, to reach a general quadrature at intervals of 4448.485 yr. Each quadrant constitutes a 556.06-yr unit that is 256 × 2.1721 yr, the accepted mean QBO. For a student interested in music, one might note that "middle C" on the piano is generally accepted to correspond to 256 Herz (cycles per second). An Earth–moon connection is provided by a 111.25-yr cycle (×5 = 556 yr, that appears also to be the precession of the lunar perigee and an approach to maximum terrestrial spin rate). The 111-yr period approximates what Wood (2001) calls the "Proxigee–Syzygy Perihelion Cycle".

The 556-yr period is precisely 50 × 11.1212 yr, the mean sunspot cycle, but because of its imprecise length, as noted above, it coincides only from time to time with the lunar cycles and likewise with the "ideal" QBO. Over progressively longer intervals, however, the SSC is found to conform to planetary motions, notably the "King-Hele Cycle" (177.9394 yr = 16 × SSC). This cycle reflects the alignments of Jupiter with the center of the sun and the center of mass (barycenter) of the solar system. A rather similar cycle of 178.73 yr (the so-called "Jose Cycle") represents the sun's inertial orbital precession. These two cycles at a ratio of 224:225 create a 40 036.36-yr period; and this in turn is exactly 18 432.0 times the "ideal" QBO. It is also nine times the 4448.485-yr quadrature cycle (2048 × QBO). Evidently, Currie's discovery of the QBO period in his SSC spectral analysis (1973) is confirmed by the general interlocking of both with the fundamental long-term planetary periods.

Studies of planetary pair beat frequencies have disclosed simple logarithmic relationships. If unity (n) = 0.0678784 yr (approximately the inner sun's sidereal spin rate), then $8^2 \times n)^{-2}$ = 2.17211 (i.e. QBO). In the same sequence can be found

$128^2 \times n = 1112.12$, and thus ½ = 556.06; or
$256^2 \times n = 4448.485$ (quadrature cycle: 224 × Saturn/Jupiter lap, = 400 SSC)

Evidently on a long-term basis the QBO appears to be eminently predictable from the astronomic constants. This, however, is not so simple. It would in fact be foolhardy, if not egregiously mistaken, to make forecasts without considering all the variables.

QBO forecasts

Schove (1983), in a volume that collects many of his earlier papers, recounts his heroic attempts to forecast climatic conditions from the "regularity" of the QBO. In England your vacation plans can often depend on the probability of a warm/dry summer versus a cold/wet one. Schove noticed that odd/even-numbered years could be used with this object. But just when it seemed to be working fine (on a biennial basis), it would break down into a triennial or chaotic interval.

Whereas the solar cycle is complicated by the low-frequency electromagnetic radiations (notably UV), equatorially focused, and the high-frequency volatility of the solar wind focused, on the magnetic poles, especially the northern one, there is also the moon, notably mediated by the lunar year (quite distinct from the solar year), the 18.6134-yr period of the nodal cycle which strongly influences the geostrophic oceanic circulation, and finally the 18.0303-yr "Saros", the principal tide cycle. Although the seasons and their radiation characteristics are well established on a 365-day basis, the lunar year introduces an eccentricity, so that no two successive years can ever be climatically the same.

As shown by Pettersson (1912, 1914), the lunar apsides cycle of 8.849 yr has a sun-directed orientation only every third return, thus creating a 26.547-yr periodicity. This is repeated at the same season approximately every 53.094 yr. If we calibrate this to the tide cycles (Wood, 2001), the key year is 1912, which was marked by the highest predicted tides in five centuries. Wood pointed out that if the skipper of the *Titanic* had known about this, history would have been different. (The higher tidal buoyancy and warmer waters greatly accelerated the spring production of icebergs that year.) The same triple apsides periodicity is matched by "strong" to "very strong" El Niñs. Of importance for the ENSO returns is the perihelion cycle (coming close to the Christmas/New Year season). The proxigee–syzygy–perihelion return coincides with the apsides every 61.943 yr, with triple apsides every 186.17 yr, e.g. AD 1912, 1726, 1540 (each with its matching El Niñ).

In summary, the QBO is found to be closely related not only to the solar cycle, but also to the lunar cycles. Inasmuch as the latter are very precisely established, in contrast to the gross variability of the solar cycle, the lunar periodicities would appear to offer the better prospect for predictive purposes.

Acknowledgments

Appreciation is expressed to Columbia University and the NASA/Goddard Institute for Space Studies (Dr James Hansen) for logistic support.

Rhodes W. Fairbridge

Bibliography

Berson, F.A., and Kulkarni, R.N., 1968. Sunspot cycles and the quasi-biennial stratospheric oscillation. *Nature*, **217**: 1133–1134.

Charvátová I., and Střeštík, J., 1991. Solar variability as a manifestation of the Sun's motion. *Journal of Atmospheric and Terrestrial Physics*, **53**: 1019–1025.

Currie, R.G., 1973. Fine structure in the sunspot spectrum – 2 to 70 years. *Astrophysics and Space Science*, **20**: 509–518.

Ebdon, R.A., 1975. The quasi-biennial oscillation and its association with tropospheric circulation patterns. *Meteorology Magazine*, **104**: 282–297.

Fairbridge, R.W. (ed.), 1967. *Encyclopedic of Atmospheric Sciences and Astrology*. New York: Reinhold.

Fairbridge, R.W., and Sanders, J.E., 1987. The sun's orbit, A.D. 750-2050: basis for new perspectives on planetary dynamics and Earth-Moon linkage. In Rampino, M.R. et al., eds., *Climate – History, Periodicity, and Predictability*. New York: Van Nostrand Reinhold, pp. 446–471 (bibliography, 475–541).

Flohn, H., and Fantechi, R. (eds.), 1984. *The Climate of Europe: past, present, and future*. Dordrecht: Reidel.

Groveman, B.S. and Landsberg, H.E., 1979. Simulated northern hemisphere temperature departures 1579–1880. *Geophys. Res. Let.*, **6**: 767–770.

Labitzke, K., and Van Loon, H., 1990. Associations between the 11-year solar cycle, the quasi-biennial oscillation and the atmosphere: a summary of recent work. *Philosophical Transaction of the Royal Society of London, A*, **330**: 577–587.

Lamb, H.H., 1977. *Climate History and the Future*. London: Methuen, 835p.

Naujokat, B., 1986. An update of the observed quasi-biennial oscillation of the stratospheric winds over the tropics. *Journal of Atmospheric Sciences*, **43**: 1875–1877.

Peixoto, J.P., and Oort, A.H., 1992. *Physics of Climate.* New York: American Institute of Physics, 520 p.
Pettersson, O., 1912. The connection between hydrographical and meteorological phenomena. *Royal Meteorological Society, Quarterly Journal*, **38**: 173–191.
Pettersson, O., 1914. Climatic variations in historic and prehistoric time. *Svenska Hydrografisk–Biologiska Kommissionensskrifter, v*, 1–26.
Reed, R.J., 1961. The present status of the 26-month oscillation. *American Meteorological Society Bulletin*, **46**: 374–387.
Ribes, E., 1990. Astronomical determinations of the solar variability. *Philosophical Transactions of the Royal Society of London, A*, **330**: 487–497.
Ribes, E. et al., 1991. The variability of the solar diameter. In Sonett, C.P. et al., eds., *The Sun in Time.* Tucson, AZ: University of Arizona Press, pp. 59–97.
Schove, D.J. (ed.), 1983. *Sunspot Cycles* (Benchmark Papers in Geology, vol. 68). Stroudsburg, PA: Hutchinson Ross.
Von Rudloff, H., 1967. Die Schwankungen und Pendelungen der Klimas im Langperiodischen Bereich. *Wiss. Mitteilungen Met. Inst. Univ. München*, no. 15.
Windelius, G., and Carlborg, N., 1995. Solar orbital angular momentum and some cyclic effects on Earth systems. *Journal of Coastal Research*, special issue 17, pp. 383–395.
Wood, F.J., 2001. *Tidal Dynamics*, vol. 2. *Journal of Coastal Research*, special issue 31.

Cross-references

Cycles and Periodicities
Determinism, Climatic
El Niñ
Krakatoa Winds
Oscillations
Seasonal Climate Prediction
Southern Oscillation
Sunspots
Teleconnections
Tree-Ring Analysis

R

RADAR, CLIMATIC APPLICATIONS

Radar Detection and Ranging (RADAR) was first developed in the late 1930s by Great Britain, Germany, and the United States for military applications. It was not until 1941 that radar was first used for storm detection and monitoring (Ligda, 1951). Since then weather surveillance radar (WSR) has been an important meteorological tool, particularly in the near real-time operational analysis of storm-scale (meso-γ scale) features. Radar has become a valuable tool in climatological research as well, since, unlike climatological station networks, there are few gaps in coverage between stations where WSR networks exist.

While near real-time satellite imagery is available for most regions of the world, radar provides a number of advantages over satellite remote sensing systems. For example, radar allows a cross-sectional as opposed to "top-down" view, and in some cases a finer temporal resolution. The cross-sectional perspective allows for better determination of internal storm structure, which is valuable when dealing with severe thunderstorms and tornadogenesis.

Radar physics

Radar is a sounding system that transmits radio waves from a rotating parabolic antenna. When each 360° scan is complete, the antenna tilts upward and a subsequent scan is performed. While the specific volume coverage pattern (VCP) applied dictates the number of tilt scans, most scanning strategies employ between five and 14 sweeps at various elevation angles.

Microwave energy emitted by the radar antenna is transmitted in a series of pulses that interact with atmospheric mass, and a portion of that energy is back-scattered to the antenna (known as an echo). The energy received is then amplified and displayed in one of two primary formats: (1) the plan position indicator (PPI), and (2) the range-height indicator (RHI). Distance to the reflecting target is determined by the time elapsed between pulse emission and return, and direction provided according to the azimuth of the radar "beam" when the target is sensed. The primary atmospheric targets are hydrometeors such as raindrops, snow crystals, and hailstones, although insects and birds may also be detected.

The magnitude of radiation reflected back to the antenna is a function of object density and a refractive index that is dependent upon target diameter relative to the wavelength of the energy transmitted (usually between 3 and 10 cm). The reflectivity factor converted for image display (Figure R1) by the amplifier is equal to the summation of the sixth power of hydrometeor diameters within the volume sampled.

Doppler radar, used to detect sub storm-scale motion, was first used on 27 May 1953 at the Cavendish Laboratory, Cambridge, England (Barratt and Browne, 1953). Doppler radar research and development continued through the 1950s with the most significant advances made by Lhermitte in France. By the 1960s the United States Air Force, Sperry Rand Corporation which had employed the services of Lhermitte, and the National Severe Storms Laboratory (NSSL) in Norman, Oklahoma, began research toward the development of operational Doppler weather radar. The United States Federal Aviation Administration (FAA) was also developing Doppler radar for

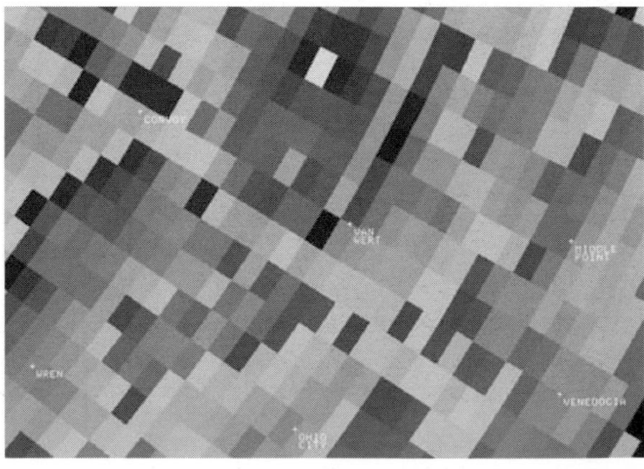

Figure R1 0.5° base reflectivity image at 2030 UTC, 10 November 2002, from WSR-88D KIWX (Northern Indiana) radar.

aviation services, and together with the NSSL, the Joint Doppler Operational Project (JDOP) was begun in 1976 to determine the potential for Doppler radar in operational meteorology (JDOP Staff, 1979). In 1979 the Severe Environmental Storms and Mesoscale Experiment (SESAME) field project (Alberty et al., 1979) was launched by NSSL to investigate the utility of Doppler radar in detecting fine-scale circulations associated with severe convective storms. As a result, by 1988 the United States National Weather Service (NWS) had developed a Doppler radar system (WSR-88D) that has become the cornerstone of the US NEXRAD Doppler radar network.

Doppler radar is based on the principle that the frequency of back-scattered radiation received at a fixed point location changes with time due to target motion. If a target is moving downrange, then the shift is to a lower frequency (red shift); when the target is moving toward the radar the frequency increases (blue shift). When opposite frequency shifts are detected between adjacent image pixels (known as "gates") the possibility of a mesocyclone (parent circulation that may generate a tornado) or microburst (severe "straight-line" winds) exists, depending on whether the rotation detected is cyclonic or anticyclonic.

In many cases a supercell thunderstorm will produce rotation (mesocyclone circulation) several minutes prior to tornadogenesis (Figure R2). Detection of a mesocyclone prior to tornadogenesis allows meteorologists the opportunity to issue timely warnings that help save lives and property. For example, since installation of the NEXRAD Doppler radar network in the United States in the early 1990s, mean tornado warning lead times have improved from 11.4 minutes prior to installation of the WSR-88Ds, to 13.8 minutes within the first 3 years of their operation (Bieringer and Ray, 1996). Probability of Detection (POD) has also improved from 75.4% before commissioning of the first WSR-88Ds in 1991, to 86.4% by 1995.

Radar imagery

Radar imagery is available in two display formats; Plan Position Indicator (PPI) and Range-Height Elevation (RHI). The PPI imagery represents a top-down view of any VCP elevation, while the RHI provides a cross-sectional view through integration of all elevation scans. Image products are either based on either reflectivity or radial velocity data.

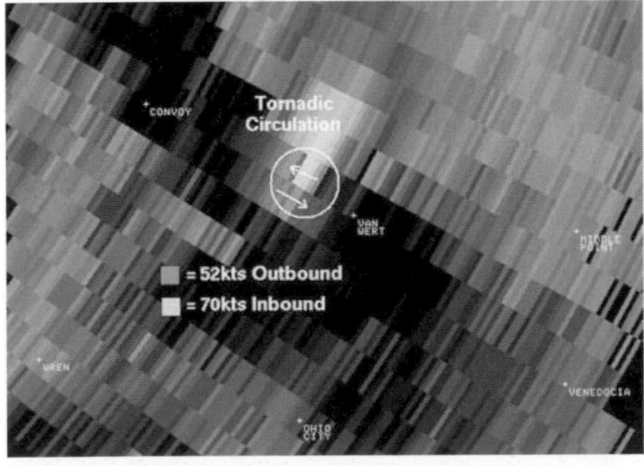

Figure R2 0.5° base reflectivity image at 2030 UTC, 10 November 2002, from WSR-88D KIWX (Northern Indiana) radar indicating a mesocyclone and tornado circulation northwest of Van Wert, Ohio. The WSR-88D radar is upradial northwest of the upper-left portion of the image.

Reflectivity imagery provides information regarding location and intensity of precipitation, while radial velocity products are used to detect storm scale motion based upon the Doppler principle. Common reflectivity imagery includes base reflectivity, composite layer reflectivity, echo top heights, and vertically integrated liquid products. Radial velocity imagery is provided in both base velocity and storm-relative velocity forms.

Reflectivity products

Reflectivity products depict the magnitude of back-scattered radiation received by the antenna from targets within the radar beam at a given location. The radiation received is amplified and color images generated to indicate strength of the reflected radar signal. With most radar systems warmer hues indicate heavier precipitation and cooler colors lighter precipitation. Radar reflectivity products are useful in determining the location and character of precipitation and meso-γ scale storm structure, which is very valuable in estimating thunderstorm intensity and character.

Base reflectivity

Base reflectivity products are perhaps the most common radar image types available (see Figure R1). While the lowest elevation scan is the most frequently used (generally 0.5°), separate base reflectivity products are available for each tilt-scan angle.

Composite reflectivity

Composite reflectivity is a display of the maximum echo intensity (reflectivity) detected at any elevation angle within a complete volume scan. When compared against base reflectivity, the composite reflectivity can reveal important storm structure and intensity trends. For example, when composite reflectivity is very large for a storm when the lowest level tilt scan reveals much weaker reflectivity, the storm is likely to produce heavy precipitation, and in many cases large hail.

Vertically integrated liquid

The vertically integrated liquid (VIL) product expresses density of water (liquid and solid) within a tropospheric column with a surface area of $1\,m^2$. Values of VIL are provided in units of $kg\,m^2$. Since wet hail produces very high reflectivity due to its large diameter, VIL products are commonly used to detect severe thunderstorms.

Echo tops

The echo tops product depicts maximum height in which reflectivity is sensed, which can reach up to 70 000 feet above ground level. Echo top information is valuable in identifying areas of strong updrafts and determining thunderstorm severity. In addition, severe weather events often coincide with the collapse of the echo top.

Precipitation

Accumulated precipitation products involve a summation of radar reflectivity over space to provide estimates of rainfall totals for specified time periods. Common accumulated time periods are 1 hour, 3 hours, and storm total. Accuracy of estimated precipitation totals derived from radar reflectivity products has improved in recent years, although convective precipitation totals are often both underestimated and overestimated. Legates (2000) has shown that weather radar systems tend to underestimate precipitation when hail is minimal, as is the case of tropical convection. When large hail occurs it is not uncommon for a 2–4 inch overestimation of accumulated liquid precipitation since the reflectivity from wet hail is very large.

Radial velocity products

Radial velocity products (see Figure R2) depict the average radial velocity of targets in the radar beam at a given elevation and range. A single Doppler radar beam (considered a "radial" outward from the radar antenna) can detect motion either directly toward or away from the radar antenna. Rarely is target motion purely radial; rather, most targets exhibit motion that is partially parallel to the radar beam. For example, hydrometeor particle motion may be westerly at $10\,m\,s^{-1}$, but if the radar antenna is located to the southwest, then only 50% of motion is considered radial. For this reason radial velocity is not typically related to particle speed. In cases where all of the target motion is perpendicular to the radar beam, radial velocity is zero.

Radial velocity values that depict outbound motion are considered positive and typically assigned warm colors (typically red) to indicate a "red shift" (decrease) in frequency, while inbound radial velocity values are considered negative and usually given cool colors such as blue or green to depict a "blue shift" (increase) in frequency. Gray or white are often used to depict zero radial velocity. Purple coloration is provided for pixels in which the radar cannot determine whether the back-scattered pulse received was from the current emission or a previous pulse. This is referred to as a "second trip" echo, which is also known as "range folding."

Base velocity

Base velocity data are the default radial velocity information as described above. The disadvantage of using base velocity data is that it is impossible to separate pure particle motion from the motion of the storm that is carrying the particle. In other words, the base radial velocity will only depict true particle motion for those particles within a stationary storm system.

Storm relative velocity

To control for the effects of storm motion, the radial velocity contributed from storm motion is subtracted from the total radial velocity values to depict true particle motion. This would be analogous to moving the radar along with the storm. Storm relative radial velocity, sometimes referred to as storm relative motion (SRM), provides the best estimate of particle motion as a result of mico-γ scale circulations within an individual storm.

Radar limitations

As with any remote sensing system, there are limitations to consider when using radar systems and products. For example, all radar systems exhibit range-height and aspect-ratio problems, and some suffer from significant signal attenuation depending on the wavelength transmitted. Other limitations include ground clutter and anomalous propagation.

Ground clutter

When a radar antenna is sited near objects that extend a significant distance above the Earth's surface, such as tall trees, buildings, hills, or mountains, a portion of the radar signal at the lower elevation tilt scans (0.5–1.5°) will be returned by those objects to the radar antenna and receiver. While algorithms are used to detect these non-meteorological returns, and filter them out of the image as much as possible, a generally concentric reflectivity ring typically appears around the radar site. This problem is enhanced when the radar is operating in high-power (clear-air) mode.

While this may seem to be a significant limitation when storms pass near the immediate radar site, the "cone of silence" that exists as a result of the upper limit of elevation scans (usually near 15°) usually does not allow the detection of the precipitating system close to the radar site regardless of ground clutter. As a result, use of higher tilt scans (1.5° or higher) to characterize the precipitation as separate from the ground clutter will only yield acceptable results until the storm passes close within the immediate vicinity of the radar site.

Range–height limitations

The range–height problem exists for all surface-based radar systems due to the Earth's spherical shape. Curvature of the Earth's surface results in a radar beam reaching greater and greater heights downrange even though its trajectory does not change. An added component of the range–height problem is refraction of the radar beam that results in a natural tendency of the beam to curve slightly upward with distance. In some cases the beam can become either sub- or super-refracted, which either enhances the upward (sub-refraction) curvature or reverses the sense of curvature so that the beam approaches the ground surface (super-refraction). To compound the problem, nearly all radar systems begin their first scan at an elevation angle of 0.5°. These factors limit the useful range of any individual radar system to about 240 nautical miles, and provide a lower elevation scan limit of 8000 feet above ground level just 60 nautical miles from the radar.

Aspect-ratio problems

As the energy is transmitted downrange along the radar beam, the beam itself widens due to the parabolic design of the transmitting antenna. By the time the pulses reach 60 nautical miles downrange the beam width has increased to greater than 1 km. This deteriorating beam width resolution with distance may result in an inability to resolve some of the smaller tropospheric circulations such as mesocyclones that can lead to tornadogenesis.

Signal attenuation

Weather radar systems transmit pulses of energy with wavelengths ranging from 3 to 10 cm, while most hydrometeors exhibit diameters much smaller than this (Rayleigh scattering). When large hail occurs the hailstone diameters the 3–10 cm wavelength range (Mie scattering). As a result shorter wavelength radar systems are more effective in detecting small particles such as cloud droplets and drizzle. However, this shorter wavelength energy is also partially absorbed by these same particles, a process referred to as attenuation. Signal attenuation makes it difficult to accurately measure the intensity of back-scattered energy for more distant small particle targets.

Radar systems that use longer wavelengths are often desired for operational purposes since absorption by the intervening particles is drastically reduced. This means that a distant thunderstorm that lies behind a closer storm will also appear on the radar screen in an accurate manner. Since detecting severe weather is one of the more important missions of weather surveillance radars, such as the US WSR-88D NEXRAD network, these radar systems typically use a greater wavelength such as 10 cm.

Anomalous propagation (AP)

The pulses of energy emitted by a radar antenna normally propagate in a slightly curved upward arch due to the decrease in atmospheric density with height. This enhances the range–height limitations typical with all surface-based radars. When thermal inversions exist, however, the radar beam can arch down toward the ground (super-refraction). In some cases the Earth's surface may reflect the signal back to the radar system, resulting in a very strong return signal at the lowest elevation tilt scans (0.5–1.5°). To the inexperienced observer the resulting image may suggest an intense thunderstorm is occurring, when in fact no storm exists at all. To check the possibility that high-intensity low-level returns are not in fact storms, the radar analyst observes imagery from successively higher tilt scans (1.5° or higher) to determine the depth of the intense reflectivity core. In cases of AP, the high reflectivity area noted at the lower scan angles would not appear at higher levels. A further check may be performed through use of an image loop to determine if the high reflectivity core is moving in a manner expected of a thunderstorm, or if the apparent core remains relatively stationary which suggests an AP echo.

Radar in climatological research

The primary advantage of using radar data in climatological research, compared with data collected from climatological and meteorological stations, is improved spatial and temporal resolution. While climatological and meteorological surface stations are in most cases several tens, if not more than 100 kilometers apart, the resolution of most radar data ranges from about 0.5 km to 2 km, with updated products available every 5–7 minutes.

Although the use of radar data in climatological research has been increasing in recent years, many climatologists have been slow to utilize this valuable resource. A review of the current literature reveals a greater number of radar climatologies being developed by meteorologists in national weather services than by climatologists. A sample of such titles includes: Topographic influences on Amarillo radar echo climatology (Marshall and Peterson, 1980); Relationships between vertically-integrated liquid and thunderstorm severity over central Florida (Kitzmiller et al., 1990); and A radar-based climatology of July convective initiation in Georgia and surrounding area (Outlaw and Murphy, 2000).

For researchers that have been involved with radar climatology, a majority use PPI reflectivity imagery to determine location of precipitation and precipitation intensities from convective systems. One of the primary reasons for this is that precipitation, especially from convective systems, is perhaps the most spatially variable of all climatological elements (Riggs and Truppi, 1957). Examples of such studies are a radar climatology for the state of South Carolina developed by Landers (1969), an investigation of diurnal precipitation patterns in Florida, by Frank and Smith (1968), a radar climatology of summertime convective clouds in the Black Hills region of the United States (Kuo and Orville, 1973), and a thundertsorm climatology for eastern Colorado (Karr and Wooten, 1976).

David L. Arnold

Bibliography

Alberty, R.L., Burgess, D.W., Hane, C.E., and Weaver, J.F., 1979. *SESAME 1979 Operations Summary*. Environmental Research Laboratories, NOAA. Boulder, CO: US Government Printing Office.

Barratt, P., and Browne, I.C., 1953. A new method of measuring vertical air currents. *Quarterly Journal of the Royal Meteorological Society*, **79**: 550.

Bieringer, P., and Ray, P.S., 1996. A comparison of tornado warning lead times with and without NEXRAD Doppler radar. *Weather and Forecasting*, **11**: 47–52.

Frank, N.L., and Smith, D.L., 1968. On the correlation of radar echoes over Florida with avrious meteorological parameters. *Journal of Applied Meteorology*, **2**: 582–593.

JDOP Staff, 1979. *Final Report of the Joint Doppler Operational Project*. NOAA Technical Memorandum. ERL NSSL-86. Norman, OK: National Severe Storms Laboratory.

Karr, T.W., and Wooten, R.L., 1976. Summer radar echo distribution around Limon, Colorado. *Monthly Weather Review*, **104**: 728–734.

Kitzmiller, D.H., Saffle, R.E., McDonald, M., Miller, R.G., Lang, J., and McGovern, W.E., 1990. Relationships between vertically-integrated liquid and thunderstorm sevrity over central Florida. *Preprint volume, Sixteenth Conference on Severe Local Storms*. Kananaskis Park, Alberta: American Meteorological Society, 22–26 October, pp. 603–606.

Kuo, J.-T., and Orville, H.D., 1973. A radar climatology of summertime convective clouds in the Black Hills. *Journal of Applied Meteorology*, **12**: 359–368.

Landers, H., 1969. *Preliminary Report on South Carolina Radar Climatology*. Clemson Climatic Research Series no. 4. Clemson, SC: South Carolina Agricultural Experiment Station.

Legates, D.R., 2000. Real-time calibration of radar precipitation estimates. *Professional Geographer*, **52**: 235–246.

Ligda, M.G.H., 1951. Radar storm observation. In Malone, T.F., ed., *Compendium of Meteorology*. Boston, MA: American Meteorological Society, pp. 1265–1282.

Marshall, T.P., and Peterson, R.E., 1980. Topographic influences on Amarillo radar echo climatology. *Preprint volume, Eighth Conference on Weather Forecasting and Analysis*. Denver, CO: American Meteorological Society, 10–13 June.

Outlaw, D.E., and Murphy, M.P., 2000. A radar-based climatology of July convective initiation in Georgia and surrounding area. Eastern Region Technical Attachment no. 2000–04. Silver Spring, MD: NOAA National Weather Service.

Riggs, L.P., and Truppi, L.E., 1957. A survey of radar climatology. *Proceedings of the Sixth Weather Radar Conference*. Boston, MA: American Meteorological Society, pp. 227–232.

Cross-references

Satellite Observations of the Earth's Climate System
Thunderstorms
Tornadoes

RADIATION CLIMATOLOGY

Radiation is the most fundamental energy available to the Earth–atmosphere system (EAS). All significant atmospheric and oceanic circulation systems and critical chemical, material and energy cycles are driven directly or indirectly by the continuous supply of radiation energy from the sun. The conversion of this energy into biochemical energy provides the foundation for life. In addition, the unequal receipt of this solar energy is primarily responsible for the Earth's climates and the seasons.

Radiation describes both a type of energy and a means of energy exchange. All objects with a temperature above absolute zero (0 Kelvin) will emit electromagnetic radiation that travels at the speed of light as a series of waves. The "quality" of this energy is related to its wavelength (λ), the distance from one crest to the next (usually measured in micrometers, μm). The entire range of potential wavelengths forms the electromagnetic spectrum which extends from gamma and X-rays ($<0.1\,\mu$m) to radio waves ($>1000\,\mu$m) and beyond. In climatology we are often interested in specific parts of the spectrum – for example, radiation at wavelengths between 0.4 and 0.7 μm corresponds to the visible spectrum for humans and the energy that vegetation employs in photosynthesis. The amount of energy associated with a radiation transfer (a quantum, Q) is inversely proportional to its wavelength,

$$Q = \frac{hc}{\lambda}$$

where, h is Planck's constant (6.626×10^{-34} J s) and c is the speed of light, $3.8\times10^{6}\,\mathrm{m\,s^{-1}}$. The loss (emission) or gain (absorption) of a quantum of radiation is associated with a change in the energy state of a molecule or atom. This change only occurs at specific, often widely separated, wavelengths known as absorption lines. If interactions between molecules occur (as is the case with liquids and solids) these lines broaden to form absorption bands.

At the outset it is worth defining some of the terms used to describe radiation exchange. A complete description of the transfer must include its energy quantity and quality, and directional properties. The radiation flux is the energy emitted per unit time by an object ($\mathrm{J\,s^{-1}}$ or Watts (W)). To compare emitting objects it is common to present the flux per unit area of the emitting surface – this is termed an emittance (irradiance refers to received radiation). More generally, this term is called a flux density ($\mathrm{W\,m^{-2}}$). The radiation emitted/received toward/from a given direction is described by the intensity, the flux passing through a unit of solid angle, a steradian ($\mathrm{W\,ster^{-1}}$). A similar term is the radiance, the intensity divided by the area of the emitting surface ($\mathrm{W\,m^{-2}\,ster^{-1}}$). Initially, it is useful to describe these properties for an ideal object (known as a blackbody) before examining the objects that are of most interest here, that is, the sun, the Earth and the atmosphere.

The emittance of a blackbody at different wavelengths is a function of surface temperature only (Planck's law). A plot of the spectral emittance has a positive skew, with peak energy output occurring at shorter wavelengths (Figure R3). The total emittance is proportional to temperature to the fourth power (the Stefan-Boltzmann law),

$$E = \sigma T^4$$

where σ is the Stefan–Boltzmann constant ($5.67 \times 10^{-8}\,\mathrm{W\,m^{-2}\,K^{-4}}$) and T is temperature on the Kelvin scale. The peak of the spectral emittance curve, which establishes the wavelength at which most radiation is emitted, is also a function of temperature (Wien's law),

$$\lambda_{\max} = \alpha / T$$

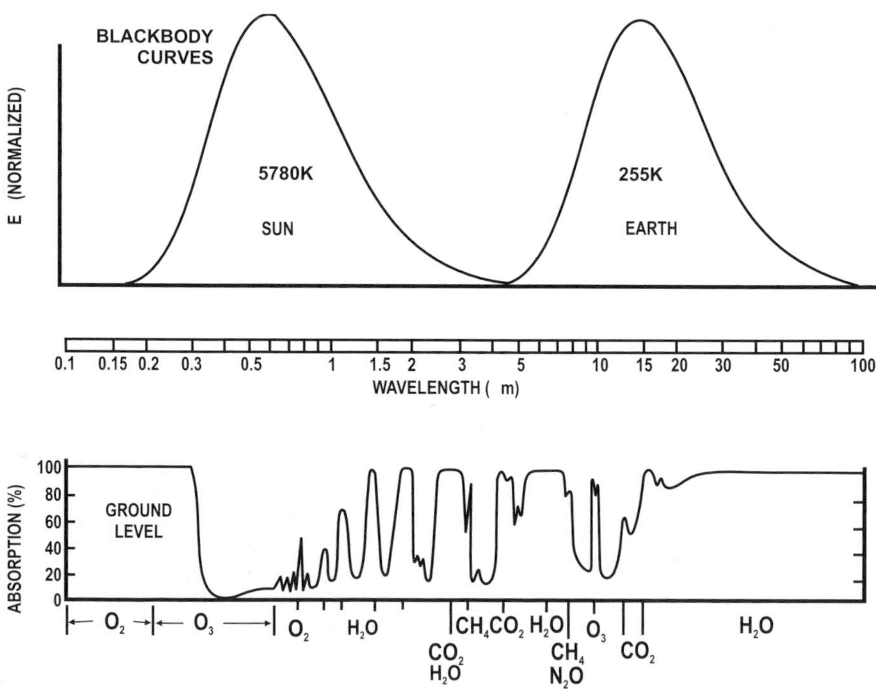

Figure R3 Blackbody spectral emmitance (E_λ) in $\mathrm{W\,m^{-2}\,\mu m^{-1}}$ for two objects at the temperature of the Sun (5780 K) and Earth (255 K). The curves have been standardized to demonstrate the difference between the radiation emitted by the Sun (shortwave) and that by the Earth (longwave). The actual area under the curves shown is given by the Stefan-Boltzmann law. The lower diagram illustrates the spectral absorption characteristics of the atmosphere and the gases responsible (modified from Wallace and Hobbs, 1977).

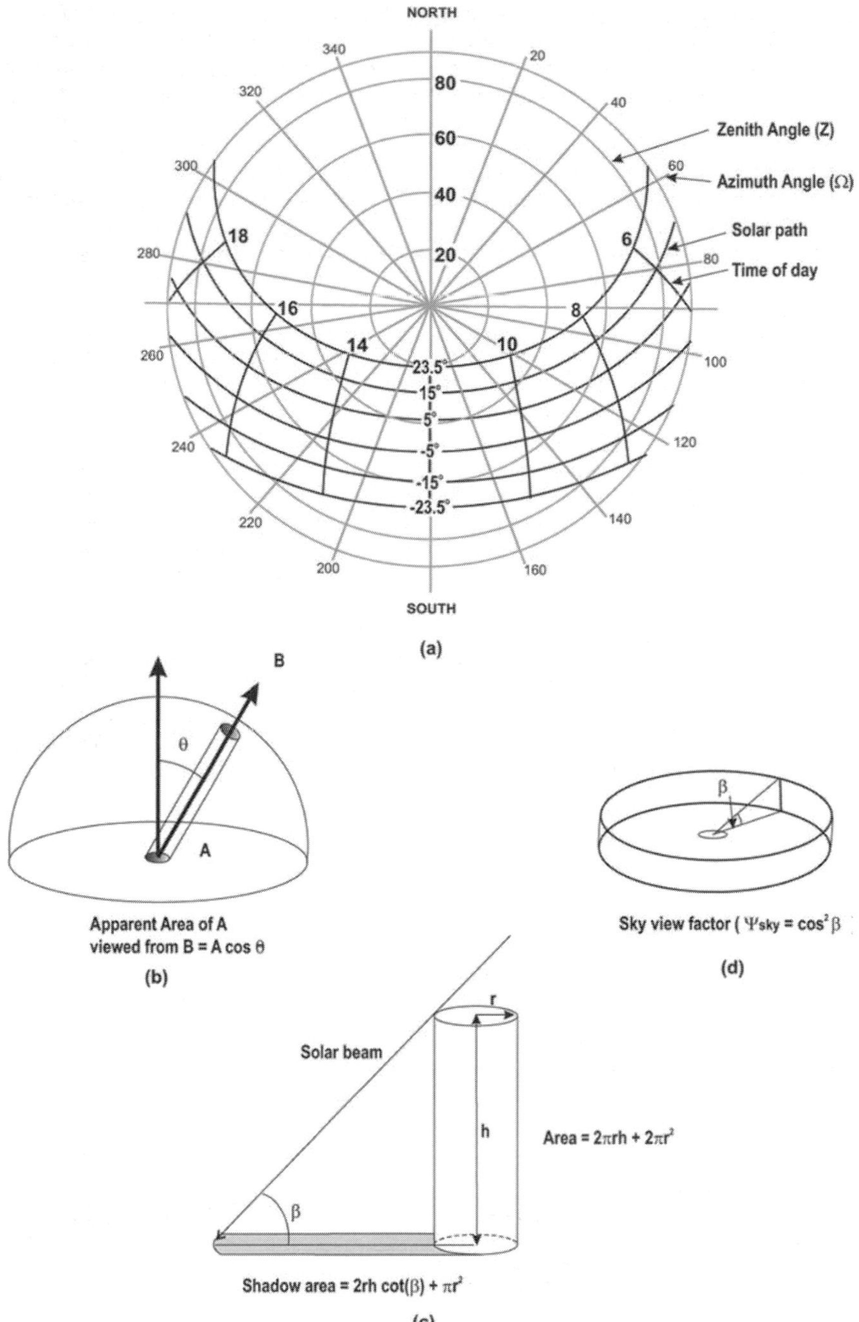

Figure R4 Radiation geometry: (**a**) The solar path diagram for 45°N illustrating the position of the sun (Z, Ω) for different times of day and times of the year (shown by solar declination marked along the center line). (**b**) For a Lambertian surface the intensity of emitted radiation is equal in all directions. The apparent area of the emitting surface declines with the cosine of the viewing angle (θ) so that the measured emittance is equal in all directions (modified from Oke, 1987). (**c**) The solar radiation intercepted by a cylinder (used to represent a human, for example) equals its shadow area. To obtain an average irradiance this quantity is divided by the area of the cylinder (after Montieth and Unsworth, 1990). (**d**) The sky view factor for a simple basin used to represent a forest clearing, for example (after Oke, 1987).

where, a equals 2898 mm K. The emission of a blackbody is isotropic; that is, the intensity of the radiant flux is equal in all directions. However, according to Lambert's cosine law, the apparent area of the emitting surface decreases as the viewing angle increases from the perpendicular (Figure R4b).

A consequence of the laws stated above is that the spectral characteristics of radiation emitted by the Earth and the sun are substantially different (see Figure R3). At an approximate temperature of 6000 K, the sun's radiation profile suggests $E \approx 7.35 \times 10^7$ W m^{-2} and $\lambda_{max} \approx 0.5$ μm – by comparison, an approximate value for the Earth's surface of 300 K yields values of 460 W m^{-2} and 10 μm. At the top of the atmosphere, the radiation spectra of the sun and the Earth effectively form two non-overlapping curves and are termed shortwave (or solar) and longwave (or terrestrial) radiation, respectively.

The spectral emittance for real objects shows significant deviations from that for a blackbody at the same temperature. Emissivity is a measure of the ability of a surface to emit as a blackbody at a given wavelength (ε_λ). An important corollary is that its absorptivity (a_λ, the ability to absorb at that wavelength) equals its emissivity (that is, $\varepsilon_\lambda = a_\lambda$). This is Kirkchoff's law.

When radiation encounters a medium it may be reflected, absorbed or transmitted. For a given wavelength these fates may be expressed as proportions,

$$r_\lambda + a_\lambda + \tau_\lambda = 1$$

where the symbols represent the spectral reflectivity (r_λ), absorptivity (a_λ) and transmissivity (τ_λ). If the intercepting surface is opaque no transmission occurs, and the radiation is either absorbed or reflected. However, if the medium is transparent some radiation will be transmitted, a proportion that will depend upon the density of absorbing molecules and the distance taken to pass through that medium (termed the path length).

In the EAS, most solid materials and water act as nearly full emitters in the longwave portion of the spectrum. As a result we can effectively use the terms reflectivity (or albedo) and emissivity to refer to the behavior of Earth materials in the shortwave and longwave portions of the spectrum, respectively. On the other hand, the atmosphere has a distinctive absorption profile (see Figure R3), which is highly dependent on wavelength. While much of shortwave radiation is transmitted (particularly at visible wavelengths), most longwave radiation is absorbed. In addition, aerosols and molecules selectively scatter radiation. The nature of this scattering depends on the physical dimensions of the scatterers and the wavelength of radiation. For example, Rayleigh scattering occurs in all directions and is inversely proportional to wavelength – as a result, blue light is preferentially scattered (producing a blue sky). On the other hand, a mixture of larger particles of smoke and dust scatters a wide spectrum of grayish light in a predominantly forward direction known as Mie scattering (Wallace and Hobbs, 1977).

Radiation exchange

The transfer of radiation between objects and surfaces depends on both the magnitude and direction of emissions and their disposition relative to each other. The directional attributes of the exchange can be categorized into parallel (along one path) and diffuse (along many paths) types. Direct solar radiation is treated as parallel (or beam) radiation while scattered solar radiation and emitted and received longwave radiation are treated as diffuse radiation.

The direct solar irradiance (S) on a surface at the base of the atmosphere is a function of atmospheric transmissivity (τ) and the angle formed between the beam and a line perpendicular to the surface. For a horizontal surface,

$$S = 1370 \cos Z \, \tau^m$$

where Z is the solar zenith angle, m is path length (equal to $1/\cos Z$) and τ is an overall measure of atmospheric transmissivity. Surface irradiance is maximized when the surface is perpendicular to the beam. For complex shapes (such as those of plants or animals), surface irradiance can be estimated by substituting simpler geometrical shapes for which the irradiance is more easily obtained (Figure R4d).

For diffuse exchanges, geometrical relations control the proportion of radiation exiting one surface ($E\uparrow_A$) that impinges on another ($E\downarrow_{A\to B}$),

$$E\downarrow_{A\to B} = E\uparrow_A \Psi_{A\to B}$$

where $\Psi_{A\to B}$ is the view factor (a value between 0 and 1). If the emittance is perfectly diffuse then the view factor is a function of geometry only. For many common configurations of shapes the view factors are readily available in engineering textbooks. In climatology it is often sufficient to distinguish between radiation derived from the sky and that derived from the terrain – hence, the sky view factor is a useful measure of the nature of exchanges. As with beam exchanges, it may be possible to employ simplified configurations to substitute for real, more complex situations (Figure R4c).

The radiation and energy budgets

In the EAS a statement of the radiation budget of a surface accounts for the nature and direction of these radiation transfers,

$$Q^* = K\downarrow - K\uparrow + L\downarrow - L\uparrow$$

where, the K and L refer to shortwave and longwave radiation, respectively, the arrows refer to receipt ($\downarrow$) and loss ($\uparrow$) and Q^* represents net radiation. The radiation budget is itself a part of a more comprehensive statement of all relevant energy exchanges. For a typical land surface,

$$Q^* = Q_H + Q_E + Q_G$$

where, Q^* is said to be partitioned into sensible (Q_H) and latent (Q_E) heat exchanges by convection with the atmosphere and sensible heat exchange by conduction (Q_G) with the substrate. Here, the non-radiative terms are treated as positive when directed away, and negative when direct toward, a surface.

Although the energy balance statement separates the non-radiative and radiation terms, they are intimately linked. For example, modifying surface albedo will directly affect shortwave absorption and the energy available for heating the surface (affecting $L\uparrow$, Q_H and Q_G) and evaporating water (Q_E). Changing these fluxes will, in turn modify near-surface air properties and affect further exchanges.

Measurement and modeling

Radiation instruments are categorized according to the spectral ranges to which they are sensitive. While the surface(s) of a radiometer absorbs radiation from both solar and terrestrial sources, those of a pyranometer and pyrgeometer absorb shortwave ($0.3 \le \lambda \le 4.0$ μm) and longwave radiation ($\lambda \ge 4$ μm), respectively. These instruments commonly consist of flat plates that absorb radiation from all directions, and must be positioned carefully to maximize exposure to the desired source of radiation. Net values are obtained by orienting two parallel sensing surfaces in opposite directions. On the other hand, a pyrheliometer is used to obtain measurements of direct (beam) solar radiation and its view must be aligned with the solar beam. Most of these instruments rely on thermometry, whereby the temperature of a spectrally sensitive surface is converted to radiation receipt (DeFelice, 1998).

Surface radiation measurements are usually made at a site that has a substantially unimpeded view of the sky vault; however, specific projects must consider the view factors of the instrument with regard to radiating surfaces to determine their individual contribution to the measured value. Downward-facing instruments on airborne or satellite platforms will have a wide field of view that incorporates many diverse radiation sources. Moreover, the spectral characteristics of the instrumentation must be such as to separate surface from atmospheric properties. While aircraft-based measurements are usually associated with specific experiments, satellite observations now perform routine observations of the global radiation budget (ERBE Science Team, 1986).

In the absence of actual measurements, models of varying sophistication are employed to determine radiation exchanges at the surface and throughout the depth of atmosphere. The simplest models are based on statistical relations established with existing measurements and use readily available meteorological parameters. For example, $L\downarrow$ from the sky may be estimated from the air temperature near the ground and lower atmosphere emissivity, which can be approximated from knowledge of humidity (Oke, 1987). More complex models simulate the radiative exchanges at several layers in the atmosphere based on knowledge of the temperature profile and the radiative properties of the layers above and below (Kiehl and Trenberth, 1997).

Table R1 Radiative properties of common surface materials (from Oke, 1987)

Surface	Remarks	Albedo (α)	Emissivity (ε)
Soil	Dark, wet Light, dry	0.05 to 0.40	0.98 to 0.98
Desert		0.2 to 0.45	0.84 to 0.91
Grass	Long (1 m) Short (0.02 m)	0.16 to 0.26	0.90 to 0.95
Forests, deciduous	Bare Leaved	0.15 to 0.20	0.97 to 0.98
Forests, coniferous		0.05 to 0.15	0.97 to 0.99
Water	Small zenith angle Large zenith angle	0.03 to 0.10 0.10 to 1.00	0.92 to 0.97 0.92 to 0.97
Snow	Old Fresh	0.40 to 0.95	0.82 to 0.99
Ice	Sea Glacier	0.30 to 0.45 0.20 to 0.40	0.92 to 0.97

Earth–sun geometry

The receipt of solar energy by the Earth is governed by its relationship to the sun that changes during the year. The Earth revolves around the sun once every 365 1/4 days and rotates upon its axis once every 24 hours. The orbit is nearly circular with an average Earth–sun distance of 150×10^6 km, and this results in an irradiance on a surface perpendicular to the beam of 1370 W m^{-2} – this value is known as the solar constant. The average irradiance at the top of the atmosphere is one-quarter of this value, 342 W m^{-2}. In reality this value will vary by about 7% during the year owing to the slight eccentricity of the Earth's orbit placing it closest to the sun on 3 January and furthest on 4 July. However, the resulting variation in intercepted radiation is not responsible for the seasons.

As the Earth revolves around the sun on its orbital plane, it rotates around an axis that is tilted at an angle of 23½° toward a fixed point outside the solar system. During its revolution the orientation of the Earth in relation to the sun is continually changing. At the equinoxes (21 March and 22 September), the shadow line on the Earth bisects both hemispheres (resulting in equal lengths of daylight and darkness at all locations). On these days the solar beam is perpendicular to the equator at noon and the zenith angle of the sun at noon is equal to the latitude of the location – thus, the tilt of the Earth with regard to the solar beam (the solar declination), is 0°. From late March to June the declination increases in latitude reaching a maximum (23½°) on 21 June (the summer solstice) and the noon sun is now directly overhead at 23½°N (Tropic of Cancer). During this transition the illuminated half of the Earth is predominantly in the northern hemisphere where each location experiences more than 12 hours of daylight (Table R2). As the Earth continues on its orbit, the solar declination returns to the equator and then into the southern hemisphere, reaching a minimum declination of −23½° (the Topic of Capricorn) on 21 December – the winter solstice.

The effect of these two movements is clearly seen in a solar path diagram that shows the path of the sun across the sky hemisphere at different times of the year (Figure R4a). The position of the sun on this hemisphere can be fixed by its zenith (Z) and

Table R2 Variation in daylength (DL, hours), solar zenith angle at noon (Z) and daily solar radiation at the top of the atmosphere (K_{EX}, in MJ day^{-1}) at three solar declinations (δ). DL is obtained from the half-daylength angle H where, $\cos(H) = -\tan(\phi)\tan(\delta)$]. K_{EX} is obtained from $86\,400/\pi\ 1370\ [H \sin(\phi)\sin(\delta) + \cos(\phi)\cos(\delta)\sin(H)]$ (Sellers, 1965)

	Equinox ($\delta = 0°$)			Winter solstice ($\delta = -23½°$)			Summer solstice ($\delta = 23½°$)		
Latitude (ϕ)	DL	Z	K_{EX}	DL	Z	K_{EX}	DL	Z	K_{EX}
0	12.0	0	37.7	12.0	23½	34.6	12.0	23½	34.6
15	12.0	15	36.5	11.1	38½	27.5	12.9	8½	39.7
30	12.0	30	32.8	10.1	53½	19.1	13.9	6½	42.6
45	12.0	45	26.9	8.6	68½	10.2	15.4	21½	43.4
60	12.0	60	19.1	5.5	83½	2.1	18.5	36½	42.8
75	12.0	75	10.1	0	—	—	24.0	51½	45.4
90	—	90	0	0	—	—	24.0	23½	47.0

azimuth angles (Ω), which are obtained from the declination of the sun (δ), latitude (ϕ) and time of day,

$$\cos Z = [\sin(\phi)\sin(\delta) + \cos(\phi)\cos(\delta)\cos(h)]$$

$$\cos \Omega = [\sin(\delta)\cos(\phi) - \cos(\delta)\sin(\phi)\cos(h)]/\sin Z$$

The azimuth (Ω) should be subtracted from 360° if the local time is after noon. In these equations the time of the day is expressed as the hour angle (h), that formed between the longitude of the position and that of the noonday sun (Oke, 1987).

The global radiation budget

Figure R5 depicts the average energy budget of the Earth using an average solar input to the system of 342 W m^{-2}. Of this value just 235 W m^{-2} is absorbed by the EAS and contributes to heating. The remainder, accounting for 30%, is reflected. In the atmosphere, 67 W m^{-2} is absorbed, causing heating in those layers where the chief absorbing gases are concentrated – in the stratosphere (ozone) and near the surface (water vapor). About 58% of the energy at the top of the atmosphere is transmitted to the surface (approximately equal amounts via direct and diffuse paths) of which 168 W m^{-2} is absorbed (Kiehl and Trenberth, 1997).

The EAS maintains thermal equilibrium by emitting 235 W m^{-2} to space as longwave radiation. This would require a blackbody temperature of 254 K. However, the surface emittance is a great deal higher at 390 W m^{-2} – implying a temperature of 287 K. The difference between these temperatures is a result of the selective absorption of the atmosphere, which creates the natural greenhouse effect. In contrast to shortwave radiation, which the atmosphere largely transmits or scatters, the gases in the atmosphere have exceptional absorption properties in the longwave range. As a result only a small proportion of the radiation emitted by the surface (<10%) is transmitted through the atmosphere directly to space. This "gap" in the absorption capacity of the gases forms the atmospheric window. Outside this narrow range the atmosphere is an excellent absorber (chiefly as a result of water vapor and carbon dioxide) and emitter. It is the "recycling" of longwave energy between the Earth's surface and the overlying atmosphere that causes the greenhouse effect – 50% of the incident longwave radiation at the surface originates in the lowest 100 m of the atmosphere (Montieth and Unsworth, 1990). The atmosphere is heated chiefly through its energy exchanges with the underlying surface. Thereafter this energy is transferred via radiation and convection through its depth.

In terms of net radiation the Earth's surface gains heat by radiation. Conversely, the atmosphere cools by radiation and it is the exchange of sensible and latent heat energy (by convection) that allows the budget of the both the Earth and the atmosphere to be balanced. At the top of the atmosphere the energy balance and the radiation balance are equivalent and Q^* is zero. Note that clouds play a very significant role in the Earth's radiation balance – they are significant reflectors of shortwave and absorbers/emitters of longwave radiation. The net effect of clouds is estimated to warm the EAS by an amount equal to 30 W m^{-2}. Each of these radiation terms shows considerable spatial and temporal variation.

Spatial variations

The broad radiation climate of the Earth can be seen in a meridional plot of Q^* for the Earth's surface, the atmosphere and the

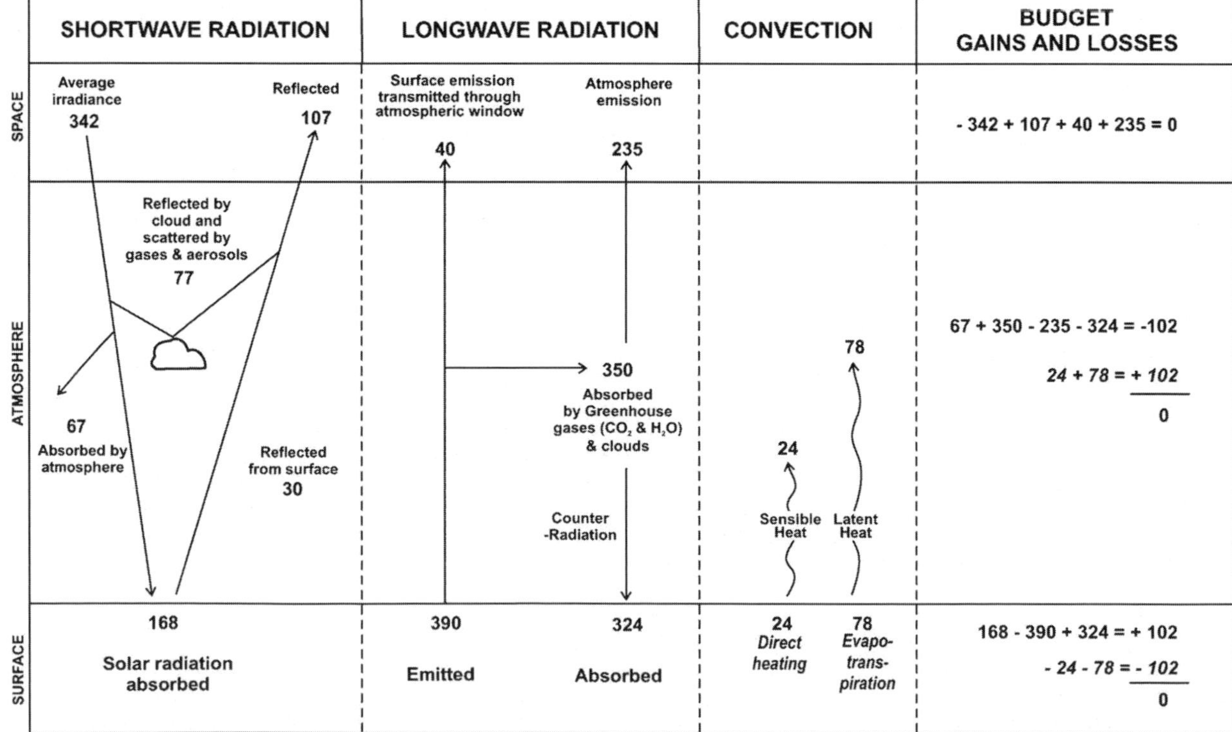

Figure R5 The Earth's average energy budget showing the radiative and convective fluxes in W m^{-2} (Kiehl and Trenberth, 1997).

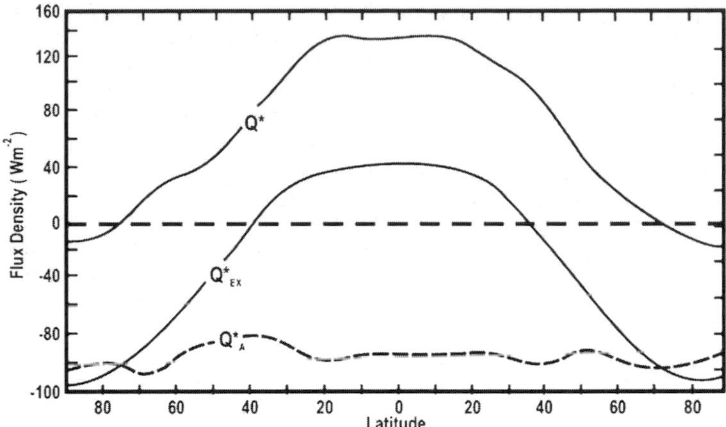

Figure R6 The meridional variation in net radiation at the top of the atmosphere (Q^*_{EX}), the Earth's surface (Q^*) and the atmosphere (Q^*_A). Modified from Sellers, 1969.

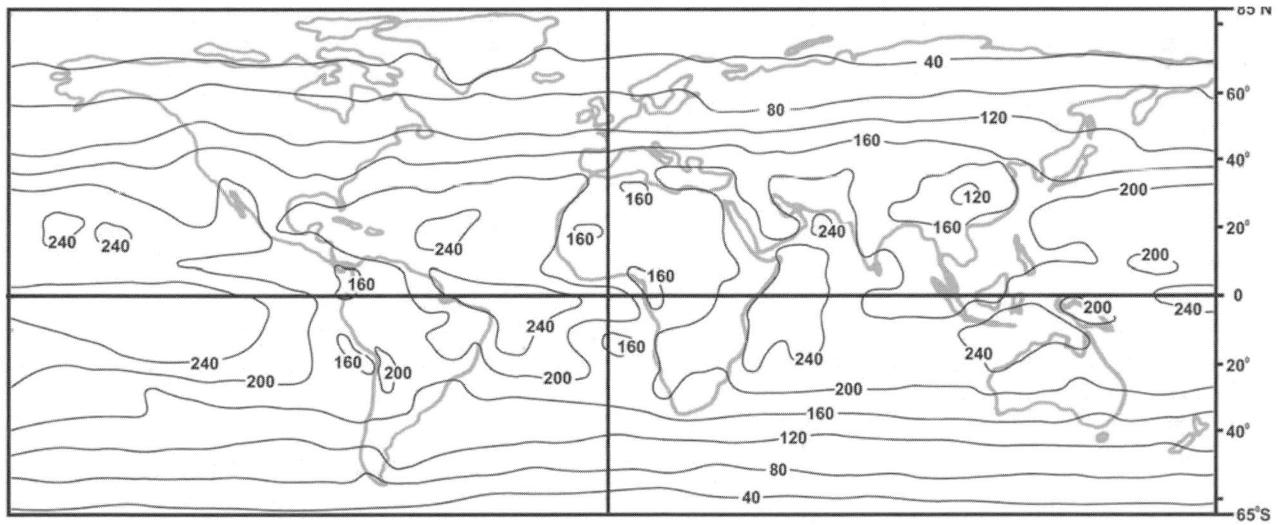

Figure R7 The average net shortwave (K^*) budget at the Earth's surface during September 1985–1989. Data are based on the ERBE experiment and is obtained from on online data library at http://ingrid.ldgo.columbia.edu/.

entire EAS (Figure R6). At the top of the atmosphere Q^* is positive within 35° of the equator and is negative at higher latitudes. Within the EAS the horizontal movement (advection) of sensible and latent energies by ocean and air currents ensures the long-term stability of climate. At the Earth's surface the pattern of Q^* is much the same but, apart from polar areas, is positive. At these high latitudes the low solar irradiance, coupled with the high albedo of ice, results in little surface absorption and radiative heating – here it is the atmosphere that transfers heat to the surface. For the atmosphere, Q^* is negative and shows little latitudinal variation.

The spatial pattern of Q^* is largely controlled by the net shortwave budget (K^*) at the Earth's surface (Figure R7). The broad pattern of decrease from the equator to poles is disrupted by significant longitudinal variations that result primarily from the different radiative properties of land and water. At any given latitude, K^* is higher over water where the irradiance is greater (low cloudiness) and the albedo is low. Over land surfaces, unexpectedly low K^* values are found in arid areas where cloud-free skies ensure high solar receipt, but a high albedo (due to lack of vegetation and bright soils) reduces absorption. The net longwave budget may enhance these differences. Cooler ocean surfaces and warmer, more humid, ocean atmospheres will suppress longwave radiation loss; conversely hot and arid land surfaces overlain by warm but dry atmospheres will enhance longwave loss (Breon et al., 1994; Gupta et al., 1999).

Temporal variations

In addition to spatial fluctuations the magnitudes of the radiation terms vary with the passage the sun across the sky hemisphere. This path varies with time of day and time of year (see Figure R4a). The movement of the solar declination (and the latitude of maximum solar irradiance) over the course of the year causes a corresponding shift in the global pattern of

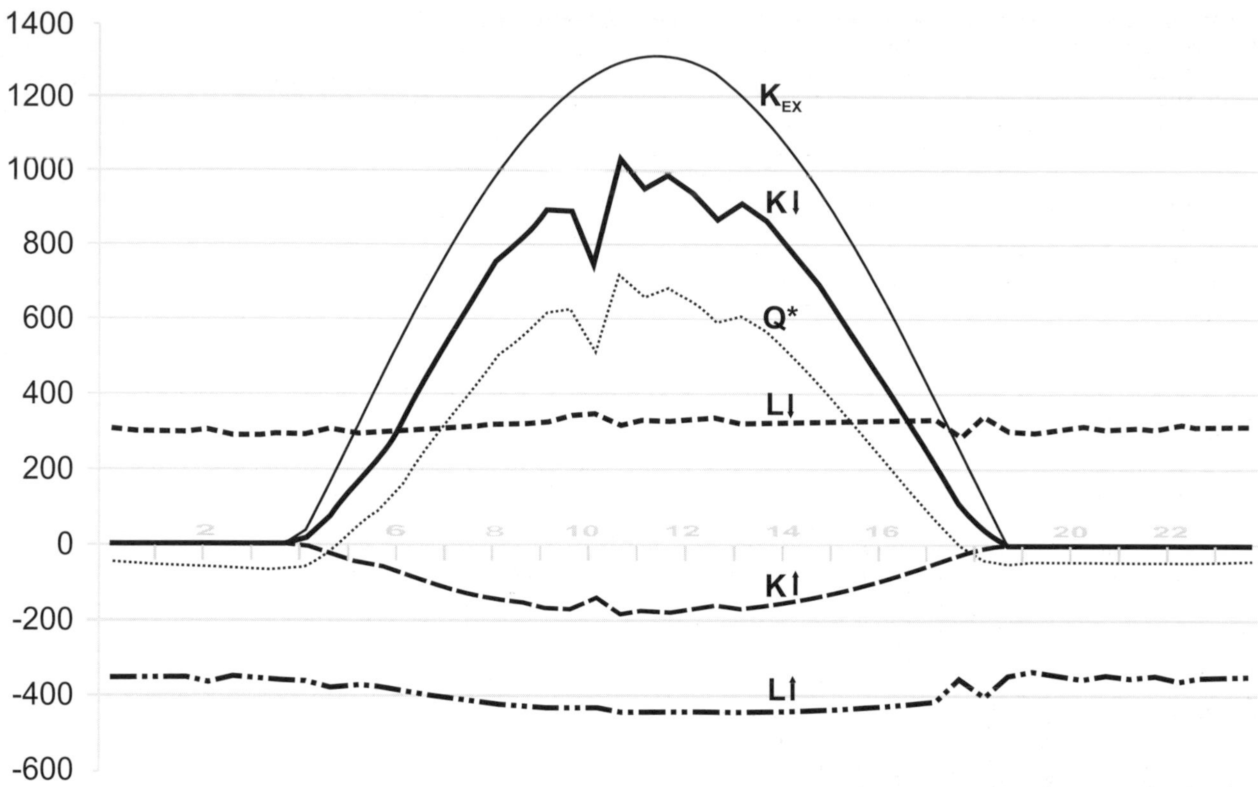

Figure R8 Measured surface radiation budget over recently grazed, fairly dry vegetation in the central US (39°06′N, 96°27′W) for 4th June, 1987 obtained during the FIFE experiment. The symbols refer to incident shortwave radiation receipt (K↓) and reflection (K↑) and, to longwave radiation receipt (L↓) and emission (L↑). K_{EX} describes solar radiation available on a horizontal surface at the top of the atmosphere.

radiation exchange. The effect of these seasonal movements is most apparent outside of the tropics where the noon sun experiences a change in zenith angle of 47° over the year and the length of daylight varies considerably (Table R2).

Superimposed on the seasonal variation is a diurnal pattern that results from the changing zenith angle of the sun during the day. For a flat surface on a clear day, Q^* has a symmetric appearance with the peak receipt occurring at local noon and negative values overnight (see Figure R8). This pattern is largely driven by access to solar energy, which is available during daylight hours and varies inversely with the cosine of the solar zenith angle. By comparison, the net longwave exchange varies little during the day and night, and is negative. The overall result is positive values for Q^* from just after sunrise to just before sunset, and negative values overnight. The effects of cloud on the surface radiation balance will greatly depend upon their character. For example, if the sky is obscured by low-level, thick cloud the solar input to the system is uniformly reduced. On the other hand, the cloud behaves as a full absorber and emitter at longer wavelengths. The net effect will be to reduce the diurnal variation in Q^* and the daily temperature range.

Climates near the ground

Distinctive near-surface climates at all scales can be linked to the radiation budget (Smith et al., 2002). In particular, distinctive local climates can be generated during calm conditions and clear skies when mixing is reduced and daytime radiation gain and nighttime radiation loss are enhanced. During the daytime such conditions often result in late afternoon convective activity as the convergence of heat in the lower atmosphere produces an unstable vertical temperature profile. At night these conditions result in radiative divergence, cooling and the development of an inversion in a thin layer of air adjacent to the surface. One of the human responses to damaging frost caused by this form of cooling is based explicitly on modifying the radiation exchange by generating a cloud using a fine water spray or smoke particulates. This will have the effect of increasing atmospheric emissivity and reducing the rate of heat loss from the near-surface air volume.

Local variations in land cover and/or topography will enhance the effects of radiation exchanges. At the boundary between land and water the radically different radiative and thermal properties generate land and sea breezes at night and during the day, respectively. Similarly, variations in the slope and aspect of surfaces will result in spatial variations in surface irradiance by creating areas of shade and sunlight in proximity. In valley configurations the surface irradiance/emittance will depend on the topographic geometry and orientation with regard to the sun's path. A slope that presents a near-perpendicular surface to the solar beam will warm and generate shallow up-slope (anabatic) winds as the adjacent air becomes warm and buoyant. At night, near-surface cooling increases toward the valley ridges (where radiative loss to the sky is enhanced) and a shallow layer of cool air slides downslope (katabatic winds). These locally generated winds may assume particular importance for human affairs, as valleys and coastal areas are preferred areas of occupation – for example, both the sea breeze and katabatic airflow produce low-level inversions that have implications for pollution control.

The climates of all living things are intimately connected to the immediate radiation environment. Over land, the type and magnitude of surface–air exchanges are partly governed by the behavior of the vegetative cover, which can vary their "resistance" to exchanges of mass and energy according to environmental conditions. In the process of photosynthesis, plant leaves absorb light (green is selectively transmitted and reflected) and exchange water and carbon dioxide with the ambient air. If there is insufficient light, or the plant is experiencing water stress, these exchanges do not occur, and an important link between the atmosphere and soil moisture is severed. If the vegetative canopy is elevated, dense and extensive an entirely different climate is created in the canopy layer between the free atmosphere and the Earth's surface (by analogy, a similar subclimate develops within urban canopy formed by buildings). Within this environment the radiation budget is entirely altered. Shortwave receipt at the soil surface is greatly reduced due to the interception of solar radiation by the canopy. Moreover, net longwave radiation is small in magnitude as the overlying leafy canopy is an excellent absorber/emitter at these wavelengths.

The relationships between radiation and the climates and the behavior of animals are most obvious in hot and arid climates where protection from the intense solar radiation is imperative. For example, cold-blooded creatures (such as lizards and snakes) manage their climates through periods of exposure to, and withdrawal from, the direct solar beam. For other creatures, including humans, survival depends on the construction of shelter that modifies the immediate radiation environment. As an example, ancient Anasazi dwellings at Mesa Verde, New Mexico, suggest a remarkable fit to a semi-arid climate. The dwellings are placed in a cliff notch formed at the juncture of two rock types. The overlying rock provides a natural canopy to shade the building group beneath during the hot summer months yet, during the cool winter months (when the solar path is lower), it allows access to the solar beam (Knowles, 1978).

Conclusion

The measurement and modeling of radiative exchanges at all scales will continue to occupy a central role in understanding climate. Many of the globally significant environmental issues of our time (e.g. global warming, stratospheric ozone depletion and urban air pollution) are fundamentally concerned with radiation exchanges. At the grandest scale the global radiation budget is a critical indicator of the state of the EAS. The role of satellites as tools at this scale will increase in importance as they begin to provide long-term records (comparable to surface measurements), improve their spatial and spectral precision and are accompanied by detailed radiation models.

Acknowledgments

The author acknowledges Professor Johan Feddema of the University of Kansas for comments on an early draft, and the work of Stephen Hannon at UCD who produced the final copies of the diagrams.

Gerald Mills

Bibliography

Breon, F.-M., Frouin, R., and Gauthier, C., 1994. Global shortwave energy budget at the earth's surface from ERBE observations. *Journal of Climate*, **7**: 309–324.

DeFelice, T.P., 1998. *An Introduction to Meteorological Instrumentation and Measurement*, NJ: Prentice-Hall.

ERBE Science Team, 1986. First data from the Earth Radiation Budget Experiment (ERBE). *Bulletin of the American Meteorological Society*, **67**(7): 818–824.

Feagle, R.G., and Businger, J.A., 1980. *An Introduction to Atmospheric Physics*. International Geophysics Series, Vol. 25, 2nd edn. New York: Academic Press.

Gupta, S.K., Ritchey, N.A., Wilber, A.C., Whitlock, C.H., Gibson, G.G., and Stackhouse, P.W., 1999. A climatology of surface radiation budget derived from satellite data. *Journal of Climate*, **12**: 2691–2710.

Kiehl, J.T., and Trenberth, K.E., 1997. Earth's annual global mean energy budget. *Bulletin of the American Meteorological Society*, **78**(2): 197–208.

Knowles, R., 1978. *Energy and Form: An ecological approach to urban growth*. MA: MIT Press.

Montieth, J.L., and Unsworth, M.H., 1990. *Principles of Environmental Physics*, 2nd edn. London: Edward Arnold.

Oke, T.R., 1987. *Boundary Layer Climates*, 2nd edn. New York: Routledge.

Sellers, W., 1965. *Physical Climatology*. Chicago, IL: University of Chicago Press.

Smith, G.L., Wilber, A.C., Gupta, S.K., and Stackhouse, P.W., 2002. Surface radiation budget and climate classification. *Journal of Climate*, **15**: 1175–1188.

Wallace, J.M., and Hobbs, P.V., 1977. *Atmospheric Science: an introductory survey*. New York: Academic Press.

Cross-references

Energy Budget Climatology
Radiation Laws
Solar Radiation
Terrestrial Radiation

RADIATION LAWS

In the quantitative study of radiation, much use is made of the concept of a *black body* or *full radiator*. This is a theoretical object that absorbs all radiant energy incident on it and that, as a consequence of Kirchoff's Law (see below), also emits the maximum amount of radiant energy a body can emit at a given temperature. In the natural environment many liquid and solid surfaces approximate black bodies, although gases show large deviations from this type of behavior. Nevertheless, black body radiation provides the starting point for the study of the radiative behavior of all bodies.

Black body radiation

Planck's Law describes the quantity of radiation emitted at each wavelength (λ) by a black body at temperature T (kelvins):

$$E_{B\lambda} = c_1\lambda^{-5}/[\exp(c_2/\lambda T) - 1] \quad (1)$$

where $E_{B\lambda}$ is the monochromatic or spectral emittance (i.e. the radiant energy emitted per unit area per unit time within a waveband of unit width centered on λ). The constants of equation (1) are $c_1 = 3.74 \times 10^8\,\mathrm{W\,m^{-2}\,\mu m^4}$ and $c_2 = 1.44 \times 10^4\,\mathrm{K\,\mu m}$.

Wien's (Displacement) Law gives the wavelength of maximum monochromatic emittance, $\lambda_{B(max)}$,

$$\lambda_{B(max)} = c_3/T$$

where $c_3 = 2897\,\mu\mathrm{m\,K}$.

The Stefan–Boltzmann Law gives the full spectrum emittance E_B,

$$E_B = \int_0^\infty E_{B\lambda} d\lambda = \sigma T^4 \qquad (2)$$

where $\sigma = 5.67 \times 10^{-8}\,\mathrm{W\,m^{-2}\,K^{-4}}$.

Radiation from real bodies

The ratio of monochromatic emittance from a real body at a specific wavelength and temperature to that of a black body at the same wavelength and temperature is termed the monochromatic (spectral) emissivity, $\varepsilon_\lambda = E_\lambda/E_{B\lambda}$. Thus, if the behavior of the function $\varepsilon_\lambda = f(\lambda)$ is known for the real body, one may compute its monochromatic emittance as

$$E_\lambda = \varepsilon_\lambda E_{B\lambda}$$

where $E_{B\lambda}$ is evaluated from equation (1). A body showing $\varepsilon_\lambda \neq f(\lambda)$ is a *gray body*. When ε_λ varies with wavelength the body is a *discontinuous emitter*.

When a monochromatic irradiance I_λ is incident on a body, the fraction absorbed is the monochromatic (spectral) absorptivity (or absorptance), a_λ. *Kirchoff's Law* states that $\varepsilon_\lambda = a_\lambda$. Thus, a body that emits only weakly (or not at all) at a given wavelength absorbs only weakly (or not at all) at that wavelength. For example, both absorption by and emission from the Earth's atmosphere are characterized by rapid changes in absorptive and emissive ability with wavelength between *lines* and *bands* (high ε_λ and a_λ) and *windows* (low ε_λ and a_λ).

The full spectrum *emissivity* is defined analogously with the monochromatic equivalent, $\varepsilon = E/E_B$, where E is the emittance of the real body. Thus, $E = \varepsilon E_B$, where E_B is evaluated from equation (2).

There are numerous other physical laws that relate to the angular attributes of radiation (e.g. Lambert's Law), to radiative transfer (e.g. Beer's Law, Schwarzschild's Equation), and to the scattering and reflection of radiation (e.g. Fresnel's formula). Details may be obtained from the bibliography.

A. John Arnfield

Bibliography

Liou, K.N., 2002. *An Introduction to Atmospheric Radiation*. New York: Academic Press.

Monteith, J.L., and Unsworth, M.H., 1990. *Principles of Environmental Physics*. New York: Edward Arnold.

Wallace, J.M., and Hobbs, P.V., 1977. *Atmospheric Science: An Introductory Survey*. New York: Academic Press.

Cross-references:

Energy Budget Climatology
Solar Radiation
Terrestrial Radiation

RAINFOREST CLIMATES

The tropical rainforest climate is given the name based on several characteristics of temperature and precipitation which result in a broad-leafed evergreen forest biome. The tropical rainforest climate is defined in terms of the high frequency of precipitation and high temperatures throughout the year. The climate is limited to tropical areas as a result of high amounts of radiation evenly distributed through the seasons. Coupled with these temperature characteristics is precipitation distributed through the year. A dense, highly diverse, evergreen forest vegetation has evolved in response to this combination of energy and precipitation.

Geographically, the climate defines a forest region found along the equator in Asia, Oceania, Africa, and South America. The three main areas are in the East Indies, the Congo Valley of Africa, and the Amazon Valley of South America. The climate is generally restricted to some 10° of latitude either side of the equator. The climate is found in small areas as far as the Tropics of Cancer and Capricorn. For example the climate is found on the island of Madagascar and in Brazil at a latitude of 23°S. It has its greatest latitudinal extent on the east sides of the continents due to the prevailing easterly trade winds and warm water along these coasts. In east Africa the trade winds penetrate inland as far as Ngorongoro Crater and the rift valley. The rainforest climate is found in limited areas in these sites. The actual distribution of rainforest climates is much smaller than generally perceived. In Africa, for example, only a relatively small region lying along the equator on the west side of the continent has this type of climate. The same is true of Asia and Oceania. While scattered over a very large area of the Pacific Ocean and the Asian continent, the climate is not found over a large amount of land in one continuous unit.

This type of climate is restricted to low elevations, usually below 1000 m. This also limits the climate to locations where temperatures do not average much below 20°C. At elevations where temperatures drop below this level the rainforest declines and an alpine ecosystem becomes dominant. There is, however, a limited amount of midlatitude rainforests which results from large amounts of precipitation evenly distributed through the year but under cooler but fairly even temperatures. An example of the latter is the rainforest found on the western slopes of the Coast Ranges on the Olympic Peninsula of the United States.

Temperature characteristics

The even distribution in temperature results from the location of these areas within some 10° of latitude either side of the equator. The vertical rays of the sun are thus never more than about 33½° from the zenith. Resulting insolation varies little as the solar beam stays so close to the zenith. The surface temperatures vary little from month to month. The monthly mean temperatures, as well as daily mean temperatures range from 18°C to 30°C,

Figure R9 Rainforest canopy in Maui, Hawaii.

Figure R10 Rainforest in Caribbean National Forest, Puerto Rico.

depending on the actual site. This climate is thus winterless in the sense of having a cold season. It is probably only in terms of sensible temperatures that this climate can be considered hot. Much higher temperatures occur in other climates, including those of the midlatitudes. Because of the location near the equator the length of daylight, and hence daily insolation, also varies only slightly from day to day. There is an average of 12 h of daylight and 12 h of darkness each day through the year. The annual range in temperature is generally less than 2.8°C. Belem, Brazil, experiences a range from the warmest to the coolest month of only

1.7°C. The combination of nearly even hours of daylight and darkness, the even intensity of radiation through the year, and the high humidity associated with the climate, keeps the diurnal range of temperature very small. Contrary to the rest of the world, diurnal temperature variation is typically greater than the annual range. The typical difference between afternoon and early morning temperatures is less than 10°C.

Precipitation is very frequent and the high frequency is one of the main distinguishing elements of the climate. The intertropical convergence zone (ITC) is present over these regions on a continuing basis. The ITC is the semipermanent low-pressure trough which is located near the equator. This trough of low pressure is consistent through time, but not particularly deep. The pressure is normally slightly below the average sea-level pressure, but the trough is sufficient to bring about convergence most of the year. The ITC shifts location with the seasons, migrating north and south over the equator. In most years it does not shift beyond 15°N or 15°S of the equator. The areas with a rainforest climate lie within the area affected by the ITC in all months. Where the ITC migrates the farthest from the equator, the climate is most restricted in area. The convergence of maritime tropical air into the ITC and the lifting along the ITC produces the precipitation.

Moisture characteristics

Humidity is high and varies between 50% and 100% depending on time of day and season. Partly cloudy skies are the rule and days with clear skies are limited. Cumulus cloud forms are most common building in the daytime hours to altocumulus and cumulonimbus when the air becomes unstable. Precipitation is probable on more than 50% of the days in each month, but there are seasonal differences in frequency in some areas. Duitzenzorg, Java, averages 322 days a year with rain. In the Americas, Cedral, Costa Rica, averages 355 days with rain. In this case the probability of rain on any given day is 97%. Average annual precipitation is quite variable, ranging anywhere from 150 to 300 cm. Belem, Brazil, averages 274 cm per year. Significantly, the variation from year to year in the annual total precipitation is relatively low, being less than 50% in most cases. The seasonal distribution tends to be fairly even in most areas. All areas receive at least 6 cm of precipitation per month and it can run much higher than this. Interestingly enough, the location of the sites which have the highest annual rainfall on each of the continents, are not found in the heart of the rainforest. They are found in other climatic regions where there is a pronounced seasonal distribution of precipitation, and enhanced by orographic lifting. As might be expected from the abundant year-round precipitation, large rivers are a part of the environment. Examples are the Congo of Africa and the Amazon of South America.

The precipitation that occurs is almost entirely rain and occurs mainly as convectional showers with thunderstorms being common. The rainfall can be extremely intense as the moisture content of the air can be very high. Associated with the thunderstorms are immense displays of thunder and lightning. Many of the animist religions found in tropical Africa have among their gods those of thunder and lightning. Among the Yoruba people of west Africa *Shango* is the god of thunder. It is also of note that the highest regional fatality rate due to lightning occurs in Kenya along the eastern escarpment of the rift valley. Tremendous thunderstorms erupt as the maritime air rises along the escarpment. The winds associated with the thunderstorms cause considerable blow-down in the forest, as well as lightning damage. The typical thunderstorm does not grow to great heights and as a result hail is rare but does occur on occasion. Due to the extremely high moisture content of the trade winds at times, and the nature of the convectional storms, record rainfalls occur in short periods of time. Cilaos, La Reunion, in the Indian Ocean, has experienced as much as 187 cm of rain in one 24-h period.

Morning temperature inversions are quite common as the ground cools slightly. Only a few degrees drop in temperature is often enough to produce the inversion. Since humidity is so high in the daytime, when cooling takes place at night the dewpoint may be reached when temperatures drop into the range of 15–20°C. Early morning fog and heavy dew are both common. Wood is the primary fuel for cooking and heating in most of the rural areas of the rainforest. These inversions tend to trap the smoke from cooking fires and early mornings are often replete with smog. It is of note that tornadoes and hurricanes are rare in these regions. The reason for this is the low rotational element needed to produce these storms.

Weak easterly waves also traverse the ITC. These systems, though weak, tend to lower the pressure and increase convergence. These systems bring prolonged periods of rain that may last for several days. Winds are predominantly from the eastern quadrants; they vary from northeast to southeast depending on actual location and the season. Velocities at standard instrument height tend to be low, as the pressure gradient is low and the rainforest further restricts air flow. Occasionally in winter the harmattan blows from the Sahara Desert bringing dust storms into the fringes of the rainforest in western Africa. These storms can reduce visibility to less than 400 m.

The weather is fairly predictable in most areas of rainforests. Day-to-day changes are relatively small. There is never much question as to how the weather will alter a person's choice of clothing. Afternoon showers are most common, although precipitation can occur during all hours of the day. The afternoon showers are initiated by the surface heating which brings about instability. At some locations the time of onset of the diurnal showers can also be predicted within minutes on most days. This regularity in timing is strong enough for it to be used in timing daily activities.

The rainforest biome

The rainforest climate has favored the evolution of a tremendous variety of plants and animals, including disease organisms. The most widespread indigenous disease today is malaria. Some 40% of the global population lives in regions where malaria persists. While not limited to the rainforest climate it is endemic in all areas where the rainforest climate is found. In parts of Africa it is estimated that nearly every male, by the time he reaches the age of 15 years, is infected by malaria. If the warming of Earth continues, the incidence of malaria is certain to increase as warmer temperatures increase the voracity of malaria-carrying mosquitoes.

Clothing requirements are thus similar throughout the year. Light cotton clothing for daytime wear and a light jacket or other overgarment for evening constitutes a basic wardrobe. Sensible temperatures reach the uncomfortable level for the majority of people when the temperature exceeds 30°C and relative humidity reaches 100%. Since relative humidity is often extremely high, the weather tends to be uncomfortable, particularly in the afternoon hours. It is for this reason that human activity, especially in rural areas without air conditioning, often ceases or is greatly reduced in the late afternoon hours. For instance, most activities may shut down from 2 to 5 p.m., and resume again after 5:00 p.m.

John J. Hidore

Bibliography

Hastenrath, S., 1991. *Climate and Circulation of the Tropics*. Dordrecht: Kluwer.
Hidore, J., 1996. *Global Environmental Change. Its Nature and Impact*. Upper Saddle River, NJ: Prentice-Hall.
McGregor, G.R., and Nieuwolt, S., *Tropical Climatology: an introduction to the climate of low latitudes*. New York: Wiley.
Oliver, J.E., and Hidore, J.J., 2001. *Climatology – An Atmospheric Science*. Upper Saddle River, NJ: Prentice-Hall.

Cross-references

Africa: Climate of
Asia, Climate of South
Doldrums
Intertropical Convergence Zone
Monsoons and Monsoon Climate
Precipitation Distribution
Vegetation and Climate
South America, Climate of
Tropical and Equatorial Climates

RAINSHADOW

An area on the leeward (downwind) side of a mountain or mountain range where precipitation is significantly less than that on the windward (upwind) side is considered to lie within the mountain's "rainshadow" (also known as precipitation shadow). The windward side of the mountain or mountain range acts as a wedge where the surface air is forced upward and is cooled adiabatically. Cooling may proceed until saturation is reached, then condensation can begin, and precipitation may result. As the air passes over the summit and begins its descent on the leeward side, it is warmed adiabatically. Descending and warming air causes a loss of saturation, and is conductive to the development of clear skies, thereby resulting in a rainshadow. This process is responsible for the creation of certain desert regions around the world. The Great Basin Desert of the southwestern United States, the Takla Makan of China, and the Patagonia Desert of Argentina are all examples of deserts found on the leeward sides of mountains.

L.M. Trapasso

Bibliography

Glickman, T., 2000. *Glossary of Meteorology*. Boston, MA: American Meteorological Society.
Lutgens, F., and Tarbuck, E., 2001. *The Atmosphere*, 8th edn. Englewood Cliffs, NJ: Prentice-Hall.

Cross-references

Adiabatic Phenomena
Local Winds
Orographic Precipitation
Precipitation Distribution

REANALYSIS PROJECTS

Reanalysis projects are the production of retroactive records of global analyses of atmospheric fields. They evolved in the 1990s from the assimilated datasets that were produced in the 1980s as an answer to the problem of providing homogeneous datasets for climatological research. The twentieth century saw a tremendous increase in the amount of climatological data. Four periods can be distinguished: the "surface" period covered the first three decades of the century during which time ever-increasing quantities of surface and some upper air data were collected; the "upper air balloon" period from the 1940s to the International Geophysical Year in 1957 when the upper air networks of both radiosondes and pilot balloons were established; the "modern rawinsonde" period from 1958 to 1978 when radiosondes were tracked by radar to obtain wind observations; and the "satellite" period after 1978. Various international efforts were made to make these diverse data compatible by specifying various standards of instruments, observing times and so on. But it became apparent in the last quarter of the twentieth century that the data could not be objectively used in some climatological analyses, particularly over the vexed question of climate variability. This was because of problems of urbanization, lack of data over the oceans and poles, changes of station site and/or methodological practices, to name but a few. In addition the development of computers and computer models required that data be evenly distributed both on the Earth's surface and vertically through the atmosphere. Consequently two strategies were developed to overcome these problems and both relied on producing "gridded" data; that is data tied to latitude–longitude grid points rather than to individual geographical locations. The first strategy was the "homogenization" of datasets; this occurred in the 1980s and was applied to both station data and gridded data (see Trenberth and Paolino, 1980; Easterling and Peterson, 1995; Jones and Bradley, 1995). The second strategy was the production of assimilated datasets of atmospheric fields (and later incorporating oceanographic fields) in the early 1990s that then evolved into reanalysis projects. This second strategy will now be described.

During the 1970s and 1980s the major operational weather prediction centers developed a variety of global analyses based on atmospheric general circulation models (AGCM). These analyses depended on assimilation systems which quality control, grid and check for and correct inconsistencies in the raw data. The basic outputs of these global analyses have been used in global climate models (see Gates et al., 1996). However, the assimilation systems and the analyses they produced were continually evolving and were consequently not strictly comparable. One answer to this was the production of "control" datasets for a series of years such as the NASA/DAO (National Aeronautics and Space Administration/Data Assimilation Office of the Goddard Laboratory for Atmospheres) assimilated dataset 1985–1989 (Shubert et al., 1993). The second answer was to produce retroactive records of global analyses of atmospheric fields using a frozen state-of-the-art analysis system – reanalysis projects.

The concept of the first reanalysis project (National Centers for Environmental Prediction/National Center for Atmospheric Research (NCEP/NCAR) 40-Year Reanalysis) was discussed at a National Meteorological Center (NMC) workshop in April 1991. It arose because there seemed to be some "climate changes" in the NMC climate assimilation system which were really the result of operational changes. For example the upper air observation times changed in June 1957 from 6-hourly intervals beginning at 0300 UTC to the current 6-hourly intervals beginning 0000 UTC. Similarly the statistical methods used in the assimilation system were continually being improved. The past data include changes in observation systems and a decision had to be made at the outset whether to use all available data at a given time or to select a subset of data that remains stable throughout the period. In the case of the NCEP/NCAR-40 it was decided to adopt the former strategy as this would provide the most accurate analysis (as opposed to the most stable climate) for the period of the reanalysis, which was 1957–1996. The data used in NCEP/NCAR-40 consisted of global rawinsonde data obtained from NCEP (1962 onwards), the US Air Force (1948–1970), China (1954–1962), the former USSR (1961–1978), Japan and several other countries; COADS (Comprehensive Ocean–Atmosphere Data Set, Woodruffe et al., 1987); aircraft and constant level balloon data from a variety of sources, mainly post-1962; surface land synoptic data since 1949; satellite sounder data since 1969; satellite cloud drift winds using geostationary meteorological satellites. Once all these data were assimilated the reanalysis output came in two main formats: the binary universal format representation (BUFR) of the World Meteorological Organization (WMO) and the gridded binary format (GRIB). The data output is given for 17 standard pressure levels on a 2.5° latitude × 2.5° longitude grid at 0000, 0600, 1200 and 1800 UTC each day. In addition a variety of monthly means of atmospheric fields are produced. Full details of NCEP/NCAR-40 are given in Kalnay et al., 1996.

Whilst NCEP/NCAR-40 was being developed it was hoped that the reanalysis could be extended backward in time to incorporate (the mainly northern hemisphere) upper air data for the 1948–1957 decade. This hope became a reality in 2001 when the NCEP-NCAR 50-year reanalysis (1948–1998) was described by Kistler et al. (2001). Information and data for both of these reanalyses can be accessed online at http://www.cdc.noaa.gov/cdc/reanalysis/reanalysis.shtml.

The NASA/DAO assimilated dataset mentioned above became the second United States reanalysis project when it was expanded to provide a 17-year reanalysis from 1979 to 1997. It was then redesignated as NCEP/DOE (Department of Energy) AMIP-II Reanalysis (or Reanalysis 2) being an improvement on the NCEP/NCAR reanalysis because it improved the parameterizations of the physical properties and fixed the errors that had been discovered. It has been running in real time since 2001 and now covers the period from 1979 to present. Details may be found on http://wesley.wwb.noaa.gov/reanalysis2/ index.html.

The European Centre for Medium-Range Weather Forecasts (ECMWF) began a 15-year reanalysis programme (ERA-15) in February 1993 to produce an assimilated dataset for the period 1979–1993. This project used its own archive of observations received from WMO, COADS, the Hadley Centre (UK) sea ice and sea surface temperature (SST) dataset as well as the NCEP SST analyses, data from the First GARP (Global Atmospheric Research Programme) Global Experiment (FGGE) of 1978–1979, the Alpine Experiment (ALPEX) of 1982, Japanese and Australian data, TOGA (Tropical Ocean Global Atmosphere) data, and the vertical sounder data from the TIROS satellites. ECMWF had its own data assimilation system and the project was completed in September 1996. A full description is available online at http://www.ecmwf.int/ research/era/ERA-15/Project/. In 2000 a new reanalysis was begun by ECMWF, known as ERA-40, which covered the period 1957–2001 and was completed in March 2003. It used the same data sources as ERA-15 but with the addition of a variety of satellite radiances (from 1972 onwards), cloud photographs and cloud motion winds, particularly those of the European Meteosat geostationary satellite from 1982 to 1988. The output will be 6-hourly analyses on a grid with spacing of about 125 km in the horizontal and with 60 levels in the vertical between the surface and 65 km (Simmons and Gibson, 2000). At the time of writing (February 2003) four sets of data are available and it is hoped that subsets of the data will be supplied through CD-ROMS, finance permitting. The current availability of ERA-40 analyses can be checked online at http://www.ecmwf.int/research/era/Data_Services/section1.html. During the process of validation some deficiencies in the reanalyses have been found, so there will be a rerun covering the latter part of the ERA-40 period. In February 2003 it had not been decided whether these reruns would begin in 1991, 1987 or 1979.

The other major reanalysis is the joint project of the Japanese Meteorological Agency (JMA) and the Japanese Central Research Institute of the Electric Power Industry called the Japanese Re-Analysis 25 years (JRA-25) which will cover the period 1979–2004. This project began in 2001 with data preparation and the assimilation procedures. The JMA will use its own historical archive of observational data as well the merged database of the NCEP and ECMWF archives. The assimilation system will utilize the latest JMA operational models. The reanalysis began in 2003 and will be completed in 2005. The web page is http://www.jreap.org/indexe.html.

International workshops on reanalysis have already occurred. The WMO World Climate Research Programme (WCRP) sponsored one in the USA in October 1997 (WCRP, 1998) and one in the UK in 1999 (WCRP, 2000). The latter consisted of 109 papers covering the development and use of the various reanalysis projects. The third workshop was held under the auspices of ECMWF at Reading, UK, in November 2001 and a printed version of the 34 papers is available at http://www.ecmwf.int/publications/library/ecpublications/proceedings/ERA40-reanalysis_workshop/index.html.

Brian D. Giles

Bibliography

Easterling, D.R., and Peterson, T.C., 1995. A new method for detecting undocumented discontinuities in climatological time series. *International Journal of Climatology*, **15**: 369–377.

Gates, W.L., Henderson-Sellers, A., Boer, G.J. et al., 1996. Climate models – evaluation'. In Houghton, J.T., Meira Filho, L.G., Callandar, B.A. et al., eds., *Climate Change 1995. The science of climate change*. Intergovernmental Panel on Climate Change. Cambridge: Cambridge University Press, pp. 229–284.

Jones, P.D., and Bradley, R.S., 1995. Climatic variations in the longest instrumental records. In Bradley, R.S., and Jones, P.D., eds., *Climate Since A.D. 1500*. London: Routledge, pp. 246–268.

Kalnay, E., Kanamitsu, M., Kistler, R., et al., 1996. The NCEP/NCAR 40-year reanalysis project. *Bulletin of the American Meteorological Society*, **77**: 437–471.

Kistler, R., Kalnay, E., Collins, W., et al., 2001. The NCEP-NCAR 50-year reanalysis: monthly means CD-ROM and documentation. *Bulletin of the American Meteorological Society*, **82**: 247–267.

Shubert, S.D., Rood, R.B., and Pfaendtner, J., 1993. An assimilated dataset for earth science applications. *Bulletin of the American Meteorological Society*, **74**: 2331–2342.

Simmons, A.J., and Gibson, J.K., 2000. *The ERA-40 Project Plan*. The ERA-40 Project Report Series No. 1, ECMWF. (available online at http://www.ecmwf.int/research/era/Products/Report_Series/ERA40PRS_1

Trenberth, K.A., and Paolino, D.A., 1980. The Northern Hemisphere sea-level pressure data set: trends, errors and discontinuities. *Monthly Weather Review*, **108**: 855–872.

WCRP, 1998. *Proceedings of the First WCRP International Conference on Reanalysis* (Silver Spring, MD, USA, 27-31 October 1997.) WCRP-104, WMO/TD-N 876. Geneva: WCRP.

WCRP, 2000. *Proceedings of the Second WCRP International Conference on Reanalysis.* (Wokefield Park, Reading, UK, 23-27 August 1999.) WCRP-109, WMO/TD-N 985. Geneva: WCRP.

Woodruffe, S.D., Slutz, R.J., Jenne, R.I., and Steurer, P.M., 1987. A comprehensive ocean–atmosphere data set. *Bulletin of the American Meteorological Society*, **68**: 1239–1250.

Cross-references

Climate Data Centers
Climate Modeling Research Centers

RELATIVE HUMIDITY

Relative humidity (or simply *humidity*) is the ratio of atmospheric vapor pressure to saturation vapor pressure. The World Meteorological Organization prefers to define relative humidity as the ratio of specific humidity (or mixing ratio) to saturation mixing ratio, expressed in percent. It can be computed from wet-bulb temperature and related psychrometric data.

Relative humidity (U, in percent) is expressed:

$$U = 100\frac{e'}{e_w'},$$

or

$$U = \frac{\text{Absolute humidity}}{\text{Saturation absolute humidity}} \times 100.$$

where e' is the actual vapor pressure of the air and e_w' the saturation vapor pressure at the same pressure and temperature. Mixing ratio (r) is defined as the ratio of mass of water vapor (m_v) to mass of dry air (m_a), thus $r = m_v/m_a$. Saturation occurs when moist air at pressure p and temperature T may exist in equilibrium with pure water or ice at the same temperature and pressure. In terms of mixing ratio (r) and saturation mixing ratio (r_w), it may be expressed:

$$U = 100\frac{r}{r_w} \times \frac{0.62197 + r_w}{0.62197 + r}.$$

Table R3 Relative humidity (percent) from temperature and temperature–dewpoint depression

Dewpoint depression (°C)	Temperature (°C) 40	35	30	25	20	15	10	5	0	−5	−10	−15	−20	−25	−30	−35	−40
1	95	95	94	94	94	94	94	93	93	93	92	92	92	91	91	91	90
2	90	89	89	89	88	88	87	87	86	86	85	85	84	83	83	82	81
3	85	85	84	83	83	82	82	81	80	79	79	78	77	76	75	74	73
4	81	80	79	78	78	77	76	75	74	73	73	71	70	69	68	67	66
5	76	75	75	74	73	72	71	70	69	68	67	66	64	63	62	60	59
6	72	71	70	69	68	67	66	65	64	63	61	60	59	57	56	54	53
7	68	67	66	65	64	63	62	61	59	58	57	55	54	52	51	49	47
8	64	63	62	61	60	59	57	56	55	53	52	50	49	47	46	44	42
9	61	60	59	57	56	55	54	52	51	49	48	46	45	43	41	39	38
10	58	56	55	54	53	51	50	48	47	45	44	42	41	39	37	35	34
11	54	53	52	50	49	48	46	45	43	42	40	39	37	35	33	32	27
12	51	50	49	47	46	45	43	41	40	38	37	35	34	32	30	28	24
13	48	47	46	44	43	42	40	38	37	35	34	32	30	29	27	25	21
14	46	44	43	41	40	39	37	36	34	32	29	29	28	26	24	23	19
15	43	42	40	39	37	36	34	33	31	30	28	27	25	23	22	20	16
16	40	39	38	36	35	33	32	30	29	27	26	24	23	21	20	16	
17	38	37	35	34	32	31	29	28	27	25	23	22	21	19	18	14	
18	36	34	33	32	30	29	27	26	24	23	21	20	19	13	16	12	
19	34	32	31	30	28	27	25	24	22	21	20	18	17	15	14	11	
20	32	30	29	28	26	25	23	22	21	19	18	16	15	14	12	09	
21	30	28	27	26	24	23	22	20	13	17	16	15	14	12	09		
22	28	27	25	24	23	21	20	19	17	16	15	13	12	11	08		
23	26	25	24	22	21	19	13	17	13	13	13	12	11	10	07		
24	25	23	22	21	19	18	17	16	14	13	12	11	10	09	06		
25	23	22	21	19	18	17	16	14	13	12	11	10	09	08	06		
26	22	20	19	18	17	16	14	13	12	11	10	09	08	05			
27	20	19	18	17	16	14	13	12	11	10	09	08	07	05			
28	19	18	17	15	14	13	12	11	10	09	08	07	06	04			
29	18	17	15	13	13	12	11	10	09	08	07	06	06	04			
30	17	16	14	13	12	11	10	03	08	03	03	06	03	03			

Note: Computed from saturation vapor pressures (with respect to water surface) given in *Smithsonian Meteorological Tables*, (6th rev. (edn), Washington, DC, 1951).

Table R3 provides relative humidity values derived, in this example, from temperature and temperature-dewpoint depression.

Relative humidity generally decreases with height like many other atmospheric characteristics, but it will increase just below clouds since in these the relative humidity is generally 100%. It also has a marked diurnal oscillation opposite in phase to temperature, with a daily maximum at dawn and a minimum in the afternoon. In addition, in certain climatic belts it has an annual variation, also in contrasting phase to temperature. In temperate regions it is generally above 60–80%; in arid regions 20–40%.

In the event of mixing of airmasses of differing relative humidity, a resultant midpoint can be established. If the midpoint falls above the 100% saturation curve, then condensation occurs.

John E. Oliver

Bibliography

Curry, J.A., and Webster, P.J., 1999. *Thermodynamics of Atmospheres and Oceans*. San Diego, CA: Academic Press.

McIntosh, D.H. (ed.), 1963. *Meteorological Glossary*. London: Her Majesty's Stationery Office.

Moran, J.M., and Morgan, M.D., 1997. *Meteorology: the atmosphere and the science of weather*, 5th edn. Upper Saddle River, NJ: Prentice-Hall.

Wallace, J.M., and Hobbs, P.V., 1997. *Atmospheric Science: an introductory survey.* San Diego, CA: Academic Press.

Goody, R., 1995. *Principles of Atmospheric Physics and Chemistry*. New York: Oxford University Press.

Cross-references

Dew/Dewpoint
Humidity

RELIGION AND CLIMATE

Religion, a set of beliefs in one or more gods or supernatural beings, has been defined as humans' attempts to explain the complexities within their milieu and to ensure survival in harsh environments or mental peace in salubrious environments. In times of peace and prosperity, as well as war and tragedy, people have paid homage to who or what they believe is the author and sustainer of life and goodness. One of the premises of many religions was the maintenance of the harmonious relationship between believers and their physical environment, particularly aspects of weather and climate that impacted food production. Three of the largest monotheistic religions, Christianity, Islam, and Judaism, originated within rural agricultural societies in dry climates. All major world religions have specific climate regions of origin, routes of diffusion, and patterns of distribution. Environmental stress invoked, and still invokes, a religious response. Reducing or eliminating critical aspects of environmental stress diminishes traditional organized religious fervor and encourages the proliferation of new religions and cults. All major religions are currently impacted by modern technology, affluence, and urbanization. Secularism has tempered or is replacing mainline religions in highly developed modern societies. The thesis presented for discussion in this item is: if natural hazards, climatic perturbations, environmental stress, and a need for a stable food supply were factors in the origins of many religions, modern technological creations and the ability to control aspects of human microenvironments could be factors in the decline of universal, mainline religious group adherence. Concomitantly, human technological advancements may induce a modification of religious codes of conduct, replacement of things worshiped, and places revered.

Climate determinism

People and cultural groups react and respond differently, physiologically and psychologically, to climatic regimes. There was, at one time in human history, a very close relationship between the physical environment, specifically climates, in which humans lived and the level of human development, cultural attributes, personality traits, and religion. In some physical environments, and in some climatic regimes, it was easier for humans in that particular stage of socioeconomic and technical development to survive and multiply. Ellsworth Huntington, an American geographer, in his works, *Civilization and Climate* and *Mainsprings of Civilization*, describes at length the role of climate determining religion. In relating religious beliefs to the environment, he suggests that the very nature of the desert environment allows comprehension of a single, omnipotent God. To those who lived in forest climates such a concept had no meaning. In the enclosed environment of the forest, with its distinct flora and fauna, it is much easier to conceive of physical site spirits or animal gods, each of whom rules his own segment of the forest. It follows that animistic beliefs were more likely to be found in the religions of the forest people. Those who lived in tropical or subtropical climates, with forest and grassland biomes (home to a wide diversity of aquatic and zoological creatures) that provided sustenance to humans without the constant threat of starvation, multifaceted deities and philosophical "isms" were recognized as gods or served as gods. Different climate realms became the originating hearths of formal and informal religious systems. The firm relationship between humans and God does not describe all religions and all that is worshiped on Earth. There are "gods of gaps" versus faith as trust in God.

Environmentalism, possibilism, and probabilism

In various works, H.T. Buckle, an English geographer, and more so Friedrich Ratzel, a German geographer, contended that the physical environment determined or molded peoples' culture and religion, as well as governments. Ratzel's carefully crafted theories dominated geographic thought for decades but were eventually rejected as being overly simplistic. Possiblism and probabilism emerged as a way to explain human cultural and religious diversity. Both stressed that the environment put restrictions on human religious expression, but de-emphasized the determining factors in the environment and climate. However, Ellen Churchill Semple, an American geographer influenced by Ratzel, credited the desert climate of the Middle East as the primary reason for the origin and dispersal of Islam. She stressed that the extreme conditions and harsh physical environment cannot help but influence people to be fierce and militant so that they can survive. She also researched aspects of the origins of Judaism and Christianity, focusing upon pastoral images taken from the lives of nomadic herding peoples that dominate their religious literature. Semple contended that the desert and steppe nomads conceived an orderly, monotheistic deity and religion through their observations of the night skies and other facets of their dry environment. Vidal de la Blanche, a French geographer, credited climate and the physical environment for the type of religion, that developed in a region, and for the diffusion of that religion. In his writings he emphasized the local climate and the physical environment of a place or region and fostered a region-specific culture from which a region-specific religion emerged.

Ellsworth Huntington was a contemporary of both Blanche and Semple. His views of climate and religion differ from theirs. Huntington examined the population decline in the Middle East and related it to a drying climate. Cultural stress created by climatic change and drought made it easier for those who lived in the Middle East to accept a new religion – Islam. Huntington believed that Islam would never have become as accepted as it was, had the weather and climate of the region not changed. Richard Hartshorne, an American geographer, questioned the idea of using climate or the physical environment to explain the development of a religion in his book, *The Nature of Geography*. He wrote that the climate and environment of a place may have led to the development of a "faith system" in primitive societies, but that they have no influence upon religion in mature and advanced cultures. Jan Broek, a Dutch geographer, and John Webb, an English geographer, in the late 1950s and 1960s taught and wrote that "elements of the physical environment have played a significant role in most religions: holy mountains and rivers, sacred caves, groves, and lakes". In 1961 A.H. Bhattacharya, an Indian geographer, wrote "each religion reflected the geographical spirit of the cradle land from which it disseminated.... Islam is cardinally a desert religion, Christianity a religion of the Mediterranean region, and the Indian religion a product of the prolific Monsoon environment."

David E. Sopher, an American geographer, wrote in 1967 that most of the world's monotheistic religions came out of a nomadic pastoral lifestyle within specific climatic regions. He credits exceptions to the actions of small groups of intellectuals in the affluent cultural/administrative center of their society. A review of the literature in the geography of religion reveals a resilient theme in connection with the type of climate of an area and the development of a distinct religion (Table R4).

Human religious decision making

Humans live in confining physical and socioeconomic environments that vary in personal impact according to population density and stage of scientific and technical development (Figure R11). Within their confines or circles, humans make judgments as to the life system and set of beliefs they will seek, follow, or reject, based upon learned cultural and perceived environmental factors. Initially their actions to ensure survival directed their probe for a set of beliefs until their emotional needs were met within the limits placed by their cultural and physical environments (Table R5). Humanity's increased numbers have altered the world's climate through direct or indirect activities or actions. Areas of extreme high population densities, large urban metropolitan areas, represent humanity's most profound influence upon natural environments, upon the climate of a place, and upon traditional cultural environments. A crude human "Survival Assurance Index" can be postulated that considers those aspects of a cultural group's mores in altered environments which turn religious zeal to religious deviation, to religion doubting, and into personalized ego-serving, anti-religious philosophies that weaken the bonds and moral glue that hold a society together.

In the past the accepted definition (understanding) of climate was the physical condition of a place or region that synthesized various phenomena of the atmosphere (temperature, wind, moisture, storms, etc.), and it daily affected the lives of humans and animals.

Today the definition of climate has evolved into the holistic physical and social milieu of a place or region that synthesizes all phenomena that affect life on Earth. The natural hazards, and specifically climate perturbations, of the past that led humans to create a set of beliefs to ensure their survival at a place have been tempered by technology and human creativity. Natural hazard fears have been replaced by fears of human actions which threaten an individual's survival daily. Fears of cultural hazards, not climatic events, have led to a lessening of the personal affinity for a climate-based universal religion and

Table R4 Selected Views on Climatic Determinism and Religion (After Siddiqi and Oliver, 1986)

Author	Date	Concept
1. Hippocrates	ca. 420 B.C.	The environment (climate) influences all human behavior, including what they worship
2. Ibn Khaldun	A.D. 1332–1406	Climate influences all aspects of life, including religion
3. Nathaniel Carpenter	A.D. 1589–1628	Human character determined by climate
4. Charles Montesquieu	A.D. 1689–1755	Climate influences politics and religion
5. Carl Ritter	A.D. 1779–1859	Strong interconnection between climate and religion
6. H. T. Buckle	A.D. 1821–1862	Climate molds religions
7. Friedrich Ratzel	A.D. 1844–1904	Climate influences religion's origins and dispersals
8. Ellen Churchill Semple	A.D. 1863–1932	Climate plays a role in the origin and diffusion of religion
9. E. G. Dexter	A.D. 1868–1938	Climate and human conduct
10. Jean Bruhes	A.D. 1869–1930	Relationship between climate and human destinies
11. Ellsworth Huntington	A.D. 1876–1947	Climatic religious determinism
12. Griffith Taylor	A.D. 1880–1963	Humans are not free agents
13. Stephen Visher	A.D. 1888–1967	Climatic influences are permissive
14. Clarence Mills	A.D. 1891–1974	'Climate makes the man'

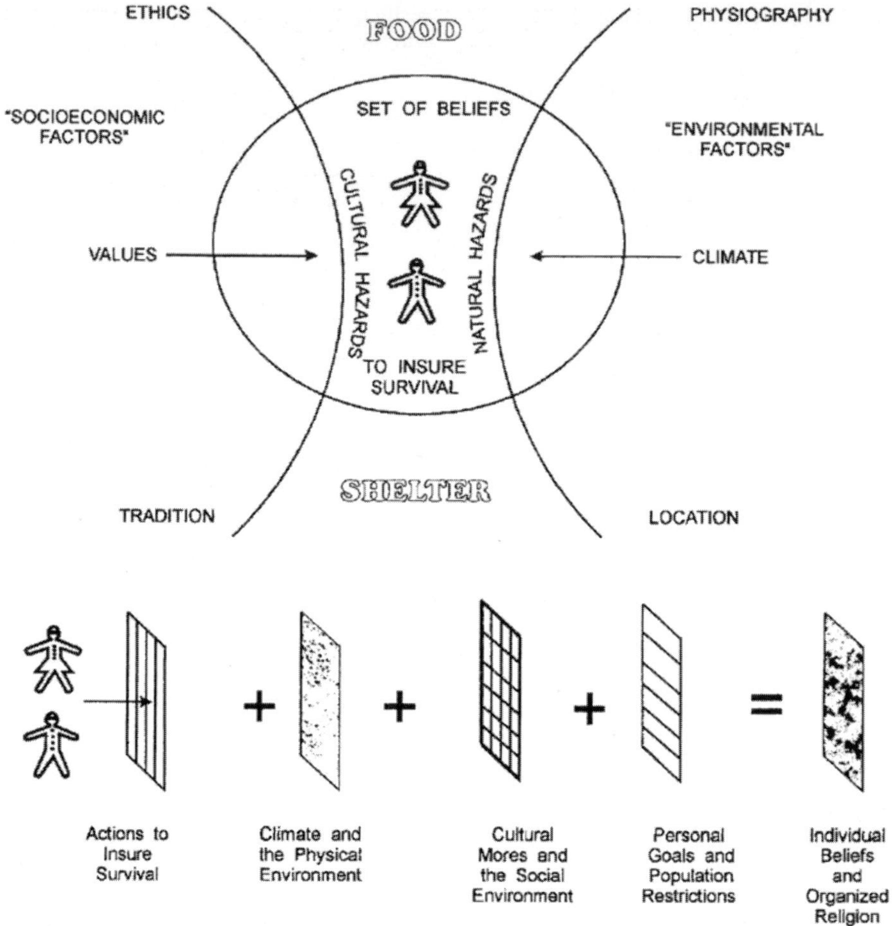

Figure R11 Human religious decision making: historic.

Table R5 Worship of the Sun: The Source of Heat and Light

1. Allure of the Sun (Sun day)
2. Fear of Darkness (fortnight)
3. Agricultural/Religious Calendars (seasons, Latin for "sowing period")
4. Sun Festivals (northern hemisphere)
5. "Sun stand" on winter solstice, December 21 (call back the Sun, Yule Log, hope for life after death, Christmas)
6. "Equal day and night" or vernal equinox, March 21 (joy for the return of the Sun, new life, or new beginning, Easter)
7. "Sun stand" on summer solstice, June 21 (bonfires and festivals to increase the Sun's heat or rekindle its diminishing flame)
8. "Equal day and night" or autumnal equinox, September 21 (lowering or disappearance of the Sun, period of maturity passing to decline, thanksgiving to the Sun that ripened the crops)

Sun worship was a reflection of human dependence upon the mysteries of the Sun's behavior, the Earth's primary source of energy, and observed relationship between the Sun, climate, and food production.

led to the creation of personalized philosophical and new-age religions, i.e. the "gods of gaps" assumption.

Jewish religious festivals and climate as a case study

An example of the influence climate has had upon the development of religion is the calendar of Jewish religious festivals and rituals. When the Hebrews brought to the Promised Land their lunar calendars and lunar festivals, characteristics of nomadic shepherds who depended upon the light of the moon for convenient movement of the flocks, they merged their festivals with ancient Canaanite agricultural festivals whose dates were determined by the position of the sun. The Hebrews had their own concept of the universe, of climate and of their world (Figure R12). The times fixed for the observances were arranged in accordance with the climate/weather and in order not to interfere with agricultural tasks or agricultural labor needs. During religious festivals, rituals were followed to invoke the cooperation and

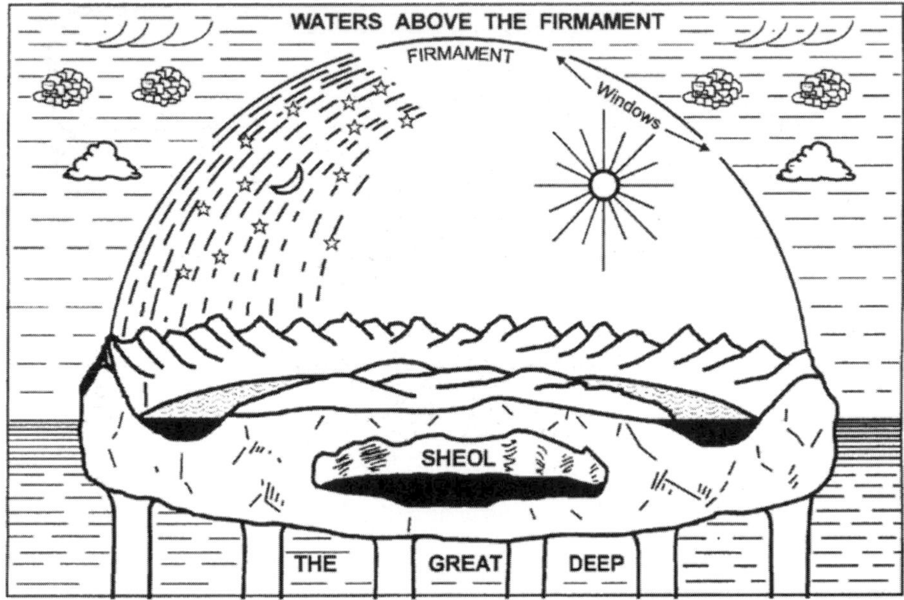

Figure R12 Climate and the Hebrew world. This diagram represents a cross-section of the Hebrew world as they imagined it and as they described it in their sacred literature. Rain came through windows in the firmanent, and climate was created by interactions of elements within the domed firmanent. *Source*: modified from Miller, M.S., and Miller, J.L., *Harper's Bible Dictionary*. London: Harper & Rowe, 1961, p. 193.

blessings of God in all aspects of weather, agricultural life, and food production. Not only did the festivals provide a welcomed respite for those who were tilling fields or tending flocks, they knit together the developing religious soul of the Hebrews/Israelites and promoted a sense of union with God. Besides daily worship, there were weekly, monthly, and yearly festivals to be observed. Festivals in ancient Israel included the activities shown in Table R6. Jewish religious festivals had an agroclimatic base. Today, these festivals are retained despite the occupational evolution of most Jews from rural shepherds thousands of years ago into modern urban professionals.

Climate, environmental and social stress and modern religious thought

Religious climatic inertia and historic precedence are evident also in the spatial dynamics of major religions. The current world distribution of major religions remains primarily in accord with the basic climate regions in which they originated (Figure R13). Islam is found in the Middle East and North Africa (BWH and BWK climates) plus Indonesia (Af Climate); Buddhism, Confucianism, Taoism, and Shintoism are found in eastern Asia (Ca climates) while animism remains in a few primal cultures located in northern Siberia, northern North America (Dd and ET climates), portions of the Amazon Basin of South America, southern Africa, New Guinea, and northern Australia (Af and Aw climates). Christianity is the dominant world religion found in all continents. It originated in ancient Israel (Cs and Ca climates) and spread northward into Europe and Eurasia (Cs, Cb, BS, BW, Da, Db, and Dc climates), then to similar climatic regions in North America, South America, and Australia. Only in the continent of Africa has the pattern of religions changed radically in the past 100 years. African religious groups are multifarious, having been impacted by colonial rulers, and changeable. Depictions of any religious patterns are of limited value. Christians are still a minority in Africa while animism is the dominant religious force. In many instances Christian doctrine is wedded to traditional beliefs.

In most Western societies, secularism has tempered the growth and development of mainline Christian denominations. Climatic stress has been replaced by social stress. Social stress, found in almost all technologically and economically developed nations, produces great strain in an individual, in a family, and in a relationship. Excessive demands by society placed upon an individual create great personal and interpersonal tension, produce a debilitating and wearing effect, and at times deform an individual's physical body and mental health. Also, militant atheism, as manifested in the anti-religious commitment and ideology of communism, has proven to be a deadly foe of religion in general and Christianity in particular. The demise of the communist dictatorship and the break-up of the Soviet Union, the emancipation of Eastern Europe, and the relaxation of some political control in China have lessened somewhat the impact of militant atheism. As militant atheism waned, traditional universal religions were revived and new religious cults emerged. The New Age Movement, for one, is based upon Eastern mysticism, Hinduism, and paganism. Devoid of an environmental or climatic-linked base, there is no one specific holy book, beliefs vary according to the individual, and God is an impersonal force or principle, not a person. Adherents believe that people have inner power but need to discover it, and that followers can contact supernatural beings through meditation, self-awareness, and "spirit guides". A cycle of reincarnations may occur. These beliefs

Table R6

I. Septenary (cycles of Sabbaths)
  A. Weekly Sabbath – to restore the human body's energy lost in the manual labor required of farmers in a dry, summer subtropical climate.
  B. Feast of the Trumpets or Seventh New Moon – a break from labor involved in sowing grain.
  C. Sabbatical Year or every seventh year – a fallow year when the land was to have a complete rest from cultivation.
  D. Year of the Jubilee or every forty-nine years – a complete year of rest from all agricultural or economic activity, reversion of property, and remission of slaves.

II. Yearly
  A. Passover – the lambing season and just before barley harvest.
  B. Pentecost or First Fruits – conclusion of the grain harvest and prior to the grape harvest.
  C. Day of Atonement – after the fruits of the ground were harvested and before plowing of the land.
  4. Tabernacles or Books – plowing and sowing time begin.

III. Jewish Religious Calendar and Agroclimatic Events

Month	Festival	Agricultural activity	Weather
1. Abib or Nisan (March–April)	Passover	Prior to barley harvest; lambing season	Warm/humid cloudy/latter or spring rains
2. Iyyar Ziv (April–May)	Little Passover	Barley harvest	Warm/low humidity/cloudy/dry
3. Sivan (May–June)	Pentecost/Feast of Weeks	Wheat harvest/lambing season	Hot/very dry/low humidity/few clouds
4. Tammuz (June–July)	—	—	Hot/very dry/low humidity/no clouds
5. Ab (July–August)	—	Grapes, figs, olives ripen and harvest	Hot/very very dry/few clouds/low humidity
6. Elul (August–September)	—	Vintage begins/pomegranates ripen	Warm/very very dry/few clouds/low humidity
7. Tishri or Ethanim (September–October)	Day of Atonement/Feast of the Tabernacles/Feast of Booths	Harvest of the fruits of the ground/ ploughing of fields	Cool/early or beginning of fall rains/some cloudiness/humid
8. Marchesvan or Bul (October–November)	Feast of the Trumpets	Sowing of barley and wheat	Cool/intermittent rain/moist-to-dry/ cloudy/humid
9. Chislev (November–December)	Hanakkah/Feast of Dedication	Trees and plants growing	Cloudy/humid
10. Tebeth (December–January)	—	Trees and plants growing	Cloudy/high humid
11. Sebat (January–February)	Arbor Day	Blossoming of fruit trees	Cold/heavy rain/very cloudy/high humidity
12. Adar (February–March)	Purim	Blossoming of nut trees	Moist-to-dry/cloudy/humid
13. Veadar or Second Adar (Intercalary Month)	—	—	—

are not linked to agroclimatic events, and there is no heaven or hell. In general, cults of modern religious thought are separated from the physical environment and employ meditation, crystals, and psychic powers. They strive for world peace and holistic health. The new religious cults and movements have had less effect upon universal religious systems than modern secularism. They are found in Climatic Realms C, D, A, and B.

Secularism and environmental change

There are many religious and semireligious movements emerging to help individuals cope with problems they face in the socioeconomic and political environments of our times. These, in most part, rely upon human reason, experiences, science, or technology, and not upon supernatural deities or revelations. Modern urban societies engender a global view of life issues devoid of old mystical, faith-based elements. Scientific achievements, social institutions, and rational thoughts have replaced religious institutions, religious beliefs, and religious values in the minds of New Age and new cult believers. Organized, mainline religions are being separated from the core institutions of society, specifically the family, education, business, and the state. Modernity, science, and urbanization have created a new environment for new religions – that of an urban climate. In these climatic regions humans, political structures, and corporations determine whether a person is successful or unsuccessful, whether a person eats or does not have food, and whether a person survives walking down a sidewalk or is harmed. Nature almost has been eliminated as a factor in choices that must be made in an urban society for human survival. Society, those who make up society, and society's idols, are the new gods that must be appeased in the world's urban centers.

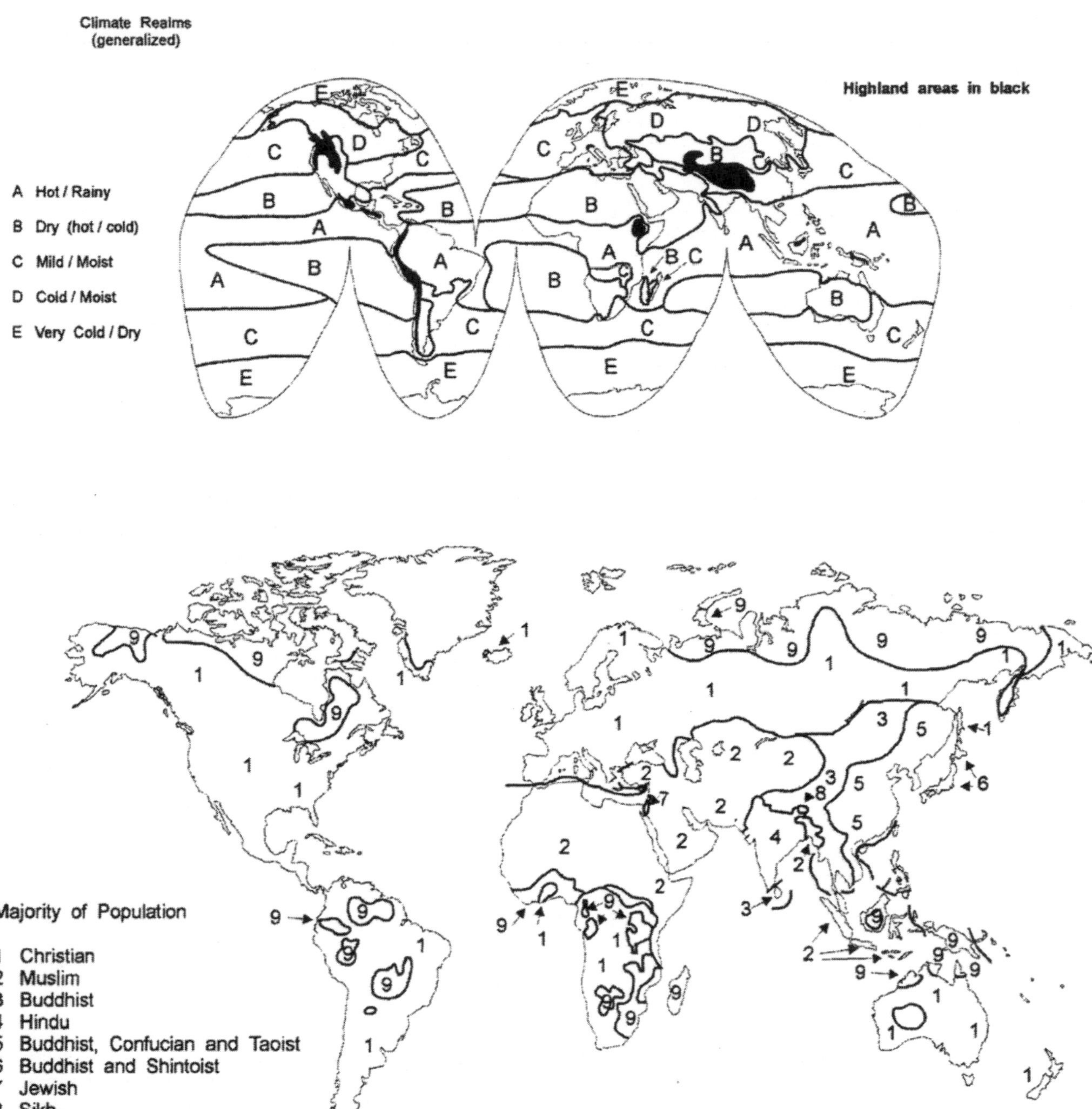

Figure R13 Climate and religious affiliation.

Religion and human-made urban and socioeconomic climates

Concerned urban area leaders are struggling to provide economic and cultural amenities for urban inhabitants as well as to create a pleasant place to live. The bulk of all urbanized land is used for human residences, not for food production or economic activities, together with extensive transportation networks needed for access to the city's economic and social functions. Migration of people and wealth to urban areas has caused much anxiety about the future of rural life in society. The shift of population from rural to urban has been too recent and too rapid to fully understand its total ramifications. Urban expansion cannot continue indefinitely. Large cities have spread and coalesced with other large cities into a formless mass in which the very function of a city has been lost. The urban habitat is where humans now live, not the area in which

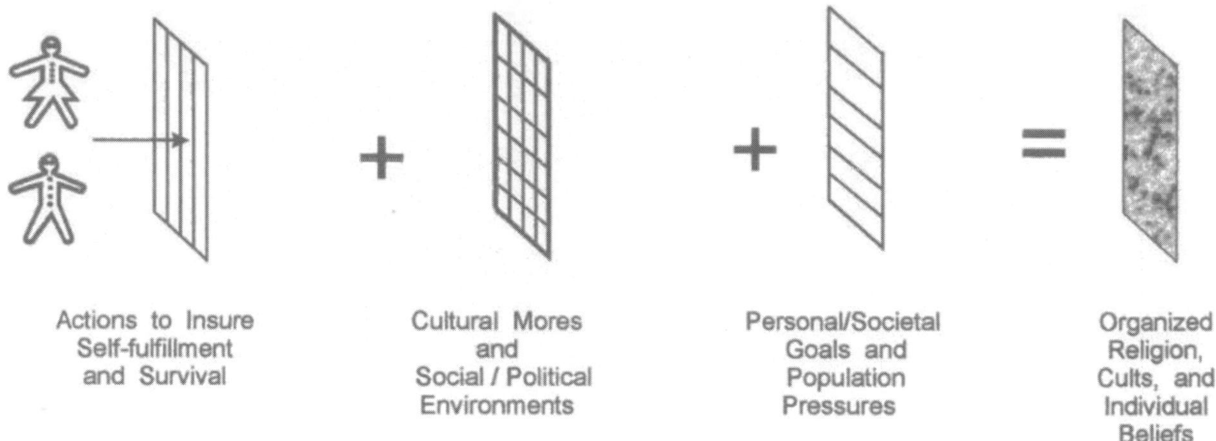

Figure R14 Human religious decision making: post-modern.

humans are biologically and socially best suited to live. Humans need space for complex biological and psychological reasons. Great distortions in human behavior occur when population density reaches certain excess thresholds. Planless urban development, accumulating refuse and scrap, and increasing pollution and crime affect people and their outlook on life. Modification of the natural environment and construction of a brick-and-cement urban habitat have created a physical environment, in most part, not conducive to human happiness and well-being. Also, it has created a new type of climate – an urban climate.

The issues and concerns facing a modern urban dweller daily are not the same issues and concerns of those who resided in the site of a religious system's origin. For the major universal religions the inhabitants of the site of origin were agriculturalists dependent upon the physical environment, particularly weather and climate, for survival. Drought, food shortages, and famine were common occurrences in the primitive societies based upon dry-land agricultural systems. Organized methods and objects of worship were contrived or revealed to temper those elements that impacted food production in many cultures of the world. In contemporary society the same fears and concerns are not manifested in urban complexes. Science and technology, along with human ingenuity and creativeness, have enabled humans to separate themselves, in part, from nature and place themselves in the hands of other humans (Figure R14).

Cultural hazards, not natural hazards, confront modern urban dwellers today. Organized universal religions are now facing new challenges in an urbanized world plagued by urban terrorism, ethnic strife, economic uncertainty, and religious leaders' indiscretions. Still, the religious impulse is extremely strong and vital in America. Two out of three Americans claim religion is very important to them and their families. One out of every two Americans attends weekly worship services. All surveys, whether professional or casual, report that a belief in a supreme being and devotion to prayer is very important to a majority of Americans. There is a deep desire for spiritual support and a hunger for a benevolent supreme being. Americans strive for and want spiritual reassurance and gain spiritual sustenance in personal experiences and private meditative practices. The Bible Geography Specialty Group of the Association of American Geographers contends that the United States is the most religiously diverse and most religiously tolerant of all highly developed nations in the world. Humans live in a mosaic of environments. In the past the physical environment, particularly climate, and now the cultural environment, serve as a catalyst for religious evaluation. In modern urban societies, where multitiered economic and complex social stratifications are the rule, microevaluation of religious thought is favored. Here, environmental change and cultural change are inextricably interwoven. Culture serves to mediate between environment and society. It also serves as the catalyst for religious innovation. Adaptation to new environments, specifically urban environments, involves concomitant cultural innovation and religious diversity. Environmental changes trigger macro cultural responses and micro or personalized religious commitments (Figure R15).

New religions, new cults, and new philosophies of life are created by humans in response to the perceived new threats to their happiness, success, and survival. The new urban environments, these human-made urban climates, are presenting issues to urban dwellers that the universal religions are not addressing or are incapable of addressing. As it was thousands of years ago, new environments and new climates (both physical and social) are engendering new religions. The old religions, the mainline religions, the universal religions are being challenged, membership captured, and resources taken by new religions, cults, and philosophies. The decline in membership of many religious systems in the world is, in part, a result of the change in human habitats and changes in human concerns. Yet physical environment-induced stress or calamities (droughts, floods, tornadoes, hurricanes) and cultural environment-induced stress (terrorism, war, loss of a job, bankruptcy) still evoke traditional religious responses and increase worship service attendance. Viability and sustainability of religious systems will depend upon their capability to adapt to the rapid changes in the climate of where people live – specifically the physical climate and the socioeconomic climate. In modern climate-controlled or climate-tempered society, people's loyalty, beliefs, and hearts are where their fears and riches are.

Wiliam A. Dando

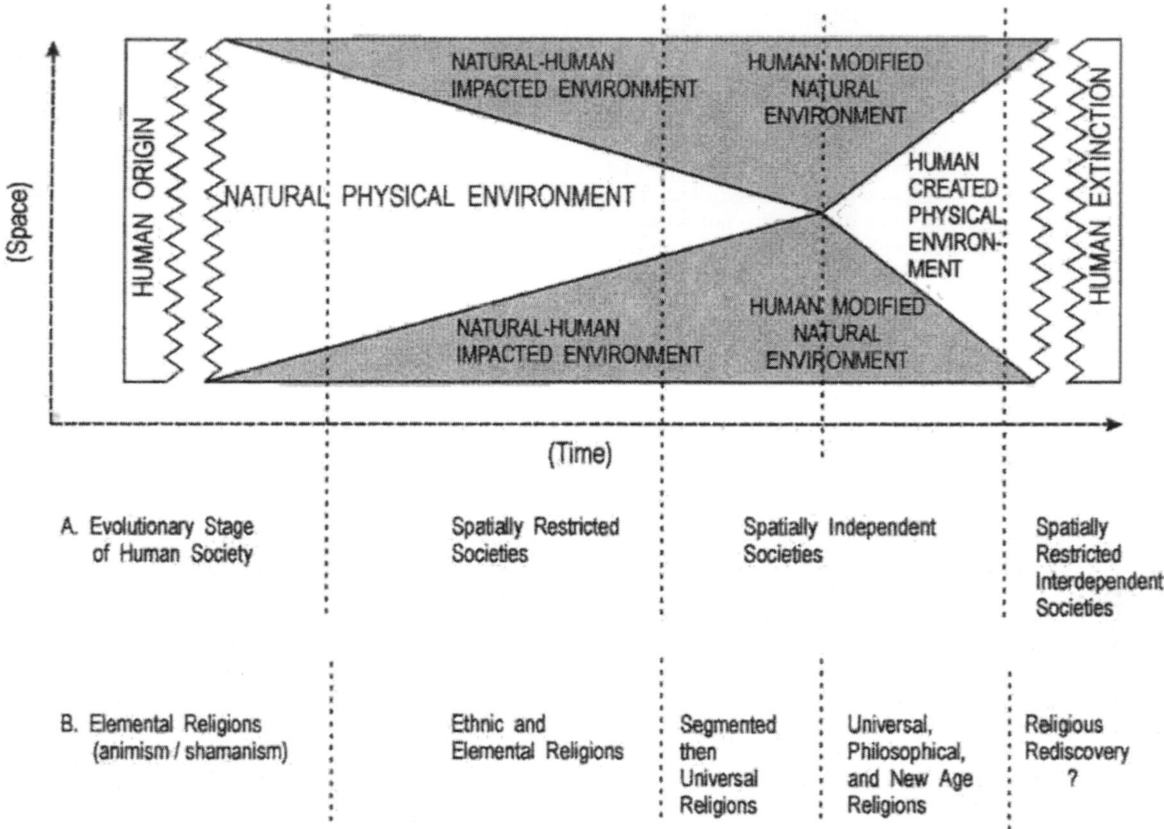

Figure R15 The religious environmental continuum.

Bibliography

Al-Faruqi, I., and Faruqi, L., 1986. *The Cultural Atlas of Islam*. New York: Macmillan, pp. xii–xiv and 3–19.

Allen, J., 2002. *The Student Atlas of World Geography*. New York: McGraw-Hill, pp. 3–28.

Cully, K., and Harper, F., 1969. *Will the Church Lose the City*? Cleveland: World Publishing Company, pp. 206–233.

Dando W. (ed.), 1988. *Earth Science Meteorology–Climatology Instructional Manual*. Grand Forks, ND: National Science Foundation and the University of North Dakota Press, Unit Nine, pp. 224–257.

Dowley, T., (ed.), 1997. *Atlas of the Bible and Christianity*. Grand Rapids, MI: Baker Book House, pp. 115–154.

Eck, D.F., 2001. *A New Religious America*. San Francisco, CA: Harper, pp. 335–386.

Endfield, G., and Nash, D., 2002. Missionaries and morals: climatic discourse in nineteenth-century central South Africa. *Annals of the Association of American Geographers*, **92**(4): 727–742.

Huntington, E., 1915. *Civilization and Climate*. New Haven, CT: Yale University Press.

Huntington, E., 1945. *Mainsprings of Civilization*. New York: Wiley.

James, P., and Martin, G., 1993. *All Possible Worlds: a history of geographical ideas*. New York: Wiley.

Miller, M., and Miller, J., 1961. *Harper's Bible Dictionary*. London: Harper & Rowe, p. 193.

Moore, R., and Day, D., 1966. *Urban Church Breakthrough*. New York: Harper & Rowe, pp. 157–179.

Siddiqi, A., and Oliver, J., 1987. Climatic determinism. In Oliver, J., and Fairbridge, R., (eds.), *The Encyclopedia of Climatology*. New York: Van Nostrand & Reinholt, pp. 382–386.

Smith, H., 1958. *The Religions of Man*. New York: Mentor Book/Harper & Brothers.

Sopher, D., 1967. *Geography of Religions*. Englewood Cliffs, NJ: Prentice-Hall.

Tatham, G., 1957. Environmentalism and possibilism. In Taylor, G., (ed.), *Geography in the 20th Century*. London: Methuen, pp. 128–162.

Voehringer, E., 1967. *Christianity and Other World Religions*. Philadelphia, PA: Lutheran Church Press, pp. 9–30.

Cross-references

Climate Classification
Determinism, Climatic
Cultural Climatology

ROSSBY WAVE/ROSSBY NUMBER

The configuration of the higher atmospheric flow, particularly in the westerlies, tends to revert to a wavelike pattern, a situation confirmed by both observation and hydrodynamic experiments (Fultz, 1949). Although these waves in the westerlies, within the middle or upper troposphere, exist in a continuum of lengths, they tend to be of either short (<50° longitude) or long (50–120° longitude) wavelength. Frequently, the situation is one of rapidly moving shortwaves superimposed on a field of slowly progressing or quasistationary longwaves.

Named after the late C.G. Rossby, a meteorologist who contributed much to their theoretical understanding, Rossby waves are the features of the greatest length within the westerlies (Rossby, 1939). Characterized by cold troughs and warm ridges, and best developed between about 700 and 200 mb, they may be quasistationary or progress slowly, depending on their thermal structure. In either case they are usually not responsible for generating the synoptic-scale systems associated with the daily weather changes so typical in the midlatitude westerlies. Rather, their primary impact is to function as steering influences on these systems.

Possible origin

Important vorticity relationships help account for the development and maintenance of longwaves (Rossby) in the westerlies. One such relationship exists between the horizontal area of airstreams in the troposphere and their absolute vorticity. For example, as the westerlies ascend a mountain barrier, such as the Rocky Mountains, the airstream is vertically compressed between the rising terrain and the tropopause; vertical compression is compensated by lateral expansion (not detectable on airflow maps). Lateral expansion necessitates a decrease of absolute vorticity (somewhat analogous to the outstretching arms of the spinning ice skater, a motion that then slows the spin rate), which is achieved by a turn to the right in the northern hemisphere, transporting the air to lower latitudes and values of earth vorticity (Figure R16). On the downslope side of the barrier, however, vertical stretching between the tropopause and the lowering terrain leads to horizontal contraction, and absolute vorticity may again increase. This increase is achieved by lessening the right-hand turning initiated earlier, and, at the base of the barrier, where the lateral area of the airstreams is similar to the initial state, the degree and direction of trajectory turning (vorticity) are similar to the original state (in the theoretical absence of other perturbations).

In this fashion, interaction with the Earth may generate torques in the atmosphere and force certain features in the tropospheric flow (Reiter, 1963). Accordingly, mean longwave patterns display geographic preferences. Other more complex forcing may be associated with less conspicuous Earth features (Namias, 1980; Walsh and Richman, 1981).

Establishment of a train of waves

Once a meridional trajectory is initiated, a second important vorticity relationship helps ensure establishment of a downstream wave pattern. Above the friction layer in the middle or upper troposphere, airstreams tend to follow trajectories of constant absolute vorticity (Petterssen, 1956). In longwaves particularly, this absolute vorticity is maintained through a seesaw of the value of the terms representing curvature vorticity and Earth vorticity. Earth vorticity (f) must be considered because the size of longwaves is such that latitude changes are of significance in the equation; shear vorticity can be ignored at this large scale, so curvature vorticity becomes the dominant relative vorticity term. Thus, in this situation, the absolute vorticity of the air particles is the sum of vorticity due to curvature and f. For example, a turn to lower latitudes where the value of f decreases will necessitate cyclonic curvature (increasing the value of x) if the sum of these turns is to remain constant (Figure R17). This cyclonic curvature then redirects the particle to higher latitudes and values of f, so the trajectory then must bend increasingly to the right (decreasing the value of x) in order for constant absolute vorticity to remain constant. Such a sustained equilibrium between the value of these two terms in the equation theoretically leads to a train of longwaves around the hemisphere that may have been originally stimulated by only one perturbation arising from Earth–atmosphere interaction. In reality, however, these interactions are continuously occurring as the westerlies circle the Earth and the observed wave train may differ substantially from the theoretical plan.

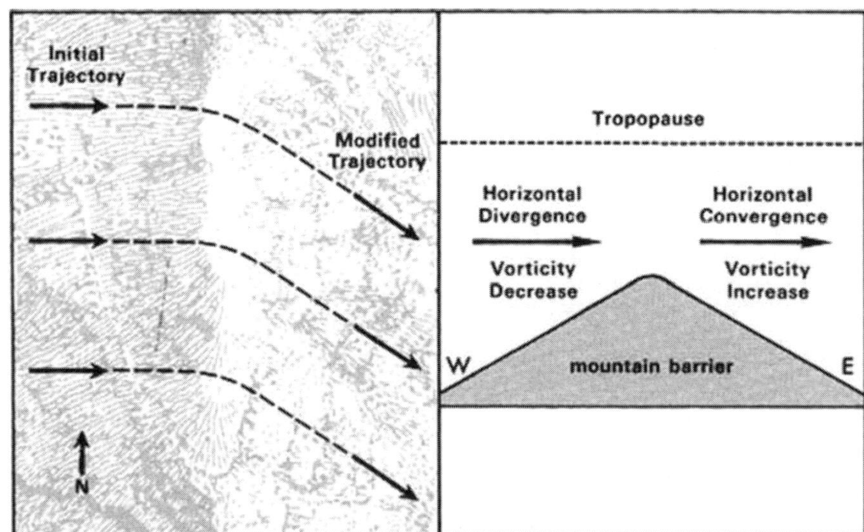

Figure R16 Schematic diagram summarizing the process in which the trajectory of an airstream in the northern hemisphere westerlies can be altered by passage over a mountain barrier.

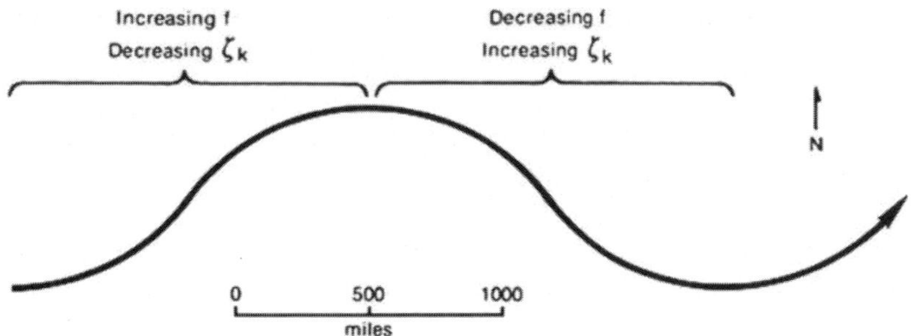

Figure R17 Diagrammatic summary of the changes of relative and Earth vorticity in the upper troposphere as air passes through a quasistationary wave in the northern hemisphere westerlies.

Figure R18 Illustration of the effect that increasing the amplitude of an initial wave (from A to B) might have on the downstream wave train in the northern hemisphere westerlies.

Teleconnection

Once established, waves within a longwave train are teleconnected because of the tendency of the air to follow constant absolute vorticity trajectories. For example, a perturbation in one wave may, by consequence, increase the amplitude of that wave. A trajectory issuing from that now-amplified feature would retain at least some of that amplitude downstream and might result in the entire wave train acquiring greater amplification (Figure R18). In this way weather events in remote corners of the hemisphere may have ramifications far downstream in less time than is needed for the events themselves to reach the affected area.

Impact of Rossby waves on weather

The impact of Rossby waves on surface weather is primarily in their function as steering influences on the path taken by midlatitude cyclones and anticyclones. This influence operates because cyclone and anticyclone paths correspond closely to jet stream positions, where vertical atmospheric motions supporting these systems are strongest. Jet streams follow the general configuration of the upper tropospheric circulation, which normally features a number of quasistationary Rossby waves. The number of such waves around the hemisphere averages three or four in winter but increases to five to seven in summer when the circulation is weaker and stationary waves are shorter.

The configuration of the upper tropospheric westerly flow is one of the most important factors contributing to the weather of specific regions (for a general discussion, see Reiter, 1963). As noted, mean troughs and ridges often have geographic preferences; by guiding airmasses into a region or diverting storm systems, these features also can lend specific character to a region's climate. Beneath a stationary ridge, for example, storminess is suppressed by subsidence or the presence of warm air aloft and the weather, perhaps for long periods, would be stable. Where ridges are often forced (perhaps by Earth interaction), such weather would recur so often as to characterize the region's climate. The weather associated with stationary troughs, on the other hand, is often cloudy and wet. Here, cold airmasses brought equatorward by the upper air flow would produce lower mean temperatures than would be expected beneath ridges at the same location. Within a given region, alternation between dominance by slowly progressive Rossby waves would produce periods of different weather, each lasting a week or two.

Importance of Rossby waves in the general circulation

Rossby waves are an integral part of the meridional heat transfer in the westerlies. By ushering cold surface air equatorward and warm tropical air poleward, they help offset the heat imbalance arising from the nonuniform distribution of solar energy between the equator and the poles. Furthermore, the weak upward vertical motions east of longwave troughs help promote cyclonic formation, moving generally northeastward as they develop. These eddy circulations further contribute to the meridional exchange of heat, moisture, and momentum necessary in the middle latitudes to maintain the global energy budget.

The Rossby number

This is a nondimensional number (Ro) that defines the ratio of intertial force to the *Coriolis force* for a given flow in a rotating fluid. With U as a characteristic velocity, f the Coriolis parameter, and L a characteristic length,

$$\mathrm{Ro} = \frac{U}{fL}$$

In experiments of the dishpan type system, which is cylindrical rather than spherical, the Rossby number will still be a basic relationship provided the Coriolis parameter is replaced by a cylindrical value that is simply twice the rotation rate of the system. An analogous but distinctive nondimensional number is the *thermal rossby number* (Ro_T), which is the inertial force due to the thermal wind (U_T) and the Coriolis force (f) in the flow of a fluid heated at the center and cooled at the rim. With L as a characteristic length,

$$\mathrm{Ro}_T = \frac{U_T}{fL}.$$

The thermal wind (U_T) is given by

$$U_T = g_\epsilon \frac{(\Delta r\theta)\delta}{f\Delta r},$$

where g is the acceleration of gravity, ε is the coefficient of thermal expansion, $\Delta\theta r/\Delta r_r$ a characteristic radial temperature gradient, and δ the depth of the fluid.

Jay R. Harman and John E. Oliver

Bibliography

Eliassen, A., and Kleinschmidt, E., Jr, 1957. Dynamic meteorology. In Flugge, E. (ed.), *Handbuch der Physik*. Berlin: Springer-Verlag, pp. 1–154.

Fultz, D., 1949. A preliminary report on experiments with thermally produced lateral mining in a rotating hemispherical shell of liquid. *Journal of Meteorology*, **6**(1): 17–33.

Fultz, D., 1951. Experimental analogies to atmospheric motions. In Malone T.F., ed., *Compendium of Meteorology*. Boston, MA: American Meteorological Society, pp. 1235–1248.

Geer, I.W. (ed.), 1996. *Glossary of Weather and Climate*. Boston, MA: American Meteorological Society.

Holton, J.R., 1992. *An Introduction to Dynamic Meteorology*, 3rd edn. San Diego, CA: Acadamic Press.

Huschke, R.E. (ed.), 1959. *Glossary of Meteorology*. Boston, MA: American Meteorological Society.

Namias, J., 1980. Causes of some extreme Northern Hemispheric climatic anomalies from summer. *Monthly Weather Review*, **108**: 1333–1346.

Petterssen, S., 1956. *Weather Analysis and Forecasting*, vol. 1. New York: McGraw-Hill.

Reiter, E.L., 1963. *Jet Steam Meteorology*. Chicago, IL: University of Chicago Press.

Rossby, C.G., 1939. Relation between variations in the intensity of the zonal circulation of the atmosphere and the displacements of the semi-permanent centers of action. *Journal of Marine Research*, **2**: 38.

Walsh, J.E., and Richman, M.B., 1981. Seasonality in the association between surface temperatures over the United States and the North Pacific Ocean. *Monthly Weather Review*, **109**: 767–783.

Cross-references

Atmospheric Circulation, Global
Coriolis Effect
Jet Streams
Middle Latitude Climates
Vorticity

S

SATELLITE OBSERVATIONS OF THE EARTH'S CLIMATE SYSTEM

The Earth's climate is established via a complex set of interactions between and within its major subsystems. Predicting the evolution of the Earth's climate is difficult as the processes that govern these interactions are connected over vast ranges of time and space. Credible climate prediction requires a deep level of understanding which in turn requires observations that are able to link processes that take place on the smaller–faster scales, such as those affecting the evolution of clouds, precipitation and other aspects of the hydrological cycle, to the global scale. Climate also requires that these observations be made over sufficiently extended time periods. The almost regular-in-time and global-in-space nature of satellite observations makes them particularly attractive for this purpose and the use of satellite data in the study of Earth's climate has grown significantly over the last two decades.

Selected examples of climate studies that use satellite data are introduced below to highlight the general issues involved in constructing climate information from such data. Other examples of programs that focus on developing satellite climate data, such as the Earth radiation budget and solar irradiance programs, are not discussed.[1] It will become apparent that a satellite observing system is much more than the instrument on a satellite and thus its accuracy is defined by more than the instrument calibration. Rather satellite observing systems are complex systems defined by mixtures of properties including the properties of the sensors, the platform including its orbit, and the properties of all sorts of other extraneous information associated with the inverse process. As such, proper characterization of these observing systems is complicated.

[1] More can be found on the radiation budget and total solar irradiance satellite programs, http://asd-www.larc.nasa.gov/ceres/ASDceres.html, http://daac.gsfc.nasa.gov/CAMPAIGN_DOCS/FTP_SITE/INT_DIS/readmes/sol_irrad.html)

Terminologies

Asynchronous sampling. This refers to the way sensors on polar orbiting satellites, in particular, observe the Earth. Because these satellites move with respect to Earth, the observations made from them mix space and time variations of properties.

Brightness temperature. An alternative way of expressing spectral radiances. It is a measure of the equivalent blackbody temperature of a hypothetical body emitting that same amount of radiation as detected by the sensor. This is not equivalent to the (true) thermodynamic temperature of the medium.

Calibration. The process by which raw instrument data (e.g. voltages) are converted to physical quantities (e.g. radiance units). Radiometric calibration ideally requires the conversion be connected in some way to a reference standard.

Climate subsystems. Atmosphere, the lithosphere, hydrosphere, cryosphere and biosphere (Piexoto and Oort, 1992).

Data inversion. The process of extracting geophysical information from sensor data. Most sensor data are in the form of (spectral) radiances.

Geostationary satellites. These satellites are high above the Earth, nominally at about 36 000 km in an orbit that lies in the equatorial plane. As a result the satellite remains stationary with respect to a fixed point on the equator. Sensors on these satellites are thus able to stare at the Earth watching the atmosphere evolve below it.

Observatory. The sum of the spacecraft, entire instrument payload and the orbit characteristics that define how the Earth below is sampled in space and time.

Observing system. The sum of the observatory, model systems used in data inversion and all other components of the inversion process including observing networks used in vicarious calibration.

Polar orbiting satellites. Satellites in low-altitude orbits, often between about 400 and 1000 km above the Earth's surface. These satellites view the Earth in a swath with a periodicity that is related to orbit altitude. The angle of inclination of the orbit dictates whether or not the swath extends over the polar regions.

Radiances. The detected amount of radiant energy over some defined spectral region confined to some acceptance angle characteristic of a radiometer.

Radiometers. Instruments for measuring the amount of energy of EM radiation integrated over a spectral region of nominal width $\Delta\lambda$ typically in a few selected spectral regions. These regions are referred to as channels as defined by the width and a nominal central wavelength λ.

Representativeness. The space–time properties of a measurement in relation to the true space–time properties of a given climate parameter.

Resolving power. The ratio $\lambda/\Delta\lambda$ Most satellite visible and infrared radiometers as used in climate research have resolving powers of order 1–10, but spectrometer systems such as being planned for the next-generation sounders have powers that are orders of magnitudes greater.

Satellite platforms. The spacecraft.

Sensor. A particular instrument flown on a satellite. By far the majority of sensors are spectral radiometers.

Vicarious calibration. The process of calibration using non-traceable and often indirect methods.

Selected traits of a climate satellite observing system

Figure S1 depicts a few characteristics of satellite observing systems that might be considered desirable for many types of climate problems. The figure is in the form of a 2×2 matrix with one dimension representing measurement accuracy and the other measurement precision. Accuracy is especially important for many studies that delve into detailed study of processes requiring quantitative information about parameters and relationships between parameters. Precision is critical for monitoring the types of climate change expected under global warming. Detection in and of itself requires precision but not necessarily accuracy. Understanding the nature of such change and attributing cause to such change requires both precision and accuracy to quantify changes to processes. The dilemma confronting the climate community is that only the operational satellite observing systems have the long-term commitments necessary to sustain observations over time-scales of interest to climate. These operational systems generally do not have a precision adequate for climate trend analysis; nor do they possess sufficient accuracy to quantify the typically small changes to processes responsible for the observed changes. In partial response to these shortcomings, concepts for dedicated satellite monitoring missions have been proposed (e.g. Goody et al., 2003).

Factors affecting both accuracy and precision of satellite observing systems are many and complex and include: (a) the platform, (b) the space–time representativeness, (c) the inversion process responsible for extracting information, (d) sensor properties and calibration.

Platforms

The most elementary factor that governs the overall statistical properties of the measurements is the type of orbit the satellite is placed into. Earth orbiting satellites are typically launched in one of two orbits – the Low Earth Orbits of polar orbiting satellites (LEO) and the geostationary orbit (GEO). LEO satellites vary most in size from microsatellites carrying payloads of just a few kilograms to large observatories with payloads exceeding thousands of kilograms. LEO also typically carry one of two types of payloads – experimental one-of-a-kind sensors or prototype operational sensors and operational payloads designed with versions of the same instrument flown over many years on multiple satellites. Examples of the experimental class of missions are NASA's Earth Observing System missions (EOS) and Earth System Science Pathfinder (ESSP) missions, Japan's ADEOS missions and ESA's Earth missions.[2] The basis for these missions is typically rationalized in terms of fulfilling climate observational needs, usually in the context of studying climate processes over finite periods of time. Further discussion of experimental satellites, including a historical perspective, can be found in Kidder and VonderHaar (1995), among other references.

Operational satellite programs of the USA include satellites in both LEO and GEO, whereas those of Europe and Japan, at present, are based solely on GEO, although the launch of the ESA-EUMETSAT Polar System (MetOp, http://www.esa.int/export/esaME/index.html) is imminent. The convergence of the US military and civil polar satellite programs has occurred forming the NPOESS program (http://www.ipo.noaa.gov/index2.html). The first launch of NPOESS is expected around 2008 and, when combined with MetOp, an internationally coordinated system of polar satellites will be established.

The GEO platforms are larger (thousands of kilograms) than most polar orbiters, and their ability to observe the same region of Earth almost continuously makes them ideal for observing the evolution of severe weather and other short-term phenomena. To date, the payloads of geostationary satellites consist of rudimentary imaging and sounding radiometers, although more advanced sensors are being planned beginning with the new imager of Eumetsat's MSG1.

Representativeness

The time–space sampling properties inherent to different satellites are fundamentally dictated by its orbit and the type and number of satellites that compose the observing system.

Geostationary satellites provide a more synoptic-like method of observing Earth; that is sensors on GEO effectively produce a snapshot of information observing large regions of Earth at the same time. Unlike the LEO polar orbiters, they do not provide information at higher latitudes. Despite the sampling limitation and the lack of advanced sensors, GEO data continue to play an important role in climate research. The International Satellite Cloud Climatology Project (ISCCP, Rossow and Schiffer, 1991; http://isccp.giss.nasa.gov), for example, collects and analyzes observations from the imager radiometers flown on operational GEO. ISCCP pieces together these GEO observations, augmenting them with data from polar orbiting satellites to produce a global distribution of cloud properties. ISCCP is one of the first climate projects to integrate multiple satellite observations of this type.

Polar orbiting satellites sample the Earth in a very different manner. These satellites observe Earth "asynoptically" such that different portions of Earth are observed at different times. Global fields are then created in such a way that the space–time properties of quantities are mixed. This creates problems when observing processes that vary irregularly on small space and short time scales, or when observing processes that possess diurnal variability. The latter model of variability is particularly troublesome given that key elements of the Earth's hydrological cycle possess pronounced diurnal variability. Undersampling of such behavior introduces aliasing on the longer time-scales of primary interest to climate studies.

[2] More information about NASA's EOS, ESA's Earth observations and Japan's ADEOS programs can be found under the respective home pages of each program, http://eospso.gsfc.nasa.gov/, http://www.eorc.jaxa.jp/ADEOS/, http://www.esa.int/esaSA/earth.html

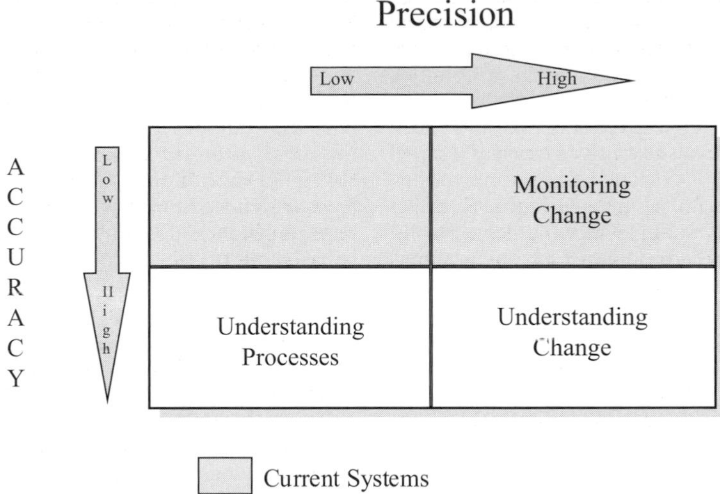

Figure S1 The climate measurement problem – understanding climate processes requires accuracy, monitoring climate change requires high precision and detection, and understanding climate change requires both.

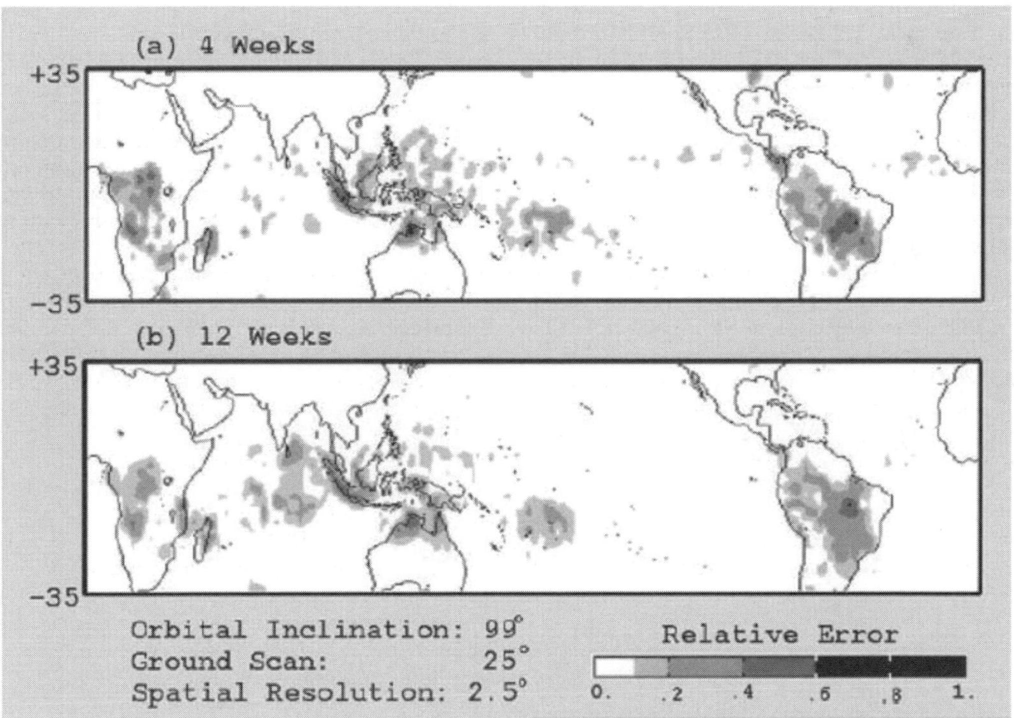

Figure S2 Illustration of the effects of asynoptic sampling on a proxy for convective precipitation. Shown is the sampling error in monthly and seasonal mean precipitation when observations from only a single polar orbiting satellite are used to sample precipitation. Large biases in excess of 20% arise because of aliasing of errors associated with the under-sampled diurnal cycle on the longer-term means (Salby and Callaghan, 1996).

The inversion process

An essential part of any remote sensing observing system is that part of the system that takes measurements and inverts them into geophysical parameters, such as temperature, moisture, and cloudiness, among others. This inversion process is complex and almost always requires other sources of data. One example of such data is the *a-priori* information often needed to constrain the inversion process. *A-priori* data typically derive from some climatological database which inescapably enters the final product. Therefore the accuracy and stability of the final product depend in part on the quality of the direct measurements and in part on *a-priori* and other data used in the inversion process. The observing system is thus more than the instrument on a satellite, being now defined as a complex system involving the satellite and its platform properties, the sensors on the satellite, the inversion process and all related data required for this process. Consequently, assessing the real quality of the observing system is difficult. An example of how these other factors associated with the inversion process affect climatological distributions of water vapor is described in Engelen and Stephens (1999). The problems faced in interpreting satellite data without proper understanding of the inversion process are also highlighted below in the discussion of trends in microwave brightness radiances.

Calibration

The suitability of space-based data for studying climate ultimately hinges on how well the sensor is calibrated. The purpose of calibration is to convert raw instrument output to appropriate physical units using known standard references in the translation. Calibration therefore establishes the absolute accuracy of the measurement. While some sensors are calibrated prior to launch, in-orbit instrument changes require that this calibration be carried out routinely on orbit. Such a requirement adds significant costs to instrument design, so often sensors are only partially calibrated in orbit, calibrated only prior to launch or not at all. Indirect methods of calibration, referred to as vicarious calibration, are needed. The approach to vicarious calibration varies from application to application and often involves comparison of a measurement to some model simulation of the measurement. We learn from the examples below that vicarious calibration is generally less accurate than direct calibration, and that it is often difficult to assign errors to the process. Despite these shortcomings, vicarious calibration, when conducted routinely and with care, serves a vital role in the use of satellite observations in climate research.

Calibration of the AVHRR

Satellite sensors degrade in orbit, initially because of the outgassing from the radiometer components and subsequently because of continuous exposure to the space environment. Thus it is generally insufficient merely to calibrate sensors before launch. Issues associated with instrument degradation are illustrated in the example of channel 1 of the Advanced Very High Resolution Radiometer (AVHRR, http://noaasis.noaa.gov/ NOAASIS/ml/avhrr.html). This channel, uncalibrated on orbit, is used with channel 2 to monitor land cover and vegetation. Post-launch calibration of these channels has been carried out using the vicarious calibration approach. One approach builds on the method developed for ISCCP and is based on: (a) the use of terrestrial targets that are radiometrically stable in time, (b) model simulations and (c) matching the performance of the AVHRR sensor to a known, calibrated sensor. The terrestrial target typically used to calibrate the visible channel 1 is a portion of the Libyan Desert which is taken to be stable. The absolute calibration scale is based on congruent aircraft/satellite radiance measurements over White Sands, New Mexico. Figure S3 shows the effects of how this vicarious calibration removes the significant and spurious downward trends of channel 1 solar radiances. The absolute accuracy of this vicarious calibration procedure is between 5% and 10%. Measure of precision, on the other hand, suggested by the rms deviation of the albedo, is significantly smaller than absolute accuracy.

WV channel calibration on Meteosat

Routine onboard calibration of operational-like sensors is carried out more as an exception than as a rule. Furthermore, the inflated size of the optics of imagers on GEO and the associated costs of calibration usually prevents direct calibration in orbit. Vicarious calibration has become a common procedure for calibrating operational GEO sensors.[3] An example of this type of calibration for the Meteosat (http://www.eumetsat.de) water vapor channel is described at length by van de Berg et al. (1995). The procedure developed is to match in time satellite radiances to radiosonde data at selected locations where the quality of these data is maintained. A radiative transfer model converts these sounding data into synthetic radiances which are

[3] Although the IR sensors of the GEO satellites are operationally calibrated in orbit, vicarious calibration is required to correct for unwanted effects.

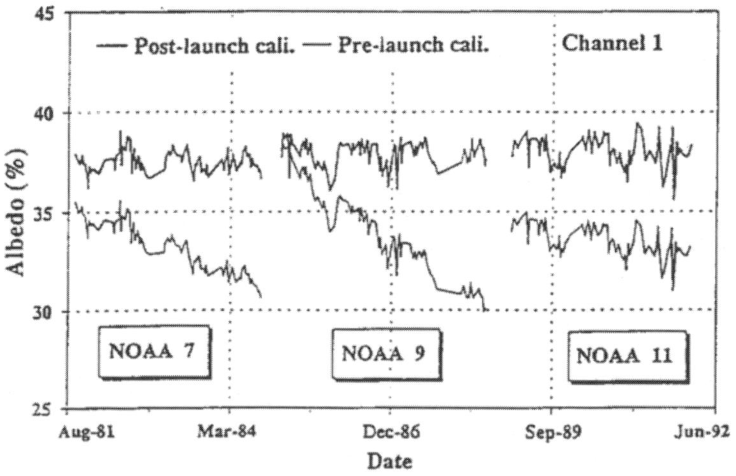

Figure S3 Albedo derived from AVHRR channel 1 over the southeastern portion of the Libyan desert (Rao, 1997).

then used to convert raw radiance counts to radiance. The absolute accuracy of this approach is difficult to estimate and requires some calibration of the model. The precision might be deduced from the variability of the calibration which is claimed to be at the 2% level or even slightly less (Schmetz, 1989; van de Berg et al., 1995).

The TOMS experience

The two examples described above refer to calibration of sensors that are either otherwise uncalibrated, and include no means of onboard calibration or whose calibration for one reason or another is incomplete. As noted in the above example, the intrinsic accuracy of this vicarious calibration is one issue, but the precision of the calibration is another. Unfortunately, the precision of the AVHRR calibration cannot be addressed when calibration scales are referenced to a single (or even limited number) of matched aircraft flights. High precision generally requires accurate description of an instrument time-dependent characteristics. Characterizing instrument stability must therefore be a major consideration in the design of monitoring systems.

Experience has shown that routine vicarious calibration has to be an essential ingredient of any monitoring system. A clear illustration of its value was provided in the example of detection of decadal ozone climate trends using TOMS data (Hilsenrath et al., 1997). The gradual degradation of optical components (chiefly the solar diffuser that served as part of the on-orbit calibration procedure) prompted the development of an *ad-hoc* vicarious calibration approach to account for instrument drifts. Like the case of water vapor radiances, the TOMS approach relied heavily on a surface observing network of Dobson spectrometers.

Constructing climate records from multiple satellites

The ISCCP, which began in 1983, was the first internationally coordinated project for integrating satellite radiance data from multiple platforms. This program involved routine collection of operational polar and GEO satellite data merged to produce homogeneous global radiances and climatological cloud information from these radiances. One of the key lessons learned from this activity concerned the difficulty associated with the maintenance of an observing network of multiple satellites for an extended period. The ISCCP notionally planned to make use of data from five geostationary satellites as well as data from a single polar orbiter in an effort to fill in the data voids over polar regions. Actual availability of data, however, fell below this hypothetical combination and the amount of data varied over the course of the project. The extent to which this has affected trends in ISCCP data is under investigation. The long-term trends evident in ISCCP data are not reproduced in other satellite cloud climatologies, such as the one PATMOS produced, based on AVHRR (Jacobowitz et al., 2003).

In addition to PATMOS, there are other examples of studies that combine data obtained from multiple versions of the same instrument flown on a series of NOAA polar orbiting spacecraft. Data from the TIROS Operational Vertical Sounder (TOVS, www2.ncdc.noaa.gov/docs/podug/html/c4/sec4-0.htm) accumulated over more than 20 years provide an important climate resource. The issues associated with combining these data concern the removal of sampling biases from the data (representativeness errors) and the need to correct for different response characteristics of the different versions of the same instrument. The NOAA spacecraft carrying the TOVS are placed in a nominal sun-synchronous orbit, which means the time at which any particular location is fixed. Unfortunately these orbits drift at a disturbing rate (Figure S4), creating time-dependent representativeness errors and thus spurious trends in climate parameters extracted from the measurements.

TOVS channel 12 water vapor radiances

Bates et al. (1996) describe an empirical method to correct for sampling biases introduced by both satellite drift and lack of resolution of the diurnal cycle in a 20-year record of the TOVS channel 12 water vapor radiances. They also examine biases that arise through differences between the particular versions of TOVS flown on the different satellites. In a crude sense the satellite-to-satellite differences can be thought of as an indication of the precision of the data. For monthly mean data this difference is about 0.25 K, except for the NOAA-9 satellite which differs from the other satellites by 0.4–0.45 K. Assigning an absolute accuracy to the brightness temperature data is more difficult, requiring post-launch vicarious calibration procedures.

MSU temperature trends

Microwave radiance data, converted to brightness temperatures, obtained from the microwave sounding unit (MSU, see http://orbitnet.nesdis.noaa.gov/crad/st/amsuclimate/amsu.html) of TOVS have been interpreted as a direct measure of atmospheric temperature. The apparent lack of any trend in the time-series of these brightness temperatures have been much debated over the past decade since they appear to be contrary to arguments for global warming. MSU data have been combined by different investigators to produce extended time-series. The details of the temperature trends, however, depend on the nature of assumption made in correcting for instrument, sampling biases, and inter-satellite sensor differences. Two notable efforts are those at the university of Alabama Huntsville (UAH, e.g. Christy et al., 1995) and those by the Remote Sensing

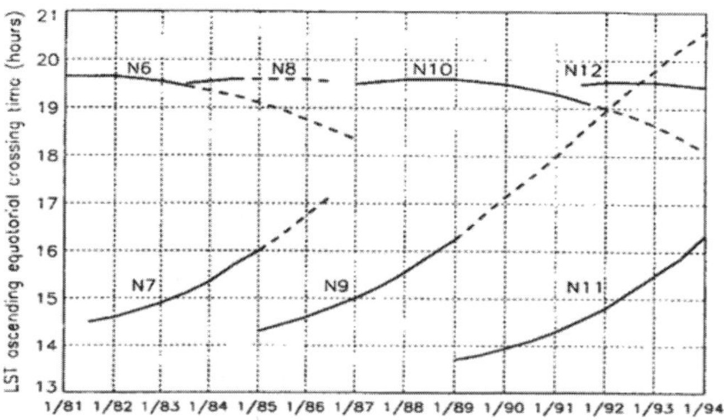

Figure S4 Ascending node equatorial crossing times for the NOAA series satellites up to 1994. The solid line is the period of operational coverage (from Bates et al., 1996).

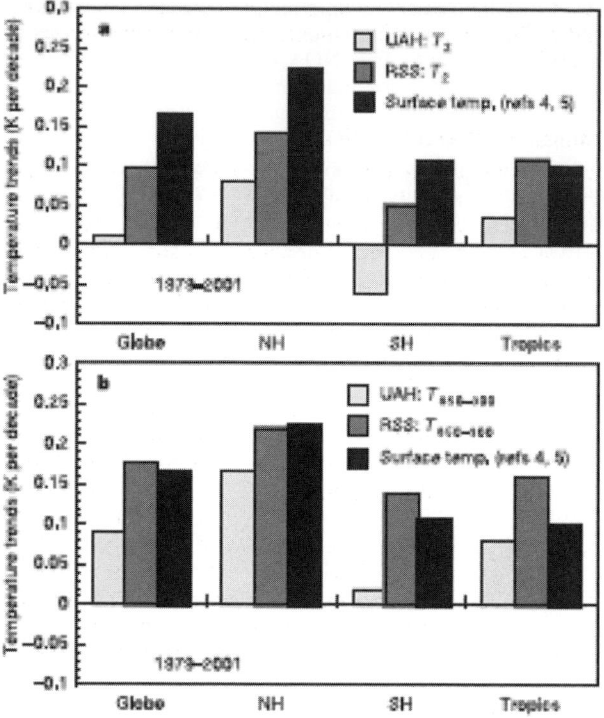

Figure S5 Trends in monthly mean temperature anomalies. (**a**) MSU channel 2 brightness temperature trend and (**b**), MSU-derived 850–300-hPa layer temperature trend for the globe, northern hemisphere (NH), southern hemisphere (SH) and tropics (30°N–30°S). The results using both UAH (version 5) and RSS data sets are shown (from Fu et al., 2004).

Systems (RSS, Wentz and Schabel, 2000), each producing different MSU temperature trends.

What tends to be lost in the debate on MSU trends is that most of the uncertainty and thus source of debate stems from the interpretation of the observations more so than the details of the MSU observations themselves. The inversion processes applied in studies that analyze these data are inappropriately primitive. The radiance data of channel 2, or simple combinations of channel data, are mostly used in trend analysis and are interpreted simply as a direct measure of the atmospheric temperature. The different channels of the MSU, however, measure emission from different portions of the atmospheric column. The problem is that this emission is smeared over almost the entire troposphere, including contributions from the surface and even portions of the stratosphere. Any trend or lack of trend in the brightness temperatures cannot be simply interpreted as a trend in actual temperature as changes to surface properties (soil moisture over land, wind over ocean and atmosphere properties, e.g. lapse rates, precipitation, among other factors) affect the interpretation of the data (e.g. Hansen et al., 1995; O'Brien, 2001). In a recent study Fu et al. (2004) develop a simple correction for the effects of changes in stratospheric temperatures on the MSU temperature trends. They difference between MSU channels 2 and 4 brightness temperatures which they empirically relate to the mean 850–300 hPa temperature.

Figure S5 shows the MSU channel 2 trends (T2, Figure S5a) and MSU-inferred T850-300 (Figure S5b) for the globe, northern hemisphere, southern hemisphere, and tropics (30°N–30°S) using both the UAH and RSS MSU datasets, as well as surface temperature trends based on *in-situ* observations. The trends of 0.01 K per decade (UAH) and 0.1 K per decade (RSS) for global mean are substantially smaller than the surface temperature trend of 0.17 K per decade. In the southern hemisphere the trend from UAH is actually negative. However, as shown in Figure S5b, the global trends of T850-300 are 0.09 K per decade (UAH) and 0.18 K per decade (RSS), which are about 0.08 K per decade larger than the corresponding channel 2 trends alone.

Graeme L. Stephens

Bibliography

Bates, J.J., Wu, X., and Jackson, D.L., 1996. Interannual variability of upper-tropospheric water vapor band brightness temperature. *Journal of Climate*, **9**: 427–438.

Christy, J.R., Spencer, R.W., and McNider, R.T., 1995. Reducing noise in the MSU daily lower-tropospheric global temperature dataset. *Journal of Climate*, **8**: 888–896.

Engelen, R., and Stephens, G.L., 1999. Characterization of water vapour from TOVS/HIRS and SSMT-2 Measurements. *Quarterly Journal of the Royal Meteorological Society*, **125**: 331–351.

Fu, Q., Johansen, C., Warren, S., and Siedel, D., 2004. Contribution of stratospheric cooling to satellite-inferred tropospheric temperature trends. *Nature*, **429**: 55–57.

Goody, R.J., Anderson, T., Karl, R., et al., 2003. Why monitor climate? *Bulletin of the American Meteorological Association*, **84**: 873–878.

Hansen, J., Wilson, H., Sato, M., Ruedy, R., Shah K., and Hansen, E., 1995. Satellite and surface temperature data at odds? *Climatic Change*, **30**: 103–117.

Hilsenrath, E., Bhartia, P.K., and Cebula, R., 1997. Calibration of BUV satellite ozone data – an example for detecting environmental trends. In Guenther, et al., eds., Workshop on Strategies for Calibration and Validation of Global Change Measurements, 10–12 May 1995. NASA reference publication 1397.

Jacobowitz, H., Stowe, L., Ohring, G., Heidinger, A., Knapp, K., and Nalli, N., 2003. The AVHRR Pathfinder atmosphere (PATMOS) climate data set. *Bulletin of the American Meteorological Association*, **84**: 785–793.

Kidder, S., and VonderHaar, T., 1995. *Satellite Meteorology: an introduction*. San Diego, CA: Academic Press.

O'Brien, D., 2001. For the dependence of MSU brightness temperature on lapse rate. *Journal of Quant Spectros. Rad. Transfer*, **69**: 159–170.

Piexoto, J.P., and Oort, A.H., 1992. *Physics of Climate*, New York: American Institute of Physics.

Rao, C.R.N., 1997. The NOAA/NASA AVHRR calibration activity. In Guenther, R. ed., Workshop on Strategies for Calibration and Validation of Global Change Measurements, 10–12 May 1995. NASA reference publication 1397.

Rossow, W.C., and Schiffer, R., 1991. ISCCP Data Products. *Bulletin of the American Meteorological Society*, **72**: 2–20.

Salby, M.L., and Callaghan, P., 1996. Sampling error in climate properties derived from satellite measurements: relationship to diurnal variability. *Journal of Climate*, **10**: 18–36.

Schmetz, J., 1989. Operational calibration of the Meteosat water vapor channel by calculated radiances. *Applied Optics*, **28**: 3030–3038.

Van de Berg, L.C.L., Schmetz, J., and Whitlock, J., 1995. On the calibration of the Meteosat water vapor channel. *Journal of Geophysics Research*, **100**: 21069–21076.

Wentz, F., and Schabel, M., 2000. Precise climate monitoring using complimentary satellite data sets. *Nature*, **403**: 414–416.

Cross-references

Electromagnetic Radiation
Radar, Climatic Applications
Units and Conversions

SAVANNA CLIMATE

The savanna climate is named for the mixed grassland and tree vegetation which is characteristic of the climate. On the margins with the rainforest, trees dominate the landscape but open areas of grass and shrubs are interspersed. In the drier areas grasses predominate with some trees scattered throughout. The distinguishing characteristics of the climate are the year-round high temperatures and pronounced wet and dry seasons. The climate is found marginal to the tropical rainforest climate and in most cases poleward from the rainforest. The climate occupies much of the area from the boundary with the rainforest climate to the Tropics of Cancer and Capricorn. Large areas of the tropics experience this climate. Most of Africa south of the Sahara Desert has a savanna climate, contrary to secular perception of Africa as a tropical jungle. Similarly, much of Southeast Asia and northern Australia has this climate.

The location of the savanna climate with respect to the general circulation is such that the area is subject to the seasonal migration of the equatorial convergence zone (ITCZ) and the subtropical high. The migration of the Hadley cell in the tropics is both pronounced and unpredictable. The ITCZ found between the northern and southern cells migrates as much as 40° in some areas. The migration is influenced by the extent of the landmasses in each hemisphere. Since the landmasses are more extensive north of the equator, the ITCZ tends to migrate farther in the northern hemisphere than in the southern. Therefore the savanna climate is generally found at higher latitudes in the northern hemisphere. In the southern hemisphere the ITCZ and the accompanying precipitation does not go much beyond 15°S. In the northern hemisphere the system moves as far as 30°N.

Temperature characteristics

The temperature is controlled by latitude and the shift of the semipermanent pressure belts. The variation in the length of the day through the year is not too significant as the photoperiod varies from 11 to 13 hours. There is substantial variation in the angle of incidence. On the poleward margins the elevation of the sun varies as much as 45° from summer to winter solstice. Annual average temperature in this climate averages very nearly that of the rainforest climate, ranging from 18°C to 21°C. Monthly averages vary from high sun to low sun season, primarily as a result of the variation in insolation. The range in temperature from month to month averages between 2.8°C and 8.3°C. Sites on the margins of the rainforest have the least variation and those sites on the poleward margins the greatest variation. The greater variation on the poleward margins is due to the incursion of dry continental air which allows greater radiation heating and cooling. There is no winter as we know it in terms of temperature. The frost-free season is continuous, although temperatures as low as the teens have been recorded on the dry margins of the climatic region.

During the rainy season diurnal temperatures vary little as a result of the high humidity and even hours of daylight and darkness. In the dry season there is a greater exchange of radiant energy between the surface and atmosphere and so the diurnal range is greater. During this part of the year the diurnal range averages between 11°C and 17°C. As in other tropical climates the change in temperature from day to night is greater than the change from month to month.

Moisture characteristics

The most significant feature of the savanna climate is the pronounced seasonal moisture pattern. The wet and dry seasons are the result of the migration of the ITCZ. In the northern

Figure S6 Savanna with mixed woodland and grass as is found on the wetter margins of the climate, Chobe National Park, Simbabwe.

Figure S7 Dry savanna dominated by grasses, Botswanna.

hemisphere the ITCZ moves north of the equator in the months of May through September. The rainy season north of the equator occurs during these months. Atmospheric humidity is high when the ITCZ brings in moisture during the high sun season. During the low sun season dry air from the subtropical high replaces the more humid air. The change in humidity is visible in terms of cloud cover. When tropical maritime air is present clouds exist much of the time and clear days are infrequent. Low and middle forms of cumulus clouds are most prevalent. At this time of year dewpoints will range from 16°C to 22°C. In the dry season clear days are common and dewpoints range from 0°C to 7°C.

Average annual precipitation totals vary a great deal, and are not used to define the region. On the dry margins annual totals may diminish to as little at 25 cm. On the wet margins of the rainforest, especially where orographic lifting occurs, totals reach over 10 m. It is probable that the greatest annual average rainfalls occur in this climate. Variability from year to year is very high. The reason is that the total precipitation in a given year is subject to the extent of the migration of the ITCZ and the length of time the migration persists. The farther a site is from the mean location of the ITCZ the lower the mean annual precipitation tends to be, and the greater the relative variability. There are exceptions to the generalization that totals decrease with distance from the ITCZ, such as in the Assam Hills of India where some of the highest totals on the planet occur.

Seasonality of weather increases poleward, as might be expected. In a traverse away from the equator, the low sun precipitation begins to decrease first. The high sun precipitation remains as high as in the rainforest climate. Midway between the ITCZ and the tropical desert the low sun precipitation drops to near zero. Continuing poleward, the high sun precipitation declines until it is no longer sufficient to support woodland growth. The tropical steppe emerges at this point.

Grasses dwindle as high sun precipitation decreases. In the zone where precipitation becomes too little for grasses to grow, true desert appears. The rainy season is concurrent with the high sun and presence of the ITCZ overhead. The heating of the lower atmosphere favors the formation of thunderstorms, with the greatest frequency of thunderstorms in the afternoon hours. Occasional incursions of low-pressure systems bring more widespread precipitation. Near coastal areas remnants of tropical cyclones can produce extensive precipitation. These storms do not penetrate very far inland. The dry season is a product of rather stable desert air originating from the subsidence and divergence in the semipermanent subtropical high-pressure zone. Extremely dry and dusty winds frequently move into and through these climatic areas. These go by a variety of names, but the Harmattan is perhaps the best known.

Seasonal variation

This climate has the most pronounced seasonality of precipitation of any of the climatic types. An example of the extremes that are reached is found in Asia. Rangoon Burma has a winter 3-month average precipitation of only 2.5 cm. In contrast there is a 3-month summer average of 190 cm. The seasonality increases northward along the coast of Burma where the seasonal averages are 3.8 cm and 432 cm. The maximum in Asia is reached at Cherrapunji, India, where in 2 winter months they receive an average of 2.5 cm and in 2 summer months 536 cm. Cherrapunji has recorded a 5-day total of 381 cm of rain, a 1-month total of 930 cm and a yearly total of 2647 cm.

Migration of the ITCZ north and south of the equator is highly variable from year to year. As a result the amount of precipitation received also varies a great deal. In addition to the marked seasonal variation in precipitation so there is great year-to-year variation in rainfall. In some years an area may receive much-above-average precipitation, and the next year practically none. Most of the inhabitants of this climate region are agriculturalists. The variability of rainfall, whether too much or too little, often causes serious social, political, and economic problems. When too much rain falls, crops are often washed out, or are subject to mold and mildew. Too little rain frequently results in drought.

Severe drought in turn results in famine. All regions of savanna climate have experienced drought and famine. Perhaps the greatest human struggles with drought-induced famine in the savanna climate have occurred in Asia and Africa.

John J. Hidore

Bibliography

Hastenrath, S., 1991. *Climate and Circulation of the Tropics*. Dordrecht: Kluwer.
Hidore, J., 1996. *Global Environmental Change: its nature and impact*. Upper Saddle River, NJ: Prentice-Hall.
Oliver, J., and Hidore, J., 1993. *Climatology: an atmospheric science*. Upper Saddle River, NJ: Prentice-Hall.
Yevjevich, V., da Cunha, L., and Vlachos, E., 1983. *Coping with Droughts*. Littleton, CO: Water Resource Publications.

Cross-references

Africa: Climate of
Doldrums
Intertropical Convergence Zone
South America, Climate of
Tropical and Equatorial Climates

SCALES OF CLIMATE

The types of climate that prevail over a region are determined by quantifying the prevailing weather elements whose periods of observation extend over a length of time. The longer the observation period the better the representation of the existing climate. The observed weather phenomena range over a broad spectrum of scales, but in reality none of the weather features is discrete, being part of a continuum. As a result, attempts to divide weather phenomena into distinct scales have resulted in disagreement with regard to the scale limits. The characteristics of the climatic spectrum depend on the nature of the exchange of heat, moisture, and momentum between the Earth's surface and the atmosphere. The physical laws that control the exchange processes between the Earth's surface and the atmosphere are the same everywhere. However, as with the problem of meteorological scale, so there is disagreement on areal scale in climatology. To illustrate this, two common approaches are outlined here. The first is a three-part division, the second includes a further category often identified in regional climatic studies.

The basic scales

The subdivision of the climatic phenomena provides a basis for dividing the climatology into micro-, meso-, and macroclimatology, and these features are revealed by having a network of observing stations spaced at intervals of a few square kilometers, about 1 km, and 20 km or so respectively (Munn, 1966). The schemes of classification use horizontal and vertical characteristics as criteria as shown in Table S1 and Figure S8.

Microclimatology

Geiger (1966) emphasized the importance of the air layer below the agreed level of about 2 m of the atmosphere dealing with the "climate near the ground". In this level the horizontal extent of the circulation could be of the order of tens of kilometers or less. The magnitude of all the meteorological elements in this air layer is subject to vertical and horizontal changes. The vertical variations are due to the nearness of the ground, whereas the horizontal differences are due to changes in the nature and moisture of the soil, differences in surface slopes, and by the type and height of vegetation cover on it. Terms such as position climate, miniature climate, and spot-climate also can be adopted for microclimate but are somewhat ambiguous. Huschke (1959) viewed the microclimate as a fine climatic structure of the air space which extends from the Earth's surface to a height where the effects of the immediate character of the underlying surface cannot be distinguished from the general local climate. Munn (1966) was of the opinion that microclimatology is restricted in depth to the lowest hundred meters of the atmosphere, the surface boundary layer.

Mesoclimatology

In mesoclimatology the configuration of the ground, type of soil, and its vegetation cover are considered as features of the locality, which are subject to only slow changes over time, and determine the climate that prevails in a particular place and can be called local climate. The horizontal extent of this type of phenomena (mesoscale feature) is of the order of hundreds of kilometers. The concept of topoclimate proposed by Thornthwaite (1953) serves as an intermediate position between macro- and micro-climates; but Scaetta (1935) argued that the word topoclimate should be changed to mesoclimate, and Utaaker (1974) felt that the term topoclimate is more appropriate for local or mesoclimates since the topography of the land is the main causal factor responsible for local climatic peculiarities. In the opinion of Konstantinov et al. (1974), the formation of microclimate and local climate occurs under the influence of the underlying surface. Barrett (1974, p. 9) described the mesoscale phenomena as "local embroideries upon a larger, more generalized climatic picture".

Table S1 Scales of meteorological motion systems

	Approximate characteristic dimensions			
Motion system	Horizontal scale (km)	Vertical scale (km)	Time scale (n)	Total energy[a]
Macroscale				
Planetary waves	5×10^3	10	2×10^2 to 4×10^2	—
Synoptic perturbations	5×10^2 to 2×10^3	10	10^2	Average depression 10^{-3}
Mesoscale phenomena	1 to 10^2	1 to 10	1 to 10	Average thunderton 10^{-8}
Microscale phenomena	$<10^{-1}$	$<10^{-2}$	10^{-2} to 10^{-1}	Average windgust 10^{-17}

[a] Base 1 = daily solar energy intercepted by the Earth.
Source: From Barry (1970).

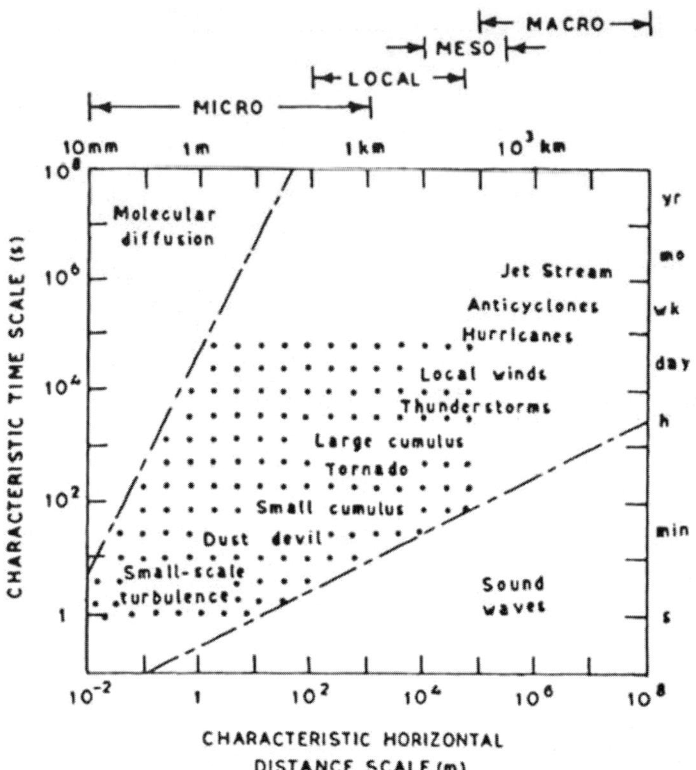

Figure S8 Time and space of various atmospheric phenomena. The shaded area represents the characteristic domain of boundary layer features (after Smagorinsky, 1974).

Table S2 Scale of climate and its corresponding meteorological phenomena

Climate	Horizontal distribution (m)	Vertical distribution (m)	Example of climatic phenomena	Lifetime of corresponding meteorological phenomena(s)
Microclimate	10^{-2}–10^{2}	10^{-2}–10^{1}	Climate of greenhouse	10^{1}–10^{1}
Local climate	10^{2}–10^{4}	10^{-1}–10^{1}	Thermal belt of slope	10^{1}–10^{4}
Mesoclimate	10^{3}–(2×10^{5})	10^{0}–(6×10^{3})	Climate of basin	10^{4}–10^{5}
Macroclimate	(2×10^{5})–(5×10^{7})	10^{0}–10^{5}	Climate zone, monsoon region	10^{5}–10^{6}

Source: From Yoshino (1975), p. 3.

Macroclimatology

This is concerned with broad features of the climates of substantial parts of the globe or most of a continent that are governed by the large-scale atmospheric circulation systems. It involves consideration of mean weather belts on a global basis. Large-scale disturbances of the order of thousands of kilometers, including the moving cyclones and anticyclones, define macroclimatic features of a continent. Regional macroclimatology deals not only with the variations in the amount, intensity, and seasonal distribution of the climatic elements but also the general tendency of synoptic weather features. The high relief partly produces changes in the geographical distribution of a meteorological phenomenon.

Scale and local climates

A knowledge of the interrelationships between the climatic patterns and Earth surface, together with the influence of major atmospheric systems on even lower levels of the atmosphere, is of vital importance in understanding the nature of the macroclimatology of a continent.

In a review of climatic scale, Yoshino (1975) suggests that the term local climate is intermediate between microscales and

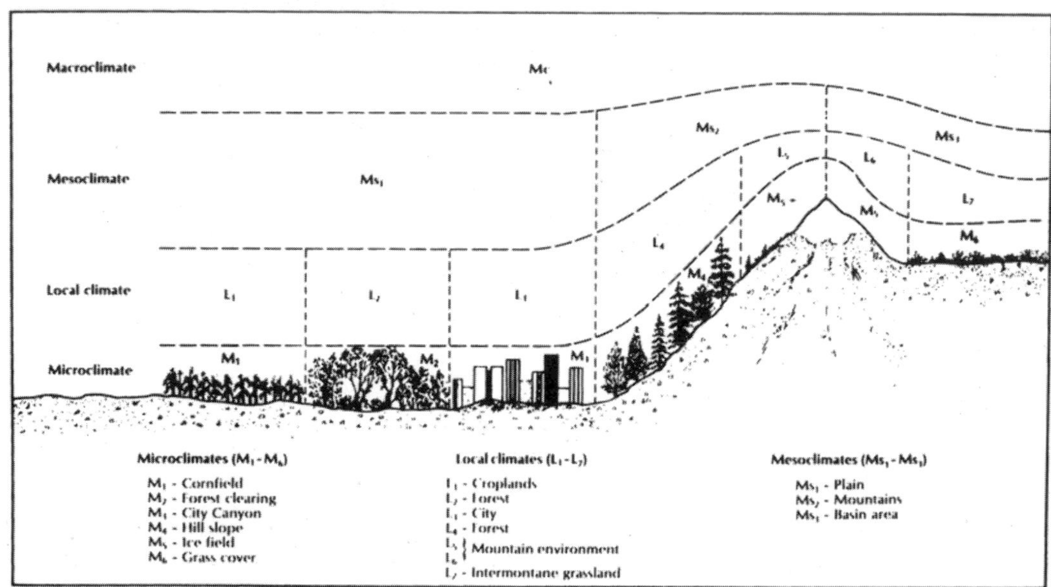

Figure S9 Area scales of climatic investigation.

mesoscales (Table S2). He suggests that the determining climatic factors in local climate are mesorelief, vegetation masses, etc. The characteristics of local climate are manifested in a layer of air observed at places tens and even hundreds of meters high. Further, mesoclimatic phenomena have been defined as those limited to the size of objects representing single independent units in geographical taxonomy, so that they are independent types of local climate. In the USSR there are two different views on scale; the first view is that local climate is a subclass between microclimate and macroclimate, and the second that local climate is nothing but a subclass of a microclimate in a broader sense in which local climate and microclimate are combined.

A graphic representation of the Yoshino scales is given in Figure S9.

A.A.L.N. Sarma

Bibliography

Barrett, E.C., 1974. *Climatology from Satellites*. London: Methuen.

Barry, R.C., 1970. A framework for climatological research with particular reference to scale concepts, *Institute of British Geographers Transactions*, **49**: 61–70.

Geer, I.W. (ed.), 1996. *Glossary of Weather and Climate*, Boston, MA: American Meteorological Society.

Geiger, R., 1966. *The Climate Near the Ground.* Cambridge, MA: Harvard University Press.

Huschke, R.E., 1959. *Glossary of Meteorology*. Boston: American Meteorological Society.

Konstantinov, A.R., Sakaly, L.I., and Diagot, L.S., 1974. The significance of the components of heat and water balances in the formation of micro and local climate. In *Proceedings of Physical and Dynamic Climatology*, WMO-no. 247. Geneva: World Meteorological Organization.

Munn, R.E., 1966. *Descriptive Micrometeorology*. New York: Academic Press.

Oliver, J.E., and Hidore, J.J., 2002. *Climatology: an atmospheric science*. Upper Saddle River, NJ: Prentice-Hall.

Scaetta, H., 1935. Terminologie climatique, bioclimatique et microclimatique. *La Met*, **11**: 342–347.

Smagorinsky, J., 1974. Global atmospheric modeling and numerical simulation of climate. In Hess, W.N., ed., *Weather and Climate Modification*. New York: Wiley.

Thornthwaite, C.W., 1953. *Topoclimatology.* Baltimore, MD: Johns Hopkins University Press.

Utaaker, K., 1974. Physics of local and microclimate. In *Proceedings of Physical and Dynamic Climatology*, WMO no. 347. Geneva: World Meteorological Organization.

Yoshino, M., 1975. *Climate in a Small Area.* Tokyo: University of Tokyo Press.

Cross-references

Climate Classification
Climatology
Local Climatology
Microclimatology

SEA ICE AND CLIMATE

Sea ice forms in those portions of the high-latitude oceans where the water temperature falls below the freezing point (approximately 271.2 K). Once it has formed, a sea ice cover generally fractures into floes with typical dimensions of 10–1000 m. The fracturing is a response to stresses imposed by the oceans and the winds, which also cause the ice to drift at speeds of up to 10 km day^{-1}.

In the Antarctic, the area covered by sea ice varies seasonally from 2.5 × 10^6 km^2 to 2.0 × 10^6 km^2. Only slightly more than 10% of the ice survives the summer melt. The average thickness is 1.5 m, and recent satellite observations indicate that typical concentrations are 60–85%. Because the ice is unconstrained by land at its equatorward boundary, it is free to interact with the open ocean. In contrast, much of the Arctic ice cover is constrained by land areas. The seasonal range is much less ($7.0 \times 10^6\,km^2$ to $14.1 \times 10^6\,km^2$) than in the Antarctic, and much of the ice has survived at least one summer melt season. The mean thickness of Arctic sea ice is 3–4 m, and concentrations are typically 90–100%.

Thus, there are fundamental differences between sea ice in the north polar and south polar regions. Obviously, the seasonal maximum is reversed because of hemispheric location. In the north polar regions the sea ice area doubles and ice extends to middle latitudes while in the south sea ice enlarges to four or five times the summer area but remains in polar regions. Southern hemisphere sea ice is restricted to the high latitudes by the Antarctic Circumpolar Current. This ocean current circles the pole, creating an even latitudinal distribution of heat in the southern ocean. The north polar area sea ice appears to be decreasing more rapidly than that of the south polar area although it is an older, thicker and more persistent covering.

Physical characteristics of sea ice

Expectations concerning a climatic role of sea ice are based on the profound surface changes associated with the formation or disappearance of an ice cover. First, the formation of sea ice increases the surface albedo (reflectivity of solar radiation) from 0.10 to 0.15 over open water to approximately 0.60 in the case of snow-free ice and values of 0.70–0.90 are typical of snow-covered ice. Second, the low thermal conductivity of sea ice results in an effective reduction of the vertical fluxes of moisture and heat at the surface, thereby insulating the atmosphere from the ocean. Third, because of the latent heat associated with melting and freezing, sea ice acts as a thermal reservoir that delays the normal seasonal temperature cycle. Finally, sea ice alters the salinity distribution and hence the density stratification of the ocean, both by the expulsion of brine during freezing and by the large-scale advective transport of low-salinity ice. Results of sea ice model studies show that the wintertime heat input to the atmosphere is one to two orders of magnitude larger over thin ice ($\leq$0.5 m) than over perennial ice.

Air–sea-ice interactions

Local effects

There is abundant evidence that fluctuations of sea ice extent are associated with *local* anomalies of air temperature. For example, annual air temperatures at the Antarctic island station, Orcadas, are highly correlated ($r = -0.77$) with the duration of ice cover at nearby Scotia Bay (Schwerdtfeger, 1970). Ice fluctuations at a single longitude generally are not, however, representative of those at other longitudes. There are also statistically significant correlations between ice extent and air temperatures observed several months later at stations near the ice edge. Sea ice may therefore be of modest predictive value at some locations.

Regional and interannual effects

In both the Arctic and the Antarctic, year-to-year fluctuations of ice extent have been shown to be associated with shifts of cyclone trajectories and frequencies. The data suggest that cyclone development is aided by anomalously large areas of open water within the ice pack. However, the associations between storm tracks and ice fluctuations are not indicative of simple causal relationships. Feedbacks undoubtedly complicate the directionality, and other factors such as remote atmospheric forcing and oceanic processes are most likely involved. In fact, statistical studies of ice–atmosphere associations on the monthly and interannual time-scales indicate that the associations are often manifestations of atmospheric effects on the ice, i.e. equatorward winds and below-normal temperatures increase the areal extent of sea ice.

The effects of sea ice on the atmosphere can be isolated in controlled experiments with three-dimensional mathematical models of the atmosphere. One such experiment, based on parallel simulations in which the prescribed Arctic ice extremes differed by approximately 10° latitude, produced statistically significant differences of atmospheric sea level pressure (up to 8 mb), 700 mb geopotential height (up to 100 gpm), and tropospheric temperature (up to 8°C at low levels). Although the largest differences occurred in the general vicinity of the ice margin, substantial effects were found over Europe and the subtropical Atlantic and Pacific. The model results therefore suggest that a general advance or retreat of the ice edge by 500–1000 km could have a significant atmospheric impact over large portions of the northern hemisphere.

Long-term changes

Satellite detection of sea ice

Polar latitudes have historically made sea ice a difficult subject for scientific study. Prior to the 1970s the extent and variability of sea ice distribution was not well understood and it remained for successful remote sensing of the polar regions by satellites to provide meaningful answers.

Sea ice, prior to the advent of passive microwave imagery, was observed by satellites detecting visible and infrared radiation. Visible and infrared radiation from sea ice was often obstructed by cloud cover and visible signals were dependent upon daylight. This produced frequent and regular gaps in the data accumulation. Microwave detection is possible during all hours of the day, and is not hampered by non-precipitating clouds (Parkinson and Gloersen, 1993). There is also greater contrast between the microwave signals of ice and open water. These properties allow microwave radiometry to detect the large-scale characteristics of polar sea ice.

Changes in sea ice

Since the late 1970s there has been a well-documented decline in Arctic sea ice, and this may be related to global warming trends. However, some studies have shown that the decline is periodically reversed with several reversals occurring about the time of a maximum North Atlantic Oscillation (NAO) index. Parkinson (2000) reports that several studies show connections with the NAO and the spatially broader Arctic Oscillation. Accordingly, the possibility of close connections between the sea ice cover and major oscillatory patterns in the atmosphere and oceans requires more research to establish the real cause of changes in the areal coverage of sea ice.

Importance of sea ice changes

Two major changes occur during the transformation of seawater into ice: salt is rejected from the ice and heat is released. Salt rejection increases the salinity and density of surface waters, which can cause them to sink. Salt rejection during the formation of sea ice is a distillation process, allowing only a fraction of the initial salt content to become captured within the ice. Because evaporation and precipitation rates in the polar regions are low, the formation and redistribution of sea ice is the only effective way that freshwater is created and distributed in the polar regions (Aagard and Carmack, 1994).

As water changes its state from liquid to solid, the energy requirements necessary to maintain the new solid form are decreased. The extra energy is then released. Conversely, sea ice absorbs heat as it melts. The release and absorption of heat results in a moderation of temperature extremes. At present it is not possible to verify the effect of this on long-term climatic changes, yet it is on the longer time-scales that sea ice is most likely one of the key climatic variables.

John E. Walsh

Bibliography

Aagard, K., and Carmack, E.C., 1994. The Arctic Ocean and climate: a perspective. In *The Polar Oceans and their Role in Shaping the Global Environment*. Geophysical Monograph 85. American Geophysical Union, Washington DC.

Alfred Wegener Institute, 1997. Sea Ice-An Antarctic Habitat, http://www.awi-bremerhaven.de/Eistour/index-e.html

Antarctic Adventure, 1997. Sea Ice, Antarctic Adventure (Tasmania), http://www.antarctic.com.au/encyclopaedia/physical/SeaIce.html

Hays, J.D., Imbrie, J., and Shackleton, N.J., 1976. Variations of the Earth's orbit: pacemaker of the ice ages. *Science*, **194**: 1121–1132.

Herman, G.F., and Johnson, W.T., 1978. The sensitivity of the general circulation to arctic sea ice boundaries. *Monthly Weather Review*, **106**: 1649–1664.

Hibler, W.D., 1989. Arctic ice–ocean dynamics. In Herman, I., ed. *The Arctic Seas*. New York: Van Nostrand Reinhold.

Kukla, G., and Gavin, J., 1981. Summer ice and carbon dioxide. *Science*, **214**: 497–503.

Manabe, S., and Stouffer, R.J., 1980. Sensitivity of a global climate model to an increase of CO_2 concentration in the atmosphere. *Journal of Geophysics Research*, **85**: 5529–5554.

Parkinson, C.L., and Gloersen, P., 1993. Global sea ice coverage. In Gurney, R., Foster, J., and Parkinson, C., eds. *Atlas of Satellite Observations Related to Global Change*. Cambridge: Cambridge University Press, pp. 371–384.

Parkinson, C.L., and Kellogg, W.W., 1979. Arctic sea ice decay simulated for a CO_2 induced temperature rise, *Climate Change*, **2**: 149–162.

Parkinson, C.L., Comiso, J.C., Zwally, H.J., Gloersen, P., Cambell, W.J., and Cavalieri, D.J., 1987. *Arctic Sea Ice, 1973–1976*. NASA SP-489. National Aeronautics and Space Administration.

Parkinson, C., 2000. Recent trend reversals in Arctic Sea ice extents: Possible connections to the North Atlantic Oscillation. *Polar Geography*, **24**: 1–12.

Schwerdtfeger, W., 1970. The climate of the Antarctic. In Orvig, S., ed., *Climates of the Polar Regions*. Amsterdam: Elsevier, pp. 253–322.

Semtner, A.J., 1984. The climatic response of the Arctic Ocean to Soviet river diversions. *Climate Change*, **6**: 109–130.

Zwally, H.J., Parkinson, C.L., and Comiso, J., 1983. Variability of Antarctic sea ice and CO_2 change. *Science*, **220**: 1005–1012.

Cross-references

Antarctic Climates
Arctic Climates
Climatic Variation: Historical
Ocean–Atmosphere Interaction
Oscillations

SEA LEVEL RISE

Sea level is a fundamental climatological parameter that integrates diverse physical phenomena of the ocean–atmosphere system. The ocean surface adjusts to diurnal, seasonal, and interannual variations in atmospheric pressure, geostrophic currents, and coastal upwelling. Long-term changes in sea level derive from deformation of the Earth's crust as well as from eustatic and steric changes associated with climate change. At present, sea level is rising at around 1–2 mm/yr (Church et al., 2001). However, rates of sea level rise (SLR) are anticipated to increase by factors of 3–5 during this century, due to greenhouse gas-induced climate warming. Future SLR arouses concern because of potentially adverse consequences to coastal populations. These include land loss due to inundation and erosion, as well as increased frequency of coastal floods, with a rising water level. A knowledge of current sea level trends is essential in establishing an accurate baseline for future projections and for coastal zone management.

Causes of sea level change

Over the last few million years, changes in the extent of polar ice sheets and glaciers have periodically removed or added water to the world's oceans, affecting global sea level. *Glacial eustatic* changes in sea level are those caused by variations in ocean water volume from growth or melting of continental ice. At the peak of the last ice age, around 20 000 years ago, global sea level was 120 m below the present level (Bard et al., 1996). As the glaciers receded, sea level rose continuously for over 10 000 years, punctuated by at least three major meltwater pulses, around 13 800, 11 300, and 7800 years ago, when rates of SLR increased by factors up to 15–20 times present values (Cronin et al., 2002; Bard et al., 1996; Blanchon and Shaw, 1995). Although most ice melting ended by 6000 years ago, some minor melting of the Antarctic ice sheet may have persisted into the late Holocene. In many areas, sea levels fell since the mid-Holocene, caused by "siphoning", due to displacement of ocean water toward coastal regions subsiding at the periphery of former ice sheets and also weighted down by added meltwater (Peltier, 2001). However, the rate of SLR over the last few millennia has been rather low: $\leq$0.3 mm/yr (Table S3).

Although the ice contained in mountain glaciers represents only ~0.5 m of global sea level, they are highly sensitive to climate change. A worldwide compilation of data from glaciers suggests that the retreat of glaciers has added 0.2–0.4 mm/yr to observed SLR during the twentieth century (Table S3). Recent rates may be increasing. Shrinkage of Alaskan glaciers alone may have added the equivalent of 0.14 $\pm$ 0.04 mm/yr SLR, between the mid-1950s and the mid-1990s, and 0.27 $\pm$ 0.10 mm/yr SLR from the mid-1990s to 2000-2001 (Arendt et al., 2002).

The Antarctic and Greenland ice sheets contain enough water to elevate global sea level by ~70 m, so that even a small change in volume could create a significant effect. The behavior of ice sheets is determined by calculation of the mass budget (i.e. net losses via melting and discharge of icebergs vs snow accumulation), measurement of elevation changes over time by satellite altimeters, and modeling studies. During the 1990s, coastal glaciers in Greenland have been thinning rapidly, in contrast to gains at higher elevations, resulting in a net loss equivalent to a sea level rise of ~0.13 mm/yr. In Antarctica, the West Antarctic Ice Sheet, especially around the Amundsen Sea, has been shrinking, adding ~0.1 mm/yr to SLR (Rignot and Thomas, 2002). In contrast, large sectors of East Antarctica are either stable or gaining mass, although the overall mass balance is still poorly constrained.

As oceans warm, water density decreases and, even if no meltwater were added from ice sheets or glaciers, the volume would increase, causing a *steric* change. Thermal expansion is a major contributor to modern SLR, adding an estimated 0.5 ± 0.2 mm/yr during the second half of the twentieth century (Table S3). (Changes in ocean salinity have a negligible effect on global sea level at present.) Because of the high heat capacity of the ocean, a considerable time lag may exist before surface warming has penetrated to depths. Thus, global sea level will continue to rise long after greenhouse gas concentrations have stabilized.

Crustal deformation introduces a major source of geographic variability into tide-gauge records. Although deglaciation ended over 6000 years ago, sea level has continued to change due to the Earth's delayed viscoelastic response to the redistribution of mass on its surface (*glacial isostasy*). The effects of these glacial isostatic adjustments (GIA) extend for thousands of kilometers from the centers of the former ice sheets (Peltier, 2001). Differential vertical motions over shorter distances are associated with tectonic deformation and fault displacements, particularly at convergent plate margins. More localized crustal motions include land subsidence due to sediment loading by major rivers, such as the Mississippi or the Ganges–Brahmaputra, compaction of peat in coastal salt marshes, and overpumping of water, oil, or natural gas.

Changes in the amount of water stored in lakes and reservoirs, groundwater, and modification in land-use patterns that alter evapotranspiration and river runoff rates may also influence sea level. These human activities may have reduced river runoff by approximately 328 km^3/yr, equivalent to 0.9 ± 0.5 mm/yr of SLR (Gornitz, 2001). Alternatively, water sequestered behind dams may largely counteract other anthropogenic effects (Sahagian, 2000). These estimates carry significant uncertainties, due to limited historical data and simplifying assumptions. In the near future, satellite radar altimetry and gravity measurements will help quantify the global water mass budget and the extent of such anthropogenic transformations.

Recent sea level trends (last 100–200 years)

Variations in sea level are measured using a tide gauge, which is usually placed on piers. The tide gauge consists of a graduated staff to measure the rise and fall of water with the tides. Older gauges use a float operating in a stilling well, to dampen the effects of waves. More modern, automated systems employ acoustical signals to record the ever-changing water levels at 6-minute intervals (Figure S10). The tide gauge measurements are linked geodetically to a local fixed benchmark on stable

Table S3 Estimated sea level rise

Source	Minimum (mm/yr)	Mid-value (mm/yr)	Maximum (mm/yr)
Thermal expansion	0.3	0.5	0.7
Mountain glaciers	0.2	0.3	0.4
Greenland	0.0	0.05	0.1
Antarctica	−0.2	−0.1	0.0
Long-term glacial eustatic	0.0	0.25	0.5
Land water storage	−1.1	−0.35	0.4
Miscellaneous	0.0	0.05	0.10
Total	−0.8	0.7	2.2
Observed	1.0	1.5	2.0

Adapted from Church et al., 2001.

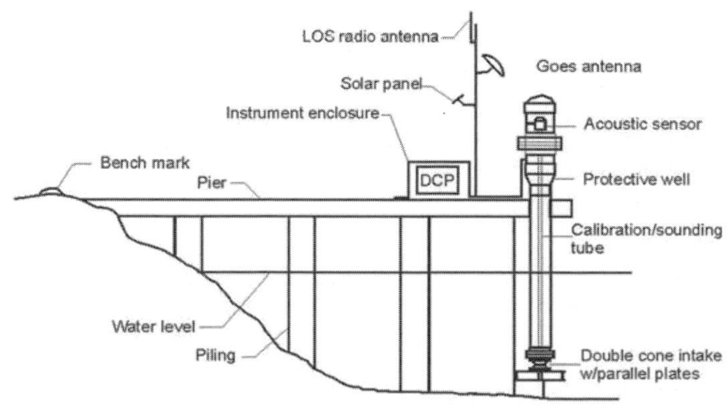

Figure S10 Schematic diagram of a tide gauge with acoustic-type sensor (after NOAA NOS Co-ops).

ground. Global sea level data are compiled by the Permanent Service for Mean Sea Level (Spencer and Woodworth, 1993; http://www.pol.ac.uk/psmsl/). However, most records are too short or are too broken to be useful for climate studies. A time-series of at least 50 years, and preferably longer, is recommended in order to minimize effects of interdecadal fluctuations and to derive a meaningful long-term sea level trend (Douglas, 2001). Only 175 stations have nearly continuous series of ≥50 years. Furthermore, most usable stations are located in the northern hemisphere, particularly in Europe and North America. Thus, selection of an optimal record length becomes a trade-off between the number of sufficiently long records and geographical coverage.

Tide gauges measure *relative* sea level change, which includes shifts in the land relative to the sea surface, in addition to the recent climate-related signal. These vertical land motions introduce considerable spatial variability in tide-gauge records. Late Holocene sea level proxies (e.g. peats, corals, wood, mollusks, etc.) have been used to construct long-term sea level curves which include glacial isostatic and other geologic trends (Gornitz, 1995). Differencing the recent and late Holocene sea level trends provides a measure of the climate-related *absolute* sea level change. Geophysical models filter the glacial isostatic component from raw sea level data by computing gravitational changes induced by the global redistribution of mass, including both ice and water, caused by deglaciation (Peltier, 2001; Lambeck and Johnston, 1998). These models are calibrated by means of radiocarbon-dated geological proxies. The geophysical method is sensitive to assumptions of viscoelastic properties of the Earth's mantle as well as the thicknesses and melting histories of the former ice sheets. Neither methodology eliminates other sources of land motion, such as neo-tectonic displacements. Continuously-operating global positioning satellite (GPS) stations at or near tide gauges will measure vertical crustal changes. Nearly a decade of observations may be required to achieve the ~1 mm/yr accuracy for absolute sea level measurements.

Radar altimeters, such as those on board the ERS-1, Geosat, and TOPEX/POSEIDON satellites, measure ocean surface heights. By precisely tracking the satellite's orbit, the satellite's distance relative to the Earth's center can be established. A difference of these two measurements yields the absolute sea level. Satellite altimetry, especially since the launch of TOPEX/POSEIDON in 1992, has become a major tool for investigating interannual ocean basin variability, such as the El Niño-Southern Oscillation (ENSO) and sea level change (Nerem and Mitchum, 2001).

Mean global sea level, determined from tide gauges, and more recently from satellite altimeters, has been increasing by 1–2 mm/yr over the last 100 years (see Table S3). Recent measurements suggest a global trend close to 1.8 mm/year (Church et al., 2004). The observed rates are significantly higher than those of the past few thousand years, derived from geological data (Gornitz, 1995; Shennan and Woodworth, 1992; Varekamp and Thomas, 1998). Precisely dating the onset of this accelerated phase is difficult because of limited instrumental data prior to 1850 and proxy sea level data for the last 1000 years.

The disparity between observed sea level trends and the sum of individual contributing sources still needs to be explained (Table S3). The sea level trend of 2.5 ± 0.7 mm/yr detected by the TOPEX/POSEIDON altimeter is largely related to ocean thermal expansion during a period (1993–1998) strongly influenced by major ENSO events, and may therefore be unrepresentative of longer-term sea level changes (Nerem and Mitchum, 2001). The longer-term (1955–1996) thermal expansion has been 0.5 ± 0.05 mm/yr (Cabanes et al., 2001). Along with mountain glacier retreat supplying close to 0.3 mm/yr during most of the twentieth century (Table S3), this accounts for less than 1 mm/yr SLR. The thermal expansion may be strongly enhanced at shallow coastal stations, which are therefore not truly representative of the world's oceans, leading to an overestimate in global sea level trends derived from tidal data (Cabanes et al., 2001). On the other hand, the eustatic contribution from small glaciers and the Greenland ice sheet may have been growing in recent years (Meier and Wahr, 2002; Rignot and Thomas, 2002).

Future sea level rise

Rates of SLR are anticipated to accelerate as future climate warms, due to increasing concentrations of atmospheric greenhouse gases. Diverse scenarios of greenhouse gas and aerosol emissions have been input into a suite of coupled Atmospheric–Oceanic General Circulation Models (AOGCM) by the Intergovernmental Panel on Climate Change (Houghton et al., 2001). The global mean sea level by 2100 is projected to rise by 9–88 cm over 1990 values, with a central value of 48 cm (for all scenarios). These values are slightly below those published previously by the IPCC, due to model improvements that show a smaller contribution from glaciers and ice sheets. However, recent observations suggest that these models may have underestimated the role of the cryosphere (Arendt et al., 2002; Meier and Wahr, 2002; Rignot and Thomas, 2002). Even after eventual stabilization of greenhouse gas concentrations, sea level will continue to rise, because of the time lag in ocean warming.

Vivien Gornitz

Bibliography

Arendt, A.A., Echelmeyer, K.A., Harrison, W.D., Lingle, C.S., and Valentine, V.B., 2002. Rapid wastage of Alaska glaciers and their contribution to rising sea level. *Science*, **297**: 382–385.

Bard, E., Hamelin, B., Arnold, M., et al., 1996. Deglacial sea-level record from Tahiti corals and the timing of global meltwater discharge. *Nature*, **382**: 241–244.

Blanchon, P., and Shaw, J., 1995. Reef drowning during the last deglaciation: evidence for catastrophic sea-level rise and ice-sheet collapse. *Geology*, **23**: 4–8.

Cabanes, C., Cazenave, A., and Le Provost, C., 2001. Sea level rise during past 40 years determined from satellite and in situ observations. *Science*, **294**: 840–842.

Church, J.A., Gregory, J.M., Huybrechts, P., et al., 2001. Changes in sea level. In Houghton, J.T., Ding, Y., Griggs, D.J., et al., eds., *Climate Change 2001: the scientific basis*. Cambridge: Cambridge University Press, pp. 639–693.

Church, J.A., White, N.J., Coleman, R., Lambeck, K., and Mitrovica, J.X., 2004. Estimates of the regional distribution of sea level rise over the 1950–2000 period. *Journal of Climatology*, **17**: 2609–2625.

Cronin, T.M., Vogt, P.R., and Willard, D.A., 2002. Rapid sea-level rise at 8.0–7.5 ka and Antarctic ice sheet thinning. *EOS Transactions AGU*, **83**(47), Fall Meeting Supplement, Abstract OS71D-0324.

Douglas, B.C., 2001. Sea level change in the era of the recording tide gauge. In Douglas, B.C., Kearney, M.S., and Leatherman, S.P., eds., *Sea Level Rise: history and consequences*. San Diego, CA: Academic Press, pp. 37–64.

Gornitz, V., 1995. A comparison of differences between recent and late Holocene sea-level trends from eastern North America and other selected region. *Journal of Coastal Research*, special issue no. 17, pp. 287–297.

Gornitz, V., 2001. Impoundment, groundwater mining, and other hydrological transformations: impacts on global sea level rise. In: Douglas, B.C., Kearney, M.S., and Leatherman, S.P., eds., *Sea Level Rise: history and consequences*. San Diego, CA: Academic Press, pp. 97–119.

Houghton, J.T., Ding, Y., Griggs, D.J., et al. (eds.), 2001. *Climate Change 2001: the scientific basis*. Cambridge: Cambridge University Press.

Lambeck, K., and Johnston, P., 1998. The viscosity of the mantle: evidence from analyses of glacial rebound phenomena. In Jackson, I., ed., *The Earth's Mantle*. Cambridge: Cambridge University Press, pp. 461–502.

Meier, M.F., and Wahr, J.M., 2002. Sea level is rising: do we know why? *Proceedings of the National Academy of Sciences*, **99**(10): 6524–6526.

Nerem, R.S., and Mitchum, G.T., 2001. Observations of sea level change from satellite altimetry. In Douglas, B.C., Kearney, M.S., and Leatherman, S.P., eds., *Sea Level Rise: history and consequences*. San Diego, CA: Academic Press, pp. 121–163.

Peltier, W.R., 2001. Global glacial isostatic adjustment and modern instrumental records of relative sea level history. In Douglas, B.C., Kearney, M.S., and Leatherman, S.P., eds., *Sea Level Rise: history and consequences*. San Diego, CA: Academic Press, pp. 65–95.

Rignot, E., and Thomas, R.H., 2002. Mass balance of polar ice sheets. *Science*, **297**: 1502–1506.

Sahagian, D., 2000. Global physical effects of anthropogenic hydrologic alterations: sea level and water redistribution. *Global and Planetary Change*, **25**: 39–48.

Shennan, I., and Woodworth, P.L., 1992. A comparison of late Holocene and twentieth-century sea level trends from the UK and North Sea region. *Geophysics Journal International*, **109**: 96–105.

Spencer, I., and Woodworth, P.L., 1993. *Data Holdings of the Permanent Service for Mean Sea Level*, PSMSL, Birkenhead: Bidston Observatory.

Varekamp, J.C., and Thomas, E., 1998. Climate change and the rise and fall of sea level over the millennium. *EOS Transactions AGU*, **79**: 69, 74–75.

Cross-references

El Niño
Greenhouse Effect and Greenhouse Gases
Ice Ages
Ocean–Atmosphere Interactions
Southern Oscillation

SEASONAL AFFECTIVE DISORDER

The decreasing daylight that heralds the coming of winter means for some people the onset of unpleasant and sometimes even debilitating symptoms. Some people experience mood swings and changes in level of activities that coincide with shifts in the seasons. In its most extreme form these changes constitute seasonal affective disorder (SAD). The most widely recognized form of SAD is the winter type (w-SAD) whose symptoms first appear in fall, gradually intensify into winter and then usually abate during spring. In addition to exhibiting the characteristic symptoms of non-seasonal anxiety and depression (hostility, anger, irritability), people with w-SAD also demonstrate atypical conditions. These different symptoms include increased rather than decreased sleep and increased rather than decreased appetite, which is often accompanied by a craving for carbohydrates. The latter behavior typically leads to weight gain. Another seasonal pattern of activity level and mood begins in late spring, intensifies in early summer and then remits in fall. In contrast to victims of w-SAD, people experiencing the summer form of SAD (s-SAD) tend to demonstrate the typical symptoms of non-seasonal depression including a lack of sleep and poor appetite, leading to weight loss. The summer form of SAD typically occurs at lower latitudes whereas the winter form prevails at middle and higher latitudes.

Since the first documentation of SAD in the early 1980s, most research on this syndrome has focused on w-SAD. The considerably greater prevalence of w-SAD and the early identification of an effective treatment have stimulated greater research interest in w-SAD than in s-SAD. In marked contrast, knowledge of the less common s-SAD, which lacks a specific treatment, has advanced little since its discovery in the late 1980s. Consequently the remainder of this item will center on w-SAD.

Epidemiological surveys suggest that 1–4% of the general population of Europe and North America experience w-SAD, while as many as 20% exhibit subsyndromal symptoms (Wehr, 2001). Seasonal changes in this latter group are not severe enough to meet all clinical criteria for w-SAD (American Psychiatric Association, 1994). Studies consistently find a strong relationship with gender. Women with w-SAD outnumber men with w-SAD by four to one. Furthermore, women experience greater increases in overeating, weight gain, and increased sleep than their male counterparts. The winter form of SAD is more prevalent among young adults than adolescents and older adults. This age-related pattern of w-SAD contrasts with that of non-seasonal depression that is most common in the early 60s.

A number of studies have sought to identify the factors that influence the occurrence of SAD. Given the considerable seasonal changes in photoperiod at mid and high latitudes, logic suggests that the prevalence of w-SAD should increase with latitude. Recent literature reviews suggest, however, that this relationship is weak. Studies of the influence of climate and social–cultural factors on patterns of prevalence also have failed to demonstrate significant linkages. This inability to find consistent epidemiological relationships may be the consequence of the relatively small number of studies, limited sample sizes, and the reliability and heterogeneous nature of the methodology used in studies of prevalence (Haggarty et al., 2001; Mersch et al., 1999). Furthermore, the relatively mild symptoms exhibited by many SAD patients may obscure identification of causative factors.

Explanations

Scientists have proposed several explanations for the onset of SAD. An early hypothesis proposed that seasonal changes in photoperiod disrupted one or more of the body's rhythmic behaviors such as the daily pattern of change in blood pressure, body temperature, and activity level (sleep/wake cycle). Such endogenous rhythms of about 24 hours are called circadian (about a day) rhythms and are driven by an internal biological clock. The human biological clock has a period of about 25 hours, but varies considerably from person to person. In the absence of external clues, circadian rhythms increasingly go out of phase with the outside world. In the case of the activity/sleep cycle, for example, people awaken later each day. When persons who have been living in isolation for an extended period are then exposed to the day/night cycle, their biological clock and their circadian rhythms soon are back in phase.

The primary biological clock in humans resides in the suprachiasmatic nucleus (SCN), a small mass of tissue located in the hypothalamus of the brain. The most potent synchronizer of this clock is bright light. A dedicated neural pathway (termed the retinohypothalanic tract) leads directly from the retina of the eye to the SCN. This tract, including the light-absorbing cells in the retina, differs from the pathway that mediates vision. The retinohypothalamic tract synchronizes the activities of the SCN with the natural cycles of day and night. In response to that cycle, the SCN coordinates via neurons the release of melatonin from the pineal gland, a pea-sized mass of nerve tissue located deep between the cerebral hemispheres. Melatonin plays a major role in the activity/sleep cycle. Darkness stimulates the release of this hormone whereas both dim and bright light inhibit its release. Melatonin levels, therefore, follow a circadian pattern, peaking around midnight and dropping to their lowest level around noon. Furthermore, the duration of nocturnal secretion of melatonin follows a distinct seasonal pattern with the period becoming longer in winter and shorter in summer. Sites that respond to the melatonin signal induce changes in physiology and behavior that vary with the duration of the signal; that is, the seasons.

Understanding of circadian rhythms led to the phase-delayed hypothesis for w-SAD. The theory proposes that w-SAD develops when circadian rhythms are phase delayed relative to the external clock or to other rhythms. Studies utilizing the most reliable measures of circadian rhythm phase change have shown evidence of a phase advance in w-SAD patients when they are exposed to light. Subsequently the atypical symptoms of w-SAD remit. There remains, however, a subset of SAD patients who do not have demonstrable phase-delayed circadian rhythms, do not require a phase shift for a response to light, or both. Therefore, other mechanisms must be playing a role.

Additional mechanisms

Another avenue of research on the cause(s) of w-SAD has focused on neurotransmitters, particularly serotonin, that are implicated in mood disorders. Like melatonin, serotonin levels exhibit a marked seasonality, albeit it in the opposite direction. Serotonin reaches its lowest levels during midwinter and its highest during midsummer. Given the role of serotonin in regulating hunger, low levels in midwinter could explain the typical tendency of patients with w-SAD to crave carbohydrates and gain weight during winter. Various lines of research, including the use of medications that enhance serotonin function in the brain, demonstrate a major role of the serotonin system in the clinical symptoms of w-SAD. Other research suggests that norepinephrine and dopamine (neurotransmitters in the same family as serotonin and implicated in mood disorders) also play some role in w-SAD.

Although preliminary in nature, emerging evidence suggests that genetic factors influence vulnerability to or protection from SAD. Studies of twins, for example, have found that the "atypical" symptoms of SAD are inheritable. A related line of study has focused on Icelanders, descendants of a population that was essentially isolated for much of the past 1000 years. Despite living at high latitudes, modern-day Icelanders have a significantly lower prevalence of SAD than populations residing at lower latitudes along the US coastline. A recent follow-up study conducted in Winnipeg, Canada, of two adult populations – one of wholly Icelandic descent and the other of non-Icelandic descent – found that the rate of SAD among the people of Icelandic descent was about half the rate of the non-Icelandic population. The authors of these studies suggest that genetic adaptation to the long winter season occurred within the native Icelandic population. Other genetic studies have begun to find associations between patients with SAD and certain alleles of genes that are associated with the functioning of the biological clock and the brain's neurotransmitters such as serotonin.

Although substantial evidence continues to accumulate for the circadian phase shift and the serotogenic hypothesis of SAD, conflicting results suggest that SAD is a complex and diverse condition. Recent progress on determining the molecular mechanisms of the human biological clock and how it regulates circadian rhythms should provide important insights into the etiology of pathophysiology of SAD (Lam and Levitan, 2000).

Bright-light therapy is the first-line option for the treatment of patients with SAD. Patients typically are exposed for 2 hours to light-boxes emitting at least 2500 lx (one lux equals one lumen/meter). Administering treatments between 6 a.m. and 10 a.m. is recommended. If treatment during that period is not feasible, however, some patients benefit from early evening exposure. (With regard to the phase-delay hypothesis, the efficacy of evening exposure, although typically less than that of morning exposure, has been surprising.) Continuing treatment throughout the winter season prevents relapses. Responses to bright-light therapy are good to excellent. When properly administered light therapy has few adverse side-effects (eye irritation, headache, and nausea), all of which are typically mild and reversible. Usually a patient will respond to bright-light therapy within only 2–4 days. A full remission, however, may require several weeks. If a patient does not respond to bright light, prescription of an antidepressant drug is an alternative. Given current research, the best choice at this time is medication that influences serotonin activity. For persons seeking a more natural treatment, simply being outdoors an hour or two can make a significant difference, particularly in combination with exercise such as walking, jogging, or skiing. When indoors, persons with SAD should opt for a room with a window or bright artificial lighting. As a point of reference, indoor lighting should be typically 100 lx at home and 300 to 500 lx at work.

Although individuals who experience seasonally related changes in activity level and mood have been known since ancient times, the first systematic description of SAD and its treatment was not published until 1984 (Rosenthal et al., 1984). Since that milestone the number of studies has grown rapidly, and the field of study now encompasses contributions from epidemiology to molecular biology. Although we continue to learn more about the characteristics of the syndrome, its patterns of prevalence, and its causes, we now know that SAD is much more complex than originally thought. Such factors as improved design of epidemiological surveys of larger populations, advanced brain-imaging techniques, and further applications of molecular biology to the workings of the biological clock and its mediation of circadian rhythms will continue to lead to a greater understanding of the SAD syndrome and consequently its diagnosis and treatment.

Michael D. Morgan

Bibliography

American Psychiatric Association, 1994. *Diagnostic and Statistical Manual of Mental Disorders: DSM IV* (4e). Washington, DC: American Psychiatric Press.

Alexsonn, J., Stefansson, J.G., Magnusson, A., Sigvaldson, H., and Karlsson, M.M., 2002. Seasonal affective disorders: relevance of Icelandic and Icelandic–Canadian evidence to etiologic hypothesis. *Canadian Journal of Psychiatry*, **47**(2): 153–159.
Haggarty, J.P., Cemovsky, C., and Husni, M., 2001. The limited influence of latitude on rate of seasonal affect disorder. *Journal of Nervous and Mental Disease*, **189**(7): 482–484.
Harmatz, M.G., Well, A.S., Overtree, C.E., Kawamura, K.Y., Rosal, M., and Ockene, I.S., 2000. Seasonal variation of depression and other moods: a longitudinal approach. *Journal of Biological Rhythms*, **15**(4): 344–350.
Lam, R.W., and Levitan, R.D., 2000. Pathophysiology of seasonal affective disorder: a review. *Journal of Psychiatry and Neuroscience*, **25**(5): 468–480.
Levitt, A.J., and Boyle, M.H., 2002. The impact of latitude on the prevalence of seasonal depression. *Canadian Journal of Psychiatry*, **47**(4): 361–367.
Magnusson, A., 2000. An overview of epidemiological studies on seasonal affective disorder. *Acta Psychiatrica Scandinavica*, **101**: 176–184.
Magnusson, A., and Stefansson, J.G., 1993. Prevalence of seasonal affective disorder in Iceland. *Archives of General Psychiatry*, **50**: 941–946.
Mersch, P.P., Middendorp, H.M., Bouhuys, A.L., Beersma, D.G.M., and van den Hoofdakker, R.H., 1999. Seasonal affective disorder and latitude: a review of the literature. *Journal of Affective Disorders*, **53**(1): 35–48.
Partonen, T., and Magnusson, A., 2001. *Seasonal Affective Disorder: practice and research.* Oxford: Oxford University Press.
Rosenthal, N.E., Sack, D.A., Gillin, I.C., et al., 1984. Seasonal affective disorder: a description of the syndrome and preliminary findings with light therapy. *Archives of General Psychiatry*, **41**(1): 72–80.
Saeed, S.A., and Bruce, T.I., 1998. Seasonal affective disorders. *American Family Physician*, **57**(6): 1340–1347.
Terman, M., and Terman, I., 1999. Bright light therapy: side effects and benefits across the symptom spectrum. *Journal of Clinical Psychiatry*, **60**(11): 799–808.
Wehr, T.A., 2001. Photoperiodism in humans and other primates: evidence and implications. *Journal of Biological Rhythms*, **16**(4): 348–367.

Cross-references

Bioclimatology
Climate Comfort Indices
Seasons

SEASONAL CLIMATE PREDICTION

Seasonal climate is generally defined as the mean temperature and precipitation and other statistics of weather over a region for a 3-month period. Seasonal climate prediction aims to predict the climate anomalies (departures from normal) of coming seasons based on current conditions of the atmosphere, ocean and land surfaces. Seasonal climate prediction is of great societal interest. For example, if the major droughts around the world could be predicted in advance, it would lead to substantial savings for the economy of the countries affected by the droughts.

Climate prediction approaches have been evolving over hundreds of years. Earlier approaches were to identify some environmental indicators which could suggest likely abnormal climate in the next season. In the early twentieth century, as more and more meteorological data became available, scientists started to construct empirical models for seasonal climate predictions. Before the 1980s, however, these empirical models were mostly based on statistical techniques, providing little understanding of the physical mechanisms responsible for relationships between current conditions and the future climate. In recent years, owing to the advances in numerical modeling and climate diagnostics in last two decades, numerical climate prediction models have emerged and statistical models have become more physically based, leading to improvement in prediction skill.

Predictability of seasonal climate

Since the equations controlling atmospheric motion are nonlinear and sensitive to initial conditions, errors in initial conditions, no matter how small, will inevitably grow and eventually lead to unskillful deterministic forecasts. All studies have led to a conclusion that the day-to-day weather fluctuation is not predictable beyond 1–2 weeks. While the predictability of the day-to-day weather is limited to such a short period, seasonal climate anomalies are found to be predictable for a much longer period in some circumstances and for some regions.

The predictability of seasonal climate comes from the slowly evolving upper oceans and land surfaces. The atmosphere is coupled with oceans and land surfaces by exchange of heat, water (liquid water or water vapor) and momentum. For example, the surface wind drives oceanic circulations, precipitation provides fresh water to oceans and land surfaces, and oceans and land surfaces in turn release water vapor and heat to the atmosphere. Because of the air–sea and air–land interactions, anomalies in oceans and land surfaces may cause anomalies in the atmosphere, and vice-versa. Since the thermal inertia of the upper oceans and land surfaces is much larger than that of the atmosphere, the fluctuations in upper oceans and land surfaces are much slower than the day-to-day weather process. The slower fluctuations can lead to longer-range predictions, hence providing a physical basis for seasonal climate predictability.

The most predictable seasonal climate anomaly is that associated with the El Niño and Southern Oscillation (ENSO). ENSO is a coupled ocean–atmosphere phenomenon. Its sea surface temperature (SST) manifestation is the episodic warming and cooling in the equatorial central and eastern Pacific at seasonal to interannual timescales (see Figure S11). These slowly evolving SST exert an influence on the tropical atmosphere by redistributing the surface heating, low-level wind fields, tropical convection and mid/upper troposphere heating. The atmospheric heating anomalies in tropics then force a planetary wave train of climate anomalies to a large area of the globe, in particular the Pacific/North America (PNA) region. Figure S12 shows the linear correlation between the tropical SST shown in Figure S11 and the global 500 hPa height for the January–February–March season, indicating the global climate impact of the ENSO phenomenon. Current researches have shown that ENSO is predictable several seasons in advance, and therefore the associated climate anomalies are also predictable to some degree with that time lead.

SST anomalies in other ocean basins are also found to be important in seasonal climate prediction. For instance, the tropical Atlantic Ocean is important for Brazil and western Africa, and the Indian Ocean is influential for eastern Africa, Southern Asia, and Australia.

As an example of the seasonal climate predictability resulting from the land surface feedbacks, the variability of the Asian summer monsoon is related to the winter snow cover over the Eurasia continent. In addition, a portion of the temperature and precipitation anomalies over the North American continent in the summer is caused by the soil moisture conditions in the spring.

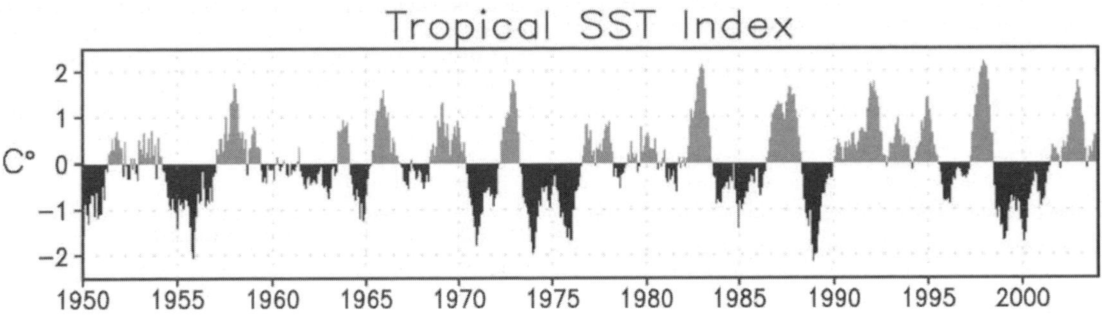

Figure S11 The time-series of monthly sea surface temperature (SST) anomalies averaged over the area (180–130°W, 5°S–5°N) for the period 1950–2003. Unit is degree Celsius. The large positive (negative) departures correspond to El Niño (La Niña).

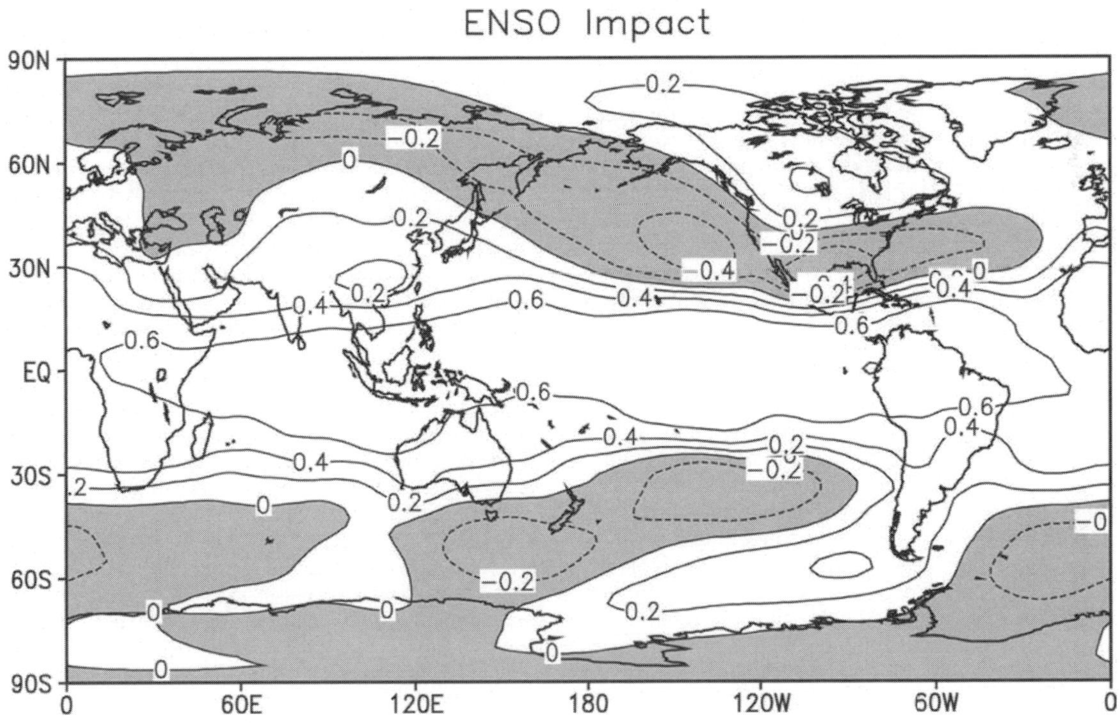

Figure S12 Linear correlation coefficient of 500 hPa height anomalies with the tropical Pacific sea surface temperature (SST) index shown in Figure S11 for the January–February–March seasons of the period 1950–2003.

Prediction methods

All the forecast models currently in practice can be grouped into two categories: statistical (or empirical) models and dynamical models.

Statistical models

Statistical forecast is based on the principle that the future state of a predictand can be predicted from antecedent and current values of predictor variables. Predictors should be selected based upon the understanding of the mechanisms of climate variability. Formulations of predictor–predictand relationships are ascertained by appropriate statistical techniques, including testing on independent data or cross-validation. The statistical techniques widely used for this purpose include canonical correlation analysis, multiple linear regression, constructed analog, Markov process, singular spectrum analysis, singular value decomposition, linear discriminant analysis, and neural networks. The quantitative predictor–predictand relationship, representing the statistical model, is then used to calculate predictand values in real-time situations.

In most cases predicators for seasonal climate predictions are boundary forcing, such as SST, soil moisture and snow cover. Although sometimes atmospheric predictors are included, these predictors, except the quasi-biennial oscillation (QBO) in the tropical stratosphere, generally are boundary forced signals, and thus do not mean that the atmosphere itself has sufficient memory to provide predictability at seasonal time-scales.

Dynamical models

There are two types of dynamical models currently used for climate prediction. One is atmosphere-only general circulation models (AGCM), the other one is coupled ocean–atmosphere general circulation models (CGCM). Both are based on the full set of the physical equations of motion. A two-tier climate prediction approach is used with AGCM, in which the SST are predicted first with statistical models or coupled dynamical models and then used to force the overlying atmosphere. In CGCM, similar to nature, both atmosphere and oceans evolve freely and influence each other. Being initialized appropriately, forward integrations of the coupled model directly result in seasonal climate predictions, in a more straightforward way than the two-tier approach.

Nevertheless, AGCM are still widely used in operational forecasting. The reason is that, in the current CGCM, SST tend to drift away from their realistic values as the integration proceeds, thus yielding unrealistic patterns of atmospheric anomalies. This problem is expected to be alleviated, and eventually solved, by continuing efforts in model development.

In both AGCM and CGCM, land surfaces are coupled with the atmosphere through land-surface process models, which determine soil moisture, surface temperature, snow cover and vegetation. Initialization of the land surface, however, remains a challenging problem because observations of soil moisture and evaporation, for example, are scarce.

Format and expressing uncertainty

Because the precise state of the atmosphere is unpredictable beyond 1–2 weeks, climate prediction is inherently probabilistic. For this reason seasonal climate predictions are usually issued in the format of a probability distribution, which indicates the chances that the temperature and precipitation will be *above*, *near* or *below* their long-term averages (climatology). For statistical forecasts, the probability distribution is estimated based on conditional probabilities indicating the relative frequency with which a particular climate pattern was observed under certain boundary forcing, such as El Niño or La Niña SST. For dynamical forecast the probability distribution is estimated by counting the ensemble members falling into the three categories. Calibration procedure as a post-process is usually applied here for reducing the influence of model biases.

Summary

Seasonal climate is potentially predictable in many regions of the world, especially the tropics. Its predictability comes from the slowly evolving underlying boundary conditions that force the atmosphere, such as SST and land surface characteristics. Skillful seasonal climate predictions, therefore, depend on the extent to which the relevant boundary conditions and their climate impact are predictable. Since the atmospheric motion is inherently chaotic, climate predictions need to be expressed probabilistically, indicating the likelihood of the climate shifting from its unconditional climatology. Currently, both statistical (empirical) and dynamical models are used in operational forecasts. These two types of model are compensatory for their different characteristics. The improvement of prediction skill in the future will depend on the enhancement of the observing system and model development, in particular the improvement of the fully coupled ocean–land–atmosphere model.

Peitao Peng

Acknowledgments

Drs Arun Kumar and Huug van den Dool at CPC/NCEP reviewed the manuscript and gave important comments and suggestions.

Bibliography

Anderson, J.L., van den Dool, H., Barnston, A.G., Chen, W., Stern, W., and Ploshay, J., 1999. Present day capabilities of numerical and statistical models for atmospheric extratropical seasonal simulation and prediction. *Bulletin of the American Meteorological Society*, **80**: 1349–1362.

Goddard, L., Mason, S.J., Zebiak, S.E., et al., 2001. Current approaches to seasonal-to-interannual climate predictions. *International Journal of Climatology*, **21**: 1111–1152.

Hastenrath S., 2003. Climate prediction (Empirical and numerical). In Holton, J.R., Pyle, J., and Curry, J.A., eds., *Encyclopedia of Atmospheric Science*. Academic Press, pp. 411–417.

Hoerling, M.P., and Kumar, A., 2003. Seasonal and interannual weather prediction. *Encyclopedia of Atmospheric Science*. In Holton, J.R., Pyle, J., and Curry, J.A., eds., New york: Academic Press, pp. 2562–2567.

Ji, M., Leetmaa, A., and Kousky, V.E., 1996. Coupled model forecasts of ENSO during 1980s and 1990s at the National Meteorology Center. *Journal of Climate*, **9**: 3105–3120.

Kumar, A., and Heorling, M.P., 1995. Prospects and limitations of seasonal atmospheric GCM predictions. *Bulletin of the American Meteorological Society*, **76**: 335–345.

Latif, M., Anderson, D., Barnett, T., et al., 1998. A review of predictability and prediction of ENSO. *Journal of Geophysics Research*, **103**: 14,375–14,393.

Palmer, T.N., and Anderson, D.L.T., 1994. The prospects for seasonal forecasting. *Quarterly Journal of the Royal Meteorological Society*, **120**: 755–793.

Peng, P., Kumar, A., Barnston, A.G., and Goddard, L., 2000. Simulation skills of the SST forced global climate variability of the NCEP-MRF9 and the Scripps-MPI ECHAMs models. *Journal of Climate*, **13**: 3657–3679.

Shukla, J., Anderson J., Baumhefner D., et al., 2000. Dynamical seasonal prediction. *Bulletin of the American Meteorological Society*, **81**: 2593–2606.

Trenberth, K.E., (ed.) 1992. *Climate System Modeling*. London: Cambridge University Press.

Cross-references

Cycles and Periodicities
Dynamic Climatology
El Niño
Ocean–Atmosphere Interactions
Synoptic Climatology
Teleconnections

SEASONALITY: ASTRONOMIC FORCING

The basic concepts of *season* are treated under that heading. However, in decadal or century-length time-series, distinctive alternations of seasonal characteristics are apparent, and seem to relate to astronomical forcing. The whole question is controversial, however, being bound up with prejudice, politics and even religion, so it must be approached with care.

The best-established systematic variables are connected with the atmospheric pressure *oscillations*, which have long been recognized on the basis of two "nodes", usually employing the data from two weather centers that have long been in operation. The usual parameters employed are mean pressure records, but closely related also are sea-level data (the "inverted barometer effect").

As established by David Thomson (1995), analysis of instrumental records since AD 1659 shows that the dominant frequency of the seasons is one cycle per *anomalistic year*; that is, perihelion to perihelion (365.25964 days), which also controls the annual temperature cycle. Fluctuations of that temperature are "probably of solar origin", but are partly by albedo feedback. False readings of seasonality are partially based on using the equinox-to-equinox standard, the *tropical year* (365.24220 days).

As customarily taken, "seasonality" is a response to the different attitudes of the Earth's surface in the reception of the solar constant, approximately 1.95 cal/cm^2/min (or 1.4 × 10 ergs/cm^2/s). It should be remembered, however, that the sun's spin rate is somewhat variable and this affects its radiation characteristics. Furthermore its emissions fall into two broad categories: (a) electromagnetic radiation, which travels at the speed of light in direct rays, so that its impact on the Earth's surface (its "perpendicularity") is limited to the equatorial belt and the boundaries of the tropics of Cancer and Capricorn (roughly 23.5°N and S). And (b) its particle or corpuscular radiation that reaches the Earth in the plasma of the solar wind, the particles being magnetized so that they are funneled in to the north magnetic pole and the auroral oval, following a spiral trajectory and with greatly variable energetic attributes (corresponding to the violence of the solar eruptions).

Another customary assumption is the alleged "11-yr" solar cycle, which is often grossly in error. Its long-term periodicity has a mean of 11.1212 yr, over some intervals being 11.17 yr, but over century intervals it has a ±6 yr variability. No climatic time-series should ever be statistically locked in to an "11-yr" standard.

A second aspect of "seasonality" is the lunar factor. Whereas the solar year is closely related to the calendar year of 365 days, or the leap year mean of 365.25 d, further corrections have to have been made over centuries. In contrast, the lunar year may be 28 d × 336 d, or 28 d × 13, thus 364 d, or 28 d × 14, thus 393 d, with further computations relating to various religious calendars. For climatic purposes several lunar cycles, beyond the 28 d standard, are helpful:

1. *Perigee–syzygy cycle* (PSC, approx. 3 or 6 yr). During the twentieth century, phases fell near the vernal equinox on 19–28 September. The ascending node and perigee (closest approach in the 8.849-yr apsides cycle) coincide in 1910, 1916, 1922, 1934 . . . 1982, 1988, 1994. They are antiphase in 1913, 1919 . . . 1985, 1991, 1997. In successive cycles they may be up to 1 week apart. Sanders (1995) remarked on the rough parallelism of the PSC with rainfall. The 6-yr period appears in many climate spectra, perhaps relating to the *ca.* 438 d. Chandler Wobble, and thus with the high-latitude atmospheric angular momentum.
2. *Quasi-biennial oscillation* (QBO, average 2.172 yr), the well-known weather cycle (Schove, 1983), but also a less well-known periodicity of solar emissions. To create a seasonality, one looks to a 13.03-yr return (×6), which is also important in climatic time-series, including ENSO events. The QBO (×96) is also in a planetary alignment (Uranus–Jupiter–Venus synod) which resonates at 208.5227-yr intervals, and which is seen in tree-ring ^{14}C flux rates (Thomson, 1990). It is linked to innumerable other planetary geometrics.
3. *Lunar nutation and declination cycle* (NDC, approx. 18.605 yr, ×5 = 93.025 yr). Coincides with tidal maximum in AD 1912 and with climatic spectra (Pettersson, 1930; Currie, 1995). Declination brings lunar zenith to more than 1100 km farther north (or south), accelerating geostrophic currents such as the Gulf Stream (Lamb, 1977).

   Tidal cycles and others are strongly seasonal, so that if the astronomical periodicity is an uneven number, such as 18.605 yr, its effect is less than is the case when its multiple epoch returns at the same season. In this case the fifth return (93.025 yr) brings the "seasonality" factor into more emphatic relief. Solar cycle warming, at times in phase with the lunar forcing, is a supplementary factor (see also below, item 4). Pettersson (1930), using tax returns from the Hanseatic League ports reflecting the herring fishery which is tidally responsive, was able to trace this 93-yr seasonal return through more than eight centuries, a truly remarkable use of economic proxies for climatic purposes.
4. *Arctic tidal/temperature resonance* (approx. 11–22–45–179–359 yr). First observed in Greenland ice core 0–18 temperature proxies, later this series was seen as beat frequency of lunar nodal/apsides (L_N/L_A) tide cycles and as a lunar–solar resonance in Hudson Bay temperature proxies (Guiot, 1987). A 185 "staircase" of isostatically uplifted beachridges on Hudson Bay, spanning about 8300 yr, was provisionally identified by Fairbridge and Hillaire-Marcel (1977) as a 44.5-yr solar cycle, but evidently a lunar resonance is also involved. A period of about 74.4 yr (4 × 18.6/7 × 10.6 yr) spans both lunar and (high-energy) SSC. Similar raised beaches are seen throughout the polar and subpolar regions (Fletcher et al., 1993; Møller et al., 2002).

All these high Arctic data are seasonally constrained by the universal presence of sea ice in winter, which inhibits wave action and evaporation, and reduces tidal effects. Care should be taken with "annual" statistics because of the seasonality asymmetry in the sub-Arctic, where a Jan.–Feb.–Mar.–Apr.–May interval is typical "winter", whereas in the midlatitudes the seasons are more closely expressed as quarter-year units (NH): Dec.–Jan.–Feb. (winter); Mar.–Apr.–May (spring); June–July–Aug. (summer); Sept.–Oct.–Nov. (fall). In equatorial and monsoon lands the seasons are best identified with rainfall (in Australia "the Wet"). The "year" of an ENSO is particularly deceptive because its precursors appear in one statistical year while its full expression usually follows the "Christ-child" anniversary and falls in January.

Global aspects

Extreme asymmetry of land and sea on planet Earth renders the northern hemisphere to be more continental in climate and the southern to be more oceanic. Because of the present axial tilt and the state of the precession, the southern hemisphere receives higher insolation at perihelion in early January, and the northern hemisphere has a milder seasonality on a 10^5-yr time scale. Thus its continentality is moderated in both summer and winter.

Over long-term intervals one may note the seasonality controls of three fundamental resonance periods involving the astronomic beat frequencies and the mean sunspot cycle (SSC). The key beat frequency is a "lap" of the two largest planets, Saturn and Jupiter (SJL: 19.8593 yr), which has been described as the "heartbeat of the solar system" (Fairbridge and Sanders, 1987). Also involved is the mean beat frequency of the inner planets (IPC), 46.3383 yr, and the mean SSC, 11.1212 yr. The three resonance events are also exact, whole number multiples of the QBO, 2.1721 yr. The three resonances are:

1. 556.06 yr, "Stacey Cycle": SJL × 28; SSC × 50; QBO × 256.
2. 278.03 yr, "Stacey Hemicycle": SJL × 14; SSC × 25; QBO × 128.
3. 139.015 yr: SJL × 7; IPC × 3; QBO × 64.

Those three resonance intervals return at close to the Earth's perihelion, at the present time in earliest January, but this date shifts over long periods. Thus the season shifts but the "seasonality" remains constant.

The ^{14}C flux variance recorded in tree rings, inversely to the global temperature patterns, was about 4–5 times greater in the early Holocene (about 8000–9000 years ago) than today. The ^{14}C maximum (temperature minimum) in 7590 BC was 8897 yr before the AD 1307 start of the Little Ice Age. Then, within one decade of 1307, the climate of Britain and western Europe swung from a medieval "medieval climatic optimum", 2–3°C warmer than today, to one that much colder. Documentary evidence shows that, even in Britain, crops were devastated, and in central Europe monastic records indicated that in places half the population died of starvation. Comparable events occurred about a dozen times during the Holocene, but require more study.

The interval 8897 yr corresponds to 16 × 5560 yr (the basic planetary quadrature unit). The 556-yr alignment corresponds to 2 × 278 yr ($64^2 \times 0.0678784$ yr, the ISSR, inner sun spin rate); and 278 yr is the basic resonance period of the two largest planets, Jupiter and Saturn (14 SJL) and the mean sunspot period (25 SSC). Over the long term of 8897 yr, other synodic pairs are involved, e.g. NUL × 52, the 317-yr double synods (USL/SJL) × 28, the 12.783-yr NJL × 697, and exactly 2 × 4448.485 yr, the planetary quadrature cycle. This period is also exactly 800 × 11.1212 yr, the mean SSC. Therefore, sunspot periodicity relates both to planetary cycles and long-term climatic change.

From the evidence of the ^{14}C flux oscillations, those of glacigene sediments in deep-sea cores in the North Atlantic and ^{18}O fluctuations in Greenland ice cores, the swing from one climate-forcing extreme to the other was achieved in less than one century; probably much less. Sampling problems prevent a more precise evaluation. Even in terms of mean temperatures, the swing from one extreme to the other was in the order of 5°C.

In summary, "seasonality" is characteristically expressed by cycles of both solar and lunar causes. Worldwide time-series analyses of all sorts of meteorological and geophysical data disclose them both (Currie, 1995; and his lifelong bibliography). While the lunar cycles, mainly expressed in oceanic tidal data and in their related atmospheric derivatives, are very regular, those dependent on solar activity are highly irregular.

Acknowledgments

Appreciation is expressed for many decades of scientific help from Clyde Stacey, John Sanders, Göran Windelius, Fergus Wood (all regrettably deceased), and others, as well as logistic help from Dr James Hansen (Goddard Institute for Space Studies, NASA).

Rhodes W. Fairbridge

Bibliography

Cerveny, R.S., and Balling, R.C., 1999. Lunar influence on diurnal temperature range. *Geophysics Research Letters*, **26**(11): 1605–1607.

Currie, R.G., 1988. Lunar tides and the wealth of nations. *New Scientist*, 5 Nov. pp. 52–55.

Currie, R.G., 1995. Variance contribution of M and S signals to Nile River data over a 30–8 year bandwidth. *Journal of Coastal Research*, special issue 17, pp. 29–38.

Fairbridge, R.W., and Hillaire-Marcel, C., 1977. An 8000 yr paleoclimatic record of the "Double-Hale" 45 yr solar cycle: *Nature*, **268**: 413–416.

Fairbridge, R.W., and Sanders, J.E., 1987. The sun's orbit, AD 750–2050: basis for new perspectives on planetary dynamics and Earth-Moon linkage. In: Rampino, M.R., Sanders, J.E., Newman, W.S., and Königsson, L.K., eds. *Climate – History, Periodicity, and Predictability*. New York: Van Nostrand Reinhold, pp. 446–471 (bibliography, 475–541).

Fletcher, C.H. III, Fairbridge, R.W., Møller, J., and Long, A.J., et al., 1993. Emergence of the Varanger Peninsula, Arctic Norway, and climate changes since deglaciation. *The Holocene*, **3**(2): 116–127.

Guiot, J., 1987. Reconstruction of seasonal temperatures in central Canada since AD 1700 and detection of the 18.6- and 22-year signals. *Climatic Change*, **10**: 249–268.

Lamb, H.H., 1977. *Climate History and the Future*. London: Methuen.

Møller, J.J., Yevzerov, V.Y., et al., 2002. Holocene raised beach ridges and sea-ice-pushed boulders on Kola Peninsula, northwest Russia: indicators of climate change. *The Holocene*, **12**(2): 169–176.

Pettersson, O., 1930. The tidal force. *Geographischer Annaler*, **12**: 261–322.

Sanders, J., 1995. Astronomical forcing functions: from Hutton to Milankovitch and beyond. *Northeastern Geology and Environmental Sciences*, **17**(3): 306–345.

Schove, D.J. (ed.), 1983. *Sunspot Cycles* (*Benchmark Papers in Geology*, vol. 68). Stroudsburg: Hutchinson Ross.

Thomson, D.J., 1990. Time series analysis of Holocene climate data. *Royal Society, London, Philosophical Transactions*, **A-330**: 601–616.

Thomson, D.J., 1995. The seasons, global temperature and precession. *Science*, **268**: 59–67.

Windelius, G., 1993. Large volcanic eruptions and earthquakes in pace with the Chandler Wobble. *Cycles*, **44**: 140–144.

Wood, F., 2001. *Tidal Dynamics*, (3rd edn.). West Palm Beach: Coastal Education and Research Foundation, 2 vols.

Cross-references

Blocking
Continental Climate and Continentality
El Niño
Ice Ages
Maritime Climate
Quasi-Biennial Oscillation
Seasons
Southern Oscillation
Sunspots

SEASONS

Season is term for a period of time during the calendar year characterized by or associated with a set of coherent climatic activities or weather phenomena. Traditionally, there are four seasons of three months each winter, spring, summer, and autumn (fall), which are reversed in each hemisphere. The seasons are terrestrial responses to solar heating inequities that are governed by the dynamic relationship that exists between the Earth and the sun at various times during the year. The single most important heat source on the surface of the Earth is the sun. Endogenic heat flow from the earth's interior is three orders of magnitude less than solar energy. The amount of solar heat intercepted by the Earth can be expressed as a fixed value referred to as the solar constant, which is approximately 1.95 cal/cm^2/min. Seasonality is a response to the different attitudes in the reception of the solar constant at locations on the Earth's spherical surface. Variations in the reception of the solar constant lead to terrestrial heating imbalances that are largely a latitudinal phenomenon. The amount of terrestrial heating accomplished by the sun depends on (1) the intensity of the sun's rays; and (2) the duration of time to which the land is exposed to the sun's rays. These two variables are regulated partly by the fact that the Earth is a sphere but mainly by the fact that the Earth is in orbit around the sun. Of greatest import are the orbital characteristics of the Earth, i.e., revolution, inclination, and parallelism.

Revolution

The Earth revolves around the sun in a slightly elliptical orbit (Figure S13). On or about 3 January the Earth is as close as 147 million km (91.5 million miles) to the sun. This closest position of the Earth to the sun is referred to as perihelion (Greek for *around the Sun*) and occurs during the winter season of the northern hemisphere. On or about 4 July the Earth is as far away as 152 million km (94.5 million miles) from the sun. This most distant position of the Earth from the sun is called aphelion (Greek for *away from the Sun*) and occurs during the summer season of the northern hemisphere. This 3% disparity in the proximity of the Earth to the sun is responsible for approximately 7% more intense insolation (incoming solar radiation) in January than in July. This is evidently not the cause of the natural seasons.

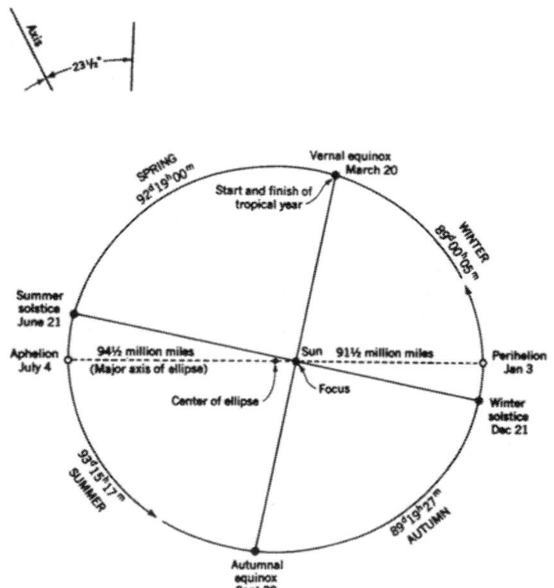

Figure S13 Dates of equinox and solstice: duration of seasons for the northern hemisphere.

Perpendicularity

Far more important than closeness to the Sun is another factor governing the intensity of insolation received at the surface of the Earth: the directness (perpendicularity) with which the rays of the sun strike the surface of the Earth. The more direct the solar beam, the greater its potential for heating. Because the Earth is a sphere, only one point on its surface can receive perpendicular (90°) rays from the sun at any one time. Stated differently, only one latitude on the Earth receives perpendicular rays from the sun on any one day of the year. When the sun has reached its zenith (highest altitude in the sky), the latitude on which perpendicular rays from the sun are received on a horizontal portion of the Earth's surface is called the declination of the sun. Figure S14 is a curiously shaped graph called an analemma that shows the declination of the sun for any day of the year. Only locations between 23.5°N and 23.5°S ever receive direct rays from the sun. The most effective solar heating is accomplished at the declination of the sun due to the directness of the solar beam there. The equator receives perpendicular rays twice a year (at the time of the crossover on the analemma) and is never farther than 23.5° of latitude from the location of perpendicular solar rays. Due to the relative directness of the solar rays at low latitudes, effective heating is accomplished annually and only slight seasonality is expressed there. On the other hand, at polar areas where the most indirect rays of the sun are received throughout the year, large seasonality is expressed. Not only a function of revolution and sphericity, perpendicularity of the sun's rays is also a function of the Earth's orbital characteristics, referred to as inclination and parallelism.

Inclination and parallelism

As the Earth revolves around the sun it describes a geometrical plane referred to as the plane of the ecliptic (Figure S15). While orbiting, the Earth rotates about its axis, which is inclined 23.5° from a vertical to the plane of the ecliptic. This tilt of the Earth's rotational axis is called inclination. The orientation of the rotational axis remains fixed in the direction of Polaris, the North Star, throughout revolution so that the Earth's axis is always parallel to itself at any moment in time (Figure S16). This orbital characteristic is called parallelism. The angle of inclination (presently 23.5° from the vertical to the plane of the ecliptic) is often referred to as the obliquity of the ecliptic. The Earth can be thought of as a gyroscope tilted 23.5° from a position normal to the plane described by its orbital path around the sun. Due to inclination and parallelism during revolution, the northern hemisphere (NH) is tilted some increment toward the sun for 6 months of the year, whereas the southern hemisphere (SH) is directed away from the sun. During the other 6 months of the year the NH is tilted away from the sun, and the SH is directed towards the sun. The hemisphere that is tilted in the direction of the sun receives more direct rays and is heated more effectively, resulting in the warm season of that

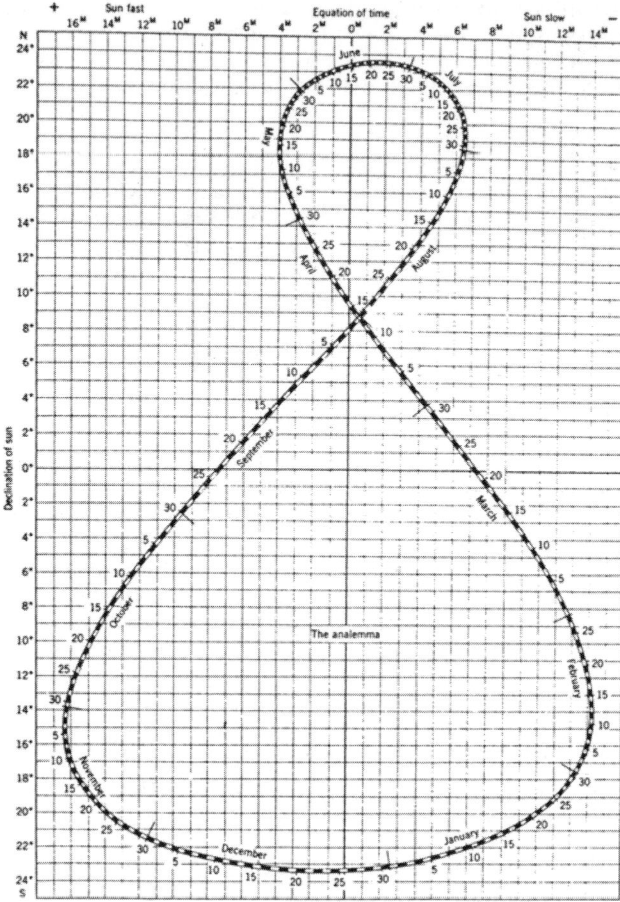

Figure S14 The analemma allows both the sun's declination and the equation of time to be estimated for any date in the year.

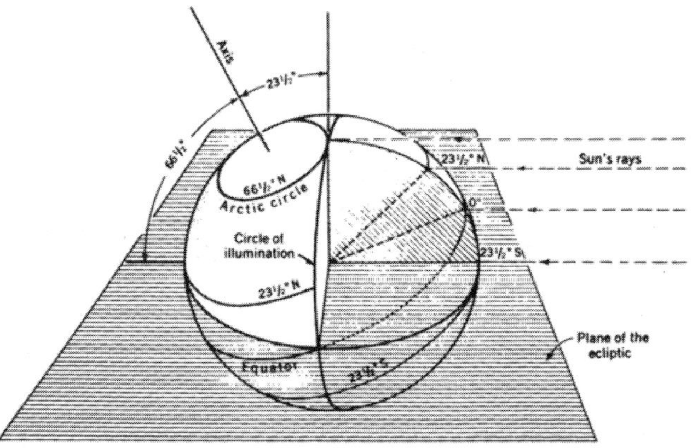

Figure S15 At winter solstice the sun's rays at noon are perpendicular at 23.5°S, the Tropic of Capricorn. All the region lying poleward of the Arctic Circle experiences night for the full 24 hours.

hemisphere. The hemisphere that is tilted away from the sun receives less direct rays and is heated less effectively, resulting in the cool season of that hemisphere. Thus, at the same period of time that the warm season occurs in the NH, the cool season occurs in the SH and vice-versa.

Duration of exposure

While a hemisphere is tilted toward the sun, the surface of the Earth in that hemisphere not only receives more direct rays from the sun but also lengthened time of exposure. Figure S17 contains sunset–sunrise diagrams from which estimates can be made of the number of daylight hours at any latitude during various times of the year. During the month of June when the NH is tilted toward the sun, sunrise occurs before 6:00 a.m. at all northern latitudes and after 6:00 a.m. at all southern latitudes. Similarly, during the month of June, sunset occurs after 6:00 p.m. at all northern latitudes and before 6:00 p.m. at all southern latitudes. During the month of December, when the NH is directed away from the sun, sunrise occurs after 6:00 a.m. at all northern latitudes and before 6:00 a.m. at all southern latitudes, whereas sunset occurs before 6:00 p.m. at all northern latitudes and after 6:00 p.m. at all southern latitudes. Earlier sunrises and later sunsets of the hemisphere that is tilted in the direction of the sun result in longer daylight hours. These lengthened time exposures, coupled with more direct rays of the sun, lead to the terrestrial heating imbalance that causes the seasons across the globe.

The march of the seasons

On or about 21 June the Earth arrives at a point in its revolution about the sun when the sun's perpendicular rays are received in the NH at 23.5°N, the Tropic of Cancer (Figures S16 and S18a). This marks the northernmost latitude to receive perpendicular solar rays (Figure S14). At this time and place the perpendicular rays of the sun are said to stand still in their northward migration. This is referred to as the June solstice (Latin for *Sun stand still*) and marks the beginning of summer in the NH and winter in the SH. Further orbital motion of the Earth in its inclined and parallel attitude results in the southward migration of the perpendicular rays from the sun so that 3 months later, on or about 21 September, the sun's perpendicular rays strike the surface of the Earth at the equator (Figure S18b). The inclination of the Earth's rotational axis is neither directed toward nor away from the sun and heating of the NH and the SH is accomplished impartially. With the declination of the sun at the equator, there are 12 hours of daylight and 12 hours of nighttime in both hemispheres. This time is referred to as the September equinox (Latin for *equal night*) and marks the beginning of autumn in the NH (autumnal equinox) and the beginning of spring in the SH. Continuation of the Earth in its orbit results in the SH being directed toward the sun. The most direct rays from the sun are received in the SH, causing preferential heating there. On or about 21 December the perpendicular rays of the sun are positioned at 23.5°S, the Tropic of Capricorn (Figure S18c). This latitude marks the southernmost location to receive perpendicular rays from the sun. This time is referred to as the December solstice, after which further orbital motion of the Earth results in the northward migration of the sun's perpendicular rays. The December solstice signals the beginning of the winter season in the NH and the summer season in the SH. With more orbital motion, the perpendicular rays of the sun do migrate equatorward where 3 months later, on or about 21 March, the perpendicular rays of the sun strike the surface of the Earth at the equator. This is the March equinox and marks the beginning of spring for the NH (vernal equinox) and fall for the SH. Thus, the seasons advance every year in an organized manner in both hemispheres with spring following winter, summer following spring, fall following summer, and winter following fall.

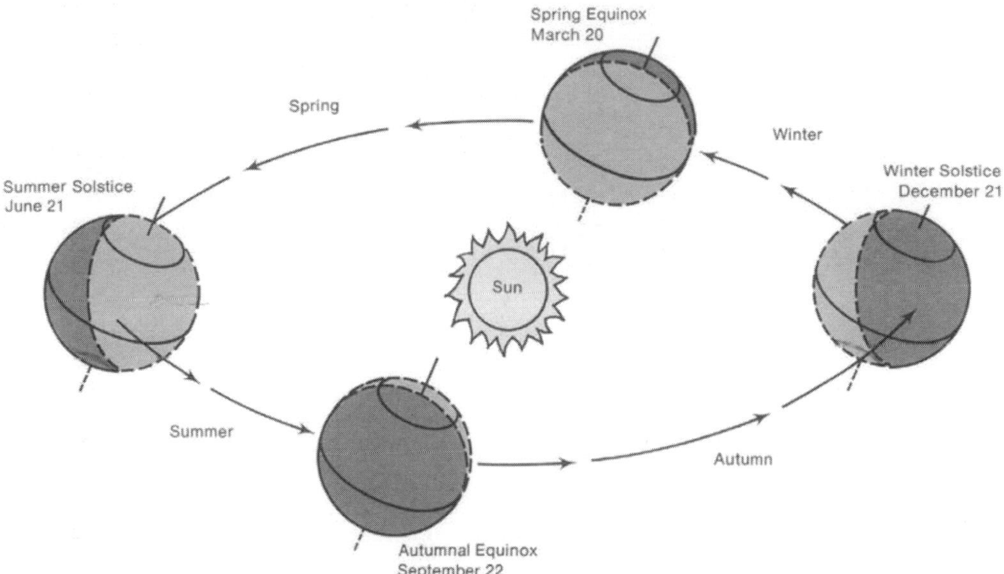

Figure S16 Revolution of the Earth about the sun. The relative variation in the inclination of the Earth's axis of rotation with respect to the sun is a principal cause of the seasons (after Strahler, 1971).

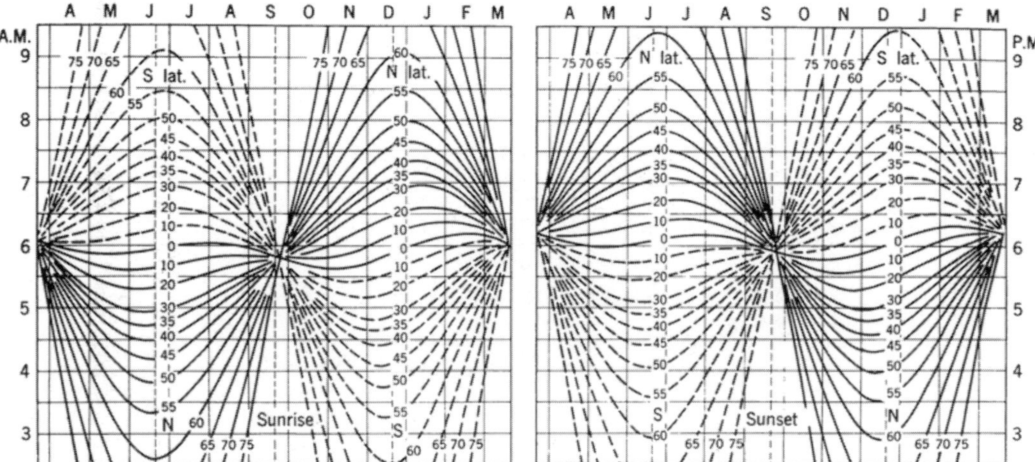

Figure S17 Sunset–sunrise diagram.

Land of the midnight sun

Nowhere are the various seasons less expressed than at the equator and more expressed than at the poles. Because the Earth is a sphere, only half of it is ever illuminated by the rays of the sun. At the sun's distance away from the Earth, its rays are approximately parallel as they are intercepted by the spherical Earth (Figure S18). The dividing line between the sunlit portion and the darkened portion of the Earth is called the circle of illumination. On the dates of the equinoxes, the circle of illumination passes through the poles resulting in 12 hours of daylight and 12 hours of nighttime at every location on the Earth (Figure S18b). Although the length of daylight and nighttime is equal everywhere across the Earth, the angle of incidence (the angle at which the rays of the sun strike the surface of the Earth) varies as a function of latitude. At the equinoxes when the angle of incidence is 90° (perpendicular) at the equator, it is zero (tangential) at the poles. On the dates of the June and December solstices the circle of illumination passes through 66.5°N, the Arctic Circle, and 66.5°S, the Antarctic Circle. This causes the two hemispheres to be disproportionately illuminated and heated. At the June solstice the area of the Earth's surface between the Arctic Circle and the North Pole is continuously illuminated during rotation, whereas the area of the Earth's surface between the Antarctic Circle and the South Pole is continuously darkened (Figure S18a). This results in 24 hours of daylight for all locations higher than 66.5°N and 24 hours of nighttime for all locations below 66.5°S. At the

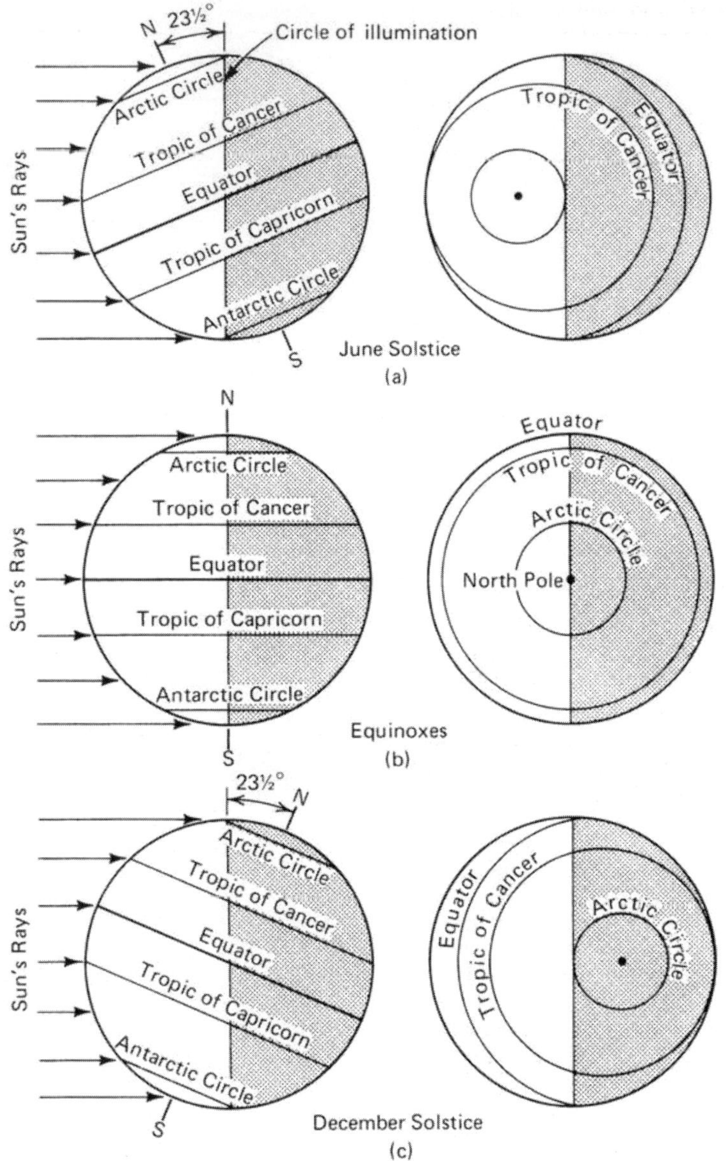

Figure S18 Characteristics of the summer and winter solstices and equinoxes (from Lutgens and Tarbuck, 1979).

December solstice it is the area of the Earth's surface below the Antarctic Circle that receives 24 hours of daylight, whereas nighttime prevails in the NH above the Arctic Circle (Figure S18c). Both poles receive 24 hours of daylight for a continuous 6 months of the year. The polar areas remain relatively cool year-round due to the low angle of incidence with which the solar rays strike the surface of the Earth.

Delineation of the seasons

With polar exceptions the identification of four periods of time with coherent climatic activities can be a difficult task. Barry and Perry (1973) suggest that adherence to just four seasons of the year is a gross oversimplification and that the actual seasons of the Earth are many more in number. Table S4 contains the average beginning dates of the natural seasons as identified by various studies. Although there is some disagreement about the average beginning dates of the seasons, each of the studies identifies more than the traditional four natural seasons. The recognition of the seasons is further complicated by regular recurrences of seasonally anomalous periods of weather called singularities. Singularities vary in length but are most frequently expressed for a week or two. In many climatic records of the NH, a general warming trend is recognized during late October or early November. This period is called the "Indian Summer" in North America and is known by other names in various parts of the world. Another popularized singularity that occurs in the NH is that of a warming period in early to middle January. This warming trend is referred to as the January Thaw,

Table S4 Natural calendars according to various studies

Reference[a]	Average beginning dates of the seasons					
	Autumn	Winter I	Winter II	Early spring	Spring	Summer
Baur (1958)	16 Aug.	16 Nov.	1 Jan.	15 Feb.	1 Apr.	17 May
Bradka (1966)	30 Aug.	1 Nov.	19 Jan.	9 March	21 Apr.	7 June
Bryson and Lahey (1958)	21 Aug.	1 Nov.			21 March	25 June
Chu (1962)[b]	8 Aug.	7 Nov.		4 Feb.		6 May
Dzerdzeevski (1957)	26 Aug.	7 Oct.	6 Dec.	13 March	19 Apr.	22 May
Multanovski (1920, in Bradka, 1966)	13 Aug.	5 Oct.	27 Dec.	13 March		18 May
Lamb (1950)	10 Sept.	20 Nov.	20 Jan.		1 Apr.	18 June
Sakata (1950)	29 Aug.	27 Nov.	27 Dec.	10 Feb.	22 March	10 June
Yoshino (1968)	21 Aug.	26 Nov.	19 Dec.	7 Feb.	17 March	18 May
Craddock (1957)	15 Sept.		4 Dec.	13 March	16 June	
Zaharova (1969)	24 Aug.	16 Nov.	25 Dec.	Early spring	10 March	7 May

Note: Some of the definitions used by different workers are not identical. Baur and Yoshino apparently include "early summer" in summer and "early autumn" in autumn.
[a] For references see Barry and Chorley (1998).
[b] These refer to some of the twenty-four "solar terms" of 15-day duration recognized in China ca. 247 BC.
Source: After Barry and Perry (1973).

although surface temperatures often do not reach the melting point of water.

For long-term global climate studies it is often convenient to designate winter as DJF; summer as JJA and so on. In annual statistics, December is often set forward to the following year. When asatronomical seasons are involved, their recurrence time may be greatly amplified in climate effect by the season. Thus, the quasibiennial oscillation (QBO) of 2.172 years, will only return at the same seasonafter 13 years (2 × 5 Biennial + 1 triennial)

Associated seasons

Complementing the above astronomically defined natural seasons are a host of related physical and cultural seasons whose activities are made possible by the expression of the natural seasons. The frost-free season refers to the time period between the last killing frost in the spring and the first killing frost of the fall. This time period is also referred to as the growing season and is often subdivided into smaller units of time, including the planting season, sprouting season, and harvesting season. Climatically, various elements of the weather are often temporally concentrated, making possible the identification of a wet–dry season, hot–cold season snow season, and a hail season, to name a few. Generally, a severe weather season is recognized in the North American continent by the occurrence of tornadoes in the late spring–early summer of the year, whereas a hurricane season occurs most often in late summer. Culturally, many human activities are related to the climatic events expressed during the natural seasons, i.e. tourist season, skiing season, construction season, and various sports-related seasons.

Ted Alsop

Bibliography

Aguado, E., and Burt, J.E., 2001. *Understanding Weather and Climate*. Upper Saddle River, NJ: Prentice-Hall.

Ahrens, D.L., 2000. *Meteorology Today: an introduction to weather, climate, and the environment*, 6th edn. Pacific Grove, CA: Brooks & Cole.

Barry, R.G., and Chorley, R.J., 1998. *The Atmosphere, Weather, and Climate*, 7th edn. London: Routledge.

Barry, R.G., and Perry, A.H., 1973. *Synoptic Climatology*. London: Methuen.

Bradka, J., 1966. Natural seasons in the Northern Hemisphere. *Geofys. Sbornich* **14**, 597–648.

Bryson, R.A., and Lahey, J.F., 1958. *The March of the Seasons*. Madison, WI: Meteorology Department, University of Wisconsin.

Cole, F.W., 1980. *Introduction to Meteorology*. New York: Wiley.

Griffiths, J.F., and Driscoll, D.M., *Survey of Climatology*. Columbus, OH: Merrill.

Lutgens, F.K., and Tarbuck, E.J., 2001. *The Atmosphere*, 8th edn. Upper Saddle River, NJ: Prentice-Hall.

Mackenzie, F.T., 1998. *Our Changing Planet*, 2nd edn. Upper Saddle River, NJ: Prentice-Hall.

Moran J.M., and Morgan, M.D., 1997. *Meteorology: the atmosphere and the science of weather*, 5th edn. Upper Saddle River, NJ: Prentice-Hall.

Oliver, J.E., and Hidore, J.J., 2002. *Climatology: an atmospheric science*. Upper Saddle River, NJ: Prentice-Hall.

Oliver, J.E., and Fairbridge, R.W., 1987. *Encyclopedia of Climatology*. New York: Van Nostrand Reinhold.

Strahler, A.N., 1971. *The Earth Sciences*. New York: Harper & Row.

Wahl, E.W., 1953. Singularities and the general circulation. *Journal of Meteorology*, **10**: 42–45.

Cross-references

Indian Summer
Kepler's Laws
Local Winds
Middle Latitude Climates
Monsoons and Monsoon Climate
Solar Radiation
Temperature Distribution

SENSIBLE HEAT

Sensible heat refers to heat that can be "sensed" by a thermometer or felt by an individual. The temperature of a parcel of air is defined as the average kinetic energy of its molecules. As more energy is added to a parcel of air, the molecules move faster (obtain more kinetic energy), which we interpret as an increase in its temperature. Consequently, the sensible heat of an air parcel is directly related to the temperature of the parcel.

In contrast to energy transfer through radiation exchange (both solar and terrestrial radiant energy) and latent heat (stored potential energy because the water is in the vapor state), sensible heat flux includes the transport of energy through conduction (molecular diffusion) and convection (moving air currents). Turbulent atmospheric motions dominate the surface boundary layer such that convection dominates the sensible heat exchange, except for within a millimeter of the surface.

Generally, interest focuses only on changes in the sensible heat and not with its absolute magnitude. Mathematically, if the pressure is held constant, the change in sensible heat (ΔH) can be related to the change in temperature (ΔT) by

$$\Delta H = c_p \Delta T$$

where c_p is the specific heat at constant pressure and, at 0°C, equals approximately 1.0 joules per gram per °C for dry air, 1.95 joules per gram per °C for water vapor, and 4.22 joules per gram per °C for liquid water. To the atmospheric physicist the sensible heat of a parcel of air is equal to its specific enthalpy, a thermodynamic property given by the First Law of Thermodynamics.

The pole-to-equator temperature gradient (warm equator and cold poles) causes a net poleward transfer of energy through sensible and latent heat fluxes as well as transport of heat through ocean currents. Of these three, however, sensible heat is the dominant transport mechanism, accounting for about half of the total poleward energy transport in the tropics and midlatitudes and virtually all of it in the higher latitudes. Tropical deserts exhibit the largest exchange of sensible heat to the atmosphere.

David R. Legates

Bibliography

Hartmann, D.L., 1994. *Global Physical Climatology*. San Diego, CA: Academic Press.

Wallace, J.M., and Hobbs, P.V., 1977. *Atmospheric Science: an introductory survey*. Orlando, FL: Academic Press.

Cross-references

Bowen Ratio
Energy Budget Climatology
Climate Comfort Indices
Latent Heat
Phase Changes

SIBERIAN (ASIATIC) HIGH

The Siberian (Asiatic) High is a semipermanent cell of high pressure centered poleward of 45°N in northeastern Siberia. It is one of the major centers of action during the winter in the northern hemisphere (Rossby, 1945).

The influence of the high is rarely felt above 2400 m (8000 ft), as it is made up of cold, dense air. The axis of the high tilts strongly southward as the intensity decreases. No indication of the sea level high is present at the 3000-m (10 000-ft) level (Petterssen, 1969).

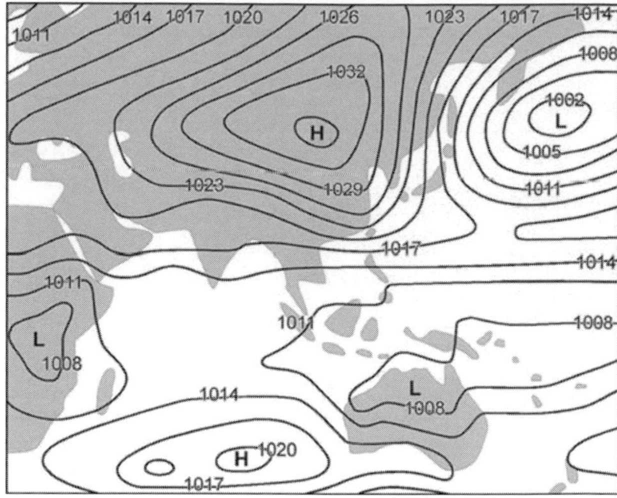

Figure S19 Mean January sea level pressure showing location of the Siberian High.

Daily synoptic weather maps for the northern hemisphere show many moving high- and low-pressure systems related to the interaction of cold and warm airmasses along the polar front. These moving disturbances can be eliminated by constructing mean charts, thereby delineating clearly the semipermanent nature of the Siberian High.

As shown in Figure S19, the Siberian High has a mean January sea level pressure of 1035 mb (30.56 in), 22 mb (0.64 in) above the mean sea level pressure of 1013 mb (29.92 in) for the world. This rather high mean pressure is attributed to the extensive continental cooling during winter of the vast Asiatic land mass (44 134 000 sq km; 17 040 000 sq miles). Indeed, the highest pressure ever recorded on Earth occurred under the Siberian High. On 31 December 1968, a pressure of 1084 mb (32.01 in) was measured at Agata, Siberia (66.83°N, 98.71°E) during an extremely cold period (Allaby, 2001). By contrast, the weaker thermal influence of the smaller North American landmass (24 063 000 sq km; 9 300 000 sq miles) results in an attenuated and divided anticyclone of 1020 mb (30.12 in) (Willett and Sanders, 1959).

The cold, dry air that develops over Asia during the winter is limited on the south by the Himalayas and other mountain ranges extending westward toward southern Europe. The mountain ranges along the eastern coast of Asia afford some restriction to the eastward spread of the cold air, whereas no major obstacles exist on the western side to prevent warmer air coming from the Atlantic. As a consequence of these topographical configurations, the Siberian High is further removed from the maritime influences of the western coast. Generally, the high covers the large area from the mountain slopes around the Caspian Sea to the Anadyr range in northeastern Siberia, centering around the Baikal region. The high is so powerful that the subtropical high-pressure belt is completely submerged (Petterssen, 1969).

The Siberian High tends to be displaced toward the Pacific side of Eurasia during periods of strong circulation. Westward displacement in the direction of Europe occurs during alternating periods of weak circulation. Weather conditions are deeply affected by these pressure changes. For example, northwest Europe will experience an outflow of extremely cold, dry continental air from Asia during the time of weak circulation (Rossby, 1945).

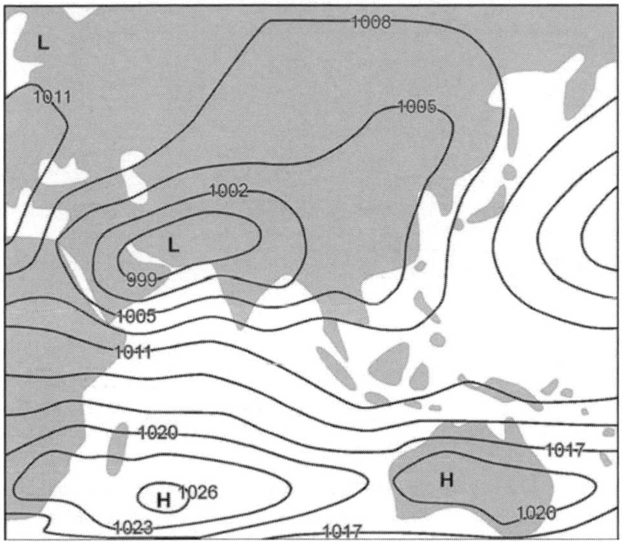

Figure S20 July mean sea level pressure over Asia.

Winds circulate clockwise around northern hemisphere anticyclones. Thus, the winter monsoon (monsoon winds being those winds whose direction reverses with the seasons) blows from the northeast in central Asia, bringing cool, dry air to the Indian subcontinent (Strahler and Strahler, 2005).

The Siberian High is associated with the coldest temperatures ever recorded in the northern hemisphere. Average January temperatures are −46.2°C (−50.1°F) for Verkhoyansk (67.7°N, 133.6°E) and −51.0°C (−60°F) for Oimekon (63.4°N, 143.2°E) in northeastern Siberia. In addition, the highest values for continentality (which by one definition is simply the temperature range between two extreme months) are found in the same region (Landsberg, 1958).

Continental heating in summer results in a low-pressure cell of thermal origin centered equatorward of 30°N. The net effect of the heating of the Asiatic landmass is to pull the tropical belt of low pressure poleward across the Tropic of Cancer (23.5°N).

As shown in Figure S20, the Asiatic Low is centered over the hot deserts and plateaus of southwest Asia during the summer. The July mean sea level pressure is 999 mb (29.5 in), or 14 mb (0.42 in) lower than normal sea level pressure of 1013 mb (29.92 in). An analogous situation develops over the arid southwestern portion of North America. A weaker low-pressure cell exists with a July mean of 1011 mb (29.86 in). The difference in the dimensions and pressure of the two continental summer lows is of course attributed to the greater land mass of Asia (Trewartha and Horne, 1980).

Winds follow the pressure gradient from high to low and flow counterclockwise around a low in the northern hemisphere. Thus, the development of the Asiatic Low results in the summer monsoons. The summer southwest monsoon winds blow across the open warm waters of the Indian Ocean. The air is forced to ascend and adiabatically cools as it strikes the high plateaus and mountains of India. This cooling lowers the air temperature below the dew point, resulting in the conspicuous rainy season of the monsoonal climates (Strahler and Strahler, 2005).

Robert M. Hordon

Bibliography

Ahrens, C.D., 2003. *Meteorology Today: An Introduction to Weather, Climate, and the Environment,* 7th edn. Brooks/Cole (Thomson Learning).

Akin, W.E., 1991. *Global Patterns: climate, vegetation, and soils*. Norman, OK: University of Oklahoma Press.

Allaby, M., 2001. *Encyclopedia of Weather and Climate*. New York: Facts on File.

Barry, R.G., and Chorley, R.J., 1998. *Atmosphere, Weather, and Climate*, 7th edn. London: Routledge.

Barry, R.G., and Chorley, R.J., 2003. *Atmosphere, Weather, and Climate*, 8th edn. London: Routledge.

Curry, J.A., and Webster, P.J., 1999. *Thermodynamics of Atmospheres and Oceans*. San Diego, CA: Academic Press.

Graedel, T.E., and Crutzen, P.J., 1993. *Atmospheric Change: an Earth system perspective.* New York: Freeman.

Hartmann, D.L., 1994. *Global Physical Climatology.* San Diego, CA: Academic Press.

Hidore, J.J.,and Oliver, J.E., 2001. *Climatology: An Atmospheric Science, Upper Saddle River: Prentice Hall.*

Landsberg, H., 1958. *Physical Climatology*, 2nd edn. DuBois, PA: Gray.

Lutgens, F.K., and Tarbuck, E.J., 2004. *The Atmosphere: An Introduction to Meterology,* 9th edn. Upper Saddle River, New Jersey: Pearson Prentice Hall.

Nese, J.M., and Grenci, L.M., 1998. *A World of Weather: fundamentals of meteorology*, 2nd edn. Dubuque, Io: Kendall/Hunt.

Oliver, J.E., and Hidore, J.J., 2002. *Climatology: An Atmospheric Science*, 2nd edn. Upper Saddle River, New Jersey: Pearson Prentice Hall.

Petterssen, S., 1969. *Introduction to Meterology,* 3rd edn. New York: McGraw-Hill.

Robinson, P.J. and Henderson-Sellers, A., 1999. *Contemporary Climatology*, 2nd edn. Harlow: Longman.

Rossby, C.G., 1945. The scientific basis of modern meteorology. In Berry, F.A., Bollay, E., and Beers, N.R., eds., *Handbook of Meteorology*, New York: McGraw-Hill, pp. 502–529.

Strahler, A., and Strahler, A.N., 2005. *Physical Geography,* 3rd edn., Hoboken, New Jersey: John Wiley.

Trewartha, G.T., and Horne, L.H., 1980. *An Introduction to Climate,* 5th edn. New York: McGraw-Hill.

Wallace, J.M., and Hobbs, P.V., 1997. *Atmospheric Science: an introductory survey.* San Diego, CA: Academic Press.

Willett, H.C., and Sanders, F., 1959. *Descriptive Meteorology*, 2nd edn. New York: Academic Press.

Cross-references

Airmass Climatology
Atmospheric Circulation, Global
Azores (Bermuda) High
Centers of Action
Continental Climate and Continentality
Monsoons and Monsoon Climate
North American (Canadian) High
Pacific (Hawaiian) High
Zonal Index

SINGULARITIES

Singularities are specific types of weather that occur fairly regularly at a specific time of the year. They are often warm or cold, wet or dry, conditions that are considered departures from the normal annual march of temperature and precipitation.

In the United State, the best-known singularities are the Indian Summer, the January Thaw, and April Showers. Clearly such features are not observed throughout all climate regions.

Table S5 Brief summary of studies of the January Thaw

Study	Geographical coverage	Period of record	Analyzed variables	Conclusions
Esten and Mason (1910)	Storrs, CT	1888–1909	Record max and min; max, min and avg. temps	Singularity
Marvin (1919)	Continental US	1778–1865	Avg. weekly temps	No strong singularity
Nunn (1927)	Northeast US	1873–1925	Avg. temps	Singularity
Slocum (1941)	Northeast US	1871–1939	Avg. temps	No conclusion
Wahl (1952)	Northeast US	1873–1952	Avg. temps, sea-level pressures	Singularity
Wahl (1953)	Boston, MA	1873–1952	Avg. temps	Singularity
Brier (1954)	N. hemisphere	1899–1939	Sea level pressure	No conclusion
Lautzenheiser (1957)	Boston, MA	1911–1950	Avg. temps	No strong singularity
Dickson (1959)	Nashville, TN	1871–1950	Avg. temps, tornadoes	Singularity
Bingham (1961)	Northeast US	1896–1956	Avg. weekly temps	No strong singularity
Duquet (1963)	Northeast US	1872–1961	Weekly avg. temps	Singularity
Newman (1965)	Boston, MA	1872–1964	Max and min temps	No strong singularity
Frederick (1966)	US and SW Canada	1897–1956	Avg., max, and min temps	Singularity
Hayden (1976)	East Coast US	1954–1970	Means of surf heights	Singularity
Logan (1982)	Portland, Maine	1965–1979	Avg. temps	No conclusion
Lanzante and Hernack (1982)	New Brunswick, NJ	1858–1981	Max temps	Singularity
Lanzante (1983)	N. America, Atlantic and Pacific Oceans	1947–1976	700-mb heights	Singularity
Kalnicky (1987)	N. hemisphere	1899–1969	Sea-level pressure	No conclusion
Guttman and Plantico (1987, 1989)	Eastern US	1951–1980	Max and min temps	Singularity
Guttman (1991)	Central Park, NY	1876–1987	Max and min temps	No strong singularity

From Godfrey et al., 2002, where references to cited articles are given.

The January Thaw, for example, is an anomalous warming that occurs in the northeastern United States in mid- to late January. As Table S5 indicates, this feature has been the topic of many studies. The conclusions are quite variable, with some indicating the existence of a singularity and others finding no consistent record.

John E. Oliver

Bibliography

Brier, G.W., 1954. A note on singularities. *Bulletin of the American Meteorological Society*, **35**: 378–379.

Bryson, R.A., and Lowry, W.P., 1955. Synoptic climatology of the Arizona summer precipitation singularity. *Bulletin of the American Meteorological Society*, **36**: 329–339.

Godfrey, C.M., Wilks, D.S., and Schultz, D.M., 2002. Is the January Thaw a statistical phantom. *Bulletin of the Meteorological Society*, **83**: 53–62.

Kalnicky, R.A., 1987. Seasons, singularities, and climate changes over the midlatitudes of the Northern Hemisphere during 1899–1969. *Journal of Climatology and Applied Meteorology*, **26**: 1496–1510.

Logan, R., 1982. The January thaw. *Weatherwise*, **35**: 263–267.

Nunn, R., 1927: The "January thaw". *Monthly Weather Review*, **55**: 20–21.

Rebman, E.J., 1953. Singularities in weather at Walla Walla, Wash., as related to the index of zonal westerlies. *Monthly Weather Review*, **81**: 386–387.

Talman, C.F., 1919. Literature concerning supposed recurrent irregularities in the annual march of temperature. *Monthly Weather Review*, **47**: 555–565.

Wahl, E.W., 1953. Singularities and the general circulation. *Journal of Meteorology*, **10**: 42–45.

Cross-reference

Indian Summer

SNOW AND SNOW COVER

Importance of snow cover

Snow cover refers to the blanket of snow covering the ground, and includes the concepts of depth and areal extent (Sturm et al., 1995). Snow cover is a key component of the global climate system through its role in modifying energy and moisture fluxes between the surface and the atmosphere, and through its role as a water store in hydrological systems. Snow is a highly reflective material; the reflectivity or ***albedo*** of new snow is 0.8–0.9, which means that 80–90% of the incident solar energy is reflected away from the surface. This property, combined with the excellent insulating characteristics of a snow cover, dramatically reduces the energy exchange between the surface and the atmosphere. Empirical studies have shown that mean surface air temperatures are typically 5°C colder when a snow cover is present (Groisman and Davies, 2001). This positive feedback is responsible for the rapid expansion of northern hemisphere snow cover extent in October–November. The larger-scale climatic significance of the snow–albedo feedback is modulated by cloud cover and by the small amount of total solar radiation received in high latitudes during winter months. Groisman et al. (1994a) observed that snow cover exhibited the greatest influence on the Earth radiative balance in the spring (April to May) period when incoming solar radiation was greatest over snow-covered areas. The impact of snow on surface reflectivity is an example of a ***direct feedback*** to the climate system. Snow is also involved in a number of ***indirect feedbacks*** such as its influence on modulating sea ice growth, and linkages to cloud cover and surface temperature through snow's role in soil moisture recharge (Cess et al., 1991; Randall et al., 1994). There is an extensive body of literature linking snow cover feedbacks to monsoon circulations (e.g. Vernekar et al., 1995; Gutzler and Preston, 1997). However, recent research suggests snow's role may be limited (Robock et al., 2003). Groisman and Davies (2001) and Armstrong et al. (2003) provide more detailed reviews of the climatic significance of snow.

Snow cover is also a critical component in natural and human systems. Snow accumulation is an important resource for drinking water, irrigation, hydroelectrical generation and natural river ecosystems. The presence of an insulating snow cover is also critical for ecological systems (Jones et al., 2001). Beneath even 30 cm of snow organisms and soil are well protected from the extreme diurnal temperature fluctuations occurring at the snow surface, and exchanges of carbon, methane and other gases between the land surface and the atmosphere can continue during the winter period (Sommerfeld et al., 1993). Snow influences on soil temperature are also important for hydrology. When soil moisture freezes, the hydraulic conductivity is reduced, leading to either more runoff due to decreased infiltration or higher soil moisture content due to restricted drainage. Knowing whether the soil is frozen or not is important in predicting surface runoff and spring soil moisture reserves (Zhang and Armstrong, 2001).

Finally, snow supports a multi-billion-dollar recreation and tourism industry in midlatitudinal mountain regions of the world (e.g. Rocky Mountains, Appalachians, Alps). But snow can also be a nuisance and a hazard. Snow can play havoc with driving conditions through reduced visibility and traction, heavy loads of snow can collapse buildings, and rapid melt of snow can cause flooding. In 1987–1988 the snow removal budget for the city of Montréal was $47 million (Phillips, 1990). In mountainous areas of the world, snow avalanches are an ever-present hazard with the potential for loss of life, property damage and disruption of transportation. McClung and Schaerer (1993) provide an excellent review of avalanche science.

Snowfall

The initial characteristics of a ***snowpack*** start with the snowfall process. While the physics of snowfall formation is highly complex (see review by Schemenauer et al., 1981), the basic ingredients are a source of moisture, a mechanism for vertical motion in the atmosphere to cause precipitation to form (e.g. orographic lift, frontal lift, convection), and air temperatures at or below 0°C. The form of the initial ice crystals (e.g. columnar, platelike, dendritic) depends on the temperature at formation, and the crystals may undergo considerable change as they fall through layers with different temperature and humidity. The shape and size of the final snowflakes thus depend on a host of factors, which can be further complicated by wind action breaking crystals into smaller fragments. The density (mass per unit volume, expressed in units of $\mathrm{kg\,m^{-3}}$) of natural snowfall varies with crystal type, size and liquid water content, but is typically in the range 50–120 $\mathrm{kg\,m^{-3}}$ (Pomeroy and Gray, 1995). Snowfall densities vary from event to event, and there are important regional differences in snowfall densities based on temperature and precipitation climates. While some data suggest snowfall density is air temperature-dependent (e.g. Pomeroy and Gray, 1995), snowfall density is not strictly a function of air temperature, and densities can vary greatly for any given air temperature (Doesken and Judson, 1997). The density and crystal structure of newly fallen snow are important in many engineering and operational fields such as avalanche risk forecasting, design of roofs and structures to withstand snow loading, the accretion of snow on transmission lines (a particular problem in Japan where wet snow is frequent), aircraft deicing, and vehicle trafficability.

Spatial extent

Snow cover is encountered over most of the northern hemisphere mid- and high-latitudes during the winter season, and over many mountainous regions of the world for extended periods. Figure S21 shows the seasonal range in snow cover over the northern hemisphere from satellite data. The temporal variability is dominated by the seasonal cycle with average snow cover extent ranging from an average minimum extent of 3.6 million $\mathrm{km^2}$ in August, to an average maximum extent of 46.8 million $\mathrm{km^2}$ in late January (Table S6). Most of the global snow cover extent is located over the northern hemisphere: with the Antarctic and Greenland landmasses excluded, the southern hemisphere mean maximum terrestrial snow cover extent is less than 2% of the corresponding winter maximum snow cover extent over the northern hemisphere. In the southern hemisphere exclusive of Antarctica, most of the seasonal snow cover extent is located over South America.

While annual snow accumulations can exceed several meters in humid mountainous regions of the world, shallow snow covers in the order 10–30 cm depth are typical for large areas of the northern hemisphere that experience relatively cold, dry winters (e.g. midlatitudinal continental and polar regions). The contrast in snow depths between regions close to winter moisture sources (i.e. Pacific and Atlantic oceans) is evident in Figure S22, which shows mean February snow depths over North America derived from surface observations.

Snow cover characteristics

Over a winter season a snowpack grows and develops as a complex layered structure reflecting the weather and climate conditions of each precipitation event, and metamorphic and melt processes in the snowpack. The international standards (units, terminology, and symbols) for describing the vertical structure of a snowpack are provided in Colbeck et al. (1990). The three basic properties used to describe a snow cover are depth, density and snow water equivalent (SWE) related as follows:

$$\mathrm{SWE\ (mm)} = 0.01\, h_s \times \rho_s$$

where h_s is the depth of snow (cm) and ρ_s is the density of snow ($\mathrm{kg\,m^{-3}}$). The conversion from a mass of snow ($\mathrm{kg\,m^{-2}}$) to a depth of water (mm) is based on the fact that 1 mm of water spread over an area of 1 $\mathrm{m^2}$ weighs 1 kg. Over a snow season a typical snowpack is characterized by an extended accumulation period that may include discrete melt events, followed by a rapid melt or ablation period (see example in Figure S23). Snowpack

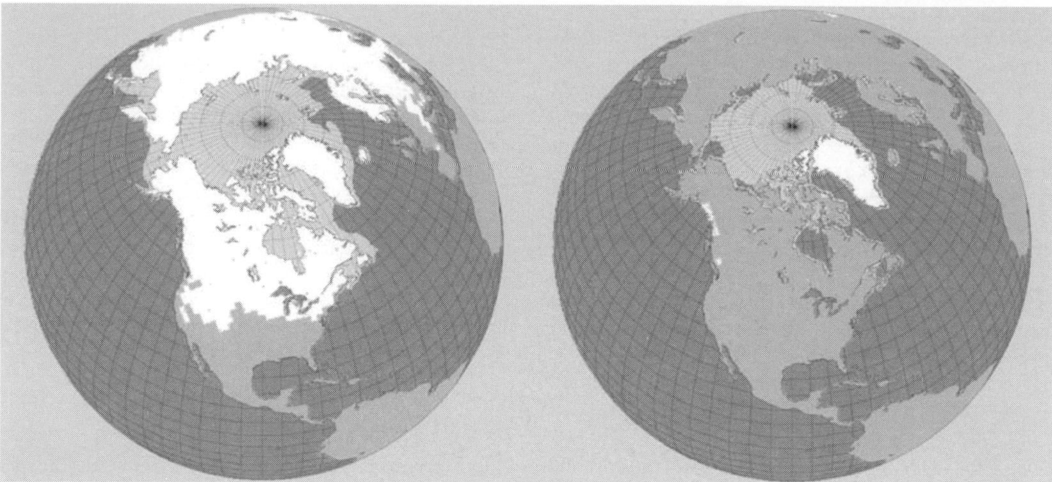

Figure S21 Mean seasonal variation in northern hemisphere snow (white) and sea ice (light grey) extent between February (left) and August (right) as derived from satellite data. Data are from Weekly Snow Cover and Sea Ice Extent, National Snow and Ice Data Center, 1996.

Table S6 Snow cover extent in million square kilometers for terrestrial portions of the globe (after Table 6.1 in Goodison et al., 1999)

Northern hemisphere	Excluding Greenland	Including Greenland
Maximum (late January)	44.8	46.8
Minimum (late August)	1.4	3.6
Southern hemisphere	Excluding Antarctica	Including Antarctica
Maximum (June–October)[b]	0.85[a]	14.5[c]
Minimum (January–March)[b]	0.07	13.7[c]
Global extent	Excluding Greenland + Antarctic	Including Greenland + Antarctic
Northern hemisphere winter	44.9	60.5
Southern hemisphere winter	2.3	18.1

[a] The main contribution is South America, which was estimated to have a mean winter snow cover extent of 0.45×10^6 km^2 over the 1988–2001 period (Foster et al., 2003). New Zealand mean winter snow cover extent is estimated to be 0.06×10^6 km^2 (Fitzharris and McAlevey, 1999).
[b] These month ranges are approximations since seasonal snow cover extent in southern hemisphere mountainous regions exhibits strong month-to-month and year-to-year variability.
[c] Seasonal snow cover variation in Antarctic snow cover was not taken into account.

density typically increases rapidly at the start of the season due to changes in the size, shape and bonding of snow crystals. This process is termed ***metamorphism***, and is caused by temperature and water vapor gradients, crystal settlement and wind packing (Pomeroy and Gray, 1995). During the melt season, snow density can also experience rapid increases from melt and refreezing. Melting snow densities typically range from 350 to 500 kg m^{-3}. In extremely cold environments with shallow snowpacks, snow density can actually decrease over time due to the formation of ***depth hoar*** in response to strong temperature and vapor gradients through the snowpack. A detailed review of the physics of snow metamorphism is provided by Langham (1981).

At a continental scale, differences in climate give rise to distinct snow cover–climate regions. Sturm et al. (1995) were able to classify global snow cover into six distinct classes with unique stratigraphic attributes (tundra, taiga, alpine, maritime, prairie and ephemeral) using wind, precipitation and air temperature data. Snow cover also exhibits extensive local scale variations due to the effects of wind redistribution, vegetation interception and sublimation, and the influence of topography on wind speed and the local energy balance. An extensive review of these processes is provided by Pomeroy and Gray (1995). Sublimation losses from blowing snow and tree canopies are important when determining the water budget of hydrological systems. Pomeroy and Gray (1995) estimated sublimation loss over prairie environments to be 15–41% of annual snowfall, and they estimated that approximately one-third of total snowfall falling on spruce and pine was lost through canopy sublimation.

Snowpack modelling

The ability to accurately model the accumulation and ablation of a snow cover is important for many applications such as flood forecasting, reservoir management, and climate system simulations. A wide range of approaches have been used to simulate the accumulation and ablation of a snowpack from simplified degree-day melt models to more complex physical models of the energy and mass balance such as Anderson (1976). In more

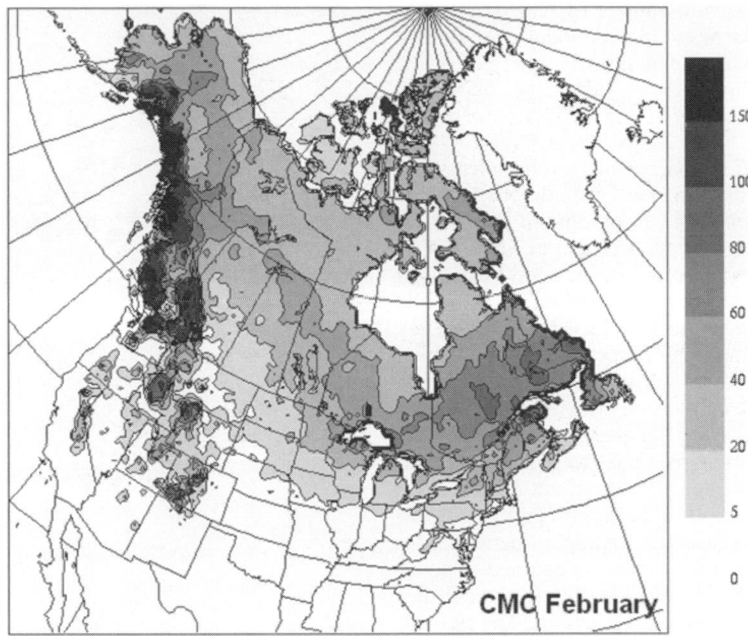

Figure S22 February mean snow depth (cm) over North America derived from an objective analysis of surface observations over the period 1979–1997 (Brown et al., 2003).

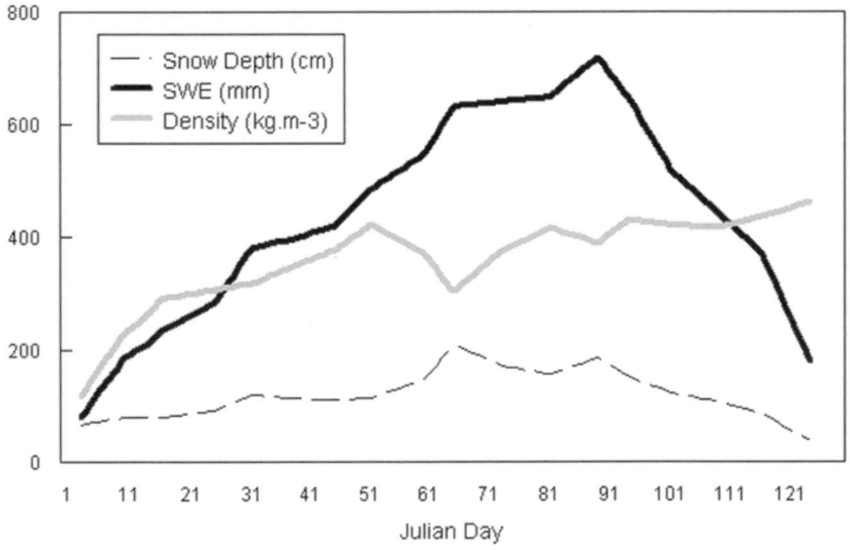

Figure S23 Temporal variation in snow depth (cm), SWE (mm) and mean snowpack density (kg. m^{-3}) from weekly snowpit observations at Col de Porte, France, 1995. Note the effect of the major snowfall event on Day 66 in decreasing the average snowpack density. Data courtesy E. Martin, Météo-France.

recent years, modelers have incorporated the physics of snow crystal size and shape evolution into physical models (e.g. Brun et al., 1992) along with improved understanding of heat transfer at the snow surface (Jordan et al., 1999), snowmelt (Marsh, 1999) and blowing snow (Pomeroy and Gray, 1995). The latest generation multilayer physical snowpack models have been shown to provide accurate simulations of point snowpack properties and melt at open locations for a range of sites and snow climates. Accurate simulation of snow cover over an area is a much more difficult problem, however, since this involves taking into account local-scale variations in topography and vegetation that are the main factors driving spatial variation in snow cover accumulation and melt. Important areas of ongoing snow modeling research include improved representation of blowing snow, vegetation and canopy processes, treatment of patchy snow, as well as development of approaches to include local-scale heterogeneity into global and regional climate models.

Measuring snow

The measurement of ***snowfall*** (the depth of freshly fallen snow that accumulates during the observing period, traditionally measured with a ruler), ***solid precipitation*** (the amount of liquid water in the snowfall intercepted by a precipitation gauge), and snow on the ground (depth, SWE) is a science in its own right. Reviews of measurement equipment and techniques are provided by WMO (1981), Goodison et al. (1981), Sevruk (1992), Pomeroy and Gray (1995), and Doesken and Judson (1997). Accurate information on the amount and solid fraction of winter precipitation is required for input to hydrological and climate models, but this is often difficult to provide in practice. For example, standard techniques have been developed for correcting precipitation gauges for wind-induced undercatch (Goodison et al., 1998) but not all observing sites have the required information to apply corrections. In addition, the available surface-based observational networks for snowfall and solid precipitation tend to be concentrated in populated lower-elevation areas, and in recent years many of these sites have been closed or converted to automated stations. The automation of precipitation measurements is difficult, and has important consequences for the homogeneity of climate data series.

Accurate information on the amount and spatial distribution of SWE is critical information for water resource management (e.g. agriculture, hydroelectric power generation, flood forecasting). Water authorities and utilities make use of surface-based snow surveys (and snow pillows in mountainous terrain) to monitor peak SWE values prior to snowmelt, as well as satellite imagery to map snow cover extent. Information on snow depth and SWE can also be derived from a variety of aircraft or satellite sensors with varying degrees of success (Hall and Martinec, 1985).

Temporal variability and climate change

Regional and continental snow cover exhibits large interannual variability in response to atmospheric circulation patterns that influence temperature and precipitation. Interannual variability in North American and Eurasian snow cover extent have been shown to be strongly correlated to the Pacific-North America (PNA) and North Atlantic Oscillation (NAO) circulation patterns respectively (Gutzler and Rosen, 1992). Snow cover exhibits large regional variability in response to ENSO events (Cayan, 1996; Clark et al., 2001), with a tendency for El-Niño events to be associated with greater snow cover over Eurasia and less snow cover over North America (Groisman et al., 1994b). A detailed review of snow cover–atmosphere relationships is provided by Groisman and Davies (2001).

Snow cover extent (SCE) exhibits significant negative correlations to air temperature in many regions of the northern hemisphere (see the "temperature response regions" in Groisman et al., 1994b), and for the hemisphere as a whole (Robinson and Dewey, 1990; Karl et al., 1993). This negative relationship (see Figure S24) reflects the positive snow–albedo feedback, and analysis of northern hemisphere snow cover has shown that the most significant decreases have occurred in the second half of the snow year when the snow–albedo feedback is strongest. Satellite records indicate that the northern hemisphere annual SCE has decreased by about 10% since 1966, largely due to decreases in spring and summer since the mid-1980s over both the Eurasian and American continents (Robinson, 1997, 1999). Analysis of reconstructed SCE information since 1915 (Brown, 2000) showed that most of the observed reduction occurred during the second half of the twentieth century. This period has been characterized by widespread trends toward less winter snow, earlier snowmelt and earlier snowmelt runoff, with important implications for water resources, e.g. reduced storage of water in the snowpack and earlier melt translate to a lower freshwater pulse for recharge of soil moisture and reservoirs, and increased potential for evaporation loss. Global Climate Model (GCM) simulations suggest widespread reductions in snow cover over the next 50–100 years in response to global warming. However, there is considerable uncertainty in model-projected regional patterns of snow cover change (Frei and Robinson, 1998), particularly in mountainous regions. A discussion of the impacts of reductions in snow cover is provided in Fitzharris (1996).

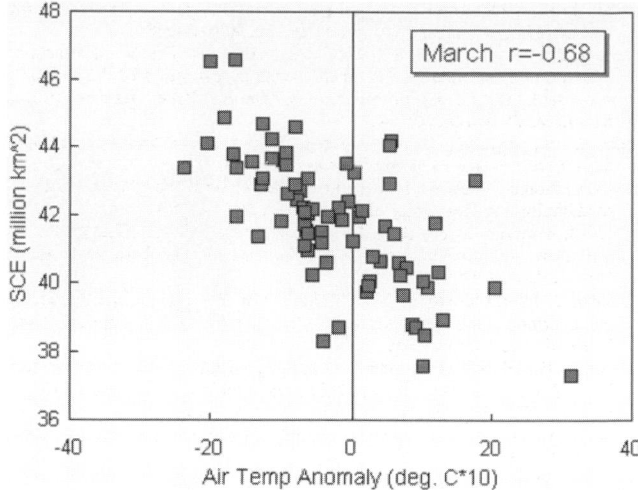

Figure S24 Scatterplot of reconstructed (1922–1971) and satellite-observed (1972–1997) northern hemisphere snow cover extent (SCE) versus northern hemisphere midlatitudinal (40–60°N) land surface air temperature anomalies for March. Air temperature anomalies were computed from the Jones (1994) gridded land temperature dataset, and the snow cover data are from Brown (2000). The inferred air temperature sensitivity for northern hemisphere March SCE is $-1.26 \times 10^6\,\text{km}^2\,°\text{C}^{-1}$.

Ross D. Brown and David A. Robinson

Bibliography

Anderson, E., 1976. *A Point Energy Balance Model of a Snow Cover*. Office of Hydrology, National Weather Service, NOAA Tech. Rep. NWS 19.

Armstrong, R.L., 1977. Continuous monitoring of metamorphic changes of internal snow structure as a tool in avalanche studies. *Journal of Glaciology*, **19**: 325–334.

Brown, R.D., 2000. Northern Hemisphere snow cover variability and change, 1915–1997. *Journal of Climate*, **13**: 2339–2355.

Brown, R.D., Brasnett, B., and Robinson, D., 2003. Gridded North American monthly snow depth and snow water equivalent for GCM evaluation. *Atmosphere–Ocean*, **41**: 1–14.

Brun, E., David, P., Sudul, M., and Brunot, G., 1992. A numerical model to simulate snow-cover stratigraphy for operational avalanche forecasting. *Journal of Glaciology*, **38**: 13–22.

Cess, R.D., Potter, G.L., Zhang M.-H. et al., 1991. Intercomparison of snow-feedback as produced by 17 general circulation models. *Science*, **253**: 888–892.

Cayan, D.R., 1996. Interannual climate variability and snowpack in the western United States. *Journal of Climate*, **9**: 928–948.
Clark, M.P., Serreze, M.C., and McCabe, G.J., 2001. Historical effects of El Nino and La Nina events on the seasonal evolution of the montane snowpack in the Columbia and Colorado River Basins. *Water Resources Research*, **37**: 741–757.
Colbeck, S., Akitaya, E., Armstrong, R., et al., 1990. *International Classification for Seasonal Snow on the Ground*. Boulder, CO: International Commission on Snow and Ice (IAHS) World Data Center-A for Glaciology, University of Colorado, CB 449.
Dery, S.J., and Yau, M.K., 2002. Large-scale mass balance effects of blowing snow and surface sublimation. *Journal of Geophysical Research (Atmospheres)*, **107**(D23): 4679.
Doesken, N.J., and Judson, A., 1997. *The SNOW Booklet: a guide to the science, climatology and measurement of snow in the United States*. Colorado State University.
Fitzharris, B., 1996. The cryosphere: changes and their impacts. In Watson, R.T., Zinyowera, M.C., Moss, R.H., and Dokken, D.K., eds., *Climate Change 1995, Impacts, Adaptations and Mitigation of Climate Change: Scientific-Technical Analyses*. Cambridge: Cambridge University Press, pp. 241–265.
Fitzharris, B.B., and McAlevey, B.P., 1999. Remote sensing of seasonal snow cover in the mountains of New Zealand using satellite imagery. *Geocarto International*, **14**: 33–42.
Foster, J.L., Chang, A.T.C., Hall, D.K., and Kelly, R., 2003. An examinaation of South American snow cover extent and snow mass from 1979–2002 using passive microwave satellite data. Proceedings, Seventh International Conference on Southern Hemisphere Meteorology and Oceanography, 24–28 March, 2003, Wellington, New Zealand.
Frei, A., and Robinson, D.A., 1998. Evaluation of snow extent and its variability in the Atmospheric Model Intercomparison Project. *Journal of Geophysical Research Atmospheres*, **103**(D8): 8859–8871.
Goodison, B.E., Ferguson, H.L., and McKay, G.A., 1981. Measurement and Data Analysis. In Grey, D.M., and Male, D.H., eds., *Handbook of Snow*. Pergamon Press, pp. 191–274.
Goodison, B.E., Brown, R.D., and Crane R.G., (eds.), 1999. Cryospheric systems. In King, M.D., ed., *EOS Science Plan*. Greenbelt, MD: NASA/Goddard Space Flight Centre, pp. 261–307.
Goodison, B.E., Louie, P.Y.T., and Yang, D., 1998. *WMO Solid Precipitation Measurement Intercomparison*. WMO Instruments and Observing Methods Report No. 67, WMO/TD No. 872, Geneva, Switzerland.
Groisman P.Ya., and Davies, T.D., 2001. Oxford: *Snow cover and the climate system*. In Jones, H.J., Pomeroy, J., Walker, D.A., and Hoham, R., eds., *Snow Ecology: an interdisciplinary examination of snow-covered ecosystems*. Cambridge: Cambridge University Press, pp. 1–44.
Groisman, P.Ya, Karl, T.R., and Knight, R.W., 1994a. Observed impact of snow cover on the heat balance and the rise of continental spring temperatures. *Science*, **263**: 198–200.
Groisman, P.Ya, Karl, T.R., and Knight, R.W., 1994b. Changes of snow cover, temperature and radiative heat balance over the Northern Hemisphere. *Journal of Climate*, **7**: 1633–1656.
Gutzler, D.S., and Preston, J.W., 1997. Evidence for a relationship between spring snow cover in North America and summer rainfall in New Mexico. *Geophysical Research Letters*, **24**: 2207–2210.
Gutzler, D.S., and Rosen, R.D., 1992. Interannual variability of wintertime snow cover across the Northern Hemisphere. *Journal of Climate*, **5**: 1441–1447.
Hall, D.K., and Martinec, J., 1985. *Remote Sensing of Ice and Snow*. New York: Chapman & Hall.
Jones, H.J., Pomeroy, J., Walker, D.A., and Hoham, R., (eds.), 2001. *Snow Ecology – an interdisciplinary examination of snow-covered ecosystems*. Cambridge: Cambridge University Press.
Jones, P.D., 1994. Hemispheric surface air temperature variations: a reanalysis and an update to 1993. *Journal of Climate*, **7**: 1794–1802.
Jordan, R.E., Andreas, E.L., and Makshtas, A.P., 1999. Heat budget of snow-covered sea ice at North Pole 4. *Journal of Geophysical Research, (Oceans)*, **104**: 7785–7806.
Karl, T.R., Groisman, P.Y., Knight, R.W., and Heim, R.R., Jr, 1993. Recent variations of snow cover and snowfall in North America and their relation to precipitation and temperature variations. *Journal of Climate*, **6**: 1327–1344.
Langham, E.J., 1981. Physics and properties of snowcover. In Gray, D.M., and Male, D.H., eds., *Handbook of Snow*. Oxford: Pergamon Press, pp. 275–337.
Marsh, P., 1999. Snowcover formation and melt: recent advances and future prospects. *Hydrological Processes*, **13**: 2519–2536.
McClung, D., and Schaerer, P., 1993. *The Avalanche Handbook*. Seattle: The Mountaineers.
Phillips, D., 1990. *The Climates of Canada*. Ottawa, Canada: Minister of Supply and Services.
Pomeroy, J.W., and Gray, D.M., 1995. *Snowcover: accumulation, relocation and management*. Saskatoon: National Hydrology Research Institute Science Report No. 7.
Randall, D.A., Cess, R.D., Blanchet, J.P., et al., 1994. Analysis of snow feedbacks in 14 general circulation models. *Journal of Geophysical Research*, **99**(D10): 20757–20771.
Robinson, D.A., 1997. Hemispheric snow cover and surface albedo for model validation. *Annals of Glaciology*, **25**: 241–245.
Robinson, D.A., 1999. Northern Hemisphere snow cover during the satellite era. *Proceedings, Fifth Conference on Polar Meteorology and Oceanography*. Dallas, TX. Boston, MA: American Meteorological Society, pp. 255–260.
Robinson, D.A., and Dewey, K.F., 1990. Recent secular variations in the extent of northern hemisphere snow cover. *Geophysical Research Letters*, **17**: 1557–1560.
Robock, A., Mu, M., Vinnikov, K., and Robinson, D.A., 2003. Land surface conditions over Eurasia and Indian summer monsoon rainfall. *Journal of Geophysical Research (Atmospheres)*, **108**(D4): 4131–4143.
Schemenauer, R.S., Berry, M.O., and Maxwell, J.B., 1981. *Snowfall Formation*. In Gray, D.M., and Male, D.H., eds., *Handbook of Snow*. Oxford: Pergamon Press, pp. 129–152.
Sevruk, B., (ed.), 1992. *Snow Cover Measurements and Areal Assessment of Precipitation and Soil Moisture*. Operational Hydrology Report 35, Publication 749, Geneva: World Meteorological Organization.
Sommerfeld, R.A., Moisier, A.R., and Musselman, R.C., 1993. CO_2, CH_4, and N_2O flux through a Wyoming snowpack and the implication for global budgets. *Nature*, **361**: 140–142.
Sturm, M., Holmgren, J., and Liston, G.E., 1995. A seasonal snow cover classification system for local to global applications. *Journal of Climate*, **8**: 1261–1283.
Taylor, P.A., Li, P.Y., and Wilson, J.D., 2002. Lagrangian simulation of suspended particles in the neutrally stratified surface boundary layer. *Journal of Geophysical Research (Atmospheres)*, **107**(D24): 4762.
Vernekar, A.D., Zhou, J., and Shukla, J., 1995. The effect of Eurasian snow cover on the Indian Monsoon. *Journal of Climate*, **8**: 248–266.
WMO, 1981. *Guide to Hydrological Practices*, 4th edn., vol. 1. WMO-No. 168. Geneva: World Meteorological Organization.
Zhang, T., and Armstrong, R.L., 2001. Soil freeze/thaw cycles over snow-free land detected by passive microwave remote sensing. *Geophysical Research Letters*, **28**: 763–776.

Cross-references

Albedo and Reflectivity
Antarctic Climates
Arctic Climates
Hydroclimatology
Taiga Climate
Tourism and Climate
Tundra Climate
Oscillations

SOLAR ACTIVITY

The sun is a variable star. A variety of transient phenomena are observed on or near its visible surface, including looping prominences and explosive flares. "Solar activity" is a general, inclusive term employed to characterize these and other phenomena, along with their variations in time. Some effects of solar activity are propagated throughout the solar system by the streaming solar wind. In the vicinity of the Earth, results of solar activity include auroras, geomagnetic storms and on occasion regional power outages (Allen et al., 1989). Telecommunications may be disrupted and the orbits of satellites perturbed. In order to understand the nature of the near-Earth and interplanetary environment it is necessary to gain an understanding of the nature and effects of solar activity.

Basic considerations

The visible surface of the sun is the photosphere. This is a layer only about 300 km thick, where the solar plasma becomes opaque to visible light. Below the photosphere is the convective zone, which is permeated by magnetic fields originating yet deeper in the sun. Above the photosphere is the chromosphere, a tenuous, normally invisible layer about 2500 km in thickness, where temperatures soar to more than 10^4 K. Above the chromosphere is the still-hotter solar corona with temperatures of about 2×10^6 K, which may be seen as a halo of light surrounding the sun during an eclipse. These three regions of the sun (photosphere, chromosphere and corona) each have characteristic forms of solar activity that will be outlined below.

Solar magnetic fields cause the phenomena of solar activity. The sun, like the Earth, has a bipolar magnetic field (like that of a bar magnet). The sun's field is rapidly and continually changing, undergoing a cycle in which the polarity reverses every 11 years or so. The magnetic lines of force of the Earth's field run roughly north–south (a poloidal field), but in the sun a combination of effects leads to fields in the equatorial and middle latitudes that trend more nearly east–west (a toroidal field). This is a consequence of the differential rotation of the sun; our star rotates more rapidly in the equatorial regions than near the poles, and the effect on deeply buried, originally poloidal field lines is to wind them around the sun's equatorial region, leading to the observed east–west pattern.

The vigorous convection in the outer portions of the sun brings internal magnetic fields up to surface regions of the sun, and leads to irregularities and disruptions of the fields. The phenomena of solar activity result from the evolution of these disturbed fields.

Sunspots and the photosphere

Sunspots are dark markings on the visible surface of the sun. They may occasionally be seen with the naked eye under conditions when the sun is partially obscured by haze or dust. Records of sunspot observations from the Orient date back many centuries (Stephenson and Wolfendale, 1988). Galileo was first to make telescopic observations of sunspots, and regular observations of high quality have been obtained since early in the nineteenth century.

Sunspots generally consist of a dark region, the umbra, surrounded by a lighter but still dark region termed the penumbra. They tend to occur in pairs, though single spots and complex groups are also seen. The leading spot of each pair has the same polarity as the other leading spots on the same hemisphere of the sun, while the trailing spots have polarities opposite to the leading spots. Thus spot pairs are thought to mark the emerging and returning locations of loops of magnetic field lines through the photosphere. The spots are dark due to their relatively low temperatures (about 2000 K lower than the surrounding photosphere). The fields are thought to block the convective motion of hot gases from below.

Bright areas in the vicinity of sunspot groups are termed faculae. These emit more radiation on average than the surrounding photosphere, approximately canceling the deficit associated with the sunspots (Chapman et al., 1986; Foukal and Lean, 1988).

The sunspot number is one of the most ancient and fundamental measures of solar activity. Introduced by R. Wolf, the sunspot number is proportional to $(10g + s)$, where g is the number of sunspot groups visible, and s is the number of individual spots counted on the surface of the sun. Following the recognition of the 11-year cycle in sunspot occurrence by Schwabe in 1843, Wolf reconstructed the cycle back to 1610. As previously noted, the polarities of leading sunspots and the polarity of the main dipolar magnetic field of the sun are reversed in the subsequent cycle, indicating that the fundamental period of activity is in fact approximately 22 years.

Many techniques have been employed to construct a record of solar activity over longer times (Stephenson and Wolfendale, 1988). Observations of auroras and of visible sunspots have been used to reconstruct solar activity cycles before 1600 (e.g. Wittman, 1978; Stothers, 1979, Siscoe, 1980). The record of atmospheric ^{14}C preserved in tree rings yields a record of solar activity spanning more than 8000 years (Sonett, 1984; Damon and Linick, 1986). Most recently records of the isotope ^{10}Be preserved in polar ice cores have been employed to obtain a high-resolution record of solar activity in prehistoric times (Attolini et al., 1988; Beer et al., 1990). We will return to a discussion of characteristics and implications of the record of solar activity in a later section.

Chromospheric solar activity

The chromosphere may be imaged using light emitted at a wavelength of 6563 Å(0.6563 μm). This wavelength is a characteristic of a transition of the electron of hydrogen atoms from the second to the first discrete level of excitation; it is an advantageous wavelength for observation (Gibson, 1972), since hydrogen is the most abundant element in the solar atmosphere. The 6563 Å line is termed "hydrogen alpha".

Solar flares are often evident in H_α images, indicating that they are an important component of chromospheric solar activity. Flares are explosive outbursts resulting from the release of magnetic energy in active regions; they emit radiation at many wavelengths (including energetic X-rays and gamma radiation) and may accelerate particles beyond solar escape velocity. Solar flares may cause nuclear reactions in the photosphere and may perturb the solar wind, triggering geomagnetic disturbances some time later at the Earth.

The corona

The solar corona is best imaged at X-ray wavelengths; here the cooler photosphere and chromosphere are dark, but hotter coronal structures show up clearly. Such images show two different types of regions: those in which the coronal loop structures are closed, emerging from and returning to the sun; and regions where the fields are open, stretching outward into interplanetary space. The former regions are identified with the same solar active regions seen in photospheric and chromospheric images, while the latter are broad regions of (predominantly) a single magnetic polarity ("unipolar regions"). These regions may also be coronal holes, which are thought to be the source of high-speed streams in the solar wind.

Coronal mass ejections involve the expulsion of vast bubbles of low-density plasma from the sun. They sometimes occur in association with solar flares, and are of undoubted importance in determining conditions in near-Earth space.

Solar activity cycle

Sunspot minimum is characterized by the presence of only a few active regions and sunspots, and by relatively large and well-developed coronal holes. The first sunspots of a cycle appear in relatively high latitudes (about 45°N and S), with leading spot polarities opposite to those of the preceding cycle. With time the sunspots and active regions appear in successively lower latitudes, until at sunspot maximum most spots appear within 10–15° of the solar equator. More complex active regions, with many sunspots, are found at higher levels of solar activity, and these complex regions generate most of the solar flares occurring in a given cycle. The large-scale solar magnetic fields are more complex during the high-activity phases of the cycle, and coronal holes may dominate the corona only near the poles.

The level of solar activity as measured by sunspot numbers increases relatively rapidly in the first few years of the cycle, with the subsequent decay to sunspot minimum requiring a longer period of time. During the decaying phase of the cycle, extensive regions of magnetic flux migrate from the sunspot zones to higher latitudes, first neutralizing the fields prevailing there and then establishing new polar caps of opposite polarity (Newkirk and Frazier, 1982; Wang, et al., 1989).

A detailed discussion of the solar cycle would require much more space than is available here. Interested readers should consult the volumes by Zirin (1988) and Foukal (1990).

As we have seen, there are many different varieties of solar activity. It is important to remember that these diverse phenomena are related, and that they are all linked with the fundamental 11-year sunspot cycle, which itself represents a harmonic of the more fundamental 22-year Hale magnetic cycle.

Periodicity and prediction

The source of the 11-year cycle of solar activity is an outstanding problem in solar physics. There appears to be nothing in the physical make-up of the sun that would give a fundamental oscillation of this period.

The time-series of sunspot numbers has been analyzed by numerous investigators, with a remarkable diversity of results. Rozelot (1994) provides an extensive review, while Kuklin (1976) summarizes much earlier work. Most investigators might agree that the 11-year period is real (Dicke, 1978), though Wilson suggests that there are in fact two modes, one of 120 ± 4 months and another of 140 ± 5 months, which when averaged yield the 11-year cycle. There is less agreement on longer periods of about 60 and about 80–90 years (the "Gleissberg cycle"). A number of investigators find a period of about 178 years (Jose, 1965; Cohen and Lintz, 1974; Fairbridge and Hameed, 1983). Still longer periods are seen in proxy data such as those obtained from radiocarbon fluctuations in tree rings (Sonett, 1984; Damon and Linick, 1986).

Fluctuations of solar activity with periods shorter than that of the basic sunspot cycle have also been described. A 155-day period in solar flare occurrence is well documented (Silverman, 1990), and a 25.6-month ("quasi-biennial") period is noted by several authors (Westcott, 1964; Apostolov, 1985). (The new science of helioseismology deals with oscillations of the solar surface with much shorter periods.)

Advance prediction of the level of solar activity is a desirable goal, since astronauts, satellites, space probes and the Earth itself are not immune to the effects of solar phenomena such as large solar flares and coronal mass ejections. The periodicity of sunspot activity forms a basis for simple predictions, and methods such as that of McNish and Lincoln (1949) have been successively improved to yield forecasts of activity levels. Unfortunately the past is not a perfectly reliable guide to the future, and none of the available methods has been an unqualified success. Other models are described in Schatten and Sofia (1987), Wilson (1988) and Butcher (1990).

Another controversial subject is the possible relationship between solar activity and weather and climate on Earth. Literally hundreds of investigations on this topic have been performed, with a diversity of results. The bibliography by Fairbridge (1987) includes dozens of such studies. The hypothesis of a relationship linking solar activity and terrestrial weather has received considerable support in the form of strong positive statistical correlations reported by Labitzke and van Loon (1989) and Friis-Christensen and Lassen (1991). The question has considerable socioeconomic significance. As previously noted, a drop in the solar output of perhaps a large fraction of a percent, occurring over several decades, would lead to significant climatic cooling on the Earth. This is presently the best explanation for the occurrence of the Little Ice Age, a period of reduced temperatures and concomitant climatic and socioeconomic stresses in the seventeenth and eighteenth centuries. This episode was contemporaneous with a prolonged minimum of solar activity now known as the Maunder minimum.

James H. Shirley

Bibliography

Allen, J., Sauer, H., Frank, L., and Reiff, P., 1989. Effects of the March 1989 solar activity. *EOS*, **70**(46): 1479.

Apostolov, E.M., 1985. Quasi-biennial oscillation in sunspot activity. *Bulletin of the Astronomical Institute, Czechoslovakia*, **36**: 97–102.

Attolini, M.R., Cecchini, S., Gastagnoli, G.C., et al., 1988. On the existence of the 11-yr cycle in solar activity before the Maunder minimum. *Journal of Geophysical Research*, **93**: 12729–12734.

Beer, J., Blinov, A., Bonani, G. et al., 1990. Use of ^{10}Be in polar ice to trace the 11-year cycle of solar activity. *Nature*, **347**: 164–166.

Butcher, E.C., 1990. The prediction of the magnitude of sunspot maxima for cycle 22 using abnormal quiet days in Sq(H). *Geophysical Research Letters*, **17**: 117–118.

Chapman, G.A., Herzog, A.D., and Lawrence, J.K., 1986. Time integrated energy budget of a solar activity complex. *Nature*, **319**: 654–655.

Charvátová I., 1990. The relations between solar motion and solar variability. *Bulletin of the Astronomical Institute, Czechoslovakia*, **41**: 56–59.

Charvátová I., and Strestik, J., 1991. Solar variability as a manifestation of the Sun's motion. *Journal of Atmospheric and Terrestrial Physics*, **53**: 1019–1025.

Cohen, T.J., and Lintz, P.R., 1974. Long term periodicities in the sunspot cycle. *Nature*, **250**: 398–400.

Damon, P.E., and Linick, T.W., 1986. Geomagnetic–heliomagnetic modulation of atmospheric radiocarbon production. *Radiocarbon*, **28**: 266–278.

Dicke, R.H., 1978. Is there a chronometer hidden deep in the Sun? *Nature*, **276**: 676–680.

Eddy, J.A., 1983. The Maunder minimum: a reappraisal. *Solar Physics*, **89**: 195–207.

Fairbridge, R.W., 1987. A comprehensive bibliography. In Rampino, M.R., et al., eds. *Climate: History, Periodicity, Predictability.* New York: Van Nostrand Reinhold.

Fairbridge, R.W., and Hameed, S., 1983. Phase coherence of solar cycle minima over two 178-year periods. *Astronomics Journal*, **88**: 867–869.

Fairbridge, R.W., and Sanders, J.E., 1987. The Sun's orbit, AD 750–2050: basis for new perspectives on planetary dynamics and Earth-Moon linkage. In Rampino, M.R., et al., eds., *Climate – History, Periodicity, and Predictability*. New York: Van Nostrand Reinhold, pp. 446–471 (bibliography, pp. 475–541).

Fairbridge, R.W., and Shirley, J.H., 1987. Prolonged minima and the 179-yr cycle of the Sun's inertial motion. *Solar Physics*, **110**: 191–210.

Feynman, J., and Gabriel, S.B., 1990. Period and phase of the 88-year solar cycle and the Maunder minimum: evidence for a chaotic Sun. *Solar Physics*, **127**: 393–403.
Foukal, P., 1990. *Solar Astrophysics*. New York: John Wiley & Sons.
Foukal, P., and Lean, J., 1988. Magnetic modulation of solar luminosity by photospheric activity. *Astrophysics Journal*, **328**: 347–357.
Friis-Christensen, E., and Lassen, K., 1991. Length of the solar cycle: an indicator of solar activity closely associated with climate. *Science*, **254**: 698–700.
Gibson, E.G., 1972. Description of solar structure and processes. *Reviews of Geophysics and Space Physics*, **10**: 395–461.
Gilman, P.A., and Howard, R., 1984. Variations in solar rotation with the sunspot cycle. *Astrophysics Journal*, **283**: 385–391.
Holweger, H., Livingston, W., and Steenbock, W., 1983. Sunspot cycle and associated variation of the solar spectral irradiance. *Nature*, **302**: 125–126.
Howard, R.W., 1981. Global velocity fields of the Sun and the activity cycle. *American Scientist*, **69**: 28–36.
Hudson, H.S., 1987. Solar variability and oscillations. *Reviews in Geophysics*, **25**: 651–662.
Jose, P.D., (1965). Sun's motion and sunspots. *Astronics Journal*, **70**: 193–200.
Kuhn, J.R., Libbrecht, K.G., and Dicke, R.H., 1988. The surface temperature of the Sun and changes in the solar constant. *Science*, **242**: 908–911.
Kuklin, G.V., 1976. Cyclical and secular variations of solar activity. In Bumba, V., and Kleczek, J., eds., *Basic Mechanisms of Solar Activity*. Dordrecht: International Astronomical Union, pp. 147–190.
Labitzke, K., and van Loon, H., 1989. Recent work correlating the 11-year solar cycle with atmospheric elements grouped according to the phase of the quasi-biennial oscillation. *Space Science Review*, **49**: 239–258.
Lean, J., 1989. Contribution of ultraviolet irradiance variations to changes in the Sun's total irradiance. *Science*, **244**: 197–200.
McNish, A.G., and Lincoln, J.V., 1949. Prediction of sunspot numbers. *Transactions of the American Geophysics Union*, **30**(5): 673–685.
Newkirk, G., Jr, and Frazier, K., 1982. The solar cycle. *Physics Today*, **April 1982**: 25–34.
Rozelot, J.P., 1994. On the stability of the 11-year solar cycle period (and a few others). *Solar Physics*, **149**: 149–154.
Schatten, K.H., and Sofia, S., 1987. Forecast of an exceptionally large even-numbered solar cycle. *Geophysical Research Letters*, **14**: 632–635.
Shirley, J.H., Sperber, K.R., and Fairbridge, R.W., 1990. Sun's inertial motion and luminosity. *Solar Physics*, **127**: 379–392.
Silverman, S.M., 1990. The 155-day solar period in the sixteenth century and later. *Nature*, **347**: 365–367.
Siscoe, G.L., 1980. Evidence in the auroral record for secular solar variability. *Reviews in Geophysics and Space Physics*, **18**: 647.
Sonett, C.P., 1984. Very long solar periods and the radiocarbon record. *Reviews of Geophysics and Space Physics*, **22**: 239–254.
Stephenson, F.R., and Wolfendale, A.W., (eds.), 1988. *Secular Solar and Geomagnetic Variations in the Last 10,000 Years*. Dordrecht: Kluwer.
Stothers, R., 1979. Solar activity cycle during classical antiquity. *Astronomy and Astrophysics*, **77**: 121–127.
Wang, Y.-M., Nash, A.G., and Sheeley, N.R., Jr, 1989. Magnetic flux transport on the Sun. *Science*, **245**: 712–718.
Westcott, P., 1964. The 25- or 26-month periodic tendency in sunspots. *Journal of Atmospheric Science*, **21**: 572–573.
White, O.R., (ed.), 1977. *The Solar Output and its Variation*. Boulder, CO: Colorado University Press.
Willson, R.C., and Hudson, H.S., 1991. The Sun's luminosity over a complete solar cycle. *Nature*, **351**: 42–44.
Wilson, R.M., 1987. On the distribution of sunspot cycle periods. *Journal of Geophysical Research*, **92**: 10101–4.
Wilson, R.M., 1988. A prediction for the size of sunspot cycle 22. *Geophysical Research Letters*, **15**: 125–128.
Wittman, A., 1978. The sunspot cycle before the Maunder minimum. *Astronomy and Astrophysics*, **66**: 93–97.
Woodard, M.F., and Libbrecht, K.G., 1993. Observations of time variation in the Sun's rotation. *Science*, **260**: 1778–1781.
Zirin, H., 1988. *Astrophysics of the Sun*. Cambridge: Cambridge University Press.

Cross-references

Maunder, Edward Walter and Maunder Mininum
Sunspots

SOLAR CONSTANT

The solar constant is defined as the total irradiance of the sun at the mean orbital distance of the Earth. It has a value of about 1368 W m^{-2}. The solar constant is a fundamental quantity in atmospheric physics since it represents the amount of solar energy arriving at the top of the atmosphere. The term remains in use even though the total solar irradiance is not constant.

The concept of the solar constant was introduced by A. Pouillet in 1837 (Pap, 1986). S.P. Langley devised a method for measuring the solar constant in the 1880s, and surface measurements of the solar constant were made routinely beginning in the early years of this century (Hoyt, 1979). Space-age measurements of the solar irradiance have been obtained by instruments such as the active cavity radiometer irradiance monitor (ACRIM) on the Solar Maximum mission and the Earth Radiation Budget (ERB) instrument on the Nimbus-7 satellite (Mechikunnel et al., 1988). Very accurate models of the solar total irradiance over the past half-century have recently become available (Tobiska et al., 2000).

It was recognized that the solar output might not be constant by Langley and others. However, in the first half of the twentieth century the greatest number of investigators favored the notion of an unchanging sun. Evidence suggesting variation of the solar output, obtained by C.G. Abbot and others, was thus very controversial. The difficulties in measuring the solar constant through the ever-changing atmosphere of the Earth precluded a resolution of this question prior to the space age.

Today there is irrefutable evidence of variations of the solar output of up to a few tenths of a percent, over time periods of days to weeks (Willson et al., 1981; Tobiska et al., 2000). The solar output varies in step with the 11-year cycle of sunspot activity (Willson and Hudson, 1991), with the highest irradiance and luminosity recorded near the maximum of the sunspot cycle. The amplitude of this effect approaches 0.1%.

If the solar output in addition experiences long-term variations, these are likely to have important implications for terrestrial climatic change (Schatten and Arking, 1990; Reid, 2000; and subsequent articles in that volume). Thus the subject of solar output variations on all timescales is an active area of research. In one study Shirley et al. (1990) found evidence of a relationship linking solar luminosity with the solar inertial motion.

By far the largest fraction of the solar output is found in the visible range of the electromagnetic spectrum. Some portions of the continuum of solar emissions at higher and lower wavelengths vary with the phase of the sunspot cycle (White, 1977; Tobiska, 2001). Variations in the solar ultraviolet flux (in particular) may play an important role in changes of the total irradiance of the Sun (Lean, 1989).

A special issue of the journal *Space Science Reviews* (volume 94, #1-2, 2000) contains 37 articles summarizing our state of knowledge of solar irradiance variations and their climatic effects as of the turn of the present century.

James H. Shirley

Bibliography

Hoyt, D.V., 1979. The Smithsonian Astrophysical Observatory solar constant program. *Reviews in Geophysics and Space Physics*, **17**: 427–453.
Lean, J., 1989. Contribution of ultraviolet irradiance variations to changes in the Sun's total irradiance. *Science*, **244**: 197–200.

Mechikunnel, A.T., Lee, R.B. III, Kyle, H.L., and Major, E.R., 1988. Intercomparison of solar total irradiance data. *Journal of Geophysical Research*, **93**: 9503–9509.
Pap, J., 1986. Variation of the solar constant during the solar cycle. *Astrophysics and Space Science*, **127**: 55–71.
Reid, G.C., 2000. Solar variability and the Earth's climate: introduction and overview. *Space Science Review*, **94**: 1–11.
Schatten, K.H., and Arking, A., (eds), 1990. *Climate Impact of Solar Variability*. Washington, DC: NASA, NASA CP-3086 .
Shirley, J.H., Sperber, K.R., and Fairbridge, R.W., 1990. Sun's inertial motion and luminosity. *Solar Physics*, **127**: 379–392.
Space Science Reviews 94, #1-2, 2000. Special issue on solar variability.
Tobiska, W.K., 2001. Variability of the solar constant from irradiances shortward of Lyman-Alpha. *Advances in Space Research*, **29**: 1969–1974.
Tobiska, W.K. et al., 2000. The SOLAR2000 Empirical Solar Irradiance Model and Forecast Tool. *Journal of Atmospheric, Solar and Terrestrial Physics*, **62**: 1233–1250.
White, O.R. (ed.), 1977. *The Solar Output and its Variation*. Boulder, CO: Colorado University Press.
Willson, R.C., and Hudson, H.S., 1991. The Sun's luminosity over a complete solar cycle. *Nature*, **351**: 42–44.
Willson, R.C., Gulkis, S., Janssen, M. et al., 1981. Observations of solar irradiance variability. *Science*, **211**: 700–702.

Cross-references

Solar Activity
Sunspots

SOLAR FLARE

Solar flares are energetic eruptions from the surface of the sun. They typically originate within active regions, which are areas where internal solar magnetic fields have emerged through the solar surface (the photosphere), forming bright faculae and dark sunspots. The release of energy in flares accelerates particles to high velocities (e.g. Chupp, 1990). In some cases these particles may later be detected at the surface of the Earth, even when the flare occurs on the opposite side of the sun. Some geomagnetic storms and auroras may result from strong solar flares, while others appear to be caused by coronal mass ejections. Large flares release tremendous amounts of energy (up to about 10^{26} J in less than an hour).

The largest flares seem to pass through three stages. The preflare (or precursor) stage involves brightening of a portion of the solar surface within an active region. Loops of plasma confined by magnetic tubes emerging from the solar surface expand, sometimes over a period of several days. The impulsive phase (or flash) lasts from a few seconds to a few minutes. This phase includes the emission of bursts of X-rays and gamma rays and other radiation; a shock wave forms. The main, or extended, phase follows. The flare becomes visible in optical wavelengths, and soft X-rays and radio emissions continue. Flares are often observed in the hydrogen alpha spectrum line, which provides an image of the structure of the solar chromosphere (a thin layer immediately above the photosphere). Rust (1993) provides a review of observations of the development of flares.

The physics of solar flares is not well understood. They are linked with collisions and twisting of magnetic strands of force generated within the solar interior. The energy is thought to derive from a magnetic tension.

There are many varieties of flares. Some are confined to closed loops, while others disrupt loop structures; some are linked with coronal mass ejections that may change the structure of the corona. Flares may be classified by the area of the chromosphere that brightens in the event; by the signal seen in solar radio wave observations (i.e. bursts); or by the enhancement in solar X-ray flux that accompanies the flare, or by other schemes (Orrall, 1991).

Flares, in common with many other solar phenomena, are more frequent at times of higher solar activity. In addition there is a well-developed cycle of flare occurrence, with a period of 155 days, whose origin is not well understood (Ichimoto et al., 1985; Bai and Sturrock, 1987).

A record of solar flare activity spanning about 10^7 years is preserved in near-surface rocks on the moon. Protons emitted from the sun in flares may produce radioactive isotopes such as ^{26}Al and ^{53}Mn in lunar rocks; radiometric dating then provides an estimate of the time of occurrence of the flare. The data suggest that the level of intensity of flare activity has not changed significantly in recent millennia.

James H. Shirley

Bibliography

Bai, T., and Sturrock, P.A., 1987. The 152-day periodicity of the solar flare occurrence rate. *Nature*, **327**: 601–613.
Chupp, E.L., 1990. Transient particle acceleration associated with solar flares. *Science*, **250**: 229–236.
Foukal, P.V., 1990. *Solar Astrophysics*. New York: John Wiley.
Ichimoto, K., Kubota, J., Suzuki, M., et al., 1985. Periodic behavior of solar flare activity. *Nature*, **316**: 422–424.
Orrall, F.Q., 1991. Solar activity. In Maras, S.P., ed., *The Astronomy and Astrophysics Encyclopedia*. New York: Van Nostrand Reinhold.
Rust, D., 1993. Solar flare prediction needed. *EOS*, **74**(47): 553–559.
Zirin, H., 1988. *Astrophysics of the Sun*. New York: Cambridge University Press.

Cross-references

Solar Constant
Solar Radiation
Solar Wind

SOLAR RADIATION

Solar radiation is the common term for the electromagnetic radiation emitted by the sun. Virtually all of the radiant energy received by Earth emanates from the sun as solar radiation. Without this input of radiant energy the Earth would be a cold, dark, lifeless planet. In combination with the Earth's rotation, oceanic and atmospheric circulation systems are driven by solar radiation absorbed by the surface and, to a lesser extent, the atmosphere that surrounds it. Through photosynthesis, energy derived from the solar radiation spectrum is used by plants to convert water and carbon dioxide into carbohydrates. Indeed, the light from the sky is scattered solar radiation.

Solar output

The amount and spectral nature of this radiant energy is a function of the emission temperature of the solar disk, which is approximately 5777 K, based on the total solar emission (LeNoble, 1993).

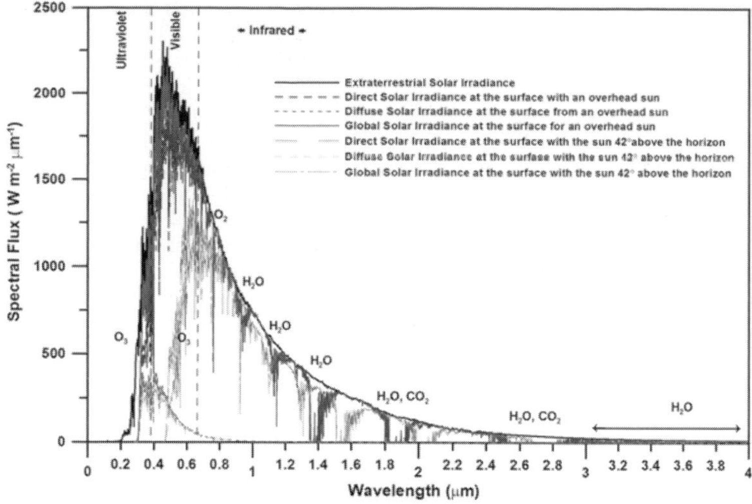

Figure S25 Extraterrestrial spectrum, direct beam spectrum and diffuse spectrum for an overhead sun (solar elevation 90°) and when the sun is at 42° elevation.

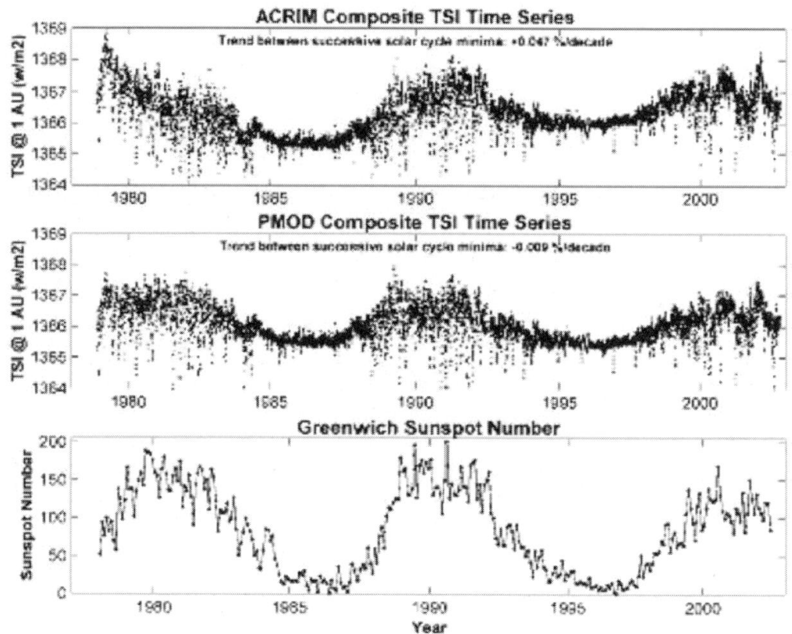

Figure S26 Variation in the total solar irriadiance as published by two laboratories. The lower panel indicates the sunspot cycle (from Wilson and Mordvinov, 2003).

Small fluctuations in the emission temperature do not greatly alter the overall output of energy, but do change the spectral (wavelength) characteristics of the radiant flux, with shorter wavelengths changing more than longer wavelengths. Less than 0.2% of the energy output of the sun is emitted at wavelengths shorter than 250 nm and a similar amount is radiated at wavelengths longer than about 6 μm. The wavelength region of greatest emission is in the visible portion of the spectrum, 400–700 nm, the peak of the irradiance distribution being at 465 nm (Figure S25). The total solar irradiance (TSI), often referred to as the solar constant, is the energy flux received by the Earth from the sun and has an average value of 1367 W m^{-2}. Figure S26 illustrates the variation of the TSI between 1979 and 2002 as measured by various satellite instruments. The fluctuations in the sun's emittance are linked to solar activities such as sunspots, auroras and geomagnetic disturbances that have an approximate 11-year cycle (see bottom panel of Figure S25). Increased solar activity leads to an increase in irradiance (Hoyt and Schatten, 1997).

Solar geometry

The actual amount of solar radiation reaching the surface of the Earth is a function of sun–Earth geometry and the absorption and scattering properties of the atmosphere that surrounds the planet.

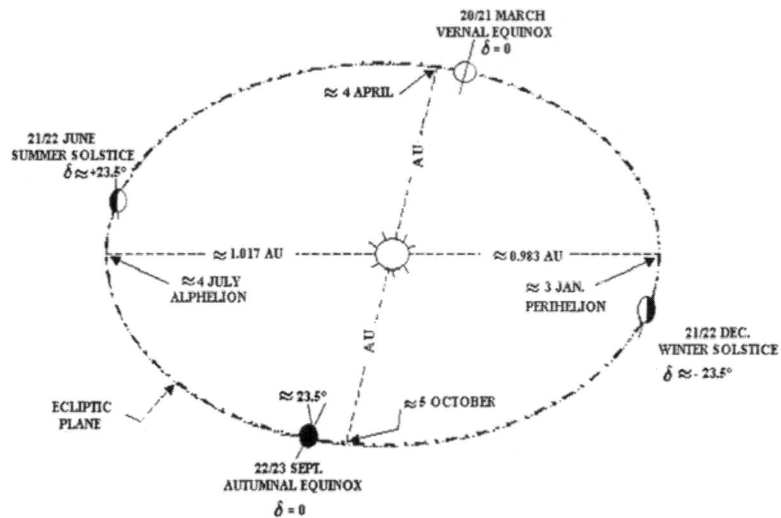

Figure S27 Orbital characteristics of the Earth about the sun (adapted from Iqbal, 1983).

The Earth's orbit about the sun is not concentric, varying between 1.47×10^8 km and 1.52×10^8 km, with an average distance of 1.496×10^8 km (known as 1 Astronomical Unit (AU)). Figure S27 is a schematic of the orbit of the Earth about the sun showing that the greatest distance between the two bodies is on 4 July and the nearest approach of the Earth to the sun is 3 January. The TSI varies by the inverse square of the distance from the sun. This variation is known as the eccentricity correction factor.

The amount of energy at the top of the atmosphere at any given location above the Earth is also affected by the tilt of the Earth's axis from the plane about which it orbits the sun (the ecliptic plane). As the Earth orbits the sun the apparent location of the sun moves south and north of the equator because of the changing angle between the sun and the Earth. The solar declination is the angular measure of this variation and varies between 23°27′ (north) at the summer solstice in the Northern Hemisphere and −23°27′ (south) at the winter solstice. At the equinoxes, when the sun is directly overhead at the equator, the solar declination is zero. Figure S28 maps the amount of solar radiation that would reach the Earth's surface based on the changing sun–Earth distance and the changing angle between the Earth and the sun.

The final geometric factor affecting the receipt of solar radiation at a given location is the daily eastward rotation of the Earth on its axis. The angular elevation between the solar disk and the plane of the surface varies between zero at sunrise and sunset, and a maximum of [90 − |latitude − solar declination|][1] when the sun is equatorward of the location. The receipt of solar radiation on a flat surface increases as the solar elevation increases, reaching a maximum at the time the sun is at its maximum height in the sky; defined as solar noon.

The Atmosphere

The amount and spectral nature of this solar radiation changes as it passes through the atmosphere (refer to Figure S25). Gases such as ozone, oxygen and water vapor absorb radiation and in so doing warm the atmosphere, while air molecules scatter radiation and particulates (aerosols) both scatter and absorb radiation. The scattering of radiation separates the total spectral flux into two components: (1) the direct beam spectral flux, which is the spectral radiation that reaches the surface without being absorbed or scattered; and (2) the diffuse spectral radiation, which is the radiation that has been scattered in any manner. In combination, the total amount of solar radiation that reaches the surface is termed global spectral radiation or global spectral irradiance. When these fluxes are integrated over the entire wavelength range for which the sun emits, the integrated fluxes are generally known as global, direct and diffuse irradiance.

Gases absorb in various spectral regions and at various levels of the atmosphere. Ozone absorbs most of the ultraviolet portion of the solar spectrum as the rays pass through the stratosphere where the largest concentration of the O_3 is located. Water vapor is more abundant in the troposphere and absorbs the longer infrared wavelengths of solar radiation. Gases such as oxygen and carbon dioxide are more uniformly mixed throughout the atmosphere (see Figure S25). Two gases associated with human activity, sulfur dioxide and nitrogen oxides, are also responsible for the absorption of solar radiation in the UV and visible wavelength regions of the solar spectrum.

Scattering of solar radiation in cloud-free skies is primarily by air molecules. Known as Rayleigh scattering, this scattering of light is by particles whose diameter is much smaller than the wavelength of the radiation being scattered. The amount of scattering is roughly proportional to the inverse of the wavelength to the fourth power (λ^{-4}) and is responsible for the blue color of the sky seen on clear days. It can be seen in Figure S25 that Rayleigh scattering becomes negligible at wavelengths greater than about 600 nm. Scattering by particles larger than the wavelength of radiation, such as smoke, volcanic aerosols, dust or water droplets, is referred to as Mie scattering. This is more directional in nature than Rayleigh scatter, and less dependent on the wavelength of the solar radiation. The loss of color contrast in the sky during hazy or polluted days is the result of Mie scatter. Clouds are the most significant scatterer of solar radiation, being capable of increasing the instantaneous amount of solar radiation reaching the Earth's surface to values above the TSI and reducing the flux to virtually zero when thick. Figure S29, a schematic of the global shortwave radiation

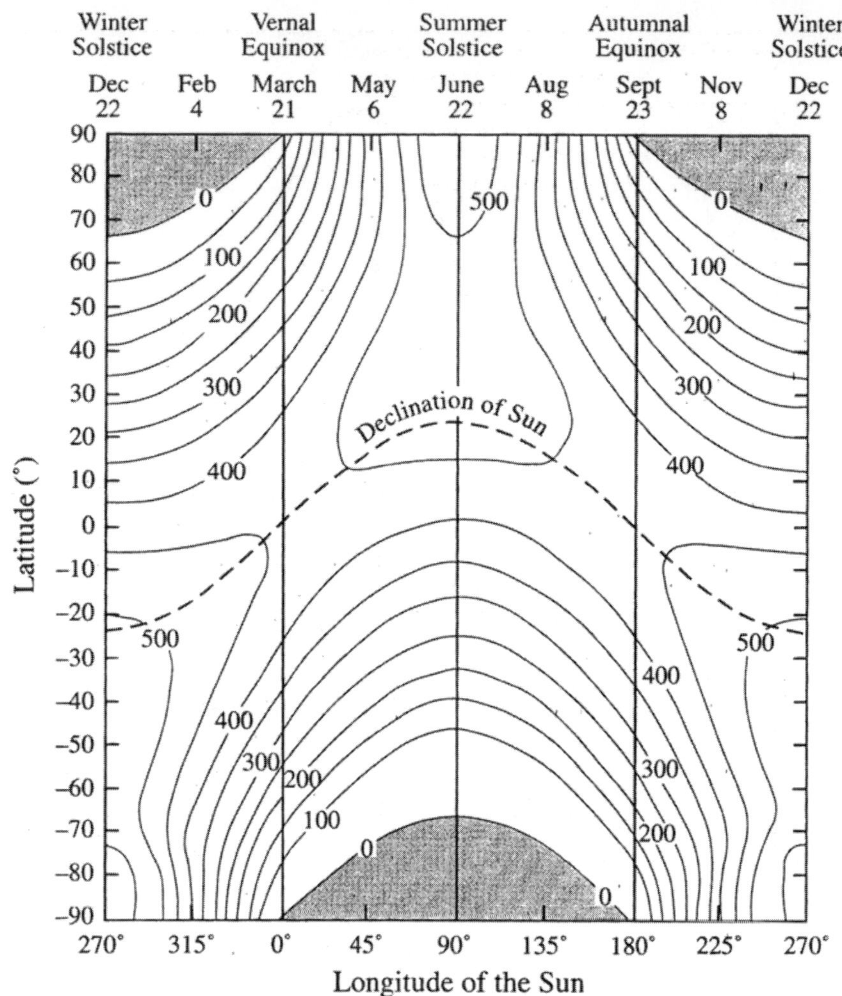

Figure S28 Geographical and temporal distribution of solar radiation (W m^{-2}) at the surface of the Earth.

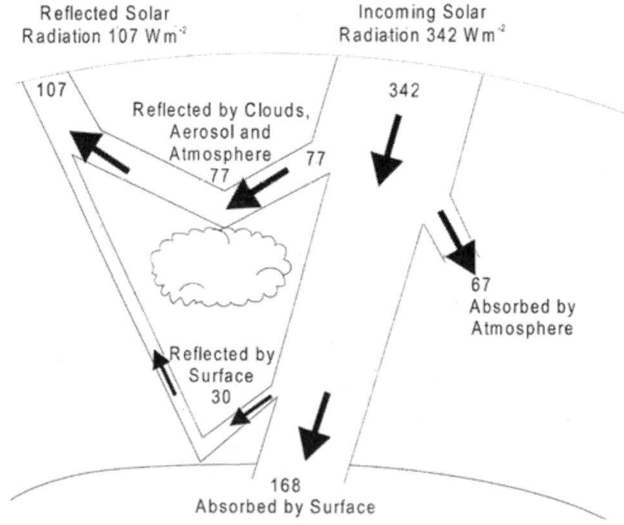

Figure S29 The global average shortwave radiation budget showing the distribution of solar radiation in the atmosphere and at the surface (adapted from IPCC, 2001).

budget, illustrates the effectiveness of clouds in intercepting and scattering radiation. Because of the effectiveness of clouds in this process, changes in global cloud cover can play a significant role in changing the Earth's climate (IPCC, 2001a). The 5 March plot on Figure S30 illustrates the effect of the passage of a single cloud between the sun and the surface, showing first a rapid decrease in the amount of solar radiation reaching the surface and then an increase substantially greater than the amount of radiation received from a clear sky when the direct beam irradiance is least attenuated.

Solar radiation at the surface

The amount of solar radiation reaching the surface of the Earth is dictated by both geometric and atmospheric factors. Absorption of this radiation at the surface, however, is related to the characteristic properties of the surface. Figure S30 shows the changes in the incoming and reflected global solar fluxes at the Bratt's Lake Observatory, part of a global surface radiation budget monitoring network (Ohmura et al., 1998). The plots on the graph show how the solar geometry alters the amount of radiation received between summer and winter with respect to the amount of solar radiation that is received at the surface and

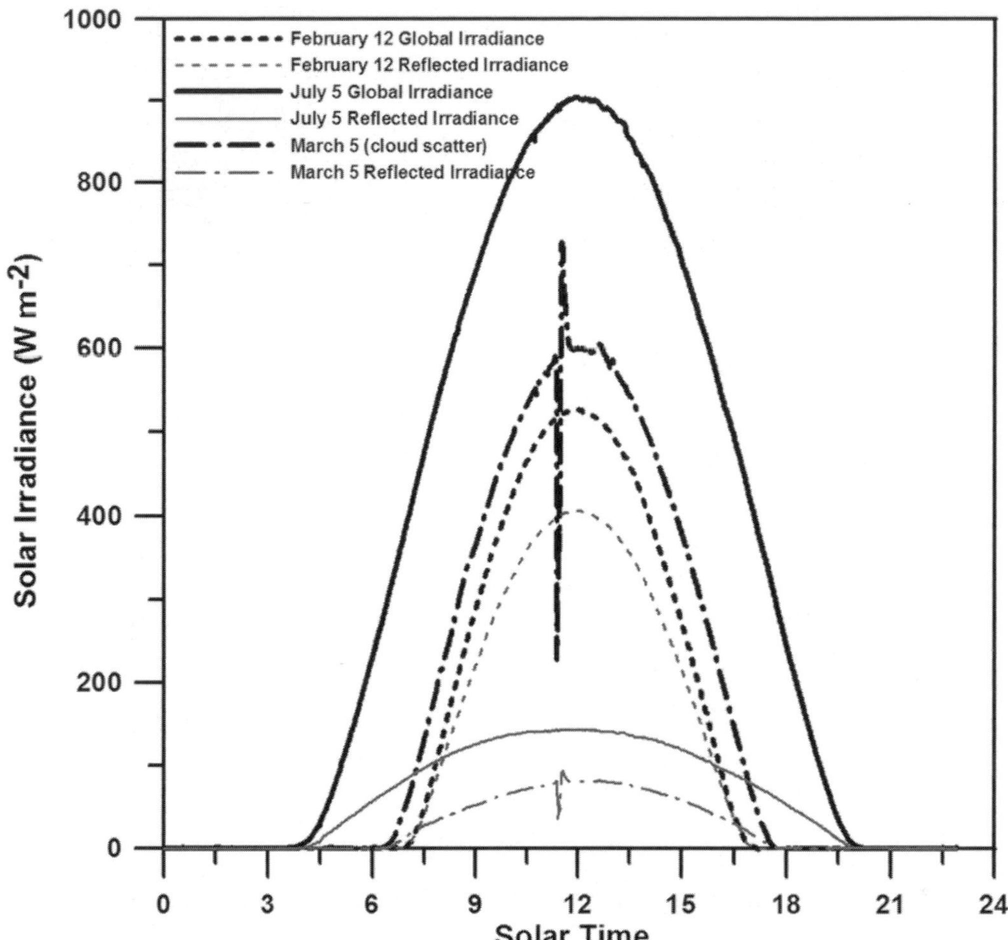

Figure S30 Global and reflected irradiance observations for three nearly clear-sky days from the Bratt's Lake Observatory (104.7°W, 50.28°N). The blue curve illustrates the effect of cloud passing between the sun and the surface. Note the increase above the normal irradiance levels caused by the scattering of radiation from the side of the cloud after reducing the surface flux by more than half. This can also be observed in the reflected irradiance. The 12 February and 5 July days show the changes in the radiation received at the surface, primarily because of solar geometry. The difference in the reflected irradiance between the two days is a function of the surface cover.

the changes that the surface plays in absorbing the radiation. The reflected fluxes in the figure are due to the differences in the land cover between summer (wheat) and winter (snow). The effect of the cloud passage on the 5 March solar irradiance curve is also apparent in the reflected flux. The energy that is not reflected is absorbed and used in a variety of processes from heating the surface, to evaporating water, to photosynthesis. The ratio between the solar radiation reflected by a surface and that incident on the surface is called the albedo. The albedo of snow ranges between 0.4 and 0.95, while the albedo of most plants ranges between 0.16 and 0.25. The albedo of man-made urban surfaces varies between 0.05 for asphalt and 0.4 for light-colored brick (Oke, 1987).

While accurate observations of the incoming and reflected solar radiation can be obtained at the surface of the Earth, satellite instrumentation is used to better understand the global distribution of the amount of solar radiation available at the surface. The global distribution of solar radiation at the surface, based on observations made by NASA/AtSR ERBE, can be seen for January and June 1991 in Figure S31. The relationships between solar geometry and cloud cover are apparent. The amount of radiation reaching the surface during the winter is zero near the North Pole, while the flux in the South Polar region is at a maximum. In the June distribution the high solar radiation values over the Sahara region are correlated with the low cloud amount. Similarly, over the southern ocean region, which is predominantly cloud-covered throughout the year, the solar radiation reaching the surface is low.

Some countries are now producing solar radiation distributions from satellite in near real-time. These include the USA (http://www.ghcc.msfc.nasa.gov/goesprod/GEIUSlatest.html) and Australia (http://www.bom.gov.au/sat/solrad. shtml). Global maps of daily and monthly solar radiation are available from http://www.osdpd.noaa.gov/PSB/EPS/RB/ RB.html.

Solar radiation and humans

Although solar radiation is plentiful and free, the utilization of this resource in most of the western world has been limited by

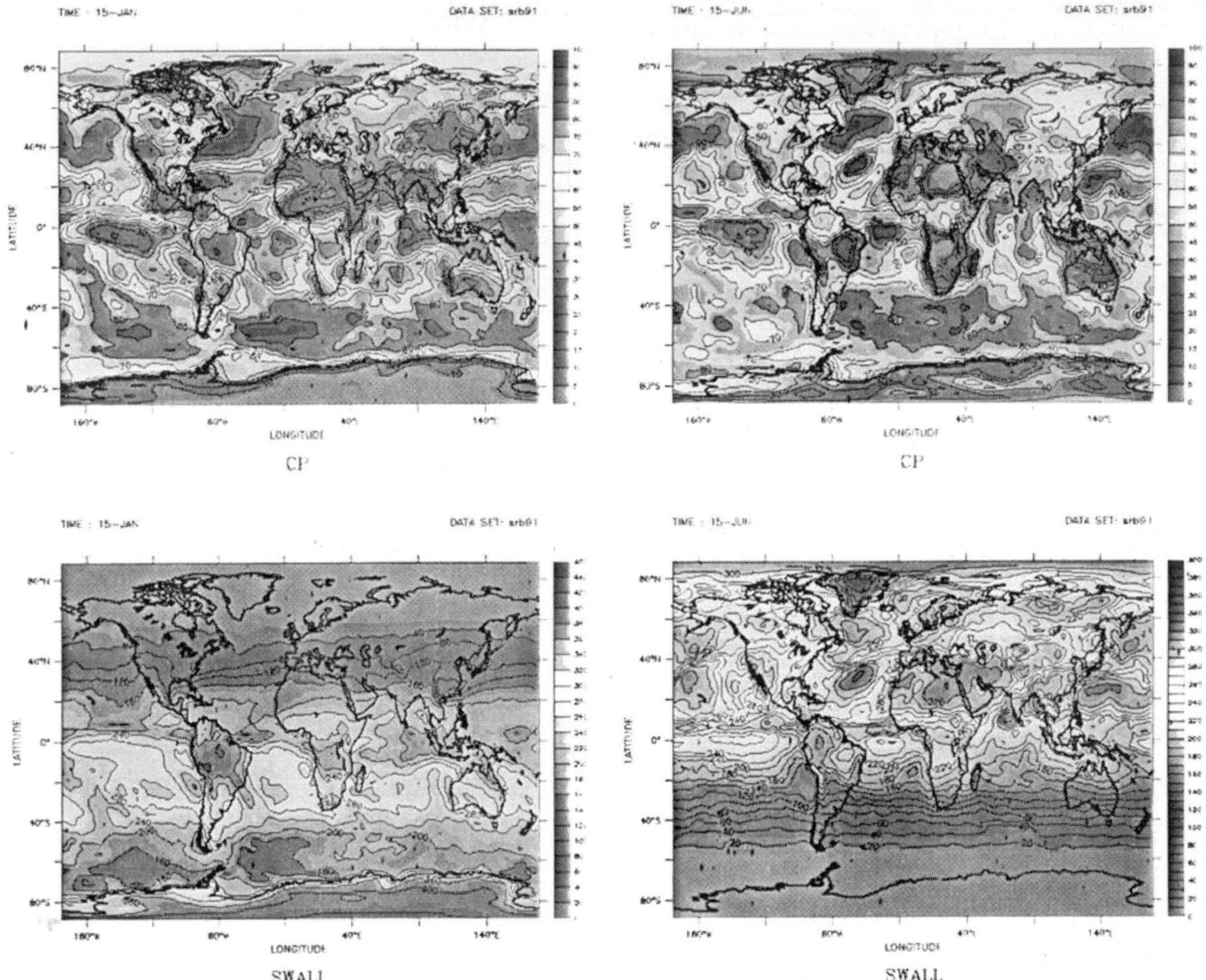

Figure S31 Surface radiation budget (SRB) distributions of monthly mean cloud amount (top) and monthly mean solar radiation at the surface (bottom) for January and June 1991 from the Earth Radiation Budge Experiment (ERBE). The cloud amount is in percent, the solar radiation is in W m^{-2} (from http://asd-www.larc.nasa.gov).

[1] Latitude is in degrees, positive north. The | | indicate the absolute value of the difference.

[2] 1 EJ = 1 × 10^{18} Joules (SI prefix exa).

the relatively high costs of products that utilize solar radiation, the relatively inexpensive costs associated with energy from fossil fuels and nuclear power, the high research and development costs associated with changing from conventional methods, and the lack of incentives to change (Interlaboratory Working Group, 2000). As the cost of conventional energy increases and the global population becomes more aware of the potential of climate change associated with the burning of fossil fuels, the use of solar radiation is expected to increase. Primarily, solar energy is used in the heating of water, the heating and to a lesser extent cooling of buildings, the production of off-grid electricity through photovoltaics, the use of thermal heating to produce electricity and the use of domestic parabolic collectors for cooking in rural areas of developing nations (Duffie and Beckman, 1991). The IPCC (2001b) indicates that the use of photovoltaics increased from 37 MW to 150 MW between 1986 and 1998, with growth estimated to be to 1000 MW by 2005. The annual worldwide growth rate in the sale of photovoltaic modules is between 15% and 20% per annum, which is expected to continue through 2020 (Interlaboratory Working Group, 2000). The IPCC (2001b) estimated that by 2005 up to 5 million square meters of solar thermal panels would be in use in Europe for domestic hot-water heating. Concentrating collectors have been used for the commercial generation of power by focusing solar radiation onto a central superheat boiler to generate steam that is passed through a turbine. An example of this type of system is the California Solar One generating plant, which was first operated at 10 MW (Duffie and Beckman, 1991), but has now been upgraded to operate at 30 MW, with the eventual goal of producing 100 MW of electricity (IPCC, 2001b).

Developing countries and those countries without ready access to other energy sources have more readily embraced the use of solar energy than developed countries. Major projects using solar radiation are planned or ongoing in Brazil, India, Japan, South Africa, Uganda and Vietnam. Both the United Nations Environment Programme and the World Bank are actively involved in increasing the use of solar energy throughout the globe, but particularly in developing countries. The amount of solar radiation estimated to be available for use around the globe is between 1575 EJ[2] per year and 49 837 EJ per year (IPCC, 2001b). In comparison, the global energy use today is approximately 425 EJ per year (IEA, 1998).

L.J. Bruce McArthur

Acknowledgments

The author acknowledges Dr C. McLinden for the provision of the data used in Figure S25, Dr D. Halliwell for the provision of data used in Figure S30 and the use of the Ferret program in the production of Figure S31. Ferret is a product of NOAA's Pacific Marine Environmental Laboratory. Ms O. Neibergall and Mr F. Abdi aided in the drafting of Figures S27 and S29.

Bibliography

Duffie, J.A., and Beckman, W.A., 1991. *Solar Engineering of Thermal Processes*, 2nd edn. New York: John Wiley.

Hoyt, D.V., and Schatten, K.H., 1997. *The Role of the Sun in Climate Change*. New York: Oxford University Press.

IEA, 1998. *World Energy Outlook – 1998 Update*. Paris: IEA/OECD.

Interlaboratory Working Group, 2000. *Scenarios for a Clean Energy Future* (Oak Ridge, TN: Oak Ridge National Laboratory; Berkeley, CA: Lawrence Berkeley National Laboratory; and Golden, CO: National Renewable Energy Laboratory), ORNL/CON-476, LBNL-44029, and NREL/TP-620-29379, November.

IPPC, 2001a. Climate Change 2001: the scientific basis. Contribution of Working Group 1 to the Third Assessment Report of the Intergovernmental Panel on Climate Change (Houghton, J.T., Ding, Y., Griggs, D.J., et al., eds.) Cambridge: Cambridge University Press.

IPPC, 2001b. *Climate Change 2001: mitigation.* Contribution of Working Group 3 to the Third Assessment Report of the Intergovernmental Panel on Climate Change (Metz, B., Davidson, O., Swart, R., and Pan, J., eds.). Cambridge: Cambridge University Press.

Iqbal, M., 1983. *An Introduction to Solar Radiation*. London: Academic Press.

LeNoble, J., 1993. *Atmospheric Radiative Transfer*. Hampton, VA: A. Deepak.

Liou, K-N., 1980. *An Introduction to Atmospheric Radiation*. San Diego, CA: Academic Press.

Ohmura, A., Dutton, E.G., Forgan, B., et al., 1998. Baseline Surface Radiation Network (BSRN/WCRP): New precision radiometry for climate research. *Bulletin of the American Meteorological Society*, **79**(10): 2115–2136.

Oke, T.R., 1987. *Boundary Layer Climates*. New York: Methuen. Wells, N., 1997. *The Atmosphere and Ocean: a physical introduction*. Chichester: John Wiley & Sons.

Willson, R.C., and Mordvinov, A.V., 2003. Secular total solar irradiance trend during solar cycles 21–23. *Geophysical Research Letters*, **30**(5): 1199, doi:10.1029/2002GL016038.

Cross-references

Electromagnetic Radiation
Energy Budget Climatology
Radiation Laws
Terrestrial Radiation
Ultraviolet Radiation

SOLAR WIND

The visible outer blanket of the sun during a total solar eclipse is called the corona. The corona does not abruptly end but extends billions of kilometers into space. The outward movement and expansion of the corona are a function of distance from the sun. Expansion close to the sun is very slow because the pull of gravity is dominant; as the distance to the sun increases, the outward flow increases. The flow of the gas stream is the solar wind, the term originally designated by Eugene N. Parker in his classic 1958 paper on the dynamics of the interplanetary gas (see Parker, 2001).

The solar wind is a stream of ionized gas blown constantly at supersonic speeds away from the sun in all directions. The component makeup of the solar wind is mainly electrons, protons, alpha particles (hydrogen and helium atomic nuclei – both positively charged) and some heavier ions.

The beautiful display of the auroras occurs when a strong solar wind collides with the ionosphere. The solar wind also impacts satellites, communication and navigation systems. For additional details see "Solar Wind" in the *Encyclopedia of Planetary Science*.

John E. Oliver

Bibliography

Brandt, J.C., 1970. *Introduction to the Solar Wind*. San Francisco, CA: W.H. Freeman.

Marsch, E., and Schwenn, R. (eds.), (1992) *Solar Wind*. New York: Pergamon Press.

Parker, E.N., 2001. A History of the Solar Wind concept. In Huber, M., et al., eds., *The Century of Space Science*, vol. 1. Dordrecht: Kluwer.

Shirley, J.H., and Fairbridge, R.W., 1997. *Encyclopedia of Planetary Sciences*. London: Chapman and Hall.

Cross-references

Solar Activity
Solar Flare
Solar Radiation

SOUTH AMERICA, CLIMATE OF

Extending from about 10°N to 55°S, South America possesses a great variety of physical geographic features. The differences are perhaps epitomized by the contrast of landscapes associated with the Andean Mountains and the great river basin of the Amazon. This variety, together with the latitudinal extent, provides the setting for climatic diversity over the continent.

Of singular importance in describing the climate of South America is its tapered shape, with much of its area lying in tropical latitudes and diminishing in higher latitudes. This causes the property of continentality, so significant in other continents that extend into middle latitudes, to be totally absent from South America. Thus the oceans are of major and immediate importance in the climates over large areas of the continent.

General characteristics

Circulation patterns

Figure S32 shows the January and July pressure and wind patterns over South America. The seasonal shift of the systems is clearly apparent, and it is these that govern the synoptic climatology of the continent. Essentially, the major controls are the locations of the Intertropical Convergence (ITC), the Pacific and Atlantic High-Pressure Cells, and the prevailing westerlies of high latitudes.

The ITC occupies its most northerly position during the period June to September when, on average, it is located between 7°N and 9°N. Its advance and retreat are not, however, a simple linear migration. During its advance south and east over Brazil in spring and summer (southern hemisphere), the movement is retarded in northeast Brazil. This causes the climate of this region to differ markedly from the expected norm. It represents one of the several climatic anomalies that occur in South America. The migration of the ITC and associated humid unstable westerlies produces a variety of monthly distribution of precipitation in the equatorial realm. Data illustrating this are listed in Table S7.

The second basic synoptic control is related to the location of the semipermanent high-pressure cells. As seen in Figure S32, these attain their most equatorward position in the low sun season. The eastern part of the Pacific High that influences the climates of the west coast produces subsiding stable air so that much of the land bordering the west coast receives minimal precipitation. In contrast, the Atlantic High provides much of the east coast (north of Patagonia) with effective precipitation.

The southern part of South America is influenced by the third major control, the strong westerly wind belt (Figure S32). The strength and relative uniformity of these winds are a function of the minimal continental area that extends into the circulation of the Southern Ocean.

Each of the features described – the ITC, the High-Pressure Cells, and the westerly wind belt – have important consequences

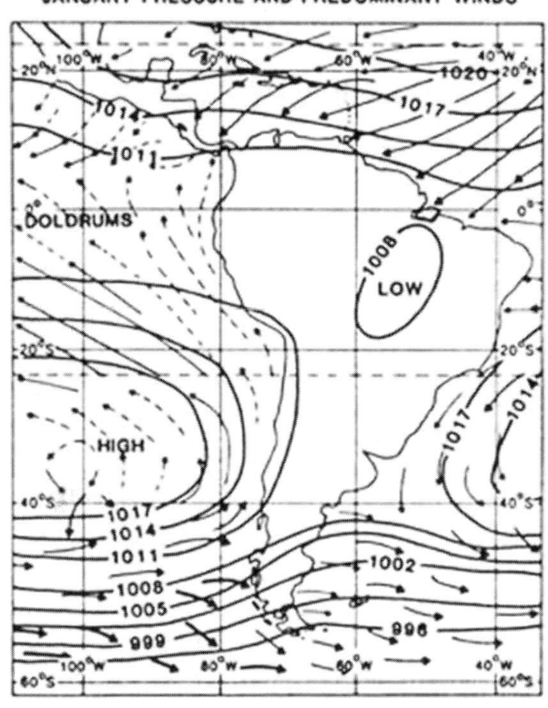

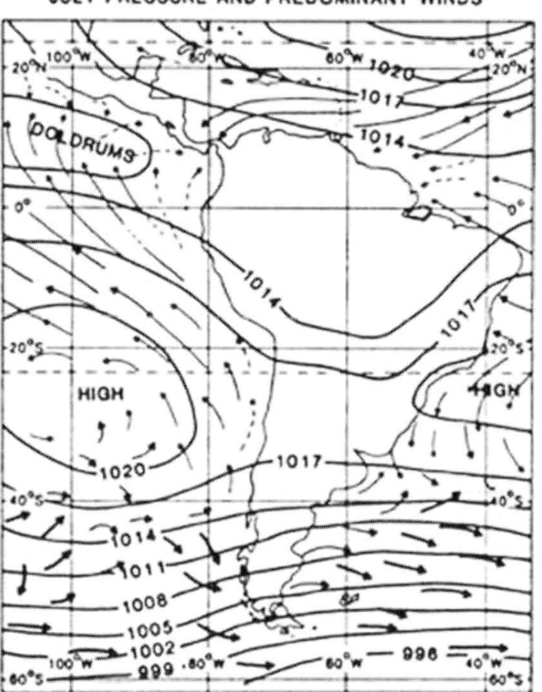

Figure S32 January and July circulation patterns over South America.

Table S7 Precipitation in cm for stations in Amazon Basin

Location	J	F	M	A	M	J	J	A	S	O	N	D	Year
Para (Lower Amazon) 1°27′S, 48°29′W, 13 m	32	36	37	32	26	17	15	11	9	8	7	15	245
Manaus (Middle Amazon) 3°08′S, 60°01′W, 44 m	25	23	26	22	17	8	6	4	5	11	14	20	181
Uapes (Upper Amazon) 0°08′S, 67°05′W, 83 m	26	19	25	27	30	23	22	18	13	18	18	26	265

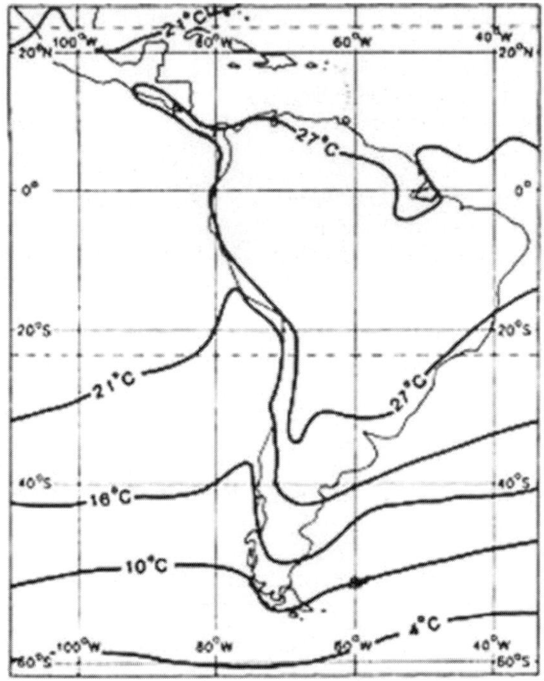

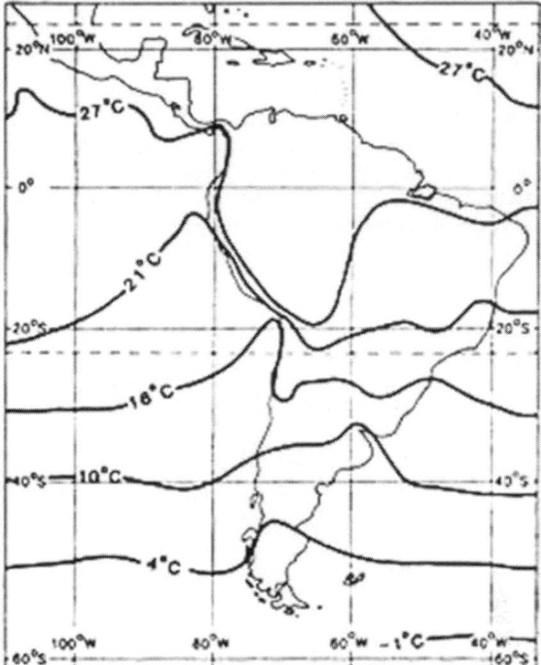

Figure S33 January and July normal temperatures for South America.

in the regional climatology of South America. Additionally, CLIVAR research (Nogué-Paegle, 2002) shows that a South American Monsoon System (SAMS) develops over a large extension of landmass crossed by the equator where conditions vary from the forests of Amazonia to the mountain desert of the Altiplano. This gives rise to a unique seasonal pattern of convection and rainfall.

Temperature and precipitation

Figure S33 provides January and July normal temperatures for South America. The general view shows dominance of high average temperatures in the equatorial and tropical regions with a fairly even decline poleward. The southern part of the continent, given the absence of a continental effect, has equable temperatures.

Major variations from the general temperature pattern occur because of the Andes Mountains and the influence of ocean currents. As Figure S34 shows, the mountains create an altitudinal variation of temperature with well-marked zones being identified. The influence of ocean currents can be seen in the shape of the isotherms along the coastal regions. The cold Peruvian Current causes lower temperatures on the west coast, whereas the warm Brazilian Current results in a poleward shift of the isotherms.

Given the very large areas over which temperature variations are minimal, it follows that much of the climatic differences in South America often are a result of precipitation distribution. The average annual precipitation illustrated in Figure S35 indicates that high amounts occur in a broad belt from the northwest coast across the Amazon Valley region to southeastern Brazil. A second wet area is in the southwest where onshore westerlies strike the mountainous coast of Chile. Between these

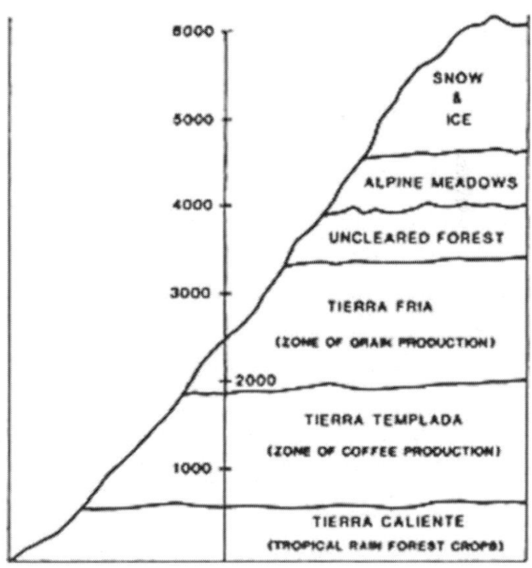

Figure S34 Altitudinal zonation of a mountain located in tropical South America.

moist areas a steep gradient to arid and subhumid regions is seen. The west coast desert and Patagonia both experience less than 250 cm of precipitation each year. The anomalous dry climates of northern Venezuela and northeast Brazil are seen on the distributional map.

Seasonality of precipitation, Figure S36, shows that, with the exception of southern Chile and the east coast area from about 25°S to 40°S, most of the continent experiences seasonal increments of precipitation.

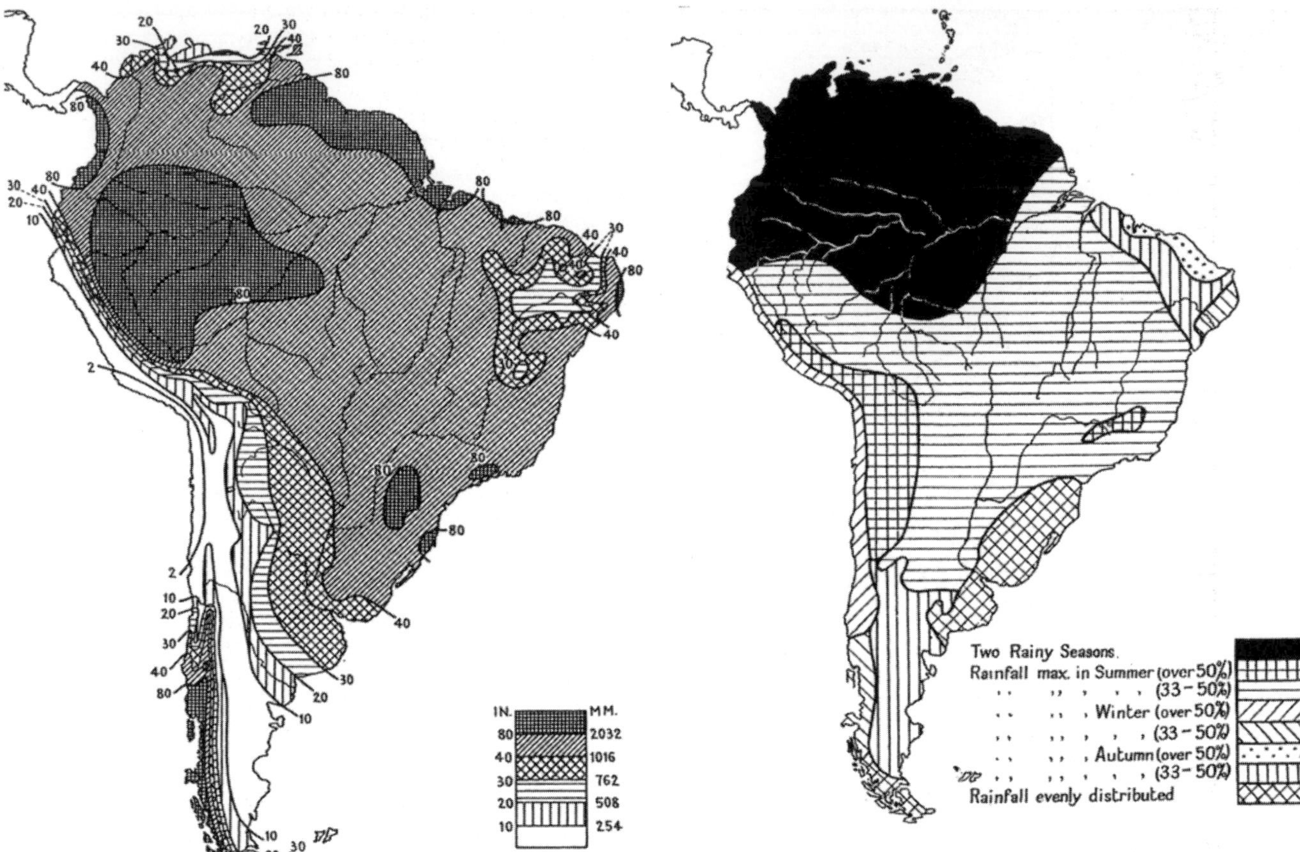

Figure S35 Average annual rainfall in South America.

Figure S36 Seasonality of precipitation (after Kendrew, 1942, p. 361).

Climatic regions

Using a standard climatic classification scheme (in this case the Köppen system as modified by Trewartha), the climates of South America generally can be located (Figure S37). The distribution, with the exception of the undifferentiated highlands, follows closely the distribution seen on hypothetical or model continental climatic regions. The extensive equatorial climate (Af) is bordered by areas receiving rain during the high sun season (Aw). These then give way to desert (BW, BS) on the west coast and humid climates (Ca) on the east. Thereafter the Mediterranean regime (Cs) and humid middle-latitude (Cf) climates extend from the desert to Tierra del Fuego. On the eastern side of the Andes, dry climates (BS, BW) occur in the lee of the mountains.

There are, of course, areas within the regions shown that are anomalous. These are discussed within the framework of the regional climates.

Regional climates

To describe the general characteristics of the regional climates of South America it proves convenient to use the following major divisions: equatorial and tropical regions; the extratropical region west of the Andes; and the extratropical region east of the Andes.

Equatorial and tropical climates

These are areas that essentially experience high average annual temperatures with copious rainfall throughout the year (Af) or with seasonal rainfall (Aw). Additionally, the dry climates of the west coast are included in this group.

The Af climate largely corresponds to the Amazon Basin. Here, moist convergent winds bring precipitation that reaches its highest amounts where air ascends the slope of the Andes. The actual distribution of rainfall in this regime is clearly not as simple as had been thought earlier, and throughout much of the basin periods of relative dryness occur. Sample data have already been presented indicating this feature (Table S7).

The data indicate that in the lower Amazon Basin, as represented by the rainfall for Para (Belem), a dry period occurs from about September to November. This region provides the typically cited classical type of climate for Af regions. Mornings often are clear, but clouds appear before noon to lead to heavy rainfall in the afternoon. Occasionally, the rain may last for a 24-hour period and thunder characteristically occurs. The vegetation of this area is lush.

In the middle of the basin the dry season also occurs (see data for Manaus) but the total amount of rain is less. This is reflected in the less luxuriant vegetation that occurs. The upper basin, as

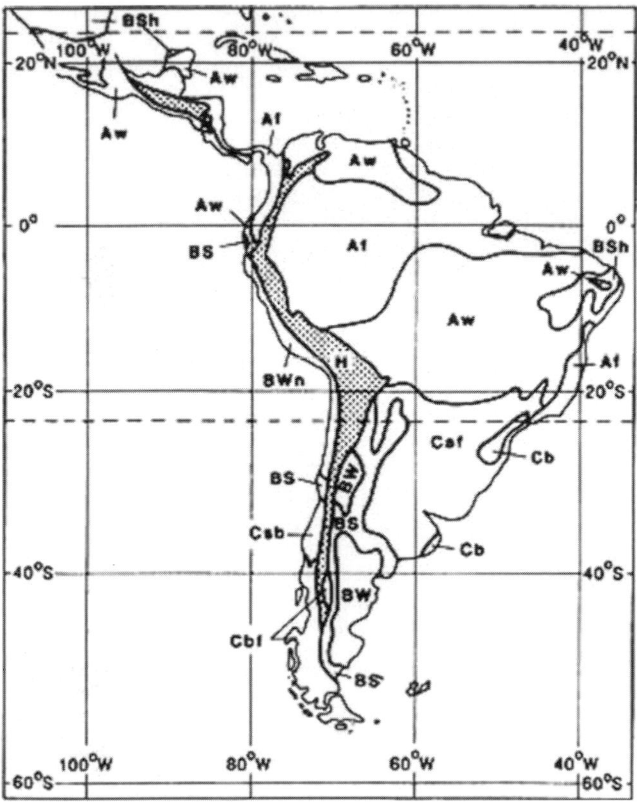

Figure S37 Climatic regions of South America according to the Köppen Classification System as modified by Trewartha. *Key*: A = tropical rainy climates: Af, Am, tropical rainforest; Aw, tropical savanna. B = dry climates: BS, steppe; BW, desert (h designates tropical and subtropical, n indicates frequent fog); C = humid mesothermal climates: Cs, dry summer; Ca, warm summer; Cb, cool summer. The stippled (H) area is undifferentiated highlands.

noted, is probably the wettest region and Uapes (Table S7) has in excess of 250 cm of rainfall annually.

The tropical wet and dry climate (Aw) borders the Amazon region. Characterized by rainfall during the high sun season, the areas are appreciably warmer than the equatorial rainforest. Typical data are shown in Table S8; at Cuyaba it is seen that the coolest month has a temperature 23°C, but in addition to this, it has an absolute minimum of 6°C.

From about the equator to the Mediterranean regime of central Chile on the west coast of South America aridity prevails along hundreds of miles of coastline. The dominant feature creating these conditions is the location of the Pacific subtropical high pressure. Stable, subsiding air resulting from this cell is directed seaward by the Andean mountains so that the air has minimal moisture content. It might be anticipated that during the seasonal shift of the high-pressure cells a weakening would occur in the system and that it would move away from the coastline. This does not happen since the high-pressure cell merely migrates northward and southward without any major change in its longitudinal location.

The stability of the air is strengthened by the cold offshore North Peruvian Current. This chills the lower levels of the air and diminishes any prospective convectional activity. The cooling of the air is so marked that the moisture does condense and fog is a characteristic of the coastal area. This is reflected in the Köppen designation of BWn, where the lower-case n indicates frequent fog. Table S9 provides illustrative data.

The circulation of waters off this coastline is at times modified by El Niño conditions. This occurs when the subtropical high-pressure cell is located south of its normal position and permits the northeast trades to cross the equator and flow northeasterly parallel to the coast of Peru. A warm current replaces the cold to lead to many dire consequences: one of these is the coming of precipitation to the normally arid area.

Within the general climatic region under discussion are climates that differ from what might be anticipated. Of particular interest in this case are the climates associated with

Table S8 Precipitation in cm and temperature in °C for tropical wet-dry stations

Location	J	F	M	A	M	J	J	A	S	O	N	D	Year
Caceres 16°04′S, 57°41′W, 118 m	26 27	19 27	17 27	9 26	4 24	3 22	1 23	1 24	5 26	11 27	18 27	21 27	135 25.6
Cuyaba 15°36′S, 56°46′W	21 26	20 26	23 26	11 26	5 24	1.4 23	0.6 23	1.2 25	4 27	13 27	16 27	19 27	135.2 26
Parana 12°26′S, 48°06′W, 260 m	29 25	24 24	24 22	10 17	1 15	0 12	0.25 11	0.5 13	3 15	13 18	23 21	31 23	158.75 18

Table S9 Precipitation in mm and relative humidity at 1300 hours for west coast desert

Location	J	F	M	A	M	J	J	A	S	O	N	D	Year
Lima 12°05′S, 77°03′W, 120 m	2.5 69	0 66	0 64	0 66	5.1 76	5.1 80	7.6 77	7.6 78	7.6 76	2.5 72	2.5 71	0 70	40.0 72

Northeast Brazil and those of the dry littoral of Venezuela and Colombia.

Northeast Brazil

This is an exceptional area in which rainfall diminishes from in excess of 100 cm along the coast to less than 40 cm inland. Rainfall that does occur reaches a maximum in summer, to give a highly unusual As climate designation of Köppen. The precise cause of the dry region is complex, involving the position of the subtropical high-pressure system over the Atlantic Ocean, the positions of the advance and retreat of the ITC, and the role of coastal and inland topographic variations. In general terms, what appears to occur is that during the southern winter almost all of Brazil south of the equator is under the influence of a greatly enlarged subtropical anticyclone, and rainfall is at a minimum over wide areas. In the summer the anticyclone weakens and equatorial air invades much of southern Brazil, with the major exception of the northeastern portion, which remains under the influence of the high pressure.

The situation in the region is complicated by the fact that even these tropical areas can be influenced by deep intrusions of polar air heralded by a cold front. During the winter months of April to October cold fronts invade Matto Grosso and the southern Amazon Basin once to three times a year. Such invasions cause a drastic fall in temperature, called friagem, and do give rise to widespread showers, often accompanied by hail.

The dry littoral regions

Along the northern margins of Venezuela and Colombia the location is such that onshore trade winds, mountains, and the anticipated presence of the ITC should provide copious rainfall. This is clearly not the case, as data in Table S10 indicate.

To explain the aridity of the area requires analysis of local divergence and subsidence over the region. In spring, the driest season, a deep easterly flow prevails with divergence up to at least 2000 m to give rise to subsidence. In summer the ITC is absent and subsidence continues, although it is not as severe as spring and some rainfall does occur. Winter, like fall, is relatively rainy but windfield analysis still indicates divergence at the surface. In all, analysis of resultant atmospheric flow indicates that the littoral area is a region in which vertical downward motion of the atmosphere, although varying in strength from season to season, is the dominant factor. It should be noted that to this explanation must be added the suggested effects of land–sea differential friction and the shape of the coastline as contributing toward the relative aridity.

Extratropical regions west of the Andes

Along the west coast of southern South America is a poleward transition from the arid desert to a humid middle-latitude climate. A Mediterranean scrub woodland (Cs), extending some 550 km south of 32°S, occurs between these two extremes.

Atmospheric circulation of the eastern Pacific Ocean is conditioned by the location of the semipermanent high-pressure cell and its associated winds. The desert area owes its origin to subsiding air of the high and enhancement of stability by the cold offshore waters. As winter approaches, the high-pressure cell contracts and decreases in intensity; it becomes centered at approximately 25°S. This permits the belt of westerlies to migrate northward, giving the Mediterranean area its winter precipitation.

The southwestern strip of the continent has been called "one of the most monotonous unpleasant marine climates in the world" (Rumney, 1968, p. 279). As shown in Figure S32, it receives year-round onshore winds from the cool South Pacific. The lack of any continental influence causes this area to have one of the most equable climates in the world (Table S11).

Extratropical regions east of the Andes

The most exceptional feature of this region is the rainfall deficiency that occurs to give the Patagonian desert. The apparent obvious reason – location on the lee side of a mountain barrier cannot totally explain the aridity. It appears that the westerly winds from the Pacific Ocean, on rising over the Andean chain, form an upper-level anticyclonic ridge over the mountains, with a trough to the east. The narrow width of the continent in this

Table S10 Precipitation in cm for Venezuelan stations

Location	J	F	M	A	M	J	J	A	S	O	N	D	Year
Caracas 10°30′N, 66°56′W, 950 m	2.2	1.0	1.5	3.3	7.8	10.1	10.9	10.9	10.6	10.9	9.4	4.6	83.2
Maracaibo 10°30′N, 71°36′W, 6 m	0.2	0	0.7	2.0	6.8	5.6	4.6	5.6	7.1	14.9	8.3	1.5	57.3

Table S11 Temperature (°C) for selected stations in southwest Chile

Location	J	F	M	A	M	J	J	A	S	O	N	D	Year
Cabo Raper 46°50′S, 75°38′W, 40 m	10.5	10.0	10.2	9.4	8.0	6.6	5.5	6.1	6.3	7.5	8.6	10.0	9.7
Los Evangelistas 52°24′S, 75°06′W, 55 m	8.3	8.3	8.0	6.9	5.5	4.4	4.1	4.1	5.0	5.5	6.1	7.2	6.1

Note: Temperature range for Cabo Raper is 5°C and for Los Evangelistas 4.2°C.

Table S12 Precipitation in cm for Argentina including Patagonian Desert

Location	J	F	M	A	M	J	J	A	S	O	N	D	Year
Bahia Blanca 38°43′S, 62°16′W, 30 m	4.3	5.6	6.3	5.8	3.0	2.3	2.5	2.5	4.0	5.6	5.3	4.8	52.0
Parana 31°44′S, 60°31′W, 65 m	7.8	7.8	9.9	12.4	6.6	3.0	3.0	4.0	6.1	7.1	9.4	11.4	88.5
Santa Cruz 50°01′S, 68°32′W, 11 m	1.5	0.7	0.7	1.5	1.0	1.2	1.0	1.2	0.7	0.7	1.0	1.7	12.9

area causes the trough to be located over the Atlantic, and it is thus ineffectual in providing rainfall to the dry region. This negates the formation of disturbances that typically provide rainfall for similar middle-latitude situations.

The east coastal region north of Patagonia receives rainfall partly because of the greater breadth of the continent at this latitude. The upper-air trough is now over the continent and causes cyclonic rather than anticyclonic conditions to prevail. Low-pressure systems that move eastward and southeastward bring appreciable precipitation to the Argentinian Pampa. Data applicable to stations on the west side of the Andes are given in Table S12.

Climate variability and change

Climate variability in South America is in many ways connected to the Southern Oscillation (SO) and the El Niño phenomena. Of particular significance are the anomalously moist conditions in Peru and southern Ecuador during an El Niño event. It has also been shown that droughts in northeastern Brazil are related to strong El Niño events.

Over the longer time period it has suggested that there is evidence for both a Medieval Warm Period and a Little Ice Age in extratropical South America (Cioccale, 1999). Offering evidence from altitudinal migration studies that reflect relative environmental suitability and glacial advance/retreat it is suggested that about AD 1320 a marked cooling occurred, representing the Little Ice Age. Other studies of paleohydrology and glacial activity in Argentina and Chile also suggest signatures representive of the Little Ice Age. Perhaps such might be expected, for Vargas et al. (1995) show that many rapid climatic changes have occurred in South America since the end of the last Ice Age.

Observations in more recent years indicate further changes. In the Andes Mountains of Peru the Oori Kalis glacier retreated approximately 4 m a year between 1963 and 1978. Since that time the rate has sharply increased. Glaciers in Patagonia have receded by an average of almost 1.5 km over the last 13 years. There has also been an increase in maximum, minimum, and average daily temperatures of some 1°C over the last 100 years in southern Patagonia, east of the Andes.

New interest concerning climate change and variability in South America has led to renewed research. At the same time, modeling studies, as illustrated by Nicolini et al., 2002, provide further information about the future climate of South America.

Norys Jiminez and John E. Oliver

Bibliography

Adams, J., 2004. *South America during the last 150,000 years*, www.esd.ornl.gov.

Boucher, K., 1975. *Global Climates*. New York: Halsted Press.

Caviedes, C.N., 1973. Secas and El Niño, two simultaneous climatic hazards in South America. *Proceedings of the Association of American Geographers*, **5**: 4449.

Cioccale, M.A., 1999. Climatic fluctuations in the central region of Argentina in the last 1000 years. *Quaternary International*, **62**: 35–47.

Kendrew, W.G., 1942 and following. *The Climates of the Continents*. New York: Oxford University Press.

Lenters, J.D., and Cook, K.H., 1999. Summertime precipitation variability in South America: role of large-scale circulation. *Monthly Weather Review*, **127**: 409–443.

Nicolini, M., Salio, P., Katzfey, J.J., McGregor, J.L., and Saulo, A.C., 2002. January and July regional climate simulation over South America. *Journal of Geophysical Research*, **107**: (D22), 10.1029/2001, JD 000736.

Nogués-Paegle, J., 2002. Progress in Pan American CLIVAR Research: Understanding the South American Monsoon. *Meteorologica*, **27**: 1–30.

Rumney, G.R., 1968. *Climatology and the World's Climates*. New York: Macmillan.

Schwerdtfeger, W., 1976. *Climates of Central and South America*, vol. 12: *World Survey of Climatology*. Landsberg, H.E., (ed.). New York: Elsevier.

Trewartha, G.T., 1981. *The Earth's Problem Climates*, 2nd edn. Madison, WI: University of Wisconsin Press.

Vargas, W.M., Minetti, J.L., and Poblete, A.G., 1995. Statistical study of climatic jump in the regional zonal circulation over South America. *Journal of the Meteorological Society of Japan*, **73**: 849–856.

Whetton, P.H., Pittock, A.B., Labraga, J.C., Mullan, A.B., and Joubert, A., 1996. Southern hemisphere climate comparing models with reality. In Henderson-Sellers, A., and Giambelluca, T., eds., *Climate Change, Developing Southern Hemisphere Perspectives*. Chichester: John Wiley & Sons.

Cross-references

Arid Climates
El Niño
Intertropical Convergence Zone
Maritime Climate
Mediterranean Climates
Ocean Circulation
Southern Oscillation
Trade Winds and the Trade Wind Inversion
Tropical and Equatorial Climates

SOUTHERN OSCILLATION

This is a term introduced by Sir Gilbert Walker (Walker 1923, 1924), then director of the Meteorological Service of India, for an alternation between high and low pressure in a seesaw pattern between two specific stations, but later extended to broad areas of the southeastern Pacific and the central to eastern Indian Ocean. These were referred to as the corresponding "nodes" (Berlage, 1957, 1959) or "core regions". A state with southeastern Pacific pressure high is said to have a *high SO index*; when pressure is low, there is a *low SO index*. A popularized account is provided by Nash (2002).

This atmospheric pressure differential sets up a "meteorologic tide", recorded by tide gauges (Fairbridge and Krebs, 1962), that creates a high average mean sea level during the low-pressure phase, and low MSL in the high-pressure phase.

A third correlation was obtained with rainfall data by Walker and Bliss (1932), and hence with lake levels and sediment cores. The earliest evidence of the pressure variation was reported in 1897 by H.H. Hildebrandsson for the inverse relations between Sydney, Australia, and Buenos Aires, although neither is within the nodal peak areas. In fact there are, it has since been found, worldwide "teleconnections" (Bjerknes, 1969). This term was introduced by the Baron Gerard de Geer (1858–1943) and his student Carl C. Caldenius (1887–1961) in Sweden for the long-distance linkages (e.g. to Argentina) suspected to exist in lake varve sediments, though they were difficult to prove. The small variations can serve as a "vernier" of year-to-year significance, so-called "wiggle-matching", provided that one knows the overall pattern and sidereal chronology (as distinct from the elastic radiocarbon indicators).

Another index of the Southern Oscillation (SO) is the sea-surface temperature (SST). In an illustration provided by P.B. Wright (in Oliver and Fairbridge, 1987, p.797), whose monthly values of each are shown for the period 1950–1983, represented by (a) the pressure anomaly for Tahiti (S.E. Pacific Node) minus that for Darwin (Indian Ocean Node), that is the "pressure gradient"; and (b) the SST anomaly in the equatorial belt roughly 180°W and 90°W. Simple inspection shows a general inverse relationship, warming with lower atmospheric pressure and vice-versa, but not in the detailed wiggles.

Wright (*op. cit.*, p. 796) pointed to the remarkable persistence of the SO, with anomalies maintaining the same magnitude and sign for many months over certain seasons. Between September and December (southern hemisphere spring) there is a 3-month lag correlation between SO and SST (0.91), but February to May it is much less (0.36). There is notable year-to-year variability, the years with the lowest SO index corresponding to El Niño (EN) years, which means warm SST in the southeastern Pacific. The intimate relationship between SO and EN years has led to their joint identification as "ENSO", although they are not always very closely linked.

Solar forcings

Inasmuch as the teleconnections show that both SO and ENSO phenomena have worldwide repercussions, their explanations must necessarily be complex and planetary in nature. Accordingly, the ultimate source must lie in the sun and effective solar radiation, which, contrary to most assumptions, is remarkably variable and closely related to the spin-rate of the sun. This variance is at least partially predictable, being related to the torques set up in the photosphere by planetary dynamics such as the Saturn–Jupiter beat (19.8593 yr), or various Earth–moon and Earth–Venus cycles (Fairbridge and Sanders, 1987; Windelius and Carlborg, 1995).

Linkage of the sun to the Earth's climatic systems follows two systems: (a) electro-magnetic radiation with the UV playing a significant role, through ozone production in the upper atmosphere; and (b) particle and plasma radiation in the solar wind. Because of the particle magnetization this influence is focused on the magnetic poles and the two auroral ovals. It plays an important role in the Atlantic Oscillation.

The first of these two solar forcings has its direct effect in the troposphere, focused on our equatorial belt, and consequently influences the Walker Circulation. The principal energy source of the latter is in the rising air over the countless islands and volcanic peaks of Indonesia and New Guinea. Its major variance is the QBO, the Quasi-biennial Oscillation, which has a mean period of 2.172 yr. But inasmuch as the Earth's climatic seasons relate closely to a 12-month frame, an astronomically defined pulse is likely to be amplified at "round-number" peaks, in this case (×6), theoretically at 13.032 yr, but in most climatic time series is simply 13 yr.

The moon is also involved in these cycles, playing an eccentric role because of the difference between the lengths of the solar and lunar years. The extreme perigee–syzygy alignment occurs at a little over 26 yr (QBO × 12). On a longer term, the ^{14}C flux determined from tree rings provides a 69.5/104.26/208.522-yr spectral series (Thomson, 1990), respectively QBO ×32/64/96.

Lunar forcings

The lunar linkage with the QBO suggests an oceanic tidal factor in the SO global role. It is controlled by the position of Indonesia and New Guinea in the Walker Circulation, which is dominated by the orographic forcing and upward flow of warm air generated by the high volcanoes and mountainous islands. This archipelago is uniquely placed on planet Earth astride the equator, and it is also midway between the world's two largest continental shelves (Sahul and Sunda), each more than 1.5 million km^2. A rising tide over such a shelf margin brings cold offshore water onto the shelf and gives a minor cooling pulse to the atmosphere. At spring tides this pulse is larger and with more important cyclical enhancements the tides assume an even greater role (Wood, 2001). When the ENSO brings higher sea levels and warmer waters to these shelves the net oceanic flow is always from east to west, although modulated by the tide cycles. According to Currie (1995) these tide cycles are maximized at the 18.6-yr nodal intervals, and then appear worldwide, appearing in the spectra of every geophysical time-series from pressure to SST, to sea level, and other important parameters.

This region at the junction of the Pacific and Indian oceans is thus critical, not only for the SO, but also for global climate. It creates a unique "Warm Pool" (sometimes referred to as the "Western Pacific Warm Pool"). Its linkage to lunar cycles introduces a boundary current relationship. If one considers geostrophic currents such as the Kuro Shio or equally well to the Gulf Stream, as they head north, their velocities increase as the moon's zenith shifts northward (by over 1100 km) during the 9.3-yr hemicycle of the 18.6-yr lunar period. Due to the Coriolis Effect the acceleration causes a fall of sea level on the left and rise on the right, when SST will rise. These relationships were first observed in the North Atlantic by Russian workers (Lamb, 1977) but apply equally in the Pacific and to both hemispheres.

Exact coherence of terrestrial climate patterns to lunar and solar cycles must not be expected because of the dominance of the 12-month cycle of sunshine (insolation). Thus the QBO is a biennial cycle for a while and then jumps to triennial. The lunar cycle of 9.3 yr only corresponds to the Earth's orbital period every 186 yr (20 × 9.3). The Earth–moon perigee–syzygy cycle, with closest approaches and alignments with the sun, is some 52 yr, but its season varies from four times per year to five, creating 13-yr and 26-yr subcycles. No regular annual cycle is seen. In phase in 1953 and 2005, it coincides with major ENSO years in 1983 and 1997. Those same years mark synchronous events of the solar synodic rotations and the moon's perigee–syzygy cycle. The latter is almost in phase with the apsides cycle of 8.849 yr.

El Niño links

Comparison of the SO with the El Niño historical series shows a close association, but also some gaps and unexpectedly missing years. Nevertheless the Quinn et al. (1978, 1983) historical base provides an excellent long-term reference. Two important aspects call for attention: (1) there is no systematic correlation with the sunspot cycle, and flare phenomena are too irregular to be considered. Nevertheless, as mentioned above, the solar influence on the Walker Circulation and the 2.1721-yr QBO needed to be taken into the models. Over 50 ENSO events in four centuries match the QBO, and of these 14 are classed as "super" or "very strong". (2) Coincidence of many lunar periodicities is seen, notably the 8.849-yr apsides cycle and the 9.3-yr declination hemicycle. These two also affect the Walker Circulation, so that both solar and lunar forcings are involved. Synergy can be deduced from the fact that at 69.508 yr there are 32 × QBO of 2.172 yr and 4 × lunar tidal beats of 17.3769 yr.

Long-term astronomic control is shown by the Quadrature Cycle (of Stacey, 1963) when all the planets are at approximately 45° to one another in the celestial sphere, and

$$4448.485 = 64 \times 69.508 = 2048 \times \text{QBO}.$$

The 69.508 period is especially significant because it appears in the dendrochronological spectra, notably the ^{14}C flux which is inversely related to solar radiation and linked not so much to the sun, but mainly to cosmic radiation. Nevertheless, the inverse ^{14}C flux pattern is present in the warm cycles of the Greenland ice cores (Stuiver and Braziunas, 1988).

The lunar cycle of 558.14 yr (Wood, 2001), which is the extreme proxigee–perigee–syzygy–perihelion peak, is marked by the resonance intervals of several planetary beat frequencies. Notable is the Saturn–Jupiter lap of 19.8593 yr which (×7) has an important resonance at 139.015 yr (=64 QBO; and 1/32 × 4448.485), the perigee–syzygy precession, but is also locked in to the lunar systems by the nutation and declination period of 18.603 yr (× 30 = 139.015 yr). Correspondingly, the 93.015-yr period that is well known in Swedish tidal history (Pettersson, 1914a,b), and is 1/6 × 558 yr, and the 31-yr tidal periods (1/18 × 558 yr) many of which (±1) are present in the ENSO records, e.g. 1976–1914–1884–1821–1791–1728–1696–1634–1574–1539/41. Observational gaps are of course present and the "old year/new year" problem causes the ±1 yr anomalies. The perihelion epoch of the 558.14-yr tide cycle also causes a juggle between December and January dates. The next major ENSO should be in 2006/7.

Linkage with climatic proxies is provided by the 208.52-yr cycle in the tree-ring ^{14}C flux (Thomson, 1990). It coincides with ENSO events in 1552–1701–1969. In the astronomical frame it corresponds with the moon's perigee–perihelion–syzygy and Jupiter/Mercury laps and its 1/8 signal is 26.06534 yr (in turn, ×12 QBO, and 2 × 13.0325 yr, the Uranus–Jupiter–Venus synod). Important El Niño dates include: 1969, **1932**, 1857, **1728**, 1708, **1671**, **1652**, **1634**, **1578**, **1541** (very strong or "super" El Niños are shown in bold face).

Evidence of "super-El-Niños" in the geological past is provided by the proxy of ancient beachridges which are well preserved (and isotopically dated) along the semiarid coasts of Peru (Ortlieb et al., 1995). Human populations were also involved by the fluctuations in climate and the effect on foodstuff supplies, so that geoarcheological evidence could be cited.

Multiples of the 31-yr lunar cycle have been found to correlate very closely with the isotopically determined dates, after sidereal correction (Ortlieb et al., 1995). We can use AD 1541 as a well-established anchor point (Table S13).

Conclusion

The range of the ENSO teleconnection can now be traced directly through the "sluice-gates" of the Indonesian archipelago to the Indian Ocean, and thence to the Indian monsoonal rains and pressure anomalies. This east–west flow weakens appreciably with the ENSO, and with the larger lunar fluctuations can lead to "monsoon failures", droughts and catastrophic starvation (although modern organization and foodstuff reserves can now mitigate the worst events). The seesaw relationship of the SO in general to the El Niño in particular (first described by Walker and Bliss, 1932), is illustrated in the sketch by Harrison and Larkin (1998), showing (a) the "positive" sea-level pressure (SLP; low MSL) over northern Australia and the adjacent Indian Ocean, favoring (b) equatorial westerlies, a warming of the eastern Pacific, and (c) negative SLP (high MSL) associated with the southeastern Pacific and the El Niño in Peru and Ecuador (with heavy rains). Rainfall anomalies around the globe are reviewed by Ropalewski and Halpert (1987). Not shown is the displacement of the cold Peru ("Humboldt") Current with weakened upwelling and SE trade winds, matched in the northern hemisphere by a similar shift of the cold, south-setting California Current and NE trades, while warm currents hug the coasts, south-setting in Peru and Ecuador, north-setting in Mexico and California. Use of coral-growth proxies for SST (Evans et al., 2002) shows that not only is the SO a major forcing, but overall Pacific SST cycles occurred over the last four centuries, that cluster around El Niño peaks (1634, 1686, 1720, 1791, 1844, 1900). The

Table S13 Peruvian beachridge dates (calibrated radiocarbon: Ortlieb et al., 1995), compared with a projection of the lunar 31-yr tidal peaks, showing long-term fluctuations of the Southern Oscillation

Beachridge dates	Tidal dates	Historical events
AD (BP)	1541 (AD)	Spanish Conquest of Peru
1500 (450)	1504	End of Columbus' last voyage
1252 (940)	1231	
600 (1565)	611	Classic Maya stage
130 (1820)	140	
95 (1855)	109	
BC		
290 (2238)	288 (BC)	
787 (2735)	784	
1180 (3130)	1187	Chira Ridge J (Peru)
1480 (3430)	1466	Chira Ridge L (Peru)
1650 (3600)	1652	

systematic monitoring and study of ENSO is thus a matter of worldwide importance for both scientific and social needs.

Acknowledgments

Appreciation is expressed to Dr James Hansen (GISS/NASA) and colleagues for helpful discussions, and especially to our deceased associates, notably J.E. Sanders and F.E. Wood.

Rhodes W. Fairbridge

Bibliography

Berlage, H.P., 1966. The Southern Oscillation and world weather. *Kon. Nederlands Meteor. Inst. Med. Verh.*, **88**: 1–152.

Bjerknes, J., 1969. Atmospheric teleconnections from the equatorial Pacific. *Monthly Weather Review*, **97**: 163–172.

Broecker, W.S., and Denton, G.H., 1990. What drives glacial cycles? *Scientific American*, **262**: 48–56.

Currie, R.G., 1995. Variance contribution of M_n and S_c signals to Nile River data over a 30.8 year bandwidth. *Journal. Coastal Research*, special issue 17, pp. 29–38.

Dunbar, R.B., et al., 1994. Eastern Pacific sea surface temperature since 1600 A.D.: the $S^{18}O$ record of climate variability in Galapagos corals. *Paleoceanography*, **9**: 291–314.

Fairbridge, R.W., and Krebs, O.A., Jr, 1962. Sea level and the Southern Oscillation. *Geophysical Journal*, **6**: 532–545.

Fairbridge, R.W., and Sanders, J.E., 1987. The Sun's orbit, AD 750–2050: basis for new perspectives on planetary dynamics and Earth-Moon linkage. *In* Rampino, M.R. et al., eds., *Climate – History, Periodicity, and Predictability*. New York: Van Nostrand Reinhold, pp. 446–471 (bibliography, pp. 475–541).

Harrison, D.E. and Larkin, N.K. 1998. El niño-Southern Oscillation sea surface temperature and wind anomalies, 1946–1993. *Reviews of Geophysics*, **36**: 353.

Lamb, H.H., 1977. *Climate History and the Future*. London: Methuen.

Nash, J.M., 2002. *El Niño*. New York: Warner Books.

Oliver, J.E. and Fairbridge, R.W. (eds.), 1987. *The Encyclopedia of Climatology*. New York: Van Nostrand Reinhold.

Ortlieb, L., Fournier, M., and Macheré J., 1995. Beach ridges and major late Holocene El Niño events in northern Peru. In Finkl, C.W., ed., *Journal of Coastal Research*, special issue 17, pp. 109–117.

Pettersson, O., 1914. On the occurrence of lunar periods in solar activity and the climate of the earth. *Svenska Hydrogr.-Biol. Komm. Skriften.*

Quinn, W.H., Zopf, D.O., Short K.S., and Kuo Yang, R.T.W., 1978. Historical trends and statistics of the Southern Oscillation, El Niño, and Indonesian droughts. *Fishery Bulletin*, **76**(3): 663–678.

Quinn, W.H., and Neal, V.T. 1983. Long-term variations in the Southern Oscillation, El Niño, and Chilean subtropical rainfall. *Fishery Bulletin*, **81**(2): 363–374.

Ropelewski, C.F., and Halpert, M.S., 1987. Global and regional scale precipitation patterns associated with El Niño/Southern Oscillation. *Monthly Weather Review.*, **115**: 1606–1626.

Stacey, C.M., 1963. Cyclical measures: some tidal aspects concerning equinoctial years. *New York Academy of Science Annals*, **105**(7): 421–460.

Stuiver, M. and Braziunas, T.F., 1989. Atmospheric ^{14}C and century-scale solar oscillations. *Nature*, **338**: 405–408.

Thomson, D.J., 1990. Time series analysis of Holocene climate data. *Philosophical Transactions, Royal Society of London*, **A330**: 601–616.

Walker, G.T., 1923. Correlation in seasonal variations of weather. VIII. A preliminary study of world-weather. *Memoirs of the Indian Meteorological Department*, **24**(Part 4): 75–131.

Walker, G.T., 1924. Correlation in seasonal variations of weather. IX. A further study of world weather. *Memoirs of the Indian Meteorological Department*, **24**(Part 9): 275–332.

Walker, G.T., and Bliss, E.W., 1932. World Weather V. *Royal Meteorological Society memoirs*, **4**: 53–84.

Windelius, G., 1993. Large volcanic eruptions and earthquakes in pace with the Chandler Wobble. *Cycles*, **44**: 140–144.

Windelius, G., and Carlborg, N., 1995. Solar orbital angular momentum and some cyclic effects on Earth systems. *Journal of Coastal Research*, special issue 17, pp. 383–395.

Wood, F.E., 2001. *Tidal Dynamics*, 2 vols. *Journal of Coastal Research*, special issue 31.

Wright, P.B., 1984. Possible role of cloudiness in the persistence of the Southern Oscillation. *Nature*, **310**: 128–130.

Wright, P.B., 1985. The Southern Oscillation: an ocean–atmosphere feedback system? *American Meteorological Society Bulletin*, **66**: 398–412.

Cross-references

Cycles and Periodicities
El Niño
Oscillations
Quasi-Biennial Oscillation
Sunspots

SPORTS, RECREATION AND CLIMATE

Recreation and sports have been vital elements in most all societies for many centuries. These activities require leisure time, and modern economies have allowed both the time and resources to make recreation and sports available to all. In the United States alone, over a *quarter of a trillion dollars* is spent yearly on recreation. Recreation consists of a myriad of activities and ranges from hiking in the wilderness to watching the World Cup Final of soccer (football) with over 100 000 other souls in attendance. Such activities have been studied by applied climatologists and meteorologists, and their work has been nicely outlined by Perry (1997). The World Meteorological Organization (1996) has published a booklet promoting the value of weather and climate information to sports and recreational activities.

Outdoor sports are, above all, subject to the vagaries of the atmosphere. Thornes (1983) characterized sports into three types with respect to the atmospheric environment. The first is specialized weather sports that must have a certain mix of conditions to take place. Thus, if one wishes to participate in snowboarding, snow must be present. Or, if kite flying is planned, sufficient wind must exist. The second type is weather interference sports. Baseball, rugby, tennis, and polo are examples of this type. Performance can be negatively affected by the atmospheric conditions and the condition of the playing surface. The third type is weather advantage sports in which performance may be enhanced by going first, last, or choosing position. In some respects snow skiing fits this category because the notoriously changeable winter conditions and the packing of the course taking place with successive participants.

Physiological considerations

The basic biometeorological considerations of sports performance have been well researched and are outlined by Tromp (1980). There are well-known effects of thermal stress and high altitude. As sports are performed, the tendency is for human heat production to exceed the ability of the body to dissipate the heat in all but the coldest weather. Therefore, the core temperature of an athlete's body tends to creep higher and higher and increase dehydration, with its concomitant decreases in athletic performance. The interaction of the atmosphere with sport has been accorded long and loving debate. At Coors Field in Denver (1600 m), much speculation has been accorded altitude's role in increasing the number of home runs, but recent data suggest the characteristics of local winds and the quality of pitching play at least as much a role as altitude (Chambers et al., 2003). At Laramie, Wyoming, the University of Wyoming's football stadium intimidates opponents with the sign "Welcome to 7220 feet". Altitudinal effects were brought to prominence during the Mexico City Olympics of 1968. The city, at 2300 m, was the highest venue of any modern Olympic Games and presented a multiplicity of effects. At this altitude the air has 30% less oxygen than at sea level, so sports that required endurance performances all saw decayed times from previous Olympics. However, another altitudinal effect is decreased density of the air resulting in less frictional resistance. Races at distances of less than 400 m, relay races, and the long jump all saw records. Caution must be used when assessing climatic influences on sport performance. Fitness, skill, and mental outlook play roles that are as least as large.

Objective climatologies of human comfort have been applied to recreation and sports. Yapp and McDonald (1978) developed a thermodynamic climatology of human recreation. They considered the combined influence of solar radiation, temperature, humidity, wind, Earth surface type, and clothing in a thermodynamic model of human comfort. Then they classified locations based on their thermodynamic suitability for recreational activities. DeFreitas (1990) has analyzed various climates as to their recreational potentials.

Cimate and the origins and diffusion of sport

Elements of climate can be seen in locational aspects of sports (Rooney, 1974). From a standpoint of logic it can be stated that as individual sports were invented and evolved, they did so in reasonable harmony with their local climatic environments. Golf, originated in Scotland, has thermodynamically optimal conditions that range in the upper teens and low twenties Celsius – conditions that mimic Scottish summers. Baseball is slow-paced and more suited to the summers of the American midlands than soccer, which requires more endurance per game.

The geographic differences in production of players with superior skills at a sport are partially attributable to climate. The childhood years see sports played at the grassroots. Although there are some ice rinks in Florida, it is apparent that a relatively small percentage of children use ice rinks. In Canada most children play hockey on the outdoor ice and snow that lasts so much of the winter season. The consequence is that, in North America, the National Hockey League talent pool is concentrated in Canada. Conversely, when examining the origins of Major League Baseball players, there is a preponderance of talent from the southern United States where mild winters allow year-round play.

Whereas it can be argued that sports have partially evolved with respect to the climate present in a society, it is also clear that climate should not be construed as defining all locational aspects of sports. The strong imprint of human society is seen in the dissemination of sports into disparate climates far beyond their climatic origins. Golf has been transplanted to much of the world and is enjoyed by enthusiasts even in many locales where summer thermodynamic conditions are dangerous. In another instance, cricket – a favorite game in the British Isles – was ported to the Indian Subcontinent during colonial rule. Although cricket is more comfortably played in the British Isles, the sport is widespread in India and Pakistan where warm-season conditions are less comfortable. In the United States, major sports evolved with noted seasonalities relating to comfort of play and spectatorship. The beginnings and ends of these seasons have become blurred as dictated by revenue considerations. American professional football was once a fall sport and finished before Christmas, but its playoffs now string all the way through January; the result is some games are played in brutally cold weather that is dangerous for fans and players alike. The Superbowl, however, is always played at a predetermined city that is climatically benign or has a domed stadium. The most popular sport in the world is soccer; it has spread over the globe to be played in venues in every inhabitable climate. No matter whether it is in climates of Amazonia or Siberia, soccer is played and excellence in performance achieved.

Some sports are played during the most extreme conditions: American football games at every level have the reputation of going on despite the weather conditions. Sports competitions are canceled or postponed because of inherent danger to the participants or spectators. Baseball, for instance, becomes dangerous when the pitcher cannot control the flight of the ball due to heavy rain. Golf tournaments are halted when lightning is in the vicinity.

Lessening the atmospheric impact

As sports have developed, two technologies have attempted to diminish the effects of weather conditions on individual events. They are the advent of artificial turf and the increased presence of domed stadia. Fans of European football (soccer) and American football are well aware of the connection between the wear-and-tear on the athletic field and the quality of play during the season. In the deteriorating conditions of fall into winter, mud and unsure footing become more common, and skill might play a lesser role. To obviate this effect, artificial turf with improved subsurface drainage has been placed at many venues at considerable expense.

Some outdoor sports have been played indoors for many decades. Ice hockey was the first major sport to make the transition. The space taken up by a rink is small enough so that the activity can be fit into many modest-sized arenas. This increases player and fan comfort and enables better control of the smoothness of the ice surface. As buildings became larger, space-extensive sports were placed inside. In the United States the Astrodome opened in 1965; this was an attempt to bring baseball inside out of the oppressive heat and humidity of the Texas Gulf Coast summers. The Astrodome began its existence with natural grass turf and semi-transparent roof panels to allow entrance of sunlight. The panels proved impossible as a backdrop for fly balls, so they were painted over and the grass died; ironically, this necessitated the first installation of artificial turf. Through time, domed stadia were put in large cities where the climate provided many inclement days during a sports season. The logic of the existence of the Kingdome, Seattle's original domed stadium, is obvious. The presence of an indoor venue made possible a full slate of home baseball games in the rain, drizzle, cool marine west coast climate of Seattle. Likewise, the dome in Minneapolis allows baseball and football to be played in comfort in a climate that provides many cold days in their respective seasons. As architectural technologies improved, domes were made retractable. In Toronto, Seattle (the newer dome), and Phoenix, hugely expensive retractable domes allow competition to be "outdoors" when conditions permit! In the building of domes, local societal concern and beliefs often supersede climate. The Green Bay Packers football team proudly plays on "the frozen tundra of Lambeau Field" to the consternation of teams from warmer climates; the cold climate is a source of pride.

Travel for recreation

The citizens of modern societies have had increasing opportunities to travel for recreation. Inherent to this translocation is the cursory climatic expectations of most travelers. Doekson (1996) states "part of the enjoyment in travel and tourism is derived from experiencing at first hand the elements of the weather that define a region's climate". Tourists, therefore, travel long distances drawn by the promise of a location's climate and seasonality.

During the Raj in India, the British Colonial administration made a yearly summer switch from the Ganges Plain to Simla (2200 m) to enjoy milder temperatures and more comfortable recreation. Today, climatology is taken into account as cruise lines reposition their ships to be in the most attractive regions by season. To wit, some of the cruise ships in the Alaskan summer trade migrate to the Caribbean for the winter season. Residents of the cloudy northern European plain make pilgrimages to the Mediterranean to bask in the plentiful sunshine. Likewise, residents of the California lowlands flock to the Sierra Nevada to enjoy the snowpacks of the high altitude winters.

Steve Stadler

Bibliography

Chambers, F., Page, B., and Zaidins, C., 2003. Atmosphere, weather, and baseball: do baseballs really fly at Denver's Coors Field? *Professional Geographer*, **55**: 491–504.

DeFreitas, C.J., 1990. Recreation climate assessment. *International Journal of Climatology*, **10**: 89–103.

Doeksen, N.J., 1996. Recreation and climate. In Schneider, S., ed., *Encyclopedia of Climate and Weather*. New York: Oxford University Press, pp. 638–689.

Perry, A., 1997. Recreation and tourism. In Thompson, R.D., and Perry, A., eds., *Applied Climatology, Principles and Practices*. London: Routledge, pp. 240–248.

Rooney, J.F., Jr, 1974. *The Geography of American Sport from Cabin Creek to Anaheim*. Reading, MA: Addison-Wesley.

Thornes, J.E., 1983. The effect of weather on attendance at sports events. In Bole, J. and C. Jenkins, eds, *Geographical Perspectives on Sport*, University of Birmingham, 201–210.

Tromp, S.W., 1980. *Biometeorology*. London: Heydon.

World Meteorological Organization, 1996. *Weather and Sports*. Geneva: World Meteorological Organization.

Yapp, R., and McDonald, S., 1978. Recreation climate model. *Journal of Environmental Management*, **7**: 243–260.

Cross-reference

Bioclimatology

SPRING GREEN WAVE

In midlatitude climates with strong seasonal contrasts, the transition from winter to spring is often abrupt, and characterized by rapid plant growth and development (Schwartz, 1994). Early satellite observations showed how this onset of vegetation activity moves poleward across the landscape like a wave, and thus it became commonly know as the "green wave" (e.g. Rouse, 1977). Despite such simple verbal imagery, the green wave is a multifaceted collection of processes and phenomena that can be observed and measured in many ways. Primarily, it is a time marked by profound changes in the interactions between the atmosphere and biosphere, as plants again become more active participants in the exchange of energy, mass, and momentum between the land surface and atmosphere. Recently, much progress has been made toward understanding these interactions better, especially in measuring the effects of climate change on the biosphere, and implications for climate modeling (Menzel, 2002; Schwartz, 2003). Research efforts have combined developments in remote sensing with phenology, the traditional study of plant and animal life stage events driven by environmental factors (primarily temperature) in spring (Reed et al., 1994; White et al., 1997; Schwartz et al., 2002).

Early modeling work showed that plant phenological development (e.g. first leaf appearance) in spring does not usually occur gradually, but very rapidly over a short period of time, driven by the occurrence of specific synoptic-scale weather events (warm sectors of low-pressure systems, Schwartz and Marotz, 1988). These findings validated phenological events as a sharp and integrative measure of the onset of the spring season. Further, spring phenology is among the most sensitive signs of plant and animal responses to changing temperatures (Chmielewski and Rötzer, 2002). Thus, conventional species-based observations can serve as valuable independent measures of change (Menzel, 2002). For example, phenological observations show that the onset of spring has gotten earlier about 5–6 days over the last 40 years in many parts of Europe and North America, but no similar changes are apparent in China (Menzel and Fabian, 1999; Schwartz and Reiter, 2000; Schwartz and Chen, 2002). Even so, most current surface phenology observation networks are independently administered, largely uncoordinated, and do not cover many places on Earth. A coordinated global monitoring network is being slowly built that combines appropriate native species in each region with widely adapted cloned "indicator" plants, such as lilacs, to help interconnect the species responses between regions (Schwartz, 1999). However, this network is far from complete, and even when finished will not provide the full-surface global and regional coverage needed for planet-wide biospheric monitoring. Some needed phenological information can be extracted from satellite-based sensors, but these satellite measures themselves are limited by cloud-cover interference and other problems. Fortunately, many of these issues will be effectively addressed by a networking scheme that combines satellite-based, phenological model output, and surface phenological observations (Schwartz, 1994, 1998; Zhao and Schwartz, 2003; Figure S38).

Besides monitoring year-to-year changes in plant and animal responses, phenological events can also be valuable indicators of the progression of seasonal changes in the surface energy balance and associated meteorological variables. Earlier long-term studies showed that phenology can highlight abrupt springtime changes in diurnal temperature range, relative humidity, visibility,

Figure S38 Average 1961–1990 spring indices (SI) model first leaf date (1 January = 1).

and other variables affected by the rapid upturn in plant transpiration and photosynthesis (e.g. Schwartz, 1992). Recent work has confirmed that these changes are ultimately driven by changes in the latent–sensible heat balance, and related to carbon flux (Fitzjarrald et al., 2001; Schwartz and Crawford, 2001; Wilson and Baldocchi, 2000). As new phenological data become available globally, and analyses continue, spring phenological studies have excellent potential to further advance global change science, through coordination with information coming from satellites and carbon flux monitoring sites around the world.

Mark D. Schwartz

Bibliography

Chmielewski, F.M., and Rötzer, T., 2002. Annual and spatial variability of the beginning of growing season in Europe in relation to air temperature changes. *Climate Research*, **19**: 257–264.

Fitzjarrald, D.R., Acevedo, O.C., and Moore, K.E., 2001. Climatic consequences of leaf presence in the eastern United States. *Journal of Climate*, **14**: 598–614.

Menzel, A., 2002. Phenology, its importance to the global change community. *Climate Change*, **54**: 379–385.

Menzel, A., and Fabian, P., 1999. Growing season extended in Europe. *Nature*, **397**: 659.

Reed, B.C., Brown, J.F., VanderZee, D., Loveland, T.R., Merchant, J.W., and Ohlen, D.O., 1994. Variability of land cover phenology in the United States. *Journal of Vegetation Science*, **5**: 703–714.

Rouse, J.W., Jr, 1977. *Applied Regional Monitoring of the Vernal Advancement and Retrogradation (Green Wave Effect) of Natural Vegetation in the Great Plains Corridor*, Final Report, Contract No. NAS5-20796. Greenbelt, MD: Goddard Space Flight Center.

Schwartz, M.D., 1992. Phenology and springtime surface layer change. *Monthly Weather Review*, **120**: 2570–2578.

Schwartz, M.D., 1994. Monitoring global change with phenology: the case of the Spring Green Wave. *International Journal of Biometeorology*, **38**: 18–22.

Schwartz, M.D., 1998. Green-wave phenology. *Nature*, **394**: 839–840.

Schwartz, M.D., 1999. Advancing to full bloom: planning phenological research for the 21st century. *International Journal of Biometeorology*, **42**: 113–118.

Schwartz, M.D. (ed)., 2003. *Phenology: An Integrative Environmental Science*. Dordrecht: Kluwer.

Schwartz, M.D., and Chen, X., 2002. Examining the onset of spring in China. *Climate Research*, **21**: 157–164.

Schwartz, M.D., and Crawford, T.M., 2001. Detecting energy balance modifications at the onset of spring. *Physical Geography*, **21**: 394–409.

Schwartz, M.D., and Marotz, G.A., 1988. Synoptic events and spring phenology, *Physical Geography*, **9**: 151–161.

Schwartz, M.D., and Reiter, B.E., 2000. Changes in North American spring. *International Journal of Climatology*, **20**: 929–932.

Schwartz, M.D., Reed, B.C., and White, M.A., 2002. Assessing satellite-derived start-of-season (SOS) measures in the conterminous USA. *International Journal of Climatology*, **22**: 1793–1805.

White, M.A., Thornton, P.E., and Running, S.W., 1997. A continental phenology model for monitoring vegetation responses to interannual climatic variability. *Global Biogeochemical Cycles*, **11**: 217–234.

Wilson, K.B., and Baldocchi, D.D., 2000. Seasonal and interannual variability of energy fluxes over a broadleaved temperate deciduous forest in North America. *Agriculture, Forestry and Meteorology*, **100**: 1–18.

Zhao, T., and Schwartz, M.D., 2003. Examining the onset of spring in Wisconsin. *Climate Research*, **24**: 59–70.

Cross-references

Frost
Middle Latitude Climates
Seasons
Vegetation and Climate

STANDARD ATMOSPHERE

The Standard Atmosphere is defined as "a hypothetical vertical distribution of atmospheric temperature, pressure and density, which, by international agreement, is taken to be representative of the atmosphere for purposes of pressure altimetry calibration, aircraft performance calculations, aircraft and missile design, etc." (Huschke, 1959). In terms of climatic utility, it represents an approximation of annual average middle-latitude conditions.

History

In his review of the Standard Atmosphere, Sissenwine (1969) suggests that the derivation of a standard for conditions through the atmosphere dates back to 1864. At that time, to aid in providing fairly realistic temperatures for ballooning and mountaineering, the French adopted a simple system. Following World War I, English-speaking countries used various standards, but it was not until 1924 that the first true *standard* was proposed by the International Committee for Air Navigation (ICAN). It used a pressure profile computed from equations and a decrease in temperature of 6.5°C/km to 11 km and thereafter isothermal conditions. A parallel development in the United States by the National Advisory Committee for Aeronautics (NACA) produced a Standard Atmosphere slightly different from that suggested by ICAN, but the NACA standard remained in use in the United States until 1952. With the extension of high-altitude commercial flights after World War II, the International Civil Aviation Organization (ICAO), a successor of ICAN, was established and in 1952 adopted a new Standard Atmosphere. Shortly thereafter, in 1962, the US Committee on Extension of the Standard Atmosphere (COESA) proposed a standard that incorporated new upper-air research based on observed data rather than those interpolated from theoretical estimates. Data represent idealized middle-latitude, year-round, mean conditions for daylight hours for the range of solar activity between maximum and minimum sunspot activity. The most recent version is the US Standard Atmosphere, 1976.

The standards

Complete tables for Standard Atmosphere are provided in many sources. Sissenwine's article (1969) is particularly useful. Not only does it supply a complete listing of the 1962 COESA Standard Atmosphere, it also provides data applicable to Supplemental Atmospheres. These include monthly (January and July) values as well as those applicable to areas other than those of middle latitudes.

Sample values of the 1976 US Standard are given in Tables S14 and S15. Up to 32 km, values are the same as those of the ICAO Standard Atmosphere and the lowest 50 km are those accepted by the International Standards Organization (ISO). Values for greater heights are derived using rocket and satellite data.

John E. Oliver

Table S14 US Standard Atmosphere to the stratopause

Height (km)	Temperature (K)	Pressure (mb)	Height (km)	Temperature (K)	Pressure (mb)
0.0	288.2	1013.2	11.0	216.8	227.0
1.0	281.7	898.8	12.0	216.6	194.0
2.0	275.2	795.0	14.0	216.6	141.7
3.0	268.7	701.2	16.0	216.6	103.5
4.0	262.2	616.6	18.0	216.6	75.65
5.0	255.7	540.5	20.0	216.6	55.29
6.0	249.2	472.2	25.0	221.6	25.49
7.0	242.6	411.0	30.0	226.5	11.97
8.0	236.2	356.5	35.0	236.5	5.746
9.0	229.7	308.0	40.0	250.4	2.871
10.0	223.3	265.0	50.0	270.6	0.798

Source: From US Committee on Extension of the Standard Atmosphere (1976).

Table S15 Tempatures of the Standard Atmosphere

Height (km)	Temperature (K)	Height (km)	Temperature (K)
0.000	288.15	150	634.39
11.019	216.65	160	696.29
20.063	216.65	170	747.57
32.162	228.65	190	825.31
47.350	270.65	230	915.78
51.413	270.65	260	950.99
71.802	214.65	300	976.01
85.638	187.65	350	990.06
86.000	186.87	400	995.83
91.000	186.87	500	999.24
100	195.08	750	999.99
110	240.00	1000	1000.00
120	360.00		

Source: From U.S. Committee on Extension of the Standard Atmosphere (1976).

Bibliography

Huschke, R.E., 1959. *Glossary of Meteorology*. Boston, MA: American Meteorological Society.

Geer, I.W. 1996. *Glossary of Weather and Climate*. Boston: Amer. Met.Soc.

International Civil Aviation Organization, 1954. *Manual of the ICAO Standard Atmosphere*. Document 7488. Washington, DC: ICAO.

Sissenwine, N., 1969. Standard and supplemental atmospheres. In Rex, D.F., ed., *World Survey of Climatology*, vol. 4. New York: Elsevier.

US Committee on Extension of the Standard Atmosphere, 1962. *US Standard Atmosphere*. Washington, DC: US Government Printing Office.

US Committee on Extension of the Standard Atmosphere, 1976. *US Standard Atmosphere*. Washington, DC: US Government Printing Office.

Cross-references

Atmospheric Nomenclature
Troposphere

STATISTICAL CLIMATOLOGY

Given the vast quantity of climatic data, statistical methods have wide applicability and utility in the study of past, present, and future climates. The data are a testament to the efforts of countless scientists and volunteer observers all over the world, and make statistical climatology both possible and powerful. Still, climatic processes and phenomena are not observed everywhere at all times and it is often difficult to detect meaningful climatic signals amidst substantial variability in both space and time. By using both traditional and innovative statistical methods, however, climatologists are able to detect and quantify rates of climatic change, estimate the probability of extreme events, and reveal uncertainties in our current understanding of climatic processes and models.

Statistical graphics

Graphical methods sometimes are not included in discussions of statistical approaches; however, they often are the most useful, powerful, and compelling component of any data analysis. The creative and innovative use of illustrations of all kinds is at the heart of many climatic analyses (see Cleveland, 1993; Schmid, 1983; Tufte, 1983 for general discussions of statistical graphics and scientific visualization). Essential descriptions of climatic data – such as histograms, time-series plots, and maps – are used to evaluate data quality and illustrate key relationships. Even more importantly, careful use of statistical graphics and maps can help to generate scientific questions and research hypotheses.

Histograms, for instance, are used to describe overall frequency distributions (Figure S39a). They are particularly important in climatology, where large amounts of data must be visualized and assessed. Percentiles and the existence of outliers also may be derived and evaluated from histograms. By spreading the weight of individual observations over a range of values, kernel-density methods (Silverman, 1986) provide a smoother alternative to the sometimes discontinuous histogram (Figure S39b).

Time-series plots of various types are critical in the visualization of trends, cyclical behavior, and other types of variability (Figure S40a). After detrending (usually by linear least-squares regression), persistence or autocorrelation within a time-series often is more evident (Figure S40b). Scatterplots of lagged relationships (Figure S40c) and autocorrelation functions (Figure S40d) further illustrate persistence or "memory" in climatological time-series. For instance, the climate system has relatively little memory from year to year when it comes to hemispherically averaged air temperature anomalies (Figure S40).

Maps of all kinds – contours of pressure, temperature, or precipitation, graduated-circle maps, etc. – are invaluable in visualizing data. In addition, it is often necessary to reduce the dimensionality of data that are sampled in two or three spatial dimensions as well as in time. By averaging one or more of these dimensions (or using other types of filtering), space–time patterns are more discernible. When global-scale air temperature is averaged over all longitudes, for instance, it becomes evident that the two warming periods during the twentieth century were very different in their latitudinal extent (Figure S41).

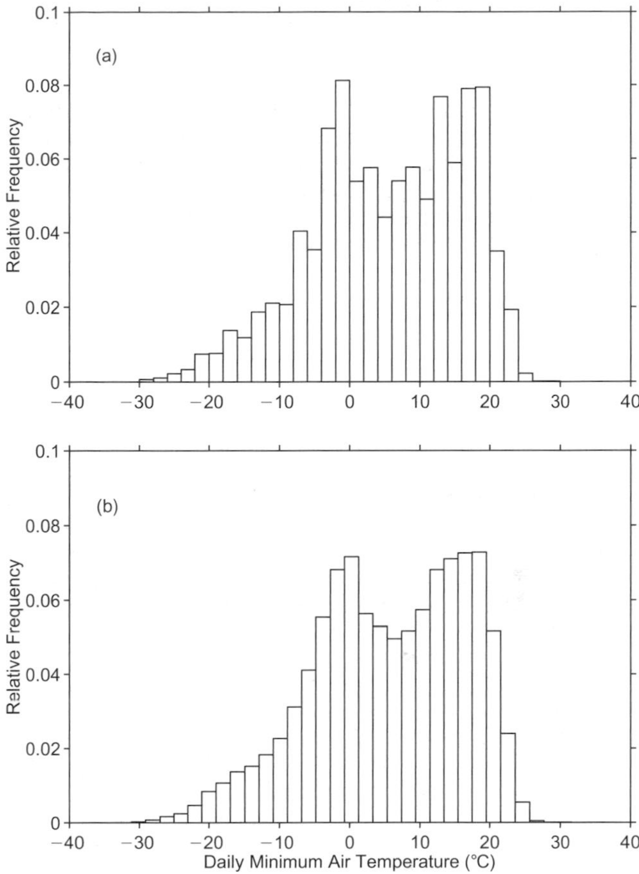

Figure S39 Histograms of daily minimum air temperature (°C) at Lincoln, Illinois, from 1961 to 1990: (**a**) bins of original data and (**b**) bins with kernel-density smoothing applied. The bimodal nature of the distribution is more apparent after the kernel-density smoothing has been applied. Data from US Daily Historical Climate Network.

Probability distributions

Probability distribution *functions* may be fit to empirical distributions (i.e. those depicted in Figure S39). While many climatological variables may be assumed to have prescribed distributions – normal distribution for air temperature, gamma distribution for precipitation, Weibull distribution for wind speed – it often is worthwhile to examine the characteristics and fit of a number of distributions. For instance, as averages, monthly and annual air temperature generally follow a normal distribution; however, daily air temperature may have more complex distributions, with extended tails or bimodality (e.g. Figure S39). Straightforward transformations of precipitation data have been shown to sometimes produce better probability models than the standard gamma distribution (Legates, 1991). For analyses where extreme events are of interest, a number of specialized distributions are available. Extreme value distributions are particularly important in determining "return periods" for extreme events, such as the 50- or 100-year precipitation amount (Katz et al., 2002). The interactions between changing parameters, such as mean and variance, and their resulting influence on the

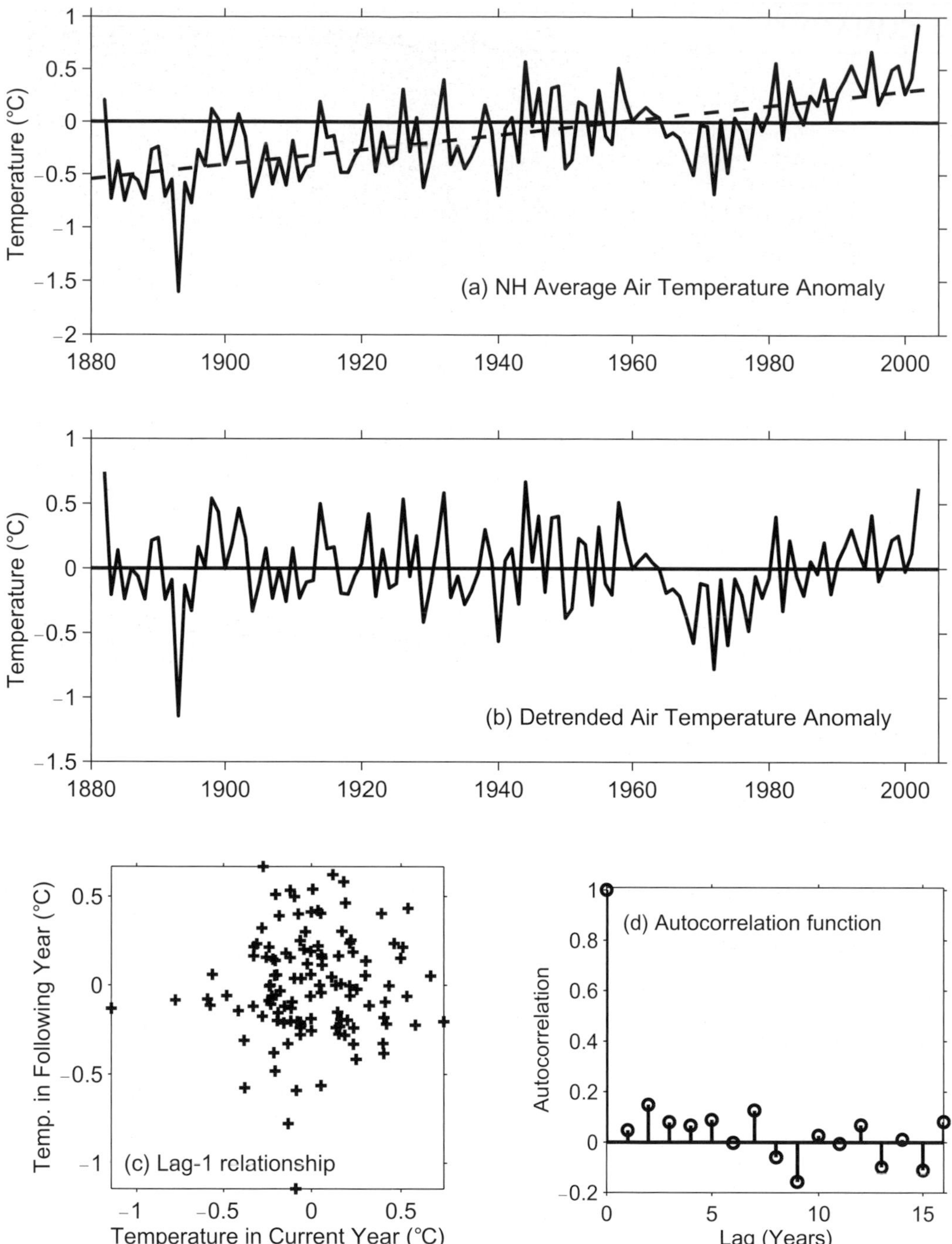

Figure S40 Visualization of northern hemisphere average air-temperature anomalies (°C) during January: (**a**) original data with linear least-squares trend (dashed line), (**b**) detrended northern hemisphere air-temperature anomalies, (**c**) lag-1 relationship between current year's anomaly and the following year's anomaly, and (**d**) autocorrelation function. Data from Jones and Moberg (2003).

probability of extreme events also is of critical importance in climatic change research (Katz and Brown, 1992; Mearns et al., 1984; Robeson, 2002a). A wide range of probability models for climatological variables, both discrete and continuous, are discussed by von Storch and Zwiers (1999) and Wilks (1995).

Regression

The estimation of bivariate and multivariate relationships via regression is one of the most common and powerful statistical tools in all of science. Regression analysis develops an equation that relates a dependent variable to a set of independent variables. As a result, regression can be used: (a) to

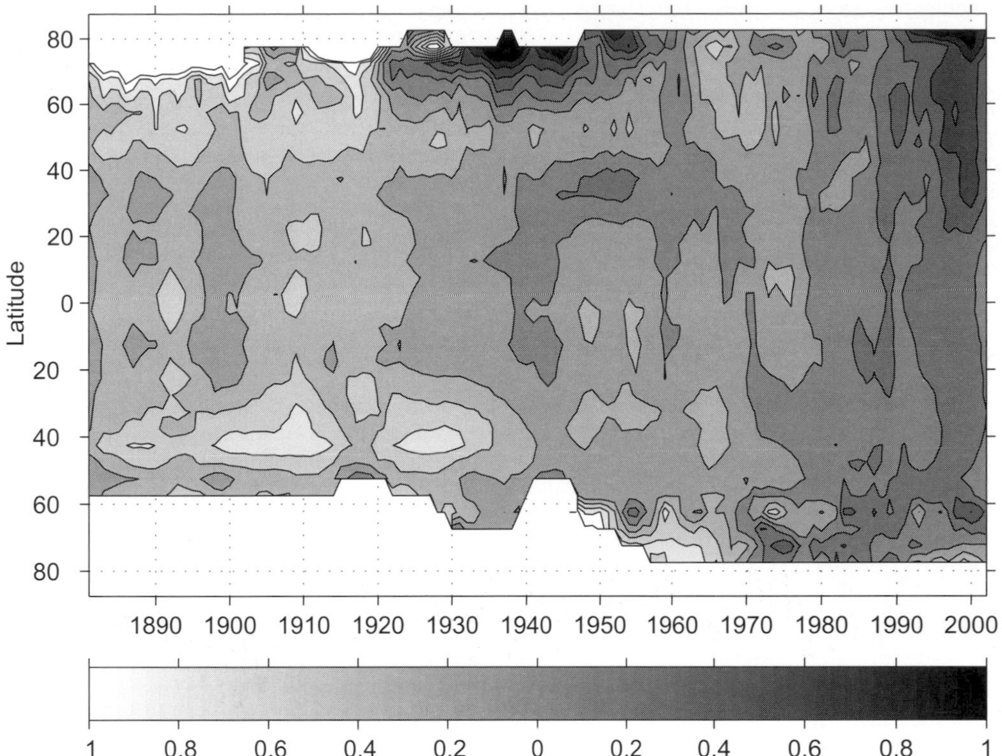

Figure S41 Zonal averages of annually averaged air-temperature anomalies (°C): an example of a time-space "slice" of climatic data (contour interval is 0.2°C). In this representation of global-scale near-surface air temperature variation, longitude has been averaged to leave latitude and time as the remaining dimensions. As a result, it becomes clear that the warming during the 1930s and 1940s occurred primarily in the high northern latitudes while the warming in the latter portion of the twentieth century was more widespread. Adapted from Delworth and Knutson (2000) using data from Jones and Moberg (2003).

evaluate the strength and sensitivity of statistical relationships between variables and (b) to estimate "missing" values of the dependent variable, which would include both interpolation and extrapolation (including forecasting). Diagnostic tools associated with regression, such as the coefficient of determination (R^2), are used to evaluate the amount of variability that can be accounted for in one variable by knowing the values of other variables.

As an example of a simple regression analysis, air temperature and atmospheric pressure data for 83 winter days near Bloomington, Indiana can be used to estimate the relationship between those variables and dewpoint temperature (Table S16). The regression analysis tells us that air temperature has a direct relationship (positive regression coefficient) with dewpoint temperature while atmospheric pressure has an inverse relationship (negative regression coefficient). The relationship is relatively strong, with air temperature and pressure "explaining" nearly 85% of the variability in dewpoint temperature (R^2 = 0.846). In addition, the sensitivity of the dependent variable to the independent variables can be estimated from the magnitude of the regression coefficients: for a 1°C increase in air temperature, dewpoint temperature is expected to increase by 0.85°C. For each 1 mb increase in pressure, dewpoint temperature is expected to decrease by 0.3°C. These relationships also make intuitive sense. Through its influence on water-vapor capacity and its covariation with evaporation, air temperature controls and limits dewpoint

Table S16 Regression analysis of daily dewpoint temperature, air temperature, and atmospheric pressure for 83 winter days near Bloomington, Indiana, during 2001

Predictor	Regression coefficient	Standard error	t	p-Value
Constant	302.34	63.29	4.78	0.000
T_a	0.85258	0.05905	14.44	0.000
P	−0.30137	0.06202	−4.86	0.000

Notes: The regression equation is $T_d = 302 + 0.853\ T_a - 0.301\ P$, where:
T_d is dewpoint temperature (°C)
T_a is air temperature (°C)
P is atmospheric pressure (mb)
$R^2 = 0.846$

temperature. Also, synoptic-scale pressure systems during winter in Indiana are such that high-pressure systems are associated with dry air of continental origin and low-pressure systems transport moist air of maritime origin.

Whenever regression is used, it is critical to consider the limitations and conditions of the data used in model development. In the example shown in Table S16, the primary limitation is that the regression was developed using only winter days for 1 year at one location. It would not be advisable to apply the regression equation at other times of year and at other locations without careful consideration of the potential errors that might occur. Also, the representativeness of the 1 year's data should be considered (i.e. whether this one winter represented the range of conditions that can occur).

Although ordinary least-squares (OLS) regression is the most widely used method for estimating regression coefficients, a number of alternatives are available. OLS is particularly sensitive to *outliers* (although outliers can be defined in many ways, they are "unusual" values; in regression, when the outliers occur at the tails of the distribution of the independent variable, the outliers also are known as *influence points*, as they can strongly influence the regression coefficients). For data that contain outliers, a number of robust (or resistant) regression methods can be evaluated, usually in concert with OLS. Whereas OLS minimizes the sum of the squared residuals (residuals are the difference between the observed and estimated values of the dependent variable), robust methods include those that: (a) minimize the sum of the absolute values of the residuals, (b) minimize the median residual, and (c) use OLS after a certain proportion of the data has been "trimmed". The book by Rousseeuw and Leroy (1987) contains detailed information on robust regression methods, whereas Lanzante (1996) has a number of examples using climatic data.

Variations through time

Trend detection

Although climatologists frequently analyze climatic data for linear trends, there are many different types of temporal variability that occur in climatic time-series. These include nonlinear trends, cyclical variability, quasiperiodic behavior, and discontinuities. The data "window" that is available is a key component of any analysis of temporal variability. It is critical to keep in mind that variations outside the window, if they were known, can change the interpretation of variations inside the window. A "trend", for instance, may be part of a long-term oscillation that is not apparent within the data window (Figure S42).

The regression methods discussed above can be applied to time-series data to estimate linear trends with time (e.g. Figure S40a). Linear trends estimates are useful for studying a variety of types of climatic change, including changes in air temperature (Easterling et al., 1997), extreme precipitation events (Zhang et al., 2001), wind speed (Pryor and Barthelmie, 2003), and growing-season length (Robeson, 2002b).

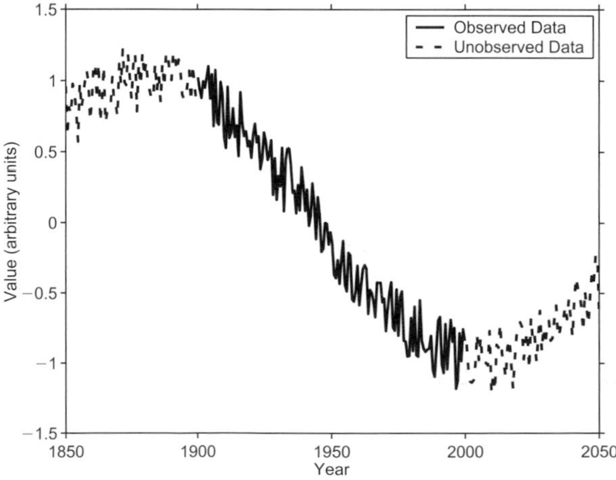

Figure S42 Hypothetical data that are sampled only during the twentieth century. The downward trend during the twentieth century is part of a long-term oscillation that is not apparent within the data window.

Smoothing and filtering

Moving averages, also known as running means, frequently are used to reveal variability in a time-series that is masked by either a prominent periodicity (such as a daily cycle) or high levels of variability ("noise"). A moving average, however, is but one special case of a large class of linear operators known as *digital filters* (see Hamming, 1983 for a general discuss of filters, and Holloway, 1958 or Panofsky and Brier, 1958 for overviews of their application in climatology). Although the filters usually used in climatology are designed to smooth a time-series (smoothing filters are known as "low-pass" filters, as they "pass" low-frequency information and remove high-frequency information), filters can be designed to have a wide range of properties (e.g. high-pass and band-pass filters; Hamming, 1983).

The general form of most digital filters is

$$y_t = \sum_{i=-k}^{k} w_i x_{t+i}$$

where x_t is the original time-series at time t, w_i are the filter weights, and y_t is the filtered time-series. The filter weights usually sum to 1, so that the mean of the filtered time-series is the same as the mean of the original time-series. The filter has length of $L = 2k + 1$ and usually is symmetric about the center weight (w_0).

For instance, a common filtering application involves daily air-quality data, which sometimes have a quasi-weekly cycle that is related to the weekday–weekend commuting pattern. To remove this cycle, a seven-point (equally weighted) moving average could be applied, where all of the w_i are 1/7 (Figure S43a). Moving averages of this type not only can be designed to remove periodicities, but they help to visualize longer-term patterns. In actuality, equally weighted moving averages have properties that make them unsuitable for many kinds of filtering (e.g. they "leak" high-frequency information). As a result, filters that have smoother shapes – such as Gaussian filters – frequently are used to remove high-frequency noise. The Matlab technical programming language has numerous tools for exploring digital filters and their properties.

Spectral analysis

Another approach to analyzing variations of climatic processes with time is to analyze data in the frequency domain. Frequency-domain approaches transform climatic data into a series of periodic (e.g. cosine) functions, each with a characteristic frequency, amplitude, and phase. Frequency represents the time scale, such as one cycle per year, while amplitude indicates how much variability is in the time-series at that time scale (e.g. Figure S43b). Phase (not represented in Figure S43b) represents the timing of the peak of the periodic function. Many climatological analyses are naturally done in the frequency domain since numerous climatic phenomena are well represented by periodic functions (e.g. daily and annual cycles of air temperature and solar radiation).

Spectral analysis is a key tool in the analysis of past climates, particularly in identifying the relationships between Earth's orbital parameters and ice-age climates (e.g. Hays et al., 1980) and in analyzing solar influences on climate (e.g. Chambers et al., 1999). In addition to describing periodic and quasi-periodic variations with time, many calculations (e.g. filtering, estimation of autocorrelation functions, etc.) are more straightforward when done in the frequency domain. See Bloomfield (2000) for further discussion of frequency-domain methods.

Variations through space

Spatial visualizations and analysis of climatic information provide highly interpretable tools for analyzing climatic data. Some of the most useful and provocative graphics in the 2001 Intergovernmental Panel on Climate Change (IPCC, 2001) volume, for instance, are the maps of air temperature and precipitation trends during the twentieth century. Well-constructed maps can be used to summarize statistical results, as well as to explore climatic data. A number of useful statistical approaches also can be used to evaluate data that are explicitly spatial, such as spatial autocorrelation, semivariance analysis, and eigenvector analysis.

Spatial dependence

Nearly all climatic variables are such that nearby locations are more similar to one another than are distant locations. This similarity represents the spatial dependence of the variable and it can be measured in a number of ways. Spatial autocorrelation measures the degree of similarity of locations while semivariance measures the degree of dissimilarity.

Spatial autocorrelation can be expressed as a single number (as in the case of Moran's I) that usually measures the strength of the similarity between contiguous neighbors. Spatial autocorrelation also can be expressed as a function, whereby the correlation of a variable with itself is calculated over a number of spatial lags and then plotted against the distance of each lag (Figure S44). Spatial autocorrelation functions can provide information on the dominant spatial scales of variability (e.g. thousands of kilometers for synoptic-scale phenomena) and also suggest appropriate weighting functions for spatial interpolation (Thiebaux and Pedder, 1987).

Climatological variables frequently have spatial autocorrelation functions that vary with direction, or are anisotropic (Thiebaux, 1976). Solar radiation, for instance, should be more similar in east–west directions than in north–south directions. This information can inform climatologists on how best to sample and interpolate spatial climatic variables – different sampling strategies or spatial weighting functions may be needed in north–south directions than in east–west directions.

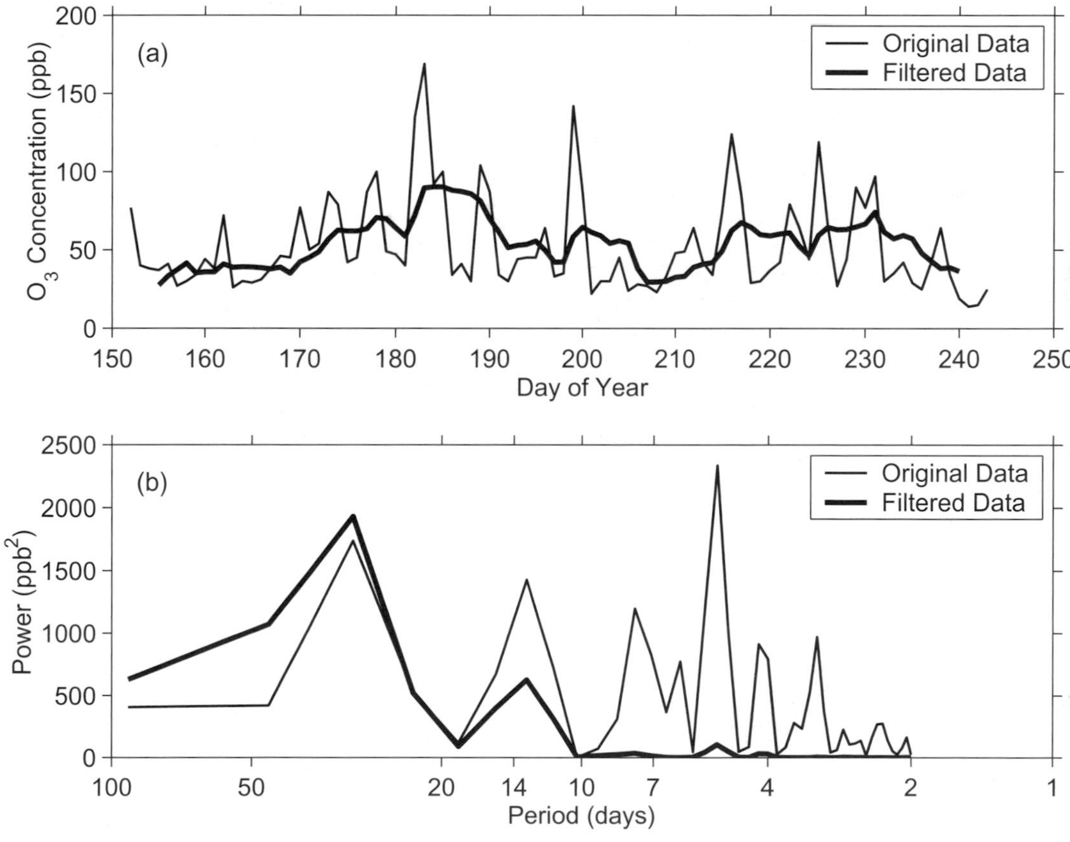

Figure S43 Ambient ozone data (maximum 8-h concentration for each day) from Narragansett, Rhode Island, during the summer of 2002: (**a**) time-series before and after applying a seven-point equally weighted moving average and (**b**) power spectrum before and after applying moving average. The moving average smooths the time-series, thereby removing much of high-frequency (short period) information, while preserving low-frequency (long period) information. Equally weighted moving averages, however, are not recommended in most filtering applications, as they "leak" at high frequencies (e.g. note the small peak between 4 and 7 days in the filtered power spectrum).

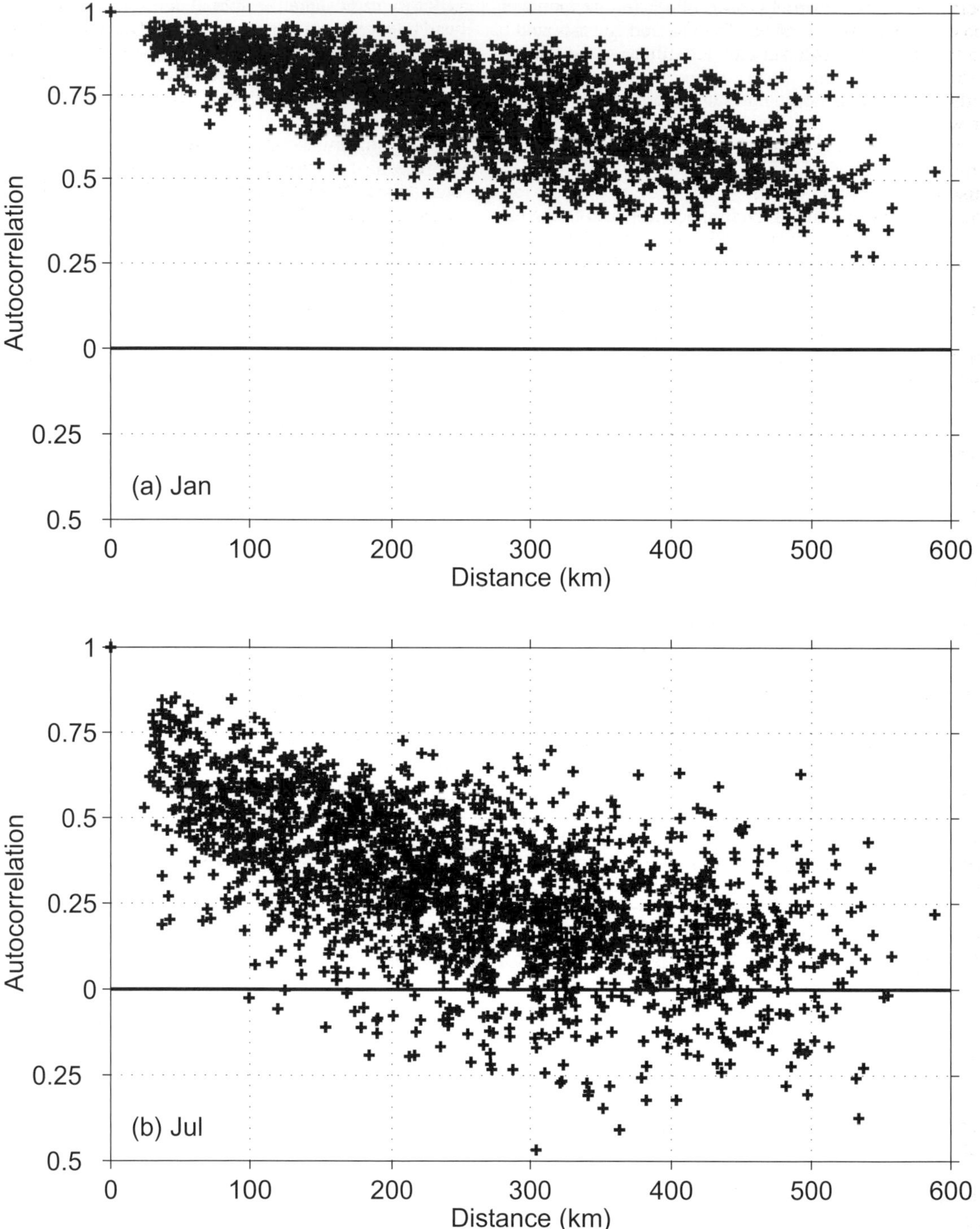

Figure S44 Spatial autocorrelation functions for January and July precipitation using 1961–1990 data from the states of Illinois and Indiana (data are monthly averages from the US Historical Climate Network stations). January precipitation shows much greater spatial coherence, with locations that are more than 500 km apart still having relatively high autocorrelations. July precipitation, on the other hand, has relatively little spatial coherence, as spatial autocorrelations drop quickly (e.g. below 0.5 at distances of ≈100 km).

Spatial interpolation

Although gridded data from computer models and satellite-based sensors are widely used in climatology, many climatological variables are sampled at irregularly spaced locations, such as networks of surface climate stations. To be compared with models or satellite data – or simply to construct a contour map – irregularly spaced data are interpolated to a regular grid. Interpolation also is needed for spatial averaging – the global average temperature is not the simple mean of all the thermometer-based measurements that are made. The data are interpolated to a grid and then averaged according to how much area each grid point represents.

All methods of spatial interpolation estimate values at locations that were not in the original sample, usually using some form of distance-weighting or spatial function fitting.

Some methods, such as objective analysis and kriging, make explicit use of spatial autocorrelation functions or semivariograms to determine optimal spatial weighting (Davis, 1986; Thiebaux and Pedder, 1987; Isaaks and Srivastava, 1991). The choice of interpolation method can be important (Lam, 1983; Robeson, 1997); however, in data-rich regions, most spatial interpolation methods produce similar results. Adding additional explanatory variables – such as elevation data, though, can substantially improve spatial interpolation results (Daly et al., 1994; Willmott and Matsuura, 1995).

Significance testing

While tests for statistical significance continue to play important roles in climatology, it is a rare occasion when classical, parametric significance tests are directly applicable. Climatic data frequently have characteristics – such as autocorrelation, nonnormality, heteroscedasticity, and excessive sample sizes – that limit the effectiveness of traditional procedures.

In many cases the sample sizes of climatic data are extremely large. At many primary weather stations, for instance, many variables are observed on an hourly basis. Statistical tests performed with samples that contain thousands of observations can produce "significant" results that are scientifically unimportant. For example, two locations where monthly average pressure data are available for 50 years ($n = 600$) would need to have a correlation of only 0.08 to produce a significant result at the 0.05 level for type-I error (i.e. $\alpha = 0.05$). In such cases statistical significance can be misinterpreted if only the significance of the test and not the correlation itself is reported.

Climatic data are nearly always autocorrelated – in both space and time. Autocorrelation results in "effective" sample sizes that are much smaller than the number of observations (n). For instance, if an air-temperature time-series that consists of 100 years of observations ($n = 100$) produces a trend estimate of 1°C/100 yr with a standard error of 0.5°C/100 yr, the trend would generally be considered "significant" (i.e. for a two-tailed test, the associated t-value would be 2.0 with a p-value of 0.048). If the time-series has positive autocorrelation (as most air-temperature time-series do; Katz and Skaggs, 1981), the number of "independent" observations is less than 100. For a time-series where first-order autocorrelation dominates, an established method (Chatfield, 1975; Mitchell et al., 1966) for estimating the effective sample size is

$$n' = n\frac{1 - r_1}{1 + r_1}$$

where n' is the effective sample size and r_1 is the first-order autocorrelation coefficient. Using the equation above with $n = 100$ and $r_1 = 0.7$ produces $n' = 23$. Applying this level of autocorrelation to the regression trend example would change the 1°C/100 yr trend to "not significant" (note that autocorrelation is estimated using the regression-trend residuals and, therefore, is not influenced by the trend itself, which is a form of nonstationary mean). While a number of procedures are available (e.g. Santer et al., 2000), most produce even more conservative estimates of n' than the one above. Regardless of which procedure is used, it is clear that consideration of autocorrelation is essential when determining if a relationship is deemed "significant".

Computer-intensive methods (Efron and Gong, 1983) also are used in climatology to evaluate statistical significance and confidence. Bootstrapping methods intensively resample the original data (with replacement) to generate empirical probability distributions. This can be particularly important for variables that do not have theoretical distributions (such as mean absolute error or MAE) or when assumptions of statistical tests are not met. Cross-validation usually entails successively removing one or more data values at a time and then estimating the removed values using the remaining data. This process is repeated in order to generate unbiased estimates of "predictability". Cross-validation has been used in climatology to evaluate climate forecasts and to evaluate spatial interpolation methods (Michaelson, 1987; Robeson, 1994).

Model evaluation

Models of all kinds should be compared with independent observations to evaluate their utility. Sometimes this procedure is called "model validation"; however, that term connotes a binary decision as to whether a model is valid or not. In reality, the evaluation of a model is a subjective procedure that utilizes a wide range of graphics and statistical measures. Essential graphics include scatterplots of model estimates (P_i) against the observations (O_i) and histograms of model error ($e_i = P_i - O_i$). Error measures such as root mean squared error and mean absolute error also are essential; however, neither preserves the sign of the error. Although correlation is useful in evaluating "fit" when developing statistical models, it is not recommended for independent evaluations of models, as models with large error can still produce high correlations (i.e. when the model estimates covary with the observations, but there is a large systematic error). See Willmott (1981) for further discussion of model evaluation.

In model evaluation, as in all areas of statistical climatology, the continued development of new methods is critical. Due to extensive sample sizes, spatial and temporal autocorrelation, and unusual data types (such as multidimensional vectors) frequently encountered in climatology, standard statistical methods frequently are not appropriate. Climatologists, therefore, must continue to explore the creative use of maps, statistical graphics, and alternative statistical methods.

Scott M. Robeson

Bibliography

Bloomfield, P., 2000. *Fourier Analysis of Time Series*, 2nd edn. New York: Wiley.

Chambers, F.M., Ogle, M.I., and Blackford, J.J., 1999. Palaeoenvironmental evidence for solar forcing of Holocene climate: linkages to solar science. *Progress in Physical Geography*, **23**: 181–204.

Chatfield, C., 1975. *The Analysis of Time Series: theory and practice*. London: Chapman & Hall.

Cleveland, W.S., 1993. *Visualizing Data*. Summit, NJ: Hobart Press.

Daly, C., Neilson, R.P., and Phillips, D.L., 1994. A statistical–topographic model for mapping climatological precipitation over mountainous terrain. *Journal of Applied Meteorology*, **33**: 140–158.

Davis, J., 1986. *Statistics and Data Analysis in Geology*, 2nd edn. New York: Wiley.
Delworth, T.L., and Knutson, T.R., 2000. Simulation of early 20th century global warming. *Science*, **287**: 2246–2250.
Easterling, D.R., Horton, B., Jones, P.D., et al., 1997. Maximum and minimum temperature trends for the globe. *Science*, **277**: 364–367.
Efron, B., and Gong, G., 1983. A leisurely look at the bootstrap, the jackknife, and cross-validation. *American Statistician*, **37**: 36–48.
Hamming, R., 1983. *Digital Filters*, 2nd edn. New York: Prentice-Hall.
Hays, J.D., Imbrie, J., and Shackleton, N.J., 1980. Variations in the Earth's orbit: pacemaker of the ice ages. *Science*, **194**: 1121–1132.
Holloway, J., 1958. Smoothing and filtering of time series and space fields. *Advances in Geophysics*, **4**: 351–389.
IPCC (Intergovernmental Panel on Climate Change), 2001. *Climate Change 2001: the scientific basis*. New York: Cambridge University Press.
Isaaks, E.H. and Srivastava, R.M., 1989. *Applied Geostatistics*. New York: Oxford University Press.
Jones, P.D., and Moberg, A., 2003. Hemispheric and large-scale surface air temperature variations: an extensive revision and an update to 2001. *Journal of Climate*, **16**: 206–223.
Katz, R.W., and Brown, B.G., 1992. Extreme events in a changing climate: variability is more important than averages. *Climatic Change*, **21**: 289–302.
Katz, R.W., and Skaggs, R.H., 1981. On the use of autoregressive–moving average processes to model meteorological time series. *Monthly Weather Review*, **109**: 479–484.
Katz, R.W., Parlange, M.B., and Naveau, P., 2002. Statistics of extremes in hydrology. *Advances in Water Resources*, **25**: 1287–1304.
Lam, N., 1983. Spatial interpolation methods: a review. *American Cartographer*, **10**: 129–149.
Lanzante, J., 1996. Resistant, robust and non-parametric techniques for the analysis of climate data: theory and examples, including applications to historical radiosonde station data. *International Journal of Climatology*, **16**: 1197–1226.
Legates, D.R., 1991. An evaluation of procedures to estimate monthly precipitation probabilities. *Journal of Hydrology*, **122**: 129–140.
Mearns, L., Katz, R., and Schneider, S., 1984. Extreme high-temperature events: Changes in their probabilities with changes in mean temperature. *Journal of Climate and Applied Meteorology*, **23**: 1601–1613.
Michaelson, J., 1987. Cross-validation in statistical climate forecast models. *Journal of Climate and Applied Meteorology*, **26**: 1589–1600.
Mitchell, Jr, J.M., Dzerdzeevskii, B., Flohn, H., et al., 1966. *Climatic Change*. WMO Technical Note No. 79. Geneva.
Panofsky, H.A., and Brier, G.W., 1958. *Some Applications of Statistics to Meteorology*. State College, PA: Pennsylvania State University Press.
Pryor, S.C., and Barthelmie, R.J., 2003. Long term variability of flow over the Baltic. *International Journal of Climatology*, **23**: 271–289.
Robeson, S.M., 1994. Influence of spatial sampling and interpolation on estimates of terrestrial air temperature change. *Climate Research*, **4**: 119–126.
Robeson, S.M., 1997. Spherical methods for spatial interpolation: Review and evaluation. *Cartography and Geographic Information Systems*, **24**: 3–20.
Robeson, S.M., 2002a. Relationships between mean and standard deviation of air temperature: Implications for global warming. *Climate Research*, **22**: 205–213.
Robeson, S.M., 2002b. Increasing growing-season length in Illinois during the 20th century. *Climatic Change*, **52**: 219–238.
Rousseeuw, P.J., and Leroy, A.M., 1987. *Robust Regression and Outlier Detection*. New York: Wiley.
Santer, B.D., Wigley, T.M.L., Boyle, J.S., et al., 2000. Statistical significance of trends and trend differences in layer-average temperature time series. *Journal of Geophysical Research*, **105**: 7337–7356.
Schmid, K., 1983. *Statistical Graphics: design principles and practices*. New York: Wiley.
Silverman, B.W., 1986. *Density Estimation for Statistics and Data Analysis*. New York: Chapman & Hall.
Thiebaux, H.J., 1976. Anisotropic correlation functions for objective analysis. *Monthly Weather Review*, **104**: 994–1002.
Thiebaux, H.J., and Pedder, M., 1987. *Spatial Objective Analysis*. New York: Academic Press.
Tufte, E., 1983. *The Visual Display of Quantitative Information*. Cheshire, CT: Graphics Press.
von Storch, H., and Zwiers, F., 1999. *Statistical Analysis in Climate Research*. New York: Cambridge University Press.
Wilks, D.S., 1995. *Statistical Methods in the Atmospheric Sciences*. New York: Academic Press.
Willmott, C.J., 1981. On the evaluation of models. *Physical Geography*, **2**: 184–194.
Willmott, C.J., and Matsuura, K., 1995. Smart interpolation of annually averaged air temperature in the United States. *Journal of Applied Meteorology*, **34**: 2577–2586.
Zhang, X., Hogg, W.D., and Mekis, E., 2001. Spatial and temporal characteristics of heavy precipitation events over Canada. *Journal of Climate*, **14**: 1923–1936.

SUNSPOTS

Sunspots are relatively cool, dark areas that are up to 50 000 km in diameter. They represent the visible manifestations of convection cells near the surface of the sun's photosphere. Marking concentrations of strong magnetic fields (2000–3000 gauss), they have lifetimes of a few days up to some weeks. At the solar surface the field is only about 1 gauss. They vary in number and size in a complex pattern of fluctuations, notably with the following periods: 27 days (the sun's rotation rate), 6–17 years, average 11.12 ($\pm$ 6) yr (solar cycle), and 178-179 years (all-planet synod). Characteristically observed in pairs of opposite magnetic polarity, individual spots appear to be connected beneath the surface by tubes of high-energy particles. Energy is released in the solar core by the transmutation of hydrogen to helium under appreciably enhanced pressure and temperature, the later estimated at about 20×10^6 K. The temperature drops near the surface (photosphere) in places to as little as 4200 K but then rises again in the surrounding corona to about 1 million K. The total solar energy output is 3.86×1033 erg/s. A minute fraction of this output is modulated by quasi-periodic stress variables, affecting the solar wind and electromagnetic radiations that seem to focus on the sunspots.

The sun and solar activities

On a clear day the sun appears as a bright, featureless orb and care must be taken to warn young people not to look at it without protective systems or blindness ensues. In ancient China and in western Europe, over the last two millennia, sunspots have been observed with the naked eye on foggy days, by twilight, through volcanic or desert dust, or with pinhole observation devices. In China, in particular, sunspots were seen to vary in number, size, shape, and distribution, sometimes in arrangements and in sequences that were found to correlate in mysterious ways with the climates, crops, harvests, and the general good or bad fortune of the emperors. The records of sunspot variability, together with those of auroras (related to the magnetic fields), were carefully preserved and used as a basis for astrologic forecasts. The unwise, or unlucky, astrologers who failed in their tasks were usually executed, which custom, perhaps, may help in part to explain the dedication of the observers. The earliest documented record of sunspots, together with auroras and the 76-yr passage of Halley's Comet, may have been in 467 and 466 BC (historical details are provided by Schove, 1983).

A single sunspot is seen as a dark area or *umbra*, surrounded by a brighter ring or *penumbra*. Over a few days a given spot may grow from a pinpoint to a large feature, typically up to 37 000 km in diameter. in very active areas several spots may merge into a group extending over as much as 1 50 000 km. A group has a mean lifetime of about 1.5 months. The spot displays a strong magnetic field, up to 3100 gauss, characterized by plasma flow in or out of the spot, according to magnetic polarity, at a velocity of up to 3 km/s. The middle of the spot is a relatively cool region having a temperature of 4600 K within the granular-patterned photosphere at about 6000 K. The photosphere is generally regarded as a gaseous layer of convection cells that pulsate up and down in a cyclic way about every 5 minutes (Shirley and Fairbridge, 1997).

In the chromosphere (the brightly colored solar atmosphere, situated above the photosphere), but visible only during eclipses, a bright region, or *plage*, with a vortex structure, rises above and surrounds the sunspot. The appearance of a plage precedes that of the spot and has a longer life span (200–300 days), being accompanied by an enhanced magnetic field. Filaments of a plage may undergo a short-lived breakout called a *flare*. A single spot may be characterized by up to 40 flares in the span of its lifetime.

A solar flare is a highly energetic event that is accompanied by the emission of *solar cosmic rays*. The latter are corpuscular or particle radiations, as distinct from and in addition to electromagnetic radiation, such as enhanced *X-rays* and *ultraviolet (UV) radiation* (especially Lyman-alpha flux) that occur at the same time. These transmissions modulate the *solar wind* to interact with the Earth's magnetic field and with stratospheric gases. Very important eruptions are characterized as *proton events;* on such occasions charged particles are carried down in large numbers into the Earth's troposphere, affecting the terrestrial climate. Ordinary solar flares do not affect total ozone in the stratosphere, but proton events can cause a drop in ozone by more than 20%, probably by the production of NO_x, which destroys the O_3. Ozone is important to climate, in that it is the stratosphere's major greenhouse gas. These solar phenomena interfere also with terrestrial electronic, radio, and television transmissions.

Although the *solar constant* is relatively stable at 2.00 ± 0.04 cal/cm^2/min (equivalent to 1395 W/m^2), satellite observations demonstrate small variations in the sun's luminosity over the 22-yr solar magnetic cycle. Longer-term periods, e.g. around 80 yr, appear to reflect small variations in solar radius (Gilliland, 1982), which in turn probably reflects spin rate (see also under Causes below). Day-to-day variations up to 0.4% occur during peaks of enhanced solar activity. The annual global average of incoming solar radiation at the top of the atmosphere is 350 Wts/m^2, of which only 51% reaches to the surface of the Earth, 16% being absorbed in the atmosphere and 35% being lost by reflection from clouds, snow, etc. (albedo). The mean surface air temperature at the Earth's surface is now 15°C but the long-term average over many millions of years is about 18 ± 5°C. In spite of astronomical arguments that the sun's radiation has changed through time (because the sun is a waning star), there is evidence of the existence of the sunspot cycle in late Precambrian time, and proxy evidence (from sediments, geochemistry, and paleontology) that there has been no gross variation of mean surface temperature on Earth for something more than 10^9 yr.

Solar activity–climatic relationships

The question of how the various solar transmissions modulate the terrestrial weather and climate is one of our ongoing problems, ranging from astrophysics to climatology. The linkages are complex and variable but fall into certain distinctive categories: (1) solar emissions (as ionized particles or so-called corpuscular emission); (2) electromagnetic radiation such as infrared, visible, UV, and radio waves, and as magnetic fields); (3) by changing radii, tilt and velocity (sun, planets, Earth, moon); and (4) by changing terrestrial parameters (geomagnetic, spin rate, atmospheric chemistry and dynamics, and oceanic processes). It is little wonder that there have been innumerable claims and counterclaims about sunspots and weather, ranging from the reckless to the ultraconservative, indeed, the one being often as foolish as the other. A reasonably balanced appraisal has been presented by Agee (1980). Short-term variations are difficult to pin down, but well-planned studies are now beginning to show results. The fluctuating height of the tropopause, for example, is found to match the sunspot cycle, with evidence of both particulate and electromagnetic radiation. We can learn more from the geological record.

Particularly useful, as a guide for monitoring long-term solar behavior, is the evidence of radiocarbon (^{14}C) fluxes preserved in tree rings that have now been counted and correlated back for more than 10 000 yr (Suess, 1978; Stuiver and Braziunas, 1989). The particulate bombardment by cosmic particles of nitrogen atoms in the stratosphere generates ^{14}C, which has a half-life of 5730 yr (although conventionally taken in dating records as 5568 yr, an earlier estimate). This low-energy radioactive isotope, combined in molecular form as CO_2, gradually sinks into the troposphere; entering trees during photosynthesis, it becomes incorporated in the wood (with a short time lag), or in the ocean it becomes incorporated in the $CaCO_3$ of invertebrate animals such as foraminifera and clams and planktonic plant life. Because of variable mixing rates and contamination, care must be taken with correlations, but in favorable places the tree-ring width (as a climatic proxy for precipitation or favorable temperature) correlates nicely with the ^{14}C values, a clear sun–climate relationship (Sonett and Suess, 1984).

Besides the direct observations of sunspots, there is also a wealth of long-term records of *auroras*, which serve as a supplementary index of solar activity (Siscoe, 1980; Schove, 1983; Suess and Tsurutani, 1998). The aurora represent a stratospheric optical display seen at high latitudes, oriented about the magnetic poles (aurora borealis and aurora australis), and active during periods of enhanced solar emissions, thus during sunspot peaks and during energetic solar flare events. The incoming solar (and galactic) cosmic rays are largely screened off by the Earth's own geomagnetic field, but this field intensity is not constant and varies with the terrestrial spin rate. It becomes weaker when the spin rate drops, and the rotation itself is modulated by solar activity. During a weakening of the main dipole field, however, the quadripolar influences increase, and sometimes auroral effects can be seen as far as 45° away from the magnetic pole. This could mean they could be visible even to latitude 35°N when the north magnetic pole is in that same quadrant. In the Middle Ages there were thus occasional auroral observations made in Constantinople, Baghdad, and Cairo (Siscoe, 1980). The reconstruction of an over-2-millennium record of sunspots by Schove (1983) has thus a powerful support from auroras, another and quasi-independent proxy resource. A third independent control of the longterm sunspot record has been provided by the ring-by-ring analyses of dendrochronological series (Stuiver and Braziunas, 1989). Over the last millennium, from the integrated evidence, protracted periods of sunspot weakness or "death" have been given identifying names: the Maunder Minimum (1645–1715), Spörer Minimum (1420–1530), Wolf Minimum (1280–1340), and Oort Minimum (1010–1050). Comparable (names) periods are traced back for over 5000 yr, always being associated with cold episodes in Earth history.

From the terrestrial climatic point of view, besides the flare effects (mentioned earlier), probably the most important stratospheric reaction to incoming pulses of solar radiation associated with the sunspot cycle is the creation of ozone (O_3) from oxygen by the solar UV flux, particularly in the shorter wavelengths. In addition to ozone being the principal greenhouse gas of the stratosphere, surges of ozone concentration are accompanied by a rise in ambient temperature. Its concentration fluctuates markedly in terms of the sunspot cycle (Callis et al., 1979). The behavior of the ozone layer is complicated, however, because of the slow sinking and diffusion of this gas into the troposphere and the liberation by humans of a large number of atmospheric pollutants. However, the chlorofluorocarbons, once thought to be a serious destructive agent of ozone, are now somewhat discounted. Ozone, nevertheless, is important for human health from several points of view; major influenza epidemics, for example, correlate with the sunspot peaks (Hope-Simpson, 1978).

The sun's rotation and orbital motions

The dynamic, changeable nature of sunspots is partly related to the *rotation of the sun*, which normally has a period of about 27 days (first determined in 1613 by Galileo). On Earth we observe a 27-day fluctuation in the cosmic-ray intensity (Forbush, 1966), the geomagnetic field, the UV radiation, stratospheric ozone, and other atmospheric parameters. The sun's rotation period is very close to that of the moon's sidereal cycle, so that they are sometimes in phase for quite a long time. Also, it has occasional accelerations. Changes of the solar spin rate are seen to correlate with the UV transmission, terrestrial ozone, and certain tropospheric (climatic) processes.

The sun does not rotate as a solid body, however, but having highly fluid (gaseous) outer layers it exhibits differential rotation (so-called Spörer Effect). The photosphere rotates much slower than its core. Sunspots near the equatorial belt are seen to rotate faster than those near the poles. In 1863 R.C. Carrington found the equatorial spin rate about 25 days, grading smoothly to a near-polar rate of about 32 days. In their life cycles the sunspots erupt first in solar midlatitudes (30° to 40°) and then slowly drift equatorward where they disappear. Maximum numbers are seen at 10° to 20° in each hemisphere. Plotted for both solar hemispheres, over an 11-yr interval, the spots make a distinctive, repeating pattern, Maunder's Butterfly Diagram.

In addition to the 27-day cycle, the sunspots vary in size and number over an approximately 11.12-yr cycle, a periodicity known as the *Sunspot Cycle*, or because of its many other ramifications, the *Solar Cycle*. It is hardly a cycle in the pure sense, inasmuch as it varies greatly in strength and in length, from around 6 to 17 yr, and is thus more accurately defined as a quasi-periodic fluctuation of about 11 ± 6 yr.

In the course of one sunspot cycle the magnetic tube on the sun's photosphere rises and erupts with the first spot, the *p*-spot (i.e., preceding), succeeded shortly after by its companion, the *f*-spot (i.e., following). The magnetic polarity of the *p*-spot matches that of the corresponding polar region, which alternates on average once every 11 yr, discovered by G. E. Hale in 1908. This *heliomagnetic (Hale) cycle* of about 22.34-yr represents a dispersed pattern of magnetic fields contrasting with a weak dipolar field that reverses every 11 yr (Babcock, 1961). Over the course of one cycle the total magnetic flux varies by a factor of three, but in the sunspot areas the flux rises by up to three orders of magnitude.

The phase and amplitude of the Sunspot Cycle also fluctuate somewhat regularly over a cycle of approximately 178–179 yr (Jose, 1965; Fairbridge and Hameed, 1983), which is equivalent in length to the All-Planet Synod (of Ren and Li, 1980) and to the variance in the sun's motion with respect to the barycenter. This long cycle also is reflected in the precession of the sun's orbit around the systemic barycenter (center of mass), as first shown by Jose (1965); it has been calculated for the last 12 centuries by Fairbridge (1984a). The mean period is 178.73376-yr.

As in the case of planet Earth, which precesses because of its ecliptic tilt and equatorial bulge being gravitationally affected by the body of the moon, the sun undergoes a similar precession. The sun is tilted to 5° from the mean plane of the planets, and as a nonrigid body it is also an oblate spheroid, so that the combined revolutions of the planets cause a precession of the solar orbit. However, because the vectoral distribution of the constantly changing collective gravitational effect of the planets, the sun, in its orbit around the barycenter, is constantly accelerating or decelerating.

A long-recognized feature of the solar system's periodicities is the phenomenon of *commensurability* (Stacey, 1967), which, following the pattern of the *Titius-Bode Law*, means that many of the longer planetary cycles associated with sunspot behavior are also *beat frequencies* of the fundamental periods (Mörth and Schlamminger, 1979). Thus, for example, the Saturn–Jupiter lap (synodic cycle), with an average length of 19.859-yr occurs exactly nine times during the 178.7-yr cycle. The sun's orbit around the barycenter (to be discussed under Causes), although variable in length, also has an average length of 19.859 yr ($= 9 \times 178.7$ yr). Alignments of Jupiter alone with the heliocenter and barycenter generate a cycle of 177.94 yr (the "King-Hele Cycle"). There are always 16 sunspot cycles per 177.94-yr period.

Wolf numbers

The commonly accepted index of sunspot numbers, the *Wolf numbers*, defines the average daily value as follows: $R = k\,(10g - f)$, where k = a calibration constant for any given observatory, f = number of spots observed on the solar disk, and g = number of groups. The relative number R is also sometimes called the Züich number, after the principal observing station (Waldmeier, 1961). Averages are provided by month and year. The cycles are numbered for identification, beginning with cycle 0 (min. 1745.0, max. 1750.3); cycle −12 began in 1610.8 and cycle 21 began in 1975.

The Wolf number is observed to fluctuate from minimums ($R\,m$), approaching 0, to maximums ($R\,M$), which display an extraordinary range of variance, from about 10 to over 200. Short cycles (10 yr or less) have high Wolf numbers, and long cycles (11 yr or more) have low ones. The Wolf number tends to be higher for the cycle of positive polarity (northern hemisphere), so that the 22-yr heliomagnetic cycle is sometimes regarded as the more fundamental signal. Correlations between Wolf numbers and features of the Earth's climate have met with varied degrees of success or failure, in part due to excessive mixing of conflicting signals. One factor, so far unexplained, is the two-stage response of the climate reaction. The double sunspot cycle responds differently from the single (Willett, 1962). Cycles with higher numbers (>80) may be the reverse of the lower ones. Most successful results so far have resulted from comparing the Wolf numbers over the centuries: protracted low peaks match the colder epochs (of several decades).

Attempts to predict the timing and amplitude of future Wolf maximums have proved to be a risky lottery that has caused much embarassment. Essentially, there are three methods employed, but so far no consensus has been reached. The methods used are: (1) spectral pattern analysis employing statistical techniques on the astronomically based sunspot-observational data, i.e., since Galileo's first use of the telescope during cycle −12, which began 1610; (2) long-term cycle analysis (employing the historical data plus that furnished by ancient proxy sources, interpreted in terms of planetary–solar cycles, e.g., Fairbridge and Hameed, 1983); (3) planetary dynamic analysis (e.g., Jose, 1965: employing our knowledge of the physics of the solar system), there being essentially two sources of energy: (a) the angular momentum of the system, which amounts to 3.148×10^{50} g/cm^2/s, with a translational kinetic energy of 1.99×10^{42} erg; and (b) the rotational energy of the planets, which is 0.7×10^{42} erg, the separate spin motions generating an interplanetary torque that in turn critically affects the sun's spin and its sunspots (Mörth and Schlamminger, 1979; Landscheidt, 1984). The mechanical effect of the sun's gaseous surface is akin to the crack of a cowboy's stock whip – a resonance crescendo effect, triggering flare eruptions and electron showers. Of the different methods, (1) is the usual one adopted but, although useful as a first approximation, it is almost doomed to fail because of the short term of the database. Methods (2) and (3) essentially coincide and therefore tend to be mutually confirmatory, but (3) offers a far more refined signal (see below).

Causes of sunspots

The cause of sunspots essentially is related to the thermal convection within the sun and to activity in the photosphere, but the dynamics appear to be exogenetically modulated. The problem has many complex and controversial aspects. The variability and periodicity of solar activity have led many investigators to consider the possibility of planetary interference (e.g. Schuster, 1911; Huntington, 1923). A correlation between the spectrum of sunspot numbers and various planetary periods and synods makes it extremely probable that some sort of gravitational dynamic process is involved (Johnson, 1946). Even the period of Mercury (88 days) is represented in the sunspot spectral analyses (Bigg, 1967), as are the periods and beat frequencies of certain other planets, notably Jupiter, Venus, and Earth. A simple vertical tide-raising mechanism, however, is ruled out because the amplitude generated by even the greatest planets is inadequate to produce a tide on the sun of more than a few cm. Horizontal tides, however, are appreciable and can be observed (Pimm and Bjorn, 1969).

The gravitational dynamic theory goes back to Sir Isaac Newton (in *Principia*, 1687 transl. 1729/1947), who pointed out that the mutual gravitational attraction between the sun and each of the planets resulted in a common center of mass (the systemic *barycenter*, analogous to the center of mass that exists in the Earth–moon pair). Because of the planetary revolutions around the barycenter, as mentioned above, the sun describes a miniorbit, which is in plan (e) an *epitrochoid* (a small loop within a large loop). The smaller one is mainly controlled by the 11.86-yr period of Jupiter, and the larger one by the 29.65-yr period of Sature. Or (b) a *cardioid* (a simple loop with a cross-over), which alternates several times with the epitrochoid (Fairbridge and Sanders, 1987; Charvatová 1997). These two major planets are responsible for 86% of the orbital angular momentum of the solar system. Additional inputs by the other planets contribute to the precession of the entire system, so that each solar miniorbit is different (sun's revolution varying 9 to 14 yr), but returning to a close alignment with most of the planets within a 90° sector once every 178 to 179 yr (so-called All-Planet Synod, APS). As Newton remarked, if "all the planets were placed on one side of the Sun, the distance of the common centre of gravity of all from the centre of the Sun would scarcely amount to one diameter of the Sun." That distance, however, varies from nearly zero (at peribac: minimum sun–barycenter radius, minimum orbital velocity) to $> 1.5 \times 10^6$ km at apobac: max. sun–barycenter radius, max. orbital velocity). Angular momentum is exchanged between sun and planets as the velocity accelerates or decelerates.

Effective solar radiation to the Earth (calculated by Borisenkov et al., 1983) varies constantly because of the changing distance between the sun and the Earth; three changeable orbital systems are involved: the sun's, the Earth's and the moon's. Furthermore the emission from the sun are in great variety.

The constantly changing planetary alignments of spinning celestial bodies generate not only a torque on one another, but also on the sun, expressed as:

$$\Delta L = \int_{t_0}^{t_1} \gamma(t)dt$$

The strongest pulses develop when Jupiter laps Saturn every 19.875-yr. This Saturn–Jupiter lap (SJL) is literally the *Pulse of the Solar System*. Numerous terrestrial climate series possess an ~20-yr periodicity. As pointed out by Landscheidt (1981), four consecutive pulses define a wave with a quasi-period of 79.46-yr that marks off the epochs of lowest orbital velocity and relative minimum distance of the sun from the barycenter. An empirical cycle of ~80 yr in the sunspot numbers has long been identified as the Gleissberg Cycle with distinctive heliomagnetic behavior (Gleissberg, 1965), and represented in such terrestrial climate series as tree-rings (Maksimov, 1952), in sedimentary varves (Anderson and Koopmans, 1963), and in ^{14}C variations (Stuiver and Braziunas, 1979). The SJL hemicycle of 9–10 yr appears also to be present in many terrestrial climatic series, especially when it coincides with short sunspot cycles (see also note below on flare periodicity).

In summary, there are two principal perturbing elements concerned with solar activity: (1) orbital alignments of planets that control the trajectory of the sun's miniorbit and its angular momentum (Jose, 1965); and (2) rotational torque of the planets. The first relates to long-term effects (the sunspot cycle and longer periodicities), the second mainly to short-term processes such as high-energy flare activity (particularly X-ray bursts). Concerning the latter, the angular acceleration of the vector of the tidal forces of Venus, Earth, and Jupiter form a cyclic pattern of about 120 days.

There is still a third factor: the sun's own spin rate. Quoting Landscheidt (1984): "The Sun, rotating on its axis, and the Sun, revolving around the center of mass (CM), could be looked at as coupled oscillators capable of internal resonance, resulting in slight positive or negative accelerations of the Sun's spin." Slowing tends to enhance activity, and speeding up decreases that activity. Historical observations disclose a speed-up just before the great seventeenth-century sunspot dearth, the Maunder Minimum, and recently a similar phenomenon has been noticed.

Jumps or jerks in the sun's spin rate were noted, for example, in 1967 and again in 1970, coinciding exactly with alignments of Jupiter, the sun's center, and the barycenter (Shirley and Fairbridge, 1998). Jupiter carries 61% of the system's angular momentum and (with the other planets) appears to transfer some of this energy to the sun, involving both the sun's revolution (velocity) and its rotation (spin rate). Jupiter's conjunctions and oppositions of this sort vary from 3 to 15 yr, but they generate a mean periodicity of 9.8 yr, epochs marked by highly energetic flare events.

From the terrestrial climatic point of view it seems to be no random coincidence that similar jerks in the Earth's motions (both angular velocity and spin rate) were observed, likewise in 1967 and 1970, for example, and again in 1974 and 1981. They are associated with alternation of zonal and blocking circulation regimes and large changes in the angular momentum of the atmosphere, which also is transmitted to the oceanic gyres. In the oceans the current velocity affects temperature (SST) and, because of both the Coriolis effect (the tilt on the sea-surface topography) and the steric effect (expansion or contraction of water body) there will be a corresponding series of sea level changes. The latter have long been a mystery, because the amplitude of these sudden changes grossly exceeds the limits explicable in terms of glacioeustasy (melting or development of ice caps), and are too rapid for tectonoeustasy (plate tectonics).

Inasmuch as the net angular momentum of any rotational system must be conserved, an acceleration of the atmospheric circulation may be compensated in part by a decease in the motion of the solid Earth's lithosphere and mantle. Below the mantle–core boundary, futhermore, there is a layer of about 100 km that behaves as a turbulent liquid, separating the outer spheres from the solid inner core. The latter tends to maintain its own angular momentum, so that the slowing of the outer layers is recorded by an apparent acceleration of the core. This acceleration can be monitored by watching the rate of drift of the geomagnetic intensity field. Normally the latter drifts westward at about 0.2°/yr (about 20 km/yr near the equator), which means that the core is turning a little faster (to the east) than the crust and mantle. Small variations in the regional distribution of mass in the core and the dynamic motions at the boundary combine to cause long-term secular changes in the value of gravity and the height of the geoid as monitored by repeated satellite observation. Short-term jerks or abrupt changes in the geomagnetic field are now regarded by Alldredge (1984), however, as exogenetic. It should be noted that specialists in geomagnetics and in the geophysics of the Earth's core have in the past tended to attribute all these dynamic phenomena to the endogenetic convective turbulence of the core. Many meteorologists, in contrast, tend to think of the atmosphere as existing in its own little in self-contained world, whereas some solar physicists place a time clock deep in the interior of the Sun (Dicke, 1978), both of which would seem to be emphatically denied by the evidence presented here. However, its seems in point of fact that all the phenomena are interconnected: sunsports, flares, spin rates. All are triggered ultimately by the energetics of the angular momentum of the solar system (3.148×10^{50} g/cm^2/s). In a basic NASA/JPL study, Dickey and Eubanks (1985) wrote: "Changes in the gravitational attraction of the Moon and Sun deform the Earth and change the Earth's moment of inertia, this, in turn causes periodic variations in Earth rotation." Corresponding changes are measured in the Earth's oblateness. Thus Earth rotation and polar studies are embarking on a new era.

Just as spring tides are generated on Earth by the moon twice during the synodic month (291/2 days), at both full and new-moon phases, the jerks in the sun's spin rate occur at alignments, both in conjunction and opposition, of Jupiter with the sun and barycenter. The mean periodicity of flares so generated is just over 2 yr (Landscheidt, 1984). This is one-quarter of the 9.8-yr flare cycle and one-fifth of the sunspot cycle, although we should recall that these are all average lengths and cannot be used casually for predictive purposes.

Quasi-biennial oscillation

On planet Earth the Quasi-biennial Oscillation Cycle (QBO of about 2.172-yr) is commonly ascribed to a self-generating dynamic relationship in the atmosphere. It was first discovered in American temperature data by Clayton in 1884 and has been the topic of innumerable subsequent studies. Correlations have been pointed out between the QBO and pressure, equatorial stratospheric winds, wine harvests, sea level, lake varves, tree rings, ozone, and other variables (Ebdon, 1975, among others). Inasmuch as terrestrial climates and their proxies respond much more vigorously to the 12-month seasonal determinism, a 26-month pulse is likely to conform to a biennial series for some time and then jump in phase to a triennial. Schove (1983, p. 230) recognized that strong but short solar cycles (~10 yr) are generally associated with 2-yr weather cycles, whereas weak but long solar epochs (often ~12 yr) are more likely to match 3-yr sequences. This characteristic has been successfully used as a vernier device by Schove (1983, p. 333 ff.) for correlating floating time series (tree rings, varves) during the Holocene.

Schove (1983, p. 319, i.e., a reprint of his 1964 paper) had recognized a pressure parameter in his weather series that could be traced back to the equatorial Walker Circulation and the Southern Oscillation. That, in turn, is found to be often linked also to the El Niño phenomenon, and indeed today they are commonly abbreviated together as ENSO. The Southern Oscillation is a classical ocean-atmosphere feedback system, with global signals, such as tropospheric pressure and temperature, cloudiness, and atmospheric CO_2. It is periodically modified either in the east (Pacific) by a weakening of the southeast trade wind, a reduction of upwelling off Peru with warm-water invasion (El Niño), or in the west (Indian Ocean) by eruption of cold air from central Asia (Monsoon fluctuation). Very roughly, three QBOs make one El Niño (~7 yr), but there is little regularity. Historians have traced El Niño back four centuries; important years recently have been 1925, 1930, 1934, 1941, 1943, 1951, 1953 (mild), 1957–1958, 1965, 1969, 1972, 1976–1977, and 1982–1983. Catastrophic flooding is associated with the El Niño in the mountainous and semiarid coastal areas of Peru (documented for example in 1701, 1720, 1728, and 1891). The El Niño of 1925 saw the greatest floods since the Spanish occupation; on the coast, off Lima, the SST rose from 15°C to 26°C in 3-months. The situation was very similar in 1982–1983.

Sunspots and the lunar connection

A series of papers by Currie (1984) and others have indicated a strong signal in terrestrial climate series from the lunar nodal period (18.61-yr), which is also the cause of a nutation of the same period in the Earth's orbital precession cycle (Borisenkov et al., 1983, showed its insolation effect). In long-term proxy records, e.g., from North American tree rings, from the Nile, and from ancient China (Currie and Fairbridge, 1985), it is evident that in certain periods the 11-yr sunspot cycle is identifiable, but in others it is not, being replaced by the 18.6-yr lunar period (Wood, 1978, 2001). Currie has demonstrated that the lunar tidal effect normally develops a triple-phase standing wave around the northern hemisphere, analogous to the simplest Rossby wave. When the maximum lunar–solar tide (which peaks 4 years later than the maximum lunar declination) is in phase with the sunspot cycle, the climatic effects are sometimes suppressed, sometimes enhanced. From time to time, however, the phase is strangely reversed, a "flip flop" of a 180° within one or two cycles (Currie, 1984). Fairbridge (1984b) compared the Nile floods and droughts (since AD 622) with both the solar and lunar cycles, finding indeed that in some centuries the droughts matched the sunspot maximums and in others the minimums, whereas in other epochs the lunar cycle was dominant.

The long-term lunar period of about 556 yr, the Precession of the Lunar Perigee (PLP), discovered by Pettersson (1914) and often called the Pettersson Cycle, is also the hemicycle of the 1112-yr Outer Planets quadrature cycle (of Stacey, 1967), and every third alignment, 1668 yr (Stacey's Zero Check) seems to correlate with the high sea levels of the Fairbridge (1961) eustatic curve.

Conclusion

One key to understanding all these terrestrial climate trends is the *amplitude* and the *timing* of the solar emissions, which have been suggested in some of the early papers. Evidently the sunspots have a message, but many more studies are needed before it can be clearly read and used as a tool for reliable prediction.

Rhodes W. Fairbridge

Bibliography

Agee, E.M., 1980. Present climatic cooling and a proposed causitive mechanism. *American Meteorological Society Bulletin*, **61**(11): 1356–1367.

Alldredge, L.R., 1984. A discussion of impulses and jerks in the geomagnetic field. *Journal of Geophysical Research*, **89**: 4403–4412.

Anderson, R.G., and Koopmans, L.H., 1963. Harmonic analysis of varve time series. *Journal of Geophysical Research*, **68**(3): 877–893.

Babcock, H.W., 1961. The topology of the Sun's magnetic field and 22-yr cycle. *Astrophysics Journal*, **133**: 572–587.

Bigg, E.K., 1967. Influence of the planet Mercury on Sunspots. *Astronomy Journal*, **72**(4): 463–466.

Borisenkov, Y.E., Tsvetkov, A.V., and Agaponov, S.V., 1983. On some characteristics of insolation changes in the past and the future. *Climate Change*, **5**: 237–244.

Bucha, V., 1977. Mechanism of solar-terrestrial relations and changes of atmospheric circulation. *Studia Geophysica ek Geodesica*, **21** (suppl. 416): 350–360.

Bucha, V., 1984. Mechanism for linking solar activity to weather-scale effects, climatic changes and glaciations in the northern hemisphere. In Mörner, N.A., and Karlen, W., eds., *Climatic Changes on a Yearly to Millennial Basis*. Dordrecht: Reidel, pp. 415–448.

Callis, L.B., Natarajan, M., and Nealy, J.E., 1979. Ozone and temperature trends associated with the 11-year solar cycle. *Science*, **204**: 1303–1306.

Charvátová, I., 1997. Solar-terrestrial and climatic phenomena in relation to solar inertial motion. *Surveys in Geophysics*, **18**: 131–146.

Currie, R.G., 1984. On bistable phasing of 18.6 year nodal-induced flood in India, *Geophysical Research Letters*, **11**: 50–53.

Currie, R.G., and Fairbridge, R.W., 1985. Periodic 18.6-year and cyclic 11-year induce drought and flood in northeastern China and some global implications, *Quaternary Science Review*, **4**: 109–134.

Dicke, R.H., 1978. Is there a chronometer hidden deep in the sun? *Nature*, **276**: 676–680.

Dickey, J.O., and Eubanks, T.M., 1985. Earth rotation and polar motion: Measurements and implications, *IEEE Transactions in Geoscience and Remote Sensing*, (special issue: Geodynamics).

Ebdon, R.A., 1975. The quasi-biennial oscillation and its association with tropospheric patterns. *Meteorology Magazine*, **104**: 282–297.

Fairbridge, R.W., 1961. Eustatic changes in sea level. In Ahrens, L.H., Press, F., et al., eds. *Physics and Chemistry of the Earth*, vol. 4. London: Pergamon Press, pp. 99–185.

Fairbridge, R.W., (ed.), 1967. *The Encyclopedia of Atmospheric Sciences and Astrogeology*. New York: Reinhold.

Fairbridge, R.W., 1984a. Planetary periodicities and terrestrial climate stress. In Mörner, N.A., and Karlen, W., eds., *Climatic Changes on a Yearly to Millennial Basis*. Dordrecht: Reidel, pp. 509–520.

Fairbridge, R.W., 1984b. The Nile floods as a glacial climatic/solar proxy. In Mörner, N.A., and Karlen, W., eds., *Climatic Changes on a Yearly to Millennial Basis*. Dordrecht: Reidel, pp. 181–190.

Fairbridge, R.W., and Hameed, S., 1983. Phase coherence of solar cycle minima over two 178-year periods, *Astronomy Journal*, **88**: 867–869.

Fairbridge, R.W., and Sanders, J.E., 1987. The Sun's orbit, AD 750–2050. In Rampino, M.R., et al., eds., *Climate*. New York: Van Nostrend Rein hold, pp. 446–471.

Forbush, S.E., 1966. Time variation of cosmic rays. In Bartels, J., ed., *Handbuch der Geophysik III*. New York: Springer-Verlag, pp. 159–247.

Gilliland, R.L., 1982. Solar, volcanic, and CO_2 forcing of recent climatic changes. *Climate Change* **4**: 111–131.

Gleissberg, W., 1965. The eighty-year cycle in auroral frequency numbers. *British Astronomical Association Journal*, **75**: 227–231.

Hope-Simpson, R.E., 1978. Sunspots and flu: a correlation. *Nature*, **275**: 86.

Huntington, E., 1923. *Earth and Sun, a Hypothesis of Weather and Sunspots*. New Haven: Yale University Press.

Johnson, M.O., 1946. *Correlation of Cycles in Weather, Solar Activity, Geomagnetic Activity, and Planetary Configurations*. San Francisco, CA: Phillips & Van Orden.

Jose, P.D., 1965. Sun's motion and sunspots. *Astronomy Journal*, **70**: 193–200.

Landscheidt, T., 1984. Cycles of solar flares and weather. In Mörner, N.A., and Karlen, W., eds., *Climatic Changes on a Yearly to Millennial Basis*. Dordrecht: Reidel pp. 473–481.

Lawrence, E.N., 1965. Terrestrial climate and the solar cycle. *Weather*, **20**: 334–343.

Maksimov, I.V., 1952. On the eighty-year cycle of terrestrial climatic fluctuations. *Doklady Akademia Nauk S.S.S.R.*, **86**(5), 917–920. (Transl. E.R.Hope, DRB, Canada).

Mörth, H.T., and Schlamminger, L., 1979. Planetary motion, sunspots and climate. In McCormac, B.M., and Seliga, T.A., eds., *Solar Terrestrial Influences on Weather and Climate*. Dordrecht: Reidel, pp. 193–207.

Newton, I., 1729. *Mathematical Principles of Natural Philosophy and System of the World*, A. Motte (trans.) and revised by F. Cajori (1947). Berkeley, CA: University of California Press.

Pettersson, O., 1914. On the occurrence of lunar periods in solar activity and the climate of the earth, *Svenska Hydrogr.-Biol. Komm. Skriften*.

Pimm, R.S., and Bjorn, T., 1969. Prediction of smoothed sunspot number using dynamic relations between the Sun and planets, Final Report NASA-21445. Washington, DC: National Aeronautics and Space Administration.

Ren, A., and Li, Z., 1980. Effect of motions of planets on climate changes in China. *Kexue Tongbao*, **25**(2): 417–422.

Schove, D.J., (ed.), 1983. *Sunspot Cycles*. Stroudsburg: Hutchinson Ross. (Note: An error occurs on p. 396, sunspot minimums AD 507–1501).

Schuster, A., 1911. The influence of planets on the formation of sunspots. *Royal Society of London Proceedings*, **A85**: 309–323.

Schuurmans, C.J.E., 1981. Solar activity and climate. In Berger, A., ed., *Climatic Variations and Variability: facts and theories*. Dordrecht: Reidel, pp. 559–575.

Shirley, J.H. and Fairbridge, R.W., eds. 1998. *Encyclopedia of Planetary Sciences*. London: Chapman Hall (now Springer).
Siscoe, G.L., 1980. Evidence in the auroral record for secular solar variability, *Reviews in Geophysics and Space Physics*, **18**: 647–658.
Sonett, C.P., and Suess, H.E., 1984. Correlation of bristlecone pine ring widths with atmospheric C-14 variations: a climate–sun relation. *Nature*, **307**: 141–143.
Stacey, C.M., 1967. Earth motions. In Fairbridge, R.W., ed., *The Encyclopedia of Atmospheric Sciences and Astrogeology*. New York: Reinhold, pp. 335–340.
Stuiver, M., and Braziunas, T.F., 1989. Atmospheric ^{14}C and century-scale solar oscillations. *Nature*, **338**: 405–408.
Suess, H.E., 1978. La Jolla measurements of radiocarbon in tree-ring dated wood. *Radiocarbon*, **20**: 1–18.
Suess, S.T., and Tsurutani, B.T., eds., 1998. *From the Sun: Auroras, Magnetic Storms, Solar Flares, Cosmic Rays*, Washington: American Geophysical Union.
Waldmeier, M., 1961. *The Sunspot Activity in the Years 1610–1960*. Zurich: Schulthess.
Willett, H.C., 1962. The relationship of total atmospheric ozone in the sunspot cycle. *Journal of Geophysical Research*, **67**: 661–670.
Willett, H.C., 1965. Solar-climatic relationships in the light of standardized climatic data. *Journal of Atmospheric Science*, **22**: 120–128.
Wood, F.J., 1978. The strategic role of Perigean spring tides. In *Nautical History and North American Coastal Flooding, 1635–1976*. Washington, DC: National Oceanic and Atmospheric Administration.
Wood, F.J., 2001. *Tidal Dynamics*, 2 vols. *Journal of Coastal Research*, special issues 30/31.

Cross-references

Carbon-14 Dating
Climate Variation: Historical
Cycles and Periodicities
Maunder, Edward Walter and Maunder Minimum
Sunspots

SYNOPTIC CLIMATOLOGY

The term synoptic climatology was adopted at the headquarters of the US Air Force in the early 1940s in reference to analyses made of past weather situations in order to assess the frequencies of different operational conditions (Jacobs, 1947). Synoptic is used by meteorologists to denote the synchronous weather conditions typically depicted on a synoptic weather map. By extension, because large-scale atmospheric circulation systems such as cyclones and anticyclones are analyzed from such weather charts, features with a horizontal dimension of *ca.* 1000 km and a lifespan of about 5–7 days are called synoptic-scale systems. In a broad sense, therefore, synoptic climatology is the study of local and regional climates in terms of the properties and behavior of the atmosphere over and around a given area; the information used in such studies is primarily that shown on synoptic weather maps (Barry and Carleton, 2001; Yarnal, 1993; Court, 1957; Hare, 1955; Jacobs, 1946). Synoptic climatology is a major branch of climatological study in parallel with dynamic climatology, microclimatology, bioclimatology, paleoclimatology, and applied climatology. The modern term climate dynamics encompasses concepts and methods of both synoptic and dynamic climatology.

The basic procedures involved in a synoptic–climatological study are, first, the determination of categories of atmospheric circulation type (or other comparable descriptors of synoptic-scale weather processes); and, second, the statistical assessment of weather conditions in relation to these patterns. It has been common practice to distinguish a small number of types of airflow or pressure pattern from daily weather maps and then to calculate average values for each type of temperature and daily precipitation totals at weather stations. Indeed, these studies began almost as soon as weather maps were first produced in the late nineteenth century by such well-known meteorologists as Abercromby (1883) and van Bebber and Köppen (1895). An illustration of the contrast between mean conditions and the aggregate effect of different airflow patterns on afternoon winter precipitation frequencies in Hokkaido, Japan, is shown in Figure S45. This emphasizes the contribution made to the interpretation of climatic regimes through use of a synoptic framework for analysis instead of the calendar intervals employed in conventional climatic statistics.

Synoptic climatological classifications

Classifications of atmospheric properties and processes can be made on the basis of various synoptic features including: the pressure pattern, the airflow direction and isobaric curvature, streamlines, the large-scale steering pattern of the midtropospheric circulation, and derived properties such as divergence and vorticity (LeDrew, 1984). The spatial scale of the analysis may be regional (*ca.* 10^6 km^2) up to hemispheric (Yarnal, 1984); the time interval considered is usually 1 day, but may be a synoptic interval of up to 5 or 6 days. Table S17 sets out the range of methods that have been used, according to the spatial scale. For convenience, approaches used in identifying synoptic types can be considered under three headings. First, the sea-level pressure map (or upper-level contour chart) can be regarded in a static sense and the map pattern over a given area at a specified time forms the basis of classification. Second, there are kinematic classifications where the large-scale movement of pressure systems is examined. The most important approach here is the Grosswetter classification scheme developed in Germany by Baur (1951; Hess and Brezowsky, 1977). Similar work was undertaken in the United States (Elliott, 1951). Third, observations of weather elements may be grouped directly into sets of weather types or complexes; Russian climatologists use the term complex climatology in this connection.

Until the 1960s, synoptic classifications were developed through subjective manual procedures. By studying numerous daily weather map sequences the researcher gains familiarity with the more commonly recurring patterns of synoptic weather map features and thereby designates a limited number of map pattern types. The recognition of numerous types will reduce the variability of weather conditions within each category but, unless a very long record is considered, most type categories will be represented by only a few cases, which makes their characterization difficult. Conversely, if few types are distinguished then each category inevitably contains a wide variability of weather conditions. From several empirical studies of the pressure patterns and associated weather over the European Alps, it appears that the variability of weather conditions (as defined by the standard deviation of individual weather elements) shows a significant decrease as the number of types distinguished in a synoptic classification increases from 10 to 30, but little change thereafter (Fliri, 1965). Modern analyses almost invariably use objective classification procedures, taking advantage of the

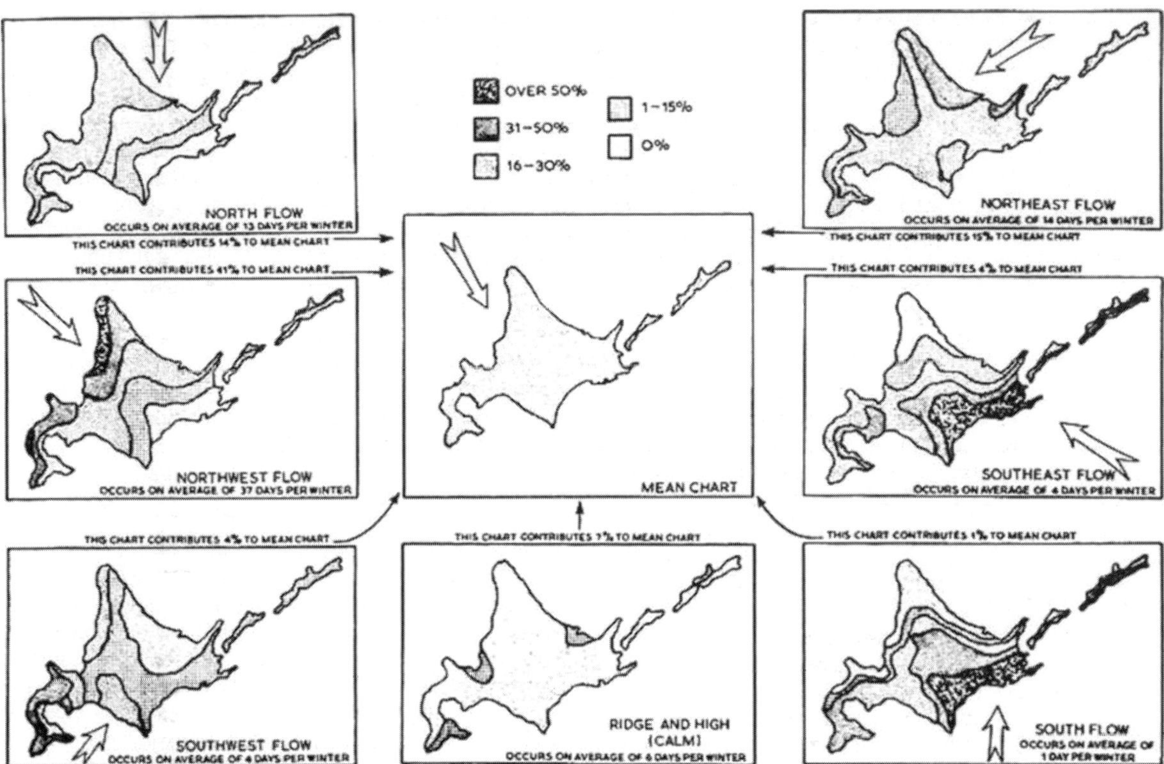

Figure S45 Percentage frequency of days at 1800 hours having precipitation amounts for preceding 6-hourly period greater than 1 mm (0.04 inch) in Hokkaido, Japan for December–February (from Jacobs, 1947).

availability of digital datasets on pressure fields or wind fields, and of meteorological observations. A variety of such objective procedures now exist. The major ones are:

1. Correlation analysis of pairs of maps with an arbitrary assigned level of similarity. Categories are then determined by some type of grouping or clustering method.
2. Objective specification of pressure or height fields by principal component (or empirical orthogonal function) analysis followed by some clustering procedure.

Correlation methods (Lund, 1963) examine the similarity of pressure patterns over an area by correlating grid point pressure values for each possible pair of maps. The pattern (key day) that has the largest number of maps correlated with it (using a threshold of, say, $r = 0.8$) is selected as type A, and these cases are abstracted and the next largest group is designated type B, and so on. On completion of this process, each case is rechecked to see that it is assigned to the key-day group with which it has the highest correlation. An analogous classification procedure using sums of squares has been developed by Kirchhofer (1973) and widely applied to develop regional classifications for the Alps, the western United States, and the Canadian Arctic. The two procedutes are essentially the same (Willmott, 1987). Both of these approaches assign classes on the basis of the shape of the pressure patterns but do not take account of the intensity of pressure systems.

The earliest studies in objective specification of isobaric or contour patterns made use of Fischer–Tschebyschev orthogonal polynomial equations (Friedman, 1955; Hare, 1958; Malone, 1958), but later workers generally use the more flexible empirical orthogonal functions (principal components or eigenvectors). These are the optimal set of mathematically determined functions that provide the most efficient representation of variance in the dataset. Each function is mutually uncorrelated (orthogonal) in space and the coefficients of the functions are orthogonal in time. The method is described in standard statistical texts and various meteorological studies (Grimmer, 1963; Joliffe, 1986; North et al., 1982). The first few principal components of pressure fields usually describe simple zonal and cellular patterns. An individual pressure map is represented by some combination of these components and a classification based on principal components is constructed by defining arbitrary ranges of the coefficient (or amplitude) of each component (Kruizinga, 1979). Typing via map patterns and principal components each has its own advantages and drawbacks. The orthogonality constraint of eigenvectors can be avoided by using obliquely rotated principal components, but examination of this technique is not yet far advanced (Richman, 1981).

The Grosswetter concept involves the identification of large-scale weather patterns over a region. Local weather conditions will differ, but overall there is a large-scale (synoptic) interrelationship. A Grosswetterlage (large-scale weather pattern) is the mean pressure distribution (at sea level) during a time interval during which the position of the stationary (steering) cyclones and anticyclones and the steering within a special circulation region remain essentially unchanged (Baur, 1951, p. 825). later work takes account of the 500-mb midtropospheric circulation patterns also. A daily catalog of *Die Grosswetterlagen*

Table S17 Classification methods used in synoptic climatology

Global scale

1.1 *Subjective schemes*
Description of seasonal changes of pressure and circulation fields.
Characterization of typical circulation regimes (zonal, meridional blocking).

1.2 *Objective schemes*
Calculation of zonal and meridional circulation indices.

Continental scale

2.1 *Subjective schemes*
Classification of pressure and circulation fields, based on the major centers of action (*Grosswetterlagen*).
Delimitation of zonal and meridional circulation types.
Assessment of weather conditions in relation to cyclone–anticyclone tracks.
Determination of the frequency of high and low centers.
Classification of airmasses and fronts.

2.2 *Objective schemes*
Classification based on derived parameters of the pressure and circulation fields (pressure gradient, relative vorticity, flow direction, etc.).
Correlation of weather conditions with typical upper-level contour patterns.
Classification based on mathematical–statistical specification of pressure fields (orthogonal polynomial functions).

Regional scale

3.1 *Subjective schemes*
Grouping similar pressure fields or airflow patterns (pressure- or airflow-pattern types).
Airmass and frontal classifications.

3.2 *Objective schemes*
Classification based on derived parameters of the pressure and circulation fields (pressure gradient, relative vorticity, flow direction, etc.).
Classification based on upper airflow at selected stations.
Correlation of weather conditions with typical upper-level contour patterns.
Classification based on mathematical-statistical specification of pressure fields (empirical orthogonal functions and clustering; self-organizing maps).

Local Scale

4.1 *Subjective schemes*
Definition of weather types according to locally observed weather elements (complex climatology).

4.2 *Objective schemes*
Intercorrelation of locally observed weather elements and statistical condensation of these into local weather types.

Source: After Wanner (1980).

Mitteleuropas is published monthly by the Deutscher Wetterdienst (1994). Russian meteorologists have extended this approach to hemispheric circulation patterns. In the classification developed at the Arctic and Antarctic Research Institute, St Petersburg, by Vangengeim (Girs, 1966) and at the Institute of Geography, Moscow by Dzerdzeevski (1962) broad categories of zonal and meridional circulation and associated cyclone tracks are distinguished, with subdivisions according to the number of sectors affected by amplified waves or blocking of the westerlies and associated meridional motion. Daily catalogs of these classifications are available from 1900 to the present.

The categorization of weather types from combinations of the joint occurrence of several weather elements (temperature, cloudiness, wind speed, etc.) has proved of limited value in interpreting climatic conditions. The selection of six classes of temperature, three of cloudiness, four of wind speed, and three of precipitation, for example, gives rise to a system with 228 possible combinations. Frequency data on some specific climatological contingencies may be of general climatological interest, however, and especially if a suitable means of synoptic interpretation is adopted. Objective weather types have been determined in one case study for Madison, Wisconsin, using principal component analysis (Christensen and Bryson, 1966). These types are shown to have a synoptically realistic spatial pattern. This approach for individual stations was also applied by Kalkstein et al. (1987) using weather data and synoptic circulation variables to obtain a temporal synoptic index. Subsequently, Kalkstein et al. (1996) developed a continental-scale analysis of weather types (spatial synoptic classification) (SSC) based on airmass characteristics. A calendar for six types based on the SSC method has been prepared for 327 stations in North America, spanning over 40 years (Sheridan, 2002).

Applications

Many regional synoptic catalogs have been developed. A large number of these are described by Barry and Perry (1973) and Barry and Carleton (2001). In several instances climatic analysis has not proceeded beyond the development of the catalog, but certain catalogs have been widely used. Illustrations are given of some of these applications. The most common application of synoptic catalogs is in the quantitative characterization of regional climatic conditions. This has included analysis of the local and regional weather characteristics of the synoptic types, their annual frequency, and typical sequences. German climatologists have used the Grosswetter catalog extensively to analyze temperature and precipitation conditions in central Europe; British climatologists have similarly made use of Lamb's catalog (1972) and objective procedures to reproduce the type categories have been developed by Jones and Kelly (1992) while El-Kadi and Smithson (1996) use the Kirchhofer approach to generate comparable sea-level pressure patterns. The Lamb catalog is currently maintained using an objective procedure, and the latter has been used back to the late nineteenth century, by the Climatic Research Unit at the University of East Anglia, UK. Such baseline characterizations should be an essential element in all climatic descriptions. For some other areas special-purpose synoptic groupings have been devised to examine such diverse questions as urban influences on air pollution, airflow patterns favoring dispersion of particular spores and insects, critical forest-fire weather conditions, and so on.

A question of general climatological significance concerns the delimitation of natural seasons. The tendency for particular calendar intervals to be characterized predominantly by a certain type of weather regime is well established. This has been shown for the British Isles, Europe, and East Asia, as well as in a broader hemispheric framework by Bradka (1966), Bryson and Lahey (1958), and Lamb (1964). In northwest Europe, for example, spring is characterized by high variability of temperature and precipitation, whereas fall weather tends to be much more persistent, with a high frequency of anticyclonic conditions. More controversial is the purported occurrence of weather singularities; i.e. the recurrence of some weather characteristic about a specified calendar date. Examples include the 20–25 January thaw in New England and the onset of summer rains in

Arizona–New Mexico about 1 July. These events are most usefully interpreted in terms of tendencies for recurrent circulation regimes, but even so it is known that long-term trends in the relative frequency of such singularities occur, presumably in relation to changes in the hemispheric circulation patterns.

Another basic topic where synoptic catalogs find application is in studies of climatic fluctuations. It is a commonly held premise that such fluctuations arise, at least in part, as a result of the changing frequency of particular types of circulation pattern. Hence, if northerly outbreaks of Arctic air in winter become more frequent, winters are likely to become more severe. However, studies suggest that changes in the frequency of identified types of pressure pattern provide an insufficient basis on their own for inferring changes in associated climatic factors (Barry et al., 1981; Yarnal, 1985). This is because, among other things, synoptic catalogs do not generally take account of the intensity or duration of the pressure patterns or airflow regimes, synoptic types are each associated with a range of possible climatic conditions, and because the internal characteristics of a type may themselves vary over time due to other external factors (such as changes in sea-surface temperatures, in atmospheric stability, in incoming solar radiation).

Originally it was hoped that synoptic catalogs would find extensive use in long-range forecasting. It was considered that recurring sequences of weather patterns could perhaps be recognized, enabling predictions to be developed when similar precursor conditions were present. Some limited applications of this approach have indeed been possible, particularly when computer searches for analog situations are made. However, there are usually only a few close analogs of any large-scale situation when this approach is applied to 15–30-day intervals, and when consideration is given not only to the surface and upper-air patterns, but also to external variables such as snow cover or soil moisture content, sea-surface temperature anomalies, and so on. Moreover, the subsequent weather developments do not always proceed in a similar direction when given apparently analogous starting points! Nevertheless, the application of a synoptic climatology can serve as an important check on numerical simulations using atmospheric general circulation models (Kidson, 1995).

Roger G. Barry

Bibliography

Abercromby, R., 1883. On certain types of British weather. *Royal Meteorological Society Quarterly Journal*, **9**: 1–25.

Barry, R.G., and Carleton, A.M., 2001. *Synoptic and Dynamic Climatology*. London: Routledge.

Barry, R.G., and Perry, A.H., 1973. *Synoptic Climatology: methods and applications*. London: Methuen.

Barry, R.G., Kiladis, G., and Bradley, R.S., 1981. Synoptic climatology of the western United States in relation to climatic fluctuations during the twentieth century. *Journal of Climatology*, **1**: 97–113.

Baur, F., 1951. Extended-range weather forecasting: In Malone, T.F., ed., *Compendium of Meteorology*. Boston, MA: American Meteorological Society, pp. 814–833.

Blasing, T.J., 1975. A comparison of map-pattern correlation and principal component eigenvector methods for analyzing climatic anomaly patterns. In *Preprint Volume: Fourth Conference on Probability and Statistics in Atmospheric Sciences*. Boston, MA: American Meteorological Society, pp. 96–101.

Bradka, J., 1966. Natural seasons in the Northern Hemisphere. *Geofysiske Sbornik*, **14**: 597–648.

Bryson, RA., and Lahey, J.F., 1958. *The March of the Seasons*. Madison, WI: Meteorology Department, University of Wisconsin.

Christensen, W.I., Jr, and Bryson, R.A., 1966. An investigation of the potential of component analysis for weather classification. *Monthly Weather Review*, **94**: 697–709.

Court, A., 1957. Climatology: complex, dynamic and synoptic. *Association of American Geographers Annals*, **47**: 125–136.

Deutscher Wetterdienst, 1994. Grosswetterlagen Europas. *Amsblatt deutsch. Wetterdiebstes*, **47**: Offenbach-am-Main.

Dzerdzeevski, B.L., 1962. Fluctuations of climate and of general circulation of the atmosphere in extratropical latitudes of the Northern Hemisphere and some problems of dynamic climatology. *Tellus*, **14**: 328–336.

El-Kadi, A.K.A., and Smithson, P.A., 1996. An automated classification of pressure patterns over the British Isles. *Transactions of the Institute of British Geographers*, n.s. **21**: 141–156.

Elliott, R.D., 1951. Extended range forecasting by weather types. In Malone, T.F., ed., *Compendium of Meteorology*. Boston, MA: American Meteorological Society, pp. 834–840.

Fliri, F., 1965. Über Signifikanzen synoptich-klimatologischer Mittelwerte in verschiedenen Alpinen Wetterlagensystemem. *Carinthia II*, Vienna, Sonderheft 24, 36–48.

Friedman, D.G., 1955. Specification of temperature and precipitation in terms of circulation patterns. *Journal of Meteorology*, **12**: 428–435.

Girs, A.A., 1966. Intra-periodical transformations of the atmosphere and their causes. In Girs, A.A., and Dydina, L.A., eds., *Contributions to Long-Range Weather Forecasting in the Arctic*. Jerusalem: Israel Program of Scientific Translations, pp. 13–45.

Grimmer, M., 1963. The space filtering of monthly surface temperature anomaly data in terms of pattern, using empirical orthogonal functions, *Royal Meteorological Society Quarterly Journal*, **89** 395–408.

Hare, FK., 1955. Dynamic and synoptic climatology. *Association of American Geographers Annals*, **45**: 152–162.

Hare, F.K., 1958. The quantitative representation of the north polar pressure field. In Sutcliffe, R.C., ed., *The Polar Atmosphere Symposium, Part 1*. Oxford: Pergamon Press, pp. 137–150.

Hess, P., and Brezowsky, H., 1977. Katalog der Grosswetterlagen Europas, 1891–1976. *Deutsch. Wetterdienstes Ber.*, **15** (113).

Jacobs, W.J., 1946. Synoptic climatology. *American Meteorological Society Bulletin*, **27**: 306–311.

Jacobs, W.J., 1947. Wartime developments in applied climatology. *Meteorological Monographs*, **1**(1): 1–52.

Joliffe, I.T., 1986. *Principal Component Analysis*. New York: Springer-Verlag.

Jones, P.D., and Kelly, P.M., 1982. Principal component analysis of the Lamb catalogue of daily weather types. Part I: Annual frequencies. *Journal of Climatology*, **2**(2): 147–158.

Kalkstein, L.S., Tan, G., and Skindov, J.A., 1987. An evaluation of three clustering procedures for use in synoptic climatological classification. *Journal of Climatology and Applied Meterology*, **25**: 717–730.

Kalkstein, L.S., Nichols, M.C., Barthel, C.D., and Greene, J.S., 1996. A new spatial synoptic classification: application to air mass analysis. *International Journal of Climatology*, **16**(9): 983–1004.

Kidson, J.W., 1995. A synoptic climatological evaluation of the changes in the CSIRO nine-level model with doubled CO_2 in the New Zealand region. *International Journal of Climatology*, **14**: 711–21.

Kirchhofer, W., 1973. Classification of European 500-mb patterns. *Schweizer Meteorol. Zentralanstalt. Arbeits* No. 43, pp. 1–16.

Knowles, H.T., and Jehn, K.G., 1975. A central Texas synoptic climatology and its use as a precipitation forecast tool. *Monthly Weather Review*, **103**: 730–736.

Kruizinga, S., 1979. Objective classification of daily 500 mbar patterns. In *Preprint Volume, Sixth Conference on Probability and Statistics in Atmospheric Sciences*. Boston, MA: American Meteorological Society, pp. 126–129.

Lamb, H.H., 1964. *The English Climate*, 2nd edn. London: English Universities Press.

Lamb, H.H., 1972. British Isles weather types and a register of the daily sequence of circulation patterns, 1861–1971. *Geophysics Memoirs*, **16**(116): 1–85.

LeDrew, E.F., 1984. The role of local heat sources in synoptic activity within the Polar Basin. *Atmosphere–Ocean*, **22**: 309–327.

Lund, I.A., 1963. Map-pattern classification by statistical methods. *Journal of Applied Meteorology*, **2**: 56–65.

Lydolph, P.E., 1959. Federov's complex method in climatology. *Association of American Geographers Annals*, **49**: 120–144.
McCutchan, M.H., 1978. A model for predicting synoptic weather types based on Model Output Statistics. *Journal of Appllied Meteorology*, **17**: 1466–1475.
Malone, T.F., 1958. *Weather Analysis and Forecasting*, vol. 2. S. Petterssen (ed.). New York: McGraw-Hill, pp. 238–256.
North, G.R., Bell, T.L., Calahan, R.F., and Moeng, F.J., 1982. Sampling errors in the estimation of empirical orthogonal functions. *Monthly Weather Review*, **110**(7): 699–706.
Richman, M.-B., 1981. Obliquely rotated principal components: An improved meteorological map typing technique? *Journal of Applied Meteorology*, **20**(10): 1145–1159.
Savina, S.S., 1987. Large-scale atmospheric processes in the Northern Hemisphere and climatic extremes in the European part of the USSR. (in Russian). *Materialy Meteorologicheskikh Issledovanii*, No. 13 Moscow: Institute of Geography, Academy of Sciences, USSR.
Sheridan, S.C., 2002. The redevelopment of a weather-type classification for North America. *International Journal of Climatology*, **22**(1): 51–68.
Sowden, I.P., and Parker, D.E., 1981. A study of climatic variability in relation to the Lamb synoptic types. *Journal of Climatology*, **1**(1): 3–10.
van Bebber, W.J., and Köppen, W., 1895. Die Isobarentypen des Nordatlantischen Ozeans und Westeuropas, ihre Beziehung zur Lage und Bewegung der Barometrischer Maxima und Minima. *Archiv deutsch. Seewarte*, **18**(4): 1–27.
Wanner, H., 1980. Grundzüge der Zirkulation der mittleren Breite und ihre Bedeutung fü die Wetterlagenalanyse im Alpenraum. In Oeschger, H., Messerli, B., and Svilar, M., eds., *Das Klimat – Analysen und Modelle, Geschichte und Zukunft*. Berlin: Springer-Verlag, pp. 117–124.
Willmott, C.J., 1987. Synoptic weather map classification: correlation versus sum-of-square. *Professional Geographer*, **30L**: 205–207.
Yarnal, B., 1984. The effect of weather map scale on the results of a synoptic climatology. *Journal of Climatology*, **4**: 481–493.
Yarnal, B., 1985. A 500 mb synoptic climatology of Pacific northwest coast winters in relation to climatic variability, 1948–1949 to 1977–1978. *Journal of Climatology*, **5**: 237–252.
Yarnal, B., 1993. *Synoptic Climatology in Environmental Analysis*. London: Belhaven Press.

Cross-references

Atmospheric Circulation, Global
Climatology
Dynamic Climatology
Jet Streams
Satellite Observations of the Earth's Climate System
Teleconnections

T

TAIGA CLIMATE

Location and definition

The vast majority of the world's people live in the middle latitudes between 30° and 60°. Using the Köppen system of climate classification there are two major climate regions in the middle latitudes. The Mild Midlatitude climate is designated by the upper-case letter "C" and is exemplified by Mediterranean, Marine West Coast, and Humid Subtropical climates. The Severe Midlatitude climates are designated by the upper-case letter "D" and include the Humid Continental climate of New York and Chicago and the Sub-Arctic climate of Moscow and Montreal. The circumpolar sub-Arctic region generally coincides with 50–70°N latitude and includes much of Alaska, Canada's Northwest Territories, as well as northern Europe and Asia. The sub-Arctic climate region generally coincides with the boreal forest and is sometimes referred to as the "Taiga", the Russian word for forest (*Taiga*, 2000).

Climatic characteristics

High latitudes have large annual variations in daylength and solar input. Additionally, albedo varies widely from a maximum in winter when snow covers the surface to a minimum in the summer, especially in the dark green boreal forest. The continental climate of Edmonton in Alberta, Canada, has an annual temperature range of 31.4°C (Figure T1). The annual temperatures in the sub-Arctic, which can range from −54°C to 21°C, exhibit the greatest range of all climate regions (Kaplan, 1996). While the summers in the Taiga are relatively cool, temperature maxima can exceed 30°C (86°F). In Verkhoyansk, Siberia, annual temperature ranges have been recorded from −70°C to 30°C. Most of the year the Taiga climate is dominated by cold, Arctic air, and this realm serves as the source region for continental Arctic airmasses.

The average annual precipitation total in the Taiga is only 12–50 cm (5–20 in), yet low evaporation rates result in a moist climate (Eugster et al., 2000). As the climograph for Edmonton shows (Figure T1), the precipitation has a marked annual cycle with the maximum amount falling as rain in the summer. Winter precipitation is in the form of snowfall and the amounts are very small; however, snow remains on the ground because the temperatures are so low (Strahler and Strahler, 2002). The mean monthly temperature in the Taiga can remain below freezing for up to 7 months, and there are only 50–100 frost-free days. Therefore, winters are long and harsh and there is a short, cool summer punctuated by a very brief fall and spring (Quayle, 1999).

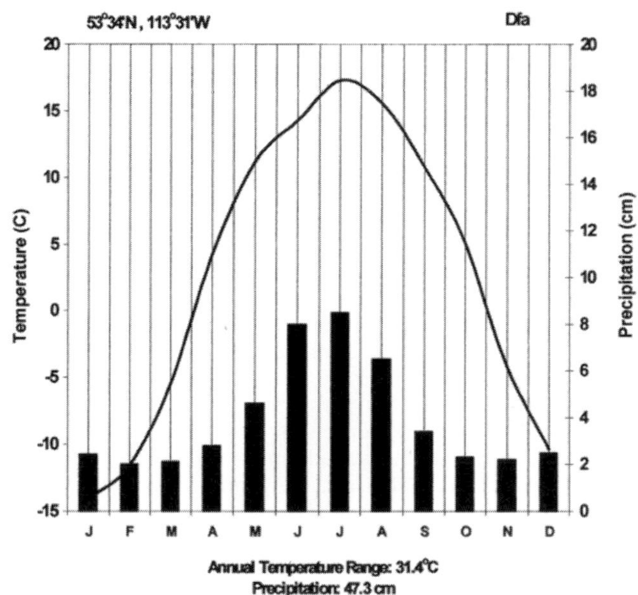

Figure T1 Climograph of Edmonton, Canada (after Eugster et al., 2000).

Flora and fauna

The boreal forest constitutes a nearly unbroken band of coniferous trees typifying the dominant vegetation of the sub-Arctic climate zone and extends across the higher latitudes of North

America and Eurasia (Figure T2). Specifically, the Taiga refers to the more open-canopy form of the boreal forest found in the northern reaches of this biome, or in the transitional ecotone between the dense coniferous forest and the frigid Tundra (Eugster et al., 2000).

In Europe and North America the boreal forest is composed mostly of evergreen species in three main genera: pine (*Pinus*), spruce (*Picea*), and fir (*Abies*). Broadleaf deciduous trees include the alder (*Alnus*), birch (*Betula*), and golden patches of aspen (*Populus*) dot the landscape in the fall (Strahler and Strahler, 2002). In northern North America widely spaced black spruce trees are common. In Siberia the deciduous larch or tamarack (*Larix*) is found in this climate zone, and in western Russia and across Scandinavia the Scots Pine is a widespread species. These forests also extend southward along mountain ranges (Oliver and Hidore, 2002). The giant sequoias are the largest living things and the Taiga is the largest biome on Earth.

Several characteristics of conifers demonstrate their ability to maximize a short growing season and winter drought conditions. The narrow needleleaf has a relatively small surface area to reduce transpiration during winter months when the ground is frozen. These needle-shaped leaves also have a thick, waxy cuticle into which the stomata are sunken, thereby escaping desiccating winds. The dark color and low albedo of conifers enables enhanced absorption of solar radiation. By retaining their foliage, evergreens can photosynthesize as soon as the short growing season begins, and the conical shape of confers allows snow to shed easily (*Taiga or Boreal Forest*, 2003).

Animals of the Taiga include the snowshoe hare, whose heavy coat of fur changes colors to blend into the background of the season. Mice and moles burrow beneath the snow in tunnels to escape the fierce winds of the sub-Arctic. Many species of birds, such as the insect-eating wood warbler, migrate to escape the harsh winter. Some birds survive year-round, such as the seed-eating finches and sparrows, and the omnivorous ravens. Other animals of the Taiga are the lynx, lemming, bobcat, bear, badger, beaver, fox, red squirrel, elk or wapiti (referred to as the red deer in Europe) and moose (known as elk in Europe) (*Taiga*, 1996). The many members of the weasel family include the fisher, pine martin, mink, ermine, sable, and wolverine.

Soils

The soils of the Taiga are of the order spodosol and are generally nutrient-poor. As a result of the acidification of the soil beneath needleleaf trees, podzolization occurs. These edaphic conditions contribute to patchy vegetation. For example, the larch forests survive in waterlogged substrate underlain by permafrost with shrubby undergrowth. The jack pine (*Pinus banksiana*) forests of North America occur on sandy outwash plains, and numerous bogs result from the poorly drained soils overlying glacial depressions. In the bogs are plant species that typify the tundra, such as mosses, cotton grass, and members of the heath family.

The great ice sheets of the most recent Ice Age scraped and shaped the landscape beneath the Taiga. Flowing ice contributed to severe erosion and exposed bedrock across vast areas. In many locations glacial debris blankets the surface. The boggy areas formed in shallow rock basins that were infilled with organic materials such as peat. The impermeable bedrock below does not permit percolation and saturated, mucky soils result.

While the growing season of the Taiga is short, there are a limited number of crops grown. In Finland, Sweden, and lands surrounding the Baltic Sea, crops include wheat, barley, rye, and oats. In conjunction with dairying, these crops provide subsistence. In the sub-Arctic areas of eastern Canada the major (nonmineral) export is lumber and pulpwood from the coniferous forests. Similarly, in Sweden, Finland, and European Russia, pine and fir are the primary plant resources as they are exported in the form of pulp, paper, cellulose, and lumber for construction.

Concerns

In recent decades public concern for the integrity of the northern needleleaf forest in conjunction with key international conferences led to a greater degree of governmental commitment to sustainable forest management. There is an enhanced awareness of the boreal forest ecosystem as a crucial component of biological diversity particularly, as well as of sustainable development in general. Also of concern is the rate of climate change that is forecast for the boreal forest. The rate of increasing temperatures that is predicted to occur over the next century is ten or more times faster than the warming that took place during the past 10000 years (*New York Times*, 1997; IPCC, 1992; Korpilahti, 2002).

Because climate and vegetation are inextricably linked, such rapid changes in climate have been shown to affect plant distributions and alter compositions of natural communities (Graham and Grimm, 1990). Many species respond individually, rather than as communities, to climatic change. There are unique species such as the bristlecone pine (*Pinus longaeva*) and the

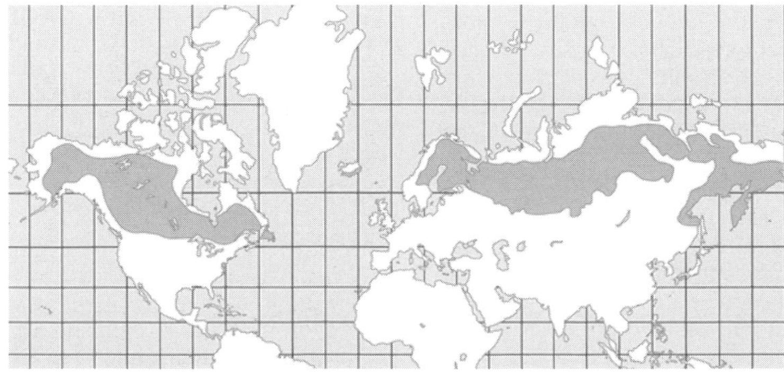

Figure T2 The taiga, cartographer T. Truckenbrod.

giant sequoia (*Sequoiadendron giganteum*) that have maintained their present locations for thousands of years, demonstrating substantial physiological tolerance to climatic variations (Kutner and Morse, 1996). However, a temperature difference of a few degrees, or slight variations in rainfall patterns, can determine whether some species survive or become extinct.

Evidence from fossil pollen records reveals migration rates of various tree species since the end of the last glacial period. According to a study by the Environmental Protection Agency (1989), beech and maple trees migrate at a rate of 10–20 km per century, hemlock migrate 20–25 km per century, and pine and oak species migrate 30–40 km per century. However, research (Schwartz, 1992) suggests that during the next century plant species may be required to shift as much as 500 km, which is well beyond the migration rates of many species.

Those species that cannot tolerate rapid climate change, and that have the ability to migrate, will likely do so, resulting in a number of new associations. In addition to differences in migration propensity and rates, community types will be altered due to changed disturbance regimes and competition as well (Davis, 1989). As atmospheric scientists are only beginning to piece together the many variables and mechanisms that are involved in climate change, the myriad repercussions of such rapid change also are just beginning to be revealed.

Mary Snow

Bibliography

Davis, M.B., 1989. Insights from paleoecology on global change. *Ecological Society of America Bulletin*, **70**: 222–228.

Environmental Protection Agency, 1989. *The Potential Effects of Global Climate Change on the United States*. Washington, DC: EPA.

Eugster, W., Rouse, W.R., Pielke, R.A., et al., 2000. Land–atmosphere energy exchange in Arctic tundra and boreal forest: available data and feedbacks to climate. *Global Change Biology*, **6**: 84–115.

Graham, R.W., and Grimm, E.C., 1990. Effects of global climate change on the patterns of terrestrial biological communities. *Trends in Ecology and Evolution*, **5**: 289–292.

Intergovernmental Panel on Climate Change, 1992. *Climate Change 1992: the supplementary report to the IPCC scientific assessment*. Houghton, J.T., Callender, B.A., and Barney, S.K., eds. New York: Cambridge University Press.

Kaplan, E., 1996. *Taiga*. Tarrytown, NY: Marshall Cavendish.

Korpilahti, E., 2002. *Climate Change, Biodiversity, and Boreal Ecosystems*. IBFRA Conference, Joensuu, Finland, 30 July–5 August 1995.

Kutner, L.S., and Morse, L.E., 1996. Reintroduction in a changing climate. In Falk, D.A., Millar, C.I., Olwell, M., eds., *Restoring Diversity: strategies for reintroduction of endangered plants*. Washington, DC: Island Press, pp. 33–37.

New York Times, 1997. Excerpts from report on warming's impact. *New York Times*, 1 December, p. 4.

Oliver, J.E., and Hidore, J.J., 2002. *Climatology: an atmospheric science*, 2nd edn. Upper Saddle River, NJ: Prentice-Hall.

Quayle, L., 1999. *Weather*. New York: Crescent Books.

Schwartz, M.W., 1992. Modeling effects of habitat fragmentation on the ability of trees to respond to climatic warming. *Biodiversity and Conservation*, **2**: 51–61.

Strahler, A., and Strahler, A., 2002. *Physical Geography: science and systems of the human environment*, 2nd edn. New York: Wiley.

Taiga, 2000. Retrieved 19 May 2003, from http://www.blueplanetbiomes.org/taiga.htm *Taiga or Boreal Forest*, 1996. Retrieved 19 May 2003, from http://www.runet.edu/~swoodwar/CLASSES/GEO235/biomes/taiga/taiga.html

Cross-references

Arctic Climates
Asia, Climates of Siberia, Central and East Asia
North America, Climate of
Tundra Climate
Vegetation and Climate

TELECONNECTIONS

"Teleconnection" refers to the tendency for atmospheric circulation patterns and their associated weather conditions to vary directly or indirectly over large and spatially discontiguous areas. Teleconnections are characterized by well-defined spatial patterns that can persist over long periods of time. Because of the persistent and recurring nature of teleconnections, they are sometimes called "modes of low-frequency circulation variability".

In the extratropics the ability of large-scale atmospheric circulation features to be teleconnected over large distances is related to the wave structure of the westerlies. At any given time the middle tropospheric flow in the westerlies can be characterized by a series of transient shortwave features superimposed upon a set of much larger quasistationary Rossby waves. The ridge and trough structure of the Rossby waves creates conditions through which geopotential heights and associated surface pressure patterns vary directly or indirectly within different parts of the wave train. Variations in the position or amplitude in one portion of a Rossby wave are teleconnected to other portions of the wave train. The quasistationary nature of Rossby waves and the tendency for ridges and troughs to be anchored in particular locations contribute to the spatial persistence of patterns.

Extratropical teleconnections

Patterns of teleconnection are often explored through the construction of simple one-point linear correlation maps or through the use of empirical orthogonal functions, such as principal components analysis (Barry and Carleton, 2001). For example, Figure T3 shows a map of the correlations between January 500-hPa geopotential heights at 55°N, 115°W and 500-hPa heights over much of the northern hemisphere. Most noteworthy in Figure T3 is the strong inverse relationships that exist between January 500-hPa geopotential heights over western North America and those over eastern North America and the north Pacific Ocean. This particular teleconnection is called the Pacific–North America (PNA) pattern (Wallace and Gutzler, 1981; Barnston and Livezey, 1987; Yarnal and Leathers, 1988). In its positive mode the PNA pattern is associated with a Rossby wave train that features amplified ridging over western North America and deep troughing over eastern North America and the north Pacific. A negative PNA is characterized by a de-amplification or reversal of these features.

The PNA pattern represents one of many extratropical teleconnections that have been identified for the northern hemisphere. In the PNA example the centers of teleconnection are aligned in a west-to-east pattern and are associated with meridional variations in the Rossby wave structure. By contrast, other patterns are characterized north–south-oriented features. The North Atlantic Oscillation (NAO) is an example of a north–south teleconnection pattern and is characterized by inversely varying pressures and geopototential heights between the regions surrounding Iceland and the Azores (Rogers, 1984).

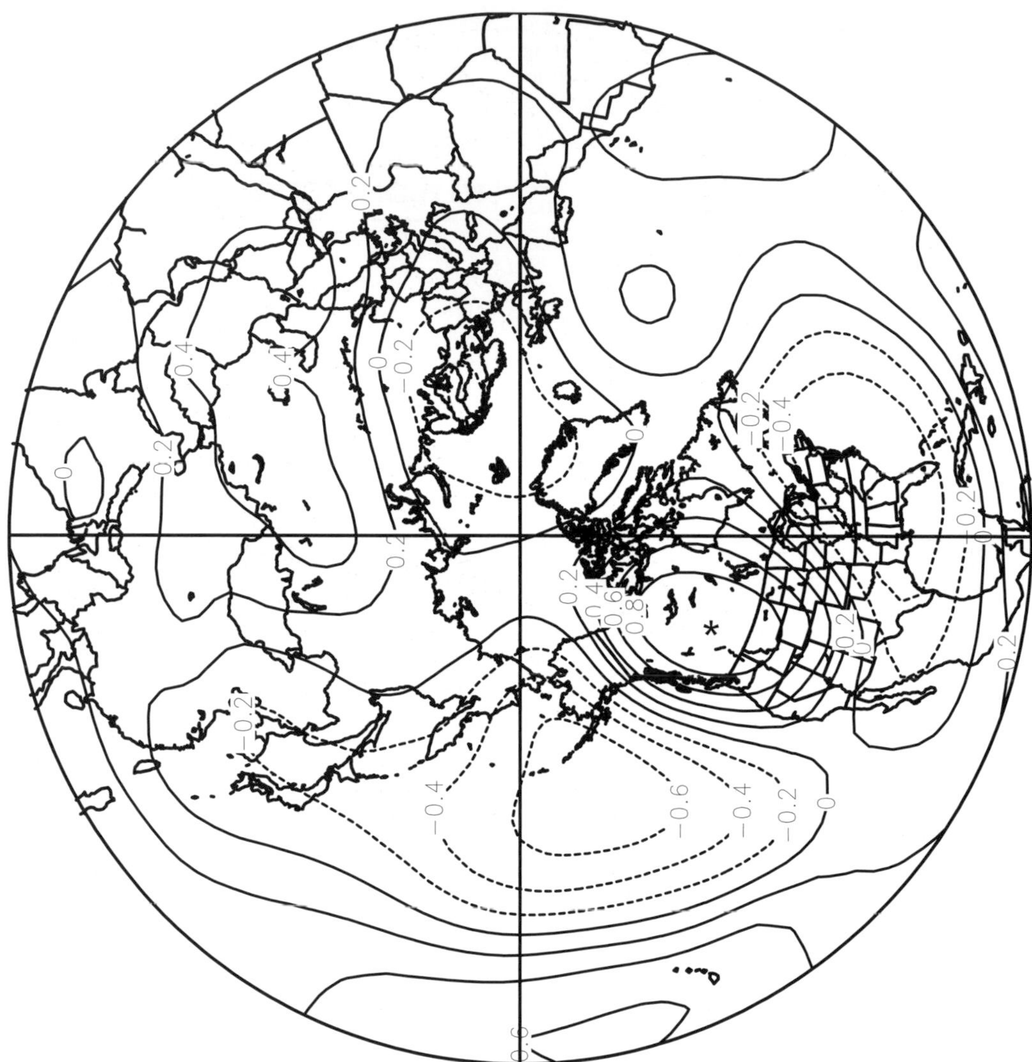

Figure T3 One-point linear correlations between average January 500-hPa geopotential heights at 55°N, 115°W (shown by an asterisk in North America) and January 500-hPa heights for the northern hemisphere (1949–2002). Data were obtained from the National Centers for Environmental Prediction/National Center for Atmospheric Research (NCEP/NCAR) reanalysis dataset.

Figure T4, which shows a map of correlations between January 500-hPa heights near Iceland and locations over the northern hemisphere, highlights the NAO teleconnection. Additional information on the PNA and NAO, including a description of all major northern hemisphere extratropical teleconnections can be found on the United States Climate Prediction Center (CPC) website (http://www.cpc.ncep. noaa.gov/). The CPC teleconnections are based on a rotated principal components analysis (RPCA) of 700-hPa geopotential heights, similar to that used by Barnston and Livezey (1987).

Temporal variability

The use of RPCA to examine teleconnections also yields important insight into the temporal variation of the major patterns. Certain patterns emerge as leading modes of circulation during particular months. For example, the CPC reports that the PNA pattern is the leading mode of northern hemisphere 700-hPa circulation during January. By contrast, the PNA is not among the leading modes of variability during June and July (http://www.cpc.ncep.noaa.gov/). The NAO is most dominant during March, April, October, and November.

Interannual variations in the strength of teleconnection patterns can be determined through the use of teleconnection indices. Different techniques have been used to create these indices. Yarnal and Leathers (1988) used standardized 700-hPa geopotential heights at three locations nearest the wave centers of the PNA pattern to develop a PNA index. Rogers (1984) calculated the difference in mean sea-level pressure between the Azores and Iceland to characterize the NAO. An alternative method of index development uses RPCA to statistically isolate the teleconnections and to calculate factor scores for each pattern (Barnston and Livezey, 1987). These scores provide an amplitude time-series for each pattern. Figure T5 shows the factor score time-series for January NAO (http://www.cpc.ncep. noaa.gov/). As can be seen in Figure T5, the NAO index exhibits a considerable

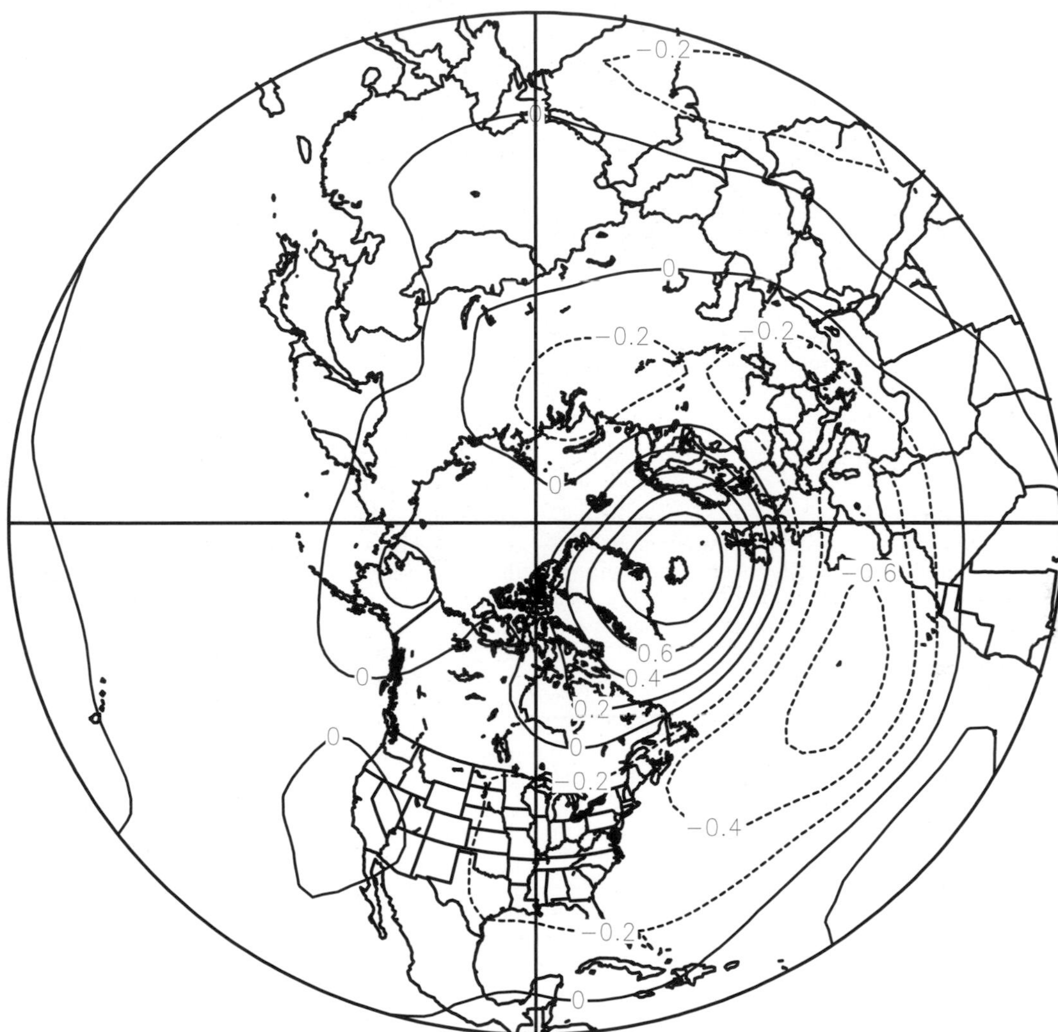

Figure T4 One-point linear correlations between average January 500-hPa geopotential heights near Iceland (shown by an asterisk) and January 500-hPa heights for the northern hemisphere (1949–2002). Data were obtained from the National Centers for Environmental Prediction/National Center for Atmospheric Research (NCEP/NCAR) reanalysis dataset.

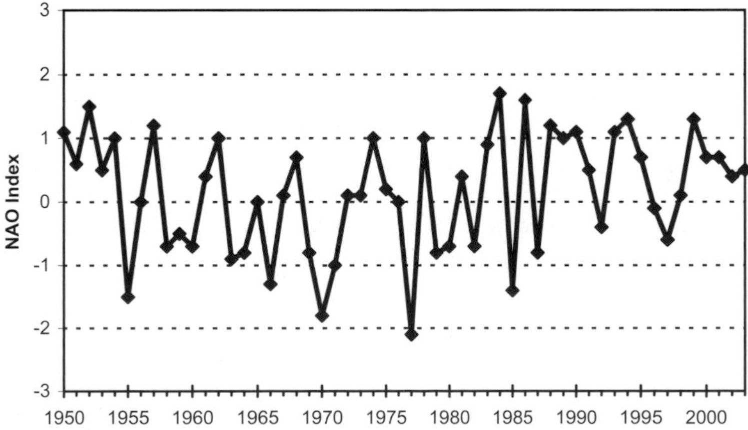

Figure T5 January NAO index for 1950–2003 as derived by the Climate Prediction Center using a rotated principal components analysis of 700-hPa geopotential height data.

amount of interannual variability. However, also apparent in the NAO record is a first-order trend toward a more positive NAO in recent decades. Temporal variability in the NAO, as with other teleconnections, exerts a strong influence on regional surface climate via its impact on airmass advective frequencies and storm track (Hurrell, 1995, 1996; Hurrell and van Loon, 1997).

Teleconnective processes

As described by Harman (1991), teleconnected wave trains result from the tendency for air flow to follow trajectories that conserve absolute vorticity. Under conditions of non-divergent flow, absolute vorticity is conserved through compensating changes in Earth vorticity (f) and relative vorticity (ζ_R). This relationship is captured in equation (1):

$$\frac{d}{dt}(\zeta_R + f) = 0 \quad (1)$$

The relative vorticity term can be broken into vorticity associated with wind shear (ζ_s) and that caused by flow curvature (ζ_k). Shear vorticity contributions are typically much smaller in spatial extent than that caused by flow curvature. As such, ζ_s can be disregarded. The conservation of absolute vorticity relationship can now be expressed by equation (2):

$$\frac{d}{dt}(\zeta_k + f) = 0 \quad (2)$$

Variations in Earth vorticity (f) are driven by latitudinal changes in Coriolis; therefore, any change in the latitude of the flow will produce a change in f and a compensating change in ζ_k. Figure T6 shows a simplified example of northern hemisphere circulation in which the flow is initially moving poleward. Under such conditions, f will increase and require a decrease in ζ_k. As ζ_k decreases, curvature will become increasingly anticyclonic and ultimately produce a ridge axis and equatorward flow. As flow moves south, f will decrease and force a compensating increase in ζ_k and the development of a trough. This relationship, which is referred to as the conservation of absolute vorticity trajectory (CAVT), yields a series of teleconnected waves.

Forcing mechanisms

A full discussion of the mechanisms through which teleconnected circulation patterns are forced is beyond the scope of this discussion. However, central among these mechanisms are the influences of diabatic heating and orography (see Barry and Carleton (2001) for a detailed description of these processes). A central component of orographic forcing involves the vertical compression of the air column as it moves over a mountain barrier. As described by Harman (1991), this vertical compression is compensated by a horizontal expansion of the column and a decrease in absolute vorticity. This decrease in vorticity occurs primarily through changes in ζ_k and a redirection of the flow. As the air flows past the orographic barrier, vertical expansion and horizontal contraction produce an increase in ζ_k. In this fashion orographic features can create geographically fixed waves that are then teleconnected downstream via CAVT processes.

The role of diabatic heating in planetary wave forcing involves a number of processes, including sensible and latent heat exchanges and radiative transfer (Barry and Carleton, 2001). For example, during winter the sea surface temperatures off the east coasts of Asia and North America are significantly warmer than the adjacent land areas. Cold airmasses flowing eastward over this warm sea surface are destabilized. The resultant instability produces vertical air motion, horizontal compression of the air column, and increased ζ_k. Changes in ζ_k, associated flow directional change and CAVT can yield a wave train geographically fixed to the surface heat source. Forcing of this type tends to vary seasonally as the strength and location of the diabatic heat source changes. Trenberth et al. (1998) provide a thorough review of the role that tropical sea surface temperature (SST) variability, such as that associated with El Niño/Southern Oscillation, plays in driving extratropical teleconnections. Of the dominant midlatitude teleconnection patterns, only a few exhibit strong linkages with tropical SST variability. Of these, the PNA pattern appears particularly sensitive to SST variations in the tropical Pacific. Ultimately, these tropical–extratropical connections rely on Rossby wave propagation. However, this process can be complicated by midlatitude influences, thus clouding the exact mechanisms through which extratropical modes of circulation are driven by tropical SST variation.

Adam Burnett

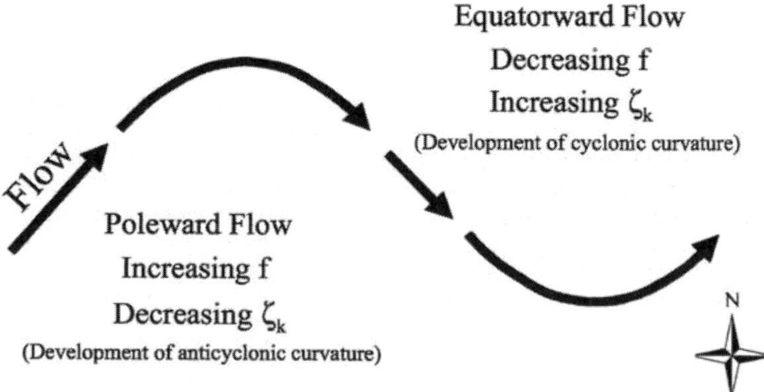

Figure T6 Simplified diagram showing the conservation of absolute vorticity trajectory. Changes in earth vorticity (f) are compensated for by changes in curvature vorticity (ζ_k).

Bibliography

Barnston, A.G., and Livezey, R.E., 1987. Classification, seasonality and persistence of low-frequency atmospheric circulation patterns. *Monthly Weather Review*, **115**: 1083–1126.

Barry, R.G., and Carleton, A.M., 2001. *Synoptic and Dynamic Climatology*. London: Routledge.

Harman, J.R., 1991. *Synoptic Climatology of the Westerlies: process and pattern*. Washington, DC: Association of American Geographers.

Hurrell, J.W., 1995. Decadal trends in the North Atlantic Oscillation: regional temperatures and precipitation. *Science*, **296**: 676–679.

Hurrell, J.W., 1996. Influence of variations in extratropical wintertime teleconnections on Northern Hemisphere temperatures. *Geophysical Research Letters*, **23**: 665–668.

Hurrell, J.W., and van Loon, H., 1997. Decadal variations in climate associated with the North Atlantic Oscillation. *Climatic Change*, **36**: 301–326.

Rogers, J., 1984. The association between the North Atlantic Oscillation and the Southern Oscillation in the Northern Hemisphere. *Monthly Weather Review*, **112**: 1999–2015.

Trenberth, K.E., Branstator, G.W., Karoly, D., Kumar, A., Lau, N.-C., and Ropelewski, C., 1998. Progress during TOGA in understanding and modeling global teleconnections associated with tropical sea surface temperatures. *Journal of Geophysical Research*, **103**(C7): 14,291–14,324.

Wallace, J.M., and Gutzler, D.S., 1981. Teleconnections in the 500 mb geopotential height field during the Northern Hemisphere winter. *Monthly Weather Review*, **109**: 784–812.

Yarnal, B., and Leathers, D.J., 1988. Relationships between interdecadal and interannual climatic variations and their effect on Pennsylvania climate. *Annals of the Association of American Geographers*, **78**: 624–641.

Cross-references

Cycles and Periodicities
El Niño
North Atlantic Oscillation
Ocean–Atmosphere Interaction
Oscillations
Vorticity

TEMPERATURE DISTRIBUTION

Each day Earth receives energy in the form of incoming solar radiation from the sun. This shortwave solar radiation ranges mostly from ultraviolet (0.2 μm wavelength) to the near-infrared (2.0 μm wavelength), but reaches its maximum at around 0.5 μm wavelength (blue–green visible light). This insolation is absorbed by Earth's surface and is converted to heat (longwave radiation). Earth's (terrestrial) longwave radiation reaches its peak intensity at the 10 μm wavelength (thermal infrared) and is responsible for heating the lower atmosphere.

Temperature is represented by a human-invented quantitative measure of heat energy emitted by or contained within a surface or material. As such, these numerical temperature values can be used to differentiate one climatic region from another. Spatial variations in temperatures (i.e. temperature distributions) occur both vertically and horizontally within the Earth's atmospheric envelope. These temperature distributions, their causes and variations, are discussed below.

Vertical distributions

The atmosphere can be divided into four distinct layers based on characteristic temperature distributions (see Figure T7). The

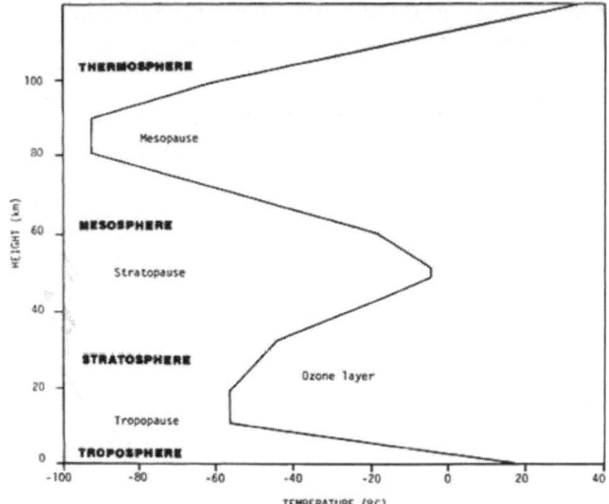

Figure T7 Vertical temperature structure of the atmosphere.

layer closest to the Earth's surface is called the *troposphere*. Almost all atmospheric water vapor and turbulent mixing exist within this layer, thus it is here that virtually all phenomena related to weather or climate take place. Temperatures tend to decrease with increasing altitude above Earth's surface (the source of terrestrial heat radiation). Cold temperatures observed at high elevations and snowcapped mountains are physical manifestations of this tropospheric temperature trend. On the average, temperatures in the troposphere will decrease at the *normal* or *environmental lapse rate* of 6.5°C/1000 m of rise.

The troposphere ends at a boundary layer called the *tropopause*. The tropopause is at its highest elevation over the tropics (approximately 16 km) but decreases in elevation poleward (9 km or less at the poles; Lutgens and Tarbuck, 2001). Temperatures in the tropopause remain constant with height and this layer is *isothermal* (same temperature) in nature.

Temperatures tend to increase with height within the *stratosphere*. This increasing temperature trend is caused by heat released during the interaction between incoming ultraviolet radiation and the ozone (O_3) layer (ozonosphere) nested within the stratosphere. It is the formation, the destruction (into molecular and atomic oxygen, O_2 and O, respectively), and the reformation ($O_2 + O \rightarrow O_3$) of ozone that shields the Earth from the harmful effects of ultraviolet radiation and, at the same time, adds heat energy to the stratosphere. The upper limit of the stratosphere is bounded by another isothermal layer called the *stratopause*.

Above the stratopause lies the next layer of the atmosphere called the *mesosphere*. The temperatures in this atmospheric shell decrease with an increase in altitude. The mesosphere (middle sphere) ascends to an elevation of around 80 km where it ends at the last isothermal layer called the *mesopause*.

Extending from the mesopause to the outer limits of the atmosphere lies the final layer, the *thermosphere*, which accounts for a minute fraction of the atmosphere's total mass. The particles (mostly ions) in this uppermost layer are energized by incoming solar radiation and move at very fast speeds, thus causing heat sensors to register an increasing temperature with height.

Factors influencing the horizontal distribution

Though it has been shown that temperatures vary vertically from the surface of the planet to the outer limits of the atmosphere, horizontal temperature distributions and the mechanisms that create them are of greater importance to climatologists and meteorologists. The causative factors and resulting temperature distributions on the surface of Earth are discussed in further detail.

Latitude and sun angle

Latitude is the single most important factor determining planetary temperature distributions. In general, as one traverses from the equator (0° latitude) to either pole (90° latitude) temperatures decrease. This equator-to-pole temperature gradient is directly related to the angle at which the sun's rays strike Earth's surface.

Solar radiation, having traveled some 150 million km, reaches Earth's spheroid surface in essentially parallel lines (see Figure T8). As such, there can be only one latitude where the sun's rays intersect the surface at a direct (90°) angle. A displacement north or south from this line of direct (maximum) radiation must result in an incident sun angle less than 90°; thus the intensity of the radiation decreases (Figure T9).

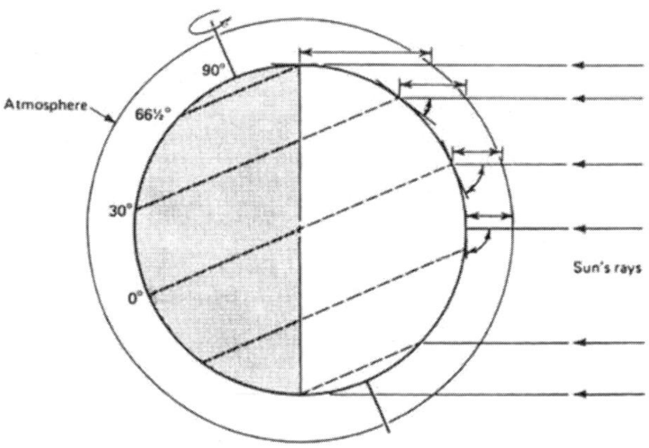

Figure T8 The Earth in the December solstice position. Parallel sun rays strike the Earth at a variety of sun angles. Rays striking the Earth at low angles, must traverse more of the atmosphere than rays striking at a high angle, and are thus subject to greater depletion by reflection and absorption (Lutgens and Tarbuck, 2001).

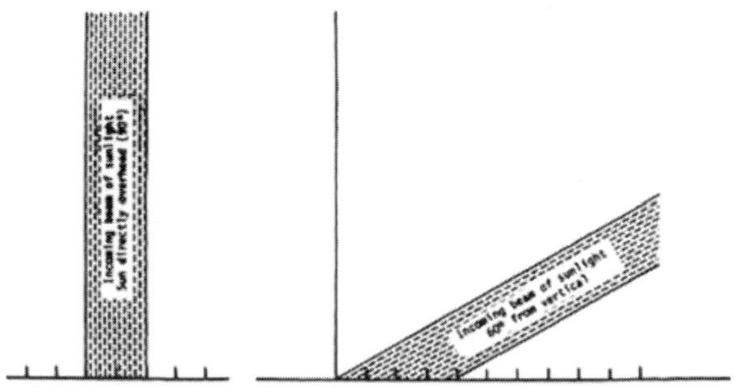

Figure T9 The sun rays intersecting Earth's surface at 60° (right) will cover twice the ground surface but with only half the intensity of a direct 90° angle (left).

The relationship between the intensity of solar radiation and sun angle is described by *Lambert's Cosine Law*: $I = I_0$ (cos γ), where I = radiant intensity of radiation reaching some point on the surface, I_0 = radiant intensity at maximum (where sun angle is 90° overhead) and γ = angle from the vertical to the direction of the radiation (zenith angle; Rosenberg, 1974). For example, when γ equals 60° from vertical, the intensity of solar radiation is half of its maximum (cos 60° = 0.5000; Figure T9).

The thickness of the atmosphere through which insolation must pass is also affected by sun angle. Low-latitude high-sun angles pass through a shorter distance of atmosphere as opposed to high-latitude oblique sun angles, which must traverse a much greater distance (see Figure T8). By traveling through a greater distance of atmospheric gases, the intensity of the solar radiation is partially extinguished by the gas molecules. This relationship is best described by *Beer's Law*:

$$I = I_0 e^{-ax}$$

where I_0 is the initial radiant intensity and I is the radiant intensity after passing through a depth x of a medium (i.e. the atmosphere) with extinction coefficient a. Extinction of radiant intensity can occur by absorption and scattering of the solar radiation (Rosenberg, 1974). Again, the low sun angles in polar regions are less intense than the tropical high sun angles.

Seasons

The migration of the maximum sun angle north and south of the equator is responsible for seasonal variations on the Earth. On 21–22 December the most direct (90°) sun angle contacts 23½°S latitude (the Tropic of Capricorn). In this orbital position – the *December Solstice* – the northern hemisphere is tilted its farthest back (23½°) from the sun and experiences the winter season. This represents the shortest day of the year in the northern hemisphere, and the North Pole is in total darkness (see Figure T8). In the southern hemisphere this is the longest day of the year, the first day of summer and the sun is visible above horizon 24 hours a day at the South Pole. During 21–22 June, radiation strikes directly upon 23½°N latitude (Tropic of Cancer). The *June Solstice* marks the first day of summer in the northern hemisphere and winter in the southern hemisphere.

The vernal and autumnal equinoxes (21–22 March and September, respectively) are the days when the sun's rays hit directly on the equator. Earth experiences 12 hours of daylight and 12 hours of darkness on these equinox dates.

During the course of a year the direct and most intense rays of the sun will migrate from 23½°N to 23½°S and back again. For this reason the tropics are always warm, energy-rich areas of the world, whereas the poles are always cold, energy-poor regions of the world. Again, the angle of incoming solar radiation will cause an equator to pole (latitude-dependent) temperature gradient found on the Earth.

Air mass circulation

As stated above, incident sun angles cause the tropics to be energy-rich, and the poles energy-deficient. This imbalance of energy has not, however, caused the tropics to become warmer nor the poles to become colder through time. The tropics and the poles maintain their climatic characteristics by exchanging energy, mass, and momentum through the middle latitudes (23½° to 66½° N or S). The weather and climates in the middle latitudes are variable in nature due to the invasion of large air masses poleward from the tropics and equatorward from Arctic and Antarctic regions.

An *air mass* is a large body of air measuring hundreds to thousands of kilometers in length and breadth, and may extend from the surface to the tropopause. Air masses have definable characteristics of temperature and humidity derived from their *source regions*; that is, the region from which the air mass originates.

The circulation of these air masses, and more importantly the frequency with which an air mass type dominates an area, will affect the temperature distribution and the climate of that area (Oliver, 1970). For example, if the midwestern United States experiences invasions of continental arctic (CA) air from northern Canada during a winter season, it may become a record cold winter. In contrast, if that same region during the winter is frequented by maritime tropical (mT) air from the Gulf of Mexico, a mild season would result. Air mass movements are variable and their variations are reflected in the temperature distributions.

Cloud cover

The effects of cloud cover on temperature are most evident in the diurnal temperature cycle (Figure T10). Cloud cover during daylight hours prevents some incoming solar radiation from reaching Earth's surface by reflecting it back to space. Thus temperatures may be cooler on cloudy days. Cloud cover at night causes overnight low temperatures to be warmer. The clouds act like a blanket, keeping longwave terrestrial (heat) radiation close to Earth's surface. This heat radiation would otherwise be lost to space by escaping through a cloudless sky.

Excessive cloud cover may cause variances in annual temperature cycles as well. Monsoonal regions are particularly susceptible. During the wet monsoon, long periods of heavy cloud cover will cause temperature values to be depressed. Temperature data for Calcutta, India (Table T1), shows April, May, and June to have the highest temperatures, whereas July and August (considered months with the highest temperatures) are cooler beneath the monsoonal cloud cover (beginning in June and extending through August).

On a global scale, areas of persistent cloud cover or clear skies will cause alterations in temperature distributions. The equatorial region of the Earth is one with abundant cloud cover. The perennial high sun angles of the tropics warm the moist equatorial air, causing it to become buoyant and rise (equatorial low pressure) and cool aloft to form ample cloud cover. By contrast, subtropical regions (30–35° N or S latitudes) are regions of stable, descending air (subtropical high pressure), which is conducive to the formation and maintenance of clear skies. Subtropical high-pressure systems and their associated clear skies are responsible for many of the world's large deserts (e.g. North Africa and Saudi Arabia).

Though the highest sun angles occur in the tropics, the clear skies and abundant solar radiation of the subtropics allow these regions to yield higher temperatures. Many all-time record high temperatures were observed at stations located between 30° and 40° latitude. Some examples are Azizia, Libya (57.8°C); Death Valley, California (56.7°C); Seville, Spain (50°C); and Tirat Tsvi, Israel (53.4°C) (National Climatic Data Center). From the microscale to the global scale, and within diurnal as well as annual temperature cycles, cloud cover remains a factor that affects temperature distributions.

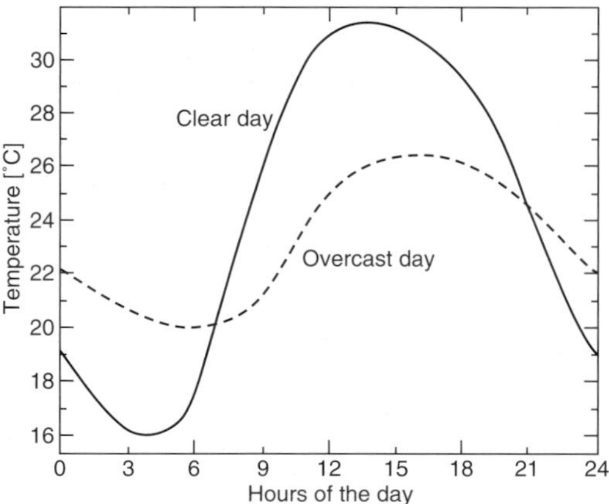

Figure T10 The daily temperature cycle on a clear day versus a cloudy day. Cloud cover will depress the daily maximum while raising the daily minimum temperature (after Lutgens and Tarbuck, 2001).

Distribution of land and sea

The surface of Earth is covered by a wide range of materials. The character of these surfaces will affect the way they accept solar radiation and emit terrestrial radiation. The largest division of surface materials lies between land and water. These two substances react differently to incoming and outgoing radiation. In general, land heats and cools considerably faster than water. This difference in the rate of their temperature gain or loss is caused primarily by their different specific heat values. *Specific heat* is the amount of heat necessary to raise the temperature of a material 1°C. Water has a specific heat of 1 calorie per gram °C, whereas most earthen materials range between 0.3 and 0.5 calories per gram °C. Thus land masses can heat and cool two to three times faster than water bodies.

Water is also transparent (or translucent) and accepts sunlight into its upper layers. In addition, water has the ability to circulate the heat energy. Unlike water, an opaque and rigid

Table T1 Mean monthly temperatures (°C), Calcutta, India

Month	J	F	M	A	M	J	J	A	S	O	N	D
	18	21	27	30	31	30	28	27	28	27	22	18

land surface reacts more quickly to heat energy exchanges with the atmosphere.

Surface temperature characteristics vary with distance from large bodies of water (e.g. large lakes or oceans). Proximity to a large body of water tends to moderate extreme temperatures. This effect is clearly demonstrated in annual temperature range value (i.e. the difference between the highest mean monthly temperature in summer and the lowest mean monthly temperature in winter). For example, choosing three stations of approximately equal latitude (32°N), we find Bermuda (surrounded by water) has a temperature range of 9–10°C. By comparison, Charleston, South Carolina (a coastal station), has an annual temperature range of 17°C. Further inland, Pine Bluff, Arkansas (a landlocked station), has an annual temperature range of 23–24°C (Tanner, 1964). Therefore, the *continentality* of a location (the degree to which the interior of a continent affects the climate) will have an influence on temperature distributions.

Albedo

Land surfaces are composed of many different materials, each with its own set of characteristics that may affect Earth's energy balance and temperature distributions. Albedo is one such characteristic and is defined as the amount of solar radiation reflected by a surface to the amount incident upon it; it is commonly expressed as a percentage (Huschke, 1959). Surfaces with high albedo (e.g. snow and ice) reflect 80–85% of the incoming solar radiation and express lower temperatures than surfaces of low albedo (e.g. green forest, albedo = 3–10%).

Cloud cover has the most variable albedo, which changes with the type and thickness of clouds (ranges from negligible values to 80%). In summary, the changing colors, textures, and substances of surface materials will determine albedo values and thus affect surface temperature distributions.

Ocean drifts and currents

Ocean water is transported by two major mechanisms: (1) frictional drag of prevailing winds on the ocean surface; and (2) differences in water density (density currents) with varying heat and salt concentrations. A slow, inconspicuous movement (3–4 km/h) over broad areas are called *drifts*, as opposed to *currents*, which are narrower, extend to greater depths and move faster (10 km/h).

If a drift or current originates in tropical (low-latitude) regions and migrates poleward, it will invade colder ocean waters and by comparison to its surroundings would appear as a warm current. Conversely, drifts and currents that stem from the higher latitudes and advect equatorward are perceived as cold (or cool) currents.

Where they exist, warm and cold currents will influence surface temperature distributions, especially along coastal regions. For instance, if we examine the mean annual temperatures of the two South American coastal locations at the same latitude, the effect of ocean currents is clearly visible. Lima, Peru, is under the dominance of the cold Peruvian current moving equatorward and has a mean annual temperature of 19°C, whereas Salvador, Brazil, under the influence of the poleward-flowing, warm Brazilian current, has a mean annual temperature of 26°C (National Climatic Data Center). Though all the atmosphere–ocean interrelationships are not fully understood, the consequences of ocean movements and climate (and more specifically temperature distributions) cannot be ignored.

Urban–rural temperatures

Surface temperature distributions tend to take on distinctive patterns in and around urban areas, designating what is called an urban heat island. In general, temperatures tend to increase as one traverses from the rural surroundings to the center of an urban area.

Observed horizontal distribution

The actual horizontal distribution of temperatures are shown in Figures T11 and T12. These maps depict mean global

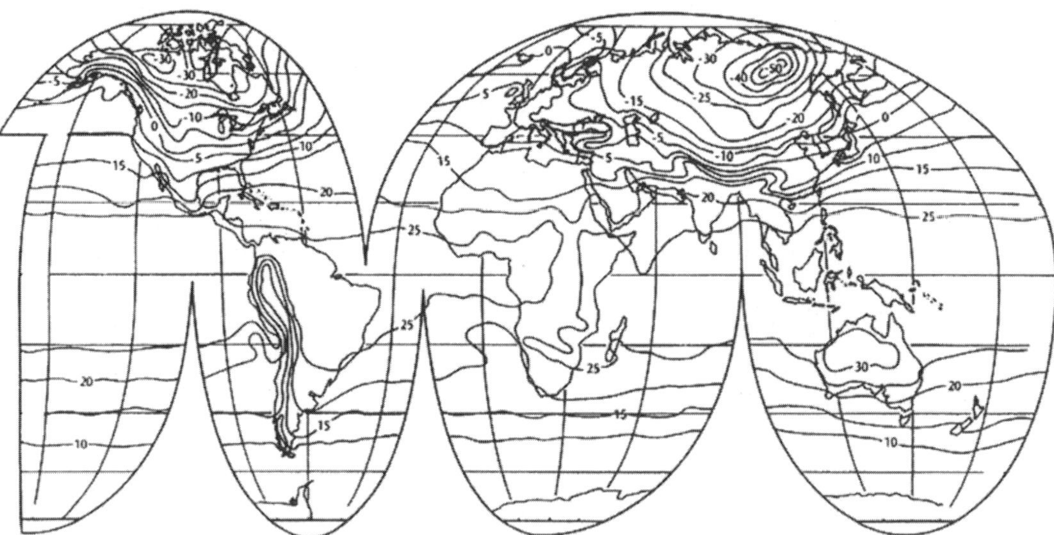

Figure T11 Map showing global distributions of mean January temperatures. The isotherm line patterns reflect some of the controls on surface temperatures.

isotherms (lines of equal temperature value) for January and July. The patterns of the isotherms tend to confirm some of the influencing factors previously mentioned.

Equator to pole gradient

The first noteworthy pattern can be seen on both figures (regardless of season). The east–west arrangement of isotherms mimics the orientation of latitude lines and reflects a changing sun angle. Here, the effects of Lambert's Cosine Law and Beer's Law work together to create an obvious equator-to-pole temperature gradient. The isotherms will deviate from their east–west alignment when passing from sea to land and back again. These disruptions in the continuity of the isotherms demonstrate the way land and sea accept solar radiation and emit terrestrial radiation. During the same season, and at comparable latitudes, land and adjacent sea may not maintain the same temperature.

Maximums and minimums

The maximum temperatures are observed in the high sun hemisphere (hemisphere that is experiencing its summer season).

Figure T12 Map showing global distributions of mean July temperatures. The isotherm line patterns reflect some of the controls on surface temperatures.

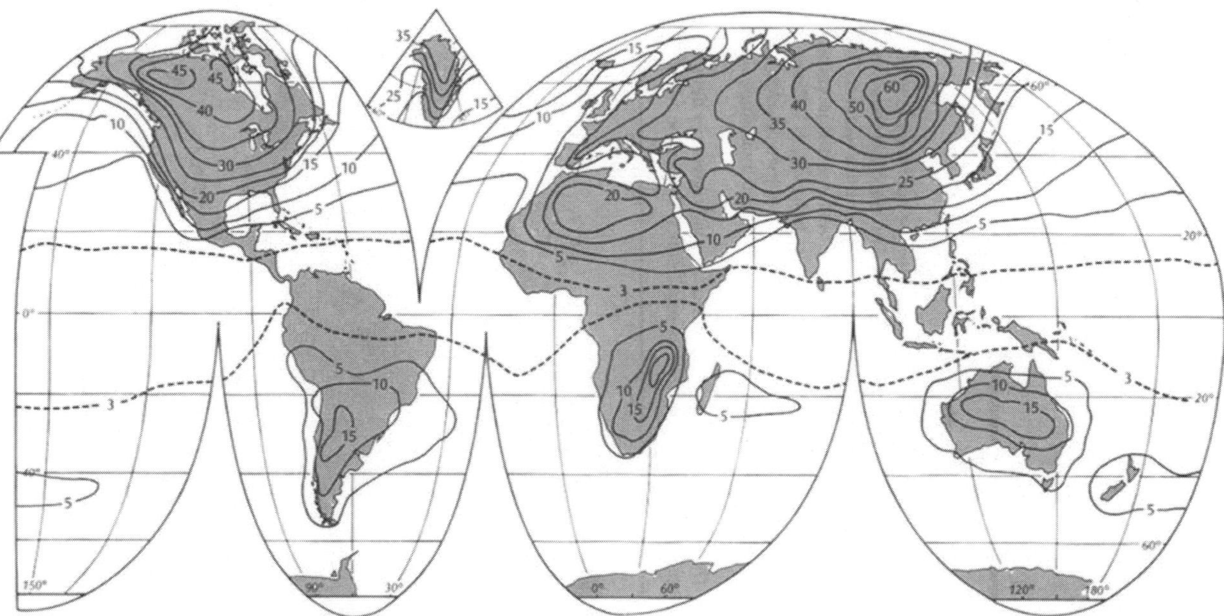

Figure T13 Map depicting global distributions of temperature range. The difference between mean January and mean July temperatures constitutes useful information to the climatologist.

This shifting of maximum temperatures from the northern hemisphere (in July) to the southern hemisphere (in January) follows the migration of the direct (90°) incident sun angle from the June solstice position (23½°N latitude) to the December solstice position (23½°S latitude). Further note that the maximum temperatures are found in the subtropical regions (approximately 30° N or S latitude) where clear skies dominate, and not in the tropics where a cloud cover prevails much of the time Finally, maximum temperature (e.g. over North Africa) and the minimum temperatures (e.g. over Antarctica and northern Siberia) are found over land bodies. Here again, a low specific heat value and other factors mentioned earlier allow land bodies to respond more rapidly to seasonal changes. This is further exemplified in Figure T13, which displays average annual temperature ranges (the difference in temperature between January and July averages). In this figure, small temperature ranges are found over the oceans where temperature changes are less dramatic and large temperature ranges (i.e. closed isotherms) are found over continental regions (effects of continentality).

Ranges and seasonality

The temperature range isoline map (Figure T13) is useful in demonstrating more than just the continentality of a particular region. Temperature ranges can further indicate the degree to which a region is affected by the seasonal changes. The small temperature range isolines that straddle the equator are showing that the tropics experience no appreciable seasonal changes. The highest sun angles within the tropics year-round; therefore the climate remains essentially the same. Near the poles, low sun angles and periods of total darkness account for seasonal variation. However, the middle-latitude temperature ranges are much higher (seasonal changes are more prominent). This is an area where sun angles, air masses, ocean currents, surface albedo, and cloud cover vary through time and space. With alterations in these and other factors come changes in temperature across the surface.

L.M. Trapasso

Bibliography

Ahrens C.D., 2005. *Essentials of Meteorology*, 4th edn. Pacific Grove, CA: Brooks/Cole.

Barry, R.G., and Chorley, R.J., 1998. *Atmosphere, Weather, and Climate*, 7th edn. London: Routledge.

Critchfield, H.J., 1983. *General Climatology*. Englewood Cliffs, NJ: Prentice-Hall.

Gabler, R.E., Petersen, J.F., and Trapasso, L.M., 2004. *Essentials of Physical Geography*, 7th edn. Pacific Grove, CA: Brooks-Cole.

Huschke (ed.), 1959. *Glossary of Meteorology*. Boston: American Meteorological Society.

Lutgens, F.K., and Tarbuck, E.J., 2001. *The Atmosphere*, 8th edn. Englewood Cliffs, NJ: Prentice-Hall.

National Climatic Data Center, Asheville, NC, www.noaa.gov/ncdc.html, <http://www.noaa.gov/ncdc.html>

Oliver, J.E., 1970. An air mass evaluation of the concept of continentality. *Professional Geographer*, **22**(2), 83–87.

Oliver, J.E., and Hidore, J.J., 2002. *Climatology: An Atmospheric Science*. Upper Saddle River, NJ: Prentice-Hall.

Robinson, P.J., and Henderson-Sellers, A., 1999. *Contemporary Climatology*, 2nd edn. Harlow: Longman.

Rosenberg, J.J., 1974. *Microclimate*. New York: Wiley.

Thompson, R.D., 1998. *Atmospheric Processes and Systems*. London: Routledge.

Trewartha, G.T., and Horn, L.H., 1980. *An Introduction to Climate*. New York: McGraw-Hill.

Cross-references

Energy Budget Climatology
Continental Climate and Continentality
Radiation Climatology
Seasons
Standard Atmosphere
Troposphere
Urban Climatology

TERRESTRIAL RADIATION

Electromagnetic radiation emitted by the Earth's surface and atmosphere is called *terrestrial* or *longwave* radiation (the latter being the preference of the World Meteorological Organization). Since these bodies exhibit temperatures in the range 200–300 K, terrestrial radiation lies in the infrared portion of the electromagnetic spectrum. At mean Earth surface temperature (288 K), nearly all the emission from a black body occurs between the wavelength limits of 4 and 100 μm, with peak monochromatic emittance at 10 μm.

The spectral distribution of emission from the Earth's surface generally approximates that from a black body. Emissivities for natural surfaces mainly lie in the range 0.90–0.97, although fresh snow exhibits very high values (>0.99) and smaller ones are associated with dry, mineral materials such as desert soils, dry rocks and building materials (0.7–0.9).

Both the amount and spectral distribution of emission by the atmosphere primarily depend on the vertical profiles of temperature and emitting gases and the presence, type, and location of clouds and aerosols. A cloudless atmosphere emits in the characteristic fashion of a gas, with prominent emission lines and bands. The major constituents of the Earth's atmosphere responsible for emission are water vapor (bands from 5–7 μm and beyond 15 μm), carbon dioxide (13–17 μm and at wavelengths <5 μm) and ozone (a strong, narrow band centered on 9.6 μm, with minor ones at 4.7 and 14.1 μm), although other gases (the so-called "greenhouse gases") also play a part (e.g. nitrous oxide, methane, carbon monoxide and chlorofluorocarbons). In the approximate wavelength range 8–13 μm (the *atmospheric window*) emission is small in the troposphere and may be attributable to water vapor *dimers*. Clouds radiate at all wavelengths, although only thick low clouds can be approximated as black bodies. Middle-layer clouds possess emissivities near 0.9, whereas cirriform clouds may have emissivities as low as 0.3. Emission of terrestrial radiation from the atmosphere is hence increased by the presence of clouds due to enhanced radiation in the window region. The role of aerosols is controversial but is probably significant in increasing emission in the window.

Following Kirchoff's Law, the atmosphere's constituents absorb terrestrial radiation in the same wavelength regions in which they emit.

At the Earth's surface, upward emission generally exceeds the irradiance from the atmosphere (the *counter-radiation*), although the difference is smaller with the presence of clouds or fog. On an annual, global basis, the counter-radiation is about 85% of the surface emittance.

Within the atmosphere, complex exchanges of terrestrial radiation occur, depending on the vertical distributions of temperature and emitting and/or absorbing constituents. Such exchanges generally result in radiative cooling of the air (up to approximately 3 K/day). The radiation emerging from the top of the atmosphere averages about 63% of that emitted by the surface.. Highest terrestrial radiation losses occur in the subtropics (where a warm surface is overlain by a relatively cloudless atmosphere), with lower values associated with the high, cold cloud tops of the *intertropical convergence zone* and with the colder regions of the high latitudes. Annually and globally averaged terrestrial radiation losses roughly balance absorbed *solar radiation* to maintain the planetary equilibrium climate.

A. John Arnfield

Bibliography

Liou, K.N., 2002. *An Introduction to Atmospheric Radiation.* New York: Academic Press.

Paltridge, G.W., and Platt, C.M.R., 1976. *Radiative Processes in Meteorology and Climatology.* New York: Elsevier.

Cross-references

Energy Budget Climatology
Greenhouse Effect and Greenhouse Gases
Intertropical Convergence Zone
Radiation Laws
Solar Radiation
Ultraviolet Radiation

THORNTHWAITE, CHARLES W. (1899–1963)

Charles Warren Thornthwaite was born in 1899, on a small farm outside of Pinconning, in central Michigan, the first of four children. His family realized that the only escape from the back-breaking farm work was through education, an understanding that drove Warren, as he preferred to be called, the rest of his life.

With no high school in Pinconning, Warren left home after grade school to attend high school and college in Mount Pleasant, Michigan. He graduated from Central Michigan Normal School (now Central Michigan University) in 1922 with a teacher's certificate and accepted a job teaching high school in Owosso, Michigan. For two summers he registered for geography courses at the University of Michigan and was then persuaded to enter into the geography graduate program at the University of California at Berkeley under the guidance of Carl Sauer.

Upon finishing his dissertation research in 1927, Thornthwaite accepted a teaching position in the Department of Geography at the University of Oklahoma where he realized that there was a need for a more rational understanding of the moisture factor in climate, long ignored by earlier climatologists. He immediately began work on what was to become one of his most significant contributions to climatology, the development of a new and more rational classification of world climates. His first attempts, published in 1931 and 1933, left Thornthwaite unsatisfied with his expression for the climatic demands for water and, over the next 11 years, he developed his concept of potential evapotranspiration (the water loss from a uniform, closed cover of vegetation that never suffers from a lack of water). This led to his expression for potential evapotranspiration in 1944 and its use, along with precipitation, in a climatic water budget in 1946. It served as the basis for his seminal paper in 1948 entitled "An approach toward a rational classification of climate".

Thornthwaite left Oklahoma in 1934 to work on a government-sponsored study of the Great Plains and, in the following year, to become Chief of the Climatic and Physiographic Division of the Soil Conservation Service in the Department of Agriculture, a position he held for the next 11 years.

Thornthwaite took a leave of absence from government service in 1946 to explore opportunities as a consulting agricultural climatologist at Seabrook Farms in southern New Jersey. During the next two years he completed his work on the climatic water budget and his new climatic classification, developed a significant improvement in the field of phenology (the relationship of climate to periodic biological activity) permitting the scheduling of planting and harvesting of crops, used his climatic water budget to develop a simple bookkeeping approach to scientific irrigation, undertook a multi-year research study of microclimatology for the US Air Force, and established a Laboratory of Climatology for the training of climatologists. The success of these activities encouraged him to resign from his government position and to undertake a new career as a private consulting climatologist.

Scientists from all over the world came to visit the Laboratory and to work with the students in training there. During the 1950s he used his climatic water budget approach to solve practical problems in soil tractionability (the movement of humans and vehicles over unpaved surfaces), the leaching of radioactive strontium through the soil, and the disposal of factory effluent by spray irrigation. He also directed the development of a series of sensitive micrometeorological instruments – both horizontal and vertical wind anemometers, net radiometers, soil heat flux transducers, and dewpoint hygrometers.

Thornthwaite was elected the first president of the UN World Meteorologic Organization Commission for Climatology in 1951 and re-elected in 1953. This provided the opportunity for world travel to attend Commission activities and to represent the US at various other UN organizational meetings. He easily became the best-known climatologist of his day and his reputation was international. He received many honors including the Cullum Medal of the American Geographical Society, the Outstanding Achievement Award from the Association of American Geographers (he was Honorary President in 1960–1961), and an Honorary Doctorate from Central Michigan University. He died of cancer at the height of his career in 1963.

Thornthwaite possessed that rare spark of genius that allowed him to rise from a poor rural farm background to become a world-renowned climatologist and geographer. Without question, he was the most outstanding American climatologist of the twentieth century and his work in applied climatology and the water budget profoundly influenced the development and direction of the modern field of climatology.

John R. Mather

Select bibliography

(Thornthwaite published more than 100 research papers and monographs)

The climates of the Earth. *Geographical Review*, **23** (1933): 433–440.

A year of evaporation from a natural land surface (with Benjamin Holzman). *Transactions of the American Geophysical Union*, **21** (1940): 510–511.

Atlas of Climatic Types in the United States 1900–1939. US Department of Agriculture Miscellaneous Publication 421. Washington, DC: Government Printing Office, 1941.

Climate and Accelerated Erosion in the Arid and Semiarid Southwest, With Special Reference to the Polacca Wash Drainage Basin (with C.F. Steward Sharpe and E.F. Dosch). US Department of Agriculture Technical Bulletin 808. Washington, DC: Government Printing Office, 1942.

Problems in the classification of climates. *Geographical Review*, **33** (1943): 233–255.

Status and prospects of climatology (with John Leighly). *Scientific Monthly*, **57** (1943): 457–465.

Report of the Committee on Transpiration and Evaporation, 1943–44 (with H.G. Wilm). *Transactions of the American Geophysical Union*, **25** (1944): 686–693.

The moisture factor in climate. *Transactions of the American Geophysical Union*, **27** (1946): 41–48.

Climate and moisture conservation. *Annals of the Association of American Geographers*, **37** (1947): 87–100.

An approach toward a rational classification of climate. *Geographical Review*, **38** (1948): 55–94.

A charter for climatology. *World Meteorological Organization Bulletin*, **2** (1952): 40–46.

Operations research in agriculture. *Journal of the Operations Research Society of America*, **1** (1952): 33–38.

Climate in relation to crops (with J.R. Mather). In "Recent studies in bioclimatology: a group." *Meteorological Monographs*, **2**, no. 8 (1954): 1–10.

Topoclimatology. In *Proceedings of the Toronto Meteorological Conference 1953*, pp. 227–32. London: Royal Meteorological Society, 1954.

Climatic classification in forestry (with F.K. Hare). *Unasylva*, **9** (1955): 51–59.

The water balance (with J.R. Mather). *Publications in Climatology*, **8** (1955): 1–104.

The water budget and its use in irrigation (with J.R. Mather). In *Water: the yearbook of agriculture 1955*, pp. 346–58. Washington, DC: US Department of Agriculture, 1955.

Microclimatic investigations at Point Barrow, Alaska (with J.R. Mather). *Publications in Climatology*, **9** (1956): 1–51.

Modification of rural microclimates. In Thomas, W.L., ed., *Man's Role in Changing the Face of the Earth*, pp. 567–583. Chicago, IL: University of Chicago Press, 1956.

Instructions and tables for computing potential evapotranspiration and the water balance (with J.R. Mather, and D.B. Carter). *Publications in Climatology*, **10** (1957): 181–311.

The task ahead in climatology. *World Meteorological Organization Bulletin*, **6** (1957): 2–7.

Estimating soil moisture and tractionability conditions for strategic planning (with C.E. Molineux, J.R. Mather, and D.B. Carter). *Air Force Surveys in Geophysics*, **94** (1958).

"Movement of radiostrontium in soils" (with J.R. Mather, and J.K. Nakamura). *Science*, **131** (1960): 1015–1019.

The measurement of vertical winds and momentum flux (with W.J. Superior, J.R. Mather, and F.K. Hare). *Publications in Climatology*, **14** (1961): 1–89.

The task ahead. *Annals of the Association of American Geographers*, **51** (1961): 345–356.

THUNDER

Thunder is the sound that accompanies and follows immediately after the passage of a lightning flash or thunderbolt. We consider the sound of thunder to include its infrasonic (inaudible) components as well as the familiar audible loud noise. A flash of lightning heats a long column of air, averaging about 6 km in length and about 15 cm in diameter. The heating is to a temperature of 12 000 K and is practically instantaneous, at least as far as the generation of sound is concerned. Shock waves start to travel radially outward from the long irregular column that makes up the air heated by the flash. The shock wave is a sharp wave front of sound pressure that gives rise to the familiar "crack" sound of a nearby thunderbolt. At distances of about 10 m or so, the shock waves change into sound waves. The listener hears the sound waves arriving at varying times from the various parts of the thunderbolt, giving rise to the familiar prolonged duration of the sound. Reflections from both the ground and the inhomogeneities in the atmosphere contribute substantially to the prolongation.

The greatest distance of audibility for thunder heard by a casual observer is about 25 km. But thunder has been heard at distances of the order of 100 km when the listener is in a quiet location out of doors. The total acoustical energy radiated by a typical thunderbolt is approximately 10^8 joules. Near the flash, the distribution of energy at various frequencies includes both ultrasonic and infrasonic components. The ultrasonic components are rapidly attenuated with distance from the lightning flash, then the audible components are attenuated although less rapidly, leaving finally only the infrasonic components at large distances. The infrasonic sound energy per unit frequency (i.e. the power spectrum), at an average distance of about 5 km from thunderbolts, is inversely proportional to frequency in the spectral range between about 1.0 and 10 hertz. The propagation of the audible parts of the sound wave is strongly influenced by the temperature of the atmosphere and by the winds between the thunderbolt and the listener. In an upwind direction from the source of sound, and with the temperature falling rapidly with altitude, a listener can be in the shadow zone and hear nothing from a thunderbolt only a few km away.

Richard K. Cook

Bibliography

Bhartendu, B., and Currie, B.W., 1963. Atmospheric pressure variations from lightning discharges. *Canadian Journal of Physics*, **41**: 1929–1933.

Eagleman, J.R., 1990. *Severe and Unusual Weather*, 2nd edn., Lenexa, KS: Trimedia.
Remillard, W.J., 1960. *The Acoustics of Thunder*. Cambridge, MA: Acoustic Research Laboratory, Harvard University.
Schonland, B.F.J., 1950. *The Flight of Thunderbolts*. London: Oxford University Press.

Cross-references

Lightning
Thunderstorms

THUNDERSTORMS

A thunderstorm is a deep convective cloud (cumulonimbus) that produces lightning and thunder, heavy rain, strong surface outflow of cool air, hail and, on rare occasions, a tornado. A characteristic feature of the thunderstorm, viewed from a distance, is its cylindrical or slightly hourglass shape capped by an anvil-shaped top; the anvil is caused by air diverging from the upper portions of the storm's updraft (Figure T14).

The vertical extent of thunderstorms varies from one climatological region to another. In moist regions, cloud base can be several hundred meters above the ground, whereas in more arid regions, cloud base can be at a height of several kilometers. Storm tops typically vary from 9 to 18 km above the ground. Within a given region, the taller the storm, the stronger is the updraft and the more severe is the surface weather associated with it.

Formation mechanisms

A thunderstorm can be thought of as a thermodynamic machine driven by the release of latent heat of condensation (water vapor converted into cloud droplets) and, to a lesser extent, latent heat of freezing (water droplets converted into ice particles) and latent heat of deposition (water vapor converted into ice crystals). When a convective cloud develops in an unstable environment (that is, one where a rising parcel of air is buoyant – warmer than its surroundings), the potential energy of latent heat is converted into kinetic energy of vertical air motion. As long as the rising mass of air remains buoyant, it accelerates upward. After passing through the equilibrium level (where the temperature of the rising air is equal to that of its surroundings), the air becomes colder than its surroundings and rapidly decelerates.

If the storm is of the ordinary type (see discussion below), it will rain out and die soon after reaching its maximum vertical extent. However, when the environmental winds exhibit a great deal of vertical shear – marked change in wind direction and increase of wind speed with height – longer-lived severe storms can occur. Another important factor is dry midlevel environmental air that helps to promote downdraft formation. In a severe thunderstorm the updraft continues to be vigorous with horizontally diverging air above the equilibrium level producing an expanding anvil cloud (Figure T14). The top of the updraft is characterized by a convective turret (overshooting top) that can extend a kilometer or two above the top of the anvil into the lower stratosphere.

How does the thermodynamic engine get started? In other words, what causes the air to rise so condensation can begin to occur? The primary mechanisms might be called buoyant lifting and dynamic lifting.

Buoyant lifting occurs when air above a warm surface (land or water) is heated until it becomes warmer than its surroundings.

Figure T14 A severe thunderstorm near Topeka, Kansas, at 1750 CST on 21 April 1961. Photographed from the US Weather Bureau's DC-6B aircraft at 6 km height by T. Fujita, University of Chicago (courtesy of the late Professor Fujita).

Successive thermals of heated air develop into cumulus clouds that grow higher and higher until a full-fledged thunderstorm forms. Through the solar heating process, this type of storm reaches maturity in the late afternoon and early evening.

Geographic features that encourage buoyant lifting are isolated mountain peaks or mountainous terrain in general. Thunderstorms typically develop in these areas during the late morning and early afternoon. These storms begin earlier than those over flatter terrain because the southeastward- and southward-facing (northeastward- and northward-facing in the southern hemisphere) slopes are more perpendicular to the morning sun's rays.

Another important geographic feature is the land–water boundary. On a summer day the heated land becomes warmer than the water, and a lake breeze or sea breeze develops with a localized circulation of air flow from water to land at low levels and return flow aloft from land to water. Through dynamic lifting, the lake or sea breeze front (leading edge of the advancing circulation) favors thunderstorm development over land. If the land cools sufficiently during the night, a land breeze can develop with thunderstorms occurring just offshore. In the fall and early winter seasons, thunderstorms or thunder-snowstorms can form over warm lakes during cold air outbreaks.

Dynamic lifting plays a major role in those thunderstorms that develop along a convergence line (e.g. cold front, mesoscale pressure trough, dry line). These storms can become organized into a squall line. The leading edge of a squall line is characterized by strong winds and heavy rain, often followed by a wide area of stratiform precipitation. Such lines develop their own propagating mechanism and live on through the night, occasionally having an overall lifetime of more than 24 hours.

Global and seasonal distribution

In general, thunderstorms occur most frequently in the tropics and somewhat less frequently in the middle latitudes. Occurrence drops off rapidly in the poleward direction with thunderstorms rarely observed beyond the Arctic and Antarctic circles. The worldwide distribution of thunderstorms by season is shown in Figure T15.

The overall distribution pattern is influenced by three primary elements: (1) the intertropical convergence zone (ITCZ); (2) solar heating of landmasses; and (3) warm ocean currents. The presence of the ITCZ in the vicinity of the equator provides a favored environment for thunderstorm development during the entire year. Most activity is over land and warm (shallow) tropical water. Since the ITCZ is located within the prevailing easterlies, thunderstorms extend to the west of the landmasses. Note that thunderstorm maximums in this zone move northward and southward with the sun.

Outside the tropics, terrestrial thunderstorm activity also follows the sun. Pronounced thunderstorm maximums occur in the summer hemisphere and minimums in the winter hemisphere. Since South America and Africa are the only continents that straddle the equator, the presence of the ITCZ and solar heating provides these two continents with the greatest thunderstorm activity.

Ocean currents also play an important role in the global distribution of thunderstorms. Generally, warm ocean currents flow poleward along the east coasts of continents and cold equatorward-moving currents flow along the west coasts. Therefore, during most seasons it is not surprising to see local thunderstorm maximums over the Brazil current east of South America, the Agulhas current east of the tip of Africa, and the East Australia current between Australia and New Zealand. Likewise, but less spectacular, there is enhanced activity over the Gulf Stream from the east coast of North America toward Europe and the Kuroshio along the east coast of Asia. Pronounced thunderstorm minimums are found along continental west coasts outside the tropics where colder ocean currents occur.

Thunderstorm structure

The following discussion is based on studies of northern hemisphere midlatitude thunderstorms. Discussed are the single cell ("ordinary") thunderstorm, the multicell thunderstorm, and the supercell thunderstorm.

Ordinary thunderstorm

The ordinary type of thunderstorm over land is one that develops owing to solar heating. It frequently is called an airmass thunderstorm because it forms within a uniform airmass far from fronts or other large-scale lifting mechanisms. The environmental wind field in which it forms does not contain marked change of wind speed or direction with height. The storm moves in the direction of, and at a slightly slower speed than, the average wind throughout storm depth.

During the roughly 1 hour of the storm's existence, it passes through three distinct stages: cumulus (growing) stage, mature stage, and dissipating stage (Figure T16). During the cumulus stage, water vapor in the rising updraft condenses on microscopic particles (that are plentiful in the atmosphere) to form cloud droplets. As condensation forms, latent heat is released to fuel the buoyancy of the cloud. When cloud droplets collide with each other and coalesce, raindrops form. Raindrops continue to grow by colliding with cloud droplets and other raindrops within the updraft.

At temperatures below freezing, water vapor within the updraft is converted directly to ice crystals through the deposition of water vapor on microscopic particles (another source of buoyant energy for the cloud). However, cloud droplets that rise above the freezing height remain in liquid form (supercooled) down to temperatures of -20°C or less before they freeze. The presence of these supercooled cloud droplets plays an important role in the formation of graupel (snow pellets), hail, and electrical discharges in the cloud.

When an ice crystal intercepts a supercooled cloud droplet, the droplet immediately freezes on the crystal; this process is called riming. As the process continues, the crystal becomes completely rimed, and then it is called graupel. If air entering the updraft is extremely moist, there will be an abundance of supercooled cloud droplets such that the graupel and coexisting frozen raindrops continue to accumulate rime ice. If the riming particles remain at midaltitudes in the updraft, where temperatures are -10°C to -30°C, the particles serve as embryos that eventually grow into hailstones.

It appears that the presence of ice crystals and riming graupel at midaltitudes within the growing storm leads to the development of an electric field. The most prevalent hypothesis is that the electric field arises owing to noninductive charging; that is, charging in the absence of a surrounding electric field. Laboratory experiments show that, under typical midaltitude within-storm conditions, when an ice crystal hits and bounces off a riming graupel particle, the ice crystal acquires a positive charge and the graupel acquires a negative charge. The

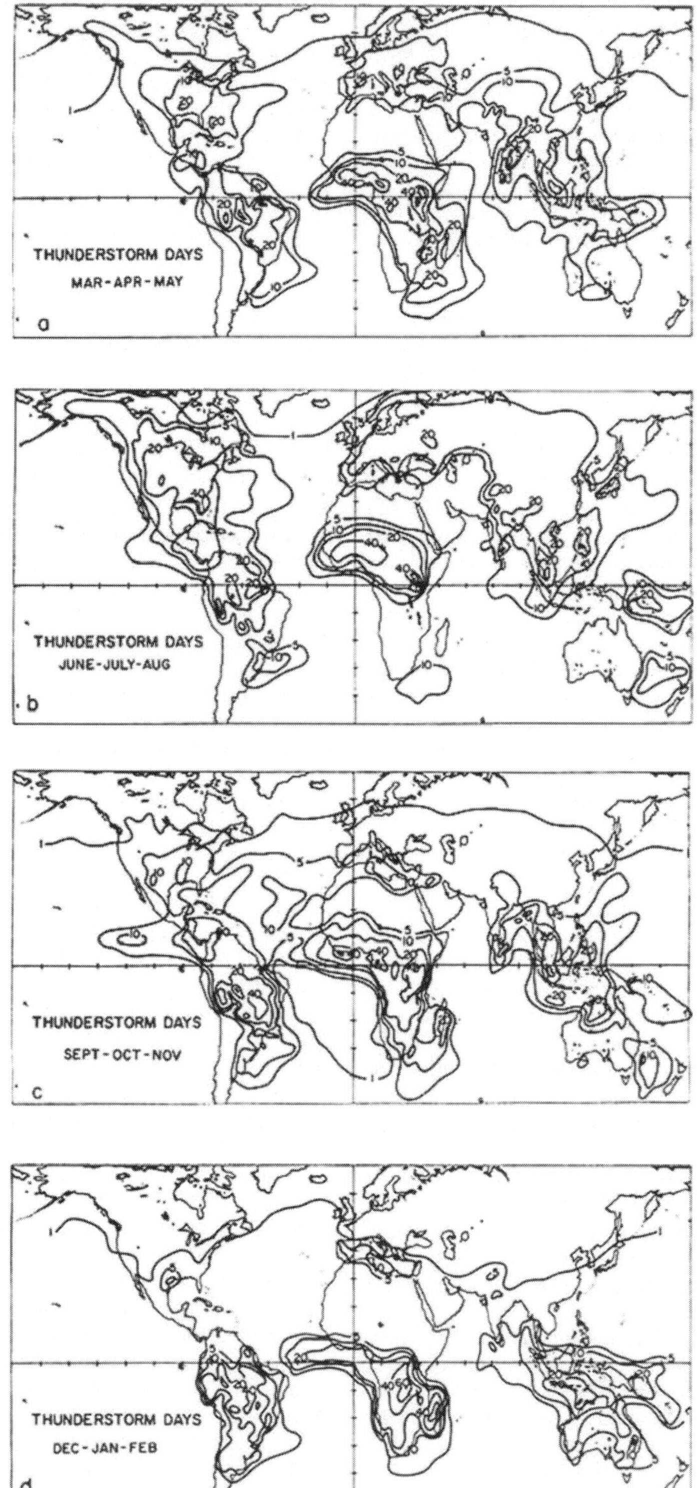

Figure T15 Geographical distribution of the average number of days on which thunderstorms occur during each season of the year. Isobronts are drawn for frequencies of 1, 5, 10, 20, 40, and 60 thunderstorm days per season (after World Meteorological Organization, 1953, 1956).

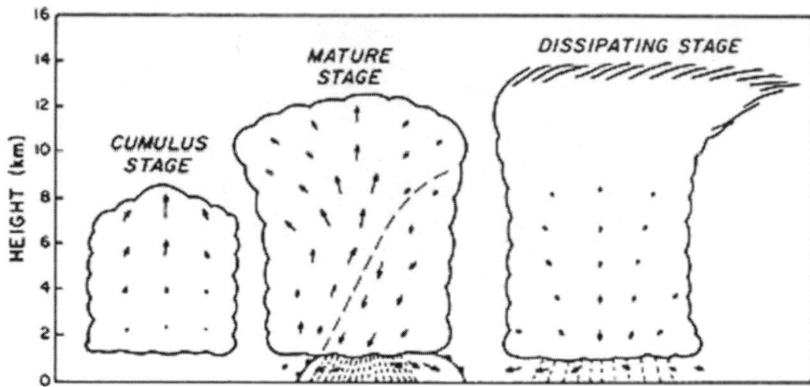

Figure T16 Vertical cross-section of three stages in the life cycle of an ordinary thunderstorm. Gust front boundaries beneath mature stage indicate leading edges of cold-air flow. Time increases as successive stages occur from left to right (after Byers and Braham, 1948).

hypothesis specifies that the lighter ice crystals are carried to the top of the cloud, while the heavier graupel particles remain suspended at midaltitudes. As a consequence of this differential transport, a positive electric field is found in the upper portions of the cloud and a negative field is found at midaltitudes, leading to the development of electrical discharges (lightning).

The beginning of the mature stage is marked by vigorous updrafts (15–25 m s^{-1}) and the start of precipitation particles (rain, graupel, hail) falling toward the ground. It is during this stage that the electric field attains its maximum strength, with frequent in-cloud and cloud-to-ground lightning. The descending precipitation particles are accompanied by a strong downdraft of air (reaching strengths of 10–15 m s^{-1}). When the evaporatively cooled air reaches the surface, it diverges outward producing a strong expanding gust front.

As the surface gust front expands, it cuts off the supply of moist air to the updraft and the updraft dies. Without an updraft available to produce more precipitation and the corresponding release of latent heat, the storm enters the dissipating stage. The downdraft weakens as precipitation decreases and consequently the storm dies.

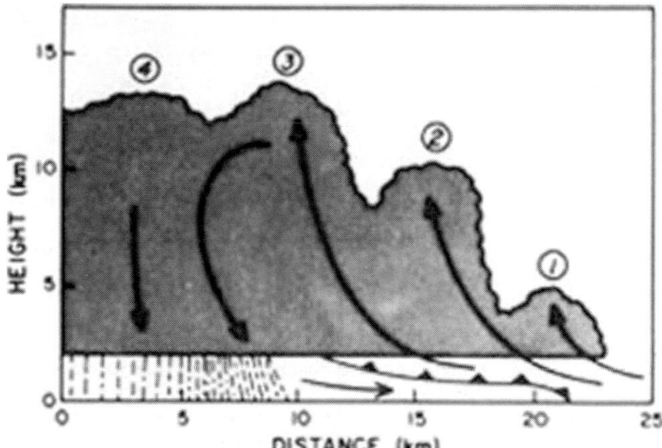

Figure T17 Vertical cross-section through a multicell storm, perpendicular to cell motion and mean environmental wind. Cell motion is away from the observer into the page; storm motion is at an angle to the right of cell motion. The gust front boundary is indicated with black triangles (modified after Kessler, 1985, p. 147).

Multicell thunderstorm

In addition to the single thunderstorm cell, as depicted in Figure T16, thunderstorms also organize themselves into a group of cells where each one is at a different developmental stage at a specific instant. This multicell thunderstorm can become severe and produce surface damage due to strong winds and hail. In Figure T17, cells 1 and 2 are in the cumulus stage, cell 3 is at the beginning of the mature stage and cell 4 is in the dissipating stage. Precipitation falling beneath cell 4 and a portion of 3 consists of both rain and hail.

The vertical cross-section in Figure T17 is perpendicular to cell motion, with the cells moving away from the viewer directly into the page. Moist low-altitude air approaching the storm from the right side (when looking in the direction of cell motion) is lifted by the outflowing cold air. Thus, the storm continues to evolve by producing its own lifting mechanism.

Although the individual cells of a multicell thunderstorm generally move with the mean wind, the cell ensemble does not move in that direction. With new cells developing at periodic intervals to the right of the existing cells, and old cells dying on the left flank, the storm as a whole moves (propagates in discrete steps) at an angle to the right of individual cell motion. This process of discrete propagation can last for hours, even though the lifetime of individual cells is about 1 hour.

Supercell thunderstorm

Severe thunderstorms that are capable of producing tornadoes develop an extraordinary structure during the mature stage. This "supercell" thunderstorm typically is an isolated storm that develops in an environment marked by extreme wind shear with height. However, not all supercell storms are isolated; some can be part of a squall line. The supercell storm differs from the ordinary storm because wind variation with height enables the precipitation downdraft to form adjacent to, and coexist with, the updraft. This feature permits the storm to have a quasi-steady mature stage that can persist for hours.

A notable characteristic of the supercell storm is its strong, rotating updraft (Figure T18); the rotating column is called the

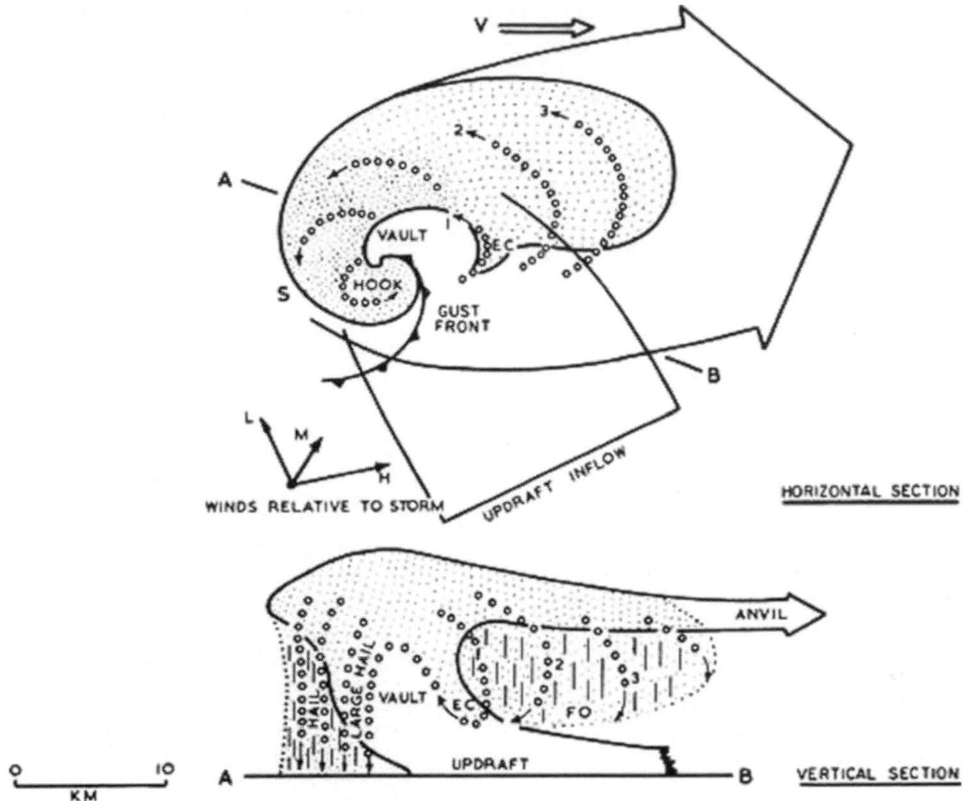

Figure T18 Horizontal and vertical cross-sections through the Browning model of a supercell thunderstorm. Stipple shading indicates the radar echo portion of the storm. (See Figure T14 for a visualization of this storm type.) Vertical section AB (lower panel) is oriented in the direction of the mean wind shear (indicated in upper panel). Environmental winds relative to the moving storm (V) are indicated at low (L), middle (M) and high (H) altitudes. The broad arrow represents low-altitude air entering the updraft, rising, and then flowing downstream with the high-altitude winds. The dotted (open circle) curves are precipitation trajectories. In horizontal section (upper panel), light and heavy stippling denotes areas of rain and hail, respectively, near the ground. Broken vertical hatching in the forward overhand (FO) and hailfall area indicates downdrafts with strong flow normal to the vertical section. Hail embryos are found in the embryo curtain (EC) (after Kessler, 1985, p. 152).

thunderstorm mesocyclone. Low-altitude air (*L*) enters the updraft, turns cyclonically (counterclockwise in the northern hemisphere) as it rises and exits through the anvil with the high-altitude storm-relative winds (*H*). Due to the updraft's strength ($30–50\,m\,s^{-1}$), cloud droplets do not have time to grow into radar-detectable precipitation particles until they have risen high in the storm. Consequently, the updraft is marked by a pronounced weak-echo vault that forms within the storm's radar echo. The fact that the updraft is rotating is indicated by descending precipitation being drawn around in the hook-shaped radar pattern in the storm's right rear quadrant. The low-altitude weak-echo inflow air beneath the vault forms the other part of the mesocyclone's circulation. Doppler radar measurements, which show the component of precipitation motion toward or away from the radar, indicate that the mesocyclone is typically about 10 km in diameter and extends vertically throughout much of the storm's depth; peak rotational velocities of $20–25\,m\,s^{-1}$ are found at an average radius of 2.5 km. If the storm produces a tornado, the tornado usually forms within the mesocyclone.

Being continuously fed with low-altitude moist air along its right flank, the storm appears to propagate continuously toward the right; this process is in contrast with the marked discrete propagation of multicell storms. The updraft within the supercell storm begins to die when there no longer is a balance of forces across the gust front. As low-altitude outflow air behind the gust front increases in relative strength, the gust front accelerates around the mesocyclone center, choking off the low-altitude moisture source that is vital for sustaining the updraft. In some supercell storms a new updraft forms along the gust front near the dying one. Since the new updraft grows in the cyclonically rotating portion of the storm, the rotation is concentrated by the updraft and it starts to rotate, producing a new mesocyclone center. On rare occasions this process can continue with the production of a series of mesocyclones over a period of several hours. This appears to be the process that takes place in storms that produce a family of sequential tornadoes, with each new tornado being associated with a new mesocyclone center.

Improving understanding of thunderstorm processes

Even though the basic characteristics of severe storms are known, complete understanding of the processes taking place within the storms is lacking. Use of conventional radars since the mid-1940s has increased our understanding of thunderstorms.

Starting in the early 1970s, Doppler radars and three-dimensional computer modeling of the thunderstorm life cycle have provided new insights into the workings of severe storms. During the 1990s, meteorologists began to use mobile instrumentation (Doppler radars, balloon launch equipment, lightning sensors, surface weather instruments) to follow thunderstorms in order to systematically measure hitherto-unknown details of the within-storm and near-storm conditions. More recently, dual-polarization Doppler radars permit discrimination among raindrops, hail, and ice crystals within evolving storms. Three-dimensional lightning detectors, that allow one to track in-cloud lightning discharges, are helping to understand the electrical component of storms. All of these facilities hold promise for solving many of the remaining mysteries about severe thunderstorms.

Rodger A. Brown

Bibliography

Byers, H.R., and Braham, R.R., Jr, 1948. Thunderstorm structure and circulation. *Journal of Meteorology*, **5**(3): 71–86.

Doswell, C.A., III (ed.), 2001. *Severe Convective Storms*. Meteorological Monograph 28(50). Boston, MA: American Meteorological Society.

Kessler, E. (ed.), 1985. *Thunderstorms: A Social, Scientific, and Technological Documentary*, vol. 2: *Thunderstorm Morphology and Dynamics*, 2nd edn. Norman, OK: University of Oklahoma Press.

Ludlam, F.H., 1980. *Clouds and Storms*. University Park, PA: Pennsylvania State University Press.

MacGorman, D.R., and Rust, W.D., 1998. *The Electrical Nature of Storms*. Oxford: Oxford University Press.

Rogers, R.R., and Yau, M.K., 1989. *A Short Course in Cloud Physics*, 3rd edn. Oxford: Pergamon Press.

World Meteorological Organization, 1953. *World Distribution of Thunderstorm Days, Part 1: Tables*. WMO No. 21. Geneva: World Meteorological Organization.

World Meteorological Organization, 1956. *World Distribution of Thunderstorm Days, Part 2: Tables of Marine Data and World Maps*. WMO No. 21. Geneva: World Meteorological Organization.

Cross-references

Airmass Climatology
Climate Hazards
Dynamic Climatology
Hail
Lightning
Tornadoes

TORNADOES

Tornadoes are zones of extremely rapid, rotating winds beneath the base of cumulonimbus clouds. Though the overwhelming majority of tornadoes rotate cyclonically (counterclockwise in the northern hemisphere), a few spin in the opposite direction. Strong tornadic winds result from extraordinarily large differences in atmospheric pressure over short distances. Over just a few tenths of a kilometer, the pressure difference between the core of a tornado and the area immediately outside the funnel can be as great as 100 mb. In what follows we adopt a northern hemisphere perspective, where "counterclockwise" is cyclonic.

Tornado characteristics and dimensions

It is difficult to generalize about tornadoes because they occur in a wide variety of shapes and sizes. While the majority have diameters of about 50 meters or so, some may extend as much as a kilometer in diameter or more. Usually they last no longer than a few minutes, but some have endured for several hours. Tornadoes normally move across the surface at speeds of about 50–65 km/h; so a typical tornado normally covers no more than a couple of kilometers from the time it touches the ground to when it dies out.

Estimates of wind speeds within tornadoes have historically been based primarily on the damage produced, and it was once believed that velocities could approach 800 km/h. The advent of weather radar has allowed meteorologists to make far more accurate estimates, and it is now understood that the weakest have wind speeds as low as 65 km/h and the most severe in excess of about 450 km/h.

Tornado formation

Tornadoes can develop in any situation that produces severe weather – frontal boundaries, squall lines, mesoscale convective complexes, supercells, and tropical cyclones. The processes that lead to their formation are not very well understood. Typically, the most intense and destructive tornadoes are those that arise from supercells.

Supercell tornado development

In a supercell storm, the first observable step in tornado formation is the slow rotation of a large segment of the cloud (up to 10 km (6 miles) in diameter). Such rotation begins deep within the cloud interior, several kilometers above the surface. The resulting large vortices, called *mesocyclones*, often precede the formation of the actual tornado by some 30 minutes or so.

The initiation of a mesocyclone depends on the presence of vertical wind shear near the surface. Figure T19 illustrates this process. Moving upward from the surface, the wind direction shifts from an easterly to westerly direction as the wind speed increases. This wind shear causes a horizontal vortex to develop. Under the right conditions, strong updrafts in the storm lift the middle of the vortex so that it bends upward with both a counterclockwise and clockwise limb. The counterclockwise rotating vortex provides the rotation needed for the development of the mesocyclone.

Intensification of the mesocyclone requires that the area of rotation decrease, which leads to an increase in wind speed. The narrowing column of counterclockwise rotating air stretches downward, and a portion of the cloud base protrudes downward to form a *wall cloud* (Figure T20). Wall clouds form where cool, humid air from zones of precipitation is drawn into the updraft feeding the main cloud. The humid air condenses at a lower height than does the air feeding into the rest of the cloud. Wall clouds most often occur on the southern or southwestern portions of supercells, near areas of large hail and heavy rainfall. *Funnel clouds* form when a narrow, rapidly rotating vortex emerges from the base of the wall cloud. A funnel cloud has all the characteristics and intensity of a true tornado; the only difference between the two is that a funnel cloud has yet to touch the ground. Funnel clouds and tornadoes are visible as the condensation that occurs as rising air cools adiabatically around the vortex. In some instances the tornado occurs without the accompanying condensation, and the vortex is not visually evident

Figure T19 The development of a mesocyclone. Mesocyclones begin with the development of a horizontally rotating tube of air due to wind shear. Convection can then lift the vortex along the middle, creating clockwise and counterclockwise vortices. The counterclockwise vortex grows into a mesocyclone. (From Klemp, 1987, Dynamics of tornadic thunderstorms, *Annual Review of Fluid Dynamics*, **19**: 369–402.)

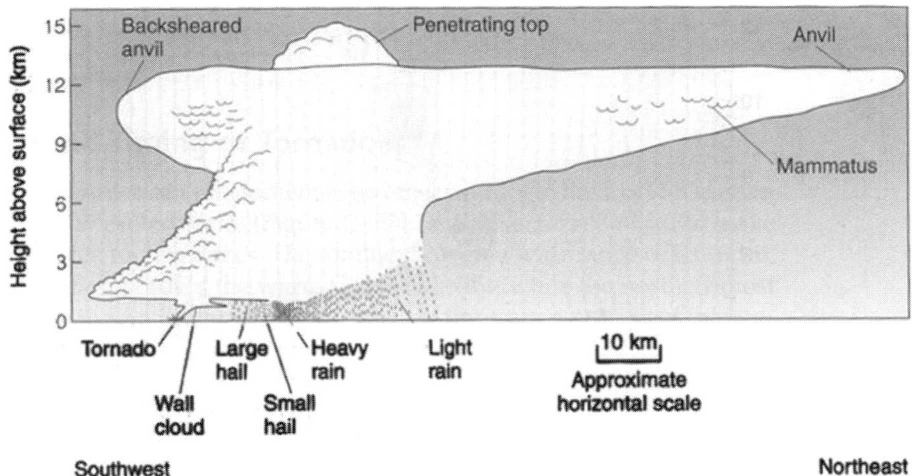

Figure T20 A generalized view of a typical supercell. A wall cloud descends from the cloud near the left side of the figure. (From Aguado and Burt, 2002. *Understanding Weather and Climate*. Used with permission of Prentice-Hall.)

except for the dust that is entrained from below. At other times condensation may cause the visual appearance of the tornado, but nearby rain obscures the event to people some distance away.

Because Doppler radar enables forecasters to observe the rotating winds of mesocyclones, the network of these radars has greatly increased lead times in the issuance of tornado warnings. Only about 50% of all mesocyclones actually spawn tornadoes, however. Exactly why some mesocyclones produce tornadoes and others do not is unknown, and for that reason forecasters cannot tell in advance which mesocyclones will produce tornadoes. Despite these uncertainties, Doppler radar has already proven itself to be the most important tool yet developed in "nowcasting" (making short-term predictions) the formation of tornadoes.

Nonsupercell tornado development

The exact mechanisms that lead to nonsupercell tornadoes are also poorly understood, though recent research indicates that

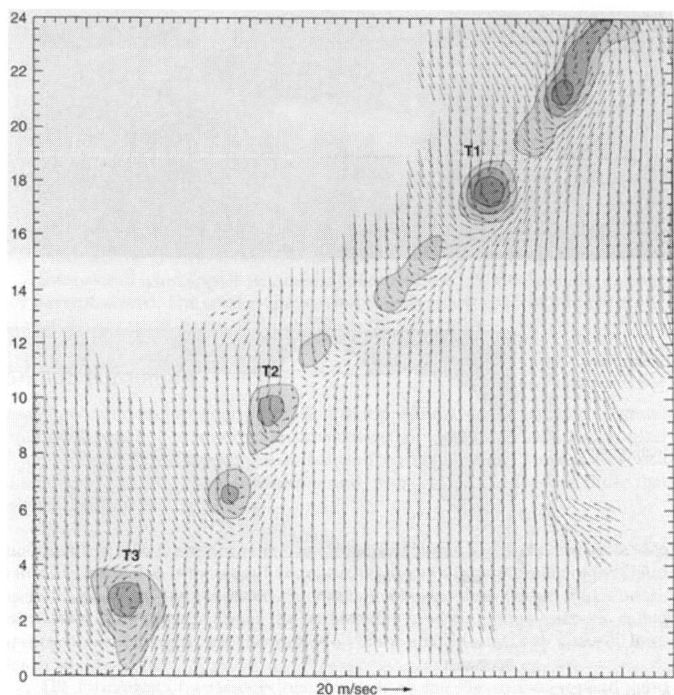

Figure T21 Convergence of two outflow boundaries can sometimes lead to the formation of tornadoes, as in this example. The vectors represent wind velocities. (From Aguado and Burt, 2002. *Understanding Weather and Climate.* Used with permission of Prentice-Hall.)

such tornadoes have their origins nearer to the surface than do those that begin as mesocyclones. Figure T21 illustrates one situation that may lead to nonsupercell tornadoes. The arrows show the outflow of air from two thunderstorm regions: at the top left and bottom right of the figure. From the bottom left to the top right of the figure is a zone of convergence between the two masses of air. At certain areas along the convergence zone (the circled areas), strong rotation develops. Another possible mechanism is shown in Figure T22, where strong convection along the convergence zone causes uplift and the formation of a cumulus cloud. In Figure T22c the cloud develops into a cumulonimbus, and the strong rotation stretches down from the cloud base to form a funnel cloud.

The location and timing of North American tornadoes

Though tornadoes occur over much of the world, no other country in the world has nearly as many as does the United States. Other areas in which tornadoes are fairly common include Japan, Australia, western Europe, South Africa and northern Argentina, but several factors combine to make North America a haven for tornadoes. First, the continent covers a wide range of latitudes; its southeastern portion borders the warm Gulf of Mexico, while the northernmost portion extends into the Arctic. Furthermore, much of the eastern portion of the continent is relatively flat and, in particular, there is no major mountain range extending in an east–west direction. Together, these features allow for a collision of northward-moving maritime tropical air from the Gulf of Mexico with southward-moving continental polar air along the polar front. This setting, coupled with the frequent presence of potential instability, provides a favorable situation for tornado development. The frequent occurrence of dry lines also contributes to the high incidence of tornadoes across much of southern Plains. Tornadoes occur at least occasionally in almost all 50 states. A great many tornadoes touch down along a wide strip running southwest to northeast between the southern Plains and the lower Great Lakes region, commonly called *Tornado Alley*.

Texas easily leads the United States in total number of tornadoes, with well over 100 annually. But when the large size of that state is accounted for, it ranks only ninth in number of tornadoes per unit area. Interestingly, the state with the greatest number of tornadoes per unit area is Florida, which lies altogether outside Tornado Alley. Unlike the twisters in Tornado Alley, many of Florida's tornadoes are embedded in passing hurricanes and tropical storms. Oklahoma is a close second to Florida in tornado density, but has about twice as many strong or violent tornadoes.

Tornadoes can occur at any time of the year; however, as shown in Figure T23, there is a strong concentration of tornadoes during the spring, when airmass contrasts are especially strong. May has the greatest number of tornadoes, with June a close second. The peak tornado season is not uniform throughout the country, because spring does not arrive everywhere at once. Louisiana and Arkansas have the greatest number of tornadoes in April, but Illinois, Wisconsin, and Nebraska are most apt to have them in June. Farther to the west, the Plains states of Texas, Oklahoma, and Kansas all share May as the peak month.

Tornadoes are far less common in Canada, with an annual average of only about 100. The greatest concentration is in the extreme southern part of Ontario, between Lake Huron and Lake Erie. The majority of Canadian tornadoes outside Ontario occur in the southern region of the Prairie provinces and southwestern

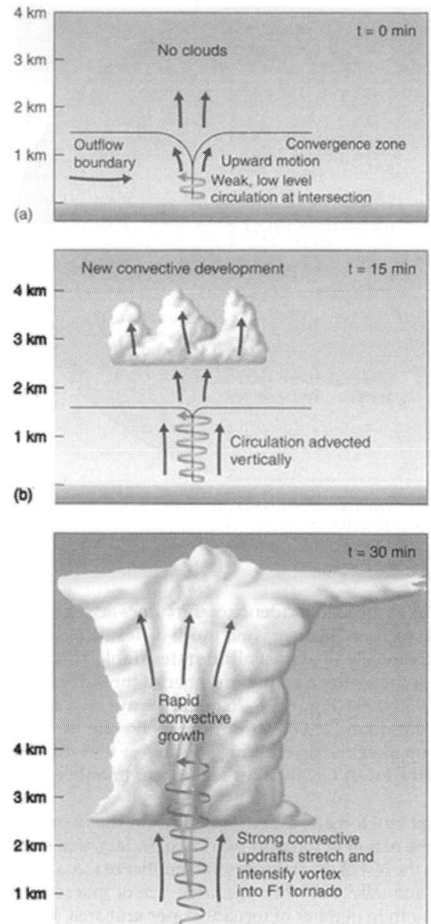

Figure T22 The development of a tornado along a convergent boundary. Spinning motions along the boundary (**a**) can be carried upward if there is sufficient convection (**b**). Once the cumulonimbus develops (**c**), the downward movement of the strong rotation can lead to tornadoes. (From Aguado and Burt, 2002. *Understanding Weather and Climate.* Used with permission of Prentice-Hall.)

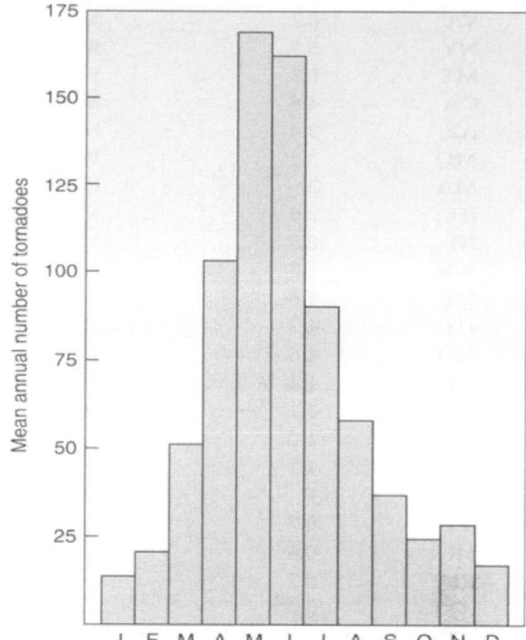

Figure T23 The monthly distribution of tornadoes in the United States, from 1950 to 1995. (From Aguado and Burt, 2002. *Understanding Weather and Climate.* Used with permission of Prentice-Hall.)

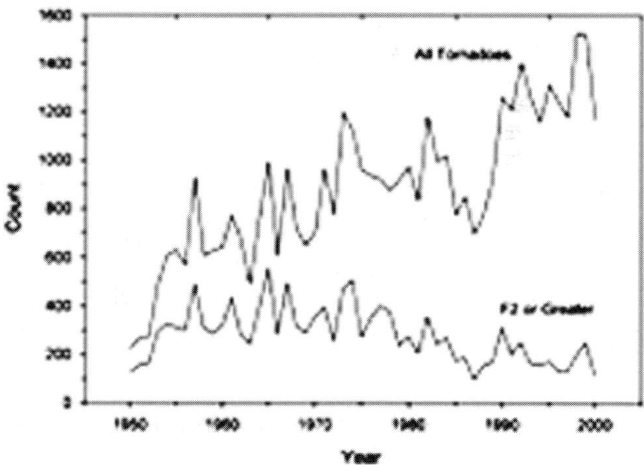

Figure T24 Observed counts of US tornadoes by year as recorded by the Storm Prediction Center. Data courtesy of Dr Matthew Menne, Climatic Data Center, NOAA.

Quebec. The Canadian tornado season extends from April through October, with the greatest frequency in June and July. Despite the fact that the greatest concentration of tornadoes in general, and of very strong tornadoes in particular, occurs in Ontario, it is interesting to note that the worst Canadian tornado outbreak in recent decades occurred in 1987 in Edmonton, Alberta, where the tornadoes of 31 July left 27 people dead, more than 300 injured, and thousands homeless. Total damages were estimated at $300 million.

Trends in US tornado occurrence

Tornado occurrences in the US, based on a log of events collected by the Storm Prediction Center of NOAA, shows a dramatic rise in tornado frequency from 1950 to the present (Figure T24). However, the same log suggests that intense storms (F2 or greater) have decreased since the 1970s, which implies that any increase in total storms has resulted from much higher frequency of moderate and weak events.

However, there is good reason to doubt the accuracy of this figure as an indicator of occurrence, because the official log does not account for increasing population (more "observers"), improved methods of detection (including Doppler radar), nor changes in the way tornadoes are classified. In contrast to Figure T24, a recent reconstruction assembled from objective evaluation of tornado environments (i.e. atmospheric conditions suitable for tornadoes) suggests a rather different trend. According to the reconstruction, strong-to-violent tornado frequency was relatively constant from 1950 through the early 1990s, then increased dramatically, roughly doubling in the ensuing 10 years.

Tornado damage

Most of the structural damage from tornadoes results from their extreme winds. It was once believed that homes were destroyed mostly by the pressure differences associated with a tornado's passage, which supposedly caused the air to push outward against the walls so violently that the house would explode. For this reason, people were advised to open their windows if they saw an approaching tornado, so that the pressure within the house could be reduced.

We now know that this was not good advice, in part because few homes actually explode. Moreover, though winds are the major factor in tornado damage, flying debris is the primary cause of tornado injuries, and opening a window increases the risk of personal injury from flying debris. (We must also suspect that opening the windows is useless in any case, because they are likely be "opened" anyway by flying objects.)

Although most tornadoes rotate around a single, central core, some of the most violent ones have relatively small zones of intense rotations (about 10 m (30 ft) in diameter) called *suction vortices*. It is these small vortices that probably allow the familiar phenomenon of one home being totally destroyed while the one next door remains relatively unscathed.

Except for those rare times when tornado-chasers make first-hand observations of passing tornadoes, it is impossible to get a precise reading on their pressure changes and wind speeds. But it is possible to classify them according to the magnitude of the damage they cause. The *Fujita scale* (named for the eminent tornado specialist Theodore Fujita) provides a widely used system for ranking tornado intensity. As shown in Table T2, documented tornadoes fall into seven levels of intensity, with each assigned a particular F-value ranging from 0 to 6 (F6 tornadoes are theoretical and have not been observed in nature). In the US the majority (69%) fall into the *weak* category, which includes F0 and F1 tornadoes. Twenty-nine percent of tornadoes are classified as *strong* (F2 and F3), which makes them capable of causing major structural damage even to well-constructed homes. Fortunately, only 2% of tornadoes are *violent* (F4 and F5). Those tornadoes are capable of wreaking incredible destruction. Cars can be picked up and carried tens of meters, pieces of straw can be driven into wooden beams, and freight cars can be carried off their tracks. Indeed, these storms are the true stars in all the movies and videos about tornadoes.

Between 1950 and 1999 there were 51 F5 tornadoes in the US and none in Canada. Thus they occur about once a year in all of North America. Texas holds the lead for the greatest number of F5 tornadoes (six) over that period, while Alabama, Iowa, Kansas, and Oklahoma each had five.

Fatalities

Because they are small and last for such a short time, the overwhelming majority of tornadoes do not inflict fatalities. Consider the fact that, between 1950 and 1994, an average of 91 people died in 760 reported tornadoes each year. This means that at least 88% of all tornadoes did not kill anybody, and the actual proportion of fatality-free tornadoes is even higher than 88%, for at least two reasons. First, tornadoes in which no one dies are preferentially undercounted, because a storm that kills is almost certain to be reported. Second, most fatalities result from a few very large storms that kill up to dozens of people. According to the National Severe Storms Laboratory (NSSL), fewer than 5% of all US tornado deaths are associated with weak (F0 or F1) tornadoes, nearly 30% with strong tornadoes (F2 and F3), and about 70% with violent tornadoes (F4 and F5). Thus, only 2% of all tornadoes are responsible for more than two-thirds of all fatalities.

A disproportionately large percentage of tornado-related deaths occur in mobile homes, which offer little protection to their occupants because they are easily blown off their foundations and tossed about by strong winds. During the 1985–2003 period there were a reported 997 tornado fatalities in the US. Of those, 403 (40%) occurred in mobile homes, in contrast to the 313 (31%) in regular residential structures.

Ten percent of the tornado fatalities during the 1985–2003 period were passengers and drivers of trucks and automobiles. Such victims often panic and make the fatal mistake of leaving the relative safety of their homes to outrun the storm in their vehicles. By far the safest place you can be when a tornado threatens is inside a well-constructed building, preferably in the

Table T2 The Fujita scale

Intensity	Wind speed (km/h)	Wind speed (mph)	Typical amount of damage
F0	<116	<72	Light: broken branches, shallow trees uprooted, damaged signs and chimneys.
F1	116–180	72–112	Moderate: damage to roofs, moving autos swept off road, mobile homes overturned.
F2	181–253	113–157	Considerable: roofs torn off homes, mobile homes completely destroyed, large trees uprooted.
F3	254–332	158–206	Severe: trains overturned, roofs and walls torn off well-constructed houses.
F4	333–419	207–260	Devastating: frame houses completely destroyed, cars picked up and blown downwind.
F5	420–512	261–318	Incredible: steel-reinforced concrete structures badly damaged.
F6	>513	>319	Inconceivable: might possibly occur in a small part of an F4 or F5 tornado. It would be difficult to identify damage done specifically by these winds, as it would be indistinguishable from that of the main body of the tornado.

F0 and F1 tornadoes are collectively called weak; F2 and F3 strong; and F4 and F5 violent.

basement and away from the windows. The notion that the southwest corner of a building offers additional protection is a commonly heard myth.

Experts cite the following safety rules for tornado safety:

1. Stay indoors and seek shelter in a basement.
2. If you are in a building with no basement, move to an interior portion of the lowest floor and crouch to the floor. If possible, cover yourself with a mattress or some other form of padding to protect you from falling or flying debris.
3. If you are in a mobile home, evacuate. If no fixed building is nearby, move away from all mobile homes and lie flat on low ground.
4. If driving a car or truck, you might be able to avoid the path of a tornado by driving at right angles to its path. Remember that most tornadoes move from southwest to northeast in the US and Canada. If you cannot avoid the tornado's path, pull off the road and seek shelter. If none is available, run to low ground away from the road.

Watches and warnings

Without question one of the most important responsibilities of the National Weather Service is to issue severe weather advisories. These take two forms: *watches* and *warnings*, either of which can be issued for severe storms or tornadoes. The declaration of a watch does not mean that severe weather has developed or is imminent; it simply tells the public that the weather situation is conducive to the formation of such activity. Most watches are issued for a period of 4–6 h, and for an area that normally encompasses several counties – about 50 000–100 000 sq km (20 000–40 000 sq miles).

The Storm Prediction Center (SPC) of the US Weather Service in Norman, Oklahoma, has responsibility for putting out severe storm and tornado watches for the entire country. Operating 24 hours a day, every day, the center constantly monitors surface weather station data, information from weather balloons and wide-bodied commercial aircraft, and satellite data for all of the US. If any particular part of the country appears vulnerable to impending severe storm activity, SPC issues a severe storm or tornado watch. The advisory then goes to the local office of the National Weather Service, which notifies local television and radio stations. The broadcast media then relay the information to the public. Watches for the entire country are also available from a number of sources, many of which are on the Worldwide Web.

Tornado warnings alert the public to the observation of an actual tornado (usually by a trained weather spotter) or the detection of tornado precursors on Doppler radar. Unlike watches, which are issued by the SPC, warnings are given by local weather forecast offices. They warn the public to take immediate safety precautions, such as finding shelter in a basement. The information is broadcast immediately by television and radio stations, and civil defense sirens are sounded.

Tornado research

Despite the inherent difficulty of studying something as complex, violent, and short-lived as a tornado, scientists can peer into their structure using a number of research tools. For many years, much of the information was obtained by films and pictures of tornadoes. Unfortunately, dust and debris kicked up by a tornado, as well as screening by trees and buildings, often made it impossible to observe the part of the tornado nearest to the ground (and of most concern to people).

In 1980 scientists developed the Totable Tornado Observatory, better known by its acronym, "Toto" (an intentional reference to the dog in *The Wizard of Oz*). Toto was a lightweight package that could be carried in the back of a pickup truck and set up by storm-chasers in 30 seconds or less. Ideally, the instrument would be placed in the path of a tornado where it could make observations of temperature, wind, and pressure. Unfortunately, getting Toto in the path of a tornado proved exceedingly difficult. Moreover, wind tunnel tests have shown that it tends to tip over at wind speeds near 50 m/s, well below those that can occur in a tornado. For these reasons storm-chasers no longer use Toto.

More recently, scientists from the University of Oklahoma have begun placing a network of smaller instruments called "turtles" in the paths of tornadoes. Though they only record temperature and pressure, turtles have an advantage in that they can be laid out in series across a road, crossing the path of a tornado. They remain a potentially useful tool for examining the environment of a tornado.

Other methods have been attempted for making direct observations within a tornado or mesocyclone. One is to release portable weather balloons with radiosonde instruments in the near-storm environment. If they are released from the right location, the wind can carry the balloons right into the heart of a severe storm. Less successfully, researchers have attempted to fire light instrument-carrying rockets (on the order of 1 kg) into severe storms from small aircraft. Unfortunately, the measurement devices have proven to be too fragile to survive the turbulence of the storm and are no longer used.

By far the most useful instrument for studying the winds of tornado-producing storms has been Doppler radar. The NEXRAD network of Doppler radar units has yielded a marked improvement in the issuance of tornado warnings. A recent study has shown that installation of the first six Doppler radar units in the heart of Tornado Alley has improved forecasters' ability to predict tornadoes. Prior to the radar deployment, 33% of the tornadoes in the study area occurred without a tornado warning; afterward, the figure dropped to 13%.

The biggest limitation of the Doppler radar network is that the farther a storm is from the transmitter/receiver unit, the less detailed is the information about its internal winds. Because tornadoes are usually quite small, they must pass within a few kilometers of the radar if precise wind information is to be obtained. For this reason, tornado researchers also use portable Doppler radar units carried on research vehicles and aircraft.

The tornado outbreak of 3–6 May 1999

In May of 1999 a devastating series of large tornadoes reminded the world of just how devastating tornadoes can be. Hardest hit was central Oklahoma, where a series of tornadoes on the first day of the outbreak killed 44 people and destroyed 2600 homes and businesses. On the same day, tornadoes killed another five people and destroyed 1100 buildings in the Wichita, Kansas area; and on 5 and 6 May the storms claimed another five lives in Texas and Tennessee. This was the deadliest outbreak of tornadoes since the record-breaking "Xenia Tornado Outbreak" of 3–4 April 1974, which spawned a total of 148 tornadoes (seven of which were rated as F5) from northern Alabama and Georgia to Windsor, Ontario, in Canada, and killed over 300 people.

Many factors contribute to the loss of life inflicted by tornadoes, including their maximum wind speed and their paths relative to large populations. In this case extremely powerful winds descended upon large population centers. However, the effectiveness of the warning system also impacts the death total and, fortunately, the system was at its very best during the 1999 outbreak. Field crews observed and reported the early development of conditions favorable for tornadoes, which complemented the forecasting efforts of meteorologists at the respective Weather Service offices. Meteorologists issued public advisories a full 2 hours in advance of the tornadoes, television and radio news crews disseminated the information, and the public generally responded wisely. Many residents sought shelter in basements, and others living in homes without basements took cover in interior hallways in time to save their lives. (People living in mobile homes received no protection at all from the storms; in the town of Bridge Creek every one of the 11 people who died were in trailers. Likewise, many fatalities occurred on highways – three of which were of people taking shelter under freeway overpasses!) Experts at the National Oceanographic and Atmospheric Agency estimate that the warning system saved as many as 7000 lives.

Waterspouts

So far we have discussed tornadoes over land. Similar features, called *waterspouts*, occur over warm-water bodies. Waterspouts are typically smaller than tornadoes, having diameters between about 5 and 100 m (17–330 ft). Though they are generally weaker than tornadoes, they can have wind speeds of up to 150 km/h (90 mph), which makes them strong enough to damage boats.

Some waterspouts originate when land-based tornadoes move offshore. The majority, however, are formed over the water itself. These "fair-weather" waterspouts develop as the warm water heats the air from below and causes it to become unstable. As the air rises within the unstable atmosphere, adiabatic cooling lowers the air temperature to the dewpoint, and the resultant condensation gives the waterspout its ropelike appearance. Waterspouts form in conjunction with cumulus congestus clouds, those having strong vertical development but not enough to form the anvil that characterizes cumulonimbus.

Contrary to what we might assume, the visible water in the waterspout is not sucked up from the ocean below; it actually comes from the water vapor in the air. Waterspouts are particularly common in the area around the Florida Keys, where they can occur several times each day during the summer.

Edward Aguado and James E. Burt

Bibliography

Bieringer, P., and Ray, P.S., 1996. A comparison of tornado warning lead times with and without NEXRAD Doppler radar. *Weather and Forecasting*, **11**: 47–52.

Bluestein, H., 1999. *Tornado Alley: monster storms of the Great Plains*. New York: Oxford University Press.

Church, C., Burgess, D., Doswell, C., and Davies-Jones, R. (eds.), 1993. *The Tornado: its structure, dynamics, prediction and hazards*. American Geophysical Union.

Davies-Jones, R.P., 1995. Tornadoes. *Scientific American*, **273**: 48–59.

Dowell, D.C., and Bluestein, H.B., 1997. The Arcadia, Oklahoma, storm of 17 May, 1981: analysis of a supercell during tornadogenesis. *Monthly Weather Review*, **125**: 2562–2582.

Glass, R.I., Craven, R.B., Bregman, D.J. et al., 1980. Injuries from the Wichita Falls tornado. *Science*, **207**: 734–38.

Grazulis, T.P., 2001. *The Tornado: nature's ultimate windstorm*. Norman, OK: University of Oklahoma Press.

Keller, D., and Vonnegut, B., 1976. Wind speed estimation based on the penetration of straws and splinters in wood. *Weatherwise* (October), pp. 228–232.

Rasmussen, et al., 1994. Verification of the origins of rotation in tornadoes experiment. *Bulletin of the American Meteorological Society*, **75**: 995–1006.

Wakimoto, R.M., and Wilson, J.W., 1989. Non-supercell tornadoes. *Monthly Weather Review*, **117**: pp. 1113–1140.

Cross-references

Climate Hazards
Synoptic Climatology
Thunderstorms
Vorticity

TOURISM AND CLIMATE

Tourism is the world's fastest-growing industry and in many countries it is a major and rapidly developing economic sector of the economy. Climate constitutes an important part of the environmental context in which recreation and tourism takes place, and climate is a resource exploited by tourism (Perry, 1997). Because tourism is a voluntary and discretionary activity, participation will often depend on perceived favorable climatic conditions. The betterment of health has been a strong motive for travel since the taking of waters at mineral and hot springs was fashionable in the seventeenth and eighteenth centuries. In the twentieth century a holiday in the sun is still perceived by many people as vital to their well-being, despite concerns over the links between skin cancer and UV-B radiation. The search for a salubrious, amenable climate was once the prerogative of the wealthy, but has now become a mass phenomenon. Leisure time is a scarce resource and spending time in a climate perceived by the individual as being most ideal has become of great importance to many people. Weather can ruin a holiday, but climate can ruin a holiday destination. Climate is a major component of the tourist destination image and the weather conditions anticipated by the holidaymaker are one of the major influences on holiday destination decisions. The Mediterranean, with its hot dry summer, is currently the world's most popular and successful tourist destinations, with 120 million visitors every year, many of them from the tourist-exporting countries of northern Europe, and 25% of the world's hotel accommodation is found in this area. This area may become less attractive to tourists in high summer in the future as it becomes too hot for comfort in the peak season with a much higher frequency of severe heat waves. (Perry, 2000). This is an example of a change in the seasonality of a tourist destination, and might have the effect of changing the length of the operating season of tourist facilities and infrastructure with implications for profitability. In northern latitudes of Europe and North America warmer weather in the future may produce a longer tourist season and encourage more people to take their holidays at home.

Areas of study

Three main areas of study have emerged in climate-tourism studies:-

1. Forecasting how weather and climate affects participation rates for different types of leisure activities and the levels of personal safety, comfort and satisfaction that ensue.
2. Improving the range and provision of weather and climate information for the leisure industry.
3. Investigating the likely impacts of climate change, particularly projected global warming, on tourism and recreational activities and enterprises.

Smith (1993) has drawn a distinction between weather-sensitive tourism, where the climate is insufficiently reliable to attract mass leisure participation, but good weather can stimulate tourist activity, and climate-dependent tourism where travel to the destination is based on the reliability of the climatic conditions Climate as a motivating factor in holiday destination choice has been examined by Maddison (2001) and Giles and Perry (1998). Predicting both spectator and participant numbers at outdoor recreational sites is important for implementing effective visitor management strategies. Activity choice, for example whether to play golf, swim or go hill-walking, is in large measure determined by perceived weather conditions. For many activities there are critical thresholds beyond which participation and enjoyment levels fall and safety may be endangered.

Tourism and leisure activities generate a large percentage of the total demand for weather information from meteorological services. However, portraying the suitability of the climatic environment for different types of leisure activity and for different groups of potential customers is a difficult task. Easily available information in holiday brochures is often misleading, incomplete, and may be distorted by commercial interests.

To present the totality of the climatic environment for tourists it is necessary to develop indices that integrate a number of climatic parameters. A tourism climate index is a useful concept but there are various facets to tourism climate, e.g. aesthetic (sunshine hours), physical (rain and wind) and thermal (combined metabolic effect), which make an effective and simple descriptive index difficult to devise. Whilst climatologists use various standard measures of comfort and discomfort these are little known or understood by the general public, and there remains the challenge of developing indices of holiday climate that can be widely used and understood by the traveling public.

Climate change

Climate change will impact upon tourism and in turn impacts, through growing greenhouse gas emissions, on the climate. Impacts may be direct, e.g. changing the environment of resorts as a result of sea-level rise, or indirect, for example raising conflicts over water resources. Tourism has a strong international dimension and is sensitive to any change of climate that alters the competitive balance of holiday destinations. Tourism is vulnerable to natural disasters and extreme events such as hurricanes, avalanches, heatwaves and droughts, whose frequency and distribution may change in the future. Infrastructure developments such as campsites are often placed in hazard-prone sites such as flood plains, to take advantage of scenic attractiveness, but are then prone to sudden-onset disasters such as flash floods.

Winter sports centers, especially those at lower altitudes, will be increasingly at risk from shorter seasons, higher temperatures (making adaptation strategies such as artificial snowmaking difficult to implement) and unreliable natural snow cover. The most vulnerable tourist resorts are a function of the likely magnitude and extent of the climate impact, the importance of tourism to the local economy and the capacity to adapt (Agnew and Viner 2001). The primary resources of sun, sea and beaches are likely to be re-evaluated in the light of expected climate change, but tourism is a continuously adapting industry and climate change will present new challenges that could lead to opportunities for tourist investment to capitalize on the changed environmental conditions.

Allen Perry

Bibliography

Agnew, M.D., and Viner, D., 2001. Potential impacts of climate change on international tourism. *Tourism and Hospitality Research*, **3**: 37–60.

Giles, A.R., and Perry, A.H., 1998. The use of a temporal analogue to investigate the possible impact of projected global warming on the UK tourist industry. *Tourism Management*, pp. 75–80.

Maddison, D., 2001. In search of warmer climates? The impact of climate change on flows of British tourists. *Climatic Change*, **49**: 193–208.

Perry, A.H., 1997. Recreation and tourism. In Thompson, R.D., and Perry, A., eds., *Applied Climatology*. London: Routledge, pp. 240–250.

Perry, A.H., 2000. Tourism and recreation. In *Assessment of Potential Effects and Adaptations for Climate Change in Europe* ACACIA Report, University of East Anglia, pp. 217–226.

Smith, K., 1993. The influence of weather and climate on recreation and tourism. *Weather*, **48**: 398–404.

Cross-references

Applied Climatology
Commerce and Climate

TRADE WINDS AND THE TRADE WIND INVERSION

Easterly winds of the tropical oceans, both north and south of the equator, have long been known as trade winds. Here, "trade" comes from a now-obsolete form, meaning "track", and refers the persistent direction of these winds. The trades are among the most important features of the atmospheric circulation, gathering energy from the ocean surface and transporting it toward the equator and controlling the climate of nearly half the Earth's surface.

The trade winds form the equatorward limb of the Hadley circulation in each hemisphere. They flow out of the subtropical high-pressure cells as northeasterlies in the northern hemisphere, and southeasterlies in the southern hemisphere. They meet at the near-equatorial convergence zone (CZ, see Ramage et al., 1981), often referred to as the intertropical convergence zone (ITCZ). The CZ is a zone of frequent convection that migrates north and south with changing sun angle, but remains in the northern hemisphere over the oceans. Tropical winds are dominantly easterly throughout the lower and middle troposphere. Rising air along the CZ begins to diverge in the upper troposphere. As this flow moves poleward in each hemisphere, Coriolis causes it to become a westerly wind. The poleward flow descends at about 30°–35° N and S, forming the subtropical high-pressure belt and completing the Hadley cell circulation.

Observations indicate that the trades accumulate both latent and sensible heat as they flow across the ocean surface. The trades transport most of this energy to the CZ area, where it contributes to the massive convection there, which in turn sets in motion the Hadley cells. In this respect the trades are seen as drivers of the atmospheric general circulation.

Trade winds form the equatorward sides of the subtropical anticyclones. The preferred positions of these large, semipermanent high-pressure systems are in the eastern portions of the ocean basins. The highs shift with the seasons, affecting the location of the trade wind zone. In each hemisphere, subtropical anticyclones shift equatorward during winter, relative to their summer positions. In the northern hemisphere, with its relatively large continents, they also shift landward (east) during winter.

Over continental areas, especially Asia, the annual cycle of heating and cooling strongly influences the patterns of tropical winds. For example, while easterlies are found across the entire tropical north Pacific in January, the Asian monsoon disrupts this pattern in summer, resulting in southwesterly winds in the sector from East Africa all the way to the western Pacific. Similarly, the monsoon trough over northern Australia during the Austral summer limits the trades of the south Pacific to the eastern two-thirds of the basin.

Trade wind inversion

Throughout much of the trade wind zone, subsidence in the subtropical anticyclones produces a persistent temperature inversion. This so-called trade wind inversion (TWI) originates on the eastern sides of the oceans, and steadily increases in height and decreases in strength moving westward and equatorward. Like any temperature inversion, the TWI is a highly stable layer which effectively caps vertical motion from below. This tends to prevent deep convection and makes thunderstorm development rather rare. The inversion also isolates the air above it from the oceanic moisture source. This upper air has been desiccated by its descent in the downward limb of the Hadley cell, and therefore air above the TWI is generally very arid. Despite inhibiting vertical exchange in the mean, Riehl (1979) has shown that the TWI is not a material boundary. There are exchanges of air across the boundary, which certainly contributes to the downstream weakening of the TWI. The TWI does not represent the upper limit of the trades, but rather separates well-mixed, humid marine air from an arid upper air. Riehl (1979) defines four layers in the trade flow, going from the surface upward: subcloud layer, cloud layer, inversion layer, and free atmosphere.

Many of the important characteristics of the TWI have been known since the Atlantic Meteor Expedition of 1925–1927, which focused on observations within the North and South Atlantic trade winds zones. In general, the strong subsidence and cold ocean currents along the eastern margins of the oceans favor the maintenance of an inversion at 500 m altitude or lower. As air flowing out of this zone moves westward and equatorward, convection in subcloud and cloud layers deepens. As a result the height of the inversion increases. In the North Pacific, for example, the inversion rises from less than 500 m at the California coast to over 2000 m at Hawaii. At the same time the strength of the TWI decreases with distance downstream. The temperature difference from the bottom to the top of the inversion layer is a good measure of its strength. An 8–10° temperature difference at the California coast weakens to about 3°C over Hawaii.

Climate of the trade wind zone

The climate of the trade wind zone is remarkably consistent, in large part due to the persistence of the trade winds. In Hawaii, trades occur approximately 90% of the time during summer, and 50% of the time during winter (Blumenstock and Price, 1967). The consistent direction and intensity of the winds, and their long trajectory over ocean, are responsible for Hawaii's agreeable climate (Giambelluca and Schroeder, 1998). The TWI strongly affects weather in Hawaii and other similarly positioned oceanic island groups. The inversion limits rainfall in the region near Hawaii to around 700 mm per year. Because of the persistent orographic lifting of trade winds on the windward slopes, some spots in Hawaii receive spectacular amounts. Mount Waiale'ale on Kaua'i averages around 11 m of rainfall annually (Giambelluca et al., 1986), making it a candidate for the wettest spot on earth. The TWI, however, inhibits upward flow beyond about 2000 m. Therefore, the upper slopes of Hawaii's highest mountains, including Haleakala on Maui, and Mauna Loa and Mauna Kea on Hawaii Island, are quite dry. This juxtaposition of persistent orographic lift and the capping effect of the TWI give rise to incredibly steep gradients in rainfall and other climatic variables, along Hawaii's upper mountain slopes.

T.W. Giambelluca

Bibliography

Blumenstock, D.I., and Price, S., 1967. Climate of Hawaii. In *Climates of the States*, No. 60–51, Climatography of the United States, US Department of Commerce.

Giambelluca, T.W., and Schroeder, T.A., 1998. Climate. In Juvik, S.P., Juvik, J.O., and Paradise, T.R. (eds.), *Atlas of Hawaii*. University of Hawaii Press.

Giambelluca, T.W., Nullet, M.A., and Schroeder, T.A., 1986. *Hawaii Rainfall Atlas*. Report R76, Hawaii Division of Water and Land Development, Department of Land and Natural Resources, Honolulu.

Ramage, C.S., Khalsa, S.J.S., and Meisner, B.N., 1981. The central Pacific near-equatorial convergence zone. *Journal of Geophysical Research*, **86**(C7): 6580–6598.

Riehl, H., 1979. *Climate and Weather in the Tropics*. New York: Academic Press.

Cross-references

Easterly Waves
Intertropical Convergence Zone
Monsoons and Monsoon Climate
Orographic Precipitation
Tropical Cyclones
Tropical and Equatorial Climates

TREE-RING ANALYSIS

Well-defined annual-growth rings can be observed in the wood (xylem) from many species of temperate forest trees throughout the world. The rings are formed by cell division of the cambium, which lies along the inner portion of the surrounding bark (Figure T25). In certain circumstances these growth rings contain useful information about varying environmental conditions affecting their growth. Dendroclimatology involves the study and use of these growth rings to reconstruct past variations in climate.

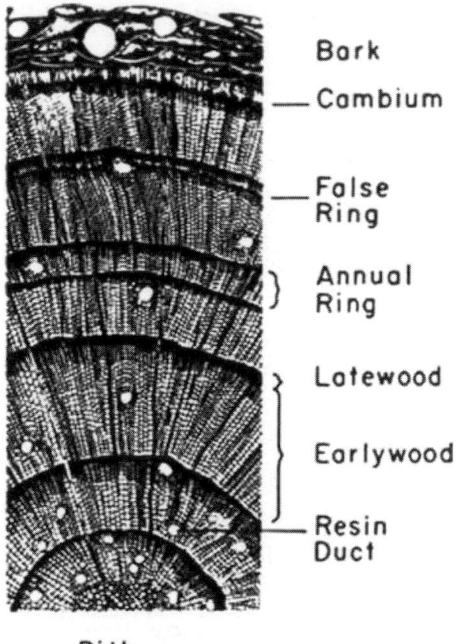

Figure T25 Drawing of cell structure along a cross-section of a young conifer stem. The earlywood consists of large and relatively thin-walled cells and the latewood is composed of small, think-walled cells. Variations in cell size and thickness within the ring can produce false rings in either earlywood or latewood. The large circular openings are resin ducts, which form a complex system of tubes throughout a coniferous trees. Note the variability in width from one ring to the next and in the proportion of earlywood versus latewood, which can be used for crossdating and reconstruction of climate.

Growth-ring formation

Initiation of cell division generally occurs throughout the cambium in the spring. The newly formed xylem cells lying inside the cambium expand into large thin-walled cells. In conifers this earlywood appears as light-colored tissue on the inside of the annual ring (Figure T25). As the season progresses, the growth slows down, the latewood cells do not expand as much as those in the earlywood and the cell walls become thicker. This forms denser, darker wood in conifers with the last-formed cells of the year usually becoming the smallest with the thickest walls. The abrupt change in color and structure between these cells and the lighter-colored earlywood of the next ring forms a distinct boundary between annual rings.

Sometimes the growing season is interrupted and more than one growth layer is apparent within the annual ring. These are referred to as intra-annual growth bands, multiple rings, or false rings (Figure T25). The outer margins of false rings are usually, but not always, less distinct than the outer margins of true annual rings. At other times the environment can be so harsh that no ring is formed in certain portions of the tree. These are called locally absent or missing rings and can usually be found elsewhere in the tree or in other nearby trees. Because of false and missing rings, the annual growth layers and thus the age of the tree cannot be reliably determined from simple counting of rings. This information must be obtained by using a dendrochronological procedure called *crossdating*.

Dendrochronology

The early Greeks were the first to note that trees form more or less annual layers and that the widths of these layers can vary as a function of local environmental conditions. Two French naturalists, Duhamel and Buffon, noted in 1737 that a frost-damaged layer occurred 20 rings in from the bark on each of several newly felled trees. A.C. Twining in 1827 was the first to grasp the possible use of tree-ring matching to establish the dates of the rings. According to Russian workers, in 1892 F.N. Shvedov was not only the first to crossdate tree rings but also the first to use variations in ring structure to infer the past history of climate.

A.E. Douglass is generally recognized as the first American worker to utilize crossdating in a systematic fashion. He is the acknowledged founder of the discipline of dendrochronology, which can be succinctly defined as the field of *systematic dating* of tree rings and their application to environmental problems, the dating of structures, or past events. He attracted international attention when he dendrochronologically dated the prehistoric pueblo dwellings of the southwestern United States (Douglass, 1936).

The crossdating procedure utilizes similarities in ring width or other morphological ring features that vary as a function of time (Figure T26). The pattern of variation is similar among trees growing throughout the same region for the same time period, presumably due to variations in macroclimatic factors affecting the trees' growing conditions. The crossdating procedure is described most fully by Stokes and Smiley (1996) and is summarized by Fritts (1976).

Crossdating includes the comparison of specimens by: (1) visual and statistical matching of features, such as width or earlywood–latewood color and thickness, which vary from one ring to the next; (2) examining these features for the presence or absence of synchrony over time; (3) identifying any inconsistent features not coinciding on all available specimens; (4) interpreting from the differences in ring structure what was the most probable cause of each discrepancy (such as failure of cambial initiation causing missing rings or interruption of cambial activity causing multiple growth layers); (5) establishing the validity of the interpretation by finding consistent structural features in the rings of different trees; and (6) finally arriving at a corrected chronological sequence that is consistent on all dated annual rings.

There is a lower limit in the number of rings (years) that must overlap in time and the number of trees (samples) necessary to establish accurate dating. Many workers would like at least 100 years of overlap and samples from ten or more trees, depending on the difficulty encountered in dating the ring sequences from the site. The majority of dendroclimatic studies incorporate two radii from at least 20 trees of the same species that are affected by similar climatic conditions.

If crossdating is done properly and carefully, with a sufficient number of suitable specimens, the troublesome ring features causing possible counting errors can be identified and corrected, and no further discrepancies will be detected no matter how many additional tree-ring materials are dated against the chronology. Literally hundreds of independently dated chronologies from southwestern North America follow the same general ring-width pattern. If there were even one error in these well-dated chronologies, the same dating error would have to have occurred in all samples of the southwestern collections, which, of course, is extremely unlikely. Because of the high accuracy of the crossdating methods, a tree-ring date can be

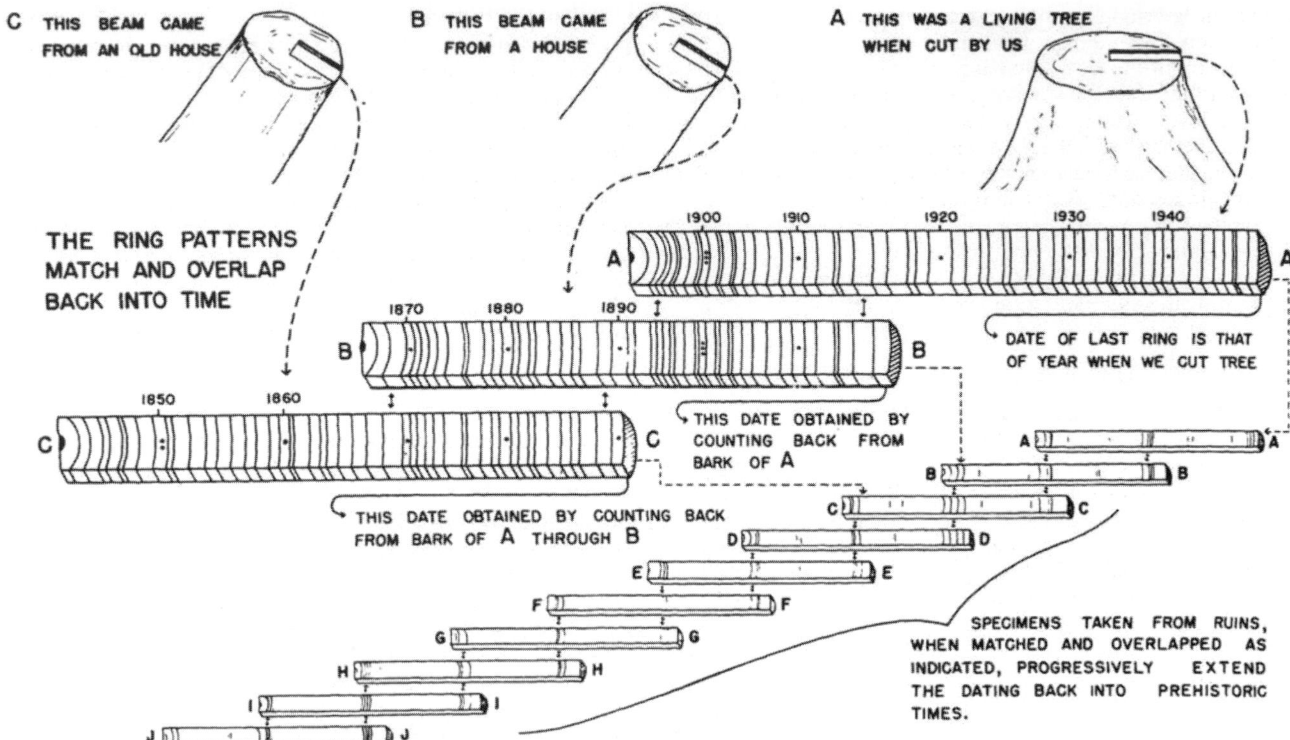

Figure T26 Crossdating is the basic principle of dendrochronology that provides annual resolution and the ability to date samples from dead wood. (Reprinted with permission from the University of Arizona Press, modified from Stokes and Smiley, 1996.)

considered for all intents and purposes absolute (Douglass, 1946; LaMarche and Harlan, 1973; Baillie, 1982). The certainty of crossdating allows one to average the measurements of the dated growth rings from many trees to obtain an important indicator of the relative effect of yearly climate on the mean yearly growth of the sampled trees.

However, tree-ring dating cannot be applied to all wood materials. Some tree species, particularly those from the tropics or subtropics, may have rings too poorly defined for crossdating. Other species may have distinct growth rings, but their ring features do not exhibit sufficient variation in common among trees to enable one to crossdate them. Still others have so many complications with both false and missing rings that synchrony of ring features is not certain, and thus dating is impossible. In addition, there are some trees that are too short-lived and have an insufficient number of rings overlapping in time to allow accurate crossdating. Therefore, in the past, dendrochronology has been applied to a restricted number of species and sites in temperate regions, but increasingly more species are being examined for their dendrochronological potential and researchers are finding unique sites in the tropics with trees that produce annual growth. Gymnosperms include the most favored species for dendrochronological investigation because they generally have the least ambiguous ring structure and most datable ring characteristics.

Dendroclimatology is considered a subfield of dendrochronology because dendrochronological procedures must be applied to date the ring materials. This assures that measurements such as ring widths from different trees of the same species are properly placed in time and can safely be averaged to produce a single chronology for the sampled site. This chronology is the time series that is used to reconstruct past climatic variations.

Tree-ring parameters reflecting variations in climate

Until recently dendroclimatic studies were largely confined to analyses of ring widths. Sometimes earlywood and latewood widths were considered, but the exact boundary between the two layers within the same ring was difficult to measure optically. There are new developments that now make these and other ring measurements practical. Wood density changes revealed by optical film density after exposure of the wood to X-rays provide information on intraseasonal climatic variations not contained in simple ring-width measurements. Also a specific density value can be chosen and used to determine the exact position of the earlywood–latewood boundary within each ring.

The area occupied by vessels in the ring can be measured optically and related to temperature and precipitation. Isotopes such as ^{18}O, deuterium, and ^{13}C in the cellulose of the ring also appear to contain a signal of past climatic conditions (Long, 1982; Wigley, 1982; Leavitt, 2002). For example, Feng and Epstein (1996) found a worldwide warming trend over the past 100 years using deuterium isotopes.

Some of these parameters offer considerably more promise than others, especially maximum wood density of the ring (Briffa et al., 1992; D'Arrigo et al., 1992; Luckman et al., 1997). Briffa et al., (1992) used a network of densitometric data from tree rings to reconstruct summer temperature across western North America. They found in their particular case that maximum latewood density was more adept at reconstructing summer temperature than was ring width.

For many of the less promising procedures the relationships to climatic variables are still not well understood. At present most of them involve more costly equipment and time-consuming procedures than the collection and processing of ring widths. However, these new parameters provide new options for dendroclimatic research. They can furnish considerably more information than ring-width analysis, especially for species on moist temperate sites where the ring widths are more complacent and contain less variation related to climatic conditions than growth rings from stressed trees in the American southwest.

The growth response to climate

The climatic signal in a tree-ring record arises from a variety of limiting environmental conditions occurring throughout the calendar year. The most obvious are those conditions affecting the rate and duration of cambial cell division during the growing period. Trees growing at high altitudes, near their polar limits, or on very arid sites may have very short growing seasons lasting for only a few weeks. Under less limiting conditions the cambium can remain active longer and growth can sometimes continue into early fall.

At other times in the year the growing tissue usually remains dormant. However, many plant processes such as photosynthesis, food storage and consumption, respiration, water uptake and loss, root growth, and ion absorption continue even through midwinter (Fritts, 1976). Whenever environmental conditions limit these processes, conditions may change within the plant that can alter the rate and timing of cambial division, the amount of cell enlargement, and cell-wall thickness in one or more of the growing seasons that follow.

The ring widths of trees growing on predominantly arid sites, for example, can be limited chiefly by conditions linked to low precipitation and high temperatures during the growing season, but at certain times during the year other factors such as low temperatures, excess moisture, heavy snow pack, and cloud cover can be limiting. Low temperatures can markedly reduce the ring widths in trees growing near their upper altitudinal limits, but low moisture and high wind also can be important limiting factors.

Different factors become limiting at different times during the year. There are also differences in response attributable to species, topography of the site, altitude, and a variety of other characteristics such as soil drainage, soil color, mineral nutrition, or shading from neighboring trees (Fritts, 1976). Any of these, as well as other factors, can modify the action of macroclimatic factors on the microenvironments of the trees and ultimately affect the tree's growth.

Modeling the climate response

Some dendroclimatological studies, especially the earlier works, assume a simple linear climate response developed in an *a-priori* fashion or inferred from a simple correlation analysis. The appropriate climatic conditions were then deduced from the associated variations in the past growth record (Schulman, 1956; Fritts, 1976). As dendroclimatological technology advanced, one type of multivariate technique called a response function (Fritts et al., 1971; Fritts, 1974) was developed to determine the overall tree-ring response to climatic factors. The generalized results of the response function were then used to infer the climatic conditions indicated by the tree-ring record.

We now know that it usually is too simplistic to assume growth is a linear response to a single climatic factor such as precipitation or temperature, even for trees growing in extremely arid or cold sites. On the other hand, the physiological processes behind the tree-growth and climate linkages are too poorly understood to construct equations for a mechanistic growth model. The only feasible approach at present is to model ring width as a lagging and autoregressive, as well as direct, response to a climate input (Figure T27) of factors that can vary throughout the year. Then one can apply a multivariate statistical model (Figure T28) to calibrate the ring-width variation with all significantly correlated climatic factors and apply calibration equations to past tree growth to reconstruct past climate.

Climate factors correlated with ring width

The most common climatic variables that have been considered are those that can be reasonably associated with environmental conditions that can limit processes affecting tree growth. Monthly mean, mean maximum or mean minimum temperature, or total monthly precipitation have been correlated or calibrated most often with ring width simply because long homogeneous records of these variables are readily available from weather stations near the tree sites. Relative humidity, sunshine, snowfall, dates of freezing as well as first and last frost, and other related measurements have also been considered (Fritts, 1976). However, the records of these types of measurements frequently are too short and are less commonly available from stations near the tree sites.

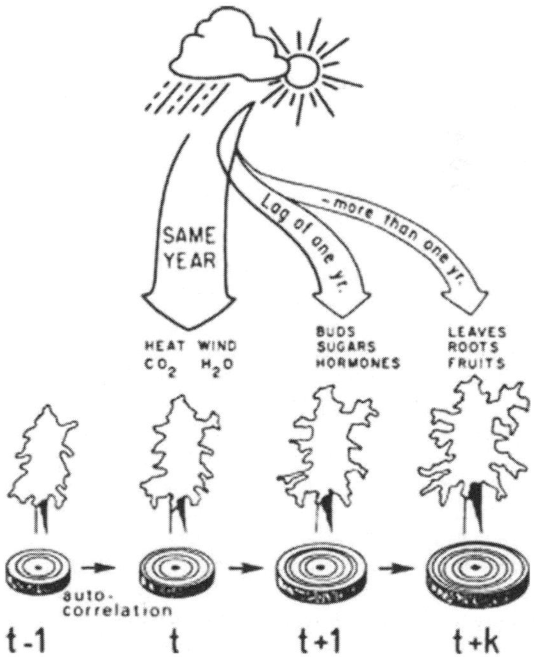

Figure T27 Climate of a given year *t* affects ring width for the same year through heat, wind, carbon dioxide, water, and other factors that ultimately impinge on tree growth. However, climate can also affect the width of rings formed in succeeding years, $t + 1$ up to $t + k$, through its effects on buds, sugar, hormones, and the amounts of leaves, roots, and fruits that in turn can influence growth. The width of the ring in year $t - 1$ is also statistically related to the ring width in year *t*. This can be modeled as autocorrelation.

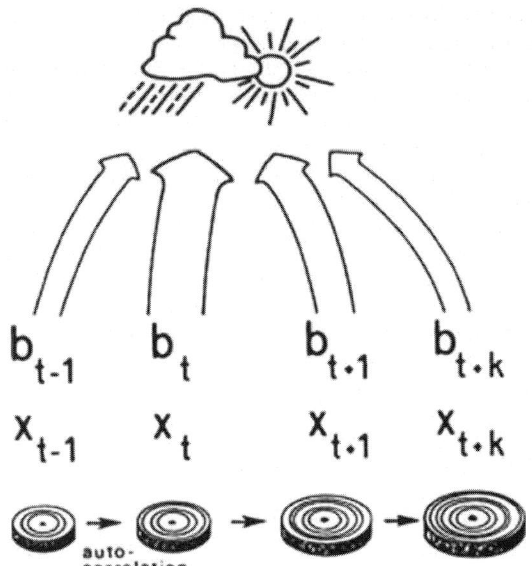

Figure T28 A multivariate transfer function can be applied to the growth model in Figure T27 to calibrate measurements of growth with measurements of climatic factors. Partial regression coefficients, or scaling factors, b_i are obtained, which are multiplied by the corresponding ring-width indices, x_i, and their products added to obtain an estimate or the calibrated variable of year *t* climate. The scaling factors are then applied to ring-width indices for early time periods to reconstruct the corresponding past climate. The scaling factor b_{t-1} is included in the model to correct the causal linkages for any autocorrelated effects.

There are some types of climatic variables such as sea-level pressure and sea-surface temperature that are known not to impinge directly on any physiological processes important to tree growth but that have been found to be significantly correlated with ring width. Apparently these variables are so closely linked with other climatic variables such as cloud cover, light levels, soil moisture, temperature, and wind, which all can influence physiological conditions affecting tree growth (Brubaker, 1980) that the linked variable can be calibrated with and reliably reconstructed from ring widths (Fritts, 1976; D'Arrigo et al., 1993).

Two or more variables such as temperature and precipitation are sometimes combined into a single drought index that appears better related to growth than either temperature or precipitation alone. Palmer Drought Severity Index (PDSI; Palmer, 1965) is the most frequently used composite variable that seems to affect tree growth (Meko et al., 1993; Cook et al., 1999; Stahle et al., 2000).

Measured stream flow can be calibrated with and reconstructed from tree growth because it is correlated with watershed soil moisture, which affects growth (Stockton and Jacoby, 1976; Stockton et al., 1985; Holmes et al., 1982). Fluctuating lake levels, affecting soil moisture and soil oxygen available to trees along the shore, have been successfully calibrated with ring width. The recorded annual catch of albacore tuna north of San Francisco has also been calibrated with and reconstructed from tree growth in western North America (Clark et al., 1975). This was possible because the schools of fish respond to sea-surface temperatures, which are in turn linked with the variations in the prevailing winds, surface pressure, and movement of storms inland, all of which influence the climatic factors impinging on tree growth.

El Niño and the Southern Oscillation (ENSO) affect tree growth in many places around the world through climatic teleconnections. Because of the annual resolution and millennial length of chronologies, tree rings are an excellent tool for reconstructing past ENSO events. Sites in the southwestern United States that are sensitive to winter precipitation have correlated significantly with ENSO events (D'Arrigo and Jacoby, 1991). This association is so robust that southwestern fire occurrence decreases following strong El Niño events (Swetnam and Betancourt, 1990; Grissino-Mayer and Swetnam, 2000).

Douglass was the first to propose that sunspots may be recorded in tree rings through a climatic influence. Since that time many researchers have pursued and identified significant association between the sunspot cycle (11 years) or the double sunspot cycle (22 years) and tree-ring width (Mitchell et al., 1979; Murphy et al., 1996; Cook et al., 1997). Although others have not found significant sunspot cycles recorded in trees rings, which has led to some discussion in the literature (LaMarche and Fritts, 1972; Briffa, 1994).

Chronology development

Modern dendroclimatologists have learned not to use ring measurements from a single tree to deduce past variations in climate because the climatic signal in a single tree's ring-width variance is too small. The measurements can be corrected for nonclimatic influences, such as changes in tree age, by indexing procedures. The indices for many trees from a given locale can be average together to form a mean chronology for the site. The more-or-less random nonclimatic noise found in different trees, including any measurement errors, cancel one another so the yearly mean index values have a much-reduced variance. The climatic signal represented by the variance common in all trees, on the other hand, is not reduced. Thus in the averaging process the ratio of signal to noise is enhanced. Using these techniques, chronologies can be developed that are responding to forcing factors at different spatial scales. By grouping trees together into stand chronologies, then aggregating stands into watershed-level chronologies, and finally combining watershed-level chronologies into regional chronologies, dendrochronologists can explore these forcing factors at multiple spatial scales (Figuer T29; Swetnam and Baisan, 1996).

Other dendroclimatic practices and procedures have been developed to facilitate data acquisition and assure that the common signal is largely attributable to macroclimatic linkages. First, a site is carefully selected to maximize the amount of climatic variance in the sampled ring widths. The more climate has been the dominant limiting factor, the greater the common variation in the ring-width variable due to the climatic signal, and the smaller the number of cores needed to obtain a given ratio of signal-to-noise. Then, so as not to harm the living tree, cylinders of wood about 4–5 mm in diameter are extracted from two or more sides of at least 10 trees growing on the same kind of site. Trees are selected that show no evidence of major nonclimatic influences such as fire, insect damage, or forest cutting.

In the laboratory the samples are mounted and the cross-section surface prepared so that the rings are clearly visible. The ring structures in the different cores are carefully crossdated to determine the exact year of ring formation (Stokes and Smiley, 1996). This technique uses the pattern of wide and narrow rings to visually determine where two samples of wood date against

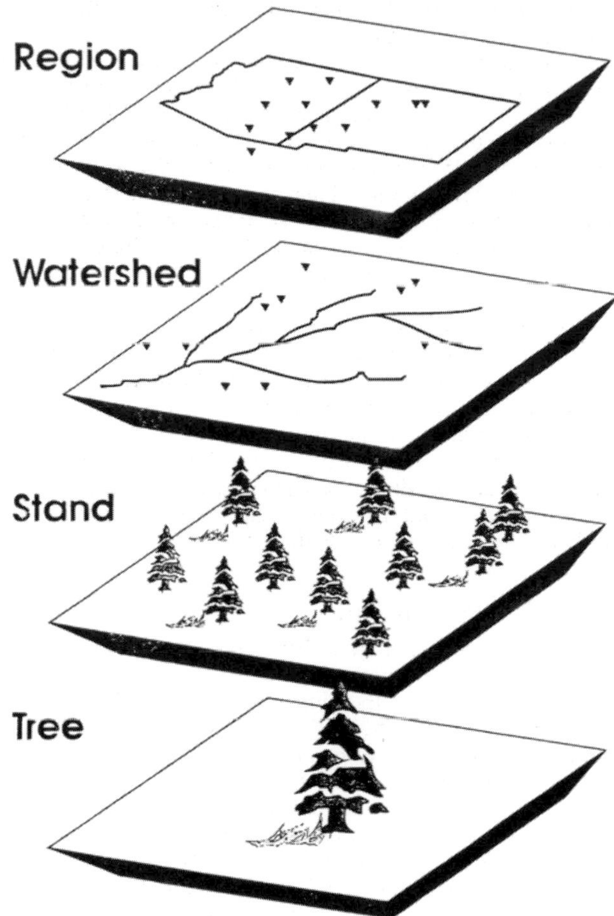

Figure T29 By collecting samples over a broad area and aggregating those samples based on topographic breaks or air mass boundaries, we can reconstruct environmental variables at multiple spatial scales. (Reprinted with permission from USDA Forest Service Swetnam and Baisan 1996.)

each other. This is the key principle of dendrochronology and provides the annual resolution needed for comparison to climatic data. Characteristics such as ring width, earlywood and latewood width, or wood density may be used for dating purposes. These characteristics are often measured and computer programs are used to provide quality control checks of the dating accuracy (Holmes, 1983) and the precision of measurement (Fritts, 1976).

As was stated earlier, many measurement parameters have substantial trends over time associated with the changing geometry of the ring, increasing tree age, and slowly changing features of the tree's immediate environment. These very low-frequency trends and changes in growth can be approximated by an exponential or polynomial curve, cubic spline, or some other similar function. In the case of ring-width measurements, an appropriate curve is fitted to the data for each core by least-squares techniques, and the ring width for each year is divided by the value from the fitted curve for that year to obtain a relative index of growth (Fritts, 1976; Graybill, 1982). This procedure is called standardization. A slightly different procedure is sometimes needed to standardize other types of measurements and recently Cook (1985) has developed a new procedure using time-series modeling techniques.

Since the changes associated with the growth function itself can contribute to a large percentage of the ring-width variance, the removal of this function by standardization enhances the climatic signal remaining in the indexed measurements. Some long-term climatic trends in growth may also be removed by this standardization procedure.

Both the mean and variance of the standardized indices for most ring-width time-series are relatively homogeneous over time. In a sample of chronologies from arid-site trees (Fritts and Shatz, 1975), roughly 90% approximated a normal distribution. However, the autocorrelation in standardized tree-ring indices is usually higher than it is for annual averages of most instrumental climatic measurements.

The signal-to-noise ratio in a typical sample of ring-width indices from a single core ranges from 0.1 to somewhat over 1.0. When the indexed values from all cores and trees collected and dated for a site are averaged to obtain a site chronology, the signal-to-noise ratio is increased. Arid-site chronologies developed from 10 or more trees, two cores per tree, typically have signal-to-noise ratios of 15 or higher (Graybill, 1982).

Generally the data are combined only from trees of one species growing on one site type because the measurements from these trees are likely to portray a consistent growth response. The chronology from such a group of trees will provide the clearest information on climate. If data are averaged from trees of different species or from sites with different characteristics, the chronology is more likely to contain a mixture of diverse climatic responses.

Climate also varies over space. Thus a reasonable sampling strategy is to treat collections from different localities, species, and site types as different chronologies in order to maintain the clearest tree-ring response. Instead of averaging the data from different chronologies, the chronologies can be entered into an array of m different chronologies of n years in length, each representing a particular species, type of site, and geographic area (Fritts and Shatz, 1975). The m elements of the array can be analyzed to extract the dendroclimatic information pertinent to each study of climate. Other ring parameters such as maximum and minimum wood density can be measured, standardized, and entered as additional chronologies to increase the size of the array and its information content.

There will be a variety of climatic signals in such a chronology array, but the different kinds of information can be factored out and used along with the information that is in common to reconstruct seasonal, annual, and decade-long variations in temperature, precipitation, drought, atmospheric pressure, and other climatic parameters. Some of the calibration procedures that can be used to accomplish this factoring will now be discussed.

Calibration

The complexities of autocorrelation and diversity in a tree-ring response can be approximated by a linear multiple regression model of m predictor variables applied at k lags to allow for m times k possible growth–climate linkages. The sign and magnitude of the multiple regression coefficients express the direction and magnitude of each linkage. A regression using empirical orthogonal functions (EOF) of monthly climatic variables as predictors and ring-width indices as predictands provides the equation for a response function. However, to interpret the equation the partial regression coefficients associated with the EOF must be transformed into coefficients of the climatic variable space by multiplying them with the eigenvectors of climate (Fritts et al., 1971; Fritts, 1976).

If variables of tree growth are the statistical predictors and a variable of climate is the predictand, the multiple regression equation becomes a transfer function that transfers the tree-ring index data into estimates of climate (see Figure T29). Multivariate transfer functions of varying complexity can be used to sort out and demodulate different tree-growth and climate linkages. Unique coefficients are calculated for each variable, season, lag, and chronology. Autoregressive moving-average time-series analyses (Meko, 1981; Cook, 1985) can also be used to identify and correct each chronology for the important autoregressive and moving-average elements. The residual from the fitted model from the time-series analysis approximates a simplified nonlagging signal of the limiting conditions of climate. This simplification appears to clarify some of the higher frequency variations in reconstructions of climate.

Verification

It is possible that large and complex transfer functions provide statistically significant calibration statistics but fail when they are applied to independent data (data not used for calibration). Although the reliability of regression estimates almost always declines for such data, it is important to assess the magnitude of the decline and to test whether the independent reconstructions are likely to have arisen by chance.

Climatic data or other related information outside the calibration period can be set aside for this purpose. Figure T30 shows an example where the expected decline in the association between the instrumental (real) data and the reconstructions from the independent compared to the dependent record is evident from the statistics. However, the first two statistics that can be tested (the correlation and the correlation of the first differences) are significantly different from random expectation ($p \leq 0.05$) for both time periods. The reduction of error terms greater than zero indicates meaningful climatic information has been reconstructed. Once the reconstructions are validated, as in the above example, the statistics from different calibration models can be compared to select the transfer function providing the most reliable climatic estimates.

Where suitable data exist, such reconstructions should be further validated for time periods outside both the calibration and verification periods. Such tests can use information from other sources such as historical data, observations on phenology, dates of lake freezing (Gordon et al., 1985) or climate reconstructions from independent proxy sources (Lough and Fritts, 1985), including independent tree-ring data. These indirect sources of paleoclimatic information must reflect climate in a similar way as the tree-ring data used in the reconstruction, have a time resolution comparable to the tree-ring data, and have a sufficiently strong signal of the reconstructed climatic variable to make an effective comparison possible. However, the inequalities between different data sources are often so great, or the errors so large in the datasets compared, that this second type of validation test can be statistically insignificant even when the climatic reconstruction is correct. For this reason, variables tightly coupled to the particular climatic variable in question and with little inherent error of their own provide the best validity checks. Much work remains to be done on finding suitable data and developing the optimum techniques for this second type of validation (see Wigley et al., 1984).

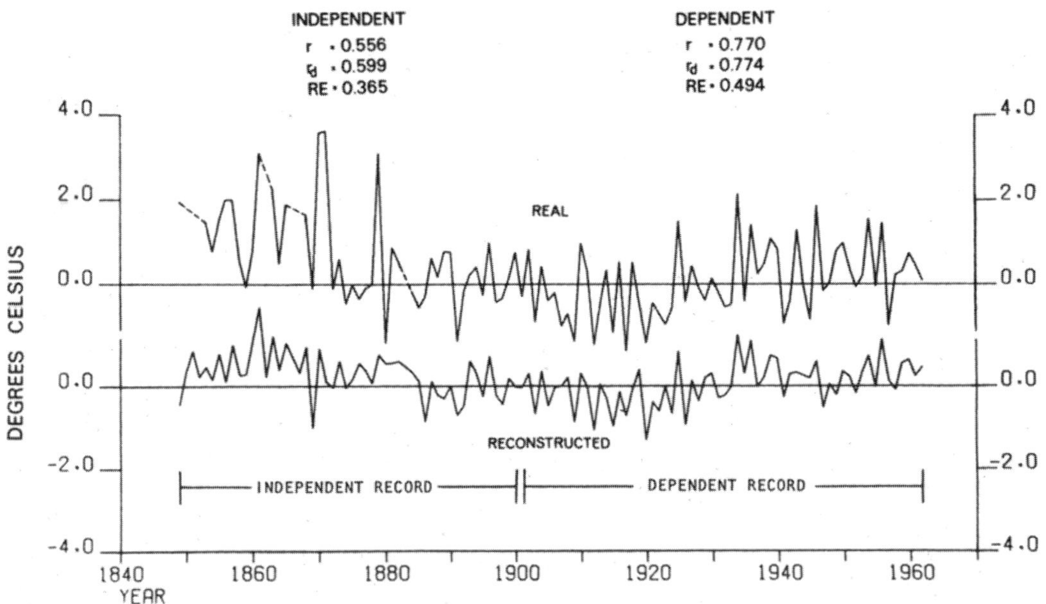

Figure T30 The mean spring temperature (real) and estimated (reconstructed) data for Santa Fe, New Mexico, plotted for the dependent record used for calibration and the independent record used for verification. Three of the six statistics of association used for verification are also shown: r = the correlation coefficient, r_d = the correlation coefficient for the first differences, and RE = the reduction of error. Temperatures and their estimates are expressed as departures from the 1901–1970 mean instrumented value.

Spatial climate reconstruction strategy

The simplest strategy is to calibrate and reconstruct one climatic record from one tree-ring chronology. However, the signal of climate in a single chronology is often too weak to obtain a reliable reconstruction. If an array of chronologies can be properly calibrated, the collective signal can be utilized while the influences of the more or less random chronology errors are averaged out. This leads to spatial reconstructions of climate on the scale of a single continent.

It has been reported (Fritts, 1982) that in some cases it may be better to calibrate climatic data for separate seasons rather than using larger intervals of time because of differences in the seasonal growth responses. Canonical regression analysis (Blasing, 1978) can be used to estimate more than one predictand. This technique has been extensively used to reconstruct the climate over different seasons and over spatial arrays of instrumented data using spatial arrays of tree-ring chronologies (Fritts et al., 1979; Fritts and Lough, 1985). Different lags in the climate response can be modeled by including tree-ring datasets at different lags as predictors and the autoregressive elements can be assessed or removed beforehand by applying time-series modeling.

Fritts (1965) first applied networks of tree-ring chronologies in climatology to reconstruct a spatial array of climate. Since that time many researchers have used a spatial array of tree-ring sites to reconstruct climate over large areas (D'Arrigo and Jacoby, 1991; Meko et al., 1993; Stahle et al. 2000). This technique has proved to be more robust than single-site climate reconstructions because the spatial continuity of the sites helps to strengthen the broad-scale climate signal (Meko et al., 1993). Stahle et al. (2000) also used a gridded network of 154 points covering a 2° × 3° latitude/longitude grid to reconstruct drought across North America from AD 1500 to AD 1978. This work was able to put the 1950s drought and 1930s dust bowl into the perspective of the past 500 years. They found that a sixteenth-century drought was by far the most severe drought during that time period and may have contributed to the fall of early settlements such as the Jamestown and Roanoke colonies in the southeastern United States (Stahle et al., 2000).

The first high-quality grids of arid-site tree-ring chronologies were assembled for western North America by Schulman (1956), Fritts (1965), Fritts and Shatz (1975), Stockton et al. and Meko (1975), Brubaker (1982), and Stockton et al. (1985). A North American grid of high-altitude chronologies from bristlecone pine was published by LaMarche (1974). Chronology information from the eastern United States was assembled first by Dewitt and Ames (1978) and then updated by Cook (1982). Chronology information for the American Arctic was first summarized in a systematic way by Cropper and Fritts (1981) and then updated by Jacoby (1982), Brubaker and Cook (1983), and Stockton et al. (1985).

Although the reliability of the climatic estimates generally declines with increasing distance from the tree sites, the estimates of variables such as atmospheric pressure and air temperature may remain relatively reliable at distances of 1000 km or more beyond the area covered by the tree sites (Fritts, 1991). However, the variances of the integrated climatic estimates are substantially lower than the variances of the unsmoothed instrumented data with which they were calibrated and verified, and they largely reflect conditions of climate averaged over long time scales and large regions.

Different strategies may be needed to obtain the most reliable climatic estimates over denser grids covering smaller areas in other regions of the world. In addition, the reliability of estimates will most certainly improve as tree-ring chronologies for new areas, species, and tree-ring parameters are added to the existing tree-ring chronology grids.

Multimillennial-length climate reconstructions

Over the past few decades of work, tree-ring research has expanded its temporal as well as spatial coverage. The efforts of many dendrochronologists from around the world are starting to produce multimillennial-length chronologies (Table T3). These chronologies enable us to examine today's climate and climate change from the perspective of over 2000 years of climatic history.

The long chronologies have aided our understanding of climatic trends. Mann et al. (1999) found a twentieth-century warming trend in the northern hemisphere that was anomalous over the past millennium, and were able to corroborate that the 1990s was the warmest decade and that 1998 was the warmest year in the past millennium. A distinctive warming trend was also reconstructed in New Zealand, demonstrating increased temperatures since the 1950s (D'Arrigo et al., 1998).

Worldwide interest

The first international workshop on dendroclimatology, held in 1974 at the University of Arizona, Tucson, marked the beginning of international interest and cooperation in dendroclimtology. It was convened for the purpose of sharing the new

Table T3 Multimillennial tree-ring chronologies

Length (years)	Species	Location	Citation
9000	*Pinus longaeva*	White Mountains, CA, USA	Hughes and Graumlich, 1996
7500	*Pinus sylvestris*	Finnish Lapland	Eronen et al., 2002
7400	*Pinus sylvestris*	Swedish Lapland	Grudd et al., 2002
7272	*Quercus petraea* and *Quercus robur*	Germany and Ireland	Pilcher et al., 1984
4000	*Larix sibirica*	Yamal Siberia	Hantemirov and Shiyatov, 2002
3620	*Fitzroya cupressoides*	South America	Lara and Villalba, 1993
3600	*Lagarostrobos franklinii*	Tasmania	Cook et al., 2000
3000	*Sequoiadendron giganteum*	California, USA	Hughes et al., 1996
2427	*Larix gmelinii*	Taimyr, Siberia	Naurzbaev et al., 2002
2129	*Psuedotsuga menziesii* and *Pinus ponderosa*	EL Malpais, NM, USA	Grissino-Mayer, 1996

technology, which at that time had been largely developed at the Laboratory of Tree-Ring Research (Fritts, 1976). This was followed by the second international workshop on global dendroclimatology, held in July 1980 at the University of East Anglia, Norwich, United Kingdom. Workers from a variety of countries described new techniques developed at their laboratories, and there was much debate on the strengths, weaknesses, and suitability of some standard procedures used in dendroclimatology. Progress in the collection of new chronologies was reviewed and plans were generated for developing both regional and global tree-ring chronology networks. Soviet scientists also have held at least two all-Soviet Union conferences to stimulate similar dendroclimatological work through eastern Europe and central Asia.

These meetings led to international dendrochronology conferences which have provided a forum for the worldwide dendrochronological community to gather and exchange fresh ideas on the science. The first conference was held in Palisades, New York, in 1986 (Jacoby and Hornbeck, 1987); the second in Lund, Sweden in 1990 (Barthilin et al., 1992); then Tucson Arizona in 1994 (Dean et al., 1996); Mendoza, Argentina, in 1998; and the most recent was held in Laval, Canada, in 2002. Along with these conferences there is an International Dendroecological Fieldweek, which has been held in Europe for 19 years, and there is also a North American Dendroecological Fieldweek that is on its thirteenth annual event. These fieldweeks provide an opportunity to novice dendrochronologists and senior researchers to gather for an intensive week-long training session in dendrochronology.

The International Tree-Ring Data Bank (ITRDB) was established in 1974 at the Laboratory of Tree-Ring Research, University of Arizona, Tucson, for preserving all basic tree-ring measurements and the derived chronologies. That databank is now managed by the World Data Center–A (WDC-A) for Paleoclimatology at the National Geophysical Data Center (NGDC) in Boulder, Colorado, which is part of the National Oceanic and Atmospheric Administration (NOAA). The ITRDB now holds nearly 12 000 tree-ring chronologies representing over 150 tree and shrub species (Grissino-Mayer and Fritts, 1997).

Dendrochronologists are greatly expanding the spatial horizons of tree-ring research by examining the potential for dendrochronological research in many areas that were previously underrepresented. New chronologies have been published for the southern hemisphere from Argentina, Chile, New Zealand, Australia, and South Africa. Some of the new collection efforts in Europe have been described by Eckstein (1972, 1982). Over the past two decades tree-ring work has been expanded into tropical regions with relative success (Jacoby, 1989; Worbes, 1995; Stahle, 1999). For more information on dendroclimatology the reader should consult Schulman (1956), Fritts (1976), articles from the *Tree-Ring Research (*formerly the *Tree-Ring Bulletin)*, and Hughes et al. (1982).

Harold C. Fritts and James H. Speer

Bibliography

Baillie, M.G.L., 1982. *Tree-Ring Dating and Archaeology*. Chicago, IL: University of Chicago Press.

Bartholin, T.S., Berglund, B.E., Eckstein, D., and Schweingruber, F.H., 1992. Tree rings and environment. *Proceedings of the International Dendrochronological Symposium*, Ystad, South Sweden, 3–9 September 1990. Lundqua Report Volume 34.

Blasing, T.J., 1978. Time series and multivariate analysis in paleoclimatology. In Shuggart, H.H., Jr, ed., *Time Series and Ecological Processes*. SIAM-SIMS Conference Series, No. 5. Philadelphia, PA: Society for Industrial and Applied Mathematics, pp. 211–226.

Briffa, K.R., Jones, P.D., and Scweingruber, F.H., 1992. Tree-ring density reconstructions of summer temperature patterns across western North American since 1600. *Journal of Climate*, **5**(7): 735–754.

Briffa, K.R., 1994. Grasping at shadows? A selective review of the search for sunspot – related variability in tree rings. In Nesme-Ribes, E., ed., *The Solar Engine and its Influence on Terrestrial Atmosphere and Climate*. Berlin: Springer Verlag, pp. 417–435.

Brubaker, L.B., 1980. Spatial patterns of tree-growth anomalies in the Pacific Northwest. *Ecology*, **61**: 798–807.

Brubaker, L.B., 1982. Western North America. In Hughes, M.K., Kelly, P.M., Pilcher, J.R., and LaMarche, V.C., eds., *Climate from Tree Rings*. Cambridge: Cambridge University Press, pp. 118–126.

Brubaker, L.B., and Cook, E.R., 1983. Tree-ring studies of Holocene environments. In Wright, H.E., ed., *Late-Quaternary Environments of the United States*, vol. 2: *The Holocene*. Minneapolis; MN: University of Minnesota Press, pp. 222–35.

Clark, N.E., Blasing, T.J., and Fritts, H.C., 1975. Influence of intra-annual climatic fluctuations on biological systems. *Nature*, **256**(5515): 302–305.

Cook, E.R., 1982. Eastern North America. In Hughes, M.K., Kelly, P.M., Pilcher, J.R., and LaMarche, V.C., eds., *Climate from Tree Rings*. Cambridge: Cambridge University Press, pp. 126–133.

Cook, E.R., 1985. A time-series analysis approach to tree-ring standardization. PhD dissertation. Tucson, AZ: Laboratory of Tree-Ring Research, University of Arizona.

Cook, E.R., Meko, D.M., and Stockton, C.W., 1997. A new assessment of possible solar and lunar forcing of the bidecadal drought rhythm in the western United States. *Journal of Climate*, **10**: 1343–1356.

Cook, E.R., Meko, D.M., Stahle, D.W., and Cleaveland, M.K., 1999. Drought reconstructions for the continental United States. *Journal of Climate*, **12**(4): 1145–1162.

Cook, E.R., Buckley, B.M., D'Arrigo, R.D., and Peterson, M.J., 2000. Warm-season temperatures since 1600 BC reconstructed from Tasmanian tree rings and their relationship to large-scale sea surface temperature anomalies. *Climate Dynamics*, **16**(2–3): 79–91.

Cropper, J.R., and Fritts, H.C., 1981. Tree-ring width chronologies from the North American Arctic. *Arctic and Alpine Research*, **13**(3): 245–260.

D'Arrigo, R.D., and Jacoby, G.C., 1991. A 1000-year record of winter precipitation from northwestern New Mexico, USA: a reconstruction from tree-rings and its relation to El Niño and the Southern Oscillation. *The Holocene*, **1**(2): 95–101.

D'Arrigo, R.D., Jacoby, G.C., and Free, R.M., 1992. Tree-ring width and maximum latewood density at the North American tree line: parameters of climate change. *Canadian Journal of Forestry Research*, **22**(9): 1290–1296.

D'Arrigo, R.D., Cook, E.R., Jacoby, G.C., and Briffa, K.R., 1993. NAO and sea surface temperature signals in tree-ring records from the North Atlantic sector. *Quaternary Science Review*, **12**(6): 431–440.

D'Arrigo, R.D., Cook, E.R., Salinger, M.J., et al.,1998. Tree-ring records from New Zealand; long-term context for recent warming trend. *Climate Dynamics,* **14**(3): 191–199.

Dean, J.S., Meko, D.M., and Swetnam, T.W., 1996. Tree-rings, environment and humanity. Proceedings of the International Conference, Tucson, Arizona, 17–21 May 1994. *Radiocarbon* (special issue), pp. 889.

Dewitt, E., and Ames, M. (eds.), 1978. *Tree-ring Chronologies of Eastern North America*, vol. 1: Laboratory of Tree-Ring Research, Chronology Series IV. Tucson, AZ: University of Arizona.

Douglass, A.E., 1936. *Climate Cycles and Tree Growth*, vol. 3. Publication No. 289. Washington, DC: Carnegie Institution of Washington.

Douglass, A.E., 1946. Precision of ring dating in tree-ring chronologies. *University of Arizona Bulletin*, **17**(3).

Eckstein, D., 1972. Tree-ring research in Europe. *Tree-Ring Bulletin*, **32**: 1–18.

Eckstein, D., 1982. Europe. In Hughes, M.K., Kelly, P.M., Pilcher, J.R., and LaMarche, V.C., eds., *Climate from Tree Rings*. Cambridge: Cambridge University Press, pp. 142–148.

Eronen, M., Zetterberg, P., Briffa, K.R., Lindholm, M., Merilainen, J., and Timonen, M., 2002. The supra-long Scots pine tree-ring record for Finnish Lapland: Part 1, chronology construction and initial inferences. *The Holocene*, **12**(6): 673–680.

Feng, X., and Epstein, S., 1996. Climate trends from isotopic records of tree rings: the past 100-200 years. *Climate Change*, **33**: 551–562.

Fritts, H.C., 1965. Tree-ring evidence for climatic changes in western North America. *Monthly Weather Review*, **93**(7): 421–443.

Fritts, H.C., 1974. Relationships of ring widths in aridsite conifers to variations in monthly temperature and precipitation. *Ecol. Mons.*, **44**(4): 411–440.

Fritts, H.C., 1976. *Tree Rings and Climate*. London: Academic Press.

Fritts, H.C., 1982. An overview of dendroclimatic techniques, procedures and prospects, In Hughes, M.K., Kelly, P.M., Pilcher, J.R., and LaMarche, V.C., eds., *Climate from Tree Rings*. Cambridge: Cambridge University Press, pp. 191–197.

Fritts, H.C., and Lough, J.M., 1985. An estimate of average annual temperature variations for North America, 1602 to 1961. *Climatic Change*, **7**: 203–224.

Fritts, H.C., and Shatz, D.J., 1975. Selecting and characterizing tree-ring chronologies for dendroclimatic analysis. *Tree-Ring Bulletin*, **35**: 31–40.

Fritts, H.C., Blasing, T.J., Hayden, B.P., and Kutzbach, J.E., 1971. Multivariate techniques for specifying tree-growth and climatic relationships and for reconstructing anomalies in paleoclimate. *Journal Applied Meteorology*, **10**(5): 845–864.

Fritts, H.C., Lofgren, G.R., and Gordon, G.A., 1979. Variations in climate since 1602 as reconstructed from tree rings. *Quaternary Research*, **12**: 18–46.

Fritts, H.C. 1991. *Reconstructing Large-Scale Climatic Patterns from Tree-Ring Data.* Tucson, AZ: University of Arizona Press, p. 286.

Gordon, G.A., Lough, J.M., Fritts, H.C., and Kelly, P.M., 1985. Comparison of sea level pressure reconstructions from western North American tree rings with a proxy record of winter severity in Japan. *Journal Climatology Applied Meteorology*, **24**: 1219–1224.

Graybill, D.A., 1982. Chronology development and analysis. In Hughes, M.K., Kelly, P.M., Pilcher, J.R., and LaMarche, V.C., eds., *Climate from Tree Rings*. Cambridge: Cambridge University Press, pp. 21–30.

Grissino-Mayer, H.D., 1996. A 2129-year reconstruction of precipitation for northwestern New Mexico, USA. In Dean, J.S., Meko, D.M., and Swetnam, T.W., eds., *Tree Rings, Environment and Humanity*. Proceedings of the International Conference, Tucson, Arizona, 17–21 May 1994, 191–204.

Grissino-Mayer, H.D., and Fritts, H.C., 1997. The International Tree-Ring Data Bank: an enhanced global database serving the global scientific community. *The Holocene*, **7**(2): 235–238.

Grissino-Mayer, H.D., and Swetnam, T.W., 2000. Century-scale climate forcing of fire regimes in the American Southwest. *The Holocene*, **10**(2): 213–220.

Grudd, H., Briffa, H.R., Karlen, W., Bartholin, T.S., Jones, P.D., and Kromer, B., 2002. A 7400-year tree-ring chronology in northern Swedish Lapland: natural climatic variability expressed on annual to millennial timescales. *The Holocene*, **12**(6): 657–666.

Hantemirov, R.M., and Shiyatov, S.G., 2002. A continuous multimillennial ring-width chronology in Yamal, northwestern Siberia. *The Holocene*, **12**(6): 717–726.

Holmes, R.L., 1983. Computer-assisted quality control in tree-ring dating and measuring. *Tree-Ring Bulletin*, **43**: 69–78.

Holmes, R.L., Stockton, C.W., and LaMarche, V.C., Jr, 1982. Extension of river flow records in Argentina. In Hughes, M.K., Kelly, P.M., Pilcher, J.R., and LaMarche, V.C., eds., *Climate from Tree Rings*. Cambridge: Cambridge University Press, pp. 168–170.

Hughes, M.K., and Graumlich, L.J., 1996. Multimillennial dendroclimatic studies from the western United States. In Jones, P.D., Bradley, R.S., and Jouzel, J., eds., *Climatic Variations and Forcing Mechanisms of the last 2000 years*. NATO ASI Series vol. 141, pp. 109–124.

Hughes, M.K., Kelly, P.M., Pilcher, J.R., and LaMarche, V.C., Jr eds., 1982. *Climate from Tree Rings*, Cambridge: Cambridge University Press.

Hughes, M.K., Touchan, R., and Brown, P.M., 1996. A multimillennial network of giant sequoia chronologies for dendroclimatology. In Dean, J.S., Meko, D.M., and Swetnam, T.W., eds., *Tree Rings, Environment and Humanity:* Proceedings of the International Conference, Tucson, Arizona, 17–21 May 1994, pp. 225–234.

Jacoby, G.C., 1982. The Arctic. Hughes, M.K., Kelly, P.M., Pilcher, J.R., and LaMarche, V.C., eds., *Climate from Tree Rings*. Cambridge: Cambridge University Press, pp. 107–118.

Jacoby, G.C., Jr., and Hornbeck, J.W., compilers 1987. Proceedings of the International Symposium on Ecological Aspects of Tree-Ring Analysis, August 17–21, 1986, Tarrytown, New York. U.S. Department of Energy, Publication CONF-8608144: 1–726.

Jacoby, G.C., 1989. Overview of tree-ring analysis in tropical regions. *IAWA Bulletin,* **10**(2): 99–108.

LaMarche, V.C., Jr., and Fritts, H.C., 1972. Tree-rings and sunspot numbers. *Tree-Ring Bulletin,* **32**: 19–33.

LaMarche, V.C., Jr, 1974. Frequency-dependent relationships between tree-ring series along an ecological gradient and some dendroclimatic implications. *Tree-Ring Bulletin*, **34**: 1–20.

LaMarche, V.C., Jr, and Harlan, T.P., 1973. Accuracy of tree-ring dating of bristlecone pine for calibration of the radiocarbon time scale. *Journal of Geophysical Research*, **78**(36): 8849–8858.

Lara, A., and Villalba, R., 1993. A 3,620-year temperature reconstruction from *Fitzroya cupressoides* tree rings in southern South America. *Science*, **260**: 1104–1106.

Leavitt, S.W., 2002. Prospects for reconstruction of seasonal environment from tree-ring delta C-13: baseline findings from the Great Lakes area, USA. *Chemical Geology*, **192**(1–2): 47–58.

Long, A., 1982. Stable isotopes in tree rings. In Hughes, M.K., Kelly, P.M., Pilcher, J.R., and LaMarche, V.C., eds., *Climate from Tree Rings*. Cambridge: Cambridge University Press, pp. 12–18.

Lough, J.M., and Fritts, H.C., 1985. The Southern Oscillation and tree rings: 1600–1961. *Journal of Climatology Applied Meteorology*, **24**: 952–966.

Luckman, B.H., Briffa, K.R., Jones, P.D., and Schweingruber, F.H., 1997. Tree-ring based reconstruction of summer temperatures at the Columbia Icefield, Alberta, Canada, AD 1073–1983. *The Holocene*, **7**(4): 375–389.

Mann, M.E., Bradley, R.S., and Hughes, M.K., 1999. Northern hemisphere temperatures during the past millennium: Inferences, uncertainties, and limitations. *Geophysical Research Letters*, **26**(6): 759–762.

Meko, D. M., 1981. Applications of Box-Jenkins methods of time-series analysis to the reconstruction of drought from tree rings. PhD. dissertation, University of Arizona, Tucson.

Meko, D.M., Cook, E.R., Stahle, D.W., Stockton, C.W., and Hughes, M.K., 1993. Spatial patterns of tree-growth anomalies in the United States and southeastern Canada. *Journal of Climate*, **6**: 1773–1786.

Mitchell, J.M., Stockton, C.W., and Meko, D.M., 1979. Evidence of a 22-year rhythm of drought in the western United States related to the Hale Solar Cycle since the 17th century. In McCormac, B.M., and Seliga, T.A., eds., *Solar-Terrestrial Influences on Weather and Climate*. Dordrecht: Reidel, pp. 125–143.

Murphy, J.O., Sampson, H., Veblen, T.T., and Villalba, R., 1996. Reconstruction of the annual variation in Zurich sunspot number from tree ring-index time series. In Dean, J.S., Meko, D.M., and Swetnam, T.W., eds., *Tree Rings, Environment and Humanity*. Proceedings of the International Conference, Tucson, Arizona, 17–21 May 1994, Radiocarbon pp. 853–869.

Naurzbaev, M.M., Vaganov, E.A., Sidorova, O.V., and Schweingruber, F.H., 2002. Summer temperatures in eastern Taimyr inferred from a 2427-year late Holocene tree-ring chronology and earlier floating series. *The Holocene*, **12**(6): 727–736.

Palmer, W.C., 1965. Meteorological drought. *USDC Weather Bureau Research Paper*, Num. 45.

Pilcher, J.R., Baillie, M.G.L., Schmidt, B., and Becker, B., 1984. A 7,272-year tree-ring chronology for western Europe. *Nature*, **312**: 150–152.

Schulman, E., 1956. *Dendroclimatic Changes in Semiarid America*. Tucson, AZ: University of Arizona Press.

Stahle, D.W., 1999. Useful strategies for the development of tropical tree-ring chronologies. *IAWA Journal*, **20**(3): 249–253.

Stahle, D.W., Cook, E.R., Cleaveland, M.K., et al., 2000. Tree-ring data document 16th century megadrought over North America. *EOS. Transactions of the American Geophysical Union*, **81**(12): 121–125.

Stockton, C.W., and Jacoby, G.C., Jr, 1976. Long-term surface water supply and stream flow trends in the Upper Colorado River Basin based on tree-ring analysis. *Lake Powell Research Project Bull.* **18**, 709.

Stockton, C.W., and Meko, D.M., 1975. A Long-term history of drought occurrence in western United States as inferred from tree rings. *Weatherwise*, **28**(6): 245–249.

Stockton, C.W., Boggess, W.R., and Meko, D.M., 1985. Climate and tree rings. In Hect, A.D., ed., *Paleoclimate Analysis and Modeling*. New York: Wiley, pp. 71–161.

Stokes, M.A., and Smiley, T.L., 1996. *An Introduction to Tree-Ring Dating*. Tucson, AZ: University of Arizona Press.

Swetnam, T.W., and Baisan, C.H., 1996. Historical fire regime patterns in the southwestern United States since AD 1700. In Allen, C.D. ed., *Fire Effects in Southwestern Forests*. Proceedings of the Second La Mesa Fire Symposium, Los Alamos, New Mexico, 29–31 March 1994, pp. 11–32.

Swetnam, T.W., and Betancourt, J.L., 1990. Fire–Southern Oscillation relations in the southwestern United States. *Science*, **249**: 1017–1020.

Wigley, T.M.L., 1982. Oxygen-18, carbon-13 and carbon-14 in tree rings. In Hughes, M.K., Kelly, P.M., Pilcher, J.R., and LaMarche, V.C., eds., *Climate from Tree Rings*. Cambridge: Cambridge University Press, pp. 18–21.

Wigley, T.M.L., Briffa, K.R., and Jones, P.D., 1984. On the average value of correlated time series, with applications in dendroclimatology and hydrometeorology. *Journal of Climatology and Applied Meteorology*, **23**(2): 201–213.

Worbes, M., 1995. How to measure growth dynamics in tropical trees: a review. *IAWA Journal*, **16**(4): 227–351.

Cross-references

Climate Variation: Historical
Vegetation and Climate

TROPICAL AND EQUATORIAL CLIMATES

The tropical region of the Earth is generally defined geographically as the area between the Tropic of Cancer, situated at 23.5°N latitude, and the Tropic of Capricorn at 23.5°S. The tropics are also sometimes said to be the latitudes that lie between, and partly include, the subtropical high-pressure regions that are centered on average at about 30–35°N and S latitudes. Tropical climates are most commonly defined as those occurring between the subtropical high-pressure regions (the climatic tropics), although some suggest that tropical climates extend between 35°N and S latitudes. Included within the tropical region is the equatorial zone, which is variably defined but most often stated to be the zone extending 3° of latitude on either side of the equator, although an equatorial climate is generally stated to be the climatic type extending 10–12°N and S of the equator (and occurring on low ground–non-highlands only). Therefore, an equatorial climate is most often noted to be a subtype of the broader category of tropical climate.

More specifically, tropical climates have been defined using several climatic criteria. Supan in 1896 defined regions of tropical climates as those having average annual temperatures exceeding 68°F (20°C), a value also adopted by Trewartha (1954). Köppen in 1918 used a criterion of all months exceeding an average temperature of 64°F (18°C), which was also adopted by Miller (1931). A classification by Trewartha in 1968 employed 65°F (18.3°C) for the average of the coolest month. Figure T31 shows the location of the 18°C isotherm for the average temperature of the coolest month. Thus the tropical and equatorial climates cover approximately half of the Earth's surface.

General characteristics

Temperatures

Tropical climates, in addition to having all months with average temperatures above about 64°F (18°C), experience no killing frosts. The terms summer and winter have essentially no meaning in terms of temperature in these climates, and these regions are sometimes described as having no winters. Greeks referred to tropical regions as the Torrid Zone. Another main temperature characteristic of tropical climates is revealed in Figure T31. Average annual temperature range is small, and day–night (diurnal) temperature ranges, although also small, often exceed the average annual range. These characteristics have led to the often-quoted statement that "night is the winter of the tropics". Temperature range, diurnally and seasonally, is generally least in the equatorial region and increases until the poleward extent of this climatic type is reached. Temperature characteristics of tropical dry climates vary from those stated here, as discussed in the section on semiarid and arid climatic types.

Pressure and wind

As noted, the poleward border of tropical climates is situated generally at the high-pressure zones located at about 30–35°N and S latitudes. Here are located the subtropical anticyclones. Near the equator is the low-pressure zone termed the Intertropical Convergence Zone (ITCZ). This is generally considered to be the region of convergence of the northeast trade winds of the northern hemisphere with the southeast trades that are primarily in the southern hemisphere (Figure T32). However, secondary convergence areas exist, and satellite-derived wind data have confirmed the occurrence of separate northern and southern ITCZ. The northern one is created by the actual convergence of the trades, and migrates north and south of the equator with the seasons and the migration of the solar declination, whereas the southern ITCZ remains in the southern hemisphere and is produced by a

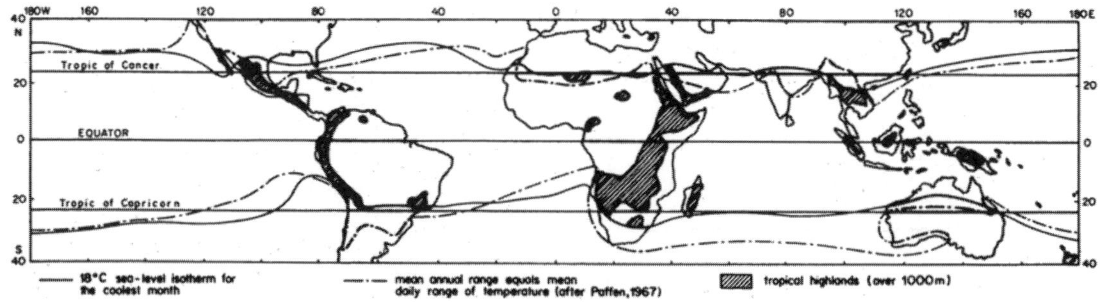

Figure T31 The 18°C sea-level isotherm and annual/daily-range isotherm (Nieuwolt, 1977).

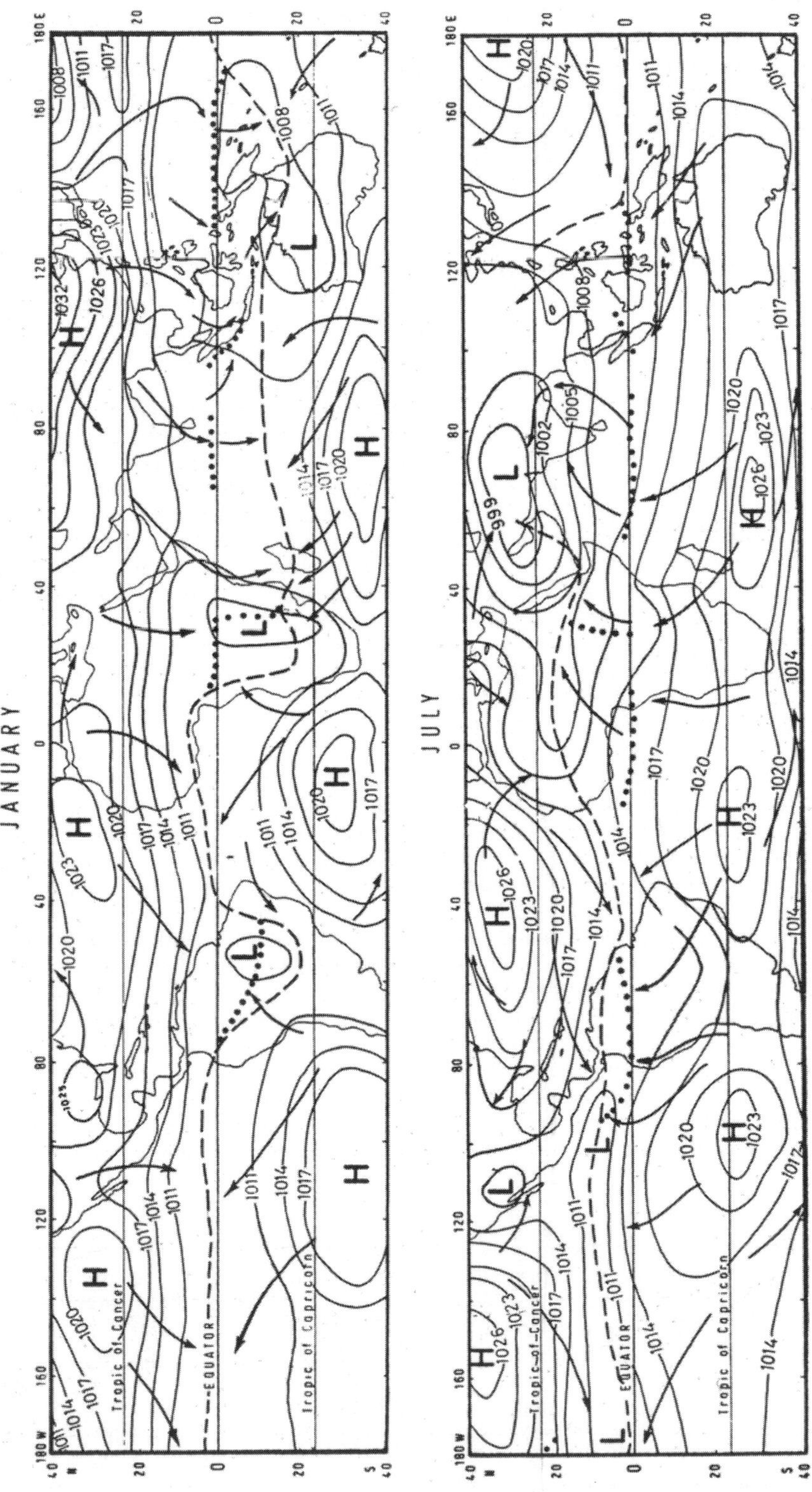

Figure T32 Average sea-level pressure and winds in the tropics. Dashed line is average position of main ITC; dotted lines indicate secondary convergence areas (Nieuwolt, 1977).

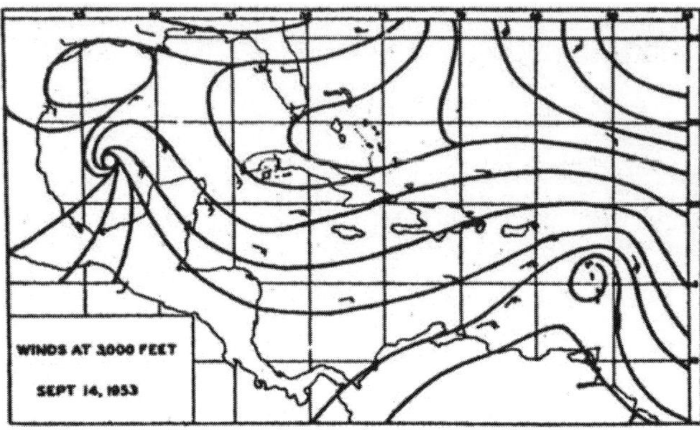

Figure T33 Representative map of streamlines in the Caribbean (after Riehl, 1954).

squeezing effect within the southeast trade winds. The southern ITCZ does not produce the significant cloud cover that is associated with its northern counterpart, nor does it produce a pronounced region of light and variable winds – the doldrums – that is characteristic of the zone of the northern ITCZ.

Pressure gradients within the tropics are relatively small, as is the Coriolis effect, which has a value of zero at the equator. For these reasons detailed pressure analysis is not as useful as wind analysis in the tropics. Instead, streamline analysis is a significant way of delineating tropical atmospheric circulation. A streamline is a line drawn parallel to the wind direction. A regional streamline chart for the Caribbean is shown in Figure T33. The streamlines indicate areas of convergence above which the air is rising and divergence above which air is subsiding. Convergent and divergent regions correspond to cyclonic and anticyclonic flow, respectively.

A surface wind characteristic that occurs with some regularity is the so-called surge of the trades, which is caused by higher-latitude disturbances, in this case Rossby waves. This leads to an acceleration of trade winds for a few days, followed by a slackening. Convergence accompanies the increase in velocity, which produces shear lines, squalls and heavy showers. Other frequent perturbations in the general surface easterly flow in the tropics are easterly waves. These are migratory wave-like disturbances that travel east to west within the trades, generally slower than the regional wind. They consist of a weak pressure trough, and are often characterized by a cluster of thunderstorms in the convergence that usually occurs east of the trough. Many originate over Africa, and usually intensify over the western portions of oceans. Several per week may influence a given location in the Caribbean from May to November.

There are several wind and related oceanic phenomena that occur periodically within the tropics that strongly influence the climate of the tropics and many parts of midlatitudes as well. El Niño/La Niña and the related Southern Oscillation (SO), together given the mnemonic ENSO, have been intensively studied and widely discussed elsewhere. Less discussed, but also an important influence on the climate of tropical regions, is the quasi-biennial oscillation (QBO). (The semiannual oscillation, abbreviated SAO, which is a tropical circulation feature that occurs in a region extending from near the stratopause to the lowermost thermosphere, i.e., primarily within the mesosphere, apparently has a less distinct influence on the surface climatology of the tropics than the QBO, and will not be discussed here.) The QBO is a periodic oscillation of winds in the equatorial lower stratosphere. Wind here is sometimes from the east and at other times from the west. The easterlies were deduced from movement of atmospheric materials injected into the stratosphere by the eruption of Krakatau (Krakatoa) in 1883. The equatorial stratospheric westerlies were discovered by von Berson, and subsequently called the Berson (Berson's) westerlies, by launching pilot balloons above Lake Victoria in 1908. Research in the early 1960s revealed these opposite-blowing winds to be the two phases of the QBO. The transition from one phase to the other, or period, varies between about 20 and 36 months, mostly 24–30 months, with an average period of 28 months, or approximately 2 years. Thus the equatorial stratospheric winds blow from the east for about 12–15 months and then reverse and blow from the west 12–15 months; another reversal occurs and the winds again blow from the east. Transition time from constant prevailing easterlies or westerlies to the opposite direction is just 2–4 months. Wind reversals occur first at higher levels in the stratosphere and descend to near the tropical tropopause. The cause(s) of the QBO are complex and not precisely known, but vertically propagating atmospheric waves and radiative diabatic cooling have been implicated. The QBO is known to be related to other atmospheric phenomena, such as high-latitude stratospheric circulation and the distribution of ozone in the stratosphere. It has an influence on surface climate, at least in the tropics, and also on the frequency of tropical cyclones.

Trade wind inversion

Riehl (1954) noted that the trade wind inversion is the most important factor about the thermal structure of the oceanic tropical atmosphere. One of the first-known accounts of the phenomenon was by Piazzi-Smyth in August 1856, when he climbed the Peak of Teneriffe in the Canary Islands, and passed from a lower region of rich agricultural slopes into an upper zone of cactus and semi-desert. The height of the base of this temperature inversion varies from about 1300 to 1650 feet (400–500 m) toward the eastern sides of the oceans, closely coinciding with the eastern extremities of the subtropical anticyclones, to about 6550–8200 feet (2000–2500 m) near the western sides and near the equator (Figure T34). However, locally the height varies due to differential surface heating and evaporation, the latter occurring from the surface and from the tops of cumulus clouds. The height of the base of the inversion has been observed to be as low as 165 feet (50 m) in the Atlantic off the coast of Africa. The strength of the inversion also varies: in the northern hemisphere it may be very weak or occasionally absent; it is usually weakest when the height of its base is highest, and vice-versa. The inversion is formed primarily by broad-scale subsidence of air from the eastern ends of the subtropical anticyclones. The subsiding air, often quite dry, is usually prevented from reaching the surface because of a surface layer of moist air. In this way the tropical troposphere is divided into an upper dry layer and a lower moist layer.

The inversion acts as a regulating valve for vertical cloud development within much of the tropics. Below the inversion there is high humidity, a steep lapse rate, and often considerable cumulus cloud cover. Above the inversion humidity values are very low and cumulus clouds seldom develop through the base of the inversion into this drier region. Only at the ITCZ does the

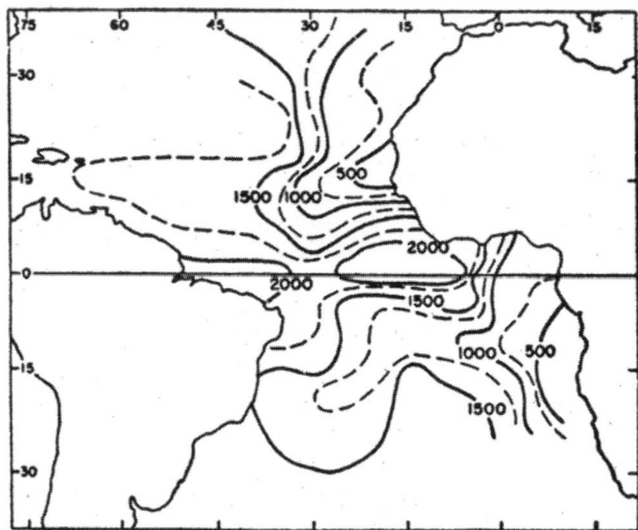

Figure T34 Height (meters) of the base of the trade wind inversion over the Atlantic Ocean.

moist air rise to great heights, accompanied by significant vertical development of cumulus clouds.

Monsoons

The monsoon is a seasonal wind reversal. Although many locations on Earth experience seasonal wind shifts, and the term monsoon has been widely applied, monsoons are best developed in southern and eastern Asia within and near the fringes of the tropics, especially over the Indian Ocean and westernmost Pacific and adjacent land. Figures T35 and T36 show that summer monsoon winds are directed from ocean to land, providing significant precipitation, and that in winter winds are the reverse direction, and usually much drier. The distribution of the major monsoon regions in the world, and a detailed discussion of the causes and characteristics of monsoons, is given in the entry on monsoon climates.

Climatic regions

Tropical and equatorial climates have been subdivided in several ways, and many names have been applied to the categories. Generally, in tropical areas climatic types are mostly determined by rainfall, especially the seasonal pattern, rather than by temperature characteristics. The simplest classification uses the following groups: (1) tropical wet, (2) tropical wet-and-dry, and (3) low-latitude dry. Most classifications employ four or five groups. A typical four-class grouping includes: (1) tropical rainforest, (2) tropical wet-and-dry, (3) tropical steppe, and (4) tropical desert. Classifications using five categories add a monsoon climatic type to the four-class grouping. A classification employing five main categories will be discussed here.

Tropical rainforest

This climatic type is also often called equatorial, humid equatorial, wet equatorial, tropical wet, and tropical rainy. It is the Af category in the Köppen classification, and Ar in Trewartha's. This climate characteristically occupies a zone extending as

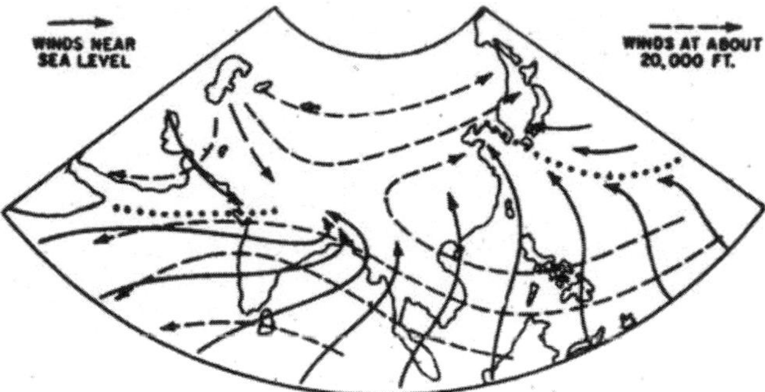

Figure T35 Surface and upper-level winds during the Asian summer monsoon (Petterssen, 1958).

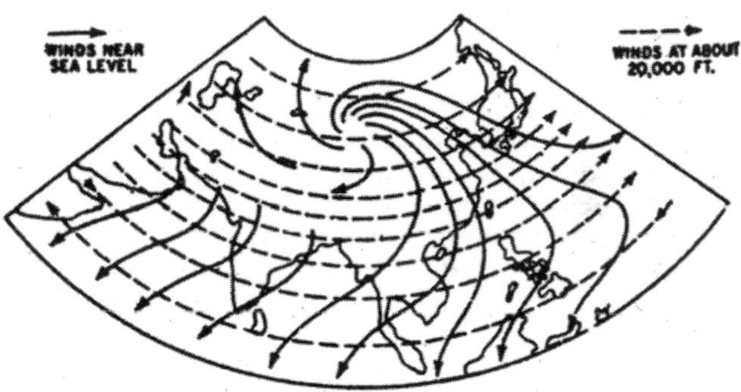

Figure T36 Surface and upper-level winds during the Asian winter monsoon (Petterssen, 1958).

much as 12° on either side of the equator, interrupted only by the monsoonal areas. It is also usually restricted to elevations less than 3300 feet (1000 m). The largest regions of this climate straddle the equator closely, coinciding with the Amazon Basin in Brazil, the Congo Basin of central Africa, and the countries of Indonesia and Malaysia.

Typically, this climatic type has consistently relatively high average monthly temperatures – about 80°F (27°C) throughout the year, although some locations have slightly lower averages. Many locations have an annual range of temperature between the month with the highest average and the month with the lowest average that is less than 8°F (4°C); such locations are designated with the letter I (for isothermal) in the Köppen system. Some locations have ranges that are less than 5°F (3°C). At these locales the monthly temperature range is often considerably less – as much as three times less – than the diurnal range, making the statement "night is the winter of the tropics" particularly *apropos.* Large annual average precipitation of about 60–100 inches (1525–2540 mm) characterizes most of this climatic type. Earth's maximum annual average rainfall typifies many stations with this type of climate. Data for representative stations with a tropical rainforest climate are given in Table T4. These temperature and precipitation conditions support Earth's major rainforests, or selva.

Many locations in the tropical rainforest climate experience double maximums of monthly temperature and rainfall, because the sun, and often the ITCZ, traverse each location twice per year. However, a double maximum does not occur in some places because passage of the ITCZ and its accompanying cloud

Table T4 Climatic data for tropical rainforest stations

	Jan	Feb	Mar	Apr	May	June	July	Aug	Sept	Oct	Nov	Dec	Year
Kisangani, Democratic Republic of the Congo (Zaire)													
Temp. (°C)	25	25	25	25	25	24	24	24	24	24	24	24	24
Precip. (mm)	84	104	176	143	155	89	112	227	191	249	171	71	1772
Manaus, Brazil													
Temp. (°C)	26	26	26	26	26	27	27	28	28	28	28	27	27
Precip. (mm)	249	231	262	221	170	84	58	38	46	107	142	202	1810
Pontianak, Indonesia													
Temp. (°C)	27	27	28	28	28	27	28	28	28	28	28	27	28
Precip. (mm)	274	208	242	277	282	221	165	203	228	366	389	323	3178

cover suppresses maximum temperatures, and because the rain-producing capacity of the ITCZ is not always geographically and temporally consistent. Also inconsistent are wind patterns. Light and variable winds associated with the ITCZ, the doldrums, predominate here, disturbed only by local winds associated with thunderstorms and by the relatively infrequent occurrence in these low latitudes of storms of tropical origin (e.g. hurricanes, typhoons, or tropical cyclones).

Tropical wet-and-dry

This climatic type is also often called tropical savanna (savannah). It is designated as Aw in the Köppen system and with the same two letters by Trewartha. It lies poleward of the tropical rainforest (equatorial) climate, quite consistently terminating near the Tropic of Cancer and Tropic of Capricorn, except north of the equator in Africa, where it extends to only about 15°N latitude, and where it extends to approximately 27°N into southern Florida. It is the primary climate of the trade wind belt. The largest occurrences of this climate are in Africa, where it covers more than one-third of the continent; South America, where it is north and south of the Amazon Basin; the Caribbean; along the west coast of Central America; India, covering much of that country; peninsular southeast Asia; and northern Australia.

The tropical wet-and-dry climate coincides closely with the occurrence of vegetation called savanna, which consists of mostly tropical grasslands, shrubs, and drought-resistant trees, the latter covering 10–40% of the land. This vegetation, and thus the climate, are transitional between the nearly constantly wet equatorial regions where the influence of the ITCZ is dominant, and the much drier regions of the subtropics. The wet-and-dry portion of the name of this climate results from its hybrid nature, being under the influence of very different climatic controls in the different seasons. Here there is no dominant double season of high insolation (high sun) and consequently no double maximums of temperature and precipitation as in the tropical rainforest climate. Instead, most regions receive significant rainfall during the high-sun season, when influenced primarily by the ITCZ coupled with rain-bearing easterly waves. Low-sun season is influenced by an equatorward shift of the subtropical high-pressure cells, which reduces rainfall, often substantially. Some areas on the eastern coasts in this climatic type have more consistent rainfall, also receiving rainfall in winter and fall from the stronger influence of the trade winds on these coastal regions.

Coastal portions of this climate typically receive 35–70 inches (890–1780 mm) of precipitation annually, most of it in the summer, and have average monthly temperatures of 65–70°F (18–21°C) during winter and 75–80°F (24–27°C) in the summer months. More continental locations generally receive 20–50 inches (510–1270 mm), with the lower amounts most commonly encountered near the poleward edges of this climatic type; much of the annual total is received in summer. Temperatures are similar to coastal locations, except that greater diurnal ranges are experienced. Data for representative stations with a tropical wet-and-dry climate are given in Table T5.

Monsoon

The monsoon climatic type is discussed in detail in the item on monsoon climates; only general characteristics will be given here. This climatic type is designated Am in the Köppen system, but is not identified as a separate climatic type in the Trewartha classification. (Am regions on maps of the Köppen system are sometimes within Ar and sometimes within Aw regions on maps depicting Trewartha's system; some consider the monsoon climate to be a variety of the tropical rainforest climate while others consider it as a subtype of tropical wet-and-dry climate.) Monsoon climates, as defined by Köppen, occur primarily in the Philippines, sections of coastal Vietnam, along the eastern coast of the Bay of Bengal, southwestern India, a small portion of western Africa (Sierra Leone and Liberia), and northeastern coast of Brazil. Some have suggested that the term monsoon climate be restricted to Asia and northern Australia, but this restriction has not been widely followed, and many consider that a true Am climate in Australia occurs in only an extremely small portion of the northeastern part of that country.

Monsoon regions are in approximately the same latitudes as the tropical rainforest (equatorial) climates, although they also extend slightly farther north in southeast Asia, and thus share some characteristics with that climatic type. Temperatures are quite similar to those of the rainforest, with all months averaging above 64°F (18°C), and average about 80°F (27°C) at many locations, but there is often a slightly greater, albeit still small, seasonal variation. Warmest conditions usually precede the arrival of the ITCZ and its associated cloud cover, and there is frequently a secondary temperature maximum following the departure of the ITCZ.

Although amount of rainfall in the monsoon regions is often similar to that in equatorial climates, the seasonal distribution is different, and it is this characteristic that primarily distinguishes

Table T5 Climatic data for tropical wet-and-dry (savanna) stations

	Jan	Feb	Mar	Apr	May	June	July	Aug	Sept	Oct	Nov	Dec	Year
Calcutta, India													
Temp. (°C)	20	23	28	30	31	30	29	29	30	28	24	21	27
Precip. (mm)	13	24	27	43	121	259	301	306	290	160	35	3	1582
Cuiaba, Brazil													
Temp. (°C)	27	27	27	26	26	24	24	26	28	28	28	27	27
Precip. (mm)	216	198	232	116	52	13	9	12	37	130	165	195	1375
Dakar, Senegal													
Temp. (°C)	21	20	21	22	23	26	27	27	28	28	26	25	24
Precip. (mm)	0	2	0	0	1	15	88	249	163	49	5	6	578
Darwin, Australia													
Temp. (°C)	28	28	28	28	27	25	25	26	28	29	27	29	28
Precip. (mm)	341	338	274	121	9	1	2	5	17	66	156	233	1562
Miami, Florida, USA													
Temp. (°C)	20	20	21	24	26	27	28	28	28	25	22	21	24
Precip. (mm)	55	50	53	91	155	229	176	171	222	208	69	42	1521

Table T6 Climatic data for monsoon stations

	Jan	Feb	Mar	Apr	May	June	July	Aug	Sept	Oct	Nov	Dec	Year
Cochin, India													
Temp. (°C)	27	28	29	30	29	27	26	27	27	27	28	27	28
Precip. (mm)	23	20	51	124	297	724	592	353	196	340	170	41	2931
Mangalore, India													
Temp. (°C)	27	27	28	29	29	27	26	26	26	27	27	27	27
Precip. (mm)	3	3	5	38	157	942	988	597	267	206	74	13	3293
Monrovia, Liberia													
Temp. (°C)	27	27	28	27	26	24	24	25	25	26	26	26	26
Precip. (mm)	5	3	112	297	341	918	616	473	760	641	209	74	4449
Rangoon (Yangon), Myanmar													
Temp. (°C)	25	27	29	31	29	27	27	27	27	28	27	26	27
Precip. (mm)	3	5	8	51	307	480	582	528	394	180	69	10	2617

the two climatic types. Monsoon climates have at least one month that receives less than 2.4 inches (61 mm), although the so-called dry season is generally just a few months in duration. This dry season is nearly always during the winter. Often the dry period is not as distinct as the dry season of the Aw climate, and the copious rainfall of the summer is sufficient to sustain tree growth during the short dry season, and thus there are no water-balance deficits in winter in the monsoon climate, as there are in Aw climates. Data for representative stations with a monsoon climate are given in Table T6.

Tropical semiarid and arid

Many of Earth's driest areas occur within tropical/subtropical regions, and are referred to generally as low-latitude or tropical dry climates. The deserts in these latitudes, particularly those at low elevations, are often called warm deserts, to distinguish them from the so-called cold deserts of higher latitudes (and elevations), although this distinction is somewhat arbitrary, as many tropical deserts experience low nighttime temperatures and some even have relatively low summertime monthly average temperatures. Tropical dry climates are situated mostly between 12° and 35°N and S, and are most commonly divided into two subgroups: tropical semiarid or steppe, supporting grasses, and tropical arid or desert, with xerophytic vegetation or no vegetative cover. Tropical semiarid is designated BSh in the Köppen and Trewartha classifications, and deserts are shown as BWh (interior tropical deserts) and BWk (coastal tropical deserts). (Some midlatitude deserts are also designated as BWk.) As further discussed in the item on arid climates, the main causes of dry regions in the tropics are subsidence of air in the subtropical anticyclones, cold ocean currents, which enhance the stability caused by anticyclones, topographic barriers blocking rain-bearing rains, and great distances from moist ocean winds. Often several of these mechanisms operate in conjunction to produce dryness in an area.

Semiarid regions, especially the large steppes of Africa and Australia, lie at the edges of major deserts, producing a transition from the extremely dry deserts to the more moist tropical savanna climates. (The term steppe is also used for some midlatitude grasslands, especially the extensive grasslands that extend along a significant portion of the southern edge of the Ukraine and Russia; here the climatic designation is BSk.) Much of the northern half of Africa reveals the progression equatorward of desert, centered along the Tropic of Cancer, to steppe – including much of the Sahel – to tropical savanna, and finally to tropical

Table T7 Climatic data for tropical semiarid (steppe) stations

	Jan	Feb	Mar	Apr	May	June	July	Aug	Sept	Oct	Nov	Dec	Year
Broome, Australia													
Temp. (°C)	30	30	29	28	25	22	21	23	25	28	29	30	27
Precip. (mm)	160	147	99	30	15	23	5	3	<3	<3	15	84	581
Kayes, Mali													
Temp. (°C)	26	28	32	34	36	32	29	27	28	29	29	29	29
Precip. (mm)	3	0	0	<3	25	97	160	241	188	43	<3	<3	757
Maracaibo, Venezuela													
Temp. (°C)	28	28	29	29	30	30	30	30	29	29	28	28	29
Precip. (mm)	3	<1	8	20	69	56	46	56	72	150	84	15	580

rainforest at the equator. This same progression occurs in the southern half of the continent. Some of the more poleward semiarid regions have a precipitation regime that is the reverse of the adjacent tropical wet-and-dry climate. Summers, not winters, are typically dry because the subtropical highs shift over these regions, whereas in winter they are more strongly influenced by migrating storms of the westerlies. Data for representative stations with a semiarid climate are given in Table T7.

The largest and driest true arid regions, or deserts (BWh and BWk), are centered approximately on the Tropic of Cancer and Tropic of Capricorn, just poleward of the semiarid tropical steppes. Although several criteria are involved in determining aridity, and many defining schemes have been proposed, a common characteristic of all deserts is meager precipitation. Tropical deserts in the Köppen system have about 11 inches (280 mm) for maximum annual precipitation; 12 inches (305 mm) under the Thornthwaite classification; and 14 inches (355 mm) in the Miller scheme. Additionally, nearly all tropical deserts experience a large annual excess of evaporation over precipitation, with the former often 15–20 times greater than the latter. Relative humidity varies greatly, from as low as 2% in parts of the Sahara Desert to 100%, with frequent fogs, in coastal deserts.

The largest of the tropical deserts are in northern and southwestern Africa, southwestern Asia, central Australia, west-central South America, and southwestern United States/northern Mexico. Deserts are often categorized into the two broad groupings: coastal and interior.

Coastal tropical deserts include the Atacama of west-central South America, Namib of southwestern Africa, and the arid coast of Lower California, i.e. the Baja peninsula of northwestern Mexico. (The Sahara Desert of Africa and the Australian Desert extend from the interior to west coasts, but they are considered to be interior deserts.) Coastal deserts are produced mostly by a combination of dry subsiding air of the subtropical anticyclones coupled with the stabilizing influence of cool ocean currents. These effectively suppress precipitation much of the time, but relative humidity may be as high as 81% in the summer, and copious fog and dew are common in several of these deserts; fog has occurred on 200 days in 1 year in the Namib Desert. The lowest average annual precipitation on Earth, 0.03 inch (0.8 mm), also often shown as 0.02 inch, occurs in the Atacama Desert at Arica, Chile. Iquique (also Iquiqui), Chile, 120 miles (193 km) south of Arica and also in the Atacama Desert, experienced 14 consecutive years with no measurable rainfall. (In some sources this 14-year record is attributed to Arica, but most cite Iquique as the South American record holder; some reports state that the longest rain-free period here was 20 years. There are also reports that Wadi Halfa, in the Sahara Desert, experienced a 19-year period with no precipitation.) Temperatures are relatively low in the coastal deserts partly because of the cool ocean current and sometimes-abundant cloud cover. Monthly averages range between 60°F and 70°F (15–21°C). Average annual temperature in the Atacama Desert is 66°F (19°C), and it is 63°F (17°C) in the Namib Desert. Data for representative stations with a coastal tropical desert climate are given in Table T8.

Interior tropical deserts, some of which may extend to coastal regions, are among the largest arid regions on Earth. The Sahara in northern Africa is approximately 3.5 million square miles (9.1 million square kilometers), which exceeds the area of the conterminous United States. The Australian Desert, which has many local names, covers 1.3 million square miles (3.4 million square kilometers), and covers such a large portion of Australia that the continent is sometimes called the "desert continent". The Arabian Desert covers 1 million square miles (2.6 million square kilometers), and the Thar (Great Indian) Desert of western India and southeastern Pakistan is most frequently stated to cover 230 000 square miles (600 000 square kilometers). The desert region of northern Mexico and southwestern United States is a large contiguous region with several local names, including Chihuahuan, Sonoran and Mojave. This vast dry region is in the subtropical/midlatitude transition zone.

The aridity of these interior tropical deserts is primarily because of their location relative to the subtropical anticyclones, as was true of the coastal tropical deserts, but also because much of their areas are far removed from moist, rain-bearing oceanic winds. In the case of the Mojave Desert in California, moisture is also blocked by the tall Panamint Mountains and Sierra Nevada to the west, producing a persistent rainshadow effect. The infrequent rains usually fall from intense summer convectional storms, and occasionally from winter cyclonic storms originating in middle latitudes. The latter storms occasionally produce snow in the deserts. Precipitation is extremely variable in amount, intensity, occurrence, and distribution.

Temperatures are generally considerably higher in these deserts than in coastal tropical deserts. Average monthly temperatures are generally between 70°F and 90°F (21–32°C), and the highest temperatures on Earth, both average annual and daily maximum, have occurred in this climate, including a record-high reading of 136°F (58°C) at El Azizia, Libya – about 50 miles (80 kilometers) south of Tripoli – on 13 September 1922 (although the validity of this value has been questioned).

Table T8 Climatic data for tropical desert stations

	Jan	Feb	Mar	Apr	May	June	July	Aug	Sept	Oct	Nov	Dec	Year
Coastal tropical deserts													
Antofagasta, Chile													
Temp. (°C)	21	21	20	18	16	15	14	14	15	16	18	19	17
Precip. (mm)	0	0	0	<3	<3	3	5	3	<3	3	<3	0	14
Iquique, Chile													
Temp. (°C)	21	21	19	18	17	16	16	16	16	17	18	20	18
Precip. (mm)	0	0	0	0	0	0	3	0	0	0	0	0	3
Walvis Bay, Namibia													
Temp. (°C)	19	20	19	18	17	16	15	15	15	15	16	18	17
Precip. (mm)	<1	5	8	3	2	<1	1	2	<1	<1	<1	<1	21
Interior tropical deserts													
Adrar, Algeria													
Temp. (°C)	12	14	18	24	31	35	37	36	31	25	18	13	25
Precip. (mm)	<1	<1	3	<1	<1	<1	<1	<1	<1	5	5	<1	13
Al-Hofuf, Saudi Arabia													
Temp. (°C)	14	16	21	25	31	34	35	34	31	27	21	16	25
Precip. (mm)	23	8	16	16	1	0	0	0	0	1	1	6	49
Alice Springs, Australia													
Temp. (°C)	29	28	25	20	15	12	12	14	18	23	26	28	21
Precip. (mm)	43	33	28	10	15	13	8	8	8	18	30	38	252
Greenland Ranch in Death Valley, California, USA													
Temp. (°C)	11	14	19	24	29	34	39	37	32	24	16	12	24
Precip. (mm)	8	5	5	3	0	0	3	3	3	3	5	5	43

Death Valley, in the Mojave Desert of southeastern California, is the hottest – as well as driest and lowest – spot in North America, and experienced a temperature of 134°F (57°C) in July 1913; this still stands as the highest temperature officially recorded in the western hemisphere. In contrast, relatively low winter temperatures give these deserts the greatest annual temperature range found in tropical regions. Diurnal temperature ranges are also extreme: in the Sahara Desert in Algeria a daily range of 100°F (56°C) has been recorded, when the temperature changed in a 24-hour period from 132°F (56°C) to 32°F (0°C); this value is generally accepted as the greatest daily range of temperature to have been recorded on Earth. (Although the range – 100°F – is the same, sometimes the high and low temperatures are presented as 126°F and 26°F, respectively.) Data for representative stations with an interior tropical desert climate are given in Table T8.

High temperatures of interior tropical deserts are sometimes accompanied by hot, dry, dust-bearing winds, especially in Africa. Many of these winds have local names. The Ghibli is a hot wind in Libya, usually blowing from the Sahara; this wind in Morocco is called the Chergui. A Khamsin wind is a hot, extremely dry wind reaching into Egypt from the south and east. Another wind sweeping northward to the Mediterranean area from the Sahara or Arabian deserts is the well-known Sirocco. The Harmattan is a reverse wind from the Sirocco, blowing mostly southward from the Mediterranean or northern Africa, covering much of northern and northwestern Africa. This wind, which is usually warm rather than very hot, and extremely dry, is considered a continental part of the trade wind system.

James Henry

Bibliography

Boucher, K., 1975. *Global Climate*. New York: John Wiley & Sons.

Byers, H.R., 1959. *General Meteorology*, 3rd edn. New York: McGraw-Hill.

Crowe, P.R., 1951. Wind and weather in the equatorial zone. *Institute of British Geographers Transactions*, **17**: 23–76.

Goudie, A., and Wilkinson, J., 1977. *The Warm Desert Environment*. Cambridge: Cambridge University Press.

Hamilton, K., 1998. Dynamics of the tropical middle atmosphere: a tutorial review. *Atmosphere–Ocean*, **36**: 319–354.

Kendrew, W.G., 1961. *The Climates of the Continents*, 5th edn. Oxford: Clarendon Press.

Krause, P.F., and Flood, K., 1997. *Weather and Climate Extremes*. Alexandria, VA: US Army Corps of Engineers.

Lydolph, P.E., 1985. *The Climate of the Earth*. Totowa, NJ: Rowman & Allanheld.

Miller, A.A., 1931. *Climatology*. London: Methuen.

Nieuwolt, S., 1977. *Tropical Climatology*. New York: John Wiley & Sons.

Oliver, J.E., and Hidore, J.J., 2002. *Climatology: an atmospheric science*. New York: Prentice-Hall.

Petterssen, S., 1958. *Introduction to Meteorology*. New York: McGraw-Hill.

Riehl, H., 1954. *Tropical Meteorology*. New York: McGraw-Hill.

Saucier, W.J., 1955. *Principles of Meteorological Analysis*. Chicago, IL: University of Chicago Press.

Thornthwaite, C.W., 1931. The climates of North America according to a new classification, *Geographical Review*, **21**: 633–655.

Trewartha, G.T., 1954. *An Introduction to Climate*, 3rd edn. New York: McGraw-Hill.

Trewartha, G.T., 1968. *An Introduction to Climate*, 4th edn. New York: McGraw-Hill.

Trewartha, G.T., 1981. *The Earth's Problem Climates*, 2nd edn. Madison, WI: University of Wisconsin Press.

Cross-references

Arid Climates
Central America and Caribbean, Climate of

Deserts
Intertropical Convergence Zone
Monsoons and Monsoon Climate
Rainforest Climates
Savanna Climate
South America, Climate of
Tropical Cyclones

TROPICAL CYCLONES

The term "tropical cyclone" is used to refer to warm-core, nonfrontal low-pressure systems of synoptic scale that develop over tropical or subtropical oceans. This definition differentiates *tropical cyclones* from *extratropical (midlatitude) cyclones* that exhibit a cold-core in the upper troposphere and often form along fronts in higher latitudes. *Subtropical cyclones* are hybrid systems that exhibit some characteristics of tropical cyclones and some characteristics of extratropical cyclones. Tropical cyclones extract much of their energy from the upper layer of the ocean, while extratropical cyclones derive much of their energy from the baroclinic temperature gradients in which they form. This classification system differentiates cyclones based on their initial sources of energy, their thermal structure in the upper troposphere and the presence or absence of fronts. However, cyclones exhibit characteristics related to their large-scale environment, and it is possible for a cyclone to transform from one type into another type when the cyclone moves away from its genesis region. For example, many tropical cyclones undergo an *extratropical transition* into extratropical cyclones with fronts when they move into higher latitudes. Tropical cyclones are observed over the North Atlantic Ocean, the eastern, central and western North Pacific Ocean, the central and western South Pacific Ocean and the northern and southern Indian Ocean. A cyclone with some of the characteristics of a tropical cyclone was observed over the southern Atlantic Ocean in March 2004.

Tropical disturbances

A *tropical disturbance* is a discrete tropical weather system with organized convection that originates in the tropics or the subtropics. Tropical disturbances are not typically associated with fronts and they may not always exhibit a detectable perturbation in the surface wind field. A tropical weather system must maintain its identity for at least 24 hours in order to be classified as a tropical disturbance. This latter requirement makes it possible to differentiate migratory tropical disturbances from the typical diurnal convection found in tropical and subtropical regions.

Tropical disturbances originate in several regions. Tropical (easterly) waves develop over the interior of Africa and translate westward over the North Atlantic Ocean. Tropical waves are elongated troughs of low pressure with a meridional extent of hundreds of kilometers. Figure T37 is a visible satellite image of a strong tropical wave as it approaches the Caribbean Sea. Clusters of thunderstorms that originate in the Intertropical Convergence Zone (ITCZ) may become separated from the ITCZ and are classified as tropical disturbances if they maintain convection for at least 24 hours. Other tropical disturbances

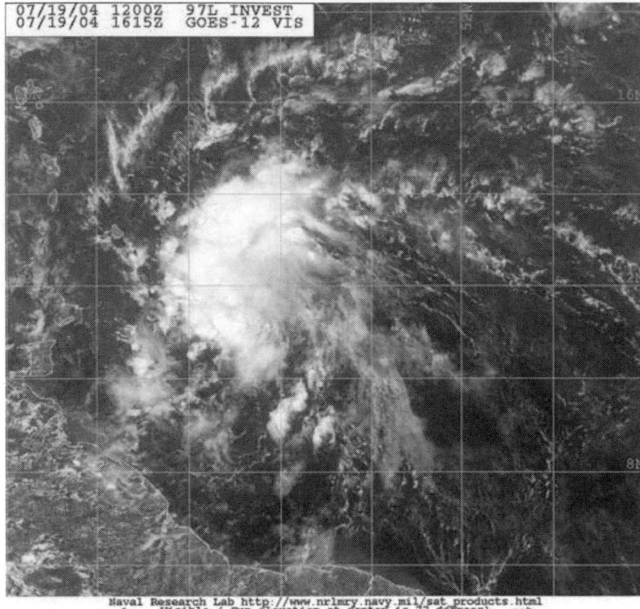

Figure T37 Visible satellite image of a strong tropical wave approaching the Caribbean Sea (from the Naval Research Laboratory's tropical cyclone web demonstration page, produced by the Marine Meteorology Division in Monterey, California).

form in the monsoon trough over the western Pacific Ocean. Tropical disturbances generally move from east to west in the prevailing flow at lower latitudes. Tropical disturbances are not tropical cyclones although the systems may evolve into tropical cyclones when certain conditions exist.

Tropical cyclogenesis

The term "*tropical cyclogensis*" refers to the formation of a tropical cyclone. Tropical cyclogenesis occurs when an existing disturbance composed of an area of convection with embedded thunderstorms organizes into a distinct surface low-pressure system with cyclonic convergent flow. Although the precise physical processes that lead to the formation of a tropical cyclone are not known, a number of conditions seem to be necessary for tropical cyclogenesis to occur. First, a disturbance must exist with significant convection, vorticity (i.e. tendency for cyclonic rotation) and convergence near the surface. Second, the disturbance must be located over a warm ocean. The sea surface temperature (SST) needs to be warmer than 26°C and the warm water needs to extend through a sufficient depth thought to be a few tens of meters. Third, the environmental lapse rate (i.e the rate at which the atmosphere cools with height) needs to be sufficiently large so that the atmosphere is conditionally unstable. Fourth, the middle troposphere (i.e. 3–7 km above the surface) must contain sufficient moisture to allow for the development of deep convection in thunderstorms. Fifth, the vertical wind shear (i.e. the difference in the winds between the lower troposphere and upper troposphere) must be relatively small. Sixth, the disturbance needs to be at least 500 km from the equator.

All tropical cyclones develop from existing disturbances. Most tropical cyclones develop from tropical disturbances.

However, some tropical cyclones develop from disturbances associated with dissipating frontal boundaries, subtropical cyclones and mesoscale convective systems (MCS). In order for a disturbance to develop into a tropical cyclone it must be located over a warm ocean and the flow must be convergent near the surface. As the air passes over the ocean, water evaporates from the surface into the atmosphere. Since water absorbs energy as it evaporates, this process represents a transfer of latent energy from the surface to the air. The air is usually slightly warmer than the water and there is an additional transfer of internal energy (formerly called sensible heat) into the atmosphere. The convergent flow near the surface brings the latent and internal energy into the region of convection associated with the disturbance, and the air begins to rise in thunderstorms. As the air rises, it cools until it becomes saturated and then the water vapor begins to condense. When the water vapor condenses, the latent energy is converted into internal energy and the temperatures in the middle and upper troposphere begin to increase. If sufficient condensation occurs in the upper troposphere, the temperatures will be higher near the center of the disturbance than they will be at the outer edges of the system and the system will develop a warm core.

The development of the warm core enhances the divergence of mass in the upper level of the disturbance. When the divergence of mass at the upper levels exceeds the convergence of mass in the lower levels the pressure at the surface will begin to decrease. As the pressure at the surface decreases, the horizontal pressure gradient increases and the wind speed at the surface increases. As the wind speed at the surface increases, the fluxes of latent and internal energy from the ocean to the atmosphere increase. Convergence also increases, more latent energy is drawn into the disturbance and the warm core in the upper levels becomes more pronounced. The convergence will also collect vorticity in the lower levels of the atmosphere and increase the tendency for the air to rotate cyclonically. If the surface pressure continues to fall and the flow of air at the surface develops a clearly identifiable cyclonic circulation, then the system is classified as a *tropical cyclone*.

The relative importance of the processes that produce tropical cyclogenesis has been debated by tropical meteorologists. Charney and Eliassen (1964) proposed a mechanism to explain tropical cyclogenesis called Conditional Instability of the Second Kind (CISK). CISK proposes that the convergence in the lower levels enhances convection and the release of latent energy, which further enhances the convergence. The positive feedback between the convection on the cumulus scale of motion and the larger-scale disturbance eventually produces a tropical cyclone. Emanuel (1987) proposed a mechanism called Wind Induced Surface Heat Exchange (WISHE) in which the increased convergence enhances the fluxes of latent and internal energy from the ocean to the atmosphere. The additional energy is redistributed into the upper levels of the disturbance by the convection. WISHE proposes that the increased fluxes of energy from the surface are critical to tropical cyclogenesis. Although both theories of tropical cyclogenesis include the effects of the surface fluxes of energy and convection, they assign differing levels of importance to the two processes. Convection is assumed to be the predominant factor in CISK, while the surface fluxes assume the predominant role in WISHE.

Most tropical disturbances do not develop into tropical cyclones. Numerous factors can inhibit or prevent the genesis of a tropical cyclone from a disturbance. If the SST are too cold, insufficient latent energy is transferred from the ocean to the atmosphere to support the formation of a warm core. Even when the SST are greater than 26°C, if the layer of warm water is too shallow the winds will mix the upper layer of the ocean and cooler water will rise to the surface. The upwelling of cooler water will reduce the transfer of latent energy and inhibit the genesis of a tropical cyclone. If the environmental lapse rate is too small, the atmosphere will be stable and only shallow convection will develop. If the middle of the troposphere is too dry, the water rising in clouds will evaporate and deep convection will not be able to develop. Stability and dry air in the middle troposphere prevent the development of a warm core in the upper troposphere. If the vertical wind shear is too strong, the winds in the upper troposphere will advect (i.e. transport horizontally) the warm core away from the center of the disturbance. Strong winds aloft also inhibit upper-level divergence over part of the disturbance and prevent the decrease of pressure at the surface. When the vertical wind shear is very strong, the upper-level winds may separate the upper portion of the disturbance from the lower part of the disturbance and the system is literally blown apart.

Classification of tropical cyclones

Observation of a clearly identifiable cyclonic circulation at the surface and a warm core aloft results in the classification of a system as a tropical cyclone. If the maximum sustained wind speed at the surface is less than $17\,\mathrm{m\,s^{-1}}$ (34 knots), then the tropical cyclone is classified as a *tropical depression*. When a tropical depression forms it is given a numerical designation by the meteorological organization with responsibility for that location of the world. For example, the first tropical depression that forms over the eastern North Pacific Ocean each season is designated Tropical Depression One-E. If the maximum sustained wind speed at the surface is $18–32\,\mathrm{m\,s^{-1}}$ (35–64 knots), then the tropical cyclone is classified as a *tropical storm*. When a system is classified as a tropical storm it is given a name from a list of names for that basin. If the maximum sustained winds at the surface are $33\,\mathrm{m\,s^{-1}}$ (65 knots) or greater, a tropical cyclone is classified as a *hurricane* in the western hemisphere (i.e. over the Atlantic Ocean and the eastern and central North Pacific Oceans), as a *typhoon* over the western Pacific Ocean, and as a *severe cyclone* over the Indian Ocean. A tropical cyclone in the western hemisphere with maximum sustained winds at the surface of $49\,\mathrm{m\,s^{-1}}$ (96 knots) or greater is classified as a *major hurricane*. A tropical cyclone over the western Pacific Ocean with maximum sustained winds at the surface of $66\,\mathrm{m\,s^{-1}}$ (130 knots) is classified as a *super typhoon*.

Structure

The convergent cyclonic flow in the lower atmosphere of a tropical cyclone tends to organize the convection into spiral bands. Figure T38 is a visible satellite image of Tropical Depression Three-E, which intensified into Tropical Storm Celia over the eastern North Pacific Ocean during July of 2004. The image depicts the central mass of convection over the center of the tropical depression, which is called the Central Dense Overcast (CDO). Strong rising motion near the center of the tropical cyclone transports mass upward. A primary *spiral band* of convection extends to the west and north of the center of the depression. Figure T39 is a visible image of the same tropical cyclone after it had intensified into Tropical Storm Celia. The strongest convection is still evident near the center of the

Figure T38 Visible satellite image of tropical depression Three-E (2004) (from the Naval Research Laboratory's tropical cyclone web demonstration page, produced by the Marine Meteorology Division in Monterey, California).

system and several spiral bands are visible to the east and south of the center. The strong rising motion near the center of the tropical storm produces divergence at upper levels of the troposphere and the *cirrus canopy* spreads out over the top of the storm. Evidence of the upper-level divergence can be seen in some of the thin, wispy streamers of cirrus radiating away from the center of the storm on its northeast side.

When tropical cyclogenesis occurs, the fastest horizontal winds often are observed 100–200 km from the center of the tropical cyclone in the spiral bands. As the tropical storm intensifies, the pressure gradient increases, the low-level convergence becomes stronger, and the Radius of Maximum Winds (which is the distance from the center of the tropical cyclone at which the fastest wind speeds are found) decreases. As the intensity of the tropical storm approaches a classification of a hurricane or typhoon, the primary spiral band wraps around the center of the circulation. A clear area may begin to appear on radar and satellite imagery. The clear area at the center of the tropical cyclone is called the *eye*. Subsidence is observed in the eye in the upper and middle troposphere and the sinking motion helps account for the clearing. In the lower troposphere near the surface of the ocean the air is moister and shallow cumulus clouds may occur below the subsidence in the eye. Figure T40 is a visible satellite image of Hurricane Alex on 4 August 2004. The eye is clearly visible in the center of the storm. The diameter of the eye may be only 10 km in small intense tropical cyclones. In large tropical cyclones and those storms undergoing an extratropical transition the diameter of the eye may be 100 km. The horizontal winds are usually relatively light near the center of the eye. Wind speeds may increase toward the outer portions of the eye. Tall, vertical cumulonimbus clouds surround the eye. The inner edge of

Figure T39 Visible satellite image of Tropical Storm Celia (2004) (from the Naval Research Laboratory's tropical cyclone web demonstration page, produced by the Marine Meteorology Division in Monterey, California).

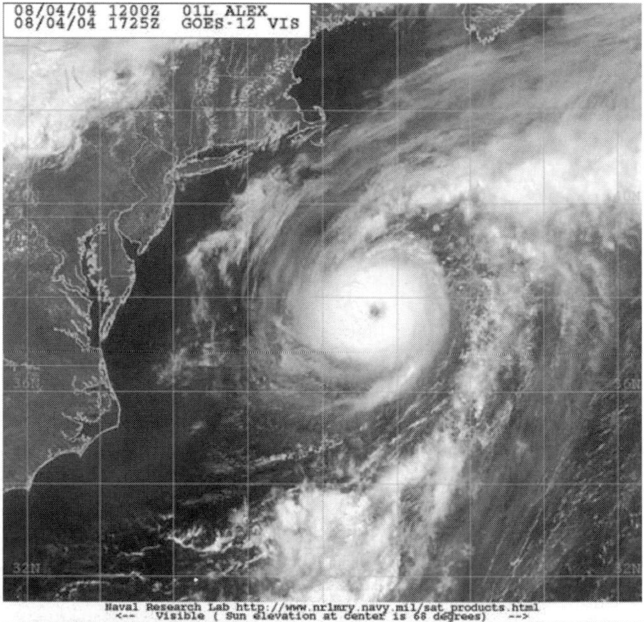

Figure T40 Visible satellite image of Hurricane Alex on 4 August 2004 (from the Naval Research Laboratory's tropical cyclone web demonstration page, produced by the Marine Meteorology Division in Monterey, California).

the vertical clouds around the eye is called the *eyewall*. The fastest horizontal winds and strongest low-level convergence in a hurricane or typhoon are often found just outside the eyewall. The highest rates of vertical motion (i.e. the strongest updrafts and downdrafts) are found in the convection that surrounds the eye. The speed of the strongest updrafts may exceed 25 m s^{-1} in the eyewalls of intense tropical cyclones (Dodge et al., 1999). The speed of the strongest downdrafts is roughly 50–60% of the strongest updrafts in the eyewall. Aerial reconnaissance has observed that the eyewalls of some intense tropical cyclones appeared to slope outward with height in the upper levels of the storms.

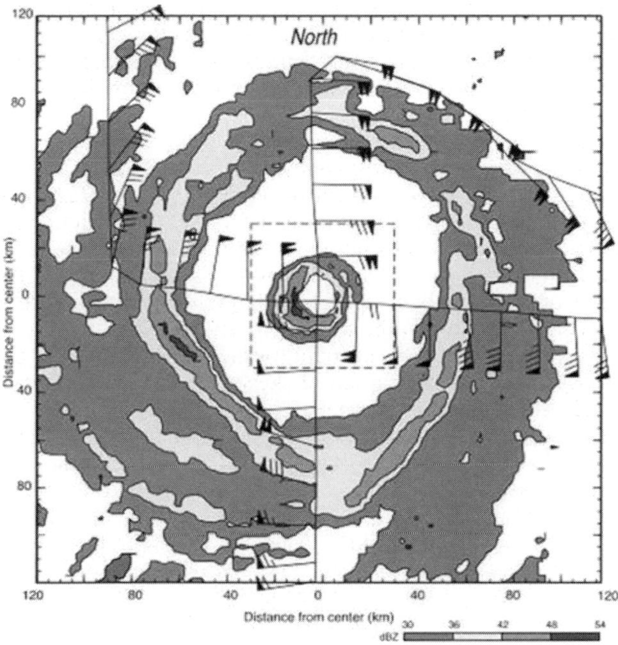

Figure T41 Radar image of concentric eyewalls in Hurricane Gilbert (1988) (from Dodge et al., 1999).

Willoughby et al. (1982) documented the existence of multiple *concentric eyewalls* and an *eyewall replacement cycle* in certain hurricanes. Many times in intense tropical cyclones the convection in the eyewall moves inward as the storm intensifies and the radius of the eye becomes smaller. In some intense tropical cyclones an outer spiral band wraps completely around the center of the storm and surrounds the original eyewall. In this case the storm exhibits multiple concentric eyewalls. Figure T41 is a radar image of Hurricane Gilbert (1999) that exhibits multiple concentric eyewalls. The inner eyewall is found 10–20 km from the center of Gilbert. The edge of the outer eyewall is 50 km from the center of the storm. Figure T42 shows a cross-section of the radar reflectivity of the concentric eyewalls. The highest reflectivity, which represents the highest rainfall rate, is found in the inner eyewall. A secondary reflectivity maximum is found in the outer eyewall. Lower reflectivities, which indicate the presence of stratiform rain, are found between the regions of strong convection in the eyewalls. Most of the convergent flow in the lower levels of the atmosphere rises in the outer eyewall and convection diminishes in the inner eyewall until it eventually dissipates. The outer eyewall becomes the primary eyewall and the tropical cyclone exhibits an eye with a much larger radius. When the outer eyewall replaces the inner eyewall, the minimum surface pressure rises and the maximum sustained wind speeds may decrease. The reduction in intensity may be only temporary as the new eyewall will begin to contract if the large-scale environment is still favorable of

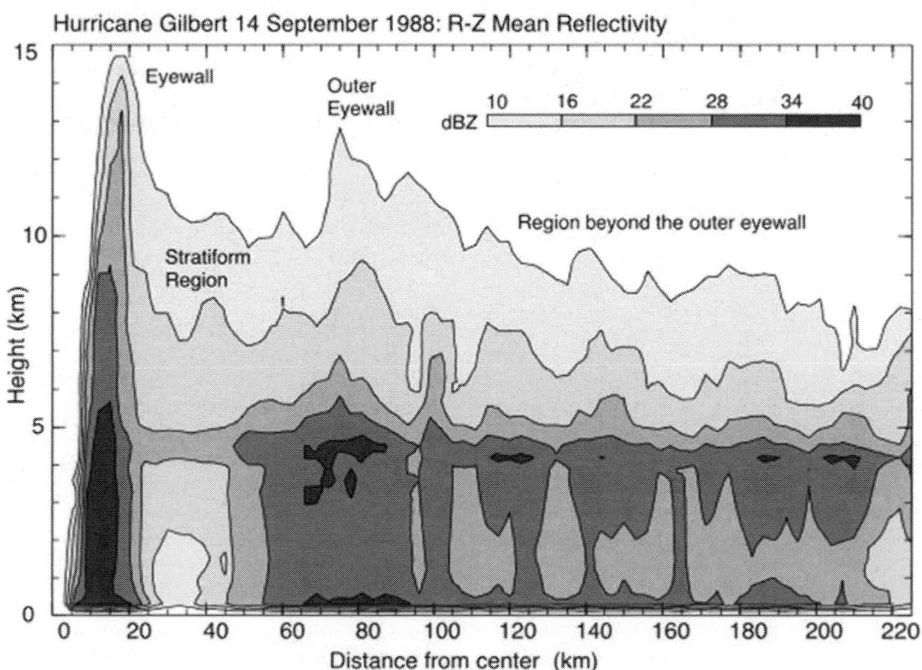

Figure T42 Radar cross-section of concentric eyewalls in Hurricane Gilbert (1988) (from Dodge et al., 1999).

intensification of the tropical cyclone. Additional spiral bands may wrap around the center of the storm and a number of eyewall replacement cycles may occur.

Tropical cyclones exhibit a wide range of sizes and their wind fields result from the interaction of the storm and its surrounding environment. For example, gale force winds ($17\,m\,s^{-1}$) extended to a radius of 1100 km in Typhoon Tip over the Northwest Pacific Ocean on 12 October 1979. In contrast the radius of gale force winds was only 50 km in Tropical Cyclone Tracy on 24 December 1974.

Climatology

Approximately 90 tropical cyclones form each year. The total number of tropical cyclones is remarkably consistent from year to year, although the global distribution of the cyclones may vary significantly. Table T9 presents the monthly frequencies of tropical cyclones for six regions of the world.

The greatest number of tropical cyclones form over the western North Pacific Ocean. Approximately 32 tropical cyclones form over the western North Pacific Ocean in an average year. Tropical cyclones can form in any month, although they are relatively infrequent during the late winter and early spring. SST are warm (>28°C) over much of the year in this region. Tropical cyclogenesis occurs most frequently during the period of July through October. During these months the monsoon trough provides a region of low-level cyclonic flow and convection favorable for the formation of tropical cyclones.

Numerous tropical cyclones also form over the southern Indian Ocean and the southwest Pacific Ocean. Although tropical cyclones may form over this region during any month, the majority of the storms occur during the austral summer. During austral summer a monsoon trough (also sometimes called the Southwest Pacific Convergence Zone) provides low-level cyclonic inflow and convection. Approximately 15 tropical cyclones form annually over the eastern North Pacific Ocean. Although this is only half the number that form over the western North Pacific Ocean, when the numbers are normalized for the size of the region the greatest frequency of tropical cyclogenesis occurs over the eastern North Pacific. In one out of every two years the first tropical cyclone occurs over the eastern North Pacific in late May. Tropical cyclogenesis is most common during July through September. Many of the tropical cyclones that form over this region appear to develop out of tropical waves that migrate across the Atlantic Ocean and central America before entering the eastern North Pacific. Abundant regions of warm SST and relatively low magnitudes of vertical wind shear create a very favorable environment for genesis of tropical cyclones from these tropical waves.

Around ten tropical storms form annually over the North Atlantic basin. Although the hurricane season in the Atlantic officially runs from the first of June through the end of November, tropical cyclogenesis occurs most frequently during August through October. Tropical cyclogenesis during the early or later part of the hurricane season often occurs along old fronts or from subtropical cyclones. Early in the hurricane season tropical cyclones most commonly form over the southern Gulf of Mexico or the western Caribbean Sea. During late summer and early autumn the SST are at their peak and wind shear is generally reduced over the lower latitudes, which allows tropical waves to develop into tropical cyclones. Tropical cyclones can form over most of the subtropical and tropical north Atlantic during this period. Toward the end of the hurricane season the preferred regions for tropical cyclogenesis move back to the southern Gulf of Mexico and the southwest Caribbean Sea.

Five tropical cyclones form on average over the northern Indian Ocean. The formation of tropical cyclones is most frequent in the fall, but there is a secondary period of formation during the early summer. Many of the tropical cyclones develop from disturbances that are associated with the monsoon trough. An average of one tropical cyclone forms each year over the central North Pacific Ocean around Hawaii. The most likely months for tropical cyclogenesis to occur are August, September and October.

The formation of tropical cyclones over ocean basins that experience relatively high frequencies of these systems is marked by periods of activity and inactivity. Maloney and Hartmann (2001) have proposed that these periods of frequent formation of tropical cyclones and quiescent periods are related to the Madden–Julian Oscillation (MJO). The MJO is associated with periods of enhanced westerly and easterly flows in the lower troposphere. It has been proposed that the enhanced westerly flows interact with the prevailing easterly winds in the tropics to generate cyclonic rotation favorable to tropical cyclogenesis. During these periods tropical cyclones form more frequently. The periods of enhanced easterly flow are thought to be associated with the periods of reduced formation of tropical cyclones.

The frequency of tropical cyclone formation also exhibits an interannual variability. Studies have examined statistical links between the frequency of tropical cyclone formation and large-scale phenomena such as El Niño–Southern Oscillation (ENSO) and the Quasi-Biennial Oscillation (QBO). Wang and Chan (2002) examined the links between ENSO and tropical cyclone activity over the western North Pacific Ocean. It was found that tropical cyclones developed farther to the south and east during an El Niño. Since the storms formed further to the

Table T9 Tropical cyclone frequencies

	Jan	Feb	Mar	Apr	May	June	July	Aug	Sep	Oct	Nov	Dec	Year
Western North Pacific	0.6	0.3	0.5	0.8	1.3	2.0	4.7	6.6	5.8	4.7	2.9	1.6	31.8
Southern Indian/Southwest Pacific	6.1	6.7	4.8	3.2	0.9	0.2	0.4	0.1	0.3	0.7	1.4	3.4	28.2
Eastern Pacific	0.0	0.0	0.0	0.0	0.5	2.0	3.6	3.4	3.4	1.7	0.2	0.0	15.0
North Atlantic	0.0	0.0	0.0	0.0	0.0	0.6	0.8	2.6	3.5	1.6	0.5	0.0	9.6
Northern Indian	0.1	0.1	0.0	0.1	0.0	0.6	0.0	0.0	0.3	1.0	1.3	0.5	5.0
Central Pacific	0.0	0.0	0.0	0.0	0.0	0.0	0.0	0.5	0.3	0.2	0.1	0.0	1.1
Total	6.8	7.1	5.3	4.1	2.7	5.4	9.5	13.2	13.6	9.9	6.4	5.5	90.7

Table T10 Saffir–Simpson scale

Category	Wind speed	Damage	Structures affected
One	33–42 m s^{-1} (64–82 knots)	Minimal	Unanchored mobile homes
Two	43–49 m s^{-1} (83–95 knots)	Moderate	Mobile homes, roofing, windows
Three	50–58 m s^{-1} (96–113 knots)	Extensive	Small residences, utility buildings
Four	59–69 m s^{-1} (114–135 knots)	Extreme	Residences and some commercial buildings
Five	>70 m s^{-1} (>135 knots)	Catastrophic	Residences and industrial buildings

south and east they typically moved further west before recurving poleward and existed for longer periods of time. Higher sea-level pressures and lower SST tend to be associated with reduced formation of tropical cyclones over the southwest Pacific Ocean around Australia during an El Niño. Although the formation of tropical cyclones over the central Pacific is relatively rare, they appear to form more frequently during an El Niño. Warmer SST and enhanced convection provide a more favorable environment during an El Niño. The frequency of the formation of the tropical cyclones over the eastern North Pacific is not strongly affected by an El Niño. The favorable conditions associated with warmer SST are somewhat offset by higher magnitudes of vertical wind shear. However, cyclones that do form over the eastern North Pacific during an El Niño exhibit higher intensities. The strongest hurricane on record in this region was hurricane Linda, which formed during the El Niño in 1997. The frequency of the formation of tropical cyclones over the Atlantic is reduced during an El Niño. The convection over the tropical Pacific Ocean generates stronger westerly winds that increase the vertical wind shear over the Atlantic Ocean and inhibit the formation of tropical cyclones.

Movement of tropical cyclones

The movement of tropical cyclones results from the interaction of the system with its surrounding environment. However, tropical cyclones over many ocean basins exhibit similar paths. Tropical cyclones often form in the lower latitudes of the tropics equatorward of the subtropical ridge. These systems are initially embedded in deep easterly flow that causes them to move generally east to west. As the systems near the western periphery of the subtropical ridge the tropical cyclones begin to experience a more poleward direction of movement. The poleward movement around the western edge of the subtropical high is called *recurvature*. Many tropical cyclones reach their peak intensity at the western edge of the subtropical high. As the tropical cyclones move poleward they begin to be more strongly affected by the midlatitude westerly flow. Eventually many tropical cyclones recurve and move rapidly from west to east in the midlatitude flow. These tropical cyclones often experience an extratropical transition or are absorbed into the circulation of a larger extratropical cyclone.

Hazards associated with tropical cyclones

The most obvious hazard associated with a tropical cyclone is the danger posed by high wind speeds. The National Hurricane Center in Miami, Florida uses the Saffir–Simpson Hurricane Scale to classify hurricanes (Table T10). Even a minimal hurricane can do damage to unanchored mobile homes and poorly maintained structures. A Category Two hurricane can do significant damage to mobile homes and certain types of roofing. The wind speeds associated with a Category Three hurricane can do extensive damage to any structures in their path. The winds in Category Four and Five hurricanes are capable of doing significant damage to many structures unless they are specifically constructed to withstand extremely high wind speeds.

In coastal communities the *storm surge* may pose the greatest hazard from tropical cyclones. On the side of the storm where the winds are blowing toward the coast friction between the high winds and the water will produce a flow of water toward the coast. This flow of water produces an increase in the mean elevation of the ocean at the shore that is called the *storm surge*. Strong wind speeds also produce high waves that may add to the water-level rise. The height of the storm surge is a function of the strength of the tropical cyclone, the bathymetry of the area near the shore and the shape of the coastline. Some of the deadliest tropical cyclones in history produced catastrophic storm surges. A tropical cyclone that struck Bangladesh in 1970 produced a storm surge that may have killed 300 000 people. A hurricane that hit Galveston, Texas, in 1900 produced a storm surge that killed 8000 people.

Once a tropical cyclone moves inland the greatest threat is from fresh water flooding caused by heavy rainfall. Generally, the amount of rain produced by a tropical cyclone is a function of its intensity, size and how fast it moves. Large slow-moving storms can produce up to a meter of rain in a day; while smaller, more rapidly moving systems may only produce a few centimeters of rainfall. Rainfall may be significantly enhanced in regions where topography interacts with the tropical cyclone to produce winds flowing up the slope. Flooding produced by Hurricane Mitch in 1998 over Central America produced catastrophic flooding that may have killed 20 000 people.

Tornadoes are another hazard posed by tropical cyclones after the move inland. Although the environment around a tropical cyclone is not particularly unstable, increased friction as the winds flow from the sea to over the land can create sufficient shear to generate tornadoes. The tornadoes are often associated with strong convection in the spiral bands. After a tropical cyclone moves farther inland, warming of the surface during the daytime may increase the instability. When outer spiral bands move into these more unstable regions, convection may produce additional tornadoes. Although most tropical cyclones produce relatively few tornadoes, a few hurricanes have produced over 100 tornadoes after making landfall in the United States.

Jay Hobgood

Bibliography

Charney, J.G., and Eliassen, A., 1964. On the growth of the hurricane depression. *Journal of Atmospheric Science*, **21**: 68–75.

Dodge, P.P., Burpee, R.W., and Marks, F.D., 1999. The kinematic structure of a hurricane with sea level pressure less than 900 mb. *Monthly Weather Reivew*, **127**: 987–1004.

Emanuel, K.A., 1987. An air–sea interaction model of intraseasonal oscillations in the tropical atmosphere. *Journal of Atmospheric Science*, **44**: 2324–2340.

Maloney, E.D., and Hartmann, D.L., 2001. The Madden–Julian Oscillation, barotropic dynamics and North Pacific tropical cyclone formation. Part I: Observations. *Journal of Atmospheric Science*, **58**: 2545–2558.

Wang, B., and Chan, J.C.L., 2002. How does ENSO regulate tropical storm activity over the western North Pacific? *Journal of Climate*, **15**: 1643–1658.

Willoughby, H.E., Clos, J.A., and Shoreibah, M.G., 1982. Concentric eye walls, secondary wind maxima, and the evolution of the hurricane vortex. *Journal of Atmospheric Science*, **39**: 395–411.

Cross-references

Climate Hazards
Coastal Climate
Easterly Wave
El Niño
The Madden–Julian Oscillation
Ocean–Atmosphere Interaction
Satellite Observations of the Earth's Climate System
Trade Winds and the Trade Wind Inversion

TROPOSPHERE

The vertical distribution of the atmosphere traditionally has been divided up into a series of concentric layers primarily based on the thermal structure of the atmosphere. The bottom layer of the atmosphere is called the *troposphere*, and it is within this layer that we find most of the atmospheric water vapor as well as most clouds and weather phenomena. It is primarily composed of nitrogen (78%) and oxygen (21%), with only small concentrations of other trace gases. Nearly all atmospheric water vapor or moisture is found in the troposphere. The upper boundary of this layer is called the tropopause and represents the upper limit of large-scale turbulence and mixing in the atmosphere. The height of the tropopause varies with time of year and latitude, ranging from an average height of 5–6 km over the polar regions to 15–16 km over the equator, and tending to be higher in summer than in winter.

The Earth's surface is the primary absorber of incoming solar radiation therefore the warmest portion of the troposphere is normally adjacent to the surface. Temperatures decrease at an average rate of 6.5°C/km with increasing height in the troposphere; within the tropopause, temperatures remain the same with increasing elevation. Sometimes the temperature does not decrease with height in the troposphere, but increases. Such a situation is known as a temperature inversion. Temperature inversions limit or prevent the vertical mixing of air. Such atmospheric stability can lead to air pollution episodes with air pollutants emitted at ground level becoming trapped beneath the temperature inversion.

One of the most important properties of our atmosphere is its ability to be compressed due to the influence of the Earth's gravitational field. Although the atmosphere extends well over 100 km in height, over 80% of the mass is concentrated in the troposphere. Therefore, atmospheric pressure decreases rapidly with height, averaging only about 20–30% of surface values at the tropopause.

K. Dewey

Bibliography

Aguado, E., and Burt, J.E., 2001. *Understanding Weather and Climate*. Upper Saddle River, NJ: Prentice-Hall.

Barry, R.G., and Chorley, R.J., 1998. *Atmosphere, Weather, and Climate*, 7th edn. London: Routledge.

McIlveen, R., 1992. *Fundamentals of Weather and Climate*. London: Chapman & Hall.

Nese, J.M., and Grenchi, L.M., 1998. *A World of Weather: fundamentals of meteorology*, 2nd edn. Dubuque, IO: Kendall/Hunt.

Robinson, P.J., and Henderson-Sellers, A., 1999. *Contemporary Climatology*, 2nd edn. Harlow: Longman.

Rex, D.F., 1969. Climate of the free atmosphere. In Landsberg, H.E., ed., *World Survey of Climatology*, vol. 4. New York: Elsevier.

Cross-references

Atmospheric Nomenclature
Inversion
Lapse Rate

TUNDRA CLIMATE

Location and definition

There are two types of polar climates that lie poleward of 70°N where the Taiga gives way to the mostly treeless landscape. The coldest of these two regions is the Ice Cap climate (EF) and the milder is the Tundra climate (ET). The word Tundra is derived from the Finnish word *tunturia* for barren land, or treeless plain (*Tundra*, 2003). While there are a few areas of Tundra in the Antarctic, as a climatic region the Tundra generally is located around the Arctic Ocean and above the timberline of high-latitude mountains (Oliver and Hidore, 2002; Figure T43).

Tundra refers to both a climate type as well as a vegetation association (Strahler and Strahler, 2002). Tundra climate is categorized as either Arctic Tundra or Alpine Tundra. The Alpine Tundra experiences a different climate than does the Arctic Tundra in terms of daylength and seasonal changes. However, the Alpine Tundra is similar in terms of the cold temperatures and poor, thin soils. Therefore, most tree species are absent from these montane regions and Tundra vegetation invades and thrives.

The Tundra biome is subdivided into the High, Middle, and Low Arctic Tundra based on latitude. The High Arctic Tundra is restricted to the islands of the Arctic Oceans. The Middle Arctic Tundra is confined to the Arctic Coastal Plain where the terrain is fairly level and the freeze/thaw cycle creates geometric rock patterns on the surface. The Low Arctic Tundra describes the vast majority of the Tundra with steeper slopes that are better drained. Along waterways, trees, such as the willow and alder, can grow to a few meters high. However, the

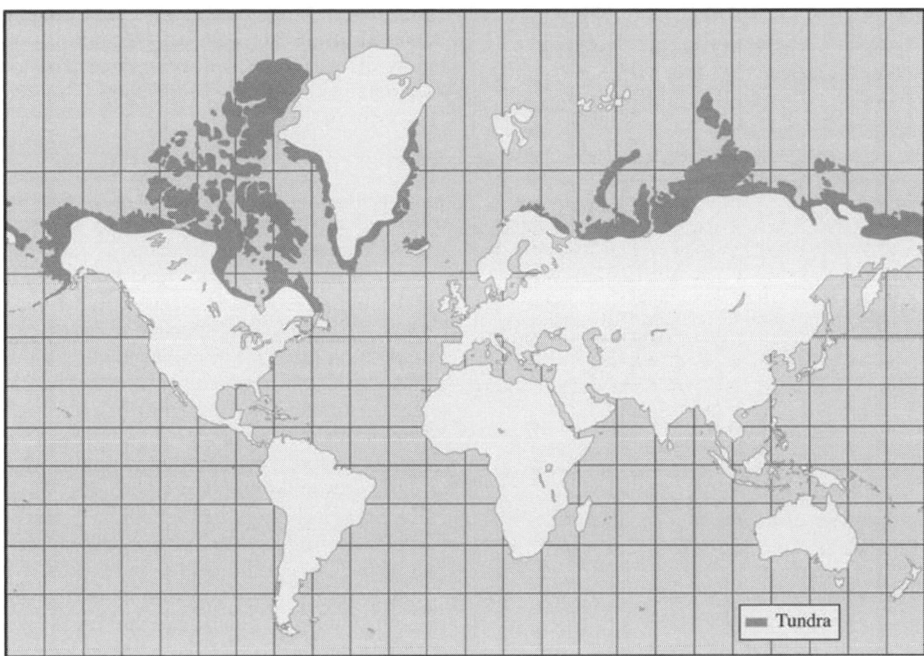

Figure T43 The Tundra; cartographer T. Truckenbrod.

Tundra climate does not favor tree growth as abrasion from wind-driven snow stunts the branches and subsurface dynamics damage the roots (*Tundra*, 1996).

Climatic characteristics

The Tundra is sometimes considered a polar desert as the precipitation is usually less than 20 cm (8 in). While at lower latitudes this scant amount of precipitation would constitute a desert, in the Tundra, evaporation rates are so low that there remains adequate moisture. In other words, the small amount of precipitation exceeds the small demand of potential evapotranspiration. According to Köppen's classification, this Tundra ET climate is under snow cover for approximately 8–10 months of the year. Snowfall actually benefits Tundra plants by providing a layer of insulation against the high-speed and desiccating winds. The amount of precipitation received by any area in the Tundra depends on proximity to large water bodies and to semi-permanent high- or low-pressure cells.

Since much of the Tundra in North America, Asia, and Greenland is located near major water bodies, there is a lesser degree of continentality than in many areas of the sub-Arctic climate region. Most of the Tundra is under the influence of high pressure; however, winter storms develop predominantly in the two areas where the barometric pressure remains low. The Aleutian Low influences the weather from eastern Siberia to the Gulf of Alaska and beyond, while the Icelandic Low generates storms from central Canada to Scandinavia and western Europe (Ritter, 2002).

The daylength of most of the Tundra, which is above the Arctic Circle, varies from 0 to 24 hours. During summer months, Tundra regions have very long days, yet the sun is never very high above the horizon. The average temperature of the warmest month in the Tundra is below 10°C and above freezing. The warmest summers are located in the interiors of Siberia, Alaska, and Canada. Along with the high albedo of snow, this low angle of the sun contributes to a very low solar radiation receipt. Consequently, the summertime daily high temperature is mild and not much greater than the daily minimum. While the daily temperature range is low, the annual temperature range is moderately high (Figure T44).

The long, cold, dark winters of the Tundra result in 6–10 months with mean monthly temperatures below 0°C. The coldest weather occurs in northeastern Siberia (Hatter et al., 2003). There are no crops grown in the Tundra, and the short growing season for the native species is only 60–80 days. When the Tundra blooms, however, the colors are varied and the landscape is quite beautiful.

Edaphic conditions

The widespread existence of permafrost, a perennially frozen layer of soil and regolith below the surface, typifies most of the Tundra. In some locations the permafrost extends to depths of several hundred meters. During long Tundra winters the surface soil layer is thoroughly frozen. In the summer the temperatures are warm enough to thaw the upper layer of permafrost, but only to a depth of 10 cm to 1 m. Rather than the extremely cold temperatures associated with the Tundra, it is the layer of permafrost that is the primary limiting factor to tree growth as roots cannot delve deeply into the soil.

Soils of the Tundra are nutrient-poor and slow to form due to the extremely cold temperatures. In the Alpine Tundra the terrain is rugged and steep. These glacially carved areas have the distinctive U-shaped hanging valleys, tallus slopes, and cirques. Where soil is present, it too tends to be poorly developed and shallow. Because thin soils of the Tundra overlie an impermeable layer of permafrost, rain and snowmelt are unable to percolate downward and the surface becomes saturated and boggy.

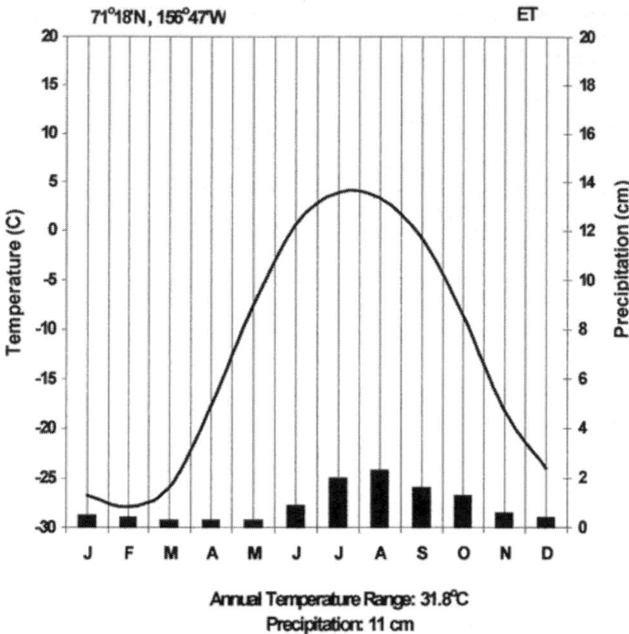

Figure T44 Climograph of Barrow, Alaska (after Eugster et al., 2000).

Mostly of the order Gelisols, Tundra soils are subject to disruption from frost churning and mixing. This freeze/thaw cycle pulls organic and partially decomposed matter from the surface down to the lower layers and brings rocky materials from the C-horizon up to the surface. This process of cryoturbation contributes to the circular, striped, and polygonal patterns of the Tundra. Together, the very thin active soil layer, the existence of permafrost, the persistent freeze/thaw cycle, and the occurrence of solifluction during warmer months create a patchwork of microhabitats and their associated plant communities on the Tundra landscape.

Vegetation

Vegetation lifeforms of the Tundra are well adapted to the soil and climate conditions and include tussock-forming graminoids; mats, or cushion plants; rosettes; and dwarf shrubs. Bunch grasses and cushion plants grow in tight clumps, enabling them to retain heat and avoid exposure to high-speed, cold winds. Many plants have dark-colored leaves that permit greater absorption of heat energy. Cotton grass has winged seeds that maximize the wind's kinetic energy to disperse offspring across the Tundra. For protection from the elements, many buds or other reproductive structures of Tundra vegetation grow close to the ground or beneath the surface as rhizomes.

On bare rock the first species to take hold are lichens. Lichens are a combination of algae and fungi and are an important food source for herbivores on the Tundra. This plant excretes an acid solution that begins the chemical dissolution of rock, and constitutes the first step in soil formation. Also present are herbaceous species such as mosses, perhaps best known is the commercially used sphagnum moss; sedges; Arctic meadow grass; perennial forbs; and members of the heath family. Dwarf shrubs and trees, such as willows and birches, are typically less than 1 m high and are found only in the most favorable growing conditions of the Tundra. Generally the root networks of the few woody species branch out laterally just below the surface because they can only penetrate to the depth of the thawed ground.

Animals

The animal life of the Tundra demonstrates remarkable adaptive mechanisms as well. Migratory species venture south when the temperatures begin to drop. These species include caribou and birds such as ravens; snow buntings; falcons; loons; sandpipers; terns; snow birds; and various species of gulls (*Tundra Biome*, 2003). Other species migrate only when their prey population crashes. For example, the snowy owl has been seen as far south as Virginia during times when lemming populations declined drastically. The lemming is a small rodent and a major herbivore in the Tundra.

Other adaptive mechanisms include morphological characteristics. Animals which reside in the Tundra year round typically have large, compact bodies, a thick fur or covering of feathers serving as insulation, or they have a plumage or pelage that turns white in the winter and brown in the summer, like the Arctic hare. Residing year-round are the reindeer, rabbit, Arctic fox, wolf, weasel, polar bear, and musk ox. The musk ox is not a true ox, but is more closely related to goats and sheep. Insects include mosquitoes, moths, blackflies, grasshoppers and Arctic bumblebees. Fish species are cod, flatfish, trout and salmon (*Tundra Biome*, 2003). Relative to other biomes, species composition and the food chain of the Tundra are simple.

The Alpine Tundra is home to mountain goats, sheep, marmots and pikas, while the insect species include springtails, beetles, and butterflies. The vegetation of the Alpine Tundra is very similar to the Arctic Tundra as it is composed of tussock grasses, species of heath, and dwarf shrubs and trees. The Arctic Tundra is an important breeding ground for waterfowl, such as geese and swan, as well as for caribou which breed in September and October, allowing the calves to be born in the milder months of May and June. Caribou feed mostly on lichens, but they also eat shrubs, grasses, the tender shoots of trees, and mushrooms.

Concerns

The Arctic Tundra is a fragile environment. While some ecosystems take fewer than 50 years to completely return to a climax community, the slow rate of ecological processes in the Tundra causes this region to require up to centuries to recover from disturbance (Oliver and Hidore, 2002). Increased development, especially in terms of oil exploration, along with climatic change, have left their mark on portions of the Tundra.

Thermal erosion results from scraping away or destroying the upper layer of plants and organic matter in a permafrost region (Strahler and Strahler, 2002). This upper layer serves as insulation, and when it is damaged or removed the summer thaw extends further downward, enhancing the melting of ice wedges and causing general surface subsidence. Clay and silt mix with meltwater, wash downstream, and form muddy trenches to create a thermokarst landscape. The evidence of this alteration of permafrost regions became obvious during World War II when military bases with their airfields and other transportation networks caused mud basins the size of small lakes to form, that engulfed surrounding buildings. Presently, engineers are required to construct buildings on pilings in the Tundra (Strahler and Strahler, 2002).

While the Alaskan oil pipeline is supported above the ground, to minimize permafrost damage, trucks and other heavy equipment used in construction and exploration for oil in the Arctic leave behind deep trenches and a scarred landscape. The debate over drilling in the Alaskan Arctic Wildlife Refuge (ANWR) is a heated one. Proponents of drilling in this vulnerable environment point to the United States' dependence on foreign oil. Opponents to drilling in ANWR say that if the US increased fuel efficiency for cars, SUV, and light trucks from 24 to 39 miles per gallon, it would save 51 billion barrels of oil over the next 10 years. Some argue that exploration would affect only a small portion of the "adesolate tundra". Others point out that legislation was passed 30 years ago to raise the fuel standards for light trucks and cars to 27.5 mpg. In the early 1980s that standard was relaxed to 24 mpg. The difference in oil consumption for a single year between the lax legislation and the 27.5 mpg standard exceeds the most optimistic predictions of the total yield of the proposed drilling in the refuge (an estimated 5.7–16 billion barrels) (Van Deerlin, 2003).

The debate over drilling in the Arctic is related to the increasing temperatures that have been recorded in that region. Warming temperatures have been especially pronounced in the Arctic Tundra (Aleksandrova, 1980; Bliss and Matveyana, 1992; Chapman and Walsh, 1993; Keeling et al., 1996). In the last 100 years, portions of the Tundra have warmed by 10°C, which is 10 times the global average, and sea ice covers 15% less of the Arctic Ocean than it did 20 years ago (McFarling, 2002). The Tundra must be completely frozen before oil exploration can occur. On the North Slope of Alaska, the number of days that developers can venture onto the Tundra has decreased from 200 to 100 days over the past 30 years as freezing occurs later and melting begins earlier in the year (Pinsker, 2003).

Vegetation and animal communities have been monitored in the Tundra and have been shown to exhibit varying responses to the warming of that region. Browne (2002) reports that more than 75% of the Canadian Arctic Tundra is predicted to disappear, taking with it one in five of the species living there. Responses to rising temperatures at the micro-scale show changes in the vessel anatomy of Tundra plants (Gorsuch et al., 2001), and at the macro-scale the overall structure of the Taiga–Tundra ecotone has been altered (Payette et al., 2001). In an analysis of the period 1901–1995 the Tundra is a statistically significant outlier regarding the rate of warming in that region (Fraedrich et al., 2001). Species and systems alike are making manifest the substantial repercussions of climate change.

Mary Snow

Bibliography

Aleksandrova, V.D., 1980. *The Arctic and Antarctic: their division into geobotanical areas*. New York: Cambridge University Press.

Bliss, L.C., and Matveyana, N.V., 1992. *Circumpolar Arctic Vegetation. Arctic ecosystems in a changing climate: an ecophysiological perspective*. New York: Academic Press, pp. 59–89.

Browne, A., 2002. How climate change is killing off rare animals: Conservationists warn that nature "crown jewels" are facing ruin. *London Observer*, 10 February, p. 150.

Chapman, W.L., and Walsh, J.E., 1993. Recent variations of sea ice and air temperature in high latitudes. *Bulletin of the American Meteorological Society*, **74**: 33–47.

Eugster, W., Rouse, W.R., Pielke, R.A., et al., 2000. Land–atmosphere energy exchange in Arctic tundra and boreal forest: available data and feedbacks to climate. *Global Change Biology*, **6**: 84–115.

Fraedrich, K., Gerstengarbe, F.W., and Werner, P.C., 2001. Climate shifts during the last century. *Climate Change*, **50**: 405–417.

Gorsuch, D.M., Oberbauer, S.F., and Fisher, J.B., 2001. Comparative vessel anatomy of Arctic deciduous and evergreen dicots. *American Journal of Botany*, **88**: 1643–1649.

Hatter, G., Howanic, J., and Hustrulid, E., 2003. *Tundra*. Retrieved 19 May, 2003, from http://ths.sps.lane.edu/biomes/tundra4/tundra4d.html

Keeling, C.D., Chin, J.E., and Whorf, T.P., 1996. Increased activity of northern vegetation infected from atmospheric CO_2 measurements. *Nature*, **382**: 146–149.

McFarling, U.L., 2002. The Arctic meltdown: quick thaw alarms natives and scientists. *Seattle Times*, 15 April, p. A3.

Oliver, J.E., and Hidore, J.J., 2002. *Climatology: an atmospheric science*, 2nd edn. Upper Saddle River, NJ: Prentice-Hall.

Payette, S., Fortin, M., and Gamache, I., 2001. The subarctic forest-tundra: the structure of a biome in a changing climate. *Bioscience*, **51**: 709–718.

Pinsker, L.M., 2003. Tales from a warming Arctic. *Geotimes*, **48**: 11–12.

Ritter, M., 2002. Arctic Tundra Retrieved 19 May 2003, from http://www.uwsp.edu/geo/ faculty/ritter/geog101/lectures/climates_ tundra.html

Strahler, A., and Strahler, A., 2002. *Physical Geography: science and systems of the human environment*, 2nd edn. New York: Wiley.

Tundra, 1996. Retrieved June 5 2003, from http://www.runet.edu/~swoodwar/CLASSES/GEOG235/biomes/tundra/tundra.html

Tundra Biome. Retrieved 5 June 2003, from http://www.creekland.org/studentpro/shainaE/tund.html

Van Deerlin, L., 2003. Bush's oil junta determined to drill in ANWR. *San Diego Union-Tribune*, 9 April, p. B7.

Cross-references

Arctic Climates
Asia, Climates of Siberia, Central and East Asia
Taiga Climate
North America, Climate of

TURBULENCE AND DIFFUSION

Analysis and forecasting of meteorological motions usually is based on observations 100 miles or more apart and, therefore, only reflect the behavior of large-scale motions. However, motions exist on all scales, down to perhaps a millimeter. Further, since the basic equations of meteorology are nonlinear, the small-scale motions interact with the large-scale features of the atmosphere by their ability to mix air with different large-scale characteristics. Sometimes meteorologists define *turbulence* as small-scale motion, i.e. motion on too small a scale to be described by ordinary weather data. On the other hand, students of fluid mechanics require their turbulence to have certain definite properties, such as important nonlinear effects, large random vorticity, and the ability to diffuse contaminants. Fortunately, most of the small-scale motion is of this character; yet certain small-scale phenomena, such as internal waves, should not be classed as turbulence.

Distribution of turbulence

Most atmospheric turbulence can be ascribed to one of two causes:

1. mechanical turbulence caused by rapid change of the large-scale wind from one point to another; and

2. heat convection, caused by heating from below or cooling from above.

There are three general regions in which turbulence is important: (1) near the ground; (2) in convective clouds (cumulus type); and (3) between 25 000 ft and 40 000 ft where turbulence often occurs in clear air (clear-air turbulence, CAT).

On a windy day, turbulence is always strong near the ground, since the wind vanishes right at the surface, and we thus have a strong wind shear. Such mechanical turbulence depends on the wind speed and is especially strong over rough terrain. Convection will increase the turbulence at daytime due to solar heating and will suppress it at night.

Cumulus clouds always indicate convection cells, and some of the most destructive turbulence occurs in cumulonimbus clouds, where the upward and downward motions may reach many tens of miles per hour. Turbulence near 30 000 ft (CAT) almost certainly is due to the rapid vertical variation of wind near the jet stream. Layers of excessive shear and turbulence often are less than 1000 ft thick and often are missed by conventional meteorological observing equipment. CAT appears to be most prominent over mountainous terrain.

These three general types of regions are characterized by strong vertical turbulent motions that might be felt by aircraft. It is possible that quasihorizontal eddies of scales of tens of miles are common throughout the atmosphere.

Structure of turbulence

In general, there are eddies of all sizes. The smallest tend to be *isotropic* – equally well developed in all directions; whereas the larger ones tend to have larger horizontal than vertical dimensions. Very close to the ground, eddies tend to be small (particularly of small vertical dimensions); therefore, these eddies can produce little vertical mixing; with increasing height, vertical mixing increases. In general, turbulence caused mechanically has smaller scales than turbulence caused by heating. Thus, when convective turbulence is added to mechanical turbulence, relatively slow fluctuations are superimposed on the fast variations previously present. Particularly, the fluctuations of wind direction are greatly amplified by adding convection to mechanical turbulence. In general, mechanical eddies near the ground are elongated in the direction of the wind, whereas convective eddies have about the same dimensions in all horizontal directions.

Effects of turbulence

Turbulence alters the distribution of the mean meteorological variables by its ability to produce vertical mixing. Since this is weakest near the ground, we find the strongest gradients immediately above the surface, with weaker gradients above. Since the mean winds are nearly horizontal, most of the vertical flux of momentum, moisture, and heat is produced by turbulence. Thus, evaporation requires turbulence to carry off the vapor.

Turbulence creates irregular patches of temperature and moisture, which are able to scatter electromagnetic waves and sound waves. For example, propagation of microwaves beyond the horizon has been ascribed to turbulence.

Turbulence affects structures such as towers, planes, bridges, missiles, etc. Not only does turbulence increase the maximum wind far above the mean wind, but fluctuations within particular periods can produce resonance with certain natural modes of vibrations of the structures, producing failure of the structures.

Turbulence produces mixing of concentrated contaminants with the surrounding air, a process called *turbulent diffusion*. This is particularly important in connection with general air pollution and with problems involving atomic debris, poisonous missile fuels, and chemical and biological warfare. The treatment of diffusion is described in the next section.

Turbulent diffusion

The coefficient of molecular diffusivity in air is so small that molecular diffusion almost always can be neglected compared to turbulent diffusion (spreading of contaminants by turbulent mixing). In most diffusion problems the source strength is given, as well as some characteristics of the atmosphere; required are either instantaneous concentrations at various points or average concentrations (dosages). Originally, this type of problem was usually attacked by writing an equation analogous to the molecular diffusion equation, but with the molecular diffusivity replaced by a much larger *turbulent* diffusivity. However, whereas the molecular diffusivity is often a constant, the turbulent diffusivity varies with height above ground and meteorological conditions. Further, the turbulent diffusivity may depend on the geometry of the contaminant (since only eddies smaller than the contaminant cloud can mix efficiently). However, the geometry of the contaminant depends on the diffusivity, so that this technique presents severe theoretical difficulties. Nevertheless, it has been used for vertical spreading from ground sources, with particular success, both in the United States and abroad. Usually, wind and diffusivity are assumed to be power laws of height so that exact mathematical solutions are possible.

Another technique of solving diffusion problems is based on the assumption that the concentration of contaminants follows a Gaussian (normal) distribution, a reasonable assumption because the turbulent wind field is very nearly Gaussian. The theoretical expressions are then made more precise by the requirement of continuity – namely, that material once emitted cannot disappear.

The resulting equations contain some unknown parameters that describe the dimensions of the cloud or plume. Usually these are related statistically to the time after emission and to the energy of the turbulence. The energy of the turbulence can be measured locally (for short-range diffusion estimates), or it can be related to the large-scale distribution of wind, temperature, and roughness in the atmosphere. Corrections have to be made for deposition rate of fall in case of large impurities, washout, and chemical interactions. In connection with air pollution regulations, a form of the Gaussian technique for pollution estimates most commonly used is the Pasquill–Gifford method. Here lateral and vertical plume spread are computed from graphs based on empirical relations between plume spread, solar radiations, and wind. Numerical approximations to these graphs are also available, which, in addition, allow for the effect of varying ground roughness on turbulent intensities.

It has been recognized that the vertical distribution of pollutants can be far from Gaussian, particularly on clear days in light winds. For example, the height of maximum concentration does not remain at the level of emission, but descends to the ground, only to rise again later. Thus, a relatively new technique has been developed for clear, convective days. New length and velocity scales are chosen: the thickness of the mixed layer h and w^*, a velocity depending on the surface heat flux and this thickness. If all quantities are normalized by these scales, universal curves for normalized concentration as function of normalized distance and normalized height can be derived for different normalized source heights, which are equally valid in the atmosphere and in laboratory models.

A distinction must be made between continuous sources discussed so far (such as a stack) and instantaneous sources (such as an atomic bomb). In the case of a continuous source, the greatest concentration occurs along the mean wind direction. Variations of horizontal and vertical wind directions on all time scales diluted the average concentration measured at a given point. For an elevated source the greatest concentrations on the ground occur with light winds and strong vertical turbulence.

In the case of an instantaneous source, only small eddies spread the contaminant initially; the large eddies just transport the whole puff in various directions. Thus, the concentration remains high near the center of the puff, the material remains concentrated, and only a relatively small volume of air is affected by the contamination initially. Larger eddies take over later and dilution increases.

Hans A. Panofsky

Bibliography

Anon, 1975. *Lectures on Air Pollution and Environmental Impact Analyses*, 1975. American Meterological Society.

Arya, S.P., 1998. *Air Pollution Meteorology and Dispersion.* New York: Oxford University Press.

Panofsky, H.A., and Dutton, J.A., 1984. *Atmospheric Turbulence.* New York: Wiley-Interscience.

Pasquill, F., and Smith, P.B., 1984. *Atmospheric Diffusion.* New York: Academic Press.

Cross-references

Air Pollution Climatology
Boundary Layer Climatology
Cloud Climatology
Lapse Rate

U

ULTRAVIOLET RADIATION

In general scientific usage, *ultraviolet radiation* refers to radiant emissions in that portion of the electromagnetic spectrum between X-rays and visible radiation, i.e. in the wavelength range 0.01–0.4 μm approximately. However, outside the Earth's atmosphere, 99% of the solar irradiance occurs in the spectral range 0.15–4 μm; hence, climatologically significant ultraviolet radiation is that lying between 0.15 μm and 0.4 μm in the spectrum, a band containing about 9% of the radiant energy of solar origin prior to atmospheric attenuation. Solar ultraviolet radiation is conventionally subdivided into three wavelength bands, based primarily on potential biological activity: UV-A (0.32–0.4 μm), UV-B (0.28–0.32 μm), and UV-C (< 0.28 μm).

Radiation with wavelengths shorter than 0.21 μm (most of the UV-C) is absorbed by oxygen and nitrogen in processes of ionization, dissociation, and heating at elevations in excess of 80 km in the thermosphere and ionosphere. However, most absorption takes place in the stratosphere, where ozone absorbs radiation in the wavelength range 0.21–0.31 μm. Collectively, these attenuation processes reduce the solar irradiance at the surface by only about 3% on a global, annual basis. Nevertheless, ozone's absorption of ultraviolet radiation to a large extent maintains the temperature structure of the stratosphere. As a result, solar radiation at the Earth's surface lacks UV-C and is deficient in UV-B, possessing little radiant energy at wavelengths shorter than 0.31 μm.

Large percentage fluctuations have been detected in the monochromatic irradiance received outside the atmosphere at ultraviolet wavelengths, primarily associated with solar activity. Because the magnitude of these fluctuations is greatest at the extreme short end of the solar ultraviolet, and decreases with increasing wavelength, the effect on the spectrally integrated solar irradiance is small (probably less than 0.1%). However, changes in stratospheric warming resulting from irradiance fluctuations in the far ultraviolet have been linked to dynamic processes occurring in the troposphere by some authors, and speculation has ensued that longer-term variations in ultraviolet emissions by the sun may be responsible for climate changes.

Concern has been expressed about the role of stratospheric ozone depletion, brought about by the introduction of chlorofluorocarbons and other byproducts of human activity into the atmosphere, in increasing UV-B irradiances at ground level. Ultraviolet radiation, particularly at wavelengths shorter than 0.3 μm, has strong actinic effects, and excessive exposure to such radiation has been linked to sunburn, skin cancer, and eye disease in humans. Deleterious effects on terrestrial and marine ecosystems also have been suggested, as well as the potential for adverse effects on air quality. Nevertheless, this radiation possesses beneficial attributes such as forming vitamin D in the skin and devitalizing bacteria and similar microorganisms of disease.

A. John Arnfield

Bibliography

Kemp, D.D., 1994. *Global Environmental Issues: A Climatological Approach*. New York: Routledge.

Liou, K.N., 2002. *An Introduction to Atmospheric Radiation*. New York: Academic Press.

Paltridge, G.W., and Platt, C.M.R., 1976. *Radiative Processes in Meteorology and Climatology*. New York: Elsevier.

United Nations Environment Programme, 2003. *Environmental Effects of Ozone Depletion and its Interactions with Climate Change*: 2002 Assessment. Nairobi: UNEP.

Cross-references

Energy Budget Climatology
Electromagnetic Radiation
Ozone
Seasonal Affective Disorder
Sunspots

UNITS AND CONVERSIONS

The strength and utility of a measurement system depends on the existence of a consistent set of units and standards around which a complete system can develop. These fundamental standards of weights and measures are maintained on a federal level in the United States by the National Bureau of Standards. On a worldwide level the International Bureau of Weights and Measures in Paris, which was established by the Treaty of Meter in 1875, sets the standard.

The Treaty of Meter also provided for an International Conference of Weights and Measures to meet at regular intervals. In 1957 this conference changed the world standard of length from the platinum–iridium meter bar to a multiple (1 650 763.73) of the wavelength of orange-red light of krypton 86. This new wavelength standard has several advantages over the metal one. It is indestructible, immutable, a constant of nature, and can be reproduced anywhere in the world with an accuracy of one part in one hundred million. In 1960 this same conference adopted an International System of Units (abbreviated SI for Système International). The SI is a metric system based on seven fundamental physical quantities in terms of which all others are to be defined so as to be consistent with the generally accepted equations of physics. These quantities and their units are mass (kilogram), length (meter), time (second), temperature (Kelvin), current (ampere), and luminous intensity (candela). The seventh unit, the mole, was adopted in 1971.

The SI supplants the older cgs and mks systems, although they are still the two most commonly used. In the United States the customary system still employed by established tradition uses the foot, pound, and second; these are defined in terms of the SI units by simple numerical ratios. This customary system, inherited from Britain, is widely used in industry and commerce in both the United States and much of the English-speaking world, but it is not inherently consistent and is little used for strictly scientific measurements. Certain units of measurement, e.g. the nautical mile, which is equivalent to one degree of latitude, have such obvious value that it will be difficult to replace them.

Metric units of measurement

In the metric system of measurement, designations of multiples and subdivisions of any unit may be arrived at by combining with the name of the unit the prefixes deka, hecto, and kilo meaning, respectively, 10, 100, and 1000, and deci, centi, and milli, meaning, respectively, one-tenth, one-hundredth, and one-thousandth. In certain cases, particularly in scientific usage, it becomes convenient to provide for multiples larger than 1000 and for subdivisions smaller than one-thousandth. Accordingly, the following prefixes have been introduced and these are now generally recognized:

yotta (Y), meaning 10^{24}
zetta (Z), meaning 10^{21}
exa (E), meaning 10^{18}
peta (P), meaning 10^{15}
tera (T), meaning 10^{12}
giga (G), meaning 10^{9}
mega (M), meaning 10^{6}
kilo (k), meaning 10^{3}
hecto (h), meaning 10^{2}
deka (da), meaning 10^{1}
deci (d), meaning 10^{-1}
centi (c), meaning 10^{-2}
milli (m), meaning 10^{-3}
micro (μ), meaning 10^{-6}
nano (n), meaning 10^{-9}
pico (p), meaning 10^{-12}
femto (f), meaning 10^{-15}
atto (a), meaning 10^{-18}
zepto (z), meaning 10^{-21}
yocto (y), meaning 10^{-24}

These prefixes increasingly are used with data exchanges. The bit (b) and byte (B) are the standard unit of data quantity, and Gb, MB and GB are commonly used.

Time

Second: the standard unit of time. It was originally defined as 1/86 400 of a mean solar day, but the standard was changed in 1960 by the Eleventh General Conference on Weights and Measures to be 1/31 556 925.9747 of the tropical year 1900. The second is the duration of 9 192 631 770 periods of the radiation corresponding to the transition between the two hyperfine levels of the ground state of the cesium 133 atom. The present standard avoids the variability caused by changes in the Earth's period of rotation about its axis.

1 minute = 60 seconds
1 hour = 3600 seconds
1 sidereal month = 27.32167 days
1 synodical month = 29.53059 days
1 tropical (ordinary) year = 31 556 925.9747 seconds
1 calendar year (common) = 31 536 000 seconds

Temperature

Temperature: a term used to express the relative intensity of heat. It is identified with the kinetic energy of translation of molecules and, in accordance with the kinetic theory of gases, has an absolute zero where all motion has ceased. It is also an aspect of matter that governs the ability to transfer heat from or to other matters. Heat will not transfer between two bodies if they are at the same temperature. There are three temperature scales currently employed: the Fahrenheit scale, the Centigrade (Celsius) scale, and the Kelvin scale. By international agreement, Celsius is favored over Centigrade, but the latter is still widely preferred.

Temperature scales – Centigrade (Celsius) to Fahrenheit

Equivalents
1 C° = 1.8 F°
1 F° = 0.5556 C°
1 K (Kelvin) = 1 C° −273°C = −459.4°F = 0°K

Conversion formulas
°F = 9/5°C + 32
°C = 5/9 (°F −32)

Note: °F is read "degrees Fahrenheit"
F° is read "Fahrenheit degrees"
°C is read "degrees Centigrade or Celsius"
C° is read "Centigrade or Celsius degrees"

1°F per foot = 0.0182269°C per centimeter
1°C per centimeter = 54.864°F per foot
1 ft per °F = 0.54864 m per °C
1 m per °C = 1.82269 ft per °F

Mass and weight

Gram: the unit of mass in the metric system that was first defined to be equal to the mass of a cubic centimeter of pure water at the temperature of its maximum density (4°C). A platinum cylinder was made, known as the kilogram, and declared to be the standard for 1000 grams.

Gram-atomic weight: the atomic weight of an element expressed in grams, customarily based on a scale on which the gram-atomic weight of oxygen is 16 000 grams, but more precisely, based on a scale on which the gram-atomic weight of carbon 12 is 12 000. The gram-atomic weight of all elements contains the same number of atoms. This number, called Avogadro's number, is 6.032×10^{23} and is termed a gram atom.

Gram-equivalent weight: the equivalent weight of an element of compound expressed in grams, customarily based on a scale on which the equivalent weight of oxygen is 8000 grams. It is the gram-atomic weight of an element (or formula weight of a radical) divided by the absolute value of its valence (oxidation state).

Gram-molecular weight: the molecular weight of an element or compound expressed in grams. The gram-molecular weight of all elements or compounds contains the same number of molecules, 6.032×10^{23} (Avogadro's number), and is termed one mole.

Ton: in the United States a ton is a unit of weight equal to 2000 pounds, commonly called a short ton. The long ton, more widely used in England, is equal to 2240 pounds. The metric ton is equal to 1000 kilograms.

1 short ton = 0.90718 metric ton = 0.892857 long ton
1 long ton = 1.12 short tons = 1.016047 metric tons
1 metric ton = 0.984207 long ton = 1.10232 short tons
1 pound (avoirdupois) = 7000 grains = 16 ounces
1 ounce (av.) = 437.5 grains
1 ounce (troy) = 480 grains
1 pound (troy) = 5760 grains
1 ounce (apothecaries) = 480 grains
1 kg = 10^3 g = 2.20462 lb
1 lb = 0.453592 kg = 453.592 g

Linear measures

Meter: the standard unit of length in the metric system. It was first defined by a platinum end standard, the meter of the archives, in Paris. This length, based on a measure of an arc from Dunkirk, France, to Montzuich, Spain, was to be one ten-millionth part of the meridional quadrant of the Earth. In 1889 at the first general conference of the International Bureau of Weights and Measures, the meter was redefined in terms of the platinum–iridium bar known as the international prototype meter at the international bureau. This standard meter is now defined as 1 650 763.73 times the wavelength in a vacuum of krypton 86.

The *micron* is a unit of length equal to 10^{-6} meters. It is commonly employed in the measure of short distances such as the wavelength of light and is represented by the Greek letter μÅThe Angstrom unit (Åis also used for wavelengths; one Å = 10^{-10} m.

Mile: the mile most commonly used in English-speaking countries is the English or statute mile of 5280 feet. This distance was supposedly designated by 1000 paces of the Roman Legions stationed in Britain. Navigators use the nautical mile, which was originally defined as the length of one minute of arc on a great circle drawn on the surface of a sphere with the same area as the Earth. This length is 6080.27 feet. However, since the Earth is an oblate spheroid flattened at the poles, the length of one minute of arc measured along a meridian varies in different latitudes. It is shortest near the poles and longest near the equator with an average length of 6077.015 feet. To resolve this confusion, an international agreement was made defining the nautical mile as 1852 meters, the present value of the international nautical mile. The United States nautical mile, equal to 1.00067387 international nautical miles, and the British nautical mile, equal to 1.00063931 international nautical miles, are occasionally used.

Foot: a unit of length used in the British system of units and employed in English-speaking countries. It was originally standardized in England in 1870 as one-third the length of a yard. In 1959 the foot was defined as 0.3048 meter in accordance with an agreement among the directors of the standards laboratories of English-speaking countries. As a linear measure the foot is of great antiquity, originally identified as the length of a man's foot.

Yard: the fundamental distance in the English measuring system is taken as the distance at 62°F between two fine lines on gold plugs in a bronze bar at Westminster, England, known as the Troughton scale. In 1893 the yard was redefined as 3600/3937 meter.

1 cm = 0.39370 in = 0.032808 ft
1 km = 10^5 cm = 0.62137 mile
1 fathom = 6 ft = 1.8288 m
1 nautical mile = 1.85325 km
1 in = 2.54001 cm
1 ft = 30 480 cm
1 statute mile = 1.60935 km = 5280 ft
1 astronomical unit = 1.496×10^8 km = 92 957 000 miles
1 light year = 9460×10^{12} km = 5.878×10^{12} miles
1 parsec = 3.085×10^{13} km = 1.917×10^{13} miles

Metric to English units – equivalents of length
1 micron (μ) = 0.001 millimeter (mm) = 0.00004 inch (in)
1 mm = 0.1 centimeter (cm) = 0.03937 in
1000 mm = 100 cm = 1 meter (m) = 39.27 in = 3.2808 feet (ft)
1 m = 0.001 kilometer (km) = 1.0936 yards (yd)
1000 m = 1 km = 0.62137 mile
1 in = 2.54 cm
12 in = 1 ft = 0.3048 m

Square measures

1 square ft = 0.00002295684 acre
1 acre = 43 560 ft^2 = 0.0015625 mile^2
1 square yard = 0.836127 m^2
1 hectare = 2.471054 acres
1 square mile (statute) = 640 acres
1 square cm = 0.1550 square in = 0.0010764 square ft
1 square km = 10^{10} square cm = 0.3861 square mile
1 square in = 6452 square cm
1 square ft = 929.0 square cm
1 square mile = 2.5900 square km

1 mm^2 = 0.00155 in^2	1 in^2 = 6.452 cm^2
1 m^2 = 10.764 ft^2	1 ft^2 = 0.09290 m^2
1 km^2 = 0.3861 mile^2	1 mile^2 = 2.5900 km^2

Cubic measures

1 gal (UK) = 4.5461 liters = 1.200956 gal (US)
1 liter = 0.219969 gal (UK) = 0.264173 gal (US)

1 gal (US) = 3.7854 liters = 0.832670 gal (UK)
1 cc = 0.0610 cu in = 0.000035314 cu ft
1 cu in = 16.387 cc
1 cu ft = 28 317 cc
1 mm^3 = 0.000061 in^3 1 in^3 = 16.387 cm^3 (cc)
1 cm^3 (cc) = 0.0610 in^3 1 ft^3 = 0.028317 m^3
1 m^3 = 35.315 ft^3 1 mile^3 = 4.1681 km^3
1 km^3 = 0.239911 mile^3

Velocity

1 knot = 51.4 cm/s
1 ft/s = 30.480 cm/s = 1.0973 km/h
1 mi/h = 1.6093 km/h = 44.704 cm/s
1 cm/s = 3.728 × 10^{-4} mile/min = 0.02237 mile/h
1 km/h = 27.7778 cm/s
1° ψ per day = 1.2863 m/s = 2.8774 mile/h = 2.4987 knots

Force

As units of force, frequency gram weight, kilogram weight, pound weight, etc., are used. We denote them here by g*, kg*, and lb*.

1 dyne = 0.0010197 g* = 2.2481 × 10^{-6} lb* = 1 g/s^2
1 megadyne = 10^6 dynes
1 g* = 980.665 dynes
1 kg* = 980.665 × 10^3 dynes
1 lb* = 4.4482 × 10^5 dynes

Pressure

The SI unit of pressure is the pascal, symbol Pa, the special name given to a pressure of one newton per square metre (N/m^2). The relationships between the pascal and some other pressure units are shown in Table U1, but note that not all are, or can be, expressed exactly.

Following the Eighth Congress of the World Meteorological Organization in 1986, the term hectopascal (hPa) is preferred to the numerically identical millibar (mb) for meteorological purposes. This choice was made, despite the fact that hecto (× 100) is not a preferred multiple in the SI system.

The so-called "manometic" pressure unit definitions such as millimeters of mercury and inches of mercury depend on an assumed liquid density and acceleration due to gravity, assumptions which inherently limit knowledge of their relationship with the pascal. In order to encourage the demise of non-SI units, whose definitions are becoming inadequate for the most precise measurement of pressure, there is international effort to exclude them from conversion tables or, in the meantime, restrict the precision of newly published conversion factors.

Table U1 Units of pressure

Unit	Symbol	No. of pascals
bar	bar	1 × 10^5
millibar	mb	100
hectopascal	hPa	100
conventional millimeter of mercury	mmHg	133.322...
conventional inch of mercury	inHg	3 386.39...
torr	torr	10 1325/760
pound-force per square inch	lbf/in^2	6 894.76...

The torr is defined as exactly 101325/760 Pa – the "760" coming from the original and arbitrary definition of standard atmosphere. Its value differs from the conventional millimeter of mercury by about 1 part in 7 million.

Other pressure units and conversions include the following.

1 barye = 1 dyne/cm^2 = 9.8692 × 10^{-7} atmosphere = 10^{-6} bar = 1.4504 × 10^{-5} lb/sq in
1 bar = 10^6 baryes = 10^6 dynes/cm^2 = 0.98692 atmosphere = 14.504 lb/sq in = pressure of 750.06 mm of Hg
1 millibar = 10^{-3} = 10^3 dynes/cm^2 = pressure of 0.75006 mmHg
1 megabar = 10^6 bars
1 kg/cm^2 = 0.980665 × 10^6 dynes/cm^2 = 14.233 lb/sq in
1 lb/sq ft = 4.7254 × 10^{-4} atmosphere = 478.80 dynes/cm^2 = 47.880 newtons/m^2
1 lb/sq in = 0.068046 atmosphere = 0.068947 bar = 6.8947 × 10^4 dynes/cm^2 = 6894.8 newtons/m^2
1 atmosphere = 1.0332 kg/cm^2 = 1.01325 bars = 14.7 lb/sq in = pressure of 760 mmHg at 0°C and g = 980.665 cm/s^2

Energy–work

Foot candle: the unit of illumination in general use in the United States. This is defined as the illumination on an area, one foot square, on which there is a luminous flux of one lumen uniformly distributed. It may also be defined as the illumination on a surface of which all points are at a distance of one foot from a uniform point source of one candle.

Foot-lambert: a unit of luminance defined as the uniform luminance of a perfectly diffusing surface either emitting or reflecting light at the rate of one lumen per square foot. It may also be defined as $1/n$ candle per square foot and is sometimes referred to as the apparent foot-candle.

Foot-pound: a unit in the English gravitational system; it may be defined in two ways: (1) a unit of energy equal to the work done when one pound of force displaces a point to which the force is applied one foot in the direction of the applied force; (2) a unit of torque equal to the time-rate of change of angular momentum produced by a pound of force acting at a perpendicular distance of one foot from the axis of rotation.

Foot-poundal: a unit in the English absolute system; it may be defined either as a unit of energy, equal to the work done by a force of magnitude one poundal when the point at which the force is applied is displaced one foot in the direction of the force, or as a unit of torque equal to the time rate of change of angular momentum produced by a force of magnitude one poundal acting at a perpendicular distance of one foot from the axis of rotation.

Calorie: used in the cgs system, this unit represents the quantity of heat required to raise 1 gram of water through 1°C (at 15°C).

Kilocalorie: equal to 10^3 calories, this is often cited as a calorie in metabolic studies. This use has led to considerable confusion.

Langley: proposed in 1942 as a solar radiation unit. Initially given as calories per sq cm per minute, the time dimension was dropped in 1947. Thus 1 Langley = 1 cal/cm^2 = 697.8 W m^{-2}. The unit was named for S.P. Langley (1834–1906) of the Smithsonian Institution.

British thermal unit (BTU): the energy required to raise the temperature of one pound of water through 1°F (39° to 40°F).
1 erg = 1 dyne centimeter = 1 g cm^2/s^2
1 joule = 10^7 ergs = 0.102 m kg* = 0.737 ft-lb

1 gram calorie corresponds to 4.19×10^7 ergs
1 watt = 10^7 ergs per second = 1 joule per second
1 horsepower = 746 watts
British thermal unit (39°F) (Btu) = 1060.4 joules (absolute)
British thermal unit (60°F) (Btu) = 1054.6 joules (absolute)
Gram calorie (mean) = 1.5593×10^{-6} horsepower

Energy flow representation

In attempting to standardize the representation of energy flow, the World Meteorological Organization (1971) suggests the following symbols:

$Q\downarrow$ ($Q\downarrow = K\downarrow + L\downarrow$)	Downward radiation
$K\downarrow$ ($K\downarrow = S + D$)	Global solar radiation
S	Vertical component of direct solar radiation
D	Diffuse solar radiation
$L\downarrow$ ($A\downarrow$)($L\downarrow = A\downarrow$)	Downward atmospheric radiation
$Q\uparrow$ ($Q\uparrow = K\uparrow + L\uparrow$)	Upward radiation
$K\uparrow$	Reflected solar radiation
R	Upward solar radiation reflected by Earth's surface alone
$L\uparrow$	Upward terrestrial radiation
Lq	Upward terrestrial radiation – surface
r	Reflected atmospheric radiation
$A\uparrow$	Upward atmospheric radiation
Q	Net radiation
K^*	Net solar radiation
L^*	Net terrestrial radiation

Given these symbols, components of the Earth's budget can be represented by equations. Of particular importance is net radiation.

Net radiation = Incoming − Outgoing

$$Q = Q\downarrow - Q\uparrow$$

$$\text{or } Q^* = K^* + L$$

Since $K^* = K\downarrow - K\uparrow$ and $L^* = L\downarrow - L\uparrow$

$$Q^* = (K\downarrow - K\uparrow) + (L\downarrow - L\uparrow)$$

alternatively

Since $Q\downarrow = K\downarrow + L\downarrow$ and $Q\uparrow K\uparrow + L\uparrow$

$$\text{then } Q^* = (K\downarrow + L\downarrow) - (K\uparrow + L\uparrow)$$

John E. Oliver

Bibliography

Brown, J.M., 1973. *Tables and Conversions for Microclimatology*. USDA Forest Service Technical Report NC-8. Washington, DC: US Government Printing Office.

Cardarelli, F., 1997. *Scientific Unit Conversion*. London: Springer-Verlag.

Carmichael, R.D., and Smith, E.R., 1962. *Mathematical Tables and Formulas*. New York: Dover.

Clark, S.E., Jr, (ed.), 1966. *Handbook of Physical Constants*, rev. edn. Geological Society of America Memorandum 97. Boulder, Co: Geological Society of America.

Clason, W.E. (comp.), 1964. *Elsevier's Lexicon of International and National Units*. New York: Elsevier.

Forsythe, W.E. (comp.), 1954. *Smithsonian Physical Tables*. Washington, DC: Smithsonian Institution.

Gilbert, J.A., 1967. Units, numbers, constants and symbols. In Fairbridge, R.W., ed., *The Encyclopedia of Atmospheric Sciences and Astrogeology*. New York: Reinhold.

Gray, D.E. (ed.), 1957. *American Institute of Physics Handbook*. New York: McGraw-Hill.

Hodgman, C.D. (ed.), 1960. *Handbook of Chemistry and Physics*. Cleveland, OH: Chemical Rubber Publishing Co.

Jeffreys, H., 1948. The figures of the Earth and of the Moon (3rd paper), *Royal Astonomical Society Monthly Notices*, **5**: 219–247.

Jerrard, H.G., and McNeill, D.B., 1980. *A Dictionary of Scientific Units*. London: Chapman & Hall.

Kayan, C.E. (ed.), 1959. *Systems of Units*, Publication No. 57. Washington, DC: American Association for the Advancement of Science.

National Physical Laboratory, United Kingdom. www.npl.co.uk

NIST References on Constants, Units and Uncertainties. http:.physics.nist.gov

McNish, A.G., 1957. Dimensions, units and standards. *Physics Today*, **10**: 19–25.

United Nations Studies in Methods, 1987. *Energy Statistics, Units of Measure and Conversion Factors*. Series F, No. 44. New York: United Nations.

World Meteorological Organization, 1971. *Guide to Meteorological Instruments and Observing Practices*, 4th edn. W.M.O. No. 8TP3. Geneva: World Meteorological Organization.

URBAN CLIMATOLOGY

Urbanization and climate

The process of urbanization significantly alters natural surface and atmospheric conditions. Oke, in Thompson and Perry (1997), suggests that urban atmospheres demonstrate the strongest evidence we have of the potential for human activities to change climate. In the twentieth century rapid urbanization has occurred on a worldwide scale and the majority of the world's population lives in cities. Rapid expansion of cities has produced concurrent alterations in the urban climatic environment (Landsberg, 1981).

In general there are many apparent anthropogenic impacts on our atmospheric environment (Changnon, 1983). These range from microscale (e.g. replacing trees with a parking lot) to macroscale (e.g. carbon dioxide effects on global climate by fossil fuel combustion and emissions). This discussion focuses on the causes of alterations in climate in urban areas. Processes are reviewed that are keys to the formation of an urban climate and the underlying energy, moisture, and air movement patterns that distinguish urban climates from their surroundings.

An extensive literature addresses the specific problem of air pollution on regional and global scales as it relates to general processes of urbanization. The field of urban climatology has developed to the level of theory and detailed investigations that we see today in part due to early interests in air quality in cities. This discussion, however, does not focus on air pollution effects *per se*. Urban climate can be understood, to a large degree, by the study of modifications which develop primarily through the effects of land-use changes and feedbacks into the energy, moisture, and local air motion systems. Air quality certainly plays a role. The following discussion is limited to micro-, local, and meso-scales. It is also restricted to processes taking place in what are called the urban canopy layer (UCL – beneath roof level) and the urban boundary layer (UBL – extending from roof level to the height at which urban influences are absent; Oke, 1998, see Figure U1).

The first real growth of urban climatology dates from the 1920s, followed by rapid increases in interest in urban climates between the 1930s and 1960s (especially in Germany, Austria, France, and North America). After World War II, and into the environmental era of the 1960s and 1970s and beyond, there was an exponential increase in urban climatic investigations. Investigations have simultaneously become less descriptive, more oriented to quantitative and theoretical modeling, and more integrative and interdisciplinary (e.g. World Meteorological Organization, 1970; de Dear et al., 2000). There are many scholarly reviews of the subject and accompanying bibliographies that illustrate the overall problem of how cities alter their climatic environment (e.g. Brazel, 1987; Chandler, 1976; Beryland and Kondratyev, 1972; Landsberg, 1981; Lee, 1984; Oke, 1974, 1979, 1979, 1980; and in Bonan, 2002). Today, urban climatology has achieved its own recognition as a subdiscipline in climatology and among allied disciplines such

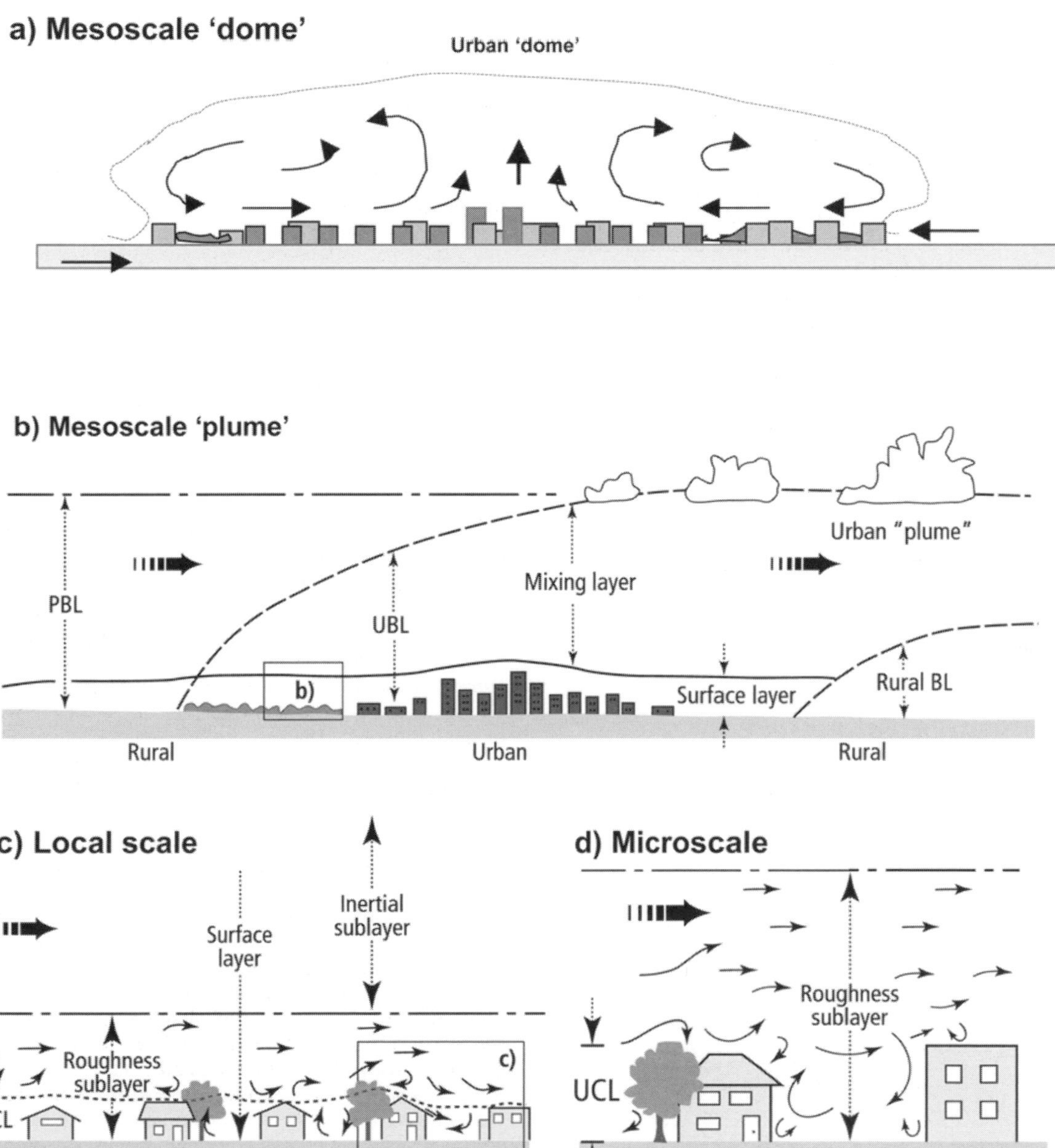

Figure U1 The idealized vertical structure of urban-modified air. (**a**) The whole city (mesoscale) in near-calm conditions with its "dome", and (**b**) in a steady regional air flow with its urban "plume". (**c**) A single urban terrain zone (local scale) showing the internal layering of the urban canopy (UCL) and lower portion of the urban boundary (UBL) layers. (**d**) A single street canyon (microscale) and building elements (after Oke, 1998).

as planning, ecology, environmental science, and meteorology (e.g. as evidenced in de Dear et al., 2000).

Factors controlling urban climates

Many aspects of urbanization change the physical environment and lead to alterations in energy exchanges, thermal conditions, moisture fluxes (evaporation, precipitation, and runoff) and wind circulation systems. These include: (1) air pollution, (2) anthropogenic heat, (3) surface waterproofing, (4) thermal properties of the surface materials, and (5) surface geometry (Oke, 1981). Other factors that must be considered relate to the setting of the city, such as relief, proximity to water bodies, size of the city, population density, and land-use distributions. Oke (1997) provides a summary of the typical resulting alterations of climatic elements in cities compared to rural environs (see Table U2).

The magnitude–frequency concept is important but less studied in urban climatology. How large are climatic alterations in a city and how often are they that large? The latter part of the question demands analysis of the linkages of microscale–mesoscale alteration magnitudes with more macroscale conditions, such as the role of synoptic climatology on urban climate variations (e.g. Unwin, 1980). Typically, cloudy and/or windy days reduce heat island magnitudes for a city.

Urbanization causes changes in the energy, moisture, and momentum systems, but few studies demonstrate pre–post urban climate conditions (Lowry, 1977). One excellent example is the experiment conducted to specifically illustrate how a city, as it developed from a rural environment, affected the local and regional climate (Landsberg, 1981). Most urban climate studies rely on geographic comparisons between the city and its surroundings in order to produce an estimate of the urban effect on local climate. Considerable attention has been given to the study of historical weather records in cities to evaluate temperature trends that are more urban in origin versus trends more attributable to global signals of change (e.g. global warming). A large number of stations with long records are needed for trend analyses, and many of the world's weather stations are in or near urban-affected locales. Thus, it is often difficult to decipher global change from such sites (e.g. Hansen, et al., 1999, 2001). Furthermore, placing typical weather stations in urban areas to study how cities alter climate is a most challenging endeavor in and of itself (Oke, 1999). This arises because of the complex structure of the urban atmosphere, processes operating at differing scales, and the resultant climate patterns in the city (see Figure U1, after Oke, 1998).

Table U2 Urban climate effects for a mid-latitude city with about 1 million inhabitants (values for summer unless otherwise noted)

Variable	Change	Magnitude/comments
Turbulence intensity	Greater	10–50%
Wind speed	Decreased	5–30% at 10 m in strong flow
	Increased	In weak flow with heat island
Wind direction	Altered	1–10 degrees
UV radiation	Much less	25–90%
Solar radiation	Less	1–25%
Infrared input	Greater	5–40%
Visibility	Reduced	
Evaporation	Less	About 50%
Convective heat flux	Greater	About 50%
Heat storage	Greater	About 200%
Air temperature	Warmer	1–3°C per 100 years; 1–3°C annual mean up to 12°C hourly mean
Humidity	Drier	Summer daytime
	More moist	Summer night, all day winter
Cloud	More haze	In and downwind of city
	More cloud	Especially in lee of city
Fog	More or less	Depends on aerosol and surroundings
Precipitation		
Snow	Less	Some turns to rain
Total	More?	To the lee of rather than in city
Thunderstorms	More	
Tornadoes	Less	

Source: Adapted after Oke (1997), p. 275.

Methods of evaluating urban climate

Many methods are used to determine how a city affects climate. Early methodologies included sampling the differences between urban–rural environments, upwind–downwind portions of the urban area, urban–regional ratios of various climatic variables, time trends of differences and ratios, time segment differences such as weekday versus weekend, and point sampling in mobile surveys throughout the urban environment (Lowry, 1977). Much of this sampling led to the discovery of the famous heat-island phenomenon.

Methodological inadequacies of many of the early studies have been pointed out in recent decades, and accurate time-series data remain a problem of analysis (Lowry, 1977; myriad studies on filtering out urban effects to reveal global trends). These inadequacies led to increased studies of processes involved, particularly fluxes of energy, moisture, and momentum in urban environments. Process studies help to provide a better characterization of how urbanization alters the surface–atmospheric system (Oke, 1979). Much attention has been given to internal variability of climate conditions within the urban environment and to the importance of the UCL (Arnfield, 1982; Grimmond, 1992; Grimmond and Oke, 1995, 1999; Johnson and Watson, 1984; Oke, 1981).

The urban heat island

Cities, no matter what their size, tend to be warmer than their surroundings. One can observe this biking through the urban landscape (Melhuish and Pedder, 1998). This fact was discovered by scientists well over a century ago (e.g. Howard, 1833), and is well known as evidenced by its mention in virtually every book on modern climatology and ecology (e.g. Thompson and Perry, 1997; Bonan, 2002). Since cities are regional agglomerations of people, buildings, and urban activities, they are spots on the broader, more rural surrounding land. These spots produce a heat-island effect on the spatial temperature distribution in an area (see Figures U7a and b in the later section on remote sensing).

The reasons for heat-island formation are many-fold (Table U3 after Oke, 1979). City size, morphology, land-use configuration, and geographic setting (e.g. relief, elevation, regional climate) dictate the intensity of the heat island, its geographic extent, orientation, and its persistence through time. Oke (1981) illustrates the effect of city size (as indicated by population size) on the maximum urban heat-island intensity (see Figure U2). Although population is only a surrogate measure for many of the causes of a heat island, there is a significant correlation indicated, and population serves as a useful reference for expected heat-island formation.

In evaluating the heat-island intensity of cities, several cautionary notes must be emphasized. The method of sampling and/or the accuracy of historical weather records from available rural and urban locales must be carefully considered (Oke, 1999). Much of the earlier work in heat-island detection is of somewhat limited value due to non-specification of probable pre-urban conditions in an area, and biases in spatial sampling of the extent of the heat island (Lowry, 1977; Landsberg, 1981). Numerical adjustments of historical temperature records must be considered by the urban researcher, since available data may be biased (e.g. observation time of various stations may be different) and influences of local changes in station locations on temperature and precipitation statistics may go undetected (e.g. Winkler and Skaggs, 1981). The more process-oriented urban climatic studies in the last two decades are not only providing measures of explanation of the heat-island phenomenon, but also are allowing for a better understanding of the efficacy of the earlier, more descriptive studies of the urban heat island.

Table U3 Mechanisms hypothesized to cause the urban heat island effect

- Urban Boundary Layer
  - Anthropogenic heat from roofs and stacks
  - Entrainment of air scoured from warmer canopy layer
  - Entrainment of heat from overlying stable air by the process of penetrative convection
  - Shortwave radiative flux convergence within polluted air
- Urban Canopy Layer
  - Anthropogenic heat from building sides
  - Greater shortwave absorption due to canyon geometry
  - Decreased net longwave loss due to reduction of sky view factor by canyon geometry
  - Greater daytime heat storage (and nocturnal release) due to thermal properties of building materials
  - Greater sensible heat flux due to decreased evaporation resulting from removal of vegetation and surface waterproofing
  - Convergence of sensible heat due to reduction of wind speed in the canopy

Source: After Oke (1979).

Radiation and energy balance of the city

Shortwave radiation

An urban area affects the exchanges of shortwave and longwave radiation by air pollution and complex changes of surface radiative characteristics. Figure U3 illustrates radiative fluxes and processes in a city atmospheric system and the profiles of shortwave radiative heating and longwave radiative cooling due to the existence of an urban aerosol layer. The attenuation of incoming shortwave radiation (fluxes 2 and 4 divided by 1; see Figure U3) has been analyzed in numerous urban climatic environments. It is thought that the attenuation in the city is 2–10% more than in the country. Generally, the ultraviolet portion of the electromagnetic spectrum (< 0.4 μm), the so-called UV, is depleted by as much as 50% (e.g. Petersen et al., 1978). However, total depletion across all solar wavelengths (0.15–4.0 μm) is much smaller, less than 10% (Petersen and Flowers, 1977; Petersen and Stoffel, 1980). The processes of scattering and absorption, which account for the attenuation, require further study to predict their relative contribution to total shortwave radiation depletion.

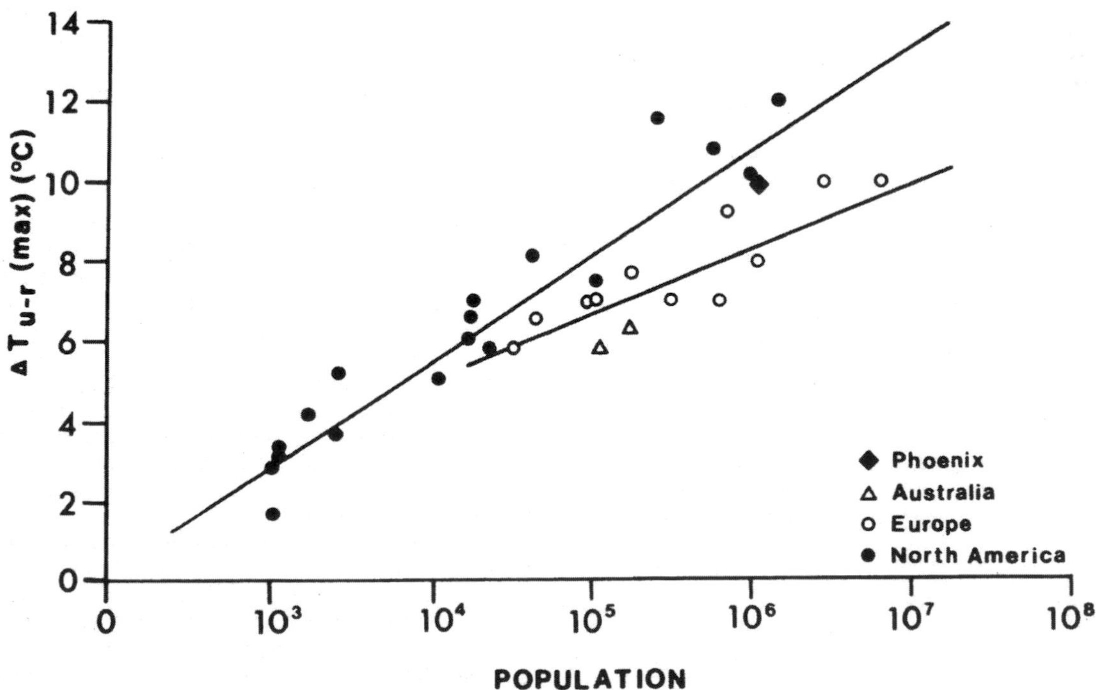

Figure U2 The maximum urban heat island intensity (ΔT_{u-r} [max]) versus population for 32 cities in Europe, North America, and Australia (data from Oke, 1981).

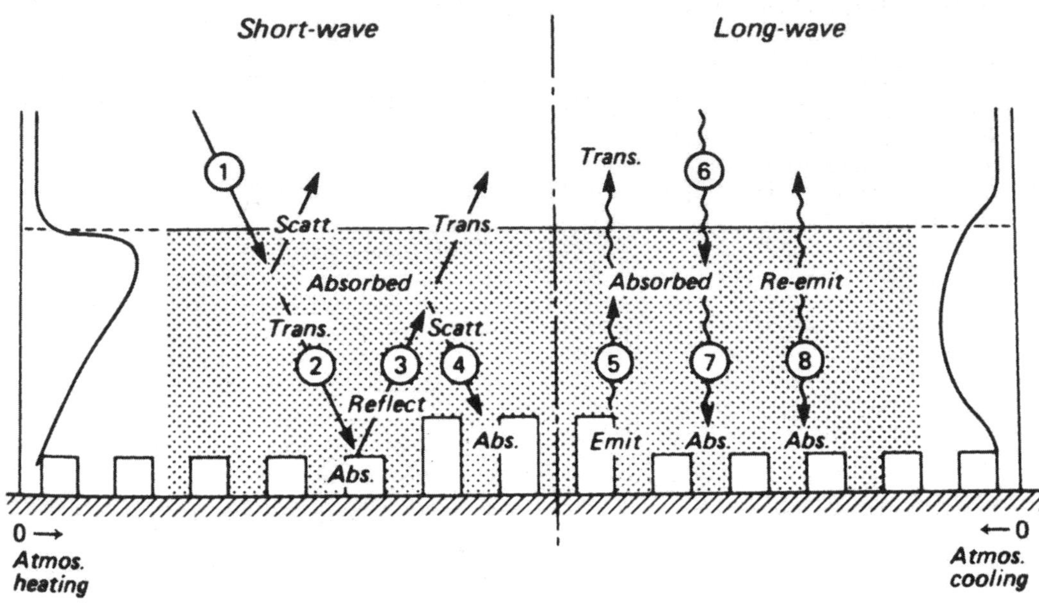

Figure U3 Radiative exchanges in a polluted urban boundary layer, including generalized profiles of shortwave radiative heating (left) and longwave cooling (right) due to the aerosol layer (shaded). Numbered fluxes are referred to in the text (after Atwater, 1971; Oke, 1982).

Albedo

The second major effect by the city is the change in the ratio of outgoing shortwave radiation (flux 3) to that of incident shortwave radiation at the complex, three-dimensional city interface (fluxes 2 and 4). This ratio, the albedo, is typically less in urban areas than in the surrounding landscape. Lower albedoes are due to darker surface materials making up the urban landscape mosaic and also due to effects of trapping of shortwave radiation by the vertical walls and the urban canyon-like morphology. There is considerable variation of albedo within the city depending on the percent vegetative cover, building material variation and roof composition, and land-use characteristics (see Table U4).

The difference in albedo between a city and its environs depends also, of course, on the surrounding terrain. A city surrounded by dark forest may experience very little albedo difference from the forest cover (both may range from 10% to 20%). In winter a midlatitude to high-latitude city with surrounding snow cover may display a much lower albedo than its surroundings. Thus, since cities receive 2–10% less shortwave radiation than their surroundings, yet have slightly lower albedoes (by less than 10%), most cities experience very small overall differences in absorbed shortwave radiation from rural locations nearby.

Table U4 Albedo values of urban surface materials

Material	Albedo (%)
Concrete	27.1
Blacktop/asphalt	10.3
Brick, red	32.0
Brick, yellow/buff	40.0
Brick, white/cream	60.0
Glass	9.0
Paint, dark	27.5
Paint, white	68.7
Roofing shingles	25.0
Snow, weathered	55.0
Stone	31.7
Tar-gravel roof	13.5
Yard (90% lawn, 10% soil)	24.0

Source: Data from Arnfield (1982), p. 104.

Longwave radiation

Longwave radiation is affected by city pollution and the fact that most urban surface areas are warmer. Warmer surfaces promote greater thermal emission of energy vertically upward from the city surface compared to rural areas, particularly at night (flux 5). Some longwave radiation is reradiated by urban aerosols back to the surface and also from the warmer urban air layer (fluxes 7 and 8). Thus, increases in incoming longwave radiation and outgoing longwave radiation are usually experienced in urban areas (that is, fluxes 7 and 8 are greater than flux 6 would be at the urban surface).

It is thought that outgoing longwave radiation increases are slightly greater than the incoming increases in the city, again especially on clear, calm nights. During daytime there is little difference between the city and its surroundings. However, surface emissivity (amount emitted relative to black-body amounts for a given temperature) can be quite different between country and city areas, and can account for considerable longwave radiation variations (Yap, 1975). A major consideration is that in the city a three-dimensional surface temperature must be characterized to accurately estimate the flux values of radiation and the energy budget (e.g. Voogt and Oke, 1997, 1998).

Net all-wave radiation

Generally, variations in the shortwave and longwave radiation fluxes between a city and its environs are relatively small, and the net all-wave radiation (the balance between incoming and outgoing shortwave and longwave radiation) may actually be less in urban areas (e.g. in St Louis there is 4% less net radiation in the city; White et al., 1978). The exact differences depend on land use, city building density and arrangement, aerosol composition of the urban atmosphere and climate of the city surroundings, as well as other factors such as artificial heat emissions and the topography of the region. In the UCL, what is called the urban sky view factor (USVF) is decreased considerably below a totally unobstructed horizon. This latter condition is more typical of the countryside. An obstructed horizon consisting of structures that make up the UCL reduces radiative loss and can account for excess heat in the city (Oke, 1981). The emission of heat from the UCL structures has been hypothesized as explaining a city's heat-island intensity. Figure U4 shows a correlation of the UCL sky view factor with the maximum heat island intensity for 31 cities plotted in Figure U4. The USVF is specifically defined as:

$$\text{USVF} = (1 - 2\text{UWVF})$$

$$\text{UWVF} = 0.5(\sin^2\theta + \cos\theta - 1)(\cos\theta)^{-1}$$

$$\theta = \tan^{-1}(H/0.5W)$$

where: H = height of UCL
W = width of the UCL.

Note the tightening of the correlation, indicating that for a given population, a European city has a more open geometry, whereas most US cities have deeper canyons in their core areas. The individual correlations of European–Australian and North American population versus maximum heat-island intensities now merge into one generalized correlation shown in Figure U4. Urban climatologists are increasingly using photographic and GIS methods to derive USVF variability within cities. Figures U5a and b illustrate results for downtown Salt Lake City for USVF and building heights (Brown and Grimmond, 2001).

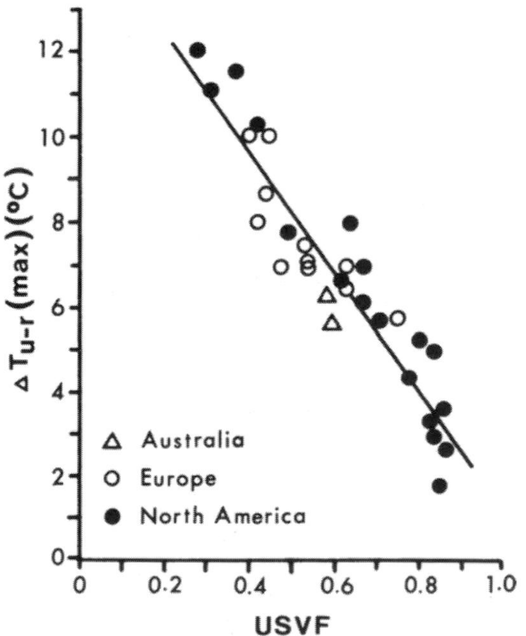

Figure U4 The maximum urban heat-island intensity (ΔT_{u-r} [max]) versus urban sky view factor (USVF) for 31 cities. The equation is ΔT_{u-r} (max) = −13.3 USVF + 14.86 (r^2 = 0.87, standard error of the estimate = 0.96°C).

Total energy balance

Not many long-term studies of a city's entire energy balance have been made (Oke, 1979). One specific reason relates to important paradigms in climatology, shifting only in the last five decades to considering physical principles of the energy, moisture and mass exchanges at the Earth–atmosphere interface (Miller, 1965). However, over the last decade, many cities worldwide have been studied through the use of tall towers with energy flux sensors, GIS, remote sensing and modeling procedures (e.g. Grimmond, 1992; Grimmond and Oke, 1995, 1999).

The city energy budget or balance can be simplified in Figure U6a to:

$$Q^* + Qf = Qe + Qh + \Delta Qs + \Delta Qa$$

where: Q^* = net all-wave radiation
= $K^* + L^*$ (net shortwave and longwave radiation)
Qf = anthropogenic heat emission ($Qfv + Qfh + Qfm$)
Qe = latent heat flux
Qh = sensible heat flux
ΔQs = net heat storage in the city
ΔQa = net advection into or out of the city.

Since Q^* is not that different between a city and its surroundings, the observed heating in cities is likely accounted for by Qf, (Qe, Qh), and ΔQs. Usually, ΔQa is negligible. The contribution of Qf to the total energy balance is highly city-dependent and seasonally variable. Qf ranges from nil to 300% of Q^*, depending on the degree of industrialization (Qf is high in more industrialized cities), latitude (Qf is high in higher-latitude cities), and season (Qf is higher in winter). Qf is composed of heat produced by combustion of vehicle fuels (Qfv), stationary source releases such as within buildings (Qfh), and heat released by metabolism (Qfm). Qfv is a function of type and amount of gasoline used, number of vehicles, distance traveled, and fuel efficiency. Qfh requires an analysis of consumer usage of fuel such as gas and electricity. Qfm can be evaluated by active and sleep rates. For people metabolic rates have been given (Oke, 1979). Bach (1970) suggests rates for animals.

Sensible, latent, and storage heat

The repartitioning of energy in urban areas among sensible, latent, and storage heat processes primarily depends on the mosaic of land uses in the city as compared to rural areas. Generally, the drier urban building and road materials induce higher Qh, less Qe, and higher ΔQs in urban areas. A term labeled Qg is specified in the energy budget equation often as part of ΔQs and is the soil heat flux. Significant Qe does occur in cities. This is theorized to be due to urban irrigation effects and vegetation in the city (Kalanda et al., 1980). Marotz and Coiner (1973) indicate that vegetation in urban areas is not as limited as supposed, and Oke (1978) shows that $\Delta Qs/Q^*$ ratios for rural areas vary by only about 0.10 from those for suburban and urban areas.

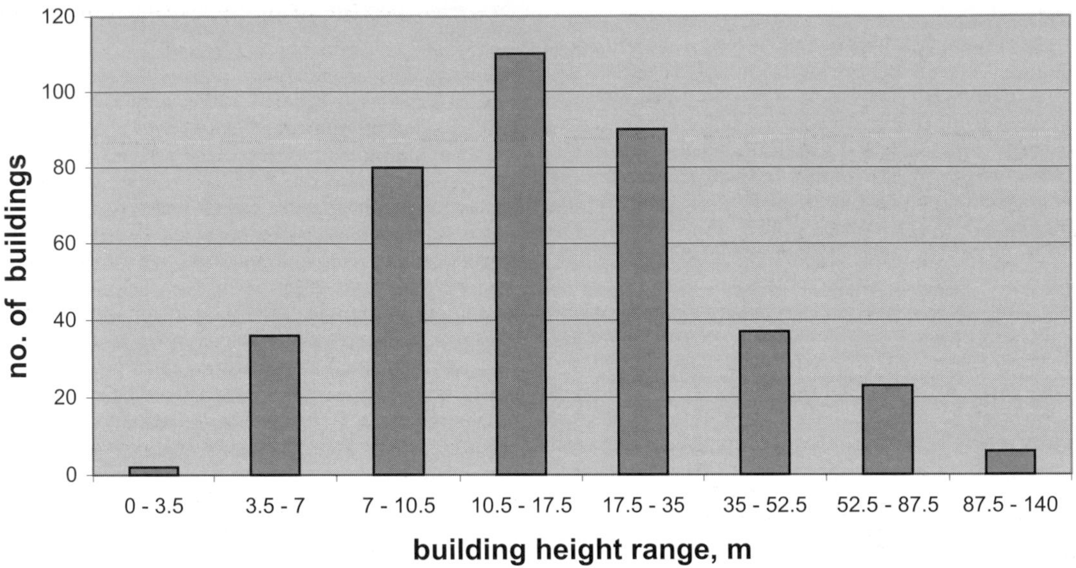

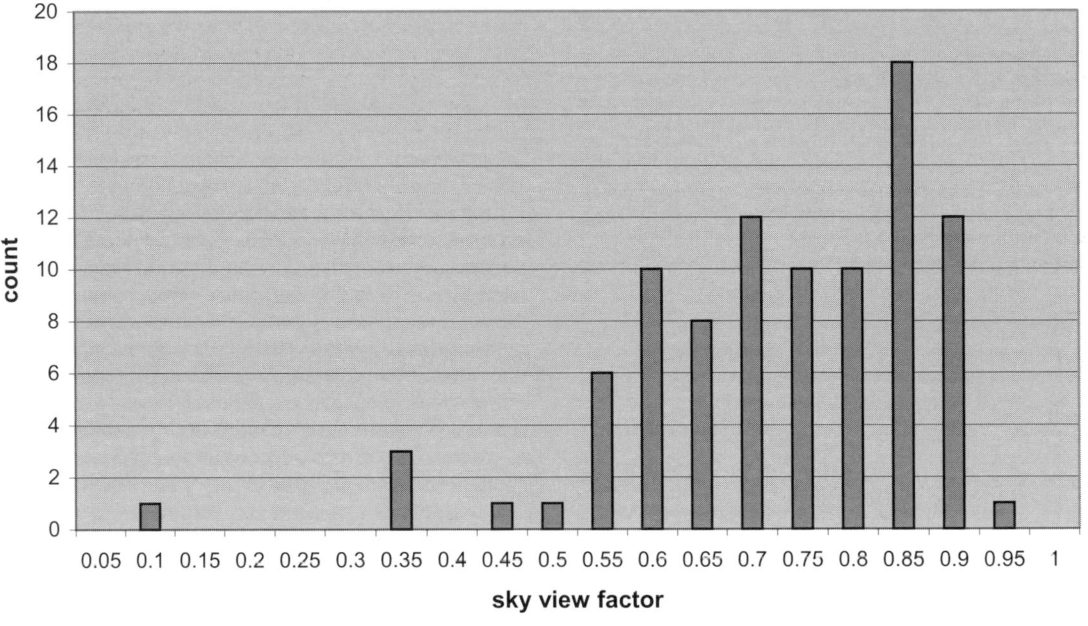

Figure U5 *Upper*: histogram of the computed sky view factor for street-level positions in downtown Salt Lake City. *Lower*: histogram of the building heights for downtown Salt Lake City (after Brown and Grimmond, 2001).

Table U5 shows some recent results for selected US cities after Grimmond and Oke (1995). Note the substantive fluxes of Qh and ΔQs, also significant Qe, especially for the more moist city of Chicago. In a detailed evapotranspiration study of nine places, Grimmond and Oke (1999) show that the ratio of Qe/Q^* ranges from 0.09 (Mexico City with very little external water use) to Chicago's ratio of 0.46 (Table U5).

Goward (1981) lists thermal properties of typical interface materials, noting that urban area materials have similar thermal properties and that urban thermal inertias are higher than dry soils but lower than wet soils (see Table U6). Differences in UCL structure and composition are the keys to explaining heat excesses in cities, rather than just the thermal properties of city materials *per se*. Much research remains on this point, and has become a focal point in the field of urban climatology (Arnfield, 1982; Barring et al., 1985; Goldreich, 1985; Goward, 1981; Grimmond and Oke, 1995, 1999; Johnson and Watson, 1984; Oke, 1981, 1982; Terjung and O'Rourke, 1980).

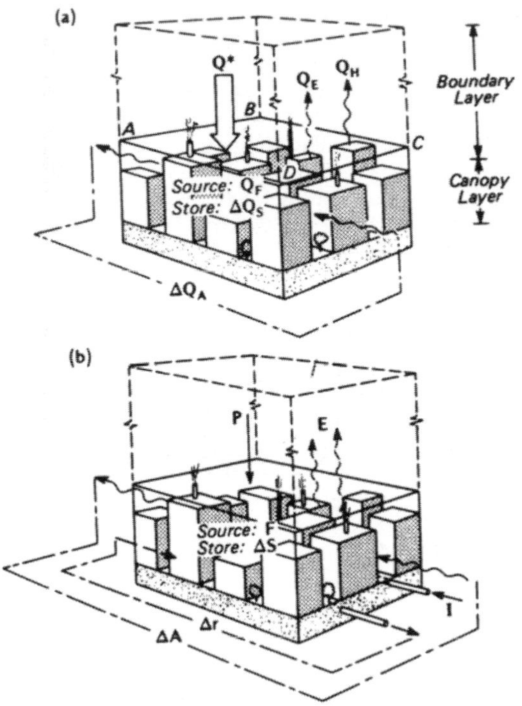

Figure U6 Schematic depiction of the fluxes involved in (**a**) the energy and (**b**) the water balance of an urban building air volume (after Oke, 1978, p. 242).

Table U5 Summary of daytime mean summer energy balance fluxes (MJ/m^2/day) for selected cities derived from tall tower observations (Grimmond and Oke, 1995).

Location	Q^*	Qh	Qe	ΔQs	Qg
Tucson, AZ	16.27	7.54	4.11	4.62	Na
Sacramento, CA	12.65	5.19	3.79	3.67	12.73
Chicago, ILl	17.20	5.58	7.11	4.51	2.65
Los Angeles, CA	16.40	5.74	4.12	6.54	1.37

Remote sensing of urban surface temperature

The extreme land cover and spatial heterogeneity of the city landscape is problematic to the synoptic collection of *in-situ* measurements of energy balance flux parameters. Longwave radiation emitted upward from the urban surface (L^*), however, can be measured as a separate flux across the city landscape using thermal infrared (TIR) remote sensing. A number of investigators have provided evidence using TIR remote sensing data that urban areas are strong daytime longwave emitters (Rao, 1972; Pease and Nichols, 1976; Auer, 1978, Carlson and Boland, 1978; White et al., 1978; Roth et al., 1989). Other researchers have used remote sensing data to observe nighttime longwave radiation characteristics of cities (Dabberdt and Davis, 1974; Roth et al., 1989). As indicated by Table U6, however, the thermal properties of surfaces typical of the urban landscape are highly variable, and this variability is compounded even further by other factors, such as inconsistencies and deterioration caused by weathering or residue formation from automobiles. Flynn (1980), therefore, suggests that the problem related to the effects of surface material types and patterns on urban climatological phenomena, including the urban heat island effect, is a spatial one. A solution is to measure the attributes of space at a scale commensurate with that where the process becomes evident. The object of study then becomes the urban space that describes the energy budget or climatic processes within an areal unit. Flynn (1980) notes that approaching the problem in this way focuses on the pattern analysis and measurement of a basic urban surface (i.e. surface material type) and may explain some of the variability in urban temperatures across the city landscape. Thus, although TIR remote sensing data obtained from satellites are useful for discerning the general temperature characteristics of urban surfaces, these data must be obtained at spatial scales at which energy budgets for discrete surfaces material types can be identified (e.g. concrete, trees, rooftops, asphalt) to adequately quantify how thermal energy emanating from the composite urban landscape forces development of the urban heat island effect.

In this respect, high spatial resolution (i.e. < 20 m) TIR remote sensing data have been shown to be extremely useful for measuring urban thermal energy balance characteristics (Quattrochi and Ridd, 1994, 1998; Lo et al., 1997; Quattrochi et al., 2000). For example, Figure U7a is a mosaic of individual

Table U6 Thermal properties of typical interface materials

Material	Thermal conductivity W m^{-1} °C^{-1}	Specific heat J kg^{-1} °C^{-1} × 10^3	Density kg m^{-3} × 10^3	Thermal admittance J m^{-2} °C^{-1} sec $^{-1/2}$ × 10^3
Asphalt	0.7454	0.92	2.114	1.204
Brick	0.6910	0.84	1.970	1.067
Concrete	0.9338	0.67	2.307	1.185
Glass	0.8794	0.67	2.600	1.213
Granite	2.7219	0.67	2.600	2.176
Limestone	0.9338	0.92	1.650	1.182
Sand (dry)	0.3308	0.80	1.515	0.633
Wood	0.2094	1.38	0.500	0.377
Soil (wet)	2.4288	1.48	2.000	2.681
Soil (dry)	0.2513	0.80	1.600	0.567
Water (20°C)	0.5988	4.15	0.998	1.579
Air (20°C)	0.0251	1.01	1.001	0.056

Source: After Goward (1981), p. 24. *Note*: Thermal admittance is the square root of the product of the other three quantities.

flight lines of airborne TIR data acquired at a 10 m spatial resolution during the day in summertime over the Atlanta, Georgia, USA, central business district (CBD). Given these are TIR data, surfaces that emit high surface temperatures (i.e. exhibit a high thermal energy response) are depicted in white to very light gray tones on Figure U7a. On the other hand, surfaces that are comparatively low emitters of surface temperatures (and subsequently, exhibit lower thermal energy responses in relation to other surfaces) are represented as dark gray to black in tone in the figure. The A–A′ line in Figure U7a represents a transect centered on the Atlanta CBD and extending 5 km to the northwest and southeast of the CBD. There is a mixture of land covers along the transect that, in turn, is reflected in the fluctuations of temperatures along the graph given in Figure U7b. It may be seen in reference to Figures U7a and b that, in general, the residential land covers that inherently have more vegetation (e.g. trees, grass) are relatively cool (~ 20°C) as opposed to the surfaces in the CBD that have considerably less vegetation and are predominantly composed of pavements, rooftops and other impervious surfaces, exhibit considerably higher surface temperatures of approximately 35°C. Hence, these high spatial resolution TIR aircraft remote sensing data can be used to identify thermal energy responses from discrete urban surfaces to assist in more accurately measuring longwave energy upwelling from the city surface as a factor in the overall city landscape regime and, ultimately, in assessing their effects on urban climatology.

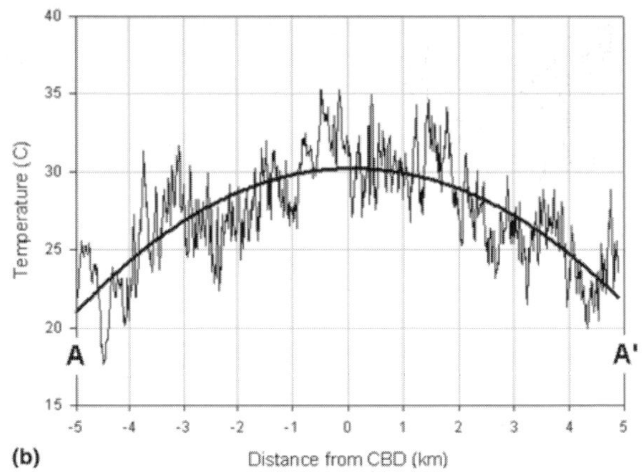

Figure U7 (**a**) Mosaic of individual flight lines of daytime airborne thermal infrared remote sensing data obtained over the Atlanta, Georgia, USA, central business district in summer. The A–A′ line represents a transect across the area to illustrate the variability in surface temperature energy responses across the Atlanta urban landscape (refer to **b**). (**b**) Graph of the mean surface temperature (solid line) across the Atlanta, Georgia, USA, central business district area as identified from daytime high-resolution TIR aircraft remote sensing data. Reproduced with permission, the American Society for Photogrammetry and Remote Sensing. Quattrochi, D.A., et al. "A Decision Support Information System for Urban Landscape Management Using Thermal Infrared Data." *Photogrammetric Engineering and Remote Sensing (PE&RS)*. October 2000: 1195–1207.

Moisture environment of cities

The processes of urbanization affect the surficial and atmospheric water budget. The water balance of an urban area is (Figure U6b):

$$p + I + F = E + \Delta r + \Delta S + \Delta A$$

where: p = precipitation
I = water supply from rivers and reservoirs
F = water released to air by combustion
E = evapotranspiration
Δr = net runoff
ΔS = moisture storage
ΔA = net moisture advection.

Precipitation

Many studies have shown that precipitation has appeared to have increased up to 30% in, and/or downwind of, large urban areas (e.g. Changnon, 1969; Dettwiller and Changnon, 1976; Atkinson, 1971). Stemming from Changnon's famous La Porte, Indiana, anomaly (1968) of increased precipitation downwind of the Chicago-Gary industrial area, much attention has been given recently to this subject. Detailed climatic investigations as part of the Metropolitan Meteorological Experiment (METROMEX) in St Louis, Missouri, have yielded estimates of urban effects on precipitation (Changnon et al., 1977; Ackerman et al., 1978; Changnon, 1981). Results indicate that summer is the time of maximum urban effect on precipitation (Figure U8) due to the dominance of convection. Analysis of rain cells, radar echo counts, detailed raingauge measurements, and temperature and wind field studies in and around St Louis have led to causal explanations of rainfall excesses that include: (1) pollution; (2) modification of the thermal structure of the UBL; and (3) deepening of the mixing layer and stronger vertical entrainment of heat and moisture in portions of the metropolitan area. A convergence zone over the city is set up by increased sensible heating. The city is aerodynamically rougher, has more pollution (condensation nuclei sources), causes more thermal and mechanical turbulence, and particularly affects convective cloud formation and precipitation. However, topographic settings of cities must be clearly understood relative to induced differences in precipitation receipt in the urban region, before causal explanations such as the above can be applied to other cities in different environments.

Sheppard et al. (2002) use the Tropical Rainfall Measuring Mission (TRMM) satellite's precipitation radar (PR) data to identify warm-season rainfall patterns in and around six selected cities in the southeast United States (Atlanta, Georgia; Montgomery, Alabama; Nashville, Tennessee; San Antonio, Waco, and Dallas, Texas). The results substantiated earlier

METROMEX findings of downwind increases in the rainfall of about 28% on average some 30–60 km downwind of the cities studied, and highlight the usefulness of satellite data to determine urban effects. Precipitation increases due to human influences may show cyclic behavior due to pollution effects, even at the weekly time scale, as suggested in a study of the regional pollution and precipitation enhancement toward end of workweeks on the Atlantic seaboard of the US (Cerveny and Balling, 1999). Table U7 illustrates results from several cities focusing on the fluxes of *Qe* relative to *Q** and the expression of water loss in millimeters per day, after Grimmond and Oke (1999).

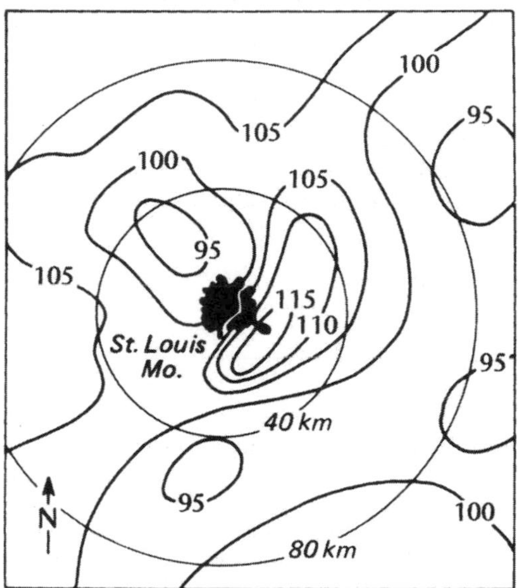

Figure U8 Average rural–urban ratios of summer rainfall in the St Louis area in the period 1949–1968. The rural–urban ratio is the ratio of the precipitation recorded at any station to that at two stations near the urban center (after Oke, 1978, p. 266).

Runoff

Since *I* and *F* are extra sources in cities, there is a net gain of moisture by urbanization. Water loss through *E* would be less in a city due to a waterproofing effect (Oke, 1980). The ΔS term would correspondingly be less. Assuming ΔA to be negligible, Δr would thus increase in cities. Indeed, many studies have indicated increased *r* in city environments, as well as changed hydrographic characteristics (e.g. Mather, 1978). There are four effects of land-use changes on the hydrology of the urban area: (1) peak flow is increased; (2) total runoff is increased; (3) water quality is lowered; and (4) hydrologic amenities – appearance of river channel and aesthetic impressions – are lowered (Mather, 1978).

Figure U9 illustrates two factors that control discharge characteristics before and after urbanization, shown here as a ratio related to percent impervious area and percent area served by storm sewers. These two factors are controls on the discharge of an area. Note that urban–preurban ratios of runoff increase as percent impervious surface area and percent storm sewers increase. Geographic information system processing of data is now used in hydrological studies, merged with simulation programs for stream flow analysis in urban areas (e.g. Brun and Band, 2000). In their work, relationships among runoff, base flow, impervious cover, and percent soil saturation for upper Gwynns Falls watershed, Baltimore, Maryland, were studied. It was possible to determine through modeling procedures the threshold percent impervious cover that would significantly alter runoff in the urbanizing watershed. In this case, current percent impervious surface cover is approximately 18%, whereas modeling suggests 20% is a threshold beyond which runoff alterations by land cover would significantly occur.

Humidity

Since *E* and ΔS are less, dewpoint is usually less in daytime in cities (e.g. Sisterson and Dirks, 1978; Brazel and Balling, 1986). At night a humidity island may result (e.g. Chandler, 1967; Kopec, 1973), due to extra *E* at night and contributions of *F*. Since recent studies have not shown *Qe* and *E* to be that

Table U7 Daily mean fluxes of Qe and water loss (E) for selected cities and conditions (after Grimmond and Oke, 1999)

Location[a]	*Q**	*Qe*	*Qe/Q**	*E* (mm.day)	*P* (mm/month)	*P* (mean monthly)	Practices[b]
C95	14.89	6.80	0.46	2.76	80.5	92.0	1
A93	13.74	4.93	0.36	2.00	0.0	0.0	2
T90	12.50	4.90	0.39	1.99	16.3	5.4	3
A94	15.58	4.70	0.30	1.94	0.0	0.0	4
Mi95	13.74	4.58	0.33	1.87	79.8	180.9	5
S91	9.72	4.38	0.45	1.77	0.0	1.8	6
Vs92	8.88	2.68	0.30	1.10	23.2	41.1	7
Sg94	12.45	3.46	0.28	1.40	0.0	0.25	8
Vl92	11.41	1.48	0.13	0.60	23.2	41.1	9
Me93	3.38	0.31	0.09	0.14	0.0	0.0	10

[a] In order, Chicago, ILl June/Aug 1995; Arcadia, CA July/Aug 1993; Tucson, AZ , June 1990; Arcadia, CA July 1994; Miami, FL, May/June 1995; Sacramento, CA, Aug 1991; Vancouver, BC, July/Sept, 1992; San Gabriel, CA, July, 1994; Vancouver, BC, Aug, 1992; Mexico City, D.F., Dec 1993.

[b] 1 = extensive irrigation; 2 = extensive irrigation (see A94); 3 = xeric landscaping; low water use vegetation, greenspace irrigated automatically at night/early a.m.; 4 = As A93. irrigation ~1.37 mm/day; 5 = frequent rainfall, also irrigation; 6 = irrigation on alternate days; 7 = irrigation ban – external use ~0.25 mm/day; 8 = As A93 but more hand watering, irrigation ~1.32 mm/day; 9 = irrigation ban in Vs92; 10 = some street washing, very little external use.

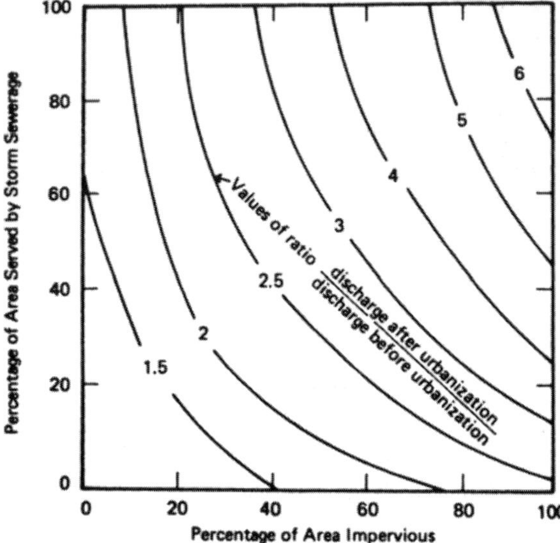

Figure U9 Two factors that cause increased discharge in the urban area as related to the ratio of discharge *after* urbanization to that *before* urbanization (after Mather, 1978).

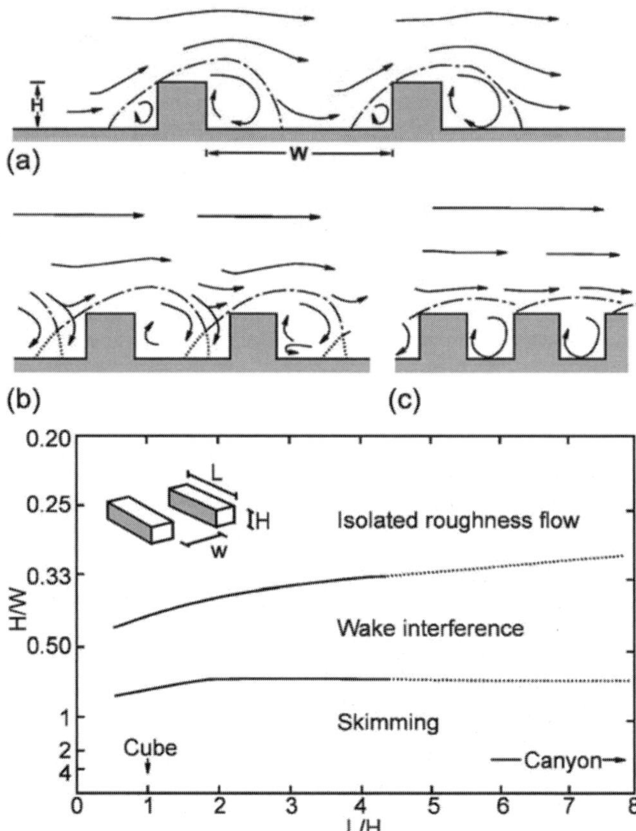

Figure U10 Threshold H/W ratios for the transition flow between flow regimes (**a**) skimming → wake interference; (**b**) wake interference → isolated roughness (from Hunter et al., 1992, used by permission of Pergamon Press Ltd).

different between city and surroundings, evaluations of F are important in the understanding of the humidity contrasts. Its absence may mean a negative humidity island in the city (drier); its presence a positive humidity island (wetter). Fog differences are not great between cities and rural areas although, due to pollution, cities have more haze, smog, and perhaps more slight fog. More cloud condensation nuclei and ice nuclei have been observed in cities, but knowledge is very unsatisfactory at this point about the effects on urban precipitation (Landsberg, 1981). Holmer and Eliasson (1999) report vapor pressure results between urban and rural areas for nine locations from previous studies plus their own study in Goteborg, Sweden. They identified an urban moisture excess, particularly at night, that also was related to the urban heat island and longwave radiative fluxes.

Air movement in cities

Urban areas affect air flow, both speed and direction, on micro- and meso-scales. The aerodynamic roughness of urban areas (level above the ground where wind speed becomes greater than zero on average) is larger by some 0.5–4.0 m than in rural areas, and the city therefore produces more frictional drag. A heat-island effect in cities influences the pressure field and also vertical stability of the air. This will affect local airflow. These effects also depend on overall gradient wind strength across the urban region. Many studies have shown that, as wind increases, the magnitude of the urban heat island diminishes.

Wind speeds

When regional air flow is strong, the aerodynamic roughness of the city is dominant and is important in inducing increases in mechanical turbulence on the order of 30–50% (Oke, 1980). Under light winds, air pressure differences and atmospheric stability effects, in addition to roughness, are critical in determining city wind speeds from place to place. The built environment affects the flow regimes of air through and over the city (Hunter et al., 1990). Generally, the length/height ratio and height/width ratio of buildings control the flow regime among isolated roughness flow, wake interference flow, and skimming flow (viewed in Figures U10a and b). Generally, under light regional flow, winds tend to accelerate in cities compared to rural areas because of strong rural atmospheric stability and the increased turbulence over the warmer, rougher city surface. Under strong flow, winds generally decelerate in the urban area due to greater frictional effects of the city on airflow. Studies have shown that there appears to be a critical regional flow speed separating the accelerating from decelerating wind response in the city (Chandler, 1965; Bornstein and Johnson, 1977; Lee, 1979). This speed, however, is quite variable and ranges from 0.8 to 5.6 m/s.

Wind direction

Wind direction differences in cities and in rural areas relate to frictional retardation effects. Decelerating winds will promote cyclonic turning toward lower pressure; accelerating winds produce anticyclonic turning (Ackerman, 1974; Angell and Bernstein, 1975; Lee, 1973). Wind direction changes induced by a city also vary diurnally in relation to the depth of the mixing layer. During the day the turning effect lessens, whereas at night it reaches a maximum.

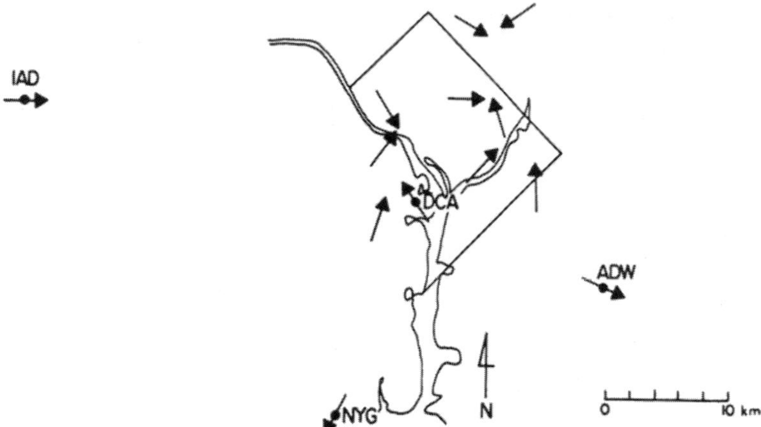

Figure U11 Wind convergence on a hot summer afternoon toward the city of Washington, DC. IAD = Dulles International Airport; DCA = Washington National Airport; ADW = Andrews Air Force Base; and NYG = Quantico Marine Base (after Landsberg, 1981, p. 135).

Urban circulations, much like the classic sea breeze, have been documented in addition to being theorized from laws of thermodynamics (Bach, 1970; Hjelmfelt, 1982; Shreffler, 1978; Vukovich and King, 1980). Landsberg (1981) illustrates the summer air circulation for Washington, DC (Figure U11). Note the convergence locally in the city boundary. Other air flow systems, such as synoptic fronts and sea breezes, also can be affected by cities. Manhattan, New York, for example, slows cold frontal passages upwind of the city and speeds them up downwind, especially when a strong heat island prevails (Loose and Bornstein, 1977). Wind conditions at street level and in the UCL are very complex and relate to physical barriers, building sizes, orientations, building densities, and general land-use patterns. As a generalization, winds are slowed in the UCL. However, effects of wind channeling and funneling can occur, depending on prevailing winds and the overall orientation and morphology of the city landscape.

Conclusions

The climate of cities will continue to be of importance to the ever-expanding urban population of the world. The application of research findings of urban climatology in building designs and urban environmental planning is beginning to emerge but is not yet widespread (Bonan, 2002). Due to the complexity of the urban landscape and the variability of dimensions, land use, morphology, and other characteristics, much research still remains on just how a city affects the surface and atmospheric climatic environment, and the city's overall urban ecology. Equally, if not more important, are the interactions of the city climate system with other elements of the entire urban ecosystem (Douglas, 1981; Bonan, 2002). The discovery of these interrelationships will eventually aid in planning solutions related to pollution, health, comfort, water supplies, and general quality of life among urban dwellers.

Anthony J. Brazel and Dale Quatrocchi

Bibliography

Ackerman, B., 1974. Wind fields over St. Louis in undisturbed weather. *American Meteorological Society Bulletin*, **55**: 93–95.

Ackerman, B.S., Changnon, S.A., Dzurisin, G., et al., 1978. *Summary of METROMEX*, vol. 2: *Causes of Precipitation Anomalies*. Bulletin 63. Urbana, IL: Illinois State Water Survey.

Angell, J.K., and Bernstein, A.B., 1975. Flow across an urban area determined from double-theodolite pilot balloon observations. *Journal of Applied Meteorology*, **14**: 1072–1079.

Arnfield, A.J., 1982. An approach to the estimation of the surface radiative properties and radiation budget of cities. *Physical Geography*, **3**(2): 97–122.

Atkinson, B.W., 1971. The effect of an urban area on the precipitation from a moving thunderstorm. *Journal of Applied Meteorology*, **10**: 47–55.

Atwater, M.A., 1971. The radiation budget for polluted layers of the urban environment. *Journal of Applied Meteorology*, **10**: 205–214.

Auer, A.H., Jr, 1978. Correlation of land use and cover with meteorological anomalies. *Journal of Applied Meteorology*, **17**: 636–643.

Bach, W., 1970. An urban circulation model, *Archiv für Meteorologie, Geophysik, und Bioklimatologie*, **B18**: 155–168.

Barring, L., Mattsson, J.O., and Lindqvist, S., 1985. Canyon geometry, street temperatures, and urban heat island in Malmo. Sweden. *Journal of Climatology*, **5**: 433–444.

Beryland, M.E., and Kondratyev, K.Ya., 1972. *Cities and the Global Climate* (A. Nurklik, trans.). Downsview, Ontario: Atmospheric Environment Service.

Bonan, G.B., 2002. *Ecological Climatology Concepts and Applications*. Cambridge: Cambridge University Press.

Bornstein, R.D., and Johnson, D.S., 1977. Urban–rural wind velocity differences. *Atmosphere and Environment*, **1**: 597–604.

Brazel, A.J., 1987. Urban climatology. In Oliver, J., and Fairbridge, R.W., eds., *Encyclopedia of Earth Sciences*, vol. XI: *Encyclopedia of Climatology*. New York: Van Nostrand Reinhold, pp. 889–901.

Brazel, S.W., and Balling, R.C., Jr, 1986. Temporal analysis off long-term atmospheric moisture levels in Phoenix, Arizona. *Journal of Climatology and Applied Meteorology*, **25**: 112–117.

Brown, M.J., and Grimmond, S., 2001. Sky view factor measurements in downtown Salt Lake City. Data report for the DOE CBNP URBAN Experiment, October 2000, Los Alamos National Laboratory, New Mexico, LA-UR-01-1424.

Brun, S.E., and Band, L.E., 2000. Simulating runoff behavior in an urbanizing watershed, Computers. *Environment and Urban Systems*, **24**: 5–22.

Carlson, T.N., and Boland, F.E., 1978. Analysis of urban-rural canopy using a surface heat flux/temperature model. *Journal of Applied Meteorology*, **17**: 998–1013.

Cerveny, R.S. and Balling, R.C., Jr, 1999. Identification of anthropogenic weekly cycles in Northwest Atlantic pollution, precipitation, and tropical cyclones. *Nature*, **394**: 561–562.

Chandler, T.J., 1965. *The Climate of London*. London: Hutchinson.

Chandler, T.J., 1967. Absolute and relative humidities in towns. *American Meteorological Society Bulletin*, **48**: 394–399.

Chandler, T.J., 1976. *Urban Climatology and Its Relevance to Urban Design*. Technical Note 149. Geneva: World Meteorological Organization.

Changnon, S.A., Jr, 1969. Recent studies of urban effects on precipitation in the United States. *American Meteorological Society Bulletin*, **50**: 411–421.

Changnon, S.A., Jr, 1981. METROMEX: a review and summary. *Meteorological Monographs* **40**: 1–181.

Changnon, S.A., Jr, 1983. Purposeful and accidental weather modification: our current understanding. *Physical Geography*, **4**(2): 126–139.

Changnon, S.A., Huff, F.A., Schickedanz, P.T., and Vogel, J.L., 1977. *Summary of METROMEX, vol. 1: Weather Anomalies and Impacts*. Bulletin 62. Urbana, IL: Illinois State Water Survey.

Dabberdt, W.F., and Davis, P.A., 1974. *Determination of Energetic Characteristics of Urban–Rural Surfaces in the Greater St Louis Area*. US Environmental Protection Agency project report, April, Washington, DC, USA.

De Dear, R.J., Kalma, J.D., Oke, T.R., and Auliciems, A., 2000. Biometeorology and urban climatology at the turn of the millennium. Selected papers from the conference ICB-ICUC '99 (Sydney, 8–12 November 1999), WCASP-50, WMO/TD-No. 1026.

Dettwiller, J.W., and Changnon, S.A., Jr, 1976. Possible urban effects on maximum daily rainfall at Paris, St. Louis, and Chicago. *Journal of Applied Meteorology*, **15**: 517–519.

Douglas, I., 1981. The city as an ecosystem. *Progress in Physical Geography*, **5**(3): 315–367.

Flynn, J.J., 1980. Point pattern analysis and remote sensing techniques applied to explain the form of the urban heat island. Doctoral dissertation, State University of New York, College of Environmental Science and Forestry, Syracuse, New York.

Goldreich, Y., 1985. The structure of the ground-level heat island in a central business district. *Journal of Climatology and Applied Meteorology*, **24**(11): 1237–1244.

Goward, S.N., 1981. Thermal behavior of urban landscapes and the urban heat island. *Physical Geography*, **2**(1): 19–33.

Grimmond, C.S.B., 1992. The surburban energy balance: methodological considerations and results for a mid-latitude west coast city under winter and spring conditions. *International Journal of Climatology*, **12**: 481–497.

Grimmond, C.S.B. and Oke, T.R., 1995. Comparison of heat fluxes from summertime observations in the suburbs of four North American cities. *Journal of Applied Meteorology*, **34**(4): 873–889.

Grimmond, C.S.B., and Oke, T.R., 1999. Evapotranspiration rates in urban areas. In *Impacts of Urban Growth on Surface Water and Groundwater Quality* (Proceedings of IUGG 99 Symposium HS5, Birmingham, July 1999), IAHS Publication no. 259, pp. 235–243.

Hansen, J., Ruedy, R., Glascoe, J., and Sato. M., 1999. GISS analysis of surface temperature change. *Journal of Geophysical Research*, **104**(D24): 30 997–31 022.

Hansen, J., Ruedy, R., Sato, M., et al., 2001. A closer look at United States and global surface temperature change. *Journal Geophysical Research*, **106**(D20): 23 947–23, 963.

Hjelmfelt, M.R., 1982. Numerical simulation of the effects of St. Louis on mesoscale boundary-layer airflow and vertical air motion: simulations of urban vs. non-urban effects. *Journal of Applied Meteorology*, **21**: 1239–1257.

Holmer, B. and Eliasson, I., 1999. Urban–rural vapour pressure differences and their role in the development of urban heat islands. *International Journal of Climatology*, **19**: 989–1009.

Howard, L., 1833. *The Climate of London*, vols. 1–3. London: Harvey & Darton.

Hunter, I., Watson, I.D., and Johnson, G.T., 1990/1991. Modelling air flow regimes in urban canyons. *Energy and Buildings*, **15–16**: 315–324.

Johnson, G.T., and Watson, I.D., 1984. The determination of view-factors in urban canyons, *Journal of Climatology and Applied Meteorology*, **23**(2): 329–335.

Kalanda, B.D., Oke, T.R., and Spittlehouse, D.L., 1980. Suburban energy balance estimates for Vancouver, B.C. using the Bowen ratio-energy balance approach. *Journal of Applied Meteorology*, **19**: 791–802.

Kopec, R.J., 1973. Daily spatial and secular variations of atmospheric humidity in a small city. *Journal of Applied Meteorology*, **12**: 639–648.

Landsberg, H.E., 1981. *The Urban Climate*. New York: Academic Press.

Lee, D.O., 1973. Urban influence on wind directions over London. *Weather*, **32**: 162–170.

Lee, D.O., 1979. The influence of atmospheric stability and the urban heat island on urban-rural wind speed differences. *Atmosphere and Environment*, **13**: 1175–1180.

Lee, D.O., 1984. Urban climates. *Progress in Physical Geography*, **8**(1): 1–31.

Lo, C.P., Quattrochi, D.A., and Luvall, J.C., 1997. Application of high-resolution thermal infrared remote sensing and GIS to assess the urban heat island effect. *International Journal of Remote Sensing*, **18**: 287–304.

Loose, T., and Bornstein, R.D., 1977. Observations of mesoscale effects on frontal movement through an urban area. *Monthly Weather Review*, **105**(5): 563–571.

Lowry, W.P., 1977. Empirical estimation of urban effects on climate: a problem analysis. *Journal of Applied Meteorology*, **16**: 129–135.

Marotz, G.A., and Coiner, J.C., 1973. Acquisition and characterization of surface material data for urban climatological studies. *Journal of Applied Meteorology*, **12**: 919–923.

Mather, J.R., 1978. *The Climatic Water Budget in Environmental Analysis*. Lexington, MA: Lexington Books.

Melhuish, E., and Pedder, M., 1998. Observing an urban heat island by bicycle. *Weather*, **53**(4): 121–128.

Miller, D.H., 1965. The heat and water budget of the Earth's surface. *Advances in Geophysics*, **11**: 175–302.

Oke, T.R., 1974. *Review of Urban Climatology 1968–1973*. Technical Note 134. Geneva: World Meteorological Organization.

Oke, T.R., 1979. *Review of Urban Climatology 1973–1976*. Technical Note 169. Geneva: World Meteorological Organization.

Oke, T.R., 1980. Climatic impacts of urbanization. In Bach, W., Pankrath, J., and Williams, J., eds., *Interactions of Energy* and *Climate*. Boston, MA: Reidel, pp. 339–356.

Oke, T.R., 1981. Canyon geometry and the nocturnal urban heat island: comparison of scale model and field observations. *Journal of Climate*, **1**: 237–254.

Oke, T.R., 1982. The energetic basis of the urban heat island. *Royal Meteorological Society Quarterly Journal*, **108**(455): 1–24.

Oke, T.R., 1987. *Boundary Layer Climates*. London: Methuen.

Oke, T.R., 1997. Urban climates and global environmental change. In Thompson, R.D., and Perry, A.H., eds. *Applied Climatology: Principles and Practice*. London: Rutledge, pp. 273–287.

Oke, T.R., 1998. Observing weather and climate. *Proceedings of the Technical Conference on Meteorology and the Environment*. Instruments and Methods of Observation, Instruments and Observing Methods Report No. 70, WMO/TD-No. 877, WMO, Geneva, pp. 1–8.

Oke, T.R., 1999. Observing urban weather and climate using ŝtandard' stations. In De Dear, R.J., Kalma, J.D., Oke, T.R., and Auliciems, A., eds., *Biometeorology and Urban Climatology at the Turn of the Millennium*. Selected papers from the conference ICB-ICUC '99 (Sydney, 8–12 November 1999), WCASP-50, WMO/TD-No. 1026, pp. 443–448.

Pease, R.W., and Nichols, D.A., 1976. Energy balance maps from remotely sensed imagery. *Photogrammetric Engineering and Remote Sensing*, **42**: 1367–1373.

Petersen, J.T., and Flowers, E.C., 1977. Interaction between air pollution and solar radiation. *Solar Energy*, **19**: 23–32.

Petersen, J.T., and Stoffel, T.L., 1980. Analysis of urban-rural solar radiation data from St. Louis, Missouri. *Journal of Applied Meteorology*, **19**: 275–283.

Petersen, J.T., Flowers, E.C., and Rudisill, J.H., 1978. Urban–rural solar radiation and atmospheric turbidity measurements in the Los Angeles basin. *Journal of Applied Meteorology*, **17**: 1595–1609.

Quattrochi, D.A., and Ridd, M.K., 1994. Measurement and analysis of thermal energy responses from discrete urban surfaces using remote sensing data. *International Journal of Remote Sensing*, **15**: 1991–2022.

Quattrochi, D.A., and Ridd, M.K., 1998. Analysis of vegetation within a semi-arid urban environment using high spatial resolution airborne thermal infrared remote sensing data. *Atmosphere and Environment*, **32**: 19–33.

Quattrochi, D.A., Luvall, J.C., Rickman, D.L., Estes, M.G., Jr, Laymon, C.A., and Howell, B.F., 2000. A decision support system for urban landscape management using thermal infrared data. *Photogrammetric Engineering and Remote Sensing*, **66**: 1195–1207.

Rao, P.K., 1972. Remote sensing of urban ĥeat islands' from an environmental satellite. *American Meteorological Society Bulletin*, **50**: 522–528.

Roth, M., Oke, T.R., and Emery, W.J., 1989. Satellite-derived urban heat islands from three coastal cities and the utilization of such data in urban climatology. *International Journal of Remote Sensing*, **10**: 1699–1720.

Shepperd, J.M., Pierce, H., and Negri, A.J., 2002. Rainfall modification by major urban areas: observations from spaceborne rain radar on the TRMM satellite. *Journal of Applied Meteorology*, **41**: 689–701.

Shreffler, J.H., 1978. Detection of centripetal heat-island circulations from tower data in St. Louis. *Boundary-Layer Meteorology*, **15**: 229–242.

Sisterson, D.L., and Dirks, B.A., 1978. Structure of the daytime urban moisture field. *Atmosphere and Environment*, **12**: 1943–1949.

Terjung, W.H., and O'Rourke, P.A., 1980. Simulating the causal elements of urban heat islands. *Boundary-Layer Meteorology*, **19**: 93–118.

Thompson, R.D., and Perry, A.H., 1997. *Applied Climatology: principles and practice*. London: Rutledge.

Unwin, D.J., 1980. The synoptic climatology of Birmingham's urban heat island 1965–1974. *Weather*, **35**: 43–50.

Voogt, J.A., and Oke, T.R., 1997. Complete urban surface temperatures. *Journal of Applied Meteorology*, **36:** 1117–1132.

Voogt, J.A,. and Oke, T.R., 1998. Radiometric temperatures of urban canyon walls obtained from vehicle traverses. *Theoretical and Applied Climatology*, **60**: 199–217.

Vukovich, F.M., and King, W.J., 1980. A theoretical study of the St. Louis heat island: comparisons between observed data and simulation results on the urban heat island circulation. *Journal of Applied Meteorology*, **19**(7): 761–770.

White, J.M., Eaton, F.D., and Auer, A.H., Jr, 1978. The net radiation budget of the St. Louis metropolitan area. *Journal of Applied Meteorology*, **17**: 593–599.

Winkler, J.A., and Skaggs, R.H., 1981. Effect of temperature adjustments on the Minneapolis–St. Paul urban heat island. *Journal of Applied Meteorology*, **20**: 1295–1300.

World Meteorological Organization, 1970. *Urban Climates*. Technical Note 108, No. 254, TP 141. Geneva: World Meteorological Organization.

Yap, D., 1975. Seasonal excess urban energy and the nocturnal heat island—Toronto. *Archiv für Meteorologie, Geophysik, und Bioklimatologie*, **B23**: 69–80.

Cross-references

Air Pollution Climatology
Bioclimatology
Energy Budget Climatology
Local Climatology
Microclimatology
Water Budget Analysis

V

VEGETATION AND CLIMATE

Descriptions of the processes that shape the world around us emphasize the roles that energy and water budgets play. Much recent work in physical geography has been concerned with understanding the interactions of the various systems and subsystems that make up our environment. Utilizing factors derived from moisture and energy budgets, we have been able to achieve a more rational understanding of the distribution of the great vegetation biomes as well as the general relation of climatic factors to vegetation.

Scales of interactions

Monteith (1981) has suggested the scales shown in Figure V1 involving the coupling of the surface vegetation to the atmosphere. On the continental scale the location and extent of the major plant biomes may be determined by the general circulation of the atmosphere as well as by the space and time variations of the principal energy and moisture factors of climate. Local climates (mesoclimates) are more influential in the development of plant communities and plant associations and their ultimate survival. It is, of course, the interactions of the mesoclimate with the plant communities that result in the microclimates (or climates near the ground) to which the individual members of the plant community must react. On an even smaller scale, the interaction of the individual plant with its microclimate results in the epiclimate (called teleoclimate by Gates, 1968), which involves what occurs in the individual air layer in direct contact with the leaves, branches, and stems of plants. This climate, in the immediate vicinity of the plant, influences the temperature, gas concentration, and moisture condition within the plant tissue. The endoclimate determines the rate of growth and development of the individual plant organs and, in this way, provides a feedback from the individual plant through the community to the biome (Monteith, 1981).

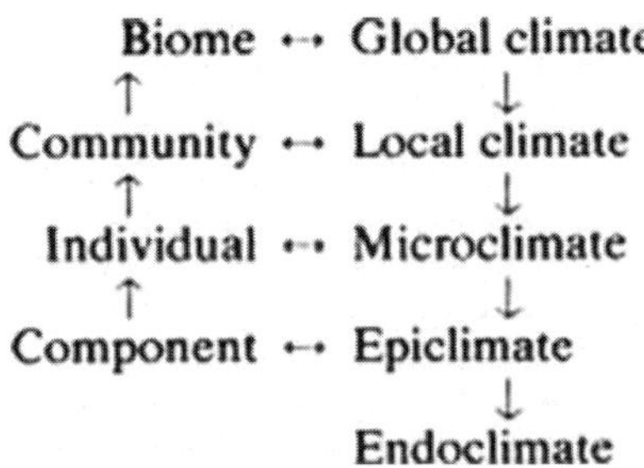

Figure V1 Monteith scales.

History

Long before instrumental observations were available, climates were often named after their characteristic vegetation. For example, forest, desert, savanna, and tundra have long been used as names of climate types as well as of vegetation types. With the development of meteorological instruments and the collection of systematic climatic observations, the possibility of expressing explicitly the relation between climate and the growth and development of plants or of vegetation covers seemed about to be realized.

The eighteenth-century French scientist Réaumur was possibly the first to make an exact determination of the quantity of heat required to bring a plant from one stage of maturity to another. He felt that the mean daily air temperatures, summed between one stage of development and another for a plant, should be approximately constant for that particular period of development from year to year. Each plant species should have its own *thermal constant* to progress from germination to maturity. Réaumur's work stimulated others to undertake more detailed observations on plant development. There was a rapid expansion of observations of the relation between climatic factors and the dates of budding, leafing, flowering, and fruiting of annuals. Such phenological studies suggested a general relation between climate and the geographical distribution of vegetation.

The early work of Wahlenberg (1813) and von Humboldt (1817) showed the relation between temperature and vegetation distribution on mountain slopes, whereas Dove (1846) and

Linsser (1869; see references in Thornthwaite and Mather, 1954) pointed out that, in many regions of the world, rainfall would exercise a greater control than temperature over the distribution of vegetation. Linsser actually developed a moisture factor that allowed him to subdivide the vegetation zones of the Earth into five major categories. The important vegetation classification of de Candolle (1855) established five basic types of plants – four temperature types including megatherms (plants needing continuously high temperatures and moisture), mesotherms (plants needing only moderate amounts of heat and moisture), microtherms (plants able to survive short summers and colder winters), hekistotherms (plants of the polar zones – beyond the tree limit), and xerophytes (plants tolerant of dryness) – in terms of their physiologic response to climate.

Schimper's world map of vegetation formations (1898) separated, for the first time, the vegetation of the tropics and temperate zones into rainforest, forest, woodland, grassland, and desert. Tundra was used to represent the one vegetation class of the polar zone. He also introduced two transitional zones, without definite boundaries – the parklike landscape and the semidesert.

Köppen (1900) was among the first to express the idea that plants could serve as meteorological instruments capable of integrating the various climatic elements and that climatic regions could be defined in terms of plant regions as long as the climatic and edaphic influences on vegetation were carefully separated. His well-known climatic classifications were based on climatic records, but the climatic boundaries were identified by study of vegetation, soil, and hydrologic features.

Köppen's first map of climate classification appeared in 1900. In it he attempted to combine certain of the details of Grisebach's early map (1866) of vegetation regions (24 regions were identified on floral content) and de Candolle's 1855 classification described earlier. Köppen accepted temperature divisions of tropical, warm temperate, sub-Arctic, and polar to accord with the four temperature classes of de Candolle, along with four moisture divisions (forest – not xerophytes; tree savannas – xerophytic trees and shrubs; steppe; and desert). Through a combination of temperature and moisture types, he identified 14 types of climate.

By the mid-1900s the standard vegetation map generally included an assemblage of specific vegetation types having spatial qualities and a strong relationship to one of the earlier climatic maps of Köppen. However, even Köppen recognized that not all his climates could be readily identified by distinct vegetation classes even though most users attempted such a direct correlation. Often, schematic diagrams were constructed suggesting the close similarities between idealized patterns of vegetation and climate (see Figure V3). Such diagrams are misleading in their simplification and generalization and their failure to identify vegetation on the basis of form, taxonomic, and cover factors.

Vegetation specialists today would question whether plant formations are entities that are only produced and delimited by climate (de Laubenfels, 1975). Many vegetation maps in current use depict ecosystems rather than vegetation formations. Correlation of such maps with the distribution of other environmental components may not truly indicate the role of the environment in the distribution and growth of vegetation formations.

De Laubenfels (1975) suggests that only four distinct undisturbed vegetation formations can be identified (rainforest, seasonal forest, woodland, and desert; see Figure V3). These are distinct groupings rather than points or regions along a continuum. In addition to the four major vegetation formations, there are three others – grassland, savanna, and brush – that might be considered as disturbed cover types. The disturbance – often fire – if repeated frequently may result in fairly permanent vegetation types.

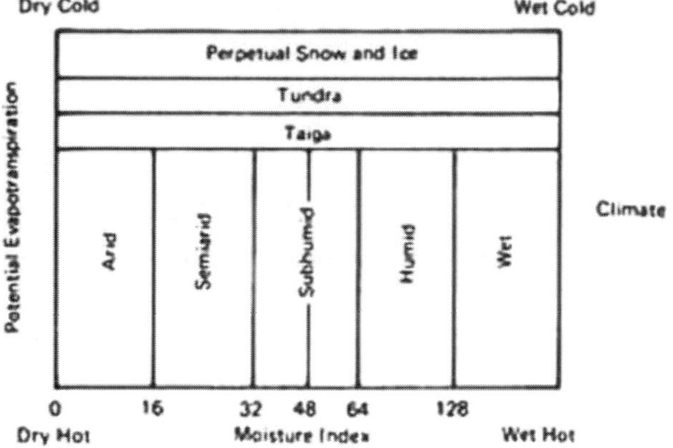

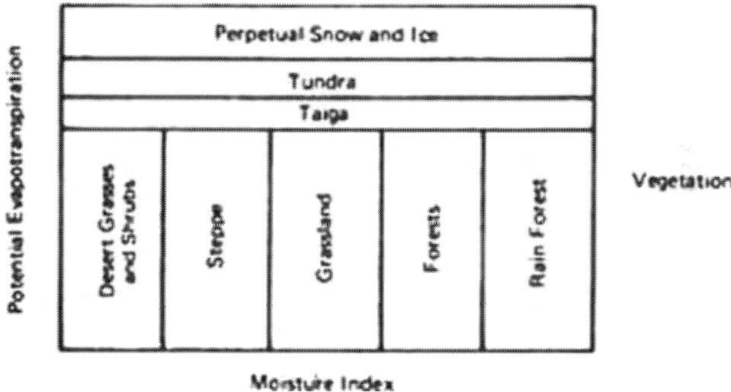

Figure V2 Schematic distribution of climate and vegetation on the basis of varying temperature (energy) and moisture indices.

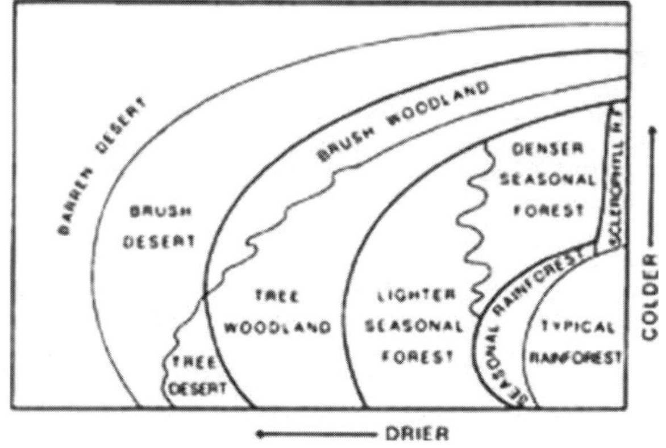

Figure V3 Schematic representation of various primary and secondary vegetation types with respect to moisture and temperature gradients. Grasslands and savanna may often replace many of them over large areas (from de Laubenfels, 1975).

Limitations of climatic observations

The standardized climatological stations used to record the climate of a country or region are neither situated nor equipped to measure temperature or humidity at the places and at the times that are critical for plants.

The climate at 5 feet above the ground in a standard weather shelter is very different from that within a few inches of the ground in the open. Nocturnal temperatures are lower and daytime temperatures are higher close to the ground, where many plants grow, than they are a few feet above it.

Climatic influences on crop scheduling and irrigation

Food producers recognize that yield and quality are higher if plants receive the proper amount of moisture throughout the growing season and are harvested at the peak of their maturity. To help in the scheduling of crops, Thornthwaite and his associates (Thornthwaite and Mather, 1954) undertook studies of the relations between climate and crops. They attempted to link the water used by plants in transpiration with the rate of plant development. This approach involved a study of the old field of phenology (the study of the timing of periodic biologic events with respect to climate and other conditions of the natural environment).

Thornthwaite was one of the first scientists to provide quantitative evaluation of the moisture factor in climate by relating precipitation to potential evapotranspiration (defined as the amount of water that will be lost from a surface completely covered with vegetation all growing to the same height that never suffers from a lack of water). Potential and actual evapotranspiration are the same only under ideal conditions of soil moisture and vegetation. Unlike actual evapotranspiration, potential is defined to be independent of soil type, kind of crop, or mode of cultivation, and is thus a function of the climate alone.

Soil tanks, covered with the same kind of vegetation that surrounds them and having a water supply fully adequate to the needs of the vegetation, have been used to measure potential evapotranspiration water losses. Various investigators have also developed physical and empirical methods to compute this quantity (for example Rosenberg, Hart, and Brown; Penman; Baier and Robertson; Blaney and Criddle; Jensen and Haise, see references in Mather, 1978).

Daily observations of plant development by Thornthwaite and his associates at the Laboratory of Climatology (Higgins, 1952) made it possible to associate definite climatic conditions with definite amounts of plant development as well as with definite amounts of plant water need (potential evapotranspiration). The daily observations showed that, as incoming energy from the sun increased, both plant development and plant water use increased. This led to the formulation of a development unit as the amount of development that will occur in a plant as a unit amount of water is being transpired. The units, given in the metric system, show that as 100 development units are accumulated in a plant's life, 10 mm depth of water is transpired by the plant. This is an unfamiliar idea but it is useful because it provides a single solution for the important agricultural problems of scheduling supplementary irrigation and crop planting and harvesting.

Active climatic factors and vegetation distribution

The influence of climate on vegetation under different environmental conditions may be quite variable. Eyre (1963) indicated that, in certain cases, chemical composition and physical characteristics of the soil and parent material could be as important as climatic factors. Thornthwaite (1952), writing as a climatologist, took a somewhat more deterministic view of the influence of climate. He felt that climatic factors almost by themselves could be used to distinguish vegetation associations. For example, he stated that the evidence for a "grassland climate" was unmistakable. Such a climate is marked by a specific balance between moisture deficit and surplus so that few periods of either very dry or moist soil conditions would exist during the year. This led Thornthwaite to conclude that grasslands are a natural phenomenon, in equilibrium with their climatic environment. Whereas Eyre considered the climatic factors of temperature and precipitation (P), Thornthwaite considered soil moisture distribution, plant water need, and the seasonal course of moisture surplus and deficit. He felt that these were climatic factors that might truly influence vegetation distribution. Soil moisture content, for example, can combine both climatic and edaphic factors. Since soil moisture storage is a climatic factor according to Thornthwaite, his reference to a definite grassland climate becomes more understandable.

Hare (1954) and later Thornthwaite and Hare (1955) attempted to correlate the factor of potential evapotranspiration (PE) and the moisture index ($I_m = 100 \times [(P/PE)-1]$) with the distribution of vegetation. They found in a temperate, humid climate, I_m can roughly distinguish different forest types, whereas the sub-Arctic, boreal forest subdivisions are better correlated with PE. Major (1963), in an alternative approach, has considered a third factor, actual evapotranspiration (AE), and has suggested that vascular plant activity and growth might be related to the actual water loss of the vegetation since this factor expresses the real water activity of the plant rather than some unrealized potential value.

Relation of vegetation to global and local climates

North America

Mather and Yoshioka (1966, 1968) have studied the relation among natural vegetation, the climatic moisture index, I_m, and potential evapotranspiration, PE, at stations in all major vegetation regions across North America (see Figure V4). Information concerning the characteristic natural vegetation at each station, as identified from the Shantz and Zon (1924) vegetation map, supplemented with data from the Kühler (1964) map, has been indicated by means of a letter symbol. (All references in this section will be found in Mather and Yoshioka, 1966). Climatic stations were located in the different vegetation regions and the data of I_m and PE obtained for each station.

By using annual potential evapotranspiration and the climatic moisture index, it has been possible to locate discrete and, in many cases, nonoverlapping areas that are related to regions of natural vegetation. If truly active factors of climate can be selected, it appears possible to obtain a relation between climatic and botanic distributions in midlatitudes. From the distributions shown in Figure V4 one might conclude that it is possible to speak not only of a forest climate as opposed to a grassland or a desert climate, but that it also may be possible to identify a birch–maple as opposed to a spruce or oak–chestnut forest climate in a region such as the United States. It does not, of course, eliminate the possibility of edaphically influenced vegetation associations within large, homogeneous climatic

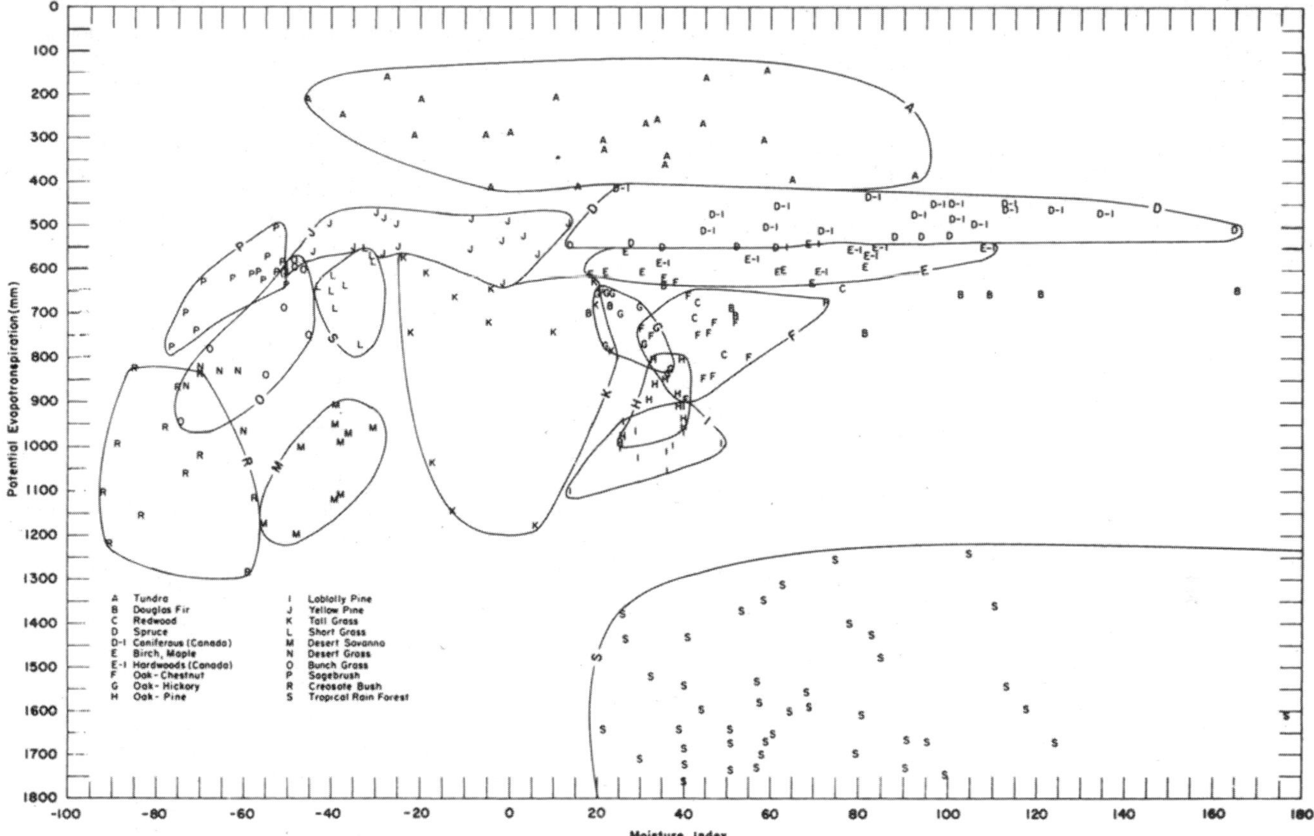

Figure V4 Relation between the climatic moisture index, potential evapotranspiration, and natural vegetation at representative locations in North America (from Mather and Yoshioka, 1968).

regions, or the possibility of more than one vegetation association being able to survive and compete successfully within a given climatic range. Edaphic and other environmental factors must also contribute to these areas of overlap, but the general lack of overlap bears witness to the responsiveness of the botanic environment to significant climatic stimuli.

South Africa

It is possible to locate 22 stations within one degree of 30°S across Africa at which climatic moisture balances can be evaluated. The moisture indices at these stations range from the very dry ($I_m = -92$) Port Nolloth on the west coast to the moist Mid Illovo ($I_m = 26$), Umbogintwine, or Durban (both $I_m = 12$) on the east coast. There is a progressive change in the moisture index at stations along 30°S as one traverses the continent. The stations used and their moisture indices are shown in Figure V5.

Vegetation varies across South Africa in a manner entirely consistent with the climatic indices. On the west coast there is a definitely arid-type vegetation consisting of succulent bush, as Adamson (1938) calls it, or a desert shrub, desert grassland type of vegetation. Adamson separates the succulent bush (I_m −80 to −90) from the arid bush (I_m −50 to −75) and indicates that slightly more moisture is found in the arid bush country. At Bethulie the vegetation is in transition from arid bush to grassland and the moisture index has dropped to −45. Grassland prevails in the region from 26°E to 29°E as the moisture index varies from −45 to −6. Another change in the vegetation occurs as the moisture index shifts from negative to positive between Kokstad (29°25′E) and Nottingham Road (30°01′E). According to Adamson, east of 30°E at this latitude the vegetation shifts to a temperate savanna and then farther east to a subtropical forest. Small but definitely positive moisture indices are found here. There, thus, appears to be a clear relationship between climatic indices and the broadscale pattern of vegetation in South Africa.

Pine-savanna vegetation association of Nicaragua

Within the tropics, and especially when tropical savanna vegetation associations are considered, the role of climate seems of less importance. Parsons (1955) has described in some detail the Miskito pine savanna of the Nicaraguan and Honduran coast. Not only is this pine savanna significant in that it marks the most tropical penetration of pines in the western hemisphere but it is one of the rainiest areas in the western hemisphere with a savanna-type vegetation. Parsons feels that it clearly contradicts the concept that there is such a thing as a "savanna climate". He feels that the limits of the pine savanna are well marked by the termination of the gravel shelf on which this vegetation association is found. Parsons also considers fire, hurricanes, and impeded subsurface drainage as possible

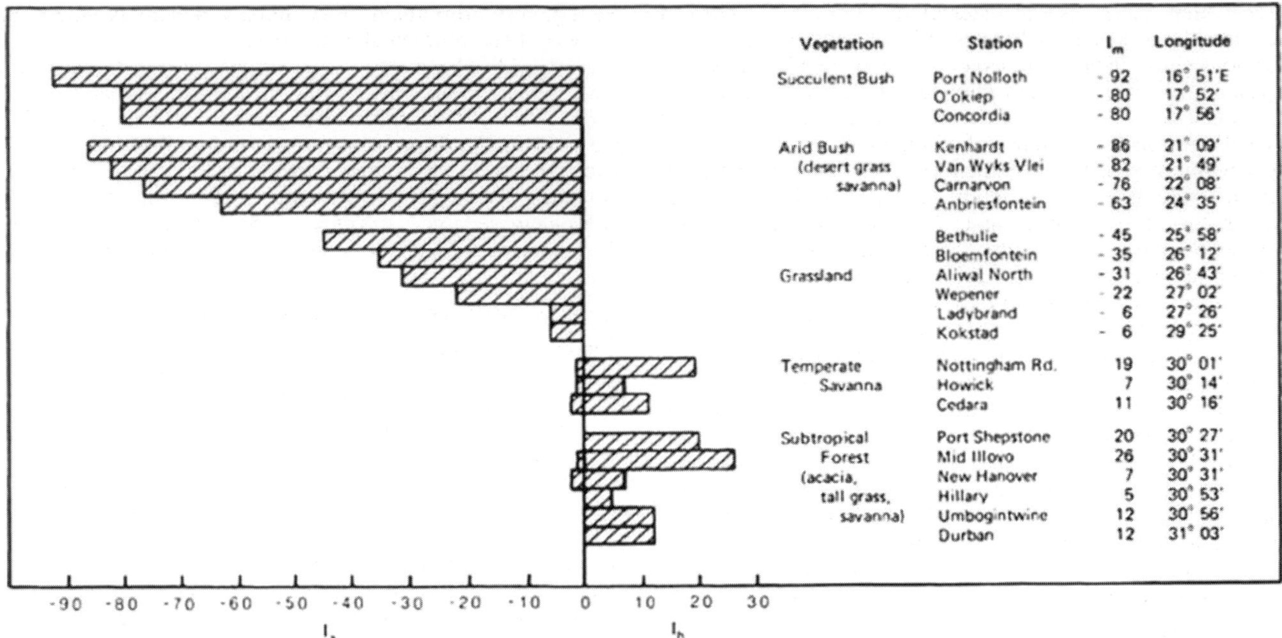

Figure V5 Relation of natural vegetation and climatic moisture indices at selected locations near latitude 30°S across South Africa (from Mather and Yoshioka, 1966).

contributing factors. There is a suggestion that hardwoods might be able to take over the area if fire and other cultural destructive agents were eliminated.

Puerto Cabezas is located on the coast of Nicaragua near the middle of the area of pine savanna. It experiences 3293 mm of precipitation on the average through the year, ranging from 458 mm in June to 50 mm in March. The moisture index for Puerto Cabezas is +106. With a potential evapotranspiration of 1545 mm, this area, by all rights, should experience a tropical rainforest type of vegetation. Climate, in this case, may be a "permissive" factor, but it certainly is not a determinant one since savanna rather than rainforest exists there.

Denevan (1961), in a study of the upland pine vegetation of Nicaragua, emphasizes the role of fire in the development of the pine savanna. He considers the frequency of burning as the determining factor. The existence of a pine savanna in this region poses two separate problems; first, there is the problem of conifers in tropical and subtropical climatic regions; second, there is the problem of the savanna vegetation and its relation, if any, to climate. Denevan feels that pines are not able to compete successfully against broadleaf vegetation in tropical and subtropical areas without the aid of fires and humans. Other pine associations in tropical areas should be restudied with this possibility in mind.

Global natural vegetation

Extending Mather and Yoshioka's climate and vegetation analyses to the entire Earth, Brewer (2001) achieved similar success. Using global long-term averages of seasonal values of PE and Thornthwaite's I_m from the spring, summer, and fall, he was able to estimate global natural vegetation from an independent dataset. Vegetation data were obtained from the Matthews' (1983) natural vegetation database, composed by merging numerous descriptions and maps of vegetation, augmented with satellite data, into a single vegetation database. Vegetation categories were defined according to United Nations Educational, Scientific and Cultural Organization (UNESCO) guidelines including lifeform, density, and seasonality (UNESCO, 1973).

Using multivariate statistical analysis, Brewer accurately determined the locations of natural vegetation cores, including the elusive "grassland climate". Discriminant analysis effectively grouped vegetation based on common responses to seasonal active climate factors. Transition zones and certain transition-stage vegetation categories, however, proved difficult to determine precisely.

Seasonal climatic values used in Brewer's analysis yielded higher correlation to natural vegetation distributions than did annual climatic values. Adding wintertime PE and I_m values provided negligible improvement in determining vegetation categories. Indeed, it appears that the winter contribution to vegetation development is masked by the other seasons. In mid- and high-latitudes there is very little or no vegetation growth during the winter while in tropical regions the winter season is potentially redundant as it merely represents one-quarter of the total growing season. Additionally, moisture index modifications such as that proposed by Willmott and Feddema (1992) and the use of monthly active climatic factors show promise in further improving global vegetation–climate relationships. Also, neighborhood classification techniques readily available in geographic information systems may lead to better understanding of vegetation in transition zones.

Relation of vegetation to microclimates and epiclimates

One has to be impressed with the great variety of responses of individual vegetation species and vegetation components (leaves, branches, trunks, and stems) to the different factors of the microclimates and epiclimates. It is difficult to make generalizations since plant reactions are so influenced by individual exposures of the plants and their leaves. These change both from plant to plant and from level to level within the same plant. Exposure of a number of different plants next to each other within a single environment will result in the measurement of quite different leaf temperatures as well as transpiration, photosynthetic respiration, and growth rates. Gates (1980) reported on three trees in Sydney, Australia, at noon on a warm summer day (air temperature 32°C) with a thin overcast of high cirrus clouds. The leaves of *Erythrina indica* were all standing in a vertical position, whereas those of a nearby cottonwood were horizontal. Leaf sizes and temperatures were essentially the same, but the *Erythrina* was regulating its temperature by the vertical exposure of its leaves (less surface exposed to radiation), whereas the cottonwood was controlling its leaf temperature by increased evapotranspiration loss. A nearby jacaranda tree maintained nearly the same leaf temperature as a result of the small size of its leaves in relation to the other two trees and a larger convection coefficient.

Under conditions of heat stress, taller-growing plants have some advantage over lower-growing plants because the higher leaves are exposed to greater convective cooling as a result of increased air movement at those levels. Temperatures of desert plants with very small leaves are closely coupled to air temperature by convection. Such plants may even drop their leaves under extremely dry conditions, whereas desert succulents have thick waxy cuticles to prevent undue water loss. Many desert plants have dense crowns of thorns to reflect sunlight, to shade the upper part of the plant, or to slow the movement of solar heat inward to the plant surface itself. Alpine or tundra plants tend to be low or dwarfed to protect them from convectional heat loss and, possibly, to protect them from the extreme cold of winter under a blanket of insulating snow. Gates has measured needle temperatures 9°C warmer on the leeward side of alpine fir trees than on the exposed windward side, which were at air temperature. He feels such a temperature gradient is partly the reason for the increased leeward growth of such trees. The physical effect of the wind itself will also help shape the tree with the prevailing wind direction.

Raunkiaer (1934), Taylor (1975) and others (see references in Gates, 1980) all concluded that leaf size and form were strong indicators of climatic or environmental conditions. To illustrate this, they pointed to the small leaves of desert, Arctic, or alpine plants, the small-lobed sun leaves on trees, the large understory leaves on tropical forest plants, and the systematic manner in which leaves graded in cross-sections from wet to dry or warm to cool.

Grace (1981) suggests that the shape and size of trees and the physiognomy of the vegetation are partially controlled by wind. Structural features of the vegetation influence the rate of transport of heat, gases, and momentum across the boundary layer between the plant and the atmosphere, with higher transport rates existing when vegetation is tall rather than short, or possessing small leaves, stems, and branches rather than large. Thus, with other factors being equal, tall plants or those with small leaves will be at a competitive disadvantage to short plants with large leaves in areas where temperatures are low and wind speeds high. The very low-growing, large-leaved plants of the Arctic and alpine environments (so-called cushion plants) illustrate this point since their low stature decreases convective heat loss and their large leaves increase solar heating, permitting them to maximize productivity under generally cold environmental conditions.

Gates (1980) has noted rapid changes in leaf temperatures as wind or solar radiation conditions change. A leaf that was 5–7°C above air temperature in clear, still air might cool by 2–3°C when clouds move across the sun or by as much as 10°C when the leaves are rapidly exposed to windy conditions. The temperature changes are almost instantaneous. Although he feels that these large changes in temperature may produce some physiological consequences, they are uncertain at present.

Climatic effects of vegetation change

Flohn (1973) has estimated that, over the last 8000 years, 11% of the land area of the globe has been changed into arable land, while more than 30% of the existing forest land is no longer in its undisturbed state. Extrapolating the present rate of removal of tropical rainforests, Peters et al. (1989) estimate that none will exist by 2030. Sixty percent of Central Europe has been converted from forest to farmland since AD 1000 (Darby, 1956).

Vegetation influences the hydrologic cycle through the receipt and disposition of precipitation. Forest vegetation will intercept and reevaporate large amounts of precipitation, whereas the deep root systems of trees will remove water from greater depths of soil and for a longer time than will grassland. Thus, the effect of changing land use or crop type within a basin is to alter interception loss, transpiration, moisture loss from the soil, and, through feedback processes, the soil moisture content and the runoff from the basin. Muller (1966) has shown that the regrowth of forests in the central New York state area has reduced runoff to the Delaware River by about one inch over the past 30 years (resulting in further upstream movement of the salt line during dry periods). Sellers and Lockwood (1981) developed a complex multilayer model to simulate the action of the hydrologic cycle in four distinct vegetation regions in Britain (pine forest, oak forest, wheatland, and grassland). The model was used with hourly climatic data for the year 1977 and showed the smallest runoff from the pine forest and the greatest runoff from the wheat field. Oak forest and grassland gave nearly the same runoff, the latter being slightly larger. Sellers and Lockwood found these differences largely to be a result of differences in the interception and evaporation properties of the different canopies (Table V1).

Thus, changing a forest to a nonforest vegetation will lead to a marked increase in runoff, as will a change from a grassland to cultivated crops. As a result, a change in vegetation cover may have significant consequences for soil hydrology, soil development, and erosion potential.

Vegetation and general circulation models (GCMs')

Encouraging large-scale links between vegetation and climate has led to the development of vegetation models that can

Table V1 Annual totals of simulated runoff, intercepted loss, and transpiration (in mm depth) for four vegetation types

Vegetation	Runoff	Interception loss	Transpiration
Wheat	417	158	134
Grass	318	62	335
Oak forest	279	172	263
Pine forest	152	253	309

Source: After Sellers and Lockwood (1981).

translate the output from a GCM into maps of past, present, or potential future vegetation. These future vegetation estimates also feed back into the GCMs' to provide more accurate estimates of possible future climate. Most vegetation models used in conjunction with GCMs' simulate an existing ecosystem rather than natural or potentially dominant vegetation.

Brewer (2001) estimated potential vegetation under a 2 × CO_2 climate change scenario using GCM model output for temperature and precipitation similar to that used by the IPCC. Likely areas of core natural vegetation change are shown in Figure V6. Most changes appear in the southern hemisphere, specifically South America and Africa, with minor changes throughout North America, Europe, and Asia. Most startling in his analysis is the global loss of deserts under possible future climate conditions, similar to conditions thought to exist nearly 8000 years ago (Bradley, 1994).

Recent Intergovernmental Panel on Climate Change (IPCC) reports are the result of a combination of biogeography and biochemistry process-based model outputs frequently linked to GCMs'. While each of these types of models has inherent limitations, efforts are under way to blend the two types of process-based models into realistic assessment tools for climate change (IPCC, 1997). Many of the models examined by the IPCC, including the biogeography-based MAPSS (Neilson, 1995) and BIOME3 (Haxeltine and Prentice, 1996) and the biochemistry-based CENTURY (Parton et al., 1987) and BIOME-BGC (Hunt and Running, 1992) (see IPCC, 1997 for a complete list of models tested) exhibited similar skill in simulating current ecosystems from climate data. The models showed, however, broad divergence in attempting to explain vegetation under alternative climates (IPCC, 1997). Similar to Brewer's (2001) multivariate method, these models also showed difficulty with transitional zones both spatially and temporally.

A continuing analysis of the aforementioned vegetation models is under way through the Vegetation/Ecosystem Modeling and Analysis Project (VEMAP members, 1995). This project addresses the response of vegetation to environmental variability and climate both temporally and spatially (Kittel et al., 1995). Future refinements in GCMs' should lead to more reliable estimates of the position and extent of vegetation change.

Conclusions

It is clear that vegetation develops in response to many different stimuli, among which we might list climatic, edaphic, and cultural conditions as possibly the most important. The response of vegetation to climate is both direct and indirect: direct through the role that the factors of temperature, radiation, moisture, and wind play on the growth and development of the vegetation, and indirect through the influence that the climatic factors have on soil conditions, disease organisms, competing botanic associations, and cultural practices. In addition, the reciprocal influence of vegetation on the microclimate of the particular area and on the other factors of the microenvironment creates another level of influence that must be considered in evaluating the factors that contribute to the distribution of vegetation.

The degree to which the distribution of vegetation can be explained on the basis of climatic conditions depends *in part* on the proper selection of climatic factors. Temperature and precipitation by themselves are poor descriptors of climate. The active factors of climate from a vegetation viewpoint are water surplus or water deficit (water supply in relation to water need), potential evapotranspiration (available energy expressed in terms of water need), actual evapotranspiration (actual plant water use), and soil moisture storage. Knowledge of these factors is available from computations of the climatic water balance, and combinations of these factors are expressed in shorthand form in the moisture index of the Thornthwaite climatic water balance.

Figure V6 Potential areas of vegetation change (shaded) under 2 × CO_2 climate change cenarios. (from Brewer, 2001).

Use of more expressive climatic parameters, or finally the use of the frequencies of some of these elements to describe distributions in environmental biology, should yield increased appreciation of the role of climate. The ability to identify some individual formation classes within the biochores provides a new and powerful tool for analysis and classification of the vast body of existing environmental data from many parts of the world. It introduces another aspect of rationality and simplification into the complex world of environmental influences and distributions.

The foregoing should not depreciate the role of edaphic, cultural, and other factors in influencing vegetation distribution. There are many examples of nonclimatic controls on vegetation such as the savanna and the low-latitude pine associations. It is, of course, only through a complete interpretation of all contributing influences that the most understandable picture of the geographic distribution of vegetation can be obtained. As more basic studies of climatic influences on vegetation growth and development are completed, we should be able to develop even more powerful tools for evaluating the botanic potential of newly developing areas and for increasing productivity. Additionally, the increased knowledge of the interaction between climate and vegetation will yield more realistic approximations of future climate in our changing world.

John R. Mather and Michael J. Brewer

Bibliography

Adamson, R.S., 1938. Notes on the vegetation of the Kamiesberg. *Memoirs of the Botanical Survey of South Africa,* **18**: 1–25.

Bradley, R.S., 1994. *Quaternary Climatology*. New York: Chapman & Hall.

Brewer, M.J., 2001. Estimating Natural Vegetation from Climatic Data. *Publications in Climatology*, **54**(1).

Darby, H.C., 1956. The clearing of the woodland in Europe. In Thomas, W.L., ed., *Man's Role in Changing the Face of the Earth*. Chicago, IL: University of Chicago Press, pp. 183–186.

de Candolle, A., 1855. *Geographie Botanique Raisonne*, 2 vols. Paris: V. Masson.

de Laubenfels, D.J., 1975. *Mapping the World's Vegetation*. Syracuse, NY: Syracuse University Press.

Denevan, W.M., 1961. The upland pine forests of Nicaragua. *University of California Publications in Geography,* **12**: 251–320.

Eyre, S.R., 1963. *Vegetation and Soils, A World Picture*. Chicago, IL: Aldine.

Flohn, H., 1973. Naturliche und Anthropogene Klimamodifikationen. *Annalen Meteorologische*, **6**: 59–66.

Gates, D.M., 1968. Energy exchange in the biosphere. In Eckardt, F.E., ed., *Functioning of Terrestrial Ecosystems at the Primary Production Level*. Paris: UNESCO, pp. 33–43.

Gates, D.M., 1980. *Biophysical Ecology*. New York: Springer-Verlag.

Grace, J., 1981. Some effects of wind on plants. In Grace, J., Ford, E.D., and Jarvis, P.G., eds., *Plants and Their Atmospheric Environment*. London: Blackwell, pp. 31–56.

Grainger, A., 1980. The state of the world's tropical forests. *Ecologist*, **10**: 6–54.

Grisebach, A.R.H., 1866. Die Vegetation-Gebiete der Erde, ubersichlich Zusammengestellt. *Petermanns Geographische Mitteilungen*, **12**: 45–53.

Hare, F.K., 1954. The boreal conifer zone. *Geographical Studies,* **1**(1): 4–18.

Haxeltine, A., and Prentice, I.C., 1996. BIOME3: an equilibrium terrestrial biosphere model based on ecophysiological constraints, resource availability and competition among plant functional types. *Global Biogeochemical Cycles*, **10**(4): 693–709.

Higgins, J.J., 1952. Instructions for making phenological observations of garden peas. *Publications in Climatology*, **5**(2): 1–8.

Hunt, E.R., and Running, S.W., 1992. Effects of climate and lifeform on dry matter yield from simulations using BIOME-BGC. *Proceedings of the Internatonal Geoscience and Remote Sensing Seminar*, vol. 2, pp. 1631–1633.

Intergovernmental Panel on Climate Change (IPCC), 1997. *IPCC Special Report on the Regional Impacts of Climate Change: an assessment of vulnerability* (Watson, R.T., Zinyowera, M.C., and Moss, R.H., eds.). Cambridge: Cambridge University Press.

Kittel, T.G.F., Rosenbloom, N.A., Painter, T.H., Schimel, D.S., and VEMAP Modeling Participants, 1995. The VEMAP integrated database for modeling United States ecosystem/vegetation sensitivity to climate change. *Journal of Biogeography*, **22**: 857–862.

Köppen, W., 1900. Versuch einer Klassification der Klimate, vorzsugsweise nach ihren Beziehungen zur Pflanzenwelt. *Geographraphische Zeitschrift*, **6**: 593–611, 657–679.

Kühler, A.W., 1964. *Potential Natural Vegetation of the Conterminous United States,* American Geographical Society, Special Publication No. 36.

Major, J., 1963. A climatic index to vascular plant activity. *Ecology*, **44**(3): 485–498.

Mather, J.R., 1978. *The Climatic Water Budget in Environmental Analysis*. Lexington, MA: Lexington Books.

Mather, J.R., and Yoshioka, G.A., 1966. The role of climate in the distribution of vegetation. *Publications in Climatology*, **19**(4): 305–395.

Mather, J.R., and Yoshioka, G.A., 1968. The role of climate in the distribution of vegetation. *Annals of the Association of American Geographers*, **58**(1): 29–41.

Matthews, E., 1983. Global vegetation and land use: new high-resolution data bases for climate studies. *Journal of Climate and Applied Meteorology*, **22**: 474–487.

Monteith, J.L., 1981. Coupling of plants to the atmosphere. In Grace, J., Ford, E.D., and Jarvis, P.G., eds., *Plants and Their Atmospheric Environment*. London: Blackwell Scientific Publications, pp. 1–29.

Muller, R.A., 1966. The effects of reforestation on water yield. *Publications in Climatology*, **19**(3): 251–304.

Neilson, R.P., 1995. A model for predicting continental scale vegetation distribution and water balance. *Applied Ecology*, **5**: 362–385.

Parsons, J.J., 1955. The Miskito pine savanna of Nicaragua and Honduras. *Annals of the Association of American Geographers,* **45**: 36–63.

Parton, W.J., Schimel, D.S., Cole, C.V., and Ojima, D.S., 1987. Analysis of factors controlling soil organic matter levels in Great Plains grasslands. *Soil Sciences Society of American Journal*, **51**: 1173–1179.

Peters, C.M., Gentry, A.H., and Mendelsohn, R.O., 1989. Valuation of an Amazonian Rainforest. *Nature*, **339**: 655–656.

Schimper, A.F.W., 1898. *Pflanzen-geographie auf Physiologischer Grundlage*. Jena.

Sellers, P.J., and Lockwood, J.G., 1981. A numerical simulation of the effects of changing vegetation type on surface hydroclimatology. *Climate Change*, **3**: 121–136.

Shantz, H.L. and Zon, R., 1924. Natural Vegetation. *Atlas of American Agriculture,* Part I, Sect. E. Washington. Government Printing Office.

Taylor, S.E., 1975. Optimal leaf form. In Gates, D.M., and Schmerl, R.B., eds., *Perspectives in Biophysical Ecology.* New York, NY: Springer, pp. 73–86.

Thornthwaite, C.W., 1952. Grassland climates. *Publications in Climatology*, **5**(6): 1–14.

Thornthwaite, C.W., and Hare, F.K., 1955. Climatic classification in forestry. *Unasylva*, **9**(12): 50–59.

Thornthwaite, C.W., and Mather, J.R., 1954. Climate in relation to crops. *Meteorological Monographs*, **2**(8): 1–10.

United Nations Educational, Scientific and Cultural Organization (UNESCO), 1973. *International Classification and Mapping of Vegetation*, Series 6: *Ecology and Conservation*. Paris: United Nations Educational, Scientific, and Cultural Organization.

VEMAP Members (J.M. Melillo, J. Borchers, J. Chaney, H. Fisher, S. Fox, A. Haxeltine, A. Janetos, D.W. Kicklighter, T.G.F. Kittel, A.D. McGuire, R. McKeown, R. Neilson, R. Nemani, D.S. Ojima, T. Painter, Y. Pan, W.J. Parton, L. Pierce, L. Pitelka, C. Prentice, B. Rizzo, N.A. Rosenbloom, S. Running, D.S. Schimel, S. Sitch, T. Smith, I. Woodward), 1995. Vegetation/Ecosystem Modeling and Analysis Project (VEMAP): Comparing biogeography and biogeochemistry models in a continental-scale study of terrestrial ecosystem responses to climate change and CO_2 doubling. *Global Biogeochemical Cycles*, **9**(4): 407–437.

Willmott, C.J., and Feddema, J.J., 1992. A More Rational Climatic Moisture Index. *Professional Geographer*, **44**(1).

Cross-references

Agroclimatology
Applied Climatology
Arid Climates
Bioclimatology
Climate Classification
Humid Climates
Tree-Ring Analysis

VOLCANIC ERUPTIONS AND THEIR IMPACT ON THE EARTH'S CLIMATE

"The bright Sun was extinguish'd, and the stars did wander darkling in the eternal space. . . ." So begins Byron's lament *Darkness*, inspired apparently by the gloom of 1816, "the year without a summer". The missing summer is linked to effects of the most explosive eruption in modern history and perhaps the last 10 000 years, that of Tambora on the Indonesian island of Sumbawa. Widespread famine in Europe and China, southward and westward migration of New England farmers are commonly quoted anthropological effects, while proposed cultural impacts range from Byron's poetry to influences on J.M.W. Turner's paintings, and even Mary Shelley's "Frankenstein".

That volcanic eruptions can affect the Earth's climate has been appreciated for some time. On a heuristic and phenomenological level this is easy to understand – powerful volcanic eruptions eject ash and gases into the atmosphere, and the resulting cloud or plume intercepts sunlight and causes changes in weather. Cooling is most often associated with volcanic eruptions. However, in detail, the response of the climate system to volcanic eruptions is quite complex.

Sulfur-rich aerosols are the primary culprit in the effect of volcanism on climate, and these set up complex radiative responses in the climate system that involve stratospheric heating, surface cooling, and loss of ozone. This review examines the origin and evolution of our ideas on volcanic effects on climate, discusses these effects in the context of eruption variables such as size of eruption and composition of the magma, and then traces the impact of volcanic input into the atmosphere with an emphasis on mechanisms. Table V2 gives some definitions of terminology.

Proposing the connection

The earliest known appreciation of the volcanic effect on weather is generally accepted to be Plutarch's writings from about AD 100. The descriptions relay events in 44 BC, the year of Julius Caesar's brutal assassination, and notes a dim sun that was "pale and without radiance" that resulted in a cold summer "owing to the feebleness of the warmth" that was responsible for shriveled crops and famine from Rome to Egypt. While the association is oblique, most workers believe that Plutarch was describing the effect of an eruption from Mount Etna in Sicily that year. A celebrated early account of the volcanic eruption–climate association is that of the great American intellectual and diplomat Benjamin Franklin, who opined that the "dry fog" and unusually cold summer of 1783 and cold winter of 1783–1784 may have been the result of "great burning balls or globes" (meteorites) or "the vast quantity of smoke, long continuing to issue" from volcanoes in Iceland. It is now known that these effects were associated with the 1783 Lakagigar or Laki fissure eruption emanating from the Grimsvotn volcano. Severe impacts of volcanic eruptions on global climate are recognized from two major eruptions in Indonesia in the 1800s. The cold summer of 1816 is part of an average global temperature drop of 0.4–0.7°C generally attributed to the Tambora eruption. On the back of the rapid spread of global communications in the late 1800s, remarkably insightful and prescient Dutch Government and Royal Society of London studies of the effects of the 1883 eruption of Krakatau produced the first widespread reports of strange atmospheric effects linked to volcanoes. This prompted others such as W.J. Humphreys in 1913 to suggest a cause and effect between volcanoes and climate. The examples above are but a few of the eruptions recorded in human history that have lent valuable insight into our understanding of the volcanic impact on climate. Current research is revealing other past eruptions that are associated with major global climatic perturbation. One such eruption, the 1600 eruption of Huaynaputina, in southern Peru, is now recognized as one of the most important in recent history.

Teasing out the volcanic signal

While it may be known that an eruption occurred at a certain time and place, the impact of that eruption on climate is quite hard to pin down against the background of meteorological variability, an issue of signal to noise. The utility of historic weather data with spotty global coverage for synoptic analysis notwithstanding, even well-known eruptions have confusing accounts. For instance, while many records suggest that the summer of 1783 was abnormally cold, some records indicate that it was exceptionally hot, leading to an early harvest. The awful 1816 summer attributed to the eruption of Tambora happened within the framework of the Little Maunder Minimum – a global temperature decline that appears to have started before the eruption. These examples illustrate that the context of volcanic eruptions within natural variability needs to be better understood.

The importance of written records in helping identify the volcanic signal and assessing the impact of eruptions in historic times is clearly demonstrated above. However, a complete picture requires a multidisciplinary approach involving the detective work of gleaning the volcanic signal contained in the deposits of the eruption and a variety of other proxy records such as ice cores and tree rings. Some of the more common proxies are described in Table V3. To help produce a coherent database of volcanic effects these data are compiled into various indices such as Volcanic Explosivity Index (VEI), the Dust Veil Index (DVI), and the Ice Core Volcanic Index (IVI). Alan Robock and co-workers have assessed the veracity and utility of these indices. They note that none of them is perfect. Any perfect index would have to include the radiative forcing associated with each eruption and this is most directly a function of the sulfur content of the stratospheric aerosol. This is difficult to measure even for recent large eruptions, and is therefore a major drawback for historic eruptions; the problems increase the further back one looks.

Despite these difficulties it is generally accepted that a statistically significant average temperature decrease of 0.2–0.5°C for 1–3 years after an eruption can be expected (Figure V7). It is generally agreed now that the key to the severity of the climate

Table V2 Glossary

Aerosol – A suspension of fine liquid or solid particles in a gas (air).
Albedo – A measure of surface or atmospheric reflectivity. The fraction of incident radiation reflected back.
Caldera volcanoes – Depression marking the sites of volcanic eruption that have collapsed due to the removal of magma, a decrease in magma pressure, or due to weakness of the rocks. Smallest are typically a few kilometers, while the largest can be ~100 km in size.
Dust Veil Index (DVI) – A quantitative measure of the amount of fine ash in the atmosphere after a volcanic eruption. Developed by H.H. Lamb it is based on historic reports, estimates of volume of ejecta, temperature data, optical phenomena and, after 1883, radiation measurements.
Effusive eruption – Passive, largely non-explosive, eruptions of lava, fire fountains, and small explosions resulting from gas release from relatively fluid magma with low silica content such as basalt (~50% SiO_2).
El Nino–Southern Oscillation (ENSO) – A climatic cycle consisting of quasi-periodic warming and cooling phases, El Niño and La Niña respectively, driven by interactions between the atmosphere and ocean in the tropical Pacific.
Eruption intensity – A measure of the mass eruption rate which controls column height, and therefore potential stratospheric injection of an eruption. High-intensity eruptions are more likely to penetrate well into the stratosphere.
Eruption magnitude – A measure of the mass (kg) of material ejected during an eruption.
Explosive eruption – Violent eruptions that involve explosive disruption of magma driven by the escape of gas at high pressure. These typically involve relatively viscous magma with high silica content such as andesite (~58–63% SiO_2) to rhyolite (>70% SiO_2). They commonly produce eruption columns of ash, gas, and pulverized rock (pyroclasts) that penetrate into stratosphere.
Flood basalt – Effusive eruptions resulting in sheet-like regionally extensive basalt lava. Often form thick plateaux such as the Columbia River (Washington/Oregon, USA) or Deccan Traps (India). Commonly associated with the break-up of continents or tectonic plates, these are the largest volcanic phenomena and can result in over a million cubic kilometers of lava being erupted.
Heterogeneous chemical reactions – Reactions that occur between reactants of different phases, i.e. species in gas with those in liquid or solid phases in an aerosol.
Ice Core Volcanic Index (IVI) – Index of volcanic aerosol loading developed by Robock and Free based on ice-core records of acidity or sulfate.
Magma – Molten silicate from the interior of the Earth discharged during a volcanic eruption. Consists of liquid silicate, crystals, and gas.
Microphysics – Processes affecting the size, number, and distribution of aerosol particles. Microphysical processes include nucleation, condensation, evaporation, coagulation, and sedimentation.
Nucleation – Process by which gas or liquid phase species form new liquid or solid particles. Heterogeneous nucleation involves the presence of a pre-existing nucleus onto which molecules can aggregate. Homogeneous nucleation occurs spontaneously. Stratospheric sulfate aerosols are thought to form from binary homogeneous nucleation of water vapor and sulfuric acid vapor.
Optical depth – Measure of the cumulative extinction of electromagnetic radiation in transit through the atmosphere.
Plinian eruption – Most explosive, widely distributed, and highest eruption column type of pyroclastic eruption.
Proxy record of volcanic activity – Written record, natural phenomenon or feature that provides some indication of a volcanic event.
Pyroclastic eruptions – Eruptions dominated by explosive production of ash, gas, and rock fragments resulting from violent fragmentation of magma. See explosive eruption, above.
Radiative effects – Pertaining to the net effect of absorption, reflection, and transmission on atmospheric dynamics.
Residence time – Longevity of particles in the atmosphere before they are removed through sedimentation or physicochemical reactions.
Sedimentation – Process of removal of particles (ash, aerosol) due to gravity.
Sulfate – The SO_4^{2-} molecule (oxidized sulfur) found in solution with liquid water; dissolved form of sulfuric acid.
Teleconnection – Broadly (globally) correlated pattern that is not necessarily continuous.
Volcanic Explosivity Index (VEI) – Classification scheme describing the size (deposit mass or volume) of a volcanic eruption.
Volcanic plumes – Eruptive mixtures of volcanic particles, gases, and air that form vertically rising clouds associated with volcanic eruptions. Plumes may spread laterally as gravity currents at neutral buoyancy levels.

Table V3 Common proxy records for volcanic eruptions

Ice cores	Volcanic aerosols deposited on permanent ice become part of the seasonal record of ice layers. Drill cores through key ice accumulations reveal acidity profiles that record volcanic deposition with accuracies of 1–3 years going back at least 10 000 years. Identifying the source volcano for a given acidity peak is a complex process requiring an understanding of eruption style, magnitude, location, and magma composition.
Dust layers	Allied to acidity profiles, some permanent ice caps can record periods of volcanic "dust" deposition as well. This typically indicates close proximity to the volcano and with knowledge of the ash compositions from various volcanoes a match can be made. An example is the Quelcaya Ice Cap in Peru, where ash from the 1600 Huaynaputina eruption is recorded.
Tree rings	Annual growth rings in trees in temperate regions provide a highly accurate record of time. Coupled to variations in the density of rings or evidence of frost damage within rings, this record provides a faithful record of eruptions large enough to perturb climate so that growing conditions are altered.
Written records	Although somewhat subjective, scholarly, historical, and religious records, as well as works of art and literature, can provide valuable insight into poorly documented eruptions in the recent past or broader impacts of "known" eruptions. Examples include Etna, Italy in 44 BC, Huaynaputina, Peru in 1600, and the "mystery cloud" of 536 AD.
Atmospheric phenomena *Coloured and dim suns and moons* *Large sunspots, coronal appearances* *Hazy and pale sky (no clouds)* *Vivid sunsets, twilight effects*	Resulting from the scattering, transmission, and absorption of the sun's radiation by volcanic dust palls and aerosols, these phenomena, often seen after major eruptions, led to the development of the Dust Veil Index (DVI) by Lamb, 1970.

effect caused by an eruption is the magnitude of the eruption, the sulfur content of the magma, and the amount of sulfur released into the stratosphere as an aerosol. A classification of eruptions based on the magnitude of an eruption, the Volcanic Explosivity Index (VEI) is presented in Table V4, while key climate-impacting parameters of important volcanic eruptions are given in Table V5.

The scale of volcanic eruptions and their effects

The foregoing should impress on the reader that the climatic effects of volcanic eruptions can be quite perturbing and severe. However, the eruptions highlighted above are puny compared to those recorded in Earth history. Some of the largest eruptions in the geological record form large caldera volcanoes such as Toba in Sumatra, and Flood Basalt plateaux such as those of the Columbia River, in Oregon and Washington, and the Deccan Traps, in Mahabaleshwar in India. These have a VEI of 8 – the highest. The 74 ka eruption of Toba, erupted almost 40–50 times as much magma as Tambora in 1815, and individual eruptions such as the Roza flow in the Columbia River plateau are at least a couple of orders of magnitude greater than the Laki eruption of 1783. It should be no surprise then that authors such as Rampino, Self, and Stothers have drawn an analogy with the "nuclear winter" scenario and offer the theory that "volcanic winters" might be a likely consequence of the largest eruptions. This concept is rooted in the effect of volcanic aerosols on the opacity of the atmosphere and reduction of sunlight. The effect appears to increase with eruption magnitude as shown in Figure V8. Based on this, the coincidence of the Toba eruption with the end of the Brørup interstadial and the Wisconsinan glacial advance has been suggested as a cause-and-effect relationship by Rampino and his colleagues. Presciently, Humphreys had already speculated in his 1913 paper the possibility that sustained volcanic activity might trigger Ice Ages. Developing this further, Stanley Ambrose has proposed that the Toba volcanic winter brought widespread famine and death to human populations around the world. This abrupt "bottleneck" in population is thought to have favored evolutionary changes, which occur much faster in small populations, to produce rapid population differentiation of modern human races only 70 000 years ago.

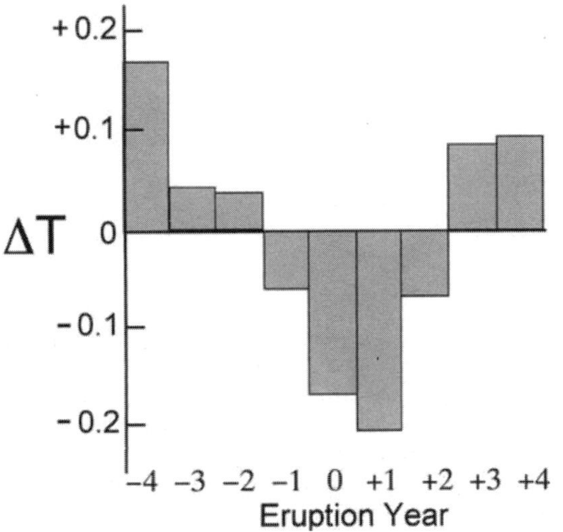

Figure V7 Northern hemisphere temperature variation averaged over several major eruptions. Eruption years are: 1815, 1875, 1883, 1902, 1947, 1956, 1963. Data suggest a 0.2°C drop in the year following the eruption (after Self et al., 1981).

In the early 1980s several studies posited a primary cause-and-effect relation between the eruption of the Deccan Trap flood basalts and the mass extinctions at the Cretaceous–Tertiary (K–T) boundary 65 million years ago. While this is now eschewed by the mainstream community in favor of the more supportable impact hypothesis, the temporal coincidence and the significant effect suggested by the results of scaling calculations of eruptions such as the 1783 Laki eruption to those of the Deccan scale beg the question of what the exact relationship is.

The effect of eruption size is not straightforward, however, and, as explained below, the size of the eruption, or its magnitude, may be less important than the sulfate aerosol production and intensity of the eruption.

Table V4 Classification of volcanic eruptions in the Volcanic Explosivity Index (VEI) – a measure of the magnitude of eruptions. Modified after Newhall and Self, 1982

VEI	Description	Plume height	Volume range	Classification	Period	Example
0	Non-explosive	< 100 m	1000 m^3	Hawaiian	Days	Kilauea, Hawaii
1	Gentle	100–1000 m	10 000 m^3	Hawaiian/Strombolian	Days	Stromboli, Italy
2	Explosive	1–5 km	1 000 000 m^3	Strombolian/Vulcanian	Weeks	Galeras, Colombia, 1992
3	Severe	3–15 km	10 000 000 m^3	Vulcanian	Years	Ruiz, Colombia, 1985
4	Cataclysmic	10–25 km	0.1 km^3	Sub-Plinian	Decades	Galunggung, 1982 Lascar, Chile, 1993
5	Paroxysmal	> 25 km	1 km^3	Sub Plinian/Plinian	Centuries	Mt St Helens, USA, 1980
6	Colossal	>25 km	10 km^3	Plinian	Centuries — Millenia	Krakatau, Indonesia, 1883 Huaynaputina, Peru, 1600 Pinatubo, Philippines, 1991
7	Super-colossal	> 25 km	100 km^3		Millenia	Tambora, Indonesia, 1815
8	Mega-colossal	> 25 km	1000 km^3	Ultra-Plinian	Megallenia	Toba, Indonesia, 74 ka Yellowstone, USA, 2 Ma

Note: Effusive eruptions are poorly represented in this scheme. For instance, flood basalt eruptions would be VEI 0–1 in terms of description, but a VEI 8 in terms of volume.

Table V5 Magnitudes, plume heights, stratospheric aerosol loadings, and climate impact of some key eruptions

Eruption latitude, year	VEI	Magma volume (km^3)	Plume height (km)	H_2SO_4 aerosol (kg)	Northern hemisphere temperature decrease (°C)
Explosive eruptions					
Toba[a] 2°S, 73.5 ka	8	>1000	~40?	10^{12-13}	3–5
Tambora 8°S 1815	7	>50	>40	2×10^{11}	0.4–0.7
Krakatau 6°S 1883	6	>10	>40	5×10^{10}	0.3
Huaynaputina[b] 16°S 1600	6	>9	33–42	7×10^{10}	0.8
St Helens 46°N 1980	5	0.35	22	3×10^{8}	0–0.1
Effusive eruptions					
Rosa Flow[c] Columbia River Plateau 46°N, 14.7 Ma	8?	1400	13	10^{13}	5–15
Laki 64°N 1783	4	14	13	10^{11}	~1.0

Data sources are all Rampino and Self, 1984 except for [a] Rampino and Ambrose, 2000; [b] de Silva and Zielinski, 1998; [c] Thordarson and Self, 1996.

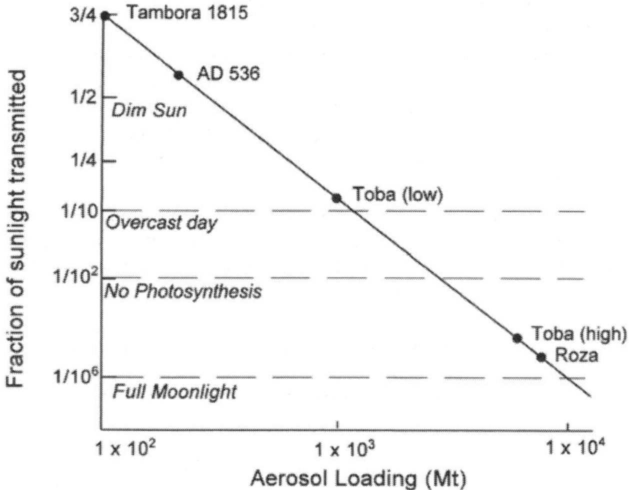

Figure V8 Volcanic aerosols commonly increase the opacity of the atmosphere and this can be quantified by the reduction in sunlight transmitted. Show here are estimates from major historic and pre-historic eruptions. All these eruptions are mentioned in the text (from Rampino et al., 1988).

Cause, effect, and mechanisms

The complex set of interactions summarized in Figure V9 constitute the volcanic impact on the Earth's climate. Volcanic eruptions inject finely comminuted magma as ash and gases into the atmosphere. Ash consists of frozen magma shards and fragments, pulverized rock fragments from the conduit of the volcano, and crystals. Volcanic gases are dominated by water vapor (~80%), and carbon dioxide (~10%), and the rest is made up of N_2, SO_2, H_2S, CO, H_2, HCl, and HBr. It is volcanic aerosols, formed by sulfur species injected into the stratosphere, rather than the ash, that are the main agent of climate effects. These stratospheric aerosols affect the global radiation by absorbing and back-scattering incoming solar radiation; the former leads to stratospheric heating, and the latter to surface cooling. Stratospheric heating results in a perturbation of the atmospheric dynamics by affecting both vertical and meridional circulation. The volcanic aerosols may also catalyze a reduction of stratospheric ozone through heterogeneous chemical reactions. This may be exacerbated by the aforementioned dynamic processes, and so lead to reduced radiative heating of the lower stratosphere.

As ash, water vapor, sulfur dioxide (SO_2), HCl and other gaseous species are injected into the atmosphere, the ash falls out of the atmosphere very rapidly – in a matter of minutes to weeks. Ash therefore has little global climate impact beyond local cooling that is often manifested in reduced amplitude of the diurnal cycle of surface temperature variations. Moreover, water vapor, and soluble gaseous components such as HCl, condense as temperatures drop in the upper troposphere and these components often rain out before they enter the stratosphere. Thus tropospheric particles are removed from the atmosphere in several weeks by frequent precipitation. The intensity of an eruption is therefore another important factor in the impact on the global climate. Since most volcanic eruptions are effusive, they have little chance to impact the global atmosphere as their eruption plumes do not penetrate the tropopause.

Predominantly effusive eruptions involving relatively low-viscosity basalt magma such as the 1783 Laki eruption, and those that occur on the islands of Hawaii, do not in general have global impacts, even though these magmas typically have a higher concentration of sulfur. Therefore it is the very

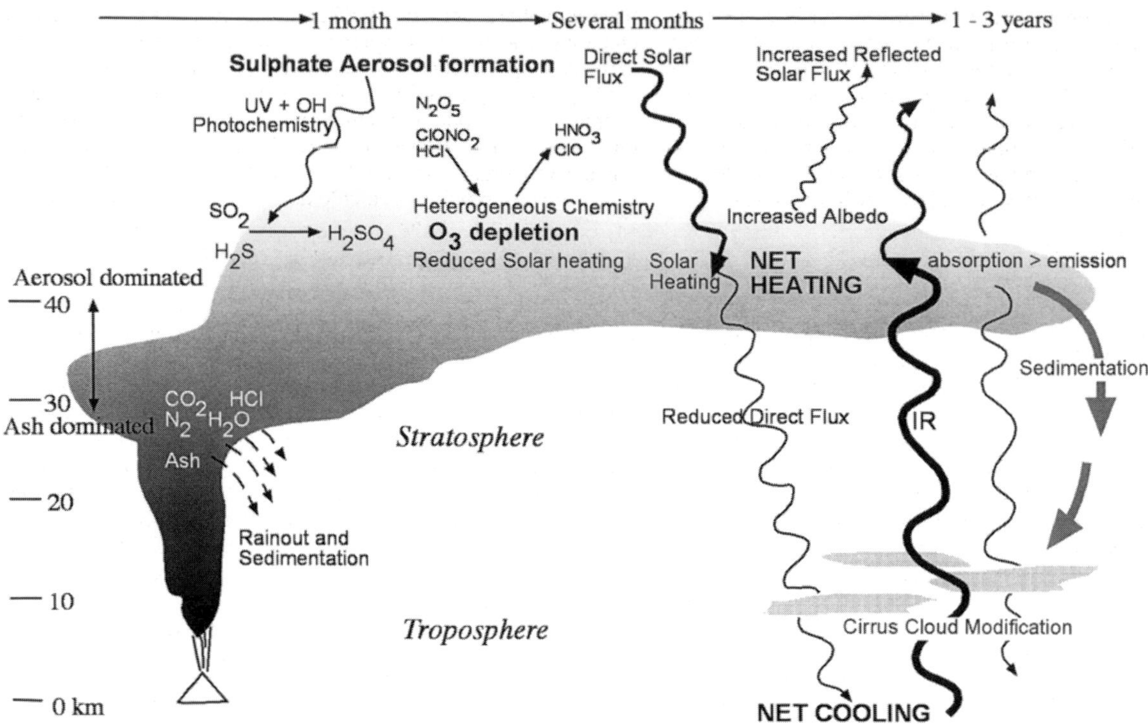

Figure V9 Evolution of the atmospheric input from a major plinian eruption (VEI 6) and the key processes that lead to climatic impact. Ash and gases are injected directly into the stratosphere but most solid and gaseous species are washed out rapidly (days to weeks). Sulfur gases have a much longer life in the atmosphere and may form an aerosol within a month or so. Over several months scattering of solar radiation may lead to cooling of the surface. In the stratosphere, absorption of the direct solar flux and infrared radiation escaping from the surface and troposphere results in stratospheric heating despite the reduction in solar heating resulting from ozone. These effects may persist for 1–3 years, after which growth and coagulation lead to sedimentation and recirculation of the aerosol into the troposphere. Here it may seed cloud formation and increase planetary albedo. Further details are in the text (adapted and modified from similar drawings by Richard Turco, Alan Robock, and others).

explosive, plinian-style pyroclastic eruptions, e.g. those at Mount St Helens, Washington, USA, in 1980; El Chichó, Mexico, in 1982; and Mount Pinatubo, Philippines, in 1991, that penetrate the tropopause and inject sulfur-rich aerosols into the stratosphere. These eruptions involve more silica-rich magmas such as andesite or rhyolite that are more viscous and volatile-rich, and hence more explosive. However, they are relatively sulfur-poor compared to basalts. This demonstrates the importance of stratospheric input, and thus magnitude of the eruption, for climate effects. Since the stratosphere is drier than the troposphere, particles are not rained out, so ash that enters the stratosphere may remain in the stratosphere for several months before sedimenting out, while aerosols, depending on latitude and altitude of the initial injection, are known to have atmospheric residence times of 1–2 years. During this time the aerosols are picked up by meridional circulation in the tropics, transported poleward in the midlatitudes, and then brought back to the troposphere at higher latitudes. Volcanic aerosols are rapidly advected around the globe in a matter of weeks, and thus circle the globe many times during their lifetime. They become more dispersed with time.

Once injected into the stratosphere the sulfur species, mainly SO_2, oxidizes within about a month to produce H_2SO_4. This combines with water vapor to give stratospheric aerosols. The aerosols act to cool the surface by reflection and/or warm the stratosphere by absorbing the direct solar flux or infrared being emitted from the surface and escaping through the troposphere. Whether cooling or warming dominates depends on particle size of the aerosol; for warming to dominate, particles must be greater than 2 μm. However, these particles would sediment out in a matter of months, and it seems unlikely that significant periods of surface warming can result from volcanic aerosols.

Particle size also affects the optical depth of aerosols, and this in turn affects the radiative response to volcanic input. Smaller particles provide greater aerosol optical depth per unit mass and thus a greater relative effect than larger particles. These size-related properties moderate the radiative and climate influences of volcanic eruptions, and this leads to the counterintuitive assertion that the influence of a volcanic eruption on global climate may be limited as the size of a volcanic eruption increases. It turns out that larger eruptions may not necessarily produce proportionally larger temperature anomalies; the greater input of volcanic SO_2 from a larger eruption leads to larger particles, rather than greater numbers, and these not only sediment out faster, but have smaller optical depth per unit mass than smaller particles.

Impact of a tropical eruption

The latitude of eruption is another important factor. Tropical eruptions have a more pronounced impact on global climate because of atmospheric circulation and the likely spread of aerosol across hemisphere. Alan Robock and his colleagues have modeled the effects of a tropical eruption and show that the stratospheric heating is larger in the tropics than in the higher latitudes. This leads to an enhanced pole-to-equator temperature gradient that is naturally more pronounced in the winter. In the northern hemisphere winter, horizontal advection may result in winter warming as the jet stream (polar vortex) is made stronger and amplifies the North Atlantic Oscillation circulation pattern.

Two years after the 1991 Pinatubo eruption, stratospheric temperatures were found to have changed from higher than average to new lows. These observations are correlated with an observed decrease in stratospheric ozone following the eruption. It is probable that the loss of ozone resulted in less ultraviolet absorption and thus reduced the heating of the lower stratosphere. As far as we know, volcanic eruptions catalyze the loss of ozone in two ways: through chemical reactions and through changes in atmospheric dynamics. Ozone-destroying Cl and Br in volcanic aerosols are rained out before reaching the stratosphere. So it is not the introduction of these species that causes ozone depletion. Instead volcanic aerosols alter stratospheric chemistry in a way that activates stratospheric Cl and Br, from anthropogenic sources. These reactions, known as heterogeneous reactions, are recognized by workers such as Susan Solomon and others as a crucial part of the aerosol–ozone connection. The reactions involve atmospheric gases and liquid and solid aerosols. They are inversely temperature-dependent, and thus lead to enhanced stratospheric ozone depletion at mid-latitudes and polar regions. Ozone depletion can also result from dynamic uplifting from convection produced by the heating effect of volcanic aerosols. In the tropics this enhanced circulation can lead to ozone "depletion" by simply pushing ozone out to higher latitudes. Michael Mills notes that the 6–8% decrease of the total equatorial ozone column after the Mount Pinatubo eruption was substantially greater than tropical ozone depletion caused by anthropogenic sources, although this effect lasted only about a year. These competing effects notwithstanding, the net effect of aerosols in the stratosphere is stratospheric heating.

Aerosols are eventually transported to the troposphere through sedimentation and atmospheric circulation and are taken up by clouds or rain. This may lead to increased planetary albedo as the incorporation of aerosols results in a decrease in the average size of droplets in cirrus clouds, which may contribute to tropospheric and surface cooling. Significant tropospheric cooling on a global scale has been identified after recent eruptions of Agung in 1964, El Chichon in 1982, and Pinatubo in 1991 after removal of natural variations in temperature in the troposphere caused by phenomena such as the ENSO and the recent global warming trend. In the case of Agung and Pinatubo, cooling of the troposphere was of the order of 0.4°C. Mills notes that "Pinatubo more than offset the climate forcing of greenhouse gases added to the atmosphere since preindustrial times, at least for a short while." For El Chichon the effect was swamped by the 1982–1983 El Niño. Krakatau, 1883, and Tambora, 1815, may have cooled the atmosphere by about 0.3°C and 0.4–0.7°C respectively.

Teleconnections and broader Earth system impacts

More abstract volcano–climate connections have been posited over the years. One intriguing connection is that the El Niño–Southern Oscillation cycle (ENSO) may be enhanced by stratospheric aerosols. ENSOs are known to be a response to changing sea-surface temperatures. Since stratospheric volcanic aerosols can influence sea-surface temperatures through surface cooling, an impact of a tropical explosive eruption may be a change in tropospheric circulation and sea-surface temperatures that leads to an ENSO event. Paul Handler originally suggested this but was challenged on the basis of weakness in his statistical analysis showing coincidence of volcanic eruptions and ENSOs, relative timing of the two phenomena, and the weakness in the physical model. A viable physical model was developed by M. Hirono through an analysis of the impact of aerosols from the 1982 eruption of El Chichon. However, Alan Robock effectively dismissed the notion that volcanic eruptions can produce El Niños, and argued that the contemporaneity of volcanic eruptions such as El Chichon 1982 and powerful El Ninos is a coincidence. The debate has come full circle as, recently, Adams and colleagues have presented a much stronger statistical case that over the past 400 years the ENSO phenomenon has been influenced by the after-effects of explosive volcanic eruptions. Volcano–climate interactions are clearly an important part of the systemic view of the Earth, and understanding their impact will lead to increased awareness of the teleconnections that abound across the various "spheres" of the Earth system.

We end the discussion of volcanic effect on climate with a broader perspective. Earth-system impacts of superexplosive eruptions or prolonged episodes of volcanism have been suggested as perturbations or forcings throughout Earth history. Numerous periods of anomalous climate from the geological record, e.g. "greenhouse" periods such as the Mesozoic Warm Period, and even the K-T Strangelove ocean, have been associated with anomalous volcanism. While these are contentious issues and are actively debated they nevertheless are part of the volcanism and climate story. In ending this review it is important to remind the reader that volcanism is a primary source of planetary atmospheres, supplying the gases that accumulated as the atmosphere or condensed into a hydrosphere after the primordial atmosphere was lost to space. Addressing these any further is beyond the scope of this article, and for more detailed treatments of these issues the reader is referred to the *Origin and Evolution of Planetary and Satellite Atmospheres* edited by Atreya et al. (University of Arizona Press, 1989) and the works of Kastings, Schubert, and others therein.

Acknowledgments

The author is grateful to Mike Rampino for his review and to John Oliver for editorial input. This article has been improved in both style and substance as a result. Any errors remain the responsibility of the author.

Further reading

Adams, J.B., Mann, M.E., and Caspar, M.A., 2003. Proxy evidence for an El Niño-like response to volcanic forcing. *Nature*, **426**: 274–278.

Alvarez, W., and Asaro, F., 1990. An extraterrestrial impact *and* Courtillot, V.E, 1990. A volcanic eruption. *Scientific American*, **256**: 44–60.

Ambrose, S.H., 1998. Late Pleistocene human population bottlenecks, volcanic winter, and the differentiation of modern humans. *Journal of Human Evolution*, **34**: 623–651.

Andres, R.J., and Kasgnoc, A.D., 1998. A time-averaged inventory of subaerial volcanic sulfur emissions. *Journal of Geophysical Research*, **103**: 25251–25261.

Angell, J.K., 1990. Variation in global tropospheric temperature after adjustment for the El Niño influence. *Geophysics Research Letters*, **17**(8): 1093–1096.

Carey, S., and Bursik, M., 2000. Volcanic plumes. In Sigurdsson et al., eds, *Encyclopedia of Volcanoes*. pp. 527–544.
de Silva, S.L., and Zielinski, G., 1998. The global impact of the 1600 eruption of Huaynaputina, Southern Peru. *Nature*, **393**: 455–458.
Francis, P.W., 1990. The golden glow of volcanic winter. In *Volcanoes: A Planetary Perspective*. Oxford: Oxford University Press, pp. 368–394.
Hamill, P., Jensen, E.J., Russell, P.B., and Baumann, J.J., 1997. The life-cycle of stratospheric aerosol properties. *Bulletin of the American Meteorological Society*, **78**(7): 1–16.
Handler, P., 1989. The effect of volcanic aerosols on global climate. *Journal of Volcanology and Geothermal Research*, **37**: 233–249.
Hirono, M., 1988. On the trigger of El Niño–Southern Oscillation by the forcing of early El Chichó volcanic aerosols, *Journal of Geophysical Research*, **93**: 5364–5384.
Hooper, P., 2000. Flood basalt provinces. In Sigurdsson et al., eds, *Encyclopedia of Volcanoes*. pp. 345–360.
Humphreys, W.J., 1913. Volcanic dust and other factors in the production of climate changes, and their possible relation to ice ages. *Bulletin of Meteorological and Weather Observations*, **6**: 1–34.
Kasting, J.F., and Toon, O.B., 1989. Climate evolution on the terrestrial planets. In: Atreya, S.L., et al., eds. *Origin and Evolution of Planetary and Satellite Atmospheres*. Tucson: University of Arizona Press, pp. 423–449.
Mass, C.F., and Portman, D.A., 1989. Major volcanic eruptions and climate: a critical evaluation. *Journal of Climate*, **2**: 566–593.
McCormick, M.P., Thomason, L.W., and Trepte, C.R., 1995. Atmospheric effects of the Mt. Pinatubo eruption. *Nature*, **373**: 399–404.
Mills, M.J., 2001. Volcanic aerosol and global atmospheric effects. In Sigurdsson et al., eds, *Encyclopedia of Volcanoes*, pp. 931–944.
Newhall, C.G., and Self, S., 1982. The Volcanic Explosivity Index (VEI): An estimate of explosive magnitude for historical volcanism. *Journal of Geophysical Research*, **87**(2): 1231–1238.
Pang, K.D., 1991. The legacies of eruption: matching traces of ancient volcanism with chronicles of cold and famine. *The Sciences*, **31**: pp. 30–35.
Rampino, M.R., and Self, S., 1984. The atmospheric effects of El Chichon. *Scientific American*, **250**: 48–57.
Rampino, M.R., and Self, S., 1992. Volcanic winter and accelerated glaciation following the Toba super-eruption, *Nature*, **359**: 50–52.
Rampino, M.R., Self, S., and Stothers, R.B., 1988. Volcanic winters. *Annual Reviews of Earth and Planetary Science*, **16**(41): 1–51.
Rampino, M.R., and Ambrose, S.H., 2000. Volcanic winter in the Garden of Eden: The Toba super-eruption and Late Pleistocene human population crash. *Geological Society of American Special Paper*, **345**: pp. 71–82.
Robock, A., 2000. Volcanic eruptions and climate. *Review of Geophysics*, **38**: 191–219.
Robock, A., and Mao, J., 1992. Winter warming from large volcanic eruptions. *Geophysics Research Letters*, **12**: 2405–2408.
Schoeberl, M.R., Bhartia, P.K., Hilsenrath, E., and Torres, O., 1993. Tropical ozone loss following the eruption of Mt. Pinatubo. *Geophysics Research Letters*, **20**: 29–32.
Schubert, G., Turcotte, D.L., Solomon, S.C., and Sleep, N., 1989. Coupled evolution of the atmospheres and interiors of planets and satellites. In Atreya, S.K., Pollack, J.B., and Matthews, M.S., editors. *Origin and Evolution of Planetary and Satellite Atmospheres*. Tucson, Arizona: University of Arizona Press, pp. 450–483.
Self, S., Rampino, M.R., and Barbera, J.J., 1981. The possible effects of large 19th and 20th century volcanic eruptions and zonal and hemispheric surface temperatures. *Journal of Volcanology and Geothermal Research*, **11**: 41–60.
Solomon, S., Portmann, R.W., Garcia, R.R., Thomason, L.W., Poole, L.R., and McCormick, M.P., 1996. The role of aerosol trends and variability in anthropogenic ozone depletion at northern mid-latitudes. *Journal of Geophysical Research*, **101**: 6713–6727.
Stommel, H., and Stommel, E., 1983. *Volcano Weather: The story of 1816, the Year without a Summer*. Newport, RI: Seven Seas Press.
Thordarson, T., and Self, S., 1996. Sulfur, chlorine and luorive degassing and atmospheric loading by the Roza eruption, Columbia River Basalt, Washington, USA. *Journal of Volcanology and Geothermal Research*, **74**: 49–73.
Thordarson, T., and Self, S., 2002. Atmospheric and environmental effects of the 1783–1784 Laki eruption: A review and reassessment. *Journal of Geophysics Research*, **108**: No. D1 10.1029/2001JD002042.
Turco, R.P., Toon, O.B., Ackerman, T.P., Pollack, J.B., and Sagan, C., 1983. Nuclear winter: global consequences of multiple nuclear explosions. *Science*, **222**(4630): 1283–1292.
Wexler, H., 1952. Volcanoes and world climate. *Scientific American*, **220**: 3–5.
Yang, F., and Schlesinger, M.E., 2002. On the surface and atmospheric temperature change following the 1991 Pinatubo volcanic eruption – a GCM study. *Journal of Geophysics Research*, **107**: 10.1029/2001JD000373.

S.L. de Silva

VORTICITY

Large-scale fluid circulations are of great significance in Earth systems as they redistribute energy, moisture and momentum over the planetary surface. Of the various motions observed in the atmosphere and hydrosphere, three-dimensional rotation, or vorticity, is commonly calculated to diagnose the behavior of weather systems and ocean circulation. Simply, vorticity is a vector description of the local rotation in fluids. Expressed as the total magnitude of change in wind components *u, v, w* along orthogonal Cartesian axes *x, y, z*, vorticity describes the complete spin experienced by a fluid parcel:

$$\nabla_2 \times V_3 = \mathbf{i}\left(\frac{\partial w}{\partial y} - \frac{\partial v}{\partial z}\right) + \mathbf{j}\left(\frac{\partial u}{\partial z} - \frac{\partial w}{\partial x}\right) + \mathbf{k}\left(\frac{\partial v}{\partial x} - \frac{\partial u}{\partial y}\right)$$

where **i**, **j**, **k** are unit vectors in *x, y, z* coordinates, *u, v, w* are wind speed components in *x, y, z* coordinates.

Vorticity is used in numerous meteorological models to describe the rotational dynamics of atmospheric phenomena (Holton, 1972; Carlson, 1991; Cushman-Roisin, 1994). For example, the generation of updraft rotation in severe thunderstorms appears to be greatly influenced by the orientation of the horizontal vorticity vector in relation to the horizontal velocity vector, i.e. streamwise and crosswise vorticity (Davies-Jones, 1984). Most regional or synoptic-scale studies of atmospheric circulation focus on the so-called vertical component of vorticity (spin about the *z*-axis) owing to the relationship between horizontal vorticity, divergence and vertical motion in the atmosphere.

Absolute and relative vorticity

The vertical component of vorticity is typically expressed as:

$$\zeta = \frac{\partial v}{\partial x} - \frac{\partial u}{\partial y}$$

Absolute vorticity is a measure of the total vertical vorticity experienced by a fluid parcel in the Earth-reference frame. Mathematically, absolute vorticity simplifies to:

$$\zeta_A = \zeta_R + f$$

the sum of ζ_R, the relative vorticity produced by wind shear and airflow through a curved path, and *f*, the Coriolis parameter, which quantifies the background rotation of the Earth. Earth vorticity, f, is a function of latitude and positive (negative) in the northern (southern) hemisphere. Thus, positive/counterclockwise (negative/clockwise) rotation is cyclonic in the northern (southern) hemisphere. Earth vorticity values range from 0 at the equator to $\pm 2\Omega$ at the poles, where Ω equals the angular rotation rate of the Earth.

Relative vorticity, ζ_R, is the sum of shear and curvature vorticity components and ranges from positive (cyclonic) to negative (anticyclonic) values; at 0, relative spin = 0. Relative vorticity is measured in relation to the ground, in contrast to absolute vorticity, where the "background" motion of the Earth's rotation is an additional portion of the total spin. The shear component of relative vorticity arises from horizontal variations in wind velocity. In the vicinity of an isotach maximum (jet streak), winds increase toward the center across transects along and normal to the current within which the jet is embedded. Air parcels on the poleward side of a jet streak experience cyclonic vorticity; conversely, parcels on the equatorward side of a jet streak occur in a zone of anticyclonic, vorticity. Thus, even in straight-line airflow, vorticity is generated by lateral shear.

The remaining source of relative vorticity derives from rotation incurred as air flows through a curved path. In the northern hemisphere, cyclonic (anticyclonic) flow trajectories exert positive (negative) torque on air parcels. In geostrophic flow patterns, therefore, curvature vorticity maxima (minima) are associated with geopotential height minima (maxima).

The sum of Earth vorticity and relative vorticity, absolute vorticity, ranges from near-zero values in regions where ζ_R and f are opposed, to large positive or negative values in regions where ζ_R and f are aligned. Absolute vorticity is usually expressed as a unitless quantity, $10^{-5}\,s^{-1}$ (length measurements cancel).

Rossby (1940) observed that air parcels flowing at a level of non-divergence (divergence terms = 0), typically near 500 mb in the troposphere, conserve their absolute vorticity through time. Earth vorticity increases (decreases) occur as air flows across parallels of latitude toward the hemispheric pole (equator). Air parcel trajectories adjust for changing values of f by curving to raise or lower relative vorticity. As continued adjustments produce oscillations of Earth and relative vorticity after an initial perturbation, conservation of absolute vorticity produces a train of standing waves, or Rossby waves.

Airflow at 500 mb gains Earth vorticity as it translates toward the hemispheric pole. In the northern hemisphere, conservation of absolute vorticity is realized through anticyclonic curvature of this airflow, which is associated with a ridge in geopotential height contours. Air flowing through the ridge returns equatorward, and begins to lose Earth vorticity. As Earth vorticity declines toward the equator, relative vorticity increases in association with cyclonic curvature in the parcel trajectory, culminating in the formation of geopotential trough. The features of this wave train are related to one another via absolute vorticity conservation in a process known as teleconnection.

Vorticity advection and vertical motion

As air flows through a pattern of vorticity (the pattern is determined by curvature and shear), vorticity is advected downstream. Because the magnitude of relative or absolute vorticity and its advecting current are large above the friction layer (in relation to near-surface values), deformations of the three-dimensional wind field caused by differential vorticity advection are most significant in the upper troposphere.

Because the rotation of a fluid and its horizontal radius are related by the conservation of angular momentum, positive vorticity advection (transfer of large vorticity toward regions of smaller vorticity), or PVA, is associated with horizontal divergence. Upper troposphere divergence attends rising motion by continuity and overlies surface convergence (a configuration known as "Dines compensation"). Similarly, upper tropospheric negative vorticity advection, or NVA, corresponds to regions of horizontal convergence, subsidence and surface divergence. Because of the co-location of geopotential minima (maxima) with vorticity maxima (minima), the region downstream of a northern hemisphere geopotential trough (ridge) is a zone of PVA (NVA). Within an isotach maximum, quadrants of PVA (NVA) are located in the right (left) entrance and left (right) exit of the jet streak.

The relationships between spin, divergence and vertical motion provide large-scale, dynamic mechanisms that drive relative humidity and static stability adjustments and deep convection. Vorticity advection patterns can provide a preliminary diagnostic assessment for the prediction of cloudcover, precipitation and severe weather.

Potential vorticity

Potential vorticity (PV) represents the balance between horizontal rotation and atmospheric thickness (or static stability). PV is defined mathematically as:

$$PV = \zeta_\theta\left(\frac{d\theta}{dp}\right)$$

where ζ_θ is the isentropic absolute vorticity, and θ is the potential temperature and p is pressure. On isentropic surfaces (surfaces of constant potential temperature), PV is conserved (Hoskins et al., 1985). Thus, following isentropic motion, static stability decreases (increases) are accompanied by absolute vorticity increases (decreases). As a result, the stratosphere, a reservoir of highly stable air, is a significant source of vorticity. A key mechanism of midlatitude cyclogenesis is the downslope flow along isentropic surfaces of stratospheric air into the upper and middle troposphere. As the stratospheric air becomes destabilized, absolute vorticity leads to positive vorticity advection, ascending motions and surface cyclone development.

The vorticity–thickness relationship defined by potential vorticity conservation also plays a role in the generation of anchor waves, perturbations of the midlatitude westerlies that generate the Rossby wave trains described above. Anchor waves are "anchored" to geographical features that induce thickness variations in the westerlies. Topographic anchor waves, such as the mountain ridge and lee trough of the Rocky Mountains of western North America, are year-round perturbations. Anchoring associated with baroclinic gradients, such as the troughs in the lee of east coast margins of Asia and North America, are strongest in winter, when cold continental surfaces are juxtaposed against warm ocean currents. The vorticity forcing of topographic and thermal anchors, in coordination with the strength of the zonal current of the westerlies, determines the number of waves and their position within the circulation. Thus, in addition to driving large-scale vertical motions, vorticity is closely related to seasonal climatology.

Greg Bierly

Bibliography

Carlson, T.N., 1991. *Mid-latitude Weather Systems*. London: Harper Collins Academic.

Cushman-Roisin, B., 1994. *Introduction to Geophysical Fluid Dynamics*. Englewood Cliffs, NJ: Prentice-Hall.

Davies-Jones, R.P., 1984. Streamwise vorticity: the origin of updraft rotation in supercell storms. *Journal of Atmospheric Science*, **41**: 2991–3006.

Holton, J.R., 1972. *An Introduction to Dynamic Meteorology*. New York: Academic Press.

Hoskins, B.J., McIntyre, M.E., and Robinson, A.W., 1985. On the use and significance of isentropic potential vorticity maps. *Quarterly Journal of the Royal Meteorological Society*, **111**: 877–946.

Rossby, C.-G., 1940. Planetary flow patterns in the atmosphere. *Quarterly Journal of the Royal Meteorological Society*, **66**: 68–87.

Cross-references

Coriolis Effect
Jet Streams
Rossby Number/Rossby Wave
Zonal Index

W

WALKER CIRCULATION

In a series of articles extending from 1923 to 1937, Sir Gilbert Walker statistically defined the concept of the Southern Oscillation. As part of the research a distinctive circulation pattern located in the equatorial Pacific Ocean was described. This subsequently has been called the Walker Circulation (Bjerknes, 1969).

The Walker Circulation is a convective cycle that owes its origin to the gradient of sea surface temperatures along the equator in the Pacific Ocean. When the cold water belt off the coast of equatorial South America is well established, the air above will be cool and will not become part of the ascending air of the Hadley Circulation. Instead, the air flows westward between the Hadley cells of the two hemispheres. Upon reaching the eastern part of the oceans, in the vicinity of 170°E, it is in contact with warmer water and becomes heated and supplied with moisture. A thermal circulation, embedded in the normal equatorial flow, results. Figure W1 provides a schematic representation of the extent and location of the Walker Circulation during a non-ENSO period.

The flow of air in the Walker Circulation is contingent on ocean temperatures and must be considered part of a larger atmospheric system. Whereas the Walker Circulation maintains an east–west exchange of air in the equatorial belt from South America to the west Pacific, the Southern Oscillation results in an exchange of mass along the complete circumference of the globe in tropical latitudes. As Bjerknes (1969) notes, the Walker Circulation is distinguished from other tropical east–west exchanges in that it taps potential energy by combining the large-scale rise of warm–moist and descent of colder dry air. Fluctuations in the Walker Circulation are therefore likely to initiate some of the major pulses of the Southern Oscillation.

John E. Oliver

Bibliography

Bjerknes, J., 1969. Atmospheric teleconnections from the equatorial Pacific. *Monthly Weather Review*, **97**(3): 163–172.

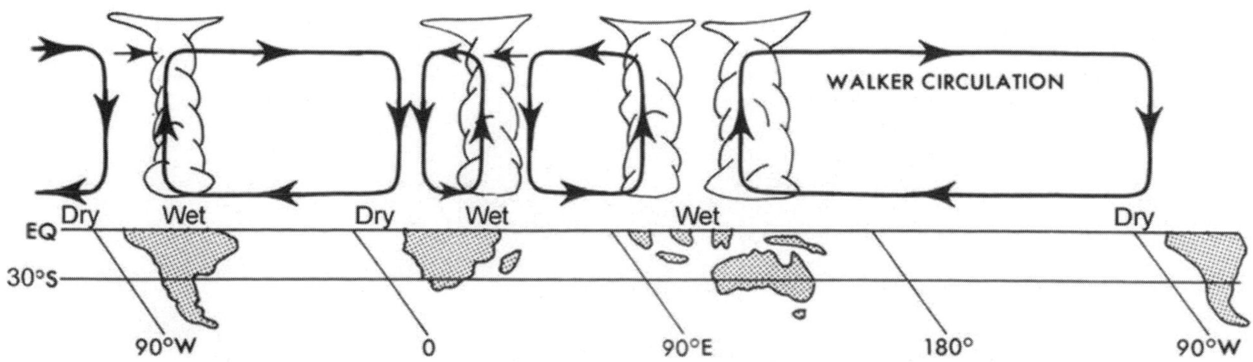

Figure W1 Schematic diagram showing the Walker Circulation along the equator in non-ENSO conditions. Ascending air results in precipitation in the areas indicated (wet). Descending air results in dry conditions in the areas shown (dry) (after Wyrtki, 1982 and Peixoto and Oort, 1992).

Lockwood, J.G., 1979. *Causes of Climate*. New York: Halstead Press.
Peixoto, J.P., and Oort, A.H., 1992. *Physcis of Climate*. New York, American Institute of Physics.
Walker, G.T., 1928. World weather, *Royal Meteorological Society Memoirs*, 2.
Walker, G.T., 1930. World weather, *Royal Meteorological Society Memoirs*, 3.
Walker, G.T., 1932. World weather, *Royal Meteorological Society Memoirs*, 4.

Cross-references

Cycles and Periodicities
El Niño
Hadley Cell
The Madden–Julian Oscillation
Ocean–Atmosphere Interaction
Southern Oscillation
Teleconnections

WATER BUDGET ANALYSIS

Water budget analyses represent an environmental systems approach to the hydrologic cycle, with emphasis on the transport, storage and utilization of water at the Earth's surface. The geographical scales of analyses range from global water budgets down to studies of the income, outgo, and storage of water from small tanks set in the soil, known as lysimeters or evapotranspirometers. Time scales range from average water budgets derived from climatic data averaged over a number of years (typically 30 or more) to continuous daily water budgets within a real-time framework. This overview of water budget methodology begins with a description of the climatic water budget, with applications organized within a time scale beginning with long-term averages and closing with daily water budgets.

Mather (1974, 1978) has published comprehensive compilations of water budget procedures, analyses, and applications; Muller (1982) has published an overview of his water budget approaches to runoff and river regimen; and Miller (1977) has published a detailed survey of interactions of water with ecosystems at the surface of the Earth.

Climatic water budget

The climatic water budget is usually developed for a place, with the data inputs based on mean monthly temperature and precipitation derived from a climatic station or array of stations. The climatic water budget was introduced into the literature by Thornthwaite and his colleagues (Thornthwaite, 1948; Thornthwaite and Mather, 1955), initially for analyses of global and regional climatic classification in terms of the interactions of energy and moisture in the various regions. Focus was directed toward the determination of humid and dry climatic realms, identification of sub-humid climatic regions neither humid nor dry, and establishment of an orderly or regular structure of climatic types based solely on climatic parameters rather than vegetation characteristics (Carter and Mather, 1966). Numerous applications for environmental monitoring and analyses were developed at about the same time, and professional interests in applications quickly surpassed the initial studies of climatic classification (Thornthwaite and Mather, 1955).

Another key feature of the climatic water budget is that it serves as an excellent instructional tool for environmental managers and decision-makers, while providing a working foundation of many climatological principles for students interested in water resources. The various components of the water budget are easy to understand, and the bookkeeping methodology provides a straightforward assessment of the distribution of moisture in the environment.

The climatic water budget for Seabrook, New Jersey (Figure W2), is used to illustrate components of the budget. Thornthwaite developed and refined many of his water budget concepts at the Laboratory of Climatology at Seabrook Farms, and this average water budget for Seabrook appeared in the initial monograph about the water budget and its applications (Thornthwaite and Mather, 1955).

Potential evapotranspiration

Potential evapotranspiration (PE), the basic building block of the water budget, was developed as a concept by Thornthwaite in the mid-1940s. Potential evapotranspiration is defined as the amount of water that would evaporate and transpire from a landscape fully covered by a homogeneous stand of vegetation without any shortage of soil moisture within the rooting zone. Another specification is that estimates of PE be determined for a large area with similar vegetation and soil moisture conditions so that advection effects are eliminated.

Thornthwaite used measurements of water use in irrigation districts (primarily from the United States) and evapotranspirometer data to derive a complex set of empirical equations for estimation of monthly PE based solely on temperature data and a latitude factor to adjust for daylength. The calculation procedures were set out in an instruction manual (Thornthwaite and Mather, 1957), so that climatologists rather quickly calculated PE and average climatic water budgets for most climatic regions.

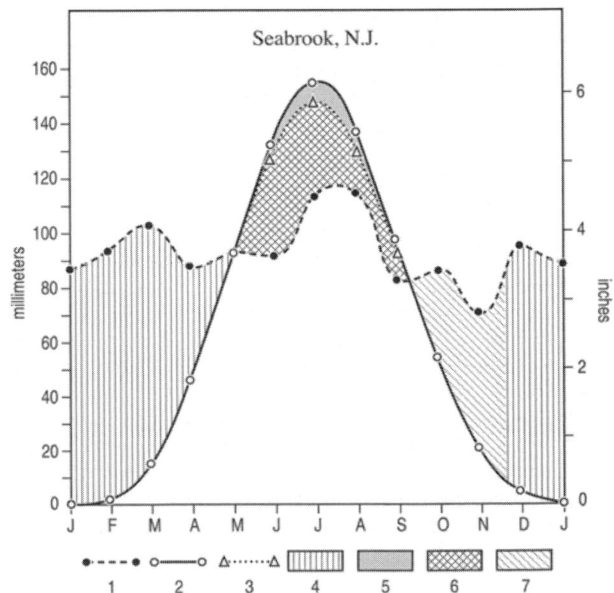

Figure W2 Traditional climatic water-budget graph. (1) P; (2) PE; (3) AE; (4) S; (5) D; (6) soil-moisture utilization; (7) soil-moisture recharge (Thornthwaite and Mather, 1955).

PE can be measured on a daily basis by means of moist evapotranspirometers or weighing lysimeters. These devices are large tanks filled with soil set into the ground with the top flush with the soil surface. Plants are grown in the evapotranspirometers, with a large buffer zone of similar plants around the tanks, to reflect natural field conditions.

The evapotranspirometer can be treated as a water budget system in terms of transport and storage of water. Precipitation or irrigation represents income, and measured percolation to subsurface drainage tanks represents one component of outflow. If the soil moisture in the tank is maintained close to capacity, the difference between total income (precipitation and irrigation water) and percolation represents an estimate of PE.

Weighing lysimeters are much more elaborate and expensive, but these instruments provide much more information. Decreases in weight represent evapotranspiration losses, after the other incomes and outflows of water are taken into account. Hence, weighing lysimeters have the capability even of estimating the diurnal regimes of PE rates, but it is difficult to maintain representativeness to the surrounding landscape. In the United States weighing lysimeters have been limited mostly to drier western states where the economy depends on availability of inexpensive water from well-watered uplands.

PE can be thought of as an energy supply term and, at the same time, as the climatic demand for water from the landscape. PE became the keystone of the water budget analysis; it represented an approximate estimate of potential or optimum water demand in the landscape that could be met by current precipitation and soil moisture utilization. It provided the basis for the budgeting procedure.

At the same time the PE concept became controversial. For some investigators PE estimates were great improvements over pan evaporation data; pans behave much like wet wicks in dry landscapes, and pan evaporation data needed to be adjusted downward by geographical coefficients to represent water loss from ponds and small lakes (Farnsworth et al., 1982). At the same time, the PE concept for a dry region is not entirely satisfying. This dilemma revolves around dimensions of irrigated buffer zones necessary to eliminate advection effects for measurement of PE in dry landscapes.

Thornthwaite also recognized that PE estimates based on temperature and daylength were necessary surrogates for measurements of surface energy budgets needed for rigorous estimates of PE, and he justified use of the empirical equations because of the unavailability of routine energy budget data. Over the ensuing decades more rigorous estimates of PE based on solar and longwave radiation, humidity, and wind have been developed, with the most frequently used estimates based on modifications of formulations originally developed by Penman (1948) in England and Jensen and Haise (1963) in the United States. Nevertheless, unavailability of energy budget data on a routine basis has restricted use of the more rigorous PE estimates mostly to scientific investigations, with the standard Thornthwaite PE estimates used for regional analyses. In general, error rates for regional PE estimates are considered to be rather minimal, particularly in most midlatitude environments; Brutsaert (1982) has published a comprehensive analysis of evaporation, including the PE concept.

Precipitation

Precipitation (P) represents the atmospheric delivery of moisture for the fundamental interactions of energy and moisture at the surface. Unfortunately, most national and regional networks of precipitation gages represent minuscule samples of precipitation. Some networks are reasonably adequate for water budget analyses, especially on an average annual basis. However, standard precipitation data are clearly inadequate in areas marked by large spatial and temporal variability in rainfall, over many sparsely settled areas and especially in mountainous regions with complex patterns of orographic precipitation and rainshadow valleys. Furthermore, there are few long-term observational series over the world's oceans, which occupy more than two-thirds of the surface areas of the Earth, and remote-sensing techniques have not as yet been refined to fill in data gaps between standard gages satisfactorily. Of particular significance is the generally agreed upon undercatch of standard rain gages, especially during windy storms, as opposed to the true precipitation delivered to global landscapes (Mather, 1974); indeed, in river-basin analyses, precipitation may be more in error than PE.

Soil moisture storage

Soil moisture storage (ST) represents water available within rooting zones of the plants for transpiration and evaporation. In classical terms it represents the differences between field capacity and the wilting point. In the water budget framework, soil moisture is generally expressed in units equivalent to precipitation. On an average worldwide basis, Thornthwaite originally estimated the available soil-moisture storage capacity to approximate 100 mm, but 300 mm later was used for global comparisons (Carter and Mather, 1966). In the United States the modern county soil surveys published by the Natural Resources Conservation Service include useful information for estimating available soil moisture capacities for combinations of geological substrate, soil types, slopes, land use, and vegetation covers.

Within the water budget model it is assumed that soil moisture will be utilized by plants to meet the demands of PE when precipitation is less than PE. For the calculation of soil-moisture depletion, Thornthwaite originally suggested that plants could draw on soil moisture equally as needed until the available soil moisture within the rooting zone was exhausted; this approach became known as the equal-availability model (Figure W3, curve A). A few years later Thornthwaite and his

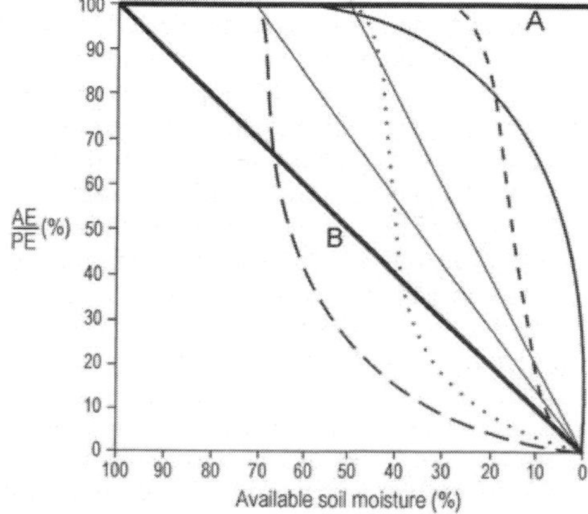

Figure W3 Selected soil-moisture depletion curves. Curve A represents equal availability, and B decreasing availability (after Mather, 1974).

colleagues introduced the decreasing-availability model (Figure W3, curve B), which stated that the plants would withdraw soil moisture to meet the PE demand in proportions relative to the availability of soil moisture within the rooting zone. For example, if available soil moisture within the rooting zone amounted to only 60% of capacity, the plants would withdraw 60% of the PE demand, but if available soil moisture were only 20% of capacity, plants would be able to withdraw only 20% of the demand (Mather, 1974). Figure W3 also illustrates various other soil-moisture depletion models either proposed or in use.

Most investigators prefer some form of the decreasing-availability model over the equal-availability model. A later innovation has been the introduction of a two-layer or even multiple-layer accounting system for soil moisture storage and depletion. The most commonly used approach is to set up a two-layer accounting system, with an upper shallow layer in which moisture is treated as equally available and given the first priorities for soil moisture depletion and recharge. Soil moisture in the thicker lower layer is treated as decreasingly available and second in the priority system of depletion and recharge. Multiple-layer systems are most commonly used with daily or weekly water budgets, and are particularly well adapted to computer modeling.

Actual evapotranspiration

Actual evapotranspiration (AE) in Thornthwaite's average water budget models represents precipitation and soil moisture withdrawals actually used by the plants to try to meet the energy demand represented by PE. Actual evapotranspiration, then, can be equal to but never greater than PE; there are, of course, places and times when AE is less than PE. Thornthwaite's adoption of the term actual evapotranspiration was unfortunate, however, because it implies empirical measurements of evapotranspiration, rather than calculations within the model framework. Most researchers in the agricultural sciences have substituted evapotranspiration, or simply ET, for actual evapotranspiration (AE), regardless of whether ET is measured or calculated.

No distinction is made between transpiration by plants and evaporation from soil surfaces in the Thornthwaite water budget models. Thornthwaite pointed out that most water loss from fully vegetated surfaces was transpiration by plants rather than small amounts of evaporation from shaded soil surfaces; linking of the two pathways of water vapor to the atmosphere as evapotranspiration, either potential or actual, was the cornerstone of water budget analyses as applied to landscapes, river basins, and regions.

In most situations ET represents water passing through plant systems. Indeed a number of measures of biological activity, plant development and growth, crop yields, and even physical and chemical processes in the soils are tied to climatic processes and are indexed effectively from a climatic perspective by AE or ET.

Moisture deficit and surplus

The moisture deficit (D), often simply termed the deficit, is defined as the difference between PE and ET. The deficit represents the additional water that would have been used by plants if it were available, and can serve as a measure of irrigation potential of places and regions; thus, D is also a measure of potential increases in plant growth and crop yields.

The moisture surplus (S), represents precipitation not used for evapotranspiration or soil moisture recharge, and, therefore, water available for surface runoff to lakes and streams or for percolation to groundwater tables. The term surplus betrays an agricultural bias; surplus becomes the input to the surface and groundwater systems studied by hydrologists and engineers. In water budget climatology, surplus water is just as much a part of regional climates as clouds and precipitation are for meteorologists and atmospheric scientists. The surplus is useful for estimating runoff, streamflow, groundwater recharge, and physical or chemical processes in the soil related to downward migration of water.

Figure W2 shows the standard graphical format of the average climatic water budget components of a place. Seabrook, New Jersey, is used here to represent typical annual regimes for a humid climate region in middle latitudes. This average monthly budget is based on an available soil moisture storage capacity of 300 mm and the decreasing-availability model of soil moisture depletion (Figure W3, curve B). Figure W2 emphasizes the seasonality of water budget components, soil moisture withdrawals and deficits during summer, soil moisture recharge during fall, and production of surplus water during winter and early spring.

The average climatic water budget model can be expressed in two interrelated fundamental equations expressing water movement and storage (1) P = ET + S; and energy demands (2) PE = ET + D, with the terms as defined in the text.

Average water budgets

In recent decades there has been considerable interest in global water budgets, in which both storage and exchanges of water in the atmosphere, on the continents in terms of surface waters, soil moisture, groundwater, snow, and glacial ice, and ocean water are evaluated. These analyses indicate, in part, the very small storage of water vapor in the atmosphere at any one time, estimated to average only about 25 mm of equivalent precipitation on a global basis, the large magnitude of evapotranspiration on the continents relative to runoff to the oceans, the importance of evaporation from the oceans, and the dominance of maritime sources of atmospheric moisture for precipitation over most regions of the continents (Baumgartner and Reichel, 1975; Mather, 1974). Figure W4 is a schematic representation of

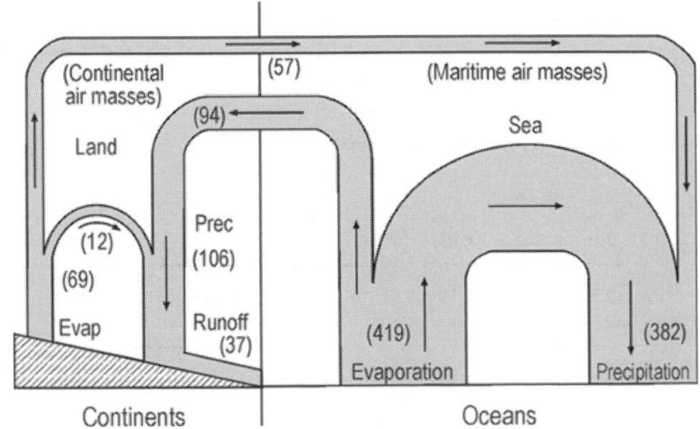

Figure W4 Schematic representation of the global hydrologic cycle. Widths are proportional to volumes of water; values are mean annual water volumes in thousands of cubic kilometers (after Mather, 1974).

the hydrologic cycle in terms of the average annual water budget of the Earth.

The water budget graph for Seabrook, New Jersey (Figure W2) is a representation of the average water budget of a place. Thornthwaite and his associates have calculated average climatic water budgets for thousands of places using a soil moisture storage capacity of 300 mm and the decreasing-availability model of soil moisture depletion; these average budgets have been published by continents in *Publications in Climatology* (Mather, 1962–1965). A moisture index (Im) was also derived from the average water budget calculations; it represents a comparative index of the degree to which mean annual precipitation meets the climatic demands for water (PE) at each place. The index was used as part of an overall climatic classification having five basic types from wettest to driest: perhumid, humid, subhumid, semiarid, and arid (Carter and Mather, 1966). Detailed maps of each of the continents are available, but there has been less interest in classification than applications.

Components of the average water budget, especially PE, AE, and Im, have been used to illustrate regional and global patterns of natural vegetation (Mather and Yoshioka, 1968) and the basic types of the United States comprehensive soil classification (Mather, 1978). Actual evapotranspiration (ET) also has been used to describe global patterns of a host of biological interactions with climate including primary productivity of land plants (Leith and Box, 1972), the decomposition of organic debris and forest-fire hazards (Meentemeyer, 1978), and plant-litter production (Meentemeyer et al., 1982). Carter et al. (1972) prepared a summary statement setting out environmental relationships and responses to components of the average budget, extending relationships to weathering and geomorphology.

Monthly water budgets

For study and management of environmental resources in most climatic regions, average water budget components suggest an unrealistic seasonal stability of moisture conditions. Although variability of energy availability from one year to another on a given month (expressed by PE) is relatively small, the variability of precipitation for a given month over a series of years can be very large. In addition, use of average monthly precipitation in average water budget calculations masks the effects of precipitation variability on evaporation, deficits, and surpluses through time.

As an example, Table W1 shows mean annual water budget components for Baton Rouge, Louisiana (based on the period 1961–1990) calculated first by means of the average water budget, and secondly by means of a continuous monthly water budget procedure. In the continuous monthly water budget procedure, components for each month are calculated before proceeding on to the following month through the period of interest. Monthly precipitation variability is taken into account, and a deficit, for example, will be calculated for a very dry month that, on the average, is very wet. When monthly means of continuous water budgets are calculated and summed on an average annual basis, Table W1 shows that average annual ET (AE) decreased at Baton Rouge from 1045 mm to 946 mm, or a decrease of nearly 100 mm. At the same time, deficits and surpluses increased to 105 mm and 601 mm, respectively. Mean annual totals based on continuous monthly water budget analyses are more representative of the inter-annual variability of environmental conditions than components of the average water budget.

Table W1 Comparison of average annual and continuous monthly water budgets, Baton Rouge, Louisiana, 1961–1990 (in millimeters)

Water budget component	Average annual budget	Means of continuous monthly budgets
PE	1064	1051
P	1546	1546
AE	1045	946
D	20	105
S	502	601

Source: Louisiana Office of State Climatology, Southern Regional Climate Center, Louisiana State University.

Figure W5 illustrates the monthly water budget components at Baton Rouge during a recent period of years. This graphical style does not show monthly precipitation, but emphasizes the seasonality of generation of surplus water for runoff in humid subtropical and midlatitude climatic regimes. Effects of climatic variability also are illustrated in Figure W6, which shows seasonal and annual values of the AE/PE ratio plotted against PE for four places on the United States beginning with New Brunswick, New Jersey, in the humid continental climatic region. Despite the designation as a humid climate, dry years at New Brunswick are much drier in terms of the AE/PE ratio than wetter years in the subhumid climate at Dodge City, Kansas, where the variability is very great. Bradford, in the mountains of Pennsylvania, is just about always wet in these terms, and Alamosa, located on a rain-shadow valley of Colorado, is always dry.

The Palmer drought index series (Palmer, 1965) and the crop-moisture index are developed from continuous monthly and weekly water budgets using a two-layer soil moisture system. These two indices represent unique adaptations of water budget components for the identification of periods of significant departures-from-normal of seasonal moisture characteristics for each place or region. The Palmer indices and the crop moisture index are both prepared on a near real-time basis by the Climate Prediction Center of the National Weather Service (NWS) for climatic monitoring across the United States; these products are available in the *Weekly Weather and Crop Bulletin*, published by Joint Agricultural Weather Facility, operated by the NWS and the US Department of Agriculture.

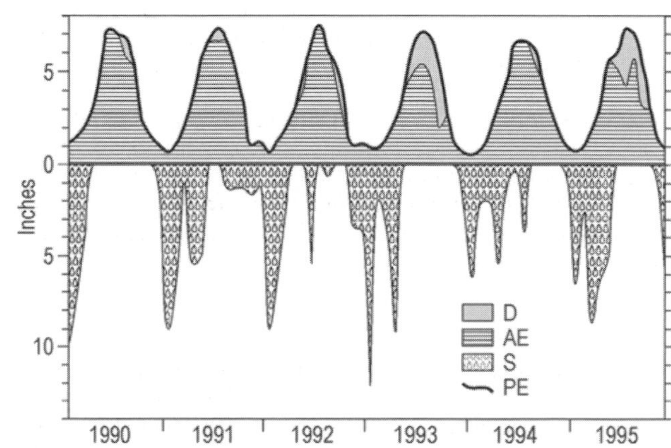

Figure W5 Modeled continuous monthly water-budget components for Baton Rouge, Louisiana, 1990–1995.

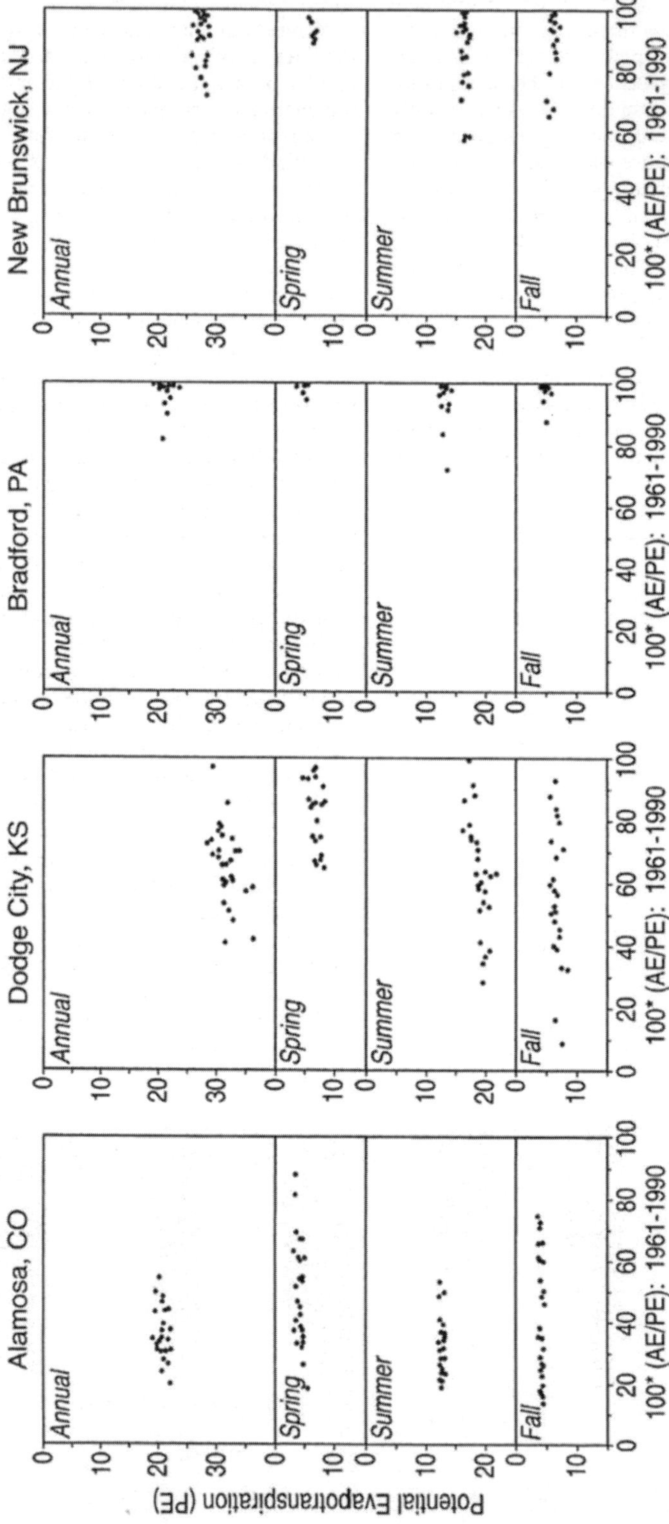

Figure W6 Seasonal and annual ratios of AE/PE plotted against PE for the humid East (New Brunswick, New Jersey, and Bradford, Pennsylvania), the subhumid Great Plains (Dodge City, Kansas), and a high rainshadow desert (Alamosa, Colorado).

An especially interesting application of continuous monthly water budget procedures has been the study of relationships of land use and land cover to runoff and streamflow. For a given climatic regime, does a parcel of land or an entire drainage basin yield more or less water now than, say, 50 years ago because of changes in the use of the land and because of engineering modifications of the drainage systems? These relationships are difficult to evaluate because the investigator is usually searching for relatively small differences among quantities that are difficult to estimate well for entire drainage basins, with climatic variation adding to the complexity of the problem.

The first water budget analysis of this problem was an investigation of the effects of reforestation on water yield from small drainage basins on the Allegheny Plateau in New York (Muller, 1966). Monthly water budget calculations were used to estimate runoff and streamflow due directly to month-by-month variations of PE and precipitation. Monthly estimates of streamflow were then compared to measured streamflow, which is the consequence of climatic variability and changes in land cover. In these drainage basins the land-use changes were mostly replacement of pastures with conifer reforestation blocks. The differences between calculated and measured runoff (Figure W7) were assumed to represent the effects of reforestation, and this study indicated that reforestation significantly reduced annual runoff for a number of years.

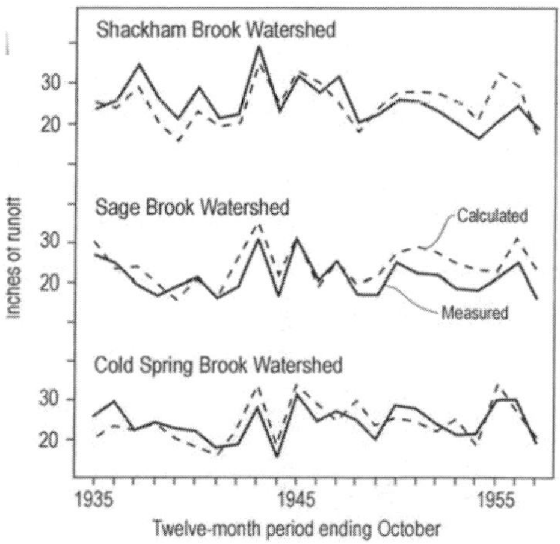

Figure W7 Modeled and measured water yield from partially reforested watersheds on the Allegheny Plateau, New York (after Muller, 1966).

Water budget analyses have also been used to show some of the effects of urbanization on runoff. Figure W8 shows the results of a simulation of the monthly surplus or runoff for undeveloped land in Middlesex County, New Jersey, and for the same land and years as if this land had been developed for typical middle-class subdivisions (Muller, 1969). The figure shows extra water coming off the subdivisions; most of the supplementary runoff is generated during summer and fall when undeveloped land normally produces little to no runoff. Examples of other hydrologic applications of continuous monthly water budget procedures include, for example, effects of land-use changes on river-basin regimen for water resources optimization (Shelton, 1981), comparisons of drainage basin modeled and measured runoff (Rohli and Grymes, 1995), and long-term trends of precipitation and modeled runoff (Keim et al., 1995).

In another study of the Mississippi River drainage basin using monthly temperature and precipitation data from 147 climate divisions for 1932–1988, the standard continuous monthly water budget accounted for 96% of the variation in annual discharge for the entire basin above Tarbert Landing, Louisiana (Hoff and Muller, 1998). Comparisons between modeled and observed runoff suggested that land use and river management had not changed precipitation and runoff relationships from the entire basin through the years, but that significant runoff changes had occurred within the Ohio–Tennessee basins.

Most recently, the water budget methodology was utilized to evaluate the sensitivity of runoff from the Cannonsville watershed in the Catskill Mountains (one of the primary sources for potable water for the City of New York) during historical periods of extreme high and low precipitation since the mid-1950s (Frei et al., 2002). A modified daily Thornthwaite water budget with snowpack and groundwater storage was developed for the study. Using four GCM model estimates of global warming by 2080, estimated average annual runoff is shown to range from between +10% and −30% of average annual water yield in recent decades, about equal to extreme wet and dry years in the 1960s and 1970s, respectively.

Other applications include monitoring salinity regimes in estuaries (Mather et al., 1972), analyses of water budget components of a large lake system (Sanderson, 1966), relationships

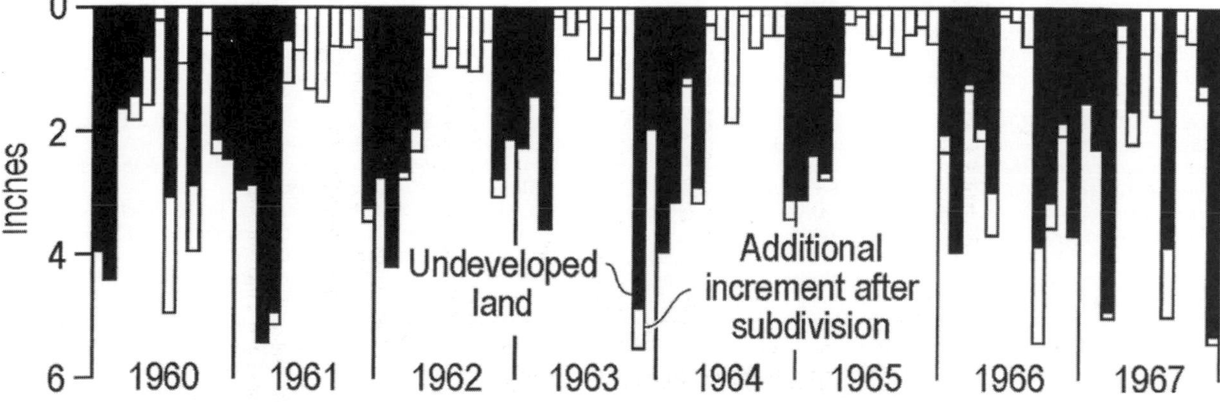

Figure W8 Water-budget simulation of surplus water generated from undeveloped land and residential subdivisions in Middlesex County, New Jersey (after Muller, 1982).

among water budget components and crop yields (Chang, 1968), and evaluations of the relationships of pine bark beetle outbreaks to deficits and surpluses (Kalkstein, 1981).

Weekly and daily water budget analyses

The annual sums of estimated evapotranspiration, deficits, and surpluses are changed if weekly or daily water budget calculations are used to replace continuous monthly calculations. In humid and subhumid climatic regions, daily and weekly variability of precipitation results in smaller seasonal or annual sums of ET and correspondingly larger sums for D and S. Weekly or daily water budget data are even more representative of real-world climate and environmental interactions than continuous monthly water budgets. As a result of the increased demand for daily and weekly water budgets, there are now a number of computer programs available for long-term analysis.

Daily water budgets, in which daily values of PE are compared to available soil moisture, are being developed extensively in the agricultural sciences. Basically, daily budgets are being used to model crop development and growth, and to study the water economies of commercial crops. Examples include quantification of relationships between ET and productivity (Hank, 1974), development of specialized water budget models for agriculture (Ritchie, 1972), and simulation models for growth of corn and sorghum (Stapper and Arkin, 1980). Daily water budget models also have been used to analyze differential stress in terms of deficits on crop production at various physiological growth stages in corn and soybeans (Sudar et al., 1981). Daily water budget models also have been incorporated by the National Weather Service in their River Forecast Centers for real-time prediction of river stages and floods on selected drainage basins (Peck, 1976).

Robert A. Muller and John M. Grymes III

Bibliography

Baumgartner, A., and Reichel, E., 1975. *The World Water Balance: mean annual global, continental and maritime precipitation, evaporation and runoff* (R. Lee, trans.). Amsterdam: Elsevier.

Brutsaert, W., 1982. *Evaporation into the Atmosphere: theory, history, and applications*. Dordrecht: Reidel.

Carter, D.B., and Mather, J.R., 1966. Climatic classification for environmental biology. *Publications in Climatology*, **19**: 305–395.

Carter, D.B., Schmudde, T.H., and Sharpe, D.M., 1972. *The Interface as a Working Environment: a purpose for physical geography*. Technical Paper No. 7. Washington, DC: Commission on College Geography, Association of American Geographers.

Chang, J.H., 1968. *Climate and Agriculture: an ecological survey*. Chicago, IL: Aldine.

Denmead, O.T., and Shaw, R.H., 1960. The effects of soil moisture stress at different stages of growth on the development and yield of corn. *Agronomy Journal*, **45**: 385–390.

Farnsworth, R.K., Thompson, W.S., and Peck, E.L., 1982. *Evaporation Atlas for the Contiguous 48 United States*. NOAA Technical Report NWS 33. Washington, DC: US Government Printing Office.

Frei, A., Armstrong, R.L., Clark, M.P., and Serreze, M.C., 2002. Catskill Mountain water resources: vulnerability, hydroclimatology, and climate-change sensitivity. *Annals of the Association of American Geographers*, **92**: 203–224.

Hank, R.J., 1974. Model for predicting plant growth as influenced by evapotranspiration and soil water. *Agronomics Journal*, **66**: 35–41.

Hoff, J.L., and Muller, R.A., 1998. Effects of land use and management on the geographical and temporal variations of runoff from the Mississippi River basin. GCIP Mississippi River Climate Conference. 8–12 June 1998, St Louis, MO, pp. 239–240.

Jensen, M.E., and Haise, H.R., 1963. Estimating evapotranspiration from solar radiation. Am. Soc. Civil Engineers Proc. *Journal of the Irrigation and Drainage Division*, **89**: 15–41.

Kalkstein, L.S., 1981. An improved technique to evaluate climate-southern pine beetle relationships. *Forest Science*, **27**: 579–589.

Keim, B.D., Faiers, G.E., Muller, R.A., Grymes, J.M., III and Rohli, R.V., 1995. Long-term trends of precipitation and runoff in Louisiana, U.S.A. *International Journal of Climatology*, **15**: 531–541.

Leith, H., and Box, E.O., 1972. Evapotranspiration and primary productivity. *Publications in Climatology*, **25**: 37–46.

Mather, J.R. (ed.), 1962–1965. Average climatic water balance data of the continents. *Publications in Climatology*, vols. 15–18.

Mather, J.R., 1974. *Climatology: fundamentals and applications*. New York: McGraw-Hill.

Mather, J.R., 1978. *The Climatic Water Budget in Environmental Analysis*. Lexington, MA: Lexington Books.

Mather, J.R., and Yoshioka, G.A., 1968. The role of climate in the distribution of vegetation. *Annals of the Association of American Geographers*, **58**: 29–41.

Mather, J.R., Swaye, F.J., Jr, and Hartmann, B.J., 1972. The influence of the climatic water balance on conditions in the estuarine environment. *Publications in Climatology*, **25**: 1–41.

Meentemeyer, V., 1978. Microclimate and lignin control of litter decomposition rates. *Ecology*, **59**: 465–472.

Meentemeyer, V., Box, E.O., and Thompson, R., 1982. World patterns and amounts of terrestrial plant litter production. *Bioscience*, **32**: 125–128.

Miller, D.H., 1977. *Water at the Surface of the Earth: an introduction to ecosystem hydrodynamics*. New York: Academic Press.

Muller, R.A., 1966. The effects of reforestation on water yield – a case study using energy and water balance models for the Allegheny Plateau, New York. *Publications in Climatology*, **19**: 251–304.

Muller, R.A., 1969. Water balance evaluations of the effects of subdivisions on water yield in Middlesex County, New Jersey. *Proceedings of the Association of American Geographers*, **1**: 121–125.

Muller, R.A., 1970a. Frequency of moisture deficits and surpluses in the humid subtropical climatic region of the United States. *Southeastern Geographers*, **10**: 30–40.

Muller, R.A., 1970b. Frequency analyses of the ration of actual to potential evapotranspiration for the study of climate and vegetation relationships. *Proceedings of the Association of American Geographers*, **2**: 118–122.

Muller, R.A., 1982. The water budget as a tool for inventory and analysis of factors affecting variability and change of river regimen. In Ma, L.J.C., and Noble, A.G., eds., *The Environment: Chinese and American views*. New York: Methuen, pp. 171–186.

Muller, R.A., and Larimore, P.B., Jr, 1975. Atlas of seasonal water budget components of Louisiana. *Publications in Climatology*, **28**: 1–19.

Palmer, W.C., 1965. *Meteorological Drought*. Weather Bureau Research Paper No. 45. Washington, DC: US Department of Commerce.

Peck, E., 1976. *Catchment Modeling and Initial Parameter Estimation for National Weather Service River Forecast System*. NOAA-NES-Hydro 31. Washington, DC: US Government Printing Office.

Penman, H.L., 1948. Natural evaporation from open water, bare soil, and grass. *Proceedings of the Royal Society of London*, **A193**: 120–145.

Ritchie, J.T., 1972. Model for predicting evaporation from a row crop with incomplete cover. *Water Resources Research*, **8**: 1204–1213.

Rohli, R.V., and Grymes, J.M., III, 1995. Differences between modeled surplus and USGS-measured discharge in Lake Pontchartrain basin, U.S.A. *Water Resources Bulletin*, **31**: 97–107.

Sanderson, M., 1966. A climatic water balance of the Lake Erie Basin. *Publications in Climatology*, **19**: 1–87.

Shelton, M.L., 1981. Runoff and land use in the Deschutes Basin. *Annals of the Association of American Geographers*, **71**: 11–27.

Stapper, M., and Arkin, G.F., 1980. *Dynamic Growth and Development Model for Maize*. Texas Agricultural Experimental Station Report 80–2.

Sudar, R.A., Saxton, K.F., and Spomer, R.E., 1981. A predictive model of water stress in corn and soybeans. *Transactions of the American Society of Agricultural Engineers*, **24**: 97–102.

Thornthwaite, C.W., 1948. An approach toward a rational classification of climate. *Geographical Review*, **38**: 55–94.

Thornthwaite, C.W., and Mather, J.R., 1955. The water balance. *Publications in Climatology*, **8**: 9–86.

Thornthwaite, C.W., and Mather, J.R., 1957. Instructions and tables for computing potential evapotranspiration and the water balance. *Publications in Climatology*, **10**: 185–311.

Cross-references

Agroclimatology
Aridity Indexes
Climate Classification
Evaporation
Evapotranspiration
Humid Climates
Hydroclimatology
Mather, John R.
Palmer Index/Palmer Drought Severity Index
Precipitation Distribution
Thornthwaite, Charles W.

WEATHER

A major component of the atmospheric sciences, weather is defined as a state or condition of the atmosphere at any particular place and time. A broad, integrated weather picture is known as a weather system. Weather is specifically distinguished from climate, which represents a regional or global synthesis of weather extended through time on the scale of years, rather than minutes or hours. Weather involves measurement of multiple parameters, so-called weather elements: temperature, pressure, wind, cloud condition, visibility, humidity, precipitation, and other hydrometeors (dew, rain, hail, snow). In relation to humans, weather often conditions personal comfort or economy and goes further to embrace related factors such as sea state, tidal hazards, avalanche liability, and river flooding. Additionally, weather study also includes optical phenomena, such as lightning, aurora, and solar phenomena.

In aviation, the word weather is used in relation to the synoptic weather observations of the immediate past and present. Weather also has a special geological meaning. *Weathering* is the process of rock disintegration (both mechanical and chemical, abrasion and corrosion) under the influences of the atmosphere.

The scientific study of weather is meteorology. This is defined by the American Meteorological Society (Geer, 1996) as the study dealing with phenomena of the atmosphere. This not only includes the physics, chemistry and dynamics of the atmosphere, but is extended to include many of the direct effects of the atmosphere upon the Earth's surface, the oceans, and life in general. The goals often ascribed to meteorology are the complete understanding, accurate prediction, and artificial control of atmospheric phenomena.

John E. Oliver

Bibliography

Aguado, E., and Burt, J.E., 2001. *Understanding Weather and Climate*.Upper Saddle River, NJ: Prentice-Hall.
Ahrens, D.L., 2000. *Meteorology Today: an introduction to weather, climate, and the environment*, 6th edn. Pacific Grove, CA: Brooks & Cole.
Barry, R.G., and Chorley, R.J., 1998. *Atmosphere, Weather, and Climate*. London: Routledge.
Geer, I.W., 1996. *Glossary of Weather and Climate*. Boston, MA: American Meteorological Society.
Lutgens, F.K., and Tarbuck, E.J., 2001. *The Atmosphere*, 8th edn. Upper Saddle River, NJ: Prentice-Hall.
Oliver, J.E., and Hidore, J.J., 2002. *Climatology: an atmospheric science*. Upper Saddle River, NJ: Prentice-Hall.
Thompson, R.D., 1998. *Atmospheric Processes and Systems*. London: Routledge.

Cross-references

Applied Climatology
Bioclimatology
Climatology

WIND CHILL

The term "wind chill" was coined by the Antarctic explorer Paul A. Siple in his dissertation, "Adaptation of the explorer to the climate of Antarctica", submitted in 1939. During their stay in Antarctica, Siple and his colleague, Charles F. Passel, conducted experiments on the time required to freeze 250 grams of water in a plastic cylinder placed outside in the wind, including a formula relating the rate of energy loss of the cylinder to the air temperature and wind speed. The formula was later used to compute the *wind chill equivalent temperature*. This parameter was supposed to take into account the exaggerated sensation of cold that one experiences when the wind is blowing at low temperatures. For example, according to the formula, an appropriately clothed person exposed to a temperature of -15°C and a wind of 12 meters per second ($m\,s^{-1}$) would feel as cold as when walking in still air at a temperature of -39°C. The US National Weather Service adopted the wind chill formula in 1973 and used it until 1 November 2001, when it was replaced.

Flaws in the formula for wind chill equivalent temperature were recognized from the start. Water has no internal heat source like the human body. A plastic bottle does not resemble human skin: the formula ignores the thermal resistance of skin, and the assumed skin temperature is unrealistically high. Wind speed, customarily measured on a 10-meter tall mast, is greater than that experienced at face level. Finally, people accustomed to cold winter weather complained that the wind chill equivalent temperature overstated the effect of wind.

In the fall of 2000, responding to the apparent inadequacies in the formula for wind chill equivalent temperature, the US Office of the Federal Coordinator for Meteorology formed the Joint Action Group for Temperature Indices (JAG/TI), whose first job was to evaluate the shortcomings of existing wind chill formulas and, if necessary, propose a new formula. The JAG/TI included government and university representatives from the US and Canada. The JAG/TI enlisted Dr Maurice Bluestein (Indiana University–Purdue University in Indianapolis) and Mr Randall Osczevski (Defense and Civil Institute of Environmental Medicine – DCIEM, Toronto, Ontario, Canada) to devise a new formula, based in part upon their recent research on wind chill and in part upon new tests with human subjects (Table W2).

DCIEM conducted clinical trials in a wind tunnel under controlled conditions of temperature and airflow. Subjects had

Table W2 Wind Chill Calculation Chart, (metric units). Table of wind chill temperatures, where T air = air temperature in °C and V_{10} = observed wind speed at 10 m elevation, in km/h; courtesy of the Meterological Service of Canada, Environmental Canada.

T air / V_{10}	5	0	-5	-10	-15	-20	-25	-30	-35	-40	-45	-50
5	4	-2	-7	-13	-19	-24	-30	-36	-41	-47	-53	-58
10	3	-3	-9	-15	-21	-27	-33	-39	-45	-51	-57	-63
15	2	-4	-11	-17	-23	-29	-35	-41	-48	-54	-60	-66
20	1	-5	-12	-18	-24	-30	-37	-43	-49	-56	-62	-68
25	1	-6	-12	-19	-25	-32	-38	-44	-51	-57	-64	-70
30	0	-6	-13	-20	-26	-33	-39	-46	-52	-59	-65	-72
35	0	-7	-14	-20	-27	-33	-40	-47	-53	-60	-66	-73
40	-1	-7	-14	-21	-27	-34	-41	-48	-54	-61	-68	-74
45	-1	-8	-15	-21	-28	-35	-42	-48	-55	-62	-69	-75
50	-1	-8	-15	-22	-29	-35	-42	-49	-56	-63	-69	-76
55	-2	-8	-15	-22	-29	-36	-43	-50	-57	-63	-70	-77
60	-2	-9	-16	-23	-30	-36	-43	-50	-57	-64	-71	-78
65	-2	-9	-16	-23	-30	-37	-44	-51	-58	-65	-72	-79
70	-2	-9	-16	-23	-30	-37	-44	-51	-58	-65	-72	-80
75	-3	-10	-17	-24	-31	-38	-45	-52	-59	-66	-73	-80
80	-3	-10	-17	-24	-31	-38	-45	-52	-60	-67	-74	-81

FROSTBITE GUIDE
Low risk of frostbite for most people
Increasing risk of frostbite for most people in 10 to 30 minutes of exposure
High risk for most people in 5 to 10 minutes of exposure
High risk for most people in 2 to 5 minutes of exposure
High risk for most people in 2 minutes of exposure or less

sensors for measuring temperature and energy transfer attached to various parts of the face. Bluestein and Osczevski calibrated their new formula for wind chill temperature against the tests with human subjects. Following JAG/TI adoption of the formula, the Meteorological Service of Canada began using it operationally on 1 October 2001; the US National Weather Service followed suit on 1 November 2001.

The new wind chill temperature (WCT) has the following features:

1. It corrects the observed wind speed, normally measured at a height of 10 meters, to the average height of the human face, about 1.5 meters.
2. It is based on a model of the human face that incorporates modern heat transfer theory.
3. The formula builds in a margin of safety in that it applies to persons whose skin has unusually high thermal resistance. For these people, interior body heat does not pass as readily to the surface of the skin as in the great majority of the population.
4. It uses a low-wind threshold of 4.8 km/h rather than the 6.4 km/h used in the Siple–Passel formula. Pedestrians on busy streets are more likely to walk at 4.8 km/h than 6.4 km/h.
5. The formula assumes nighttime conditions with clear skies (no incoming solar radiation and minimal infrared radiation received from the sky).
6. It gives values higher than those generated by the old Siple–Passel formula.

In metric units the equation for wind chill temperature (°C) is

$$\mathrm{WCT} = 13.12 + 0.6215T - 11.37V^{0.16} + 0.3965TV^{0.16}$$

where T is the air temperature in °C, and V is the wind speed in kilometers per hour measured at a height of 10 meters (standard practice).

In English units the equation for wind chill temperature (°F) is

$$\mathrm{WCT} = 35.74 + 0.6215T - 35.75V^{0.16} + 0.4275TV^{0.16}$$

where T is the air temperature in °F and V is the wind speed in miles per hour measured at a height of 33 feet.

The JAG/TI agreed to delay incorporating the effects of solar radiation on the formula until researchers determine the correct adjustments for solar zenith angle and cloudiness. In addition, the JAG/TI wants to refine estimates relating exposure time at a given WCT to the onset of frostbite. Finally, they may devise a formula that takes into account the chilling effects of water on the skin.

Thomas W. Schlatter

Bibliography

Henson, R., 2002. Cold rush: scientists search for an index that fits the chill. *Weatherwise*, **55**(1): 14–19.

Websites

http://www.msc.ec.gc.ca/education/windchill/index_e.cfm (Canada)
http://www.nws.noaa.gov/om/windchill/index.shtml (US)

Cross-references

Applied Climatology
Bioclimatology
Climate Comfort Indices
Heat Index

WIND POWER CLIMATOLOGY

Humans have long used the wind to ease workloads in a wide variety of applications. Sailing ships have been in use for at least three millennia and wind-driven gristmills were invented more than two millennia ago. At the beginning of the twentieth century, wind took on a new importance in powering electrical devices that revolutionized the quality of life. In the United States and in other developed countries, wind power generation became virtually moribund after rural electrification efforts started in the 1930s. However, the relatively high costs of fuel in Europe and environmental concerns revitalized the wind power industry at the end of the twentieth century. Key to this wind revolution was improvement of the machines. As of the beginning of the twenty-first century, wind power represented less than 1% of world electrical generation, but was the fastest-growing source of alternative energy.

The successful use of wind power is largely dependent on climatic knowledge. The problem concerning productive use is that winds do not blow evenly over space and time over Earth. Witness the global circulation components known as the trade winds and the doldrums. The former was closely associated with dependable winds that made for dependable sailing schedules. The latter was associated with light and variable winds presenting a challenge to their productive use. As humans attempt to tap the kinetic energy of winds for generating electricity we are, ironically, still subject to basic limitations of an area's climatology.

Turbine technology

The wind has been used to turn turbines ever since the first small turbines were attached to windmills in Denmark in the 1890s. The fossil energy crises of the 1970s drove investors to fund wind power generation, notably in California. These machines were much smaller and less reliable than today's versions. As government tax credits declined, so did the wind power industry. Modern wind turbines used in utility-scale wind power farms represent a quantum leap in technology from that era. The late twentieth century witnessed considerable improvement in turbine technology including use of strong, lightweight carbon fiber materials in the blades, blade designs employing lift instead of drag, and computerized controls. Large turbines have the ability to continuously adjust blade pitch so as to maximize the wind power harvest or halt the revolution of the blades. However, the most noticeable change has been in their size. Modern turbines represent some of the largest machines ever built.

As turbine sizes have increased, they become more cost-effective in electrical production. In the late 1970s the cost of utility-scale wind power per kilowatt hour (kWh) was nearly 40¢. By the first decade of the twenty-first century the cost had decreased to on the order of 3–5¢ per kWh. Although there has been continued argument over the appropriateness of state and federal production tax credits in the United States, it is clear that the cost of utility-scale wind power has declined to become competitive with fossil-fuel-derived power.

A typical utility-scale turbine of today has a height of 50 meters or more at the hub connecting three blades that may be larger than 30 meters. Newer turbines produce more than a megawatt (one megawatt is enough to power in excess of 250 US households). Most installations will have many turbines strung along a ridgeline in order to take advantage of the strongest winds in the area and to make the most of the economies of scale. Figure W9 illustrates a typical utility-scale

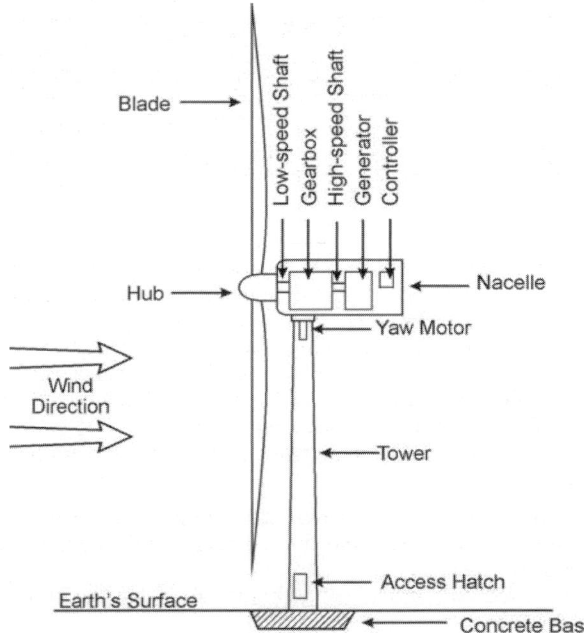

Figure W9 Utility-scale wind turbine.

turbine designed for a 20-year lifetime. An anemometer feeds wind speed and direction data to the controller. When wind speeds exceed 3 m/s the blades start to turn. An onboard yaw motor keeps the hub turned toward the wind direction and the controller adjusts blade pitch in order to maximize the power derived from the wind. Blade rotation speeds are moderate on large turbines, with 30–60 rpm being typical. At the hub the blades are connected to a shaft connected to a gearbox in the nacelle. The purpose of the gearbox is to convert rotational speeds up to 1200–1500 rpm on a high-speed shaft; these higher speeds are suitable for use in generators. The generators in the nacelles produce electricity at high voltage (e.g. 960 kV) and the electricity is fed to transformers at an electrical substation in order to be put on a power grid at 25 kV. At wind speeds of 25+ m/s the turbine's energy production will be cut out so that neither blades nor surroundings are damaged.

Small turbines are used for on-site power. A typical small turbine installation might have blades sweeping an area 10 m across while mounted on a 15 m tower. These have outputs less than 50 kW. A small turbine is shown in Figure W10. Though small turbines tend to cost more to produce each kilowatt hour than do large turbines, their relatively small up-front costs and greater flexibility in siting make them attractive in some home, small industrial, and cooperative ownership settings.

Figure W10 A small wind turbine (courtesy Bergey Windpower Company).

Turbine technology continues to improve and has been fostered by government-led research and development. For instance the US Department of Energy's National Wind Technology Center runs a low-speed turbine program. By increasing electrical output at progressively lower wind speeds, the aim is to make lower wind areas commercially viable for wind development, thus increasing the usable area by 20 times. In Europe, Denmark's Risø National Energy Laboratory is a leading organization in wind power research.

Wind power calculation

The key to economic success in developing wind power is in understanding of the wind power characteristics of a proposed site. Calculation of available wind power relies on knowledge of the physics of kinetic energy. The kinetic energy (KE) of wind with total mass (M) and velocity (V) is given by the expression:

$$\mathrm{KE} = \tfrac{1}{2} \times M \times V^2 \tag{1}$$

For purposes of wind turbine power calculation, consider the air moving onto a turbine's blades (Figure W9) as a parcel with the shape of a huge cylinder. In a given time, the turbine's blades sweep out cross-sectional area (A) and the cylinder has a depth (D) in the dimension horizontal to Earth's surface. So, the volume (Vol.) of this cylinder is given by the equation:

$$\mathrm{Vol} = A \times D \tag{2}$$

The density of air (ρ) in the cylindrical parcel can be represented by:

$$\rho = M/\mathrm{Vol} \tag{3}$$

Restated, this becomes:

$$M = \rho \times \mathrm{Vol} \tag{4}$$

If a time (T) is required for the cylinder of air with depth (D) to pass through the plane of the wind turbine blades, then the air parcel's velocity (V) can be expressed as:

$$V = D/T \tag{5}$$

In Equation (4) can be reworked as:

$$D = V \times T \tag{6}$$

Equation (1) can be restated using the substitutions from the other equations. First, substitute for M using equation (4):

$$\mathrm{KE} = \tfrac{1}{2} \times (\rho \times \mathrm{Vol}) \times V^2 \tag{7}$$

Next, substitute for Vol. using equation (2):

$$\mathrm{KE} = \tfrac{1}{2} \times (\rho \times A \times D) \times V^2 \tag{8}$$

Next, substitute for D using equation (6):

$$\mathrm{KE} = \tfrac{1}{2} \times (\rho \times A \times V \times T) \times V^2 \tag{9}$$

Equation (9) can be restated as:

$$\mathrm{KE} = \tfrac{1}{2} \times (\rho \times A \times V^3 \times T) \tag{10}$$

Power is defined as kinetic energy divided by time. Substituting the right side of equation (10) for KE, power is expressed as:

$$\mathrm{Power} = (\tfrac{1}{2} \times \rho \times A \times V^3 \times T)/T \tag{11}$$

Equation (11) can be reduced to:

$$\text{Power} = \tfrac{1}{2} \times \rho \times A \times V^3 \quad (12)$$

In reality, the density of air (ρ) of the wind does not vary as widely as wind velocity (V) so that equation (12) means that power approximately varies as a cubic function of wind speed: the power that can be derived from the wind increases greatly with increases in wind speed.

The wind industry commonly uses wind power density (WPD) as a standard measure. WPD is independent of turbine blade size and efficiency. It is calculated by dividing power by the cross-sectional area (A) of the air parcel. Thus, WPD is dependent only on the velocity and density of the air parcel:

$$\text{WPD} = \text{Power}/A \quad (13)$$

Wind power density is expressed in W/m^2. The US Department of Energy defines a site as economically suitable for large turbines if it has a wind power density greater 400 W/m^2 at 50 m above the surface; this equates to a wind of approximately 7 m/s and above with turbine sites close to sea level. The Department of Energy's wind power map of the continental United States is given in Figure W11.

The large capital investment in wind turbines dictates caution, so utility-scale projects are thoroughly researched with onsite meteorological instrumentation on tall towers mimicking turbine heights. Unfortunately, these specially instrumented usually have short lengths of record (several months) and are not of true climatological proportions.

A site's wind resource must also be evaluated on its time variability. Equation (13) is an average over time. Extreme care must be used with using equation (13). Almost always, the cube of long-term mean wind speed is less than the mean of the cubes of the individual wind speed observations. This is because wind speed/power plots approximate Rayleigh functions. An example is given in Figure W12. To accurately estimate wind power density, the calculations should be a summation of data taken frequently (e.g. every few minutes) over long time periods (years). The preferred form of the WPD equation becomes:

$$\text{WPD} = 0.5 \times 1/n \times \sum_{i=1}^{n} (\rho_i \times V_i^3) \quad (14)$$

In equation (14) n is the number of wind speed observations and ρ_j and V_j are, respectively, the ith (first, second, third, etc.) observations of air density and wind speed. Wind speed and air density vary between observation times so that the time variability of wind's kinetic energy is accounted for.

Few instrumented tall towers exist by which to service the data requirements of equation (14); therefore the more ubiquitous 10 m tower data area are usually used and faster wind speeds at higher levels occupied by large turbines inferred by a one-seventh power law.

Wind power mapping

In that long-term wind measurements are not present in all possible places, especially far from large cities, wind power mapping has been employed as a first approximation of the wind resource. Typically, the wind map is generated by using long-term wind observations from existing wind observation stations. Worldwide, wind reporting sites with observations an hour or less apart are much less common than temperature or precipitation measurement sites. Usually, relatively few sites can be employed in wind power mapping of large areas. Moreover, the World Meteorological Organization 10 m standard height for wind observations applies directly to small turbines; calculation of long-term wind power density for utility applications must be interpolated by a power law unless using rare tall tower data.

There are various ways to spatially interpolate the wind between measurement sites. The most common involve computer models that incorporate atmospheric and land surface physics. By using generalized physical relationships between wind speed, terrain, and vegetation, these models adjust the wind power density between observation sites. An example output of this type of model is the US Department of Energy (USDOE) map in Figure W11. This model calculated wind power density for 1/4° latitude by 1/3° longitude grid blocks. The USDOE acknowledges that, because of local terrain features, their estimates might be off by 50–100% at individual sites. The USDOE and state entities have generated maps at scales ranging down to grid cells a few hundred meters across. Empirical models have also been employed as the basis for wind density mapping. By plotting the relationship between wind speed, wind density, altitude, and surface roughness, empirical functions are derived and then applied to individual grid cells over a large area. Both types of models (physical and empirically) require substantial spatially referenced data, and are limited by the quality of those data. In any case, model-based wind power density maps serve as only the background context by which to identify areas where wind power development might be feasible. Typically, developers erect tall instrument towers at proposed wind farm sites in order to examine wind speeds and directions over a series of months. The placement and juxtaposition. of the turbines is a carefully planned balance of topography and wind characteristics.

Wind power climatologies

A wind power climatology is a data time-series in which the diurnal, seasonal, and interannual characteristics of wind are considered. Wind speed, direction, and variation (expressed as standard deviations) are commonly used. Wind power climatologies include wind power roses as diagnostic tools. Figure W13 is an example of a monthly wind power climatology derived from several years of data. In this example the average wind power is quite good, but the summer low in available wind power is of interest because it coincides with the peak electrical demands of the hot summer at this location.

Wind power climatology wind roses are variations on wind rose data usually seen. Figure W14 includes the usual wind rose data; that is, it shows the percentage of time that the wind comes from each of 16 directions. Additionally, Figure W14 shows the percentage of the wind power derived from various directions. The implication of this figure is that large turbines should be lined along a west–east axis on high ground so as to take advantage of the additional lift provided as the air ascends from northerly and southerly quadrants.

From wind power climatologies, the potential yield of a specific turbine type can be modeled. Using data such as Figure W12 the projected output capacity factor of specific turbines can be estimated and cost/benefits calculated.

Siting

Economies of scale suggest that large turbines be placed in proximity to each other. This lessens the amount of infrastructure

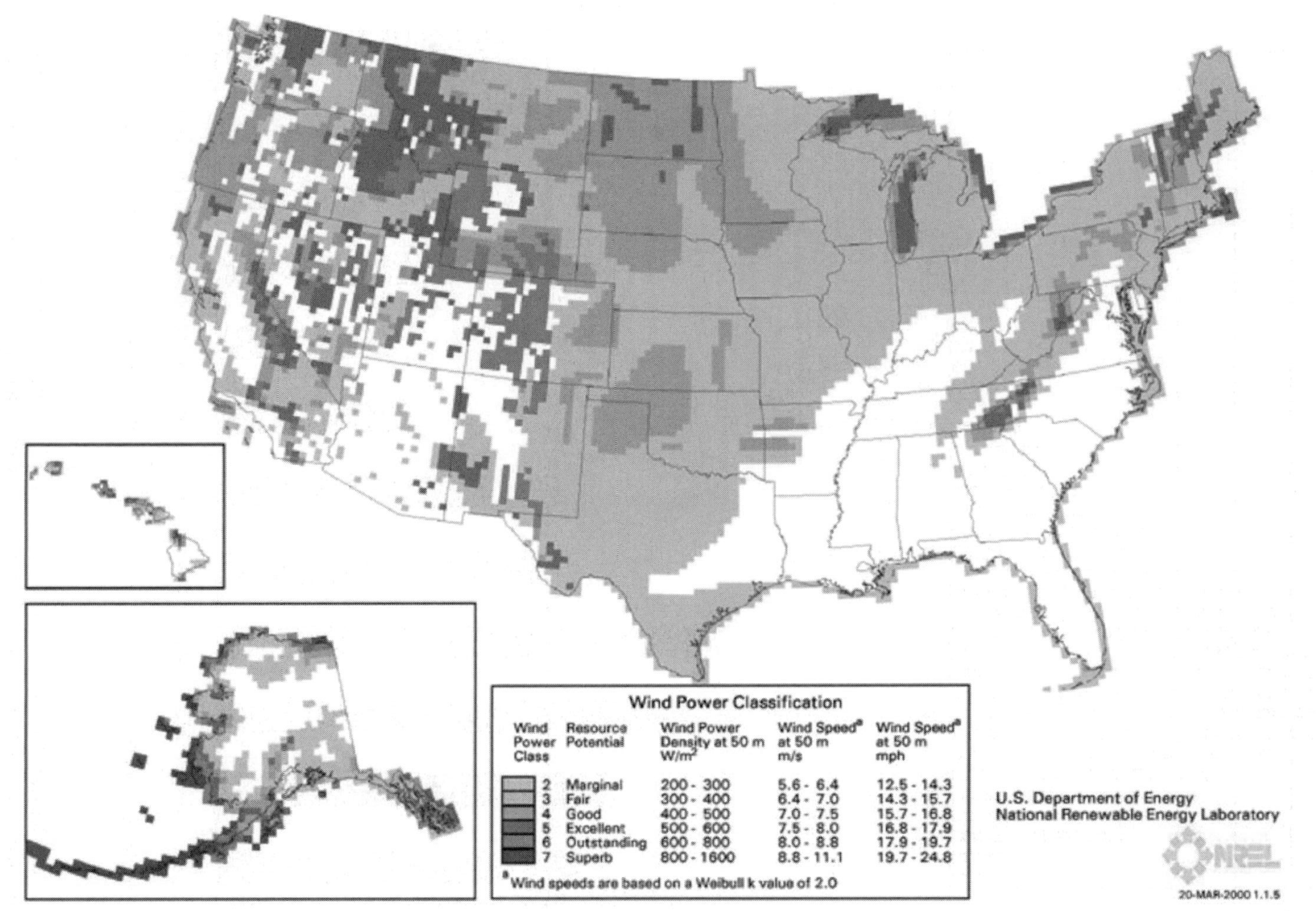

Figure W11 Wind power density in the continental United States. (US Department of Energy, National Renewable Energy Laboratory, Wind Energy Resource Atlas of the United States, 1987).

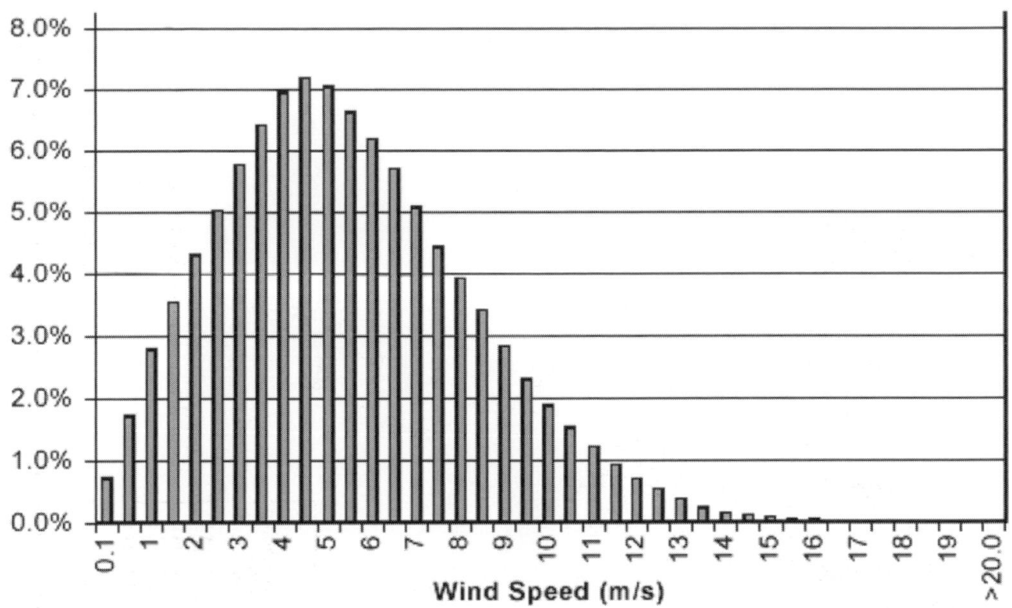

Figure W12 Wind speed distribution over time at 10 m. The original data were in 5-min increments over 9 years. The curve is a Rayleigh function (courtesy Oklahoma Wind Power Initiative).

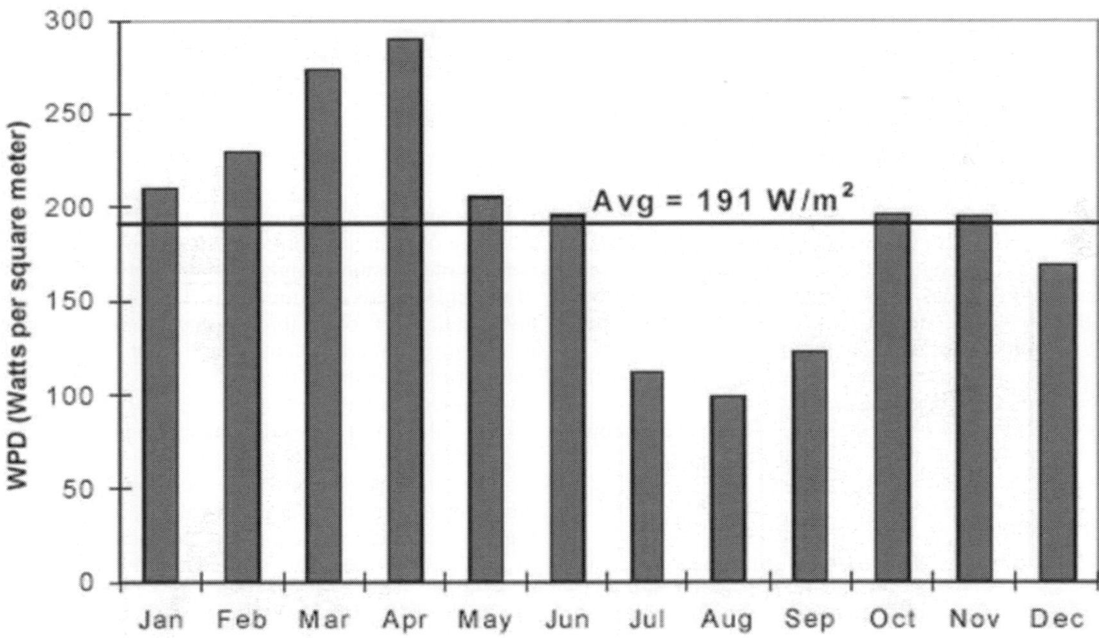

Figure W13 Monthly 10-m wind power climatology for Cheyenne, Oklahoma. The original data were in 5-min increments over 9 years (courtesy Oklahoma Wind Power Initiative).

(roads, buried power lines, leases) that must be developed. However, individual turbines must not be too close to each other so as to minimize the rotor-caused turbulence that is inherited by the wind. Thus, wind turbines are usually found arranged in a line or generously separated multiple lines. For any given turbine the goal is to have it turn at capacity all the time. The temporal variability of wind precludes this, but modern turbines can be profitable if the actual wind power is greater than 0.3 of the capacity over long time periods.

In that wind turbines do not turn at capacity at all times, the most rational economic decision is to site them where the winds blow the fastest and longest. In mountain areas this is typically on sharp ridge crests and in passes. In the former case higher altitudes are associated with faster wind speeds. In the

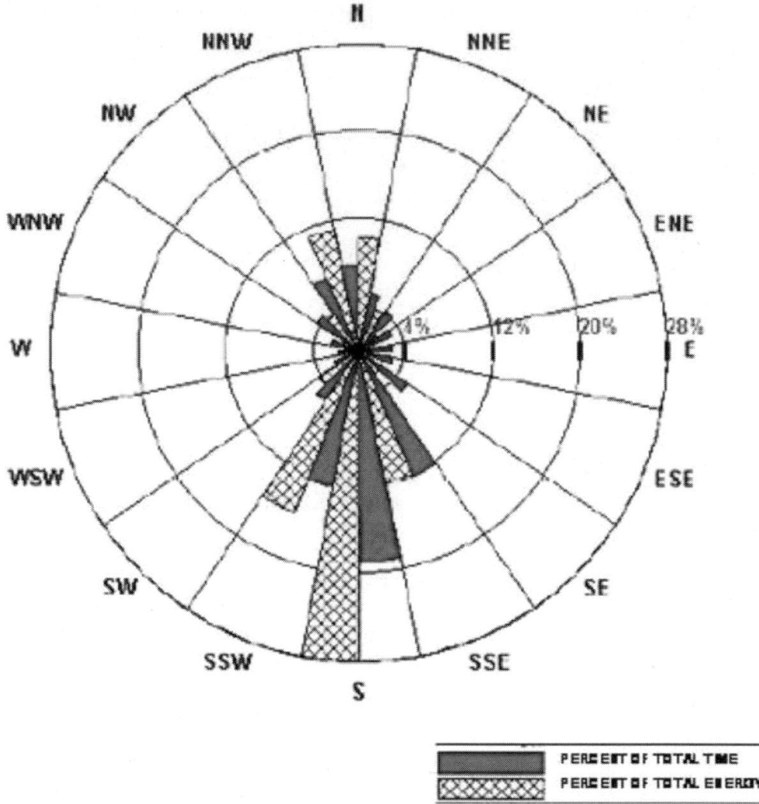

Figure W14 Wind speed and wind power rose at 10 m for Cheyenne, Oklahoma. The original data were in 5-min increments over 9 years (courtesy Oklahoma Wind Power Initiative).

latter case the Venturi effect causes local maxima of winds. Non-mountainous areas also require attention to siting detail. In general, lower areas on subdued terrain have lower wind power densities than higher areas, and forested areas are avoided because of the increased friction. For a line of turbines, preferred siting is on unforested ridges with map contour lines at right angles to the wind direction(s) providing the most wind power. The effect of a ridge – even a subdued ridge – is to increase the vertical speed (z) of the wind. This assists turbine rotor blades to turn because they move the blade edge.

Small wind turbines

The above discussion has focused on utility-scale wind, but small wind turbines (50 kW or less) are increasingly being used. Many times smaller than utility-scale turbines, the small turbines represent less up-front capital cost and great flexibility of siting. They are not as cost-efficient per kWh as large turbines, but they are able to provide electricity to places where there are no other sources or supplement power purchased from electrical utilities. There has been a large deployment of turbines in rural areas and, in particular, underdeveloped countries.

Downsides of wind power use

Two major downsides have slowed the development of the wind as an alternative to fossil fuels. They are the intermittency of the wind and the locational disconnect between windy sites and population centers. The intermittency of the wind power source has been of great concern to power providers. Over very short periods of time the boundary layer wind is composed of gusts of multiple speeds and directions. This short-term intermittency has been largely solved through engineering of turbines that are geared to provide constant power output. Over scales of days and seasons, however, there are times that the wind will not blow enough to turn turbines to their capacity, if at all. In these cases a utility must have other energy to place on the power grid and such ability incurs costs.

A second sort of problem is the situation in which the best wind power sites are not proximal to population centers. The transmission of power entails loss of power over distance. For example, the Great Plains of the United States have large areas of excellent wind resources but much of that area is hundreds of kilometers from large cities. Some rural sites might be near transmission lines but the lines might not have enough capacity to accommodate the additional power. The building of new transmission lines is expensive and requires time to acquire rights of way and permitting.

The future

Non-fossil-fuel-derived electricity with little negative environmental impact is a great gift of the atmosphere. The beginning of the present century has seen a huge increase in wind-generated electricity, and it is clear that much development will ensue given current levels of technology. Also clear is the fact that the temporal variability of wind power limits its use on power grids. As a small percentage of generation capacity placed on a power grid, the wind can be compensated for if it does not blow. When wind becomes a substantial part of the capacity (greater than 10% or so) calm conditions can create a shortage of power to the grid as a whole: this circumstance is unacceptable. Thus, utilities are wary of having too much wind in their generation portfolios and electrical wind generation regulated to a small minority of the power mix.

The future holds promise in the development of large turbines that can turn at lower wind speeds than present, and significantly decrease the amount of time the kinetic energy of wind can be tapped. Also the use of wind energy in hydrogen fuel cell production and the storage of converted wind energy in compressed fluids and next-generation batteries has the potential to open many avenues to increased use of the wind.

Steve Stadler and Tim Hughes

Bibliography

American Wind Energy Association, 1986. *Standard Procedures for Meteorological Measurements at a Potential Wind Turbine Site.* Washington, DC: American Wind Energy Association.

American Wind Energy Association, 1988. *Standard Performance Testing of Wind Energy Conversion Systems.* Arlington, VA: American Wind Energy Association.

American Wind Energy Association, 1992. *Wind Energy for Sustainable Development.* Washington, DC: American Wind Energy Association.

American Wind Energy Association, 1993. *Recommended Practice for the Siting of Wind Energy Conversion Systems.* Washington, DC: American Wind Energy Association.

American Wind Energy Association, 1994a. *Understanding Your Wind Resource.* Washington, DC: American Wind Energy Association.

American Wind Energy Association, 1994b. *Wind Energy Regulators Handbook.* Washington, DC: American Wind Energy Association.

Baker, T.L., 1985. *A Field Guide to American Windmills.* Norman, OK: University of Oklahoma Press.

Elliott, D.L., and Schwartz, M.N., 1993. *Wind Energy Potential in the United States.* PNL-SA-23109. Richland, WA: Pacific Northwest Laboratory.

Elliott, D.L., Holladay, C.G., Barchet, W.R., Foote, H.P., and Sandusky, W.F., 1987. *Wind Energy Resource Atlas of the United States.* Richland, WA: Pacific Northwest Laboratory.

Frost, W., and Asplide, C., 1994. Characteristics of the wind. In Spera, D., ed., *Fundamental Concepts of Wind Turbine Engineering.* New York: American Academy of Mechanical Engineers, pp. 371–445.

Hiester, T., and Pennell, W., 1981. *The Meteorological Aspects of Siting Large Wind Turbines.* Report No. PNL-2522. Richland, WA: Pacific Northwest Laboratory.

Koeppl, G.W., 1982. *Putnam's Power From the Wind.* New York: Van Nostrand Reinhold.

Landsberg, H.E. (editor in chief), 1969–1984. *World Survey of Climatology*, vols 1–15. Amsterdam: Elsevier Science.

National Wind Technology Center, www.nrel/gov/wind

Pacific Northwest Laboratory. 1993. *Shaded Relief Elevation Maps for Wind Prospectors.* Washington, DC: American Wind Energy Association.

Pennell, W., 1983. *Siting Guidelines for Utility Application of Wind Turbines.* Palo Alto, CA: Electric Power Research Institute.

Petersen, E.L., Mortensen, N.G., Landberg, L., Højstrup, J. and Frank, H.P., 1997. Wind power meteorology. Part I: climate and turbulence. *Wind Energy*, **1**: 2–22.

Risø National Laboratory, http://www.enero.dk/presentation/countries/risoe.htm

Shepherd, D.G., 1990. *Historical Development of the Windmill.* DOE/NASA-5266-2. Washington, DC: US Department of Energy.

Spera, D.A., 1994. *Wind Turbine Technology: fundamental concepts of wind turbine engineering.* Fairfield, NJ: American Society of Mechanical Engineers.

Thomas, N., 1995. *Understanding Your Wind Resource.* Washington, DC: American Wind Energy Association.

Van Wijk, A.J.M., and Coelingh, J.P., 1993. *Wind Power Potential in the OECD Countries.* Utrecht: Utrecht University.

Walker, S.N., 1988. *Local Wind Measurements for Micrositing.* Golden, CO: Solar Energy Research Institute.

Wegley, H., Orgill, M., and Drake, R., 1978. *A Siting Handbook for Small Wind Energy Conversion Systems.* Hanford, WA: Battelle Pacific Northwest National Laboratory.

Cross-references

Commerce and Climate
Pressure, Surface
Winds and Wind Systems

WINDS AND WIND SYSTEMS

Causes of wind

Wind is simply the movement of air. On a global scale the world's winds act as a primary thermal equilibrium mechanism helping to offset the persistent latitudinal energy imbalance of the Earth. This imbalance is caused by the equatorial region absorbing more radiation than it can effectively lose and the polar regions losing more radiation than they receive from the sun. The large-scale motion of the atmosphere – atmospheric circulation – transports a great amount of the excess heat from the equatorial regions into the colder high latitudes. Although the radiative imbalance of the Earth–atmospheric system drives the large-scale motions of the atmosphere, a variety of other mechanisms, such as topography, create the winds at smaller scales.

Five basic forces are primarily responsible for producing the world's wind systems; however, the relative importance of each force varies with scale. For example, a force that might be significant at the global circulation scale may be relatively unimportant at the scale of tornadoes. Therefore, to achieve an understanding of winds and wind systems, a consideration of both the forces and the specific scale of study is required.

Pressure gradient force

The initial force controlling most of the motions of the atmosphere is that induced by variations in the horizontal pressure field over the Earth. If pressure varies in a horizontal plane, the forces acting on the plane would not balance and air would flow from high to low pressure. This movement is created by a force referred to as the pressure gradient force (PGF). Variations in pressure on a horizontal plane could be created by surplus heating at certain locations and/or deficit heating (cooling) at other areas. Because density varies inversely with temperature and directly with pressure, a rise in temperature will cause a decrease in density and therefore a fall in pressure. In the east–west direction, Δx, PGF can be expressed as:

$$\text{PGF} = \frac{1}{\rho}\frac{p}{\Delta x}$$

where ρ is density and p is pressure.

Gravity

The only other force that can initiate atmospheric motion is gravity, which operates only in the vertical. All particles that make up the atmosphere, regardless of height or size, experience a downward acceleration due to the mass of the Earth. This acceleration per unit mass, called gravity, would collapse the atmosphere to the Earth's surface in the absence of some counterbalancing force. Fortunately, the decrease of pressure with height produces a strong vertical pressure gradient force that offsets the downward force of gravity. This counterbalancing relationship is commonly expressed in the hydrostatic equation as:

$$-\rho g = \frac{\Delta p}{\Delta z}$$

where g is the acceleration due to gravity and z is the vertical coordinate.

Coriolis force

An apparent force operating on atmospheric motion arises due to the rotation of the Earth. This force is a product of the acceleration observed by using the rotating Earth as a fixed frame of reference. It is not a true force since no mass is directly involved. As an example of the Coriolis effect, consider an air particle moving freely from the North Pole, P, with a velocity, v, in the direction of some point A, which the particle should reach in time dt (Figure W15). Given that the Earth's surface is rotating with counterclockwise angular velocity, ω, while the particle is traveling the distance vdt, point A rotates to point B while point A′ rotates to position A. Consequently, the moving particle arrives at point A and appears to have followed the curved path PCA′ relative to the Earth's surface or it was deflected from its original path. This force, postulated to account for this deflection, is termed the Coriolis effect after the physicist G.G. de Coriolis (1792–1843) who first treated it quantitatively.

The acceleration produced by the horizontal component of the Coriolis force is $2v\Omega \sin\Phi$ where Ω is the Earth's angular velocity, v is the velocity of the particle, and Φ is latitude (the quantity $2\Omega\sin\Phi$ is termed the Coriolis parameter, f). The Coriolis force is strongest at the poles and equal to 0 at the equator, and at all latitudes away from the equator it is directly related to the wind speed.

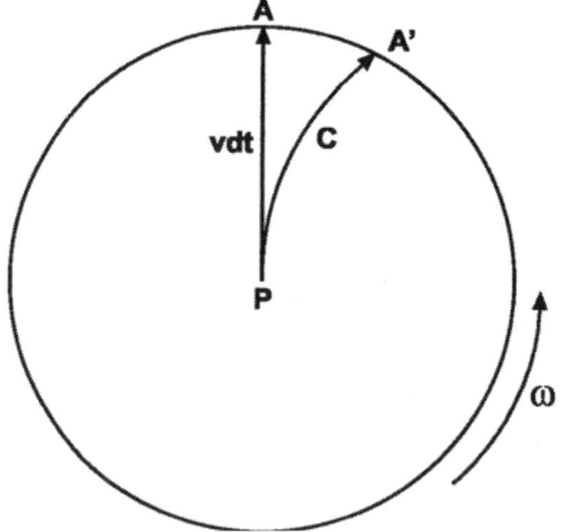

Figure W15 Coriolis effect due to Earth's rotation.

Centrifugal force

Another apparent force is related to the inertia built up by the air particle. As an example, consider a ball attached to a string and whirled at a constant angular velocity. If the ball is observed from a coordinate system rotating with the ball, it appears that the ball is stationary. However, there is still the force of the string acting on the ball, and an additional force is needed to balance the force of the string. The force of the string is balanced by the centrifugal force, which acts to move the ball in a direction tangent to the circle of motion. Simplistically for the atmosphere, the centrifugal force can be expressed as $\omega^2 r$, where r is the radius of curvature. Generally, this force is small compared to other major forces operating on large-scale atmospheric flows.

Friction

The last of the major forces influencing air movement is friction or viscosity. Normally, friction is important only throughout the planetary boundary layer, which may range in altitude from 30 m to 3 km, depending on atmospheric stability. Friction between moving air and the surface of the Earth produces a retarding stress with a direction opposite to the air motion. Its magnitude is a function of both the roughness of the terrain over which the air is passing and the actual wind speed. Frictional stresses also can occur between two adjacent layers of air moving in different directions and speeds; however, these stresses are generally less in magnitude than those between the surface and air.

Interaction between forces

These five forces – the pressure gradient force, the centrifugal force, the Coriolis effect, gravity, and friction – achieve a balance in our atmosphere and therefore all influence atmospheric motion. However, depending on the height in the atmosphere and/or the scale of motion, some of these forces might be far less important than others. For example, in the levels above the planetary boundary layer, an approximate balance can occur between the horizontal pressure gradient force and the Coriolis effect. If an exact balance of these two factors is achieved, the resulting air motion is called the geostrophic wind, v_g, defined in the east–west direction as:

$$v_g = \frac{1}{fp}\left(\frac{\Delta p}{\Delta x}\right)$$

Because the horizontal pressure gradient force operates from high to low pressure so that air is directed perpendicular to the isobars, the Coriolis effect also must be perpendicular to the isobars and directed from low to high pressure (Figure W16). In the northern hemisphere, the Coriolis force deflects to the right of the apparent direction of wind. The geostrophic wind velocity must exist such that low pressure is found to the left of the direction of motion. In the southern hemisphere low pressure is found to the right of the direction of motion because the Coriolis force acts to deflect motion to the left. This principle is known as Buys–Ballot's Law.

Below 1000 meters, in the planetary boundary layer, friction becomes important and a three-way balance between the pressure gradient force, Coriolis force, and friction is achieved. Friction acts opposite to the wind direction, slowing the wind. This decreases the Coriolis effect and disrupts the geostrophic flow parallel to the isobars. As a result of the interplay of these three forces, the actual wind blows at an angle across the isobars, in a direction from higher to lower pressure (Figure W17). This creates

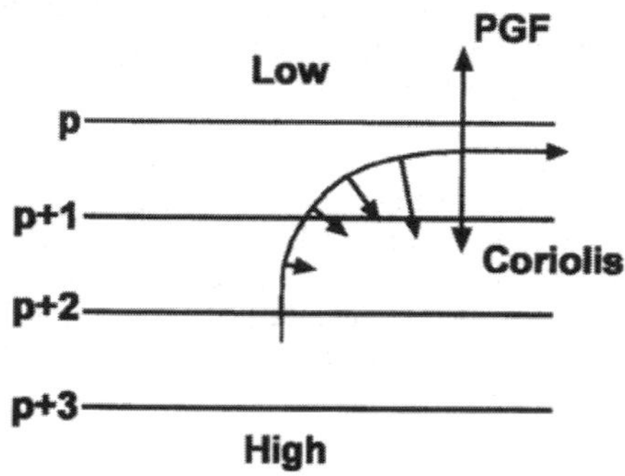

Figure W16 Approach to geostrophic equilibrium by a parcel initially at rest.

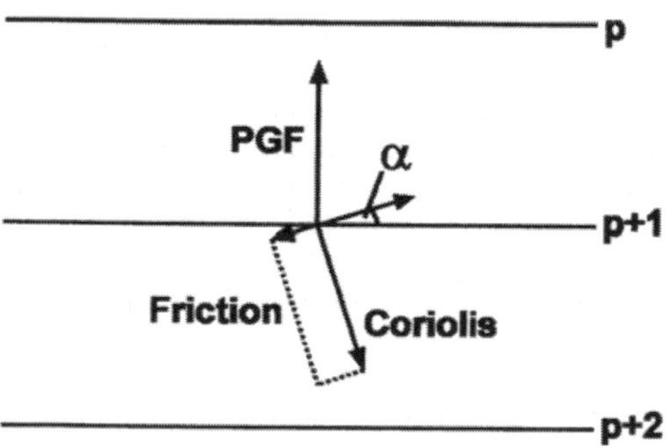

Figure W17 Balance among pressure gradient force, Coriolis force, and frictional force.

surface winds that flow clockwise and out of high-pressure areas (anticyclones) and counterclockwise and into low-pressure areas (cyclones).

Because pressure and temperature are directly related to one another, thermal patterns provide an alternative method of discussing winds. Through manipulation of the geostrophic wind equation and the quantitative relationship between temperature and pressure, the change in geostrophic wind speed with height to a first approximation can be expressed in the east–west direction as:

$$\frac{\Delta v}{\Delta x} = \frac{g}{fT}\left(\frac{\Delta T}{\Delta x}\right)$$

where $\Delta v/\Delta x$ is the change in north–south component of wind with height and T is temperature. This equation shows us that greater temperature changes in the horizontal will force larger wind speed changes with height.

The preceding discussion represents a simplistic explanation of the fundamental principles of dynamic meteorology and climatology. A solid mastery of these principles is a necessary prerequisite to a full understanding of the nature of the actual winds and wind systems that are found in the world's atmosphere.

The world's wind systems

Atmospheric circulations may range in size from a small gust or swirl of wind to the global circulation cells that give rise to the trade winds and the westerlies. The specific winds of a given location are often shaped by local conditions.

Planetary wind systems

The mean planetary wind pattern, also referred to as the general circulation of the atmosphere, represents the averaging of wind observations over a long period of time and thus depicts the largest of scales of atmospheric motion. The actual circulation at any moment can differ significantly from this time-averaged pattern. A very generalized model of the mean wind patterns (Figure W18)

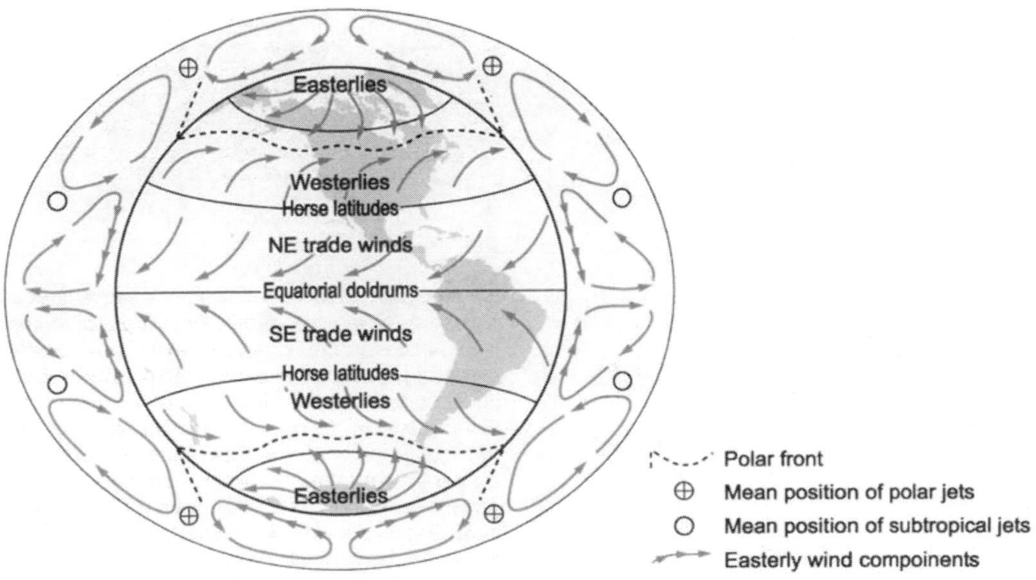

Figure W18 General circulation of the atmosphere. Double-headed arrows in cross-section indicate easterly wind components.

employs the assumption that the Earth is covered only by oceans and has normal rotational characteristics.

Under these conditions the equatorial zone would be the region of most consistent receipt of solar radiation (the noon sun is always within 23½° of being directly overhead). The resulting intense surface heating produces an enormous vertical transport of air, leading to the development of a broad band of lower surface pressure straddling the equator. This heated, ascending air begins to cool and diverge at upper levels as it flows toward both of the polar zones. At one time (1735) the English philosopher and meteorologist George Hadley proposed that the poleward-moving air ultimately reached the poles and then the airstreams returned to the equator region along the Earth's surface. However, he noted, if the air at the Earth's surface flowed toward the equator from the polar zones, the Coriolis effect would eventually cause the air to be flowing from east to west in both hemispheres. This type of flow would be in direct opposition to the west to east rotation of the Earth. Hence, in later years a three-cell model was developed to explain the basic patterns of vertical airflow. The three-cell model has the poleward-flowing air beginning to descend, finally reaching the surface in the area around 30° latitude in both hemispheres.

Many of the surface circulation features have traditional names dating back to the period of sailing vessels during the eighteenth and nineteenth centuries. This zone of descending air near 30° latitude produces a region of relatively higher pressure dominated by fair weather and lack of strong surface horizontal airflow. Legend has it that sailing vessels were often becalmed as they attempted to pass through these areas and the horses were the first to go overboard in an attempt to lighten the vessels, hence the term Horse Latitudes was given to this region, which also is traditionally referred to as the subtropical high-pressure belt.

Because of this accumulation of air near the Earth's surface, airstreams begin to diverge from the subtropical high-pressure belt and flow toward both the polar region and the equator. Out of respect to Hadley, scientists today refer to this convectively driven cell immediately on either side of the equator as the Hadley Cell. Under the influence of the Coriolis force, the airstreams diverging from each of the hemispheric subtropical high-pressure zones are deflected to the right in the northern hemisphere and to the left in the southern hemisphere. The equatorward-moving air, termed the trade winds, is of a persistent nature and takes on the characteristics of southeasterly winds in the southern hemisphere and northeasterly winds in the northern hemisphere. These trade winds of each hemisphere converge along a zone called the Intertropical Convergence Zone (ITCZ). The exact location of the ITCZ varies with the time of year and is related to the migration of the direct rays of the sun. Generally, the ITCZ is located north of the equator during the northern hemisphere's summer months and to the south of the equator during the summer season of the southern hemisphere. Because this equatorial zone is both a region with little airmass contrast and dominated by persistent clouds and precipitation, it is often referred to as the doldrums.

Poleward of the subtropical zone in each hemisphere is an area of winds known as the westerlies. In reality, the mean airflow in this zone in the northern hemisphere is from the southwest and in the southern hemisphere from the northwest. The characteristic feature of the westerlies is the net transport of airmasses from west to east. Along the poleward margin of this region is an area referred to as the disturbed westerlies, which has the traveling cyclones or low-pressure cells moving along frontal boundaries that separate the warm tropical airstreams from the colder polar airstreams. These cold polar airstreams (northeasterly winds in the northern hemisphere and southeasterly winds in the southern hemisphere) originate in another region dominated by high pressure, the polar zones. It should be noted, though, that this region of high pressure is not as extensive as the subtropical high-pressure region. These polar airstreams, also called polar easterlies, collide with the westerlies along what is known as the polar front. Considerable uplift occurs along this frontal zone and, as the air ascends, it diverges toward both the polar and subtropical regions completing the two circulation cells that are poleward of the equatorial Hadley Cell. This middle cell is termed the Ferrel Cell. The polar front zone is meteorologically the most active region of the world and is examined in more detail below in the section entitled "Traveling Systems".

One of the most dramatic features of the upper-air global circulation is the presence of jet streams. The two primary jet streams are the polar jet located above the polar front (at approximately 10–15 km above sea level) and the subtropical jet located at approximately the same elevation along the northern margin of the Hadley Cell. Each of these jet streams is hundreds of km wide and several km thick. Known as steering currents, they direct the low-pressure cells and provide quasi-barriers to the movement of the major air masses. Surface weather conditions such as droughts, heat waves, and storminess have been related to the positioning of the jet stream and the related attendant upper-air flow patterns.

One of the interesting aspects of the jet stream and all of the upper-air flow is the variation in wind velocities during the year. Using the concept of the thermal wind, the greater the contrast in thermal energy across a region, the greater the resulting wind velocity. Therefore, the jet stream (as well as most upper-air flow) reaches its maximum velocities during the winter season and early spring when the latitudinal thermal gradient is at a maximum. Conversely, during the summer season of either hemisphere, the latitudinal thermal gradient diminishes to a minimum along with the velocity of the upperair wind velocities. The jet stream is in a wavelike form and assists in the exchange of heat and momentum poleward in an attempt to maintain the hemispheric and global heat balance. This poleward and corresponding equatorward undulation in the upper-air flow is often called meridional flow, and the amplitude of this meridional energy exchange is also related to the latitudinal thermal gradient. The undulations observed in the upper-air flow reach the highest amplitudes, crossing the largest amount of latitude, during the winter season. During summer the high-amplitude wave pattern becomes one of greatly reduced amplitude as the result of a reduced latitudinal thermal gradient.

The concept of seasonality can be built into this model just as it was for the interpretation of annual variation in the velocity of the winds. During the northern hemisphere winter season, all of the circulation regimes migrate southward from their normal positions illustrated in the above model; conversely, these circulation features migrate northward with the advancement of summer into the northern hemisphere. This seasonal shift in zones of circulation results, for some cases, in dramatic reversals in prevailing wind patterns. For example, in India, the prevailing northeast winter winds gradually change over to southwesterly winds as the summer monsoon and zones of prevailing winds migrate northward with the advance of summer. In the United States, much of the country is dominated by northerly flow during the winter season; however, as the circulation belts migrate northward with the direct rays of the sun, the polar circulation gives way to the dominant summertime pattern of westerlies. The northward movement of the subtropical high-pressure belt through California during summer is notable for the dramatic seasonality of precipitation, with almost all of the precipitation in this region occurring during winter while under the influence of the westerlies. As mentioned earlier, the migration of the ITCZ is also related to the seasonal migration of these generalized circulation features.

There are two major limitations in the use of this model of generalized circulation patterns. First, the three vertical cells are far too simplified to express the true complexity of upper-air flow. Many climatologists do not adhere to the three-cell model, preferring instead to illustrate only the Hadley Cell and a single more complex midlatitude-polar cell. Second, the horizontal patterns are so highly generalized that the actual circulation features at any given time would vary considerably from this model. However, it is generally agreed that this schematic diagram is a useful starting point for the much more detailed flow characteristics that evolve once the actual geographic distribution of landmasses, sea surfaces, and topographic features are included.

Traveling wind systems

Of all the regions of the world, the midlatitudes of each hemisphere are by far the most active in terms of moving atmospheric systems. Although the subtropics do have tropical cyclones (specific regional names include hurricanes, typhoons, cyclones, and Baguios), these are not daily occurrences and are in fact rare features of the day-to-day weather pattern. The primary features of the midlatitudes that produce variations in surface windflow characteristics are the cold, warm, and occluded portions of the polar front, the low-pressure cells traveling along the polar frontal system, and the high-pressure cells that originate in either the polar or the subtropical source region.

The low-pressure cells that originate along the polar front undergo a life cycle that was first described by J. Bjerknes in 1918. The first step in the stages of development is cyclogenesis. It is at this point that advection of northerly and southerly airstreams begins. The airstreams in the northern hemisphere move in a counterclockwise circulation (southern hemisphere, clockwise) as the airstreams ascend and converge toward the low center. As the pressure at the center of the low continues to fall, the warm and cold airstreams continue to advect over large regions and with increasing velocities. Eventually, the cold air sweeps around the low-pressure cell, eliminating the wedge of northward-moving air in a process called occlusion. As the occlusion process advances, the wind velocities drop off, and the latitudinal transport of polar and tropical air diminishes. Another interesting aspect of these traveling cyclones is their tendency to occur in a sequence known as the family of wave cyclones. Occasionally, near the region of cyclogenesis, the cold air breaks through the polar front in the form of a rapidly moving cold front known as a polar outbreak. On either side of the polar front are large high-pressure systems that make up distinctive airmasses. These airmasses take on the characteristics of their source region, extend over thousands of square kilometers, and have similar temperature and moisture structures. The basic airflow associated with these high-pressure cells is a descending and clockwise spiraling motion (southern hemisphere, counterclockwise).

Local winds

The surface winds associated with this scale are often shaped by local conditions such as topography, location, and differential heating. This scale of winds may be classified into five main groups: (a) diurnal winds, including those winds which are associated the diurnal cycle; (b) jet-effect winds, including those winds which are strongly influenced by the local topography; (c) antitriptic winds, including those winds arising from pressure and thermal gradients (such as land/sea breezes or Chinook/föhn winds) or by gravity (such as fall winds); (d) local winds created local heating or instability created by the overrunning of cold air, such as dust storms or haboobs; and (e) winds created by strong pressure gradients over a relatively small area, the uninterrupted flow over a level surface, or both sets of conditions (such as the desert khamsin wind or winds associated with blizzard events).

Differential heating at the Earth's surface can result in several local wind circulations. The sea (and lake) breeze and corresponding land breeze are examples of local wind circulations that reverse direction on a diurnal basis. Mountain and valley breezes also typically occur with a diurnal reversal in circulation. It should be stressed that these circulation systems can be completely obscured by the larger-scale synoptic systems and are therefore most pronounced during clear and relatively weak synoptic weather conditions. During the winter months cold dense air can spill out of intermontane regions, descending into surrounding valleys in the form of strong cold airstreams. Throughout the world, literally hundreds of local winds have been identified and labeled with a labyrinth of vernacular names.

Small-scale or local winds have a direct impact on many human activities. Microbursts and the winds associated with thunderstorms pose serious concerns to the aviation industry. Sudden dust storms in desert environments such as the southwestern United States have caused death and injury on highways through the region. Chinook/foehn winds in many alpine regions of the world have caused rapid melting of snow that has contributed to flash flooding in some areas. Additionally, architects have begun to consider the impact on building design of air flow within the urban landscape. Such interrelationships between human activity and the climatology of local winds will undoubtedly continue to be expanded and researched in the future.

Observation and measurement of winds

Many practical and theoretical problems in meteorology and climatology require accurate measurements of wind speed, direction, and/or gustiness. From a theoretical perspective the study of the forces, laws, and effects of air motion lies near the heart of the atmospheric sciences. From a practical standpoint, examples of problems requiring wind observations include the siting of emission sources of noxious atmospheric pollutants, design of homes for optimum ventilation, studies of insect or disease migration, or analysis of the aeolian landforms. Increasingly, engineers, architects, scientists, and planners are requiring wind data for assessing both the useful and the destructive potential of atmospheric motions.

The science that deals with the study of wind measurements is called anemometry. The measurement techniques described below represent a limited sample of the many methods used to accurately measure or estimate wind parameters on various spatial and temporal scales.

Wind speed estimations

In 1804 British Navy Admiral Francis Beaufort created a system for estimating wind speeds for observations of wind effects. Originally, the system was designed for use at sea; later it was modified for use on land. In 1874 the Beaufort system was officially adopted for international use, and in 1955 the US Weather Service expanded the scale by adding additional force numbers to the original system.

The first known anemometer was a swinging-plate type instrument, which was described in literature found dating back to the 1450s. By hanging a plate with some string from a horizontal rod, the intensity of the wind speed was measured by the extent to which the angle of the plate deviated from the perfectly perpendicular position it took on during still wind conditions. These types of anemometers are known as pressure-type anemometers since they rely upon the pressure of the wind against the surface of the plate in positioning the plate in a static position away from its perpendicular position.

Theoretically, sound estimates of wind speed and direction can be derived from pressure and/or temperature measurements made throughout the atmosphere. Using the dynamical principles that govern the flow of a thermally active fluid over a spherical surface, remarkably precise estimates of wind parameters can be generated, particularly for heights above the planetary boundary layer where frictional influences can be ignored. However, the effects of friction and response time on instrumentation limit the effectiveness and precision of wind sensor as wind speeds approach zero and the structural support limits the sensor's survival at high speeds.

The World Meteorological Organization (WMO) has set standards for measurement of wind speed and direction. "Sustained" wind speeds are measured as 10-minute averages, in order to be compatible with most standard software (and literature on the subject) while gusts are the maximum speeds recorded above the sustained wind speed average within the 10-minute average period. Wind direction is measured in degrees in a clockwise direction from north. The WMO standard exposure height for surface winds is 10 km and the anemometer must have good exposure in all directions within about three km.

Langrangian measurements

Langrangian measurements of winds are taken from within the moving body of air. Observations of movements of clouds, smoke puffs, or visible plumes have long been used to calculate winds above the various levels of the atmosphere. Satellite estimates of wind speed, for example, have long been made particularly over oceanic areas. Alternative methods have employed scatterometers, altimeters, and synthetic aperture radars to measure the state of the ocean surface, e.g. wave height, for determining surface wind speeds.

More direct Langrangian measurements of the wind field may be made by tracking balloons that rise through the atmosphere. In clear conditions a pilot balloon (pibal) of known ascent rate may be released and tracked with an instrument called a theodolite. Altitude and azimuth angles measured by the theodolite at specific time intervals allow wind speed and direction at different heights to be determined trigonometrically using the pibal method. Similar techniques can be employed in cloudy conditions using balloons carrying radar targets or radio transmitters. The rawinsonde (rawin) method tracks the ascending target with radar, and wind speed and direction are trigonometrically determined. The rawinsonde and/or rabal methods for wind determination also use radar for establishing altitude and azimuth angles, but the height of the balloon is determined from the radiosonde data. Above the 30-km ceiling for most balloons, various targets released from rockets can be tracked to calculate wind velocities in the high atmosphere.

Doppler radar and wind profiler now commonly obtained detailed wind information for lower tropospheric wind measurements on a regular basis. Doppler radars measure the radial velocity of targets (cloud droplets, dust or precipitation) in relation to the station to give a basic inflow/outflow speed determination in relation to the station. Wind profilers, in essence upward-oriented Doppler radars, conversely give vertical profile of wind speeds as a function of time. Doppler radars have even been mounted on vehicles for detailed observation of tornadic winds.

Eulerian measurements

Eulerian measurements of wind are taken by various instruments positioned at some fixed point near the ground. Four basic Eulerian methods are used in anemometry to measure wind speeds.

1. Pressure anemometers, such as pitot-static tubes, operate on the principle that the square of wind speed, v is functionally related to the pressure exerted by the airflow. The pressure types are very accurate in measuring the fine structure of the winds and, accordingly, are often called gust recorders.
2. Mechanical anemometers, such as the popular three-cup rotation anemometer, measure wind speeds by recording rotations per unit time interval, by generating electrical current, or by applying pressure on a set of springs. Air motion causes atmospheric pressure differences between the concave and convex sides of the cups, thereby forcing rotation about a vertical axis. The propeller, or windmill, mechanical anemometers operate on a similar principle, but the axis of rotation is generally parallel to the wind direction. Today the most common type of anemometer is a rotational device known as a cup anemometer. This consists of three cups made of lightweight material which are equidistantly attached with spokes to a rotating center rod. The speed of rotation varies with the magnitude of the wind speed. This is known as a "drag" device because its rotation is caused by the drag of the cups against the wind. Another increasingly popular type of anemometer is the propeller anemometer. This anemometer is simply a very small two- or three-blade propeller, calibrated so that the revolutions per minute (rpm) corresponds to specific wind speeds. This anemometer is known as a "lift" device because it relies upon the rational lift caused by the wind on the propellers to measure the wind speed.
3. Thermoelectric anemometers such as hot-wire anemometers are often required to measure very light wind speeds. Because the cooling rate of an object is a function of the flow rate of a fluid, the energy required to maintain a constant temperature of a probe can be directly related to the wind speed.
4. Sonic anemometers capitalize on the principle that true sound velocity is proportional to wind speed in the direction of air motion. Using a set of emitters and receivers, the true sound speed can be used to accurately determine existing wind speeds.

Wind direction is usually measured by various types of wind vanes. The most popular types are asymmetrically shaped arrows with the center of gravity mounted on a vertical axis. The tail fins provide resistance, counterweight and move to the downwind position. Aerovanes, or bivanes, record both speed and direction by combining the propeller-type anemometer with a streamlined shape to keep the instrument facing the wind. Due to their exceptional visibility, wind socks are still widely used at smaller airports to display the existing wind direction.

Actual wind data

The great variability of surface winds through both time and space, caused in large part by significant frictional effects, poses problems in accurate and representative wind measurement. To standardize surface wind measurements, all instruments are to be exposed to freely flowing air at least 10 m away from any obstacle. However, practical considerations often lead to poor site selections for equipment, resulting in systematic errors in direction and speed measurement.

In the lowest portions of the atmosphere, wind speeds are retarded by the friction associated with the rough surfaces. Accordingly, in the first few hundred meters above the ground, wind speeds dramatically increase with height, but speeds increase more gradually with height above the lower friction layer. Because Coriolis force is directly related to wind speed, winds in the northern hemisphere tend to veer to the right with increasing altitude, resulting in the well-known Ekman spiral.

Most atmospheric measurements (i.e. temperature, precipitation, pressure, etc.) can be represented by using scalar quantities. However, because wind has both a speed and a direction, it is often represented as a vector quantity. Consequently, a variety of techniques are unique to the presentation of wind data.

Following WMO convention, wind direction is given according to compass headings expressed nominally as N, NNE, NE, NNW, or numerically such that 90° is east, 180° is south, and so on. In either case, wind direction always gives the directional origin of the wind; for example, a southerly wind blows from the south to the north. On maps, actual wind directions are often depicted by arrows that fly with the wind (Figure W18), whereas curved streamlines generally are used to show general air trajectories or theoretical flow patterns.

Wind speed is often given in knots, although miles per hour and meters per second are also popular units of measure. Speeds can be illustrated using barbs or feathers on the arrows of wind direction or they can be depicted using isotachs drawn through points with equal speed or wind force values.

The vectorial nature of wind allows many other diagrammatic techniques to be employed. Many types of wind roses can be drawn, often with the relative length, or thickness, of lines radiating from the center indicating frequencies of principal wind directions.

The northerly and westerly components of the wind, calculated as $v \cos\theta$ and $-v \sin\theta$, respectively, with v equal to wind speed and θ equal to wind direction, allow scalar treatments of vectorial wind data. Resultant winds, calculated as vector summations of the winds, have proved to be particularly useful in studies on atmospheric diffusion processes.

Many Internet sites, both public and private, now offer realtime wind observations for an incredible number of sites around the world. Climatological compilations of surface and upper-air wind measurements are available for hourly, 3-hourly, daily, monthly, annual, and long-term averages for periods extending back more than 50 years for upper-air measurements and more than a century for surface winds. For the United States the primary archive of wind information is the National Climatic Data Center (NCDC) located in Asheville, North Carolina and accessible at www.ncdc.noaa.gov. Radiosonde and rawinsonde reports for daily winds at many levels of the atmosphere have been compiled for hundreds of land- and sea-based stations around the world.

Robert C. Balling, Jr and Randall S. Cerveny

Bibliography

Barry, R.G., and Chorley, R.J., 1998. *Atmosphere, Weather and Climate*, 7th ed. London: Routledge.

Holton, J., 1992. *An Introduction to Dynamic Meteorology*. New York: Academic Press.

Peixoto, J.P., and Oort, A.H., 1992. *Physics of Climate*. New York: American Institute of Physics.

Thompson, R.D., 1998. *Atmospheric Processes and Systems*. London: Routledge.

Wallace, J.M., and Hobbs, P.V., 1997. *Atmospheric Science: an introductory survey*. San Diego, CA: Academic Press.

Cross-references

Atmospheric Circulation, Global
Katabatic (Gravity) Winds
Land and Sea Breezes
Local Winds
Mountain and Valley Winds
Pressure, Surface
Trade Winds
Vorticity
Wind Power Climatology

Z

ZONAL INDEX

The zonal index is one of the two standard circulation index conditions, specifically, a measure of midlatitude westerlies (in both hemispheres) expressed as the horizontal pressure difference between latitudes 33° and 55° or as the corresponding geostrophic wind. A high-index weather situation (over 8 mb) represents a strong westerly component and a low index (less than 3 mb) relative weakening, i.e. implying a stronger meridional component; in extreme low-index conditions the usual poleward pressure gradient may be reversed and expressed with a negative figure. In Cartesian coordinates, the x axis is eastwest and the y axis northsouth, so that the zonal wind (x) may be expressed as positive if it blows from west to east, and vice-versa. It may be compared with the meridional wind (y axis).

The index cycle

A roughly cyclic variation in the zonal index is called the index cycle. This cycle has a period that varies from 3 to 8 weeks, although extremely high- or low-index conditions generally do not persist for more than 2 weeks.

The typical index cycle can be divided into a number of stages beginning with a high zonal index with a long wavelength pattern (Figure Z1). The wavelength pattern aloft is shortened, leading to a complete breakup of the zonal winds at both the surface and aloft. Closed cells are formed, which are

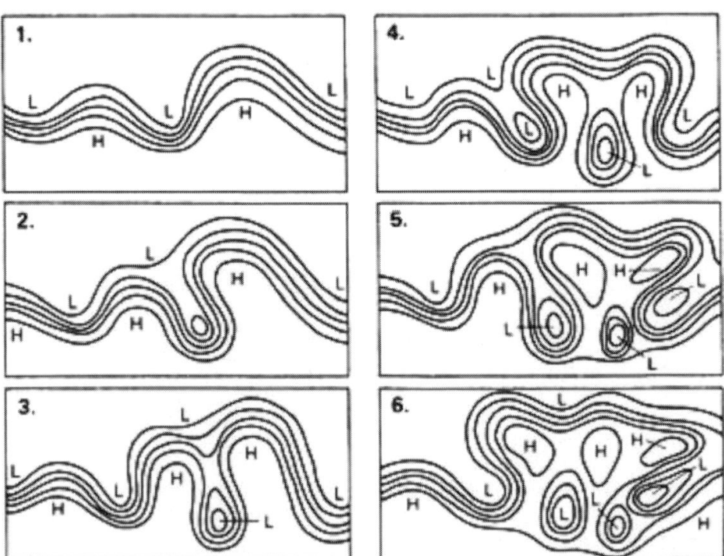

Figure Z1 Schematic representation of the stages in the breakdown of zonal flow at the 500 mb level. Time period involved in the breakdown may be 8–10 days and depicted waves may extend from the central United States to Eastern Europe (from Lockwood, 1979).

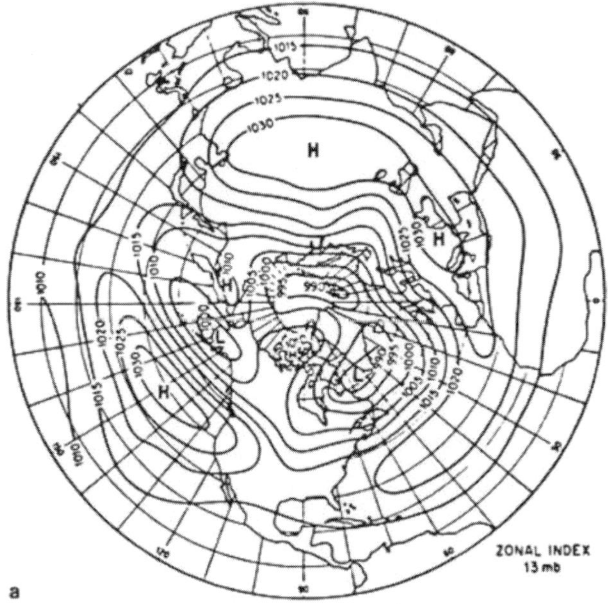

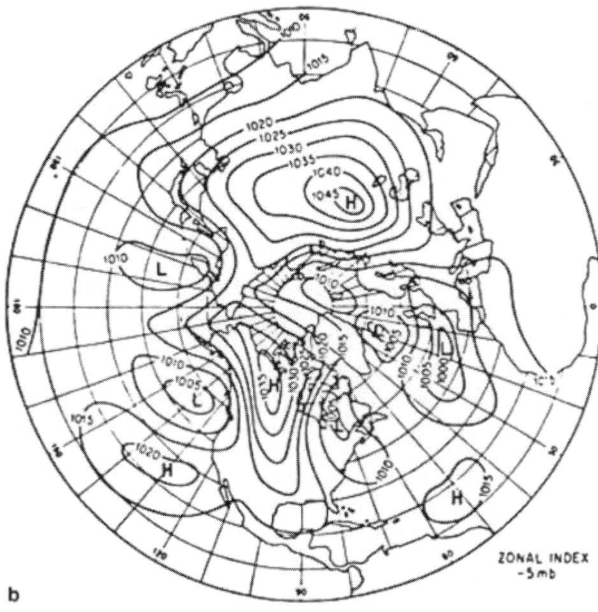

Figure Z2 Examples illustrating mean pressure distribution at sea level during one week of (**a**) high zonal index and (**b**) low zonal index.

eventually cut off to permit the re-establishment of the open wave pattern in the upper air.

During northern hemisphere high-index conditions:

1. Icelandic and Aleutian lows are strongly developed, often with double center in east–west trough.
2. Atlantic and Pacific subtropical highs strongly developed, somewhat north of usual winter latitude, in east–west orientation, extending over neighboring continental margins. Continental highs at 3540 °N.
3. Polar continental highs missing at high latitudes over Siberia and Canada, and replaced by lows, though polar anticyclone may be strong.
4. Frontal zones oriented more east–west than usual, with lows migrating rapidly eastward at high latitude.
5. Midlatitude westerly weather is mainly fair, airmass contrasts being rather small.
6. Subpolar weather in the low-pressure belts is stormy with rapid succession of lows but with mild temperatures.

An example is given in Figure Z2a.

Northern hemisphere low-index conditions are characterized by

1. Icelandic and Aleutian lows subdivided into separate weak cells, one generally to the southwest and the other to the southeast of usual positions. Develop a north–south orientation.
2. Atlantic and Pacific subtropical highs are weak, displaced to low latitudes and subdivided into secondary cells by the north–south extensions of lows.
3. Polar continental highs are well developed over Siberia and Canada and are quite separate from the subtropical highs. They may form a continuous high-pressure belt around the Arctic.
4. Frontal zones are oriented more north–south than usual, situated east of each low-pressure trough, generated between strong northern flows of tropical air on the west sides of highs and a spilling of polar air into the low-pressure troughs. Cyclogenesis reaches much farther toward the equator than usual.
5. Midlatitude weather marked by alternate very warm and very cold contrasts with high precipitation. Migrating lows move eastward into continents south of continental highs, thus from southern California to lower Mississippi and from Bay of Biscay to Mediterranean and South Russia. Maximum north–south heat exchange results.
6. High-latitude weather is not stormy but sees rapid influxes of warm air.

An example of such low-index conditions is shown in Figure Z2b.

John E. Oliver

Bibliography

Aguado, E., and Burt, J.E., 2001. *Understanding Weather and Climate*. Upper Saddle River, NJ: Prentice-Hall.

Barry, R.G., and Chorley, R.J., 1998. *Atmosphere, Weather, and Climate*, 7th edn. London: Routledge.

Boucher, K., 1975. *Global Climates*. New York: Halstead Press.

Glickman, T., 2000. *Glossary of Meteorology*, 2nd edn. Boston, MA: American Meteorological Society.

Holton, J., 1992. *An Introduction to Dynamic Meteorology*, 3rd edn. New York: Academic Press.

Lockwood, J.G., 1979. *Causes of Climate*. New York: Wiley.

McIlveen, R., 1992. *Fundamentals of Weather & Climate*. London: Chapman & Hall.

Namias, J., and Clapp, P.F., 1951. Observational studies of general circulation patterns. In Malone, T.F., ed., *Compendium of Meteorology*. Boston, MA: American Meteorological Society, pp. 551–567.

Nese, J.M., and Grenchi, L.M., 1998. *A World of Weather: fundamentals of meteorology*, 2nd edn. Dubuque, IO: Kendall/Hunt.

Petterssen, S., 1956. *Weather Analysis and Forecasting*, 2nd edn. New York: McGraw-Hill.

Rossby, C.G., 1959. Current problems in meteorology. In Bolin, B., ed., *The Atmosphere and Sea in Motion.* New York: Rockefeller Institute Press, pp. 9–50.
Thompson, R.D., 1998. *Atmospheric Processes and Systems.* London: Routledge.
Wallace, J.M., and Hobbs, P.V., 1997. *Atmospheric Science: an introductory survey.* San Diego, CA: Academic Press.
Willett, H.C., 1944. *Descriptive Meteorology.* New York: Academic Press.

Cross-references

Atmospheric Circulation, Global
Centers of Action
Global Environmental Change: Impacts
Pressure, Upper Air
Rossby Wave/Rossby Number

Index of Authors Cited

Subject Index

Note: page numbers in **bold** indicate main articles; those in *italics* refer to figures and tables.